Ophiolites and Oceanic Crust: New Insights from Field Studies and the Ocean Drilling Program

Edited by

Yildirim Dilek
Department of Geology
114 Shideler Hall, Miami University
Oxford, Ohio 45056 USA

and

Eldridge M. Moores
Department of Geology
University of California
Davis, California 95616 USA

and

Don Elthon
Department of Chemistry
University of Houston
Houston, Texas 77204 USA

and

Adolphe Nicolas
CNRS/Université Montpellier II
Laboratoire de Tectonophysique
34095 Montpelier cedex 05
France

SPECIAL PAPER

349

Geological Society of America
3300 Penrose Place
P.O. Box 9140
Boulder, Colorado 80301-9140
USA
2000

Published by The Geological Society of America, Inc.
3300 Penrose Place, P.O. Box 9140, Boulder, Colorado 80301, USA
www.geosociety.org

Printed in U.S.A.

GSA Books Science Editor Abhijit Basu
GSA Books Managing Editor Rebecca Herr
Cover design by Margo Good

Library of Congress Cataloging-in-Publication Data

Ophiolites and oceanic crust: New insights from field studies and the Ocean Drilling Program / edited by Yildirim Dilek . . . [et al.].
p.cm.—(Special paper ; 349)
Includes bibliographical references and index.
ISBN 0-8137-2349-3
1. Ophiolites. 2. Ocean bottom. 3. Earth—Crust. I. Dilek, Yildirim, 1957- II. Special papers (Geological Society of America) ; 349.

QE462.O6 O635 2000
552′.3—dc21 00–030831

Front Cover:
Dike-gabbro contact in the Kizildag ophiolite (Turkey) along the Karaçay Valley in Hatay. The light-colored rock in the foreground is a fine-grained isotropic gabbro, which contains basaltic dikes and dikelets of irregular size and shape. Steeply southeast-dipping sheeted dikes in the background are emanating from this contact and the plutonic rocks, suggesting that the dike-gabbro transition zone here is the root zone of the sheeted dike complex. Irregular plagiogranite intrusions are common both in the gabbro and the sheeted dikes across this igneous contact. Photograph by Yildirim Dilek.

Back Cover:
Participants of the Geological Society of America Penrose Conference on Ophiolites and Oceanic Crust: New Insights from Field Studies and the Ocean Drilling Program, in front of a blueschist block in the Franciscan Complex at Tiburón Peninsula, northern California, September 16, 1998.

10 9 8 7 6 5 4 3 2 1

Contents

Preface vii

Section I. Ophiolites, Oceanic Crust, and Global Tectonics: Introduction 1

1. *Tethyan ophiolites, mantle convection, and tectonic "historical contingency": A resolution of the "ophiolite conundrum"* 3
   Eldridge M. Moores, Louise H. Kellogg, and Yildirim Dilek

2. *Evidence for subduction-related contamination of the mantle beneath the southern Chile Ridge: Implications for ambiguous ophiolite compositions* 13
   Marnie E. Sturm, Emily M. Klein, Jill L. Karsten, and Jeffrey A. Karson

3. *The ophiolite–oceanic lithosphere analogue: New insights from the Northern Apennines (Italy)* 21
   Elisabetta Rampone and Giovanni B. Piccardo

4. *Significance of serpentine mud volcanism in convergent margins* 35
   Patricia Fryer, John P. Lockwood, Nathan Becker, Steven Phipps, and Clifford S. Todd

5. *Spatial and temporal relationships between ophiolites and their metamorphic soles: A test of models of forearc ophiolite genesis* 53
   John Wakabayashi and Yildirim Dilek

Section II. Oceanic Lower Crust and Upper Mantle: Introduction 65

6. *Large mantle upwellings and related variations in crustal thickness in the Oman ophiolite* 67
   Adolphe Nicolas and Françoise Boudier

7. *Lower oceanic crust formed at an ultra-slow-spreading ridge: Ocean Drilling Program Hole 735B, Southwest Indian Ridge* 75
   Paul T. Robinson, Henry J.B. Dick, James H. Natland, and the ODP Leg 176 Shipboard Party

8. *Magmatic and tectonic controls on the evolution of oceanic magma chambers at slow-spreading ridges: Perspectives from ophiolitic and continental layered intrusions* 87
   Peter Thy and Yildirim Dilek

9. *Syntexis and the genesis of lower oceanic crust* 105
   Jean H. Bédard, Réjean Hébert, Alain Berclaz, and Véronika Varfalvy

10. *Can downward flow of dense cumulate slurry through mushy upper gabbros produce lower gabbros at a fast-spreading center?* 121
    W. Roger Buck

11. *Evolution of gabbroic rocks of the Northern Apennine ophiolites (Italy): Comparison with the lower oceanic crust from modern slow-spreading ridges* 129
    Riccardo Tribuzio, Massimo Tiepolo, and Riccardo Vannucci

12. *New insights concerning the influence of water during the formation of podiform chromitite* 139
    Stephen J. Edwards, Julian A. Pearce, and Jonathan Freeman

13. *Gauging stress from mantle chromitite pods in the Oman ophiolite* 149
    Benjamin Holtzman

Section III. Structure and Physical Properties of Upper Oceanic Crust: Introduction 159

14. *Dependence of crustal accretion and ridge-axis topography on spreading rate, mantle temperature, and hydrothermal cooling* 161
    Yongshun John Chen

15. *Comparison of volcanic construction in the Troodos ophiolite and oceanic crust using paleomagnetic inclinations from Cyprus Crustal Study Project (CCSP) CY-1 and CY-1A and Ocean Drilling Program (ODP) 504B drill cores* 181
Hans Schouten and Charles R. Denham

16. *On the boundary between seismic layers 2 and 3: A stress change?* 195
Philippe A. Pezard

17. *Characterization of the upper oceanic crust using high-resolution seismic amplitude modeling* 203
Yue-Feng Sun and David Goldberg

Section IV. Hydrothermal Processes: Introduction 211

18. *A synthesis of geological and geochemical investigations of the TAG hydrothermal field: Insights into fluid-flow and mixing processes in a hydrothermal system* 213
Susan E. Humphris and Margaret K. Tivey

19. *Volatiles in mid-ocean ridge environments* 237
Deborah S. Kelley and Gretchen L. Früh-Green

20. *Lower-crustal cracking front at fast-spreading ridges: Evidence from the East Pacific Rise and the Oman ophiolite* 261
Craig E. Manning, Christopher J. MacLeod, and Patricia E. Weston

21. *Hydrothermal alteration and fluid fluxes in ophiolites and oceanic crust* 273
Jeffrey C. Alt and Damon A.H. Teagle

22. *Hydrothermal alteration patterns in supra-subduction zone ophiolites* 283
Kathryn M. Gillis and Neil R. Banerjee

Section V. Pacific Rim Ophiolites: Introduction 299

23. *Macquarie Island: Its geology, structural history, and the timing and tectonic setting of its N-MORB to E-MORB magmatism* 301
Rick Varne, Anthony V. Brown, and Trevor Falloon

24. *The Timor ophiolite, Indonesia: Model or myth?* 321
Ron Harris and Timothy Long

25. *Proterozoic–early Paleozoic ophiolites of the Andean basement of southern South America* 331
Victor A. Ramos, Mónica Escayola, Diana I. Mutti, and Graciela I. Vujovich

26. *Prospecting for ophiolites along the California continental margin* 351
Robert G. Coleman

27. *Mesozoic assimilation of oceanic crust and island arc into the North American continental margin in California and Nevada: Insights from geophysical data* 365
Nicola J. Godfrey and Yildirim Dilek

28. *Tectonostratigraphic significance of sedimentary strata occurring within and above the Coast Range ophiolite (California Coast Ranges) and the Josephine ophiolite (Klamath Mountains), northwestern California* 383
Emile A. Pessagno, Jr., Donna M. Hull, and Clifford A. Hopson

29. *Models for origin and emplacement of Jurassic ophiolites of northern California* 395
Raymond V. Ingersoll

30. Geology and geochemistry of the ophiolitic Trinity terrane, California: Evidence of middle Paleozoic depleted supra-subduction zone magmatism in a proto-arc setting 403
Rodney V. Metcalf, E. Timothy Wallin, Keith R. Willse, and Edward R. Muller

31. Sulfide minerals as an indicator for petrogenesis and serpentinization of peridotites: An example from the Hayama-Mineoka belt, central Japan 419
Hiroshi Sato and Yujiro Ogawa

Section VI. Ophiolites from the Iapetus, Rheic-Pleionic, Neotethyan, and Indian Oceans: Introduction 431

32. Contrasting mantle sequences exposed in the Lewis Hills massif: Evidence for the early, arc-related history of the Bay of Islands ophiolite 433
Günter Suhr and Stephen John Edwards

33. Petrology, tectonics, and hydrothermal alteration of a fossil backarc oceanic crust: Solund-Stavfjord ophiolite complex of the western Norwegian Caledonides—A review 443
Harald Furnes, Kjell Petter Skjerlie, and Yildirim Dilek

34. Early Devonian boninites from the Magnitogorsk arc, southern Urals (Russia): Implications for early development of a collisional orogen 461
Piera Spadea and Jane H. Scarrow

35. Constraints on the formation and emplacement age of western Greek ophiolites (Vourinos, Pindos, and Othris) inferred from deformation structures in peridotites 473
Annie Rassios and Alan G. Smith

36. Petrology and geochemistry of the Shahr-Babak ophiolite, central Iran 485
A. Mohamad Ghazi and A.A. Hassanipak

37. Aswad massif (United Arab Emirates): Archetype of the Oman-UAE ophiolite belt 499
Adolphe Nicolas, Françoise Boudier, K. Michibayashi, and L. Gerbert-Gaillard

38. A new structural profile along the Muscat-Ibra transect, Oman: Implications for emplacement of the Samail ophiolite 513
David R. Gray, Robert T. Gregory, and John McL. Miller

39. Formation and evolution of the western Indian Ocean as evidenced by the Masirah ophiolite: A review 525
Tjerk Peters

Index 537

Acknowledgments

We are indebted to the following reviewers for their timely and thorough reviews of the manuscripts; their efforts greatly improved the content of the volume and are gratefully acknowledged.

Jeffrey C. Alt, *University of Michigan*
Asish R. Basu, *University of Rochester*
Keir Becker, *University of Miami*
Jean H. Bédard, *Geological Survey of Canada, Québec*
Ron F. Berry, *University of Tasmania, Australia*
Donna Blackman, *Scripps Institution of Oceanography*
Sherman Bloomer, *Oregon State University*
Françoise Boudier, *Université Montpellier II, France*
C. Kent Brooks, *University of Copenhagen, Denmark*
Pamela C. Burnley, *Georgia State University*
David A. Butterfield, *National Oceanic and Atmospheric Administration/Pacific Marine Environmental Laboratory, Seattle*
Mathilde Cannat, *Université Pierre et Marie Curie, France*
Yongshun J. Chen, *Oregon State University*
Millard Coffin, *University of Texas, Austin*
Robert G. Coleman, *Stanford University*
Brian Cousens, *Carleton University, Canada*
Darrel S. Cowan, *University of Washington*
Anthony J. Crawford, *University of Tasmania, Australia*
Susan M. DeBari, *Western Washington University*
William R. Dickinson, *University of Arizona*
Yildirim Dilek, *Miami University*
Robert Duncan, *Oregon State University*
Stephen J. Edwards, *University of Greenwich, United Kingdom*
Don Elthon, *University of Houston*
W. Gary Ernst, *Stanford University*
Russell C. Evarts, *U.S. Geological Survey, Menlo Park*
Martin F.J. Flower, *University of Illinois, Chicago*
Patricia Fryer, *University of Hawaii*
Jeff Gee, *Scripps Institution of Oceanography*
Mario J. Giaramita, *California State University, Stanislaus*
Kathryn M. Gillis, *University of Victoria, Canada*
Robert T. Gregory, *Southern Methodist University*
Bradley Hacker, *University of California, Santa Barbara*
Robert Hall, *Royal Holloway London University, United Kingdom*
Gregory D. Harper, *State University of New York, Albany*
Ron Harris, *Brigham Young University*
James W. Hawkins, *Scripps Institution of Oceanography*
Clifford A. Hopson, *University of California, Santa Barbara*
Susan E. Humphris, *Woods Hole Oceanographic Institution*
Stephen Hurst, *University of Illinois, Urbana*
Raymond V. Ingersoll, *University of California, Los Angeles*
Rebecca Jamieson, *Dalhousie University, Canada*
David Jousselin, *University of Oregon*
Thierry Juteau, *Université de Bretagne Occidentale, France*
Suzanne M. Kay, *Cornell University*
Pavel Kepezhinskas, *University of South Florida*
Yves Lagabrielle, *Université de Bretagne Occidentale, France*
Craig E. Manning, *University of California, Los Angeles*
Martin A. Menzies, *Royal Holloway London University, United Kingdom*
Rodney V. Metcalf, *University of Nevada, Las Vegas*
Catherine Mével, *Université Pierre et Marie Curie, France*
Peter J. Michael, *University of Tulsa*
Kate C. Miller, *University of Texas, El Paso*
Rachel Mills, *University of Southampton, United Kingdom*
Eldridge Moores, *University of California, Davis*
Karlis Muehlenbachs, *University of Alberta, Canada*
H. Richard Naslund, *State University of New York, Binghamton*
Adolphe Nicolas, *Université Montpellier II, France*
Simon Peacock, *Arizona State University*
Michael R. Perfit, *University of Florida*
Tjerk Peters, *Universität Bern, Switzerland*
Hazel Prichard, *University of Cardiff, United Kingdom*
Stephen Roberts, *University of Southampton, United Kingdom*
Paul T. Robinson, *Dalhousie University, Canada*
Kent Ross, *University of Houston*
Matthew H. Salisbury, *Geological Survey of Canada, Nova Scotia*
David Sandwell, *Scripps Institution of Oceanography*
Peter Schiffman, *University of California, Davis*
John W. Shervais, *University of South Carolina*
Steven B. Shirey, *Carnegie Institution of Washington*
Norman Sleep, *Stanford University*
Jonathan E. Snow, *Max-Planck Institute für Chemie, Germany*
John G. Spray, *University of New Brunswick, Canada*
Debra S. Stakes, *Monterey Bay Aquarium Research Institute, Moss Landing, California*
Hubert Staudigel, *Scripps Institution of Oceanography*
Joann M. Stock, *California Institute of Technology*
Christopher A. Suczek, *Western Washington University*
Rupert Sutherland, *Institute of Geological & Nuclear Sciences, New Zealand*
Stephen Swift, *Woods Hole Oceanographic Institution*
Damon A.H. Teagle, *University of Southampton, United Kingdom*
George Thompson, *Stanford University*
Peter Thy, *University of California, Davis*
David Vanko, *Georgia State University*
Robert J. Varga, *Wooster College*
John Wakabayashi, *Hayward, California*
Richard J. Walker, *University of Maryland*
Elisabeth Widom, *Miami University*
Roy H. Wilkens, *University of Hawaii*
Richard Wilson, *University of Aarhus, Denmark*
Mei-Fu Zhou, *University of Hong Kong, China*
Robert Zierenberg, *University of California, Davis*

Preface

This volume contains 39 papers resulting from the Penrose Conference on Ophiolites and Oceanic Crust: New Insights from Field Studies and the Ocean Drilling Program, which was convened September 13–17, 1998, in Marshall, California. The meeting brought together 86 earth scientists with a diverse range of scientific backgrounds that include structural geology, tectonics, geophysics, petrology, and geochemistry. It was a lively and engaging meeting, with numerous active debates on both the current state of knowledge and future directions for innovative research. We hope that the vigor, quality, and depth of these discussions are apparent in the papers of this volume.

Ophiolites have been of particular importance in the reconstruction of ancient plate boundaries ever since their recognition as on-land fragments of oceanic lithosphere. Furthermore, the internal architecture of well-preserved ophiolite complexes shows that ophiolites are good structural analogues for oceanic crust. Such complexes provide three-dimensional exposures and age relationships that can be used to study the nature of extensional tectonics and magmatic construction in oceanic spreading environments. It is clear that properly interpreting the paleotectonic environment in which an ophiolite formed, as well as the magmatic and tectonic processes involved, is much more complicated than was once widely believed. Likewise, it is also clear that, in any given tectonic environment, the oceanic crust exhibits compositional heterogeneities that are far greater than earth scientists realized only a few years also.

Panel discussions at the conference and comparative studies of magmatic and structural processes operative during formation of modern oceanic crust and ophiolites, as presented in the papers of this volume, indicate that the original Penrose Conference definition of *ophiolite* has been very useful and remains effective in ophiolite and oceanic-crust comparisons, as long as the term is used independently of its origin of formation and/or tectonic significance. The Penrose Conference definition needs to be expanded, however, to include more information about the geologic context of individual complexes as revealed in the underlying and overlying rock units.

The scientific understanding of both ophiolites and oceanic crust has undergone a remarkable transformation since the original Penrose Conference on Ophiolites, held in 1972. This transformation has been made possible by the continuous interactions among the scientific communities studying ophiolites and marine geology and geophysics, by many detailed studies of well-developed ophiolites around the world, by enhanced understanding of the tectonic implications of ophiolites, and by greatly improved sampling, observation, and exploration of the modern seafloor. It is within this context that we organized the 1998 conference to bring together a multidisciplinary group of geoscientists with backgrounds in ophiolites and/or marine geology and geophysics. The expressed aims of the conference were to reevaluate existing models of oceanic-crust generation and ophiolite formation, to reassess the ophiolite–oceanic-crust analogy, to attempt to reach a new consensus on the architecture of oceanic lithosphere, and to discuss the significance of ophiolites and oceanic crust in plate-tectonic processes.

Many of the most important scientific questions regarding the formation and evolution of ophiolites and oceanic crust can, in our view, be most suitably addressed by the mutual integration of data, observations, and models from both environments. In the ocean basins, questions regarding processes of mantle flow, mantle melting, and melt segregation are currently being addressed by large integrated geophysical investigations (e.g., MELT, or Mantle Electromagnetic and Tomography Experiment), U-series isotope studies in erupted basalts, petrologic studies of abyssal peridotites, and various theoretical and experimental approaches. Similar studies of ophiolite complexes are equally important in providing key structural and petrologic information and context for these oceanographic studies. Although recent successful deep drilling of the oceanic crust and upper mantle at the Hess Deep (Ocean Drilling Program Leg 147), the Mid-Atlantic Ridge (Leg 153), and the Southwest Indian Ridge (Legs 118 and 176) has allowed more rigorous assessment of the formation of the plutonic foundation of oceanic crust, studies of ophiolites and mafic layered intrusions are crucial to placing studies of these oceanographic suites in context. The forte of on-land ophiolite studies is the capability for detailed sampling and observation, the ability to map field relationships in four dimensions, and the ability to place the oceanic rocks in a detailed stratigraphic and structural context generally unobtainable from marine studies.

It is also clear that the most productive marine geology and geophysics field programs are those that represent truly multidisciplinary studies, in which geophysics, petrology, structural geology, tectonics, geochemistry, and other disciplines are integrated to address an important scientific problem. The marine geology and geophysics community has developed, on an international scale, a scientific culture in which important problems are identified through extensive community planning, and large, multidisciplinary groups of investigators are typically assembled to address these problems. The scale of field investigations of modern oceanic lithosphere is typically much grander than studies of ophiolites in terms of the number of scientists involved, the breadth of their disciplines, and the level of funding. Although generally more modest in the level of funding, and more individually carried out, studies of many ophiolites have also been multiyear, multinational, multidisciplinary efforts that have been highly productive in shedding light on the nature of individual complexes in particular and their significance for ocean-lithosphere formation in general.

It is breathtaking to contemplate how much progress has been made in understanding ophiolites and the oceanic crust in the past 25 years since the first Penrose Field Conference on Ophiolites. Since 1972, the profound importance of hydrothermal activity in the formation and evolution of the oceanic crust has been recognized, as has the presence of exemplary correlative hydrothermal systems in ophiolites. Geophysical, structural, and petrologic studies of ophiolites and oceanic crust have led to the development of an entirely new generation of geologic models for the formation of oceanic crust. The counterparts to the structural and tectonic characteristics of slow-, intermediate-, and fast-spreading ridges in the oceans have been found in ophiolite complexes. The petrologic and tectonic segmentation of mid-ocean ridges, at a variety of scales, has been established, and representatives of these features have been discovered in ophiolites. The use of mid-oceanic ridge basalts to probe mantle processes at a local and global scale has produced a remarkable improvement in understanding mantle melting, history, and structure. The incomplete and disrupted nature of oceanic crust produced at slow-spreading ridges has become well established, and counterparts have been found in ophiolites. The variability of internal structures and geochemical characteristics of ophiolites has become separable into features related to their formation at spreading centers and to subsequent emplacement-related tectonics.

This volume contains many contributions to these and other related topics of inquiry. We have organized the volume thematically. There are sections on ophiolites, oceanic crust, and global tectonics; oceanic lower crust and upper mantle; structure and physical properties of upper oceanic crust; hydrothermal processes; Pacific Rim ophiolites; and ophiolites from the Iapetus, Rheic-Pleionic, Neotethyan, and Indian Oceans. In creating this structure, we hope that the reader will find the information accessible and informative and will easily get a good grasp of the state of present knowledge and where some future directions of research might lie.

Progress in the next 25 years could even surpass the progress of the past 25. Deep drilling in the oceanic lithosphere may be successful and provide a wealth of petrological, geochemical, and structural constraints on how the oceanic crust evolves in terms of magmatic, structural, and hydrothermal alteration processes. Advances in geochemical instrumentation and computational facilities could produce their own revolutionary changes in the development and refinement of sophisticated models of the crust and upper-mantle evolution. Deep-ocean observatories may well lead to key time-series studies that define the processes of ocean-floor evolution with unprecedented clarity. Simultaneously, on-land multidisciplinary studies of ophiolites and their tectonic settings should shed more light upon both the nature of the emplacement mechanism and the changes in oceanic crustal structure and composition through time. More widespread application of satellite imagery should significantly enhance our knowledge of less well-known complexes. We hope that we can use these combined oceanic and on-land results to begin to refine our extrapolation of oceanic processes backward in time to the 96% of Earth history not reflected in the modern oceans.

We thank the meeting participants for the synergy and excitement they brought to the meeting and for sharing their knowledge and observations with each other, and we are grateful to the contributors to this volume for their time and effort expended. We also thank Lois Elms of the Western Experience, whose indefatigable efforts in organizing the logistics of the conference contributed so much to its success. We extend our sincere appreciation to Abhijit Basu, Geological Society of America Books Science Editor, for his enthusiastic support and suggestions that helped to bring about and to enhance the volume, and to the many reviewers of manuscripts that led to their improvement. We are particularly indebted to Rebecca Herr, Maria Mullins, and the GSA Publications Office staff for their invaluable assistance in copy editing, formatting, and finalizing this book. We gratefully acknowledge the grants from the Marine Geology and Geophysics Program of the National Science Foundation, the Joint Oceanographic Institutions–U.S. Science Advisory Committee, and the Geological Society of America to support travel by conference participants and the preparation of this volume.

Yildirim Dilek, *Miami University*
Eldridge Moores, *University of California, Davis*
Don Elthon, *University of Houston*
Adolphe Nicolas, *Université Montpellier II*

Section I
Ophiolites, Oceanic Crust, and Global Tectonics: Introduction

Three fifths of the surface of the solid Earth is made of oceanic hosphere, all of which has formed at oceanic spreading centers during e past 170 m.y. Hence understanding the structure and evolution of eanic lithosphere and mid-ocean ridges is particularly significant to tter understand the mantle processes and the kinematics of plate tec- nics. Ophiolites—which contain crustal and mantle rocks that are ry similar to the rocks sampled by drilling and dredging in the ocean sins—display the only evidence for structural, magmatic, and hydro- ermal processes associated with seafloor spreading prior to 170 Ma. nce the advent of plate tectonics and establishment of the concept of origin of ophiolites at oceanic spreading centers, ophiolites have ayed a major role in deciphering the paleogeography of ancient ocean sins, the tectonic evolution of orogenic belts, and the nature of mid- ean ridge processes. The crustal and mantle rocks within ophiolites, eir overlying sedimentary cover, and the commonly underlying tec- nic basement and/or mélange complexes collectively record essential formation about structural, magmatic, and metamorphic processes as- ciated with the formation of oceanic crust at spreading centers and ctonic displacement and emplacement of this fossil crust into conti- ntal margins as part of active-margin tectonics. Thus the ophiolites d associated rock units preserve evidence for the first-order global ctonic events in Earth's history. This section includes papers that dress some of the global tectonic aspects of the ophiolite–oceanic- ust analogy and evolution of ophiolites.

It has been recognized since the late 1970s that many ophiolites ve petrologic and geochemical characteristics that are unlike mod- n mid-ocean ridge crust and upper mantle. Because of these char- teristics, ophiolites increasingly have been linked to supra-subduc- n zone environments ranging from nascent arcs and forearcs to ra-arc or backarc basins. In this book, the papers by Moores et al., urm et al., and Wakabayashi and Dilek examine the enigmatic con- pt of the supra-subduction zone origin of ophiolites from different rspectives and provide new insights and interpretations on the com- sitional variability in oceanic crust, as well as on the tectonic sig- ficance of ophiolites and their emplacement. The kinematics of con- ental rifting and mantle response to this rifting event may strongly ntrol the mode of seafloor spreading and the generation of oceanic hosphere, as suggested in the study of the Ligurian ophiolites by mpone and Piccardo. Documentation of the formation of serpen- e mud volcanoes in the Mariana forearc by Fryer et al. is also portant from the global tectonic perspective for better understand- g fluid flux in convergent-margin settings.

Moores et al. consider at a global scale the ophiolite conundrum splayed by the apparent conflict between geochemical-petrologic and uctural evidence from the Tethyan ophiolites and suggest that the emical compositions of these ophiolites may reflect derivation from eviously subducted mantle, rather than a supra-subduction zone ori- n at the time of their igneous formation. They argue that mantle- urce regions are heterogeneous at all scales and that the composition the mantle tapped by oceanic spreading centers is nonuniform, in rt as a function of the Earth's tectonic history prior to melting. Thus e history of subduction in and around any oceanic basin and the nount of involvement of previously subducted lithosphere in mantle elting beneath spreading centers might have strongly controlled the agma compositions of so-called "supra-subduction zone ophiolites."

Sturm et al. cite the recent discovery of supra-subduction zone geochemical signatures in basalts from the southern Chile Ridge spreading center as evidence that the distinctions between mid-ocean ridge and supra-subduction zone environments are not as clear and well defined as once believed. They propose that the existence of supra-subduction-zone-like geochemical signatures in a particular ophiolite complex does not eliminate the possibility that the ophiolite actually formed at a mid-ocean ridge. Particularly for those ophiolites in which the overall geochemical and geologic character is similar to that of mid-ocean ridge environments, but which have a subtle supra-subduction-zone-like signature, Sturm et al. suggest that caution is warranted in interpreting their origin.

Rampone and Piccardo review the recent petrologic and isotopic data from the Ligurian ophiolites in the Northern Apennines (Italy) and report that the atypical oceanic mantle character of these ophiolites is a result of their primary association with the subcontinental lithospheric mantle of Hercynian origin. Rampone and Piccardo argue that the mantle section of the ophiolites was developed through a passive extension of the continental lithosphere during which the subcontinental mantle (Proterozoic age?) underwent tectonic exhumation and uplift, analogous to the history of mantle peridotites exposed in the Red Sea (Zabargad) and the Galicia margin of Iberia. They show that the Jurassic extrusive rocks of mid-oceanic-ridge basalt (MORB) type in the Ligurian ophiolites are not linked to the peridotites by a genetic melt-residua relationship. This study shows that the Ligurian ophiolites represent the remnants of a Red Sea–type embryonic ocean basin, rather than a typical mid-oceanic ridge of a mature ocean basin (e.g., Atlantic Ocean).

Fryer et al. describe the occurrence of serpentine mud volcanoes in the Mariana forearc and compare their observations to on-land serpentinite bodies exposed in ancient convergent-margin settings. By using high-resolution side-scan sonar imagery, Fryer et al. investigate the spatial and temporal relationships between serpentinite protrusions and faults and conclude that episodic mud volcanism in the Mariana forearc is facilitated by deep-seated faults, along which dehydration-generated fluids from the downgoing plate and remobilized fault-gouge material rise to form large submarine serpentine mud seamounts. Blueschist clasts found in serpentine mud suggest derivation of material from as deep as 30 km. Some of the sedimentary serpentinite deposits in the northern California Coast Ranges and the Koryak fold-and-thrust belt in Kamchatka may be the remnants of such serpentine mud volcanoes developed in the forearc setting of former subduction zones.

Wakabayashi and Dilek examine the spatial and temporal relationships between ophiolites and their metamorphic soles from several well-studied examples around the world in order to test the models of forearc ophiolite genesis. They discuss the timing and the mechanism(s) of formation of metamorphic soles beneath the ophiolites and state that most metamorphic soles form at the inception of a subduction zone and that they are hence slightly younger than their ophiolitic counterparts. Wakabayashi and Dilek show that the ophiolite-sole pairs of inferred forearc origin consistently have older ophiolites over younger soles, incompatible with the idea of the generation of these ophiolites in a forearc setting after the initiation of subduction beneath them. They thus conclude that the apparent forearc setting of the ophiolites in question may be an artifact of their emplacement tectonics, rather than their igneous environment of formation.

Geological Society of America
Special Paper 349
2000

Tethyan ophiolites, mantle convection, and tectonic "historical contingency": A resolution of the "ophiolite conundrum"

Eldridge M. Moores and Louise H. Kellogg
Department of Geology, University of California, Davis, California 95616, USA
Yildirim Dilek
Department of Geology, Miami University, Oxford, Ohio 45056, USA

ABSTRACT

The "ophiolite conundrum" describes the conflict in many ophiolite complexes between (1) structural and stratigraphic evidence for seafloor spreading in a non-island-arc environment and (2) geochemical evidence for derivation of magmas from highly depleted mantle similar to that found at present over subduction zones ("supra-subduction zone" settings). Tethyan ophiolites provide many excellent examples of the conundrum. These complexes crop out in four major belts from the western Mediterranean to the Himalaya. Many bodies display a complete ophiolite sequence (e.g., Troodos, Cyprus, and Semail, Oman), whereas others display only a partial sequence, characteristically with lavas and/or sedimentary rocks overlying serpentinized peridotite (e.g., western Mediterranean complexes, parts of Kizildag, Turkey). Chemical compositions of the lavas in these ophiolites range from standard E-MORBs (enriched mid-oceanic-ridge basalts) to highly depleted magmas (e.g., low in Nb, Ta, Ti, and some other trace elements), reminiscent of modern volcanic-arc and backarc basin (supra-subduction zone) settings. Geologic or stratigraphic evidence for any extensive arc edifice is generally lacking, and most complexes are overlain by pelagic sedimentary rocks.

Modern mid-ocean ridges show compositional variations reflecting incomplete mixing of mantle components. Some mid-oceanic ridge lavas display trace element signatures (i.e., incompatible element enrichment) unlike those of MORB affinity but reminiscent of the supra-subduction zone field. These data and a new model for a two-layer mantle of depleted material in the upper 1500 km and denser, enriched mantle below suggest a resolution of the ophiolite conundrum. In this model, the composition of mantle tapped by oceanic spreading centers for the past 1100 m.y. is nonuniform or "historically contingent," i.e., it depends upon the prior history of spreading and subduction. Spreading-center magma compositions may have varied just as the patterns of continental assembly and fragmentation have altered the tectonic arrangement of Earth's surface.

INTRODUCTION

For about a century since the pioneering work of Steinmann (1906, 1927), ophiolite complexes have been recognized as major components in orogenic belts (e.g., Hess, 1939, 1955). The recognition of ophiolite sequences as the products of spreading in an oceanic environment was one of the cardinal ingredients in the application of the "New Global Tectonics" to mountain belts some 25–30 years ago (e.g., Gass, 1968; Dewey and Bird, 1970; Coleman, 1971; Moores, 1970; Moores and Vine, 1971). The early 1970s also brought recognition that as no oceanic crust older than 200 m.y. is preserved in the oceans themselves, ophiolite complexes (and greenstone belts) preserve the only evidence available about oceanic spreading processes for the 96% of Earth history prior to 200 Ma. This recognition led to early attempts to relate oceanic crustal structure to that of ophiolites in order to make estimates of the spreading rates that formed an individual ophiolite (e.g., Moores and Jackson, 1974; Kidd and Cann, 1977).

The tectonic environment of many ophiolite complexes, as revealed in the rocks surrounding these complexes, provides important clues as to their environment of origin. On the basis of their tectonic environment, ophiolite complexes are separable into "Tethyan" and "Cordilleran" types (Moores, 1982). Tethyan-type ophiolite complexes tectonically overlie continental crust and shallow-water continental-margin or platform sediments, and they characteristically are in turn overlain by pelagic sedimentary rocks with little or no volcaniclastic material. Cordilleran-type complexes display no clear relationship with old continental basement or platform-miogeoclinal sedimentary rocks, are spatially and temporally associated with tectonic mélanges commonly containing high-grade metamorphic rocks, and commonly include arc-type volcaniclastic deposits in their extrusive sections. In some Cordilleran ophiolites, extrusive rocks, sheeted

Moores, E.M., Kellogg, L.H., and Dilek, Y., 2000, Tethyan ophiolites, mantle convection, and tectonic "historical contingency": A resolution of the "ophiolite conundrum," *in* Dilek, Y., Moores, E.M., Elthon, D., and Nicolas, A., eds., Ophiolites and Oceanic Crust: New Insights from Field Studies and the Ocean Drilling Program: Boulder, Colorado, Geological Society of America Special Paper 349, p. 3–12.

dikes, and hypabyssal intrusions have commonly been deposited on and intruded into an older and multiply deformed oceanic crust, suggesting a polygenetic evolutionary history of these island-arc–related ophiolites (i.e., Dilek et al., 1991; Saleeby, 1992).

Much discussion of the relationship between ophiolites and modern oceanic lithosphere has centered on possible modern analogues. Modern oceanic environments of lithosphere formation include fast-, intermediate-, and slow-spreading mid-ocean ridges, incipient rifts, such as the Red Sea or Gulf of California, backarc basins (e.g., Lau Basin), intra-arc rifts, and possibly incipient arc development (Bloomer et al., 1995). Dilek et al. (1998) reviewed the internal structures of oceanic crust from fast-, intermediate-, and slow-spreading environments and compared the findings with ophiolites in which the structural and lithologic features could be established confidently as having formed during spreading, rather than subsequently during emplacement. Fast-spreading ridges, exemplified by Hess Deep, exhibit a complete ophiolite sequence with minimal crustal attenuation, similar to that of the Semail complex. Slow-spreading ridges, exemplified by the MARK area (the intersection of the Mid-Atlantic Ridge and the Kane Fracture Zone), have incomplete, highly faulted sections, with surface exposure of serpentinite and gabbro, similar to ophiolites in the western Mediterranean (Alps and Apennines; Lagabrielle and Cannat, 1990). Intermediate-spreading ridges, exemplified by modern oceanic crust drilled in Ocean Drilling Program (ODP) Hole 504B south of the Costa Rica Rift, resemble sections of the Ordovician ophiolites of Norway. The obvious inference is that the Semail complex, the Norwegian ophiolites, and the western Mediterranean ophiolites formed at fast-, intermediate-, and slow-spreading ridges, respectively. If so, the structure of well-preserved ophiolite complexes can be used as an indicator of past spreading rates and would become a powerful tectonic tool.

Since Miyashiro (1973) pointed out the arc-like chemistry of the Troodos ophiolite, most geochemical and petrologic studies of ophiolites have centered around the chemical compositions of many well-developed examples, such as Troodos and Semail, that contain highly depleted, non-MORB lavas, crystallized from mantle that was already depleted of its MORB constituents. These compositions occur in the modern oceans chiefly in arc-related, backarc, or supra-subduction zone environments (e.g., Robinson et al., 1983; Pearce et al., 1984; Bloomer et al., 1995; Robinson and Malpas, 1990), and such environments have been invoked for the ophiolites as well. This geochemical and petrologic evidence stands in strong contrast with evidence—from both the structure within Tethyan ophiolite complexes and the paleogeographic environments inferred from surrounding and overlying sedimentary deposits—that these ophiolites originated well away from any type of subduction-related activity.

This seemingly irreconcilable conflict between geochemical-petrologic evidence and structural-stratigraphic evidence has led to the so-called ophiolite conundrum (e.g., Dilek and Moores, 1990). How can one reconcile these two different lines of evidence? What if anything can ophiolite complexes tell us about spreading processes and environments in the past? Can the structural differences from fast- to slow-spreading environments be applied to ophiolite complexes, or can the latter offer little light to shed on past spreading processes?

We present a new attempt at resolution of this conflict based upon recent information acquired from mid-oceanic spreading environments, from isotope evidence of mantle-mixing rates, from new interpretations of convection and structure in Earth's mantle, and from ophiolite complexes from the Tethyan region (Fig. 1). We propose a model wherein the nature of mantle tapped at mid-oceanic ridges has varied in the past in response to prior tectonic history of a region and/or the mantle. This view potentially resolves the conundrum and also leads to new possible insight about the structure in the deep mantle in Earth's past.

TETHYAN OPHIOLITES

These bodies are prime illustrations of features that have led to the conundrum. They crop out in four principal belts, the Jurassic Alpine-Apennine belts of the western Mediterranean (e.g., Lagabrielle and Cannat, 1990), the Jurassic Dinaride-Hellenide complexes of the western Balkan peninsula (e.g., Smith, 1993), the Jurassic–Cretaceous bodies of eastern Greece and central Turkey (Smith, 1993; Dilek and Moores, 1990), and the Late Cretaceous bodies of southern Turkey and Cyprus and extending around the "peri-Arab crescent" to Oman (e.g., Dilek and Thy, 1998; Dilek et al., 1998; Dilek and Moores, 1990; Peters et al., 1991; and references therein).

The Alpine-Apennine complexes include those in the Alps themselves, mostly highly deformed and metamorphosed, and the better-preserved complexes of the Apennines and Corsica (Lagabrielle and Cannat, 1990). These complexes generally have MORB compositions, but their sections are incomplete. The least deformed complexes charactistically have tectonite peridotites overlain unconformably by breccias (ophicalcites), pillow lavas, and pelagic sediments. Plutonic mafic rocks are rare, and sheeted dikes are virtually absent. These complexes have been ascribed to formation in oblique, slow-spreading Red Sea–like openings in continental crust that was consolidated in the 300 Ma Hercynian orogeny (the eastward continuation of the southern Appalachians).

The western Balkan complexes include those of Albania (e.g., Shallo et al., 1990), Bosnia-Herzegovina, and Greece (Smith, 1993). These ophiolites generally exhibit a complete sequence, from harzburgite and/or lherzolite tectonite at the base to overlying cumulate mafic-ultramafic rocks, sheeted dikes, and pillow lavas (Smith, 1993; Jackson et al., 1975; Rassios et al., 1983). Some complexes in the eastern Balkans, however, display less complete sequences in which mantle tectonites and sheeted dikes are lacking. These complexes range from MORB-like in the west to depleted in the east, but the compositional transition occurs within a few tens of kilometers across strike (Bortolotti et al., 1996). These complexes could have formed by spreading above a southwest-dipping subduction zone (Smith, 1993) or by narrow opening of a continent previously consolidated during the Hercynian orogeny (e.g., Dewey et al., 1973).

In central Anatolia, Jurassic–Cretaceous ophiolites crop out as dismembered bodies in the Izmir-Ankara-Erzincan suture zone, an east-trending belt of mélange extending some 1000 km in northern Turkey (Tankut, 1990). Dismembered ophiolitic rocks are associated with both MORB- and arc-like dikes, lavas, and volcaniclastic rocks and attest to a complex oceanic paleogeography of the presuture ocean basin (Tankut et al., 1998). The chemistry of some dike intrusions within the gabbros and serpentinized peridotites of dismembered ophiolites in the Ankara mélange shows subalkalic characteristics with incompatible element abundances characteristic of N-MORB affinity (i.e., normal MORB). Some dike intrusions, on the other hand, display typical island-arc tholeiite signatures with large ion lithophile element (LILE) enrichment and high field strength element (HFSE) depletion compared to N-MORB (Tankut et al., 1998). On the basis of these chemical features and crosscutting relationships of the dike intrusions, Tankut et al. (1998) suggested that the remnants of the Neotethyan oceanic lithosphere preserved in the Ankara mélange developed at a mid-ocean ridge (MOR) system within the northern branch of the Neotethys and that this MOR-generated oceanic lithosphere was subsequently placed in a supra-subduction zone setting during regional oblique convergence. Intrusions of dikes with

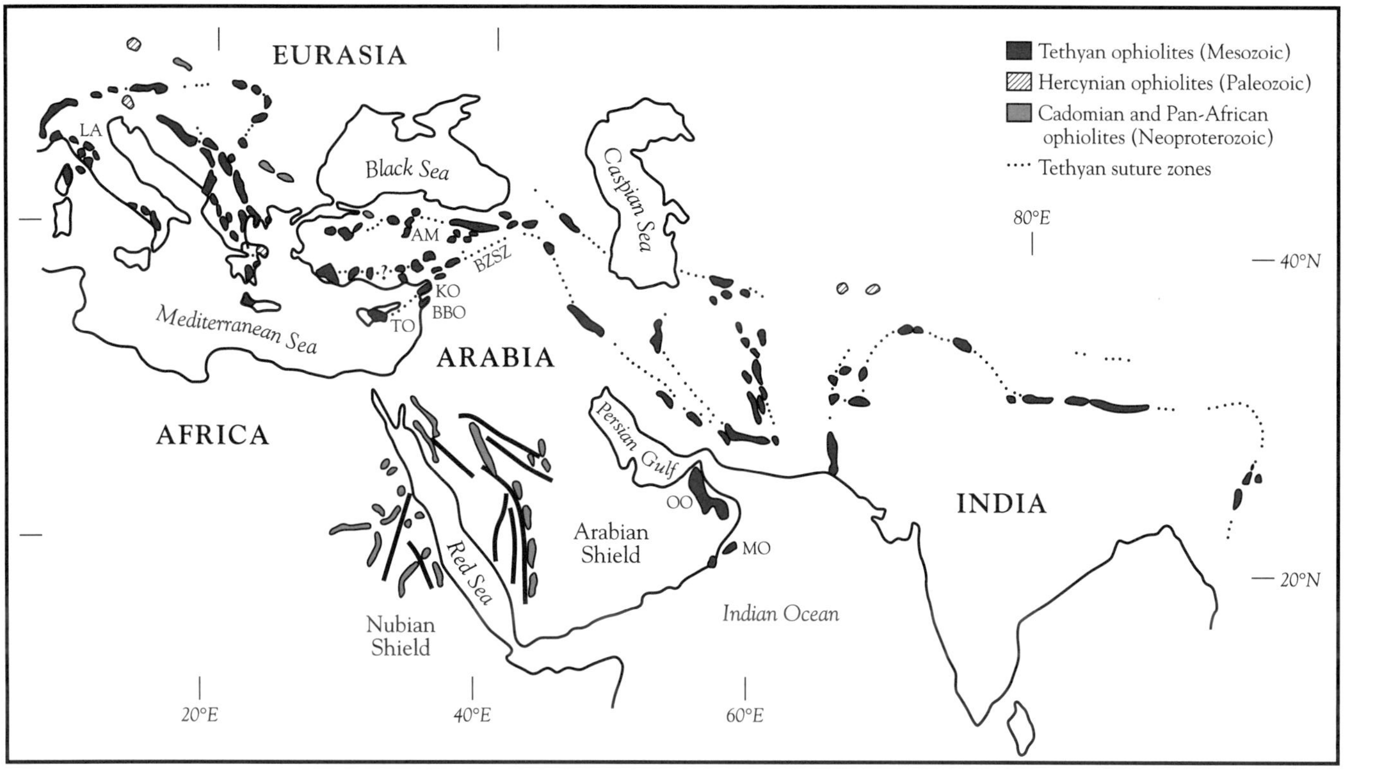

Figure 1. Distribution of major Neotethyan, Hercynian, and Cadomian–Pan-African ophiolites (data are from Dilek and Moores, 1990; Lippard et al., 1986; Stern et al., 1990). LA—Ligurian ophiolites, AM—Ankara mélange, TO—Troodos ophiolite, BBO—Baër-Bassit ophiolite, KO—Kizildag ophiolite, BZSZ—Bitlis-Zagros suture zone, OO—Oman ophiolite, MO—Masirah ophiolite.

an island-arc signature were thus a result of the subduction-related magmatic plumbing that crosscut the Neotethyan oceanic lithosphere in the upper plate.

The Late Cretaceous ophiolites of southern Turkey, Cyprus, and farther east display many fairly complete sequences (e.g., Dilek and Moores, 1990; Dilek and Thy, 1998; Dilek et al., 1999). These complexes mostly tectonically overlie microcontinental or continental blocks of African-Arabian (Pan-African; Dewey et al., 1973; Lippard et al., 1986) affinity. They formed at spreading centers in early Late Cretaceous time, and most were emplaced some 2 m.y. later over continental blocks (with the notable exception of the Troodos complex, which is being emplaced today). Compositional relationships indicate derivation of ophiolitic rocks from mantle sources that had undergone previous melt extraction (Dilek and Thy, 1998). These compositions are generally ascribed to a supra-subduction zone environment, and most tectonic models of this region have the complexes formed over a north-dipping subduction zone (e.g., Moores et al., 1984; Pearce et al., 1984; Robinson and Malpas, 1990). However, the internal structure and petrology of these bodies indicate a complex tectonic environment. Ophiolites in southern Turkey include serpentinized peridotites, cumulate and isotropic gabbros, and discontinuous exposures of thin extrusive rocks, all intruded by isolated mafic dike swarms (Dilek et al., 1999). The Troodos and Kizildag (Turkey) complexes display, on the other hand, a complete pseudostratigraphy and no isolated dike swarms, but their dikes and pillow lavas lack the Ta-Nb and LREE (light rare earth element) enrichment seen in the Tauride dike swarms (Turkey). These differences may be the result of variations in either the mantle-source composition or the nature and extent of melt extractions and subsequent metasomatism by LILE- and LREE-enriched fluids during formation of ophiolitic magmas. Major LREE pattern differences between the Kizildag and Troodos ophiolites, only 300 km apart, suggest that the Troodos oceanic crust originated from a more prolonged and complete melting compared to that which created Kizildag.

Dilek et al. (1999) argued that a hydrous component associated with a subduction zone is insufficient by itself to explain the evolution of supra-subduction zone ophiolites. A slight positive correlation between the extent of melting and LREE enrichment may have resulted from infiltration by fluids or melts derived from a more fertile source. There seems to be no compelling geochemical evidence for the operation of an active subduction zone during melting. Mixing of the depleted mantle and metasomatizing fluids might have occurred previously during the earlier episodes of orogenic events and associated subduction in the region (i.e., Hercynian, Paleotethys, and/or Pan-African; Dilek and Moores, 1990). The apparent "arc signature" of the dikes and extrusive rocks may have arisen from an asthenospheric window associated with interactions along the complex plate boundary, from preexisting, previously subducted, subcontinental lithosphere, from crustal contamination in a narrow ocean basin, or from a mantle wedge over a subducting slab.

In addition, the supra-subduction zone chemistry contrasts with the general lack of any arc-like volcanic deposits associated with these ophiolites. Virtually all complexes are overlain by pelagic sedimentary rocks. There is simply no evidence in the field for robust development of western Pacific–like island arcs in the Tauride region of southern and central Anatolia or in the Pelagonian platform in Greece. Rather, the regional structural data indicate spreading of small oceanic basins separated by several microcontinents derived from the northern margin of Africa-Arabia followed in quick succession by development of mantle-rooted thrust faults (nascent subduction zones) within these small oceans and subsequent emplacement of the ophiolites. Metamorphic soles beneath these ophiolites

TABLE 1. MANTLE DOMAINS AND POSSIBLE ORIGINS

Mantle type	Characteristics	Proposed origin	Possible mantle location
DMM	high $^{143}Nd/^{144}Nd$ (~0.5132–0.5133) low $^{87}Sr/^{86}Sr$ (~0.702) intermediate $^{3}He/^{4}He$ (~8 times atmospheric value) low $^{206}Pb/^{204}Pb$ (~17–18)	extraction of continents from primitive mantle?	upper mantle (0–2000 km?), deep-mantle graveyards? (Engebretson et al., 1992)
HMU	high $^{206}Pb/^{204}Pb$ (~21–22) relative to DMM: slightly lower $^{143}Nd/^{144}Nd$ (~0.5129) higher $^{87}Sr/^{86}Sr$ (~0.7030) lower $^{3}He/^{4}He$ (~5–6 times atmospheric value)	separation of Pb to core; subduction and metasomatism of suboceanic lithosphere	Same as for DMM
EM I	low $^{143}Nd/^{144}Nd$ (~0.5121–0.5123) intermediate $^{87}Sr/^{86}Sr$ (0.7050) high $^{3}He/^{4}He$ (20–30 times atmospheric value) low $^{206}Pb/^{204}Pb$ (~17.5)	recycled ancient pelagic sedimentary rocks or ancient subcontinental lithosphere or relatively undepleted mantle (Hofmann, 1997)	1500–2900 km (Kellogg et al., 1999)
EM II	low $^{143}Nd/^{144}Nd$ (~0.5120–0.5122) high $^{87}Sr/^{86}Sr$ (>0.707) low $^{3}He/^{4}He$ (~0) low $^{206}Pb/^{204}Pb$ (~19)	subducted and recycled ocean crust with some continent and/ or sedimentary rocks	throughout mantle; possibly concentrated in thermal boundary layers
FOZO			throughout mantle?

Note: Modified after Hart (1988) and Hofmann (1997).

display inverted metamorphic field gradients suggestive of their formation under high-temperature conditions at or near spreading axes (Hacker, 1990; Dilek et al., 1999). There are no extensive accretionary-prism complexes of subduction zone origin (i.e., like the Franciscan Complex) associated with the Tethyan ophiolites.

MANTLE VARIATIONS BENEATH MODERN OCEANS

The geological and geochemical data from the eastern Mediterranean region thus illustrate the "ophiolite conundrum." Regional geologic evidence that ophiolites formed in a series of small ocean basins without the presence of any island-arc sequences contrasts with the evidence from geochemistry that ophiolites were derived from mantle more depleted than that thought generally to represent the present oceans.

In contrast to conventional dogma, however, heterogeneous mantle is present within the modern oceans, as well, and it has given rise to magmatic rocks of different compositions, in addition to N-MORB and E-MORB at many localities within the modern oceans.

Hofmann (1997), Hart (1988), and Zindler and Hart (1986) have summarized the evidence for mantle heterogeneity from oceanic-island basalts. They recognized four types of mantle domains or isotopically defined end members: DMM, a depleted mantle component thought to represent the source of MORB lavas; HIMU, a mantle component with high U/Pb ratios (= μ); and EM I and EM II, two "enriched-mantle" types (Table 1).

Plotting isotope data for various oceanic islands produces an approximately linear array, converging on a focal zone (FOZO), which has been identified as a separate component by Hart et al. (1992). Table 1 summarizes the different mantle types, their possible origins and source regions.

These mantle isotope heterogeneities require time to develop. Zindler and Hart (1986) reported that kilometer-sized heterogeneities can survive in the mantle for billions of years, even in the presence of a melt. These long-surviving heterogeneities present a difficulty, because numerical models of mantle mixing indicate that convection may rapidly stir different components (Christensen, 1989, 1990; Davies, 1990; Kellogg, 1992; van Keken and Ballentine, 1998, and many others). Although most workers agree that the primary DMM source resides in the shallow mantle, a number of possible locations have been proposed for the HIMU, EM I and EM II mantle sources, including a mesospheric boundary layer at 670 km (Allègre and Turcotte, 1985) and mantle plumes possibly originating at the core-mantle boundary (Schilling, 1975). (An alternative view, presented by Anderson [1995], holds that the DMM source lies beneath a heterogeneous, shallow, enriched-mantle region.) Entrainment of different components and mixing of recycled material results in a marble-cake structure (Allègre and Turcotte, 1986) with the scales of the heterogeneities ranging from very small (10 m) to very large (>1000 km) (Zindler and Hart, 1986).

Additional evidence for multiple chemical components in the mantle comes from Earth's heat-flow budget (Stein, 1995), which requires that some part of the mantle has a higher rate of heat production than that of the DMM source (Richter, 1985). This geochemical evidence suggesting that the mantle is poorly mixed, combined with rapid mixing (or stirring) seen in numerical models of whole-mantle convection, contrasts with recent high-resolution seismological models showing subducted slabs penetrating into the lower mantle (e.g., Fukao et al., 1992; van der Hilst et al., 1997; Kennett et al., 1998). Recently, Kellogg et al. (1999) have suggested a model, discussed later in this paper, that reconciles the heat-flow, petrological, and geochemical evidence for an undepleted mantle source with seismic evidence for deep penetration of mantle slabs into the lower mantle (Fig. 2). In their model, a reservoir of essentially undepleted or possibly enriched mantle extends from ~1500 km depth to the core-mantle boundary. This material is slightly denser than the depleted mantle above, so it can persist without mixing for billions of years, accounting for the incomplete degassing of Earth's mantle through time (Allègre et al., 1983). The dense layer must be enriched in heat-producing elements to account for Earth's overall heat budget. Because this layer would be hot and not much more intrinsically dense than the overlying depleted mantle, the interface between the dense layer and the overlying mantle would be highly irregular and would vary with time, depending on the amount and age of subducted material piled on top of it. The interface in regions where a large amount of lithosphere had been subducted relatively recently would be depressed (the "mantle graveyards" of Engebretson et al., 1992). Such regions would alternate with regions having a raised interface over dense material that had been displaced

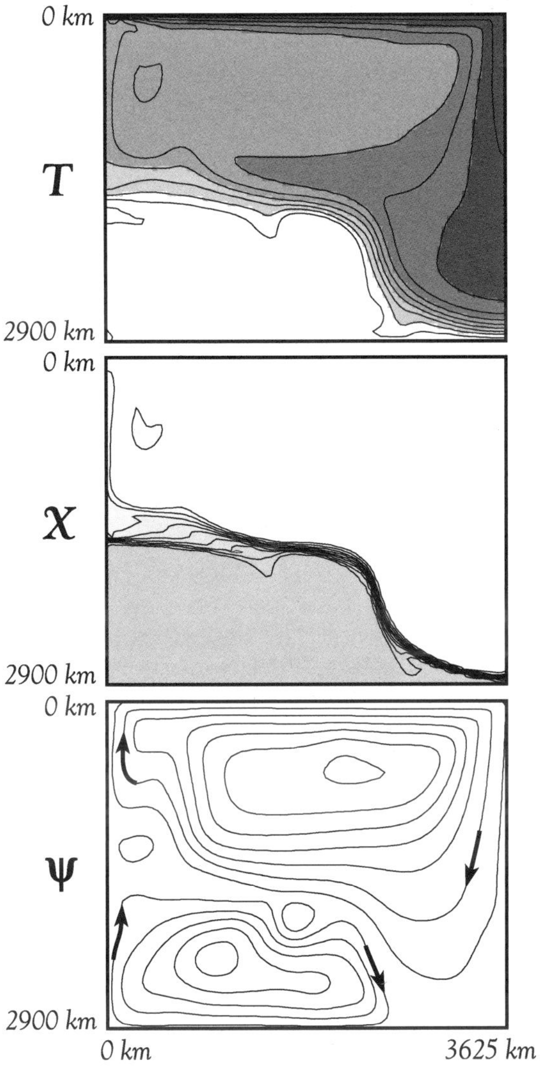

Figure 2. Cross section of a single time "snapshot" of a computer model of mantle convection. A layer of dense material enriched in heat-producing elements is in the lower 1000 km of the mantle. Each box represents the mantle from the surface (0 km depth) to the core-mantle boundary (2900 km depth). The top cross section shows temperature, with white representing hot and gray cooler mantle. The cold boundary layer descends along the right side. The middle cross section shows composition; shaded material is from the deep layer. Contours of temperature and composition are at intervals of 0.1 on an arbitrary scale. The bottom frame shows streamlines. The large cells in the upper and lower layer both move clockwise.

to form domes from which would originate plumes to produce hotspots or possibly "infant Pacific" or Easter Island microplates. This model also explains recent seismic evidence for compositional heterogeneity in the deep mantle (below about 1500 km) (Kennett et al., 1998; van der Hilst and Karason, 1999).

Thus geochemical and theoretical considerations of mantle composition and structures point to a complex process of changes in the internal architecture of mantle through time. Depleted and undepleted mantle components occupy different regions at various times, depending on the prior history of surface plate activity, the internal mantle convective patterns, and also, no doubt, the interactions between the lower mantle and the core. Numerical models of convection in a two-layer mantle with elevated internal heating in an intrinsically dense lower layer show complex, time-dependent flow patterns with thousands of kilometers of topography on the interface between layers. Downgoing slabs can push aside the dense layer, but stirring of the two primary components is incomplete. Domes of dense material form away from these slab graveyards, and a thermal boundary layer at the interface results in upwelling plumes rising from the high spots. Large-scale mantle upwellings may be associated with formation of large igneous provinces (LIPs) or with continental rifting. Entrainment of dense material plus subducted material in this thermal boundary layer results in heterogeneous isotopic signatures in the resulting igneous rocks.

EVIDENCE FROM BACKARC BASINS AND MID-OCEANIC RIDGES

The isotopic nature of supra-subduction zone mantle is known principally from the western Pacific. Okamura et al. (1998) suggested that arc lavas of the Japan Sea region may have resulted from a mixture of DMM and EM II mantle types. Cousens et al. (1994) reported that some backarc basalts from the Japan Sea include an enriched component derived from subducted sediment. Trull et al. (1990) reported lavas from the Andaman Sea that were derived from mantle with a "subduction component."

Geochemical and petrologic studies from several mid-oceanic ridge regions also point toward the existence of persistent chemical heterogeneities. In the northeast Pacific, Cousens (1996) and Cousens et al. (1995) summarized geochemical evidence from lavas from the Juan de Fuca Ridge for their derivation from a mixture of mantle types, possibly DMM and HIMU, i.e., involving a component of previously subducted oceanic crust. Sherman et al. (1997) similarly reported lavas derived from mantle with a "subduction component" from the southern Chile Ridge.

In the Atlantic, Ross and Elthon (1993) reported on cumulate plagioclase lherzolite, gabbronorite, olivine gabbronorite, and noritic anorthosite from DSDP (Deep Sea Drilling Project) Hole 334, 104 km west of the Mid-Atlantic Ridge; the rocks were interpreted to have been derived from a highly depleted mantle source similar to that of supra-subduction zone lavas. Ross and Elthon (1997) also described gabbros from Sites 921–924 from ODP Leg 153 on the western wall of the Mid-Atlantic Ridge median valley south of the Kane Fracture Zone (MARK area); the gabbros were derived from two parent magmas, one of MORB composition and the other relatively depleted in incompatible trace elements.

In the Indian Ocean, Mahoney et al. (1998) documented the existence of mantle sources for old and young volcanic rocks that differ considerably from those of modern Pacific-Atlantic rocks. Mahoney et al. ascribed these Indian Ocean rocks to derivation from isotopically distinct mantle, including small amounts of old continental lithosphere, or from old, previously subducted material. They also reported that ophiolites from the Neotethyan suture, from Semail and Masirah, Oman, and from Xigaze, Tibet, also show similar differences compared to rocks derived from Atlantic-Pacific mantle.

MODEL FOR RESOLUTION OF THE OPHIOLITE CONUNDRUM

These data taken together suggest a possible resolution to the ophiolite conundrum. To reiterate briefly, the data suggest the following:

- Mantle heterogeneities are widespread and range in scale from a few centimeters to very large (e.g., Indian Ocean vs. Pacific-Atlantic).

- These heterogeneities can exist for tens to hundreds of million years or longer.
- Modern ridges display depleted or "subduction-component" lavas in addition to "MORB-type" lavas.
- Ophiolite complexes exhibit structural features reminiscent of fast-, intermediate-, and slow-spreading ridges.
- Many Tethyan ophiolites show no evidence of any nearby Pacific-type island arc.

These data suggest a model whereby the chemistry of lavas in ophiolites reflects the composition of the mantle from which the magmas have been derived, but that mantle composition does not necessarily indicate a concurrently active tectonic environment. Instead, these mantle-source compositions are *historically contingent* (e.g., Cushing, 1994; Gould, 1989) in that they may indicate the tectonic history prior to the oceanic spreading that produced the ophiolite rather than the modern tectonic environment. In the case of the Tethyan ophiolites, the complexes in the Balkan and central Turkey ophiolites may have been derived from depleted mantle subducted during the ~300-m.y.-long Hercynian-Appalachian episode of orogenesis. Similarly, the southern Turkey and Oman ophiolites, including Troodos, may have originated from mantle previously subducted during the 600–500 Ma Pan-African episode of orogenesis. Both the Hercynian and Pan-African episodes involved amalgamation of several western Pacific–like oceanic island arcs—and presumably abundant subducted mantle—during convergence of continental masses to form a supercontinent (Pangea and Gondwana, respectively).

NEW MODEL FOR MANTLE STRUCTURE

Kellogg et al. (1999) have postulated the existence of a layer in the deep mantle that is distinct from the depleted-mantle source of MORB. This hypothesis is quite different from the commonly posed end-member models of whole-mantle convection or convection layered with a boundary at the 670-km-depth seismic discontinuity. Recent seismic models of the mantle indicate that the 670 km boundary is not currently a significant barrier to mass exchange between the upper and lower mantle. But estimates of the volumes of different geochemical reservoirs (see Hofmann, 1997, for a review) and constraints provided by heat flow (Kellogg et al., 1999) suggest that at least one quarter of the mantle must be enriched in heat-producing elements relative to the MORB source region. Kellogg et al. (1999) proposed a dense layer, or *hot abyssal layer,* in the lowermost mantle that is enriched in heat-producing elements and remains unmixed from the overlying mantle. Such a layer is consistent with seismic observations of increased heterogeneity beginning at a depth of about 1500 km (Kennett et al., 1998; Van der Hilst and Karason, 1999).

Figure 2 shows a computer simulation of mantle convection with a layer enriched in heat-producing elements in the lower 1000 km of the mantle. The heat production in the upper layer corresponds to a depleted-mantle source with U concentration of 7 ppb (i.e., 7 in 10^9). The lower layer is enriched by a factor of 3.7. Although the cold subducted slab penetrates to the core-mantle boundary (Fig. 2A), the dense material does not overturn, and there is no whole-mantle mixing with the MORB source. Instead, a thermal boundary layer develops at the interface between layers, and instabilities form and rise from this boundary. The hot upwellings rise from high spots on the interface where they have entrained some components that will "flavor" future volcanic rocks with a distinctive signature: a mix of the ancient material in the dense layer (Fig 2B, shaded region) and slab material from its graveyard in the mid-mantle boundary layer. Complete penetration of the slab to the core-mantle boundary is likely to be a rare event; several factors will determine whether this occurs, primarily the age of the lithosphere at the time of subduction.

SPECULATIVE RECONSTRUCTIONS OF THE EARTH'S SURFACE AND INTERIOR

Through the use of this new model of mantle structure, we make some speculations on the possible crust-mantle tectonic history of Earth. Schematic cartoon cross sections (Fig. 3, A–C) through Earth at ca. 250 Ma, 165 Ma, and 100 Ma represent a first attempt to speculate on the structure of the mantle in the light of surface information.

At 250 Ma (Fig. 3A), Pangea had formed by the by-then extinct Appalachian-Hercynian-Altaid subductions and resultant collisions, depressing the 2000 km boundary. Sonomia had just collided in the region of the western United States and added a swath of crust several hundred kilometers wide to Pangea's western margin. Cimmeride island arcs derived from Gondwana were present in Panthalassa (Paleotethys), as were island arcs represented by the Stikinia and Alexander terranes of North America. Formation of the Triassic hotspot lavas of Wrangellia are shown over a "bump" on the interface between the depleted mantle and the hot abyssal layer.

At 165 Ma (Fig. 3B), Stikinia had collided with the western margin of Laurasia (e.g., Moores, 1998). Plumes rising from a bump in the hot abyssal layer produced the large igneous provinces of the central Atlantic rifted margin (Bahamas hotspot, etc.). Separation of Laurasia and Gondwana along the Atlantic highly magmatic rifted margins led to "leaky-transform" volcanic-rock–poor Neotethyan rifts that formed from previously subducted upper mantle, giving a supra-subduction zone chemical signature to their oceanic magmas, as discussed earlier in this paper. A push-up in the lower mantle underlay the Pacific microplate forming along the Farallon-Izanagi spreading center in Panthalassa.

At 100 Ma (Fig. 3C), the Atlantic Ocean, by then about 30° wide along the equator, separated North America from Eurasia and Gondwana. Narrow Neotethys rifts continued to form from previously subducted upper mantle. Subductions along the western North American and eastern Asian margins depressed the hot abyssal layer beneath those regions. Complementary double push-ups in the mantle beneath the Izanagi-Pacific and Pacific-Farallon spreading centers gave rise to the Cretaceous "superplume epoch."

These reconstructions imply that large upwellings may be produced when subducted slabs push enriched material in the lower mantle into large piles at some distance from the subduction zone. These reconstructions further imply that the production of large igneous provinces (LIPs) is likely to follow epochs of rapid subduction.

The reconstructions of Figure 3 are clearly highly speculative, generalized, and incomplete, but they emphasize that the structure of the mantle and the resultant source regions for spreading-center magmas would have been quite different at various times in the past. Thus the compositional nature of the mantle source of oceanic-crust magmas preserved in ophiolites is historically contingent, a function of its prior history, of the nature of previous subduction episodes linked at least in part to those of continental assembly and fragmentation, and plate-tectonic patterns and rates. Some ophiolites having supra-subduction zone chemistry may indeed have formed in a supra-subduction zone environment, but others, despite having the same chemistry, formed in a mid-oceanic ridge setting that was tapping into magma sources derived from previously subducted mantle. In this sense, the relationship between composition and tectonic environment is nonuniformitarian. Chemical compositions alone cannot serve as a reliable indicator of tectonic environment of ophiolite or-

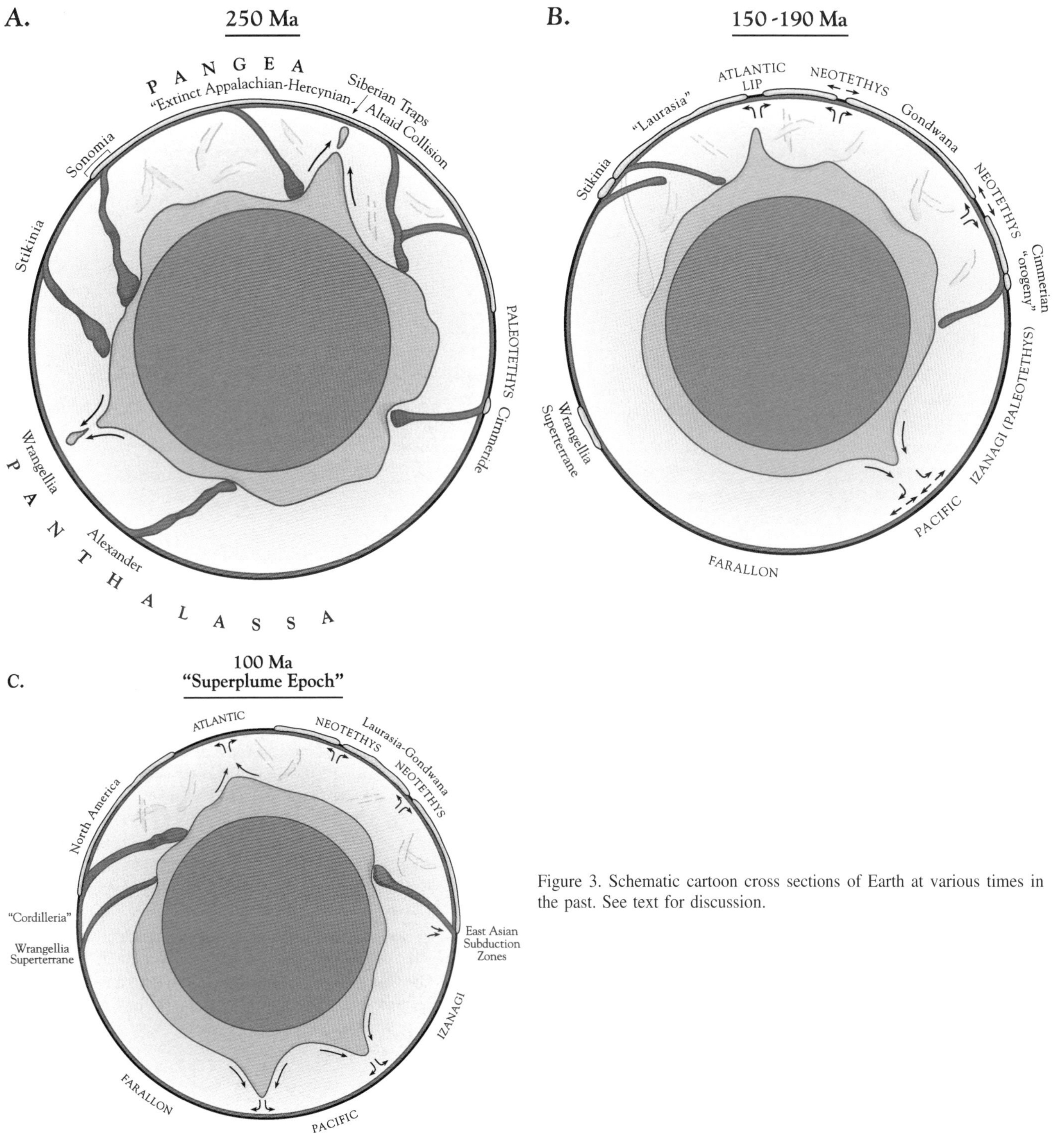

Figure 3. Schematic cartoon cross sections of Earth at various times in the past. See text for discussion.

igin; they must be used in concert with other data from structural geology and regional stratigraphy.

POSSIBLE LINK BETWEEN HIGHLY MAGMATIC RIFTED MARGINS AND CONTINENTAL-ASSEMBLY EPISODES

Approximately half of Earth's Mesozoic rifted continental margins exhibit thick igneous sequences present as dikes, marginal plutonic complexes, and submarine seaward-dipping reflectors (e.g., Coffin and Eldholm, 1994). These highly magmatic rifted margins imply a role for plume-generated rifting. By contrast, Neoproterozoic and Paleozoic continental-margin sequences preserved in the continents show no such concentrations of igneous rocks. Volcanic rocks are present along Paleozoic rifted margins, such as the North American Cordillera (Stewart, 1972) and those related to the Iapetus Ocean (the proto-Atlantic; Rankin et al., 1989), but they do not begin to compare in volume to the igneous rocks in the Mesozoic rifted margins. Abundant volcanic, possibly plume-related rocks were produced at ca. 1100 Ma, however, at least in North America (e.g., Hoffman,

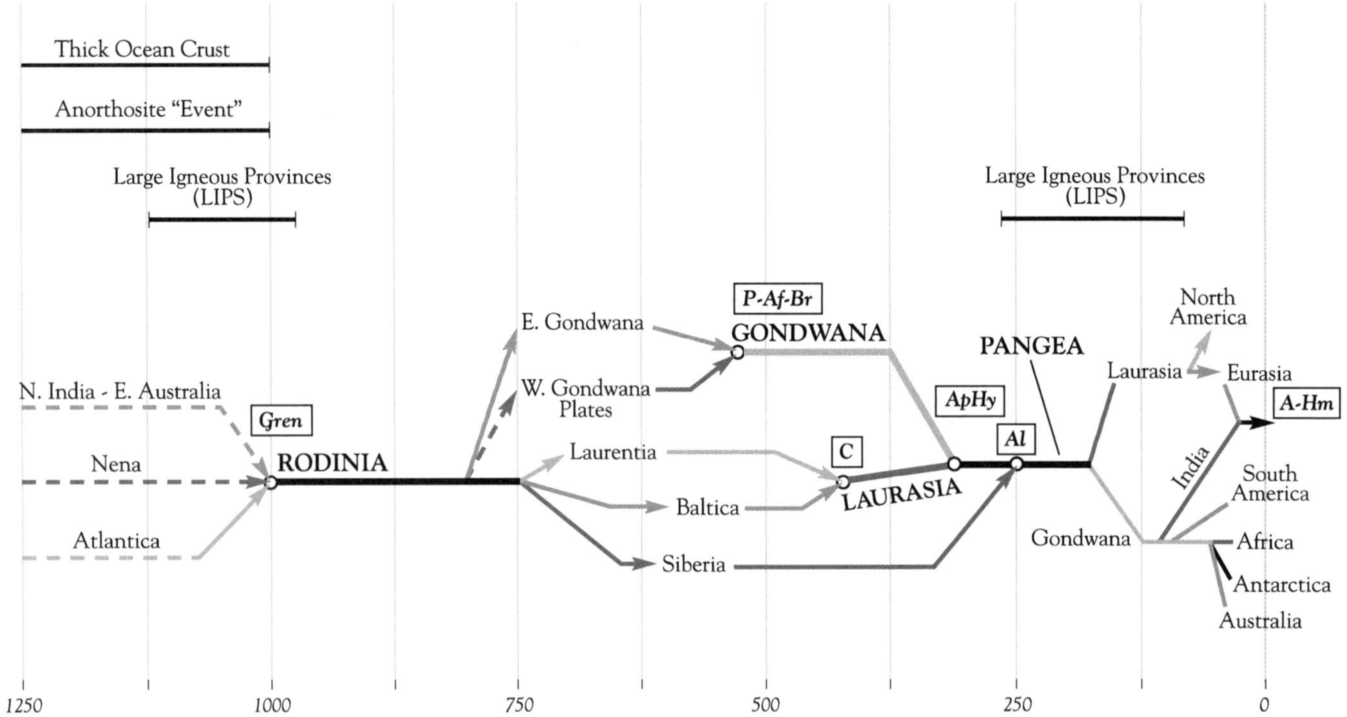

Figure 4. Diagram showing schematic evolution of continental assembly and dispersal for 1250 Ma to present. Modified after Rodgers (1996). Al—Altaid-Uralian orogenies of central Asia. A-Hm—Alpine-Himalayan orogenies. ApHy—Appalachian-Hercynian orogenies. Atlantica—composite continent formed of 2 Ga cratons within South America and Africa. C—Caledonide orogeny. Gren—Grenville and related orogenies. Nena—composite continent formed of Baltica, Laurentia, Eastern Siberia, and Eastern Antarctica. P-Af-Br—Pan-African–Brasiliano orogenies.

1989). This time apparently correlates with the assembly of a previous supercontinent, Rodinia. Figure 4 shows a diagrammatic history of major continental convergence and dispersal and times of formation of major large igneous provinces. To a first approximation, there seems to be an association between times of maximum continental convergence and large igneous provinces preserved on the continents or along continental margins. Perhaps periods of major convergence lead to large depressions of the depleted-mantle layer (between 2000 and 2900 km depth) beneath major subduction zones and complementary rises elsewhere to form plume-generated large igneous provinces. Periods of continental dispersal such as the Neoproterozoic and Paleozoic may have seen a relative quiescence of plume activity. A major exception to this scenario, of course, is the Cretaceous superplume episode, which occurred during the most recent episode of relatively great continental dispersal.

Figure 4 also displays schematically a correlation between major igneous events and continental dispersal, convergence, and major orogenic events. The diagram shows the presence of anorthosites and anorogenic granites, the possible thick oceanic crust prior to 1000 Ma (Moores, 1993), and the existence of the large igneous province in North America, as indicated by the Midcontinent rift, the MacKenzie dike swarm, and the Coppermine River basalts and Muskox intrusion of northwestern Canada (e.g., Hoffman, 1989; Van Schmus and Hinze, 1985). Convergence of continents during the Grenville orogeny gave rise to the supercontinent Rodinia. This large land mass fragmented by 750 Ma into East Gondwana, several separate cratons in South America–Africa, Laurentia, Baltica, and Siberia. Amalgamation of these cratonic fragments in the ca. 600–500 Ma Pan-African–Brasiliano orogeny produced the supercontinent of Gondwana. Convergence and amalgamation of Laurentia and Baltica along the Caledonide orogenic belt gave rise by ca. 400 Ma to Laurussia. Amalgamation of Laurussia first with Gondwana along the Appalachian-Hercynian orogens at ca. 300 Ma and then with Siberia along the Altaid orogen at ca. 250 Ma led to the supercontinent Pangea. Fragmentation of Pangea into Laurasia and Gondwana by the opening of the central Atlantic and early Neotethys rifts began at ca. 150 Ma and was followed by the breakup of Gondwana, separation of Europe and North America, and collision of India and Africa with Eurasia in the Alpine-Himalayan orogenies.

The correlation between the pattern of continental assembly and fragmentation and the presence of large igneous provinces provides an intriguing comparison. Times of major continental convergence must represent times of enhanced supply of lithospheric slabs to the mantle. To the extent that continental assembly and breakup patterns through time represent "punctuations" on a uniform process, these times may also correspond to major changes in mantle-source regions and resultant magmatism.

CONCLUSIONS

- The chemical composition of mantle-source regions is historically contingent, i.e., it is at least in part a function of Earth's tectonic history prior to melting.
- Tethyan ophiolites are of widely varying tectonic settings, but their chemical compositions may reflect derivation from previously subducted mantle, rather than a supra-subduction zone environment at the time of their formation.
- The amount of involvement of previously subducted lithosphere in mantle melting beneath mid-ocean ridges has varied in the past in concert with subduction maxima and minima. Times of major continental convergence may approximately correlate with major ep-

isodes of hotspot activity, formation of large igneous provinces, and derivation of mid-oceanic ridge magmas from previously subducted mantle.

- The "ophiolite conundrum" is resolved. This resolution will make possible the understanding of sea-floor spreading processes in pre-Mesozoic time.

ACKNOWLEDGMENTS

Moores and Dilek thank the participants of the 1998 Ophiolite Penrose Conference on Ophiolites and Oceanic Crust: New Insights from Field Studies and the Ocean Drilling Program. The Marine Geology and Geophysics Program at the National Science Foundation (NSF) and the Joint Oceanographic Institutions–U.S. Science Advisory Committee (JOI-USSAC) are thanked for financial support of the conference.

Kellogg thanks the participants of the 1998 Massachusetts Institute of Technology–Harvard Seminar on mantle convection for valuable discussions. This work was supported by a Presidential Faculty Fellowship from the National Science Foundation (to Kellogg).

Dilek thanks the NSF for grants (EAR-9219064 and EAR-9796011) in support of his studies of the Neotethyan ophiolites in Turkey and JOI-USSAC for grants for his involvement in ODP Legs 148 and 153. Moores acknowledges support from many NSF grants over the past 35 years for his studies of ophiolite complexes.

We also thank our colleagues R.J. Twiss and J.S. McClain for many helpful discussions and the GSA reviewers, A.R. Basu, B. Cousens, and D. Elthon, for many helpful suggestions.

REFERENCES CITED

Allègre, C.J., and Turcotte, D.L., 1985, Geodynamic mixing in the mesosphere boundary layer and the origin of oceanic islands: Geophysical Research Letters, v. 12, p. 207–210.

Allègre, C.J., and Turcotte, D.L., 1986, Implications of a two-component marble-cake mantle: Nature, v. 323, p. 123–127.

Allègre, C.J., Staudacher, T., Sarda, P., and Kurz, M., 1983, Constraints on evolution of Earth's mantle from rare gas systematics: Nature, v. 303, p. 762–766.

Anderson, D.L., 1995, Lithosphere, asthenosphere, and perisphere: Reviews of Geophysics, v. 33, p. 125–149.

Bloomer, S.H., Taylor, B., MacLeod, C.J., Stern, R.J., Fryer, P., Hawkins, J.W., and Johnson, L., 1995, Early arc volcanism and the ophiolite problem: A perspective from drilling in the western Pacific: American Geophysical Union Monograph 88, p. 1–30.

Bortolotti, V., Kodra, A., Marroni, M., Mustafa, F., Pandolfi, L., Principi, G., and Saccani, E., 1996, Geology and petrology of ophiolitic sequences in the Mirdita region (northern Albania): Ofioliti, v. 21, p. 3–20.

Christensen, U., 1989, Mixing by time-dependent convection: Earth and Planetary Science Letters, v. 95, p. 382–394.

Christensen, U., 1990, Mixing by time-dependent mantle convection, Reply: Earth and Planetary Science Letters, v. 98, p. 408–410.

Coffin, M.F., and Eldholm, O., 1994, Large igneous provinces: Crustal structure, dimensions, and external consequences: Reviews of Geophysics, v. 32, p. 1–36.

Coleman, R.G., 1971, Plate tectonic emplacement of upper mantle peridotites along continental edges: Journal of Geophysical Research, v. 76, p. 1212–1222.

Cousens, B.L., 1996, Depleted and enriched upper mantle sources for basaltic rocks from diverse tectonic environments in the northeast Pacific Ocean: The generation of oceanic alkaline vs. tholeiitic basalts, *in* Basu, A., and Hart, S., eds., Earth processes: Reading the isotopic code: American Geophysical Union Geophysical Monograph 95, p. 207–231.

Cousens, B.L., Allan, J.F., and Gorton, M.P., 1994, Subduction-modified pelagic sediments as the enriched component in backarc basalts from the Japan Sea: Ocean Drilling Program Sites 797 and 794: Contributions to Mineralogy and Petrology, v. 114, p. 421–434.

Cousens, B.L., Allan, J.F., Leybourne, M.I., Chase, R.L., Van Wagoner, N., 1995, Mixing of magmas from enriched and depleted mantle sources in the Northeast Pacific: West Valley segment, Juan de Fuca Ridge: Contributions to Mineralogy and Petrology, v. 120, p. 337–357.

Cushing, J.T., 1994, Quantum mechanics, historical contingence, and the Copenhagen Hegemony: Chicago, University of Chicago Press, 317 p.

Davies, G.F., 1990, Mixing by time-dependent convection: Comment: Earth and Planetary Science Letters, v. 98, p. 405–407.

Dewey, J.F., and Bird, J.M., 1970, Mountain belts and the new global tectonics: Journal of Geophysical Research, v. 75, p. 2625–2647.

Dewey, J.F., Pitman, W.C., III, Ryan, W.B.F., and Bonnin, J., 1973, Plate tectonics and the evolution of the Alpine system: Geological Society of America Bulletin, v. 84, p. 3137–3180.

Dilek, Y., and Moores, E.M., 1990, Regional tectonics of the Eastern Mediterranean ophiolites, *in* Malpas, J., Moores, E.M., Panayiotou, A., and Xenophontos, C., eds., Ophiolites, oceanic crustal analogues: Proceedings of the Symposium "Troodos 1987": Nicosia, Cyprus, The Geological Survey Department, p. 295–309.

Dilek, Y., and Thy, P., 1998, Structure, petrology, and seafloor spreading tectonics of the Kizildag ophiolite (Turkey), *in* Mills, R., and Harrison, K., eds., Modern ocean floor processes and the geological record: Geological Society of London Special Publication 148, p. 43–69.

Dilek, Y., Thy, P., and Moores, E.M., 1991, Episodic dike intrusions in the northwestern Sierra Nevada, California: Implications for multistage evolution of a Jurassic arc terrane: Geology, v. 19, p. 180–184.

Dilek, Y., Moores, E.M., and Furnes, H., 1998, Structure of modern oceanic crust and ophiolites and implications for faulting and magmatism at oceanic spreading centers, *in* Buck, R., Karson, J., Delaney, P., and Lagabrielle, Y., eds., Faulting and magmatism at mid-ocean ridges: American Geophysical Union Monograph 106, p. 219–266.

Dilek, Y., Thy, P., Hacker, B., and Grundvig, S., 1999, Structure and petrology of Tauride ophiolites and mafic dike intrusions (Turkey): Implications for the Neotethyan ocean: Geological Society of America Bulletin, v. 111, p. 1192–1216.

Engebretson, D., Kelley, K.P., Cashman, H.J., and Richards, M.A., 1992, 180 million years of subduction: GSA Today, v. 2, no. 5, p. 93–95, 100.

Fukao, Y., Obayashi, M., Inoue, H., Nenbai, M., 1992, Subducting slabs stagnant in the mantle transition zone: Journal of Geophysical Research, v. 97, p. 4809–4822.

Gass, I.G., 1968, Is the Troodos massif of Cyprus a fragment of Mesozoic ocean floor?: Nature, v. 221, p. 926–930.

Gould, S.J., 1989, Wonderful life: The Burgess Shale and the nature of history: Stephen, New York, W.W. Norton, 347 p.

Hacker, B.R., 1990, Simulation of the metamorphic and deformational history of the metamorphic sole of the Oman ophiolite: Journal of Geophysical Research, v. 95, p. 4895–4907.

Hart, S.R., 1988, Heterogeneous mantle domains: Signatures, genesis, and mixing chronologies: Earth and Planetary Science Letters, v. 90, p. 273–296.

Hart, S.R., Hauri, E.H., Oschmann, L.A., and Whitehead, J.A., 1992, Mantle plumes and entrainment: Isotopic evidence: Science, v. 256, p. 517–520.

Hess, H.H., 1939, Island arcs, gravity anomalies, and serpentinite intrusions: A contribution to the ophiolite problem: International Geological Congress, Moscow 1937, Report 17, v. 2, p. 263–283.

Hess, H.H., 1955, Serpentines, orogeny and epeirogeny: Geological Society of America Special Paper 62, p. 391–408.

Hoffman, P.F., 1989, Speculations on Laurentia's first gigayear (2.0–1.0 Ga): Geology, v. 17, p. 135–138.

Hofmann, A.W., 1997, Mantle geochemistry: The message from oceanic volcanism: Nature, v. 385, p. 219–229.

Jackson, E.D., Green, H.W., II, and Moores, E.M., 1975, The Vourinos ophiolite, Greece: Cyclic units of lineated cumulates overlying harzburgite tectonite: Geological Society of America Bulletin, v. 86, p. 390–398.

Kellogg, L.H., 1992, Mixing in the mantle: Annual Reviews of Earth and Planetary Sciences, v. 20, p. 365–388.

Kellogg, L.H., Hager, B.H., and van der Hilst, R.D., 1999, Compositional stratification in the deep mantle: Science, v. 283, p. 1881–1884.

Kennett, B.L.N., Widiyantoro, S., and van der Hilst, R.D., 1998, Joint seismic tomography for bulk sound and shear wave speed in the Earth's mantle: Journal of Geophysical Research, v. 103, p. 12469–12493.

Kidd, R.G.W., and Cann, J.R., 1977, Chilling statistics indicate an ocean-floor spreading origin for the Troodos complex, Cyprus: Earth and Planetary Science Letters, v. 24, p. 151–155.

Lagabrielle, Y., and Cannat, M., 1990, Alpine Jurassic ophiolites resemble the modern central Atlantic basement: Geology, v. 18, p. 319–322.

Lippard, S.J., Shelton, A.W., and Gass, I.G., 1986, The ophiolite of Northern Oman: Geological Society of London Memoir 11, 178 p.

Mahoney, J.J., Frei, R., Tejada, M.L.G., Mo, X.X., Leat, P.T., and Nägler, T.F., 1998, Tracing the Indian Ocean mantle domain through time: Isotopic results from old West Indian, East Tethyan, and south Pacific seafloor: Journal of Petrology, v. 39, p. 1285–1306.

Miyashiro, A., 1973, The Troodos complex was probably formed in an island arc: Earth and Planetary Science Letters, v. 19, p. 218–224.

Moores, E.M., 1970, Ultramafics and orogeny, with models for the U.S. Cordillera and the Tethys: Nature, v. 228, p. 837–842.

Moores, E.M., 1982, Origin and emplacement of ophiolites: Reviews of Geophysics and Space Physics, v. 20, p. 735–760.

Moores, E.M., 1993, Neoproterozoic oceanic crustal thinning, emergence of continents, and origin of the Phanerozoic ecosystem: A model: Geology, v. 21, p. 5–8.

Moores, E.M., 1998, Ophiolites, the Sierra Nevada, "Cordilleria", and orogeny along the Pacific and Caribbean margins of North and South America: International Geology Review, v. 40, p. 40–54.

Moores, E.M., and Jackson, E.D., 1974, Ophiolites and oceanic crust: Nature, v. 250, p. 136–138.

Moores, E.M., and Vine, F.J., 1971, Troodos massif, Cyprus, and other ophiolites as oceanic crust: Evaluation and implications: Philosophical Transactions of the Royal Society London, v. 268A, p. 443–466.

Moores, E.M., Robinson, P.T., Malpas, J., and Xenophontos, C., 1984, Model for the origin of the Troodos Massif, Cyprus, and other mideast ophiolites: Geology, v. 12, p. 500–503.

Okamura, S., Arculus, R.J., Martynov, Y.A., Kagami, H., Yoshida, T., and Kawano, Y., 1998, Multiple magma sources involved in marginal-sea formation: Pb, Sr, and Nd isotopic evidence from the Japan Sea region: Geology, v. 26, p. 619–622.

Pearce, J.A., and Lippard, S.J., and Roberts, S., 1984, Characteristics and tectonic significance of supra-subduction ophiolites, *in* Kokelaar, B.P., and Howels, M.F., eds., Marginal basin geology: Geological Society of London Special Publication 16, p. 77–94.

Peters, Tj., Nicolas, A., and Coleman, R.G., 1991, Ophiolite genesis and evolution of the oceanic lithosphere: Proceedings of the Ophiolite Conference, Muscat, Oman, 7–18 January, 1990: Dordrecht, Netherlands, Kluwer Academic Publishers, 903 p.

Rankin, D.W., Hull, L.M., Drake, A.A., Goldsmith, J.R., Ratcliffe, N.M., and Stanley, R.S., 1989, Proterozoic evolution of the rifted margin of Laurentia, *in* Rankin, D.W., and Hatcher, R.D., Jr., eds., The Appalachian–Ouachita Orogen in the United States: Boulder, Colorado, Geological Society of America, Geology of North America, v. F-2, p. 10–41.

Rassios, A., Moores, E.M., and Green, H.W., II, 1983, Magmatic structure and stratigraphy of the Vourinos ophiolite cumulate zone, northern Greece: Ofioliti, v. 8, p. 377–410.

Richter, F.M., 1985, Models for the Archean thermal regime: Earth and Planetary Science Letters, v. 73, p. 350–360.

Robinson, P.T., and Malpas, J., 1990, The Troodos ophiolite of Cyprus: New perspectives on its origin and emplacement, *in* Malpas, J., Moores, E.M., Panayiotou, A., and Xenophontos, C., eds., Ophiolites, oceanic crustal analogues: Proceedings of the Symposium "Troodos 1987": Nicosia, Cyprus, The Geological Survey Department, p. 13–26.

Robinson, P.T., Melson, W.G., O'Hearn, T., and Schmincke, H.-U., 1983, Volcanic glass compositions of the Troodos ophiolite, Cyprus: Geology, v. 11, p. 400–404.

Rodgers, J.J.W., 1996, A history of continents in the past three billion years: Journal of Geology, v. 104, p. 91–107.

Ross, K., and Elthon, D., 1993, Cumulates from strongly depleted mid-ocean-ridge basalt: Nature, v. 367, p. 826–828.

Ross, K., and Elthon, D., 1997, Cumulus and postcumulus crystallization in the oceanic crust: Major- and trace-element geochemistry of Leg 153 gabbroic rocks, *in* Karson, J.A., Cannat, M., Miller, D.J., and Elthon, D., eds., Proceeding of the Ocean Drilling Program, Scientific Results, v. 153, p. 333–350.

Saleeby, J., 1992, Petrotectonic and paleogeographic settings of U.S. Cordilleran ophiolites, *in* Burchfiel, B.C., Lipman, P.W., and Zoback, M.L., eds., The Cordilleran orogen: Conterminous U.S.: Boulder, Colorado, Geological Society of America, Geology of North America, v. G-3, p. 653–682.

Shallo, M., Kodra, A., and Gjata, K., 1990, Geotectonics of Albanian ophiolites, *in* Malpas, J., Moores, E.M., Panayiotou, A., and Xenophontos, C., eds., Ophiolites, oceanic crustal analogues: Proceedings of the Symposium "Troodos 1987": Nicosia, Cyprus, The Geological Survey Department, p. 265–271.

Sherman, S.B., Karsten, J.L., and Klein, E.M., 1997, Petrogenesis of axial lavas from the southern Chile Ridge: Major element constraints: Journal of Geophysical Research, v. 102, p. 14963–14990.

Smith, A.G., 1993, Tectonic significance of the Hellenic–Dinaric ophiolites: Geological Society of London Special Publication 76, p. 213–245.

Stein, C., 1995, Heat flow of the Earth, *in* Ahrens, T.J., ed., Global earth physics: A handbook of physical constants: Washington, D.C., American Geophysical Union, p. 144–158.

Steinmann, G., 1906, Geologische Beobachtungen in den Alpen, II, Die schard'tsche Überfaltungstheorie und die geologischen Bedeutung der Teefseeabsätze und der ophiolitische Massengesteine: Berichte der Naturforschenden Gesellschaft zu Freiburg I. BR. C. A. Wagners Universitäts-Buchdruckerei, Freiburg I. BR, p. 1–66.

Steinmann, G., 1927, Die ophiolitischen Zonen in der mediterranean Kettengebirgen: International Geological Congress, 14th, Madrid, v. 2, p. 638–667.

Stern, R.J., Nielsen, K.C., Best, E., Sultan, M., Arvidson, R.E., and Kröner, A., 1990, Orientation of Late Precambrian sutures in the Arabian–Nubian Shield: Geology, v. 18, p. 1103–1106.

Stewart, J.H., 1972, Initial deposits in the Cordilleran geosyncline: Evidence of a late Precambrian (<850 m.y.) continental separation: Geological Society of America Bulletin, v. 83, p. 1345–1360.

Tankut, A., 1990, Geochemical implications for tectonic setting of the ophiolitic rocks from the ophiolite mélange belt of the Ankara mélange: Bulletin of the Mineral Research and Exploration Institute of Turkey (MTA), v. 110, p. 17–28.

Tankut, A., Dilek, Y., and Önen, P., 1998, Petrology and geochemistry of the Neo-Tethyan volcanism as revealed in the Ankara mélange, Turkey: Journal of Volcanology and Geothermal Research, v. 85, p. 265–284.

Trull, T.W., Perfit, M.R., and Kurz, M.D., 1990, He and Sr isotopic constraints on subduction contributions to Woodlark Basin volcanism, *in* V.M. Goldschmidt conference, Baltimore, Maryland, May 2–4, 1990: Program and Abstracts, Geochemical Society, p. 87.

van der Hilst, R.D., and Karason, H., 1999, Compositional heterogeneity in the bottom 1000 km of the Earth's mantle: Towards a hybrid convection model: Science, v. 283, p. 1885–1888.

van der Hilst, R.D., Widayantoro, S., and Engdahl, E.R., 1997, Evidence for deep mantle circulation from global tomography: Nature, v. 386, p. 578–584.

van Keken, P.E., and Ballentine, C.J., 1998, Whole-mantle versus layered mantle convection and the role of a high-viscosity lower mantle in terrestrial volatile evolution: Earth and Planetary Science Letters, v. 156, p. 19–32.

van Schmus, W.R., and Hinze, W.J., 1985, The midcontinent rift system: Annual Reviews of Earth and Planetary Sciences, v. 13, p. 345–383.

Zindler, A., and Hart, S., 1986, Chemical geodynamics: Annual Reviews of Earth and Planetary Sciences, v. 14, p. 493–571.

MANUSCRIPT ACCEPTED BY THE SOCIETY FEBRUARY 23, 2000

Printed in the U.S.A.

Geological Society of America
Special Paper 349
2000

Evidence for subduction-related contamination of the mantle beneath the southern Chile Ridge: Implications for ambiguous ophiolite compositions

Marnie E. Sturm and Emily M. Klein
Division of Earth and Ocean Sciences, Nicholas School of the Environment, Duke University, Durham, North Carolina 27708, USA
Jill L. Karsten
School of Ocean and Earth Science and Technology, University of Hawaii, Honolulu, Hawaii 96822, USA
Jeffrey A. Karson
Division of Earth and Ocean Sciences, Nicholas School of the Environment, Duke University, Durham, North Carolina 27708, USA

ABSTRACT

The recent discovery of convergent-margin geochemical signatures along the southern Chile Ridge spreading center may have implications for the interpretation of ophiolite provenance. The location of the ridge near a ridge-trench-trench (RTT) triple junction suggests that the convergent-margin signatures may relate to processes associated with spreading-center subduction. The possibility that ophiolite obduction may also occur near RTT triple junctions where a spreading center is being subducted implies that mid-ocean ridge basalts (MORBs) with supra-subduction zone (SSZ) signatures may have been preferentially emplaced as ophiolite complexes in the past. There is evidence that this process may have occurred immediately south of the Chile triple junction in the region of the Taitao Peninsula. Controversy surrounding the provenance of some well-studied ophiolites (e.g., Bay of Islands, Semail massif) has recently tended to favor the view that these ophiolites formed above subduction zones, primarily on the basis of their trace element chemistry. The discovery of MORB with convergent-margin geochemical signatures near ophiolitic material with similar chemical systematics points to a modern example of a tectonic setting in which ocean crust with anomalous convergent-margin-like compositions is formed and may be emplaced as an ophiolite. As a result, our southern Chile Ridge findings suggest that the geochemical distinction between MORB and SSZ lavas may be less well defined than previously thought.

INTRODUCTION

It is generally accepted that ophiolite complexes formed in a variety of tectonic settings (e.g., fast- and slow-spreading mid-ocean ridges, island arcs, backarc basins, "leaky" transforms) although we rarely have the opportunity to observe modern examples of ophiolite emplacement. Ideally, the provenance of an ancient ophiolite complex may be ascertained by studying its present geologic setting, its internal structure, the timing of emplacement, the nature and provenance of overlying sedimentary rocks, and its igneous geochemistry (e.g., Pearce et al., 1984; Dewey and Bird, 1971; Moores and Jackson, 1974; Robertson, 1981; Gealey, 1977; Pearce and Cann, 1973; Moores, 1982; Searle and Cox, 1999). Geochemistry, in particular, has been widely used to determine the provenance of ophiolites around the world because processes within subduction zones are believed to impart a diagnostic geochemical signature to supra-subduction zone (SSZ) lavas that are distinct from those of mid-ocean ridge basalts (MORBs) (e.g., Jenner et al., 1991; Pearce et al., 1984; Elthon, 1991; Wood et al., 1979). Compared to MORBs, lavas formed near a subduction zone are selectively enriched in alkali metal, alkaline earth, and other incompatible elements (e.g., Sr, K, Rb, Ba, Th) and relatively depleted in high field strength elements (e.g., Ta, Nb, Hf, Zr, Ti) (Pearce et al., 1984; Gill, 1981; Plank and Langmuir, 1993). These chemical signatures have been variously attributed to modification of the mantle-source region of arc basalts by a subduction component—such as ocean sediments, altered ocean crust, or melts or fluids derived therefrom—and/or selective retention of high field strength elements in the subducting lithosphere (Best, 1975; Morris et al., 1990; Hawkesworth et al., 1977; Saunders and Tarney, 1979; Plank and Langmuir, 1998).

Although there is abundant evidence that some ancient ophiolite complexes formed above a subduction zone, the provenance of others remains ambiguous. For example, the discovery of ophiolites with MORB-like major element systematics but with SSZ trace element systematics (e.g., Bay of Islands, Elthon, 1991) has led to controversy surrounding their provenance. The recent identification of a suite of modern MORBs from the southern Chile Ridge with arc-like geochemical signatures may shed light on a possible tectonic environ-

Sturm, M.E., Klein, E.M., Karsten, J.L., and Karson, J.A., 2000, Evidence for subduction-related contamination of the mantle beneath the southern Chile Ridge: Implications for ambiguous ophiolite compositions, *in* Dilek, Y., Moores, E.M., Elthon, D., and Nicolas, A., eds., Ophiolites and Oceanic Crust: New Insights from Field Studies and the Ocean Drilling Program: Boulder, Colorado, Geological Society of America Special Paper 349, p. 13–20.

ment of formation for ophiolites with both MORB-like characteristics and SSZ geochemical affinities.

TECTONIC SETTING OF THE SOUTHERN CHILE RIDGE

The Nazca, Antarctic, and South American plates meet at the Chile triple junction in a ridge-trench-trench (RTT) configuration (Cande and Leslie, 1986). The Chile Ridge, an intermediate-rate (~6 cm/yr full rate) spreading ridge, forms one limb of the triple junction and separates the Nazca and Antarctic plates. The Chile Ridge was first subducted beneath the westward-migrating South American plate at ca. 14 Ma (Cande and Leslie, 1986) and is currently being subducted at a rate of 9 cm/yr near 46°20′S (Fig. 1; Cande and Leslie, 1986). During the Pliocene, the Chile triple junction was located at the present location of the Taitao Peninsula near 46°50′S, about 50 km south of the present-day triple junction (Cande and Leslie, 1986).

Lithologies located on the subaerial Taitao Peninsula include an ophiolite complex and two volcanic and shallow intrusive rocks of variable chemical characteristics. The main units are the Bahia Barrientos ophiolite, the later-stage bimodal dike complex, the associated volcanic-sedimentary sequences (known as the Chile margin unit and the main volcanic unit) and several plutonic intrusions (Guivel et al., 1996, 1999; Forsythe et al., 1986; Keading et al., 1990; Mpodozis et al., 1985; Lagabrielle et al., 1994; Le Moigne et al., 1996; Bourgois et al., 1996). These magmatic suites are characterized by a wide range of chemical affinities and display a large number of possible sources including MORB, contaminated MORB, slab-melting–derived magmas, and calc-alkalic magmas (Guivel et al., 1996).

The Bahia Barrientos ophiolite lacks a complete continuous section and consists of mantle peridotites, gabbros, and doleritic dikes of normal MORB (N-MORB) with depleted rare earth element (REE) patterns (Guivel et al., 1996). Lavas from the bimodal dike complex consist of imbricated mafic dolerite and andesite-dacite-rhyolite and display REE patterns typical of calc-alkalic series lavas (Guivel et al., 1996, 1999; Nelson et al., 1993). The precise age of the ophiolite and the bimodal dike complex has not been determined and cannot yet be conclusively linked to processes associated with the nearby Chile triple junction (Guivel et al., 1999). Although both units were deposited in a shallow-water environment, the Chile margin unit consists of interbedded sedimentary and volcanic material whereas the main volcanic unit contains pillow lavas and associated sedimentary rocks with a distinct greenschist metamorphic overprint (Guivel et al., 1996). No primary contact of the volcanic-sedimentary units with the ophiolite has been located, and they unconformably overlie pre-Jurassic continental basement with either a tectonic (main volcanic unit) or sedimentary (Chile margin unit) contact (Lagabrielle et al., 1994; Bourgois et al., 1993). Chemical similarities among some of the lavas from the Chile margin unit, the main volcanic unit, and the southern Chile Ridge suggest that they may have originated from the buried active spreading center as it was subducted at shallow depths (Guivel et al., 1996, 1999).

The submarine Taitao Ridge (a morphologic feature and not a spreading center) extends into the trench, immediately south of the triple junction (Fig. 1; Forsythe et al., 1995; Guivel et al., 1999) and consists of at least three superposed tectonic slices. The upper slice consists of mafic to felsic volcanic rocks sampled during Ocean Drilling Program Leg 141 (Site 862) and during subsequent field studies; the compositions of the mafic lavas range from MORB with arc-like trace element characteristics similar to some southern Chile Ridge lavas to calc-alkalic lavas clearly derived from an arc environment

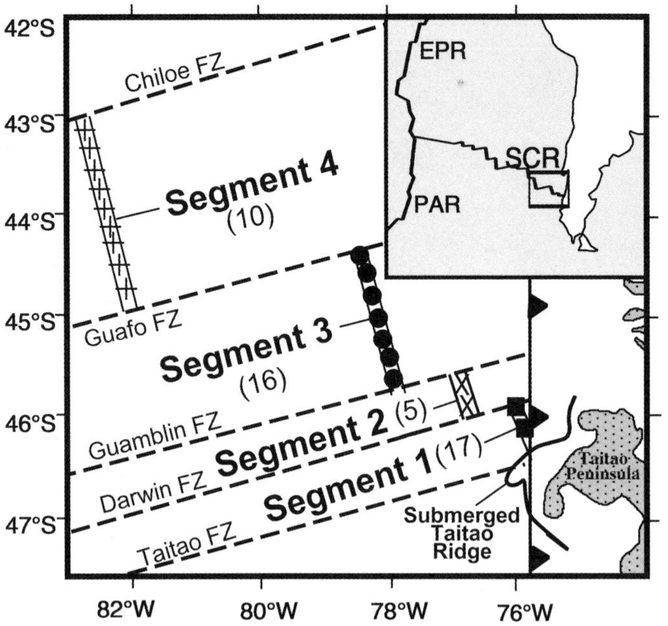

Figure 1. Tectonic elements of the southern Chile Ridge (SCR) near its intersection with the Chile Trench (barbed vertical line). Ridge segments are labeled 1–4 with increasing distance from the trench. Fracture zones (FZ) are shown as dashed lines. Dredging and wax coring sites along each segment are as follows: filled squares—segment 1, Xs—segment 2, filled circles—segment 3, crosses—segment 4. The number of sampling sites along each ridge segment is indicated in parentheses. Heavy line surrounding Taitao Peninsula is the contour showing the bathymetry at 2000 m below sea level; the contour's westward deflection indicates the submerged Taitao Ridge. Inset shows regional location of the study area (box); EPR—East Pacific Rise, PAR—Pacific-Antarctic Ridge.

(Forsythe et al., 1995; Kurnosov et al., 1995; Guivel et al., 1999). The compositions of the lower tectonic slices are not known.

The four segments of the southern Chile Ridge closest to the subduction zone were sampled on the Southern Cross Expedition in 1993 (R/V *Knorr* 138–11); all sites reported here are located within the axial valley (Sherman et al., 1997; Sturm et al., 1999). For convenience, the segments are numbered 1–4 stepping away from the trench (Fig. 1). All analyses from the southern Chile Ridge were performed on fresh, hand-picked glasses by direct-current-plasma and inductively-coupled-plasma mass spectrometry for major and trace elements, respectively, and thermal-ionization mass spectrometry for isotopes; representative analyses are reported in Klein and Karsten (1995), Sherman et al. (1997), and Sturm et al. (1999).

GEOCHEMISTRY OF THE SOUTHERN CHILE RIDGE

Major element data for southern Chile Ridge lavas reported by Sherman et al. (1997) show that the ridge-axis samples are tholeiitic basalts typical of lavas erupted along mid-ocean ridges worldwide. The major element systematics suggest that the southern Chile Ridge lavas were derived from parental magma of diverse compositions and underwent variable extents of fractionation (Mg# = 0.42–0.65), although most are relatively undifferentiated (Mg# > 0.6). [Mg# is the Mg number, defined as 100 × Mg/(Mg + Fe).]

In contrast to the major element compositions, the trace element signatures in the southern Chile Ridge basalts are unlike the majority of MORBs sampled previously (Klein and Karsten, 1995). Specifi-

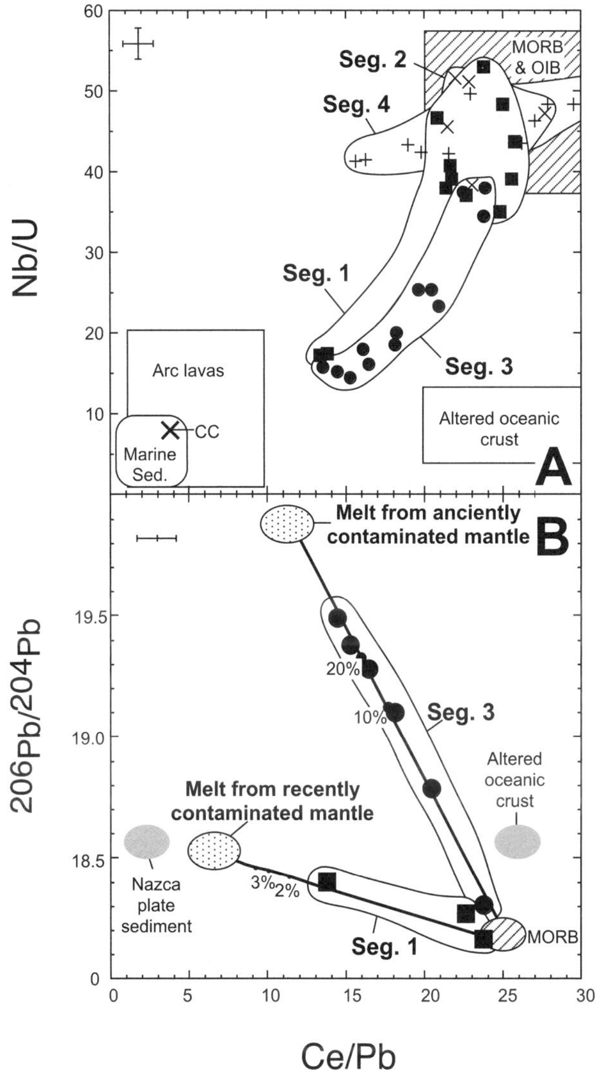

Figure 2. Ce/Pb vs. (A) Nb/U and (B) $^{206}Pb/^{204}Pb$ for Chile Ridge glasses (symbols as in Fig. 1). A, Fields are shown for mid-oceanic-ridge basalt (MORB) and oceanic-island basalt (OIB) (Sun and McDonough, 1989; Hofmann et al., 1986, 1988; Hickey et al., 1986), marine sediments (Ben Othman et al., 1989; Plank and Langmuir, 1993; Plank and Ludden, 1992), arc lavas (Hickey et al., 1986), altered oceanic crust (estimate modified from Hart and Staudigel, 1989; Chauvel et al., 1992; Plank, 1993; and Staudigel et al., 1995) and average upper continental crust (point X labeled CC; Taylor and McLennan, 1985). Lavas from segments 1 and 3 extend from the MORB field toward compositions commonly associated with arc lavas; this fact suggests contamination of the mantle source with a small amount of sediment and altered crust (Klein and Karsten, 1995). B, Solid bold lines represent calculated trends that show mixing (% indicated) between a 10% melt derived from mantle contaminated with previously recycled material (segment 3) or recently recycled material (segment 1) (shaded ovals) and N-MORB melts (obliquely ruled oval) also produced by 10% melting. Contaminated mantle is composed of 2% recently subducted contaminant (consisting of 20% local sediment and 80% altered oceanic crust) for segment 1 and 15% anciently (2.2 Ga) subducted contaminant (consisting of 3% terrigenous sediment and 97% altered ocean crust) for segment 3. For details of model calculations, see Sturm et al. (1999). Also shown in Figure 2B are fields for Nazca plate sediments and altered oceanic crust (Unruh and Tatsumoto, 1976; Dasch, 1981; Sun and McDonough, 1989; Stolper and Newman, 1994). Analytical uncertainties ($\pm 2\sigma$) that exceed symbol size are shown by barred crosses in upper left of each plot.

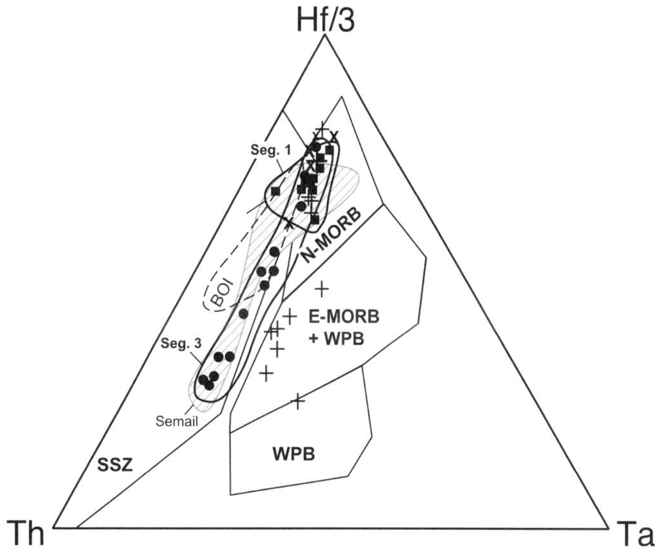

Figure 3. Th-Hf-Ta discrimination diagram of Wood (1980) showing Chile Ridge glasses (symbols as in Fig. 1). Fields are shown for depleted MORB (i.e., normal MORB or N-MORB), enriched MORB (E-MORB) + within-plate basalts (WPB), within-plate basalts, supra-subduction zone (SSZ) lavas, Bay of Islands ophiolite (BOI; enclosed by dashed line; Elthon, 1991), and Semail massif (oblique rules; Alabaster et al., 1982; Pearce et al., 1984). Note that lavas from segments 1 and 3 extend from the N-MORB field well into the SSZ field.

cally, samples from segments 1 and 3 (Fig. 1) extend well outside the MORB field toward low Ce/Pb, Nb/U (Fig. 2A), and Rb/Cs ratios and high Th/La ratios and thus approach compositions commonly associated with arc lavas. On the most commonly used discrimination diagram in the ophiolite literature, Th-Hf-Ta (Wood et al., 1979; Wood, 1980), lavas from segment 1 extend from the MORB field into the SSZ field with increasing extent of incompatible element enrichment; the most enriched segment 1 lavas, plotting well within the SSZ field, are those recovered nearest the trench (Fig. 3). In a similar but more dramatic way, segment 3 lavas exhibit a regular north-to-south trend of increasing enrichment, extending from the MORB field far into the SSZ field (Fig. 3). Like segment 3 lavas, segment 4 basalts also trend toward low Ce/Pb and high Th/La ratios, but at relatively high, MORB-like, Nb/U ratios (Fig. 2A), and extend from the MORB field into the "within-plate basalt" field on a Th-Hf-Ta diagram (Fig. 3). Segment 2 lavas have N-MORB compositions (Figs. 2A and 3). Here we focus on the systematics of segments 1 and 3 because they display the most robust convergent-margin signatures.

ORIGIN OF ANOMALOUS MORB SIGNATURES ALONG THE SOUTHERN CHILE RIDGE

Klein and Karsten (1995) interpreted the unusual trace element signatures of segments 1 and 3 as evidence of contamination of the subridge mantle beneath the southern Chile Ridge by a subduction component. Sturm et al. (1999) further constrained the timing of the recycling events for segments 1 and 3 by using combined isotope and trace element modeling. The uniformly MORB-like $^3He/^4He$ ratios and modest variations in Pb, Sr, and Nd isotope ratios of segment 1 suggest recent (since 20 Ma) introduction of a contaminant into its source, consistent with recycling of terrigenous sediment and altered oceanic crust (in bulk or in part) from the adjacent subduction zone

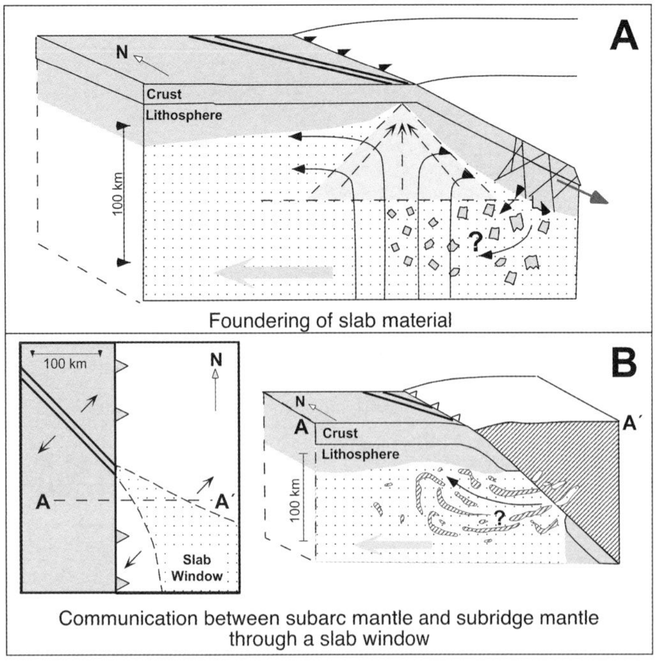

Figure 4. Schematic diagrams of a subducting spreading center showing the possible modes of contamination of the mantle in the area corresponding to segment 1 of the southern Chile Ridge. A, Foundering of slab-derived components following breakup of hot, young, buoyant crust in the subduction zone (van den Beukel, 1990). B, A plan view (left panel) of development of a "slab window" (Thorkelson, 1996) and a cross-sectional view (right panel) that illustrate communication between the metasomatized subarc mantle and the subridge mantle. Large gray arrows represent shallow asthenospheric mantle flow postulated for this region (Parmentier and Oliver, 1979).

(Fig. 2B). In contrast, variations in the segment 3 lavas extend to highly radiogenic Pb and Sr isotopic compositions (e.g., $^{206}Pb/^{204}Pb = 19.5$) and the lowest $^3He/^4He$ yet measured in MORB ($3.5R_A$, where R_A is Earth's atmospheric ratio of $^3He/^4He$). These systematics are most readily explained by ancient (ca. 2 Ga) recycling of terrigenous sediment and altered oceanic crust similar to origins suggested for the Society Islands although there are no hot spots near the southern Chile Ridge (Fig. 2B; Sturm et al., 1999; Chauvel et al., 1992; Devey et al., 1990; Hémond et al., 1994).

Because isotope systematics suggest that the segment 1 source was contaminated recently (Fig. 2B), the most likely scenario is that the segment 1 mantle was contaminated by material derived from the adjacent subduction zone (Sturm et al., 1999). Several mechanisms related to the unusual tectonic setting of spreading-ridge subduction may have allowed a subduction component (with its distinctive trace element signature) to be incorporated into the segment 1 mantle source. These may include incorporation of slab-derived components into the melting regime beneath segment 1 as a result of foundering after slab breakup (Fig. 4A; van den Beukel, 1990). Alternatively, others have proposed that ridge subduction may lead to the development of a "slab window" (Thorkelson, 1996) that may allow communication between metasomatized subarc mantle and the mantle beneath segment 1 (Fig. 4B). This communication may lead to transport of material through the slab window or may take the form of migration and focusing of contaminated fluids and melts toward the trench in the channel created by continual lithospheric spreading. In these scenarios, transport of contaminants may be facilitated by the shallow westward asthenospheric mantle flow (large gray arrows; Fig. 4B) postulated for this region (Parmentier and Oliver, 1979).

In contrast to segment 1, the segment 3 SSZ signature likely reflects ancient recycling of subducted material and storage of that material at depth (e.g., McKenzie and O'Nions, 1983; McCulloch, 1993; Hanan and Graham, 1996). Thus, the relationship of the anomalous segment 3 chemical signatures to processes occurring at the nearby subduction zone is less clear. Sturm et al. (1999) postulated, however, that pulsation from the nearby subducting plate may have initiated disturbance of a boundary layer at depth in which the segment 3 chemical signature was stored (Larsen et al., 1996). As a result, this discrete mantle reservoir may have risen diapirically through the upper mantle in the relatively recent past and emerged beneath segment 3 to be passively sampled at the ridge (Sturm et al., 1999). Therefore, it is possible that both the segment 1 and 3 signatures are linked to processes associated with spreading-ridge subduction, although the connection of the segment 3 signature to modern subduction is more tenuous.

SIGNIFICANCE FOR OPHIOLITES

The southern Chile Ridge is unusual from a tectonic perspective because it is an active mid-ocean spreading center that intersects a subduction zone, an uncommon setting in the current plate configuration but one known to have occurred fairly frequently in the geologic past as an inevitable consequence of plate tectonics (e.g., Delong and Fox, 1977; Karsten et al., 1996; van den Beukel, 1990). On the basis of plate reconstructions (e.g., Bohannon and Parsons, 1995), thermal and mechanical models (e.g., van den Beukel, 1990), and regional geologic evidence (e.g., Gealey, 1977; Mitchell, 1985), processes associated with spreading-ridge subduction may have played a significant role in ophiolite emplacement throughout Earth's history (e.g., Moores, 1982; Casey and Dewey, 1984; van den Beukel, 1990; Mitchell, 1985; Forsythe et al., 1986; Coleman, 1981). The observation that many ophiolites appear to have relatively thin lithosphere (<10 km) and young ages compared to the timing of subduction suggests that they were created at a spreading center shortly before emplacement and therefore close to a subduction zone (Dewey and Casey, 1979; Pearce et al., 1984). As a result, both eruption of MORB with unusual arc-like compositions along a spreading ridge in advance of a subduction zone and the potential emplacement of parts of the oceanic lithosphere as ophiolite complexes appear to be closely tied to spreading-ridge subduction. This connection implies that MORB with SSZ signatures may have been preferentially emplaced as ophiolite complexes in the past.

It follows that the use of diagnostic trace-element signatures to discriminate petrogenetic provenance (e.g., Pearce et al., 1984; Elthon, 1991) could point to the somewhat misleading conclusion that an ophiolite was accreted above a subduction zone in a forearc, arc, or backarc environment when it actually accreted at a mid-ocean ridge on the downgoing plate. For example, if oceanic lithosphere with SSZ trace element signatures similar to the southern Chile Ridge segment 1 or 3 lavas were to be emplaced as an ophiolite, it would have the structure, lithology, and major element composition of MORB but trace element systematics more characteristic of arc (or backarc or forearc) lavas (Figs. 2 and 3). As a result, the current definition of SSZ ophiolites must be expanded to include MORBs with convergent-margin-like trace element (and isotopic) chemical signatures that formed by processes associated with nearby subduction (Fig. 4; Sturm et al., 1999).

Examination of lavas from the Taitao Ridge and the Taitao Pen-

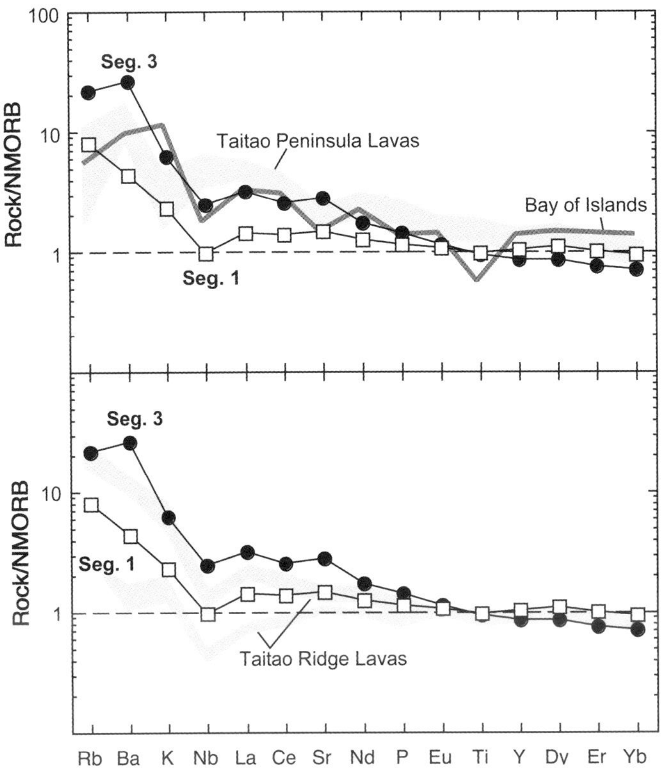

Figure 5. Trace element abundance patterns for representative enriched southern Chile Ridge glasses from segments 1 and 3 (symbols as in Fig. 1) normalized to the composition for N-MORB (Klein and Karsten, 1995; Sturm et al., 1999; Sun and McDonough, 1989). Lavas (also normalized) from the Taitao Peninsula (bimodal dike complex) and the Taitao Ridge (both indicated by shaded fields; Guivel et al., 1999) and an average of compositions from the Bay of Islands ophiolite (bold line; Jenner et al., 1991) are shown for comparison.

insula in relationship to the southern Chile Ridge provides some evidence that MORB with convergent-margin-like trace element characteristics may have been emplaced into the forearc region as part of an ophiolite suite. Lavas and shallow intrusive rocks from the bimodal dike complex, the Chile margin unit, and the main volcanic unit on the Taitao Peninsula and from young lavas from the Taitao Ridge (Fig. 1) display trace element and isotope signatures similar to those found along segments 1 and 3 of the southern Chile Ridge (Fig. 5). This fact has led Lagabrielle et al. (1994) and Guivel et al. (1996, 1999) to conclude that these lavas and shallow intrusive rocks most likely originated from the actively subducting southern Chile Ridge spreading center. The formation of some Taitao Peninsula rocks and Taitao Ridge rocks was coeval with subduction of two segments of the southern Chile Ridge beneath the South American margin at ca. 6.9 and 1.5 Ma, respectively (Guivel et al., 1999). In the case of the Taitao Ridge, Behrmann et al. (1994) argued that the volcanic pile probably formed at a ridge-transform intersection along the mid-ocean ridge and represents a possible candidate to become an ophiolite body emplaced into the South American forearc. Seismic profiles collected during a cruise to study the Chile triple junction in 1997, showing that the Taitao Ridge is composed of tectonic slices, support this view. It is unclear, however, whether the Taitao Peninsula lavas resulted from (1) direct eruption of magma from the contaminated suboceanic mantle through the thin forearc (Guivel et al., 1996), (2) emplacement of oceanic lithosphere into the forearc region following slab breakup and underthrusting (van den Beukel, 1990), or (3) both processes. In any case, lavas erupted in the Chile triple junction region (southern Chile Ridge, Taitao Ridge, and Taitao Peninsula) raise the possibility that the interaction of an active spreading center with a trench is associated with the eruption of both N-MORB and anomalous MORB with SSZ trace element characteristics followed by emplacement of this oceanic lithosphere as part of an ophiolite complex.

Although some ancient ophiolites show clear evidence that they formed above a subduction zone (e.g., Troodos; Schminke et al., 1983; Mehegan and Robinson, 1985; Pearce et al., 1984; Desmons et al., 1980), until now, no explanation has been widely accepted for the relatively common occurrence of ophiolitic material with MORB-like major element systematics but with SSZ trace element affinities. The Chile triple junction region may shed light on the origin of ancient ophiolites such as the Bay of Islands Complex and the Semail massif of Oman, whose provenance has remained ambiguous owing to competing geologic and geochemical evidence pointing to a mid-ocean ridge or an SSZ environment of formation.

There has been considerable controversy, for example, regarding the origin of one the world's best-studied ophiolites, the Ordovician Bay of Islands Complex in western Newfoundland. Although many of the internal structural relationships, major element systematics, and some trace element systematics suggest that the Bay of Islands Complex formed at a mid-ocean ridge spreading center (e.g., Malpas, 1977; Casey et al., 1981, 1985; Karson and Elthon, 1987; Komor et al., 1987; Beard, 1986; Coish, 1989; Pearce et al., 1984), the mineral chemistry and regional geology of the Bay of Islands Complex have led others to suggest that it may have SSZ affinities (Bédard, 1991; Searle and Stevens, 1984). The recent discovery of Nb and Ta depletions (and Th enrichment), considered to be diagnostic of SSZ environments, led some workers to reinterpret the Bay of Islands ophiolite as having formed in an SSZ environment (Elthon, 1991; Jenner et al., 1991) because, like the southern Chile Ridge lavas, Bay of Islands lavas show strong SSZ affinities on the Th-Hf-Ta diagram (Figs. 3 and 5; Elthon, 1991).

Similarly, studies of the regional geology and internal structure and petrology of the Semail ophiolite led to the widely held view that it formed in a mid-ocean ridge environment (e.g., Gass, 1968; Moores and Vine, 1969, 1971; Coleman, 1981; Pallister and Hopson, 1981; Smewing et al., 1984; Nicolas and Boudier, 1991; Ceuleneer, 1991; Boudier et al., 1988). In contrast, geochemical studies suggested a SSZ setting (Pearce et al., 1984; Alabaster et al., 1982) and were ultimately used to argue that the Semail massif formed in this environment. In fact, both interpretations may be correct in that some lavas from the Bay of Islands Complex and Semail, like those from the Chile triple junction region (Fig. 5), may have formed at true mid-ocean spreading centers but near RTT intersections where trace elements were affected by processes associated with nearby subduction.

The southern Chile Ridge data suggest that the mere existence of convergent-margin-like geochemical signatures in a particular ophiolite complex does not eliminate the possibility that the ophiolite formed at a mid-ocean ridge. Thus, such ophiolites may actually reveal mid-ocean ridge crustal structure and processes rather than arc processes as several authors have inferred. Although unambiguously determining the tectonic setting in which an ophiolite formed may not always be possible, we have identified a modern example of a historically common tectonic setting that apparently produces and emplaces ophiolitic material with MORB-like lithologic and major element chemical characteristics but SSZ trace element affinities, a relatively common combination found among ancient ophiolite complexes.

ACKNOWLEDGMENTS

This paper benefited from many thought-provoking discussions with the participants and conveners of the 1998 Penrose Conference on Ophiolites and Ocean Crust, particularly Julian Pearce and Eldridge Moores. We especially thank Yves Lagabrielle, Don Elthon, and an anonymous reviewer for their insightful and constructive comments. This work was supported by grants from the National Science Foundation (OCE9116169, OCE9258631, OCE9504280, and OCE9618736 to E. Klein and OCE9116430 and OCE9503617 to J. Karsten).

REFERENCES CITED

Alabaster, T., Pearce, J.A., and Malpas, J., 1982, The volcanic stratigraphy and petrogenesis of the Oman ophiolite complex: Contributions to Mineralogy and Petrology, v. 81, p. 168–213.

Beard, J.S., 1986, Characteristic mineralogy of arc-related cumulate gabbros: Implications for the tectonic setting of gabbroic plutons and for andesite genesis: Geology, v. 14, p. 848–851.

Bédard, J.H., 1991, Cumulate recycling and crustal evolution in the Bay of Islands ophiolite: Journal of Geology, v. 99, p. 225–249.

Behrmann, J.H., Leslie, S.D., Cande, S.C., and ODP Leg 141 Scientific Party, 1994, Tectonics and geology of spreading ridge subduction at the Chile Triple Junction: A synthesis of results from Leg 141 of the Ocean Drilling Program: Geologische Rundschau, v. 83, p. 832–853.

Ben Othman, D., White, W., and Patchett, J., 1989, The geochemistry of marine sediments, island arc magma genesis, and crust-mantle recycling: Earth and Planetary Science Letters, v. 94, p. 1–21.

Best, M.G., 1975, Amphibole-bearing cumulate inclusions, Grand Canyon, Arizona, and their bearing on undersaturated hydrous magmas in the upper mantle: Journal of Petrology, v. 16, p. 212–236.

Bohannon, R.G., and Parsons, T., 1995, Tectonic implications of post-30 Ma Pacific and North American relative plate motions: Geological Society of America Bulletin, v. 107, p. 937–959.

Boudier, F., Ceuleneer, G., and Nicolas, A., 1988, Shear zones, thrusts, and related magmatism in the Oman ophiolite: Initiation of thrusting on an oceanic ridge: Tectonophysics, v. 151, p. 275–296.

Bourgois, J., Lagabrielle, Y., Le Moigne, J., Urbina, O., Janin, M.C., and Beuzart, P., 1993, Preliminary results of a field study of the Taitao ophiolite (southern Chile): Implications for the evolution of the Chile triple junction: Ofioliti, v. 18, p. 113–129.

Bourgois, J., Martin, H., Lagabrielle, Y., Le Moigne, J., Frutos-Jara, J., 1996, Subduction erosion related to spreading-ridge subduction: Taitao Peninsula (Chile margin triple junction area): Geology, v. 24, p. 723–726.

Cande, S.C., and Leslie, R.B., 1986, Late Cenozoic tectonics of the southern Chile trench: Journal of Geophysical Research, v. 91, p. 471–496.

Casey, J.F., and Dewey, J.F., 1984, Initiation of subduction zones along transform and accreting plate boundaries, triple-junction evolution, and forearc spreading centres: Implications for ophiolitic geology and obduction, *in* Gass, I.G., Lippard, S.J., and Shelton, A.W., eds., Ophiolites and oceanic lithosphere: Geological Society of London Special Publication 13, p. 269–290.

Casey, J.F., Dewey, J.F., Fox, P.J., Karson, J.A., and Rosencrantz, E., 1981, Heterogeneous nature of oceanic crust and upper mantle: A perspective from the Bay of Islands Ophiolite Complex, *in* Emiliani, C., ed., The oceanic lithosphere: The sea, Volume 7: New York, John Wiley and Sons, p. 305–338.

Casey, J.F., Elthon, D.L., Siroky, F.X., Karson, J.A., and Sullivan, J., 1985, Geochemical and geological evidence bearing on the origin of the Bay of Islands and Coastal Complex ophiolites of western Newfoundland, *in* Sharman, G.F., and Francheteau, J., eds., Oceanic lithosphere: Tectonophysics, v. 116, p. 1–40.

Ceuleneer, G., 1991, Evidences for paleo-spreading center in the Oman ophiolite: Mantle structures in the Maqsad area, *in* Peters, Tj., Nicolas, A., and Coleman, R.J., eds., Ophiolite genesis and evolution of the oceanic lithosphere: Norwell, Massachusetts, Kluwer Academic Publishers, p. 147–173.

Chauvel, C., Hofmann, A., and Vidal, P., 1992, HIMU-EM: The French-Polynesian connection: Earth and Planetary Science Letters, v. 110, p. 99–119.

Coish, R.A., 1989, Boninitic lavas in Appalachian ophiolites: A review, *in* Crawford, A.J., ed., Boninites: London, Unwin Hyman, p. 264–287.

Coleman, R.G., 1981, Tectonic setting for ophiolite obduction in Oman: Journal of Geophysical Research, v. 86, p. 2497–2508.

Dasch, E.J., 1981, Lead isotopic composition of metalliferous sediments from the Nazca Plate, *in* Kulm, L.D., et al., eds., Nazca Plate: Crustal formation and Andean convergence: Geological Society of America Memoir 154, p. 199–209.

Delong, S.E., and Fox, P.J., 1977, Geological consequences of ridge subduction, *in* Taleani, M., and Pitman, W.C., eds., Island arcs, deep sea trenches, and back-arc basins: Washington, D.C., American Geophysical Union, p. 221–246.

Desmons, J., Delaloye, M., Desmet, A., Gagny, C., Rocci, G., and Voldet, P., 1980, Trace and rare earth element abundances in Troodos lavas and sheeted dikes, Cyprus: Ofioliti, v. 5, p. 35–56.

Devey, C.W., Albarède, F., Cheminéeè, J.-L., Michard, A., Mühe, R., and Stoffers, P., 1990, Active submarine volcanism on the Society hotspot swell (west Pacific): A geochemical study: Journal of Geophysical Research, v. 95, p. 5049–5066.

Dewey, J.F., and Bird, J., 1971, Origin and emplacement of the ophiolite suite: Appalachian ophiolites in Newfoundland: Journal of Geophysical Research, v. 76, p. 3179–3206.

Dewey, J.F., and Casey, J.F., 1979, Nucleation of subduction zones along transform faults and ophiolite emplacement: Geological Society of America Abstracts with Programs, v. 11, p. 413.

Elthon, D., 1991, Geochemical evidence for formation of the Bay of Islands ophiolite above a subduction zone: Nature, v. 354, p. 140–143.

Forsythe, R.D., Nelson, E.P., Carr, M.J., Kaeding, M.E., Herve, M., Mpodozis, C., Soffia, J.M., and Harambour, S., 1986, Pliocene near-trench magmatism in southern Chile: A possible manifestation of ridge collision: Geology, v. 14, p. 23–27.

Forsythe, R.D., Meen, J.K., Bender, J.F., and Elthon, D., 1995, Geochemical data on volcanic rocks and glasses recovered from Site 862: Implications for the origin of the Taitao Ridge, Chile Triple Junction region: Proceedings of the Ocean Drilling Program, Scientific Results, v. 141, p. 331–348.

Gass, I.G., 1968, Is the Troodos massif of Cyprus a fragment of Mesozoic ocean floor?: Nature, v. 220, p. 39–42.

Gealey, W.K., 1977, Ophiolite obduction and the geologic evolution of the Oman Mountains and adjacent areas: Geological Society of America Bulletin, v. 88, p. 1183–1191.

Gill, J.B., 1981, Orogenic andesites and plate tectonics: New York, Springer-Verlag, 385 p.

Guivel, C., Lagabrielle, Y., Bourgois, J., Maury, R.C., Martin, H., Arnaud, N., and Cotten, J., 1996, Magmatic responses to active spreading ridge subduction: Multiple magma sources in the Taitao Peninsula region (46°–47°S, Chile Triple Junction): Third International Symposium on Andean Geodynamics, St. Malo, France, p. 575–578.

Guivel, C., Lagabrielle, Y., Bourgois, J., Maury, R.C., Fourcade, S., Martin, H., and Arnaud, N., 1999, New constraints for the origin of ridge-subduction-related plutonic and volcanic suites from the Chile Triple Junction (Taitao Peninsula and Site 862, ODP Leg 141): Tectonophysics, v. 311, p. 83–111.

Hanan, B.B., and Graham, D.W., 1996, Lead and helium isotope evidence from oceanic basalts for a common deep source of mantle plumes: Science, v. 272, p. 991–995.

Hart, S.R., and Staudigel, H., 1989, Isotopic characterization and identification of recycled components, *in* Hart, S.R., and Gullen, L., eds., Crust/mantle recycling at convergence zones: Norwell, Massachusetts, Kluwer Academic Publishers, p. 15–28.

Hawkesworth, C.J., O'Nions, R.K., Pankhurst, R.J., Hamilton, P.J., and Evensen,

N.M., 1977, A geochemical study of island-arc and back-arc tholeiites from the Scotia Sea: Earth and Planetary Science Letters, v. 36, p. 253–262.

Hémond, C., Devey, C.W., and Chauvel, C., 1994, Source compositions and melting processes in the Society and Austral plumes (South Pacific Ocean): Element and isotope (Sr, Nd, Pb, Th) geochemistry: Chemical Geology, v. 115, p. 7–45.

Hickey, R., Frey, F.A., Gerlach, D.C., and Lopez-Escobar, L., 1986, Multiple sources for basaltic arc rocks from the southern volcanic zone of the Andes (34°–41°S): Trace element and isotopic evidence for contributions from subducted oceanic crust, mantle, and continental crust: Journal of Geophysical Research, v. 91, p. 5963–5983.

Hofmann, A.W., Jochun, K., Seufert, M., and White, W.M., 1986, Nb and Pb in basalts: New constraints on mantle evolution: Earth and Planetary Science Letters, v. 79, p. 33–45.

Jenner, G.A., Dunning, G.R., Malpas, J., Brown, M., and Brace, T., 1991, Bay of Islands and Little Port complexes, revisited: Age, geochemical and isotopic evidence confirm supra-subduction-zone origin: Canadian Journal of Earth Science, v. 28, p. 1635–1652.

Kaeding, M., Forsythe, R.D., and Nelson, E.P., 1990, Geochemistry of the Taitao ophiolite and near-trench intrusions from the Chile Margin Triple Junction: Journal of South American Earth Sciences, v. 3, p. 161–177.

Karson, J.A., and Elthon, D., 1987, Evidence for variations in magma production along oceanic spreading centers: A critical appraisal: Geology, v. 15, p. 127–131.

Karsten, J.L., Klein, E.M., and Sherman, S.B., 1996, Subduction zone geochemical characteristics in ocean ridge basalts from the southern Chile Ridge: Implications of modern ridge subduction systems for the Archean: Lithos, v. 37, p. 143–161.

Klein, E.M., and Karsten, J.L., 1995, Ocean-ridge basalts with convergent margin geochemical affinities from the Chile Ridge: Nature, v. 374, p. 52–57.

Komor, S.C., Elthon, D., and Casey, J.F., 1987, Petrology of a leucogabbroic interval within basal layered gabbros at North Arm Mountain, Bay of Islands Ophiolite: Contributions to Mineralogy and Petrology, v. 95, p. 278–300.

Kurnosov, V., Forsythe, R., Lindsley-Griffen, N., Zolotarev, B., Kashinvez, G., Eroshchev-Shak, V., Artamonov, A., and Chudaev, O., 1995, Comparison of the alteration and petrology of the Taitao Ridge to the Taitao Ophiolite: Proceedings of the Ocean Drilling Program, Scientific Results, v. 141, p. 349–360.

Lagabrielle, Y., Le Moigne, J., Maury, R.C., Cotton, J., and Bourgois, J., 1994, Volcanic record of the subduction of an active spreading ridge, Taitao Peninsula (southern Chile): Geology, v. 22, p. 515–518.

Larsen, T.B., Yuen, D.A., Smedsmo, J.L., and Malevsky, A.V., 1996, Thermomechanical modeling of pulsation tectonics and consequences on lithospheric dynamics: Geophysical Research Letters, v. 23, p. 217–220.

Le Moigne, J., Lagabrielle, Y., Whitechurch, H., Girardeau, J., Bourgois, J., and Maury, R.C., 1996, Petrography and geochemistry of the ophiolitic and volcanic suites of the Taitao Peninsula, Chile Triple Junction area: Journal of South American Earth Sciences, v. 9, p. 43–58.

Malpas, J., 1977, Petrology and tectonic significance of Newfoundland ophiolites with examples from the Bay of Islands, *in* Coleman, R.G., and Irwin, W.P., eds., North American ophiolites: State of Oregon Deparment of Geology and Mineral Resources Bulletin, v. 95, p. 13–23.

McCulloch, M.T., 1993, The role of subducted slabs in an evolving Earth: Earth and Planetary Science Letters, v. 115, p. 89–100.

McKenzie, D., and O'Nions, R.K., 1983, Mantle reservoirs and ocean island basalts: Nature, v. 301, p. 229–231.

Mehegan, J.M., and Robinson, P.T., 1985, Lava compositions of the Troodos Ophiolite, Cyprus: Eos, Transactions of the American Geophysical Union, v. 66, p. 1123.

Mitchell, A.H.G., 1985, Ophiolite detachment and emplacement related to spreading ridge subduction: Ofioliti, v. 10, p. 355–362.

Moores, E.M., 1982, Origin and emplacement of ophiolites: Reviews of Geophysics, v. 20, p. 735–760.

Moores, E.M., and Jackson, E.D., 1974, Ophiolites and oceanic crust: Nature, v. 250, p. 136–139.

Morris, J.D., Leeman, W.P., and Tera, F., 1990, The subducted component in island arc lavas: Constraints from Be isotopes and B-Be systematics: Nature, v. 344, p. 31–36.

Mpodozis, C.M., Hervé, M., Nasi, C., Soffia, M.J., Forsythe, R.D., and Nelson, E.P., 1985, El magmatismo plioceno de Peninsula Tres Montes y su relación con la evolución del punto triple de Chile austral: Revista Geológica de Chile, v. 25–26, p. 13–28.

Nelson, E., Forsythe, R., Diemer, J., Allen, M., and Urbina, O., 1993, Taitao ophiolite: A ridge collision ophiolite in the forearc of southern Chile (46°S): Revista Geológica de Chile, v. 20, no. 2, p. 137–165.

Nicolas, A., and Boudier, F., 1991, Rooting of the Sheeted Dike Complex in the Oman Ophiolite, *in* Peters, Tj., Nicolas, A., and Coleman, R.J., eds., Ophiolite genesis and evolution of the oceanic lithosphere: Norwell, Massachusetts, Kluwer Academic Publishers, p. 39–54.

Pallister, J.S., and Hopson, C.A., 1981, Samail Ophiolite plutonic suite: Field relations, phase variation, cryptic variation and layering, and a model of a spreading ridge magma chamber: Journal of Geophysical Research, v. 86, p. 2593–2644.

Parmentier, E.M., and Oliver, J.E., 1979, A study of shallow global mantle flow due to the accretion and subduction of lithospheric plates: Geophysical Journal of the Royal Astronomical Society, v. 57, p. 1–21.

Pearce, J.A., and Cann, J.R., 1973, Tectonic setting of basic volcanic rocks determined using trace element analyses: Earth and Planetary Science Letters: v. 19, p. 290–300.

Pearce, J.A., Lippard, S.J., and Roberts, S., 1984, Characteristics and tectonic significance of supra-subduction zone ophiolites, *in* Kokelaar, B.P., and Howells, M.F., eds., Marginal basin geology: Geological Society of London Special Publication 16, p. 77–94.

Plank, T., 1993, Mantle melting and crustal recycling in subduction zones [Ph.D. thesis]: New York, Columbia University, 439 p.

Plank, T., and Langmuir, C.H., 1993, Tracing trace elements from sediment input to volcanic output at subduction zones: Nature, v. 362, p. 739–742.

Plank, T., and Langmuir, C.H., 1998, The chemical composition of subducting sediment and its consequences for the crust and mantle: Chemical Geology, v. 145, p. 325–394.

Plank, T., and Ludden, J.N., 1992, Geochemistry of sediments in the Argo abyssal plain at Site 765: A continental margin reference section for sediment recycling in subduction zones: Proceedings of the Ocean Drilling Program, Scientific Results, v. 123, p. 167–189.

Robertson, A.H.F., 1981, Metallogenesis on a Mesozoic passive continental margin, Antalya Complex, southwest Turkey: Earth and Planetary Science Letters, v. 54, p. 323–345.

Schmincke, H.-U., Rautenschlein, M., Robinson, P.T., Mehegan, J.M., 1983, Troodos extrusives series of Cyprus: A comparison with oceanic crust: Geology, v. 11, p. 405–409.

Searle, M.P., and Cox, J., 1999, Tectonic setting, origin, and obduction of the Oman ophiolite: Geological Society of America Bulletin, v. 111, p. 104–122.

Searle, M.P., and Stevens, R.K., 1984, Obduction processes in ancient, modern, and future ophiolites, *in* Gass, I.G., Lippard, S.J., and Shelton, A.W., eds., Ophiolites and oceanic lithosphere: Geological Society of London Special Publication 13, p. 303–320.

Sherman, S.B., Karsten, J.L., and Klein, E.M., 1997, Petrogenesis of axial lavas from the southern Chile Ridge: Major element constraints: Journal of Geophysical Research, v. 102, p. 14963–14990.

Staudigel, H., Davies, G.R., Hart, S.R., Marchant, K.M., and Smith, B.M., 1995, Large scale isotopic Sr, Nd, and O isotopic anatomy of altered oceanic crust: DSDP/ODP sites 417/418: Earth and Planetary Science Letters, v. 130, p. 169–185.

Stolper, E., and Newman, S., 1994, The role of water in the petrogenesis of Mariana trough magmas: Earth and Planetary Science Letters, v. 121, p. 293–325.

Sturm, M.E., Klein, E.M., Graham, D.W., and Karsten, J.L., 1999, Age constraints on crustal recycling to the mantle beneath the southern Chile Ridge: He-Pb-Sr-Nd isotope systematics: Journal of Geophysical Research, v. 104, p. 5097–5114.

Sun, S.-S., and McDonough, W.F., 1989, Chemical and isotopic systematics of oceanic basalts: Implications for mantle composition and processes, *in* Saunders, A.D., and Norry, M.J., Magmatism in the ocean basins: Geological Society of London Special Publication 42, p. 313–345.

Taylor, S.R., and McLennan, S.M., 1985, The continental crust: Its composition and evolution: Oxford, England, Blackwell Scientific, 312 p.

Thorkelson, D.J., 1996, Subduction of diverging plates and the principles of slab window formation: Tectonophysics, v. 255, p. 47–63.

Unruh, D.M., and Tatsumoto, M., 1976, Lead isotopic composition and uranium, thorium, and lead concentrations in sediments and basalts from the Nazca plate: Deep Sea Drilling Project, Initial Reports, v. 34, p. 341–347.

van den Beukel, J., 1990, Breakup of young oceanic lithosphere in the upper part of a subduction zone: Implications for the emplacement of ophiolites: Tectonics, v. 9, p. 825–844.

Wood, D.A., 1980, The application of a Th-Hf-Ta diagram to problems of tectonomagmatic classification and to establish the nature of crustal contamination of basaltic lavas of the British Tertiary Volcanic Province: Earth and Planetary Science Letters, v. 50, p. 11–30.

Wood, D.A., Joron, J.-L., and Treuil, M., 1979, A reappraisal of the use of trace elements to classify and discriminate between magma series erupted in different tectonic settings: Earth and Planetary Science Letters, v. 45, p. 326–336.

Manuscript Accepted by the Society February 23, 2000

Printed in the U.S.A.

Geological Society of America
Special Paper 349
2000

The ophiolite–oceanic lithosphere analogue: New insights from the Northern Apennines (Italy)

Elisabetta Rampone*
Dipartimento di Scienze della Terra, Università di Milano, Via Botticelli 23, 20133 Milano, Italy; e-mail: betta@r10.terra.unimi.it
Giovanni B. Piccardo
Dipartimento per lo Studio del Territorio e delle sue Risorse, Università di Genova, Corso Europa 26, 16132 Genova, Italy

ABSTRACT

In this paper, we discuss the results of recent petrologic and isotopic studies on the Northern Apennine ophiolites, which are remnants of the oceanic lithosphere of the Jurassic Ligurian Tethys Ocean. In the Northern Apennines, ophiolites crop out in two distinct geologic units, the External and Internal Ligurides, which have been related, respectively, to pericontinental and intraoceanic settings of the Ligurian Sea sector of the Tethys. In the External Ligurides, subcontinental-lithosphere peridotites (presumably Proterozoic) are associated with Hercynian continental crust. Both mantle and crustal rocks display a composite subsolidus retrograde evolution, related to their uplift toward the seafloor, where they are associated with Jurassic mid-oceanic-ridge basalts (MORBs). The Internal Liguride ophiolites comprise the spatial association of depleted peridotites (of Permian age) and younger (mostly Jurassic) MORB-type magmatism, not linked to the peridotites by a genetic melt and residua relationship. Structural and petrologic features recorded by the Northern Apennine ophiolites thus indicate that the Jurassic Ligurian Sea sector of the Tethys was mostly underlain by older (Proterozoic and Permian), subcontinental lithospheric mantle and younger, unrelated, MORB seafloor. This peculiar lithologic association cannot be produced at mid-ocean ridges of mature oceans; rather, it most likely originated after the breakup of the continental crust in response to passive extension of the continental lithosphere. The exposure of subcontinental mantle peridotites on the ocean floor is a unique feature testifying to the early stages of inception of an oceanic basin by means of passive lithospheric extension. Modern oceanic analogues of the Ligurian Sea sector of the Tethys represented by the Northern Apennine ophiolites are currently found in embryonic oceans (e.g., the Red Sea) and at the transition zone between continental and oceanic lithosphere of mature oceanic basins (e.g., the Galicia margin, western Spain).

INTRODUCTION

In the past 20 years, a large number of geophysical and petrologic studies devoted to the oceanic basins have provided clear evidence that the nature and structure of oceanic lithosphere and its mechanism of formation can be quite different in settings that reflect distinct stages of evolution of an ocean (i.e., embryonic oceans, passive margins, mid-ocean ridges).

A large divergence from the classical model of oceanic lithosphere generated at mid-ocean ridges has been discovered in oceanic settings along passive continental margins (e.g., the Iberian margin, western Spain) and in embryonic oceans (e.g., the Red Sea). The ocean-continent boundary along the Iberian margin is characterized by large exposures of fertile-mantle peridotites that have been tectonically denuded and are now in direct contact with continental crust material (Boillot et al., 1988; Kornprobst and Tabit, 1988). Embryonic oceans like the Red Sea are mainly floored by thinned continental lithosphere (gneisses and fertile-mantle lherzolites) injected by MORB (mid-oceanic-ridge basalt) dikes, as exposed on Zabargad Island.

Oceanic lithosphere formed at mid-ocean ridges is characterized by the existence of a cogenetic relationship between the magmatic crust and the associated residual peridotites, as demonstrated by Snow et al. (1994). In contrast, such a cogenetic link cannot exist in marginal oceanic settings or in embryonic oceans, where old, subcontinental lithospheric mantle is associated with young, MORB-type magmatism.

Ophiolites display highly variable structural and petrologic features that are related to their original oceanic settings. Most of the available petrologic and isotopic studies on ophiolites have provided evidence of a cogenetic relationship between mantle peridotites and the associated magmatic rocks, as is expected in mature oceanic lith-

*Present address: Dipartimento di Scienze Geologiche e Geotecnologie, Università di Milano Bicocca, Via Emanueli, 20100 Milano, Italy.

Rampone, E., and Piccardo, G.B., 2000, The ophiolite–oceanic lithosphere analogue: New insights from the Northern Apennines (Italy), *in* Dilek, Y., Moores, E.M., Elthon, D., and Nicolas, A., eds., Ophiolites and Oceanic Crust: New Insights from Field Studies and the Ocean Drilling Program: Boulder, Colorado, Geological Society of America Special Paper 349, p. 21–34.

osphere, but a few studies have pointed to the lack of a genetic mantle-crust link. Thus, the question arises whether ophiolites can indeed represent different stages of evolution of an oceanic basin.

In this paper, we review recent results of the investigations on Northern Apennine ophiolites and show that their petrologic and isotopic features are not consistent with oceanic lithosphere formed at a mid-ocean ridge. Indeed, the Northern Apennine ophiolites consist of the spatial association of older (Proterozoic and Permian), subcontinental lithospheric mantle and younger (mostly Jurassic), unrelated, MORB-type magmatism. These ophiolites are interpreted as ancient oceanic lithosphere formed during early stages of oceanization, following a passive extension of continental lithosphere leading to lithospheric mantle denudation. Modern oceanic analogues for this tectonic setting are represented by passive margins (e.g., the Galicia Bank) and embryonic oceans (e.g., the Red Sea). The main goals of this contribution are, therefore, to provide some geologic and isotopic evidence for the inadequacy of interpreting all ophiolites as analogues to mid-ocean ridges and to highlight the peculiar nature of an ophiolite developed close to a continental margin.

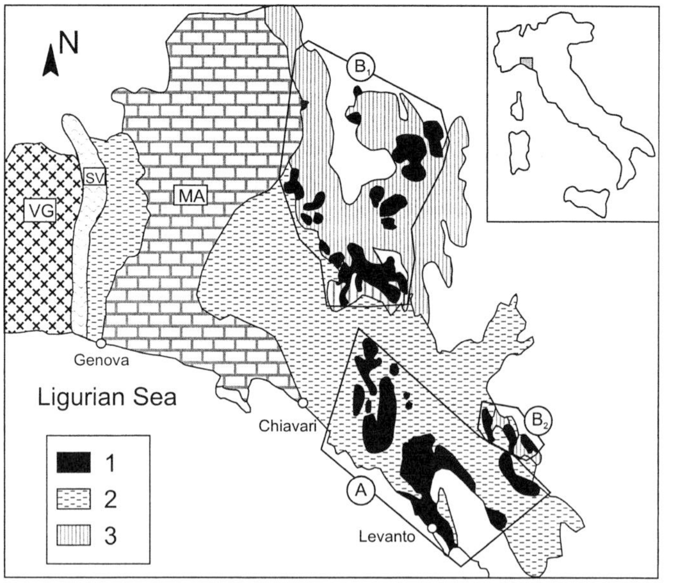

Figure 1. Tectonic sketch map of Eastern Liguria (Northern Apennines). VG—Voltri Group. SV—Sestri-Voltaggio units. MA—Mount Antola Formation and Mount Caio flysch (calcareous sandy turbidites with interbedded shales). (1) Ophiolites (peridotites, gabbros, and basalts); field A—Internal Liguride ophiolites; fields B_1 and B_2—External Liguride (Mount Ragola–Mount Aiona and Suvero, respectively) ophiolites. (2) Internal Liguride sedimentary sequences—radiolarian cherts, *Calpionella* limestones, Palombini, Lavagna, and Giariette shales, and Gottero sandstones. (3) External Liguride flyschoid sequences. The white areas are disregarded.

STRUCTURAL REMARKS ON THE NORTHERN APENNINE OPHIOLITES

Ophiolite sequences cropping out in the Alpine-Apennine belt are the remnants of oceanic lithosphere of the Jurassic Ligurian Tethys. The Ligurian Ocean was formed during the late Middle or early Late Jurassic from rifting of the Apulian continental block away from Europe (Lemoine et al., 1987, and references therein).

The large number of recent structural and petrologic studies devoted to the Alpine-Apennine ophiolites have demonstrated that these ophiolites show some peculiar and atypical characteristics when compared with oceanic lithosphere produced at mid-ocean ridges in mature oceans. Some of the striking features include (1) overall abundance of serpentinized mantle peridotites, directly overlain by volcanic and sedimentary rocks; (2) occurrence of gabbroic rocks as discrete dikes and bodies intruded into the peridotites; (3) absence of a basaltic sheeted-dike complex; (4) thinness and/or scarceness of submarine basaltic cover; (5) presence of ophiolitic breccias and radiolarian cherts (i.e., the oldest oceanic sediments) commonly lying directly on top of the peridotite-gabbro basement. These features led to the generally accepted interpretation that the Ligurian Ocean domain was floored by a serpentinitic and gabbroic basement before basaltic extrusion and chert deposition. Field observations and age determinations have shown that the two magmatic events (early gabbro intrusion and late basalt extrusion) were separated by a long series of tectonic, metamorphic, and sedimentary events.

The diversity of the Jurassic oceanic lithosphere, which floored the Jurassic Ligurian Tethys, is shown by the Northern Apennine ophiolites, occurring in two distinct geologic units, the so-called Internal and External Ligurides (Fig. 1). The terms "External" and "Internal" refer to the inferred paleogeographic position (i.e., pericontinental vs. intraoceanic settings, respectively) in the Jurassic oceanic basin.

In the Internal Ligurides, ophiolitic rocks show structural relationships similar to those described for the whole Ligurian basin; they consist of a peridotite-gabbro basement stratigraphically covered by ophiolitic breccias, pillowed basaltic lava flows, and marine sedimentary rocks (Abbate et al., 1994, and references therein). Field and structural evidence indicates that gabbroic rocks constitute discrete intrusive bodies within mantle peridotites. Subsequently, the peridotite-gabbro association underwent a tectonic-metamorphic decompressional evolution as testified by deformation and recrystallization fabrics seen along shear zones; this evidence indicates progressive uplift and final exposure at the seafloor, where the peridotites were extensively serpentinized. The uppermost serpentinites were intensively fractured with development of tectonic breccias (ophicalcites), which were partially covered by sedimentary ophiolitic breccias. Ophicalcites and sedimentary breccias were discontinuously covered by MORB pillow-lava flows and the Oxfordian-Callovian radiolarian cherts, i.e., the oldest pelagic sediments (Marcucci and Passerini, 1991). Discrete basaltic dikes, temporally related to the basaltic extrusions, commonly crosscut serpentinized peridotites and foliated gabbros, as well as the overlying tectonic and sedimentary breccias.

The Apennine ophiolites of the External Ligurides crop out in a peculiar geologic context, where they are spatially associated with continental crust (Marroni et al., 1998, and references therein). These ophiolites mostly consist of mantle lherzolites and pillowed basaltic lavas that occur as large blocks (olistoliths) within the Cretaceous–Eocene flysch sequences. Because of the association with continental crust, the source area for the External Liguride lherzolites and basalts has been located close to the Adria continental margin (Piccardo et al., 1990, and references therein). Accordingly, the External Ligurides have been regarded as a fossil example of the ocean to continent transition (Marroni et al., 1998). Peridotites record a composite tectonic-metamorphic evolution, which indicates progressive upwelling of the upper-mantle rocks toward the seafloor prior to widespread serpentinization (Piccardo et al., 1990, 1992; Rampone et al., 1995). Abundant chilled basaltic dikes intrude peridotites and crosscut all mantle structures. The continental crust consists of late Hercynian granitoids, mafic and felsic granulites, and paragneisses (Eberhardt et

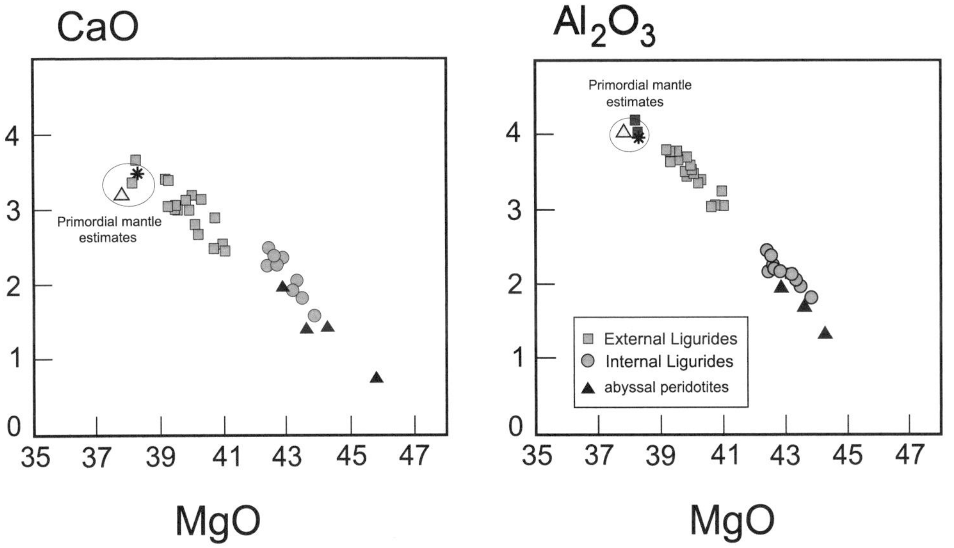

Figure 2. Whole-rock abundances of CaO and Al_2O_3 vs. MgO for the Internal and External Liguride peridotites (data from Rampone et al., 1995, 1996; all data on anhydrous basis in wt%). Primordial mantle estimates are from Hofmann (1988) and Jagoutz et al. (1979). The representative compositions of abyssal peridotites are from Dick (1989).

al., 1962; Marroni et al., 1998, and references therein). In some larger olistoliths, the primary relationships between granitoids, ophiolitic basalts, and radiolarian cherts are preserved, and, in places, granite remnants are interposed between overlying pillowed basalts and mantle peridotites.

PETROLOGY OF THE NORTHERN APENNINE OPHIOLITES

As just outlined, the predominant lithotype in the Northern Apennine ophiolites is mantle peridotite; however, the peridotites of the External and Internal Ligurides display distinctly different petrologic and isotopic features. Gabbros are more common in the Internal Ligurides whereas basalts occur in both. The External Liguride ophiolites also show relationships with rocks from the lower continental crust.

The External Liguride Peridotites

The External Liguride peridotites crop out as kilometer-scale bodies that largely preserve mantle textures and assemblages despite widespread serpentinization. They are dominantly fertile spinel lherzolites with isotropic, granular textures. Pyroxenite bands, mainly consisting of spinel-bearing Al-augite clinopyroxenites and Cr-diopside websterites, are common within the peridotites. They represent high-pressure cumulates from basaltic melts (Rampone et al., 1995, and references therein).

The External Liguride peridotites are characterized by a fertile composition and have only slight depletion in fusible components, as indicated by

- lherzolitic modal compositions (clinopyroxene up to 10–15 vol%),
- relatively high Al_2O_3 (2.86–4.00 wt%) and CaO (2.33–3.39 wt%) contents (Fig. 2); a few samples even approach primordial-mantle values (Jagoutz et al., 1979),
- clinopyroxene REE (rare earth element) spectra, showing only moderate light REE (LREE) depletion (Ce_N/Sm_N = 0.6–0.8) and absolute concentrations at 10–16 times C1 chondrite (Fig. 3A),
- the high Na, Ti, Sr, and Zr contents in clinopyroxenes and whole rocks (Ernst and Piccardo, 1979; Beccaluva et al., 1984; Ottonello et al., 1984; Rampone et al., 1993, 1995).

The External Liguride lherzolites display a complete static equilibrium recrystallization under spinel-facies conditions. Disseminated kaersutitic to Ti-pargasitic amphiboles occur in equilibrium with the spinel-bearing assemblage: they show LREE-depleted REE spectra (Fig. 3A) and very low Sr, Zr, and Ba contents. The occurrence of disseminated, LREE-depleted, Ti-rich amphibole is a ubiquitous feature of many subcontinental lithospheric peridotites (Vannucci et al., 1995, Ionov et al., 1997, and references therein). The stability of pargasitic amphibole fixes the spinel-facies re-equilibration to temperatures lower than 1100 °C; this result is in agreement with thermometric estimates on the spinel-facies assemblage, which yield temperatures in the range of 1000–1100 °C. The equilibrium spinel-facies recrystallization in a rather "cold" ($T < 1100$ °C) thermal regime has been interpreted as a stage of annealing recrystallization after accretion of the External Liguride mantle to the conductive lithosphere (Piccardo et al., 1994; Rampone et al., 1995). Information about the timing of lithospheric accretion has been derived from Rb-Sr and Sm-Nd isotope studies (Rampone et al., 1995).

Present-day Sr and Nd isotope ratios for the External Liguride peridotites are shown in Figure 4. As a whole, they plot within the depleted end of the MORB field, similar to many subcontinental orogenic spinel lherzolites from the western Mediterranean area (e.g., the Pyrenees and Lanzo North peridotites). One sample, in particular, displays extremely depleted Sr and Nd isotopic compositions ($^{87}Sr/^{86}Sr$ = 0.701736; $^{143}Nd/^{144}Nd$ = 0.513543). Similar, exceptionally low Sr ratios have been also measured in the Lanzo North and Pyrenean peridotite massifs (see Fig. 4) and have been considered to reflect their isolation from the convective mantle a long time ago and incorporation in the subcontinental lithosphere. This interpretation is supported by model-age calculations (Fig. 5A). Most External Liguride peridotites display Nd model ages (assuming a CHUR [chondritic uniform reservoir] mantle source) in the range of 1.9–1.7 Ga (Fig. 5A), and consistent results are obtained with Rb-Sr systematics (Rampone et al., 1995). Moreover, the single sample with extremely depleted Sr and Nd isotopic compositions yields Sr and Nd model ages of 1.2 and 2.1 Ga, respectively (a Nd model age of 780 Ma is obtained if an extremely depleted mantle source is assumed; see Rampone et al., 1995, for more details): these can be considered as minimum ages of differentiation from the asthenospheric mantle. The coupling of high $^{143}Nd/^{144}Nd$ (>0.51310) and rather low $^{147}Sm/$

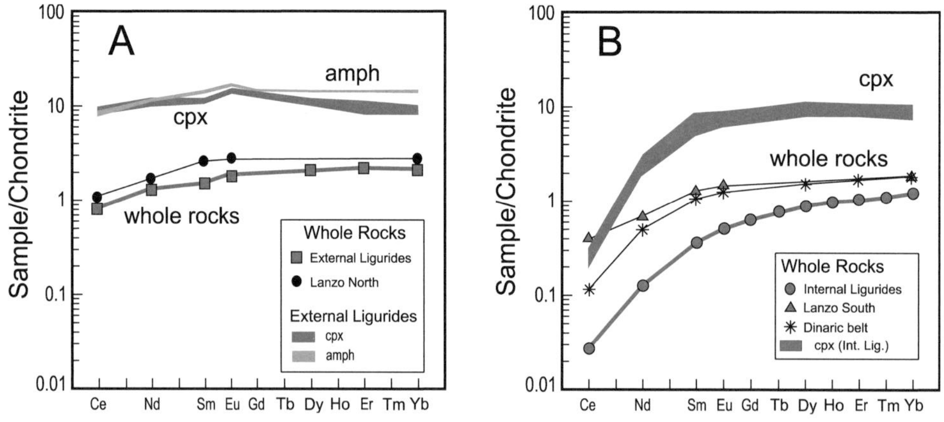

Figure 3. Representative chondrite-normalized REE compositions. A, Whole rocks, clinopyroxenes (cpx), and pargasitic amphiboles (amph) from the External Liguride peridotites (data from Rampone et al., 1995); we have also reported, for comparison, a representative whole-rock composition for the Lanzo North peridotites from the Western Alps (Bodinier, 1988). B, Whole rocks and clinopyroxenes from the Internal Liguride (Int. Lig.) peridotites; the Internal Liguride data are compared with those of the Lanzo South (Bodinier, 1988) and Central Dinaric ophiolite belt (Lugovic et al., 1991) peridotites. Normalizing values are from Anders and Ebihara (1982).

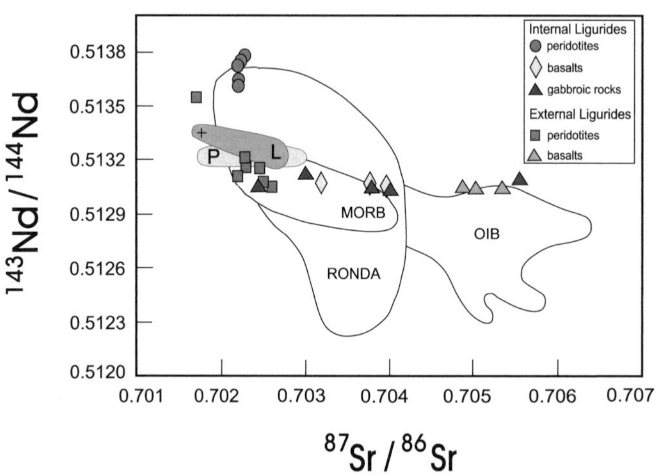

Figure 4. Present-day $^{143}Nd/^{144}Nd$ vs. $^{87}Sr/^{86}Sr$ diagram for the Northern Apennine (Internal and External Liguride) peridotites and basalts and for the Internal Liguride gabbroic rocks (data from Rampone et al., 1995, 1996, 1998). Also shown are the fields for MORB and OIB (oceanic-island basalt) (Hofmann, 1997), as well as for the Ronda peridotites (Reisberg et al., 1989, and references therein). Field L refers to the Lanzo (Western Alps) peridotites (data from Bodinier et al., 1991); the cross inside the field indicates the composition of the sample with the lowest $^{87}Sr/^{86}Sr$ ratio. Field P refers to the Pyrenean peridotites (data from Downes et al., 1991).

^{144}Nd (0.22–0.30) ratios is a peculiar feature of many subcontinental peridotites (see the data for Lanzo North, Ronda, and Pyrenees in Fig. 5A) and records ancient (usually Proterozoic) depletion events.

In the case of the Ronda and Pyrenean peridotites, Proterozoic ages of lithospheric accretion have also been obtained by their striking correlations between $^{187}Os/^{188}Os$ and Al_2O_3 contents (Reisberg and Lorand, 1995; Burnham et al., 1998) (see Fig. 5B). Such correlations between an isotopic ratio and a whole-rock element abundance are solely recorded by the Re-Os systematics; they yield old, Proterozoic model ages that reflect the time since the peridotites became isolated from the convective mantle (Reisberg and Lorand, 1995). The External Liguride peridotites also display a correlation between the Al_2O_3 contents and the $^{187}Os/^{188}Os$ ratios (Fig. 5B) that is consistent with the trends defined by the Ronda and Pyrenean peridotites. This correlation provides further evidence for a Proterozoic age of lithospheric accretion of the External Liguride mantle, although the External Liguride compositional range is too limited to determine a robust Os model age (Snow et al., 2000).

The External Liguride peridotites and associated pyroxenites show the effects of subsequent tectonic and metamorphic evolution in terms of widespread partial recrystallization from spinel- to plagioclase-bearing assemblages, and progressive deformation leading to the development of porphyroclastic textures and tectonite to mylonite fabrics in kilometer-wide shear zones (Beccaluva et al., 1984; Piccardo et al., 1990; Rampone et al., 1993, 1995). The spinel- to plagioclase-facies transition is reflected by peculiar microstructures: (1) large exsolution lamellae of orthopyroxene and plagioclase within spinel-facies clinopyroxene, (2) plagioclase rims around spinel and, (3) granoblastic domains consisting of olivine + plagioclase + neoblastic pyroxenes. This recrystallization is accompanied by significant within-mineral major and trace element redistribution that, according to mass-balance calculations, occurred in a closed system without concomitant partial melting (Rampone et al., 1993). This inference is consistent with both the fertile composition of the External Liguride peridotites and the relatively low temperatures (T = 900–950 °C) obtained for the plagioclase-facies metamorphic recrystallization (Beccaluva et al., 1984; Rampone et al., 1995). Sm-Nd data on plagioclase-clinopyroxene pairs have yielded ages of ca. 165 Ma for this metamorphic recrystallization (Rampone et al., 1995). The subsequent retrograde evolution, which ended with the seafloor emplacement, is documented by (1) the development of amphibole-bearing assemblages and later widespread serpentinization and (2) the intrusion of chilled basaltic dikes crosscutting the mantle tectonite and mylonite fabrics (Piccardo et al., 1990).

The evolution recorded by the External Liguride peridotites can be summarized as follows: (1) old (presumably Proterozoic) accretion of the peridotites to the lithosphere and annealing recrystallization (at $T < 1100$ °C) under spinel-facies conditions, (2) Early Jurassic decompressional evolution, and (3) shallow-level emplacement. The

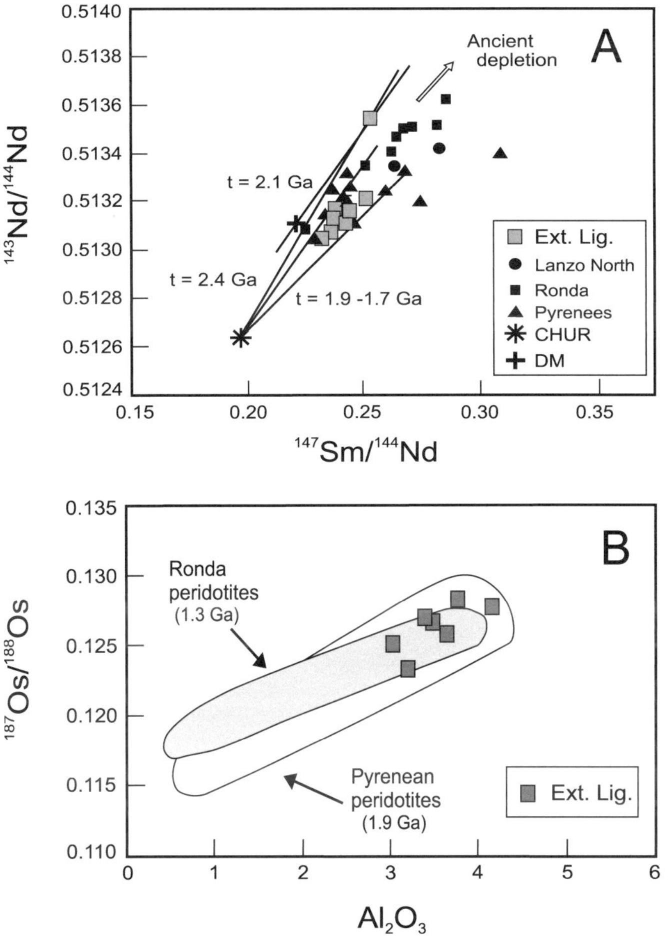

Figure 5. A, $^{143}Nd/^{144}Nd$ vs. $^{147}Sm/^{144}Nd$ diagram for the External Liguride (Ext. Lig.) peridotites (data from Rampone et al., 1995), compared with data from other subcontinental orogenic peridotites (Ronda, Reisberg et al., 1989; Pyrenees, Downes et al., 1991, Mukasa et al., 1991; Lanzo North, Bodinier et al., 1991). All data are from clinopyroxene separates. The DM (depleted mantle) and CHUR (chondritic uniform reservoir) source ratios are, respectively, $^{143}Nd/^{144}Nd = 0.513114$, $^{147}Sm/^{144}Nd = 0.222$ and $^{143}Nd/^{144}Nd = 0.512638$, $^{147}Sm/^{144}Nd = 0.1967$ (see text and Rampone et al., 1995, for more explanation). B, $^{187}Os/^{188}Os$ vs. Al_2O_3 diagram for the External Liguride peridotites (whole-rock data; from Snow et al., 2000). Also shown are the fields for the compositions of the Ronda (shaded; Reisberg and Lorand, 1995) and Pyrenees (unshaded; Reisberg and Lorand, 1995; Burnham et al., 1998) peridotites. They both define strong positive correlations that yield ages of 1.9 ± 0.3 Ga (Pyrenees) and 1.3 Ga (Ronda) for the isolation of these peridotites from the convective mantle and their incorporation into the subcontinental lithosphere.

External Liguride lherzolites therefore represent a slice of subcontinental lithospheric mantle that underwent progressive tectonic unroofing during early stages of rifting and oceanic opening of the Ligurian Tethys.

The Internal Liguride peridotites

The Internal Liguride peridotites consist of clinopyroxene-poor (5–10 vol%) spinel lherzolites. As a whole, the Internal Liguride peridotites show compositions depleted in fusible components: their MgO, CaO, and Al_2O_3 contents are comparable to those of abyssal peridotites (see Fig. 2). Their REE spectra are strongly fractionated ($Ce_N/Yb_N = 0.015–0.028$) and display extremely low LREE absolute concentrations: these compositions are among the most depleted of ophiolitic peridotites from the Western Mediterranean area (see Fig. 3B). Geochemical modeling indicates that the Internal Liguride peridotite compositions are consistent with mantle residua left after low degrees (<10%) of fractional melting, presumably initiated in the garnet-facies stability field (Rampone et al., 1996).

In terms of major and trace element chemistry, the Internal Liguride ultramafic rocks thus bear strong similarities to present-day oceanic mantle. They could therefore represent depleted oceanic mantle that produced the Jurassic MORB-type crust of the Ligurian Tethys. A Middle to Late Jurassic age of shallow emplacement for the Internal Liguride peridotites is fixed by the Oxfordian–Callovian age of the overlying radiolarian cherts (Marcucci and Passerini, 1991). The question is whether the partial melting event responsible for their residual signature occurred during the Jurassic uplift. In this respect, isotopic investigations have provided new and unexpected insights.

The present-day Sr and Nd isotope ratios of the Internal Liguride peridotites are shown in Figure 4. Their $^{87}Sr/^{86}Sr$ ratios (0.702203–0.702285) are consistent with a MORB-type mantle, but the $^{143}Nd/^{144}Nd$ ratios (0.513619–0.513775) are very high and plot significantly above the MORB field. Their high Nd ratios are coupled to very high $^{147}Sm/^{144}Nd$ ratios (0.54–0.56), as shown in the $^{147}Sm/^{144}Nd$ versus $^{143}Nd/^{144}Nd$ diagram (Fig. 6A), where the Internal Liguride ultramafic rocks are compared with other depleted peridotites of the Western Alpine belt (e.g., the Lanzo South peridotite, Bodinier et al., 1991, and Balmuccia peridotite, Ivrea zone, Voshage et al., 1988). It is noteworthy that the Internal Liguride samples display the most "extreme" Nd isotope compositions. As discussed in Rampone et al. (1996, 1998), such compositions are not consistent with a Jurassic partial-melting event of a MORB-type asthenosphere: Nd model ages calculated for an average Internal Liguride peridotite composition and assuming a depleted mantle (DM) source (see Fig. 6A) yield a Permian time of depletion ($t = 275$ Ma). A Permian age (ca. 270 Ma) of partial melting has been also inferred for the Balmuccia peridotites (Voshage et al., 1988), which plot on the trend defined by the Internal Liguride peridotites and the DM source in Figure 6A. It is interesting that the Lanzo South peridotites also plot on this trend; however, they have been interpreted as "MORB material that rose as an asthenospheric diapir in the early Mesozoic, shortly before having been emplaced at shallow level" (Bodinier et al., 1991, p. 207). Furthermore, asthenospheric mantle upwelling and melting during the Permian is well documented by the widespread occurrence, in the Alpine belt, of Permian gabbroic bodies intruded beneath or within thinned continental crust (see discussions in Piccardo et al., 1994, Rampone et al., 1996, 1998, and references therein). Thus, the Permian age of depletion recorded by the Internal Liguride ultramafic rocks is surprising as they represent the upper mantle of the Jurassic oceanic lithosphere, but it is reasonable in the frame of the Permian extension-related mantle partial melting and magma production in the Alpine realm (Dal Piaz, 1993, and references therein).

The Internal Liguride peridotites, after the melting event, were completely recrystallized under spinel-facies conditions; the spinel-facies assemblages record equilibration temperatures in the range of 1150–1250 °C (Beccaluva et al., 1984; Rampone et al., 1996). The peridotites subsequently underwent an extended subsolidus decompressional history. They were first partially re-equilibrated under plagioclase-facies conditions (Beccaluva et al., 1984; Rampone et al., 1996) and then were intruded by gabbroic bodies. Both the peridotites and gabbros then underwent a tectono-metamorphic retrograde evolution, from upper-amphibolite-facies to greenschist-facies condi-

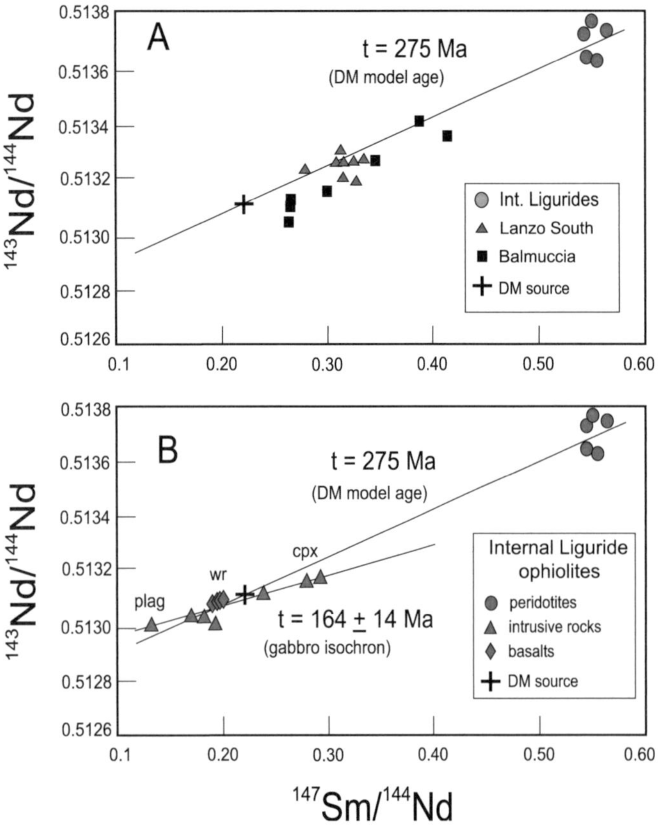

Figure 6. $^{143}Nd/^{144}Nd$ vs. $^{147}Sm/^{144}Nd$ diagrams. A, Internal Liguride peridotites (data from Rampone et al., 1996), compared with data from the Balmuccia (Ivrea zone, Voshage et al., 1988) and Lanzo South (Bodinier et al., 1991) peridotites. All data refer to clinopyroxene separates (see Fig. 5 for definition of the DM source and text for explanation on the model-age calculation). B, Internal Liguride peridotite data together with data from basalts (whole rocks, wr) and gabbroic rocks (whole rocks, clinopyroxene [cpx] and plagioclase [plag] separates) of the Internal Liguride ophiolites (data from Rampone et al., 1998).

tions, which ended with the seafloor emplacement documented by the widespread serpentinization and intense tectonization causing the formation of ophicalcites.

Briefly, the main evolution stages of the Internal Liguride peridotites are (1) Permian partial melting and subsequent equilibration under spinel-facies conditions, (2) Middle Jurassic gabbroic intrusion and progressive decompressional uplift, and (3) Late Jurassic seafloor emplacement.

Gabbroic rocks

Gabbroic rocks mostly occur as intrusive bodies and dikes in the peridotites and are more commonly found in the Internal Liguride sequences. As a whole, their compositions range from ultramafic cumulates to plagiogranites. The predominant rock types are ultramafic cumulates (plagioclase-bearing dunite), Mg-rich gabbros (troctolite, olivine gabbro, minor orthopyroxene-bearing gabbro, and rare anorthosite), intermediate gabbros (gabbronorite), Fe- and Ti-rich gabbroids (Fe-gabbro and diorite), and plagiogranites (diorite, quartz diorite, and trondjemite) (Serri, 1980; Hebert et al., 1989).

The Northern Apennine ophiolitic gabbroids represent the cumulate products at different stages of low-pressure crystal fractionation from tholeiitic MORB-type parental magmas (Beccaluva and Piccardo, 1978; Serri, 1980; Hebert et al., 1989; Tiepolo et al., 1997; Marroni et al., 1998; Tribuzio et al., 1999). This conclusion is mainly documented by

1. the crystallization order (olivine + Cr-Mg spinel, plagioclase, clinopyroxene ± orthopyroxene, Fe-Ti oxides), typical of fractionation of MORB tholeiites under low to moderate pressures;

2. the mineral-chemistry trends defined by the main compositional parameters of the constituent minerals (e.g., the Fo content of olivine, the An content of plagioclase, and the Mg-number [Mg/(Mg + Fe)] of pyroxenes) (Piccardo, 1995), which are consistent with trends recorded by oceanic cumulates (Elthon et al., 1992); and

3. the trace element composition of computed liquids in equilibrium with clinopyroxene from olivine gabbros, which supports a MORB affinity (Piccardo, 1995; Tiepolo et al., 1997; Tribuzio et al., 1999) (Fig. 7A).

A MORB signature is also denoted by available Sm-Nd isotope data on some plagioclase-bearing dunites and an olivine gabbro from the Internal Liguride ophiolites that define an isochron giving an age of 164 ± 14 Ma and an initial ϵ_{Nd} value (at 164 Ma) of 8.6 (see Fig. 6B and Rampone et al., 1998). As clearly documented in Figure 6B, the Internal Liguride gabbroic rocks and peridotites do not fall on the same linear array, thus indicating that peridotites and associated gabbroic rocks are not linked by a simple residua and melt relationship, as expected in a mature oceanic lithosphere; these features have been discussed by Rampone et al. (1998).

As a whole, available geochronologic data on the Northern Apennine ophiolitic gabbroids yield intrusive ages older than the age of basaltic extrusion and chert deposition. In particular, relatively younger ages are recorded by the Internal Liguride intrusive rocks (160–167 Ma; Bigazzi et al., 1972, 1973; Rampone et al., 1998) with respect to gabbroic rocks from the External Ligurides (185 Ma; Bigazzi et al., 1973). Triassic ages of magmatic emplacement are also documented for gabbroic rocks from the Montgenevre ophiolites in the Western Alps (192–212 Ma, Carpena and Caby, 1984; 185 Ma, Costa and Caby, 1997). These data therefore indicate that the gabbroic rocks were mostly intruded in a preoceanic subcontinental environment, prior to the Late Jurassic oceanic breakup (Dal Piaz and Lombardo, 1985; Piccardo et al., 1994).

As already outlined for the peridotites, the Northern Apennine ophiolitic gabbros underwent a retrograde tectono-metamorphic evolution, from upper-amphibolite- to greenschist- to low-grade-facies conditions, which was mainly recorded along localized ductile shear zones (Molli, 1994, Marroni et al., 1998, and references therein). Similar decompressional metamorphic histories are documented for many Alpine ophiolitic gabbros and have been related to the exhumation of these rocks, in response to the Triassic–Jurassic rifting of the Ligurian Sea sector of the Tethys Ocean (Lemoine et al., 1987, and references therein).

Basaltic rocks

Basaltic rocks are abundant in both the External and Internal Liguride ophiolites; they mostly occur as pillow or massive lava flows and as discrete dikes intruding deformed gabbros and mantle peridotites. Petrologic and geochemical studies of the Alpine-Apennine ophiolitic basalts have provided clear evidence of their overall mid-oceanic-ridge tholeiite affinity (Ferrara et al., 1976; Venturelli et al., 1981; Beccaluva et al., 1984; Ottonello et al., 1984; Vannucci et al., 1993a; Rampone et al., 1998). As a whole, the Internal and External Liguride basaltic rocks display a large degree of differentiation;

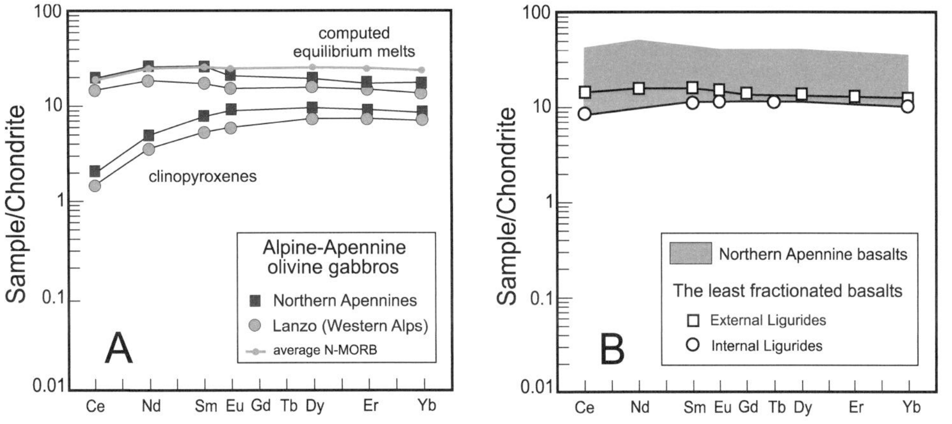

Figure 7. Chondrite-normalized REE abundances. A, Representative clinopyroxenes in ophiolitic olivine gabbros from Lanzo in the Western Alps (Bodinier et al., 1991) and from the Northern Apennines (Rampone et al., 1997; Tribuzio et al., 1999). Also shown are the compositions of computed equilibrium melts (calculated by using $D_{cpx/melt}$ partition coefficients from Hart and Dunn, 1993) and an average composition for N-MORB (from Hofmann, 1988). B, Representative whole-rock data for the least fractionated Internal and External Liguride basalts (data from Ottonello et al., 1984, and Vannucci et al., 1993a, respectively). The field represents the entire range of compositions defined by the Northern Apennine basalts (in both External and Internal Ligurides) (data from Ottonello et al., 1984; Venturelli et al., 1981; Vannucci et al., 1993a; Marroni et al., 1998; Rampone et al., 1998).

this fact is demonstrated by their REE compositions, which range from ~10 times C1 chondrite to >40 times C1 chondrite (see the gray field in Fig. 7B; data from Venturelli et al., 1981, Beccaluva et al., 1984, Ottonello et al., 1984, Marroni et al., 1998). The most primitive Internal Liguride basalts show moderate LREE fractionation (Ce_N/Sm_N = 0.6) and heavy REE (HREE) abundances at ~10 times C1 chondrites, whereas even the least differentiated External Liguride basalts display almost flat or slightly LREE-enriched REE spectra (Ce_N/Sm_N = 0.9–1) (Fig. 7B). On the basis of a combined investigation of whole-rock and clinopyroxene trace element chemistry, it has been demonstrated that the compositions of both the External and Internal Liguride basalts are consistent with magmas generated by varying degrees of fractional melting (totaling no more than 10%) of a MORB-type asthenospheric spinel-facies mantle source (Vannucci et al., 1993a). The MORB affinity of the Northern Apennine ophiolitic basalts has been recently confirmed by a Sr-Nd isotope study that has shown that, in spite of a large $^{87}Sr/^{86}Sr$ variability caused by seawater alteration, the Internal and External Liguride basalts have fairly homogeneous $^{143}Nd/^{144}Nd$ ratios ranging from 0.513046 to 0.513098 (Rampone et al., 1998; see Fig. 4).

Information concerning the age of basaltic activity in the Ligurian Tethys has been mostly derived from U-Pb dating of zircons in silicic differentiates (plagiogranites and albitites), which occur as veins or dikes crosscutting the contacts between the serpentinite-gabbro basement and the overlying breccias and intruding the massive basalts. They are therefore considered to be almost contemporaneous with the basaltic volcanism. These data have yielded ages in the range of 150–160 Ma for the basaltic volcanism (Bortolotti et al., 1991, 1995; Borsi et al., 1996; Ohnenstetter et al., 1981; Costa and Caby, 1997). This age is consistent with paleontologic data on the coeval radiolarian cherts, which indicate timing of deposition not older than 150–160 Ma (De Wever and Caby, 1981; Lemoine, 1983; Marcucci and Passerini, 1991).

External Liguride continental crust

Ophiolitic rocks (pillow basalts and fertile lherzolites) cropping out in the External Ligurides are spatially associated with lower continental crust (mafic and felsic granulite) and granitoid upper crust (Eberhardt et al., 1962; Marroni et al., 1998, and references therein). On the basis of petrologic and geochronologic studies, the mafic granulites have been inferred to have been derived from gabbroic protoliths (crystallized from tholeiitic melts) that were emplaced at low-crustal levels (P = ~8 kbar) during Early Permian time (290 Ma; Marroni and Tribuzio, 1996). These mafic granulites subsequently underwent a retrograde metamorphic evolution from granulite- to amphibolite- to greenschist-facies conditions. Radiometric data (Ar-Ar amphibole age of 228 Ma; Meli et al., 1996) suggest that the granulite- to amphibolite-facies recrystallization occurred between Permian and Middle Triassic time (Marroni et al., 1998). A similar retrograde evolution has been also inferred for the quartzo-feldspatic granulites that are primarily associated with mafic granulites (Marroni et al., 1998).

The granitoids mostly consist of leucogranites; there are subordinate granodiorites, tonalites, and diorites. Petrologic and radiometric data indicate that the leucogranites were emplaced at shallow levels (P < 3 kbar) (Marroni et al., 1998) during late Hercynian time (310–280 Ma) (Eberhardt et al., 1962; Ferrara and Tonarini, 1985). After the Middle Triassic, the granitoids were cataclastically deformed and were finally brought to shallow crustal levels during the Jurassic, as testified to by the intrusion of basaltic dikes and by the occurrence of primary stratigraphic contacts between granitoids and Oxfordian–Callovian radiolarian cherts (Marroni et al., 1998, and references therein).

ORIGINAL TECTONIC SETTING OF THE NORTHERN APENNINE OPHIOLITES

In the past, different interpretations of the main structural and/or petrologic features of the Alpine-Apennine Jurassic ophiolites have led to the development of different models for their original tectonic setting. Besides the common interpretation suggesting that the Northern Apennine ophiolites represent sections of a MORB-type fossil oceanic lithosphere, mainly based on the MORB affinity of the magmatic rocks, the inferred widespread occurrence of peridotites exposed on the paleo-seafloor has led some authors to suggest that these ophiolites were formed in a transform-zone setting (Gianelli and Principi, 1977; Lemoine, 1980; Weissert and Bernoulli, 1985).

Investigations on the structure of present-day oceanic lithosphere have provided increasingly strong evidence that serpentinized mantle peridotites are extensively exposed on the seafloor in slow-spreading ridge environments. Serpentinized peridotites are intruded by discrete gabbroic bodies and only partially covered by basaltic lava flows. On the basis of these features, it has been argued by some authors that Alpine-Apennine ophiolites represent mature oceanic lithosphere formed in a slow-spreading ridge setting (Barrett and Spooner, 1977; Lagabrielle and Cannat, 1990; Lagabrielle and Lemoine, 1997).

Other contributions, however, stressed the probable subcontinental origin of many mantle rocks flooring the Ligurian Ocean (Decandia and Elter, 1969, 1972; Piccardo, 1976) and opened the discussion on the diversity of the Alpine-Apennine ophiolites compared with mature oceanic lithosphere. On the basis of the atypical association of fertile subcontinental-type lherzolitic mantle and MORB magmatism (as recognized in the External Ligurides), it was argued that these ophiolites formed during early stages of opening of an oceanic basin, subsequent to a continental rifting episode. This interpretation implies a marginal (pericontinental) position of this oceanic basin (Beccaluva and Piccardo, 1978). Further petrologic investigations on the Northern Apennine ophiolitic ultramafic rocks demonstrated the depleted compositions of the Internal Liguride peridotites and implied the existence of a residua and melt relationship between the peridotites and associated MORB magmatism (Ottonello et al., 1984). It was therefore asserted that the Internal Liguride ophiolites represent oceanic lithosphere produced during a more evolved stage of development of the Ligurian Tethys (Beccaluva et al., 1984). The comparison with peridotites from different settings in modern oceanic basins emphasized close similarities between the External Liguride peridotites and the marginal-basin and prerift lherzolites as well as between the Internal Liguride ultramafic rocks and mid-ocean ridge peridotites (Piccardo et al., 1990).

However, recent petrologic and isotope investigations of ultramafic rocks of the Northern Apennine ophiolites (as reported previously in this chapter) have shown that none of the Ligurian peridotites can be considered as typical oceanic mantle and that a simple mantle-residua and basaltic-melt genetic relationship does not exist even in the Internal Liguride ophiolites.

It is clear that the External Liguride peridotites consist of fertile subcontinental lithospheric mantle (presumably Proterozoic), whereas the Internal Liguride mantle ultramafic rocks are composed of depleted peridotites that underwent MORB-type partial melting during the Permian, well before production of the associated Jurassic basaltic crust. Both the External and Internal Liguride peridotites display a composite subsolidus retrograde evolution, which reflects their uplift from lithospheric mantle depths, and subsequent emplacement on the ocean floor.

The Northern Apennine ophiolites therefore represent the spatial association of older (Proterozoic and Permian), subcontinental mantle peridotites, which are partly still linked to continental-crust material, and younger (mostly Jurassic), unrelated, MORB-type igneous rocks. This peculiar association is unlike what is observed in the present-day mature oceanic lithosphere, in which the mantle and the associated mafic crust are linked by a direct genetic relationship. In contrast, the Northern Apennine ophiolites most likely represent lithologic associations that are expected to develop after the breakup of continental crust in response to passive extension of the lithosphere.

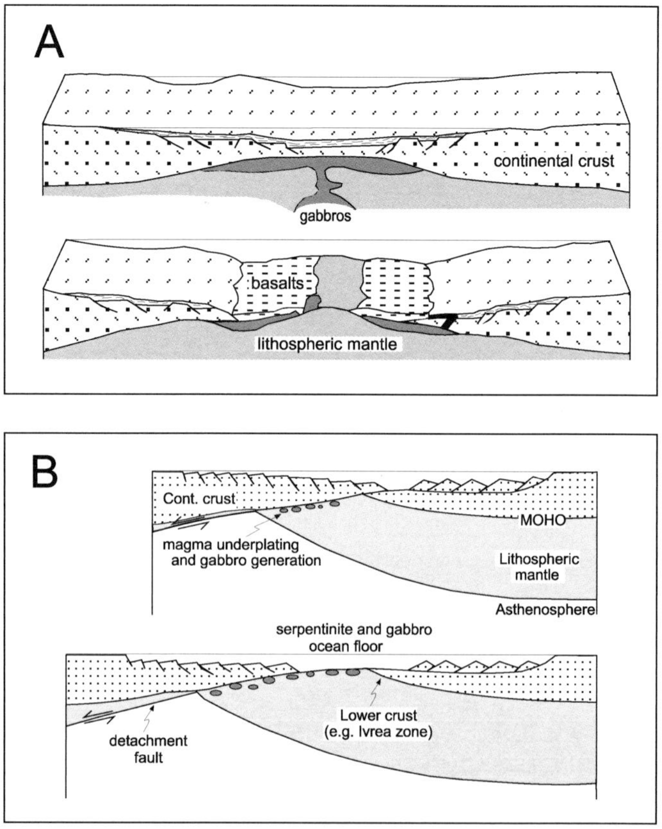

Figure 8. Models of mantle denudation and oceanic inception of the Jurassic Ligurian Tethys by passive extension of the continental lithosphere; left-side and right-side continental margins are, respectively, Europe and Adria. A, Symmetric extension according to Elter (1972). Upper section: Triassic–Dogger interval; note thinning of the continental crust and the underplating of the gabbroic bodies. Lower section: Configuration at Malm after the breakup of the continental crust, the unroofing of the mantle lherzolites, and the extrusion of the pillowed basalts (redrawn after Elter, 1972). B, Asymmetric extension according to Lemoine et al. (1987). Upper section: Liassic–Middle Jurassic rifting; note the underplating of gabbroic bodies. Lower section: Late Middle Jurassic spreading and seafloor exposure of lithospheric subcontinental mantle with gabbroic intrusions (redrawn after Lemoine et al., 1987).

This latter geodynamic process is the most suitable to account for the tectonic denudation of large sectors of subcontinental mantle. The exhumation of lithospheric mantle in response to passive lithospheric extension has been confirmed by analogue geophysical modeling (Brun and Beslier, 1996) performed to study the modes of passive-margin formation and related mantle exhumation at the continent-ocean boundary.

Mechanisms of passive extension of the lithosphere were already invoked in pioneering studies on the Northern Apennine ophiolites to explain the peculiar structural features of the seafloor of the Ligurian Tethys (Decandia and Elter, 1969, 1972; Fig. 8A). More recently, models of asymmetric passive extension of the lithosphere by means of normal simple-shear mechanisms (according to the model of Wernicke, 1985) have been proposed to account for some asymmetrical features recognized on both sides of the preoceanic continental rift of the Ligurian Tethys. On the basis of the contrast between synextensional (Liassic–Dogger) subsidence of the Adria margin and uplift of the Briançonnais ridge of the European margin, Adria and Europe are interpreted as the lower- and upper-plate conjugate margins, respectively, during simple-shear extension (Lemoine et al., 1987; Hoogerduijn Strating et al., 1990, 1993; Vissers et al., 1991) (see Fig. 8B).

Other contributions have proposed an eastward-dipping detachment zone such that Adria would form the upper plate of the extensional system. This interpretation is based on structural, metamorphic, and magmatic features recorded by the Austroalpine (Sesia-Lanzo) and South Alpine (Ivrea-Verbano) units (i.e., the marginal units of the Adria plate), which indicate their involvement in the preoceanic extensional regime (Dal Piaz, 1993; Trommsdorff et al., 1993; Piccardo et al., 1994). According to these authors, the occurrence of post-Hercynian gabbroic bodies in the Austroalpine units of the Western and Central Alps indicates the early intrusion into the extending Adria lithosphere of basaltic magmas produced by partial melting of the passively upwelling asthenosphere. Moreover, the original location of the gabbroic bodies indicates that asthenospheric partial melting started asymmetrically with respect to the future axis of the continental breakup and that the melting zone shifted progressively toward the location of the future oceanic basin (Fig. 9) (see Piccardo et al., 1994, Rampone et al., 1998, and references therein).

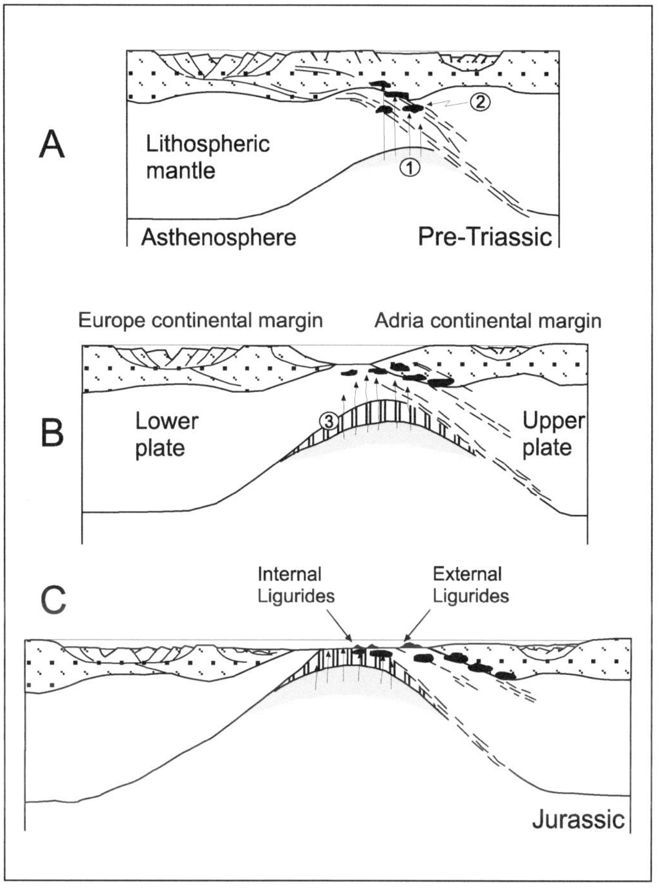

Figure 9. Model for the evolution of the Ligurian Sea sector of the Tethys Ocean from post-Hercynian (pre-Triassic) to Jurassic continental breakup and seafloor spreading (redrawn after Piccardo et al., 1994). The model only considers the Permian basaltic magmatism and the intrusion of mantle-derived melts (in black) into the extending Adria lithosphere. Late Variscan crustal magmatism is not considered. A, Inception of the asymmetric passive extension of the continental lithosphere. The upwelling asthenospheric mantle (1) undergoes partial melting, producing (2) primary melts for the Permian gabbroic intrusions and (3) leaving Permian residual mantle subsequently accreted to the subcontinental lithosphere. B, Jurassic breakup of the continental crust and seafloor exposure of the subcontinental lithospheric mantle (formation of External Liguride ophiolites). C, Seafloor exposure of the depleted Permian mantle and formation of the Internal Liguride ophiolites. More evolved stages of ocean spreading are not recorded in the Northern Apennine ophiolites.

MODERN OCEANIC ANALOGUES

Modern analogues for the original oceanic setting of the Alpine-Apennine ophiolites are represented by peculiar oceanic settings located along passive continental margins of mature oceans (e.g., the western Galicia margin, Spain), and by embryonic oceans (e.g., the Red Sea).

The western Galicia margin

The western Galicia margin formed from the divergence between the Iberian and North American plates during the Late Jurassic–Early Cretaceous and is characterized by the occurrence of mantle peridotites that are directly overlain by thinned continental upper crust at the ocean-continent boundary (Fig. 10). Peridotites mainly consist of spinel lherzolites whose compositional features (bulk rock and mineral chemistry) form a rather fertile signature, not comparable to the composition of typical mid-oceanic ridge peridotites (Evans and Girardeau, 1988; Boillot et al., 1988; Girardeau et al., 1988; Kornprobst and Tabit, 1988). The Galicia peridotites then record a complex retrograde evolution, characterized by progressive deformation (from coarse-grained to mylonite-type fabrics) and recrystallization (from spinel- to plagioclase-bearing assemblages) that occurred under decreasing P and T conditions and led to the peridotites' exposure (Kornprobst and Tabit, 1988). This evolution is quite consistent with that already described for the External Liguride peridotites.

Thus, petrologic and structural evidence suggests that the Galicia peridotites represent a piece of deep subcontinental lithospheric mantle, rather than the oceanic mantle of the Atlantic Ocean. This subcontinental mantle was tectonically denuded and emplaced on the ocean floor in response to passive extension and thinning of the lithosphere that occurred during the rifting stage of the North Atlantic (Boillot et al., 1980, 1998; Kornprobst and Tabit, 1988; Brun and Beslier, 1996).

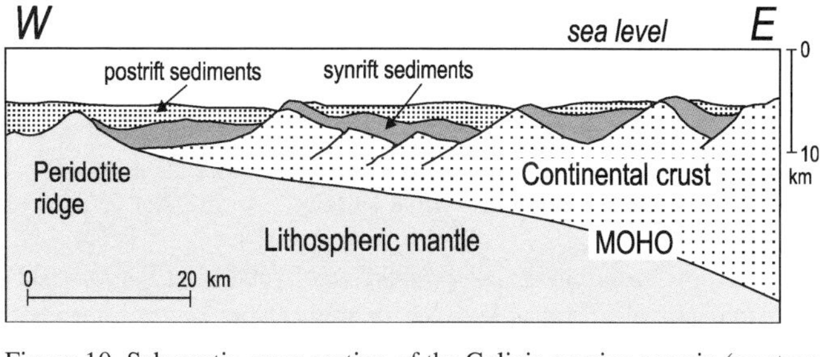

Figure 10. Schematic cross section of the Galicia passive margin (western Spain), as inferred from seismic data and direct sampling (after Boillot et al., 1988).

The embryonic ocean of the northern Red Sea

The Red Sea is believed to have formed by asymmetric passive extension of the Nubian-Arabian continental lithosphere during the

Cenozoic (Bohannon et al., 1989, Voggenreiter et al., 1988, and references therein). Geological and geophysical data indicate that the southern part of the Rea Sea is characterized by a well-developed axial rift valley, which is floored by MORB. By contrast, oceanic crust is not yet developed in the northern Red Sea, which consists of thinned and extended continental lithosphere, most probably intruded by abundant basaltic dikes (Bonatti et al., 1986, and references therein). The unique opportunity for the direct investigation of the lithosphere underneath this embryonic ocean is provided by the small island of Zabargad (St. John's; northern Red Sea), which represents an uplifted sliver of the sub–Red Sea lithosphere (Styles and Gerdes, 1983, Bonatti et al., 1981, and references therein).

Zabargad Island exposes mantle peridotites (northern, central, and southern bodies) spatially associated with polymetamorphic Pan-African continental basement rocks, sedimentary sequences of Cretaceous–Paleocene and Miocene age, and recent reef limestones (Bonatti et al., 1981, and references therein). The Zabargad peridotites have been interpreted as (1) lithospheric subcontinental mantle of Pan-African age (Bonatti et al., 1986; Brueckner et al., 1988; Piccardo et al., 1988, 1990) or (2) asthenospheric mantle that ascended during the Red Sea rifting (Nicolas et al., 1987). The latter interpretation has been subsequently restricted to the southern body (Dupuy et al., 1991). On the basis of the results of detailed structural-petrologic and isotopic investigations on the three peridotite bodies, the subcontinental origin of Zabargad peridotites is considered more reliable (see Piccardo et al., 1993, for a thorough discussion). The peridotites display compositional and structural features that are strikingly similar to those described for the External Liguride lherzolites. They are mostly fertile spinel lherzolites, whose undepleted signatures are indicated by the high contents of fusible components in both bulk rocks and constituent minerals. Whole-rock REE spectra show slight negative LREE fractionation and MREE (middle REE) to HREE values at ~2 times C1 chondrite (Bonatti et al., 1986; Piccardo et al., 1988, 1993). Peridotites are cut by Al-diopside pyroxenite dikes that crystallized from tholeiitic melts at relatively high pressures and temperatures (Piccardo et al., 1988, 1993; Vannucci et al., 1993b). After this event, Zabargad lherzolites and pyroxenites underwent complete equilibrium recrystallization to spinel-facies conditions at temperatures lower than 1100 °C. LREE-depleted amphiboles (kaersutite and Ti-pargasite) are widespread and in structural and chemical equilibrium with the spinel-facies assemblage. This spinel-facies equilibration has been interpreted as the stage of annealing to the regional geotherm after accretion to the lithosphere (Piccardo et al., 1993).

Zabargad lherzolites show isotope compositions similar to those of mantle xenoliths from the Cenozoic volcanic fields of the western Arabian margin (type 1A of Henjes-Kunst et al., 1990), which represent the subcontinental lithospheric mantle of the area (McGuire, 1988; Henjes-Kunst et al., 1990). When plotted on a $^{147}Sm/^{144}Nd$ versus $^{143}Nd/^{144}Nd$ diagram, Zabargad and Arabian spinel lherzolites define a linear array consistent with a 700 Ma reference isochron (Fig. 11). This age corresponds to the timing of Pan-African magmatic and tectono-metamorphic activity, well documented in the continental crust of the Nubian-Arabian shield and in the basement crustal rocks exposed on Zabargad (Lancelot and Bosch, 1990; Brueckner et al., 1995).

On this basis, the Zabargad lherzolites have been interpreted as mantle material that was accreted to the subcontinental lithosphere during the Pan-African orogeny, long before their Cenozoic involvement in the Red Sea rifting (Brueckner et al., 1988; Piccardo et al., 1993). An even older (Proterozoic) age of lithospheric accretion has

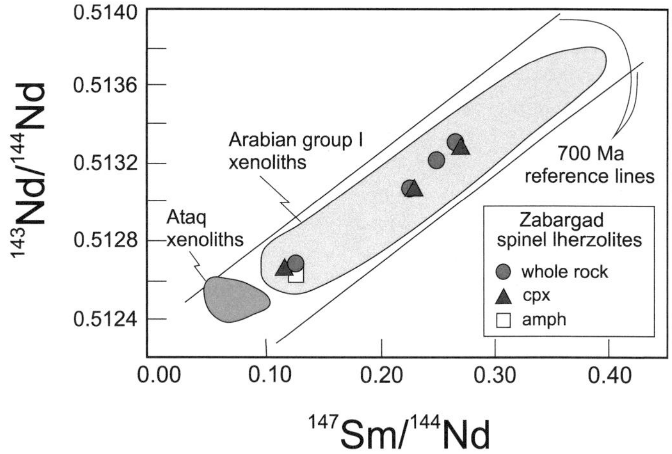

Figure 11. $^{143}Nd/^{144}Nd$ vs. $^{147}Sm/^{144}Nd$ diagram for the Zabargad (Red Sea) spinel lherzolites. The data refer to whole rocks, clinopyroxene (cpx), and pargasitic amphibole (amph) separates from peridotites from the central and southern bodies (all data from Brueckner et al., 1988). Also shown are the fields for the Arabian group I and Ataq xenoliths (Henjes-Kunst et al., 1990; Menzies and Murthy, 1980), which represent the lithospheric mantle underneath the western Arabian peninsula. These data define a positive correlation, almost parallel to 700 Ma reference lines (see text and Henjes-Kunst et al., 1990, for further explanation).

been recently inferred for the Zabargad peridotites, on the basis of Os isotope data (Snow and Schmidt, 1998). Zabargad peridotites then underwent a decompressional evolution marked by the development along shear zones of tectonite-to-mylonite fabrics and by the partial recrystallization to plagioclase-bearing and later amphibole-bearing assemblages (Piccardo et al., 1988, 1993). Subsequent sea-water alteration occurred as a consequence of the Cenozoic Red Sea rifting, which exposed the lherzolites at the seafloor (Bonatti et al., 1986).

The peridotites are spatially associated with granulite-facies mafic and felsic gneisses. On the basis of isotope investigations, it is inferred that the granulite-facies metamorphism might have occurred during the Pan-African event. They are therefore considered as a fragment of continental crust of the Arabian-Nubian shield (Lancelot and Bosch, 1990; Brueckner et al., 1995). The gneisses record a retrograde metamorphic recrystallization from amphibolite- to greenschist-facies assemblages (Boudier et al., 1988; Seyler and Bonatti, 1988) that parallels the retrograde history recorded by the peridotites. This evolution has been related to the progressive uplift of the peridotite-gneiss association during the Red Sea rifting and crustal thinning (Brueckner et al., 1995, and references therein).

Mantle rocks and continental crustal rocks are cut by various generations of basaltic dikes of Quaternary age (5 Ma), which are related to the final stages of the Red Sea rifting (Petrini et al., 1988, and references therein). Major and trace element chemistry, and Nd isotope ratios (0.51290–0.51307) indicate that the basalts are olivine tholeiites with transitional MORB affinity.

Geophysical and direct petrologic and geochemical data from the sub–Red Sea lithosphere exposed on Zabargad Island thus suggest that the northern Red Sea consists of subcontinental lithospheric mantle, which is associated with continental lower crust and younger MORB-type dike intrusions. These findings collectively indicate the onset of partial melting of the rising asthenosphere beneath the northern Red Sea. This peculiar association represents the early stages of inception of an oceanic basin developed by means of passive extension of the continental lithosphere (Fig. 12).

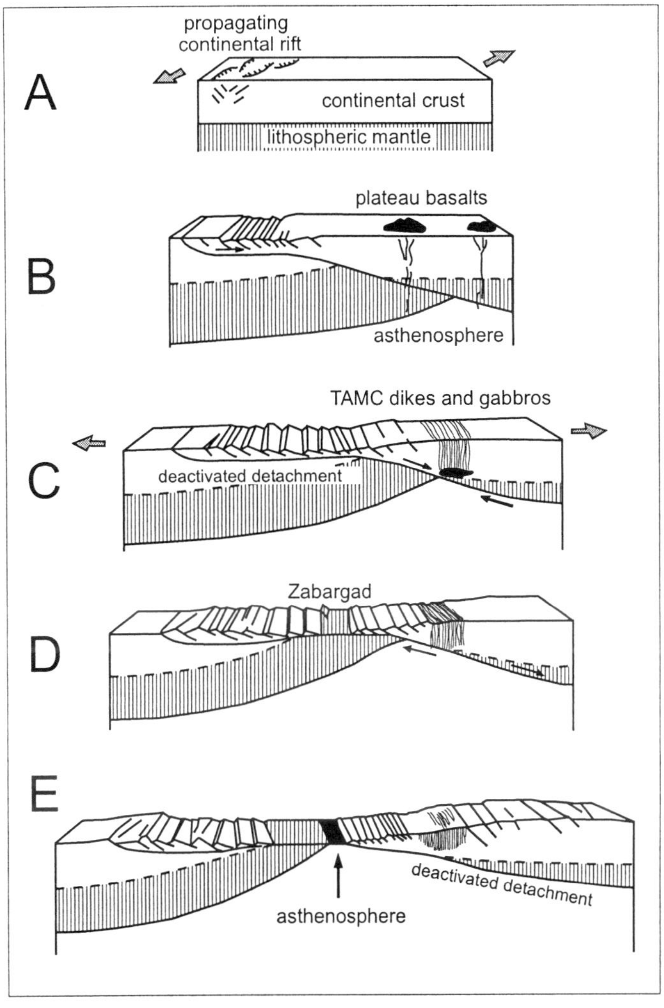

Figure 12. Model for the evolution of the Red Sea, according to an asymmetric extensional process, from early transtensional stages (propagating continental rift) to late oceanization (redrawn after Voggenreiter et al., 1988, and Stampfli et al., 1991). A, Transtensional stage (lower Oligocene). B, Simple shear initiation. C, Asthenosphere passive upwelling, partial melting, and emplacement of layered gabbroic bodies and basaltic dikes (Tihama Asir Magmatic Complex, TAMC). D, Tectonic denudation of the subcontinental mantle (e.g., the Zabargad peridotites). E, Beginning of oceanization (from 5 Ma to the present), as recognized for the southern sector of the Red Sea.

CONCLUSIONS

Petrologic and isotopic studies of the Northern Apennine ophiolites have provided clear evidence that none of the Liguride peridotites can be considered as typical oceanic mantle. In the External Ligurides, subcontinental lithospheric peridotites (presumably Proterozoic) are associated with Hercynian continental crust. Both mantle and crustal rocks display a composite subsolidus retrograde evolution, related to their uplift toward the seafloor, where they became associated with Jurassic MORB. This peculiar lithologic association is typically recorded in embryonic oceans (e.g., the Red Sea) and in the transition zone between continental and oceanic lithosphere of mature oceanic basins (e.g., the Galicia margin). The Internal Liguride ophiolites, classically related to more evolved oceanic stages of the Jurassic Ligurian Tethys consist of the spatial association of depleted peridotites (of Permian age) and younger (mostly Jurassic), unrelated, MORB-type magmatism. Thus, even in this case, peridotites and associated magmatic rocks are not linked by a simple genetic relationship, as is expected for the oceanic lithosphere generated at mid-ocean ridges.

The current data and interpretations suggest that the Jurassic Ligurian Tethys was mostly floored by older (Proterozoic and Permian) subcontinental lithospheric mantle. The most suitable geodynamic process to account for the tectonic denudation of large sectors of subcontinental mantle is passive extension of the continental lithosphere. The exposure of subcontinental mantle peridotites on the ocean floor is a unique feature testifying to the early stages of oceanization by means of passive lithosphere extension. This process is currently recorded in embryonic oceans, such as the Red Sea, and at passive margins, such as the Galicia Bank (western Spain).

The Northern Apennine ophiolites therefore cannot be interpreted as remnants of fossil mature oceanic lithosphere. By contrast, they most likely represent the lithologic associations that develop after the breakup of the continental crust in response to the passive extension of the continental lithosphere.

ACKNOWLEDGMENTS

The research was financially supported by Italian MURST (Ministero dell'Università e della Ricerca Scientifica e Tecnologica) and CNR (Consiglio Nazionale delle Ricerche). G. Ernst, M. Menzies, and J. Snow are greatly thanked for their constructive reviews. We also thank G.V. Dal Piaz, P. Elter, and V. Trommsdorff for their helpful comments on an early draft of the manuscript. We are indebted with Y. Dilek for his careful handling of the paper.

REFERENCES CITED

Abbate, E., Bortolotti, V., Passerini, P., Principi, G., and Treves, B., 1994, Oceanisation processes and sedimentary evolution of the Northern Apennine ophiolite suite: A discussion: Memorie della Società Geologica Italiana, v. 48, p. 117–136.

Anders, E., and Ebihara, M., 1982, Solar system abundances of the elements: Geochimica et Cosmochimica Acta, v. 46, p. 2363–2380.

Barrett, J.J., and Spooner, E.F.C., 1977, Ophiolitic breccias associated with allochtonous oceanic crustal rocks in the East Ligurian Apennines, Italy: A comparison with observations from rifted oceanic ridges: Earth and Planetary Science Letters, v. 35, p. 79–91.

Beccaluva, L., and Piccardo, G.B., 1978, Petrology of the Northern Apennine ophiolites and their significance in the Western Mediterranean area, *in* Closs, H., Roeder, D., and Schmidt, K., eds., Alps, Apennines, Hellenides: Inter-Union Commission on Geodynamics Scientific Report 38, E. Schweizerbart'sche Verlagsbuchhandlung, Stuttgart, p. 243–254.

Beccaluva, L., Macciotta, G., Piccardo, G.B., and Zeda, O., 1984, Petrology of lherzolitic rocks from the Northern Apennine ophiolites: Lithos, v. 17, p. 299–316.

Bigazzi, G., Ferrara, G., and Innocenti, F., 1972, Fission track ages of gabbros from Northern Apennines ophiolites: Earth and Planetary Science Letters, v. 14, p. 242–244.

Bigazzi, G., Bonadonna, F.P., Ferrara, G., and Innocenti, F., 1973, Fission track ages of zircons and apatites from Northern Apennine ophiolites: Fortschritte der Mineralogie, v. 50, p. 51–53.

Bodinier, J.L., 1988, Geochemistry and petrogenesis of the Lanzo peridotite body, western Alps: Tectonophysics, v. 149, p. 67–88.

Bodinier, J.L., Menzies, M.A., and Thirwall, M.F., 1991, Continental to oceanic mantle transition: REE and Sr-Nd isotopic geochemistry of the Lanzo Lherzolite Massif, *in* Menzies, M.A., Dupuy, C., and Nicolas, A., eds., Orogenic lherzolites and mantle processes: Journal of Petrology, Special Lherzolite Issue, p. 191–210.

Bohannon, R.G., Naeser, C.W., Schmidt, D.L., and Zimmermann, R.A., 1989, The timing of uplift, volcanism, and rifting peripheral to the Red Sea: A case for passive rifting?: Journal of Geophysical Research, v. 94, p. 1683–1701.

Boillot, G., Grimaud, S., Mauffret, A., Mougenot, D., Mergoil-Daniel, J., Kornprobst, J., and Torrent, G., 1980, Ocean-continent boundary off the Iberian margin: A serpentinite diapir west of the Galicia bank: Earth and Planentary Science Letters: v. 48, p. 23–34.

Boillot, G., Girardeau, J., and Kornprobst, J., 1988, Rifting of the Galicia margin: Crustal thinning and emplacement of mantle rocks on the seafloor, *in* Boillot, G., Winterer, E.L., et al., Proceedings of the Ocean Drilling Program, Scientific Results: College Station, Texas, Ocean Drilling Program, v. 103, p. 741–756.

Bonatti, E., Hamlyn, P.R., and Ottonello, G., 1981, The upper mantle beneath a young oceanic rift: Peridotites from the island of Zabargad (Red Sea): Geology, v. 9, p. 474–491.

Bonatti, E., Ottonello, G., and Hamlyn, P.R., 1986, Peridotites from the Island of Zabargad (St. John's), Red Sea: Petrology and geochemistry: Journal of Geophysical Research, v. 21, p. 559–631.

Borsi, L., Scharer, U., Gaggero, L., and Crispini, L., 1996, Age, origin, and geodynamic significance of plagiogranites in lherzolites and gabbros of the Piedmont-Ligurian ocean basin: Earth and Planetary Science Letters, v. 140, p. 227–241.

Bortolotti, V., Cellai, D., Vaggelli, G., and Villa, I.M., 1991, $^{40}Ar/^{39}Ar$ dating of Apenninic ophiolites: 2. Basalts from the Aiola sequence, Southern Tuscany: Ofioliti, v. 16, p. 37–42.

Bortolotti, V., Cellai, D., Chiari, M., Vaggelli, G., and Villa, I.M., 1995, $^{40}Ar/^{39}Ar$ dating of Apenninic ophiolites: 3. Plagiogranites from Sasso di Castro, Northern Tuscany, Italy: Ofioliti, v. 20, p. 55–65.

Boudier, F., Nicolas, A., Kiénast, J.R., and Mével, C., 1988, The gneiss of Zabargad Island: Deep crust of a rift: Tectonophysics, v. 150, p. 209–227.

Brueckner, H.K., Zindler, A., Seyler, M., and Bonatti, E., 1988, Zabargad and the isotopic evolution of the sub–Red Sea mantle and crust: Tectonophysics, v. 150, p. 163–176.

Brueckner, H.K., Elhaddad, M.A., Hamelin, B., Hemming, S., Kröner, A., Reiberg, L., and Seyler, M., 1995, A Pan African origin and uplift for the gneisses and peridotites of Zabargad Island, Red Sea: A Nd, Sr, Pb, and Os study: Journal of Geophysical Research, v. 100 (B11), p. 22283–22297.

Brun, J.P., and Beslier, M.O., 1996, Mantle exumation at passive margins: Earth and Planetary Science Letters, v. 142, p. 161–173.

Burnham, O.M., Rogers, N.W., Pearson, D.G., Van Calsteren, P.W., and Hawkesworth, C.J., 1998, The petrogenesis of the Eastern Pyrenean peridotites: An integrated study of their whole-rock geochemistry and Re-Os isotope composition: Geochimica et Cosmochimica Acta, v. 62, p. 2293–2310.

Carpena, J., and Caby, R., 1984, Fission-track evidence for Late Triassic oceanic crust in the French Occidental Alps: Geology, v. 12, p. 108–111.

Costa, S., and Caby, J., 1997, Evolution of the Ligurian Tethys in the Western Alps: REE study and Sm-Nd and U-Pb geochronology of the Montgenevre ophiolites (Southeast of France): Terra Nova 9, Abstract Supplement 1, p. 510.

Dal Piaz, G.V., 1993, Evolution of Austroalpine and upper Pennine basement in the North-Western Alps from Variscan convergence to post-Variscan extension, *in* Von Raumer, J.F., and Neubauer, F., eds., Pre-Mesozoic geology in the Alps: Berlin, Heidelberg, and New York, Springer-Verlag, p. 327–344.

Dal Piaz, G.V., and Lombardo, B., 1985, Review of radiometric dating in the Western Italian Alps: Rendiconti della Società Italiana di Mineralogia e Petrologia, v. 40, p. 125–138.

Decandia, F.A., and Elter, P., 1969, Riflessioni sul problema delle ofioliti nell'Appennino Settentrionale (nota preliminare): Atti della Società Toscana di Scienze Naturali, v. 75, p. 1–9.

Decandia, F.A., and Elter, P., 1972, La zona ofiolitifera del Bracco nel settore compreso fra Levanto e la Val Graveglia (Appennino Ligure): Memorie della Società Geologica Italiana, v. 11, p. 503–530.

De Wever, P., and Caby, R., 1981, Datation de la base des Schistes Lustrés postophiolitiques par des radiolaires (Oxfordien Supérieur–Kimmeridgien Moyen) dans les Alpes Cottiennes (Saint-Veran, France): Comptes Rendus de l'Académie des Sciences, Paris, v. 292, p. 467–472.

Dick, H.J.B., 1989, Abyssal peridotites, very slow spreading ridge and ocean ridge magmatism, *in* Saunders, A.D., and Norry, M.J., eds., Magmatism in the ocean basins: Geological Society of London Special Publication 42, p. 71–106.

Downes, H., Bodinier, J.L., Thirwall, M.F., Lorand, J.P., and Fabries, J., 1991, REE and Sr-Nd isotopic geochemistry of the Eastern Pyrenean peridotite massifs: Subcontinental lithospheric mantle modified by continental magmatism, *in* Menzies, M.A., Dupuy, C., and Nicolas, A., eds., Orogenic lherzolites and mantle processes: Journal of Petrology, Special Lherzolite Issue, p. 97–116.

Dupuy, C., Mevel, C., Bodinier, J.-L., and Savoyant, L., 1991, Zabargad peridotite: Evidence for multistage metasomatism during Red Sea rifting: Geology, v. 19, p. 722–725.

Eberhardt, P., Ferrara, G., and Tongiorgi, E., 1962, Determination de l'age de granites allochtones de l'Apennin septentrional: Bulletin de la Societé Geologique de France, v. 4, p. 666–667.

Elter, P., 1972, La zona ofiolitica del Bracco nel quadro dell'Appennino Settentrionale: Introduzione alla geologia delle Liguridi: Pacini, Pisa, 66° Congresso della Società Geologica Italiana, Guida alle Escursioni, p. 5–35.

Elthon, D., Stewart, M., and Ross, D.K., 1992, Compositional trends of minerals in oceanic cumulates: Journal of Geophysical Research, v. 97, p. 15189–15199.

Ernst, W.G., and Piccardo, G.B., 1979, Petrogenesis of some Ligurian peridotites: I. Mineral and bulk rock chemistry: Geochimica et Cosmochimica Acta, v. 43, p. 219–237.

Evans, C., and Girardeau, J., 1988, Galicia margin peridotites: Undepleted abyssal peridotites from the North Atlantic, *in* Boillot, G., Winterer, E.L., et al., eds., Proceedings of the Ocean Drilling Program, Scientific Results: College Station, Texas, Ocean Drilling Program, v. 103, p. 195–207.

Ferrara, G., and Tonarini, S., 1985, Radiometric geochronology in Tuscany: Results and problems: Rendiconti della Società Italiana di Mineralogia e Petrologia, v. 40, p. 111–124.

Ferrara, G., Innocenti, F., Ricci, C.A., and Serri, G., 1976, Ocean-floor affinity of basalts from Northern Apennine ophiolites: Geochemical evidence: Chemical Geology, v. 17, p. 101–111.

Gianelli, G., and Principi, G., 1977, Northern Apennine ophiolite: An ancient transcurrent fault zone: Bollettino della Società Geologica Italiana, v. 96, p. 53–58.

Girardeau, J., Evans, C., and Beslier, M.O., 1988, Structural analyses of plagioclase-bearing peridotites emplaced at the end of continental rifting: Hole 637A, ODP Leg 103 on the Galicia margin, *in* Boillot, G., Winterer, E.L., et al., eds., Proceedings of the Ocean Drilling Program, Scientific Results, v. 103: College Station, Texas, Ocean Drilling Program, v. 103, p. 209–223.

Hart, S.R., and Dunn, T., 1993, Experimental cpx/melt partitioning of 24 trace elements: Contributions to Mineralogy and Petrology, v. 113, p. 1–8.

Hebert, R., Serri, G., and Hekinian, R., 1989, Mineral chemistry of ultramafic tectonites and ultramafic to gabbroic cumulates from the major oceanic basins and Northern Apennine ophiolites (Italy): A comparison: Chemical Geology, v. 77, p. 183–207.

Henjes-Kunst, F., Altherr, R., and Baumann, A., 1990, Evolution and composition of the lithospheric mantle underneath the Western Arabian peninsula: Constraints from Sr-Nd isotope systematics of mantle xenoliths: Contributions to Mineralogy and Petrology, v. 105, p. 460–472.

Hofmann, A.W., 1988, Chemical differentiation of the earth: The relationships between mantle, continental crust, and oceanic crust: Earth and Planetary Science Letters, v. 90, p. 297–314.

Hofmann, A.W., 1997, Mantle geochemistry: The message from oceanic volcanism: Nature, v. 385, 219–229.

Hoogerduijn Strating, E.H., Piccardo, G.B., Rampone, E., Scambelluri, M., Vissers, R.L.M., 1990, The structure and petrology of the Erro-Tobbio peridotite (Voltri Massif, Ligurian Alps): A two day excursion with emphasis on processes in the upper mantle: Ofioliti, v. 15, p. 119–184.

Hoogerduijn Strating, E.H., Rampone, E., Piccardo, G.B., Drury, M.R., and Vissers, R.L.M., 1993, Subsolidus emplacement of mantle peridotites during incipient oceanic rifting and opening of the Mesozoic Tethys (Voltri Massif, NW Italy): Journal of Petrology, v. 34, p. 901–927.

Ionov, D., Griffin, W.L., and O'Reilly, S., 1997, Volatile-bearing minerals and lithophile trace elements in the upper mantle: Chemical Geology, v. 141, p. 153–184.

Jagoutz, E., Palme, H., Baddenhausen, H., Blum, K., Cendales, M., Dreibus, G.,

Spettel, B., Lorenz, V., and Wanke, H., 1979, The abundance of major, minor, and trace elements in the earth's mantle derived from ultramafic nodules: Proceedings of Lunar and Planetary Science Conference, 10th, p. 2031–2050.

Kornprobst, J., and Tabit, A., 1988, Plagioclase-bearing ultramafic tectonites from the Galicia margin (Leg 103, Site 637): Comparison of their origin and evolution with low pressure ultramafic bodies in western Europe, *in* Boillot, G., Winterer, E.L., et al., eds., Proceedings of the Ocean Drilling Program, Scientific Results: College Station, Texas, Ocean Drilling Program, v. 103, p. 253–263.

Lagabrielle, Y., and Cannat, M., 1990, Alpine Jurassic ophiolites resemble the modern Central Atlantic basement: Geology, v. 18, p. 319–322.

Lagabrielle, Y., and Lemoine, M., 1997, Alpine, Corsican and Apennine ophiolites: The slow-spreading ridge model: Comptes Rendus de l'Académie des Sciences, Paris, v. 325, p. 909–920.

Lancelot, J., and Bosch, D., 1990, A Pan-African age for the lower continental crust on Zabargad Island and implications for the development of the Red Sea rift, Colloquium of African Geology, 15th, Nancy, France, Abstract, p. 154.

Lemoine, M., 1980, Serpentinites, gabbros, and ophicalcites in the Piemont-Ligurian domain of the Western Alps: Possible indicators of oceanic fracture zones and of associated serpentinite protrusions in the Jurassic-Cretaceous Tethys: Geneva, Archives des Sciences, v. 33, 103–115.

Lemoine, M., 1983, Serpentinite oceanic bottom in south Queyras ophiolite (French Western Alps): Record of the incipient oceanic opening of the Mesozoic Ligurian Tethys: Eclogae Geologicae Helvetiae, v. 76, p. 611–629.

Lemoine, M., Tricart, P., and Boillot, G., 1987, Ultramafic and gabbroic ocean floor of the Ligurian Tethys (Alps, Corsica, Apennines): In search of a genetic model: Geology, v. 15, p. 622–625.

Lugovic, B., Altherr, R., Raczek, I., Hofmann, A.W., and Majer, V., 1991, Geochemistry of peridotites and mafic igneous rocks from the Central Dinaric Ophiolite Belt, Yugoslavia: Contributions to Mineralogy and Petrology, v. 106, p. 201–216.

Marcucci, M., and Passerini, P., 1991, Radiolarian-bearing siliceous sediments in the Mesozoic of the Northern and Central Apennines: Ofioliti, v. 16, p. 121–126.

Marroni, M., and Tribuzio, R., 1996, Gabbro-derived granulites from External Liguride Units (Northern Apennine, Italy): Implications for the rifting processes in the western Tethys: Geologische Rundschau, v. 85, p. 239–249.

Marroni, M., Molli, G., Montanini, A., and Tribuzio, R., 1998, The association of continental crust rocks with ophiolites in the Northern Apennines (Italy): Implications for the continent-ocean transition in the Western Tethys: Tectonophysics, v. 292, p. 43–66.

McGuire, A.V., 1988, The mantle beneath the Red Sea margin: Xenoliths from western Saudi Arabia: Tectonophysics, v. 150, p. 101–119.

Meli, S., Montanini, A., Thoni, M., and Frank, W., 1996, Age of mafic granulite blocks from the External Liguride Units (Northern Apennine, Italy): Padova, Memorie di Scienze Geologiche, v. 48, p. 65–72.

Menzies, M.A., and Murthy, V.R., 1980, Nd and Sr isotope geochemistry of hydrous mantle nodules and their host alkali basalts: Implications for local heterogeneities in metasomatically veined mantle: Earth and Planetary Science Letters, v. 46, p. 323–334.

Molli, G., 1994, Microstructural features of high temperature shear zones in gabbros of the Northern Apennine ophiolites: Journal of Structural Geology, v. 16, p. 1535–1541.

Mukasa, S.B., Shervais, J.W., Wilshire, H.G., and Nielson, J.E., 1991, Intrinsic Nd, Pb and Sr isotopic heterogeneities exhibited by the Lherz alpine peridotite massif, French Pyrenees, *in* Menzies, M.A., Dupuy, C., and Nicolas, A., eds., Orogenic lherzolites and mantle processes: Journal of Petrology, Special Lherzolite Issue, p. 97–116.

Nicolas, A., Boudier, F., and Montigny, R., 1987, Structure of Zabargad Island and the early rifting of the Red Sea: Journal of Geophysical Research, v. 92, p. 461–474.

Ohnenstetter, M., Ohnenstetter, D., Vidal, Ph., Cornichet, J., Hermitte D., and Mace, J., 1981, Crystallization and age of zircon from Corsican ophiolitic albitites: Consequences for oceanic expansion in Jurassic times: Earth and Planetary Science Letters, v. 54, p. 397–408.

Ottonello, G., Piccardo, G.B., and Joron, J.L., 1984, Rare earth and 3rd transition element geochemistry of peridotitic rock. II. Ligurian peridotites and associated basalts: Journal of Petrology, v. 25, p. 379–393.

Petrini, R., Joron, J.L., Ottonello, G., Bonatti, E., and Seyler, M., 1988, Basaltic dykes from Zabargad Island, Red Sea: Petrology and geochemistry: Tectonophysics, v. 150, p. 229–248.

Piccardo, G.B., 1976, Petrologia del massiccio lherzolitico di Suvero (La Spezia): Ofioliti, v. 1, p. 279–317.

Piccardo, G.B., 1995, Ophiolites, *in* Ranalli, G., ed., Plate tectonics: The first twenty-five years: Proceedings of the VIII Summer School Earth and Planetary Sciences, Siena, Italy, Tipografica Senese, p. 267–296.

Piccardo, G.B., Messiga, B., and Vannucci, R., 1988, The Zabargad peridotite-pyroxenite association: Petrological constraints on its evolution: Tectonophysics, v. 150, p. 135–162.

Piccardo, G.B., Rampone, E., and Vannucci, R., 1990, Upper mantle evolution during continental rifting and ocean formation: Evidence from peridotite bodies of the Western Alpine–Northern Apennine system: Memoires de la Société Geologique de France, v. 156, p. 323–333.

Piccardo, G.B., Rampone, E., Vannucci, R., 1992, Ligurian peridotites and ophiolites: From rift to ocean formation in the Jurassic Ligure-Piemontese basin: Acta Vulcanologica, v. 2, p. 313–325.

Piccardo, G.B., Rampone, E., Vannucci, R., Shimizu, N., Ottolini, L., Bottazzi, P., 1993, Mantle processes in the subcontinental lithosphere: The case study on the rifted sp-lherzolites from Zabargad (Red Sea): European Journal of Mineralogy, v. 5, p. 1039–1056.

Piccardo, G.B., Rampone, E., Vannucci, R., and Cimmino, F., 1994, Upper mantle evolution of ophiolitic peridotites from the Northern Apennine: Petrological constraints to the geodynamic processes: Memorie della Società Geologica Italiana, v. 48, p. 137–148.

Rampone, E., Piccardo, G.B., Vannucci, R., Bottazzi, P., and Ottolini, L., 1993, Subsolidus reactions monitored by trace element partitioning: The spinel- to plagioclase-facies transition in mantle peridotites: Contributions to Mineralogy and Petrology, v. 115, p. 1–17.

Rampone, E., Hofmann, A.W., Piccardo, G.B., Vannucci, R., Bottazzi, P., and Ottolini, L., 1995, Petrology, mineral, and isotope geochemistry of the External Liguride peridotites (Northern Apennine, Italy): Journal of Petrology, v. 36, p. 81–105.

Rampone, E., Hofmann, A.W., Piccardo, G.B., Vannucci, R., Bottazzi, P., and Ottolini, L., 1996, Trace element and isotope geochemistry of depleted peridotites from an N-MORB type ophiolite (Internal Liguride, N. Italy): Contributions to Mineralogy and Petrology, v. 123, p. 61–76.

Rampone, E., Piccardo, G.B., Vannucci, R., and Bottazzi, P., 1997, Chemistry and origin of trapped melts in ophiolitic peridotites: Geochimica et Cosmochimica Acta, v. 61, p. 4557–4569.

Rampone, E., Hofmann, A.W., and Raczek, I., 1998, Isotopic contrasts within the Internal Liguride ophiolite (N. Italy): The lack of a genetic mantle-crust link: Earth and Planetary Science Letters, v. 163, p. 175–189.

Reisberg, L., and Lorand, J.-P., 1995, Longevity of sub-continental mantle lithosphere from osmium isotope systematics in orogenic peridotite massifs: Nature, v. 376, p. 159–162.

Reisberg, L., Zindler, A., and Jagoutz, E., 1989, Further Sr and Nd isotopic results from peridotites of the Ronda Ultramafic Complex: Earth and Planetary Science Letters, v. 96, p. 161–180.

Serri, G., 1980, Chemistry and petrology of gabbroic complexes from the Northern Apennine ophiolites, *in* Panayiotou, A., ed., Ophiolites, Nicosia, Cyprus Geological Survey Department, p. 296–313.

Seyler, M., and Bonatti, E., 1988, Petrology of a gneiss-amphibolite lower crustal unit from Zabargad Island, Red Sea: Tectonophysics, v. 150, p. 177–208.

Snow, J.E., and Schmidt, G., 1998, Pre-Pan African melting in the Zabargad peridotites: Os isotopic evidence: Terra Nova, v. 11, p. 45–50.

Snow, J.E., Hart, S.R., and Dick, H.J.B., 1994, Nd and Sr isotope evidence linking mid-ocean ridge basalts and abyssal peridotites: Nature, v. 371, 57–60.

Snow, J.E., Schmidt, G., and Rampone, E., 2000, Os isotopes and highly siderophile elements (HSE) in the Ligurian ophiolites, Italy: Earth and Planetary Science Letters, v. 175, p. 119–132.

Stampfli, G.M., Marcoux, J., and Baud, A., 1991, Tethyan margins in space and time: Palaeogeography, Palaeoclimatology, Palaeoecology, v. 87, p. 373–409.

Styles, P., and Gerdes, K.D., 1983, St. John's Island (Red Sea): A new geophysical model and its implications for the emplacement of ultramafic rocks in fracture zones and at continental margins: Earth and Planetary Science Letters, v. 65, p. 353–368.

Tiepolo, M., Tribuzio, R., and Vannucci, R., 1997, Mg- and Fe-gabbroids from Northern Apennine ophiolites: Parental liquids and igneous differentiation processes: Ofioliti, v. 22, p. 57–69.

Tribuzio, R., Tiepolo, M., Vannucci, R., and Bottazzi, P., 1999, Trace element distribution within the olivine-bearing gabbros from the Northern Apennine ophiolites (Italy): Evidence for post-cumulus crystallization in MOR-type gabbroic rocks: Contributions to Mineralogy and Petrology, v. 134, p. 123–133.

Trommsdorff, V., Piccardo, G.B., and Montrasio, A., 1993, From magmatism through metamorphism to sea-floor emplacement of sub-continental Adria lithosphere during pre-Alpine rifting (Malenco, Italy): Schweizerische Mineralogische und Petrographische Mitteilungen, v. 73, p. 191–203.

Vannucci, R., Rampone, E., Piccardo, G.B., Ottolini, L., and Bottazzi, P., 1993a, Ophiolitic magmatism in the Ligurian Tethys: An ion microprobe study of basaltic clinopyroxenes: Contributions to Mineralogy and Petrology, v. 115, p. 123–137.

Vannucci, R., Shimizu, N., Piccardo, G.B., Ottolini, L., and Bottazzi, P., 1993b, Distribution of trace elements during breakdown of mantle garnets: An example from Zabargad: Contributions to Mineralogy and Petrology, v. 113, p. 437–449.

Vannucci, R., Piccardo, G.B., Rivalenti, G., Zanetti, A., Rampone, E., Ottolini, L., Oberti, R., Mazzucchelli, M., and Bottazzi, P., 1995, Origin of LREE-depleted amphiboles in the subcontinental mantle: Geochimica et Cosmochimica Acta, v. 59, p. 1763–1771.

Venturelli, G., Thorpe, R.S., and Potts, P.J., 1981, Rare earth and trace element characteristics of ophiolitic metabasalts from the Alpine-Apennine belt: Earth and Planetary Science Letters, v. 53, p. 109–123.

Vissers, R.L.M., Drury, M.R., Hoogerduijn Strating, E.H., and van der Wal, D., 1991, Shear zones in the upper mantle: A case study in an Alpine lherzolite massif: Geology, v. 19, p. 990–993.

Voggenreiter, W., Hotzl, H., and Mechie, J., 1988, Low-angle detachment origin for the Red Sea rift system?: Tectonophysics, v. 150, p. 51–75.

Voshage, H., Sinigoi, S., Mazzucchelli, M., Demarchi, G., Rivalenti, G., and Hofmann, A.W., 1988, Isotopic constraints on the origin of ultramafic and mafic dykes in the Balmuccia peridotite (Ivrea Zone): Contributions to Mineralogy and Petrology, v. 100, p. 261–267.

Weissert, H.J., and Bernoulli, D., 1985, A transform margin in the Mesozoic Tethys: Evidence from the Swiss Alps: Geologische Rundschau, v. 74, p. 665–679.

Wernicke, B., 1985, Uniform sense normal simple shear of the continental lithosphere: Canadian Journal of Earth Sciences, v. 22, p. 108–125.

Manuscript Accepted by the Society February 23, 2000

Printed in the U.S.A.

Geological Society of America
Special Paper 349
2000

Significance of serpentine mud volcanism in convergent margins

Patricia Fryer
Hawaii Institute of Geophysics and Planetology, University of Hawaii, 2525 Correa Road, Honolulu, Hawaii 96822, USA
John P. Lockwood
Geohazards, P.O. Box 479, Volcano, Hawaii 96785, USA
Nathan Becker
Hawaii Institute of Geophysics and Planetology, University of Hawaii, 2525 Correa Road, Honolulu, Hawaii 96822, USA
Steven Phipps
Department of Geology, University of Pennsylvania, Philadelphia, Pennsylvania 19104-6316, USA
Clifford S. Todd
Department of Geology and Geophysics, University of Hawaii, 2525 Correa Road, Honolulu, Hawaii 96822, USA

ABSTRACT

Serpentinites are distinctive components of on-land exposures of former convergent-margin terranes throughout the world, commonly mingled with high-pressure, low-temperature blueschist-facies metamorphic rocks in complex mélange assemblages that have long perplexed geologists. Serpentinite blocks within these mélanges commonly retain vestiges of original igneous textures and were once widely considered to be igneous intrusive rocks, altered in situ. These chaotic assemblages never show contact-metamorphic aureoles, but usually do show extensive internal shearing and are commonly found along fault zones; thus an alternative explanation was that they formed as tectonically intrusive (i.e., protrusive) bodies. Early workers suggested that these serpentinites were emplaced during or after deformation and metamorphism. Particularly puzzling were occurrences of sedimentary serpentinite, masses of originally unconsolidated serpentine enclosing clasts of serpentinized peridotite and metabasites (some reaching blueschist-grade metamorphism), showing evidence of sedimentary structures and occasionally incorporating cherts and marine fossils.

Recent discoveries of large submarine serpentine mud volcanoes containing blueschist materials in the forearc of the active Mariana convergent plate margin show that serpentine mud, crystals, and clasts can be emplaced directly on the seafloor by protrusive processes. Mudflows from these serpentine seamounts are incorporated in the forearc strata before subsequent emplacement on land and before the deformation and associated metamorphism of the materials enclosing them. The Mariana forearc mud volcanism is episodic and brings up materials derived both from the descending plate and from the mantle and crustal sections of the overriding plate. These materials rise along deep-seated faults above the subduction zone and probably represent fault gouge mobilized by seismicity and rising fluids released during dehydration of materials on the downgoing plate. The serpentine mud volcanoes of the Mariana forearc are the only ones known to be currently active and to be venting slab-derived fluids. These mud volcanoes contain a wide variety of metamorphosed rocks either as clasts or as disaggregated crystals disseminated throughout the muds. Blueschist materials discovered in these seamounts indicate depths of origin of ≤30 km. Probably these materials derive from the top of the descending Pacific plate. Other, unmetamorphosed rocks of plate origin have also been recovered from several of these seamounts. Many examples of sedimentary serpentinite exist throughout the world in ancient convergent-margin settings; thus the phenomenon of serpentine mud volcanism is likely a more important agent of recycling of subducted constituents than previously thought.

INTRODUCTION

Convergent margins constitute a critical element in the study of global geochemical cycling and the evolution of continental crust. Serpentinite masses and high-pressure, low-temperature metamorphic rocks were long ago recognized as typical of such terranes where they are exposed on land (e.g., Hess, 1955) well before the concepts of "seafloor-spreading" and continental accretion were postulated. The serpentinite bodies were generally interpreted as emplaced into these highly deformed terranes either during or after tectonism. In a series of three papers in the early 1970s, Lockwood (1971a, 1971b, 1972) brought attention to the worldwide existence of sedimentary serpentinites. He proposed (1972) that many serpentinite bodies found in the margins of continental areas ("accreted terranes" in modern terms) were emplaced well before tectonism by sedimentary processes and that their sources were seafloor protrusions of unconsolidated serpentinite material that formed in forearc regions.

The discovery of serpentine mud volcanoes in the Mariana con-

Fryer, P., et al., 2000, Significance of serpentine mud volcanism in convergent margins, *in* Dilek, Y., Moores, E.M., Elthon, D., and Nicolas, A., eds., Ophiolites and Oceanic Crust: New Insights from Field Studies and the Ocean Drilling Program: Boulder, Colorado, Geological Society of America Special Paper 349, p. 35–51.

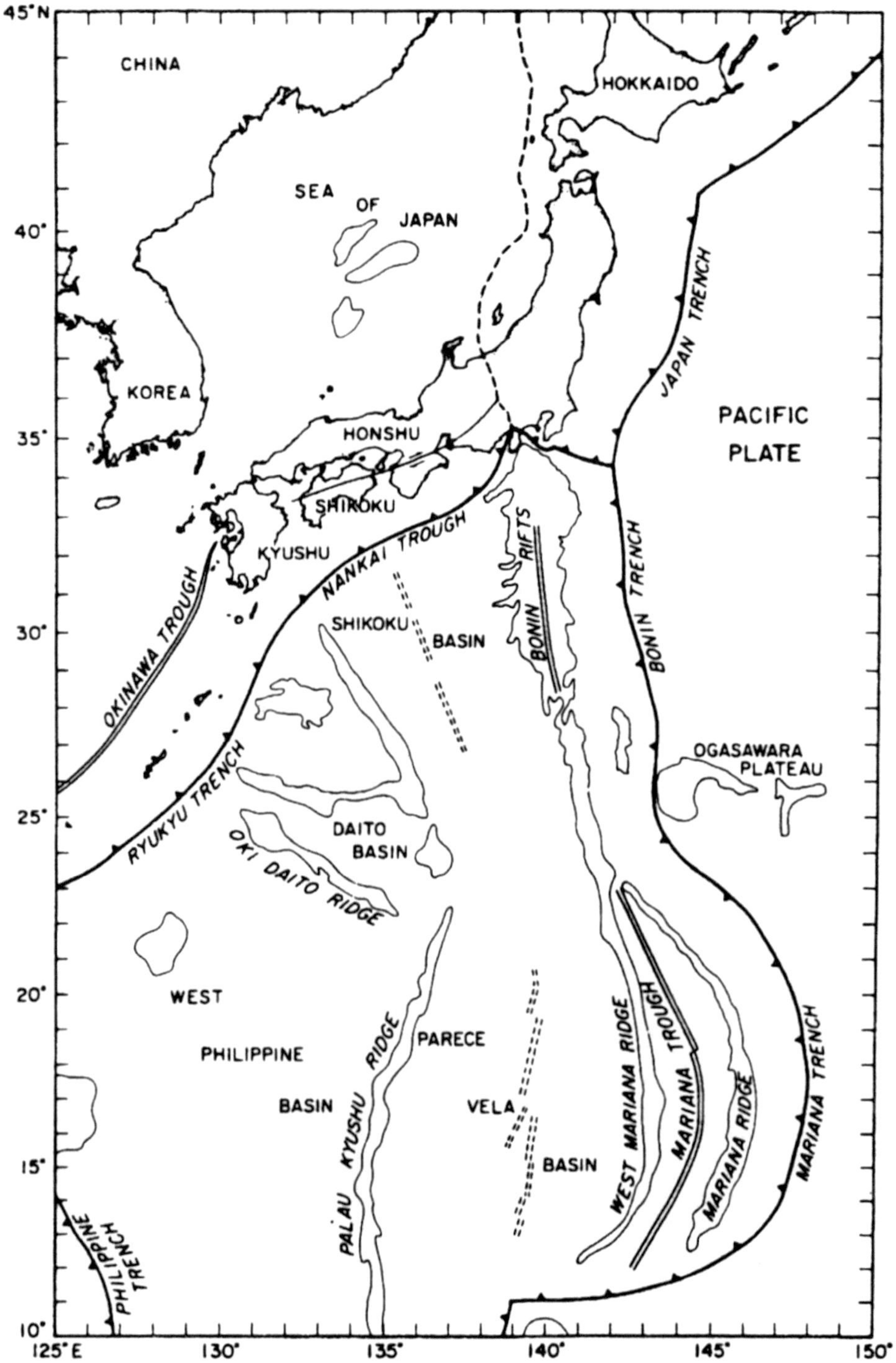

Figure 1. Bathymetry and geologic features of the central western Pacific region including the Mariana convergent-margin system. Ridges and basins are outlined by the 4 km bathymetric contour, except for the Izu-Bonin arc, West Mariana Ridge, and Mariana arc, which are outlined by the 3 km bathymetric contour. Barbed lines locate axes of trenches, medium double lines locate active spreading centers, and dashed lines locate relict spreading centers. A close-up view of the southeastern quarter of the Mariana system is shown in Figure 2 (after Fryer, 1992a).

vergent margin (Fig. 1) (Fryer et al., 1985; Fryer, 1996a) has shown that the process envisioned by Lockwood (1972) is currently taking place in an active convergent plate margin. The subducting plate releases volatiles and associated constituents through a series of dewatering events and dehydration reactions as it descends beneath the Mariana convergent margin (Fryer and Fryer, 1987; Peacock, 1990, 1993; Mottl, 1992; Liu et al., 1996). The slab-derived fluids affect the physical properties of the décollement and interact with the lithosphere of the overriding plate. Faulting of the overriding plate facilitates escape of slab-derived fluids and provides routes for both fluids and mobilized fault gouge to move upward toward the seafloor. Fault gouge generated within the fault zones mixes with slab-derived fluids to produce a gravitationally unstable slurry of fluid-laden mud and rock that rises along a conduit within the fault plane to the seafloor (Fryer, 1992a, 1996a). When this rising slurry reaches the seafloor, it protrudes to form discrete edifices—mud volcanoes—that can reach 20–30 km in diameter and grow as high as 2 km (Fryer et al., 1985, 1999; Fryer and Fryer, 1987; Fryer and Pearce, 1992; Fryer and Mottl, 1992; Fryer, 1992a, 1996a). These serpentine mud volcanoes appear to be long-lived (Fryer et al., 1995). Sediments as old as Eocene contain serpentine in drill cores from near the base of one of them (Big Blue Seamount) (Desprairies, 1981). Fryer (1992a) suggested that the mud volcanoes are active episodically, likely triggered by seismic activity on the faults beneath them. The mudflows can contain rock clasts derived from anywhere along the route to the seafloor. These may include materials from the subducted slab, e.g., blueschist-grade (high-pressure, low-temperature) rocks (Maekawa et al., 1992, 1993, 1995; Fryer et al., 1999; Fryer and Todd, 1999; Todd and Fryer, 1999) and various types of mafic rocks (Johnson, 1992). The serpentine muds also contain hemipelagic cherts, likely also slab derived (Johnson et al., 1991). Johnson et al. (1991) first interpreted dredge hauls from near one of the mud volcanoes (Conical Seamount) that contained such cherts, mid-oceanic-ridge basalt (MORB), and oceanic-island basalt (OIB) (as well as several other lithologies) as evidence of accretion of slivers of the Pacific plate onto the Mariana forearc. A subsequent *Shinkai 6500* submersible dive to the site of the dredge hauls showed the fault scarp from which they were recovered to be entirely composed of flow upon flow of serpentine muds containing numerous clasts of various lithologies. Thus, we now think it likely that the cherts and the other slab-derived rocks were entrained in serpentine muds rising from the décollement. The blueschist-facies rocks that Maekawa et al. (1992, 1993, 1995) described were recovered from ODP (Ocean Drilling Program) drill samples on Conical Seamount. These are the first blueschist-grade rocks to have been discovered in an active convergent margin and are unequivocal proof of deep-seated origin of the muds. The pressure range for the drill core samples is 5–7 kbar (~16–20 km), and the temperature range is 150 to 250 °C (Maekawa et al., 1993, their Fig. 3). The muds cored at these seamounts also contain pore fluids that have a distinct slab signature (Mottl, 1992; Fryer et al., 1999).

Note that this type of mud volcano is distinct from its much smaller cousins formed by the remobilization of sedimentary sequences in convergent-margin settings elsewhere, typically in many accretionary sediment wedges. Some examples are the mud volcanoes described from the Costa Rica margin (Shipley et al., 1990), those outboard of the Barbados forearc (Henry et al., 1990), or the mud volcanoes exposed on the Mediterranean Ridge (Robertson, 1996). Mud volcanoes formed by remobilization of sediments and associated with either expulsion of overpressured formation fluids or with expulsion of hydrocarbons attain maximum sizes that are approximately an order of magnitude smaller than those of the serpentine mud volcanoes of the Mariana system. Of course, the deposits formed by the protruded materials from sedimentary mud volcanoes are also mineralogically and geochemically distinct from those of serpentine mud volcanoes.

This paper presents detailed descriptions of several recently surveyed Mariana serpentine mud volcanoes and relates them to deposits and settings described from several localities within subaerial, ancient convergent-margin terranes. We postulate that they formed in a manner similar to that which we observe on the Mariana forearc region today. The relationships between forearc structures and the occurrence of this type of mud volcano provide a general model for the possible conditions under which such features may develop in convergent-margins settings and be preserved during obduction events.

TECTONIC CONTROL OVER DISTRIBUTION

The serpentine mud volcanoes of the Mariana forearc are associated with major fault zones in the highly disrupted, outer 100 km of the convergent margin (Fig. 2). The muds and the fluids emanating from these mud volcanoes indicate pressure and temperature conditions at depth and composition of the lithosphere in the overriding plate and along the décollement. The Mariana system is nonaccretionary (lacks a wedge of sediments accreted from the plate), and the area between the trench and the volcanic arc (the forearc) is pervasively faulted (Mrozowski et al., 1981; Fryer, 1992b; Wessel et al., 1994). Stern and Smoot (1998) suggested the reason that the seamounts are confined to the outer half of the system may be related to gradients in degree of serpentinization across the forearc, decreasing from the trench toward the volcanic arc. Fryer et al. (1985) suggested that the variation in rheology of the forearc wedge may contribute to the degree of faulting of the outer forearc. The outer half of the wedge lies within a zone of brittle lithosphere where subduction of seamounts may prompt vertical tectonism and thus facilitate a high degree of deep-penetrating fracturing of the forearc that provides routes for egress of rising serpentine muds (Fryer et al., 1985). We agree with Stern and Smoot (1998) that serpentinization in the forearc mantle may decrease toward the arc, but are of the opinion that the distribution of serpentine mud volcanism is more likely controlled by the distribution of faulting and volume of fluid flux in the supra-subduction zone environment. Likely the regions adjacent to faults are highly serpentinized, but it is possible that intervening parts of forearc mantle may be little affected by metamorphic hydration. Understanding the complexity of the structural settings in which these serpentine mud volcanoes form is of critical importance in recognizing possible on-land analogues of these structures.

Comprehensive surveys of the forearc to map fault traces are lacking; however, data compiled from several studies show that the Mariana forearc was subjected to along-strike and across-strike extension, as well as vertical tectonic movement. The active volcanic arc and the forearc region have increased in curvature with time as the convergent margin underwent several episodes of arc rifting and backarc extension (Wessel et al., 1994; Fryer, 1996a). These various extensional stresses produced extensive conjugate fault sets (Wessel et al., 1994; Mrozowski et al., 1981; Fryer and Fryer, 1987; Fryer, 1996a). Focal mechanisms for relocated earthquakes (Engdahl et al., 1998) show strike-slip as well as dip-slip mechanisms for faulting in the forearc wedge (Fig. 3). The vertical tectonism, related in part to subduction, resulted in development of large, discrete fault blocks and grabens on the outer forearc seamounts (Fryer and Smoot, 1985).

Side-scan sonar and bathymetry data (Fig. 2) collected from the southern half of the Mariana forearc demonstrate several structural settings and numerous stages of formation of active serpentine and blueschist mud volcanoes. The mud volcanoes occur within forearc grabens at the base of horst blocks (e.g., Conical Seamount), along the upper edges of horst blocks (e.g., Celestial Seamount), along the trace of major strike-slip fault lineaments (Pacman and Big Blue Seamounts), or at the upper edge of the inner trench slope (the Chamorro Seamounts) (Fig. 4). Young, active edifices have high-backscatter intensities of side-scan sonar energy whereas older or less active mud volcanoes have a more uniform backscatter character (Figs. 5 and 6). Multiple overlapping lobes of mudflows on several serpentine seamounts show variation in backscatter intensity (proportional to amount of sediment cover) that indicates the episodic nature of eruptions (see Figs. 5 and 6). The faulted surfaces of horsts contain smaller, lobate mudflow features, protrusive masses of various sizes, and possible sites of fluid egress (Figs. 5 and 7). The distribution of the smaller protrusions of serpentine of apparently varying ages around a main edifice suggests that in highly faulted regions, the egress routes for the serpentine protrusions may vary, possibly becoming centralized with time. In other areas (Figs. 6 and 8), the mud volcanoes are aligned along the tops of major breaks in slope of the inner wall of the trench and lie collinearly with major fault traces that have no or very little apparent throw (strike-slip faulting?). Our recent geophysical survey data from the southern half of the Mariana forearc demonstrate many of these structural relationships.

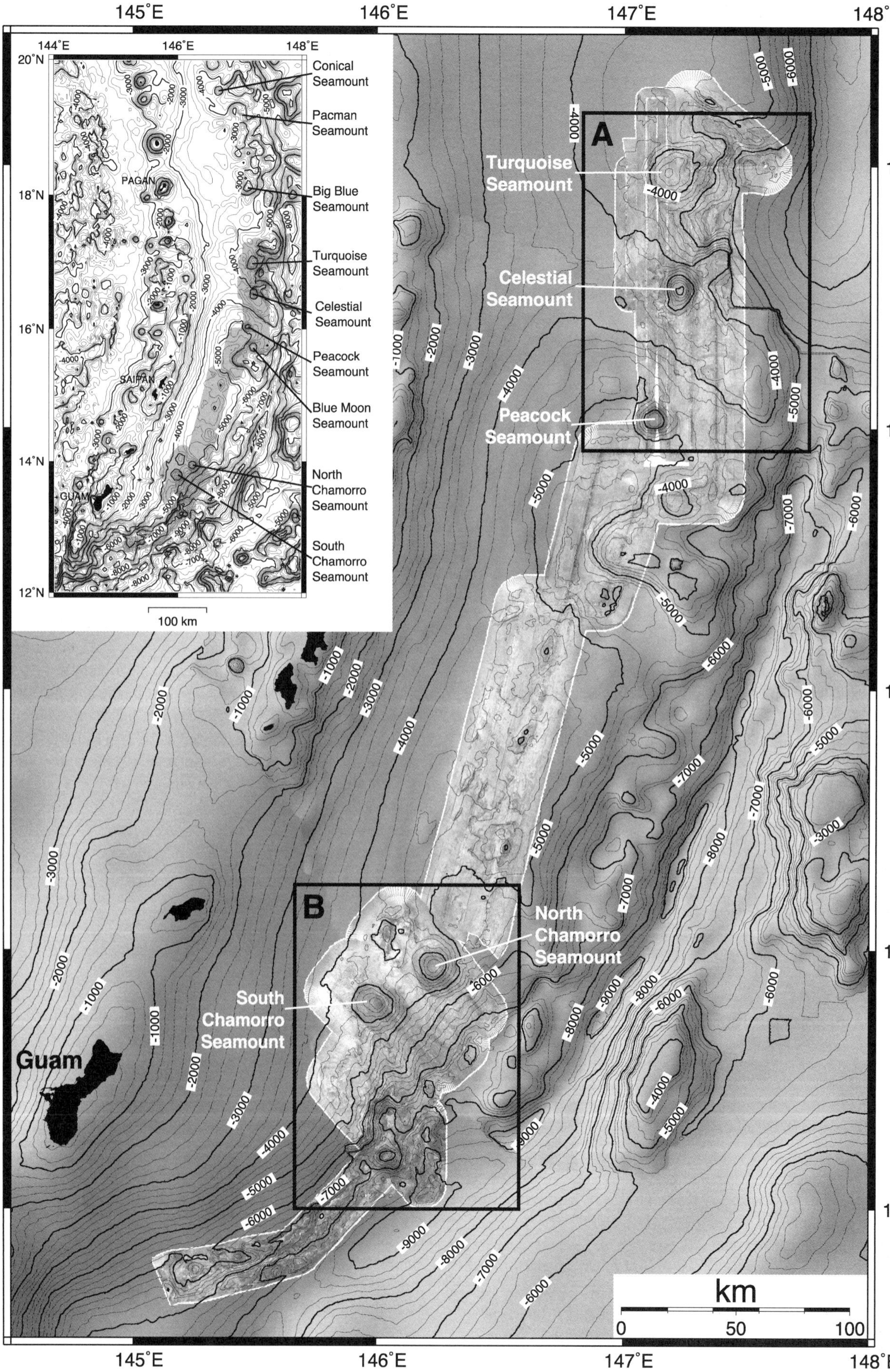
145°E
146°E
147°E
148°E
17°N
16°N
15°N
14°N
13°N
144°E
20°N
18°N
16°N
14°N
12°N
Conical Seamount
Pacman Seamount
Big Blue Seamount
Turquoise Seamount
Celestial Seamount
Peacock Seamount
Blue Moon Seamount
North Chamorro Seamount
South Chamorro Seamount
PAGAN
SAIPAN
GUAM
100 km
A
B
Turquoise Seamount
Celestial Seamount
Peacock Seamount
North Chamorro Seamount
South Chamorro Seamount
Guam
km
0
50
100

Area A mud volcanoes

The main edifices associated with the fault block at 16°30′N (Figs. 5 and 7) show some of the types of possible relationships of the serpentine mud volcanoes with regional fault structures. Peacock Seamount is forming at the base of a large horst block whereas Celestial Seamount is growing at the upper edge of the block. Turquoise Seamount lies within a fault-controlled channel between two horsts, but also is controlled by faulting that is apparently conjugate to that which bounds the horst to the south. The detailed surveys of these seamounts that were completed recently demonstrate several aspects of the stages of growth and the nature of fault control over these seamounts.

Peacock Seamount. Peacock Seamount lies at the base of the northeast-trending horst block in the center of the area (Figs. 5 and 7). It has a high-backscatter surface. The edifice is nearly symmetrical with a relatively simple morphology. Sediment cores from the summit of the seamount recovered vitric silt about 1.3 m thick overlying serpentine mud. Some of the sediment is semi-indurated. We suggest that the mud volcano is forming by the rise of fault gouge along the southern bounding fault of the horst. The presence of semi-indurated sediment as the primary lithology recovered in the cores from the summit and the lack of serpentine mud except at the base of the core catcher in most of the cores suggest that the carapace of semi-indurated sediment represents material from the deeper forearc sedimentary section. Possibly less well indurated sediments have been sloughed off the summit region during up-doming of the seamount as serpentine materials began protruding from below. Serpentine muds apparently have not reached the seafloor at the summit, at least not where we cored. If this is so, then the seamount may be in the earliest stage of development of a serpentine mud volcano. The high backscatter suggests deformation of the surface of the edifice with roughness on the order of 10 cm. This deformation may be related to disruption of the surface of the seamount in response to tumescence precursory to the eruption of serpentine muds.

Celestial Seamount. Celestial Seamount is located at the northern edge of the horst block, where a major northeast-trending lineament intersects the northern bounding fault of the block (Figs. 4 and 6). The northeast-trending lineament forms the northern boundary of a small horst that projects northeastward from the northern edge of the main horst (Figs. 5 and 7). Celestial Seamount partially drapes the northern edge of the main horst (best seen in Fig. 7). Low-backscatter lobes, probably of older serpentine protrusions, can be seen at the base of the edifice on the south and east sides (Figs. 5 and 7). The lower backscatter suggests that these lobes are covered with sediment. The seafloor on the upper surface of the horst is broken by northwest-trending lineaments, probably faults (Figs. 5 and 7). A series of northwest-trending lineaments also stair-steps down the northern flank of the horst. There is a large normal fault that trends northeast across the western edge of the survey area (best seen in Fig. 7). This fault lines up with a fault trace that cuts through the seamount to the north (i.e., Turquoise Seamount, described next). On the top of the horst to the west of the main edifice are patches of high-backscatter mounds (0.5 to 5 km in diameter) that may represent small protrusions of serpentine muds. These patches are located in a region that is fractured by the small, northwest-trending faults. The area to the west of the edifice also has a generally higher-backscatter character than the surrounding seafloor. The edges of the higher-backscatter regions are lobate, suggesting that they may represent mudflows or regions of fluid egress associated with the northwest-trending fault lineaments. The location of the main edifice is most likely controlled by the intersection of the north bounding fault with the northeast-trending lineament (see area A in Fig. 2 and perspective view in Fig. 7). Sediment cored from the summit depression of the Celestial Seamount is dark-blue, serpentine-rich, sandy silt with needles of aragonite and accessory magnetite, chlorite, and a few small pebbles of mica and amphibole schist. The high backscatter of the main edifice, the recovery of dark-blue (reduced) serpentine muds at the seafloor with virtually no sediment cover, and the low chlorinity (M. Mottl, 1999, personal communication) of the pore fluids in the sediment all suggest that the seamount is currently active.

Turquoise Seamount. Turquoise Seamount is located at the northern base of the horst. It lies in a forearc basin between the horst at 16°30′N and another horst at about 17°15′N. A core taken at this seamount recovered 1.6 m of foraminiferal sand with about 1 cm of green serpentine mud in the core catcher. The lithology of the muds from this seamount is very unusual by comparison with that from other mud volcanoes that we have sampled. The serpentine mud contains serpentine, aragonite, glass shards, and grains of fresh olivine. It is possible that mudflows from this seamount entrained volcaniclastic turbidites deposited in the fault-controlled forearc channel that this mud volcano partially blocks. An apparent fault trace, trending northeast (best seen in Fig. 7), projects through the summit region and northeast flank of the edifice. The northeastern quarter of the seamount appears to have been disturbed by movement along this fault. There is a prominent notch in the northeast side of the edifice. Movement on the northeast-trending fault also appears to have disturbed the lower southern flank of the seamount. A few of the northwest-trending faults that lie collinearly with the stair-step faults on the north flank of the horst to the south have cut through the southern flank of the seamount. A subtle radial fracturing, centered to the southeast of the topographic summit of the seamount, can be seen in the side-scan imagery (best seen in Fig. 5). The low-backscatter character of the seamount, the fault deformation, and the recovery of >1.5 m of foraminiferal sand in the summit core suggests that the seamount is not currently active.

Figure 2. Bathymetry of southeastern part of Mariana forearc with side-scan sonar imagery superimposed where available. Contour interval is 250 m on the main map. The bathymetry is presented in shaded relief in the areas outside the regions of side-scan imagery. Illumination is from the west. Areas outlined and marked A and B are locations for Figures 5 and 7 and for Figures 6 and 8, respectively. Inset shows the bathymetry of southeastern part of Mariana system from 12°N to 20°N. Contour interval for inset is 200 m. Serpentine and blueschist mud volcanoes of Mariana forearc that have been sampled are labeled on inset map. The shaded area in inset shows location of the side-scan sonar survey data given in the main map.

Area B mud volcanoes

The two main serpentine mud volcanoes located near 14°N (Fig. 2, area B) are associated with major fault trends that cut through otherwise normal forearc structure. These seamounts lie at the top of the inner slope of the trench, at the break in slope of the forearc. In part, the formation of the forearc break in slope is fault controlled. Also, the fact that major fault traces with no apparent vertical offset lie along the trend of the mud volcanoes and crosscut them suggests that it may be easier for rising serpentine muds to find egress to the seafloor where major fault traces intersect.

South Chamorro Seamount. South Chamorro Seamount lies at

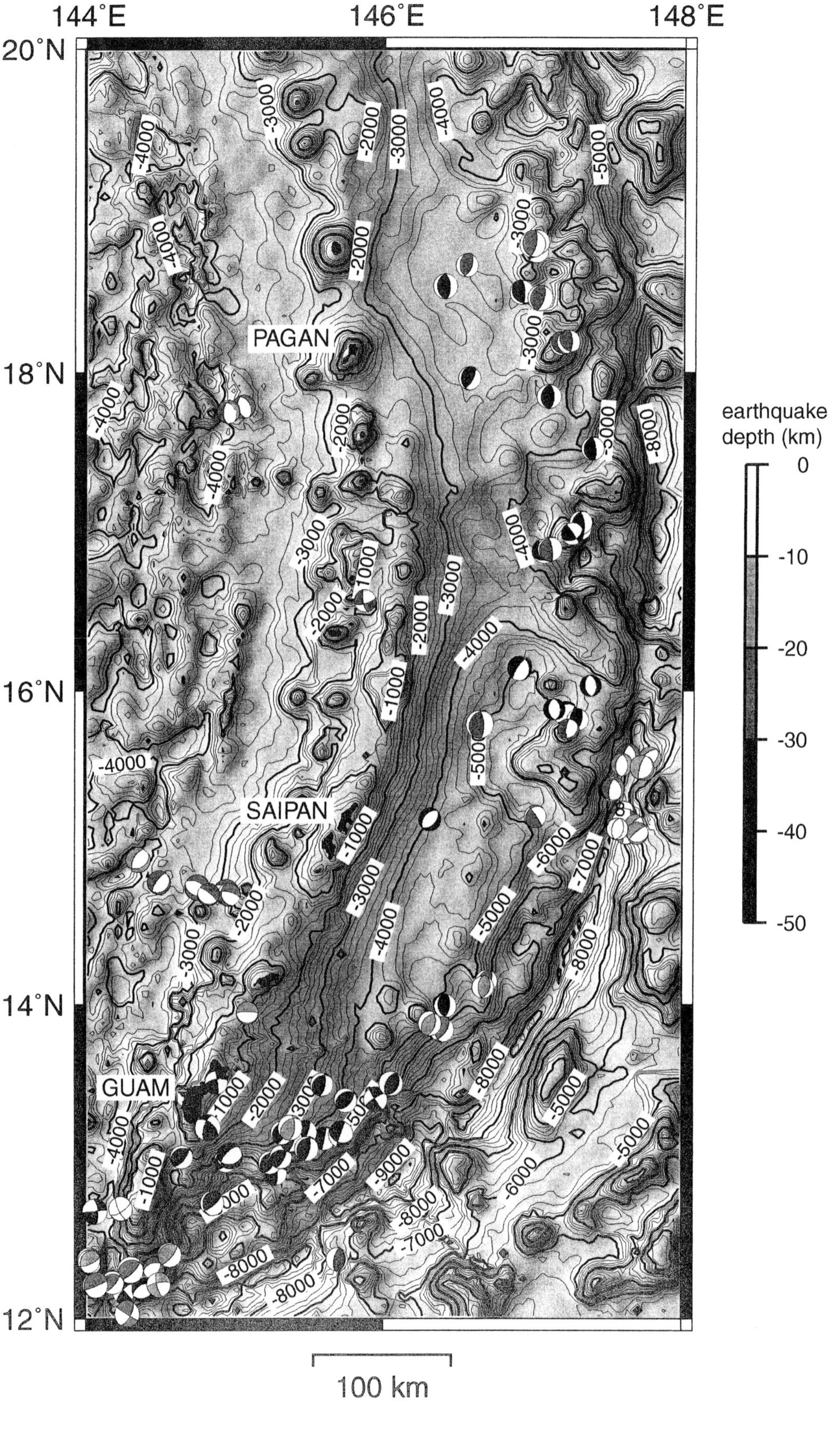

Figure 3. The bathymetry of the southeastern part of the Mariana system from 12°N to 20°N. Contour interval is 200 m. Labels for the various serpentine and blueschist mud volcanoes of the Mariana forearc are given on the inset map in Figure 1. The map shows shaded relief with illumination from the west. Positions of relocated (Engdahl et al., 1998) earthquakes are shown by their fault mechanisms (from Harvard centroid moment catalogue for earthquakes with moment magnitudes greater than 5.5). Depths of earthquakes are indicated by increasing darkness of the focal-mechanism solutions.

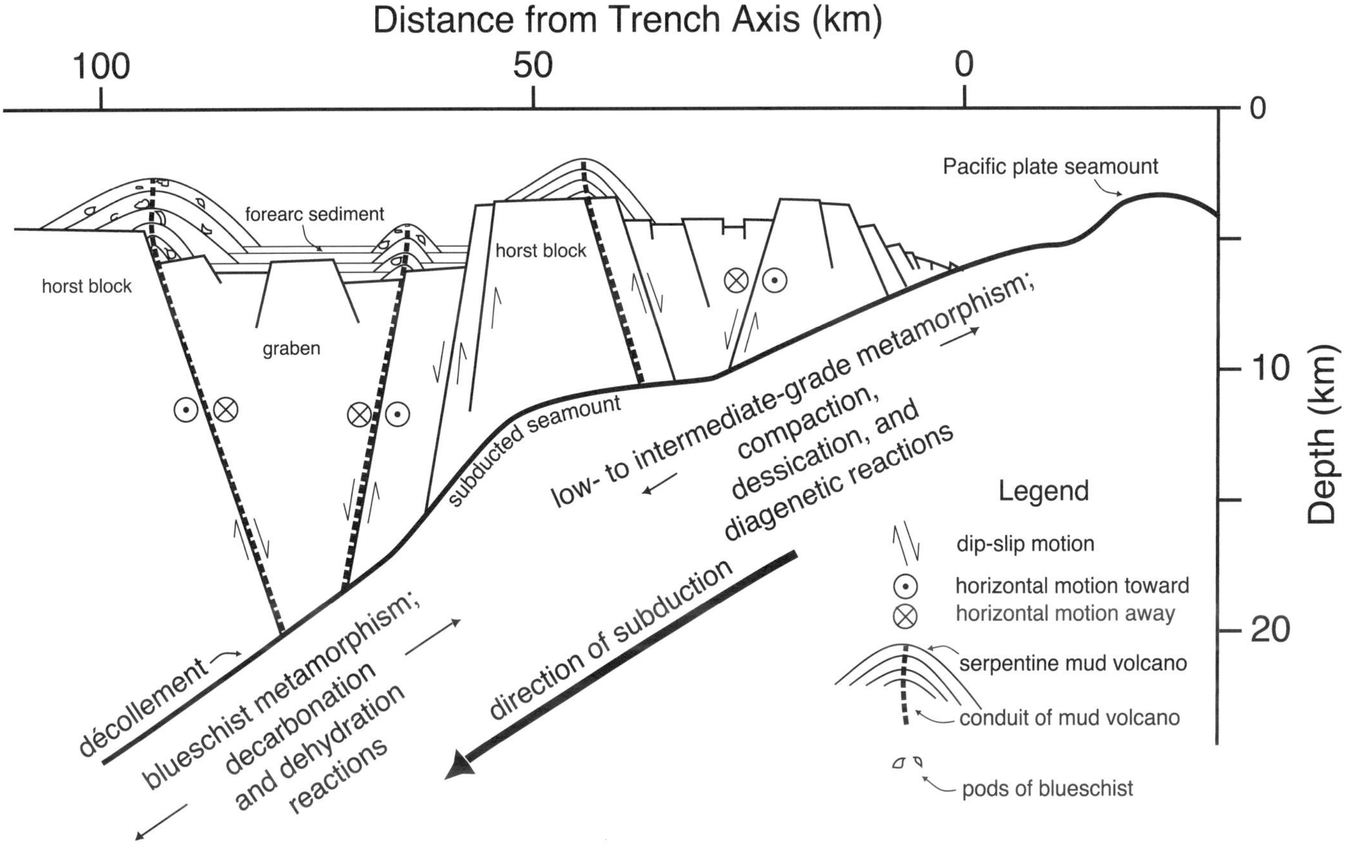

Figure 4. Schematic sketch of the types of structural settings in which serpentine and blueschist mud volcanoes form on the Mariana forearc. Muds rise along fault-controlled conduits, and volcanoes form mainly at tops or bases of horst blocks or along strike-slip faults. Muds and included rock clasts may derive from the subducted slab or from the walls of the conduit within any part of the overriding plate through which the conduit passes. Modified after Fryer et al. (1999, their Fig. 2).

the break in slope between the forearc basin and the inner slope of the Mariana Trench (Figs. 6 and 8). The strike of the fault block north of the seamount lies collinearly with lineaments that project southward from the southern base of the mud volcano. A series of small lineaments also parallels the strike of the inner slope of the trench and probably represents small normal faults that stair-step down the face of the inner trench slope. The southeastern quadrant of the edifice has slumped toward the southeast and created a scarp-outlined collapse sector. The inner slope of the trench immediately southeast of the seamount is littered with debris flows from this collapse. The debris-flow area was investigated with the *Shinkai 6500* submersible in 1995 (Fryer, 1996b). The debris flows are blocks of serpentinized ultramafic rocks and clasts of carbonate chimney material set in a matrix of fine serpentine mud. Accessory materials in the muds include aragonite, brucite, amphiboles, and garnets. The debris flows can readily be seen in both the map and perspective views of area B (Figs. 6 and 8).

A small knoll has formed at the apex of the sector-collapse feature, on the southeastern side of the summit. This knoll was investigated with the *Shinkai 6500* submersible on three dives in 1995 and 1996 (Fryer, 1996b; Fryer and Mottl, 1997). The knoll is a tumescent region exposing blue and green serpentine muds enclosing clasts of serpentinized ultramafic rocks, blocks of semi-indurated foraminiferal sand, and small pebbles of blueschist set in a matrix dominated by serpentine. Accessory minerals in the muds include aragonite, magnetite, and brucite and minor crossite, phengite, hydrated garnets, and possibly other high-pressure phases (e.g., XRD [X-ray diffraction] analysis has shown the presence of two major peaks of jadeite, but we have not been able to confirm its presence optically or by microprobe analysis) (Fryer et al., 1999; Fryer and Todd, 1999; Todd and Fryer, 1999). Some small pebbles of blueschists have been recovered from this site. For one of these, we have been able to suggest a lower pressure and temperature *(P-T)* limit for formation. To apply typical geobarometers, one needs an appropriate equilibrium mineral assemblage and mineral-composition information for several phases. We cannot attain this for the muds, as the mineral assemblages are chaotic and possibly representative of various depths along the conduit. Pressure and temperature conditions can, however, be determined for individual rock clasts. Even if the entire assemblage is not present in a rock, *P-T* limits can be calculated in some cases. It can be shown that white mica in the assemblage Mu + Bt + Qtz + Chl + H_2O must have a higher phengite content than in the assemblage Mu + Chl + H_2O, present in one of the rock samples from South Chamorro Seamount. Therefore, thermobarometric calculations based on the larger assemblage (Massone and Szpurka, 1997) put a lower bound on *P-T* conditions of formation for this sample (Fig. 9). Abundant large euhedral aragonite crystals with inclusions of the matrix mud and various discrete matrix minerals (including crossite) are present in the muds from many of the seamounts. They would likely not have formed outside the stability field of aragonite. Therefore, the presence of this mineral also puts a lower bound on *P-T* conditions (Fig. 9; Johannes and Puhan, 1971). Our data indicate that the rock

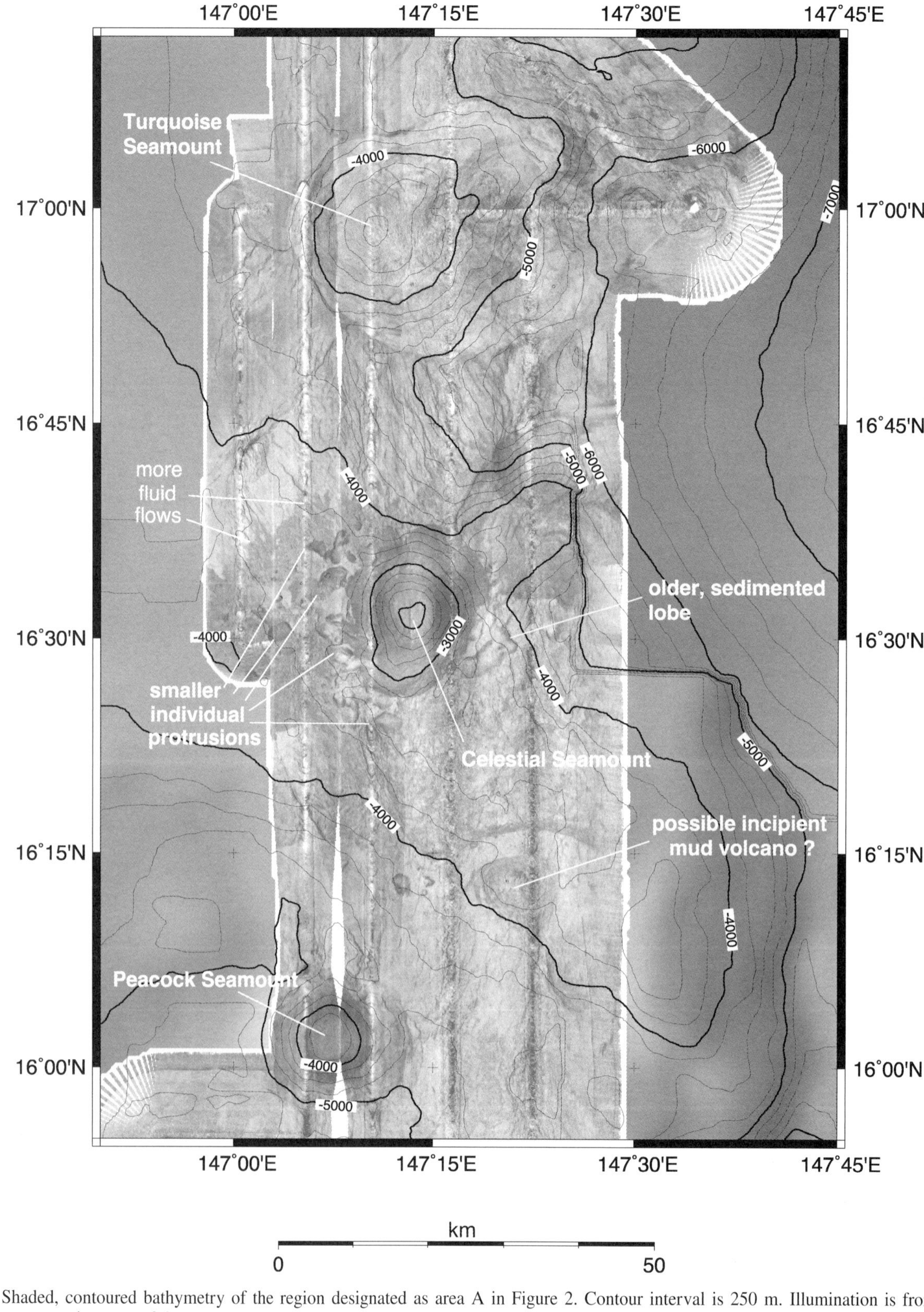

Figure 5. Shaded, contoured bathymetry of the region designated as area A in Figure 2. Contour interval is 250 m. Illumination is from the west. The side-scan sonar imagery of the survey region is superimposed on the bathymetry. Areas of high backscatter (rough surfaces or high-relief regions) are printed as dark areas. White areas indicate gaps in side-scan coverage. Nadir cross-talk artifacts show up as narrow speckled stripes along track lines. Note fault lineaments trending northwest in the area to the west of Celestial Seamount and stair-stepping down the slope north of the edifice. These are essentially parallel to the major bounding faults that form the northeast and southwest edges of the block upon which Celestial Seamount has formed. Note also older lobes (more sediment-covered and thus lower-backscatter regions) of serpentine mud protrusions on the lower east flank of Celestial Seamount (immediately east of the 3000 m contour). Protrusions of serpentine muds are probably responsible for the clusters of small (~0.5–5 km in diameter), higher-backscatter areas on the western side of Celestial Seamount between 147°E and 147°15′E. Note northeast-trending faults on the western side of the survey area between 16°45′N and 17°N. A perspective view from the northeast corner of this map is presented in Figure 7.

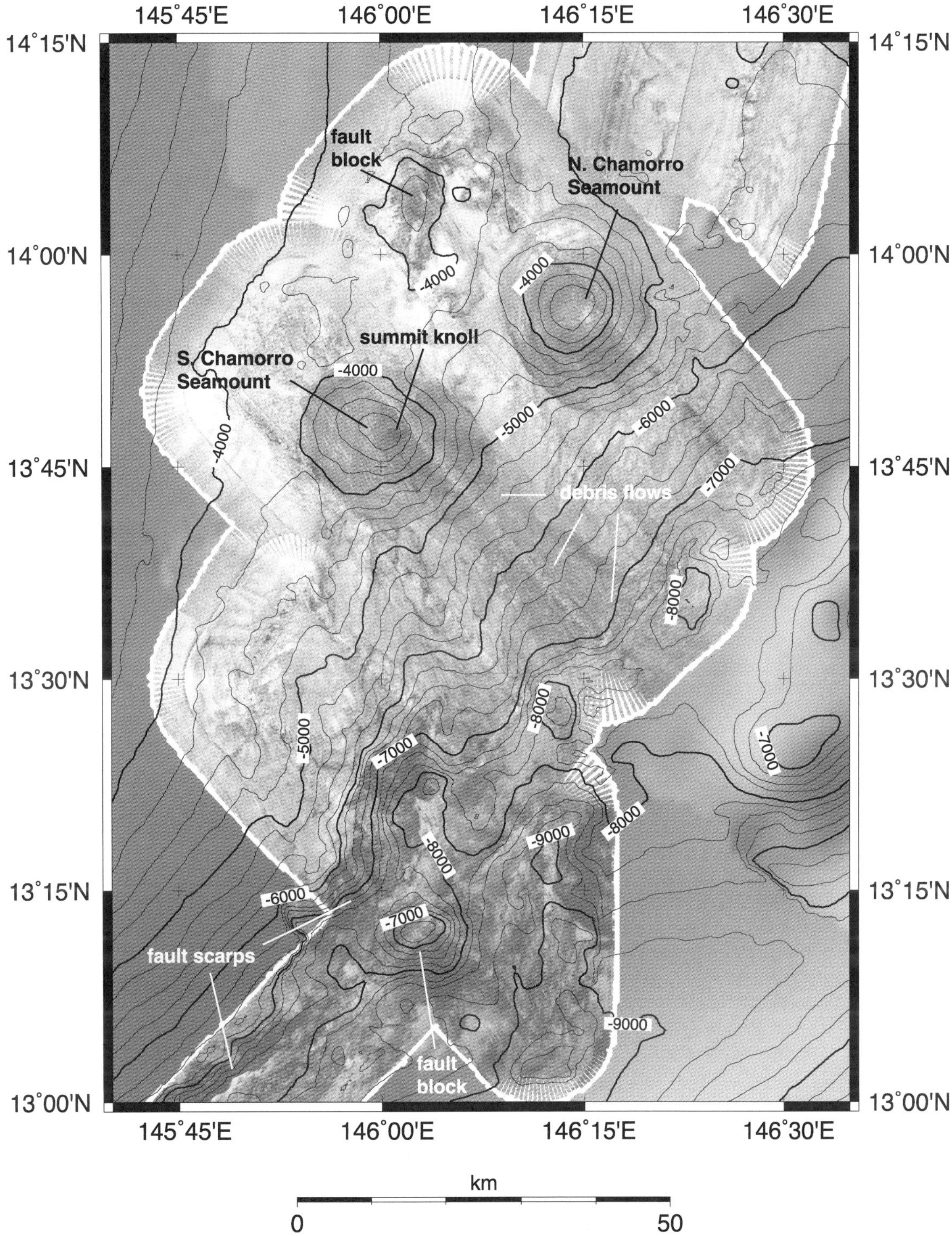

Figure 6. Shaded, contoured bathymetry of the region designated as area B in Figure 2. Contour interval is 250 m. Illumination is from the west. The side-scan sonar imagery of the survey region is superimposed on the bathymetry. Areas of high backscatter (rough surfaces or high-relief regions) are printed as dark areas. Note the small (2 km in diameter) area of high backscatter at the summit of South Chamorro Seamount. This is the summit knoll site of the most recent protrusive event at this seamount and the location of the megafauna community and fluid venting. A sector-collapse feature can be seen in the southeastern quadrant of the edifice, outlined by two radiating scarps, and debris flows cover a swath ~10 km wide reaching from the apex of the collapse feature near the summit for at least 40 km to depths of 8000 m on the inner slope of the trench. A perspective view from the northeast corner of this map is presented in Figure 8.

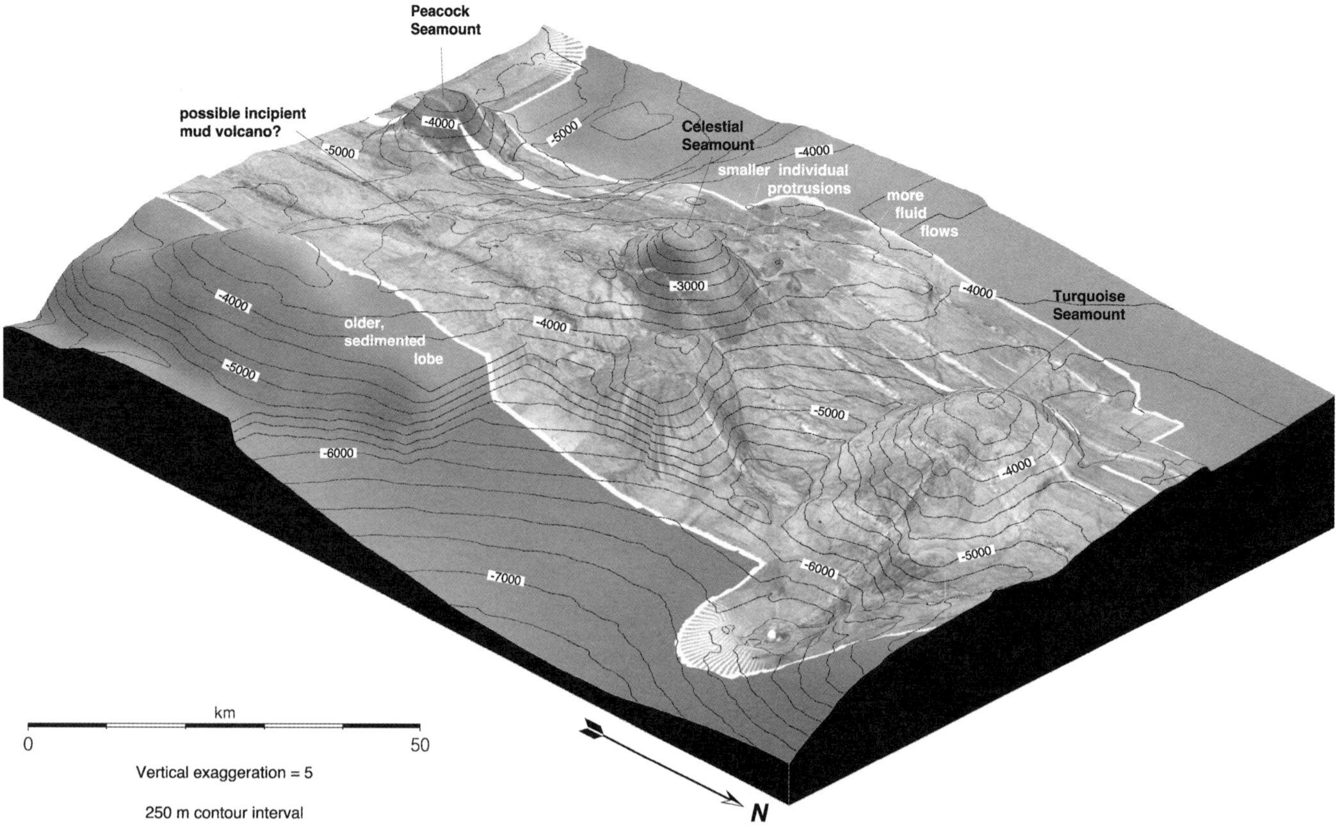

Figure 7. Perspective view of area A in Figure 2 looking from the northeast. Bathymetry is shown in shaded relief illuminated from the west and is contoured at 250 m intervals, labeled every 1000 m. The side-scan imagery is overlain on the bathymetric surface. North direction is indicated by arrows. Vertical exaggeration is 5×.

must come from depths approaching 30 km (Fig. 9) (Fryer and Todd, 1999). Depth to the slab under South Chamorro Seamount is ~27 km.

The individual crystals of sodic amphibole contained in the mudflow matrix have sodic rims around more calcic cores. The fact that the grains are euhedral and lack evidence of retrograde metamorphism suggests that they may have been emplaced by rapid rise from depth. In order to bring materials up from the depths required for equilibrium assemblages of blueschist facies (20 to 30 km) without permitting retrograde reactions, some sort of driving mechanism must be invoked. If protrusive events are triggered by earthquakes along the décollement or within the overriding wedge, release of slab-derived fluids might facilitate such a rapid rise. Fluids related to dehydration of the slab would find more ready access to the overlying forearc wedge and egress along well-established, fault-controlled conduits beneath the mud volcanoes at times of seismic activity. Once triggered, the muds could rise within the conduit in response to gravitational instability and not require further seismic activity. We suspect, however, that serpentine mud volcanoes that are actively undergoing protrusion events (e.g., South Chamorro Seamount) might display microseismicity associated with the rise of muds within the conduit or with flank deformation (sector collapse on South Chamorro Seamount). Determining the patterns of such microseismicity could help to establish the pluming system of these edifices in much greater detail and might provide some constraints on rates of rise of material within the seamount.

The *Shinkai 6500* submersible dives at the summit also collected carbonate chimney structures and carbonate armoring of fissure edges. A biological community consisting of mussels, snails, tube worms, and crabs is active at the summit knoll, particularly in the fissures that are evidently seeps for slab-related fluids (Fryer and Mottl, 1997). The summit region immediately west of the knoll is covered with foraminiferal sands. South Chamorro Seamount is the only site at which megafaunal communities have been observed on the serpentine seamounts, although small limpets and high-spire gastropods were observed grazing on bacterial mats on active carbonate chimney structures on Conical Seamount (Fryer et al., 1990b).

Fryer and Mottl (1997) suggested that the summit knoll on South Chamorro Seamount formed during a very recent or still active protrusion event (Fryer and Mottl, 1997). We suggest that seismic activity on the subduction zone or along the faults (antithetic to the décollement) beneath the two Chamorro seamounts may trigger protrusion events. In the case of South Chamorro Seamount, seismic activity may have initiated the sector-collapse events and reactivated protrusion at this site.

North Chamorro Seamount. North Chamorro Seamount lies in a position similar to that of South Chamorro Seamount, except that it is centered more over the edge of the break in slope of the inner trench (Figs. 6 and 8). Therefore, the body of the edifice drapes the inner slope of the trench to a greater extent than does South Chamorro Seamount. There are no obvious debris flows mantling the inner slope of the trench to the southeast of this seamount, although the southeastern edge of the seamount is somewhat oblate in the downslope direction. North Chamorro Seamount has a relatively flat to slightly indented summit region. Cores taken at the summit yielded as much as 2.06 m of foraminiferal sands with blue-green serpentine mud in

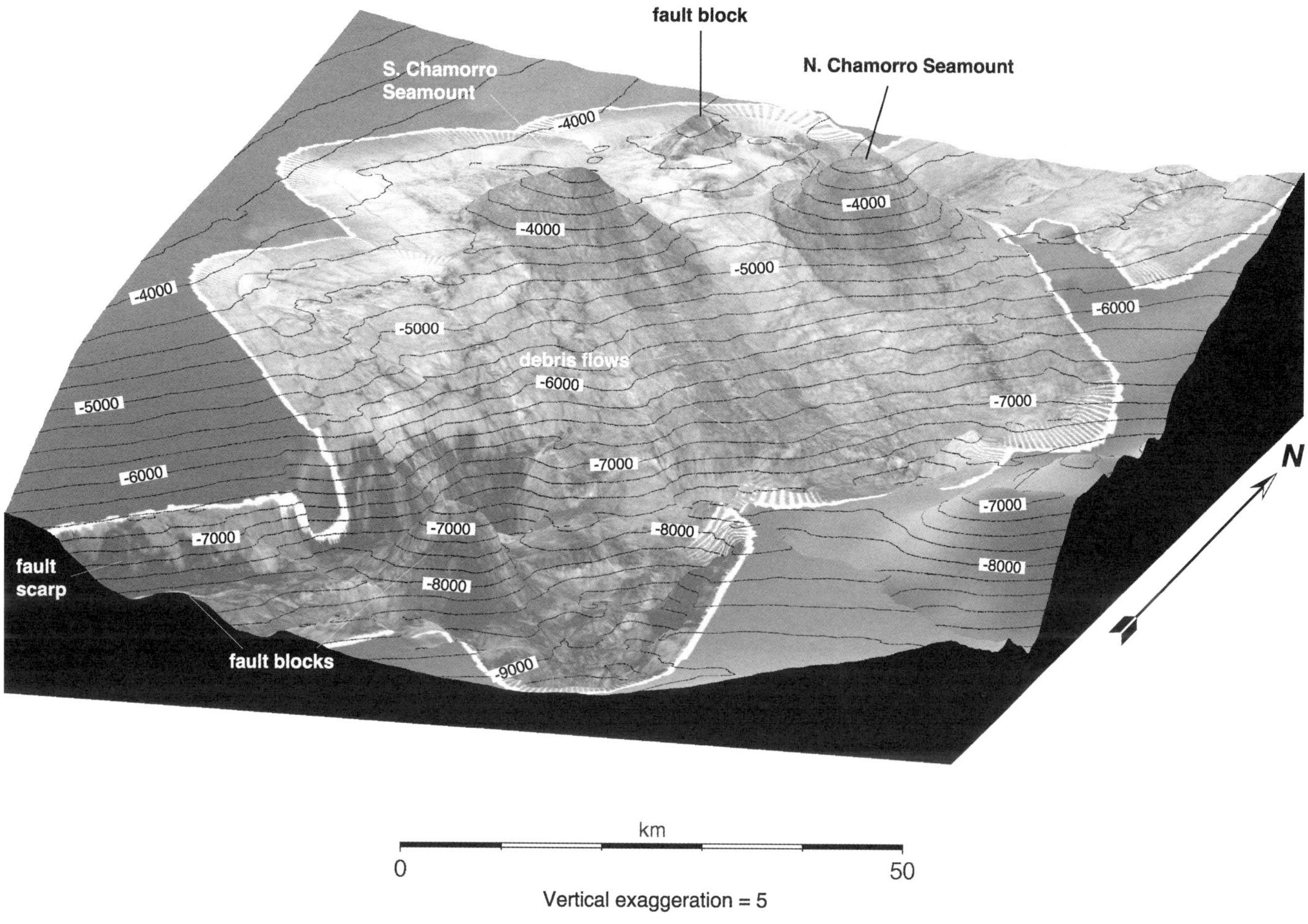

Figure 8. Perspective view of area B in Figure 2 looking from the southeast. Bathymetry is shown in shaded relief illuminated from the west and is contoured at 250 m intervals, labeled every 1000 m. North direction is indicated by arrows. Vertical exaggeration is 5×. These views show two serpentine mud volcanoes. They are (informally named) North Chamorro Seamount and South Chamorro Seamount. The third high-relief feature in the image is a fault block. Cores from the horst recovered siltstone and manganese nodules. Fault scarps on the lower part of the inner trench slope have very high backscatter because of their high relief and likely exposure of hard rock.

the bottom of the cores. The edifice appears to have been formed by at least two major episodes of protrusions. There is a basal ring of slightly lower backscatter material that surrounds the base of the seamount and that is probably a slightly more sediment-covered series of older serpentine mud protrusions. A distinct demarcation occurs about one third of the way up the flank where a higher backscatter begins and characterizes the upper flanks. The summit depression of the seamount has very low backscatter probably because of the flat foraminiferal sand cover. This seamount is probably currently inactive.

MODES OF ORIGIN

Fryer (1992a) stressed that a careful distinction must be made between the modes of origin of some of the edifices that occur in the Mariana and Izu-Bonin forearc regions in order to avoid confusion in terminology associated with these edifices. There are horst blocks of massive, but highly faulted serpentinite that represent uplifted fragments of forearc mantle (e.g., some of the local highs along a serpentinite ridge that forms the inner trench slope of the Izu-Bonin system) and edifices that probably have originated exclusively by mud volcanism (e.g., Conical, Celestial, Peacock, and the two Chamorro Seamounts). In addition, there are probably seamounts that have formed by a combination of both mechanisms (e.g., Pacman [see Fryer, 1996a] and Big Blue Seamounts).

A ridge that forms the outer 50 km of the Izu-Bonin forearc area (the continuation of the Mariana convergent-margin system to the north) is composed of serpentinized peridotite (Fryer and Fryer, 1987; Horine et al., 1990). The serpentinite ridge may represent metamorphically expanded supra-subduction zone mantle (Fryer and Fryer, 1987). The hydration of the toe of the overriding plate could account for the formation of this ridge, but superimposed on it are discrete apparent mud-volcano edifices. Although none of these edifices shows signs of current activity, it is clear that conduits for the egress of serpentine muds have formed along the serpentinite ridge in the past. Recent geochemical studies of ultramafic clasts in drill samples of mudflows from one of the Izu-Bonin serpentine seamounts (Torishima Forearc Seamount, ODP Leg 125 drilling) define provenance and relate isotopic anomalies (Hf and Nd) to cycling of subduction and mantle components in arc-basin systems (Parkinson and Pearce, 1998; Pearce et al., 2000). These workers have determined that the peridotites of the Izu-Bonin forearc seamount formed by subduction-

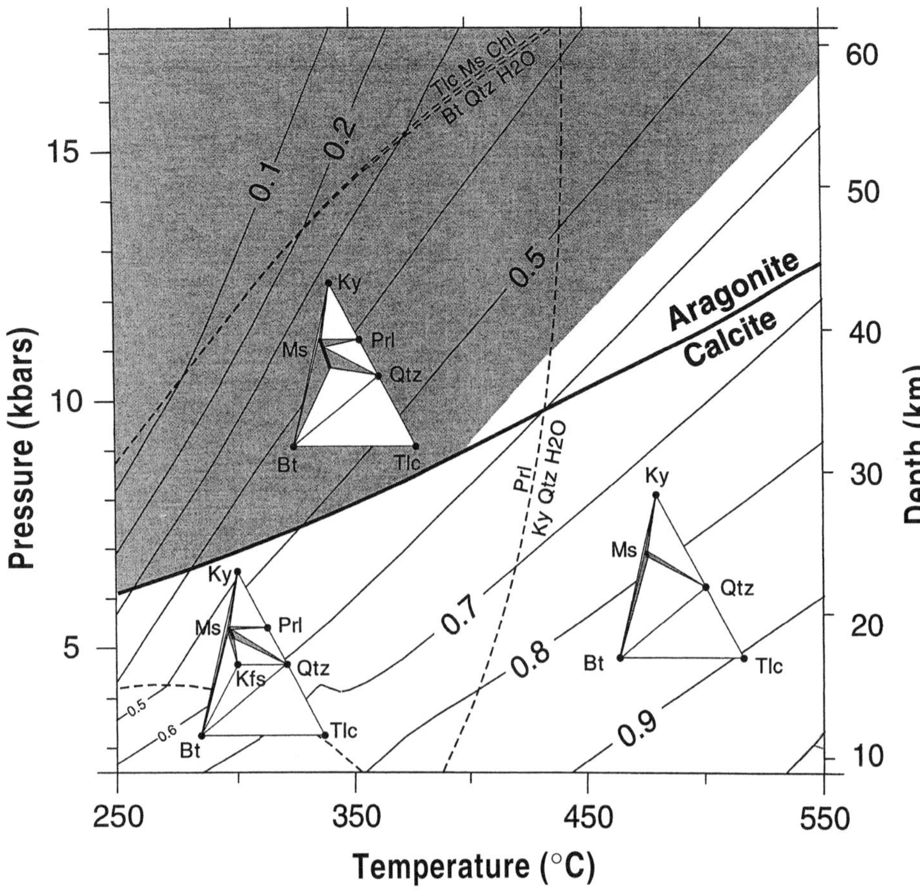

Figure 9. Pressure vs. temperature diagram. Shaded area shows limits on the conditions of formation of a rock sample from South Chamorro Seamount based on the presence of aragonite and thermobaric calculations using the phengite component of muscovite as described in the text. Numbered lines indicate the muscovite end-member content of white mica. The chemographic triangles are projected from chlorite onto the SiO_2-Al_2O_3-$KAlO_2$ plane.

related melting, whereas the peridotites recovered in drilling on Conical Seamount in the Mariana system are related to trapped or accreted oceanic lithosphere. Parkinson and Pearce (1998) showed that both types of mantle lithosphere likely underwent extensive interaction with subduction-derived magmas. Such would not be the case for serpentinites of mid-ocean ridge fracture zones or exposures at the edges of rifted passive margins. These geochemical provenance characteristics plus the presence of high-pressure, low-temperature metamorphic minerals and rocks in the serpentinite masses could serve to distinguish serpentine mudflow deposits formed in convergent-margin environments from debris-flow deposits formed elsewhere.

An area of serpentinized peridotite horsts, ~180 km long by 60 km wide, occurs in the Mariana forearc between 17.8°N and 19.4°N. Discrete serpentine mud volcanoes are superimposed on the horst blocks in at least two localities that we have sampled by coring (Pacman Seamount and Big Blue Seamount; see Fig. 2). This region bears a resemblance in form, extent, and perhaps in tectonic origin to the New Idria serpentinite—a fault bounded, sheared serpentinite mass that makes up part of the Diablo Range in California (Fryer et al., 1995). In fact, in the southern end of the New Idria mass there are several discrete cylindrical bodies of highly sheared chrysotile with clasts of serpentinized ultramafic rocks that are currently being mined. These were suggested by Fryer et al. (1995) to represent the eroded conduits of serpentine mud volcanoes, like those envisioned for the Mariana forearc. If this interpretation is correct, both the fault-bounded blocks of serpentinite in the central Mariana forearc and the elongate ridge in the outermost part of the Izu-Bonin forearc may be in situ examples of what has previously been considered "Alpine-type" serpentinite masses. Because these features also contain serpentine mud volcanoes, the distinction between types of serpentinite deposits will have to depend on a careful examination of field relationships, bulk and mineral geochemistry, and mineral constituents of the clasts enclosed in the serpentinite.

SEDIMENTARY SERPENTINITE DEPOSITS ON LAND

Sedimentary serpentinites, with ages ranging from early Paleozoic to late Cenozoic, are found throughout the world in zones of continental accretion or intercratonic collision (Lockwood, 1971a). These deposits are characterized by having nearly monolithologic content of serpentinized peridotite. Clasts of other lithologies are not voluminous, but can include eclogites, blueschists, various metagabbros or metabasalts, and a wide variety of sedimentary rocks and marine fossils. These were originally interpreted as submarine landslides. The high-pressure, low-temperature clasts, when present, are clear indications of formation in a subduction environment. The internal structures of these sedimentary serpentinites are commonly chaotic, typically made up of poorly sorted angular to subrounded blocks of serpentinized peridotite mixed with sand- to clay-sized serpentinite fragments. Grading and cross-bedding may be preserved. The deposits commonly form planar bodies of widely varying thickness with great lateral extents and are normally conformable to the bedding of the enclosing strata. Mineralogically, the matrix fragments dominantly consist of chrysotile, but lizardite may develop during metamorphism (Lockwood, 1971b). The matrix of detrital serpentinites differs in bulk composition from massive Alpine-type serpentinites and from the enclosed serpentinized ultramafic clasts. It typically

Figure 10. Photograph of a block of sedimentary serpentinite with an unidentified marine pelecypod from near Wilbur Springs, northern California (from Lockwood, 1971b).

Figure 11. Photograph of an exposure of sedimentary serpentinite along the Pambak River, in the Lesser Caucasus, showing numerous clasts embedded in a sheared serpentine matrix with variations in size of clasts in distinct zones (bedding variations) (from Lockwood, 1971b).

contains higher amounts of SiO_2, Al_2O_3, F_2O_3, FeO, CaO, and CO_2 (Lockwood, 1971b), suggesting either incorporation of terrigenous-derived detrital clay during submarine transport or incorporation of mafic minerals or lithic clasts derived from the subducted slab or from sequences in the overriding plate wedge (or both). Carbonates found as acicular crystals in the mudflow matrix can contribute to the CaO and CO_2 enrichment. Fossils are rarely preserved in these deposits, but where found invariably demonstrate a marine origin (Fig. 10). The sedimentary serpentinite deposits were suggested to have been emplaced by some sort of protrusive mechanism as early as 1964 (Bailey et al., 1964, p. 164). Bailey et al. (1964) and Gresens (1970) suggested that piercements of serpentinite may occur on the seafloor and thus provide sources for the sedimentary serpentinite deposits and gravity-slide deposits, although a possible relationship to tectonism of convergent margins was not discussed. Lockwood (1971b, 1972) suggested that protrusion of serpentinite onto the seafloor, followed by sedimentary redistribution, might explain the deposits in many localities. He suggested that they may represent protrusive bodies, citing protrusive serpentinite piercement structures described by others in several areas: parts of the Franciscan Formation of California (Bailey et al., 1964), the northeastern coast of Lake Sevan, Armenia (Knipper and Kostanyan, 1964), Table Mountain serpentinite extrusion in California (Dickinson, 1966), and serpentinite bodies in the Diablo Range, California (Oakeshott, 1968). Lockwood offered the suggestion that the sedimentary serpentinite deposits in these and other localities (e.g., the Appalachian ultramafic belt, the New Zealand ultramafic belt, the Lesser Caucasus, and the Scandinavian Caledonides and numerous areas in central Norway and Sweden) may have been emplaced simultaneously with the enclosing strata (not as later intrusions). He suggested that structurally weakened serpentinized peridotite in a region of dehydration (such as a convergent margin) could protrude above the seafloor and provide the source for submarine landslides, mudflows, or turbidity currents (Fig. 11). Lockwood (1972) noted that the source areas of these deposits must provide a large volume of serpentinite of varying grain sizes and associated lithic clasts very rapidly to the adjacent sites of deposit. These contrast with allochthonous, gravity-slide, ophiolitic masses that include relatively undeformed spilitic rocks and chert (as in the Apennines). The mechanism Lockwood (1972) suggested for emplacement of these protrusions required that faulting in forearc environments would provide a pathway for the rise of metamorphosed ultramafic materials and that fluids derived from dehydration reactions could provide a driving mechanism to complement the inherent gravitational instability caused by serpentinization of the ultramafic rocks.

Several localities in California, e.g., the Knoxville Formation in the Wilbur Springs area, expose layers (ranging from a few centimeters to over 500 m thick) of fossiliferous sedimentary serpentinite interlayered with sedimentary horizons (Carlson, 1984). The Mysterious Valley Formation has many similarities to Mariana mud-volcano deposits. The formation, lying in a fossil Mesozoic forearc in the California Coast Ranges of Napa County, is a 1-km-thick accumulation of chaotic serpentinite-rich olistostromes that may represent deposits from a Middle to Upper Jurassic serpentine mud volcano constructed on forearc crust. The Mysterious Valley Formation lies structurally above the Franciscan subduction complex and thus overlies rocky "bedrock" serpentinite that locally represents the Middle Jurassic Coast Range ophiolite (the crust of a fossil oceanic forearc) (Phipps, 1992; Dickinson et al., 1996; Giaramita et al., 1998). The Mysterious Valley Formation underlies bedded, arc-derived subsea-fan rocks of the Great Valley Group forearc-basin fill (Ingersoll and Dickinson, 1981; Phipps, 1984, 1992; MacPherson and Phipps, 1988). The formation consists of serpentinite debris flows and breccias, mafic breccias and associated finer-grained clastic rocks, and abundant pebble- to mountain-sized blocks of serpentinite, greenstone, mafic breccia, chert, clastic sedimentary rocks, phyllites, actinolitic greenschists, amphibolite, and garnet amphibolites (Phipps, 1984, 1992). Rare clasts of incipient blueschist-grade rocks were found in the matrix (MacPherson and Phipps, 1988). The basalt clasts in the Mysterious Valley Formation are mostly MORB compositions with rare OIB clasts (MacPherson and Phipps, 1988). The compositional range of the MVF clast population is virtually identical to that of the mafic clasts from Conical Seamount (Johnson, 1992; Phipps and MacPherson, 1999). The clasts are enclosed in a variably sheared matrix of sedimentary serpentinite, serpentine-bearing mudstone, and clay mudstone. The matrix of the sedimentary serpentinites contains marine fossils of Late Jurassic age (MacPherson and Phipps, 1988).

Cuban sedimentary serpentinite deposits include clasts of serpentinized ultramafic rocks, porphyritic volcanic rocks, gabbro, and limestone that have been interpreted by some early workers to represent detrital accumulations emplaced directly on the seafloor, derived from adjacent protrusive masses that had risen along regional thrust faults (Bucher, 1956; Kozary, 1968). These workers described field exposures that show the contact where the sheared serpentinite masses within the fault zones become the seafloor protrusions of sedimentary serpentinite that these fault-controlled masses produce. These deposits may preserve cross sections of a serpentine mud volcano complete with conduit and proximal mudflows. In the Guajira Peninsula of Colombia, sedimentary serpentinite sequences of varying thickness are intercalated with flysch (Lockwood, 1971a). These serpentinite deposits include well-preserved original sedimentary structures, including graded bedding. It is not likely that these were derived from local mass wasting, but were possibly emplaced in seafloor channels within more massive "mudflow" deposits. Other deposits in north-central Colombia also show sedimentary structures such as graded bedding (Zimmerle, 1968).

Within the Koryak fold-and-thrust belt on the east side of the Penshinskaya Gulf in Kamchatka there are extensive deposits of sedimentary serpentinites, reaching at least 2800 m thick (some of this thickness may be due to imbricate thrusting). These serpentinites preserve sedimentary structures (graded bedding and cross-bedding) (Bogadayeva and Matveyenko, 1960; Mikhailov, 1962). The deposits, which are part of the Kuyul ophiolite terrane, are fossiliferous and contain clasts of rock types from several sources (Khudoley and Sokolov, 1998). Some of the included clasts are fragments of sandstone and siltstone derived from the adjacent Ainyn terrane, interpreted to be the landward slope of an accretionary prism (Sokolov et al., 1996). Other rock clasts were derived from MORB and supra-subduction zone sources (Khudoley and Sokolov, 1998). The deposits also include hemipelagic cherts (Sokolov et al., 1996), likely derived from a subducted plate. Several aspects of the Kuyul terrane make it an excellent analogue for Mariana forearc serpentine seamounts. The serpentinite mélanges within it are juxtaposed with forearc-basin sequences. Rock clasts found in the serpentinite mélanges are similar in diversity to those identified in deposits from the Mariana forearc mud volcanoes. They include supra-subduction zone rocks, MORB, and cherts. The sedimentary serpentinite mélange deposits lie adjacent to sections of more massive serpentinite (possibly similar to the horst blocks underlying Pacman and Big Blue Seamounts?). Recent detailed mapping by Sokolov et al. (1996) and Khudoley and Sokolov (1998) shows the interrelationships of these sequences of terranes and the juxtaposition of multiple accretionary terranes of different ages in the regional Koryak fold-and-thrust belt on the east side of the Penshinskaya Gulf. Although the Kuyul terrane is complex and deformed, these Kamchatka deposits appear to be excellent analogues for Mariana forearc serpentine seamounts and to have preserved many of the types of serpentinite sequences we would expect to see were the Mariana outer forearc uplifted and obducted onto a continental margin.

A REVISED MODEL FOR EMPLACEMENT OF SERPENTINITE MASSES IN CONVERGENT MARGINS

We conclude that the model for formation of sedimentary serpentinite masses suggested by Lockwood (1972) is essentially correct, but requires some modifications to fit the observations made in the Mariana region. Several workers have suggested that subduction may be initiated along former transform faults at times of major plate reorganizations (e.g., Uyeda and Ben-Avraham, 1972; Stern and Bloomer, 1992). If transform faults become subduction zones, then it is possible that exposures of uplifted serpentinized mantle within the fault zones (megamullions, e.g., Tucholke et al., 1998) could become sites of serpentinite seamounts in convergent margins. Metamorphic expansion of supra-subduction zone ultramafic rocks in the outer edge of an overriding plate at a convergent margin can facilitate exposure of large tracts of serpentinized forearc mantle (Fryer and Fryer, 1987). Either mechanism could expose large blocks of metamorphosed mantle of the overriding plate in a convergent-margin setting. Faulting in forearc regions can uplift regions of serpentinized forearc mantle and may occur as a consequence of vertical tectonism related to seamount or ridge subduction (Fryer and Fryer, 1987). These regions of massive serpentinite are often themselves highly faulted by regional extensional stresses imposed on forearcs. Extensional faulting in forearcs may be related to trench rollback or increasing curvature of a forearc with time. Faults provide egress for rising serpentinized mantle, slab-derived fluids, and fragments of subducted plate. The resulting mud volcanism of serpentine material is thus most often superimposed on these faulted serpentinized blocks or found along major fault zones. The serpentine that makes up the bulk of Mariana forearc serpentine mud volcanoes is formed in response to hydration metamorphism of supra-subduction zone mantle and rises because of density contrasts along fault-controlled conduits in the overriding plate. Protrusion events may be triggered by seismic activity in the subduction zone or in the overriding plate. Such events are likely episodic and may provide a mechanism for rapid flushing of fault gouge through the conduit.

As the material rising to form mud volcanoes has been derived from anywhere along the path of the rising fluids, from within either the subducted slab or the overriding plate, the mud volcanoes' compositions have a wide range. The mud volcanoes' deposits range in thickness from several centimeters to at least several tens of meters. They can be intercalated between and are conformable with forearc sediments. The deposits are made up of a matrix of clay- to sand-sized serpentine particles that can contain a wide variety of additional mineral grains and minute lithic fragments (carbonates, mafic minerals, chlorites, garnets, and other high-pressure metamorphic minerals). The matrix encloses rock clasts with a wide range in sizes and lithology. It is possible that the rising material includes coarser clasts than commonly observed at the seafloor on these seamounts. Larger clasts have greater density than the finer fractions and may sink back into the muds or along the conduit and may not be exposed at the surface. Clasts, like the muds, may be derived from any lithology along the conduit path. They are dominantly serpentinized peridotite, but can include metabasalts and metagabbros of blueschist and greenschist grade (Johnson, 1992; Maekawa et al., 1992, 1993, 1995) and may also include various types of sedimentary rocks (some derived from the subducted slab). We have seen igneous clasts of such varied origins as OIB, island-arc boninitic suites, arc tholeiites, and MORB (Johnson, 1992).

The materials are deposited on the seafloor as either thick lobes of varying extent (see Figs. 5 and 6) or radiating thin flows of muds (e.g., Conical Seamount, Fryer et al., 1990a; Fryer, 1996a) probably depending on the amount of slab-derived fluids that accompany the protrusive events. The deposits thus formed build large mud volcanoes and may be reworked during later deformation of the seamount, as is clearly the case at South Chamorro Seamount (Fig. 8). The edifice may deform under its own weight over time as the lobes of serpentinite masses are subjected to gravitational instability. Concentric pressure ridges on the flanks of Conical Seamount (Fryer et al., 1990a) and the radial fracturing apparent on the summit of Turquoise Seamount (Fig. 5) suggest that such internal deformation is not unusual in these seamounts.

Preservation of serpentinite horsts in forearcs and of mud volcanoes and their deposits during emplacement on land in collision events may depend on their setting in the forearc. It is unlikely that mud volcanoes forming on the upper edges of forearc horsts would survive intact, such as Celestial, Blue Moon, or the Chamorro Seamounts. In these settings, perhaps the only remaining evidence of their existence would be the presence of the conduit of the edifice and possibly distal parts of mudflows that happen to be preserved in adjacent forearc basins. The conduit, if preserved, might appear as a more or less cylindrical body of sheared unconsolidated serpentine with embedded clasts of various lithologies including some metabasites of high-pressure, low-temperature origin. This conduit would likely be enclosed in variably serpentinized peridotite, but may also be associated with shallower-level forearc sequences. The body of the edifice may be entirely removed by erosion. Serpentine mud volcanoes forming at the edges of forearc grabens would be far more likely to be preserved, e.g., Conical, Turquoise, or Peacock Seamounts. Seamounts at the edges of these grabens may be partially buried by forearc sediments so that uplift and erosion during exhumation may expose the edifice more or less intact, preserving contacts with the upper forearc volcaniclastic sediments. It is possible that the Mysterious Valley Formation is an example of such an exposure. Exposures in Cuba may be similar. Deposits from serpentine mud volcanoes that have been redistributed after protrusion and have partially infilled bathymetric lows adjacent to the mud volcanoes may be preserved intact, but without an obvious connection to the major edifice. The debris flows from the sector collapse on the southeastern side of South Chamorro Seamount are an example. These debris flows are draping the inner slope of the trench and are ponded in a deep basin near the trench axis. Such reworked serpentine mudflows may eventually be incorporated in a wedge of accreted sediment from the subduction plate or be drawn down the subduction zone to become part of a subduction complex. It is possible that some of the sedimentary serpentinite material within the Franciscan Formation in California could have formed by such a mechanism. Seamounts such as Pacman and Big Blue are composite edifices (Fryer et al., 1995, 1999). They represent horst blocks of uplifted forearc mantle that have been faulted and subsequently partially draped by serpentine mudflows. Such edifices may be partially preserved during emplacement. It is possible that only part of the serpentine mudflows associated with these edifices might be preserved during collision and emplacement on land. The mudflows that in-fill graben structures within these seamounts might be preserved (e.g., the debris flows draping the east-flank graben on Pacman Seamount; see Fryer et al., 1995), and the conduits for the muds may also be preserved in relationship to the graben-filling mudflows. Exposures of such edifices could preserve sedimentary serpentinite masses within surrounding blocks of more massive serpentinite.

Of course, when forearc horst blocks of massive serpentinite or mud-volcano edifices are emplaced on land during obduction events, the serpentinite masses may be deformed. In particular, it may be difficult to recognize serpentine mudflow deposits as sedimentary serpentinites if they have been subjected to postemplacement deformation and metamorphism. It is possible, therefore, that sedimentary serpentinites of mud-volcano origin are undercounted in the geologic record and may be more important than previously recognized. After emplacement, these deposits may be further sheared and remobilized. Sedimentary remobilization of the serpentine masses may give rise to the sorts of sedimentary structures (cross-bedding and graded bedding) seen in some exposures on land; however, it is also possible for such sedimentary structures to be formed in situ during protrusive events on the flanks of the mud volcanoes. A careful comparison of detailed structures of the in situ deposits with those studied previously on land will help clarify this point. We think that it is possible with careful mapping and mineral and geochemical studies to distinguish sedimentary serpentinite deposits formed by forearc mud volcanism from serpentinite deposits formed by other means.

Our model suggests that in form, texture, composition, and structure, the deposits formed in situ as mud volcanoes on the Mariana forearc are very similar to those of many chaotic sedimentary serpentinite deposits observed in former convergent-margin terranes on land. We contend that the mechanism of serpentine mud volcanism observed in the Mariana forearc is a more common phenomenon in convergent margins than has been previously recognized.

ACKNOWLEDGMENTS

We thank the Captain and crew of the R/V *Moana Wave* for assistance in obtaining the side-scan data presented in this paper. We particularly thank the Hawaii Mapping Research Group members who sailed with us, B. Applegate, F. Martinez, L. Peterson, and S. Tatori. We greatly appreciate the critical reviews and helpful comments on the manuscript by reviewers S. Peacock and J. Wakabayashi. The side-scan and geophysical data were obtained with support of National Science Foundation grant OCE9633501 and an NSF–ODP Site Augmentation grant from the Ocean Drilling Program (JSC-8–97) through the U.S. Science Support Program. The U.S. Science Support Program associated with the Ocean Drilling Program is sponsored by the National Science Foundation and Joint Oceanographic Institutions (JOI), Inc. Any opinions, findings, and conclusions or recommendations expressed in this publication are those of the authors and do not necessarily reflect the views of the National Science Foundation or JOI. ODP is funded by the National Science Foundation and research agencies in Australia, Belgium, Canada, Chinese Taipei, Denmark, Finland, France, Germany, Iceland, Italy, Japan, Korea, Netherlands, Norway, Spain, Sweden, Switzerland, Turkey, and the United Kingdom. This is SOEST contribution no. 5259 and HIGP contribution no. 1113.

REFERENCES CITED

Bailey, E.H., Irwin, W.P., and Jones, D.L., 1964, Franciscan and related rocks and their significance in the geology of Western California: California Division of Mines and Geology, Bulletin, v. 183, 177 p.

Bogadayeva, M.V., and Matveyenko, V.T., 1960, Osnovnyye I ul'traosnovnyye intruzii Koryaksko-Anadyrskogo rayona (Severo-Vostok SSSR) [Basic and ultrabasic intrusions of the Koryak-Anadyr region (northeast USSR)]: All-Union Science-Research Institute for Gold and Rare Metals (Magadan), Trudy, Vyp. 59, p. 1–72.

Bucher, W.H., 1956, Role of gravity in orogenesis: Geological Society of America Bulletin, v. 67, no. 10, p. 1295–1318.

Carlson, C., 1984, Stratigraphic and structural significance of foliate serpentine breccias, Wilbur Springs, *in* Society of Economic Paleontologists and Mineralogists, Field Trip Guidebook no. 3, p. 108–112.

Desprairies, A., 1981, Authigenic minerals in volcanigenic sediments cored during Deep Sea Drilling Project Leg 60, *in* Hussong, D.M., Uyeda, S., eds., Initial Reports of the Deep Sea Drilling Project, Washington, D.C., U.S. Govt. Printing Office, v. 60, p. 437–453.

Dickinson, W.R., 1966, Table Mountain serpentinite extrusion in California Coast Ranges: Geological Society of America Bulletin, v. 77, p. 451–472.

Dickinson, W.R., Hopson, C.A., Saleeby, J.B., Schweickert, R.A., Ingersoll, R.V., Pessagno, E.A., Jr., Mattinson, J.M., Luyendyk, B.P., Beebe, W., Hull, D.M., Munoz, I.M., and Blome, C.D., 1996, Alternate origins of the Coast Range ophiolite (California): Introduction and implications: GSA Today, v. 6, no. 2, p. 1–10.

Engdahl, E.R., van der Hilst, R., Buland, R., 1998, Global teleseismic earthquake relocation with improved travel times and procedures for depth determination: Bulletin of the Seismological Society of America, v. 88, no. 3, p. 722–743.

Fryer, P., 1992a, A synthesis of Leg 125 drilling of serpentine seamounts on the Mariana and Izu-Bonin forearcs, Ocean Drilling Program Leg 125, *in* Proceedings of the Ocean Drilling Program, Scientific Results Leg 125: College Station, Texas, Ocean Drilling Program, p. 593–614.

Fryer, P., 1992b, Egress of slab-related fluids on the Mariana Forearc: Implications for fluid flux in a nonaccretionary convergent margin setting [abs.]: Eos, Transactions of the American Geophysical Union, v. 73, no. 43, p. 638.

Fryer, P., 1996a, Tectonic evolution of the Mariana convergent margin: Reviews of Geophysics, v. 34, no. 1, p. 89–125.

Fryer, P., 1996b, An actively venting serpentine seamount on the southeastern Mariana forearc: *Shinkai 6500* dives 280 and 281: Proceedings, Japanese Science and Technology Center, Deep Sea Research, v. 12, p. 247–256.

Fryer, P., and Fryer, G.J., 1987, Origins of nonvolcanic seamounts in a forearc environment, *in* Keating, B., Fryer, P., and Batiza, R., eds., Seamounts, islands, and atolls: Washington D.C., American Geophysical Union Monograph Series, v. 43, p. 61–69.

Fryer, P., and Mottl, M.J., 1992, Lithology, mineralogy, and origin of serpentine muds drilled at Conical Seamount and Torishima Forearc Seamount, Ocean Drilling Program Leg 125, *in* Proceedings of the Ocean Drilling Program, Scientific Results Leg 125: College Station, Texas, Ocean Drilling Program, p. 343–362.

Fryer, P., and Mottl, M.J., 1997, *Shinkai 6500* investigations of a resurgent mud volcano on the southeastern Mariana forearc: Proceedings, Japanese Science and Technology Center, Deep Sea Research, v. 13, p. 103–114.

Fryer, P., and Pearce J.A., 1992, Introduction to the scientific results of Leg 125: Ocean Drilling Program Leg 125, Scientific Results Leg 125: College Station, Texas, Ocean Drilling Program, p. 3–11.

Fryer, P., and Smoot, N.C., 1985, Processes of seamount subduction in the Mariana and Izu-Bonin trenches: Marine Geology, v. 64, p. 77–90.

Fryer, P., and Todd, C., 1999, Mariana Blueschist mud volcanism sampling the subducted slab [abs.]: Eos, Transactions, American Geophysical Union, v. 80, no. 17, p. S349.

Fryer, P., Ambos, E.L., and Hussong, D.M., 1985, Origin and emplacement of Mariana forearc seamounts: Geology, v. 13, p. 774–777.

Fryer, P., Pearce, J.A., et al., 1990a, Proceedings of the Ocean Drilling Program: Initial Reports, Leg 125, College Station, Texas, Ocean Drilling Program, 1092 p.

Fryer, P., Saboda, K.L., Johnson, L.E., MacKay, M.E., Moore, G.F., and Stoffers, P., 1990b, Conical Seamount: SeaMARC II, Alvin submersible, and seismic reflection studies, *in* Proceedings of the Ocean Drilling Program: Initial Reports, Leg 125, College Station, Texas, Ocean Drilling Program, p. 5–14.

Fryer, P., Mottl, M., Johnson, L.E., Haggerty, J.A., Phipps, S., and Maekawa, H., 1995, Serpentine bodies in the forearcs of Western Pacific convergent margins: Origin and associated fluids, *in* Taylor, B., and Natland, J., eds., Active margins and marginal basins of the Western Pacific, American Geophysical Union, Monograph Series, v. 88, p. 259–279.

Fryer, P., Wheat, C.G., and Mottl, M.J., 1999, Mariana Blueschist mud volcanism: Implications for conditions within the subduction zone: Geology, v. 27, no. 2, p. 103–106.

Giaramita, M., MacPhearson, G.J., and Phipps, S.P., 1998, Petrologically diverse basalts from a fossil ocean forearc in California: The Llanda and Black Mountain remnants of the Coast Range ophiolite: Geological Society of America Bulletin, v. 110, no. 5, p. 553–571.

Gresens, R.L., 1970, Serpentinites, blueschists, and tectonic continental margins?: Geological Society of America Bulletin, v. 81, p. 307–310.

Henry, P., Le Pichon, X., Lallemant, S., Foucher, J.-P., and Westbrook, G., 1990, Mud volcano field seaward of the Barbados accretionary complex: A deep-towed scan sonar: Journal of Geophysical Research, v. 95, no. B6, p. 8917–8929.

Hess, H.H., 1955, Serpentinites, orogeny, and epeirogeny: Geological Society of America Special Paper 62, p. 391–408.

Horine, R.L., Moore, G.F., and Taylor, B., 1990, Structure of the outer Izu-Bonin forearc from seismic-reflection profiling and gravity modeling: Ocean Drilling Program Leg 125, Initial Results Leg 125, College Station, Texas, Ocean Drilling Program, p. 81–94.

Ingersoll, R.V., and Dickinson, W.R., 1981, Great Valley Group (sequence), Sacramento Valley, California, *in* Frizzell, V., ed., Upper Mesozoic Franciscan rocks and Great Valley sequence, central Coast Ranges, California: Pacific Section, Society of Economic Palenotologists and Mineralogists, Field Trip Guide, p. 1–33.

Johannes, W., and Puhan, D., 1971, The calcite-aragonite transition, reinvestigated: Contributions to Mineralogy and Petrology, v. 31, p. 28–38.

Johnson, L.E., 1992, Mafic clasts in serpentine seamounts: Petrology and geochemistry of a diverse crustal suite from the outer Mariana forearc, *in* Proceedings of the Ocean Drilling Program: Scientific Results Leg 125, College Station, Texas, Ocean Drilling Program, p. 401–413.

Johnson, L.E., Fryer, P., Taylor, B., Silk, M., Sliter, W.V., Jones, D.L., Itaya, T., and Ishii, T., 1991, New evidence for crustal accretion in the outer Mariana forearc: Cretaceous radiolarian cherts and MORB-like lavas: Geology, v. 19, p. 811–814.

Khudoley, A.K., and Sokolov, S.D., 1998, Structural evolution of the northeast Asian continental margin: An example from the western Koryak fold and thrust belt (northeast Russia): Geological Magazine, v. 135, no. 3, p. 311–330.

Knipper, A.L., and Kostanyan, Yu.L., 1964, Vozrast giperbazitov severovostochnogo poberezh'ya ozera Sevan [The age of ultramafics of the northeastern coast of Lake Sevan]: Akademii Nauk SSSR, Izvestiia Seriia Geologicheskaia, no. 10, p. 67–69.

Kozary, M.T., 1968, Ultramafic rocks in thrust zones of northwestern Oriente Province, Cuba: American Association of Petroleum Geologists Bulletin, v. 52, no. 12, p. 2298–2317.

Liu, J., Bohlen, S.R., and Ernst, W.G., 1996, Stability of hydrous phases in subducting oceanic crust: Earth and Planetary Science Letters, v. 143, p. 161–171.

Lockwood, J.P., 1971a, Sedimentary and gravity-slide emplacement of serpentinite: Geological Society of America Bulletin, v. 82, no. 4, p. 919–936.

Lockwood, J.P., 1971b, Detrital serpentinite from the Guajira Peninsula, Colombia, *in* Donnelly, T.W., ed., Caribbean geophysical, tectonic, and petrologic studies: Geological Society of America Memoir 130, p. 5–75.

Lockwood, J.P., 1972, Possible mechanisms for the emplacement of alpine-type serpentinite, *in* Shagam, R., et al., eds., Studies in Earth and Space Sciences: Geological Society of America Memoir 132 (Hess Volume), p. 273–287.

MacPherson, G.J., and Phipps, S.P., 1988, Geochemistry and petrology of mafic volcanic rocks from olistostromes in the basal Great Valley Group, northern California Coast Ranges: Geological Society of America Bulletin, v. 100, p. 1770–1779.

Maekawa, H., Shozui, M., Ishii, T., Saboda, K.L., and Ogawa, Y., 1992, Metamorphic rocks from the serpentinite seamounts in the Mariana and Izu-Ogasawara forearcs: Ocean Drilling Program Leg 125, Scientific Results Leg 125, College Station, Texas, Ocean Drilling Program, p. 415–430.

Maekawa, H., Shozui, M., Ishii, T., Fryer, P., Pearce, J.A., 1993, Blueschist metamorphism in an active subduction zone: Nature, v. 364, no. 6437, p. 520–523.

Maekawa, H., Fryer, P., Ozaki, M., 1995, Incipient blueschist-facies metamorphism in the active subduction zone beneath the Mariana Forearc, *in* Taylor, B., and Natland, J., eds., Active margins and marginal basins of the Western Pacific, American Geophysical Union Monograph Series, v. 88, p. 281–290.

Massone, H.J., and Szpurka, Z., 1997, Thermodynamic properties of white micas on the basis of high-pressure experiments in the system K_2O-MgO-Al_2O_3-SiO_2-H_2O and K_2O-FeO-Al_2O_3-SiO_2-H_2O: Lithos, v. 41, no. 1–3, p. 229–250.

Mikhailov, A.F., 1962, Geologo-petrologicheskaya kharakteristika giperbazitiv i giperbazitovykh brekchiy yugo-zapadnoy chasti Penzhinskogo Kraiya [Geologic-petrologic character of ultramafics and ultramafic breccias of the southwestern part of the Penzhinskiy Range]: All-Union Scientific Research Geologic Institute (VSEGEI) Trudy, v. 73, no. 4, p. 111–132.

Mottl, M.J., 1992, Pore waters from serpentinite seamounts in the Mariana and Izu-Bonin forearc serpentinite: Leg 125, *in* Proceedings of the Ocean Drilling

Program: Scientific Results Leg 125, College Station, Texas, Ocean Drilling Program, p. 373–386.

Mrozowski, C.L., Hayes, D.E., and Taylor, B., 1981, Multichannel seismic reflection surveys of Leg 60 Sites, Deep Sea Drilling Project, *in* Hussong, D.M., and Uyeda, S., eds., Initial Reports of the Deep Sea Drilling Project, Volume 60: Washington, D.C., U.S. Government Printing Office, p. 57–71.

Oakeshott, G.B., 1968, Diapiric structures in Diablo Range, California, *in* Diapirism and diapirs: A symposium: American Association of Petroleum Geologists Memoir, p. 228–243.

Parkinson, I.J., and Pearce, J.A., 1998, Peridotites from the Izu-Bonin-Mariana forearc (ODP Leg 125): Evidence for mantle melting and melt-mantle interaction in a supra-subduction zone setting: Journal of Petrology, v. 39, no. 9, p. 1577–1618.

Peacock, S.M., 1990, Fluid processes in subduction zones: Science, v. 248, p. 329–336.

Peacock, S.M., 1993, Metamorphism, dehydration, and the importance of the blueschist to eclogite transition in subducting oceanic crust: Geological Society of America Bulletin, v. 105, p. 684–694.

Pearce, J.A., Kempton, P.D., Nowell, G.M., and Noble, S.R., 1999, Hf-Nd element and isotope perspective on the nature and provenance of mantle and subduction components in western Pacific arc-basin systems: Journal of Petrology, v. 40, no. 11, p. 1579–1611.

Phipps, S.P., 1984, Ophiolotic olistostromes in the basal Great Valley sequence, Napa County, northern California Coast Ranges, *in* Raymond, L., ed., Mélanges: Their nature, origin, and significance: Geological Society of America Special Paper 198, p. 103–125.

Phipps, S.P., 1992, Late Cenozoic tectonic wedging and blind thrust faulting beneath the western Sacramento Valley and northern Coast Ranges, California, Day 2, *in* Bartow, A., Erskine, M. and Unrugh, J.R., eds., Field guide to the tectonics of the boundary between the California Coast Ranges and the Great Valley of California: Pacific Section, Society of Economic Paleontologists and Mineralogists, Guide Book 70, p. 63–83.

Phipps S.P., and MacPherson, G.J., 1999, The formation of Franciscan ophiolitic melanges: Actualistic models from the western Pacific: Geological Society of America Abstracts with Programs, v. 31, no. 6, p. A-85.

Robertson, A., 1996, Mud volcanism on the Mediterranean Ridge: Initial results of Ocean Drilling Program Leg 160: Geology, v. 24, no. 3, p. 239–242.

Shipley, T.H., Stoffa, P.L., and Dean, D.F., 1990, Underthrust sediments, fluid migration paths and mud volcanoes associated with the accretionary wedge off Costa Rica: Middle America Trench: Journal of Geophysical Research, v. 95, no. B6, p. 8743–8752.

Sokolov, S.D., Peive, A.A., Krylov, K.A., Crigor'yev, V.N., Batanova, V.G., Golozubov, V.V., Luchitskaya, M.V., and Aleksyutin, M.V., 1996, Elementy veshshestrennoi i strukturnoi uporyadochennosti v serpentinitorom melanzhye [Elements of material and structural order in serpentinite mélange]: Geotektonika, v. 1, p. 47–62.

Stern, R.J., and Bloomer, S.H., 1992, Subduction zone infancy: Examples from the Eocene Izu-Bonin-Mariana and Jurassic California arcs: Geological Society of America Bulletin, v. 104, p. 1621–1636.

Stern, R.J., and Smoot, N.C., 1998, A bathymetric overview of the Mariana forearc: The Island Arc, v. 7, no. 3, p. 525–540.

Todd, C.S., and Fryer, P., 1999, Blueschist mud volcanism in the Mariana forearc: Sampling the subducted slab: Geological Society of America Abstracts with Programs, v. 31, no. 7, p. A-103.

Tucholke, B.E., Lin, J., and Kleinrock, M.C., 1998, Megamullions and mullion structure defining oceanic metamorphic core complexes on the Mid-Atlantic Ridge: Journal of Geophysical Research, v. 103, no. B5, p. 9857–9866.

Uyeda, S., and Ben-Avraham, Z., 1972, Origin and development of the Philippine Sea: Nature, Physical Science, v. 240, p. 176–178.

Wessel, J., Fryer, P., Wessel, P., and Taylor, B., 1994, Extension in the northern Mariana forearc: Journal of Geophysical Research, v. 99, no. B8, p. 15,181–15,203.

Zimmerle W., 1968, Serpentine greywacke from the north Coast basin Colombia and their geotectonic significance: Neues Jahrbuch fuer Mineralogie Abhandlungen, v. 109, p. 156–182.

Manuscript Accepted by the Society February 23, 2000

Printed in the U.S.A.

Geological Society of America
Special Paper 349
2000

Spatial and temporal relationships between ophiolites and their metamorphic soles: A test of models of forearc ophiolite genesis

John Wakabayashi
1329 Sheridan Lane, Hayward, California 94544, USA
Yildirim Dilek
Department of Geology, Miami University, Oxford, Ohio 45056, USA

ABSTRACT

Thin sheets of high-grade metamorphic rocks, called metamorphic soles, are found structurally beneath many ophiolites, apparently formed at the initiation of intraoceanic subduction, and may be useful in determining the tectonic setting of ophiolite genesis. Many ophiolites are interpreted to have formed in a supra-subduction zone setting largely on the basis of their petrology and geochemistry. Of these supra-subduction zone ophiolites, most have been interpreted to have formed in a forearc setting. Generation of ophiolites in a forearc setting indicates that such ophiolites must postdate the initiation of subduction beneath them and must therefore be younger than their metamorphic soles. However, the examples of pairs of ophiolites and their metamorphic soles reviewed herein show that the ophiolites are not younger than their soles. The spatial and temporal relationships reviewed herein may be incompatible with the generation of ophiolites in forearcs, although volumetrically subordinate igneous rocks that postdate the formation of the soles have been found in some of the examples. The apparent forearc setting of the ophiolites may be an artifact of their emplacement tectonics, not their igneous environment of formation. Although we know of no geochronologic confirmation of an older subduction zone structurally beneath purported forearc ophiolites, detailed geochronology has yet to be conducted on most ophiolite and metamorphic sole pairs. For the scenario in which an ophiolite is formed above one subduction zone but emplaced over a different one, either backarc formation or intra-arc formation of the ophiolite can be compatible with its regional geochronologic and structural relationships.

INTRODUCTION

Subophiolitic soles

Subophiolitic soles are thin (<500 m thick), fault-bounded sheets of highly strained high-grade metamorphic rocks that are found structurally beneath many ophiolite complexes (e.g., Williams and Smyth, 1973, Jamieson, 1986). The higher-grade parts of the metamorphic soles are composed mainly of metamorphosed mafic rocks of oceanic affinity, with minor metamorphosed pelagic sedimentary rocks. The pressure-temperature (*P-T*) conditions of metamorphism for subophiolitic soles are consistent with high-temperature metamorphism beneath hot—and thus necessarily young—suboceanic mantle. On the basis of their oceanic protoliths, the presence of an ophiolite structurally above them, and their metamorphic petrology, metamorphic soles are thought to form at the inception of oceanic subduction underneath the hot subophiolitic mantle of the hanging wall (e.g., Williams and Smyth, 1973, Malpas, 1979, Nicolas and Le Pichon, 1980, Spray, 1984, Jamieson, 1986) (Fig. 1). The inverted-temperature anomaly responsible for the high-grade metamorphism of the sole decays quickly (in <2 m.y.) as subduction continues (Peacock, 1988; Hacker, 1990, 1991; Hacker et al., 1996). Thus, the high-grade metamorphism of the sole can occur only at the inception of subduction because the hanging wall would thereafter be too cold to cause high-grade metamorphism. As a result of thermal insulation from continuing subduction, the metamorphic rocks of the sole cool rapidly through the blocking temperature of commonly applied isotopic systems such as $^{40}Ar/^{39}Ar$. Because of this rapid cooling, the metamorphic age of the aureole may be a close approximation to the inception of intraoceanic subduction (e.g., Spray, 1984, Peacock, 1988) that predates the "obduction" or emplacement of the ophiolite in the upper plate onto a continental margin, island arc, or subduction complex (Moores, 1982). Thus, the timing of formation of metamorphic soles beneath ophiolites provides a critical age constraint for the tectonic evolution of fossil oceanic crust. Geochronologic data, where available, and the metamorphism of continental-margin materials in the lower parts of some soles indicate that emplacement of some ophiolites onto a continental margin (common for Tethyan ophiolites—such ophiolites will be called "Tethyan type" herein; illustrated in Fig. 2) took place within less than 5 m.y. after the inception of subduction, indicating that less than 500 km of oceanic crust was subducted prior to emplacement of the ophiolite and cessation of

Wakabayashi, J., and Dilek, Y., 2000, Spatial and temporal relationships between ophiolites and their metamorphic soles: A test of models of forearc ophiolite genesis, *in* Dilek, Y., Moores, E.M., Elthon, D., and Nicolas, A., eds., Ophiolites and Oceanic Crust: New Insights from Field Studies and the Ocean Drilling Program: Boulder, Colorado, Geological Society of America Special Paper 349, p. 53–64.

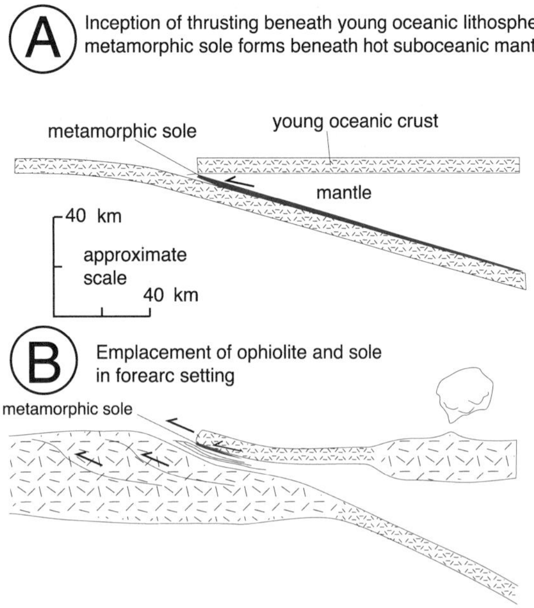

Figure 1. Schematic cross-sectional cartoons illustrating subophiolitic sole development and emplacement of ophiolites in forearc settings. A, Metamorphic sole develops beneath hot, young oceanic lithosphere at inception of subduction. Note that the oceanic crust has already been formed. No setting of ophiolite formation is specified in this cartoon. B, Emplacement traps the ophiolite and its sole in a forearc setting. The thickness of the sole is greatly exaggerated in the cartoons; the true thickness would be too thin to be visible on the diagram. Note that upward tectonic movement of the sole relative to the crustal part of the ophiolite takes place some time after metamorphism of the sole, either before or during emplacement.

subduction (e.g., Hacker et al., 1996). Other ophiolites were emplaced over subduction complexes (common for ophiolites of the North American Cordillera—such ophiolites will be called "Cordilleran type" herein; illustrated in Fig. 3) as subduction continued following development of the metamorphic sole. How much time elapsed between sole development and emplacement of Cordilleran-type ophiolites is dependent on how the term "emplacement" is defined for such ophiolites. If emplacement is defined as the time at which rocks of the subduction complex were first underplated beneath the ophiolite, then the emplacement time for such an ophiolite would be defined to be synchronous with sole development, because the sole is essentially the structurally highest part of the subduction complex (e.g., Platt, 1975, Peacock, 1988). Other definitions of emplacement for Cordilleran-type ophiolites may be formulated for different degrees of exhumation of the ophiolite; such definitions would yield different timing between sole development and emplacement ranging from a few million years to tens of million years.

Forearc ophiolites and subophiolitic soles

The tectonic setting of ophiolite generation has long been a subject of debate (e.g., Miyashiro, 1973, Hynes, 1975, Gass et al., 1975, Moores, 1982, Coleman, 1984, Pearce et al., 1984). Many ophiolites, including the largest and most complete complexes, have geochemical characteristics suggesting formation over a subduction zone ("supra-subduction zone" or SSZ ophiolites of Pearce et al., 1984). A forearc setting is one of the most widely cited supra-subduction zone environments regarding ophiolite formation because the geochemical characteristics and mineral assemblages of these ophiolites are similar to those of igneous rocks in young forearcs and because of the evidence for extension in ophiolites and forearcs (Bloomer et al., 1995, and references therein).

Any tectonic model suggesting a forearc origin for ophiolite generation—i.e., forearc ophiolite models—requires the existence of an active subduction zone for the evolution of ophiolitic magmas (Fig. 2). Formation of magmas over a young subduction zone has been proposed for the development of igneous rocks in the Marianas forearc, which is considered an analogue for ophiolite generation in a nascent forearc setting (Bloomer et al., 1995). Because subophiolitic soles form at the inception of subduction, the ages of subophiolitic soles compared to the ophiolites that structurally overlie them can be used to evaluate the plausibility of the forearc ophiolite model. Note that the spatial and temporal relationships of forearc ophiolites to their metamorphic soles also apply to all other types of supra-subduction zone ophiolites that were emplaced above the same subduction zones over which they formed (Fig. 2, A and B); these other types of supra-subduction zone ophiolites include intra-arc and backarc ophiolites that evolved to an apparent forearc setting as a result of subduction erosion. In this paper, the term "forearc ophiolite" will also apply to these other types of supra-subduction zone ophiolites.

The presence of a subophiolitic sole indicates that, in order to produce the metamorphism, some type of oceanic crust must have existed structurally above the downgoing slab prior to development of the sole because the *P-T* conditions of metamorphism, as well as the types of structurally adjacent rocks, are not compatible with other tectonic settings (e.g., Williams and Smyth, 1973, Jamieson, 1986). Jamieson (1986) noted that, in most cases, the overlying oceanic crust was apparently hot and young (<20 m.y.) at the time of the generation of the subjacent metamorphic sole. The youth of both the overlying oceanic crust and the subducting oceanic plate at the time of sole formation is consistent with thermal models for subophiolitic soles as well as ophiolite preservation models (Hacker, 1991; Hacker et al., 1996), and the regional structural and geochronologic data from the Neotethyan ophiolites (Dilek et al., 1999).

Because emplacement of ophiolites onto a continental margin or over a subduction complex involves a subduction zone that is dipping beneath them, essentially all ophiolites are emplaced into forearc settings (Figs. 1, 2A, 2B, and 3). A key issue is whether the apparent forearc setting of ophiolites is (1) a reflection of both their tectonic setting of origin and subsequent emplacement or (2) a tectonic artifact of their postmagmatic emplacement in a forearc setting following changes in the regional tectonic regime and plate interactions (Dilek and Moores, 1990; Hacker et al., 1996).

This chapter will (1) examine ophiolite and metamorphic sole

Figure 2. Cross-sectional cartoons illustrating different settings of formation of supra-subduction zone ophiolites and relationships with metamorphic soles. Scale is approximate; dimensions of marginal basins and depth of sole metamorphism have been minimized for space considerations. All patterned parts of the figure depict oceanic or continental crust, and the unpatterned area below depicts mantle. A and B, Settings in which emplacement of the ophiolites is above the same subduction zones over which they were formed. In such cases, the ophiolite should always be younger than its sole. In A, the ophiolite forms shortly after the inception of subduction. In this scenario, subduction begins, and the metamorphic sole that will underlie the future ophiolite is formed. Partial melting of the high-grade rocks of the sole and partial melting in the

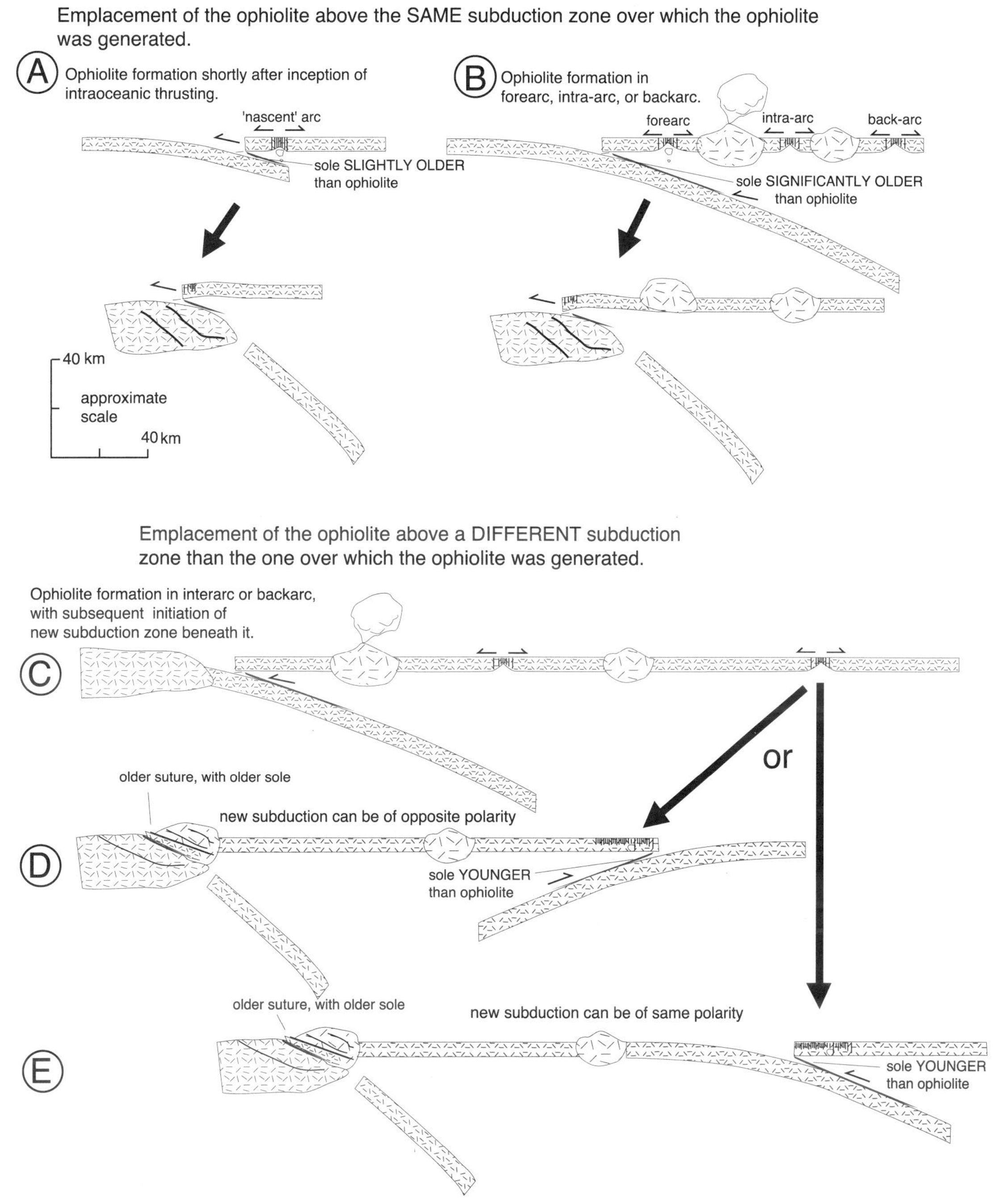

mantle hanging wall above the sole lead to the development of a nascent arc and spreading. Supra-subduction zone magmas formed by the spreading intrude and erupt over the preexisting oceanic crust; the resulting rocks constitute the bulk (if not all) of the crust found in the future ophiolite. Thus metamorphism of the sole predates formation of the ophiolite. In B, the ophiolite forms in association with previously developed volcanic arcs, in forearc, intra-arc, and backarc settings. As in A, supra-subduction zone magmas intrude or erupt over preexisting crust. In this scenario, the future sole, formed at the inception of subduction, will be much older than the ophiolite. For both A and B, emplacement follows some time after inception of subduction, and ophiolites can be emplaced by failed subduction of a continental margin ("Tethyan type"), as illustrated, or over a subduction complex ("Cordilleran type"; see Fig. 3 for example of this type of emplacement for the Coast Range ophiolite of California). C–E, Development of supra-subduction zone ophiolites in intra-arc and backarc settings, but with subsequent development of a new subduction zone beneath the ophiolites as a result of regional or global plate reorganization; in D and E, the new subduction zone has the opposite polarity or the same polarity, respectively. The backarc example is illustrated in D and E, but the scenarios could also apply to intra-arc ophiolites as well. Emplacement (not illustrated) of the ophiolite would follow the stages shown in D and E and can result in ophiolites of either the Tethyan or the Cordilleran type. In C, D, and E, the development of the new subduction zone results in the eventual emplacement of the supra-subduction zone ophiolite above a different subduction zone than the one over which the ophiolite formed; the formation of the metamorphic sole postdates formation of the ophiolite. This scenario requires the existence of a suture and arc remnants; both the suture and the arc remnants predate the formation of the ophiolite. Some late melting may occur during the inception of new subduction beneath the young ophiolite, and the resultant igneous rocks would postdate both the ophiolite and the sole, but the volumes of these rocks would be minor.

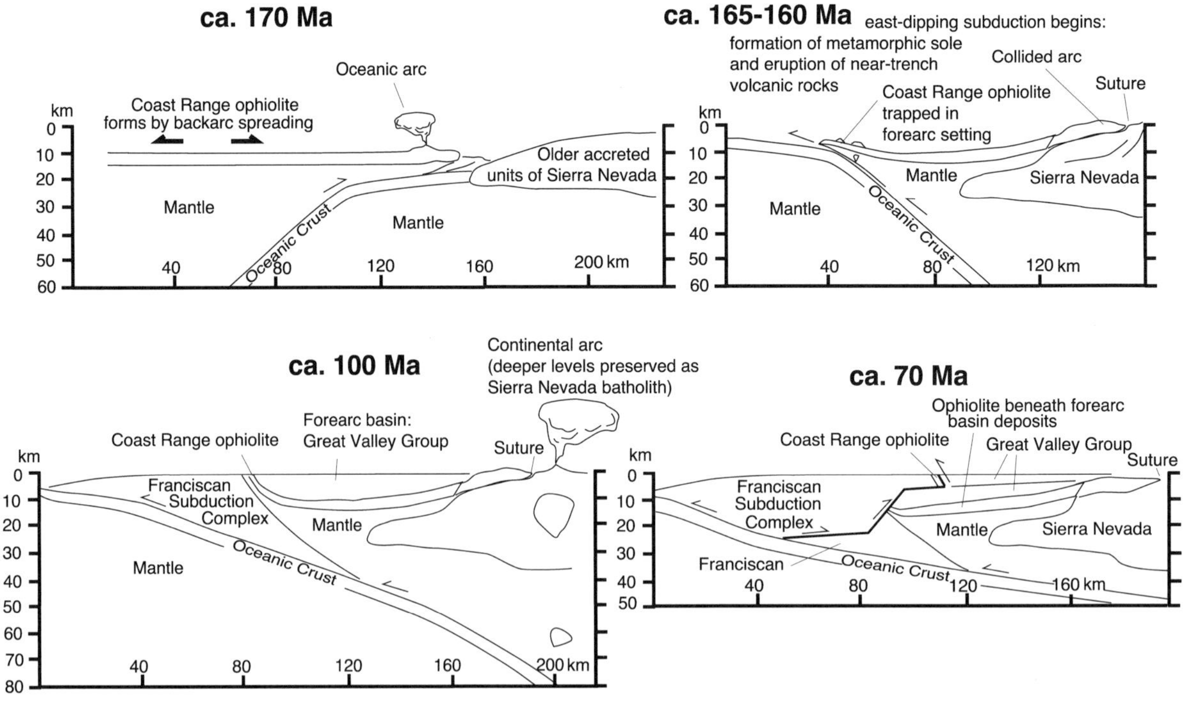

Figure 3. Development of a supra-subduction zone ophiolite in a backarc setting and subsequent emplacement in a forearc setting above a subduction zone that is different from the one over which the ophiolite formed. The sequence of cartoons shown uses the evolution of the Coast Range ophiolite of California as an example, illustrating a model similar to that proposed by Moores (1970), Schweickert and Cowan (1975), Godfrey and Klemperer (1998), and others (Godfrey and Dilek, this volume). The present-day geometry resembles the 70 Ma cross section, except for the absence of a subducting plate and the slivering of the Franciscan subduction complex by strike-slip faulting. No vertical exaggeration; 70 Ma cross section simplified from Wakabayashi (1999).

relationships predicted by forearc ophiolite models, (2) discuss the significance of geochronologic studies of ophiolites and metamorphic soles as tests for the plausibility of forearc ophiolite genesis, (3) review some well-documented ophiolite and metamorphic sole pairs from around the world, particularly from the standpoint of geochronologic relationships, and (4) evaluate and summarize the relationships in these examples in the context of ophiolite generation models. For most of the examples, ophiolite and metamorphic sole pairs have been selected that have been dated with U-Pb zircon geochronology to determine the igneous age of the ophiolite and $^{40}Ar/^{39}Ar$ hornblende geochronology to determine age of the metamorphic sole.

OPHIOLITE–METAMORPHIC SOLE AGE RELATIONSHIPS PREDICTED BY THE FOREARC OPHIOLITE MODEL AND GEOCHRONOLOGIC TESTS OF THE MODEL

Predicted age relationships

The forearc ophiolite model predicts specific relationships between ophiolites and their metamorphic soles. In order to generate an ophiolite above the subduction zone that the ophiolite will be emplaced over (the tectonic relationship demanded by a forearc setting of ophiolite generation), the ophiolite should be *younger* than its metamorphic sole, because the metamorphic sole is formed at the inception of subduction. The earliest possible generation of melt (thus representing the smallest possible age gap between sole and ophiolite) from the subjacent subduction zone should correspond to the high-temperature peak of metamorphic sole formation. This type of setting would be equivalent to the nascent arc model proposed by some for ophiolite generation (e.g., Bloomer et al., 1995). Partial melting of the metamorphosed downgoing plate would be followed by magma migration and crystallization.

Geochronologic tests and problems

The presence of a metamorphic sole indicates the preexistence of young oceanic crust overlying the sole at the time of its formation; the sole *must* be younger than such crust. A critical issue is whether *most* of the ophiolitic rocks that occur above metamorphic soles (1) are part of the young oceanic crust that predated the formation of the sole or (2) formed later, intruding into and erupting on top of this preexisting crustal and mantle material (Figs. 1 and 2). This problem may be at least partly addressed with geochronologic data.

A persistent problem with ophiolite geochronology is that zircons, which may be dated by the U-Pb method, are commonly restricted to silicic differentiates, such as plagiogranites. These intrusions may have formed late relative to the rest of the ophiolite (e.g., Hacker, 1990; Hacker et al., 1996; Moore et al., 1998). Ages of plagiogranites, and hence most U-Pb ages from ophiolites, may represent minimum ages of ophiolite formation. Hornblendes, which can be dated by the $^{40}Ar/^{39}Ar$ step-heating method, may be more widespread in plutonic ophiolitic rocks than zircons. However, most of these hornblendes are the result of high-grade hydrothermal metamorphism of ophiolitic plutonic rocks and thus commonly postdate the igneous crystallization age of the rocks in which they occur (e.g., Liou and Ernst, 1979; Evarts and Schiffman, 1983). In addition, the closure temperature for Ar in hornblende is well below the igneous crystallization temperature of the rocks. Conventional K-Ar dating exploits the same minerals (generally hornblende) in ophiolitic rocks as $^{40}Ar/$

TABLE 1. AGES OF OPHIOLITE-METAMORPHIC SOLE PAIRS

Ophiolite	Ophiolite age	Sole age
Coast Range, California, USA	156–164 Ma, U/Pb zircon*; ~170–172 Ma, U/Pb zircon†	159–163 Ma, $^{40}Ar/^{39}Ar$ hornblende§
Brooks, Range, Alaska, USA	184–187 Ma, $^{40}Ar/^{39}Ar$ hornblende#; 170–162 Ma, U/Pb zircon**	163–169 Ma, $^{40}Ar/^{39}Ar$ hornblende#
Trinity, California, USA	556–579, 469–475 Ma, 403–431 Ma, U/Pb zircon††	186–407 Ma, K/Ar hornblende§§
Semail, Oman	93.5–97.3 Ma, U/Pb zircon##	92.9–94.9 Ma, $^{40}Ar/^{39}Ar$ hornblende***
Bay of Islands, Newfoundland, Canada	486 Ma, U/Pb zircon†††	463–469 Ma, $^{40}Ar/^{39}Ar$ hornblende§§§

Sources: * Hopson et al. (1981). † Hopson et al. (1996), Mattinson and Hopson (1992). § Ross and Sharp (1986, 1988). # Wirth et al. (1993). ** Moore (1998). †† Wallin and Metcalf (1998), Wallin et al. (1995), Mattinson and Hopson (1972a, b). §§ Lanphere et al. (1968), Hotz (1974). ## Tilton et al. (1981). *** Hacker et al. (1996). ††† Dunning and Krogh (1985). §§§ Dallmeyer and Williams (1975).

^{39}Ar dating, but, unlike the $^{40}Ar/^{39}Ar$ step heating method, conventional K-Ar dating lacks a means of evaluating excess Ar (that may make ages older than they should be) or Ar loss (that makes ages younger). Because of the shortcomings of conventional K-Ar dates, our analysis of ophiolite and subophiolitic sole geochronology focuses primarily on U-Pb and $^{40}Ar/^{39}Ar$ dates.

Existing methods for dating ophiolitic rocks thus tend to exclude the older crystallization ages and bias the interpreted age of ophiolites toward younger ages. For subophiolitic soles, $^{40}Ar/^{39}Ar$ hornblende dates provide the closest approximation to metamorphic crystallization ages. Metamorphic temperatures estimated for higher-grade rocks of the metamorphic soles commonly exceed the closure temperature for Ar in hornblende by as much as several hundred degrees Celsius; thus $^{40}Ar/^{39}Ar$ hornblende ages from metamorphic soles are considered cooling ages (e.g., Hacker et al., 1996). However, cooling in metamorphic soles is apparently extremely rapid, such that the hornblende ages are probably less than 5 m.y. younger than their true crystallization age and may be within 1 or 2 m.y. of the true metamorphic ages in most cases, on the basis of field examples and thermal modeling (e.g., Hacker et al., 1996, Hacker, 1990, 1991).

Considering the limitations imposed by existing geochronologic methods, it is apparent that most ages for both ophiolites (U-Pb and $^{40}Ar/^{39}Ar$) and metamorphic soles ($^{40}Ar/^{39}Ar$) are probably systematically younger than their true crystallization ages. However, the bias toward younger ages for ophiolites may be equal to or greater than that of the soles. Thus, for an ophiolite to have been formed in a forearc setting relative to the subduction zone over which the ophiolite is emplaced, the igneous ages should be notably younger than the subophiolitic sole ages. This age difference should be several million years even for ophiolites associated with "nascent arcs."

AGE RELATIONSHIPS OF EXAMPLES OF OPHIOLITE–METAMORPHIC SOLE PAIRS

The age relationships discussed in the following examples are summarized in Table 1.

Semail ophiolite, Oman

The Semail ophiolite is the largest, best-preserved, and best-exposed ophiolite in the world. Ophiolitic rocks form an outcrop belt that is more than 600 km long by 150 km wide. The ophiolite, estimated to be 16–20 km thick, has an 8–12-km-thick section of peridotite tectonite overlain by a crustal section that includes gabbro (0.5–6.5 km thick), sheeted dikes (1–1.5 km thick), and extrusive rocks (0.5–2 km thick) (Boudier and Coleman, 1981; Pallister and Hopson, 1981; Lippard et al., 1986). The high-grade part of the metamorphic sole beneath the peridotite tectonite is <300 m thick; metamorphic conditions have been estimated at 700–900 °C and 9–13 kbar (Gnos, 1998). The Semail ophiolite was emplaced over a continental margin within less than 5 m.y. after the initiation of subduction beneath it (Hacker et al., 1996). The Semail ophiolite has been interpreted to have formed either at a fast-spreading mid-ocean ridge (Nicolas, 1989) or in a supra-subduction zone setting (Lippard et al., 1986). supra-subduction zone models for the Semail ophiolite generally interpret the ophiolite as having formed in a nascent forearc setting (Lippard et al., 1986; Searle and Cox, 1999).

Plagiogranites from the ophiolite crystallized in the interval 93.5–97.3 Ma on the basis of U-Pb zircon ages (Tilton et al., 1981), and the ophiolite cooled to ~550 °C by ca. 93.8 Ma on the basis of $^{40}Ar/^{39}Ar$ hornblende ages (Hacker et al., 1996, and references therein). The metamorphic sole cooled to equivalent temperatures by ca. 93.5 Ma (range 92.9–94.9 Ma; the initiation of subduction may have been slightly different in different areas), based on $^{40}Ar/^{39}Ar$ hornblende ages (Hacker et al., 1996). The ophiolite ages appear to be slightly older to indistinguishable from the sole ages. If the ophiolite formed in the forearc setting relative to the subduction zone over which the ophiolite was emplaced, then the ophiolitic ages would be expected to be younger.

Bay of Islands ophiolite, Newfoundland

The Bay of Islands ophiolite, the largest of the ophiolites in the Appalachian orogen, is exposed over an area of ~100 km by 25 km in Newfoundland (Dunning and Krogh, 1985). The ophiolite, ~12 km thick, has a peridotite tectonite section as thick as 6 km and a crustal section that reaches thicknesses of slightly greater than 6 km. The layered gabbros and massive gabbro section are as thick as 4 km, and the remainder of the crustal section comprises a sheeted-dike complex and overlying basaltic lavas (e.g., Church and Stevens, 1971). The Bay of Islands ophiolite was emplaced over the eastern North American continental margin. Thus, the Bay of Islands ophiolite is similar to Tethyan ophiolites in its mode of emplacement and field relationships. The ophiolite has been considered to have a mid-ocean ridge (e.g., Karson, 1984) or backarc affinity (e.g., Jenner et al., 1991). Although a forearc origin has not been proposed for the Bay of Islands ophiolite, it is discussed here because of the similarity of its field relationships to those of the Tethyan ophiolites for which forearc origins have been commonly proposed.

The metamorphic sole of the Bay of Islands ophiolite is as thick as 150 m, including as much as a 70 m thickness of amphibolite or higher-grade metamorphosed mafic rocks (Malpas, 1979; McCaig, 1983). The conditions of metamorphism for the sole have been estimated as 720–850 °C and 7–12 kbar (Malpas, 1979; McCaig, 1983).

Although older U-Pb zircon age determinations attributed to the Bay of Islands ophiolite yield ages of 504–508 Ma (Mattinson, 1975, 1976), subsequent dating and reinterpretation of the field relationships and isotopic systematics associated with older dates indicate a 486

Ma age for the ophiolite (Dunning and Krogh, 1985). The metamorphic sole yields $^{40}Ar/^{39}Ar$ hornblende ages of 463–469 Ma (Archibald and Farrar, 1976; Dallmeyer and Williams, 1975; recalculated to new decay constants of Steiger and Jäger, 1977). The Bay of Islands ophiolite is apparently older than the sole and is unlikely to have formed in a forearc setting.

Brooks Range ophiolites, Alaska

The western Brooks Range ophiolites of Alaska occur as fault-bounded sheets exposed over a distance of >300 km; the individual sheets are as much as 40 km by 20 km in lateral extent (Wirth et al., 1993). The ophiolites consist of tectonized peridotite, layered and massive gabbros, and late-stage intrusions as well as sheeted dikes and massive and pillowed basaltic rocks. Late granitic dikes intrude ultramafic and gabbroic rocks at one locality. Although similar to other ophiolites in the North American Cordillera in terms of igneous and inferred emplacement age, the Brooks Range ophiolites bear more similarity to Tethyan-type ophiolites in that they were apparently emplaced over a continental margin instead of over an arc or other accreted terrane (Wirth et al., 1993). In this regard, the tectonic setting of this ophiolite is analogous to the Semail and Bay of Islands ophiolites. Some researchers have proposed an intra-arc origin for the Brooks Range ophiolites (Wirth et al., 1987, 1989; Harris, 1989), and a nascent intra-arc setting has been proposed subsequently (Harris, 1998).

The metamorphic sole of the Brooks Range ophiolites is less than 500 m thick (Harris, 1998). Samples 5 to 50 m structurally beneath the ophiolite formed under *P-T* conditions of 560 °C at 5 kbar, but structurally higher samples show evidence of much higher temperatures of metamorphism (Harris, 1998).

Wirth et al. (1993) reported $^{40}Ar/^{39}Ar$ hornblende ages for two ophiolites as 184–187 Ma; $^{40}Ar/^{39}Ar$ hornblende ages from their metamorphic soles range from 163 to 169 Ma. Hacker et al. (1996) reinterpreted the ages and suggested that the metamorphic sole and igneous rock $^{40}Ar/^{39}Ar$ ages are indistinguishable, indicating emplacement of young, hot ophiolite, as in Semail. Harris (1998) reported a 165 Ma $^{40}Ar/^{39}Ar$ age from the sole, and Moore et al. (1993) reported a 170 Ma U-Pb zircon age from plagiogranite in the overlying ophiolite. Moore et al. (1998) reported several U-Pb zircon ages from the Brooks Range ophiolites ranging from 162 to 170 Ma. These dates were from intrusive rocks that may significantly postdate the main stage of ophiolite generation (Moore et al., 1998). In summary, the Brooks Range ophiolite may have been very young at the time of the formation of the sole, but it does not appear to postdate the formation of the sole. Thus, geochronologic relationships between the Brooks Range ophiolite and its sole are incompatible with the generation of the ophiolite in a forearc setting (or any other supra-subduction zone setting in which emplacement occurred above the same subduction zone over which the ophiolite formed).

Trinity ophiolite, California

The Trinity ophiolite is a composite terrane exposed over an area of 100 km by >50 km; its thickness reaches at least 5 km (e.g., Irwin, 1981). The emplacement relationships of the Trinity ophiolite are of the Cordilleran type; the ophiolite is emplaced over a subduction complex or accreted terranes (Irwin, 1981). The ophiolite is made up primarily of ultramafic tectonites and mafic intrusions. A small part of the terrane yields U-Pb zircon ages of 556–579 Ma (Wallin et al., 1995), and the largest part of the ophiolite yields U-Pb zircon ages of 469–475 Ma (Mattinson and Hopson, 1972a, 1972b; Wallin and Metcalf, 1998). The second largest part of the assemblage is considerably younger and has yielded U-Pb zircon ages of 404–431 Ma (Wallin et al., 1995; Wallin and Metcalf, 1998). The oldest part of the Trinity ophiolite is interpreted as being of mid-ocean ridge origin (Wallin and Metcalf, 1998); however, the origin of the younger rocks is controversial, and interpretations range from mid-ocean ridge (Boudier et al., 1989), to backarc basin (Brouxel and Lapierre, 1988), to forearc setting (Saleeby, 1990; Wallin and Metcalf, 1998).

The ophiolite is structurally underlain by the Central Metamorphic belt, which is interpreted to be the metamorphic sole of the Trinity ophiolite, in part on the basis of the inverted metamorphic field gradient at the structurally higher levels beneath the ophiolite (Peacock, 1988; Peacock and Norris, 1989). However, it is not a typical metamorphic sole in that the metamorphic terrane is as thick as 5 km (e.g., Irwin, 1981) compared to typical sole thicknesses of less than 500 m (e.g., Jamieson, 1986). It is likely that much of the metamorphism of the Central Metamorphic belt, except for the structurally highest part of the belt, is a consequence of a later thermal event or events that are not related to heat derived from initial thrusting beneath a hot ophiolitic hanging wall (Peacock, 1988). Peak metamorphic conditions for the structurally highest part of the Central Metamorphic belt are estimated as 650 ± 50 °C at 5 kbar (Hacker and Peacock, 1990). Structurally beneath the Central Metamorphic belt is a collage of accreted allochthonous units including accretionary-prism, arc, and ophiolitic terranes, and all subduction complexes structurally beneath the Central Metamorphic belt are interpreted to be younger than the Central Metamorphic belt (e.g., Irwin, 1981). Hornblende K-Ar ages of the Central Metamorphic belt range from 186 to 407 Ma, probably indicating variable amounts of Ar loss (Lanphere et al., 1968; Hotz, 1974; recalculated to decay constants of Steiger and Jäger, 1977) consistent with the premise of significant metamorphism that postdates initial thrusting beneath the ophiolite. The oldest ages (398 and 407 Ma, Hotz, 1974) may come closest to dating the initiation of subduction beneath the Trinity ophiolite. Most of the youngest group of Trinity terrane igneous rocks are significantly older than the ages from the Central Metamorphic belt that apparently marks the initiation of subduction beneath the Trinity terrane. Thus, a forearc origin for the Trinity terrane appears to be inconsistent with regional field relationships and existing geochronologic data. It should be noted, however, that the K-Ar ages for the Central Metamorphic belt, even the older ones, may have been partially reset by later heating. This terrane in particular may provide a good test of forearc ophiolite models because the younger intrusive activity spans more than 25 m.y. Thus, if these rocks indeed formed in a forearc setting, subduction had to be ongoing for at least that long prior to the eruption of the youngest of these rocks. This amount of time is more than long enough for the structurally highest part of the young subduction complex, which includes the metamorphic sole, to have cooled (e.g., Hacker, 1990, 1991). If the structurally highest part of the Central Metamorphic belt yields $^{40}Ar/^{39}Ar$ hornblende ages of at least 425 Ma, then the forearc model for the younger Trinity ophiolite rocks will gain considerable plausibility.

Coast Range ophiolite, California

The Coast Range ophiolite forms scattered exposures over a distance of 900 km in coastal California, a distance that extends to 1300 km or more if slip on the dextral San Andreas fault system is restored (Hopson et al., 1981). Most ophiolite remnants are less than 10 km in across-strike width, and the thicknesses of the crustal section of the ophiolite are generally less than 5 km. The ophiolite remnants fall into three groups (Hopson et al., 1981): (1) Exposures along the western margin of the northern Sacramento Valley generally consist of ultramafic rocks overlain by mafic breccia and volcanic rocks; plutonic rocks are relatively rare, and there are no sheeted dike intrusions. (2) Ophiolite remnants in the western northern Coast Ranges and central Coast Ranges contain ultramafic rocks, gabbro, basalt, and, commonly, overlying tuffs or volcanic breccias; sheeted intrusive rocks are present in only a few of the remnants. (3) A nearly complete ophiolite sequence with ultramafic rocks, cumulate and massive gabbros, sheeted intrusive rocks, and basalts is present west of the Salinian block in south-central coastal California; tuffaceous cover strata are lacking on such exposures.

The Coast Range ophiolite structurally overlies the Franciscan subduction complex (e.g., Bailey et al., 1970). Thus, like the Trinity ophiolite, the Coast Range ophiolite is a typical Cordilleran-type ophiolite in that the ophiolite was emplaced over a subduction complex or other accreted terranes.

The Coast Range ophiolite has been ascribed to a forearc origin (Shervais, 1990; Saleeby, 1996; Giaramita et al., 1998), an intra-arc or backarc origin (e.g., Moores, 1970, Schweickert and Cowan, 1975, Evarts, 1977, Dickinson et al., 1996, Godfrey and Klemperer, 1998), or a mid-ocean ridge origin (Hopson et al., 1981, 1996). The igneous age of the ophiolite ranges from ca. 155 to 164 Ma, on the basis of U-Pb zircon geochronology (Hopson et al., 1981); however, more recent U-Pb zircon dating, and reinterpretation of older U-Pb zircon data, indicates that the igneous age of the ophiolite is 165 to 170 Ma (Hopson et al., 1996; Mattinson and Hopson, 1992).

An intact subophiolitic sole is not found beneath the Coast Range ophiolite, with the possible exception of a few thrust sheets, several hundred meters to several kilometers in longest dimension, consisting of amphibolite, eclogite, and coarse blueschist, present at the structurally highest level of the Franciscan Complex. An intact sole is associated with similar Cordilleran subduction complexes, the Shuksan suite (Brown et al., 1982), and the Catalina schists (Platt, 1975). Amphibolite, eclogite, and coarse-grained blueschist tectonic blocks (termed "high-grade blocks") are common in some Franciscan Complex mélanges and closely resemble the rocks of the rare thrust sheets noted above. Most high-grade blocks display early high-temperature, moderate- to high-pressure metamorphic assemblages (amphibolite and, rarely, granulite facies) that are variably overprinted with low- to moderate-temperature, high-pressure assemblages (blueschist and eclogite facies) typically associated with subduction-zone metamorphism (Moore and Blake, 1989; Wakabayashi, 1990). These rocks consistently give the oldest metamorphic ages in the Franciscan Complex; the $^{40}Ar/^{39}Ar$ hornblende ages fall in a comparatively small range of 159 to 163 Ma (Ross and Sharp, 1986, 1988). The *P-T* conditions estimated for high-temperature metamorphism in the high-grade blocks are similar to *P-T* conditions estimated from some subophiolitic soles (~600–800 °C, 5–10 kbar, Wakabayashi, 1987, 1990), consistent with previous interpretations of their origin (e.g., Platt, 1975, Cloos, 1985). Moreover, the anticlockwise *P-T-t* (pressure-temperature-time; pressure plotted on the positive *y*-axis) path of the high-grade blocks suggests that these rocks formed during the early stages of subduction (Wakabayashi, 1990). Similar anticlockwise *P-T* paths have been determined from metamorphic soles of other ophiolites by thermobarometry (Dilek and Whitney, 1997; Searle and Cox, 1999) and thermal modeling (Hacker, 1991). The high-grade blocks apparently represent a dismembered, overprinted subophiolitic sole, and the $^{40}Ar/^{39}Ar$ hornblende ages from these rocks can be regarded as the $^{40}Ar/^{39}Ar$ hornblende ages for the subophiolitic sole of the Coast Range ophiolite. Thus, the metamorphic sole age (159–163 Ma) is younger than the igneous age (ca. 170–172 Ma) of the Coast Range ophiolite, a relationship that is inconsistent with the generation of the ophiolite in a forearc.

Tuffs and volcanic breccias that overlie pelagic sedimentary rocks that rest on the basaltic volcanic sequences of some of the ophiolite remnants, as well as felsic intrusions into these rocks, almost certainly postdate the inception of subduction and are therefore genuine forearc igneous rocks (Hopson et al., 1981). Because these rocks postdate the development of the mafic intrusive and basaltic rocks of the ophiolite, as well as the pelagic sedimentary rocks, Hopson et al. (1981) considered them to be a postophiolitic unit rather than part of the Coast Range ophiolite itself. However, at the Del Puerto remnant of the Coast Range ophiolite, felsic volcanic rocks occur stratigraphically below pelagic sedimentary rocks (Evarts et al., 1999, and references therein). In addition, geochronologic data, including a 155 Ma U-Pb zircon age from a plagiogranite dike in the gabbroic section (Hopson et al., 1981) and 156 and 157 Ma $^{40}Ar/^{39}Ar$ hornblende ages from hornblende diorites (Evarts et al., 1992), indicate that part of this ophiolite remnant may have formed in a forearc setting after the initiation of subduction. Analogous igneous rocks that postdate metamorphic sole development are associated with Neotethyan ophiolites (Dilek et al., 1999). Although most of the 170–172 Ma Coast Range ophiolite is incompatible with a forearc setting of formation because it is older than its sole, some parts of the ophiolite, particularly felsic and intermediate volcanic and plutonic rocks with demonstrably younger ages, are best explained by generation in a forearc.

DISCUSSION

How young must oceanic crust be to form a metamorphic sole?

The essential correspondence of ophiolite and metamorphic sole ages for the Semail ophiolite has been cited for evidence of initiation of subduction at a spreading center, and the question has been raised as to whether this relationship may be universal for the formation of metamorphic soles (Hacker et al., 1996). This issue is important in evaluating the validity of the existing data set of ophiolite-sole age pairs. The Bay of Islands ophiolite-sole pair was once regarded as problematic because of the apparent difference in ophiolite and sole ages of nearly 50 m.y. (508 Ma for the ophiolite vs. 460 Ma for the sole) (e.g., discussion in Malpas, 1979). New dating of the ophiolite and application of more recently accepted decay constants have yielded ages of 486 Ma for the ophiolite versus 463 to 469 Ma for the sole and thus narrowed the age gap to 17 to 23 m.y. Within the limits of the uncertainties in the ages, the minimum age gap is smaller by an additional 5 m.y.

There are two aspects of the metamorphic geothermometry and geobarometry that reduce the need for ultrayoung oceanic crust (less than 1 m.y.) to produce the high-grade metamorphism in the sole. Nearly all metamorphic temperature determinations of the soles are based on the garnet-clinopyroxene geothermometer of Ellis and Green (1979). New experimental data and subsequent recalibration of this geothermometer (Pattison and Newton, 1989) yield consistently lower temperatures than the Ellis and Green geothermometer, with estimated temperature differences of as much as 160 °C. The estimated temperature for the highest-temperature sample of McCaig (1983) from the Bay of Islands sole, recalculated on the basis of the

Pattison and Newton (1989) geothermometer, is ~680 °C versus 844 °C as originally estimated. A different recalibration of the garnet-clinopyroxene thermometer by Krogh (1988) also yields lower temperatures of metamorphism than those estimated by the Ellis and Green (1979) geothermometer, but higher than those estimated by the Pattison and Newton (1989) geothermometer.

In addition to temperature considerations, the estimated pressure of metamorphism of the Bay of Islands metamorphic sole is 7 to 12 kbar (McCaig, 1983). A greater depth of metamorphism indicates more heating of the sole rocks than estimated in two-dimensional models of the Semail sole (the pressure estimate for metamorphism of the Bay of Islands metamorphic sole indicates depths of ~22 to 37 km vs. the 18 km depth used for the Semail sole in models of Hacker, 1990, 1991), because (1) a higher peak temperature is attained for the hanging wall (for a given age of hanging-wall crust) and (2) a longer transit time is required for the metamorphosed rocks to reach their depth of peak metamorphism as they are being subducted and then underplated, allowing for a longer period of heating. The possibility of a lower temperature of metamorphism for the sole and a greater depth of metamorphism may allow a somewhat older (but still relatively young) oceanic crust at the time of sole formation. The relationship of the Coast Range ophiolite and its "sole" (Franciscan high-grade blocks and related thrust sheets) may be similar to the Bay of Islands relationship in that temperatures of metamorphism have been overestimated (the temperature range cited above was based on the Ellis and Green, 1979, geothermometer) and the metamorphism of the highest-temperature rocks may have taken place at considerable depth.

In summary, it appears that formation of metamorphic soles by initiation of subduction at a spreading center is not necessarily a universal relationship. Thus, the age differences between ophiolites and their soles summarized above are plausible.

Are metamorphic soles "welded" to the base of ophiolites, or are they in fault contact?: Similarities to field relationships of high-pressure, low-temperature metamorphic rocks

The terminology "metamorphic sole" implies that these rocks became essentially part of the ophiolite structural block, once they were metamorphosed and underplated. This concept is important to tectonic interpretations that link metamorphic soles with the ophiolites that are found structurally above them. Because of difficulties in estimating metamorphic pressure from the mineral assemblages of metamorphic soles, many estimates of pressure have relied on a model that related the depth of metamorphism of the sole to the structural thickness of the overlying ophiolite (e.g., Searle and Malpas, 1980); such a model assumed the linkage discussed above. Where quantitative geobarometry has been conducted—such as the for the Bay of Islands (e.g., McCaig, 1983), St. Anthony Complex (Jamieson, 1986), Ballantrae (Treloar et al., 1980), or Semail (Gnos, 1998) metamorphic soles or the dismembered sole of the Coast Range ophiolite (Wakabayashi, 1990)—metamorphic pressure estimates for the soles far exceed burial metamorphic pressures of the crustal section of the overlying ophiolite. The structural thickness of the structurally intervening ultramafic rocks is insufficient to account for this difference in metamorphic pressure. Ultramafic rocks present above the metamorphic mafic sole show evidence of high-pressure formation, consistent with the depth of metamorphism displayed in the sole rocks (summarized in Spray, 1984; Jamieson, 1986), so the metamorphic pressure gap may be between the crustal section of the ophiolite and the high-pressure ultramafic rocks. However, mantle flow may have occurred subsequent to the stabilization of mineral compositions (from which P and T were estimated in the ultramafic rocks) and prior to the initiation of intraoceanic thrusting and sole development; the ultramafic rocks may have been at shallower crustal levels (than those indicated by their mineral assemblages) at the time of sole formation. The low-P on high-P metamorphic relationship of the ophiolite over its sole is analogous to field relationships of high-P, low-T metamorphic rocks (e.g., Platt, 1986) and may indicate the presence of normal faulting between the crustal section of the ophiolite and the sole. The detailed work of Gnos (1998) also indicates an inverted pressure gradient within the Semail sole. If such a relationship is common to other soles, then the high-grade part of the sole (and structurally adjacent ultramafic rocks) is sandwiched between the low-P crustal part of the ophiolite above and the lower-P part of the sole below. Such a relationship also is similar to field relationships of many high-P, low-T rocks (Maruyama et al., 1996). The similarity in field relationships between high-P metamorphic rocks and metamorphic soles is not surprising, because metamorphic soles commonly make up the structurally highest (and oldest) part of subduction complexes (Peacock, 1988).

This discussion indicates that a significant amount of vertical tectonic transport of the sole relative to the crustal part of the ophiolite apparently occurred after high-grade metamorphism of the sole. Although this tectonic scenario indicates that ophiolites are allochthonous with respect to the soles they tectonically overlie, the general genetic relationships between ophiolites and soles and predicted age relationships discussed herein should still be valid, because the sole and ophiolite are still interpreted to be associated with the same subduction zone.

Structural and geochronologic relationships are difficult to reconcile with forearc ophiolite formation

Subophiolitic metamorphic soles are a consequence of the initiation of subduction beneath oceanic crust. If ophiolites indeed formed in forearc settings, they must postdate the inception of subduction beneath them and therefore be younger than the soles beneath them. The previously reviewed examples of ophiolite-sole pairs show that, in those examples, the ophiolites are not younger than the soles. Forearc igneous rocks that postdate metamorphic sole formation are associated with the Coast Range ophiolite and Neotethyan ophiolites and may be a feature of many other ophiolites. However, these rocks constitute a subordinate part of the ophiolites with which they are associated. We are aware of no examples in the world for which these younger forearc rocks constitute the predominant part of an ophiolite. Furthermore, although examples of active mid-ocean ridge, backarc, and intra-arc basin seafloor spreading are well documented and young forearc volcanic rocks have been identified, no example of an active forearc spreading center has been found (this statement includes spreading over a very young subduction zone). The geochronologic record of ophiolites and their soles is far from complete. It is thus possible that future geochronologic studies may identify ophiolites younger than their soles.

Backarc settings as a supra-subduction zone alternative to forearc ophiolite generation: The Coast Range ophiolite of California as an example

Backarc or intra-arc settings for ophiolite formation may be compatible with the regional geologic and geochronologic relationships for supra-subduction zone ophiolites, if emplacement of the ophiolite took place above a different subduction zone than the one over which the ophiolite formed. If the ophiolites were generated by backarc spreading, initiation of new subduction under the ophiolites and generation of the subophiolitic sole may require a subduction polarity flip (e.g., Moores, 1970) (Figs. 2D and 3). However, the constraints imposed by ophiolite and metamorphic sole relationships only require that the ophiolite be emplaced above a different subduction zone than the one over which it formed, whether the subduction zone is of the same or different polarity (Fig. 2, D and E). It is probable that forearc settings have been proposed for many ophiolites partly on the basis on their being trapped in a forearc setting during the final emplacement process (Figs. 1, 2, and 3). Late Cenozoic subduction polarity flips have occurred in the southwest Pacific, including the ongoing subduction polarity flip in eastern New Guinea (Cooper and Taylor, 1987) and the ca. 10 Ma to 5 Ma polarity flip in northern Sulawesi (Hall, 1996), yet a single polarity of subduction that persists for the entire history of an active plate margin is more commonly adopted as a model of the evolution of active plate margins (e.g., Hamilton, 1969; Saleeby, 1981, 1996).

Uniformly east-dipping subduction has been proposed for the Mesozoic tectonic evolution of California (e.g., Hamilton, 1969); a variant of this hypothesis includes the development of forearc ophiolites (Saleeby, 1981, 1996). Alternatively, the tectonic evolution of the active plate margin of California included one or more subduction polarity flips and the concomitant development of ophiolites in backarc basins (Moores, 1970; Schweickert and Cowan, 1975; Cannat and Boudier, 1985; Ingersoll and Schweickert, 1986; Ingersoll, this volume). For the Coast Range ophiolite, these different types of models result in two contrasting tectonic settings for its formation as a supra-subduction ophiolite: a forearc setting for the models of uniform, east-dipping subduction (Saleeby, 1996), contrasted to a backarc setting for models that include subduction polarity flips (Dickinson et al., 1996). Note that a third competing model for the generation of the Coast Range ophiolite is the formation of the ophiolite at a mid-ocean ridge, followed by large-scale tectonic transport prior to emplacement (Hopson et al., 1981, 1996). However, our discussion will compare the relative plausibility of the competing supra-subduction zone models, while acknowledging the possibility that the supra-subduction zone setting interpretation for the generation of the Coast Range ophiolite may be incorrect.

As summarized previously, the Coast Range ophiolite (170–172 Ma U-Pb age) is apparently older than rocks that have been interpreted as remnants of its metamorphic sole (159–163 Ma $^{40}Ar/^{39}Ar$ hornblende ages), a relationship incompatible with its generation in a forearc setting. The Coast Range ophiolite may have formed in a backarc setting relative to a west-dipping subduction zone, now represented by the buried west-dipping zone of ultramafic rocks (Godfrey and Klemperer, 1998) (Fig. 3). When the continental margin blocked the west-dipping subduction zone, a new, east-dipping, subduction zone formed beneath the Coast Range ophiolite, emplacing the ophiolite over the developing Franciscan subduction complex and trapping the ophiolite in a forearc setting. The initiation of subduction beneath the oceanic crust resulted in the development of a metamorphic sole and in near-trench forearc volcanism that is represented by the felsic intrusions, tuffs, and volcaniclastic rocks that are associated with many of the Coast Range ophiolite remnants. The west-dipping suture zone interpreted by Godfrey and Klemperer (1998) may be entirely buried by the Upper Jurassic and younger clastic sedimentary rocks of the Great Valley Group, or the suture may crop out in the Sierra Nevada. If the suture is exposed, it may coincide with high-pressure epidote and garnet amphibolites that are found structurally beneath ca. 200 Ma ophiolitic rocks in the northern part of the western Sierra Nevada (Wakabayashi and Dilek, 1987). These rocks may be related to high-pressure garnet amphibolites of the Tuolumne Complex that have yielded $^{40}Ar/^{39}Ar$ and K/Ar hornblende ages of 178 to 200 Ma (Sharp and Leighton, 1987). These garnet amphibolites may have formed as part of a subophiolitic sole as west-dipping subduction began.

Mid-ocean ridge settings as an alternative to forearc ophiolite genesis

A backarc basin setting of ophiolite genesis may not be compatible with field relationships for ophiolites that lack regional association with an older subduction zone or older arc. Such ophiolites include many of the Tethyan ophiolites (such as Semail) and the Brooks Range ophiolites that have been noted for their similarity to Tethyan ophiolites (Wirth et al., 1993; Harris, 1998). For such ophiolites, a mid-ocean ridge setting would fit ophiolite-sole relationships and regional geology better than a backarc setting.

A mid-ocean ridge setting of ophiolite generation is compatible with ophiolite-sole relationships in all cases, mainly because the constraint of preexisting subduction is removed. However, it is beyond the scope of this paper to debate the relative merits of supra-subduction zone versus mid-ocean ridge ophiolites.

CONCLUSIONS

Although the forearc setting of supra-subduction zone ophiolite generation is consistent with the petrologic and geochemical characteristics of many ophiolites, the apparent incompatibility of forearc ophiolite generation with the structural and geochronologic relationships between the ophiolites and their metamorphic soles should be addressed. Alternative settings for supra-subduction zone ophiolite generation, such as backarc or intra-arc settings, are more compatible with regional structural and geochronologic relationships. Further geochronologic studies of different ophiolite complexes, particularly involving methods with high closure temperatures in both metamorphic and igneous rocks, will better test the plausibility of forearc or alternative supra-subduction zone settings of ophiolite generation.

ACKNOWLEDGMENTS

We thank R. Jamieson, R. Harris, and E. Moores for helpful, constructive reviews of the manuscript.

REFERENCES CITED

Archibald, D.A., and Farrar, E., 1976, K-Ar ages of amphiboles from the Bay of Islands ophiolite and Little Port Complex, western Newfoundland, and their geological implications: Canadian Journal of Earth Sciences, v. 13, p. 520–529.

Bailey, E.H., Blake, M.C., Jr., and Jones, D.L., 1970, On-land Mesozoic ocean crust in California Coast Ranges: U.S. Geological Survey Professional Paper 700-C, p. 70–81.

Bloomer, S.H., Taylor, B., MacLeod, C.J., Stern, R.J., Fryer, P., Hawkins, J.W., and Johnson, L., 1995, Early arc volcanism and ophiolite problem: A perspective from drilling in the western Pacific, *in* Taylor, B., and Natland, J., eds., Active margins and marginal basins of the western Pacific: American Geophysical Union, Geophysical Monograph 88, p. 1–30.

Boudier, F., and Coleman, R.G., 1981, Cross section through the peridotite in the Samail ophiolite, southeastern Oman Mountains: Journal of Geophysical Research, v. 86, p. 2573–2592.

Boudier, F., Le Sueur, E., and Nicolas, A., 1989, Structure of an atypical ophiolite: The Trinity complex, eastern Klamath Mountains, California: Geological Society of America Bulletin, v. 101, p. 820–833.

Brouxel, M., and Lapierre, H., 1988, Geochemical study of an early Paleozoic island-arc-back-arc basin system: Part 1: The Trinity complex, eastern ophiolite (northern California): Geological Society of America Bulletin, v. 100, p. 1111–1119.

Brown, E.H., Wilson, D.L., Armstrong, R.L., and Harakal, J.E., 1982, Petrologic, structural, and age relations of serpentinite, amphibolite, and blueschist in the Shuksan Suite of the Iron Mountain-Gee Point area, North Cascades, Washington: Geological Society of America Bulletin, v. 93, p. 1087–1098.

Cannat, M., and Boudier, F., 1985, Structural study of intra-oceanic thrusting in the Klamath Mountains, northern California: Implications on accretion geometry: Tectonics, v. 4, p. 435–452.

Church, W.R., and Stevens, R.K., 1971, Early Paleozoic ophiolite complexes of the Newfoundland Appalachians as mantle-ocean crust sequences: Journal of Geophysical Research, v. 76, p. 1460–1466.

Cloos, M., 1985, Thermal evolution of convergent plate-margins: Thermal modeling and re-evaluation of isotopic Ar-ages for blueschists in the Franciscan Complex of California: Tectonics, v. 4, p. 421–433.

Coleman, R.G., 1984, The diversity of ophiolites: Geologie en Mijnbouw, v. 62, p. 141–150.

Cooper, P., and Taylor, B., 1987, Seismotectonics of New Guinea: A model for arc reversal following arc-continent collision: Tectonics, v. 6, p. 53–67.

Dallmeyer, R.D., and Williams, H., 1975, $^{40}Ar/^{39}Ar$ ages from the Bay of Islands metamorphic aureole: Their bearing on the timing of Ordovician ophiolite: Canadian Journal of Earth Sciences, v. 12, p. 1685–1690.

Dickinson, W.R., Schweickert, R.A., and Ingersoll, R.D., 1996, Coast Range ophiolite as backarc/interarc basin lithosphere: GSA Today, v. 6, no. 2, p. 2–3.

Dilek, Y., and Moores, E.M., 1990, Regional tectonics of the eastern Mediterranean ophiolites, *in* Malpas, J., Moores, E., Panayiotou, A., and Xenophontos, C., eds., Ophiolites, Oceanic crustal analogues: Proceedings of the Symposium "Troodos 1987": Nicosia, Cyprus, Geological Survey Department, Ministry of Agriculture and Natural Resources, p. 285–309.

Dilek, Y., and Whitney, D.L., 1997, Counterclockwise P-T-t trajectory from the metamorphic sole of a Neo-Tethyan ophiolite (Turkey): Tectonophysics, v. 280, p. 295–310.

Dilek, Y., Thy, P., Hacker, B., and Grundvid, S., 1999, Structure and petrology of Tauride ophiolites and mafic dike intrusions (Turkey): Implications for the Neotethyan ocean: Geological Society of America Bulletin, v. 111, p. 1192–1216.

Dunning, G.R., and Krogh, E., 1985, Geochronology of ophiolites of the Newfoundland Appalachians: Canadian Journal of Earth Sciences, v. 22, p. 1659–1670.

Ellis, D.J., and Green, D.H., 1979, An experimental study of the effect of Ca upon garnet-clinopyroxene Fe-Mg exchange equilibria: Contributions to Mineralogy and Petrology, v. 71, p. 13–22.

Evarts, R.C., 1977, The geology and petrology of the Del Puerto ophiolite, Diablo Range, central California, *in* Coleman, R.G., and Irwin, W.P., eds., North American ophiolites: Oregon Department of Geology and Mineral Industries Bulletin 95, p. 121–139.

Evarts, R.C., and Schiffman, P., 1983, Submarine hydrothermal alteration of the Del Puerto ophiolite, California: American Journal of Science, v. 283, p. 289–340.

Evarts, R.C., Sharp, W.D., and Phelps, D.W., 1992, The Del Puerto Canyon remnant of the Great Valley ophiolite: Geochemical and age constraints on its formation and evolution: American Association of Petroleum Geologists Bulletin, v. 76, p. 418.

Evarts, R.C., Coleman, R.G., and Schiffman, P., 1999, The Del Puerto ophiolite: Petrology and tectonic setting, *in* Wagner, D.L., and Graham, S.A., eds., Geologic field trips in northern California: California Division of Mines and Geology Special Publication 119, p. 136–149.

Gass, I.G., Smith, A.G., and Vine, F.J., 1975, Origin and emplacement of ophiolites: Geodynamics Today, a review of Earth's Dynamic Processes: London, Royal Society of London, p. 54–65.

Giaramita, M., MacPherson, G.J., and Phipps, S.P., 1998, Petrologically diverse basalts from a fossil oceanic forearc in California: The Llanada and Black Mountain remnants of the Coast Range ophiolite: Geological Society of America Bulletin, v. 110, p. 553–571.

Gnos, E., 1998, Peak metamorphic conditions of garnet amphibolites beneath the Semail ophiolite: Implications for an inverted pressure gradient: International Geology Review, v. 40, p. 281–304.

Godfrey, N.J., and Klemperer, S.L., 1998, Ophiolitic basement to a forearc basin and implications for continental growth: The Coast Range/Great Valley ophiolite: Tectonics, v. 17, p. 558–570.

Hacker, B.R., 1990, Simulation of the metamorphic and deformational history of the metamorphic sole of the Oman ophiolite: Journal of Geophysical Research, v. 95, p. 4895–4907.

Hacker, B.R., 1991, The role of deformation in the formation of metamorphic field gradients: Ridge subduction beneath the Oman ophiolite: Tectonics, v. 10, p. 455–473.

Hacker, B.R., and Peacock, S.M., 1990, Comparison of the Central Metamorphic Belt and Trinity terrane of the Klamath Mountains with the Feather River terrane of the Sierra Nevada, *in* Harwood, D.S., and Miller, M.M., eds., Paleozoic and early Mesozoic paleogeographic relations: Sierra Nevada, Klamath Mountains and related terranes: Geological Society of America Special Paper 255, p. 75–92.

Hacker, B.R., Mosenfelder, J.L., and Gnos, E., 1996, Rapid emplacement of the Oman ophiolite: Thermal and geochronologic constraints: Tectonics, v. 15, p. 1230–1247.

Hall, R., 1996, Reconstructing Cenozoic Southeast Asia, *in* Hall, R., and Blundell, D., eds., Tectonic evolution of Southeast Asia, Geological Society of London Special Publication 106, p. 153–184.

Hamilton, W.B., 1969, Mesozoic California and underflow of the Pacific mantle: Geological Society of America Bulletin, v. 80, p. 2409–2430.

Harris, R.A., 1989, Process of allochthon emplacement with special reference to the Brooks Range ophiolite, Alaska and Timor, Indonesia [Ph.D. dissert.]: University of London, 480 pp.

Harris, R.A., 1998, Origin and tectonic evolution of the metamorphic sole beneath the Brooks Range ophiolite, Alaska, *in* Oldow, J.S., and Avé Lallemant, H.G., eds., Architecture of the central Brooks Range fold and thrust belt, arctic Alaska: Geological Society of America Special Paper 324, p. 293–312.

Hopson, C.A., Mattinson, J.M., and Pessagno, E.A., Jr., 1981, The Coast Range ophiolite, western California, *in* Ernst, W.G., ed., The geotectonic development of California: Englewood Cliffs, New Jersey, Prentice-Hall, p. 419–510.

Hopson, C.A., Pessagno, E.A., Jr., Mattinson, J.M., Luyendyk, B.P., Beebe, W., Hull, D.M., Munoz, I.M., and Blome, C.D., 1996, Coast Range ophiolite as paleoequatorial mid-ocean lithosphere: GSA Today, v. 6, no. 2, p. 3–4.

Hotz, P.E., 1974, Preliminary geologic map of the Yreka Quadrangle, California: U.S. Geological Survey Miscellaneous Field Studies Map MF-568, scale 1: 62 500.

Hynes, A., 1975, Comment on "The Troodos ophiolitic complex was probably formed in an island arc" by A. Miyashiro: Earth and Planetary Science Letters, v. 25, p. 213–216.

Ingersoll, R.V., and Schweickert, R.A., 1986, A plate tectonic model for ophiolite genesis, Nevadan orogeny, and forearc initiation, northern California: Tectonics, v. 5, p. 901–912.

Irwin, W.P., 1981, Tectonic accretion of the Klamath Mountains, *in* Ernst, W.G., ed., The geotectonic development of California: Englewood Cliffs, New Jersey, Prentice-Hall, p. 29–49.

Jamieson, R.A., 1986, PT paths from high temperature shear zones beneath ophiolites: Journal of Metamorphic Geology, v. 4, p. 3–22.

Jenner, G.A., Dunning, G.R., Malpas, J., Brown, M., and Brace, T., 1991, Bay of Islands and Little Port complexes, revisited: Age, geochemical, and isotopic evidence confirm suprasubduction zone origin: Canadian Journal of Earth Sciences, v. 28, p. 1635–1652.

Karson, J.A., 1984, Variations in structure and petrology of the Coastal Complex, Newfoundland: Anatomy of an oceanic fracture zone, *in* Gass, I.G., Lippard, S.J., and Shelton, A.W., eds., Ophiolites and oceanic lithosphere: London, Blackwell Scientific, p. 131–144.

Krogh, E.J., 1988, The garnet-clinopyroxene Fe-Mg geothermometer: A reinterpretation of existing experimental data: Contributions to Mineralogy and Petrology, v. 99, p. 44–48.

Lanphere, M.A., Irwin, W.P., and Hotz, P.E., 1968, Isotopic age of the Nevadan orogeny and older plutonic and metamorphic events in the Klamath Mountains, California: Geological Society of America Bulletin, v. 79, p. 1027–1052.

Liou, J.G., and Ernst, W.G., 1979, Oceanic ridge metamorphism of the East Taiwan ophiolite: Contributions to Mineralogy and Petrology, v. 65, p. 335–348.

Lippard, S.J., Shelton, A.W., and Gass, I.G., 1986, The ophiolites of northern Oman: Geological Society of London Memoir 11, p. 1–178.

Malpas, J., 1979, The dynamothermal aureole of the Bay of Islands ophiolite suite: Canadian Journal of Earth Sciences, v. 16, p. 2086–2101.

Maruyama, S., Liou, J.G., and Terabayashi, M., 1996, Blueschists and eclogites of the world and their exhumation: International Geology Review, v. 38, p. 485–594.

Mattinson, J.M., 1975, Early Paleozoic ophiolite complexes of Newfoundland: Isotopic ages of zircons: Geology, v. 3, p. 181–183.

Mattinson, J.M., 1976, Ages of zircons from the Bay of Islands ophiolite complex, western Newfoundland: Geology, v. 4, p. 393–394.

Mattinson, J.M., and Hopson, C.A., 1972a, Paleozoic ages of ophiolite complexes in Washington and northern California: Eos, Transactions of the American Geophysical Union, v. 53, p. 543.

Mattinson, J.M., and Hopson, C.A., 1972b, Paleozoic ophiolite complexes in Washington and northern California: Carnegie Institution of Washington Yearbook, v. 71, p. 578–583.

Mattinson, J.M., and Hopson, C.A., 1992, U-Pb ages of the Coast Range ophiolite: A critical reevaluation based on new high-precision Pb-Pb ages: American Association of Petroleum Geologists Bulletin, v. 76, p. 425.

McCaig, A.M., 1983, P-T conditions during emplacement of the Bay of Islands ophiolite complex: Earth and Planetary Science Letters, v. 63, p. 459–473.

Miyashiro, A., 1973, The Troodos ophiolitic complex was probably formed in an island arc: Earth and Planetary Science Letters, v. 19, p. 218–224.

Moore, D.E., and Blake, M.C., Jr., 1989, New evidence for polyphase metamorphism of glaucophane schist and eclogite exotic blocks in the Franciscan Complex, California and Oregon: Journal of Metamorphic Geology, v. 7, p. 211–228.

Moore, T.E., Aleinikoff, J.N., and Walter, M., 1993, Middle Jurassic U/Pb crystallization age for Siniktanneyak Mountain ophiolite: Geological Society of America Abstracts with Programs, v. 25, p. 124.

Moore, T.E., Aleinoff, J.N., and Wirth, K.R., 1998, Middle Jurassic U/Pb ages for the Brooks Range ophiolites, Alaska: Eos, Transactions of the American Geophysical Union, v. 79, no. 45, p. F807–F808.

Moores, E.M., 1970, Ultramafics and orogeny, with models of the U.S. Cordillera and the Tethys: Nature, v. 228, p. 837–842.

Moores, E.M., 1982, Origin and emplacement of ophiolites: Reviews of Geophysics and Space Physics, v. 20, p. 735–760.

Nicolas, A., 1989, Structures of ophiolites and dynamics of oceanic lithosphere: Dordrecht, Netherlands, Kluwer Academic Publishers, 367 p.

Nicolas, A., and Le Pichon, X., 1980, Thrusting of young lithosphere in subduction zones with special reference to structures in ophiolitic peridotites: Earth and Planetary Science Letters, v. 46, p. 397–406.

Pallister, J.S., and Hopson, C.A., 1981, Samail ophiolite plutonic suite: Field relations, phase variation, cryptic variation, and layering, and a model of a spreading ridge magma chamber: Journal of Geophysical Research, v. 86, p. 2673–2745.

Pattison, D.R.M., and Newton, R.C., 1989, Reversed experimental calibration of the garnet-clinopyroxene Fe-Mg exchange thermometer: Contributions to Mineralogy and Petrology, v. 101, p. 87–103.

Peacock, S.M., 1988, Inverted metamorphic gradients in the westernmost Cordillera, *in* Ernst, W.G., ed., Metamorphism and crustal evolution of the western United States, Rubey Volume VII, New Jersey, Prentice-Hall, p. 954–975.

Peacock, S.M., and Norris, P.J., 1989, Metamorphic evolution of the Central Metamorphic Belt, Klamath province, California: An inverted metamorphic gradient beneath the Trinity peridotite: Journal of Metamorphic Geology, v. 7, p. 191–209.

Pearce, J.A., Lippard, S.J., and Roberts, S., 1984, Characteristics and tectonic significance of supra-subduction zone ophiolites, *in* Kokelaar, B.P., and Howells, M.F., eds., Marginal basin geology: Volcanic and associated sedimentary and tectonic processes in modern and ancient marginal basins: Geological Society of London Special Publication 16, p. 77–96.

Platt, J.P., 1975, Metamorphic and deformational processes in the Franciscan Complex, California: Some insights from the Catalina Schist terrane: Geological Society of America Bulletin, v. 86, p. 1337–1347.

Platt, J.P., 1986, Dynamics of orogenic wedges and the uplift of high-pressure metamorphic rocks: Geological Society of America Bulletin, v. 97, p. 1037–1053.

Ross, J.A., and Sharp, W.D., 1986, $^{40}Ar/^{39}Ar$ and Sm/Nd dating of garnet amphibolite in the Coast Ranges, California: Eos, Transactions of the American Geophysical Union, v. 67, no. 44, p. 1249.

Ross, J.A., and Sharp, W.D., 1988, The effects of sub-blocking temperature metamorphism on the K/Ar systematics of hornblendes: $^{40}Ar/^{39}Ar$ dating of polymetamorphic garnet amphibolite from the Franciscan Complex, California, Contributions to Mineralogy and Petrology, v. 100, p. 213–221.

Saleeby, J.B., 1981, Ocean floor accretion and volcanoplutonic arc evolution of the Mesozoic Sierra Nevada, *in* Ernst, W.G., ed., The geotectonic development of California: Englewood Cliffs, New Jersey, Prentice-Hall, p. 133–181.

Saleeby, J.B., 1990, Geochronological and tectonostratigraphic framework of the Sierran-Klamath ophiolitic assemblages, *in* Harwood, D.S., and Miller, M.M., eds., Paleozoic and early Mesozoic paleogeographic relations: Sierra Nevada, Klamath Mountains, and related terranes: Geological Society of America Special Paper 255, p. 93–114.

Saleeby, J.B., 1996, Coast Range ophiolite as parautochthonous forearc lithosphere: GSA Today, v. 6, no. 2, p. 6–10.

Schweickert, R.A., and Cowan, D.S., 1975, Early Mesozoic tectonic evolution of the western Sierra Nevada, California: Geological Society of America Bulletin, v. 86, p. 1329–1336.

Searle, M., and Cox, J., 1999, Tectonic setting, origin, and obduction of the Oman ophiolite: Geological Society of America Bulletin, v. 111, p. 104–122.

Searle, M.P., and Malpas, J., 1980, Structure and metamorphism of rocks beneath the Semail ophiolite of Oman and their significance in ophiolite obduction: Transactions of the Royal Society of Edinburgh, Earth Sciences, v. 71, p. 247–262.

Sharp, W.D., and Leighton, C.W., 1987, Accretion of the Foothills ophiolite, western Sierra Nevada foothills: Geological Society of America Abstracts with Programs, v. 19, no. 6, p. 450.

Shervais, J.W., 1990, Island arc and ocean crust ophiolites: Contrasts in the petrology, geochemistry and tectonic style of ophiolite assemblages in the California Coast Ranges, *in* Malpas, J., Moores, E., Panayiotou, A., and Xenophontos, C., eds., Ophiolites, oceanic crustal analogues, Proceedings of the Symposium "Troodos 1987": Nicosia, Cyprus, Geological Survey Department, Ministry of Agriculture and Natural Resources, p. 507–520.

Spray, J.G., 1984, Possible causes and consequences of upper mantle decoupling and ophiolite displacement, *in* Gass, I.G., Lippard, S.J., and Shelton, A.W., eds., Ophiolites and oceanic lithosphere: Geological Society of London Special Publication 13, p. 255–268.

Steiger, R.H., and Jäger, E., 1977, Subcommission on Geochronology: Convention on the use of decay constants in geo- and cosmochronology: Earth and Planetary Science Letters, v. 36, p. 359–362.

Tilton, G.R., Hopson, C.A., and Wright, J.E., 1981, Uranium-lead isotopic ages of the Samail ophiolite, Oman, with application to Tethyan ocean ridge tectonics: Journal of Geophysical Research, v. 86, p. 2763–2775.

Treloar, P.J., Bluck, B.J., Bowes, D.R., and Dudek, A., 1980, Hornblende-garnet metapyroxenite beneath serpentinite in the Ballantrae ophiolite of southwest Scotland and its bearing on the depth provenance of oceanic lithosphere: Transactions of the Royal Society of Edinburgh, Earth Sciences, v. 71, p. 201–212.

Wakabayashi, J., 1987, Amphibolite grade metamorphism of Franciscan rocks from the San Francisco Bay Area, California: Geological Society of America Abstracts with Programs, v. 19, no. 6, p. 460.

Wakabayashi, J., 1990, Counterclockwise P-T-t paths from amphibolites, Franciscan Complex, California: Relics from the early stages of subduction zone metamorphism: Journal of Geology, v. 98, p. 657–680.

Wakabayashi, J., 1999, Subduction and the rock record: Concepts developed in the Franciscan Complex, California, *in* Sloan, D., Moores, E.M., and Stout, D., eds., Classic Cordilleran concepts: A view from California: Geological Society of America Special Paper 338, p. 123–133.

Wakabayashi, J., and Dilek, Y., 1987, An alpine-style collision in the northern Sierra

Nevada, California: Structural and metamorphic evidence: Eos, Transactions of the American Geophysical Union, no. 44, v. 68, p. 1474.

Wallin, E.T., and Metcalf, R.V., 1998, Supra-subduction zone ophiolite formed in an extensional forearc: Trinity terrane, Klamath Mountains, California: Journal of Geology, v. 106, p. 591–608.

Wallin, E.T., Coleman, D.S., Lindsley-Griffin, N., and Potter, A.W., 1995, Silurian plutonism in the Trinity terrane (Neoproterozoic and Ordovician), Klamath Mountains, California, USA: Tectonics, v. 14, p. 1007–1013.

Williams, H., and Smyth, R., 1973, Metamorphic aureoles beneath ophiolite suites and Alpine peridotites: Tectonic implications with west Newfoundland examples: American Journal of Science, v. 273, p. 594–621.

Wirth, K.R., Harding, D.J., and Bird, J.M., 1987, Basalt geochemistry, Brooks Range, Alaska: Geological Society of America Abstracts with Programs, v. 19, no. 6, p. 464.

Wirth, K.R., Bird, J.M., and Cheatham, M.M., 1989, Geochemistry of western Brooks Range basalt, Alaska: Geological Society of America Abstracts with Programs, v. 21, no. 5, p. 160–161.

Wirth, K.R., Bird, J.M., Blythe, A.E., Harding, D.J., and Heizler, M.T., 1993, Age and evolution of the Brooks Range ophiolites, Alaska: Results from $^{40}Ar/^{39}Ar$ thermochronometry: Tectonics, v. 12, p. 410–432.

Manuscript Accepted by the Society February 23, 2000

Printed in the U.S.A.

Geological Society of America
Special Paper 349
2000

Section II
Oceanic Lower Crust and Upper Mantle

The mode and nature of magmatic and tectonic processes operating in the lower oceanic crust and upper mantle remain controversial and the subject of intense debate. Limited sampling of the lower crust and upper mantle in the oceanic basins has certainly been a hindrance to further progress in this field, but it has also become increasingly clear that lower-crustal and upper-mantle rocks are very complex, both petrologically and structurally. Because of limited sampling of the lower crust and upper mantle from the oceanic basins, many of the inferences about processes operating at this level have been derived from studies of ophiolite complexes and mafic layered intrusions. Although ophiolites and mafic layered intrusions are generally considered to be good analogues for bona fide oceanic crust, enough differences exist to raise concerns about how rigorously these inferences can be extended to understanding magmatic and tectonic processes in the oceanic basins.

Two papers in this section describe the large-scale geologic and petrologic characteristics of ophiolites and/or oceanic crust. Nicolas and Boudier describe the thick cumulate section (Moho transition zones) of the Semail ophiolite (Oman) and several broad areas that locally have steep foliation and lineation trajectories that crosscut the flat-lying Moho; these areas within the ophiolite are considered to reflect mantle diapirs rising below a ridge segment. In addition to these structural characteristics, the thick dunite transition zones are characterized by abundant dunites and wehrlites, which are inferred to result from intense melt circulation through these areas, coupled with melt and rock interactions and compaction. Nicholas and Boudier conclude that these mantle diapirs and their ultramafic rocks are spatially associated with thin gabbroic crust and that the diapirs are the major sites for mantle upwelling and melt delivery to the crustal environment. Melt delivery occurs at these localized centers and is then distributed somewhat evenly along the ridge segments. Robinson et al. summarize results from two Ocean Drilling Program legs that have penetrated ~1.5 km of gabbroic rocks on the Southwest Indian Ridge. The cored rocks are dominantly olivine gabbro and troctolite, with lesser amounts of gabbro, microgabbro, and ferrogabbro. These rocks can be divided into five major sequences of cumulates that individually show increasing magmatic differentiation upward; in the simplest sense, these major sequences could represent the remnants of crystallized magma from five small and separate magma chambers or, alternatively, five major inputs of primitive magma into a larger, evolving magma chamber. The bulk composition of the gabbroic section is approximately that of a primitive to moderately evolved basalt, but mass balance indicates that a considerable thickness of primitive cumulates must be present, although currently not sampled.

At a smaller spatial scale, two papers in this section present data and/or interpretations on how complicated the interactions can be for basaltic liquids and crystal mushes in tectonically active magma-chamber environments. Thy and Dilek examine the evidence bearing on various physical and chemical processes occurring in magma chambers associated with mafic layered intrusions, ophiolites, and mid-ocean ridges. They argue that the processes that occur in magma chambers in mid-ocean ridges are similar to those inferred to have formed ophiolites and mafic layered intrusions. In particular, they note the presence of thick crystal-mush zones in magma-chamber models for mid-ocean ridges, ophiolites, and mafic layered intrusions, and they present evidence that compaction of crystal mushes during deformation and solidification can produce some of the petrologic and textural features of gabbroic rocks, particularly when potential effects of the expelled intercumulus liquid are considered. Bédard et al. describe petrographic and field observations on how complicated and intricate the interactions might be between basaltic liquids, crystal mushes, and preexisting cumulates in the Bay of Islands ophiolite. They argue that basaltic magmas routinely react with and partially assimilate or infiltrate preexisting rocks or crystals to such a degree that most so-called "cumulate" gabbroic and ultramafic rocks are actually complex hybrids and reaction or assimilation products, without any clear linkages to traditional crystal-accumulation processes. Bédard et al. further suggest that the lower oceanic and ophiolitic crust might act as a reactive filter to such a strong degree that the compositions of most erupted basalts in these environments have been strongly modified by the reaction and assimilation processes in the lower crust.

Regardless of uncertainties in how individual gabbroic rocks form, it is widely accepted that thick sequences of gabbroic rocks are formed from relatively small magma lenses in mid-ocean ridge environments. Differences in perspective have spawned a large number of models to explain how this process might occur (Sinton and Detrick, 1992; Nicolas et al., 1993; Phipps Morgan and Chen, 1993; Quick and Denlinger, 1993), but none is successful at adequately explaining the full range of constraints. Buck presents a variant on existing models in which a small and shallow melt lens is stable immediately below the sheeted-dike complex at a mid-ocean ridge. Buck argues that early-formed cumulates from a crystallizing basaltic liquid are, broadly speaking, olivine gabbros and are more dense than evolved gabbro cumulates. These early-formed olivine gabbros could sink through less dense gabbroic rocks until they reach equally dense rocks. This process would produce a density stratification of the gabbroic part of the crust such that the more primitive cumulates would finally reside at the base of the gabbroic section, thereby producing a compositionally stratified oceanic crust such as is observed in ophiolites and modern oceanic crust.

In spite of lingering questions about how the plutonic foundation of the ocean crust is formed, it is generally accepted that these plutonic rocks contain valuable chemical information about the magmatic environment in which they formed. Tribuzio et al. report on the major and trace element geochemical variations in gabbroic rocks from the Northern Apennine ophiolites, which are very similar to rocks from modern slow-spreading ridges such as the Southwest Indian Ridge, the Mid-Atlantic Ridge, and the Mid-Cayman Rise. Northern Apennine ophiolite gabbros have mineral compositions, in terms of major and trace elements, that are essentially identical to gabbroic rocks from these slow-spreading ridges.

Podiform chromitites are one of the more enigmatic features of upper-mantle outcrops, including ophiolites. Edwards et al. examine the issue of how to dissolve as much Cr in a basaltic melt as possible, so that, from a mass-balance perspective, it is possible to precipitate the large amounts of Cr-spinel found in podiform chromites. Edwards et al. suggest that dissolved water in the basaltic melt is the key factor because it will promote melting of refractory peridotite and greater Cr solubility in the basaltic melt. Their model suggests that Cr is retained in the melt until melt and peridotite reaction raises the Si content of the melt to the extent that the Cr solubility decreases and Cr-spinel is precipitated as podiform chromite. Holtzman reports that the podiform chromites potentially could be used to provide constraints on the mantle stress regime. He presents numerical models of the mechanical interaction between a podiform chromite and a dunite matrix, suggesting that a fracture in the chromite requires a low strength contrast between the chromite and the matrix. These models also indicate that two thresholds are observed: a minimum strain rate produces a single midpoint fracture and a higher strain rate produces a more distributed deformation. These numerical experiments are not currently intended to offer meaningful strain rate estimates.

REFERENCES CITED

Nicolas, A., Freydier, C., Godard, M., and Vauchez, A., 1993, Magma chambers at oceanic ridges: How large?: Geology, v. 21, p. 53–56.

Phipps Morgan, J., and Chen, Y.J., 1993, The genesis of oceanic crust: Magma injection, hydrothermal circulation, and crustal flow: Journal of Geophysical Research, v. 98, p. 6283–6297.

Quick, J.E., and Denlinger, R.P., 1993. Ductile deformation and the origin of layered gabbro in ophiolites: Journal of Geophysical Research, v. 98, p. 14015–14027.

Sinton, J.M., and Detrick, R.S., 1992, Mid-ocean ridge magma chambers: Journal of Geophysical Research, v. 97, p. 197–216.

Geological Society of America
Special Paper 349
2000

Large mantle upwellings and related variations in crustal thickness in the Oman ophiolite

Adolphe Nicolas and Françoise Boudier
CNRS/UMII, Laboratoire de Tectonophysique, place Eugène Bataillon, CC049, 34095 Montpellier cedex 05, France

ABSTRACT

Small mantle diapirs structurally defined and mapped in the southern part of the Oman ophiolite are associated with a thick Moho transition zone (MTZ) and copious wehrlitic intrusions in the overlying crust. The presence of an abundance of wehrlites in the crust has been used as a first indicator of focused melt delivery to the crust that is related to mantle diapirism. The second indicator of melt activity is the thickness of the MTZ, which has previously been shown to sharply decrease away from mapped diapirs. Both indicators converge, but the second allows a finer definition of the presumed upwelling areas. Throughout this 500-km-long ophiolite, three broad areas, thus defined, may represent the loci of mantle upwelling below the paleoridge at which the ophiolite originated. The northernmost area, 70 km long, is located in the United Arab Emirates part of the ophiolite; another area, 30 km long, is located in the northern Fizh massif in Oman. The southernmost area covers 130 km of a 150-km-long propagating-ridge segment that opened into very young lithosphere in the central and southern part of the ophiolite. In contrast with the situation of the central area, where the ophiolite represents a piece of lithosphere that was drifted away from the oceanic ridge before obduction, the spreading axis of the northern and southern propagating-ridge segments and thus the underlying mantle diapirs were incorporated into the ophiolite during its oceanic detachment. The small individual diapirs, 10–20 km across, which have been mapped within the propagating-ridge segment, represent distinct heads of a much larger mantle upwelling. If the criterion of MTZ thickness is used, we see that these three upwelling zones encompass one to three small structurally defined diapirs. In a preceding study, we showed that the gabbro section of the crust was significantly thinner above the small structurally defined diapirs than away from them. We extend the previous study with 17 new, more complete sections and confirm the earlier result, as well showing that it applies to the wider upwelling domains. The thickness of the gabbro section is, on average, 2.5 km above these domains versus 3.7 km away from the upwelling areas. The thickness of the sheeted-dike section may also be less above the upwellings (on average, 1 km thick there versus 1.2 km away from the upwellings). If to these thicknesses is added the 0.7 km estimate for the so-called *Geotimes* or V1 lavas directly associated with the ophiolite, the total thickness of the crust is only 4.2 km above the presumed mantle upwellings compared to 5.6 km away from these areas.

INTRODUCTION

After several years of structural mapping in the mantle section of the Oman ophiolite, we have identified localized areas with steep foliation and lineation trajectories in the mantle below a rather flat lying Moho. These are interpreted as mantle diapirs rising below a ridge segment (Ceuleneer et al., 1988). Further mapping revealed that the diapirs are aligned in the southern and central part of the ophiolite (Fig. 1) along a propagating-ridge segment that had been incorporated into the ophiolite when the detachment took place in the immediate vicinity of the ridge (Nicolas and Boudier, 1995). Finally, three of these diapirs are now mapped in great detail (Jousselin et al., 1998).

This mapping has confirmed that the size of the diapir is small (10–20 km) and that the rising mantle flow has rotated to the horizontal within a Moho transition zone (MTZ) that is a few hundred meters thick. This rotation was possible because the MTZ was so rich in melt below the ridge that the solid dunite was locally dispersed as an olivine-melt suspension (Boudier and Nicolas, 1995), thus drastically reducing its viscosity. The wehrlite mush generated by the melt flood has been injected either as wehrlite sills within the crystallizing gabbros located above the mantle diapirs or as large and discordant wehrlite intrusions, injected at a short distance from the ridge axis (Benn et al., 1988; Juteau et al., 1988).

We have concluded that mantle diapirs are revealed by a structural signature (steep foliations and lineations) if they have been entrapped in the ophiolite during its detachment and, in any case, by the presence of a thick MTZ and of abundant wehrlite sills and intrusions into the overlying crust. These criteria were used to tentatively locate diapirs along the Oman ophiolite belt (MacLeod and Rothery, 1992; Nicolas et al., 1988).

In this paper, we present a map (Fig. 2) of compiled data on wehrlite occurrence in the lower crust and thickness of MTZs in the

Nicolas, A., and Boudier, F., 2000, Large mantle upwellings and related variations in crustal thickness in the Oman ophiolite, *in* Dilek, Y., Moores, E.M., Elthon, D., and Nicolas, A., eds., Ophiolites and Oceanic Crust: New Insights from Field Studies and the Ocean Drilling Program: Boulder, Colorado, Geological Society of America Special Paper 349, p. 67–73.

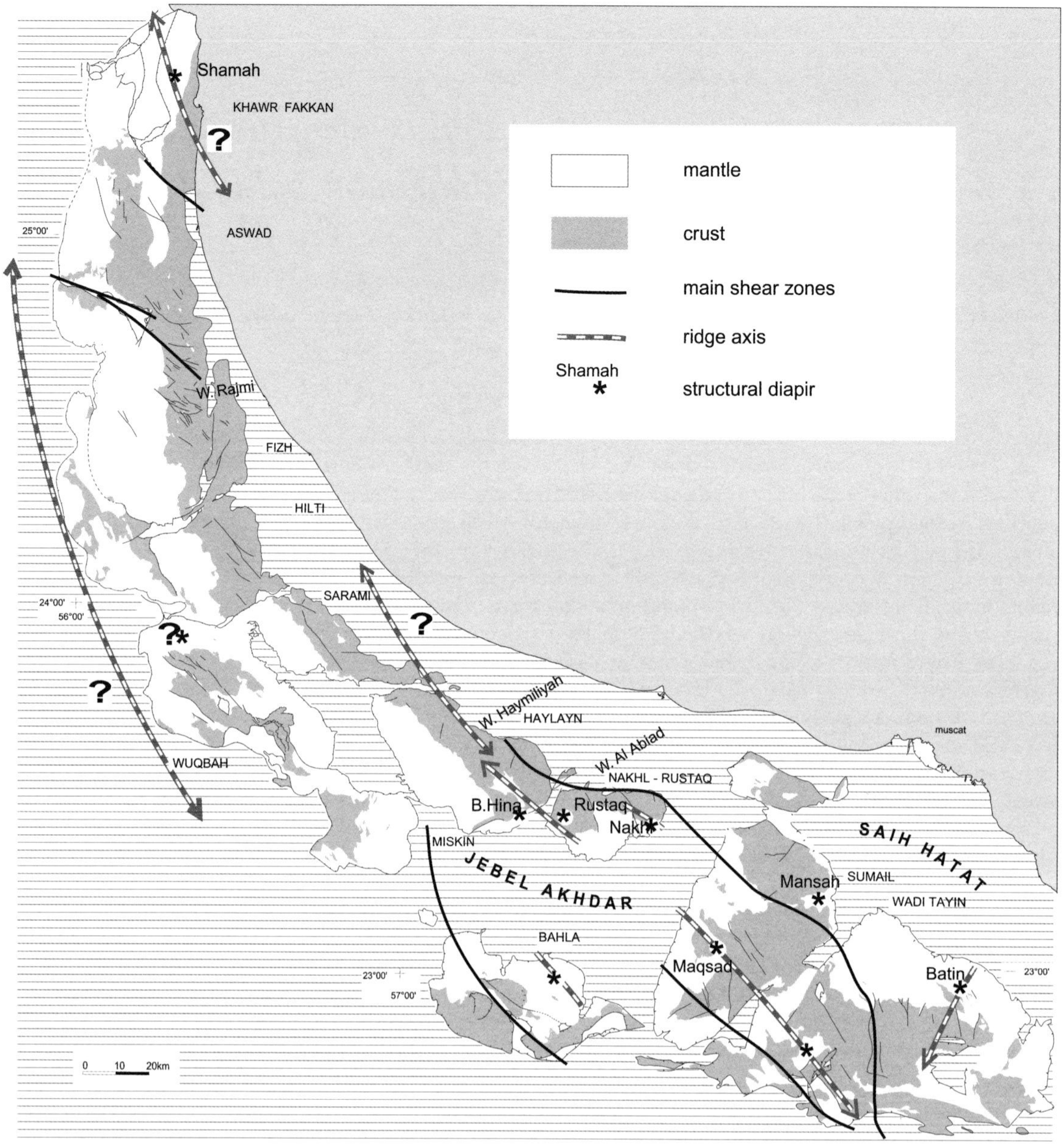

Figure 1. Simplified map of the Oman ophiolite belt illustrating the present state of understanding of the ridge system in which the ophiolite originated (Nicolas et al., in press).

ophiolite belt in order to locate areas of mantle upwelling independently of the mapped, structurally defined diapirs. Extending our earlier estimates of crust and MTZ thicknesses and how they relate to diapirs in the underlying mantle (Nicolas et al., 1996), we also investigate how crustal thickness relates to the newly defined mantle-upwelling areas. The map (Fig. 2) is based on two distinct sets of data: mapping of the MTZ by our group and integration of the cumulate peridotite gabbro (CPG) as that unit appears on the geologic maps of the Sultanate of Oman (Ministry of Petroleum and Minerals, 1986 and 1987).

MOHO TRANSITION ZONE

Mapping the MTZ and estimating its thickness are generally straightforward as this zone is limited by sharp lithologic changes, downward to harzburgites and upward to layered gabbros, the latter limit being identified here as the Moho. In most occurrences, these two limits are parallel to both the foliation in the immediately underlying harzburgites and the layering in the lowermost gabbros. The limits of this zone are more difficult to delineate above mantle diapirs such as Maqsad (Semail massif) where they have been mapped in great detail (Jousselin et al., 1998). There, MTZ rooting into the harzburgites becomes irregular through a network of dunite veins. Invasion of the MTZ by tholeiitic melt is responsible for the development of troctolites and wehrlites that are difficult to distinguish from the crustal wehrlites as, above diapirs, wehrlites are especially abundant in the lower crust as sills and intrusions. The tholeiitic melt also feeds gabbro sills that are intruded into the MTZ; these sills are locally so thick and abundant that it is equally difficult to trace the MTZ limit with the crustal layered gabbros (Boudier and Nicolas, 1995). Above diapirs there is, indeed, a remarkable petrologic and structural continuity between the MTZ and the lower gabbro unit.

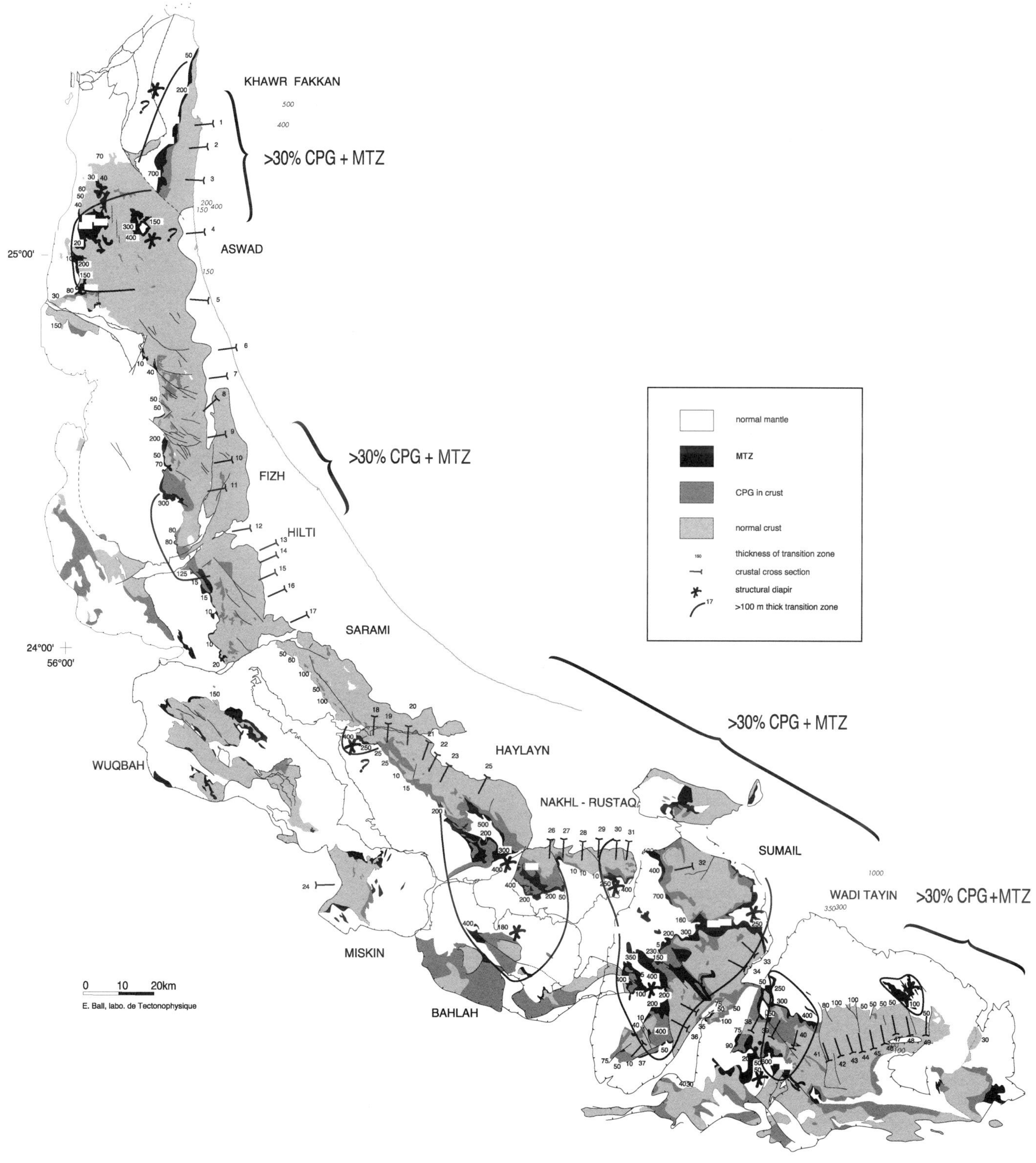

Figure 2. Map of the dunite (i.e., the Moho transition zone, MTZ; black) and wehrlite (i.e., part of the cumulate peridotite gabbro, CPG; dark gray) occurrences in the Oman ophiolite. The crustal section is represented in lighter gray. Large and small numbers refer to the thickness of the MTZ and the order of cross sections in Figure 3, respectively. Also represented are the small, structurally defined diapirs (stars), the areas where the MTZ is thicker than 100 m, and the olivine-rich areas. This map is the best fit of the southern and central massifs of the Oman belt assuming that they have been scattered by gravity sliding (see text).

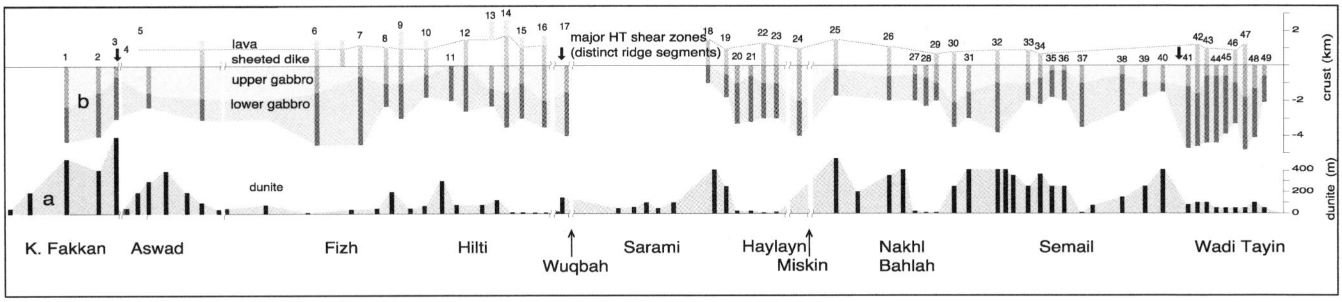

Figure 3. A, Dunite (of the MTZ) and B, crustal-thickness estimates based on the 49 cross sections located and numbered in Figure 2. This dual plot is a projection of the belt onto a vertical plane, from north to southeast, maintaining the relative distances between the cross sections.

The difficulty is increased further in areas such as Wadi Haymiliyah in the Haylayn massif and Wadi Rajmi in the Fizh massif where dunite sills, a couple of hundred meters thick, are inserted in the layered gabbros. The question of the origin is, however, still open. One piece of evidence in favor of a crustal origin is the absence of crystallographic fabrics that are generally present in mantle dunites; another one is that along contacts with the gabbros, we have locally observed a pyroxenite breccia with a dunite matrix and dunite veining into the pyroxenite; obviously it is not a solid dunite that has intruded the pyroxenite, but an olivine mush, which since has been compacted. Such evidence suggests that these dunites result from the complete compaction of a former wehrlite sill, rather than from a tectonic insertion of a MTZ sliver into the lower crust.

WEHRLITES IN THE CRUSTAL SECTION

It is probably to avoid the difficulty met in tracing the Moho in areas of intense melt circulation that the geologic maps of Oman (Ministry of Petroleum and Minerals, 1986 and 1987) have grouped as cumulate peridotite gabbros (CPGs) the impregnated dunites belonging to the MTZ and the gabbros rich in wehrlite sills and discordant intrusions. These maps do not distinguish between wehrlite sills and intrusions, a distinction that could be introduced in a discussion about the origin of the considered areas with respect to diapirs. Only the large wehrlite intrusions appear in the 1986 and 1987 Ministry maps. In spite of the difficulties we have mentioned, our detailed structural mapping allows us, in most situations, to draw the Moho limit inside this CPG unit and thus to distinguish the compacting-mantle MTZ from the lower crust intruded by the wehrlitic-rich mush.

A map combining these Ministry maps and our own sources as in Figure 2 is a good proxy to identify zones of intense melt circulation and reaction, presumably on top of mantle upwellings, as will be independently checked by comparing these petrologically defined areas with the structurally defined mantle diapirs.

Unfortunately, we do not have similar geologic maps in the Emirates part of the belt; as a consequence, our estimation of the wehrlite content in the crust there is restricted to the mapped large wehrlite intrusions. Olivine content in the crust is therefore underestimated.

OLIVINE-RICH ZONES

To investigate the relationships, at the scale of the belt, among the small, structurally defined diapirs, the thick MTZs, and the olivine-rich zones, we have compiled the map in Figure 2. The central and southern massifs have been fit together to remove the separation proposed (according to our observations and by S. Hanna, personal commun.) to have occurred with the uplift of Jebel Akhdar. Voids and overlaps between the massifs are minimized, and the continuity of internal structures from one massif to the next (Nicolas et al., in press) is accounted for.

At the scale of the entire belt (Fig. 2), it is possible to distinguish (1) large areas where the dunite-wehrlite (MTZ + CPG) association constitutes over 30% of the surface of the crust from (2) surrounding areas where these rock units are less represented. The separation into olivine-rich and olivine-poor areas has been achieved by making multiple transverse scans of the map in Figure 2. In areas where the dunite-wehrlite association is below 30%, the thickness of the MTZ does not exceed 100 m and is commonly no thicker than a few tens of meters (Figs. 2 and 3). Olivine-rich rocks form nearly all of the northernmost part of the ophiolite (i.e., the Emirates part—the Khawr Fakkan and Aswad massifs), extending 70 km along the ridge direction as defined by the sheeted-dikes azimuth. Another area of olivine-rich rocks, 30 km long, forms the southern part of the Fizh massif, and a third area stretches for 130 km and makes up most of the large propagating-ridge segment mapped in the central to southern part of the belt (Nicolas and Boudier, 1995).

THICKNESS OF MTZ

Special attention has been paid to the thickness of the MTZ because this parameter reflects melt influx below the ridge of origin. Once dunites (being the end member of residual mantle after partial melting) have been created, they are preserved despite further melt and rock interactions and plastic deformation. The only real difficulty is to distinguish mantle dunites generated by reaction with melt in the MTZ from crustal dunites produced by total compaction of the wehrlite mush as already described.

Recent detailed mapping of the MTZ in the vicinity of diapirs has revealed sharp variations in structure and thickness. The MTZ that is several hundred meters thick above diapirs and dominantly deformed in the magmatic state (Boudier and Nicolas, 1995) reduces to ten to a few tens of meters of plastically deformed dunites at distances of only 10 km from the diapir (Fig. 2). At a small scale, the change from MTZ thicknesses of <20–100 m at the periphery of the Maqsad diapir to thicknesses of 150–450 m inside the diapir has been traced over a lateral distance of 3 km (Fig. 2). Also, in Wadi Al Abiad, in the Nakhl-Rustaq massif, a very thin MTZ (1 to 10 m) is observed less than 10 km away from the Nakhl diapir (Nicolas and Boudier, 1995), which is itself associated with a 250–350-m-thick MTZ. Finally, in the Aswad massif, to the north, transitions from MTZs >200 m thick to MTZs only 10–50 m thick occur within a 5 km distance from the diapir.

CRUSTAL THICKNESS

In our previous study, we compared the thickness of the gabbro units to that of the MTZs. We concluded that diapiric areas having

MTZs thicker than 100 m were capped by gabbro units 1–1.5 km thinner, on average, than the gabbro units that are located away from diapirs and associated with MTZs thinner than 100 m. Because new data derived from further mapping are available, the number of cross sections yielding the information on MTZ and crust thickness has been increased from the original number of 32 to 49 (Figs. 2 and 3). The definitions used and the limitations presented in the Nicolas et al. (1996) paper are still valid here. In particular, with the exception of the Miskin section, no attempt has been made to estimate the crustal thickness in the western massifs because of the shear and thrust tectonics related to their position near the front of the ophiolite nappe. We stress again that estimation of the thickness of the gabbro unit relies on two main guides: (1) the general dips of the top (lavas and metalliferous sedimentary rocks) and bottom (metamorphic sole and detachment-related foliations in the mantle sequence) surfaces of the ophiolite and (2) the attitudes of the formations immediately surrounding the gabbro units, namely, the Moho surface and the sheeted-dike orientation (assumed to have been initially nearly vertical). In contrast, minor attention is paid to the attitudes of layering and magmatic foliation in the gabbros because of the common occurrence of magmatic to submagmatic folds at all scales and because the magmatic foliation steepens up-section. For these reasons, earlier estimations of crustal thickness that did not account for these effects tend to yield larger results than ours (Lippard et al., 1986; Pallister and Hopson, 1981; Reuber, 1988). Cross sections where shear zones, a few hundred meters thick and generally transcurrent, have been mapped were eliminated because the gabbro unit there is abnormally thin or, less commonly, abnormally thick. For examples, the two cross sections in the gabbros of the Sarami massif, which is cut by a major transcurrent fault, are only 1.5 km and 2 km thick; similarly, the Wadi Haylayn gabbro section in the Haylayn massif, cut by a flat-lying shear zone, is 1.8 km thick. In contrast, the Rajmi gabbro section in the Fizh massif, cut by numerous shear zones both flat and steep, is 5 km thick, possibly as a result of tectonic thrusting. Finally, a few sections described in our 1996 study (Nicolas et al., 1999) have been reevaluated on the basis of new data: Fujayrah (no. 3 in the 1996 study) has been reevaluated from 2.3 to 3 km, Halhal (no. 9), from 2.5 km to 3.3 km, and Mayah (no. 10) from 2.2 to 2.6 km.

With respect to the 1996 study in which it was assumed that the overlying basaltic lid (sheeted dikes and extrusive rocks) was uniformly 1.5–2 km thick, we present here detailed estimates for the sheeted-dike unit (27 measurements) and the extrusive rocks (10 measurements). For the latter, the proposed thickness is that of the *Geotimes* V1 volcanic rocks that lie on top of the sheeted dikes and below the metalliferous sedimentary rocks (umbers). When these typical sediments are absent, it can be difficult to distinguish these ophiolite-related lavas from the overlying Lasail (Lippard et al., 1986; Ernewein et al., 1988) V2 ones.

DISCUSSION

Relationship between olivine-rich areas and diapirs

Detailed study of the Maqsad diapir (Jousselin et al., 1998), overlain by a thick MTZ, has shown that this MTZ is the level where mantle flow rotates abruptly from vertical to actively horizontal and divergent at rates one order of magnitude faster than the spreading rate. This large deformation taking place at the outskirts of diapirs alone cannot explain the sharp and drastic reduction in MTZ thicknesses that have been already described herein. A mass-balance estimation shows that the largest part of the MTZ is expelled upward, as a mush crystallizing in the crustal section as wehrlite sills and intrusions, and probably downward, as compacted dunite lenses (Jousselin and Nicolas, in press). If this interpretation is correct, an important conclusion regarding the location of mantle diapirs can be derived (Fig. 4). Domains where the MTZ section is a few hundred meters thick should be located above in situ mantle upwelling, even if the mantle structure is not typically characterized by steep foliations and lineations. Other domains should have a thin MTZ, whether they are still on the paleo–ridge axis but located away from a diapir or they have drifted away from the ridge of their origin.

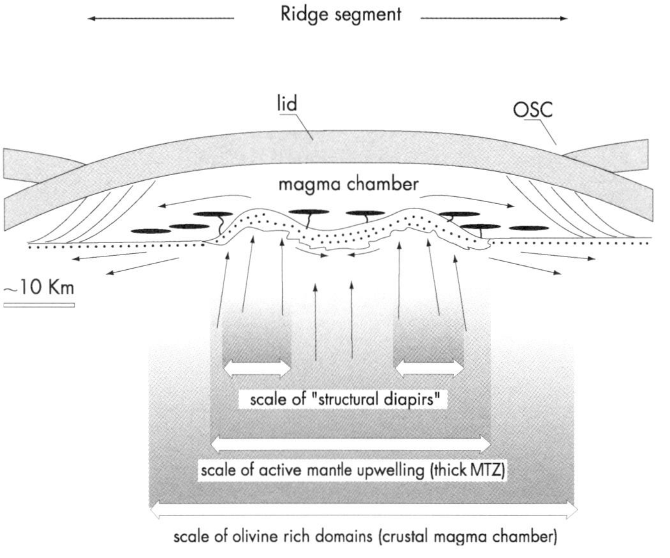

Figure 4. Possible deep structures of a ridge segment, based on the data reported here, and indication of their respective scales. Dots, MTZ (Moho transition zone); black lenses, wehrlite sites; OSC—overlapping spreading center.

This key can be applied to tentatively interpret the variations in the thickness of the MTZ shown in Figure 2 in which the massifs of the ophiolite belt have been organized in the best-fit reconstruction. We have fixed the limit between thin and thick MTZs at 100 m and have contoured this limit. Although arbitrary, choosing this limit is justified by the fact that it clearly separates the internal domain where the average thickness of MTZs is 290 m (65 measurements) from the outer domain where it is 45 m (68 measurements). Several conclusions can be derived from the contoured map in Figure 2:

1. The small mantle diapirs revealed by structural mapping are associated with both olivine-rich crust and thick Moho transition zones, confirming earlier results (Nicolas et al., 1996). This conclusion is verified for the Batin, Samad, Maqsad, Mansah, Nakhl, Rustaq–Bani Hina, Bahla, and Aswad diapirs, although in the last case, the direct structural evidence is limited because the mantle is largely concealed below the flat-lying crust (Nicolas et al., this volume). The steep structures mapped in the Shamah area of the Khawr Fakkan massif may also relate to a diapir, as suggested by our earlier mapping (Ceuleneer et al., 1988). The western tip of Haylayn massif, another area where steep mantle structures and a thick MTZ coincide, seems to represent a mantle diapir, despite the fact that this area is only 7 km across. The situation in the Wuqbah and Miskin massifs where steep structures have been mapped is more ambiguous. These mantle structures are associated with a 150-m-thick MTZ and a few large wehrlite intrusions. Unfortunately these massifs are sliced by large thrusts because of their position close to the front of the ophiolite

TABLE 1. THICKNESS OF MTZ AND CRUST UNITS ALONG THE BELT

Crust and dunite thickness						
Cross section	Lower gabbro (km)	Upper gabbro (km)	Sheeted dike (km)	Lava (km)	Dunite (m)	Area
1	2	2.3			200	Khawr Fakkan
2	2.5	1.5			500	Khawr Fakkan
3	2.4	0.6			700	Khawr Fakkan
4		1.5			350	Aswad
5	2	1.5	1.1	0.8	100	Aswad
6	3	1.5	1	0.5		Fizh
7	3.9	0.6	1.2		50	Fizh
8	1.3	1	1.1	1	50	Fizh
9	2	1	1	1	70	Fizh
10	1.3	0.5	1	0.5	300	Fizh
11	2				100	Fizh
12	2.6		1.5		100	Fizh
13	1	1.3		1	100	Hilti
14	2	1.5	1.6	1	15	Hilti
15	2	1	1	0.5	15	Hilti
16	1.5	2	1.1	0.5	10	Hilti
17	2.5	1.5			10	Hilti
18	1		1.5		400	Haylayn
19	1.3	0.5	1		250	Haylayn
20	2.3	1			25	Haylayn
21	2.6	0.6			25	Haylayn
22	2	1	1.3		10	Haylayn
23	2	1	1.2		15	Haylayn
24	2	2	1			Miskin
25	1.5	0.2	1.5		500	S. Haylayn
26	1.4	0.6	1.1		400	Nakhl
27	1.5	0.5			10	Nakhl
28	1.6	0.7			10	Nakhl
29	1	1	0.7		10	Nakhl
30	1.4	2.1	0.8		250	Nakhl
31	1.5	1.5			400	Nakhl
32	2.8	1	0.9		400	Semail
33	1	1	0.9		250	Semail
34	1.5	0.7	0.7		360	Semail
35	1.5	0.3			250	Semail
36	1.7	0.3			250	Semail
37	2.5	1			30	Semail
38	2.1	0.5			150	Wadi Tayin
39	0.9	0.9			250	Wadi Tayin
40	0.5	1			400	Wadi Tayin
41	3.5	1.2			80	Wadi Tayin
42	3.8	1.6	1.2		100	Wadi Tayin
43	3.8	0.6	1		100	Wadi Tayin
44	3.8	0.6			50	Wadi Tayin
45	3.3	0.6			50	Wadi Tayin
46	2.3	1	0.9		50	Wadi Tayin
47	3	1.8	1.2		50	Wadi Tayin
48	2.8	1.3			100	Wadi Tayin
49	1.5	0.6			50	Wadi Tayin

Note: Locations are shown in Figure 2. The 132 measurements of MTZ thickness have been averaged to create the 49 cross sections presented in Figure 3.

nappe, a tectonic complexity that may be blurring their original structure. In contrast, the absence of thick dunites in the western part of the Samad area (Wadi Tayin massif), where steep structures have been mapped, suggests that there is no diapir there and that foliation is steepened along the tectonic contact with the Hawasina Formation.

2. Conversely, the prediction that away from in situ diapirs the MTZ must be thin because of compaction (Jousselin and Nicolas, in press) is verified everywhere except in the Bani Umar Gharbi area in the Fizh massif where a thick MTZ is not related to steep mantle structures. The flat-lying mantle foliations and the regular trend of mineral lineations suggest that this area drifted away from the ridge, which we have located outside the belt to the west (Fig. 1) (Nicolas et al., in press). Why the presumed MTZ was neither compacted nor highly strained when it drifted away from a diapir is not understood. One possibility is that this olivine-rich zone may not be a real MTZ, but a particularly large wehrlite sill intruded in the lower crust that, as already described herein for two other such intrusions, would have been totally compacted.

3. The contouring of the 100 m MTZ thickness allows a finer analysis of mantle upwelling within the three large olivine-rich domains that we have defined. This approach is mainly applicable to the southern domain, which covers most of the northwest–southeast ophiolite segment studied. In the northern part of this segment, one can define a single zone of associated steep structures and thick MTZs covering the Bani Hina, Rustaq, and Bahla structurally defined diapirs and another zone including the Nakhl, Maqsad, and Mansah structurally defined diapirs, the latter being an off-axis diapir (Jousselin et al., 1998). East of Samad in the Wadi Tayin massif is an upwelling zone that includes the Samad diapir. This upwelling zone extends farther north than the area of observable steep structures in the mantle.

TABLE 2. THICKNESS OF CRUSTAL UNITS DEPENDING ON WHETHER THE ASSOCIATED MTZ IS THICKER OR THINNER THAN 100 m

Thickness of DTZ	Lower gabbros (km)	Upper gabbros (km)	Sheeted dikes (km)	V1 lavas (km)	Total (km)
>100 m (19 cross sections)	1.6	0.9	1	0.7	4.2
≤100 m (30 cross sections	2.4	1.3	1.2	0.7	5.6

Crustal thickness and mantle diapirs

Table 1 presents detailed information on the thickness of the various components of the crust and also illustrates, together with Figure 3, how crustal thickness relates to the large dunite- and wehrlite-rich areas, presumably ascribed to large mantle upwellings. The earlier relationship noted between thin gabbro sections and small diapirs (Nicolas et al., 1996) is now extended to these larger areas. With the exception of the *Geotimes* V1 lavas (for which we have only nine estimates yielding an average thickness of 0.7 km), all other components of the crust display reduced thicknesses in olivine-rich areas in comparison to their thicknesses in olivine-poor areas (Table 2).

The slight reduction in thickness of the sheeted dikes above olivine-rich areas suggests that, because they are located above mantle diapirs, these zones may have had a higher heat budget that pushed isotherms upward. The effect is illustrated by the upward migration of the melt lens with spreading rate in currently active oceanic ridges [Purdy et al., 1992; Phipps Morgan and Chen, 1993]. Finally, Figures 2 and 3 suggest that there is a tendency for the crust, both in olivine-rich and olivine-poor areas, to be thinner in the central part of the belt and thicker at both extremities. No explanation for this tendency is attempted.

CONCLUSIONS

Along the Oman ophiolite belt, three areas are remarkable by the abundance of olivine in their lower-crust and upper-mantle sections due to a large fraction (≥30%) of wehrlites in the crustal section and the exceptional thickness of the MTZs. This olivine abundance is ascribed to intense melt circulation through these areas with important melt-rock reactions followed by compaction. From north to south, three such areas extend along the sheeted-dike direction for 70 km, 30 km, and 130 km. They incorporate all the 10–20-km-wide structurally defined mantle diapirs mapped so far on the basis of the presence of steep structures (Ceuleneer et al., 1988; Nicolas and Boudier, 1995; Jousselin et al., 1998). For these two reasons, they are regarded as representing large zones of mantle upwelling below the ridge system of their origin. The structurally defined diapirs are considered as local heads protruding at shallow depth in the mantle above the large upwellings.

A finer analysis of areas where the thickness of the MTZs is in excess of 100 m within the large olivine-rich domains demonstrates their close association with the small mantle diapirs. This association confirms the sharp and massive reduction in thickness of the MTZs for parts of the ophiolite that drifted away from the diapirs (Jousselin et al., 1998). In the eastern part of the belt, the association also allows us to subdivide the large upwelling area defined by the abundance of olivine, which nearly covers the entire ophiolite segment studied, into three more localized areas, extending 20–60 km along the ridge trend and comprising between one and three structurally defined diapirs. This finer analysis shows that individual mantle-upwelling zones along the Oman belt range in size from 10 to 60 km. Along fast- to superfast-spreading oceanic ridges, it is still debated whether the mantle upwelling is active or passive (Lin and Phipps Morgan, 1992; Sparks et al., 1993). The results on which the active models are based assign to mantle diapirs dimensions of 20–50 km (Barth and Mutter, 1996; Dunn and Toomey, 1997; Wang et al., 1996), which are more in agreement with the size of the mantle upwellings now described in Oman than with that of the small diapirs described in our earlier papers.

The earlier conclusion that mantle diapirs were associated both with thick MTZs and thin gabbroic crust, compared to areas located away from diapirs (Nicolas et al., 1996), is confirmed and made more precise by the new sections. It applies well to the larger upwelling zones now recognized. The average difference in crustal thickness between upwellings and other zones along the ridge axis reaches 1.4 km, comparing well with results obtained at 9°N along the East Pacific Rise (Barth and Mutter, 1996; Wang et al., 1996).

ACKNOWLEDGMENTS

This work relies on field and laboratory assistance by J.L. Bouchez, M. Misseri, S. Crambert, G. Ceuleneer, I. Reuber, K. Benn, C. MacLeod, F. Quatrevaux, E. Gnos, B. Ildefonse, D. Jousselin, and K. Michibayashi and on technical assistance by C. Nevado, E. Ball, B. Marie, B. Ildefonse, and M. Peyret.

Constructive review was provided by Dona Blackman and an unidentified reviewer. The hospitality of M. Kassim and H. Al Azri from the Ministry of Petroleum and Minerals of Oman is acknowledged as well as continuous financial support from the CNRS over nearly 20 years.

REFERENCES CITED

Barth, G.A., and Mutter, J.C., 1996, Variability in oceanic crustal thickness and structure: Multichannel seismic reflexion results from the northern East Pacific Rise: Journal of Geophysical Research, v. 101, p. 17951–17975.

Benn, K., Nicolas, A., and Reuber, I., 1988, Mantle-crust transition zone and origin of wehrlitic magmas: Evidence from the Oman ophiolite: Tectonophysics, v. 151, p. 75–85.

Boudier, F., and Nicolas, A., 1995, Nature of the Moho transition zone in the Oman ophiolite: Journal of Petrology, v. 36, p. 777–796.

Ceuleneer, G., Nicolas, A., and Boudier, F., 1988, Mantle flow patterns at an oceanic spreading centre: The Oman peridotite record: Tectonophysics, v. 151, p. 1–26.

Dunn, R.S., and Toomey, D.R., 1997, Seismological evidence for three-dimensional melt migration beneath the East Pacific Rise: Nature, v. 388, p. 259–262.

Ernewein, M., Pflumio, C., and Whitechurch, H., 1988, The death of an accretion zone as evidenced by the magmatic history of the Sumail ophiolite (Oman): Tectonophysics, v. 151, p. 247–274.

Jousselin, D., Nicolas, A., and Boudier, F., 1998, Detailed mapping of mantle diapir below a paleo-spreading center in the Oman ophiolite: Journal of Geophysical Research, v. 103, p. 153–170.

Jousselin, D., and Nicolas, A., 2001, The Moho Transition Zone in the Oman ophiolite: Relation with wehrlites in the crust and dunites in the mantle: Marine Geophysical Researches, special issue (in press).

Juteau, T., Beurrier, M., Dahl, R., and Nehlig, P., 1988, Segmentation at a fossil spreading axis: The plutonic sequence of the Wadi Haymiliyah area (Haylayn block, Sumail Nappe, Oman): Tectonophysics, v. 51, p. 167–197.

Lin, J., and Phipps Morgan, J., 1992, The spreading rate dependence of three-dimensional mid-ocean ridge gravity structure: Geophysical Research Letters, v. 19, no. 1, p. 13–16.

Lippard, S.J., Shelton, A.W., and Gass, I.G., 1986, The ophiolite of Northern Oman: Geological Society of London Memoir 11, 178 p.

MacLeod, C.J., and Rothery, D.A., 1992, Ridge axial segmentation in the Oman ophiolite: Evidence from along-strike variations in the sheeted dyke complex, *in* Parson, L.M., Murton, B.J., and Browning, P., eds., Ophiolites and their modern oceanic analogues: London, Geological Society Special Publication 60, p. 39–63.

Ministry of Petroleum and Minerals of Sultanate of Oman, 1986 and 1987, Geological Map of Oman, scale: 1:100 000.

Nicolas, A., and Boudier, F., 1995, Mapping oceanic ridge segments in Oman ophiolites: Journal of Geophysical Research, v. 100, p. 6179–6197.

Nicolas, A., Boudier, F., and Ceuleneer, G., 1988, Mantle flow patterns and magma chambers at ocean ridges: Evidence from Oman ophiolite: Marine Geophysical Researches, v. 9, p. 293–310.

Nicolas, A., Boudier, F., and Ildefonse, B., 1996, Variable Moho depth in the Oman ophiolite: Implication for oceanic ridges: Journal of Geophysical Research, v. 101, p. 17941–17950.

Nicolas, A., Boudier, F., Ildefonse, B., and Ball, E., 2001, Accretion of Oman ophiolite in a microplate system: Discussion of a new structural map: Marine Geophysical Researches, special issue.

Pallister, J.S., and Hopson, C.A., 1981, Samail ophiolite plutonic suite: Field relations, phase variation, cryptic variation and layering, and a model of a spreading ridge magma chamber: Journal of Geophysical Research, v. 86, p. 2593–2644.

Phipps Morgan, J., and Chen, Y.J., 1993, The genesis of oceanic crust: Magma injection, hydrothermal circulation, and crustal flow: Journal of Geophysical Research, v. 98, p. 6283–6297.

Purdy, J.M., Kong, L.S.L., Christeson, G.S., and Solomon, S.C., 1992, Relationship between spreading rate and the seismic structure of mid-ocean ridges: Nature, v. 355, p. 815–817.

Reuber, I., 1988, Complexity of the crustal sequence in northern Oman ophiolite (Fizh and southern Aswad block): The effect of early slicing?: Tectonophysics, v. 151, p. 137–165.

Sparks, D.W., Parmentier, E.M., and Phipps Morgan, J., 1993, Three-dimensional convection beneath oceanic spreading center: Implications for along-axis variations in crustal thickness and gravity: Journal of Geophysical Research, v. 98, p. 21977–21995.

Wang, X.J., Cochran, J.R., and Barth, G.A., 1996, Gravity anomalies, crustal thickness, and the pattern of mantle flow at the fast spreading East Pacific Rise, 9°–10° N: Evidence for three-dimensional upwelling: Journal of Geophysical Research, Solid Earth, v. 101, p. 17927–17940.

Manuscript Accepted by the Society February 23, 2000

Printed in the U.S.A.

Geological Society of America
Special Paper 349
2000

Lower oceanic crust formed at an ultra-slow-spreading ridge: Ocean Drilling Program Hole 735B, Southwest Indian Ridge

Paul T. Robinson*
Institute of Geochemistry, GeoForschungsZentrum, Telegrafenberg, D-14473 Potsdam, Germany
Henry J.B. Dick
Department of Geology and Geophysics, Woods Hole Oceanographic Institution, Woods Hole, Massachusetts, USA
James H. Natland
Rosenstiel School of Marine and Atmospheric Science, Miami, Florida, USA
ODP Leg 176 Shipboard Party†

ABSTRACT

Ocean Drilling Program (ODP) Hole 735B, drilled on Legs 118 and 176, penetrated 1508 m of oceanic layer 3 on a transverse ridge adjacent to the Atlantis II Fracture Zone, Southwest Indian Ridge. The cored sequence consists predominantly of olivine gabbro and troctolite and lesser amounts of gabbro, microgabbro, and gabbronorite rich in Fe-Ti oxides. The section contains five major blocks of relatively primitive olivine gabbro and troctolite, composed of many smaller igneous bodies. Each of these composite blocks shows a small upward decrease in Mg# [defined as 100 × Mg/(Mg + Fe^{2+})] and contains more fractionated Fe- and Ti-rich gabbros near the top. Small, crosscutting bodies of olivine gabbro and troctolite with diffuse boundaries may represent conduits through crystal mushes for melts migrating upward and feeding individual intrusions. Oxide gabbros and oxide gabbronorites are commonly associated with shear zones of intense deformation, which crosscut the section at all levels. However, oxide-rich rocks decrease in abundance downward and are nearly absent in the lower 500 m of the section. The oxide gabbros and gabbronorites appear to have formed from late-stage, Fe- and Ti-rich, intercumulus melts that were expelled out of fractionating olivine gabbros into the shear zones.

The fabrics of the recovered gabbros are consistent with synkinematic cooling and extension of the crustal section in a mid-ocean ridge environment. However, thick intervals of the core have only a weak magmatic foliation. The magmatic foliation is commonly overprinted by a weak, parallel, deformational fabric probably reflecting the transition from a largely magmatic to a largely crystalline state. Deformation in this crustal section decreases markedly downward.

Metamorphism and alteration also decrease downward, and much of the core has less than 5% background alteration. Major zones of crystal-plastic (ductile flow by dislocated creep) deformation in the upper part of the core probably formed under conditions equivalent to granulite-facies conditions when there was little or no melt present. Late-magmatic and hydrothermal fluids produced a variety of plagioclase, amphibole, and diopside veins. Late-stage, low-temperature veins of calcite, smectite, zeolite, and prehnite are present in a few intervals.

The fact that the cored section is unlike ophiolites as defined by the 1972 Penrose Conference Participants suggests that no ophiolite representing an ultra-slow-spreading-ridge environment like the Southwest Indian Ridge may be preserved.

INTRODUCTION

During the past 30 years of ocean drilling, a great deal has been learned about the upper ocean crust. Numerous holes have penetrated the upper few hundred meters of oceanic layer 2 and have yielded large amounts of data on the petrology, geochemistry, magnetic character, and physical properties of the basement. One hole, ODP (Ocean Drilling Program) Hole 504B, penetrated slightly >2000 m into basement, sampling all of the volcanic layer and the upper part of the sheeted-dike sequence. Some workers have suggested that this hole penetrated the layer 2–layer 3 boundary (Detrick et al., 1994); if so, it is the only hole in the oceans to have done so.

Despite many efforts to penetrate and sample the lower ocean crust, we still have very little information on oceanic layer 3. Gabbros interpreted as layer 3 rocks have been dredged from slow-spreading ridges, such as the Mid-Atlantic Ridge and the Southwest Indian

*Permanent address: Centre for Marine Geology, Dalhousie University, Halifax, Nova Scotia, Canada.

†Alt, J.C., Bach, W., Bideau, D., Gee, J.S., Haggis, S., Hertogen, J., Hirth, G., Holm, P.M., Ildefonse, B., Iturrino, G.J., John, B.E., Kelley, D.S., Kikawa, E., Kingdon, A., LeRoux, P.J., Maeda, J., Meyer, P.S., Miller, D.J., Naslund, H. R., Niu, Y.-L., Snow, J., Stephen, R.A., Trimby, P.W., Worm, H.-U., and Yoshinobu, A.

Robinson, P.T., Dick, H.J.B., Natland, J.H., and ODP Leg 176 Shipboard Party, 2000, Lower oceanic crust formed at an ultra-slow-spreading ridge: Ocean Drilling Program Hole 735B, Southwest Indian Ridge, *in* Dilek, Y., Moores, E.M., Elthon, D., and Nicolas, A., eds., Ophiolites and Oceanic Crust: New Insights from Field Studies and the Ocean Drilling Program: Boulder, Colorado, Geological Society of America Special Paper 349, p. 75–86.

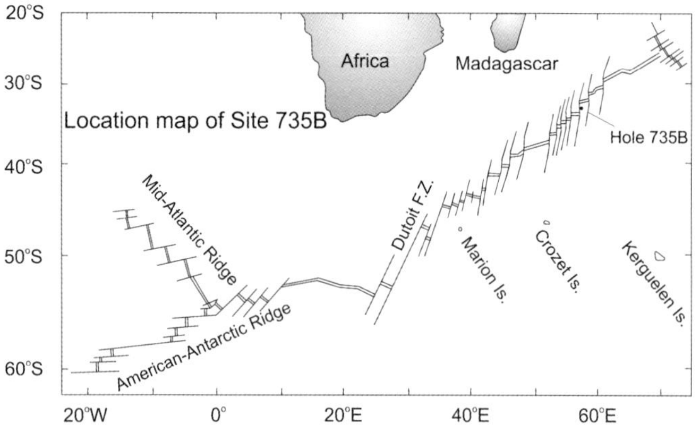

Figure 1. Southwest Indian Ridge showing location of ODP Hole 735B on the Atlantis II Fracture Zone.

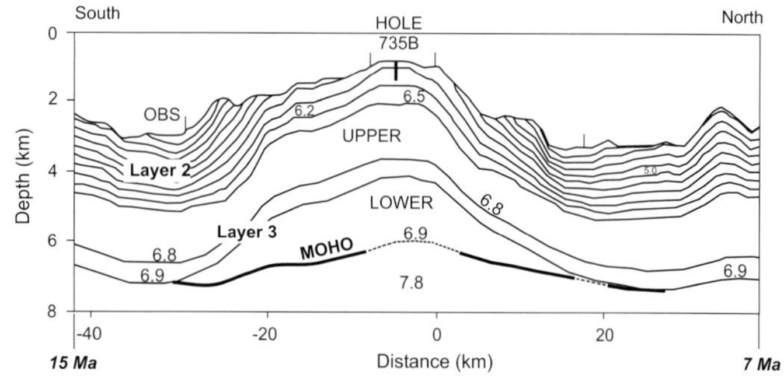

Figure 2. P-wave seismic velocity model for the Atlantis Bank and surrounding area on the north-south Line CAM 101 (Muller et al., 1997). The velocity contour interval is 0.3 km/s, and the OBS (ocean-bottom seismograph) locations are shown on the seafloor. The position of Hole 735B has been projected to the line from 1 km to the west.

Ridge (Fisher et al., 1986) and were sampled on a few DSDP (Deep Sea Drilling Project) legs (37, 45, 83) prior to ODP Leg 118. However, available samples are sparse and gave a very incomplete picture of the lower crust.

During ODP Leg 118, the *JOIDES Resolution* spudded Hole 735B on the Atlantis Bank on the eastern side of the Atlantis II Fracture Zone and drilled to a depth of 500.7 m with 87% recovery (Robinson, Von Herzen, et al., 1989). The hole penetrated a sequence of gabbros, gabbronorites, oxide gabbros, and troctolites with the seismic characteristics and physical properties of oceanic layer 3. In 1997, the *JOIDES Resolution* returned to this site and deepened the hole to 1508 mbsf (meters below seafloor), again with very high core recovery (Shipboard Scientific Party, 1999). The recovered core represents the first significant sample of in situ lower ocean crust, and its study has produced many new insights into crustal-accretion processes at ultra-slow-spreading ridges.

TECTONIC SETTING OF SITE 735

ODP Site 735 is located on the east side of the Atlantis II Fracture Zone, which offsets the Southwest Indian Ridge at long ~57°E (Fig. 1). The ridge originated with the breakup of Gondwana in the Mesozoic and, just prior to 80 Ma, joined the Central and Southeastern Indian Ridges to form the Indian Ocean triple junction (Fisher and Sclater, 1983). Once formed, the triple junction migrated steadily to the northeast, creating a number of new ridge segments and fracture zones, including the Atlantis II Fracture Zone (Sclater et al., 1981). Both the fracture zone and the adjacent crust are entirely oceanic in origin and have been unaffected by processes associated with continental breakup. Spreading on the Southwest Indian Ridge has been relatively constant over the past 30 m.y. at a half rate of 8 mm/yr (Fisher and Sclater, 1983); this half rate places the ridge at the very low end of the spreading-rate spectrum.

The Atlantis II Fracture Zone is a 199-km-long, left-lateral ridge offset that grew at a rate of ~4 mm/yr over the past 20 m.y. (Dick et al., 1991b, 1991c). The transform valley is a deep feature with relief just over 5500 m; the valley extends from 6480 mbsl (meters below sea level) at its deepest point to 694 mbsl on the shallowest part of the Atlantis Bank. The walls are very steep; the slopes range from ~25° to 40° and are typically covered with abundant talus. A turbidite sequence as much as 500 m thick covers the valley floor, which is bisected by a median ridge 1.5 km high. On the transform-valley walls, 37 dredges were carried out during the 1986 site survey and recovered a large volume of rock composed of 38% peridotite, 20% gabbro, 13% diabase, and 29% pillow lava. Peridotite was dredged from 5675 to 2548 mbsl and gabbro from 6050 to 1500 mbsl; peridotite and gabbro occurred together in eight dredges from 5652 to 3200 mbsl (Dick et al., 1991c). These lithology distributions suggest that the crust-mantle boundary lies at ~2500 mbsl on the fracture-zone wall.

Hole 735B is located on the Atlantis Bank, a flat-topped, steep-sided platform ~9 km long in a north-south direction and 4 km wide. The bank is the shallowest of a series of uplifted blocks that form a linear transverse ridge parallel to the transform. The platform has a wave-cut, commonly bare-rock, surface exposing foliated and massive gabbro, crisscrossed by a network of joints and veins. Locally, it is mantled with a thin layer of carbonate sediment. After truncation by wave action, it subsided to its present depth, presumably by normal lithospheric cooling (Dick et al., 1992).

On the basis of well-defined magnetic anomalies that cross the Atlantis Bank (anomaly 5r.2n, 11.75 Ma), this crustal segment was formed at the Southwest Indian Ridge at ca. 11.5 Ma. A U-Pb isotopic age on zircon from trondhjemite recovered in Hole 735B confirms the magnetic anomaly age (Stakes et al., 1991). The Atlantis Bank gabbros are believed to have formed beneath the ridge axis ~15–19 km from the current ridge-transform intersection (Dick et al., 1991c, 1992).

A recent survey of the area (Muller et al., 1997) suggests that the crustal thickness ranges from 5 to 6 km and that the maximum thickness is directly beneath the Atlantis Bank (Fig. 2). Despite the absence of both sheeted dikes and pillow lavas, Muller et al. (1997) postulated a 2-km-thick seismic layer 2 beneath the Atlantis Bank; they placed the Moho at a depth of 6 km below sea level. In this ultra-slow-spreading environment near the distal end of a ridge segment, it is unlikely that the lower igneous crust would be 5–6 km thick. Partially serpentinized peridotites, dredged from the west wall of the bank at water depths of ~2500 m, suggest that the gabbro-peridotite boundary is significantly higher than the seismic Moho postulated by Muller et al. (1997). Thus, the low seismic velocities beneath the Atlantis Bank probably reflect extensive serpentinization of the upper mantle.

The emplacement and unroofing of the Atlantis Bank are thought to reflect detachment faulting during a period of amagmatic spreading in the rift valley (Dick et al., 1991c, 1992). During the initial stages of faulting, the section was thinned as the upper sequence of lavas and sheeted dikes was removed, thus exposing the lower crust. This tectonic erosion was followed by block uplift as the crustal section migrated out of the rift valley and into the inside-corner high at the

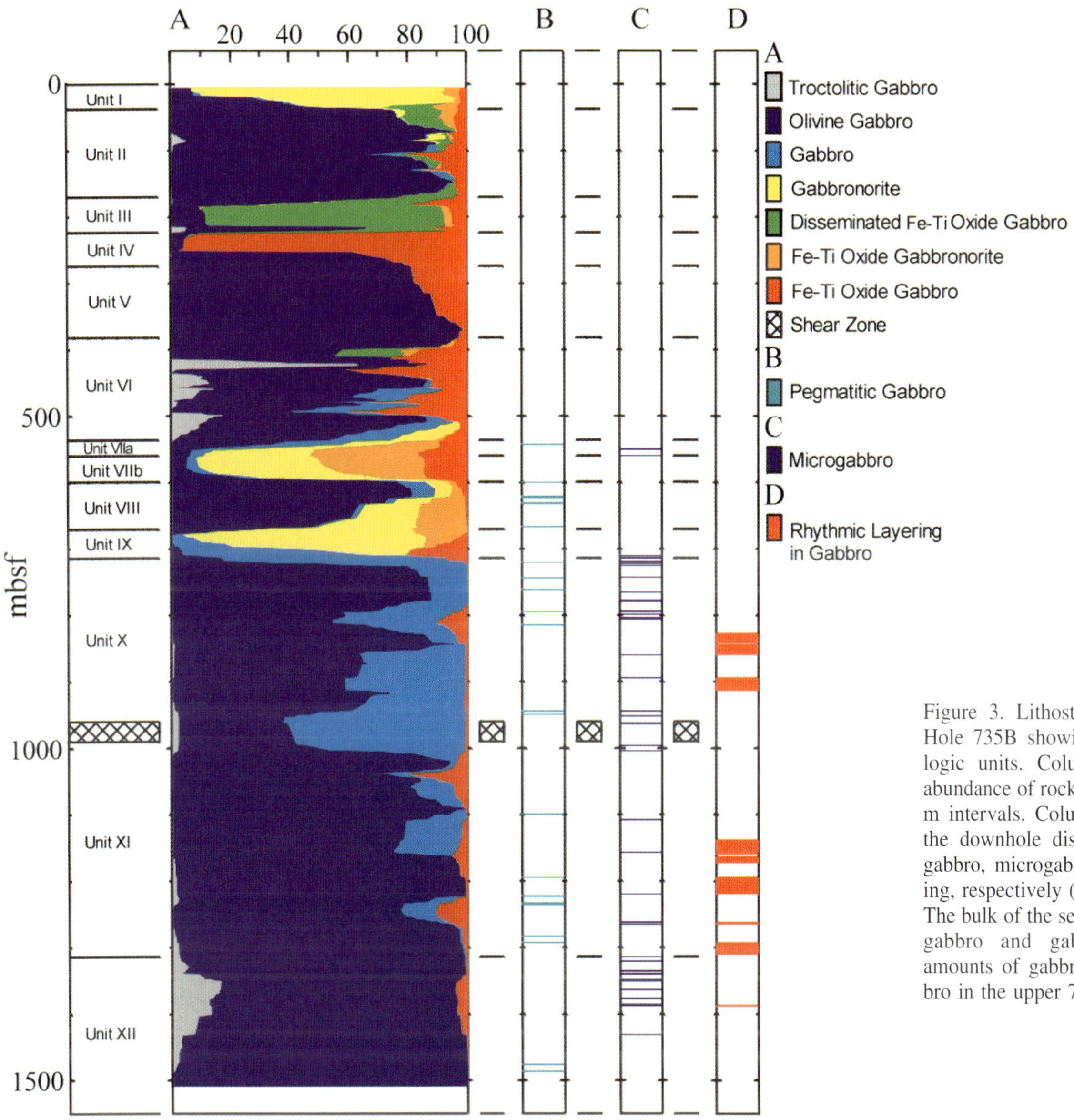

Figure 3. Lithostratigraphic column for Hole 735B showing the 12 major lithologic units. Column A is the relative abundance of rock types averaged over 20 m intervals. Columns B, C, and D give the downhole distribution of pegmatitic gabbro, microgabbro, and igneous layering, respectively (after Dick et al., 1999). The bulk of the section consists of olivine gabbro and gabbro with significant amounts of gabbronorite and oxide gabbro in the upper 700 m.

ridge-transform intersection. Uplift continued until the section reached wave base where it was eroded to form the smooth platform observed today (MacLeod et al., 1999). After truncation by wave action, the section then subsided to its current level.

In 1998, another ODP Hole (1105A) was drilled on the Atlantis Bank ~1.2 km from 735B as part of an engineering leg (179) to test the newly developed hammer-in casing. This hole penetrated 158 m and recovered 118.43 m of gabbro, olivine gabbro, oxide gabbro, and felsic rocks, similar to those in Hole 735B (Pettigrew, Casey, et al., 1999). Study of these rocks will provide valuable information on the three-dimensional structure of the bank.

LITHOLOGY OF THE LOWER OCEAN CRUST

During two legs of drilling, Hole 735B penetrated 1508 m into the lower oceanic crust with an average of 87% recovery. However, low recovery occurred primarily in the upper 100 m of the hole and in narrow shear zones at several depths; thus there were long sections in which core recovery was close to 100% (Robinson, Von Herzen, et al., 1989; Shipboard Scientific Party, 1999).

Twelve major lithologic units have been recognized in Hole 735B; they range in thickness from 44 to 354 m (Fig. 3). The bulk of the section consists of olivine gabbro and lesser amounts of gabbro and troctolitic gabbro, although gabbronorite is abundant in lithologic units I, VII, and IX. Fe-Ti oxide gabbro, Fe-Ti oxide gabbronorite, and disseminated Fe-Ti oxide gabbro are common in the upper 700 m of the section where they are typically associated with zones of crystal-plastic deformation (Dick et al., 1991a) (Fig. 4). These relatively evolved rocks are much less abundant in the lower parts of the section where deformation is also less intense. The average modal compositions of the different rock types are given in Table 1.

Most of the recovered rocks are equigranular with an average grain size between 5 and 15 mm. The principal minerals are similar in size and follow the same general trends. Where there is a disparity among the minerals, the size order is usually pyroxene > plagioclase > olivine. The grain sizes in varitextured zones range from fine (<1

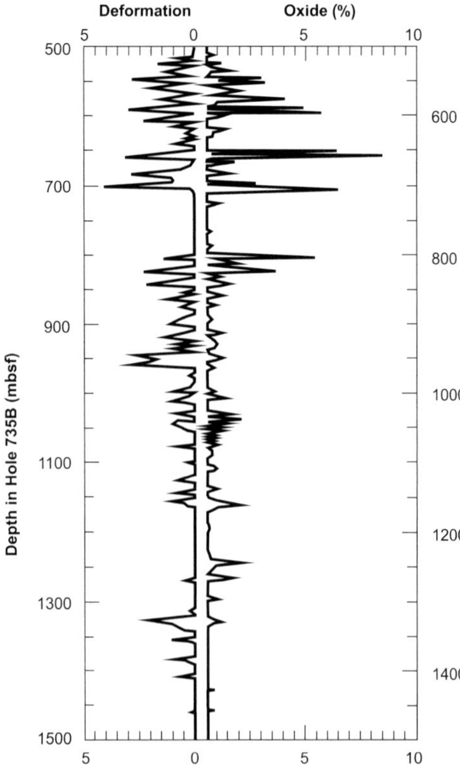

Figure 4. Downhole profile of the lower 1000 m of Hole 735B showing the relationship between intensity of crystal-plastic deformation and abundance of oxides in the core (Shipboard Scientific Party, 1999). Deformation intensity is on an arbitrary scale of 0–5.

mm) to pegmatitic (>30 mm); such zones are scattered throughout the core; grain-size peaks occur at 510, 635, 825, 1100, 1215, 1300, 1425, and 1480 mbsf (Fig. 3).

Grain-size or modal layering is rare, being present in only ~12% of the core (Fig. 3). Grain-size variations may be sharp or gradational over a few tens of centimeters whereas modal layering is typically defined by distinct changes in the proportions of olivine, pyroxene, and plagioclase. Layering may also be defined by textural changes or by weak magmatic foliations.

The dominant coarse-grained olivine gabbro is cut by numerous fine- to medium-grained bodies of microgabbro, gabbronorite, olivine gabbro, and troctolite with either sharp or gradational contacts (Bloomer et al., 1991). Sharp contacts are typically planar with distinctly finer grain sizes whereas gradational contacts swirl and curve through the host rock. These crosscutting bodies may represent channelways along which hot, relatively primitive melts fed upward into the overlying intrusions, or the bodies could be protodikes. One definite, fine-grained diabase dike with planar, chilled margins cuts the sequence at 105 mbsf, and a 0.5-m-thick interval of recrystallized diabase intercalated with highly deformed gneiss at 25 mbsf is interpreted as a dike.

Oxide gabbro, Fe-Ti oxide olivine gabbro, and Fe-Ti oxide gabbronorite make up ~15% of the recovered core, but they are abundant only in the upper 600 m of the section (Fig. 3). Fe-Ti oxides are most abundant in lithologic unit III (primarily disseminated-oxide olivine gabbro) and unit IV (massive-oxide olivine gabbro). Most of these rocks have a weak to strong penetrative foliation due to crystal-plastic deformation. They consist primarily of clinopyroxene, plagioclase and olivine, locally accompanied by hypersthene or pigeonite. Fe-Ti oxides range from <1 to 50 mod% (modal percent) and consist chiefly of ilmenite and magnetite (Natland et al., 1991; Ozawa et al., 1991). The oxides are typically anhedral and occur in thin bands or anastomosing layers that surround the silicate minerals and commonly define a foliation in the rock. They also fill what appear to be pull-apart cracks in large pyroxene and plagioclase grains. Apatite and sphene are closely associated with the oxide bands and form euhedral to subhedral crystals up to several millimeters across.

Although not voluminous (<0.5 vol%), felsic rocks are common and widely distributed in the core. These are mostly leucodiorites, accompanied by lesser amounts of trondhjemite, tonalite, and rare granite. They typically form narrow veins or small, irregular patches of uncertain origin. Many felsic veins are either hydrothermal in origin or show a hydrothermal overprint, but some of the veins and patches have sharp chilled boundaries and retain igneous textures. Most of them consist of sodic plagioclase and minor amphibole, but some contain variable amounts of quartz, diopside, sphene, and zircon.

METAMORPHISM AND ALTERATION

The cored sequence in Hole 735B preserves a complex record of metamorphism and alteration spanning a range from near-solidus to near-ambient temperatures. Granulite- and amphibolite-facies metamorphism is recorded in pyroxene and amphibole gneisses at the top of the section, and lower-temperature background alteration is widespread, particularly in the upper 1000 m. Hydrothermal veins are also abundant in the upper parts of the section and range from high-temperature diopside-rich veins to low-temperature carbonate and zeolite varieties. Background alteration generally ranges from <1% to over 40%, although it reaches nearly 100% in some narrow intervals (Fig. 5). The most abundant and pervasive background alteration occurs in the interval between 500 and 600 mbsf. In general, both metamorphism and alteration decrease downward, and below 1000 m there are long intervals of nearly fresh rock.

Five principal styles of metamorphism and alteration are recorded in the core (Robinson et al., 1991): (1) high-temperature metamorphism associated with zones of crystal-plastic deformation that produced pyroxene and amphibole gneisses; (2) static replacement of olivine and locally orthopyroxene by talc, Mg-amphibole, and secondary magnetite and marginal replacement of pyroxene by brown

TABLE 1. AVERAGE PRIMARY MODAL COMPOSITIONS (%) OF THE MAJOR ROCK TYPES IN HOLE 735B

Rock type	Olivine	Orthopyroxene	Clinopyroxene	Amphibole	Plagioclase	Opaques	Apatite
Troctolite	30	0	3	0	67	<1	0
Troctolitic Gabbro	16	<1	11	<1	73	<1	0
Olivine Gabbro	10	<1	31	<1	59	<1	0
Gabbro	3	<1	38	<1	59	<1	tr
Gabbronorite	3	4	26	1	65	1	tr
Oxide Gabbronorite	1	4	34	<1	52	9	<1
Oxide Gabbro	2	<1	40	<1	52	6	<1

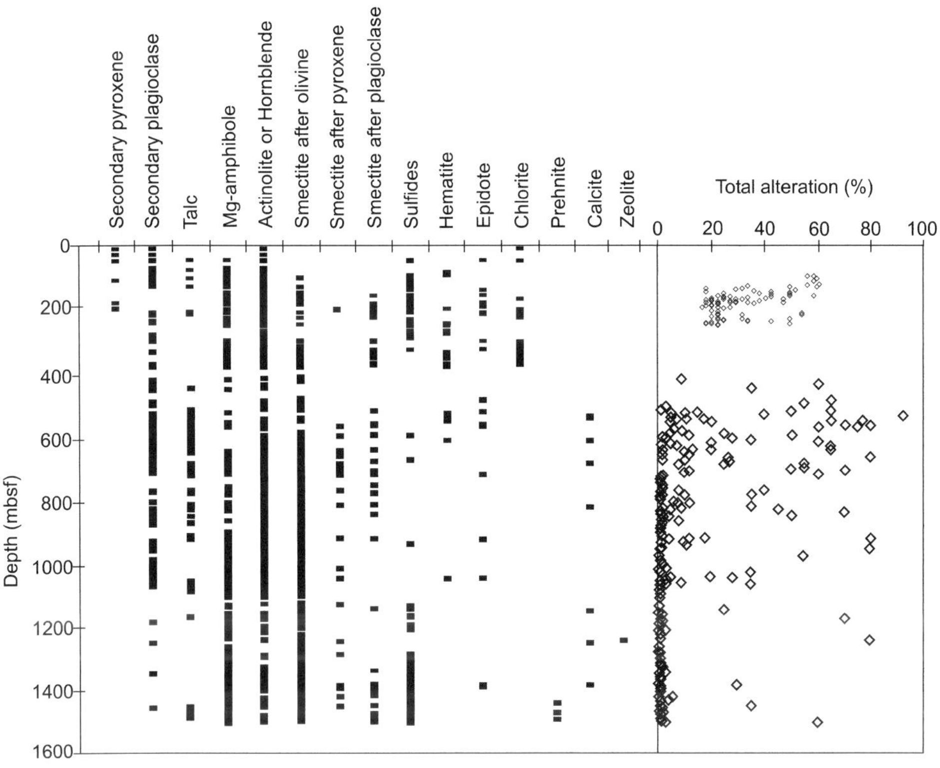

Figure 5. Intensity of background alteration and distribution of secondary minerals in Hole 735B (data from Robinson, unpublished data, and Shipboard Scientific Party, 1999).

and green amphibole; (3) infilling of hydrothermal veins formed during brittle deformation; (4) late-stage, low-temperature replacement of olivine and some pyroxene by carbonate and Fe oxides; and (5) deposition of smectite in subvertical fractures.

Granulite-facies metamorphism (>800–950 °C) is preserved in narrow zones of crystal-plastic deformation where intense shearing has produced bands of olivine and pyroxene neoblasts that alternate with plagioclase-rich bands. A few of these zones are impregnated with Fe-Ti oxides, and many have abundant brown hornblende, indicating that deformation continued as the temperature decreased and the water/rock ratios increased. Amphibole gneisses are particularly well developed in the upper 50 m of the section (lithologic unit I) but also occur in lithologic unit II and in thin shear zones lower in the section. Most of these rocks have a well-developed foliation and consist chiefly of green or brown amphibole and plagioclase, with lesser amounts of Fe-Ti oxides and brown mica. A few samples contain small amounts of actinolitic amphibole and intermediate to sodic plagioclase. Porphyroclasts of pyroxene, many of which are also replaced by amphibole, are preserved in some of the less deformed rocks.

In weakly deformed or undeformed rocks, the background alteration is much less intense and chiefly involves partial replacement of (1) pyroxene by brown and green amphibole; (2) olivine by talc, amphibole, and secondary magnetite; and (3) plagioclase by minor amphibole, smectite, or sodic feldspar. Background alteration decreases downward, and long intervals in the lower part of the core are nearly pristine.

More than 3000 separate veins were identified in the 1508-m-long section, and these make up ~2.5% of the recovered core. Most of the veins occur in the upper 500 m of the section; below this level they decrease rapidly (Fig. 6) Only a few low-temperature carbonate and smectite veins occur in the lower 700 m of the sequence.

Five major categories of veins are recognized on the basis of their mineral constituents: (1) felsic, including plagioclase and plagioclase + amphibole veins; (2) diopside and diopside + plagioclase veins; (3) amphibole veins; (4) carbonate and smectite veins; and (5) chlorite, zeolite, and prehnite veins (Fig. 6). In addition, there are a few small veins of epidote and quartz, but these are very rare. This is a rather arbitrary division because many veins are zoned with high-temperature minerals along the margins and lower-temperature minerals in the centers.

Felsic material that does not exhibit a strong hydrothermal overprint typically occurs as anastomosing veinlets and/or irregular patches as wide as 15 cm. By number, these veins constitute only ~4% of the total, but they account for 45 vol%. Most have sharp boundaries with the host rock, but some, particularly those with a hydrothermal overprint, have alteration halos 1–2 cm wide. They tend to be associated with oxide gabbros and, thus, decrease markedly in abundance in the lower parts of the section.

Dioritic veins consist chiefly of intermediate plagioclase and small amounts of amphibole and 1%–2% of accessory minerals such as apatite, zircon, titanite, and magnetite. Those with more granitic compositions contain small amounts of potassium feldspar and quartz or myrmekite and trace amounts of biotite. Veins with a hydrothermal overprint typically contain strongly zoned plagioclase, abundant brown or green amphibole, and chlorite. In some of these veins, quartz grains are partly replaced with brown smectite.

Plagioclase and plagioclase + amphibole veins are typically associated with the felsic veins and are commonly difficult to distinguish from them. In many cases, the plagioclase veins have splayed off larger felsic veins, and the plagioclase veins nearly always display a strong hydrothermal overprint. Plagioclase is the dominant mineral in these veins, and it commonly shows significant hydrothermal alteration to sodic feldspar, chlorite, zeolite, carbonate, smectite, and prehnite. Most of these veins contain at least small amounts of green amphibole, partly altered to colorless amphibole, chlorite, and secondary magnetite.

Monomineralic amphibole veins are abundant only in the upper 200 m of the section (Fig. 6) where they are closely associated with highly deformed amphibole gneisses in lithologic units I and II. A

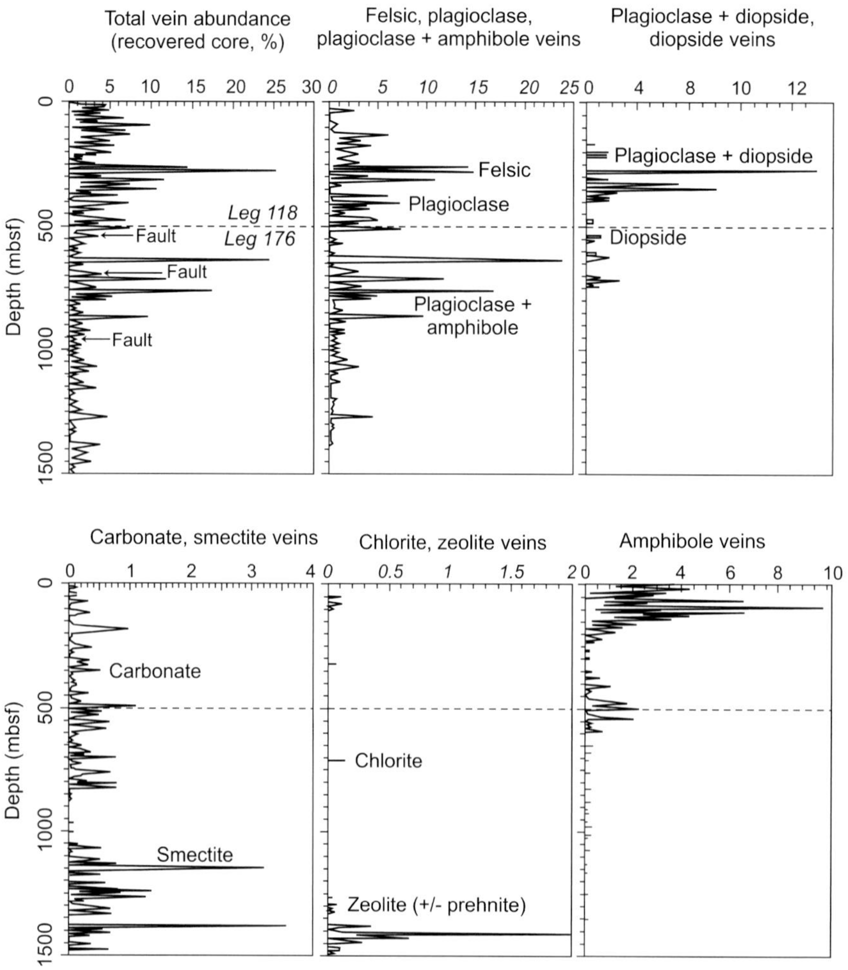

Figure 6. Downhole distribution of veins in ODP Hole 735B by the percentage of core. Note the predominance of relatively high temperature veins in the upper 750 m and the downward decrease in total vein abundance (Leg 118 data from Dick, et al., 1991; Leg 176 data from Shipboard Scientific Party, 1999).

smaller cluster occurs between ~350 and 550 mbsf; amphibole veins virtually disappear in the lower parts of the section. These veins are typically 1–2 mm wide and consist almost entirely of green or brownish-green amphibole, locally accompanied by small amounts of plagioclase, titanite, and chlorite. Most of them show little or no hydrothermal alteration although locally the amphibole is rimmed by chlorite.

Diopside and plagioclase + diopside veins are restricted to the interval between ~200 and 750 mbsf where they form several small clusters (Fig. 6). Typical examples have relatively large (up to 2 mm), euhedral to subhedral crystals of diopside along the vein margins, followed inward by mixtures of amphibole and plagioclase. A few have rims of plagioclase along the vein wall. Many of the diopside crystals are strongly zoned, filled with small, dark inclusions, and rimmed by amphibole. Late-stage, lower-temperature minerals such as prehnite, carbonate, smectite, chlorite, and various zeolites are commonly present in the vein groundmass.

Carbonate and smectite veins are irregularly distributed through the section except for a 200-m-thick interval between ~850 and 1050 mbsf (Fig. 6). Most of these are monomineralic smectite veins, ranging from hairline cracks to ~1 mm in width, although mixed smectite + carbonate veins are locally present. The carbonate veins tend to occur in narrow clusters and are almost always associated with zones of late-stage oxidation of olivine and orthopyroxene.

Zeolite veins occur only near the base of the section between ~1300 and 1500 mbsf (Fig. 6). They are typically 1–2 mm wide and consist chiefly of natrolite, prehnite, and smectite.

GEOCHEMISTRY

The rocks recovered from Hole 735B span a wide compositional range from troctolite to granite; olivine gabbro and gabbro are the most voluminous. Felsic and silicic rocks are present in very small quantities and are restricted to net veins and irregular groundmass patches. Fe-Ti oxide gabbros, on the other hand, are common and voluminous and display the most extreme lithologic and geochemical variations in the section. Magnesium numbers [defined as 100 × Mg/(Mg + Fe^{2+})] range from 87 in the most primitive troctolites to 24 in the most evolved felsic units (Fig. 7).

Troctolites, particularly those between 500 and 520 mbsf, are the most primitive rocks recovered. They are characterized by high Mg#s (>80), low SiO_2, relatively high Al_2O_3, low Na_2O and CaO (Table 2), and very low incompatible elements such as Ti, Zr, and Y. Locally, the troctolites pass imperceptibly into troctolitic gabbros and then olivine gabbros with slightly increasing SiO_2, CaO, TiO_2, Zr, and Y and decreasing Al_2O_3 and MgO. The troctolites between 419 and 520 m, however, are clearly intrusive into the olivine gabbros and form a distinct group of extremely primitive rocks.

Gabbros are almost indistinguishable geochemically from olivine gabbros except for having slightly lower MgO and slightly higher CaO, which together reflect the decrease in olivine and the increase in plagioclase.

Oxide gabbros are characterized by high Fe_2O_3 and TiO_2, relatively high Na_2O and P_2O_5, and low MgO, Al_2O_3, and CaO (Table 2). These are relatively evolved rocks with Mg#s generally between

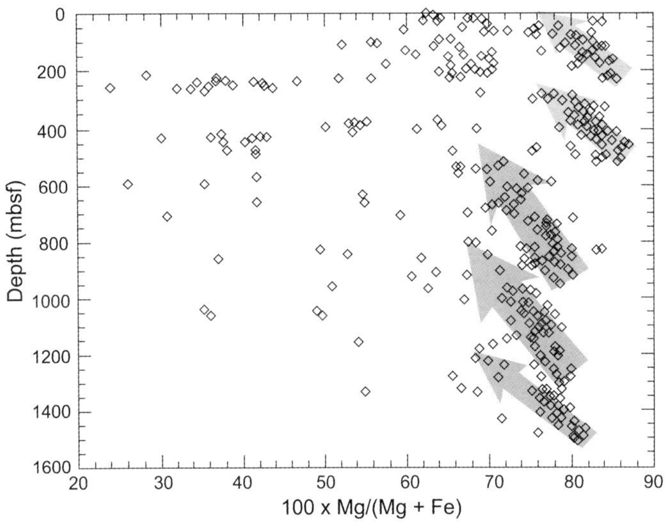

Figure 7. Mg# [defined as 100 × Mg/(Mg + Fe^{2+})] vs. depth for samples from Hole 735B. Mg#s are calculated by assuming that FeO equals 85% of the total Fe. Shaded arrows define five magmatic sequences that become more evolved upward. Each sequence is tentatively interpreted as an intrusive body made up of many smaller igneous units (after Dick et al., 1999, and Shipboard Scientific Party, 1999).

26 and 70. In the upper 500 m of the section, oxide olivine gabbros have an average Mg# of 29.6. Felsic samples have notably high SiO_2, Al_2O_3, Na_2O, and K_2O and low Fe_2O_3, MgO, and CaO. They have somewhat higher Mg#s than the oxide gabbros and the lowest Ca#s [defined as 100 × Ca/(Ca + Al)] of all the recovered rocks.

Overall, the section appears to have been constructed from as many as five major intrusions, each of which has differentiated upward from primitive to evolved compositions (Fig. 7). Boundaries between the five units lie at ~225 m, ~560 m, ~950 m, and ~1350 m depth. Although Mg#s are imperfect differentiation indices in Fe-Ti oxide gabbros, there are fairly clear fractionation patterns with depth. The very low Mg#s (<40) reflect, in part, addition of Fe-Ti oxides to the host gabbros, although a few represent highly evolved felsic rocks. However, even when these late-stage, oxide-rich rocks, which intrude or crosscut the section along shear zones, are removed from consideration, the main body of gabbro, disseminated-oxide gabbro, olivine gabbro, and troctolite shows a cyclic geochemistry (Fig. 7). In the main body of rocks, Mg#s range from a high of 87 in troctolites to a low of 63 in disseminated-oxide gabbros. It is noteworthy that the initial melts in the lower three cycles are less primitive than those in the upper two cycles (Fig. 7), a feature consistent with the observed variations in bulk-rock compositions.

The geochemical boundary at 560 m corresponds to a major fault and that at 960 m to a major shear zone (Shipboard Scientific Party, 1999).

STRUCTURAL GEOLOGY

The Hole 735B section exhibits a wide range of structural features, with considerable variation in the style and intensity of deformation. Highly deformed amphibole gneisses and mylonites alternate with essentially undeformed sections, but there is a general decrease downward in the intensity and distribution of deformation. Faults range from major features to narrow shear zones and show both normal and reverse sense of shear.

Primary magmatic foliations are locally present but are usually overprinted by crystal-plastic deformation.

Most of the rocks from Hole 735B are coarse- to medium-grained with hypidiomorphic-granular, intergranular, or subophitic textures. Where present, magmatic foliations are normally weak and are defined by the parallel alignment of platy minerals, particularly plagioclase or, more rarely, clinopyroxene. The magmatic foliations have no preferred orientation although most dip from 20° to 50° in the core (Shipboard Scientific Party, 1999).

Crystal-plastic deformation is highly localized and is most pronounced in the intervals between 0 and 50 mbsf and between 450 and 600 mbsf (Fig. 8). Most of the rocks (77%) have no crystal-plastic deformation, and only 7% have more than a weak foliation.

TABLE 2. AVERAGED MAJOR OXIDE COMPOSITIONS OF SAMPLES FROM HOLE 735B AS A FUNCTION OF ROCK TYPE AND DEPTH IN THE DRILL CORE

Rock type	Depth (mbsf)	N†	SiO_2	TiO_2	Al_3O_2	Fe_2O_3§	MnO	MgO	CaO	Na_2O	K_2O	P_2O_5
*Felsic**		3	64.06	1.62	18.49	1.80	0.05	1.27	4.93	6.24	1.43	0.11
Oxide gabbro	520–705	5	46.55	2.97	12.24	17.69	0.27	6.06	9.99	3.27	0.10	0.86
	800–960	6	47.33	2.84	12.80	14.81	0.22	7.36	11.20	2.86	0.06	0.53
	960–1320	2	48.30	2.31	15.01	13.38	0.21	6.13	10.06	3.40	0.08	1.12
Gabbro	520–705	4	52.21	0.52	16.72	7.40	0.13	7.38	12.13	3.43	0.06	0.02
	705–800	2	51.24	0.47	16.37	6.78	0.13	9.03	13.16	2.76	0.03	0.02
	800–960	7	51.04	0.43	16.38	6.93	0.14	8.97	13.21	2.84	0.03	0.03
	960–1320	6	51.75	0.38	16.37	6.45	0.13	8.77	13.15	2.97	0.03	0.01
	1320–1503	7	51.55	0.35	17.94	5.16	0.11	8.37	13.52	2.95	0.03	0.01
Olivine gabbro	520–705	12	51.50	0.44	16.18	7.78	0.14	8.85	11.91	3.13	0.06	0.02
	705–800	11	50.65	0.34	16.46	6.93	0.13	10.16	12.67	2.63	0.02	0.02
	800–960	22	50.74	0.44	15.91	7.10	0.13	10.23	12.71	2.69	0.03	0.03
	960–1320	40	51.04	0.42	16.37	7.16	0.13	9.41	12.56	2.85	0.03	0.02
	1320–1503	18	51.14	0.36	16.99	5.80	0.12	9.29	13.62	2.65	0.02	0.02
Troctolitic gabbro	800–960	2	50.95	0.36	17.29	6.19	0.12	9.26	13.05	2.68	0.08	0.02
	1320–1503	3	49.29	0.28	17.90	7.84	0.13	10.70	11.11	2.72	0.02	0.01
Troctolite	500–520	4	46.99	0.13	18.21	6.52	0.10	15.52	10.65	1.81	0.06	0.01

Note: Depth intervals correspond to the chemical-stratigraphy units (see Shipboard Scientific Party, 1999). Major element compositions are all normalized to 100%. (Data are from Shipboard Scientific Party, 1999).

* Average of 3 felsic samples.

† N = number of samples used to calculate the average; mafic rock samples have been excluded.

§ Total Fe as Fe_2O_3.

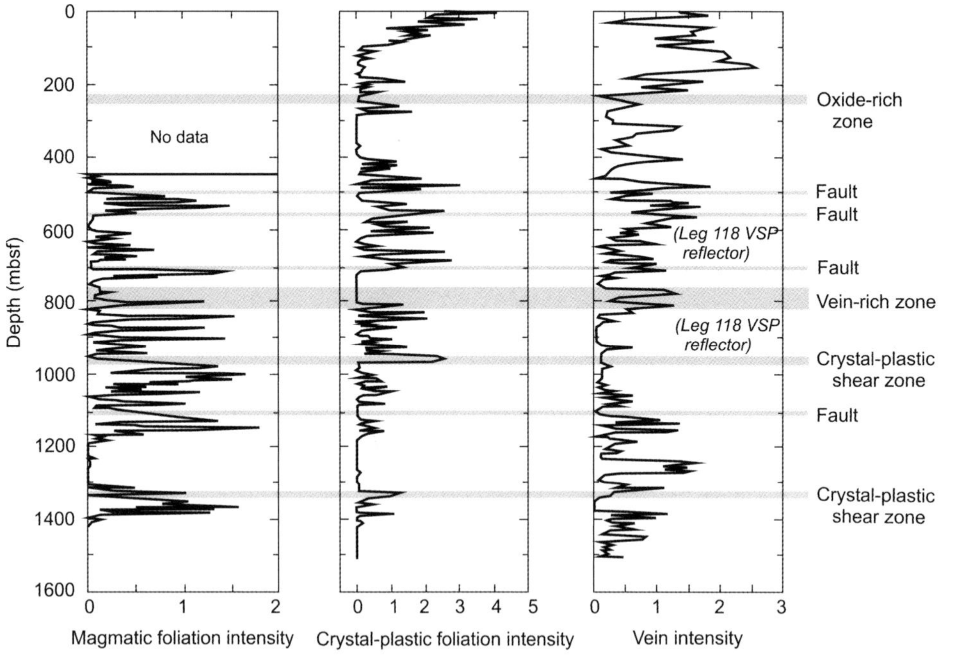

Figure 8. Downhole variations in the intensity of magmatic foliation, crystal-plastic deformation, and vein intensity in Hole 735B (Shipboard Scientific Party, 1999). VSP—vertical seismic profile.

Long intervals of core have no deformation whatsoever, and these intervals increase in number and length with depth.

More than 150 zones of intense crystal-plastic deformation have mylonitic fabrics. These zones are typically narrow bands a few millimeters to a few centimeters thick, although thicker intervals are present in the upper 50 m of the core. In most cases, the transition from mylonitic fabric to undeformed rock takes place over a few centimeters. Like the magmatic foliations, crystal-plastic foliations show no systematic variations in dip with depth in the hole; however, there is a strong preferred orientation to the deformation fabric. When oriented relative to the remanent magnetic vector, the crystal-plastic foliations dip north toward the ridge axis (Shipboard Scientific Party, 1999).

In general, the intensities of the crystal-plastic and magmatic fabrics show a good positive correlation except in areas where magmatic foliations have been completely overprinted. There are also a few zones with strong magmatic foliations that show little or no overprinting by tectonic fabrics. Shear zones with crystal-plastic deformation fabrics may show either a normal or reverse sense of shear in the core reference frame.

Both high- and low-temperature brittle deformation features are also present in the Hole 735B rocks. High-temperature brittle deformation is recorded by numerous felsic veins and magmatic breccias. Although many of these veins and breccias have a hydrothermal overprint, their original mineral assemblage (plagioclase ± clinopyroxene ± amphibole ± oxides ± biotite ± quartz ± alkali feldspar ± titanite ± apatite ± zircon) suggests a magmatic origin. Typically, the felsic material forms anastomosing vein networks and irregular patches in which individual veins have no preferred orientation.

Lower-temperature brittle deformation features are mostly discrete faults and associated cataclasites. Major faults of unknown displacement occur at 560 and 690–700 mbsf, and smaller zones of cataclasis occur at 490, 1076, and 1100–1120 mbsf. The major faults overprint granulite- and amphibolite-facies shear zones and were active during greenschist-facies conditions (Shipboard Scientific Party, 1999).

In addition to the major faults, there are more than 700 discrete "microfaults" or zones of cataclasis with associated breccia and gouge in the section. These are irregularly distributed in the core: most occur in the upper 750 m, and there are almost none below 1400 mbsf. Commonly, these zones of brittle deformation are filled with calcite, amphibole, and/or smectite, and many have small slickensides.

As noted above, there is commonly an association between the occurrence of crystal-plastic shear zones and the distribution of Fe-Ti oxides (Fig. 4) although some deformed intervals have no oxides and some oxide-rich zones are undeformed. This association between the shear zones and oxides suggests that crystal-plastic deformation tends to occur in zones rich in fractionated melt. However, once deformation is initiated, Fe-Ti oxide–rich melts may migrate into the shear zones, causing increased oxide concentration. Continued deformation may cause shearing and fracturing of the oxide-rich bands. An absence of annealing in the sheared and fractured oxides suggests that the late-stage deformation occurred in the solid state, well below the gabbro solidus (cf. Agar and Lloyd, 1997).

PALEOMAGNETISM

The transverse ridge on the east side of the Atlantis II Fracture Zone is crossed by a number of well-defined magnetic anomalies with the Atlantis Bank lying at anomaly 5r.2n (Dick et al., 1991c). The presence of these anomalies in an area underlain by gabbro and peridotite indicates that the lower ocean crust and (perhaps) the upper mantle are significant sources of marine magnetic anomalies.

The average natural remanent magnetization (NRM) of minicores from Hole 735B is 2.5 A/m (Shipboard Scientific Party, 1999), and there is no significant decrease downhole (cf. Pariso and Johnson, 1993) despite a significant downward decrease in the proportion of oxide gabbros. There is, however, considerably less scatter of NRM values in the lower 1000 m of the hole (Fig. 9). Downhole logs of magnetic susceptibility show numerous spikes in the record, most of which correlate closely with thin bands of oxide gabbro or felsic veins (Fig. 10).

It is clear that the gabbros sampled in Hole 735B can easily account for the observed seafloor magnetic anomalies on the Atlantis Bank (Kikawa and Ozawa, 1992). A gabbro layer 3–5 km thick with

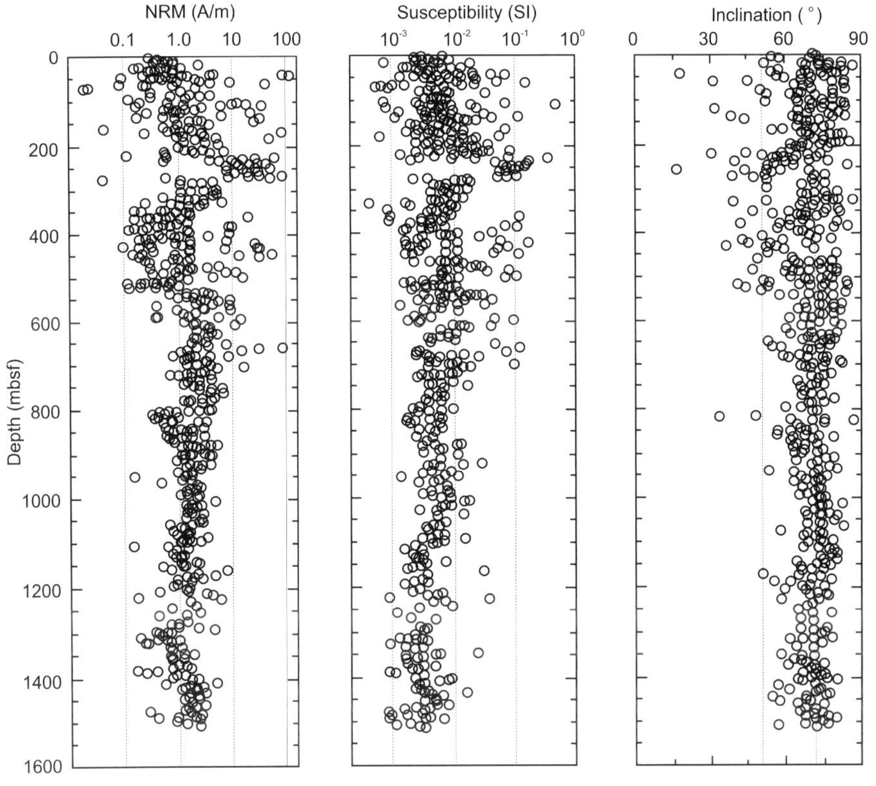

Figure 9. Downhole variations in stable thermal remanent magnetization (NRM), magnetic susceptibility, and stable inclinations in Hole 735B (Shipboard Scientific Party, 1999) (Leg 118 data after Kikawa and Pariso, 1991).

an average magnetization of 2.5 A/m could account for most seafloor anomalies observed elsewhere. Thus, it seems clear that gabbroic crust can contribute significantly to marine magnetic anomalies and locally may even dominate over the contribution of lavas and sheeted dikes. However, the rapid acquisition of thermal remanence of the Hole 735B gabbros resulting from their uplift and unroofing at the ridge-transform intersection may not be typical of the Southwest Indian Ridge as a whole. Farther from the ridge-transform intersection, the gabbros would have cooled beneath a 1–2-km-thick carapace of lavas and sheeted dikes and thus would have acquired their remanence more slowly.

Hole 735B was spudded into a negative magnetic anomaly (5r.2n), and the primary remnant magnetism has an average inclination of −71°, uncorrected for any deviation of the hole (Shipboard Scientific Party, 1999). Although there is some scatter in the upper 500 m of the section, most values cluster about the average, and there are no significant downhole trends (Fig. 9). This result indicates that the section has not been tectonically disrupted since it cooled below the Curie point at ~580 °C. However, the average inclination is ~20° steeper than that expected for the site (51°), suggesting some rotation of the section as a whole.

PHYSICAL PROPERTIES

Because the Atlantis Bank has been unroofed by some unspecified amount, its exact position in the ocean crust is not known. However, both the compressional-wave velocities and densities of the plutonic rocks are characteristic of oceanic layer 3. Compressional-wave velocities of minicores ranged from a low of 5.53 to a high of 7.83 km/s (Robinson, Von Herzen, et al., 1989). However, the average minicore velocity for the entire section is 6.78 ± 0.29 km/s, which corresponds closely to the average in situ velocity measured by vertical seismic profiling (6.4–6.5 km/s; Robinson, Von Herzen, et al., 1989). In situ velocities are normally somewhat lower than minicore velocities because of the presence of faults, fractures, breccia zones, and highly altered zones that would not normally be represented in minicores. However, both the minicore and in situ velocities are typical for gabbro and metagabbro and well within the range for oceanic layer 3 (6.5–7.0 km/s).

Bulk densities also show considerable scatter, particularly in the upper 700 m of the section (Fig. 11). The mean bulk density is 2.98 ± 0.10 g/cm^3, and the mean grain density is 2.99 ± 0.11 g/cm^3, very close to the density of typical olivine gabbro (Shipboard Scientific Party, 1999). The highest densities (up to 3.21 g/cm^3) are from oxide gabbros with abundant ilmenite and magnetite, and the lowest are from olivine gabbros with high proportions of plagioclase. The observed densities are characteristic of layer 3 gabbros.

SUMMARY AND DISCUSSION

Because it is normally covered by 1–2 km of lavas and sheeted dikes, the lower part of the ocean crust is largely inaccessible. Attempts to penetrate the volcanic carapace by drilling have proved expensive and fruitless. Because the Atlantis Bank is a segment of lower ocean crust that has been tectonically unroofed, it is readily accessible to dredging and drilling. Thus far, seven research cruises have visited the area including ODP Legs 118, 176, and 178, two survey cruises, a seismic survey cruise, and a diving cruise.

During ODP Legs 118 and 176, *JOIDES Resolution* drilled Hole 735B to a combined depth of 1508 mbsf with an average core recovery of 87%. For the first time, a significant part of the lower ocean crust has been sampled in situ; this success has allowed us to determine the stratigraphy of the crust and to assess the tectonic, magmatic, and metamorphic processes involved in crustal construction at an ultra-slow-spreading ridge.

The ocean crust is believed to be relatively thin and discontinuous at slow-spreading ridges (<10 mm/yr half rate) or near large transform faults (Reid and Jackson, 1981; Mutter and Detrick, 1984;

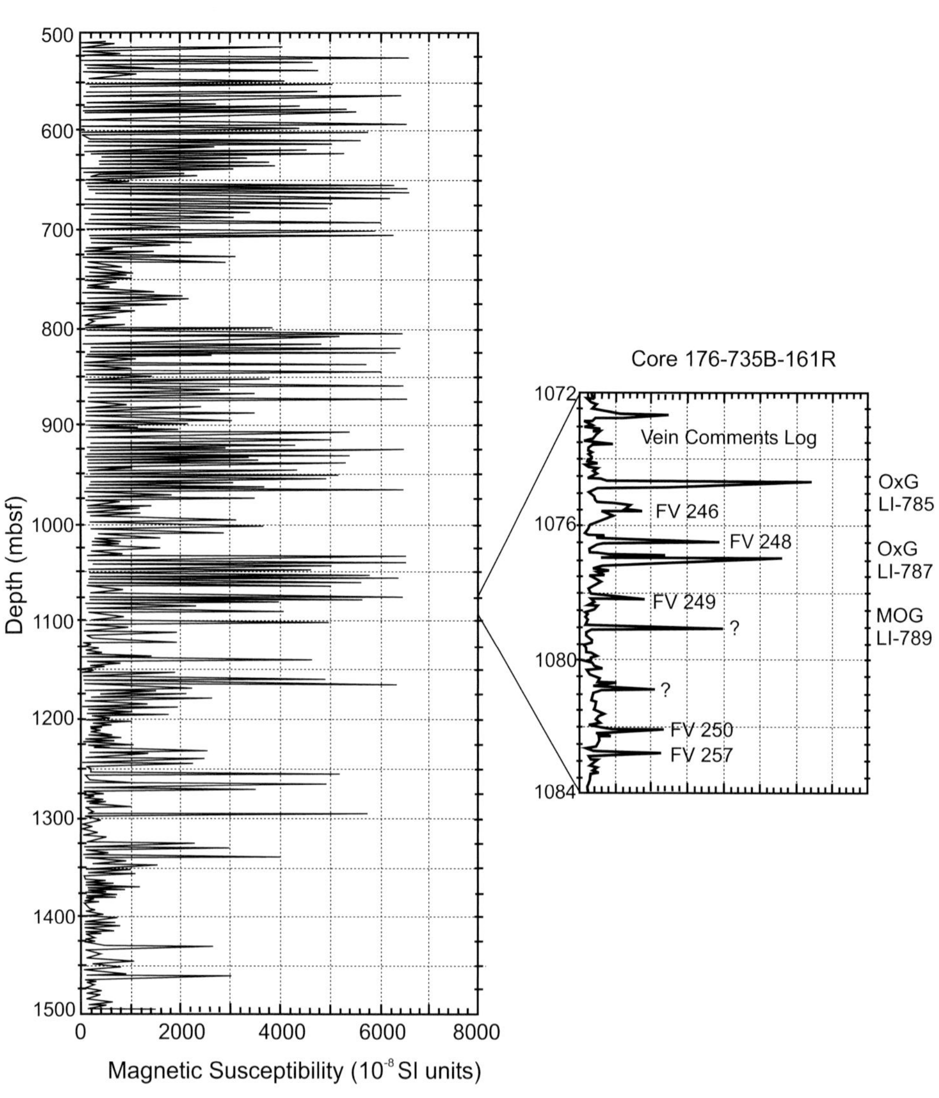

Figure 10. Magnetic susceptibility log for Hole 735B between 500 and 1508 mbsf. Enlarged part of the column on the right shows the close correlation between magnetic susceptibility and oxide layers and felsic veins. FV—felsic vein, OxG—oxide gabbro, MOG—mixed olivine gabbro (Shipboard Scientific Party, 1999).

Whitehead et al., 1984; Dick, 1989; Cannat, 1993; Bown and White, 1994; Tucholke and Lin, 1994). Large, steady-state magma chambers are absent at slow-spreading ridges (Whitehead et al., 1984; Detrick et al., 1990; Dick et al., 1991a, 1992; Bloomer and Meyer, 1992; Sinton and Detrick, 1992), and the internal structure of the crust is controlled as much by dynamic processes of tectonism and alteration as by igneous processes (Robinson, Von Herzen, et al., 1989; Dick et al., 1992; Cannat, Karson, et al., 1995).

On the basis of the seismic velocity structure beneath the Atlantis Bank, the Moho discontinuity has been placed at a depth of ~6 km below sea level (Fig. 2) (Muller et al., 1997). However, the fact that the gabbro-peridotite transition lies at a depth of ~2500 m on the western edge of the platform suggests that the crust-mantle boundary is much shallower than the Moho. If so, the low velocities below the Atlantis Bank probably reflect extensive serpentinization of upper-mantle peridotite. This inference suggests that serpentine can be a significant component of oceanic layer 3 and that the crust-mantle boundary in the oceans may not always correspond to the seismic Moho.

The bulk of the crustal section consists of olivine gabbro and gabbro with lesser amounts of troctolite and gabbronorite. Although there are many small-scale igneous units, the section appears to have been constructed from five major intrusions, each of which shows an upward fractionation from troctolite to gabbro, gabbronorite, and disseminated-oxide gabbro (Fig. 7). A striking feature of the section is the abundance of oxide-rich rocks in the upper part of the sequence and their common association with zones of crystal-plastic deformation.

The calculated bulk composition of the 1508-m-thick section indicates that the gabbros are close in composition to primitive or moderately differentiated basalts (Dick et al., 1999). However, calculated compositions for the upper, middle, and lower 500-m-thick sections show significant differences, with strong enrichment in TiO_2 and total Fe in the upper section, accompanied by a decrease in Mg#. However, except for an upward increase in Cr, trace element contents remain essentially constant through the 1508-m-thick section. This decoupling of major and trace elements appears to preclude formation of the oxide gabbros by simple upward fractionation of a large basaltic intrusion (Dick et al., 1999). Rather they must be of hybrid origin formed by intrusion of late-magmatic melts into a crystal mush of olivine gabbro. These liquids could have come from lower in the section or have been intruded laterally along the ridge axis. Mass balancing of these late-magmatic liquids with reasonable parental-magma compositions (Dick et al., 1999) suggests that there must be a considerable thickness of primitive cumulates that have yet to be sampled. These could lie deeper in the crustal section or in the upper mantle, they could have been removed by faulting, or they could be

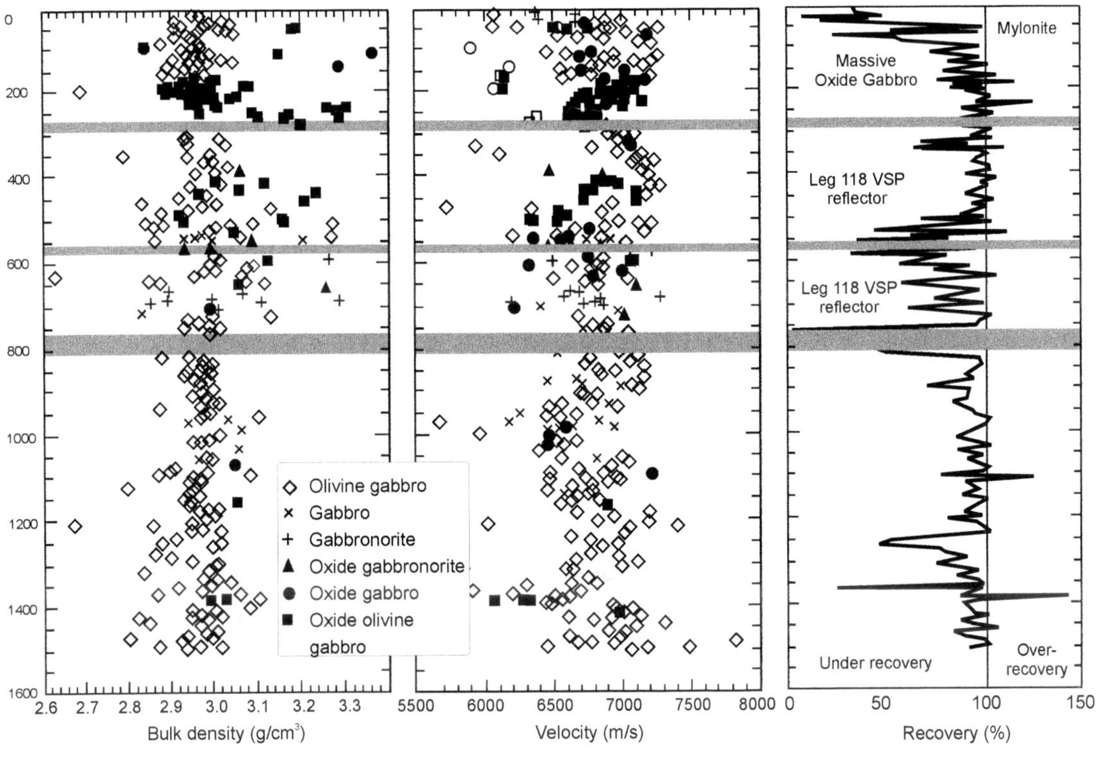

Figure 11. Downhole variations in bulk density and compressional-wave velocity for minicore samples from Hole 735B. Percent recovery is also given. Horizontal bands are oxide-rich or vein-rich zones (Shipboard Scientific Party, 1999) (Leg 118 data after Iturrino et al., 1991). VSP—vertical seismic profile.

located closer to the midpoint of the paleo–volcanic segment if the oxide-rich melts migrated along the axis.

The absence of any eruptive equivalents to the oxide gabbros (ferrobasalts) on Southwest Indian Ridge suggests that these melts represent late-magmatic liquids trapped in the crystallizing olivine gabbros. This interpretation is supported by the presence of local patches of oxide gabbro in undeformed olivine gabbro deep in the hole that may be local pockets of such melt.

The close association of oxide gabbros and crystal-plastic shear zones suggests that the crystal-plastic deformation developed preferentially in zones rich in fractionated melt. However, the abundance of oxides in some zones strongly suggests that once deformation was initiated, melt migrated into and along these zones, causing increased oxide concentrations.

Metamorphism and alteration also appear to have been strongly influenced by deformation. High-temperature (750 to 850 °C) granulite- and amphibolite-facies metamorphism is closely associated with zones of intense shearing. Because the brittle-ductile deformation and associated metamorphism took place at high temperatures, it must have occurred shortly after the rocks crystallized beneath the ridge axis. As the section migrated away from the ridge axis and cooled, brittle deformation became dominant, leading to formation of abundant veins (Robinson et al., 1991). High-temperature diopside, amphibole, and felsic veins are concentrated in the upper parts of the sequence where crystal-plastic deformation was most intense; low-temperature carbonate, zeolite, and smectite veins occur chiefly in the lower, least-deformed parts of the core. The zones of oxidative alteration are invariably associated with these low-temperature veins and must reflect migration of seawater along late-stage cracks formed after emplacement of the Atlantis Bank into the transverse ridge.

It is clear that Hole 735B sampled a section of lower oceanic crust with the seismic characteristics and physical properties characteristic of layer 3. The presence of well-defined magnetic anomalies on the Atlantis Bank clearly demonstrates that lower-crustal gabbros can contribute significantly to seafloor magnetic anomalies. Indeed, a crustal section 3–5 km thick with the average magnetization measured in Hole 735B gabbros (2.5 A/m) could account for most seafloor anomalies measured elsewhere.

The crustal section sampled in Hole 735B differs significantly from the crustal sections in ophiolites such as described by the Penrose Conference Participants (1972). Notable differences include the paucity of grain-size or phase layering in Hole 735B, the abundance of high-temperature shear zones with well-developed amphibolite gneisses, and the abundance of oxide gabbros, most of which are distributed along zones of intense shear. Wehrlites, which are common in many classic ophiolites, are absent, and the bulk geochemistry of the section is distinct from that of most ophiolites, which typically have an arc or supra-subduction zone signature. Some of the features observed in Hole 735B gabbros, particularly the abundance of high-temperature shear zones, have been reported in Alpine and Ligurian ophiolites, and more similarities may be revealed by careful comparison of the two (Ishiwatari, 1985; Lagabrielle and Cannat, 1990). However, the likelihood seems small that sections of cold ocean lithosphere formed by slow spreading—like the lithosphere at 735B—would be emplaced on land as ophiolites. Thus, there may be no direct on-land equivalent of the Atlantis Bank crust.

REFERENCES CITED

Agar, S.M., and Lloyd, G.E., 1997, Deformation of Fe-Ti oxides in gabbroic shear zones from the MARK area, *in* Karson, J.A., Cannat, M., Miller, D.J., and Elthon, D., eds., Proceedings of the Ocean Drilling Program, Scientific Results: College Station, Texas, Ocean Drilling Program, v. 153, p. 123–141.

Bloomer, S.H., and Meyer, P.S., 1992, Mid-ocean ridges: Slimline magma chambers; discussion: Nature, v. 357, p. 117–118.

Bloomer, S.H., Meyer, P.S., Dick, H.J.B., Ozawa, K., and Natland, J.H., 1991, Textural and mineralogic variations in gabbroic rocks from Hole 735B, *in* Von Herzen, R.P., Robinson, P.T., et al., Proceedings of the Ocean Drilling

Program, Scientific Results: College Station, Texas, Ocean Drilling Program, v. 118, p. 21–39.

Bown, J.W., and White, R.S., 1994, Variation with spreading rate of oceanic crustal thickness and geochemistry: Earth and Planetary Science Letters, v. 121, p. 435–449.

Cannat, M., 1993, Emplacement of mantle rocks in the seafloor at mid-ocean ridges: Journal of Geophysical Research, v. 98, p. 4163–4172.

Cannat, M., Karson, J.A., et al., 1995, Proceedings of the Ocean Drilling Program, Initial Reports: College Station, Texas, Ocean Drilling Program, v. 153, 798 p.

Detrick, R.S., Mutter, J.C., Buhl, P., and Kim, I.I., 1990, No evidence from multichannel seismic reflection data for a crustal magma chamber in the MARK area on the Mid-Atlantic Ridge: Nature, v. 347, p. 61–64.

Detrick, R.S., Collins, J., Stephen, R., and Swift, S., 1994, In situ evidence for the nature of the seismic layer 2–3 boundary in ocean crust: Nature, v. 370, p. 288–290.

Dick, H.J.B., 1989, Abyssal peridotites, very slow spreading ridges, and ocean ridge magmatism, *in* Saunders, A.D., and Norry, M.J., eds., Magmatism in the ocean basins: Geological Society of London Special Publication 42, p. 71–105.

Dick, H.J.B., Meyer, P.S., Bloomer, S., Kirby, S., Stakes, D., and Mawer, C., 1991a, Lithostratigraphic evolution of an in-situ section of oceanic layer 3, *in* Von Herzen, R.P., Robinson, P.T., et al., Proceedings of the Ocean Drilling Program, Scientific Results: College Station, Texas, Ocean Drilling Program, v. 118, p. 439–538.

Dick, H.J.B., Schouten, H., Meyer, P.S., Gallo, D.G., Berg, H., Tyce, R., Patriat, P., Johnson, K., Snow, J., and Fisher, A., 1991b, Bathymetric map of the Atlantis II Fracture Zone, Southwest Indian Ridge, *in* Von Herzen, R.P., Robinson, P.T., et al., Proceedings of the Ocean Drilling Program, Scientific Results: College Station, Texas, Ocean Drilling Program, v. 118, foldout map, scale: 1:24000.

Dick, H.J.B., Schouten, H., Meyer, P.S., Gallo, D.G., Berg, H., Tyce, R., Patriat, P., Johnson, K.T.M., Snow, J., and Fisher, A., 1991c, Tectonic evolution of the Atlantis II Fracture Zone, *in* Von Herzen, R.P., Robinson, P.T., et al., Proceedings of the Ocean Drilling Program, Scientific Results: College Station, Texas, Ocean Drilling Program, v. 118, p. 359–398.

Dick, H.J.B., Robinson, P.T., and Meyer, P.S., 1992, The plutonic foundation of a slow-spreading ridge, *in* Duncan, R.A., Rea, D.K., Weissel, J.K., von Rad, U., and Kidd, R.B., eds., The Indian Ocean: A synthesis of results from the Ocean Drilling Program: American Geophysical Union, Geophysical Monograph 70, p. 1–50.

Dick, H.J.B, Natland, J.H., Alt, J.C., Bach, W., Bideau, D., Gee, J.S., Haggas, S., Hertogen, J.G.H., Hirth, G., Holm, P.M., Ildefonse, B., Iturrino, G.J., John, B.E., Kelley, D.S., Kikawa, E., Kingdon, A., LeRoux, P.J., Maeda, J., Meyer, P.S., Miller, D.J., Naslund, H.R., Niu., Y., Robinson, P.T., Snow, J., Stephen, R.A., Trimby, P.A., Worm, H.-U., and Yoshinobu, A., 1999, A long in-situ section of the lower ocean crust: Results of ODP Leg 175 Drilling at the Southwest Indian Ridge: Earth and Planetary Science Letters, v. 179, p. 31–51.

Fisher, R.L., and Sclater, J.G., 1983, Tectonic evolution of the southwest Indian Ocean since the mid-Cretaceous: Plate motions and stability of the pole of Antarctica/Africa for at least 80 My: Geophysical Journal of the Royal Astrological Society, v. 73, p. 553–576.

Fisher, R.L., Dick, H.J.B., Natland, J.H., and Meyer, P.S., 1986, Mafic/ultramafic suites of the slowly spreading southwest Indian Ridge: Protea exploration of the Antarctic Plate Boundary, 24°47′E: Ophioliti, v. 11, p. 147–178.

Ishiwatari, A., 1985, Alpine ophiolites: Product of low-degree mantle melting in a Mesozoic transcurrent rift zone: Earth and Planetary Science Letters, v. 76, p. 93–108.

Iturrino, G.J., Christensen, N.I., Kirby, S., and Salisbury, M.H., 1991, Seismic velocities and elastic properties of oceanic gabbroic rocks from Hole 735B, *in* Von Herzen, R.P., Robinson, P.T., et al., Proceedings of the Ocean Drilling Program, Scientific Results: College Station, Texas, Ocean Drilling Program, v. 118, p. 227–244.

Kikawa, E., and Ozawa, K., 1992, Contribution of oceanic gabbros to sea-floor spreading magnetic anomalies: Science, v. 258, p. 796–799.

Kikawa, E., and Pariso, J.E., 1991, Magnetic properties of gabbros from Hole 735B, Southwest Indian Ridge, *in* Von Herzen, R.P., Robinson, P.T., et al., Proceedings of the Ocean Drilling Program, Scientific Results: College Station, Texas, Ocean Drilling Program, v. 118, p. 285–307.

Lagabrielle, Y., and Cannat, M., 1990, Alpine Jurassic ophiolites resemble the modern central Atlantic basement: Geology, v. 18, p. 319–322.

MacLeod, C.J., Dick, H.J.B., Allerton, S., Robinson, P.T., and Shipboard Scientific Party, 1998, Geologic mapping of slow-spread lower ocean crust: A deep-towed video and wireline rock drilling survey of Atlantis Bank, ODP Site 735, SW Indian Ridge: InterRidge News, v. 7, p. 39–43.

Muller, M.R., Robinson, C.J., Minshull, R.S., and Bickle, M.J., 1997, Thin crust beneath ocean drilling program borehole 735B at the Southwest Indian Ridge?: Earth and Planetary Science Letters, v. 148, p. 93–108.

Mutter, J.C., and Detrick, R.S., 1984, Multichannel seismic evidence for anomalously thin crust at Blake Spur fracture zone: Geology, v. 12, p. 534–537.

Natland, J.H., Meyer, P.S., Dick, H.J.B., and Bloomer, S.H., 1991, Magmatic oxides and sulfides in gabbroic rocks from Hole 735B and the later development of the liquid line of descent, *in* Von Herzen, R.P., Robinson, P.T., et al., Proceedings of the Ocean Drilling Program, Scientific Results: College Station, Texas, Ocean Drilling Program, v. 118,, p. 75–111.

Ozawa, K., Meyer, P.S., and Bloomer, S.H., 1991, Mineralogy and textures of iron-titanium oxide gabbros and associated olivine gabbros from Hole 735B, *in* Von Herzen, R.P., Robinson, P.T., et al., Proceedings of the Ocean Drilling Program, Scientific Results: College Station, Texas, Ocean Drilling Program, v. 118, p. 41–73.

Pariso, J.E., and Johnson, H.P., 1993, Do lower crustal rocks record reversals of the Earth's magnetic field? Magnetic petrology of gabbros from Ocean Drilling Program Hole 735B: Journal of Geophysical Research, v. 98, p. 16,013–16,032.

Penrose Conference Participants, 1972, Penrose field conference on ophiolites: Geotimes, v. 17, p. 24–25.

Pettigrew, T.L., Casey, J.F., et al., 1999, Proceedings of the Ocean Drilling Program, Initial Reports 179 [CD-ROM]: College Station, Texas, Ocean Drilling Program.

Reid, I., and Jackson, H.R., 1981, Oceanic spreading rate and crustal thickness: Marine Geophysical Research, v. 5, p. 165–172.

Robinson, P.T., Von Herzen, R., et al., 1989, Proceedings of the Ocean Drilling Program, Initial Reports 118: College Station, Texas, Ocean Drilling Program, 823 p.

Robinson, P.T., Dick, H.J.B., and Von Herzen, R.P., 1991, Metamorphism and alteration in oceanic layer 3: Hole 735B, *in* Von Herzen, R.P., Robinson, P.T., et al., Proceedings of the Ocean Drilling Program, Scientific Results: College Station, Texas, Ocean Drilling Program, v. 118, p. 541–552.

Sclater, J.G., Fisher, R.L., Patriat, P., Tapscott, C., and Parsons, B., 1981, Eocene to recent development of the Southwest Indian Ridge: A consequence of the evolution of the Indian Ocean Triple Junction: Geophysical Journal of the Royal Astronomical Society, v. 64, p. 587–604.

Shipboard Scientific Party, 1999, Leg 176 Summary *in* Dick, H.J.B., Natland, J.H., Miller, D.J., et al., Proceedings of the Ocean Drilling Program, Initial Reports 176: College Station, Texas, Ocean Drilling Program, p. 1–70.

Sinton, J.M., and Detrick, R.S., 1992, Mid-ocean ridge magma chambers: Journal of Geophysical Research, v. 97, p. 197–216.

Stakes, D., Mevel, C., Cannat, M., and Chaput, T., 1991, Metamorphic stratigraphy of Hole 735B, *in* Von Herzen, R.P., Robinson, P.T., et al., Proceedings of the Ocean Drilling Program, Scientific Results: College Station, Texas, Ocean Drilling Program, v. 118, p. 153–180.

Tucholke, B.E., and Lin, J., 1994, A geological model for the structure of ridge segments in slow-spreading ocean crust: Journal of Geophysical Research, v. 99, p. 11937–11958.

Whitehead, J.A., Dick, H.J.B., and Schouten, H., 1984, A mechanism for magmatic accretion under spreading centers: Nature, v. 312, p. 146–148.

MANUSCRIPT ACCEPTED BY THE SOCIETY FEBRUARY 23, 2000

Geological Society of America
Special Paper 349
2000

Magmatic and tectonic controls on the evolution of oceanic magma chambers at slow-spreading ridges: Perspectives from ophiolitic and continental layered intrusions

Peter Thy
Department of Geology, University of California, One Shields Avenue, Davis, California 95616, USA;
e-mail: thy@geology.ucdavis.edu
Yildirim Dilek
Department of Geology, Miami University, Oxford, Ohio 45056, USA

ABSTRACT

Oceanic lithosphere formed at slow-spreading ridges shows pronounced lateral and vertical variations in crustal thickness and architecture as a result of a complex interplay between magmatic and tectonic processes. Emplacement of magma at slow-spreading centers is dominantly controlled by injection of sills, dikes, and ephemeral chambers. The internal evolution of such magma reservoirs is controlled by processes similar to those inferred to dominate the evolution of magma chambers at fast-spreading ridges. Magma chambers crystallize and undergo differentiation along their margins and floors by forming crystal-mush zones. Such zones solidify by compaction and vertical melt migration and by recycling of magma into the main reservoir. Compaction and plastic flow in the crystal mush result in banding and foliation and intensify preexisting modal layering and lamination. The presence of thick crystal-mush zones and their solidification by compaction and interstitial melt migration are common features of many continental, ophiolitic, and oceanic magma chambers. The magma-chamber development at slow-spreading centers, such as the Southwest Indian Ridge, may significantly be affected by tectonic uplift and withdrawal of the bulk of magma from the melt chambers. These processes collectively may result in cessation of compaction and a high concentration of trapped interstitial liquid in the partially solidified gabbro mush near the former segregation and crystallization front. This interstitial trapped liquid will crystallize to form evolved ferrogabbros partially as a result of lateral migration and syn- and post-tectonic channeled melt movement into uplift-induced pressure-release zones. Such fossil compaction profiles are not commonly recognized from layered gabbros of stable continental cratons and margins, but may be expected to occur more commonly in ophiolitic complexes. In general, magma-chamber processes that operate to develop gabbroic complexes at slow-spreading ridges are similar in many respects to those inferred from continental layered intrusions and from gabbros in fossil oceanic crust as preserved in ophiolites.

INTRODUCTION AND BACKGROUND

Considerable efforts have been directed toward investigating the nature of lower oceanic crust and magma-chamber processes at oceanic spreading centers over the past two decades (e.g., Phipps Morgan et al., 1994; Kelemen and Aharanov, 1998). Models regarding the internal structure, stratigraphy, and petrogenesis of modern oceanic crust have been delineated mainly by the way magma chambers and magma-chamber processes have been viewed. The seismic velocity and density layering of in situ oceanic crust was correlated with the pseudostratigraphy found in ophiolites, and a layered structure of oceanic lithosphere has been widely accepted (i.e., Penrose Conference Participants, 1972; Christensen, 1978; Christensen and Smewing, 1981). In these models, the oceanic seismic layer 3 corresponds to the thick masses of gabbroic and ultramafic cumulates seen in ophiolitic complexes (Cann, 1970; Coleman, 1977). This association resulted in the idea that large, steady-state magma chambers, which are continuously replenished by mafic or ultramafic magma, are located beneath oceanic spreading centers (Fig. 1A) and that the magma chambers grow as a result of magmatic spreading and deposition of solid or semisolid cumulates on chamber walls (Cann, 1974; Smewing, 1981; Pallister and Hopson, 1981). Consequently, the idea of the existence at mid-ocean ridges of laterally steady-state mafic or ultramafic magma chambers originated from the early ophiolite studies and became widely accepted.

Another perspective for the early models of oceanic magma-chamber evolution has come from studies of continental layered intrusions (Wager and Deer, 1939; Hess, 1960; Jackson, 1961; Morse, 1969; Irvine, 1974; Wager and Brown, 1967; McBirney, 1995; Cawthorn, 1996; Irvine et al., 1998). Internal compositional variations

Thy, P., and Dilek, Y., 2000, Magmatic and tectonic controls on the evolution of oceanic magma chambers at slow-spreading ridges: Perspectives from ophiolitic and continental layered intrusions, *in* Dilek, Y., Moores, E.M., Elthon, D., and Nicolas, A., eds., Ophiolites and Oceanic Crust: New Insights from Field Studies and the Ocean Drilling Program: Boulder, Colorado, Geological Society of America Special Paper 349, p. 87–104.

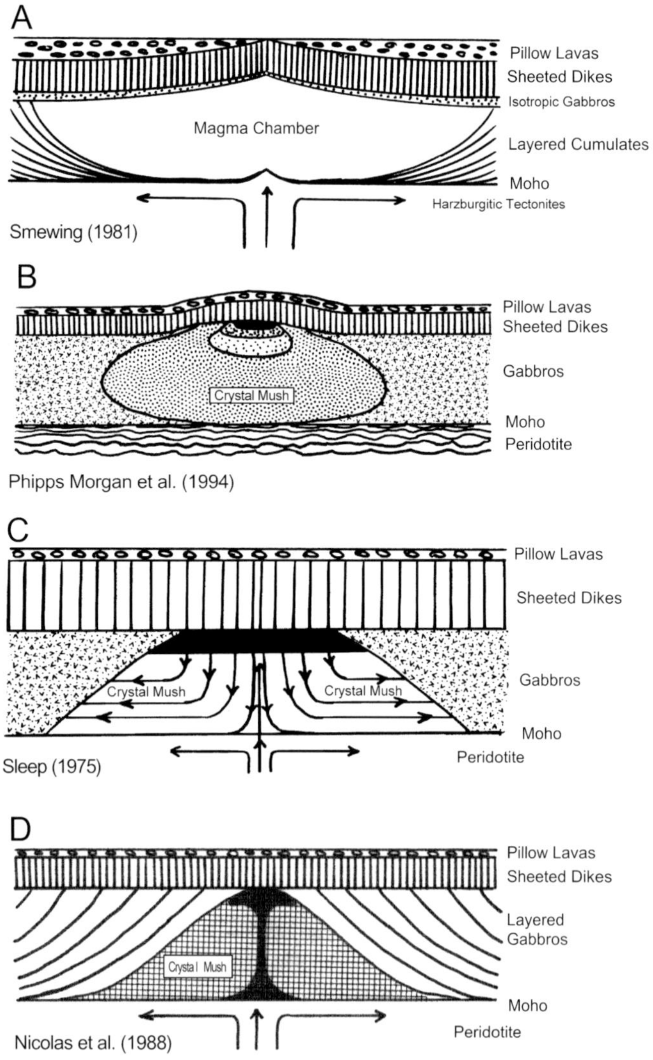

Figure 1. Schematic models for magma chambers at fast-spreading ridges. A, Model based on the studies of the Semail ophiolite by Smewing (1981), illustrating a large steady-state magma chamber. B, Magma-chamber structure at the East Pacific Rise based on seismic imaging (Phipps Morgan and Chen, 1993; Phipps Morgan et al., 1994). A large crystal-mush zone (or low-velocity zone) extends to the crust-mantle transition and shows increasing melt fraction toward the ridge crest. A sill-like magma lens (black) occurs below the crest. C, Modeling of an axial magma chamber by Sleep (1975) based on thermal balancing between magma injection and hydrothermal cooling. A sill-shaped magma chamber (black) occurs under the ridge crest from which gabbroic (and ultramafic) crystal mush moves down and outward, forming a stratified complex. Lateral flow is terminated by cooling from above. D, Model based on studies of the Semail ophiolite by Nicolas et al. (1988), showing a dome-shaped, melt-rich zone (squares) located above a mantle upwelling and a narrow feeder to a magma lens (black) beneath the ridge. Convection occurs in the mush with the result that melt is centrally located. Lateral movement, magmatic flow in the crystal-mush complex, and cooling from above collectively result in layered gabbros.

and observations on the development of layering in many continental gabbroic intrusions have been used to argue for fractional crystallization either by gravitational and current separation or by boundary-layer crystallization. The orientation of layering has been interpreted to reflect the floor and original shape of magma chambers. Because plutonic complexes in many ophiolites commonly display layering, it was concluded that similar processes were operative in oceanic magma chambers. These early views of solidification processes in magma chambers were uniformly based on observations from continental layered intrusions because of the lack of direct observations from modern oceanic spreading centers. For the same reason, ophiolitic studies were used to provide the crucial stratigraphic and structural guidelines for understanding the evolution of oceanic crust.

Thermal modeling of ridge structures (Sleep, 1975, 1978; Lister, 1983; Morton and Sleep, 1985; Phipps Morgan and Chen, 1993; Buck, this volume) and seismic imaging of axial magma chambers (e.g., Detrick et al., 1987; Harding et al., 1989) have played major roles in rectifying the simplistic early models of magma chambers. In contrast to the models proposing large, melt-dominated, oceanic magma chambers, the results of theoretical and geophysical studies have suggested the occurrence of only small molten zones beneath fast-spreading ridges that are equated with axial magma chambers ranging in size from a few hundred meters to 1 km in width and tens to hundreds of meters in thickness (Fig. 1B). Currently accepted models (e.g., Phipps Morgan and Chen, 1993) suggest that elongated melt lenses composed of molten magma, less than 1 km wide, are perched at the base of sheeted-dike complexes at fast-spreading centers (Fig. 1, B and C). Asymmetrically surrounding such lenses and extending to the base of seismic layer 3 are partially molten gabbroic rocks with only small amounts of melt (<10%; Fig. 1B; Nicolas et al., 1988; Detrick, 1991; Sinton and Detrick, 1992; Phipps Morgan and Chen, 1993; Phipps Morgan et al., 1994). The layering observed in gabbroic and ultramafic sections in ophiolites, and also believed to exist in oceanic layer 3, is thought to originate by viscous flow in the crustal section away from a frequently replenished shallow-level magma lens; this flow is interpreted to be mechanically coupled to plastic flow in the mantle section, rather than to gravitational processes in the chamber (Nicolas, 1989; Phipps Morgan et al., 1994). Development of melt lenses is believed to result from a balance between hydrothermal cooling and episodic magma injection into steady-state chambers (Sleep, 1978). This seismic model of oceanic magma chambers at fast-spreading ridges is compatible with field observations, principally from the Semail ophiolite (Fig. 1D; Browning, 1984; Nicolas et al., 1988, 1993; Nicolas, 1989; Nicolas and Ildefonse, 1996; Boudier et al., 1996).

The interpretations of magma chambers at fast-spreading ridges may not be directly applicable to magma-starved, slow-spreading ridges (Karson, 1990; Solomon and Toomey, 1992). Seismic imaging of slow-spreading ridges has largely failed to detect axial magma chambers (Detrick et al., 1990; Kent et al., 1990) and has indicated that the evolution of oceanic crust along such ridges is principally controlled by a complex interplay between magmatic accretion and amagmatic extension. Structural models for slow-spreading centers suggest widespread plastic and brittle deformation, isostatic uplift, listric normal faulting, and crustal exhumation (Harper, 1985; Karson et al., 1987; Cannat et al., 1991). These processes collectively result in pronounced lateral and vertical variations in crustal thickness along the axis and in structurally controlled, intermittent emplacement of magma in fissures, dikes, and small, rapidly solidifying chambers at all crustal levels (Fig. 2; Karson, 1990). However, the internal evolution of such ephemeral magma chambers at slow-spreading centers must be controlled by the same physical processes as those that control the evolution of steady-state magma chambers at fast-spreading centers, such as cooling, crystallization, compaction, and fractionation. The differences between the slow- and fast-spreading regimes lie in the sizes and cooling rates of magma chambers and in the interplay between ductile or brittle deformation and solidification processes in semisolid or solid chambers. Such views of slow-spreading oceanic crust have been confirmed by studies of ophiolites of inferred

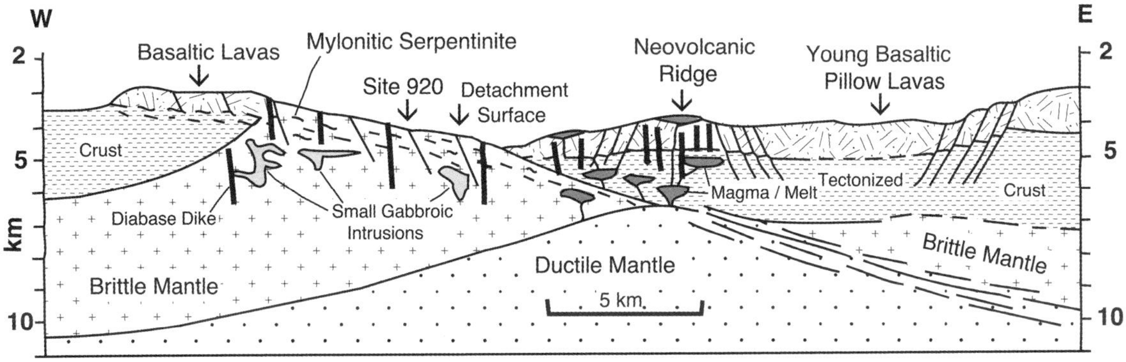

Figure 2. Interpretive general cross section across the slow-spreading Mid-Atlantic Ridge near the Kane Fracture Zone (23°20′N, MARK area; Cannat, Karson, et al., 1995; Cannat, 1996; Karson and Lawrence, 1997; Dilek et al., 1998; Lagabrielle et al., 1998). The markedly asymmetric architecture of the ridge-axis environment is due to tectonic extension and exhumation of lower-crustal and upper-mantle rocks along a major detachment fault system dipping toward the neovolcanic ridge. Thin, discontinuous exposures of basaltic rocks capping the serpentinized peridotites could be allochthonous, hanging-wall fragments and/or lava flows erupted directly onto the detachment surface. Oceanic lower crust beneath the spreading center, which is marked by the neovolcanic ridge, is tectonized as a result of high- to low-angle normal faulting, brittle to ductile shearing, and cataclastic to mylonitic deformation. A network of small magma reservoirs is located above the perturbed upper mantle, which is upwelling in the footwall of the detachment fault. These ephemeral magma reservoirs feed dike and sill intrusions in the upper crust and basaltic lava eruptions (including the neovolcanic ridge) on the seafloor and form composite nested gabbroic intrusions in the lower crust. This asymmetric crustal architecture and the inferred spatial and temporal relationships between magmatism and tectonism from the MARK area are similar to those observed and interpreted from the Alpine-Apennine (Lagabrielle and Lemoine, 1997) and Neotethyan ophiolites (Dilek et al., 1990; Dilek and Thy, 1998) and from the transverse ridge and Hole 735B at the Atlantis II Fracture Zone on the Southwest Indian Ridge (Dick et al., 1991).

slow-spreading origin (Harper, 1985; Malpas et al., 1989; Malpas, 1990; Dilek and Eddy, 1992; Dilek and Thy, 1998).

The only direct confirmation of the nature of oceanic magma chambers has come from the results of deep drilling by the Ocean Drilling Program into the plutonic foundation of in situ oceanic crust near active spreading centers (ODP Legs 118, 147, 153, 176, and 179). Drilling at Site 735 at the Atlantis II Fracture Zone of the Southwest Indian Ridge has provided the most comprehensive sampling of oceanic gabbros available for structural and petrologic studies (Robinson, Von Herzen, et al., 1989; Dick et al., 1998; Natland et al., 1998). ODP Legs 118 and 176 recovered a near-continuous section of gabbros to a depth of ~1500 mbsf (meters below seafloor) from Hole 735B. Most recently, ODP Leg 179 drilled an ~160-m-deep hole offset by a few kilometers from Site 735 (Hole 1105A; Casey et al., 1998). At the time of writing, only the results from the upper part of Hole 735B recovered during Leg 118 were available. More recently, the ODP initial results volumes from Leg 176 (Dick, Natland, Miller, et al., 1999) and Leg 179 (Pettigrew, Casey, Miller, et al., 1999) have been released. See also Robinson et al. (chapter 7, this volume) for a summary of the Leg 176 results. Plutonic rocks recovered from Hole 735B contain evolved Fe-Ti oxide gabbros intercalated with relatively primitive olivine gabbros (Bloomer et al., 1991; Dick et al., 1991, 1992; Hébert et al., 1991; Natland et al., 1991; Ozawa et al., 1991). This relationship was interpreted by Dick et al. (1991) to have resulted from crystallization combined with lateral and vertical migration of evolved intercumulus ferrobasaltic melt in a partially consolidated crystal mush now represented by the host olivine gabbros. Textural and compositional characteristics of these gabbros have led to the suggestion that melt migration was in part aided by localized deformation in the crystal mush (Dick et al., 1991, 1992). This interpretation suggests that interaction between tectonism and magma-chamber processes strongly affects the evolution of magma at slow-spreading centers.

The studies of core samples from in situ oceanic gabbros have contributed significantly to our understanding of the dynamics of magma chambers. Consequently, it is widely accepted that continental layered intrusions and ophiolites are no longer the only measure by which oceanic magma chambers should be interpreted. In fact, the ideas developed for modern oceanic magma chambers are now being applied to better define the models for the evolution of continental layered intrusions, where more attention is being directed toward melt migration, compaction, viscous flow, and deformation during solidification and consolidation of cumulates (Irvine, 1980; McBirney, 1995; McBirney and Nicolas, 1997). However, these recent developments in our understanding of oceanic magma chambers appear to have led to the view that gabbroic complexes at modern oceanic spreading centers are very different from their ophiolitic and continental counterparts and that observations from ophiolitic and continental intrusions therefore offer little help for the understanding magma-chamber processes in an oceanic environment (Dick et al., 1992; Sinton and Detrick, 1992). In this paper, we present some data, observations, and interpretations indicating that the gabbros from the ultra-slow-spreading Southwest Indian Ridge show igneous and structural features that are similar to those documented from continental and ophiolitic gabbroic intrusions. This result suggests that magma-chamber processes responsible for the development of gabbros in modern oceanic crust can be defined, in part, by understanding processes commonly used to explain the evolution of continental and ophiolitic gabbroic complexes. First, we briefly review the current information from well-known gabbroic complexes, and then we use this information to discuss the formation of oceanic gabbros.

GABBROIC COMPLEXES

Here we summarize the available information on the nature of layering, mode of differentiation, and compositional variability in cumulus phases and cumulate rocks of gabbroic complexes from modern oceanic lithosphere developed at the Atlantis II Fracture Zone (Southwest Indian Ridge), the Troodos ophiolite, and the Skaergaard layered intrusion. Common to the origin of these three gabbroic complexes is the apparent slow-spreading rate in their tectonic settings at the time of their intrusion. The Skaergaard intrusion was emplaced at the continental margin of East Greenland during the early stages of Tertiary continental rifting and the initial formation of the North Atlantic Ocean. The Troodos ophiolite formed at a slow-spreading center in a Cretaceous Neotethyan seaway between the African-Arabian plate and the Eurasian plate. The gabbros from the Atlantis II Fracture Zone formed along the ultra-slow-spreading Southwest Indian Ridge with a half spreading rate of ~8 mm/yr.

Southwest Indian Ridge

The holes drilled by ODP Legs 118, 176, and 179 are situated on a wave-cut platform at the crest of a transverse ridge on the east side of the Atlantis II Fracture Zone, which offsets the slow-spreading Southwest Indian Ridge. Sites 735 and 1105 sampled gabbroic massifs formed ca. 11 Ma ago beneath the median valley of the spreading center and currently located only 15–20 km away from the Atlantis II transform intersection with the ridge. The upper part of Hole 735B was drilled during Leg 118 and penetrated to a depth of 500 m in medium- to coarse-grained gabbroic rocks (Robinson, Von Herzen, et al., 1989; Von Herzen, Robinson, et al., 1991). The Shipboard Scientific Party (1989) divided the recovered core into six lithological units (Fig. 3) based principally on the predominant gabbro types. The gabbro complex is characterized by an association of evolved Fe-Ti oxide gabbros with relatively primitive olivine gabbros. The Fe-Ti oxide gabbro forms a major constituent of some units or a minor component of layers and vein networks in dominant olivine gabbro units (Fig. 3). Volumetrically minor intervals of gabbro, gabbronorite, Fe-Ti oxide gabbronorite, and troctolitic gabbro also occur in the sequence. The whole section is variously intruded by plagiogranitic veins and small bodies of microgabbro.

The olivine gabbros are adcumulate to poikilitic textured and include plagioclase, augite, and olivine cumulus phases. Mineral zoning and interstitial crystallization of evolved mineral assemblages (e.g., Fe-Ti oxides) are not commonly found in these dominantly adcumulate-textured gabbros. The Fe-Ti oxide gabbros are relatively evolved and include plagioclase, augite, olivine, ilmenite, magnetite, low-Ca pyroxene, and apatite as cumulus and intercumulus phases. Grain-size variations from fine- to coarse-grained gabbros are commonly seen on the scale of centimeters to meters. Modal layering occurs much less commonly and is defined mostly by gradational fluctuations principally in the abundances of olivine, pyroxenes, and Fe-Ti oxide minerals. The contacts between the gabbros and the Fe-Ti gabbros are, when preserved, mostly gradational. Only rare examples of intrusive and crosscutting relationships between the two gabbro types have been reported (Dick et al., 1991). Hole 735B was deepened during Leg 176 to a total depth of ~1500 mbsf in dominantly coarse-grained olivine gabbro with subordinate amounts of Fe-Ti oxide gabbro (Dick et al., 1998; Natland et al., 1998). Hole 1105A was drilled during Leg 179 at a position offset by 1–2 km to the north-northwest from Hole 735B and recovered gabbro sequences quite similar to those from the upper part of Hole 735B (Casey et al., 1998).

The gabbros from the Southwest Indian Ridge underwent various degrees of brittle and plastic deformation to the extent that in some places it is difficult to distinguish between magmatic and metamorphic planar features, such as foliation, lamination, banding, and layering. The Fe-Ti oxide–rich intervals generally show strong plastic to brittle deformation fabrics and grain-size reduction as a result of strain localization in these lithologies. The contacts between the olivine gabbros and the Fe-Ti oxide gabbros are therefore commonly obliterated by shearing and grain-size reduction. The uppermost ~40 m of the core is made of a strongly foliated metagabbronorite with

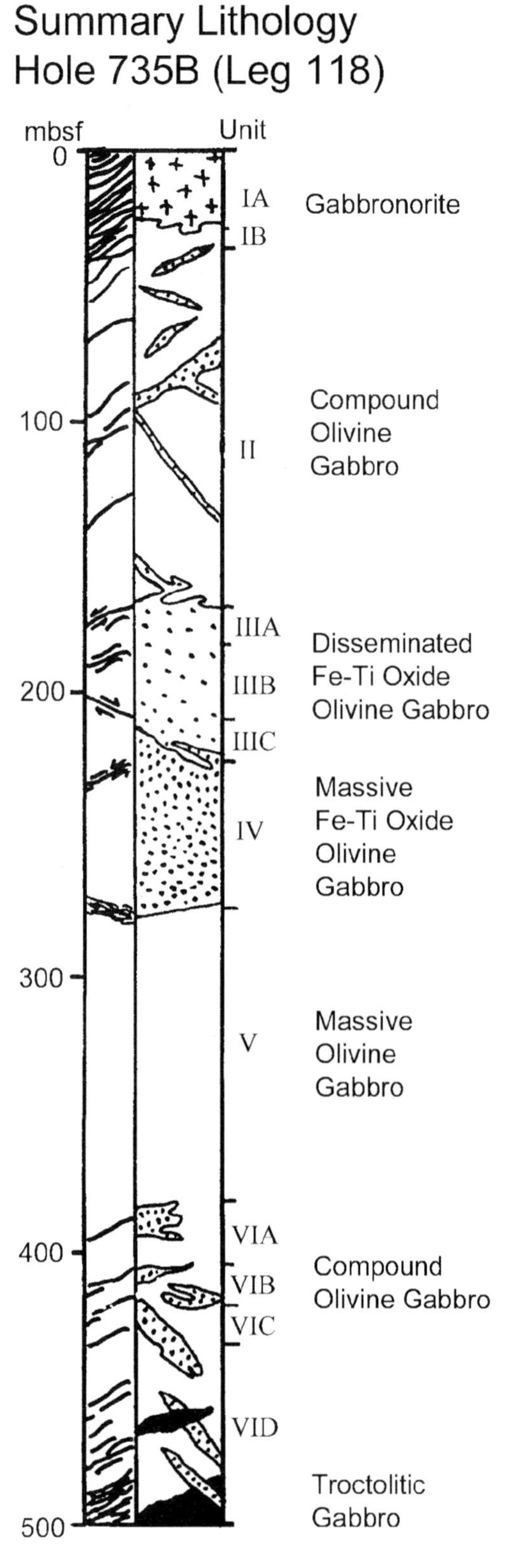

Figure 3. Summary of the visual synthesis of lithostratigraphic units of ODP Hole 735B to 500 mbsf. Left column illustrates the distribution and development of subsolidus ductile deformation (Dick et al., 1991). Right column shows the main lithologies as illustrated by Bloomer et al. (1991). The petrography of the gabbros is described by their dominant lithologies. Compound olivine gabbro refers to intercalated or mixed Fe-Ti oxide olivine gabbro and olivine gabbro.

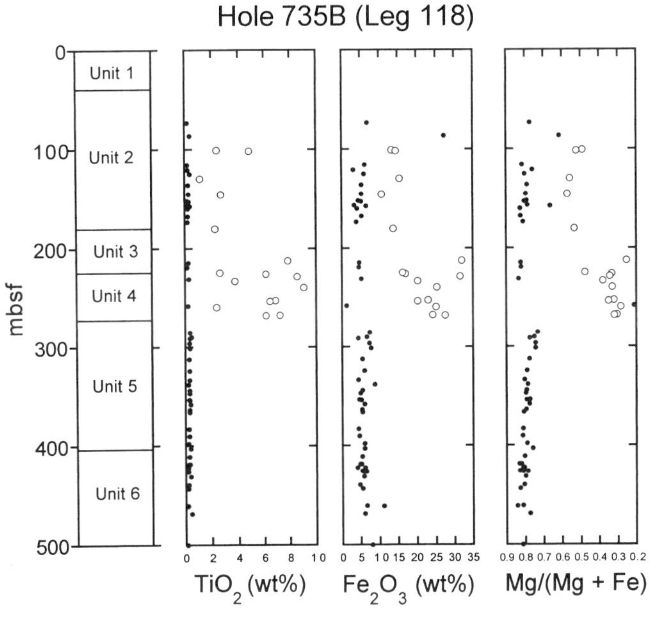

Figure 4. The compositional variation of selected elements in the cumulates from Hole 735B (Leg 118; data from Dick et al., 1991). For simplicity, the cumulates have been differentiated on the basis of their TiO_2 content. A group with $TiO_2 > 0.8$ wt% is represented by Fe-Ti oxide–bearing gabbros (unfilled symbols); a group with $TiO_2 < 0.8$ wt% is represented by Fe-Ti oxide–free gabbros (filled symbols). Fe is given as total Fe_2O_3, and the Mg/(Mg + Fe) ratio has been calculated by assuming that all Fe occurs as Fe^{2+}. The summary units are as defined by the Shipboard Scientific Party (1989); mbsf is meters below sea floor.

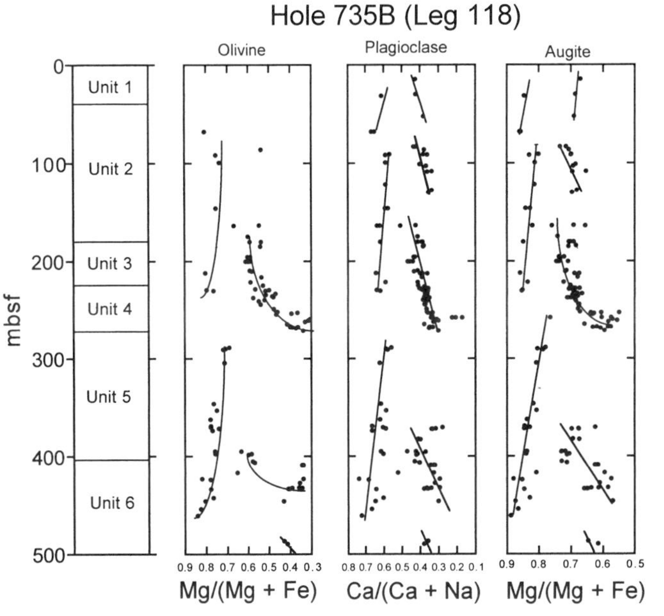

Figure 5. Cryptic variation in the upper part of Hole 735B (Leg 118) as revealed by cumulus olivine, plagioclase, and augite. Fe in olivine and augite has been calculated as total Fe^{2+} (data from Ozawa et al., 1991, and Natland et al., 1991). The curves and lines are the best visual fits to the data.

porphyroblastic to mylonitic textures, but thinner brittle to cataclastic shear zones occur at several locations in the core (Dick et al., 1991).

The compositional variation of the gabbroic cumulates from Hole 735B can be illustrated by the shipboard XRF analyses. The available results, shown for selected elements as a function of depth (Fig. 4), confirm the petrographic observations that there are two compositional groups. The volumetrically dominant and relatively primitive group of olivine gabbros has low TiO_2 (<1%) and Fe_2O_3 (total Fe) contents and high Mg/(Mg + Fe_{total}) ratios. This group is represented by gabbro, olivine gabbro, and troctolitic gabbro cumulates. The other group has very variable and high TiO_2 (>1%) and Fe_2O_3 contents and low Mg/(Mg + Fe_{total}) ratios and is represented by evolved Fe-Ti oxide gabbro, Fe-Ti oxide olivine gabbro, gabbronorite, and Fe-Ti oxide gabbronorite cumulates. In addition to Fe-Ti oxide minerals, cumulus apatite is petrographically identified in some evolved gabbros (Bloomer et al., 1991). This evolved cumulus assemblage is represented by high TiO_2, P_2O_5, and Fe_2O_3 contents. The differences between the two compositional groups do not necessarily record distinct melt compositions and are principally related to the absence or presence of cumulus Fe-Ti oxide minerals in the cumulates.

The interpretation of the compositional variation in cumulates is complicated because such rocks represent mixtures of the constituent cumulus phases as well as intercumulus crystallization and overgrowth. The problems with sampling the two gabbro components that are closely intermixed in the rocks also contribute to the complications. The relative proportions of cumulus and intercumulus phases control the bulk cumulate compositions; hence, volumetric combinations determined by the accumulative or segregative processes of up to seven cumulus and intercumulus phases (olivine, plagioclase, augite, low-Ca pyroxene, ilmenite, magnetite, and apatite) dominate cumulate compositions. Therefore, we cannot assume a priori that the two predominant cumulate groups identified by differences in the major oxide abundance also represent distinct liquid compositions and their evolutions. It is plausible, for example, that the variation in TiO_2 and Fe_2O_3 reflects variations in crystallization and accumulation processes, notably of Fe-Ti oxide minerals, and not the existence of two distinct liquid lines of descent. We can test these possibilities by examining the compositional variation with the stratigraphic height (i.e., the cryptic variation) of the three main cumulus phases (olivine, plagioclase, and augite) based on the data given by Ozawa et al. (1991) and Natland et al. (1991). Figure 5 shows that the two groups of gabbros are retained when cumulus-phase compositions are considered. The low-TiO_2 group of cumulates originated from a primitive-melt composition, whereas the high-TiO_2 group originated from a more evolved group of melts with wide-ranging compositions (see Hébert et al., 1991, and Bloomer et al., 1991). The apparently bimodal variation in the core—indicated for all cumulus phases—suggests limited mixing, fractionation, and compositional re-equilibration between the two magmatic components. This suggestion may then indicate that one cumulate component was solidified, or partially solidified, before the other component was formed or was introduced into the preexisting cumulus pile. These observations and arguments, however, do not exclude the possibility that the two groups of magmas are related by a common liquid line of descent.

These fundamental first-order observations, based on the compositions of the cumulates, concur in general with the conclusions of Dick et al. (1991, 1992) in that the main part of the Leg 118 core represents olivine gabbro intrusion(s) that display limited cryptic variation. The partially molten gabbro bodies were affected by compaction and syntectonic ductile deformation; both processes caused interstitial melt to migrate upward and laterally. Crystallization of Fe-Ti oxides and reactions with the host olivine gabbro may be responsible for the formation of the Fe-Ti oxide gabbros intercalated with the olivine gabbro throughout much of the gabbro sections.

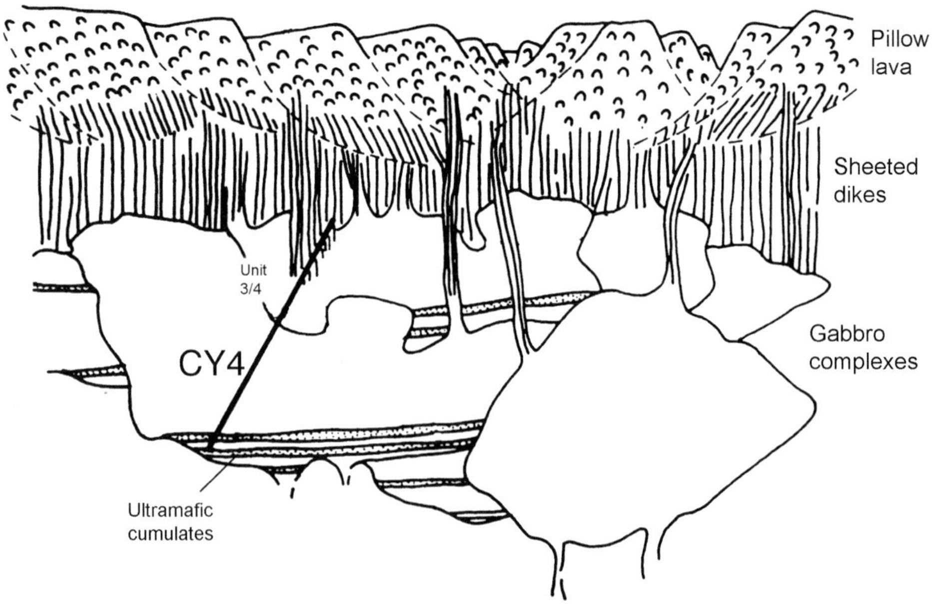

Figure 6. Schematic cross section through the upper crust of the Troodos ophiolite near the location of drill site CY4 of the Cyprus Crustal Study Project. Shown are pillow lavas, sheeted dikes, and composite nested gabbro complexes with local ultramafic cumulates at the bottom. The fossil upper oceanic crust, represented by the extrusive rocks and sheeted dikes in the Troodos ophiolite, is deformed by high-angle to listric normal faults, resulting in block rotation and formation of localized brittle-ductile to ductile shear zones in these rocks and near the dike-gabbro boundary. The line marked "CY4" illustrates the inferred penetration of the drill hole CY4. (Modified from Malpas, 1990.)

Troodos ophiolite

The main section of the plutonic complex of the Troodos ophiolite (Fig. 6) was drilled by the Cyprus Crustal Study Project (Hole CY4). This hole penetrated into the transition from sheeted dikes to gabbro at ~630 m, continued through gabbroic cumulates to a depth of 1750 m, and was terminated in ultramafic cumulates at a depth of 2260 m without reaching tectonized peridotites (Fig. 7; Gibson et al., 1989; Thy et al., 1989; Malpas et al., 1989). Fine- to medium-grained gabbroic screens and small, high-level gabbro bodies are present in the sheeted-dike complex from ~230 m depth downward, but gabbroic rocks first become dominant below 630 m where fine- to medium-grained gabbronorite rocks are developed. These rocks are mostly isotropic and show adcumulate to mesocumulate textures; they are composed of plagioclase, augite, and orthopyroxene cumulus phases. Fe-Ti oxides are not present as early-crystallizing phases, except in a short interval between 1196 and 1213 m (Thy et al., 1989). Weakly developed lamination as well as feldspathic and pyroxenitic layers occur sporadically. Irregular patches and layers showing a coarse-grained to pegmatitic texture occur locally and contain interstitial igneous amphibole.

The units defined in the CY4 drill core are mainly based on the observed compositional variation of pyroxene (Fig. 7). The contact at 1300 m (between units 3 and 4) is intrusive: the lower gabbro shows grain-size fining against the upper gabbro. The lower gabbros down to a depth of 1750 m are medium to coarse grained and commonly show grain-size and modal layering of gabbronorite, gabbro, olivine gabbro, and websterite on a scale ranging from 1 to 10 m. Olivine locally joins the cumulus assemblage. The gabbros are texturally isotropic adcumulates and show limited mineral zoning. The lowermost part of the core is composed dominantly of websterites with minor amounts of clinopyroxenites, wehrlites, lherzolites, and gabbronorites. Olivine is consistently a minor cumulus phase in the websterites. The lower cumulates are adcumulates to mesocumulates and display well-developed modal and grain-size layering on scales from a few centimeters to several meters. Medium-grained to pegmatitic gabbroic rocks are locally present as layers and patches in the websteritic cumulates (Thy et al., 1989; Browning et al., 1989).

The dike and cumulate rocks above 850 m depth are pervasively altered, as seen by the replacement of pyroxenes by hornblende. The intensity of alteration decreases downward, but zones of hornblende formation and meter-thick amphibolized gabbro are irregularly present throughout most of the gabbro sections. A major shear zone occurs at ~715 m, but otherwise the gabbros show little evidence of ductile deformation. The websterite section shows a stronger sign of ductile deformation marked by porphyroblastic texture and granulation in centimeter-thick bands. The contact between the gabbros and the websterites at 1750 m is faulted (Gibson et al., 1989); however, this brittle deformation did not affect the cryptic variation (Fig. 7).

The crystallization order in the lower gabbro and websterite cumulates are olivine, augite, orthopyroxene, and plagioclase, whereas the upper gabbros consist of plagioclase, augite, and orthopyroxene cumulates as well as the local appearance of Fe-Ti oxide minerals. These "boninitic" mineral assemblages and crystallization order—including the early appearance of low-Ca pyroxene—are unknown from mid-oceanic spreading ridges (see also Greenbaum, 1972; Allen, 1975; Thy, 1987; Thy and Moores, 1988; Thy and Xenophontos, 1991; Sobolev et al., 1993; Portnyagin et al., 1997), although the structural evidence suggests that the Troodos magma chambers formed and evolved at a spreading center.

The cryptic variation for the main cumulus phases is illustrated in Figure 7. The lower gabbroic to websteritic cumulates (units 4 and 5) show a systematic upward increase in Fe for all three mafic phases. In the upper part of the lower gabbros (unit 4), a reversal to more primitive compositions (and olivine saturation) is followed by a decrease in the Fe content of pyroxene. Plagioclase shows a very weak compositional variation in albite content that mimics the variation in the mafic phases in the gabbroic cumulates. The unsystematic variation of plagioclase composition in the websteritic cumulates is related to the intercumulus origin of plagioclase in these rocks. These first-order observations from the lower cumulate series led Thy (1987) to suggest that the variations were in general consistent with fractional crystallization in a closed magma chamber.

The intrusive contact between units 3 and 4 is a major structural as well as compositional boundary in the core. The gabbros below this boundary are more primitive in terms of major element ratios as well as minor elements. The more primitive nature of the minor elements, in particular, is seen for the lower-gabbro pyroxenes, which have markedly lower Al and Ti contents compared to the upper-gabbro pyroxenes (Thy, 1987; Malpas et al., 1989). The upper gabbros

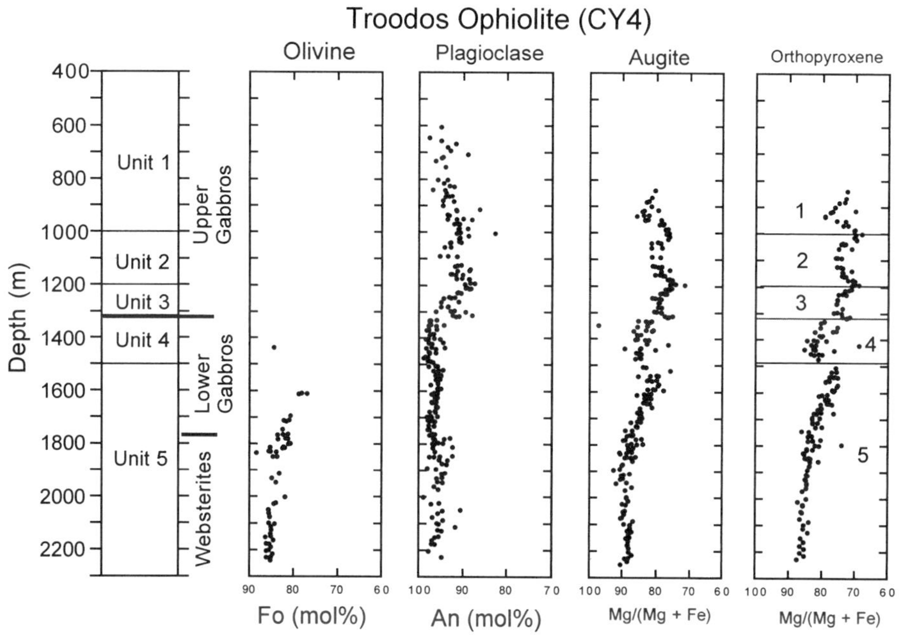

Figure 7. Cryptic variation in the plutonic complex of the Troodos ophiolite penetrated by Hole CY4 of the Cyprus Crustal Study Project (data from Thy et al., 1989). Shown are cumulus olivine (Fo in mol%), plagioclase (An in mol%), augite [in % Mg/(Mg + Fe)], and orthopyroxene [in % Mg/(Mg + Fe)]. Petrographically, the core is divided into two main groups: websterite (below 1760 m) and gabbro. The unit divisions are based on identified systematic variations and discontinuities in the cryptic variation (see column for orthopyroxene). The lower unit 5 is composed dominantly of websterites, but also includes subordinate amounts of gabbros. Only the contact between units 3 and 4 is interpreted as intrusive (Thy et al., 1989). Abbreviations: Fo—forsterite, An—anorthite. The Mg/(Mg + Fe) ratios for the pyroxenes were calculated with Fe as total Fe^{2+}.

show systematic variations that allow three units to be defined (units 1–3); each shows a systematic upward slow regression in Mg contents of the pyroxenes, followed by a trend toward increasing Fe contents. Fe-Ti oxide minerals occur intermittently in the most evolved gabbros at the top of unit 3 just prior to initiation of the regression marking the beginning of unit 2. These units were interpreted (Thy et al., 1989) to reflect an open-system magma chamber with three cycles (or three nested magma chambers), each showing gradational filling and mixing between resident magma and a new influx of a relatively more primitive magma before closed-system fractionation was resumed.

These simple views were challenged by Browning et al. (1989) who examined selected intervals in the lower websterite cumulates in more detail. They observed small-scale cyclic variations related to the layering in both modes and mineral compositions and argued that these variations reflected an open-system variation; hence, they precluded the possibility of large, well-mixed magma chambers. However, Dilek et al. (1992) emphasized that the variations observed by Browning et al. (1989) were related to the modal layering in the websterite cumulates. Qualitatively similar modal and compositional variations are found to be related to fine-scale modal layering in cumulus complexes that are interpreted to be controlled dominantly by closed-system crystallization (e.g., McBirney and Noyes, 1979; Thy et al., 1988; Conrad and Naslund, 1989). However, it is correct to say, as perhaps was intended by Browning et al. (1989), that locally the Troodos cumulates record boundary-layer development and fractionation. The systematic cryptic variation with increasing Fe content of the mafic cumulus minerals of the upper parts of unit 5 is difficult not to attribute to fractional crystallization in a partially closed magma chamber.

Field studies have elaborated on the spatial and temporal relationships between intrusive episodes and the high-temperature deformation in the Troodos plutonic complex (Moores and Vine, 1971; Malpas et al., 1989; Malpas, 1990; Dilek and Eddy, 1992). Malpas et al. (1989) mapped structural details in the Mount Olympus area of the Troodos complex and observed two major plutonic suites on the basis of relative ages of deformation and magmatism. An early suite consists mainly of harzburgite, dunite, layered olivine pyroxenite, and gabbro and displays a penetrative high-temperature deformation fabric. This tectonic fabric is generally parallel to the compositional layering in the plutonic suite. Malpas (1990) suggested that these structures of the early suite reflect the intrusion of plutons into upwelling-mantle structures at, or near, a spreading center, and that they might have resulted from upward and lateral solid flow. A late suite of plutons includes gabbroic and websteritic cumulates and displays locally well-preserved primary cumulus textures containing xenolithic inclusions of the early suite (Malpas, 1990). These field observations thus illustrate the complex relationships between the gabbroic and ultramafic plutons of the Troodos ophiolite and strongly support an evolution in multiple magma chambers (e.g., Allen, 1975) at a variety of crustal levels associated with intermittent episodes of magmatic and tectonic spreading (Malpas, 1990). The upper parts of the CY4 drill core sampled the early plutonic suite, which records a relatively open magmatic system (Thy et al., 1989; Malpas, 1990). The lower part of the drill core sampled a late-stage pluton, which may represent one of several episodes of plutonism that constituted the late suite (Malpas, 1990).

Skaergaard intrusion

The Skaergaard intrusion was emplaced during an early stage of rifting of the East Greenland continental margin. Since its discovery by L.R. Wager in 1930, this intrusion has remained an important source for testing the effects of crystallization processes in a closed basaltic magma chamber. The Skaergaard magma solidified in a small, upper-crustal chamber (Wager and Deer, 1939; Wager and Brown, 1967) by initially forming chilled margins and marginal gabbros and by subsequently crystallizing inward from the margins (Fig. 8). The marginal and upper border series (MBS, UBS) formed as a result of crystallization from the walls and the roof of the chamber (respectively), while the layered series (LS) developed concurrently by accumulating upward from the floor (Wager, 1960; Wager and Brown, 1967; McBirney, 1989, 1995; Naslund, 1984; Hoover, 1989). The crystallization fronts met at the sandwich horizon (SH). The solidification processes resulted in development of a stratigraphically differentiated cumulate section, which allows detailed assessments of the modes of solidification.

The principal manifestation of the evolving nature of the paren-

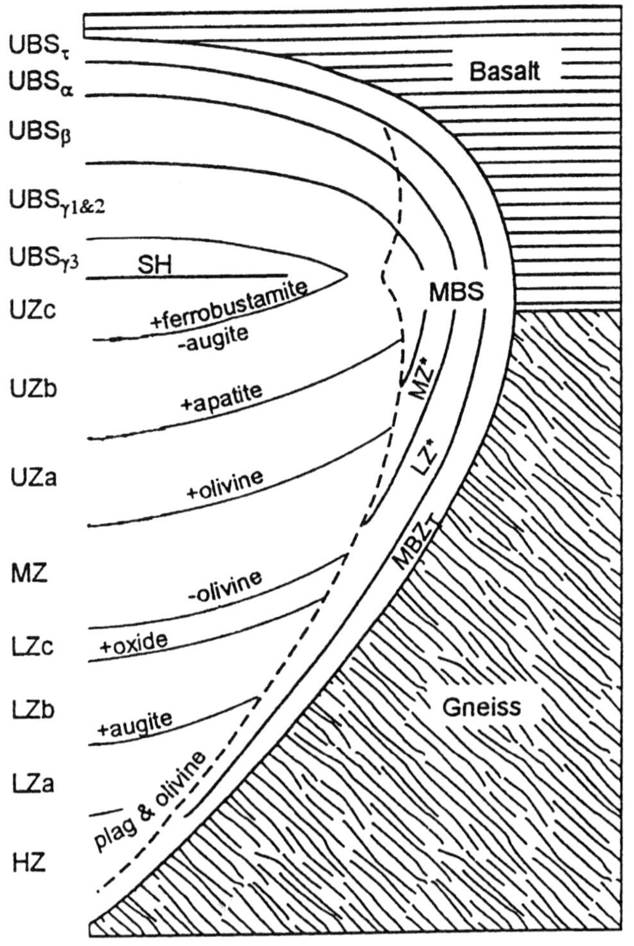

Figure 8. Schematic representation of the margins and the interior of the Skaergaard intrusion. Illustrated are the relative extent of the marginal and upper border series (MBS, UBS) and layered series (LS) with zone divisions and incoming and exiting cumulus mineral phases (Wager and Brown, 1967; Naslund, 1984; Hoover. 1989). Augite appears in the LZa and HZ as intercumulus grains. Note that ferrobustamite was described by Wager and Brown (1967) as wollastonite. (Modified from McBirney, 1995.)

tal magma lies in the mineral compositions: There is a systematic enrichment in the low-temperature components, such as albite and fayalite, of the cumulus phases toward the center of the intrusion (SH). An equally important manifestation is a systematic appearance and/or disappearance of minerals in the magmatic stratigraphy; the appearances and disappearances (zone boundaries) can be related to crystallization behavior with falling temperatures in experimental and natural basaltic systems. This feature allows the layered and border series to be divided into cumulus zones depending on the principal mineral constituents that provide convenient reference points for the magmatic evolution (Fig. 8). It is important to note that although the solid products of the Skaergaard magma are well exposed, the nature of the corresponding liquid and its evolution can only be inferred (Wager, 1960; Hunter and Sparks, 1987).

The layered series of the Skaergaard intrusion was sampled and measured along a profile by Wager and Deer (1939) in the central part of the complex. This profile is nearly 2700 m thick and includes a small hidden zone (HZ) mostly penetrated by drilling (Maaløe, 1976a). The zone divisions in the layered series are illustrated in Figure 8. Plagioclase is a cumulus phase throughout the intrusion. Olivine is a cumulus phase in the lowermost cumulates, but disappears as a cumulus phase in the middle zone (MZ). Pigeonite (inverted) was defined as a cumulus phase from lower LZb to upper UZa by Wager and Brown (1967). However, recent work has suggested that pigeonite throughout the layered series is of postcumulus origin (McBirney, 1989). Cumulus olivine reappears in the upper part of the intrusion (UZ). Augite appears in LZa (and HZ) as intercumulus poikilitic grains, but otherwise appears throughout the rest of the intrusion as cumulus grains (except for being replaced by ferrobustamite in the uppermost subzone, UZc). Cumulus Fe-Ti oxide minerals appear in the upper part of the lower zone, and their appearance defines a zone boundary (LZb-LZc). Apatite is present as a cumulus phase in the upper zone, where its first appearance defines a zone boundary (UZa-UZb). In general, the upper and marginal border series show corresponding cumulus zones from the roof and the walls and downward and inward, respectively (Fig. 8). The crystallization order of the parental Skaergaard magma was olivine and plagioclase, then augite, then Fe-Ti oxides, and finally apatite; there is a late-stage gap in olivine crystallization.

The available information on the composition of cumulus silicate phases in the layered series is illustrated in Figure 9, where the compositional values are plotted relative to the scale of Wager's stratigraphic column. Cumulus plagioclase shows a systematic increase in its albite content with increasing stratigraphic height from An_{67} at the base to An_{30} at the top. The cumulus olivine similarly shows an increase in its fayalite content with increasing stratigraphic height from Fo_{65} at the base to fayalite (Fo_0) at the top. Cumulus augite shows the most marked trend toward increasing Fe content in the uppermost cumulates of the layered series.

The variation in bulk cumulate compositions (i.e., whole rock; Fig. 10) is a strong reflection of the cumulus assemblages and their cryptic variations. Although the variations do not completely coincide with the zone divisions based on cumulus phases, the TiO_2 and FeO contents are clearly controlled by the presence of cumulus Fe-Ti oxide minerals. Whereas the $Mg/(Mg + Fe_{total})$ ratio and the TiO_2 content show systematic decreases in MZ and UZ, reflecting the cumulus assemblages and the concurrent depletion of the liquid in Fe, the total Fe content of the cumulates remains relatively high throughout the middle and upper zones.

The observed cryptic variation in the mafic cumulus assemblage of the Skaergaard intrusion is in general consistent with perfect or near-perfect fractional crystallization (Maaløe, 1976b). The exception is plagioclase for which a terminal composition of An_{30} cannot be attributed to fractional crystallization alone. It is possible, as originally suggested by Maaløe (1976b), that plagioclase primocrysts in the upper zone remained in the liquid as a result of a reversal in density contrast and increasing liquid viscosity. The consequence, as observed, may have been a retarded evolution in plagioclase composition.

Rhythmic layering characterizes the gabbros of the Skaergaard intrusion (Wager and Deer, 1939; Wager and Brown, 1967; Maaløe, 1978, 1987; Irvine, 1982, 1987; Irvine et al., 1998). The most prominent type of layers in the layered series is described as modally graded with mafic minerals concentrated at the base. Their appearance may vary greatly depending on the scale of layering (5–50 cm), sharpness of contacts, mineral modes, and mineral sorting (Irvine, 1987). Thick uniform layers that are only weakly stratified with gradational contacts alternate with rhythmically layered units throughout the layered series. The marginal border series rocks display an irregular, coarser-grained banding that is distinctly different from the layered series. This banding is generally defined by grain-size variations, colloform banding directed into the center of the intrusion, and local development of perpendicular feldspars (crescumulates; Wager et al.,

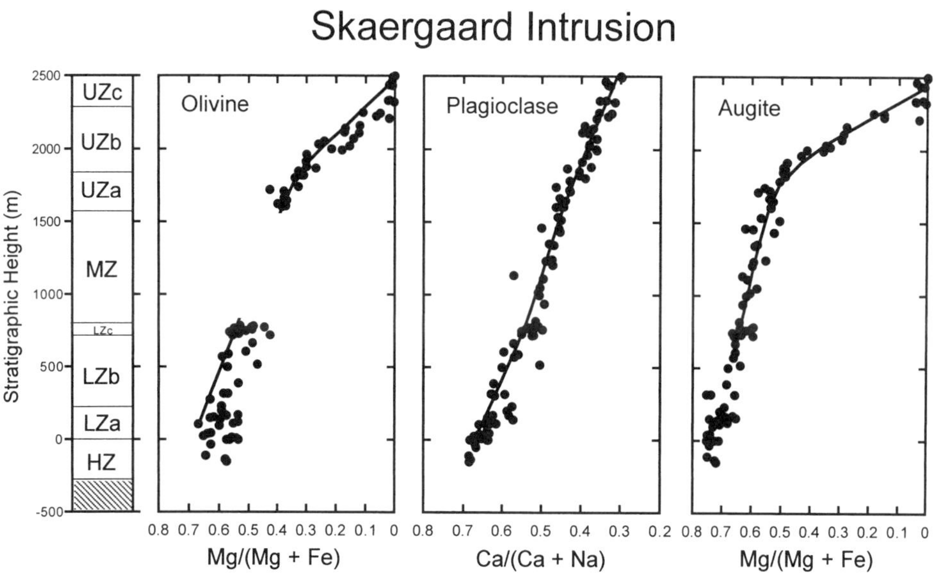

Figure 9. Cryptic variations of major silicate cumulus phases of the layered series of the Skaergaard intrusion shown as a function of height in Wager's (1960) central profile. The zone divisions are from Figure 8. The shaded part of the column is the basement. Data are from Wager and Brown (1967, connected by solid line), McBirney (1989), Maaløe (1976a), and Nwe (1976). All compositional cumulate data have been scaled to Wager's (1960) stratigraphic column, assuming an HZ of ~300 m. All Fe in augite has been calculated as total Fe^{2+}. Except for LZ augites, all analyses are of cumulus phases.

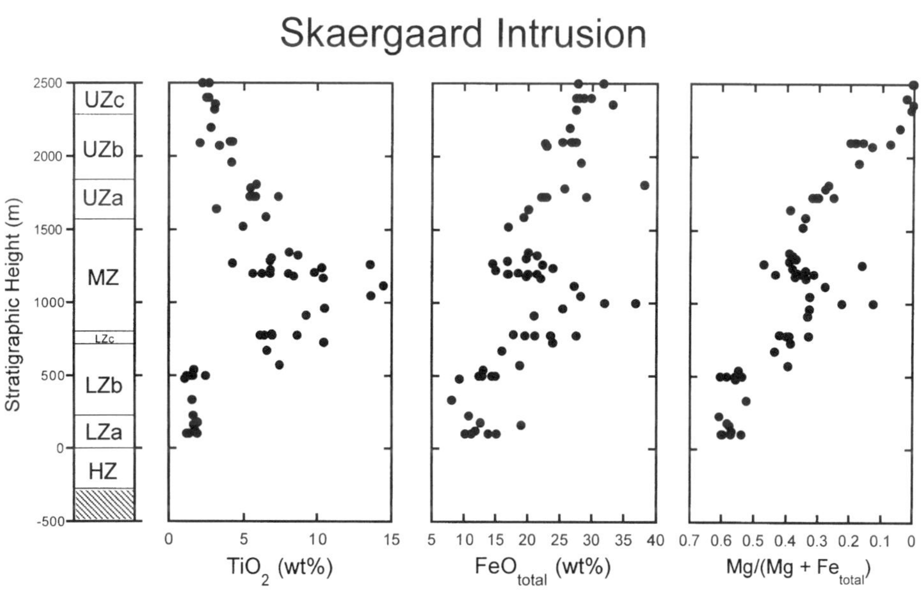

Figure 10. The compositional variation of selected elements in cumulates of the Skaergaard intrusion. The zone divisions and the stratigraphic column are as depicted in Figure 9 (data from McBirney, 1989, and Sørensen, 1995).

1960). The banding in the marginal border series is a strong indication of in situ crystallization in response to cooling by heat loss through the margins of the intrusion (Wager and Brown, 1967).

The mode of formation of the layered rocks and differentiation of the intrusion have been interpreted as being controlled mainly by gravitational segregation and density currents (Wager and Deer, 1939; Wager and Brown, 1967; Irvine, 1987). More recent investigations have questioned this classical view and suggested complexities involving in situ boundary-layer crystallization (Maaløe, 1978, 1987; McBirney and Noyes, 1979; Boudreau and McBirney, 1997). In particular, McBirney and coworkers have advocated that compaction, plastic flow, and melt migration largely controlled the internal differentiation of the complex and the layer formation (McBirney, 1995; McBirney and Hunter, 1995; McBirney and Nicolas, 1997). These views, which have been strongly opposed by Irvine et al. (1998), suggest that "the prominent layering that characterizes the main, interior part of the Skaergaard intrusion was developed largely by mineralogical recrystallization and reorganization of an initially unstratified, relatively stagnant crystal mush" and that "the crystal mush was fractionated and aggregated simply by gravitational compaction" (quotes from Irvine et al., 1998, p. 1400). Despite disagreements about the origin of layering, it is generally agreed that migration of interstitial melts played a central role in the development and solidification of the Skaergaard crystal mush.

OCEANIC MAGMA-CHAMBER PROCESSES

In this section we discuss magma-chamber processes as inferred from studies of gabbros from the Southwest Indian Ridge in the light of observations and interpretations provided by studies of ophiolitic and continental layered gabbros.

Structural and stratigraphic considerations

Repeated intrusion of batches of magma were involved in the construction of ocean crust at the Southwest Indian Ridge (Dick et al., 1991), similar to the development of the Troodos gabbros as inferred from crustal drilling and field mapping in plutonic rocks of the ophiolite (Allen, 1975; Malpas et al., 1989; Thy et al., 1989; Malpas, 1990). Fluctuations in intrusion frequency and volume of the Troodos magma system (Malpas, 1990) and temporal relocations of spreading centers (Varga and Moores, 1985; Eddy et al., 1998) resulted in the development of nested chambers of limited vertical and lateral extent that were intruded at various structural levels in the preexisting crust (Fig. 6). The Southwest Indian Ridge gabbro complexes therefore show strong structural similarities to the plutonic rocks of the Troodos ophiolite.

The interpretation of the structural setting of the gabbros drilled at Site 735 is complicated because of the significant amount of block uplift (5–6 km) and crustal exhumation near the Atlantis II transform. Thus the stratigraphic position of the drill core in the gabbroic massif cannot readily be deduced from the available information. This complication affects the formulation of structural models and makes lithologic correlations difficult (Casey et al., 1998). The olivine gabbros recovered from Hole 735B display no evidence of internal chilled margins or marginal gabbros. The olivine gabbros appear to constitute several internal coherent blocks that may either have formed in a single solidifying and replenished magma chamber or as individual and partially nested chambers; however, internal intrusive relationships might have been obliterated as a result of shearing and strain localization (e.g., at the contact between units 4 and 5).

The problems encountered in the Site 735 drill cores are not fundamentally different in these respects from those experienced in similar studies of tectonically and magmatically disrupted plutonic sequences in ophiolites and continental layered intrusions. The fact is that the stratigraphic position and the volumetric relationships of disrupted fractionated plutonic sequences can rarely be identified and scaled to allow identification of relatively open and closed system evolution.

Crystallization orders and zone divisions

Although the gabbros from Hole 735B do not allow zones based on cumulus phases to be defined, crystallization orders of the mineral phases and the petrogenesis of the rocks can still be inferred from careful petrographic observations. Troctolitic gabbros were recovered from Hole 735B at the base of unit 6 (Fig. 3); whether these troctolitic gabbros are intrusive into the olivine gabbros has been debated (Dick et al., 1991, 1992; Bloomer et al., 1991). The troctolitic layers are found dominantly in the most primitive gabbros. This is an indication, but not a conclusive evidence, that the primitive liquids were saturated, or nearly saturated, with respect to olivine and plagioclase.

The dominant lithology of the gabbros indicates mutual equilibrium of olivine, plagioclase, and augite. Cumulus low-Ca pyroxene (orthopyroxene or inverted pigeonite) occurs both in gabbronorites and Fe-Ti oxide gabbronorites. This finding suggests that low-Ca pyroxene appears in the crystallization order prior to Fe-Ti oxide minerals (Bloomer et al., 1991). Fe-Ti oxides coexisting with cumulus low-Ca pyroxene show that these minerals were the next phases to appear. The Fe-Ti oxide minerals are dominantly ilmenite, but magnetite also appears in significant amounts in textures suggesting that both of these phases are primary in origin. Finally, apatite is present in considerable amounts in some Fe-Ti oxide gabbros (Bloomer et al., 1991). This petrographic information suggests the following crystallization order: olivine, plagioclase; augite; low-Ca pyroxene; ilmenite, magnetite; apatite. It is possible that more primitive plutonic rocks (troctolites, dunites, anorthosites) exist below the deepest penetration level of Hole 735B. It is also possible that a gap exists in olivine crystallization, approximately coinciding with the appearance of low-Ca pyroxene, as indicated by a decrease in abundance or the lack of olivine in gabbros that contain low-Ca pyroxene (Bloomer et al., 1991). Regardless, this inferred crystallization order is in general similar to the crystallization order (or zone divisions) seen in the Skaergaard intrusion, as well as to the order predicted from low-pressure anhydrous experimental work on tholeiitic basalts. This similarity was also pointed out by Meyer et al. (1989) in a study of dredge samples from the Southwest Indian Ridge (54°S, 7°16′E). The complication, however, is that the rocks did not accumulate or accrete on the walls of the chamber in a systematic fashion, which could allow identification of cumulus zones, the volume relationships, and the extent of fractionation. Therefore, we are unable to relate the magmatic evolution to any simple stratigraphic and volumetric scale (Natland et al., 1991) as can be done to a certain extent for the Skaergaard intrusion (Wager, 1960).

Layering

Layering is an integral part of most gabbroic intrusions mainly in the form of modally graded and rhythmically repeated layers (Irvine, 1982, 1987). Layering is interpreted to be synonymous with crystal fractionation and is thought to reflect mechanical and density sorting, compaction and flow, fluctuations in nucleation and growth rates, or multiple injection of melt batches. However, since the early view by Wager and Deer (1939) that the layering in the Skaergaard intrusion formed from deposition and sorting from gravitationally driven magma flows, we have made few advances in understanding how and why such structures develop (cf. Boudreau and McBirney, 1997; McBirney and Nicolas, 1997; Irvine et al., 1998). Layering under certain conditions may be discordant to cryptic variation and zone divisions and therefore unrelated to differentiation (Wilson and Larsen, 1982, 1985).

Layering observed in the gabbros from Hole 735B is in general indistinct and caused by size variations and, to a lesser extent, by modal variations (Bloomer et al., 1991). As such, this layering differs from the typical modally graded rhythmic layering in the layered series of the Skaergaard intrusion and in many other intrusions. The gabbros of the Troodos ophiolite show layering that varies from well-defined size and modal variations in the lower parts to indistinct or nonexisting in the upper gabbros. The grain-size layering in the gabbros from Hole 735B has been attributed to fluctuations in nucleation and growth rates in a boundary layer along the margins of the chamber (Bloomer et al., 1991). In many respects, the Hole 735B style of layering, therefore, may be more akin to the layering observed in the border series of the Skaergaard intrusion (Naslund, 1984; Hoover, 1989; Irvine et al., 1998). However, an important consideration for the Hole 735B gabbros is that it is commonly difficult to distinguish between primary igneous structures and secondary features that resulted from metamorphic recrystallization and deformation. The presence or absence of igneous lamination and layering is therefore insufficient evidence with which to identify fractionation processes.

Gabbro and olivine gabbro differentiation

The shipboard studies of whole-rock compositions of Fe-Ti oxide–free gabbros from Site 735 reveal systematic up-hole variations particularly in the Mg/(Mg + Fe) ratios. The results from ODP Leg 118 show two segments, or blocks, composed of relatively primitive gabbro, each with a small but systematic, upward decrease in the Mg/(Mg + Fe) ratios (Fig. 4). The results from the lower part of Hole 735B, drilled during ODP Leg 176, show at least two additional, relatively coherent segments of gabbro, each with a systematic upward decrease in the Mg/(Mg + Fe) ratios (Natland et al., 1998; Dick et al., 1998). The general evolution trend within each segment is an upward decrease in Mg/(Mg + Fe) ratios with repeated reversals to

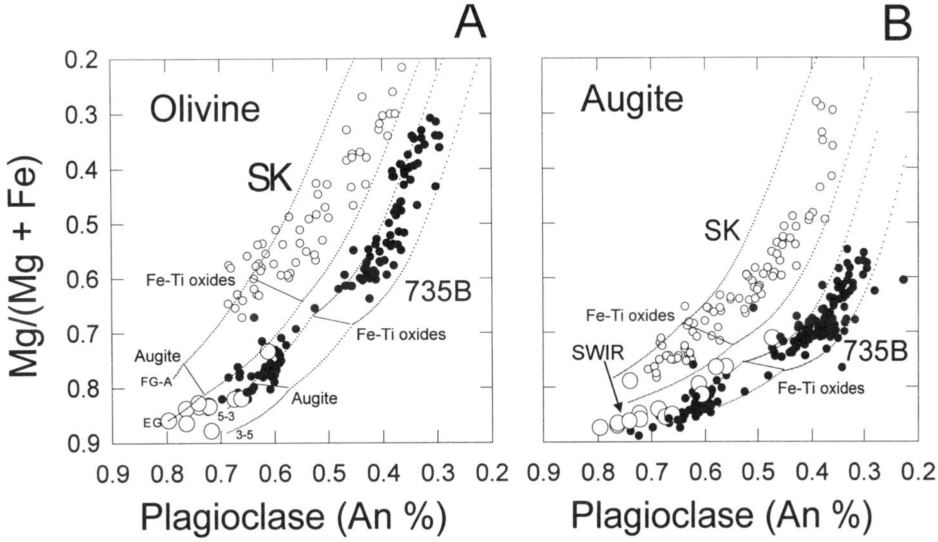

Figure 11. Mineral covariance of Skaergaard (SK) and Hole 735B (Leg 118) cumulus phases. A, An content of plagioclase and Fo content of olivine. B, An content of plagioclase and Mg/(Mg + Fe) ratio of augite. Data for Hole 735B are from Ozawa et al. (1991), Bloomer et al. (1991), and Natland et al. (1991); data from Skaergaard are from Wager and Brown (1967), McBirney (1989), Maaløe (1976a), and Nwe (1976). The large unfilled symbols labeled SWIR designate gabbros dredged from along the Southwest Indian Ridge and described by Meyer et al. (1989) and Bloomer et al. (1989). Observed variation is compared to forward fractionation modeling for potential Skaergaard parental liquids shown in steps of 1% of the remaining liquid (EG, Wager and Brown, 1967; FG-A, Brooks and Nielsen, 1990) and representative Atlantis II Fracture Zone basalt glasses (5–3 and 3–5, Natland et al., 1991). The incoming of augite and Fe-Ti oxides are indicated. Fe in pyroxenes has been calculated assuming all Fe as Fe^{2+}.

more primitive compositions between individual segments. It is not an easy task to demonstrate conclusively that the compositional variables observed in whole-rock compositions also record the coexisting liquid evolution. So far this has been demonstrated only for the upper part of Hole 735B (Bloomer et al., 1991; Hébert et al., 1991; Natland et al., 1991; Ozawa et al., 1991). The available data are illustrated in Figure 5 and show a systematic upward decrease, particularly in the Mg/(Mg + Fe) ratios in augite and the Ca/(Ca + Na) ratios in plagioclase of the relatively primitive olivine gabbros. This trend can be related to the whole-rock variations (Fig. 4) and shows that it is possible to use the Mg/(Mg + Fe) ratios of whole-rock cumulate compositions as a proxy for cryptic variation in the mafic phases normally used as an indicator of differentiation in layered complexes. Similar conclusions can be reached for the Skaergaard intrusion in which the gross variation in cumulates reflects, in a general sense, the zone divisions and the cryptic variations of the mineral chemistries (Figs. 9 and 10).

Solidification of the Skaergaard magma occurred initially by forming a chilled margin and then proceeded by crystallizing inward from the margins of the chamber. The marginal series formed by crystallization from the walls and the roof, while the layered series accumulated upward from the floor (Naslund, 1984; Hoover, 1989). The cryptic variation in the layered series of the intrusion is an upward decrease in Mg/(Mg + Fe) ratios, whereas in the upper marginal series it is a downward decrease in Mg/(Mg + Fe) ratios. The systematic upward decrease seen for all olivine gabbro cumulate segments at Site 735, therefore, most likely records an accumulated series that crystallized as the Mg/(Mg + Fe) ratio of the residual liquid decreased during solidification (and subsequent isolation) of earlier-formed cumulates advancing upward from a floor or marginal zone.

Crystallization models using the crystallization orders observed in the gabbros and simulating perfect fractional crystallization are illustrated in Figure 11 for two bracketing, potential parental magma compositions associated with the Atlantis II Fracture Zone (Natland et al., 1991, their Table 1, analyses 5–3 and 6–1); also shown in Figure 11 is a similar model developed for the Skaergaard intrusion. The compositional variation in coexisting plagioclase, olivine, and augite can be relatively well duplicated by the modeling, and the compositional differences between the Skaergaard and the Hole 735B gabbros are related to compositional differences in the parental magmas. The simulation of fractional crystallization shows that the gabbro and olivine gabbro differentiation can be accounted for by less than 35% fractionation and that the residual 65% liquid is not yet saturated in Fe-Ti oxide minerals. The presence of liquid in the cumulates would change these values, dependent on the amount of trapped liquid, but would not change the observation that the interstitial liquid was not saturated in the Fe-Ti oxide minerals.

Very few gabbroic intrusions crystallized as a closed system, like the Skaergaard intrusion. Many intrusions in tectonically and magmatically active environments record in their cumulates the evidence of replenishments by fresh or relatively primitive magma (Jackson, 1961; Irvine, 1980; Wilson et al., 1981; Wilson and Larsen, 1985; Thy et al., 1989). These replenishments result in cryptic variations that show repeated reversals to more primitive compositions followed by subsequent evolutions toward less primitive compositions. Such compositional reversals may be abrupt or continuous, and their frequency may reflect tectonic activity. Smewing (1981) suggested that very fast spreading rates and development of replenished steady-state magma chambers would result in fundamentally constant cumulus compositions without visible fractionation revealed by the cumulate successions. The results from Hole 735B record several reversals to relatively more primitive compositions that may reflect replenishment and mixing of the residual magma with more primitive magma. In general, this pattern is identical to the internal evolution inferred for many large, layered intrusions (Irvine, 1980; Wilson and Larsen, 1985) and for the upper gabbro units of the Troodos ophiolite (Thy et al., 1989). The observation that the Site 735 reversals do not always return to the same primitive composition suggests some extent of mixing with a relatively evolved residual magma.

Some layered intrusions and ophiolitic gabbros contain small-scale cyclic variations in cumulus compositions (Jackson, 1961; Irvine, 1980; Browning, 1984; Elthon et al., 1984) that are interpreted to signal frequent replenishments of magma and are used to infer limitations to chamber height (Browning, 1984). Indications from the Semail ophiolite suggest a maximum chamber height of a few hundred meters (Browning, 1984). The observation that the cryptic variations return to approximately similar compositions after each cycle, and therefore show little depth dependence in the cumulate stratigraphy, is seen as an indication that the gabbro (and ultramafic) cumulate sections were emplaced through the same size-limited magma lens (Quick and Denlinger, 1993; Phipps Morgan et al., 1994). Such limitations in chamber dimensions and emplacement modes cannot easily be made for the Site 735 gabbros. The repetitions recorded in the Site 735 gabbros suggest much larger magma chambers considerable deeper than 400 m (assuming the simplest geometric configuration and interpretations of the stratigraphic column). These values are more similar to those that can be inferred for the synorogenic layered Fongen-Hyllingen complex (Wilson et al., 1981; Wilson and Larsen, 1985) and for the plutonic complex of the Troodos ophiolite (Thy et al., 1989). The chamber dimensions inferred for the slow-spreading, Site 735 gabbros do not suggest that they were emplaced through the floor of a thin subaxial magma lens.

Fe-Ti oxide gabbro differentiation

The Fe-Ti oxide gabbros occur as zones and veins intercalated with the host gabbros and olivine gabbros with few clear indications of intrusive relationships. The compositional and petrographic evidence from the host gabbros suggests that these rocks formed in a cooling magma chamber by accumulation or crystallization along its the margins. The petrology and spatial distribution of the Fe-Ti oxide gabbro veins are distinctly different from their host rocks and suggest a different mode of origin. There are several important observations that can be made about the Fe-Ti oxide gabbros. First, despite a large short-order variability, there is a general increase in the volumetric abundances of Fe-Ti oxide gabbros in the upper parts of Hole 735B (Fig. 5; Dick et al., 1998). Second, olivine gabbros and Fe-Ti oxide gabbros appear to represent a continuum of parental-melt compositions (Figs. 5 and 11). Finally, certain intervals show downward decreases in Mg/(Mg + Fe) ratios of olivine and augite and Ca/(Ca + Na) ratios of plagioclase. These trends are the reverse of the normal fractionation trend seen in the host cumulate segments. The most prominent of these trends appears to be related to high concentrations of Fe-Ti oxide gabbros (units 3 and 4) and a compositional reversal in the host cumulates (units 4 to 5). Ozawa et al. (1991) identified three such cycles in the Fe-Ti oxide gabbros. These observations suggest that the Fe-Ti oxide gabbros increase volumetrically upward and that they developed partly independently of the host cumulates. Despite the wide ranges in the compositions of cumulates and cumulus phases, the two groups of gabbros (olivine gabbros and Fe-Ti oxide gabbros) may be related to a common liquid line of descent (Fig. 11).

Several well-established processes operative during the solidification of layered intrusions may in general explain some of these first-order observations.

Cumulate compaction and melt migration

The gabbros from Site 735 are texturally and chemically typical poikilitic to granular cumulates (adcumulates to heteradcumulates; Wager et al., 1960). This fact signifies either extensive equilibration between the cumulus mush and the adjacent magma or compaction and melt migration toward the main reservoir. The lack of evidence for crystallization of trapped melt in the olivine gabbros indicates that only small amounts of liquid existed in the cumulus pile at the point when interstitial liquid ceased to migrate. The strong adcumulate nature of the gabbros is also supported by studies of minor and trace element concentrations that suggest a final porosity of ~3% or less (Natland et al., 1991). The planar fabric and deformation textures seen in most of the gabbros point toward compaction as a factor in development of the cumulate textures as observed in many layered intrusions (Shirley, 1986; Hunter, 1996; Meurer and Boudreau, 1998a). The final porosity of consolidated crystal mushes is not easily predicted or estimated (Irvine, 1980; Meurer and Boudreau, 1998b). Estimates of trapped liquid in the lower zone of the Skaergaard intrusion have suggested values from 10% to 15% in the hidden zone and an upward decrease to at least the base of the middle zone (Wager, 1963; Henderson, 1975). It can only be assumed that the porosity increased throughout the middle and upper zones (Irvine, 1980; Shirley, 1986; Meurer and Boudreau, 1996). This assumption means that at the point when the middle zone of the Skaergaard intrusion was accumulating, the lower zone constituted ~1000 m of crystal mush, of which at least the upper part was unconsolidated and contained variable and upwardly increasing amounts of interstitial liquid.

The effects of load on a crystal mush are to compact, reorganize, deform, and recrystallize the solid framework and to move the interstitial liquid upward toward lower-pressure regimes (Irvine, 1980; Hunter, 1996). Irvine (1980) estimated that as much as 50% of an interstitial liquid may continuously be returned to the main magma reservoir. The amount of liquid in a solidified crystal mush would depend on many factors, among which accumulation and solidification processes and cooling rates are particularly important (Sparks et al., 1985). An initial porosity of 50%, or higher, for a crystal mush in the magma chamber at Site 735 may be a reasonable estimate based on theoretical predictions (Irvine, 1980). The "final" porosity of the Site 735 cumulates may be estimated at well above 20% from the porosity of the olivine gabbros (<3%) and the average amount of Fe-Ti oxide gabbros (~20%), assuming that the Fe-Ti oxide gabbro component represents redistributed and segregated interstitial liquid (Dick et al., 1992). The preliminary data from the lower part of Hole 735B (Dick et al., 1998) show that the amount of Fe-Ti oxides decreases significantly in the lower parts of the core to a few percent. More than 20% trapped liquid near the top of a compacting crystal mush appears reasonable and is certainly much less than a likely initial porosity. The distribution of the estimated trapped liquid is generally consistent with theoretically predicted distributions showing increasing porosity through a cumulate pile toward an active accumulation front (Irvine, 1980; Shirley, 1986; Meurer and Boudreau, 1996).

The olivine gabbro sequences at Site 735 suggest an extent of solidification similar to the middle part of the lower zone (LZb) of the Skaergaard intrusion (35% to 50% depending on the trapped liquid). However, the LZb of the Skaergaard reflects a much higher extent of compaction (<5% trapped liquid; Wager, 1963; Henderson, 1975) than that indicated by Hole 735B gabbros. This high extent of compaction is despite the fact that both cumulate packages would predict the presence of comparably thick masses of cumulates and a magma reservoir amounting to a total of at least 50% by volume (the mass fraction may be higher because of Fe enrichments in the upper cumulates and liquid). The preservation of a zone with high porosity and high trapped liquid in the upper parts of the Site 735 gabbros may reflect sudden removal of magma from the chamber, thus effectively terminating melt migration and trapping the residual liquid in the crystal mush. Such removal may be caused by tectonic uplift and melt extraction, possibly by injection into a subaxial dike and sill system. It is possible that the trapped liquid distribution in the Site 735 gabbros represents a "fossil" compaction profile in an interrupted or decapitated magma chamber.

The liquid density configuration in a crystal-mush zone formed from an evolving magma would display a systematic and unstable upward increase prior to Fe-Ti oxide crystallization. As a result of compaction caused by the "collapse" of the solid framework, this liquid would then move upward with the stable density gradient toward lower-pressure regimes (Irvine, 1980). Pipes of pegmatitic rocks described from the Skaergaard intrusion and other layered intrusions might have resulted from upward channeled flow of interstitial liquid as a result of compaction (Tegner et al., 1994; McBirney, 1995; Meurer et al., 1997). Previous interpretations based on the observations from Hole 735B have argued for a dominantly lateral migration of evolved ferrobasaltic melt in a partially consolidated crystal mush (Dick et al., 1991, 1992; Ozawa et al., 1991; Natland et al., 1991; Bloomer et al., 1991). Lateral migration is supported by the apparent retention of the trapped interstitial liquid in the cumulus package without large-scale vertical migrations. The lack of strong modal layering in the crystal mush does not suggest density stratification, which may restrain melt migration and impose an uneven porosity distribution (Meurer et al., 1997). It is therefore not clear why the residual liquids would migrate laterally during compaction instead of vertically as predicted for a compacting crystal mush. One reason may be that tectonic disturbances of the crystal mush (Dick et al., 1991, 1992) result in intermittent lateral or sublateral reconfiguration of the pressure regime.

The migrating melt is initially in compositional stable equilibrium with the host solid cumulates. Fractionation of the liquid is only possible if the residual liquid is isolated or removed from its solid crystallization products. Cooling and crystallization of the cumulate pile during liquid migration would change the stable assemblage and eventually saturate the migrating liquid first in Fe-Ti oxide minerals (~1100 °C) and subsequently in apatite (~1030 °C). Continued flux through the same channels may result in modal Fe-Ti oxide and silicate proportions that exceed or greatly differ from cotectic proportions, leaving behind a trail of "secondary" cumulus zones either intermingled with the host or as narrow bands. The inferred higher modal proportions of Fe-Ti oxide minerals in the uppermost gabbros may therefore reflect a higher ratio of liquid to host rock. The observed limited re-equilibration between the migrating and crystallizing liquid (Bloomer et al., 1991) is likely to have been caused by relatively fast cooling rates of the crystal and liquid mush.

The flow distance and the efficiency of fractionation can be predicted to show a positive correlation with differentiation of the migrating liquid, as measured in the mineral assemblages and the mineral compositions. The high Fe contents of both olivine and augite at the transition zone between units 4 and 5 (Fig. 5) may be a simple reflection of high porosity and channeled flow at this horizon. Ozawa et al. (1991) suggested that the downward increase in the Fe content at this horizon may reflect downward migration of dense liquids. However, the production of extended Fe enrichment well past Fe-Ti oxide saturation, as suggested by Natland et al. (1991), has no experimental confirmation. As a result of Fe-Ti oxide crystallization, the Fe content would fall markedly, and the density of the residual liquid would decrease significantly, resulting in a highly unstable density configuration in the interstitial liquid. At this point, however, the low temperature and increased viscosity may have prevented upward migration. This solidification model for the trapped liquid explains several features observed in the Fe-Ti oxide gabbros (Dick et al., 1991, 1992; Natland et al., 1991), such as the cumulus nature, strong compaction of the olivine gabbro host, mineral assemblage and mineral compositional variability, bimodal mineral compositions, restricted re-equilibration and reactions, and gradational to sharp contacts with little evidence for intrusive relationships.

Liquid fractionation and zoned magma chamber

An alternative idea for the origin of the Fe-Ti oxide gabbros may be derived from the results of recent studies of the Skaergaard intrusion. Migration of ferrobasaltic melts has been proposed to account for what is perceived as an overabundance of Fe-rich material in the middle part of the layered series (Fig. 10; McBirney, 1995). Preferential crystallization along the margins and the roof of a chamber could result in the formation of dense Fe-rich melts that may repeatedly have plunged down the walls to pond at the top of the cumulus pile. Such a scenario for the Site 735 gabbros requires local variations in the crystallization conditions such that, at any specific time, the melt that formed at the roof or the walls differed from the melt present at the top of the cumulus pile. For the Skaergaard intrusion, Hoover (1989) argued that crystals forming along the margins were comparatively less evolved than those formed on the floor and along the roof, and he related this difference to a less efficient fractionation along the margins. Studies of the cumulates of the roof zone by Naslund (1984) and Hoover (1989) found that the crystallization orders at the margins differed from those inferred for the layered series. In particular, saturation with Fe-Ti oxides and apatite occurred much earlier along the roof than along the floor. This earlier saturation would imply that, at any time, liquids were relatively more evolved along the roof than on the floor. Hoover (1989) pointed out that the relatively fractionated liquids in the roof zone remained largely in situ despite the tendency of such liquids to descend along the walls; this scenario would require that the roof zone constituted an independently convecting and probably compositionally zoned cell above the main magma body. Naslund (1984) and later McBirney (1995) have proposed that the upper, downward-propagating cell exchanged material with the main chamber through sinking of mafic minerals into the main chamber and/or the percolation of felsic differentiates upward into the upper border zone during its late-stage evolution. McBirney (1995) recently suggested that dense, crystal-laden liquids produced along the margins and the roof descended or sank into the main magma and ponded on the floor of the Skaergaard intrusion.

It is possible that the growth of the cumulus pile on the floor or margins of a magma chamber at Site 735 was concurrent with the migration of dense melts downslope along the marginal magma-solid contacts, as suggested for the Skaergaard intrusion. Upward and inward growth of the layered and perhaps marginal series may have overtaken and prevented complete re-equilibration and mixing between ferrobasaltic "ponds and layers." Such a model may explain some features observed in the cores, such as the relatively sharp to gradational contacts, incomplete re-equilibration, and the lack of mutual intrusive relationships between the two melt types. The reason why dense liquids remained unequilibrated and preserved in gabbros from Holes 735B and 1105A may be due to a fast cooling rate and the rapid segregation of the dense liquids into deformationally controlled channelways and high-porosity zones. This contrasts with the Skaergaard intrusion for which such dense liquids may have been completely re-equilibrated or mixed with their host cumulates. Although it cannot be completely excluded as a possibility, the migration of dense liquids from the roof zone of a chamber is perhaps not likely considering that the Fe-Ti oxide gabbros are dominantly found in the uppermost few segments of host gabbros and not in the lower segments despite the compositional similarities of the host gabbros.

Remelted and partially re-equilibrated metabasaltic xenoliths

Basaltic rafts detached from the roof zone of a magma chamber may form an interconnected network that can be incorporated into the cumulate sequence (Habekost and Wilson, 1989). The raft complex described from the synorogenic Fongen-Hyllingen intrusion of the central Norwegian Caledonides forms a concordant to subconcordant, dominantly metabasaltic network with individual rafts as thick as 100 m and with lateral extents of 1–2 km. This raft complex is believed to have formed from largely in situ blocks that were detached from the roof zone as a consequence of the upward migration of the chamber (Habekost and Wilson, 1989). The individual rafts were embedded in the cumulates, as the crystallization and accumulation front migrated upward in response to chamber expansion and magma replenishments. The result is a network of highly re-equilibrated, fine-grained metabasaltic xenoliths intermixed with medium-grained cumulates down to the scale of a few centimeters that may be difficult to distinguish from magmatic layered cumulates (Thy and Esbensen, 1982). Because the Fongen-Hyllingen intrusion cooled slowly, the original mineral assemblages and mineral chemistry of the basalts re-equilibrated with the host cumulates; as a result, they are no longer easily distinguished, except on the basis of structural and textural grounds (Thy and Esbensen, 1982; Habekost and Wilson, 1989). It is conceivable that the Fe-Ti oxide gabbro layers and vein networks in the Site 735 gabbros originated in a fashion similar to that suggested by Habekost and Wilson (1989) for the Fongen-Hyllingen intrusion. Partial melting of a hydrated basaltic protolith may have produced Fe-Ti–rich melts that may have segregated into certain horizons, migrated, or crystallized largely in situ. Relatively fast cooling may have prohibited complete re-equilibration of the partially molten xenoliths. The main drawback for this idea is that it is uncertain whether partial melting of hydrated basalt would produce Fe-Ti–rich melts in sufficient amounts to explain the observed volume of Fe-Ti oxide gabbros at Site 735 (except perhaps for near-complete melting of ferrobasaltic protoliths). Most experimental evidence points toward partial melts of dacitic or rhyolitic compositions. It is also a problem that restitic compositions and textures are not widely observed in association with the Fe-Ti oxide gabbros from Hole 735B.

Replenishment and incomplete mixing with dense ferrobasaltic liquids

As a final possibility, we consider the effects of replenishment and injection of a dense ferrobasaltic liquid into a partially solidified magma chamber. If an incoming magma is relatively dense, it may pool on the floor of the chamber or mix with the resident magma depending on several factors, such as injection rate and density contrast (Campbell and Turner, 1989). Wiebe and Snyder (1993) have argued that the low injection rate of a dense, replenishing magma into a relatively evolved chamber of the Newark Island intrusion prohibited fountaining and resulted in pooling at the mush-liquid interface. In order to inhibit turbulent mixing and to pool along the floor, the injection rate must be low, and the resident liquid should be less dense than the replenishing magma. An evolving tholeiitic magma shows increasing density until the point of saturation with Fe-Ti oxide minerals, after which the density rapidly decreases (Stolper and Walker, 1980). The resident magma in the Site 735 chamber was saturated in olivine, augite, and plagioclase and evolved toward increasing density. Because the incoming magma was likely saturated in Fe-Ti oxide minerals, it may have been significantly denser than the resident melt and could have been able to pool at the mush-melt interface. It is also plausible that certain influx rates and density contrasts, in combination with fast cooling, may result in fountaining and incomplete mixing of the two components. This line of reasoning requires that coexisting basaltic and ferrobasaltic melts be present and interacting with each other during the early stages of magma-chamber evolution. There is no evidence, however, to suggest the coexistence of two basaltic melts at Site 735 or in its surroundings.

TECTONISM AND MAGMA-CHAMBER EVOLUTION

The structural evidence from the Site 735 gabbros suggests that crustal accretion at the ultra-slow-spreading Southwest Indian Ridge was controlled by an interplay of magmatism, tectonism, and hydrothermal activity (Dick et al., 1991). The intrusion at all crustal levels of small, rapidly crystallizing magma bodies into preexisting crystalline and semisolid gabbros was controlled by the temporal and spatial variability in the tectonic activity. Active ridge-axial and listric normal faults could have penetrated crystal-mush zones beneath the axial rift and resulted in shearing and ductile deformation of partially solid, rigid gabbro cumulates. This process may have created irregular zones of enhanced permeability and pressure release into which the late intercumulus melt could migrate from the compacting crystal mush. The result of these magmatic and tectonic processes is a complex intercalation of undeformed olivine gabbros and bands of sheared ferrogabbro. Ductile deformation, shearing, and recrystallization appear to have continued into the subsolidus regime, as demonstrated by a continuum between plastic and brittle behavior and the associated medium- to low-grade hydrothermal metamorphism of the deformed rocks. Magmatic layering and lamination in the gabbros is commonly overprinted by a weakly parallel plastic fabric that may have resulted from compaction and shearing.

Central to our understanding of crustal-accretion processes is to resolve the relative timing of magmatic and tectonic processes and their interaction in a cooling crystal mush (Dick et al., 1991). The relatively sharp lithologic boundaries between the two gabbro components at Site 735 have been taken to indicate penetration of a late liquid into an olivine gabbro mush (Bloomer et al., 1991; Natland et al., 1991). However, there are relatively few direct indications in the cores for discordant intrusive relationships between the Fe-Ti oxide and olivine gabbros (Dick et al., 1991). Because of the restricted lateral resolution in the cores, as well as the uniform nature of the host, intrusive relationships may not be easily detected. Intrusive relationships or channeling (upward or downward) of dense, crystal-laden Fe-rich liquids into mushes have been documented from layered intrusions (Tegner et al., 1994; McBirney, 1995). The Fe-Ti oxide gabbro from Hole 735B is suggested to have formed from evolved interstitial liquids as they migrated and cooled along localized fissures and cracks in a near-rigid gabbro mush (Bloomer et al., 1991; Dick et al., 1991; Natland et al., 1991). The very low final porosity of both the olivine and the Fe-Ti oxide gabbros suggests that compaction occurred simultaneously with cooling and that migration of the melt was driven by compaction and deformation. Deformation may have contributed to, and accelerated, compaction and subsolidus recrystallization and led to the development of penetrative foliation preferentially in the Fe-Ti oxide gabbros.

The mineral assemblages and mineral chemistry of various gabbros, tectonites, and veins provide a guide for the relative timing of their formation during cooling of Site 735 rocks. Consideration of phase assemblages and the experimental results of relatively similar melt compositions (Toplis and Carroll, 1996) indicate that the formation temperatures of olivine gabbros are well above 1100 °C. The crystallization of the intercumulus liquid would have proceeded with the precipitation of Fe-Ti oxides at ~1100 °C and apatite at ~1030 °C. Re-equilibration during cooling and elimination of the interstitial liquid were likely to have continued to much lower temperatures. Subsolidus deformation is interpreted to have occurred at temperatures as high as 1000 °C (Stakes et al., 1991). Therefore, it is plausible that ductile deformation and the development of metamorphic fabric overlapped with igneous hypersolidus to solidus cooling and that interstitial melt may still have been present although in sufficiently low amounts to allow the gabbros to accommodate shear stress (Stakes et al., 1991; Dick et al., 1992). The close spatial association between ductile shear zones and the concentrations of Fe-Ti oxides is also attributed to the oxides having a strong preferred susceptibility to ductile deformation and, hence, may not necessarily imply melt-assisted grain-boundary sliding (Agar and Lloyd, 1997).

Another argument for migration of Fe-Ti–rich melts into shear zones in the gabbros is that these zones contain modally very high concentrations of Fe-Ti oxide minerals far exceeding the cotectic proportions. Toplis and Carroll (1996) conducted experimental studies of ferrobasalts at various ranges of oxygen pressures and showed that the total amounts of Fe-Ti oxides for perfect fractional crystallization may reach 50% and that they then modally decrease significantly with falling temperatures as Fe and Ti are depleted in the melt. Such amounts are sufficient to explain the modal proportions seen in many of the strongly sheared Fe-Ti oxide gabbros and do not require sustained liquid flow into the shear zones. However, there is little doubt that interstitial melt migrated in the crystal mush or that this migration may have in part been channeled along weak zones; nevertheless, there is no compelling evidence suggesting that the late interstitial melt exclusively migrated into actively deforming brittle shear zones. Subsolidus recrystallization, alteration, and hydrothermal veining apparently occurred until much lower temperatures (Stakes et al., 1991; Dick et al., 1991, 1992).

Features related to penetrative ductile deformation and metamorphism are commonly observed in gabbroic complexes intruded and solidified in tectonically active settings and orogenic belts (e.g., Wilson et al., 1981). Synmagmatic faulting and slumping in crystal mushes are observed in well-exposed layered intrusions and reflect synchronous tectonic activity. There are ample indications for a late-stage migration of interstitial melt in layered intrusions commonly seen as concordant and discordant pegmatitic veins. These occurrences may have formed from upward channeling of liquid and interaction with semi-impermeable modal layering in the mush. There is little support, however, for large-scale melt migration along synmagmatic fault and shear zones in layered intrusions despite the fact that these may have consolidated in tectonically active environments (e.g., Wilson et al., 1981).

CONCLUSIONS

Observations from drilling at Sites 735 and 1105 near the Atlantis II Fracture Zone into the gabbros developed at the ultra-slow-spreading Southwest Indian Ridge suggest magma-chamber processes that in many respects are similar to those inferred from continental and ophiolitic counterparts. Magma chambers solidified by forming thick crystal-mush zones along the margins and floors in response to cooling. The remaining melt slowly changed its composition as previously crystallized minerals were removed and the interstitial liquid was recycled back into the main melt volume. Also similar to many layered intrusions and ophiolitic cumulate complexes, the oceanic gabbros show signs of repeated injection of new magma into the chamber that mixed with or displaced the resident magma. The crystal mush solidified by compaction and recrystallization and by migration of the trapped liquid upward in the pile across modal layering and boundaries of replenishment. Compaction and plastic flow in the semisolid mush may have resulted in, or enhanced the preexisting, modal layering and lamination. The presence of thick crystal-mush zones and their solidification by compaction and interstitial melt migration are common features of many continental, ophiolitic, and oceanic magma chambers.

The prominent difference between oceanic gabbro complexes and continental layered intrusions is the presence in the former of subordinate amounts of veins and lenses of evolved ferrogabbros in dominantly relatively primitive olivine gabbros. The juxtaposition of such different gabbro types is a characteristic feature of the Site 735 gabbros that is not widely encountered in continental gabbros. The distribution of the ferrogabbros in Hole 735B may record a fossil compaction profile with increasing amounts of evolved trapped liquid toward a former crystallization front. Tectonic uplift and the vertical and lateral drainage of the residual melt from the magma lens may have resulted in termination of compaction and vertical melt migration in the cumulus mush. The trapped melt in the cumulus mush was moderately affected by syn- and post-tectonic, channeled melt movement and uplift-induced pressure release and crack propagation. Thus vertical migration of the interstitial melt was largely arrested, resulting in dominantly lateral migration and segregation of the remaining melt facilitated by uplift-induced ruptures in the crystal mush. Such fossil compaction profiles are not commonly recorded in gabbros of stable continental cratons and margins, but may occur more commonly in ophiolitic complexes.

ACKNOWLEDGMENTS

This work was supported by grants from the U.S. Science Advisory Committee of the Joint Oceanographic Institutions to P. Thy and Y. Dilek for their participation in ODP Legs 179 (Hammer Drilling and NERO) and 153 (MARK area), respectively. Thy's study of the Troodos ophiolite was partially supported by SnF (Denmark and the Carlsberg Foundation); additional support was provided by National Science Foundation grants to E. M. Moores. The manuscript benefited from helpful comments and reviews by H.R. Naslund, S.H. Bloomer, C.K. Brooks, and J.R. Wilson.

REFERENCES CITED

Agar, S.M., and Lloyd, G.E., 1997, Deformation of Fe-Ti oxides in gabbroic shear zones from the MARK area, *in* Karson, J.A., Cannat, M., Miller, J.D., and Elthon, D., eds., Proceedings of the Ocean Drilling Program, Scientific Results 153: College Station, Texas, Ocean Drilling Program, p. 123–141.

Allen, C.R., 1975, The petrology of a portion of the Troodos plutonic complex, Cyprus [Ph.D. thesis]: Cambridge University, 161 p.

Bloomer, S.H., Natland, J.H., and Fisher, R.L., 1989, Mineral relations in gabbroic rocks from fracture zones of Indian Ocean ridges: Evidence for extensive fractionation, partial diversity, and boundary-layer recrystallization, *in* Saunders, A., and Norrey, M.J., eds., Magmatism in the ocean basins: Geological Society of London Special Publication 42, p. 107–124.

Bloomer, S.H., Meyer, P.S., Dick, H.J.B., Ozawa, K., and Natland, J.H., 1991, Textural and mineralogical variations in gabbroic rocks from Hole 735B, *in* Von Herzen, R.P., Robinson, P.T., et. al., eds., Proceedings of the Ocean Drilling Program, Scientific Results 118: College Station, Texas, Ocean Drilling Program, p. 21–39.

Boudier, F., Nicolas, A., and Ildefonse, B., 1996, Magma chambers in the Oman ophiolite: Fed from the top and the bottom: Earth and Planetary Science Letters, v. 144, p. 239–250.

Boudreau, A.E., and McBirney, A.R., 1997, The Skaergaard layered series: Part III. Non-dynamic layering: Journal of Petrology, v. 38, p. 1003–1020.

Brooks, C.K., and Nielsen, T.F.D., 1990, A discussion of Hunter and Sparks (Contrib Mineral Petrol 95:451–461): Contributions to Mineralogy and Petrology, v. 104, p. 235–247.

Browning, P., 1984, Cryptic variation within the cumulate sequence of the Oman ophiolite: Magma chamber depth and petrological implications, *in* Gass, I.G., Lippard, S.J., and Shelton, A.W., eds., Ophiolites and oceanic lithosphere: Geological Society of London Special Publication 13, p. 71–82.

Browning, P., Roberts, S., and Alabaster, T., 1989, Fine-scale modal layering and cyclic units in ultramafic cumulates from the CY-4 borehole, Troodos ophiolite: Evidence for open system magma chamber, *in* Gibson, I.L., Malpas, J., Robinson, P.T., and Xenophontos, C., eds., Cyprus Crustal Study Project, Hole CY-4: Geological Survey of Canada, Paper 88-9, p. 193–220.

Campbell, I.H., and Turner, J.S., 1989, Fountains in magma chambers: Journal of Petrology, v. 30, p. 885–923.

Cann, J.R., 1970, New model for the structure of oceanic crust: Nature, v. 226, p. 928–930.

Cann, J.R., 1974, A model for oceanic crustal structure developed: Geophysical Journal of the Royal Astronomical Society, v. 39, p. 169–187.

Cannat, M., 1996, How thick is the magmatic crust at slow spreading oceanic ridges?: Journal of Geophysical Research, v. 101, p. 2847–2857.

Cannat, M., Karson, J.A., et al., 1995, Proceedings of the Ocean Drilling Program, Initial Reports 153: College Station, Texas, Ocean Drilling Program, 798 p.

Cannat, M., Mevel, C., and Stakes, D., 1991, Stretching of the deep crust at the slow-spreading Southwest Indian Ridge: Tectonophysics, v. 190, p. 73–94.

Casey, J.F., Miller, J., and Scientific Party, 1998, Gabbroic complex drilled during ODP Leg 179, Southwest Indian Ridge: A comparison with Hole 735B [abs.]: Eos, Transactions of the American Geophysical Union, v. 79, p. F942.

Cawthorn, R.G., ed., 1996, Layered intrusions: New York, Elsevier, 531 p.

Christensen, N.L., 1978, Ophiolites, seismic velocities, and oceanic crustal structure: Tectonophysics, v. 47, p. 131–157.

Christensen, N.L., and Smewing, J.D., 1981, Geology and seismic structure of the northern section of the Oman ophiolite: Journal of Geophysical Research, v. 86, p. 2545–2555.

Coleman, R.G., 1977, Ophiolites: Berlin, Springer-Verlag, 229 p.

Conrad, M.E., and Naslund, H.R., 1989, Modally graded rhythmic layering in the Skaergaard intrusion: Journal of Petrology, v. 30, p. 251–269.

Detrick, R.S., 1991, Ridge crest magma chambers: A review of results from recent marine seismic experiments, *in* Peters, Tj., Nicolas, A., and Coleman, R., eds., Ophiolites: Genesis and evolution of oceanic lithosphere, Dordrecht, Kluwer Academic Publishers, p. 7–20.

Detrick, R.S., Buhl, P., Vera, E., Mutter, J., Orcutt, J., Madsen, J., and Brocher, T., 1987, Multi-channel seismic imaging of a crustal magma chamber along the East Pacific Rise: Nature, v. 326, p. 35–41.

Detrick, R.S., Mutter, J.C., Buhl, P., and Kim, I.I., 1990, No evidence from multichannel reflection data for a crustal magma chamber in the MARK area on the mid-Atlantic ridge: Nature, v. 347, p. 61–64.

Dick, H.J.B., Meyer, P.S., Bloomer, S., Kirby, S., Stakes, D., and Mawer, C., 1991, Lithostratigraphic evolution of an in situ section of oceanic layer 3, *in* Von Herzen, R.P., Robinson, P.T., et al., eds., Proceedings of the Ocean Drilling Program, Scientific Results 118: College Station, Texas, Ocean Drilling Program, p. 439–538.

Dick, H.J.B., Natland, J.H., Miller, D.J., et al., 1999, Proceedings of the Ocean Drilling Program, Initial Reports 176 (CD-ROM): College Station, Texas, Ocean Drilling Program.

Dick, H.J.B., Natland, J.H., Robinson, P.T., MacLeod, C.J., and Allerton, S.S., ODP Leg 176 and JCR Scientific Parties, 1998, Accretion of the lower ocean crust [abs.]: Eos, Transactions of the American Geophysical Union, v. 79, p. F40.

Dick, H.J.B., Robinson, P.T., and Meyer, P.S., 1992, The plutonic foundation of a slow-spreading ridge, *in* Duncan, R.A., Rea, D.K., Kidd, R.B., von Rad, U., and Weissel, J.K., eds., Synthesis of results from scientific drilling in the Indian Ocean: Washington, D.C., American Geophysical Union, Geophysical Monograph 70, p. 1–39.

Dilek, Y., and Eddy, C.A., 1992, The Troodos (Cyprus) and Kizildag (S. Turkey) ophiolites as structural models for slow-spreading ridge segments: Journal of Geology, v. 100, p. 305–322.

Dilek, Y., and Thy, P., 1998, Structure, petrology, and seafloor spreading tectonics of the Kizildag ophiolite, Turkey, *in* Mills, R.A., and Harrison, K., eds., Modern ocean floor processes and the geological record, Geological Society of London, Special Publication 148, p. 43–69.

Dilek, Y., Moores, E.M., and Furnes, H., 1998, Structure of modern oceanic crust and ophiolites and implications for faulting and magmatism at oceanic spreading centers, *in* Buck, R., Karson, J., Delaney, P., and Lagabrielle, Y., eds., Faulting and magmatism at mid-ocean ridges: Washington, D.C., American Geophysical Union, Geophysical Monograph 106, p. 219–266.

Dilek, Y., Thy, P., Moores, E.M., and Ramsden, T.W., 1990, Tectonic evolution of the Troodos ophiolite within the Tethyan framework: Tectonics, v. 9, p. 811–823.

Dilek, Y., Thy, P., Moores, E.M., and Ramsden, T.W., 1992, Reply to discussion: Tectonics, v. 11, p. 916–923.

Eddy, C.A., Dilek, Y., Hurst, S., and Moores, E.M., 1998, Seamount formation and associated caldera complex and hydrothermal mineralization in ancient oceanic crust, Troodos Ophiolite (Cyprus): Tectonophysics, v. 292, p. 189–210.

Elthon, D., Casey, J.F., and Komor, S., 1984, Cryptic mineral chemistry variations in a detailed traverse through the cumulate ultramafic rocks of the North Arm Mountain massif of the Bay of Islands ophiolite, Newfoundland, *in* Gass, I.G., Lippard, S.J., and Shelton, A.W., eds., Ophiolites and oceanic lithosphere, Geological Society of London Special Publication 13, p. 269–290.

Gibson, I.L., Malpas, J., Robinson, P.T., and Xenophontos, C., eds., 1989, Cyprus Crustal Study Project, Hole CY-4: Geological Survey of Canada, Paper 88-9, 393 p.

Greenbaum, D., 1972, Magmatic processes at ocean ridges and evidence from the Troodos massif, Cyprus: Nature Physical Sciences, v. 238, p. 18–21.

Habekost, E.M., and Wilson, J.R., 1989, Raft-like metabasaltic inclusions in the Fongen-Hyllingen layered complex, and their implication for magma chamber evolution: Journal of Petrology, v. 30, p. 1415–1441.

Harding, A.J., Orcutt, J.A., Kappus, M.E., Vera, E.E., Mutter, J.C., Buhl, P., Detrick, R.S., and Brocher, T.M., 1989, The structure of the young oceanic crust at 13°N on the East Pacific Rise from expanding spread profiles: Journal of Geophysical Research, v. 94, p. 12163–12196.

Harper, G.D., 1985, Tectonics of slow-spreading mid-ocean ridges and consequences of a variable depth to the brittle/ductile transition: Tectonics, v. 4, p. 395–409.

Hébert, R., Constantin, M., and Robinson, P.T., 1991, Primary mineralogy of Leg 118 gabbroic rocks and their place in the spectrum of oceanic mafic igneous rocks, *in* Von Herzen, R.P., Robinson, P.T., et. al., eds., Proceedings of the Ocean Drilling Program, Scientific Results 118: College Station, Texas, Ocean Drilling Program, p. 3–20.

Henderson., P., 1975, Geochemical indicators of the efficiency of fractionation of the Skaergaard intrusion, East Greenland: Mineralogical Magazine, v. 40, p. 285–291.

Hess, H.H., 1960, Stillwater Igneous Complex, Montana: A quantitative mineralogical study: Geological Society of America Memoir 80, 230 p.

Hoover, J.D., 1989, Petrology of the marginal border series of the Skaergaard intrusion: Journal of Petrology, v. 30, p. 399–439.

Hunter, R.H., 1996, Texture development in cumulate rocks, *in* Cawthorn, R.G., ed., Layered intrusions: Amsterdam, Elsevier, p. 77–101.

Hunter, R.H., and Sparks, R.S.J., 1987, The differentiation of the Skaergaard intrusion: Contributions to Mineralogy and Petrology, v. 95, p. 451–461.

Irvine, T.N., 1974, Petrology of the Duke Island ultramafic complex, southeastern Alaska, Geological Society of America Memoir 138, 240 p.

Irvine, T.N., 1980, Magmatic infiltration metasomatism, double diffusive fractional crystallization, and adcumulus growth in the Muskox intrusion and other layered intrusions, *in* Hargraves, R.B., ed., Physics of magmatic processes: Princeton University Press, p. 325–383.

Irvine, T.N., 1982, Terminology for layered intrusions: Journal of Petrology, v. 23, p. 127–162.

Irvine, T.N., 1987, Layering and related structures in the Duke Island and Skaergaard intrusions: Similarities, differences, and origins, *in* Parsons, I., ed.,

Origins of igneous layering: Dordrecht, D. Reidel Publishing Co., p. 185–245.

Irvine, T.N., Andersen, J.C.Ø., and Brooks, C.K., 1998, Included blocks (and blocks within blocks) in the Skaergaard intrusion: Geological relations and the origins of rhythmic modally graded layers: Geological Society of America Bulletin, v. 110, p. 1398–1447.

Jackson, E.D., 1961, Primary textures and mineral associations in the ultramafic zone of the Stillwater complex, Montana: U.S. Geological Survey Professional Paper 358, 106 p.

Karson, J.A., 1990, Sea floor spreading on the mid-Atlantic ridge: implications for the structure of ophiolites and oceanic lithosphere produced in slow-spreading environments, *in* Malpas, J., Moores, E.M., Panayiotou, A., and Xenophontos, C., eds., Ophiolites, oceanic crustal analogues: Nicosia, Cyprus, Geological Survey Department, p. 125–130.

Karson, J.A., and Lawrence, R.M., 1997, Tectonic setting of serpentinite exposures on the western median valley wall of the MARK area in the vicinity of Site 920, Proceedings of the Ocean Drilling Program, Scientific Results 153: College Station, Texas, Ocean Drilling Program, p. 3–21.

Karson, J.A., Thompson, G., Humpris, S.E., Edmond, J.M., Bryan, W.B., Brown, J.B., Winters, A.T., Pockalny, R.A., Casey, J.F., Campbell, A.C., Klinkhammer, G.P., Palmer, M.R., Kinzler, R.J., and Sulanowska, M.M., 1987, Along-axis variations in seafloor spreading in the MARK area: Nature, v. 328, p. 681–685.

Kelemen, P.B., and Aharanov, E., 1998, Periodic formation of magma fractures and generation of layered gabbros in the lower crust beneath oceanic spreading ridges, *in* Buck, W.R., Delaney, P.T., Karson, J.A., and Lagabrielle, Y., eds., Faulting and magmatism at mid-ocean ridges: Washington, D.C., American Geophysical Union, Geophysical Monograph 106, p. 267–289.

Kent, G.M., Harding, A.J., and Orcutt, J.A., 1990, Evidence for a smaller magma chamber beneath the East Pacific Rise at 9°30′ N: Nature, v. 344, p. 650–653.

Lagabrielle, Y., and Lemoine, M., 1997, Alpine, Corsican and Apennine ophiolites: The slow-spreading ridge model: Comptes Rendus des Seances de l'Academie des Sciences, v. 325, p. 909–920.

Lagabrielle, Y., Bideau, D., Cannat, M., Karson, J.A., and Mével, C., 1998, Ultramafic-mafic plutonic rock suites exposed along the Mid-Atlantic Ridge (10°N–30°N): Symmetrical-asymmetrical distribution and implications for seafloor spreading processes, *in* Buck, W.R., Delaney, P.T., Karson, J.A., and Lagabrielle, Y., eds., Faulting and magmatism at mid-ocean ridges: Washington, D.C., American Geophysical Union, Geophysical Monograph 106, p. 153–176.

Lister, C.R.B., 1983, On the intermittency and crystallization mechanism of subseafloor magma chambers: Geophysical Journal of the Royal Astronomical Society, v. 73, p. 351–365.

Maaløe, S., 1976a, The zoned plagioclase of the Skaergaard intrusion, East Greenland: Journal of Petrology, v. 17, p. 398–419.

Maaløe, S., 1976b, Quantitative aspects of fractional crystallization of major elements: Journal of Geology, v. 84, p. 81–96.

Maaløe, S., 1978, The origin of rhythmic layering: Mineralogical Magazine, v. 42, p. 337–345.

Maaløe, S., 1987, Rhythmic layering of Skaergaard intrusion, *in* Parsons, I., ed., Origins of igneous layering: Dordrecht, D. Reidel Publishing Co., p. 247–262.

Malpas, J., 1990, Crustal accretionary processes in the Troodos ophiolite, Cyprus: Evidence from field mapping and deep crustal drilling, *in* Malpas, J., Moores, E.M., Panayiotou, A., and Xenophontos, C., eds., Ophiolites, oceanic crustal analogues: Nicosia, Cyprus, Geological Survey Department, p. 65–74.

Malpas, J., Brace, T., and Dunsworth, S.M., 1989, Structural and petrological relationships of the CY-4 drill hole of the Cyprus Crustal Study Project, *in* Gibson, I.L., Malpas, J., Robinson, P.T., and Xenophontos, C., eds., Cyprus Crustal Study Project, Hole CY-4: Geolocial Survey of Canada Paper 88-9, p. 39–67.

McBirney, A.R., 1989, The Skaergaard layered series: I. Structure and average compositions: Journal of Petrology, v. 30, p. 363–397.

McBirney, A.R., 1995, Mechanisms of differentiation in the Skaergaard intrusion: Journal of Geological Society of London, v. 152, p. 421–435.

McBirney, A.R., and Hunter, R.H., 1995, The cumulate paradigm reconsidered: Journal of Geology, v. 103, p. 114–122.

McBirney, A.R., and Nicolas, A., 1997, The Skaergaard layered series. Part II. Magmatic flow and dynamic layering: Journal of Petrology, v. 38, p. 569–580.

McBirney, A.R., and Noyes, R.M., 1979, Crystallization and layering of the Skaergaard intrusion: Journal of Petrology, v. 20, p. 487–554.

Meurer, W.P., and Boudreau, A.E., 1996, Compaction of density-stratified cumulates: Effects on trapped liquid distribution: Journal of Geology, v. 104, p. 115–120.

Meurer, W.P., and Boudreau, A.E., 1998a, Compaction of igneous cumulates. Part II. Compaction and the development of igneous foliations: Journal of Geology, v. 106, p. 293–304.

Meurer, W.P., and Boudreau, A.E., 1998b, Compaction of igneous cumulates. Part I. Geochemical consequences for cumulates and liquid fractionation: Journal of Geology, v. 106, p. 281–292.

Meurer, W.P., Klaber, S., and Boudreau, A.E., 1997, Discordant bodies from olivine-bearing zones III and IV of the Stillwater complex, Montana: Evidence for postcumulus fluid migration and reaction in layered intrusions: Contributions to Mineralogy and Petrology, v. 130, p. 81–92.

Meyer, P.S., Dick, H.J.B., and Thompson, G., 1989, Cumulate gabbros from the Southwest Indian Ridge, 54°S–7°16′E: Implications for magmatic processes at a slow spreading ridge: Contributions to Mineralogy and Petrology, v. 103, p. 44–63.

Moores, E.M., and Vine, F.J., 1971, The Troodos massif, Cyprus, and other ophiolites as oceanic crust: evaluation and implications: Philosophical Transactions of the Royal Society of London, Series A, v. 268, p. 443–466.

Morse, S.A., 1969, The Kiglapait Layered Intrusion, Labrador: Geological Society of America Memoir 112, 204 p.

Morton, E.M., and Sleep, N.H., 1985, A mid-ocean ridge thermal model: Constraints on the volume of axial hydrothermal flux: Journal of Geophysical Research, v. 90, p.11345–11353.

Naslund, H.R., 1984, Petrology of the upper border series of the Skaergaard intrusion: Journal of Petrology, v. 25, p. 185–212.

Natland, J.H., Meyer, P.S., Dick, H.J.B., and Bloomer, S.H., 1991, Magmatic oxides and sulfides in gabbroic rocks from Hole 735B and the later development of the liquid line of descent, *in* Von Herzen, R.P., Robinson, P.T., et al., eds., Proceedings of the Ocean Drilling Program, Scientific Results 118: College Station, Texas, Ocean Drilling Program, p. 75–111.

Natland, J.H., Dick, H.J.B., and Leg 176 Scientific Party, 1998, Plutonic assemblages of transverse ridges near transform faults of the central and Southwest Indian Ridge [abs.]: Eos, Transactions of the American Geophysical Union, v. 79, p. F894.

Nicolas, A., 1989, Structures of ophiolites and dynamics of oceanic lithosphere: Dordrecht, Kluwer Academic Publishers, 367 p.

Nicolas, A., and Ildefonse, B., 1996, Flow mechanism and viscosity in basaltic magma chambers: Geophysical Research Letters, v. 23, p. 2013–2016.

Nicolas, A., Reuber, I., and Benn, K., 1988, A new magma chamber model based on structural studies in the Oman ophiolite: Tectonophysics, v. 151, p. 87–105.

Nicolas, A., Freydier, Cl., Godard, M., and Vauchez, A., 1993, Magma chamber at oceanic ridges: How large?: Geology, v. 21, p. 53–56.

Nwe, Y.Y., 1976, Electron-probe studies of the earlier pyroxenes and olivines from the Skaergaard intrusion, East Greenland: Contributions to Mineralogy and Petrology, v. 55, p. 105–126.

Ozawa, K., Meyer, P., and Bloomer, S.H., 1991, Mineralogy and textures of iron-titanium oxide gabbros and associated olivine gabbros from Hole 735B, *in* Von Herzen, R.P., Robinson, P.T., et. al., eds., Proceedings of the Ocean Drilling Program, Scientific Results 118: College Station, Texas, Ocean Drilling Program, p. 41–73.

Pallister, J.S., and Hopson, C.A., 1981, Semail ophiolite plutonic suite: Field relations, phase variations, cryptic variation and layering, and a model of a spreading ridge magma chamber: Journal of Geophysical Research, v. 86, p. 2593–2644.

Penrose Conference Participants, 1972, Penrose field conference on ophiolites: Geotimes, v. 17, p. 24–25.

Pettigrew, T.L., Casey, J.F., Miller, D.J., et al., 1999, Proceedings of the Ocean

Drilling Program, Initial Reports 179 (CD-ROM): College Station, Texas, Ocean Drilling Program.

Phipps Morgan, J., and Chen, Y.J., 1993, The genesis of oceanic crust: Magma injection, hydrothermal circulation, and crustal flow: Journal of Geophysical Research, v. 98, p. 6283–6297.

Phipps Morgan, J., Harding, A., Orcutt, J., Kent, G., and Chen, Y.J., 1994, An observational and theoretical synthesis of magma chamber geometry and crustal genesis along a mid-ocean ridge spreading center, *in* Ryan, M.P., ed., Magmatic systems: San Diego, Academic Press, p. 139–178.

Portnyagin, M.V., Danyushevsky, L.V., and Kamenetsky, V.S., 1997, Coexistence of two distinct mantle sources during formation of ophiolites: A case study of primitive pillow-lavas from the lowest part of the volcanic section of the Troodos ophiolite, Cyprus: Contributions to Mineralogy and Petrology, v. 128, p. 287–301.

Quick, J.E., and Denlinger, R.P., 1993, Ductile deformation and the origin of layered gabbro in ophiolites: Journal of Geophysical Research, v. 98, p. 14015–14027.

Robinson, P.T., Von Herzen, R., et al., 1989, Proceedings of the Ocean Drilling Program, Initial Reports 118: College Station, Texas, Ocean Drilling Program, 823 p.

Shipboard Scientific Party, 1989, Site 735, *in* Robinson, P.T., Von Herzen, R., et al., Proceedings of the Ocean Drilling Program, Initial Reports 118: College Station, Texas, Ocean Drilling Program, p. 89–222.

Shirley, D.N., 1986, Compaction of igneous cumulates: Journal of Geology, v. 94, p. 795–809.

Sinton, J.M., and Detrick, R.S., 1992, Mid-ocean ridge magma chambers: Journal of Geophysical Research, v. 100, p. 2211–2237.

Sleep, N.H., 1975, Formation of oceanic crust: Some thermal constraints: Journal of Geophysical Research, v. 80, p. 4037–4042.

Sleep, N.H., 1978, Thermal structure and kinematics of mid-oceanic ridge axis: Some implications to basaltic volcanism. Geophysical Research Letters, v. 5, p. 426–428.

Smewing, J.D., 1981, Mixing characteristics and compositional differences in mantle-derived melts beneath spreading axes: Evidence from cyclically layered rocks in the ophiolite of north Oman: Journal of Geophysical Research, v. 86, p. 2645–2659.

Sobolev, A.V., Portnyagin, M.V., Dimitriev, L.V., Tsameryan, O.P., Danyushevsky, L.V., Kononkova, N.N., Shimizu, N., and Robinson, P.T., 1993, Petrology of ultramafic lavas and associated rocks of the Troodos massif, Cyprus: Petrology, v. 1, p. 379–412.

Solomon, S.C., and Toomey, D.R., 1992, The structure of mid-ocean ridges: Annual Review of Earth and Planetary Sciences, v. 20, p. 329–364.

Sørensen, K.B., 1995, Textural and chemical evolution of the Fe-Ti oxide minerals during late- and post magmatic cooling of the Skaergaard intrusion, East Greenland [MS thesis]: Copenhagen, University of Copenhagen, 113 p.

Sparks, R.S.J., Huppert, H.E., Kerr, R.C., McKenzie, D.P., and Tait, S.R., 1985, Postcumulus processes in layered intrusions: Geological Magazine, v. 122, p. 555–568.

Stakes, D., Mevel, C., Cannat, M., and Chapu, T., 1991, Metamorphic stratigraphy of Hole 735B, *in* Von Herzen, R.P., Robinson, P.T., et. al., eds., Proceedings of the Ocean Drilling Program, Scientific Results 118: College Station, Texas, Ocean Drilling Program, p. 153–180.

Stolper, E., and Walker, D., 1980, Melt density and the average composition of basalt: Contributions to Mineralogy and Petrology, v. 74, p. 7–12.

Tegner, C., Wilson, J.R., and Cawthorn, R.G., 1994, The dunite-clinopyroxenite pegmatoidal pipe, Tweefontain, eastern Bushveld complex, South Africa: South African Journal of Geology, v. 97, p. 415–430.

Thy, P., 1987, Petrogenetic implications of mineral crystallization trends of Troodos cumulates, Cyprus: Geological Magazine, v. 124, p. 1–11.

Thy, P., and Esbensen, K.H., 1982, Origin of fine-scale graded layers in layered intrusions: Geological Magazine, v. 119, p. 405–412.

Thy, P., and Moores, E.M., 1988, Crustal accretion and tectonic setting of the Troodos ophiolite, Cyprus: Tectonophysics, v. 147, p. 211–245.

Thy, P., and Xenophontos, C., 1991, Crystallization orders and phase chemistry of glassy lavas from the pillow sequences, Troodos ophiolite, Cyprus: Journal of Petrology, v. 32, p. 403–428.

Thy, P., Jakobsen, N.N., and Wilson, J.R., 1988, Fine-scale graded layers in the Fongen-Hyllingen gabbroic complex, Norway: Canadian Mineralogist, v. 26, p. 235–243.

Thy, P., Schiffman, P., and Moores, E.M., 1989, Igneous mineral stratigraphy and chemistry of the Cyprus Crustal Study Project drill core in the plutonic sequences of the Troodos ophiolite, *in* Gibson, I.L., Malpas, J., Robinson, P.T., and Xenophontos, C., eds., Cyprus Crustal Study Project, Hole CY-4: Geological Survey of Canada, Paper 88–9, p. 147–185.

Toplis, M.J., and Carroll, M.R., 1996, Differentiation of ferro-basaltic magmas under conditions open and closed to oxygen: Implications for the Skaergaard intrusions and other natural systems: Journal of Petrology, v. 37, p. 837–858.

Varga, R.J., and Moores, E.M., 1985, Spreading structure of the Troodos ophiolite, Cyprus: Geology, v. 13, p. 846–850.

Von Herzen, R.P., Robinson, P.T., et al., 1991, Proceedings of the Ocean Drilling Program, Scientific Results 118: College Station, Texas, Ocean Drilling Program.

Wager, L.R., 1960, The major element variation of the layered series of the Skaergaard intrusion and a re-estimation of the average composition of the hidden layered series and of the successive residual magmas: Journal of Petrology, v. 1, p. 364–398.

Wager, L.R., 1963, The mechanism of adcumulus growth in the layered series of the Skaergaard intrusion: Mineralogical Society of America Special Paper 1, p. 1–9.

Wager, L.R., and Brown, G.M., 1967, Layered igneous rocks: San Francisco, Freeman, 588 p.

Wager, L.R., and Deer, W.A., 1939, Geological investigations in East Greenland. Part III. The petrology of the Skaergaard intrusion, Kangerrdlugssuaq, East Greenland: Meddelelser om Grønland, v. 105 (4) (reissued 1962), 352 p.

Wager, L.R., Brown, G.M., and Wadsworth, W.J., 1960, Types of igneous cumulates: Journal of Petrology, v. 1, p. 73–85.

Wiebe, R.A., and Snyder, D., 1993, Slow, dense replenishments of a basic magma chamber: The layered series of the Newark Island layered intrusion, Nain, Labrador: Contributions to Mineralogy and Petrology, v. 113, p. 59–72.

Wilson, J.R., and Larsen, S.B., 1982, Discordant layering relations in the Fongen–Hyllingen basic intrusion: Nature, v. 299, p. 625–626.

Wilson, J.R., and Larsen, S.B., 1985, Two-dimensional study of a layered intrusion: The Hyllingen series, Norway: Geological Magazine, v. 122, p. 97–124.

Wilson, J.R., Esbensen, K.H., and Thy, P., 1981, Igneous petrology of the synorogenic Fongen–Hyllingen layered basic complex, south-central Scandinavian Caledonides: Journal of Petrology, v. 22, p. 584–627.

Manuscript Accepted by the Society February 23, 2000

Printed in the U.S.A.

Geological Society of America
Special Paper 349
2000

Syntexis and the genesis of lower oceanic crust

Jean H. Bédard
Geological Survey of Canada, Centre Géoscientifique de Québec, 2700 Einstein, C.P. 7500, Ste-Foy, Québec G1V 4C7, Canada; e-mail: Bedard@gsc.nrcan.gc.ca
Réjean Hébert, Alain Berclaz, Véronika Varfalvy
Département de Géologie et de Génie Géologique, Université Laval, Ste-Foy, Québec G1K 7P4, Canada

ABSTRACT

We propose that the term "syntexis" be revived and expanded to encompass all of the processes, mechanical and chemical, by which magmas react with and assimilate their host rocks. Synkinematic sills (≤50 m thick) make up the dominant proportion of the crust of the North Arm Mountain massif of the Bay of Islands ophiolite, Newfoundland. Field, textural, and geochemical evidence implies that at North Arm, most near-monomineralic rocks (chromitites, anorthosites, and pyroxenites) are products of syntectic assimilation reactions, and a substantial proportion of wehrlites, feldspathic peridotites, and olivine gabbros are syntectic hybrids. Chromitites formed by incongruent dissolution of pyroxene or feldspar into olivine + chromite–saturated melts. Such reaction-derived chromites associated with gabbroic protoliths are systematically more aluminous than those associated with pyroxenitic protoliths; this fact suggests that the composition of the Cr-spinel in crustal chromitites at North Arm is controlled by the nature of the phases being assimilated, not by the composition of the magma. Pyroxenite and anorthosite reaction rims are interpreted to have formed through pore-scale hybridization between invading primitive magmas and partial melts of gabbroic hosts. Most feldspathic peridotites and olivine gabbros display textures suggesting that they are mixtures of invading primitive magmas and disaggregated gabbroic hosts and do not represent simple cotectic cumulates. The data from North Arm suggest that the lower ophiolitic (and oceanic) crust may act as a reactive filter and that the chemical evolution of primary magmas owes as much to assimilation and reaction with older cumulates as it does to fractional crystallization.

INTRODUCTION

Geochemical variations among suites of mid-oceanic-ridge basalts (MORBs) are typically interpreted in terms of differing degrees of melting of a variably depleted or enriched mantle (e.g., Klein and Langmuir, 1987; McKenzie and O'Nions, 1991; Langmuir et al., 1992). However, most MORBs are too evolved geochemically (high FeO/MgO ratios, low Cr and Ni contents) to represent primary mantle melts (Bender et al., 1978; Fujii and Bougault, 1983; Falloon and Green, 1988; Hess, 1992). The cumulates that constitute the lower half of the oceanic crust commonly contain significant proportions of evolved rocks, such as gabbronorites and ferrogabbros (Dick et al., 1991; Natland et al., 1991; Sinton and Detrick, 1992; Mével et al., 1991; Constantin et al., 1996; Natland and Dick, 1996), supporting the idea that MORBs differentiate in the crust. The differentiation process is generally assumed to be some form of closed-system fractional crystallization (see review in Shen and Forsyth, 1995), with additional scatter caused by magma mixing in long-lived, replenished magma chambers (Dungan and Rhodes, 1978; Walker et al., 1979; O'Hara and Mathews, 1981). Recent attention has focused on the possibility that magmatic reactions between ascending melts and the mantle may also play a role in basalt petrogenesis (Richter, 1986; Navon and Stolper, 1987; Kelemen, 1990; Kelemen et al., 1990; Bodinier et al., 1991; Nielson and Wilshire, 1993; Navon et al., 1996; Varfalvy et al., 1996; Batanova et al., 1998). We do not question the existence or importance of intramantle processes, but wish to point out an apparent gap in reasoning. If reactions between primitive basaltic melts and shallow peridotitic mantle are plausible, then such melts should also react strongly with lower-crustal rocks, rocks that are composed of minerals with which the melts are in gross chemical disequilibrium. Oddly enough, the possible importance of reactions between ascending primitive melts and crustal rocks in oceanic environments has received little attention.

Since the publication of *The Evolution of the Igneous Rocks* (Bowen, 1928), fractional crystallization has been considered the dominant postsegregation magmatic differentiation process, whereas the potential effects of syntexis—the melting and dissolution of xenoliths (Lœwinson-Lessing, 1899; Daly, 1914, 1933; Green, 1994)—have been almost completely neglected. Interest in syntexis as a petrogenetically significant process in the oceanic crust was sparked by field studies from ophiolites (Pedersen, 1986; Bédard, 1988, 1991, 1993, 1994b; Malpas, 1990; Bédard and Constantin, 1991; Bédard and Hébert, 1996, 1998) and intracontinental or rifted-margin layered intrusions (Bédard et al., 1988; Bernstein et al., 1992, 1996; Tegner and Robins, 1996; Sørensen and Wilson, 1996). These studies imply that primitive replenishments to the base of active or fossil magma

Bédard, J.H., et al., 2000, Syntexis and the genesis of lower oceanic crust, *in* Dilek, Y., Moores, E.M., Elthon, D., and Nicolas, A., eds., Ophiolites and Oceanic Crust: New Insights from Field Studies and the Ocean Drilling Program: Boulder, Colorado, Geological Society of America Special Paper 349, p. 105–119.

chambers may react with and assimilate older cumulates. The assimilation of continental crust leaves distinct elemental and isotopic fingerprints on contaminated magmas. Assimilation of mafic cumulates in oceanic environments, on the other hand, would be difficult to detect in erupted lavas because of (1) the lack of isotopic contrasts between reactants and (2) the generally low abundance of incompatible trace elements in the cumulates. This paper reviews field, textural, and geochemical evidence from the Bay of Islands ophiolite that documents the action of a variety of syntectic processes. It is our contention that syntectic hybrids and reaction products constitute a more important proportion of the lower oceanic and continental crust than has generally been assumed and that syntexis may significantly affect the geochemistry of MORBs and continental flood basalts.

DEFINITIONS OF SYNTEXIS AND SYNTECTIC PROCESSES

Syntexis was once considered one of the principal processes by which magmas differentiate in the crust (Lœwinson-Lessing, 1899; Daly, 1914, 1933), but the concept fell into disfavor after Bowen (1928) quantified the heat of dissolution constraints on the process, i.e., that there are limits to the amount of material that can be assimilated without causing the magma to solidify completely. Bowen (1928) also showed that xenoliths and xenocrysts generally react with the magma, rather than simply dissolving into it, and that assimilation tends to leave the liquid line of descent unchanged, since multiply-saturated magmas are constrained to follow cotectic boundaries. Although all of these points are valid, the cumulative geochemical evidence from continental basalts is that primitive magmas do tend to assimilate their wall rocks, though not to the extent originally assumed by Daly (1914, 1933). Furthermore, cotectic boundaries do not buffer the isotopic and trace element signatures of contaminated residual melts, and melts saturated only in olivine will not be buffered to the same extent as multiply-saturated melts.

In this paper we describe relationships among the plutonic rocks of the North Arm Mountain massif of the Bay of Islands ophiolite complex (Fig. 1), in order to document some of the ways in which syntexis has proceeded in the lower oceanic crust. We also make reference to figures from earlier publications. Before the data are presented, however, it will be useful to define some terms.

The usage that is proposed here is that "syntexis" should be considered a general term for a family of processes by which host rocks, xenoliths, and xenocrysts move toward equilibrium with invading magmas. The geochemical processes that we consider syntectic are (1) congruent and incongruent dissolution of unstable phases into invading magma; (2) magmatic reactions between the invading magma and host rocks, xenoliths, and xenocrysts; and (3) diffusional re-equilibration of major and trace elements between the invading melt and host rocks, xenoliths, and xenocrysts. By this definition, syntexis includes assimilation. By using the terms "solute" and "solvent" as semantic analogues, we refer to phases that are being assimilated as the "assimilate" and the melt into which the assimilate dissolves as the "assimilant."

We suggest that the term "syntexis" should also be extended to cover the physical processes by which material is transferred from host rocks to the invading magma to form xenoliths, xenocrysts, and hybrids. These physical transfer processes include (1) magma mixing, (2) stoping, (3) disaggregation, (4) brecciation, and, in some cases, (5) heterogeneous shear (shear stoping). We propose that intimate mixtures of invading magmas and voluminous, partially re-equilibrated xenocrysts—and the rocks resulting from such mixtures—should be called "syntectic hybrids." We also recognize that syntectic hybrids may be difficult to distinguish from the products of magma mixing or mingling if the blended magmas were strongly porphyritic.

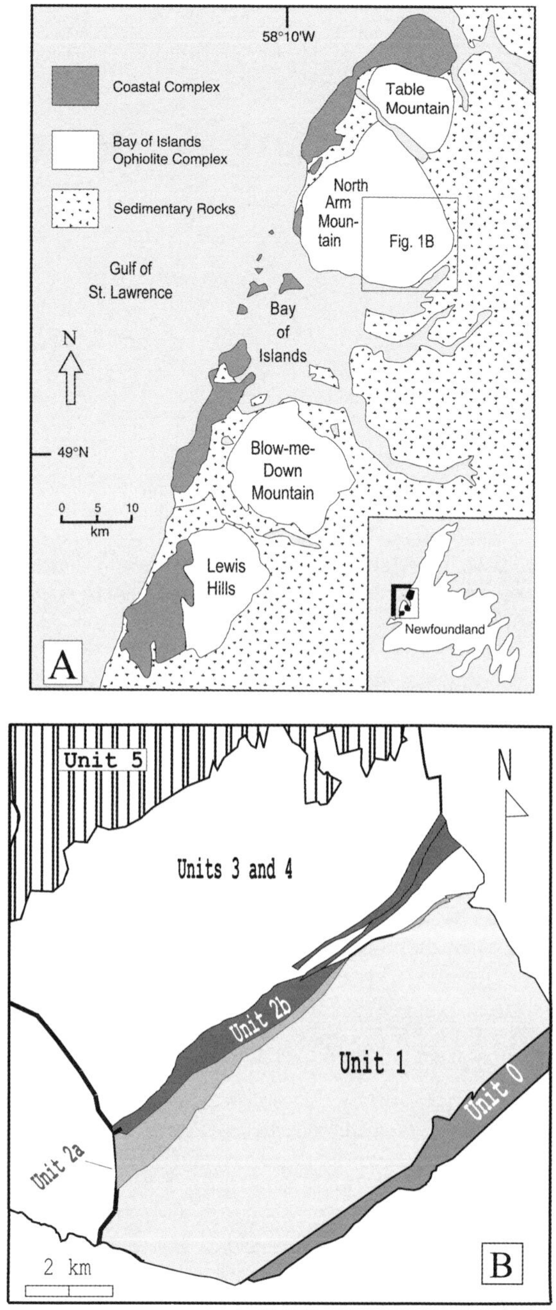

Figure 1. A, Map showing the major components of the Bay of Islands ophiolitic complex, western Newfoundland, Canada, and the location of the area mapped in B. B, Map of part of the North Arm Mountain massif adapted from Bédard and Hébert (1998). Units are as follows: 0—metamorphic sole, 1—mantle, 2a—ultramafic-dominated lower crust, 2b—mixed mafic and ultramafic lower crust, 3 and 4—layered and variably textured gabbros, 5—sheeted dikes. Heavy lines are late block faults.

The penetration of host rocks by minor intrusions related to the invading magmas could perhaps be included within the family of syntectic processes, but only in cases where such intrusions represent the first of a continuum of events leading to more complete digestion of the host rocks. For example, consider the intrusion of a swarm of

Figure 2. A, Layered olivine gabbro and troctolite cumulates in unit 3, containing numerous anorthosite boudins. B, Close-up of anorthosite boudins from Figure 2A. Note the internal "swirls." Lens cap for scale, 55 mm.

small dikes at the edges of a larger intrusion. If dike intrusion is followed by anatexis of host rocks, disaggregation of the restites into the invading magma, and partial dissolution of xenocrysts, then in such a case we suggest that dike intrusion also be considered a syntectic process. Conversely, intrusion of an isolated dike into a distal facies of a host rock should not be considered a syntectic process.

NORTH ARM MOUNTAIN MASSIF, BAY OF ISLANDS OPHIOLITE

Several units are recognized in the part of the North Arm Mountain massif that was studied (Fig. 1). (1) Mantle tectonites are principally restites from basalt genesis, but with complex, arc-related metasomatic overprints (Varfalvy et al., 1996, 1997; Batanova et al., 1998). (2) A mixed ultramafic and mafic lower cumulate zone consists of, according to our interpretation, synkinematic boninite sills formed of magmas that underplated the ophiolite complex and their reaction and differentiation products (Bédard, 1991; Bédard and Hébert, 1996). (3) Layered gabbros and subordinate peridotites that we also interpret as synkinematic sills, and which record interactions between primitive parental magmas and older cumulate rocks hosting the sills (Bédard and Constantin, 1991; Bédard, 1993). (4) A varitextured gabbro-hornblendite-trondhjemite unit (Rosencrantz, 1983; Berclaz et al., 1994). The hornblendites overprint or intrude rocks of unit 3 and are commonly associated with prominent shear zones. Porphyritic microgabbros appear to feed up into the sheeted dikes. (5–7) Sheeted dikes, lavas, and sedimentary rocks form the autochthonous superstructure of the ophiolitic complex (Rosencrantz, 1983). Preliminary geochemical modeling suggests that the cumulates of unit 3 are comagmatic with the dikes and lavas. Weak arc signatures in the geochemical data of the dikes and lavas (Siroky et al., 1985; Elthon, 1991) suggest that most of units 3, 4, 5, and 6 formed through normal seafloor-spreading processes in a mature back-arc environment (e.g., Swinden et al., 1997).

Most of the rocks in units 1 to 3 are deformed into isoclinal and sheath folds (Bédard, 1991; Bédard and Constantin, 1991). Deformation began well above the solidus (viscous), continued through the rheologic locking point and below the solidus (plastic), and on into the brittle regime. Structural analysis (Berclaz et al., 1994, 1998) indicates that most of unit 3 was affected by synspreading, extensional shear with a sinistral sense and dominant dip-slip motion, albeit with a significant ridge-parallel shear component. In contrast, most of unit 2 was deformed in a younger dextral compressional regime.

Essentially all of the rocks in units 2 and 3 are cumulates in the geochemical sense, i.e., concentrations of crystals from which most of the coexisting melt has escaped (Bédard, 1994a). Our terminology is entirely descriptive, however, and so we refer to shallowly discordant tabular intrusions as peridotitic sills because morphologically, they are sills, whereas lithologically, they are peridotites. Discussion of postemplacement crystal-liquid segregation mechanisms in the crust is deferred to a later contribution.

Layered gabbro unit (unit 3)

Unit 3 exhibits marked lateral and vertical facies variations. The layered gabbros thin toward the northeast, and the proportion of ferrogabbro, hornblende gabbro, and feldspar-phenocrystic microgabbro increases markedly eastward. The thickest section is found in the southwest (Fig. 1), where the base of unit 3 is dominated by isomodally layered, foliated, leuco– to mela–olivine gabbros (Komor et al., 1987; Bédard and Constantin, 1991) that locally contain swarms of strongly deformed anorthosite boudins (Figs. 2 and 3). Isoclinal and sheath folds are common (Bédard and Constantin, 1991; Bédard, 1993). The fact that the mineral foliation is discordant to layering in fold noses implies an origin through plastic deformation, not soft-sediment slumping or viscous magmatic flow (Bédard and Constantin, 1991; Berclaz et al., 1994, 1998). Isoclinal folding and transposition of preexisting lithologic contacts produces pseudo–rhythmic layering. Detailed mapping in the lower half of the section (Bédard and Constantin, 1991) shows that late- to postkinematic sills and dikes of olivine gabbro (2–25 m thick), lesser peridotite (1–5 m thick), and subordinate feldspar-phenocrystic gabbro (2–25 m) or Fe-oxide–bearing olivine gabbro (<1–20 m) together constitute about

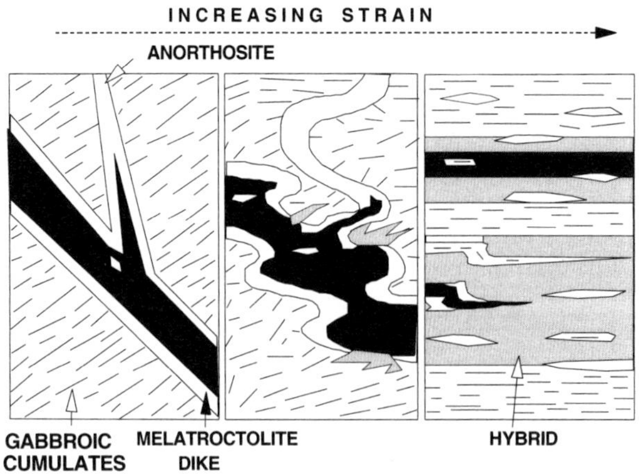

Figure 3. Sketch of contact relationships attendant upon intrusion of a synkinematic melatroctolite dike into an olivine gabbro host that is already above its solidus. Pore-scale hybridization of the two melts at the interface causes supersaturation of feldspar and forms anorthosite reaction rims. Anorthosite veinlets may be rooted at the contact. As deformation proceeds, the more rigid anorthosite tends to break up into boudin structures, and local hybridization of the intrusion and host occurs. With more extensive deformation, everything is transposed into near-parallelism, with only the anorthosite boudins remaining as clues to the original process.

half of the section. The late- to postkinematic intrusions are hosted by a complex of previously deformed and hydrothermally altered synkinematic sills, most of which are composed of olivine gabbro and troctolite, and which were themselves emplaced into an earlier series of strongly deformed leuco–olivine gabbros. These oldest rocks may also represent intrusions of sills, though we cannot prove this interpretation.

The peridotites of unit 3 are typically wehrlites, commonly feldspathic and grading to melatroctolite and olivine gabbro; they are interpreted to have crystallized from intracumulate injections of replenishing magma (Bédard and Constantin, 1991). The high-Ti clinopyroxenes (my own unpublished data) that occur in most of the peridotite and melatroctolite injections suggest that they are cogenetic with unit 3 gabbros. Discordant melatroctolite dikes may develop anorthosite reaction rims against common olivine gabbros (Fig. 7 in Bédard, 1993). Anorthosite dikelets extend from the anorthosite reaction rims into the olivine gabbro hosts. In the field, one can see progressively more deformed equivalents of the anorthosite dikelets and reaction rims that record every stage in their disruption (Fig. 3). As strain increases, the anorthosites break up into elongated boudins (Fig. 2). Presumably this deformation style reflects the higher rigidity of the anorthosite in comparison to the surrounding gabbros and troctolites. Since feldspar is not markedly weaker than pyroxene or olivine, it seems plausible to deduce that the olivine gabbro host rocks and melatroctolite dikes still contained melt when they were deformed, while the anorthosite had almost no porosity and hence was more rigid. Deformation of these intrusions and hosts continued into the subsolidus, generating a rather homogeneous-looking sequence of isoclinally folded (pseudo–rhythmically layered) troctolites and olivine gabbros that contain isolated, extremely stretched anorthosite boudins (Fig. 2).

Least deformed (late-kinematic) gabbros commonly have a leopard texture, with amoeboid clots and angular rafts of layered gabbro in an olivine-rich matrix (Fig. 4; see also Bédard, 1993). The textures suggest that the gabbroic clots are xenoclasts, whereas the layered rafts represent olivine gabbro cumulate layers (or reaction rims) that were more completely solidified and, hence, more resistant to disaggregation. The implication is that these heterogeneous olivine gabbros and feldspathic peridotites represent a slurry containing gabbroic xenoclasts produced by the disaggregation of a loosely consolidated gabbroic cumulate into a primitive, olivine-saturated magma that intruded the cumulate mass. More deformed facies are rather homogeneous-looking olivine gabbros that could be interpreted as normal cotectic cumulates. However, even the most homogeneous examples still contain scattered anorthosite boudins (Fig. 2), which are interpreted to represent disrupted and stretched anorthosite reaction rims and veins (Fig. 3). The presence of these relict reaction rims suggests that the more deformed facies are also syntectic hybrids (Bédard and Constantin, 1991; Bédard, 1993). This suggestion implies that much of the gabbroic crust (~50%) is composed of hybrid olivine gabbros, troctolites, and feldspathic peridotites; most of the rest of the crust is composed of cumulates derived from sill-like intracumulate intrusions (~40%), with subordinate anorthositic to leucogabbroic reaction products (~10%).

Mixed mafic and ultramafic cumulates (unit 2)

Rocks of unit 2 generally occur between the mantle (unit 1) and the layered gabbroic cumulates (unit 3), but intrusions of unit 2 affinity also occur within the layered gabbros, particularly in the northeastern part of the massif (Fig. 1). Cumulate peridotites and pyroxenites dominate the base of unit 2. The abundance of orthopyroxene and the low Ti contents of pyroxenes (Bédard, 1991; Bédard and Hébert, 1996), together with trace element data (Bédard, 1994a), imply a boninitic, arc affinity for the ultramafic rocks. Synmagmatic to postmagmatic deformation is nearly ubiquitous (Bédard, 1991), but where deformation is less severe, peridotite bodies may be seen to contain xenoliths of, and have contacts discordant to, layered and foliated gabbros, pyroxenites, and peridotites (Bédard, 1991; Bédard and Hébert, 1996, 1998). These contact relationships led Bédard (1988, 1991, 1993) to infer an intrusive origin for most lower-crustal peridotites. Intrusive contacts become harder to recognize in the more deformed facies, but because strongly deformed peridotites are petrologically, mineralogically, and geochemically indistinguishable from the less deformed equivalents (Bédard and Hébert, 1996) and every gradation can be seen in the field between undeformed intrusions and progressively deformed equivalents, we interpret the deformed peridotites also to be intrusions.

Two or more magmatic lineages are present among the ultramafic rocks (Bédard and Hébert, 1996): a dominant wehrlitic series (dunite-wehrlite-clinopyroxenite-gabbro) and a subordinate harzburgitic series (dunite-harzburgite-orthopyroxenite-websterite-gabbronorite). Gabbroic rocks become more abundant upward in the section and toward the northeast. Gabbroic rocks are dominantly gabbronorites and olivine gabbros as well as subordinate norites and amphibole ferrogabbros.

The process of stoping or disaggregation of host rocks into intruding magmas is commonly documented in unit 2. Such processes are particularly evident where there are mineral-assemblage contrasts between intrusion and host. For example, harzburgite-series intrusions emplaced into wehrlite-series hosts (Figs. 5 and 6) typically have olivine websterite or lherzolite boundary zones that contain abundant websterite xenoliths and xenocrystic clinopyroxene derived from the adjacent hosts (Figs. 5 and 10 in Bédard, 1991). Clinopyroxene-rich xenoliths in these boundary zones are typically rimmed

Figure 4. A, Lit-par-lit intrusions of feldspathic peridotite into layered olivine leucogabbro in unit 3. B, Close up of contact of angular leuco–olivine gabbro raft of Figure 4A. Note how layering is truncated by the intrusive contact. C, Hybrid leopard-textured olivine gabbro in unit 3. Note the extremely heterogeneous texture, with gabbroic schlieren in a melagabbroic matrix. Lens cap for scale, 55 mm.

by orthopyroxenite reaction products (Figs. 2 and 3 in Bédard and Hébert, 1996). Flow of magma within the intrusion may entrain the reaction products from the rims, forming trains of porphyroclasts (Figs. 5 and 6, A and B).

Feldspathic peridotites and olivine gabbros in unit 2 occur as halos or heterogeneous border facies (Fig. 7) separating feldspar-free intrusive peridotites from olivine-free host rocks (Bédard, 1991). The halos commonly contain angular to ragged xenoliths of layered gabbro that are obviously derived from the adjacent host rocks (Figs. 15, 19, 21, and 22 in Bédard, 1991). Furthermore, these olivine gabbro and feldspathic peridotite border facies commonly contain an abundance of aluminous Cr-spinel filling dissolution pits in feldspar (Fig. 3 in Bédard and Hébert, 1998). The textures and compositions of these aluminous Cr-spinels were interpreted to signify that they formed by incongruent dissolution of plagioclase (with most of the Cr being derived from the liquid) (Bowen, 1928; Smith, 1958; Bédard, 1991; Bédard and Hébert, 1998), which implies that the feldspars are unstable xenocrysts. The textures and spatial associations suggest to us that the olivine gabbros of unit 2 are syntectic hybrids, formed by disaggregation of gabbronoritic or ferrogabbroic hosts into primitive-magma intrusions. These hybrid olivine gabbros and feldspathic peridotites constitute about 20% of unit 2.

Olivine- and pyroxene-rich schlieren are extremely common in gabbronorites that are within a few meters of large peridotite intrusions (Fig. 17 in Bédard, 1991; Fig. 2b in Bédard and Hébert, 1996; and Fig. 7 herein). Every gradation has been observed from undeformed peridotite veins that are still rooted in the main peridotite intrusion to progressively more deformed equivalents (V in Fig. 7). As the peridotite veins are sheared and folded, they break up into schlieren; the process yields a heterogeneous, wispy-layered border facies composed of olivine gabbronorite (H in Fig. 7).

Within the lower crust, chromitites typically occur near contacts between intrusive dunites and their host rocks (Bédard, 1991; Bédard and Hébert, 1998). Thin, in situ reaction rims of high-Cr chromite are locally associated with pyroxenite hosts (Fig. 2 in Bédard and Hébert, 1998), though deformation has typically led to the development of pseudo-rhythmic chromitite-pyroxenite-dunite layering. Aluminous chromitite schlieren are associated with gabbroic hosts and hybrid feldspathic peridotites and have spinels with compositions similar to those of the spinels filling dissolution pits in feldspars. Chromitites that contain Cr-spinels enriched in Fe^{2+}, Fe^{3+}, Ti, and V occur in the vicinity of ferrogabbros. Contact relationships and the clear correspondence between the composition of the chromite and the nature of the host rocks led to the suggestion that crustal chromitite schlieren at North Arm formed by incongruent dissolution of pyroxenes, feldspars, and locally magnetite into primitive magmas saturated in olivine + chromite (Bédard, 1991; Bédard and Hébert, 1998).

Rims of pyroxenite that are 1–10 cm thick line many peridotite-gabbronorite contacts. The pyroxenite rims finger into the gabbronorite substrate and commonly contain anorthositic relicts (Bédard, 1991). Bédard (1991) suggested that pyroxenite rims result from pore-scale hybridization between primitive melts of the intrusion and anatectic melts of the host (Fig. 8). Field relationships suggest that

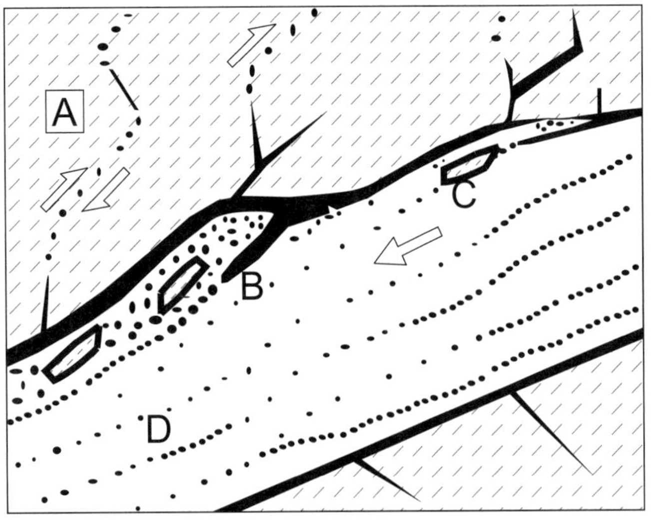

Figure 5. Sketch illustrating the contact relationships of a harzburgitic sill (white) emplaced into deformed wehrlitic rocks (dashed lines). The contacts of the sill are lined by orthopyroxenite (black), which is being stripped off (at B) to form trains of porphyroclasts (black spots labeled D; see also Fig. 6, A and B) within the harzburgite sill. Clinopyroxene-rich xenoliths (such as at C) are rimmed by orthopyroxenite. Veins of orthopyroxenite cut the wehrlite host; subsequent increments of deformation break the veins up into trains of porphyroclasts (at A; see also Fig. 6C).

these pyroxenite reaction rims commonly detach from their substrates during deformation (Figs. 5, 8, 16, and 18 in Bédard, 1991); and Bédard (1991) suggested that much of the pyroxenite constituting unit 2 represents partly digested gabbroic host rocks.

DISCUSSION

Can primary mantle melts cross a mushy lower crust?

A currently unresolved question is the mechanism by which mantle-derived magmas pass through the lower oceanic crust in order to erupt at the surface. The Moho is a major discontinuity in terms of structure, lithology, rheology, and density, and so it represents a potentially significant obstacle to melt ascent (Parsons et al., 1992). Observations from North Arm and other ophiolites (Abrajano et al., 1989; Malpas, 1990; Dilek and Eddy, 1992; Umino et al., 1990; Orberger et al., 1995) and ocean ridges (Garmany, 1989; Garmany et al., 1994; Dunn et al., 1994) suggest that magmatic underplating may be more common than previously appreciated.

The presence of a significant partial-melt fraction in the lower crust would also represent an obstacle to dike propagation, since melt fractions >5% greatly reduce bulk viscosities (Hirth and Kohlstedt, 1995) and might reduce the elasticity of the medium to the point at which crack propagation is inhibited. If the ≤10% lower-crustal melt fraction estimated by Sato et al. (1989) or ~20% melt fraction estimated by Sinha et al. (1998) is correct, cracks may not be able to propagate through the mushy lower crust. Petrologic studies on gabbroic rocks from Ocean Drilling Program Site 735B (Dick et al., 1991; Natland et al., 1991) show a correlation between the presence of ferrogabbro and the degree of deformation. This result was interpreted to imply that residual melts segregate into shallowly dipping shear zones, and similar Fe-Ti–rich subhorizontal shear zones occur in the North Arm (Berclaz et al., 1994, 1998). If a vertically propagating, magma-driven crack intersects a subhorizontal shear zone or a layer with a high melt fraction (a recent sill or a melt-rich cumulate layer), vertical crack propagation will probably terminate there, and the magma will form an intracumulate sill. Consequently, unless there are favorable structures (faults or shear zones) that facilitate ascent of primitive melts (e.g., Reuber, 1988; Stakes and Taylor, 1992), primitive melts from the mantle will tend to be trapped in the lower crust, where they will simultaneously crystallize and assimilate their host rocks.

Heat transport and anatexis

Geophysical data suggest that fast-spreading ocean ridges are characterized by a long-lived, high-level melt lens and a lower crust composed of mushy cumulates with ≤20% melt (Vera et al., 1990; Sato et al., 1989; Detrick et al., 1990). Thus, at fast-spreading ridges, the lower crust should remain at or above its solidus until seafloor spreading carries that segment several kilometers away from the ridge axis. This scenario implies that the lower-crustal rocks in the vicinity of fast-spreading ridges will already be at or above their solidus temperatures when they are intruded by the next batch of primitive magmas to ascend from the mantle. The addition of heat from the new influxes of primitive magma should increase the proportion of melt in the host cumulates, which should make them even better traps for ascending magma.

Thermal models and the presence of deep earthquakes suggest that magma chambers at slow-spreading oceanic ridges are ephemeral and that seawater penetrates deeply into the crust (Sleep, 1975; Lewis, 1983; Lister, 1983; Harper, 1985; Toomey et al., 1988; Natland et al., 1991; Kong et al., 1992; Solomon and Toomey, 1992). Thus, at slow-spreading ridges, only magmas ascending from the mantle during periods of high magmatic flux would encounter suprasolidus lower-crustal rocks (e.g., Sinha et al., 1998). Magmas that ascend from the mantle during periods of low magmatic flux would probably intrude colder crustal rocks that have been affected by hydrothermal alteration. Field evidence from North Arm implies that many primitive intrusions assimilated rocks that had already been affected by amphibolite- to greenschist-facies hydrothermal metamorphism (Fig. 19 in Bédard, 1991; Fig. 3 in Bédard and Hébert, 1996) and thus suggests that a comparison with slow-spreading ridges would be most appropriate (Berclaz et al., 1998).

Primitive magmas ascending from the mantle will generally have temperatures higher than the solidus of the crustal host rocks above them. Whether an invading magma chills and crystallizes when it is injected into host rocks or causes them to react or melt will depend on the relative volumes and temperatures of intrusions and hosts and on the solidus and liquidus temperatures of magma and hosts rocks.

Hot magmas evidently have more potential to assimilate host rocks than do low-temperature magmas (e.g., Huppert and Sparks, 1985; Sparks, 1986), whereas cold (or refractory) host rocks will tend to be more difficult to melt and may chill the magma, so preventing reaction entirely. However, even a low ambient temperature does not preclude syntexis, since xenocrysts can dissolve into (or react with) a melt so long as the magma in the intrusion remains above its solidus. On short time scales, the local proportion of intrusive to host (or xenocrystic) material (Marshall and Sparks, 1984) is as important as the preintrusion ambient temperature in determining whether an intrusion solidifies quickly or not. This proportion is likely to vary with time and in space, depending on how intrusions are distributed and how quickly they are emplaced. To summarize, an evolved, low-

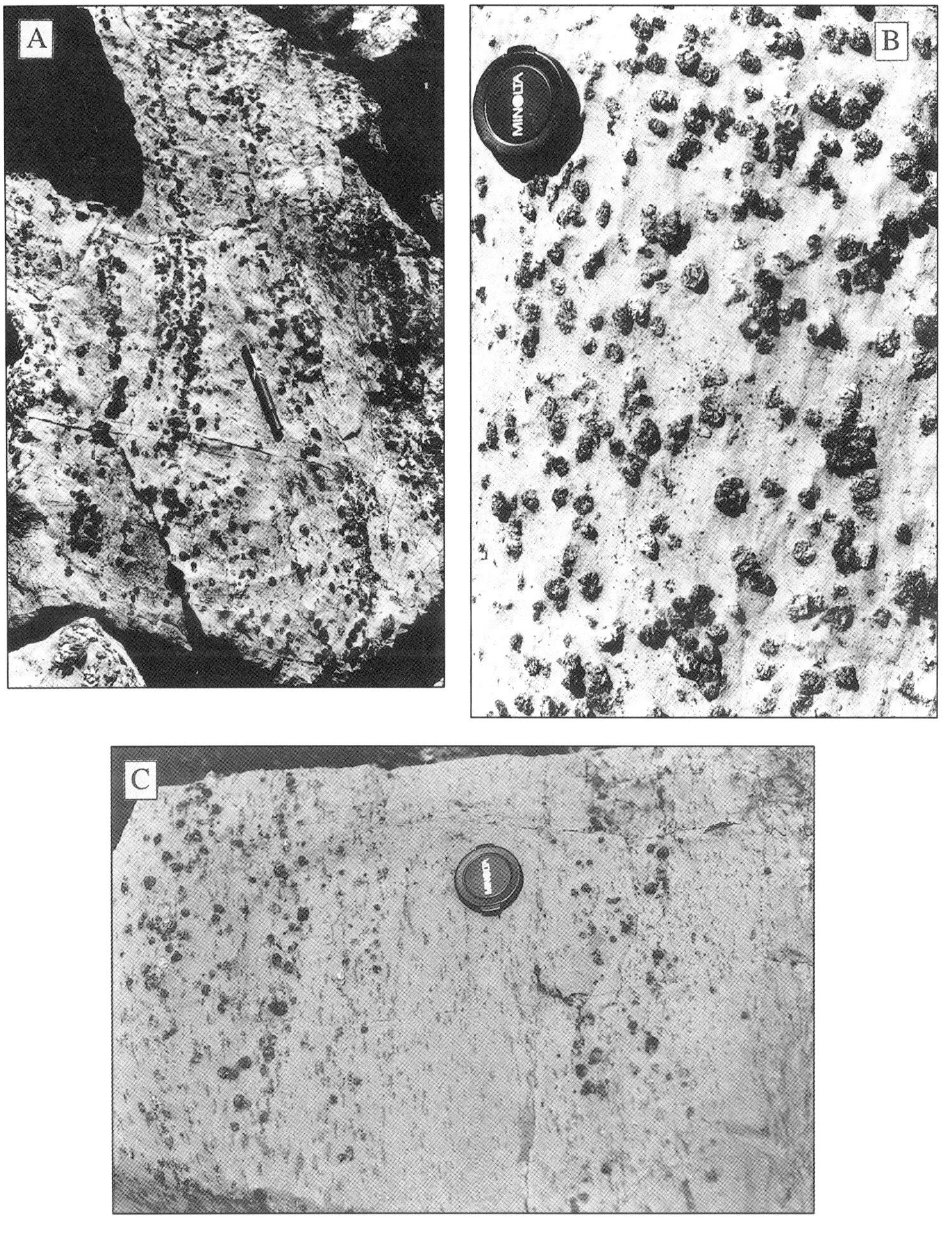

Figure 6. A, Trains of orthopyroxene porphyroclasts in core of a 20-m-thick harzburgite sill emplaced near the base of unit 2. Alignment is probably due to magmatic flow. Magic marker for scale. B, Close-up of orthopyroxene porphyroclasts in the same sill as A. C, Foliated lherzolite hosting the harzburgite sill of Figures 5 and 6, A and B. Note the extremely flattened aspect of clinopyroxenes (dark gray), which contrasts with the rounded orthopyroxene porphyroclasts (black), which we interpret to represent disaggregated orthopyroxenite veinlets emplaced into an already-deformed wehrlite.

temperature magma such as a ferrobasalt that is intermittently injected as small, isolated dikes into a cold, refractory host rock such as a pyroxenite will almost certainly freeze without having interacted with the host rocks to any significant degree. The circumstance of a voluminous sill swarm of hot basalt or picrite that is injected rapidly into evolved host rocks like ferrogabbros that are already at or above their solidus is quite a different matter.

Ascending melts will evolve as they cool, crystallize, and assimilate; therefore, they will progressively lose the capacity to assimilate their host rocks. Replenishments at fast-spreading ridges should encounter mushy lower crust that is already above its solidus, and one would predict extensive lower-crustal syntexis and reaction at depth and progressively decreasing extents of reaction as the magmas ascend. At slow-spreading ridges, the presence of hydrous minerals formed by prior hydrothermal alteration would perhaps compensate for the lower ambient temperatures of the host rocks by fluxing anatexis, because fluids released by the breakdown of amphibole, epidote, chlorite, or prehnite would lower the solidus temperature. This process also provides a mechanism by which Na, Cl, H_2O, S, and other water-soluble elements may be transferred into magmas.

The lower continental crust is generally inferred to be of basaltic composition (Rudnick and Fountain, 1995; Rudnick et al., 1998), which in petrologic terms implies gabbroic to gabbronoritic assemblages that are not radically different from the types of rocks that characterize the lower oceanic crust. Temperatures at the base of the continental crust in regions of active volcanism should be at the granulite grade, which implies near-solidus temperatures (~900 °C) for such assemblages. Thus, most of the principles that have been deduced from the study of the oceanic crust and ophiolitic rocks should also apply to a lower continental crust that is underplated by basaltic magmas during periods of intense flood-basalt volcanism.

Dikes and syntexis

The margins of large sills or plugs may be sites of intense dike intrusion that may constitute an important precursor to syntexis. The

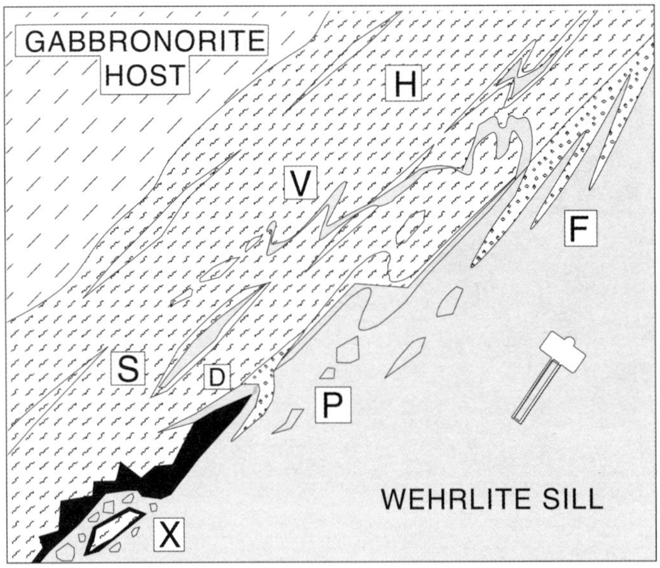

Figure 7. Sketch of relationships at the contact of a feldspar-free wehrlite sill and an olivine-free gabbronorite host. The heterogeneous contact zone (labeled H and marked with wavy lines) is dominated by olivine gabbronorite. Pyroxenite reaction rims (black) form at the contact between gabbro and wehrlite. The rims break up to form clasts (labeled P) in the sill, some of which still have gabbro in their cores (clast labeled X). When the pyroxenite rims spall off into the sill, they expose partially molten gabbro, which disaggregates to form hybrid feldspathic peridotite as at F or D. A small veinlet (labeled V) rooted in the main wehrlite is injected into the host rock, where it cuts older, more deformed, more re-equilibrated veinlets. The vein V is folded and begins to break up into pyroxenitic schlieren (labeled S), some of which still have peridotitic cores. Injection of peridotite into the contact zone may assist in detaching pyroxenite rims from their substrates as at D.

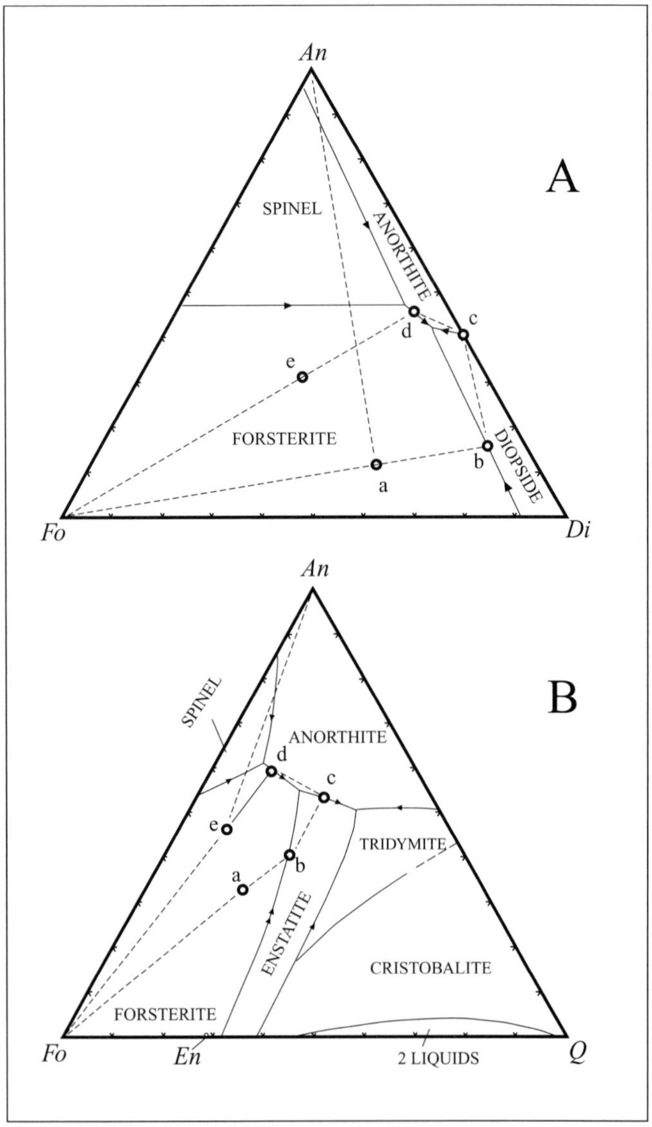

Figure 8. A, Fo-An-Di phase diagram at 1 atm, adapted from Onuma and Tohara (1983). Solid lines are phase boundaries for system containing 0.2 wt% $MgCr_2O_4$. Dashed lines are fractionation or mixing curves. Primitive magma *a* fractionates olivine and becomes cosaturated in diopside at *b*. Primitive magma *e* fractionates olivine and becomes cosaturated in anorthite at *d*. Heating of olivine-free gabbro produces anatectic magma of composition *c*. Mixtures of *c* with *b* are in the diopside-only field and will produce pyroxenite rims. Mixtures of *c* with *d* are in the anorthite-only field and will produce anorthosite. Dissolution of anorthite into melts *a* or *e* produces spinel by incongruent dissolution. B, Fo-An-Q phase diagram at 1 atm, adapted from Andersen (1915). Solid lines are phase boundaries. Dotted lines are fractionation or mixing curves. Primitive magma *a* fractionates olivine and becomes saturated in enstatite at *b*. Primitive magma *e* fractionates olivine and becomes cosaturated in anorthite at *d*. Heating of olivine-free gabbro produces anatectic magma of composition *c*. Mixtures of *c* with *b* are in the enstatite-only field and will produce pyroxenite rims. Mixtures of *c* with *d* are in the anorthite-only field and will produce anorthosite. Addition of anorthite to melt *e* produces spinel by incongruent dissolution.

proportion of dike to host is an important factor, since it will affect the rheology, temperature, and stable mineral assemblages of the resultant mixture. High dike/host ratios will allow advection of a lot of heat into the local system. In extreme cases, this process will lead to anatexis of the host rocks and perhaps collapse (by stoping and/or brecciation) of the marginal host and dike mass into the main intrusion. Smaller dike/host ratios will still reduce the viscosity and may lead to strain localization and development of a shear zone, which may, in turn, localize further intrusions. Deformation of contact zones laced with marginal dikes and re-equilibration of dikes with the host will produce mixed, hybrid rocks with mineral proportions that do not record cotectic crystallization processes.

Identification of characteristic marginal-dike assemblages provides important clues for an attempt to reconstruct the multiple intrusion history of synkinematic intrusion complexes like those at North Arm. For example, a harzburgite sill emplaced into an already-deformed wehrlitic host in unit 2 has prominent orthopyroxenite reaction rims along its contacts (Fig. 5; cf. Fig. 2a in Bédard and Hébert, 1996). Orthopyroxenitic veins are common in the wehrlites that occur near the harzburgite sill and appear to represent offshoots from it. Nearby, there are some unusual lherzolite occurrences characterized by extremely flattened and stretched clinopyroxene grains and relatively undeformed orthopyroxene porphyroclasts (Fig. 6C). Some of the orthopyroxene porphyroclasts form rough trains discordant to the dominant foliation, and we interpret them to have originally been orthopyroxenite veins, subsequently dismembered by deformation (A in Fig. 5). Thus, this lherzolite does not represent an equilibrium cotectic assemblage, but instead is a hybrid rock.

Similarly, gabbronorites that are in contact with large peridotite intrusions typically contain abundant olivine- and pyroxene-rich schlieren near their margins (Fig. 7; cf. Fig. 2 in Bédard and Hébert, 1998). These mafic schlieren are interpreted to be re-equilibrated veins or dikelets derived from the main peridotitic intrusions. As veins are deformed and recrystallized, the mineral assemblage within them will tend to equilibrate with the minerals in the dominant host lithology. When subordinate peridotite veins re-equilibrate with large volumes of gabbronorite, olivine will react out, leaving wisps and dikes rich in orthopyroxene (S in Fig. 7). Without the constraint provided by the detailed field observations of crosscutting relationships, the gradation from peridotite to olivine gabbro to gabbronorite in this intrusive contact zone might be misinterpreted as the result of fractional crystallization acting on a single magma.

Stoping, disaggregation of cumulates, and the formation of syntectic hybrids

As we have defined it, the term "syntexis" also includes the physical transfer processes of stoping, disaggregation, brecciation, and heterogeneous shear. It seems self-evident that host rocks and xenocrysts will dissolve into an invading melt only if they are exposed to it. A smooth wall presents the least efficient geometry for assimilation, both because this geometry minimizes reactant surface area and because the buildup of assimilate components in the boundary layer decreases the chemical-potential gradient that drives assimilation reactions. Thus, any mechanism that acts to increase surface area by mechanically mixing reactants will increase the bulk reaction and dissolution rates. Stoping and brecciation of host rocks are the classical mechanisms invoked (Paterson et al., 1996; Clarke et al., 1998). The field relationships at North Arm also suggest that heterogeneous shear at an angle to contacts may be an important factor, because synmagmatic deformation may have triggered disaggregation of poorly consolidated crystal aggregates (i.e., cumulates) into intruding melts and facilitated the detachment of reaction rims from their substrates (e.g., Figs. 16C, 17, and 18 in Bédard, 1991). We suggest the term "shear stoping" for this deformation-assisted mixing mechanism. We believe that the scale and significance of these different mechanical mixing mechanisms, particularly shear stoping, has hitherto been underestimated, especially for synkinematic intrusions. Mechanical mixing of components is exemplified at North Arm by the marginal clinopyroxene-bearing zones of harzburgite-series intrusions emplaced into wehrlite-series hosts and by the presence of feldspathic peridotites and olivine gabbros as halos separating feldspar-free intrusive peridotites from gabbroic hosts.

Dissolution or reaction?

Dissolution consumes heat, so that, in general, "inclusions pass into solution by precipitating their heat equivalent of the phase with which the liquid is saturated" (Bowen, 1928). This statement implies that the capacity of an intrusion to assimilate foreign matter is finite and will generally be accompanied by net crystallization. However, Kelemen (1990) showed that assimilation of orthopyroxene by olivine-saturated melts may be nearly isenthalpic. The progress of isenthalpic reactions would be limited only by accessibility of reactants and the solubility limits of melts. The rates of dissolution of most minerals into silicate melts are typically rapid (micrometers per hour), and the dissolution rate is controlled by the efficiency with which dissolved components are swept away from the dissolution surface (Kuo and Kirkpatrick, 1985; Brearley and Scarfe, 1986; Donaldson, 1990; Kerr, 1995; Edwards and Russell, 1996). In practice, it seems safe to assume that small amounts of xenocrysts (<30%) are probably readily assimilated, even where dissolution reactions are strongly endothermic, but that catastrophic disaggregation of a host rock that dumps 60% xenocrysts into a melt will probably overload the magma's capacity to dissolve them.

In practical terms, the first question that springs to mind when considering the fate of a xenocryst suddenly immersed in an invading magma is whether or not that mineral is stable in that melt. If it is not, then such a xenocryst will either dissolve or react with that melt (e.g., Bowen, 1928; McBirney, 1979). If solid-solution series are involved, a xenocrystic phase may react with a melt even if it is nominally stable in it (Bowen, 1928). This behavior was investigated experimentally by Tsuchiyama and Takahashi (1983) and Tsuchiyama (1985, 1986), who showed that on a short time scale, xenocrysts that have MgO/FeO ratios (olivine or pyroxene) or anorthite contents (plagioclase) that are higher than the normal "equilibrium" values of these phases in their new host magmas will either act as nuclei for subsequent growth or react with release of heat (Bowen, 1928). In contrast, xenocrysts that have MgO/FeO ratios that are too low or that are too albitic for their new host magmas will produce a partial melt as they react toward equilibrium compositions, but will consume heat and lower the temperature. In natural cases, recognizing that dissolution reactions have occurred can be difficult, in part because syntexis almost certainly operates in tandem with fractional crystallization (Bowen, 1928). Congruent dissolution reactions that go to completion are likely to be entirely cryptic in petrologic terms, though they will have a geochemical impact on the assimilant melt. Incongruent dissolution reactions have more potential to leave petrologically visible traces, because they produce by-products. In a sense, these by-products are restite phases, which provide important clues to the genesis of the rocks in which they are found. Perhaps the least ambiguous situation is where the by-products of incongruent dissolution occur as in situ reaction rims against the walls of an intrusion or surrounding xenoliths. In unit 2 at North Arm, orthopyroxenite rims commonly line the contacts between harzburgite intrusions and wehrlitic wall rocks, suggesting that clinopyroxene dissolves incongruently into orthopyroxene-saturated magmas. Thin chromitite rims lining some pyroxenite-dunite or gabbro-peridotite contacts at North Arm also appear to be incongruent dissolution products (Bédard, 1991; Bédard and Hébert, 1998).

There are two ways to conceptualize incongruent dissolution reactions. The first makes use of classical phase-equilibrium phase diagrams (Figs. 8 and 9), whereas the second uses the concept of the solubility limit. First imagine a feldspar xenocryst that is plunged into a primitive, olivine-saturated melt such as *a* in Figure 8A or *e* in Figure 8B. In these cases, the mixing line intercepts the spinel field, which implies that the feldspar cannot dissolve without also producing a spinel by-product. At moderate Cr contents, the spinel field expands greatly, and olivine and plagioclase may not even constitute an equilibrium assemblage. Cr-rich pyroxenes also tend to dissolve incongruently, yielding chromite as a by-product (Dickey et al., 1971; Dickey and Yoder, 1972; Dickey, 1975; Brearley and Scarfe, 1986). For example, consider a Cr-rich orthopyroxene (>0.3% Cr_2O_3) that is ingested into a melt saturated in olivine + chromite (Fig. 9). When it heats up, the orthopyroxene xenocryst "bulk" composition intersects the chromite field, and so orthopyroxene breaks down incongruently to yield a chromite by-product.

Alternatively, one might consider the solubility limit of Cr in

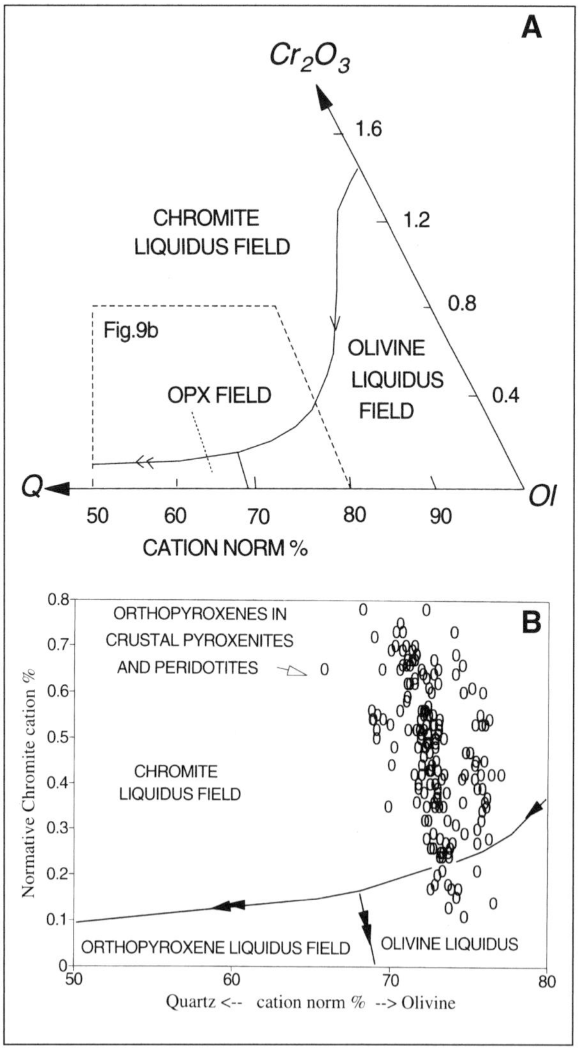

Figure 9. Part of cation-normative equivalent SiO_2-olivine-Cr_2O_3 phase diagram (Irvine, 1977), adapted from Bédard and Hébert (1998). Note that the vertical axis is exaggerated in Figure 9A and that Figure 9B is plotted in Cartesian rather than triangular coordinates. Orthopyroxene analyses are from lower-crustal pyroxenites and peridotites of unit 2, North Arm. Many of these orthopyroxenes plot within the chromite liquidus field and so would melt incongruently, generating chromite by-products.

primitive melts as the determining constraint. The solubility of Cr depends significantly on temperature, bulk composition, and oxygen fugacity (Murck and Campbell, 1986). For example, ~500 ppm Cr can dissolve in a 1250 °C basalt at the quartz-fayalite-magnetite oxygen buffer, with up to 2000 ppm entering solution at 1350 °C. So, if an orthopyroxene xenocryst that contains 0.5 wt% Cr_2O_3 (3420 ppm Cr) attempts to dissolve in such a melt, then the excess Cr will be insoluble, and a chromite restite will be left behind.

In the ideal case, reaction products will have "equilibrium" compositions and so be identical to the composition of phenocrysts in the melt. In the real world, however, most of the digestion will occur in boundary layers where melt compositions may depart significantly from the bulk of the magma in an intrusion. At North Arm, Bédard and Hébert (1998) proposed that high-Cr chromitites formed by assimilation of pyroxenites, high-Al chromitites formed by assimilation of gabbros, and high–Fe^{2+}-Fe^{3+}-Ti-V chromitites formed by assimilation of Fe-Ti oxide–rich gabbros. Thus, the compositions of the chromitites at North Arm seem to depend more on the nature of the rock being assimilated than on the composition of the magma emerging from the mantle. Consequently, we infer that the compositions of mineral phases produced by syntexis are unreliable guides to magma composition and, by extension, are unreliable paleotectonic indicators (Bédard and Hébert, 1998).

The influence of phase-equilibrium topology during magma mixing

In contrast to the simple incongruent-dissolution mechanisms discussed previously, the common occurrence of narrow pyroxenite and anorthosite reaction rims separating intrusive peridotites from host gabbros at North Arm was interpreted to have formed as a result of magma mixing at the interface between hosts and multiply-saturated intrusions (Bédard, 1991, 1993, 1994b). The nature of the reaction product in this scenario is essentially controlled by the topology of the phase diagram, a subject that we now explore.

The major element composition of multiply-saturated melts is determined by phase relationships (cotectics, eutectics, and peritectics), and so the major element composition of contaminated melts will be relatively insensitive to magma mixing or assimilation processes (Bowen, 1928). In contrast, magma mixing has potentially catastrophic impacts on (1) the trace element and isotopic chemistry of hybrid melts, (2) the major element chemistry of melts that are not multiply saturated, and (3) the crystallization products (cumulates) of mixed magmas.

Consider the consequences of emplacing a primitive, high-temperature melt into a gabbroic host. Assuming that the host is already close to its solidus temperature, then addition of heat from the intrusion will probably trigger anatexis. If the host is an olivine-free gabbro, gabbronorite, or norite, then partial melts will have eutectoid compositions such as *c* in Figure 8, A and B. At the interface between the intrusion and the host, the anatectic melt may mix with the intrusion's melt. If melt *c* is added to a magma saturated only in olivine such as *a,* then the hybrid melt will be compositionally undersaturated and will dissolve all phases with which it comes into contact, contributing to ablation of the contact. In contrast, addition of the same component *c* to a melt such as *b,* which is cosaturated in clinopyroxene and olivine (Fig. 8A) or orthopyroxene and olivine (Fig. 8B), will inevitably pull the hybrid into the pyroxene-only field. As a result, the hybrid melt will dissolve plagioclase or olivine and precipitate pyroxene until the melt composition returns to a multiply-saturated cotectic. This pyroxenite-forming mechanism has been demonstrated experimentally by Dunn (1991). At North Arm, the host gabbro immediately adjacent to the pyroxenite rims is nearly anorthositic, has high-MgO pyroxenes and extremely anorthitic plagioclases, and very low contents of all incompatible trace elements except Sr and Eu (Bédard, 1993, 1994b). All of these features are consistent with the extraction from the reaction-zone gabbro of a partial melt and with the enrichment of feldspar in the restite reflecting progressive exhaustion of pyroxene from the protolith. Note that the major element composition of the contaminated melt will tend to stay on the cotectic. Incompatible trace elements, on the other hand, would be free to vary, whereas the solid products (reaction rims) will be radically different from the crystallization products of simple fractional crystallization.

We propose that anorthositic rims at the contacts of primitive melatroctolite dikes in unit 3 at North Arm have formed in an analogous manner. In this case, however, the intrusion is saturated in troctolite of composition *d* and mixes with interstitial melts of cotectic composition *c* (Fig. 8, A and B), leading to supersaturation in plagioclase and resorption of pyroxene and olivine.

It seems evident that a monomineralic rim formed in this way (e.g., pyroxenite or anorthosite) will have a low porosity (or else it would not be monomineralic) and so will have low permeability (Wark and Watson, 1998). As reaction rims thicken, they will armor the substrate against further reaction and will eventually halt this intrapore magma-mixing process entirely. However, heat will continue to diffuse across the reaction rim into the host rocks, which may continue to melt, thus weakening the attachment of the rims to their original substrate. Either because of gravitational instability or cross-contact shear, loosened rims would detach from their substrate and fall (or be entrained) into the main, still-liquid intrusion (Fig. 7). Removal of a reaction rim would expose fresh surfaces to reaction and produce a new crop of reaction products. Repeated detachment and brecciation of reaction rims explain why pyroxenites in unit 2 at North Arm are concentrated along strike from gabbroic lenses and why so many peridotites look like breccias, containing pyroxenite clasts within a peridotitic matrix (see Fig. 8 in Bédard, 1991). This shear-enhanced stoping mechanism provides an attractive mechanism to effect near-complete digestion of host rocks. At North Arm, there are thick (100–300 m) concentrations of pyroxenite near the Moho (Fig. 2 in Bédard, 1991; Fig. 3 in Elthon et al., 1982). These pyroxenites are riddled with synkinematic peridotite sills and locally contain relict gabbroic slivers. We interpret them to have originated as gabbroic septa (now partly digested) that foundered into and reacted with the boninitic magmas underplating the crust and intruding as sills. Similarly, recycling of anorthosite reaction rims can account for the ubiquitous anorthosite boudins in unit 3 (Figs. 2 and 3) and might also explain the unusually feldspathic average mode of unit 3 (Komor et al., 1987).

Significance of modal ratios in cumulates: Cotectic or syntectic?

In classical studies of cumulate rocks (Bowen, 1928; Irvine, 1970), it is generally assumed that the proportions of minerals in cumulates were determined by the geometry of phase boundaries and that deviations from cotectic mineral ratios were caused either by mechanical effects (buoyancy contrasts, sorting by currents, e.g., Irvine et al., 1998) or by diffusion- controlled oscillations about the cotectic (e.g., McBirney and Noyes, 1979). Most of the plutonic rocks at North Arm are indeed cumulates of some kind (rather than frozen magmas), representing segregations of crystals; but graded beds attributable to sorting mechanisms are entirely absent, and most of the layering seen in the rocks is structural in origin. What is commonly seen at North Arm, however, are alternations of relatively massive, isomodal layers, each with a radically different ratio of olivine/plagioclase or of olivine/pyroxene (e.g., Fig. 6C; see also Irvine and Findlay, 1972), and relatively massive bodies within which mineral proportions show extreme variations from place to place on a hand-specimen scale (e.g., Figs. 4C and 6C). Although such variations in mineral proportions are difficult to interpret in terms of a cotectic control (Fig. 8), they are entirely consistent with a syntectic model, in which monomineralic reaction rims and/or xenocrystic materials derived from disaggregation of host rocks are mechanically mixed into the invading magma. Indeed, it could be argued that an extreme diversity of mineral proportions in a sequence of unsorted and poorly layered cumulates is itself an indication that the rocks do not represent equilibrium cotectic crystal segregates.

The crust as a reactive filter

The data from North Arm imply that the plutonic part (units 2 and 3) of the crust formed as networks of thin (<50 m) synkinematic sills emplaced into variably consolidated cumulates (Bédard, 1988, 1991, 1993; Bédard and Constantin, 1991; Bédard and Hébert, 1996, 1998). These data imply that partial-assimilation reactions between near-solidus gabbroic cumulate hosts and ascending primitive intrusions generated higher-temperature reaction products in the lower crust near the mantle—the source of the primitive magmas—and lower-temperature reaction products with increasing height in the crust. In principle, this concept is analogous to a cationic exchange column, in which lower-crustal "cumulates" represent near-total re-equilibration of host rocks with the ascending melt, whereas the uppermost gabbroic rocks represent the distilled essences of the entire process. In between these two extremes are the complex products of syntexis, which represent intermediate, metastable stages in the equilibration of more evolved host rocks with primitive melts. In effect, the mechanics of the plumbing system put Bowen's "reaction series" into reverse.

As ascending primary magmas react and equilibrate with their wall rocks while they wend their way toward the surface, the melt composition will be driven toward lower-temperature cotectic boundaries. In such a model, most MORBs would be multiply saturated because the geometry and dynamics of the conduit system forces parental melts to equilibrate with gabbroic cumulates en route to the surface (Bédard, 1993), not because they formed as eutectic mantle melts. The fractionation-assimilation path by which this process happens is unlikely to be unique, and the significance of such syntectic cumulates will be difficult to interpret because the proportions of phases in the end product will be a complex mixture of (1) phases originally present in the intrusion as phenocrysts, (2) xenocrysts and melts derived from the host, (3) reaction products, and (4) phases that crystallize from hybrid or interstitial melts. In addition, at some juncture, the melt in the pores of the cumulus framework may be expelled to some degree (Bédard, 1994a), and there may be additional complexities introduced through postcumulus metasomatic effects (Bédard et al., 1988). Thus, the complex assimilation-fractionation history of an individual intrusion or eruption will be difficult to reconstruct, even if the syntectic cumulates are well preserved and available for study. It will be even more difficult to determine the relative contributions of assimilation versus fractionation when working backward from lava compositions, especially if the magmas have been homogenized in high-level holding chambers before eruption.

Implications for the geochemical evolution of MORBs and continental flood basalts

Given that melts emerging from the mantle at ocean ridges typically have to pass through 5 km or so of near-solidus-temperature gabbroic oceanic crust before erupting, is it reasonable to think that none of these magmas will ever react with the crust en route to the surface? On the contrary, the preceding discussion suggests to us that syntexis of lower-crustal cumulates at ocean ridges should be an extremely common phenomenon. Are there any features of modern MORBs that would allow us to detect the action of syntexis? It turns out that there are several paradoxes and inconsistencies in currently accepted models of MORB genesis that could readily be explained if MORB parental magmas assimilated part of the cumulate lower crust in the course of differentiation. We do not mean to imply that the accepted explanations for all of the features enumerated below are incorrect, but it is always useful to consider a unifying and simplifying hypothesis.

MORBs define an extensive distribution along the three-phase (olivine + plagioclase + clinopyroxene) cotectic, a distribution that appears to require extensive clinopyroxene fractionation, even though clinopyroxene phenocrysts are rarely observed in MORB (Rhodes et al., 1979; Clague et al., 1981; Bryan et al., 1981). This paradox is generally explained by proposing a polybaric crystallization history, in which pyroxene fractionated at higher pressures. The common pyroxenite and anorthosite reaction rims seen at North Arm suggest an alternative explanation. Plagioclase or pyroxene may be assimilated or overproduced preferentially during syntexis of host gabbros and might never have formed free phenocrysts in the melt; thus no phenocrysts would exist in MORB lavas. In such a model, the observed range in clinopyroxene content along the cotectic may reflect assimilation, rather than crystallization, of clinopyroxene.

MORBs are quite aluminous and may contain abundant plagioclase phenocrysts (Bryan et al., 1981; Bryan, 1983). Assimilation of plagioclase at depth could provide the enrichment in Al_2O_3, CaO, Sr, Eu, and Na_2O needed to explain subsequent supersaturation in feldspar (Elthon, 1984). Aluminous spinels and partly resorbed calcic plagioclase grains are also common in MORBs (Rhodes et al., 1979; Fisk and Bence, 1980; Clague et al., 1981; Allan et al., 1988; Roeder and Reynolds, 1991). Re-equilibration and incongruent dissolution of plagioclase xenocrysts, or enrichment of boundary layers in Al_2O_3 due to dissolution of feldspar, could readily explain these features.

Diverse populations of melt inclusions in individual olivine or chromite phenocrysts (Nielsen et al., 1995; Kamenetsky et al., 1998) have been interpreted in terms of magma mixing during melt aggregation in the mantle. A simpler explanation might be that these olivines crystallized in a contaminated boundary layer as the magma was assimilating the walls of an intracrustal conduit and that the diverse melt populations represent incomplete mixing in these contaminated boundary layers. The low contents of incompatible trace elements observed in some of these inclusions could be explained by assimilation of extremely depleted cumulate rocks. A numerical study (O'Hara, 1998) suggests that assimilation of oceanic lithosphere may have a marked impact on melt trace element systematics.

Assimilation of hydrothermally altered rocks in the lower crust, as documented at North Arm, would contribute H_2O, S, alkalis, and Cl to the magma and might explain the enrichment in these elements seen in MORBs by Michael and Schilling (1989).

When considered together, data from MORB fall along the global array in terms of Fe8 versus Na8 (projected to 8% MgO), which has been interpreted as being due principally to variations in temperature acting on a relatively homogeneous mantle (Klein and Langmuir, 1987; Langmuir et al., 1992). However, when individual slow-spreading ridge segments are considered, the data show marked local cross-trends (Fig. 10), the origin of which remains problematic (Klein and Langmuir, 1989; Langmuir et al., 1992). We performed numerical calculations by using the Nielsen CHAOS 5 program (Nielsen, 1988, 1991) to test the hypothesis that these Atlantic Ridge local cross-trends might be the result of syntexis. The calculations are simplified to the extent that we assumed equilibrium throughout, whereas observations from North Arm suggest that syntexis is commonly influenced by kinetics. The calculations used a primitive MORB parent from Langmuir et al. (1992, Table C2). We assumed a conservative mass assimilated/mass crystallized ratio of 0.5. Detailed exposition of these models is beyond the scope of this paper, but the results (Fig. 10) suggest that assimilation of diverse cumulate rocks produces significant deviations from the closed-system evolution of the melt, deviations that are of the right magnitude and sense to account for the local geochemical cross-trends seen at the Mid-Atlantic Ridge.

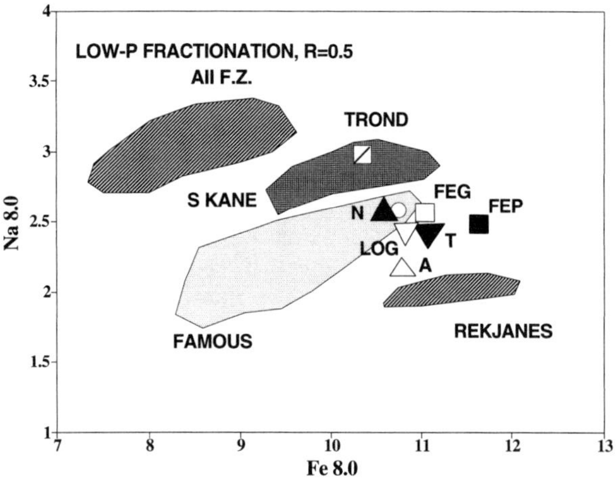

Figure 10. Na8 versus Fe8 diagram adapted from Langmuir et al. (1992), showing low-pressure fractionation starting with a conservative mass assimilated/mass crystallized ratio (*R*) of 0.5. Atlantic MORB fields (Atlantis II Fracture Zone, south part of Kane Fracture Zone, FAMOUS area, and Reykjanes peninsula lavas of Iceland) (from Fig. 21 in Langmuir et al., 1992). Parental MORB from Table C2 in Langmuir et al. (1992). Closed-system residue at 8% MgO is open circle. Results of coupled fractionation-assimilation calculations using the Nielsen (1988, 1991) program at 8% MgO are shown as various symbols, corresponding to different assimilates from North Arm, Bay of Islands (my own unpublished data): N—gabbronorite, TROND—trondhjemite, FEG—ferrogabbro, FEP—ferropyroxenite, T—troctolite, LOG—leuco-olivine gabbro, A—anorthosite.

CONCLUSIONS

The sophisticated melt-inversion models applied to MORB and to continental flood basalts are used to infer very fine scale variations in parameters such as mantle-source composition, degree of melting, and the dynamics of the melting process. Some of the parameters that enter into these calculations are potentially very sensitive to postsegregation processes like syntexis. We consider it unjustifiable to assume, a priori, that simple, closed-system fractional crystallization is the only mechanism by which basalts differentiate and to attribute all remaining geochemical diversity solely to intramantle processes. The evidence from ophiolites and layered intrusions shows that, on the contrary, syntectic assimilation, reaction, and hybridization with mafic lower-crustal host rocks may have a significant impact on the petrologic and geochemical evolution of basaltic magmas.

ACKNOWLEDGMENTS

We thank Henry Dick, Peter Kelemen, Stuart McCallum, Alexander McBirney, Alan Boudreau, William Meurer, Yildirim Dilek, Sven Maaløe, and many others for discussions that contributed to the ideas presented here. Peter Thy and Kent Ross provided constructive reviews. This is Geological Survey of Canada Contribution no. 1999049.

REFERENCES CITED

Abrajano, T.A., Pasteris, J.D., and Bacuta, G.C., 1989, Zambales ophiolite, Philippines. I. Geology and petrology of the critical zone of the Acoje massif: Tectonophysics, v. 168, p. 65–100.

Allan, J.F., Sack, R.O., and Batiza, R., 1988, Cr-rich spinels as petrogenetic indicators: MORB-type lavas from the Lamont Seamount chain, eastern Pacific: American Mineralogist, v. 73, p. 741–753.

Andersen, O., 1915, The system anorthite-fosterite-silica: American Journal of Sciences, v. 39, 4th series, p. 407–454.

Batanova, V.G., Suhr, G., and Sobolev, A.V., 1998, Origin of geochemical heterogeneity in the mantle peridotites from the Bay of Islands ophiolite, Newfoundland, Canada: Ion probe study of clinopyroxenes: Geochimica et Cosmochimica Acta, v. 62, p. 853–866.

Bédard, J.H., 1988, Magma chamber dynamics and recycling of crustal cumulates by the mantle: Evidence from the Bay of Islands ophiolite [abs.]: Eos, Transactions of the American Geophysical Union, v. 69, p. 1476.

Bédard, J.H., 1991, Cumulate recycling and crustal evolution in the Bay of Islands ophiolite: Journal of Geology, v. 99, p. 225–249.

Bédard, J.H., 1993, Oceanic crust as a reactive filter: Synkinematic intrusion, hybridization, and assimilation in an ophiolitic magma chamber, western Newfoundland: Geology, v. 21, p. 77–80.

Bédard, J.H., 1994a, A procedure for calculating the equilibrium distribution of trace elements in the minerals of cumulate rocks, and the concentration of trace elements in the coexisting liquids: Chemical Geology, v. 118: p. 143–153.

Bédard, J.H., 1994b, Boninitic cumulates and pyroxenitic digestion products from the lower crust of the Bay of Islands ophiolite: Geological Association of Canada/Mineralogical Association of Canada Meeting, Program with Abstracts, v. 19, p. A8.

Bédard, J.H., and Constantin, M., 1991, Syn- and post-kinematic intrusions of troctolite, gabbro, and peridotite into layered gabbroic cumulates in the Bay of Islands ophiolite: Genesis of anorthosite by partial assimilation, and troctolite by hybridization: Geological Survey of Canada, Current Research Paper 91-1, part D, p. 79–88.

Bédard, J.H., and Hébert, R., 1996, The lower crust of the Bay of Islands ophiolite, Canada: Petrology, mineralogy, and the importance of syntexis in magmatic differentiation in ophiolites and at ocean ridges: Journal of Geophysical Research, v. 101, p. 25105–25124.

Bédard, J.H., and Hébert, R., 1998, Formation of chromitites by assimilation of crustal pyroxenites and gabbros into peridotitic intrusions: North Arm Mountain massif, Bay of Islands ophiolite, Newfoundland, Canada: Journal of Geophysical Research, v. 103, p. 5165–5184.

Bédard, J.H., Sparks, R.S.J., Renner, R., Cheadle, M.J., and Hallworth, M.A., 1988, Peridotite sills and metasomatic gabbros in the Eastern Layered Series of the Rhum complex: Journal of the Geological Society of London, v. 145, p. 207–224.

Bender, J.F., Hodges, F.N., and Bence, A.E., 1978, Petrogenesis of basalts from the project FAMOUS area: Experimental study from 0 to 15 kbars: Earth and Planetary Science Letters, v. 41, p. 277–302.

Berclaz, A., Bédard, J.H., Hébert, R., Varfalvy, V., Mélançon, B., Yao, K.R., and el Bilali, L., 1994, Tectonic evolution of the plutonic section of the Bay of Islands ophiolite, Newfoundland [abs.]: Eos, Transactions of the American Geophysical Union, v. 75, p. 650–651.

Berclaz, A., Hébert, R., and Bédard, J.H., 1998, Predominance and significance of deformation processes in the genesis of deep oceanic crust at a slow-spreading ridge [abs.]: Evidence from the Bay of Islands ophiolite, Newfoundland: Geological Association of Canada/Mineralogical Association of Canada Meeting, Program with Abstracts, v. 23, p. A15–16.

Bernstein, S., Rosing, M.T., Brooks, C.K., Bird, D.K., 1992, An ocean-ridge type magma chamber at a passive volcanic, continental margin: The Kap Edvard Holm layered gabbro complex, East Greenland: Geological Magazine, v. 129, p. 437–456.

Bernstein, S., Kelemen, P.B., Brooks, C.K., 1996, Evolution of the Kap Edvard Holm Complex: A mafic intrusion at a rifted continental margin: Journal of Petrology, v. 37, p. 497–519.

Bodinier, J., Menzies, M.A., and Thirwall, M.F., 1991, Continental to oceanic mantle transition: REE and Sr-Nd isotopic geochemistry of the Lanzo Lherzolite Massif: Journal of Petrology, Special Lherzolites Issue, p. 191–210.

Bowen, N.L., 1928, The evolution of the igneous rocks: Princeton, New Jersey, Princeton University Press, 334 p.

Brearly, M., and Scarfe, C.M., 1986, Dissolution rates of upper mantle minerals in an alkali basalt melt at high pressure: An experimental study and implications for ultramafic xenolith survival: Journal of Petrology, v. 27, p. 1157–1186.

Bryan, W.B., 1983, Systematics of modal phenocryst assemblages in submarine basalts: Petrologic implications: Contributions to Mineralogy and Petrology, v. 83, p. 62–74.

Bryan, W.B., Thompson, G., and Ludden, J.N., 1981, Compositional variation in Normal MORB from 22°–25°N: Mid-Atlantic Ridge and Kane Fracture Zone: Journal of Geophysical Research, v. 86, p. 11815–11836.

Clague, D.A., Frey, F.A., Thompson, G., and Rindge, S., 1981, Minor and trace element geochemistry of volcanic rocks dredged from the Galapagos spreading center: Role of crystal fractionation and mantle heterogeneity: Journal of Geophysical Research, v. 86, p. 9469–9482.

Clarke, D.B., Henry, A.S., and White, M.A., 1998, Exploding xenoliths and the absence of "elephants graveyards" in granite batholiths: Journal of Structural Geology, v. 20, p. 1325–1343.

Constantin, M., Hékinian, R., Bideau, D., and Hébert, R., 1996, Construction of the oceanic lithosphere by magmatic intrusions: Petrological evidence from plutonic rocks formed along the fast-spreading East Pacific Rise: Geology, v. 24, p. 731–734.

Daly, R.A., 1914, Igneous rocks and their origin: New York, McGraw Hill, 312 p.

Daly, R.A., 1933, Igneous rocks and the depths of the Earth: New York, McGraw Hill, 598 p.

Detrick, R.S., Mutter, J.C., Buhl, P., and Kim, I.I., 1990, No evidence from multichannel reflection data for a crustal magma chamber in the MARK area on the Mid-Atlantic ridge: Nature, v. 347, p. 61–64.

Dick, H.J.B., Meyer, P., Bloomer, S., Kirby, S., Stakes, D., and Mauwer, C., 1991, Lithostratigraphic evolution in an in situ section of oceanic layer 3: Proceedings of the Ocean Drilling Program, Scientific Results v. 118, College Station, Texas, Ocean Drilling Program, p. 439–538.

Dickey, J.S., Jr., 1975, A hypothesis of origin for podiform chromite deposits: Geochimica et Cosmochimica Acta, v. 39, p. 1061–1074.

Dickey, J.S., Jr., and Yoder, H.S.J., 1972, Partitioning of chromium and aluminium between clinopyroxene and spinel: Carnegie Institute of Washington Yearbook, v. 71, p. 384–392.

Dickey, J.S., Jr., Yoder, H.S.J., and Schairer, J.F., 1971, Chromium in silicate-oxide systems: Carnegie Institute of Washington Yearbook, v. 70, p. 118–122.

Dilek, Y., and Eddy, C.A., 1992, The Troodos (Cyprus) and Kizildag (S. Turkey) ophiolites as structural models for slow-spreading ridge segments: Journal of Geology, v. 100, p. 305–322.

Donaldson, C.H., 1990, Forsterite dissolution in superheated basaltic, andesitic and rhyolitic melts: Mineralogical Magazine, v. 54, p. 67–74.

Dungan, M.A., and Rhodes, J.M., 1978, Residual glasses and melt inclusions in basalts from DSDP Legs 45 and 46: Evidence for magma mixing: Contributions to Mineralogy and Petrology, v. 67, p. 417–433.

Dunn, T., 1991, Experimental assimilation of two-pyroxene granulite into primitive basalt [abs.]: Eos, Transactions of the American Geophysical Union, v. 72, p. 294.

Dunn, R.A., Toomey, D.R., Solomon, S.C., and Purdy, G.M., 1994, Seismic structure of the lower crust and Moho beneath the East Pacific Rise at 9°30′N [abs.]: Eos, Transactions of the American Geophysical Union, v. 75, p. 657.

Edwards, B.R., and Russell, J.K., 1996, A review and analysis of silicate mineral dissolution experiments in natural silicate melts: Chemical Geology, v. 130, p. 233–245.

Elthon, D.L., 1984, Plagioclase buoyancy in oceanic basalts: Chemical effects: Geochimica et Cosmochimica Acta, v. 48, p. 753–768.

Elthon, D., 1991, Geochemical evidence for formation of the Bay of Islands ophiolite above a subduction zone: Nature, v. 354, p. 140–143.

Elthon, D., Casey, J.F., and Komor, S., 1982, Mineral chemistry of ultramafic cumulates from the North Arm Mountain massif of the Bay of Islands ophiolite: Evidence for high-pressure crystal fractionation of oceanic basalts: Journal of Geophysical Research, v. 87, p. 8717–8734.

Falloon, T.J., and Green, D.H., 1988, Anhydrous partial melting of peridotite from 8 to 35 kb and the petrogenesis of MORB: Journal of Petrology, Special Lithosphere Issue, p. 379–414.

Fisk, M.R., and Bence, A.E., 1980, Experimental crystallization of chrome spinel in FAMOUS basalt 527-1-1: Earth and Planetary Science Letters, v. 48, p. 111–123.

Fujii, T., and Bougault, H., 1983, Melting relations of a magnesian abyssal tholeiite and the origin of MORBs: Earth and Planetary Science Letters, v. 62, p. 283–295.

Garmany, J., 1989, Accumulations of melt at the base of young oceanic crust: Nature, v. 340, p. 628–632.

Garmany, J.D., Nakamura, Y., and Hello, Y., 1994, Mapping the occurrence of melt intrusions at the base of young ocean crust near 9°30′N on the East Pacific Rise [abs.]: Eos, Transactions of the American Geophysical Union, v. 75, p. 602.

Green, N.L., 1994, Mechanism for middle to upper crustal contamination: Evidence from continental-margin magmas: Geology, v. 22, p. 231–234.

Harper, G.D., 1985, Tectonics of slow spreading mid-ocean ridges and consequences of a variable depth to the brittle/ductile transition: Tectonophysics, v. 4, p. 395–409.

Hess, P.C., 1992, Phase equilibria constraints on the origin of ocean floor basalts, *in* Phipps Morgan, J., Blackman, D., and Sinton, J.L., eds., Mantle flow and melt generation at mid-ocean ridges: Washington D.C., American Geophysical Union, Geophysical Series Monograph 71, p. 67–102.

Hirth, G., and Kohlstedt, D.L., 1995, Experimental constraints on the dynamics of the partially molten upper mantle: Deformation in the diffusion creep regime: Journal of Geophysical Research, v. 100, p. 1981–2001.

Huppert, H.E., and Sparks, R.S.J., 1985, Cooling and contamination of mafic and ultramafic magmas during ascent through continental crust: Earth and Planetary Science Letters, v. 74, p. 371–386.

Irvine, T.N., 1970, Crystallization sequences in the Muskox intrusion and other layered intrusions. I. Olivine-pyroxene-plagioclase relations: Geological Society of South Africa, Symposium on the Bushveld and other layered intrusions, Special Publication 1, p. 441–476.

Irvine, T.N., 1977, Origins of chromite layers in the Muskox intrusion and other stratiform intrusions: A new interpretation: Geology, v. 5, p. 273–277.

Irvine, T.N., and Findlay, T.C., 1972, Alpine-type peridotite with particular reference to the Bay of Islands igneous complex: Canadian Department of Energy, Mines and Resources, Special Publication 42, p. 77–128.

Irvine, T.N., Andersen, J.C.O., and Brooks, C.K., 1998. Included blocks (and blocks within blocks) in the Skaergaard intrusion: Geologic relations and the origins of rhythmic modally graded layers: Geological Society of America Bulletin, v. 110, p. 1398–1447.

Kamenetsky, V.S., Eggins, S.M., Crawford, A.J., Green, D.H., Gasparon, M., and Falloon, T.J., 1998, Calcic melt inclusions in primitive olivine at 43°N MAR: Evidence for melt-rock reaction/melting involving clinopyroxene-rich lithologies during MORB generation: Earth and Planetary Science Letters, v. 160, p. 115–132.

Kelemen, P.B., 1990, Reaction between ultramafic rock and fractionating basaltic magma. I. Phase relations, the origin of calc-alkaline magma series, and the formation of discordant dunite: Journal of Petrology, v. 31, p. 51–98.

Kelemen, P.B., Johnson, K.T.M., Kinzler, R.J., and Irving, A.J., 1990, High field-strength element depletions in arc basalts due to mantle-magma interaction: Nature, v. 345, p. 521–524.

Kerr, R.C., 1995, Convective crystal dissolution: Contributions to Mineralogy and Petrology, v. 121, p. 237–246.

Klein, E.M., and Langmuir, C.H., 1987, Global correlations of ocean ridge basalt chemistry with axial depth and crustal thickness: Journal of Geophysical Research, v. 92, p. 8089–8115.

Klein, E.M., and Langmuir, C.H., 1989, Local versus global variations in ocean ridge basalt composition: A Reply: Journal of Geophysical Research, v. 94, p. 4241–4252.

Komor, S.C., Elthon, D., and Casey, J.F., 1987, Petrology of a leucogabbroic interval within basal layered gabbros at North Arm Mountain, Bay of Islands ophiolite: Contributions to Mineralogy and Petrology, v. 95, p. 278–300.

Kong, L.S.L., Solomon, S.C., and Purdy, G.M., 1992, Microearthquake characteristics of a mid-ocean ridge along-axis high: Journal of Geophysical Research, v. 97, p. 1659–1685.

Kuo, L.C., and Kirkpatrick, R.J., 1985, Dissolution of mafic minerals and its implications for the ascent velocities of peridotite-bearing basaltic magmas: Journal of Geology, v. 93, p. 697–700.

Langmuir, C.H., Klein, E.M., and Plank, T., 1992, Petrological constraints on melt formation and migration beneath mid-ocean ridges, *in* Phipps Morgan, J., Blackman, D., and Sinton, J.L., eds., Mantle flow and melt generation at mid-ocean ridges: Washington, D.C., American Geophysical Union, Geophysical Series Monograph 71, p. 183–280.

Lewis, B.T.R., 1983, The East Pacific Rise and the thermal model: Journal of Geophysical Research, v. 88, p. 3348–3354.

Lister, C.R.B., 1983, On the intermittency and crystallization mechanisms of sub-seafloor magma chambers: Geophysical Journal of the Royal Astronomical Society, v. 73, p. 351–365.

Lœwinson-Lessing, F., 1899, Compte Rendu du Congrès Geologique International, 7e Session, St. Petersburg, Russia, p. 379.

Malpas, J., 1990, Crustal accretionary processes in the Troodos ophiolite, Cyprus: Evidence from field mapping and deep crustal drilling, *in* Malpas, J., Moores, E.M., Panayiotou, A., and Xenophontos, C., eds., Ophiolites, ocean crustal analogues: Proceedings of the Symposium "Troodos 1987", Nicosia, Cyprus, Geological Survey Department, Ministry of Agriculture and Natural Resources, p. 65–74.

Marshall, L.A., and Sparks, R.S.J., 1984, Origin of some mixed-magma and net-veined ring intrusions: Journal of the Geological Society of London, v. 141, p. 171–182.

McBirney, A.R., 1979, Effects of assimilation, *in* Yoder, H.S. ed., The evolution of the igneous rocks: Fiftieth anniversary perspectives: Princeton, New Jersey, Princeton University Press, p. 307–338.

McBirney, A.R., and Noyes, R.M., 1979, Crystallization and layering of the Skaergaard Intrusion: Journal of Petrology, v. 20, p. 487–554.

McKenzie, D.P., and O'Nions, R.K., 1991, Partial melt distributions from inversion of rare earth element concentrations: Journal of Petrology, v. 32, p. 1021–1091.

Mével, C., Cannat, M., Gente, P., Marion, E., Auzende, J.M., and Karson, J.A., 1991, Emplacement of deep crustal and mantle rocks on the West Median Valley wall of the MARK area (MAR, 23°N): Tectonophysics, v. 190, p. 31–53.

Michael, P.J., and Schilling, J.G., 1989, Chlorine in mid-ocean ridge magmas: Evidence for assimilation of seawater-influenced components: Geochimica et Cosmochimica Acta, v. 53, p. 3131–3143.

Murck, B.W., and Campbell, I.H., 1986, The effect of temperature, oxygen fugacity, and melt composition on the behaviour of chromium in basic and ultrabasic melts: Geochimica et Cosmochimica Acta, v. 50, p. 1871–1887.

Natland, J.H., and Dick, H.J.B., 1996, Melt migration through high-level gabbroic cumulates of the East Pacific Rise at Hess Deep: Inferences from rock textures and mineral compositions, *in* Mével, C., Gillis, K., Allan, J.F., and Meyer, P.S., eds., Proceedings of the Ocean Drilling Project, Scientific Results 147, College Station, Texas, Ocean Drilling Program, p. 21–58.

Natland, J.H., Meyer, P.S., Dick, H.J.B., and Bloomer, S.H., 1991, Magmatic oxides and sulfides in gabbroic rocks from Hole 735B and the later development of the liquid line of descent: Proceedings of the Ocean Drilling Program, Scientific Results 118, College Station, Texas, Ocean Drilling Program, p. 75–111.

Navon, O., and Stolper, E., 1987, Geochemical consequences of melt percolation: The upper mantle as a chromatographic column: Journal of Geology, v. 95, p. 285–307.

Navon, O., Frey, F.A., and Takazawa, E., 1996, Magma transport and metasomatism in the mantle: A critical review of current geochemical models, Discussion: American Mineralogist, v. 81, p. 754–759.

Nielsen, R.L., 1988, A model for the simulation of combined major and trace element liquid lines of descent: Geochimica et Cosmochimica Acta, v. 52, p. 27–38.

Nielsen, R.L., 1991, CHAOS 5 (FORTRAN) program to model petrologic processes: College of Oceanography, Oregon State University, Corvallis, Oregon.

Nielsen, R.L., Crum, J., Bourgeois, R., Hascall, K., Forsythe, L.M., Fisk, M.R., and Christie, D.M., 1995, Melt inclusions in high-An plagioclase from the Gorda Ridge: An example of the local diversity of MORB parent magmas: Contributions to Mineralogy and Petrology, v. 122, p. 34–50.

Nielson, J.E., and Wilshire, H.G., 1993, Magma transport and metasomatism in the mantle: A critical review of current geochemical models: American Mineralogist, v. 78, p. 1117–1134.

O'Hara, M.J., 1998, Volcanic plumbing and the space problem: Thermal and geochemical consequences of large scale assimilation in ocean island development: Journal of Petrology, v. 39, p. 1077–1089.

O'Hara, M.J., and Mathews, R.E., 1981, Geochemical evolution in an advancing, periodically replenished, periodically tapped, continuously fractionated magma chamber: Journal of the Geological Society of London, v. 138, p. 237–277.

Onuma, K., and Tohara, T., 1983, Effect of chromium on phase relations in the join forsterite-anorthite-diopside in air at 1 Atm: Contributions to Mineralogy and Petrology, v. 84, p. 174–181.

Orberger, B., Lorand, J.P., Girardeau, J., Mercier, J.C.C., and Pitragool, S., 1995, Petrogenesis of ultramafic rocks and associated chromitites in the Nan Uttaradit ophiolite, northern Thailand: Lithos, v. 35, p. 153–182.

Parsons, T., Howie, J.M., and Thompson, G.A., 1992, Seismic constraints on the nature of lower crustal reflectors beneath the extending southern transition zone of the Colorado Plateau, Arizona: Journal of Geophysical Research, v. 97, p. 12391–12407.

Paterson, S.R., Fowler, T.K., and Miller, R.B., 1996, Pluton emplacement in arcs: A crustal-scale exchange process: Transactions of the Royal Society of Edinburgh, Earth Sciences, v. 87, p. 115–123.

Pedersen, R.B., 1986, The nature and significance of magma chamber margins in ophiolites: Examples from the Norwegian Caledonides: Earth and Planetary Science Letters, v. 77, p. 100–112.

Reuber, I., 1988, Complexity of the crustal sequence in the northern Oman ophiolite (Fizh and southern Aswad blocks): The effect of early slicing?: Tectonophysics, v. 151, p. 137–165.

Rhodes, J.M., Dungan, M.A., Blanchard, D.P., and Long, P.E., 1979, Magma mixing at mid-ocean ridges: Evidence from basalts drilled near 22°N on the Mid-Atlantic Ridge: Tectonophysics, v. 55, p. 35–61.

Richter, F.M., 1986, Simple models for trace element fractionation during melt segregation: Earth and Planetary Science Letters, v. 77, p. 333–344.

Roeder, P.L., and Reynolds, I., 1991, Crystallization of chromite and chromium solubility in basaltic melts: Journal of Petrology, v. 32, p. 909–934.

Rosencrantz, E., 1983, The structure of the sheeted dikes and associated rocks in the North Arm massif, Bay of Islands ophiolite complex, and the intrusive process at oceanic spreading centers: Canadian Journal of Earth Sciences, v. 20, p. 787–801.

Rudnick, R.L., and Fountain, D.M., 1995, Nature and composition of the continental crust: A lower crustal perspective: Reviews of Geophysics, v. 33, p. 267–309.

Rudnick, R.L., McDonough, W.F., and O'Connell, R.J., 1998, Thermal structure, thickness, and composition of continental lithosphere: Chemical Geology, v. 145, p. 395–411.

Sato, H., Sacks, I.S., and Murase, T., 1989, The use of laboratory velocity data for estimating temperature and partial melt fraction in the low-velocity zone: Comparison with heat flow and electrical conductivity studies: Journal of Geophysical Research, v. 94, p. 5689–5704.

Shen, Y., and Forsyth, D.W., 1995, Geochemical constraints on initial and final depths of melting beneath mid-ocean ridges: Journal of Geophysical Research, v. 100, p. 2211–2237.

Sinha, M.C., Constable, S.C., Peirce, C., White, A., Heinson, G., MacGregor, L.M., and Navin, D.A., 1998, Magmatic processes at slow spreading ridges: Implications of the RAMESSES experiment at 57°45′N on the Mid-Atlantic Ridge: Geophysical Journal International, v. 135, p. 731–745.

Sinton, J.M., and Detrick, R.S., 1992, Mid-ocean ridge magma chambers: Journal of Geophysical Research, v. 97, p. 197–216.

Siroky, F.X., Elthon, D.L., Casey, J.F., and Butler, J.C., 1985, Major element variations in basalts and diabases from the North Arm Mountain Massif, Bay of Islands Ophiolite: Implications for magma chamber processes at mid-ocean ridges: Tectonophysics, v. 116, p. 41–61.

Sleep, N.H., 1975, Formation of oceanic crust: Some thermal constraints: Journal of Geophysical Research, v. 80, p. 4037–4042.

Smith, C.H., 1958, Bay of Islands Igneous Complex, Western Newfoundland: Geological Survey of Canada Memoir 290, 132 p.

Solomon, S.C., and Toomey, D.R., 1992, The structure of mid-ocean ridges: Annual Review of Earth and Planetary Science, v. 20, p. 329–364.

Sørensen, H.S., and Wilson, J.R., 1996, Petrology of the Treknattan intrusion in the Fongen–Hyllingen complex, Trondheim region, Norway: A late intrusion into an evolved layered complex: Lithos, v. 38, p. 109–127.

Sparks, R.S.J., 1986, The role of crustal contamination in magma evolution through geologic time: Earth and Planetary Science Letters, v. 78, p. 211–223.

Stakes, D.S., and Taylor, H.P., Jr., 1992, The northern Samail Ophiolite: An oxygen isotope, microprobe, and field study: Journal of Geophysical Research, v. 97, p. 7043–7080.

Swinden, H.S., Jenner, G.A., and Szybinski, Z.A., 1997, Magmatic and tectonic evolution of the Cambrian–Ordovician Laurentian margin of Iapetus: Geochemical and isotopic constraints from the Notre Dame subzone, Newfoundland, *in* Sinha, A.K., Whalen, J.B., and Hogan, J.P., eds., The nature of magmatism in the Appalachian Orogen: Geological Society of America Memoir 191, p. 337–365.

Tegner, C., and Robins, B., 1996, Picrite sills and crystal-melt reactions in the Honningsvag Intrusive Suite, northern Norway: Mineralogical Magazine, v. 60, p. 53–66.

Toomey, D.R., Solomon, S.C., and Purdy, G.M., 1988, Microearthquakes beneath the median valley of the Mid-Atlantic Ridge near 23°N: Tomography and tectonics: Journal of Geophysical Research, v. 93, p. 9093–9112.

Tsuchiyama, A., 1985, Dissolution kinetics of plagioclase in the melt of the system diopside-albite-anorthite, and origin of dusty plagioclase in andesites: Contributions to Mineralogy and Petrology, v. 89, p. 1–16.

Tsuchiyama, A., 1986, Experimental study of olivine-melt reaction and its petrological implications: Journal of Volcanology and Geothermal Research, v. 29, p. 245–264.

Tsuchiyama, A., and Takahashi, E., 1983, Melting kinetics of plagioclase feldspar: Contributions to Mineralogy and Petrology, v. 84, p. 345–354.

Umino, S., Yanai, S., Jaman, A.R., Nakamura, Y., and Iiyama, J.T., 1990, The transition from spreading to subduction: Evidence from the Semail ophiolite, northern Oman Mountains, *in* Malpas, J., Moores, E.M., Panayiotou, A., and Xenophontos, C., eds., Ophiolites, ocean crustal analogues: Proceedings of the Symposium "Troodos 1987", Nicosia, Cyprus, Geological Survey Department, Ministry of Agriculture and Natural Resources, p. 375–384.

Varfalvy, V., Hébert, R., and Bédard, J.H., 1996, Melt-upper mantle interactions in the North Arm Massif, Bay of Islands ophiolite, Newfoundland: Implications for the genesis of boninitic magmas: Chemical Geology, v. 129, p. 71–90.

Varfalvy, V., Hébert, R., Bédard, J.H., and Laflèche, M.R., 1997, Petrology and geochemistry of pyroxenite dykes in upper mantle peridotites of the North Arm Massif, Bay of Islands ophiolite, Newfoundland: Implications for the genesis of boninitic and related magmas: Canadian Mineralogist, v. 35, p. 543–570.

Vera, E.E., Mutter, J.C., Buhl, P., Orcutt, J.A., Harding, A.J., and Kappus, M.E., et al., 1990, The structure of 0- to 0.2 m.y.-old oceanic crust at 9°N on the East Pacific Rise from expanded spread profiles: Journal of Geophysical Research, v. 95, p. 15529–15556.

Walker, D., Shibata, T., and DeLong, S.E., 1979, Abyssal tholeiites from the Oceanographer Fracture Zone II. Phase equilibria and mixing: Contributions to Mineralogy and Petrology, v. 70, p. 111–125.

Wark, D.A., and Watson, E.B., 1998, Grain-scale permeabilities of texturally equilibrated, monomineralic rocks: Earth and Planetary Science Letters, v. 164, p. 591–605.

MANUSCRIPT ACCEPTED BY THE SOCIETY FEBRUARY 23, 2000

Printed in the U.S.A.

Geological Society of America
Special Paper 349
2000

Can downward flow of dense cumulate slurry through mushy upper gabbros produce lower gabbros at a fast-spreading center?

W. Roger Buck
Lamont-Doherty Earth Observatory of Columbia University, Palisades, New York 10964, USA; email: buck@ldeo.columbia.edu

ABSTRACT

A variant on the model of a shallow magma lens as the site of gabbro crystallization at fast-spreading ridges is presented; the model has the following new feature: Periodically, interstitial melt–bearing cumulate that has built up below fractionated magma in the magma lens descends as a slurry out of the lens by density-driven flow.

The model requires a viscosity of less than ~10^{10} Pa·s for the still partially molten upper gabbros just below the magma lens. This is about the viscosity for a magma lens several tens of meters thick at the critical Rayleigh number for thermal convection. Such a magma lens could convect when refreshed with an injection of primary melt, but the magma in its most common state would be marginally subcritical for thermal convection. A dense cumulate of olivine gabbro builds up on the bottom of the lens as the refreshed magma body begins to cool. When a sufficient mass is achieved, the cumulate plus its interstitial melt drops through the lens's floor of partially molten upper gabbros and spreads out in thin layers in the lowermost crust to augment the lower gabbros.

The fractionated liquid in the lens adds to the upper crust during diking events. Remaining fractionate cools to form a gabbro crystal mush, which is less dense than cumulate but more dense than melt. Subsequently, fresh primary melt again fills the shallow lens at the base of the brittle upper crust and pushes the gabbro crystal mush down and laterally out of the magma lens to cooler locations where crystallization is completed. In this model, the upper gabbros of the oceanic crust are made from such later-crystallizing gabbro additions.

INTRODUCTION

There has long been great interest in how and where melt cools and crystallizes to form the oceanic crust. A major issue has been the size of magma chambers where primary melts crystallize and fractionate to make the magma that forms the upper crust. One view, based partly on ophiolite studies, was that the largely liquid magma chamber is as thick as the entire gabbroic lower crust (several kilometers thick), and possibly much wider (Pallister and Hopson, 1981). Another view is that only a small, thin magma chamber exists at the top of the gabbro section of a mid-ocean ridge.

This model involving small magma chambers, first put forward by Sleep (1975), gained a great boost from the seismic detection of small, axial magma-chamber (AMC) reflections along mid-ocean ridges spreading at fast and intermediate velocities (Morton and Sleep, 1985; Detrick et al., 1987). The idea got further support from thermal models based explicitly on the idea that all the lower crust was crystallized in a small, shallow, magma lens. These models successfully predict deepening of the magma lens with decreasing spreading rate and decreasing crustal thickness (Quick and Denlinger, 1993; Henstock et al., 1993; Phipps Morgan and Chen, 1993). This trend is seen in seismic data on a variety of ridges (e.g., Purdy et al., 1992). For fast-spreading ridges, these models predict that the complete gabbro section, making up the bulk of the crust, is crystallized at shallow depths.

A great appeal of this model is its simplicity. The upward motion of buoyant melt (by whatever means) is halted by cold, effectively solid upper crust. Cooling of the melt occurs by a combination of hydrothermal circulation through the fractured shallow upper crust, conduction, and convection in the magma chamber. The upper-crustal sheeted dikes and lava flows are made from magma extracted from the chamber, whereas the gabbro section is made from the residual magma of successively filled magma lenses. The residual magma is thought to be a hot crystal mush that thus can flow in a viscous manner over a broad (10–20 km wide) region (Sleep, 1975; Dewey and Kidd, 1977; Sinton and Detrick, 1992; Nicolas et al., 1993). The viscous gabbros flow down and away from the magma lens much as the ice in a glacier flows away from the surface where the glacier is fed by snowfall.

The Oman ophiolite, however, shows that the gabbro section is not at all uniform, as predicted by the "gabbro glacier" model (Fig. 1). The lower gabbros are compositionally different from—and, in fact, appear to be the cumulate complement to—the fractionated upper gabbros (Kelemen and Aharanov, 1998). The dense, cumulate-textured lower gabbros are sometimes modally layered, with small-scale graded bedding (Boudier et al., 1996). The upper gabbros in

Buck, W.R., 2000, Can downward flow of dense cumulate slurry through mushy upper gabbros produce lower gabbros at a fast-spreading center?, *in* Dilek, Y., Moores, E.M., Elthon, D., and Nicolas, A., eds., Ophiolites and Oceanic Crust: New Insights from Field Studies and the Ocean Drilling Program: Boulder, Colorado, Geological Society of America Special Paper 349, p. 121–127.

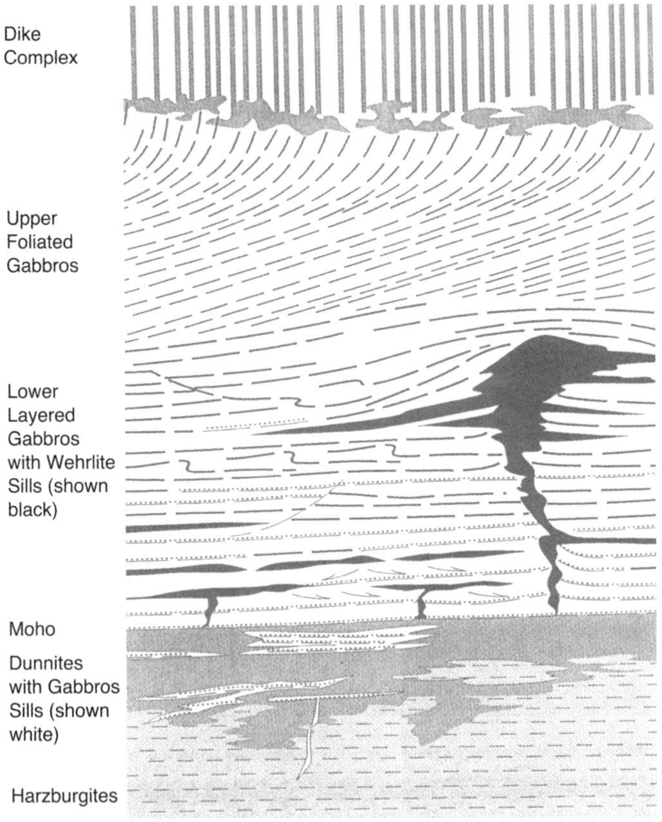

Figure 1. Schematic section through the Oman ophiolite from Boudier et al. (1996). The isotropic and foliated upper gabbros are similar in bulk composition to the dikes and pillows (not shown). The lower, layered gabbros appear to be formed of cumulates that are complementary to the upper rocks. The black area represents wehrlites, thought to be intruded from the mantle.

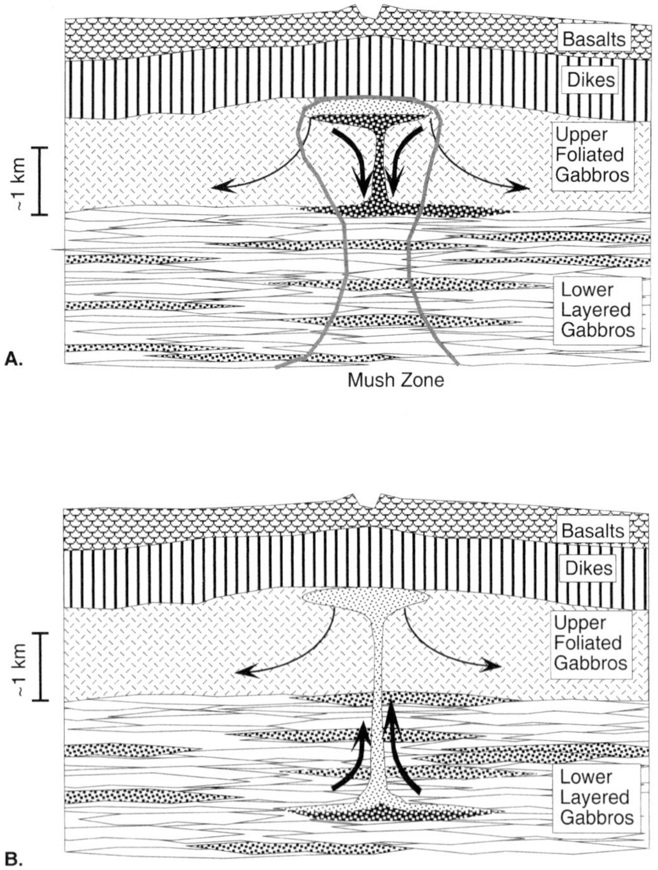

Figure 2. Illustrations of two contrasting models for formation of the gabbro sections of fast-spreading oceanic lower crust. A, The present model—cumulate descent—in which most primary melt migrates to a shallow magma lens where the melt is cooled. Dense cumulate slurry formed there flows down through the partially molten upper gabbros to form the lower gabbros. B, One of the models for crystallization near the base of the crust—the cumulite "restite" model (Korenaga and Keleman, 1997)—that contrasts most directly with the present model. Primary melts are trapped in multiple sills throughout the crust. Cumulate in these sills is left behind as "restite" when the fractionated liquid moves to a shallow magma chamber.

contrast have a more fractionated composition, essentially the same composition as the upper-crustal dikes and pillow basalts (Korenaga and Kelemen, 1997).

The ophiolite data, coupled with seismic data indicating that the base of the crust is partially molten at the axis of the lat 9°N section of the East Pacific Rise (Garmany, 1989; Dunn and Toomey, 1997), have led to new, more complex models with multiple magma lenses at different depths (e.g., Schouten and Denham, 1995; Boudier et al., 1996; Korenaga and Kelemen, 1997; Chenevez and Nicolas, 1997; Kelemen and Aharanov, 1998). A common feature of this new generation of models is that all or part of the lower gabbros, which have cumulate textures, is crystallized near the base of the crust (Fig. 2B), whereas the upper gabbros, which have isotropic and foliated textures, are crystallized at the base of the upper crust.

A major unresolved problem with these "crystallization at the bottom of the crust" models for fast-spreading ridges is heat transport. How does the very considerable latent heat of crystallization get transported from the base of the crust out into the ocean? The models for crystallization on top of the gabbro section do this by hydrothermal circulation through the brittle and fractured upper crust. There is no evidence for hydrothermal circulation through the base of fast-spreading crust close to the ridge axis. One expects earthquakes associated with such cooling, but lower-crustal earthquakes are not observed at the axis of fast-spreading ridges (e.g., Solomon and Toomey, 1992). Also, seismic results are interpreted in terms of partial melt beneath the axial magma chamber (Vera et al., 1990; Harding et al., 1989). Chen (1998) has shown that even moderate amounts of crystallization of the lower gabbros at the base of the crust could not occur in the absence of penetrating hydrothermal circulation. There is evidence for off-axis hydrothermal circulation through the entire crust (e.g., Gregory and Taylor, 1981). This observation, along with other data, led Boudier et al. (1996) to propose that some lower-crustal gabbros may crystallize a few kilometers off-axis when melt intrudes from the mantle.

This paper is an attempt to reconcile the old simple model of crystallization in a small, shallow, magma lens with the ophiolite-inspired view of a compositionally stratified gabbroic lower crust (at least for crust formed at fast-spreading centers). I will not review the debate about whether ophiolites represent "normal" oceanic crust or whether the Oman ophiolite represents fast-spreading oceanic lithosphere. I also will not review the extensive literature on magma-chamber processes, but I will discuss some simple physical arguments about crystallization of magma and segregation of cumulate crystals away from the lower-density, later-crystallized mush. My

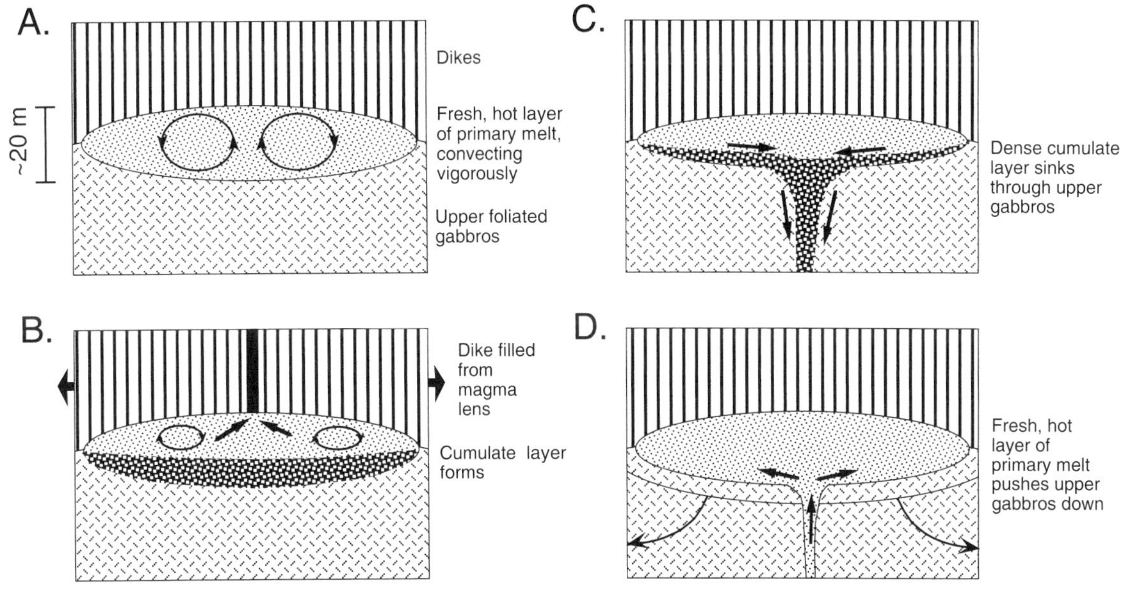

Figure 3. Detailed illustration of magma-chamber processes in the proposed model. A, Fresh, hot primary melt flows into the chamber, resulting in a brief period of rapid convection and cooling. B, The first-formed, dense crystals settle to the bottom of the magma lens. Overlying dikes and flows are fed from the fractionated magma lens. C, The mafic cumulate at the bottom of the magma lens undergoes a Rayleigh-Taylor–type instability and flows as a crystal slurry down through the less dense upper-gabbro section, which floors the magma lens. D, The complementary fractionated magma has cooled and partially crystallized. When a new packet of magma flows up, it pushes the gabbro magma mush down and out to the side to add to the upper gabbros.

goal is to consider ways that crystallization, driven by shallow-level hydrothermal cooling, could lead to the pattern of gabbros seen in the Oman ophiolite.

CUMULATE FORMATION IN A SHALLOW MAGMA LENS

Assume that mantle-derived melt migrates through the crust until its ascent is halted by cold, strong upper crust that will not flow out of the way (Fig. 2A). Support for this assumption comes from seismic observations of magma chambers at 1–3 km depth below the seafloor at the axis of ridges spreading at fast to intermediate velocities (Morton and Sleep, 1985; Detrick et al., 1987; Mutter et al., 1995; Carbotte et al., 1996). The thickness of the lens and the melt fraction in it are not well established, but the lens is likely to be ~10–100 m thick and only partially molten (Ponce-Correa et al., 1999; Singh et al., 1998; Hussenoeder et al., 1996; Kent et al., 1993).

When batches of melt form or add to a magma lens (or magma chamber), the viscosity of the melt-rich layer may be low enough to allow cumulate settling and thermal convection (Fig. 3A). The first crystals to form, such as olivine, are denser than the residual liquid, and so they settle through low-viscosity magma to form a cumulate layer (e.g., Bowen, 1922; Stolper and Walker, 1980; Naslund and McBirney, 1996). Later-formed crystals, such as plagioclase, have densities close to that of the melt and so stay in suspension (Daly et al., 1966). However, the cumulate layer is likely to be a combination of several minerals plus a large fraction of liquid melt. The viscosity of the crystal-liquid combination depends mainly on the volume fraction of crystals present (Murase and McBirney, 1973).

CUMULATE FLOW

The substrate of the magma chamber is upper gabbro, which though still partially molten is much more viscous than the primary melt that periodically enters the chamber. Thus, cumulate (plus interstitial melt) may form a layer on the bottom of the chamber (Fig. 3B). There are several ways that the dense cumulate could move down into the lower crust. Here, I look at one of the simplest ways for this sinking to happen: diapiric flow through the viscous, but less dense, upper gabbros.

The initial phase of such sinking can be treated in terms of a Rayleigh-Taylor instability. If the cumulate layer has a viscosity comparable to that of the surrounding material, then it is mathematically easy to look at conditions allowing it to sink. The time constant, τ, for the growth of such an instability is $13\mu/\Delta\rho g b$ (Turcotte and Schubert, 1982), where μ is the viscosity, $\Delta\rho$ is the density contrast, g is the acceleration due to gravity, and b is the thickness of the dense layer. The first-formed cumulate layer is likely to be of order 100 kg/m^3 denser than the material above and below it (i.e., the density contrast $\Delta\rho$ = 100 kg/m^3), and a reasonable cumulate layer thickness b is 10 m. Thus, the diapir should form on a time scale much less than that for magma-chamber replenishment, which is on the order of 10–100 yr (as discussed subsequently). If I assume that τ = 1 week (~6.5 × 10^5 s), then the viscosity μ must be ~10^9 Pa·s. If a region from the shallow magma chamber down to the bottom of the upper gabbros has a similar viscosity, then a blob of the mafic cumulate could sink into the lower gabbros on a reasonable time scale (estimated by using the formula for flow of a Stokes' sphere, as given in Turcotte and Schubert, 1982).

The sinking of the cumulate diapir could stop when it encounters denser material such as mantle or more dense parts of the lower crust, as shown in Figure 2A. When the downward motion stops, the cumulate slurry probably will spread out in a thin, nearly horizontal layer. The slurry consists of high-density minerals such as olivine, as well as low-density minerals such as plagioclase, plus entrained melt. As a layer spreads out, the minerals segregate according to density, possibly producing graded bedding. Gradually, filter pressing leads to expulsion of part of the interstitial melt, leaving the crystalline residue sufficiently viscous to prevent thermal convection and thus preserve the layering.

Figure 3 shows a cartoon version of the sequence of events suggested here for the region around a magma lens. The lower gabbros consist of the denser cumulate from each batch of primary melt. I assume that the upper gabbros are formed when new batches of primary melt fill the magma lens and push the partially crystallized remnants of the older magma out and/or down. The time interval between density-flow events depends on the size of the cumulate bodies that flow down, the thickness of the lower gabbros, and the spreading rate. If the dense cumulate starts out 10 m thick and 1 km wide, then one flowing down every 50–100 yr could build the lower gabbros at a fast-spreading ridge.

VISCOSITY OF THE MAGMA LENS AND UPPER GABBROS

Diapiric sinking of dense cumulate can only work if the viscosity of the upper gabbros is low enough to allow a dense cumulate layer to sink through it. Thus an estimate is needed of the viscosity of those gabbros when they are formed in a cooling magma lens. Convection and cooling should occur in the magma lens as long as the viscosity there is low and the temperature drop across the layer is moderate. As convectively enhanced cooling continues, the crystallinity and thus the viscosity of the layer increase markedly. When the layer viscosity increases to the point where thermal convection is no longer possible, the layer stops convecting and then cools by the much slower process of thermal conduction.

It is easy to show that a lens with pure melt in it could not be a steady-state feature of a ridge. Such a lens should transport heat to the hydrothermally cooled upper crust much faster than heat is supplied. The temperature drop across such a chamber is a hotly debated topic (e.g., Marsh, 1989; Huppert and Sparks, 1988). However, even if the temperature drop ΔT across a layer of melt were only a few degrees, then the convective heat transport should be much faster than the rate of supply of heat into crust.

To estimate the rate of cooling in a shallow magma lens, I assume that all the lower crust of thickness h is crystallized and liberates its latent heat in the lens (Wilson et al., 1988; Chenevez and Nicolas, 1997). For a total plate-spreading rate v, the rate of heat liberated in the sill is

$$q_{in} = \rho L h v, \tag{1}$$

where L is the latent heat and ρ is density. Taking v = 10 cm/yr, h = 4 km, L = 400 kJ/kg, and ρ = 3000 kg/m^3 results in q_{in} = 1.5 × 10^4 W/m.

I assume that in a steady-state situation, the convecting-magma lens (no matter what its thickness) would melt the top of the chamber until the conductive (and hydrothermal) heat flow matched the heat transported by convection. The rate of convective heat transport through a lens of thickness l and width W is

$$q_{out} = WK\mathrm{Nu}\Delta T/l, \tag{2}$$

where K is conductivity and Nu is the layer Nusselt number (the ratio of the rate of heat transported convectively to what would be carried by conduction alone). To get the convective heat transport through a magma lens, I assume a standard relationship between the layer Rayleigh number and the Nusselt number describing heat transport (Turcotte and Schubert, 1982):

$$\mathrm{Nu} = 0.12\mathrm{Ra}^{0.3}, \tag{3}$$

where Ra is the Rayleigh number given by

$$\mathrm{Ra} = \frac{\rho g \alpha \Delta T l^3}{\kappa \mu} \tag{4}$$

I assume a melt viscosity of 10 Pa·s (Murase and McBirney, 1973), thermal diffusivity of 10^{-6} m^2/s, and $\rho g \alpha$ = 1 where ρ is melt density, α is thermal expansion, and g is the acceleration of gravity. Then for ΔT = 10 K and layer thickness l = 50 m, the Rayleigh number is 1.3 × 10^{11} and Nu = 260. Therefore, the rate of convective heat transport q_{out} = 3 × 10^5 W/m. This result is about 20 times the heat input estimated by equation 1.

The convecting-magma lens would transport far more heat than could be supplied in steady state, and this finding implies that a pure magma lens is not a steady-state feature. As noted above, seismic results do not require pure melt in the AMC. Thus, I expect that when a magma lens is filled with a batch of melt, it cools and crystallizes very fast. At the point at which the viscosity of the layer increases enough to stop thermally driven convection, the rate of heat transport goes down abruptly. This effect should occur when the layer Rayleigh number reaches the critical value (Ra_{cr}) for convection (on the order of 10^3). The value of the critical viscosity (μ_{cr}) can be estimated because when Ra = Ra_{cr},

$$\mu_{cr} = \frac{\rho g \alpha \Delta T l^3}{\kappa \mathrm{Ra}_{cr}}. \tag{5}$$

Taking ΔT = 10 °C and layer thickness l = 50 m leads to a critical viscosity of ~10^9 Pa·s. The viscosity of the lens should vary greatly between times of magma replenishment, but its average value should be less than the critical viscosity so that the convective heat transport matches the average needed for crystallizing the lower crust. This critical viscosity for magma-chamber convection is about what is needed for dense cumulate to descend through the upper gabbros on a reasonable time scale.

Each new melt layer will first crystallize olivine-rich, high-density cumulates. These may be so dense as to settle to the bottom of the magma lens and then through the upper gabbros. Later-crystallizing phases will be less dense and less mafic and may be able to participate in the convection until the system "seizes up" as the crystallinity increases and the viscosity approaches the critical value. It is those less dense phases that will become the upper gabbros as they flow out in the "gabbro glacier" fashion.

Thus, I expect that each new packet of melt that forces its way to the top of the lower crust should cool on top of material with roughly the critical viscosity. The temperature-dependent viscosity of the crystal mush should not change much because conductive cooling should be slow compared to the rate of downward advection. Since the Rayleigh number depends on layer thickness, one might expect that a thick layer of mushy gabbro would convect. There is no ophiolite data indicating that the gabbro layer thermally convects outside a small magma lens (Chenevez et al., 1998). There are several possible reasons why low-viscosity upper-gabbro mush might not thermally convect even if it is at the critical viscosity for convection in a thin lens: (1) the hottest temperatures would be on top of that layer (at

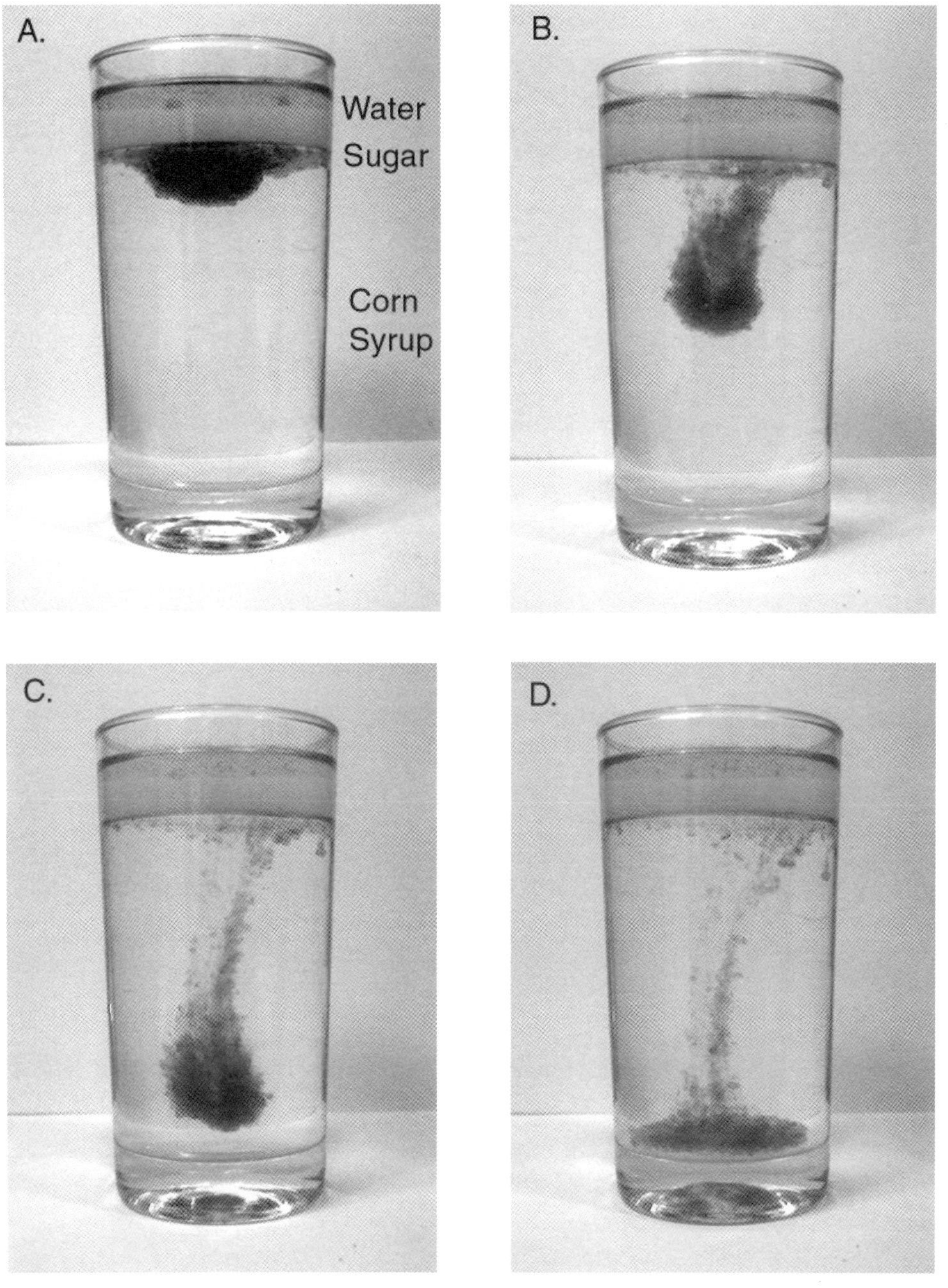

Figure 4. Photographs showing the results of a kitchen tabletop experiment of a Rayleigh-Taylor instability. In the experiment, water represents the fractionated magma of a magma lens. The sugar crystals represent cumulate in such a lens that flows down through less dense "upper gabbroic" corn syrup to form a layer of "lower gabbro."

the base of the magma lens) and so would not drive convection; (2) compositionally denser material might flow deeper, leading to stable stratification; and (3) the crystal-mush zone may be too narrow to convect. Outside a narrow region of upwelling of melt and descent of cumulate, melt may migrate out of the partially molten crystal mush and thus make the residual mush stronger.

ANALOGUE MODEL

A crude analogue experiment illustrating the idea for part of this process is pictured in Figure 4, which shows a layer of corn syrup (Karo) in the bottom half of a glass, overlain by a layer of water. The syrup, water, and dark sugar crystals represent the mushy part of the upper gabbros, a batch of melt, and early-formed cumulate, respectively. A layer of cumulate is slowly built up by sprinkling the sugar into the water. When the layer is sufficiently thick, it sinks through the (upper gabbro) syrup. A cumulate sugar (lower gabbro) layer is made when the sugar spreads out as it hits the bottom of the glass. The experiment even shows evidence of filter pressing: the interstitial water that was swept down with the sugar crystals slowly seeps out of the sugar and forms small water diapirs that float up through the syrup, although these are not shown clearly in Figure 4.

DISCUSSION

This "cumulate descent" model circumvents the thermal problem of the "crystallization at the base of the crust" models, while still predicting a compositionally stratified lower oceanic crust. Most of the cooling to form the cumulate lower gabbros happens close to the upper crust where heat transport by hydrothermal circulation is known to occur. The present model also may be consistent with the increase in temperature with depth inferred from studies of the phase equilibria of the gabbroic section of the Oman ophiolite. Korenaga and Kelemen (1997) determined that the upper gabbros crystallized at ~1175 °C whereas the lower gabbros may have formed at temperatures up to ~1225 °C. The cumulates forming the lower gabbros would have crystallized when the magma lens was hotter, and the upper gabbros would have crystallized when it had cooled. Perhaps the cumulates passed through the upper gabbros fast enough that there was little heat exchange. Hot mantle below it and hot melt passing through it could heat the lowest part of the layered gabbro section. The seismically observed low-velocity region near the Moho under the axis of fast-spreading ridges (e.g., Garmany, 1989; Dunn and Toomey, 1997) may indicate the presence of melt at those depths, but does not require that the melt is crystallizing there.

The subvertical foliation of some of the upper gabbros may be related to the downward flow of the cumulate predicted by this model. Also the graded bedding seen in some of the lower gabbros may be consistent with gravity-driven flow of a cumulate slurry. However, detailed discussion of such features will have to be left to others more conversant with interpretation of igneous processes.

In this model, the primary melt moves up through the hottest part of the crust to fill a shallow magma lens. This process may occur through a variety of mechanisms such as flow through porous channels, fracture propagation, or diapiric rise. Whatever the means, the upward flow is driven by the density contrast between melt and the rock through which it passes. In developing this model, I looked at diapiric sinking of cumulate slurry that forms because of cooling and crystallization in the magma lens. Only a narrow region of the axial crust need be weak for dense cumulate to flow down through it. The narrower the "conduit," the lower its viscosity must be to allow such flow. Passage of melt moving up from the mantle may keep such a conduit hot and partially molten while adjacent lower crust might be much stronger. However, the cumulate also may move down in other ways. A pathway for descent of low-viscosity cumulate slurry might even be opened as a hydrofracture in the lower crust when the brittle upper crust was split by a dike. One hopes that studies of the composition and fabrics of gabbros in ophiolites may show how buoyancy might drive the crust to become chemically stratified.

ACKNOWLEDGMENTS

I thank Adolphe Nicolas and John Chen for constructive reviews and Hannes Brueckner, Laurence Coogan, and Anjana Shah for many helpful suggestions. Supported by National Science Foundation grant OCE 98–19866.

REFERENCES CITED

Boudier, F., Nicolas, A., and Ildefonse, B., 1996, Magma chambers in the Oman ophiolite: Fed from the top or the bottom?: Earth and Planetary Science Letters, v. 144, p. 239–250.

Bowen, N.L., 1922, The reaction principle in petrogenesis: Journal of Geology, v. 30, p. 177–198.

Carbotte, S.M., Mutter, C.Z., Mutter, J., Spiegelman, M., Correa, G., McNutt, M., Bhat, V., and Cruz-Orozco, R., 1996, Influence of spreading rate and magma supply on crustal magma bodies: Results from a recent seismic study of the shallow East Pacific Rise north of the Orozco transform fault: RIDGE Events Newsletter, v. 7, p. 1–4.

Chen, Y.J., 1998, A second melt lens at fast spreading mid-ocean ridges: EOS, Transactions of the American Geophysical Union, v. 79, no. 45, p. F837.

Chenevez, J., and Nicolas, A., 1997, Crustal feeding in the Oman ophiolite: From the top and from the bottom? A thermal and a mass balance model: Geophysical Research Letters, v. 24, p. 1811–1814.

Chenevez, J., Machetel, P., and Nicolas, A., 1998, Numerical models of magma chambers in the Oman ophiolite: Journal of Geophysical Research, v. 103, no. B7, p. 15443–15455.

Daly, R.A., Manger, G.E., and Clark, S.P., 1966, Density of rocks: Geological Society of America Memoir 97, p. 19–26.

Detrick, R.S., Buhl, P., Vera, E.E., Mutter, J.C., Orcutt, J.A., Madsen, J.A., and Brocher, T.M., 1987, Multichannel seismic imaging of the axial magma chamber along the East Pacific Rise between 9°N and 13°N: Nature, v. 326, p. 35–41.

Dewey, J.F., and Kidd, S.F., 1977, Geometry of plate accretion: Geological Society of America Bulletin, v. 88, p. 960–968.

Dunn, R.A., and Toomey, D.R., 1997, Seismological evidence for three-dimensional melt migration beneath the East Pacific Rise: Nature, v. 388, p. 259–262.

Garmany, J., 1989, Accumulations of melt at the base of the oceanic crust: Nature, v. 340, p. 628–632.

Gregory, R.T., and Taylor, H.P., Jr., 1981, An oxygen isotope profile in a section of Cretaceous oceanic crust, Samail Ophiolite, Oman: Evidence of $\delta^{18}O$ buffering of the oceans by deep (>5 km) seawater-hydrothermal circulation at mid-ocean ridges: Journal of Geophysical Research, v. 86, no. B4, p. 2737–2755.

Harding, A.J., Orcutt, J.A., Kappus, M.E., Vera, E.E., Mutter, J.C., Buhl, P., Detrick, R.S., and Brocher, T.M., 1989, Structure of young oceanic crust at 13°N on the East Pacific Rise from expanding spread profiles: Journal of Geophysical Research, v. 94, p. 12163–12196.

Henstock, T.J., Woods, A.W., and White, R.S., 1993, The accretion of oceanic crust by episodic sill intrusion: Journal of Geophysical Research, v. 98, p. 4143–4161.

Huppert, H.E., and Sparks, R.S.J., 1988, Melting the roof of a chamber containing hot turbulently convecting fluid: Journal of Fluid Mechanics, v. 188, p. 107–131.

Hussenoeder, S.A., Tivey, M.A., Schouten, H., and Searle, R.C., 1996, Near-bottom magnetic survey of the Mid-Atlantic Ridge axis, 24°–24°40′N: Implications for crustal accretion at slow spreading ridges: Journal of Geophysical Research, v. 101, p. 22051–22069.

Kelemen, P.B., and Aharonov, E., 1998, Periodic formation of magma fractures and generation of layered gabbros in the lower crust beneath oceanic spreading ridges, *in* Buck, W.R., Delaney, P.T., Karson, J.A., and Lagabrielle, Y., eds., Faulting and magmatism at mid-ocean ridges: American Geophysical Union, Geophysical Monograph 106, p. 267–290.

Kent, G.M., Harding, A.J., and Orcutt, J.A., 1993, Distribution of magma beneath the East Pacific Rise between the Clipperton transform and the 9°17′N deval from forward modeling of common depth point data: Journal of Geophysical Research, v. 98, p. 13945–13969.

Korenaga, J., and Kelemen, P.B., 1997, The origin of gabbro sills in the Moho transition zone of the Oman ophiolite: Implications from magma transport in the oceanic lower crust: Journal of Geophysical Research, v. 102, p. 27729–27749.

Marsh, B.D., 1989, Magma chambers: Annual Review of Earth and Planetary Sciences, v. 17, p. 439–474.

Morton, J.L., and Sleep, N.H., 1985, A mid-ocean ridge thermal model: Constraints on the volume of axial hydrothermal flux: Journal of Geophysical Research, v. 90, p. 11345–11353.

Murase, T., and McBirney, A.R., 1973, Properties of some common igneous rocks and their melts at high temperatures: Geological Society of America Bulletin, v. 84, p. 3563–3592.

Mutter, J.C., Cormier, M.-H., Wang, L., Buhl, P., Detrick, R.S., Kent, G., Orcutt, J., and Harding, A., 1995, Asymmetrical crustal structure of the East Pacific Rise at 14°15′S from Expanded Spread Profiles: Eos, Transactions of the American Geophysical Union, v. 76, p. S274.

Naslund, H.R., and McBirney, A.R., 1996, Mechanisms of formation of igneous layering, *in* Cawthorne, R.G., ed., Layered intrusions: New York, Elsevier, p. 1–43.

Nicolas, A., Freydier, C., Godard, M., and Vauchez, A., 1993, Magma chambers at oceanic ridges: How large?: Geology, v. 21, p. 53–56.

Pallister, J.S., and Hopson, C.A., 1981, Samail ophiolite plutonic suite: Field relations, phase variation, cryptic variation and layering, and a model of a spreading ridge magma chamber: Journal of Geophysical Research, v. 86, p. 2593–2644.

Phipps Morgan, J., and Chen, Y.J., 1993, The genesis of oceanic crust: Magma injection, hydrothermal circulation, and crustal flow: Journal of Geophysical Research, v. 98, p. 6283–6297.

Ponce-Correa, G.M., Mutter, J.C., and Carbotte, S., 1999, Fresnel zone: A pitfall in seismic imaging of mid-ocean ridge magma lenses: Geophysical Research Letters, v. 26, p. 3021–3024.

Purdy, G.M., Kong, L.S.L., Christeson, G.L., and Solomon, S.C., 1992, Relationship between spreading rate and the seismic structure of mid-ocean ridges: Nature, v. 355, p. 815–817.

Quick, J.E., and Denlinger, R.P., 1993, Ductile deformation and the origin of layered gabbro in ophiolites: Journal of Geophysical Research, v. 98, p. 14015–14027.

Schouten, H., and Denham, C., 1995, Virtual ocean crust: Eos, Transactions of the American Geophysical Union, v. 76, p. S48.

Singh, S.C., Kent, G.M., Collier, J.S., Harding, A.J., and Orcutt, J.A., 1998, Melt to mush variations in crustal magma properties along the ridge crest at the southern East Pacific Rise: Nature, v. 394, p. 874–878.

Sinton, J.M., and Detrick, R.S., 1992, Mid-ocean ridge magma chambers: Journal of Geophysical Research, v. 97, p. 197–216.

Sleep, N.H., 1975, Formation of oceanic crust: Some thermal constraints: Journal of Geophysical Research, v. 80, p. 4037–4042.

Solomon, S.C., and Toomey, D.R., 1992, The structure of mid-ocean ridges: Annual Review of Earth and Planetary Sciences, v. 20, p. 329–364.

Stolper, E., and Walker, D., 1980, Melt density and the average composition of basalt: Contributions to Mineralogy and Petrology, v. 74, p. 7–12.

Turcotte, D., and Schubert, G., 1982, Geodynamics: Application of continuum physics to geological problems: New York, John Wiley & Sons, p. 450.

Vera, E.E., Mutter, J.C., Buhl, P., Orcutt, J.A., Harding, A.J., Kappus, M.E., Detrick, R.S., and Brocher, T.M., 1990, The structure of 0- to 0.2-m.y.-old oceanic crust at 9°N on the East Pacific Rise from expanded spread profiles: Journal of Geophysical Research, v. 95, p. 15529–15556.

Wilson, D.S., Clague, D.A., Sleep, N.H., and Morton, J.L., 1988, Implications of magma convection for the size and temperature of magma chambers at fast spreading ridges: Journal of Geophysical Research, v. 93, p. 11974–11984.

Manuscript Accepted by the Society February 23, 2000

Printed in the U.S.A.

Geological Society of America
Special Paper 349
2000

Evolution of gabbroic rocks of the Northern Apennine ophiolites (Italy): Comparison with the lower oceanic crust from modern slow-spreading ridges

Riccardo Tribuzio
Dipartimento di Scienze della Terra, Università di Pavia, Via Ferrata 1, 27100 Pavia, Italy; e-mail: tribuzio@crystal.unipv.it, and *Consiglio Nazionale delle Ricerche—Centro di Studio per la Cristallochimica e la Cristallografia, Via Ferrata 1, 27100 Pavia, Italy*
Massimo Tiepolo
Dipartimento di Scienze della Terra, Università di Pavia, Via Ferrata 1, 27100 Pavia, Italy
Riccardo Vannucci
Dipartimento di Scienze della Terra, Università di Pavia, Via Ferrata 1, 27100 Pavia, Italy, and *Consiglio Nazionale delle Ricerche—Centro di Studio per la Cristallochimica e la Cristallografia, Via Ferrata 1, 27100 Pavia, Italy*

ABSTRACT

Field and petrographic relationships together with major and trace element mineral chemistry have allowed us to determine the igneous to high-temperature metamorphic evolution of the gabbroic rocks of the Northern Apennine ophiolites. Gabbroic rocks formed by the intrusion of liquids of normal mid-oceanic-ridge (N-MORB) type in a heterogeneous mantle section under low-pressure conditions. These liquids underwent an igneous-differentiation process controlled by fractional crystallization, most likely associated with late percolation in the gabbroic cumulate pile of a volatile-bearing igneous agent, possibly a trondhjemite-type liquid. Such an igneous-differentiation process produced highly evolved liquids that gave rise to Fe-rich rocks (mostly Fe-Ti oxide–bearing diorites). The gabbroic rocks underwent high-temperature ($T \approx 900$ °C) recrystallization in ductile shear zones in the absence of seawater-derived fluids. The gabbroic rocks of the Northern Apennine ophiolites bear striking similarities to those recovered from modern slow-spreading ridges, such as the Southwest Indian Ridge, the Mid-Atlantic Ridge (at its intersection with the Kane Fracture Zone, i.e., the MARK area), and the Mid-Cayman Rise. The Northern Apennine ophiolites are interpreted to represent a fossil analogue of a magma-poor slow-spreading center that formed as a result of continental lithospheric extension.

INTRODUCTION AND GEOLOGIC SETTING

The ophiolites exposed in the Northern Apennines, Western Alps, and Corsica are lithospheric remnants of the Jurassic Western Neotethyan Ocean (Ligurian Tethys). In the Northern Apennine ophiolites, the orogenic collisional tectonics, which led to the closure of the Ligurian Tethys, was only locally accompanied by substantial metamorphic recrystallization, and therefore the primary lithostratigraphic features are generally well preserved. Since the early 1970s (Decandia and Elter, 1972; Abbate et al., 1980), it has been recognized that the Northern Apennine ophiolites have an "anomalous" pseudostratigraphy such that (1) the gabbroic rocks show intrusive relationships with ultramafic mantle rocks; (2) a true sheeted-dike complex and a continuous thick basaltic layer are lacking; (3) ophiolitic breccias are locally widespread, commonly interlayered with basaltic lavas; and (4) the mantle and gabbro association is in places directly overlain by pelagic sedimentary rocks.

In the northernmost sector of the Apennines (Eastern Liguria), two main ophiolite-bearing units, the Internal and External Ligurides, are distinguished (Fig. 1). In the Internal Ligurides, the ophiolites are stratigraphically associated with a thick turbiditic sequence (Upper Cretaceous to lower Paleocene) evolving through time from distal to proximal. In the External Liguride units, the ophiolites are present as gravity-slide blocks in Upper Cretaceous sedimentary mélanges and are characterized by the scarcity of gabbroic rocks; these mélanges also include blocks of continental crustal rocks, mainly granitoids and granulite-facies rocks. On the basis of the lithostratigraphic and structural features, the ophiolites of the Internal and External Ligurides are considered to represent intraoceanic and pericontinental domains of the Ligurian Tethys, respectively (Cortesogno et al., 1987; Marroni

Tribuzio, R., Tiepolo, M., and Vannucci, R., 2000, Evolution of gabbroic rocks of the Northern Apennine ophiolites (Italy): Comparison with the lower oceanic crust from modern slow-spreading ridges, *in* Dilek, Y., Moores, E.M., Elthon, D., and Nicolas, A., eds., Ophiolites and Oceanic Crust: New Insights from Field Studies and the Ocean Drilling Program: Boulder, Colorado, Geological Society of America Special Paper 349, p. 129–138.

et al., 1998). The ophiolites cropping out in southern Tuscany are commonly included in Upper Cretaceous to Eocene flyschoid sequences (Bortolotti, 1983); these ophiolites contain a significant amount of gabbroic rocks and are not associated with continental crustal rocks. The paleogeographic domain of the southern Tuscany ophiolites, although they are volumetrically important, has not been determined accurately.

The petrologic features of the mantle and gabbro association are well known for the Internal Liguride ophiolites. The mantle ultramafic rocks are mainly undeformed, clinopyroxene-poor, depleted lherzolites that are the residue after the extraction in Permian time of MORB-type liquids (i.e., mid-oceanic-ridge basalt) (Rampone et al., 1996, 1998). These lherzolites show incipient re-equilibration under plagioclase-facies conditions and were interpreted as fragments of oceanic-type mantle. The associated gabbroic complex formed from normal-MORB (N-MORB) parental liquids in Middle to Late Jurassic time (Hébert et al., 1989; Borsi et al., 1996; Tiepolo et al., 1997; Rampone et al., 1998). Gabbroic rocks show a widespread metamorphic overprint, characterized by a progressive temperature decrease under high thermal gradients in intraoceanic conditions (Cortesogno et al., 1975; Messiga and Tribuzio, 1991; Molli, 1995; Tribuzio et al., 1995). Rocks affected by the early high-temperature metamorphic event are localized along ductile shear zones and include an assemblage of clinopyroxene, plagioclase, titanian pargasite, and ilmenite. This event was followed by a retrograde event at amphibolite- to subgreenschist-facies conditions, most likely related to circulation of seawater-derived fluids and locally accompanied by the development of brittle deformation structures.

The mantle ultramafic rocks of the External Liguride ophiolites differ from those that crop out in the Internal Liguride units in that they dominantly consist of fertile spinel lherzolites with pyroxenite bands (Rampone et al., 1995). These lherzolites display a partial recrystallization under plagioclase-facies conditions, generally accompanied by ductile deformation, and are considered as slices of subcontinental mantle, probably accreted to the lithosphere in the Proterozoic. Little is known about the igneous and metamorphic features of the gabbroic rocks of the External Liguride ophiolites (Marroni et al., 1998) or about the mantle ultramafic rocks of the southern Tuscany ophiolites. By contrast, the petrology of the gabbroic rocks of the southern Tuscany ophiolites has been extensively investigated (Cortesogno et al., 1975; Serri, 1980; Hébert et al., 1989; Tiepolo et al., 1997; Tribuzio et al., 1999). The available petrologic data for the gabbroic rocks of the southern Tuscany and External Liguride ophiolites indicate that they share similar igneous petrologic features, which closely resemble those of the Internal Liguride ophiolites, and that they were similarly affected by high-temperature metamorphism in ductile shear zones.

In this chapter, new ion-microprobe trace element analyses of clinopyroxenes from the gabbroic rocks of the Internal and the External Liguride ophiolites have been assembled with recently published data. Field and petrographic relationships, as well as major and trace element mineral compositions, have allowed us to give insights into (1) the compositions of the parental liquids, (2) the igneous-differentiation processes, and (3) the nature of the fluid phase that was possibly involved in the high-temperature shear zones. In particular, we show that the gabbroic rocks of the Northern Apennine ophiolites show striking petrologic similarities to gabbroic rocks exposed in modern slow-spreading environments. The reconstructed igneous to high-temperature metamorphic evolution is thus believed to be relevant to the understanding of the formation of lower oceanic crust.

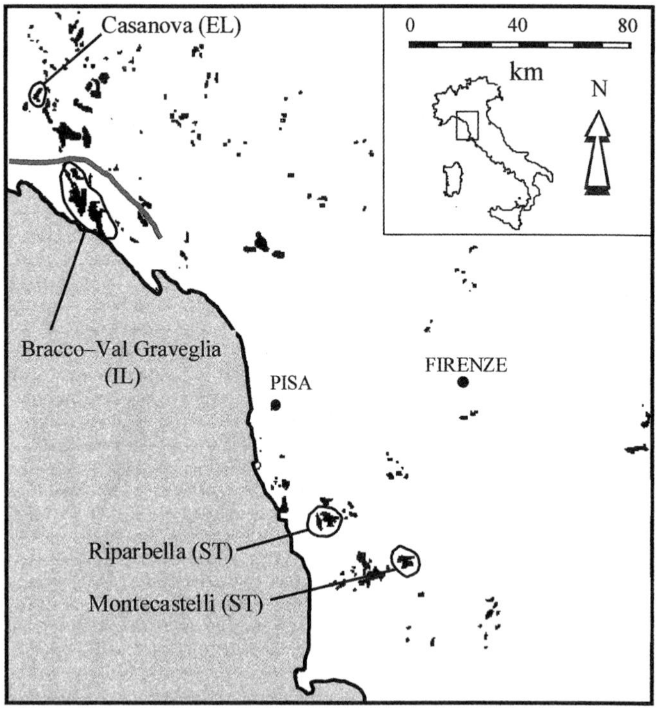

Figure 1. Location of the ophiolitic massifs in the Northern Apennines considered in this chapter. IL—Internal Liguride ophiolites, EL—External Liguride ophiolites, ST—southern Tuscany ophiolites. The thick gray line in the northernmost sector of the Apennines separates the Internal from the External Ligurides.

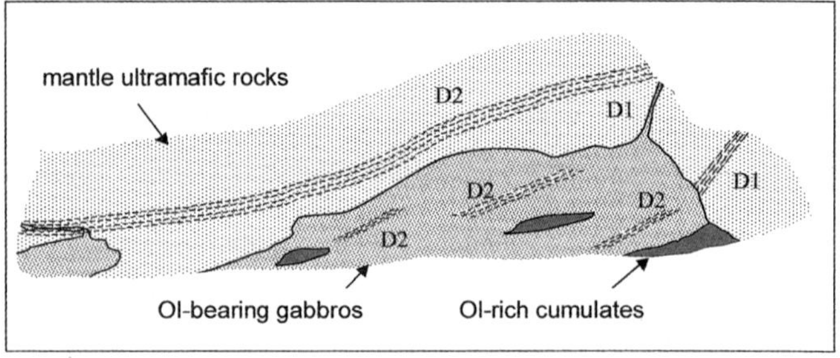

Figure 2. Schematic representation of the main mesostructural features of the mantle and gabbro association on the basis of the interpretation of field data in the Bracco area (redrawn after Molli, 1996). In gabbroic rocks, high-temperature shear zones (D2) are slightly oblique to the igneous layering, which is defined by the lenticular bodies of olivine-rich cumulates. The earliest ductile shear zones in mantle ultramafic rocks (D1) predate the intrusion of gabbroic rocks.

MESOSTRUCTURAL FEATURES OF GABBROIC ROCKS

Detailed field investigations of the gabbroic rocks have been carried out (Decandia and Elter, 1972; Cortesogno et al., 1987; Molli, 1995, 1996) in the Bracco massif (Internal Liguride ophiolites), which is volumetrically the most representative (~4 × 5 km wide and 0.9 km thick). However, the following features have also been observed in the gabbroic rocks of the External Liguride and southern Tuscany ophiolites. In general, the gabbroic rocks are intrusive into the mantle ultramafic rocks (Fig. 2). Within the peridotites, gabbroic

TABLE 1. MAJOR ELEMENT COMPOSITION OF CLINOPYROXENES

Sample	SC3	N1A	AP1/2	MC2	GEL1/9
Rock type	Melatroctolite	Fe-Ti oxide-bearing diorite	Fe-Ti oxide-bearing diorite	Fe-Ti oxide-bearing diorite	Ol-bearing gabbro
Location	IL	IL	IL	IL	EL
SiO_2	51.64	52.31	51.20	52.48	52.3
TiO_2	0.50	0.83	0.43	0.53	0.55
Al_2O_3	4.39	1.80	1.81	1.67	4.11
Cr_2O_3	1.37	B.D.L.*	B.D.L.*	B.D.L.*	1.04
FeO†	3.36	13.00	11.74	11.16	5.06
MnO	0.16	0.51	0.39	0.41	0.15
MgO	15.93	12.60	12.16	13.66	16.8
CaO	22.73	19.11	20.23	19.42	21
Na_2O	0.90	1.07	0.39	0.62	0.64
Total	100.98	101.23	98.35	99.95	101.65
Mg#§	90.0	63.7	64.5	68.0	84.8

Note:
* B.D.L.—below detection limits.
† Total Fe as FeO.
§ Mg# = 100 × Mg/(Mg + Fe_{total}).

dikes (up to 1 m thick) are relatively common; some show variations in modal compositions and grain size parallel to the peridotite walls (Tribuzio et al., 1997). The gabbroic dikes locally crosscut a tectonite foliation in mantle peridotites possibly developed under spinel-facies conditions.

The gabbroic rocks are volumetrically dominated by olivine-bearing (5%–15%) gabbros, which are commonly coarse-grained, nearly isotropic and with a subophitic texture. These rocks show meter-scale pegmatoid patches, which are generally relatively rich in plagioclase (up to 80%). The olivine-bearing gabbros contain lenticular bodies (tens of meters in size) of troctolites to melatroctolites (Bezzi and Piccardo, 1971; Cortesogno et al., 1987; Molli, 1995). Troctolites commonly exhibit an igneous foliation, produced by the preferred orientation of plagioclase and olivine grains. These rocks may also display modal layering (up to a decimeter in scale), shown by transitions from leucocratic, plagioclase-rich rocks to melatroctolites and by the presence of olivine-bearing gabbro layers. Melatroctolites have olivine (commonly in the range of 80%–90%) and accessory spinel as cumulus phases and minor plagioclase and clinopyroxene with a poikilitic texture. Melatroctolites locally include thin layers of chromitite, anorthosite, and troctolite and commonly show modal and/or grain-size layering.

Fe-rich gabbroic rocks are volumetrically subordinate and do not display direct relationships with the olivine-bearing gabbros (Cortesogno et al., 1975; Serri, 1980; Tiepolo et al., 1997). They mainly crop out as clasts in ophiolitic breccias, and they are mostly represented by Fe-Ti oxide–bearing (up to 15%) diorites, locally containing relatively high amounts (up to 5%) of titanian pargasite and apatite. Rare gabbronorites and hornblende-bearing plagioclase-rich rocks (diorites to albitites) are also present. The most representative outcrop of Fe-rich rocks is located in the Bartolina quarry in southern Tuscany in which the Fe-rich rocks are commonly intruded by trondhjemite dikelets. Similar relationships have also been observed in the clasts of the ophiolitic breccias. In the Internal Liguride ophiolites, Fe-rich gabbroic dikes have been found intruding the mantle ultramafic rocks (Tribuzio et al., 1997).

The olivine-bearing gabbros exhibit high-temperature recrystallization along ductile shear zones (Cortesogno et al., 1975; Molli, 1995; Tribuzio et al., 1995). The width of the shear zones generally ranges from a few decimeters to several meters; the high-temperature foliation overprints the magmatic layering slightly oblique to it. These shear zones locally occur in groups, where they form anastomosing patterns of deformation areas around lenses of undeformed or less-deformed host rocks (Molli, 1995). Geometric and kinematic studies of the shear zones indicate a dominantly normal, extensional displacement and noncoaxial deformation in general. According to Molli (1996), these high-temperature structures have in places been affected by later ductile shear zones that developed under conditions of lower-amphibolite to upper-greenschist facies.

The high-temperature foliation is locally crosscut by undeformed MORB dikes with chilled margins (Cortesogno et al., 1975, 1987; Molli, 1995), and less commonly, by trondhjemite dikelets. The gabbroic rocks in places contain small, elongated bodies of (1) fine-grained melanocratic gabbros and (2) diorites to albitites, but there are no clear relationships between these bodies and the high-temperature foliation. In the Internal Liguride ophiolites, the high-temperature foliation and the bodies of microgabbros and diorites to albitites are commonly crosscut at high angles by hornblende-bearing (± plagioclase) veins (width commonly less than a few millimeters) occurring as parallel swarms. These veins are in general extensional, with opening perpendicular to the walls of fractures, and resulted in coronal hornblende growth in the gabbroic rocks. The hornblende-bearing veins are locally present in the gabbroic rocks of the southern Tuscany ophiolites, but they have not been found in the External Liguride ophiolites.

IGNEOUS PETROGENETIC CONSTRAINTS FROM MAJOR ELEMENT MINERAL COMPOSITIONS

Field and petrographic relationships coupled with whole-rock and mineral-chemistry variations indicate that the gabbroic rocks were derived from an igneous-differentiation process mainly controlled by fractional crystallization (Serri, 1980; Hébert et al., 1989; Tiepolo et al., 1997). The forsterite component in olivine gradually decreases from melatroctolites to olivine-bearing gabbros (from 89 to 68 mol%). Likewise, the Mg# in clinopyroxene [Mg# = 100 × Mg/(Mg + Fe_{total}); Table 1] decreases from 0.90 in melatroctolites to 0.64 in Fe-Ti oxide–bearing diorites; such decrease is coupled with a decrease of Cr and Al and an increase of Mn. Furthermore, there is a rough positive correlation between the Mg# of clinopyroxene and the amount of anorthite component in plagioclase, which ranges from about 70 mol% in Mg-rich rocks to 28 mol% in Fe-Ti oxide–bearing

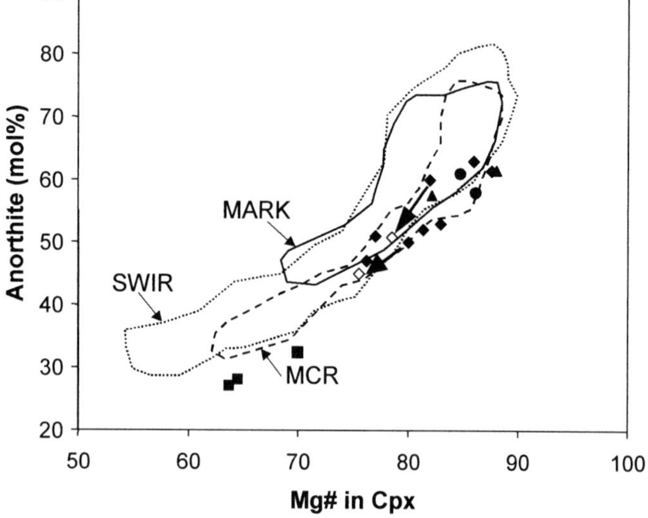

Figure 3. Anorthite content of plagioclase (mol%) vs. Mg# [i.e., 100 × Mg/(Mg + Fe_{total})] of coexisting clinopyroxene (Cpx). Data refer to cores of grains; the large arrows show the zoning in plagioclase-clinopyroxene pairs from olivine-bearing gabbros of Southern Tuscany (data after Tiepolo et al., 1997, Hébert et al., 1989, Marroni et al., 1998, Rampone et al., 1998, Tribuzio et al., 1999). Field of gabbroic rocks from (1) Southwest Indian Ridge (SWIR, Leg 118, Hole 735B) after Ozawa et al. (1991) and Hébert et al. (1991); (2) Mid-Cayman Rise (MCR) after Elthon (1987); and (3) Mid-Atlantic Ridge at its intersection with the Kane Fracture Zone (MARK area, Leg 153) after Ross and Elthon (1997). Diamonds—olivine-bearing gabbros of southern Tuscany (full and empty symbols refer to core and rim compositions, respectively), triangles—olivine-bearing gabbros of the Internal Liguride ophiolites, circles—olivine-bearing gabbros of the External Liguride ophiolites, squares—Fe-Ti oxide–bearing diorites of the Internal Liguride ophiolites.

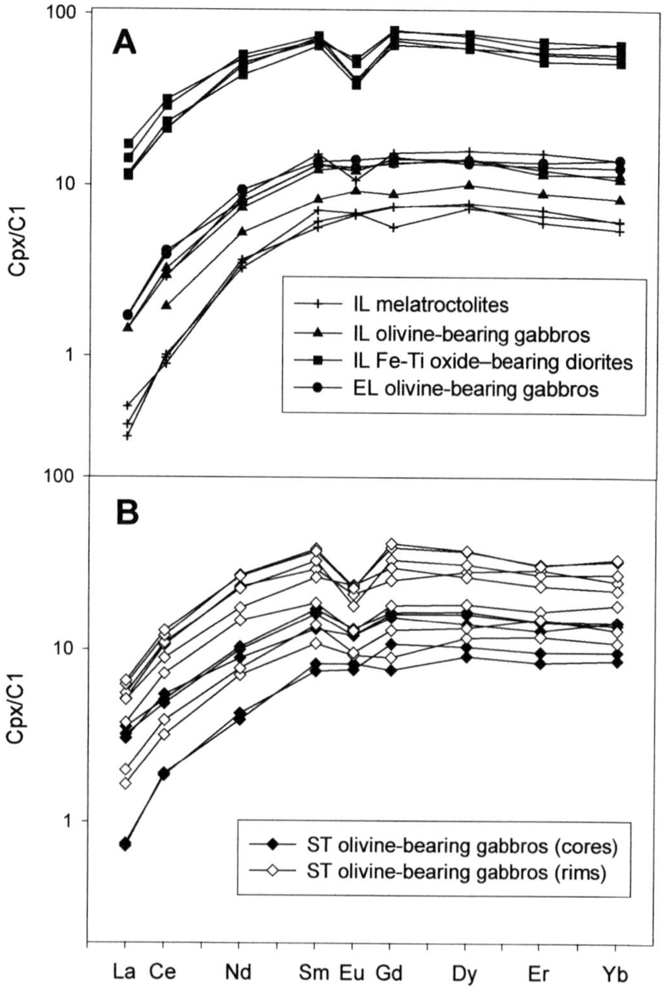

Figure 4. REE abundances in clinopyroxene; normalization to C1 chondrites (Anders and Ebihara, 1982) A, Clinopyroxenes from gabbroic rocks of the Internal (IL) and the External Liguride (EL) ophiolites. B, Olivine-bearing gabbros of the southern Tuscany (ST) ophiolites. Data are reported in Table 2, Tribuzio et al. (1995, 1999), and Rampone et al. (1997).

diorites (Fig. 3). A similar covariation has been observed for modern MOR-type gabbroic rocks from slow-spreading settings (Elthon, 1987; Ozawa et al., 1991; Hébert et al., 1991; Ross and Elthon, 1997).

Serri (1980) reconstructed a general fractional-crystallization order for the gabbroic sequence of the Northern Apennine ophiolites that has been confirmed by more recent studies (Hébert et al., 1989; Tiepolo et al., 1997). The fractionation process started with the crystallization of olivine and Cr-spinel, followed by the replacement of Cr-spinel by plagioclase at the liquidus, thus giving rise to melatroctolites and troctolites, respectively. The subsequent appearance of clinopyroxene at the liquidus resulted in the crystallization of olivine-bearing gabbros. Concerning the Fe-rich rocks, the major element variation trends are indicative of a continuous differentiation process, controlled by fractional crystallization, from melatroctolites to the Fe-Ti oxide–bearing diorites. According to Serri (1980), the Fe-rich rocks were derived from isolated magma injections that, because of closed-system fractional crystallization, became progressively markedly differentiated, resulting in the segregation of Fe-Ti oxides.

Estimation of the depth at which the gabbroic rocks formed is affected by a great uncertainty. The above crystallization order indicates low- to moderate-pressure conditions (Elthon, 1993). The recent geobarometer of Nimis and Ulmer (1998), based on the compositions of clinopyroxene crystallized from anhydrous basalts, can be applied to the Mg-rich rocks (olivine-bearing gabbros to melatroctolites). Results obtained on clinopyroxenes from about 25 rock samples mostly fall within the 0.2–0.4 GPa range; a few clinopyroxenes from the melatroctolites yield slightly higher values, which in any case do not exceed 0.5 GPa.

IGNEOUS-DIFFERENTIATION PROCESS AS REVEALED BY TRACE ELEMENT CLINOPYROXENE COMPOSITIONS

The clinopyroxene from gabbroic rocks of the Northern Apennine ophiolites has a chondrite-normalized REE (rare earth element) pattern characterized by LREE (light REE) depletion, a flat HREE (heavy REE) distribution, and a weak to negligible negative Eu anomaly (Fig. 4). The concentrations of REEs, Y, and Zr are highly variable, whereas Sr, V, Sc, and Ti abundances are within a relatively narrow range (Fig. 5; see also Table 2). It is noteworthy that different ophiolitic massifs considered in this study have olivine-bearing gabbros containing clinopyroxene with quite similar trace element compositions, thus confirming the close similarities indicated by petrographic, major element chemistry, and whole-rock chemistry data (Serri, 1980; Hébert et al., 1989; Tiepolo et al., 1997; Marroni et al., 1998). These similarities imply that the gabbroic rocks of different

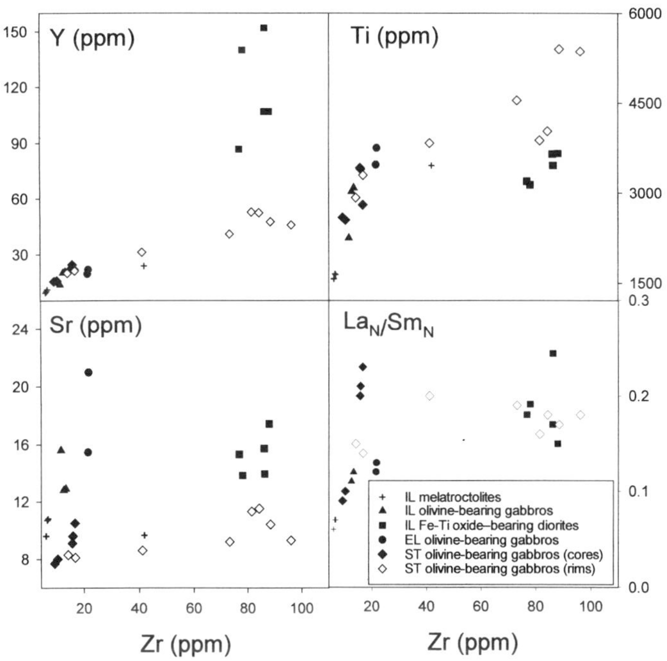

Figure 5. Y, Ti, Sr, and La_N/Sm_N vs. Zr in clinopyroxenes from the gabbroic rocks of the Northern Apennine ophiolites. IL and EL—Internal and External Ligurides, respectively; ST—southern Tuscany.

ophiolite massifs can be reconciled with the same type of parental magma and the same type of igneous-differentiation process.

The variations in trace element compositions of clinopyroxene are generally consistent with the fractional-crystallization process reported in the previous section. For instance, the relatively constant Sr contents are in good agreement with the substantial segregation of plagioclase. Likewise, the clinopyroxenes from Fe-Ti oxide–bearing diorites have extremely high REE, Y, and Zr contents and show a more marked negative Eu anomaly than those from the olivine-bearing gabbros, consistent with crystallization from evolved liquids. Furthermore, the clinopyroxenes from Fe-Ti oxide–bearing diorites do not have Ti and V concentrations that are significantly higher than those from the olivine-bearing gabbros, thus suggesting that the Ti and V contents in the evolved liquids were buffered by the segregation of Fe-Ti oxides. However, the relatively large variations in La_N/Sm_N of clinopyroxenes (Fig. 5) suggest that some other differentiation process besides simple fractional crystallization was involved. On the basis of the mineral/liquid partition coefficients reported in the current literature, it is actually inferred that a fractional-crystallization process controlled by olivine, clinopyroxene, and plagioclase (± Fe-Ti oxides) does not result in significant changes in the La_N/Sm_N ratio.

We have recently shown (Tribuzio et al., 1999) that clinopyroxene and plagioclase from the olivine-bearing gabbros are zoned and that these rocks are characterized by the occurrence of interstitial mineral accessories (Fe-Ti oxides, apatite, and titanian pargasite). The clinopyroxene rims show a slight Mg# and Cr decrease, a slight enrichment in Ti and V, and a marked increase in Zr, REEs, and Y relative to the cores. The REE pattern of the clinopyroxene rims differs in a more pronounced negative Eu anomaly and locally in a significant outward increase in the La_N/Sm_N ratio. A similar clinopyroxene zoning was also reported by Ross and Elthon (1997) for the gabbroic rocks from the MARK area (Leg 153; the MARK area is at the intersection between the Mid-Atlantic Ridge and the Kane Fracture Zone). In gabbroic rocks of the Northern Apennine ophiolites, the plagioclase rims have slightly lower anorthite component and higher LREEs, Ba, and K than the cores. Likewise, a similar

TABLE 2. TRACE ELEMENT COMPOSITIONS OF CLINOPYROXENES

Sample	SC3	SC3	SC3	N1A	N1A	AP1/2	MC2	MC2	GEL 1/9	GEL 1/9
Rock type	Mela-troctolite	Mela-troctolite	Mela-troctolite	Fe-Ti oxide-bearing diorite	Fe-Ti oxide-bearing diorite	Fe-Ti oxide-bearing diorite	Fe-Ti oxide-bearing diorite	Fe-Ti oxide-bearing diorite	Ol-bearing gabbro	Ol-bearing gabbro
Location	IL	IL	IL	IL	IL	IL	IL	IL	EL	EL
ppm										
Sc	83.	85.	82.	167.	165.	149.	152.	140.	109.	103.
Ti	1570.	1640.	1650.	3660.	3650.	3200.	3460.	3140.	3750.	3470.
V	214.	226.	233.	357.	303.	347.	N.A.*	N.A.*	374.	314.
Cr	7870.	8260.	8550.	B.D.L.†	B.D.L.†	B.D.L.†	B.D.L.†	B.D.L.†	5960.	7560.
Sr	9.6	10.8	10.7	17.4	15.7	15.3	13.9	13.8	21.0	15.4
Y	9.5	11.2	10.2	107.	107.	86.7	90.8	79.2	21.9	19.5
Zr	5.9	6.6	6.3	87.9	86.0	76.8	86.2	78.0	21.7	21.5
La	0.08	0.09	0.12	2.66	2.71	2.75	4.06	3.36	0.41	0.40
Ce	0.63	0.60	0.55	13.10	13.12	14.32	19.40	17.80	2.54	2.41
Nd	1.49	1.67	1.60	22.70	23.68	20.01	25.10	26.30	3.65	4.29
Sm	0.92	0.84	1.06	10.90	10.30	9.72	10.50	11.10	1.97	2.05
Eu	0.38	0.38	0.39	2.89	3.07	2.15	2.27	2.30	0.71	0.79
Gd	1.49	1.47	1.12	15.60	15.86	13.79	13.00	14.20	2.66	2.88
Dy	1.87	1.93	1.80	18.80	18.24	15.47	15.80	16.70	3.44	3.29
Er	0.97	1.15	1.07	11.10	10.19	9.58	8.52	9.35	2.18	2.07
Yb	0.88	0.97	0.99	10.60	10.57	9.31	8.31	8.92	2.25	2.02
La_N/Sm_N	0.06	0.07	0.07	0.15	0.17	0.18	0.24	0.19	0.13	0.12
Eu/Eu*	1.00	1.03	1.08	0.68	0.73	0.57	0.59	0.56	0.95	0.99
Zr/Y	0.62	0.59	0.61	0.82	0.80	0.89	0.95	0.98	0.99	1.10
Ti/Zr	264.	248.	263.	42.	42.	42.	40.	40.	173.	162.

Note: Analyses were carried out using a Cameca IMS 4f ion microprobe. IL—Internal Ligurides, EL—External Ligurides.
* N.A.—not analyzed.
† B.D.L.—below detection limit.

plagioclase zoning was reported by Meen and Elthon (1996) for a troctolite from the Mid-Cayman Rise.

The zoning of clinopyroxene and plagioclase, as well as the precipitation of interstitial accessory minerals, has been ascribed to a process of postcumulus fractional crystallization of a volatile-bearing interstitial liquid (Tribuzio et al., 1999; 2000). In particular, the process of entrapment of interstitial liquid is most likely responsible for the high concentrations of incompatible trace elements in the rims of clinopyroxene and plagioclase. Such a process was probably active not only in the olivine-bearing gabbros but also in the olivine-rich cumulates, as suggested by the relatively high concentrations of Zr, REEs, and Y in poikilitic Mg-rich clinopyroxene from the melatroctolite BR2 of the Internal Liguride ophiolites (Rampone et al., 1997).

Because of the LREE variations in clinopyroxenes, we have argued that the interstitial liquid had an exotic component such that the formation of the olivine-bearing gabbros did not result from closed-system crystallization. The exotic component could be related to the upward migration—as a result of buoyancy in the gabbroic cumulate pile—of (1) trondhjemite-type, Si-Al-Na–rich liquids (Dick et al., 1991; Natland and Dick, 1996; Pedersen et al., 1996) or (2) exsolved igneous fluids (Kelley and Malpas, 1997; Kelley, 1996). In the ophiolites of the Northern Apennines and Western Alps, the trondhjemite-type rocks that locally crosscut the gabbroic rocks show slightly LREE-enriched patterns (La_N/Sm_N = 2.1–3.7, for Sm_N = 50–90; Borsi et al., 1996). Therefore, although the origin of these trondhjemite-type rocks is not completely established, the formation of the postcumulus system in the olivine-bearing gabbros can be reconciled with the percolation through the gabbroic crystal mush of trondhjemite-type, volatile-bearing liquids. The resulting mixing relationships could produce an interstitial liquid that evolved in situ through fractional crystallization, thus yielding the postcumulus system.

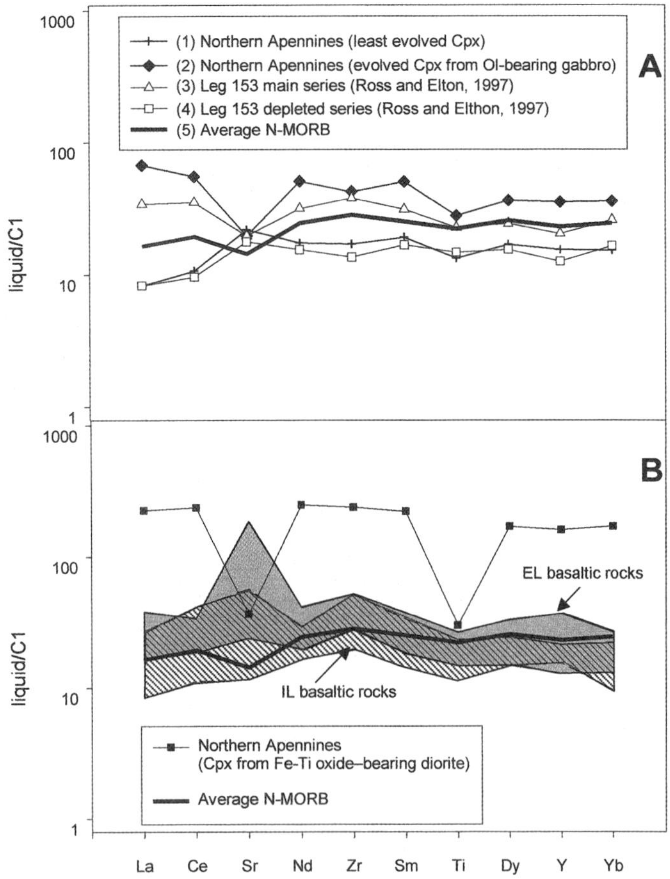

Figure 6. Trace element compositions of liquids in equilibrium with clinopyroxene, calculated through the clinopyroxene/liquid partition coefficients reported by Skulski et al. (1994) for alkali olivine basalt at 1.0 GPa, which are as follows: La = 0.05, Ce = 0.09, Sr = 0.06, Nd = 0.2, Zr = 0.1, Sm = 0.33, Ti = 0.28, Dy = 0.45, Y = 0.43, and Yb = 0.39. A, Liquids in equilibrium with (from top to bottom in key) (1) the least evolved clinopyroxenes from the gabbroic rocks of the Northern Apennine ophiolites (melatroctolite SC3), (2) the clinopyroxene core from an evolved olivine-bearing gabbro of the Northern Apennine ophiolites (sample CB17), (3) clinopyroxenes from the main gabbroic series from Leg 153 (Ross and Elthon, 1997; sample 15R-2 114–118), (4) clinopyroxenes from the depleted gabbroic series from Leg 153 (Ross and Elthon, 1997, sample 12R-1 24–28). Average N-MORB compositions (after Hofmann, 1988) are also reported. B, Liquids in equilibrium with clinopyroxene from Fe-Ti oxide–bearing diorite (sample N1A) and fields of basaltic rocks of the Internal (IL) and External Liguride (EL) ophiolites (data after Ottonello et al., 1984; Cortesogno and Gaggero, 1992; Marroni et al., 1998; Rampone et al., 1998). The high Sr contents in the basaltic rocks of the External Liguride ophiolites are most likely the result of the ocean-floor metamorphism (Marroni et al., 1998; Rampone et al., 1998).

COMPOSITIONS OF THE PARENTAL LIQUIDS

In the preceding section, we have shown that the clinopyroxenes are characterized by large trace element variations, which are controlled by fractional crystallization and by a postcumulus process that results in an overall increase of incompatible trace elements and a slight LREE enrichment. Therefore, the effects of the igneous-differentiation process may obliterate the geochemical signature of the primary melt. This possibility implies that the trace element compositions of the parental liquids of the gabbroic rocks, calculated by dividing clinopyroxenes compositions by clinopyroxene/liquid partition coefficients, may not give accurate information about the primary melt compositions.

Compared to clinopyroxenes from other gabbroic rocks of the Northern Apennine ophiolites, the clinopyroxene from melatroctolite SC3 has the highest Mg#, Cr, and Al values, the lowest amounts of Mn and incompatible trace elements, the deepest LREE depletion, and no Eu anomaly. The effects of the igneous-differentiation process in the SC3 clinopyroxene should therefore be negligible. Figure 6A shows the compositions of the liquid in equilibrium with such a least evolved clinopyroxene, calculated by applying the clinopyroxene/liquid partition coefficients for a basaltic system (Skulski et al., 1994). The resulting liquid is similar to that of average N-MORB (Hofmann, 1988) and is also consistent with that of the least evolved basalts of the Northern Apennine ophiolites. The average N-MORB indeed differs in the slightly higher contents of incompatible trace elements and in the presence of a weak Sr depletion relative to neighboring REEs, thus suggesting that the parental liquids of the sample SC3 had a slightly less evolved character than average N-MORB.

The least evolved liquid calculated for the gabbroic rocks of the Northern Apennine ophiolites bears striking similarities to that calculated for the depleted gabbroic series from ODP Leg 153 in the MARK area (Ross and Elthon, 1997). On the basis of clinopyroxene trace element compositions, Ross and Elthon (1997) found two different gabbroic series—named the "main series" and the "depleted series"—in samples gathered during Leg 153 (Sites 921 and 923), which were ascribed to distinct primary basalts. Computed liquid of the Leg 153 main series is within the compositional field defined by the liquids in equilibrium with the least and most evolved clinopyroxenes from the Mg-rich gabbroic rocks of the Northern Apennine ophiolites. The data presented in this work show a gradual transition between the least and most evolved clinopyroxene, thus suggesting that such a wide compositional field is related to the igneous-differentiation process.

The concentrations of REEs, Y, and Zr in the parental liquids of Fe-Ti oxide–bearing diorites, calculated through the same set of clinopyroxene/liquid partition coefficients used for the Mg-rich gabbroic rocks, are unrealistically high (Fig. 6B). Similarly high REE concentrations are obtained by applying a set of partition coefficients for a basaltic system to the apatite (REE compositions reported in Tribuzio et al., 1996). Experimental data show that the clinopyroxene/liquid partition coefficients for REEs, Y, and Zr, as well as the apatite/liquid partition coefficients for REEs, markedly increase with SiO_2 contents in the liquid (Watson and Green, 1981; Sisson, 1991). It thus seems plausible that the high concentrations of REEs, Y, and Zr in clinopyroxene from Fe-Ti oxide–bearing diorite are controlled not only by the processes of fractional-crystallization and entrapment of interstitial liquid, but also by an increase in the value of the clinopyroxene/liquid partition coefficient. This likely circumstance is consistent with the extensive separation of Fe-Ti oxides, which probably resulted in a significant SiO_2 increase in the residual liquids (see also Toplis and Carroll, 1996). Most likely, Fe-Ti oxide–bearing diorites did not develop from basaltic liquids. This implies that Fe-Ti oxide–bearing diorites probably developed from liquids that were SiO_2 richer than basalts (andesites to dacites).

ORIGIN OF THE FLUID PHASE IN THE HIGH-TEMPERATURE SHEAR ZONES

In weakly deformed olivine-bearing gabbros, textural and paragenetic relicts of igneous minerals are generally preserved (Cortesogno et al., 1975; Molli, 1994, 1995; Tribuzio et al., 1995). The igneous clinopyroxene is recrystallized into aggregates of neoblastic clinopyroxene, which is slightly depleted in Al relative to the igneous relict and commonly associated with subordinate titanian pargasite (Fig. 7). The latter is locally interstitial to the neoblastic clinopyroxene, but triple junctions at 120° among these two minerals are commonly present. Igneous plagioclase is replaced by aggregates of neoblastic plagioclase, which shows a barely appreciable decrease in anorthite component relative to the igneous relict. Lenticular aggregates of chlorite and tremolite are most likely a pseudomorphic low-temperature replacement of deformed olivines or of aggregates of neoblastic olivine. Accessory ilmenite is locally recrystallized and rimmed by aggregates of red-brown amphibole.

The amphibole-plagioclase geothermometer (Holland and Blundy, 1994) gives temperature estimates by assuming pressure conditions of 0.2–0.4 GPa, mostly in the range of 880–920 °C. These estimates agree with those reported by Tribuzio et al. (1995) on the basis of major element compositions of neoblastic clinopyroxene and titanian pargasite. From a textural and major element chemistry viewpoint, the deformed ophiolitic gabbros of the Northern Apennines bear close similarities to high-temperature deformed gabbros recovered from modern slow-spreading ridges (e.g., Ito and Anderson, 1983; Dick et al., 1991; Mével and Cannat, 1991).

Neoblastic clinopyroxene and titanian pargasite have high contents of REEs, Y, Zr, and Ti (Fig. 8). These high concentrations are not consistent with crystallization in the presence of seawater-derived fluids (see also Gillis, 1996), because seawater and hydrothermal vent fluids from modern oceanic basins have extremely low concentrations of these elements. For instance, the REE concentrations in hydrothermal vent fluids at ~350 °C from the Mid-Atlantic Ridge (Michard, 1989) are about four orders of magnitude lower than the REE concentrations in N-MORB. Moreover, it is difficult to reconcile the relatively high F/Cl ratios of the titanian pargasite (Tribuzio et al., 1995) with the compositions of seawater and hydrothermal vent fluids, which are characterized by marked enrichments of Cl over F (e.g., Campbell et al., 1988). Neoblastic titanian pargasite is actually geochemically distinct from the metamorphic amphiboles postdating the high-temperature ductile events that are clearly related to the interaction of the gabbroic rocks with seawater-derived fluids (Riccardi, 1994). Both major and trace element analyses show that the neoblastic titanian pargasites are compositionally indistinguishable from those occurring as interstitial accessory phases of likely late magmatic origin (Tribuzio et al., 1999, 2000) in undeformed rocks. Likewise, the compositional variations between porphyroclastic and neoblastic clinopyroxenes are generally similar to those occurring between core and rim in clinopyroxenes from the undeformed rocks.

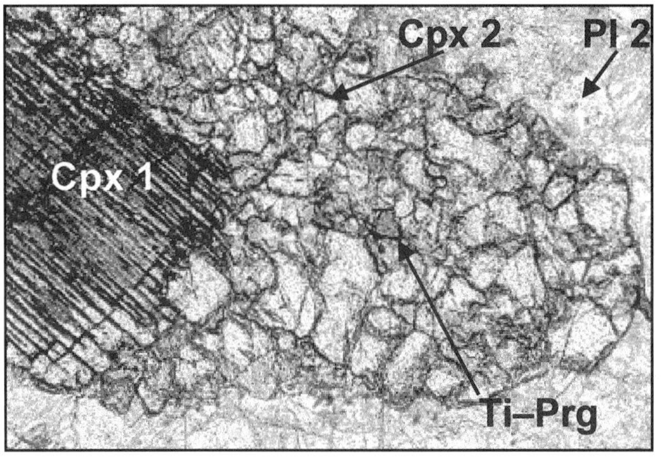

Figure 7. Photomicrograph showing a relevant microtextural feature of the weakly deformed olivine-bearing gabbros. A porphyroclast of igneous clinopyroxene (Cpx 1) with thin exsolution lamellae of Ca-poor pyroxene is partially replaced by aggregates of unexsolved neoblastic clinopyroxene (Cpx 2) and minor titanian pargasite (Ti-Prg). These aggregates are at the contact with polygonal aggregates of neoblastic plagioclase (Pl 2). Width of field of view represents ~0.5 mm.

Before being subjected to the high-temperature recrystallization, deformed olivine-bearing gabbros could have contained a significant amount of interstitial titanian pargasite. Otherwise, the high-temperature metamorphic event involved a phase of igneous origin that had significant amounts of volatiles and led to the crystallization of titanian pargasite. This igneous agent could be a trondhjemite-type liquid, as suggested by the following arguments: (1) experimental work indicates that deformation of a crystal mush with a low amount of interstitial liquid results in plastically deformed crystals (Nicolas, 1992, and references therein), and (2) mobility of these highly evolved liquids probably occurred in the gabbroic cumulate pile (see the preceding section) and in any case postdated the high-temperature metamorphic event. We conclude that the recrystallization in high-temperature shear zones developed either in the absence of any fluid phase, or in the presence of small amounts of a volatile-bearing trondhjemite-type liquid. Nevertheless, there is no evidence for the involvement of seawater-derived fluids.

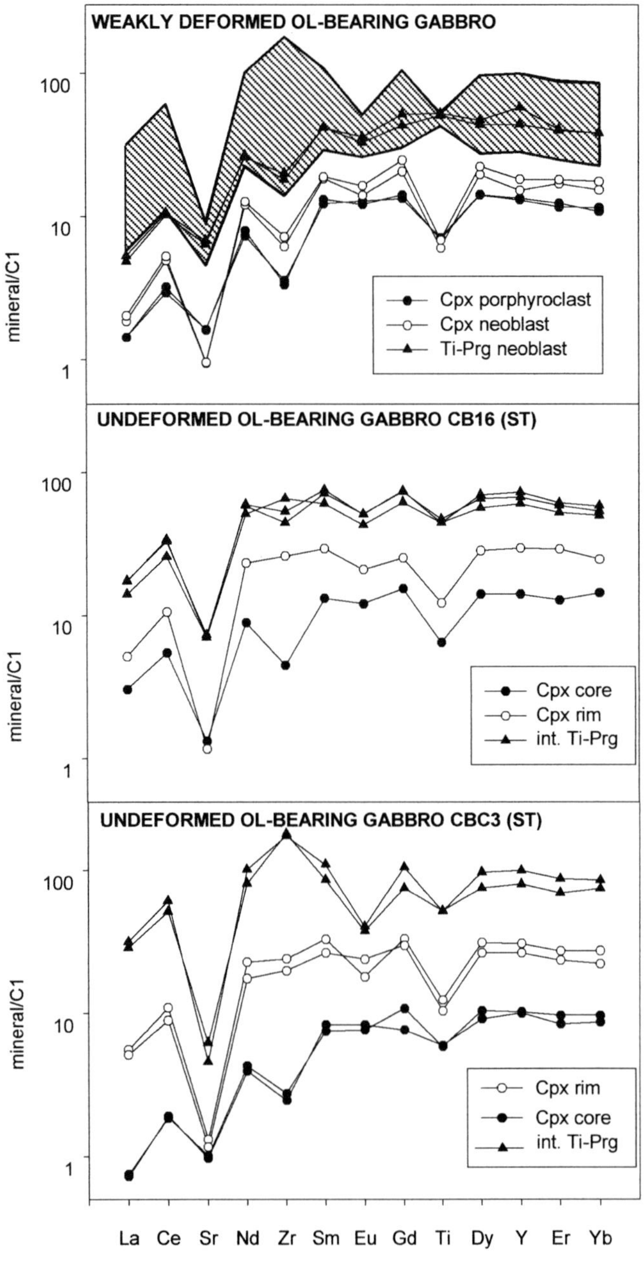

Figure 8. Incompatible trace elements in clinopyroxenes and associated titanian pargasites (Ti-Prg) in weakly deformed to undeformed olivine-bearing gabbros. The field marked by oblique lines refers to the interstitial (int.) titanian pargasites from undeformed olivine-bearing gabbros. Data after Tribuzio et al. (1995, 1999, 2000). IL—Internal Liguride ophiolites; ST—southern Tuscany ophiolites.

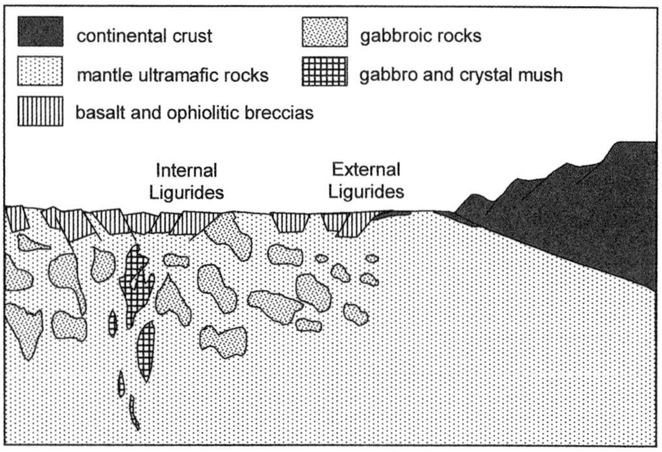

Figure 9. Sketch depicting an interpretive geologic cross section of the lithosphere of the Jurassic Ligurian Tethys. The basin probably formed as a result of continental lithospheric extension. The development of a magma-poor slow-spreading center is inferred on the basis of the close petrologic similarities between the gabbroic rocks of the Northern Apennine ophiolites and those from modern slow-spreading ridges (see text for further details). The ocean-continent transition of this simplified picture is mostly based on the studies of the External Liguride ophiolites (Marroni et al., 1998). The sketch is not to scale and does not show deformation features, except for a few normal faults in the upper levels.

EVOLUTION OF GABBROIC ROCKS AND GEODYNAMIC IMPLICATIONS FOR THE NORTHERN APENNINE OPHIOLITES

Field and petrographic relationships together with major and trace element mineral chemistry have allowed us to determine a detailed igneous to high-temperature metamorphic evolution of the gabbroic rocks of the Northern Apennine ophiolites. The gabbroic rocks formed by the intrusion of N-MORB-type liquids in a heterogeneous mantle section under low-pressure conditions. These liquids underwent an igneous-differentiation process controlled by fractional crystallization, most likely associated with late percolation in the cumulate pile of a volatile-bearing igneous agent, possibly a trondhjemite-type liquid. Gabbroic rocks underwent high-temperature ($T \approx 900$ °C) metamorphic recrystallization in ductile shear zones, in the absence of seawater-derived fluids. These shear zones were locally crosscut by MORB dikes and trondhjemite dikelets, as well as by hornblende-bearing veins indicating the later development of an active, high-temperature hydrothermal system.

Gabbroic rocks of the Northern Apennine ophiolites bear striking petrologic similarities to those recovered from modern slow-spreading ridges. Likewise, Lagabrielle and Cannat (1990) and Cannat et al. (1992) reported close lithostratigraphic similarities between the Northern Apennine ophiolites and the modern slow-spreading ridges. In particular, the mantle and gabbro association of the Northern Apennine ophiolites fits well with the slow-spreading ridge model of discontinuous magmatic crust (Cannat, 1993), characterized by the lack of a continuous gabbroic layer and by a large ultramafic component (Fig. 9). In the Northern Apennine ophiolites, the available structural data indicate that the igneous and high-temperature metamorphic evolution occurred in an extensional tectonic regime. In particular, the high-temperature shear zones most likely developed as low- to medium-angle fault zones (Molli, 1995). This probability not only agrees with the model of ridge development reported by Cannat (1993) for magma-poor slow-spreading centers, but also with a model of basin development through passive rifting of continental lithosphere. The latter geodynamic mechanism was indicated by the structural evolution recorded by the Erro-Tobbio mantle rocks (Western Alps ophiolites; Vissers et al., 1991).

According to the passive rifting model, the subcontinental mantle was unroofed along low-angle detachment faults and injected by basaltic liquids resulting from the decompressed and melted asthenospheric mantle underneath (see also Piccardo et al., 1990). This model is also consistent with the occurrence of remnants of pericontinental basin sections in which the gabbroic rocks are scarce (i.e., the External Liguride ophiolites; Marroni et al., 1998), but it does not conflict with the development a slow-spreading ridge in the Ligurian Ocean. The Northern Apennine ophiolites are thus interpreted to represent a fossil analogue of a magma-poor slow-spreading center that formed as a result of continental lithospheric extension.

ACKNOWLEDGMENTS

We are grateful to M. Cannat, Y. Dilek, and D. Elthon for their thorough reviews. This work was supported by Ministero dell' Università e della Ricerche Scientifica e Tecnologica and Consiglio Nazionale della Ricerche funds.

REFERENCES CITED

Abbate, E., Bortolotti, V., and Principi, G., 1980, Apennine ophiolites: A peculiar oceanic crust: Ofioliti, v. 1, p. 59–96.

Anders, E., and Ebihara, M., 1982, Solar system abundances of the elements: Geochimica et Cosmochimica Acta, v. 46, p. 2363–2380.

Bezzi, A., and Piccardo, G.B., 1971, Structural features of the Ligurian ophiolites: Petrologic evidence for the "oceanic" floor of the Northern Apennines geosyncline; A contribution to the problem of the alpine-type gabbro-peridotite association: Memorie della Società Geologica Italiana, v. 10, p. 53–63.

Borsi, L., Sharer, U., Gaggero, L., and Crispini, L., 1996, Age, origin, and geodynamic significance of plagiogranites in lherzolites and gabbros of the Piedmont-Ligurian ocean basin: Earth and Planetary Science Letters, v. 140, p. 227–241.

Bortolotti, V., 1983, Stratigrafia, tettonica ed evoluzione geodinamica delle ofioliti in Toscana: Memorie della Società Geologica Italiana, v. 25, p. 63–74.

Campbell, A.C., Palmer, M.R., Kinkhammer, G.P., Bowers, T.S., Edmond, J.M., Lawrence, J.R., Casey, J.F., Thompson, G., Humpris, S., Rona, P., and Karson, J.A., 1988, Chemistry of hot springs on the Mid-Atlantic Ridge: Nature, v. 335, p. 514–519.

Cannat, M., 1993, Emplacement of mantle rocks in the seafloor at mid-ocean ridges: Journal of Geophysical Research, v. 98, p. 4163–4172.

Cannat, M., Bideau, D., and Bougault, H., 1992, Serpentinized peridotites and gabbros in the Mid-Atlantic Ridge axial valley at 15°37′N and 16°52′N: Earth and Planetary Science Letters, v. 109, p. 87–106.

Cortesogno, L., and Gaggero, L., 1992, The basaltic dykes in the Bracco gabbroic massif: Petrology of the earliest phases of basalt activity in the Northern Apennines ophiolites: Ofioliti, v. 17, p. 183–198.

Cortesogno, L., Gianelli, G., and Piccardo, G.B., 1975, Preorogenic metamorphic and tectonic evolution of the ophiolite mafic rocks (Northern Apennine and Tuscany): Bollettino della Società Geologica Italiana, v. 94, p. 291–327.

Cortesogno, L., Galbiati, B., and Principi G., 1987, Note alla "Carta geologica delle ofioliti del Bracco" e ricostruzione della paleogeografia Giurassico-Cretacica: Ofioliti, v. 12, p. 261–342.

Decandia, F.A., and Elter, P., 1972, La "zona" ofiolitifera del Bracco, nel settore compreso tra Levanto e la Val Graveglia: Memorie della Società Geologica Italiana, v. 11, p. 503–530.

Dick, H.J.B., Meyer, P.S., Bloomer, S., Kirby, S., Stakes, D., and Mawer, C., 1991, Lithostratigraphic evolution of an in-situ section of oceanic layer 3, *in* Von Herzen, R.P., Robinson, P.T., et al., eds., Proceedings of the Ocean Drilling Program, Scientific Results 118: College Station, Texas, Ocean Drilling Program, p. 439–537.

Elthon, D., 1987, Petrology of gabbroic rocks from the Mid Cayman Rise Spreading Centre: Journal of Geophysical Research, v. 92, p. 658–682.

Elthon, D., 1993, The crystallization of mid-ocean ridge basalts at moderate and high pressures: European Journal of Mineralogy, v. 5, p. 1025–1037.

Gillis, K.M., 1996, Rare earth element constraints on the origin of amphibole in gabbroic rocks from Site 894, Hess Deep, *in* Mével, C., Gillis, K.M., Allan, J.F., and Meyer, P.S, eds., Proceedings of the Ocean Drilling Program, Scientific Results 147: College Station, Texas, Ocean Drilling Program, p. 59–75.

Hébert, R., Serri, G., and Hekinian, R., 1989, Mineral chemistry of ultramafic tectonites and ultramafic to gabbroic cumulates from the major oceanic basins and Northern Apennines ophiolites (Italy): A comparison: Chemical Geology, v. 77, p. 183–207.

Hébert, R., Costantin, M., and Robinson, P.T., 1991, Primary mineralogy of Leg 118 gabbroic rocks and their place in the spectrum of oceanic mafic igneous rocks, *in* Von Herzen, R.P., Robinson, P.T., et al., eds., Proceedings of the Ocean Drilling Program, Scientific Results 118: College Station, Texas, Ocean Drilling Program, p. 3–20.

Hofmann, A.W., 1988, Chemical differentiation of the Earth: The relationship between mantle, continental crust, and oceanic crust: Earth and Planetary Science Letters, v. 90, p. 297–314.

Holland, T., and Blundy, J., 1994, Non-ideal interactions in calcic amphiboles and their bearing on amphibole-plagioclase thermometry: Contributions to Mineralogy and Petrology, v. 116, p. 433–447.

Ito, E., and Anderson, A.T., Jr., 1983, Submarine metamorphism of gabbros from the Mid-Cayman Rise: Petrographic and mineral constraints on hydrothermal processes at low-spreading ridges: Contributions to Mineralogy and Petrology, v. 82, p. 371–388.

Kelley, D.S., 1996, Fluid evolution in slow-spreading environments, *in* Karson, J.A., Cannat, M., Millet, D.J., and Elthon, D., eds., Proceedings of the Ocean Drilling Program, Scientific Results 153: College Station, Texas, Ocean Drilling Program, p. 399–415.

Kelley, D.S., and Malpas, J., 1997, Melt-fluid evolution in gabbroic rocks from Hess Deep, *in* Mével, C., Gillis, K.M., Allan, J.F., and Meyer, P.S., eds., Proceedings of the Ocean Drilling Program, Scientific Results 147: College Station, Texas, Ocean Drilling Program, p. 213–226.

Lagabrielle, Y., and Cannat, M., 1990, Alpine Jurassic ophiolites resemble the modern central Atlantic basement: Geology, v. 18, p. 319–322.

Marroni, M., Molli, G., Montanini, A., and Tribuzio, R., 1998, The association of continental crust rocks with ophiolites in the Northern Apennines (Italy): Implications for the continent-ocean transition in the Western Tethys: Tectonophysics v. 292, p. 43–66.

Meen, J.K., and Elthon, D., 1996, Anatomy of a Mid-Cayman Rise gabbro [abs.]: Journal of Conference Abstracts, Cambridge Publications, v. 1, p. 393.

Messiga, B., and Tribuzio, R., 1991, The reaction between olivine and plagioclase as a consequence of fluid-rock interactions during sub-seafloor metamorphism (Al-Mg-gabbros, Northern Apennine Ophiolites, Italy): Schweizerische Mineralogische und Petrografische Mitteilungen, v. 71, p. 405–414.

Mével, C., and Cannat, M., 1991, Lithospheric stretching and hydrothermal processes in oceanic gabbros from slow spreading ridges, *in* Peters, Tj., et. al., eds., Ophiolite genesis and evolution of the oceanic lithosphere: Dordrecht, Netherlands, Kluwer Academic Publishers, p. 293–312.

Michard, A., 1989, Rare earth element systematics in hydrothermal fluids: Geochimica et Cosmochimica Acta, v. 53, p. 745–750.

Molli, G., 1994, Microstructural features of high temperature shear zones in gabbros of the Northern Apennine Ophiolites: Journal of Structural Geology, v. 16, p. 1535–1541.

Molli, G., 1995, Pre-orogenic high temperature shear zones in an ophiolite complex (Bracco massif, Northern Apennines, Italy), *in* Vissers, R.L.M., and Nicolas, A., eds., Mantle and lower crust exposed in oceanic ridges and in ophiolites: Dordrecht, Netherlands, Kluwer Academic Publishers, p. 147–161.

Molli, G., 1996, Pre-orogenic tectonic framework of the northern Apennine ophiolites: Eclogae Geologicae Helvetiae, v. 89, p. 163–180.

Natland, J., and Dick, H.J.B., 1996, Melt migration through high-level gabbroic cumulates of the East Pacific Rise at Hess Deep: Inferences from rock textures and mineral compositions, *in* Mével, C., Gillis, K.M., Allan, J.F., and Meyer, P.S, eds., Proceedings of the Ocean Drilling Program, Scientific Results 147: College Station, Texas, Ocean Drilling Program, p. 21–58.

Nicolas, A., 1992, Kinematics in magmatic rocks with special reference to gabbros: Journal of Petrology, v. 33, p. 891–915.

Nimis, P., and Ulmer, P., 1998, Clinopyroxene geobarometry of magmatic rocks. Part 1: An expanded structural geobarometer for anhydrous and hydrous, basic and ultrabasic systems: Contributions to Mineralogy and Petrology, v. 133, p. 122–135.

Ottonello, G., Joron, J.L., and Piccardo, G.B., 1984, Rare earth and 3d transition element geochemistry of peridotitic rocks. II. Ligurian peridotites and associated basalts: Journal of Petrology, v. 25, p. 373–393.

Ozawa, K., Meyer, P.S., and Bloomer, S.H., 1991, Mineralogy and textures of iron-titanium oxide gabbros and associated olivine gabbros from hole 735B, *in* Von Herzen, R.P., Robinson, P.T., et al., eds., Proceedings of the Ocean Drilling Program, Scientific Results 118: College Station, Texas, Ocean Drilling Program, p. 41–73.

Pedersen, R.B., Malpas, J., and Falloon, T., 1996, Petrology and geochemistry of gabbroic and related rocks from Site 894, Hess Deep, *in* Mével, C., Gillis, K.M., Allan, J.F., and Meyer, P.S., eds., Proceedings of the Ocean Drilling Program, Scientific Results 147: College Station, Texas, Ocean Drilling Program, p. 3–19.

Piccardo, G.B., Rampone E., and Vannucci, R., 1990, Upper mantle evolution during continental rifting and ocean formation: Evidences from peridotite bodies of the western Alpine-northern Apennine system: Mémoires de la Société Géologique de France, v. 156, p. 323–333.

Rampone, E., Hofmann, A.W., Piccardo, G.B., Vannucci, R., Bottazzi, P., and Ottolini, L., 1995, Petrology, mineral, and isotope geochemistry of the External Liguride peridotites (northern Apennine, Italy): Journal of Petrology, v. 36, p. 81–105.

Rampone, E., Hofmann, A.W., Piccardo, G.B., Vannucci, R., Bottazzi, P., and Ottolini, L., 1996, Trace element and isotope geochemistry of depleted peridotites from an N-MORB type ophiolite (Internal Liguride Peridotites, N. Italy): Contributions to Mineralogy and Petrology, v. 123, p. 61–76.

Rampone, E., Piccardo, G.B., Vannucci, R., and Bottazzi, P., 1997, Chemistry and origin of trapped melts in ophiolitic peridotites: Geochimica et Cosmochimica Acta, v. 61, p. 4557–4569.

Rampone, E., Hofmann, A.W., and Raczek, I., 1998, Isotopic contrasts within the Internal Liguride ophiolite (N. Italy): The lack of a genetic mantle-crust link: Earth and Planetary Science Letters, v. 163, p. 175–189.

Riccardi, M.P., 1994, Amphiboles related to ocean-floor metamorphism from ophiolitic metagabbros of Northern Apennine: Compositional and crystal-chemical features [Ph.D. thesis]: Pavia, Università di Pavia, 138 p.

Ross, K., and Elthon, D., 1997, Cumulus and postcumulus crystallization in the ocean crust: Major- and trace-element geochemistry of Leg 153 gabbroic rocks, *in* Karson, J.A., Cannat, M., Millet, D.J., Elthon, D., eds., Proceedings of the Ocean Drilling Program, Scientific Results 153: College Station, Texas, Ocean Drilling Program, p. 333–350.

Serri, G., 1980, Chemistry and petrology of gabbroic complexes from Northern Apennines ophiolites: Proceedings, International Ophiolite Symposium, Cyprus 1979, p. 296–311.

Sisson, T.W., 1991, Pyroxene-high silica rhyolite trace element partition coefficients measured by ion microprobe: Geochimica et Cosmochimica Acta, v. 55, p. 1575–1585.

Skulski, T., Minarik, W., and Watson, E.B., 1994, High-pressure experimental trace-element partitioning between clinopyroxene and basaltic melts: Chemical Geology, v. 117, p. 127–147.

Tiepolo, M., Tribuzio, R., and Vannucci, R., 1997, Mg- and Fe-gabbroids from Northern Apennine ophiolites: Parental liquids and igneous differentiation processes: Ofioliti, v. 22, p. 57–69.

Toplis, M.J., and Carroll, M.R., 1996, Differentiation of ferro-basaltic magmas under conditions open and closed to oxygen: Implications for the Skaergaard Intrusion and other natural systems: Journal of Petrology, v. 37, p. 837–858.

Tribuzio, R., Riccardi, M.P., and Ottolini, L., 1995, Trace element redistribution in high temperature deformed gabbros from East Ligurian ophiolites (Northern Apennines, Italy): Constraints on the origin of syndeformation fluids: Journal of Metamorphic Geology, v. 13, p. 367–377.

Tribuzio, R., Messiga, B., Vannucci, R., and Bottazzi, P., 1996, Rare earth element redistribution during high-pressure/low-temperature metamorphism in ophiolitic Fe-gabbros (Liguria, northwestern Italy): Implications for light REE mobility in subduction zones: Geology, v. 24, p. 711–714.

Tribuzio, R., Riccardi, M.P., and Messiga, B., 1997, Amphibolitization of Mg- and Fe-rich gabbroic dykes within mantle-derived serpentinites from Northern Apennine ophiolites: Evidence for high-temperature hydration of the oceanic lithosphere: Ofioliti, v. 22, p. 71–80.

Tribuzio, R., Tiepolo, M., Vannucci, R., and Bottazzi, P., 1999, Trace element distribution within olivine-bearing gabbros from the Northern Apennine ophiolites (Italy): Evidence for post-cumulus crystallization in MOR-type gabbroic rocks: Contributions to Mineralogy and Petrology, v. 134, p. 123–133.

Tribuzio, R., Tiepolo M., and Thirlwall, M.F., 2000, Titanian pargasite in gabbroic rocks from the Northern Apennine ophiolites (Italy): Evidence for late percolation of igneous fluids in a MOR-type cumulate pile: Earth Planetary Science Letters, v. 176, p. 281–293.

Vissers, R.L.M., Drury, M.R., Hoogerdujing Strating, E.H., and van der Wal, D., 1991, Shear zones in the upper mantle: A case of study in an Alpine lherzolite massif: Geology, v. 19, p. 990–993.

Watson, E.B., and Green, T.H., 1981, Apatite/liquid partition coefficients for the rare earth elements and strontium: Earth and Planetary Science Letters, v. 56, p. 405–421.

MANUSCRIPT ACCEPTED BY THE SOCIETY FEBRUARY 23, 2000

Printed in the U.S.A.

Geological Society of America
Special Paper 349
2000

New insights concerning the influence of water during the formation of podiform chromitite

Stephen J. Edwards
Lithospheric Processes and Resources Group, School of Earth and Environmental Sciences, University of Greenwich, Chatham Maritime, Kent ME4 4TB, England; e-mail: s.j.edwards@greenwich.ac.uk
Julian A. Pearce and Jonathan Freeman
Department of Geological Sciences, University of Durham, Durham DH1 3LE, England

ABSTRACT

Podiform chromitites are pod-like bodies dominated by chromian spinel and enveloped by dunite. Most major bodies are Cr rich and occur in the harzburgitic mantle section and the mantle-crust transition zone of ophiolites formed in supra-subduction zone environments. Current models for the origin of podiform chromitites require a combination of crystal fractionation and melt-peridotite reaction as basaltic melt migrates through the upper mantle. Although such models in general account for large accumulations of chromian spinel and the field and textural relationships in and around podiform bodies, they do not adequately deal with the problem of the low solubility of Cr in basaltic melts, which begs the questions: How does Cr enter melt and why is it transported to a restricted zone in a melt conduit where accumulation of chromian spinel takes place? The present study suggests that the answer may lie with water and its ability to depolymerize the silica network of the melt. The formation of most, if not all, melts capable of forming podiform chromitites requires the involvement of water in the zone of melt generation. Water not only promotes partial melting of refractory peridotites, it also dissolves in the resulting melt. This hydrous melt will have a greater abundance of octahedral sites than its anhydrous equivalent, and this modification promotes greater solubility of Cr in the hydrous melt owing to the high octahedral site preference energy of Cr^{3+}—the major ion controlling Cr solubility. Cr is held in the melt until the melt and peridotite reaction raises the Si content and extent of polymerization of the melt (so that the concentration of octahedral sites is reduced), at which point the Cr becomes insoluble and chromian spinel precipitates. Growth of chromian spinel takes place in an open system in which there is a continuous supply of Cr-rich melt during a melting episode. An examination of the partial-melting history of peridotites preserved in the Troodos ophiolite, Cyprus, indicates that Cr behaves as an incompatible element during the hydrous melting of refractory peridotites. The highest Cr contents of melts that separate from these peridotites are rarely seen in the extrusive sequences of the Troodos ophiolite, suggesting that these melts fractionated substantial volumes of chromian spinel prior to eruption.

INTRODUCTION

At the 1998 Geological Society of America Penrose Conference on Ophiolites and Oceanic Crust, substantial debate focused on the origin of mantle dunite, which reflects the current interest in the processes of melt migration and reaction in the uppermost mantle. Despite the ubiquitous association of podiform chromitite orebodies with mantle dunite, this aspect of mantle mineralization received little attention. In an attempt to reconcile this shortfall, the present study reviews the formation of these enigmatic and economically important orebodies and uses the findings to develop new insights on the importance of water and the extent of melt polymerization throughout the ore-forming process. As a review, the study relies heavily on previously published data.

Podiform chromitites (orebodies dominated by chromian spinel) present a problem to petrologists because they represent large accumulations of chromian spinel, $(Mg,Fe^{2+})(Cr,Al,Fe^{3+})_2O_4$, that has largely unfractionated Cr/Al ratios and that crystallized from basaltic melt in which Cr has a low solubility (Roeder and Reynolds, 1991). Numerous processes—such as fractional crystallization in convecting melt pockets (Lago et al., 1982), melt mixing and mingling (Paktunc, 1990; Ballhaus, 1998), and melt and peridotite reaction (Zhou et al., 1994, 1996)—have been invoked to explain this shortcoming, but still there are problems with understanding how Cr enters melt, how it is transported in melt, and how such large volumes of chromian spinel can precipitate over a short distance in a melt conduit. The presence of water in the ore-forming system may provide a key to solving these problems.

Edwards, S.J., Pearce, J.A., and Freeman, J., 2000, New insights concerning the influence of water during the formation of podiform chromitite, *in* Dilek, Y., Moores, E.M., Elthon, D., and Nicolas, A., eds., Ophiolites and Oceanic Crust: New Insights from Field Studies and the Ocean Drilling Program: Boulder, Colorado, Geological Society of America Special Paper 349, p. 139–147.

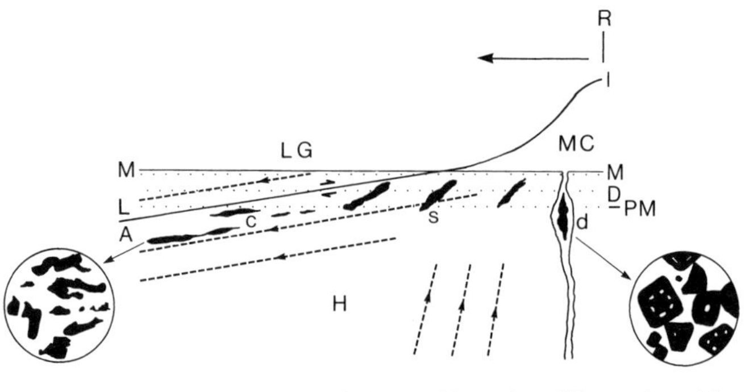

Figure 1. Simplified diagram of the position of podiform chromitites (black) that are discordant (d), subcordant (s), and concordant (c) with respect to mantle flow lines (dashed arrows) in uppermost mantle beneath an active spreading ridge (with spreading direction indicated) (adapted from Lago et al., 1982, and Nicolas, 1989). Microtextures of discordant and concordant chromitite pods are shown to demonstrate the differences in morphology of their chromian spinel and the presence and absence, respectively, of silicate inclusions in this spinel. The 1000 °C isotherm marks the top of the magma chamber and the boundary between lithosphere and asthenosphere. Dunitic rocks of the transition zone occur between the Moho (M) and the petrologic Moho (PM) and are predominantly underlain by harzburgite and overlain by layered gabbro. LG—layered gabbro; MC—magma chamber; R—Ridge; H—Harzburgite; D—Dunitic rocks; L—Lithosphere; A—Asthenosphere.

THE PETROLOGY OF PODIFORM CHROMITITE

Podiform chromitites are restricted almost entirely to sections of uppermost mantle and the mantle-crust transition zone preserved in ophiolites (including alpine-type peridotites) (Pearce et al., 1984; Boudier and Nicolas, 1985). The majority of these orebodies lie within harzburgite, rather than lherzolite, and are a key component of the supra-subduction zone ophiolites of Pearce et al. (1984) and Roberts (1988), and the harzburgite ophiolite types of Boudier and Nicolas (1985). The predominance of podiform chromitites in supra-subduction zone ophiolites is reflected in the chemistry of their volcanic sequences. Relative to volcanic rocks produced at mid-ocean ridges, those formed in supra-subduction zone settings preserve a chemical record explicable by extensive early crystal fractionation dominated by olivine and chromian spinel, as exemplified by the Cr-Y plots of Pearce et al. (1984).

Within supra-subduction zone ophiolites, podiform chromitites exhibit a range of compositions from Al-rich to Cr-rich varieties. This reflects differences in the conditions under which they formed, such as geodynamic setting, compositions of melt and host peridotite, crystallization conditions, and extent of melt and peridotite reactions (Leblanc and Violette, 1983; Zhou et al., 1994, 1996; Leblanc, 1995, 1997; Zhou and Robinson, 1997; Malpas et al., 1997; Robinson et al., 1997).

Location, structure, and texture

Chromitite pods are always enveloped by dunite (unless the dunite has been removed during deformation) and are located either at depth (near the harzburgite-lherzolite transition) or between 1 and 2 km beneath the layered gabbros (Thayer, 1960) near to the petrologic Moho (Fig. 1). The pods occur as concordant, subcordant, and discordant bodies with respect to mantle flow lines preserved in tectonized peridotites of the mantle and mantle-crust transition zone (Cassard et al., 1981; Fig. 1). The orientation of a pod with respect to these flow lines depends on the original position of formation of the pod and its subsequent transposition during mantle flow. Primary textures in discordant pods are coarse, very coarse, or even pegmatitic (Panayiotou et al., 1986; Georgiou, 1987) and reflect magmatic crystallization processes. These textures become progressively overprinted in subcordant and concordant pods by textures produced during high-temperature deformation related mainly to asthenospheric flow (Fig. 1). Lorand and Ceuleneer (1989) showed that chromian spinel in discordant pods is usually rich in silicate mineral inclusions, whereas, in concordant pods, these inclusions are rare or absent because of expulsion during strain-induced recrystallization and grain-boundary migration associated with asthenospheric deformation (Fig. 1).

Formation

Podiform chromitite orebodies are thought to form during chromian spinel–dominated crystallization of basaltic melt migrating through the upper mantle (Brown, 1980; Augé and Roberts, 1982; Lago et al., 1982; Leblanc and Ceuleneer, 1992). Lago et al. (1982) modeled the accumulation of chromian spinel within dikes of basaltic melt crossing through mantle diapirs below a spreading ridge. Their results suggested that the melt and peridotite reaction produces dunite and that active convection in cavities promotes the crystallization, growth, and accumulation of chromian spinel. This model gives an indication of what may be happening, but it does not adequately address the problem that fractionating basaltic melts do not contain sufficient Cr to enable large volumes of chromian spinel to form before the onset of silicate-dominated crystallization (Murck and Campbell, 1986; Roeder and Reynolds, 1991; Campbell and Murck, 1993). In stratiform intrusions, this problem is overcome by mixing highly magnesian melt on the olivine–chromian spinel cotectic with evolved, Al-rich tholeiitic melt in the orthopyroxene stability field; this produces a hybrid melt that lies in the chromian spinel stability field (Irvine, 1977; Irvine and Sharpe, 1986).

Mixing processes present two potential ways to form podiform chromitite. The first involves the mixing of refractory (boninitic affinity) and fertile (mid-oceanic-ridge basalt affinity) melts produced by multistage melting and melt segregation in the mantle (Paktunc, 1990; Ballhaus, 1998). Although an appealing mechanism, field relationships in ophiolites demonstrate that it is highly unlikely that melts produced by multistage melting will coexist and mix in a conduit, as earlier melt generally leaves a crystalline product, rather than melt, in the conduit (Edwards, 1991, 1995). However, melt mixing and formation of podiform chromitite is highly probable during a continuous magmatic event where a conduit remains open and is filled by melt at different stages in its chemical evolution (Leblanc and Ceuleneer, 1992). It was estimated by Augé (1987) and Leblanc and Ceuleneer (1992) that in this open system, substantial accumulation of chromian spinel will only occur if the melt/spinel volume ratio is ~300 to 500, implying that there must be a substantial melt flow to sustain the supply of Cr.

A second possible way to form podiform chromitite by mixing processes involves melt and peridotite reaction, as has been proposed by Arai and Yurimoto (1994), Zhou et al. (1994, 1996), and Edwards (1995). Such reaction increases the Si/Al ratio of melt, moving the melt into the chromian spinel stability field in much the same way as melt mixing. As a general rule, pods of Al-rich chromian spinel form where Al-rich (tholeiitic) melt reacts with lherzolite or harzburgite, whereas Cr-rich pods form during reaction between harzburgite and Al-poor (boninitic) melt (Zhou et al., 1994; Leblanc, 1995).

Field evidence for the formation of dunite by the dissolution of pyroxene in harzburgite (and rarely lherzolite) at the margins of chromitites is the most conclusive evidence in support of this second process. In summary, current models for the origin of podiform chromitite require a combination of crystal fractionation and melt and peridotite reaction in an open system as basaltic melt migrates through upper mantle during a melting episode in supra-subduction zone environments.

Silicate inclusions and evidence for water in melt

Detailed studies of podiform chromitite samples from sections of ophiolite preserved in New Caledonia (Johan, 1986), the Maqsad area of Oman (Lorand and Ceuleneer, 1989), the Troodos massif in Cyprus (McElduff and Stumpfl, 1991), and the Kempirsai district in Kazakhstan (Melcher et al., 1997) have described an array of silicate inclusions in chromian spinel, including hornblende, pargasitic hornblende, pargasite, the pure-Na analogue of phlogopite, nepheline, clinopyroxene, and rare orthopyroxene and olivine. The hydrous inclusions contrast sharply with the general absence of primary hydrous minerals in the intergranular silicates in the chromitite pod, although the primary amphibole that coats chromian spinel grains in the podiform deposits of Oman is a notable exception (Augé, 1987; Leblanc and Ceuleneer, 1992).

The hydrous inclusions suggest that water was an integral component of the chromitite-forming melt. Additional evidence for this suggestion comes from the types of inclusions in chromian spinel in dunite and harzburgite adjacent to chromitite pods. Inclusions in dunite formed by melt and peridotite reaction are comparable to those in chromitite, whereas those in harzburgite are anhydrous and comprise olivine and orthopyroxene (McElduff and Stumpfl, 1991). Furthermore, these associations suggest that the inclusions did not form during wholesale alteration of the mantle sequence or during downward percolation of seawater. The isotopic study of Dunlop and Fouillac (1986) demonstrates that the primary water trapped in chromian spinel is magmatic in origin and ultimately derived from subducted oceanic lithosphere.

In samples studied from the New Caledonia, Oman, and Troodos ophiolites, inclusions are randomly distributed and vary in abundance. Inclusion-bearing and inclusion-free chromian spinel grains may coexist in the same sample; within the former, inclusions may occur in isolation, in well-defined clusters, or in zones parallel to the crystal faces of euhedral grains. These patterns imply that silicates and/or melt were entrapped at discrete intervals during growth of chromian spinel. However, the extreme contrast between the mineral assemblage in the inclusions and that of the silicate matrix is difficult to explain. The contrast suggests that the inclusions are unlikely to simply represent entrapped melt or intercumulus silicate minerals (Augé, 1987; McElduff and Stumpfl, 1991). Instead, the peculiar mineral assemblage of the inclusions may result from reaction between anhydrous silicates and fluid-rich melt during and after entrapment (Lorand and Ceuleneer, 1989; McElduff and Stumpfl, 1991). Conversely, chromitite pods may record the involvement of two chemically distinct melts: a hydrous melt trapped in the inclusions and a more anhydrous melt that crystallized in the interstitial matrix (Schiano et al., 1997). However, this possibility again raises the question of how such compositionally distinct melts could coexist. A favorable explanation that deals with many of the mineral-assemblage discrepancies is that only one melt is involved, but it evolves along two different paths. In this case, the inclusions form by entrapment and closed-system evolution of hydrous basaltic melt, whereas the intergranular silicates, dominated by anhydrous mineral phases, are products of open-system evolution of the same melt (Schiano et al., 1997). The open system causes dominantly anhydrous silicates to crystallize around grains of chromian spinel while enabling volatiles to be transported away from the site of chromitite formation in the residual melt. This residual melt fraction is then capable of crystallizing pyroxenites containing late-stage hydrous magmatic minerals and having very coarse to pegmatitic grain sizes (Edwards, 1990, 1995).

It is clear that the melts from which podiform chromitites form are hydrous and that a component of this hydrous melt is trapped during growth of chromian spinel. An argument for the subsolidus entrapment of the inclusions is difficult to build because of the extreme contrast between the mineral assemblage of the inclusions and that of the intercumulus silicates (McElduff and Stumpfl, 1991), despite evidence that there may be substantial subsolidus growth of chromian spinel (Waters and Boudreau, 1996).

DISSOLVED WATER IN BASALTIC MELT

It is widely accepted that water is necessary for partial melting of supra-subduction zone mantle to produce melts capable of subsequent formation of podiform chromitites with largely unfractionated Cr/Al ratios (e.g., Pearce et al., 1984; Roberts, 1988; Zhou and Robinson, 1997; Robinson et al., 1997). This mantle contains an estimated <0.5 wt% water, which contrasts sharply with the 0.01–0.04 wt% water estimated to be contained in fertile mantle beneath mid-ocean ridges (Pearce et al., 1992). During hydrous melting of mantle, water is present in the zone of melting either as hydrous minerals in the melting zone or as hydrous fluid flushing this zone. Regardless of its state, water, being incompatible during mantle melting, enters the melt. Evidence for this statement is the moderate to high pre-eruptive water contents of many supra-subduction zone melts; for example, up to 1.6–1.9 wt% has been reported for back-arc basin basaltic melts (Danyushevsky et al., 1993; Stolper and Newman, 1994) and ≤3.3 wt% has been reported for basaltic andesite melt at Goosenest volcano, Oregon (Sisson and Layne, 1993). For comparison, Jambon (1994) has estimated the mean water content of mid-oceanic-ridge basalts to be 0.33 wt%, whereas that of back-arc basin basalts is 0.71 wt%.

For the purpose of the present study, it is necessary to ascertain whether, at the depth of formation of podiform chromitite, water is dissolved in melt or exists as a separate vapor phase. Figure 2 has been constructed for this purpose and portrays the approximate water-saturation curves for hydrous basaltic and andesitic melts crystallizing at a depth of chromitite formation of 7 km (0.2 GPa). Podiform chromitites form during the very early stages of fractional crystallization of melt, and Figure 2 demonstrates that a separate water vapor phase is unlikely to exist at this time for initial water concentrations in the melt of at least 4.5 wt%. For example, consider a melt with an initial water content of ~2 wt%, similar to that which formed the Troodos upper pillow lavas (Muenow et al., 1990). At 0.2 GPa, the weight fraction of this melt that has to crystallize for water saturation to occur is ~58% (basalt) to 68% (andesite) (Fig. 2). This result implies that the coarse to pegmatitic grain sizes in podiform chromitite (and pyroxenite dikes) are not the product of "true" pegmatite-forming processes, such as those that take place at and above the water-saturation limit in granitic systems, but instead reflect high diffusion rates in the melt brought about through depolymerization of the silica network by dissolved, dissociated water. In summary, we predict that water involved in melting will become dissolved in the

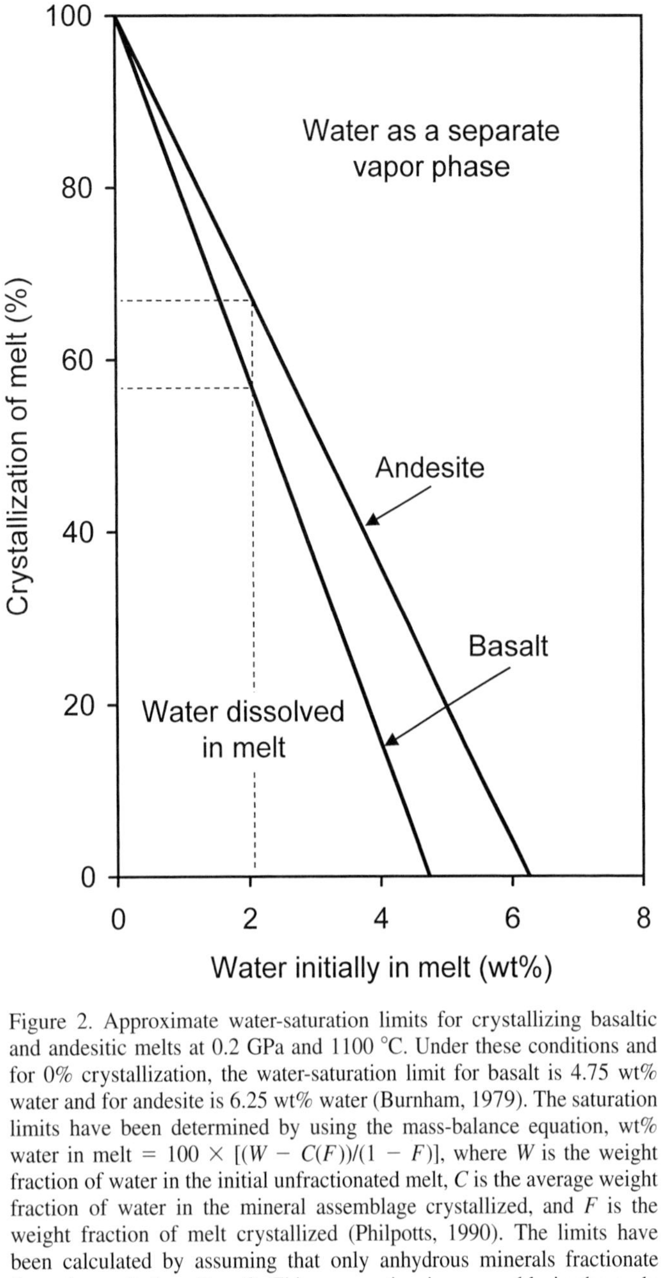

Figure 2. Approximate water-saturation limits for crystallizing basaltic and andesitic melts at 0.2 GPa and 1100 °C. Under these conditions and for 0% crystallization, the water-saturation limit for basalt is 4.75 wt% water and for andesite is 6.25 wt% water (Burnham, 1979). The saturation limits have been determined by using the mass-balance equation, wt% water in melt = $100 \times [(W - C(F))/(1 - F)]$, where W is the weight fraction of water in the initial unfractionated melt, C is the average weight fraction of water in the mineral assemblage crystallized, and F is the weight fraction of melt crystallized (Philpotts, 1990). The limits have been calculated by assuming that only anhydrous minerals fractionate from the melt, i.e., $C = 0$. This assumption is reasonable in the early stages of fractional crystallization when C is estimated to be $<<0.1$ wt%, but is likely to introduce error in the late stages of fractional crystallization. The present study is concerned with the early stages of crystallization, for which the calculated water-saturation limits are meaningful. The dashed lines indicate the percentages of crystallization needed to create water saturation in basaltic and andesitic melts with an initial water content of 2 wt%, such as that found in some Troodos upper pillow lavas investigated by Muenow et al. (1990).

melt and will remain so until after the crystallization of podiform chromitite and pyroxenite.

Cr SOLUBILITY IN BASALTIC MELT

All studies of the formation of podiform chromitite have been hampered by the problem of precipitating large volumes of chromian spinel from basaltic melts in which Cr has a very low solubility (e.g., Hill and Roeder, 1974; Fisk and Bence, 1980; Sharpe and Irvine, 1983; Roeder and Reynolds, 1991). The solubility of Cr is sensitive to changes in oxygen fugacity and temperature (Hill and Roeder, 1974; Barnes, 1986; Murck and Campbell, 1986; Roeder and Reynolds, 1991), pressure (Dick and Bullen, 1984), and melt composition (Irvine, 1967, 1976; Dick and Bullen, 1984). Numerous mechanisms involving changes in one or more of these variables have been invoked to bring about precipitation of chromian spinel. However, three key questions still require consideration: (1) How does Cr enter melt? (2) How is Cr transported from the zone of melt generation to the point at which massive precipitation of chromian spinel occurs? (3) How is continuous precipitation of chromian spinel with unfractionated Cr/Al ratios achieved over a short distance in a melt conduit? Several studies have attempted to answer these questions by suggesting that changes in the degree of polymerization of the silica network of the melt, which result from changes in Si activity, control the solubility of Cr (Dick and Bullen, 1984; Murck and Campbell, 1986; Peck and Keays, 1990; Melcher et al., 1997). We concur with this suggestion, but wish to further examine the questions with respect to the polymerization state of melt containing dissolved, dissociated water.

Roeder and Reynolds (1991) demonstrated that Cr^{2+}, Cr^{3+}, and Cr^{6+} may be present in basaltic melt and that it is the concentration of Cr^{3+} which controls chromian spinel stability and the composition of precipitated chromian spinel. Trivalent Cr has a very high octahedral site preference energy (Burns, 1973), and its behavior during mantle melting and crystal fractionation is controlled by the availability of appropriate sites in minerals and melt. Murck and Campbell (1986) predicted that Cr^{3+} will not exist in tetrahedral co-ordination in melts. It is clear, therefore, that Cr^{3+} will be most soluble in melts with high concentrations of octahedral sites, i.e., in melts with a low degree of polymerization of the silica network (Hess, 1971, 1980). Factors that are known to decrease melt polymerization and viscosity at a fixed pressure are a decrease in Si content, an increase in temperature (Murase and McBirney, 1973) and, most significantly for the present study, an increase in dissolved water content (review by Wood and Fraser, 1984; Lange, 1994; McMillan, 1994; Richet et al., 1996).

Compatibility of Cr during partial melting

Progressive partial melting of the mantle is known to increase the Ni content of peridotite residues, as would be expected for a compatible element, whereas Cr concentrations vary in a manner suggesting either compatible (Frey et al., 1985) or incompatible (Kurat et al., 1980) behavior, or both types of behavior (Roberts and Neary, 1993; Figs. 3 and 4). These variations in the geochemical behavior of Cr may be a reflection of the abundance of chromian spinel in peridotite and the partitioning of Cr into this mineral during partial melting (Dickey and Yoder, 1972; Jaques and Green, 1980; Dick and Bullen, 1984; Arai, 1992). With the near-constant input of Cr into the melt that is suggested by the peridotite data in Figure 3, the dramatic increase in Cr content of melts derived from progressively more refractory mantle sources (600 ppm Cr and 15 wt% Al_2O_3 in tholeiitic melt, but 1200 ppm Cr and 9 wt% Al_2O_3 in boninitic melt as stated by Malpas et al., 1997) must be explained by the eventual melting of chromian spinel (Roberts and Neary, 1993) and/or by changes in the Cr-carrying capacity of the melt.

Podiform chromitites form from hydrous melts that are produced by partial melting of mantle in the presence of water. This water

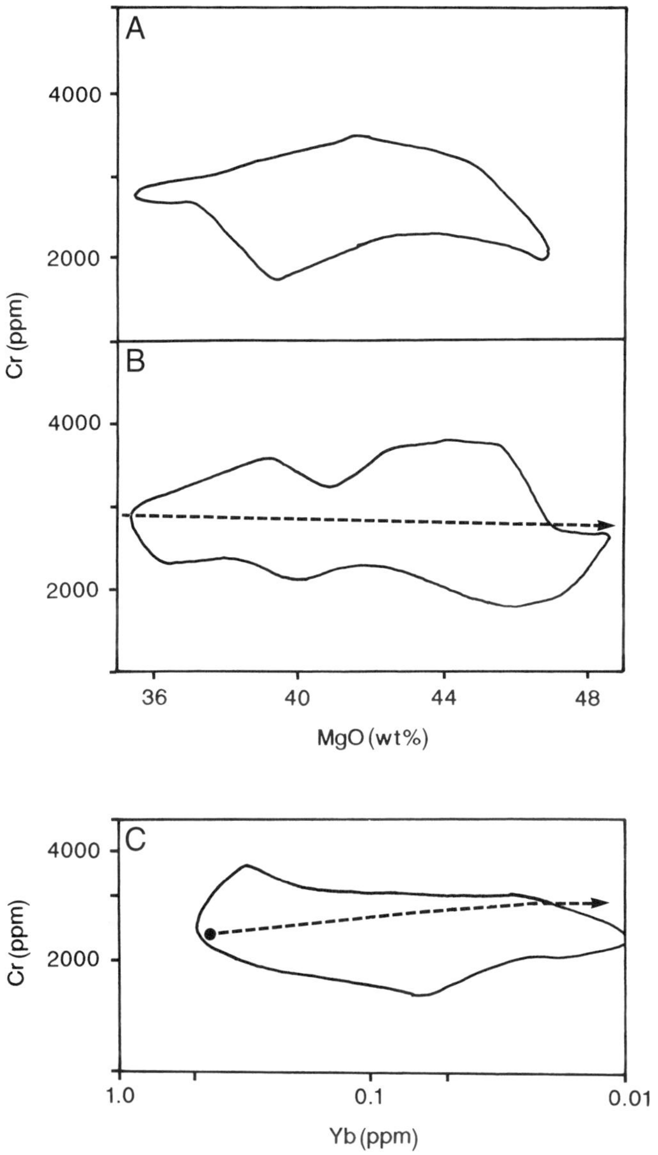

Figure 3. Whole-rock peridotite compositions. A, Data for Tinaquillo, Zabargad, Ronda, and the Western Alps taken from Liang and Elthon (1990). B, Data for xenoliths from alkali basalts taken from Liang and Elthon (1990), with the dashed arrow from Maaløe and Aoki (1977) indicating the partial-melting trend. C, Data for abyssal and Bonin-Mariana inner-trench-wall samples, with the dashed arrow indicating the partial-melting trend (Pearce and Parkinson, 1993).

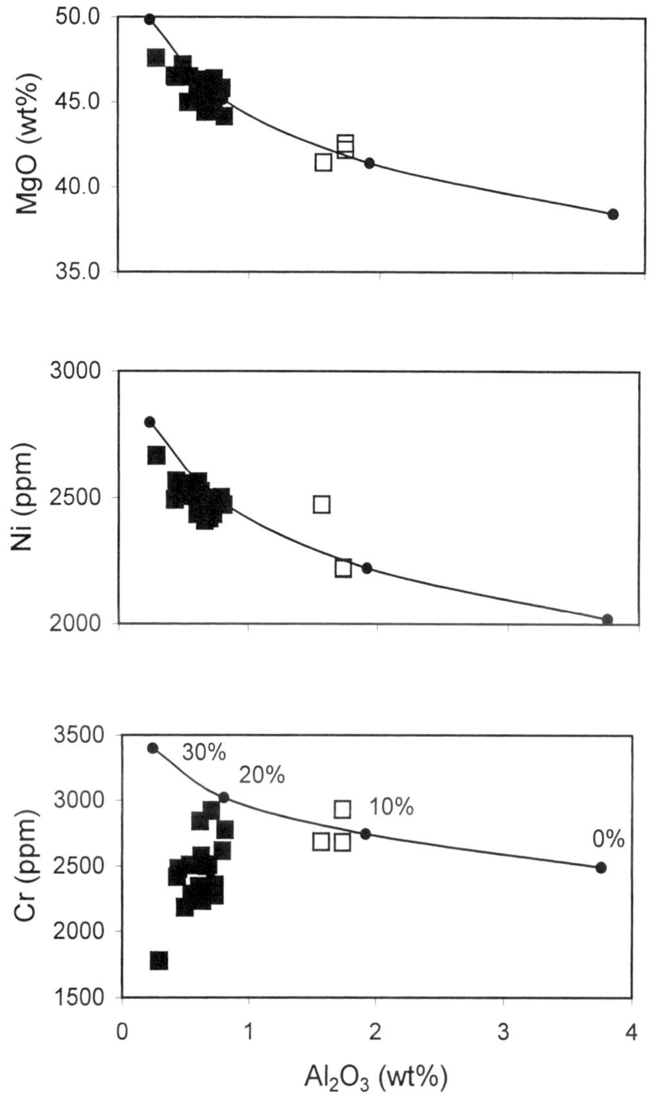

Figure 4. Variation of MgO, Ni, and Cr with Al_2O_3 exhibited by Troodos peridotites. Open squares—lherzolitic residues in the serpentinite diapir; solid squares—the main body of harzburgitic residues. Calculated fractional-melting curves illustrate compositional changes in the residue for 0% to 30% partial melting. All data are reproduced from Freeman (1996).

promotes melting to beyond the clinopyroxene-out phase boundary, so that extensive melting of lherzolite ultimately leaves a residue of harzburgite containing no free clinopyroxene. Although Cr is partitioned into residual chromian spinel during this melting (Dick and Bullen, 1984), the overall modal abundance of Cr-bearing phases in the residue decreases. Roberts (1992) suggested that these mineral and chemical changes may affect the bulk distribution coefficient of Cr, and Freeman (1996) found that the residual peridotites of the Troodos ophiolite may actually record this effect (Fig. 4). Figure 4 implies that the coefficient lies close to unity when melting leaves a lherzolitic residue, but that it drops significantly after melting proceeds beyond the lherzolite-harzburgite transition; from here on orthopyroxene, and presumably chromian spinel, must play the important role in melting—a conclusion also reached by Roberts (1992). It is the melts derived from melting of refractory peridotite, during which the behavior of Cr is incompatible, that are expected to be capable of forming podiform chromitites (Roberts, 1992; Freeman, 1996). In support of this requirement, Freeman (1996) calculated the Cr content of melts produced by fractional melting of the highly refractory Troodos harzburgites to be in the range of 870–3250 ppm. Such concentrations have been found in refractory lavas erupted in supra-subduction zone environments, although the majority of analyses give values at the lower end of this range (e.g., Crawford, 1989, and references therein), implying that very Cr-rich melts are not erupted without first fractionating appreciable concentrations of chromian spinel. This inference is consistent with the observations of Pearce et al. (1984) and Roberts (1988) that lavas erupted in supra-subduction zone environments undergo greater amounts of chromian spinel fractionation during their ascent through the mantle than those erupted in mid-ocean ridge environments.

Once Cr has entered melt, it must be transported away from the melting zone quickly and efficiently to maximize the potential for subsequent formation of podiform chromitite. The water in the zone of partial melting will enter the melt and have a profound influence on the ability of melt to transport Cr because it will depolymerize the silica network of the melt, thus countering the effect that the high Si content of the melt will have on maintaining a high degree of polymerization and so rendering Cr^{3+} largely insoluble. It has not been possible to quantify the effects of this process, but a few weight percent of water in melt will have up to an order of magnitude greater effect at the molecular level because of the low molecular weight of water in comparison to that of the other major element oxides. In terms of polymerization potential, the increase in Si content of melt arising from hydrous mantle melting is reduced or nullified by water; for example, recast as molecular proportions, 1 wt% H_2O is equivalent to 3.5 wt% SiO_2. Water is, therefore, essential to enable partial melting of highly refractory mantle peridotite to generate melt with a depolymerized silica network and, thus, to liberate appreciable amounts of Cr from the mantle and to transport this Cr in the resulting melt.

Melt and peridotite reaction and crystallization of chromian spinel

As outlined earlier, it is generally accepted that a melt and peridotite reaction may be largely responsible for bringing about massive precipitation of chromian spinel to form podiform chromitite. The reaction requires the dissolution of wall-rock pyroxene and formation of olivine to drive the melt into the chromian spinel stability field. For effective reaction to take place, melt must be able to migrate through the peridotite matrix. For an interconnected melt network to form along olivine-olivine grain boundaries, the melt fraction may be as low as 0.5 wt% (Daines and Richter, 1988), whereas olivine-pyroxene and pyroxene-pyroxene grain boundaries are unlikely to be wetted by melt fractions even as high as 5 wt% in the absence of water (Toramaru and Fujii, 1986; von Bargen and Waff, 1988). Water, again, is essential, this time for initiating melt migration through, and reaction with, pyroxene-bearing peridotite. Because dissolved water reduces the permeability threshold of this peridotite and expands the primary phase volumes of olivine and spinel at the expense of pyroxene (e.g., Bowen and Tuttle, 1949, Nicholls and Ringwood, 1973, Roberts, 1988), a positive feedback cycle will be established involving dunite formation and enhanced permeability (cf. Kelemen et al., 1995). In addition, the presence of water will greatly increase diffusion rates, enhancing the dissolution of unstable minerals and growth of stable ones (e.g., Watson, 1994). Consequently, migrating melt will become progressively more polymerized as it moves to lower pressure, cools, and Si enters the melt because of the conversion of wall-rock pyroxene to olivine. As a result, Cr^{3+} will become insoluble, and nucleation of chromian spinel will take place (Fig. 5). Through a combination of these reactions and fractional crystallization, the melt will become saturated to supersaturated, first with respect to chromian spinel and olivine and then with respect to pyroxene. This step is seen clearly in cogenetic dikes in the mantle sections of the Oman (Ceuleneer and Nicolas, 1985), Zambales (Hock et al., 1986), and Bay of Islands (Edwards, 1990, 1995) ophiolites, where pyroxenite dikes formed after pods of chromitite and dikes of chromian spinel–bearing dunite.

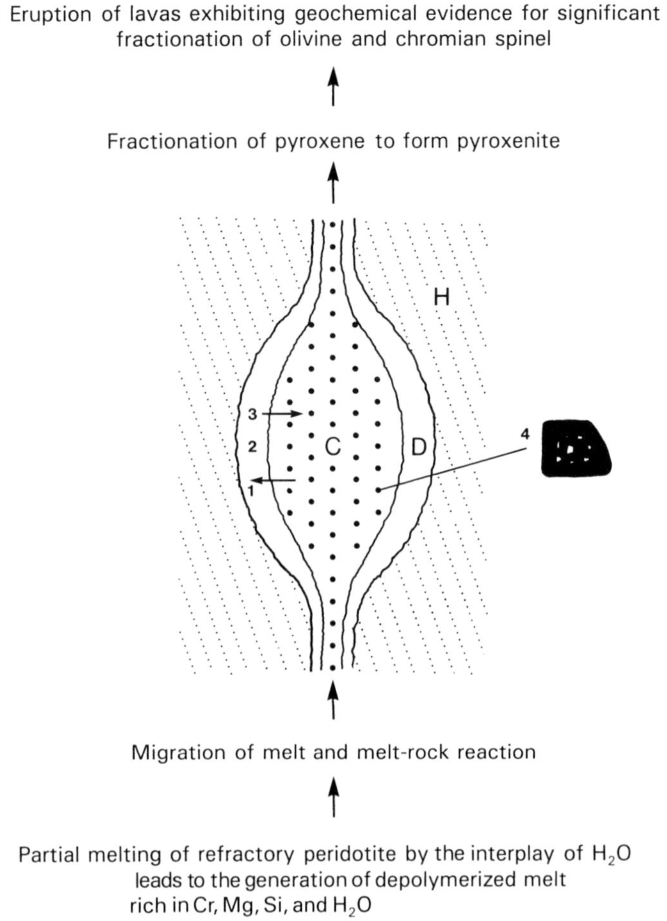

Figure 5. Simplified flow diagram emphasizing some of the key processes involved in the initial formation of a chromitite (C) and dunite (D) assemblage in harzburgitic mantle (H). After formation of hydrous melt, important steps involving water occur at the following numbered locations: (1) Hydrous melt enters peridotite wall rock and wets grain boundaries. (2) Dissolution of pyroxene and growth and recrystallization of olivine and chromian spinel. (3) Addition of Si-richer melt to the melt in the conduit, causing increased melt polymerization and crystallization of chromian spinel nuclei in the conduit. (4) Growth of chromian spinel. The diagram represents a snapshot of the ore-forming system. In reality, the system is more complex because it is open during a melting and melt-extraction event.

Melt focusing

Despite the very high chromian spinel/melt partition coefficient of Cr (600 at 1200 °C as reported by Pearce and Parkinson, 1993) and a Cr content of melt up to and exceeding 1200 ppm, sufficient segregation of chromian spinel to form a podiform body requires an open system, as suggested earlier by the melt/spinel ratio of 300–500. Achievement of this ratio requires the effective focusing of melt in the shallow mantle below a spreading axis (Roberts and Neary, 1993; Zhou et al., 1996). Melt focusing during a melting event is also required to provide a continuous supply of Cr and to maximize the likelihood of mixing between melts of differing compositions (particularly in terms of their Si and water contents).

It is possible that the cold walls of fracture zones play an important role in focusing melt, and indeed hydrous fluid, to permit or enhance the formation of podiform chromitite. As an example, the only noteworthy podiform chromitites in the Bay of Islands ophiolite in western Newfoundland are located in refractory upper-mantle peridotites where boninitic melts were channeled to shallow depth at a ridge and fracture zone intersection in a back-arc spreading environment (Edwards, 1995). High concentrations of podiform chromitite are also found along a paleotransform fault in New Caledonia (Leblanc, 1995).

FUTURE RESEARCH

The present study has demonstrated that future models examining the formation of podiform chromitite should consider the role of water dissolved in melt and the effect this water has on the extent of melt polymerization. With respect to these considerations, two key aspects require much greater study. The first is the need to quantify the processes outlined and discussed in this study. With this achievement may come the answer to vexing questions, such as why do some supra-subduction zone ophiolites contain podiform chromitites and others do not and why do some dunites contain podiform chromitite and others do not? Second, it is necessary to determine the precise composition of the hydrous component of melt and the effect of network-forming, network-modifying, and charge-balancing cations in hydrous melt. In particular, the role of Na, K, and Ca needs to be better understood, because these elements are intimately associated with the precipitation of chromian spinel as evidenced by the compositions of silicate inclusions.

ACKNOWLEDGMENTS

This paper has greatly benefited from discussions with Iain McDonald and from the thoughtful and constructive reviews by Stephen Roberts, Paul Robinson, and an anonymous referee. The editors are thanked for their enthusiasm and efficiency. The participation of Stephen Edwards at the 1998 Geological Society of America Penrose Conference on ophiolites and oceanic lithosphere was made possible by funds from the School of Earth and Environmental Sciences at the University of Greenwich.

REFERENCES CITED

Arai, S., 1992, Chemistry of chromium spinel in volcanic rocks as a potential guide to magma chemistry: Mineralogical Magazine, v. 56, p. 173–184.

Arai, S., and Yurimoto, H., 1994, Podiform chromitites from the Tari-Misaka ultramafic complex, southwestern Japan, as mantle-melt interaction products: Economic Geology, v. 89, p. 1279–1288.

Augé, T., 1987, Chromite deposits in the northern Oman ophiolite: Mineralogical constraints: Mineralium Deposita, v. 22, p. 1–10.

Augé, T., and Roberts, S., 1982, Petrology and geochemistry of some chromitiferous bodies within the Oman ophiolite: Ofioliti, v. 2/3, p. 133–154.

Ballhaus, C., 1998, Origin of podiform chromite deposits by magma mingling: Earth and Planetary Science Letters, v. 156, p. 185–193.

Barnes, S.J., 1986, The distribution of chromium among orthopyroxene, spinel, and silicate liquid at atmospheric pressure: Geochimica et Cosmochimica Acta, v. 50, p. 1889–1909.

Boudier, F., and Nicolas, A., 1985, Harzburgite and lherzolite subtypes in ophiolitic and oceanic environments: Earth and Planetary Science Letters, v. 76, p. 84–92.

Bowen, N.L., and Tuttle, O.F., 1949, The system MgO-SiO_2-H_2O: Geological Society of America Bulletin, v. 60, p. 439–460.

Brown, M., 1980, Textural and geochemical evidence for the origin of some chromite deposits in the Oman ophiolite, *in* Panayiotou, A., ed., Ophiolites: Proceedings of the International Ophiolite Symposium, Cyprus 1979: Nicosia, Cyprus, Geological Survey Department, p. 714–721.

Burnham, C.W., 1979, The importance of volatile constituents, *in* Yoder, H.S., Jr., ed., The evolution of the igneous rocks: Princeton, Princeton University Press, p. 439–482.

Burns, R.G., 1973, The partitioning of trace transition elements in crystal structures: A provocative review with applications to mantle geochemistry: Geochimica et Cosmochimica Acta, v. 37, p. 2395–2403.

Campbell, I.H., and Murck, B.W., 1993, Petrology of the G and H chromite zones in the Mountain View Area of the Stillwater Complex, Montana: Journal of Petrology, v. 34, p. 291–316.

Cassard, D., Nicolas, A., Rabinovitch, M., Leblanc, M., and Prinzhofer, A., 1981, Structural classification of chromite pods in southern New Caledonia: Economic Geology, v. 76, p. 805–831.

Ceuleneer, G., and Nicolas, A., 1985, Structure in podiform chromite from the Maqsad district (Sumail ophiolite, Oman): Mineralium Deposita, v. 20, p. 177–185.

Crawford, A.J., editor, 1989, Boninites and related rocks: London, Unwin Hyman, 465 p.

Daines, M.J., and Richter, F.M., 1988, An experimental method for directly determining the interconnectivity of melt in a partially molten system: Geophysical Research Letters, v. 15, p. 1459–1462.

Danyushevsky, L.V., Falloon, T.J., Sobolev, A.V., Crawford, A.J., Carroll, M., and Price, R.C., 1993, The H_2O content of basalt glasses from southwest Pacific back-arc basins: Earth and Planetary Science Letters, v. 117, p. 347–362.

Dick, H.J.B., and Bullen, T., 1984, Chromian spinel as a petrogenetic indicator in abyssal and alpine-type peridotites and spatially associated lavas: Contributions to Mineralogy and Petrology, v. 86, p. 54–76.

Dickey, J.S., and Yoder, H.S., 1972, Partitioning of chromium and aluminium between clinopyroxene and spinel: Carnegie Institution of Washington Year Book, v. 71, p. 384–392.

Dunlop, H.M., and Fouillac, A.M., 1986, Isotope geochemistry of Oman basic-ultrabasic rocks and chromite deposits, *in* Gallagher, M.J., Ixer, R.A., Neary, C.R., and Prichard, H.M., eds., Metallogeny of basic and ultrabasic rocks: London, Institution of Mining and Metallurgy, p. 291–304.

Edwards, S.J., 1990, Harzburgites and refractory melts in the Lewis Hills massif, Bay of Islands ophiolite complex: The base-metals and precious-metals story: Canadian Mineralogist, v. 28, p. 537–552.

Edwards, S.J., 1991, Magmatic and fluid processes in the upper mantle: A study of the Bay of Islands ophiolite complex, Newfoundland [Ph.D. thesis]: St. John's, Memorial University of Newfoundland, 368 p.

Edwards, S.J., 1995, Boninitic and tholeiitic dykes in the Lewis Hills mantle section of the Bay of Islands ophiolite: Implications for magmatism adjacent to a fracture zone in a back-arc spreading environment: Canadian Journal of Earth Sciences, v. 32, p. 2128–2146.

Fisk, M.R., and Bence, A.E., 1980, Experimental crystallization of chromite spinel in FAMOUS basalt 527–1–1: Earth and Planetary Science Letters, v. 48, p. 111–123.

Freeman, J., 1996, Mantle-melt and mantle-fluid interactions in supra-subduction zones: Evidence from the Troodos massif, Cyprus [Ph.D. thesis]: Durham, University of Durham, 323 p.

Frey, F.A., Suen, C.J., and Stockman, H.W., 1985, The Ronda high temperature peridotite: Geochemistry and petrogenesis: Geochimica et Cosmochimica Acta, v. 49, p. 2469–2491.

Georgiou, E., 1987, Geology, petrology, and mineral chemistry of the Akapnou Forest plutonic complex and the associated chromite occurrences, Cyprus [Ph.D. thesis]: Orléans, l'Université d'Orléans, 281 p.

Hess, P.C., 1971, Polymer model of silicate melts: Geochimica et Cosmochimica Acta, v. 35, p. 289–306.

Hess, P.C., 1980, Polymerization model for silicate melts, *in* Hargraves, R.B., ed., Physics of magmatic processes: Princeton, Princeton University Press, p. 3–48.

Hill, R., and Roeder, P.L., 1974, The crystallisation of spinel from basaltic liquid as a function of oxygen fugacity: Journal of Geology, v. 82, p. 709–729.

Hock, M., Friedrich, G., Pluger, W.L., and Wichowski, A., 1986, Refractory and metallurgical-type chromite ores, Zambales ophiolite, Luzon, Philippines: Mineralium Deposita, v. 21, p. 190–199.

Irvine, T.N., 1967, Chromian spinel as a petrogenetic indicator. Part 2, petrologic applications: Canadian Journal of Earth Sciences, v. 4, p. 71–103.

Irvine, T.N., 1976, Chromite crystallization in the join Mg_2SiO_4-$CaMgSi_2O_6$-$CaAl_2Si_2O_8$-$MgCr_2O_4$-SiO_2: Carnegie Institution of Washington Year Book, v. 76, p. 465–472.

Irvine, T.N., 1977, Origin of chromite layers in the Muskox intrusion and other intrusions: A new interpretation: Geology, v. 5, p. 273–277.

Irvine, T.N., and Sharpe, M.R., 1986, Magma mixing and the origin of stratiform oxide ore zones in the Bushveld and Stillwater complexes, *in* Gallagher, M.J., Ixer, R.A., Neary, C.R., and Prichard, H.M., eds., Metallogeny of basic and ultrabasic rocks: London, Institution of Mining and Metallurgy, p. 183–198.

Jambon, A., 1994, Earth degassing and large-scale geochemical cycling of volatile elements, *in* Carroll, M.R., and Holloway, J.R., eds., Volatiles in magmas: Mineralogical Society of America Reviews in Mineralogy, v. 30, p. 479–517.

Jaques, A.L., and Green, D.H., 1980, Anhydrous melting of peridotite at 0–15 kb pressure and the genesis of tholeiitic basalts: Contributions to Mineralogy and Petrology, v. 73, p. 287–310.

Johan, Z., 1986, Chromite deposits in the Massif du Sud ophiolite, New Caledonia: Genetic considerations, *in* Chromites, UNESCO's IGCP-197 Project on Metallogeny of Ophiolites: Athens, Theophrastus Publications, p. 311–339.

Kelemen, P.B., Whitehead, J.A., Aharonov, E., and Jordahl, K.A., 1995, Experiments on flow focusing in soluble porous media, with applications to melt extraction from the mantle: Journal of Geophysical Research, v. 100, p. 475–496.

Kurat, G., Palme, H., Spettel, B., Baddenhausen, H., Hofmeister, H., Palme, C., and Wänke., H., 1980, Geochemistry of ultramafic xenoliths from Kapfenstein, Austria: Evidence for a variety of upper mantle processes: Geochimica et Cosmochimica Acta, v. 44, p. 45–60.

Lago, B.L., Rabinowicz, M., and Nicolas, A., 1982, Podiform chromite ore bodies: A genetic model: Journal of Petrology, v. 23, p. 103–125.

Lange, R.A., 1994, The effect of H_2O, CO_2, and F on the density and viscosity of silicate melts, *in* Carroll, M.R., and Holloway, J.R., eds., Volatiles in magmas: Mineralogical Society of America Reviews in Mineralogy, v. 30, p. 331–369.

Leblanc, M., 1995, Chromite and ultramafic rock compositional zoning through a paleotransform fault, Poum, New Caledonia: Economic Geology, v. 90, p. 2028–2039.

Leblanc, M., 1997, Chromite and ultramafic rock compositional zoning through a paleotransform fault, Poum, New Caledonia: reply: Economic Geology, v. 92, p. 503–504.

Leblanc, M., and Ceuleneer, G., 1992, Chromite crystallization in a multicellular magma flow: Evidence from a chromitite dike in the Oman ophiolite: Lithos, v. 27, p. 231–257.

Leblanc, M., and Violette, J.F., 1983, Distribution of aluminium-rich and chromium-rich chromite pods in ophiolite peridotites: Economic Geology, v. 78, p. 293–301.

Liang, Y., and Elthon, D., 1990, Evidence from chromium abundances in mantle rocks for extraction of picrite and komatiite melts: Nature, v. 343, p. 551–553.

Lorand, J.P., and Ceuleneer, G., 1989, Silicate and base-metal sulfide inclusions in chromites from the Maqsad area (Oman ophiolite, Gulf of Oman): A model for entrapment: Lithos, v. 22, p. 173–190.

Maaløe, S., and Aoki, K., 1977, The major element composition of the upper mantle estimated from the composition of lherzolites: Contributions to Mineralogy and Petrology, v. 63, p. 161–173.

Malpas, J., Robinson, P.T., and Zhou, M.-F., 1997, Chromite and ultramafic rock compositional zoning through a paleotransform fault, Poum, New Caledonia: discussion: Economic Geology, v. 92, p. 502–503.

McElduff, B., and Stumpfl, E.F., 1991, The chromite deposits of the Troodos Complex, Cyprus: Evidence for the role of a fluid phase accompanying chromite formation: Mineralium Deposita, v. 26, p. 307–318.

McMillan, P.F., 1994, Water solubility and speciation models, *in* Carroll, M.R., and Holloway, J.R., eds., Volatiles in magmas: Mineralogical Society of America Reviews in Mineralogy, v. 30, p. 131–156.

Melcher, F., Grum, W., Simon, G., Thalhammer, T.V., and Stumpfl, E.F., 1997, Petrogenesis of the ophiolitic giant chromite deposits of Kempirsai, Kazakhstan: A study of solid and fluid inclusions in chromite: Journal of Petrology, v. 38, p. 1419–1458.

Muenow, D.W., Garcia, M.O., Aggrey, K.E., Bednarz, U., and Schmincke, H.U., 1990, Volatiles in submarine glasses as a discriminant of tectonic origin: Application to the Troodos ophiolite: Nature, v. 343, p. 159–161.

Murase, T., and McBirney, A.R., 1973, Properties of some common igneous rocks and their melts at high temperatures: Geological Society of America Bulletin, v. 84, p. 3563–3592.

Murck, B.W., and Campbell, I.H., 1986, The effects of temperature, oxygen fugacity, and melt composition on the behavior of chromium in basic and ultrabasic melts: Geochimica et Cosmochimica Acta, v. 50, p. 1871–1887.

Nicholls, I.A., and Ringwood, A.E., 1973, Effect of water on olivine stability in tholeiites and the production of silica-saturated magmas in the island-arc environment: Journal of Geology, v. 81, p. 285–300.

Nicolas, A., 1989, Structures of ophiolites and dynamics of oceanic lithosphere: Dordrecht, Kluwer Academic Publishers, 367 p.

Paktunc, A.D., 1990, Origin of podiform chromite deposits by multistage melting, melt segregation and magma mixing in the upper mantle: Ore Geology Reviews, v. 5, p. 211–222.

Panayiotou, A., Michaelides, A.E., and Georgiou, E., 1986, The chromite deposits of the Troodos ophiolite complex, Cyprus, *in* Chromites, UNESCO's IGCP-197 Project on Metallogeny of Ophiolites: Athens, Theophrastus Publications, p. 161–198.

Pearce, J.A., and Parkinson, I.J., 1993, Trace elements models for mantle melting: Application to volcanic arc petrogenesis, *in* Prichard, H.M., Alabaster, T., Harris, N.B.W., and Neary, C.R., eds., Magmatic processes and plate tectonics: London, Geological Society Special Publication, v. 76, p. 373–403.

Pearce, J.A., Lippard, S.J., and Roberts, S., 1984, Characteristics and tectonic significance of supra-subduction zone ophiolites, *in* Kokelaar, B.P., and Howells, M.F., eds., Marginal basin geology: London, Geological Society Special Publication, v. 16, p. 77–94.

Pearce, J.A., Van der Laan, S.R., Arculus, R.J., Murton, B.J., Ishii, T., Peate, D.W., and Parkinson, I.J., 1992, Boninite and harzburgite from Leg 125 (Bonin-Mariana forearc): A case study of magma genesis during the initial stages of subduction, *in* Fryer, P., Pearce, J.A., Stokking, L.B., et al., eds., Proceedings of the Ocean Drilling Program, Scientific Results, Leg 125: Texas, College Station, p. 623–659.

Peck, D.C., and Keays, R.R., 1990, Geology, geochemistry, and origin of platinum-group element: Chromitite occurrences in the Heazlewood River Complex, Tasmania: Economic Geology, v. 85, p. 765–793.

Philpotts, A.R., 1990, Principles of igneous and metamorphic petrology: New Jersey, Prentice-Hall, 498 p.

Richet, P., Lejeune, A.-M., Holtz, F., and Roux, J., 1996, Water and the viscosity of andesite melts: Chemical Geology, v. 128, p. 185–197.

Roberts, S., 1988, Ophiolitic chromitite formation: A marginal basin phenomenon?: Economic Geology, v. 83, p. 1034–1036.

Roberts, S., 1992, Influence of the partial melting regime on the formation of ophiolite chromite, *in* Parson, L.M., Murton, B.J., and Browning, P., eds., Ophiolites and their modern oceanic analogues: London, Geological Society Special Publication, v. 60, p. 203–217.

Roberts, S., and Neary, C., 1993, Petrogenesis of ophiolite chromite, *in* Prichard, H.M., Alabaster, T., Harris, N.B.W., and Neary, C.R., eds., Magmatic processes and plate tectonics: London, Geological Society Special Publication, v. 76, p. 257–272.

Robinson, P.T., Zhou, M.-F., Malpas, J., and Bai, W.-J., 1997, Podiform chromitites: Their composition, origin, and environment of formation: Episodes, v. 20, p. 247–252.

Roeder, P.L., and Reynolds, I., 1991, Crystallisation of chromite and chromium solubility in basaltic melts: Journal of Petrology, v. 32, p. 909–934.

Schiano, P., Clocchiatti, R., Lorand, J.-P., Massare, D., Deloule, E., and Chaussidon, M., 1997, Primitive basaltic melts included in podiform chromites from the Oman ophiolite: Earth and Planetary Science Letters, v. 146, p. 489–497.

Sharpe, M.R., and Irvine, T.N., 1983, Melting relations of two Bushveld chilled margin rocks and implications for the origin of chromitite: Carnegie Institution of Washington Year Book, v. 82, p. 295–300.

Sisson, T.W., and Layne, G.D., 1993, H_2O in basalt and basaltic andesite glass inclusions from four subduction-related volcanoes: Earth and Planetary Science Letters, v. 117, p. 619–635.

Stolper, E., and Newman, S., 1994, The role of water in the petrogenesis of Mariana trough magmas: Earth and Planetary Science Letters, v. 121, p. 293–325.

Thayer, T.P., 1960, Some critical differences between alpine-type and stratiform peridotite-gabbro complexes, *in* Sorgenfrei, T., ed., International Geological Congress Report on the 21st Session, Copenhagen: Det Berlingske Bogtrykkeri, p. 247–259.

Toramaru, A., and Fujii, N., 1986, Connectivity of melt phase in a partially molten peridotite: Journal of Geophysical Research, v. 91, p. 9239–9252.

von Bargen, N., and Waff, H.S., 1988, Wetting of enstatite by basaltic melt at 1350 °C and 1–25 GPa pressure: Journal of Geophysical Research, v. 93, p. 1153–1158.

Waters, C., and Boudreau, A.E., 1996, A reevaluation of crystal-size distributions in chromite cumulates: American Mineralogist, v. 81, p. 1452–1459.

Watson, E.B., 1994, Diffusion in volatile-bearing magmas, *in* Carroll, M.R., and Holloway, J.R., eds., Volatiles in magmas: Mineralogical Society of America Reviews in Mineralogy, v. 30, p. 371–411.

Wood, B.J., and Fraser, D.G., 1984, Elementary thermodynamics for geologists: Oxford, Oxford University Press, 303 p.

Zhou, M.-F., and Robinson, P.T., 1997, Origin and tectonic environment of podiform chromite deposits: Economic Geology, v. 92, p. 259–262.

Zhou, M.-F., Robinson, P.T., and Bai, W.-J., 1994, Formation of podiform chromitites by melt/rock interaction in the upper mantle: Mineralium Deposita, v. 29, p. 98–101.

Zhou, M.-F., Robinson, P.T., Malpas, J., and Li, Z., 1996, Podiform chromitites in the Luobusa ophiolite (southern Tibet): Implications for melt-rock interaction and chromite segregation in the upper mantle: Journal of Petrology, v. 37, p. 3–21.

MANUSCRIPT ACCEPTED BY THE SOCIETY FEBRUARY 23, 2000

Printed in the U.S.A.

Geological Society of America
Special Paper 349
2000

Gauging stress from mantle chromitite pods in the Oman ophiolite

Benjamin Holtzman
Department of Geology and Geophysics, University of Minnesota, 310 Pillsbury Drive, Minneapolis, Minnesota 55455, USA; e-mail: holtz007@tc.umn.edu

ABSTRACT

Chromitite pods form in the uppermost mantle at mid-ocean ridges, in locations of large volumes of migrating melt, and deform within the peridotite flowing away from the ridge. At some time after chromite has crystallized, it begins to deform by fracture and possibly creep mechanisms in a matrix deforming mainly by dislocation creep. Numerical modeling of the mechanical interaction of an elastic-plastic chromite inclusion and a viscoelastic dunite matrix suggests that fracture formation in the mono- or polycrystalline chromite inclusion requires a low strength contrast between the matrix and the inclusion. In numerical models, two thresholds of strain rate define two regimes of fracture geometry: below the first strain-rate threshold, the inclusion does not fracture; in the first regime, above the first threshold, a single mid-point fracture occurs; above the second threshold, the deformation is more distributed through the inclusion, because the strength contrast between inclusion and matrix is minimized. Melt present during deformation would lower the effective strengths of both the chromite inclusion and the dunite. The modeling approach may be used to better understand these effects. The existence of thresholds of fracture geometry with varying strength contrast may lend insight to many analogous geologic problems.

INTRODUCTION

The southern part of the Oman ophiolite contains an array of structurally defined diapirs, believed to represent a young ridge propagating into slightly older lithosphere (Boudier et al., 1997). The diapirs deliver large volumes of melt that pond in the Moho transition zone, the predominantly dunite boundary layer of highly variable thickness between the gabbroic lower crust and the peridotite mantle (Jousselin et al., 1998; Nicolas et al., 1996). Thus melt may be delivered to localized centers and then distributed relatively evenly along ridge segments, driven by isostatic and mechanical interactions between the crust and mantle (e.g., Wang and Cochran, 1993). The mechanical interactions are poorly understood because it is difficult to define the rheology and magnitude of magmatic deformation in the lower gabbroic crust as well as the rheology and dynamics of the melt-rich uppermost asthenosphere. This paper describes the initial stages of a study of podiform chromitite bodies as mechanical inhomogeneities and rheological markers in the uppermost asthenosphere; the study aims at improved comprehension of the rheology of this boundary layer.

Chromitite pods form in channels of high melt flow beneath the ridge axis and deform in the outward flow of peridotite (Cassard et al., 1981; Ceuleneer and Nicolas, 1985). As chemical heterogeneities, the chromitite pods provide opportunities for insight to mechanical and petrologic processes of melt migration in the uppermost mantle. Among the only strain markers in peridotite, the chromitite pods record the kinematics of corner flow and the rheological transition from asthenosphere to lithosphere. As mechanical heterogeneities deforming with—but differently from—the surrounding dunite and harzburgite, the chromitite pods are gauges of stress and strain in their immediate physical environment. This paper focuses on the rheological information one can extract from mechanical interactions at a microstructural scale. The largest concentration of chromitite pods (~27) in Oman exists in the Maqsad diapir, which is the source of samples and focus of questions in this study.

The numerical experiments in this study are designed to address the rheological conditions at the very beginning of observable brittle deformation of chromitite pods, which occurs soon after their formation in asthenospheric flow (Ceuleneer and Nicolas, 1985). The juxtaposition of brittle fracture in chromite and dislocation creep in olivine provides rheological constraints on stress levels during deformation. In these small-strain numerical experiments at a microstructural scale, a single elastic-plastic inclusion (composed of mono- or polycrystalline chromite) is embedded in a viscoelastic matrix (dunite). These models are based on the simplest conditions, designed to test the tool, illuminate questions, and guide field observations.

THE CHROMITITE PODS

Pod formation

Pods range in size from ~1 to 300 m; many are ~50 m in length (Nicolas, 1989). They are usually surrounded by dunite envelopes of the same origin as the reactive dunite "veins" believed to represent asthenospheric dikes (Nicolas, 1989) and/or channels of focused reactive melt flow (Kelemen et al., 1997). The low solubility of chromite in silicate liquids requires a large melt to chromite volume ratio, ~300–400:1 (Leblanc and Ceuleneer, 1992). The high flux

Holtzman, B., 2000, Gauging stress from mantle chromitite pods in the Oman ophiolite, *in* Dilek, Y., Moores, E.M., Elthon, D., and Nicolas, A., eds., Ophiolites and Oceanic Crust: New Insights from Field Studies and the Ocean Drilling Program: Boulder, Colorado, Geological Society of America Special Paper 349, p. 149–158.

requirement implies that pod formation is closely related to the dominant mechanism of melt migration through the upper asthenosphere. Lago et al. (1982, based on studies in the New Caledonian ophiolite) and Leblanc and Ceuleneer (1992, from studies in Maqsad) proposed that the pods form in vigorous convection in irregular cavities in asthenospheric mantle dikes. Examples of magmatic textures found in chromitite pods are shown in Figures 1G and 2 and illustrated extensively in Nicolas (1989). There are several, more geochemically oriented hypotheses for pod formation. Irvine (1977) proposed that the mixing of two magmas of different sources could drive melt bulk composition off the olivine-chromite cotectic into the chromite stability field, causing chromite to crystallize as the melt composition returns to the cotectic. This hypothesis was demonstrated experimentally by Ballhaus (1998). Zhou et al. (1994) proposed a similar mechanism caused by melt-rock interaction instead of magma mixing. Pods probably form by a range of chemical-mechanical processes, producing the range of observed textures and morphologies.

Pod deformation

Soon after formation, the pods begin to deform. In the New Caledonian peridotite massifs, Cassard et al. (1981) recognized that pods situated at a high angle to the peridotite foliation ("discordant") are generally undeformed, whereas more deformed pods lie closer to the foliation ("concordant"). Thus the pods form in an orientation determined by the melt-migration channel and then progressively rotate into alignment with asthenospheric shearing, which is defined by peridotite foliations and olivine fabrics (Nicolas, 1989). A similar interpretation of chromite pods in the Maqsad diapir was made by Ceuleneer and Nicolas (1985). In general, the diapir has weaker olivine fabrics and more discordant pods in the center. Fabrics strengthen and pods become more concordant with radial distance (Jousselin et al., 1998; Ceuleneer and Rabinowicz, 1992). Jousselin et al. (1998) refined the interpreted location of the paleoridge axis, providing a geodynamic reference frame for the chromite pod deformation.

The association between olivine fabric strength and fracture in chromite also appears at smaller scales. Ceuleneer and Nicolas (1985) described a particular diffuse and discordant pod that shows well-defined deformation gradients in both the chromitite and the dunite. At the top of the pod, the chromitite is undeformed, and the surrounding dunite has no lattice-preferred orientation (LPO). At the base, the chromitite is moderately deformed by pull-apart fractures, and the dunite has a well-developed (010)[100] dominant slip system, representing a high-temperature deformation (~1200 °C, Nicolas, 1989). This pod illustrates an essential association between high-temperature dislocation creep in olivine and pervasive brittle deformation in chromite. The stress and strain gradients in the surrounding dunite and peridotite describe the mechanical interaction of the pod and its matrix.

The dominant visible deformation mechanism in the chromite grains deformed at high temperature is homogeneously distributed pull-apart fracturing (modes I and II); the fractures form normal to the stretching direction and progressively fill with olivine and/or melt (Cassard et al., 1981; Secher, 1981; Nicolas, 1989; Christiansen, 1985). As illustrated in Figure 2B, these pull-apart fractures appear to follow straight or angular grain boundaries. This style contrasts with fractures associated with serpentinization, which generally cut through grains (Fig. 2A). Temperature-dependent behavior has been observed in fracture experiments of a similar spinel from 25 to 1400 °C (Ghosh et al., 1991), which suggests that the tensile strength of the grain boundary at high temperature is the strength of interest for the numerical models. The experimental constraints on this parameter are described in the next section.

An important question concerns the amount of plastic deformation occurring in the chromite grain preceding or in conjunction with fracturing. Any creep mechanism would relax the stress loading in the inclusion; such relaxation would require compensation in some other parameter, such as an increase in strain rate or melt pressure, to enable the tensile fracture strength to be reached. Mechanisms of plastic deformation in chromian spinel may include diffusion creep (Ozawa, 1989; Christiansen, 1985), dissolution followed by precipitation in the presence of melt, and/or dislocation creep (Secher, 1981). These mechanisms would likely change during the pod's deformation history and may or may not temporally coexist with the dominant fracture mechanism.

As fracture progresses, olivine and/or crystallized melt (mainly plagioclase and clinopyroxene) fills the spaces in the pull-apart fractures preventing void spaces from opening. Whether the olivine enters by solid-state creep or is diffused through the melt is unknown. If melt is present during the initiation of fracture, it could lower the effective confining pressure, aiding fracture, and/or relax stresses at the crack tip by providing a fast diffusion path, deterring fracture. Because the onset of chromite deformation is spatially and temporally associated with dislocation creep in the olivine matrix, the melt percentage present when chromite begins to fracture must be low enough to permit olivine dislocation creep to dominate over melt-aided grain-boundary sliding. For a grain size of 5 mm, a strain rate of 10^{-12}, and a stress of 0.125 MPa, the melt fraction at the dislocation-creep–diffusion creep transition would be about 17%, calculated by equating the general flow laws for dislocation and diffusion creep in Zimmerman (1999).

EXPERIMENTAL DESIGN

The experiments presented in this paper treat the simplest and most definable case of interaction between chromite and dunite: melt plays no mechanical role, and the effective confining pressure is simply lithostatic. The aim is to see whether chromite inclusions could fracture under such conditions and, if so, then how. The finite-element method used in this study allows a juxtaposition of a rate-independent (elastic-plastic "chromite") inclusion with a rate-dependent (visco-elastic "dunite") matrix. When the rheologies of the two phases are defined, stress and strain rate are the two main dependent variables in the system.

Stress loading in composites

Eshelby theory (Eshelby, 1957) and shear-lag theory (Cox, 1952; Ji and Zhao, 1993) quantify stress loading in a rigid inclusion in a weaker deforming matrix and have been applied to mainly elastic materials (both theories are described in Clyne and Withers, 1993). Stress loading occurs when the inclusion is stronger than the matrix; higher stress builds up in the inclusion, and strain is concentrated in the matrix. The shear-lag theory is built on the premise that all the stress loading in the inclusion occurs by shear stresses transferred across the interface. This theory predicts that the tensile deviatoric stresses ought to be inhomogeneous and highest in the middle; thus, inclusions should fracture in their centers, and subsequent fractures should form in the middle of the new segments (for inclusions of uniform thickness). In contrast, the Eshelby theory relies on the assumption and observation that during elastic loading, the stress in an ellipsoidal inclusion is homogeneous. Stress is transferred as a result

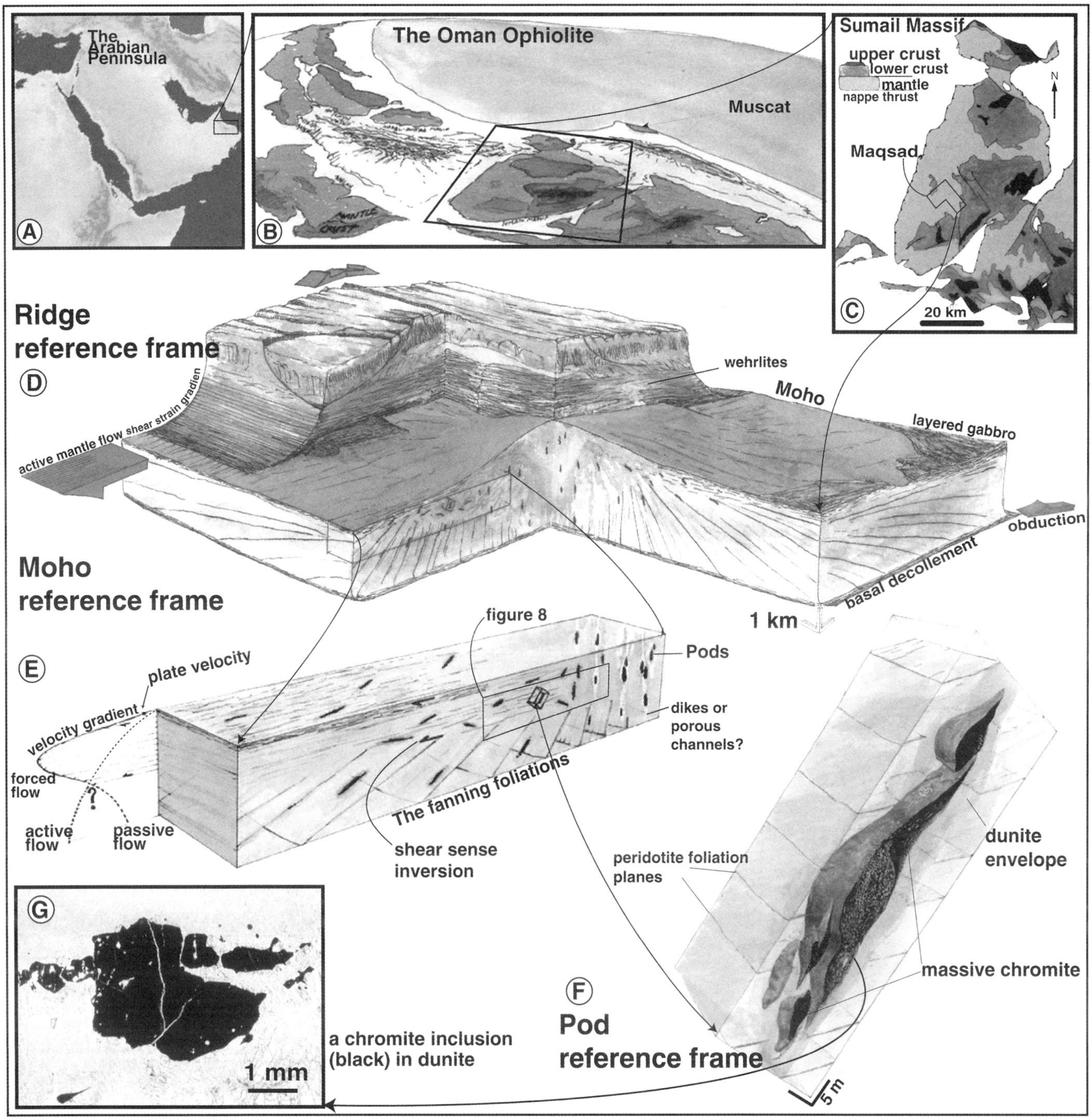

Figure 1. A, The Arabian Peninsula. B, The Oman ophiolite; lighter gray is mantle; darker gray is crust. C, The Semail massif; simplified from the "Oman Ophiolite Lineation Trajectories" map (A. Nicolas and F. Boudier, 1998, personal commun.). The most recent structural map of the Maqsad area may be found in Jousselin et al. (1998), upon which the block diagram was based. D, Schematic block diagram of the structure of the diapir and its relationship to the overlying crust. The diapir is defined by (1) tightly spaced concentric rings of isodip lines, connecting strike lines of peridotite foliation planes with equal dip, steepest in the inner ring and shallowest outward, all dipping away from the center, like nested cones; and (2) lineations radiating from the center of the isodip lines. The sense of shear in peridotites in the diapir just beneath the Moho generally indicates that the mantle material is moving faster than the crust, but switches to passive flow after some distance. The chromitite pods (27 known) are strain markers in this diapiric flow. The topographic bulge of the Moho above the thick transition zone follows Nicolas et al. (1996) and is in accord with seismic observations of thin crust above mantle gravity anomalies on the East Pacific Rise (e.g., Wang and Cochran, 1993). At the diapir scale, active mantle flow may impart strain on the lower crust, across the Moho. The amount of strain depends on relative velocities and the rheology of the transition zone forming the boundary layer between the lower crust and uppermost mantle, where chromitite pods may provide rheology constraints. E, A channel of flow just beneath the Moho. The "forced flow" exists where the shear-sense inversion is recognized; the "active flow" gradient exists near the diapir creating the fanning foliations, but switches to passive flow farther from the diapir (Jousselin et al., 1998). F, The chromitite pod reference frame, a schematic picture of an undeformed discordant chromitite pod. G, A common texture in pods is a single grain or a polycrystalline nodule in dunite. This texture is ideal for the small-scale modeling presented in this chapter.

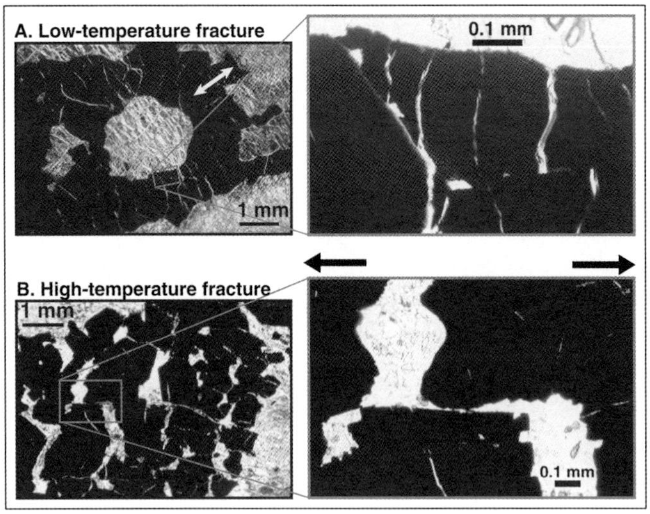

Figure 2. Chromite fracture properties in samples from Maqsad. A, An inclusion of olivine within a chromite inclusion (black) in dunite (partially serpentinized), from an undeformed discordant pod. The white arrow shows the inferred serpentine expansion direction, which is normal to the fractures in the chromite grains. These fractures are caused by the serpentine expansion, a relatively low temperature process. At higher magnification, it appears that the fractures generally cut through the chromite grains, until they hit grain boundaries, suggesting that the lattice and grain-boundary strengths are not very different at low temperature. B, A deformed inclusion of chromite from a moderately deformed pod. The high-temperature deformation predominantly occurs by well-distributed pull-apart fractures that fill with olivine as they open. The fracture walls generally appear angular and smooth, suggesting that they formed on grain boundaries. This suggestion implies that the grain boundary is significantly weaker than the lattice at high temperature. Olivine grains immediately adjacent to the fractured chromite grains shown in B are large (~5 mm), suggesting low stress. However, they have well-defined linear subgrains ~10–30 μm wide, suggesting sustained high stress on the order of 50 MPa, though strains were too low for significant recrystallization and grain-size reduction to occur (Karato et al., 1980).

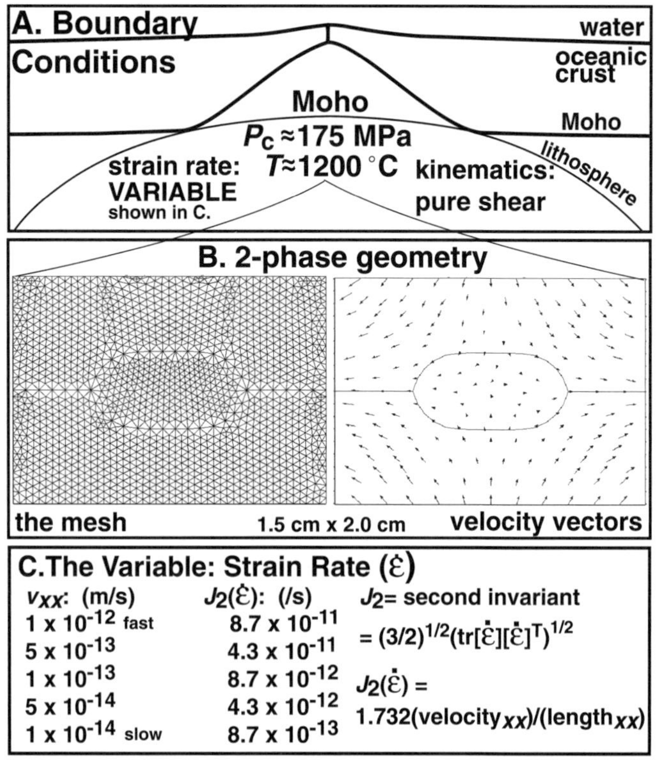

Figure 3. Experimental design. A, Diagram showing the simple boundary conditions used in this experiment. B, The mesh contains ~3000 elements. The velocity vectors reflect pure-shear plane-strain kinematics applied to the walls at varying strain rates shown in C.

of strain misfit or incompatibility, which is a function of shape and elastic strength contrast. Several numerical studies of linear and nonlinear viscous materials have also addressed loading of and interactions between an inclusion and its matrix (e.g., Ten and Yuen, 1999; Kenkmann and Dresen, 1998; Masuda and Ando, 1988). The studies presented here differ in that they juxtapose rate-dependent (viscous) and rate-independent rheologies. The results of numerical experiments presented here agree with both contrasting analytical theories.

The finite-element method

The mechanical analysis is performed by using the geodynamical software ADELI. The finite-element technique corresponds to a material description of the geologic medium originally derived from FLAC code (Cundall and Board, 1988). Meshing is performed with triangular constant-strain elements assembled in pairs to form quadrilateral elements. The main feature of this technique is the use of the dynamic relaxation method, which damps out-of-balance forces and enables an estimation of the quasi-static solution of the mechanical problem. An overview of this method can be found in Board (1989) and Hassani et al. (1997). The boundary conditions are applied to the two-phase system, comprising two simple rheologies in the geometric relationship shown, producing stress and strain gradients in the system.

The boundary conditions (Fig. 3). The confining pressure (P_c = 175 MPa) is assumed to be lithostatic, summing 2 km of water (density $\rho = 1$ kg/m^3), 5 km of crust ($\rho = 2.8$ kg/m^3), and 500 m of mantle ($\rho = 3.3$ kg/m^3). Although thermal calculations are not made in these experiments, the temperature is considered to be 1200 °C at depths of a few kilometers below the Moho for the purpose of defining rheological parameters.

A constant velocity (Voight condition) for each experimental run is applied in pure shear, plane strain to a rectangular box whose contents are described below. Plane strain is used as a first-order simplification and is in general agreement with olivine fabrics and flow models of Oman peridotites (Nicolas, 1989). The velocities, v, yield equal strain rates in the principal directions, as shown in the following description of the applied strain rate for two-dimensional pure shear:

$$[\varepsilon] = \begin{bmatrix} \dot{\varepsilon}_{xx} & 0 \\ 0 & \dot{\varepsilon}_{yy} \end{bmatrix}, \qquad \dot{\varepsilon}_{xx} = v_{xx}/l_x, \qquad \dot{\varepsilon}_{yy} = v_{yy}/l_y, \quad (1)$$

where the x is horizontal and y is vertical in the model reference frame and l is length (l_x = 20 mm, l_y = 15 mm). Though the velocities should change as the length of the walls change in order to conserve area and strain rate, the strains are so small that the area change with pressure is not significant. At the scales of these models, the kinematic boundary conditions (as opposed to applied stresses) are sensible because there are no rigid walls, only moving particles.

The two-phase geometry and size (Fig. 3). The mesh contains a single ellipsoidal inclusion with an axial ratio of ~2:1 (~10 × 5

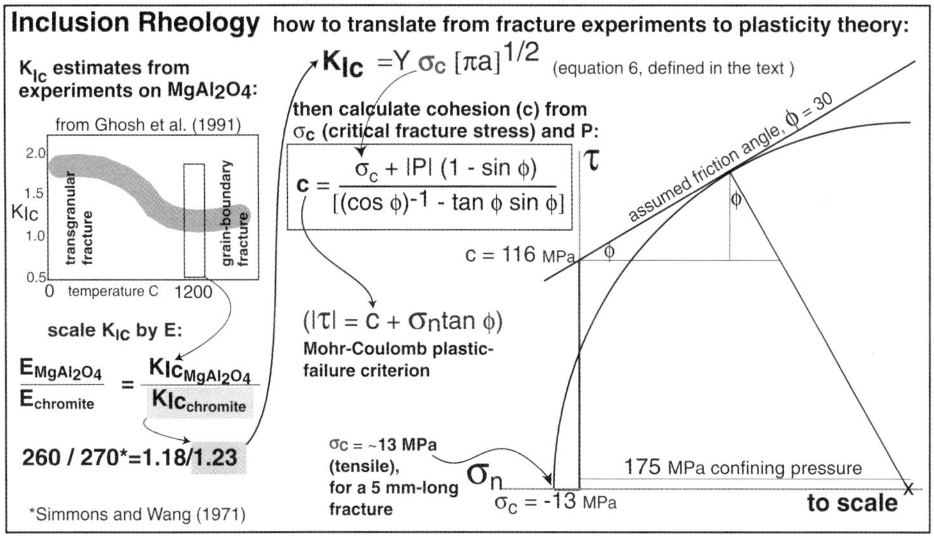

Figure 4. Definition of inclusion cohesion, based on an estimate of fracture strength from experiments, a scaling of K_{Ic} (critical stress intensity factor for mode I cracks) and E (Young's modulus) of the Mg- and Cr-spinel, and a geometric relationship of tensile strength in Mohr space to cohesion via an assumption of confining pressure.

mm), surrounded by matrix in a 20 × 15 mm box. The matrix-inclusion interface is fully coupled, allowing no slip (interfacial slip is a mechanism of stress relaxation, e.g., Kenkmann and Dresen, 1998). The shape is chosen because it represents a common average ratio for unfractured chromite nodules, commonly found at the millimeter to centimeter scale. Shape and size are important in controlling loading: a round inclusion is the strongest whereas a longer, thinner inclusion will fracture more easily; and if two inclusions of different sizes were subject to the same flow field, the larger one would be subjected to a slower strain rate.

Rheology of dunite. The matrix is defined as a linear viscoelastic (Maxwell) material:

$$\sigma = \eta\dot{\varepsilon} - (\eta/E)(d\sigma/dt) \tag{2}$$

where σ = stress (in MPa), t = time (in s), and an estimated viscosity (η) of 5×10^{19} Pa·s is used for dry olivine deforming by dislocation creep at 1200 °C at pressures of the uppermost mantle (D.L. Kohlstedt, 1998, personal communication). The elastic parameters E (Young's modulus) and ν (Poisson's ratio) are estimated from an average of values from Simmons and Wang (1971): E = 200 GPa and ν = 0.28. For every experiment, an equivalent "blank" is run, a box of matrix without the inclusion, as a control for values disturbed by the inclusion and to measure the non-steady-state "background" stress.

Rheology of the "chromite" inclusion. The inclusion has an elastic-plastic pressure-dependent rheology that allows simulation of brittle fracture as localized deformation. The elastic part of the loading path is described by Hooke's law. The elastic parameters used for chromite are E = 275 GPa and ν = 0.28 (averaged from Simmons and Wang, 1971). The Drucker-Prager plastic yield surface is identical at a point in time to the Mohr-Coulomb failure criterion (τ),

$$\tau = -\tan(\phi)\sigma_n + c, \tag{3}$$

where ϕ = angle of internal friction, σ_n = normal stress, and c = cohesion. The plastic flow law is

$$G(\sigma) = J_2'(\sigma) + \alpha_p J_1(\sigma), \tag{4}$$

where $\alpha_p = (6 \sin \psi)/(3 - \sin \psi)$ and J_1 is the mean stress or pressure. J_2' is the second invariant of the deviatoric stress tensor:

$$J_2' = (3/2)^{1/2}\|\mathrm{dev}(\sigma)\|, \tag{5}$$

where $\|\mathrm{dev}(\sigma)\| = (3/2)^{1/2}(\mathrm{tr}[\mathrm{dev}(\sigma)][\mathrm{dev}(\sigma)]^T)^{1/2}$, tr = trace, the superscript T is the transposed matrix, and

$$\mathrm{dev}(\sigma) = \begin{bmatrix} \sigma_1 - P & 0 \\ 0 & \sigma_3 - P \end{bmatrix},$$

where P = pressure. A nonassociated plastic flow law is used, meaning that the dilatancy angle ψ (60°) does not equal ϕ (30°). Physically, ψ represents the amount of volumetric strain that results from a shear strain applied to a material. The unusually large dilatancy angle enables the formation of localized shear bands in the continuum mesh, in the orientation of a tensile (mode I) fracture, but produces strain hardening in the dilating elements in the shear band.

The cohesion is defined from experiments of fracture strength. The connection of fracture mechanics and loading theory is based on Eidelman and Reches (1992) and Y. Leroy and F. Lehner (1998, personal communication). Fracture mechanics defines a critical stress at which fracture occurs (Lawn, 1993):

$$K_{Ic} = Y\sigma_c(\pi a)^{1/2}, \tag{6}$$

where K_{Ic} (in units of MPa·m$^{1/2}$) is the "fracture toughness" for mode I fractures, σ_c is the critical stress required to open a given fracture, and a is the half length of the fracture. Y is a constant added to correct for stress changes associated with certain geometries of initial notches or crack paths. (Y = 1.12 for a throughgoing crack across a body; Lawn, 1993). If the K_{Ic} value and the crack length are known, then equation 6 gives the critical stress. At high temperature and slow strain rates, it is possible that strain-rate–dependent subcritical fracture could occur and thus lower the tensile failure strength. For simplicity, critical fracture is assumed. Figure 4 shows a simple relationship between σ_c and the cohesion value in the Mohr-Coulomb fracture criterion. This relationship creates a necessary quantitative link between experiments and the numerical models.

K_{Ic} in this model is estimated from values of a similar spinel, $MgAl_2O_4$, a material that is well studied owing to its applications in high-temperature optics and electronics. Structurally, it is very similar to chromite, which has Cr as well as Al in the Al site and Fe as well as Mg in the Mg site (Leblanc and Ceuleneer, 1992). Ghosh et al. (1991) tested the fracture behavior of $MgAl_2O_4$ at room temperature to 1400 °C, measuring K_{Ic} of a fine-grained (average 35 μm) hot-pressed transparent aggregate, with a Young's modulus of 258 GPa.

Ghosh et al. observed a nonlinear descent of K_{Ic} from 0 to ~800 °C, flattening between 800 and 1400 °C, shown qualitatively in Figure 4. They attributed the drop in K_{Ic} to a decrease in the Young's modulus with increasing temperature and the flattening to nonlinear inelastic processes in the wake region of the crack. Such processes may include dislocation activity or microfractures splaying off the main crack, which would relax stress and stabilize the fracture.

The decrease in K_{Ic} accompanies a gradual transition from mostly transgranular (through the lattice) fractures at lower temperatures to intergranular (grain-boundary) fractures at higher temperatures, similar to the transition in chromite shown in Figure 2. Therefore, the K_{Ic} value given for 1200 °C (~1.2 MPa·m$^{1/2}$) reflects the strength of the grain boundaries. This value is applied to the model chromite, scaling the ratio of K_{Ic} values to the ratio of the Young's modulus for spinel to that of chromite, simply by assuming that both moduli reflect the same atomic-scale properties [$(K_{chromite} = K_{Mg\text{-}spinel}(E_{chromite}/E_{Mg\text{-}spinel})$]. As shown in Figure 4, the scaled value for chromite is 1.23 MPa·m$^{1/2}$, small relative to material and experimental variability), which gives a tensile yield stress (σ_c) of 13 MPa for the 10-mm-long inclusion (with a 5-mm-long fracture). A confining pressure of 175 MPa yields a cohesion of 116 MPa.

RESULTS

A range of strain rates is applied to the system, each strain rate constituting a different experiment. From a given experiment, the recorded parameters include pressure, deviatoric stress, and number of elements in the inclusion that have reached the plastic yield surface. In general, the stress in the inclusion is homogeneous, so most of the elements reach the plastic yield surface at the same time. Then, after some duration (a plateau), dilatancy begins to occur in some elements; stress builds in those elements and relaxes in others, and the deformation localizes. Where the localization occurs reflects the stresses in the inclusion; it is not predetermined in the code. When localization occurs, stress values become less meaningful and are ignored, for reasons described after equation 5. For simplicity, the critical stress is considered to be reached when the bulk (>80%) of the elements in the inclusion reach the failure envelope. However, each experiment continues until the deformation localizes. To describe the effect of strain rate on the system, the curves in Figure 6 illustrate qualitatively the deformation history for the range of strain-rate experiments, followed by the more quantitative "amplification" curves.

Experiment at a strain rate of 4.3 × 10^{-12} s^{-1}

Failure occurs in the inclusion at a minimum strain rate of 4.3 × 10^{-12} s^{-1}. Figure 5 illustrates the deviatoric stress gradients in the mesh, velocity vectors, and deviatoric stress vectors produced in this experiment, at a time step just before plastic failure begins. The highest deviatoric stress in the matrix is coincident with the principal stress and strain axes of the system, orthogonal to the long and short axes of the inclusion. Matrix stress is also highest in contact with the "corners" of the inclusion where the velocity vectors are tangent to the interface and the shear stress contribution to the deviatoric stress tensor is largest. Pressure gradients exist, lowest at the ends of the inclusion and highest on the long faces, as expected from the inclusion's disturbance of the particle paths. (See Ozawa, 1989, for a possible compositional expression of these gradients.) However, the volumetric strains are negligible. With progressive deformation, stress increases in the inclusion and the matrix until the plastic yield surface

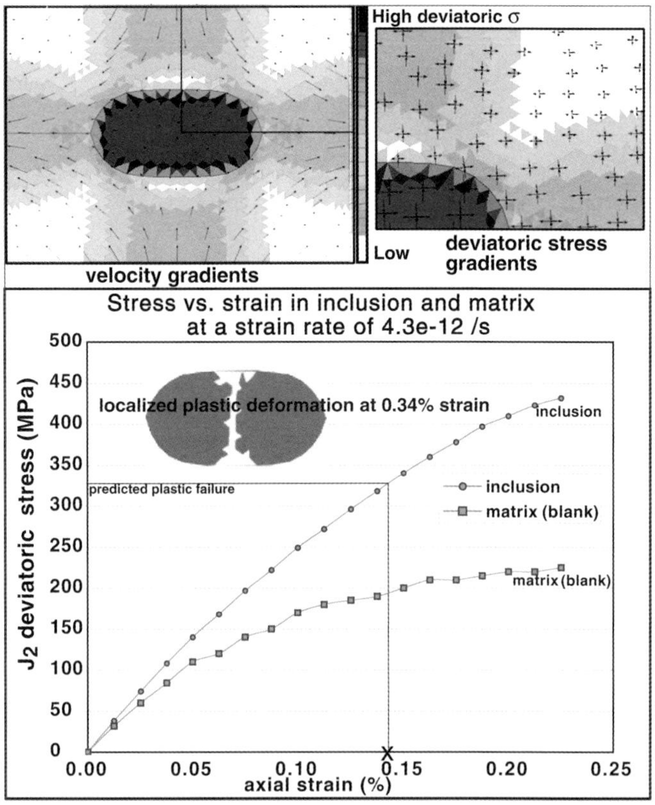

Figure 5. The results from the experiment run at a strain rate of 4.3 × 10^{-12} s^{-1}. These curves stop before localization begins at a strain corresponding to the peak number of elements in which the stress has reached the plastic yield surface. However, the geometry of localization, a single mid-point band, is shown. The x marks the onset of plastic deformation (see Fig. 6), and the corresponding stress provides an analytical test explained in the Results section.

is reached and plastic deformation begins. After about 80% of the elements in the inclusion are deforming plastically, the deformation localizes into a single shear band with the orientation of a mode I, or tensile, fracture.

As a simple test for numerical reliability, the onset of plastic deformation in the inclusion should agree with an analytical prediction. The J_2 of the deviatoric stress at which an individual element should reach plastic failure is 326 MPa (at 175 MPa confining pressure), calculated analytically. This value occurs as the average stress in the inclusion at a strain applied to the whole system of ~0.14% (marked with an x in Figs. 5 and 6). As shown in Figure 6, the first element in the inclusion fails plastically at ~0.16% strain in the 4.3 × 10^{-12} s^{-1} experiment, in agreement with the prediction.

The plastic yield in the inclusion (Fig. 6)

The curves in Figure 6 illustrate the percent of elements in the inclusion that is deforming plastically at a given system strain (= strain rate × time). The geometries of "fracture" (localized plastic strain) are also shown, for a time step after the curve dives down. In the 8.7 × 10^{-13} s^{-1} experiment, run to 25% strain, no elements ever reach the yield surface, well after the system has come to a steady state. In the next two higher strain rates (4.3 × 10^{-12} and 8.7 × 10^{-12} s^{-1}), the inclusions reach their peak number of plastified elements and then quickly localize single "mode I" shear bands in the center, with rela-

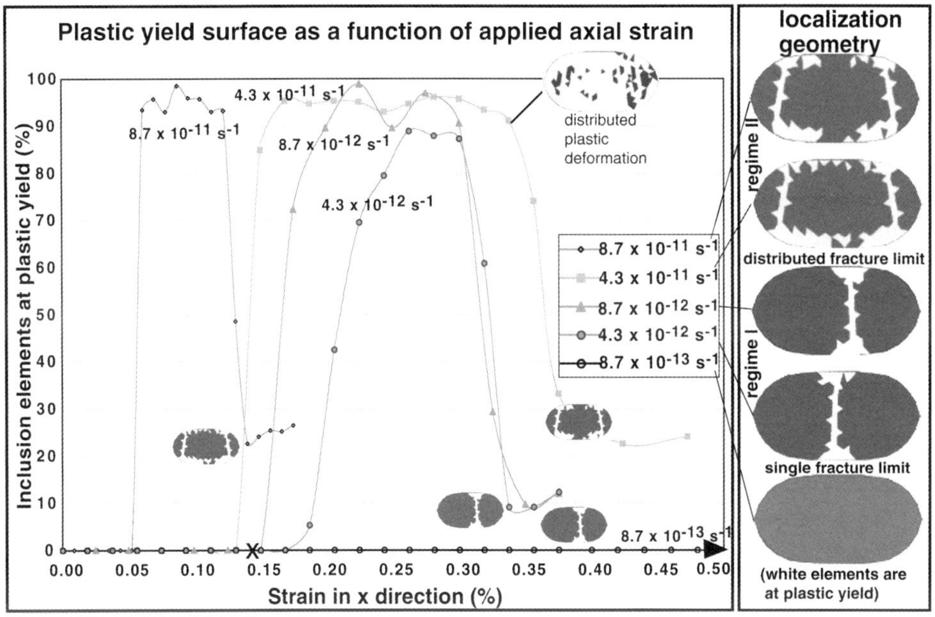

Figure 6. These curves illustrate the percent of elements in the inclusion that is deforming at the plastic yield surface as a function of strain, for each strain rate experiment. They describe qualitatively the effect of strain rate on deformation history of the inclusion. The x corresponds to a stress shown in Figure 5.

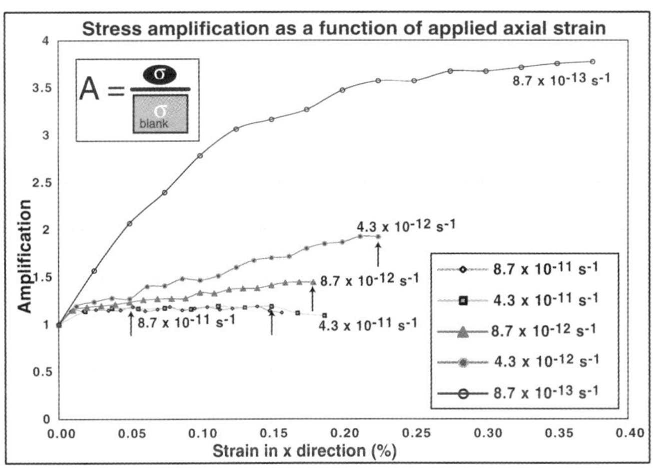

Figure 7. The amplification curves. Amplification, *A*, is the ratio of the homogeneous deviatoric stress in the inclusion to the deviatoric stress in a matrix (the "blank") deforming at the same strain rate without an inclusion, the "background" stress. The vertical arrows mark where the plastification curve reaches its plateau, where almost all the elements have reached the critical failure stress and a real inclusion would have fractured.

tively short plateaus of well-distributed plastic deformation. This result corresponds to the prediction of the shear-lag theory of mid-point fracture and represents a state of matrix-inclusion interaction in which the inclusion controls the geometry of its stresses and, thus, of its failure. The mid-point "fracture" determines a minimum strain rate capable of causing fracture, corresponding to the first limit shown in Figure 6.

At strain rates of 4.3×10^{-11} and higher, another limit is defined. The inclusions undergo relatively long (compared to the elastic loading time) and well-defined plateaus of distributed deformation and then localize suddenly. The fractures occur toward the ends of the inclusions and correspond to the locations of highest deviatoric or shear stress in the matrix, where the velocity vectors are tangential to the side of the inclusion. At these stress and strain-rate levels, the matrix appears to control the location of fracture in the inclusion. With diminishing strength contrast, the two materials behave more similarly, and the deformation becomes more evenly distributed through the inclusion, expressed as long plastification plateaus and double fractures. The system becomes more mechanically homogeneous, even when the rheologies are very different in the two materials. This behavior appears to correspond to the deformation style in brittlely deformed chromite.

The amplification curves (Fig. 7)

The strength contrast is quantified by the "amplification" (*A*) factor, which is the ratio of inclusion stress to the stress in the blank (i.e., the stress that would exist if there were no inclusion). The slowest-strain-rate experiment (8.7×10^{-13} s^{-1}) reached a steady state at an amplification of ~4. This experiment was run to 25% strain, and the amplification did not rise after the blank reached steady-state stress. A strain rate of 4.3×10^{-12} s^{-1} amplifies the stress in the inclusion by a factor of ~1.9 at the time of fracture. Thus, the maximum amplification value occurs at which the inclusion will fracture. A 50% faster strain rate (8.7×10^{-12} s^{-1}) produces similar behavior, reaching an amplification of ~1.5. The two strain rates of 10^{-12} magnitude produce fracturing after the relaxation time, t_r, of the matrix at ~10 yr for 8.7×10^{-12} s^{-1} and ~23 yr for 4.3×10^{-12} s^{-1} ($t_r = \eta/E = (5 \times 10^{19}\ \text{Pa·s})/(2 \times 10^{11}\ \text{Pa}) = \sim 7.9$ yr.)

At increasing strain rates, the two fastest-deforming systems reach amplifications of ~1.2 and then descend toward 1. These curves descend as the plastic deformation begins to relax stress in the inclusion and possibly unloads into the matrix. (After fracture, the unloading would probably be much larger in nature than in a dilating mesh, so the amplification would drop below 1.) For these two rates, yielding occurs more quickly (~1–5 yr) than the matrix relaxation time. Thus the elastic properties of the matrix dominate its behavior; the fact that the matrix *E* is smaller than the inclusion *E* explains why the amplification is positive. These higher strain rates represent a state in which the strength contrast is negligible and in a natural system could dynamically fluctuate about $A = 1$.

DISCUSSION

The meaning of two regimes

These experiments show that there is a minimum strain rate that must be exceeded for fracturing to occur. At the slowest experimental strain rate, the flowing matrix cannot transmit enough stress or strain to the inclusion to reach the plastic yield stress. Above the first strain-rate threshold, a single midpoint "fracture" occurs (regime I) suggesting that the stress is distributed within the inclusion as proposed in the shear-lag theory. At higher strain rates, a second threshold is reached, marking a transition to more distributed deformation (regime II). It is useful to deconstruct these results into a qualitative discussion of fracture simulation and deformation geometry and a quantitative discussion of stress gradients.

On the simulation of fracture. The numerical method employed in these experiments is an unusual use of nonassociated plasticity theory. Defining the dilatancy angle (60°) to be larger than the friction angle (30°) causes a shearing element to dilate more than it shears. This property causes significant strain hardening in the dilating shear zone, the exact opposite of what should happen in a natural mode I fracture. For this reason, stress values are meaningful only until the onset of plastic deformation and are plotted only up to the very beginning of localization. However, this method does produce a geometry of localization that is more true to natural processes than the concurrent magnitude of stress. The dilation vector of each element aligns with the principal tensile stress, and the localized band generally aligns with that of a true mode 1 fracture, reflecting the consistent behavior of each element. Therefore, the transition from single midpoint band to double end bands probably represents a natural process of distributed deformation increasing with diminishing strength contrast.

These numerical results may be qualitatively understood in terms of the Eshelby and shear-lag theories. In regime I (single midpoint fracture), the stress in the inclusion is homogeneous until the onset of plastic deformation. Thus, the fundamental assumption of Eshelby theory that stress is homogeneous in an elastic ellipsoid is reproduced numerically. The shear-lag theory predicts that inhomogeneous tensile stresses will be highest in the center, a prediction that is reproduced after the plastic flow law dominates the behavior and localization begins. Thus, in stages through the deformation history, these experiments agree with both apparently contrasting theories.

Regime II is a little less straightforward in terms of these theories. The double-end bands essentially connect the points of highest deviatoric shear stress and strain in the flowing matrix, traversing the inclusion. Thus, the matrix is determining where the strain localizes in the inclusion. This result is not predicted by the shear-lag theory, because the assumption that the inclusion is "stronger" is less valid as the amplification decreases toward unity. However, this behavior is consistent with the Eshelby concept of "strain misfit": the inclusion is behaving in such a way as to minimize the misfit between the two materials.

A more quantitative view of stress gradients

This study predicts that stresses in the matrix should rise above background values in the vicinity of the stronger inclusion. This prediction deviates from a common assumption for mantle mechanics, which is that stress is relatively constant and strain rate varies. The important question is how the magnitudes and shapes of stress gradients compare to stress estimates from natural microstructures. First I discuss the discrepancies in model and natural background stresses, then the stress gradients around the inclusion.

The model prediction for the minimum background deviatoric stress in the matrix required to cause plastic failure is 127 MPa ($J_2{}' =$ 220 MPa = 1.225 [2(dev σ)]$^{1/2}$), at a strain rate of 4.3×10^{-12} s^{-1}. For regime II (distributed deformation), this background stress would be higher. If this strain rate were applied to a matrix with $n = 3.5$ under the same conditions, the steady-state background stress would be ~7 MPa, at 1200 °C and 175 MPa pressure, according to the dry dislocation-creep flow law in Hirth and Kohlstedt (1996). The background stress in the upper asthenosphere at the ridge is estimated at <10 MPa, from deformation-mechanism maps in Karato et al. (1986) if mantle strain rates are slower than 1×10^{-10} s^{-1} and <1 MPa from grain-size paleopiezometers applied to Maqsad peridotite by Ceuleneer and Rabinowicz (1992). This suggests that the strain rate is reasonable, and the stress exponent is very important in studying stress loading.

If failure were to occur in the inclusion in a numerical experiment with a power-law matrix rheology (stress exponent $n = 3.5$) at the same strain rate, then the stress-amplification factor would be ~18, as opposed to 1.9 for the Newtonian matrix. In a natural sample from Maqsad, as described in the caption to Figure 2B, stress estimates from subgrain wall spacing in olivine immediately adjacent to chromite are high, on the order of 50 MPa (using the paleopiezometers defined by Karato et al., 1980). This implies that dislocation creep was the dominant deformation mechanism, so $n = 3.5$ is reasonable, and the amplification from the background stress was ~5–10. The discrepancy between model predictions and natural estimates calls for studies that explore loading behavior as a function of the stress exponent in the matrix.

The shape of the stress gradients, i.e., the curvature and the distance from the border of the pod to where the matrix is no longer subject to the pod, will depend highly on the stress exponent in the flow law that best describes the natural deformation mechanism. Numerical experiments with power-law matrixes have shown that deformation is more localized around the inclusion for non-Newtonian than for Newtonian materials (Ten and Yuen, 1999; Masuda and Ando, 1988). The kinematics of matrix flow, i.e., the degree of noncoaxiality, and also interactions with other inclusions will change the shape of stress gradients.

If a discrepancy is defined between power-law matrix models and paleopiezometric estimates of stress gradients in natural samples, an obvious culprit is intergranular melt. Such melt would lower the effective confining pressure, lowering the deviatoric stress that the matrix must load into the inclusion. If melt played a role, then strain rate could not be estimated independently, because the effective confining pressure is not known. Furthermore, reducing the effective confining pressure increases the relative importance of chromite tensile-strength variations caused by plastic creep, subcritical fracture, stoichiometric and nonstoichiometric (defect) lattice properties, melt-assisted crack-tip blunting, and other mechanisms of stress relaxation. Finally, if melt were always present as films on grain boundaries (if basaltic melt on chromite is highly wetting), then the tensile strength of those boundaries would be minuscule.

Implications and further studies

A rearrangement of the assumed knowns and unknowns enables a potential estimate of melt pressure (G. Hirth, 1998, personal commun.). The approach begins with observations of natural microstructures and inferences of local stress using experimentally defined pa-

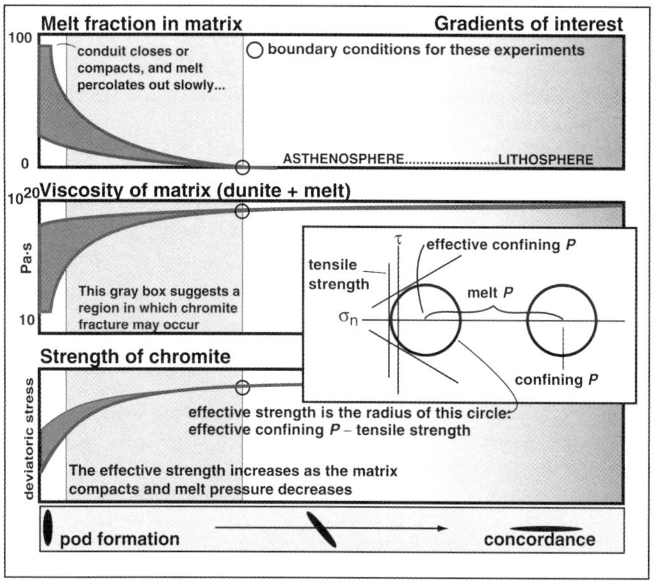

Figure 8. This figure illustrates several speculative imposed gradients that affect the deformation of chromite. The horizontal axis, marked by the formation and deformation and rotation of a pod, is the equivalent to strain, time, or distance from the center of the diapir, quantities that will vary with radial direction. Progressing from left to right (in a direction normal to the ridge axis) also signifies the transition from asthenosphere to lithosphere. The vertical gray box defines a hypothetical region in which conditions would favor the initiation of chromite fracture. At the left of the box, melt has the largest influence, weakening the matrix and reducing the effective pressure on the chromite. At the right, melt fraction has significantly decreased. These experiments are designed for the conditions at the far right of the box marked by the circles. Defining these gradients may aid in the extrapolation of future pod studies to larger-scale geodynamic questions.

leopiezometers to determine a stress gradient around a chromite body. With an estimate of the stress gradient, an inference of the deformation mechanisms to determine flow-law parameters, and an assumption or estimate of temperature, the flow law will provide an estimate of strain rate. Numerical models run at that strain rate should be able to match the observed stress gradient. Then, the homogeneous deviatoric stress that the model produces in the inclusion is taken to be the stress at which it fractured. The tensile strength of the inclusion is calculated analytically. In Mohr space, when the deviatoric stress circle is pinned at σ_3 to this critical tensile-fracture stress, the mean stress is given. As shown in Figure 8, the difference between this mean stress in the inclusion and the lithostatic pressure is the melt pressure. Thus, a tool intended for estimating strain rate could become a tool for estimating melt pressure.

Any estimate of melt pressure in the upper mantle would be of interest in melt-migration models. The differences among the debated dominant mechanisms of melt migration, dikes, focused porous flow, and/or porosity waves, are essentially related to the relationships between the strength of the solid matrix and the melt driving forces. The approach outlined herein offers a possibility, albeit complex and indirect, of defining these relationships at a moment in a dynamic process.

Chromitite pods provide many opportunities to estimate rheological conditions in the upper mantle and Moho transition zone in the context of ridge dynamics. The changing physical conditions to which a pod is subjected from formation to concordance may be described by a series of imposed and related gradients (Fig. 8). These curves are conceptual and speculative, intended to show the processes of which the pods are spatial and temporal markers. The rheological evolution of the system may be grossly imagined as follows: If the pod forms in an asthenospheric melt conduit, its initial matrix may be a melt suspension in a state of overpressure. The chromite may undergo granular flow and/or plastic deformation in a near-subsolidus state, flowing during compaction of the conduit around the chromitite body. Trapped melt probably still remains, but the matrix viscosity is rapidly increasing to that of a melt-rich peridotite. For the solid matrix to remain intact, the melt fraction should be less than ~20% (McKenzie, 1984). At <10% melt fraction, matrix viscosity can be estimated from the work of Hirth and Kohlstedt (1995a, 1995b).

CONCLUSIONS

These numerical experiments on the mechanical interaction of an elastic-plastic inclusion and a viscoelastic matrix show that such systems are potentially useful stress gauges because the inclusion failure geometry provides a constraint on stress loading. The numerically produced geometries of localized plastic deformation suggest that there may be two regimes: if mid-point fracture occurs, then the lower strain-rate limit provides the estimate; if the deformation is evenly distributed throughout the inclusion, then the upper strain-rate limit corresponding to regime II is the minimum estimate. The fact that the fractures in a chromite nodule or pod are evenly distributed suggests that the strength contrast between inclusion and fracture was minimal (~1). The relationship of increasingly distributed deformation with decreasing strength contrast has conceptual value for a very wide range of geologic situations: fractured pebbles in an unlithified matrix (e.g., Eidelman and Reches, 1992), rigid porphyroblasts in mylonite (e.g., Masuda and Kuriyama, 1988; Kenkmann and Dresen, 1998), or rigid blocks surrounded by networks of weaker material at any scale.

These experiments are not intended to offer meaningful strain-rate estimates for the following reasons: (1) The model matrix uses a linear stress-strain relationship, but the observed dislocation creep is known to be nonlinear. (2) The method of simulating fracture does not produce meaningful stress estimates for the two regimes. The most valuable result of these experiments is the existence of the transition of single to distributed "fracture" with diminishing strength contrast. (3) Melt probably plays an important role in the chromite fracture process, introducing a second independent variable. Melt pressure may be responsible for diminishing the strength contrast between olivine and chromite by lowering the effective confining pressure and thus the deviatoric stress required to fracture the inclusion. To further explore the role of melt, models may be used to simulate natural stress gradients in olivine around chromite inclusions, estimated by paleopiezometers. Then, rheological constraints from the small-strain models may be applied to melt-migration questions and larger-scale, larger-strain problems.

ACKNOWLEDGMENTS

I thank the people who have guided this study: Jean Chery, an author of the finite-element code ADELI, for greatly contributing to the experimental design; Christian Teyssier, Shun Karato, David Kohlstedt, and Mark Zimmerman for criticism and guidance; the members of the Laboratoire de Tectonophysique in Montpellier, France, for the initial ideas, motivation, and pod samples; Yves Leroy, Florian Lehner, Greg Hirth, Peter Kelemen, Jan Tullis, Alexei

Poliakov, Lamaar Halverson, Basil Tikoff, and others for discussion and advice; and David Jousselin and two anonymous reviewers for thoughtful and thorough reviews. The numerical experiments were performed at the Minnesota Supercomputer Institute under National Science Foundation grant CDA-9502979, and financial support was generously provided by the Department of Geology and Geophysics, University of Minnesota.

REFERENCES CITED

Board, M., 1989, FLAC (Fast Lagrangian Analysis of Continua) Version 2.20: Software Summary: Minneapolis, Itasca Consulting Group, Inc., v. 1.

Boudier, F., Nicolas, A., Ildefonse, B., and Jousselin, D., 1997: EPR microplates, a model for the Oman Ophiolite: Terra Nova, v. 9, p. 79–82.

Cassard, D., Nicolas, A., Rabinovitch, M., Moutte, J., Leblanc, M., and Prinzhofer, A., 1981, Structural classification of chromite pods in southern New Caledonia: Economic Geology, v. 76, p. 805–831.

Ceuleneer, G., and Nicolas, A., 1985, Structures in podiform chromite from the Maqsad district (Sumail ophiolite, Oman): Mineralium Deposita v. 20, p. 177–185.

Ceuleneer, G., and Rabinowicz, M., 1992. Mantle flow and melt migration beneath oceanic ridges: Models derived from observations in ophiolites, *in* Phipps Morgan, J., Blackman, D.K., and Sinton, J.M., eds., Mantle flow and melt generation at mid-ocean ridges: Washington, D.C., American Geophysical Union, Geophysical Monograph 71, 361 p.

Christiansen, F.G., 1985, Deformation fabric and microstructures in ophiolitic chromitites and host ultramafics, Sultanate of Oman: Geologische Rundschau v. 74, n.1, p. 64–76.

Clyne, T.W., and Withers, P.J., 1993, An introduction to metal matrix composites: Cambridge, UK, Cambridge University Press, 509 p.

Cox, H.L., 1952, The elasticity and strength of paper and other fibrous materials: British Journal of Applied Physics, v. 3, p. 73–79.

Cundall, P.A., and Board, M., 1988, A microcomputer program for modeling large-strain plasticity problems: Proceedings of the 6th International Congress on Numerical Methods in Geomechanics, Rotterdam, Swoboda, p. 2101–2108.

Eidelman, A., and Reches, Z., 1992, Fractured pebbles: A new stress indicator: Geology, v. 20, p. 307–310.

Eshelby, J.D., 1957, The determination of the elastic field of an ellipsoidal inclusion, and related problems: Proceedings of the Royal Society of London, v. A241, p. 376–396.

Ghosh, A., White, K.W., Jenkins, M.G., Kobayashi, A.S., and Bradt, R.C., 1991, Fracture resistance of a transparent magnesium aluminate spinel: Journal of the American Ceramic Society, v. 74, n. 7, p. 1624–1630.

Hassani, R., Jongmans, D., and Chery, J., 1997, Study of plate deformation and stress in subduction processes using two-dimensional numerical models: Journal of Geophysical Research, v. 102, n. B8, p. 17951–17965.

Hirth, G., and Kohlstedt, D.L., 1995a, Experimental constraints on the dynamics of the partially molten upper mantle: Deformation in the diffusion creep regime: Journal of Geophysical Research, v. 100, p. 1981–2001.

Hirth, G., and Kohlstedt, D.L., 1995b. Experimental constraints on the dynamics of the partially molten upper mantle. 2. Deformation in the dislocation creep regime: Journal of Geophysical Research, v. 100, n. B8, p. 15441–15449.

Hirth, G., and Kohlstedt, D.L., 1996, Water in the oceanic upper mantle: Implications for rheology, melt extraction, and the evolution of the lithosphere: Earth and Planetary Science Letters, v. 144, p. 93–108.

Irvine, T.N., 1977, Origin of chromitite layers in th Muskox intrusion and other intrusions: a new interpretation: Geology, v. 5, p. 273–277.

Ji, S., and Zhao, P., 1993, Location of tensile fractures within rigid-brittle inclusions in a ductile flowing matrix: Tectonophysics, v. 220, p. 23–31.

Jousselin, D., Nicolas, A., and Boudier, F., 1998, Detailed mapping of a mantle diapir below a paleo-spreading center in the Oman ophiolite: Journal of Geophysical Research, v. 103, n. B8, p. 18153–18170.

Karato, S.-I., Toriumi, M., and Fujii, T., 1980, Dynamic recrystallization of olivine single crystals during high temperature creep: Geophysical Research Letters, v. 7, n. 9, p. 649–652.

Karato, S., Paterson, M.S., and Fitzgerald, J.D., 1986, Rheology of synthetic olivine aggregates: Influence of grain size and water: Journal of Geophysical Research, v. 19, n. B8, p. 8151–8176.

Kelemen, P.B., Hirth, G., Shimizu, N., Speigelman, M., and Dick, H.J.B., 1997, A review of melt migration processes in the adiabatically upwelling mantle beneath oceanic spreading ridges: Philosophical Transactions of the Royal Society of London, v. A355, p. 283–318.

Kenkmann, T., and Dresen, G., 1998. Stress gradients around porphyroclasts: Palaeopiezometric estimates and numerical modelling: Journal of Structural Geology, v. 20, n. 2/3, p. 163–173.

Lago, B.L., Rabinowicz, M., and Nicolas, A., 1982, Podiform chromite ore bodies: A genetic model: Journal of Petrology, v. 23, n. 1, p. 103–125.

Lawn, B., 1993, Fracture of brittle solids, Cambridge, UK, Cambridge University Press, 378 p.

Leblanc, M., and Ceuleneer, G., 1992, Chromite crystallization in a multicellular flow: Evidence from a chromitite dike in the Oman ophiolite: Lithos, v. 27, p. 231–257.

Masuda, T., and Ando, S., 1988, Viscous flow around a rigid spherical body: A hydrodynamical approach: Tectonophysics, v. 148, p. 337–346.

Masuda, T., and Kuriyama, M., 1988, Successive "mid-point" fracturing during micro-boudinage: An estimate of the stress-strain relation during a natural deformation: Tectonophysics, v. 147, p. 171–177.

McKenzie, D.P., 1984, The generation and compaction of partially molten rock: Journal of Petrology, v. 25, n. 3, p. 713–765.

Nicolas, A. 1989, Structures of ophiolites and dynamics of oceanic lithosphere: Dordrecht, Netherlands, Kluwer Academic Publishers, 367 p.

Nicolas, A., Boudier, F., and Ildefonse, B., 1996, Variable crustal thickness in the Oman ophiolite: Implications for oceanic crust: Journal of Geophysical Research, v. 101, n. B8, p. 17941–17950.

Ozawa, K., 1989, Stress-induced Al-Cr zoning of spinel in deformed peridotites: Nature, v. 338, no. 6211, p. 141–144.

Ramberg, H., 1955, Natural and experimental boudinage and pinch and swell structures: Journal of Geology, v. 63, p. 512–526.

Secher, D., 1981, Les lherzolites ophiolitiques de Nouvelle Caledonie et leurs gisements de chromite: Déformation de la chromite [Thèse 3eme cycle]: l'Universite de Nantes, France, 250 p.

Simmons, G., and Wang, H., 1971, Single crystal elastic constants and calculated aggregate properties: A handbook: Cambridge, Massachusetts, MIT Press, 370 p.

Ten, A., and Yuen, D.A., 1999, Time-dependent formation of mantled inclusion structures for different rheologies under simple shear: Earth and Planetary Science Letters, v. 165, p. 25–35.

Wang, X., and Cochran, J., 1993. Gravity anomalies, isostasy, and mantle flow at the East Pacific Rise crest: Journal of Geophysical Research, v. 98, n. B11, p. 19505–19531.

Zhou, M.-F., Robinson, P.T., and Bai, W.-J., 1994, Formation of podiform chromitites by melt/rock interaction in the upper mantle: Mineralium Deposita, v. 29, p. 98–101.

Zimmerman, M.E., 1999, The structure and rheology of partially molten mantle rocks [Ph.D. thesis]: University of Minnesota, 159 p.

Manuscript Accepted by the Society February 23, 2000

Geological Society of America
Special Paper 349
2000

Section III
Structure and Physical Properties of Upper Oceanic Crust

Remarkable progress has been made in the 1990s in developing a better understanding of how the upper part of the oceanic crust forms and evolves. Detailed geophysical, petrologic, and geologic investigations of active spreading centers have resulted, among other things, in the observation and sampling of lava flows erupted within the past few years, in studies of the birth and demise of active hydrothermal systems, and in the seismic imaging of shallow magma lenses that feed eruptive events. Drilling within the upper oceanic crust further provides the opportunity to measure directly or to infer the physical properties of oceanic crust with greater accuracy than ever before. As a consequence of these more detailed observations and new data, it is becoming possible to test and refine some of the generalized assumptions and models that have been proposed over the years.

An important objective is to integrate these new observations and data into models that explain as many of the features of mid-ocean ridges as possible. Chen has developed a crustal-genesis model with a strong focus on the effects of anomalous mantle temperature on crustal accretion and ridge-axis topography along spreading centers. His model incorporates the effects of melt production, melt migration, and melt injection with seafloor observations such as seismic data, bathymetry, and gravity. Chen predicts several features, including the proposal that there is a critical mantle temperature at which small variations in temperature cause a transition from crust with a magma lens to one without, resulting in significant changes in ridge-axis thermal structure and topography.

Although there has been a general consensus in the scientific community regarding how the upper crust's volcanic carapace forms by injection of subvertical sheeted dikes that feed eruptive pillowed lavas or sheet flows, details of this process have never been particularly well defined. Schouten and Denham report that paleomagnetic inclinations of dikes and lavas from the Troodos ophiolite and ODP (Ocean Drilling Program) Hole 504B have consistent patterns that indicate postdepositional rotation about an axis parallel to the axis of seafloor spreading. Schouten and Denham show that this postdepositional rotation might have resulted from both block rotation and a vertically variable amount of rotation due to lava burial. The resulting angular unconformity between the sheeted dikes and the lowermost lavas is interpreted to represent a horizontal decoupling horizon that partitions deformation between the lavas and the dikes. The strong similarities in the structure of the upper part of the Hole 504B crust and Troodos ophiolite crust suggest a new framework for further comparative studies of structural similarities of ophiolites and oceanic crust.

The nature of the boundary between seismic layers 2 and 3 has been the subject of numerous investigations over the years, with many proposed origins for this seismic discontinuity. The boundary between these seismic layers has been detected regionally and has been interpreted to lie within the sheeted-dike complex that was drilled in ODP Hole 504B (Detrick et al., 1994). This hole, therefore, is suitable for a more detailed examination of the nature and cause of this seismic boundary. Toward this objective, Pezard summarizes a series of downhole geophysical experiments undertaken in Hole 504B that suggest a gradual change in the stress regime with depth in the sheeted-dike complex. This change from a compressional regime above to a strike-slip regime below appears to coincide with (or is close to) the location of the boundary between layers 2 and 3. This stress change is inferred to result from a depth-dependent opening of cracks and fractures. These results could explain why the seismic boundary appears to be dynamic, becoming shallower as the crust ages.

In an effort to further define the physical state of the upper crust, Sun and Goldberg use high-resolution velocity and density borehole profiles from ODP Hole 504B to construct synthetic seismograms; they then compare these results with field measurements. They conclude that variations in the seismic amplitude result mostly from porosity differences, which dominate the changes in physical properties in the upper crust. They further suggest that the seismic amplitude may be used to map porosity and lithologic variations within the upper crust.

REFERENCE CITED

Detrick, R.S., Collins, J., Stephen, R., and Swift, S., 1994, In situ evidence for the nature of the seismic layer 2/3 boundary in oceanic crust: Nature, v. 370, p. 288–290.

Geological Society of America
Special Paper 349
2000

Dependence of crustal accretion and ridge-axis topography on spreading rate, mantle temperature, and hydrothermal cooling

Yongshun John Chen
College of Oceanic and Atmospheric Sciences, Oregon State University, Corvallis, Oregon 97331, USA; e-mail: chen@oce.orst.edu

ABSTRACT

New observations indicate that mantle temperature and hydrothermal cooling, in addition to spreading rate, play important roles in controlling the production of oceanic crust and ridge-axis topography at mid-ocean ridges. A complete crustal-genesis model is presented in this paper with the focus on the effects of anomalous mantle temperatures on crustal accretion and ridge-axis topography at mid-ocean ridges. Anomalous mantle temperatures are associated with either hotspots or regions of relatively cold mantle ("cold spots"). The contribution of this new model is that it integrates all three processes—melt production, melt migration, and melt injection—of oceanic crustal genesis into one simple, self-consistent model, whereas previous models have focused on only one of these processes. Calculations show that there exists a critical mantle temperature at which a small variation in mantle temperature causes a rapid transition from the presence of a crustal melt lens to the absence of such a melt lens beneath the ridge axis, resulting in a significant change in ridge-axis thermal structure and topography. Calculations also show that the sensitivity to the mantle-temperature anomaly is greatest at ridges having intermediate spreading rates, in agreement with observations along several such ridges. Since many ridges spreading at slow and intermediate rates are influenced by either a nearby hotspot or a relatively cold mantle anomaly (e.g., the Australian-Antarctic discordance at lat 49°S, long 127°E, so named because of its deviation from the global trend of seafloor topography), interaction between ridges and mantle having an anomalous temperature is important in controlling the crustal accretion and ridge-axis topography at mid-ocean ridges.

INTRODUCTION

Mid-ocean ridges represent the longest plate boundaries on Earth's surface, and their persistent volcanism has produced over half of Earth's crust (Fig. 1). In contrast to inferred large magma chambers of continental intrusions, only small magma chambers have been detected at mid-ocean ridges such as at the fast-spreading East Pacific Rise (Detrick et al., 1987, 1993a; Mutter et al., 1995), at the intermediate–spreading rate Juan de Fuca Ridge (Morton et al., 1987; Rohr et al., 1988) and at the Vala Fa Ridge (Collier and Sinha, 1990, 1992). Mid-ocean ridge magma chambers are extremely small (~1 km wide and ~100 m thick) and lens shaped (Kent et al., 1990), and their depths below the seafloor increase (Fig. 2A) from ~1 km at a fast spreading rate to ~3 km at an intermediate spreading rate (Purdy et al., 1992). However, during pre–1990 seismic surveys (Detrick et al., 1990), such magma lenses were not detected along the slow-spreading Mid-Atlantic Ridge (its half–spreading rate ranges from 10 mm/yr to 25 mm/yr). This first-order inverse relationship between the magma-chamber depth and spreading rate could reflect the dependence of the ridge-axis thermal structure on spreading rate (solid line in Fig. 2A), if the ascending melt ponds at a horizon governed by solidus temperature as proposed in crustal-genesis models (Phipps Morgan and Chen, 1993a, 1993b; Henstock et al., 1993; Quick and Denlinger, 1993).

One of the fundamental observations of the global mid-ocean ridges is the contrasting ridge-axis morphology between the slow-spreading Mid-Atlantic Ridge (10–15 mm/yr, half–spreading rate) with an axial rift valley and rough flank topography and the fast-spreading East Pacific Rise (50–75 mm/yr) with an axial high and smooth flank topography (Macdonald, 1982, 1986). Detailed analysis of the Geosat altimetry gravity profiles across mid-ocean ridges revealed an abrupt change at spreading rates of 60–70 mm/yr from a variable high-amplitude gravity trough at slow ridges to a uniform low-amplitude gravity high at fast ridges (Small and Sandwell, 1989). These observed transitions in ridge-axis morphology and gravity have been attributed to the first-order dependence of the ridge-axis thermal structure on spreading rate and the resulting weak (ductile) zone in the lower crust at the fast– and intermediate–spreading rate ridges (Chen and Morgan, 1990a, 1990b). It is shown that these observed transitions are well correlated to the transition from the presence of a crustal magma lens to the absence of such a magma lens predicted from crustal-genesis models (Phipps Morgan and Chen, 1993b).

Petrologic and geophysical evidence has shown that a large part of the global mid-ocean ridge system has been influenced by on-axis

Chen, Y.J., 2000, Dependence of crustal accretion and ridge-axis topography on spreading rate, mantle temperature, and hydrothermal cooling, *in* Dilek, Y., Moores, E.M., Elthon, D., and Nicolas, A., eds., Ophiolites and Oceanic Crust: New Insights from Field Studies and the Ocean Drilling Program: Boulder, Colorado, Geological Society of America Special Paper 349, p. 161–179.

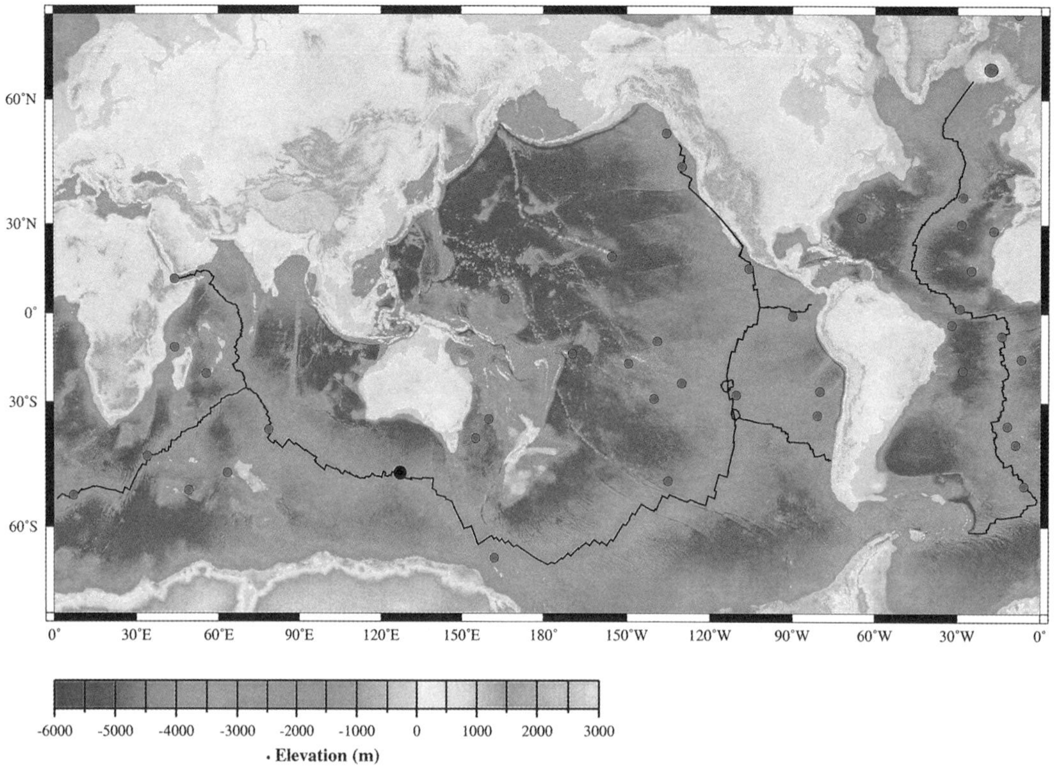

Figure 1. Map of the global mid-ocean ridge system (thick lines) and the locations of oceanic hotspots (solid dots). Also shown is the recently published global seafloor topography derived from satellite altimetry and ship depth soundings (Smith and Sandwell, 1997). The two larger solid dots show the locations of the Iceland hotspot and the Australian-Antarctic discordance (AAD) cold mantle anomaly, which represent the two extreme end members of hotspot–ridge interactions. RR—the Reykjanes Ridge south of Iceland.

or near-axis hotspots (Fig. 1), and the ridge-hotspot interactions have been the focus of extensive studies (Morgan, 1978; Schilling et al., 1985; Schilling, 1991; Sleep, 1992, 1996; Ito and Lin, 1995; Ito et al., 1996; Ribe et al., 1995; Small, 1995). The influence of hotspots on mid-ocean ridges is usually associated with a mantle-temperature anomaly due to the asthenospheric flow from the hotspot. However, some observations at these hotspot-influenced ridges show the exceptions to the global correlation between spreading rate and ridge-axis morphology. One notable example is the Reykjanes Ridge, which is located south of the Iceland hotspot in the North Atlantic. Another extreme example is the Australian-Antarctic discordance (lat 49°S, long 127°E; so named because of its deviation from the global trend of seafloor topography) in the Indian Ocean where various observations indicate a "cold" mantle beneath the ridge. These observations indicate that the mantle temperature also plays an important role in controlling the styles of crustal accretion and ridge-axis morphology at these anomalous mid-ocean ridges.

In this study, I present a complete crustal-genesis model that explores the effects of mantle temperature and hydrothermal cooling, in addition to spreading rate, on crustal accretion and ridge-axis topography. A conceptual model is presented first, and it is followed by a detailed description of the numerical procedure for solving the temperature and crust and mantle flow fields. The model results are, in general, consistent with field observations, and therefore, for the first time, a model has been devised that quantitatively describes the dependence of crustal accretion and axial morphology of mid-ocean ridges on mantle temperature as well as on spreading rate and hydrothermal cooling. Detailed model applications to the two extreme end members of ridge interactions with mantle having an anomalous temperature—the Reykjanes Ridge near the on-axis Iceland hotspot and the Australian-Antarctic discordance over a relatively cold mantle—are not reported here; instead, I focus on the sensitivity of the model results to mantle temperature and hydrothermal cooling at intermediate spreading rates. What is more important is that the specific testable model predictions are presented.

NEW OBSERVATIONS

Since 1990, significant efforts have been devoted to survey the more remote mid-ocean ridges including the south Mid-Atlantic Ridge (40°S–52°S) (Schilling et al., 1994), the Antarctic–East Pacific Rise (Lonsdale, 1994), the Southeast Indian Ridge (Cochran et al., 1997; Sempéré et al., 1997), the Southwest Indian Ridge (Grindlay et al., 1996; Madsen et al., 1996), and the Galapagos spreading center (Canales et al., 1997). In particular, extensive surveys including geophysical mapping, dredging, and seismic experiments have been conducted along the intermediate–spreading rate Southeast Indian Ridge, where parameters other than the spreading rate should have a large influence as suggested from previous studies (e.g., Chen and Morgan, 1990a, 1990b; Small and Sandwell, 1989, 1994; Shaw and Lin, 1996). These surveys have revealed detailed information about the rifting processes and crustal accretion along this ridge, which has an almost constant spreading rate. The systematic variations in the seafloor observations (such as deepening in zero-age seafloor depth and transition from an axial high to a rift valley) from the St. Paul and Amsterdam hotspot (at lat 38°S, long 78°E) eastward toward the Australian-Antarctic discordance (Fig. 1) have provided strong evidence that mantle temperature plays an important role in controlling the

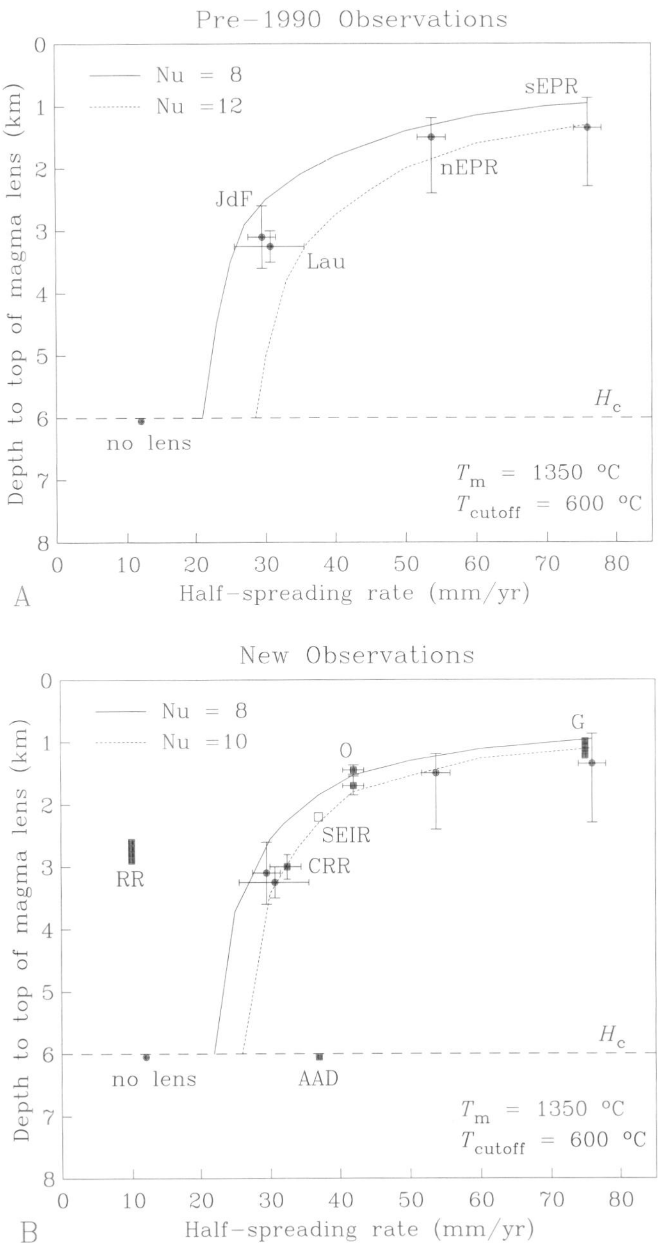

Figure 2. A, Depth to the top of a magma lens as a function of spreading rate, as determined by pre-1990 seismic observations. The two curves show the results of a suite of numerical experiments from previous studies (Phipps Morgan and Chen, 1993a) with Nu = 8 (solid curve) and 12 (dotted curve), where all other parameters are held constant. For these hydrothermal cooling parameters, a steady-state magma lens can exist within the crust only at half-spreading rates greater than ~22 mm/yr. Filled circles (and associated uncertainties) show multichannel seismic observations of the depth of the magma lenses along ridges spreading at intermediate and fast rates (Purdy et al., 1992) and the absence of such magma lenses at slow-spreading ridges (Detrick et al., 1990). JdF—Juan de Fuca Ridge, Lau—spreading ridge in Lau backarc basin, nEPR—northern East Pacific Rise, sEPR—southern East Pacific Rise. (Plot is modified from Fig. 7 of Phipps Morgan and Chen, 1993a.) B, Similar plot with more recent seismic observations added. New data of magma-lens observations are shown in filled squares. AAD—Australian-Antarctic discordance (Tolstoy et al., 1995); CRR—Costa Rica Rift (Mutter, 1995); G—East Pacific Rise south of Garrett transform (Tolstoy et al., 1997); O—East Pacific Rise north of Orozco transform (Carbotte et al., 1998); RR—Reykjanes Ridge at 57°43′N (Sinha et al., 1997); SEIR—Southeast Indian Ridge (Tolstoy et al., 1995). The lens depth at the Southeast Indian Ridge (shown as an open square) is set to be the layer 2 thickness because an accurate determination is not currently available. Model predictions of this study are shown for two cases of Nu = 8 and Nu = 10. There is a good agreement between the model predictions of the depth dependence of the magma lens with spreading rate and both the pre-1990 and newer seismic observations, except at the two anomalous ridges influenced either by a hotspot (Reykjanes Ridge) or a cold spot (Australian-Antarctic discordance). Note that the two symbols just beneath the dashed line at 6 km, which is the global average of the crustal thickness (Chen, 1992), show the two observations of no crustal magma lens; the left one is at the Mid-Atlantic Ridge (Detrick et al., 1990) and the right one is at the Australian-Antarctic discordance (Tolstoy et al., 1995).

crustal accretion and rifting at this ridge. Similar along-axis variations have been observed by a survey along the intermediate–spreading rate Galapagos spreading center that also indicated a strong influence by the nearby Galapagos hotspot (Canales et al., 1997).

There have been a number of new seismic observations of magma chambers at several mid-ocean ridges since 1990 that are shown as solid squares in Figure 2B (pre-1990 observations are shown in solid circles). A multichannel seismic investigation of a robust segment of the intermediate–spreading rate Costa Rica Rift has found a magma chamber at a depth of ~3 km (CRR in Fig. 2B) (Mutter, 1995). Seismic reflection data collected from the fast– to intermediate–spreading rate East Pacific Rise, north of the Orozco transform, have revealed a 300 m difference in magma-chamber depths between the two segments surveyed (O in Fig. 2B) (Carbotte et al., 1998). Detailed analysis of the seismic reflection data has revealed a systematic deepening of the magma-chamber depth (~260 m over a 60 km distance) along the ultra-fast-spreading southern East Pacific Rise toward the Garrett transform (G in Fig. 2B) (Tolstoy et al., 1997).

These three new observations are in agreement with previous observations reported in Purdy et al. (1992) and verify model predictions by Phipps Morgan and Chen (1993a) as shown in Figure 2B. The 300 m difference in magma-chamber depths between the two Orozco segments was interpreted to reflect the influence of magma supply (Phipps Morgan and Chen, 1993b) because the two segments are associated with markedly different magmatic budgets as indicated by the difference in axial morphology and crustal structures (Carbotte et al., 1998).

However, two new seismic observations are clearly out of this general spreading-rate–dependent global trend of ridge-axis thermal structure and topography. First, a multichannel seismic experiment was conducted at the Australian-Antarctic discordance and the adjacent Southeast Indian Ridge in the Indian Ocean (Tolstoy et al., 1995). Although both ridges (the Australian-Antarctic discordance and the Southeast Indian Ridge) are spreading at an intermediate rate of 37 mm/yr, they have rather contrasting ridge-axis topography: a well-defined rift valley at the Australian-Antarctic discordance versus an axial high at the adjacent Southeast Indian Ridge. As expected, a magma chamber was detected at the magmatically robust segment of the Southeast Indian Ridge as well as a thicker (7.2 km) crust (Tolstoy et al., 1995). Although the exact depth of the magma chamber is not available yet, a 2.2 km thickness of layer 2 (shown as an open square labeled SEIR in Fig. 2B) is consistent with the global trend for magma-chamber depth if the magma lens is located at or above the boundary of layers 2 and 3. Such a magma chamber was not observed at the Australian-Antarctic discordance survey site (indicated by

AAD and a solid square beneath the dashed line designating the Moho in Fig. 2B). The absence of a magma chamber at the intermediate–spreading Australian-Antarctic discordance could have resulted from a significantly reduced magma supply, a clear indication of the influence of a relatively cold mantle beneath the area that is consistent with seafloor observations and other geophysical and geochemical observations.

Second, a magma chamber has been detected for the first time beneath a slow-spreading ridge—the Mid-Atlantic Ridge. Sinha et al. (1997) reported the results of a multicomponent geophysical survey on the Reykjanes Ridge at 57°43′N, where a magma chamber was observed at a depth of ~2.5–3 km below a magmatically robust axial volcanic ridge (RR and a solid bar in Fig. 2B). Such magma chambers have not been detected along other parts of the Mid-Atlantic Ridge, as indicated by the solid dot beneath the Moho (dashed line) at the slow spreading rate in Figure 2B (Detrick et al., 1990).

These two new observations at the Australian-Antarctic discordance and the Reykjanes Ridge, which are clearly out of the global trend (Fig. 2B), constitute a strong indication that mantle-temperature variations play an important role in controlling the styles of crustal accretion, ridge-axis topography, regional ridge-axis depth, and geochemical anomalies. These signatures have up to now been attributed solely to the influence of mantle plumes.

A COMPLETE CRUSTAL-GENESIS MODEL

A complete crustal-genesis model is presented here that is built upon the assumption that the interplay between heat supplied by the new magma and hydrothermal cooling controls the crustal-accretion process, ridge thermal structure, and ridge-axis topography. Complete crustal genesis at the mid-ocean ridges consists of three processes: melt production at depths of ~20–60 km over a broad region in the upper mantle, melt migration to the shallow level of ~1–3 km, and melt injection into the magma lens and the resulting crustal flow.

Conceptual model

The basic model geometry is shown in Figure 3A, and it includes the effects of all the three processes of crustal genesis at a mid-ocean ridge. The basic modeling concepts of both the melt production and the melt injection resulting in crustal flow are not particularly new, as they have been presented by Chen (1996) and Phipps Morgan and Chen (1993a). What is new here is that all three mechanisms are combined into a "complete" crustal-genesis model for the mid-ocean ridges whereas most of the previous models were developed just for the melt-production part of crustal genesis (McKenzie, 1984; Klein and Langmuir, 1987; Forsyth, 1993; Shen and Forsyth, 1995; Su et al., 1994; Chen, 1996), for the melt injection into the crust (Phipps Morgan and Chen, 1993a, 1993b), or for the melt-migration process (e.g., Phipps Morgan, 1987; Cordery and Phipps Morgan, 1994; Sparks and Parmentier, 1993).

Melt production. Several studies have been published on the calculation of melt-production rate beneath the mid-ocean ridges or on the relationship between mantle temperature and crustal thickness (e.g., Klein and Langmuir, 1987; Shen and Forsyth, 1995; Su et al., 1994). All these models show a positive correlation between the melt-production rate (crustal thickness) and the mantle temperature; a hotter upwelling mantle results in a thicker crust. In this study, the total melt-production rate is calculated on the basis of a numerical model for a passive spreading center (Chen, 1996). It is a fractional melting model that includes the effect of decreasing melt production by melt

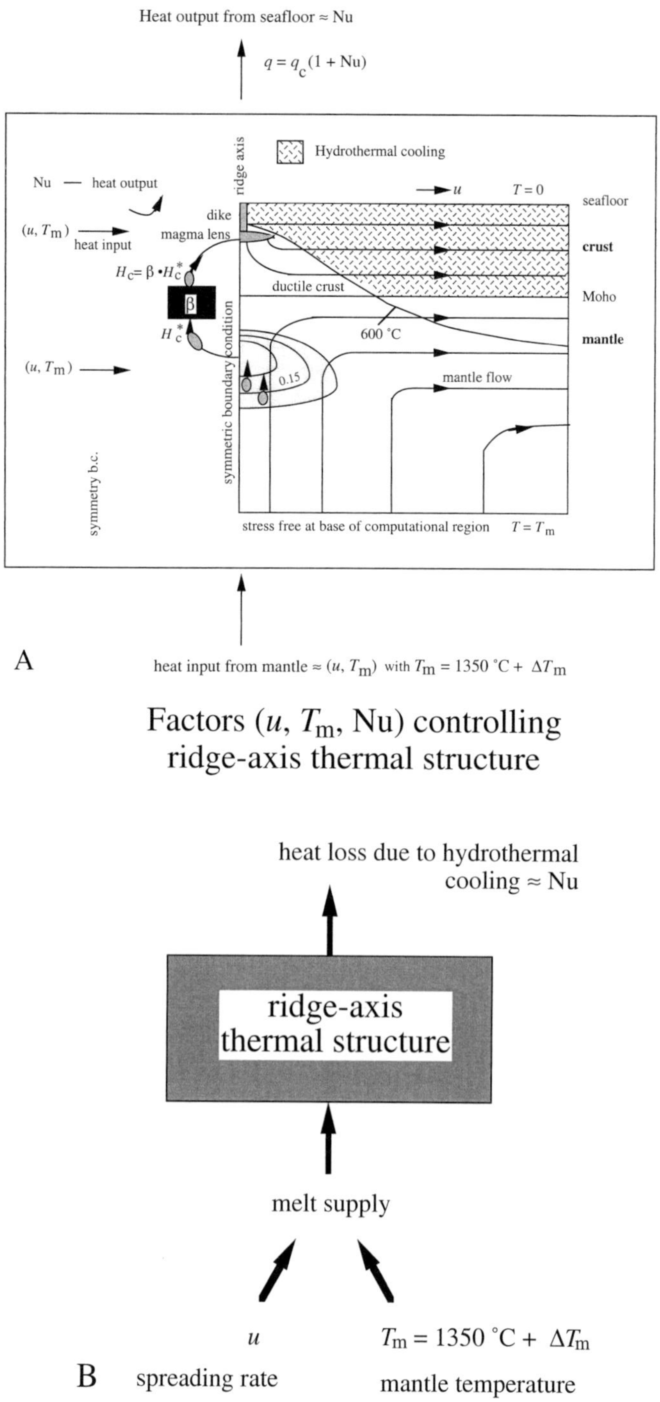

Figure 3. A, Sketch of a complete crustal-genesis model of mid-ocean ridges. Melt is extracted from the partial-melting region beneath the ridge (Chen, 1996). It is then transported to the ridge axis at shallow crustal level, where it is injected into a magma lens and a dike (Phipps Morgan and Chen, 1993a). Gabbros are formed by the downward flow of melt plus crystals from the lens. A sheeted-dike complex forms from a "steady-state" dike. Although the geometry of the lens is kept constant, the depth to the top of the lens is determined by the axial depth of the melt solidus temperature of 1200 °C. The melt-migration mechanism is modeled here by the efficiency factor β that relates the output from the partial-melting region to the melt input into the crust as $H_c = \beta(u)H_c^*$. The values of β are chosen to produce a 6-km-thick crust for a normal mantle temperature of 1350 °C at a given spreading rate. Symmetry b.c.—symmetry boundary condition. B, Chart summarizing the three-parameter control on ridge-axis thermal structure.

TABLE 1. VALUES OF THE EFFICIENCY FACTOR β OF THE MELT-MIGRATION MODEL AT DIFFERENT HALF-SPREADING RATES: $H_c = \beta \times H_c^*$

u (mm/yr)	10	20	25	30	37	42	50	60	75
β (%)	89	68	63	60	58	58	57	56	56

extraction as the residual mantle matrix becomes increasingly difficult to melt. The output of this model is the maximum crustal thickness H_c^* (it equals the total melt-production rate divided by the spreading rate), which is a function of spreading rate and mantle temperature (Fig. 3A).

Melt migration. Melt migration is not yet well understood because of its complexity and the lack of observational constraints, although limited information has been gained from several previous studies (e.g., Phipps Morgan, 1987; Cordery and Phipps Morgan, 1994; Sparks and Parmentier, 1993). On the other hand, it is certain that not all the melt produced within the upwelling mantle will reach the ridge axis and contribute to the formation of new oceanic crust within the neovolcanic zone. The percentage of the melt that does make it to the ridge axis will probably vary among ridges with different spreading rates and will also depend on other parameters that control the melt-migration processes. According to this notion, the effect of the melt-migration mechanism is modeled here by introducing an efficiency factor β. It is defined as the percentage of the total melt (H_c^*) extracted in the partial-melting region in the upper mantle that eventually reaches the shallow magma-injection zone at the ridge axis and forms the new oceanic crust (H_c^*). By this definition, the crustal thickness is related to the maximum crustal thickness, defined by the total melt production, as

$$H_c = \beta \times H_c^*. \qquad (1)$$

One important seismic observation over the past two decades is that, on average, the oceanic crust has a constant thickness of ~6 km that is independent of the spreading rate (Chen, 1992). By using this value as a constraint, the values of β as shown in Table 1 are chosen at a given spreading rate so as to produce a 6-km-thick crust for a normal mantle temperature of 1350 °C. For the particular melt-production model of Chen (1996) used here, the overall efficiency of the melt migration varies from ~90% at 10 mm/yr to ~55% at >60 mm/yr. Because there are no observational constraints on how β varies with mantle temperature, it is kept constant at a given spreading rate. In Equation 1 crustal thickness is a function of mantle temperature since the maximum crustal thickness (a measure of the total melt production) depends strongly on mantle temperature (Chen, 1996). It is worth pointing out that all the complexity of the melt-migration mechanisms is included in the β factor, and the effects of the uncertainties related to different melt-production and melt-migration models on the resulting crustal thickness are significantly reduced by calibrating the crustal thickness as 6 km at the normal mantle temperature.

Melt injection and crustal flow. When the melt reaches the crust at the ridge axis, it is injected into a dike and a magma lens (Phipps Morgan and Chen, 1993a, 1993b). The geometry of the magma lens is assumed to be ≤500 m wide (half-width) and ≤150 m thick. The depth of the magma lens is determined by the temperature of the solidus (1200 °C) beneath the ridge axis, which is solved for iteratively. If there is no crustal region beneath the ridge axis that is hotter than 1200 °C, a steady-state crustal magma lens is not permitted. In this case, a crustal-size dike is assumed for the crustal accretion at the ridge axis.

The hydrothermal cooling is treated as an enhanced thermal conductivity (K_c^*) within the region where hydrothermal activity occurs. Here K_c^* is parameterized by the Nusselt number (Nu): $K_c^* = (1 + \mathrm{Nu}) \times K_c$; Nu is defined as the ratio of hydrothermal heat transport within a permeable layer to heat transport by heat conduction alone.

Three controlling parameters (u, T_m, Nu). To summarize, the model presented in this study is based on the current understanding that it is the interplay between heat input into the crust by melt supply—which depends on spreading rate (= *u*) and mantle temperature (= T_m)—and heat loss to the seafloor due to hydrothermal cooling (∝ Nu) that controls the ridge-axis thermal structure, especially within the crust (Fig. 3B). The mantle temperature (in degrees Celsius) can be expressed as

$$T_m = 1350 + \Delta T_m, \qquad (2)$$

where ΔT_m is the perturbation relative to the normal mantle temperature of 1350 °C, which was used in previous model calculations (Chen and Morgan, 1990a, 1990b; Phipps Morgan and Chen, 1993a, 1993b). Therefore, all the model predictions of the presence or absence of a steady-state crustal magma lens, the lens depth beneath a ridge axis, and the resulting thermal structure depend on these three important parameters: spreading rate (*u*), mantle temperature (T_m or ΔT_m), and the Nusselt number (Nu).

Near a hotspot such as the Iceland hotspot or a "cold spot" such as the Australian-Antarctic discordance, the mantle beneath the ridge axis could be either hotter or cooler than the normal mantle temperature because of interactions between the ridge and the hotspot or cold spot. It is the effect of this mantle-temperature anomaly, ΔT_m in equation 2, that is the focus of this study.

Model formulation

Governing equations. The system of equations governing the resulting steady-state heat and mass transport is presented next. The equation for conservation of energy is

$$\frac{\partial(u_i T)}{\partial x_i} = \kappa\nabla^2 T + L\dot{\psi}, \qquad (3)$$

where $\dot{\psi}$ is the magma-injection rate and is zero outside the magma-intrusion zone (dike and lens), *L* is the magma's latent heat, κ is thermal diffusivity, u_i is velocity, *x* is distance to ridge axis, and *T* is temperature. The equation for conservation of mass is

$$\frac{\partial u_i}{\partial x_i} = \dot{\psi} \qquad (4)$$

The equation for conservation of momentum is

$$\frac{\partial p}{\partial x_i} + \frac{\partial \tau_{ij}}{\partial x_j} = 0, \qquad (5)$$

where

$$\tau_{ij} = \eta\left(\frac{\partial u_i}{\partial x_j} + \frac{\partial u_j}{\partial x_i}\right) + \eta\delta_{ij}\frac{\partial u_i}{\partial x_i} \qquad (6)$$

and *p* is the pressure. The "effective viscosity" η is defined by the laboratory-determined power-law rheology and depends on both temperature and strain rate (Chen and Morgan, 1990b):

$$\eta = B\dot{e}_{\max}^{(1-n)/n}\exp\left(\frac{Q}{nRT}\right), \qquad (7)$$

TABLE 2. MODEL PARAMETERS WITH POSSIBLE RANGES

Parameter name	Symbol	Probable range
Activation energy (crust)	Q_C	260 kJ/mol
Activation energy (mantle)	Q_m	520 kJ/mol
Power-law number (mantle & crust)	n	3
Spreading half-rate	u	10–80 mm/yr
Crustal thickness	H_C	3–10 km
Mantle temperature	T_m	1350 ± 100 °C
Cut-off temperature limiting the hydrothermal circulation	T_{cutoff}	600 °C
Nusselt number	Nu	2–10

where B = a constant, n = number of moles of melt, Q = the activation energy, $\dot{e}^{max}$ = the maximum shear-strain rate, R = the ideal-gas constant, and T = the temperature in kelvins.

The range of the material parameters (B, n, and Q) for the oceanic crust and mantle rocks can be found in experimental studies (e.g., Brace and Kohlstedt, 1980; Kirby, 1983), and their values used in this study are listed in Table 2. The maximum shear-strain rate ($\dot{e}_{max}$) is related to the second invariant of the strain rate tensor ($\dot{e}_{II}$) by

$$\dot{e}_{II} = (\dot{e}_{kl}\dot{e}_{kl})^{1/2} = \sqrt{2}\,\dot{e}_{max}. \tag{8}$$

Melt injection. The amount of melt entering the lens and dike ($\dot{\psi}$) is determined by mass balance. The magma-injection rate ($\dot{\psi}$) within the lens or the dike is calculated according to the total crustal thickness and the depth of the lens (Phipps Morgan and Chen, 1993a), that is, the total magma flowing at a steady-state rate into the lens and dike regions balances the amount of crust per year leaving the model domain at the right-side boundary (= $H_c u$) because of plate spreading. The same principle of mass balance also determines how much asthenospheric material enters the model domain from below. For magma-lens accretion, as observed at ridges spreading at intermediate and fast rates, all the latent heat of cooling for the crust that lies below the depth of the magma lens is released at the top of the magma lens.

The geometry of the magma lens is determined by seismic studies as being ~500 m wide (half-width) and 150 m thick, and the depth to the top of the lens is determined by the axial depth of the calculated 1200 °C isotherm, which is solved for during each iteration.

Here, it is assumed that all the melt is injected into either a magma lens or a dike where the latent heat is released before the melt flows out of the injection region. However, seismic observations on the East Pacific Rise (Garmany, 1989; Dunn and Toomey, 1997), ophiolite observations (Kelemen and Dick, 1995), and melt-inclusion studies at the Gorda Ridge (Nielsen et al., 1995) have suggested that not all the lower crust was formed by the downward flow from a shallow magma chamber. Sill intrusion could also contribute to the formation of the lower crust according to ophiolite studies (Boudier and Nicolas, 1995; Kelemen et al., 1997). I ran calibration tests on the models of Phipps Morgan and Chen (1993a) and Chen (1998); the tests showed that allowing as much as 30% of the lower crust to be added outside of the magma lens would not change the main conclusion of the previous model calculations (Phipps Morgan and Chen, 1993a; Chen, 1998). These results indicated that the simple magma lens and dike model has captured the basic physics of magma-chamber formation and crustal flow at mid-ocean ridges.

Hydrothermal cooling. Most of the heat input to the magma lens and dike is removed by hydrothermal cooling. This effect of hydrothermal cooling is parameterized by the Nusselt number (Nu) within the region penetrated by seawater circulation, and Nu is assumed to be zero elsewhere. Nu is defined as the ratio of the hydrothermal heat transport to the heat transport by heat conduction alone; therefore, it describes the efficiency of hydrothermal cooling. Thus, the total heat flux ($\mathbf{q}_{total}$) has the form

$$\mathbf{q}_{total} = -k(1 + \mathrm{Nu})\nabla T \tag{9}$$

In the model, the depth of hydrothermal circulation is limited by an isotherm of ~600 °C within the crust.

Boundary conditions. Boundary conditions similar to those in the previous models (Chen and Morgan, 1990b; Phipps Morgan and Chen, 1993a) are used here, and heat and mass transport are solved for within a model domain (200 km × 200 km) on one side of a symmetric ridge axis (Fig. 3A). For solving temperature field, the mantle is flowing into the model region from below at a constant mantle temperature, and the top surface (seafloor) is kept at 0 °C. Heat is free to move out of the right side (off-axis boundary), but no heat is allowed to move across the left side (ridge axis). For the viscous flow, the vertical velocity and horizontal shear stress are zero at the top surface, and the bottom boundary is assumed to be a shear and normal stress-free surface. At the right side, passive plate spreading is assumed to occur beneath the rigid part of the oceanic lithosphere, and symmetric conditions are adopted at the ridge axis (the left side). Further discussion of these boundary conditions appears in Chen and Morgan (1990b).

Similar to previous studies (Phipps Morgan and Chen, 1993a, 1993b), the Moho is assumed to be a material discontinuity, and the crust is assumed to form by the flow out of the melt lens and dike. The effect of these assumptions is implemented by introducing the boundary condition at the Moho as

$$v = 0 \qquad y = H_c \tag{10}$$

—that is, the vertical velocity (v) is zero at the Moho—which prevents mantle material from flowing across. This view is, in general, supported by seismic observations suggesting that the Moho reflector can be traced within 1 km of the ridge axis (Detrick et al., 1993a; Harding et al., 1993; Kent et al., 1994) and indicating therefore that crustal accretion is generally completed within the narrow neovolcanic zone, although in some cases, crust grows by off-axis volcanism (e.g., Batiza et al., 1990, 1996; Shen et al., 1993; Niu and Batiza, 1997) or by sill intrusion into the lower crust as shown by ophiolite studies (Boudier and Nicolas, 1995; Kelemen et al., 1997).

Brittle failure. The yield-strength envelope of the oceanic lithosphere is determined by brittle failure within the top rigid part of the lithosphere and by creep-flow laws at depths. Experimental studies have shown that brittle failure can be described by frictional resistance that has a simple linear relationship over a range of normal stresses, known as the Byerlee's law (Byerlee, 1978). Following the derivation of Chen and Morgan (1990b), the yield strength (S) of the upper brittle part of the oceanic lithosphere, defined as the maximum shear stress (τ) that the lithosphere can sustain (i.e., $S = \tau_{max}$), has the form

$$S = 0.4(1 - \lambda)\rho g y, \tag{11}$$

where g is the acceleration due to gravity and λ is ?. It is a "viscous-plastic" model in that viscous shear stress beneath the seafloor is limited by a depth-dependent yield strength, defined in equation 11. During numerical calculation, if the calculated maximum shear stress within a finite element (τ_e) exceeds the yield strength in that element (S_e), then the viscosity of the element is adjusted according to a "predictor-corrector method" (Chen and Morgan, 1990b),

$$\eta_{new} = \frac{S_e}{\tau_e}\eta_{old} \tag{12}$$

which guarantees $\tau_e - S_e \leq 0$ within the element.

Melt production within the upper mantle

One of the new components of this model is the inclusion of a mechanism to calculate the total melt-production rate within the upwelling upper mantle. A simple melt-production model of Chen (1996) is adopted here and is briefly described.

First, the melt-production rate within the upper mantle is calculated for the case of batch melting, that is, all the melt is retained within the solid mantle matrix so that the upwelling mantle keeps a constant composition. The following analytical expressions for the solidus T_s and liquidus T_l temperatures as a function of pressure for the peridotite mantle were obtained by McKenzie and Bickle (1988):

$$P = [(T_s - 1100)/136] + 4.968 \times 10^{-4}\exp[1.2 \times 10^{-2}(T_s - 1100)] \tag{13}$$

$$T_l = 1736.2 + 4.343P + 180 \tan^{-1}(P/2.2169), \tag{14}$$

where P is the pressure (in GPa) and T_s and T_l are the solidus and liquidus temperatures (in °C), respectively. The results of the expressions agree well with experimental observations. By defining a dimensionless temperature T',

$$T' = \frac{T - (T_s + T_l)/2}{T_l - T_s} \tag{15}$$

McKenzie and Bickle (1988) obtained an expression for the fraction of melt F of the peridotite at a temperature T and pressure P:

$$F = 0.5 + T' + (T'^2 - 0.25)\,(0.4256 + 2.988T'). \tag{16}$$

In order to compute the amount of melt separated from the residual mantle for fractional melting, additional considerations have to be made, as described in the following.

Figure 4A shows a schematic cross section of a mid-ocean ridge in which a broad partial-melting region in the upper mantle exists beneath the ridge as the consequence of the rapid mantle upwelling. For a given temperature distribution, the fraction of melt $F(x, y)$ within the upwelling mantle can be calculated through equations 13–16 for the case of batch melting and is illustrated as contours in Figure 4A.

For this calculation, I focus on a small mass per unit time $\dot{W}_0 = \rho_m u_y dx$ passing through point $A(x, y_0)$ and determine the amount of melt $d\dot{W}_m(x) = D(x)\, d\dot{W}_0$ that can be extracted from this mass after it goes through the partial-melting region (past point D). Here ρ_m is the mantle density, u_y the upwelling velocity at point A, and y_0 is chosen such that the point A is always below the melting region. The melt depletion $D(x)$ is defined as the percentage of melt extracted from the initial mass. The fraction of melt $F(x, y)$ along the flow line from A to E (Fig. 4A) is shown in Figure 4B. Following Chen (1996), the amount of melt extracted from this small mass $d\dot{W}_m(x)$ after it goes through the partial-melting region can be calculated in terms of the maximum fraction of melt (F_{max}) along its entire flow line from A to E:

$$d\dot{W}_m(x) = D(x)\dot{W}_0 = [1 - e^{-F_{max}}]\dot{W}_0. \tag{17}$$

Once $d\dot{W}_m(x)$ is known, for a steady state, the total melt-production rate from the upwelling mantle within the entire melting region is simply

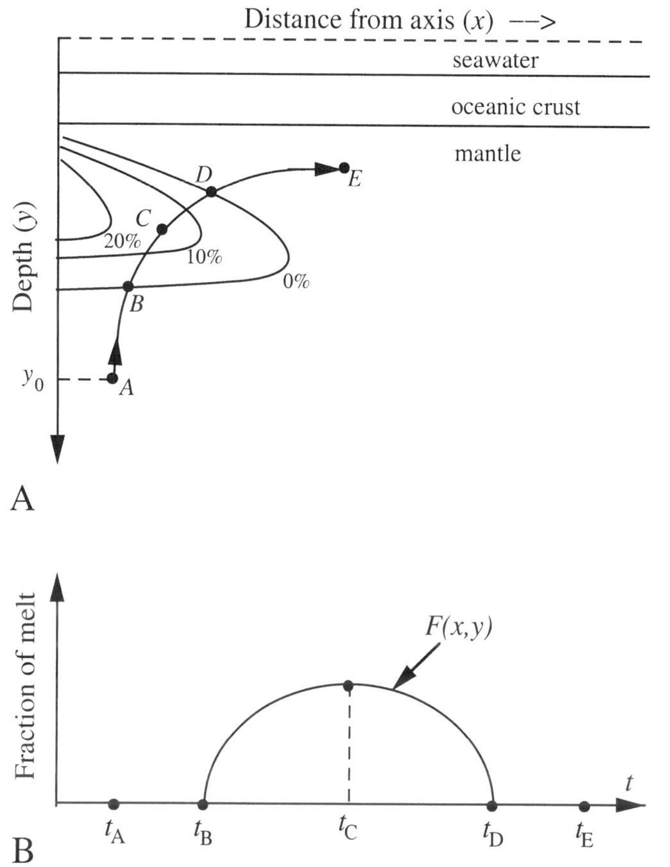

Figure 4. A, Sketch of a decompression-driven melting process beneath a mid-ocean ridge axis. Fractions of melt for the mantle rock are shown as contours. The curve from A to E represents a flow line of the upwelling mantle flow. B, The fraction of melt $F(S)$ for a no-melt-separation case, that is, all the melt is retained in the system shown as a function of S, the distance along the state path from point A in A. (After Chen, 1996.)

$$\dot{W}_m = \int_0^\infty d\dot{W}_m(x) = \int_0^\infty \rho_m u_y(x, y_0) D(x)\, dx. \tag{18}$$

If all the melt would reach the ridge axis and form oceanic crust (an ideal melt migration), the maximum thickness (H_c*) of the oceanic crust could be calculated by equating the accretion rate of new crust per unit length of ridge at the ridge axis ($\rho_c H_c{*}u$) to equation 18. This approach gives

$$H_c^* = \frac{\rho_m}{\rho_c}\int_0^\infty \frac{u_y(x, y_0)}{u} D(x)\, dx, \tag{19}$$

where ρ_c is the density of the crust (in kg/m^3) and u is the half-spreading rate (in mm/yr).

Numerical procedure and solution algorithm

During each numerical experiment with the model parameters chosen as listed in Table 2, solutions of the coupled temperature and velocity fields are solved for by using a finite-element code (RIDGE2M), which is an extension of the early code of RIDGE2D

TABLE 3. VARIABLE MODEL PARAMETERS DURING EACH ITERATION

Parameter name	Symbol	Probable range	In each iteration
Depth to the top of the magma lens	y_{top}	1.5–3.5 km	varies with $y_{solidus}$ (T)
Magma-injection rate	ψ	0–$u{*}h$	is $u * (h - y_{top})$

developed for a two-dimensional ridge setting (Chen, 1989; Chen and Morgan, 1990b; Phipps Morgan and Chen, 1993a). Details of the RIDGE2D code and its numerical procedure and algorithm can be found in Chen and Morgan (1990b) and Phipps Morgan and Chen (1993a), such as the iteration procedure for dealing with the nonlinear viscosity and the coupling between temperature and flow fields. Model parameters that are calculated during each iteration are included in Table 3, and the iteration procedure adopted in this new model is described as follows:

1. For the first iteration, a magma lens is placed at y_{top} = 1.5 km depth with a half-width (W_{ml}) of 500 m; ψ is calculated accordingly. The mantle temperature is set to be the normal value of 1350 °C, and the crustal thickness is set at 6 km. A solution vector (which has three nodal values of u, v, and T) is calculated by assuming a uniform viscosity of η = 1020 Pa·s.

2. For iterations 2–5, the approximate solutions associated with the initial geometry of crustal accretion are computed. Viscosity is updated in each iteration according to equation 7 by using the new solutions.

3. For subsequent iterations (>5), the depth of the magma lens is determined by the axial depth of the 1200 °C isotherm, and therefore it varies during each iteration. If the entire crust is cooler than 1200 °C (i.e., no magma lens), then the crustal accretion is replaced by a dike extending from the seafloor to the Moho (at 6 km depth). Consequently, ψ is adjusted accordingly (Table 3).

4. The solution converges after a number of iterations (N_1), that is, variations in the crustal-accretion geometry (presence or absence of a magma lens and the lens depth) and the resulting temperature and flow fields are satisfactorily small in each iteration. (Model runs show that $N_1 \approx 20$.)

5. The total melt-production rate and the maximum crustal thickness $H_c{*}$ are computed. The efficiency factor β is set to 6 km ÷ $H_c{*}$, according to equation 1.

6. If the input $\Delta T_m = 0$, the computation and output results are stopped; otherwise, T_m is set to 1350 °C + ΔT_m, and the solution vector is computed.

7. The total melt-production rate and the maximum crustal thickness $H_c{*}$ are then computed. The crustal thickness H_c is calculated according to equation 1 by using the β value obtained at step 5. The boundary condition (from equation 10) is reset at the new Moho depth of H_c.

8. The new solution is computed.

9. The final step is a check for convergence to an acceptable accuracy toward a steady-state solution. If convergence is not satisfied, steps 7–9 are repeated; otherwise, the computation stops and the results are output.

The RIDGE2M code implemented according to the described procedure has proved to be consistent and rapidly converging. In the following computations, I used a grid mesh of 80 × 120 elements with a variable grid spacing: the size of the element increases from 125 m × 75 m within the melt lens to 15 km × 20 km at the lower right corner (x = 200 km; y = 200 km). For this grid mesh, most of the computation converges after 40 iterations and takes ~10 minutes on a Sun Ultra 10 workstation.

MODEL RESULTS

Spreading-rate dependence

The predicted dependence of the magma-lens depth (the depth of 1200 °C beneath the ridge axis) is shown in Figure 2B for a "normal" mantle temperature of 1350 °C and two cases of Nu = 8 (solid line) and Nu = 10 (dotted line). Similar to previous model results, the predicted trend of the lens depth matches the seismic observations quite well: the lens depth deepens from ~1–1.8 km at fast ridges (40–75 mm/yr) to ~3–3.5 km at intermediate ridges (~30 mm/yr), and the sharp transition from a deeper lens to no crustal magma lens occurs at ~25 mm/yr. Note that the new observations of the magma-lens depths south of Garrett transform (G in Fig. 2B) (Tolstoy et al., 1997), north of the Orozco transform (O in Fig. 2B) (Carbotte et al., 1998), at the Southeast Indian Ridge (SEIR in Fig. 2B) east of the Australian-Antarctic discordance (AAD in Fig. 2B) (Tolstoy et al., 1995), and at the Costa Rica Rift (CRR in Fig. 2B) (Mutter, 1995) are also fitted well by the model. Therefore, it seems that values of Nu = 8–10 could be representative of the hydrothermal cooling that predicts the "correct" trend of the spreading-rate dependence of the magma-lens depth. The slight difference in the model predictions between Figure 2A and Figure 2B is because of the power-law rheology of the new model presented in this study. Two-layered rheology representing rigid lithosphere and creeping asthenosphere was employed in previous models (Phipps Morgan and Chen, 1993a), and the layer boundary was defined by the isotherm of 750 °C. With the power-law rheology that is strongly temperature dependent, the crustal flow in the "hot" wedge of lower crust beneath the magma lens is affected by the crustal thermal structure. This sensitivity of crustal flow on thermal structure tends to make a sharper transition from a magma lens to no-magma lens at slow spreading rates.

Five model crustal thermal structures and flow fields away from the ridge axis are shown in Figure 5 for spreading rate u = 60, 50, 35, 20, and 10 mm/yr. Although a narrow magma lens is present at ridges forming at both fast and intermediate rates, the lens depth at 35 mm/yr (Fig. 5C) is almost doubled from that at 60 mm/yr (Fig. 5A). In contrast, a magma lens is not permitted in the crust by the steady-state thermal structure at slow-spreading ridges (Fig. 5, D and E). Almost the entire crust is cooled below 400 °C by hydrothermal cooling at 10 mm/yr. As a matter of fact, the axial thermal structures of fast- and slow-spreading ridges would not have differed dramatically as shown here if the heat had been lost to the seafloor by heat conduction alone. After Clive Lister's (Lister, 1974) first recognition of the important role of hydrothermal systems in cooling the new ocean crust, it is now well-accepted that pervasive hydrothermal circulation would rapidly solidify magma bodies and prevent the existence of large crustal magma chambers, even at the fastest-spreading ridges (note the narrow region of the elevated temperature in Fig. 5, A and B).

Model calculations of the partial-melting region in the upper mantle beneath the ridge axis are shown in Figure 6, A–C, as shaded contours for three spreading rates of 10, 35, and 60 mm/yr, respec-

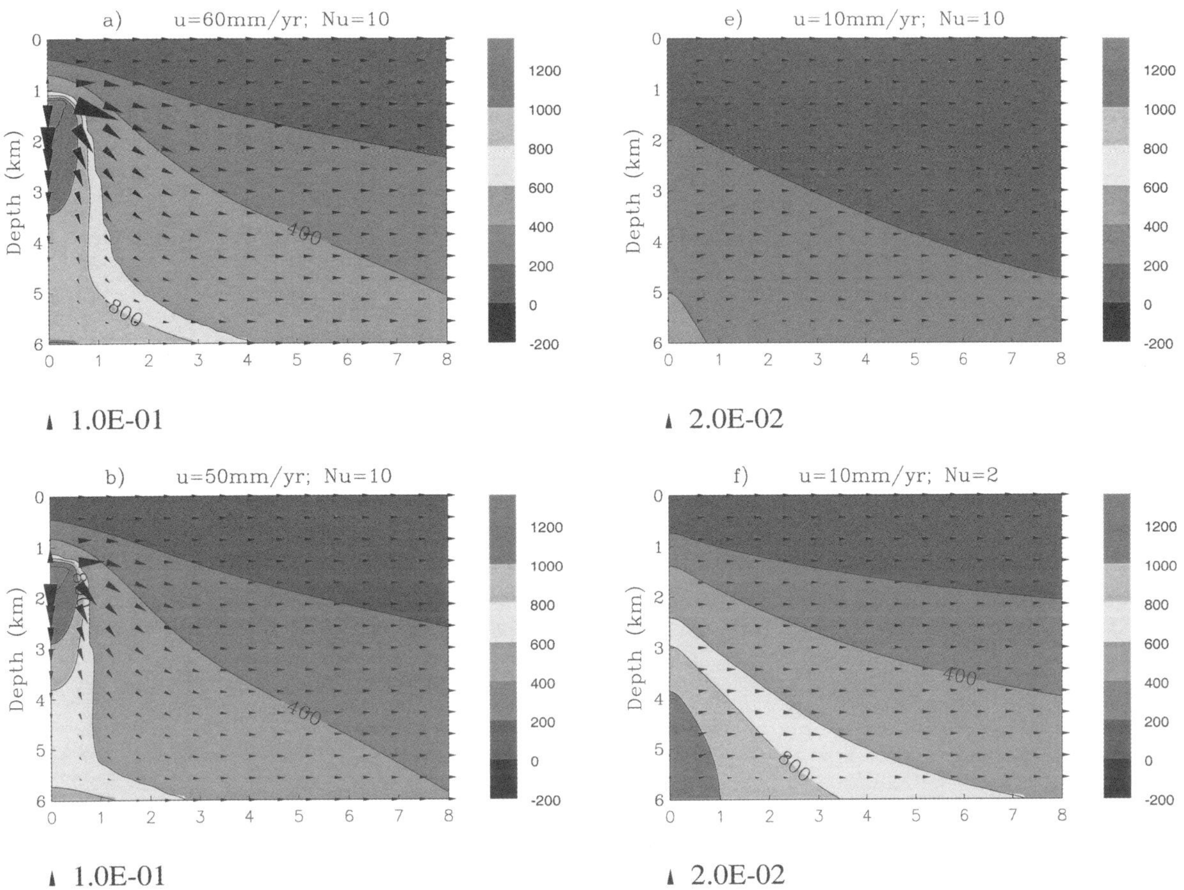

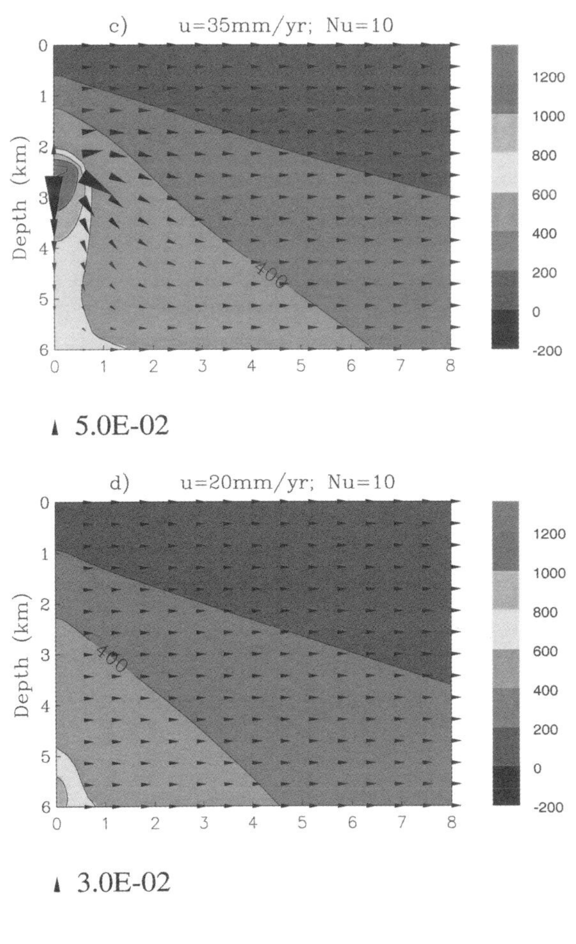

Figure 5. Crustal thermal structures (shown by isotherms and shaded contours) and flow vectors (black triangles scaled as shown) of a suite of ridge models with Nu = 10 and T_m = 1350 °C but at different half–spreading rates: A, 60 mm/yr; B, 50 mm/yr; C, 35 mm/yr; D, 20 mm/yr; and E, 10 mm/yr. Although a magma lens is present at both fast– and intermediate–spreading ridges, the lens is at a deeper depth (2.4 km) for ridges spreading at the intermediate spreading rate than those spreading at the fast spreading rate (1.2 km). F, Crustal thermal structure of a slow-spreading ridge (u = 10 mm/yr) with a normal mantle temperature (T_m = 1350 °C) and less hydrothermal cooling (Nu = 2). See text for detailed descriptions.

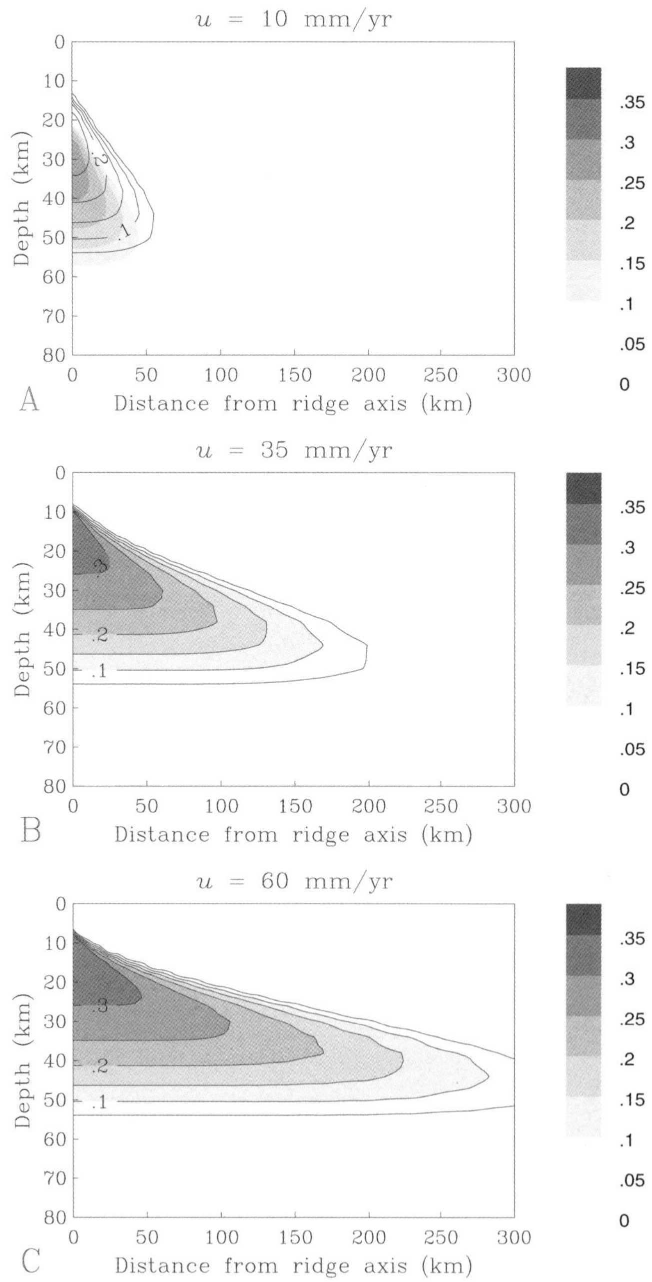

Figure 6. Partial-melting region shown as contours of melt fraction in the upper mantle beneath the ridge axis at three spreading rates: A, 10 mm/yr, B, 35 mm/yr, and C, 60 mm/yr. The melt-fraction contours are calculated from the model thermal structure by using equations 12–15 and are shown at an interval of 0.05 melt fraction (scale shading is on the right). The effect of increasing spreading rate is mainly to increase the horizontal extent of these melt-fraction contours. At a slow-spreading rate, however, the melt-production region becomes not only narrow but also thinner from the top, the effect of a cold lithosphere at a slow-spreading ridge axis.

tively. The noticeable change with spreading rate is the horizontal extent (half-width) of the partial-melting region, which varies from 50 km at 10 mm/yr to over 300 km at 60 mm/yr. The vertical extent is basically unchanged, except for being depressed from above by a cold lithosphere at the slow spreading rate (Fig. 6A), because the initial melting depth is mainly controlled by mantle temperature and chemical compositions of the mantle (Langmuir et al., 1992; Shen and Forsyth, 1995), which were kept constant in these calculations. Intuitively, it is clear that it is more difficult to focus all the melt produced within the partial-melting region with a wider horizontal extent laterally into the rather narrow neovolcanic zone (~2 km wide). On the other hand, the global seismic observations of a constant average thickness of the oceanic crust (Chen, 1992) lead one to the conclusion that the amount of crustal production at the seafloor relative to the total melt production within the upper mantle decreases with increasing spreading rate. This ratio was defined in the previous section as the β factor, and its values vary from ~90% at 10 m/yr to 56% at 60 mm/yr (Table 1).

Effects of mantle temperature

To address the effect of mantle temperature, a suite of model runs was carried out for a range of mantle-temperature variations from −70 °C to +70 °C at different spreading rates. The thermal structure of the ridge and the size of the partial-melting region are quite sensitive to the variations in mantle temperature. Figure 7 shows the variation of the partial-melting region from a negative mantle-temperature anomaly of −20 °C to a positive anomaly of +20 °C at a fast-spreading ridge. The region of partial melting is stretched in both downward and horizontal directions with increasing mantle temperature, similar to the trend with increasing spreading rate shown in Figure 6. Therefore, the total melt production beneath a mid-ocean ridge increases with either spreading rate (Fig. 6) or mantle temperature (Fig. 7).

It is clear from Figures 6 and 7 that although an increase in both spreading rate and mantle temperature significantly increases the horizontal extent of the melt-production region, the effect of spreading rate and mantle temperature on the vertical dimension of the partial-melting region is slightly different. The melt-production region is depressed from above because of a cold lithosphere at slow spreading rates (Fig. 6), whereas it is thickened at the bottom at a higher mantle temperature (Fig. 7).

The predicted crustal thicknesses, defined in equation 1 as the product of the β factor and the total melt production, at three spreading rates of 10, 35, and 60 mm/yr are shown in Figure 8; the crust becomes thicker with higher mantle temperature. Although the slopes of the three curves differ slightly, they generally fall within a quite narrow band. The relatively larger slope (more sensitivity to mantle temperature) at the slow spreading rate (10 mm/yr) is because a slow-spreading ridge has more of the melting at greater depth relative to ridges spreading at intermediate and fast rates (Fig. 6). Therefore, its melt production would be more affected by mantle-temperature variations, which tend to change the bottom extent of the melting region (Fig. 7). Given the uncertainties of the model, one can conclude that the dependence of the crustal thickness on mantle temperature is not very sensitive to spreading rate. On average, therefore, crustal thickness increases from 3 km at $\Delta T_m = -70$ °C to 9 km at $\Delta T_m = +70$ °C.

The predicted magma-lens depth is shown in Figure 9 as a function of the mantle-temperature anomaly for spreading rates of 60, 35, and 10 mm/yr. An important result is that there exists a "threshold" mantle temperature above which small changes in mantle temperature can cause a dramatic change in the axial thermal structure—a sharp transition from the presence of a magma lens to the absence of such a lens in the crust. For example, as shown in Figure 9A, a change in mantle-temperature anomaly from −35 °C to −45 °C at an intermediate spreading rate (35 mm/yr) leads to a deepening of the 1200

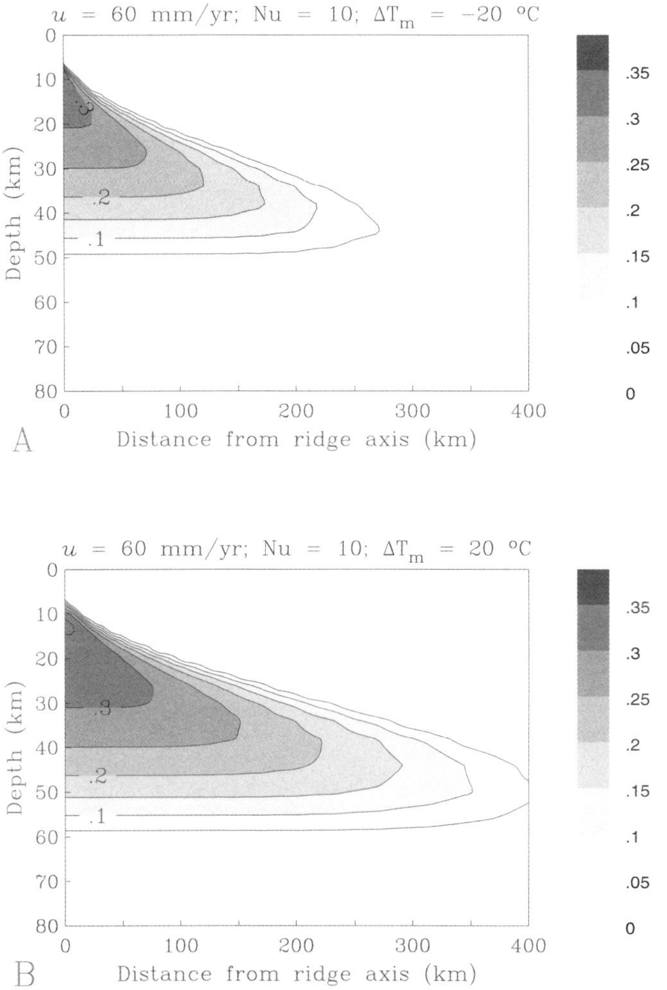

Figure 7. Partial-melting region shown as contours of melt fraction for two model mantle-temperature anomalies from the normal mantle temperature of 1350 °C: A, −20 °C and B, +20 °C. The effect of increasing mantle temperature is not only to increase the horizontal extent of the melt-fraction contours but also to move the contours downward, the effect of a hotter mantle. Therefore, the melt-production region becomes wider off-axis and thicker at the bottom with higher mantle temperature.

°C isotherm from 3 km in the lower crust (a crustal magma lens) to 7 km in the mantle (absence of a crustal magma lens).

Is this sharp transition from the presence of a magma lens to the absence of such a lens in the crust real? Does this ~4 km drop in the depth of the 1200 °C isotherm really occur at a ridge with an intermediate spreading rate and a decrease of 10 °C in mantle temperature? To answer these questions, it is necessary to understand the physics behind these numerical simulations.

The answer to the first question is yes. The increase of the slope with decreasing mantle-temperature anomaly toward the threshold mantle temperature shown in Figure 9 is real. The effect happens because all the latent heat of the magma-to-cumulate phase change for all of the "cumulate gabbro" crust beneath the magma lens is released at the lens; a similar positive feedback effect was discussed in a previous publication (Chen and Phipps Morgan, 1996). At 1350 °C and 35 mm/yr half rate, a steady-state magma lens is predicted at ~2.5 km depth (Fig. 9A) for Nu = 10, as observed from multichannel seismic reflection lines across the Costa Rica Rift (CRR in Fig. 2B) (Mutter, 1995). As mantle temperature decreases, the total magmatic heat supplied to the magma lens is reduced, and the hydrothermal cooling pushes the lens to a greater depth. This change, in turn, reduces the latent heat released at the magma lens, which further enables the hydrothermal circulation to cool the lens. This positive feedback process is the primary reason for the rapid deepening of the magma-lens depth as the threshold mantle temperature is approached.

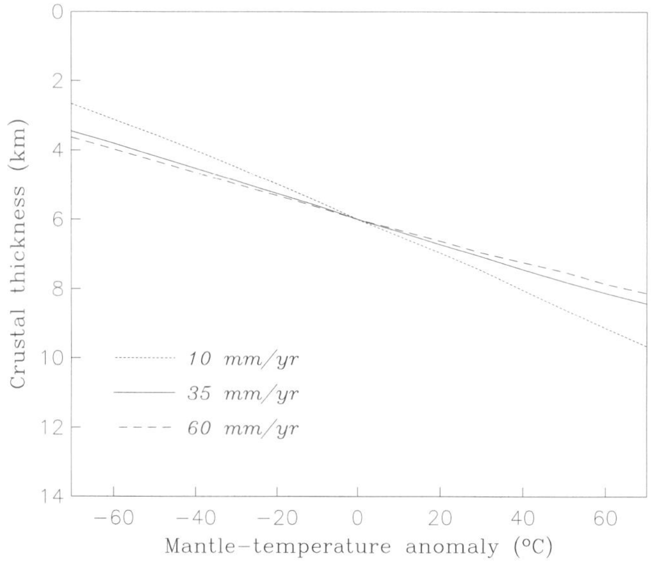

Figure 8. Calculated crustal thickness as a function of mantle temperature at three spreading rates of 10, 35, and 60 mm/yr. The mantle-temperature anomaly (ΔT_m) of 0 °C corresponds to T_m = 1350 °C, which produces a 6-km-thick crust at all spreading rates. Although each curve has a slightly different slope, all three curves fall within a narrow band, indicating that the dependence of crustal thickness on mantle temperature is quite similar at all spreading rates, especially at intermediate and fast rates. The slightly higher sensitivity to the mantle temperature at the slow spreading rate of 10 mm/yr is due to the presence of a cold lithosphere at the ridge axis that decreases the melt production near the top of the partial-melting region, as shown in Figure 6A.

The answer to the second question seems obvious—this must be an artifact of the numerical calculations; in the real world, the depth of the 1200 °C isotherm cannot have such a large drop (it is a staggering ~10 km drop for the case of 10 mm/yr as shown in Fig. 9B). The truth is that one crucial fact is ignored when this question is raised. That is, there are two 1200 °C isotherms beneath the ridge axis when a magma lens is present in the crust, and what is plotted in Figure 9 is the depth of the shallower 1200 °C isotherm in the crust. One example is shown in Figure 10 for a ridge spreading at an intermediate rate (35 mm/yr) with a normal mantle temperature and Nu = 10. In this case the 1200 °C isotherm intersects the vertical axis (x = 0) at two depths, one at 2.5 km in the crust and one at 8.5 km in the mantle. This point is not clearly shown in previous plots of crustal thermal structures (Fig. 5 in this study and Fig. 3 in Phipps Morgan and Chen, 1993a).

The situation of having two 1200 °C isotherms beneath the ridge axis stems from the fact that the crust and the mantle have separate heat sources. The thermal structure of the crust is mainly controlled by the heat input from the melt supply, whereas heat is brought into the mantle by upwelling from below. Because of the mantle's zero vertical velocity at the Moho (equation 15), the mantle loses heat to the crust only by conduction across the Moho, which is an only small contribution to the total heat supplied to the crust. As discussed for the first question, the 1200 °C isotherm in the crust deepens rapidly toward the threshold mantle temperature. When mantle temperature further decreases, the entire crust is cooled below 1200 °C by hydro-

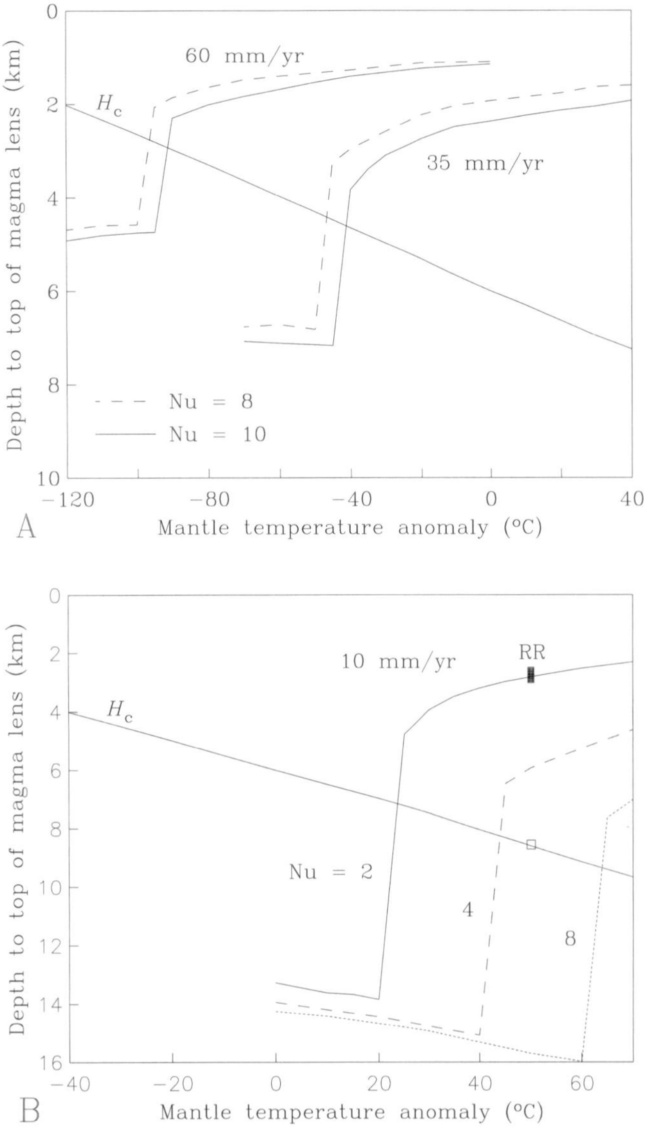

Figure 9. A, Magma-lens depth plotted as a function of mantle temperature at two spreading rates of 35 mm/yr and 60 mm/yr for two Nusselt numbers of 8 and 10. Also shown is the crustal thickness as a function of mantle temperature. B, Similar plot at a slow-spreading ridge of 10 mm/yr for three Nusselt numbers of 2, 4, and 8. Filled bar (labeled RR) shows the depth range of the magma chamber at 57°43′N beneath the Reykjanes Ridge (Sinha et al., 1997). The mantle-temperature anomaly (ΔT_m) of 0 °C corresponds to T_m = 1350 °C. There exists a "threshold" mantle temperature above which small changes in mantle temperature can cause a dramatic change in axial thermal structure and a sharp transition from the presence of a magma lens to the absence of such a lens in the crust. See text for detailed descriptions.

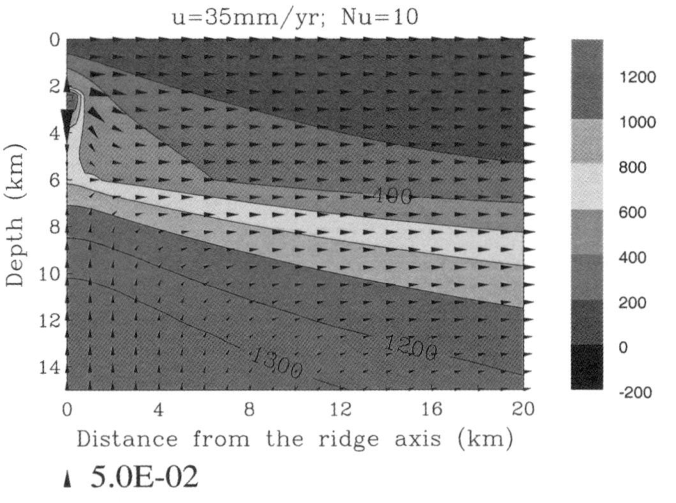

Figure 10. Thermal structure (shaded contours) and flow vectors of a ridge model with an intermediate spreading rate, Nu = 10, and T_m = 1350 °C. Both the crust and part of the upper mantle are included in this plot, whereas only the crustal fields are shown in Figure 5. Notice the narrow region of the elevated temperatures within the crust beneath the axis. The 1200 °C isotherm intersects the vertical axis (x = 0) at two depths, one at 2.5 km in the crust and one at 8.5 km in the mantle. These two 1200 °C isotherms result from two separate heat sources to the crust and the mantle.

thermal cooling (leading to absence of a magma lens); then the depth of the 1200 °C isotherm plotted in Figure 9 would be the 1200 °C isotherm in the mantle. Therefore, the drop in Figure 9 represents the "artificial" shift from the 1200 °C in the crust to the 1200 °C in the mantle.

It is interesting to point out from Figure 9 that the magma-lens depth is not very sensitive to positive mantle-temperature anomalies ($\Delta T_m > 0$) at fast-spreading ridges. A magma lens deepens less than 100 m when ΔT_m changes from 0 to −20 °C at u = 60 mm/yr (Fig. 9A). Significant changes in magma-lens depth occur only near or at the threshold mantle temperature (Fig. 9). A similar trend is also seen in the spreading-rate dependence shown in Figure 2: there is very little change in the magma-lens depth between the northern East Pacific Rise (u = 55 mm/yr) and the southern East Pacific Rise (u = 75 mm/yr) compared with the significant deepening in the magma-lens depth with decreasing spreading rate toward the transition at intermediate spreading rate.

Influence of hydrothermal cooling

The thermal structure of a mid-ocean ridge also strongly depends on hydrothermal cooling, modeled here by the Nusselt number (Nu), which describes the efficiency of heat removal by hydrothermal cooling. A higher Nu represents a more efficient hydrothermal cooling system; therefore, more heat will be brought to the seafloor for a given mantle temperature and spreading rate. This effect is clearly demonstrated in Figure 9 where the predicted magma-lens depth as a function of mantle-temperature anomaly is also shown for different values of Nu. For example as shown in Figure 9A, at 35 mm/yr, ~400 m deepening of the magma lens on average occurs for $\Delta T_m > -30$ °C when Nu increases from 8 to 10. Significant variations in hydrothermal activity have been found at different parts of the Mid-Atlantic Ridge, which has led to the suggestion that both tectonic ridge segmentation and hotspot influence are important in controlling hydrothermal venting at slow-spreading ridges (German and Parson, 1998).

To further illustrate the effect of hydrothermal cooling, Figure 5F shows one model run for u = 10 mm/yr, Nu = 2, and ΔT_m = 0 °C. Although a steady-state magma lens is not permitted in the crust, the crust is much hotter than the ridge with Nu = 10 (Fig. 5E), indicating the strong dependence of crustal thermal structure on hydrothermal cooling, as discussed by Chen and Phipps Morgan (1996).

Seismic topographic inversions of combined earthquake and explosive traveltimes have revealed a seismic low-velocity zone (LVZ) at >3 km depth beneath the seafloor on the Mid-Atlantic Ridge at 26°N. The LVZ was interpreted as a zone of elevated temperature and partial melt beneath the ridge axis (Kong et al., 1992). Although the seismic experiment could have captured an intermittent magma-intrusion event with partial melt at that depth, the LVZ could also represent a zone of elevated temperature caused by a temporary (local) reduction in hydrothermal cooling as shown in Figure 5C.

Chen and Phipps Morgan (1996) reported that although axial heat flux is primarily a function of spreading rate, it also depends on both the magma supply (crustal thickness) and hydrothermal cooling represented by the Nusselt number (Nu). Chen and Phipps Morgan argued that although magma chambers have typically been found along the ridges spreading at intermediate and fast rates, a similar magma chamber could also occur at slow-spreading ridges if the hydrothermal cooling were of low intensity (small Nu). Their prediction was confirmed by two field observations. First, a detailed along-axis survey has found only one hydrothermal vent along 600 km of the Reykjanes Ridge, which corresponds to a value of 0.014 for the "plume incidence" factor (German et al., 1994; German and Parson, 1998). That intensity of hydrothermal activity is about a factor of four less than the intensity found in normal segments of the Mid-Atlantic Ridge. On the other hand, a multicomponent geophysical survey at a magmatically active axial volcanic ridge at 57°43′N on the Reykjanes Ridge has detected a melt body (interpreted as a magma chamber) at a depth of ~2.5–3 km (filled bars in Fig. 7B) below the seafloor (Sinha et al., 1997). Model calculations show that in order to match both the observed shallow magma chamber (filled bar, Fig. 9B) and a slightly thicker crust at the southern Reykjanes Ridge (open square, Fig. 9B), the model calculation must use a smaller Nu number, which corresponds to less hydrothermal cooling. The model predictions for Nu = 2 match the observed shallow magma-chamber depth at a reasonable mantle-temperature anomaly of ΔT_m = 50 °C.

One interesting result coming out of Figure 9 is that although positive mantle-temperature anomalies ($\Delta T_m > 0$) would cause little variation in the magma-lens depth at ridges spreading at intermediate and fast rates, the magma-lens depth is quite sensitive to the change in hydrothermal cooling at slow and intermediate spreading rates. At an intermediate–spreading rate ridge (35 mm/yr) with a normal mantle temperature, a 400 m shallowing in magma-lens depth would require large positive mantle-temperature anomalies ($\Delta T_m > 50$) whereas the same amount of shallowing can take place if the Nu is reduced from 10 to 8.

Sensitivity to variations in mantle temperature and hydrothermal cooling

The effects of the three important parameters (u, ΔT_m, Nu) summarized in Figure 3B can be illustrated in Figure 2B. First, on the global scale, the control by spreading rate is shown by the deepening of the magma chamber as spreading rate slows and by the sharp transition to the absence of such a crustal magma chamber at ~25 mm/yr. Second, the influence of ΔT_m is to shift this global spreading-rate–dependence curve either upward or downward (effectively shift the lens-to-no-lens transition toward a slower spreading rate or a higher spreading rate), corresponding to either a positive ΔT_m or a negative ΔT_m anomaly. Similar effects can be achieved by either subdued hydrothermal cooling (a smaller Nu) or enhanced hydrothermal cooling (a larger Nu).

It is clear from Figure 2B that the sensitivity of the magma-chamber depth on either ΔT_m or Nu is highest at intermediate spreading rates, indicated by the steep slope of the curves of the magma-lens depth versus spreading rate. Therefore, a small perturbation in either mantle temperature or hydrothermal cooling would lead to a large variation in the magma-lens depth at intermediate–spreading ridges.

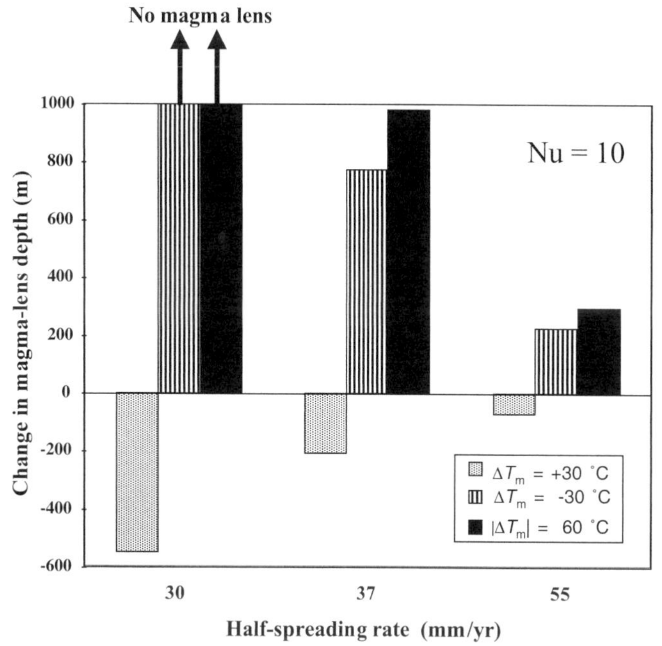

Figure 11. Sensitivity of magma-lens depth to mantle temperature at ridges spreading at intermediate and fast rates.

This point is further illustrated in Figure 11 where the changes of the magma-lens depth relative to the lens depth at the normal mantle temperature are plotted at three spreading rates. At each spreading rate, the left vertical bar shows the shallowing (negative lens-depth change) of the magma lens for ΔT_m = 30 °C, the middle bar shows the deepening (positive lens-depth change) for $\Delta T_m = -30$ °C, and the right bar shows the total lens-depth change corresponding to the variation of ΔT_m from 30 °C to −30 °C. Two major results are well shown in Figure 11.

First, for the same variation in mantle temperature, the deepening due to a negative ΔT_m is significantly larger than shallowing due to a positive ΔT_m at all three spreading rates. This result is expected because vigorous hydrothermal cooling would prevent significant shallowing of the 1200 °C isotherm for $\Delta T_m > 0$; however, it would help to push the isotherm farther downward for $\Delta T_m < 0$ (Fig. 9).

Second, it shows that ridges with intermediate spreading rates have the highest sensitivity to the mantle-temperature variations. The change of magma-chamber depth at u = 37 mm/yr is about a factor of 3 larger than that at u = 55 mm/yr. Further, the magma lens shallows by 550 m at u = 30 mm/yr, compared to by 200 m at u = 37 mm/yr and by 70 m at u = 55 mm/yr. In particular, the 1200 °C isotherm is pushed downward below the Moho (absence of a crustal magma chamber) at u = 30 mm/yr if the mantle is 30 °C cooler than the normal 1350 °C. The lens-to-no-lens transition occurs at $\Delta T_m = -20$ °C at this intermediate spreading rate of 30 mm/yr. Note that the slope of the curve of the magma-lens depth versus spreading rate for Nu = 10 in Figure 2B is steepest at 30 mm/yr among the three spreading rates shown in Figure 11; therefore, a small perturbation

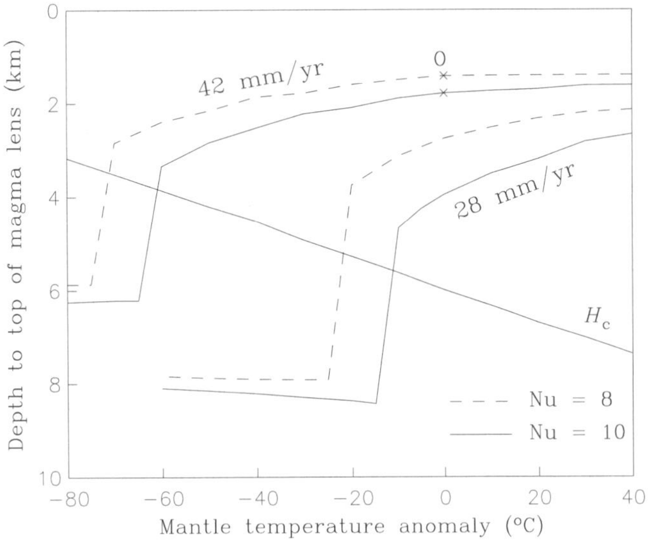

Figure 12. Magma-lens depth plotted as a function of mantle temperature at two spreading rates of 28 mm/yr and 42 mm/yr for two Nusselt numbers of 8 and 10. Also shown is the crustal thickness (H_c) as a function of mantle temperature. The two magma-lens depths observed at two East Pacific Rise segments north of Orozco transform (Carbotte et al., 1998) are shown as crosses (labeled O). The mantle-temperature anomaly (ΔT_m) of 0 °C corresponds to T_m = 1350 °C.

in either mantle temperature or hydrothermal cooling would lead to a large variation in the magma-lens depth.

The conclusion from these model calculations that intermediate–spreading rate ridges have the highest sensitivity to the mantle-temperature variations (or to the magma-supply fluctuations) is consistent with field observations and previous studies along several such ridges including the Southeast Indian Ridge (Tolstoy et al., 1995; Cochran et al., 1997; Sempéré et al., 1997), the Juan de Fuca and Gorda Ridges (Hooft and Detrick, 1995), and the Galapagos spreading center (Canales et al., 1997). Along these ridges, noticeable transitions in ridge-axis topography from an axial high to a rift valley were observed and were attributed to the variations in mantle temperature due to a nearby hotspot or a relatively cold mantle anomaly.

For example, the contrasting axial morphologies across the Blanco transform between the Juan de Fuca and Gorda Ridges, where both ridges are spreading at the same intermediate rate of 28 mm/yr, were attributed to a small difference in mantle temperature (ΔT_m = 10–15 °C) on the basis of the analysis of available gravity and bathymetry data (Hooft and Detrick, 1995). This difference in mantle temperature was due to the presence of the Cobb thermal anomaly at the Axial Seamount along the Juan de Fuca Ridge. At the slightly higher but still intermediate spreading rate of 42 mm/yr, the 300 m difference in the observed magma-chamber depths of the two East Pacific Rise segments north of Orozco (O in Fig. 2B) was attributed to the increased melt supply to the robust south segment from a small hotspot indicated by the off-axis seamount chain, which strikes roughly parallel to Pacific-hotspot–determined plate motion (Carbotte et al., 1998). Both the broad and inflated ridge axis and thicker crust suggested by gravity are consistent with the hypothesis of a hotspot influence on the south segment compared to the lack of such influence on the northern segment, where the axial high is much narrower and ~300 m deeper, similar to that of the East Pacific Rise elsewhere (Carbotte et al., 1998).

Predicted magma-lens depth is shown in Figure 12 as a function of mantle-temperature anomaly at these two spreading rates of 28 mm/yr and 42 mm/yr for Nu = 10. Although the 300 m difference in magma-chamber depth between the two Orozco segments corresponds to at least a 30 °C difference in their mantle temperature, the same amount of variation in mantle temperature would lead to a transition from a magma chamber to no crustal magma chamber at the Juan de Fuca and Gorda Ridges (28 mm/yr). In fact, the model calculations show that a small difference of 10–15 °C in mantle temperature inferred from analysis of bathymetry and gravity (Hooft and Detrick, 1995) can indeed lead to the transition from a magma chamber at the Juan de Fuca Ridge with an axial high to the absence of a long-lived crustal magma chamber (which needs to be confirmed by future seismic observations) with a rift valley at the Gorda Ridge. Again, these two calculations show that intermediate–spreading rate ridges have the highest sensitivity to the mantle-temperature variations.

Figure 12 shows that the magma lens in ridges spreading at 42 mm/yr could be 300 m shallower if the Nu decreases from 10 to 8, whereas the same amount of lens-depth change would require a large positive mantle-temperature anomaly. Therefore, these model calculations suggest that the observed 300 m shallowing in the magma-lens depth of the south Orozco segment relative to the north Orozco segment (Carbotte et al., 1998) would be mostly due to reduced hydrothermal cooling and less likely directly due to a positive mantle-temperature anomaly at the southern segment. However, the reality may be more complicated than this proposal since a positive mantle-temperature anomaly might lead to reduced hydrothermal cooling because of its effect on permeability structure. Significantly reduced hydrothermal cooling has been observed along the Reykjanes Ridge and was related to the permeability change by the adjacent Iceland hotspot (German and Parson, 1998). Thus I favor the interpretation that the observed 300 m shallowing in the magma-lens depth of the south Orozco segment relative to the north Orozco segment is caused by a small hotspot (Carbotte et al., 1998); however, this hotspot effect on the magma-lens depth is probably directly reflected in the reduced hydrothermal cooling and indirectly related to the increased magma supply from a hotter mantle. On the other hand, a negative mantle-temperature anomaly at the north Orozco segment as indicated by its starved magma supply could lead to a slightly deeper magma lens, which therefore could contribute partly to the observed 300 m difference in magma-lens depths between the magmatically robust south segment (influenced by a nearby small hotspot) and the magma-starved north segment near the Orozco transform. Similarly, the large shallowing of the magma lens at 17°S might also be related to variations in hydrothermal cooling as caused by changes in local permeability structures due to locally increased magma supply (temporal variability) (Mutter et al., 1995; Hooft et al., 1997).

DISCUSSION

Expanded model calculations (Fig. 13) have shown that the thermal and dynamic model of mid-ocean ridges with three parameters (u, T_m, Nu) presented in this study are consistent with various observations at different mid-ocean ridges, including several seismic observations at the Australian-Antarctic discordance and the Reykjanes Ridge. It is shown that most of the ridges with an axial-high morphology (solid squares, Fig. 13) are correlated with a hot crustal thermal structure having a magma lens and the ridges with a rift-valley morphology (open squares, Fig. 13) are associated with a cold crustal thermal structure having no magma lens in the crust. The exception involving the presence of a rift valley at the Australian-

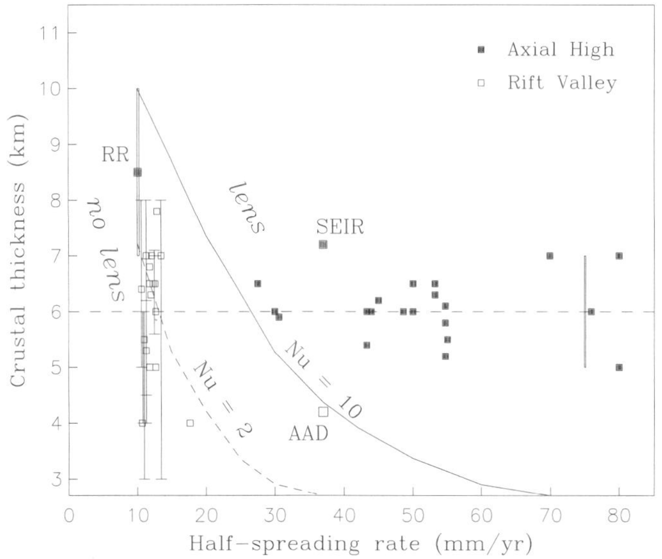

Figure 13. The boundary between a steady-state magma lens (in the upper right corner) and no magma lens in the crust (in the lower left corner) as a function of (u, T_m, Nu) for two cases of Nu = 10 (solid curve) and Nu = 2 (dashed curve). The vertical axes are the mantle temperature on the right and the corresponding crustal thicknesses are on the left. The crustal thickness is related to the mantle temperature by the average trend in Figure 8. Also plotted are the axial morphology (axial high vs. rift valley) at various ridge segments and corresponding crustal thickness determined from seismic measurements. Squares connected by solid lines show the range of crustal-thickness variations seen along several fracture zones (Detrick et al., 1993b). All data are a near-ridge subset of the global data set (Chen, 1992) except for those data annotated by symbols: RR—Reykjanes Ridge, AAD—Australian-Antarctic discordance, SEIR—Southeast Indian Ridge. For Nu = 10, the model boundary separates the axial-high morphology with a magma lens from the rift-valley morphology with no magma lens in the crust for the mid-ocean ridges with a normal mantle temperature. Observations of a rift-valley morphology, no magma lens, and a thin crust of 4.2 km at the Australian-Antarctic discordance (Tolstoy et al., 1995) are well predicted with the model for Nu = 10 and $\Delta T_m = -50$ °C. On the other hand, the observation of a magma lens at 57°43′N of the Reykjanes Ridge (Sinha et al., 1997) requires a smaller Nusselt number (Nu = 2), which is consistent with vent field studies along the Reykjanes Ridge (German and Parson, 1998). The higher mantle temperature (>20 °C) associated with a thicker crust (7–10 km) along most of the Reykjanes Ridge permits a magma lens in the crust (with an axial-high morphology) if the hydrothermal activity is a factor of four less than that along the normal Mid-Atlantic Ridge.

Antarctic discordance is well explained by a negative mantle-temperature anomaly (–40 °C). However, the anomalous axial-high morphology at the Reykjanes Ridge cannot be explained by a positive mantle-temperature anomaly alone; such an anomaly would produce a thicker crust than the seismic observations show. Because the seismic (and, to a lesser extent, gravity) data indicate that the crustal thickness is 7–10 km, the axial-high morphology can only be explained by both a hotter (>35 °C) mantle and a less efficient hydrothermal cooling regime (Nu < 3) than in a normal segment of the Mid-Atlantic Ridge.

The model calculations discussed herein support a "threshold" mechanism to explain mid-ocean ridge dynamics and the characteristic tectonic styles of rifting processes and resulting ridge-axis morphology, ridge-axis gravity, and crustal structures. A threshold mechanism is based on two relationships: (1) A sharp transition in the ridge-axis morphology is related to a crustal thermal structure and a weaker crustal rheology than the mantle at a given temperature (Chen and Morgan, 1990a, 1993b). (2) A sharp transition in the crustal thermal structure is related to the lens-to-no-lens transition, which is controlled by three parameters—u, T_m, and Nu—as shown in this study. There exists a threshold value of any one of these three parameters for a sharp transition in the ridge-axis thermal structure (Figs. 2, 9, and 13) and, therefore, the ridge-axis morphology if the other two parameters are fixed. This threshold mechanism, which could be responsible for the sharp transition in the ridge-axis morphology, has been strongly supported by detailed seafloor bathymetry observations along the Southeast Indian Ridge (Cochran et al., 1997) and along the intermediate–spreading rate Galapagos spreading center (Canales et al., 1997).

The model described herein directly relates seafloor observations such as seismic data, bathymetry, and gravity to the model parameters of spreading rate, mantle temperature, and hydrothermal cooling. Therefore, the validity of the model will ultimately be tested by observations, or, vice versa, reasonable model predictions could also guide future field programs. For example, the model prediction of the presence of a shallow magma lens (<3 km) along most of the Reykjanes Ridge (north of 58°50′N) remains to be tested by future seismic field experiments.

The simple model presented here is a significant first step toward constructing a complete model for oceanic crustal genesis and toward relating various seafloor observations to the crustal-accretion processes and to the mantle upwelling systems. Many aspects of these processes have not been dealt with in this first-step study.

First, the current model does not include the effects of both the presence of melt and melt extraction on viscosity of the upper mantle as has been discussed in mantle-flow models.

Second, buoyancy effects due to density variations in the crust and mantle are not explicitly included in the model. The mantle-flow models (Parmentier and Phipps Morgan, 1990; Rabinowicz et al., 1993; Sparks and Parmentier, 1993; Sparks et al., 1993; Su and Buck, 1993; Jha et al., 1994) take into account various buoyancy effects in the upwelling mantle such as those due to melt depletion, melt retention, and other parameters. These effects have not been incorporated into the simple melt-production model (Chen, 1996) used here and are beyond the scope of this study. On the other hand, their effects would probably be significantly reduced by the modeling approach of including the efficiency factor β. Applying β for calculating the crustal thickness would calibrate the melt-production rate from these mantle-flow models against that from the simple model of this study at the normal mantle temperature (producing the 6-km-thick crust); therefore, it would significantly reduce the disagreement in the model output of the crustal thickness between the mantle-flow models and the model used in this study.

Third, there are two basic assumptions built into the model presented in this study: spreading-rate dependence of melt migration and a steady-state magma chamber. The efficiency factor β introduced in this study to describe the melt-migration process is assumed to depend only on spreading rate. Once it is chosen at the normal mantle temperature of 1350 °C for a given spreading rate, it will be kept constant for investigating the effects of mantle temperature and hydrothermal cooling. Although β would decrease for hotter mantle temperatures, similar to the argument for its spreading-rate dependence, a lower percentage of the melt produced in the mantle would make it to the ridge axis to form new crust because the partial-melting region becomes wider at hotter mantle temperatures. However, this subtle variation would be hard to quantify owing to the lack of either direct or indirect observations of the geometry of the mantle partial-melting region. Therefore, it would not be wise to introduce the dependence of β on mantle temperature; if one did, the results would probably not be very meaningful without any observational constraints. On the other hand, this effect would not change the general trend presented here.

The steady-state nature of the present model prevents investigation of time-dependent crustal accretion at slow-spreading ridges and the transient behavior of interactions between magma intrusion or eruption and rifting dynamics at mid-ocean ridges. Extension of the current model into a time-dependent model would enable the exploration of what time scales can appropriately be used to parameterize dike intrusion above the gabbro sections of the crust as a time-averaged, steady-state process despite the inherently time-dependent nature of individual dike-intrusion episodes. In addition, the magma chamber might be fed by episodic melt intrusion from underlying mantle, or the magma might be supplied laterally by along-axis flow within connected magma chambers that as a whole are fed by concentrated mantle upwelling at discrete ridge-segment centers as suggested from three-dimensional gravity studies at the slow-spreading Mid-Atlantic Ridge (Kuo and Forsyth, 1988; Lin et al., 1990; Lin and Phipps Morgan, 1992). The current two-dimensional model limits understanding of how the crustal accretion occurs along a segment if mantle upwelling is focused and if along-axis crustal flow results.

Finally, the most poorly understood model parameter is Nu, which describes the efficiency of the hydrothermal cooling at a ridge. However, with more seismic data of the crustal thickness and crustal structure and other data defining the mantle temperature becoming available, theoretical studies such as this one can be used to estimate the Nu for different ridges and thus provide important constraints on theoretical studies of hydrothermal circulation systems at the mid-ocean ridges.

Improvements in the model that are not explicitly part of this study but are important aspects

Episodic melt injection and temporal variability. By extending the models to be time dependent (by adding an additional iteration in the code), episodic melt-injection events and the related temporal variability in the seafloor observations could be investigated. A time-dependent model can be used to explore the following questions. To what extent and over what time scales would an intermittent process change the temperature structure of oceanic crust from the steady-state temperature structure? What is the life span t_0 of a magma-intrusion event, that is, how long it would take for the intruded dike or magma lens to be cooled significantly? Applying the model at different spreading rates, one can quantitatively determine the relationship of t_0 to spreading rate and compare these predictions with the suggested variation of the episodic nature of magma intrusions with spreading rate.

Additional melt-injection region. It would be beneficial to explore the effect of a second magma injection region either above the Moho or within the mush zone beneath the magma lens, as suggested by seismic observations of PmS on the East Pacific Rise (Garmany, 1989), ophiolite observations (Boudier and Nicolas, 1995; Kelemen et al., 1997), and melt-inclusion studies at the Gorda Ridge (Nielsen et al., 1995).

Various buoyancy effects. The buoyancy effects in the upwelling mantle such as those due to melt depletion and melt retention should be included to examine their effects on the mantle-flow field and the resulting total melt-production rate.

Extension to three-dimensional simulation. A three-dimensional model can be used to investigate along-axis variations in crustal accretion on a segment scale. In particular, it would enable testing of whether the along-axis crustal flow at fast-spreading ridges is sufficient to smooth the gravity, topography, crustal thickness, and other features as observed at the fast-spreading East Pacific Rise south of Garrett Fracture Zone at 14°S if mantle upwelling at fast-spreading ridges is three-dimensional.

CONCLUSION

A complete crustal-genesis model is presented in this study with a strong focus on investigating how anomalous mantle temperatures due to hotspot (or cold spot) and ridge interactions affect the crustal accretion and ridge-axis topography at mid-ocean ridges. The major progress of this new model is that it integrates all three processes (melt production, melt migration, and melt injection) of oceanic crustal genesis at mid-ocean ridges into one simple, self-consistent model, whereas previous models have only focused on one of these processes.

It is concluded that mantle temperature and hydrothermal cooling in addition to spreading rate play important roles in controlling crustal accretion and ridge-axis morphology at mid-ocean ridges. Model calculations show that there exists a critical mantle temperature and, for ridges at that temperature, a small temperature variation causes a lens-to-no-lens transition, which produces the significant change in ridge-axis thermal structure and topography through temperature-dependent rheologies. The sensitivity to the mantle-temperature anomaly is greatest at ridges spreading at intermediate rates as has been observed along the intermediate–spreading rate Southeast Indian Ridge, Galapagos spreading center, and the Juan de Fuca and Gorda Ridges. Because large parts of the slow-spreading Mid-Atlantic Ridge and these intermediate–spreading rate ridges are influenced either by a nearby hotspot or by relatively cold mantle as at the Australian-Antarctic discordance, it is shown in this study that ridge interactions with hotspots and anomalously cold mantle are equally important in controlling the crustal accretion and ridge-axis topography along these oceanic spreading centers.

In this study, the seafloor observations such as seismic surveys, bathymetry, and gravity are related to model parameters of spreading rate, mantle temperature, and hydrothermal cooling. The validity of the model would ultimately be tested by these observations. What is more important, specific model predictions could be forecast such as (1) a magma chamber would be present both at the inflated Reykjanes Ridge with an axial high and at these intermediate–spreading rate ridges with an axial high and (2) such a magma chamber would be absent at either the Australian-Antarctic discordance or the previously listed intermediate–spreading rate ridges with a well-developed rift valley. For example, it is unlikely that a magma chamber would be detected at the Gorda Ridge, which has a pronounced rift valley and which might be associated with a slightly colder mantle. On the other hand, given the observations of a thicker crust (7–10 km) associated with a higher mantle temperature (>30 °C) and little hydrothermal activity along the Reykjanes Ridge, the model predicts the presence of a shallow magma lens (<3 km) along most of the Reykjanes Ridge (north of 58°50′N) (Figs. 9B and 13). Although these model predictions would ultimately be tested by future field observations, they could help geoscientists to digest the currently available data and also to design future hypothesis-testing field experiments.

ACKNOWLEDGMENTS

I thank Suzanne Carbotte, Chris German, Roger Searle, and Martin Sinha for sending me their preprints and reprints and sharing their recently collected data. I thank Paul Wessel and Walter Smith for their graphics package GMT and Bob Parker for his graphics programs that are used to create the figures in this paper. Water Smith and Jochen Braunmiller helped to create Figure 1. David Sandwell and Norman Sleep provided constructive reviews that improved the manuscript. This work was partly supported by the College of Oceanic and Atmospheric Sciences, Oregon State University.

REFERENCES CITED

Batiza, R., Niu, Y., and Zayac, W.C., 1990, Chemistry of seamounts near the East Pacific Rise: Implications for the geometry of subaxial mantle flow: Geology, v. 18, p. 1122–1125.

Batiza, R., Niu, Y., Karsten, J.L., Boger, W., Potts, E., Norby, L., and Burtler, R., 1996, Steady and non-steady state magma chambers below the East Pacific Rise: Geophysical Research Letters, v. 23, p. 221–224.

Boudier, F., and Nicolas, A., 1995, Nature of the Moho Transition Zone in the Oman ophiolite: Journal of Petrology, v. 36, p. 777–796.

Brace, W.F., and Kohlstedt, D.L., 1980, Limits on lithosphere stress imposed by laboratory experiments: Journal of Geophysical Research, v. 85, p. 6248–6252.

Byerlee, J.D., 1978, Friction of rocks: Pure and Applied Geophysics, v. 116, p. 615–626.

Canales, J.P., Danobeitia, J.J., Detrick, R.S., Hooft, E.E.E., Bartolome, R., and Naar, D.F., 1997, Variations in axial morphology along the Galapagos spreading center and the influence of the Galapagos hotspot: Journal of Geophysical Research, v. 102, p. 27341–27354.

Carbotte, S., Mutter, C., Mutter, J., and Ponce-Correa, G., 1998, Influence of magma supply and spreading rate on crustal magma bodies and emplacement of the extrusive layer: Insights from the East Pacific Rise at lat 16°N: Geology, v. 26, p. 455–458.

Chen, Y., 1989, Dynamics of mid-ocean ridge systems [Ph.D. thesis]: Princeton University, 221 p.

Chen, Y., 1992, Oceanic crustal thickness versus spreading rate: Geophysical Research Letters, v. 19, p. 753–756.

Chen, Y.J., 1996, Constraints on the melt production rate beneath the mid-ocean ridges based on passive flow models: Pure and Applied Geophysics, v. 146, p. 590–620.

Chen, Y.J., 1998, A second melt lens at fast spreading mid-ocean ridges: EOS, Transactions of the American Geophysical Union, v. 79, p. F657.

Chen, Y., and Morgan, W.J., 1990a, Rift valley/no rift valley transition at mid-ocean ridges: Journal of Geophysical Research, v. 95, p. 17571–17581.

Chen, Y., and Morgan, W.J., 1990b, A nonlinear-rheology model for mid-ocean ridge axis topography: Journal of Geophysical Research, v. 95, p. 17583–17604.

Chen, Y.J., and Phipps Morgan, J., 1996, The effect of magma emplacement geometry, spreading rate, and crustal thickness on hydrothermal heat flux at mid-ocean ridge axes: Journal of Geophysical Research, v. 101, p. 11475–11482.

Cochran, J.R., Sempéré, J.-C., and SEIR Scientific Team, 1997, The Southeast Indian Ridge between 88°E and 118°E: Gravity anomalies and crustal accretion at intermediate spreading rates: Journal of Geophysical Research, v. 102, p. 15463–15487.

Collier, J., and Sinha, M., 1990, Seismic images of a magma chamber beneath the Lau Basin back-arc spreading centre: Nature, v. 346, p. 646–648.

Collier, J.S., and Sinha, M.C., 1992, Seismic mapping of a magma chamber beneath the Valu Fa Ridge, Lau Basin: Journal of Geophysical Research, v. 97, p. 14031–14053.

Cordery, M.J., and Phipps Morgan, J., 1992, Melting and mantle flow beneath a mid-ocean spreading center: Earth and Planetary Science Letters, v. 111, p. 493–516.

Detrick, R.S., Madsen, J.P., Buhl, P.E., Vera, J., Mutter, J., Orcutt, J., and Brocker, T., 1987, Multichannel seismic imaging of an axial magma chamber along the East Pacific Rise between 4°N and 13°N: Nature, v. 326, p. 35–41.

Detrick, R.S., Mutter, J., Buhl, P.E., and Kim, I.I., 1990, No evidence from multichannel reflection data for a crustal magma chamber in the Mid-Atlantic Ridge K area on the Mid-Atlantic Ridge: Nature, v. 347, p. 61–64.

Detrick, R.S., Harding, A.J., Kent, G.M., Orcutt, J.A., Mutter, J., and Buhl, P., 1993a, Seismic structure of the southern East Pacific Rise: Science, v. 259, p. 499–503.

Detrick, R.S., White, R.S., and Purdy, G.M., 1993b, Crustal structure of North Atlantic fracture zones: Reviews of Geophysics, v. 31, p. 439–458.

Dunn, R.A., and Toomey, D.R., 1997, Seismological evidence for three-dimensional melt migration beneath the East Pacific Rise: Nature, v. 388, p. 259–262.

Forsyth, D.W., 1993, Crustal thickness and the average depth and degree of melting in fractional melting models of passive flow beneath mid-ocean ridges: Journal of Geophysical Research, v. 98, p. 16073–16079.

Garmany, J., 1989, Accumulations of melt at the base of young oceanic crust: Nature, v. 340, p. 628–632.

German, C.R., and Parson, L.M., 1998, Distributions of hydrothermal activity along the Mid-Atlantic Ridge: Interplay of magmatic and tectonic controls: Earth and Planetary Science Letters, v. 160, p. 327–341.

German, C.R., Briem, J., Chin, C., Danielsen, M., Holland, S., James, R., Jonsdottir, A., Ludford, E., Moser, C., Olafsson, J., Palmer, M.R., and Rudnicki, M.D., 1994, Hydrothermal activity on the Reykjanes Ridge: The Steinaholl vent-field at 63°06′N: Earth and Planetary Science Letters, v. 121, p. 647–654.

Grindlay, N.R., Madsen, J., Rommevaus, C., and Sclater, J., 1996, Segmentation of the Southwest Indian Ridge 15°E–35°E: Implications for crustal accretion at ultra-slow spreading rates: Eos, Transactions of the American Geophysical Union, v. 77, p. F672.

Harding, A.J., Kent, G.M., and Orcutt, J.A., 1993, A multichannel seismic investigation of upper crustal structure at 9°N on the East Pacific Rise: Implications for crustal accretion: Journal of Geophysical Research, v. 98, p. 13925–13944.

Henstock, T.J., Woods, A.W., and White, R.S., 1993, The accretion of oceanic crust by episodic sill intrusion: Journal of Geophysical Research, v. 98, p. 4143–4154.

Hooft, E.E., and Detrick, R.S., 1995, Relationship between axial morphology, crustal thickness, and mantle temperature along the Juan de Fuca and Gorda Ridges: Journal of Geophysical Research, v. 100, p. 22499–22508.

Hooft, E.E., Detrick, R.S., and Kent, G.M., 1997, Seismic structure and indicators of magma budget along the southern East Pacific Rise: Journal of Geophysical Research, v. 102, p. 27319–27340.

Ito, G.T., and Lin, J., 1995, Oceanic spreading center-hotspot interactions: Constraints from along-isochron bathymetric and gravity anomalies: Geology, v. 23, p. 657–660.

Ito, G.T., Lin, J., and Gable, C., 1996, Dynamics of mantle flow and melting at a ridge-centered hotspot: Iceland and the Mid-Atlantic Ridge: Earth and Planetary Science Letters, v. 144, p. 543–574.

Jha, K., Parmentier, E.M., and Phipps Morgan, J., 1994, The role of mantle-depletion and melt-retention in spreading center segmentation: Earth and Planetary Science Letters, v. 125, p. 221–234.

Kelemen, P.B., and Dick, H.J.B., 1995, Focused melt flow and localized deformation in the upper mantle: Juxtaposition of replacive dunite and ductile shear zones in the Josephine peridotite, SW Oregon: Journal of Geophysical Research, v. 100, p. 423–438.

Kelemen, P.B., Koga, K., and Shimizu, N., 1997, Geochemistry of gabbro sills in the crust/mantle transition zone of the Oman ophiolite: Implications for the origin of the oceanic lower crust: Earth and Planetary Science Letters, v. 146, p. 475–488.

Kent, G.M., Harding, A.J., and Orcutt, J.A., 1990, Evidence for a smaller magma chamber beneath the East Pacific Rise at 9°30′N: Nature, v. 344, p. 650–653.

Kent, G.M., Harding, A.J., Orcutt, J.A., Detrick, R.S., Mutter, J.C., and Buhl, P., 1994, Uniform accretion of oceanic crust south of the Garrett transform at 14°15′S on the East Pacific Rise: Journal of Geophysical Research, v. 99, p. 9097–9116.

Kirby, S.H., 1983, Rheology of the lithosphere: Reviews of Geophysics, v. 21, p. 1458–1487.

Klein, E.M., and Langmuir, C.H., 1987, Global correlations of ocean ridge basalt chemistry with axial depth and crustal thickness: Journal of Geophysical Research, v. 92, p. 8089–8115.

Kong, L.S., Solomon, S.C., and Purdy, G.M., 1992, Microearthquake characteristics of a mid-ocean ridge along-axis high: Journal of Geophysical Research, v. 97, p. 1659–1685.

Kuo, B.-Y., and Forsyth, D.W., 1988, Gravity anomalies of the ridge-transform system in the South Atlantic between 31 and 34.5°S: Upwelling centers and variations in crustal thickness: Marine Geophysical Research, v. 10, p. 205–232.

Langmuir, C.H., Klein, E.M., and Plank, T., 1992, Petrological systematics of mid-ocean ridge basalts: Constraints on melt generation beneath ocean ridges, *in* Phipps-Morgan, J., Blackman, D.K., and Sinton, J.M., eds., Mantle flow and melt generation at mid-ocean ridges: Washington, D.C., American Geophysical Union, Geophysical Monograph v. 71, p. 183–280.

Lin, J., and Phipps-Morgan, J., 1992, The spreading rate dependence of three-dimensional mid-ocean ridge gravity structure: Geophysical Research Letters, v. 19, p. 13–16.

Lin, J., Purdy, G.M., Schouten, H., Sempéré, J.-C., and Zervas, C., 1990, Evidence from gravity data for focused magmatic accretion along the mid-Atlantic ridge: Nature, v. 344, p. 627–632.

Lister, C.R.B., 1974, On the penetration of water into hot rock: Royal Astronomical Society Geophysical Journal, v. 39, p. 465–509.

Lonsdale, P., 1994, Geomorphology and structural segmentation of the crest of the southern (Pacific-Antarctic) East Pacific Rise: Journal of Geophysical Research, v. 99, p. 4683–4702.

Macdonald, K.C., 1982, Mid-ocean ridges: Fine scale tectonic, volcanic, and hydrothermal processes within the plate boundary zone: Annual Reviews of Earth and Planetary Sciences, v. 10, p. 155–190.

Macdonald, K.C., 1986, The crest of the Mid-Atlantic Ridge: Models for crustal generation processes and tectonics, *in* Vogt, P., and Tucholke, B., eds., The Geology of North America: The Western North Atlantic region: Boulder, Colorado, Geological Society of America, v. M, p. 51–68.

Madsen, J.A., Grindlay, N.R., Rommevaus, C., and Sclater, J., 1996, Morphologic characteristic of ridge segments along the Southwest Indian Ridge, 15–35° E: Comparison between no major offset vs. transform dominated segments: Eos, Transactions of the American Geophysical Union, v. 77, p. F671.

McKenzie, D.P., 1984, The generation and compaction of partially molten rocks: Journal of Petrology, v. 25, p. 713–765.

McKenzie, D.P., and Bickle, M.J., 1988, The volume and composition of melt generated by extension of the lithosphere: Journal of Petrology, v. 29, p. 625–679.

Morgan, W.J., 1978, Rodriguea, Darwin, Amsterdam, . . . , A second type of hotspot island: Journal of Geophysical Research, v. 83, p. 5355–5360.

Morton, J.L., Sleep, N.H., Normark, W.R., and Tomkins, D.H., 1987, Structure of the southern Juan de Fuca Ridge from seismic reflection records: Journal of Geophysical Research, v. 92, p. 11315–11326.

Mutter, C.Z., 1995, Seismic and hydrosweep study of the western Costa Rica Rift: EOS, Transactions of the American Geophysical Union, v. 76, p. F595.

Mutter, J.C., Carbotte, S.M., Su, W., Xu, L., Buhl, P., Detrick, R.S., Kent, G.M., Orcutt, J.A., and Harding, A.J., 1995, Seismic images of active magma systems beneath the East Pacific Rise between 17°05′ and 17°35′S: Science, v. 268, p. 391–395.

Nielsen, R.L., Crum, J., Bourgouis, R., Forsythe, L.M., Fisk, M.R., and Christie, D.R., 1995, Melt inclusions in high-An plagioclase from the Gorda Ridge: An example of local diversity of MORB parent magmas: Contributed Papers of Mineralogical Petrology, v. 122, p. 34–50.

Niu, Y., and Batiza, R., 1997, Trace element evidence from seamounts for recycled oceanic crust in the eastern equatorial Pacific mantle: Earth and Planetary Science Letters, v. 148, p. 471–483.

Parmentier, E.M., and Phipps-Morgan, J., 1990, Spreading rate dependence of three-dimensional structure in oceanic spreading centres: Nature, v. 348, p. 325–328.

Phipps-Morgan, J., 1987, Melt migration beneath mid-ocean spreading centers: Geophysical Research Letters, v. 14, p. 1238–1241.

Phipps-Morgan, J., and Chen, Y.J., 1993a, The genesis of oceanic crust: Magma injection, hydrothermal circulation, and crustal flow: Journal of Geophysical Research, v. 98, p. 6283–6297.

Phipps-Morgan, J., and Chen, Y.J., 1993b, Dependence of ridge-axis morphology on magma supply and spreading rate: Nature, v. 364, p. 706–708.

Purdy, G.M., Kong, L.S.L., Christeson, G.L., and Solomon, S., 1992, Relationship between spreading rate and the seismic structure of mid-ocean ridges: Nature, v. 355, p. 815–817.

Quick, J.E., and Denlinger, R.P., 1993, Ductile deformation and the origin of layered gabbro in ophiolites: Journal of Geophysical Research, v. 98, p. 14015–14027.

Rabinowicz, M., Rouzo, S., Sempéré, J.-C., and Rosemberg, C., 1993, Three-dimensional mantle flow beneath mid-ocean ridges: Journal of Geophysical Research, v. 98, p. 7851–7869.

Ribe, N.M., Christensen, U.R., and Theissing, J., 1995, The dynamics of plume-ridge interaction. 1. Ridge-centered plumes: Earth and Planetary Science Letters, v. 134, p. 155–168.

Rohr, K.M.M., Milkereit, B., and Yorath, C.J., 1988, Asymmetric deep crustal structure across the Juan de Fuca Ridge: Geology, v. 16, p. 533–537.

Schilling, J.-G., 1991, Fluxes and excess temperatures of mantle plumes inferred from their interaction with migrating mid-ocean ridges: Nature, v. 352, p. 397–403.

Schilling, J.-G., Thompson, G., Kingsley, R.H., and Humphris, S., 1985, Hotspot-migrating ridge interaction in the South Atlantic: Nature, v. 313, p. 187–191.

Schilling, J.-G., Hanan, B.B., McCully, B., and Kingsley, R.H., 1994, Influence of the Sierra Leone mantle plume on the equatorial Mid-Atlantic Ridge: A Nd-Sr-Pb isotopic study: Journal of Geophysical Research, v. 99, p. 12005–12028.

Sempéré, J.-C., Cochran, J.R., and SEIR Scientific Team, 1997, The Southeast Indian Ridge between 88°E and 118°E: Variations in crustal accretion at constant spreading rates: Journal of Geophysical Research, v. 102, p. 15489–15505.

Shaw, P.R., and Lin, J., 1996, Models of ocean ridge lithospheric deformation: Dependence on crustal thickness, spreading rate, and segmentation: Journal of Geophysical Research, v. 101, p. 17977–17993.

Shen, Y., and Forsyth, D.W., 1995, Geochemical constraints on initial and final depths of melting beneath mid-ocean ridges: Journal of Geophysical Research, v. 100, p. 2211–2237.

Shen, Y., Forsyth, D.W., Scheirer, D.S., and Macdonald, K., 1993, Two forms of volcanism: Implications for mantle flow and off-axis crustal production on the west flank of the southern East Pacific Rise: Journal of Geophysical Research, v. 98, p. 17875–17899.

Sinha, M.C., Navin, D.A., MacGregor, L.M., Constable, S., Peirce, C., White, A., Heinson, G., and Inglis, M.A., 1997, Evidence for accumulated melt beneath the slow-spreading Mid-Atlantic Ridge: Royal Society Philosophical Transactions Series A, v. 355, p. 233–253.

Sleep, N.H., 1992, Hotspot volcanism and mantle plumes: Annual Reviews of Earth and Planetary Sciences, v. 20, p. 19–43.

Sleep, N.H., 1996, Lateral flow of hot plume material ponded at sublithospheric depths: Journal of Geophysical Research, v. 101, p. 28065–28083.

Small, C., 1995, Observations of ridge-hotspot interactions in the Southern Ocean: Journal of Geophysical Research, v. 100, p. 17931–17946.

Small, C., and Sandwell, D.T., 1989, An abrupt change in ridge axis gravity with spreading rate: Journal of Geophysical Research, v. 94, p. 17383–17392.

Small, C., and Sandwell, D.T., 1994, Imaging mid-ocean ridge transitions with satellite gravity: Geology, v. 22, p. 123–126.

Smith, W.H.F., and Sandwell, D.T., 1997, Global sea floor topography from satellite altimetry and ship depth soundings: Science, v. 277, p. 1956–1962.

Sparks, D.W., and Parmentier, E.M., 1993, The structure of three-dimensional convection beneath oceanic spreading centers: Geophysical Journal International, v. 112, p. 81–91.

Sparks, D.W., Parmentier E.M., and Phipps Morgan, J., 1993, Three-dimensional convection beneath a segmented spreading center: implications for along-axis variations in crustal thickness and gravity: Journal of Geophysical Research, v. 98, p. 21977–21996.

Su, W., and Buck, W.R., 1993, Buoyancy effects on mantle flow under mid-ocean ridge: Journal of Geophysical Research, v. 98, p. 12191–12205.

Su, W., Mutter, C.Z., Mutter, J.C., and Buck, W.R., 1994, Some theoretical predictions on the relationships among spreading rate, mantle temperature, and crustal thickness: Journal of Geophysical Research, v. 99, p. 3215–3227.

Tolstoy, M., Harding, A.J., Orcutt, J.A., and Phipps Morgan, J., 1995, Crustal thickenss at the Australian Antarctic discordance and neighboring South East Indian Ridge: Eos, Transactions of the American Geophysical Union, v. 76, p. F570.

Tolstoy, M., Harding, A.J., and Orcutt, J.A., 1997, Deepening of the axial magma chamber on the southern East Pacific Rise toward the Garrett Fracture Zone: Journal of Geophysical Research, v. 102, p. 3097–3108.

Manuscript Accepted by the Society February 23, 2000

Printed in the U.S.A.

Geological Society of America
Special Paper 349
2000

Comparison of volcanic construction in the Troodos ophiolite and oceanic crust using paleomagnetic inclinations from Cyprus Crustal Study Project (CCSP) CY-1 and CY-1A and Ocean Drilling Program (ODP) 504B drill cores

Hans Schouten
Department of Geology and Geophysics, Woods Hole Oceanographic Institution, Woods Hole, Massachusetts 02543, USA; e-mail: hschouten@whoi.edu
Charles R. Denham
U.S. Geological Survey, Woods Hole, Massachusetts 02543, USA

ABSTRACT

Paleomagnetic inclinations of lavas and dikes in CY-1 and CY-1A (Troodos ophiolite) and 504B (Costa Rica Rift) drill cores show remarkably similar patterns of postdepositional rotation about a horizontal axis. We interpret these patterns in terms of architecture of upper ophiolite and oceanic crust, lava deposition, and deformation at the paleo–spreading center. We recognize two components of postdepositional rotation about an axis parallel to the spreading axis: a variable rotation with depth due to lava-flow loading (zero at the top of the lava pile and generally increasing with depth) and a depth-invariant block rotation. At both drill sites, maximum rotation due to lava-flow loading dominates over estimated block rotation. Rotation (toward the spreading axis) due to lava-flow loading increases from zero at the top of the lava pile to a maximum of ~50°–70° at the base. Block rotation (away from the axis of spreading) is ~0° in 504B and ~20° in CY-1 and CY-1A. The component due to lava-flow loading predicts a narrower zone of lava deposition at the Costa Rica paleoridge (~3 km) than at the Troodos paleoridge (~5 km). This difference suggests that the Troodos spreading center had a wider rift valley or was more like the East Pacific Rise where lavas can flow several kilometers down the flanks unobstructed by large inward-facing fault scarps.

The fact that the sheeted dikes in Hole 504B and in the surface geology down section from Holes CY-1 and CY-1A are unrotated relative to the top of the lava pile implies an abrupt change in dip of 50°–70° between the sheeted dikes and the dikes in the transition zone from sheeted dikes to overlying lava flow dikes (where most dikes are orthogonal to the lava flows). We suggest that this angular discontinuity near the base of the transition zone at both sites represents a horizontal decoupling horizon that partitions deformation between the lavas and sheeted dikes. This interpretation is an alternative to the traditional views that the discontinuity is evidence of a normal fault that cuts 504B and the axis of an abandoned spreading center that crosses the Akaki River section of the Troodos ophiolite at this stratigraphic level. We speculate that the contrasting kinematic histories recorded in the lava section and the sheeted dikes in Hole 504B and in and near Holes CY-1 and CY-1A reflect their contrasting mechanical response to subsidence due to lava-flow loading, i.e., the lava pile deformed primarily by bending and the sheeted dike section deformed by vertical slip along closely spaced planes parallel to the dikes.

INTRODUCTION

Despite several decades of scientific drilling, the architecture of upper oceanic crust remains poorly understood. This statement is not surprising because there are few holes deeper than a few hundred meters into oceanic basement and a sufficiently detailed framework for the interpretation of the drilling observations is only now beginning to emerge. We now have a critical volume of high-resolution geophysical and geologic observations of many spreading centers—the seats of construction of upper oceanic crust—that provide the appropriate framework for the interpretation of the drilling data. The most relevant component of this framework is that volcanic construction at mid-oceanic ridge spreading centers typically occurs over a broad zone, ~4–8 km wide, which predicts that the basic structure

Schouten, H., and Denham, C.R., 2000, Comparison of volcanic construction in the Troodos ophiolite and oceanic crust using paleomagnetic inclinations from Cyprus Crustal Study Project (CCSP) CY-1 and CY-1A and Ocean Drilling Program (ODP) 504B drill cores, *in* Dilek, Y., Moores, E.M., Elthon, D., and Nicolas, A., eds., Ophiolites and Oceanic Crust: New Insights from Field Studies and the Ocean Drilling Program: Boulder, Colorado, Geological Society of America Special Paper 349, p. 181–194.

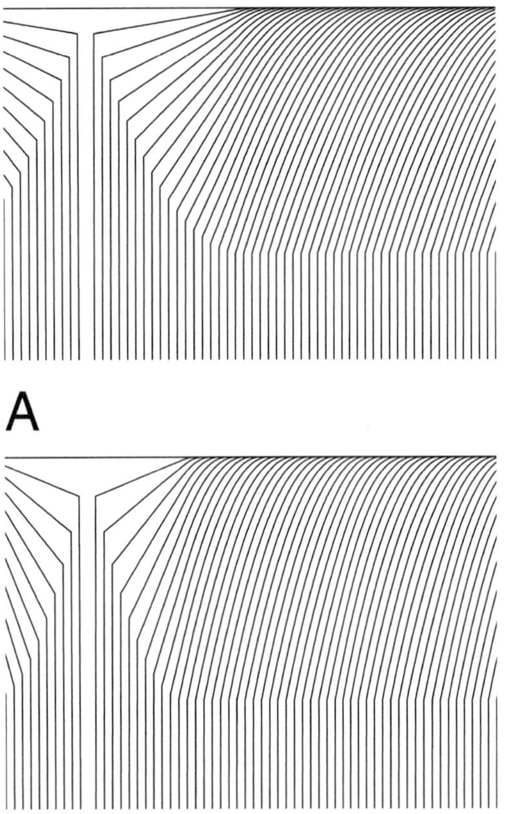

A

B

Figure 1. The effect of varying flow shape at constant cross-sectional area. Perfect ocean-floor spreading (zero width to the zone of dike injection) and vertical compensation are assumed. A, Assemblage of multiple dike-flow units with thinner, longer flows. B, Assemblage of thicker, shorter flows. Note how the flows in B have a steeper dip and are more deformed than those in A (after Cann, 1974).

of upper oceanic crust consists of overlapping lava flows dipping toward the axis (Fig. 1). Further, if this prediction is correct, then postdepositional rotation due to lava burial or lava-flow loading (e.g., Bodvarsson and Walker, 1964; Cann, 1970, 1974; Pålmason, 1973, 1980; Kidd, 1977; Rosencrantz, 1982) may be an important process shaping the architecture of upper oceanic crust.

Off-axis volcanism on the upper flanks of the East Pacific Rise has been invoked to explain the distribution of anomalously young ages (e.g., Goldstein et al., 1994) and the occurrence of fresh lava flows and constructional volcanic features well outside the axial trough (e.g., Gregg et al., 1996; Macdonald et al., 1996). However, these observations are too few and local in nature to assess the general magnitude of the lava-deposition process. On the other hand, multichannel seismic experiments have shown a ubiquitous pattern of layer 2A thickening along the East Pacific Rise (e.g., Toomey et al., 1990; Vera and Diebold, 1994; Harding et al., 1993; Kent et al., 1994; Carbotte et al., 1997). Instead of reaching its full thickness at the axis, layer 2A increases to its full thickness over a finite spatial interval, usually within ~2–4 km of the spreading axis. This pattern is generally attributed to off-axis lava deposition resulting from overflow of lava from the axial summit trough and/or off-axis eruptions (e.g., Christeson et al., 1994; Perfit et al., 1994). A recent analysis (Schouten et al., 1999) of near-bottom magnetic measurements of the CAMH (central anomaly magnetization high) and layer 2A thickening in the East Pacific Rise at its intersection with lat 9°–10°N concludes that seismic layer 2A in young oceanic crust represents the highly magnetic lava layer (Fig. 2C compares magnetic layer thickness and seismic layer 2A). This correlation between the magnetic lava layer and seismic layer 2A has several important implications: (1) the base of layer 2A in young oceanic crust is not an alteration front (e.g., Karson, 1998), but the transition from lavas to sheeted dikes, and (2) construction of the lava layer takes place over the full width of layer 2A thickening, which on the East Pacific Rise typically extends 2–4 km from the axis (Schouten et al., 1999).

Whether this lava is fed uniquely from the axial summit trough and transported over the 2–4 km through lava tubes or channels, or is erupted from widely distributed off-axis fissures and deposited locally, is not known. The relatively smooth thickening of layer 2A and its fairly constant full thickness off axis on the East Pacific Rise suggests a steady component to the process of lava deposition. We hypothesize that the steady component of the lava-deposition process (how is lava deposition distributed about the spreading center on the average?) is information we can extract from drill cores. This information will give us an average architecture of the lava layer as well as spatial constraints on the paleo–ridge environment where the lava originally was deposited. In this paper, we consider postdepositional rotation about a horizontal axis of the lava, estimated from rotations of the paleomagnetic vector measured in drill cores, to estimate the volcanic architecture of oceanic crust and ophiolite.

POSTDEPOSITIONAL ROTATION IN UPPER OCEANIC CRUST ABOUT A HORIZONTAL AXIS

At a seafloor spreading center, magma erupts out of the ocean floor through fissures that cut across previously erupted lavas. The result is a layer composed of overlapping lava flows and crosscutting feeder dikes. The structure of this layer depends on how the lava deposition is distributed about the spreading center, how dike injection is distributed, and how the lava pile subsides in response to lava-flow loading (also known as lava burial). If the subsidence is by vertical slip along closely spaced vertical planes (vertical isostatic compensation [Cann, 1974]), then there will be no rotation. If the compensation is by bending, then the postdepositional rotation of the lavas will be related to the way lava deposition was distributed about the spreading center. This rotation should be zero at the top of the lava pile and increase with depth of burial as a function of the distribution of lava deposition (see Fig. 1).

As the seafloor moves away from the spreading center, the upper crust undergoes extensional faulting and associated block rotation. Block rotation is by definition constant throughout the block and therefore depth invariant for the thickness of the block. We assume that the blocks include at least the lava pile and the sheeted dikes (Agar and Klitgord, 1995). We postulate that the postdepositional rotation measured in upper oceanic crust is the sum of two rotations about an axis parallel to the spreading center: a depth-dependent rotation due to lava loading and a depth-invariant block rotation. Block rotation equals the postdepositional rotation in the top of the lava pile, where the rotation due to lava-flow loading is zero, by definition.

In models for the structure of oceanic crust (e.g., Bodvarsson and Walker, 1964; Cann, 1970, 1974; Pålmason, 1973, 1980; Kidd, 1977; Rosencrantz, 1982), differential subsidence due to lava-flow loading is dependent on two spatial distributions, i.e., lava deposition and dike injection. For models in which the dike-injection width is much less than the lava-deposition width, which appears to be the

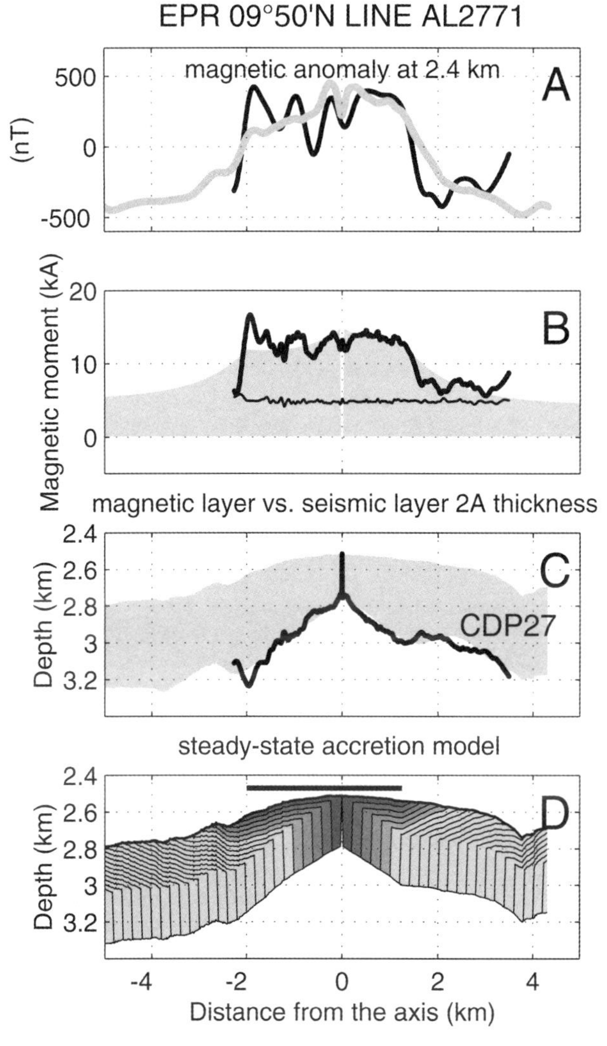

Figure 2. Example of deconvolution and modeling of observed magnetic data to estimate the architecture of the magnetic source beneath near-bottom magnetic profile AL2771 across the East Pacific Rise (EPR). A, Observed magnetic anomaly (bold line) and model anomaly (shaded line), calculated at 2.4 km water depth. B, Magnetic moment (magnetization × 0.33 km) (bold line), annihilator × 0.33 km × 15 (thin line), and the model magnetic moment (shaded). C, Nearby seismic layer 2A thickness from CDP 27 (common depth point; shaded), as determined by Harding et al. (1993), and the base of the magnetic layer (bold line) obtained by deconvolution of the observed magnetic moment for lava distribution, which was then integrated for magnetic layer thickness. Minimum rms (root mean square) misfit between seismic layer 2A thickness and magnetic thickness is 31 m. Minimization of this misfit yielded an independent estimate of the lava magnetization function with a decay constant of ~20 k.y., which is identical to the function proposed by Gee and Kent (1994). D, A forward model of the magnetic layer for a steady-state lava distribution. Isochrons are every 4 k.y. at 50 km/m.y. half rate. Darker shades reflect higher magnetization of younger lava flows. Horizontal bar shows extent of lava deposition and thickening of the magnetic lava layer in the model, which matches the width of the central anomaly magnetic high (CAMH) anomaly in A (after Schouten et al., 1999).

case on the East Pacific Rise (e.g., Hooft et al., 1996), this subsidence becomes dependent on lava deposition only (see Fig. 1). We assume that this differential subsidence occurs only by bending (rotation about a horizontal axis parallel to the ridge crest). Then, if we know the rotation down a drill core, we should be able to derive lava-deposition distribution from it. Alternatively, we can derive the lava-deposition distribution from the horizontal derivative of the thickening of layer 2A and predict the postdepositional rotation due to lava loading down an imaginary core. This option exists because, for the dike-in-dike injection case, lava-layer thickness is the integral (starting at the axis) of the lava-deposition distribution (e.g., Schouten et al., 1999).

In Figure 3 we show an example of extrusive-layer architecture derived from seismic layer 2A thickness of multichannel seismic line 1107 which crosses the East Pacific Rise at 17°25′S (Carbotte et al., 1997). Note the variable isochron dips in Figure 3, plots D and E, which are directly related to postdepositional rotation (if subsidence is by bending only), layer 2A thickening, and lava-deposition distribution. As in Figure 2, the width of the (synthetic) CAMH anomaly in Figure 3B equals the width over which seismic layer 2A thickens, i.e., the width of the lava-deposition (neovolcanic) zone.

Postdepositional rotation of lava samples from drill cores in oceanic crust and ophiolites can be estimated from the stable paleomagnetic inclination if we know the paleomagnetic field vector and the orientation of the rotation axis. This axis of deformation at a seafloor spreading center is postulated to parallel the axis of spreading. Because drill cores typically provide no azimuthal information, there are four possible postdepositional-rotation solutions for a given inclination, which depend on whether the paleofield was normal or reversed and whether the paleomagnetic declination in the sample location in the hole is pointing to the left or right of the postdepositional rotation axis. Furthermore, as pointed out by Kidd (1977), the postdepositional rotation of core samples is not necessarily the same as the dip of the isochrons. In an extreme case, differential subsidence due to lava-flow loading can be accommodated exclusively by downdip slip along vertical faults (vertical compensation), "bending" the isochrons but without actually rotating the rocks (see Fig. 1). Thus, in the presence of vertical faulting, estimates of postdepositional rotation in drill-core samples will underestimate the differential subsidence and provide, consequently, a maximum estimate of the length scales of lava deposition.

Differential subsidence along subvertical faults is more likely to occur near the rifting axis than far from it. Seismic observations of a finite layer 2A thickness at the axis require that several hundred meters of extrusive rocks accumulate within an axial zone narrower than a seismic wavelength (a few hundred meters) (Hooft et al., 1996). This narrow peak of lava deposition at the axis (model lava distribution, Fig. 3) requires differential subsidence of as much as 2 or more, which is probably accommodated by a combination of vertical faulting and rotation rather than exclusively by rotation. Consequently, in the lower half of the lava layer, which was deposited nearest to the axis where vertical faulting is presumed rife, the postdepositional rotation determined from core paleomagnetic inclinations will likely underestimate the differential subsidence due to lava-flow loading. In the upper half of the lava layer, deposited farther from the axis where lava-flow loading is gentler and bending may be the dominant subsidence mode, the postdepositional rotation determined from the drill core will provide a more accurate estimate of the subsidence, lava-deposition distribution, and architecture.

The examples in Figures 2 and 3 provide a first-order estimate of the type (length scale, order of magnitude, character) of extrusive

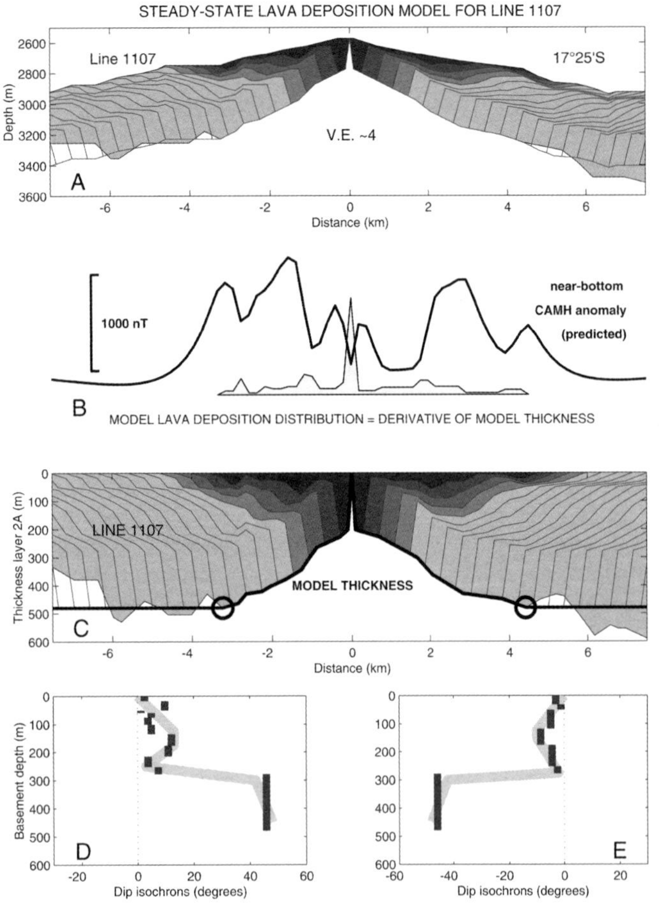

Figure 3. A, Architecture of magnetic lava layer predicted from layer 2A thickness of multichannel seismic line 1107 crossing the East Pacific Rise at 17°25′S (Carbotte et al., 1997), shown in grays. B, Model lava-deposition distribution (shaded) is derivative of model thickness (shown with bold line in C). Near-bottom CAMH magnetic anomaly (bold line) calculated at 200 m above top of model, for half spreading rate of 75 km/m.y. and assuming that lava magnetization decreases 5.5-fold with a decay time of 20 k.y. (darker shades indicate higher magnetization in A and C). C, Isochrons (drawn every ~2.5 k.y.) based on the model distribution show the predicted architecture of the extrusive layer. Steep ~45° dip of isochrons in lower 250 m reflects the narrow peak of on-axis lava deposition, which is responsible for the almost instantaneous growth of ~200 m lava layer at the axis, required by the seismic observations; gentle <10° dip of isochrons in upper layer reflects the broad off-axis lava deposition, which is responsible for the gentle 2A thickening observed in the seismic record out to ~4 km from the axis. Model thickness (bold line) follows observed thickness until open circles indicating limits of 2A thickening estimated by Carbotte. D and E, Dip of model isochrons vs. basement depth.

architecture that may be associated with a fast-spreading mid-oceanic ridge. At slower-spreading mid-oceanic ridges, seismic layer 2A thickening is much harder to observe because of topographic roughness. However, it is reasonable to assume that most spreading centers will have a spatial distribution of lava deposition characterized by large amounts of lava deposited near the axis and less lava deposited with distance away from the axis. Off the axis of slow-spreading centers and those spreading at intermediate rates, lava may pond in flanking lows or may be truncated against normal-fault scarps. Off the axis of fast-spreading centers such as the East Pacific Rise, lavas may drape the axial high's gentle upper slopes (see Figs. 2 and 3). These differences in lava deposition and the resultant architecture of the extrusive layer must, to some degree, be reflected in the postdepositional rotation, which we can estimate from the paleomagnetic inclinations in drill cores.

EXTRUSIVE ARCHITECTURE FROM CCSP HOLES CY-1 AND CY-1A (FIG. 4)

We show how drill-core data, and paleomagnetic inclinations in particular, can be used effectively to predict architecture of upper oceanic crust. We describe here the results of a pilot study of the extrusive architecture of upper crust from Cyprus Crustal Study Project (CCSP) Holes CY-1 and CY-1A in the Troodos ophiolite, Cyprus.

Holes CY-1 and CY-1A were drilled in Upper and Lower Pillow Lavas exposed along the Akaki River canyon located on the northern flank of the Troodos ophiolite. The holes are ~1 km apart, and together they constitute an almost complete section of the extrusive Troodos upper crust (Gibson et al., 1991). Stable paleomagnetic inclinations (Jim Hall, 1997, personal communication), dike fractions (based on the lithologic log of Malpas and Barnable, 1991), and a steady-state solution for extrusive architecture based on the averaged inclination in the CY-1 and CY-1A cores are shown in Figure 4. Paleomagnetic inclinations and dike fraction are plotted against composite depth, where CY-1A is suspended below CY-1 from 500 m (Vine and Smith, 1990). The data points marked by x in Figure 4A denote possible dikes or sills (i.e., intrusive rocks); the open circles denote probable pillows (i.e., extrusive rocks). Despite the scatter of the individual measurements, the averages of the pillow inclinations (heavy dots with error bars in Fig. 4A) display a sinusoidal change in inclination with composite depth in the core. We do not use the magnetic inclinations of the intrusive dikes or sills, which are later and therefore always less rotated than the pillows they intrude. Inclination maxima are located at the top and at ~500–600 m depth in the hole; minima are at ~100–400 m and at ~900–1100 m. The second maximum coincides with the transition (from 515–665 m depth) from relatively depleted low-Ti lavas (LTL) to undepleted high-Ti lavas (HTL) in the core. These chemistries loosely correlate with the Troodos Upper and Lower Pillow Lavas, respectively (e.g., Robinson et al., 1983). The Upper Pillow Lavas were deposited later than the Lower Pillow Lavas. In a seafloor-spreading environment, later deposition of lavas is equivalent to lava deposition farther from the spreading axis; earlier deposition is equivalent to deposition closer to the axis. We interpret the sinusoidal inclination variation to reflect a postdepositional rotation of the core samples due to lava-flow loading.

The paleo–field vector for the Troodos ophiolite is normal (Cretaceous super chron) and plunges 36°W (36°/274°W [Clube et al., 1985]). The orientation of lava-flow surfaces and dike margins near Holes CY-1 and CY-1A (Fig. 6) defines a structural axis that dips gently north (related to doming during emplacement of the ophiolite), so that vector and the postdepositional rotation axis are approximately orthogonal. This geometry implies that a 1° rotation about this axis equals an ~1° inclination change in the CY-1 and CY-1A core samples. Because the cores do not provide azimuthal information but the paleopolarity (normal) is known, there are now two possible postdepositional rotation solutions for the inclinations in the CY-1/1 core samples, i.e., the declination of the paleomagnetic vectors in the CY-1 and CY-1A points either approximately west or east. The mean inclination of the pillows in the CY-1 core is 26° ± 10° (Fig. 4A). Assuming the westward declination of the paleomagnetic vectors, this

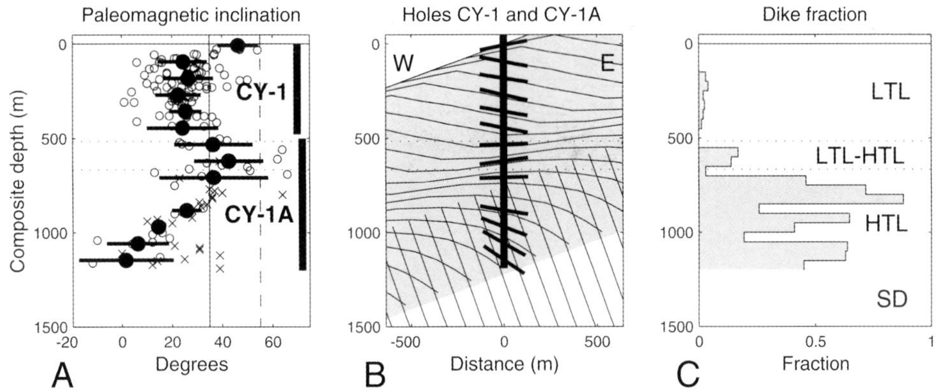

Figure 4. Extrusive architecture from CCSP Holes CY-1 and CY-1A, Troodos ophiolite, Cyprus. A, Paleomagnetic inclination and dike fraction in CCSP Holes CY-1 and CY-1A. CY-1A is suspended beneath CY-1 starting at 500 m composite depth (see bold vertical bars). Circles—stable paleomagnetic inclinations of lavas, x—those of dikes (Jim Hall, 1997, personal communication). Heavy dots with horizontal limits are 87.5 m core-length averages of the lava inclinations with their standard deviation. Paleo–field inclination is 36°W (vertical thin line). Vertical dashed line at 56° represents the paleo–field inclination shifted by block rotation (estimated at 20° to the west). Horizontal dotted lines at 515 and 665 m bracket transition from low-Ti lava (LTL) in upper part to high-Ti lava (HTL) in lower part of section. B, Model of architecture of upper ophiolite crust based on drill cores CY-1 and CY-1A. Steady-state lava-deposition isochrons are compared with postdepositional rotation (short lines crossing bold vertical line in center) based on the averaged inclinations shown in A. Shading indicates the lava pile. Steeply dipping parallel lines schematically mark transition zone from lava flows to dikes and underlying sheeted dikes (SD). This architecture indicates that the Troodos paleo–ridge crest was to the east of CY-1 and CY-1A. C, Dike fraction based on lithologic log compiled by Malpas and Barnable (1991). High dike fraction in lower part of section is indicative of transition zone from lava flows to dikes.

inclination predicts a mean postdepositional rotation toward the east of 10° ± 10°. This result is consistent with field studies in the upper Akaki section (e.g., Schmincke and Rautenschlein, 1987; Schmincke and Bednarz, 1990; Varga et al., 1998) that show an average ~20°N dip of lava-flow surfaces, which means approximately zero postdepositional rotation about a ~20°N-plunging rotation axis (see also Fig. 6B). (The alternative solution of the eastward declination of the paleomagnetic vectors in the hole produces an unacceptably high mean postdepositional rotation of the core samples of 117° ± 12° toward the west.) Consequently, we assume that the declination of the paleomagnetic vectors in Holes CY-1 and CY-1A is to the west and that the postdepositional rotation of the core samples approximately equals the paleo–field inclination (36°, thin vertical line in Fig. 4A) minus the paleomagnetic inclination (a positive postdepositional rotation implies rotation toward the east).

In Figure 5A, we compare the postdepositional rotations based on CY-1 and CY-1A lava-inclination data (small dots with error bars) with some field observations (numbered open circles) in the Akaki River canyon. Approximate location of these observations is shown with the corresponding number on the geologic map in Figure 5B. Observations 1–4 are converted to postdepositional rotation using an axis plunging 17° towards 352° (see Fig. 6B), an ophiolite emplacement rotation of 17°N, and assuming the lava flow surfaces were horizontal at deposition. Observation 1: Chalks at sediment-lava interface in Akaki canyon dip 15°NW (Schmincke and Rautenschlein, 1987, Stop 1). Observation 2: UPL lava surfaces dip ~20° towards 100° (Schmincke and Rautenschlein, 1987, Stop 2). Observation location 3: Average sheet flow dips 17° towards 100° (Varga et al., 1998, Kamara Potamos, Fig. 5B). Observation 4: Average sheet flow dips 18° towards 330° (central Akaki, this study, Fig. 6B). Observation 5: Average sheet flow dips 24° towards 261° (Klirou bridge, this study, Fig. 6A) and 32° towards 273° (Akaki canyon, Varga et al., 1998, Fig. 5A). Observation 6: Tops of lobate pillows dip 50° towards the east and cross-cutting dikes dip 40° towards the west (Xenophontos and Malpas, 1987, Stop 15). Observation 7: Average sheeted dike dips 72° ± 9° towards 108° (Malpas et al., 1989). Observation 3 is at the top of the LTL-HTL transition mapped along the Akaki river and Observation 4 is in the middle of the transition (e.g. Schmincke and Rautenschlein, 1987). The comparison between drill-core data and field observations in Figure 5A shows that predictions based on paleomagnetic inclinations in the CY-1 and CY-1A cores agree quite well with rotations based on dips of sheet flows, pillow surfaces, and dikes observed in the Akaki River canyon.

We postulate that the postdepositional rotation in CY-1 and CY-1A has two principal components: a depth-invariant rotation due to block faulting and a variable (with depth) rotation due to lava-flow loading that is zero at the top of the lava pile and that increases with depth. In a seafloor-spreading lava-deposition environment, the very top of the lava section should be unaffected by lava loading and therefore record only the component due to block rotation. The inclination in the top 50 m of CY-1 and CY-1A is 46° ± 8°; thus the top was rotated 10° ± 8° to the west. We estimate a block rotation of 20° to the west to avoid modeling problems caused by negative postdepositional rotation due to lava loading and consequent negative lava loads.

Figure 4B displays a steady-state solution for extrusive architecture based on the averaged inclination data shown in Figure 4A. The solution consists of a depth-invariant 20° westward block rotation plus a variable (with depth) eastward rotation, which we interpret as being due to lava-flow loading. The overlapping isochron structure (thin lines) reflects the postdepositional rotation derived from the averaged inclinations (the short lines crossing the heavy vertical line that represents the drill core). The sense of overlap of isochrons (i.e., lava flows) implies that the spreading center was east of CY-1 and CY-1A. The average lava-deposition distribution (i.e., the horizontal derivative of the isochrons in Fig. 4B) is shown in Figure 7A. The fact that this lava-deposition distribution is distinctly bimodal shows that the HTL section was deposited within ~1 km of the axis and the LTL section was deposited in a separate depocenter ~2 km from the axis. The LTL-HTL transition (bracketed by the vertical dotted

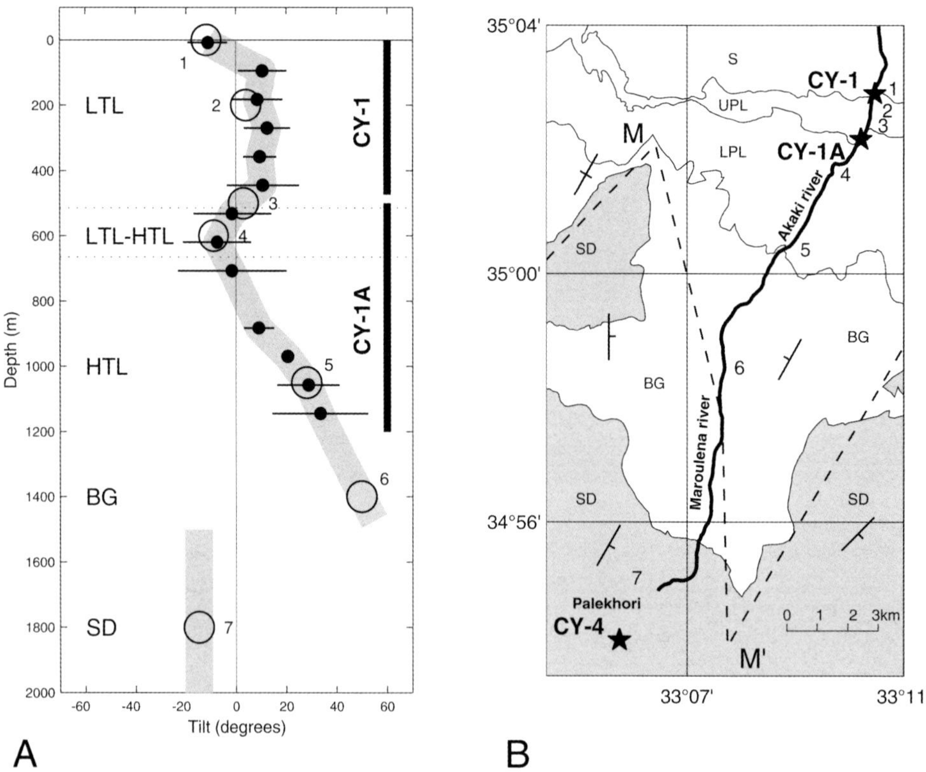

Figure 5. A, Average postdepositional rotation about a gently north-dipping axis from CY-1 and CY-1A core inclinations (small dots with horizontal error bars) compared with average rotations from field observations along the Akaki River section (open circles). 1, 2—Schmincke and Rautenschlein, (1987); 3, 5—Varga et al., (1998); 4, 5—this study (Fig. 6); 6—Xenophontos and Malpas (1987); 7—Malpas et al., (1989). Observations 1–5 are shown in stereonets in Figure 6. LTL, LTL-HTL, HTL—see Figure 4 caption. BG—Basal Group (~75%–95% dikes). SD—Sheeted dikes (>95% dikes). Field rotations are based on dips of sheet flows, pillow surfaces, and dikes, assuming that rotation axis plunges gently north (related to doming during the emplacement of the ophiolite). Field observations generally confirm average rotation structure predicted from the drill cores. B, Generalized geologic map of the Akaki and Maroulena river system with average strike and dip direction of the dikes indicated. S—sediments; UPL—upper pillow lava; LPL—lower pillow lava. Numbers indicate location of field observation shown in A. Stars indicate drill sites CY-1, CY-1A, and CY-4. Lowest unit (SD) is shaded. Dashed line is northern limit of east- to southeast-dipping dikes. M-M′ is axis of Mitsero graben. We suggest the dashed line represents a decoupling horizon that partitions deformation between the lavas and the sheeted dikes.

lines in Fig. 7A) represents a slight overlap of a proximal HTL depocenter and a distal LTL depocenter. Lava deposition occurs out to several kilometers from the axis (total width, ~5 km), which is a length scale quite similar to what is observed at the present East Pacific Rise.

EXTRUSIVE ARCHITECTURE FROM DSDP-ODP HOLE 504B (FIG. 8)

Deep Sea Drilling Project–Ocean Drilling Program (DSDP-ODP) Hole 504B was drilled ~200 km south of the Costa Rica Rift in 6.8 Ma crust (chron 3Ar) formed at an intermediate spreading rate near the paleomagnetic equator (zero inclination) at a time when Earth's magnetic field was reversed (e.g., Cann et al., 1983). Paleomagnetic north at 504B equals the geographic north, and we assume that postdepositional rotations due to lava-flow loading and block rotation are about a horizontal, east-west–oriented axis (parallel to the magnetic lineations and the active Costa Rica Rift). Thus we can equate degrees of paleomagnetic inclination in the drill core with degrees of rotation subsequent to magnetization, and, if the lavas were deposited on a horizontal surface—which for simplicity we assume—then paleomagnetic inclination equals dip of the lava horizons in this drill core. On the basis of the average paleomagnetic inclination in the upper 550 m of pillow lavas, Furuta and Levi (1983) speculated that the entire upper pillow section might be tilted toward the axis by ~30°. On the other hand, near-zero paleomagnetic inclinations in the underlying sheeted dikes (below ~1050 mbsf [meters below seafloor]) suggested no tilt at all (e.g., Smith and Banerjee, 1986).

In a manner similar to that described in the previous example, we derive the architecture of upper oceanic crust from paleomagnetic inclinations of the lavas in Hole 504B (Fig. 8). The geometry is simple. The paleo–field vector is reversed equatorial, and the postdepositional rotation axis is oriented east-west, parallel to the Costa Rica Rift to the north of Hole 504B. Consequently, inclination approximately equals rotation about the horizontal east-west–oriented axis (negative inclination implies rotation toward north). Averaged inclinations show progressive steepening from subhorizontal in the top of the core at the interface between the sediment and basement (275 mbsf) to −50° at 900–1000 mbsf within the transition zone from lava flows to dikes. The fact that the declination of the paleomagnetic vectors in the hole is also known (180°; Kinoshita et al., 1989) implies that the top of the crust is unrotated or subhorizontal. Thus the postdepositional rotation of the lavas increases with depth from zero

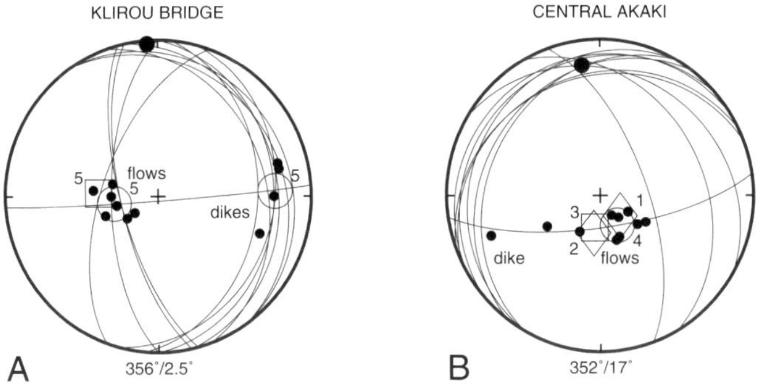

Figure 6. Equal-area stereonets showing poles to (small dots) and great circles for lava-flow surfaces and dike margins measured in the immediate vicinity of (A) Klirou bridge (location 5 in Fig. 5) and (B) central Akaki (location 4 in Fig. 5). Open circles numbered 4 and 5—average pole to the lava-flow surfaces and dike margins. Partial girdle to poles to lava flow surfaces and dike margins defines structural axis (heavy dot). Strike and plunge of structural axis are indicated. At Klirou bridge (A), average postdepositional rotation about the local axis is ~25° to the east. Note orthogonal relationship of flows and dikes at Klirou bridge. In central Akaki (B), average postdepositional rotation about the local axis is ~8° to the west. Open diamonds—average pole to chalks and lava flow surfaces measured by Schmincke and Rautenschlein (1987) at sediment-lava interface (location 1 in Fig. 5) and in the UPL (location 2 in Fig. 5), respectively. Open squares—average pole to lava-flow surfaces measured by Varga et al. (1998) in Kamara Potamos (location 3 in Fig. 5) and in Akaki River canyon (location 5 in Fig. 5). Postdepositional rotation derived from open symbols is plotted in Figure 5A.

to ~50° northward at 900–1000 mbsf (i.e., just above the sheeted dikes [>1050 mbsf]). This result is consistent with rotation due to lava-flow loading with the spreading center to the north and insignificant block rotation. The horizontal derivative of the isochrons equals the lava-deposition distribution shown in Figure 7B. This distribution indicates lava deposition out to ~1.5 km from the axis, which is distinctly narrower than the distribution derived for CY-1 and CY-1A (Fig. 7A).

Paleomagnetic inclinations of the pillows and dikes in 504B indicate an angular unconformity at ~1000 mbsf near the base of the transition zone from lava flows to dikes (Fig. 8A). Immediately above this level (750–1000 mbsf), the mean pillow inclination is −49° ± 18°, whereas the last pillows sampled below this level (1000–1050 mbsf) have an inclination of −8° ± 11° or <10° rotation. The dikes below ~1000 mbsf are subvertical (<10° rotation) on the basis of paleomagnetic inclinations (e.g., Smith and Banerjee, 1986; Pariso and Johnson, 1989; Allerton et al., 1995) and dips of chilled margins of the dikes in the cores (e.g., Dick et al., 1992; Allerton et al., 1995). This finding suggests that pillows and dikes below ~1000 mbsf are virtually unrotated (subvertical) whereas pillows immediately above this level are rotated by as much as ~50°. Pillows at the top of the 504B lava pile are also unrotated, which is interpreted to indicate zero block rotation.

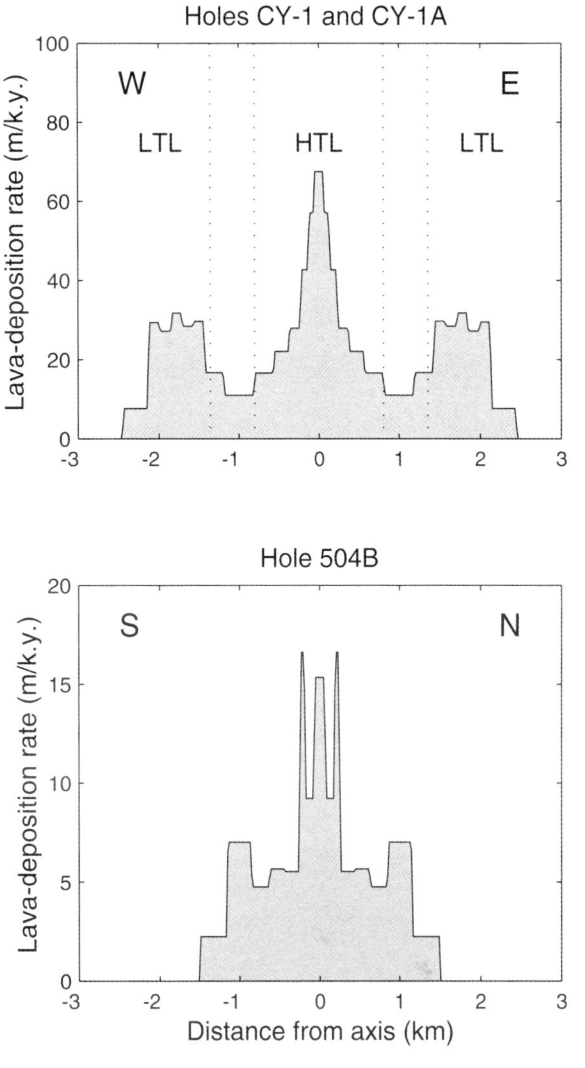

Figure 7. A, Lava-deposition thickness distribution derived from the averaged paleomagnetic inclinations in CY-1 and CY-1A and plotted symmetrically about a paleo–spreading axis. The clearly bimodal lava-deposition thickness distribution shows that the HTL section is formed within ~1 km of the axis and the LTL section is formed in a distal depocenter ~2 km from the axis (vertical dotted lines bracket the LTL-HTL transition). Total width of lava deposition is ~5 km, which suggests that the Troodos paleo–spreading center had a wide rift valley or was more like the East Pacific Rise where lavas can flow several kilometers down the flanks unobstructed by large inward-facing fault scarps. Lava-deposition rate calculated for 100 km/m.y. total spreading rate. B, Lava-deposition distribution derived from the averaged inclinations in 504B and plotted symmetrically about a paleo–spreading axis. The lava-deposition thickness distribution shows that the extrusive section is deposited over a total width of ~3 km, which is consistent with the narrower confines of a rift valley spreading at an intermediate rate. Lava-deposition rate calculated for 60 km/m.y. total spreading rate.

DISCUSSION

The two examples in Figures 4A and 8A are a graphic demonstration of a one-dimensional (1-D) architecture of upper ophiolite and oceanic crust that can be derived from paleomagnetic inclinations in drill cores. We have shown the presence of large, depth-dependent, postdepositional rotations in both cores, which we attribute to the effect of lava-flow loading or burial. In 504B, this rotation is toward the north, consistent with lava loading from the north and with the location of the hole on the southern flank of the Costa Rica Rift. In CY-1 and CY-1A, the rotation is toward the east, suggesting that CY-1 and CY-1A are located on the western flank of the Troodos paleorift. Our 1-D interpretation of these rotations differs from the 2-D interpretation of most recent authors (e.g., Pézard et al., 1997; Dilek,

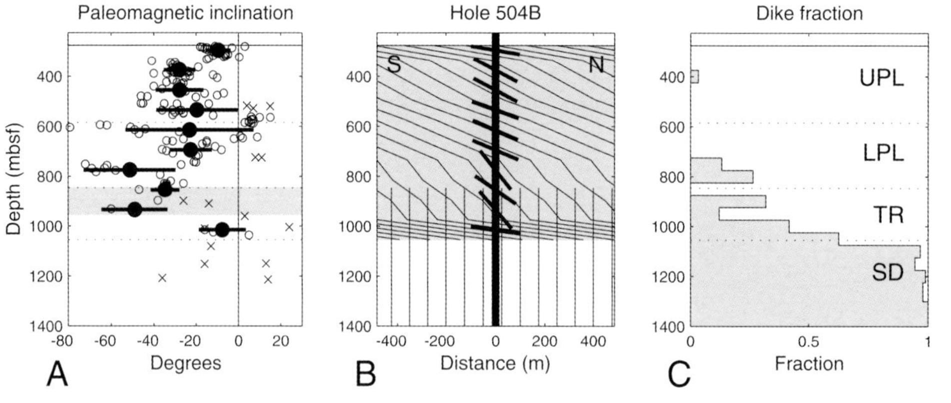

Figure 8. Extrusive architecture from DSDP-ODP Hole 504B, Costa Rica Rift. A, Paleomagnetic inclinations (Furuta and Levi, 1983; Smith and Banerjee, 1985) in ODP Hole 504B. Dotted lines separate upper pillow lavas (UPL), lower pillow lavas (LPL), transition (TR) from lava flows to dikes, and sheeted dikes (SD). Shaded band (~840–960 mbsf) is zone of maximum deformation in the lava pile (e.g., Agar, 1991) Symbols as in Figure 4. Lava inclinations averaged over 80 m core lengths. Negative inclination approximately equals north rotation about a horizontal axis. Averaged rotation of the lavas increases with depth from approximately zero at the top to ~50° rotation to the north in the zone of maximum deformation, near the base of the transition zone from lava flows to dikes. Deepest lava samples in core near ~1050 mbsf are virtually unrotated, similar to the sheeted dikes below, which have paleomagnetic inclinations and dike-margin orientations consistent with minimal postdepositional rotation about a horizontal, east-west–oriented axis. B, Model of architecture of upper oceanic crust based on 504B drill cores. Isochrons are derived from postdepositional rotation (short lines crossing bold vertical line in center) based on the averaged inclinations shown in top left panel. Shading indicates the lava pile. Vertical parallel lines schematically mark transition zone from lava flows to dikes as well as the underlying sheeted dikes. This architecture indicates that the ridge axis is to the north. Estimated block rotation is zero. C, Dike fraction (Gillis, 1987) in ODP Hole 504B.

1998) in that we recognize that rotation due to lava-flow loading (e.g., Bodvarsson and Walker, 1964; Cann, 1974) is an identifiable and important component of the postdepositional rotation process. The crustal architectures we obtain as a result make sense geologically, and we can use them effectively as a framework for the interpretation of existing drill-core and downhole information as it pertains to structure. These architectures provide a viable 1-D alternative to existing 2-D models for 504B and CY-1 and CY-1A based on tectonic rotation due to faulting.

The 1-D architecture of the Troodos Upper and Lower Pillow Lavas, based on paleomagnetic inclinations in the CY-1 and CY-1A drill cores (Fig. 4B), provides us with a new framework for the detailed evaluation of the drill-core information (e.g., dike density, chemistry, lithology) as it varies with depth in the core. The framework indicates that most of the high-Ti lava (HTL) is deposited <1 km from the axis, whereas most of the low-Ti lava (LTL) is deposited separately ~2 km from the axis (Fig. 7A). LTL and HTL loosely correlate with the Upper and Lower Pillow Lavas, respectively (e.g., Robinson et al., 1983). HTL contains an average of 50% dikes whereas LTL contains <5% (Fig. 4B), which indicates that almost all the HTL was deposited inside the dike-injection zone (or spreading zone) and the LTL was deposited outside this zone. Because both HTL and LTL were produced in on-axis eruptions in the dike-injection zone (Baragar et al., 1990), we conclude that, after eruption, almost all LTL must have flowed away from the eruption site whereas most HTL stayed behind. This conclusion is supported by calculations of viscosity of the Troodos lavas (Schmincke et al., 1990) showing that, at similar temperatures, LTL has a viscosity one to two orders of magnitude lower than that of the HTL. This difference would allow low-Ti lavas to flow one to two orders of magnitude faster or farther from their eruption center than the high-Ti lavas. Therefore, the chemical layering of the Troodos pillow lavas may be the result of a process of lava segregation by viscosity occurring on the upper flanks of the Troodos spreading center.

The 1-D architecture of the Costa Rica Rift upper oceanic crust based on paleomagnetic inclinations in the 504B drill cores (Fig. 8B) is not as distinctly bimodal as that of the Troodos upper crust. Postdepositional rotation of the lavas increases rather monotonically to ~50° immediately above the top of the sheeted dikes, suggesting less of a separation between proximal and distal lava deposition (Fig. 7B). This effect may be related to the fact that the compositions of the 504B lavas are much more uniform than those of the Troodos lavas, and so the 504B lavas probably had more uniform physical properties at the time of eruption.

Postdepositional rotation due to lava-flow loading increases more rapidly with depth in 504B than in CY-1 and CY-1A. This discrepancy predicts a narrower total width of the lava-deposition zone estimated for the Costa Rica paleoridge (~3 km, Fig. 7B) than for the Troodos paleoridge (~5 km, Fig. 7A). The Costa Rica Rift at ca. 6 Ma was, like today, an intermediate-rate spreading center and was probably not very different from intermediate-spreading-rate rift valleys today. The intermediate Costa Rica, Galapagos, and Juan de Fuca rift valleys have typical widths of <~2–3 km, consistent with the estimated 504B lava-deposition width. These data suggest that the narrower width of lava deposition (compared to the situation at Troodos) inferred from 504B was probably controlled by the width of the rift valley. The greater width of lava deposition inferred from CY-1 and CY-1A then suggests that the Troodos paleo–spreading center had a wider rift valley or was more like that of the fast-spreading East Pacific Rise (Figs. 2 and 3), where lavas can flow several kilometers down the upper flanks, unobstructed by the large inward-facing fault scarps that are characteristic of rift valleys spreading at intermediate and slow rates.

Our analysis suggests that most of the postdepositional rotation of the lavas in the 504B and CY-1 and CY-1A cores can be attributed to rotation due to lava loading (compensation by bending) with only minor block rotation. Rotation due to lava loading increases from zero in the top of the lava pile to ~50° near its base in 504B and in

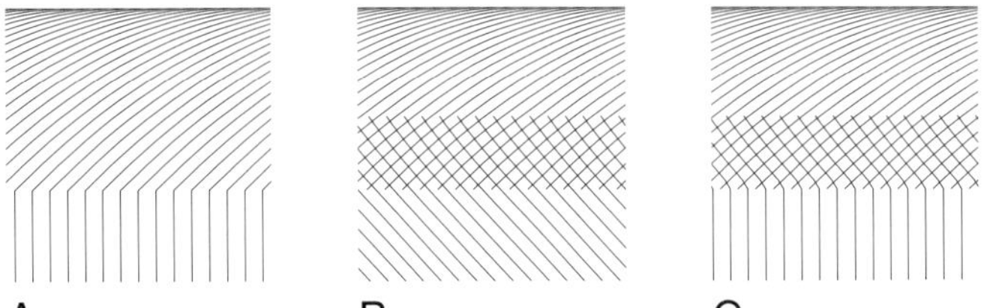

Figure 9. Nature of isostatic compensation for lava-flow loading (modified after Cann, 1974). A, Vertical isostatic compensation for lava-flow loading (i.e., involving vertical slip along closely spaced vertical planes). This situation is the same as shown in Figure 1. B, Compensation by bending involves bending the crust downward and outward under the superimposed load. Such movement inclines all preexisting dikes in the transition zone from lava flows to dikes as well as dikes in the sheeted complex. C, Compensation by bending of the lava pile (which includes the transition zone from lava flows to dikes) while vertical compensation in the sheeted complex maintains the vertical position of the dikes. This hybrid compensation model, with the decoupling of the two mechanisms near the top of the sheeted dikes, is suggested by the postdepositional rotations determined from the 504B drill cores and the Akaki River section (CY-1 and CY-1A drill cores to Palekhori).

the bottom of CY-1 and CY-1A. Block rotation is estimated at zero for 504B and a 20° back tilt for CY-1 and CY-1A. These estimates of block rotation are model dependent and not well determined. Our model assumes that block rotation occurs subsequent to lava deposition so that postdepositional rotation due to lava loading should be zero at the top of the lava pile (i.e., no load) and positive below (i.e., no negative loads). For practical purposes, we have chosen block rotations that are ~10° larger than the average rotation measured in the top 50 m of the lava pile. This choice assures lava-deposition distributions of reasonable width and aesthetic spacing of the isochrons in Figures 4 and 8.

Before we can compare these 1-D models with existing 2-D models for the postdepositional rotation in 504B and CY-1 and CY-1A, we must first consider the rotation of the deeper levels of the Costa Rica oceanic crust and the Troodos ophiolite (i.e., the sheeted dikes >1050 mbsf in 504B and the Basal Group and Sheeted Complex down section from CY-1 and CY-1A). Cann (1974) proposed that the lava pile and underlying sheeted dikes deform as one unit in response to lava-flow loading, either by vertical slip along closely spaced planes (vertical compensation) or by rotation about a horizontal axis (compensation by bending). Vertical compensation involves slip only, and there should be no rotation in the lava pile or of the sheeted dikes (see Figs. 1 and 9A). Compensation by bending inclines all preexisting dikes (Cann, 1974); the dikes in the Sheeted Complex should be rotated with the deepest (and most rotated) part of the lava pile to maintain the general orthogonal relationship between lava horizons and crosscutting dikes observed in ophiolites (Fig. 9B). Our 1-D interpretation of the pattern of postdepositional rotation in 504B and in the Akaki River section of the Troodos ophiolite suggests a different response to lava-flow loading, i.e., the lava pile (which includes the dikes in the transition zone from lava flows to dikes) bends in response to lava loading, but the underlying sheeted dikes remain vertical and therefore undergo only vertical compensation (Fig. 9C).

In both 504B and in the Akaki River section, the maximum postdepositional rotation due to lava loading is reached in the transition zone from lava flows to dikes near the base of the lava pile, whereas the sheeted dikes immediately below remained orthogonal (no postdepositional rotation) relative to the top of the lava pile. In 504B, the maximum rotation (inclination −49° ± 18°) occurs in the lower half of the lava pile (750–1000 mbsf) in the transition zone from lava flows to dikes (Fig. 8A). By contrast, the deepest pillows in 504B (1015–1036 mbsf; inclination −8° ± 11°) and the sheeted dikes below them are virtually unrotated (the mean inclination in the sheeted dikes between 1050 and 1550 mbsf is −1° [Pariso and Johnson, 1989]).

In the Akaki River section, a similar pattern of postdepositional rotation of upper crust can be recognized (see Fig. 5), where the dikes in the Basal Group dip 40°W and the dikes in the underlying Sheeted Complex dip 70°E. This situation represents an abrupt flip in the dips of dikes occurring near the top of the sheeted dikes. Similar to the pattern in 504B, the 70°E-dipping sheeted dikes are approximately orthogonal to the top of the lava pile, which dips ~20°W. In CY-1 and CY-1A, postdepositional rotation increases from −10° ± 8° in the top 50 m to 27° ± 11° in the 900–1200 m interval of CY-1 and CY-1A (Fig. 4A). The fact that the mean dike fraction in this deepest interval is ~50% (see Fig. 4C) indicates that CY-1A did not penetrate to the Basal Group (~75%–95% dikes) that typically separates Pillow Lavas from Sheeted Complex in the Troodos ophiolite. Down section of the CY-1 and CY-1A sites, field observations in the Basal Group at location 6 in Figure 5B (Xenophontos and Malpas, 1987) show that dikes and pillows of the Basal Group are rotated ~50° to the east (Fig. 5A). These field data suggest postdepositional rotation due to lava-flow loading of up to ~60°–70° (relative to the top of the Akaki River section).

The dikes in the Basal Group near Ayia Koroni are rotated 50° to the east (dipping 40°W), but dikes in the underlying Sheeted Complex near Palekhori (location 7 in Fig. 5B) are rotated 10°–30° to the west (dipping 80°–60°E) (Malpas et al., 1989). Thus there is a ~70° difference between the east-dipping sheeted dikes and the west-dipping dikes in the lowest part of the transition zone from lava flows to dikes; furthermore, the sheeted dikes near Palekhori are unrotated relative to the top of the lava pile in the Akaki River canyon (see Fig. 5A). This pattern of postdepositional rotation is very similar to that in 504B. In a 1-D interpretation of the two sites, it suggests that the compensation for lava loading is by bending in the transition zone from lava flows to dikes and that the compensation is vertical in the sheeted dikes (see Fig. 9C).

Existing models for 504B and CY-1 and CY-1A attribute the rotation of the lava pile not to lava loading but to block rotation along normal faults or volcanic growth faults; we call these the 2-D models (see Fig. 10 for a comparison of the 2-D and 1-D models). In 504B, the rotation of the lava pile above 1000 mbsf is suggested to occur along a south-dipping normal fault (e.g., Pariso and Johnson, 1989) or a south-dipping volcanic growth fault (e.g., Dilek, 1998) that crosses the hole in the transition zone from lava flows to dikes (see 2-D model in Fig. 10). The west-dipping dikes of CY-1 and CY-1A and Ayia Koroni and the east-dipping dikes of Palekhori are interpreted to lie on opposite flanks of the Mitsero graben (Varga and Moores, 1985), an abandoned spreading center crossed by the CY-1 and CY-1A to Palekhori section (see M-M′ in Fig. 5B and 2-D model in Fig. 10).

The interpretation of the presence of a fault that intersects 504B in the transition zone from lava flows to dikes (2-D 504B model in Fig. 10) is inspired by downhole measurements and core data. Maximum deformation in the core occurs in the interval between 840 and 958.5 mbsf (see shaded band in Fig. 8A), which includes a brecciated stockwork-like alteration zone at 910–928.5 mbsf (Agar, 1991).

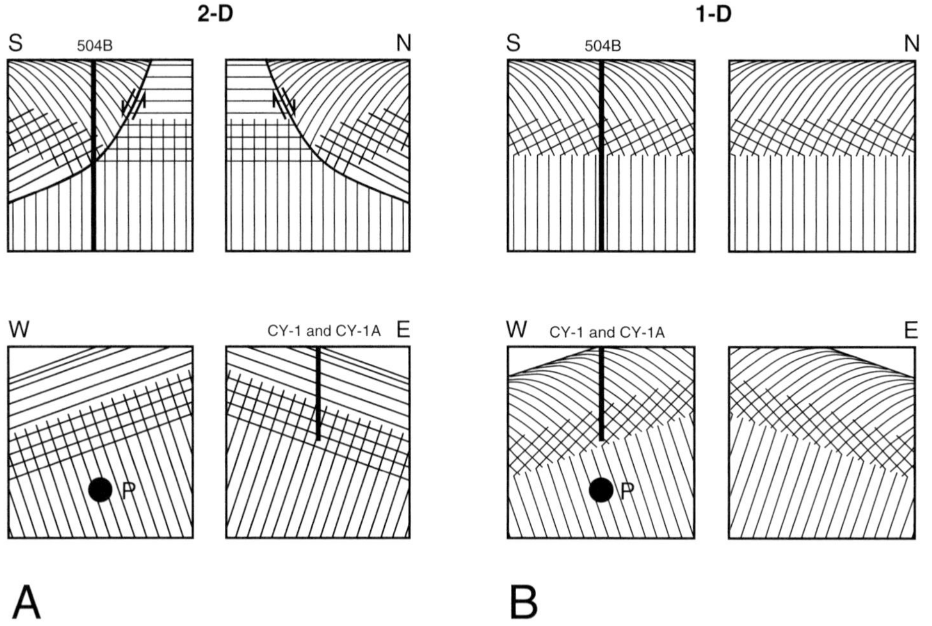

Figure 10. Comparison of (A) 2-D and (B) 1-D models to fit the postdepositional rotations in Hole 504B, Costa Rica Rift, and the Akaki River section, Troodos ophiolite, Cyprus, consisting of Holes CY-1 and CY-1A in the lava pile and transition zone from lava flows to dikes as well as the underlying sheeted dikes north of Palekhori (P). Matching panels show schematic upper-crustal architecture of lava flows, transition zone from lava flows to dikes, and sheeted dikes, on either side of the spreading axis. The 2-D model for 504B (top of A) postulates an outward-facing volcanic growth fault that intersects the hole near the top of the sheeted dikes (e.g., Dilek, 1998). The 2-D model for the Akaki River section (bottom of A) postulates the axis of an abandoned spreading center (Mitsero graben) that crosses the section near the top of the sheeted dikes north of Palekhori (e.g., Varga and Moores, 1985). This interpretation places CY-1 and CY-1A in a back-tilted block on the eastern flank of the (abandoned) spreading center and the sheeted dikes of Palekhori in a back-tilted block on the spreading center's western flank. The 1-D models (in B) postulate that the rapid change in dike orientation near the top of the sheeted dikes in 504B and in the Akaki River section is a characteristic of upper oceanic crust and ophiolite. The 1-D model places CY-1 and CY-1A and the sheeted dikes of Palekhori in the same back-tilted block on the western flank of the Troodos paleorift.

Downhole measurements suggest the presence of a main fault zone between 800 and 1100 mbsf (Pézard et al., 1997). However, this maximum deformation or fault zone is also the transition zone from lava flows to dikes, which extends from 800 to 1050 mbsf and is defined by dike fraction (Fig. 8). This correlation, plus the observation that the lavas in the maximum deformation zone are also the most rotated in the section ($-48° \pm 11°$), suggests to us that the 1-D model of compensation for lava loading by bending of the lava pile provides an attractive alternative interpretation of this deformation zone in 504B.

Varga and Moores (1985) based their identification of the Mitsero graben on the assumption that all rotations of the lavas and dikes in the Troodos ophiolite are due to block faulting. Before block rotation, all lava-flow surfaces are assumed to be approximately horizontal, and the dikes are assumed to be mostly vertical. Consequently, Varga and Moores (1985) made no distinction between dikes in the transition zone from lava flows to dikes (Basal Group) and dikes in the Sheeted Complex. Thus the rotation (dip) of dikes becomes a measure of block rotation. The Mitsero graben is defined by the east-dipping dikes on its western flank and the west-dipping dikes on its eastern flank (see M-M′ in Fig. 5B).

Our 1-D interpretation of the Mitsero graben axis is that it is not a domain boundary but a pseudostratigraphic horizon between lava pile and sheeted dikes (Figs. 9C and 10). The domains of opposing dike attitude in the Mitsero graben are stratigraphically controlled; the eastern flank of the grabon exposes mostly Basal Group in which the dikes are rotated east (dipping west) due primarily to lava-flow loading (rotation east due to lava loading > rotation west by block rotation); the western flank of the graben exposes mostly Sheeted Complex in which the dikes are rotated west (dipping east) by block rotation. Several other factors make it worthwhile to entertain this alternative solution: (1) The progressive rotation with depth of paleomagnetic inclinations in 504B and CY-1 and CY-1A suggests that rotation due to lava-flow loading, rather than block rotation, may account for up to ~50°–60° postdepositional rotation of the units in the transition zone from lava flows to dikes. (2) In this part of the ophiolite, most dikes in the Sheeted Complex dip east (Malpas et al., 1989; Baragar et al., 1990; Agar and Klitgord, 1995) whereas most dikes in the Basal Group dip west (e.g., Bear, 1960). (3) The dashed northern boundary of east-dipping dikes in Figure 5B (of which the Mitsero graben is only a part) roughly correlates with the 75% dike fraction contour mapped by Hall and Yang (1995). South of this contour, dike fraction is >75%, indicating deeper structural levels. The 75% dike-fraction contour, like the top of the sheeted dikes (95% dike fraction), is a pseudostratigraphic horizon in seafloor-spreading oceanic crust.

The foregoing description strongly suggests that the axis of the Mitsero graben may not be the vertical symmetry plane of block rotations in an abandoned spreading center (e.g., Varga and Moores, 1985) but a pseudostratigraphic horizon that roughly separates the lava pile from the underlying sheeted dikes. This analysis would mean there is no abandoned spreading center and lends support to our 1-D interpretation of the CY-1 and CY-1A to Palekhori section in Figures 10 and 11. It further suggests that, in response to lava-

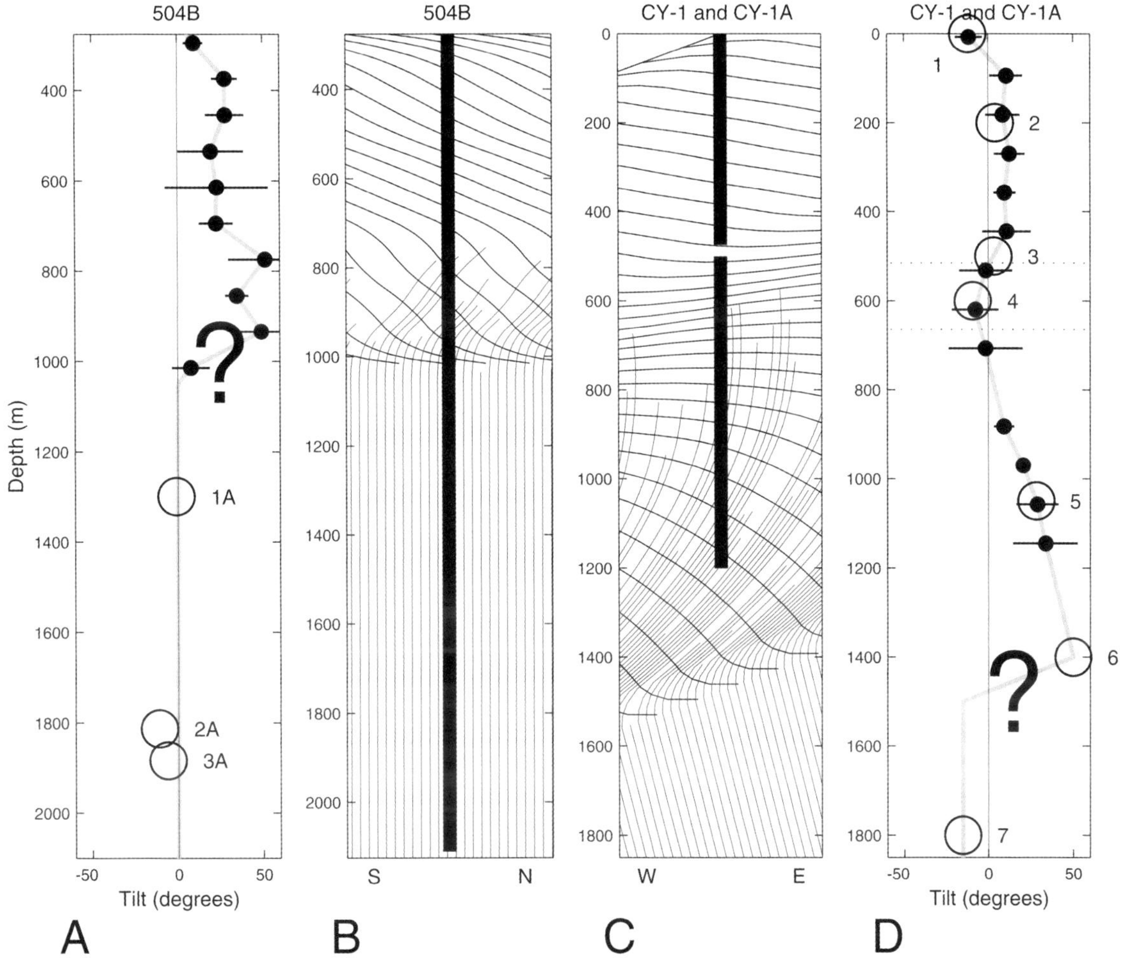

Figure 11. Postdepositional rotation of lavas and dikes, and architecture of upper oceanic crust in Holes 504B, CY-1, and CY-1A. A and D, Postdepositional rotation of the lavas derived from stable paleomagnetic inclinations in the drill cores (small-filled circles with 1-sigma error bars from Figs. 4A and 8) and model rotation (gray line). Bold question marks identify an apparent rotational disconformity at similar stratigraphic levels near the base of the lava-dike transition in 504B, CY-1, and CY-1A. A, Open circle 1A—rotation based on mean paleomagnetic inclination in sheeted dikes of 504B (1050–1550 m; Pariso and Johnson, 1989); open circles 2A and 3A—southward tilt of paleomagnetically reconstructed dike margins (Allerton et al, 1995). D, Open circles (1–7)—field observations in the vicinity and south-southwest of the CY-1 and CY1-A drill holes (also shown in Fig. 5). Dotted lines at 515 m and 665 m bracket the transition from low-Ti lava (LTL, above) to high-Ti lava (HTL, below). B and C, Architecture of upper oceanic crust. Rotation of originally horizontal lava flows and vertical dikes is according to the gray line in the adjacent panel. Model block rotation, based on rotation of uppermost lavas, is zero for 504B (B) and 20° toward the west for CY-1 And CY-1A (C). Dikes in the transition zone from lava flows to dikes are orthogonal to and rotated with the lava flows (later dikes are not shown), but the dikes in the sheeted dikes section below the lava pile are unrotated relative to the top of the lava pile. We suggest that this angular discontinuity at the base of the transition zone from lava flows to dikes (bold question mark) represents a decoupling horizon that partitions deformation between the lava pile and sheeted dikes (e.g., Agar and Klitgord, 1995). We speculate that the contrasting kinematic histories recorded in the lava pile and sheeted dikes reflect their contrasting mechanical response to subsidence due to lava-flow loading, i.e., the lava pile deforms primarily by bending and the sheeted-dike section deforms by vertical slip along closely spaced planes parallel to the dikes.

flow loading, the Troodos lava pile deformed primarily by bending (up to 70°) while the underlying sheeted dikes deformed by vertical slip along closely spaced planes parallel to vertical dikes. Coherent blocks of the deformed lava pile and Sheeted Complex were subsequently tilted westward in a sequence of domino fault blocks (Agar and Klitgord, 1975).

If we adopt the 1-D solution, then the architecture of Costa Rica Rift upper oceanic crust sampled in Hole 504B and the architecture of Troodos upper ophiolite crust sampled in CY-1 and CY-1A plus observations in the field are remarkably similar. Before block rotation, lava-flow horizons are subhorizontal at the top, dip progressively steeper with depth in the crust, and reach a maximum dip of 50°–70° near the base of the transition zone from lava flows to dikes. Most of the dikes in the transition zone are bent by the same amount as the lavas, so that the general orthogonal relationship observed in ophiolites between lava horizons and crosscutting dikes is maintained. Before block rotation, dikes in the Sheeted Complex are vertical. This fact suggests the existence of a horizontal decoupling zone that partitions deformation between the rotated lavas and the unrotated sheeted dikes somewhere near the base of the transition zone from lava flows to dikes. The existence of a decoupling zone between the lava pile and sheeted dikes was proposed by Agar and Klitgord

(1995) in their study of a similar decoupling horizon between the sheeted dikes and plutonic rocks in the Agros window, Troodos ophiolite, Cyprus.

The transition zone from lava flows to dikes is one of the most deformed intervals in 504B (e.g., Agar, 1991); it is also the interval with the largest amount of rotation. The underlying sheeted dikes appear less deformed, and paleomagnetism indicates minimal rotation. Deformation of sheeted dikes in ophiolites is typically along fault planes parallel to the dikes (e.g., Agar and Klitgord, 1995). We speculate that the contrasting kinematic histories recorded in the lava pile and sheeted dikes reflect their contrasting mechanical response to subsidence due to lava-flow loading, i.e., the lava pile deforms primarily by bending and the sheeted dike section deforms by vertical slip along closely spaced planes parallel to the dikes (Fig. 11). This interpretation predicts deformation and rotation increasing with depth of burial in the lava section and uniform fracturing without rotation throughout the sheeted-dike section. This concept would further suggest that the main fault or deformation zone in 504B is the lowest stratigraphic unit in the lava pile and not a normal-fault zone that happens to cut the hole at this level. It would also suggest that the Mitsero graben—one of the graben structures interpreted on the basis of domains of characteristic dike attitude in the Troodos ophiolite (e.g., Varga and Moores, 1985)—does not exist.

CONCLUSIONS

1. Paleomagnetic inclinations are reliable indicators of postdepositional rotation of lavas in the CY-1 and CY-1A drill cores, Troodos ophiolite, Cyprus.

2. The variation with depth of postdepositional rotation of the lavas in 504B, Costa Rica rift, and in the CY-1 and CY-1A drill cores shows a similar architecture of upper oceanic crust and ophiolite related to their construction and deformation at the spreading center.

3. Postdepositional rotation is the sum of two rotations about an axis parallel to the spreading axis: a variable rotation with depth due to lava-flow loading (zero at the top of the lava pile and maximum at the base) and a depth-invariant block rotation. At both sites, postdepositional rotation due to lava loading dominates over block rotation (assuming block faulting postdates lava loading), resulting in a net rotation of up to 50° toward the axis of the lavas and dikes in the transition zone from lava flows to dikes (where most dikes are orthogonal to the lava flows).

4. Sheeted dikes in Hole 504B and in the surface geology down section from Holes CY-1 and CY-1A are not rotated (relative to the top of the overlying pillow lavas); this lack of rotation implies an abrupt change in dip of 50°–70° between sheeted dikes and the dikes in the transition zone from lava flows to dikes.

5. We suggest that the angular discontinuity near the base of the lava pile in both the 504B and Troodos sections represents a horizontal decoupling horizon that partitions deformation between the lava pile and sheeted dikes (Agar and Klitgord [1995] documented a similar decoupling horizon between sheeted dikes and plutonic rocks in the Agros window, Troodos ophiolite, Cyprus). This interpretation is in contrast to the earlier interpretations of a normal fault that cuts 504B at this level and an abandoned spreading center—the axis of the Mitsero graben—that crosses the Akaki River section near the base of the lava pile.

6. We speculate that the contrasting kinematic histories recorded in the lava pile and the sheeted dikes in Hole 504B and in and near Holes CY-1 and CY-1A reflect their contrasting mechanical response to subsidence due to lava-flow loading, i.e., the lava pile deforms primarily by bending and the sheeted dikes deform by vertical slip along closely spaced planes parallel to the dikes.

7. This study suggests that differential subsidence due to lava-flow loading, rather than normal faulting, is primarily responsible for the large postdepositional rotations typically found lower in the oceanic lava pile. The proposed similarity between 504B and CY-1 and CY-1A upper-crustal architecture provides a new and alternative framework for further comparative studies of ODP drill cores and ophiolites.

ACKNOWLEDGMENTS

Schouten was supported by National Science Foundation grant OCE 9419472 and the Woods Hole Oceanographic Institution. A WHOI Mellon Independent Study Award provided funds for an exploratory field study on Cyprus to test predictions from paleomagnetic inclinations in the CY-1 and CY-1A drill cores. We especially thank Joe Cann, Kathy Gillis, and Costas Xenophontos for introducing us to the Troodos ophiolite and providing guidance, Jim Hall for providing the paleomagnetic inclinations for CY-1 and CY-1A, and Henry Dick and the W.M. Keck Foundation for supporting the stimulating environment of the W.M. Keck Geodynamics Program at WHOI. We also wish to acknowledge Simon Allerton, John Collins, Greg Hirth, Susan Humphris, Peter Kelemen, Kim Klitgord, Philippe Pézard, Ralph Stephen, and Maurice Tivey for helpful discussions. We thank Yildirim Dilek, Jeff Gee, and Stephen Hurst for constructive reviews that significantly improved this manuscript. This is WHOI Contribution No. 10,111.

REFERENCES CITED

Agar, S.M., 1991, Microstructural evolution of a deformation zone in the upper oceanic crust: Evidence from DSDP Hole 504B: Journal of Geodynamics, v. 13, p. 119–140.

Agar, S.M., and Klitgord, K.D., 1995, A mechanism for decoupling within the oceanic lithosphere revealed in the Troodos ophiolite: Nature, v. 374, p. 232–238.

Allerton, S.A., McNeill, A.W., Stokking, L.B., Pariso, J.E., Tartarotti, P., Marton, F.C., and Pertsev, N.N., 1995, Structures and magnetic fabrics from the lower sheeted dike complex of Hole 504B reoriented using stable magnetic remanence, *in* Erzinger, J., Becker, K., Dick, H.J.B., and Stokking, L.B., eds., Proceedings of the Ocean Drilling Program, Scientific Results, v. 137/140, p. 245–252.

Baragar, W.R.A., Lambert, M.B., Baglow, N., and Gibson, I.L., 1990, The sheeted dyke zone in the Troodos ophiolite, *in* Malpas, J., Moores, E.M., Panayiotou, A., and Xenophontos, C., eds., Ophiolites, oceanic crustal analogues; Proceedings of the Symposium "Troodos 1987", Nicosia, Cyprus, Geological Survey Department, p. 37–51.

Bear, L.M., 1960, The geology and mineral resources of the Akaki-Lythrodondha area: Nicosia, Cyprus, Geological Survey Department, Memoir no. 3, p. 1–122.

Bodvarsson, G., and Walker, G.P.L., 1964, Crustal drift in Iceland: Geophysical Journal of the Royal Astronomical Society, v. 9, p. 285–300.

Cann, J.R., 1970, New model for the structure of ocean crust: Nature, v. 226, p. 928–930.

Cann, J.R., 1974, A model for oceanic crustal structure developed: Geophysical Journal of the Royal Astronomical Society, v. 39, p. 169–187.

Cann, J.R., Langseth, M.G., Honnorez, J., Von Herzen, R.P., White, S.M., et al., 1983, Initial Reports of the Deep Sea Drilling Project: Washington, D.C., U.S. Government Printing Office, v. 69.

Carbotte, S.M., Mutter, J.C., and Xu, L., 1997, Contribution of volcanism and tectonism to axial and flank morphology of the southern EPR, 17°10′-17°40′S, from a study of layer 2A geometry: Journal of Geophysical Research, v. 102, p. 10165–10184.

Christeson, G.L., Purdy, G.M., and Fryer, G.J., 1994, Seismic constraints on shallow crustal emplacement at the fast spreading East Pacific Rise: Journal of Geophysical Research, v. 99, p. 17957–17973.

Clube, T.M.M., Creer, K.M., and Robertson, A.H.F., 1985, Paleorotation of the Troodos microplate, Cyprus: Nature, v. 317, p. 522–525.

Dick, H.J.B., Erzinger, J., Stokking, L.B., et al., 1992, Proceedings of the Ocean Drilling Program, Initial Reports: College Station, Texas, Ocean Drilling Program, v. 140.

Dilek, Y., 1998, Structure and tectonics of intermediate-spread oceanic crust drilled at DSDP/ODP Holes 504B and 896A, Costa Rica Rift, *in* Cramp, A., MacLeod, C.J., Lee, S.V., and Jones, E.J.W., eds., Geological evolution of ocean basins: Results from the Ocean Drilling Program: Geological Society of London Special Publications, v. 131, p. 179–197.

Furuta, T., and Levi, S., 1983, Basement paleomagnetism of Hole 504B, *in* Cann, J.R., Langseth, M.G., Honnorez, J., Von Herzen, R.P., White, S.M., et al., eds., Initial Reports of the Deep Sea Drilling Project: Washington, D.C., U.S. Government Printing Office, v. 69, p. 697–703.

Gee, J., and Kent, D.V., 1994, Variations in layer 2A thickness and the origin of the central anomaly magnetic high: Geophysical Research Letters, v. 21, p. 297–300.

Gibson, I.L., Malpas, J., Robinson, P.T., and Xenophontos, C., 1991, Cyprus Crustal Study Project: Initial Report, Holes CY-1 and 1A: Geological Survey of Canada, Paper 90–20, p. 1–283.

Gillis, K.M., 1987, Multistage alteration of the extrusive sequence, Troodos ophiolite, Cyprus [PhD Thesis]: Halifax, Dalhousie University, p. 1–387.

Goldstein, S.J., Perfit, M.R., Batiza, R., Fornari, D.J., and Murrell, M.T., 1994, Temporal variations in East Pacific Rise magmatic activity based on U-series dating of basalts: Nature, v. 367, p. 157–159.

Gregg, T.K.P., Fornari, D.J., Perfit, M.R., Haymon, R.M., and Fink, J.H., 1996, Rapid emplacement of a mid-ocean ridge lava flow: The East Pacific Rise at 9°46′–51′N: Earth and Planetary Science Letters, v. 144, p. E1–E7.

Hall, J.M., and Yang, J.-S., 1995, Constructional features of Troodos type oceanic crust: Relationships between dike density, alteration, magnetization, and ore body distribution and their implications for in situ oceanic crust: Journal of Geophysical Research, v. 100, p. 19973–19989.

Harding, A.J., Kent, G.M., and Orcut, J.A., 1993, A multichannel seismic investigation of upper oceanic crustal structure at 9°N on the East Pacific Rise: Implications for crustal accretion, Journal of Geophysical Research, v. 98, p. 13925–13944.

Hooft, E.E., Schouten, H., and Detrick, R.S., 1996, Constraining crustal emplacement processes from the variation in seismic layer 2A thickness at the East Pacific Rise: Earth and Planetary Science Letters, v. 142, p. 289–309.

Karson, J.A., 1998, Internal structure of oceanic lithosphere: A perspective from tectonic windows, *in* Buck, W.R, et al., eds., Faulting and magmatism at mid-ocean ridges: American Geophysical Union Geophysical Monograph 106, p. 177–218.

Kent, G.M., Harding, A.J., and Orcut, J.A., 1994, Uniform accretion of oceanic crust south of the Garrett transform at 14°15′S on the East Pacific Rise: Journal of Geophysical Research, v. 99, p. 9097–9116.

Kidd, R.G.W., 1977, A model for the process of formation of the upper oceanic crust: Geophysical Journal of the Royal Astronomical Society, v. 50, p. 149–183.

Kinoshita, H., Furuta, T., and Pariso, J., 1989, Down-hole magnetic field measurements and paleomagnetism, Hole 504B, Costa Rica Ridge, *in* Becker, K., Sakai, H., et al., eds., Proceedings of the Ocean Drilling Program, Scientific Results: College Station, Texas, Ocean Drilling Program, v. 111, p. 147–156.

Macdonald, K.C., Fox, P.J., Alexander, R.T., Pockalny, R., and Gente, P., 1996, Volcanic growth faults and the origin of abyssal hills on the flanks of the East Pacific Rise: Nature, v. 380, p. 125–129.

Malpas, J., and Barnable, G., 1991, Lithologic log of Holes CY-1 and CY-1A, *in* Gibson, I.L., Malpas, J., Robinson, P.T., and Xenophontos, C., eds., Cyprus Crustal Study Project: Initial Report, Holes CY-1 and 1A: Geological Survey of Canada, Paper 90–20, p. 1–283.

Malpas, J., Case, G., and Moore, P., 1989, The geology of the area immediately surrounding the CY-4 borehole of the Cyprus Crustal Study Project, *in* Gibson, I.L., Malpas, J., Robinson, P.T., and Xenophontos, C., eds., Cyprus Crustal Study Project: Initial Report, Hole CY-4: Geological Survey of Canada, Paper 88–9, p. 31–37.

Moores, E.M., Varga, R.J., Verosub, K.L., and Ramsden, T., 1990, Regional structure of the Troodos dyke complex, *in* Malpas, J., Moores, E.M., Panayiotou, A., and Xenophontos, C., eds., Ophiolites, oceanic crustal analogues, Proceedings of the Symposium "Troodos 1987": Nicosia, Cyprus, Geological Survey Department, p. 27–35.

Pålmason, G., 1973, Kinematics and heat flow in a volcanic rift zone, with application to Iceland: Geophysical Journal of the Royal Astronomical Society, v. 33, p. 451–481.

Pålmason, G., 1980, A continuum model of crustal generation in Iceland: kinematic aspects: Journal of Geophysics, v. 47, p. 7–18.

Pariso, J.E., and Johnson, H.P., 1989, Magnetic properties and oxide petrography of the sheeted dike complex in Hole 504B, *in* Becker, K., Sakai, H., et al., eds., Proceedings of the Ocean Drilling Program, Scientific Results: College Station, Texas, Ocean Drilling Program, v. 111, p. 159–167.

Perfit, M.R., Fornari, D.J., Smith, M.C., Bender, J.F., Langmuir, C.H., and Haymon, R.M., 1994, Small-scale spatial and temporal variations in mid-ocean ridge crest magmatic processes: Geology, v. 22, p. 375–379.

Pézard, P.A., Ayadi, M., Revil, A., Bronner, G., and Wilkens, R.H., 1997, Detailed structure of an oceanic normal fault: A multi-scalar approach at DSDP/ODP site 504B: Geophysical Research Letters, v. 24, p. 337–340.

Robinson, P.T., Melso, W.G., O'Hearn, T., and Schmincke, H.-U., 1983, Volcanic glass compositions of the Troodos ophiolite, Cyprus: Geology, v. 1, p. 400–404.

Rosencrantz, E., 1982, Formation of uppermost oceanic crust: Tectonics, v. 1, p. 471–494.

Schmincke, H.-U., and Bednarz, U., 1990, Pillow, sheet flow and breccia flow volcanoes and volcano-tectonic hydrothermal cycles in the extrusive series of the northeastern Troodos ophiolite (Cyprus), *in* Malpas, J., Moores, E.M., Panayiotou, A., and Xenophontos, C., eds., Ophiolites, oceanic crustal analogues, Proceedings of the Symposium "Troodos 1987": Nicosia, Cyprus, Geological Survey Department, p. 185–206.

Schmincke, H.-U., and Rautenschlein, M., 1987, Troodos extrusive series (Akaki River canyon) and the sheeted diabase, *in* Xenophontos, C., and Malpas, J.G., eds., Field excursion guidebook, Troodos 87: Ophiolites and Oceanic Lithosphere: Nicosia, Cyprus, Geological Survey Department, p. 260–285.

Schouten, H., Tivey, M.A., Fornari, D.J., and Cochran, J.C., 1999, The central anomaly magnetic high: Constraints on the volcanic construction and architecture of seismic layer 2A at a fast-spreading mid-ocean ridge, the EPR at 9°30′-9°50′N: Earth and Planetary Science Letters, v. 169, p. 37–50.

Smith, G.M., and Banerjee, S.K., 1985, Magnetic properties of basalts from Deep Sea Drilling Leg 83: The origin of remanence and its relation to tectonic and chemical evolution, *in* Anderson, R.N., Honnorez, J., Becker, K., et al., eds., Initial Reports of the Deep Sea Drilling Project: Washington, D.C., v. 83, U.S. Government Printing Office, p. 347–357.

Smith, G.M., and Banerjee, S.K., 1986, Magnetic structure of the upper kilometer of the marine crust at DSDP Hole 504B, eastern Pacific Ocean: Journal of Geophysical Research, v. 91, p. 10337–10354.

Toomey, D.R., Purdy, G.M., Solomon, S.C., and Wilcock, W.S.D., 1990, The three-dimensional seismic velocity structure of the East Pacific Rise near latitude 9°30′N: Nature, v. 347, p. 639–645.

Varga, R.J., and Moores, E.M., 1985, Spreading structure of the Troodos ophiolite, Cyprus: Geology, v. 13, p. 846–850.

Varga, R.J., Schiffman, P., Smith, B.M., Moores, E.M., and Hurst, S., 1987, Structure and hydrothermal metamorphism of the Solea graben, Troodos ophiolite, *in* Xenophontos, C., and Malpas, J.G., eds., Field excursion guidebook, Troodos 87: Ophiolites and oceanic lithosphere: Nicosia, Cyprus, Geological Survey Department, p. 182–213.

Varga, R.J., Gee, J.S., Staudigel, H., and Tauxe, L., 1998, Dike surface lineations as magma flow indicators within sheeted dike complex of the Troodos Ophiolite, Cyprus: Journal of Geophysical Research, v. 103, p. 5241–5256.

Vera, E.E., and Diebold, J.B., 1994, Seismic imaging of oceanic layer 2A between 9°30′N and 10°N on the East Pacific Rise from two-ship wide-aperture profiles: Journal of Geophysical Research, v. 99, p. 3031–3041.

Vine, F.J., and Smith, G.C., 1990, Structure and physical properties of the Troodos crustal section at ICRDG drill holes CY1, 1a and 4, *in* Malpas, J., Moores,

E.M., Panayiotou, A., and Xenophontos, C., eds., Ophiolites, oceanic crustal analogues, Proceedings of the Symposium "Troodos 1987": Nicosia, Cyprus, Geological Survey Department, p. 113–124.

Xenophontos, C., and Malpas, J.G., 1987, Sheeted dikes and their relationships to pillow lavas and plagiogranite, *in* Xenophontos, C., and Malpas, J.G., eds., Field excursion guidebook, Troodos 87: Ophiolites and oceanic lithosphere: Nicosia, Cyprus, Geological Survey Department, p. 82–91.

MANUSCRIPT ACCEPTED BY THE SOCIETY FEBRUARY 23, 2000

Printed in the U.S.A.

Geological Society of America
Special Paper 349
2000

On the boundary between seismic layers 2 and 3: A stress change?

Philippe A. Pezard
CEREGE (CNRS), Geophysics Department, BP80, 13545 Aix-en-Provence cedex 4, France; e-mail: pezard@cerege.fr

ABSTRACT

Over the past 30 years, the boundary between seismic layers 2 and 3 in modern oceanic crust has been successively explained by changes in porosity, fracturing, lithology, and metamorphic grade. This transition was most recently proposed to lie within the sheeted-dike complex of a deep oceanic borehole (Hole 504B of the Deep Sea Drilling Project and the Ocean Drilling Program). Both models that promote this interpretation rule out the lithologic hypothesis, which suggests that the boundary between seismic layers 2 and 3 corresponds to the downward transition from sheeted dikes to gabbro. Downhole measurements in the same hole reveal a gradual change in stress regime within the sheeted-dike complex (1.3 to 1.7 km into basement), confirming the previous findings and allowing the proposal of a common cause to explain some of the earlier hypotheses. From compressional above to strike slip below the seismic transition, this stress change is proposed to induce a depth-dependent reopening of cracks and fractures. With a nearly constant porosity throughout the sheeted-dike complex, this variable reopening of cracks can have a significant impact on acoustic and hydraulic properties of the crust and hence on the mode and nature of fluid circulation and crustal alteration. Horizontal fluid movement and heat transfer are favored in the upper part of the crust under a compressional regime, whereas vertical mining of hot fluids and associated fluid-rock interactions are enhanced at the base of the dolerites under a strike-slip regime. As the crust ages and subsides, the transition between these two regimes (where the vertical load equals the minimum horizontal stress) migrates upward, which provides an explanation for the upward migration of the alteration front in dikes with time. More generally, this result indicates that seismic methods might allow one to map subsurface changes of the stress field with depth in the upper oceanic crust.

INTRODUCTION

The nature of the boundary between seismic layers 2 and 3 in the oceanic crust has been the matter of a long-standing debate (Raitt, 1963; Christensen and Salisbury, 1975; Fox et al., 1972; Houtz and Ewing, 1976; Salisbury and Christensen, 1978; White et al., 1978; Casey et al., 1981; Spudich and Orcutt, 1982; Swift and Stephen, 1992). The debate has led more recently to elimination of the lithologic hypothesis as a possible explanation for this characteristic and systematic seismic velocity increase in the upper oceanic crust (Detrick et al., 1994; Swift et al., 1998a). The textbook explanation relating layer 3 velocities to the presence of gabbro was ruled out as the location of the boundary was proposed to be located either within or at the top of the sheeted-dike complex of the only really deep hole ever drilled into modern oceanic crust, Deep Sea Drilling Project (DSDP) and Ocean Drilling Program (ODP) Hole 504B (Cann, Langseth, et al., 1983; Anderson, Honnorez, et al., 1985; Becker, Sakai, et al., 1988; Dick, Erzinger, et al., 1992; Alt, Kinoshita, et al., 1993).

This result was obtained by combining a series of constraints on the velocity structure of the upper crust (Alt, Kinoshita, et al., 1993; Hobart et al., 1985; Collins et al., 1989; Swift et al., 1998a; Detrick et al., 1998). As a consequence, the boundary was found to stand at least 400 m, possibly up to 800 m, above yet unsampled gabbroic rocks previously believed to correspond to layer 3 seismic velocities. This gradual transition was interpreted as the product of progressive changes encountered in the character of the sheeted dikes observed in the core (Alt, Kinoshita, et al., 1993; Detrick et al., 1994). The gradual change with depth was earlier related jointly to an increase in grain size in the dolerites, a decreasing number of chilled margins of the dikes, and changes in the relative abundances of alteration minerals (decreasing abundance of chlorite and increasing abundance of amphibole). More recently, a slow transition in physical and mineralogical properties of the dolerites over several hundred meters (Laverne et al., 1995; Vanko and Laverne, 1998) was proposed as an explanation (Salisbury et al., 1996; Dilek, 1998). The possibility that the boundary between seismic layers 2 and 3 corresponds to a change in the nature of the stress field at depth into the oceanic crust is investigated here. Implications in terms of changes with depth of the physical and chemical characteristics of the ocean crust are discussed.

Although most earthquakes in young (<15 Ma) and shallow oceanic crust (with centroid depth <7 km) indicate a compressional or strike-slip stress regime (Bergman and Solomon, 1985; Bratt et al., 1985; Wiens and Stein, 1990), the faulting regime that forms the main structure of the upper oceanic crust at the kilometer scale is generally described to be normal (Sauter et al., 1991; Edwards, 1991; Kent et al., 1996; Macdonald et al., 1996; Pezard et al., 1997; Swift et al., 1998b). DSDP-ODP Hole 504B penetrates a crustal section created at an intermediate spreading rate along the Costa Rica Rift (Fig. 1). Regional seismic mapping (Kent et al., 1996; Swift et al., 1998b) and borehole-wall images (Ayadi et al., 1998) suggest mod-

Pezard, P.A., 2000, On the boundary between seismic layers 2 and 3: A stress change?, *in* Dilek, Y., Moores, E.M., Elthon, D., and Nicolas, A., eds., Ophiolites and Oceanic Crust: New Insights from Field Studies and the Ocean Drilling Program: Boulder, Colorado, Geological Society of America Special Paper 349, p. 195–202.

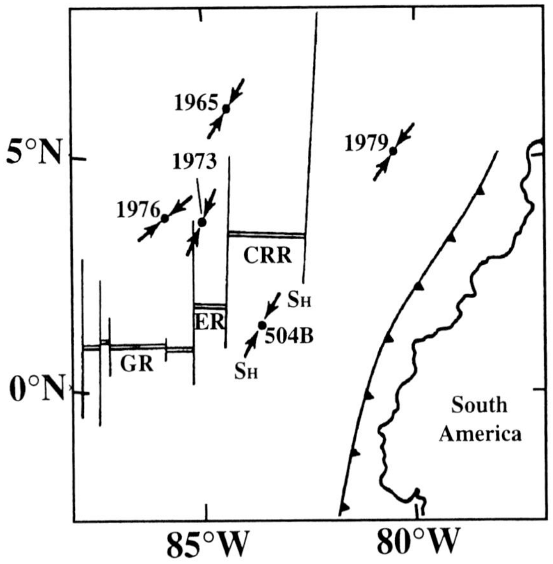

Figure 1. Location of DSDP-ODP Site 504, 200 km south of the Costa Rica Rift (CRR) in the eastern equatorial Pacific. The orientation of maximum horizontal stress is inferred from focal mechanisms of intraplate earthquakes (locations shown with occurrence dates) and hole ellipticity analysis from BHTV images of the borehole walls (Hole 504B). Double lines indicate the Cocos-Nazca spreading system; toothed line indicates the Peru-Chile Trench (other abbreviations: ER—Ecuador Rift, GR—Galapagos Rift; adapted from Morin et al., 1990).

erate off-axis tilting of the upper crust, despite the presence of active (Kent et al., 1996) normal faults (Kinoshita et al., 1989; Becker et al., 1989; Pezard, 1990; Swift et al., 1998b) with intermediate dips (Pezard et al., 1997; Ayadi et al., 1998), such as that drilled at ~800 m below seafloor (mbsf) in ODP Hole 504B.

In situ borehole instabilities, the fracture distribution mapped from images of borehole walls, and rock mechanics properties have been shown in the past to support the presence of a compressional to strike-slip stress regime (Morin et al., 1990; Moos and Zoback, 1990). This apparent conflict with active normal faulting does not necessarily lead to a contradiction as normal faults are created close to the spreading axis, under a different stress regime. The normal reactivation of these faults (Kent et al., 1996) with intermediate dips (Pezard et al., 1997) in a currently compressional environment is possible if the crust is strongly heterogeneous because of, for example, the presence of decoupling horizons at depth (Agar, 1991; Villemin et al., 1994; Agar and Klitgord, 1995).

Evidence from drilling, coring, downhole geophysical experiments and high-resolution images recorded in this deep hole during the two most recent ODP legs (Dick, Erzinger, et al., 1992; Alt, Kinoshita, et al., 1993) are presented here to illustrate and discuss this apparent conflict. Stress-induced ruptures of borehole walls are commonly described as stress-induced breakouts (Bell and Gough, 1979; Newmark et al., 1985; Plumb and Hickman, 1985; Zoback et al., 1985) when compressive in origin, but can originate from a number of different situations, either compressive or tensile (Moos and Zoback, 1990; Maury, 1993). These ruptures are either detected from orthogonal mechanical calipers of a dipmeter sonde such as the formation micro scanner (FMS) or imaged from acoustic waveforms generated with a borehole televiewer (BHTV) and reflected off the borehole wall. Also, the stress component magnitudes may be computed either from the identification of combined modes of rupture observed in a borehole (Maury, 1993) or from tensile ruptures in inclined boreholes (Peskas and Zoback, 1995).

In this study, the directions of borehole ruptures due to stress release, small-scale fault movement associated with drilling, or simply drill-pipe wear have been analyzed as a function of depth in the intrusive section of this near-vertical crustal hole. The stress tensor magnitudes in the middle and at the base of the sheeted-dike complex are computed from borehole-stability analyses. This investigation suggests that the boundary between seismic layers 2 and 3 coincides with a gradual change in stress-field characteristics with depth.

DATA ANALYSIS

DSDP-ODP Hole 504B

Located 200 km to the south of the Costa Rica Rift (Fig. 1) (Cann, Langseth, et al., 1983; Anderson, Honnorez, et al., 1985), Hole 504B remains as the reference section for the structure of modern oceanic crust. The hole penetrates 275 m of sediment and ~650 m of eruptive products (Fig. 2A), which is less than the average thickness of volcanic rocks commonly observed in ophiolites (typical ophiolite volcanic sections can be as thick as 1.0–1.5 km). The seismic data recorded in the vicinity of Hole 504B (Detrick et al., 1998) do not provide an explanation for the reduced volcanic thickness. The extrusive thickness at Site 504 appears in fact similar to that measured regionally (Swift et al., 1998b), in spite of the presence of a kilometer-scale, block-bounding fault drilled at ~800 mbsf in Hole 504B (Kinoshita et al., 1989; Pezard, 1990; Pezard et al., 1997; Ayadi et al., 1998; Swift et al., 1998b). A thick sheeted-dike complex extends >1000 m below the extrusive rocks. This intrusive section is the site of a gradual change in seismic velocity interpreted as the boundary between seismic layers 2 and 3 from seismic and borehole experiments and located either in the middle of the sheeted dikes (Detrick et al., 1994) or at the top of the dolerites (Swift et al., 1998a).

In drilling Hole 504B, operational problems related to the shearing of the drill pipe as a result of active faulting—often expressed as torque increase during coring and sometimes resulting in fishing operations—were encountered from 1060 mbsf downward. Drilling problems were found to increase with depth in the upper part of the sheeted-dike complex, with particular difficulties between 1400 and 1600 mbsf (sanding, fishing, and/or milling at 1300, 1444, 1520, 1547, 1562, and 1600 mbsf). At the beginning of ODP Leg 140, after the recovery of a 18-m-long piece of drilling assembly previously left in the hole (near 1621 mbsf), the coring proceeded uninterruptedly down to 2003 mbsf. Coring resumed with ODP Leg 148, extending down to 2111 mbsf where the drill pipe got stuck after entering 7 m into a fast-drilling fault zone (Alt, Kinoshita, et al., 1993; Dilek, 1998). Similar problems related to hole-shape changes with stress and time were encountered during downhole measurements; mechanical calipers had a particular tendency to get caught by ledges resulting from shearing of the borehole along seismically active planes (Héliot et al., 1990; Cornet et al., 1997; de Larouzière et al., 1999).

Downhole measurements and images

Downhole measurements and images of rock physical properties provide a continuous and meter- to centimeter-scale description of

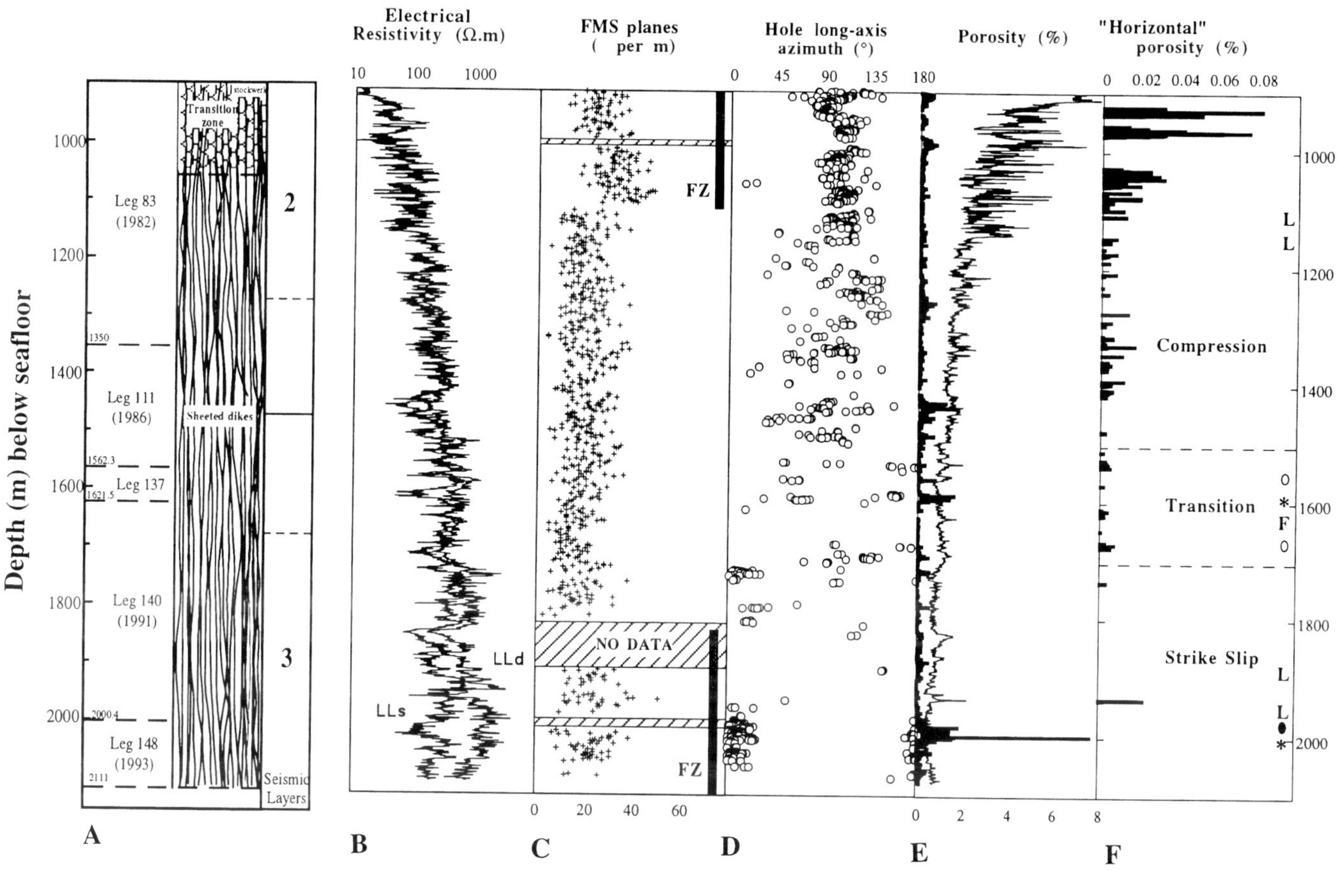

Figure 2. A, Drilling history and lithostratigraphy in the lower part of DSDP-ODP Hole 504B, showing location of the boundary between seismic layers 2 and 3 after Detrick et al. (1994). B, Electrical resistivities measured with the dual laterolog (DLL) sensor. LLd and LLs stand for laterolog deep and shallow measurements, respectively. C, A total of 20,900 planes (~20/m on average) were mapped from electrical images of the borehole walls in the lower half (transition zone and sheeted-dike complex) of Hole 504B. The local absence of data (1800–1884 mbsf, for example) is due to interruptions during recording, as the FMS tool became stuck in the hole because of ledges. FZ stands for fault zone, as determined by Ayadi et al. (1998). D, Azimuth of borehole elongation plotted when the difference between orthogonal FMS caliper readings ($C_1 - C_2$) is larger than 1.5 in. (3.8 cm). E, Computed porosity derived from Archie's formula (thin line) and minimum porosity estimate for vertical fractures (heavy line). F, Left side: Minimum porosity estimate for horizontal fractures. Right side: Evidence for the change in stress regime from compression to strike slip at depth in the hole results from computation of stress values; S_h is nearly equal to S_v (marked by open circles) in the transition zone, and, near the base of the hole, $S_h < S_v < S_H$ (marked by solid circles). Asterisk—the location of drilling-induced, near-vertical fractures associated with extensive hole cooling; F—the level of drilling-induced stress release that produced shearing of the borehole wall in a strike-slip mode and along a near-vertical fault plane; L—the locations of ledges in the borehole wall encountered during FMS recording.

crustal structures. The geophysical data discussed here were recorded during ODP Legs 140 and 148 (Dick, Erzinger, et al., 1992; Alt, Kinoshita, et al., 1993). The electrical resistivity data were recorded with a lateral sensor, the dual laterolog (DLL). In the sheeted dikes of DSDP-ODP Hole 504B, the electrical resistivity increases continuously in basement down to ~1425 mbsf; there is a more irregular pattern at greater depths (Fig. 2B). Several measurements under 100 Ω·m near the base of the hole are attributed to a local increase in fracture density, as imaged from FMS scans (Ayadi et al., 1998).

The FMS creates a picture of the borehole wall by mapping its electrical conductance from four orthogonal arrays of pad-mounted electrodes (Luthi and Souhaité, 1990; Pezard et al., 1990). The sensor can detect features as thin as a few tens of micrometers, provided that they have sufficient resistivity contrast with the surrounding matrix. Each electrode is oriented and located in space with triaxial fluxgate magnetometers, making it possible to derive the strike and dip of geologic features cut by the hole. FMS images were recorded in Hole 504B during ODP Leg 148 (Alt, Kinoshita, et al., 1993; Ayadi et al., 1998), although the recording was interrupted by ledges (Fig. 2F) encountered in the hole (1980, 1885, 1150, and 1120 mbsf). No data were recorded with the FMS over short sections above these ledges (Fig. 2C), while passing through restricted intervals with the sensor closed and inoperative.

A total of 20,900 planes were identified and mapped in the transition zone (Fig. 2A) and sheeted-dike complex of Hole 504B (from 900 to 2075 mbsf; Ayadi et al., 1998), yielding a fairly uniform average of 20 planes per m, with a slight decrease in the number of planes with increasing depth (Fig. 2C; 1100–1800 mbsf). This slight trend is in agreement with generally increasing electrical resistivity data (Fig. 2B), and increasing pressure with depth, hence decreasing aperture of the cracks. Changes in the content of fractures (Vanko and Laverne, 1998) may also contribute to this gradual change in electrical resistivity signal with depth. At a smaller scale, sharp variations in the two data sets also tend to be inversely correlated (Fig. 2, B and C; 1935 mbsf), although this is not a general rule. For example, a local drop in electrical resistivity without change in fracture density occurs near the proposed boundary between seismic layers 2 and 3 of Detrick et al. (1994; Fig. 2, B and C).

At a meter scale, a porosity profile has been derived from DLL electrical resistivity (Fig. 2E) with Archie's formula (Archie, 1942; Brace et al., 1965; Becker, 1985) and by using appropriate clay-mineral conductivities to obtain the relative volume of the conductive fraction (Pezard, 1990; Revil et al., 1996). Also, a minimum estimate of the conductive volume may be derived in electrically anisotropic settings from the difference between the two DLL measurements as detailed in Pezard (1990) and Pezard et al. (1996). The computation assumes here that clay minerals constitute the conductive element in the rock. This method provides a qualitative means to evaluate the degree of anisotropy of the formation, here mostly due to fracturing. For example, this model is blind to the simultaneous presence of horizontal and vertical fracture sets at a given depth. Also, the presence of drilling-induced, deep-reaching, and drill water–filled fractures can yield a strong electrical anomaly (Pezard et al., 1996), as seawater is over ten times more conductive than basalt alteration phases found near the base of Hole 504B. In such a case, the conductive fraction related to clay-filled fractures is computed as much larger than the total porosity (e.g., at 1450, 1600, and 2000 mbsf; Fig. 2E), which is impossible. Such a result demonstrates indirectly the presence of drilling-related, seawater-filled fractures. These near-vertical fractures imaged with the FMS (Pezard et al., 1996; Ayadi et al., 1998) are created by intensive cooling of the hole during drilling operations (Morin et al., 1990; Pezard et al., 1996). The conductive fraction related to horizontal fracturing decreases with increasing depth (Fig. 2F), with the exception of a sharp anomaly identified both in DLL and FMS data near 1935 mbsf (Fig. 2B, C, E, and F). This trend can be related to the gradual closure of horizontal cracks with increasing depth and overburden (Fig. 2F).

Downhole stress analysis

A dominant breakout direction of N122° (Fig. 3A) was derived from BHTV image analysis (Newmark et al., 1985; Morin et al., 1990) in the upper part of Hole 504B (from basement top to 1530 mbsf). This direction was interpreted as the direction of minimum horizontal stress S_h, in agreement with regional seismicity away from lithospheric fracture zones (Fig. 1). A minor mode of rupture, orthogonal to the previous one, was also identified in the upper part of the hole. This secondary mode was attributed to tensile fracturing from thermal stresses (Morin et al., 1990). It was directly associated with intense fluid circulation during specific operations such as "fishing" for pieces of drill pipe fallen in the hole during drilling.

On the basis of linear elastic theory, the combined presence of these two orthogonal modes can be explained by a highly anisotropic horizontal stress field (Guénot, 1989; Maury, 1993), which is supported by seismic data in the vicinity of the hole (Stephen and Harding, 1983). Further analytical treatment of this data set indicated the presence of a compressive stress environment near the top of the basement (Moos and Zoback, 1990), with the minimum horizontal stress close to the stress due to overburden in the upper part of the sheeted dikes. Extensional deformation modes were ruled out as basalt compressive strengths lower than 75.0 MPa were required to explain the observed failures, nearly three times less than that measured on core (Bauer and Handin, 1985; Moos and Pezard, 1996). In spite of the extremely low and possibly selective core recovery in Hole 504B (8% on average at the base of the sheeted-dike complex, for ODP Leg 148), this assumption is supported by high values of Young's modulus determined from acoustic log data (Pezard et al., 1997).

A bimodal character similar to that obtained by Morin et al. (1990) is derived from the analysis of FMS borehole enlargements in the sheeted-dike complex (Figs. 2D and 3B) with a counterclockwise shift of 10° with respect to BHTV results. This shift is due to the spontaneous mechanical behavior of the FMS sensor in the borehole; the sensor rotates to the left to compensate for cable torque and thus has a tendency to lock the FMS pads counterclockwise of the symmetry axis of enlargements. The value of the shift is related to hole size and breakout width. Although the "main" mode of Morin et al. (1990) is obtained in seismic layer 2 (Fig. 3C), it is found attenuated in the upper part of the boundary between seismic layers 2 and 3 (Fig. 3D) proposed by Detrick et al. (1994); it then disappears completely in the lower half (Fig. 3F). Below 1700 mbsf, within seismic layer 3 of Detrick et al. (1994), the secondary mode obtained from BHTV by Morin et al. (1990) is aligned with S_H (the maximum horizontal stress). A 90° rotation of the hole long-axis orientation (Fig. 2D), i.e., of the deformation mode of this deep hole, is hence detected from mechanical calipers through the sheeted dikes, from nearly east-west near the top (above 1200 mbsf) to north-south at the base (below 1700 mbsf).

Although FMS calipers are useful to detect borehole enlargements, the better azimuthal coverage of the BHTV images is needed to clearly reveal the rupture mechanisms leading to borehole enlargement (Maury, 1993). A reappraisal of images recorded during ODP Leg 140 (Dick, Erzinger, et al., 1992) was consequently undertaken. Although numerous tensional ruptures (Fig. 4) and borehole enlargements (Fig. 5) were identified, there was no typical breakout, as no rupture in compression was recognized from the BHTV images that were recorded in the bottom part of the hole (1497–1715 mbsf and 1868–1990 mbsf). Over most of these intervals, the hole is nearly cylindrical (Fig. 5A). A great majority of borehole enlargements originated in tensional ruptures (Fig. 5B), especially near the base of the hole (see Maury, 1993). Fewer were found to be related to stress release related to the shearing of an active fault plane (de Larouzière et al., 1999), as a consequence of drilling through it (Fig. 5C). A right-lateral strike-slip mode of deformation imaged at 1583 mbsf is presented here as an illustration of this behavior (Fig. 5D).

On the basis of the observations of vertical, horizontal (Fig. 4), and tangential (Fig. 5B) ruptures in tension and the absence of compressional ruptures, the magnitudes of the stress tensor were evaluated from forward modeling at two locations in the deepest section of the sheeted-dike complex (1655 and 1980 mbsf). This evaluation was made possible by the recording of compressional and shear velocities throughout basement during ODP Leg 148 (Alt, Kinoshita, et al., 1993) and by laboratory evaluation of the physical properties including mechanical resistance in compression of selected dolerite samples (Iturrino, 1995; Moos and Pezard, 1996).

The following values were used for the simulation at 1980 mbsf: 110 GPa for Young's modulus, 300 MPa for compressive strength, 0.27 for Poisson's ratio, and 0 for tensile strength. S_v (the vertical stress) was assumed to be equal to lithostatic (90 ± 3 MPa at 1980 mbsf) and vertical, that is colinear to the borehole axis as the deviation from vertical remains under 6.0° in Hole 504B. Also, the cooling necessary to produce the tensile ruptures was estimated from numerical modeling of drilling the lower part of Hole 504B with the *JOIDES Resolution.* A temperature difference between borehole fluid and borehole wall resulting in a cooling of 75 °C was used at 2000 mbsf to simulate the reentry at the begining of ODP Leg 148. A thermal stress component equal to 62 MPa during cooling was ob-

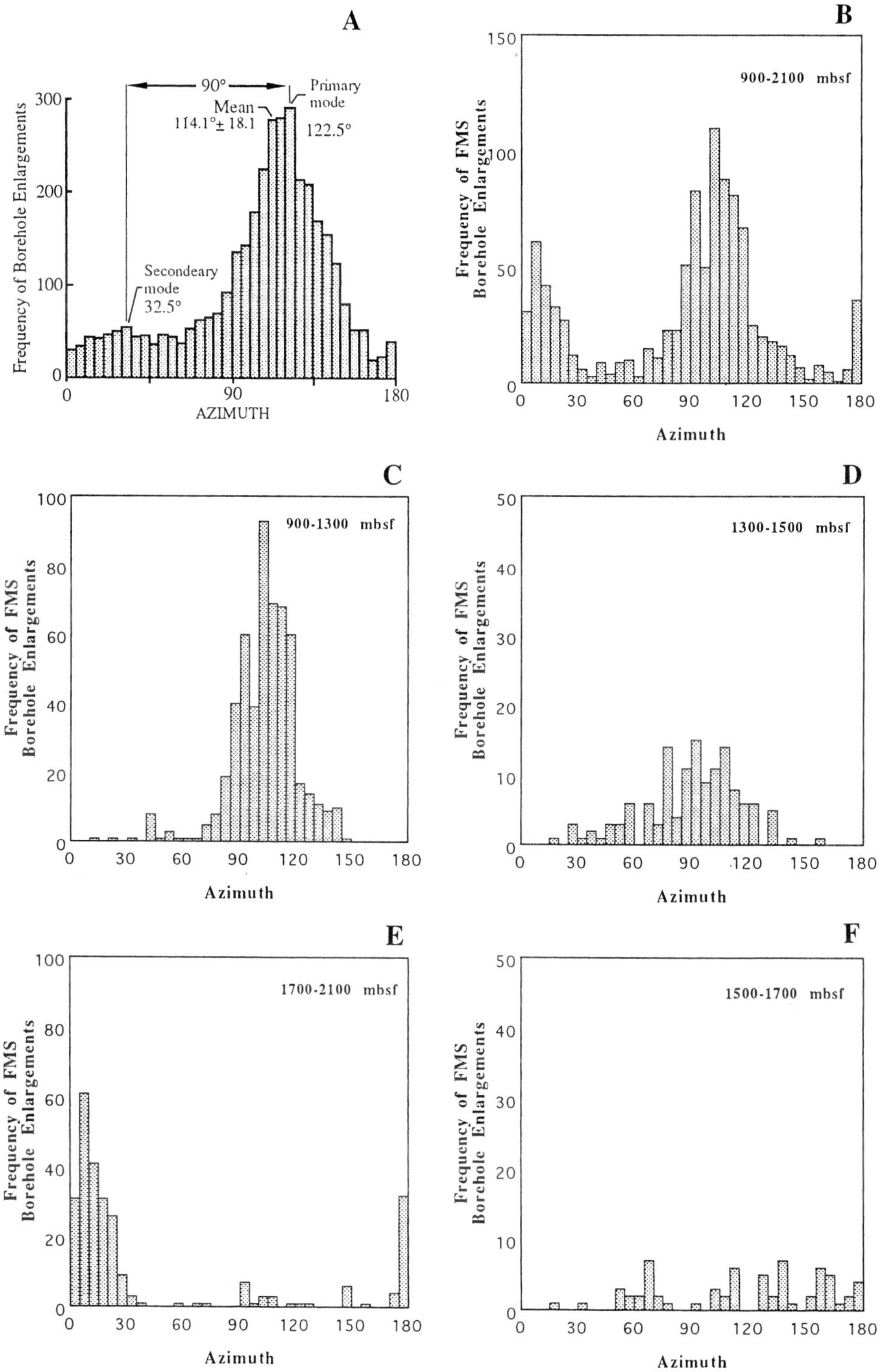

Figure 3. Histogram of borehole-enlargement frequency vs. azimuth in Hole 504B derived from (A) BHTV images between 700 and 1530 mbsf, showing azimuth of population means and modes (after Morin et al., 1990), and (B–F) FMS calipers over different depth intervals, where the difference between orthogonal caliper readings ($C_1 - C_2$) is larger than 1.5 in. (3.8 cm). The borehole-enlargement data set from FMS calipers was segmented according to the location of the boundary between seismic layers 2 and 3 in Hole 504B (chosen after Detrick et al., 1994).

tained and observed to generate the associated near-vertical and horizontal tensile ruptures.

A compressional to strike-slip regime, with S_v equal to S_h = 80 (±3) MPa and (S_H/S_h) = 1.65 is obtained at 1650 mbsf, in agreement with prior estimates (Morin et al., 1990; Moos and Zoback, 1990). The simulation indicated a strike-slip regime deeper in the hole (1980 mbsf), where S_v = 90 (±3) MPa, S_h = 85 (±3) MPa, and S_H/S_h = 1.45 (a slightly less deviatoric situation). The presence of near-vertical, seawater-filled, drilling-induced, and deep-reaching fractures near 2000 and 1600 mbsf (Fig. 2E; Pezard et al., 1996) also supports these results. Whereas the fractures near 1600 mbsf are due to repeated cooling of the hole during fishing operations at the beginning of ODP Leg 140, those near 2000 mbsf also originated in cooling operations designed to reduce wear on the coring assembly at the begining of ODP Leg 148 (Alt, Kinoshita, et al., 1993). In this case, a cooling of as much as 100 °C may actually have developed from fluid circulation (resulting in a thermal stress of ~80 MPa) at a depth where the stress due to overburden is on the order of 90 MPa. The choice of a cooling magnitude of 75 °C for the forward simulation at 1980 mbsf is thus a conservative one. Thermal stresses reduce

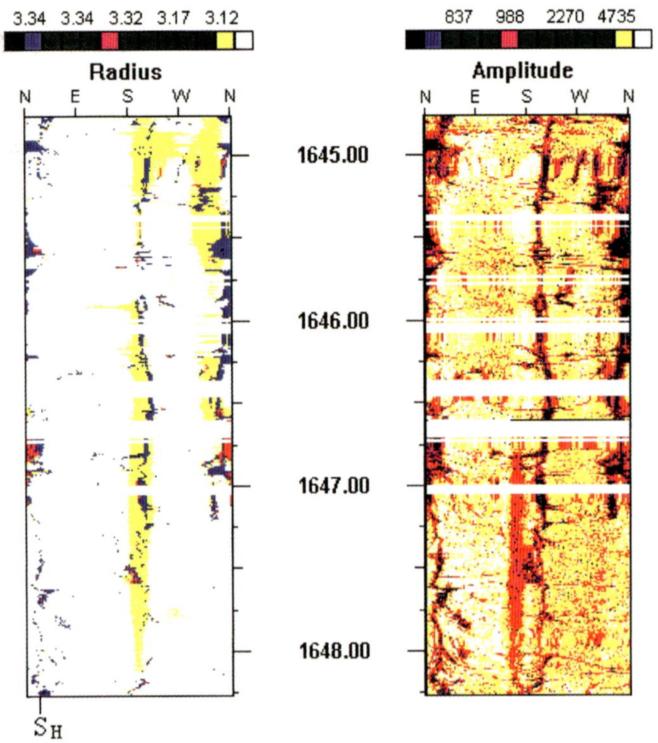

Figure 4. Example of uncalibrated BHTV acoustic images of the borehole wall within the sheeted-dike complex of Hole 504B. Left side: Transit-time data showing variations in hole size and shape vs. depth. Right side: Amplitude data relate to changes in acoustic reflectivity of the borehole wall. Near-vertical extensional ruptures, as well as some subhorizontal ones, are identified from both images in the S_H direction determined by Morin et al. (1990). The local absence of data is due to the telemetry system of the downhole sensor, as episodic transmission deficiencies occurred between the downhole tool and the surface computer during recording.

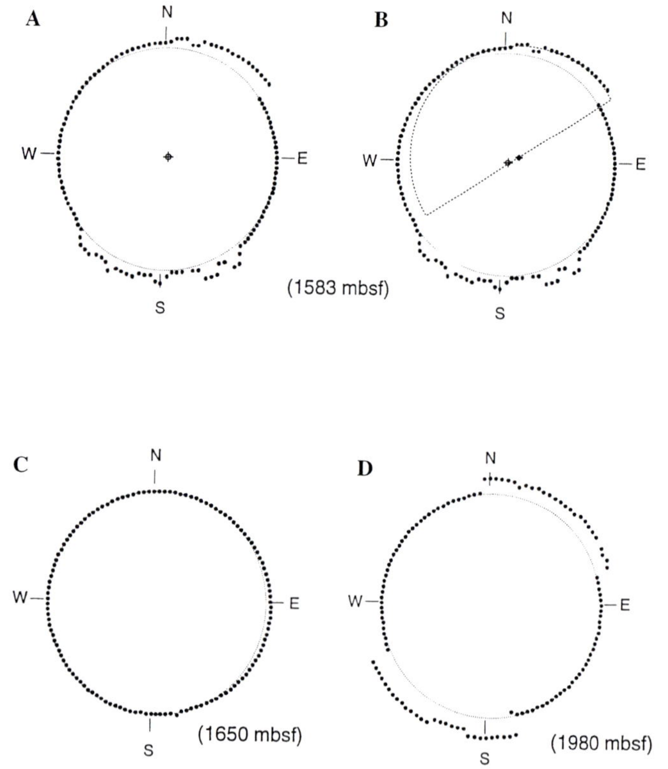

Figure 5. Borehole cross sections derived from BHTV transit-time data. The nominal size of the borehole is shown by a thin circular line. A and B, Data and interpretation showing the right-lateral shearing of the borehole along a northeast-striking fault, followed by reaming of the western part of the hole (opposite hole drift, here to the northeast in Hole 504B). C, Stable borehole with slight reaming to the east-southeast owing to pipe wear. D, Extensional ruptures of the borehole wall (radial, subvertical, and, possibly, subhorizontal) in the S_H direction determined by Morin et al. (1990).

borehole stability in such a high-temperature environment, which explains the 90° shift in borehole elongation obtained below 1700 mbsf (Figs. 2 and 3), where enlargements are due almost exclusively to tensional ruptures from thermal origin.

DISCUSSION

A series of downhole measurements recorded in the intrusive section of ODP Hole 504B document a gradual transition of the stress field from compression (above) to strike-slip (below) between 1500 and 1700 mbsf. This interval coincides with the boundary between seismic layers 2 and 3 proposed by Detrick et al. (1994) and Goldberg and Sun (1997) at this site, but appears to be located below that proposed by Swift et al. (1998a). This result has several implications. The main finding is that the debated seismic boundary could be a dynamic one, rather than a static or structural feature, getting shallower with time as the crust ages and thermally subsides. S_v at a given depth below seafloor increases with time by ~1.0 MPa in 1.0 m.y. only because of loading from the deepening water column, as postulated by Detrick et al. (1994). As hot crust is produced on-axis in relatively shallow water, the boundary between seismic layers 2 and 3 could indeed be located at a deeper level beneath the speading center than anywhere off-axis and possibly close to the dike-gabbro transition because of the relatively small overburden.

As subhorizontal structures tend to remain open under lithospheric compression, a change from a compressional to a strike-slip regime at depth in the crust has also implications for the closure of subhorizontal cracks in an elsewhere homogeneous sheeted-dike complex, as inferred from geophysical data, such as electrical resistivity below 1200 mbsf (Fig. 2). Such a selective closure should limit fluid circulation, and hence crack sealing, within the crust. A strike-slip stress regime would facilitate the extraction of higher-temperature fluids with subvertical ruptures, yielding more intense alteration and sealing of the edifice. As a consequence, attributing the boundary between seismic layers 2 and 3 to a stress change is compatible with early results implying progressive crack sealing and, hence, gradual porosity and metamorphic grade changes at depth (Christensen and Salisbury, 1975; Salisbury and Christensen, 1978; Spudich and Orcutt, 1982; Detrick et al., 1994; Laverne et al., 1995; Vanko and Laverne, 1998; Dilek, 1998), as well as more recent work relating the boundary to a change in acoustic attenuation (Goldberg and Sun, 1997).

CONCLUSION

Results of a downhole geophysical experiment in ODP Hole 504B suggest that the sheeted-dike complex is the site of a gradual change in stress regime with depth. This change from compressional above to strike-slip below coincides in depth with the transition zone between seismic layers 2 and 3 as proposed by Detrick et al. (1994). This model provides an explanation for a number of hypotheses formulated in the past to explain the presence of this seismic transition, at the time that the lithologic hypothesis was disproved. Also, subsurface changes of the stress field within the upper oceanic crust could be mapped indirectly with existing seismic terminology.

ACKNOWLEDGMENTS

This manuscript was improved from informal discussions with Christine Laverne, Dan Moos, Vincent Maury, Jeff Alt, Yildirim Dilek, and Bob Detrick, as well as from detailed reviews by Joann Stock, Steve Swift, and Keir Becker. The forward simulation routine for determining stress magnitudes from geologic observations based on borehole-wall images was written by Christophe Vergniault. This work was supported by the "Ocean" program of INSU-CNRS.

REFERENCES CITED

Agar, S.M., 1991, Microstructural evolution of a deformation zone in the upper ocean crust: Evidence from the Deep Sea Drilling Program Hole 504B: Journal of Geodynamics, v. 13, p. 119–140.

Agar, S.M., and Klitgord, K.D., 1995, A mechanism for decoupling within the oceanic lithosphere revealed in the Troodos ophiolite: Nature, v. 374, p. 232–238.

Alt, J.C., Kinoshita, H., et al., 1993, Proceedings of the Ocean Drilling Program, Initial Reports, v. 148: College Station, Texas, Ocean Drilling Program, p. 5–24.

Anderson, R.N., Honnorez, J., et al., 1985, Proceedings of the Deep Sea Drilling Program, Washington, D.C., U.S. Government Printing Office, v. 83, p. 5–11.

Archie, G.E., 1942, The electrical resistivity log as an aid in determining some reservoir characteristics: Journal of Petroleum Technology, v. 5, p. 1–8.

Ayadi, M., Pezard, P.A., Bronner, G., Tartarotti, P., and Laverne, C., 1998, Multiscalar structure at the Deep Sea Drilling Program/Ocean Drilling Program Site 504, Costa Rica Rift, III: Faulting and fluid circulation constraints from integration of FMS images, downhole logs, and core data: *in* Harvey, P.K., and Lovell, M.A., eds., Core-log integration, Geological Society of London Special Publication 136, p. 311–326.

Bauer, S.J., and Handin, J., 1985, Mechanical properties of basalt cores from the Deep Sea Drilling Program Hole 504B: Proceedings of the Deep Sea Drilling Program, Washington, D.C., U.S. Government Printing Office, v. 83, p. 371–378.

Becker, K., 1985, Large-scale electrical resistivity and bulk density of the oceanic crust, Deep Sea Drilling Program Hole 504B, Costa Rica Rift, *in* Anderson, R.N., Honnorez, J., et al., Proceedings of the Deep Sea Drilling Program, Washington, D.C., U.S. Government Printing Office, v. 83, p. 419–427.

Becker, K., Sakai, H., et al., 1988, Proceedings of the Ocean Drilling Program, Initial Reports, v. 111: College Station, Texas, Ocean Drilling Program, 360 p.

Becker, K., et al., 1989, Drilling deep into young oceanic crust, Hole 504B, Costa Rica Rift: Reviews of Geophysics, v. 27, p. 79–102.

Bell, J.S., and Gough, D.I., 1979, Northeast-southwest compressive stress in Alberta: Evidence from oil wells: Earth and Planetary Science Letters, v. 45, p. 475–482.

Bergman, E.A., and Solomon, S.C., 1985, Source mechanisms of earthquakes near mid-ocean ridges from body waves: Implications for the early evolution of the oceanic lithosphere: Journal of Geophysical Research, v. 89, p. 11415–11441.

Brace, W.F., Orange, A.S., and Madden, T.R., 1965, The effect of pressure on the electrical resistivity of water-saturated crystalline rocks: Journal of Geophysical Research, v. 70, p. 5669–5678.

Bratt, S.R., Bergman, E.A., and Solomon, S.C., 1985, Thermoelastic stress: How important as a cause of earthquakes in young oceanic lithosphere?: Journal of Geophysical Research, v. 90, p. 10249–10260.

Cann, J.R., Langseth, M.G., Honnorez, J., Von Herzen, R.P., White, S.M., et al., 1983, Proceedings of the Deep Sea Drilling Program: Washington, D.C., U.S. Government Printing Office, v. 69, p. 5–29.

Casey, J.F., Dewey, J.F., Fox, P.J., Karson, J.A., and Rosencrantz, E., 1981, Heterogeneous nature of oceanic crust and upper mantle: A perspective from the Bay of Islands ophiolite complex, *in* Emiliani, C., ed., The Sea, v. 7, p. 305–338.

Christensen, N.I., and Salisbury, M.H., 1975, Structure and constitution of the lower oceanic crust: Reviews of Geophysics, v. 13, p. 57–86.

Collins, J.A., Purdy, G.M., and Brocher, T.M., 1989, Seismic velocity structure at the Deep Sea Drilling Program Site 504, Panama basin: Evidence for thin oceanic crust: Journal of Geophysical Research, v. 94, p. 9283–9302.

Cornet, F., Helm, J., Poitrenaud, H., and Etchecopar, A., 1997, Seismic and aseismic slips induced by large-scale fluid injections: Pure and Applied Geophysics, v. 150, p. 563–583.

de Larouzière, F.D., Pezard, P.A., Ayadi, M., and Becker, K., 1999, Structure and tectonic stresses in metamorphic basement, the Ocean Drilling Program Site 976, Alboran Sea: Proceedings of the Ocean Drilling Program, Scientific Results, v. 161: College Station, Texas, Ocean Drilling Program, p. 319–329.

Detrick, R., Collins, J., Stephen, R., and Swift, S., 1994, In situ evidence for the nature of the seismic layer 2/3 boundary in oceanic crust: Nature, v. 370, p. 288–290.

Detrick, R.S., Toomey, D.R., and Collins, J.A., 1998, Three-dimensional upper crustal heterogeneity and anisotropy around Hole 504B from seismic tomography: Journal of Geophysical Research, v. 103, p. 30,485–30,504.

Dick, H.J.B., Erzinger, J., Stokking, L.B., et al., 1991, Proceedings of the Ocean Drilling Program, Initial Reports, v. 140: College Station, Texas, Ocean Drilling Program, p. 5–33.

Dilek, Y., 1998, Structure and tectonics of intermediate-spread ocenic crust drilled at the Deep Sea Drilling Program/the Ocean Drilling Program Holes 504B and 896A, Costa Rica Rift, *in* Cramp, A., MacLeod, C.J., Lee, S.V., and Jones, J.W., eds., Geological evolution of ocean basins: Results from the Ocean Drilling Program, Geological Society of London Special Publications 131, p. 179–197.

Edwards, M.H., 1991, The morphotectonic fabric of the East Pacific Rise: Implications for fault generation and crustal accretion [Ph.D. thesis], Columbia University, New York.

Fox, P.J., Schreiber, E., and Peterson, J.J., 1972, The geology of the oceanic crust: Compressional wave velocities of oceanic rocks: Journal of Geophysical Research, v. 78, p. 5155–5172.

Goldberg, D., and Sun Y.-F., 1997, Seismic structure of the upper oceanic crust revealed by in situ Q logs in DSDP/ODP Hole 504B: Geophysical Research Letters, v. 24, p. 333–336.

Guénot, A., 1989, Borehole breakouts and stress field: International Journal of Rock Mechanical and Mineralogical Science, v. 26, p. 185–195.

Héliot, D., Etchecopar, A., Cheung, P., 1990, New developments in fracture characterization from logs, *in* Proceedings of the meeting, Rock at great depth, v. 3, Maury, V., and Fourmaintraux, D., eds., Rotterdam, A.A. Balkema, p. 1471–1478.

Hobart, M.A., Langseth, M.G., and Anderson, R.N., 1985, A geophysical and geothermal survey on the south flank of the Costa Rica Rift: Sites 504 and 505: Proceedings of the Deep Sea Drilling Program, Washington, D.C., U.S. Government Printing Office, v. 83, p. 379–404.

Houtz, R., and Ewing, J., 1976, Upper crustal structure as a function of plate age: Journal of Geophysical Research, v. 81, p. 2490–2498.

Iturrino, G.J., 1995, A study of the nature of oceanic crustal reflectivity using laboratory and log measurements based on results from the Deep Sea Drilling Program and the Ocean Drilling Program sites in the Pacific and Indian Oceans [Ph.D. thesis], Rosentiel School of Marine and Atmospheric Sciences, University of Miami, Miami, p. 289.

Kent, G.M., Swift, S.A., Detrick, R.S., Collins, J.A., and Stephen, R.A., 1996, Evidence for active normal faulting on 5.9 My old crust near Hole 504B on the southern flank of the Costa Rica Rift: Geology, v. 24, p. 83–86.

Kinoshita, H.T., Furuta, T., and Pariso, J., 1989, Downhole magnetic field measurements and paleomagnetism, Hole 504B, Costa Rica Ridge: Proceedings of the Ocean Drilling Program, Initial Reports, v. 111: College Station, Texas, Ocean Drilling Program, p. 147–156.

Laverne, C., Vanko, D.A., Tartorotti, P., Alt, J.C., 1995, Chemistry and geothermometry of secondary minerals from the deep sheeted dike complex, Hole 504B: Proceedings of the Ocean Drilling Program, Scientific Results, vol. 140: College Station, Texas, Ocean Drilling Program, p. 167–189.

Luthi, S.M., and Souhaité, P.A., 1990, Method for fracture extraction and width determination from electrical borehole scans: Geophysics, v. 55, p. 821–833.

Macdonald, K.C., Fox, P.J., Alexander, R.T., Pockalny, R., and Gente, P., 1996, Volcanic growth fault and the origin of Pacific abyssal hills: Nature, v. 380, p. 125–129.

Maury, V., 1993, An overview of tunnel, underground excavation and boreholes collapse mechanisms: *in* Comprehensive Rock Engineering, Principles, Practice and Projects, Amsterdam, Pergamon Publishing, v. 4, Ch. 14, p. 369–412.

Moos, D., and Pezard, P.A., 1996, Relationships between strength and physical properties of samples recovered during Legs 137, 140, and 148 from Hole 504B: Proceedings of the Ocean Drilling Program, Scientific Results, v. 148: College Station, Texas, Ocean Drilling Program, p. 401–412.

Moos, D., and Zoback, M.D., 1990, Utilization of observations of well-bore failure to constrain the orientation and magnitude of crustal stresses: Application to continental Deep Sea Drilling Project and Ocean Drilling Program boreholes: Journal of Geophysical Research, v. 95, p. 9305–9325.

Morin, R.H., Newmark, R.L., Barton, C.A., and Anderson, R.N., 1990, State of lithospheric stress and borehole stability at Deep Sea Drilling Project Site 504B, Eastern Equatorial Pacific: Journal of Geophysical Research, v. 95, p. 9293–9303.

Newmark, R.L., Anderson, R.N., Moos, D., and Zoback, M.D., 1985, Sonic and ultrasonic logging of Hole 504B and its implications for the structure, porosity, and stress regime of the upper 1 km of the oceanic crust, *in* Proceedings of the Deep Sea Drilling Program, Washington, D.C., U.S. Government Printing Office, v. 83, p. 479–510.

Peskas, P., and Zoback, M.D., 1995, Compressive and tensile failure of inclined well bores and determination of in situ stress and rock strength: Journal of Geophysical Research, v. 110, p. 12,791–12,811.

Pezard, P.A., 1990, Electrical properties of MORB, and implications for the structure of the upper oceanic crust at the Deep Sea Drilling Program Site 504: Journal of Geophysical Research, v. 95, p. 9237–9264.

Pezard, P.A., Lovell, M.A., and the Ocean Drilling Program Leg 126 Shipboard Scientific Party, 1990, Downhole images: Electrical scanning reveals the nature of subsurface oceanic crust: Eos, Transactions of the American Geophysical Union, v. 71, p. 709.

Pezard, P.A., Becker, K., Revil, A., Ayadi, M., and Harvey, P.K., 1996, Fractures, porosity and stress in the dolerites of Hole 504B, Costa Rica Rift: Proceedings of the Ocean Drilling Program, Scientific Results, v. 148: College Station, Texas, Ocean Drilling Program, p. 317–329.

Pezard, P.A., Ayadi, M., Revil, A., Bronner, G., and Wilkens, R., 1997, Detailed structure of an oceanic normal fault: A multi-scalar approach at the DSDP/ODP Site 504: Geophysical Research Letters, v. 24, p. 337–340.

Plumb, R.A., and Hickman, S.H., 1985, Stress-induced borehole elongation: A comparision between the four-arm dipmeter and the borehole televiewer in the Auburn geothermal well: Journal of Geophysical Research, v. 90, p. 5513–5521.

Raitt, R.W., 1963, The crustal rock, *in* Hill, M.N., ed., The sea. 3. The earth beneath the sea: New York, Wiley-Interscience, p. 85–102.

Revil, A., Darot, M., Pezard, P.A., and Becker, K., 1996, Electrical conductivity in oceanic dikes, the DSDP-ODP Hole 504B: Proceedings of the the Ocean Drilling Program, Scientific Results, v. 148: College Station, Texas, Ocean Drilling Program, p. 297–305.

Salisbury, M.H., and Christensen, N.I., 1978, The seismic velocity structure of a traverse through the Bay of Islands ophiolite complex, Newfoundland: An exposure of oceanic crust and upper mantle: Journal of Geophysical Research, v. 83, p. 805–817.

Salisbury, M.H., Christensen, N.I., and Wilkens, R.H., 1996, Nature of the layer 2/3 transition from a comparison of laboratory and logging velocities and petrology at the base of Hole 504B: Proceedings of the Ocean Drilling Program, Scientific Results v. 148: College Station, Texas, Ocean Drilling Program, p. 409–416.

Sauter, D., Whitechurch, H., Munschy, M., and Humler, E., 1991, Periodicity in the accretion process on the Southeast Indian Ridge at 27°40′S: Tectonophysics, v. 195, p. 47–64.

Spudich, P., and Orcutt, J.A., 1982, Petrology and porosity of an oceanic crustal site: Results from waveform modeling of seismic refraction data: Reviews of Geophysics and Space Physics, v. 18, p. 627–645.

Stephen, R.A., and Harding, A.J., 1983, Travel time analysis of borehole seismic data: Journal of Geophysical Research, v. 88, p. 8289–8298.

Swift, S., and Stephen, R., 1992, How much gabbro is in oceanic seismic layer 3?: Geophysical Research Letters, v. 19, p. 1871–1874.

Swift, S.A., Lizarralde, D., Stephen, R.A., and Hoskins, H., 1998a, Velocity structure in upper ocean crust at Hole 504B from vertical seismic profiles: Journal of Geophysical Research, v. 103, p. 15361–15376.

Swift, S.A., Kent, G.M., Detrick, R.S., Collins, J.A., and Stephen, R.A., 1998b, Oceanic basement structure, sediment thickness, and heat flow near Hole 504B: Journal of Geophysical Research, v. 103, p. 15377–15391.

Vanko, D., and Laverne, C., 1998, Hydrothermal anorthitization of plagioclase within the magmatic/hydrothermal transition at mid-ocean ridges: Examples from a deep sheeted dike (Hole 504B, Costa Rica Rift) and a sheeted dike root zone (Oman ophiolite): Earth and Planetary Science Letters, v. 162, p. 27–43.

Villemin, T., Bergerat, F., Angelier, J., and Lacasse, C., 1994, Brittle deformation and fracture patterns on oceanic rift shoulders: the Esja peninsula, SW Iceland: Journal of Structural Geology, v. 16, p. 1641–1654.

White, R.S., McKenzie, D., and O'Nions, R.K., 1978, Oceanic crustal thickness from seismic measurements and rare earth element inversions: Journal of Geophysical Research, v. 97, p. 19683–19716.

Wiens, D.A., and Stein, S., 1990, Intraplate seismicity and stresses in young oceanic lithosphere: Journal of Geophysical Research, v. 89, p. 11442–11464.

Zoback, M.D., Moos, D., Mastin, L., and Anderson, R.N., 1985, Well bore breakouts and in-situ stress: Journal of Geophysical Research, v. 90, p. 5523–5530.

Manuscript Accepted by the Society February 23, 2000

Printed in the U.S.A.

Geological Society of America
Special Paper 349
2000

Characterization of the upper oceanic crust using high-resolution seismic amplitude modeling

Yue-Feng Sun and David Goldberg
Lamont-Doherty Earth Observatory, Columbia University, Palisades, New York 10964, USA;
e-mail: sunyf@ldeo.columbia.edu, goldberg@ldeo.columbia.edu

ABSTRACT

We examine the links between structural and lithological changes in the upper oceanic crust and seismic amplitude changes by applying modeling techniques in order to extend the conventional uses and limits of seismic data. Using high-resolution velocity and density borehole profiles from Ocean Drilling Program Hole 504B, we evaluate modeling methods and study the mechanisms of seismic attenuation in this layered porous environment. Good agreement of full-waveform synthetic seismograms with field seismic data is observed in travel time as well as in waveform shape and amplitude variations. Large seismic energy losses in the upper oceanic crust are observed where the formation consists of cyclic sequences of high- and low-porosity units, which demonstrates that energy loss in the upper oceanic crust is dominated by the effects of scattering alone. A fault zone and the boundary between seismic layers 2 and 3 are observed by amplitude changes on both synthetic and observed reflection records. The seismic amplitude, therefore, may indeed be used to map porosity and other lithologic variations.

INTRODUCTION

The seismic attributes of the oceanic crust are critical to understanding its structure and petrophysical properties. Seismic wave travel time and velocity data have been successfully used in the past to infer the geological structures at mid-ocean ridges and in the upper and lower crust. The layered structure of the oceanic crust was, in fact, discovered using seismic refraction and reflection methods. Emphasis on refraction and ocean-bottom seismic methods has revealed considerable information about the compressional-wave (P-wave) velocity structure within the limited resolution of the seismic wavelength. Raitt (1963) and then Houtz and Ewing (1976) described the structure of the oceanic crust by discriminating in situ refraction velocities and velocity gradients among seismic layers in the crust. Detrick et al. (1994) interpret velocity data in the equatorial Pacific and show that the seismic layer 2 to seismic layer 3 boundary is a gradational transition in elastic properties within the sheeted dikes, not a lithological transition from sheeted dikes to gabbro. Based on ophiolite studies, the boundary between layer 2 and layer 3 that occurs within the sheeted dike section is related to the increase in velocity associated with progressive metamorphism (Salisbury and Christensen, 1978; Salisbury et al., 1996). Exploring another important seismic attribute—the seismic attenuation—Goldberg and Sun (1997) computed a high-resolution attenuation profile at the same Pacific location. Their results suggest that changes within layers 2 and 3 are mainly due to variations in porosity and pore structure in addition to the effects of the cementation and chemical alteration of basaltic rocks. Such insights and understanding of the relationships between the seismic response and physical and structural properties of the upper oceanic crust rely heavily on the use of core samples and in situ downhole logs from Ocean Drilling Program (ODP) Hole 504B. Without ophiolite-scale mapping, however, the correlation between lithological changes and seismic parameters around this site still remains vague.

In order to refine our understanding at this site, emphasis has shifted from structural mapping to quantitative estimation of the physical properties such as porosity and pore aspect ratio and, ultimately, the interstitial fluids of the formation (Jacobson, 1992; Johnston and Christensen, 1997; Sun and Goldberg, 1997a). In the oil industry, simple velocity methods have been recently extended beyond structure and stratigraphy to determine such characteristics of a reservoir (e.g., Rasmussen and Maver, 1996; Sun and Goldberg, 1997b, 1998). These classical seismic methods using travel time and velocities alone are limited, however, because they generally neglect the intrinsic properties of the pore fluid, such as viscosity. They, in principle, cannot be used to distinguish hydrothermal fluids or oil from water, for example. On the other hand, attenuation measuring changes in waveform shape and frequency is much more sensitive to the physical and chemical state of the medium than the velocity. In the past 20 years, both laboratory experiments on attenuation and field seismic studies have accumulated convincing evidence of this greater sensitivity (e.g., Toksoz and Johnston, 1981; Bourbie et al., 1987; Wepfer and Christensen, 1991; Swift and Stephen, 1992). For example, Goldberg and Sun (1997) report an empirical relationship

Sun, Y.-F., and Goldberg, D., 2000, Characterization of the upper oceanic crust using high-resolution seismic amplitude modeling, *in* Dilek, Y., Moores, E.M., Elthon, D., and Nicolas, A., eds., Ophiolites and Oceanic Crust: New Insights from Field Studies and the Ocean Drilling Program: Boulder, Colorado, Geological Society of America Special Paper 349, p. 203–210.

between attenuation and in situ porosity of the uppermost oceanic crust for porosities >15%. Within practical constraints, such a relationship may be used to predict formation porosity from the measured seismic attenuation in similar environments.

In this paper, we review the concept of seismic energy loss and attenuation and then outline our method of separating their different components. We demonstrate this method using well-log and seismic data from the upper oceanic crust near ODP Hole 504B and compare the results with different numerical models. This study illustrates new uses of seismic data for discriminating physical attributes of the upper oceanic crust, enabling quite subtle lithologic changes to be observed.

SEISMIC ENERGY LOSS AND ATTENUATION

Attenuation is a measure of the total energy loss as a wave propagates through a medium and is inherent in all Earth materials. Seismic waves propagating in an infinite, homogeneous, and isotropic viscoelastic medium lose energy only by intrinsic attenuation and geometric spreading. The geometric spreading represents uniform spherical divergence of wave energy and is independent of the propagating medium. The definition of intrinsic attenuation explicitly refers to the "microscopic" interactions between the host solids and/or the pore fluids that generate energy loss due to viscoelasticity or anelasticity. This process mechanically converts a part of wave energy into heat.

When the medium is not homogeneous and isotropic, heterogeneities in the host medium scatter part of the incoming coherent energy in different directions and cause additional loss of energy. If these heterogeneities line up, the back-scattered energy also becomes coherent. If the heterogeneity has a thickness, for example, a bedding layer, then the back-scattered energy is usually called the reflected energy. This part of the scattered energy is, in principle, not converted into heat. The difference between intrinsic attenuation and the scattering attenuation is thus due to the heterogeneous "layered" structure of the medium at a regional scale much larger than the individual pore spaces. The knowledge of such a difference could be useful for many geologic applications.

The distinction between intrinsic attenuation and scattering (or extrinsic) attenuation is more important when the scale length of heterogeneity is comparable to or larger than the propagating wavelength (Richards and Menke, 1983; Wu, 1985). This scale dependency is only phenomenological, however, and does not obscure the distinct physical mechanisms involved with each type of attenuation. The intrinsic attenuation is due to the properties of solid framework and saturating fluid of a rock, whereas scattering attenuation is due to the contrast of properties between layers and is often caused by lithologic change. If this fact is understood, then most seismic measurements made in the field estimate the total attenuation, i.e., the sum of the intrinsic and extrinsic (scattering) components (Goldberg and Yin, 1994; Wu, 1985).

To accurately measure the total attenuation, both the intrinsic and scattering components must be identified (Jacobson, 1987). If the scattering attenuation can be estimated independently, for example from well logs or seismic data, then subtracting it from the total provides an estimate of the intrinsic attenuation. The scattering attenuation may be derived at high vertical resolution from sonic logs and/or core information or with less accuracy from synthetic well logs (Wu, 1982; Schoenberger and Levin, 1974, 1979; Spencer et al., 1982). The total or apparent attenuation can be evaluated by using vertical seismic profiles (Hauge, 1981; Stewart et al., 1984; Swift et al., 1998). This multiscale approach may be used to separate the components of attenuation at seismic frequencies and enable petrophysical and lithologic interpretations to be made from the measured total attenuation (Goldberg and Yin, 1994).

Wu (1982) was the first to give an explicit formula to calculate scattering attenuation from a known velocity profile. According to Wu's model, the energy loss of a forward-propagating wave is calculated by using back-scattered waves from multiple layer boundaries. The forward-scattered energy from multiple back-scattered waves (transmission loss of upgoing rays) is neglected as it is a small, second-order quantity (the so-called Born approximation). Menke (1983) applied a similar theory to estimate scattering attenuation from earthquake seismology. Goldberg and Yin (1994) used this method to estimate the scattering attenuation in oceanic crust from borehole logs and to compare the results with core, VSP (vertical seismic profiling), and seismic estimates. Goldberg and Sun (1997) further extend the same method to calculate an attenuation–depth profile in ODP Hole 504B and then to compare the results successfully with other seismic and laboratory estimates.

Seismic modeling using constraints from drilling information is an optimal way to integrate these methods and make lithologic interpretations. In this effort, we use the same in situ velocity and density logs from Hole 504B and produce a high-resolution discrete physical model, namely, the input to our numerical models. We are then able to investigate the extent to which the attenuation observed in field seismic data in the vicinity of ODP Hole 504B can be attributed to scattering effects and how the attenuation and energy loss are affected by in situ porosity variations. We first use a multiple back-scattering model (MBS) to extract the effects of scattering from the multichannel seismic data. This approach does not replicate a reflectivity method (or other full-waveform methods) of seismic modeling; it is solely used to assess how back-scattering attenuation affects the field seismic data. Second, we use the finite difference method (FDM) to assess the magnitude of second-order quantities that are assumed to be negligible in the MBS model. Third, we calculate scattering attenuation from inverted high-resolution seismic data and compare it qualitatively with previously published borehole estimates. The scattering components dominate the total energy loss in the upper oceanic crust, and the intrinsic component is small in this environment. The implications of these results for geoacoustic modeling may also contribute to more general studies of wave propagation and seabottom-ocean interface conditions.

DATA

Downhole data

DSDP/ODP Hole 504B is located on the Costa Rica Rift in the Panama Basin, at lat 1°13.6′N, long 83°43.8′E. This hole was drilled into the oceanic crust to a depth of ~2100 m below the seafloor (mbsf) during several expeditions over the past 20 years and is the deepest hole ever drilled in the oceanic crust. Figure 1 shows the in situ properties recorded through the crustal basalts penetrated at this site (for more detailed lithologic descriptions, see Alt, Kinoshita, et al., 1993). The sonic velocity, density, and porosity logs recorded in Hole 504B and the corrections for environmental effects are described by Goldberg and Sun (1997). We use these in situ geophysical logs for our computational modeling because they have fine vertical resolution of 0.1524 m. This resolution improves the computation of the average properties in the propagating medium. Although travel time in low-frequency seismic wave propagation may not be sensitive to changes at much less than one wavelength, the scattering attenuation

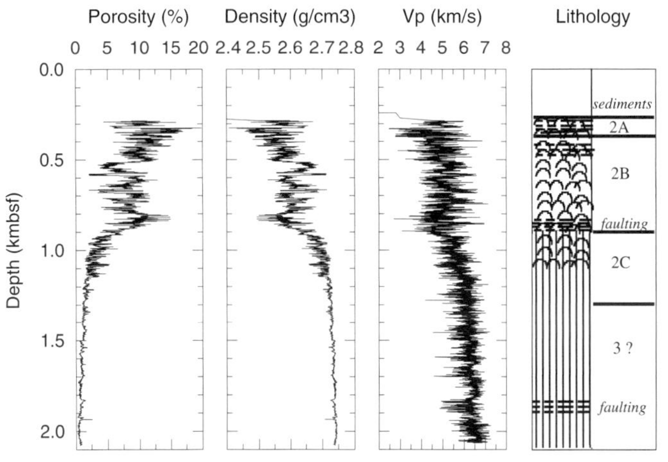

Figure 1. Basalt properties in DSDP/ODP Hole 504B on the Costa Rica Rift, Panama Basin. Porosity is derived from the deep-resistivity log. Density is a composite of the downhole density log and values derived from the resistivity log where borehole conditions were poor. Velocity was obtained from a compressional-wave sonic log. The general lithology is based on core sample interpretations (after Alt, Kinoshita, et al., 1993). Layer 2A: mostly fractured pillow lavas. Layer 2B: mostly pillow lavas. Layer 2C: sheeted dikes.

is mostly caused by layered features that are much finer than a wavelength (O'Doherty and Anstey, 1971; Jacobson, 1987; Resnick, 1990; Gist, 1994; Sams, 1995). Accurate high-resolution formation models are therefore essential for seismic amplitude and energy studies.

We divide the upper oceanic crust and the overlying sedimentary units into N_z = 13,550 layers; each layer's thickness Δz = 0.1524 m, the log's vertical resolution. Each layer i is defined by specifying its compressional-wave velocity Vp_i and density ρ_i, $i = 1, \ldots, N_z$. Adjacent model layers may have the same velocity and density parameters if the actual thickness of the physical layer is greater than Δz. The Vp and ρ profiles from 280 mbsf to 2065 mbsf are shown in Figure 1. In the overlying sediments (0–280 mbsf), discrete laboratory measurements from Costa Rica Rift samples (from Wilkens and Langseth, 1983) were used because no log data are available. Because all these velocities were measured at room temperature, a small perturbation of about +0.04 km/s was added to correct for the porosity rebound effect on these samples (Hamilton, 1976; Fisher et al., 1994). This sedimentary sequence can be divided into the three lithologic units that are summarized in Table 1 (after Collins et al., 1989).

Seismic data

We selected the multichannel seismic (MCS) reflection line (line 485, collected by the R/V *Robert D. Conrad* in May 1985) that is nearest to Hole 504B (Collins et al., 1989). The MCS data were processed into 24-fold common-mid-point (CMP) gathers, with a CMP spacing of 25 m. Line 485 crosses Hole 504B at CMP 7700 (shown in Figs. 2 and 3 for comparisons with synthetic seismograms). The seismic data processing, including stacking and migration, is described by Collins et al. (1989). The original sampling interval is 4 ms for stacked CMP data and 8 ms for migrated data. For accuracy of comparison with our synthetic results, all the field seismic data used have been resampled at a time interval of 1 ms. The center frequency of the seismic data is ~30 Hz, and typical wavelengths range from ~100 m near the seafloor to 220 m at a depth of 2 km from the seafloor. The effect of geometric spreading can be neglected if the seismic data are collected in water depths significantly greater than the depth of investigation into the crust. In general, the effect of geometric spreading should be eliminated prior to estimation of scattering and/or intrinsic attenuation.

RESULTS AND DISCUSSION

Based on the MBS model and the two-dimensional (2-D) finite-difference method (FDM), two sets of synthetic seismograms were generated by using the borehole-corrected density and sonic Vp logs in Hole 504B (see Appendix 1). One stacked CMP trace and one migrated trace from the field seismic data at the location nearest the hole are compared with these model results in Figure 2. Both the MBS and FDM synthetic records are shifted by a constant offset so as to make the peaks associated with the seafloor reflection align in two-way travel time (1140 s). Good matches are observed in travel time for the major interfaces within the sedimentary section at 1340 and 1440 ms (Fig. 2). Both synthetic traces show remarkable agreement in waveform shape and travel time with the field trace at the interface between the sedimentary section and basalt (1470 ms) and from deeper reflections to 1740 ms. Because the sedimentary rock layer that consists of limestone and chert overlying the basalts is relatively thin compared to the seismic wavelength, the reflections from the interfaces between sediment and limestone-chert and those between limestone-chert and basalt are mixed together and result in a rather complicated composite waveform.

Below 1740 ms, the agreement is less clear, even though the overall magnitudes are similar. Just below 1640 ms, an observed anomalous polarity reversal may indicate that a high-porosity zone, fluid channel, or reservoir is present. This reversal coincides with a fault zone identified by drilling at 800 mbsf (e.g., Pezard, 1990). The relatively strong reflection at ~1740 ms coincides with the depth of the boundary between layers 2 and 3 as identified by Detrick et al. (1994). We attribute some of the minor mismatches in both travel time and amplitude to the effects of multiples, noise, and other lateral variations in the crust. Another explanation is that the basalt dikes may not be vertical as assumed in these models, nor do they consider the effects of off-plane scattering.

The fact that the field trace and the FDM results agree better than the MBS model is likely because the FDM contains all of the forward-scattered and back-scattered waves and second-order multiple reflections neglected in the MBS model. A major part of the energy loss, however, can be calculated by using the MBS model. In Figure 3, the computed synthetic traces from these two different mod-

TABLE 1. SEDIMENT PROPERTIES AT DSDP/ODP HOLE 504B

Depth (mbsf)	Unit	Lithology	V_p (km/s)	ρ (g/cm³)	References
0–145	1	nannofossil oozes	1.55	1.32	Wilkens and Langseth, 1983
145–240	2	chalk	1.57	1.48	Collins et al., 1989
240–280	3	limestone/chert	2.8–4.6	1.6–2.5	Wilkens and Langseth, 1983

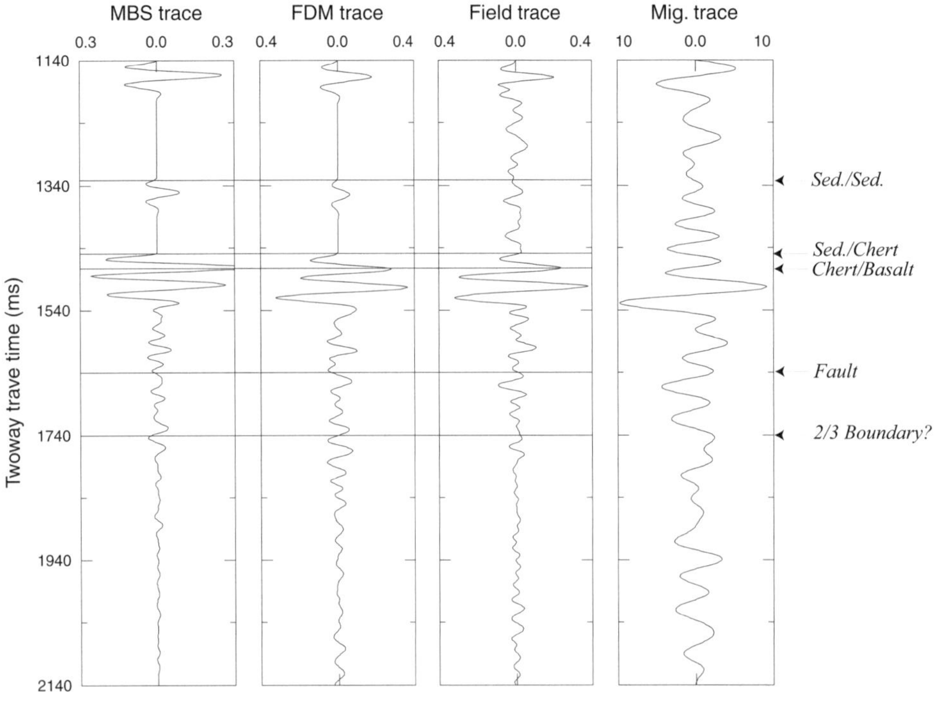

Figure 2. Amplitude comparison of synthetic traces (MBS trace and FDM trace) with real seismic data (field trace and migrated [Mig.] trace). Note that the migrated data are stretched and distorted and may only be appropriate for time-stratigraphic interpretations. Sed.—sediments.

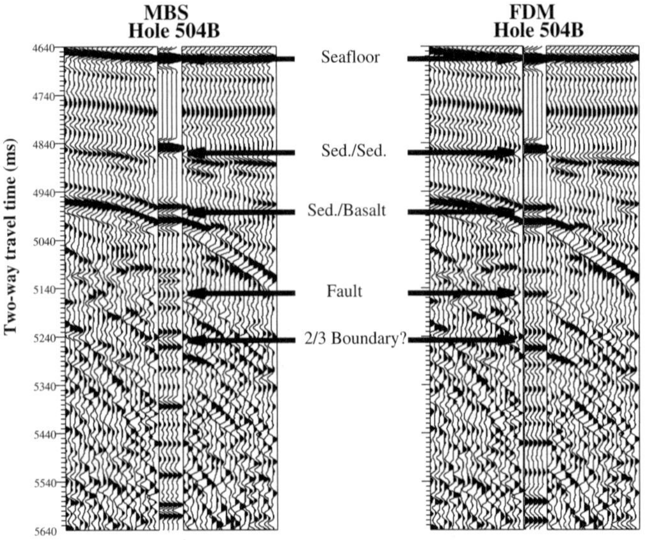

Figure 3. Comparison of synthetic results with CMP stacked traces from MCS line 485 crossing Hole 504B. MBS results show comparison of synthetic seismograms from the MBS convolution model. FDM results show comparison of synthetic seismograms from full-waveform 2-D finite-difference modeling of the wave equation. Sed.—sedimentary rocks.

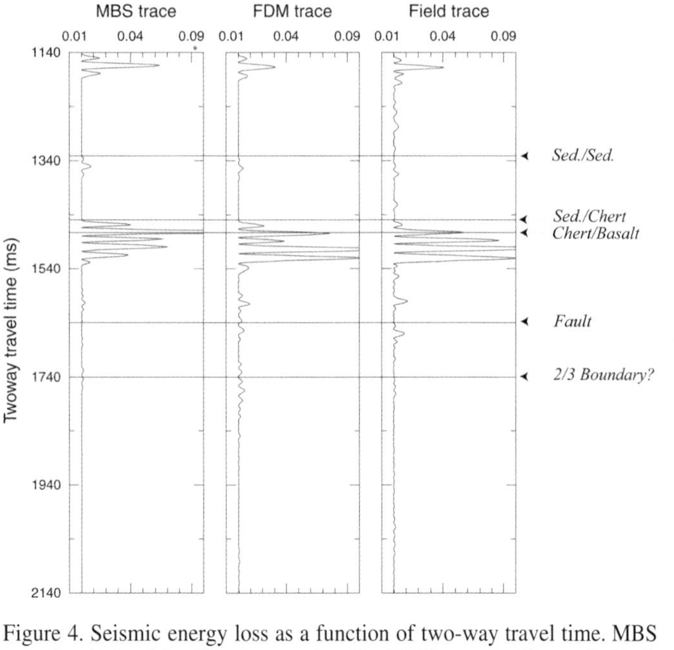

Figure 4. Seismic energy loss as a function of two-way travel time. MBS trace is obtained by directly squaring the MBS trace in Figure 2. Similarly, FDM and field-based traces are obtained from their corresponding traces in Figure 2, respectively. Note that the FDM result agrees well with the field-based seismic trace, including two small back-scattered events near 1640 ms. The horizontal scales have the SI unit $J \cdot s^{-1} \cdot m^{-2}$.

eling methods are compared with the adjacent CMP-stacked traces. Dynamic equalization has also been used on both the synthetic and field data to illustrate both strong and weak seismic events. Both the FDM and MBS synthetic traces fit well with the 2-D seismic section above 5240 ms, which, when the water depth is taken into account, corresponds to 1740 ms in Figure 2. The seismic reflection from the fault at 800 mbsf occurs at ~5140 ms (1640 ms in Fig. 2) and indicates that this is a dipping fault. The event at ~5240 ms is also dipping and corresponds to what was previously identified as the boundary between layers 2 and 3.

In Figure 4, we present the synthetic and observed seismic energy loss as a function of two-way travel time (see Appendix 2). Overall, the match between field seismic energy and the FDM energy is excellent (R = 90%) over the entire section. The greatest energy loss in the downgoing seismic wave (e.g., largest reflections) is observed between 1440 and 1540 ms where the upper oceanic crust is characterized by an alternating sequence of low- and high-porosity

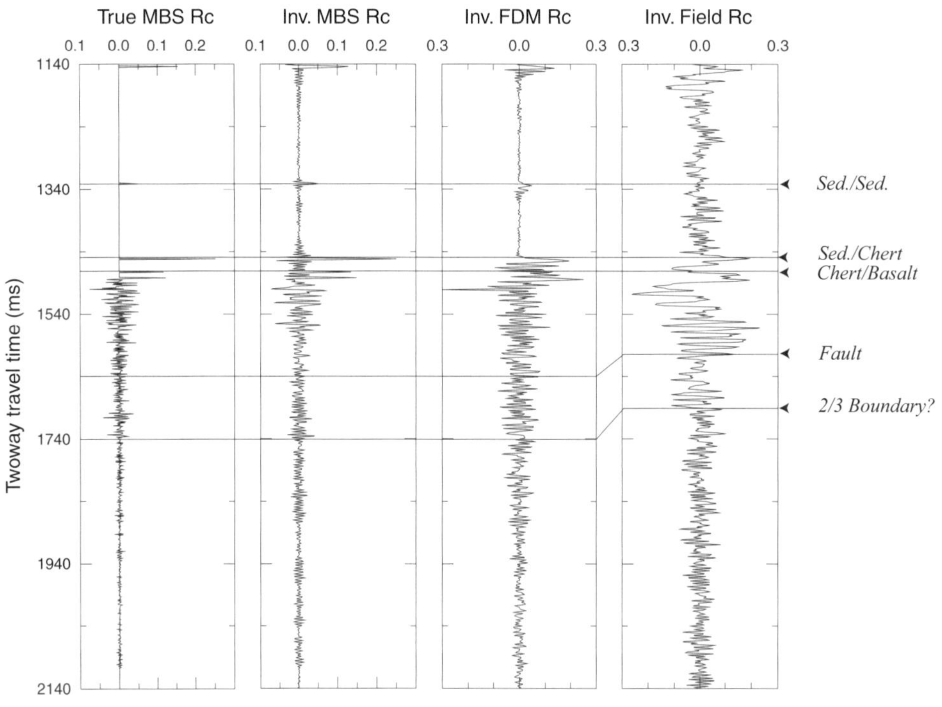

Figure 5. High-resolution reflection series. The true R_c series was generated directly from log data by using the MBS model that does not include multiples. The inverse-filtered (Inv.) MBS, FDM, and field R_c series were obtained by inverse filtering.

basalt. Geophysical logs indicate porosity of 5% to 15% over this interval (Goldberg and Sun, 1997). Relatively low attenuation occurs between 1540 and 1740 ms. The fault zone appears to influence both the field and model energy loss at ~1640 ms.

To increase the vertical resolution of these profiles, we deconvolve the seismic source effects from the MBS synthetic seismograms. The deconvolution has been achieved by using inverse filtering (Robinson and Treitel, 1980), and the results compare well with the seismic reflection coefficient series directly calculated from the density and velocity logs. In Figure 5, large seismic reflection coefficient amplitudes correlate with those in the log-derived reflection series, although several of the smaller events are contaminated by numerical errors. Above some large-amplitude threshold, therefore, the numerical inversion scheme is robust and sufficient for this analysis. We apply the same inverse filter to the FDM and field traces. The application of band-pass filters, noise, and mismatch of the source and synthetic wavelets also introduce some numerical errors in Figure 5. The major characteristics of the MBS inversion correlate well with the FDM results. The FDM inversion agrees better with the field data, however, particularly in the interval between 1440 and 1940 ms where the reflections toward the top of the hole are dominated by large porosity fluctuations. The reflections from the interfaces between sediment and sedimentary rock, and between sedimentary rock and basalt that are not resolved on the low-frequency seismic traces in Figure 2 can be clearly identified on both the inverted synthetic and field traces. The top and bottom of the fault zone are observed in the field data at 1625 and 1650 ms. Decreases in amplitude below 1740 ms may correspond to the boundary between layers 2 and 3. In the reflection coefficient series based on field data, the arrival times of both the fault and the boundary are shifted earlier, however. Neither of these features is evident on the low-resolution seismic trace in Figure 2. Their absence may be due to the stratigraphic dipping of the fault and the boundary as seen in the 2-D seismic section (Fig. 3).

We further compute high-resolution profiles of the seismic energy loss. In Figure 6, we qualitatively compare a smoothed profile of the energy loss estimated from the inverted field data with the energy loss (Q^{-1}) computed directly from borehole-log data. The Q^{-1} log was computed by using an analytical model derived from the MBS theory and converted to two-way travel time for this comparison (after Goldberg and Sun, 1997). The character of the seismic energy loss based on field data is strikingly similar to that of the computed Q^{-1} log. The relative amplitudes and shapes of the profiles can be correlated between the two profiles, as can be seen for the fault zone near 1700 ms, although a depth-time mismatch may be introduced by the stratigraphic dip of the fault. The good overall correlation of the curves suggests that seismic energy loss in the upper oceanic crust could be dominated by scattering effects, because the Q^{-1} log is computed simply from a 1-D MBS model. In the upper oceanic crust, scattering rather than intrinsic attenuation may be responsible for large seismic attenuation in the field. In principle, both scattering and intrinsic attenuation affect the field data; thus the energy loss due to intrinsic attenuation may be estimated by subtracting the effects of scattering from field seismic amplitudes.

CONCLUDING REMARKS

Our comparison of synthetic and field seismic results demonstrates that energy loss in the upper oceanic crust may be accounted for primarily by the scattering component of attenuation that is caused by variations in seismic impedance. The cause of these impedance variations is primarily the porosity that dominates the physical changes in the upper oceanic crust. Intrinsic attenuation and geometric spreading are not required to accurately match the observed field seismic data. Intrinsic attenuation would become more important, however, if the propagation distance is much longer than that measured during these logging and seismic experiments. Close agreement between our 1-D multiple back-scattering model and a 2-D finite-difference model indicates that multiple reflections from discontinuities are not significant in this environment and that structural and lithologic changes in the upper oceanic crust can be identified and interpreted in terms of the amplitudes of field seismic reflections. In

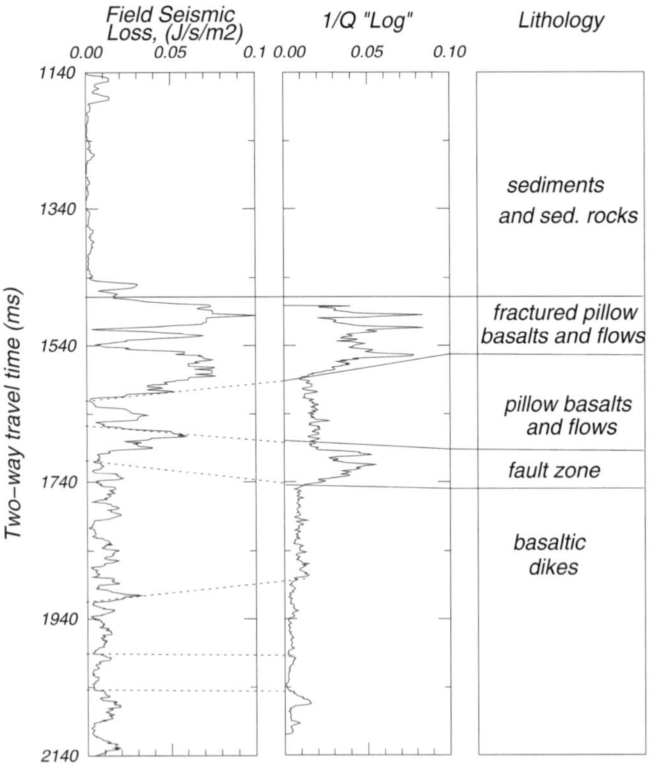

Figure 6. Interpreted seismic energy loss and in situ Q^{-1} log in DSDP/ODP Hole 504B. Trace for field seismic loss (in units of $J{\cdot}s^{-1}{\cdot}m^{-2}$) consists of a 10-point moving average of the squared R_c trace in Figure 5 (Inv. Field R_c), normalized by an arbitrary factor of 1500 $J{\cdot}s^{-1}{\cdot}m^{-2}$. The horizontal scale is dimensionless. The 1/Q "log" is the in situ Q^{-1} log from Goldberg and Sun (1997). Q^{-1} is dimensionless. Schematic lithology derived from core samples (after Alt et al., 1993). The solid lines show depth ties between log and core measurements. The dotted lines show time ties between log and seismic results. The overall similarity in character between "Field Seismic Loss" and the 1/Q "Log" suggests that seismic energy loss in the upper oceanic crust near this site is dominated by scattering attenuation rather than by intrinsic attenuation.

future investigations, the 3-D finite-element method (FEM) and MBS models may be combined to evaluate amplitude effects and distinguish lithologic changes, vertical structure, and anisotropy in the oceanic crust.

APPENDIX 1. SEISMIC MODELING

We consider the upper oceanic crust as a layered elastic medium without intrinsic losses in our numerical model. We model the compressional-wave propagation and neglect mode conversion by considering only normal-incidence plane waves. This assumption is sufficient for comparison of vertical-incidence synthetic seismograms with the stacked CMP data.

Multiple back-scattering (MBS) convolution model

Scattering attenuation is the result of heterogeneity in the formation and strongly depends on the propagating wavelength. We consider a layered medium that is transversely isotropic, such as the undeformed oceanic crust, with the symmetry axis being perpendicular to the lithostratigraphic layering. Suppose that a plane wave with frequency ω and incident energy I_0 propagates along the symmetry axis into the medium. As the plane wave hits a layer boundary, part of the incident energy will be reflected, or back-scattered, and the forward energy will be proportionally reduced. Considering the energy loss of the forward-propagating wave due to back-scattering from multiple layer boundaries and assuming that all layers have equal thickness, the forward-propagating energy, I, decays exponentially with distance. Wu (1982) gave an approximation for this scattering loss:

$$I(z) = I_0 e^{-2\alpha_s z}, \qquad (1)$$

where α_s is the amplitude-attenuation coefficient due to scattering. In this model, the forward-scattered energy from multiple back-scattered waves is neglected as it is a small, second-order quantity (Born approximation). As for intrinsic attenuation, an equivalent quality factor Q_s for scattering attenuation can be defined as follows:

$$\alpha_s = \frac{\varpi}{2\nu_0 Q_s}, \qquad (2)$$

where ν_0 is the average velocity for a given averaging interval of thickness $\gg \Delta z$.

Equations 1 and 2 are fundamental for quantifying scattering attenuation. They render it possible to treat complex scattering attenuation mathematically, as for intrinsic attenuation, although they characterize a quite different physical process of propagating waves. In order to verify this back-scattering model, we calculate a synthetic seismogram for a 1-D (vertical) medium consisting of N_z = 13,550 layers. Given a compressional-wave velocity V_{pi} and a density ρ_i for the ith layer, the reflection coefficient of the interface between layer i and layer i + 1 is

$$R_{ci} = \frac{\rho_{i+1}V_{pi+1} - \rho_i V_{pi}}{\rho_{i+1}V_{pi+1} + \rho_i V_{pi}}, \qquad i = 1, \ldots, N_z, \qquad (3)$$

and the downgoing transmission coefficient is

$$T_{di} = 1 - R_{ci}, \qquad i = 1, \ldots, N_z. \qquad (4)$$

These coefficients can be easily derived from the continuity conditions of displacement and traction at the interfaces, which hold for both forward- and back-scattered waves with normal incidence (Aki and Richards, 1980). Note that the displacement reflection coefficient R_c defined in equation 3 represents the amplitude of the displacement vector. The corresponding displacement can be obtained by multiplying by a factor of −1, enabling it to be compared with the field seismograms discussed in the next section. Assuming that all back-scattered energy at any interface between adjacent layers is totally lost, i.e., does not regenerate any forward-propagating energy, then the back-scattered coefficients for sequential interfaces, 1, 2, . . . , i, . . . , are

$$\text{Interface 1:} \quad R_{f1} = R_{c1},$$

$$\text{Interface 2:} \quad R_{f2} = R_{c2}R_{d1},$$

. . .

$$\text{Interface i:} \quad R_{fi} = R_{ci}\prod_{j=1}^{i-1} T_{dj}, \qquad (5)$$

. . .

The series R_{fi}, i = 1, . . . , N_x, can be easily converted from a depth to a time series by using the velocity model. The resulting synthetic reflection series $R_f(t)$ consists only of back-scattered energy from multiple interfaces. For a plane-wave source function W(t) with finite duration L_t, a synthetic seismogram can be computed by the convolution of the waveform W(t) with $R_f(t)$:

$$S(t) = W(t) \times R_f(t). \qquad (6)$$

In this paper, we use a Ricker wavelet with a center frequency of 30 Hz for W(t) matched to the first reflection from the boundary between seawater and the top of the sediments at Hole 504B.

Two-dimensional finite-difference model (FDM)

Seismic wave modeling based upon the complete wave equation in continuous media ensures that all complexities of wave phenomena are taken into account. In this sense, dynamic full-waveform modeling using the wave equation in the time domain is superior to other methods. We use a 2-D finite-difference method to model the wave equation and produce results sufficient for comparison with stacked compressional-wave CMP data.

Consider the following simplified equation of motion:

$$\frac{\partial^2 \mathbf{u}}{\partial t^2} = -\frac{1}{\rho}\nabla p, \tag{7}$$

where **u** is displacement, p pressure, and ρ the density of the medium. After applying the divergence operator, equation 7 reduces to

$$\frac{\partial^2 \nabla \cdot \mathbf{u}}{\partial t^2} = -\nabla \cdot \left(\frac{1}{\rho}\nabla p\right). \tag{8}$$

Let

$$p = -v^2 \rho \nabla \cdot \mathbf{u}, \tag{9}$$

where the compressional-wave velocity, $v = \sqrt{(k + 4/3\mu)/\rho}$, and k and μ are the bulk and shear modulus, respectively. Note that we have used equation 9 instead of the usual expression $p = -k\nabla \cdot \mathbf{u}$ because the compressional-wave velocity in solids is always a combined effect of the elastic bulk and shear moduli and is different from that in fluids. Substituting equation 9 into equation 8 for $\nabla \cdot \mathbf{u}$ gives

$$\frac{\partial^2 p}{\partial t^2} = v^2\left(\nabla^2 p - \frac{\nabla \rho}{\rho} \cdot \nabla p\right). \tag{10}$$

The wave equation (equation 10) can be reduced to a discrete model by using a stable central-difference scheme (Kelly et al., 1976). The 2-D FDM model size in this report is $N_z \times N_x = 3700 \times 50$ with spatial interval $\Delta z = \Delta x = 0.6$ m, where the z-axis is depth and the x-axis is horizontal direction. Because any numerical model has a finite size in space, absorbing boundary conditions have to be used to avoid artificial boundary effects from the four edges of the model space. We simulate waves that propagate out of the model space through the edges without causing artificial inward reflections by using

$$\frac{\partial p}{\partial t} \pm v\frac{\partial p}{\partial y} = 0, \qquad y = z, x, \tag{11}$$

where a + sign is used to simulate waves propagating in the positive directions of the z- and x-axes, and a − sign is used to simulate waves propagating in the negative directions of the axes. Mur (1981) developed a first-order absorbing boundary condition that results in a simple, but efficient, central-difference formula for equation 11. In addition, we place the plane-wave source 6 m below the top edge to further avoid artificial effects, and the first interface from the physical model (Vp and density logs) is placed 180 m below the top edge of the model space. This latter caution ensures high accuracy of the model for amplitude and waveform studies and further reduces the possible, even small, effects from the top edge. As in the MBS convolution model, we again use a Ricker wavelet with a center frequency of 30 Hz to simulate the plane-wave source. We use a time-sampling interval of 50 μs to satisfy the numerical stability, or the Courant-Friedrickson-Lewy conditions (e.g., Marfurt, 1984), so that

$$\Delta t = \frac{h}{\sqrt{2}v}, \tag{12}$$

where $h = min\{\Delta z, \Delta x\}$.

APPENDIX 2. SEISMIC ENERGY LOSS

The total energy of an elastic medium in motion is the sum of its kinetic energy and elastic strain energy. The total energy density, i.e., the energy per unit volume, is therefore defined as

$$H = \frac{1}{2}\rho\|\dot{\mathbf{u}}\|^2 + \frac{1}{2}\tau : \mathbf{e}, \tag{13}$$

where ρ is the mass density, **u** is the displacement, and τ and **e** are the stress and strain tensors, respectively. The energy flux rate in a plane wave, which is the energy transmitted per unit time across a unit area normal to the direction of propagation, can be derived from

$$I = \rho v \|\dot{\mathbf{u}}\|^2 \tag{14}$$

where v is the wave velocity. We define the energy reflection coefficient R_e to be the ratio of the reflected or back-scattered energy to the incident energy upon an interface and the energy transmission coefficient T_e to be the ratio of the transmitted or forward-scattered energy to the incident energy upon the interface. By using equation 14, we can obtain for an interface between layer i and layer i +1

$$R_{ei} = R_{ci}^2, \quad \text{and} \tag{15}$$

$$T_{ei} = 1 - R_{ei}, \qquad i = 1, \ldots, N_x. \tag{16}$$

That is, the energy flux rate is conserved, such that

$$R_{ei} + T_{ei} = 1, \qquad i = 1, \ldots, N_x. \tag{17}$$

Similarly, the reflection series of the back-scattered energy can be derived so that

$$R_{fei} = R_{ei}\prod_{j=1}^{i-1} T_{ej}, \quad \text{and} \tag{18}$$

$$R_{fei} = \frac{\rho_i v_i}{\rho_1 v_1} R_{fi}^2. \tag{19}$$

The MBS energy loss in terms of a displacement reflection series may then be given by the flux

$$I_r(t) = I_{inc}(0)\rho(t)v(t)R_f^2(t). \tag{20}$$

For a given plane-wave source function W(t) with finite duration L_t (or wavelet), we define an anti-wavelet $W^+(t)$ such that the convolution of the anti-wavelet W^+ with W reduces W to an impulse response:

$$W^+(t) \times W(t) = \delta(t). \tag{21}$$

The total MBS energy loss can be estimated from the seismogram by

$$I_r(t) = I_{inc}(0)\rho(t)v(t)[W^+(t) \times S(t)]^2, \tag{22}$$

where $I_r(t)$ is the instantaneous energy (flux) loss of a propagating seismic wave at two-way travel time t.

For FDM model results, we use a similar derivation in terms of pressure, such that

$$I_r(t) = I_{inc}(0)\frac{1}{\rho(t)v(t)}[W^+(t) \cdot P(t)]^2, \tag{23}$$

where $I_r(t)$ is the instantaneous energy (flux) loss and P(t) is the pressure field.

ACKNOWLEDGMENTS

We thank J. Alsop for providing the MCS data, C. Broglia for the log data, and G. Myers, D. Shearer, and R. Bookbinder for computation assistance with wave modeling. The 2-D FDM code was developed from a preliminary version written by N.M. Chen. We thank greatly the two reviewers, M.H. Salisbury and R.H. Wilkens, for their careful reviews. We also appreciate comments from Y.C. Teng, N.I. Christensen, and T.M. Brocher on early drafts. Y.-F. Sun and D. Goldberg were supported by the U.S. National Science Foundation. Lamont-Doherty Earth Observatory Contribution Number 5971.

REFERENCES CITED

Aki, K., and Richards, P., 1980, Quantitative seismology: Theory and methods: San Francisco, W.H. Freeman and Company, 932 p.

Alt, J.C., Kinoshita, H., Stokking, L.B., et al., 1993, Proceedings of the Ocean Drilling Program, Initial Reports 148: College Station, Texas, Ocean Drilling Program, 352 p.

Bourbie, T., Coussy, O., and Zinszner, B., 1987, Acoustics of porous media: Houston, Texas, Gulf Publishing Company, 334 p.

Collins, J.A., Brocher, T.M., and Purdy, G.M., 1989, Seismic-reflection structure of the upper oceanic crust: Implications from DSDP/ODP Hole 504B, Panama Basin, *in* Becker, K., Sakai, H., et al., Proceedings of the Ocean Drilling Program, Scientific Results, v. 111: College Station, Texas, Ocean Drilling Program, 344 p.

Detrick, R., Collins, J., Stephen, R., and Swift, S., 1994, In situ evidence for the nature of the seismic layer 2/3 boundary in oceanic crust: Nature, v. 370, p. 288–290.

Fisher, A.T., Fischer, K., Davoie, D., Langseth, M., and Xu, J., 1994, Geotechnical and hydrogeological properties of sediments from Middle Valley, Northern Juan De Fuca Ridge, *in* Mottl, M.J., Davis, E.E., Fisher, A.T., and Slack, J.F., eds., Proceedings of the Ocean Drilling Program, Scientific Results, v. 139: College Station, Texas, Ocean Drilling Program, p. 627–647.

Gist, G.A., 1994, Seismic attenuation from 3-D heterogeneities: A possible resolution of the VSP attenuation paradox: SEG Expanded Abstract, Society of Exploration Geophysicists, p. 1042–1045.

Goldberg, D., and Yin, C.-H., 1994, Attenuation of P-waves in oceanic crust: Multiple scattering from observed heterogeneities: Geophysical Research Letters, v. 21, p. 2311–2314.

Goldberg, D., and Sun, Y.-F., 1997, Seismic structure of the upper oceanic crust revealed by in-situ Q logs: Geophysical Research Letters, v. 24, p. 333–336.

Hamilton, E.L., 1976, Variations of density and porosity with depth in deep-sea sediments: Journal of Sedimentary Petrology, v. 46, p. 280–300.

Hauge, P.S., 1981, Measurements of attenuation from vertical seismics profiles: Geophysics, v. 46, p. 1548–1558.

Houtz, R., and Ewing, J., 1976, Upper crustal structure as a function of plate age: Journal of Geophysical Research, v. 81, p. 2490–2498.

Jacobson, R.S., 1987, An investigation into the fundamental relationships between attenuation, phase dispersion, and frequency using seismic refraction profiles over sedimentary structure: Geophysics, v. 52, p. 72–87.

Jacobson, R.S., 1992, Impact of crustal evolution on changes of the seismic properties of the uppermost ocean crust: Review of Geophysics, v. 30, p. 23–42.

Johnston, J.E., and Christensen, N.I., 1997, Seismic properties of layer 2 basalts: Geophysical Journal International, v. 128, p. 285–300.

Kelly, K.R., Ward, R.W., Treitel, S., and Alford, R.M., 1976, Synthetic seismograms: A finite-difference approach: Geophysics, v. 41, p. 2–27.

Marfurt, K.J., 1984, Accuracy of finite-difference and finite-element modeling of the scalar and elastic wave equations: Geophysics, v. 49, p. 533–549.

Menke, W., 1983, A formula for the apparent attenuation of acoustic waves in randomly layered media: Geophysical Journal of Royal Astronomical Society, v. 75, p. 541–544.

Mur, G., 1981, Absorbing boundary conditions for finite-difference approximation of the time-domain electromagnetic-field equations: IEEE Transactions of Electromagnetic Compatibility, v. 23, p. 377–382.

O'Doherty, R.F., and Anstey, N.A., 1971, Reflections on amplitudes: Geophysical Prospecting, v. 19, p. 430–458.

Pezard, P.A., 1990, Electrical properties of mid-ocean ridge basalt and implications for the structure of the upper oceanic crust in Hole 504B: Journal of Geophysical Research, v. 95, p. 9237–9264.

Raitt, R.W., 1963, The crustal rock, *in* Hill, M.N., ed., The sea: New York, Interscience, v. 3, p. 85–102.

Rasmussen, K.B., and Maver, K.G., 1996, Direct inversion for porosity of post stack seismic data, Paper SPE35509 presented at the European 3-D Reservoir Modelling Conference, Stavanger, Norway, 16–17, April.

Resnick, J.R., 1990, Stratigraphic filtering: Pure and Applied Geophysics, v. 32, p. 49–65.

Richards, P.G., and Menke, W., 1983, The apparent attenuation of a scattering medium: Bulletin of the Seismological Society of America, v. 73, p. 1005–1021.

Robinson, E.A., and Treitel, S., 1980, Geophysical signal analysis: Englewood Cliffs, New Jersey, Prentice-Hall, 466 p.

Salisbury, M.H., and Christensen, N.I., 1978, The seismic velocity structure of a traverse through the Bay of Islands Ophiolite Complex, Newfoundland, an exposure of oceanic crust and upper mantle: Journal of Geophysical Research, v. 83, p. 805–817.

Salisbury, M.H., Christensen, N.I., and Wilkens, R.H., 1996, Nature of the layer 2/3 transition from a comparison of laboratory and logging velocities and petrology at the base of Hole 504B, *in* Alt, J.C., Kinoshita, H., Stokking, L.B., and Michael, P.J., eds., Proceedings of the Ocean Drilling Program, Scientific Results, v. 148: College Station, Texas, Ocean Drilling Program, p. 409–414.

Sams, M.S., 1995, Attenuation and anisotropy: The effect of extra fine layering: Geophysics, v. 60, p. 1646–1655.

Schoenberger, M., and Levin, F.K., 1974, Apparent attenuation due to intrabed multiples: Geophysics, v. 39, p. 278–291.

Schoenberger, M., and Levin, F.K., 1979, The effect of subsurface sampling on one-dimensional synthetic seismograms: Geophysics, v. 44, p. 1813–1829.

Spencer, T.W., Sonnad, J.R., and Butler, T.M., 1982, Seismic Q-stratigraphy or dissipation: Geophysics, v. 47, p. 16–24.

Stewart, R.R., Huddleston, P.D., and Kan, T.K., 1984, Seismic versus sonic velocities: A vertical seismic profiling study: Geophysics, v. 49, p. 1153–1168.

Sun, Y.-F., and Goldberg, D., 1997a, Estimation of aspect ratio changes with pressure from seismic velocities, *in* Lovell, M.A., and Harvey, P.K., eds., Development in petrophysics: Geological Society [London] Special Publication 112, p. 131–139.

Sun, Y.-F., and Goldberg, D., 1997b, Effects of aspect ratio on wave velocities in fractured rocks: SEG Expanded Abstracts, Society of Exploration Geophysicists, v. 67, p. 925–928.

Sun, Y.-F., and Goldberg, D., 1998, Hydrocarbon signatures from high-resolution attenuation profiles: SEG Expanded Abstracts, Society of Exploration Geophysicists, v. 68, p. 996–999.

Swift, S.A., and Stephen, R.A., 1992, How much gabbro is in oceanic seismic layer 3?: Geophysical Research Letters, v. 19, p. 1871–1874.

Swift, S.A., Lizarralde, D., Stephen, R.A., and Hoskins, H., 1998, Velocity structure in upper ocean crust at Hole 504B from vertical seismic profiles: Journal of Geophysical Research, v. 103, p. 15361–15376.

Toksoz, M.N., and Johnston, D.H., 1981, Seismic attenuation: Society of Exploration Geophysicists, reprint series no. 2, 459 p.

Wepfer, W.W., and Christensen, N.I., 1991, Q structure of the oceanic crust: Marine Geophysical Research, v. 13, p. 227–237.

Wilkens, R.H., and Langseth, M.G., 1983, Physical properties of sediments of the Costa Rica Rift, Deep Sea Drilling Project Sites 504 and 505, *in* Cann, J.R., Langseth, M.G., Honnorez, J., Von Herzen, R.P., White, S.M., et al., Initial Reports of Deep Sea Drilling Project: v. 69, Washington, D.C., U.S. Government Printing Office, p. 659–674.

Wu, R.-S., 1982, Attenuation of short period seismic waves due to scattering: Geophysical Research Letters, v. 9, p. 9–12.

Wu, R.-S., 1985, Multiple scattering and energy transfer of seismic waves: Separation of scattering effect from intrinsic attenuation: Theoretical modeling, Geophysical Journal of the Royal Astronomical Society, v. 82, p. 57–80.

Manuscript Accepted by the Society February 23, 2000

Printed in the U.S.A.

Geological Society of America
Special Paper 349
2000

Section IV
Hydrothermal Processes

At the time of the first Penrose Conference on Ophiolites (Penrose Conference Participants, 1972), the significance of hydrothermal activity in the formation of oceanic and ophiolitic rocks was greatly underappreciated, at least by today's standards. Although it was widely recognized that essentially all ophiolitic rocks had been metamorphosed, the metamorphic event(s) were generally interpreted to have occurred at the time of emplacement of oceanic crust on continental margins, or even later during orogeny or interaction with meteoric water. A series of pioneering studies on oceanic and ophiolitic rocks in the middle to late 1970s demonstrated that hydrothermal activity and associated mineralization are intrinsic to the formation of submarine basaltic crust. The discovery of active hydrothermal vents and associated biota on the East Pacific Rise (Corliss et al., 1979) ushered in the modern era of hydrothermal vent studies in which the chemistry of hydrothermal fluids, dissolved chemical species, precipitates, altered rocks, and associated biota are, under ideal circumstances, studied in an integrated manner.

One of the largest and best-studied oceanic hydrothermal vent fields is the TAG (Trans-Atlantic Geotraverse) region at 26°N on the Mid-Atlantic Ridge. Humphris and Tivey provide a comprehensive overview of numerous studies of the hydrothermal vent fluids and mineral precipitates (anhydrite, quartz, and sulfides) from this locality. As a consequence of the overlapping and nested studies of both active and relict deposits, it is now possible to construct, in some detail, the temporal evolution of this key locality over the past 140,000 years. Humphris and Tivey demonstrate the critical importance of localized faulting and seawater entrainment in the evolution of hydrothermal vent fields and the chemistry of vent fluids, but unresolved questions remain about how the vent fields and mineral precipitates are linked to the alteration of deeper-level basaltic and gabbroic rocks.

Another productive approach to obtaining a better understanding of the deeper levels of submarine hydrothermal systems is the study of fluid inclusions in oceanic and ophiolitic rocks. Kelly and Früh-Green present a general overview of volatile evolution in oceanic and ophiolitic magma systems; they focus particularly on the C-O-H-NaCl components of these fluids. They show a general pattern for the evolution of magmatic and hydrothermal fluids during progressive cooling that involves an early degassing of CO_2-rich fluids, followed by release of immiscible, phase-separated high-salinity metal-rich brines and low-salinity vapors, and eventually a brine that did not form in equilibrium with an immiscible vapor phase. In other cases, methane-rich hydrothermal fluids are produced by interaction with olivine and pyroxene at high temperatures, possibly indicating that the lower crust and upper mantle might be large sources for abiogenic methane and could be an important energy source for subsurface microbial activities.

The circulation of hydrothermal fluids at depth within the crust is, at least theoretically, associated with a "cracking front" that marks the onset of brittle failure as heated rocks and water interact (Lister, 1974). Manning et al. have examined the earliest fracturing events in gabbroic rocks from the Hess Deep and the Semail ophiolite (Oman), and they show that fracturing events occur at high temperature (700–825 °C), in short-lived, pervasively distributed microfracturing events near the ridge axis. Although some details of how the distributed network of cracks forms differ from the original proposal by Lister, the results of Manning et al. provide strong support for models in which crustal cooling and penetration by seawater are intimately linked to self-generated "cracking fronts" progressively migrating deeper into the crust.

In contrast to the active and ongoing processes described in the previously noted papers, studies of those fragments of oceanic crust that have migrated completely through the zone of hydrothermal alteration can place constraints on the accumulated effects of these various processes and provide critical information on geochemical fluxes associated with hydrothermal alteration. Alt and Teagle have compiled composite oceanic and ophiolitic crustal sections for Sr, O, and S isotopes. They note that ophiolitic crust is more intensely recrystallized and that these isotopic tracers in ophiolitic crust exhibit greater compositional shifts than their oceanic crustal counterparts. They argue that the general similarities in alteration patterns for oceanic and ophiolitic crust indicate that the ophiolite–oceanic crust analogy remains invaluable in developing a better understanding of ocean ridge processes, but they caution that the isotopic differences necessarily reflect important differences in tectonic setting, primary chemistry, or heat sources. Similarly, Gillis and Banerjee focus on the geological and alteration characteristics of 11 ophiolites inferred to have formed in supra-subduction zone settings. They note the greater diversity of alteration patterns in these ophiolites, particularly the occurrence of high-temperature alteration zones in the volcanic rocks of supra-subduction zone ophiolites. They also note that sheeted dikes in supra-subduction zone ophiolites are commonly more pervasively altered than their mid-ocean ridge counterparts, as exemplified by the common occurrence of epidosite dikes in supra-subduction zone localities.

REFERENCES CITED

Corliss, J.B., Dymond, J., Gordon, L.I., Edmond, J.M., von Herzen, R.P., Ballard, R.D., Green, K.L., Williams, D., Bainbridge, A.L., Crane, K., and van Andel, T.H., 1979, Submarine thermal springs on the Galapagos Rift: Science, v. 203, p. 1073–1083.

Lister, C.R.B., 1974, On the penetration of water into hot rock: Geophysical Journal of the Royal Astronomical Society, v. 39, p. 465–509.

Penrose Conference Participants, 1972, Ophiolites: Penrose Field Conference: Geotimes, v. 17, no. 12, p. 24–25.

Geological Society of America
Special Paper 349
2000

A synthesis of geological and geochemical investigations of the TAG hydrothermal field: Insights into fluid-flow and mixing processes in a hydrothermal system

Susan E. Humphris
Department of Geology and Geophysics, Woods Hole Oceanographic Institution, Woods Hole, Massachusetts 02543, USA; e-mail: shumphis@whoi.edu
Margaret K. Tivey
Department of Marine Chemistry and Geochemistry, Woods Hole Oceanographic Institution, Woods Hole, Massachusetts 02543, USA

ABSTRACT

The TAG (Trans-Atlantic Geotraverse) hydrothermal field (26°08′N on the Mid-Atlantic Ridge) is one of the largest and best-studied sites of high-temperature hydrothermal activity and mineralization that has been found to date on the seafloor. The assemblage of active and relict deposits, with ages ranging from 0 to 140 ka, attests to a long, complex and episodic history of hydrothermal activity that appears to be controlled dominantly by large-scale faulting associated with the formation of the eastern wall of the median valley. The alternation between periods of activity and quiescence may reflect either episodic magmatism and the replenishment of the heat source or the occurrence of faulting events that reactivate fluid-flow pathways. The construction of large circular mounds, rather than the elongate deposits commonly seen in ophiolites, requires focusing of fluids along discrete pipe-like up-flow zones.

Geochemical and isotopic studies of the TAG active mound and underlying shallow up-flow zone provide strong evidence that variable amounts of entrainment of seawater play a key role in the generation of distinct fluid chemistries and the minerals that form the precipitates, the alteration of the basalts, and the growth and structural evolution of the entire deposit. We present a model for the formation and growth of the TAG active mound in which the current activity is dominated by seawater entrainment and precipitation of anhydrite due to the vigorous and focused nature of the current black smoker activity. We suggest that the pervasive, high-temperature alteration of the basement in the shallow up-flow zone likely occurs during either waxing or waning of activity, when subsurface fluid-flow rates are substantially lower and seawater entrainment is not as prevalent.

INTRODUCTION

Since the discovery of low-temperature hydrothermal activity and mineralization during the NOAA Trans-Atlantic Geotraverse (TAG) project in 1972–1973 (Scott et al., 1974a, 1974b; Rona et al., 1975), the TAG hydrothermal field at 26°08′N on the Mid-Atlantic Ridge has been the subject of numerous investigations. The magnitude of the efforts in this region increased dramatically with the discovery in 1985 of the first high-temperature hydrothermal activity to be found in the Atlantic Ocean (Rona et al., 1986). Through international and multidisciplinary studies that have included extensive geophysical surveys, detailed observations and sampling, monitoring efforts, and drilling, the active TAG hydrothermal mound is one of the best-studied sites of high-temperature hydrothermal activity and mineralization on the seafloor.

In this paper, we synthesize the results of geologic and geochemical investigations of the TAG hydrothermal field—and, in particular, the active TAG hydrothermal mound—to gain insights into the geologic and hydrologic evolution of the hydrothermal system. First, we examine the geotectonic setting and ages of all the deposits mapped within the TAG hydrothermal area to determine the controls on fluid up-flow on the scale of the entire vent field. Then we turn our attention to the active TAG hydrothermal mound and use its mineral assemblages and geochemistry to decipher its history of formation and the nature of subsurface processes on the scale of an individual hydrothermal deposit.

GEOLOGIC SETTING OF THE DEPOSITS WITHIN THE TAG HYDROTHERMAL FIELD

Geology of the TAG segment

The TAG hydrothermal field is located at the midpoint of a 40-km-long ridge segment that is bounded to the north and south by

Humphris, S.E., and Tivey, M.K., 2000, A synthesis of geological and geochemical investigations of the TAG hydrothermal field: Insights into fluid-flow and mixing processes in a hydrothermal system, *in* Dilek, Y., Moores, E.M., Elthon, D., and Nicolas, A., eds., Ophiolites and Oceanic Crust: New Insights from Field Studies and the Ocean Drilling Program: Boulder, Colorado, Geological Society of America Special Paper 349, p. 213–235.

nontransform discontinuities at 26°17′N and 25°55′N, respectively (Purdy et al., 1990; Sempere et al., 1990). The median valley of the segment has an hourglass shape centered around the midpoint at 26°10′N where depths of the inner valley floor shoal to less than 3550 m (Fig. 1). A microearthquake and seismic refraction study detected a low-velocity zone at depths as shallow as 3 km near the segment center; this zone has been attributed to a largely solid, but hot, magmatic intrusion (Kong et al., 1992). Seafloor spreading rates have been asymmetric over the past 10 m.y.; the half–spreading rates have been 11 mm/yr to the west and 13 mm/yr to the east (McGregor and Rona, 1975; McGregor et al., 1977). Asymmetry is also evident in the structural morphology of the rift valley. To the west of the neovolcanic zone, the accumulated relief of a series of uplifted terraces, covered by thick sediment and separated by well-developed rift-parallel faults, creates the western valley wall. However, the inner valley floor to the east of the neovolcanic zone is characterized by numerous ridge-parallel fissures and faults; the total extension becomes increasingly concentrated on selected faults that sum to form the steeper and higher eastern valley wall (Karson and Rona, 1990; Kleinrock and Humphris, 1996a; Zonenshain et al., 1989).

The TAG hydrothermal field

The TAG hydrothermal field is located at 26°08′N at the base of the eastern rift-valley wall where it projects into the median valley as a broad salient (Rona, 1980; Temple et al., 1979). Zones of low- and high-temperature activity, as well as two major areas of inactive sulfide deposits, have been mapped in a 5 × 5 km area (e.g., Rona et al., 1975, 1986, 1993a, 1993b; Thompson et al., 1985). The known active and inactive zones in the TAG hydrothermal field are summarized in Table 1, and their locations are shown in Figure 1.

The TAG active mound (26°08.2′N, 44°49.6′W). High-temperature hydrothermal activity is focused at the TAG active mound, which lies 2.4 km east of the neovolcanic zone and 1 km from the east wall on crust estimated from seafloor spreading rates to be ~100 k.y. old. The floor of the rift valley in this area is characterized by the intersection of actively developing, ridge-parallel (north-northeast–trending) faults and fissures with a series of preexisting, obliquely oriented (east-northeast–trending) faults, which likely provide permeable pathways through fault breccias for focusing of the upwelling hydrothermal fluids (Fig. 2) (Kleinrock and Humphris, 1996b; Kleinrock et al., 1996). The seafloor increases in depth from 3670 m on the southern and eastern sides of the active mound to more than 3680 m on its western and northern sides. On the basis of gravity data that indicate a sharp change in the thickness of sulfides near the center of the mound, from ~50 m to the north to ~15 m to the south (Evans, 1996), and the presence of faults and fissures in the sediment-covered lavas in the vicinity of the mound, Humphris and Kleinrock (1996) interpreted the change in seafloor depth to be due to faulting beneath the mound rather than being merely indicative of gently sloping terrain. In addition, the alignment of a recent graben-like depression on the mound with the ridge-parallel faults and fissures on the proximal seafloor suggests that extension is ongoing (Humphris and Kleinrock, 1996; Kleinrock and Humphris, 1996b).

On the basis of geochronologic studies, the TAG hydrothermal mound is believed to have been active episodically every few thousand years over at least the past 20 ka, and possibly the past 50 ka, although this older age was obtained from a dredged sample so its location is not well determined (Lalou et al., 1990, 1993, 1998; You and Bickle, 1998). The current activity is estimated to have begun ~80 yr ago (Lalou et al., 1998).

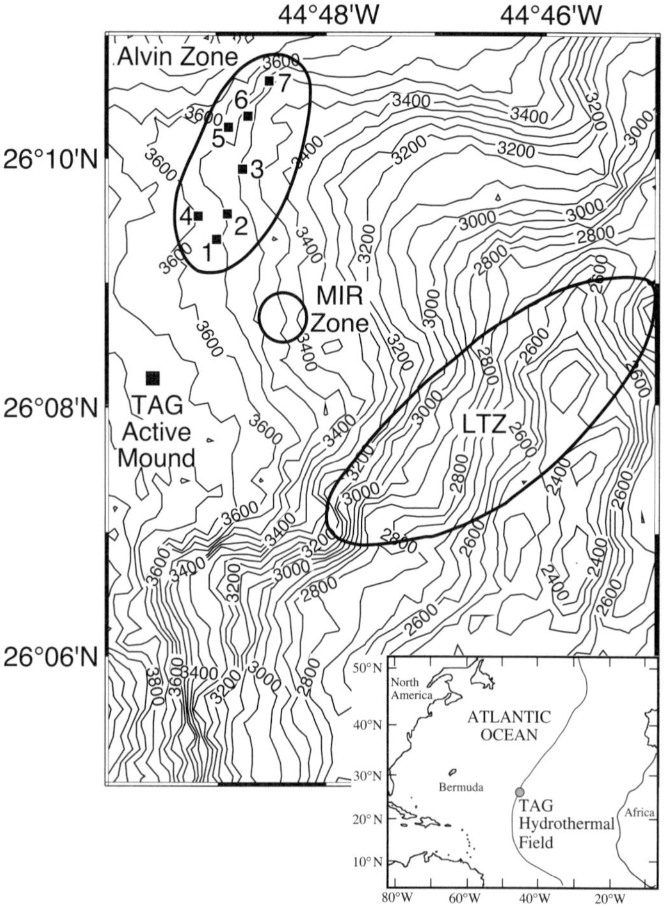

Figure 1. SeaBeam bathymetry of the TAG segment (Purdy et al., 1990), showing the locations of all the identified hydrothermal mounds and zones in the TAG hydrothermal field. Contour interval is 50 m. Numbers in the Alvin zone correspond to mound locations listed in Table 1. LTZ—low-temperature zone. Inset: Location of TAG segment on the Mid-Atlantic Ridge.

The Alvin zone. The axis-parallel faults at the TAG active mound extend north-northeast (Fig. 2) and continue into the Alvin zone (White et al., 1998). Discontinuous sulfide outcrops, standing and toppled chimneys, and large mounds have been mapped along a strike length of 2.5 km at water depths of 3400–3600 m (Rona et al., 1993a, 1993b; White et al., 1998). To date, six relict sulfide mounds similar in size to the TAG active mound have been identified (Table 1), all of which appear to be undergoing oxidation, mass wasting, and dissection by faulting (Rona et al. 1993a, 1993b, 1998; White et al., 1998). In addition, an active mound (~150 m in diameter and up to 68 m high) with low-temperature (22.5 °C) fluids discharging through manganese and iron hydroxides and silicates was discovered in 1998 at the northern end of the Alvin zone and is referred to as the Shimmering Mound (Rona et al., 1998).

Age dates, which are available for only one relict sulfide mound (the Alvin, or Double, Mound), suggest that it was hydrothermally active at ca. 48 ± 5 ka. However, a dredged manganese oxide sample attests to an earlier low-temperature episode at ca. 75 ka (Lalou et al., 1995). A sediment core collected 0.5 km west of the Alvin zone contains high concentrations of hydrothermally derived material dating from 6 and 8.5 ka that suggests more recent activity (Metz et al.,

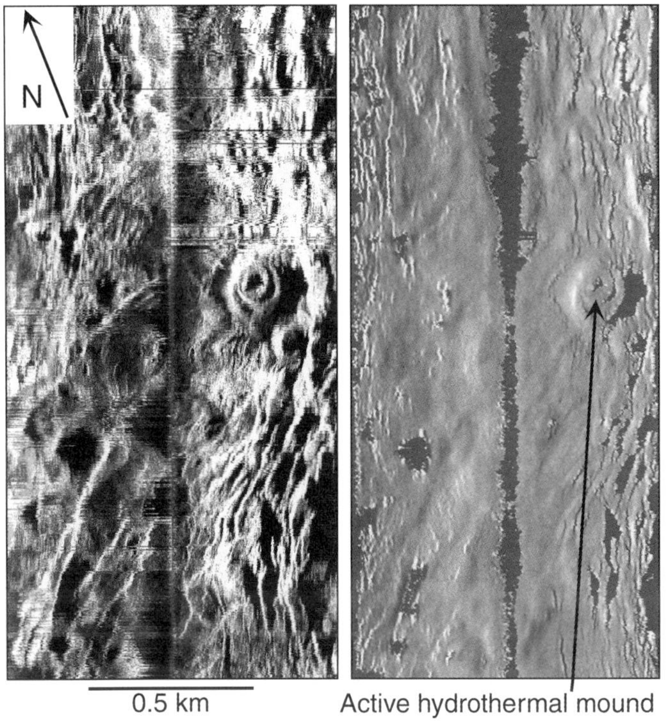

Figure 2. Sidescan-sonar backscatter image (left) and shaded-relief bathymetry (right) of the TAG active mound. Data were collected using a 120 kHz phase-difference, split-beam sidescan-sonar operated by the Deep Submergence Lab of Woods Hole Oceanographic Institution. The circular mound is ~200 m in diameter, and the two platforms and the cone of the black smoker complex can be distinguished. Note the series of faults to the north of the mound that continue into the Alvin zone (for locations, see Fig. 1). (Modified from Kleinrock et al., 1996.)

1988). The age of the currently active mound in the Alvin zone is unknown.

The Mir zone. About 2 km east-northeast of the active TAG mound and 4 km from the ridge axis lies the Mir zone. It is located on normal-fault blocks on the lower part of the eastern valley wall at 3430–3575 m water depth and covers an area ~1 km in diameter. Descriptions of the morphology of the Mir zone vary. Rona et al. (1993a, 1993b) defined three zones based on the characteristics of the hydrothermal deposits: (1) a 200-m-wide western margin where faulting has exposed step-like outcrops of hydrothermal breccias beneath a partial cover of weathered sulfide debris and metalliferous sediments, (2) a zone of semicontinuous sulfides up to 400 m wide and 700 m long with numerous standing and toppled chimneys on top of earlier sulfide talus, and (3) an area extending about 200 m east of the sulfides that contains low-temperature hydrothermal deposits, including iron oxyhydroxides, nontronite, and manganese oxide crusts. In contrast, Stepanova et al. (1996) described the Mir zone as a single mound ~400 m in diameter and up to 70 m high with a similar variety of sample types. However, a recent sidescan-sonar and coregistered bathymetric survey of the TAG hydrothermal field indicates that the Mir zone lacks large, distinctive structures, although a small (~100 m in diameter and ~10 m high) mound can be identified (White et al., 1998). A magnetic anomaly consistent with the presence of reduced crustal magnetization that may be caused by hydrothermal alteration is associated with this feature (Tivey et al., 1996).

Geochronologic studies (Lalou et al., 1993, 1995) suggest that activity in the Mir zone began with low-temperature venting at ca. 140 ka and resulted in the formation of the hydrothermal precipitates in the eastern part of the zone. High-temperature activity began at ca. 100 ka and continued episodically until ca. 0.6 ka; there was a major thermal event at 50 ka. However, high heat-flow values in the Mir zone close to an adjacent volcanic center indicate that this area continues to be thermally active, and suggest that hydrothermal activity may have ceased only recently (Rona et al., 1996).

The low-temperature zone. The first site of hydrothermal precipitates to be found in the TAG hydrothermal field was the low-temperature zone, located at mid-depth (3100–2300 m) on the eastern median-valley wall about 4 km east of the TAG active mound. Although no fluids were observed venting from the deposits, near-bottom temperature anomalies in the water column and conductive heat-flow anomalies in the sediments were interpreted to suggest ongoing, low-temperature (<20 °C) diffuse flow (Rona et al., 1984, 1993b). Hydrothermal precipitates—including surficial metal-rich staining of the carbonate ooze, as well as massive layered deposits of nontronite, manganese oxides, and iron oxides and oxyhydroxides—were mapped within a 2 × 3 km area (Rona et al., 1984; Thompson et al., 1985).

On the basis of radiometric dating of five manganese oxide samples, Lalou et al. (1986) suggested that the bulk of the deposits were formed from ca. 16 to 4 ka, although dates of 100 and 125 ka for discrete layers within two of the samples suggest there may have been an earlier episode of low-temperature hydrothermal activity.

The role of faulting in the development of the TAG hydrothermal field

The geotectonic setting and assemblage of deposits with ages ranging from 140 ka to the present attest to a long and complex history for the TAG hydrothermal field that appears to be controlled dominantly by faulting associated with the formation of the eastern median-valley wall. Age dates for manganese oxide deposits provide evidence that low-temperature activity preceded episodes of high-temperature activity in the three zones (Mir, Alvin, and low-temperature) where data are available (Table 1). On the basis of inferred seafloor-spreading flow lines, the onset of high-temperature activity at the Mir zone at ca. 100 ka appears to coincide with the time when this site was located at the slope break between the relatively flat valley floor and the bounding fault of the eastern wall (Lalou et al., 1995). This is also the case for the Alvin zone and the active mound, which reached the same location at ca. 50 ka (Lalou et al., 1995). Hence, it seems likely that the onset of high-temperature activity resulted from large-scale faulting that enhanced permeability and allowed seawater to penetrate deeper into the crust and mine sufficient heat (either from the ridge axis or from local intrusions) to sustain active circulation over long periods of time. Since that time, hydrothermal activity has been episodic at all sites on time scales of thousands of years. The alternation between periods of activity and quiescence may reflect either episodic magmatism and the replenishment of the heat source or the occurrence of faulting events that reactivate fluid-flow pathways. The construction of large mounds requires that the same near-surface fluid-flow pathways be maintained over long periods of time. At the TAG active mound, intersecting fault series provide such a focusing mechanism to produce an isolated mound (Kleinrock and Humphris, 1996b). In the Alvin zone, the extensive, along-strike sulfide outcrops suggest widespread hydrothermal discharge localized along faults; however, there were also clearly pre-

TABLE 1. COMPILATION OF CURRENTLY KNOWN ACTIVE AND INACTIVE HYDROTHERMAL DEPOSITS IN THE TAG HYDROTHERMAL FIELD

Feature*	Location	Seafloor depth (m)	Size	Tectonic setting	Brief description	Age[†] and related description (ka)
TAG active mound						
Active mound[§]	26°08.21′N 44°49.57′W	3670	200 m diameter; 40–50 m high; two platforms	Intersection of active axis parallel faults and fissures with preexisting oblique (east–northeast) fault series	Zoned sulfide + anhydrite deposit; Cu rich, 366 °C black smokers at summit of chalcopyrite + anhydrite conical edifice; Zn-rich white smokers (260–300 °C) on southeast edge	0–50; episodic; upper platform <5; present activity <0.1
Alvin zone						
Discontinuous sulfides**	26°09.33′N 44°48.99′W to 26°10.62′N 44°48.50′W	3400–3600	~2.5 km along strike	Along same set of axis-parallel faults as the TAG active mound	Weathered sulfide outcrops with oxidized surfaces; standing and toppled chimneys; iron-manganese oxides	75 based on manganese oxide; no sulfide ages
Southern Mound[††] (inactive)	26°09.33′N 44°48.99′W (1)[#]	~3580	200 m diameter; 25 m high	Mound faulted parallel to ridge axis	Weathered massive sulfide blocks that are covered with sediment	—
Alvin (or Double) Mound** (inactive)	26°09.54′N 44°48.89′W (2)[#]	3512–3540	80 × 150 m; ~30 m high	—	Weathered massive sulfide blocks	52.5–41 based on sulfide ages
Mound[§§] (no name; inactive)	26°09.09′N 44°48.75′W (approx.) (3)[#]	~3450	150–200 m diameter; 30–50 m high	—	—	—
New (or Shinkai) Mound[##]	26°09.52′N 44°49.15′W (4)[#]	3545–3615	150–200 m diameter; 35–70 m high	—	Horizontally layered massive sulfides and standing chimneys on top; some collapse in center of summit; weathered massive sulfides with oxidized surfaces; apron of metalliferous sediments	—
Shimmering Mound*** (active; low temperature)	26°10.25′N 44°48.88′W (5)[#]	3430–3504	50 m diameter; 33–64 m high	Faulted, lower valley wall	Southeast margin: fluids (22.5 °C) venting through iron-bearing silicates, iron-manganese oxides and iron oxyhydroxides.	—
Northern Mound[††] (inactive)	26°10.33′N 44°48.70′W (6)[#]	3400–3440	~150 m diameter; ~20–30 m high	—	Northwest margin: inactive; layered, gray deposits. Weathered massive sulfides; light dusting of pelagic sediments	—
Mound[§§] (no name; inactive)	26°10.62′N 44°48.50′W (7)[#]	~3520	Diameter unknown; ~20–30 m high	—	Northern extent of Alvin zone	—
Mir zone						
Discontinuous sulfides and low-temperature deposits**	Centered at 26°08.70′N 44°48.40′W	3430–3575	1 km diameter	On normal-fault blocks on lower eastern, valley wall	Hydrothermal breccias on western margin; semi-continuous weathered sulfides; iron-manganese oxides, nontronite & iron oxyhydroxides on eastern margin	140—low T; 100-0.6—high T; episodic
Low Temperature Zone						
Low temperature deposits[†††] (active?)	26°07.00′N 44°48.00′W to 26°09.00′N 44°45.00′W	2300–3100	2 × 3 km area	On top of fault scarps at mid-depth of eastern valley wall	Surficial metal-rich staining of carbonates; massive layered deposits of nontronite, iron-manganese oxides, and iron oxyhydroxides	125—low T; 16–4—low T formed bulk of deposit

Notes: This compilation represents a careful analysis and cross-correlation of all reported occurrences of hydrothermal deposits in the TAG hydrothermal field, and reconciliation with the detailed sidescan and bathymetric survey discussed by White et al. (1998).

* Where a mound has been referred to by different names by different authors, both names are given.

† Radiometric ages of hydrothermal precipitates from Lalou et al. (1986, 1990, 1993, 1995).

§ Rona et al. (1986), Thompson et al. (1988), Tivey et al. (1995), Kleinrock et al. (1996).

\# Numbers correspond to locations shown in Figure 1.

** Rona et al. (1993a, b), White et al. (1998).

†† White et al. (1998).

§§ The location and characteristics (both of which are approximate) of this mound were extracted from interpretation of a camera tow presented in Rona et al. (1993a, b).

\## White et al. (1998), Rona et al. (1998).

*** Rona et al. (1998).

††† Rona et al. (1984, 1993b), Thompson et al. (1988).

ferred fluid-flow pathways that resulted in the construction of mounds.

The other important component for maintenance of the hydrothermal system is a long-term heat source, the nature of which remains enigmatic. The TAG hydrothermal field is located at the midpoint of the segment, which is the shallowest and, hence, probably the most magmatically robust region. In addition, it is situated where the rift-valley's eastern wall protrudes into the neovolcanic zone. Hence, the geologic setting appears favorable for the extraction of heat associated with axial magmatism by circulation of seawater through faults (Temple et al., 1979; Thompson et al., 1985). However, the neovolcanic zone to the west of the TAG hydrothermal field does not appear to be particularly robust as the zone is characterized by a series of small and discontinuous volcanic regions (Kleinrock and Humphris, 1996a). Sea-surface magnetic data suggest that south of the valley's eastern-wall protrusion, the neovolcanic zone curves to the east and becomes considerably more robust (M.A. Tivey, 1998, personal commun.). Hence, the neovolcanic zone south of the TAG hydrothermal field could represent a heat source if magmatic or volcanic structures continue along strike to the north, or if there is hydrothermal circulation along strike. Finally, some workers have reported pillow mounds close to the active and relict sulfide mounds of the TAG hydrothermal field, which could be related to local heat sources (Rona et al., 1993a, 1993b; Zonenshain et al., 1989).

Comparison with the geologic settings of sulfide deposits in ophiolites

The range of tectonic settings of oceanic-crust formation exposed in ophiolite sequences (i.e., mid-ocean or back-arc spreading centers; suprasubduction zones) has been debated for many years (e.g., Coleman, 1981; Ernewein et al., 1988; Miyashiro, 1973; Moores and Vine, 1971; Pearce et al., 1981, 1984). Although very few ophiolite sequences were probably formed at true mid-ocean ridges, the massive sulfide deposits in these sections are the products of the same geologic and geochemical processes that are occurring in seafloor hydrothermal systems, and hence they represent useful analogues.

Tectonic, rather than magmatic, control on the distribution of massive sulfide deposits in ophiolite sequences is indicated by their association with large-scale fault systems (Harper, 1985; Varga and Moores, 1985). In the Troodos ophiolite in Cyprus, massive sulfide deposits are clustered into five districts, each containing many deposits within an area of about 10 km × 10 km (Hannington et al., 1998; Galley and Koski, 1999). Although large-scale fault systems control the distribution of the deposit clusters (Varga and Moores, 1985), localization within the clusters is controlled by ridge-axis–parallel normal faults (Adamides, 1987). In addition, the deposits tend to occur in areas where high-level intrusions are common (Constantinou, 1980; Varga and Moores, 1985). In Oman, the massive sulfide deposits are also clustered in groups 25–50 km apart, possibly around off-axis seamounts along the 500 km strike length of the Semail ophiolite (Alabaster et al., 1980). Mineralization is structurally controlled and is subparallel to the presumed spreading axis (Hannington et al., 1998; Galley and Koski, 1999). Deposits in both Cyprus and Oman tend to be elongate along faults with strike lengths of several hundred meters to 1 km, thereby more closely resembling the areas of discontinuous sulfide outcrops in the Mir and Alvin zones of the TAG hydrothermal field, rather than the individual circular mounds (Hannington et al., 1998).

In summary, the geotectonic setting of mineral deposits both at the TAG hydrothermal field and in many ophiolites suggests that large-scale faulting is a key element in the development of extensive massive sulfide deposits. Such faulting not only enhances the permeability of the seafloor, thereby allowing more seawater to penetrate and extract heat from magmatic intrusions at depth, but also focuses the up-flowing hydrothermal fluid so that sulfides are precipitated within a localized area.

THE TAG ACTIVE MOUND AND THE UNDERLYING SHALLOW UP-FLOW ZONE

General morphology

The morphology of the TAG active mound and its underlying stockwork has been described in detail elsewhere (e.g., Humphris and Kleinrock, 1996; Kleinrock et al., 1996; Rona et al., 1993b; Thompson et al., 1988; Tivey et al., 1995) and is summarized only briefly here. It is a circular feature, about 200 m in diameter, composed of two discrete platforms (Fig. 3); the upper platform is asymmetrically superposed on the north-northwest part of the lower platform (Kleinrock and Humphris, 1996b). A sample of sulfide talus collected by submersible from the eastern slope of the lower platform gave an age of 9.4 ka, which suggests that construction of the upper platform probably began relatively recently in the history of venting at this site (Lalou et al., 1990, 1993).

The entire mound is surrounded by a bright red-brown apron of metalliferous sediment and hydrothermal detritus that grades outward into pelagic ooze. Large slump blocks of sulfide and pyrite sand are common proximal to the mound and partially bury the flanks of the lower platform. Fine-grained iron oxides and oxyhydroxides, derived from weathering of the sulfides on the flanks and from fallout from the large, neutrally buoyant hydrothermal plume above the active mound, are widespread out to distances of up to 60 m from the mound (Mills and Elderfield, 1993; Mills, 1995).

Beneath the mound, the volcanic basement is intensely altered and veined with sulfides. This "stockwork" represents the up-flow zone of the hydrothermal system (Fig. 3) (Humphris et al., 1995). The lateral extent of the cylindrical alteration pipe, which was originally predicted to be 80–100 m in diameter on the basis of a magnetic low directly beneath the mound (Tivey et al., 1993), has now been confirmed to be 80–90 m by drilling (Humphris et al., 1995). Based on the thickness of the sulfides and the extent of the underlying up-flow zone, it is estimated that there are ~2.7 million tonnes (Mt) of massive sulfides within the mound and ~1.2 Mt in the up-flow zone, with a total Cu content of 30–60 kt (Humphris et al., 1995). These values fall well within the range for typical ophiolite-hosted massive sulfide deposits (Franklin et al., 1981) and within the largest 30% of massive sulfide deposits in Cyprus (Hannington et al., 1998).

Major types of hydrothermal solids

The surface of the mound. Detailed accounts of the mineral assemblages of the major sample types recovered from the surface of the mound have been presented by Thompson et al. (1988) and Tivey et al. (1995) and are summarized in Table 2 and Figure 3. Focused, high-temperature, black smoker activity is localized at the summit of a 10–15-m-high conical structure—the black smoker complex—located just to the west of the center of the upper platform. The cone is ~30 m in diameter at its base, and its surface is a platy layer (recovered samples are 2–10 cm thick) of massive chalcopyrite and pyrite and/or marcasite with blocks of corroding massive anhy-

TABLE 2. MAJOR SOLID TYPES RECOVERED FROM THE TAG ACTIVE MOUND AND THE UNDERLYING STOCKWORK

Sample type	Depth below seafloor (m)	Lithology	Mineral assemblage*	Geochemistry	Other comments
Black smoker complex					
Black smoker chimneys	Surface	Multiple conduits (0.5–5 cm)	Inner layer of cp; outer layer of anh + minor py and/or mc; trace sph., bn, iron-oxide, si	Cu: up to 23.5 wt%; Co: 47–1900 ppm; Se: 34–630 ppm	Large chimneys on summit black smoker complex; smaller chimneys on upper platform
Massive chalcopyrite	0–5.50	Massive; 1–2 mm tubular structures; bladed, euhedral cp	60% cp; 30% py; 10% mc with trace anh, sph	Cu: 32 wt%; Co: 120 ppm; Se: 850 ppm	Probably fragments of black smoker chimney; drilled at black smoker chimney
Massive sulfide crust	Surface	Platelike layer (2–10 cm); undersides have high porosity (50%); some small (few millimeters wide) conduits	Upper 1 mm layer of am. FeO(OH); massive cp + py and/or mc; trace sph and py	Cu: 2–23 wt%; Co: 189–1190 ppm; Se: 34–52 ppm	Surface cone of black smoker chimney
Massive anhydrite	Surface	Blocks (up to 0.5 m) with corroded edges (due to dissolution)	Anh (up to 4 mm) dominant; interstitial and includes cp + py aggregates; trace gy + si	Ca: 10–26 wt%; SO_4: 25–63 wt%; Co: 13–337 ppm; Se: 10–58 ppm	Flanks of cone of black smoker chimney; fluid inclusion *T*: ~350 °C
White smoker area					
White smoker chimneys	Surface	Multiple tortuous conduits (1.5–3 mm); low porosity (<5%) conduit linings; outer high porosity (20%–50%) layer	Low-iron sph dominant; minor mc and/or py, cp, am, si; outer FeO(OH) + si layer	Zn: 53–62 wt%; Cd: 250–2260 ppm; Pb: 300–1100 ppm; Ag: 70–1920 ppm; Au: 1–44 ppm	Field of 1–2 m spires on southeast side of platform
Mound					
Sulfide-rich mound samples	Surface	Bulbous blocks of massive sulfides with no flow conduits	Variable; from sph-rich to sph + py and/or mc; to cp + py and/or mc. Minor bn + am. si; outer FeO(OH)-si layer	Variable, from Zn rich to Fe-Zn rich, to Cu rich	Formed from fluids more similar to white smoker fluid than black smoker fluid
Iron-rich massive sulfides	Surface; mound flanks	Large (10s of cm) size blocks	Py + mc; minor cp; trace sph + am. si or qz; outer iron oxide layer with atacamite, jarosite	Fe: 36–40 wt%; Cu: 1–2 wt%; Zn: 0–6 wt%; Si: <0.5–2.1 wt%	Exposed on mound flanks similar to samples from inactive zones + ophiolites
Massive pyrites and pyrite breccias	0 to 10–15	Clasts of py or massive py (+ rare cp and/or sph) in a sandy py matrix	Py; minor cp, sph; qz; anh	Fe: 35 wt% (avg); SiO_2: 4 wt% (avg); Cu: 4 wt% (avg)	—
Chert/Fe oxides	Surface to 5	Surface: porosity >50% with small vertical conduits; banded layers of am.si and am. FeO (OH) At depth: Porosity 2%–10%; microcystalline qz and filamentous FeO (OH); traces of am. banded precursor	Am.si + FeO (OH); minor grains to iron, copper-iron, and zinc sulfides + iron oxide	SiO: 34–60 wt%; Fe: 16–35 wt%; S: 0–6 wt%; Pb: <20–22 ppm	—
Ocherous iron oxide	Surface; mound flanks	Iron oxide clasts with coating of am.si in matrix of iron oxides; clasts exhibit textures of sulfide grains	Goeth, hem, am. FeO (OH), am.si and/or qz; rare py; atacamite, jarosite	SiO_2: 4–53 wt%; Fe: 39–74 wt%; S: 0–0.2 wt%; Pb: 35–440 ppm	Similar to chert and iron oxide samples, but more Fe and Pb rich
Pyrite-anhydrite breccias	10–15 to ~30	Clasts of pyrite ± siliceous material in anh matrix; large anh veins (up to 45 cm wide)	Py + anh; minor cp; siliceous clasts; par, qz, py	Fe: 36–40 wt%; Ca: 2–20 wt%; Si: 1–25 wt%	Only beneath the center of the mound (TAG-1 and TAG-5)
Pyrite-silica-anhydrite breccias	~30 to 35–40	Similar to py + anh breccias but with qz + anh matrix	Py + qz + anh; minor cp; siliceous clasts as above	Fe: 23–39 wt%; Ca: 3–20 wt%; Si: 1–25 wt%	Only beneath the center of the mound (TAG-1 and TAG-5)
Stockwork					
Pyrite-silica breccia	35–40 to 50–60	Py, si and altered basalt clasts in matrix of qz + disseminated sulfides; porosity 2%–15%	20%–80% py; 10%–70% qz; minor anh, cp, sph	Fe: 30 wt% (avg); SiO_2: 27 wt% (avg)	—
Silicified wallrock breccias	50–60 to 100–110	Fragments of altered basalt veined and cemented by qz + py	Altered basalt clasts: qz, par, py, rutile (?); Trace chl Veins: qz + py ± rare massive cp	Fe: 29 wt% (avg); SiO_2: 23 wt% (avg); Cd: <0.5 ppm (avg); Pb: 19 ppm (avg)	Gradation from py + si breccia into silicified wall-rock breccias
Chloritized basalt breccias	100–110 to 125	Fragments of chloritized basalt, veined and cemented with py + qz + py	30%–85% chl; 10%–45% qz; 5%–15% py	Clasts; comparable to silicified wall-rock breccias but more Mg (1–6 wt%)	—

* Data from Tivey et al. (1995); Knott et al. (1998); Herzig et al. (1998b); Fouquet et al. (1998); Hannington et al. (1998); Humphris et al. (1989).
Abbreviations: cp—chalcopyrite; anh—anhydrite; py—pyrite; mc—marcasite; sph—sphalerite; bn—bornite; si—silica; FeO(OH)—iron oxyhydroxides; am.—amorphous; gy—gypsum; qz—quartz; goeth—goethite; hem—hematite; par—paragonite; chl—chlorite.

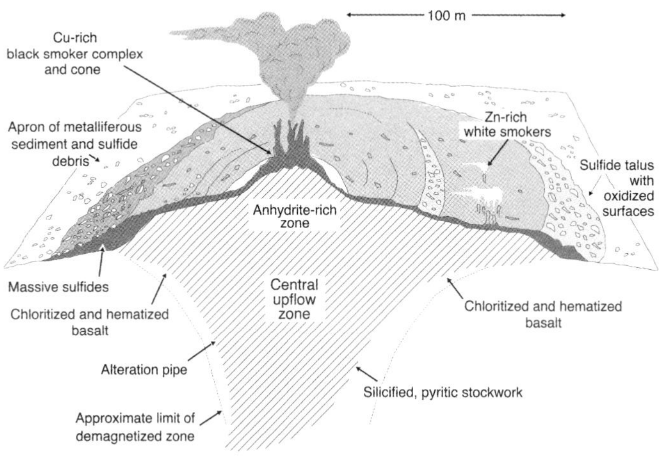

Figure 3. Schematic diagram of the structure of the TAG active mound and underlying up-flow zone showing its general morphology and the distribution of the major rock types. (Modified from Hannington et al., 1998.)

drite. On its summit, a cluster of black smoker chimneys, composed dominantly of anhydrite and chalcopyrite, vigorously discharges high-temperature (363 °C), Cu-rich fluids (Thompson et al., 1988; Tivey et al., 1995; Edmond et al., 1995). Over the eastern part of the upper platform, recent surveys have also documented less vigorous widespread black smoker activity emanating from point sources and seeping from small, rift-parallel fissures (Humphris and Kleinrock, 1996).

On the southeast edge of the lower platform, a field (~20–50 m diameter) of numerous active and inactive white smoker chimneys, known as the Kremlin area, is built on top of older sulfide deposits (Thompson et al., 1988). The 1–2-m-tall spires with bulbous tops are porous (porosity up to 50%), exhibit multiple, millimeter-sized conduits, and are composed dominantly of low-Fe sphalerite with minor amounts of amorphous silica, marcasite, or pyrite (Tivey et al., 1995). In 1990, when they were first successfully sampled, the white smoker chimneys were discharging Zn-rich fluids with temperatures of 260–300 °C. However, between 1986 and 1994, activity at this site waned until only a few chimneys were discharging fluids (Humphris and Kleinrock, 1996). By 1995, no active chimneys could be located for sampling, even after excavation of their tops (Edmonds et al., 1996).

The remainder of the surface of the mound is composed of blocks of massive sulfides, pieces of collapsed black and white smoker chimneys, and debris derived from other hydrothermal deposits originally formed in the shallow subsurface but now exposed at the surface (Table 2). Diffuse flow of low-temperature (up to 50 °C) fluids occurs over much of the surface; this flow has resulted in extensive cementation and hydrothermal recrystallization. Bulbous blocks of massive sulfides are sphalerite rich, sphalerite and pyrite and/or marcasite rich, or chalcopyrite and pyrite and/or marcasite rich; their mineral assemblages reflect the fluid temperatures and compositions from which they formed (Tivey et al., 1995).

Deep orange to dark red fragments of amorphous silica and iron oxyhydroxides occur on the rims of the mound (Tivey et al., 1995). These samples exhibit banding and likely formed by precipitation from conductively cooled (to <100 °C) black smoker fluid from which sulfides had precipitated within the mound (Tivey et al., 1995). They are similar to chert and iron oxide recovered from as deep as 5 mbsf (meters below the seafloor) within the mound, which are now composed of microcrystalline quartz and filamentous iron oxyhydroxides, but show evidence for a banded precursor of amorphous silica (Knott et al., 1998). This chert-like layer just below the surface is analogous to the silica caps overlying many ophiolite-hosted massive sulfide deposits (Franklin et al., 1981) and may prevent rapid and widespread discharge of hydrothermal fluids; subsurface cooling and precipitation of sulfides are thereby enhanced.

The flanks of the deposit are steep-sided slopes up to 20 m high that expose large blocks of massive sulfides. The surfaces of these blocks are extensively oxidized to iron oxides, oxyhydroxides, and other secondary minerals including atacamite (a hydrated copper chloride) and jarosite (a hydrated iron and potassium sulfate) (Hannington, 1993; Thompson et al., 1988). The interiors are dominantly pyrite and marcasite in various proportions, and texturally, they range from being banded, to being completely recrystallized, to exhibiting subangular clasts of pyrite in a higher-porosity, pyrite matrix (Tivey et al., 1995). The latter textural type resembles the pyrite breccias recovered within the mound down to ~10–15 mbsf (Humphris, Herzig, et al., 1996).

The interior of the mound. Five areas of the TAG active mound were drilled during Ocean Drilling Program (ODP) Leg 158; the maximum penetration of 125 mbsf was achieved close to the black smoker complex (the TAG-1 area). Recovery of rock core was poor, ranging from <1% to 63% and averaging 12%, and hence a great deal of textural, structural, and boundary information is absent. However, the stratigraphy of several holes drilled in proximity (10–15 m) at each location was combined to produce composite sections for each area (Fig. 4 shows the individual locations with letter labels; Humphris, Herzig, et al., 1996). The vertical zonation of different rock types through the mound and underlying stockwork has been described by Humphris et al. (1995) and Humphris, Herzig et al. (1996) and has been detailed mineralogically and geochemically in many papers in the ODP *Scientific Results* volume for Leg 158 (Herzig, Humphris, et al., 1998a); hence, only a brief summary, based on these references, is provided here.

One of the most striking features of the entire section is the dominance of breccias composed of mixtures of clasts of various types (e.g., sulfide, silicified wall-rock) that formed by different processes at different depths (Humphris, Herzig, et al., 1996). These breccias likely reflect the episodic nature of the hydrothermal venting: collapse and breakdown of structures occurred during periods of inactivity, and recrystallization and cementation of old material, as well as growth of new deposits, took place during periods of activity.

Massive pyrite and pyrite breccias dominate the upper 10–20 m and consist of clasts of pyrite in a matrix of pyrite with minor quartz ± anhydrite. These breccias are underlain by an anhydrite-rich zone that is best developed in the center of the mound near the black smoker complex (at TAG-1). Here, it can be divided into two breccia types: (1) matrix-supported pyrite + anhydrite breccias (clasts of pyrite in a matrix of >10% anhydrite) down to ~30 mbsf and (2) pyrite + silica + anhydrite breccias (pyrite, quartz + pyrite, and silicified basalt clasts in a matrix of >10% anhydrite + quartz) that extend to the bottom of the anhydrite-rich zone at ~35–40 mbsf. Complex, multistage fracture fillings and cavity linings of anhydrite are extremely well developed throughout this zone, where the thickness of some composite veins reaches 45 cm. To the north (TAG-5), pyrite + anhydrite breccias are only vein related, and to the east (TAG-2), anhydrite is present only as thin veins and filling vugs. Its virtual absence on the western side of the mound (TAG-4) implies that the temperatures within the mound at this location are below 150 °C—the temperature at and above which extensive precipitation of anhydrite from heating of seawater is expected (Bischoff and Seyfried,

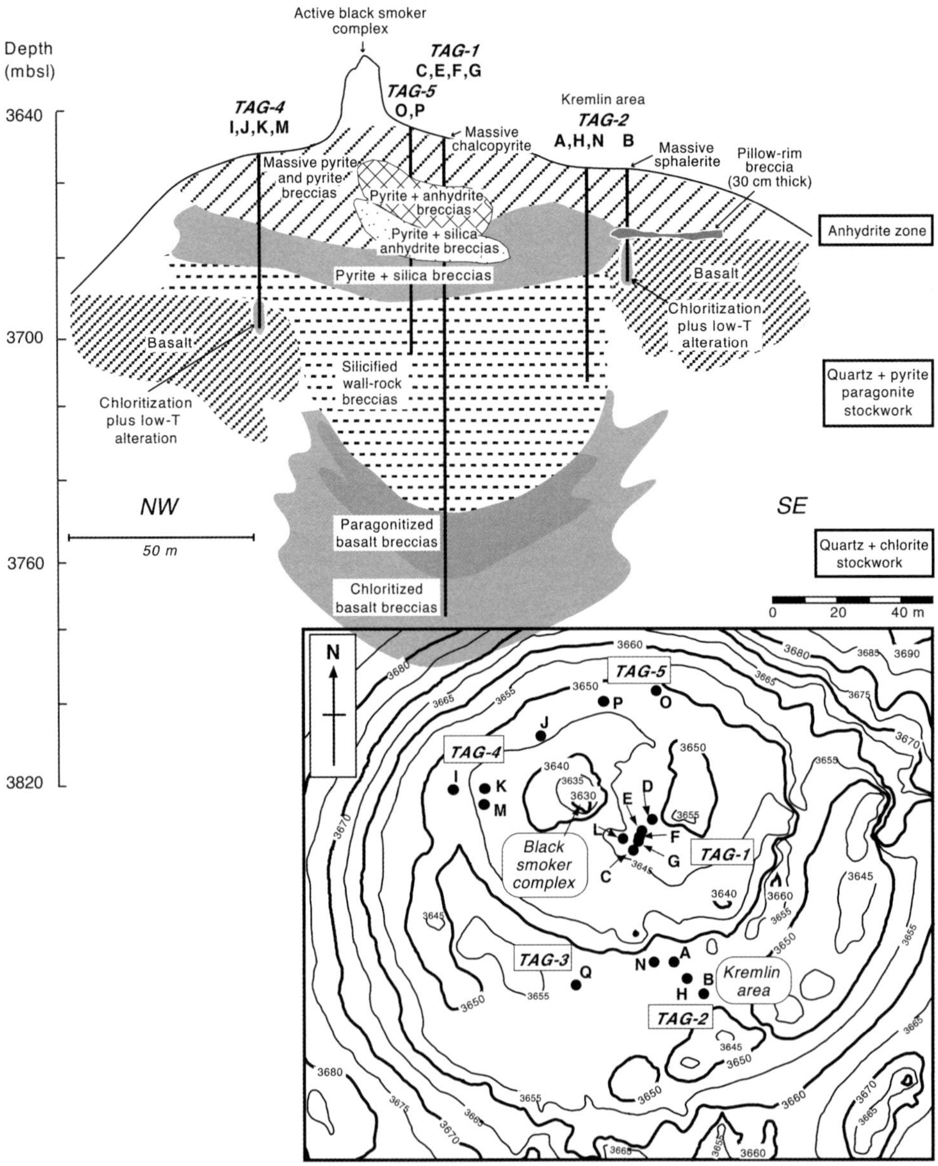

Figure 4. Northwest to southeast cross section showing an interpretation of the generalized internal structure of the TAG active mound; mbsl—meters below sea level. (Modified from Alt and Teagle, 1998.) Inset: High-resolution bathymetry (5 m contour interval) of the TAG active mound, showing the locations (solid circles with letter labels) of the holes drilled during ODP Leg 158. (After Humphris et al., 1995.)

1978). This region coincides with one of low conductive heat flow interpreted to be an indication of inflow of seawater into the mound (Becker et al., 1996). The thickness of the massive sulfide breccias and anhydrite-supported breccias corresponds closely to the present relief of the mound (~40 m).

The shallow up-flow (or stockwork) zone. The transition from pyrite + anhydrite + silica breccias to pyrite + silica breccias that represent the top of the stockwork zone is accompanied by an increase in the abundance of clasts of both pyrite + quartz and silicified basalt and an increase in quartz (to >10 vol%) in the matrix (Humphris, Herzig, et al., 1996; Honnorez et al., 1998; Alt and Teagle, 1998). Anhydrite veins are still common in this zone at TAG-1 and also occur in the pyrite + silica breccias at TAG-2. The pyrite + silica breccias then grade into silicified and paragonitized wall-rock breccias, which are distinguished from the overlying pyrite + silica breccias in that they contain significantly less pyrite (<50 vol%) and are dominantly clast supported. The breccias consist of clasts of basalt totally replaced by quartz, paragonite (a sodium mica), pyrite, and a Ti-bearing phase in a matrix of quartz and pyrite (Honnorez et al., 1998). This quartz + paragonite + pyrite complex forms the core of the stockwork.

Below ~100 mbsf at TAG-1, the breccias become less silicified and more paragonitized, and at 110 mbsf, they grade into chloritized basalt breccias consisting of chloritized basaltic fragments (1–5 cm in size) cemented by quartz and pyrite, and crosscut by veins of pyrite, quartz, and quartz + pyrite (Humphris, Herzig, et al., 1996; Honnorez et al., 1998). The fact that anhydrite, although only a minor component, was recovered on broken clast surfaces and in veins and vugs throughout the shallow up-flow zone indicates seawater penetration to at least 120 mbsf (Humphris, Herzig, et al., 1996).

A similar gradation from the central quartz + paragonite + pyrite stockwork into chloritized basalt breccias at depth in TAG-1 also occurs at the margins of the up-flow zone. On the western side (TAG-4), the boundary between the sulfide section and basement is sharp and occurs at a depth of 42 mbsf. The basaltic material recovered from this site is slightly to moderately altered in the interior, but many pieces exhibit green, chloritized alteration halos and minor veins of quartz ± chlorite ± pyrite. In addition, some of the basalts also show

TABLE 3. END-MEMBER CHEMICAL COMPOSITIONS OF TAG BLACK AND WHITE SMOKER FLUIDS: 1986–1995

	Black smoker complex				New black smoker	White smokers		Model results[#]	Diffuse flow**
	1986*	1990†	1993§	1995§	(1995)§	1990†	1994§		
Temp (°C)	290–320	360–366	363–364	369	—	273–301	270	290	—
pH (25 °C)	—	3.35	—	3.8	3	3	<3	<3	—
$Si(OH)_4$ (mmol/kg)	22	21	—	21	21	19	18	21	16.6
H_2S (mmol/kg)	—	2.5–3.5	—	3	3	0.5	3	0.88††	<0.04
Cl (mmol/kg)	659	636	633.5	645	643	—	363	636	640
Na (mmol/kg)	584	557	543	553	549	—	549	544	—
Li (μmol/kg)	411	—	368	370	367	—	352	—	366
K (mmol/kg)	17	17	18	20	20	17	20	17	—
Ca (mmol/kg)	26	31	30	31	30	27	27	27.4	21.5
Sr (μmol/kg)	99	103	99	99	100	91	95	—	71
Fe (μmol/kg)	1640	5590	5180	5040	5450	3830	3840	4106††	3260
Mn (μmol/kg)	1000	680	689	677	666	750	762	—	635
Zn (μmol/kg)	—	46	—	>36	>36	300–400	—	400	62
Cu (c)	—	120–150	—	>83	>91	3	—	65.5††	<3
$^{87}Sr/^{86}Sr$	0.7029	0.70348	0.7034§§	—	—	0.7046	0.7034§§	—	0.70304

* Data from Campbell et al. (1988).
† Data from Edmond et al. (1995) with corrected $^{87}Sr/^{86}Sr$ ratio according to James and Elderfield (1997).
§ Data from Edmonds et al. (1996).
[#] From Tivey et al. (1995), extrapolated to Mg = 0.
** Data from James and Elderfield (1996).
†† Black smoker fluid (Mg = 0) values for H_2S, Fe, and Cu of 3.68 mmol/kg, 5680 μmol/kg, and 240 μmol/kg, respectively, were used in model calculations (see Tivey et al., 1995).
§§ Gamo et al. (1996) for 1994 samples.

red outer alteration halos (as wide as 2 mm) indicative of later lower-temperature oxidation (Humphris, Herzig, et al., 1996). On the eastern side at the edge of the lower platform (TAG-2), the contact between the massive sulfide section and the underlying basaltic basement is marked by a 30 cm section of pillow-rim breccia. Fragments of altered crystalline and glassy basalt occur in a matrix of red chlorite + hematite + quartz mud. The crystalline fragments are completely replaced by yellow phyllosilicates and hematite, and the glassy fragments are blue and completely altered to chlorite (Humphris, Herzig, et al., 1996; Honnorez et al., 1998). Alteration within this layer suggests local hydrothermal flow along the contact at the base of the sulfides (Honnorez et al., 1998). The section of basalt recovered from below the pillow-rim breccia exhibits typical low-temperature alteration in which smectite ± iron hydroxides replaces olivine and fills vesicles. However, many basaltic fragments have more intensely altered red-brown alteration halos (<5 mm wide) in which the rock is totally replaced by chlorite and stained by iron oxides and oxyhydroxides. The chloritized halos present on basalts on both the western and eastern margins of the mound suggest that higher temperature (>200 °C) reactions between fluids and rocks occurred at least on a very local scale, although the heterogeneous nature of the alteration indicates that hydrothermal flow was neither pervasive nor sustained (Humphris, Herzig, et al., 1996). The relatively unaltered nature of much of the basement suggests that there is an abrupt transition from pervasive high-temperature alteration and recrystallization within the up-flow zone to low-temperature alteration more typical of shallow oceanic crust at the margins (Humphris, Herzig, et al., 1996).

Fluids discharging from the TAG active mound

In 1990, two distinct high-temperature fluid compositions were identified exiting the TAG active mound within ~70 m of one another: one at the black smoker complex and the other at the Kremlin (white smoker) area (Edmond et al., 1995). More recently, lower-temperature diffuse fluids have been sampled and exhibit distinct compositional differences relative to black and white smoker fluids (James and Elderfield, 1996).

Black smoker fluids. A time series of the end-member (Mg = 0 mmol/kg) chemistry of high-temperature (~366 °C) fluids exiting the black smoker complex since 1986 (Table 3) demonstrates that their composition has been remarkably stable, even though the 1995 samples were collected only a few months after drilling (the low iron concentration in the 1986 sample has been attributed to a sampling artifact by Campbell et al., 1988). It should be noted that there has been some confusion in the literature regarding the $^{87}Sr/^{86}Sr$ ratio of end-member black smoker fluid; this problem was discussed by James and Elderfield (1997). Although the first $^{87}Sr/^{86}Sr$ value reported for 1986 samples was 0.7029 ± 0.0013 (Campbell et al., 1988), data from 1990 and 1994 have defined end-member $^{87}Sr/^{86}Sr$ values of 0.70348 ± 0.00005 (not 0.7038 as reported in Edmond et al., 1995) and 0.7034 ± 0.0001, respectively (Edmond et al., 1995; Gamo et al., 1996; James and Elderfield, 1997); thus no significant variations have been measured. Such temporal stability has also been observed at other well-developed and mature hydrothermal systems, such as at 21°N on the East Pacific Rise and in the Guaymas Basin in the Gulf of California (Campbell et al., 1988). In 1995, fluids venting from a new cluster of black smoker chimneys ~30 m to the east of the black smoker complex were also sampled (Table 3) and are similar in their end-member fluid composition to the hydrothermal fluids of the black smoker complex (Edmonds et al., 1996).

White smoker fluids. On the southeast edge of the lower platform in the Kremlin area (TAG-2), white smoker chimneys were active between 1986 and 1994 (Humphris and Kleinrock, 1996). In 1990, when they were first successfully sampled, discharging fluids had lower temperatures (260–300 °C), a lower pH (3 at 23 °C), lower concentrations of Ca, Sr, Fe, Cu, and H_2S, and significantly higher concentrations of Zn compared to the black smoker fluids (Table 3). Similarities in other element concentrations, such as K, led Edmond et al. (1995) to speculate that the two fluids were related, and James and Elderfield (1996) used the $^{87}Sr/^{86}Sr$ ratios of both black and white smoker fluids to deduce an end-member hydrothermal-fluid $^{87}Sr/^{86}Sr$ value of 0.70343 ± 0.00006. In 1994, only a few chimneys were observed venting fluids that exhibited the same compositions as previously (Table 3).

Diffuse flow. Diffuse flow of low-temperature (up to 50 °C) fluids occurs over much of the rest of the irregular surface of the mound, as well as through isolated areas of metalliferous sediments and hydrothermal detritus that form an apron surrounding the mound (Becker et al., 1996; Mills et al., 1996). Diffuse-flow fluid compositions extrapolated to Mg = 0 mmol/kg are depleted in H_2S, Fe, Cu, Si, Ca, and Sr relative to black smoker and white smoker fluids (James and Elderfield, 1996; Table 3). However, unlike the white smoker fluids, the diffuse-flow fluid compositions, when extrapolated to Mg = mmol/kg, exhibit little or no enrichment in Zn concentrations relative to black smoker fluids. Time-series measurements of diffuse flow have demonstrated a correlation between the velocity and temperature of the fluids and tidal components; the correlation suggests that tidal variations may influence the subsurface hydrology of the hydrothermal system (Schultz et al., 1996).

Relationships between different fluids currently discharging from the TAG active mound. Although the chemical compositions of the black smoker, white smoker, and diffuse-flow fluids are each distinct, they can be related through processes of mixing of black smoker fluids with entrained seawater, precipitation and dissolution of various mineral phases within the mound, and conductive cooling.

Large differences in pH and in the Fe, H_2S, and Cu concentrations in black and white smoker fluids have been attributed to subsurface deposition of iron and copper-iron sulfides triggered by entrainment of seawater (Edmond et al., 1995). The formation of these mineral phases results in acid generation and remobilization of Zn and thus explains the elevated Zn concentrations in white smoker fluids. Edmond et al. (1995) noted that the lowest Mg value of sampled white smoker fluid was 12 mmol/L, consistent with addition of no more than 23% seawater to the hydrothermal fluid + seawater mix. Geochemical modeling has quantitatively demonstrated that the white smoker fluid can be reproduced by a combination of (1) mixing 14% seawater and 86% black smoker fluid, dissolving sphalerite to account for the order of magnitude Zn enrichment in the white smoker fluid, and (2) conductively cooling the mixture to achieve the observed temperature (Tivey et al., 1995). The lower concentrations of Ca and Sr in both the white smoker and diffuse fluids relative to black smoker fluid provide evidence for (1) mixing-induced reactions at temperatures in excess of ~150 °C between seawater and hydrothermal fluid and (2) precipitation of anhydrite ($CaSO_4$) within the mound.

Mixing of black smoker fluids with significantly greater amounts of entrained seawater can explain the composition of the diffuse fluids (James and Elderfield, 1996). The lack of significant enrichments in Zn in diffuse fluids relative to black smoker fluids may reflect deposition of zinc sulfide close to the mound-seawater interface before the diffuse fluids exit the mound (James and Elderfield, 1996). Alternatively, Zn may already have been depleted from parts of the mound involved in determining the diffuse fluid compositions. Depletions of Si in the diffuse fluids relative to black smoker fluids are consistent with precipitation of amorphous silica, chalcedony, or quartz within the mound. The U concentrations in the diffuse fluids are lower than expected from conservative mixing of seawater and hydrothermal fluid and, instead, are consistent with extraction of seawater U into mound materials (e.g., into anhydrite and oxidized crusts of sulfides) (James and Elderfield, 1996). Pore fluids from a core recovered from the surface of the mound ~110 m south of the black smoker complex also show depletions in U and Si. On the basis of chemical mass-balance calculations, Mills et al. (1996) estimated that the pore fluids were derived from mixing of high-temperature fluids (15%–8%) with seawater in the subsurface, together with extensive conductive cooling, removal of U, and removal of Si through the formation of amorphous silica, quartz, and low-temperature clay minerals.

Relationships between vent fluids and minerals precipitated within the mound and underlying up-flow zone

Calculations that reproduce the compositions of white smoker and diffuse fluids provide information on processes occurring in the mound. For example, generation of white smoker fluid from a mix of seawater and black smoker fluid would result in deposition of anhydrite, pyrite, and chalcopyrite within the mound in proportions of 19:8:1 (Tivey et al., 1995). This type of information is very powerful in defining fluid-flow and mixing processes, especially when combined with data from minerals precipitated beneath the mound surface. At the TAG active mound, the presence of abundant anhydrite (in the central part of the upwelling zone) and quartz (throughout the mound and the underlying stockwork) in veins, filling pore spaces, and cementing breccias has enabled fluid temperatures, salinities, and mixing proportions to be investigated by using microthermometric and isotopic studies. The geochemistry and isotopic composition of sulfides within the mound provide additional information on interactions among seawater, hydrothermal fluid, and previously deposited mound materials.

Anhydrite. The current phase of hydrothermal activity includes significant entrainment of seawater and the formation of abundant anhydrite veins. Analyses of anhydrite recovered at three of the five drill sites (TAG-1, TAG-2, and TAG-5) have allowed documentation of temperatures and seawater/hydrothermal fluid mixing ratios to depths of 125 mbsf. Studies of fluid inclusions provide evidence for (1) consistently high temperatures (>337 °C) in shallow parts of the mound and (2) temperatures in excess of 380 °C at depths of 55 mbsf at TAG-5 and greater than 90 mbsf at TAG-1 (Fig. 5) (Petersen et al., 1998; Tivey et al., 1998). Trapping temperatures (T_t) of <337 °C were recorded only at TAG-5 (where T_t ranged from 187 to 383 °C) and at depths less than 10 mbsf at TAG-2 (where T_t ranged from 266 to 321 °C). At TAG-1 and TAG-2, a trend of increasing trapping temperature with depth is observed; trapping temperatures near the mound surface are close to the highest measured temperatures of fluid venting at those two sites. It should be noted that there are concerns associated with analyzing fluid inclusions in anhydrite, requiring careful investigation of the reliability of fluid-inclusion data. Tivey et al. (1998) addressed such concerns, noting that the consistency of homogenization temperatures of inclusions along single arrays or in clusters argues against any factor other than stretching being a major problem for homogenization temperatures in the samples analyzed. They further noted that stretching was also not likely to be a major problem in the samples analyzed given (1) the consistency of temperatures of inclusions of a range of sizes and shapes and (2) the similarity of temperatures of inclusions to temperatures of vent fluids exiting the mound and to temperatures of fluid inclusions in quartz from similar depths in the mound.

For all fluid inclusions in anhydrite, final ice-melting temperatures (T_m) ranged from −3.8 to −0.7 °C, corresponding to salinities of 6.2–1.2 wt% NaCl equivalent—well within the range of measured salinities in mid-ocean ridge vent fluids (Petersen et al., 1998; Tivey

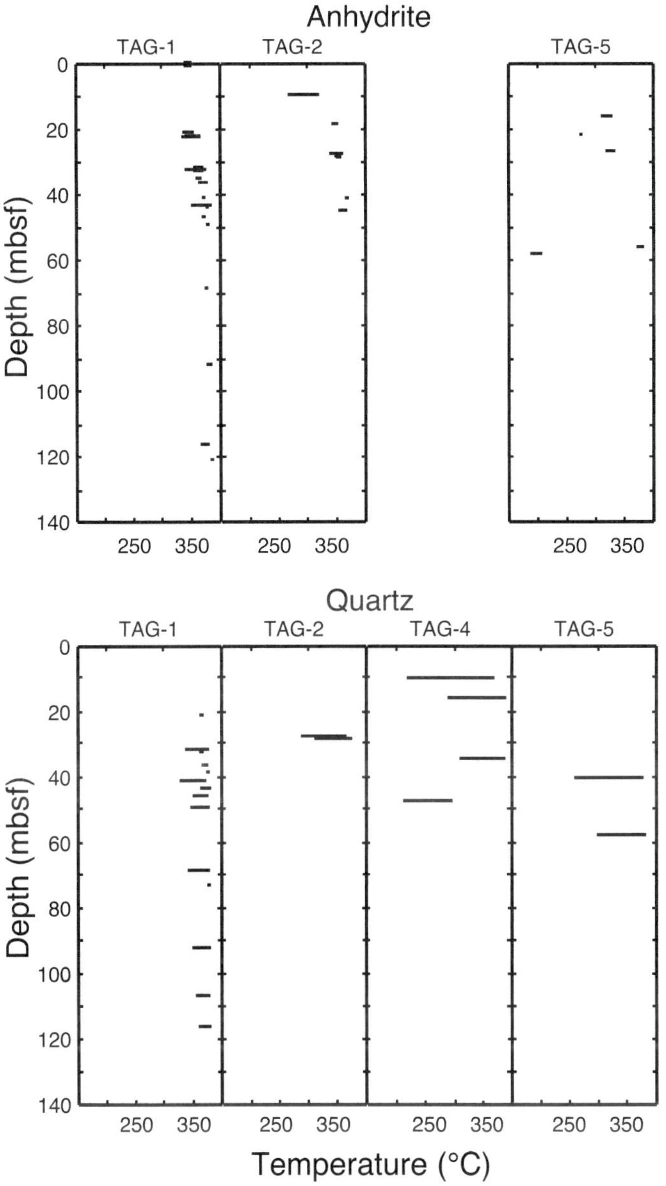

Figure 5. Trapping temperatures (pressure corrected) for fluid inclusions in anhydrite and quartz and corresponding depths of host samples from the TAG-1, TAG-2, TAG-4, and TAG-5 areas. (Data from Petersen et al., 1998, and Tivey et al., 1998.)

et al., 1998). This range in salinities likely reflects phase separation of seawater-derived hydrothermal fluids at depth. The minimum temperature at which phase separation could occur at pressures in excess of those at the surface of the mound (~36.7 MPa) and produce the lowest-salinity fluid observed is ~440–450 °C (Petersen et al., 1998). For example, phase separation at a pressure of 38 MPa (corresponding to ~20 m below the depth of deepest penetration at TAG-1) and a temperature of 450 °C would produce a vapor of salinity 0.78 wt% NaCl equivalent and a liquid of 25.42 wt% NaCl equivalent (Bischoff, 1991). The narrower range in observed salinities may in part be explained by remixing of the vapor and liquid phases during ascent and by the fact that the fluids trapped within the inclusions are a mix of hydrothermal fluid and seawater (Petersen et al., 1998). Another factor that complicates estimating the probable depth and temperature of phase separation from the fluid-inclusion data is that small amounts of anhydrite may be dissolving from the walls of the inclusions, which results in increases in the salinities of fluids within the inclusions of ~0.65 wt% since entrapment (Tivey et al., 1998). Despite these complications, the range in salinities and the increase in temperatures—from a maximum of 366 °C at the surface to maximum trapping temperatures of ~390 °C at depths of ~100 to 120 mbsf—suggest that phase separation occurred at depth beneath the TAG active mound sometime in the recent past and may be occurring now.

Sr isotope ratios of anhydrites range from 0.706370 to 0.709128 and reflect a broad range in the relative proportions of seawater and TAG black smoker fluid in mixtures from which anhydrite precipitated (Mills et al., 1998; Teagle et al., 1998a, 1998c). As shown in Table 4, calculated temperatures resulting from mixing seawater and black smoker fluid in the proportions indicated by the Sr isotope ratios are sometimes lower than temperatures at which anhydrite should precipitate (~150 °C) and are significantly less than the anhydrite formation temperatures indicated from microthermometric measurements. This finding suggests that significant conductive heating of seawater and/or black smoker fluid and seawater mixtures has occurred within the mound (Mills et al., 1998; Tivey et al., 1998).

Values of $\delta^{18}O$ in anhydrite range from 6.5‰ to 13.6‰ and thus give temperatures of 305 to 147 °C, calculated on the basis of oxygen isotope equilibrium between sulfate and black smoker fluid (Chiba et al., 1998; Teagle et al., 1998a, 1998c), which are significantly lower than the temperatures of anhydrite precipitation indicated by the fluid-inclusion data. Chiba et al. (1998) attributed these lower calculated temperatures to lack of attainment of oxygen isotope equilibrium; therefore, anhydrite must have precipitated rapidly and flow rates of sulfate-bearing fluid into the mound must have been rapid. Alternatively, Teagle et al. (1998a, 1998c) proposed that the $\delta^{18}O$ values reflect precipitation of anhydrite in isotopic equilibrium with mixtures of TAG black smoker fluid and seawater that has been previously conductively heated to 100–180 °C. However, in order to explain the anhydrite formation temperatures in excess of 337 °C (as indicated by the fluid-inclusion data), an additional step of rapid conductive heating prior to anhydrite precipitation with no oxygen isotope exchange would be required. Our preferred explanation is that the range in $\delta^{18}O$ values of anhydrite results from shifts in oxygen isotope compositions away from that of seawater sulfate (~9.5‰; Lloyd, 1967; Longinelli, 1989; Longinelli and Craig, 1967), as explained by Teagle et al. (1998a, 1998c), but that isotopic equilibrium was not attained. As noted by Chiba et al. (1998), the $\delta^{18}O$ of precipitating anhydrite becomes lower than that of seawater sulfate at temperatures greater than ~230 °C, but becomes higher at temperatures less than ~230 °C. As seawater and seawater + hydrothermal fluid mixtures are conductively heated to >337 °C, there is some isotope exchange that results in shifts to lower $\delta^{18}O$ values. Anhydrite with $\delta^{18}O$ values higher than ~9.5‰ formed from mixtures that had spent enough time at temperatures less than ~230 °C to undergo some oxygen isotope exchange prior to being heated relatively rapidly to the high temperatures indicated by microthermometric measurements (Fig. 6).

S isotope analyses of anhydrite recovered from within the TAG mound indicate a narrow range in $\delta^{34}S$ from 19.2‰ to 23.0‰, consistent with the isotopic composition of seawater sulfate, with higher values possibly resulting from partial reduction of seawater sulfate during mixing of seawater and hydrothermal fluid (Chiba et al., 1998; Gemmell and Sharpe, 1998; Knott et al., 1998). Although these values are in equilibrium with H_2S in black smoker fluid (and H_2S is assumed to be in equilibrium with pyrite with $\delta^{34}S$ of 7‰) at temperatures

TABLE 4. STRONTIUM AND OXYGEN ISOTOPE COMPOSITIONS OF ANHYDRITE AND ESTIMATED TEMPERATURES OF ANHYDRITE FORMATION

Sample no.	Depth (mbsf)	$^{87}Sr/^{86}Sr$*	$\delta^{18}O$†	HF:SW ratio	Temperature (°C) Based on $^{87}Sr/^{86}Sr$	Based on $\delta^{18}O$†	Maximum T_t§	Gain required**
TAG-1								
ALV1677-2	0	0.707214	—	29:71	205	—	353	148
216R-3	0	0.707061	8.5	35.65#	220	231	—	—
158-957C-								
11N-1, 131–134	32.01	0.706370	8.8	48:52#	275	270	363	88
11N-1, 129–129	32.01	0.70733	9.1	41:59#	245	240	—	—
11N-1, 126–126	32.01	0.706734	8.2	31:69#	210	218	—	—
11N-1, 125–125	32.01	0.706352	7.0	48:52#	275	231	—	—
11N-3, 38–40	33.99	0.706695	8.2	42:58#	250	240	—	—
11N-3, 66–68	34.25	0.706695	9.0	42:58#	250	225	—	—
11N-3, 121–123I	34.82	0.707283	10.0	31:69#	210	203	369	159
11N-3, 121–123V	34.82	0.706846	9.5	39:61#	235	214	368	133
158-957E-								
15R-1, 15–19	106.75	0.708634	13.6	9:91#	65	147	—	—
17R-1, 8–15	116.18	0.708155	—	15:85	110	—	381	271
18R-1, 19–25	120.89	0.709128	—	0.5:99.5	5	—	388	383
18R-1, 20–24	120.90	0.708628	13.6	9:91#	65	152	—	—
TAG-2								
158-957H-								
1N-1, 38–46	9.08	0.707251		29:71	205	—	313	108
5N-2, 10–12	27.75	0.707648	10.9	25:75#	175	187	—	—
5N-2, 38–40	27.85	0.707948	10.2	20:80#	140	196	—	—
5N-2, 79–84V	28.08	0.707249	—	29:71	205	—	359	154
5N-2, 79–84	28.39	0.707272	10.1	32:68#	215	201	—	—
9X-1, 15–22	44.85	0.706866	—	35:65	220	—	369	149
TAG-5								
158-957P-								
12R-4, 86-93	57.99	0.708847		4.5:95.5	32		207	175

Note: Proportions of hydrothermal fluid and seawater (HF:SW) were calculated by using Sr-isotope data and concentrations of Sr in hydrothermal fluid and seawater (see Mills et al., 1998). Temperatures from mixing were calculated by considering the heat capacities of 3.2 wt% NaCl solutions at 40 MPa and 2 °C and 370 °C (4200 $J \cdot kg^{-1} \cdot K^{-1}$ and 6300 $J \cdot kg^{-1} \cdot K^{-1}$, respectively); Bischoff and Rosenbauer, 1985) and the relative volumes of the two fluids.

* From Mills et al. (1998) and Teagle et al. (1998a).

† From Teagle et al. (1998a). Temperatures calculated from the measured oxygen-isotope composition by assuming equilibrium exchange between fluid and anhydrite.

§ From Tivey et al. (1998). Max T_t values are maximum trapping temperatures for fluid inclusions in anhydrite for corresponding depths.

Assumes a $^{87}Sr/^{86}Sr$ ratio in black smoker fluid of 0.7038. Other calculations were made by assuming a $^{87}Sr/^{86}Sr$ ratio in black smoker fluid of 0.70319.

** Temperature increase that must be achieved by conductive heating.

indicated by fluid-inclusion data, Chiba et al. (1998) reasoned that S isotope equilibrium is unlikely to have been attained, given that isotope-exchange reactions should be faster for O than for S.

Quartz. Microthermometric measurements on fluid inclusions in quartz indicate similar trapping temperatures and increasing maximum trapping temperatures with depth at each site, as observed for anhydrite (Fig. 5) (Petersen et al., 1998). However, unlike anhydrite, in which both primary and secondary fluid inclusions exhibit similar trapping temperatures (Petersen et al., 1998; Tivey et al., 1998), secondary fluid inclusions in quartz commonly exhibit lower trapping temperatures than primary fluid inclusions in the same phase (Petersen et al., 1998). Final ice melting temperatures (T_m) of fluid inclusions in quartz range from −3.7 °C to −2.2 °C, corresponding to salinities of 6.0 to 3.7 wt% NaCl equivalent (Petersen et al., 1998). The range in salinities in excess of seawater salinity is similar to that observed in fluid inclusions in anhydrite; however, the fluid inclusions in quartz lack fluids exhibiting salinities less than that of seawater. Of note is that the range of temperatures and salinities in fluid inclusions in quartz at TAG-4, where no anhydrite was recovered, indicates temperatures and salinities similar to fluid inclusions in quartz from the TAG-2 and TAG-5 sites (Petersen et al., 1998).

Temperatures for quartz formation have also been calculated from oxygen isotope ratios measured in quartz separates, assuming that the hydrothermal quartz precipitated in equilibrium with black smoker fluid having a $\delta^{18}O$ value of 1.7‰ (Shanks et al., 1995). If the empirical isotopic fractionation between quartz and water of Sharp and Kirschner (1994) is used, temperatures are similar to those documented by microthermometry, ranging from 170 to 360 °C with most calculated temperatures between 260 °C and 360 °C (Alt and Teagle, 1998). Alt and Teagle (1998) attributed the cooler temperatures to conductive cooling of hydrothermal fluids within the mound, in part reflecting cooling of the mound from seawater entrainment. Petersen et al. (1998) proposed that the lower temperatures recorded by secondary inclusions in quartz in the TAG-2, TAG-4, and TAG-5 areas may represent waning stages of hydrothermal activity. They concluded that anhydrite recovered from within the mound probably represents recent hydrothermal activity, whereas quartz probably formed during both recent and past times of activity. This conclusion is supported by textural observations that anhydrite always postdates quartz when they are observed together and by data from vent fluids sampled at the surface that indicate subsurface deposition of calcium (as anhydrite) and silica (as amorphous silica, chalcedony, or quartz).

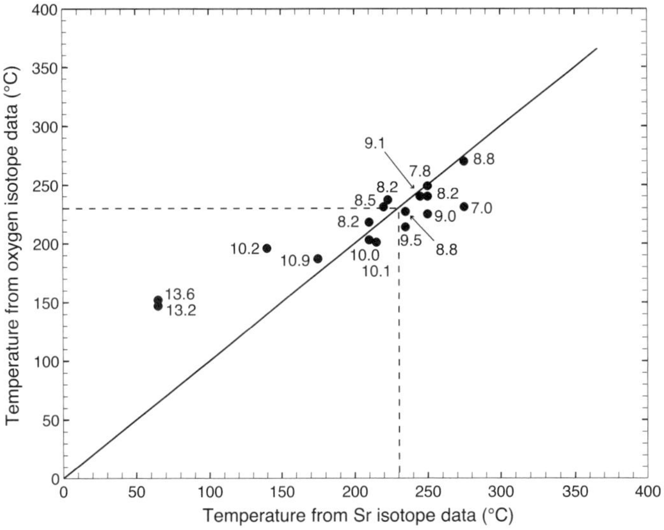

Figure 6. Comparison of temperatures calculated from the measured oxygen isotope compositions by assuming equilibrium exchange between fluid and anhydrite (from Teagle et al., 1998a) and temperatures calculated from mixing of hydrothermal fluid and seawater in proportions calculated by using the Sr isotope data (see Table 4). Numbers indicate measured $\delta^{18}O$ values (from Teagle et al., 1998a). Values higher than ~9.5‰ can be explained as the result of mixtures having spent enough time at temperatures less than ~230 °C to undergo some oxygen isotope exchange prior to being heated rapidly to the high temperatures indicated by microthermometric measurements (>337 °C). Values less than ~9.5‰ can result from incomplete exchange at the higher temperatures (see text).

However, recent U-Th isotope analyses of anhydrite from the TAG active mound have suggested that at least some of the anhydrite might have ages of ca. 10.5 ka (You and Bickle, 1998), although the mechanism by which it can be preserved during periods of inactivity is unclear.

Sulfides. The geochemical signatures of sulfide samples recovered from within the mound differ substantially from those recovered from the surface of the mound. Surface samples recovered from the black smoker complex are Cu rich and contain high concentrations of Co and Se (Table 2). Within the mound, concentrations of Cu (<0.1–12.7 wt%), Co (20–1020 ppm), and Se (<1–92 ppm) vary widely but are generally lower than at the surface and show no pattern of Cu enrichment with depth (Fouquet et al., 1998; Herzig et al., 1998b). The presence of moderately high Se, Co, and Cu concentrations at different levels and locations within the mound reflects deposition from high-temperature fluids (Fouquet et al., 1998). The absence of a Cu-rich core within the mound and the enrichment of Cu at the black smoker complex may result from remobilization of Cu within the up-flow zone (Herzig et al., 1998b). Supporting evidence includes (1) the significantly higher concentrations of Cu in TAG black smoker fluids compared with most MOR (mid-ocean ridge) vent fluids (120–150 μmol/kg vs. a maximum of 35 μmol/kg for MOR vent fluids from the Juan de Fuca Ridge and Lau Basin; Edmond et al., 1995; Von Damm, 1995) and (2) thermodynamic calculations that indicate that the TAG black smoker fluid is undersaturated with respect to chalcopyrite at 360 bars and 350 °C (Tivey et al., 1995). Thus chalcopyrite may be dissolving in fluids flowing up through the mound and reprecipitating in the black smoker chimneys at the surface.

Surface samples from the Kremlin area (TAG-2) and from other parts of the lower platform are Zn rich and contain high concentrations of Cd, Ag, Sb, Pb, and Au (Table 2), reflecting precipitation at cooler temperatures (260–300 °C) from white smoker fluids that are enriched in Zn, Cd, Ag, Sb, Pb, and Au relative to black smoker fluids (Hannington et al., 1995; Tivey et al., 1995). In contrast, Zn concentrations, as well as Cd, Ag, Sb, Pb, and Au concentrations, are uniformly low within the mound at TAG-1 and TAG-5 and at depths greater than 10.5 mbsf at TAG-2 (Fouquet et al., 1998; Herzig et al., 1998b). Highest concentrations of these elements in drill core are found in the top 10 m of the mound beneath the Kremlin area (TAG-2). The lack of Zn, Cd, Ag, Sb, Pb, and Au within the mound reflects mobilization of trace elements and their concentration in the upper 10 m of the eastern part of the mound (Fouquet et al., 1998; Herzig et al., 1998b).

$\delta^{34}S$ values of sulfide recovered from the mound surface range from 4.4‰ to 7.6‰ (Lein et al., 1991; W.C. Shanks et al., 1998, personal communication), and $\delta^{34}S$ values in H_2S of TAG black smoker fluid samples range from 6.6‰ to 7.5‰ (W.C. Shanks, 1998, personal communication). $\delta^{34}S$ values of sulfide recovered from within the mound range from 4.4‰ to 10.27‰; the lowest values are associated with anhydrite veins and quartz + pyrite stringer veins, and the higher values are associated with disseminated pyrite in pyrite + quartz breccias and altered basalt (Gemmell and Sharpe, 1998; Herzig et al., 1998b; Knott et al., 1998). (One lower value was reported [$\delta^{34}S$ = 0.35‰] and interpreted to be the result of local biogenic processes near the surface of the mound [Gemmell and Sharpe, 1998]). These S isotope values are high (by ~3‰) compared to values for sulfides from other MOR vent sites not covered by sediment (Gemmell and Sharpe, 1998; Herzig et al., 1998b; Knott et al., 1998). High $\delta^{34}S$ values are also characteristic of massive sulfides from the relict Mir and Alvin zones within the TAG hydrothermal field (Herzig et al., 1998b).

Two scenarios have been presented to explain the higher $\delta^{34}S$ values. Gemmell and Sharpe (1998) assumed that the fluid at depth beneath the mound has a S isotope composition similar to that of other MOR vent fluids at depth, i.e., $\delta^{34}S$ values of ~0‰ to 1‰. They proposed that this fluid mixes with entrained, partially reduced seawater in the upper parts of the stockwork, resulting in a modified hydrothermal fluid with a $\delta^{34}S$ value of 6‰ to 7‰. Higher $\delta^{34}S$ values result from interaction of this fluid with previously formed sulfide and sulfate followed by precipitation of sulfide from the resultant fluid that has higher $\delta^{34}S$. In contrast, Knott et al. (1998) proposed that the $\delta^{34}S$-enriched signature is likely attained deep in the system and that fluids rising beneath the mound have $\delta^{34}S$ values close to 5.5‰. They hypothesized that the higher values reflect differences in the deep plumbing of a system that has a strong "point source." They noted that massive sulfide deposits, stockwork feeder zones, and epidosites in deep reaction zones in the Troodos ophiolite (Cyprus) all have relatively uniform $\delta^{34}S$ values (Alt, 1994). Furthermore, $\delta^{34}S$ values of sulfides from Cyprus are similar to those of sulfides from TAG (Alt, 1994; Herzig et al., 1998b). Sulfides deposited at the surface and in veins reflect the $\delta^{34}S$ value of this deep hydrothermal fluid. Varying amounts of sulfate reduction from shallow entrainment of seawater and reaction with seawater sulfate results in the higher $\delta^{34}S$ values of sulfides in pyrite + quartz breccias and altered basalt. The similarity between $\delta^{34}S$ values of the black smoker fluid and those in the veins within the TAG mound are consistent with the latter scenario. Mixing of end-member or reaction-zone fluid with entrained, partially reduced seawater would result in fluids that contain some Mg (if no precipitation of Mg-rich phases is assumed); however, the fluids exiting the black smokers contain negligible Mg (Edmond et al., 1995). An alternative is that the S isotope compo-

TABLE 5. FLOW RATES, DISTANCES, AND TIMES REQUIRED TO ACHIEVE SUFFICIENT CONDUCTIVE HEATING

T_F (°C)	T_i (°C)	T_0 (°C)	L (m)	r_2/r_1	F^* (kg/s)	V^* (m/s)	t (min)
388†	2	390	100	100	0.040	0.15	11.1
388†	2	390	100	10	0.080	0.31	5.4
369	200	370	50	100	0.017	0.04	20.8
369	200	370	50	10	0.034	0.08	10.4
369	200	370	100	100	0.034	0.08	20.8
369	200	370	100	10	0.067	0.16	10.4

Note: Temperature, T_F, of fluid heated by conduction as it flows through a hollow cylinder (Tivey et al., 1998): of length L, and inner and outer radii of r_2 and r_1 respectively, was calculated by using equation 1 from Tivey et al. (1998).

$$T_F = T_i - T_0 \exp\left[\frac{2\pi\kappa_L}{FC_p \ln(r_2/r_1)}\right] + T_0$$

T_i—the initial temperature of the fluid. T_0—temperature of the outer side of the cylinder, assumed constant. L—length of pipe. k—thermal conductivity of cylinder wall (8 W m^{-1} K^{-1}; Tivey, 1998). V—fluid velocity. F—fluid flow rate = $V\pi r_1^2\rho$ (ρ = 670 km m^{-3} for seawater at ~370 °C and 40 MPa; Bischoff and Rosenbauer, 1985). r_1—inner radius of cylinder. r_2—outer radius of cylinder. C_p—heat capacity of seawater (6300 J kg^{-1} K^{-1} at ~370 °C and 40 MPa; Bischoff and Rosenbauer, 1985). t—time = L/V.

* Flow rates or velocities of greater than these values will not result in sufficient conductive heating to achieve T_F.

† Average values of C_p and ρ were used (5200 J kg^{-1} K^{-1} and 830 kg m^{-3}; Bischoff and Rosenbauer, 1985).

sition of the black smoker fluid rising through the mound is being modified by dissolution of previously deposited anhydrite and partial reduction of sulfate.

Implications for fluid flow within the mound. A consistent story emerges from the many different analyses carried out on anhydrite, quartz, and sulfide minerals recovered from within the TAG active mound. Data obtained from quartz provide information on conditions that existed within the mound in the past. High trapping temperatures for primary fluid inclusions in quartz from the TAG-4 area indicate that this part of the mound was, in the past, part of a high-temperature up-flow zone, whereas cooler temperatures recorded by oxygen isotope compositions and/or by trapping temperatures for secondary inclusions in quartz from the TAG-2, TAG-4, and TAG-5 areas provide evidence for waning stages of hydrothermal activity (Petersen et al., 1998). Data obtained from anhydrite provide information on recent activity. The combination of Sr isotope analyses of anhydrite and microthermometric analyses of fluid inclusions in anhydrite provides evidence for significant conductive heating of seawater + hydrothermal fluid mixtures within the mound (Mills et al., 1998; Tivey et al., 1998). This conductive heating can occur as fluids are transported through veins bounded by conductive sulfide-rich breccias at flow rates of up to 0.02–0.08 kg/s (Table 5) (Tivey et al., 1998). At these flow rates, which correspond to velocities of 0.04–0.31 m/s in conduits of radius 0.01 m, the time for fluid to traverse from outside the mound to sites where anhydrite has been deposited could be as little as 5–20 min (Mills and Tivey, 1999; Tivey et al., 1998). These rapid flow rates and short times necessary for conductive heating are consistent with the interpretation of Chiba et al. (1998) that oxygen isotope ratios in anhydrite are far from equilibrium.

Data obtained from anhydrite can also be combined with TAG black smoker fluid compositions to calculate the amounts of anhydrite that should precipitate from the mixtures of seawater and hydrothermal fluid indicated by Sr isotope ratios. For example, thermodynamic calculations carried out, with precipitation of Mg-bearing phases suppressed (because they have not been identified in drill-core samples; Humphris, Herzig, et al., 1996), indicate that 1.43 × 10^{-2} mol anhydrite, 2.66 × 10^{-5} mol chalcopyrite, and 3.7 × 10^{-4} mol hematite should precipitate per kilogram of a mixture of 20% black smoker fluid + 80% seawater (Tivey et al., 1998). This information can be used to estimate the time it would take to precipitate the present inventory of anhydrite within the mound. If we assume that between 1% and 10% of the 225 MW convective heat flux from the black smoker complex (Rona et al., 1993b) is available to heat entrained seawater from 2 °C to 360 °C, then 1.2 to 12 kg of seawater can be heated per second (Tivey et al., 1998). This rate corresponds to 24 to 240 separate channels each carrying ~0.05 kg/s. If 0.0143 mol anhydrite is assumed to precipitate per kilogram of heated fluid, it would take ~80 to 800 yr to precipitate the existing 20,000 m^3 of anhydrite present within the TAG active mound. This convective process of entraining and heating seawater is also the most plausible explanation for cooling black smoker fluids from >380 °C at 125 mbsf to the temperature of 366 °C measured at orifices of TAG black smoker chimneys (Tivey et al., 1998).

Differences in the geochemistry of sulfides recovered from the surface and from within the mound are easily explained as a result of the prevalent seawater entrainment into the mound. Mixing of seawater and hydrothermal fluid combined with conductive heating of fluid mixtures results in efficient precipitation of anhydrite, chalcopyrite, pyrite, and hematite, generation of more acidic fluids, and mobilization of metals. Geochemical calculations indicate that, in the absence of seawater entrainment, notable amounts of minerals are not saturated in TAG black smoker fluid until significant conductive cooling has occurred and temperatures are low (<230 °C; Tivey et al., 1995). At TAG, it is the addition of seawater and resultant changes in pH and redox conditions that triggers precipitation of anhydrite and significant amounts of sulfides, and generation of a more acidic fluid. Over time, this process has created a zoned deposit. The distribution of Zn, Cd, Ag, Sb, Pb, and Au throughout the mound indicates that zone refinement is currently restricted to the upper 10 m beneath the eastern part of the mound. These elements have already been leached from most of the rest of the mound, but there are areas of the western part of the mound (at TAG-4) where trace elements have not been as extensively remobilized.

Finally, S isotope values of sulfides from the TAG mound also appear to have been affected by interaction with entrained seawater. The distribution of $\delta^{34}S$ values observed within the mound—i.e., sulfides in veins have lower values and those in pyrite + quartz breccias and altered basalts have higher values—is consistent with reaction of hydrothermal fluid with seawater sulfate. Partial reduction of sulfate in entrained seawater, or of sulfate from previously deposited anhydrite, can explain the highest $\delta^{34}S$ values observed in the shallow up-flow zone.

ALTERATION OF BASALTS WITHIN THE SHALLOW UP-FLOW ZONE

Sequence of alteration and the geochemical consequences

Studies of the minerals in the hydrothermally altered basalts and basaltic clasts recovered from the sulfide mound and underlying stockwork at TAG have resulted in the development of a paragenesis of alteration and the associated changes in the composition of the basement (Honnorez et al., 1998; Alt and Teagle, 1998; Humphris, Herzig, et al., 1996) (Fig. 7). In terms of the internal structure of the deposit, these stages of alteration are observed in a vertical sequence,

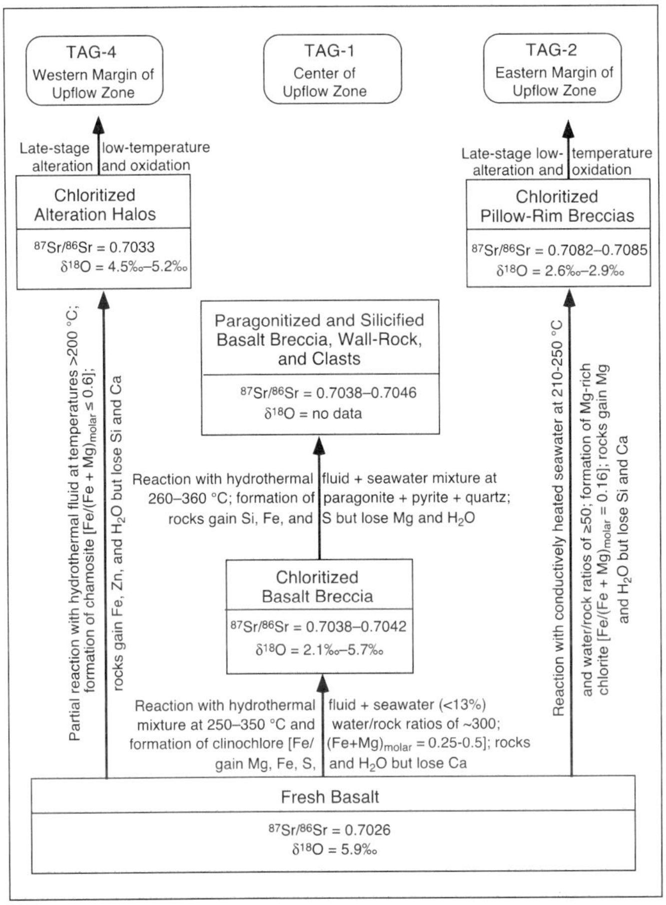

Figure 7. Sequence and conditions of alteration of basalts in the shallow up-flow zone beneath the TAG active mound based on petrographic and mineralogical observations and geochemical and isotopic data from samples recovered on ODP Leg 158. (Adapted from Teagle et al., 1998b, 1998c.)

with the green chloritized basalts at the greatest depths drilled in the stockwork, grading into paragonitized basalt breccias and then up into the zone of silicified wall-rock breccias.

Table 6 presents average chemical compositions of the various types of hydrothermally altered basalts, as well as of the fresh precursor (Humphris et al., 1998). Fresh basalt in the up-flow zone was initially converted to a chlorite ± quartz ± pyrite assemblage that resulted in uptake of Al_2O_3, Fe_2O_3, MgO, H_2O^+, S, V, and Co. In addition, essentially all of the CaO, Na_2O, and Sr were lost from the rock during alteration of plagioclase. The trace metals Cu, Ni, and Zn were leached from the rock, and almost all of the Cu was removed. SiO_2 fluxes were variable in direction and in general quite small. This process was followed by replacement of the chlorite-rich assemblage by paragonite ± quartz ± pyrite. Alteration of the chloritized basalts resulted in additional uptake of SiO_2, Fe_2O_3, S, and Co as well as small amounts of Na_2O, K_2O, Sr, Ba, and Zn. Other components were lost from the rock, including MgO, H_2O^+, V, and small amounts of Al_2O_3. Further silicification resulted in continued loss of MgO and H_2O^+ and almost complete loss of V. This stage of the alteration sequence is also responsible for the largest increases in the SiO_2, Fe_2O_3, S, and Co contents of the altered rocks, some of which showed gains in Cu and Zn, presumably in association with the pyrite. Small gains in Na_2O, K_2O, Ba, and Sr in a few samples may be related to the continued replacement of the rock by paragonite. In addition to this sequence of water-rock reactions, entrainment and heating of seawater beneath the mound resulted in precipitation of anhydrite in veins and pore spaces down to depths of at least 120 mbsf (Humphris, Herzig, et al., 1996). Anhydrite veining crosscuts the quartz + pyrite stockwork and appears to be related to the current phase of high-temperature activity.

The chloritization of basalts on the margins of the central up-flow zone suggests that reactions between seawater and rocks at high temperatures extended in the subsurface at least to the periphery of the mound at some time. However, the timing of this chloritization event relative to the chloritization envisaged as the first stage of alteration within the up-flow zone is unknown. In addition, differences in the magnitudes and directions of the MgO and Fe_2O_3 exchanges at these sites resulted from the formation of chlorites of different compositions and thus suggest that temperatures and the composition of the reacting fluids were different at the periphery of the stockwork zone.

Similar sequences of chloritization followed by silicification have been described in some other modern seafloor hydrothermal deposits. At the Sea Cliff hydrothermal field on the northern Gorda Ridge, where mineralization is developed in fault talus, Zierenberg et al. (1995) have documented an alteration sequence in which Mg metasomatism at temperatures of 220 °C converts the rock to smectites and smectite + chlorite, followed by silicification of the basalt and cementation by amorphous silica at temperatures below 100 °C. Geochemically, these alteration reactions resulted in near-total silicification of the basaltic fragments with removal of almost all cations. Within the exposed stockwork underlying sulfide mounds of the extinct hydrothermal system on the Galapagos Ridge, the alteration assemblage is dominated by smectite, silica, pyrite, and chlorite (Embley et al., 1988). Reactions between rock and water resulted in strong depletion of Ca, Na, K, and Mn, enrichment in S and Fe, and variable Si and Mg concentrations.

Alteration zones beneath some ancient, volcanic-hosted, massive sulfide deposits preserved in ophiolites also show similar alteration sequences with chloritization of basalts preceding the development of mineral assemblages of quartz + chlorite + sericite, followed by complete replacement by quartz + sulfide. Although there are some mineral compositional differences, the basic mineral and chemical changes in alteration zones of deposits in Cyprus (e.g., the Mathiati deposit) and in Oman (e.g., the Bayda deposit) are similar, having undergone loss of alkalis, precipitation of large quantities of silica, and the formation of chlorite (Collinson, 1986; Lydon and Galley, 1986; Richards et al., 1989). Of particular note is the similarity between the vertical zonation of alteration-mineral assemblages observed in the Turner-Albright deposit in the Josephine ophiolite (Zierenberg et al., 1988) and that seen at TAG. As in the case of the active TAG mound, basalt alteration and silicification within this ophiolite-based massive sulfide deposit resulted in the removal of all cations, including Al and Ti.

In the following sections we use mineral compositions and oxygen and Sr isotope data to determine the temperatures and hydrothermal fluid compositions for each of these steps in the alteration sequence.

Alteration of basalt within the central up-flow zone (TAG-1)

Chloritization of fresh basalt. The chloritized basalts recovered from >120 mbsf are typically composed of 80%–85% chlorite, 10%–15% quartz, and ~5% pyrite. The composition of the chlorite is cli-

TABLE 6. AVERAGE COMPOSITIONS OF FRESH AND ALTERED BASALTS FROM THE UPWELLING ZONE OF THE TAG ACTIVE MOUND*

	Fresh basalts and basalts interior		Paragonitized-silicified basalts and clasts: Silicified wall rock and paragonitized-silicified basalt breccias		Paragonitized-silicified basalts and clasts: Altered clasts from the mound		Chloritized basalts: Chloritized basalts from TAG-1	Chloritized basalts: Altered pillow breccias (TAG-2)	Chloritized basalts: Alteration halos (TAG-4)
	Avg.	*s*	Avg.	*s*	Avg.	*s*	range	range	range
No.	6		28		14		4	2	3
SiO_2	49.52	0.29	44.90	11.90	47.72	8.45	22.96–50.07	30.99–31.88	35.57–38.85
Al_2O_3	14.76	0.25	2.80	2.05	2.89	2.00	4.82–13.82	17.09–17.41	15.75–16.06
$Fe_2O_3^T$	10.94	0.46	24.04	5.89	24.94	3.91	22.34–35.45	12.66–12.84	20.73–24.60
MnO	0.17	0.02	<0.01	0.00	<0.01	0.00	0.02–0.06	0.04–0.07	0.07–0.08
MgO	8.30	0.20	0.17	0.12	0.11	0.10	2.46–9.82	22.98–23.46	7.88–9.01
CaO	10.41	0.31	0.20	0.30	0.53	1.08	0.03–0.15	0.05–0.21	3.22–4.23
Na_2O	2.59	0.06	0.43	0.30	0.42	0.29	0.08–0.37	0.33–0.38	1.52–1.86
K_2O	0.14	0.03	0.07	0.05	0.08	0.04	0.01–0.04	0.06–0.14	0.10–0.17
TiO_2	1.58	0.01	0.32	0.25	0.33	0.23	0.41–1.13	2.00	1.66–1.71
P_2O_5	0.15	0.00	<0.01	0.01	0.02	0.01	<0.01–0.02	0.02–0.03	0.13–0.15
S	<0.1–0.11	n.d.	21.77	5.79	21.98	4.89	7.44–28.93	<0.10	<0.10–0.58
H_2O^+	1.48	0.69	1.26	0.53	0.75	0.36	2.43–7.73	11.98–12.03	6.64–7.25
Co	70	23	213	87	212	67	150–405	55–58	47–55
Ni	145	8	<5–96	n.d.	<5–79	n.d.	22–62	200–204	145–174
Cu	73	10	<5–117	n.d.	<5–413	n.d.	<5–60	<5–14	89–148
Zn	179	148	<5–238	n.d.	<5–888	n.d.	13–75	105–233	1016–7238
$^{87}Sr/^{86}Sr$	0.702609–0.702728		0.703816–0.705676 (8 samples)		(10 samples)		0.703856–0.704181 (3 samples)	0.708225–0.708464 (2 samples)	0.703260–#0.703342
$\delta^{18}O$ (rock)	5.9–6.8 (5 samples)		no data		no data		2.1–5.7 (3 samples)	2.6–2.9	4.5–5.2 (2 samples)

Note: Avg.—average, *s*—standard deviation, No.—number of samples, n.d.—not determined. Oxides are reported in weight percent; trace elements in parts per million.

* Geochemical data from Humphris et al. (1998); isotopic data from Alt and Teagle (1998) and Teagle et al. (1998).

nochlore with $Fe/(Fe + Mg)_{molar}$ of 0.25–0.5 (Alt and Teagle, 1998; Honnorez et al., 1998), and comparison of the normalized bulk-rock MgO contents with those of the fresh basalt precursor indicates that the rocks gained 4–5 wt% MgO during chloritization (Humphris et al., 1998).

$^{87}Sr/^{86}Sr$ ratios for these chloritized rocks range from 0.70386 to 0.70418 compared with values of 0.70260–0.70272 for the relatively fresh precursors (Teagle et al., 1998b, 1998c). These values, together with the uptake of MgO by the rocks during alteration, suggest that the reacting fluids were a mix of hydrothermal fluid and entrained seawater. In order to conduct mixing calculations, the isotopic signatures of the end-member fluids must be known. The $^{87}Sr/^{86}Sr$ ratio of seawater is 0.70918 and is well determined (Hodell et al., 1991); however, the $^{87}Sr/^{86}Sr$ ratio of the end-member hydrothermal fluid that was a component of the fluid that chloritized the basalts is not well established. Measurements on black smoker fluids currently discharging at TAG give $^{87}Sr/^{86}Sr$ ratios of 0.70348 (Gamo et al., 1996; James and Elderfield, 1997). However, on the basis of the observation that none of the hydrothermal assemblages that are 100% recrystallized has $^{87}Sr/^{86}Sr$ ratios less than 0.7038, Teagle et al. (1998b, 1998c) suggested that the end-member hydrothermal fluid involved in the alteration had a $^{87}Sr/^{86}Sr$ ratio of ~0.7038. They then used this value to calculate that the reacting fluid contained <10% seawater (using the present-day $^{87}Sr/^{86}Sr$ ratio of hydrothermal fluid has a small effect and only changes this percentage to <13% seawater). If all the Mg is assumed to have been removed from a fluid containing 10% seawater during chloritization, then the integrated water/rock mass ratio for the chloritized rocks is ~300 (Alt and Teagle, 1998).

Oxygen isotope values for the bulk chloritized rocks ($\delta^{18}O$ = 2.1‰–5.7‰) have been used to determine the $\delta^{18}O$ values of chlorites; it was assumed that the samples can be considered two-component mixtures of chlorite and quartz, which are essentially the only two oxygen-bearing phases in the chloritized samples (Alt and Teagle, 1998). By using this method, Alt and Teagle (1998) calculated a range of $\delta^{18}O$ values for chlorite of 1‰–3‰ and estimated temperatures of alteration of 250–350 °C (this calculation assumes equilibrium with hydrothermal fluids having $\delta^{18}O$ = 1.7‰ as measured in the TAG black smoker vent fluids; Shanks et al., 1995). Hence, in summary, it appears that chloritization of basalts proceeded at high water/rock mass ratios via reaction of a mixture dominated by hydrothermal fluid but with a small component (<13%) of seawater at temperatures of 250–350 °C.

Paragonitization, pyritization, and silicification of the chloritized basalts. Chloritization of basalts within the main part of the upflow zone was followed by replacement by paragonite, quartz, and pyrite. In general, quartz is the most abundant mineral (40–85 vol%), and pyrite generally varies between 10 and 35 vol%. Paragonite, which comprises 3%–30% of the altered basaltic clasts, is present in much higher concentrations (up to 60 vol%) at the base of the paragonitized-silicified breccias where it grades into the chloritized basalts (Honnorez et al., 1998; Alt and Teagle, 1998). This phase of alteration was intense and pervasive and resulted in the total replacement of the earlier-formed clay minerals (e.g., chlorite, smectites, etc.), as well as all relict primary minerals, by paragonite. At the same time, the rocks were silicified and pyritized, the primary basaltic textures were progressively erased and replaced, and veins and fractures were filled with pyrite and quartz. The end products of this alteration phase are rocks composed essentially of Fe, S, and Si and minor Na and Al (reflecting the various proportions of quartz, paragonite, and pyrite) and almost complete loss of Mg and H_2O (Humphris et al., 1998). $\delta^{18}O$ measurements in quartz separates indicate temperatures of 260–360 °C for the quartz within these samples (Alt and Teagle, 1998).

The white mica more typically associated with alteration of the host rocks of ophiolite-based massive sulfide deposits is K rich. However, the formation of paragonite at TAG rather than a more potassic mica may reflect the high Na/K ratios (34) of TAG black smoker hydrothermal fluids compared with those from other seafloor hydrothermal systems (e.g., 21 at Snakepit; 15–18 at South Cleft on the Juan de Fuca Ridge; 13–26 at the Galapagos spreading center vents; Von Damm, 1995). These ratios are high, not because of high concentrations of Na (in fact the Na/Cl ratios in the black smoker hydrothermal fluids are close to the Na/Cl ratio in seawater), but because of low K concentrations in the fluid (Von Damm, 1995). These low K concentrations may be an indication that the fluid reacted with basalts from which K had been leached during earlier high-temperature alteration reactions.

$^{87}Sr/^{86}Sr$ ratios of the paragonitized and silicified basalts and clasts show a wide range from 0.70382 to 0.70706, although 13 of those samples fall within a much narrower range of 0.70382–0.70461 (Teagle et al., 1998b, 1998c). On the basis of mixing calculations, Teagle et al. (1998b) attributed the high $^{87}Sr/^{86}Sr$ ratios exhibited by five samples to contamination by late-stage anhydrite. Anhydrite has been reported as a minor component in all five samples (Humphris, Herzig, et al., 1996), and the bulk-rock compositional data indicate higher CaO and Sr concentrations in these samples than others from the same sites (Humphris et al., 1998), thus supporting Teagle et al.'s conclusion. The range in $^{87}Sr/^{86}Sr$ ratios of 0.70382–0.70461 for those samples that lack anhydrite is strikingly similar to the range exhibited by the chloritized basalts from deeper in the stockwork.

On the basis of the geochemical and isotopic data, it is clear that paragonitization occurred through reaction with a fluid that had already undergone significant chemical changes. One mechanism by which the fluid could attain a composition resulting in reactions that produce paragonite would be to have undergone a previous chloritization event earlier in its flow path. Figure 8 illustrates the phase relationships in the FeO-Fe_2O_3- and MgO-Na_2O-SiO_2-Al_2O_3-H_2O systems at a variety of temperatures (Saccocia and Seyfried, 1994). Chloritization results in a decrease in pH and in the concentration of Mg in the fluid, thereby efficiently driving the activity ratio of Mg^{2+}/H^+ lower. These reactions—as well as any conductive cooling of the fluid as proposed by Alt and Teagle (1998) to explain the temperatures of quartz formation—drive the fluid composition toward the paragonite stability field, and if the log activity ratio of Na^+/H^+ is between 3 and ~4.5 at 300 °C, paragonite will form. Since the integrated water/rock mass ratios are high, the $^{87}Sr/^{86}Sr$ ratios of the fluid will not change significantly. In the TAG shallow up-flow zone, there is clear mineralogical evidence that the rocks were chloritized prior to paragonitization and have undergone (at least) two discrete alteration events. Hence, one scenario is that after the initial phase of chloritization of the stockwork, the altered rocks were subsequently exposed to fluids that had already undergone chloritization reactions earlier in their flow path.

Alteration of basalt on the margins of the mound

Chloritization of pillow-rim breccia in the Kremlin area (TAG-2). Completely chloritized glass and basalt samples from the section of pillow-rim breccia at the contact between massive sulfide and the underlying, slightly altered basaltic basement are composed of the most Mg-rich chlorites reported from the seafloor [$Fe/(Fe + Mg)_{molar} = 0.16$; Honnorez et al., 1998]. Comparison of the normalized bulk-rock

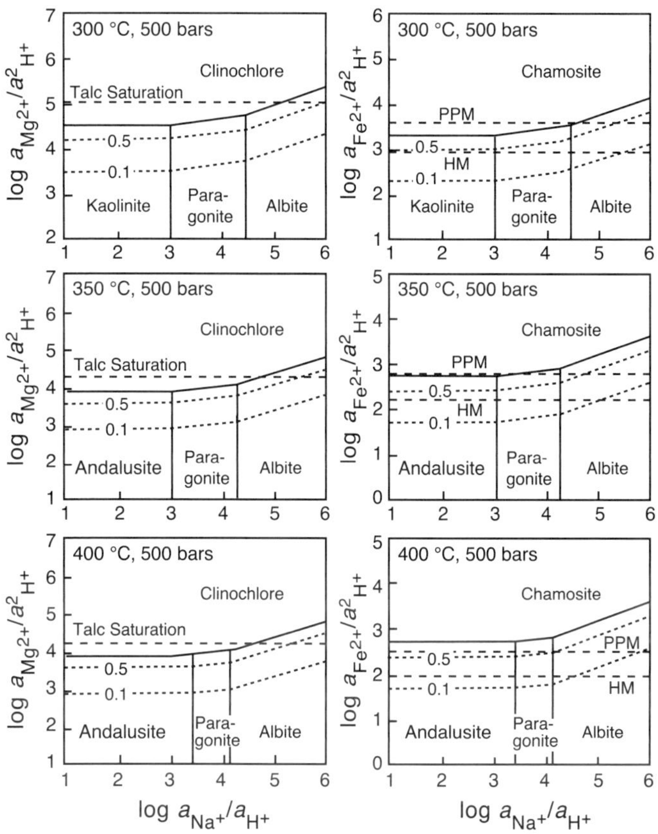

Figure 8. Phase relationships in the FeO-Fe_2O_3- and MgO-Na_2O-SiO_2-Al_2O_3-H_2O systems at 300, 350, and 400 °C and 500 bars at quartz saturation. Dotted lines represent contours of mole fraction clinochlore in the MgO system and chamosite in the FeO system. PPM—pyrite-pyrrhotite-magnetite, HM—hematite-magnetite. (After Saccocia and Seyfried, 1994.)

MgO contents with those of the fresh basalt precursor indicates that the rocks have gained ~10 wt% MgO during chloritization (Humphris et al., 1998). In addition, the $^{87}Sr/^{86}Sr$ ratios of these completely recrystallized samples are high (0.70823 and 0.70846; Teagle et al., 1998b, 1998c) and approaching the ratio in seawater. These samples have low $\delta^{18}O$ values of 2.6‰ and 2.9‰ that yield calculated temperatures of formation between 210 and 250 °C (Alt and Teagle, 1998). The large bulk-rock gains in MgO, the low $\delta^{18}O$ values, and the $^{87}Sr/^{86}Sr$ ratios approaching the seawater ratio all point to reaction of the pillow-rim breccias with conductively heated seawater (Teagle et al., 1998b, 1998c). An increase in MgO of 10 wt% during chloritization of the halos implies a minimum integrated seawater/rock ratio of ~50–100 if it is assumed that (1) all the Mg is removed from seawater and (2) the seawater had not previously lost some Mg (Alt and Teagle, 1998).

Chloritization of basalts on the western side of the mound (TAG-4). On the western side of the mound, chloritization is evident in the form of halos that rim fragments of fresher (only 5%–15% altered) basalt. The chloritized halos are about 1 cm in width and are 60%–80% altered, although some plagioclase microlites and iron-titanium oxides are still present. Chlorite and mixed-layer chlorite-smectite replace olivine, pyroxene, and interstitial areas and also fill vesicles. The chlorite in the halos is chamosite [$Fe/(Fe + Mg)_{molar} \leq 0.6$], which is more enriched in Fe and Si than the chlorite from the deep parts of the up-flow zone (Honnorez et al., 1998). The normalized bulk-rock contents of the halos relative to the precursors indicate that Fe is taken up by the rock, whereas Mg changes very little. In addition, the halos are enriched in Zn by an order of magnitude over the Zn concentration in the precursors (Humphris et al., 1998).

The $^{87}Sr/^{86}Sr$ ratios of the halos (0.70326–0.70334) are low compared to those of the other chloritized basalts at TAG and to the $^{87}Sr/^{86}Sr$ ratio of TAG black smoker hydrothermal fluids, but are slightly higher than the $^{87}Sr/^{86}Sr$ ratios of the fresher basalt interior (Teagle et al., 1998b). The halos have $\delta^{18}O$ values of 4.5‰–5.2‰ compared with 5.9‰–6.8‰ for the fresh to slightly low-temperature–altered basalts (Alt and Teagle, 1998). The uptake of both Fe and Zn during chloritization indicates that the rocks reacted with an upwelling hydrothermal fluid rather than a seawater-dominated fluid to produce the 60%–80% alteration observed in the halos (Alt and Teagle, 1998; Teagle et al., 1998b). This finding is consistent with the prediction by Saccocia and Seyfried (1994) that conductive cooling of upwelling Fe-rich hydrothermal fluids leads to more Fe-rich chlorites in the shallow crust. On the basis of mixing trajectories between hydrothermal fluids and seawater, Teagle et al. (1998b, 1998c) suggested that the chloritized halos may represent an intermediate between fresh basalt and the chloritized breccias and, hence, may have formed from a similar hydrothermal fluid. However, it is also possible that the relatively low $^{87}Sr/^{86}Sr$ ratios and high $\delta^{18}O$ values are indicative of incomplete reactions and the halos did not equilibrate with the fluids. The extent of the reactions is likely to be limited because chemical exchange between the rock and fluid away from the fractures is controlled by diffusion of elements in the pore fluids of the halos.

Implications for conditions of alteration within the shallow up-flow zone

Geochemical and isotopic data from altered basalts are consistent with two major styles of high-temperature alteration occurring in the shallow up-flow zone. The first is halo-type alteration, which is observed in the basement from margins of the mound (the TAG-2 and TAG-4 areas), and is characterized by chloritization of selvages along veins and fragment surfaces. On both the eastern and western margins of the mound, these chloritization halos provide evidence that high-temperature (>200 °C) reactions between fluid and rock occurred at least on a very local scale, although the heterogeneous nature of the alteration indicates that hydrothermal flow was neither pervasive nor sustained. Lower-temperature oxidation appears to be the most recent type of alteration and suggests a waning of activity. This suggestion is supported by (1) the decrease in activity of the white smokers on the eastern margin of the mound and (2) the low measurements of conductive heat flow on the western side of the upper platform. However, it should be noted that the chloritization of the pillow-rim breccia through reaction with conductively heated seawater at temperatures of 210–250 °C suggests local hydrothermal flow along the contact at the base of the sulfides at TAG-2. It is interesting to note that high heat-flow values (from 3.7 to >25 W/m^2) were recorded at stations 20–30 m away from the TAG-2 site in carbonate ooze, and may be the result of lateral advection of this hot, hydrothermal fluid (Becker et al., 1996).

The second style of alteration is pervasive alteration that has completely replaced the basalt in the central part of the up-flow zone. Two phases of high-temperature reactions with hydrothermal fluids (and a small component of seawater) are recorded. The rocks were initially altered to chlorite + quartz + pyrite at high water/rock mass ratios via reaction of a mixture dominated by hydrothermal fluid but with a small component (<15%) of seawater at temperatures of 250–350 °C. This phase was followed by alteration to a paragonite + quartz + pyrite assemblage at temperatures of 260–360 °C by a fluid that may have undergone previous chloritization reactions at depth. The pervasive nature of this alteration requires more prolonged interaction with the hydrothermal fluid and seawater mixture than is required for the observed halo-type alteration or than is envisaged within the currently active system.

MODEL FOR THE FORMATION OF THE TAG ACTIVE MOUND

The morphology and internal structure of the TAG active mound bear a striking resemblance to ancient deposits in ophiolites, as has been well documented by Hannington et al. (1998) for the volcanic-hosted massive sulfide deposits of Cyprus and Oman. Metal-enriched zones similar to those in the upper few meters of the TAG active mound have been observed close to the paleo–sulfide-seawater interface, and the chert and iron oxide occurrences are analogous to the red cherts (jasper) associated with many of the Cyprus-type deposits (Franklin et al., 1981). Vertical zonation of sulfide and wall-rock breccias is common: coarse massive sulfide conglomerates and quartz-cemented breccias overlie an extensive pipe-like quartz + pyrite stockwork of chloritized and silicified basalt breccias (Constantinou, 1980; Constantinou and Govett, 1973; Franklin et al., 1981). The abundance of breccia ores and sulfide conglomerates in the Cyprus deposits was previously explained by postdepositional weathering and destruction of the mounds (Constantinou, 1976, 1980). However, the observations at TAG suggest that brecciation is not a post-depositional phenomenon, but rather is an active process during construction of the hydrothermal deposit. Although anhydrite may once have cemented the Cyprus breccias, it is now absent from those deposits because it dissolved after high-temperature hydrothermal activity had ceased.

The complex and brecciated structure implies that the TAG active mound has undergone multiple stages of development, which are reflected in the sequences of alteration and veining events that can be distinguished in both the sulfide breccias and in the silicified and chloritized basalt breccias. The mound's construction is interpreted to be a combination of accumulation of sulfides at the seafloor and hydrothermal replacement and mineralization in the up-flow zone during periods of activity, followed by collapse, mass wasting, and brecciation during periods of inactivity (Humphris, Herzig, et al., 1996).

Figure 9 presents a simplified schematic model for the construction and evolution of the TAG active mound through cycles of activity and inactivity (Hannington et al., 1998; Humphris, Herzig, et al., 1996). During periods of activity, high-temperature hydrothermal fluids flow up through basement; the fluids are focused along zones of higher permeability and react with the rocks. Where flow is vigorous and well focused, clusters of chimneys are constructed. Sulfides accumulate and a mound begins to form (Fig. 9A). As was pointed out by Tivey et al. (1995), fluid flowing rapidly upward from a narrow pipe-like feeder zone will draw seawater into the structure. The seawater is heated, both conductively and by mixing with upwelling hydrothermal fluids, and anhydrite precipitates within the central part of the mound and underlying up-flow zone, cementing fragments of sulfides and thereby providing supporting structure. Other subsurface processes include precipitation of sulfides and generation of more acidic fluids due to mixing of hydrothermal fluids with seawater, dissolution and recrystallization of previously deposited material and remobilization of metals, and precipitation of a siliceous "cap" by conductive cooling of upwelling diffuse hydrothermal fluids. As the system begins to wane and up-flow becomes less vigorous, less seawater is entrained into the mound. Fluids circulating within the mound are composed of proportionally greater amounts of hydrothermal fluid, but the lower flux of high-temperature fluids and slower fluid-flow rates result in conductive cooling, deposition of quartz, and pervasive basalt alteration.

During periods of inactivity, temperatures within the mound decrease, and dissolution of anhydrite causes collapse of chimneys and disintegration of the anhydrite-supported breccias in the central part of the mound (Fig. 9B). This process results in a poorly sorted pile of talus consisting of broken chimney fragments, massive sulfides, and chert on the seafloor underlain by a pipe-like zone of altered basalts.

During periods of renewed activity (Fig. 9C), fluid flow is refocused along slightly different conduits, resulting in a shift in the position of the main vigorous discharge. This kind of change was likely the case when the upper platform began to form, possibly at ca. 5 ka, and also during the recent reactivation about 80 yr ago that shifted the black smoker complex to the west of the center of the upper platform (Lalou et al., 1998). Until fluid flow once again becomes focused, the renewed circulation may be accompanied by pervasive basalt alteration; thus the full process results in repeated overgrowth by later generations of deposits and reworking, replacement, and recrystallization of the sulfides. Mobilization of metals from dissolution and/or recrystallization of sulfides within the mound and their transport to the surface account for zone refinement of the deposit. The fact that breccias are first cemented with anhydrite and then later by quartz results in the observed vertical zonation of the deposit.

The paragenetic sequence of alteration that has been identified beneath the TAG active mound suggests that during active cycles, different types of fluid-flow and mixing processes occur as the hydrothermal discharge becomes focused and then wanes. We envisage an active phase beginning with up-flow of hydrothermal fluids into the shallow subsurface, where they encounter a collapsed pile of sulfide and breccia talus. Flow at this initial stage will likely be diffuse, and subsurface mixing with seawater will result in precipitation of anhydrite. As the anhydrite cements breccias and parts of the mound become less permeable, the high-temperature up-flow will become more focused, and a black smoker complex will form. This vigorous flow then induces higher volumes of seawater to become entrained into the mound, where it mixes with hydrothermal fluids in varying proportions and is conductively heated. The TAG mound is currently at this stage of activity. Under these conditions, high-temperature reactions with the basement are limited to selvages along veins and fragment surfaces, leading to the halo-type alteration, similar to that observed at TAG-4 that was produced during an earlier phase of activity. Evidence from fluid-inclusion data, textures that indicate overprinting of the chloritization by lower-temperature alteration, and lower values of conductive heat flow on this part of the mound all suggest that activity at TAG-4 is waning. As hydrothermal activity throughout the mound wanes, fluid flow becomes less vigorous, and less seawater is entrained. Fluid flow also becomes less focused and more diffuse, and there is more conductive cooling. This change leads

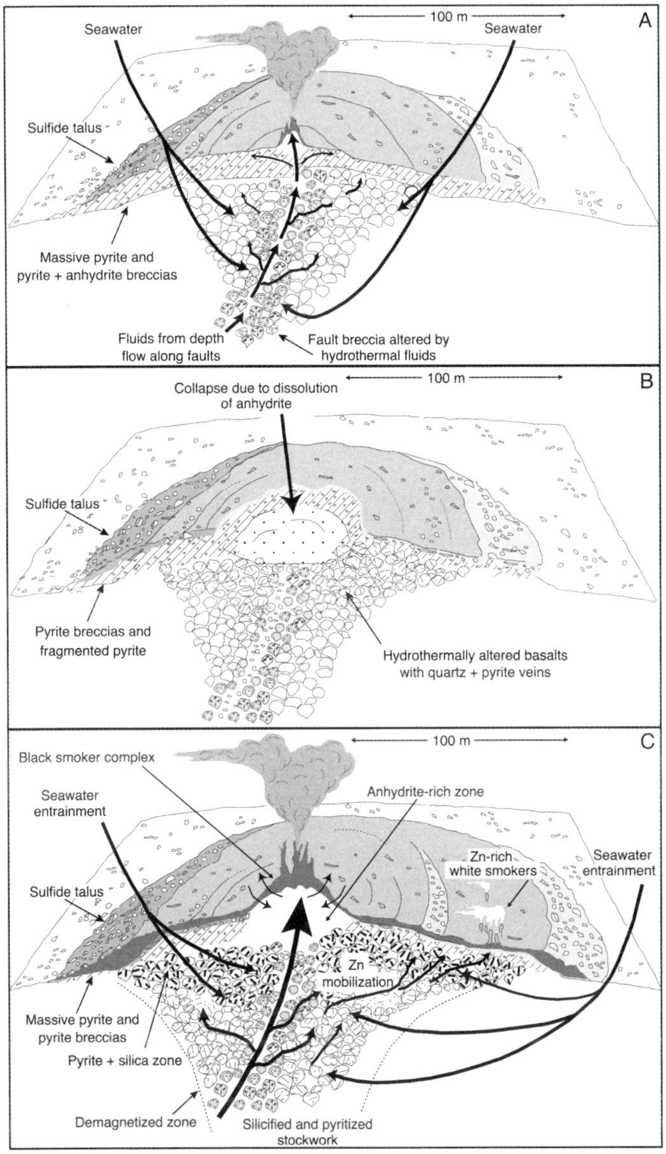

Figure 9. Schematic model of the episodic growth of the TAG active mound. (Adapted from Hannington et al., 1998.) A, During periods of activity, high-temperature hydrothermal fluids flow up through the basement; the fluids are focused along zones of higher permeability and react with the rocks. Where flow is vigorous and well focused, clusters of chimneys are constructed; elsewhere there is diffuse flow. Sulfides accumulate and a mound begins to form. Subsurface processes include entrainment of seawater, precipitation of sulfides and anhydrite, dissolution and recrystallization of previously deposited material and remobilization of metals, and precipitation of a siliceous "cap" by conductive cooling of upwelling diffuse hydrothermal fluids. B, During periods of inactivity, dissolution of anhydrite causes collapse of chimneys and disintegration of the anhydrite-supported breccias in the central part of the mound. These processes result in a poorly sorted pile of talus consisting of broken chimney fragments, massive sulfides, and chert on the seafloor underlain by a pipe-like zone of altered basalts. C, Present state of the TAG active mound schematically illustrating seawater entrainment and fluid flow. During periods of renewed activity, fluid flow is refocused along slightly different conduits; this change results in a shift in the position of the main vigorous discharge and continuation of the subsurface processes described in A. Fluid flowing rapidly upward from a narrow pipe-like feeder zone draws large volumes of seawater into the structure, where it mixes with hydrothermal fluids in varying proportions and is conductively heated. Anhydrite precipitates within the central part of the mound and underlying up-flow zone; the anhydrite cements the fragments of sulfides and thereby provides supporting structure.

to dissolution of anhydrite either by cooler fluids or by hot, sulfate-poor fluids and to precipitation of silica in pore spaces, veins, and as a cap on the deposit. Continued waning and cooling lead to the eventual dissolution of the anhydrite and collapse of anhydrite-supported lithologies.

An important question relates to the timing of the pervasive chloritization and paragonitization observed in the shallow up-flow zone. This alteration requires more prolonged interaction with hydrothermal fluids (and small amounts of seawater) than is envisaged within the currently active system. We suggest that there are two alternative scenarios by which chloritization and paragonitization can be achieved. If these alteration processes occur during the most vigorous hydrothermal activity when there is a well-developed black smoker complex and highly focused discharge, then high-temperature and pervasive water-rock reactions would be limited to areas that have a cap and are sufficiently isolated from the rapid entrainment of seawater. However, this scenario should lead to an up-flow zone in which some parts are pervasively altered whereas others show only halo-type alteration. Drilling has demonstrated that the basement rocks and fragments that comprise the breccias in the up-flow zone (except at its margins) are extensively altered and totally replaced, and thus a more pervasive process is suggested. An alternative scenario, and our preferred one, is that chloritization and paragonitization of the shallow up-flow zone either precedes, but most likely succeeds, the period of most vigorous venting. During these times when there is less vigorous, more diffuse flow and less seawater entrainment, there is more time for extensive water + rock reactions to occur (i.e., to "cook" the rock) with mixtures that are dominated by hydrothermal fluid but contain small amounts of seawater.

SUMMARY AND OUTSTANDING PROBLEMS

The TAG hydrothermal field is one of the largest hydrothermal deposits that has been found to date on the seafloor and provides an excellent example of a modern seafloor system for understanding the formation of ancient massive sulfide deposits. The presence of a large field of old massive sulfide deposits developed over a long period of time, combined with a currently active hydrothermal system, allows the controls on fluid flow to be investigated on different spatial and temporal scales. There is strong evidence that faulting plays a key role in maintaining fluid-flow pathways at TAG, as has been demonstrated for volcanogenic massive sulfide deposits in ophiolites. These pathways enable heat to be extracted either from the ridge axis or from local intrusions and also focus upwelling fluids along zones of enhanced permeability kept open, or periodically reopened, over long periods of time. The episodic nature of the activity may reflect either episodic magmatism and the replenishment of the heat source or the occurrence of faulting events that reactivate fluid-flow pathways. Unlike many ophiolitic massive sulfide deposits that are generally elongate features, the construction of large circular or quasi-circular mounds, as well as discontinuous but extensive, along-strike sulfide outcrops that result from fault control, imposes the additional requirement that, at certain sites, fluids are focused within discrete, pipe-like up-flow zones at least into the shallow subsurface, where more tortuous pathways may be followed. The intersection of faults with differing strikes provides one mechanism to produce conduits of permeable fault breccia, and such structures have been mapped on the seafloor adjacent to the TAG active mound.

Outstanding problems that remain to be addressed on the scale of a large hydrothermal field are (1) the nature of the heat source and plumbing system that maintain activity localized in one area over periods of thousands of years and (2) the causes of the episodic nature of the activity and whether they are related to magma-replenishment cycles or to the frequency of faulting events.

Although up-flow of hydrothermal fluids from depth is likely fault-controlled on the local scale at the TAG active mound, one of the key findings has been the critical role that seawater entrainment has played in the generation of distinct fluid chemistries, efficient precipitation of minerals, metal remobilization, variations in the styles of alteration of basalts, and the growth and structural evolution of the entire deposit. Despite the fact that the mineralization at TAG is very similar to the ancient analogues, the magnitude of seawater entrainment could not have been predicted from studies of sulfide deposits in ophiolites alone. In fact, an important question is whether such vigorous seawater entrainment is common in other large seafloor hydrothermal systems. If not, then we could conclude that styles of formation of seafloor massive sulfide deposits show a wide range of variability.

A great benefit of studying modern seafloor hydrothermal systems is being able to geochemically model them since it is possible to sample and analyze the hydrothermal fluids. Such studies have significantly advanced our understanding of the growth and evolution of massive sulfide deposits over the past 20 years and have emphasized the complexity of reactions within hydrothermal systems. We have only begun to investigate reactions in the very shallowest part of the up-flow zone and their relationship (if any) to fluids currently discharging from the overlying accumulation of hydrothermal precipitates. We still have not been able to completely unravel the relative timing of the different types of basement alteration and hydrothermal mineral precipitation that are present in the samples that have been collected to date. Future drilling to obtain fluids and samples from deeper within the hydrothermal system, coupled with further studies of the deep reaction zones in ophiolites, will be needed to address the many questions that still remain.

ACKNOWLEDGMENTS

In writing this paper, we have drawn on collaborations with numerous colleagues, including G. Thompson, P. Rona, M. Hannington, M. Kleinrock, J. Edmond, J. Alt, R. Mills, D. Teagle, the shipboard scientific party of ODP Leg 158, and many others over the past 15 years. We thank all of them for many hours of discussions that have shaped our thoughts about processes operating within submarine hydrothermal systems. We also wish to acknowledge W. Bach, S. White, and K. Von Damm for helpful discussions during the preparation of this manuscript and M. Sulanowska for help in preparation of the figures. We thank R. Mills and D. Teagle for helpful reviews that significantly improved this manuscript and Y. Dilek for his patience and editorial handling. Work presented here was funded by numerous grants from the National Science Foundation (including recent grant OCE-9314697 to Humphris and grant OCE-9712150 to Tivey), as well as Joint Oceanographic Institutions–U.S. Science Advisory Committee grants to both authors. This is WHOI Contribution no. 10009.

REFERENCES CITED

Adamides, N.G., 1987, Diverse modes of occurrence of Cyprus sulphide deposits and comparison with recent analogs, *in* Robinson, P.T., Gibson, I.L., and Panayiotou, A., eds., Cyprus Crustal Study Project: Initial Report, Holes CY-2 and 2a: Ottawa, Ontario, Canada, Geological Survey of Canada, p. 153–168.

Alabaster, T., Pearce, J.A., Mallick, D.I.J., and Elboushi, I.M., 1980, The volcanic stratigraphy and location of massive sulphide deposits in the Oman ophiolite, *in* Panayiotou, A., ed., Ophiolites: Proceedings of the International Ophiolite Symposium: Nicosia, Cyprus, Cyprus Ministry of Agricuture and Natural Resources, Geology Survey Department, p. 751–757.

Alt, J.C., 1994, A sulfur isotopic profile through the Troodos ophiolite, Cyprus: Primary composition and the effects of seawater hydrothermal alteration: Geochimica et Cosmochimica Acta, v. 58, p. 1825–1840.

Alt, J.C., and Teagle, D.A.H., 1998, Probing the TAG hydrothermal mound and stockwork: Oxygen isotopic profiles from deep ocean drilling, *in* Herzig, P.M., Humphris, S.E., et al., eds., Proceedings of the Ocean Drilling Program, Scientific Results, v. 158: College Station, Texas, Ocean Drilling Program, p. 285–295.

Becker, K., Von Herzen, R.P., Kirklin, J., Evans, R., Kadko, D., Kinoshite, M., Matsubayashi, O., Mills, R., Schultz, A., and Rona, P., 1996, Conductive heat flow at the TAG active hydrothermal mound: Results from 1993–1995 submersible surveys: Geophysical Research Letters, v. 23, p. 3463–3466.

Bischoff, J.L., 1991, Densities of liquids and vapors in boiling NaCl-H_2O solutions: A PVTX summary from 300°–500° C: American Journal of Science, v. 291, p. 309–338.

Bischoff, J.L., and Rosenbauer, R.J., 1985, An empirical equation of state for hydrothermal seawater (3.2 percent NaCl): American Journal of Science, v. 285, p. 725–763.

Bischoff, J.L., and Seyfried, W.E., 1978, Hydrothermal chemistry of seawater from 25° to 350° C: American Journal of Science, v. 378, p. 838–860.

Campbell, A.C., Palmer, M.R., Klinkhammer, G.P., Bowers, T.S., Edmond, J.M., Lawrence, J.R., Casey, J.F., Thompson, G., Humphris, S., Rona, P., and Karson, J.A., 1988, Chemistry of hot springs on the Mid-Atlantic Ridge: Nature, v. 335, p. 514–519.

Chiba, H., Uchiyama, N., and Teagle, D.A.H., 1998, Stable isotope study of anhydrite and sulfide minerals at the TAG hydrothermal mound, Mid-Atlantic Ridge, 26°N, *in* Herzig, P.M., Humphris, S.E., et al., eds., Proceedings of the Ocean Drilling Program, Scientific Results, v. 158: College Station, Texas, Ocean Drilling Program, p. 85–90.

Coleman, R.G., 1981, Tectonic setting for ophiolite obduction in Oman: Journal of Geophysical Research, v. 86, p. 2497–2508.

Collinson, T.B., 1986, Hydrothermal mineralization and basalt alteration in stockwork zones of the Bayda and Lasail massive sulfide deposits, Oman ophiolite [M.S. thesis]: University of California, Santa Barbara, 164 p.

Constantinou, G., 1976, Genesis of the conglomerate structure, porosity, and collomorphic textures of the massive sulphide ores of Cyprus: Geological Association of Canada Special Paper, v. 14, p. 187–210.

Constantinou, G., 1980, Metallogenesis associated with the Troodos ophiolite, *in* Panayiotou, A., ed., Ophiolites: Proceedings, International Ophiolite Symposium, Cyprus, 1979: Nicosia, Cyprus, Geological Survey Department, p. 663–674.

Constantinou, G., and Govett, G.J.S., 1973, Geology, geochemistry, and genesis of Cyprus sulfide deposits: Economic Geology, v. 68, p. 843–858.

Edmond, J.M., Campbell, A.C., Palmer, M.R., Klinkhammer, G.P., German, C.R., Edmonds, H.N., Elderfield, H., Thompson, G., and Rona, P.A., 1995, Time series studies of vent fluids from the TAG and MARK sites (1986, 1990) Mid-Atlantic Ridge: A new solution chemistry model and mechanism for Cu/Zn zonation in massive sulfide orebodies, *in* Parson, L.M., Walker, C.L., and Dixon, D.R., eds., Hydrothermal vents and processes: Geological Society of London Special Publication 87, p. 77–86.

Edmonds, H.N., German, C.R., Green, D.R.H., Huh, Y., Gamo, T., and Edmond, J.M., 1996, Continuation of the hydrothermal fluid chemistry time series at TAG, and the effects of ODP drilling: Geophysical Research Letters, v. 23, p. 3487–3489.

Embley, R.W., Jonasson, I.R., Perfit, M.R., Franklin, J.M., Tivey, M.A., Malahoff, A., Smith, M.F., and Francis, T.J.G., 1988, Submersible investigation of an extinct hydrothermal system on the Galapagos Ridge: Sulfide mounds, stockwork zone, and differentiated lavas: The Canadian Mineralogist, v. 26, p. 517–539.

Ernewein, M., Pflumio, C., and Whitechurch, H., 1988, The death of an accretion zone as evidenced by the magmatic history of the Sumail ophiolite (Oman): Tectonophysics, v. 151, p. 247–274.

Evans, R.L., 1996, A seafloor gravity profile across the TAG hydrothermal mound: Geophysical Research Letters, v. 23, p. 3447–3450.

Fouquet, Y., Henry, K., Knott, R., and Cambon, P., 1998, Geochemical section of the TAG hydrothermal mound, *in* Herzig, P.M., Humphris, S.E., et al., eds., Proceedings of the Ocean Drilling Program, Scientific Results, v. 158: College Station, Texas, Ocean Drilling Program, p. 363–387.

Franklin, J.M., Lydon, J.W., and Sangster, D.F., 1981, Volcanic-associated massive sulfide deposits: Economic Geology, v. 75, anniversary volume, p. 485–627.

Galley, A.G., and Koski, R.A., 1999, Setting and characteristics of ophiolite-hosted volcanogenic massive sulfide deposits, *in* Barrie, C.T., and Hannington, M.D., eds., Volcanic-associated massive sulfide deposits: Processes and examples in modern and ancient settings: Reviews in Economic Geology, v. 8, p. 221–246.

Gamo, T., Chiba, H., Masuda, H., Edmonds, H.N., Fujioka, K., Kodama, Y., Nanba, H., and Sano, Y., 1996, Chemical characteristics of hydrothermal fluids from the TAG mound of the Mid-Atlantic Ridge in August 1994: Implications for spatial and temporal variability of hydrothermal activity: Geophysical Research Letters, v. 23, p. 3483–3486.

Gemmell, J.B., and Sharpe, R., 1998, Detailed sulfur-isotope investigation of the TAG hydrothermal mound and stockwork zone, 26°N Mid-Atlantic Ridge, *in* Herzig, P.M., Humphris, S.E., et al., eds., Proceedings of the Ocean Drilling Program, Scientific Results, v. 158: College Station, Texas, Ocean Drilling Program, p. 71–83.

Hannington, M.D., 1993, The formation of atacamite during weathering of sulfides on the modern seafloor: The Canadian Mineralogist, v. 31, p. 945–956.

Hannington, M.D., Tivey, M.K., Larocque, A.C.L., Petersen, S., and Rona, P.A., 1995, The occurrence of gold in sulfide deposits of the TAG Hydrothermal Field, Mid-Atlantic Ridge: The Canadian Mineralogist, v. 33, p. 1285–1310.

Hannington, M.D., Galley, A.D., Herzig, P.M., and Petersen, S., 1998, Comparison of the TAG mound and stockwork complex with Cyprus-type massive sulfide deposits, *in* Herzig, P.M., Humphris, S.E., et al., eds., Proceedings of the Ocean Drilling Program, Scientific Results, v. 158: College Station, Texas, Ocean Drilling Program, p. 389–415.

Harper, G.D., 1985, Tectonics of slow spreading mid-ocean ridges and consequences of a variable depth to the brittle/ductile transition: Tectonics, v. 4, p. 395–409.

Herzig, P.M., Humphris, S.E., et al., eds., 1998a, Proceedings of the Ocean Drilling Program, Scientific Results, v. 158: College Station, Texas, Ocean Drilling Program, 427 p.

Herzig, P.M., Petersen, S., and Hannington, M.D., 1998b, Geochemistry and sulfur isotopic composition of the TAG hydrothermal mound, Mid-Atlantic Ridge, 26°N, *in* Herzig, P.M., Humphris, S.E., et al., eds., Proceedings of the Ocean Drilling Program, Scientific Results, v. 158: College Station, Texas, Ocean Drilling Program, p. 47–70.

Hodell, D.A., Mueller, P.A., and Garrido, J.R., 1991, Variations in the strontium isotopic composition of seawater during the Neogene: Geology, v. 19, p. 24–27.

Honnorez, J.J., Alt, J.C., and Humphris, S.E., 1998, Vivisection and autopsy of active and fossil hydrothermal alterations of basalt beneath and within the TAG hydrothermal mound, *in* Herzig, P.M., Humphris, S.E., et al., eds., Proceedings of the Ocean Drilling Program, Scientific Results, v. 158: College Station, Texas, Ocean Drilling Program, p. 231–254.

Humphris, S.E., and Kleinrock, M.C., 1996, Detailed morphology of the TAG active hydrothermal mound: Insights into its formation and growth: Geophysical Research Letters, v. 23, p. 3443–3446.

Humphris, S.E., Herzig, P.M., Miller, D.J., Alt, J.C., Becker, K., Brown, D., Brugmann, G., Chiba, H., Fouquet, Y., Gemmell, J.B., Guerin, G., Hannington, M.D., Holm, N.G., Honnorez, J.J., Iturrino, G.J., Knott, R., Ludwig, R., Nakamura, K., Petersen, S., Reysenbach, A.-L., Rona, P.A., Smith, S., Sturz, A.A., Tivey, M.K., and Zhao, X., 1995, The internal structure of an active sea-floor massive sulphide deposit: Nature, v. 377, p. 713–716.

Humphris, S.E., Herzig, P.M., et al., 1996, Proceedings of the Ocean Drilling Program, Initial Reports, College Station, TX, Ocean Drilling Program, 384 p.

Humphris, S.E., Alt, J.C., Teagle, D.A.H., and Honnorez, J.J., 1998, Geochemical changes during hydrothermal alteration of basement in the stockwork beneath the active TAG hydrothermal mound, *in* Herzig, P.M., Humphris, S.E., et al., eds., Proceedings of the Ocean Drilling Program, Scientific Results, v. 158: College Station, Texas, Ocean Drilling Program, p. 255–276.

James, R.H., and Elderfield, H., 1996, Chemistry of ore-forming fluids and mineral formation rates in an active hydrothermal sulfide deposit on the Mid-Atlantic Ridge: Geology, v. 24, p. 1147–1150.

James, R.H., and Elderfield, H., 1997, Chemistry of ore-forming fluids and mineral formation rates in an active hydrothermal sulfide deposit on the Mid-Atlantic Ridge: Correction: Geology, v. 25, p. 480.

Karson, J.A., and Rona, P.A., 1990, Block-tilting, transfer faults, and structural control of magmatics and hydrothermal processes in the TAG area, Mid-Atlantic Ridge 26°N: Geological Society of America Bulletin, v. 102, p. 1635–1645.

Kleinrock, M.C., and Humphris, S.E., 1996a, Structural asymmetry of the TAG rift valley: Evidence from a near-bottom survey for episodic spreading: Geophysical Research Letters, v. 23, p. 3439–3442.

Kleinrock, M.C., and Humphris, S.E., 1996b, Structural controls on the localization of seafloor hydrothermal activity at the TAG active mound, Mid-Atlantic Ridge: Nature, v. 382, p. 149–153.

Kleinrock, M.C., Humphris, S.E., and Deep -TAG Team, 1996, Detailed structure and morphology of the TAG active hydrothermal mound and its geotectonic setting, *in* Humphris, S.E., Herzig, P.M., et al., eds., Proceedings of the Ocean Drilling Program, Initial Reports, v. 158: College Station, Texas, Ocean Drilling Program, p. 15–21.

Knott, R., Fouquet, Y., Honnorez, J., Petersen, S., and Bohn, M., 1998, Petrology of hydrothermal mineralization: A vertical section through the TAG mound, *in* Herzig, P.M., Humphris, S.E., et al., eds., Proceedings of the Ocean Drilling Program, Scientific Results, v. 158: College Station, Texas, Ocean Drilling Program, p. 5–26.

Kong, L.S.L., Solomon, S.C., and Purdy, G.M., 1992, Microearthquake characteristics of a mid-ocean ridge along-axis high: Journal of Geophysical Research, v. 97, p. 1659–1685.

Lalou, C., Thompson, G., Rona, P.A., Brichet, E., and Jehanno, C., 1986, Chronology of selected hydrothermal Mn oxide deposits from the transatlantic geotraverse "TAG" area, Mid-Atlantic Ridge 26°N: Geochimica et Cosmochimica Acta, v. 50, p. 1737–1743.

Lalou, C., Thompson, G., Arnold, M., Brichet, E., Druffel, E., and Rona, P.A., 1990, Geochronology of TAG and Snake Pit hydrothermal fields, Mid-Atlantic Ridge: Witness to a long and complex hydrothermal history: Earth Planetary Science Letters, v. 97, p. 113–128.

Lalou, C., Reyss, J.L., Brichet, E., Arnold, M., Thompson, G., Fouquet, Y., and Rona, P.A., 1993, New age data for Mid-Atlantic Ridge hydrothermal sites: TAG and Snakepit geochronology revisited: Journal of Geophysical Research, v. 98, p. 9705–9713.

Lalou, C., Reyss, J.L., Brichet, E., Rona, P.A., and Thompson, G., 1995, Hydrothermal activity on a 10^5-year scale at a slow-spreading ridge, TAG hydrothermal field, Mid-Atlantic Ridge 26°N: Journal of Geophysical Research, v. 100, p. 17855–17862.

Lalou, C., Reyss, J.L., and Brichet, E., 1998, Age of sub-bottom sulfide samples at the TAG active mound, *in* Herzig, P.M., Humphris, S.E., et al., eds., Proceedings of the Ocean Drilling Program, Scientific Results, v. 158: College Station, Texas, Ocean Drilling Program, p. 111–117.

Lein, A.Y., Ul'yanova, N.V., Grinenko, V.A., and Lisitsyn, A.P., 1991, Geochemistry of the hydrothermal sulfide ores of the Mid-Atlantic Ridge (26°N): Geochemistry International, v. 28, p. 1–13.

Lloyd, R.M., 1967, Oxygen-18 composition of oceanic sulfate: Science, v. 156, p. 1228–1231.

Longinelli, A., 1989, Oxygen-18 and sulphur-34 in dissolved oceanic sulphate and phosphate, *in* Fritz, P., and Fontes, J.C., eds., The marine environment, handbook of environmental isotope geochemistry: Amsterdam, Elsevier, p. 219–225.

Longinelli, A., and Craig, H., 1967, Oxygen-18 variations in sulfate ions in seawater and saline lakes: Science, v. 156, p. 56–59.

Lydon, J.W., and Galley, A., 1986, The chemical and mineralogical zonation of Mathiati alteration pipe, Cyprus, and its genetic significance, *in* M.J. Gallagher, M.J., Ixer, R.A., Neary, C.R., and Prichard, H.M., eds., Metallogeny of basic and ultrabasic rocks: London, Institute of Mining and Metallurgy, p. 49–71.

McGregor, B.A., and Rona, P.A., 1975, Crest of Mid-Atlantic Ridge at 26°N: Journal of Geophysical Research, v. 80, p. 3307–3314.

McGregor, B.A., Harrison, C.G.A., Lavelle, J.W., and Rona, P.A., 1977, Magnetic anomaly pattern on the Mid-Atlantic Ridge crest at 26°N: Journal of Geophysical Research, v. 82, p. 231–238.

Metz, S., Trefry, J.H., and Nelsen, T.A., 1988, History and geochemistry of a metalliferous sediment core from the Mid-Atlantic Ridge at 26°N: Geochimica et Consmochimica Acta, v. 52, p. 2369–2378.

Mills, R.A., 1995, Hydrothermal deposits and metalliferous sediments from TAG, 26°N Mid-Atlantic Ridge, *in* Parson, L.M., Walker, C.L., and Dixon, D.R., eds., Hydrothermal vents and processes: Geological Society of London Special Publication 87, p. 121–132.

Mills, R., and Elderfield, H., 1993, A dual origin for the hydrothermal component in a metalliferous sediment core from the Mid-Atlantic Ridge: Journal of Geophysical Research, v. 98, p. 9671–9681.

Mills, R.A., and Tivey, M.K., 1999, Seawater entrainment and fluid evolution within the TAG hydrothermal mound: Evidence from analyses of anhydrite, *in* Cann, J., Elderfield, H., and Laughton, A., eds., Mid-ocean ridges: Dynamics of processes associated with the creation of new ocean crust: Cambridge, Cambridge University Press, p. 225–248.

Mills, R.A., Clayton, T., and Alt, J.C., 1996, Low-temperature fluid flow through sulfidic sediments from TAG: Modification of fluid chemistry and alteration of mineral deposits: Geophysical Research Letters, v. 23, p. 3495–3498.

Mills, R.A., Teagle, D.A.H., and Tivey, M.K., 1998, Fluid mixing and anhydrite precipitation within the TAG mound, *in* Herzig, P.M., Humphris, S.E., et al., eds., Proceedings of the Ocean Drilling Program, Scientific Results, v. 158: College Station, Texas, Ocean Drilling Program, p. 119–127.

Miyashiro, A., 1973, The Troodos ophiolitic complex was probably formed in an island arc: Earth and Planetary Science Letters, v. 19, p. 218–224.

Moores, E.M., and Vine, F.J., 1971, The Troodos Massif, Cyprus and other ophiolites as oceanic crust: Evaluation and implications: Philosophical Transactions of the Royal Society of London, A, v. 268, p. 443–466.

Pearce, J.A., Alabaster, T., Shelton, A.W., and Searle, M.P., 1981, The Oman ophiolite as a Cretaceous arc-basin complex: Evidence and implications: Philosophical Transactions of the Royal Society of London, A, v. 300, p. 299–317.

Pearce, J.A., Lippard, S.J., and Roberts, S., 1984, Characteristics and tectonic significance of supra-subduction zone ophiolites, *in* Kokellar, B.P., and Howell, M.F., eds., Marginal basin geology: Geological Society of London Special Publication 16, p. 77–94.

Petersen, S., Herzig, P.M., and Hannington, M.D., 1998, Fluid inclusion studies as a guide to the temperature regime within the TAG hydrothermal mound, 26°N, Mid-Atlantic Ridge, *in* Herzig, P.M., Humphris, S.E., et al., eds., Proceedings of the Ocean Drilling Program, Scientific Results, v. 158: College Station, Texas, Ocean Drilling Program, p. 163–176.

Purdy, G.M., Sempere, J.-C., Schouten, H., Dubois, D.L., and Goldsmith, R., 1990, Bathymetry of the Mid-Atlantic Ridge, 24°–31°N: A map series: Marine Geophysical Researches, v. 12, p. 247–252.

Richards, H.G., Cann, J.R., and Jensensius, J., 1989, Mineralogical zonation and metasomatism of the alteration pipes of Cyprus sulfide deposits: Economic Geology, v. 84, p. 91–115.

Rona, P.A., 1980, TAG hydrothermal field: Mid-Atlantic Ridge crest at latitude 26°N: Journal of the Geological Society of London, v. 137, p. 385–402.

Rona, P.A., McGregor, B.A., Betzer, P.R., Bolger, G.W., and Krause, D.C., 1975, Anomalous water temperatures over Mid-Atlantic Ridge crest at 26°N: Deep Sea Research, v. 22, p. 611–618.

Rona, P.A., Thompson, G., Mottl, M.J., Karson, J.A., Jenkins, W.J., Graham, D., Mallette, M., Von Damm, J., and Edmond, J.M., 1984, Hydrothermal activity at the TAG hydrothermal field, Mid-Atlantic Ridge crest at 26°N: Journal of Geophysical Research, v. 89, p. 11365–11377.

Rona, P.A., Klinkhammer, G., Nelsen, T.A., Trefry, J.H., and Elderfield, H., 1986, Black smokers, massive sulphides, and vent biota at the Mid-Atlantic Ridge: Nature, v. 321, p. 33–37.

Rona, P.A., Bogdanov, Y.A., Gurvich, E.G., Rimski-Korsakov, A., Sagalevitch, A.M., Hannington, M.D., and Thompson, G., 1993a, Relict hydrothermal zones in the TAG hydrothermal field, Mid-Atlantic Ridge 26°N, 45°W: Journal of Geophysical Research, v. 98, p. 9715–9730.

Rona, P.A., Hannington, M.D., Raman, C.V., Thompson, G., Tivey, M.K., Humphris, S.E., Lalou, C., and Petersen, S., 1993b, Active and relict seafloor hydrothermal mineralization at the TAG hydrothermal field, Mid-Atlantic Ridge: Economic Geology, v. 18, p. 1989–2017.

Rona, P., Petersen, S., Becker, K., Von Herzen, R.P., Hannington, M.D., Herzig, P.M., Naka, J., Lalou, C., and Thompson, G., 1996, Heat flow and mineralogy of TAG relict high-temperature zones: Mid-Atlantic Ridge, 26°N, 45°W: Geophysical Research Letters, v. 23, p. 3507–3510.

Rona, P.A., Fujioka, K., Ishihara, T., Chiba, H., Masuda-Nakaya, H., Oomori, T., Kleinrock, M.C., Tivey, M.A., Watanabe, M., and Lalou, C., 1998, An active, low temperature hydrothermal mound and a large inactive sulfide mound found in the TAG hydrothermal field, Mid-Atlantic Ridge 26N, 45W: Eos, Transactions of the American Geophysical Union, v. 79, p. F920.

Saccocia, P.J., and Seyfried, W.E.J., 1994, The solubility of chlorite solid solutions in 3.2 wt% NaCl fluids from 300–400 °C, 500 bars: Geochimica et Cosmochimica Acta, v. 58, p. 567–585.

Schultz, A., Dickson, P., and Elderfield, H., 1996, Temporal variations in diffuse hydrothermal flow at TAG: Geophysical Research Letters, v. 23, p. 3471–3474.

Scott, R.B., Rona, P.A., and McGregor, B.A., 1974a, The TAG hydrothermal field: Nature, v. 251, p. 301–302.

Scott, M.R., Scott, R.B., Rona, P.A., Butler, L.W., and Nalwalk, A.J., 1974b, Rapidly accumulating manganese deposit from the median valley of the Mid-Atlantic Ridge: Geophysical Research Letters, v. 1, p. 355–358.

Sempere, J.-C., Purdy, G.M., and Schouten, H., 1990, Segmentation of the Mid-Atlantic Ridge between 24°N and 30° 40′N: Nature, v. 344, p. 427–431.

Shanks, W.C., Bohlke, J.K., and Seal, R.R., 1995, Stable isotopes in mid-ocean ridge hydrothermal systems: Tracers of interactions between fluids, minerals and organisms, *in* Humphris, S.E., Zierenberg, R.A., Mullineaux, L.S., and Thomson, R., eds., Seafloor hydrothermal systems: Physical, chemical, biological and geological interactions: American Geophysical Union Monograph, v. 91, p. 85–114.

Sharp, Z.D., and Kirschner, D.L., 1994, Quartz-calcite oxygen isotope geothermometry: A calibration based on natural isotopic variations: Geochimica et Cosmochimica Acta, v. 58, p. 4491–4502.

Stepanova, T.V., Krasnov, S.G., and Cherkashev, G.A., 1996, Mineralogy, chemical composition and structure of the MIR mound, TAG hydrothermal field: Geophysical Research Letters, v. 23, p. 3515–3518.

Teagle, D.A.H., Alt, J.C., Chiba, H., and Halliday, A.N., 1998a, Dissecting an active hydrothermal deposit: The strontium and oxygen isotopic anatomy of the TAG hydrothermal mound: Anhydrite, *in* Herzig, P.M., Humphris, S.E., et al., eds., Proceedings of the Ocean Drilling Program, Scientific Results, v. 158: College Station, Texas, Ocean Drilling Program, p. 129–141.

Teagle, D.A.H., Alt, J.C., Humphris, S.E., and Halliday, A.N., 1998b, Dissecting an active hydrothermal deposit: The strontium and oxygen isotopic anatomy of the TAG hydrothermal mound: Whole rock and silicate minerals, *in* Herzig, P.M., Humphris, S.E., et al., eds., Proceedings of the Ocean Drilling Program, Scientific Results, v. 158: College Station, Texas, Ocean Drilling Program, p. 297–309.

Teagle, D.A.H., Alt, J.C., Chiba, H., Humphris, S.E., and Halliday, A.N., 1998c, Strontium and oxygen isotopic constraints on fluid mixing, alteration, and mineralization in the TAG hydrothermal deposit: Chemical Geology, v. 149, p. 1–24.

Temple, D.G., Scott, R.B., and Rona, P.A., 1979, Geology of a submarine hydrothermal field, Mid-Atlantic Ridge, 26°N latitude: Journal of Geophysical Research, v. 84, p. 7453–7466.

Thompson, G., Mottl, M.J., and Rona, P.A., 1985, Morphology, mineralogy, and chemistry of hydrothermal deposits from the TAG area, 26°N Mid-Atlantic Ridge: Chemical Geology, v. 49, p. 243–257.

Thompson, G., Humphris, S.E., Schroeder, B., Sulanowska, M., and Rona, P.A., 1988, Active vents and massive sulfides at 26°N (TAG) and 23°N (SNAKEPIT) on the Mid-Atlantic Ridge: The Canadian Mineralogist, v. 26, p. 697–711.

Tivey, M.A., Rona, P.A., and Schouten, H., 1993, Reduced crustal magnetization beneath the active sulfide mound, TAG hydrothermal field, Mid-Atlantic Ridge 26°N: Earth and Planetary Science Letters, v. 115, p. 101–115.

Tivey, M.A., Rona, P.A., and Kleinrock, M.C., 1996, Reduced crustal magnetization beneath relict hydrothermal mounds: TAG hydrothermal field, Mid-Atlantic Ridge, 26°N: Geophysical Research Letters, v. 23, p. 3511–3514.

Tivey, M.K., Humphris, S.E., Thompson, G., Hannington, M.D., and Rona, P.A., 1995, Deducing patterns of fluid flow and mixing within the active TAG mound using mineralogical and geochemical data: Journal of Geophysical Research, v. 100, p. 12527–12555.

Tivey, M.K., Mills, R.A., and Teagle, D.A.H., 1998, Temperature and salinity of fluid inclusions in anhydrite as indicators of seawater entrainment and heating in the TAG active mound, *in* Herzig, P.M., Humphris, S.E., et al., eds., Proceedings of the Ocean Drilling Program, Scientific Results, v. 158: College Station, Texas, Ocean Drilling Program, p. 179–190.

Varga, R.J., and Moores, E.M., 1985, Spreading structure of the Troodos ophiolite, Cyprus: Geology, v. 13, p. 846–850.

Von Damm, K.L., 1995, Controls on the chemistry and temporal variability of seafloor hydrothermal fluids, *in* Humphris, S.E., Zierenberg, R.A., Mullineaux, L.S., and Thomson, R., eds., Seafloor hydrothermal systems: Physical, chemical, biological and geological interactions: Washington, D.C., American Geophysical Union Monograph, v. 91, p. 222–247.

White, S.N., Humphris, S.E., and Kleinrock, M.C., 1998, New observations on the distribution of past and present hydrothermal activity in the TAG area of the Mid-Atlantic Ridge (26°08′N): Marine Geophysical Researches, v. 20, p. 41–56.

You, C.-F., and Bickle, M.J., 1998, Evolution of an active sea-floor massive sulfide deposit: Nature, v. 394, p. 668–671.

Zierenberg, R.A., Shanks, W.C., III, Seyfried, W.E., Jr., Koski, R.A., and Strickler, M.D., 1988, Mineralization, alteration, and hydrothermal metamorphism of the ophiolite-hosted Turner-Albright sulfide deposit, southwestern Oregon: Journal of Geophysical Research, v. 93, p. 4657–4674.

Zierenberg, R.A., Schiffman, P., Jonasson, I.R., Tosdal, R., Pickthorn, W., and McClain, J., 1995, Alteration of basalt hyaloclastite at the off-axis Sea Cliff hydrothermal field, Gorda Ridge: Chemical Geology, v. 126, p. 77–99.

Zonenshain, L.P., Kuzmin, M.I., Lisitsin, A.P., Bogdanov, Y.A., and Baranov, B.V., 1989, Tectonics of the Mid-Atlantic rift valley between the TAG and MARK areas (26–24°N): Evidence for vertical tectonism: Tectonophysics, v. 159, p. 1–23.

Manuscript Accepted by the Society February 23, 2000

Printed in the U.S.A.

Geological Society of America
Special Paper 349
2000

Volatiles in mid-ocean ridge environments

Deborah S. Kelley
School of Oceanography, Box 357940, University of Washington, Seattle, Washington 98195, USA; e-mail: kelley@ocean.washington.edu
Gretchen L. Früh-Green
Institute für Mineralogie und Petrographie, ETH-Zurich, CH-8092, Zurich, Switzerland; e-mail: gretli@erdw.ethz.ch

ABSTRACT

This paper focuses on the evolution of volatiles and hydrothermal fluids in high-temperature plutonic and shallow-crustal rocks from mid-ocean ridge environments. The thermal and compositional evolution of fluids preserved in samples from the ocean basins and from ophiolite suites reflects complicated histories involving multiple magma and hydrothermal fluid pulses during cooling and transport of the crust away from the zone of crustal accretion. Analyses of primary fluid inclusions in plutonic samples show that with progressive melt fractionation, magmatic fluids evolve from CO_2-rich vapors, to immiscible (CO_2 + H_2O)–rich vapors, and finally to metal-rich, CO_2 + H_2O + NaCl brines that are exsolved at temperatures in excess of 700 °C. In more water-rich systems, such as backarc environments, the last fluids to be exsolved from the melts may involve direct exsolution of 30–60 wt% NaCl brines.

Paths of magmatic fluid evolution and compositions may vary significantly in different spreading environments. Magmatic fluids in gabbroic samples from the Southwest Indian Ridge show a continuum between CO_2-rich fluids and (CO_2 + CH_4 + H_2O)–rich fluids; the latter contain 30–50 mol% CO_2 and 43 mol% CH_4. Phase equilibria and isotopic data show that the (CO_2 + CH_4 + H_2O)–rich fluids reflect Rayleigh distillation of evolved magmatic CO_2, subsequent closed-system respeciation, and attendant graphite precipitation at temperatures of ~500–800 °C.

In all spreading systems studied to date, the transition from magmatic to hydrothermal seawater–dominated conditions in plutonic rocks is marked by circulation of fluids at temperatures >400 °C, with salinities from <0.1% to 300% of seawater values. The compositions of these fluids likely reflect condensation of seawater-derived fluids or boiling of NaCl-enriched hydrothermal fluids. Similar fluids in quartz breccia samples are interpreted to represent the 300 to >400 °C conduits for systems venting on the seafloor. Plutonic samples from the Mid-Atlantic Ridge and the Southwest Indian Ridge indicate that 400 °C circulation systems may also involve (CH_4 + H_2O ± H_2)–rich fluids formed during high-temperature seawater and rock interactions within the plutonic sequences. In zones of low magmatic supply, fluid migration along deeply penetrating fault systems may intensify hydration of deep-seated mafic and/or mantle material and lead to significant CH_4 (+ H_2 + hydrocarbon) generation.

The processes by which these volatiles are transported to the overlying ocean are complex. Some of these volatiles may be released during diking-eruptive events in the course of which there are catastrophic and voluminous releases of buoyant hydrothermal plumes enriched in H_2, ^3He, and CH_4 above ocean-water background levels. The degassing of volatiles during the eruption and their release during high-temperature interaction of seawater and dike extrusive material may provide energy for subseafloor biotopes and subsequent microbial blooms.

INTRODUCTION

Volatiles play a profound role in the evolution of any tectonically active planet. Degassing of volatiles is responsible for formation of the early oceans and atmosphere and strongly influences the geochemical, deformational, and eruptive history of continental and submarine volcanoes. Both active and passive degassing of volatiles (CO_2, H_2, CH_4, H_2S, and He) and their entrainment into venting hydrothermal fluids provide a key tracer of hydrothermal sites on the seafloor (Lilley et al., 1995; Lupton, 1995). More important, it is becoming increasingly clear that migration of volatiles through the crust (either by direct degassing or through leaching by hydrothermal fluids) plays a critical role in sustaining diverse microbial communities in subsurface and near-vent environments (Deming and Baross, 1993; Fisher, 1995; Karl, 1995; Baross and Deming, 1995; Delaney et al., 1998). Examination of seafloor dike intrusion and eruptive events indicates that magmatic volatiles released during emplacement of melt into the water-saturated, porous oceanic crust provide nutrients for the massive microbial blooms that occur subsequent to the eruptions (Embley et al., 1995; Fornari and Embley, 1995; Holden et al., 1998; Delaney et al., 1998; Summit and Baross, 1998; Kelley et al., 1998). Many of these microorganisms are Archeae, believed

Kelley, D.S., and Früh-Green, G.L., 2000, Volatiles in mid-ocean ridge environments, *in* Dilek, Y., Moores, E.M., Elthon, D., and Nicolas, A., eds., Ophiolites and Oceanic Crust: New Insights from Field Studies and the Ocean Drilling Program: Boulder, Colorado, Geological Society of America Special Paper 349, p. 237–260.

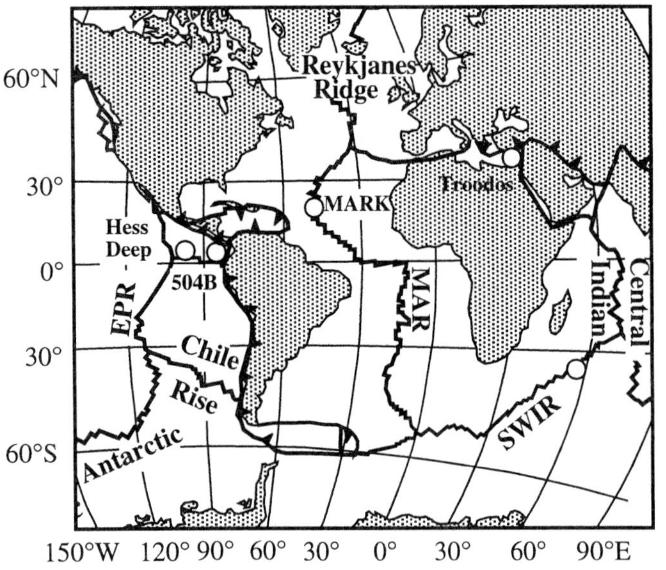

Figure 1. Locations of samples from diverse spreading systems, described in this study. EPR—East Pacific Rise, MARK—Mid-Atlantic Ridge and Kane Fracture Zone intersection, MAR—Mid-Atlantic Ridge, SWIR—Southwest Indian Ridge.

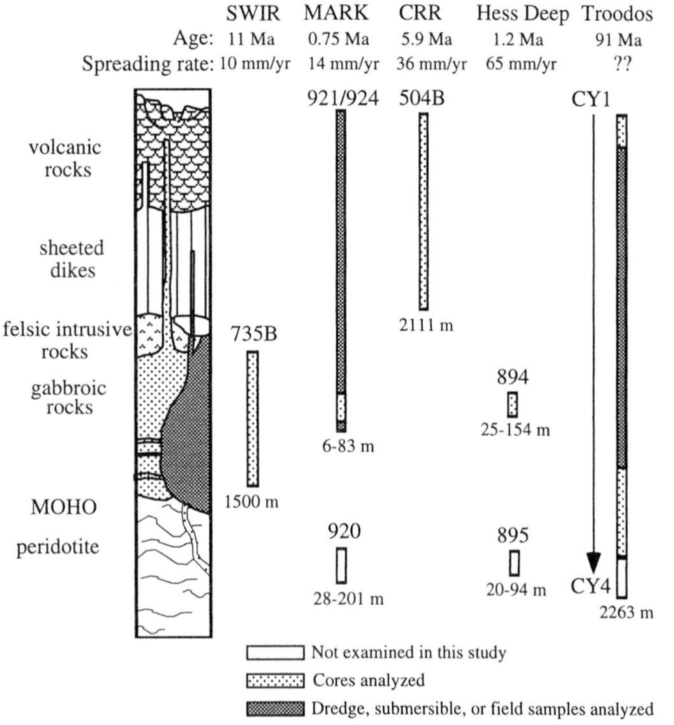

Figure 2. Simplified stratigraphic column of the oceanic crust showing schematic depth and lithology of samples used in this study in relationship to age of the crustal section and spreading rate. The light shaded rectangles reflect analyzed cores, the darker shaded rectangles correspond to samples studied from dredges and by submersibles, and the open rectangles indicate cores recovered but not yet analyzed. The numbers above each rectangle indicate the Ocean Drilling Program Hole or Site number (or the drill core designation on Troodos), and the numbers at the bottom of each section show the depth of each hole, e.g., ODP Hole 735B was drilled to a depth of 1500 m below the seafloor. SWIR—Southwest Indian Ridge, MARK—Mid-Atlantic Ridge and Kane Fracture Zone intersection, CRR—Costa Rica Rift.

to be the most ancient of extant hyperthermophilic organisms (Baross and Hoffman, 1985; Summit and Baross, 1998). The results from these and similar studies are directing scientific research to a profound new planetary paradigm—that the water-saturated outer brittle crust of volcanically active planets may harbor a substantial microbial biomass that is at least partially supported by the flux of volatiles from the mantle to the hydrosphere (Delaney et al., 1998).

This paper focuses on new insights into the evolution of volatiles and hydrothermal fluids in high-temperature plutonic and shallow-crustal systems in mid-ocean ridge environments. These insights have been gained largely from the successful recovery by the Ocean Drilling Program (ODP) of in situ rocks from the crystallized parts of submarine magma chambers and by examining tectonically exposed sections of oceanic crust. This paper explores the compositional differences in volatile chemistry and evolution that may be imposed by the spreading and tectonic environments in which fluid circulation occurs. Because of current database limitations, the discussion of volatile species focuses on the C-O-H-NaCl system. Finally, a discussion is presented summarizing recent results from programs that allow seagoing rapid response to submarine eruptions; these programs provide insight into the early release of volatiles during magmatic eruptions and the development of microorganisms within these spreading systems. The paper is not an exhaustive literature review, but presents a general overview of volatile evolution from supersolidus to hydrothermal vent conditions. From this work it is clear that our understanding of these dynamic systems must be viewed as preliminary; much work remains to fully characterize the distribution, composition (e.g., sulfur), and concentration of volatiles at depth in different tectonic settings and to evaluate the importance and magnitude of geologic and biologic interactions that affect the flux of volatiles in submarine environments.

SAMPLE LOCATIONS

The sample suites described in this study are from a diverse set of spreading environments (Figs. 1 and 2) and include variably altered and deformed gabbroic and intercalated trondhjemitic to dioritic rocks recovered from (1) the ultra-slow-spreading Southwest Indian Ridge; (2) near the eastern intersection of the Kane Fracture Zone and the Mid-Atlantic Ridge (the MARK area; this suite also includes diabasic rocks, quartz breccia, and metabasalt); (3) the Hess Deep—a rift propagating into 1 Ma crust formed at the fast-spreading East Pacific Rise, and (4) the Troodos ophiolite. The samples also include diabasic rocks recovered from ODP Hole 504B near the Costa Rica Rift; these rocks formed in a ridge environment spreading at intermediate to fast rates.

Southwest Indian Ridge

The Southwest Indian Ridge has been the site of intense dredging, submersible, and drilling studies. Drilling at Site 735B, located on a fault-bounded, wave-cut plutonic terrace ~93 km south of the present-day axis of the ridge and ~18 km inboard of the Atlantis II Fracture Zone, resulted in a 1.5-km-deep hole into the plutonic foundation of the crust with an 87% recovery (Figs. 1 and 2) (Shipboard Scientific Party, 1999). The ~1300 m plutonic section recovered during ODP Legs 118 and 176 is considered to represent crust formed at ca. 11–12 Ma beneath the median valley of the ultra-slow-spreading (1.6 cm/yr total rate) Southwest Indian Ridge and records a complex history of intrusion, brittle-ductile deformation, and penetration by hydrothermal fluids under granulite-facies metamorphic conditions (Dick et al., 1991). Intrusive relationships and the geochemistry of these rocks indicate that the crust in this region was built by small, discrete magma bodies modified by syntectonic differentiation involving migration of late iron-rich melts along shear zones (Bloomer et al., 1991; Dick et al., 1991; Meyer and Sapp, 1994). The iron oxide–rich zones commonly exhibit strong crystal-plastic and brittle-ductile deformation. Crosscutting veins and sills of microgabbros, which range from troctolite through olivine gabbro and oxide gabbronorite, are common throughout the sequence. These veins and sills may represent the periodic migration and intrusion of melts as the main gabbroic bodies cooled and solidified (Bloomer et al., 1991; Dick et al., 1991; Stakes et al., 1991; Meyer and Sapp, 1994). Discrete zones of trondhjemite veins and intrusion breccias represent the last phase of igneous activity. In the upper 500 m of the core, these zones are extensively replaced by secondary minerals formed under greenschist-facies metamorphic conditions and contain partially resorbed xenoliths of gabbro. Except for the trondhjemite veins and shear zones, much of the core exhibits little alteration, and below ~500 m there is a striking decrease in alteration down-section where much of the crust is altered <5%.

Mid-Atlantic Ridge–Kane Fracture Zone (MARK)

The MARK area (Fig. 1) is one of the best-characterized regions of the Mid-Atlantic Ridge (Cormier et al., 1984; Purdy and Detrick, 1986; Schulz et al., 1988; Detrick et al., 1984; Karson et al., 1987; Pockalny et al., 1988; Kong et al., 1988; Gao et al., 1998) and has been extensively sampled by dredging (Karson and Dick, 1983), submersible dives (Karson and Dick, 1983; Mével et al., 1991; Gente et al., 1991), and two ODP studies (Legs 109 and 153). The ridge-transform intersection is marked by a 6000-m-deep nodal basin. Westward of this basin, within a distance of 15 km, extreme uplift of the crust and mantle section has produced a lithologically complex massif that shallows to 1400 m depth and exposes ca. 500–900 ka crust (Karson and Dick, 1983; Gillis et al., 1993; Karson and Lawrence, 1997). Variably altered and deformed plutonic rocks collected from the MARK area range from olivine gabbro, to ferrogabbro, to trondhjemite (Gillis et al., 1993; Kelley et al., 1993; Kelley, 1997). Drilling at ODP Sites 921–924 successfully penetrated into tectonically exposed windows of deep to shallow-crustal rocks at the massif (Fig. 2). Intrusive relationships and the geochemical composition of these rocks indicate that the crust in this region was formed by a series of mafic intrusions that, similar to plutonic rocks from ODP Hole 735B cores, commonly show evidence of syn- to post-magmatic high-temperature deformation (Agar and Lloyd, 1997).

The plutonic rocks from the MARK area record a complicated alteration and deformational history, starting at temperatures in excess of 700 °C (Kelley and Delaney, 1987; Mével and Cannat, 1991; Gillis et al., 1993; Kelley et al., 1993; Kelley, 1997). In some of the metagabbroic rocks, this high-temperature deformation and recrystallization has been interpreted to represent ductile shear zones that may reflect initiation of faulting in the near-axis environment (Karson and Dick, 1983; Gillis et al., 1993; Mével et al., 1991; Karson and Lawrence, 1997). The relative timing and penetrative depth of these extensional features are not clear, however. The shear zones allowed enhanced penetration of fluids, which locally resulted in pervasive high-temperature alteration at temperatures of >500 °C (Kelley and Delaney, 1987; Mével and Cannat, 1991; Gillis et al., 1993; Kelley et al., 1993). The plutonic samples commonly have undergone brittle deformation in the form of microfractures to millimeter-wide fractures that are filled with lower-amphibolite-facies to greenschist-facies mineral assemblages. The degree of alteration within individual samples may be quite heterogeneous; the most pervasive alteration is adjacent to fractures. Quartz-cemented, sulfide-bearing breccias collected in this same area are interpreted to represent the 300 °C upflow conduits of deep-seated paleo–hydrothermal systems (Delaney et al., 1987; Kelley et al., 1993; Saccocia and Gillis; 1995). Metabasalts collected from this area are variably altered under greenschist- to zeolite-facies metamorphic conditions (Gillis et al., 1993).

East Pacific Rise–Hess Deep

Hess Deep is the deepest part of a rift valley that is propagating westward into the eastern flank of the East Pacific Rise ahead of the Cocos-Nazca spreading center. Hess Deep was drilled by ODP at Sites 894 and 895 during Leg 147 (Figs. 1 and 2). At a distance of some 70 km east of the East Pacific Rise axis, uplift attendant with propagation of the Cocos-Nazca spreading center has exposed a complete but structurally dissected crustal section of the East Pacific Rise including peridotite, gabbro, diabase, and basalt. Drilling at Site 894 in this area recovered ~50 m of gabbronorite, gabbro, olivine gabbro, olivine gabbronorite, and basalt, whereas at Site 895, ~9 km southeast of Site 894, a 64-m-thick discontinuous sequence of dunites and harzburgites with associated troctolites, gabbros, and basalts was recovered (Fig. 2). The mineral assemblages and textures of the gabbroic samples recovered from Site 894 indicate that the rocks crystallized predominantly from tholeiitic magma in a position immediately below the axial magma lens (Gillis, 1995; Natland and Dick, 1996; Pedersen et al., 1996). However, in isolated pockets, the chemistry and occurrence of accessory minerals such as apatite, zircon, oxide minerals, and some amphiboles are consistent with crystallization of these phases under relatively high $f(O_2)$ (oxygen fugacity) conditions and from a hydrous magma (Pedersen et al., 1996). The patches of micropegmatite represent the most evolved compositions sampled at Site 894 and may reflect re-equilibration with evolved, volatile-rich fluids that percolated through the gabbroic sequence late in the magmatic history (Pedersen et al., 1996; Kelley and Malpas, 1996). The complex suite of dunites, harzburgites, and gabbroic rocks recovered from Site 895 are thought to represent late-stage melt migration through residual mantle tectonites at the transition zone between the crust and mantle (Gillis, 1995; Dick and Natland, 1996).

At least 80% of the gabbroic rocks at site 894 are moderately altered to upper-amphibolite-facies to greenschist-facies mineral assemblages, whereas the shallow-mantle sequences at Site 895 predominantly record greenschist-facies metamorphic conditions (Früh-Green et al., 1996a, 1996b; Manning and MacLeod, 1996; Manning et al., 1996). In contrast to slow-spreading environments, there is an absence of ductile deformation at Hess Deep. With the exception of local cataclastic shear zones at shallow depths in the Site 894 gabbroic sequence, alteration is typically static and is reflected by pseudomorphic textures after primary igneous phases (Gillis, 1995; Früh-Green et al., 1996b). Metamorphism occurred through several stages of magma injection and hydrothermal alteration during progressive transport away from the axis of the East Pacific Rise and ultimate intersection with the Cocos-Nazca spreading center (Früh-Green et al., 1996a, 1996b; Kelley and Malpas, 1996; Manning and MacLeod, 1996; Manning et al., 1996; MacLeod et al., 1996).

Troodos ophiolite

The Troodos complex on the island of Cyprus has been taken as a type example for ophiolites, and studies of this complex and related ophiolites have affected many of the ideas concerning crustal evolution and the geometry of magma chambers at mid-ocean ridges (Fig. 1) (Casey and Karson, 1981; Pallister and Hopson, 1981; Malpas et al., 1989; Gillis and Robinson, 1990; Malpas, 1990). The ophiolite forms an elongate uplifted dome in which the plutonic rocks constitute the core of the complex. Mutually intrusive relationships between and within the intrusive suites provide strong evidence for the presence of multiple magma chambers beneath the Troodos paleo–spreading axis at ca. 91 Ma (Malpas, 1990). Drilling within the complex successfully penetrated 2263 m into the sheeted-dike, gabbroic, and ultramafic section (Fig. 1 and 2). In Troodos, the upper-level, generally undeformed, plutonic rocks are overlain by an extensive sheeted-dike complex that passes upward into the extrusive section, forming the flanks of the massif. The high-level gabbros are massive, coarse-grained rocks that commonly contain large pegmatitic patches. Plagiogranites and diorites occur as a series of discontinuous veins, screens, and cupolas that form part of a roof assemblage to the plutonic complex. Alteration within the plutonic rocks is extremely variable and heterogeneous. In the most highly altered samples, coarse-grained quartz and epidote almost completely replace primary minerals, resulting in epidosites, which are concentrated in shear zones or localized pods (Malpas, 1990; Kelley and Robinson, 1990; Kelley et al., 1992). Near the base of the sheeted-dike complex are localized zones of pervasive alteration in which the dikes have been completely transformed into assemblages of quartz + epidote ± chlorite (Richardson et al., 1987; Schiffman and Smith, 1988). These zones, characterized by extreme metal depletion, are believed to reflect intense high-temperature (300–400 °C) alteration at high water/rock ratios (Richardson et al., 1987; Schiffman and Smith, 1988) and have been interpreted as feeders for major conduits of hydrothermal up-flow which fed the higher-level stockwork and vent systems. Thus, the interface between the sheeted dikes and gabbro has been inferred to represent the base of the seawater-dominated hydrothermal systems (Richardson et al., 1987; Richards and Cann, 1989).

Costa Rica Rift, Hole 504B

Ocean Drilling Program Hole 504B, located 200 km south of the Costa Rica Rift, extends to a depth of 2111 m into 5.9 Ma crust formed in a ridge environment spreading at intermediate to fast rates (Fig. 1). The hole—which penetrates 274.5 m of sediments, 571.5 m of pillow lavas and minor flows, a 209 m transition zone of mixed pillow and massive lavas and dikes, and 1056 m of dikes—is believed to penetrate into crustal rocks near the boundary between seismic layers 2 and 3 (Fig. 2). The basaltic rocks recovered are geochemically homogeneous olivine- to slightly quartz-normative tholeiites and include aphyric basalts and sparsely phyric to moderately phyric basalts, which contain variable amounts of olivine, clinopyroxene, and plagioclase as phenocryst phases. Hydrothermal alteration in the volcanic and diabasic sequence was variable and complex, involving multiple pulses that included downwelling of cold seawater, mixing of hydrothermal fluids with cooler seawater, and at least two high-temperature stages involving 350–380 °C fluids (Alt et al., 1986, 1989; Kelley et al., 1995; Alt et al., 1996; Gu and Vanko, 1996).

FLUID INCLUSIONS

The evolution of volatiles during the degassing and crystallization history of mid-ocean ridge magma chambers is poorly characterized because (1) few experimental data exist on fluids other than CO_2 and H_2O in mafic systems; (2) most compositional data come from analyses of submarine basaltic glasses that are not generally highly evolved, and (3) sampling the deep ocean crust is inherently difficult. However, by investigating tectonically exposed sections of the lithosphere through drilling, submersible, and ophiolite studies and attendant examination of small capsules of fluids (fluid inclusions) entrapped during mineral growth and fracturing events, new insights into the geochemical history and the distribution of volatiles in the near solidus to subsolidus environment have been forthcoming. Fluid inclusions provide the only opportunity to directly measure the concentrations and evolution of volatile species in magmatic and seawater-derived fluids at depth. In addition, they generally provide the most accurate geothermometric and geobarometric information on these systems because minerals that form submarine plutonic rocks typically record the integrated effects of multiple magmatic and hydrothermal pulses, as evidenced by overprinting secondary mineral phases and the common development of disequilibrium textures. By analyzing multiple populations of inclusions entrapped in magmatic and successive generations of alteration mineral phases in host, enclave, and veined samples, a history of the spatial, temporal, and chemical evolution of these systems may be developed. Because inclusions are unique in the type of information they can yield, they have been used extensively in submarine research to examine fluid flow in deep and shallow hydrothermal systems (Table 1). Care must be taken when doing these types of analyses, however, as it may be difficult to determine an unequivocal origin of the inclusions. Inclusions of unquestionable primary origin are those that form arrays that mimic the habit of the mineral host, whereas those that are clearly secondary in origin form linear arrays that are not aligned with mineral faces (Roedder, 1984). In samples that have been affected by multiple hydrothermal pulses, inclusions may be so abundant that it is impossible to discern either a primary or secondary origin. Freezing-point depression and halite-dissolution measurements provide information on fluid salinities and are expressed as weight percent NaCl equivalent. Because of the low K contents in many of these systems and because many submarine magma and hydrothermal fluid systems evolve to more Na-rich compositions, NaCl equivalent is a reasonable approximation for fluid salinities. Low eutectic temperatures measured in inclusions from the Oceanographer Fracture Zone on the Mid-Atlantic Ridge, however, indicate that some fluids may contain significant amounts of Ca (Vanko et al., 1992). In the following discussion, a subset of fluid-inclusion data is presented that provides an overview of our current understanding of deep-seated volatiles in mid-ocean ridge environments.

EVOLUTION OF MAGMATIC FLUIDS

C-O-H-S *speciation*

Few data exist on the speciation of C-O-H-S fluids in mafic systems during degassing reactions associated with melt evolution, but speciation depends on the relative solubilities and abundances of the gases, on the oxidation state of the melt, and on the pressure and temperature of degassing. Analyses of submarine basaltic glasses, and experimental data on CO_2 and H_2O in mafic systems indicate that owing to the low solubility of CO_2 in basaltic melts, mid-oceanic-

TABLE 1. FLUID INCLUSION SUMMARY

Location and spreading rate*	Rock type	Mineral†	P or S	T_h (°C)	NaCl equiv. (%)	T_m	Composition
Troodos-Cyprus—slow or arc? (6)	Plagiogranite-gabbro	Plag	?	183–264 (7)	1.8–2.4		NaCl–H_2O
		Quartz	P-S	302–537 (143)	35–61		NaCl + H_2O ± KCl
		Quartz	P-S	184–390 (222)	1.7–17.2		NaCl + H_2O
		Cpx†	?	256–394 (12)	2.4–4.6		NaCl + H_2O
	Epidosite	Quartz	?	305–411 (55)	35–47		NaCl + H_2O
		Quartz	?	255–402 (68)	1.1–17.2		NaCl + H_2O
		Epidote		391–413 (19)	2.2–4.7		NaCl + H_2O
	Diabase	Quartz	?	260–377 (50)	2.0–5.4		NaCl + H_2O
Semail—fast? (5)	Plagiogranites to diorites dikes + veins	Quartz	P-S	106–436 (142)	1.7–8.9		NaCl + H_2O
		Epidote	P	284–388 (25)	2.9–4.0		NaCl + H_2O
		Quartz	P-S	224–376 (78)	1.4–13.1		NaCl + H_2O
		Epidote	P	345–402 (38)	2.7–4.0		NaCl + H_2O
	Extrusive rocks	Quartz	P-S	128–409 (221)	1.4–10.1		NaCl + H_2O
MARK—slow (2), (3), (7), (9), (13)	Gabbro-grabbronorite-oxide gabbro	Apatite	P-S	>700* (20)	42–47		± CO_2
		Apatite	P-S	364–398 (29)	0.9–2.6		± CO_2
		Apatite	P-S			−56.6 (25)	CO_2
		Apatite	S	245–296 (6)	4.2–4.7		NaCl + H_2O
		Plag	S	240–349 (270)	0.4–7.6		NaCl + H_2O
		Epidote	P-S	381–435 (9)	4.4–5.9		NaCl + H_2O
		Cpx	P	387–433 (38)	3.5–6.5		NaCl + H_2O
		Cpx	?	251–326 (8)	7.3–8.9		NaCl + H_2O
	Trondhjemite	Quartz	S?	305–330 (30)	3.5–7.6		NaCl + H_2O
		Quartz	?	262–340 (18)	35–41		NaCl + H_2O
	Trondhjemite veins in gabbro	Quartz	P-S	295–495 (69)	39–43		NaCl + H_2O
		Quartz	P-S	219–400 (148)	1.3–20.2		NaCl + H_2O
		Plag	S	273–320 (29)		+0.2 to +0.6 8	CH_4 + H_2O
	Oxide gabbro	Apatite	P	22 (20)		−56.6	CO_2
	Olivine gabbro	Apatite	P-S	301–378 (10)		+4.0 to +8.2	CO_2 + H_2O ± CH_2
		Apatite					
		Plag	S	349–365 (25)		+9.4 to +10.7	CH_4 + H_2O
	Breccias	Quartz	P-S	161–343 (205)	3.5–6.5		NaCl + H_2O
	Metabasalt	Plag	?	266–303 (15)	3.8–4.8		NaCl + H_2O
		Quartz		174–313 (34)	4.6–6.9		NaCl + H_2O
Hess Deep—fast (9), (14)	Gabbronorites	Plag	P-S	209–289 (120)	0.8–13.4		NaCl + H_2O
		Plag	S	248–308 (11)	34–39		NaCl + H_2O
		Apatite	?	181–347 (22)	29–33		NaCl + H_2O
		Apatite	S	216–372 (20)	0.2–19.3	up to +9.9	NaCl + H_2O ± CO_2
	Breccias	Quartz	P-S	162–370 (193)	0.0–4.8		NaCl + H_2O
		Epidote	P-S	173–353 (15)	2.4–3.4		
SWIR—ultra-slow (11), (14)	Olivine gabbro-gabbro-gabbronorite	Apatite	P-S	330–373 (29)		+3.5 to + 7.7	CO_2 + H_20
		Apatite	S	330–395 (56)	1.4–4.5		NaCl + H_2O
		Plag	S	234–292 (50)	0.3–3.2		NaCl + H_2O
		Plag	S	211–370 (150)		+1.6 to +15.5	CH_4 + H_2O
Hole 504B—intermediate-fast (1), (4), (7), (10)	Stockwork	Quartz	P	138–452			
		Calcite	P	203–215			
	Diabase	Plag	P-S	125–202 (460)	0.0–11.7		NaCl + H_2O

Note: P or S—Primary or secondary fluid inclusion; ?—classification unclear; T_h—homogenization temperature; NaCl equiv.—equivalent fluid salinities obtained from freezing measurements; T_m—ice melting temperature of CO_2 or clathrate melting temperature for CH_4 + H_2O inclusions.
* Numbers below locations refer to reference used: (1) Honnorez et al., 1985; (2) Kelley and Delaney, 1987; (3) Delaney et al., 1987; (4) Schops and Herzig, 1990; (5) Nehlig, 1991; (6) Kelley et al., 1992; (7) Kelley et al., 1993; (8) Kelley et al., 1995; (9) Saccocia and Gillis, 1995; (10) Gu and Vanko, 1996; (11) Kelley, 1996; (12) Kelley and Malpas, 1996; (13) Kelley, 1997; (14) Kelley and Früh-Green, 1999.
† Plag—plagioclase, Cpx—clinopyroxene.

ridge basalts (MORBs) are saturated or oversaturated with respect to CO_2 and have typical concentrations of <400 ppm (Delaney et al., 1978; Mathez and Delaney, 1981; Pineau and Javoy, 1983; Fine and Stolper, 1986; Dixon et al., 1988; Stolper and Holloway, 1988; Pawley et al., 1992; Dixon and Stolper, 1995). In contrast, the incompatible nature of H_2O, coupled with its low concentration, means that basalts are generally undersaturated with respect to H_2O, with concentrations in glasses of typically <5000 ppm (Moore, 1977; Delaney et al., 1978; Dixon et al., 1988, 1995).

At deeper levels in the crust, the compositional evolution of fluids within a crystallizing magma chamber has not been well studied, but compositions are likely to vary significantly depending on the T vs. $f(O_2)$ (temperature vs. oxygen fugacity) path and whether minerals such as graphite precipitate (Gerlach and Nordlie, 1975; Mathez, 1989a, 1989b; Mathez et al., 1989). Characterization of the $f(O_2)$ values of mafic magmas shows that some early basalts may reflect relatively reduced conditions near the iron-wüstite (IW) buffer (IW to MW [magnetite-wüstite]) (Sato and Valenza, 1980; Christie et al., 1986). Under these conditions, modeling suggests that the plutonic rock–hosted fluids are dominated by CO_2, H_2O, and CO and reduced species such as H_2S, H_2, and CH_4 may be stable (Mathez, 1984). Progressive degassing of fluids in a shallow-crustal environ-

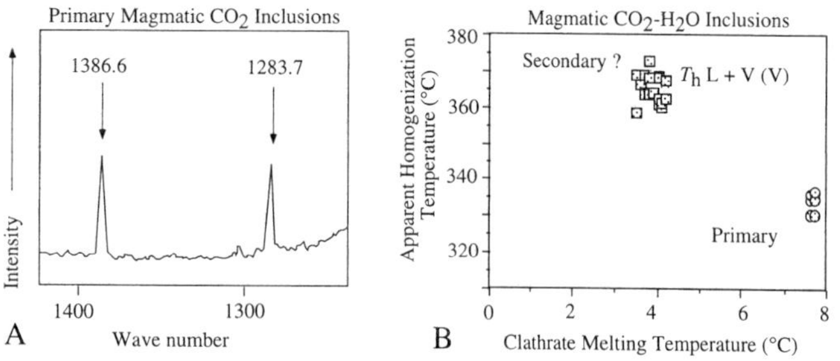

Figure 3. A, Micro-Raman analyses of a primary, apatite-hosted fluid inclusion in a gabbroic sample from MARK (Mid-Atlantic Ridge and Kane Fracture Zone intersection). Peaks at 1386.6 and 1283.7 indicate that the inclusion contains CO_2. B, Corresponding clathrate melting temperatures and apparent homogenization temperatures (T_h) for CO_2-rich primary and secondary inclusions in a gabbroic sample from the Southwest Indian Ridge. A solid melting event at −56.6 °C (not shown) and clathrate formation indicate that the inclusions are dominated by CO_2 and H_2O. The CO_2 inclusions represent the earliest generation of fluids commonly trapped in submarine plutonic samples.

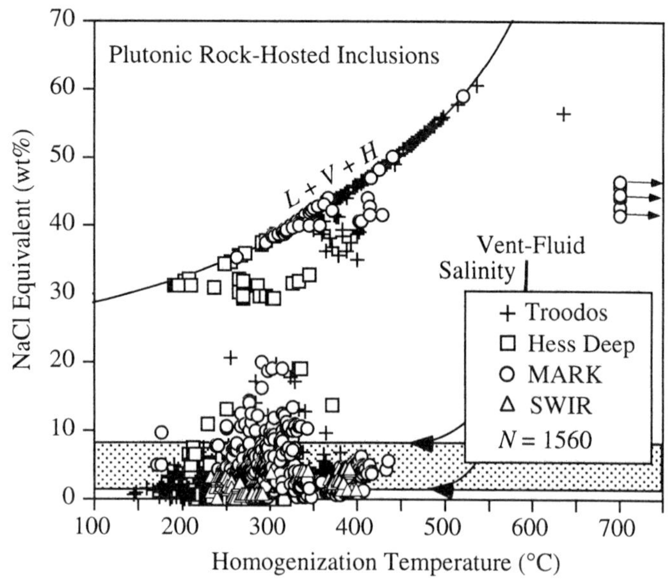

Figure 4. Corresponding homogenization temperatures and equivalent fluid salinities of plutonic rock–hosted fluid inclusions in submarine environments. Primary and secondary inclusions in gabbroic to trondhjemitic rocks from systems spreading at diverse rates exhibit similar trends in fluid evolution. The inclusions that fall along the liquid + vapor + halite trend (L + V + H) homogenize by halite dissolution and were likely entrapped in the absence of a vapor phase. These inclusions typically occur in the most evolved samples and in late-stage felsic veins. Primary, high-salinity, apatite-hosted inclusions remain unhomogenized at temperatures of 700 °C (arrows) and coexist with vapor-dominated $H_2O \pm CO_2$ vapor-rich inclusions with 1–2 wt% NaCl equivalent salinities. These relationships indicate that the inclusions trapped immiscible magmatic brines and vapors formed during supercritical phase separation at pressures of 1–1.2 kbar and at temperatures greater than 700 °C. The origin of ~25–45 wt% NaCl equivalent brines that homogenize by vapor-bubble disappearance (those inclusions just to the right of the L + V + H curve) is equivocal. They commonly occur in the most evolved plutonic samples and are believed to represent conductively cooled magmatic brines. Also shown are compositions of lower-salinity, seawater-derived inclusions. MARK—Mid-Atlantic Ridge and Kane Fracture Zone intersection, SWIR—Southwest Indian Ridge.

ment is likely to drive the system to more oxidizing conditions near or above the quartz-fayalite-magnetite (QFM) buffer where CO_2, H_2O, and SO_2 dominate (Gerlach and Nordlie, 1975; Mathez, 1984).

Within the plutonic foundation of the crust, CO_2-rich inclusions (Fig. 3) typically are the earliest generation of fluid preserved, consistent with solubility data and modeling of basaltic systems (Kelley et al., 1993; Kelley, 1996; Kelley and Früh-Green, 1999). Such inclusions are commonly entrapped in olivine and apatite during mineral growth and along healed microfractures during subsequent fracturing events. However, very few measurements of magmatic CO_2 have been made of inclusions in gabbroic and late-stage felsic differentiates (Table 1).

Fluids bearing $CO_2 + H_2O$ + NaCl

In many oceanic gabbros and more felsic rocks, CO_2 inclusions show a continuum between CO_2-rich and CO_2-poor, H_2O-rich inclusions, which occur in apatite, quartz, and more rarely plagioclase (Table 1). The ($H_2O \pm CO_2$)–rich inclusions are vapor-rich and are associated with brines containing 40–50 wt% NaCl ± $CaCl_2$ ± KCl (Fig. 4) that commonly contain daughter phases of chalcopyrite and or pyrite (Kelley and Delaney, 1987; Vanko, 1988; Vanko et al., 1992; Kelley et al., 1993; Kelley and Malpas, 1996; Kelley, 1997). In the MARK gabbros, homogenization temperatures indicate that the magmatic H_2O-rich brines were entrapped at temperatures in excess of 700 °C in association with a low-salinity $H_2O \pm CO_2$ vapor phase (Fig. 4) (Kelley and Delaney, 1987; Kelley et al., 1993). The occurrence of high-salinity inclusions with sulfide daughter phases demonstrates that the brines were capable of dissolving relatively high concentrations of metal and sulfur. Volumetric estimates of the sulfide minerals indicate that Fe and Cu concentrations in the brines approach 7000 ppm and 8000 ppm, respectively (Kelley and Delaney, 1987). The brine and vapor inclusions are remarkably similar to those found in porphyry copper systems (Gustafson and Hunt, 1975; Bodnar and Beane, 1980; Reynolds and Beane, 1985; Cline and Bodnar, 1991).

Phase separation of magmatic fluids

Conjugate brine and vapor-rich fluids in oceanic plutonic rocks most commonly occur in micropegmatitic to pegmatitic pods and in Fe-Ti oxide gabbros that contain abundant trace-mineral phases such as zircon, apatite, and magnetite (Kelley and Delaney, 1987; Kelley and Malpas, 1996; Kelley, 1997) and more rarely in quartz veins (Vanko, 1988; Nehlig, 1991; Vanko et al., 1992). The coexistence of vapors and brines as primary inclusions and as inclusions along healed microfractures requires that they were exsolved in the two-phase field under immiscible conditions (Fig. 5). Under shallow-crustal conditions, compositionally evolved interstitial melts, as represented by the pegmatitic pods, may become saturated with respect to water when ~5%–7% residual melt remains (Burnham, 1979). Experimental and theoretical considerations indicate that such conditions are probably achieved at temperatures of ~700–800 °C (Wyllie, 1977; Burnham, 1979), consistent with inclusion homogenization temperatures (Kelley and Delaney, 1987). The salinity of the exsolved aqueous phase is controlled by the Cl content of the melt and the partition coefficient between the melt and the volatile phase (Holland, 1972; Kilinc and Burnham, 1972; Cline and Bodnar, 1991). Under conditions corresponding to the two-phase field, fluids will exsolve as a highly expanded vapor phase and fine droplets of brine (Fig. 5). Compositions of the vapors and brines formed during supercritical

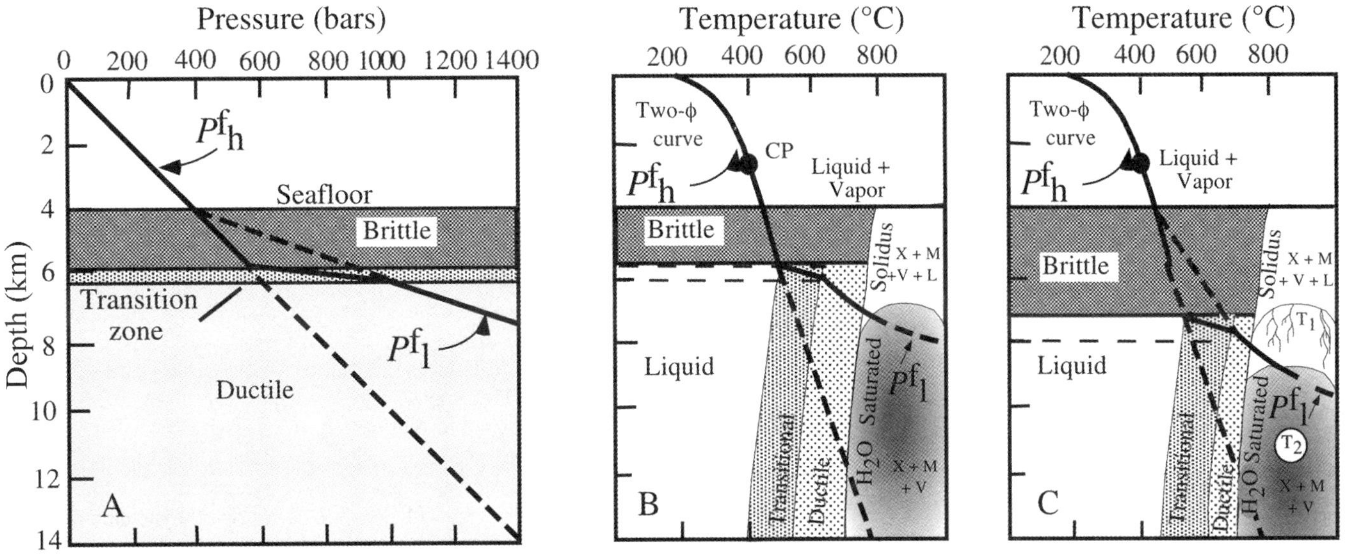

Figure 5. A, Schematic relationship between pore-fluid pressure and depth as a function of the deformational regime within the oceanic crust (after Kelley and Delaney, 1987). The upper part of the crust is depicted as brittle (i.e., well fractured; dark shading); the lower part (on axis) is represented as being partially molten or too ductile to sustain significant stress (light shading). The two regions are separated by a transition zone (medium shading) in which fracturing begins. The line with the 45° slope labeled P^f_h represents the pressure vs. depth curve for hydrostatic conditions. The more gently sloping line labeled P^f_l represents pressure vs. depth relationships for lithostatic conditions assuming a seafloor depth of 4.0 km and a rock density of 3.0 g/cm^3. The gently sloped pressure vs. depth curve connecting the two idealized hydrostatic and lithostatic curves represents the pore-fluid pressure gradient expected in the vicinity of the transition zone. B, Generalized two-phase (ϕ), liquid + vapor curve at various temperatures for a dilute saline solution under pressure-depth conditions identified in A. The part of the curve labeled P^f_h is the position in *P-T* space of the two-phase curve under hydrostatic conditions. The part of the curve labeled P^f_l is the position of the two-phase curve under lithostatic conditions. At temperatures and pressures less than the critical point (CP), circulating fluids that intersect the two-phase curve will undergo boiling. At temperatures and pressures greater than the critical point, fluids in the two-phase region will separate into a vapor and a brine. The gentle slope between the end-member curves reflects the pore-fluid pressure gradient expected in the transition zone between the brittle and ductile deformational regimes. In this idealized model, at depths of <8 km and at temperatures greater than the solidus, exsolved magmatic fluids would be composed of coexisting vapors and brines [e.g., crystals (X) + melt (M) + vapor (V) + liquid (L)]. Under higher-pressure conditions, the fluids would be exsolved as a single phase. C, As the brittle-ductile boundary shifts with cooling of a pluton, the cracking front may progressively propagate downward. If these cracks intersect pore fluids trapped under lithostatic conditions, phase separation may result.

phase separation (condensation) are dependent on the temperature and pressure at which immiscibility occurs and are independent of the parental fluid's salinity (see Fournier, 1987, for a more detailed discussion). Experimental work by Frantz et al. (1992), which examined the compositional limits of immiscibility in the $CO_2 + H_2O +$ NaCl system from 500 °C to 700 °C at pressures of 1–3 kbar, shows that exsolution of late-stage volatiles under immiscible conditions should be a common phenomenon in shallow submarine magma chambers.

Exsolution of volatiles during late-stage solidification of hydrous melt may have a profound effect on the magma and hydrothermal fluid system and subsequent degassing behavior (Fig. 5). In these high-temperature environments, the position of the two-phase field will be determined by the composition of the exsolving fluid and by the pressure gradient within the magma-chamber margins (Fig. 5), which is determined by the fracture-permeability distribution and by the magnitude of volatile exsolution (Kelley and Delaney, 1987; Fournier, 1987). In any active magma and hydrothermal fluid system, the position of the brittle-ductile boundary will not be stable, but will fluctuate with migration of the crystallization front and with changes in volatile exsolution (Fig. 5) (Fournier, 1987). In a degassing system, volatile buildup may overpressure the chamber carapace, resulting in hydraulic fracturing of the roof material (Knapp and Knight, 1977; Sparks et al., 1994). A downward shift in the brittle region due to fracturing events either within the carapace or during faulting of the host rock will result in substantial decompression of fluids within the chamber, which are at lithostatic pressure (Fig. 5C). If the initial temperature and pressure conditions of the fluid are in the one-phase field, the decompression from local fracturing will likely induce two-phase separation (Kelley and Delaney, 1987; Kelley, 1997). Because of the large positive volume change of fluids attendant with condensation, this process may lead to a cascading sequence of enhanced fracturing and volatile exsolution (Burnham, 1979; Kelley and Delaney, 1987; Kelley et al., 1992). Pervasively fractured wall rocks above intrusions in the Troodos ophiolite may attest to this process (Kelley et al., 1992). In the MARK gabbros, the coexistence of 40–50 wt% brines in association with 1–2 wt% NaCl vapors that were entrapped at temperatures of >700 °C yields pressures of ~1.0–1.2 kbar and requires minimum formation depths of 2–3 km below the seafloor (Kelley and Delaney, 1987; Kelley et al., 1993).

Fluids bearing NaCl + H_2O

Although formation of brines by condensation is widely regarded as the dominant mechanism for generating the extreme salinity variations of some fluids observed in submarine plutonic rocks and in vent fluids (Stakes and Vanko, 1986; Delaney et al., 1987; Kelley and Delaney, 1987; Vanko, 1988; Bischoff and Rosenbauer, 1989; Butterfield et al., 1990, 1994; Nehlig, 1991; Von Damm et al., 1992, 1997), accumulating evidence suggests that the exsolution of brines directly from late-stage melts may also play a significant role in for-

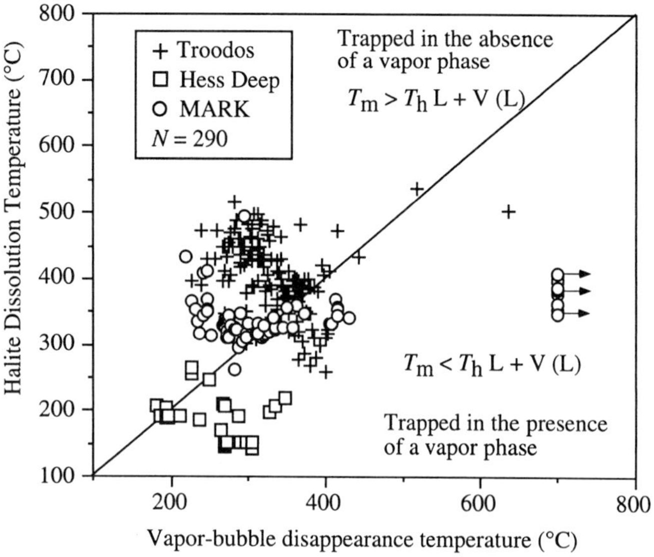

Figure 6. Corresponding vapor-bubble disappearance temperatures and halite dissolution temperatures for halite-bearing inclusions in gabbroic and trondhjemitic rocks. Fluids that homogenize by halite dissolution [$T_m > T_h$, L + V (L)] were entrapped in the absence of a vapor phase. Inclusions that homogenize by vapor-bubble disappearance [$T_m < T_h$, L + V (L)] were formed under supercritical conditions in the presence of a vapor. T_m = temperature of halite dissolution; T_h = homogenization temperature; L = liquid; V = vapor; N = number of inclusions measured; MARK—Mid-Atlantic Ridge and Kone Fracture Zone intersection.

mation of plutonic rock–hosted fluids. In highly evolved and H_2O-rich systems, such as those in backarc spreading environments and Troodos, the latest fluids may involve direct exsolution of brines formed in the absence of a vapor phase (Figs. 4 and 6) (Cline and Bodnar, 1991; Kelley et al., 1993; Kelley, 1997). At pressures of 0.5–1.0 kbar, exsolving fluids are incapable of removing substantial amounts of Cl from the melt; therefore, Cl concentrations in the melt increase as crystallization proceeds (Cline and Bodnar, 1991). Assuming isothermal crystallization of silicic melts at 700 °C and an initial Cl/H_2O ratio of 0.1, fluid salinities evolve from 1–7 wt% NaCl, as the first fluids exsolve, to 80–90 wt% NaCl, at high degrees of crystallization (Cline and Bodnar, 1991). Because of the partitioning of H_2O and Cl in the melt, at pressures of 2 kbar or greater, this evolution is reversed, and fluids evolve from highly saline brine to low-salinity fluids. The exact composition of these fluids depends on the initial volatile and Cl contents of the melts, as well as on the T vs. $f(O_2)$ path, but the general trend in fluid composition should also hold true for late-stage magmatic fluids in submarine settings. From these data, it may be predicted that the progressive degassing of fluids in submarine shallow-crustal environments will result in the following general compositional evolution with decreasing temperature: (1) CO_2, (2) $(H_2O \pm CO_2)_{vapor}$ + $(H_2O \pm CO_2 + \text{brine})_{liquid}$, and (3) brine.

This model of direct brine exsolution from late-stage melts is supported by the ubiquitous occurrence of halite-bearing inclusions, which are not associated with vapor-rich inclusions and which consistently occur in the compositionally most evolved submarine plutonic suites. Halite-bearing inclusions are especially abundant in trondhjemitic to dioritic rocks from the Troodos ophiolite (Figs. 4 and 6) (Kelley et al., 1992, 1993; Nehlig, 1991; Kelley, 1997). These inclusions have salinities of ~30–60 wt% NaCl equivalent and were entrapped at minimum temperatures of 200–500 °C; these data indicate that the circulating magmatic fluids most likely underwent significant cooling prior to inclusion entrapment (Figs. 4 and 6) (Kelley, 1997).

In many of these systems, however, it is not possible to determine whether the brine inclusions represent magmatic fluids that were exsolved under conditions corresponding to the two- or one-phase field. As previously discussed, in an evolving magma and hydrothermal fluid system, rapid stress changes and thermal pulses associated with magma emplacement, volatile exsolution, and spreading likely induce episodic fracture propagation in the crystalline host rocks (Fig. 5). Fluctuations in the brittle-ductile boundary and progressive phase-separation events can dramatically alter pore-fluid compositions. For example, a cooled, *phase-segregated* magmatic fluid at 600 °C and 800 bars pressure would exist as a 40 wt% brine and a 2 wt% NaCl vapor. If these trapped pore fluids were isothermally decompressed to 650 bars during a fracturing event, the brine phase would separate into a brine of ~52 wt% NaCl and a vapor of ~9 wt% NaCl; the vapor phase would separate into a brine of ~52 wt% NaCl and a vapor of ~0.45 wt% NaCl. In a submarine hydrothermal environment, periodic intrusions of melt may result in reheating and phase separation of pore fluids, processes that would further increase the brine content of the residual fluids. Pressure decreases associated with these fracturing events may promote formation of brines by supercritical phase separation, whereas more quiescent periods may be associated with direct exsolution of brines (Fig. 5) (Kelley et al., 1993; Kelley, 1997). Preservation of brines in the absence of a vapor phase may reflect either of these two processes. During fluid migration along anastomosing systems of microfracture networks in the gabbroic rocks, density stratification of the brine and vapor phases (Fournier, 1987; Bischoff and Rosenbauer, 1989; Fox, 1990), coupled with preferential wetting of the fracture walls with brines (Goldfarb and Delaney, 1988), may allow preferential entrapment of the brines at depth as the vapors migrate to shallower-crustal levels. Whatever the mechanism of brine formation is in hydrothermal systems, the general restriction of halite-bearing inclusions to primary apatite and quartz crystals within micropegmatitic patches, compositionally evolved veins, and upper-level intrusions indicates that percolation of late-stage magmatic brines is common to all spreading environments. It should be noted, however, that due to the incompatible nature and low concentrations of H_2O in MORB melts, it is unlikely that brines generated either by supercritical phase separation of magmatic fluids or by direct exsolution are volumetrically significant in most mid-ocean ridge systems.

Fluids bearing $CO_2 + CH_4 + H_2O$

Recent work shows that in slow-spreading environments, the evolution of magmatic fluids may follow a dramatically different evolutionary path than that already described by the sequence (1) CO_2, (2) $(H_2O \pm CO_2)_{vapor}$ + $(H_2O \pm CO_2 + \text{brine})_{liquid}$, and (3) brine. Detailed micro-Raman, microthermometric, isotopic, and mass spectrometry analyses of fluids in the upper 500 m of gabbroic rocks from the Southwest Indian Ridge, ODP Hole 735B, indicate that the magmatic fluids are rich in CO_2, CH_4, and H_2O with CH_4 concentrations of up to 50 mol% (Kelley, 1996; Kelley and Früh-Green, 1999) (Figs. 7 and 8). The $(CO_2 + CH_4 + H_2O)$–rich fluids are common in olivine and plagioclase throughout the plutonic suite and are believed to reflect respeciation of magmatic CO_2 during cooling and attendant graphite precipitation at ~500–800 °C and $f(O_2)$ values from −3 log units below to close to QFM conditions. These fluids show average $\delta^{13}C$ (CO_2) values of −7.6‰ ± 6.1‰ and a corresponding general decrease in CO_2 concentrations (7–81 ppm),

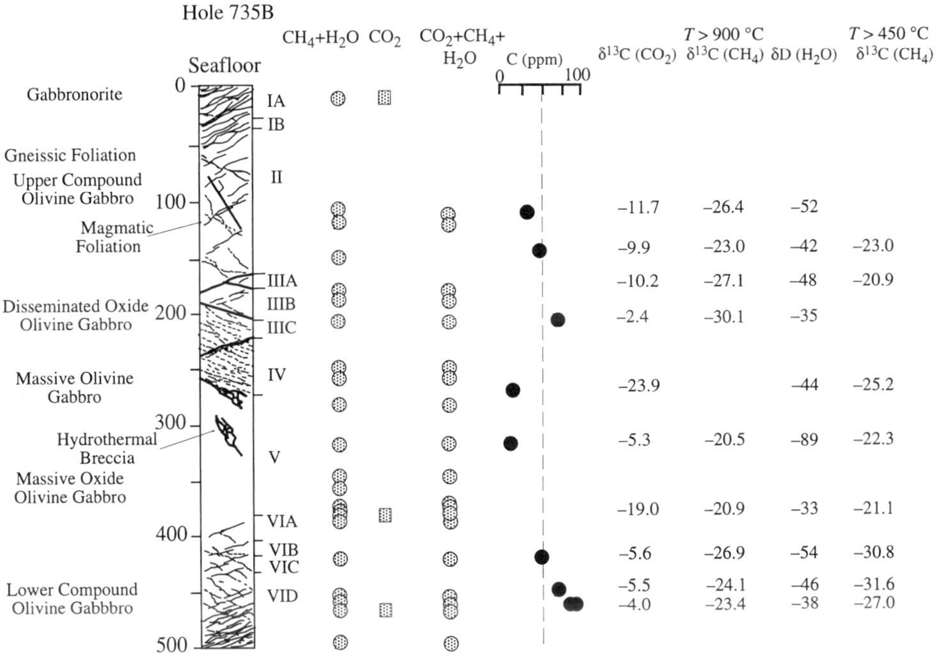

Figure 7. Location of C-O-H–bearing fluid samples measured by microthermometry, micro-Raman, mass spectrometric, and isotopic analyses (light gray circles and squares) in relationship to the stratigraphic plutonic and deformational relationships observed in core 735B (section and units designated by roman numerals are after Dick et al., 1991). CH_4-rich fluid inclusions are entrapped along healed microfractures that crosscut all lithologic units. Also shown are total C concentrations (black symbols) and average $\delta^{13}C$ and δD values downhole. The average isotope values are shown for (CO_2 + CH_4 + H_2O)–rich fluids released during step-heating experiments at temperatures of >900 °C [$\delta^{13}C$ (CO_2), $\delta^{13}C$ (CH_4), and δD (H_2O)] and for CH_4 + H_2O fluids released at temperatures >450 °C [$\delta^{13}C$ (CH_4)].

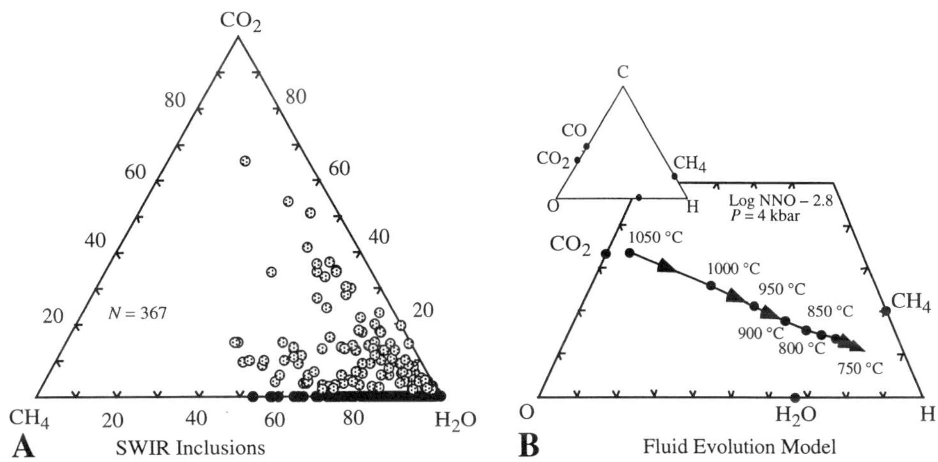

Figure 8. Fluid-inclusion data and evolution model for the fluids in the Southwest Indian Ridge plutonic suite. A, Calculated mole fractions of fluid inclusions in olivine and plagioclase mineral separates from mass spectrometric analyses (after Kelley, 1996). Fluids released at temperatures of ~300–500 °C are characterized by variable amounts of CH_4 + H_2O + H_2 ± C (black dots); those released at 600–1000 °C contain (CO_2 + CH_4 + H_2O)–rich fluids (light gray dots). N = number of inclusions measured. B, Magmatic fluid evolution model (after Mathez et al., 1989) in the C-O-H system at 4 kbar. The fluid evolution path represents the trace with cooling of the intersection of the graphite stability curve and the isopleths of log $f(O_2)$ = log NNO − 2.8. Because of the shallow intersection of the graphite stability curve with oxygen isopleths in the C-O-H system, vapors that are initially dominated by CO_2 + CO (as would be expected in a MORB system) rapidly evolve to more (CH_4 + H_2O)–rich compositions if cooling is accompanied by graphite precipitation. Geochemical, petrologic, and structural studies of the 735B plutonic suites (e.g., Dick et al., 1991; Stakes et al., 1991; Kelley, 1996) record a remarkably complex history of progressive melt fractionation, devolatilization, melt-crystal-vapor interaction, and synmagmatic deformation during cooling through the high-temperature subsolidus regime. We believe that the (CO_2 + CH_4 + H_2O)–rich fluids formed during cooling and condensation of CO_2-rich magmatic gases, in conjunction with graphite precipitation, as the plutonic sequence cooled and solidified.

consistent with formation during degassing of magmatic fluids (Figs. 7–9) (Kelley and Früh-Green, 1999). The $\delta^{13}C$ (CO_2) values, however, are too positive (Fig. 9) to be explained solely by degassing reactions and require that respeciation processes occurred during cooling of the fluids. Calculations based on C isotope fractionations between CO_2, CH_4, and graphite predict that initial Rayleigh distillation of evolved magmatic fluids and subsequent closed-system respeciation attendant with graphite precipitation during cooling of the fluids through the subsolidus regime can produce CO_2 and CH_4 with $\delta^{13}C$ values similar to those in the Southwest Indian Ridge fluids (Kelley and Früh-Green, 1999). This model is also consistent with the hydrogen isotope data, which show an enrichment in deuterium [δD (H_2O) = −33‰ to −89‰] relative to compositions considered to represent primary δD values of magmatic water in MORB melts (–80‰ to −65‰ e.g., Craig and Lupton, 1976; Kyser and O'Neil, 1984; Pineau and Javoy, 1994). Preliminary measurements of gab-

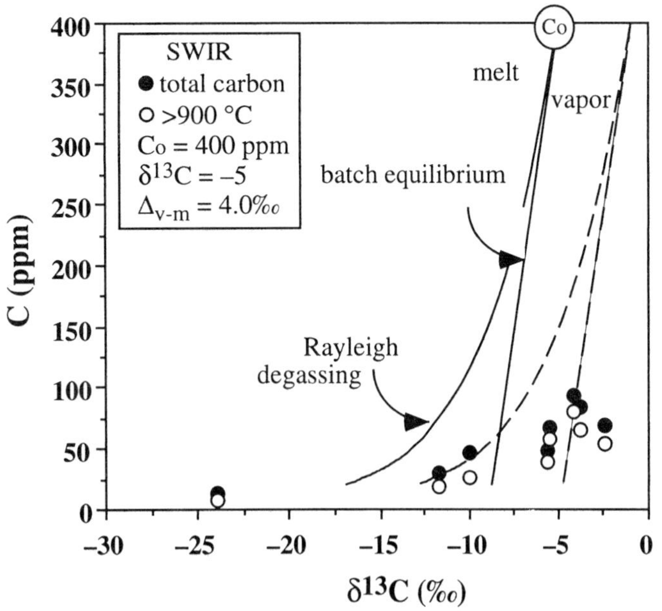

Figure 9. Southwest Indian Ridge (SWIR) data on C concentrations (ppm) vs. $\delta^{13}C$ (‰) values for C released at temperatures of >900 °C. Open circles indicate concentrations of CO_2 released at >900 °C; filled circles show values for the total C measured from the plutonic samples during both the 500 °C and >900 °C intervals. Curves indicate the calculated degassing paths for open-system Rayleigh distillation and closed-system batch devolatilization, under the assumptions of an initial C concentration (C_o) of 400 ppm, an initial $\delta^{13}C$ value of −5‰, and a vapor-melt fractionation factor ($\Delta_{v\text{-}m}$) of 4.0‰. The initial C concentration is unknown, although whole-rock measurements indicate values of 150–400 ppm (Alt and Anderson, 1991). This is a minimum value as the melts most likely have undergone significant degassing prior to solidification. The solid lines indicate the concentration and $\delta^{13}C$ value of C dissolved in the melt; dashed lines show the $\delta^{13}C$ of CO_2 in equilibrium with a given C concentration of the melt. In general, the $\delta^{13}C$ (in CO_2) decreases with decreasing C concentration and is roughly parallel to the Rayleigh distillation vapor curve, indicative of varying degrees of degassing. The values are more enriched in ^{13}C than would be predicted by this model, consistent with graphite formation during cooling. The maximum total C contents of less than 100 ppm suggests that the mafic magmas had exsolved at least 75%–95% of their original volatiles.

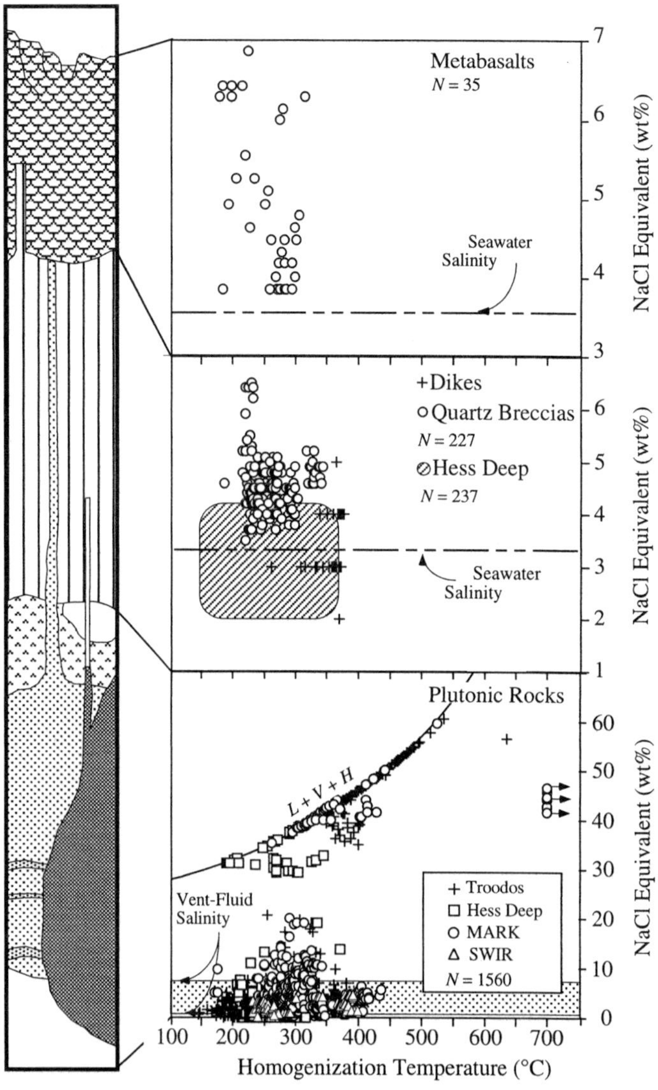

Figure 10. Uncorrected homogenization temperatures and corresponding salinities for low- to moderate-salinity inclusions hosted in gabbro, vein, quartz breccia, and metabasalt samples. Fluids in the breccia, dikes, and metabasalt samples generally show an enrichment in salinity over seawater values, but overlap salinities measured in hydrothermal vents. *N* = number of inclusions measured.

broic samples from the MARK area and the Cayman Rise yield C isotope ratios and C concentrations similar to those determined on the Southwest Indian Ridge gabbros (Früh-Green and Kelley, 1998). However, it is not yet known if (CO_2 + CH_4 + H_2O)-rich fluids are common in mid-ocean ridge environments, nor what the contribution of these fluids is to the flux of C to the hydrosphere.

HYDROTHERMAL PLUMBING SYSTEMS

Low- to moderate-salinity fluids

The transition from magmatic- to hydrothermal seawater-dominated conditions in mid-ocean ridge systems is likely to occur over a continuum of temperature and pressure conditions. Geochemistry of amphiboles that fill early microfractures and veins and oxygen isotope data indicate that the onset of brittle failure and penetration of hydrothermal fluids into the plutonic foundation of the oceanic crust may occur at temperatures up to 600–750 °C under variable strain conditions (Stakes, 1991; Stakes et al., 1991; Gillis, 1995; Manning et al., 1996). This subsolidus phase of fluid circulation has yet to be unambiguously documented by fluid inclusions. However, fluid inclusions that record temperatures of ~250–450 °C (temperatures corrected for pressure effects are up to ~500 °C) in microfractures and secondary minerals are ubiquitous in many plutonic samples from a variety of submarine environments (Table 1; Figs. 10 and 11) (Jehl et al., 1977; Gallinatti, 1984; Kelley and Delaney, 1987; Richardson et al., 1987; Vanko, 1988; Nehlig and Juteau, 1988; Cowan and Cann, 1988; Richards and Cann, 1989; Nehlig, 1991; Vanko and Stakes, 1991; Kelley et al., 1992, 1993; Vanko et al., 1992; Kelley and Malpas, 1996; Kelley, 1997). The fluid inclusions, which typically occur along microfractures in plagioclase, apatite, and quartz and as primary inclusions in epidote and hydrothermal clinopyroxene, record multiple fracturing events marked by discrete fluid compositions and temperatures. It is likely that these inclusions reflect the progressive cooling of solidified magma bodies and brittle

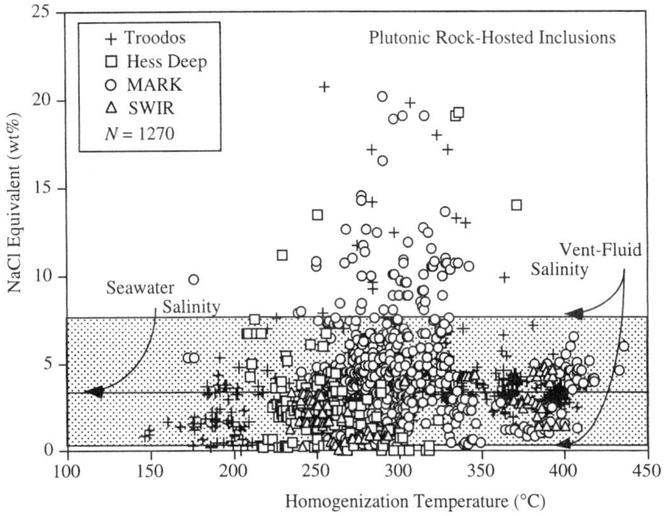

Figure 11. Uncorrected homogenization temperatures and corresponding fluid salinities in plutonic rock–hosted inclusions from Troodos, MARK (Mid-Atlantic Ridge and Kane Fracture Zone intersection), Hess Deep, and the SWIR (Southwest Indian Ridge). Most inclusions overlap the range of salinities measured for hydrothermal vents (e.g., <1 to 7 wt% NaCl). Under reasonable entrapment pressures, corrected homogenization temperatures would be ~50 °C higher than shown. The inclusion salinities likely reflect condensation processes as the sample depths are too deep to be influenced by boiling.

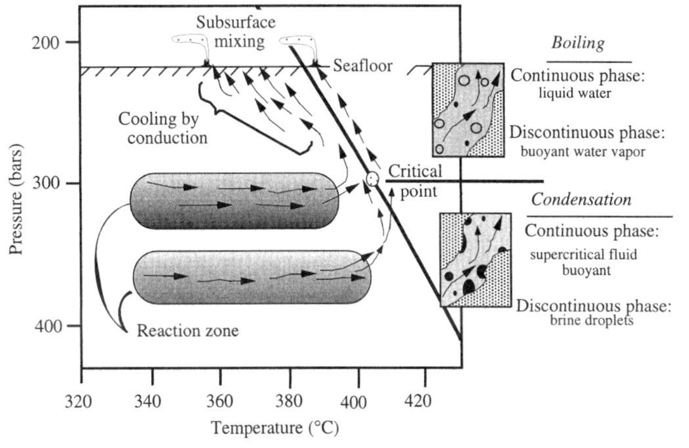

Figure 12. Idealized two-phase curve showing the critical point of seawater (407 °C, 298 bars) and hypothetical paths of seawater circulation. Fluids that circulate under shallow-depth (low-pressure), low-temperature conditions that intersect the two-phase curve will undergo boiling and separate water vapor from a continuous liquid phase. At deeper-crustal depths, and at higher temperatures, fluids that intersect the two-phase curve at temperatures and pressures greater than the critical point will condense droplets of brine from a continuous vapor phase. Both processes are operative in submarine hydrothermal systems, and many vent systems show evidence of conductive cooling of the upwelling hydrothermal fluids prior to their exiting the black smoker chimneys.

fracturing, perhaps in response to downward propagation of a cracking front (Lister, 1983). During this period, brittle failure facilitates enhanced fluid flow under greenschist-facies to upper-amphibolite-facies metamorphic conditions and leads to intense alteration adjacent to vein networks and in cataclastically deformed zones (Kelley and Delaney, 1987; Nehlig, 1991; Vanko et al., 1992; Gillis et al., 1993; Gillis, 1995; Früh-Green et al., 1996b; Manning and MacLeod, 1996; Manning et al., 1996). Petrologic data on amphibole veinlets within the Hess Deep rocks suggest that short-lived fracturing of the crystallized carapace and fluid flow may occur within the gabbroic rocks at temperatures of ~690–750 °C (Manning et al., 1996). Theoretical modeling indicates that these high-temperature events may be transient, lasting no more than several thousand years (Manning et al., 1996).

One of the most striking characteristics of the low-salinity inclusions in plutonic rocks is their temperature and compositional overlap with respect to fluids measured in active submarine vent systems (Fig. 11). This relationship is particularly evident for gabbro-hosted inclusions, where uncorrected inclusion homogenization temperatures cluster at ~250–400 °C, and fluid salinities are commonly from 0.1% to 200% of seawater values (Butterfield et al., 1990, 1997; Von Damm, 1990, 1995). The similarity is remarkable considering that the plutonic rock–hosted inclusions are along networks of microfractures that presumably represent rock-dominated systems, in contrast to the high-flow-through environments represented by the hydrothermal vents.

Several processes have been invoked to account for the range in salinities measured in active vent and deeper-crustal systems; they include supercritical phase separation of seawater-derived fluids, segregation of brines and vapors, variable mixing of hydrothermal seawater with phase-separated brines and vapors, and boiling. The development of low-salinity vapors and brines by condensation, in addition to being an important process in the high-temperature magmatic environments already described, has also been inferred to be an active process in lower-temperature, seawater-dominated systems operative beneath hydrothermal vents at 9°N on the East Pacific Rise (Von Damm et al., 1992, 1997; Haymon et al., 1993; Fornari et al., 1998) and on the Juan de Fuca Ridge within Axial Seamount and at the Endeavor Segment (Delaney et al., 1992, 1997; Butterfield et al., 1990, 1994; Lilley et al., 1993).

In an evolving magma and hydrothermal fluid system, as seawater-derived fluids penetrate into the oceanic crust and approach the heat source, they must traverse a series of compositionally dependent condensation curves (Figs. 12 and 13). As hydrothermal fluids migrate into the high-temperature reaction zone and obtain their end-member fluid compositions (Seyfried et al., 1991; Berndt and Seyfried, 1990; Seyfried and Ding, 1995; Alt, 1995), the fluid properties undergo dramatic changes: viscosity and density decrease, heat capacity increases, the coefficient of thermal expansion increases, and near the two-phase curve, there is a maximum in quartz solubility (Norton, 1984). These changes translate into a zone of maximum transport efficiency and promote the rapid, buoyant rise of end-member hydrothermal fluids within the fracture networks. Circulating seawater-like fluids that intersect the part of the curve at conditions less than the critical point (i.e., 298 bars and 407 °C for seawater; Bischoff and Rosenbauer, 1988) will undergo boiling (Fig. 12). At temperatures and pressures greater than the critical point, seawater-like fluids that intersect the condensation part of the curve will separate out droplets of brine from a low-salinity vapor (see Fournier, 1987, for additional discussion).

The geometry of the up-flow zone at depth is not known, but may be similar to an artery system in which a fine-capillary fracture network progressively feeds fluids into larger, more throughgoing fractures at shallow-crustal levels. Whatever the geometry is, however, it is clear that most end-member hydrothermal fluids pass into the zone of supercritical phase separation during their transport to the surface, resulting in the formation of vapor and brines (Figs. 11 and

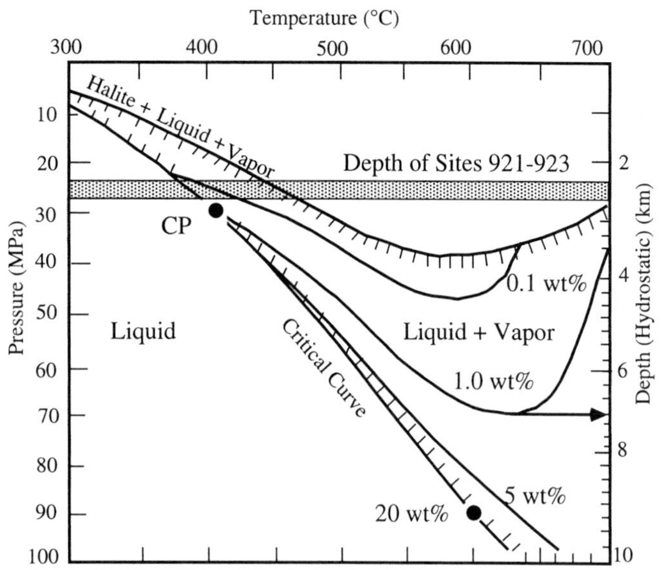

Figure 13. Isothermal (*P-X*) projection for the NaCl-H_2O system (after Bischoff and Rosenbauer, 1988). Isopleths (solid lines) of the solubility of NaCl in a vapor phase are shown for 0.1, 1.0, and 5.0 wt% NaCl solutions, as is the critical point for a 20 wt% NaCl solution. Also shown is the halite saturation curve (H + L + V) and critical curve under hydrostatic conditions. The critical curve separates the one-phase field (Liquid) from the two-phase (Liquid + Vapor) field (bounded by hatch marks). The condensation model for generation of the ~1 wt% NaCl equivalent salinity fluids in submarine plutonic rocks described in this study requires that fluid circulation occurred at temperatures of >407 °C (the critical point of seawater) and that the inclusions were trapped at pressures less than ~700 bars. This pressure estimate is a maximum in that at pressures greater than 700 bars, condensation would cause vapors with >1 wt% NaCl to separate.

12) (Delaney et al., 1987; Kelley and Delaney, 1987; Vanko, 1988; Kelley et al., 1993; Saccocia and Gillis, 1995; Von Damm, 1995; Butterfield et al., 1997). As in plutonic environments, it is likely that the volume of vapor generated by this process is much greater than for brines.

Under appropriate conditions, the salinities of the brine and vapor phases are likely to change as the fluids migrate up the conduits. Segregation due to density differences and differences in wetting characteristics will promote segregation of the immiscible vapor and brine, and with a decrease in pressure, the segregated phases will either continue to undergo condensation or boiling (Fig. 12) (Delaney et al., 1987; Goldfarb and Delaney, 1988; Fox, 1990). That both of these processes occurs is well documented in hydrothermal systems at Axial Seamount, at hydrothermal fields along the Endeavour Segment of the Juan de Fuca Ridge, and at venting sites along the East Pacific Rise (Butterfield et al., 1990, 1994; Delaney et al., 1992; Haymon et al., 1993; Von Damm, 1995; Lilley et al., 1995; Fornari et al., 1998). The preservation of inclusions with 5–20 wt% NaCl equivalent along healed microfractures within oceanic plutonic rocks suggests that segregation of the vapors and brines occurs deep within these system as well (Figs. 10 and 11) (Delaney et al., 1987; Kelley and Delaney, 1987; Fournier, 1987; Goldfarb and Delaney, 1988; Bischoff and Rosenbauer, 1989; Fox, 1990; Von Damm, 1995; Von Damm et al., 1997).

The preservation of very low salinity fluids within the gabbroic rocks provides a useful tool in estimating maximum depths at which entrapment of the fluids occurred (Fig. 13) (Kelley, 1997). In gabbroic rocks from ODP Sites 921–924 in the MARK area, at Hess Deep, and also at the Southwest Indian Ridge, fluid inclusions with homogenization temperatures of ~250–350 °C commonly record NaCl equivalent salinities of less than 1 wt% (Fig. 12). These salinities require a maximum fluid-entrapment pressure of ~700 bars because at pressures greater than this, condensation will cause endmember vapors with >>1 wt% NaCl to separate owing to the strong curvature of the isotherms in the NaCl-H_2O system (Fig. 13) (Kelley, 1997). Some of the fluid inclusions may reflect condensation of a magmatic aqueous phase (e.g., Cline and Bodnar, 1991); however, if this is the case, then the fluids must have undergone significant conductive cooling, and unreasonable pressure corrections of several kilobars are required to increase the homogenization temperatures of these fluids (250–350 °C) to trapping temperatures consistent with magmatic conditions. Whatever the source of these fluids is, however, the preservation of fluids with 1 wt% NaCl salinities requires that the vapor and brine phases maintain segregation subsequent to phase separation and that little, if any, mixing with seawater occurred.

The anastomosing microfracture systems in the plutonic rocks may provide a favorable environment for such segregation processes because in large-aperture, high-flow-through fractures, flow is likely to be turbulent, and brine and vapor phases may not remain immiscible. In contrast, in small-aperture, less continuous fracture networks such as those present in the plutonic rocks, flow is more likely to occur under laminar conditions, and preferential wetting of the fracture walls with brines may allow the vapor and brine to coexist as discrete phases (Goldfarb and Delaney, 1988; Fox, 1990). Intermittent release of such brine- and vapor-rich pools at depth during episodic fracturing events appears to be a common process affecting the compositions of fluids venting on the seafloor where salinities change from <1 wt% NaCl to 7 wt% NaCl over the lifetime of a vent system (Von Damm et al., 1992, 1997; Von Damm, 1995; Butterfield et al., 1997).

Fluids bearing $CH_4 + H_2O \pm H_2$

CH_4 has rarely been identified in fluids hosted in rocks from submarine environments, but it is a ubiquitous component of submarine hydrothermal vent fluids (Lilley et al., 1983, 1992, 1993; Welhan and Craig, 1983; Welhan, 1988; Von Damm, 1990) and has been detected as a trace component in inclusions in mid-ocean ridge glasses (Pineau et al., 1976; Welhan and Lupton, 1987; Welhan, 1988; Javoy and Pineau, 1991). Recent studies of rocks from slow-spreading ridges show that ($CH_4 + H_2O \pm H_2$)–rich fluids may also be common within high-temperature submarine plutonic environments (Table 1). ($CH_4 + H_2O$)–rich inclusions, which do not contain CO_2, are a ubiquitous feature of the 735B gabbroic suite (Vanko and Stakes, 1991; Kelley, 1996; Kelley and Früh-Green, 1999) and occur in isolated gabbroic rocks from the MARK area (Figs. 8 and 14) (Kelley, 1997). The CH_4-bearing inclusions are vapor dominated to vapor rich and occur along healed microfractures in plagioclase and less commonly in quartz. Raman spectroscopic analyses and petrographic studies indicate that the ($CH_4 + H_2O$)–bearing inclusions contain molecular H_2 and opaque, fine-grained, hexagonal daughter minerals that are believed to be graphite; carbonate minerals were observed by Vanko and Stakes (1991).

Mass spectrometric analyses show that the Southwest Indian Ridge fluids contain up to 42 mol% CH_4, which is significant in that occluded gases and vent fluids from sediment-starved environments typically contain CH_4 concentrations of <1 mol% (Fig. 8) (Kelley, 1996). Step-heating sampling of the Southwest Indian Ridge rocks at

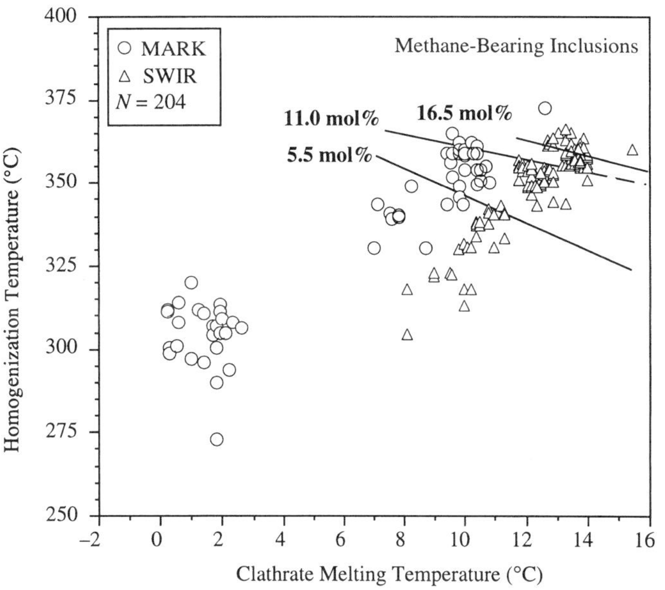

Figure 14. Corresponding clathrate melting and homogenization temperatures for CH_4-bearing inclusions from MARK (Mid-Atlantic Ridge and Kane Fracture Zone intersection) and SWIR (Southwest Indian Ridge). Melting and homogenization temperatures of the secondary inclusions from Sites 921–923 indicate that the inclusions contain up to ~11 mol% CH_4. These inclusions are similar to those in gabbroic rocks from the Southwest Indian Ridge recovered from Hole 735B and to those in gabbroic rocks previously collected from MARK (i.e., Alvin sample ALV1012–9). Experimentally derived isopleths are after Zhang and Frantz (1992). Inclusions generally contain 1–2 wt% NaCl equivalent salinities. N = number of inclusions measured.

500 °C yield CH_4 concentrations of 0.13 to 0.29 μmol/g, $\delta^{13}C$ (CH_4) values from −33.8‰ to −20.4‰, and hydrogen isotope values (δD) of −149‰ to −99‰ (Kelley and Früh-Green, 1999). The CH_4 in these fluids is slightly depleted in ^{13}C, but has hydrogen isotope values similar to end-member compositions of hydrothermal fluid from the East Pacific Rise and to fluids from the southern Juan de Fuca Ridge [$\delta^{13}C$ (CH_4) = −20.8‰ to −15.0‰ and δD (CH_4) = −126‰ to −102‰ (N = 2); Welhan and Craig, 1983; Kim et al., 1984; Merlivat et al., 1987; Evans et al., 1988; Shanks et al., 1995]. Uncorrected homogenization temperatures indicate that these fluids were entrapped at minimum temperatures of ~350–370 °C; at 1000 bars pressure, their entrapment temperatures would be close to 400 °C (Fig. 14) (Kelley, 1996). These data provide strong evidence that many of the (CH_4 + H_2O)-rich inclusions were trapped under equilibrium conditions at T and $f(O_2)$ conditions very near the QFM buffer. The isotopic compositions of these fluids do not support either a biogenic or thermogenic origin for the CH_4, but are consistent with its formation involving seawater alteration of the plutonic rocks at high temperatures (Kelley and Früh-Green, 1999).

These data in concert with accumulating evidence from water-column analyses of ocean-bottom waters and field studies indicate that both the mantle and gabbroic layer 3 may be a significant source for CH_4 in submarine hydrothermal systems. Intense CH_4 degassing has been observed along the Mid-Atlantic Ridge in attenuated crustal zones in which serpentinized peridotite bodies crop out on the seafloor (Charlou et al., 1992; Rona et al., 1992; Charlou and Donval, 1993; Bougault et al., 1993), pore waters sampled during ODP Leg 125 at a serpentinized seamount in the Mariana forearc yielded high concentrations of CH_4 and C_2H_6 (Haggerty, 1989; Mottl and Haggerty, 1989), and CH_4 seeps have been observed emanating from serpentinized bodies in the Zambales ophiolite in the Philippines (Abrajano et al., 1988). Experiments of Berndt et al. (1996) show that alteration of olivine at 300 °C and 500 bars leads to elevated H_2 contents and the reduction of dissolved CO_2 to form reduced gaseous species, such as methane, ethane, propane, and a significant amount of amorphous carbon or graphite. Continuation of this study by Allen et al. (1998) demonstrates that the H_2 reacts with the CO_2-bearing fluids to produce large quantities of formic acid and smaller concentrations of alkane and alkene gases.

Conditions under which CH_4- and H_2-rich fluids form in these deep-seated hydrothermal systems involve complex reactions that require the infiltration of water and may involve a series of metastable reaction steps during serpentinization (Holloway, 1984; Shock, 1990, 1992; Allen et al., 1998; Kelley and Früh-Green, 1999). These reactions are commonly accompanied by conditions of extremely low $f(O_2)$ at the serpentinization front (Frost, 1985; Früh-Green et al., 1996a). Reduction accompanies the formation of FeNi alloys in serpentinites, and these alloys and magnetite may act as significant catalysts for Fischer-Tropsch–type reactions to form CH_4 and other hydrocarbons (e.g., Yoshida et al., 1993; Berndt et al., 1996; Allen et al., 1998). The composition of any C-bearing fluid produced will be greatly affected by the $f(O_2)$ such that during migration of the reduction front, any CO_2 in the initial fluid will undergo reduction to CH_4 and C (graphite).

Our data together with water-column data, field studies, and experimental results suggest that the production of CH_4 and other hydrocarbon species should be a common process in mid-ocean ridge systems where high-temperature fluids interact with mafic mineral phases. Preliminary studies indicate that up to 800 ppm C is trapped in oceanic peridotites (Früh-Green et al., 1996a); thus, the presence of CO_2 and reduced-C species in oceanic gabbros and peridotites is a potential source of C in hydrothermal fluids. In addition to Fischer-Tropsch–type reactions during alteration processes, the degradation of reduced C in mantle peridotites and its subsequent incorporation in marine hydrothermal fluids during fluid + rock reactions (i.e., serpentinization) are potential sources of C and CH_4 in hydrothermal fluids. If the total CH_4 concentrations (0.3–0.6 mmol/kg) in the Southwest Indian Ridge samples are representative of those in gabbroic rocks in other submarine settings and if plutonic layer 3 is assumed to be 6 km thick and to represent ~60% of the oceanic crust, then the plutonic section of the oceanic crust may contain ~10^{19} g of abiogenic CH_4 (Kelley and Früh-Green, 1999). On the basis of preliminary studies (Früh-Green et al., 1996a), the shallow mantle may be an even larger reservoir for abiogenically produced CH_4. If these C-bearing fluids are released during subsequent hydrothermal events, they may be an important energy source for subsurface and vent-associated chemosynthetic microbial communities.

FLUIDS IN THE SHALLOW OCEANIC CRUST

Sheeted dikes

There is no direct evidence to indicate whether hydrothermal events in the plutonic rocks influence the physicochemical signature of fluids within the sheeted dikes or at sites of seafloor venting. The striking thermal and compositional overlap of low-temperature fluids in gabbroic rocks and felsic differentiates with fluids at shallower-crustal levels, however, suggests that these two systems may be connected (Fig. 10) (Kelley et al., 1993). Nevertheless, with the present data it is not possible to define their relationship. Fluid inclusions in diabasic samples from ophiolites and Hole 504B indicate temperatures of 160–400 °C and fluid salinities of <1 to 8 wt% NaCl equivalent for the circulating fluids (Table 1) (Honnorez et al., 1985; Nehlig and Juteau, 1988; Schops and Herzig, 1990; Nehlig; 1991; Kelley et al., 1992, 1995; Gu and Vanko, 1996). These values are similar to those observed in the deeper-crustal rocks. Similar to plutonic rocks from slow-spreading systems, the production of CH_4 may also be important in layer 2 of the oceanic crust. (CH_4 + CO_2)–rich inclusions in Hole 504B occur ~6 m above a zone in which olivine is significantly replaced by talc, and these inclusions may reflect alteration processes similar to those that are believed to occur in the gabbros (Fig. 15) (Kelley et al., 1995).

Investigation of Hole 504B inclusions also shows that care must be taken in interpreting low-temperature inclusions because they may re-equilibrate during migration of the crust off-axis (Kelley et al., 1995; Gu and Vanko, 1996). Fluid inclusions in samples from the sheeted-dike sequence exhibit corrected homogenization temperatures from 150 °C to 200 °C and are in apparent equilibrium with present conductive downhole temperatures (Kelley et al., 1995; Gu and Vanko, 1996).

Up-flow zones

Successive generations of fluid inclusions in the MARK, Hess Deep, and Troodos quartz breccias provide a record of fluid evolution in rocks believed to represent the ~300–400 °C up-flow conduits of submarine paleo–hydrothermal systems (Figs. 10 and 16) (Spooner and Bray, 1977; Delaney et al., 1987; Kelley et al., 1993; Saccocia and Gillis, 1995). The quartz breccias contain mineral phases similar to those described for many ore deposits, including variable amounts of chlorite, epidote, pyrite, chalcopyrite, quartz, titanite, and anatase; they can be either matrix or clast dominated. In the breccia samples, liquid-dominated inclusions are pervasive in matrix and vein quartz, occurring along crisscrossing networks of microfractures, along well-defined, individual arrays, and as primary inclusions that mimic grain boundaries. The inclusions exhibit homogenization temperatures of 170–370 °C and equivalent fluid salinities of 0.0–9.8 wt% NaCl (Figs. 10 and 16) (Gallinatti, 1984; Delaney et al., 1987; Kelley et al., 1993; Saccocia and Gillis, 1995). Textural relationships of chlorite-rich clasts and quartz indicate that in the MARK system there were at least five distinct hydrothermal pulses localized along the fault network(s) represented by the breccias (Fig. 16) (Kelley et al., 1993). In some samples there is a progression in temperature and salinity in which the earliest, highest-temperature fluids are the lowest in salinity (Fig. 16). As the clasts were mobilized along fault systems, cooler fluids precipitated quartz overgrowths on the chlorite-rich breccia fragments, and the late-stage fluids show salinity enrichments of up to three times the salinity of seawater (Delaney et al., 1987; Kelley et al., 1993).

This evolution from high-temperature, low-salinity fluids to lower-temperature brines is also recognized in nascent and well-established hydrothermal systems on the seafloor (Butterfield et al., 1997; Haymon et al., 1993; Von Damm et al., 1997) and presumably represents the progressive flushing of residual brines formed during the vigorous stage of hydrothermal venting. Fluctuations between lithostatic and hydrostatic conditions near magma-chamber margins in response to migration of the brittle-ductile boundary likely result in ephemeral phase-separation events and the episodic release of brines into the overlying system (Kelley, 1997). Work by Saccocia and Gillis (1995) showed that there is a correspondence between the mineral assemblage of the breccias and fluid salinity: high-salinity, H_2S-poor fluids correspond to chlorite-bearing breccias, and lower-salinity, H_2S-enriched fluids correspond to epidote-bearing breccias. As pointed out by Saccocia and Gillis (1995), these compositional trends are identical to those observed in present-day venting systems, and they suggest a direct link between alteration assemblages and fluids in the up-flow zone. Phase separation of the seawater-derived fluids is believed to occur at temperatures of 450–625 °C (Delaney et al., 1987; Kelley et al., 1993; Saccocia and Gillis, 1995).

Metabasalts

Fluid-inclusion data indicate that conditions of hydrothermal circulation within metabasalt samples are similar to those that exist at deeper-crustal levels (Fig. 10). Metabasalt-hosted fluid inclusions from the MARK area record somewhat lower fluid temperatures but fluid salinities similar to those observed in the quartz breccia samples (Fig. 10) (Kelley et al., 1993). Primary(?) inclusions within altered plagioclase grains indicate fluid salinities of 3.8–6.9 wt% NaCl and temperatures of 193–313 °C. The highest temperatures in the metabasalts are recorded by inclusions in vein quartz and by primary(?) inclusions in altered groundmass plagioclase. Fluid inclusions in altered basalt samples from Oman yield temperatures of 138–409 °C and equivalent fluid salinities of 1.4–10.1 wt% NaCl (Table 1) (Nehlig and Juteau, 1988; Nehlig, 1991). Many of these samples likely preserve fluids whose compositions were determined at greater depths. However, the metabasalts generally exhibit much greater degrees of alteration than the plutonic or diabasic suites (Gillis and Thompson, 1993; Gillis et al., 1993); therefore, inclusions with high salinities within altered zones of plagioclase grains may represent hydration processes involving sinking fluids under rock-dominated conditions, rather than mixing of phase-separated brines and seawater (Kelley et al., 1993; Kelley and Malpas, 1996).

SPREADING ENVIRONMENT

Although few data exist on the dependence of fluid evolution and spreading rate within magma and hydrothermal fluid systems, fluid-inclusion results suggest that there may be significant differences in fluid compositions between different spreading environments. At magmatic conditions, the primary concentration of volatiles such as H_2O vary significantly between backarc and mid-ocean ridge systems; backarc environments show H_2O enrichments of 10–20 times those of mid-ocean ridge samples (e.g., analyses from Troodos glasses indicate primary H_2O concentrations of 2.1–2.3 wt% in contrast to 0.2 wt% H_2O measured for many MORBs) (Delaney et al., 1978; Dixon et al., 1988, 1995; Muenow et al., 1990; Johnson et al., 1994; Dixon and Stolper, 1995). This water enrichment is important because at a given pressure, melts with higher water concentrations will reach water saturation earlier in their crystallization history, and it is likely that most of these magmatic fluids will be exsolved as supercritical vapors and brines (Fig. 5) (Frantz et al., 1992; Kelley et al., 1993; Kelley, 1997). Therefore, magmatically exsolved brines are much more likely to be volumetrically significant in backarc spreading environments, and it is also likely that the direct exsolution of brines in the absence of a vapor phase is important (Fig. 5 and 6) (Cline and Bodnar, 1991). In contrast, in mid-ocean ridge systems, exsolution of water from the melt only occurs within the last ~3%–5% of the residual melt. These small, remaining pockets of melt are commonly represented by micropegmatitic to pegmatitic pods that contain abundant apatite, zircon, and magnetite, which are found in most gabbroic rocks sampled to date.

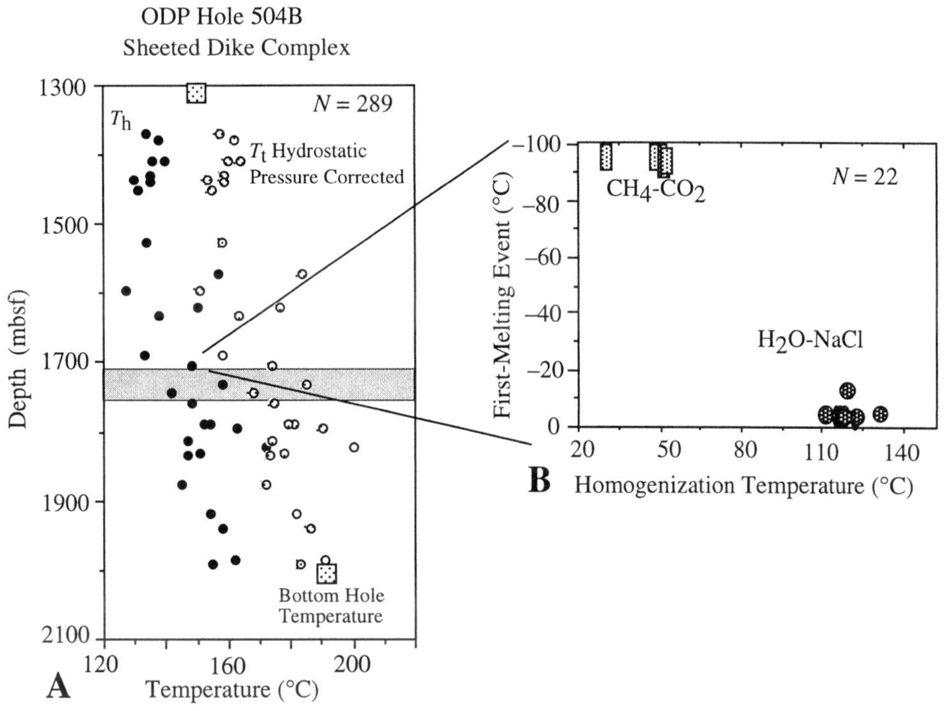

Figure 15. Fluid-inclusion data from the Costa Rica Rift (ODP Hole 504B). A, Average uncorrected homogenization temperatures (solid circles) and average pressure-corrected temperatures (gray-filled circles) as a function of depth (meters below seafloor) for plagioclase-hosted fluid inclusions in diabase recovered during ODP Legs 111, 137, and 140. Trapping temperatures (T_t) were calculated by assuming an average fluid salinity of 3.2 wt% NaCl and that hydrostatic pressures apply. Trapping temperatures exhibit a gradient from 159 °C at a depth of 1370 mbsf to 181 °C at a depth of 1992 mbsf. Corrected homogenization temperatures are in apparent equilibrium with present conductive bottom hole temperatures (gray-filled boxes) and may record off-axis, passive circulation of seawater into the sheeted-dike complex as the crust was transported away from the Costa Rica Rift. Homogenization temperatures for inclusions in diabase samples at depths shallower than 1525 m are after Nehlig (1991). *N* denotes number of inclusions measured. Gray band indicates where talc becomes an important mineral phase after olivine. B, Homogenization temperatures and first-melting temperatures for CH_4 + CO_2 and for H_2O + NaCl inclusions in plagioclase. The CH_4 + CO_2 inclusions, which occur ~6 m above a zone in which olivine is significantly replaced by talc (as shown in A), are believed to have formed from alteration of olivine.

The differences in initial volatile contents between water-poor, mid-ocean ridge systems and water-enriched environments such as backarcs may have a profound affect on ore depositional processes as well. Because chloride forms the dominant complex for metals in aqueous magmatic fluids (and lower-temperature hydrothermal fluids) (Barnes, 1979; Seyfried and Ding, 1995), the formation of significant volumes of brine is likely to be important in the transport of metals and in ore deposition—particularly in areas with melts having initial high water contents (Kelley et al., 1992). In backarc systems such as the Okinawa Trough and in the Manus Basin, large active seafloor hydrothermal systems are developed that host significant sulfide deposits, some of which are analogous to Kuroko-type deposits (Ohmoto and Skinner; 1983; Halbach et al., 1989; Yang and Scott, 1996). The sulfide deposits are associated with felsic intrusions, and limited fluid-inclusion data on Manus suggest that metal-enriched magmatic fluids were important in formation of the sulfide deposits (Yang and Scott, 1996). Similarly, limited observations based on fluid-inclusion compositions in upper-level gabbros and plagiogranites from Troodos show that the fluids are remarkably similar in composition to those of porphyry copper deposits where there is abundant data indicating that the contribution of magmatic metal-rich brines to ore deposition is important (Gustafson and Hunt, 1975; Bodnar and Beane, 1980; Titley and Beane; 1981; Reynolds and Beane, 1985; Kelley et al., 1992). It is also important to note that in addition to brine and metal transport phenomena, significant exsolution of magmatic fluids likely results in enhanced deuteric alteration of the surrounding host rock (Kelley et al., 1992; Kelley, 1997). Therefore, exposed crustal sequences like the Troodos ophiolite may not provide good analogues for examining many questions concerning rock-alteration processes with specific reference to mid-ocean ridge environments. Because of limited data on magmatic fluid compositions in mid-ocean ridge and subduction-zone–related settings, however, there is still much debate about the relative contribution of magmatic fluids to ore depositional processes, and significant research is required to address this problem.

Comparison of gabbroic rocks from Hess Deep with those from MARK and the Southwest Indian Ridge suggest that in addition to brine generation, there may be other differences in the compositional evolution of magmatic fluids particularly in slowly spreading plutonic environments. In the Hess Deep suite, (CO_2 + CH_4)–rich fluids have not yet been observed. In contrast they are ubiquitous within the Southwest Indian Ridge rocks, and similar inclusions are observed in the MARK suite, although they do not appear to be as common (Vanko and Stakes, 1991; Kelley, 1996). The CH_4-rich fluids are especially abundant in iron oxide gabbros and in gabbronorites,

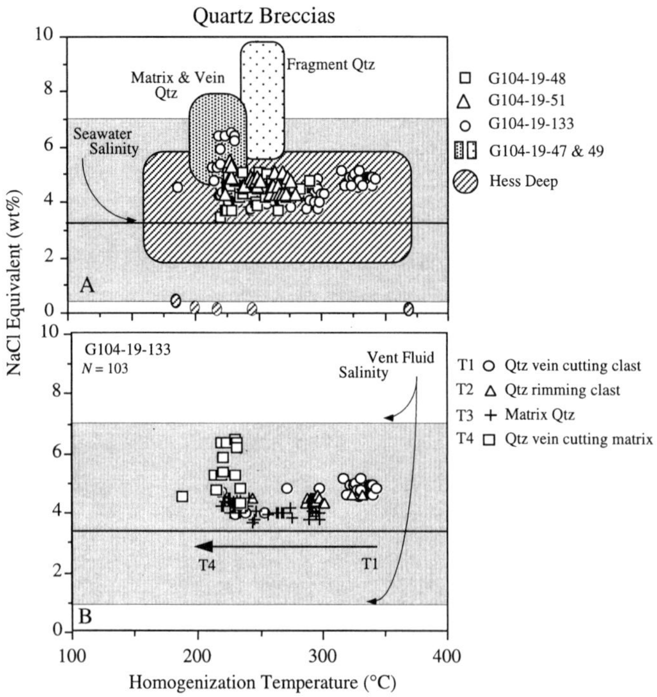

Figure 16. Corresponding temperatures of homogenization and fluid salinities for inclusions hosted in quartz breccia samples from MARK (Mid-Atlantic Ridge and Kane Fracture Zone intersection) (Delaney et al., 1987; Kelley et al., 1993) and Hess Deep (Saccocia and Gillis, 1995). A, Fluid salinities in breccia-hosted inclusions are 130%–300% of seawater values and exhibit uncorrected homogenization temperatures of 180–370 °C. Textural relationships of the chlorite- and quartz-rich fragments and matrix and vein quartz in MARK samples indicate at least five distinct hydrothermal pulses along the fault network. B, Fluid inclusions in four successive generations of quartz (T1 to T4) record initial fluid temperatures of 343 °C, down to 222 °C during late-stage veining. The high salinities of these fluids may reflect the culmination of complex cycles of fluid interaction involving wall-rock brecciation, mobilization, and rebrecciation of the fragments and periodic flushing of deeper-seated brines into the overlying circulation system.

where the fluids occur as primary inclusions in quartz and are associated with a high modal percent of trace minerals. These oxide-rich rocks are especially abundant in the Southwest Indian Ridge suite, particularly within a 100-m-thick zone at the top of the core. Results from Hess Deep suggest that these rocks may not be common in fast-spreading systems. In slow-spreading environments where melt transport to the crust is likely to be discontinuous, a closed system may be approached in late-stage differentiates allowing volatiles to evolve toward (CO_2 + CH_4 + H_2O)–rich fluids (Mathez, 1989a, 1989b, Mathez et al., 1989). Clearly, however, additional investigations are needed into processes that affect volatile compositions deep within the crust.

In all spreading systems studied to date, the transition from magmatic to hydrothermal seawater–dominated conditions in crustal layer 3 rocks is marked by migration of fluids at minimum temperatures of ~400 °C, with salinities from <0.1% to 300% of seawater values (Fig. 11) (Jehl et al., 1977; Kelley and Delaney, 1987; Richardson et al., 1987; Vanko, 1988; Nehlig and Juteau, 1988; Cowan and Cann, 1988; Richards and Cann, 1989; Nehlig, 1991; Vanko and Stakes, 1991; Kelley et al., 1992, 1993; Vanko et al., 1992; Kelley and Malpas, 1996; Kelley, 1997). Results from the Southwest Indian Ridge and MARK work show that in slow-spreading environments, the low-temperature hydrothermal systems also involve circulation of fluids with very high CH_4 concentrations at temperatures of 250–400 °C (Figs. 8 and 14) (Vanko and Stakes, 1991; Kelley, 1996, 1997; Kelley and Früh-Green, 1999). The ubiquitous occurrence of fluids with CH_4 concentrations of 15–40 mol% throughout the plutonic section from the Southwest Indian Ridge, their occurrence in gabbroic rocks from the MARK area (Figs. 8 and 14), and the common occurrence of CH_4-rich plumes that exhibit low Mn/CH_4 ratios above outcropping peridotite bodies along the Mid-Atlantic Ridge (Charlou et al., 1992; Rona et al., 1992; Charlou and Donval, 1993; Bougault et al., 1993) suggest that CH_4-rich fluids are of particular importance in deep-seated hydrothermal systems in slow-spreading environments.

In zones of low magmatic supply, where serpentinization and unroofing of mantle and deep, hot crustal rocks are significant processes in crustal development, migration of fluids along deeply penetrating fault systems may intensify hydration of deep-seated mafic and mantle material, leading to significant CH_4 (+ H_2 + hydrocarbons) generation. In crust formed at intermediate spreading rates, this process may also be important during the waning stages of magmatic activity where the brittle-ductile boundary has undergone significant migration into crustal layer 3 (Wilcock and Delaney, 1996; Fornari et al., 1999). Limited sampling of plutonic rocks from fast-spreading environments precludes evaluation of the importance of CH_4 generation during fluid-rock interactions. If this process is significant, however, it is likely to occur only in off-axis settings because the shallow nature of the brittle-ductile boundary at the ridge axis is likely to prevent fluid penetration into the on-axis gabbroic sequence. In any case, the production of CH_4 and other hydrocarbon species should be a common process in mid-ocean ridge systems where high-temperature fluids interact with mafic mineral phases.

VOLATILES RELEASED DURING DIKE INTRUSION AND ERUPTIVE EVENTS

The processes by which volatiles are transported to the overlying ocean are complex and occur over varied time scales. Beginning with the discovery and sampling of hydrothermal vents, it was known that the high-temperature fluids were enriched in volatiles—such as CO_2, CH_4, H_2, and ^3He—above ocean-water background levels (Lilley et al., 1995; Lupton, 1995). It has only been shown recently, however, that over short periods of time, significant concentrations of volatiles may also be both released and generated during seafloor eruptions (Sansone et al., 1991, 1996; Kelley et al., 1998). The detection of event plumes in 1986–1987 and 1993 on the Cleft and CoAxial Segments of the Juan de Fuca Ridge, respectively, and subsequent investigation of event plumes on the Gorda Ridge led to the realization that submarine dike intrusion and eruptive events result in the catastrophic and voluminous release of volatile-enriched buoyant hydrothermal plumes that rise up to 1000 m or more above the seafloor (Baker et al., 1987a, 1989, 1995a, 1995b; Lupton et al., 1989; Lupton, 1995; Delaney et al., 1998; Baker, 1998). These and follow-on investigations have led to the profound discovery that dike intrusion and eruptive events are associated with the massive effusion of microbially produced material (floc) from the subsurface, that extensive blooms of microorganisms both within and on the seafloor follow shortly after the eruptions, and that microorganisms thrive within the transient volatile-rich plumes (Fornari and Embley, 1995; Embley et al., 1995; Holden et al., 1998; Delaney et al., 1998; McLaughlin, 1998; Summit and Baross, 1998).

The biological responses to submarine eruptions are not well

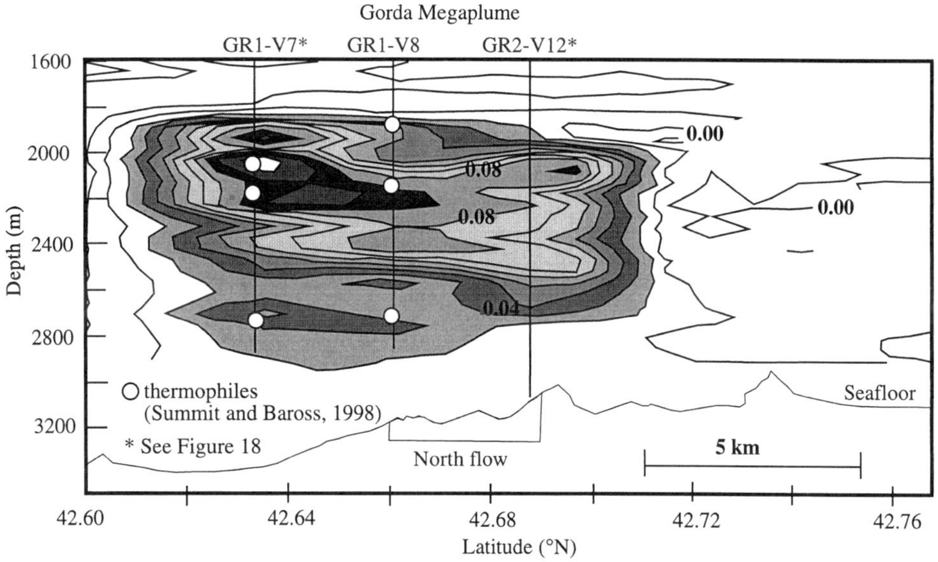

Figure 17. Event plume as defined by along-axis temperature anomalies ($\Delta\theta$ values, in bold) mapped during the first rapid-response cruise in March 1996 just a few weeks after the eruption of the Gorda Ridge megaplume (after Baker, 1998). The plume was mapped (shaded, contoured areas) and sampled with a 20 L bottle-rosette package that included a high-precision Sea Bird 911plus CTD (conductivity-temperature-depth measuring instrument) and a Sea Tech light-scattering sensor (Baker, 1998). Contour intervals are 0.01 °C. The locations of a subset of vertical casts done during the first cruise in March 1996 (GR1) and a follow-on cruise in April (GR2), as well as the locations where thermophiles were sampled within the megaplume (EP96A), are indicated by vertical lines (GR1-V7, GR1-V8, GR2-V12) and open circles, respectively. Thermophiles were not cultured from background ocean-water samples, but were cultured from samples taken within the plume (Summit and Baross, 1998). The north lava flow associated with the formation of the megaplume is marked by the dark box (labeled "North flow"). See Figure 18 for CH_4, H_2, and ^3He measurements made on discrete samples from vertical casts GR1-V7 and GR2-V12.

known, but it is believed that the blooms at least partially result from an increase in nutrients supplied by magmatic volatiles (CO_2,CH_4, H_2) released during these events (Delaney et al., 1998; Summit and Baross, 1998). Because these hydrothermal plumes or "megaplumes" integrate the heat and mass flux of fluids that saturate the porous sections of oceanic crust and because they entrain hyperthermophilic and thermophilic microorganisms that were flushed from the seafloor during these events, they are useful for studying a variety of hydrothermal and biological processes (Baker et al., 1995a, 1995b; Lupton, 1995; Lilley et al., 1995; Delaney et al., 1998; Cowan et al., 1998; Juniper et al., 1998; Summit and Baross, 1998; Kelley et al., 1998; Chadwick et al., 1998).

In studies of both the Cleft and CoAxial event plumes, real-time characterization of volatiles such as H_2 and CH_4 was not possible because of the lag in response times, the transient nature of the plumes, and the extremely labile nature of many of the volatile species (de Angelis et al., 1993; Lilley et al., 1995; McLaughlin et al., 1996; McLaughlin, 1998; Cowen et al., 1998). However, in 1996, the rapid response to an intense period of seismicity detected by the Navy's Sound Surveillance System (SOSUS) on the northernmost segment of the Gorda Ridge (Chadwick et al., 1998; Fox and Dziak, 1998) allowed examination of the plume within days following the eruption, the first comeasurement of ^3He, H_2, and CH_4 in the associated event plumes, and examination of the chemical transformation of these volatile species during aging of the plumes (Kelley et al., 1998). Although additional field work on eruptions is required, results from Gorda and from examining other basaltic systems indicate that most eruptive plumes should be enriched in volatiles such as CO_2, CH_4, and H_2 with respect to ocean-water background levels (Delaney et al., 1978; Gerlach and Taylor, 1990; Dixon and Stolper, 1995; Sansone et al., 1996; Kelley et al., 1998).

The dike intrusion and eruptive event at the Gorda Ridge generated an extensive event plume that was roughly 1 km thick and extended 15 km along axis and at least 8 km across axis, with temperature anomalies ($\Delta\theta$) up to ~0.12 °C (Baker, 1999) and a low ^3He/heat ratio of $\sim 0.4 \times 10^{-12}$ $cm^3/(g{\cdot}°C)$ at standard temperature and pressure (Figs. 17 and 18). Hydrogen concentrations in the initial event plume reached a maxiumum value of 47 nmol/L, 235 times that of the ocean-water background concentration (0.2 nmol/L) (Kelley et al., 1998). In contrast, roughly one month later, H_2 concentrations, although still elevated, had drastically dropped to 40 times the ocean-water background value. CH_4 concentrations were also elevated and similar to stable hydrothermal systems in ridge environments without sediment cover, reaching a maximum value of 7 nmol/L (Fig. 18) (Lilley et al., 1992, 1995; Kelley et al., 1998).

The highly elevated and variable H_2 concentrations in the plume, which correlate with well-defined maxima in ^3He and temperature (Fig. 18), are interpreted to reflect the intense and rapid interaction of seawater with the basaltic flows and very high temperature seawater and rock interactions in newly developed subseafloor fractures generated during dike intrusion (Kelley et al., 1998). These H_2-generating reactions likely involve reactions similar to those decribed by Berndt et al. (1996) and Allen et al. (1998). This interpretation is consistent with field data from sea-surface plume fluids generated during direct interaction of Kilauea lava with seawater where H_2 concentrations of up to 2 μM were measured (Sansone et al., 1991). H_2

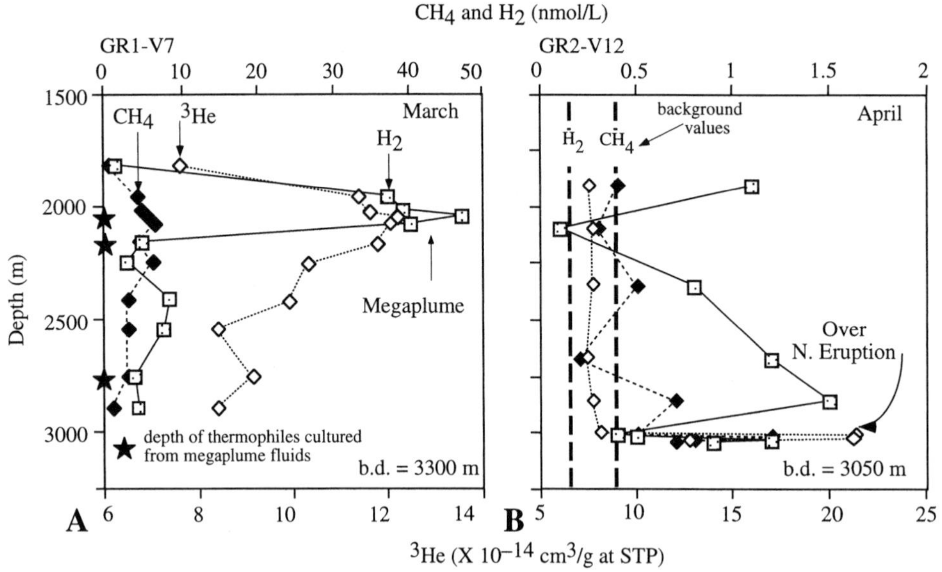

Figure 18. Profiles of CH_4, H_2, and ^3He for discrete samples taken on vertical casts through the Gorda Ridge megaplume. The location of each profile with respect to the eruption is shown in Figure 17. STP—standard temperature and pressure. A, Results from vertical cast GR1-V7 in March 1996. The most intense plume signature is marked by fluids with very high H_2 concentrations (47 nmol/L) that are 235 times the ocean-water background level (0.2 nmol/L), high ^3He concentrations, and elevated CH_4 concentrations. Also shown are the locations of samples where thermophiles and hyperthermophiles were cultured from discrete plume samples (Summit and Baross, 1999). B, Results from vertical cast GR2-V12 in April 1996. There is a sharp decrease in the concentration of H_2, indicating cooling of the flow and a lack of high-temperature water and rock interactions. The sharp increase in ^3He concentrations in deep fluids that are associated with a weak plume sampled during GR2-V12 is located 10 m off the bottom and directly over the northern eruption. The high ^3He concentrations, without a corresponding increase in H_2, likely indicate that although there was continued degassing of ^3He from the lava flow, the flow was cool enough so that H_2 was no longer being produced by high-temperature water and rock interactions (b.d.—bottom depth).

increases have also been measured in hydrothermal vents at 9°N on the East Pacific Rise, perturbed by recent volcanic eruptions. At one of the vents, H_2 showed enriched values of at least 40 mM three weeks after the 1991 eruption (Lilley et al., 1992). Plume fluids with H_2 concentrations of 20 nmol/L were also measured following an eruption at Loihi Seamount in 1996 (Sansone et al., 1996). This burst of hydrogen associated with intense water + rock reactions is likely short-lived because of the rapid cooling of the lava flows. Low H_2 concentrations were measured in a weakly developed plume associated with the northern Gorda eruption two months after the eruption (Fig. 18), and at 9°N, H_2 values had declined to 1 mM in a perturbed hydrothermal vent eleven months following the eruption (Lilley et al., 1992; Kelley et al., 1998).

CH_4 concentrations in the Gorda event plume fluids reached concentrations of 16 times the ocean-water background value and are similar to those from stable venting systems in ridge environments without sedimentary cover, where concentrations typically range from ~2 nmol/L to 100 nmol/L (Lilley et al., 1995). The source for this CH_4 is not clear owing to a lack of isotopic data on these fluids. However, methanogens have been isolated from vents at numerous vent localities (Karl, 1995); therefore, microbial activity must be considered as a potential CH_4 source in hydrothermal environments. The elevated concentration of CH_4 in the event plume fluids, coupled with field data from Kilauea (Sansone et al., 1991), which indicates that CH_4 is not produced during transient high-temperature alteration episodes associated with new surface eruptions, suggests that the source fluids for event plumes are from within the deeper crust and not from the lava flows (Kelley et al., 1998).

It is postulated that injection of melt into the water-saturated shallow crust most likely results in the short-term superheating of the surrounding pore fluids and in the vigorous flushing of both interstitial pore fluids and microbial material from the crustal reservoir (Delaney et al., 1998). The organisms that inhabit the pore spaces are released into the water column within hours to days following the eruption. The most convincing evidence for this process comes from the discovery of anaerobic thermophiles that were cultured from 17 out of 22 water samples from the Gorda event plume, but thermophiles were not detected in any of the 12 ocean-water background samples (Fig. 17 and 18) (Summit and Baross, 1998). The cultured microorganisms included novel heterotrophic Archaea that are nutritionally and phylogenetically distinct from previously described marine species. Both thermophilic and hyperthermophilic microorganisms were isolated from fluids at the Floc Site after the CoAxial eruption and included sulfur-reducers and methanogens (Holden et al., 1998). Temperatures of the venting fluids were more than 50 °C lower than the minimum growth temperature of some of the hyperthermophiles cultured. The diffuse sites at CoAxial harbored high concentrations of thermophiles when sampled three weeks and three months following the eruption (Holden et al., 1998).

Although investigation of these systems is relatively young, data collected during these rapid-response studies support a model in which subseafloor biotopes are periodically perturbed by dike intrusion and eruptive events that temporarily destroy, or enrich, thermophilic zones within the subsurface (Delaney et al., 1998). The correspondence between the massive effusion of microbial material from the seafloor, the bloom of microorganisms following an eruption, and the enriched volatile concentrations in megaplume fluids and in hydrothermal vents perturbed by eruptions (Lilley et al., 1992; Haymon et al., 1993; Embley et al., 1995; Holden et al., 1998; Delaney et al., 1998; Baker, 1999; Kelley et al., 1999; Summit and Baross, 1999) suggests that some of the supporting energy for these microbes comes from the degassing of volatiles associated with the eruption and from volatiles released during high-temperature interaction of seawater and dike-extrusive material. As volcanic activity was certainly more robust during the period of early Earth and ocean formation, dike intrusion and eruptive events and the degassing of volatiles were likely significant processes in supporting early life.

CONCLUSIONS

This paper presents an overview of volatile evolution in magma and hydrothermal fluid systems within submarine environments, and we have outlined some of the processes that determine the composition of gas species in these dynamic systems. The compositional and thermal evolution of magmatic fluids in submarine environments follows a path remarkably similar to those documented in many continental porphyry copper deposits. This path is marked by early degassing of CO_2-rich fluids followed by degassing of highly saline, metal-rich brines in concert with low-salinity vapors. The final fluids exsolved from the melt are brines formed in the absence of an immiscible vapor phase. Phase separation and segregation during cooling of the fluids result in entrapment of ~30–70 wt% NaCl brines at depth in these high-temperature plutonic environments. At shallower-crustal conditions, phase-separated and segregated, seawater-derived hydrothermal fluids with salinities of <1 wt% to ~7 wt% circulate at temperatures of >400 °C, in places pervasively altering the host rocks. Crosscutting arrays of inclusions in quartz breccias, which are interpreted to represent clogged up-flow conduits, show that the up-flowing fluids become progressively enriched in salinity within the shallow stockwork systems. Recent data show, however, that not all systems follow this evolutionary path and that fluids with >40 mol% CH_4 may form both under magmatic and lower-temperature hydrothermal conditions in some slow-spreading environments. Under hydrothermal conditions, generation of fluids bearing CH_4, H_2O, and hydrocarbons is likely to be a common process in all spreading environments where high-temperature (~400 °C) seawater-derived fluids interact with mafic-bearing rocks. Analyses of rocks from the Southwest Indian Ridge suggest that the plutonic foundation of the oceanic crust may be an immense reservoir (~10^{19} g CH_4) for abiogenic CH_4 formed in a mid-ocean ridge hydrothermal system.

There is still much that we do not know about volatiles during crustal formation and aging, however. Future research must focus on obtaining a better understanding of the origin and distribution of volatiles within the mantle and crust if we are to accurately establish geochemical fluxes (e.g., of CO_2, CH_4, H_2S, H_2) in ridge-crest systems. In addition, more detailed studies of volatiles associated with subduction-zone environments are required. If generation of CH_4 (+ H_2 + hydrocarbons) is common in the deep crust during hydration of deep-seated mafic and mantle material, then it is likely that hydration of the upper mantle also produces significant amounts of CH_4. Particular emphasis should be placed on obtaining a greater knowledge about the C cycle in shallow- and deeper-crustal environments and the biological interactions that contribute and modify C during fluid, rock, and microbial interactions. Success will require quantification and characterization of microorganisms within the seafloor. Data from recent submarine eruptive events supports the hypothesis that subseafloor habitats host a significant biomass, but to date very little is known about the nature of this biotope and the depth or extent to which microorganisms exist within the seafloor.

ACKNOWLEDGMENTS

We thank D. Butterfield, C. Manning, and D. Vanko for their thoughtful reviews and comments, which aided in refinement of this manuscript. This work was supported by National Science Foundation grant OCE-9529833 to the University of Washington (to Kelley), contribution 2211, and Swiss National Science Foundation grant 2-7-016-95 (to Früh-Green).

REFERENCES CITED

Abrajano, T.A., Sturchio, N.C., Bohlke, J.K., Lyon, G.L., Poreda, R.J., and Stevens, C.M., 1988, Methane-hydrogen gas seeps, Zambales ophiolite, Philippines: Deep or shallow origin?: Chemical Geology, v. 71, p. 211–222.

Agar, S.M., and Lloyd, E. 1997, Deformation of Fe-Ti oxides in gabbroic shear zones from the MARK area, *in* Karson, J.A., Cannat, M., Miller, D.J., and Elthon, D., eds., Proceedings of the Ocean Drilling Program, Scientific Results, v. 153: College Station, Texas, Ocean Drilling Program, p. 123–141.

Allen, D.A., Berndt, M.E., Seyfried, W.E., Jr., and Horita, J., 1998, Inorganic reduction of CO_2 to HCOOH, CH_4, and other reduced carbon compounds with application to subseafloor hydrothermal systems [abs.]: Eos, Transactions of the American Geophysical Union, v. 79, p. 58–59.

Alt, J.C., 1995, Sulfur isotopic profile through the oceanic crust: Sulfur mobility and seawater-crustal sulfur exchange during hydrothermal alteration: Geology, v. 23, p. 585–588.

Alt, J.C., Honnorez, J., Laverne, C., and Emmermann, R., 1986, Alteration of a 1 km section through the upper oceanic crust, DSDP Hole 504B: The mineralogy, chemistry, and evolution of seawater-basalt interactions: Journal of Geophysical Research, v. 91, p. 10309–10355.

Alt, J.C., Anderson, T.F., Bonnell, L., and Muehlenbachs, K., 1989, The mineralogy, chemistry, and stable isotopic composition of hydrothermally altered sheeted dikes, DSDP Hole 504B, Leg 111, *in* Becker, K., Sakai., H., et al., eds., Proceedings of the Ocean Drilling Program, Scientific Results, v. 111: College Station, Texas, Ocean Drilling Program, p. 27–40.

Alt, J.C., and Anderson, T.F., 1991, Mineralogy and isotopic composition of sulfur in layer 3 gabbros from the Indian Ocean, Hole 735B, *in* Von Herzen, R.P., Fox, J., Palmer, A., and Robinson, P.T., eds., Proceedings of the Ocean Drilling Program, Scientific Results, v. 118: College Station, Texas, Ocean Drilling Program, p. 113–125.

Alt, J.C., Laverne, C., Vanko, D.A., Tartarotti, P., Teagle, D.A.H., Bach, W., Zuleger, E., Erzinger, J., Honnorez, J., Pezard, P.A., Becker, K., Salisbury, M.H., and Wilkens, R.H., 1996, Hydrothermal alteration of a section of upper oceanic crust in the eastern equatorial Pacific: A synthesis of results from Site 504 (DSDP Legs 69, 70, and 83 and ODP Legs 111, 137, 140, and 148), *in* Alt, J.C., Kinoshita, H., Stokking, L.B., and Michael, P.J., eds., Proceedings of the Ocean Drilling Program, Scientific Results, v. 148: College Station, Texas, Ocean Drilling Program, p. 417–434.

Baker, E.T., 1998, Patterns of event and chronic hydrothermal venting following a magmatic intrusion: New perspectives from the 1996 Gorda Ridge eruption: Deep Sea Research II, Event Detection and Response: Gorda Ridge, v. 45, p. 2599–2618.

Baker, E.T., Massoth, G.J., and Feeley, R.A., 1987, Cataclysmic hydrothermal venting on the Juan de Fuca Ridge: Nature, v. 329, p. 149–151.

Baker, E.T., Lavelle, J.W., Feeley, R.A., Massoth, G.J., Walker, S.L., and Lupton, J.E., 1989, Episodic venting of hydrothermal fluids from the Juan de Fuca Ridge: Geophysical Research Letters, v. 94, p. 9237–9250.

Baker, E.T., German, C.R., and Elderfield, H., 1995a, Hydrothermal plumes over spreading-center axes: Global distributions and geological inferences, *in*

Humphris, S.E., Zierenberg, R.A., Mullineaux, L.S., and Thompson, R.E., eds., Seafloor hydrothermal systems: Physical, chemical, biological, and geological interactions: Washington, D.C., American Geophysical Union Monograph, v. 91, p. 47–71.

Baker, E.T., Massoth, G.J., and Feeley, R.A., 1995b, Hydrothermal event plumes from the CoAxial seafloor eruption site, Juan de Fuca Ridge: Geophysical Research Letters, v. 22, p. 147–150.

Barnes, H.L., 1979, Geochemistry of hydrothermal ore deposits: New York, John Wiley and Sons, 798 p.

Baross, J.A., and Deming, J.W., 1995, Growth at high temperatures: Isolation and taxonomy, physiology, and ecology, *in* Karl, D.M., ed., The microbiology of deep-sea hydrothermal vents: Baton Rouge, CRC Press, p. 169–217.

Baross, J.A., and Hoffman, S.E., 1985, Submarine hydrothermal vents and associated gradient environments as sites for the origin and evolution of life: Origins of Life, v. 15, p. 327–345.

Berndt, M.E., and Seyfried, W.E., Jr., 1990, Boron, bromine, and other trace elements as clues to the fate of chlorine in mid-ocean ridge vent fluids: Geochimica et Cosmochimica Acta, v. 54, p. 2235–2245.

Berndt, M.E., Allen, D.E., and Seyfried, W.E., Jr., 1996, Reduction of CO_2 during serpentinization of olivine at 300 °C and 500 bar: Geology, v. 24, p. 351–354.

Bischoff, J.L., and Rosenbauer, R.J., 1988, Liquid-vapor relations in the critical region of the system NaCl-H_2O from 380 °C to 415 °C: A refined determination of the critical point and two-phase boundary of seawater: Geochimica et Cosmochimica Acta, v. 52, p. 2121–2126.

Bischoff, J.L., and Rosenbauer, R.J., 1989, Salinity variation in submarine hydrothermal systems by layered double-diffusive convection: Journal of Geology, v. 97, p. 613–623.

Bloomer, S.H., Meyer, P.S., Dick, H.J.B., Ozawa, K., and Natland, J.H., 1991, Textural and mineralogical variations in gabbroic rocks from Hole 735B, *in* Von Herzen, R.P., Fox, J., Palmer, A., and Robinson, P.T., eds., Proceedings of the Ocean Drilling Program, Scientific Results, v. 118: College Station, Texas, Ocean Drilling Program, p. 21–39.

Bodnar, R.J., and Beane, R.E., 1980, Temporal and spatial variations in hydrothermal fluid characteristics during vein filling in preore cover overlying deeply-buried porphyry copper-type mineralization at Red Mountain, Arizona: Economic Geology, v. 75, p. 876–893.

Bougault, H., Charlou, J.-L., Fouquet, Y., Needham, H.D., Vaslet, N., Appriou, P., Baptiste, P.J., Rona, P.A., Dmitried, L., and Silanties, S., 1993, Fast- and slow-spreading ridges: Structure and hydrothermal activity, ultramafic topographic highs, and CH_4 output: Journal of Geophysical Research, v. 98, p. 9643–9651.

Burnham, C.W., 1979, Magma and hydrothermal fluids, *in* Barnes, H.L., ed., Geochemistry of hydrothermal ore deposits: New York, John Wiley and Sons, p. 71–133.

Butterfield, D.A., Massoth, G.J., McDuff, R.E., Lupton, J.E., and Lilley, M.D., 1990, Geochemistry of hydrothermal fluids from Axial Seamount hydrothermal emissions study vent field, Juan de Fuca Ridge: Sub-seafloor boiling and subsequent fluid-rock alteration: Journal of Geophysical Research, v. 95, p. 12895–12921.

Butterfield, D.A., McDuff, R.E., Mottl, M.J., Lilley, M.D., Lupton, J.E., and Massoth, G.J., 1994, Gradients in the composition of hydrothermal fluids from the Endeavour segment vent field: Phase separation and brine loss: Journal of Geophysical Research, v. 99, p. 9561–9583.

Butterfield, D.A., Jonasson, I.R., Massoth, G.J., Feely, R.A., Roe, K.K., Embley, R.E., Holden, J.F., McDuff, R.E., Lilley, M.D., and Delaney, J.R., 1997, Seafloor eruptions and evolution of hydrothermal fluid chemistry: Philosophical Transactions of the Royal Society of London, A, v. 355, p. 369–386.

Casey, J.F., and Karson, J.A., 1981, Magma chamber profiles from the Bay of Islands ophiolite complex: Nature, v. 292, p. 295–301.

Chadwick, W.W., Jr., Embley, R.W., and Shank, T.M., 1998, The 1996 Gorda Ridge eruption: Geologic mapping, sidescan sonar, and Seabeam comparison results: Deep Sea Research II, Event Detection and Response: Gorda Ridge, v. 45, p. 2547–2570.

Charlou, J.L., and Donval, J.-P., 1993, Hydrothermal methane venting between 12°N and 26°N along the Mid-Atlantic Ridge: Journal of Geophysical Research, v. 98, p. 9625–9642.

Charlou, J.L., Bougault, H., Appriou, P., Nelsen, T., and Rona, P., 1992, Different TDM/CH_4 hydrothermal plume signatures: TAG site at 26°N and serpentinized ultrabasic diapir at 15°05 on the Mid-Atlantic Ridge: Geochimica et Cosmochimica Acta, v. 55, p. 3209–3223.

Christie, D.M., Carmichael, I.S.E., and Langmuir, C.H., 1986, Oxidation states of mid-ocean ridge basalt glasses: Earth and Planetary Science Letters, v. 79, p. 397–411.

Cline, J.S., and Bodnar, R.J., 1991, Can economic porphyry copper mineralization be generated by a typical calc-alkaline melt?: Journal of Geophysical Research, v. 96, p. 8113–8126.

Cormier, M.H., Detrick, R.S., and Purdy, G.M., 1984, Anomalously thin crust in oceanic fracture zones, new seismic constraints from the Kane Fracture Zone: Journal of Geophysical Research, v. 89, p. 10249–10266.

Cowan, J., and Cann, J., 1988, Supercritical two-phase separation of hydrothermal fluids in the Troodos ophiolite: Nature, v. 333, p. 259–261.

Cowen, J.P., Bertram, M.A., Baker, E.T., Feeley, R.A., Massoth, G.J., and Summit, M., 1998, Geomicrobial transformation of manganese in Gorda Ridge event plumes: Deep Sea Research II, Event Detection and Response: Gorda Ridge, v. 45, p. 2713–2738.

Craig, H., and Lupton, J.E., 1976, Primordial neon, helium, and hydrogen in oceanic basalts: Earth and Planetary Science Letters, v. 31, p. 369–385.

de Angelis, M.A., Lilley, M.D., Olson, E.J., and Baross, J.A., 1993, Methane oxidation in deep-sea hydrothermal plumes of the Endeavour Segment of the Juan de Fuca Ridge: Deep-Sea Research, v. 40, p. 1169–1186.

Delaney, J.R., Muenow, D.W., and Garcia, M.O., 1978, Abundance and distribution of water, carbon, and sulfur in the glassy rims of submarine pillow basalts: Geochimica et Cosmochimica Acta, v. 42, p. 581–594.

Delaney, J.R., Mogk, D.W., and Mottl, M.J., 1987, Quartz-cemented sulfide-bearing greenstone breccias from the Mid-Atlantic Ridge: Samples of a high-temperature hydrothermal upflow zone: Journal of Geophysical Research, v. 92, p. 9175–9192.

Delaney, J.R., Robigou, V., McDuff, R.E., and Tivey, M.K., 1992, Geology of a vigorous hydrothermal system on the Endeavour Segment, Juan de Fuca Ridge: Journal of Geophysical Research, v. 97, p. 19663–19682.

Delaney, J.R., Kelley, D.S., Lilley, M.D., Butterfield, D.A., McDuff, R.E., Baross, J.A., Deming, J.W., Johnson H.P., and Robigou, V., 1997, The Endeavour Hydrothermal System I: Cellular circulation above an active cracking front yields large sulfide structures, "fresh" vent water, and hyperthermophilic archae: RIDGE Events, v. 8, n. 2, p. 11–19.

Delaney, J.R., Kelley, D.S., Lilley, M.D., Butterfield, D.A., Baross, J.A., Embley, R.W., and Summit, M., 1998, The quantum event of oceanic crustal accretion: Impacts of diking at mid-ocean ridges: Science, v. 281, p. 222–230.

Deming, J.W., and Baross, J.A., 1993, Deep-sea smokers: Windows to a subsurface biosphere: Geochimica et Cosmochimica Acta, v. 57, p. 3219–3230.

Detrick, R.S., Fox, P.J., Kastens, K., Bryan, W.B.F., Mayer, L., Karson, J., and Pockalny, R.A., 1984, Seabeam survey of the Kane Fracture Zone and adjacent Mid-Atlantic ridge rift valley [abs.]: Eos, Transactions of the American Geophysical Union, v. 65, p. 1006.

Dick, H.J.B., and Natland, J.H., 1996, Late-stage melt evolution and transport in the shallow mantle beneath the East Pacific rise, *in* Mével, C., Gillis, K.M., Allan, J.F., and Meyer, P.S., eds., Proceedings of the Ocean Drilling Program, Scientific Results, v. 147: College Station, Texas, Ocean Drilling Program, p. 103–134.

Dick, H.J.B, Meyer, P.S., Bloomer, S., Kirby, S., Stakes, D., and Mawer, C., 1991, Lithostratigraphic evolution of an in-situ section of oceanic crust, *in* Von Herzen, R.P., Fox, J., Palmer, A., and Robinson, P.T., eds., Proceedings of the Ocean Drilling Program, Scientific Results, v. 118: College Station, Texas, Ocean Drilling Program, p. 439–538.

Dixon, J.E., and Stolper, E.M., 1995, An experimental study of water and carbon dioxide solubilities in mid-ocean ridge basaltic liquids. Part II. Applications to degassing: Journal of Petrology, v. 36, p. 1633–1646.

Dixon, J.E., Stolper, E., and Delaney, J.R., 1988, Infrared spectroscopic measurements of CO_2 and H_2O in Juan de Fuca Ridge basaltic glasses: Earth and Planetary Science Letters, v. 90, p. 87–104.

Dixon, J.E., Stolper, E.M., and Holloway, J.R., 1995, An experimental study of water and carbon dioxide solubilities in mid-ocean ridge basaltic liquids. Part I. Calibration and solubility models: Journal of Petrology, v. 36, p. 1607–1631.

Embley, R.W., Chadwik, W.W., Jr., Jonasson, I.R., Butterfield, D.A., and Baker, E.T., 1995, Initial results of the rapid response to the 1993 CoAxial event: Relationships between hydrothermal and volcanic processes: Geophysical Research Letters, v. 22, p. 143–146.

Evans, E.C., White, L.D., and Rapp, J.B., 1988, Geochemistry of some gases in hydrothermal fluids from the Southern Juan de Fuca Ridge: Journal of Geophysical Research, v. 93, p. 15305–15313.

Fine, G.J., and Stolper, E.M., 1986, Dissolved carbon dioxide in basaltic glasses: Concentrations and speciation: Earth and Planetary Science Letters, v. 76, p. 263–278.

Fisher, C.R., 1995, Toward an appreciation of hydrothermal-vent animals: Their environment, physiological ecology, and tissue stable isotope values, *in* Humphris, S.E., Zierenberg, R.A., Mullineaux, L.S., and Thompson, R.E., eds., Seafloor hydrothermal systems: Physical, chemical, biological, and geological interactions: Washington, D.C., American Geophysical Union Monograph, v. 91, p. 297–316.

Fornari, D.J., and Embley, R.W., 1995, Tectonic and volcanic controls on hydrothermal processes at the mid-ocean ridge: An overview based on near-bottom and submersible studies, *in* Humphris, S.E., Zierenberg, R.A., Mullineaux, L.S., and Thompson, R.E., eds., Seafloor hydrothermal systems: Physical, chemical, biological, and geological interactions: Washington, D.C., American Geophysical Union Monograph, v. 91, p. 1–46.

Fornari, D.J., Shank, T., Von Damm, K.L., Gregg, T.K.P., Lilley, M.D., Levai, G., Bray, A., Haymon, R.M., Perfit, M.R., and Lutz, R., 1998, Time-series temperature measurements at high-temperature hydrothermal vents, East Pacific Rise 9°49′–51′N: Evidence for monitoring a crustal cracking event: Earth and Planetary Science Letters, v. 160, p. 419–431.

Fournier, R.O., 1987, Conceptual models of brine evolution in magmatic-hydrothermal systems: U.S. Geological Survey Professional Paper, v. 1350, p. 1487–1506.

Fox, C.G., 1990, Consequences of phase separation on the distribution of hydrothermal fluids at ASHES vent field, Axial Volcano, Juan de Fuca Ridge: Journal of Geophysical Research, v. 95, p. 12923–12926.

Fox, C.G., and Dziak, R.P, 1998, Hydroacoustic detection of volcanic activity on the Gorda Ridge, February–March, 1996: Deep Sea Research II, Event Detection and Response: Gorda Ridge, v. 45, p. 2513–2530.

Frantz, J.D., Popp, R.K., and Hoering, T.C., 1992, The compositional limits of fluid immiscibility in the system H_2O-NaCl-CO_2 as determined with the use of synthetic fluid inclusions in conjunction with mass spectrometry: Chemical Geology, v. 98, p. 237–256.

Frost, B.R., 1985, On the stability of sulfides, oxides, and native metals in serpentinite: Journal of Petrology, v. 26, p. 31–63.

Früh-Green, G.L., Plas, A., and Lécuyer, C., 1996a, Petrologic and stable isotope constraints on hydrothermal alteration and serpentinization of the EPR shallow mantle at Hess Deep (Site 895), *in* Mével, C., Gillis, K.M., Allen, J.F., and Meyer, P.S., eds., Proceedings of the Ocean Drilling Program, Scientific Reports, v. 147: College Station, Texas, Ocean Drilling Program, p. 255–292.

Früh-Green, G.L., Plas, A., and Dell'Angelo, L.N., 1996b, Mineralogical and stable isotope record of polyphase alteration of upper crustal gabbros of the East Pacific Rise (Hess Deep, Site 894), *in* Mével, C., Gillis, K.M., Allen, J.F., and Meyer, P.S., eds., Proceedings of the Ocean Drilling Program, Scientific Reports, v. 147: College Station, Texas, Ocean Drilling Program, p. 235–254.

Früh-Green, G.L., and Kelley, D.S., 1998, Volatiles in Mid-Ocean Ridges [abs.]: Eos, Transactions of the American Geophysical Union, v. 79, p. 45.

Gallinatti, B.C., 1984, Initiation and collapse of active circulation in a hydrothermal system at the Mid-Atlantic Ridge, 23°N: Journal of Geophysical Research, v. 89, p. 3275–3289.

Gao, D.L., Hurst, S.D., Karson, J.A., Delaney, J.R., and Spiess, F.N., 1998, Computer-aided interpretation of side-looking sonar images from the eastern intersection of the Mid-Atlantic Ridge with the Kane Transform: Journal of Geophysical Research, v. 103, p. 20997–21014.

Gente, P., Mével, C., Auzende, J., Karson, J.A., and Fouquet, Y., 1991, An example of recent accretion on the Mid-Atlantic Ridge: The Snake Pit neovolcanic ridge (MARK area, 22°22′N): Tectonophysics, v. 190, p. 1–29.

Gerlach, T.M., and Nordlie, B.E., 1975, The C-O-H-S gaseous system. 1. Composition limits and trends in basaltic cases: American Journal of Science, v. 275, p. 353–376.

Gerlach, T.M., and Taylor, B.E., 1990, Carbon isotope constraints on degassing of carbon dioxide from Kilauea volcano: Geochimica et Cosmochimica Acta, v. 54, p. 2051–2058.

Gillis, K.M., 1995, Controls on hydrothermal alteration in a section of fast-spreading oceanic crust: Earth and Planetary Science Letters, v. 134, p. 473–489.

Gillis, K.M., and Robinson, P.T., 1990, Patterns and processes of alteration in the lavas and dykes of the Troodos ophiolite, Cyprus: Journal of Geophysical Research, v. 93, p. 21523–21548.

Gillis, K.M., and Thompson, G., 1993, Metabasalts from the Mid-Atlantic Ridge: New insights into hydrothermal systems in slow-spreading crust: Contributions to Mineralogy and Petrology, v. 113, p. 502–523.

Gillis, K.M., Thompson, G., and Kelley, D.S., 1993, A view of the lower crustal component of hydrothermal systems at the Mid-Atlantic Ridge: Journal of Geophysical Research, v. 98, p. 19597–19619.

Goldfarb, M.S., and Delaney, J.R., 1988, Response of two-phase fluids to fracture configurations within submarine hydrothermal systems: Journal of Geophysical Research, v. 93, p. 4585–4594.

Gu, C., and Vanko, D.A., 1996, Nature of fluid inclusions in samples of the deep sheeted dikes cored during Leg 148 (Hole 504B), *in* Alt, J.C., Kinoshita, H., Stokking, L.B., and Michael, P.J., eds., Proceedings of the Ocean Drilling Program, Scientific Results, v. 148: College Station, Texas, Ocean Drilling Program, p. 87–95.

Gustafson, L.B., and Hunt, J.P., 1975, The porphyry copper deposit at El Salvador, Chile: Economic Geology, v. 70, p. 857–912.

Haggerty, J.A., 1989, Fluid inclusion studies of chimneys associated with serpentinite seamounts in the Mariana Forearc [abs.], Proceedings, Second Pan-American Conference on Fluid Inclusions, p. 29.

Halbach, P., Nakamura, Ko-ichi, Wahsner, M., Lange, J., Sakai, H., Kaselitz, L., Hansen, R.D., Yamano, M., Post, J., Prause, B., Seifert, R., Michaelis, W., Teichmann, F., Kinoshita, M., Marten, A., Ishibashi, J., Czerwinski, S., and Blum, N., 1989, Probably modern analogue of Kuroko-type massive sulfide deposits in the Okinawa Trough back-arc basin: Nature, v. 338, p. 496–499.

Haymon, R.M., Fornari, D.J., Von Damm, K.L., Lilley, M.D., Perfit, M.R., Edmond, J.M., Shanks, W.C., Lutz, R.A., and Grebmeier, J.M., 1993, Volcanic eruption of the mid-ocean ridge along the East Pacific Rise crest at 9°45–52′N: Direct submersible observations of seafloor phenomena associated with an eruption event in April, 1991: Earth and Planetary Science Letters, v. 119, p. 85–101.

Holden, J.F., Summit, M., and Baross, J.A., 1998, Thermophilic and hyperthermophilic microorganisms in 3–30 °C hydrothermal fluids following a deep-sea volcanic eruption: FEMS Microbiology Ecology, v. 25, p. 33–41.

Holland, H.D., 1972, Granites, solutions, and base metal deposits: Economic Geology, v. 67, p. 231–235.

Holloway, J., 1984, Graphite-CH_4-H_2O-CO_2 equilibria at low-grade metamorphic conditions: Geology, v. 12, p. 455–458.

Honnorez, J., Alt, J.C., Honnorez-Guerstein, B.-M., Laverne, C., Muehlenbachs, K., Ruiz, J., and Sultzman, E., 1985, Stockwork-like sulfide mineralization in young oceanic crust, *in* Anderson, R.N., Honnorez, J., Becker, K., et al., eds., Deep Sea Drilling Project Hole 504B, Initial Reports of the Deep Sea Drilling Program: Washington, D.C., U.S. Government Priniting Office, v. 83, p. 263–282.

Javoy, M., and Pineau, F., 1991, The volatile record of a "popping" rock from the Mid-Atlantic Ridge at 14°N: Chemical and isotopic composition of gas trapped in the vesicles: Earth and Planetary Science Letters, v. 107, p. 598–611.

Jehl, V., Poty, B., and Weisbrod, A., 1977, Hydrothermal metamorphism of the oceanic crust in the North Atlantic Ocean, Bulletin de la Societe Geologique de France, v. 19, n. 6, p. 1213–1221.

Johnson, M.C., Anderson, A.T., Jr., and Rutherford, M.J., 1994, Pre-eruptive volatile contents of magmas, *in* Carroll, M.R., and Holloway, J.R., eds., Volatiles in magmas: Mineralogical Society of America, Reviews in Mineralogy, v. 30, p. 281–330.

Juniper, S.K., Bird, D.F., Summit, M., Vong, M.P., and Baker, E.T., 1998, Bacterial and viral abundances in hydrothermal event plumes over northern Gorda Ridge: Deep Sea Research II, Event Detection and Response: Gorda Ridge, v. 45, p. 2739–2750.

Karl, D.M., 1995, Ecology of free-living, hydrothermal vent microbial communities, *in* Karl, D.M., ed., The microbiology of deep-sea hydrothermal vents: Baton Rouge, Louisiana, CRC Press, p. 35–124.

Karson, J.A., and Dick, H.J.B., 1983, Tectonics of ridge-transform intersections at the Kane Fracture Zone: Marine Geophysical Research, v. 6, p. 51–98.

Karson, J.A., et al., 1987, Along-axis variations in seafloor spreading in the MARK area: Nature, v. 328, p. 681–685.

Karson, J.A., and Lawerence, R., 1997, Tectonic setting of serpentinite exposures on the western median valley wall of the MARK area in the vicinity of Site 920, *in* Karson, J.A., Cannat, M., Miller, D.J., and Elthon, D., eds., Proceedings of the Ocean Drilling Program, Scientific Results, v. 153: College Station, Texas, Ocean Drilling Program, p. 5–21.

Kelley, D.S., 1996, Methane-rich fluids in the oceanic crust: Journal of Geophysical Research, v. 101, p. 2943–2962.

Kelley, D.S., 1997, Fluid evolution in slow-spreading environments, *in* Karson, J.A., Cannat, M., Miller, D.J., and Elthon, D., eds., Proceedings of the Ocean Drilling Program, Scientific Results, v. 153: College Station, Texas, Ocean Drilling Program, p. 399–415.

Kelley, D.S., and Delaney, J.R., 1987, Two-phase separation and fracturing in mid-ocean ridge gabbros at temperatures greater than 700 °C: Earth and Planetary Science Letters, v. 83, p. 53–66.

Kelley, D.S., and Früh-Green, G.L., 1999, Abiogenic methane in deep-seated mid-ocean ridge environments: Insights from stable isotope analyses: Journal of Geophysical Research, v. 104, p. 10439–10460.

Kelley, D.S., and Malpas, J., 1996, Melt-fluid evolution in gabbroic rocks from the Hess Deep, Ocean Drilling Leg 147, *in* Mével, C, Gillis, K.M., Allan, J.F., and Meyer, P.S., eds., Proceedings of the Ocean Drilling Program, Scientific Results, v. 147: College Station, Texas, Ocean Drilling Program, p. 213–227.

Kelley, D.S., and Robinson, P.T., 1990, Development of a brine-dominated hydrothermal system at temperatures of 400–500 °C in the upper level plutonic sequence, Troodos ophiolite, Cyprus, Geochimica et Cosmochimica Acta, v. 54, p. 653–661.

Kelley, D.S., Robinson, P.T., and Malpas, J.G., 1992, Processes of brine generation and circulation in the oceanic crust: Fluid inclusion evidence from the Troodos ophiolite, Cyprus: Journal of Geophysical Research, v. 97, p. 9307–9322.

Kelley, D.S., Gillis, K.M., and Thompson, G., 1993, Fluid evolution in submarine magma-hydrothermal systems at the Mid-Atlantic Ridge: Journal of Geophysical Research, v. 98, p. 19579–19596.

Kelley, D.S., Vanko, D.A., and Gu, C., 1995, Fluid evolution in oceanic crustal layer 2: Fluid inclusion evidence from the sheeted dike complex, Hole 504B, Costa Rica Rift, *in* Stokking, L.B., Dick, H.J.B., and Erzinger, J., eds., Proceedings of the Ocean Drilling Program, Scientific Results: College Station, Texas, Ocean Drilling Program, v. 140, p. 191–198.

Kelley, D.S., Lilley, M.D., Lupton, J.E., and Olson, E.J., 1998, Enriched H_2, CH_4 and ^3He concentrations in hydrothermal plumes associated with the 1996 Gorda Ridge eruptive event: Deep Sea Research II, Event Detection and Response: Gorda Ridge, v. 45, p. 2665–2682.

Kilinc, I.A., and Burnham, C.W., 1972, Partitioning of chloride between a silicate melt and coexisting aqueous phase from 2–8 kilobars: Economic Geology, v. 67, p. 231–235.

Kim, K.R., Welhan, J.A., and Craig, H., 1984, The hydrothermal vent fields at 13°N and 11°N on the EPR: Alvin 1984 results [abs.]: Eos, Transactions of the American Geophysical Union, v. 65, p. 973.

Knapp, R.B., and Knight, J.E., 1977, Differential thermal expansion of pore fluids: Fracture propagation and microearthquake production in hot pluton environments: Journal of Geophysical Research, v. 82, p. 2515–2522.

Kong, L.S., Detrick, R.S., Fox, P.J., Mayer, L.A., and Bryan, W.F.B., 1988, The morphology and tectonics of the MARK area from Sea Beam and Sea MARK I observations (Mid-Atlantic Ridge 23°N): Marine Geophysical Research, v. 10, p. 59–90.

Kyser, T.K., and O'Neil, J.R., 1984, Hydrogen systematics of submarine basalts: Geochimica et Cosmochimica Acta, v. 48, p. 2123–2133.

Lilley, M.D., Baross, J.A., and Gordon, L.I., 1983, Reduced gases and bacteria in hydrothermal fluids: The Galapagos spreading center and 21°N East Pacific Rise, *in* Rona, P.A., Bostrom, K., Laubier, L., and Smith, K.L., Jr., eds., Hydrothermal processes at seafloor spreading centers: New York, Plenum Press, p. 411–447.

Lilley, M.D., Olson, E.J., Lupton, J.E., and Von Damm, K.L., 1992, Volatiles in the 9°N hydrothermal system: A comparison of 1991 and 1992 data [abs.]: Eos, Transactions of the American Geophysical Union, v. 73, p. 524.

Lilley, M.D., Butterfield, D.A., Olson, E.J., Lupton, J.E., Macko, S.A., and McDuff, R.E., 1993, Anomalous CH_4 and NH_4^+ concentrations at an unsedimented mid-ocean ridge hydrothermal system: Nature, v. 364, p. 45–47.

Lilley M.D., Feely, R.A., and Trefry, J.H., 1995, Chemical and biological transformations in hydrothermal plumes, *in* Humphris, S.E., Zierenberg, R.A., Mullineaux, L.S., and Thompson, R.E., eds., Seafloor hydrothermal systems: Physical, chemical, biological, and geological interactions: Washington, D.C., American Geophysical Union Monograph, v. 91, p. 369–391.

Lister, C.R.B., 1983, The basic physics of water penetration into hot rock, *in* Rona, P.A., Bostrom, K., Laubier, L., and Smith, K.L., Jr., eds., Hydrothermal processes at seafloor spreading senters: New York, Plenum Press, p. 141–176.

Lupton, J.E., 1995, Hydrothermal plumes: Near and far field, *in* Humphris, S.E., Zierenberg, R.A., Mullineaux, L.S., and Thompson, R.E., eds., Seafloor hydrothermal systems: Physical, chemical, biological, and geological interactions: Washington, D.C., American Geophysical Union Monograph, v. 91, p. 317–346.

Lupton, J.E., Baker, E., and Massoth, G., 1989, Variable ^3He/heat ratios in submarine hydrothermal systems: Evidence from two plumes over the Juan de Fuca Ridge: Nature, v. 337, p. 161–164.

MacLeod, C.J., Celerier, B., Früh-Green, G.L., and Manning, C.E., 1996, Tectonics of Hess Deep: A synthesis of drilling results from Leg 147, *in* Mével, C., Gillis, K.M., Allan, J.F., and Meyer, P.S., eds., Proceedings of the Ocean Drilling Program, Scientific Results, v. 147: College Station, Texas, Ocean Drilling Program, p. 461–476.

Malpas, J., 1990, Crustal accretionary processes in the Troodos ophiolite, Cyprus: Evidence from field mapping and deep crustal drilling, *in* Malpas, J., Moores, E., Panayiotou, A., and Xenophontos, C., eds., Ophiolites, oceanic crustal analogues: Proceedings of the Symposium "Troodos 1987": Nicosia, Cyprus, Geological Survey Department, Ministry of Agriculture and Natural Resources, p. 65–74.

Malpas, J.G., Brace, T., and Dunsworth, S.M., 1989, Structural and petrological relationships of the CY-4 drill hole of the Cyprus Crustal Study Project, *in* Gibson, I.L., Malpas, J., Robinson, P.T., and Xenophontos, X., eds., Cyprus Crustal Study Project: Initial Report, Hole CY-4: Geological Survey of Canada Paper (88–9), p. 39–106.

Manning, C.E., and MacLeod, C.J., 1996, Fracture-controlled metamorphism of Hess Deep gabbros, Site 894: Constraints on the roots of mid-ocean ridge hydrothermal systems at fast-spreading centers, *in* Mével, C., Gillis, K.M., Allan, J.F., and Meyer, P.S., eds., Proceedings of the Ocean Drilling Program, Scientific Results, v. 147: College Station, Texas, Ocean Drilling Program, p. 189–212.

Manning, C., Weston, P.E., and Mahon, K.I., 1996, Rapid high temperature metamorphism of the East Pacific Rise gabbros from Hess Deep: Earth and Planetary Science Letters, v. 144, p. 123–132.

Mathez, E.A., 1984, Influence of degassing on oxidation states of basaltic magmas: Nature, v. 310, p. 371–375.

Mathez, E.A., 1989a, Vapor associated with mafic magma and controls on its composition, *in* Whitney, J.A., and Naldrett, A.J., eds., Ore deposition associated with magmas: Reviews in Economic Geology, v. 9, p. 21–31.

Mathez, E.A., 1989b, Interactions involving fluids in the Stillwater and Bushveld complexes: Observations from the rocks, *in* Whitney, J.A., and Naldrett, A.J., eds., Ore deposition associated with magmas: Reviews in Economic Geology, v. 9, p. 167–179.

Mathez, E.A., and Delaney, J.R., 1981, The nature and distribution of primary carbon in submarine basalts and peridotite nodules: Earth and Planetary Science Letters, v. 56, p. 217–232.

Mathez, E.A., Dietrich, V.J., Holloway, J.R., and Boudreau, A.E., 1989, Carbon distribution in the Stillwater Complex and evolution of vapor during crystallization of Stillwater and Bushveld magmas: Journal of Petrology, v. 30, p. 153–173.

McLaughlin, E.A., 1998, Microbial hydrogen oxidation in hydrothermal systems [Ph.D. thesis]: University of Washington, 125 p.

McLaughlin, E.A., Lilley, M.D., Olson, E.J., and Rust, T., 1996, Microbial hydrogen oxidation in the Loihi event plumes: Microbial rapid response [abs.]: Eos, Transactions of the American Geophysical Union, v. 46, p. 398.

Merlivat, L., Pineau, F., and Javoy, M., 1987, Hydrothermal vent waters at 13°N on the East Pacific Rise: Isotopic composition and gas concentration: Earth and Planetary Science Letters, v. 84, p. 100–108.

Mével, C., and Cannat, M., 1991, Lithospheric stretching and hydrothermal processes in oceanic gabbros from slow-spreading ridges, *in* Peters, T.J., Nicolas, A., and Coleman, R.G., eds., Ophiolite genesis and evolution of the oceanic lithosphere: Hingham, Massachusetts, Kluwer Academic Publishers, p. 293–312.

Mével, C., Cannat, M., Gente, P., Marion, E., Auzende, J.M., and Karson, J.A., 1991, Emplacement of deep crustal and mantle rocks on the west median valley wall of the MARK area (MAR 23°N): Tectonophysics, v. 190, p. 31–53.

Meyer, P.S., and Sapp, K., 1994, Crystal size distribution, mineral chemistry, and crystallization history of ODP Site 735B gabbros [abs.]: Eos, Transactions of the American Geophysical Union, v. 75, p. 627.

Moore, J.G., 1977, Water content of basalt erupted on the ocean floor: Contributions to Mineralogy and Petrology, v. 28, p. 272–279.

Mottl, M.J., and Haggerty, J.A., 1989, Upwelling of Cl-poor, S- and C-rich waters through a serpentinite seamount, Mariana Forearc, ODP Leg 125 [abs.]: Eos, Transactions of the American Geophysical Union, v. 70, p. 1382.

Muenow, D.W., Garcia, M.O., Aggrey, K.E., Bednarz, U., and Schmincke, H.U., 1990, Volatiles in submarine glasses as a discriminant of tectonic origin: Application to the Troodos ophiolite: Nature, v. 343, p. 159–161.

Natland, J.H., and Dick, H.J.B., 1996, Melt migration through high-level gabbroic cumulates of the East Pacific Rise at Hess Deep: The origin of magma lenses and the deep crustal structure of fast-spreading ridges, *in* Mével, C., Gillis, K.M., Allan, J.F., and Meyer, P.S., eds., Proceedings of the Ocean Drilling Program, Scientific Results, v. 147: College Station, Texas, Ocean Drilling Program, p. 21–58.

Nehlig, P., 1991, Salinity of oceanic hydrothermal fluids: A fluid inclusion study: Earth and Planetary Science Letters, v. 102, p. 310–325.

Nehlig, P., and Juteau, T., 1988, Flow porosities, permeabilities, and preliminary data on fluid inclusions and fossil geothermal gradients in the crustal sequence of the Sumail ophiolite (Oman): Tectonophysics, v. 151, p. 199–221.

Norton, D., 1984, Theory of hydrothermal systems: Annual Review, Earth and Planetary Science Letters, v. 12, p. 155–157.

Ohmoto, H., and Skinner, B.J., 1983, The Kuroko and related volcanogenic massive sulfide deposits: Introduction and summary of new findings: Economic Geology Monograph, v. 5, p. 1–8.

Pallister, J.S., and Hopson, C.A., 1981, Semail ophiolite plutonic suite: Field relations, phase variation, cryptic variation and layering, and a model of a spreading ridge magma chamber: Journal of Geophysical Research, v. 86, p. 2593–2644.

Pawley, A.R., Holloway, J.R., and McMillan, P.F., 1992, The effect of oxygen fugacity on the solubility of carbon-oxygen fluids in basaltic melt: Earth and Planetary Science Letters, v. 110, p. 213–225.

Pedersen, R.B., Malpas, J., and Fallon, T., 1996, Petrology and geochemistry of gabbroic and related rocks from Site 894, Hess Deep, *in* Mével, C., Gillis, K.M., Allan, J.F., and Meyer, P.S., eds., Proceedings of the Ocean Drilling Program, Scientific Results, v. 147: College Station, Texas, Ocean Drilling Program, p. 3–20.

Pineau, F., and Javoy, M., 1983, Carbon isotopes and concentrations in mid-oceanic ridge basalts: Earth and Planetary Science Letters, v. 62, p. 239–257.

Pineau, F., and Javoy, M., 1994, Strong degassing at ridge crests: The behavior of dissolved carbon and water in basalt glasses at 14°N, Mid-Atlantic Ridge: Earth and Planetary Science Letters, v. 123, p. 179–198.

Pineau F., Javoy, M., and, Bottinga, Y., 1976, $^{13}C/^{12}C$ ratios of rocks and inclusions in popping rocks of the Mid-Atlantic Ridge: Earth and Planetary Science Letters, v. 29, p. 413–421.

Pockalny, R.A., Detrick, R.S., and Fox, P.J., 1988, Morphology and tectonics of the Kane Transform from Seabeam bathymetric data: Journal of Geophysical Research, v. 93, p. 3179–3194.

Purdy, G.M., and Detrick, R.S., 1986, The crustal structure of the Mid-Atlantic Ridge at 23°N from seismic refraction studies: Journal of Geophysical Research, v. 91, p. 3739–3762.

Reynolds, T.J., and Beane, R.E., 1985, Evolution of hydrothermal fluid characteristics at the Santa Rita, New Mexico, porphyry copper deposit: Economic Geology, v. 80, p. 1328–1347.

Richards, H.G., and Cann, J.R., 1989, Mineralogical zonation and metasomatism of the alteration pipes of Cyprus sulfide deposits: Economic Geology, v. 84, p. 91–115.

Richardson, C.J., Cann, J.R., Richards, H.G., and Cowan, J.G., 1987, Metal-depleted root zones of the Troodos ore-forming hydrothermal systems, Cyprus: Earth and Planetary Science Letters, v. 84, p. 243–253.

Roedder, E., 1984, Fluid inclusions: Mineralogical Society of America, Reviews in Mineralogy, v. 12, p. 664.

Rona, P.A., Bougault, H., Charlou, J.L., Appriou, P., Nelsen, T.A., Trfry, T.H., Eberhart, G.L., Barone, A., and Needham, H.D., 1992, Hydrothermal circulation, serpentinization, and degassing at a rift-valley fracture zone intersection: Mid-Atlantic Ridge near 15°N, 45°W: Geology, v. 20, p. 783–786.

Saccocia, P.J., and Gillis, K.M., 1995, Hydrothermal upflow zones in the oceanic crust: Earth and Planetary Science Letters, v. 136, p. 1–16.

Sansone, F.J., Resing, J.A., Tribble, G.W., Sedwick, P.N., Kelley, K.M., and Hon, K., 1991, Lava-seawater interactions at shallow-water submarine lava flows: Geophysical Research Letters, v. 18, p. 1731–1734.

Sansone, F.J., Wen, X., Holmes, M.E., Resing, J.A., Rust, T.M., and McLaughlin, E.A., 1996, Dissolved gases in Loihi hydrothermal plumes following the 1996 seismic event [abs.]: Eos, Transactions of the American Geophysical Union, v. 77, p. 398.

Sato, M., and Valenza, M., 1980, Oxygen fugacities of the layered series of the Skaergaard intrusion East Greenland: American Journal of Science, v. 280-A, p. 134–158.

Schiffman, P., and Smith, B.M., 1988, Petrology and oxygen isotope geochemistry of a fossil seawater hydrothermal system within the Solea Graben, northern Troodos ophiolite, Cyprus: Journal of Geophysical Research, v. 93, p. 4612–4624.

Schops, D., and Herzig, P.M., 1990, Sulfide composition and microthermometry of fluid inclusions in the Leg 111 sheeted dike section of Ocean Drilling Program Hole 504B, Costa Rica Rift: Journal of Geophysical Research, v. 95, p. 8405–8418.

Schulz, N.J., Detrick, R.S., and Miller, S.P., 1988, Two and three-dimensional inversions of magnetic anomalies in the MARK area (Mid-Atlantic Ridge 23°N): Marine Geophysical Research, v. 10, p. 41–58.

Seyfried, W.E., Jr., and Ding, K., 1995, Phase equilibria in subseafloor hydrothermal systems: A review of the role of redox, temperature, pH, and dissolved Cl on the chemistry of hot spring fluids at mid-ocean ridges, *in* Humphris, S.E., Zierenberg, R.A., Mullineaux, L.S., and Thompson, R.E., eds., Seafloor hydrothermal systems: Physical, chemical, biological, and geological interactions: Washington, D.C., American Geophysical Union Monograph, v. 91, p. 248–272.

Seyfried, W.E., Ding, K., and Berndt, M.E., 1991, Phase equilibria constraints on the chemistry of hot spring fluids at mid-ocean ridges: Geochimica et Cosmochimica Acta, v. 55, p. 3559–3580.

Shanks, W.C., III, Bohlke, J.K., and Seal, R.R., II, 1995, Stable isotopes in mid-ocean ridge hydrothermal systems: Interactions between fluids, minerals, and organisms, *in* Humphris, S.E., Zierenberg, R.A., Mullineaux, L.S., and Thompson, R.E., eds., Seafloor hydrothermal systems: Physical, chemical, biological, and geological Interactions: Washington, D.C., American Geophysical Union Monograph, v. 91, p. 194–221.

Shipboard Scientific Party, 1999, Initial Reports, Hole 735B, Southwest Indian Ridge, *in* Dick, H.J.B., Natland, J., and Miller, D.J., eds., Proceedings of the Ocean Drilling Program, Initial Results, v. 176: College Station, Texas, Ocean Drilling Program, p. 1–70.

Shock, E.L., 1990, Geochemical constraints on the origin of organic compounds in hydrothermal systems: Origins of Life, Evolution and the Biosphere, v. 20, p. 331–367.

Shock, E.L., 1992, Chemical environments of submarine hydrothermal systems: Origins of Life, Evolution and the Biosphere, v. 22, p. 67–107.

Sparks, R.S.J., Barclay, J., Jaupart, C., Mader, H.M., and Phillips, J.C, 1994, Physical aspects of magma degassing. I. Experimental and theoretical constrains on vesiculation, *in* Carroll, M., Hollowasy, J.R., and Ribbe, P.H., eds., Volatiles in magma: Reviews in Mineralogy, v. 30, p. 413–445.

Spooner, E.T.C., and Bray, C.J., 1977, Hydrothermal fluids of seawater salinity in ophiolitic sulfide ore deposits in Cyprus: Nature, v. 266, p. 808–812.

Stakes, D.S., 1991, Oxygen and hydrogen isotope compositions of oceanic plutonic rocks: high-temperature deformation and metamorphism of oceanic layer 3, *in* Taylor, H.P., Jr., O'Neil, J.R., and Kaplan, I.R., eds., Stable Isotope Geochemistry: A tribute to Samuel Epstein: San Antonio, Texas, Geochemical Society Special Publication 3, p. 77–90.

Stakes, D.S., and Vanko, D.A., 1986, Multistage hydrothermal alteration of gabbroic rocks from the failed Mathematician Ridge: Earth and Planetary Science Letters, v. 79, p. 75–92.

Stakes, D., Mével, C., Cannat, M., and Chaput, T., 1991, Metamorphic stratigraphy of Hole 735B, *in* Von Herzen, R.P., Fox, J., Palmer, A., and Robinson, P.T., Proceedings of the Ocean Drilling Program, v. 118: College Station, Texas, Ocean Drilling Program, p. 153–180.

Stolper, E., and Holloway, J.R., 1988, Experimental determination of the solubility of carbon dioxide in molten basalt at low pressure: Earth and Planetary Science Letters, v. 87, p. 397–408.

Summit, M., and Baross, J.A., 1998, Thermophilic subseafloor microorganisms from the 1996 North Gorda Ridge eruption: Deep Sea Research II, Event Detection and Response: Gorda Ridge, v. 45, p. 2751–2766.

Titley, S.R., and Beane, R.E., 1981, Porphyry copper deposits, Part I. Geologic setting, petrology, and tectogenesis: Economic Geology, Seventy-fifth Anniversary Volume, p. 214–234.

Vanko, D.A., 1988, Temperature, pressure, and composition of hydrothermal fluids, with their bearing on the magnitude of tectonic uplift at mid-ocean ridges, inferred from fluid inclusions in oceanic layer 3 rocks: Journal of Geophysical Research, v. 93, p. 4595–4611.

Vanko, D.A., and Stakes, D.S., 1991, Fluids in oceanic layer 3: Evidence from veined rocks, Hole 735B, Southwest Indian Ridge, *in* Von Herzen, R.P., Fox, J., Palmer, A., and Robinson, P.T., Proceedings of the Ocean Drilling Program, v. 118: College Station, Texas, Ocean Drilling Program, p. 181–215.

Vanko, D.A., Griffith, J.D., and Erickson, C.L., 1992, Calcium-rich brines and other hydrothermal fluids in fluid inclusions from plutonic rocks, Oceanographer Transform, Mid-Atlantic Ridge: Geochimica et Cosmochimica Acta, v. 56, p. 35–47.

Von Damm, K.L., 1990, Seafloor hydrothermal activity: Black smoker chemistry and chimneys: Annual Reviews of Earth and Planetary Science, v. 18, p. 173–204.

Von Damm, K.L., 1995, Controls on the chemistry and temporal variability of seafloor hydrothermal fluids, *in* Humphris, S.E., Zierenberg, R.A., Mullineaux, L.S., and Thompson, R.E., eds., Seafloor hydrothermal systems: Physical, chemical, biological, and geological interactions: Washington, D.C., American Geophysical Union Monograph, v. 91, p. 222–247.

Von Damm, K.L., Colodner, D.C., and Edmonds, J., 1992, Hydrothermal fluid chemistry at 9–10°N EPR: Big changes and still changing [abs.]: EOS, Transactions of the American Geophysical Union, v. 73, p. 524.

Von Damm, K.L., Buttermore, L.G., Oosting, S.E., Bray, A.M., Fornari, D.J., Lilley, M.D., and Shanks, W.C., III, 1997, Direct observation of the evolution of a seafloor 'black smoker' from a vapor to brine: Earth and Planetary Science Letters, v. 149, p. 101–111.

Welhan, J.A., 1988, Origins of methane in hydrothermal systems: Chemical Geology, v. 71, p. 183–198.

Welhan, J.A. and Craig, H., 1983, Methane, hydrogen, and helium in hydrothermal fluids of 21°N on the East Pacific Rise, *in* Rona, P.A., Bostrom, K., Laubier, L., and Smith, K.L., Jr., eds., Hydrothermal processes at seafloor spreading centers: New York, Plenum Press, p. 391–409.

Welhan, J.A. and Lupton, J.E., 1987, Light hydrocarbon gases in Guaymas Basin hydrothermal fluids: Thermogenic versus abiogenic origin: American Association of Petroleum Geologists Bulletin, v. 71, p. 215–223.

Wilcock, S.D., and Delaney, J.R., 1996, Mid-Ocean ridge sulfide deposits: Evidence for heat extraction from magma chambers of cracking fronts?: Earth and Planetary Science Letters, v. 145, p. 49–64.

Wyllie, P.J., 1977, Crustal anatexis: An experimental view: Tectonophysics, v. 43, p. 41–71.

Yang, K., and Scott, S.D., 1996, Possible contribution of a metal-rich magmatic fluid to a sea-floor hydrothermal system: Nature, v. 383, p. 420–423.

Yoshida, T., Nishizawa, K., Tabata, M., Abe, H., Kodama, T., Tsuji, M., and Tamaura, Y., 1993, Methanation of CO_2 with H_2-reduced magnetite: Journal of Material Science, v. 28, p. 1220–1226.

Zhang, Y., and Frantz, J.D., 1992, Hydrothermal reaction involving equilibrium between minerals and mixed volatiles. II. Investigations of fluid properties in the CO_2-CH_4-H_2O system using synthetic fluid inclusions: Chemical Geology, v. 100, p. 51–72.

Manuscript Accepted by the Society February 23, 2000

Printed in the U.S.A.

Geological Society of America
Special Paper 349
2000

Lower-crustal cracking front at fast-spreading ridges: Evidence from the East Pacific Rise and the Oman ophiolite

Craig E. Manning
Department of Earth and Space Sciences, University of California, Los Angeles, California 90095-1567, USA
Christopher J. MacLeod
Department of Earth Sciences, Cardiff University, Cardiff CF1 3YE, UK
Patricia E. Weston
Department of Earth and Space Sciences, University of California, Los Angeles, California 90095-1567, USA

ABSTRACT

The earliest fracturing of lower-crustal gabbros in the Oman ophiolite and the East Pacific Rise occurred in a distributed, microscopic, semibrittle crack network that formed primarily at grain boundaries. The newly created permeability provided conduits that first delivered fluid to the lower crust and drove fluid + rock reactions. The abundance of metamorphic amphibole and modified magmatic plagioclase so produced ranges from trace quantities to >50 modal percent. The initial crack network differs in many respects from previous models. Amphibole and plagioclase compositions show that temperatures of initial cracking are high, ranging from ~700 °C near the dike-to-gabbro transition to ~825 °C near the petrologic Moho. An increase in temperatures of initial cracking with depth implies that strain rate increased with depth or that mineralogic controls on rheology or hydrolytic-weakening behavior varied through the crust. Comparison with thermal models shows that the semibrittle cracking front moves away from the ridge axis as it penetrates downward in the oceanic crust. For example, models involving rapid lower crustal cooling imply that the cracking front penetrates the dike-gabbro transition within ~1 km of the ridge and reaches the Moho within ~6 km. We suggest that microcracking occurred episodically because continuous advance of a cracking front over these distances implies velocities that are slow compared to reaction rates. The microscopic fracturing scale and inferred low fluid flux suggest that associated heat flow was conductive. These observations provide new insights into the brittle-plastic transition and fluid-rock interaction in the lower oceanic crust formed at fast-spreading ridges.

INTRODUCTION

The onset of fracturing is an important event in the evolution of the lower oceanic crust. In fast-spreading environments, where ductile shear zones are rare (Mével and Cannat, 1991; Gillis 1995; Manning and MacLeod, 1996), fracturing provides the initial permeability for the penetration of seawater into the lower crust. The onset of this mode of deformation also is associated with incipient seismicity. Models for the cooling of the oceanic lower crust suggest that fracturing begins at a downward-propagating "cracking front" that is associated with the initiation of brittle failure (Lister, 1974; Mével and Cannat, 1991). The earliest hydrothermal veins in ophiolitic gabbros have been interpreted as the fossil remnants of such a cracking front (e.g., Nehlig et al., 1988a, 1988b). However, these models have never been rigorously tested in ophiolitic or oceanic gabbros.

We have examined the earliest fracturing event in gabbros from two localities: Hess Deep, where East Pacific Rise lower crust has been exposed at the seafloor by normal faulting, and the Oman ophiolite, where lower-crustal sections are subaerially exposed. Lower crust at both localities was formed at high spreading rates so their comparison provides insights into the nature of the cracking front in mid-ocean ridge environments in which magma supply rates are high. Here, we show that the earliest brittle deformation in crust generated by fast spreading occurs at high temperatures during short-lived, distributed microfracturing events near the ridge axis.

GEOLOGIC BACKGROUND

Hess Deep

Hess Deep is a submarine rift valley in ca. 1 Ma crust that formed at the East Pacific Rise. Hess Deep lies at the western end of the Cocos-Nazca spreading center ~50–100 km east of the present East Pacific Rise spreading axis (Lonsdale, 1988; Francheteau et al., 1990). Exposed on the rift-valley walls and on fault blocks within the rift are ultramafic and gabbroic rocks, sheeted dikes, and pillow lavas, which formed at the East Pacific Rise at an inferred half spreading rate of 65 mm/yr (Francheteau et al., 1990). We examined Hess

Manning, C.E., MacLeod, C.J., and Weston, P.E., 2000, Lower-crustal cracking front at fast-spreading ridges: Evidence from the East Pacific Rise and the Oman ophiolite, *in* Dilek, Y., Moores, E.M., Elthon, D., and Nicolas, A., eds., Ophiolites and Oceanic Crust: New Insights from Field Studies and the Ocean Drilling Program: Boulder, Colorado, Geological Society of America Special Paper 349, p. 261–272.

Deep gabbroic rocks from two localities: Ocean Drilling Program (ODP) Hole 894G (2°18′N, 102°32′W), which was sited at an intrarift ridge; and samples collected during *Alvin* submersible dives on the northern scarp of the rift valley (2°21′N, 101°16′W; Karson et al., 1992; Gillis, 1995). ODP Hole 894G penetrated 154.5 m of gabbro, gabbronorite, and olivine gabbronorite (Gillis et al., 1993). The north-scarp samples include basaltic gabbro and gabbronorite. (All gabbroic rocks are referred to as gabbros for brevity.) Hole 894G gabbros are interpreted to have crystallized immediately beneath the typical stratigraphic position of the axial magma lens (Natland and Dick, 1996). The north-scarp rocks were collected a maximum of 300–700 m below the base of the sheeted dikes and are thought to have crystallized and cooled within or above the axial melt lens (Karson et al., 1992; Natland and Dick, 1996).

The Oman ophiolite

The middle Cretaceous Oman ophiolite has an internal structure that is similar to East Pacific Rise lithosphere exposed at Hess Deep. Numerous lines of evidence point to formation of Oman oceanic crust at a fast rate of spreading. U-Pb radiometric ages are similar throughout the ophiolite (Tilton et al., 1981). The crustal sequence is thick and continuous, and the restored orientation of the crust-mantle transition is perpendicular to that of the sheeted dikes. Plastic deformation, which is typical of amagmatic, axial disruption of the lower crust at slow-spreading ridges (Mével and Cannat, 1991), is rare in Oman gabbros (Pallister and Hopson, 1981; Lippard et al., 1986; Nicolas, 1989; MacLeod and Rothery, 1992). Finally, ubiquitous L-S crystal shape fabrics in the gabbros imply movement of crystal-liquid suspensions throughout the lower crust (Nicolas et al., 1988, 1994; Benn and Allard, 1989; Nicolas, 1989, 1992; Boudier et al., 1996), similar to magmatic flow fabrics in East Pacific Rise gabbros at Hess Deep (MacLeod et al., 1996).

For this study, we chose the Wadi Abyad section in the central part of the Oman ophiolite. This plutonic section was selected because it was affected less by spreading-ridge complexities (MacLeod and Rothery, 1992; Nicolas and Boudier, 1995) or obduction-related deformation than most areas in Oman. The sharp petrologic Moho and ~2.6-km-thick lower-crustal section are tilted ~30°N. The top of the plutonic sequence grades upward into the lowermost sheeted dikes, of which several hundred meters crop out. No higher levels are exposed in this area.

Detailed descriptions of the gabbroic rocks of Wadi Abyad are given by Browning (1982, 1984), Beurrier (1988), and MacLeod and Yaouancq (2000). Modally layered gabbros and olivine gabbros comprise the lower two-thirds of the plutonic section. Layering and layer-parallel foliation are parallel to the petrologic Moho. The layered gabbros pass up into foliated, massive, unlayered, medium-grained gabbro and olivine gabbro (~660 m thick). The dip of magmatic foliation increases within ~100–200 m of the base of the foliated gabbros, becoming nearly perpendicular to the Moho, and a steeply plunging lineation has developed. The foliated gabbros pass upward into 140 m of variably textured gabbro displaying extreme, meter-scale variability in texture and chemical composition. Subtle gradations among foliated, isotropic, and pegmatitic gabbro, microgabbro, and dolerite are observed. The upper boundary of the variably textured gabbros is gradational. It is marked by the absence of medium- to coarse-grained gabbros and by reduced grain size, such that microgabbros and dolerites predominate. The level corresponds to the dike-rooting zone described by previous authors elsewhere in the Oman ophiolite (Rothery, 1983; MacLeod and Rothery, 1992). This zone is the lowest level at which dikes have chilled margins.

Evolved, silica-rich plagiogranites and late wehrlites are volumetrically insignificant in the Wadi Abyad area. Rare north-northwest–trending to north-trending diabase dikes crosscut the section. Their depleted compositions suggest affinities to the younger "V2" lava series that postdates MORB-type "V1" lavas (Ernewein et al., 1988). The dikes are parallel to several dextral strike-slip faults and locally display mylonitic margins having dextral strike-slip offsets. Greenschist-facies alteration assemblages associated with dikes, faults, and parallel veins imply that this minor disruption of the section occurred at low temperatures, after the crust moved off the ridge axis. This later event may have been a manifestation of a slightly younger ridge segment or fossil propagating-ridge tip to the northwest (Juteau et al., 1988; Reuber et al., 1991; MacLeod and Rothery, 1992; Nicolas and Boudier, 1995).

MICROFRACTURING AND METAMORPHISM

In all samples studied from Hess Deep, the earliest gabbro metamorphism was associated with the development of a microfracture network (Manning and MacLeod, 1996; Manning et al., 1996). Gabbros from Wadi Abyad, Oman, display a similar, early microfracture network (Fig. 1). This event was the earliest in which secondary minerals formed because crosscutting relationships require that all other metamorphic events postdate the microfractures (e.g., Plate 2–1 in Manning and MacLeod, 1996).

The dominant microcrack morphologies are of the grain-boundary, intergrain, and intragrain type, in the terminology of Simmons and Richter (1976). Transgranular microcracks are rare. Intragrain microcracks are primarily evident as thin, planar arrays of healed fluid inclusions and micrometer-scale green to blue-green hornblende inclusions in plagioclase grains (Fig. 1, A and C). Grain-boundary microfractures are up to 40 μm wide, and porosity is entirely filled by green, blue-green, or brown hornblende (Fig. 1, A and B). On the scale of individual thin sections, the microcracks are evenly distributed throughout (Fig. 1, A and B). Shear offsets are rare (Fig. 1), pointing to an extensional, mode I origin for the crack network. Localization of deformation into fault zones or extensional fractures was not observed (later, crosscutting fractures do display such deformation localization as described subsequently). At larger scales of meters or more, the volumetric abundance of the microcracks, as expressed by changes in amphibole modal abundance, is quite variable in Hole 894G over ~150 m (Manning and MacLeod, 1996). However, a general decrease in the maximum volume of secondary-mineral formation occurs with increasing depth below the dike-gabbro transition in Wadi Abyad. This finding agrees with observations elsewhere in the Oman ophiolite (e.g., Gregory and Taylor, 1981; Stakes and Taylor, 1992).

Modification of primary magmatic minerals is spatially associated with microcracks. At the margins of microfractures, secondary calcic plagioclase rims occur on magmatic plagioclase (Fig. 1, C and D), and sporadic secondary calcic clinopyroxene replaces magmatic augite (Gillis, 1995; Manning and MacLeod, 1996). The extent of replacement of primary magmatic plagioclase is lower than that of mafic phases, resulting in larger amphibole patches where the latter are involved (Fig. 1). Figure 1D illustrates that on the microscopic scale, plagioclase recrystallization and grain-boundary modification are observed in Oman gabbros, similar to textures reported by Manning et al. (1996) for gabbros from Hess Deep. In samples from the north scarp of Hess Deep, chlorite locally postdates amphibole. It

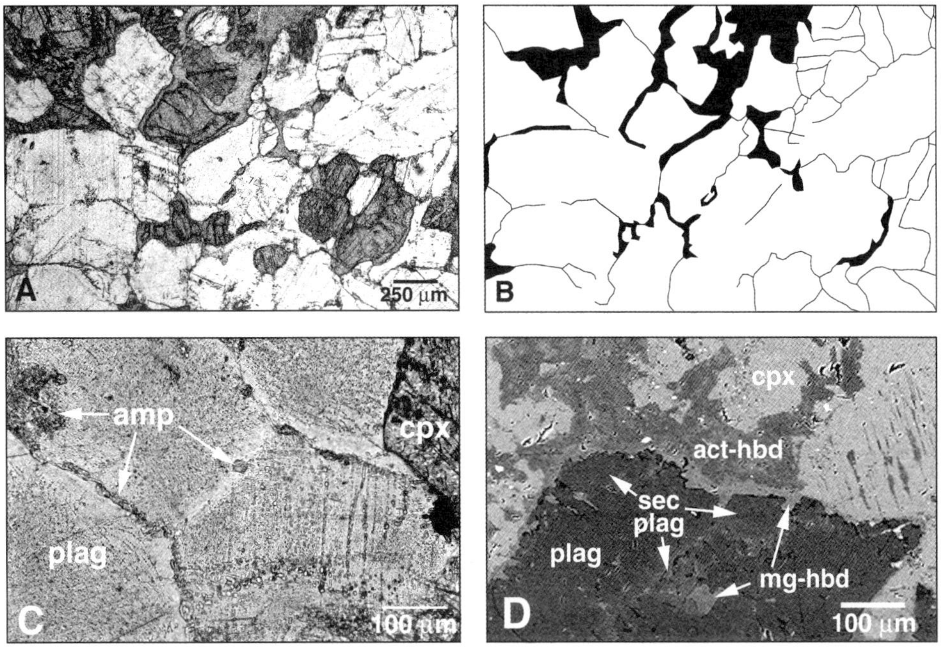

Figure 1. Textural characteristics of microscopic amphibole veins and associated alteration in gabbros from Wadi Abyad, Oman. A, Microscopic amphibole veins in foliated gabbro, sample Om75, plane-polarized light. Plagioclase grains are light, clinopyroxene, dark, and amphibole, intermediate. B, Line drawing of A at same scale showing amphibole-filled microcracks and alteration patches. Amphibole alteration is limited to narrow crack-filling zones along plagioclase-plagioclase grain boundaries, but is more extensive where clinopyroxene is involved owing to partial replacement of this phase. C, Healed fluid- and mineral-inclusion planes and grain-boundary microcracks, sample Om44, plane-polarized light. Abbreviations: amp—amphibole, cpx—clinopyroxene, plag—plagioclase. D, Backscattered-electron photomicrograph, sample Om75. Abbreviations: act-hbd—actinolitic hornblende, mg-hbd—magnesio-hornblende, cpx—clinopyroxene, plag—primary magmatic plagioclase, sec plag—secondary plagioclase. Higher intensity of backscattered electrons from magnesio-hornblende results from higher Fe content.

occurs both in grain-boundary regions and replacing plagioclase with little associated albitization. These areas were avoided in our analyses.

The coexistence of hornblende with calcic plagioclase and secondary clinopyroxene is diagnostic of the middle- to upper-amphibolite facies in mafic bulk compositions (Spear, 1981). The presence of chlorite with hornblende in the Hess Deep north-scarp samples (Gillis, 1995; Weston, 1998) suggests that metamorphism there continued to the greenschist-amphibolite transition.

Later fracture-related metamorphism occurred in both Hess Deep and in Wadi Abyad. In each locality, macroscopic fracturing followed microcracking. Macroscopic fractures are visible in outcrop and hand specimen as veins filled by secondary minerals. Macroscopic cracks are wider (millimeters to centimeters) than the microfractures. In addition, they are more continuous on the scale of decimeters to meters, and they cut across magmatic crystals rather than separating grain boundaries. In Hess Deep, progressively younger vein types are macroscopic amphibole veins, chlorite-bearing veins, and zeolite-calcite veins (Manning and MacLeod, 1996). In Wadi Abyad, later veins include macroscopic amphibole veins and epidote-bearing veins. In both localities, the later macroscopic fractures show that fracture deformation evolved from early distributed microcracking to later fracturing in which strain was localized into more widely spaced, throughgoing structures.

The evolution of fractures in Oman gabbros has been studied previously by Nehlig and Juteau (1988a, 1988b) and Nehlig (1994) in the Haylayn block to the northwest of Wadi Abyad. They described a macroscopic vein history similar to that which we observed in Wadi Abyad. However, they did not report alteration associated with microfractures. The intrusive history in the Haylayn area is more complex than that in Wadi Abyad. There is voluminous plagiogranite, and a large, late wehrlite intrusion cuts the section. The earliest fractures identified by Nehlig and Juteau (1988a, 1988b) are represented by voluminous actinolite veins. In light of the very high vein densities and the complex intrusive history in the Haylayn area, it is possible that early alteration related to microfractures has been overprinted and destroyed. In any event, because the macroscopic veins are younger than the earliest fracturing where we have observed crosscutting relationships (e.g., Manning and MacLeod, 1996), we assume that these veins were temporally distinct from the microcrack network that is the focus of this paper.

MINERAL COMPOSITIONS

Electron-microprobe analyses were performed on ten samples from Wadi Abyad and three samples from the north scarp of Hess Deep. Results were compared to those of Manning et al. (1996) on Hole 894G gabbros. The depth of the Oman samples ranged from the dike-gabbro transition to the Moho. Following Manning et al. (1996), amphibole and plagioclase compositions were determined at mutual grain edges such that analysis spots were <20 µm apart. Electron-microprobe operating conditions were 8 nA beam current, 15 kV accelerating potential, and 20 s counting times. A slightly defocused electron beam ($\sim$5 µm) was used for plagioclase analyses.

Standards were well-characterized natural and synthetic minerals. Oxide concentrations were derived by using a PAP correction algorithm (Pouchou and Pichoir, 1986).

Representative amphibole and plagioclase compositions are given in Tables 1 and 2, and key compositional parameters are plotted in Figure 2. The early amphiboles from Wadi Abyad and Hess Deep display a wide range in ^{IV}Al (Fig. 2). All amphiboles are calcic (Fig. 2C). In Hess Deep, X_{Mg} ranges from ~0.9 to ~0.6 in Hole 894G (Manning et al., 1996), but is lower (~0.7 to ~0.3) in the north-scarp samples because of higher bulk-rock Fe content (Fig. 2A). In Wadi Abyad, amphiboles are more magnesian (X_{Mg} = 0.7–0.9) at all levels except the dike-gabbro transition. As at the north scarp of Hess Deep, the gabbros at the highest levels are Fe rich, and this difference results in lower X_{Mg} values (0.6–0.7) of the secondary amphiboles. Amphibole in all samples displays an increase in A-site occupancy with ^{IV}Al (Fig. 2B). The substitution of ^{IV}Al for Si at low A-site occupancy (<0.2 atoms per formula unit) is tschermakitic, but at higher values is increasingly pargasitic.

The wide range of amphibole compositions illustrated in Figure 2, A–C, results from variations in bulk Al content on the grain scale (Manning and MacLeod, 1996). Where microcracks cut clinopyroxene, low bulk Al yielded actinolitic hornblende; where plagioclase was cut, more aluminous amphiboles grew. This difference can be seen in Figure 1D, in which magnesio-hornblendes (high-BSE [backscattered-electron] intensity) are in contact with plagioclase, and actinolitic hornblendes (low-BSE intensity) replace clinopyroxene. No compositional differences were discerned between amphiboles found within plagioclase crystals and those that occur at plagioclase-plagioclase grain contacts.

The compositional variability displayed by the amphiboles may seem inconsistent with growth at the high temperatures of the amphibolite facies. It is interesting that the compositional range reported here is not unique, but rather appears to be characteristic of amphibolite-facies amphiboles from oceanic gabbros (cf. Stakes and Vanko, 1986; Stakes et al., 1991; Stakes and Taylor, 1992; Gillis et al., 1993). This feature likely results from the combination of low strain and rapid reaction rates. Thus, the time scale for pore-filling reactions is faster than that required for transport-controlled development of homogeneous compositions of secondary minerals, consistent with formation at the high temperatures of the upper-amphibolite facies (discussed subsequently). It is important to note that growth of low-^{IV}Al actinolitic hornblende—and, locally, actinolite—at high temperatures is not inconsistent with this interpretation. Experimental studies show that Al-poor tremolitic amphibole is stable to >800 °C in Mg-rich, Al-poor bulk compositions (e.g., Jenkins, 1987), such as those represented on the microscopic scale by clinopyroxene and olivine grain-boundary environments in the gabbros.

Plagioclase compositions range widely in the sample suite (Fig. 2D). In samples from high levels of the gabbro section (i.e., both Hess Deep localities and the Oman samples from the dike-gabbro transition), average plagioclase compositions are $X_{An} \approx 0.55$. By contrast, samples from deeper in the gabbro section in Oman are significantly more calcic, with mean $X_{An} \approx 0.85$.

METAMORPHIC TEMPERATURES

Temperatures were calculated for compositionally appropriate amphibole-plagioclase pairs by using the extent of the $[NaSi]_{-1}[CaAl]$ exchange, as calibrated by Holland and Blundy (1994). Results are summarized in Table 3. Mean temperatures for amphibole-plagioclase equilibration in individual samples range from 711 to 829 °C. These temperatures are consistent with the inference from the mineral assemblage that metamorphism occurred in the upper-amphibolite facies (Spear, 1981).

Figure 3A shows temperatures as a function of stratigraphic position relative to the dike-gabbro transition. At Hess Deep, both sample suites are in high-level gabbros, near the dike-gabbro transition zone. Temperatures of initial fracturing in Hole 894G and the north scarp are statistically indistinguishable: 716 ± 5 °C and 711 ± 7 °C, respectively (uncertainties are standard errors in the mean). The samples from the same level in Wadi Abyad record nearly identical temperatures (Fig. 3A, Table 3). Samples from the foliated gabbros in Wadi Abyad yield somewhat higher temperatures of 754 ± 11 °C to 777 ± 10 °C. The two Oman samples from the deepest part of the lower crust near the Moho yield the highest temperatures (810 ± 5 °C and 829 ± 14 °C).

Two factors must be considered in assessing the accuracy of the inferred temperatures: equilibrium and thermometer applicability. First, it must be established that exchange equilibrium was closely approached at the time of fracturing and fluid flow, and that little subsequent exchange occurred. Manning et al. (1996) noted that metamorphism during fluid inflow after fracturing does not produce equilibrium textures because there is no penetrative deformation or grain-size modification. However, exchange equilibrium is likely to have occurred at the time of fracturing for three reasons: First, plagioclase shows textural evidence for re-equilibration during reaction with the fluid phase (Fig. 1D). Second, fluid + mineral reaction rates in the amphibolite facies result in equilibration on the time scale of days (Spear, 1981; Wood and Walther, 1983; Walther and Wood, 1984). Finally, $\delta^{18}O$ values of amphibole and plagioclase in Hess Deep and Oman gabbros are consistent with high-temperature equilibration with a fluid phase (e.g., Gregory and Taylor, 1981; Stakes and Taylor, 1992; Lécuyer and Gruau, 1996). Equilibration was local, on the scale of tens of micrometers, because amphibole compositions vary systematically as a function of grain-boundary mineral composition (Manning and MacLeod, 1996). Weston (1998) argued that diffusive re-equilibration during cooling is unlikely because of the sluggish diffusion rates in plagioclase and amphibole (e.g., Liu and Yund, 1992; Giletti and Casserly, 1994; Brabander and Giletti, 1995; Giletti and Shanahan, 1997).

The second factor influencing the accuracy of inferred temperatures is the compositional overlap between the data set used for geothermometer calibration and the gabbro data sets. All amphibole and plagioclase compositions in this study lie within the range of those used for calibration of the thermometer (see Manning et al., 1996). Compositional factors that might cause problems for the thermometer would be high concentrations of Mn, Ti, or halogens, as well as elevated cummingtonite substitution. Tables 1 and 2 show that the cation concentrations are low. Weston (1998) found that F and Cl contents are near detection limits in all studied samples. Likewise, cummingtonite substitution is low, as indicated by high Ca and Na in M4 sites (Tables 1 and 2).

In light of the apparent increase in temperature with depth exhibited in Wadi Abyad (Fig. 3), it is important to examine in more detail the plagioclase and amphibole compositions that suggest high temperature. Experimental investigations of the wet solidus of basalt by Helz (1973) equilibrated amphibole and plagioclase at 750–850 °C. Relevant compositional parameters of amphiboles are shown in Figure 2 for comparison with the Oman amphiboles. At oxygen fugacities of Ni-NiO, similar ^{IV}Al, ^{M4}Na, and A-site occupancy were observed. Minor differences in X_{Mg} are probably attributable to bulk compositional differences (the layered gabbros in Oman are cumu-

TABLE 1. REPRESENTATIVE AMPHIBOLE AND PLAGIOCLASE ANALYSES FROM HESS DEEP

	ODP Hole 894G–Interrift Ridge										*Alvin* dive samples–North Scarp					
Sample	4R-02-7		8R-01-10		9R-04-11		13R-01–11B		17R-02-1B		2213-1125		2218-1323		2213-1053	
analysis	50	51	16A2	17P2	20	21	48A1	49P1	8A	8P	68	69	52	51	8	9
phase	amp	plag	amp	plag	amp	plag	amp	plag	amp	plag	amp	plag	amp	plag	amp	plag
SiO_2	43.92	56.26	47.73	55.67	43.98	49.98	47.11	53.90	50.65	50.68	49.71	52.48	45.46	58.36	52.66	55.94
TiO_2	00.56	00.04	00.82	00.05	00.27	00.02	00.29	00.00	00.33	00.04	0.40	0.05	1.17	0.00	0.33	0.05
Al_2O_3	10.46	26.92	06.74	27.30	12.66	30.95	09.10	28.10	05.08	31.40	4.64	30.03	6.63	26.18	3.82	27.24
Cr_2O_3	00.02	0.00	00.04	00.02	00.01	00.00	00.08	00.02	00.00	00.05	0.08	0.05	0.00	0.00	0.00	0.00
FeO	17.69	00.33	15.91	00.56	11.79	00.67	13.41	00.58	12.32	00.39	17.54	0.39	21.66	0.31	13.85	0.69
MnO	00.19	00.07	00.24	00.07	00.27	00.05	00.26	00.01	00.18	00.01	0.52	0.03	0.41	0.02	0.24	0.00
MgO	10.20	00.01	12.11	00.00	14.43	00.15	13.62	00.00	15.02	00.00	12.68	0.00	8.50	0.00	14.75	0.06
CaO	11.80	09.74	11.44	09.97	10.95	14.48	11.32	10.95	11.94	14.40	10.59	12.77	11.20	7.64	11.47	10.59
Na_2O	01.97	05.85	01.31	06.01	02.56	03.30	01.34	05.19	00.84	03.62	0.65	4.12	1.06	7.23	0.50	5.21
K_2O	00.20	00.46	00.18	00.04	00.12	00.00	00.16	00.03	00.11	00.01	0.03	0.04	0.17	0.06	0.07	0.05
Total	97.01	99.69	96.52	99.69	97.04	99.60	96.69	98.78	96.47	100.60	96.83	99.94	96.26	99.80	97.67	99.82
Si	6.574	2.546	7.072	2.523	6.348	2.298	6.851	2.470	7.350	2.302	7.349	2.384	6.943	2.617	7.587	2.529
IV_{Al}	1.426	1.436	0.928	1.458	1.652	1.678	1.149	1.518	0.650	1.681	0.651	1.608	1.057	1.384	0.413	1.452
VI_{Al}	0.420		0.249		0.502		0.411		0.219		0.158		0.137		0.236	
Ti	0.063	0.001	0.091	0.002	0.029	0.001	0.032	0.000	0.036	0.001	0.044	0.002	0.134	0.000	0.036	0.002
Cr	0.002	0.000	0.005	0.001	0.001	0.000	0.009	0.001	0.000	0.002	0.009	0.002	0.000	0.000	0.000	0.000
Fe^{3+}	0.485	0.006	0.407	0.011	0.782	0.013	0.597	0.011	0.345	0.007	0.369	0.007	0.590	0.006	0.141	0.013
Fe^{2+}	1.730		1.564		0.642		1.034		1.150		1.799		2.176		1.528	
Mn	0.024	0.003	0.030	0.003	0.033	0.002	0.032	0.000	0.022	0.000	0.065	0.001	0.053	0.001	0.029	0.000
Mg	2.275	0.001	2.674	0.000	3.104	0.010	2.952	0.000	3.248	0.000	2.794	0.000	1.935	0.000	3.167	0.004
Ca	1.893	0.472	1.816	0.484	1.693	0.713	1.764	0.538	1.857	0.701	1.678	0.622	1.833	0.367	1.771	0.513
Na	0.572	0.513	0.376	0.528	0.716	0.294	0.378	0.461	0.236	0.319	0.186	0.363	0.314	0.629	0.140	0.457
K	0.038	0.027	0.034	0.002	0.022	0.000	0.030	0.002	0.020	0.001	0.006	0.002	0.033	0.003	0.013	0.003
Total	15.502	5.005	15.246	5.011	15.524	5.009	15.239	5.001	15.133	5.014	15.108	4.992	15.205	5.007	15.061	4.973
T (°C)	724		737		818		718		727		721		733		673	

Note: Amphibole formulae were normalized to 23 anhydrous oxygens, and then Fe^{3+} contents were calculated by following Holland and Blundy (1994); values given are the midpoints between maximum and minimum required by charge balance (Holland and Blundy, 1994). All Fe assumed to be Fe^{3+} in plagioclase. Oxide concentrations listed as 0.00 wt% are below detection.

TABLE 2. REPRESENTATIVE AMPHIBOLE AND PLAGIOCLASE ANALYSES FROM OMAN OPHIOLITE

	Wadi Abyad											
Sample	Om1		Om7		Om41		Om68		Om72		Om86a	
analysis	12	10	59		27	26	84	85	104	105	33	34
phase	amp	plag	amp	58	amp	plag	amp	plag	amp	plag	amp	plag
SiO_2	55.54	48.57	44.75	49.29	46.76	46.64	52.44	47.95	55.03	48.70	44.93	53.51
TiO_2	0.33	0.02	0.10	0.00	0.32	0.00	0.30	0.03	0.36	0.03	0.34	0.08
Al_2O_3	2.65	31.35	12.54	32.42	10.09	34.37	5.23	33.17	2.71	33.24	10.06	28.47
Cr_2O_3	0.37	0.00	0.09	0.00	0.01	0.00	0.16	0.03	0.11	0.00	0.01	0.07
FeO	4.95	0.18	11.35	0.30	12.32	0.13	8.47	0.37	6.08	0.32	17.29	0.66
MnO	0.11	0.01	0.20	0.00	0.17	0.02	0.06	0.04	0.11	0.00	0.21	0.06
MgO	20.67	0.06	13.99	0.00	14.25	0.00	17.97	0.00	20.23	0.00	10.85	0.00
CaO	12.43	16.80	11.96	15.73	12.42	18.13	12.31	16.88	12.45	16.44	11.27	11.44
Na_2O	0.66	2.41	2.46	2.74	2.08	1.61	1.01	2.22	0.49	2.50	1.53	5.17
K_2O	0.01	0.00	0.07	0.00	0.04	0.00	0.01	0.00	0.04	0.00	0.13	0.12
Total	97.72	99.41	97.51	100.48	98.46	100.89	97.97	100.70	97.61	101.23	96.62	99.57
Si	7.692	2.244	6.459	2.246	6.705	2.131	7.350	2.191	7.660	2.209	6.671	2.443
^{IV}Al	0.308	1.708	1.541	1.742	1.295	1.851	0.650	1.787	0.340	1.777	1.329	1.532
^{VI}Al	0.125		0.592		0.411		0.214		0.105		0.432	
Ti	0.034	0.001	0.011	0.000	0.035	0.000	0.032	0.001	0.038	0.001	0.038	0.003
Cr	0.041	0.000	0.010	0.000	0.001	0.000	0.018	0.001	0.012	0.000	0.001	0.003
Fe^{3+}	0.102	0.004	0.482	0.006	0.430	0.002	0.348	0.007	0.150	0.006	0.680	0.013
Fe^{2+}	0.471		0.888		1.047		0.645		0.558		1.467	
Mn	0.013	0.001	0.024	0.000	0.021	0.001	0.007	0.001	0.013	0.000	0.026	0.002
Mg	4.266	0.004	3.009	0.000	3.045	0.000	3.754	0.000	4.197	0.000	2.401	0.000
Ca	1.845	0.832	1.850	0.768	1.908	0.887	1.849	0.826	1.857	0.799	1.793	0.559
Na	0.177	0.216	0.688	0.242	0.578	0.142	0.274	0.197	0.132	0.220	0.440	0.457
K	0.002	0.000	0.013	0.000	0.007	0.000	0.002	0.000	0.007	0.000	0.025	0.007
Total	15.076	5.009	15.567	5.004	15.483	5.015	15.143	5.012	15.069	5.012	15.303	5.019
T (°C)	780		752		788		773		760		734	

Note: See Table 1 for analytical notes.

lates and, therefore, not of basaltic bulk composition). Mole fractions of anorthite are also comparable in the experimental charges and the Oman gabbros. Thus, the high temperatures inferred for the base of the gabbro section in Wadi Abyad are interpreted as an accurate assessment of the onset of fracture-related metamorphism.

DISCUSSION

Origin of microcracks by semibrittle failure

A distributed microfracture network is found in all Hess Deep and Oman gabbros studied. Failure by distributed microcracking therefore appears to be a volumetrically significant deformation mechanism for the lower crust at fast spreading rates. Macroscopic failure modes may be of three types (e.g., Rutter, 1986; Evans et al., 1990): brittle, semibrittle, and plastic. Petrographic characteristics of brittle failure include localization of strain into throughgoing fracture surfaces, microcracking, and textural indications of frictional sliding. Evidence for deformation by plastic flow includes crystal-plastic features such as twinning and undulatory extinction, and deformation is macroscopically homogeneous. Semibrittle flow involves deformation by a combination of microfracturing and crystal plasticity, leading to a macroscopically homogeneous distribution of deformation. On the basis of the dilatant, microscopic, distributed nature of the microfracture networks in Oman and Hess Deep gabbros, we infer that they represent deformation in the semibrittle field. Although we have not carried out a systematic study of the crystal-plastic textures that should accompany the microcracks, we note that patchy, undulose extinction and minor deformation lamellae in plagioclase and augite occur in the Oman gabbros (Wilks and Carter, 1986). As the formation of metamorphic amphibole in the gabbros was associated with the microcracks, it is evident that hydrothermal hydration of the lower crust occurred during semibrittle deformation, after the gabbros had cooled below the brittle-plastic transition.

Microcracking temperatures and their variation with depth

Given that metamorphism was associated with microcracking, metamorphic temperatures can be used to study the pattern of development of semibrittle deformation in the oceanic lower crust. Detailed statistical analyses have been carried out for the metamorphic temperatures associated with microfracturing in both Hess Deep localities (Hole 894G, Manning et al., 1996; north scarp, Weston, 1998). These studies reveal that metamorphism related to microfracturing in each locality very likely occurred in a single event. This is because *F*-tests show the small standard deviations of each temperature population to be consistent with a single underlying distribution at the 90% confidence level. Moreover, χ^2 and Kolmogorov-Smirnov tests indicate that the north-scarp and Hole 894G populations are not statistically distinct (Weston, 1998). This result implies that despite their geographic separation, the early metamorphic event occurred over the same temperature interval, which was no more than 66 °C (748–682 °C) at the 90% confidence level.

Statistical analyses of similar detail have not been carried out for the Oman sample suite. However, examination of the standard deviations permits preliminary conclusions about early metamorphism in Wadi Abyad. Table 3 shows that the standard deviations of the temperature populations in each sample are, with one exception (Om65), similar to or less than that of the thermometer-calibration data set (39 °C). This result implies that, like Hess Deep, the Oman samples record a single, narrow temperature distribution.

Figure 3A shows that metamorphism of the Oman gabbros was

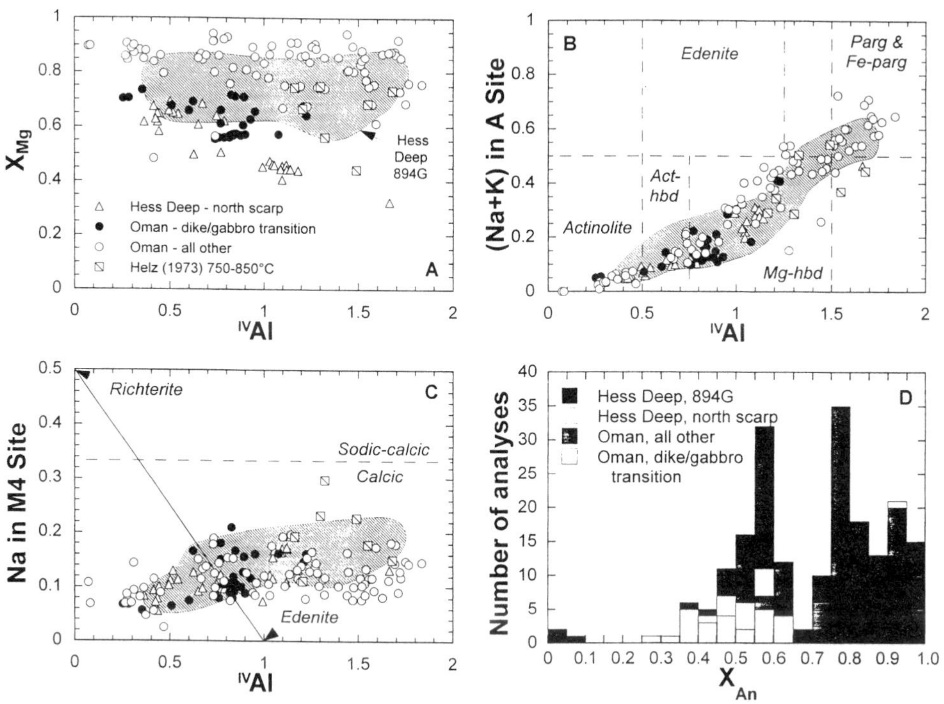

Figure 2. Amphibole and plagioclase compositional parameters. Amphibole nomenclature after Leake (1978). A, Mole fraction of Mg [X_{Mg} = Mg/(Mg + Fe^{2+})] vs. tetrahedrally coordinated Al (^{IV}Al) in amphiboles. B, Alkali cations in A site vs. ^{IV}Al in amphiboles; symbols as in A. Act-hbd—actinolitic hornblende, Mg-hbd—magnesio-hornblende, Parg—pargasitic hornblende, Fe-parg—ferropargasitic hornblende. C, Na in M4 site vs. ^{IV}Al in amphiboles with richterite-edenite exchange trajectory used in temperature determination. Near-linear increase in ^{M4}Na indicates that extent of richterite-edenite exchange is similar for all amphiboles, consistent with narrow calculated temperature interval. Symbols as in A. D, Histogram of plagioclase compositions in terms of mole fraction of anorthite [X_{An} = Ca/(Ca + Na)].

TABLE 3. TEMPERATURES OF INITIAL CRACKING

Location and sample	Depth below dike-gabbro transition (km)	Number of analyses	Mean T (°C)	Standard deviation	Standard error in mean	Notes
Hess Deep						
Hole 894G	0–1	52	716	36	5	Manning et al. (1996)
North Scarp	0–0.3	29	711	37	7	
Oman						
Om86	0.15	11	708	31	9	
Om75	0.42	9	715	36	12	
Om72	0.59	9	777	31	10	
Om68	0.67	8	754	29	11	
Om65	0.83	6	756	59	24	
Om07	1.05	9	769	31	10	
Om53	1.08	8	—	—	—	$X_{an} > 0.9$
Om48	2.22	3	—	—	—	$X_{an} > 0.9$
Om41	2.34	4	810	9	5	
Om01	2.60	8	829	38	14	

Note: Temperatures not calculated for samples in which $X_{an} > 0.9$ because this composition is outside the calibration range of the thermometer (Holland and Blundy, 1994). Pressure was assumed to be 1 kbar.

comparable to that at Hess Deep, in that similar temperatures are recorded at the same levels. In addition, the deeper levels exposed in Wadi Abyad imply that the metamorphic event caused by microfracturing occurred at progressively higher temperatures with increasing depth in the crust. Taking temperature to be the independent variable, a second-order polynomial yields an equation for the variation in depth below the dike-gabbro transition (z, in km) with temperature (T, in °C): $z = 76 - 0.22T + (1.5 \times 10^{-4})T^2$, and $R^2 = 0.93$ (Fig. 3B). (Hole 894G data were omitted from this fit because of their large depth uncertainty; however, they are consistent with it.) A linear fit gives lower R^2 (0.85) and systematic deviations at low and high z; a higher-order polynomial is probably unwarranted for the size of the data set and the precision of individual determinations. The polynomial is strictly applicable only to Wadi Abyad, because crustal thickness may vary from locality to locality. Nevertheless, if Hess Deep and Wadi Abyad gabbros are representative, the fit implies that there is a nonlinear increase in temperature of the onset of semibrittle deformation in the lower crust of fast-spreading ridges.

Constraints on the cracking front in the lower crust

The Lister model. Lister (1974, 1982) developed a conceptual model for the penetration of water into cooling crust near oceanic ridges. In this model, fluid influx occurs via permeability generated by a cracking front that propagates downward into fresh, uncracked rock. Fracturing occurs when tensile stress due to cooling and contraction slightly exceeds overburden stress. Lister envisioned that the cracking front could affect the entire crustal section and the upper mantle, though he recognized that it would penetrate below the dike-gabbro transition only after the lithospheric section had moved off

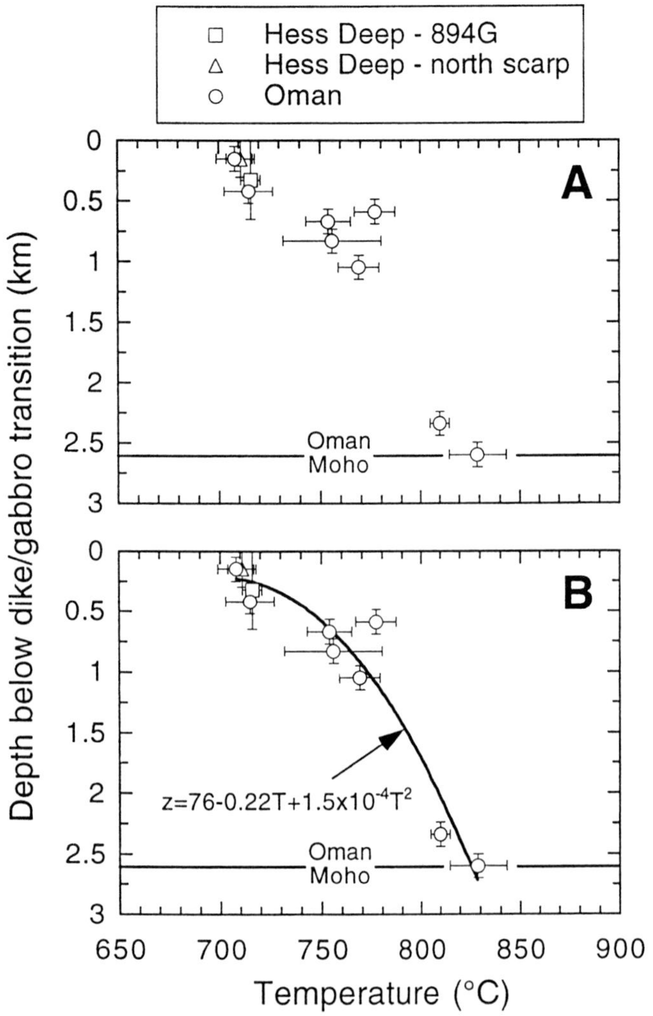

Figure 3. A, Variation in mean temperature of amphibole-plagioclase equilibration in microcracks with depth below dike-gabbro transition in Hess Deep and Oman. Temperature uncertainties are standard error of the mean (Table 3). Hole 894G mean is plotted at midpoint of an assumed possible depth range of 0–0.75 km below dike-gabbro transition. B, Same as A, with least-squares fit.

axis. Mével and Cannat (1991) elaborated on this model by pointing out that spreading rate should play a role in determining how water enters the lower crust. At slow-spreading ridges, high-temperature ductile shear zones predate a cracking front, whereas at fast-spreading rates, ductile shear zones are absent, and permeability should initially be generated solely by the cracking front. Mével and Cannat (1991) assumed a constant temperature of 500 °C for the inception of fracturing, regardless of depth in the crust.

Our observations of the earliest fracturing in Hess Deep and Wadi Abyad gabbros produced by fast spreading support the idea that water penetration and initial brittle failure are related and that this process can affect the entire crustal section. In both gabbro localities, the earliest metamorphic fluid + rock reactions are texturally related to fractures. However, the Oman and East Pacific Rise gabbros require revision of the Lister model and its subsequent modifications in several fundamental ways, including cracking scale, temperature, and nature of the front propagation. These differences provide constraints on the nature of heat transport and fracture evolution in the lower oceanic crust in fast-spreading environments.

Scale of fracturing. A key feature of the Lister model is that initial fracturing is postulated to occur as widely spaced cracks with significant lateral extents and spacings of centimeters to meters; that is, within the brittle regime of cataclastic faulting and strain localization (e.g., Rutter, 1986; Evans et al., 1990). In the model, water first penetrates "a semi-infinite expanse of hot rock at a planar 'cracking front,' it being understood that the cracks are *macrocracks* on a scale of several centimeters and nothing like the 'microfracturing' associated with brittle creep" (Lister, 1974, p. 470). Lister (1974) was interested in the possibility of fracture-controlled advective heat transport, so he focused his attention on the brittle-faulting regime where permeability is likely to be higher for longer times. Although he may not have intended his analysis to be applied strictly to the entire subsolidus deformation history, it has nevertheless gained wide currency in the literature as a guide to the links between deformation and metamorphism of the oceanic crust. Thus it is important to emphasize that our observations demonstrate that initial fracture deformation of lower-crustal oceanic rocks is actually semibrittle in character. There is minimal strain localization. Instead, the structures are distributed, dilatant microcracks found predominantly in grain-boundary regions.

Lister (1974) also proposed that the fracture system should coalesce with time via "pattern enlargement." Although different in scale, the fracture systems do show evidence for increasing degree of strain localization, which we take to be synonymous with Lister's "pattern enlargement." In Hess Deep, microfractures are crosscut by the earliest set of macroscopic fractures, which are filled by the same minerals (Manning and MacLeod, 1996). The macroscopic fractures are less abundant and more continuous than their microscopic counterparts. This finding is consistent with the channeling of fluids into fewer and larger conduits. Weston (1998) showed that the macroscopic amphibole fractures formed at only slightly lower temperatures (704–674 °C), a distinction that is statistically significant at the 90% confidence level. This result implies that the fracture system associated with the cracking front can locally evolve from pervasively distributed microcracks to larger, throughgoing fractures over a narrow temperature interval, though these larger fractures are comparatively rare (Manning and MacLeod, 1996). This textural evolution is consistent with a progressive change from semibrittle, cataclastic flow to brittle, cataclastic faulting with time and declining temperature.

Cracking temperature. The second difference between the observed initial fracture system and the Lister model is the temperature of cracking and its distribution with depth. Lister (1974) assumed that failure by transient creep would occur at a strain rate of $\sim 3 \times 10^{-8}$ s^{-1} after build up of 1%–2% plastic strain during cooling. As lithostatic pressure increases, greater strains—and, hence, lower temperatures—occur before cracking. Figure 4 illustrates Lister's (1974) estimate of the cracking temperature by using his parameter choices, a water-column depth of 2.5 km, 2 km of basalts and dikes overlying gabbros, and rock density of 3000 kg/m^3. The curve in Figure 4 corresponds to the depth at which stress at failure is equal to confining pressure (the Goetze criterion for the brittle-plastic transition). It therefore represents the maximum model cracking temperature at any depth. At the dike-gabbro transition, the maximum cracking temperature in this scenario would be 650 °C; at the Oman Moho, 550 °C.

Our calculated temperatures for initial microfracturing at any position within the gabbros are higher than those predicted by Lister (1974) by ~65 to ~275 °C (Fig. 4), or ~10%–50%. What is more important, calculated temperatures increase instead of decrease with depth. This discrepancy is independent of assumed flow law or strain rate. For example, Hirth et al. (1998) used various flow laws at the

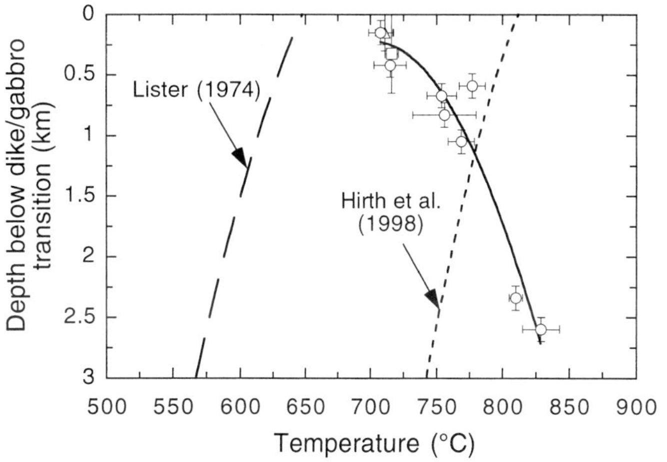

Figure 4. Comparison of observed temperature distribution with predicted position of the cracking front of Lister (1974) and the brittle-plastic transition of Hirth et al. (1998) based on the Mackwell et al. (1998) flow law for dry Maryland diabase at a strain rate of 10^{-14} s^{-1}.

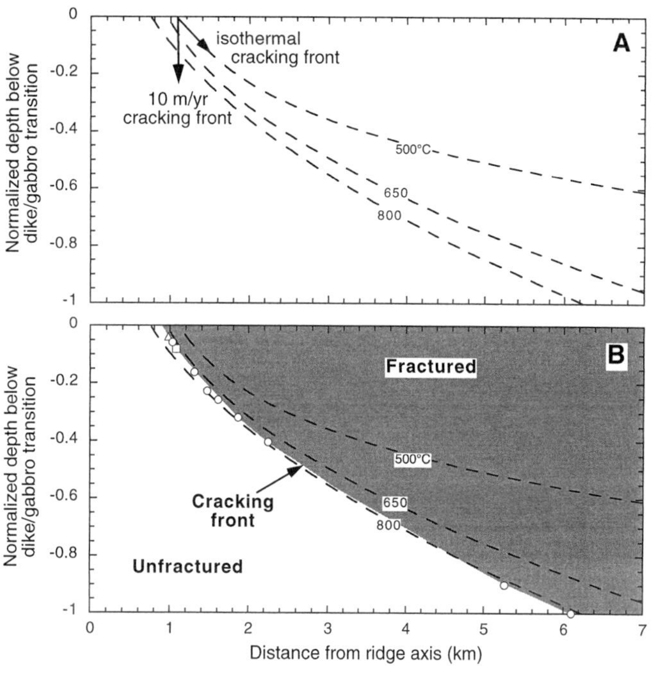

Figure 5. Variation in temperature of cracking front with depth and distance from ridge axis. Isotherms are taken from Phipps Morgan and Chen (1993) model results for a spreading rate of 50 mm/yr. Thicknesses of Oman gabbros (2.6 km), Hess Deep gabbros (assumed to be 4 km), and the model (4 km) were normalized to unity for comparison. A, Arrows show that propagation velocity of 10 m/yr at a specific locality will result in metamorphic temperatures spanning many hundreds of degrees, whereas isothermal cracking requires that the front moves downward and away from the ridge axis at a shallow angle. B, Results of this study compared to the thermal model show that the cracking front in crust formed in a fast-spreading environment occurs at progressively higher temperatures with increasing depth in the crust, requiring that fracturing occurs over a distance of several kilometers between 1 and 6 km from the ridge axis. The region that has been affected by a high-temperature cracking front is shaded. Symbols as in Figure 3.

lower strain rate of 10^{-14} s^{-1} to predict the temperature of the brittle-plastic transition in the oceanic lower crust. All are similar in form to the Lister curve. Figure 4 shows the brittle-plastic transition for the dry-Maryland-diabase flow law of Mackwell et al. (1998). Our calculated cracking temperatures lie within the brittle field from 0 to 1 km below the dike-gabbro transition; however, near the Oman Moho, calculated cracking temperatures are above the brittle-plastic transition of Hirth et al. (1998). Figure 4 illustrates that the increase in cracking temperature with depth is not consistent with failure along a front defined by the boundary between brittle and plastic deformation mechanisms for the Mackwell et al. (1998) flow law and a constant strain rate.

There are two possible explanations for the observed increase in the temperature of semibrittle deformation with depth. First, fracturing may not correspond to the brittle-plastic transition at all depths. In this scenario, the maximum stress difference encountered by the rocks was below their failure strength at the brittle-plastic transition and remained so until the yield strength was exceeded at lower temperatures. However, this scenario should produce a more random distribution of cracking temperature with depth, not the observed systematic change that is evidently independent of gabbro locality. Alternatively, the metamorphic temperatures correspond very nearly to the brittle-plastic transition, but their variation with depth indicates rheologic differences corresponding to changes in modal abundance of magmatic minerals such as olivine, increasing extent of hydrolytic weakening with depth, or increasing strain rate with depth. Because a strain-rate increase with depth is a necessary consequence of viscous flow of crust away from the ridge axis (e.g., Chen and Morgan, 1990; Quick and Denlinger, 1993; Chenevez et al., 1998), it is a strong candidate for explaining the observed variation between temperature and depth, though the other alternatives cannot be ruled out.

Propagation of the cracking front. Lister (1974, 1982) suggested that the cracking front can advance downward at the rate of tens of meters per year at all depths. Even at fast spreading rates, such rapid propagation velocities would allow penetration through the entire crustal section over very short distances normal to the spreading direction. For example, at 65 mm/yr (the spreading rate of the East Pacific Rise at the latitude of Hess Deep; Francheteau et al., 1990), a front velocity of 10 m/yr would result in penetration to the base of a hypothetical 6 km crustal section in 600 yr while the crust spread laterally 39 m. Figure 5A shows that these propagation velocities would effectively mean instantaneous penetration of the entire crust. If fracturing of the lower crust were to occur in a single such event, temperatures of cracking and metamorphism would increase with depth in the crust by many hundreds of degrees across an entire crustal section. Because Lister (1974) inferred a narrow temperature interval for fracturing, he proposed that the rapid front velocities required episodic cracking events progressively farther from the ridge axis. For an isothermal front (Mével and Cannat, 1991), the distance from ridge axis at which fracturing occurred could exceed 10 km if temperatures were ≤500 °C (Fig. 5A).

Initial fracturing in Oman and Hess Deep gabbros displays a slight increase in temperature with depth, requiring a small but significant (~100 °C) transgression of isotherms and cracking progressively more distant from the ridge axis. The rate of spreading at which Oman crust was produced, although likely rapid, is unknown in detail. However, if a rate similar to that inferred for Hess Deep is assumed, then a schematic sense of the implications of our observations can be gained through integration with thermal models, such as those of Phipps Morgan and Chen (1993). Figure 5B illustrates that the fracturing near the dike-gabbro transition would occur at ~1 km from the spreading axis, but at ~6 km at Moho depths. If cracking proceeded downward along a continuously propagating front, the penetration velocity would have been 34 mm/yr. This is slower than the spreading rate by a factor of nearly two and slower than the velocity inferred by Lister (1974) by about three orders of magnitude. When combined with the rapid reaction rates at the high temperatures inferred for cracking, the low apparent propagation velocity and the wide range in distance from the ridge axis suggest that initial semibrittle microcracking is episodic.

The inference that initial microfracturing is episodic has important implications for the subsolidus evolution of deformation and metamorphism in the lower crust. If the first microfractures at any depth are connected to shallower fracture systems, then the cumulative effect of the successive cracking required to deepen the front would be to produce a virtually continuous record of metamorphism from the upper-amphibolite facies to lower metamorphic grades. Such a record may be shown to some extent by the later chlorite observed in the north-scarp gabbros of Hess Deep (e.g., Gillis, 1995). But late chlorite is not a general feature of the deeper gabbro metamorphism in either locality. Instead, the early microfractures form at high temperatures, and locally coalesce into larger fracture systems now represented by hornblende veins. The crack arrays are open for short times over narrow temperature intervals (Manning and MacLeod, 1996; Manning et al., 1996; Weston, 1998). The rocks then cool over significant temperature intervals during which little metamorphism occurs, until fractures filled by greenschist-facies minerals are generated. Thus, the continuous record of metamorphism that should exist if multiple cracking events of substantial vertical extent were to cut the gabbros has not been observed. This finding leads to two possibilities for fracture evolution in the lower crust of fast-spreading environments: (1) cooling through the interval of initial cracking into the greenschist facies (~500 °C) is so rapid (Fig. 5) that the structures formed at these conditions are rare and have gone unrecognized, or (2) there is limited vertical connectivity to the initial microfracture systems. Distinguishing between these two possibilities is not possible on the basis of currently available data.

Heat transport and later, macroscopic fracture systems. The nature of heat transport during semibrittle microfracturing can be determined by considering the temperature of metamorphism and the compositions of secondary minerals. Statistical analysis of temperatures in Hess Deep samples implies that metamorphism occurred over a narrow temperature interval, below which no further amphibolite-facies metamorphism took place (Manning et al., 1996). The standard deviations of the temperatures for each Oman sample suggest that this statement is true for that suite as well. Given the rapid cooling rates expected for gabbros formed in fast-spreading environments (Fig. 5), metamorphism over a limited temperature range implies that fluid flow associated with the cracking front was short-lived. Moreover, the limited degree of homogenization of amphibole compositions in Hess Deep (Manning and MacLeod, 1996) and Oman (this study) suggests that cracks sealed rapidly compared to the rate of transport of chemical components. These lines of evidence imply that fluid flux is low and that the lower crust therefore cools conductively through the amphibolite facies (down to 500 °C). This result contrasts with the Lister (1974) model of the cracking front, where the onset of fracturing brings the crust into a new regime of hydrothermal cooling, in which heat is transported advectively by fluids flowing in long-lived cracks.

Although the earliest fracturing in Oman and East Pacific Rise gabbros differs in many respects from the idealized crack system postulated by Lister, later vein sets may be more similar. In both localities, later veins are longer (meters to tens of meters), are lower temperature (predominantly greenschist facies), and have spacings of centimeters to meters. If these fractures were open for long times, they may have allowed advection of thermal energy out of the lower crust. However, our observations suggest that such fracture systems developed in already cracked and hydrated crust.

CONCLUSIONS

The earliest fracturing in gabbros from fast-spreading environments such as at Hess Deep and Wadi Abyad, Oman, is defined by a semibrittle, distributed microcrack-network that is filled by hornblende. Our observations show that initial tensile failure is the event that first allows water penetration into the lower oceanic crust. This process can affect the entire crustal section. The temperatures of initial fracturing are high and increase with depth, from ~700 °C at the dike-gabbro transition to ~825 °C at the Oman Moho. Microfracturing occurred at increasing distance from the ridge axis with progressively greater depths in the gabbros. The actual distance depends on thermal structure and spreading rate. If the models of Phipps Morgan and Chen (1993) are representative, then initial fracturing occurred at ~1 km from the ridge axis at the dike-gabbro transition and at ~6 km at the Oman Moho. Microfracturing is probably episodic because the velocity required for continuous downward propagation is slow relative to expected reaction rates. The temperature data suggest that cracking occurred at increasing strain rate with depth, that the rheologic properties of the layered, foliated, and variably textured gabbros varied with depth, or that the effects of water on strength led to progressive weakening with depth. The microscopic scale and heterogeneity of amphibole compositions imply limited fluid flow, which suggests that heat flow was conductive during early metamorphism.

ACKNOWLEDGMENTS

Supported by National Science Foundation grants OCE-9405999 and 9529602 (to Manning). Reviews by K. Gillis, R. Ingersoll, D. Stakes, and R. Newton improved the manuscript.

REFERENCES CITED

Benn, K., and Allard, B., 1989, Preferred mineral orientations related to magmatic flow in ophiolite layered gabbros: Journal of Petrology, v. 30, p. 925–946.

Beurrier, M., 1988, Geologie de la nappe ophiolitique de Semail dans les parties orientale et centrale de l'Oman: Paris, Université de Paris Bureau de Recherches Geologiques et Minieres, v. 128, 412 p.

Boudier, F., Nicolas, A., and Ildefonse, B., 1996, Magma chambers in the Oman ophiolite: Fed from the top and the bottom: Earth and Planetary Science Letters, v. 144, p. 239–250.

Brabander, D.J., and Giletti, B.J., 1995, Strontium diffusion kinetics in amphiboles and significance to thermal history determination: Geochimica et Cosmochimica Acta, v. 59, p. 2223–2238.

Browning, P., 1982, The petrology, geochemistry, and structure of the plutonic rocks of the Oman Ophiolite [Ph.D. thesis]: Milton Keynes, The Open University, 404 p.

Browning, P., 1984, Cryptic variation within the cumulate sequence of the Oman ophiolite: Magma chamber depth and petrological implications, *in* Gass, I.G., Lippard, S.J., and Shelton, A.W., eds., Ophiolites and oceanic lithosphere: Geological Society of London Special Publication 14, p. 71–82.

Chen, Y., and Morgan, W.J., 1990, A nonlinear rheology model for mid-ocean ridge axis topography: Journal of Geophysical Research, v. 95, p. 17583–17604.

Chenevez, J., Machetel, P., and Nicolas, A., 1998, Numerical models of magma chambers in the Oman ophiolite: Journal of Geophysical Research, v. 103, p. 15443–15455.

Ernewein, M., Pflumio, C., and Whitechurch, H., 1988, The death of accretion zone as evidenced by the magmatic history of the Sumail ophiolite (Oman): Tectonophysics, v. 151, p. 247–274.

Evans, B., Frederich, J.T., and Wong, T.-F., 1990, The brittle-ductile transition in rocks: Recent experimental and theoretical progress, *in* Duba, A.G., Durham, W.B., Handin, J.W., and Wang, H.F., eds., The brittle-ductile transition in rocks: New York, American Geophysical Union Monograph, v. 56, p. 1–20.

Francheteau, J., Armijo, R., Cheminée, J.L., Hekinian, R., Lonsdale, P., and Blum, N., 1990, 1 Ma East Pacific Rise oceanic crust and uppermost mantle exposed by rifting in Hess Deep (equatorial Pacific Ocean): Earth and Planetary Science Letters, v. 101, p. 281–295.

Giletti, B.J., and Casserly, J.E.D., 1994, Strontium diffusion kinetics in plagioclase feldspars: Geochimica et Cosmochimica Acta, v. 58, p. 3785–3793.

Giletti, B.J., and Shananan, T.M., 1997, Alkali diffusion in plagioclase feldspar: Chemical Geology, v. 139, p. 3–20.

Gillis, K.M., 1995, Controls on hydrothermal alteration in a section of fast-spreading oceanic crust: Earth and Planetary Science Letters, v. 134, p. 473–489.

Gillis, K.M., Mével, C., Allan, J., and the Leg 147 Scientific Party, 1993, Proceedings of the Ocean Drilling Program, Initial Reports, v. 147: College Station, Texas, Ocean Drilling Program, 366 p.

Gillis, K.M., Thompson, G., and Kelley, D.S., 1993, A view of the lower crustal component of hydrothermal systems at the Mid-Atlantic Ridge: Journal of Geophysical Research, v. 98, p. 19597–19619.

Gregory, R.T., and Taylor, H.P., Jr., 1981, An oxygen isotope profile in a section of Cretaceous oceanic crust, Samail ophiolite, Oman: Evidence for $\delta^{18}O$ buffering of the oceans by deep (>5 km) seawater-hydrothermal circulation at mid-ocean ridges: Journal of Geophysical Research, v. 86, p. 2737–2755.

Helz, R.H., 1973, Phase relations of basalts in their melting range at P_{H_2O} = 5 kb as a function of oxygen fugacity: Journal of Petrology, v. 14, p. 249–302.

Hirth, G., Escartin, J., and Lin, J., 1998, The rheology of the lower oceanic crust: Implications for lithospheric deformation at mid-ocean ridges, *in* Buck, W.R., Delaney, P.T., Karson, J.A., and Lagabrielle, Y., eds., Faulting and magmatism at mid-ocean ridges: New York, American Geophysical Union Monograph, v. 106, p. 291–303.

Holland T.J.B., and Blundy, J.D., 1994, Non-ideal interactions in calcic amphiboles and their bearing on amphibole-plagioclase thermometry: Contributions to Mineralogy and Petrology, v. 116, p. 433–447.

Jenkins, D.M., 1987, Synthesis and characterization of tremolite in the system H_2O-CaO-MgO-SiO_2: American Mineralogist, v. 72, p. 707–715.

Juteau, T., Beurrier, M., Dahl, R., and Nehlig, P., 1988, Segmentation at a fossil spreading axis: The plutonic sequence of the Wadi Haymiliayah area (Haylayn Block, Sumail Nappe, Oman): Tectonophysics, v. 151, p. 167–197.

Karson, J.A., Hurst, S.D., and Lonsdale, P.F., 1992, Tectonic rotations of dikes in fast-spread oceanic crust exposed near Hess Deep: Geology, v. 20, p. 685–688.

Leake, B.E., 1978, Nomenclature of amphiboles: Canadian Mineralogist, v. 16, p. 501–520.

Lécuyer, C., and Gruau, G., 1996, Oxygen and strontium isotope compositions of Hess Deep gabbros (Holes 894F and G): High-temperature interaction of seawater with the oceanic crust layer 3: Proceedings of the Ocean Drilling Program, Scientific Results, v. 147: College Station, Texas, Ocean Drilling Program, p. 227–234.

Lippard, S.J., Shelton, A.W., and Gass, I.G., 1986, The ophiolite of northern Oman: Geological Society of London Memoir 11, 178 p.

Lister, C.R.B., 1974, On the penetration of water into hot rock: Geophysical Journal of the Royal Astronomical Society, v. 39, p. 465–509.

Lister, C.R.B., 1982, "Active" and "passive" hydrothermal systems in the oceanic crust: Predicted physical conditions, *in* Fanning, K.A., and Mannheim, F.T., eds., The dynamic environment of the ocean floor: Lexington, Massachusetts, D.C. Heath, p. 441–459.

Liu, M., and Yund, R.A., 1992, NaSi-CaAl interdiffusion in plagioclase: American Mineralogist, v. 77, p. 275–283.

Lonsdale, P., 1988, Structural pattern of the Galapagos Microplate and evolution of the Galapagos triple junctions: Journal of Geophysical Research, v. 93, p. 13551–13574.

Mackwell, S.J., Zimmerman, M.E., and Kohlstedt, D.L., 1998, High-temperature deformation of dry diabase with application to tectonics of Venus: Journal of Geophysical Research, v. 103, p. 975–984.

MacLeod, C.J., and Rothery, D.A., 1992, Ridge axial segmentation in the Oman ophiolite: Evidence from along-strike variations in the sheeted dyke complex, *in* Parson, L.M., Murton, B.J., and Browning, P., eds., Ophiolites and their modern oceanic analogues: Geological Society of London Special Publication 60, p. 39–63.

MacLeod, C.J., and Yaouancq, G., 2000, A fossil melt lens in the Oman ophiolite: Implications for magma chamber processes at fast-spreading ridges: Earth and Planetary Science Letters, v. 176, p. 357–373.

MacLeod, C.J., Boudier, F., Yaouancq, G., and Richter, C., 1996, Gabbro fabrics from ODP Site 894, Hess Deep: Implications for magma chamber processes at the East Pacific Rise: Proceedings of the Ocean Drilling Program, Scientific Results, v. 147: College Station, Texas, Ocean Drilling Program, p. 317–328.

Manning, C.E., and MacLeod, C.J., 1996, Fracture-controlled metamorphism of Hess Deep Gabbros, Site 894: Constraints on the roots of mid-ocean ridge hydrothermal systems at fast spreading centers: Proceedings of the Ocean Drilling Program, Scientific Results, v. 147: College Station, Texas, Ocean Drilling Program, p. 189–212.

Manning, C.E., Weston, P.E., and Mahon, K.I., 1996, Rapid high-temperature metamorphism of East Pacific Rise gabbros from Hess Deep: Earth and Planetary Science Letters, v. 144, p. 123–132.

Mével, C., and Cannat, M., 1991, Lithospheric stretching and hydrothermal processes in oceanic gabbros from slow-spreading ridges, *in* Peters, T., Nicolas, A., and Coleman, R.G., eds., Ophiolite genesis and evolution of the oceanic lithosphere: Dordrecht, Netherlands, Kluwer Academic Publishers, p. 293–312.

Natland, J.H., and Dick, H.J.B., 1996, Melt migration through high-level gabbroic cumulates of the East Pacific Rise at Hess Deep: Inferences from rock textures and mineral compositions: Proceedings of the Ocean Drilling Program, Scientific Results, v. 147: College Station, Texas, Ocean Drilling Program, p. 21–58.

Nehlig, P., 1994, Fracture and permeability analysis in magma-hydrothermal transition zones in the Samail ophiolite (Oman): Journal of Geophysical Research, v. 99, p. 589–601.

Nehlig, P., and Juteau, T., 1988a, Flow porosities, permeabilities, and preliminary data on fluid inclusions and fossil thermal gradients in the crustal sequence of the Sumail ophiolite (Oman): Tectonophysics, v. 151, p. 199–221.

Nehlig, P., and Juteau, T., 1988b, Deep crustal seawater penetration and circulation at ocean ridges: Evidence from the Oman ophiolite: Marine Geology, v. 84, p. 209–228.

Nicolas, A., 1989, Structures of ophiolites and dynamics of oceanic lithosphere: Boston, Kluwer Academic Publishers, 367 p.

Nicolas, A., 1992, Kinematics in magmatic rocks with special reference to gabbros: Journal of Petrology, v. 33, p. 891–915.

Nicolas, A., and Boudier, F., 1995, Mapping oceanic ridge segments in Oman ophiolite: Journal of Geophysical Research, v. 92, p. 6179–6197.

Nicolas, A., Reuber, I., and Benn, K., 1988, A new magma chamber model based on structural studies in the Oman ophiolite: Tectonophysics, v. 151, p. 87–105.

Nicolas, A., Boudier, F., and Ildefonse, B., 1994, Evidence from the Oman ophiolite for active mantle upwelling beneath a fast-spreading ridge: Nature, v. 370, p. 51–53.

Pallister, J.S., and Hopson, C.A., 1981, Samail plutonic suite: Field relations, phase variations, cryptic variation and layering, and a model of a spreading ridge magma chamber: Journal of Geophysical Research, v. 86, p. 2593–2644.

Phipps Morgan, J., and Chen, Y.J., 1993, The genesis of oceanic crust: Magma injection, hydrothermal circulation, and crustal flow: Journal of Geophysical Research, v. 98, p. 6283–6297.

Pouchou, J.L., and Pichoir, J.F., 1986, Very high elements X-ray-microanalysis—recent models of quantification: Journal de Microscopie et de Spectroscopie Electroniques, v. 11, p. 229–250.

Quick, J.E., and Denlinger, R.P., 1993, Ductile deformation and the origin of layered gabbro in ophiolites: Journal of Geophysical Research, v. 98, p. 14015–14027.

Reuber, I., Nehlig, P., and Juteau, T., 1991, Axial segmentation at a fossil oceanic spreading centre in the Haylayn block (Semail nappe, Oman): Off-axis mantle diapir and advancing ridge tip: Journal of Geodynamics, v. 13, p. 253–278.

Rothery, D.A., 1983, The base of a sheeted dyke complex, Oman ophiolite: Implications for magma chambers at oceanic spreading axes: Journal of the Geological Society of London, v. 140, p. 287–296.

Rutter, E.H., 1986, On the nomenclature of mode of failure transitions in rocks: Tectonophysics, v. 122, p. 381–387.

Simmons, G., and Richter, D., 1976, Microcracks in rocks, *in* Strens, R.G.J., ed., The physics and chemistry of minerals and rocks: London, Wiley, p. 105–137.

Spear, F.S., 1981, An experimental study of hornblende stability and compositional variability in amphibolite: American Journal of Science, v. 281, p. 697–734.

Stakes, D.S., and Taylor, H.P., Jr., 1992, The northern Samail ophiolite: An oxygen isotope, microprobe, and field study: Journal of Geophysical Research, v. 97, p. 7043–7080.

Stakes D.S., and Vanko, D.A., 1986, Multistage hydrothermal alteration of gabbroic rocks from the failed Mathematician Ridge: Earth and Planetary Science Letters, v. 79, p. 75–92.

Stakes, D.S., Mével, C., Cannat, M., and Chaput, T., 1991, Metamorphic stratigraphy of Hole 735B: Proceedings of the Ocean Drilling Program, Scientific Results, v. 118: College Station, Texas, Ocean Drilling Program, p. 153–180.

Tilton, G.R., Hopson, C.A., and Wright, J.E., 1981, Uranium-lead isotopic ages of the Samail ophiolite, Oman, with application to Tethyan ocean ridge tectonics: Journal of Geophysical Research, v. 86, p. 2763–2776.

Walther, J.V., and Wood, B.J., 1984, Rate and mechanism in prograde metamorphism: Contributions to Mineralogy and Petrology, v. 88, p. 246–259.

Weston, P.E., 1998, Temperature, timing, and fluid composition constraints for high temperature fracturing and hydrous mineral formation in gabbros from Hess Deep [M.S. thesis]: Los Angeles, University of California at Los Angeles, 152 p.

Wilks, K.R., and Carter, N.L., 1986, Ductile and semi-brittle deformations of layer 3 gabbros: Zambales, Bay of Islands, and Samail ophiolites: Eos, Transactions of the American Geophysical Union, v. 67, p. 1202.

Wood, B.J., and Walther, J.V., 1983, Rates of hydrothermal reactions: Science, v. 222, p. 413–415.

Manuscript Accepted by the Society February 23, 2000

Printed in the U.S.A.

Geological Society of America
Special Paper 349
2000

Hydrothermal alteration and fluid fluxes in ophiolites and oceanic crust

Jeffrey C. Alt and Damon A.H. Teagle
Department of Geological Sciences, University of Michigan, Ann Arbor, Michigan 48109-1063, USA; e-mail: jalt@umich.edu

ABSTRACT

We have compiled a composite oceanic crustal section for strontium, oxygen, and sulfur isotope tracers of seawater interaction, including new oxygen isotope data for a 1-km-thick section of lower oceanic crust, and in this paper, we summarize the main similarities and differences between hydrothermal alteration in ophiolites and that in oceanic crust. Ophiolitic crust is consistently more intensely recrystallized, and isotope tracers exhibit greater shifts in ophiolites than in situ oceanic crust. These differences require greater time-integrated fluid fluxes through the ophiolites (by a factor of ~3–6), as well as penetration of larger fluid fluxes to greater depths in some ophiolites. The reasons for this fundamental difference must be related in some way to differences in tectonic setting, primary chemical compositions, or heat sources, but the exact reasons remain problematic. The greater fluid fluxes and tracer exchange in ophiolites point out the need for caution when applying data from ophiolites to the global mid-ocean ridge system and estimation of the effects on oceanic and crustal chemistry. Despite significant differences between hydrothermal effects in ophiolites and oceanic crust, many important similarities exist, and the ophiolite analogy remains invaluable to understand the structure of oceanic crust and processes at mid-ocean ridges. A need remains, however, for additional and more continuous sections through ocean and ophiolitic crust in different tectonic settings, in particular through the critical lithologic transitions that coincide with changes in fluid circulation and alteration.

INTRODUCTION

The presence of sheeted-dike complexes in ophiolites provides evidence for their formation via seafloor spreading (Gass, 1968; Moores and Vine, 1971); consequently, ophiolites have been used as analogues for interpreting magmatic, tectonic, and hydrothermal processes attending the formation of oceanic crust at mid-ocean ridges. The relative ease of access to ophiolites and field mapping and sampling in three dimensions have made the ophiolite analogue invaluable to understanding in situ oceanic crust; in particular, the ophiolite analogue has provided such fundamental concepts as sheeted dikes and the basic layered model of oceanic crust. Despite formation in seafloor-spreading environments, however, most ophiolites did not form at typical mid-ocean ridges in large ocean basins. Some fragmented ophiolites may have petrologic and geochemical similarities to mid-ocean ridge basalt, but most do not. The latter include the large, relatively intact, and well-studied ophiolites such as Troodos in Cyprus and Semail in Oman, as well as those in Newfoundland and California. These ophiolites formed in supra-subduction settings, as shown by their trace element and volatile compositions and the presence of boninites (high-Mg, low-Ti basalts) and differentiated calc-alkalic rocks, as well as basalts, in contrast to the depleted tholeiites erupted at mid-ocean ridges (Miyashiro, 1973; Pearce et al., 1984; Rautenschlein et al., 1985; Harper et al., 1988; Muenow et al., 1990; Elthon, 1991; Schiffman et al., 1991; Bloomer et al., 1995).

Chemical exchange between seawater and oceanic crust is an important control on the composition of the oceans, but also significantly affects the crust, which in turn influences metasomatism and recycling in subduction zones. Data from ophiolites have been used to model the chemical and isotope fluxes between seawater and oceanic crust during hydrothermal alteration as well as the effect these have on seawater and the crust (Gregory and Taylor, 1981; Bickle and Teagle, 1992; Alt, 1994). Recent studies, however, have pointed out differences between hydrothermal effects in ophiolites and oceanic crust, particularly that fluid fluxes through ophiolites must have been greater than those that occur in modern oceanic crust formed at mid-ocean ridges (Schiffman et al., 1991; Bickle and Teagle, 1992; Alt, 1994; Alt et al., 1996, 1998). We have compiled composite crustal sections through oceanic crust for strontium, oxygen, and sulfur isotope tracers, including new oxygen isotope data for a 1-km-thick section of lower oceanic crust, and in this paper, we summarize the main similarities and differences between hydrothermal alteration in ophiolites and that in oceanic crust. We show that hydrothermal fluid fluxes are consistently greater through ophiolitic crust, which has implications for calculations of global fluid and chemical fluxes at mid-ocean ridges, and that many ophiolites exhibit metamorphic effects that vary from current models for oceanic crust. Despite these differences, ophiolites remain important on-land analogues that provide invaluable insights into processes affecting oceanic crust formed at mid-ocean ridges.

Alt, J.C., and Teagle, D.A.H., 2000, Hydrothermal alteration and fluid fluxes in ophiolites and oceanic crust, *in* Dilek, Y., Moores, E.M., Elthon, D., and Nicolas, A., eds., Ophiolites and Oceanic Crust: New Insights from Field Studies and the Ocean Drilling Program: Boulder, Colorado, Geological Society of America Special Paper 349, p. 273–282.

OCEANIC CRUST AND SUBMARINE HYDROTHERMAL SYSTEMS

An idealized section of oceanic crust consists of a volcanic section, several hundred meters thick, underlain by ~1 km of sheeted dikes, and then a plutonic section, for an average thickness of ~7 km (White et al., 1992). Crust generated at fast-spreading ridges (e.g., the East Pacific Rise) reasonably conforms to this model (Gillis, 1995), whereas crust formed at slow-spreading ridges (e.g., Mid-Atlantic Ridge, Southwest Indian Ridge) is commonly structurally disrupted in response to variations in tectonic activity and magma budget (Tucholke and Lin, 1994).

Convection in the oceanic crust can be conveniently divided into active axial hydrothermal systems and passive, off-axis systems (Lister, 1982). In this classification, active systems are restricted to mid-ocean ridge axes with the heat source provided by a magma chamber or cooling intrusive rocks. Maximum fluid temperatures at depth are high (up to ~400 °C), and discharge of these high-temperature fluids is rapid. Moreover, hydrothermal systems on slow- and fast-spreading ridges can differ. Fluid circulation at fast-spreading ridge axes is restricted to the upper crust (lavas, dikes, uppermost gabbros); fluid penetration is limited by the presence of a melt lens in the uppermost plutonic section at 1–2 km depth (Gillis, 1995). Magmatism at slow-spreading ridges is intermittent, so their high-temperature hydrothermal systems go through cycles of waxing and waning activity (Lalou et al., 1995), although little is known of how these hydrothermal cycles affect the crust at depth. The presence of earthquakes down to mantle depths on slow-spreading ridges indicates that at times the entire crust is locally cooled to the brittle realm (Huang and Solomon, 1988) and that fluids may circulate to great depths at the axis of spreading. In fast-spreading crust, penetration of small volumes of fluids to progressively greater depths occurs as the melt lens solidifies and the lower crust cools and fractures while it moves away from the axis (Manning et al., 1996).

In the simplified classification of Lister (1982) passive, off-axis convection systems are driven by cooling of the lithosphere, and circulation is thought to be restricted to the volcanic section (Fisher et al., 1994). Temperatures are low (<100 °C down to near 0 °C), flow is sluggish (Mottl and Wheat, 1994), and circulation can continue over large areas for several tens of million years (Stein and Stein, 1994). Missing from this classification is the poorly understood transition from active high-temperature convection cells at the axis to the well-developed shallow and cooler passive systems that occur much farther off axis. This transition must take place as the crust moves slightly (a few kilometers) off axis, where fluids penetrate to greater depths to cool the lower crust (plutonic section) and cooling of the upper crust (dikes and lavas) continues.

Sections of oceanic and ophiolitic crust integrate the effects of these different hydrothermal regimes. In this paper we describe these integrated effects in terms of hydrothermal processes and their implications for fluid fluxes.

HYDROTHERMAL ALTERATION IN OCEANIC CRUST AND OPHIOLITES

Mineralogy

Ophiolites. Despite abundant outcrop, it is commonly difficult to construct a stratigraphy for many ophiolites. For the purposes of this paper, we consider data from well-preserved ophiolites, principally the Troodos ophiolite, Cyprus, and the Semail ophiolite, Oman, where good stratigraphy has been established and profiles of the most useful tracers of seawater interaction have been analyzed. Hydrothermal alteration effects in the Troodos ophiolite are summarized here. The upper 20–300 m, or "seawater zone," of the volcanic section is characterized by pervasive low-temperature (~0–100 °C) alteration by seawater, with rocks slightly to highly altered to smectite ± iron oxides ± K-feldspar ± calcite ± zeolites (Gillis and Robinson, 1990; Bednarz and Schmincke, 1989; Alt, 1994). The underlying, variably altered, "low-temperature zone" extends to near the base of the volcanic section, and variably altered rocks contain secondary smectite ± zeolites ± celadonite ± iron oxyhydroxides ± sepiolite ± calcite ± opal. An ~100-m-thick "transition zone" in the uppermost dikes marks the downward mineral-assemblage change to hydrothermally metamorphosed rocks that have been totally recrystallized to albite + chlorite + quartz ± actinolite ± epidote ± titanite at temperatures above ~250 °C. Sheeted dikes are essentially 100% recrystallized to greenschist-facies minerals, and the lower dikes contain amphibole (actinolitic hornblende, actinolite) + minor quartz + chlorite + epidote (Schiffman and Smith, 1988; Baragar et al., 1989; Gillis and Robinson, 1990; Alt, 1994).

The upper ~200 m of the Troodos plutonic section is intensely hydrothermally altered (70%–100% recrystallized); the alteration intensity decreases below this depth (Allen, 1975; Alt, 1994). The underlying gabbroic rocks and cumulates are less altered: they are generally only 5%–50% recrystallized. Secondary minerals in these rocks are characteristic of amphibolite to greenschist facies and include brown and green amphiboles, actinolite, magnetite, titanite, albite, epidote, chlorite, talc, and retrograde prehnite, zeolites, calcite, and smectite.

Other ophiolites exhibit broadly similar alteration profiles; their metamorphic grade reaches a maximum within the sheeted dikes and uppermost plutonic section, and alteration intensity decreases in the overlying lavas and underlying gabbroic rocks and cumulates (e.g., Spooner and Fyfe, 1973; Spooner et al., 1977; Coish, 1977; Stern and Elthon, 1979; Liou, 1979; Cocker et al., 1982; Ishizuka, 1985; Harper et al., 1988; Pflumio, 1991; Schiffman et al., 1991; Agrinier et al., 1988; Lecuyer et al., 1990).

Oceanic Crust. The problems of constructing a continuous stratigraphy in ophiolites are compounded in oceanic crust, where critical lithologic transitions are rarely sampled and available lower-crustal sections are complicated by tectonic exposure. Numerous Ocean Drilling Program (ODP) holes penetrate variable depths in the volcanic section, but Hole 504B in the eastern Pacific provides the only intact lava plus dikes section (Alt et al., 1986a). The faulted exposures sampled by submersible at Hess Deep in the eastern Pacific cross the dikes to gabbro transition and provide a transect through the plutonic section (Agrinier et al., 1995; Gillis, 1995). Other drilled sections have provided isolated samples of lower crust from tectonic exposures: ODP Hole 735B in the Southwest Indian Ridge provided the longest lower-crustal section (1.5 km; Stakes et al., 1991; Dick, Natland, et al., 1998), Hole 894 at Hess Deep provided a 120 m section of the uppermost plutonic rocks (Lecuyer and Gruau, 1996), and Sites 921–923 in the MARK area (the intersection of the Mid-Atlantic Ridge with the Kane Fracture Zone) yielded a short (<100 m) section of middle lower crust (Barling et al., 1997).

Another problem with sampling oceanic crust by deep-ocean drilling is incomplete recovery. Although all lithologies and alteration types are sampled by coring, less competent material in secondary mineral veins and breccias may be preferentially lost during drilling. This effect is greatest in the volcanic section, where fresh and altered glassy interpillow material and breccias may be lost. Sheeted dikes

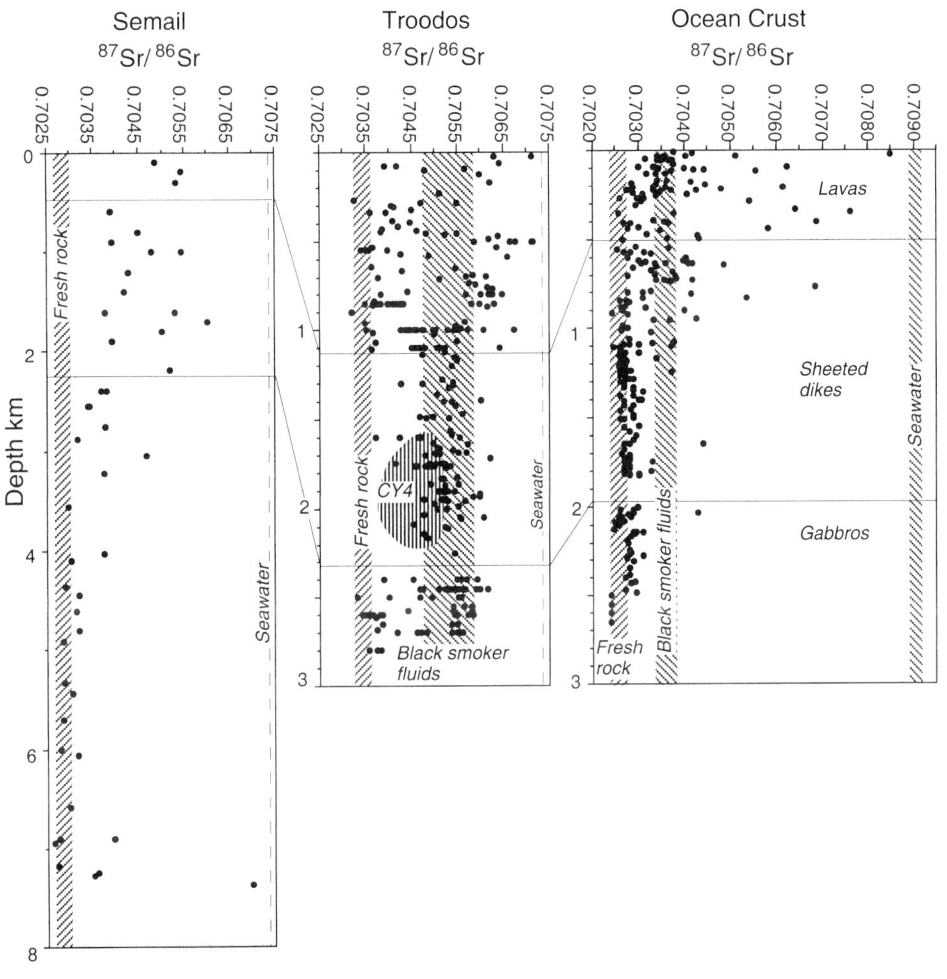

Figure 1. Profiles of bulk-rock Sr isotope data for composite sections through the Semail and Troodos ophiolites and oceanic crust. Reference lines for fresh basalt and coeval seawater (Burke et al., 1982) are shown, as well as compositions of deep hydrothermal fluids ("black smoker fluid"). Compositions of black smoker fluids are inferred from analyses of bulk rocks and secondary minerals in the Troodos ophiolite (Bickle and Teagle, 1992) and oceanic crust (Teagle et al., 1998b). Data for Troodos compiled in Bickle and Teagle (1992) and from Bickle et al. (1998). The field for dikes from the Troodos CY4 drill core (shaded) reflects restricted local variations in isotope exchange, in contrast to the greater and more uniform shifts displayed by representative dike sections from throughout the rest of the ophiolite. Semail data from McCulloch et al. (1980), Lanphere et al. (1981), and Regba et al. (1991). Oceanic crust data (see text) compiled in Teagle et al. (1998b) (Holes 504B and 896A), and from Kempton et al. (1991) (Hole 735B), Muehlenbachs (1980) (Holes 417D and 418A), Staudigel et al. (1981b, 1995) (Holes 417D and 418A), Gillis and Thompson (1993) (MARK), Lecuyer and Gruau (1996) (Hess Deep), and Barling et al. (1997) (MARK).

are more massive and coherent and lack such material, so despite low recovery during drilling, preferential loss of material may be less of a problem. Moreover, the similarity of dikes sampled by different techniques (dredging, drilling, and submersible) from several locations in different oceans suggests that the cored dikes are representative (Gillis and Thompson, 1993; Gillis, 1995; Agrinier et al., 1995; Alt et al., 1996). Coring in oceanic gabbro sections generally has very high recovery rates, close to 100%, and metamorphic effects are quite similar among several different sections sampled by various techniques, indicating that the sections are representative and that sampling bias is limited (Ito and Clayton, 1983; Kempton et al., 1991; Stakes et al., 1991; Gillis and Thompson, 1993; Agrinier et al., 1995; Lecuyer and Gruau, 1996; Barling et al., 1997; Dick, Natland, et al., 1998).

Like ophiolitic sections, oceanic crust exhibits an increase in metamorphic grade downward from lavas into sheeted dikes and uppermost gabbros, then a rapid decrease in alteration intensity downward into the deeper plutonic section. A major difference, however, is the overall much lower intensity of recrystallization in oceanic crust compared to ophiolites.

Lavas in oceanic sections are typically only slightly recrystallized (5–15%) to low-temperature (0–100 °C) phases (smectite, celadonite, iron oxyhydroxides, carbonates, and minor zeolites), but interpillow hyaloclastites and breccias can be locally highly altered (up to 100%; see review in Alt, 1999). The alteration minerals and chemistry of 6 Ma rocks from ODP Holes 504B and 896A in the eastern Pacific (Figs. 1–3) are generally similar to those at older sites (e.g., Holes 417D and 418A in the western Atlantic, Figs. 1 and 2); fluids still circulate in the uppermost basement in young crust, however, and C and perhaps other elements may be taken up from seawater for tens of millions of years (Alt and Teagle, 1999). At the base of the volcanic section in Hole 504B, there is a sharp (70 m thick) transition to mineralized rocks intensely hydrothermally metamorphosed to chlorite + albite + titanite + quartz (Alt et al., 1986a). The latter rocks were altered at 250–380 °C in a subsurface mixing zone, where hydrothermal fluids in the dikes mixed with cooler seawater circulating in the overlying volcanic section. The volcanic rock to dike transition at Hess Deep is marked by similar changes, although the position of the downward change from low-temperature alteration to higher-grade effects varies somewhat depending upon local permeability and fluid circulation (Gillis, 1995).

Oceanic dike sections from several different locations are only partly recrystallized (Alt et al., 1986a, 1996; Gillis and Thompson, 1993; Gillis, 1995). On average, dikes in Hole 504B are ~25%–35% recrystallized; most are only slightly recrystallized (10%–30%), but more intense alteration (30%–100%) is restricted to isolated patches and halos along veins (Alt et al., 1986a, 1996). Secondary minerals in the upper dike section reflect formation under greenschist- to subgreenschist-facies conditions and include albite + chlorite-smectite ± chlorite ± actinolite ± quartz ± titanite (Alt et al., 1986a, 1996; Gillis and Thompson, 1993; Gillis, 1995). The lower dikes are somewhat more intensely recrystallized than the upper dikes and contain the same secondary minerals, as well as higher-grade minerals of the upper-greenschist to amphibolite facies: e.g., Mg-hornblende, sec-

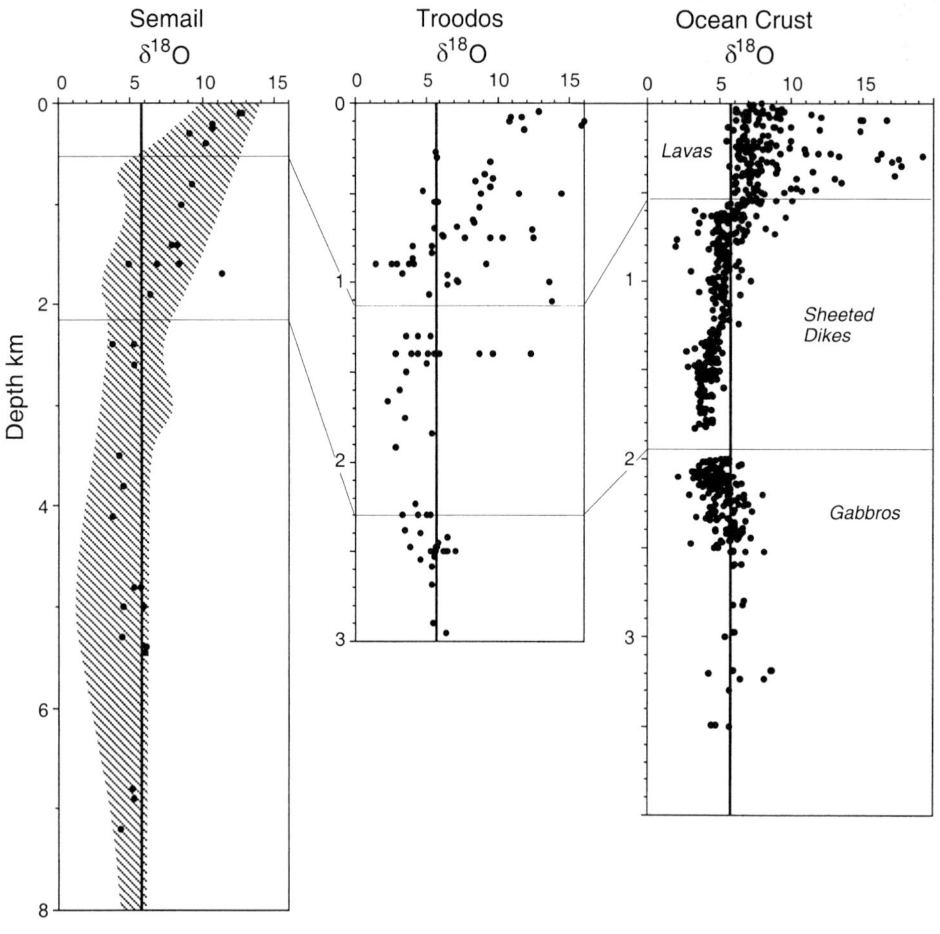

Figure 2. Profiles of whole-rock O isotope data for composite sections through the Semail and Troodos ophiolites and oceanic crust (see text). Reference line given for fresh basalt. Data for Troodos compiled in Bickle and Teagle (1992). Individual data points for Semail are from Gregory and Taylor (1981), and shaded area encloses data from multiple profiles from Stakes and Taylor (1992). Ocean-crust data from this work (Hole 735B) Alt et al. (1986b, 1996) (Hole 504B), Agrinier et al. (1995) (Hess Deep), Teagle et al. (1996) (Hole 896A), Staudigel et al. (1981a, 1981b, 1995) (Holes 417D and 418A), Muehlenbachs (1980) (Holes 417D and 418A), Stakes et al. (1991) (Hole 735B), and Lecuyer and Gruau (1996) (Hess Deep).

ondary clinopyroxene, and calcic plagioclase (Alt et al., 1986a, 1996; Gillis and Thompson, 1993; Gillis, 1995; Vanko and Laverne, 1998).

The uppermost ~100 m of the plutonic section in crust at Hess Deep (where the crust was formed at a fast-spreading mid-ocean ridge) was altered at amphibolite and transitional greenschist to amphibolite facies via intense hydrothermal circulation overlying an axial magma lens (Gillis, 1995). The underlying gabbronorites are less intensely recrystallized, with alteration mainly restricted to the formation of amphibole in microfractures that formed as hydrothermal fluids penetrated downward through the solidified rocks while they cooled and moved off axis. (Gillis, 1995; Manning et al., 1996). Metamorphism of plutonic sections from crust formed at slow-spreading ridges is broadly similar to that formed at fast-spreading ridges: metamorphic intensity is high in the uppermost few hundred meters of gabbros and decreases dramatically with depth (Ito and Clayton, 1983; Mevel and Cannat, 1991; Gillis and Thompson, 1993; Dick, Natland, et al., 1998). Most of the deeper parts of oceanic plutonic sections are essentially unaltered (Gillis, 1995; Dick, Natland, et al., 1998).

Isotope Profiles

Oceanic Crust. Profiles of whole-rock Sr, O, and S isotope compositions for a composite section of oceanic crust and for the Troodos and Semail ophiolites are shown in Figures 1–3. These isotope tracers are convenient tools to quantify the effects of recrystallization of the rocks and interaction with seawater-derived hydrothermal fluids.

In the oceanic profiles, the volcanic section is shifted toward the $^{87}Sr/^{86}Sr$ ratio of seawater and to elevated $\delta^{18}O$ values as the result of low-temperature exchange with seawater. Scatter to low $\delta^{34}S$ values for sulfide (Fig. 3) is the result of fractionation produced during partial oxidation of igneous sulfides at low temperatures (Alt et al., 1989). Sr and O isotope shifts generally decrease with depth in the oceanic sections, consistent with other geochemical and mineralogical data that indicate decreasing seawater effects downward (Alt et al., 1986a, 1996; Staudigel et al., 1995). In 6 Ma rocks from Holes 504B and 896A, Sr exchange is limited mainly by the slight (5%–15%) recrystallization of the majority of rocks. Secondary minerals have higher $^{87}Sr/^{86}Sr$ ratios and $\delta^{18}O$ values than the bulk rocks, a fact that reflects precipitation of minerals in equilibrium with seawater-like fluids; consequently, breccias and highly altered interpillow hyaloclastites exhibit greater exchange (Teagle et al., 1996, 1998b). Weighted mean bulk Sr and O isotope values for the volcanic section of Holes 504B and 896A, including veins and breccias, are 0.7040 and 7.7‰, respectively; these values indicate ~15% exchange with seawater (Alt et al., 1998).

Also shown in Figures 1 and 2 are data for the volcanic sections of 120 Ma crust drilled at Holes 417D and 418A in the western Atlantic. These data significantly overlap with those of the younger sites but exhibit scatter to higher values. Lavas from Hole 417A, adjacent to Holes 417D and 418A, are much more intensely altered, but these samples are from a basement hill affected by long-lived, focused, low-temperature hydrothermal outflow, and we do not consider these unique rocks to be representative of the uppermost oceanic crust (Muehlenbachs, 1980; Staudigel et al., 1981a, 1995; Alt and Honnorez, 1984; Teagle et al., 1996). Staudigel et al. (1995) used

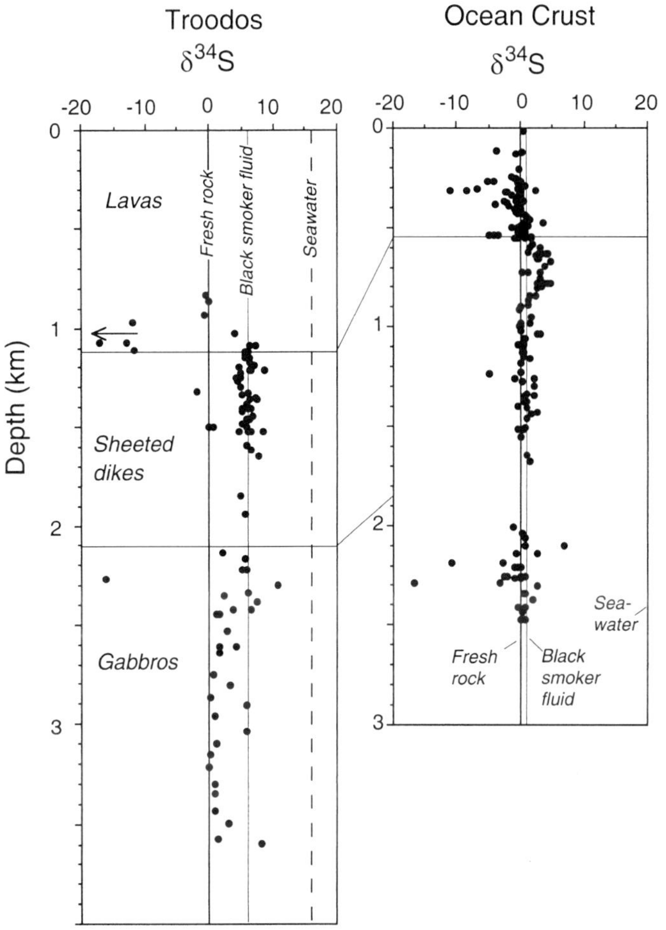

Figure 3. Profiles of whole-rock S isotope compositions for composite sections of the Troodos ophiolite and oceanic crust (Holes 504B and 735B). Data from Alt (1994, 1995). Reference lines given for MORB mantle and for modern and Cretaceous seawater (Claypool et al., 1980), as well as compositions of deep hydrothermal fluids ("black smoker fluid"). Compositions of black smoker fluid inferred from compositions of bulk rocks and secondary minerals in the Troodos ophiolite (Alt, 1994) and oceanic crust (Alt et al., 1989; Alt, 1995).

composite samples from Holes 417A, 417D, and 418A to estimate the bulk composition of the altered volcanic section at these sites and obtained Sr and O values of 0.7046 and 10‰. The isotope shifts at these sites suggest greater alteration than in the younger extrusive rocks from Holes 504B and 896A. This difference reflects in part the higher core recovery (~80%) at the older sites and a greater number of analyses of highly altered breccias and interpillow hyaloclastites (Muehlenbachs, 1980; Staudigel et al., 1981a, 1995). Some of the apparent greater alteration at the older sites may also be attributed to heterogeneity of alteration effects in uppermost crust and to ingrowth of ^{87}Sr from secondarily enriched Rb in the older sections. It is also possible that the crust at the older sites has a longer history of alteration and aging, although this view remains controversial.

Rb-Sr dating of vein smectites in oceanic lavas yields ages mostly within 10–15 m.y. of crustal generation (Staudigel and Hart, 1985; Waggoner, 1993); therefore, alteration may not yet be complete in the 6 Ma crust drilled at sites 504 and 896. The Rb-Sr dates represent "closure" ages of the minerals, however, and do not necessarily represent thc timing or duration of fluid circulation or alteration (Hart and Staudigel, 1986). Moreover, with the exception of continued addition of carbonate, all the alteration effects observed in older crust (>100 Ma) have also been reported from 3 Ma oceanic crust (Andrews, 1977; Staudigel et al., 1981b; Alt and Honnorez, 1984; Alt and Teagle, 1999); this finding suggests that most alteration may be complete within a few million years of crustal accretion. Heat-flow data indicate that, on average, fluids can circulate and remove heat for ~65 m.y. (Stein and Stein, 1994), with the implication that alteration may continue for several tens of million years. The chemical exchange resulting from such prolonged circulation is essentially unquantified, however, and the main effect of this circulation may be precipitation of carbonate within the upper crust (Staudigel et al., 1981b; Alt and Teagle, 1999).

The transition from lavas to dikes and the uppermost dikes in Hole 504B exhibit scatter to high ^{87}Sr/^{86}Sr ratios and a sharp decrease in δ^{18}O to values that are ^{18}O depleted relative to fresh MORB (mid-oceanic-ridge basalt; Figs. 1 and 2). These changes reflect the generally intense recrystallization of these rocks at temperatures of 250–380 °C in a zone of mixing between seawater and hydrothermal fluids (Alt et al., 1986a, 1996; Teagle et al., 1998b). S isotope compositions also reflect mixing in that there is a shift of sulfide δ^{34}S values toward seawater from hydrothermal values of 0‰–1‰ (Alt et al., 1989). Simple isotope mass balance indicates that, on average, 14% of the S and 20% of the Sr in the transition zone and uppermost dikes are derived from seawater.

The Sr and S isotope compositions of the major parts of oceanic sheeted-dike and plutonic sections are at most only slightly shifted from primary values. Multiple sections from several different areas sampled by various techniques (drilling, dredging, and submersible) all exhibit similar effects, indicating that these are systematic and representative of the oceanic crust (this work; Alt et al., 1996; Ito and Clayton, 1983; Stakes et al., 1991; Agrinier et al., 1995; Teagle et al., 1998b; Kempton et al., 1991; Gillis and Thompson, 1993; Lecuyer and Gruau, 1996; Barling et al., 1997). Further aging and alteration of these relatively young sections (<15 Ma) is insignificant because of their very low Rb contents and because these deep sections are effectively sealed to seawater circulation. The limited shifts in ^{87}Sr/^{86}Sr ratios and δ^{34}S values in these sections reflect in part the generally small amount of recrystallization of these rocks but also the highly evolved compositions of hydrothermal fluids (Alt et al., 1996; Teagle et al., 1998b). The isotope compositions of anhydrite in these rocks record the evolution of seawater compositions during recharge, from seawater-like in the volcanic section to highly evolved, rock-dominated fluids in the dikes (Teagle et al., 1998a).

The δ^{18}O values decrease with depth in the dikes and are low in the uppermost few hundred meters of gabbros, then increase back to igneous values at greater depths in the plutonic section (Fig. 2). This pattern is primarily the result of the higher alteration temperatures in the lower dikes and uppermost gabbros, but also the somewhat greater intensity of recrystallization of these rocks (Alt et al., 1996; Ito and Clayton, 1983; Stakes et al., 1991; Agrinier et al., 1995). Local ^{18}O enrichments of deep plutonic rocks from Hole 735B reflect late, low-temperature, smectitic alteration related to uplift and exposure of the large tectonic block penetrated by this hole. The ^{87}Sr/^{86}Sr ratios of vein amphiboles and intensely recrystallized alteration halos from the deep sheeted dikes in Hole 504B are only slightly shifted toward seawater, indicating strongly rock-dominated compositions of deep hydrothermal fluids (Teagle et al., 1998b). The upper 200 m of gabbros from Hole 735B are significantly more altered than deeper rocks (Stakes et al., 1991; Dick, Natland, et al., 1998), yet fact that the ^{87}Sr/^{86}Sr ratios of these altered gabbros are only slightly shifted toward seawater values documents the rock-dominated compositions of the hydrothermal fluids (Kempton et al., 1991).

Ophiolites. Alteration processes affecting ophiolitic crust are broadly similar to those affecting oceanic crust, so the isotope profiles from ophiolites in Figures 1–3 are generally comparable to those for oceanic crust. Important differences between the oceanic and ophiolitic profiles, however, reflect the overall more intense recrystallization and greater fluid fluxes in the ophiolites.

Although rocks in the volcanic sections of the ophiolites exhibit wide scatter, they are shifted somewhat farther from primary values for Sr and O isotope compositions than are oceanic lavas (Figs. 1 and 2). That overall average bulk-rock values for ophiolitic lavas are high ($^{87}Sr/^{86}Sr = 0.7050$, $\delta^{18}O = 10‰$) reflects significant alteration to secondary minerals. Adding vein minerals and highly altered interpillow material will only increase the net isotope exchange, so these alteration estimates are minima. The isotope data thus reflect greater alteration to secondary assemblages in the ophiolitic extrusive rocks than in oceanic lavas, as is borne out by petrographic observations (Gillis and Robinson, 1990; Alt, 1994). The greater seawater exchange documented in the ophiolites reflects the higher overall intensity of recrystallization and closer approach to isotope equilibrium with low-temperature seawater fluids in the ophiolitic sections than in oceanic lavas.

Rough decreases in Sr and O isotope values with depth in the volcanic sections are similar to trends seen in oceanic crust and likewise suggest decreasing seawater circulation and more evolved fluid compositions with depth. S isotope effects in the ophiolitic lavas are similar to those in oceanic crust and are the result of loss of S and isotope fractionation during partial oxidation by seawater fluids at low temperatures (Alt, 1994).

Sheeted-dike complexes in the ophiolites are shifted to higher, more seawater-like $^{87}Sr/^{86}Sr$ ratios and $\delta^{34}S$ values compared to those of oceanic dikes (Figs. 1 and 3). The data for the dikes of the Troodos CY4 drill core suggest restricted local variations in Sr isotope exchange, but data for dikes and plutonic rocks from throughout the rest of the Troodos ophiolite exhibit greater and more uniform isotope shifts (Fig. 1). The greater isotope shifts in the ophiolites relative to oceanic crust are the results of the overall greater intensity of recrystallization of the ophiolitic dikes and consequent closer approach to equilibrium with hydrothermal fluids, but also reflect the presence of a greater seawater component in deep hydrothermal fluids in the ophiolites. The essentially complete recrystallization of Troodos dikes and the relatively uniform shifts of dikes and epidosites (the deep-seated, upwelling feeder zones for massive sulfide deposits) throughout the ophiolite to high $^{87}Sr/^{86}Sr$ ratios and $\delta^{34}S$ values indicate equilibrium with hydrothermal fluids that had high $^{87}Sr/^{86}Sr$ ratios and $\delta^{34}S$ values (~0.7054 and ~6‰; Bickle and Teagle, 1992; Alt, 1994). Even if dikes from modern oceanic crust were completely recrystallized, the more rock-dominated Sr and S isotope compositions of hydrothermal fluids at mid-ocean ridges would result in relatively low Sr and S isotope values of these rocks (0.7028–0.7038 and ~1‰, respectively; Alt, 1994; Teagle et al., 1998b).

Sr and S isotope shifts in the ophiolites persist into the uppermost gabbros, but then decrease downward into the deeper plutonic section (Figs. 1 and 3). This trend results from decreasing penetration of hydrothermal fluids and consequent decreased recrystallization, as well as progressive tracer exchange and fluid evolution with increasing depth. The trend of decreasing hydrothermal effects downward in the plutonic section, particularly in the data for Troodos, is comparable to effects in oceanic crust. The O isotope data for Troodos are also generally similar to those of oceanic crust: ^{18}O depletions in the dikes and uppermost gabbros reflect the intensity of recrystallization and high alteration temperatures in these rocks. In contrast to oceanic crust and the Troodos and some other ophiolites (e.g., Sarmiento, Macquarie Island; Elthon et al., 1984; Cocker et al., 1982), the sheeted-dike complex in Oman is generally enriched in ^{18}O (Fig. 2). This ^{18}O enrichment reflects alteration by ^{18}O-enriched hydrothermal fluids derived from high-temperature reactions at depth, plus superimposed alteration at lower temperatures (Gregory and Taylor, 1981; Stakes and Taylor, 1992). Similar effects of ^{18}O enrichment are seen in other ophiolites (Agrinier et al., 1988; Harper et al., 1988; Barrett and Friedrichsen, 1989; Schiffman et al., 1991; Lecuyer and Fourcade, 1991).

Determining the shift in $^{87}Sr/^{86}Sr$ ratio in the deep plutonic sections is complicated somewhat by slight variations in the primary values for different MORB sections and by uncertainty in the igneous values for the ophiolites. It appears, however, that a seawater signature extends to greater depth in Oman than in Troodos or oceanic crust. Similarly, $\delta^{18}O$ values for Oman are depleted throughout the plutonic section for many profiles, indicating exchange with seawater oxygen down to the mantle (Gregory and Taylor, 1981; Stakes and Taylor, 1992). Similar deep exchange and ^{18}O depletions of plutonic rocks are observed in other ophiolites (Sarmiento, Macquarie Island; Cocker et al., 1982; Elthon et al., 1984).

DISCUSSION

Fluid Fluxes

The volume of fluid that has passed through a hydrothermal system is a key parameter for understanding heat and chemical transport, and various approaches have been used to estimate the amounts of fluid reacting with oceanic and ophiolitic rocks. The water/rock ratio, originally formulated for stable isotope exchange by Taylor (1974), is commonly used (Spooner et al., 1977; Gregory and Taylor, 1981; Elthon et al., 1984; Stakes and O'Neil, 1982; Alt et al., 1986b; Agrinier et al., 1988; Harper et al., 1988; Lecuyer and Fourcade, 1991). This estimate is based on the relative mass, volume, or molar quantities of fluid and rock required to cause the observed isotope compositions of rock and fluid. For O isotopes, this calculation assumes equilibrium and requires knowledge of the proportions of secondary minerals present in order to calculate the temperature-dependent rock/water isotope fractionation. Such calculations may be applicable to many rocks from ophiolites where recrystallization is complete (e.g., Harper et al., 1988), but can be problematic in oceanic crust characterized by partial recrystallization and disequilibrium among secondary phases. Although providing some insight into hydrothermal processes, this water/rock ratio does not represent the actual process of fluids percolating through and progressively exchanging chemical tracers with a porous medium.

Similar calculations can be made for other tracers (e.g., Sr and S isotopes, oxidation of Fe, uptake of Mg, alkalis, or P, etc.; e.g., Spooner et al., 1977). A major problem here is that these calculations assume seawater fluid compositions and hence yield seawater/rock ratios, whereas hydrothermal fluid compositions vary significantly with depth in the crust. As seawater fluids penetrate deeper and react with a progressively greater cumulative amount of rock, the fluids evolve and become more rock dominated, e.g., the fluids are depleted in Mg and have lower $^{87}Sr/^{86}Sr$ ratios than seawater (Teagle et al., 1998a, 1998b).

In order to take into account such depth variations in hydrothermal processes and evolution in fluid compositions, Bickle and Teagle (1992) modeled fluid infiltration and progressive reaction in the Troodos ophiolite. There, the wide range in $^{87}Sr/^{86}Sr$ ratios of the volcanic section are interpreted to have resulted from kinetically limited exchange between basalt and seawater fluids, whereas the more uniformly elevated ratios of the sheeted-dike complex and uppermost plutonic section essentially reflect total recrystallization and equilibrium with hydrothermal fluids having $^{87}Sr/^{86}Sr = 0.7047–0.7059$. Bickle and Teagle calculated that a time-integrated fluid flux of 2.9×10^7 kg·m^{-2} is required to account for the Sr isotope shifts observed through an ~1 km section of sheeted dikes plus uppermost plutonic rocks. A similar evaluation of Sr exchange in oceanic crust yields a significantly lower fluid flux of 5×10^6 kg·m^{-2} (Bickle and Teagle, 1992). Uncertainties in the mean bulk isotope shifts of the oceanic volcanic section used in this model (Staudigel et al., 1995; Alt et al., 1998; Teagle et al., 1998b) do not affect the results of the model calculation (Bickle and Teagle, 1992). The difference in fluid and rock exchange and greater fluid flux for Troodos is consistent with the much greater component of seawater Sr present in upwelling hydrothermal fluids inferred for Troodos compared to those sampled at modern mid-ocean ridges: e.g., $^{87}Sr/^{86}Sr = 0.7054$, or ~50% contemporaneous seawater component for Troodos, versus ~0.7034 and 15% seawater component for mid-ocean ridges (Bickle and Teagle, 1992; Von Damm, 1995; Teagle et al., 1998b).

Although calculations like these for Troodos still need to be carried out for other ophiolites, the fact that the Sr isotope profiles through these ophiolites are generally similar to the Troodos profile (Fig. 1; McCulloch et al., 1980; Lecuyer et al., 1990) implies that time-integrated fluid fluxes through these other ophiolites were similar and were significantly greater than the fluxes through modern oceanic crust. Only limited Sr data exist for parts of other ophiolites, but their $^{87}Sr/^{86}Sr$ ratios are likewise shifted further toward contemporaneous seawater values than are $^{87}Sr/^{86}Sr$ ratios for modern oceanic crust, implying consistently greater fluid fluxes through the ophiolites and less evolved or less rock-dominated hydrothermal fluid compositions (Alexander et al., 1993; Valsami-Jones and Cann, 1994).

S isotope data for the Troodos ophiolite and for oceanic crust are consistent with inferences made from the relative intensities of hydrothermal recrystallization of these sections and from Sr isotope data. The sheeted-dike complex and upper gabbros in oceanic crust contain only ~7% of a seawater S component, versus ~35% seawater S in the comparable section of the Troodos ophiolite (Fig. 3; Alt, 1994, 1995). The implications here are greater time-integrated fluid fluxes through the ophiolitic dike complex and less rock-dominated hydrothermal fluids in the ophiolite.

There is a much smaller proportion of Sr compared to O in a given mass of fluid relative to a given mass of rock. During fluid flow and interaction with rock, therefore, the Sr isotopes in the fluid become rock dominated much sooner than the O isotopes, and the seawater signature affecting the $\delta^{18}O$ value of the rock will penetrate much farther along the fluid + rock reaction pathway compared to the effects of Sr and resulting shifts in the $^{87}Sr/^{86}Sr$ ratio of the rock (Bickle and Teagle, 1992). The integrated fluid flux for the Troodos ophiolite calculated from Sr data can thus more than account for the observed shifts in O isotope compositions in the ophiolite. This difference in tracer content of the fluid can also account for the fact that the plutonic sections of the Semail ophiolite exhibit significant shifts in $\delta^{18}O$ but little or no change in $^{87}Sr/^{86}Sr$ (Figs. 1 and 2). These shifts in $\delta^{18}O$ of the Semail plutonic section indicate penetration of hydrothermal fluids down to the mantle (Gregory and Taylor, 1981; Stakes and Taylor, 1992). Similar $\delta^{18}O$ shifts in plutonic rocks from the Sarmiento and Macquarie Island ophiolites suggest penetration of seawater hydrothermal fluids deep into the plutonic sections (Cocker et al., 1982; Elthon et al., 1984). In contrast, recrystallization of the deep plutonic section of oceanic crust is slight, resulting in little or no change in $\delta^{18}O$ of the deep crust (Fig. 2) and implying only very limited pervasive deep penetration of hydrothermal fluids (this paper; Gillis, 1995; Manning et al., 1996; Dick, Natland, et al., 1998).

In summary, the available data for recrystallization and isotope tracers in ophiolites and oceanic crust consistently indicate greater integrated fluid fluxes through ophiolitic crust than through mid-ocean ridge crust. A consequence of this difference is less rock-dominated compositions of hydrothermal fluids in the ophiolites. The more intense recrystallization of the plutonic sections of some ophiolites and the persistence of isotope shifts to great depths in these sections suggest deeper penetration of large fluxes of hydrothermal fluids and resultant greater seawater tracer exchange than in mid-ocean ridge crust. The substantial heat budget required by these greater fluid fluxes through ophiolites remains a problem, however (Bickle and Teagle, 1992). Another consequence of the greater fluid fluxes and tracer exchange measured in ophiolites is that caution must be used when inferring effects on seawater composition from application of data from ophiolites to the entire world mid-ocean ridge system, for example, the buffering of the O isotope composition of the oceans (Muehlenbachs and Clayton, 1976; Gregory and Taylor, 1981; Agrinier et al., 1988, 1995; Lecuyer et al., 1991) and the influence on the Sr isotope and Mg budgets of the oceans (Bednarz and Schmincke, 1989; Lecuyer et al., 1990; Bickle and Teagle, 1992). Conversely, the greater hydrothermal effects in supra-subduction basins (ophiolites) compared to those in oceanic crust indicate that inferences on ocean chemistry based on data for oceanic crust are minima and would be enhanced by additional hydrothermal exchange with seawater in ophiolites, especially during periods of Earth history when rifting of arcs and formation of backarc basins were widespread.

In some cases, the specific geologic setting of an ophiolite can provide a basis for explanation of the discrepancies in alteration, tracer exchange, and fluid fluxes (e.g., in a rifted arc; Schiffman et al., 1991), but elsewhere such an explanation is problematic. Alt et al. (1998) examined the volcanic section of an in situ profile through oceanic forearc crust and suggested that factors such as (1) the greater primary volatile contents of subduction lavas, (2) the greater abundance of easily altered glass and mafic phases in the arc-related lavas, and (3) the higher silica contents of the arc glasses may contribute to greater alteration of supra-subduction (ophiolitic) lavas. It is unknown, however, whether such factors would influence alteration at greater depths in ophiolitic crust.

Other Differences

Other significant differences exist between hydrothermal effects in many ophiolites and those characteristic of oceanic crust, particularly in the extrusive and plutonic sections. In contrast to oceanic crust and the Troodos ophiolite, the volcanic sections in many ophiolites are more intensely recrystallized at higher metamorphic grades (prehnite-pumpellyite, greenschist), and alteration assemblages exhibit more gradual changes downward (Spooner and Fyfe, 1973; Spooner et al., 1977; Coish, 1977; Stern and Elthon, 1979; Liou, 1979; Cocker et al., 1982; Ishizuka, 1985; Harper et al., 1988; Pflumio, 1991; Schiffman et al., 1991). Such differences reflect generally more uniform or continuous thermal gradients in these settings compared to the discontinuous or stepped gradients inferred for Troodos and mid-ocean ridges (Alt et al., 1986a; Gillis and Robinson, 1990; Schiffman et al., 1991; Beiersdorfer, 1993; Alt, 1999). More uniform thermal gradients can result from burial of ophiolitic lavas by thick sequences of volcaniclastic sediments, which seal the basement from free access of seawater and convective heat removal (Schiffman et al., 1991). A lack of deep normal faults to focus fluid flow has been suggested to result in more pervasive fluid flow and alteration in some ophiolites (Schiffman et al., 1991), but modeling of Sr isotope data and heat transport suggest that significant focusing of hydrothermal fluids by faults does not occur at modern mid-ocean ridges (Cann et al., 1985; Bickle and Teagle, 1992; Bickle et al., 1998). Multiple episodes of intrusion and volcanic activity can produce superimposed and more intense alteration and metamorphic effects in ophiolitic volcanic sequences (Pflumio, 1991), and in some cases the effects of superimposed regional metamorphism may be important.

In contrast to the plutonic sections of oceanic crust and some ophiolites (e.g., Troodos, Semail, Macquarie Island, Sarmiento; Gregory and Taylor, 1981; Cocker et al., 1982; Elthon et al., 1984; Bickle and Teagle, 1992), which are depleted in ^{18}O or exhibit only slight changes, the gabbroic sections in many ophiolites are enriched in ^{18}O (Agrinier et al., 1988; Harper et al., 1988; Barrett and Friedrichsen, 1989; Schiffman et al., 1991; Lecuyer and Fourcade, 1991). The latter results from lower-temperature reactions, in some cases involving ^{18}O-enriched fluids, that are superimposed on earlier high-temperature effects that deplete the rocks in ^{18}O. In the Trinity and Xigaze ophiolites, this effect has been related to formation of these ophiolites at slow-spreading ridges, where the crust is tectonically disrupted so that fluids are allowed to penetrate into the plutonic section at temperatures of less than ~250 °C (Agrinier et al., 1988; Lecuyer and Fourcade, 1991). Renewed magmatism following tectonic extension at a slow-spreading ridge would provide a heat source for hydrothermal alteration of cooled and fractured plutonic rocks, perhaps also contributing to lower-temperature alteration and ^{18}O enrichment. We have not yet recognized the effects of this type of renewed magmatic and hydrothermal activity on modern oceanic crust at depth, however. Schiffman et al. (1991) discussed ^{18}O enrichments of plutonic rocks and metamorphism in Coast Range ophiolites in terms of a rifted arc, where burial of the system by sediments and a long duration of magmatic activity resulted in more uniform geothermal gradients, prolonged hydrothermal circulation at lower temperatures (<250 °C), and possible ^{18}O enrichment of fluids via reaction with pelagic sediments. Clearly, additional and more complete recovery of sections of oceanic crust in different tectonic environments and continued feedback between studies of ophiolites and oceanic crust are required in order to understand the origin and significance of the differences outlined here.

SUMMARY AND CONCLUSIONS

Ophiolitic crust is consistently more intensely recrystallized than in situ oceanic crust, and isotope tracers (Sr, O, S) of interaction between seawater and the crust exhibit greater shifts in ophiolites. This difference requires greater time-integrated fluid fluxes through the ophiolites than through oceanic crust, by a factor of ~3–6, as well as penetration of larger fluid fluxes to greater depths in some ophiolites. The reasons for this fundamental difference are likely related in some way to differences in tectonic setting, primary chemical compositions, or heat sources, but the specific explanations remain problematic. The greater fluid fluxes and tracer exchange in ophiolites point out the need for appropriate caution when extrapolating data from ophiolites to the global mid-ocean ridge system and effects on oceanic and crustal chemistry.

Despite significant differences between hydrothermal effects in ophiolites and oceanic crust, many important similarities also exist, and the ophiolite analogy has been invaluable in understanding the structure of oceanic crust and processes at mid-ocean ridges. A feedback between study of ophiolites and oceanic crust is essential to understand the origin and significance of similarities and differences in these crustal settings. Despite the best obtainable composite section presented here, there is a need for additional and more continuous sections through oceanic crust, in particular through the critical lithologic transitions that coincide with changes in fluid circulation and alteration: from lavas to dikes, from dikes to gabbros, from upper to lower gabbros, and from gabbros to peridotites. More continuous and well-studied sections through ophiolites are also needed. Likewise, integrated studies of oceanic crust and ophiolites are required to link petrographic and structural information and quantification of alteration types to comprehensive elemental and isotope analyses of whole rocks and secondary minerals in order to understand fluids and their effects on the crust.

ACKNOWLEDGMENTS

This work was supported by National Science Foundation grant OCE-9633353 and a grant from JOI/USSAC (Joint Oceanographic Institutions–U.S. Science Advisory Committee). The manuscript benefited from helpful reviews by Karlis Muehlenbachs, Hubert Staudigel, and Robert Zierenberg.

REFERENCES CITED

Agrinier, P., Javoy, M., and Girardeau, J., 1988, Hydrothermal activity in a peculiar oceanic ridge: Oxygen and hydrogen isotope evidence in the Xigaze ophiolite (Tibet, China): Chemical Geology, v. 71, p. 313–335.

Agrinier, P., Hekinian, R., Bideau, D., and Javoy, M., 1995, O and H stable isotope compositions of oceanic crust and upper mantle rocks exposed in the Hess Deep near the Galapagos Triple Junction: Earth and Planetary Science Letters, v. 136, p. 183–196.

Alexander, R.J., Harper, G.D., and Bowman, J.R., 1993, Oceanic faulting and fault-controlled subseafloor hydrothermal alteration in the sheeted dike complex of the Josephine ophiolite: Journal of Geophysical Research, v. 98, p. 9731–9759.

Allen, C.R., 1975, The petrology of a portion of the Troodos plutonic complex, Cyprus [Ph.D. thesis]: Cambridge, Cambridge University, 161 p.

Alt, J.C., 1994, A sulfur isotopic profile through the Troodos Ophiolite, Cyprus: Primary composition and the effects of seawater hydrothermal alteration: Geochimica et Cosmochimica Acta, v. 58, p. 1825–1840.

Alt, J.C., 1995, Sulfur isotopic profile through the oceanic crust: Sulfur mobility and seawater-crustal sulfur exchange during hydrothermal alteration: Geology, v. 23, p. 585–588.

Alt, J.C., 1999, Very low grade hydrothermal metamorphism of basic igneous rocks, *in* Frey, M., and Robinson, D., eds., Very low grade metamorphism: Oxford, Blackwell Scientific, p. 169–201.

Alt, J.C., and Honnorez, J., 1984, Alteration of the upper oceanic crust: DSDP Site 417 Mineralogy and Chemistry: Contributions to Mineralogy and Petrology, v. 37, p. 149–169.

Alt, J.C., and Teagle, D.A.H., 1999, Uptake of carbon during alteration of oceanic crust: Geochimica et Cosmochimica Acta, v. 63, p. 1527–1535.

Alt, J.C., Honnorez, J., Laverne, C., and Emmermann, R., 1986a, Hydrothermal alteration of a 1 km section through the upper oceanic crust, DSDP Hole 504B: The mineralogy, chemistry, and evolution of seawater-basalt interactions: Journal of Geophysical Research, v. 91, p. 10309–10335.

Alt, J.C., Muehlenbachs, K., and Honnorez, J., 1986b, An oxygen isotopic profile through the upper kilometer of oceanic crust, DSDP Hole 504B: Earth and Planetary Science Letters, v. 80, p. 217–229.

Alt, J.C., Anderson, T.F., and Bonnell, L., 1989, The geochemistry of sulfur in a

1.3 km section of hydrothermally altered oceanic crust, DSDP Hole 504B: Geochimica et Cosmochimica Acta, v. 53, p. 1011–1023.

Alt, J.C., Laverne, C., Vanko, D., Tartarotti, P., Teagle, D.A.H., Bach, W., Zuleger, E., Erzinger, J., Honnorez, J., Pezard, P.A., Becker, K., Salisbury, M.H., and Wilkens, R.H., 1996, Hydrothermal alteration of a section of upper oceanic crust in the eastern equatorial Pacific: A synthesis of results from Site 504 (DSDP legs 69, 70, and 83, and ODP legs 111, 137, 140, and 148), *in* Alt, J.C., Kinoshita, H., Stokking, L., and Michael, P., eds., Proceedings of the Ocean Drilling Program, Scientific Results, v. 148: College Station, Texas, Ocean Drilling Program, p. 417–434.

Alt, J.C., Teagle, D.A.H., Brewer, T.S., Shanks, W.C., and Halliday, A.N., 1998, Alteration and mineralization of an oceanic forearc and the ophiolite–ocean crust analogy: Journal of Geophysical Research, v. 103, p. 12365–12380.

Andrews, A. J., 1977, Low temperature fluid alteration of oceanic layer 2 basalts: Canadian Journal of Earth Science, v. 14, p. 911–926.

Baragar, W.R., Lambert, M.B., Baglow, N., and Gibson, I.L., 1989, Sheeted dikes from CY4 and surface sections: Troodos ophiolite, *in* Gibson, I.L., et al., eds., Cyprus Crustal Study Project, Initial Report, Hole CY4: Geological Survey of Canada Paper 88–9, p. 69–106.

Barling, J., Hertogen, J., and Weis, D., 1997, Whole-rock geochemistry and Sr, Nd, and Pb isotopic characteristics of undeformed, deformed, and recrystallized gabbros from Sites 921, 922, and 923 in the MARK area, *in* Karson, J.A., Cannat, M., Miller, D.J., and Elthon, D., eds., Proceedings of the Ocean Drilling Program, Scientific Results, v. 153: College Station, Texas, Ocean Drilling Program, p. 351–362.

Barrett, T.J., and Friedrichsen, H., 1989, Stable isotopic compositions of atypical ophiolitic rocks from east Liguria, Italy: Chemical Geology, v. 80, p. 71–84.

Bednarz, U., and Schmincke, H.U., 1989, Mass transfer during sub-seafloor alteration of the upper Troodos crust (Cyprus): Contributions to Mineralogy and Petrology, v. 102, p. 93–101.

Beiersdorfer, R.E., 1993, Metamorphism of a Late Jurassic volcano-plutonic arc, northern California, USA: Journal of Metamorphic Geology, v. 1, p. 415–428.

Bickle, M.J., and Teagle, D.A.H., 1992, Strontium alteration in the Troodos ophiolite: Implications for fluid fluxes and geochemical transport in mid-ocean ridge hydrothermal systems: Earth and Planetary Science Letters, v. 113, p. 219–237.

Bickle, M.J., Teagle, D.A.H., Beynon, J., and Chapman, H. J., 1998, The structure and controls on fluid-rock interactions in ocean ridge hydrothermal systems: Constraints from the Troodos ophilite, *in* Mills, R.A., and Harrison, K., eds., Modern ocean floor processes and the geological record: Geological Society of London Special Publication 148, p. 127–152.

Bloomer, S.H., Taylor, B., MacLeod, C.J., Stern, R.J., Fryer, P., Hawkins, J.W., and Johnson, L., 1995, Early arc volcanism and the ophiolite problem: A perspective from drilling in the western Pacific, *in* Taylor, B., and Natland, J., eds., Active margins and marginal basins of the western Pacific: American Geophysical Union Geophysical Monograph 88, p. 1–10.

Burke, W.H., Denison, R.E., Hetherington, E.A., Koepnick, R.B., Nelson, H.F., and Otto, J.B., 1982, Variation of seawater $^{87}Sr/^{86}Sr$ throughout Phanerozoic time: Geology, v. 10, p. 516–519.

Cann, J.R., Strens, M.R., and Rice, A., 1985, A simple magma-driven thermal balance model for the formation of volcanogenic massive sulphides: Earth and Planetary Science Letters, v. 76, p. 123–134.

Claypool, G.E., Holser, W.T., Kaplan, I.R., Sakai, H., and Zak, I., 1980, The age curves of sulfur and oxygen isotopes in marine sulfate and their mutual interpretation: Chemical Geology, v. 28, p. 199–260.

Cocker, J.D., Griffin, B.J., and Muehlenbachs, K., 1982, Oxygen and carbon isotope evidence for seawater-hydrothermal alteration of the Macquarie Island ophiolite: Earth and Planetary Science Letters, v. 61, p. 112–122.

Coish, R.A., 1977, Ocean floor metamorphism in the Betts Cove ophiolite, Newfoundland: Contributions to Mineralogy and Petrology, v. 60, p. 225–270.

Dick, H.J.B, Natland, J.H., et al., 1999, Proceedings of the Ocean Drilling Program, Initial Reports, v. 176 [CD-ROM]: College Station, Texas, Ocean Drilling Program.

Elthon, D., 1991, Geochemical evidence for formation of the Bay of Islands ophiolite above a subduction zone: Nature, v. 354, p. 140–143.

Elthon, D., Lawrence, J.R., Hanson, R.E., and Stern, C., 1984, Modeling of oxygen isotope data from the Sarmiento ophiolite complex, Chile, *in* Gass, I.G., Lippard, S.J., and Shelton, A.W., eds., Ophiolites and oceanic lithosphere: Geological Society of London Special Publication 13, p. 185–197.

Fisher, A.T., Becker, K., and Narasimhan, K., 1994, Off-axis hydrothermal circulation: Parametric tests of a refined model of processes at DSDP/ODP Site 504: Journal of Geophysical Research, v. 99, p. 3097–3123.

Gass, I.G., 1968, Is the Troodos Massif of Cyprus a fragment of Mesozoic ocean floor?: Nature, v. 220, p. 39–42.

Gillis, K.M., 1995, Controls on hydrothermal alteration in a section of fast spreading oceanic crust: Earth and Planetary Science Letters, v. 134, p. 473–489.

Gillis, K.M., and Robinson, P.T., 1990, Patterns and processes of alteration in the lavas and dikes of the Troodos ophiolite, Cyprus: Journal of Geophysical Research, v. 95, p. 21523–21548.

Gillis, K.M., and Thompson, G., 1993, Metabasalts from the Mid-Atlantic Ridge: New insights into hydrothermal systems in slow-spreading crust: Contributions to Mineralogy and Petrology, v. 113, p. 502–523.

Gregory, R.T., and Taylor, H.P., 1981, An oxygen isotope profile in a section of Cretaceous oceanic crust, Samail ophiolite, Oman: Evidence for $\delta^{18}O$ buffering of the oceans by deep (>5 km) seawater-hydrothermal circulation at mid-ocean ridges: Journal of Geophysical Research, v. 86, p. 2737–2755.

Harper, G.D., Bowman, J.R., and Kuhns, R., 1988, A field, chemical, and stable isotope study of subseafloor metamorphism of the Josephine ophiolite, California–Oregon: Journal of Geophysical Research, v. 93, p. 4625–4656.

Hart, S.R., and Staudigel, H., 1986, Ocean crust vein mineral deposition: Rb/Sr ages, U-Th-Pb geochemistry, and duration of circulation at DSDP Sites 261, 462, and 516: Geochimica et Cosmochimica Acta, v. 50, p. 2751–2761.

Huang, P.Y., and Solomon, S.C., 1988, Centroid depths of mid-ocean ridge earthquakes: Dependence on spreading rate: Journal of Geophysical Research, v. 93, p. 13445–13477.

Ishizuka, H., 1985, Prograde metamorphism of the Horokanai ophiolite in the Kamuikotan zone, Hokkaido, Japan: Journal of Petrology, v. 26, p. 391–417.

Ito, E., and Clayton, R.N. 1983, Submarine metamorphism of gabbros from the Mid-Cayman rise: An oxygen isotopic study: Geochimica et Cosmochimica Acta, v. 47, p. 535–546.

Kempton, P.D., Hawkesworth, C.J., and Fowler, M., 1991, Geochemistry and isotopic composition of gabbros from Layer 3 of the Indian Ocean Crust, Leg 118, Hole 735B, *in* Von Herzen, R.P., Robinson, P.T., et al., Proceedings of the Ocean Drilling Program, Scientific Results, v. 118: College Station, Texas, Ocean Drilling Program, p. 113–125.

Lalou, C., Reyss, J. L., Brichet, E., Rona, P.A., and Thompson, G., 1995, Hydrothermal activity on a 10^5-year time scale at a slow spreading ridge, TAG hydrothermal field, Mid-Atlantic Ridge, 26°N: Journal of Geophysical Research, v. 100, p. 17855–17862.

Lanphere, M.A., Coleman, R.G., and Hoson, C.A., 1981, Sr isotopic tracer study of the Samail ophiolite, Oman: Journal of Geophysical Research, v. 86, p. 2709–2720.

Lecuyer, C., and Fourcade, S., 1991, Oxygen isotope evidence for multi-stage hydrothermal alteration at a fossil slow-spreading center: The Silurian Trinity ophiolite (California, USA): Chemical Geology, v. 87, p. 231–246.

Lecuyer, C., and Gruau, G., 1996, Oxygen and strontium isotope compositions of Hess Deep Gabbros (Holes 894F and 894G): High-temperature interaction of seawater with the oceanic crust layer 3, *in* Mével, C., Gillis, K.M., Allan, J.F., and Meyer, P.S., eds., Proceedings of the Ocean Drilling Program, Scientific Results, v. 147: College Station, Texas, Ocean Drilling Program, p. 227–234.

Lecuyer, C., Brouxel, M., and Albarede, F., 1990, Elemental fluxes during hydrothermal alteration of the Trinity ophiolite (California, USA) by seawater: Chemical Geology, v. 89, p. 87–115.

Liou, J.G., 1979, Zeolite facies metamorphism of basaltic rocks from the east Taiwan ophiolite: American Mineralogist, v. 64, p. 1–14.

Lister, C.R.B., 1982, "Active" and "passive" hydrothermal systems in the ocean crust: Predicted physical conditions, *in* Fanning, K.A., and Manheim, F.T., eds., The dynamic environment of the ocean floor: Lexington, Massachussetts, D.C. Heath, p. 441–470.

Manning, C.E., Weston, P.E., and Mahon, K.I., 1996, Rapid high-temperature metamorphism of the East Pacific Rise gabbros from Hess Deep: Earth and Planetary Science Letters, v. 144, p. 123–132.

McCulloch, M.T., Gregory, R.T., Wasserburg, G.J., and Taylor, H.P., 1980, Neodymium, strontium, and oxygen isotopic study of the Cretaceous Semail ophiolite and implications for the petrogenesis and seawater-hydrothermal alteration of oceanic crust: Earth and Planetary Science Letters, v. 46, p. 201–211.

Mével, C., and Cannat, M., 1991, Lithospheric stretching and hydrothermal processes in oceanic gabbros from slow spreading ridges, *in* Peters, T., Nicolas, A., and Coleman, R.G., eds., Ophiolite genesis and evolution of the oceanic lithosphere: Dordrecht, Netherlands, Kluwer Academic Publishers, p. 293–312.

Miyashiro, A., 1973, The Troodos ophiolitic complex was probably formed in an island arc: Earth and Planetary Science Letters, v. 19, p. 218–224.

Moores, E.M., and Vine, F.J., 1971, The Troodos massif, Cyprus and other ophiolites as oceanic crust: Evaluation and implications: Royal Society of London Philosophical Transactions, ser. A, v. 268, p. 443–466.

Mottl, M.J., and Wheat, C.G., 1994, Hydrothermal circulation through mid-ocean ridge flanks: Fluxes of heat and magnesium: Geochimica et Cosmochimica Acta, v. 58, p. 2225–2237.

Muehlenbachs, K., 1980, The alteration and aging of the basaltic layer of the seafloor: Oxygen isotope evidence from DSDP/IPOD Legs 51, 52, and 53, *in* Donnelly, T., Francheteau, J., Bryan, W., Robinson, P., Flower, M., and Salisbury, M., eds., Initial Reports, Deep Sea Drilling Project, Washington, D.C., U.S. Government Printing Office, v. 51–53, p. 1159–1167.

Muehlenbachs, K., and Clayton, R.N., 1976, Oxygen isotope composition of the oceanic crust and its bearing on seawater: Journal of Geophysical Research, v. 81, p. 4365–4369.

Muenow, D.W., Garcia, M.O., Aggrey, K.E., Bednarz, U., and Schmincke, H.U., 1990, Volatiles in submarine glasses as a discriminant of tectonic origin: Application to the Troodos ophiolite: Nature, v. 343, p. 159–161.

Pearce, J.A., Lippard, S.J., and Roberts, S., 1984, Characteristics and geotectonic significance of supra-subduction zone ophiolites: Geological Society of London Special Publication 16, p. 77–96.

Pflumio, C., 1991, Evidences for polyphased oceanic alteration of the extrusive sequence of the Semail ophiolite from the Salahi block (Northern Oman), *in* Peters, T., Nicolas, A., and Coleman, R.G., eds., Ophiolite genesis and evolution of the oceanic lithosphere: Dordrecht, Netherlands, Kluwer Academic Publishers, p. 313–352.

Rautenschlein, M., Jenner, G.A., Hertogen, J., Hofmann, A.W., Kerrich, R., Schmincke, H.U., and White, W.M., 1985, Isotopic and trace element composition of volcanic glasses from the Akaki Canyon, Cyprus: Implications for the origin of the Troodos ophiolite: Earth and Planetary Science Letters, v. 75, p. 369–383.

Regba, M., Agrinier, P., Pflumio, C., and Loubet, M., 1991, A geochemical study of a fossil oceanic hydrothermal discharge zone in the Oman ophiolite (Zuha sulphide prospect): Evidence for a polyphased hydrothermal history, *in* Peters, T., Nicolas, A., and Coleman, R.G., eds., Ophiolite genesis and evolution of the oceanic lithosphere: Dordrecht, Netherlands, Kluwer Academic Publishers, p. 353–384.

Schiffman, P., and Smith, B.M., 1988, Petrology and oxygen isotope geochemistry of a fossil seawater hydrothermal system within the Solea Graben, northern Troodos ophiolite, Cyprus: Journal of Geophysical Research, v. 93, p. 4612–4624.

Schiffman, P., Evarts, R.C., Williams, A.E., and Pickthorn, W.P., 1991, Hydrothermal metamorphism in oceanic crust from the Coast Range Ophiolite of California: Fluid-rock interaction in a rifted arc, *in* Peters, T., Nicolas, A., and Coleman, R.G., eds., Ophiolite genesis and evolution of the oceanic lithosphere: Dordrecht, Netherlands, Kluwer Academic Publishers, p. 399–426.

Spooner, E.T.C., and Fyfe, W.S., 1973, Sub-seafloor metamorphism, heat, and mass transfer: Contributions to Mineralogy and Petrology, v. 42, p. 278–304.

Spooner, E.T.C., Beckinsdale, R.D., England, P.C., and Senior, A., 1977, Hydration, ^{18}O-enrichment, and oxidation during ocean floor metamorphism of ophiolitic metabasic rocks from E. Liguria, Italy: Geochimica et Cosmochimica Acta, v. 41, p. 857–871.

Stakes, D.S., and O'Neil, J.R., 1982, Mineralogy and stable isotope geochemistry of hydrothermally altered oceanic rocks: Earth and Planetary Science Letters, v. 57, p. 285–304.

Stakes, D.S., and Taylor, H.P., 1992, The northern Samail ophilite: An oxygen isotope, microprobe, and field study: Journal of Geophysical Research, v. 97, p. 7043–7080.

Stakes, D., Mével, C., Cannat, M., and Chaput, T., 1991, Metamorphic stratigraphy of Hole 735B, *in* Von Herzen, R.P., Robinson, P.T., et al., eds., Proceedings of the Ocean Drilling Program, Scientific Results, v. 118: College Station, Texas, Ocean Drilling Program, p. 153–181.

Staudigel, H., and Hart, S.R., 1985, Dating of ocean crust hydrothermal alteration: Strontium isotope ratios from Hole 504B carbonates and the re-interpretation of Sr isotope data from Deep Sea Drilling Project Sites 105, 332, 417, and 418, *in* Anderson, R.N., Honnorez, J., and Becker, K., eds., Initial Reports, Deep Sea Drilling Project, v. 83, Washington, D.C., U.S. Government Printing Office, p. 297–303.

Staudigel, H., Muehlenbachs, K., Richardson, S.H., and Hart, S.R., 1981a, Agents of low temperature ocean crust alteration: Contributions to Mineralogy and Petrology, v. 77, p. 150–157.

Staudigel, H., Hart, S.R., and Richardson, S.H., 1981b, Alteration of the oceanic crust: Processes and timing: Earth and Planetary Science Letters, v. 52, p. 311–327.

Staudigel, H., Davies, G.R., Hart, S.R., Marchant, K.M., and Smith, B.M., 1995, Large scale Sr, Nd, and O isotopic anatomy of altered oceanic crust: DSDP Sites 417/418: Earth and Planetary Science Letters, v. 130, p. 169–185.

Stein, C.A., and Stein, S., 1994, Constraints on hydrothermal heat flux through the oceanic lithosphere from global heat flow: Journal of Geophysical Research, v. 99, p. 3081–3096.

Stern, C., and Elthon, D., 1979, Vertical variations in the effects of hydrothermal metamorphism in Chilean ophiolites: Their implications for ocean floor metamorphism: Tectonophysics, v. 55, p. 179–213.

Taylor, H.P., 1974, The application of oxygen and hydrogen isotope studies to problems of hydrothermal alteration and ore deposition: Economic Geology, v. 69, p. 843–883.

Teagle, D.A.H., Alt, J.C., Bach, W., Halliday, A.N., and Erzinger, J., 1996, Alteration of upper oceanic crust in a ridge flank hydrothermal upflow zone: Mineral, chemical and isotopic constraints from ODP Hole 896A, *in* Alt, J.C., Kinoshita, H., Stokking, L.B., and Michael, P.J., eds., Proceedings of the Ocean Drilling Program, Scientific Results, v. 148: College Station, Texas, Ocean Drilling Program, p. 119–150.

Teagle, D.A.H., Alt, J.C., and Halliday, A.N., 1998a, Tracing the chemical evolution of fluids during hydrothermal recharge: Constraints from anhydrite recovered in ODP Hole 504B: Earth and Planetary Science Letters, v. 155, p. 167–182.

Teagle, D.A.H., Alt, J.C., and Halliday, A.N., 1998b, Tracing the evolution of hydrothermal fluids in the upper oceanic crust: Sr isotopic constraints from DSDP/ODP Holes 504B and 896A, *in* Mills, R.A., and Harrison, K., eds., Modern ocean-floor processes and the geological record: Geological Society of London Special Publication 148, p. 81–97.

Tucholke, B.E., and Lin, J., 1994, A geological model for the structure of ridge segments in slow-spreading ocean crust: Journal of Geophysical Research, v. 99, p. 11937–11958.

Valsami-Jones, E., and Cann, J.R., 1994, Controls on the Sr and Nd isotopic compositions of hydrothermally altered rocks from the Pindos Ophiolite, Greece: Earth and Planetary Science Letters, v. 125, p. 39–54.

Vanko, D.A., and Laverne, C., 1998, Hydrothermal anorthitization of plagioclase within the magmatic/hydrothermal transition at mid-ocean ridges: Examples from deep sheeted dikes (Hole 504B, Costa Rica Rift) and a sheeted dike root zone (Oman ophiolite): Earth and Planetary Science Letters, v. 162, p. 27–43.

Von Damm, K.L., 1995, Controls on the chemistry and temporal variability of fluids, *in* Lupton, J., Mullineaux, L., and Zierenberg, R., eds., Physical, chemical, biological, and geological interactions within submarine hydrothermal systems, Washington, D.C., American Geophysical Union, Geophysical Monograph 91, p. 222–247.

Waggoner, D.G., 1993, The age and alteration of central Pacific oceanic crust near Hawaii, Site 843, *in* Wilkens, R.H., Firth, J., and Bender, J., eds., Proceedings of the Ocean Drilling Program, v. 136: College Station, Texas, Ocean Drilling Program, p. 119–132.

White, R.S., McKenzie, D., and O'Nions, R.K., 1992, Oceanic crustal thickness from seismic measurements and rare earth element inversions: Journal of Geophysical Research, v. 97, p. 19683–19715.

Manuscript Accepted by the Society February 23, 2000

Printed in the U.S.A.

Geological Society of America
Special Paper 349
2000

Hydrothermal alteration patterns in supra-subduction zone ophiolites

Kathryn M. Gillis and Neil R. Banerjee
School of Earth and Ocean Sciences, P.O. Box 3055, University of Victoria, Victoria, British Columbia V8W 3P6, Canada

ABSTRACT

The geologic and alteration characteristics of 11 supra-subduction zone (SSZ) ophiolites have been compiled in order to assess the variability of hydrothermal alteration patterns. All SSZ ophiolites show evidence for high-temperature hydrothermal discharge in the form of polymetallic sulfide deposits and/or metalliferous sedimentary rocks. Metamorphic zonations are commonly subparallel to the igneous stratigraphy, and metamorphic grade increases with depth, but varies laterally in response to local magmatic and tectonic activity. The degree of alteration is more pervasive in sheeted-dike complexes than in volcanic or plutonic sequences. The greatest variability in the style of alteration is in volcanic sequences, where alteration occurred at low-temperature to greenschist-facies conditions. The prevalence of high porosity and permeability within volcanic sequences in modern oceanic crust, which maintains low temperatures, indicates that special conditions are required to develop zeolite- or greenschist-facies zones away from sites of hydrothermal discharge. Recent study of sediment-covered ridges show that early burial of oceanic crust, at or near a spreading center, does not lead to the widespread development of high-temperature alteration zones in the uppermost volcanic rocks. Alteration patterns in the Izu-Bonin and Tonga forearcs show many similarities to SSZ ophiolites interpreted to have formed in a forearc, such as the presence of epidosites and the nature of alteration within plutonic sequences. However, modern forearc volcanic sequences only record low-temperature alteration. Much more detailed documentation of most ophiolites is required in order to fully understand the effects of geologic setting on the evolution of hydrothermal systems.

INTRODUCTION

Alteration patterns documented for ophiolites and modern oceanic crust record the time-integrated effects of fluid-rock interaction related to oceanic hydrothermal systems and, commonly, the obduction onto land or tectonic exposure onto the seafloor. The tectonic setting of a spreading center does not strongly influence the basic components of hydrothermal systems. Heat from magmatic intrusion drives convection, and extensional processes (i.e., dike intrusion and faulting) create permeability for fluid flow. Seawater reacts with the rocks through which it is passing and promotes chemical exchange and recrystallization of the oceanic crust to distinct metamorphic assemblages. Hydrothermal systems are conceptually described in terms of recharge zones where seawater enters the crust and flows toward the heat source, reaction zones (also called root zones) where hydrothermal fluid compositions become fixed prior to their ascent to the seafloor, and discharge zones that bring metal-enriched hydrothermal fluids to the seafloor. We have learned from studies of modern oceanic crust that reactions do not proceed at the same rates in all lithologies (see Alt, 1995, for review). High-temperature reactions within sheeted dike and plutonic sequences occur at or near the ridge where geothermal gradients are steep, whereas low-temperature alteration within volcanic sequences occurs dominantly off-axis.

Our knowledge and understanding of oceanic hydrothermal systems within ophiolites have been strongly influenced by the extensive literature associated with the so-called classic ophiolites—Troodos and Semail. These ophiolites have served as our guides for studying the anatomy of hydrothermal systems and, in particular, the evolution of ore deposits and metalliferous sedimentary rocks and the nature of hydrothermal root zones. Early study of these ophiolites demonstrated the role of seawater (e.g., Heaton and Sheppard, 1976; McCulloch et al., 1981; Spooner et al., 1977) and influence of ophiolite pseudostratigraphy and structure on patterns of hydrothermal alteration (e.g., Gass and Smewing, 1973). Subsequent study of hydrothermal alteration patterns in the modern oceans has shown that ophiolites and crust generated at a mid-ocean ridge (MOR) share many common features. A compelling example is the similarity between massive sulfide deposits hosted in supra-subduction zone (SSZ) ophiolites and those actively forming at mid-ocean ridges. Indeed, Cyprus-type massive sulfide deposits hosted in the Troodos ophiolite are similar in terms of mineral assemblages and internal structure to the active TAG hydrothermal field (TAG is the Trans-Atlantic Geotraverse across the Mid-Atlantic Ridge; Humphris et al., 1995; Eddy et al., 1998). Key differences exist, however, that indicate that some aspect(s) of hydrothermal systems may be fundamentally different in SSZ and MOR settings (see Alt, 1995, for review). At present, it is difficult to evaluate the causes of these differences as the modern record provides a very incomplete view of the spatial relationships within hydrothermal systems.

Ophiolites are generally classified as either mid-ocean ridge or supra-subduction zone (SSZ) based on their geotectonic setting. MOR ophiolites are formed at spreading centers uninfluenced by subduction

Gillis, K.M., and Banerjee, N.R., 2000, Hydrothermal alteration patterns in supra-subduction zone ophiolites, *in* Dilek, Y., Moores, E.M., Elthon, D., and Nicolas, A., eds., Ophiolites and Oceanic Crust: New Insights from Field Studies and the Ocean Drilling Program: Boulder, Colorado, Geological Society of America Special Paper 349, p. 283–297.

zones and include ridges in incipient ocean basins, major ocean basins, and back-arc basins well behind any subduction zone (Pearce and Peate, 1995). Supra-subduction zone ophiolites are formed at ridges directly associated with subduction, such as incipient back-arc basins, ridges associated with subduction initiation, and ridge-trench intersections (Pearce et al., 1984). SSZ ophiolites are distinguished from MOR ophiolites by using magma chemistry, igneous petrography, and regional geologic constraints. The chemistry of SSZ magmas shows evidence for the addition of mobile elements and water to the mantle source during the course of subduction and the more common availability of depleted mantle sources (Pearce and Peate, 1995). There is much on-going debate on precisely which type of geotectonic setting many SSZ ophiolites represent.

The purpose of this paper is to compile and synthesize our current knowledge of hydrothermal alteration within SSZ ophiolites. Our intent is to highlight aspects of hydrothermal systems that are common to SSZ ophiolites and MOR crust and to identify features that are influenced by geotectonic setting. We demonstrate that there is much more diversity in hydrothermal alteration patterns within SSZ ophiolites than is currently known for MOR crust. This diversity is related in part to the range of geotectonic settings in which SSZ ophiolites may evolve. In many cases, however, much more information is required in order to evaluate the role of geologic setting and the effects of emplacement. Classification of an ophiolite as "SSZ" was based on the geochemical classification presented in Pearce et al. (1984). In this compilation, we have attempted to include SSZ ophiolites that developed at different time intervals, have diverse tectonic histories, and have fairly intact pseudostratigraphies. Spreading rate was not considered as a variable as there are no consistent criteria used in the literature to assign spreading rate.

HYDROTHERMAL CHARACTERISTICS OF SSZ OPHIOLITES

Eleven SSZ ophiolites that span a range of age and tectonic setting, and for which there is at least one journal article describing some aspect of hydrothermal alteration, are included in this review (Tables 1 and 2; Fig. 1). Information on significant aspects of the hydrothermal systems is lacking for most of these ophiolites. It is therefore only possible to compare general characteristics, including the hydrothermal mineral assemblages, degree of alteration (i.e., the extent to which igneous phases are replaced by hydrothermal minerals), the presence or absence of polymetallic sulfide deposits, epidosites, and metalliferous sedimentary rocks, and conditions such as temperature and fluid/rock ratio. Available data do not permit a meaningful review of mineral compositions, geochemical fluxes, and physicochemical conditions that goes beyond previously published reviews (e.g., Alt, 1995; Schiffman et al., 1991).

Before the hydrothermal characteristics are reviewed, a few terms need to be defined. *Low-temperature alteration* describes assemblages of clay minerals (smectite, celadonite), Na-K zeolites (e.g., phillipsite, clinoptilolite, analcite), and calcite that form at temperatures of <100 °C. The zeolite, prehnite-pumpellyite, greenschist, and amphibolite facies are identified on the basis of characteristic mineral assemblages and compositions (e.g., Beiersdorfer and Day, 1995; Frey et al., 1991; Liou et al., 1987; Schiffman and Day, 1999). We assigned a metamorphic facies for the major lithologic units of each ophiolite based on published mineral assemblages (Table 2); in doing so, we recognized that these assemblages represent the cumulative metamorphic history and not solely the peak-metamorphic assemblage (from which metamorphic facies are assigned). We used the following low-pressure, common mineral assemblage(s) to assign a metamorphic facies. The *zeolite facies* is typified by the assemblages analcite + heulandite + quartz + mixed-layer smectite/chlorite or chlorite and laumontite + albite + quartz + mixed-layer smectite/chlorite or chlorite ± pumpellyite (e.g., Coombs et al., 1959; Liou et al., 1991). The mineral assemblage of the *prehnite-pumpellyite facies* is strongly dependent upon bulk-rock composition and, to a lesser extent, the pressure and temperature conditions of metamorphism (Beiersdorfer and Day, 1995; Frey et al., 1991). Some common assemblages include the following (bulk compositions of the protolith are given in parentheses): pumpellyite + chlorite + epidote + albite + quartz (Fe-rich); pumpellyite + chlorite + actinolite + epidote + albite + quartz (intermediate); prehnite + actinolite + chlorite + epidote + albite + quartz (low Al, high ferric iron); and prehnite + pumpellyite + actinolite + epidote + albite + quartz (Beiersdorfer and Day, 1995). It is important to note that Mg-rich rocks may not stabilize prehnite-pumpellyite assemblages, accounting for the absence of the prehnite-pumpellyite facies in some localities. The common assemblages for the *greenschist* and *amphibolite facies* are: chlorite + actinolite + epidote + albite + quartz ± titanite and hornblende + plagioclase (An_{20-50}) ± quartz ± titanite, respectively (Eskola, 1920; Liou et al., 1987).

The characteristic sequence of metamorphic facies with increasing temperature at low pressures is zeolite → ± prehnite-pumpellyite → greenschist → amphibolite. Note that the prehnite-pumpellyite facies is not common in MOR oceanic crust. Facies transitions are marked by a series of continuous reactions that occur over a range of temperature and pressure and are dependent upon bulk compositions and oxygen fugacity (e.g., Liou et al., 1974). Temperature estimates for facies boundaries are based on experimental and theoretical constraints, in conjunction with mineralogical and geochemical data for natural systems. At low pressures, the transitions from zeolite to prehnite-pumpellyite facies and from zeolite to greenschist facies occur at ~100–150 °C and ~200–250 °C, respectively. The transitions from prehnite-pumpellyite to greenschist facies and from greenschist to amphibolite facies occur at 200–300 °C and 450–550 °C, respectively (e.g., Apted and Liou, 1983; Beiersdorfer and Day, 1995; Frey et al., 1991). In the following sections, stratigraphic intervals with mineral assemblages that are typical for metamorphic facies are referred to as metamorphic zones (e.g., zeolite zone).

Volcanic rocks

Volcanic sequences show a broad range in the style and degree of alteration. Away from hydrothermal discharge sites, four metamorphic zones are recognized in SSZ ophiolites: low-temperature, zeolite, prehnite-pumpellyite, and greenschist. In all zones, igneous textures are preserved, plagioclase is variably altered to K-feldspar or albite, and clinopyroxene remains unaltered. Alteration assemblages are vertically zoned with metamorphic grade increasing with depth; however, many zone sequences have been recognized (Fig. 2). The Troodos ophiolite is the only ophiolite in which the low-temperature zone comprises the entire volcanic sequence (Bednarz and Schmincke, 1989; Gass and Smewing, 1973; Gillis and Robinson, 1990), with the most pervasive alteration concentrated in the uppermost volcanic rocks. Mineral assemblages change with depth, although their distributions are not laterally continuous. The vertical distribution of metamorphic zones in the volcanic sequence of the Semail ophiolite is subparallel to the boundaries separating the eruptive units V1, V2, and V3 (Pflumio, 1991), such that from top to bottom, the sequence is low-temperature (V3) → zeolite (V2) → prehnite-pumpellyite (V1). A similar zonation is observed in Kizildag

TABLE 1. GEOLOGIC CHARACTERISTICS OF SUPRA-SUBDUCTION ZONE OPHIOLITES INCLUDED IN REVIEW

	Age (Ma)	Tectonic setting	Lava chemistry*	Sedimentary cover: Basement contact	Sedimentary cover: Sedimentary-rock type
Tethyan Ophiolites					
Pindos	Middle-Jurassic (165 ± 3, Ar-Ar, metamorphic sole) (1)	Intraoceanic forearc, initiated within a small marginal basin (2, 3)	MORB, IAT, Bon (2; 3)	Unconformable	Metalliferous sedimentary rocks, mudtones, siltstones, sandstones (3)
Kizildag (Hatay)	Late Cretaceous (73–99, K-Ar) (4)	Forearc (prearc) (5); SSZ (6, 7)	IAT, Bon (5)	Unconformable	Upper Campanian intervolcanic sedimentary rocks; upper Maastrichtian conglomerate cover (8)
Troodos	Late Cretaceous (90–92, U-Pb); (9); Turonian (10)	Incipient subduction (forearc) (11)	IAT, Tran Bon (12, 13)	Conformable	Turonian umbers and mudstones to Paleocene chalks (14)
Semail	Late Cretaceous (~94–97, U-Pb) (15)	Mid-ocean ridge (15, 16, 17) to SSZ (18)	MORB, IAT (18, 19, 20)	Conformable	Coniacian–Santonian pelagic sedimentary rocks at V2-V3 contact (21)
Caledonian Ophiolites					
Karmøy	Early Ordovician (500–470, U-Pb) (22)	Island arc, arc-splitting, basin development (23, 24, 25)	IAT, MORB, Bon to Calc-Alk to Alk (24)	Unconformable	Upper Ordovician coarse clastic rocks (26)
Leka	Early Ordovician (497 ± 2) (22)	Island arc, arc-splitting, basin development (27)	IAT, MORB, Bon to Calc-Alk to Alk (25, 28)	Unconformable	Upper Ordovician coarse clastic rocks (26)
Betts Cove	Early Ordovician (488.6 + 3.1, −1.8, U-Pb) (29)	Forearc (30)	Bon (30)	Conformable	Ordovician volcaniclastic-sedimentary sequence (31)
Western U.S. Ophiolites					
Josephine	Late Jurassic (161–164) (32)	Backarc basin (33)	IAT (33)	Conformable	Siliceous argillite and pelagic chert grades up into metagraywacke (33) (34)
Del Puerto	Late Jurassic (155–165, U-Pb) (35)	Forearc or rifted island arc (36)	IAT, MORB (36)	Conformable	Upper Jurassic turbiditic, volcaniclastic sedimentary rocks (37)
Trinity	Silurian (435 ± 21, Sm-Nd) (38)	Backarc basin in vicinity of island arc (39)	IAT (39)	Unconformable	Carboniferous shales, mudstones, sandstones (39)
Western Pacific Ophiolite					
Zambales	Late Eocene (40)	Island-arc–backarc system (41, 42); rifted subduction-related marginal basin (43)	MORB, IAT, Bon (41, 42)	Conformable	Upper Eocene pelagic limestone with tuff interbeds (44)
Modern SSZ					
Izu-Bonin	Middle to late Eocene (41.3 ± 0.5) (45)	Forearc	IAT, Bon (46)	Conformable	Volcaniclastic and hemipelagic sedimentary rocks (47)
Tonga	Middle Eocene (48)	Forearc	MORB, IAT, Bon (48, 49)	Conformable	Volcaniclastic sedimentary rocks, pelagic chalks, carbonate debris, up to 600 m thick (50)

Note: References are indicated by number in parentheses. Only the most recent or relevant references are given here; see text and cited references for complete listing of related references. 1, Spray et al. (1984); 2, Kostopoulos (1989); 3, Jones and Robertson (1991); 4, Delaloye and Wagner (1984); 5, Lytwyn and Casey (1993); 6, Dilek and Moores (1990); 7, Dilek et al. (1991); 8, Selcuk (1981), 9; Mukas and Ludden (1987); 10, Blome and Irwin (1985); 11, Robertson and Xenophontos (1993); 12, Bednarz and Schmincke (1994); 13, Rogers et al. (1989); 14, Robertson and Hudson, (1974); 15, Tilton et al. (1981); 16, Pallister and Hopson (1981); 17, Nicolas (1989); 18, Pearce et al. (1981); 19, Alabaster et al. (1982); 20, Ernewein et al. (1988); 21, Tippit et al. (1981); 22, Dunning and Pedersen (1988); 23, Pedersen et al. (1988); 24, Pedersen and Hertogen (1990); 25, Pedersen and Furnes (1991); 26, Sturt et al. (1985); 27, Furnes et al. (1988); 28, Furnes et al. (1992); 29, Dunning and Krogh (1985); 30, Bédard et al. (1998); 31, Jenner and Fryer (1980); 32, Wright and Wyld (1986); 33, Harper (1984); 34, Pinto-Auso and Harper (1985); 35, Hopson et al. (1981); 36, Shervais (1990); 37, Evarts and Schiffman (1983); 38, Jacobsen et al. (1984); 39, Brouxel and Lapierre (1988); 40, Amato (1964); 41, Hawkins and Evans (1983); 42, Geary et al. (1989); 43, Yumul (1996); 44, Schweller et al. (1984); 45, Mitchell et al. (1992); 46, Murton et al. (1992); 47, Shipboard Party (1990); 48, Bloomer et al. (1998); 49, Falloon et al. (1997); 50, Clift et al. (1998).

* IAT—island-arc tholeiite, Bon—boninite, Tran Bon—transitional boninite, MORB—mid-ocean–ridge basalt, Calc-Alk—calc-alkalic, Alk—alkalic.

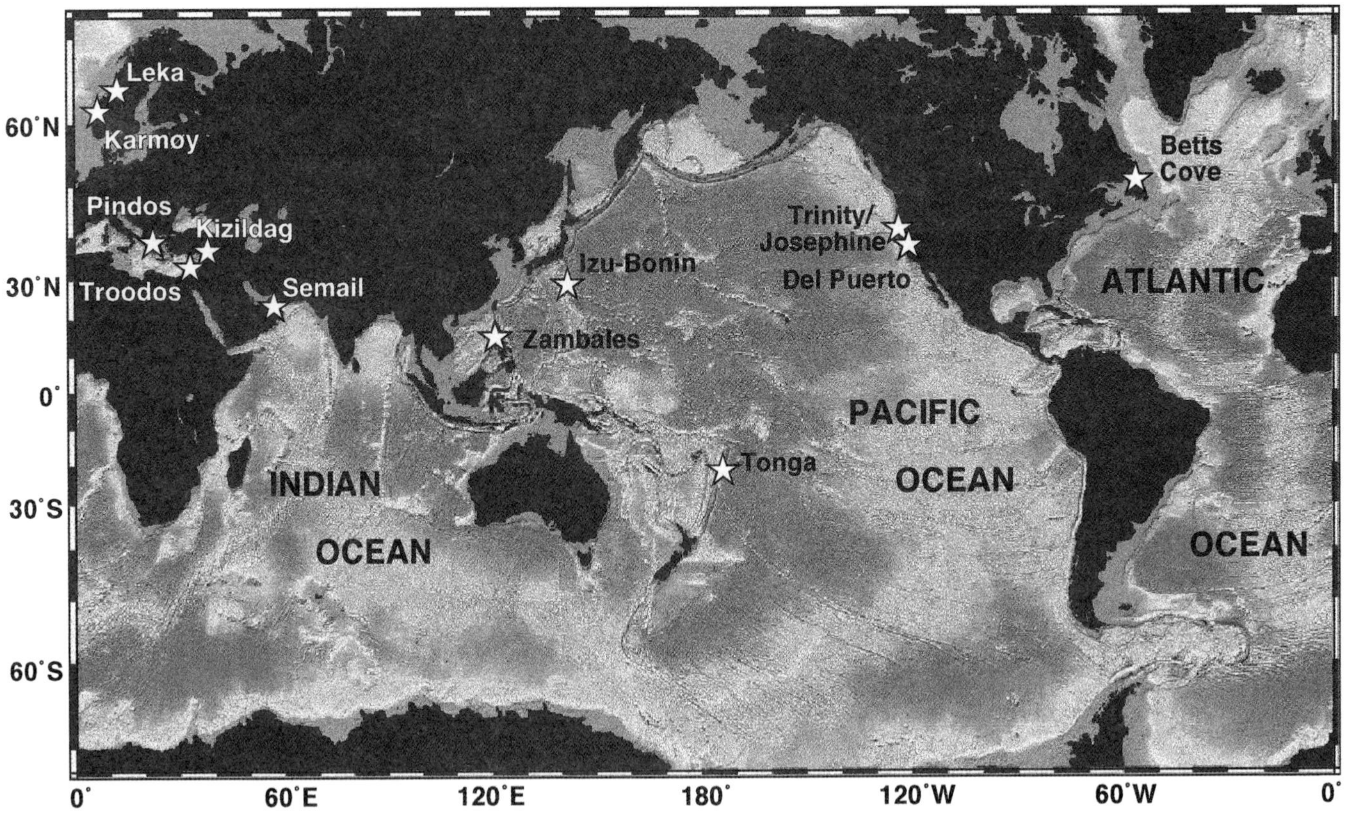

Figure 1. Locations of SSZ ophiolites and forearcs included in this review (modified from Smith and Sandwell, 1997).

whereas the Josephine volcanic rocks are altered to assemblages of the prehnite-pumpellyite to greenschist facies. The volcanic sequences of five other ophiolites are dominantly composed of one metamorphic zone, and mineral assemblages vary with depth. The Karmøy, Leka, Betts Cove, and Trinity volcanic rocks are altered to greenschist-facies assemblages, whereas the Pindos ophiolite contains dominantly prehnite-pumpellyite–facies assemblages (Table 2). Alteration within the Pindos and Josephine volcanic rocks is laterally heterogeneous with patches of albite epidosite. The stratigraphic divisions of the Del Puerto and Zambales ophiolites differ from the other ophiolites in that they each have a thick section of volcanic rocks, dikes, and sills (~2.5 km) and lack sheeted-dike complexes. The zonation in their volcanic sequences is prehnite-pumpellyite → greenschist (Evarts and Schiffman, 1983; Geary and Kay, 1983).

What factors contribute to the differences within volcanic sequences? Studies of modern oceanic crust and ophiolites show that volcanic sequences initially have high porosity and permeability relative to subjacent sheeted dike complexes (Fisher, 1998; Gillis, 1997; Nehlig and Juteau, 1988; and references therein). It follows that circulation of a large volume of ambient seawater would maintain low temperatures (<50 °C) within volcanic sequences. Indeed, modern volcanic rocks recovered away from zones of hydrothermal up-flow on ridges without sediment cover contain only low-temperature mineral assemblages (Alt, 1995). It is interesting to note that similar low-temperature assemblages have been documented only for the Troodos ophiolite (Gillis and Robinson, 1988).

One situation that may cause higher temperatures in volcanic sequences is the early sealing of the oceanic crust by the deposition of clay-rich sediments. Recent drilling by the Ocean Drilling Program (ODP Legs 139, 168, 169) at the sediment-covered Juan de Fuca and Gorda Ridges provides some insights as to how a sedimentary cover influences basement alteration. Sediment-hosted, high-temperature hydrothermal systems at Middle Valley and Escanaba Trough were drilled during ODP Legs 139 and 169 (Davis et al., 1992; Fouquet et al., 1998). Volcanic rocks recovered away from sites of active discharge were initially altered to low-temperature assemblages and subsequently overprinted by higher-temperature phases. At Site 1037 (Escanaba Trough), vesicles and vugs filled with chlorite, clay, and calcite are connected by thin veinlets containing fine needles of Mg-rich hornblende that formed at temperatures greater than 350 °C (Shipboard Scientific Party, 1998). Along the eastern flank of the Juan de Fuca Ridge, young volcanic basement (≥0.6 Ma) was buried by turbiditic sediments within 18 km of the ridge (Davis et al., 1997). Cores recovered from the upper few tens of meters of the volcanic basement, which ranges in age from 0.6 to 3.6 Ma, are dominantly altered to low-temperature assemblages (Davis et al., 1997). The youngest sites contain mainly low-temperature clays, carbonate, iron oxyhydroxides, and zeolites. Sites farther from the ridge exhibit decreased evidence of oxidative alteration, and mineral assemblages suggest restricted or closed circulation of fluids (Davis et al., 1997; Hunter et al., 1999). Calculated temperatures along the interface between the basement and the sedimentary rocks increase from 16 °C in the 0.6 Ma rocks (~18 km from the ridge axis) to ~60 °C in the 3.6 Ma rocks (100 km from the ridge axis); the temperatures may increase to >100 °C farther off axis where the thickness of the sedimentary rock increases (A. Fisher, 1999, personal commun.). Minor chlorite veinlets in the youngest cores suggest an early oxidative high-temperature event at the ridge axis (Hunter et al., 1999). These studies suggest that early burial of oceanic crust, at or near a spreading center, does not lead to the widespread development of high-temperature alteration zones in the uppermost volcanic rocks, but does not determine the nature of alteration in the lower volcanic rocks. It is interesting to note that a small proportion of samples dredged from the Mid-Atlantic Ridge that have pillow morphologies are altered to greenschist-facies assemblages, showing that high temperatures can be achieved on a local scale in permeable volcanic sequences (Humphris et al., 1978).

The characteristics of modern oceanic crust suggest that special circumstances are required in order to attain high temperatures in highly permeable crust. SSZ ophiolites that formed adjacent to a continental margin and/or volcanic arc could be quickly buried by turbiditic and volcaniclastic sedimentary rocks (e.g., Del Puerto ophiolite in California, Evarts and Schiffman, 1983). In this setting, the metamorphic zonations may develop in response to the slow cooling of plutons if extensional faulting is not widespread, similar to continental geothermal systems (Schiffman et al., 1991) and oceanic islands (e.g., La Palma in the Canary Islands, Schiffman and Staudigel, 1994). The magmatic evolution of a section of oceanic crust may also influence the distribution of metamorphic zones. For example, the Semail ophiolite was constructed by three magmatic events (V1 $\rightarrow$ V3) (Ernewein et al., 1988). In this case, the metamorphic zonation developed by the superposition of hydrothermal phases during successive magmatic and hydrothermal events (Pflumio, 1991). There is no straightforward way to explain why other volcanic sequences are altered to the zeolite or greenschist facies. Obduction-related metamorphism has been ruled out for the Josephine ophiolite (Harper, 1995; Harper et al., 1988). The sedimentary cover is missing for the Pindos and Kizildag ophiolites so its role cannot be evaluated. The Betts Cove ophiolite is conformably overlain by a volcanic-sedimentary sequence, which may have affected the thermal regime in the upper crust. Hence, more information concerning the effects of obduction, the sedimentary cover, and other geologic variables is required to evaluate how high temperatures are achieved within permeable volcanic sequences.

An interesting characteristic of several volcanic sequences altered to zeolite or higher-temperature facies is the presence of hematite and K-bearing phases and/or K enrichment (e.g., Betts Cove, Coish, 1977; Trinity, Lécuyer et al., 1990; Josephine, Harper et al., 1988; Zambales, Geary and Kay, 1983). K- and Fe^{3+}-bearing phases (smectite, celadonite, K-feldspar, iron oxyhydroxides) are common in volcanic sequences generated at modern MORs (e.g., Holes 417 and 418 [western Atlantic], Staudigel et al., 1981; Hole 504B [eastern Pacific], Alt et al., 1986), forearcs (Izu-Bonin, Alt et al., 1998; Tonga, Kelman, 1998), and the volcanic sequence of the Troodos ophiolite (Bednarz and Schmincke, 1989; Gillis and Robinson, 1990) and are indicative of interaction with oxidative seawater at low temperatures. Recrystallization of these phases at higher temperatures would lead to the formation of hematite, K-feldspar, and possibly muscovite. Their occurrence suggests that incipient alteration in volcanic sequences in some ophiolites occurred at low temperatures, followed by a higher-temperature event. Indeed, the presence of a high-temperature, K-bearing phase (stilpnomelane) in core from the Escanaba Trough indicates that low-temperature K-enriched assemblages were recrystallized during a later hydrothermal event (Shipboard Scientific Party, 1998). High bulk-rock $\delta^{18}O$ values for the Trinity and Josephine volcanic rocks are also consistent with alteration at low temperatures, although published studies for these ophiolites do not advocate this view (Harper, 1988; Lécuyer et al., 1990; Lécuyer and Fourcade, 1991). Sr and O isotope data for the Trinity volcanic rocks indicate alteration at high fluid/rock ratios (Lécuyer and Fourcade, 1991). The dominant metamorphic mineral assemblage indicates temperatures of >300 °C, whereas the presence of celadonite in the uppermost volcanic rocks suggests temperatures of <50 °C (Lécuyer et al., 1990). Fluid-inclusion data for interpillow quartz indicate alteration at temperatures between 100 and 150 °C. High $\delta^{18}O$ values for the Josephine volcanic rocks indicate alteration at high fluid/rock ratios, and the metamorphic mineral assemblages indicate temperatures of <200 °C (Harper, 1988). The O isotope data for both the Trinity and Josephine ophiolites overlap the range of $\delta^{18}O$ values for modern volcanic rocks altered at low temperatures (e.g., Hole 504B, $\delta^{18}O$ = 5.5‰–12‰, Alt et al., 1996b). An alternative history of alteration that may reconcile the high $\delta^{18}O$ values (indicating low-temperature alteration, i.e., <100 °C) with the high-temperature mineral assemblages is an early stage of low-temperature alteration at high fluid/rock ratios, followed by a higher-temperature alteration event at low fluid/rock ratios.

Volcanic sequence–sheeted dike transition

Metamorphic zones at the volcanic sequence–sheeted dike transition indicate both gradational and stepped thermal gradients. In the Troodos ophiolite, the fact that the metamorphic assemblage jumps from low-temperature to greenschist facies over tens to hundreds of meters suggests a discrete step in the geothermal gradient zonation across this lithologic boundary (Fig. 2) (Gillis and Robinson, 1990). However, in most ophiolites that have higher-temperature metamorphic zones within their volcanic sequences, the transition to a higher metamorphic grade is gradational. Hence, the volcanic sequence–sheeted dike transition does not coincide with a discrete step in the geothermal gradient (e.g., the Josephine ophiolite, Harper et al., 1988) (Fig. 2).

Sheeted dikes

All sheeted dike complexes are altered to the greenschist facies, and, in some ophiolites, the basal sheeted dikes are altered to the amphibolite facies (Fig. 2; Table 2). Most ophiolites show a change in mineral assemblage with depth, which is indicative of an increase in temperature and the degree of replacement of igneous phases, although there is very limited documentation of this trend. On a regional scale, many ophiolites display a lateral variation in their mineral assemblages that reflect fluctuations in temperature and/or fluid/rock ratios (e.g., Alexander et al., 1993; Stakes and Taylor, 1992). Higher-temperature zones extend to shallower stratigraphic depths where sheeted dikes are intruded by high-level and, commonly, off-axis plutons (e.g., Semail, Alabaster and Pearce, 1985; Haymon et al., 1989; Troodos, Eddy et al., 1998). Lower-temperature zones extend to deeper crustal levels in areas that have been extensively faulted (e.g., Josephine, Alexander et al., 1993; Troodos, Gillis and Robinson, 1990). On an outcrop scale, metamorphic grade is also variable where multiple generations of dikes are identified, such that the oldest dikes are altered more pervasively and at higher grades than younger dikes (e.g., Kizildag, Erendil, 1983; Josephine, Alexander et al., 1993).

Plutonic sequences

Plutonic sequences are variably altered to the greenschist and/or amphibolite facies (Table 2). The degree of alteration is less than in the overlying dikes and typically diminishes with depth toward the Moho. Early O isotope studies show that the incipient penetration of hydrothermal fluids into the plutonic crust occurs at high tempera-

TABLE 2. HYDROTHERMAL CHARACTERISTICS OF SUPRA-SUBDUCTION ZONE OPHIOLITES INCLUDED IN REVIEW

	Ore deposits	Stock-works	Metalliferous sedimentary rocks	Epidosite	Background alteration: Volcanic rocks*	Background alteration: Sheeted dikes*	Background alteration: Plutonic rocks*	Meta-morphic overprint
Tethyan Ophiolites								
Pindos	Fe-Cu (1)	Yes (1, 2)	Intercalated metalliferous sedimentary rocks, red chert (1)	Small (centimeters to hundreds of meters); hosted in extrusive and intrusive rocks (2)	*Prehnite-pumpellyite*: Chl + Alb + Prh + Ana (2)	*Greenschist*: Alb + Chl + Qtz + Act + Epi (2)	No published description	None reported
Kizildag (Hatay)	None reported	None reported	Intercalated metalliferous sedimentary rocks (3); Mn-rich cherts (4)	None reported	*Zeolite*: UV: Na-Ca zeo + Cla *Prehnite-pumpellyite*: LV: Prh + Zeo + Epi + Chl + Alb + Qtz + Hem (4)	*Prehnite-pumpellyite*: UD: Alb + Zeo + Chl + Pmp + Prh + Epi + Qtz + Cal *Greenschist*: LD: Act + Qtz + Pmp + Chl + Tit; Act + Epi in basal dikes (4)	*Amphibolite to greenschist*: Amp + Chl + Pla (5)	None reported
Troodos	Fe-Cu (6–9)	Yes (10, 11)	Umbers, ochers, metalliferous sedimentary rocks (12–15)	Large (hundreds of meters to 8 km^2); hosted in intrusive rocks (11, 16–18)	*Low-temperature alteration*: UV: Sme + K-fel + Na-K zeo + Cal + Fe oxy LV: Sme + Cel + Na-K zeo ± Cal (19–22)	*Greenschist*: UD: Sme/Chl + Chl + Alb + Qtz ± Epi ± Ca zeo ± Pyr ± Cha LD: Act + Qtz + Chl + Tit ± Hor ± Epi ± Ca zeo ± Mag ± Pyr ± Cha (21, 23–25)	*Amphibolite to granulite*: Hor + Int pla ± Chl ± Epi ± Mag ± Pyr ± Cha ± Qtz ± Opx (26–29)	None (21)
Semail	Fe-Cu-Zn (30–32)	Yes (31–33)	Intercalated metalliferous sedimentary rocks (34, 35)	Small (centimeters to hundreds of meters); hosted in intrusive rocks (36)	*Zeolite*: UV (V3): Alb + Chl + Cel + Fe pmp + Tit + Sco *Zeolite to prehnite-pumpellyite*: MV (V2): Alb + Chl + Prh + Epi + Qtz + Alb + K-fel + Cell + Fe pmp + Tit + Cal ± Ca zeo ± Sul *Prehnite-pumpellyite*: LV (V1): Qtz + Alb + Chl + Epi + Tit + Prh + Pmp + Pyr + Cha (33, 37)	*Greenschist*: Act + Chl + Epi + Qtz + Alb ± Prh ± Pmp ± Tit ± Sul (36–38)	*Amphibolite to greenschist*: Ca pla + Hor + Act ± Epi ± Chl ± Prh ± tit (36, 39–41)	None (37)
Caledonian Ophiolites								
Karmøy	Fe-Cu-Zn (42, 43)	None reported	None reported	Patchy zones in sheeted dikes, plagiogranites (43, 44)	UV: No published description *Greenschist*: LV: Epi + Qtz + Chl + Qtz + Tit + Pyr (43)	*Greenschist*: Alb + Epi + Qtz + Chl + Tit + Pyr ± Act ± Cal (43)	*Amphibolite to greenschist*: Hor + Act + Alb + Epi + Qtz + Tit + Pyr + Mag (43)	Greenschist (45)
Leka	None reported	None reported	None reported	None reported	*Greenschist*: UV: Bio + Cal + Act + Epi + Qtz + Chl (46) LV: no published description	No published description	No published description	Greenschist (45)
Betts Cove	Cu-Zn ± Au ± Ag (47)	Yes (48)	Interpillow red chert (47)	None reported	*Greenschist*: UV: Alb + Chl + Qtz + Cal + Mus + Hem LV: Act + Alb + Epi + Qtz + Chl + Cal (48, 49)	*Greenschist*: Act + Alb + Epi + Qtz + Chl + Cal (48, 49)	*Greenschist*: Alb + Act + Epi + Chl + Mus + Prh + Pargnt (49)	None reported

TABLE 2. HYDROTHERMAL CHARACTERISTICS OF SUPRA-SUBDUCTION ZONE OPHIOLITES INCLUDED IN REVIEW (*continued*)

	Ore deposits	Stock-works	Metalliferous sedimentary rocks	Epidosite	Background alteration: Volcanic rocks*	Background alteration: Sheeted dikes*	Background alteration: Plutonic rocks*	Meta-morphic overprint
Western U.S. Ophiolites								
Trinity	None reported	None reported	None reported	None reported	*Greenschist*: UV: Act + Chl + Epi + Alb + Qtz ± Cel ± Cal LV: Act + Chl + Epi + Alb + Qtz (50)	*Greenschist*: Act + Chl + Epi + Alb + Qtz; Hor in basal dikes (50)	*Amphibolite*: UP: Act + Epi + Tre ± Alb ± Qtz LP: Hor + Tre + Tlc + Ser (50)	None (50)
Jose-phine	Zn-Cu-Ag-Au-Co (51, 52)	Yes (52)	Iron- and manga-nese-bearing argillites and cherts (53)	Albite epidosites in volcanic rocks; patchy epidosite zones in upper dikes (53, 54)	*Prehnite-pumpelly-ite*: UV: Alb + Chl + Epi + Qtz + Tit + Sul + Pmp ± Prh ± Hem ± Mag ± Mus ± Cal *Greenschist*: LV: Alb + Chl + Epi + Qtz + Tit + Sul ± Pmp ± Hem ± Mag ± Mus ± Cal ± K-fel (53, 54, 55)	*Greenschist*: UD: Alb + Chl + Epi + Qtz + Tit ± Act ± Pyr LD: Int pla + Hor + Mag ± Chl ± Tit, or Alb + Act + Chl + Epi + Qtz ± Prh ± Tit ± Cal (53, 55, 56)	*Amphibolite*: Int pla + Hor + Mag ± Chl ± Epi (53, 55, 56, 57)	Prehnite-pumpel-lyite to green-schist (53)
Del Puerto	None (58)	None (58)	Manganese oxide deposits, iron-bear-ing cherts (58)	Patchy zones in sheeted dikes (58)	*Prehnite-pumpelly-ite*: UV: Fe pmp + Alb + Cor + Qtz + Hem + Tit + Mag ± Ca zeo ± Prh ± Sme/Chl ± Cel ± Cal *Greenschist*: LV: Chl + Epi + Alb + Qtz + Tit + Hem + Mag ± Act ± Ca zeo ± Prh ± Cal (58)	No sheeted dike complex	Dike/Plutonic Complex *Amphibolite to greenschist*: Act + Epi + Chl + Alb + Qtz + Hem + Mag ± Prh ± Cal (58)	None (58)
Western Pacific Ophiolite								
Zam-bales	Fe-Cu (59)	None reported	Thin beds of red mudstone (60)	None reported	*Greenschist*: Alb + Sme/Chl ± Qtz ± Cal ± Epi ± Fe pmp ± Pyr ± Hem ± K cla (59)	Dike-Sill Complex *Greenschist*: UD-S: Alb + Chl + Act + Qtz + Cal + Pyr ± Epi ± Tit LD-S: Alb + Chl + Act ± Epi ± Tit ± Qtz ± Pyr (59)	*Amphibolite to greenschist*: Int pla + Act + Hor ± Chl ± Epi ± Tit ± Qtz ± Alb (59)	None reported
Modern SSZ								
Izu-Bonin	None reported	None reported	None reported	None reported	*Low-temperature alteration*: Sme + Cel + Fe oxy + Cal + Na-K zeo ± Pyr ± Cha (61)	*Greenschist*: UD: Sme/Chl + Ser + Alb + K-fel + Cal + Pyr + Cha ± Chl (61)	No published description	Not applica-ble
Tonga	None reported	None reported	None reported	Patches within diabase and plagiogranite host (62)	*Low-temperature alteration*: Na-K zeo + Sme + Cel + Cal + K-fel + Fe oxy (dredge samples) (63)	*Greenschist*: Chl + Sme/Chl + Epi + Alb (dredge sam-ples) (63)	*Amphibolilte to greenschist*: Ca pla + Hor ± Chl ± Epi	Not applica-ble

Note: References are indicated in the table by numbers in parentheses: 1, Robertson and Varnavas (1993); 2, Valsami and Cann (1992); 3, Robertson (1986); 4, Erendil (1983); 5, Dilek and Thy (1998); 6, Constantinou and Govett (19792); 7, Constantinou and Govett (1973); 8, Constantinou (1976); 9, Adamides (1990); 10, Lydon and Galley (1986); 11, Richardson et al. (1987); 12, Robertson (1975); 13, Robertson (1976); 14, Boyle and Robertson (1984); 15, Boyle (1990); 16, Schiffman et al. (1987); 17, Schiffman and Smith (1988); 18, Bettison-Varga et al. (1992); 19, Gass and Smewing (1973); 20, Gillis and Robinson (1985); 21, Gillis and Robinson (1990); 22, Bednarz and Schmincke (1989); 23, Baragar et al. (1989); 24, Baragar et al. (1990); 25, Bettison-Varga et al. (1995); 26, Malpas (1990); 27, Kelley et al. (1992); 28, Kelley and Robinson (1990); 29, Gillis and Roberts (1999); 30, Ixer et al. (1984); 31, Alabaster and Pearce (1985); 32, Haymon et al. (1989); 33, Regba et al. (1990); 34, Fleet and Robertson (1980); 35, Robertson and Fleet (1986); 36, Nehlig and Juteau (1988); 37, Pflumio (1991); 38, Nehlig et al. (1994); 39, Gregory and Taylor (1981); 40, McCulloch et al. (1981); 41, Stakes and Taylor (1992); 42, Vokes and Gale (1976); 43, Scott (1991); 44, Pedersen and Malpas (1984); 45, Pedersen (1986); 46, Tveit et al. (1993); 47, Strong and Saunders (1988); 48, Saunders and Strong (1986); 49, Coish (1977); 50, Lécuyer et al. (1990); 51, Kuhns and Baitis (1987); 52, Zierenberg et al. (1988); 53, Harper et al. (1988); 54, Harper (1995); 55, Beiersdorfer (1993); 56, Alexander et al. (1993); 57, Kimball (1988); 58, Evarts and Schiffman (1983); 59, Geary and Kay (1983); 60, Schweller et al. (1984); 61, Alt et al. (1998); 62, Banerjee et al. (1998); 63, Kelman (1998). (Table abbreviations appear in footnote continued on p. 290.)

tures (Gregory and Taylor, 1981) which is consistent with recent geothermometric data for the Troodos (Gillis and Roberts, 1999) and Semail (Manning et al., this volume) ophiolites. In the Semail ophiolite, detailed study of vein systems (Nehlig and Juteau, 1988) and regional-scale O isotope data (Stakes and Taylor, 1992) show that amphibolite-facies assemblages are commonly overprinted by lower-grade assemblages, especially in areas that have been extensively deformed.

Discharge zones: Stockworks, ore deposits, metalliferous sedimentary rocks, epidosites

The discharge of hot, hydrothermal fluids is manifested in many different forms. Epidosites, mineralized fault and/or vein systems, and stockworks mark discharge zones subjacent to sulfide deposits. Exhalative massive sulfide deposits occur at all depths within volcanic sequences, from the sheeted dike–volcanic sequence transition up to the volcanic sequence–sedimentary rock interface; therefore, high-temperature hydrothermal systems may be active during all stages of crustal accretion. Lenses of metalliferous sedimentary rocks intercalated with or overlying volcanic rocks are the oxidation products of massive sulfides or the fallout of hydrothermal plume particles, mixed with biogenic and detrital sediment. Their composition and mineral assemblages vary widely, depending on their origin, proximity to the hydrothermal source, and the background sedimentation (see Mills and Elderfield, 1995, for review).

All the SSZ ophiolites included in this compilation show some evidence for high-temperature hydrothermal activity (Table 2). Metalliferous sedimentary rocks have been described for most ophiolites and include ochers (ferruginous sedimentary rocks associated with sulfides), umbers (iron-manganese oxide–bearing sedimentary rocks), iron-bearing cherts, and metal-enriched mudstones. Most ophiolites are host to one or more massive sulfide deposits that are underlain by mineralogically and chemically zoned stockworks. Details for each deposit vary in terms of sulfide assemblage, metal grade, and size, as well as the characteristics of their stockwork zones. A synthesis of ophiolite-hosted massive sulfides and their stockwork zones is presented in Galley and Koski (1999).

Epidosites have been identified in six SSZ ophiolites, all of which contain sulfide deposits. Epidosites are metal depleted and characterized by the hydrothermal replacement of igneous phases by granular assemblages of epidote + quartz ± chlorite ± titanite ± magnetite ± sulfides (Richardson et al., 1987; Schiffman et al., 1987). Epidosites form at temperatures of up to 450 °C and high cumulative fluid/rock ratios (Bickle and Teagle, 1992; Bickle et al., 1999; Richardson et al., 1987; Schiffman and Smith, 1988), which suggest that they represent hydrothermal fluid conduits. The size of epidosite zones within sheeted dike complexes is quite variable. In the Josephine (Harper, 1995; Harper et al., 1988), Pindos (Valsami and Cann, 1992), and Semail (Nehlig and Juteau, 1988) ophiolites, epidosite occurs as patchy zones within a single dike or dike outcrop. By contrast, epidosites in the Troodos ophiolite comprise areas up to 8 km^2 (Bettison-Varga et al., 1992; Richardson et al., 1987; Schiffman and Smith, 1988). Epidotized plagiogranites are described for the Troodos (Kelley and Robinson, 1990; Kelley et al., 1992; Malpas, 1990), Semail, (Stakes and Taylor, 1992), and Karmøy (Pedersen, 1986) ophiolites.

When and why does epidotization occur? Schiffman and Smith (1988) and Bettison-Varga et al. (1992) argued, on the basis of regional-scale petrologic and $\delta^{18}O$ data, that epidotization in the Troodos ophiolite occurred off axis. More recently discussed field, paleomagnetic, and textural relationships show that epidotization proceeds rapidly after dike intrusion, prior to significant dike rotation (Varga et al., 1999). Epidotized dikes are cut by nonepidotized dikes in both the Troodos (Richardson et al., 1987) and Josephine (Alexander et al., 1993) ophiolites. If epidotization occurs during or shortly after a dike or dike swarm is intruded, magmatic degassing may play a role in the formation of the epidosites. SSZ ophiolites with epidosites have magma compositions that range from mid-oceanic-ridge basalt (MORB) to island-arc tholeiite (IAT) to boninite (Table 1). An aspect of magma composition that may be significant is the high volatile content of IATs and boninites relative to MORBs (e.g., Muenow et al., 1990). To date, however, a magmatic signature has not been recognized in the geochemistry of epidosites hosted in sheeted dike complexes. Epidotized plagiogranite and trondhjemite bodies contain fluid inclusions whose characteristics are best explained by a magmatic fluid source (Kelley et al., 1992; Nehlig, 1991). Such epidotized plutons have not been linked to epidosite zones within sheeted-dike complexes.

Discharge zones within the Pindos and Josephine ophiolites are also marked by localized zones of albite epidosites within the volcanic sequences, where pillows, breccia, and interpillow zones were recrystallized at 200–250 °C and high fluid/rock ratios (Harper et al., 1988; Valsami and Cann, 1992). In the Del Puerto ophiolite, manganese oxide deposits and laumontite mounds mark discharge zones (Evarts and Schiffman, 1983).

DO SSZ OPHIOLITES RESEMBLE MODERN SSZ SETTINGS?

Our knowledge of hydrothermal alteration patterns in modern SSZ settings is limited to recent studies of the Izu-Bonin (Alt et al., 1998) and Tonga (Banerjee et al., 2000; Kelman, 1998) forearcs. The forearcs of these intraoceanic arc systems developed during the initiation of subduction, prior to establishment of volcanic arcs (Bloomer et al., 1995). Forearc or "pre-arc" crust was built on, or displaced, preexisting oceanic crust during a short-lived period of extension. Subsequent development of volcanic arcs led to the deposition of volcaniclastic and hemipelagic sediments (Table 1). At the Izu-Bonin forearc, a section through the volcanic sequence and the transition into the sheeted dike complex was recovered at ODP Site 786 (Shipboard Scientific Party, 1990a). Secondary-mineral assemblages within the volcanic sequence are indicative of low-temperature alteration

(Table 2 footnote, continued from p. 289.)
* Abbreviations for rock subdivisions: UV—upper volcanic rocks; MV—middle volcanic rocks; LV—lower volcanic rocks; UD—upper dikes; LD—lower dikes; UP—upper plutonic rocks; LP—lower plutonic rocks; S—sills. Mineral abbreviations: At—actinolite; Alb—albite; Amp—amphibole; Ana—analcime; Bio—biotite; Ca pla—calcic plagioclase; Ca zeo (e.g. laumontite, stilbite); Cal—calcite; Cel—celadonite; Cha—chalcopyrite; Chl—chlorite; Cla—clay; Cor—corrensite; Epi—epidote; Fe oxy—iron oxyhydroxides; Fe pmp—Fe pumpellyite; Hem—hematite; Hor—hornblende; Int pla—intermediate plagioclase; K cla—potassic clay; K-fel—K-feldspar; K zeo—potassic zeolite; Mag—magnetite; Mus—muscovite; Na-K zeo (e.g., phillipsite, clinoptilolite, chabazite, gmelinite); Opx—orthopyroxene; Pargnt—paragamet; Prh—prehnite; Pmp—pumpellyite; Pla—plagioclase; Pyr—pyrite; Qtz—quartz; Ser—sericite; Sme—smectite; Sme/Chl—mixed layer smectite/chlorite; Sco—scolecite; Sul—sulfides; Tlc—talc; Tit—titanite; Tre—tremolite; Zeo—zeolite.

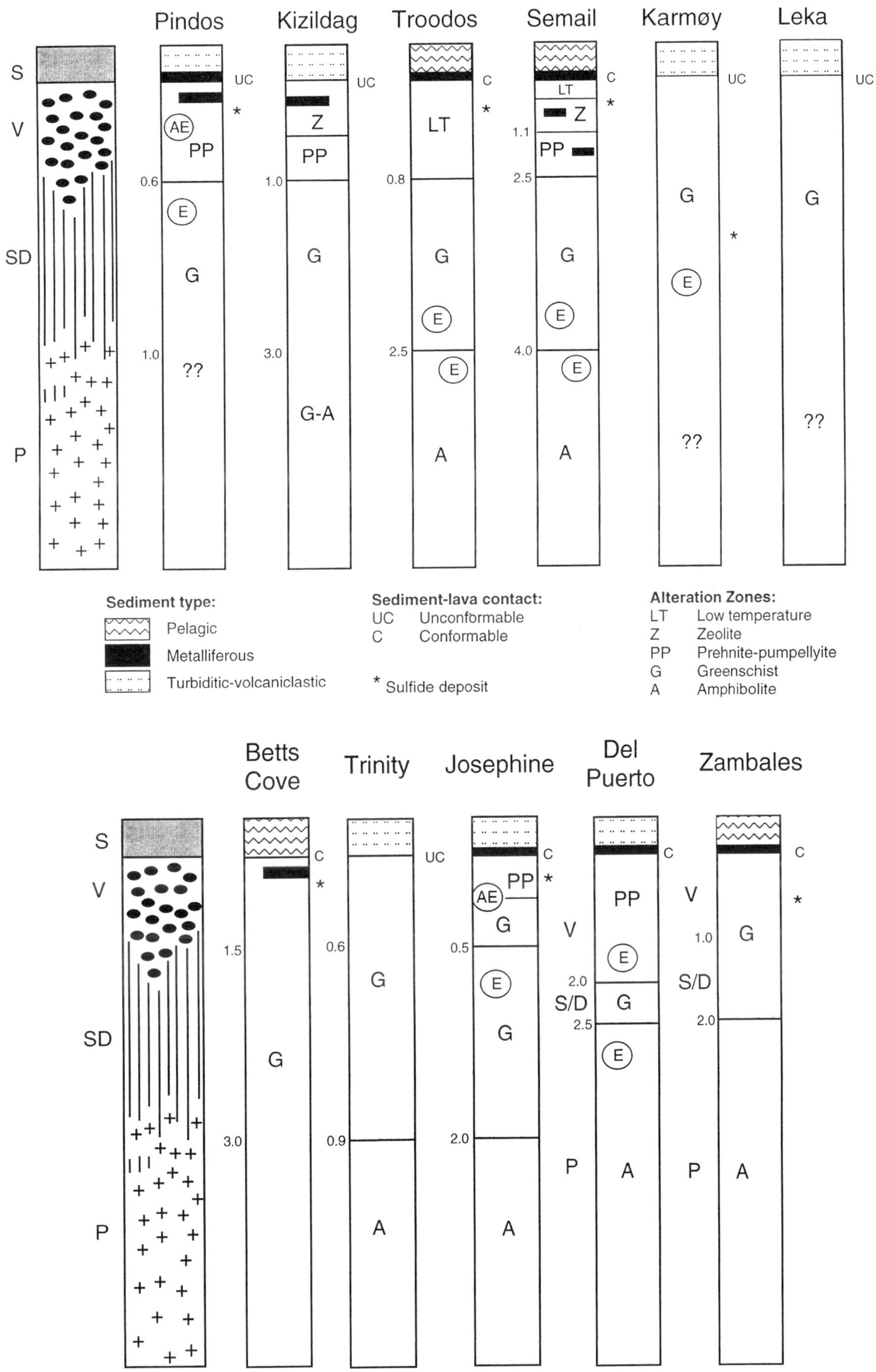

(Figure 2. See caption, following page.)

(<100 °C) (Table 2) (Alt et al., 1998). A change in the mineral assemblage and degree of alteration at the transition downward into the sheeted dike complex reflects an increase in temperature and interaction with chemically evolved hydrothermal fluids (Alt et al., 1998). A comprehensive suite of basaltic, diabasic, and plutonic rocks was dredged along the trench-facing slope of the Tonga Trench, where subduction erosion and trench rollback have exposed basement rocks (Bloomer et al., 1996). Samples with basaltic textures have low-temperature alteration assemblages that are similar to the volcanic sequence at ODP Site 786 (Table 2) (Kelman, 1998). Radiometric and biostratigraphic ages for the basement rocks and overlying sedimentary rocks, respectively, for the Tonga and Izu-Bonin forearcs indicate that the igneous basement was buried shortly after formation (McDougall, 1994; Mitchell et al., 1992; Shipboard Scientific Party, 1992; Stabell et al., 1992).

The western Pacific intraoceanic forearcs are considered reasonable modern analogues for the Troodos, Kizildag, and Pindos ophiolites, although these ophiolites developed in subduction-zone systems that did not develop volcanic arcs (Bloomer et al., 1995). Alteration of the volcanic sequences of the Izu-Bonin and Tonga forearcs is broadly similar to the low-temperature zones documented for the Troodos ophiolite and the upper part of the volcanic sequence of the Kizildag ophiolite (Table 2). The Troodos ophiolite is conformably overlain by pelagic sedimentary rocks whereas an unconformable sedimentary sequence overlies the Kizildag ophiolite (Table 1). Similarities between the Troodos ophiolite and the Izu-Bonin and Tonga forearcs suggest that volcanic-arc development and, in particular, the accumulation of volcaniclastic sediments do not affect the thermal structure of the upper forearc crust. Although the Pindos and Betts Cove ophiolites also formed in a forearc setting, higher-grade alteration is recorded in their volcanic sequences. The Pindos volcanic rocks are altered to the prehnite-pumpellyite facies with localized zones of albite epidosite (Valsami and Cann, 1992). Tectonic dismemberment and absence of a conformable sedimentary cover (Jones and Robertson, 1991) make it impossible to evaluate why higher temperatures prevailed in the upper crust of this forearc. The Betts Cove volcanic rocks are altered to the greenschist facies and are conformably overlain by volcanic and sedimentary sequences (Snooks Arm Group) (Bédard et al., 1998). Although it is plausible that the thermal structure of the Betts Cove volcanic rocks was influenced by burial and intrusion by this Ordovician arc–backarc sequence, it is not possible to rule out the role of obduction-related metamorphism.

The metamorphic evolution of the intrusive crust is most sensitive to magmatic and tectonic processes associated with crustal formation and is less affected by postaccretion events. Preliminary study of plutonic rocks recovered from the Tonga forearc shows that incipient hydrothermal alteration was focused along microfracture arrays at amphibolite-facies conditions, followed by locally pervasive greenschist-facies alteration associated with brittle deformation (Banerjee et al., 1999). Calculated temperatures for incipient alteration (>600 °C) are consistent with temperatures calculated for the Semail (Manning et al., this volume) and Troodos (Gillis and Roberts, 1999) ophiolites. Moreover, the association of greenschist-facies assemblages with zones of brittle deformation is common in many ophiolites (e.g., Harper et al., 1988; Nehlig and Juteau, 1988). An exciting find in the Tonga suite was the recovery of five epidosite samples with diabase and plagiogranite protoliths. Mineral-composition, fluid-inclusion, and stable isotope data show that their evolution was very similar to that of epidosites hosted in SSZ ophiolites (Banerjee et al., 2000).

Figure 2. Schematic diagram showing key geologic and hydrothermal characteristics of each of the 11 SSZ ophiolites considered in this study. The column on the far left is a schematic profile through a section of oceanic crust, including sedimentary cover (S), volcanic rocks (V), sheeted dikes (SD), sill-dike complexes (S-D), and plutonic rocks (P). The column for each SSZ ophiolite shows the type of sedimentary cover and the nature of its contact with the volcanic basement, the presence of sulfide deposit(s), metalliferous sedimentary rocks, and epidosites (E) or albite epidosites (AE). The distribution of metamorphic zones is also shown. The numbers on the left side of each column indicate the depth (in km) below the paleoseafloor for the volcanic sequence–sheeted-dike and sheeted-dike–plutonic rock transitions. See Tables 1 and 2 for references.

HOW DO SSZ OPHIOLITES COMPARE WITH MOR CRUST?

Many investigators have highlighted the similarities and differences in the alteration patterns in SSZ ophiolites and modern MOR crust (e.g., Alt, 1995; Pflumio, 1991; Schiffman et al., 1991). Outcrop-scale observations of MOR crust, obtained by submersible and ROV (remotely operated vehicle) surveys, are limited to Hess Deep (fast-spreading East Pacific Rise), MARK (slow-spreading Mid-Atlantic Ridge where it intersects the Kane Fracture Zone), the Atlantis Bank (slow-spreading Southwest Indian Ridge), and the Blanco Fracture Zone (intermediate-rate–spreading Juan de Fuca Ridge). Significant sections of sheeted dike and plutonic sequences have been recovered by drilling only at five locations (Holes 504B, 735B [Southwest Indian Ridge], 894 [Hess Deep], MARK, Izu-Bonin). In general, MOR crustal sequences also show an increase in metamorphic grade with depth, and metamorphic zones generally subparallel the igneous stratigraphy (e.g., Hole 504B, Alt et al., 1996a; Hess Deep, Gillis, 1995). Incipient cracking and hydrothermal flow into the upper plutonic crust at MORs occur at the same high temperatures recorded in SSZ plutonic sequences (e.g., Manning et al., this volume).

Three key differences between MOR and SSZ oceanic crust have been cited in the literature. The first is that the ophiolitic sheeted dike complexes are more pervasively altered than MOR crust. This characteristic is best exemplified by monitoring the degree of alteration of igneous clinopyroxene to actinolite (± hornblende, ± chlorite). In MOR crust, relict clinopyroxene is preserved throughout sheeted dike complexes, with localized zones of complete replacement in the basal dikes (Alt et al., 1996a; Gillis, 1995). By contrast, relict clinopyroxene is common in the upper dikes and rare in the lower dikes of the Troodos ophiolite (Baragar et al., 1989; Baragar et al., 1990). Studies of other ophiolites mention anecdotally that igneous clinopyroxene is more pervasively altered; however, documentation is generally lacking. The bulk-rock geochemistry of hydrothermally altered sheeted dikes has also been used to monitor the pervasiveness of alteration. Comparison of Sr and O isotope data for sheeted dikes from Hole 504B (Alt et al., 1996b) and Hess Deep (Agrinier et al., 1995) with the Troodos (Bickle and Teagle, 1992; Bickle et al., 1999; Schiffman and Smith, 1988; Spooner et al., 1977), Semail (Gregory and Taylor, 1981; McCulloch et al., 1981; Stakes and Taylor, 1992), and Karmøy (Menuge et al., 1989) ophiolites clearly shows that SSZ ophiolites were more pervasively altered at much higher cumulative fluid/rock ratios than MOR crust.

A second key difference between MOR and SSZ crust is the paucity of epidosites in MOR environments. As described above, epidosites are documented for six SSZ ophiolites and are rare in

MOR rock collections. This difference may simply reflect the limited nature of our MOR sample collections from the middle to lower crust. Of particular note, however, is that epidosites develop at higher cumulative fluid/rock ratios than is commonly inferred for the root zones of MOR hydrothermal systems (Von Damm, 1995). Hence, the end products of high-temperature hydrothermal systems are similar, as evidenced by the ore deposits, whereas the root zones appear to be fundamentally different.

The final key difference between MOR crust and SSZ ophiolites is the common occurrence of high-temperature alteration within SSZ volcanic sequences. MOR volcanic sequences are dominated by low-temperature alteration that developed in the off-axis setting (Alt, 1995; Gillis and Robinson, 1988). Only the Troodos volcanic sequence bears any similarity to MOR crust, although many ophiolites show some evidence for an early stage of low-temperature alteration prior to reaching peak-metamorphic temperatures.

SUMMARY

More diversity has been documented for hydrothermal alteration patterns within SSZ ophiolites than within MOR crust. In all cases, more information concerning the nature of hydrothermal alteration and its relationship to local- and regional-scale geology is required in order to adequately address how the geologic setting of SSZ ophiolites influences hydrothermal systems.

1. All SSZ ophiolites show evidence for high-temperature hydrothermal activity. Alteration assemblages are vertically zoned such that metamorphic grade increases with depth. Although metamorphic zones generally subparallel the igneous pseudostratigraphy, the distribution of these zones varies laterally in response to local magmatic and tectonic activity. The degree of alteration does not progressively increase with depth and is generally most intense within the basal sheeted dikes.

2. The thermal structure of the upper crust in SSZ ophiolites is quite variable. SSZ ophiolites that are spatially and temporally associated with volcanic arcs established high temperatures within their volcanic sequences (e.g., Del Puerto, Zambales, Betts Cove). By contrast, low temperatures were maintained within the volcanic sequences of SSZ ophiolites that formed in a subduction-zone system that did not develop a volcanic arc (e.g., Troodos, Kizildag).

3. Volcanic sequences of the Izu-Bonin and Tonga forearcs, which are covered by contemporaneous, thick sections of volcaniclastic sediments, are altered to low-temperature assemblages. This finding suggests that burial by volcaniclastic sediments cannot solely account for the high temperatures recorded in many SSZ volcanic sequences.

4. MOR volcanic sequences are altered to low-temperature mineral assemblages and are less pervasively altered than SSZ crust.

5. SSZ and MOR sheeted dike complexes are altered at similar temperatures; however, SSZ sheeted dikes are more pervasively altered at higher cumulative fluid/rock ratios than those in MOR crust. This difference is particularly evident in epidosite zones where dikes underwent extreme metasomatism.

ACKNOWLEDGMENTS

The observations and ideas presented in this paper come from a rich history of hydrothermal research that evolved from the early study of ophiolites. Gillis's views and biases about ophiolites and mid-ocean ridges have evolved through interactions with numerous colleagues, including J. Cann, D. Kelley, J. Malpas, G. Thompson, and C. Xenophontos. Our research into hydrothermal systems is supported by the Natural Sciences and Engineering Research Council.

REFERENCES CITED

Adamides, N.G., 1990, Hydrothermal circulation and ore deposition in the Troodos ophiolite, Cyprus, *in* Malpas, J., Moores, E.M., Panayiotou, A., and Xenophontos, C., eds., Ophiolites: Oceanic crustal analogues: Nicosia, Cyprus, Geological Survey Department, p. 685–704.

Agrinier, P., Hekinian, R., Bideau, D., and Javoy, M., 1995, O and H stable isotope compositions of oceanic crust and upper mantle rocks exposed in the Hess Deep near the Galapagos Triple Junction: Earth and Planetary Science Letters, v. 136, p. 183–196.

Alabaster, T., and Pearce, J.A., 1985, The interrelationship between magmatic and ore-forming hydrothermal processes in the Oman ophiolite: Economic Geology, v. 80, p. 1–16.

Alabaster, T., Pearce, J.A., and Malpas, J., 1982, The volcanic stratigraphy and petrogenesis of the Oman ophiolite complex: Contributions to Mineralogy and Petrology, v. 81, p. 168–183.

Alexander, R.J., Harper, G.D., and Bowman, J.R., 1993, Oceanic faulting and fault-controlled subseafloor hydrothermal alteration in the sheeted dike complex of the Josephine Ophiolite: Journal of Geophysical Research, v. 98, p. 9731–9759.

Alt, J.C., 1995, Subseafloor processes in mid-ocean ridge hydrothermal systems, *in* Humphris, S.E., Zierenberg, R.A., Mullineaux, L., and Thompson, R.E., eds., Seafloor hydrothermal systems: Washington, D.C., American Geophysical Union Monograph, v. 91, p. 85–114.

Alt, J.C., Honnorez, J., Laverne, C., and Emmermann, R., 1986, Hydrothermal alteration of a 1 km section through the upper oceanic crust, Deep Sea Drilling Project Hole 504B: Mineralogy, chemistry, and evolution of seawater-basalt interactions: Journal of Geophysical Research, v. 91, p. 10309–10335.

Alt, J.C., Laverne, C., Vanko, D.A., Tartarotti, P., Teagle, D.A.H., Bach, W., Zuleger, E., Erzinger, J., Honnorez, J., Pezard, P.A., Becker, K., Salisbury, M.H., and Wilkens, R.H., 1996a, Hydrothermal alteration of a section of upper oceanic crust in the eastern equatorial Pacific: A synthesis of results from Site 504 (DSDP Legs 69, 70, and 83, and ODP Legs 111, 137, 140, and 148), *in* Alt, J.C., Kinoshita, H., Stokking, L.B., Michael, P.J., et al., Proceedings of the Ocean Drilling Program, Scientific Results, Leg 148: College Station, Texas, Ocean Drilling Program, p. 417–434.

Alt, J.C., Teagle, D.A.H., Bach, W., Halliday, A.N., and Erzinger, J., 1996b, Stable and strontium isotopic profiles through hydrothermally altered upper oceanic crust, Hole 504B, *in* Alt, J.C., Kinoshita, H., Stokking, L.B., and Michael, P.J., et al., eds., Proceedings of the Ocean Drilling Program, Scientific Results, Leg 148: College Station, Texas, Ocean Drilling Program, p. 57–69.

Alt, J.C., Teagle, D.A.H., Brewer, T., Shanks, W.C., III, and Halliday, A.N., 1998, Alteration and mineralization of an oceanic forearc and the ophiolite-ocean crust analogy: Journal of Geophysical Research, v. 103, p. 12365–12380.

Amato, F.L., 1964, Stratigraphic paleontology in the Philippines: Philippine Geology, v. 19, p. 1–24.

Apted, M.J., and Liou, J.G., 1983, Phase relations among greenschist, epidote-amphibolite, and amphibolite in a basaltic system: American Journal of Science, v. 283A, p. 328–354.

Banerjee, N.R., Gillis, K.M., and Muehlenbachs, K., 2000, Discovery of epidosites in a modern oceanic setting, the Tonga forearc: Geology, v. 28, p. 151–154.

Banerjee, N.R., Gillis, K.M., and Muehlenbachs, K., 1999. Patterns of hydrothermal alteration in the Tonga forearc crust [abs.]: Eos, Transactions of the American Geophysical Union, v. 80, p. 1137.

Baragar, W.R.A., Lambert, M.B., Baglow, N., and Gibson, I.L., 1989, Sheeted dykes from CY-4 and surface sections: Troodos ophiolite, *in* Gibson, I.L., Malpas, J., Robinson, P.T., and Xenophontos, C., eds., Cyprus Crustal Study Project: Initial Report, Hole CY-4: Ottawa, Geological Survey of Canada Paper 88–9, p. 69–106.

Baragar, W.R.A., Lambert, M.B., Baglow, N., and Gibson, I.L., 1990, The sheeted dyke zone in the Troodos ophiolite, *in* Malpas, J., Moores, E.M., Panayiotou, A., and Xenophontos, C., eds., Ophiolites: Oceanic crustal analogues: Nicosia, Cyprus, Geological Survey Department, p. 37–51.

Bédard, J.H., Lauziere, K., Tremblay, A., and Sangster, A., 1998, Evidence for forearc seafloor-spreading from the Betts Cove ophiolite, Newfoundland: Oceanic crust of boninitic affinity: Tectonophysics, v. 284, p. 233–245.

Bednarz, U., and Schmincke, H.U., 1989, Mass transfer during sub-seafloor alteration of the upper Troodos crust (Cyprus): Contributions to Mineralogy and Petrology, v. 102, p. 93–101.

Bednarz, U., and Schmincke, H.U., 1994, Petrological and chemical evolution of the northeastern Troodos extrusive series, Cyprus: Journal of Petrology, v. 35, p. 489–523.

Beiersdorfer, R.E., 1993, Metamorphism of a Late Jurassic volcano-plutonic arc, northern California, USA: Journal of Metamorphic Geology, v. 11, p. 415–428.

Beiersdorfer, R.E., and Day, H.W., 1995, Mineral paragenesis of pumpellyite in low-grade mafic rocks, *in* Schiffman, P., and Day, H.W., eds., Low-grade metamorphism of mafic rocks: Boulder, Colorado, Geological Society of America Special Paper 296, p. 5–28.

Bettison-Varga, L., Varga, R.J., and Schiffman, P., 1992, Relation between ore-forming hydrothermal systems and extensional deformation in the Solea graben spreading center, Troodos ophiolite, Cyprus: Geology, v. 20, p. 987–990.

Bettison-Varga, L., Schiffman, P., and Janecky, D.R., 1995, Fluid-rock interaction in the hydrothermal upflow zone of the Solea graben, Troodos ophiolite, Cyprus, *in* Schiffman, P., and Day, H.W., eds., Low-grade metamorphism of mafic rocks: Boulder, Colorado, Geological Society of America Special Paper 296, p. 81–100.

Bickle, M.J., and Teagle, D.A.H., 1992, Strontium alteration in the Troodos ophiolite: Implications for fluid fluxes and geochemical transport in mid-ocean ridge hydrothermal systems: Earth and Planetary Science Letters, v. 113, p. 219–237.

Bickle, M.J., Teagle, D.A.H., Beynon, J., and Chapman, H.J., 1998, The structure and controls on fluid-rock interactions in ocean ridge hydrothermal systems: Constraints from the Troodos ophiolite, *in* Mills, R.A., and Harrison, K., eds., Modern ocean floor processes and the geological record: Geological Society of London Special Publication 148, p. 127–152.

Blome, C.D., and Irwin, W.P., 1985, Equivalent radiolarian ages from ophiolitic terranes of Cyprus and Oman: Geology, v. 13, p. 401–404.

Bloomer, S., Taylor, B., MacLeod, C.J., Stern, R.J., Fryer, P., Hawkins, J.W., and Johnson, L., 1995, Early arc volcanism and the ophiolite problem: A perspective from drilling in the western Pacific, *in* Taylor, B., and Natland, J., eds., Active margins and marginal basins of the western Pacific: Washington, D.C., American Geophysical Union, Geophysical Monograph 88, p. 1–30.

Bloomer, S.H., Wright, D., MacLeod, C.J., Tappin, D., and Shipboard Scientific Party, 1996, Geology of the Tonga forearc: A supra-subduction zone ophiolite [abs.]: Eos, Transactions of the American Geophysical Union, v. 77, p. 325.

Bloomer, S.H., Falloon, T.J., Petcovic, H., Nielson, R.L., and Duncan, R.A., 1998, Petrology and geochemistry of volcanic and plutonic rocks from the Tonga forearc, southwest Pacific [abs.]: Eos, Transactions of the American Geophysical Union, v. 79, p. 420.

Boyle, J.F., 1990, The composition and origin of oxide metalliferous sediments from the Troodos ophiolite, Cyprus, *in* Malpas, J., Moores, E.M., Panayiotou, A., and Xenophontos, C., eds., Ophiolites: Oceanic crustal analogues: Nicosia, Cyprus, Geological Survey Department, p. 705–717.

Boyle, J.F., and Robertson, A.H.F., 1984, Evolving metallogenesis at the Troodos spreading axis, *in* Gass, I.G., Lippard, S.J., and Shelton, A.W., eds., Ophiolites and oceanic lithosphere: Geological Society of London Special Publication 13, p. 169–181.

Brouxel, M., and Lapierre, H., 1988, Geochemical study of an early Paleozoic island-arc–back-arc basin system. Part 1: The Trinity ophiolite (northern California): Geological Society of America Bulletin, v. 100, p. 1111–1119.

Clift, P.D., MacLeod, C.J., Tappin, D.R., Wright, D.J., and Bloomer, S.H., 1998, Tectonic controls on sedimentation and diagenesis in the Tonga Trench and forearc, southwest Pacific: Geological Society of America Bulletin, v. 110, p. 483–496.

Coish, R.A., 1977, Ocean floor metamorphism in the Betts Cove Ophiolite, Newfoundland: Contributions to Mineralogy and Petrology, v. 60, p. 255–270.

Constantinou, G., 1976, Genesis of the conglomerate structure, porosity, and collomorphic textures of the massive sulphide ores of Cyprus: Geological Association of Canada Special Paper, No. 14, p. 187–210.

Constantinou, G., and Govett, G.J.S., 1972, Genesis of sulphide deposits, ochre, and umber of Cyprus: Transactions of the Institution of Mining and Metallurgy, v. 81, p. B34-B46.

Constantinou, G., and Govett, G.J.S., 1973, Geology, geochemistry, and genesis of Cyprus sulfide deposits: Economic Geology, v. 68, p. 843–858.

Coombs, D.S., Ellis, A.J., Fyfe, W.S., and Taylor, A.M., 1959, The zeolite facies, with comments on the interpretation of hydrothermal syntheses: Geochimica et Cosmochimica Acta, v. 17, p. 53–107.

Davis, E.E., Mottl, M.J., Fisher, A.T., et al., 1992, Proceedings of the Ocean Drilling Program, Initial Reports, Leg 139: College Station, Texas, Ocean Drilling Program, 1026 p.

Davis, E.E., Fisher, A.T., Firth, J.V., et al., 1997, Proceedings of the Ocean Drilling Program, Initial Reports, Leg 168: College Station, Texas, Ocean Drilling Program, 470 p.

Delaloye, M., and Wagner, J.J., 1984, Ophiolites and volcanic activity near the western edge of the Arabian plate, *in* Dixon, J.F., and Robertson, A.H.F., eds., The geological evolution of the eastern Mediterranean: Geological Society of London Special Publication 17, p. 225–233.

Dilek, Y., and Moores, E.M., 1990, Regional tectonics of the eastern Mediterranean ophiolites, *in* Malpas, J., Moores, E., Panayiotou, A., and Xenophontos, C., eds., Ophiolites: Oceanic crustal analogues: Nicosia, Cyprus, Geological Survey Department, p. 295–309.

Dilek, Y., and Thy, P., 1998, Structure, petrology, and seafloor spreading tectonics of the Kizildag Ophiolite, Turkey, *in* Mills, R.A., and Harrison, K., eds., Modern ocean floor processes and the geological record: Geological Society of London Special Publication 148, p. 43–69.

Dilek, Y., Moores, E.M., Delaloye, M., and Karson, J.A., 1991, Amagmatic extension and tectonic denudation in the Kizildag Ophiolite, southern Turkey: Implications for the evolution of Neotethyan oceanic crust, *in* Peters, T., Nicolas, A., and Coleman, R.G., eds., Ophiolite genesis and evolution of the oceanic lithosphere: London, Kluwer Academic Publishers, p. 485–500.

Dunning, G.R., and Krogh, T.E., 1985, Geochronology of ophiolites of the Newfoundland Appalachians: Canadian Journal of Earth Science, v. 22, p. 1659–1670.

Dunning, G.R., and Pedersen, R.B., 1988, U/Pb ages of ophiolites and arc-related plutons of the Norwegian Caledonides: Implications for the development of Iapetus: Contributions to Mineralogy and Petrology, v. 98, p. 13–23.

Eddy, C.A., Dilek, Y., Hurst, S., and Moores, E.M., 1998, Seamount formation and associated caldera complex and hydrothermal mineralization in ancient oceanic crust, Troodos ophiolite (Cyprus): Tectonophysics, v. 292, p. 189–210.

Erendil, M., 1983, Petrology and structure of the upper crustal units of the Kizildag ophiolite, *in* Tekeli, D., and Goncüoglu, M.C., eds., Geology of the Taurus Belt: Ankara, Turkey, Maden Tetkik ve Arama Enstitüsü, p. 269–284.

Ernewein, M., Pflumio, C., and Whitechurch, H., 1988, The death of an accretion zone as evidenced by the magmatic history of the Sumail ophiolite (Oman): Tectonophysics, v. 151, p. 247–274.

Eskola, P., 1920, The minerals facies of rocks: Norsk Geologisk Tidsskrift, v. 6, p. 143–194.

Evarts, R.C., and Schiffman, P., 1983, Submarine hydrothermal metamorphism of the Del Puerto ophiolite, California: American Journal of Science, v. 283, p. 289–340.

Falloon, T.J., Bloomer, S.H., and Crawford, A.J., 1997, Boninites recovered from the Tonga forearc [abs.]: Eos, Transactions of the American Geophysical Union, v. 78, p. 815.

Fisher, A.T., 1998, Permeability within basaltic oceanic crust: Reviews of Geophysics, v. 36, p. 143–182.

Fleet, A.J., and Robertson, A.H.F., 1980, Ocean-ridge metalliferous and pelagic sediments of the Semail Nappe, Oman: Journal of the Geological Society of London, v. 137, p. 403–422.

Fouquet, Y., Zierenberg, R.A., Miller, D.J., et al., 1998, Proceedings of the Ocean Drilling Program, Initial Reports, Leg 169: College Station, Texas, Ocean Drilling Program, 592 p.

Frey, M., de Capitani, C., and Liou, J.G., 1991, A new petrogenetic grid for low-grade metabasites: Journal of Metamorphic Geology, v. 9, p. 497–509.

Furnes, H., Pedersen, R.B., and Stillman, C.J., 1988, The Leka ophiolite complex, central Norwegian Caledonides: Field characteristics and geotectonic significance: Journal of the Geological Society of London, v. 145, p. 401–412.

Furnes, H., Pedersen, R.B., Hertogen, J., and Albrektsen, B.A., 1992, Magma development of the Leka ophiolite complex, central Norwegian Caledonides: Lithos, v. 27, p. 259–277.

Galley, A.G., and Koski, R.A., 1999, Setting and characteristics of ophiolite-hosted volcanogenic massive sulfide deposits, *in* Hannington, M.D., and Barry, C.T., eds., Volcanic-associated massive sulfide deposits: Processes and examples in modern and ancient settings: Denver, Colorado, Economic Geology Publications Company, p. 215–236.

Gass, I.G., and Smewing, J.D., 1973, Intrusion, extrusion, and metamorphism at constructive margins: Evidence from the Troodos Massif, Cyprus: Nature, v. 242, p. 26–29.

Geary, E.E., and Kay, R.W., 1983, Petrological and geochemical documentation of ocean floor metamorphism in the Zambales Ophiolite, Philippines, *in* Hayes, D.E., ed., The tectonic and geologic evolution of Southeast Asian seas and islands, Part 2: Washington, D.C., American Geophysical Union, Geophysical Monograph 27, p. 139–156.

Geary, E.E., Kay, R.W., Reynolds, J.C., and Kay, S.M., 1989, Geochemistry of mafic rocks from the Coto Block, Zambales ophiolite, Philippines: Trace element evidence for two stages of crustal growth: Tectonophysics, v. 168, p. 43–63.

Gillis, K.M., 1995, Controls on hydrothermal alteration in a section of fast-spreading oceanic crust: Earth and Planetary Science Letters, v. 134, p. 473–489.

Gillis, K.M., 1997, Distribution of porosity in a section of upper oceanic crust exposed in the Troodos ophiolite: Journal of Geophysical Research, v. 102, p. 10133–10149.

Gillis, K.M., and Roberts, M., 1999, Cracking at the magma-hydrothermal transition: Evidence from the Troodos ophiolite: Earth and Planetary Science Letters, v. 169, p. 227–244.

Gillis, K.M., and Robinson, P.T., 1985, Low-temperature alteration of the extrusive sequence, Troodos ophiolite, Cyprus: Canadian Mineralogist, v. 23, p. 431–441.

Gillis, K.M., and Robinson, P.T., 1988, Distribution of alteration zones in the upper oceanic crust: Geology, v. 16, p. 262–266.

Gillis, K.M., and Robinson, P.T., 1990, Patterns and processes of alteration in the lavas and dykes of the Troodos ophiolite, Cyprus: Journal of Geophysical Research, v. 95, p. 21523–21548.

Gregory, R.T., and Taylor, H.P., 1981, An oxygen isotope profile in a section of Cretaceous oceanic crust, Samail ophiolite, Oman: Evidence for $\delta^{18}O$ buffering of the oceans by deep (>5km) seawater-hydrothermal circulation at mid-ocean ridges: Journal of Geophysical Research, v. 86, p. 2737–2755.

Harper, G.D., 1984, The Josephine ophiolite, northwestern California: Geological Society of America Bulletin, v. 95, p. 1009–1026.

Harper, G.D., 1988, Episodic magma chambers and amagmatic extension in the Josephine ophiolite: Geology, v. 16, p. 831–834.

Harper, G.D., 1995, Pumpellyosite and prehnitite associated with epidosite in the Josephine ophiolite: Ca metasomatism during upwelling of hydrothermal fluids at a spreading axis, *in* Schiffman, P., and Day, H.W., eds., Low-grade metamorphism of mafic rocks: Boulder, Colorado, Geological Society of America Special Paper 296, p. 101–122.

Harper, G.D., Bowman, J.R., and Kuhns, R., 1988, A field, chemical, and stable isotope study of subseafloor metamorphism of the Josephine ophiolite, California-Oregon: Journal of Geophysical Research, v. 93, p. 4625–4656.

Hawkins, J.W., and Evans, C.A., 1983, Geology of the Zambales Range, Luzon, Philippine Islands: Ophiolite derived from an island arc–back arc basin pair, *in* Hayes, D.E., ed., The tectonic and geologic evolution of southeast Asian seas and islands, Part 2: Washington, D.C., American Geophysical Union, Geophysical Monograph 27, p. 95–123.

Haymon, R.M., Koski, R.A., and Abrams, M.J., 1989, Hydrothermal discharge zones beneath massive sulfide deposits mapped in the Oman ophiolite: Geology, v. 17, p. 531–535.

Heaton, T.H.E., and Sheppard, S.M.F., 1976, Hydrogen and oxygen isotope evidence for sea-water-hydrothermal alteration and ore deposition, Troodos complex, Cyprus, *in* Volcanic processes in ore genesis: London, Institution of Mining and Metallurgy, p. 42–57.

Hopson, C.A., Mattinson, J.M., and Pessagno, E.A., 1981, Coast Range Ophiolite, western California, *in* Ernst, W.G., ed., The tectonic development of California, v. 1: Englewood Cliffs, Prentice-Hall, p. 418–510.

Humphris, S.E., Morrison, M.A., and Thompson, R.N., 1978, Influence of rock crystallisation history upon subsequent lanthanide mobility during hydrothermal alteration of basalts: Chemical Geology, v. 23, p. 125–137.

Humphris, S.E., Herzig, P.M., Miller, D.J., et al., 1995, The internal structure of an active sea-floor massive sulphide deposit: Nature, v. 377, p. 713–716.

Hunter, A.G., Kempton, P.D., and Greenwood, P., 1999, Low-temperature fluid-rock interaction: An isotopic and mineralogical perspective of upper crustal evolution, eastern flank of the Juan de Fuca Ridge (JdFR), ODP Leg 168: Chemical Geology, v. 155, p. 3–28.

Ixer, R.A., Alabaster, T., and Pearce, J.A., 1984, Ore petrography and geochemistry of massive sulfide deposits within the Semail Ophiolite: Transactions of the Institute of Mining and Metallurgy, v. 93, p. 114–124.

Jacobsen, S.B., Quick, J.E., and Wasserburg, G.J., 1984, A Sr and Nd isotopic study of the Trinity Peridotite: Implications for mantle evolution: Earth and Planetary Science Letters, v. 68, p. 361–378.

Jenner, G.A., and Fryer, B.J., 1980, Geochemistry of the Upper Snooks Arm Group, Burlington Peninsula, Newfoundland: Canadian Journal of Earth Science, v. 17, p. 888–900.

Jones, G., and Robertson, A.H.F., 1991, Tectono-stratigraphy and evolution of the Mesozoic Pindos ophiolite and related units, northwestern Greece: Journal of the Geological Society of London, v. 148, p. 267–288.

Kelley, D.S., and Robinson, P.T., 1990, Development of a brine-dominated hydrothermal system at temperatures of 400–500 °C in the upper level plutonic sequence, Troodos ophiolite, Cyprus: Geochimica et Cosmochimica Acta, v. 54, p. 653–661.

Kelley, D.S., Robinson, P.T., and Malpas, J.G., 1992, Processes of brine generation and circulation in the oceanic crust: Fluid inclusion evidence from the Troodos Ophiolite, Cyprus: Journal of Geophysical Research, v. 97, p. 9307–9322.

Kelman, M.C., 1998, Hydrothermal alteration of a supra-subduction zone ophiolite analog: Tonga, southwest Pacific [M.Sc. thesis]: Corvallis, Oregon State University, 193 p.

Kimball, K.L., 1988, High-temperature hydrothermal alteration of ultramafic cumulates from the base of the sheeted dikes in the Josephine ophiolite, NW California: Journal of Geophysical Research, v. 93, p. 4675–4687.

Kostopoulos, D.K., 1989, Geochemistry, petrogenesis, and tectonic setting of the Pindos Ophiolite, NW Greece [Ph.D. thesis]: Newcastle Upon Tyne, England, University of Newcastle Upon Tyne, 528p.

Kuhns, R.J., and Baitis, H.W., 1987, Preliminary study of the Turner-Albright Zn-Cu-Ag-Au-Co massive sulfide deposit, Josephine County, Oregon: Economic Geology, v. 81, p. 1362–1376.

Lécuyer, C., and Fourcade, S., 1991, Oxygen isotope evidence for multi-stage hydrothermal alteration at a fossil slow-spreading center: The Silurian Trinity ophiolite (California, USA): Chemical Geology, v. 87, p. 231–246.

Lécuyer, C., Brouxel, M., and Albarede, F., 1990, Elemental fluxes during hydrothermal alteration of the Trinity ophiolite (California, USA) by seawater: Chemical Geology, v. 89, p. 87–115.

Liou, J.G., Kuniyoshi, S., and Ito, K., 1974, Experimental studies of the phase relations between greenschist and amphibolite in a basaltic system: American Journal of Science, v. 274, p. 613–632.

Liou, J.G., Maruyama, S., and Cho, M., 1987, Very low-grade metamorphism of volcanic and volcaniclastic rocks, *in* Frey, M., ed., Low temperature metamorphism: Glasgow, Blackie and Sons, p. 59–113.

Liou, J.G., de Capitani, C., and Frey, M., 1991, Zeolite equilibria in the system $CaAlSi_2O_8$-SiO_2-H_2O: New Zealand Journal of Geology and Geophysics, v. 34, p. 293–301.

Lydon, J.W., and Galley, A.G., 1986, The geochemical and mineralogical zonation of the Mathiati alteration pipe and its genetic significance, *in* Gallagher, M.J., Ixer, R.A., Neary, C.R., and Prichard, H.M., eds., Metallogeny of basic and ultrabasic rocks: London, London Institute of Mining and Metallurgy, p. 46–68.

Lytwyn, J.N., and Casey, J.F., 1993, The geochemistry and petrogenesis of volcanics and sheeted dikes from the Hatay (Kizildag) ophiolite, southern Turkey: Possible formation with the Troodos ophiolite, Cyprus, along fore-arc spreading centers: Tectonophysics, v. 223, p. 237–272.

Malpas, J., 1990, Crustal accretionary processes in the Troodos ophiolite, Cyprus: Evidence from field mapping and deep crustal drilling, *in* Malpas, J., Moores, E.M., Panayiotou, A., and Xenophontos, C., eds., Ophiolites: Oceanic crustal analogues: Nicosia, Cyprus, Geological Survey Department, p. 65–74.

McCulloch, M.T., Gregory, R.T., Wasserburg, G.J., and Taylor, H.P., Jr., 1981, Sm-Nd, Rb-Sr, and $^{18}O/^{16}O$ isotopic systematics in an oceanic crustal section: Evidence from the Samail ophiolite: Journal of Geophysical Research, v. 86, p. 2721–2735.

McDougall, I., 1994, Dating of rhyolitic glass in the Tonga forearc (Hole 841B), *in* Hawkins, J., Parson, L., Allan, J., et al., eds., Proceedings of the Ocean Drilling Program, Scientific Results, Leg 135: College Station, Texas, Ocean Drilling Program, p. 923.

Menuge, J., Pedersen, R.B., and Furnes, H., 1989, Seawater alteration of the Karmøy ophiolite Complex, SW Norway: Nd and Sr isotopic evidence: Norsk Geologisk Tidsskrift, v. 69, p. 191–200.

Mills, R.A., and Elderfield, H., 1995, Hydrothermal activity and the geochemistry of metalliferous sediment, *in* Humphris, S.E., Zierenberg, R.A., Mullineaux, L.S., and Thomson, R.E., eds., Seafloor hydrothermal systems: Washington, D.C., American Geophysical Union., Geophysical Monograph 91, p. 392–407.

Mitchell, J.G., Peate, D.W., Murton, B.J., Pearce, J.A., Arculus, R.J., and van der Laan, S., 1992, K-Ar dating of samples from Sites 782 and 786 (Leg 125): The Izu-Bonin forearc region, *in* Fryer, P., Pearce, J.A., Stokking, L.B., et al., eds., Proceedings of the Ocean Drilling Program, Scientific Results, Leg 125: College Station, Texas, Ocean Drilling Program, p. 203–210.

Muenow, D.W., Garcia, M.O., Aggrey, K.E., Bednarz, U., and Schmincke, H.U., 1990, Volatiles in submarine glasses as a discriminant of tectonic origin: Application to the Troodos ophiolite: Nature, v. 343, p. 159.

Mukasa, S.B., and Ludden, J.N., 1987, Uranium-lead isotopic ages of plagiogranites from the Troodos ophiolite, Cyprus and their tectonic significance: Geology, v. 15, p. 825–828.

Murton, B.J., Peete, D.W., Arculus, R.J., Pearce, J.A., and van der Laan, S., 1992, Trace element geochemistry of volcanic rocks from Site 786: The Izu-Bonin forearc, *in* Fryer, P., Pearce, J.A., Stokking, L.B., et al., eds., Proceedings of Ocean Drilling Program, Scientific Results, Leg 125: College Station, Texas, Ocean Drilling Program, p. 211–236.

Nehlig, P., 1991, Salinity of oceanic hydrothermal fluids: A fluid inclusion study: Earth and Planetary Science Letters, v. 102, p. 310–325.

Nehlig, P., and Juteau, T., 1988, Deep crustal seawater penetration and circulation at ocean ridges: Evidence from the Oman ophiolite: Marine Geology, v. 84, p. 209–228.

Nehlig, P., Juteau, T., Bendel, V., and Cotten, J., 1994, The root zones of oceanic hydrothermal systems: Constraints from the Samail ophiolite (Oman): Journal of Geophysical Research, v. 99, p. 4703–4713.

Nicolas, A., 1989, Structures of ophiolites and dynamics of oceanic lithosphere: Dordrecht, Netherlands, Kluwer Academic Publishers, 367 p.

Pallister, J.S., and Hopson, C.A., 1981, Samail ophiolite plutonic suite: Field relations, phase variation, cryptic variation and layering, and a model of a spreading ridge magma chamber: Journal of Geophysical Research, v. 86, p. 2593–2644.

Pearce, J.A., and Peate, D.W., 1995, Tectonic implications of the composition of volcanic arc magmas: Annual Reviews of Earth and Planetary Science, v. 23, p. 251–285.

Pearce, J.A., Alabaster, T., Shelton, A.W., and Searle, M.P., 1981, The Oman ophiolite as a Cretaceous arc-basin complex: Evidence and implications: Philosophical Transactions of the Royal Society of London, v. 300, p. 299–317.

Pearce, J.A., Lippard, S.J., and Roberts, S., 1984, Characteristics and tectonic significance of supra-subduction zone ophiolites, *in* Kokelaar, B.P., and Howells, M.F., eds., Marginal basin geology: Volcanic and associated sedimentary and tectonic processes in modern and ancient marginal basins: Geological Society of London Special Publication 16, p. 77–94.

Pedersen, R.B., 1986, The nature and significance of magma chamber margins in ophiolites: Examples from the Norwegian Caledonides: Earth and Planetary Science Letters, v. 77, p. 100–112.

Pedersen, R.B., and Furnes, H., 1991, Geology, magmatic affinity, and geotectonic environment of some Caledonian Ophiolites in Norway: Journal of Geodynamics, v. 13, p. 183–203.

Pedersen, R.B., and Hertogen, J., 1990, Magmatic evolution of the Karmøy ophiolite Complex, SW Norway: Relationships between MORB-IAT-boninitic-calc-alkaline and alkaline magmatism: Contributions to Mineralogy and Petrology, v. 104, p. 277–293.

Pedersen, R.B., and Malpas, J., 1984, The origin of oceanic plagiogranites from the Karmøy ophiolite, Western Norway: Contributions to Mineralogy and Petrology, v. 88, p. 36–52.

Pedersen, R.B., Furnes, H., and Dunning, G.R., 1988, Some Norwegian ophiolite complexes reconsidered: Progress in studies of the lithosphere in Norway: Norges Geologiske Undersoekelse Special Publication, v. 3, p. 80–85.

Pflumio, C., 1991, Evidences for polyphased oceanic alteration of the extrusive sequence of the Semail ophiolite from the Salahi Block (Northern Oman), *in* Peters, T., Nicolas, A., and Coleman, R.G., eds., Ophiolite genesis and evolution of the oceanic lithosphere: Proceedings of the Ophiolite Conference, Muscat, Oman, January 1990, Volume 5, Petrology and Structural Geology: Muscat, Kluwer Academic Publishers, p. 313–351.

Pinto-Auso, M., and Harper, G.D., 1985, Sedimentation, metallogenesis, and tectonic origin of the basal Galice Formation overlying the Josephine ophiolite, northwestern California: Journal of Geology, v. 93, p. 713–725.

Regba, M., Agrinier, P., Plumio, C., and Loubet, M., 1990, Evidences for polyphased oceanic alteration of the extrusive sequence of the Semail ophiolite from the Salahi Block (Northern Oman), *in* Peters, T., Nicolas, A., and Coleman, R.G., eds., Ophiolite genesis and evolution of the oceanic lithosphere: Proceedings of the Ophiolite Conference, Muscat, Oman, January 1990, Volume 5, Petrology and Structural Geology: Muscat, Kluwer Academic Publishers, p. 353–383.

Richardson, C.J., Cann, J.R., Richards, H.G., and Cowan, J.G., 1987, Metal-depleted root zones of the Troodos ore-forming hydrothermal systems, Cyprus: Earth and Planetary Science Letters, v. 84, p. 243–253.

Robertson, A.H.F., 1975, Cyprus umbers: Basalt-sediment relationships on a Mesozoic ocean ridge: Journal of the Geological Society of London, v. 131, p. 511–531.

Robertson, A.H.F., 1976, Origins of ochres and umbers: Evidence from Skouriotissa, Troodos Massif, Cyprus: Transactions of the Institute of Mining and Metallurgy, v. 85, p. B245-B251.

Robertson, A.H.F., 1986, Geochemistry and tectonic implications of metalliferous and volcaniclastic sedimentary rocks associated with Late Cretaceous ophiolitic extrusives in the Hatay area, southern Turkey: Ofioliti, v. 11, p. 121–140.

Robertson, A.H.F., and Fleet, A.J., 1986, Geochemistry and paleo-oceanography of metalliferous and pelagic sediments from the late Cretaceous Oman ophiolite: Marine and Petroleum Geology, v. 3, p. 315–337.

Robertson, A.H.F., and Hudson, J.D., 1974, Pelagic sediments in the Cretaceous and Tertiary history of the Troodos Massif, Cyprus: Special Publication of the International Association of Sedimentologists, v. 1, p. 403–436.

Robertson, A.H.F., and Varnavas, S.P., 1993, The origin of hydrothermal metalliferous sediments associated with the Early Mesozoic Othris and Pindos ophiolites, mainland Greece: Sedimentary Geology, v. 83, p. 87–113.

Robertson, A.H.F., and Xenophontos, C., 1993, Development of concepts concerning the Troodos ophiolite and adjacent units in Cyprus, *in* Prichard, H.M., Alabaster, T., Harris, N.B.W., and Neary, C.R., eds., Magmatic processes and plate tectonics: Geological Society of London Special Publication 76, p. 85–119.

Rogers, N.W., MacLeod, C.J., and Murton, B.J., 1989, Boninitic lavas from the Limassol Forest Complex, Cyprus, *in* Crawford, A.J., ed., Boninites: London, Unwin Hyman, p. 288–313.

Saunders, C.M., and Strong, D.F., 1986, Alteration-zonation related to variations in water/rock ratio at the Betts Cove ophiolite massive sulphide deposit, Newfoundland, Canada, *in* Gallagher, M.J., Ixer, R.A., Neary, G.R., and Prichard, H.M., eds., Metallogeny of basic and ultrabasic rocks: London, Institute of Mining and Metallurgy, p. 161–175.

Schiffman, P., and Day, H.W., 1999, Petrological methods for the study of very low-grade metabasites, *in* Frey, M., and Robinson, D., eds., Low-grade metamorphism: Oxford, Blackwell Science Ltd., p. 108–142.

Schiffman, P., and Smith, B.M., 1988, Petrology and oxygen isotope geochemistry of a fossil seawater hydrothermal system within the Solea graben, Northern Troodos Ophiolite, Cyprus: Journal of Geophysical Research, v. 93, p. 4612–4624.

Schiffman, P., and Staudigel, H., 1994, Hydrothermal alteration of a seamount complex on La Palma, Canary Islands: Implications for metamorphism of accreted terranes: Geology, v. 22, p. 151–154.

Schiffman, P., Smith, B.M., Varga, R.J., and Moores, E.M., 1987, Geometry, conditions, and timing of off-axis hydrothermal metamorphism and ore-deposition in the Solea graben: Nature, v. 325, p. 423–425.

Schiffman, P., Evarts, R.C., Williams, A.E., and Pickthorn, W.J., 1991, Hydrothermal metamorphism in oceanic crust from the Coast Range ophiolite of California: Fluid-rock interaction in a rifted island arc, *in* Peters, T., Nicolas, A., and Coleman, R.G., eds., Ophiolite genesis and evolution of the oceanic lithosphere: Proceedings of the Ophiolite Conference, Muscat, Oman, January 1990, Volume 5, Petrology and Structural Geology: Muscat, Kluwer Academic Publishers, p. 399–425.

Schweller, W.J., Roth, P.H., Karig, D.E., and Bachman, S.B., 1984, Sediment history and biostratigraphy of ophiolite-related Tertiary sediments, Luzon, Philippines: Geological Society of America Bulletin, v. 95, p. 1333–1342.

Scott, J.L., 1991, Mineralization in the Karmøy ophiolite, southwest Norway [M.Sc. thesis]: St. John's, Memorial University of Newfoundland, 221p.

Selcuk, H., 1981, Etude geologique de la partie meridionale du Hatay (Turquie) [Ph.D. thesis]: Geneve, Université de Geneve, 116 p.

Shervais, J.W., 1990, Island arc and ocean crust ophiolites: Contrasts in the petrology, geochemistry, and tectonic style of ophiolite assemblages in the California Coast Ranges, *in* Malpas, J., Moores, E.M., Panayiotou, A., and Xenophontos, C., eds., Ophiolites: Oceanic crustal analogues: Nicosia, Cyprus, Geological Survey Department, p. 507–520.

Shipboard Scientific Party, 1990a, Site 786, *in* Fryer, P., Pearce, J.A., et al., Proceedings of the Ocean Drilling Program, Initial Reports, Leg 125: College Station, Texas, Ocean Drilling Program, p. 313–363.

Shipboard Scientific Party, 1990b, Site 782, *in* Fryer, P., Pearce, J.A., et al., Proceedings of the Ocean Drilling Program, Initial Reports, Leg 125: College Station, Texas, Ocean Drilling Program, p. 197–252.

Shipboard Scientific Party, 1992, Site 841, *in* Parson, L.M., Hawkins, J.W., Allan, J.F., et al., Proceedings of the Ocean Drilling Program, Initial Reports, Leg 135: College Station, Texas, Ocean Drilling Program, p. 571–660.

Shipboard Scientific Party, 1998, Escanaba Trough (Site 1037), *in* Fouquet, Y., Zierenberg, R.A., Miller, D.J., et al., Proceedings of the Ocean Drilling Program, Initial Reports, Leg 169: College Station, Texas, Ocean Drilling Program, p. 205–252.

Smith, W.H.F., and Sandwell, D.T., 1997, Global sea floor topography from satellite altimetry and ship depth soundings: Science, v. 277, p. 1956–1962.

Spooner, E.T.C., Chapman, H.J., and Smewing, J.D., 1977, Strontium isotopic contamination and oxidation during ocean floor hydrothermal metamorphism of the ophiolitic rocks of the Troodos Massif, Cyprus: Geochimica et Cosmochimica Acta, v. 41, p. 873–890.

Spray, J.G., Bébien, J., Rex, D.C., and Roddick, J., 1984, Age constraints on the igneous and metamorphic evolution of the Hellenic-Dinaric ophiolites, *in* Dixon, J.E., and Robertson, A.H.F., eds., The geological evolution of the eastern Mediterranean: Geological Society of London Special Publication 17, p. 619–627.

Stabell, B., Ali, J., Ciampo, G., Milner, G., Wang, Y.-J., and Xu, Y., 1992, Biostratigraphic summary, Leg 125, *in* Fryer, P., Pearce, J.A., Stokking, L.B., et al., Proceedings of the Ocean Drilling Program, Scientific Results, Leg 125: College Station, Texas, Ocean Drilling Program, p. 615–622.

Stakes, D.S., and Taylor, H.P., 1992, The Northern Samail Ophiolite: An oxygen isotope, microprobe and field study: Journal of Geophysical Research, v. 97, p. 7043.

Staudigel, H., Hart, S.R., and Richardson, S.H., 1981, Alteration of the oceanic crust: Processes and timing: Earth and Planetary Science Letters, v. 52, p. 311–327.

Strong, D.F., and Saunders, C.M., 1988, Ophiolitic sulfide mineralization at Tilt Cove, Newfoundland: Controls by upper mantle and crustal processes: Economic Geology, v. 83, p. 239–255.

Sturt, B.A., Andersen, T.B., and Furnes, H., 1985, The Skei Group, Leka: An unconformable clastic sequence overlying the Leka ophiolite, *in* Gee, D.G., and Sturt, B.A., eds., The Caledonide Orogen–Scandinavia and related areas: New York, Wiley, p. 395–405.

Tilton, G.R., Hopson, C.A., and Wright, J.E., 1981, Uranium-lead isotopic ages of the Semail ophiolite, Oman, with implications to Tethyan ocean ridge tectonics: Journal of Geophysical Research, v. 86, p. 2763–2776.

Tippit, P.R., Pessagno, E.A.J., and Smewing, J.D., 1981, The biostratigraphy of sediments in the volcanic unit of the Semail ophiolite: Journal of Geophysical Research, v. 86, p. 2756–2762.

Tveit, R., Furnes, H., and Pedersen, R.B., 1993, Geological and geochemical development of the submarine volcanic sequence on Storøya, Leka ophiolite Complex, North Trøndelag: Norsk Geologisk Tidsskrift, v. 73, p. 81–94.

Valsami, E., and Cann, J.R., 1992, Mobility of rare earth elements in zones of intense hydrothermal alteration in the Pindos ophiolite, Greece, *in* Parson, L.M., Murton, B.J., and Browning, P., eds., Ophiolites and their modern ocean analogues: Geological Society of London Special Publication 60, p. 219–232.

Varga, R.J., Gee, J.S., Bettison-Varga, L., Anderson, R.S., and Johnson, C.L., 1999, Early formation of seafloor hydrothermal systems during structural extension: Paleomagnetic evidence from the Troodos Ophiolite: Earth and Planetary Science Letters, v. 171, p. 221–235.

Vokes, F.M., and Gale, G.H., 1976, Metallogeny related to global tectonics in Southern Scandinavia: Geological Association of Canada Special Paper 14, p. 413–441.

Von Damm, K., 1995, Controls on the chemistry and temporal variability of seafloor hydrothermal fluids, *in* Humphris, S.E., Zierenberg, R.A., Mullineaux, L.S., and Thomson, R.E., eds., Seafloor hydrothermal systems: Washington, D.C., American Geophysical Union, Geophysical Monograph 91, p. 222–247.

Wright, J.E., and Wyld, S.J., 1986, Significance of xenocrystic Precambrian zircon contained within the southern continuation of the Josephine ophiolite: Devils Elbow ophiolite remnant, Klamath Mountains, northern California: Geology, v. 14, p. 671–674.

Yumul, G.P., 1996, Varying mantle sources of supra-subduction zone ophiolites: REE evidence from the Zambales ophiolitic complex, Luzon, Philippines: Tectonophysics, v. 262, p. 243–262.

Zierenberg, R.A., Shanks, W.C., III, Seyfried, W.E., Jr., Koski, R.A., and Strickler, M.D., 1988, Mineralization, alteration, and hydrothermal metamorphism of the ophiolite-hosted Turner-Albright sulfide deposit, southwestern Oregon: Journal of Geophysical Research, v. 93, p. 4657–4674.

Manuscript Accepted by the Society February 23, 2000

Printed in the U.S.A.

Geological Society of America
Special Paper 349
2000

Section V
Pacific Rim Ophiolites

This section includes a diverse collection of papers from widely separated parts of the Pacific Rim. Many of these bodies are "Cordilleran-type" ophiolites whose relationship to the crystalline continental basement is not so clear. As such, the majority of the Pacific Rim ophiolites are polygenetic in origin and are situated within the accreted oceanic terranes and/or in trench-forearc tectonic settings characteristic of active continental margins, in contrast to "Tethyan-type" ophiolites that are commonly underlain by passive continental margins (i.e., the Ligurian and Tethyan ophiolites discussed in Section 4). The first three papers are about occurrences in the Southern Hemisphere. Varne et al. report on the geology and evolution of Macquarie Island, an ophiolite formed ca. 9.5–12 Ma by slow spreading on a short ridge segment along the Australian plate–Pacific plate boundary and subsequently elevated above sea level by transpression as this plate boundary became obliquely convergent. This young ophiolite displays a Pacific MORB signature, but it includes more primitive, highly enriched, silica-undersaturated compositions, which Varne et al. argue was not derived from a MORB source.

The ophiolite-like rocks of Timor are the focus of the paper by Harris and Long. Most of these bodies include lherzolite blocks in a mélange that display geochemical characteristics of undepleted peridotites commonly found in abyssal and passive-margin settings. These lherzolites are thus thought to be derived from the Australian continental margin, rather than from a forearc basement of the Banda arc in the upper plate. Volcanic rocks and lherzolite from a single area (the Ocussi region) may represent an ophiolite of SSZ (supra-subduction zone) composition, emplaced soon after formation.

Ramos et al. present information on little-known Proterozoic–early Paleozoic ophiolitic rocks in the central Andes of Argentina, principally in the Cordillera Frontal, Precordillera, and the Pampean ranges. These ophiolites are highly deformed and metamorphosed, but are thought to represent a series of sutures that formed during the late Precambrian to early Paleozoic assembly of the basement rocks of west-central South America.

Farther north in the western United States are exposed the most extensive occurrences of Jurassic ophiolites in the North American Cordillera; these have played a critical role in development of the ideas and tectonic models for the Mesozoic evolution of California and vicinity. The section includes four papers on the Mesozoic ophiolites of California and environs. Coleman presents new data on ophiolites within the California margin, arguing for five separate tectonic settings for ophiolite accretion: tectonic underplating by downgoing oceanic crustal slabs, incorporation of abyssal peridotite into the subduction complex, supra-subduction zone ophiolites associated with offshore intraoceanic arcs, magmatic underplating along the incipient Sierra Nevada arc, and incorporation of far-traveled ophiolites and seamounts of oceanic spreading center origin into the continental margin. Coleman states that ophiolite obduction may be a rare event after all, whereas magmatic underplating and wedging of oceanic crustal slabs along convergent margins are clearly significant processes for addition of oceanic crust into continental margins.

Godfrey and Dilek present a synthetic model to integrating geophysical work on the ophiolitic basement beneath the Great Valley with field data from the Sierra Nevada and western Nevada in a comprehensive view of the structure of the U.S. Pacific margin. Their model suggests that the geophysically well-defined, thick mafic slab beneath the Great Valley basin represents the oceanic basement of an ensimatic Jurassic arc terrane, which included the Coast Range ophiolite and the Sierra Nevada foothills ophiolite belt. This intraoceanic terrane is thought to have consisted of volcanic archipelagoes, intra-arc basins, and a deformed ophiolitic substratum and to have been accreted into the North American continental margin during a major orogenic event.

Pessagno et al. describe sedimentary strata found within and above the Jurassic Coast Range and Josephine ophiolites. They argue that the Coast Range ophiolite formed at a spreading ridge in low latitudes and subsequently was displaced northward prior to emplacement on the North American continental margin, but that the Josephine ophiolite formed at a backarc or forearc spreading center.

Finally, Ingersoll presents models for formation and emplacement of the California Jurassic ophiolites by backarc spreading behind an east-facing arc (Sierra Nevada and Coast Range ophiolites) or a west-facing arc (Josephine). The tectonics of the Taiwan collision zone, the Ryukyu intraoceanic arc, and the Okinawa Trough may be considered as a modern analogue for the evolution of the Jurassic ophiolites in California, according to Ingersoll, who argues for the implausibility of forearc rifting as a mechanism to form these ophiolites. These papers together demonstrate that there still seems to be room for disagreement on the nature of igneous development and emplacement of the western U.S. ophiolites and their tectonic significance.

Metcalf et al. present new data on the Trinity ophiolite complex in the Klamath Mountains, California. They argue that this body, the areally largest ophiolite in North America, is polygenetic and compositionally of SSZ type and that it may have formed within a primitive Mariana-Izu-Bonin–like island arc. The Trinity complex has few equivalents on the western North American margin. Rocks of similar age are found only in the Eastern belt of the northern Sierra Nevada and in central Alaska. Thus its tectonic significance with respect to North America remains uncertain.

The chemical compositions of minor phases in peridotites are often critical in determining the tectonic setting and petrogenetic history of the rocks. Sato and Ogawa document mineral-chemistry variations in ophiolitic peridotites (Hayama-Mineoka belt) from central Japan, in which they particularly focus on sulfide and native-metal phases in the rocks. They conclude that the primary sulfide phases (pentlandite) are residual after partial melting (15% to 30%) of the mantle in a mid-ocean ridge environment and that subsequent serpentinization has produced awaruite, heazlewoodite, native copper, and magnetite. On the basis of these data and interpretations and regional correlations, they further argue that the ophiolitic rocks of the Hayama–Mineoka belt are the remnants of a separate oceanic plate (Mineoka plate), rather than belonging to the Izu-Bonin-Mariana forearc setting as previously suggested.

Geological Society of America
Special Paper 349
2000

Macquarie Island: Its geology, structural history, and the timing and tectonic setting of its N-MORB to E-MORB magmatism

Rick Varne
School of Earth Sciences, University of Tasmania, GPO Box 252-79, Hobart, Tasmania, Australia 7001;
e-mail: rick.varne@utas.edu.au
Anthony V. Brown
Mineral Resources Tasmania, GPO Box 56, Rosny Park, Tasmania, Australia 7018;
e-mail: abrown@mrt.tas.gov.au
Trevor Falloon
School of Earth Sciences, University of Tasmania, GPO Box 252-79, Hobart, Tasmania, Australia 7001

ABSTRACT

Macquarie Island is an exposure above sea level of the Macquarie Ridge Complex, on the boundary between the Australian and Pacific plates south of New Zealand. Geodynamic reconstructions show that at ca. 12–9.5 Ma, oceanic crust of the Macquarie Island region was created at this plate boundary within a system of short spreading-ridge segments linked by large-offset transform faults. At this time, the spreading rate was slowing (<10 mm/yr half–spreading rate) and magmatism was waning. Probably before 5 Ma, and possibly before the extinct spreading ridge had subsided, the plate boundary became obliquely convergent, and crustal blocks were rotated, tilted, and uplifted along the ridge to form the island. Planation by marine erosion has exposed sections through the oceanic crust.

The magmatism that built the oceanic crust produced melts similar in composition to the widespread normal to enriched mid-oceanic-ridge basalt (N- to E-MORB) suite found in many spreading ridges, but the melts ranged beyond E-MORB to primitive, highly enriched, and silica-undersaturated compositions. These compositions form one end member of a continuum from MORB but seem not to have been derived from a MORB-source mantle, despite sharing a Pacific MORB isotopic signature. The survival of these primitive melts may be due to their origin in a slow-spreading system that must have been closing down as extension along the plate boundary gave way to transpression, putting a stop to the upwelling of asthenosphere and decompression melting. In a more energetic, faster-spreading system, mixing would have been more efficient, the presence of this end member could not easily have been inferred from its isotopic composition, and the igneous rocks would have resembled a typical N- to E-MORB suite. Macquarie Island may therefore provide a type example of magmatism at a very slow spreading ridge and a clue to the origins of E-MORB.

INTRODUCTION

Macquarie Island is ~37 km long and less than 5 km wide; it lies in the Southern Ocean ~1500 km south-southeast of Tasmania, near 54°35′S and 158°55′E (Figs. 1 and 2). The island is an exposure above sea level of part of the Macquarie Ridge Complex. Groups of small islets occur on the ridge 14 km north (Judge and Clerk Islands) and 33 km south (Bishop and Clerk Islands) of Macquarie Island, but otherwise the whole island group is surrounded by thousands of kilometers of open ocean.

Macquarie Island is unique, among islands of the major ocean basins, in being an exposure of oceanic crust (Varne et al., 1969). Reconnaissance mapping (Varne et al., 1969; Varne and Rubenach, 1972; Christodoulou et al., 1984) showed that Macquarie Island is composed of fault-bounded blocks derived from different layers of oceanic lithosphere originally formed at a mid-oceanic spreading center. Planation by marine erosion of the tilted and rotated blocks of oceanic lithosphere as they were uplifted has exposed various sections through the oceanic crust, including a near-complete crustal section in the northwest quarter of the island (Griffin and Varne, 1980).

Geochemical data support the identification of Macquarie Island as uplifted oceanic crust. Macquarie Island basaltic and doleritic rocks are compositionally very like ocean-floor basalts from the Mid-Atlantic Ridge (Griffin and Varne, 1980; Varne, 1989) in their petrographic, major and trace element, and normative compositions.

Recent studies have improved our understanding of the geologic evolution of Macquarie Island in three ways and have prompted the writing of this progress report and synthesis.

First, Geosat gravity, magnetic, bathymetric, swath, and seismic

Varne, R., Brown, A.V., and Falloon, T., 2000, Macquarie Island: Its geology, structural history, and the timing and tectonic setting of its N-MORB to E-MORB magmatism, *in* Dilek, Y., Moores, E.M., Elthon, D., and Nicolas, A., eds., Ophiolites and Oceanic Crust: New Insights from Field Studies and the Ocean Drilling Program: Boulder, Colorado, Geological Society of America Special Paper 349, p. 301–320.

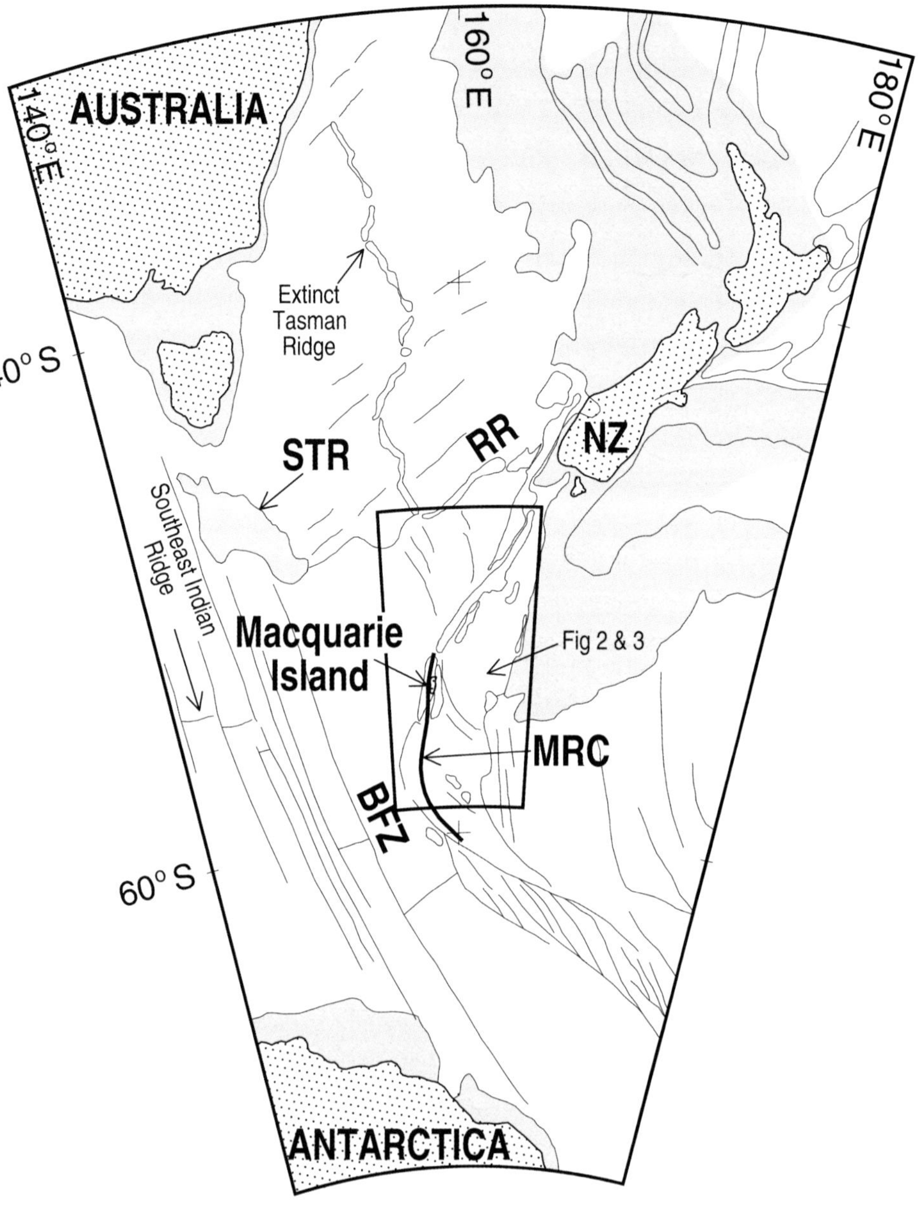

Figure 1. Location diagram, after Sutherland (1995). Dot-patterned areas represent land; shaded areas represent inferred continental crust. STR—South Tasman Rise, RR—Resolution Ridge, MRC—Macquarie Ridge Complex, BFZ—Balleny Fracture Zone, NZ—New Zealand.

data recently acquired for the oceanic crust south of New Zealand have stimulated the development of a new geodynamic model that shows how the boundary between the Pacific and Australian plates has evolved progressively from a spreading system in Eocene time, to a system of transform faults linked by short spreading centers in the late Miocene (Fig. 3), and now into a right-lateral strike-slip plate boundary (Sutherland, 1995; Wood et al., 1996; Lamarche et al., 1997; Massell et al., 2000). We believe that the long-standing discrepancy between magnetostratigraphic and radiometric ages for the rocks of the island, discussed by Duncan and Varne (1988), is resolved by this new geodynamic modeling.

Second, recent detailed regional mapping of Macquarie Island by the Geological Survey of Tasmania has been compiled and produced as a 1:10,000 scale map (seven sheets) covering the island (Goscombe and Everard 1998). The data have then been generalized to produce a 1:25 000 scale map (two sheets) and a 1:50 000 scale map (one sheet) of the island and further simplified to produce Figures 4–7 of this paper. Details of the field work and laboratory studies of samples collected during the survey are now being compiled for publication as a *Geological Survey of Tasmania Bulletin.* The new field studies and geodynamic modeling suggest an explanation for the first appearance and continued existence of the island.

Third, Kamenetsky et al. (2000) have provided new major, trace, and isotopic analyses of a set of Macquarie Island basaltic glasses. These analyses reveal a seriate variation within the usual compositional spectrum of MORB and extend beyond MORB to enriched compositions that represent a new, primitive, enriched MORB end member.

This paper starts with a review of the geographic and tectonic setting of the island and goes on to analyze the geodynamic history of the Macquarie Island region and to summarize the geologic relationships, drawing on the results of the new regional mapping. It finishes with a summary of the petrology of the island's rocks and a discussion of the petrogenesis of E-MORB.

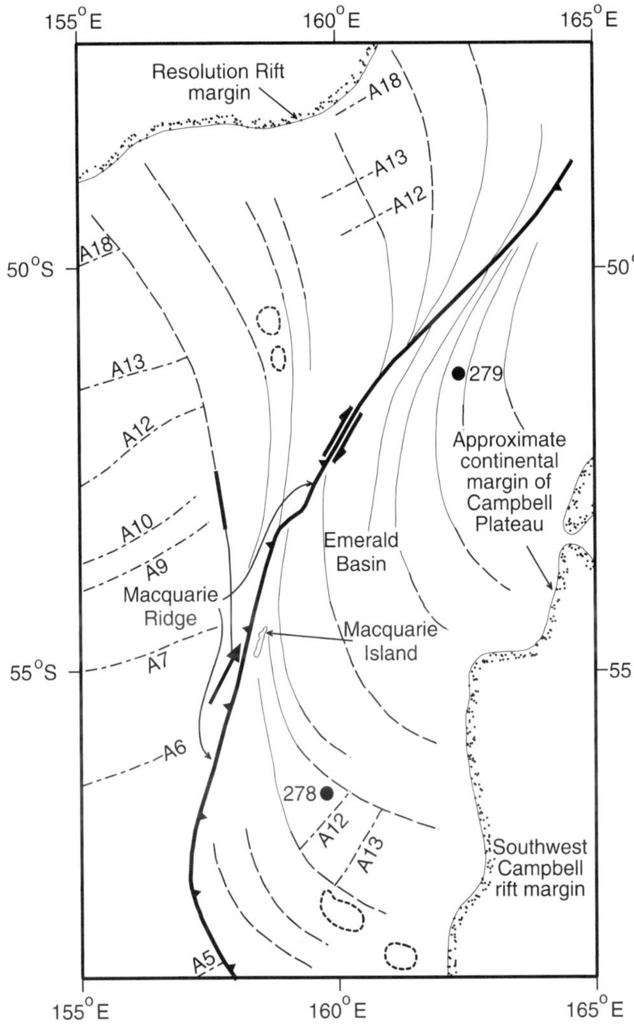

Figure 2. Sketch map showing the positions of rift margins after Sutherland (1995); magnetic anomalies (lines with long and short dashes, labeled A12, A13, etc.) from Weissel et al. (1977); curving fracture zones (solid lines) from Massell et al. (2000) continued as dashed lines following Geosat gravity anomalies identified by Sutherland (1995). DSDP sites also indicated (solid circles). Dashed-line ovoids are gravity highs. Gray arrow indicates the Australian-Pacific plate-motion vector at the Macquarie segment of the Macquarie Ridge Complex near Macquarie Island, from DeMets et al. (1994).

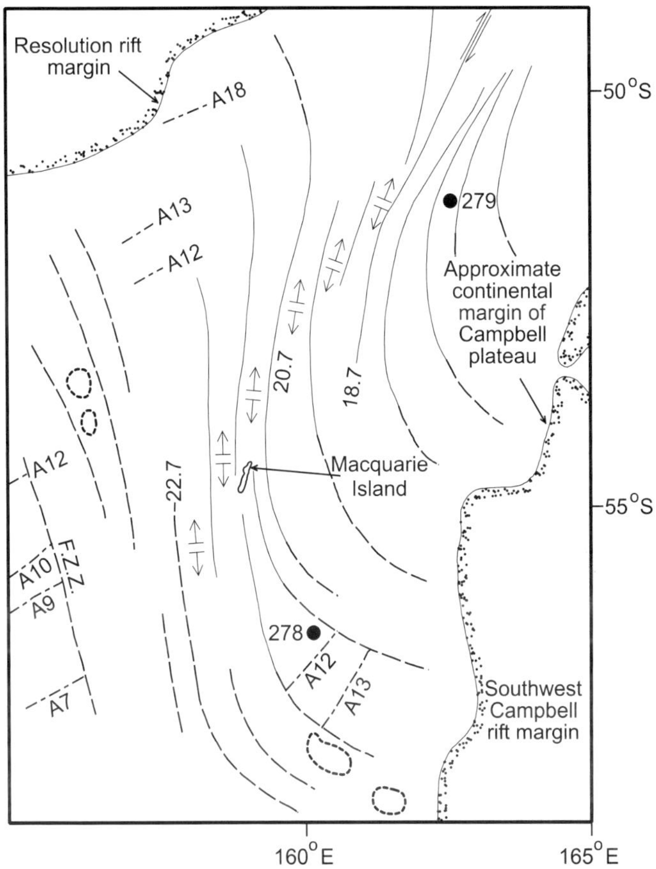

Figure 3. Reconstruction of the Macquarie Island region at ca. 11 Ma, sketched by using the best-fit reconstruction (Pacific plate fixed) and positioning of the rift margins of Sutherland (1995). The curving fracture zones, lineaments, and anomalies are relocated from Figure 2; the position and segmentation of the Australian plate–Pacific plate boundary and calculated full spreading rates in millimeters per year are from Lamarche et al. (1997). FZZ—fracture zone Z (see text for discussion).

GEOGRAPHIC AND TECTONIC SETTING OF MACQUARIE ISLAND

The Macquarie Ridge Complex marks the boundary between the Australian and Pacific plates south of New Zealand. The ridge runs south-southwest for 1200 km to join the Southeast Indian Ridge and the Pacific-Antarctic Ridge at the triple junction at lat ~57°S (Fig. 1). The Macquarie Ridge Complex is divided into four discrete segments (Hayes and Talwani, 1972; Massell et al., 2000) by marked breaks in its continuity and changes in strike near lat 51°S, 53°S, and 56°S. It is flanked by trenches, which lie to its west in the northern and southern segments, and to its east in the two central segments. The Macquarie Ridge segment (lat 53°–56°S) that includes the Macquarie Island region rises to shallow depths (<120 m) along much of its length and, in places, has a flat-topped morphology that suggests that much of the Macquarie segment might have been at sea level at times in the past, perhaps during the Pliocene (Griffin and Varne, 1980; Massell et al., 2000). The Macquarie Trench lies immediately to its east, mostly at 5000–6000 m water depth.

The Macquarie segment has a double crest with a central valley; Massell et al. (2000) have located the active plate boundary within this central valley, coincident with a zone of multiple, ridge-parallel lineaments. Macquarie Island is an exposure of the eastern ridge crest (Summerhayes, 1969; Massell et al., 2000) and therefore lies on the Pacific plate at this time. The ridge is seismically very active (average return period is 1 yr for earthquakes of magnitude >6.2; an earthquake of magnitude 7.8 was recorded in 1943 (Jones and McCue, 1988). The seismicity is shallow and probably limited to the crust (Ruff et al., 1989), and earthquake fault-plane solutions for the Macquarie segment suggest subhorizontal, east-west–oriented principal stress axes and right-lateral strike-slip movement along fault planes parallel to the ridge (Jones and McCue, 1988). Australian-Pacific plate-motion vectors calculated from modeling of global plate motions (DeMets et al., 1994) intersect the active plate boundary along the Macquarie segment of the Macquarie Ridge Complex near Macquarie Island at ~14° (gray arrow, Fig. 2). Such vectors imply transpressional relative motion between the Australian and Pacific plates and counterclockwise rotation of the Pacific plate, relative to the Australian plate.

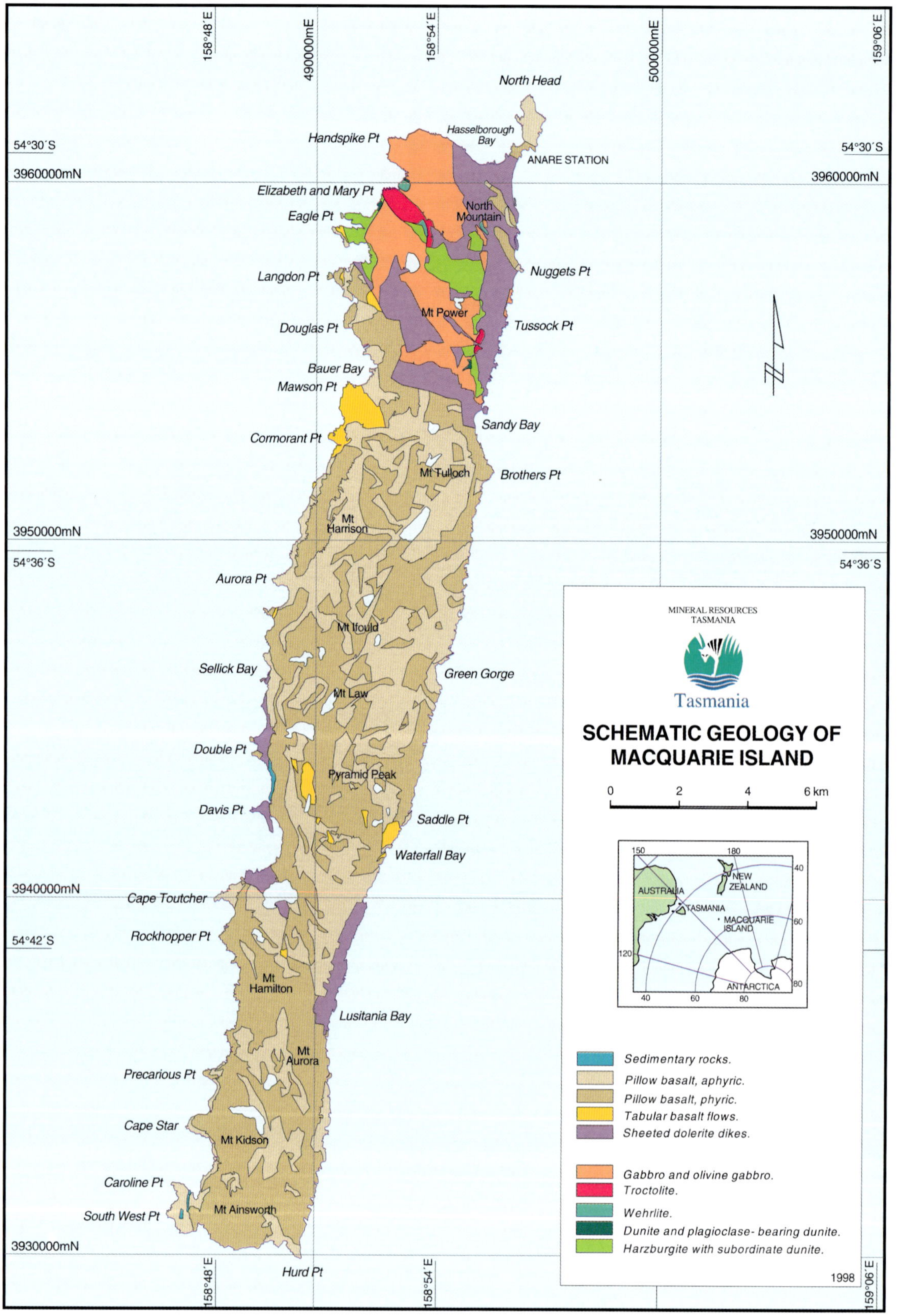

Figure 4. Generalized schematic geologic map of Macquarie Island with superficial deposits omitted, after Goscombe and Everard (1998).

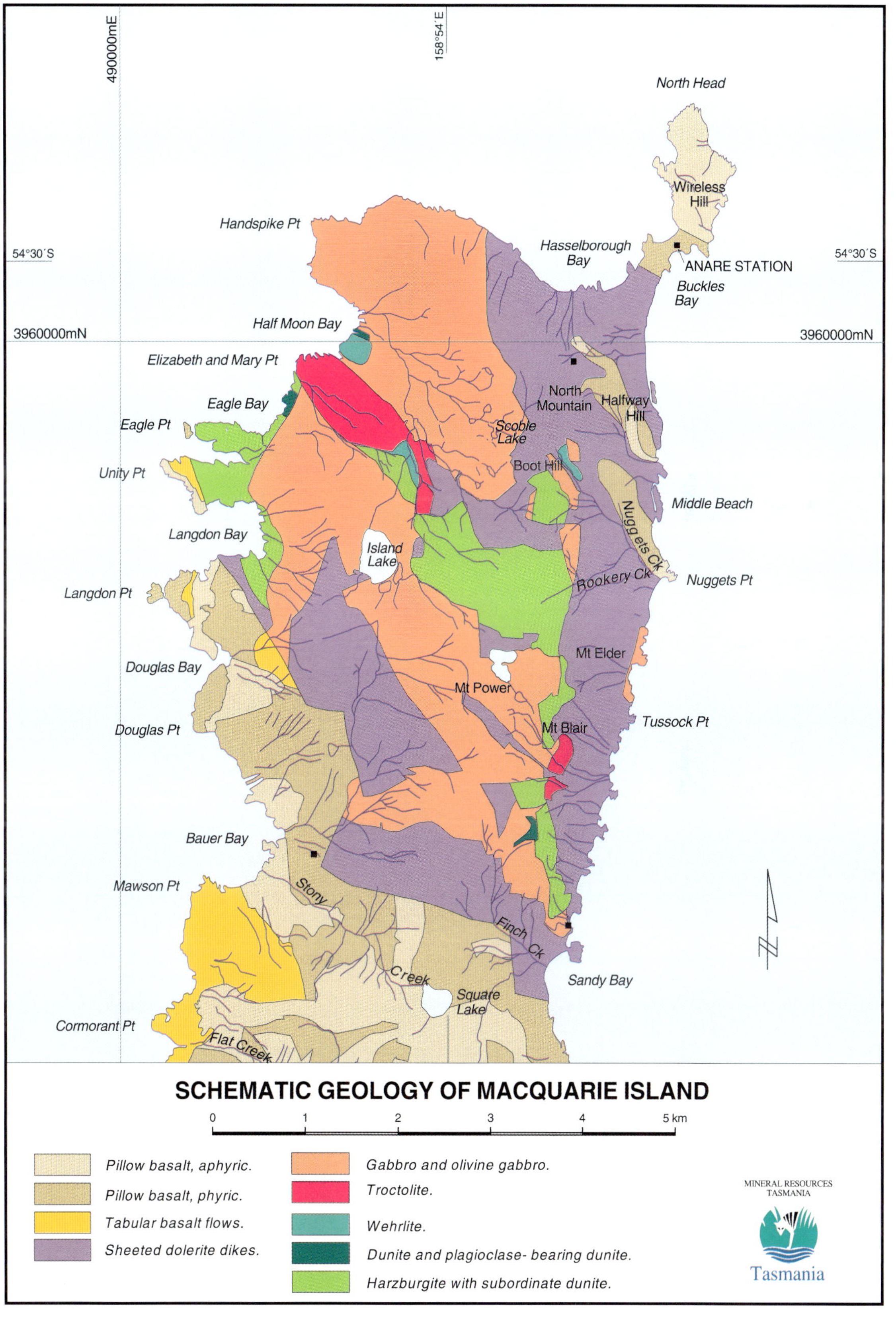

Figure 5. Geologic map of the northern part of Macquarie Island with superficial deposits omitted, after Goscombe and Everard (1998).

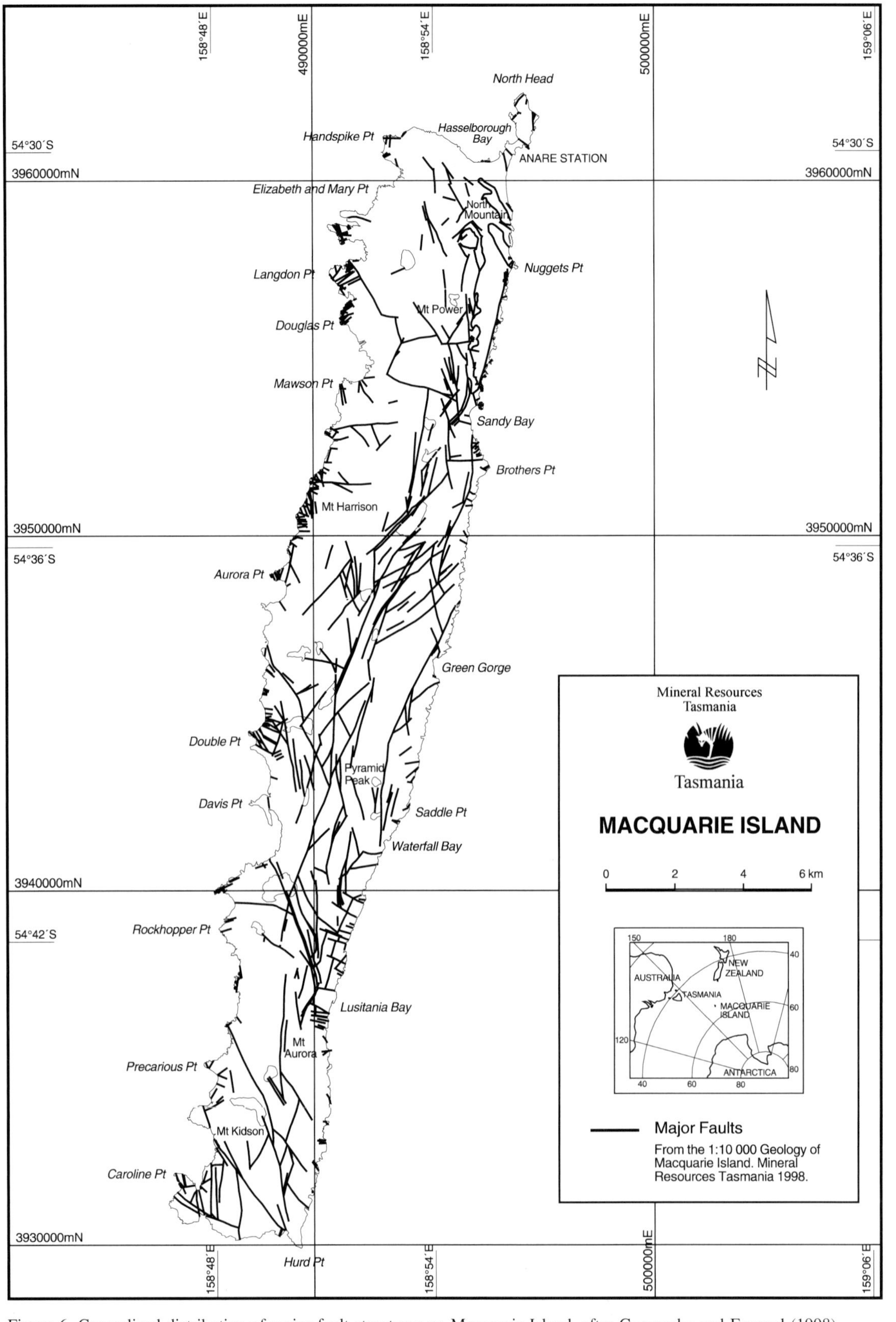

Figure 6. Generalized distribution of major fault structures on Macquarie Island, after Goscombe and Everard (1998).

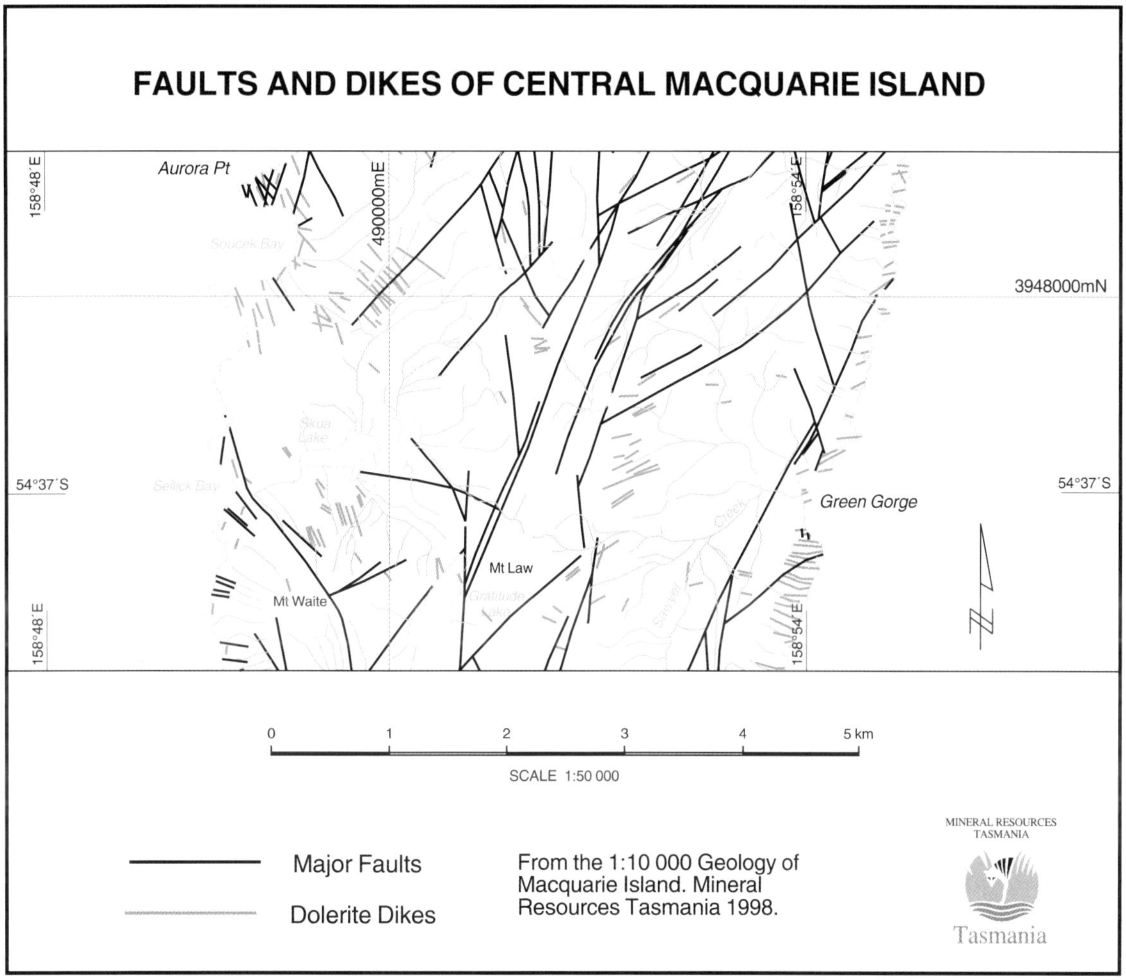

Figure 7. Central part of Macquarie Island showing the orientation of dikes cutting the volcanic sequences. Major fault structures from Figure 6 are also shown.

GEODYNAMIC HISTORY OF THE MACQUARIE ISLAND REGION

The Macquarie Ridge Complex is built of oceanic crust belonging to the south Tasman Sea and the Emerald Basin, a triangular region of the ocean floor southwest of New Zealand that is bordered by the Resolution Ridge and Resolution Rift to the north and the western margin of the continental Campbell Plateau to the east (Figs. 1, 2, 3).

The Resolution Ridge and Rift structures run westward from the southwest corner of South Island, New Zealand (Figs. 1, 2), and continue as the bathymetric structures of the Nella Dan Rift toward the southern end of the submerged continental South Tasman Rise, south of Tasmania (Fig. 1). The location of the Resolution Ridge and Nella Dan Rift bathymetric structures also marks the angular magnetic unconformity between the magnetic anomalies of the north Tasman Sea and the younger anomalies of the oceanic crust of the Balleny spreading corridor of the Southeast Indian Ridge (Cande et al., 1998). The now extinct Tasman spreading center (Figs. 1, 2) can be easily identified on gravity anomaly maps (see Fig. 2 of Sutherland, 1995).

The Balleny spreading corridor of the Southeast Indian Ridge is composed of the oceanic crust between the southern part of the Macquarie Ridge Complex and the Balleny Fracture Zone (Cande et al., 1998) in which spreading began in Cretaceous time and has continued until the present (Fig. 2). The crust of the Balleny spreading corridor merges eastward with the crust of the south Tasman Sea, also termed Southeast Tasman oceanic crust by Wood et al. (1996), which is the triangular region between the Resolution Ridge and the Macquarie Ridge Complex.

East of the Macquarie Ridge Complex and Macquarie Island, the Emerald Basin lies on the Pacific plate between the complex and the western margin of the submerged continental Campbell Plateau south of New Zealand (Fig. 2). The crust of the basin was drilled at DSDP (Deep Sea Drilling Project) Sites 278 and 279 (Kennett, Houtz, et al., 1974). At Site 278, well-dated fossiliferous middle Oligocene (ca. 28 Ma) sedimentary rocks overlie pillow basalt, and at Site 279, early middle Miocene (ca. 20 Ma) sedimentary rocks overlie basalt that may be part of a massive flow. The petrographic characteristics and major and rare earth element compositions of basalts from both sites (Ovenshine et al., 1974; Schilling and Ridley, 1974) resemble those of Macquarie Island rocks.

Weissel et al. (1977) tentatively identified linear magnetic anomalies 12 (31 Ma according to the Cande and Kent [1992] time scale) and 13 (34 Ma) near Site 278 (Fig. 2), and, presciently, a north-northeast–trending magnetic fabric in this part of the Emerald Basin. Equally presciently, Molnar et al.'s (1975) reconstructions showed how the southern part of the Tasman Sea floor could have formed by slow spreading at the Australian plate–Pacific plate boundary, near the present Macquarie Ridge Complex. Molnar et al. (1975) based their tectonic analysis on closure of the Australian plate–Pacific plate–Antarctic plate circuit, but their data set and conclusions are consistent with Sutherland's (1995) new results.

Sutherland (1995) recognized that the Resolution Ridge and the western margin of the Campbell Plateau plus the eastern margin of the Emerald Basin (Fig. 2) are conjugate margins formed in Eocene time by the rifting of continental and older oceanic crust when the Australian plate–Pacific plate boundary developed into a spreading ridge south of New Zealand. Sutherland (1995) also used Geosat data to identify a remarkable system of curving gravity-anomaly lineaments within the new ocean crust, which run parallel to fracture zones identified by side-scan data (Coffin et al., 1994; Collot et al., 1995).

Sutherland (1995) assumed that the gravity-anomaly lineaments preserved the directions of old transform faults formed during spreading at the plate boundary and employed the new constraints in plate-motion modeling to show that as the rift opened, the Australian plate–Pacific plate relative pole of rotation migrated southeast away from the New Zealand landmass during late Eocene–late Miocene time. The pole remained close to the plate boundary south of New Zealand, and its closeness implies slow spreading at the plate boundary (Lamarche et al., 1997). The pole then migrated southwest during Pliocene–Pleistocene time (Stock and Molnar, 1982).

New bathymetric, magnetic, swath, and seismic data acquired for this triangle of oceanic crust support and amplify Sutherland's modeling of the tectonic evolution of the region in general. The Geosat gravity lineaments identified by Sutherland (1995) correspond with a subparallel family of fracture zones 40–50 km wide that, near the bathymetric Macquarie Ridge, parallel the plate boundary but bend away from the ridge as they approach the conjugate rift margins (i.e., the Campbell Plateau; Lamarche et al., 1997; Massell et al., 2000). Extrapolated fractures approach rift boundaries orthogonally (Fig. 2).

The fracture zones are interpreted as old transform faults and the bathymetric ridge of the Macquarie Ridge Complex as corresponding approximately to the relict Australian plate–Pacific plate spreading center. Magnetic anomalies 18–11 in the northeastern part of the south Tasman Sea, which were identified by Wood et al. (1996), show that the crust becomes younger to the southeast away from the Resolution Ridge system, as do magnetic anomalies 18–12 identified by Weissel et al. (1977) in the spreading channel north of 50°S, 160°E (Fig. 2).

Seafloor reflectivity patterns are identified with tectonic spreading fabric and strike orthogonally to the fracture zones. Away from spreading ridges, oceanic crust typically displays a tectonic spreading fabric, usually oriented parallel to the spreading ridge at which the crust was generated (Lonsdale, 1977; Laughton and Searle, 1979). The tectonic spreading fabric resembles horst- and graben-like structures formed by closely spaced ridge-parallel normal faulting. Near transforms that offset the spreading axis, the tectonic spreading fabric may curve in the direction of the ridge offset, oblique to the spreading axis.

North of 50°S, 160°E, in the spreading channel where Weissel et al. (1977) located east-northeast–trending magnetic anomalies 18–12 (Fig. 2), Massell et al. (2000) have shown that the seafloor reflectivity patterns also strike east-northeast, validating their interpretation as tectonic spreading fabric.

Analogously, near DSDP Site 278 (Fig. 2) the Geosat gravity anomaly lineament interpreted by Sutherland (1995) is orthogonal to the strike of the linear magnetic anomalies 12 and 13 (Weissel et al., 1977).

In general, then, the strike of the tectonic fabric in spreading channels in the crust of the south Tasman Sea swings progressively to remain orthogonal to the curved confining fractures (Lamarche et al., 1997; Massell et al., 2000). For example, the strike of the tectonic spreading fabric rotates progressively clockwise along the curved spreading channel running through 50°S, 160°E, from east-northeast near the rifted margin to southeast near the western flanks of the Macquarie Ridge Complex at ~53°S, 159°30′E, where the spreading fabric is striking approximately orthogonal to the local strike of the Macquarie Ridge Complex and the fracture zone has swung parallel and close to the complex.

PLATE-BOUNDARY GEOMETRY AT 11 Ma AND CALCULATED MODEL AGES

Fracture zones and tectonic spreading fabric within the triangle of new oceanic crust south of New Zealand record the post-Eocene evolution of the Australian plate–Pacific plate boundary, as it changed from a spreading system, into a system of large-offset transform faults linked by short spreading segments, and now into a right-lateral strike-slip plate boundary (Sutherland, 1995; Wood et al., 1996; Lamarche et al., 1997; Massell et al., 2000). The gradual change in the orientation of fracture zones and spreading fabric implies a similarly gradual reorientation of the spreading segments rather than rift jumps (Lamarche et al., 1997).

Lamarche et al. (1997) used Sutherland's (1995) reconstruction of the plate boundary through time and determinations of the Australian plate–Pacific plate relative pole of rotation to investigate the geometry and spreading rates at three stages (chron 13 [35–30 Ma], chron 6 [25–20 Ma], and chron 5 [20–11 Ma]). Fracture zones were taken as fossil transform faults that offset spreading-axis segments when they were active and theoretically followed small circles that were lines of latitude relative to the rotation pole. Spreading ridges lie along great circles passing through the rotation pole and were located symmetrically to the conjugate margins.

At 11 Ma, the Lamarche et al. (1997) reconstruction shows a large-offset transform to the north and a series of short spreading segments displaced left along transform faults in the southern part of the region. The calculated full spreading rate at the plate boundary increases from north to south.

We have sketched a possible plate-boundary geometry in the Macquarie Island region at this time (Fig. 3). Our reconstruction is based on Lamarche et al.'s (1997) Figure 6C, to which we have added the fracture zones of Massell et al. (2000) augmented with gravity anomaly lineaments sketched from Sutherland's (1995) Figure 2 and the linear magnetic anomalies of Weissel et al. (1977). We note that this reconstruction of the Australian plate–Pacific plate boundary at 11 Ma resembles a simplified mirror image of the Australia plate–Pacific plate boundary east of the triple point to the south (Fig. 1).

Model ages for Macquarie Island crust

The reconstruction at 11 Ma places Macquarie Island close to the modeled position of the Australian plate–Pacific plate boundary and within the region of newly formed oceanic crust (Fig. 3). This observation alone suggests a spreading age for Macquarie Island crust of ca. 11–12 Ma. The transform orientations imply that the grain of the spreading fabric formed at 11 Ma could have run approximately east-west.

We are also able to estimate the age of Macquarie Island crust in another way, by measuring the amount of ocean crust between Macquarie Island and anomaly 12 (31 Ma, Cande and Kent, 1992), along a line parallel to the fracture zones bounding the spreading corridor, and then calculating the time required to form this amount of crust at the relevant spreading rate.

The distance between the center of the island and anomaly 12 is ~270 km. The calculated full spreading rates (Fig. 3) in the Macquarie Island spreading corridor are 29.5 mm/yr (35–30 Ma), 26.0 mm/yr (25–20 Ma), and 20.7 mm/yr (20–11 Ma). If we adopt 13–14 mm/yr as a plausible average half-spreading rate for the 30–11 Ma time frame, creation of the new seafloor in the spreading corridor between central Macquarie Island and anomaly 12 would have taken ~19–21 m.y., yielding a model spreading age for the crust of the center of the island of ca. 10–12 Ma. (The estimated model spreading age for the center of the island using this rate and the distance of ~240 km between Macquarie Island and DSDP Site 278 [mid-Oligocene, ca. 28 Ma] is similar, 11–9.5 Ma.) Wood et al. (1996) calculated a spreading rate of ~15 mm/yr, similar to the spreading rate adopted here, from direct measurements between magnetic anomalies 18 and 11 (40–30 Ma) for crust in the southeast Tasman Sea that is approximately conjugate to the crust of the Macquarie Island spreading channel (R. Sutherland, 1999, personal communication). They noted that spreading rates probably slowed in the mid-Oligocene.

These estimates of ca. 12–9.5 Ma for the model spreading age of Macquarie Island crust are in agreement with radiometric and paleontological ages for volcanic rocks of the island (see the next section).

Williamson (1979, 1988) estimated the age of the island in yet another way and arrived at a different result. Macquarie Island pillow lavas have magnetic properties that correspond well with those of ocean-floor basalts (Banerjee et al., 1974; Butler et al., 1976; Levi and Banerjee, 1977; Levi et al., 1978). Williamson (1979, 1988) measured a north-south magnetic profile on land along the island for 20 km within volcanic rocks. His model for the distribution of normal and reversed directions of remanent magnetism has a broad section of reversed polarity followed to the south by a broad normal polarity section, interrupted in the center of the island by a narrow zone of reversal (Fig. 11 of Williamson, 1988).

Weissel et al. (1977) had identified magnetic anomaly 7 in the Balleny spreading corridor of the Southeast Indian Ridge to the west of Macquarie Island (Fig. 1). Williamson (1988) argued that this anomaly could be traced eastward, across Macquarie Island to the Macquarie Trench. This interpretation would imply that the ocean crust exposed on Macquarie Island was formed at the Southeast Indian Ridge by late Oligocene spreading (at ca. 25 Ma), that the island remains part of the Australian plate, and that the Macquarie Trench marks the plate boundary.

Massell et al. (2000) have now shown that the present plate boundary runs west of Macquarie Island. In the new reconstructions based on the Geosat gravity data and swath mapping of the seafloor, the ocean crust of the Macquarie Island region had not formed by anomaly 7 time (Fig. 6B of Lamarche et al., 1997). Volcanic rocks near the location of Williamson's (1988) magnetic profile along the island have original crystallization ages of 11.5 ± 0.3 Ma (Duncan and Varne, 1988), within anomaly 5 time.

At 11 Ma, the plate vector was oblique to the Australian plate–Pacific plate boundary (Lamarche et al., 1997), and only a small amount of spreading crust was being created. The grain of the tectonic spreading fabric of the region is likely to have run roughly east-west, similar to the orientation of the magnetic anomalies sketched by Williamson (1988) west of the Macquarie Ridge Complex. We suggest that if a coherent pattern of magnetic anomalies has survived the recent transpressional tectonism in the vicinity of the plate boundary, then Williamson (1988) might have located part of anomaly 5. However, we also note that there may be enough uncertainty in the reconstructions to allow the possibility that the crust of Macquarie Island might have been located near the zone where the oceanic crust of the fast-spreading Balleny spreading corridor of the Southeast Indian Ridge merged with the oceanic crust of the slow-spreading south Tasman Sea.

Postspreading history of the Macquarie Island region

Kamp (1986) pointed out that the early Tertiary spreading rate in the south Tasman Sea was slower than that within the Balleny spreading corridor and suggested that the Southeast Indian Ridge may have been offset from its eastern extension along a fracture zone that he termed "fracture zone Z," sketched as a transform fault. Weissel et al. (1977) compared the along-strike lengths of magnetic anomalies in the Balleny spreading corridor with their counterparts on the Antarctic plate and noted that only anomalies 9 and 10 are the same length as their Antarctic counterparts; older anomalies are longer, and younger anomalies are shorter. The zone on the Australian plate where the lengths of anomalies 9 and 10 match those of their Antarctic counterparts occurs along the fracture zone that trends N20°W near 158°E in Figure 2. We have located Kamp's fracture zone Z (Fig. 3) along this fracture zone and sketched its merger with the Macquarie Ridge Complex south of lat 56°S. Fracture zone Z marks our proposed western limit for south Tasman seafloor.

Molnar et al.'s (1975) reconstructions showed how the southern part of the Tasman Sea floor could have formed by slow spreading at the Australian plate–Pacific plate boundary, near the present Macquarie Ridge Complex. Molnar et al. (1975) suggested that, in the north, the boundary was characterized by right-lateral strike-slip and, in the south, by a component of opening. Their geometrical construction shows oblique spreading continuing since 10 Ma along the southern part of the plate boundary. Overlap in the north between the new oceanic crust and the older rifted continental crust implies some subduction along the boundary near New Zealand (see also Collot et al., 1995).

The rate of formation of south Tasman seafloor was slowing with time, and the widths of the spreading segments were diminishing as the Australian plate–Pacific plate relative pole of rotation migrated south. The leftward-stepping system of shortening, slow-spreading ridge segments linked by transform faults continued curving southward into parallelism with fracture zone Z, which merges with the Macquarie Ridge Complex south of lat 56°S.

As spreading was waning, the fracture zones converged into parallelism along the plate boundary. Oblique spreading along the boundary probably ceased before 5 Ma, when increases in spreading rate occurred on both the Southeast Indian and Pacific-Antarctic Ridges, and the Australian plate–Pacific plate rotation pole moved to its present position (Sutherland, 1995). Our sketch implies at least 200 km of transcurrent right-lateral movement (cf. Figs. 2, 3) along the southern part of the boundary since 11 Ma.

Australian plate–Pacific plate relative motion seems not to have changed significantly since 5 Ma (Sutherland, 1995). The transcurrent to compressional relative motion along the boundary accommodates mismatch in the Southeast Indian Ridge and Pacific-Antarctic Ridge spreading rates (Johnson and Molnar, 1972; Ruff et al., 1989).

GEOLOGY OF MACQUARIE ISLAND

Macquarie Island is an exposure of part of the eastern crest of the double ridge system in this segment of the Macquarie Ridge Complex (Fig. 2). Groups of small islets occur on this eastern ridge 14 km north (Judge and Clerk Islands) and 33 km south (Bishop and Clerk Islands) of Macquarie Island. The whole island group is surrounded by thousands of kilometers of open ocean.

Mawson (1943) pointed out that Macquarie Island may be conveniently regarded as a mountain range rising abruptly from the sea. The island is topographically defined by an elevated and undulating, lake-studded summit plateau, standing ~300 m above sea level and cut off seaward by steep cliffs. A wide coastal platform occurs in places at the base of the cliffs on both sides of the island, with its greatest development in the northwest, where it is 1 km wide, studded by remnant sea stacks surrounded by quaking bogs (called "featherbed"). Remnant raised beaches and associated deposits are common in the northern part of the island.

The geomorphology and Quaternary history of the island have been described by Selkirk et al. (1990) who showed that its landscape is a product of the interplay among faulting, uplift, sea-level changes, erosion, and periglacial processes. Since the first appearance of this part of the Macquarie Ridge Complex above the sea surface, possibly in middle to late Pleistocene time (Selkirk et al., 1990), the rocks of the oceanic crust have been assaulted by aerial and subaerial forces that have produced surface deposits ranging from alluvial, lacustrine, swamp, and slope deposits to beach, eolian, and glaciogenic deposits.

Faults near the east coast determine coastline and cliff lines, the orientations of many valleys, and the locations of most lakes, either in lineament-bounded depressions or fault-dammed drainage (Ledingham and Peterson, 1984; Selkirk et al., 1990). The west coast of the island is more embayed than the east, and traces of coastline-parallel faulting are less evident (Figs. 4, 5, and 6).

The complex fault system that determines the island's eastern coastline, termed the Brothers fault-line system by Ledingham and Peterson (1984), parallels the strike of the Macquarie Ridge Complex and is almost certainly still active. Selkirk et al. (1990) also observed that Judge and Clerk Islands, to the north, and Bishop and Clerk Islands, to the south, fall on this same alignment, so its influence may extend beyond the shores of the island.

The Brothers fault-line system probably also includes the whole system of north-trending faults and lineaments that cuts the eastern half of the island (Figs. 4, 5, and 6). Between Nuggets Point and Sandy Bay, the coastal strip is composed of gabbro, which finishes seaward at a lineament that defines the straight eastern margin of the rock platform. To the west, gabbro is thrown against sheeted dolerite dikes along a coast-parallel fault that controls the cliff line that forms the eastern margin of the summit plateau. The coastal platform thus appears to be a narrow fault block within the Brothers fault-line system that has been planated by marine erosion, exposing deep crustal material. Inland, rising above the summit plateau are three hills (North Mountain, Mount Elder, and Mount Blair) with a raised beach at their base; the three hills must have been a small island in the past (Ledingham and Peterson, 1984). The fault zone that runs inland north from Sandy Bay to North Mountain has been the site of recent movement and is likely to be a branch of the Brothers fault-line system (cf. Figs. 5 and 6).

In the central third of the island, between Brothers Point and Saddle Point, where the Brothers fault line comes ashore, upthrown wedges between the fault and the east coast form the eastern edge of the island plateau. The western (inland) slopes are covered by fault-derived scree (Ledingham and Peterson, 1984; Selkirk et al., 1990) and farther inland, recent movement on north- to northeast-trending faults (Fig. 6) has impounded lakes on the plateau. Farther south, on the east coast north of Lusitania Bay, the coastal strip is composed of sheeted dolerite dikes in faulted contact with extrusive volcanic rocks inland, to the west. The geologic relationships along the east coast of the island suggest a component of reverse faulting associated with the Brothers fault-line system.

Selkirk et al. (1990) inferred the presence of three major fault zones from the topography and the juxtaposition of rocks of different lithologies.

The northernmost of these zones crosses the isthmus that separates North Head from the main part of the island (Figs. 4, 5). Near the ANARE (Australian National Antarctic Research Expedition) Station, little-altered basaltic rocks on North Head abruptly give way to greenschist-facies basaltic pillow lavas and dikes. A basalt from North Head yielded a well-defined $^{40}Ar/^{39}Ar$ plateau age of 9.7 ± 0.3 Ma, which Duncan and Varne (1988) interpreted as its original crystallization age. Poorly preserved coccoliths from oozes associated with pillow lavas from North Head were thought by Quilty et al. (1973) to be of early or middle Miocene age (11–24 Ma) and foraminifera from oozes suggest a late Miocene (5–11 Ma) or younger age (Varne et al., 1969), both broadly in keeping with the radiometric dating. The inferred faults across the isthmus may be related to the Brothers fault-line system or could be older.

A second topographical low across the island runs between Bauer Bay, on the west coast, and Sandy Bay, on the east, and roughly follows the complex boundary between the sheeted dolerite dikes, gabbros, and ultramafic rocks of the northwest quarter of the island and the extrusive basaltic and interbedded sedimentary rocks that compose the southern three-quarters of the island. This boundary is likely to be following crustal features formed during seafloor spreading (see the following section on "Spreading and postspreading events").

To the south, the central third of the island is composed of extrusive basaltic and intermixed sedimentary rocks. Williamson's (1979, 1988) 20-km-long north-south magnetic profile, discussed earlier, was measured in this part of the island. A hornblende separate from a massive flow or near-surface intrusion at Pyramid Peak, near Williamson's line of section, yielded a K-Ar age of 11.5 ± 0.3 Ma, and a basalt from Mawson Point, south of Bauer Bay, yielded a well-defined $^{40}Ar/^{39}Ar$ plateau age of 11.5 ± 0.3 Ma (Duncan and Varne, 1988). Both ages are interpreted as original crystallization ages.

The third major fault zone crosses the island from north of Cape Toutcher, on the west, to the east coast north of Lusitania Bay, where sheeted dolerite dikes are juxtaposed with extrusive sequences (Fig. 4). This map feature across the island also appears to mark a metamorphic (see the subsequent discussion on Metamorphism of Macquarie Island Crust) and geochemical transition in the volcanic rocks. Griffin (1982) showed that in the central part of the island, between Bauer and Sandy Bay in the north, and Sandell Bay (north of Cape Toutcher; see Fig 4) and Lusitania Bay in the south, 29 of the analyzed volcanic rocks are alkalic and 11 are tholeiitic, whereas in the southern part of the island, 6 are alkalic and 10 are tholeiitic.

New mapping

New mapping by the Geological Survey of Tasmania has revealed numerous complex relationships between the different igneous rock successions that constitute this exposed part of the Macquarie Ridge Complex. These relationships, when fully analyzed, will allow a better understanding of processes and products during spreading and transpression at the Australian plate–Pacific plate boundary.

Figures 4 and 5 are simplified geologic maps of Macquarie Island, with superficial deposits omitted, derived from new 1:10 000–scale geologic maps (seven sheets) covering the island (Goscombe and Everard, 1998). In general, outcrop is poor inland on the island, and mapping is therefore more detailed along the coasts. Faults and lineaments are shown in Figure 6.

Following the new mapping, the igneous and associated rocks have been grouped into eight main igneous rock associations that exhibit a wide range of igneous relationships and deformational features. The rock associations correspond to groupings normally accepted as parts of an idealized oceanic crustal section and are presented in order of increasing depth of formation. As the new work records (Goscombe and Everard, 1998), the rocks exhibit a far greater range of features and rock associations than previously realized.

The following are the main assemblages: extrusive basaltic and sedimentary rocks; sheeted dolerite dikes; a microgabbro transition group; a massive, coarse-grained, gabbro assemblage; a troctolite association; a mixed plagioclase wehrlite and dunite association; a mixed dunite and harzburgite association; and a harzburgite association. Dolerite dikes intrude all of these rock associations. Areas of a single rock association are typically fault bounded, but some transitional boundaries are observed.

1. Extrusive basaltic and sedimentary rocks. Extrusive basaltic and intermixed sedimentary rocks are the dominant rock association exposed on Macquarie Island and constitute 65.6% of the island (Goscombe and Everard, 1998). Pillow basalt is common, and volcanic breccias, hyaloclastite, and massive lavas are also present. Pillow basalts are commonly amygdaloidal; aphyric to very densely phyric (>30% plagioclase phenocrysts) varieties are randomly distributed over the island. Tabular basalt flows are much less common and occur mainly near Mawson Point. Rare picritic plugs and flows or shallow intrusions of hornblende-bearing basalt occur in a narrow zone down the center of the island.

Thin lenses of volcaniclastic sedimentary rocks tend to be associated with volcanic sequences that contain massive lavas. Mixed clastic turbiditic units consist of typically laminated, red mudstone and siltstone; typically gray, greenish-gray, or red sandstone and pebbly sandstone with muddy to silty matrix; and typically clast-supported conglomerate consisting of subrounded to subangular cobble- to boulder-sized clasts in a mudstone to sandstone matrix. Dolerite and basalt dominate rock clasts, but in one area near Douglas Point, the clasts consist of gabbro, indicating that rocks from lower-crustal depths were exposed to submarine erosion. Calcareous oozes occur between pillows and are widely distributed although not abundant.

Unconformities have been recognized or inferred at only seven localities. All of these, with the exception of Mawson Point, are local unconformities, where no significant time difference between the sequences above and below the unconformities has been inferred. At Mawson Point, an angular discordance may occur where steeply dipping metamorphosed pillow basalts exposed around Bauer Bay could be overlain to the south by a sequence of shallowly dipping tabular basalt flows that contain a basalt dated at 11.5 ± 0.3 Ma (Duncan and Varne, 1988). The contact between the different sequences is not exposed.

Dips in extrusive sequences range from shallow to steep and are variable even in areas of constant strike. Along the west coast and in the southern third of the island, the dominant strike trends of extrusive sequences and of basaltic dikes cutting these sequences are northwest, about N40°W. Along the east coast, in the central third of the island, the dominant strike trends are east, averaging about N80°E. East of Aurora Point, trends across the center of the island seem to swing, but outcrops are poor.

South of Brothers Point, a series of narrow, parallel, basaltic dikes cuts the extrusive rocks and forms as much as 40% of the coastal outcrop. The dikes intrude the extrusive rocks at high angles to the bedding, no matter how steep the dips of the bedding planes. The strike of the dikes is also parallel or nearly parallel to the strike of the intruded extrusive rocks. These relationships imply that the basaltic dikes were originally steeply inclined intrusions into relatively flat-lying extrusive sequences and that the axes around which the later tilting of the rocks took place must have been approximately horizontal and parallel to the intersections of the dikes and bedding planes (Varne and Rubenach, 1972).

Inland outcrops are generally poor, but ocean-floor topography little touched by erosion occurs near Pyramid Peak, where downslope from a summit formed of rounded pillows, a ridge is composed of radially-outward-dipping elongate pillow lobes (cf. Fig. 16 of Ballard and Moore, 1977).

The central part of the island may be composed of a coherent crustal block of extrusive rocks dipping shallowly east. The rocks are disturbed by recent faulting, but the east-dipping bedding, estimates of stratigraphic thickness, exposures of sheeted dikes along the west coast, and the distribution of metamorphic assemblages (see the subsequent discussion on "Metamorphism of Macquarie Island crust") are consistent with this simple picture.

2. Sheeted dolerite dikes. The sheeted dolerite dike complexes may be as much as 1500 m thick and mainly occur in the northern part of the island, associated with ultramafic and gabbroic rocks (Fig. 5). The dike complexes are composed almost entirely of subparallel dolerite dikes, typically with less than 5% gabbro screens, and no basalt or ultramafic screens have yet been found. Varne et al. (1969) reported that in the north of the island, the dikes are distributed around a near-horizontal axis trending N60°W, with most dipping ~50°SW.

In the southern part of the island, two other complexes of sheeted dolerite dikes occur in contact with extrusive rocks along the coasts north of Cape Toutcher and Lusitania Bay (Fig. 4). Screens of massive, coarse-grained gabbro are found within the east coast outcrops, but on the west coast, gabbro screens are absent and screens of pillow basalt are rare.

Structurally, sheeted dolerite dikes underlie extrusive rocks and overlie gabbro bodies, but lateral transitions between gabbros and sheeted dikes occur over zones 100–300 m wide in which sheeted dolerite dikes make up 30%–70% of the outcrop (Varne and Rubenach, 1972; Goscombe and Everard, 1998). This association indicates that the sheeted dikes formed following rifting of gabbro bodies to give elongated, sheeted, dike swarms and associated transition zones. These mid-crustal sheeted dolerite dikes are distinct from (1) the series of dikes that intrude the extrusive rocks that have basaltic textures, (2) the individual dikes that occur at all crustal levels below the sheeted dikes, and (3) the packets of sheeted dolerite dikes of limited width (5–100 m) that intrude ultramafic successions and that are dominantly aphyric or contain olivine.

3. Microgabbro transition association. The microgabbro transition succession is transitional between coarse massive gabbro and sheeted dolerite dikes. This transition differs from the lateral transition between gabbros and sheeted dolerite dikes that cuts across the microgabbro transition and so postdates this group of rocks. The fine-grained nature of the sheeted dolerite dikes suggests that the microgabbro is the transitional rock type at the top of the coarse gabbros and immediately below the sheeted dolerite dikes proper.

*4. **Coarse-grained, massive and layered gabbroic rocks with ultramafic screens.*** Massive gabbro occupies 5.9% of Macquarie Island, and layered gabbro occupies 0.1% (Goscombe and Everard, 1998). The bulk of the gabbro can therefore be considered to be compositionally and mineralogically uniform from a field-mapping perspective. Compositionally layered gabbro is known only at Handspike Point where grain-size variation of the gabbro is on a centimeter to meter scale.

Ultramafic rocks are found as steeply dipping (70° to 80°), planar, tabular and laterally continuous screens within the coarse gabbro units. Most of the screens immediately north of Half Moon Bay are plagioclase wehrlite and troctolite. The proportion of screens decreases over a short distance to the north, away from the plagioclase wehrlite association. Elsewhere, screens mostly consist of harzburgite. The highest density of ultramafic screens is in transition zones adjacent to areas of ultramafic rocks, where harzburgite has been intruded by gabbro bodies (Goscombe and Everard, 1998). Flaser gabbro around Island Lake and on Langdon Point has a fabric striking approximately N60°W. Hornblende replaces clinopyroxene along cracks cutting clinopyroxene augen.

*5. **Troctolite association.*** The main troctolite outcrop occurs as a southeast-trending body inland from Elizabeth and Mary Point. Troctolite, commonly with compositional layering, and dominantly plagioclase-bearing massive wehrlite with minor dunite also occur to the north of the harzburgite outcrops on the northwestern side of the island. To the northeast, small areas of dominantly plagioclase-bearing massive wehrlite and minor dunite crop out. In places these outcrops are also cut by gabbro and dolerite dikes. Troctolitic rocks also occur in the central fault-bounded block on the eastern escarpment. The layers range from a few centimeters to meters in thickness and are composed of melatroctolite, troctolite, olivine gabbro, laminated gabbro, and anorthosite.

Outcrops of troctolitic rocks are highly variable because of compositional layering and range from anorthosite to plagioclase wehrlite. According to Christodoulou (1994), in the southern part of Half Moon Bay, layering strikes north, dips variably east-northeast, and can be traced for a few meters. The layers range from a few centimeters to meters in thickness and are composed of melatroctolite, troctolite, olivine gabbro, laminated gabbro, and anorthosite.

Goscombe and Everard (1998) recognized two scales of compositional layering, 10–50 mm and 120–500 mm. Different mineral proportions define both. The thicker compositional layers are mainly gradational, and most become progressively richer in plagioclase upward. The gradational compositional layering indicates that the sequence is right way up. Layering is mostly planar but also gently curviplanar, almost always parallel and laterally extensive, although some thin anorthosite layers that pinch out have been observed.

Plagioclase wehrlite occurs dominantly as centimeter-scale layers within the layered troctolite succession. There is a continuous spectrum between plagioclase wehrlite and troctolite; plagioclase wehrlite typically forms the basal part of the compositional layers that then grade up to troctolite with increasing plagioclase and decreasing olivine. Plagioclase wehrlite also rarely occurs as 10–50-m-wide units within the troctolite association.

Scour surfaces with gently concave scallops have eroded preexisting layers at a few localities. The fact that the scour surface is not a sharp plane on the mineral-grain scale indicates subsequent annealing. Immediately above all scour surfaces, a 15–80-mm-wide layer of fine-grained (0.25–0.80 mm), anhedral, granular troctolite occurs. There is no igneous fabric, such as grain alignment, or ductile deformation associated with scour-surface layers, but they commonly underlie coarse-grained troctolite whose compositional layering is concordant to the scour surfaces.

*6. **Mixed plagioclase wehrlite and dunite succession with gabbro and troctolite screens.*** According to Goscombe and Everard (1998), this succession is dominated by plagioclase wehrlite with dunite and plagioclase dunite units and minor screens of gabbro and troctolite and has only been recognized at Half Moon Bay. Harzburgite is absent. In places, the fact that dolerite dikes constitute up to 50% of the rock indicates that the succession is preserved as a highly dissected suite of isolated screens of the original rock types.

Dunite and plagioclase dunite occur in four locations as tabular or elongate lenses, 2–25 m wide, within plagioclase wehrlite. No discordant contacts have been observed, and the relationship between the two rock types is unknown.

Minor troctolite and gabbro occur as screens between dolerite dikes within the plagioclase wehrlite association, mostly at its northern and southern limits. These screens imply a transitional contact to the troctolite and gabbroic successions. There are, however, abrupt changes in orientation of compositional layering across the boundaries.

*7. **Mixed dunite, wehrlite, and harzburgite succession.*** This mixed ultramafic rock association consists of significant proportions of dunite and plagioclase dunite (30%–40%), plagioclase wehrlite (30%–40%), and up to 20%–30% harzburgite.

Plagioclase wehrlite is mostly massive with ~20% being well layered (10–300 mm). Three varieties of compositional layering have been mapped. Layering is mainly defined by troctolitic layers of 20–50 mm thickness. Plagioclase-rich versus plagioclase-poor wehrlite bands, 50–300 mm wide, form layers that are not laterally continuous and pinch out. All varieties of layering are subparallel to parallel. Discordances are not apparent.

Dunite is massive with no compositional layering apparent. Its textures are adcumulate to orthocumulate. It occurs both as up to 30-m-wide tabular units apparently concordant with the harzburgite and plagioclase wehrlite units and as elliptical bodies up to 100 m long by 50 m wide. No discordant contacts with adjacent ultramafic rocks have been recognized, but discordance cannot be discounted because of poor outcrops.

Coarse-grained gabbroic veins, pegmatites, and dikes intrude all the ultramafic successions. Gabbro veins are most common in dunitic and plagioclase wehrlitic rocks. Gabbro also occurs as wider, discontinuous, lens-shaped intrusions. Some of these are pegmatitic with crystals of clinopyroxene up to 40 mm.

*8. **Harzburgite association.*** The harzburgite association consists almost entirely of massive to locally layered harzburgite with varying proportions of (1) gabbro dikes, pegmatites, and veins, (2) rare troctolite, plagioclase wehrlite, and phlogopite-plagioclase veins, and (3) in places, rare dunite lenses (Goscombe and Everard, 1998).

There are six bodies of harzburgite at Eagle and Unity Points, Boot Hill (located in Fig. 5), two regions in the central northern plateau, and two blocks (north and central) within the ultramafic thrust sliver on the eastern escarpment. There is also an area on the plateau where harzburgite occurs with minor dunite and pyroxenite units.

The maximum total thickness of exposed harzburgite, perpendicular to layering, is 700 m at Eagle Point. The massive gabbro association cuts across harzburgite and is younger.

Compositional layering is recognized in harzburgite at 14 localities, and each area has a reasonably consistent orientation throughout.

There are no discordant relationships associated with this layering, which is defined by varying proportions of olivine and orthopyroxene. Orthopyroxene commonly poikilitically encloses olivine, a texture that may be cumulate in origin. Dunite layers and lenses are concordant to compositional layering in the harzburgite in all but one exposure, and the margins appear gradational with the host harzburgite. Consequently, dunite units within the harzburgite succession are interpreted as lenses, layers, and pods that crystallized at the same time as the harzburgite and not as later intrusions. The harzburgite underwent some minor deformation before serpentinization, but no tectonite fabric has been recorded.

Gabbro and olivine gabbro dikes, pegmatites, and veins are very common in the harzburgite association, but are scarce within associations 6 and 7 and absent from association 5. Gabbro intrusions predate serpentinization, faults, retrogressive veins and all dolerite dikes; therefore, the gabbroic dikes and veins are early formed but clearly intruded a cool, already brittle, harzburgite.

Dolerite dikes (<5%–8%) are steeply dipping (70°–90°) and at present trend south-southeast. They are much less common in this association than in associations 3 to 7.

Metamorphism of Macquarie Island crust

The distribution of secondary assemblages in lavas and dikes has been used to map metamorphism on Macquarie Island (Griffin, 1982). Griffin (1982) showed that the various assemblages form a series that is essentially temperature dependent; the grades of metamorphism span a range from very low temperature ocean-floor weathering (0–20 °C), through zeolite facies (20–220 °C), lower-greenschist facies (220–320 °C), upper-greenschist facies, and finally lower-amphibolite facies (~450 °C). Pressure effects are small; pressures at the base of the dike swarms range up to ~0.15 GPa. Griffin (1982) concluded that the assemblages representing the various grades of metamorphism are widely distributed in the extrusive sequences and that this regional nature of the distribution demonstrates the regularity of the oceanic-crust metamorphic process.

K-Ar measurements imply that ocean-floor weathering was occurring from 9.6 to 5.8 Ma (Duncan and Varne, 1988) whereas $^{40}Ar/^{39}Ar$ experiments on metamorphosed rocks from the lower pillow lavas and sheeted dikes gave an age range of 6.5–7.2 Ma, which records the time of closure of these rocks to diffusion of Ar and the end of greenschist metamorphism conditions (Duncan and Varne, 1988).

Regional distribution of metamorphic assemblages in extrusive rocks. On North Head (Fig. 5), in the northern part of the island, the extrusive rocks are relatively fresh, include unaltered glass, and have undergone only the smectite-carbonate alteration that is characteristic of ocean-floor weathering. (Rocks of lower-greenschist facies occur in a narrow zone near the Anare Station, in faulted contact with the North Head alkalic volcanic rocks.)

Elsewhere the extrusive rocks were generally metamorphosed at higher grades, zeolite facies and above. However, fresh glass and rocks that have only undergone ocean-floor weathering occur together with the zeolite-bearing assemblages, suggesting patchy metamorphic conditions near the lower limit of zeolite-facies metamorphism. The intensity of alteration increased and became more uniform as the grade of metamorphism increased.

The zone of zeolite-bearing assemblages covers the eastern part of the central third of the island and extends from west of Mount Tulloch, Mount Law, and Pyramid Peak (Fig. 4) to the east coast between Brothers Point and Waterfall Bay.

Griffin (1982) suggested that a structurally concordant, east-dipping section could be followed westward from Pyramid Peak to the coast around Davis Point (Fig. 4). This extrusive section appears to have a thickness of ~1400 m and ranges from little-altered volcanic assemblages at its top, around Pyramid Peak, to lower-greenschist–facies assemblages at its base, on the west coast; therefore, a temperature gradient of ~200 °C/km is implied at the time of the metamorphism. Sheeted dikes underlie the extrusive section.

Along the west coast of the island, lower-greenschist–facies extrusive rocks intruded by dikes extend in a zone down the west coast from south of Cormorant Point, along the coast around Davis Point, to the major fault north of Cape Toutcher. The dikes dominantly strike northwest, approximately N40°W, and the dips on bedding are to the southeast.

Along the east coast, in the central third of the island, the dominant strike trends are east, averaging about N80°E. Outcrops are poor inland, and the extrusive rocks are cut by curving lineaments, fractures, and faults (Fig. 6). The Brothers fault line cuts off most of the coastal section (Ledingham and Peterson, 1984), but southwest of Brothers Point and west of the Brothers fault line, mapping at 1: 10 000 (map sheet 3 of Goscombe and Everard, 1998) shows that the general strike of lens-shaped outcrops of volcaniclastic rocks within pillow lavas swings from approximately N70°–80°E near the coast, through due east, to N70°W ~1 km inland.

Griffin's (1982) stratigraphic and metamorphic transition from Pyramid Peak to the west coast suggests that the central third of Macquarie Island can be treated as a block of fairly coherent oceanic crust. Near the base of the section, metamorphic assemblages lie within the lower-greenschist facies. Across much of the poorly exposed center of the island, and along the east coast, the widespread distribution of fresh and little-altered glass and zeolite-facies assemblages shows that shallow crustal levels are widely exposed. Similarly, the lack of reports of greenschist-facies assemblages from the center of the island, coupled with the radiometric constraints on the timing of the cooling of the crust, again suggests that only shallow crustal levels are exposed.

Regionally, these findings imply that the section may dip shallowly east, that large (≥0.5 km) vertical movements along the many faults that cut the crust are unlikely, and that little erosion has occurred. The change in regional strike across the island seems to be an accumulation of small rotations of crustal blocks.

In the southernmost third of the island, south of the major fault zone between Cape Toutcher and Lusitania Bay (Figs. 4 and 6), the strike of the tholeiitic extrusive volcanic rocks is generally northwest, and these are almost all metamorphosed to lower-greenschist–facies assemblages (Griffin, 1982), except in the southeast, where lower-grade assemblages occur.

The sheeted dolerite dikes are compositionally similar to the extrusive basaltic rocks (Griffin and Varne, 1980), but tend to be more altered, with widespread replacement of primary mafic minerals by actinolite. Plagioclase survives little changed, as do the doleritic textures (Varne and Rubenach, 1972; Banerjee et al., 1974; Griffin, 1982). Locally, the metamorphic grade lay within the amphibolite facies. Gabbroic rocks are little altered, with minor alteration to secondary greenschist-facies and lower-amphibolite–facies assemblages.

Cocker et al. (1982) used oxygen isotope compositions to divide the metamorphosed rocks into two groups. The extrusive zeolite-facies to lower-greenschist–facies metabasalts are enriched in ^{18}O relative to fresh basaltic glass, whereas the intrusive rocks—including greenschist-facies and lower-amphibolite–facies sheeted dolerite dikes, gabbro, and serpentinized harzburgite—are depleted in ^{18}O. Oxygen and carbon isotope evidence and calcite measurements suggest seafloor weathering temperatures of <20 °C (Cocker et al., 1982) and also show that apparently little-altered gabbros deep in the crustal section exchanged some oxygen with seawater.

Spreading and postspreading events in the Macquarie Island region: Possible records in the geology of Macquarie Island

The Macquarie Ridge Complex has evolved progressively from a spreading ridge into the active Australian plate–Pacific plate right-lateral strike-slip plate boundary. The location of the island just to the east of the plate boundary suggests that spreading slowed and ceased soon after Macquarie Island crust was formed (between 12 and 9.5 Ma). On Macquarie Island, the geologic record of the spreading process is therefore likely to have been modified by later rotation of crustal blocks and strike-slip faulting parallel to the current plate boundary.

Geologic results of spreading. The broad layered structure of the crust of Macquarie Island, all of its rock associations, its metamorphism, and the compositions of its rocks are characteristic products of seafloor spreading. However, seafloor spreading involves mechanical extension as well as magmatism, the later tilting of crustal blocks in rift walls, and the imprinting of the tectonic spreading fabric.

Reconstruction (Fig. 3) suggests that in the Macquarie Island region, the half–spreading rate, at ~10 mm/yr, was slow. Traces of tectonic spreading fabric imaged by Massell et al. (2000) strike between east-southeast and east-northeast near the Macquarie Island segment of the Macquarie Ridge Complex, in agreement with the approximately east-west orientation of the modeled spreading axes at 11 Ma. Macquarie Island crust seems likely to have formed at a short segment or segments of slow-spreading ridge between large-offset transforms.

Oceanic crust that has formed at slow-spreading ridges, particularly near intersections with large-offset transform faults, seems structurally more complex than crust formed at fast-spreading ridges. Studies of the MARK area (Mid-Atlantic Ridge–Kane Fracture Zone; e.g., Karson et al., 1987; Mutter and Karson, 1992; Gao et al., 1998), where the slow-spreading Mid-Atlantic Ridge is offset along the Kane transform, show that within the median valley, a contact between young pillow lavas of the rift floor and gabbroic outcrops along the base of the western wall can be traced for several kilometers. Greenstone, metagabbro, and ultramafic rocks, which were formed several kilometers within the crust, are commonly exposed in ridge-parallel escarpments along low-angle normal-slip detachment faults and in the elevated flanks of the inside corner high. Much of the southern wall of the eastern Kane transform is composed of gabbro covered by talus and pelagic sediment.

The geologic relationships around and inland from Bauer Bay resemble those of the MARK area. Extrusive sequences are in faulted contact for several kilometers with gabbro and sheeted dolerite dikes separated by gabbro screens or interleaved with harzburgite at Langdon Bay, where high-temperature deformation zones also occur, marked by flaser gabbro. West of the contact on Douglas and Langdon Points, tabular basalt flows are underlain by sandstone and conglomerate containing gabbro and dolerite clasts. This succession (Fig. 5) unconformably overlies pillow lavas (Goscombe and Everard, 1998). To the south, around Bauer Bay, steeply dipping, metamorphosed pillow basalts may underlie the sequence of little-altered, shallowly dipping tabular basalt flows at Mawson Point. Strikes of dikes within the sheeted-dike complexes largely lie parallel to the contact between extrusive and intrusive rocks, about N60°W. We conclude that this part of the crust of Macquarie Island was formed in a slow-spreading setting and retains geologic relationships similar to those of the MARK area.

At the risk of overinterpreting the data, we note that the geologic discontinuities across the island at North Head, between Bauer Bay and Sandy Bay, and between Cape Toutcher and the east coast north of Lusitania Bay, are subparallel, and all marked by compositional and metamorphic changes. The rocks of North Head, which scanty data suggest are 1.8 Ma younger than those in the central part of the island, were originally thought of as possible representatives of younger off-axis volcanism (Duncan and Varne, 1988). New analytical data (Kamenetsky et al., 2000) show that glass compositions from North Head form part of the same distinctive compositional series as glasses from the rest of the island.

If we suppose that North Head magmatism was fed by similar sources in the same spreading system, at the same spreading center, the implied half–spreading rate of ~5 mm/yr fits the regional geodynamic reconstruction, which implies that spreading was slowing to a stop at the Australian plate–Pacific plate boundary about this time.

Later traces of the effects of transpression at the plate boundary. Massell et al. (2000) have located the plate boundary along a zone of multiple, ridge-parallel lineaments. The island is a few kilometers to the east, within a zone of diffuse faulting at least 20 km wide, marked by ridge-parallel submarine lineaments traced along the submarine ridge slopes east of Macquarie Island (Massell et al., 2000).

Transpression along the plate boundary has caused crustal thickening and overall east-west crustal shortening (Williamson, 1988). Transpression has probably also caused the system of close-spaced, north- and northwest- trending fault sets and lineaments (Fig. 6) that cut the eastern half of the island, some with a component of reverse faulting.

We suggest that Macquarie Island began as a long, narrow block of oceanic crust. The block was bounded by north-trending transforms, and its seafloor-spreading fabric trended approximately east (Fig. 3).

Now, the block lies within a dextral shear zone aligned parallel to the plate boundary. The northeast block of intrusive rocks and the eastern part of the central block of extrusive rocks have been rotated little with respect to the original orientation of the spreading fabric. The extrusive rocks of the southern part of the island and along the western coast of the central block may have been rotated clockwise through 50°–60°, similar to the amount and sense of rotation of the northwest intrusive block. In Figure 7, we show the changes in orientation of dikes cutting the volcanic sequences across the central part of Macquarie Island. Major fault structures, taken from Figure 6, are also shown. Furthermore, though not illustrated here, the strike of dikes in gabbro to the west of Mount Power in the northern part of the island is about northwest, compared with the east-northeast strike of the dikes to the north-northwest of Mount Blair (see Fig. 5 for locations; Goscombe and Everard, 1998, sheet 2).

From east to west across the island, the strike therefore now curves from N80°E along the east coast to N60°W in the west, implying that the fabric has been progressively rotated across the island. On this broad scale, the main blocks that compose Macquarie Island resemble in size and shape the rhomb-shaped blocks of oceanic crust imaged within the dextral Sovanco shear zone (Cowan et al., 1986). Cowan et al. (1986) suggested that older crust has been disrupted into blocks that have rotated within a large-scale shear zone ("bookshelf tectonics"). The faults bounding the blocks are typically oblique to the boundaries of the enclosing shear zone. Following Cowan et al.'s (1986) simple model, we suggest that blocks of Macquarie Island crust were similarly rotated within a dextral shear zone bounded to its east by the Brothers fault-line system.

Transpression and overall east-west shortening, reverse faulting, and crustal thickening could have caused some eastward tilting and elevation of the crustal blocks of which the island is composed. Elevation above sea level of the rotated and tilted northwestern block was followed by marine planation, possibly more than once, which exposed the section through Macquarie Island ocean crust. The western limits and thicknesses of the crustal blocks are not known, but the northern block contains a near-complete tilted crustal section, through to harzburgite. Marine planation of the northern part of the island is more pronounced, and raised beaches are more common, than in the south, and elevation of the central part of the island may have been relatively rapid as well as more recent.

Goscombe and Everard (1999), in a preliminary note, identified three tectonic phases—D1 (extension), D2 (transtension), and D3 (transpression)—from field relationships on the island. Their D1 phase includes the crust and structures formed during or soon after spreading, discussed here earlier as "Geologic results of spreading," and their D3 phase includes the effects of the current tectonic regime along the plate boundary, discussed here as "Later traces of the effects of transpression at the plate boundary." They have ascribed the eruption of subhorizontal lavas devoid of D1 dikes around Bauer Bay, volcanic centers near Pyramid Peak, and the intrusion of some late-stage dolerite dikes, to their intermediate, D2, phase. We suggest instead that the subhorizontal lavas are devoid of D1 dikes because they were late eruptions on the floor of the spreading rift and that the volcanic centers near Pyramid Peak share a similar origin. We see no strong evidence as yet for D2.

Goscombe and Everard (1999) also suggested total clockwise block rotations ranging from 38° on the east coast, to ~77° for the northern intrusive associations, and to 106° in the south and west of the island. They seem to have derived these rotations with respect to an assumed original orientation of approximately N40°E for the spreading fabric, whereas new information shows that the fabric was more likely to have been oriented roughly east-west when Macquarie Island oceanic crust was formed (see Fig. 3, discussion in the text, and Massell et al., 2000).

PETROGENESIS OF THE MAGMATISM OF THE ISLAND: N-MORB TO E-MORB AND BEYOND

Mawson (1943), Cameron et al. (1981), Griffin and Varne (1980), Griffin (1982), Varne (1989), and Christodoulou (1994), among others, have described the petrography and chemical compositions of Macquarie Island basalts, dolerites, and gabbros. Here, we briefly summarize their findings before we focus on the petrogenesis of the magmatism of the island and search for a causative link between its petrogenesis and its geodynamic setting.

Macquarie Island basaltic and doleritic rocks range from coarsely porphyritic to aphyric, from tholeiitic to alkalic, and from almost wholly crystalline to almost wholly glassy. As discussed earlier, although fresh glass is widespread, many of the basalts and virtually all of the dolerites have been altered and metamorphosed to varying extents. According to Griffin (1982), tholeiitic basalts and dolerites are more commonly porphyritic than alkalic varieties. Plagioclase is the most common phenocryst phase, followed by olivine, spinel, and rarely clinopyroxene. Where alteration is slight, the crystalline groundmass is mainly composed of clinopyroxene, plagioclase, opaque minerals, and recrystallized glass; there are pseudomorphs after olivine and rare amphibole in relatively alkalic varieties (Griffin and Varne, 1980).

Phenocryst assemblages and major element compositions of the minerals of the basalts and dolerites resemble those of the gabbros, particularly in the layered rocks, showing that the igneous rocks are related (Griffin and Varne, 1980; Christodoulou, 1994). Compositions of troctolitic clinopyroxene (Christodoulou, 1994) resemble those of tholeiitic clinopyroxene (Griffin, 1982).

In chemical composition, most of the basalts and dolerites resemble MORB (Varne et al., 1969; Griffin and Varne, 1980; Varne, 1989). Their Mg/(Mg + Fe) values range from ~0.72 to 0.51. The rocks range normatively from *ol*- and *hy*-bearing tholeiitic varieties, similar to N-MORB, to *ne*-bearing varieties similar to E-MORB but ranging to unusual, more strongly alkalic and *ne*-normative compositions. The normative transition from *ol*- and *hy*-bearing to *ne*-bearing compositions corresponds approximately with the appearance in the rocks of groundmass amphibole and olivine. Compared with lavas, the dikes tend to be less rich in *ol* and *ne,* include the few *Q*-normative rocks, and have lower Mg/(Mg + Fe) values than the lavas; these facts imply that dikes are more differentiated than the lavas.

When minor and trace element data (Ti, Rb, Sr, Y, Zr, and Nb) were first reported long ago for Macquarie Island basalts and dolerites (Varne and Rubenach, 1972), these data showed that the group of Macquarie Island rocks resembled MORB much more closely than representative groups of basalt from orogenic and oceanic island settings. However, no representatives of the extremely depleted N-MORB were found, and some Nb-rich alkalic basalt from Macquarie Island fell outside the MORB volume and into a compositional volume termed "unlike typical ocean-floor basalt" which was also unlike oceanic island basalt. Within the decade, geochemically enriched types of MORB with high Nb contents (E-MORB) had been described from the Mid-Atlantic Ridge (Griffin and Varne, 1980).

There is also a clear tendency for tholeiitic compositions to occur in the southernmost part of the island and in the northern third (excluding North Head), whereas the alkalic types are more common in the central part of the island (Varne and Rubenach, 1972; Griffin, 1982). If the crust of Macquarie Island is viewed as representing a time sequence, younging from south to north, then the oldest magmatism was more tholeiitic than alkalic, the magmatism of the central part of the island was more alkalic than tholeiitic, the magmatism of the northern block was tholeiitic, and the magmatism at North Head was alkalic. If the composition of the igneous rocks is compared with metamorphic grade, and by implication to crustal depth, then the magmatism becomes more tholeiitic with increasing depth in the crust, although the range from tholeiitic to alkalic persists through the section.

In either case, the magmatism of Macquarie Island forms a continuum which ranges from N-MORB, through E-MORB, to compositions more alkalic than E-MORB (Griffin and Varne, 1980; Kamenetsky et al., 2000). The Macquarie Island enriched glasses have compositions that have many features in common with other E-MORB suites (Kamenetsky et al., 2000), but range to alkalic compositions that seem to be variants of MORB but more enriched than any reported elsewhere.

In view of the extreme compositional variation of the MORB magmatism of Macquarie Island, we have taken advantage of Kamenetsky et al.'s (2000) comprehensive and high-quality data set to

TABLE 1. ANALYSES OF MACQUARIE ISLAND GLASS END MEMBERS

Glass	Utas47979 (Analysis 1)*	Utas47963 (Analysis 6)*
SiO_2	49.56	48.18
TiO_2	0.97	1.91
Al_2O_3	17.09	18.17
FeO	8.06	6.81
MnO	0.14	0.08
MgO	8.75	5.90
CaO	12.82	10.85
Na_2O	2.37	4.24
K_2O	0.14	1.76
P_2O_5	0.08	0.66
Cr_2O_3	0.04	0.02
H_2O	0.25	1.49
Total	100.27	100.06

* Analyses 1 and 6 of Table 1 of Kamenetsky et al. (2000). Analyses by electron microprobe for major elements, except H_2O by Fourier transform infrared spectroscopy. All oxides in weight percent. All iron reported as FeO.

delineate the origins of its end members. We then speculate about the causes of compositional variability and enrichment in MORB.

Glass compositions

Kamenetsky et al. (2000) have presented the results of petrologic and geochemical studies of hyaloclastites and pillow-rim glasses sampled from North Head and from the central part of the island. They concentrated their attention on the compositions of glasses with high Mg/(Mg + Fe) values and microphenocrysts of forsteritic olivine ($Fo_{86-89.5}$) and Cr-rich spinel [mean Cr# = Cr/(Cr + Al) = 35]. These "primitive" compositions define a seriate variation that starts within the compositional spectrum of MORB (Table 1, analysis UTas47979) and extends to enriched compositions (Table 1, analysis UTas47963) that represent a new primitive enriched MORB end member (Kamenetsky et al., 2000).

Compared with N-MORB, this new end member is characterized by relatively low contents of MgO, FeO, SiO_2, and CaO, coupled with high contents of Al_2O_3, TiO_2, Na_2O, P_2O_5, K_2O, and incompatible trace elements. MgO abundances range from low (5.9 wt%) in the most enriched to high (up to 8.8 wt%) in the least enriched "primitive" glass compositions, similar to many E- to N-MORB suites. Enrichment in incompatible elements correlates positively with $^{87}Sr/^{86}Sr$ (0.70255–0.70275) and Pb isotope ratios ($^{206}Pb/^{204}Pb$ = 18.951–19.493; $^{207}Pb/^{204}Pb$ = 15.528–15.589; $^{208}Pb/^{204}Pb$ = 38.523–38.979) and negatively with $^{143}Nd/^{144}Nd$ (0.51310–0.51304). The Sr, Nd, and Pb isotope characteristics of Macquarie Island glasses are those of Pacific MORB, but in isotopic space, they define a short linear array aligned between DMM (Depleted MORB mantle) and HIMU [high μ = (U + Th)/Pb] magmas (Kamenetsky et al., 2000).

Significance of Macquarie Island "primitive" glass compositions. Geochemically enriched types of MORBs are found in some segments of ridges characterized by slow spreading rates or by closely spaced large-offset transforms; off-axis seamounts in the vicinity of spreading centers are also locations for occurrences of such MORBs. Their formation is commonly attributed to the influence of near-ridge plumes or hotspots, or alternatively to the presence of blobs or veins of enriched material in depleted peridotite mantle (Kamenetsky et al., 2000). Such heterogeneities seem necessary to account for the correlation between trace element and radiogenic isotope characteristics reported in many MORB suites (the so-called coupled MORB of Niu et al., 1996).

However, current models of MORB melt generation, segregation, and extraction (e.g., McKenzie and Bickle, 1988; Natland, 1989) predict that discrete primary melts will be extremely rare or even absent in the oceanic lithosphere, and traces of source heterogeneities and of small amounts of mantle melting are unlikely to survive within relatively long-lived and robust magmatic systems located beneath oceanic spreading centers.

Macquarie Island glass compositions are therefore potentially of great petrogenetic significance. First, the glasses come from a small, geologically well-constrained and closely sampled area of ocean crust. Second, they include a range of compositions that define a seriate variation from average N-MORB to unusual, highly enriched compositions. Third, this variation is characterized by strong bivariate correlations ("coupling") between major, trace, and volatile element abundances and isotope ratios. Finally, the range of compositions includes "primitive" (near-primary) compositions, modified by olivine crystallization alone.

Experimental constraints on the origins of the end members. The existence of the "primitive" glasses provides a rare opportunity to delineate the likely conditions under which the melts might have been generated. Our approach is to compare Macquarie Island glass compositions with those of experimentally determined mantle melts of appropriate MORB-source mantle compositions. Figure 8 shows the location of experimentally determined Ol + Opx + Cpx ± Sp + L cotectics for MORB-source mantle, at 1.5 and 2.0 GPa, within Falloon and Green (1988)'s CIPW molecular normative [*Jd* + *CaTs* + *Lc*]–*Di*–*Qz*–*Ol* "basalt tetrahedron" (see insets in Fig. 8).

Figure 8A is a projection from normative *Ol* onto the face [*Jd* + *CaTs* + *Lc*]–*Di*–*Qz* of the "basalt tetrahedron." Figure 8B is a projection from Di onto its base [*Jd* + *CaTs* + *Lc*]–*Qz*–*Ol.* With increasing pressure, the cotectic shifts toward the *Ol* apex. With increasing temperature and degree of melting, indicated by arrows, the compositions of partial melts of MORB-source mantle become poorer in *JCL* (= normative *Jd* + *CaTs* + *Lc*) components and richer in *Ol* and *Di.*

A field enclosing the compositions of primitive MORB glasses is shown in Figure 8, A and B. These compositions fall around the low-pressure cotectic (not shown). Six representative "primitive" Macquarie Island glass compositions (analyses 1–6 of Table 1 in Kamenetsky et al., 2000) are also projected and plotted. They define a low-pressure array from the MORB field to highly undersaturated compositions.

The most magnesian olivine in these "primitive" glasses ranges from $Fo_{88.2}$ in high-K glasses to $Fo_{89.5}$ in low-K glasses. These olivine compositions are less magnesian than those of the most magnesian olivine in Macquarie Island low-K picrite ($Fo_{90.5}$), peridotite ($Fo_{90.5}$), and harzburgite ($Fo_{91.5}$) as well as in abyssal mantle peridotites worldwide.

The compositions of the host glasses were adjusted by Kamenetsky et al. (2000) for the effects of olivine-only fractionation by adding olivine component until calculated equilibrium with olivine ($Fo_{90.5}$) was achieved. The amount of olivine added during the correction procedure did not exceed 5–8 wt%. It should be noted that the analyzed compositions of the glasses cannot represent the compositions of the erupted melts, even allowing for the loss of fugitive components, because the glass-forming melts were already modified by the crystallization of up to 5 vol% (~6 wt%) microphenocrysts of olivine. Thus the compositional path retraced by the correction for olivine crystallization is likely to pass through the actual liquidus

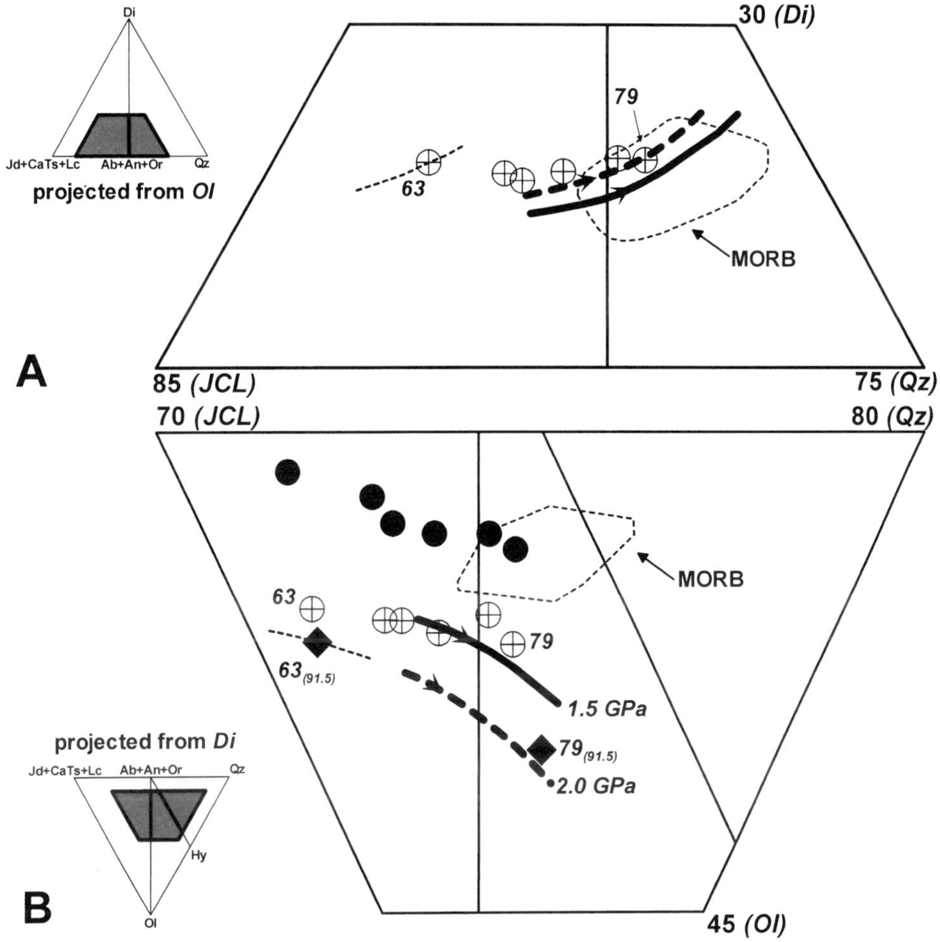

Figure 8. Comparison of Macquarie Island basaltic glass compositions with experimentally determined Ol + Opx + Cpx ± Sp + L cotectics for MORB-source mantle at 1.5 and 2.0 GPa and with the compositions of primitive MORB glasses, within the CIPW molecular normative "basalt tetrahedron" of Falloon and Green (1988) see small inset). A, Projection from *Ol* on to part of the face [*Jd* + *CaTs* + *Lc*]–*Di*–*Qz* of the "basalt tetrahedron." B, Projection from *Di* on to part of the base [*Jd* + *CaTs* + *Lc*]–*Qz*–*Ol* (see small inset). The thick line labeled 1.5 GPa delineates the Ol + Opx + Cpx ± Sp + L cotectic for MORB-source peridotite MM-3 at 1.5 GPa (data from Falloon et al., 1999); the thick dashed line labeled 2.0 GPa delineates the Ol + Opx + Cpx ± Sp + L cotectic for peridotite MM-3 at 2.0 GPa (Falloon et al., 1999). The MORB field was constructed to enclose 50 primitive MORB glasses (data points not shown) from the database of Melson and O'Hearn (1986, and references contained therein). The filled circles show the projected locations in the basalt tetrahedron of a representative set of six primitive Macquarie Island basaltic glasses (analyses 1–6 from Table 1 of Kamenetsky et al., 2000). The circles with crosses are the same six compositions, following adjustment by olivine addition to be in equilibrium with $Fo_{90.5}$ (see Kamenetsky et al., 2000). The end-member compositions of the set are labeled 63 and 79. The filled diamonds are these same end-member compositions adjusted by olivine addition to be in equilibrium with $Fo_{91.5}$ (see text for discussion). Note that in A (the projection from Ol), the projected locations of the glass compositions are insensitive to the effects of this olivine addition.

composition of the melt, which will lie somewhere between the analyzed and recalculated "near-primary" melt composition.

Kamenetsky et al. (2000) have suggested that the excellent correlation among major, trace, and isotopic compositions of the "primitive" glasses could be explained by variable degrees of partial melting of a source that was originally more enriched than MORB-source mantle, but was progressively depleted during melting toward the composition of the mantle source for typical N-MORB.

Aside from the failure of this suggestion to account for the systematic, although small, variations in the Sr, Nd, and Pb isotopc compositions, comparison of the recalculated "near-primary" melt compositions with the experimentally derived cotectics shows that none of these compositions seems to have a composition appropriate to that of a primary MORB-source mantle melt. This result implies that the melts having the recalculated "near-primary" compositions might have had a prior history of olivine fractionation that has not so far been identified in the compositions of olivine phenocrysts of the Macquarie Island glasses but possibly survives in rare picrite flows.

The "near-primary" melt compositions plot as approximately horizontal linear arrays both when viewed in projection from *Ol* (Fig. 8A), owing to almost constant normative *Di,* and when viewed in projection from *Di* (Fig. 8B), owing to almost constant normative *Ol.*

These arrays are not consistent with a simple model of isobaric melting of a spinel lherzolite source, because the progressive isobaric melting of a MORB-source mantle will produce melts that increase in normative *Ol* and *Di* with melting.

The most enriched compositions are also richer in *Di* and *JCL* components than the likely composition of any liquids produced by small degrees of melting of MORB-source mantle. To be capable of yielding the major element compositions of the enriched melts, their source mantle needs to be richer in *Di* + *JCL* components.

We have investigated this problem further by seeking conditions and compositions where primary melt compositions might exist that could be parental to the Macquarie Island "primitive" glasses.

We selected two melt compositions (Table 1) that define the two extremes of the "primitive" series, UTas47979 (henceforth, composition 79) and UTas47963 (henceforth composition 63). We recalculated these two compositions by adding olivine component (Fo_{100}), following the procedure of Kamenetsky et al. (2000), until we achieved melt compositions in equilibrium with $Fo_{91.5}$, the composition of olivine in Macquarie Island harzburgite. To attain these compositions from the glass compositions as analyzed (Table 1) required the addition of 9.34 mol% olivine for composition 63 and 8.90 mol% for composition 79, not a particularly large amount, particularly bearing in mind that the original compositions have up to 5 mol% olivine microphenocrysts unaccounted for in this calculation. The small amount of olivine that must be added bears out the particularly "primitive" compositions of the glasses.

The recalculated picritic composition 79 (filled diamond labeled "79" in Fig. 8B) now falls on the 2.0 GPa cotectic for MORB-source mantle and could have been the product of ~10–15 wt% melting of N-MORB-source mantle at pressures of 1.5–2 GPa, similar to implied melting conditions under mid-ocean ridges (Falloon and Green, 1988), whether the melt is viewed as an accumulation of melt fractions produced at various points throughout the melting regime or a single batch melt. If the model MORB-source mantle was more "refractory," i.e., depleted in melt components, the cotectics in Figure 8A would shift away from the (*Di*) and (*JDL*) apexes of the normative tetrahedron, and the amount of melting would be less.

The phase relationships also imply that the model mantle source for the recalculated picritic composition 63 (filled diamond labeled "63" in Fig. 8B) would necessarily be richer in (*Di* + *JCL*) components than MORB-source mantle. However, few experiments have been made involving similar compositions, and we also note that composition 63 includes 1.49 wt% H_2O; so, we have sketched a possible cotectic, though it is not defined by experiment (Fig. 8).

Magmatism in a waning spreading system

N-MORB magmatism is believed to be the result of extensive polybaric decompression melting within rising asthenospheric mantle, accompanied by accumulation and mixing of melt fractions, followed by crystal fractionation and magmatic differentiation in shallow magma chambers. In the energetic magmatic systems of long-lived active spreading centers, isotopic traces of source heterogeneities and small melt aliquots would rarely survive these processes that first homogenize and then differentiate the melts. Conditions promoting mixing and the obliteration of differences between small melt fractions are less likely to dominate the magmatic systems of slow-spreading ridges or of ridges segmented by large-offset transforms—so that the magmatic systems must also be segmented—whether spreading rates are fast or slow (e.g., Batiza, 1996).

Geodynamic reconstructions show that ca. 12–9.5 Ma oceanic crust of the Macquarie Island region was created at the Australian plate–Pacific plate boundary within a slow-spreading system of short, left-stepping ridge segments linked by large-offset transform faults (Fig. 3). The spreading was slowing below 20 mm/yr full spreading rate, and magmatism was shutting down as the plate boundary changed into a right-lateral strike-slip transpressional boundary when asthenospheric mantle upwelling ceased and, with it, decompression melting.

Theoretical models predict that at very slow spreading ridges, conductive heat loss from the upwelling asthenospheric mantle will reduce the temperature in the melting region; this effect results in smaller degrees of melting than at faster-spreading ridges and the production of thinner crust (Bown and White, 1994). Depth to the top of the melting zone will increase, and the degree of melting will decrease, as the spreading rate falls below 20 mm/yr.

Summary of Macquarie Island magmatism

We propose that the magmatism of Macquarie Island is the outcome, possibly inevitable, of a waning spreading system.

We recall that the magmatism formed a continuum that ranges from N-MORB through E-MORB and beyond to compositions that are extreme variants of MORB, more enriched than any reported elsewhere.

There is a high proportion of "primitive" glasses with high Mg/(Mg + Fe) values and microphenocrysts of Fo-rich olivine and Cr-rich spinel. The "primitive" glasses comprise a series of discrete, homogeneous samples with a range of compositions from N-MORB to E-MORB and beyond, which defines the compositional spectrum, step by step.

We have used experimental data to argue that the compositions defining the extremes of the compositional range—N-MORB composition 79 and enriched composition 63—came from two distinct mantle sources. The major element composition of sample 79 could have been derived by olivine fractionation from a parental melt produced by ~10–15 wt% melting of N-MORB-source mantle, at pressures of 1.5–2 GPa. The major element composition of sample 63 came from a mantle much more enriched in the felsic normative components, *Jd* ($NaAlSi_2O_6$), *CaTs* ($CaAl_2SiO_6$), and *Lc* ($KAlSi_2O_6$), which is likely to have been melted at similar pressures. Yet the Sr, Nd, and Pb isotope compositions of N-MORB sample 79 and E-MORB sample 63 both resemble Pacific MORB (Kamenetsky et al., 2000).

These inhomogeneities may not have survived in a spreading system where the products of decompression melting of rising asthenosphere were abundant and mixing was easy. The existence of a mantle rich in felsic components implied by the existence of "primitive" glasses with major and trace element compositions like sample 63 could have been unsuspected because the isotopic signature of that sample is MORB-like.

Yet, paradoxically, the existence of the spectrum of glass compositions is also a testament to the efficiency of the mixing process that produced the spectrum of compositions in the melts or their sources, or both, but failed to mix the variation away. If melting and mixing, either of sources or melts, had been more efficient, the magmatism would have resembled a typical N- to E-MORB suite with some limited variation, and evidence for the highly enriched end-member source would have gone unnoticed.

ACKNOWLEDGMENTS

We thank Rupert Sutherland for his help and constructive criticism, Steve Cande for some help with this article that he might have forgotten giving us, Eldridge Moores and Bob Duncan for their critical reviews, Mike Coffin for discussion and permission to quote from Massell et al. (2000), and Yildirim Dilek for his patience. Our thanks also to the Data Management Group, Mineral Resources Tasmania, for the production of the diagrams. Brown publishes with the permission of the Tasmanian Director of Mines.

REFERENCES CITED

Ballard, R.D., and Moore, J.G., 1977, Photographic atlas of the Mid-Atlantic Ridge Rift Valley: New York, Springer-Verlag, 114 p.

Banerjee, S.K., Butler, R.F., and Stout, J.H., 1974, Magnetic properties and mineralogy of exposed ocean crust on Macquarie Island: Zeitshrift für Geophysik, v. 40, p. 537–548.

Batiza, R., 1996, Magmatic segregation at mid-ocean ridges: A review, *in* MacLeod, C.J., Tyler, P.A., and Walker, C.L., eds., Tectonic, magmatic, hydrothermal, and biological segmentation of mid-ocean ridges: Geological Society of London Special Publication 19, p. 103–130.

Bown, J.W., and White, R.S., 1994, Variation with spreading rate of oceanic crustal thickness and geochemistry: Earth and Planetary Science Letters, v. 121, p. 435–449.

Butler, R.F., Banerjee, S.K., and Stout, J.H., 1976, Magnetic properties of oceanic pillow basalts: Evidence from Macquarie Island: Geophysical Journal of the Royal Astronomical Society, v. 47, p. 179–196.

Cameron, W.E., Nisbet, E.G., and Dietrich, V.J., 1981, Petrographic dissimilarities between ophiolitic and ocean-floor basalts, *in* Panayiotou, A., ed., Ophiolites (Proceedings of the International Ophiolite Symposium, Cyprus, 1979): Nicosia, Cyprus Geological Survey Department, p. 182–192.

Cande S.C., and Kent, D.V., 1992, A new geomagnetic polarity time scale for the Late Cretaceous and Cenozoic: Journal of Geophysical Research, v. 97, p. 13917–13951.

Cande, S.C., Stock, J., Raymond, C., and Müller, R.D., 1998, New constraints on plate tectonic puzzle of the SW Pacific: Eos, Transactions of the American Geophysical Union, v. 79, p. 81–82.

Christodoulou, C., 1994, Mineralogy of the Macquarie Island plutonic suite: Fractionation processes at shallow level magma chambers and controls on basalt major element chemistry: Ofioliti, v. 19, p. 217–245.

Christodoulou, C., Griffin B.J., and Foden J., 1984, The geology of Macquarie Island: Australian National Antarctic Research Expeditions Research Notes, v. 21, 15 p.

Cocker, J.D., Griffin, B.J., and Muehlenbachs, K., 1982, Oxygen and carbon isotope evidence for seawater-hydrothermal alteration of the Macquarie Island ophiolite: Earth and Planetary Science Letters, v. 61, p. 112–122.

Coffin, M.F., Karner, G.D., and Falvey, D.A., 1994, Research cruise yields new details of Macquarie Ridge Complex: Eos, Transactions of the American Geophysical Union, v. 75, p. 561–564.

Collot, J.-Y., Lamarche, G., Delteil, J., Wood, R., Sosson, M., Coffin, M., and Lebrun, J.-F., 1995, Morphostructure of an incipient subduction zone along a transform plate boundary: Puysegur Ridge and Trench: Geology, v. 23, p. 519–522.

Cowan, D.S., Botros, M., and Johnson, H.P., 1986, Bookshelf tectonics: Rotated crustal blocks within the Sovanco Fracture Zone: Geophysical Research Letters, v. 13, p. 995–998.

DeMets, C., Gordon, R.G., Argus, D.F., and Stein, S., 1994, Effect of recent revisions to the geomagnetic time scale on estimates of current plate motions: Geophysical Research Letters, v. 21, p. 2191–2194.

Duncan, R.A., and Varne, R., 1988, The age and distribution of the igneous rocks of Macquarie Island: Papers and Proceedings of the Royal Society of Tasmania, v. 114, p. 45–50.

Falloon, T.J., and Green, D.H., 1988, Anhydrous partial melting of peridotite from 8 to 35 kbars and the petrogenesis of MORB, *in* Menzies, M.A., and Cox, K.G., eds., Oceanic and continental lithosphere: Similarities and differences: Journal of Petrology, Special Volume, Oxford University Press, p. 379–414.

Falloon, T.J., Green, D.H., Danyushevsky, L.V., and Faul, U.H., 1999, Peridotite melting at 1 and 1.5 GPa: An experimental evaluation of techniques using diamond aggregates and natural mineral mixes for determination of near-solidus melts: Journal of Petrology, v. 40, p. 1343–1375.

Gao, D., Hurst, S.D., Karson, J.A., Delaney, J.R., and Spiess, F.N., 1998, Computer-aided interpretation of side-looking sonar images from the eastern intersection of the Mid-Atlantic Ridge with the Kane Transform: Journal of Geophysical Research, v. 103, p. 20997–21014.

Goscombe, B.D., and Everard, J.L., 1998, Geology of Macquarie Island; Mineral Resources of Tasmania, Hobart, Australia, Geological Atlas 1:10 000 Series, 7 sheets.

Goscombe, B.D., and Everard, J.L., 1999, Macquarie Island mapping reveals three tectonic phases: Eos, Transactions of the American Geophysical Union, v. 80, p. 50 and 55.

Griffin, B.J., 1982, Igneous and metamorphic petrology of lavas and dykes of the Macquarie Island ophiolite complex [Ph.D. thesis]: Hobart, University of Tasmania, 220 p.

Griffin, B.J., and Varne, R., 1980, The Macquarie Island ophiolite complex: Mid-Tertiary oceanic lithosphere from a major ocean basin: Chemical Geology, v. 30, p. 258–308.

Hayes, D.E., and Talwani, M., 1972, Geophysical investigation of the Macquarie ridge complex, *in* Hayes, D.E., ed., Antarctic oceanology II: The Australian-New Zealand sector: Washington, D.C., American Geophysical Union, Antarctic Research Series, v. 19, p. 211–234.

Johnson, T., and Molnar, P., 1972, Focal mechanisms and plate tectonics of the southwest Pacific: Journal of Geophysical Research, v. 77, p. 5000–5032.

Jones, T.D., and McCue, K.F.,1988, The seismicity and tectonics of the Macquarie Ridge: Papers and Proceedings of the Royal Society of Tasmania, v. 114, p. 51–57.

Kamenetsky, V., Everard, J.L., Crawford, A.J., Varne, R., Eggins, S.M., Lanyon, R., 2000, Enriched end-member of primitive MORB melts: Petrology and geochemistry of glasses from Macquarie Island (SW Pacific): Journal of Petrology, v. 41, p. 411–430.

Kamp, P.J.J., 1986, Late Cretaceous-Cenozoic tectonic development of the southwest Pacific region: Tectonophysics, v. 121, p. 225–251.

Karson, J.A., Thompson, G., Humphris, S.E., Edmond, J.M., Bryan, W.B., Brown, J.R., Winters, A.T., Pockalny, R.A., Casey, J.F., Campbell, A.C., Klinkhammer, G., Palmer, M.R., Kinzler, R.J., and Sulanowska, M.M., 1987, Along-axis variations in seafloor spreading in the MARK area: Nature, v. 328, p. 681–685.

Kennett, J.P., Houtz, R.E., et al., 1974, Initial Reports of the Deep Sea Drilling Project, Volume 29: Washington, D.C., U.S. Government Printing Office, 1197 p.

Lamarche, G., Collot, J.-Y., Wood, R.A., Sosson, M., Sutherland, R., and Delteil, J., 1997, The Oligocene-Miocene Pacific-Australia plate boundary, south of New Zealand: Evolution from oceanic spreading to strike-slip faulting: Earth and Planetary Science Letters, v. 148, p. 129–139.

Laughton, A.S., and Searle, R.C., 1979, Tectonic processes on slow spreading ridges, *in* Talwani, M., Harrison, C.G., and Hayes, D.E., eds., Deep drilling results in the Atlantic Ocean: Ocean crust: New York, American Geophysical Union, Maurice Ewing Series, no.2, p. 15–32.

Ledingham, R., and Peterson, J.A., 1984, Raised beach deposits and the distribution of structural lineaments on Macquarie Island: Papers and Proceedings of the Royal Society of Tasmania, v. 118, p. 223–235.

Levi, S., and Banerjee, S.K., 1977, The effects of alteration on the natural remanent magnetization of three ophiolite complexes: Possible implications for the oceanic crust: Journal of Geomagnetism and Geoelectricity, v. 29, p. 421–439.

Levi, S., Banerjee, S.K., Beske-Diehl, S., and Moskowitz, B., 1978, Limitations of ophiolite complexes as models for the magnetic layer of the oceanic lithosphere: Geophysical Research Letters, v. 5, p. 473–476.

Lonsdale, P., 1977, Structural geomorphology of a fast-spreading rise crest: Marine Geophysical Research, v. 3, p. 251–293.

Massell, C., Coffin, M.F., Mann, P., Mosher, S., Frohlich, C., Duncan, C.S., Karner, G., Ramsay, D., and Lebrun, J.F., 2000, Neotectonics of the Macquarie

Ridge Complex, Australia-Pacific plate boundary: Journal of Geophysical Research, v. 105, p. 13457–13480.

Mawson, D., 1943, Macquarie Island: Its geography and geology, Australasian Antarctic Expedition Science Report Series A, *in* Australasian Antarctic Expedition Scientific Reports, Series A, Volume 5: Sydney, Australia, Government Printing Office, 193 p.

McKenzie, D., and Bickle, M.J., 1988, The volume and composition of melt generated by extension of the lithosphere: Journal of Petrology, v. 29, p. 625–679.

Melson, W.G., and O'Hearn, T., 1986, Zero-age variations in the composition of abyssal volcanic rocks along the axial zone of the Mid-Atlantic Ridge: Boulder, Colorado, Geological Society of America, The Geology of North America, v. M, p. 117–136.

Molnar, P., Atwater, T., Mammerickx, J., and Smith, S.M., 1975, Magnetic anomalies, bathymetry and the tectonic evolution of the South Pacific since the Late Cretaceous: Geophysical Journal of the Royal Astronomical Society, v. 40, p. 383–420.

Mutter J.C., and Karson, J.A., 1992, Structural processes at slow-spreading ridges: Science, v. 257, p. 627–634.

Natland, J.H.,1989, Partial melting of a lithologically heterogeneous mantle: Inferences from crystallization histories of magnesian abyssal tholeiites from the Siqueiros Fracture Zone, *in* Saunders, A.D., and Norry, M.J., eds., Magmatism in the ocean basins: Geological Society of London Special Publication 42, p. 41–70.

Niu, Y., Waggoner, D.G., Sinton, J.M., and Mahoney, J.J., 1996, Mantle source heterogeneity and melting processes beneath seafloor spreading centers: The East Pacific Rise, 18°-19°S: Journal of Geophysical Research, v. 101, p. 27711–27733.

Ovenshine, A.T., Winkler, G.R., Andrews, P.B., and Gostin, V.A., 1974, Chemical analyses and minor element composition of Leg 29 basalts, *in* Kennett, J.P., Houtz, R.E., et al., 1974, Initial Reports of the Deep Sea Drilling Project: Washington, D.C., U.S. Government Printing Office, v. 29, p. 1097–1102.

Quilty, P.G., Rubenach, M.J., and Wilcoxon, J.A., 1973, Miocene ooze from Macquarie Island: Search, v. 4, p. 163–164.

Ruff, L.J., Given, J.W., Sanders, S.O., and Sperber, C.M., 1989, Large earthquakes in the Macquarie Ridge Complex: Transitional tectonics and subduction initiation: Pure and Applied Geophysics, v. 129, p. 71–129.

Schilling, J.-G., and Ridley, W.I., 1974, Volcanic rocks from DSDP Leg 29: Petrography and rare-earth abundances, *in* Kennett, J.P., Houtz, R.E., et al., Initial Reports of the Deep Sea Drilling Project: Washington, D.C., U.S. Government Printing Office, v. 29, p. 1103–1107.

Selkirk, P.M., Seppelt, R.D., and Selkirk, D.R., 1990, Subantarctic Macquarie Island: Environment and biology: Cambridge, Cambridge University Press, 285 p.

Stock, J., and Molnar, P., 1982, Uncertainties in the relative positions of the Australia, Antarctica, Lord Howe, and Pacific plates since the Late Cretaceous: Journal of Geophysical Research, v. 87, p. 4697–4714.

Summerhayes, C.P., 1969, Marine geology of the New Zealand Subantarctic seafloor: New Zealand Department of Science and Industry Research Bulletin, v. 190, 92 p.

Sutherland, R., 1995, The Australia-Pacific boundary and Cenozoic plate motions in the SW Pacific: Some constraints from Geosat data: Tectonics, v. 14, p. 819–831.

Varne, R., 1989, Macquarie Island, *in* Burrett, C.F., and Martin, E.L., eds., Geology and mineral resources of Tasmania: Geological Society of Australia Special Publication, v. 15, p. 389–402.

Varne, R., and Rubenach, M.J., 1972, Geology of Macquarie Island and its relationship to oceanic crust, *in* Hayes, D.E., ed., Antarctic oceanology II: The Australian-New Zealand sector. Washington, D.C., American Geophysical Union, Antarctic Research Series, v. 19, p. 251–266.

Varne, R., Gee, R.D., and Quilty, P.J., 1969, Macquarie Island and the cause of oceanic linear magnetic anomalies: Science, v. 166, p. 230–232.

Weissel, J.K., Hayes, D.E., and Herron, E.M., 1977, Plate tectonics sythesis: The displacements between Australia, New Zealand, and Antarctica since the Late Cretaceous: Marine Geology, v. 25, p. 231–277.

Williamson, P.E., 1979, The palaeomagnetism of outcropping oceanic crust on Macquarie Island: Journal of the Geological Society of Australia, v. 25, p. 387–394.

Williamson, P.E., 1988, Origin, structural, and tectonic history of the Macquarie Island region: Papers and Proceedings of the Royal Society of Tasmania, v. 114, p. 27–43.

Wood, R., Lamarche, G., Herzer, R., Delteil, J., and Davy, B., 1996, Paleogene seafloor spreading in the southeast Tasman Sea: Tectonics, v. 15, p. 966–975.

MANUSCRIPT ACCEPTED BY THE SOCIETY FEBRUARY 23, 2000

Geological Society of America
Special Paper 349
2000

The Timor ophiolite, Indonesia: Model or myth?

Ron Harris
Department of Geology, Brigham Young University, Provo, Utah 84602-4606, USA; e-mail: rharris@byu.edu
Timothy Long
Department of Geology, West Virginia University, Morgantown, West Virginia 26506 USA

ABSTRACT

Geologic and geochemical associations of ophiolite-like rocks in Timor provide conflicting evidence for claims that the collision between the Banda arc and the Australian continent is a modern analogue for the formation and emplacement of Oman-type ophiolites or that an ophiolite emplacement event took place in Timor. The only parts of an ophiolite sequence that are present in Timor are small bodies of spinel lherzolite and various types of volcanic rocks. The lherzolite bodies are mostly found as blocks within a mélange (Bobonaro mélange) derived primarily from the underthrust Australian continental margin. Mineral and whole-rock geochemical properties of the lherzolite bodies are similar to mostly undepleted and unprocessed peridotites found in abyssal and passive-margin settings.

In East Timor, lherzolite bodies are closely associated with the Aileu complex, which is considered part of the Permian–Triassic basement of the Australian passive continental margin. Mesozoic prograde metamorphism of the Aileu complex increases regionally toward the lherzolite bodies via a Barrovian zonation. Both the Aileu complex and the lherzolite bodies were similarly affected by late Neogene collisional (retrograde) metamorphism.

The Atapupu and Nefomasi lherzolite bodies in West Timor are petrologically and geochemically indistinguishable from those of East Timor. The same is true with other lherzolite blocks found within the Bobonaro mélange. The structural position of these lherzolite bodies indicates an affinity to thrust sheets accreted from the distal edge of the Australian continental margin (lower plate) rather than from forearc basement of the upper plate.

Lherzolite bodies and volcanic units in the Ocussi region are different and may represent parts of a young, SSZ (supra-subduction zone) ophiolite that was emplaced within a few million years of its birth.

INTRODUCTION

The Banda arc–Australian continent collision zone of eastern Indonesia features the closure of a Tethys-like seaway composed of marginal basins and arc terranes caught between the converging continents of Australia and Asia (Fig. 1). The region displays a tectonic scenario commonly attributed to the generation and emplacement of Oman-type ophiolites (i.e., Moores, 1982; Searle and Stevens, 1984). From ca. 35 Ma to the present, the passive continental margin of northern Australia has shouldered into and under a series of marginal basins, and the result has been the emplacement of several mafic and ultramafic bodies. The largest body, which is in eastern Sulawesi, is the size of the Oman ophiolite. Other mafic and ultramafic bodies are found throughout the outer islands of the Banda arc, which formed during the Neogene. However, little is known about the origin and tectonic affinity of these rocks.

Most models for the tectonic evolution of the outer Banda arc interpret the mafic and ultramafic rocks incorporated into the forearc as fragments of young oceanic lithosphere derived from the Banda arc upper plate (Hamilton, 1979; Earle, 1980, 1981; Brown and Earle, 1983; Sopaheluwakan et al., 1989; Linthout and Helmers, 1994). Associated metamorphic rocks are interpreted as a metamorphic sole that formed during tectonic emplacement while the "ophiolite" was hot. Volcanic rocks in West Timor of arc affinity and late Neogene age are consistent with this interpretation (Harris, 1991). However, Berry (1981), Berry and Grady (1981), and Berry and McDougall (1986) showed that the Hili Manu lherzolite of East Timor is compositionally akin to peridotite found in oceanic and continental-margin settings. They also showed that the lherzolite is associated with 70 Ma or older prograde metamorphism of Permian–Triassic Australian affinity sequences. These relationships imply more of an association with the lower, continental plate rather than the Neogene Banda arc.

Distinguishing between an upper- or lower-plate origin for the ultramafic bodies of Timor is critical for tectonic reconstruction of the Banda arc–Australian continent collision, which is commonly used as a modern analogue for the emplacement of SSZ (supra-subduction zone) ophiolites. In this paper, we present data from studies of the mineral and whole-rock chemistry and geologic associations of several ultramafic bodies found along the suture zone of the modern arc-continent collision on the island of Timor. The purpose of the study is to test whether the ultramafic bodies are parts of the relatively

Harris, R., and Long, T., 2000, The Timor ophiolite, Indonesia: Model or myth?, *in* Dilek, Y., Moores, E.M., Elthon, D., and Nicolas, A., eds., Ophiolites and Oceanic Crust: New Insights from Field Studies and the Ocean Drilling Program: Boulder, Colorado, Geological Society of America Special Paper 349, p. 321–330.

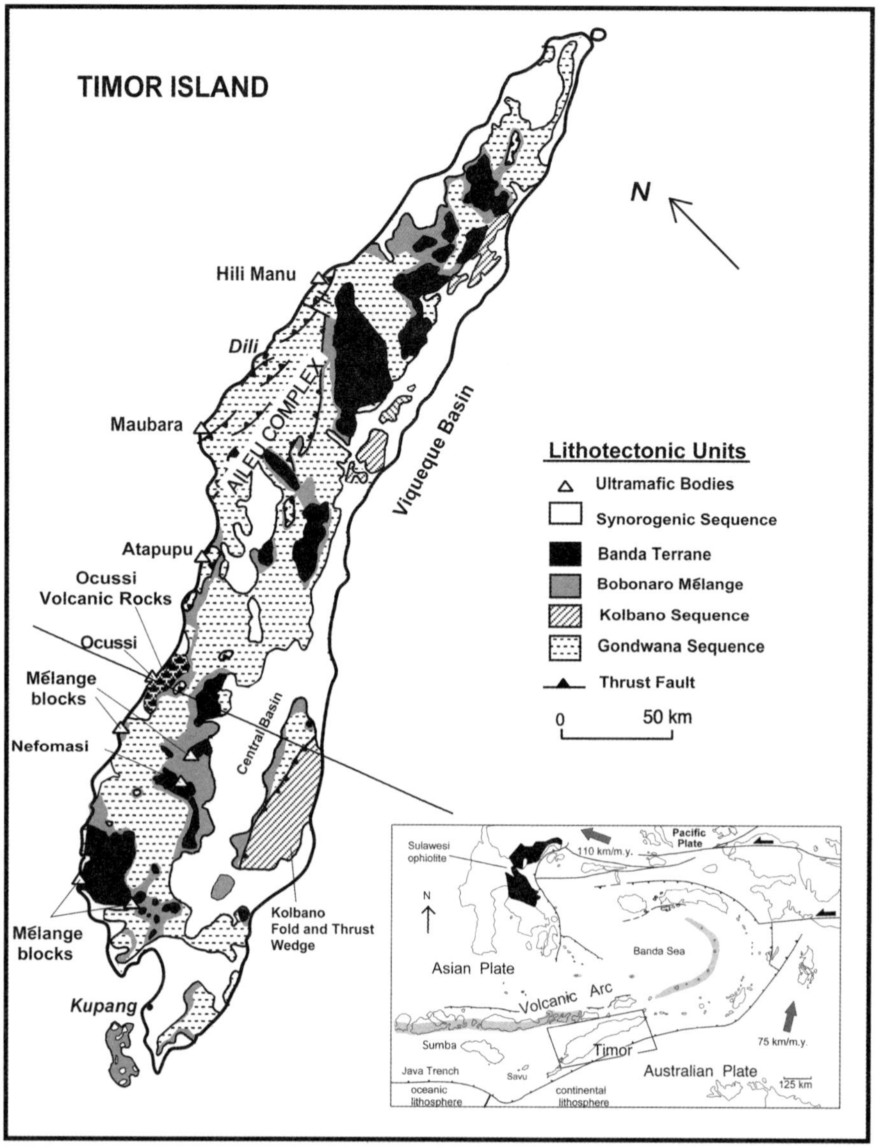

Figure 1. Map of the lithotectonic units of the island of Timor. Nappes of Banda terrane (black) are erosional remnants of the uppermost structural unit of Timor, considered by many as associated with an ophiolite. Ultramafic bodies found throughout the island and sampled as part of this study are labeled and shown (open triangles). The Banda terrane is structurally underlain by the Sonnebaite disruption zone (Bobonaro mélange, gray), a mélange zone that separates units of Asian and Australian affinity. Most of the ultramafic bodies are found in the mélange or hosted by the uppermost nappe of the Gondwana sequence of Australian affinity. Northwest-southeast line of cross section in Figure 2 is shown. Inset: Reference map shows plate motion vectors and the island of Timor in reference to the Banda arc–Australian continent collision (gray is active zone of volcanism). The Banda Sea is trapped by the northward-moving Australian plate and the westward-moving Pacific plate that impinge upon the Sunda Shelf of the Asian plate. The Sulawesi ophiolite may represent a fragment of oceanic crust from the Banda Sea that was emplaced over the Sunda Shelf during the Oligocene (Parkinson, 1998).

young Banda arc upper plate or of Mesozoic suboceanic or subcontinental mantle that was sheared from and shoved back over the lower-plate continental margin during arc-continent collision.

REGIONAL GEOLOGIC SETTING

The island of Timor, Indonesia, occurs above a Miocene to Holocene subduction zone into which the northwest Australian continental margin has entered obliquely (Fig. 1). The resulting arc-continent collision began in central Timor during the late Miocene and has propagated westward at a rate of ~110 km/m.y. (Harris, 1991) to its current position south of the island of Sumba.

Timor is part of the nonvolcanic, outer Banda arc, a contractional wedge formed mostly by mechanical accretion of underthrust Australian continental-margin cover sequences (Gondwana and Kolbano Sequences, Figs. 1 and 2). Most ophiolite-like rock units are found in the hinterland of the wedge, which consists of klippen of distal parts of the underthrust Australian continental margin and the Banda terrane of Asian affinity. The Banda terrane is made up of volcanic and sedimentary sequences associated with the Sunda and Banda volcanic arcs that overlie a polyphase metamorphic complex (Audley-Charles and Harris, 1990). The metamorphic rocks yield K-Ar and $^{40}Ar/^{39}Ar$ cooling ages of 32–35 Ma (Sopaheluwaken et al., 1989; Dropkin et al., 1993). The Bobonaro mélange is commonly

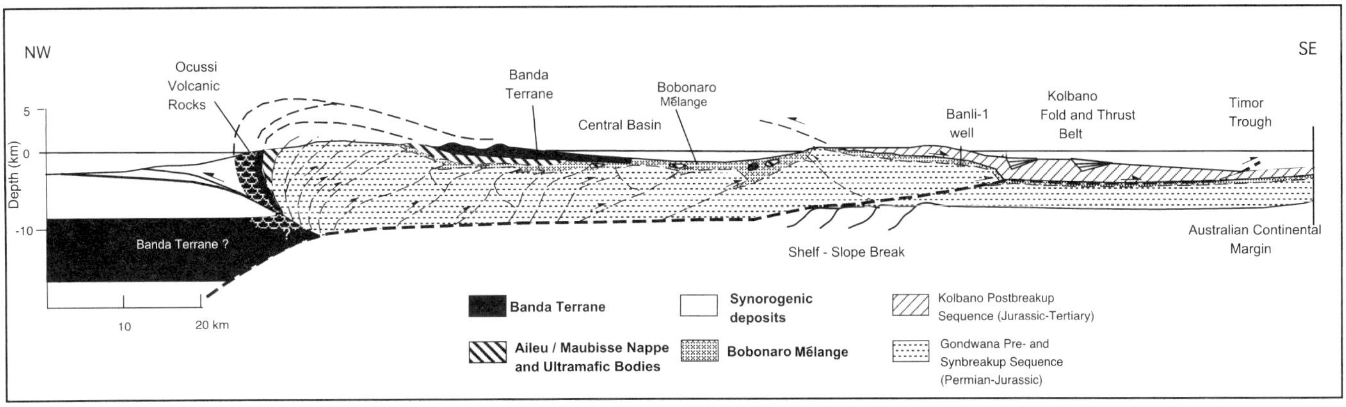

Figure 2. Northwest-southeast structural section across West Timor (modified from Harris et al., 1998). The underthrust and tectonically stacked Australian continental margin is structurally overlain by the Aileu-Maubisse nappe, Bobonaro mélange, and Banda terrane. The collisional suture between units of the Australian and Asian plates is at the base of the Banda terrane. The roots of the Banda terrane are inferred as the footwall to the retrowedge thrust system. The Ocussi volcanic rocks are the only part of the Banda terrane that shows any SSZ ophiolitic affinity. The underthrusting of continental-margin material beneath the roots of the Ocussi body may represent a model for the emplacement of SSZ ophiolites.

found at the structural base of Banda terrane klippen and is composed mostly of material derived from structurally underlying Australian continental-margin sequences (Harris et al., 1998). The mélange marks the collisional suture between the Australian lower plate and the Asian upper plate. Most of the ultramafic rocks in Timor are found as blocks or thrust sheets within this mélange or at the same structural level as the Aileu-Maubisse thrust nappe. This nappe is composed of mostly Permian and Triassic carbonate rocks, alkalic pillow basalt, and psammite that accumulated in an intracratonic rift basin. Seafloor spreading began in the basin during the Late Jurassic, which left the Aileu and Maubisse units at the edge of the northwest Australian passive continental margin. In this position, these units were some of the first ones to accrete to the Banda arc during Neogene arc-continent collision (Audley-Charles and Harris, 1990).

Many different interpretations exist concerning the structural evolution of the collisional event that formed Timor (Hamilton, 1979; Carter et al., 1976; Chamalaun and Grady, 1978, Harris, 1991). However, all conclude that Timor occupies a suture zone between the Asian and Australian plates and that Australian continental crust extends to the northern edge of the island where most of the lherzolite bodies are found. Spinel lherzolite has also been dredged from a large seamount or ridge at the edge of the western Australian passive continental margin near the transition from thinned continental crust to initial (earliest) oceanic crust (Nicholls et al. 1981). Lherzolite bodies uplifted by extensional denudation and serpentinite diapirism are not uncommon in passive-margin settings (i.e., Eldholm and Talwani, 1981; Boillot et al., 1987; Bonatti and Michael, 1989).

GEOLOGIC ASSOCIATIONS AND INTERPRETATIONS

De Roever (1940) and De Waard (1954) were the first to refer to an ophiolite or "Alpine-type" peridotite in Timor. Their interpretations were based on the discovery in central West Timor of peridotite, serpentinite, and pillow basalt that were spatially associated with a multiphase metamorphic sequence. Hamilton (1979), with no additional data, reinterpreted these units as parts of a young ophiolite sequence derived from the Banda forearc, as was predicted from studies of ancient arc-continent collision zones (i.e., Dewey and Bird, 1971). Field studies by Earle (1980, 1981), Brown and Earle (1983), and Sopaheluwakan et al. (1989) represented the mafic and ultramafic rocks of West Timor as a reversely ordered ophiolite that structurally overlies an inversely zoned metamorphic sole of Oligocene age.

Hili Manu lherzolite

Berry (1981) analyzed the Hili Manu lherzolite body of East Timor as part of his study of the Aileu complex schist belt, a mostly pelitic and mafic volcanic metamorphic belt along the north coast of Timor. He found that the petrology and geochemistry of the Hili

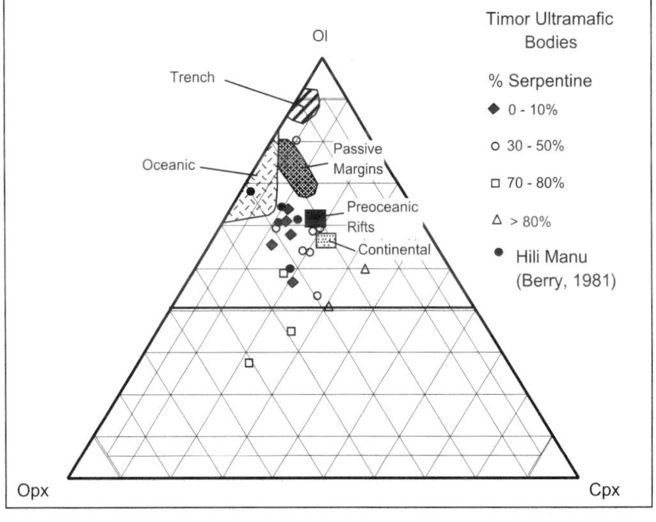

Figure 3. Modal abundance of olivine, orthopyroxene, and clinopyroxene for the ultramafic bodies of Timor compared to ultramafic rocks from known tectonic settings. Rock classification according to Streckeisen (1976). Symbols for the Timor samples correspond to degree of serpentinization, according to the amounts of (Ol + serpentinized Ol) and (Opx + Cpx + serpentinized Px). Samples with <75% serpentinization had roughly equal abundance of orthopyroxene and clinopyroxene or were slightly to moderately enriched in orthopyroxene. Also plotted are modal abundances from the Hili Manu lherzolite of Berry (1981). Fields are taken from Bonatti and Michael (1989). Timor samples cluster with relatively undepleted tectonic settings.

TABLE 1. WHOLE-ROCK GEOCHEMISTRY OF TIMOR ULTRAMAFIC BODIES

	Atapupu*	Atapupu*	Hili Manu*	Hili Manu*	W. Atapupu*	Mélange*	Mélange*	Mélange*	Mélange*	Nefo-masi†	Nefo-masi†	Nefo-masi†	Nefo-masi†	Ocussi†
Latitude	−8.97°S	−8.97°S	−8.49°S	−8.49°S	−9.09°S	−9.62°S	−9.62°S	−9.62°S	−9.46°S	−9.52	−9.52	−9.49	−9.49	−9.21
Longitude	124.9°E	124.9°E	125.93°E	125.96°E	124.8°E	123.68°E	123.68°E	123.68°E	123.91°E	124.25	124.23	124.26	124.26	124.36
SiO_2	38.21	38.24	38.2	42.99	38.69	40.47	40.91	41	38.77	39.91	38.96	39.12	39.03	38.7
TiO_2	0.04	0.01	0.02	0.06	0.04	0.04	0.08	0.07	0.03	0.11	0.03	0.05	0.06	0.01
Al_2O_3	1.62	0.77	1.47	2.4	1.74	2.03	2.67	2.65	1.77	3.46	1.02	1.51	2.1	0.94
Fe_2O_3	7.47	8.01	7.89	8.47	7.26	7.78	8.23	8.11	7.86	7.76	8.49	8.08	8.05	7.94
MnO	0.09	0.07	0.01	0.13	0.08	0.12	0.12	0.12	0.12	0.1	0.11	0.1	0.1	0.1
MgO	35.1	37.05	36.88	39.43	36.52	35.81	36.37	36.15	36.77	36.38	36.64	36.54	36.67	36.67
CaO	0.27	0.13	0.45	2.69	0.21	1.96	1.94	2.48	1.55	2.95	0.69	2.09	0.92	0.92
Na_2O	0.02	0	0.01	0.11	0.27	0.07	0.13	0.13	0.05	0.12	0.02	0.03	0.02	0.02
K_2O	0.01	0.03	0	0.01	0.03	0.01	0.03	0.01	0.04	0.03	0.02	0.02	0.02	0.02
P_2O_5	0.013	0.01	0.012	0.007	0.016	0.008	0.007	0.008	0.005	0.011	0.009	0.01	0.008	0.008
LOI	14.54	13.22	13	3.35	14.76	10.18	9.56	7.72	12.51	N/A	N/A	N/A	N/A	N/A
Total	97.38	97.54	97.94	99.65	99.62	98.75	100.05	98.45	99.48	90.83	85.99	87.55	85.33	85.33
Calc'd FeO	6.72	7.21	7.1	7.62	6.53	7	7.41	7.3	7.07	6.98	7.64	7.27	7.24	7.14
Calc'd Fe_2O_3	0.75	0.8	0.79	0.85	0.73	0.78	0.82	0.81	0.79	0.78	0.85	0.81	0.81	0.8
Mg#	89.86	89.72	89.82	89.78	90.47	89.67	89.29	89.37	89.82	89.84	89.06	89.51	89.55	89.71
Sc	10	8.4	10.6	14.5	10	13	13.5	12.4	11.9	11.8	11.2	11.8	12.2	11.2
V	51	38	50	68	43	65	61	58	56	77	36	59	59	33
Cr	2239	2445	2650	2768	2380	2540	2855	2364	2584	2687	2393	2420	2500	2530
Ni	2159	2266	2238	1909	2119	1940	1946	2014	1861	1832	2117	1908	1968	2117
Cu	13	5	13	17	17	19	26	21	22	18	15	19	17	16
Zn	37	31	34	38	37	39	36	38	36	37	37	37	35	33
Y	1.5	0.4	0.9	2.3	1.1	2.3	1.4	2.2	1.2	1.4	1.4	1.5	1.4	1.3
Sr	3.1	4.8	5	2.9	3.8	3.6	4.7	3.9	20.6	5.8	5.8	5.8	5.8	5.7

Note: Oxide contents are in weight percent; trace elements are in parts per million. FeO is partitioned into FeO and Fe_2O_3 based on methods of Ragland (1989).
* ICP analyses
† XRF analyses

Manu body is characteristic of lherzolite found at mid-ocean ridge diapirs and along continental margins.

Studies of the metamorphic rocks of the Aileu complex associated with the Hili Manu lherzolite by Berry and Grady (1981) and Berry and McDougall (1986) revealed a Barrovian prograde metamorphic zonation that decreases in metamorphic grade westward away from the Hili Manu lherzolite body for over 40 km. The prograde assemblage decreases from upper-amphibolite facies within 10 km from the lherzolite to upper-greenschist facies as much as 40 km away. The prograde mineral assemblage is overprinted by a lower-temperature retrogressive assemblage that decreases from middle- to lower-amphibolite facies to lower-greenschist facies toward the south. $^{40}Ar/^{39}Ar$ analysis of the metamorphic rocks has yielded a minimum cooling age of 70 Ma for the prograde assemblage and 5–8 Ma for the retrogressive assemblage. Berry and McDougall (1986) suggested that the prograde phase of metamorphism may be associated with Mesozoic rifting and development of the Australian passive continental margin, whereas the retrograde phase is a product of Neogene arc-continent collision. An apatite fission-track study by Harris et al. (2000) yields similar results. These age relationships and the scale and metamorphic history of the Aileu complex are not characteristic of subophiolite metamorphic aureoles.

Further studies of the Aileu complex by Prasetyadi and Harris (1996) demonstrate that highly serpentinized parts of the Hili Manu lherzolite and some smaller serpentinite and hornblende diorite bodies near Maubara (Fig. 1) have locally intruded into the Aileu schist. These contact relationships indicate a close association between serpentine diapirism and igneous intrusions with the Aileu complex that is also supported by the decrease in prograde metamorphism away from the peridotite bodies over several kilometers. The pattern of retrograde metamorphism, which is most likely related to arc-continent collision, is much more diffuse and shows no correlation with the ultramafic bodies. Transitional stratigraphic relationships between the Aileu complex and Permian and Triassic pillow lavas, shale, and crinoidal limestone of the Maubisse Formation (Barber and Audley-Charles, 1976; Prasetyadi and Harris, 1996) strengthen correlations of the lherzolite with the Australian continental margin.

Atapupu lherzolite

Small lherzolite bodies are found along the north coast of West Timor near Atapupu (Fig. 1) that are similar in composition to the Hili Manu lherzolite. Maps of the region by Simons (1940) and Helmers et al. (1989) show tectonically interlayered lenses of amphibolite, metapelite, hornblende diorite, and serpentinite that is intruded locally by granite and aplite dikes. Fresh lherzolite bodies are found between massive and brecciated serpentinite with mylonitic laminations and lenses of amphibolite. Many of the lherzolite bodies occur as blocks in the Bobonaro mélange along with fragments of the Maubisse Formation.

Thermobarometric studies of the Atapupu lherzolite by Helmers et al. (1989) indicate decompressional ascent from early equilibration of pyroxene porphyroclasts at 1030–1060 °C above 8 kbar. Neoblasts that formed along the rims of the porphyroclasts during mylonitization equilibrated at 940 and 850 °C at 6–8 kbar. Serpentinization became dominant below 550 °C. Pressure-temperature estimates in associated metapelites show initial isobaric prograde recrystallization at 850 to 800 °C at 6–7 kbar, which was followed by decompression and cooling that parallels the trend found in lherzolite bodies. These relatively high equilibration temperature estimates are identical to

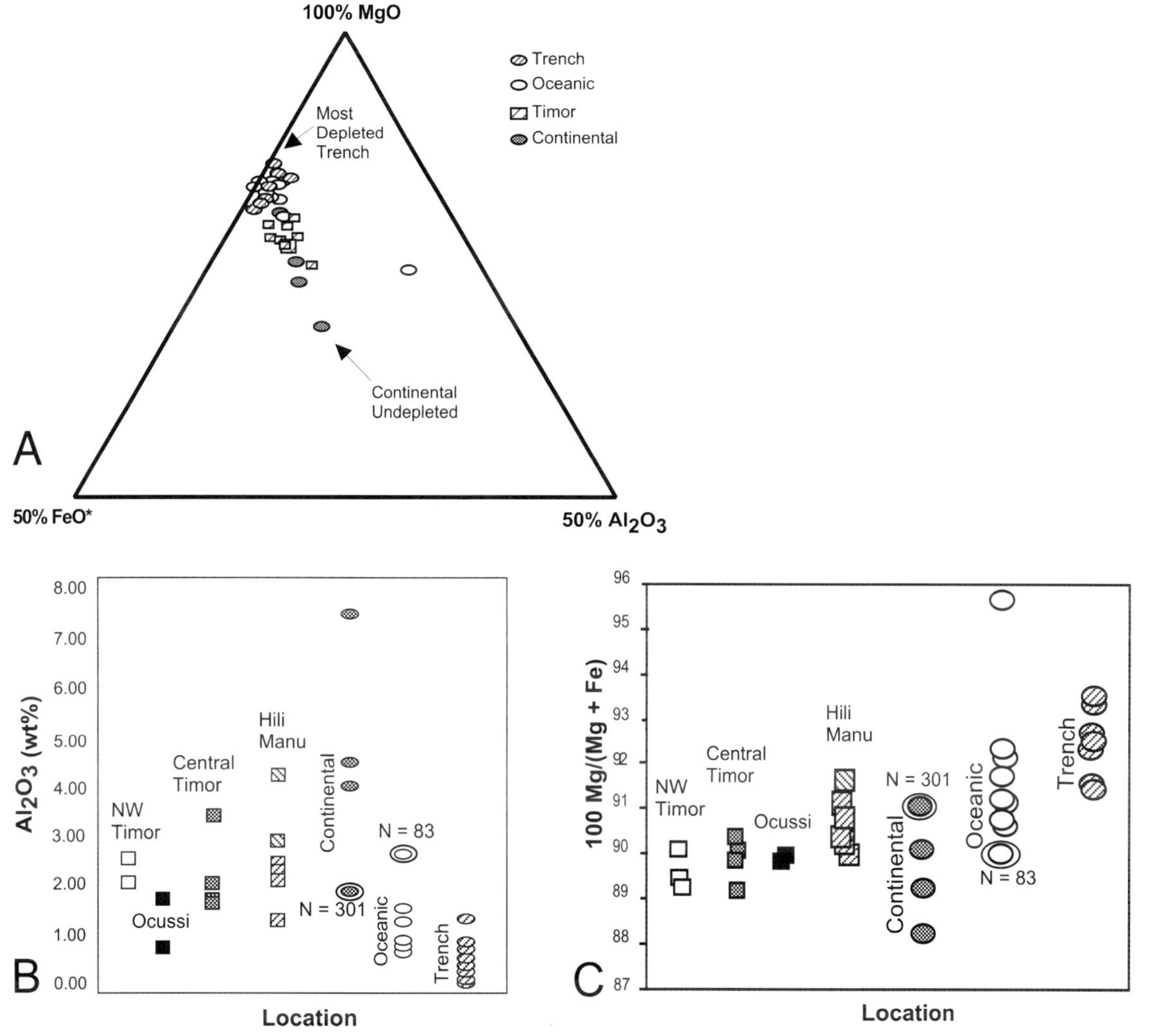

Figure 4. Whole-rock discriminant diagrams comparing Timor samples (<10% serpentine and Hili Manu lherzolite from Berry, 1981) with those from other tectonic settings (see text). A, Mg:Fe:Al ratio shows a depletion trend of decreasing FeO and Al_2O_3. Most Timor samples plot with data from preoceanic-rift upper mantle and continental undepleted and oceanic estimates. B, Al_2O_3 contents (in weight percent) of all Timor samples show some scatter, but mostly overlap with data from continental and oceanic upper mantle. C, Mg# values of Timor samples overlap data from preoceanic-rift upper-mantle samples, continental undepleted samples, and oceanic estimates.

those from lherzolite dredged from the edge of the western Australian continental margin (Nicholls et al., 1981) and require steep geothermal gradients of 40–50 °C/km. Nicholls et al. (1981) associated the steep geotherm with continental rift-valley volcanism prior to seafloor spreading.

Although the composition and pressure-temperature estimates for the Atapupu lherzolite and associated metamorphic assemblage are similar to those found at Hili Manu, Helmers et al. (1989) offered the Hamilton (1979) hypothesis of an obducted arc-type ophiolite and subophiolite metamorphism to explain their observations. They drew attention to the close association of late Neogene volcanic arc rocks with peridotite bodies at Atapupu and in Seram, suggesting that the volcanic arc was the heat source for the high-grade metamorphism during late Neogene arc-continent collision. A similar interpretation is offered for ultramafic bodies found in central Timor near metamorphic rocks of the Banda terrane (Earle, 1980, 1981; Brown and Earle, 1983; Sopaheluwakan et al., 1989).

Ocussi volcanic rocks

Helmers et al. (1989) used the close association between the Atapupu lherzolite and nearby Ocussi volcanic pile to support their claim of a young (hot), upper-plate origin for the lherzolite. The Ocussi volcanic rocks are a 3–4-km-thick sequence of interlayered pillow lavas and sheet flows that are tilted to near vertical along the north coast of West Timor (Figs. 1 and 2). The volcanic units have island-arc–basalt geochemical affinities, such as LILE (large ion lithophile element) enrichment, Nb depletion, and high Cr and low Ti contents (Harris, 1992). $^{40}Ar/^{39}Ar$ plateau ages of basalt are 3–5 Ma, and fauna from interlayered calcilutite are late Neogene age (Harris, 1992). Blocks of lherzolite are found in mélange near the volcanic pile, but no other parts of an ophiolite sequence are found. The geochemistry and age of the Ocussi volcanic rocks indicate that they formed in a SSZ setting probably as part of the upper plate of the Banda arc immediately before tectonic emplacement onto the edge

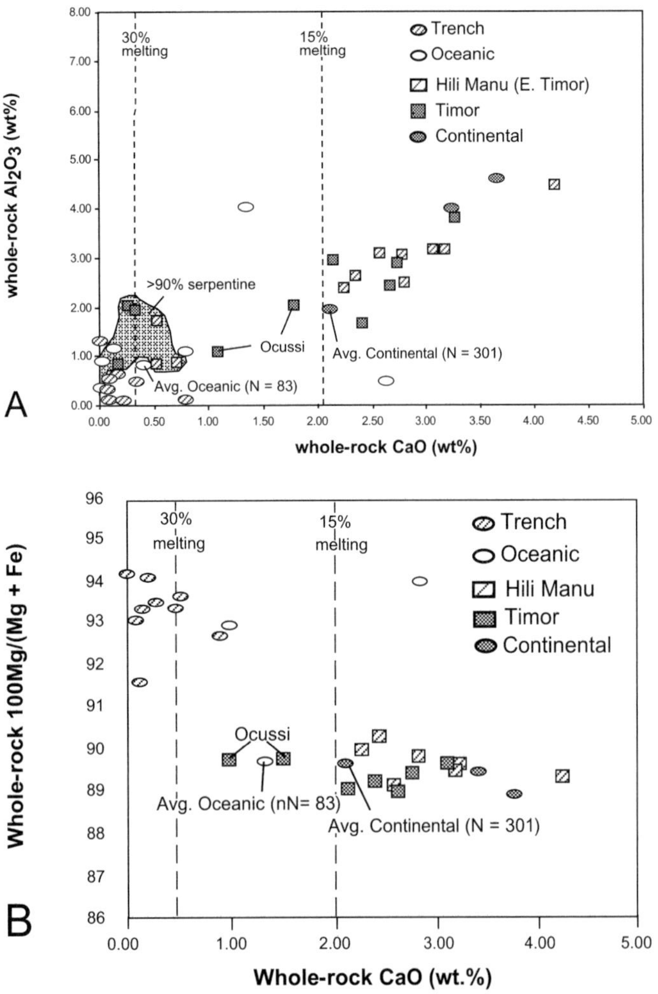

Figure 5. Partial-melting estimates after Mysen and Kushiro (1977) and Ishiwatari (1985). These plots depict approximate locations of 15% and 30% partial melting of previously undepleted upper mantle. A, CaO vs. Al_2O_3. B, CaO vs. Mg#. Except for cases of extreme serpentinization, all Timor samples plot at <30% depletion, and the majority plot at <15%.

of Australia during late Neogene arc-continent collision (Harris, 1992). These characteristics are very similar to those of most Oman-type ophiolites. Yet, little is known about the structural relationships between the Ocussi volcanic rocks and other parts of the Banda arc–Australian continent collision. It is possible that the roots of the Ocussi volcanic unit are structurally underlain by continental-margin rocks, as shown in Figure 2. This interpretation is supported by similar age and geochemical results from samples of basalt dredged from the forearc basin to the north of the Ocussi region (Harris, 1992).

Other ultramafic bodies of Timor

In addition to the Hili Manu and Atapupu lherzolite there are several other ultramafic bodies found throughout Timor. Many occur as blocks in the Bobonaro mélange that is widely exposed throughout recently emergent and less exhumed parts of northwestern Timor (Fig. 1). Below the mélange is a thrust stack composed mostly of accreted Australian continental-margin cover sequences, which are the source for most of the mélange (Fig. 2). Structurally overlying the mélange is the Banda terrane. The close spatial relationships between the Banda terrane and the ultramafic blocks in the Bobonaro mélange have led to correlations between the two.

Earle (1980) and Brown and Earle (1983) associated the metamorphism of the Banda terrane with obduction of a young backarc basin that formed at the edge of the Asian margin. They suggested that the two units were both part of the Banda terrane when it collided with the Australian margin in the Miocene. Sopaheluwakan et al. (1989) presented a similar model, but had the "ophiolite" obducted onto continental fragments hundreds of kilometers outboard of the Australian margin during the Oligocene that later collide with northwest Australia. However, thermochronological studies of Permian–Triassic rocks structurally underlying the ultramafic bodies show no indication of any pre–late Neogene event (Harris et al., 2000).

The Nefomasi lherzolite body of central West Timor is one of the largest on the island. It structurally overlies a mélange and broken formation derived mostly from underlying thrust sheets of Permian–Triassic Australian affinity sedimentary sequences (Fig. 2), as confirmed at two different exposures of the basal thrust (Harris et al., 1998). Contacts found between the Nefomasi body and the Banda terrane are associated with serpentine diapirs intruding Neogene carbonate units. These contacts are similar to those produced by serpentinite diapirs that pierce the Banda terrane throughout Timor (Harris et al., 1998). These observations are consistent with contact relationships first mapped by Rosidi et al. (1979) that show the Nefomasi body structurally underlying the Mutis metamorphic complex and other units of the Banda terrane. However, this interpretation was reversed in the geologic map of the Nefomasi region presented by Sopaheluwakan et al. (1989). Harris et al. (1998) confirmed that the Nefomasi body does not overlie the Banda terrane.

Summary

The close association of ultramafic bodies, volcanic rocks, and metamorphic sequences in Timor has been used to support claims for obduction of hot ophiolite slabs in the Banda arc–Australian continent collision (Hamilton, 1979; Earle, 1980, 1981; Brown and Earle, 1983; Sopaheluwakan et al., 1989; Linthout and Helmers, 1994). However, the lack of an ophiolite sequence, the close affinity of ophiolite-like rocks with the Aileu and Maubisse units, and age and structural relationships of the volcanic and metamorphic rocks are difficult to reconcile with claims that there is an ophiolite in Timor, that the various metamorphic complexes are associated with hot emplacement of an ophiolite, or that ophiolite-like rocks were derived from the forearc upper plate (with the exception of the Ocussi volcanic rocks). We present a new interpretation here that the ophiolite-like units of Timor—particularly the lherzolite bodies and associated pillow lavas and metamorphic rocks of Aileu and Maubisse units—are derived from near the edge of the Australian continental margin.

One way we have tested this hypothesis is by analyzing the mineral and whole-rock chemistry of ultramafic bodies throughout Timor to determine if the mantle material has a within-plate (continental or oceanic) or a SSZ chemical affinity. We assume that these two different settings are distinguishable chemically on the basis of comparisons of results compiled by Bonatti and Michael (1989) and ourselves from ultramafic rocks found in (1) the upper plate of subduction zones throughout the western Pacific (Bowin et al., 1966; Fisher and Engel, 1969; Ishii, 1985; Bloomer and Fisher, 1987; Ishii et al., 1992; Arai, 1994); (2) ocean basins, such as along the Mid-Atlantic Ridge at 43°N (Shibata and Thompson, 1986) and the Owens Fracture Zone (Hamlyn and Bonatti, 1980); (3) passive continental margins of the Red Sea (Bonatti et al., 1986), western Australia

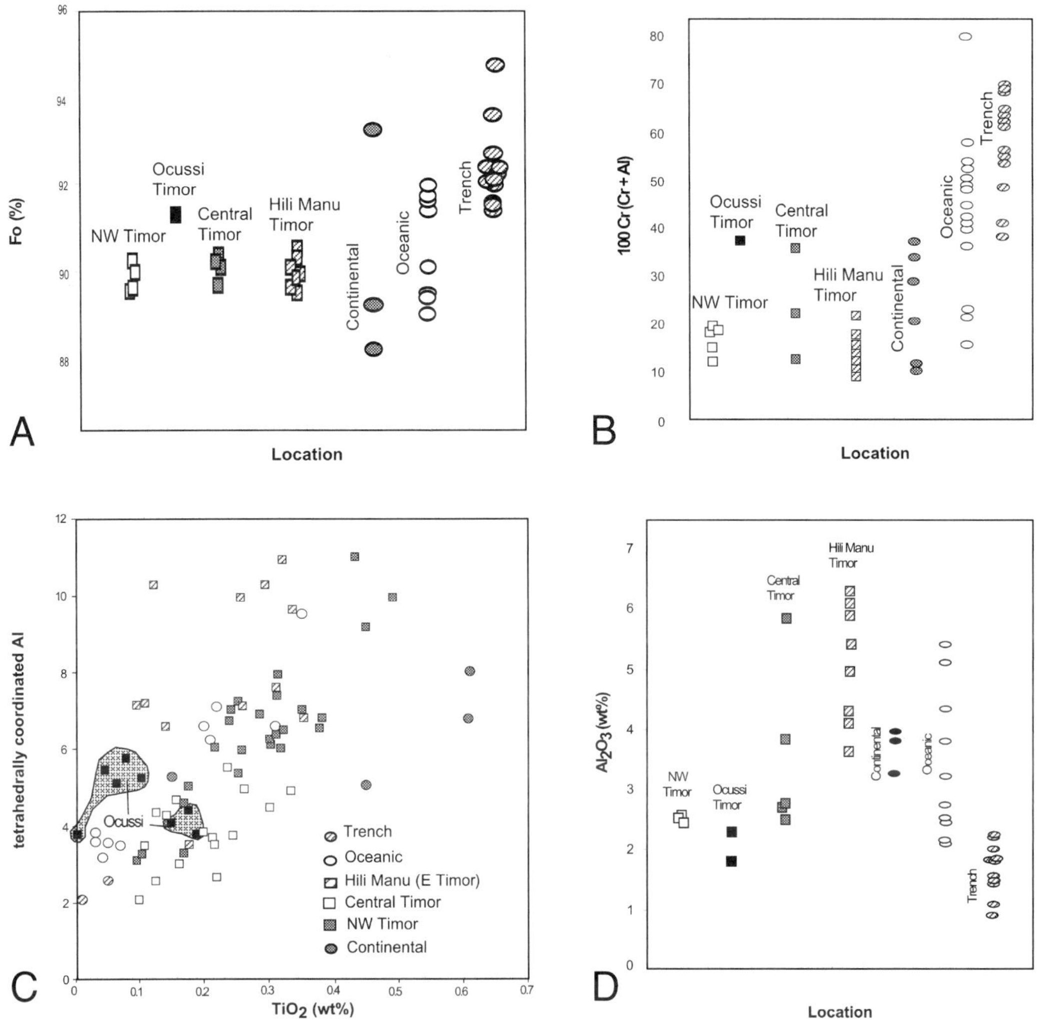

Figure 6. Mineral chemistry of Timor ultramafic bodies compared to those from known tectonic settings, listed from least to most depleted. Analyses obtained with a Cameca electron microprobe at 15 kV acceleration voltage. Points represent the average of five analyses per mineral per thin section. The reported relative abundances of Fe_2O_3 and FeO are a function of the abundance of Al_2O_3, which also occupies the tetrahedral site. Olivine FeO* was not recalculated by the analysis program, as it is assumed that no oxidized Fe was present during olivine crystallization. If only FeO* was reported in data from other sources, we recalculated for Fe_2O_3 and FeO by using the method of Droop (1987). A, Forsterite content (% Fo) of olivine. B, Spinel Cr#. C, Clinopyroxene Al:Ti ratio. D, Al_2O_3 (weight percent) in orthopyroxene.

(Nicholls et al., 1981), and Iberia (Boillot et al., 1980); and (4) nodules from subcontinental settings (Carter, 1970; Frey and Prinz, 1978; Herzberg, 1993). We acknowledge that there are exceptions to these trends, such as depleted harzburgite xenoliths found in volcanic rocks in cratonal settings (i.e., Herzberg, 1993; Bernstein et al., 1998) and relatively undepleted lherzolite found in the Japan arc (Arai, 1994).

PETROGRAPHY

All of the ultramafic rocks we found throughout East and West Timor (Fig. 1) are spinel- or spinel-bearing lherzolite (Fig. 3). The average mineral assemblage of samples from 13 ultramafic bodies consists of relatively large interstitial and subhedral crystals of deformed orthopyroxene with clinopyroxene exsolution lamellae, deformed interstitial clinopyroxene, highly dissected and kink-banded olivine, interstitial spinel, and associated opaque minerals. Some samples contain discrete mylonite zones of grain-size reduction. The majority of the samples are made of 30%–50% serpentine (Fig. 3). Three locations have <10% serpentine.

Comparisons of modal mineral contents of Timor samples with data compiled by Bonatti and Michael (1989) show a depletion trend from undepleted continental settings, preoceanic rifts, passive margins (oceanic to continental transitional type), oceanic fracture zones, and subduction-related active margins. Most Timor samples cluster on the pyroxene-rich side of the trend in the relatively undepleted to partially depleted range, well away from subduction-related active-margin peridotite.

WHOLE-ROCK GEOCHEMISTRY

Whole-rock chemical analyses were performed on samples from 11 different lherzolite bodies throughout East and West Timor (Table 1). Nine samples were analyzed at Texas Tech by using inductively-coupled plasma (ICP) and five others at West Virginia University by using X-ray fluorescence (XRF).

The most discriminating whole-rock geochemical parameters of ultramafic chemical evolution are Al_2O_3 (in weight percent), Mg/(Mg + Fe) = Mg#, and the MgO:FeO*:Al_2O_3 proportions (Ishiwatari,

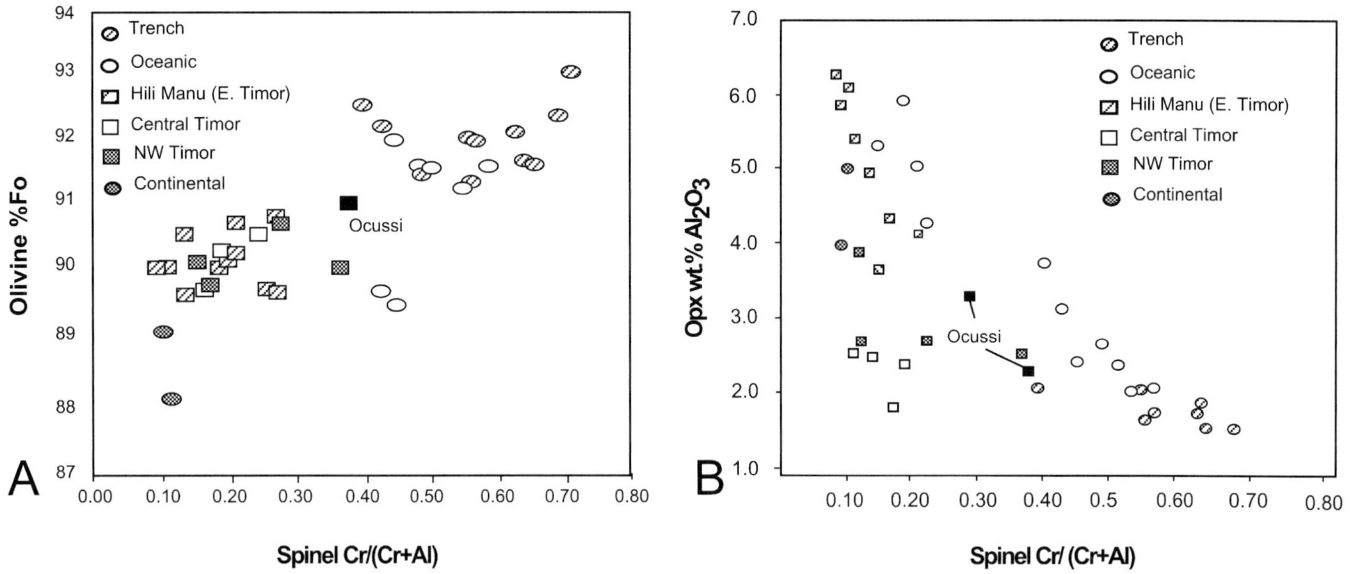

Figure 7. Covariation diagrams showing loss of magmaphile elements Fe and Al from mantle depletion, and complementary enrichment in refractory elements Mg and Cr. Most Timor samples cluster with data from estimated continental undepleted and preoceanic-rift upper-mantle rocks. A, Spinel Cr# vs. Fo content (%) in olivine. B, Spinel Cr# vs. Al_2O_3 (weight percent) in orthopyroxene.

1985; Bonatti and Michael, 1989). Al, Ca, and Fe depletion and Mg enrichment reflect increasing degrees of mantle depletion.

With the exception of lherzolite blocks found near the Ocussi volcanic rocks, most plots show that Timor ultramafic bodies overlap with continental undepleted and preoceanic rift settings (Figs. 4 and 5). There is also some overlap with the mean of oceanic upper mantle. The closest associations are with Red Sea–type peridotite (Zabargad) and peridotite from the Australian and Iberian passive continental margins (Figs. 4 and 5).

MINERAL CHEMISTRY

Samples from eight different ultramafic bodies were analyzed by electron microprobe at Pennsylvania State University. Mineral-chemistry parameters that may indicate the degree of partial melting of a lherzolitic mantle source include forsterite content of olivine (percent Fo) (Fig. 6A), spinel Cr# [Cr/(Cr + Al)] (Fig. 6B), tetrahedrally coordinated Al in clinopyroxene versus weight percent TiO_2 in clinopyroxene (Fig. 6C), and weight percent Al_2O_3 in orthopyroxene (Fig. 6D). Progressive partial melting of relatively undepleted continental upper-mantle lherzolite produces an increase in spinel Cr# and percent Fo and a decrease in orthopyroxene weight percent Al_2O_3 (Fig. 7, A and B). As with the whole-rock chemistries, Al and Fe will show magmaphile tendencies, whereas Mg and Cr are more refractory.

Fo in olivine ranges from 89% to 90.5% (Fig. 6A). Spinel compositions range from 33% to 53% chromite, with low Cr# values from 10 to 37 (Fig. 6B). X-ray diffraction studies also found edenite and magnesiohornblende, and lizardite and clinochrysotile.

In summary, the Timor lherzolite bodies have a mineral chemistry most similar to those from preoceanic rifts and passive continental margins (Figs. 6 and 7). Ultramafic rocks in the Ocussi region are the only samples that consistently show relatively more depletion.

DEGREE OF PARTIAL MELTING

Ishiwatari (1985) estimated the degree of partial fusion based on whole-rock chemistry of residual peridotite by graphing covariations of weight percent CaO with Mg# and weight percent CaO with weight percent Al_2O_3. Relative to an initial pyrolite composition, Bonatti and Michael (1989) estimated degrees of depletion ranging from near zero in Red Sea–type preoceanic-rift settings to ~10%–15% in passive-margin and transitional-crust settings, 10%–25% in oceanic peridotite, and 30% and greater in subduction-zone settings.

Plots of whole-rock weight percent CaO vs. Mg# and weight percent CaO vs. weight percent Al_2O_3 (Fig. 5) show that Timor samples mostly record less than 20% partial melting. Whole-rock Mg# and Mg:Fe:Al indicate similarities with preoceanic-rift, passive-margin, and oceanic peridotite.

Mineral-chemistry comparisons reflect the same trends. Most Fo contents in olivine of Timor lherzolites plot near the overlap between preoceanic-rift and oceanic peridotite (Fig. 6A). The same is found with comparisons of spinel Cr# (Fig. 6B), which fall into the type I abyssal peridotite category of Dick and Bullen (1984). The presence of diopside also reflects limited depletion, as clinopyroxene is typically the first major mineral lost during partial melting.

Samples collected near the Ocussi volcanic rocks display anomalously high values for spinel Cr# and percent Fo in olivine and low values for whole-rock and orthopyroxene weight percent Al_2O_3 (Figs. 6 and 7). However, the geologic association between the ultramafic and mafic rocks in the Ocussi region remains ambiguous.

ORIGIN AND TECTONIC EVOLUTION

Geologic associations with the Aileu and Maubisse units in the Hili Manu region, near Maubara, and throughout West Timor demonstrate a close affinity between ophiolite-like rocks and the Australian lower plate. Geochemical evidence supports this conclusion, but is more equivocal as to whether the lherzolite is from a suboceanic or subcontinental setting. The high temperatures of equilibration associated with Atapupu lherzolite require a geotherm most likely associated with rifting. Pressure estimates on the other hand indicate depths of at least 22–30 km, which is more consistent with the depth of continental rather than oceanic rifts. The lack of plagioclase development is also inconsistent with an oceanic rift zone origin for the spinel lherzolite bodies.

We speculate that the spinel lherzolite bodies originated from near an oceanic to continental lithospheric transition and were initially exhumed by extension during Mesozoic passive-margin development. Later these bodies were thrust with the Aileu-Maubisse nappe over parts of the continental margin more proximal to the shelf during late Neogene arc-continent collision. This hypothesis is consistent with the structural position and lack of an ophiolite sequence associated with the lherzolite bodies, age differences between prograde and retrograde metamorphism of the Aileu complex, and the composition of the lherzolite bodies.

CONCLUSION

Although many features of the Banda arc–Australian continent collision present an attractive modern analogue for the emplacement of Oman-type ophiolites, there remains a paucity of evidence of any such event. Large klippen of forearc crust, represented by the Banda terrane, do structurally overlie much of the orogenic wedge in a manner similar to Oman-type ophiolites, yet the Banda terrane is not an ophiolite. A similar tectonic scenario is found in the Hellenic subduction system on the island of Crete. Although there are many Oman-type ophiolites found along the orogenic strike of the Hellenic accretionary wedge, the "uppermost tectonic unit" or "Asteroussia nappe" of Crete (Bonneau, 1984) is more akin to the Banda terrane of Timor, and no ophiolite is found there.

The Ocussi region is different and may represent the emergent tip of a Banda arc–related SSZ ophiolite. It is also possible that other such bodies have been emplaced over the edge of the Australian continental margin in submerged parts of the orogen that have undergone less uplift. A modern analogue of an Oman-type forearc ophiolite, from an active arc-continent collision, remains to be found.

ACKNOWLEDGMENTS

Funds were provided in part by the National Science Foundation (EAR 9118151). Field studies in Timor were sponsored by the Geological Research and Development Center (GRDC) and Universitas Pembangunan Nasional (UPN "Veteran"). Field assistance was provided by Sahat Tobing and Carolus Prasetyadi. We thank Jack Renton for providing XRF analyses.

REFERENCES CITED

Arai, S., 1994, Characterization of spinel peridotites by olivine-spinel compositional relationships: Review and interpretation: Chemical Geology, v. 113, p. 191–204.

Audley-Charles, M.G., and Harris, R.A., 1990, Allochthonous terranes of the Southwest Pacific and Indonesia: Philosophical Transactions of the Royal Society of London, v. 331, p. 571–587.

Barber, A.J., and Audley-Charles, A.C., 1976, The significance of metamorphic rocks of Timor in the development of the Banda arc, eastern Indonesia: Tectonophysics, v. 30, p. 119–128.

Bernstein, S., Kelemen, P.B., and Brooks, C.K., 1998, Depleted spinel harzburgite xenoliths in Tertiary dykes from East Greenland: Restites from high degree melting: Earth and Planetary Science Letters, v. 154, p. 221–235.

Berry, R.F., 1981, Petrology of the Hili Manu lherzolite, East Timor: Journal of the Geological Society of Australia, v. 28, p. 453–469.

Berry, R.F., and Grady, A.E., 1981, Deformation and metamorphism of the Aileu Formation, north coast, East Timor and its tectonic significance: Journal of Structural Geology, v. 3, p. 143–167.

Berry, R.F., and McDougall, I., 1986, Interpretation of $^{40}Ar/^{39}Ar$ and K/Ar dating evidence from the Aileu Formation, East Timor, Indonesia: Chemical Geology, v. 59, p. 43–58.

Bloomer, S.H., and Fisher, R.L., 1987, Petrology and geochemistry of igneous rocks from the Tonga Trench: A non-accreting plate boundary: Journal of Geology, v. 95, p. 469–495.

Boillot, G., Grimaud, S., Mauffret, A., Mougenot, D., Kornprobst, J., Mergoil-Daniel, J., and Torrent, G., 1980, Ocean-continent boundary off the Iberian margin: A serpentinite diapir west of the Galicia Bank: Earth and Planetary Science Letters, v. 48, p. 23–34.

Boillot, G., Recq, M., Winterer, E.L, Meyer, A.W., Applegate, J., Baltuck, M., Bergen, J.A., Comas, M.C., Davies, T.A., Dunham, K., Evans, C.A., Girardeau, J., Goldberg, G., Haggerty, J., Jansa, L.F., Johnson, J.A., Kasahara, J., Loreau, J.P., Luna-Sierra, E., Moullade, M., Ogg, J., Sarti, M., Thurow, J., Williamson, M., 1987, Tectonic denudation of the upper mantle along passive margins: A model based on drilling results: Tectonophysics, v. 132, p. 335–342.

Bonatti, E., and Michael, P.J., 1989, Mantle peridotites from continental rifts to ocean basins to subduction zones: Earth and Planetary Science Letters, v. 91, p. 297–311.

Bonatti, E., Ottonello, G., and Hamlyn, P.R., 1986, Peridotites from the island of Zabargad (St. John), Red Sea: Petrology and geochemistry: Journal of Geophysical Research, v. 91, p. 599–631.

Bonneau, M., 1984, Correlation of the Hellenic Nappes in the southeast Aegean and their tectonic reconstruction: Geological Society of London Special Publication 17, p. 517–527.

Bowin, C.O., Nalwalk, A.J., and Hersey, J.B., 1966, Serpentinized peridotite from the north wall of the Puerto Rico Trench: Geological Society of America Bulletin, v. 77, p. 257–270.

Brown, M., and Earle, M.M., 1983, Cordierite-bearing schists and gneisses from Timor, eastern Indonesia: P-T conditions of metamorphism and tectonic implicatons: Journal of Metamorphic Geology, v. 1, p. 183–203.

Carter, J.L., 1970, Mineralogy and chemistry of the Earth's upper mantle based on the partial fusion-partial crystallization model: Geological Society of America Bulletin, v. 81, p. 2021–2034.

Carter, D.J., Audley-Charles, M.G., and Barber, A.J., 1976, Stratigraphical analysis of island arc–continental margin collision in eastern Indonesia: Journal of the Geological Society of London, v. 132, p. 179–198.

Chamalaun, F.H., and Grady, A.E., 1978, The tectonic development of Timor: A new model and its implications for petroleum exploration: Australian Petroleum Exploration Association Journal, v. 18, p. 102–108.

De Roever, W.P., 1940, Geological investigations in the southwestern Moetis region (Netherlands Timor): Geological expedition to the Lesser Sunda Islands, University of Amsterdam, v. 2, p. 97–344.

De Waard, D., 1954, Contributions to the geology of Timor. VI. The Second Geological Timor Expedition, Preliminary Results: Indonesian Journal for Natural Science, v. 110, p. 154–160.

Dewey, J.F., and Bird, J.B., 1971, Origin and emplacement of the ophiolite suite: Appalachian ophiolite in Newfoundland: Journal of Geophysical Research, v. 76, p. 3179–3206.

Dick, H.J.B., and Bullen, T., 1984, Chromian spinel as a petrogenetic indicator in abyssal and alpine-type peridotites and spatially associated lavas: Contributions to Mineralogy and Petrology, v. 86, p. 54–76.

Droop, G.T.R., 1987, A general equation for estimating Fe (super 3+) concentrations in ferromagnesian silicates and oxides from microprobe analyses, using stoichiometric criteria, Mineralogical Magazine, v. 361, p. 431–435.

Dropkin, M.J., Harris, R.A., and Zeitler, P.K., 1993, An Oligocene forearc crustal flake exposed in a contemporary arc-continent collision: Timor, Indonesia: Geological Society of America Abstracts with Programs, v. 25, no. 6, p. 482.

Earle, M.M., 1980, A study of Boi and Molo, two metamorphic massifs on Timor, eastern Indonesia [Ph.D. dissert.]: Chelsea College, University of London, 228 p.

Earle, M.M., 1981, The metamorphic rocks of Boi, Timor, eastern Indonesia, *in* Barber, A.J., and Wiiryosiujono, S., ed., The geology and tectonics of Eastern Indonesia: Geological Research and Development Centre Special Publication, no. 2, p. 239–251.

Eldholm, O., and Talwani, M., 1981, The passive margins of northern Europe and East Greenland, *in* Scrutton, R.A., ed., Dynamics of passive margins: American Geophysical Union, Geodynamic Series, v. 6, p. 30–44.

Fisher, R.L., and Engel, C.G., 1969, Ultramafic and basaltic rocks dredged from the nearshore flank of the Tonga Trench: Geological Society of America Bulletin, v. 80, p. 1373–1378.

Frey, F.A., and Prinz, M., 1978, Ultramafic inclusions from San Carlos, Arizona: Petrologic and geochemical data bearing on their petrogenesis: Earth and Planetary Science Letters, v. 38, p. 129–176.

Hamilton, W., 1979, Tectonics of the Indonesian region: United States Geological Survey Professional Paper, no. 1078, 345 p.

Hamlyn, P.R., and Bonatti, E., 1980, Petrology of mantle-derived ultramafics from the Owen Fracture Zone, northwest Indian Ocean: Implications for the nature of the oceanic upper mantle: Earth and Planetary Science Letters, v. 48, p. 65–79.

Harris, R.A., 1991, Temporal distribution of strain in the active Banda orogen: A reconciliation of rival hypotheses: Journal of Southeast Asian Earth Sciences, v. 6, p. 373–386.

Harris, R.A., 1992, Peri-collisional extension and the formation of Oman-type ophiolites in the Banda Arc and Brooks Range, *in* Parson, L.M., Murton, B.J., and Browning, P., eds., Ophiolites and their modern oceanic analogues: Geological Society of London Special Publication 60, p. 301–325.

Harris, R.A., Sawyer, R.K., and Audley-Charles, M.G., 1998, Collisional mélange development: Geologic associations of active mélange-forming processes with exhumed mélange facies in the western Banda orogen, Indonesia: Tectonics, v. 17, p. 458–480.

Harris, R.A., Kaiser, J.S., Hurford, A.J., and Carter, A., 2000, Thermal history of Australian passive margin cover sequences accreted to Timor during Neogene arc-continent collision, Indonesia: Journal of Asian Earth Sciences, v. 18, p. 47–69.

Helmers, J., Sopaheluwaken, J., Tjokrosapoetro, S., and Surya Nila, E., 1989, High grade metamorphism related to peridotite emplacement near Atapupu, Timor, with reference to the Kaibobo peridotite on Seram, Indonesia: Netherlands Journal of Sea Research, v. 24, p. 357–371.

Herzberg, C.T., 1993, Lithosphere peridotites of the Kaapvall craton: Earth and Planetary Science Letters, v. 120, p. 13–29.

Ishii, T., 1985, Dredged samples from the Ogasawara fore-arc seamount or "Ogasawara Paleoland" fore-arc ophiolite, *in* Nasu, N., ed., Formation of active ocean margins: Tokyo, Terra Scientific Publishing Company, p. 307–342.

Ishii, T., Robinson, P.T., Maelawa, H., and Fiske, R., 1992, Petrological studies of peridotites from diapiric serpentinite seamounts in the Izu-Ogaswara-Mariana forearc, Leg 125, *in* Fryer, P., Pearce, J.A., and Stokking, L.B., eds., Proceedings of the Ocean Drilling Program, Scientific Results, v. 125: College Station Ocean Drilling Program, p. 445–485.

Ishiwatari, A., 1985, Igneous petrogenesis of the Yakuno Ophiolite (Japan) in the context of the diversity of ophiolites: Contributions to Mineralogy and Petrology, v. 89, p. 155–167.

Linthout, K., and Helmers, H., 1994, Pliocene obducted, rotated, and migrated ultramafic rocks and obduction-induced anatectic granite, SW Seram and Ambon, Eastern Indonesia: Journal of Southeast Asian Earth Sciences, v. 9, p. 95–109.

Moores, E.M., 1982, Origin and emplacement of ophiolites: Reviews in Geophysics and Space Physics, v. 20, p. 735–760.

Mysen, B.O., and Kushiro, I., 1977, Compositional variations of coexisting phases with degree of melting of peridotite in the upper mantle, American Mineralogist, v. 62, p. 843–856.

Nicholls, I.A., Ferguson, J., Jones, H., Marks, G.P., and Mutter, J.C., 1981, Ultramafic blocks from the ocean floor southwest of Australia: Earth and Planetary Science Letters, v. 56, p. 362–374.

Parkinson, C., 1998, Emplacement of the East Sulawesi Ophiolite: Evidence from subophiolite metamorphic rocks: Journal of Asian Earth Sciences, v. 16, p. 13–28.

Prasetyadi, C., and Harris, R.A., 1996, Hinterland structure of the active Banda arc–continent collision, Indonesia: Constraints from the Aileu complex of East Timor: Proceedings, 25th Convention of the Indonesian Association of Geology, p. 144–173.

Rosidi, H.M.O., Suwitopiroyo, K., and Tjokrosapoetro, S., 1979, Geological map Kupang-Atambua Quadrangle, Timor: Bandung, Indonesia, Geological Research and Development Centre, scale 1:250,000.

Searle, M.P., and Stevens, R.K., 1984, Obduction processes in ancient, modern and future ophiolites, *in* Gass, I.G., Lippard, S.J., and Shelton, A.W., eds., Ophiolites and oceanic lithosphere: Geological Society of London Special Publication 13, p. 303–320.

Shibata, T., and Thompson, G., 1986, Peridotites from the Mid-Atlantic Ridge at 43° N and their petrogenetic relation to abyssal tholeiites: Contributions to Mineralogy and Petrology, v. 93, p. 144–159.

Simons, A.L., 1940, Geological investigations in northeast Netherlands Timor: Geological Expedition of the Lesser Sunda Islands, v. 1, p. 107–213.

Sopaheluwakan, J., Helmers, H., Tjokrosapoetro, S., and Surya Nila, E., 1989, Medium pressure metamorphism with inverted thermal gradient associated with ophiolite nappe emplacement in Timor: Netherlands Journal of Sea Research, v. 24, p. 333–343.

Streckeisen, A., 1976, To each plutonic rock its proper name: Earth Science Reviews, v. 12, p. 1–33.

Manuscript Accepted by the Society February 23, 2000

Printed in the U.S.A.

Geological Society of America
Special Paper 349
2000

Proterozoic–early Paleozoic ophiolites of the Andean basement of southern South America

Victor A. Ramos
Departamento de Geología, Universidad de Buenos Aires, Ciudad Universitaria, (1428) Buenos Aires, Argentina; e-mail: andes@tango.gl.fcen.uba.ar
Mónica Escayola
Departamento de Geología, Universidad Nacional de Salta, Buenos Aires 177, (4400) Salta, Argentina, and *Consejo Nacional de Investigaciones Científicas y Técnicas*
Diana I. Mutti
Departamento de Geología, Universidad de Buenos Aires, Buenos Aires, Argentina
Graciela I. Vujovich
Departamento de Geología, Universidad de Buenos Aires, Ciudad Universitaria, (1428) Buenos Aires, Argentina, and *Consejo Nacional de Investigaciones Científicas y Técnicas*

ABSTRACT

We describe a series of Middle Proterozoic to early Paleozoic ultramafic and mafic rocks in the basement of the Central Andes that have different geochemical attributes and paleogeographic settings. Several ophiolitic belts are identified on the basis of lithologic assemblages and geochemical characteristics, but only the most important belts are described here: the Western and Eastern Córdoba (Early Cambrian), Pie de Palo (Middle Proterozoic), Western Precordillera (Middle Ordovician), and eastern Cordillera Frontal (Proterozoic) belts. Most ophiolites in the Central Andean basement belong to one of two types: the harzburgite type and the lherzolite type. For example, in the metamorphic basement in the Western Córdoba belt of the Eastern Sierras Pampeanas, a harzburgite-type ophiolite is associated with important podiform chromite deposits whereas a lherzolite-type ophiolite of the Eastern Córdoba belt was emplaced in a backarc setting. The most ancient ophiolites at these latitudes are exemplified by the Middle Proterozoic Pie de Palo ophiolite, part of the Cuyania terrane, which is a Laurentian-derived block accreted to the protomargin of Gondwana during early Paleozoic time. The ophiolitic assemblage of the Western Precordillera includes serpentinized peridotites, ultramafic cumulates, layered gabbros, diabases, and basaltic pillow lavas. The ophiolites of the eastern Cordillera Frontal are emplaced within three distinctive belts in Precambrian metamorphic rocks. The dominant assemblage of the eastern Cordillera Frontal includes serpentinized peridotites, gabbros, basaltic dikes, and pillow lavas. As a whole, these ophiolite belts represent a complex system of sutures that record the accretionary history of the basement of the Central Andes in several tectonic episodes.

INTRODUCTION

The basement of the Central Andes of Argentina, situated within the flat segment of subduction between lat 27°S and 33°S, exposes a series of ophiolitic belts. Some of these belts were described in the early work of Borrello (1969) and in the review of ultramafic and mafic rocks of Villar (1975). The ophiolitic belts have received increasing attention in recent years as they have been interpreted as marking possible sutures of a collage of terranes accreted to the protomargin of Gondwana during Proterozoic–early Paleozoic time (Ramos et al., 1984, 1986). The current distribution of these possible sutures and the outline of the suspected allochthonous terranes are shown in Figure 1, along with the main geologic provinces of the Central Andes.

Present information on these belts is inhomogeneous; some of them are known at a reconnaissance level, whereas others are better studied. The objective of this overview is to describe these poorly known belts in order to call attention to their lithologic assemblages, ages, distribution, and local features and thus to encourage future studies.

ULTRAMAFIC AND MAFIC BELTS AND OPHIOLITE ASSEMBLAGES

The early reviews of Villar (1975, 1985) focused on the petrographic attributes and the Alpine-type settings of these ultramafic-mafic belts. The lithology, the complex tectonic emplacement, and the belts' general association with mylonitic rocks prompted her to

Ramos, V.A., et al., 2000, Proterozoic–early Paleozoic ophiolites of the Andean basement of southern South America, *in* Dilek, Y., Moores, E.M., Elthon, D., and Nicolas, A., eds., Ophiolites and Oceanic Crust: New Insights from Field Studies and the Ocean Drilling Program: Boulder, Colorado, Geological Society of America Special Paper 349, p. 331–349.

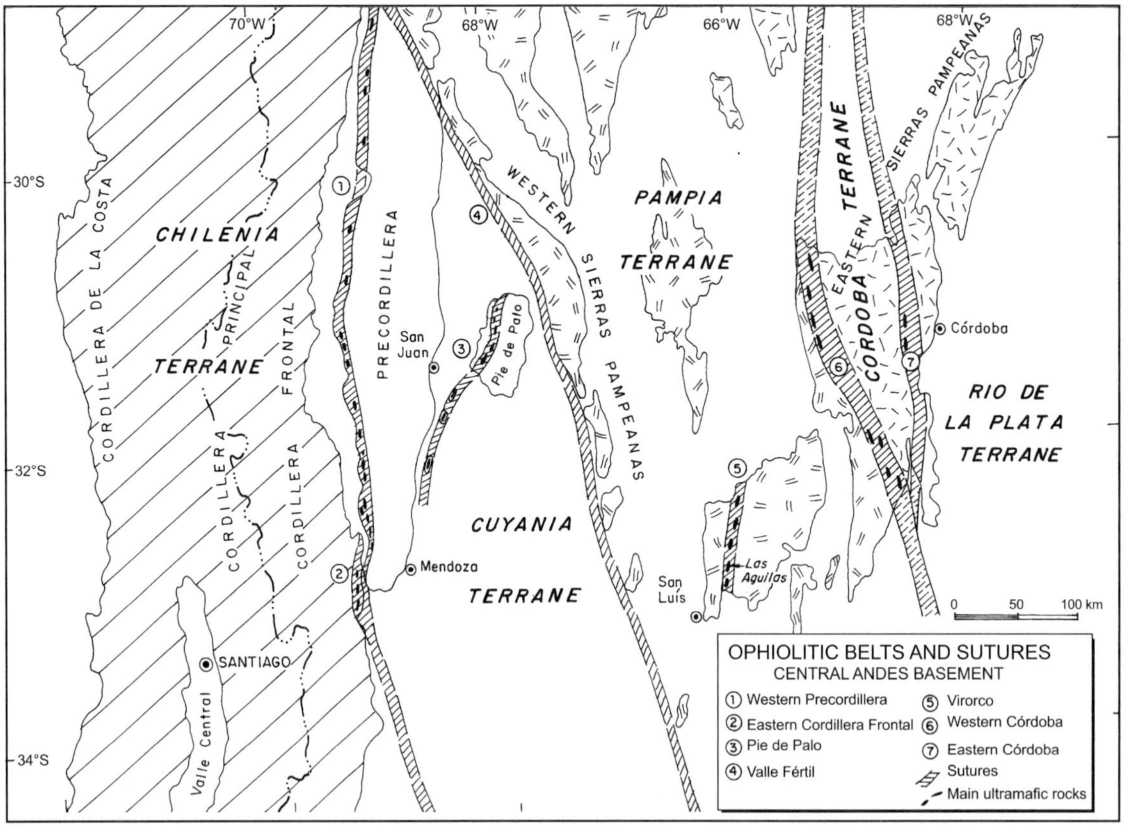

Figure 1. Ultramafic and mafic rocks interpreted as ophiolitic belts, possible sutures among Proterozoic–early Paleozoic allochthonous suspect terranes, and main geologic provinces within the flat subduction segment of the Central Andes (modified from Ramos, 1988, 1994, 1995).

interpret some of them as ophiolitic assemblages. These belts are interpreted to have been obducted onto the Río de La Plata craton successively from east to west (Ramos, 1988). We describe them in this order.

1. The Precambrian–early Paleozoic basement of the Sierras Pampeanas is exposed in a series of mountain blocks uplifted during the Andean orogeny. These mountains expose two belts of ultramafic and mafic rocks of Late Proterozoic age: the Eastern Córdoba belt and Western Córdoba belt (Fig. 1) within the Eastern Sierras Pampeanas, which separate the Córdoba terrane from Pampia terrane and the Río de La Plata craton (Kraemer et al., 1995). These belts are well developed along the uplifted basement of the east-central part of the Andes. Several ophiolitic thrust sheets have been recognized within the Late Proterozoic–early Paleozoic metamorphic basement. Since the early work of Villar (1985), the existence of two separate belts in the Eastern Sierras Pampeanas has been recognized, but Kraemer et al. (1995) were the first to focus on their tectonic importance.

2. Another tract of ultramafic and mafic rocks known as the Virorco belt is developed within the Pampia terrane, north of the city of San Luis (Fig. 1). Elongated bodies of mafic and ultramafic rocks, up to 3.5 km in length and up to 500 m width, crop out within a 50-km-long, north-northeast–trending belt. These rocks, first described by González Bonorino (1962), are characterized by ultramafic and mafic bodies metamorphosed to granulitic facies. Dominant rock types include dunites, pyroxenites, hornblendites, and amphibolites (Bjerg et al., 1996). A felsic segregation in the ultramafic rocks at Las Aguilas dated by U-Pb in zircon separates yielded a value of 478 ± 6 Ma that was interpreted by Sims et al. (1997) as a crystallization age. The extensively boudinaged rocks display a foliation parallel to the stretching lineation in the enclosing pelitic gneisses. These mafic-ultramafic bodies are associated with platinum-group minerals in Las Aguilas (Gervilla et al., 1992; Mogessie et al., 1995) and have been interpreted as a series of linear intrusions unrelated to an ophiolite assemblage (Villar, 1985), as a suture of a backarc basin (Ramos, 1988), or more recently, as a series of mafic, ultramafic, and amphibolite rocks emplaced synchronously with regional deformation in a backarc or frontal island-arc setting (Sims et al., 1997). Although geologic studies are scarce and the precise nature of this belt is unknown, it is worth mentioning that Kostadinoff et al. (1999) have shown a large positive magnetic anomaly, which coincides with the trace of this belt that has been interpreted as a major crustal discontinuity.

3. A major lineament, known as the Valle Fértil megashear, coincides with a belt of mylonitic rocks that show a conspicuous ductile deformation of Middle Ordovician age (Schmidt et al., 1995; Ramos et al., 1998). There are some ultramafic and mafic rocks along this possible suture (Mirré, 1976). Harzburgites, norites, and pyroxenites are described by Villar (1985) in the Sierra de Valle Fértil, along with other peridotites and amphibolites in the Sierra de La Huerta (Vujovich et al., 1994; Castro de Machuca et al., 1996). These units are better developed in the Sierras de Maz, Umango, and Espinal (see Villar, 1985) to the north outside of the study area. These mafic and ultramafic rocks consist of partially serpentinized peridotite associated with asbestos mines, amphibolites, and metagabbros. Protoliths

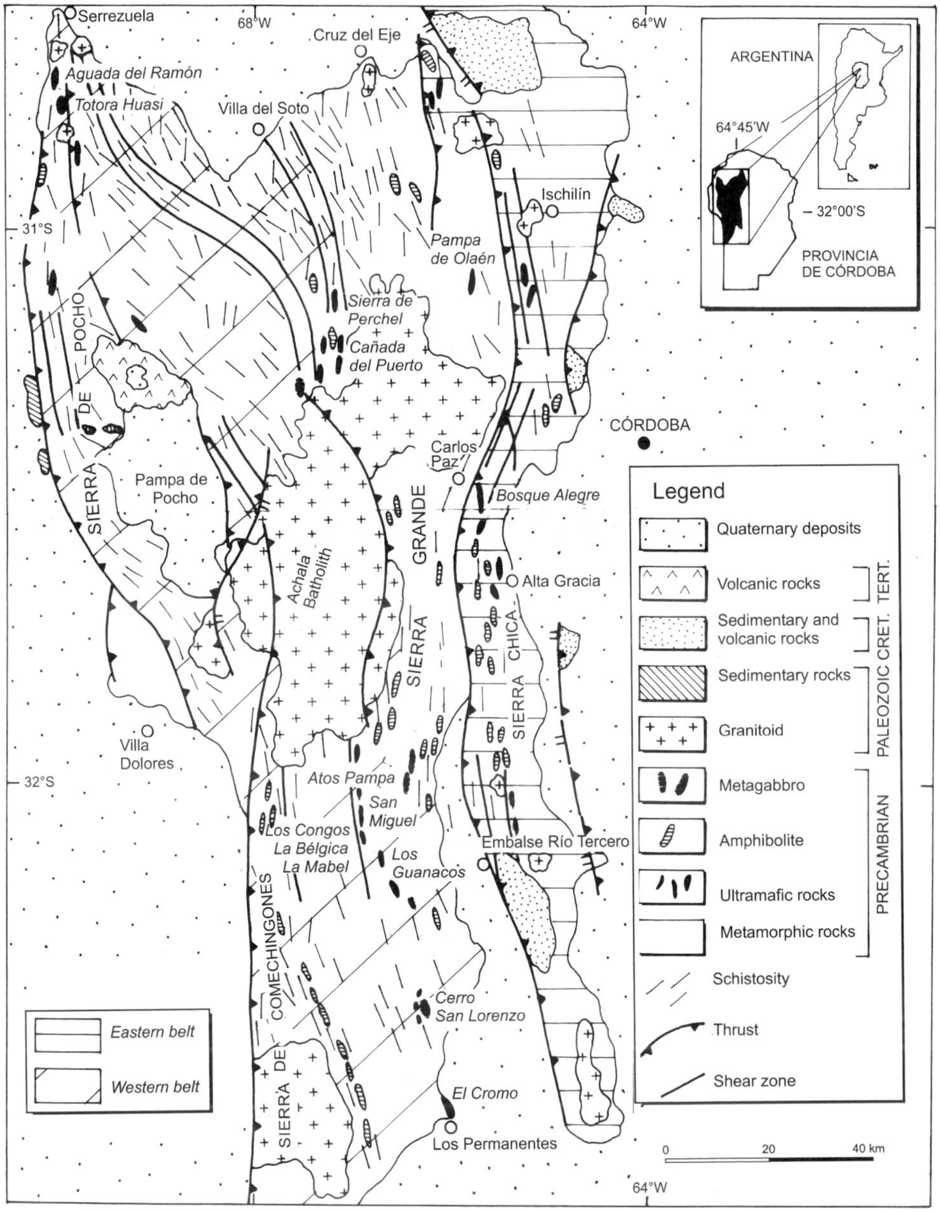

Figure 2. Generalized geologic map of the Sierras de Córdoba, showing the location of the Western and Eastern Córdoba belts of ultramafic and mafic rocks (modified from Mutti, 1992, 1997, and Escayola, 1994).

of most of these amphibolites and metagabbros are interpreted to have formed in an island-arc or backarc setting (Vujovich, 1993; Vujovich and Kay, 1996). Within the study area, the Valle Fértil lineament coincides with major gravity and magnetic anomalies that have been modeled as large ultramafic bodies emplaced in the middle to upper crust (Giménez, 1997; Martínez, 1997). This belt coincides with the proposed suture between the Cuyania and Pampia terranes (Ramos, 1995; Kay et al., 1996; Astini et al., 1996; Rapela et al., 1998).

4. Another belt of ophiolites separates the Precordillera and Pie de Palo terranes, within the Cuyania composite terrane (Fig. 1). Vujovich and Kay (1998) suggested that this belt represents a Middle Proterozoic ophiolite.

5. The westernmost tract of ultramafic and mafic rocks between lat 30°S and 34°S is represented by the Western Precordillera–eastern Cordillera Frontal belt that separates the Chilenia and Cuyania terranes. The belt was emplaced during Devonian time (Ramos et al., 1984).

The paucity of structural and petrologic data precludes a detailed description of some of these belts. We focus our discussion on the Western and Eastern Córdoba, Pie de Palo, and Western Precordillera–eastern Cordillera Frontal belts, for which critical data exist in support of their origin as ophiolitic assemblages.

EASTERN CORDOBA BELT

Eastern Córdoba ophiolites

This belt—exposed along a reverse fault that uplifted the Sierra Chica de Córdoba during the Tertiary (Fig. 2)—extends for 100 km from Ischilín in the north to Embalse Río Tercero in the south. The main exposures included the Pampa de Olaen district, the Bosque Alegre and Alta Gracia bodies, and several other minor bodies of ultramafic and mafic rocks (Mutti, 1987).

Villar (1985) described the ultramafic and mafic rocks in this belt as a typical Alpine-type assemblage. Ramos (1988) interpreted this ultramafic assemblage as an ophiolitic belt emplaced in the suture between the Río de La Plata craton and the allochthonous terranes located to the west. Structural studies of Kraemer et al. (1995) demonstrate the west vergence of the ophiolite emplacement.

The Bosque Alegre ophiolite, southwest of Córdoba city on the western slope of Sierra Chica (Mutti, 1992) constitutes the main exposure in this belt. It is characterized by an elongate body up to 2 km long with an average width of 0.8 km and is bounded on the west by a ductile shear zone associated with a highly schistose amphibolite a few centimeters thick. Ophiolitic rocks are associated with mylonitic belts in a series of west-vergent thrust sheets (Bonalumi and Gigena, 1987). The country rock is composed of schists and tonalite gneiss with a mineral assemblage of plagioclase + quartz + K-feldspar + biotite + garnet, indicating an amphibolite metamorphic facies. Dominant rocks types are lherzolite, subordinate harzburgite, pyroxenite, and gabbro, intruded by leucocratic rocks. The lherzolite has a layered structure and consists of olivine (Fo_{88} to Fo_{90}), orthopyroxene (En_{87} to En_{90}), diopside, amphibole, magnetite, ilmenite, and chromite. The harzburgite displays cumulate textures and consists of olivine, orthopyroxene, spinel, magnetite, ilmenite, chromite, and phlogopite and is intruded by numerous plagiogranite, olivine and hornblende pyroxenite, and gabbro dikes (Mutti, 1992).

Geochemical characteristics of the Eastern Córdoba ophiolites

The peridotites of the Eastern Córdoba belt have SiO_2 contents between 41.99% and 46.80% and MgO contents between 42% and 46%. In the $MgO-Al_2O_3-CaO$ diagram, the peridotites plot in the cumulate peridotitic field, whereas the amphibolites plot in the mafic cumulate field. In the FeO/MgO vs. SiO_2 diagram, the amphibolites plot as calc-alkalic rocks (Mutti, 1992). Among the trace elements when compared to chondrites (Wood et al., 1979b), the Rb and Ba have high relative values (10 × chondrites), as well as Th and Nb. Most of the Eastern Córdoba belt amphibolites show strong affinities with backarc basalts (see Table 1 and Fig. 3B), and hence we interpret them as backarc ophiolites (Mutti, 1992; Escayola et al., 1996).

Tectonic setting

Escayola (1994) identified the Eastern Córdoba belt as a typical lherzolite ophiolite in the sense of Boudier and Nicolas (1985), distinct from the typical Western Córdoba belt of harzburgitic type.

The inferred back-arc origin of the Eastern Córdoba belt led Kraemer et al. (1995) to interpret the basement of the Córdoba terrane as part of the Río de La Plata craton. On the basis of this interpretation, Ramos (1995) recognized the Córdoba terrane as a parautochthonous unit that has been incorporated into the craton as a result of the closure of a backarc basin during the latest Proterozoic to Early Cambrian.

WESTERN CORDOBA BELT

Western Córdoba ophiolites

The ultramafic and mafic rocks exposed along the western flank of Sierra Grande de Córdoba delineate a 150-km-long discontinuous belt that crops out in the Proterozoic metamorphic basement of Sierras Pampeanas (Fig. 2). This belt extends from Cruz del Eje in the north to Los Permanentes in the south.

The host rocks of the ultramafic and mafic units consist of medium- to high-grade metamorphic rocks that broadly correspond to the Brasiliano cycle (900 to 500 Ma). The metamorphism is characterized by two prograde episodes (Escayola et al., 1996) that are associated with an intense deformation that produced an axial-planar foliation. Ages obtained by Rb/Sr and K/Ar methods range from 550 to 640 Ma (Cingolani and Varela, 1975; Caminos et al., 1982). Tonalite and granodiorite stocks and granitic veins and pegmatites intrude these units. The latest granitic intrusion of the Brasiliano cycle has been dated between 530 and 520 Ma by U-Pb in zircons (Rapela et al., 1998). This date indicates a Cambrian age for the syncollisional intrusions. Mylonitization and retrograde metamorphism followed late high-grade metamorphism associated with migmatization. Dominant types are quartz + feldspar, quartz + mica + feldspar, and quartz + garnet gneisses and schists, along with migmatites and subordinate amphibolites, marbles, and kinzigites. Their protoliths have been interpreted as quartzofeldspathic and quartz-rich clastic rocks, shales, graywackes, shelf carbonates, and basaltic rocks (Mutti et al., 1997). Ultramafic rocks are bounded by reverse to dextral strike-slip faults marked by ductile shear zones (Martino et al., 1995).

In the northern segment of the Western Córdoba ophiolites, Villar (1985) recognized a series of mafic to ultramafic bodies such as Totora Huasi, Agua de Ramón, Sierra del Perchel, and Cañada del Puerto (Fig. 2). There are no detailed studies of these rocks.

The Western Córdoba ophiolites south of the Achala batholith occur in four distinctive ways: (1) metamorphic peridotites that are composed mainly of depleted harzburgite with concordant to subconcordant podiform chromite ore bodies; (2) transition-zone harzburgites, lherzolites, and websterites intruded by sills and pods of spinel dunites; (3) a layered complex composed mainly of cumulate harzburgite and gabbros; and (4) a cordierite + garnet + anthophyllite association that could represent the basaltic pillow-lava section. These rocks have been recognized in Atos Pampa–San Miguel, Los Congos, La Bélgica, Los Guanacos, Cerro San Lorenzo, El Cromo, and Los Permanentes (see locations in Fig. 2) (Fernández Gianotti, 1979; Escayola, 1994; Mutti 1994, 1997).

Metamorphic peridotites

Metamorphic peridotites occur as elongated lenses that were emplaced along ductile shear zones traceable for more than 40 km (Massabie et al., 1994). The main foliation azimuth ranges from N40°W to N40°E, and the regional structure is characterized by a series of west-vergent thrust sheets that dip between 50° and 70° to the east (Dalla Salda, 1984). Individual peridotite bodies reach up to 1700 m in length and 800 m in width. The dominant rock type as inferred from mineral relicts and pseudomorphic textures is harzburgite with Mg# [Mg/(Mg + Fe)] between 0.87 to 0.84, alumina content <2%, and alkalis content <0.2% (Mutti, 1992; Escayola, 1994; Escayola et al., 1996). The metamorphic assemblage in these rocks includes antigorite + brucite + tremolite + lizardite + chrysotile. The lack of layered textures indicates that these bodies are not cumulate rocks (Escayola, 1994).

Most of the ultramafic rocks are associated with podiform chromite bodies within the basal units, which have been interpreted as mantle tectonites (Mutti et al., 1997). In most cases, the chromitite is enclosed in an intensely sheared dunitic or orthopyroxenitic host rock. A few Fo_{94} to Fo_{92} olivine relicts occur in the serpentinized dunite mesh. Three types of chromite deposits are present: concordant, subconcordant, and, rarely, discordant with penetrative foliation and lineation. The latter is strongly deformed (Mutti and Fernández, 1999). The chromite grains vary from subhedral to anhedral and range from 20 μm to 5 mm. Microscopic examination of the samples shows dark brown spinels with exsolution textures and a wide variety of mineral

TABLE 1. REPRESENTATIVE ANALYSES OF EASTERN AND WESTERN CORDOBA BELTS

	Eastern Córdoba ophiolite backarc amphibolites*					Western Córdoba ophiolite							Western Córdoba ophiolite					
						Arc amphibolites							Harzburgites†		Spinel dunites†		Gabbroic dikes†	
Sample	104	M311	16	M262	M259	44	M220	M200	M241	SE6	SU4	SE8						
SiO_2	46.77	40	48.65	45.52	46.62	52.02	46.68	48.35	47.74	43	40.43	45.36	46.46	43.92	22.89	27.61	41.12	39.83
TiO_2	1.93	0.63	1.58	1.47	1.15	0.6	0.66	0.49	0.61	0.85	0.62	0.71	0.41	0.86	5.23	1.63	0.72	0.74
Al_2O_3	13.86	14.74	14.25	14.87	12.24	13.57	12.84	18.38	13.12	13.45	14.74	15.7	0.73	2.71	15.79	18.76	15.55	17.3
FeO	13.1	10.54	10.01	12.45	12.72	12.15	8.83	7.3	11.91	11.76	11.4	12.66	8.06	10.07	35.83	17.94	11.27	12.17
MnO	0.23	0.16	0.17	0.16	0.21	0.01	0.18	0.13	0.19	0.2	0.3	0.18	0.13	0.13	0.33	0.2	0.2	0.21
MgO	8.61	11.78	5.81	7.56	8	8.15	8.72	6.55	7.26	8.82	11.21	12	43.8	42.09	19.73	33.37	13.57	12.05
CaO	12.73	15.86	13.03	14.22	13.9	9.11	18.9	14.66	13.26	18.06	16.84	10.14	0.28	0.11	0.1	0.19	16.49	16.56
Na_2O	1.81	2	2.78	1.65	1.86	1.76	0.84	1.36	1.78	0.91	0.41	1.13	0.1	0.07	0.05	0.16	0.77	0.78
K_2O	0.37	1.07	0.53	0.5	0.65	1.2	0.21	0.28	0.24	0.22	0.19	0.23	0.01	0.01	0.03	0.02	0.27	0.3
P_2O_5	0.2	0.06	0.13	0.13	0.15	0.04	0.05	0.03	0.06	0.07	0.06	0.05	0.02	0.02	0.02	0.11	0.03	0.05
LOI	2.5	2	2.81	1.45	2.5	b.l.	1.58	1.27	2.5	2.5	3.49	1.34						
Total	102.11	98.84	99.75	99.98	100.00	98.61	99.49	98.8	98.67	99.84	99.69	99.5	100.00	99.99	100.00	99.99	99.99	99.99
Cr	321	850	170	360	200		300	289	400	280	320	480	2734	1647	68		372	308
Ni	93	341	73	216	51	500	131	103		117	154	209	2225	1235	203	775	114	96
Co	49	60	46	56	37	57	39	40	52	50	54	59	112	86	125	180	53	61
Sc	50	38	44	40	45	61	33	31	49	46	47	58	8	31	86	91	53	47
V	401	200	299	270	292	55	257	128	257	336	280	293	29	160	1376	78	282	226
Rb	15	15	33	20	15	259	20	20	15	15	18	20	2	2	16	b.l.	7	5
Cs	1	1	1	1.2	1	58	1	1	1	1	1	1	0.52	0.2	1.67	b.l.	0.09	0.01
Ba	50	370	79	70	270		102	122	50	50	19	17	9	38	111	0.43	38	23
Sr	190	81	252	416	163	470	53	116	44	65	41	79	b.l.	b.l.	8	30	418	493
Ga						95										98		
Ta	0.32	<0.50	<1	<1	<0.50	<0.50	<0.50	<0.50	0.56	0.5	0.5	0.5	0.04	0.01	0.04	0.12	0.08	0.07
Nb	8.3												25.5	21.7	33	b.l.	15.7	16.8
Hf	2.6	2	3.4	1.6	2	1	0.7	0.8	1	1	0.5	1.1	0.08	0.15	0.52	1.94	1.17	1.1
Zr	102		112	73			23	31			20	25	b.l.	b.l.	b.l.	85	96	51
Y	32	24	35	24	30	18	25	15	23	35	20	30	20	6	6	b.l.	11	13
Th	0.95	1.2	0.5	0.9	0.2	<0.50	<0.50	0.5	0.5	0.5	0.5	0.5	0.04	b.l.	0.17	b.l.	0.1	b.l.
U	0.42	0.5	0.5	0.5	1.7	0.5	0.5	0.5	0.5	0.5	0.5	0.5	0.1	0.38	0.15	0.32	b.l.	0.06
La	7.8	4.1	5.8	7.3	7.4	1.3	1.3	1	2.2	1.0	0.9	0.8	0.2	0.22	0.13	0.82	1.61	1.47
Ce	19	13	15	19	23	3	4	3	3	5	5	7	0.95	1.26	0.81	2.13	4.44	4.58
Nd	5	5	10.16	8	18	5	5	5	5	5	5	5	0.73	0.63	0.35	3.45	5.16	4.19
Sm	4	2.3	3.68	3	5.1	1	1.2	1	1.6	1.8	1.1	1.9	0.03	0.07	0.12	1.84	2	1.87
Eu	1.9	0.7	1.28	1.1	1.4	0.5	0.5	0.4	0.8	0.6	0.7	0.7	0.02	0.02	0.05	0.26	0.78	0.76
Gd													0.05	b.l.	b.l.	3.37	2.9	2.69
Tb	1.1	0.5	0.82	0.5	0.9	0.5	0.5	0.5	0.5	0.5	0.5	0.9	0.01	0.03	0.16	0.67	0.54	0.52
Ho													b.l.	b.l.	0.33	1.12	0.92	0.8
Tm													b.l.	b.l.	0.21	0.5	0.43	0.4
Yb	4.1	2.9	3.45	2.5	3.5	2.9	2.7	1.6	3.5	3.5	2.6	3.8	0.08	0.36	1.49	3.06	2.69	2.59
Lu	0.63	0.48	0.51	0.34	0.49	0.4	0.39	0.25	0.6	0.08	0.39	0.51	0.01	0.06	0.26	0.47	0.4	0.37

Note: Blank—not determined; b.l.—below limit of detection. Values are weight percent for oxides, parts per million for trace elements.

* This paper.

† Escayola et al. (1996).

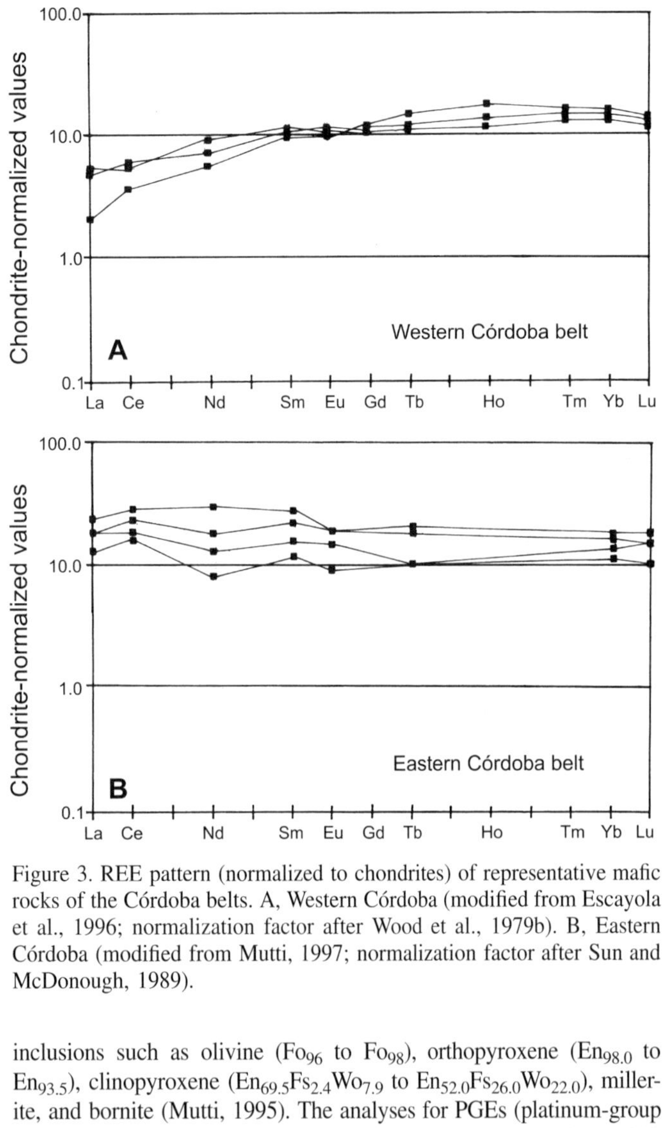

Figure 3. REE pattern (normalized to chondrites) of representative mafic rocks of the Córdoba belts. A, Western Córdoba (modified from Escayola et al., 1996; normalization factor after Wood et al., 1979b). B, Eastern Córdoba (modified from Mutti, 1997; normalization factor after Sun and McDonough, 1989).

inclusions such as olivine (Fo_{96} to Fo_{98}), orthopyroxene ($En_{98.0}$ to $En_{93.5}$), clinopyroxene ($En_{69.5}Fs_{2.4}Wo_{7.9}$ to $En_{52.0}Fs_{26.0}Wo_{22.0}$), millerite, and bornite (Mutti, 1995). The analyses for PGEs (platinum-group elements) show a concentration in the chromite dunite of 30.7 to 980 ppb, whereas the concentration in the harzburgites ranges from 1.7 to 4.9 ppb. The chondrite-normalized patterns range from highly fractionated to flat PGE profiles (Mutti and Fernández, 1999).

Transition-zone peridotite

The transition zone between the mantle and crustal units is well represented in the Western Córdoba belt of the Sierras de Córdoba and locally contains cumulate tectonized peridotites with thick, layered, gabbro sills and numerous small gabbro dikes and veins. Dominant lithologies include harzburgite, dunite, pyroxenite, and gabbro as well as intensively serpentinized amphibolites. The gabbros show relict layering ranging in thickness from 0.01 to 20 m. Transition-zone peridotites have a characteristic Mg# between 0.77 and 0.70 (though some values fall between 0.64 and 0.50) and higher alumina (>2%) and alkalis (>0.2%) contents than the metamorphic peridotites. Olivine ($Fo_{91.0}$ to $Fo_{86.8}$) and orthopyroxene ($En_{88.0}$ to $En_{86.5}$) are richer in Fe than the basal peridotites. Intercumulus clinopyroxene grains and clinopyroxene lamellae in orthopyroxene range from $En_{67.9}Fs_{2.0}Wo_{30.1}$ to $En_{50.5}Fs_{40.0}Wo_{45.5}$. Hornblende, ferrotschermakite, pargasite, and edenitic hornblende are also common (Mutti and Fernández, 1999).

In some other areas, metamorphosed websteritic dikes in the harzburgites, up to more than 4 m wide, have an antigorite + diopside + tremolite assemblage with a relict igneous texture and display diffuse contacts with the peridotite, in places with interlayered dunites. The harzburgitic rocks are also intruded by sills and pods, arranged as boudins of lherzolitic rocks with the metamorphic association of olivine + tremolite. Spinel dunite dikes, 0.5 to 3 m thick, intrude the harzburgites subconcordantly and are composed of olivine + spinel + chromite ± orthopyroxene.

Layered complex

The mafic layered section is well represented in Cerro San Lorenzo (Fig. 2) and in a few other localities as isolated lenses up to 2700 m long and 1600 m thick. In addition to the layered facies, several concordant, ~1-m-thick, lensoid dikes are observed. These quartz-bearing gabbro dikes are composed of plagioclase, clinopyroxene, quartz, hornblende, apatite, titanite, and opaque minerals. Layers containing three types of cumulate facies are common (Chincarini et al., 1996).

Plagioclase-olivine cumulate. This is the most abundant lithology of the gabbroic complex and is characterized by 2-mm-thick layers of plagioclase and subordinate olivine. It is a fine- to medium-grained rock; the anhedral olivine is generally rimmed by orthopyroxene (bronzite coronas), hornblende, and spinel, together with subhedral plagioclase and interstitial clinopyroxene and minor hypersthene surrounding olivine grains. Bronzite-hornblende coronas are interpreted as metamorphic, whereas the hypersthene coronas could be relict igneous textures.

Plagioclase-clinopyroxene cumulate. These gabbros consist of subhedral plagioclase and clinopyroxene with a well-developed layering up to 50 cm thick, interbedded with hornblende-rich layers up to 7 cm thick.

Plagioclase-clinopyroxene-orthopyroxene cumulate. These gabbronorites are locally present and are less common than the other two facies. Although these rocks underwent an amphibolite-facies metamorphism, they still show well-preserved igneous textures.

Mineral chemistry. The plagioclase composition ranges from An_{83} to An_{78} in the olivine gabbros and from An_{78} to An_{63} in the gabbro dikes. The Mg# of the cumulate rocks ranges from 0.80 to 0.65.

Basaltic section

Amphibolites, mafic schists, and gneiss constitute a widespread assemblage in the Western Córdoba belt (Fig. 2). Mutti (1997) has interpreted them as the metamorphic equivalent of the upper section of the ophiolite. Metapelites and metavolcanic rocks are the dominant protoliths, whereas the quartz + plagioclase + almandine + biotite + K-feldspar + sillimanite + staurolite + cordierite assemblage characterizes the upper-amphibolite metamorphic facies in these rocks. East of the peridotite exposures of La Mabel (see location in Fig. 2) are cordierite + garnet + anthophyllite rocks that have been interpreted as the metamorphosed basaltic section of the sequence (Escayola, 1994).

Geochemical characteristics of the Western Córdoba ophiolites

Harzburgites of the western belt are rich in MgO (42.09%–43.8%), higher than spinel harzburgites (31.3%) and spinel dunites (19.73%–33.37%). MgO contents of cumulate gabbros range from 6.04% to 16.13%. Scarce aluminous spinel–bearing dunites and harzburgites with Al_2O_3 contents of 14.0%–17.6% and FeO contents of 25.1%–31.7% are between the metamorphic and transition-zone peridotites (Escayola, 1994; Escayola et al., 1996). TiO_2 is less than 0.12% in harzburgites and lherzolites (Mutti, 1997) and 0.80% in gabbros. Escayola (1994) found higher TiO_2 contents than in the aluminous spinel facies and gabbros (Mutti, 1997).

Trace elements normalized (after Wood et al., 1979b) either to chondrites or MORB (mid-oceanic-ridge basalt) show Nb, Sr, Rb, and Ba positive anomalies (Escayola, 1994) with REE (rare earth element) patterns characteristic of mid-ocean ridges (Table 1, Fig. 3A). In the harzburgites, REEs show patterns between 0.5 and 2 times chondrites with relative depletion in light REEs (LREEs). The La/Sm ratio >1 is similar to other basal tectonites associated with mid-ocean ridges (Sun and McDonough, 1989). On the other hand, spinel dunites and spinel harzburgites are depleted in LREEs relative to heavy REEs (HREEs) and show a typical negative Eu anomaly (Escayola, 1994).

The southern sector of this western belt has somewhat different geochemical attributes according to Mutti and DiMarco (1998). In this sector, a series of northwest-trending thrust sheets with southwest vergence repeats mafic and ultramafic bodies. The amphibolites have REE contents (see Table 1) similar to those of the arc tholeiites of the South Sandwich Islands described by Pearce et al. (1984, 1995). On the other hand, the ultramafic rocks of Los Guanacos and Cerro San Lorenzo areas show some suprasubduction affinities (Mutti and DiMarco, 1998).

These characteristics as a whole show mid-ocean ridge affinities for the Western Córdoba belt, although the southern sector appears to have a more complex suprasubduction setting, possibly an intraoceanic arc (Mutti, 1995; Escayola et al., 1996).

Tectonic setting

Kraemer et al. (1995) and Ramos (1995) interpreted this belt as a mid-ocean ridge–derived ophiolite emplaced during the Brasiliano (Pampean) deformation that occurred from the end of the Proterozoic to Early Cambrian time, as a result of the accretion of the Pampia terrane to the Río de La Plata craton. In the northern and central sector, the ophiolites are associated with pelitic gneisses derived from accretionary-prism turbidites (Escayola et al., 1996). The southern sector is related to orthoamphibolites derived from a metavolcanic protolith with intraoceanic arc affinity (Mutti and DiMarco, 1998).

PIE DE PALO BELT

This Middle Proterozoic belt is exposed along the western flank of Sierra de Pie de Palo, one of the westernmost ranges of Sierras Pampeanas (Fig. 1). The belt continues farther south in the Cerros Barboza and Valdivia (Ramos and Vujovich, 1995), reaching more than 90 km in length (Fig. 4). The outcrops of Sierra de Pie de Palo and the Cerros Barboza and Valdivia collectively make up the Pie de Palo complex, characterized by schists and gneisses that include a belt of mafic and ultramafic rocks identified as an Alpine-type belt (Villar, 1985). The metamorphic grade has been determined as greenschist to amphibolite facies by Dalla Salda and Varela (1984), Castro de Machuca and Conte-Grand (1994), and Ramos and Vujovich (1995). Vujovich and Kay (1998) presented geochemical data showing that the Pie de Palo Complex is one of the best exposed juvenile Middle Proterozoic oceanic arc or backarc complexes on Earth.

The primary igneous characteristics are obliterated by the extreme deformation of the basement associated with greenschist- to amphibolite-facies metamorphism. In the southernmost part of the range, granulite-facies metamorphism is reported (Dalla Salda and Varela, 1984). The country rock in which the ophiolites are emplaced constitutes most of the central part of the Sierra de Pie de Palo and is composed of biotite + muscovite + garnet + oligoclase schists and gneisses. The protoliths of these country rocks are inferred to have been mainly silicic magmas and immature graywackes (Ramos et al., 1993; Vujovich and Kay, 1998). The protoliths of the mafic-ultramafic belt were ultramafic and mafic cumulates and lava flows. Dominant rock types include serpentinite, metagabbro, metadiorite, amphibolite, and various mafic schists. The majority of the outcrops are chlorite, talc + chlorite, and amphibolite schists with nematoblastic to decussate textures. The most important minerals are tremolite-actinolite, talc, chlorite, albite, and epidote. Cataclasis and protomylonitization are common.

Most of the mafic-ultramafic units are exposed in almost continuous belts between the Quebrada de Guayaupa and Quebrada Piedras Pintadas (Fig. 4). The mafic-ultramafic rocks are part of imbricate thrust sheets that override the Pie de Palo Complex on the siliciclastic and carbonate platform facies of the Caucete Group (Ramos et al., 1996). The northwest-vergent thrust sheets are preserved in a series of klippen along the Quebrada de Pirquitas such as the one seen in Figure 5.

The Middle Proterozoic Grenville ages for the Pie de Palo Complex were established by McDonough et al. (1993) on the basis of U-Pb zircon ages in gneisses. Different varieties of gneisses range in age from 1090 to 1060 Ma, whereas zircons from metarhyolites yield ages near 1080 Ma. Metamorphic zircons from amphibolites gave ages of 1066 Ma (McDonough et al., 1993). A Rb/Sr reference isochron age of 1027.6 Ma from Varela and Dalla Salda (1993) also suggests a Grenvillian age. Some new dates from Cerro Barboza indicate a preliminary $^{207}Pb/^{206}Pb$ age of 1118.2 Ma (Vujovich and Kay, 1998). Minimum $^{40}Ar/^{39}Ar$ hornblende ages ranging from 1030 to 650 Ma from Cerro Valdivia are reported by Ramos et al. (1998). All these data are consistent with a Grenvillian age for the igneous and metamorphic episodes that formed the mafic and ultramafic rocks in this belt.

Geochemical characteristics of the Sierra de Pie de Palo ophiolites

The ultramafic group is composed primarily of serpentinite and metapyroxenite bodies, mainly exposed in imbricated thrust sheets in Quebrada del Gato and Quebrada Piedras Pintadas. These units were interleaved with massive metagabbros, metadiorites, amphibolites, and garnet amphibolite bodies. Chemical analyses of these rocks support an ultramafic cumulate protolith on the basis of the presence of chromite, relict olivine, and pyroxene grains. They have high Cr (4700 ppm) and Ni (1800 ppm) and low REE concentrations (see Table 2). In Cerro Valdivia, ultramafic and mafic cumulate protoliths display compositional and grain-size layering with orthopyroxene, clinopyroxene, olivine, and Cr-rich spinel as cumulate phases. The REE patterns are consistent with intercumulus liquid and support an arc origin (Vujovich and Kay, 1998; Fig. 6A). Similarly, the high La/Ta ratios (>98) of these rocks are also consistent with an arc environment. On the basis of its chemical characteristics, Cerro Barboza amphibolite is interpreted as an amphibole + plagioclase + iron-titanium oxide cumulate (Vujovich and Kay, 1998). Thus an origin as an island-arc cumulate section is well established for the Quebrada

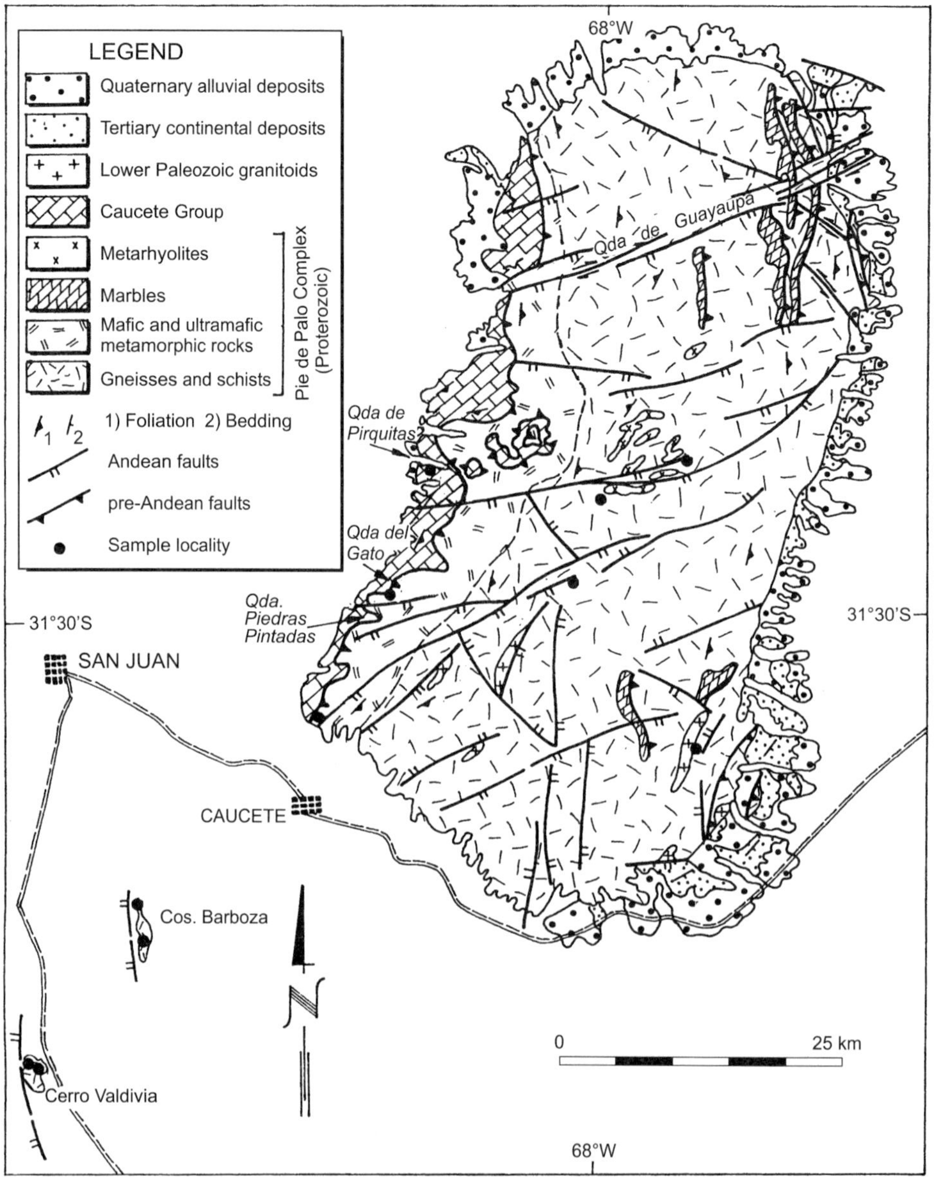

Figure 4. Generalized geology of Sierra de Pie de Palo, showing the mafic and ultramafic belt along its western side and its continuation in Cerros Barboza and Valdivia (modified from Ramos and Vujovich, 1995).

Figure 5. Ophiolitic rocks in a klippe at Quebrada de Pirquitas (view to the south). The oceanic rocks are thrust over quartzites of the Caucete Group platform.

del Gato, Cerros Barboza, and Cerro Valdivia ultramafic rocks (Vujovich and Kay, 1998).

A metamorphosed mafic sequence occurring in Las Pirquitas displays the characteristic textural field relationships of a possible sheeted-dike complex (Vujovich and Kay, 1998). These amphibolites have a distinctive pattern like N-MORB (normal MORB), characterized by LREE depletion and Ta and Th abundance (Fig. 6B). Cr (80–310 ppm) and Ni (45–80 ppm) concentrations are consistent with fractionation of olivine under the low-pressure crystallization conditions typical of N-MORB. The gabbros have typical cumulate-dominated trace element patterns with low REE concentrations and positive Eu anomalies. Vujovich and Kay (1998) correlated these rocks on geochemical grounds with the Lau Basin sequences described by Hawkins (1995a, 1995b).

In the central part of Sierra de Pie de Palo, the amphibolites are interpreted as originating in an arc or backarc environment. The country rocks consist of biotite + muscovite + oligoclase + garnet schists and appear to have a siliciclastic origin, although a magmatic origin for their protolith cannot be ruled out (Ramos et al., 1993; Vujovich and Kay, 1998).

TABLE 2. REPRESENTATIVE ANALYSES OF PIE DE PALO BELT

	Cerros Barboza and Valdivia mafic-ultramafic cumulates*					Quebradas Las Pinquitas and El Quemado amphibolites*			
Sample	V42	V43	V44	V45	B23		QQ2	PR-3	PR4a
SiO_2	53.18	51.94	51.30	52.05	42.31				
TiO_2	1.01	1.22	0.75	1.14	4.87				
Al_2O_3	3.43	4.46	13.60	15.01	12.67				
FeO	9.27	9.15	7.98	8.20	17.92	10.09	11.23	11.95	10.22
MnO	0.21	0.14	0.15	0.12	0.16				
MgO	19.52	17.79	11.51	10.53	6.87				
CaO	12.83	14.00	10.79	7.54	10.50				
Na_2O	0.33	0.69	2.40	3.55	2.50	4.64	2.54	2.63	2.49
K_2O	0.07	0.20	0.72	0.47	0.22				
P_2O_5	0.16	0.36	0.26	0.26	0.45				
Total	100.01	99.95	99.46	98.87	98.47				
Cr	2462	3172	916	584	11	311	138	77	309
Ni	403	245	130	142	24	79	55	44	72
Co	61	42	42	44	80	47	48	49	43
Sc	53.8	46.8	34.3	31.6	51.6	40.4	41.5	42.5	43.4
Cs	b.l.	0.02	0.10	0.12	0.06	0.06	b.l.	0.46	0.72
Ba	13	50	268	143	73	10	32	41	83
Sr	257	309	859	847	858	139	112	145	223
Ta	b.l.	0.11	0.05	0.10	0.26	0.05	0.06	0.07	0.06
Hf	1.1	2.1	1.7	1.6	1.2	1.1	1.4	1.6	1.3
Th	0.02	0.3	0.4	0.4	0.1	b.l.	0.1	0.1	0.1
U	0.01	0.26	0.27	0.30	0.14	0.03	b.l.	0.18	0.09
La	3.97	10.6	9.13	10.0	5.75	1.24	1.51	1.77	1.92
Ce	15.1	33.3	24.0	26.7	16.7	4.48	4.84	5.74	6.61
Nd	12.0	20.8	12.8	15.9	13.0	4.36	4.52	5.09	5.57
Sm	3.82	5.45	3.35	4.02	3.97	1.72	2.26	2.31	1.93
Eu	1.18	1.90	1.35	1.44	2.10	0.651	0.810	0.874	0.778
Tb	0.571	0.737	0.498	0.555	0.698	0.583	0.749	0.797	0.660
Yb	0.920	1.06	1.19	1.22	1.41	2.66	3.33	3.53	2.87
Lu	0.133	0.144	0.157	0.182	0.198	0.390	0.472	0.534	0.430

Note: Blank—not determined; b.l.—below limit of detection. Values are weight percent for oxides, parts per million for trace elements.
* Vujovich and Kay (1998).

As a whole, the ophiolitic rocks of Pie de Palo exhibit a typical suprasubduction affinity in the sense of Pearce et al. (1984) as depicted in Figure 7.

Tectonic setting

The petrologic, geochemical, isotopic, and geochronologic similarities between the basement of Sierra de Pie de Palo and the basement of the Precordillera led Kay et al. (1996) and Vujovich and Kay (1998) to correlate both basements and interpret them as having been derived from Laurentia. These terranes collectively constitute the Cuyania composite terrane that docked against the protomargin of Gondwana during the Middle Ordovician (Ramos et al., 1998). In this scenario, the Pie de Palo ophiolites are interpreted as the suture zone between the Middle Proterozoic Precordillera and the Pie de Palo terranes, which were amalgamated prior to the collision with the Gondwana protomargin (Vujovich and Kay, 1998). This last Ordovician collision uplifted and exposed the basement and the old suture as evidenced by the $^{40}Ar/^{39}Ar$ crystallization history (Ramos et al., 1998).

On the basis of the dominant west vergence of the ductile deformation preserved in these rocks, as well as the geochemical and petrologic characteristics, the Middle Proterozoic subduction is interpreted as east dipping (present coordinates). The Pie de Palo ultramafic-mafic assemblage is therefore explained as a disrupted ophiolite that evolved in an intraoceanic arc and/or backarc setting (Vujovich and Kay, 1998).

WESTERN PRECORDILLERA OPHIOLITE

The ophiolitic nature of the mafic and ultramafic belt of Western Precordillera has been established since the early work of Borrello (1963, 1969). This belt, extending over 900 km from south of lat 26°S to north of lat 33°S, was recognized by Haller and Ramos (1984) as the Famatinian ophiolites emplaced during the early Paleozoic.

The Western Precordillera ophiolite is represented by a series of discontinuous exposures of mafic and ultramafic rocks preserved in the western slope facies of the carbonate platform of the Precordillera (Fig. 8). The ultramafic rocks are more abundant toward the south, whereas pillow lavas and mafic and ultramafic sills are preserved in the northern sector together with distal limestone turbidites, or pelagic limestones, and locally with chert.

The northern sequences

The structural relationships of the clastic strata with the mafic and ultramafic igneous rocks are well exposed in a section along Río Jáchal, near lat 30°15′S (Ramos et al., 1984). In the western part of the section, a massive sequence of basaltic lavas crops out along an unconformity at the base of the Tertiary strata. The basalts occur as pillow lavas and columnar-jointed flows that are steeply folded with a westward vergence. The eastern part of this section consists of graywackes, shales, and local conglomerates containing angular sedimentary clasts, intruded by sills of mafic and ultramafic composition. The eastern unit is tightly folded, and on the basis of graded beds, a

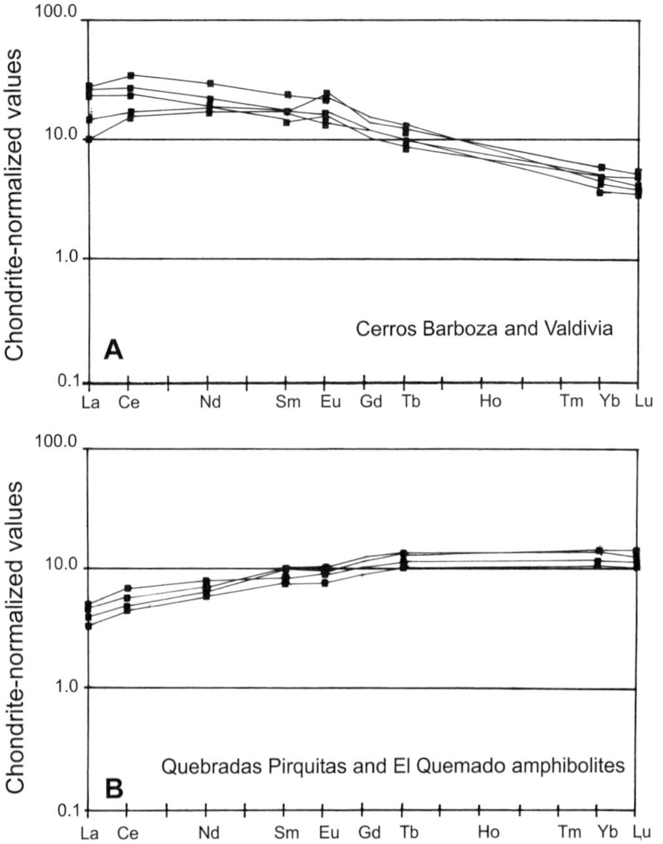

Figure 6. REE pattern normalized to chondrites from representative mafic rocks of the Pie de Palo ophiolite. A, Southern sector. B, Northern sector. (Modified from Vujovich and Kay, 1998; normalized to the Leedy chondrite.)

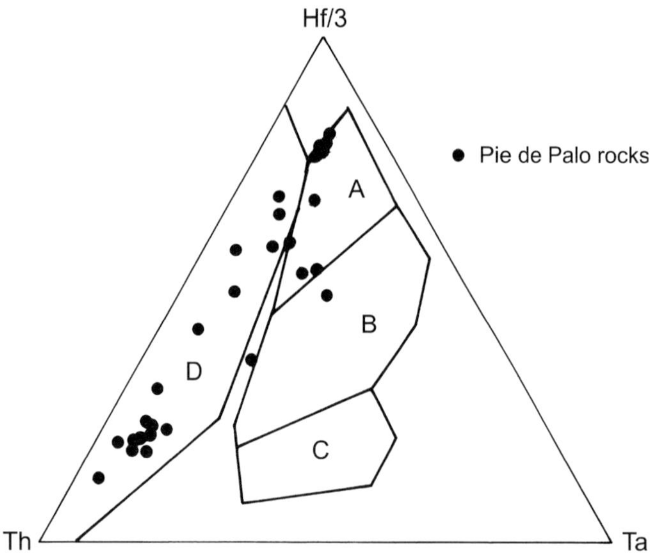

Figure 7. Hf-Th-Ta diagram of mafic rocks of Pie de Palo (from Wood et al., 1979a). Most of the rocks plot in the arc-backarc or N-MORB fields (simplified from Vujovich and Kay, 1998). Fields: A—N-MORB, B—E-MORB, C—alkalic within-plate, and D—arc.

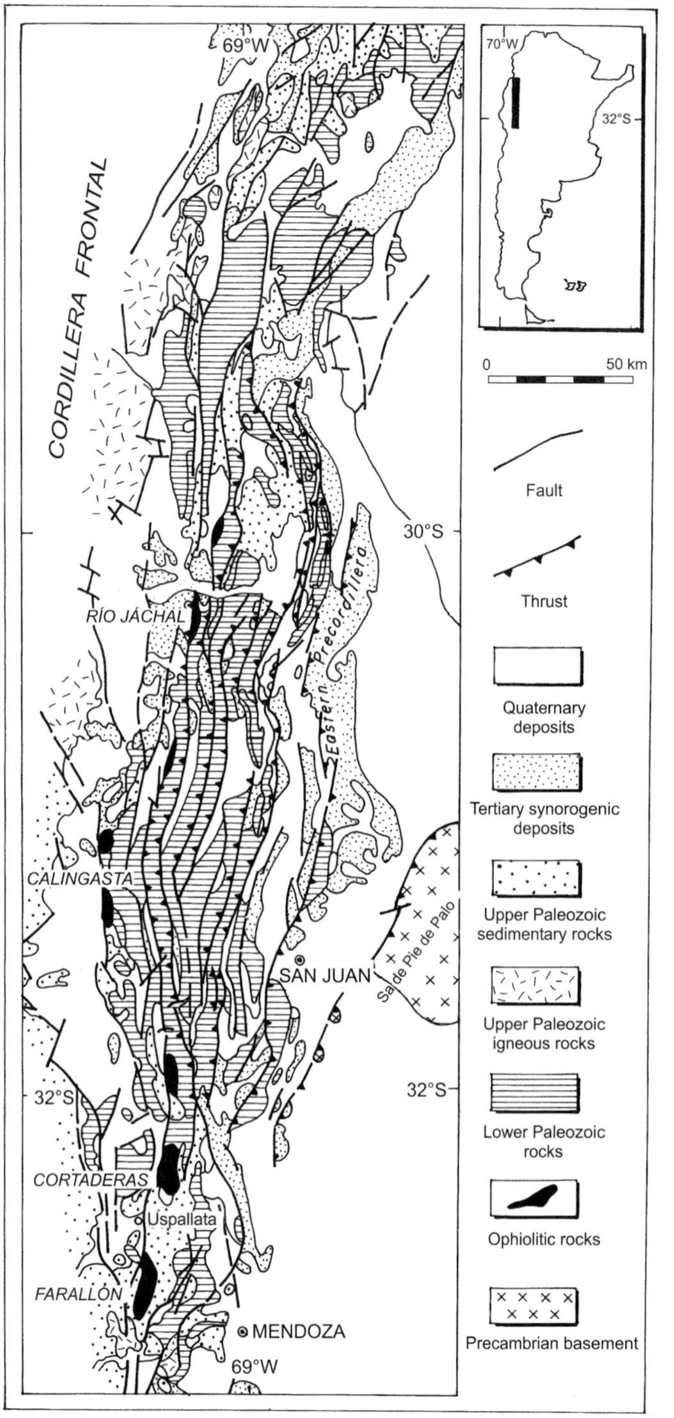

Figure 8. Geologic map of Precordillera fold-and-thrust belt, central western Argentina, showing main exposures of ophiolite assemblages (after Ramos et al., 1986).

west vergence was interpreted. Ultramafic sills occur parallel to the stratification, and one of them occupies the core of a westward-overturned syncline. Their concordance with the structure and the fact that they are genetically related to thick pillow lavas indicate that the ultramafic sills were intruded prior to folding (Ramos et al., 1986).

Similar structural and stratigraphic relationships are seen in the Calingasta section (lat 31°15′) where thick sequences of pillow lavas are interbedded with slope-deposited clastic rocks along the valley of

TABLE 3. REPRESENTATIVE ANALYSES OF WESTERN PRECORDILLERA OPHIOLITE

	Northern sequences: Ordovician basalts*								Southern sequences: Devonian basalts†				
Sample	3-31-3	3-12-2	3-12-5	3-12-6	3-12-7	3-15-1	3-15-4	3-18-2	14	21	24	43	106
SiO_2	49.21	48.99	49.20	48.30	48.43	47.85	48.89	49.04					
TiO_2	1.78	1.78	2.30	2.28	2.28	2.38	2.67	2.01					
Al_2O_3	14.53	16.36	14.08	14.29	14.76	15.19	13.55	14.40					
FeO	11.89	10.29	12.60	12.48	12.31	12.96	14.01	12.59	11.17	11.30	11.69	11.12	11.33
MnO	0.24	0.21	0.20	0.20	0.20	0.21	0.23	0.20					
MgO	7.74	6.58	7.33	6.59	6.30	6.19	5.91	7.65					
CaO	10.66	11.29	10.59	12.89	12.39	11.39	10.64	10.59					
Na_2O	2.86	3.19	2.48	1.35	1.42	2.95	2.46	2.06	4.17	2.73	3.27	4.15	1.7
K_2O	0.05	0.60	0.23	0.30	0.18	0.23	0.40	0.74					
P_2O_5	0.14	0.14	0.18	0.21	0.21	0.19	0.29	0.16					
Total	99.10	99.43	99.19	98.89	98.48	99.54	99.05	99.44					
Cr									436	213	198	316	78
Ni									104	87	82	83	63
Co									54	50	50	50	50
Sc	44.6	42.7	41.6	39.9		39.7	46.9	39.1	49.2	42.8	39.0	47.5	41.6
Cs	0.438	0.169	1.4	0.661		0.111	0.339	0.813	0.03	0.09	0.85	0.56	0.09
Ba	52.4	120.6	33.8	143.5		61.4	66.8	117.0	63	107	250	144	21
Sr									232	229	531	215	319
Ta	b.l.	0.488	b.l.	1606		0.865	b.l.	0.783	0.30	0.54	0.61	0.43	0.83
Hf	2.96	2.83	3.61	3.58		3.87	4.80	3.19	2.3	2.7	2.9	2.5	4.4
Th	0.507	0.457	0.700	0.709		0.837	1.10	0.783	0.31	0.65	0.61	0.42	0.96
U	0.194	0.169	0.220	0.223		0.578	0.350	0.259	b.l.	0.24	0.25	0.22	0.42
La	6.89	6.58	10.2	9.52		10.8	13.4	8.93	4.43	7.22	7.91	5.63	11
Ce	18.5	17.8	25.1	22.9		26.6	31.2	22.5	13.5	19	21.2	15.8	28.3
Nd	11.1	11.9	16.9	15.7		19.3	24.7	14.9	10.2	13.2	14.2	11.7	18.8
Sm	3.65	3.78	4.92	4.71		5.16	6.63	4.35	3.15	3.86	4.04	3.5	5.54
Eu	1.35	1.32	1.77	1.56		1.72	2.13	1.46	1.07	1.27	1.28	1.25	1.76
Tb	0.809	0.721	0.98	0.801		0.862	1.37	0.760	0.732	0.876	0.832	0.805	1.25
Yb	2.70	2.37	2.75	2.49		2.71	4.21	2.12	2.48	2.72	2.80	2.70	3.64
Lu	0.405	0.343	0.367	0.350		0.373	0.564	0.328	0.343	0.390	0.391	0.384	0.507

Note: Blank—not determined; b.l.—below limit of detection. Values are weight percent for oxides, parts per million for trace elements.
* Kay et al. (1984).
† Cortés and Kay (1994).

the Río San Juan. A few kilometers south of Calingasta, massive gabbro and diabase underlie pillow lavas that are several hundred meters thick.

Age constraints for the mafic sequence are derived from the age of the clastic sequence interfingered with the uppermost part of the lava flows. The occurrence of Caradocian graptolites indicates a Late Ordovician age for the pillow lavas (Blasco and Ramos, 1976).

Geochemical characteristics of the northern sequences

The chemistry of the mafic and ultramafic rocks from the Western Precordillera is presented in Kay et al. (1984) and Haller and Ramos (1984). The basalts from all the studied localities represent evolved oceanic tholeiites that formed in a similar tectonic environment (Table 3, Fig. 9). Distinctive characteristics include the presence of clinopyroxene and plagioclase phenocrysts, high TiO_2 contents (1.7%–2.4%), LREE enrichment, and a restricted compositional range. Basalts from the Sierra del Tigre and Río Jáchal areas have a greater range of TiO_2 contents and FeO/MgO ratios than basalts from the other localities. A wehrlite from the Río Jáchal section and basalts collected between lat 28°30′ and 30°36′S have similar trace element distributions and are more enriched in the LREEs than the basalts farther south (lat 31°30′S–32°48′S). On chemical grounds, the Precordillera basalts are not similar to N-MORBs, oceanic intraplate magmas, or arc-derived magmas, but they do have similarities to basalts from unevolved backarc basins and transitional to plume ridge segments such as the Reykjanes Ridge south of Iceland. The basalts plot in the enriched MORB (E-MORB) field of the Wood et al. (1979a) diagram (Fig. 10). Eruption through continental crust is unlikely, both on the basis of field information and low Ba contents. Possible origins consistent with the chemical data include an oceanic ridge in an early stage of development, a transitional ridge segment, or a backarc basin.

The southern sequences

The southwest Precordillera contains the best exposed and most complete sections of mafic and ultramafic rocks of the Western Precordillera. The Cortaderas section, the type locality of these Famatinian ophiolites, has been studied by Haller and Ramos (1984, 1993), Cortés and Kay (1994), and Davis et al. (1994, 1995a, 1995b, 1999).

Dominant rock types include serpentinized peridotites, ultramafic cumulates, layered gabbros, coarse-grained gabbros to microgabbros, and diabase. All these units are in tectonic contact and heavily deformed. Their country rocks consist of metasedimentary and metavolcanic rocks consisting of medium-bedded metasandstones and metasiltstones with interlayered intermediate-composition to mafic metavolcanic flows and tuffs. The inferred protoliths of the gray metasiltstones were fine-grained sandstones, siltstones, and rare chert (Davis et al., 1999).

The depositional age of the metasedimentary rocks is unknown. K/Ar whole-rock ages of 370 to 358 Ma from the metasedimentary rocks have been interpreted as representing the age of metamorphism (Cucchi, 1971; Caminos et al., 1979). Buggish et al. (1994) estab-

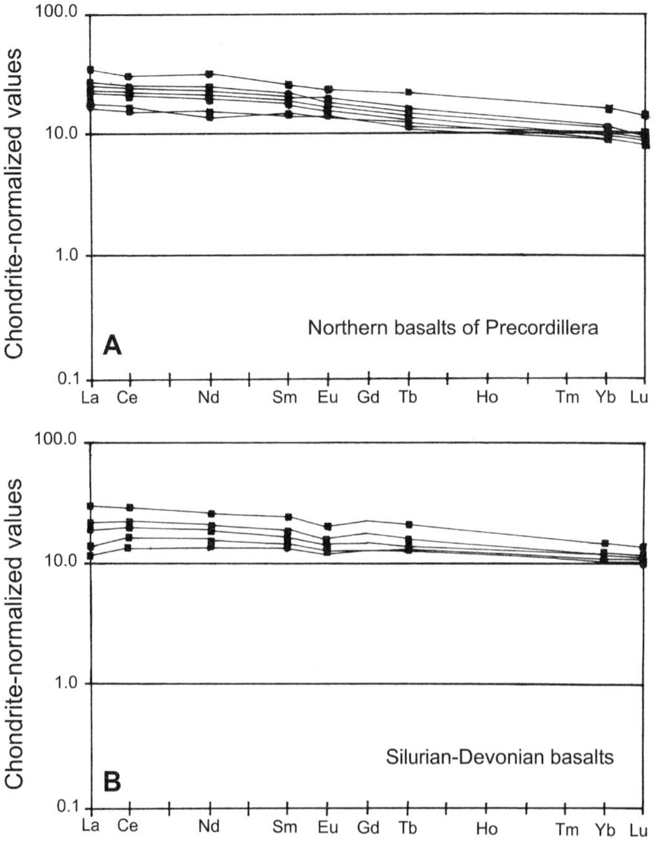

Figure 9. REE pattern normalized to chondrites from representative basalts of Precordillera. A, Ordovician basalts from the northern sector (Kay et al., 1984). B, Silurian–Devonian basalts from the southern sector (modified from Cortés and Kay, 1994). REEs in both plots were normalized to the Leedy chondrite.

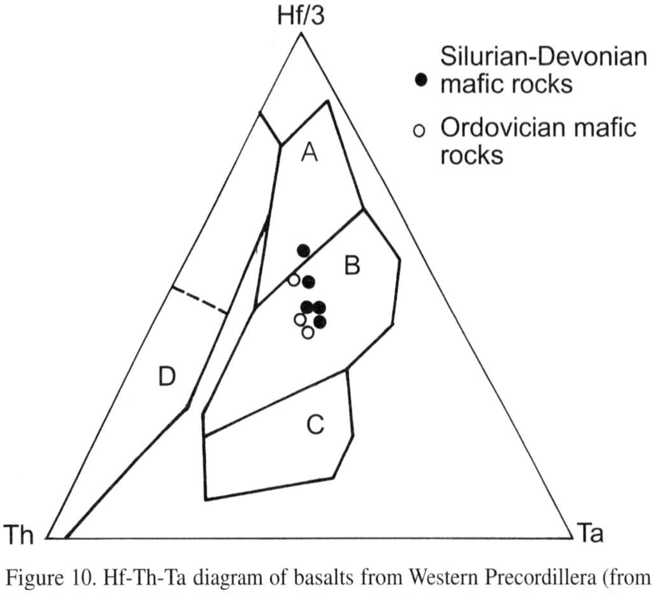

Figure 10. Hf-Th-Ta diagram of basalts from Western Precordillera (from Wood et al., 1979a). Both Ordovician and Silurian–Devonian rocks have E-MORB affinities (modified from Cortés and Kay, 1994). Fields: A—N-MORB, B—E-MORB, C—alkalic within-plate, and D—arc.

lished an age range from 353.1 to 399.6 Ma on the basis of K/Ar dating of 2–6 μm interlayered illite/muscovite separated from the metasedimentary rocks. These ages were interpreted as dating the neocrystallized phase of white mica associated with syndeformational metamorphism. On the basis of new $^{40}Ar/^{39}Ar$ and structural data, Davis et al. (1999) interpreted the metamorphic age of the metasedimentary sequence as 384.5 ± 0.5 Ma. These ages are all consistent with interpreting the main phase of deformation and regional metamorphism as Early Devonian, as was proposed by Astini et al. (1996).

Rock types in the ophiolitic units constitute a complete ophiolite pseudostratigraphy, as established by Coleman (1977), Moores (1982), and Nicolas (1989). The ultramafic rocks consist of highly serpentinized bodies, with a few relict peridotites. Each thrust sheet is composed of serpentinized ultramafic rocks, ultramafic cumulates, and layered gabbros that represent the mantle and lower-crustal part of a classic ophiolite sequence (Davis et al., 1995a, 1995b). Despite extensive serpentinization, these authors were able to distinguish relict clinopyroxene, chromite, spinel, and minor amounts of olivine and orthopyroxene. They used these data to identify wehrlite, harzburgite, lherzolite websterite, and veins of dunite crosscutting wehrlite. In some areas, mantle tectonite textures were recognized in the ultramafic rocks (Haller and Ramos, 1984).

Coarsely crystalline ultramafic cumulates are associated with the base of the layered gabbro section, which ranges from a few meters up to 150 m in thickness. High-grade metamorphism of the layered gabbro complex converted the mineral assemblage to mafic garnet granulite (Davis et al., 1999).

Microgabbros and fine-grained diabase in tectonic contact with the cumulates and layered gabbros reach several hundred meters in thickness. In some areas, rusty-orange ophicalcites are exposed. They consist of pervasively sheared and cataclastic serpentinite and mafic garnet granulite with a calcitic to dolomitic matrix and extensive carbonate veining.

West of the Cortaderas section, basalt flows up to 20 m thick are interfingered with Silurian–Devonian sandstones and shales. There are pillow lavas and sheet flows that are heavily deformed with conspicuous boudinage and pinch-and-swell structures (Cortés, 1992). These units are interpreted as the upper part of a complex accretionary prism developed during the Ordovician and Middle Devonian.

Geochemical characteristics of the southern sequences

Preliminary trace and REE analyses of samples from the mafic rocks of the Cortaderas section indicate an E-MORB or within-plate origin (Davis et al., 1995a, 1995b). Geochemical analyses from the gabbros of Cortaderas show the P-MORB signatures of anomalous oceanic crust formed in a plume or plateau setting (Haller, 1995).

The most complete geochemical study is that of Cortés and Kay (1994) in the Silurian–Devonian pillow lavas west of Cortaderas (see Table 3). The Eu anomaly and the low concentration of Ni (<104 ppm) indicate low-pressure crystallization of olivine and plagioclase, characteristic of mid-ocean ridges. The REE, Ta, Th, and Hf contents indicate a nonarc oceanic source for these mafic rocks. The La/Ta ratios (13 to 15) are within the range of those in mid-ocean ridges, but much lower than those in island arcs (La/Ta > 20 to 25). In the Wood et al. (1979a) diagram, the samples are typical of N-MORB and E-MORB (Fig. 10), and the REE patterns are consistent with a mid-ocean ridge or enriched source origin (Fig. 9B).

The isotopic study of Cortés and Kay (1994) confirmed the oceanic nature of these Silurian–Devonian lavas as well as that of the

Ordovician basalts of the northern sequences. The ϵ_{Nd} for these lavas ranges from +7.1 to +8.6. These values fall within the range of the early Paleozoic depleted mantle (Cortés and Kay, 1994). The $^{87}Sr/^{86}Sr$ ratios (0.7051 to 0.7095) of the Silurian–Devonian lavas, although higher than normal oceanic crust, are similar to the Bay of Islands ophiolite of Nova Scotia described by Jacobsen and Wasserburg (1979). These wide ranges have been interpreted in both cases as produced by seawater addition.

Tectonic setting

The Western Precordillera mafic and ultramafic belt with its almost complete pseudostratigraphic sequence is interpreted as a disrupted ophiolite (Haller and Ramos, 1984, 1993) emplaced along a suture zone between the Chilenia and Precordillera terranes (Ramos et al., 1984, 1986).

Davis et al. (1999) challenged this interpretation on the basis of different ages for the ultramafic and lower mafic sequence (Middle Proterozoic) compared to the upper mafic sequence (early Paleozoic) and suggested the occurrence of two different ophiolite assemblages along the suture. However, recent studies in the Ligurian ophiolites show similar contrasts in ages within different sections of an ophiolite assemblage (see Rampone and Piccardo, this volume). This situation can be explained as a result of the continental upper mantle and lower crust interleaving with the upper oceanic crust trapped in the collision.

The east vergence of the early ophiolite emplacement proposed by Davis et al. (1995a) could be compatible with a late, west-verging emplacement if flake tectonics in the sense of Oxburgh (1972) is applicable. Several ophiolites in the Variscan sutures of the southwestern Iberian massif recorded such a change in vergence, as proposed by Ferreira da Fonseca (1995).

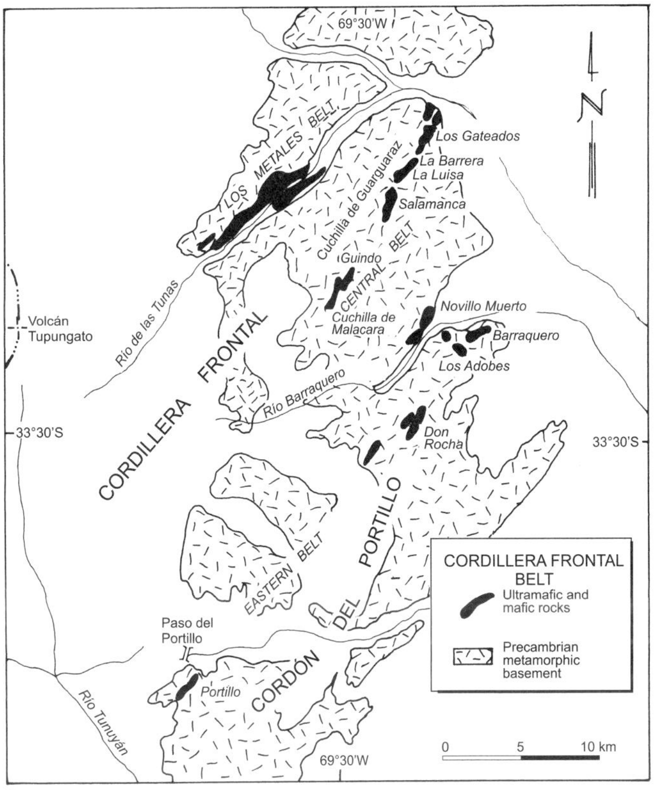

Figure 11. Different ophiolitic belts in the eastern slope of Cordillera Frontal (modified from Polanski, 1964; Haller and Ramos, 1993; and Gregori and Bjerg, 1997).

EASTERN CORDILLERA FRONTAL OPHIOLITES

Along the eastern slope of the Cordillera Frontal, a series of ophiolite assemblages, emplaced in Precambrian metamorphic rocks, is preserved south of lat 33°S (Polanski, 1964). Villar (1975, 1985) has described these rocks as ultramafic Alpine-type belts. The age of their country rock is poorly defined. U-Pb data from zircons from Cordón del Portillo yielded an age of 1069 ± 36 Ma (Ramos and Basei, 1997), indicating a basement of possible Grenvillian age.

Three belts have been recognized from west to east—Los Metales, the Central, and the Eastern belt (Fig. 11)—that could represent either a thrust-sheet repetition of a single ophiolite or different ophiolites.

Los Metales belt

These rocks are exposed along the Río de Las Tunas and have been recently described by Gregori and Bjerg (1997). This belt, 8 km long and 1.5–2 km wide, is composed of serpentinized peridotites, altered gabbros, basaltic dikes, and pillow lavas. The ultramafic and mafic rocks on the Precambrian metamorphic basement have been emplaced along a west-dipping conspicuous shear zone.

An extensive serpentinized peridotite exposure interpreted as a basal tectonite is in contact with the metamorphic rocks. Pyroxene-bearing gabbros and gabbroid rocks are scarce. The pillow lavas are in tectonic contact with the ultramafic rocks. There are two types of dikes: deformed metabasalts in greenschist metamorphic facies and basalts with no deformation and slight metamorphism. The metamorphosed mafic dikes have a regional trend parallel to the fabric of the metamorphic rocks, they are up to 20 m thick, and some are folded and faulted. The country rocks consist of amphibolites, biotite + garnet, and quartz + feldspar schists. Dikes of a second type are up to 50 m thick, have chilled margins, and preserve their original textures and mineral assemblages.

The chemical analyses of the basaltic rocks (Table 4, Fig. 12A) indicate that the metabasites have an E-MORB affinity, whereas the basaltic dikes have geochemical signatures typical of ocean-island basalts (OIBs) (Gregori and Bjerg, 1997).

Central belt

This belt is exposed along the eastern flank of Cuchilla de Guarguaráz, and consists of a series of ultramafic bodies, such as La Barrera, Salamanca, Los Gateados, and other lenses up to the Cuchilla de Malacara bodies (Fig. 11). These bodies are mainly composed of serpentinite with scarce relicts of wehrlites, lherzolites, and dunites, associated with rare bodies of gabbros and lenses of amphibolites of variable sizes. The country rocks are metasedimentary rocks. Serpentinites contain talc, concentrated along shear zones, and include carbonate-rich bodies with brucite and epidote. In some areas like Los Gateados, the ultramafic rocks are heavily deformed in an antiformal structure and contain sulfides and Cr-rich spinel (Villar and Donnari, 1989a, 1989b). Layered gabbros are commonly metamorphosed to amphibolite facies. The gabbros have geochemical affinities with E-MORB with a slight N-MORB trend (Table 4, Fig. 12B). Villar (1998) has interpreted these rocks as part of an ophiolitic assemblage.

TABLE 4. REPRESENTATIVE ANALYSES OF CORDILLERA FRONTAL OPHIOLITE

	Los Metales belt: Metabasites*							Central and eastern belts: Metagabbros†						
Sample	1.1E+07	1010394	3100293	5100293	4120293	9140293	5270294	113-1	114-3	115-4	116-4E	117-5	118-6	119-7
SiO_2	41.45	47.74	49.48	49.35	42.08	52.52	47.00	49.81	49.46	48.23	36.95	48.83	45.94	48.33
TiO_2	1.57	1.93	1.50	1.43	1.30	1.62	3.09	1.06	1.11	1.19	1.59	1.04	1.42	1.29
Al_2O_3	12.69	13.30	14.71	15.49	15.88	15.57	14.60	15.16	15.28	14.61	19.01	15.29	17.11	14.86
FeO	8.72	12.70	8.43	9.19	13.04	9.66	15.40	8.33	10.34	10.98	13.6	9.48	10.35	11.71
MnO	0.13	0.22	0.17	0.16	0.22	0.17	0.31	0.22	0.22	0.18	0.29	0.2	0.21	0.21
MgO	10.84	6.44	6.42	6.12	9.44	6.36	3.81	6.05	6.59	8.54	10.38	7.63	6.93	7.06
CaO	16.06	11.34	11.10	12.82	8.80	8.30	7.87	10.01	11.8	10.42	12.1	11.32	10	12.09
Na_2O	1.56	1.78	3.30	2.56	1.17	3.86	3.14	3.63	3.33	3.92	0.77	3.01	3.07	3.07
K_2O	0.61	0.25	0.35	0.42	2.23	0.69	0.55	0.71	0.48	0.57	0.69	0.42	0.54	0.37
P_2O_5	0.25	0.19	0.16	0.15	0.11	0.21	0.34	0.05	0.07	0.1	0.13	0.06	0.09	0.08
LOI	6.11	4.11	4.35	2.29	5.64	1.02	3.85	3.24	1.97	0.9	4.72	2.43	4.18	1.64
Total	99.99	100	99.97	99.98	99.91	99.98	99.96	98.27	100.65	99.64	100.23	99.71	99.84	100.71
Cr	196	110	175	209	236	193	15							
Ni	92	108	104	126	192	91	61	306	332	259	552	354	381	372
Co	64	45	30	33	51	33	34	91	71	21	188	85	116	80
Sc	32	42	33	33	37	34	36	56.6	49.5	237.7	99.9	116.4	72.7	59.1
V	211	287	161	153	117	156	129	41	41	42	42	40	52	43
Rb	19	5	8	7	56	8	37	240	242	281	300	222	301	265
Cs	0.25	0.17	0.11	0.15	3.7	0.07	b.l.	26.16	15.44	3.11	27.15	12.84	18.64	10.36
Ba	149	24	145	304	3206	299	76	1.07	0.7	0.08	1.73	0.57	2.02	0.93
Sr		222	296	376	261	349	207	124.1	129.99	171.2	166.49	76.18	117.8	49.36
Ga	19	18	18	19	23	12	16	386.99	413.05	165.13	1030.9	282.91	320.87	266.71
Ta	2.90	0.80	0.09	1.00	0.80	0.50	b.d.	0.439	0.223	1.947	0.744	1.043	0.534	0.427
Nb	7.00	19.00	12.00	20.00	16.00	15.00	8.00	2.96	4.93	2.02	6.38	2.6	3.39	3.72
Hf	2.40	3.00	3.10	1.90	1.90	2.80	3.80	1.95	1.7	2.05	2.8	1.69	2.46	2.57
Zr	104	124	121	89	62	100	209	64.02	69.11	56.33	95.81	55.17	78.45	76.77
Y	21	35	33	25	29	27	49	24	28.9	24.5	28.9	26.2	35	33.7
Th	3.99	0.80	0.90	0.50	0.08	0.90	1.00	0.369	0.231	0.304	0.524	0.15	0.249	1.036
U	0.93	0.09	0.50	0.10	0.07	0.06	b.l.	0.834	0.195	0.133	0.736	0.26	0.655	0.268
La	11.90	8.40	9.90	6.80	4.20	10.10	10.80	2.07	2.33	3.67	5.94	2.04	3.38	3.25
Ce	26.10	21.00	22.00	17.00	11.00	23.00	24.00	6.33	6.8	9.67	14.57	6.51	9.88	9.34
Pr								0.869	0.973	1.27	1.794	0.896	1.329	1.302
Nd	14.60	14.00	14.00	11.00	8.00	14.00	14.00	5.96	6.17	8.06	10.74	6.35	8.74	8.15
Sm	3.80	4.33	3.81	3.25	2.65	3.74	4.80	2.27	2.43	2.84	3.53	2.4	3.24	3.06
Eu	1.40	1.48	1.27	1.23	1.02	1.34	1.70	0.831	0.832	0.951	1.16	0.955	1.447	1.509
Gd								3.04	3.4	3.78	4.23	3.42	4.69	4.92
Tb	b.d.	0.90	0.80	0.70	0.80	0.80	b.l.	0.57	0.63	0.71	0.76	0.63	0.89	0.86
Dy								3.95	4.25	4.83	5.13	4.41	6.12	5.74
Ho								0.9	0.94	1.12	1.11	0.97	1.35	1.22
Er								2.81	2.94	3.43	3.58	2.97	4.14	3.93
Tm								0.428	0.396	0.492	0.511	0.423	0.595	0.558
Yb	b.d.	3.19	3.26	2.63	3.30	3.07	3.40	2.76	2.82	3.3	3.27	2.96	3.9	3.4
Lu	0.29	0.42	0.44	0.37	0.43	0.41	0.47	0.445	0.431	0.531	0.552	0.501	0.647	0.578

Note: Blank—not determined; b.l.—below limit of detection. Values are weight percent for oxides, parts per million for trace elements.
* Gregori and Bjerg (1997).
† Villar (1998).

Another ultramafic-mafic body, exposed in the Salamanca mine, consists mainly of a serpentinized peridotite in the core of a synformal structure and includes actinolite + tremolite + muscovite + biotite + garnet schists (Maiza et al., 1982). On the basis of the mineral assemblages of the mafic and ultramafic rocks, an amphibolite-epidote metamorphic grade has been estimated with pressures lower than 5 kbar and temperatures of ~593 °C (Ruviños et al., 1997). The serpentinites in La Barrera and La Luisa mines have relicts of dunite with a typical cumulate structure.

All these ultramafic rocks are associated with numerous dike swarms that are metamorphosed to amphibolite facies and contain hornblende, actinolite, sodic plagioclase, garnet, epidote, titanite, and biotite (Ruviños et al., 1997).

Eastern belt

This belt comprises the mafic and ultramafic bodies of Novillo Muerto–Barraquero, Don Rocha, Los Adobes, Guindo, and Cordón del Portillo. The main body corresponds to Novillo Muerto, which is composed of dunites, harzburgites, wehrlites, and orthopyroxenites associated with ophicalcites (Villar, 1969). The peridotites have a cumulate structure.

Geochemical analyses of the harzburgites of Novillo Muerto show a depleted mantle source because of the presence of PGEs. The REE patterns exhibit a relative enrichment of LREEs. On the basis of petrologic and geochemical considerations, Villar (1998) interpreted this sequence to be a typical cumulate peridotite-gabbro section of an ophiolitic assemblage.

Tectonic setting

These belts of ophiolitic rocks were emplaced in Devonian time, as inferred from K/Ar ages and Rb/Sr isochrons on the metamorphic rocks (Basei et al., 1998). Following the proposal of Caminos (1965), the Eastern Cordillera Frontal belts have been correlated with the

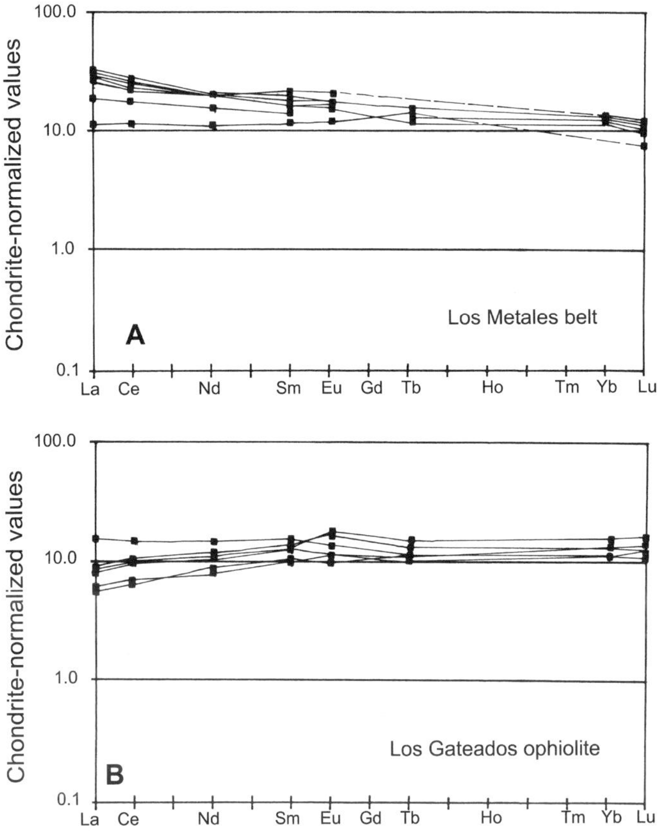

Figure 12. REE patterns normalized to chondrites from representative mafic rocks from Cordillera Frontal. A, Mafic rocks from the Los Metales belt (modified from Gregori and Bjerg, 1997; normalization factor after Sun and McDonough, 1989). B, Mafic rocks from the Central belt (data from Villar, 1998; normalization factor after Taylor and McLennan, 1985).

Famatinian ophiolites of the Western Precordillera (Ramos et al., 1986; Astini et al., 1996).

Both sets of ophiolites share a Devonian metamorphism bracketed between 375–358 Ma (Cordillera Frontal) and 399–353 Ma (Precordillera), which is interpreted as the time of the final emplacement of the ophiolites. This event is associated with the docking of Chilenia against the Precordillera terrane in the Middle to Late Devonian (Ramos et al., 1984). The dominant vergence of the Cordillera Frontal basement is to the east, opposite to the pre-Carboniferous vergence in the Precordillera. Most structural studies in the Western Precordillera indicate a west-vergent ductile deformation (Keidel, 1939; Ramos et al., 1986; Von Gosen, 1995). However, Davis et al. (1999) demonstrated a west-vergent brittle deformation superimposed on a preexisting east-vergent ductile deformation related to the ophiolite emplacement. According to these authors, the west-vergent deformation is Early Permian in age, which conflicts with the angular unconformity between plastically deformed early Paleozoic rocks and undeformed Carboniferous beds in the Western Precordillera (Ramos et al., 1986).

The isotopic studies performed by Basei et al. (1998) have shown that the mafic rocks of the Western Precordillera have latest Proterozoic Nd model ages (577 Ma) coherent with an early Paleozoic crystallization age of the basaltic rocks. On the other hand, Nd model ages on amphibolites and metamorphic rocks of the Eastern Cordillera Frontal belt range between 1700 and 1400 Ma, typical for the Grenvillian basement.

The Nd model ages are based on only a few samples; therefore, either the differences could be insignificant or indicate a complex fractionation from the mantle. However, it is possible that the difference in age is meaningful. If so, the Eastern Cordillera Frontal belt represents older ophiolites emplaced in the Precambrian basement and subsequently remobilized during an early Paleozoic deformation.

CONCLUDING REMARKS

The characterization of four distinct and poorly known ophiolitic belts widely exposed in the basement of the Central Andes provides an opportunity to examine different features through time (see Table 5). These belts represent different crystallization and emplacement periods: Middle Proterozoic (Pie de Palo), latest Proterozoic–Early Cambrian (Córdoba ophiolites), and Ordovician–Devonian (Western Precordillera and, possibly, Cordillera Frontal ophiolites).

Although the petrologic data set is far from complete, it is evident that the main units and features of these ophiolites are well preserved as in a typical ophiolite pseudostratigraphy (Moores, 1969, 1982; Nicolas, 1989). The absence of sheeted dikes, with the exception of the possible section in Cortaderas (Western Precordillera), is mainly related to the difficulty of recognizing such features in multiply deformed ancient ophiolites.

Some ophiolites are characterized by the combination of layered peridotites and basaltic pillow lavas, without the existence of gabbros, as in Los Metales belt. These assemblages were originally interpreted as a result of tectonic disruption during emplacement. However, observations from modern oceanic analogues show that the apparent lack of some crustal units and/or the incomplete pseudostratigraphy seen in these ophiolites might reflect the primary architecture and tectonic setting (Dilek et al., 1998). Furthermore, there are segments that consist of large bodies of layered gabbro, such as in San Lorenzo, Western Córdoba belt, that may also reflect the primary setting.

The dominant peridotite in the basal assemblages consists of harzburgite in most of the belts, with the exception of the Eastern Córdoba, Pie de Palo, and the Central Cordillera Frontal belts. Although the geochemical characteristics of the Eastern Córdoba belt suggest a suprasubduction setting, the lherzolites and associated mafic rocks of the Central Cordillera Frontal belt show E-MORB affinities.

As a whole, geochemical properties vary from N-MORB affinities, as in the northern Pie de Palo, Central Cordillera Frontal, and Western Córdoba belts, to enriched intraoceanic settings. Some of the belts—the southernmost segment of the Western Córdoba and the southern Pie de Palo belts—have arc affinities. These combinations indicate that the paleogeographic settings of ancient oceanic crusts as preserved in these ophiolites are complex, involving intraoceanic ridges, arcs, and/or backarc basins.

The mechanism of obduction and emplacement of different ophiolites is complex, but in all the cases, it is manifested in strong deformation and metamorphism. Some areas as the Cortaderas ophiolite may have undergone some sort of flake tectonics, with reversal of vergence during emplacement, whereas others have more straight and simple west vergence as in Pie de Palo.

In spite of vast exposures of these Middle Proterozoic to early Paleozoic ophiolites, detailed analytical work is still lacking in order to define their igneous, metamorphic, and tectonic evolution through time. There is no doubt that the ophiolites preserved in the basement of the Central Andes provide one of the most complete records of crustal evolution in Earth's history.

TABLE 5. MAIN OPHIOLITE BELTS OF SOUTHERN SOUTH AMERICA

Ophiolite	Age	Country rocks	Remarks	Tectonic setting
Eastern Córdoba	Late Proterozoic	Schists and tonalite gneiss. Amphibolite metamorphic facies	Lherzolite, subordinate harzburgite, pyroxenite, gabbros. Intruded by leucocratic rocks	Backarc basin
Western Córdoba	Early Cambrian	Quartz + feldspar, quartz.mica + feldspar, quartz + garnet gneisses and schists; migmatites, amphibolites, marbles, and kinzigites. Amphibolite to granulite metamorphic facies	*Metamorphic peridotites*: harzburgites, metamorphosed ultramafic rocks. Chromite bodies	Mid-ocean ridge ophiolite
			Transition zone: Cumulate tectonized peridotites (harzburgites, dunites, pyroxenites, gabbros). Gabbro sills, dikes, and veins. Websteritic and spinel + dunite dikes in the harzburgites. Harzburgite boudins in lherzolitic rocks *Layered complex*: Gabbroic cumulate complex with plagioclase-olivine, plagioclase-clinopyroxene, and plagioclase-clinopyroxene- orthopyroxene cumulates. Quartz-bearing gabbro dikes *Basaltic section*: Amphibolites, mafic schists, and gneisses derived from basaltic lavas	
	Ordovician?	Metapelitic schists and gneisses and amphibolites	Dunites, pyroxenites, hornblendites, amphibolites. Felsic segregation associated with ultramafic bodies	Backarc or frontal island arc setting
Virorco		Amphibolite to granulite metamorphic facies		
Valle Fértil	Middle Ordovician Middle Proterozoic	Metapelitic gneisses. Upper-amphibolite to granulite metamorphic facies	Pyroxenites, peridotites, gabbros, hornblendites, amphibolites	Island arc setting
Pie de Palo	Middle Proterozoic	Metapelitic schists and gneisses, graphite and amphibolite schists. Greenschist to amphibolite metamorphic facies	Sepentinized peridotites, pyroxenites, gabbros, diorites, amphibolites	Island arc and backarc setting
Western Precordillera				
Northern sequences	Middle Ordovician to Devonian	Greywackes and shales with local conglomerates	Clinopyroxene-bearing basaltic lavas, mafic to ultramafic sills	Anomalous enriched mid-ocean ridge
Southern sequences	Middle Ordovician to Devonian	Metasandstones and metasiltstones interlayered with metavolcanic flows and tuffs	Serpentinized peridotites (wehrlites, harzburgites, lherzolites, websterites, and veins of dunite crosscutting wehrlite), ultramafic cumulates, layered gabbros to microgabbros, diabase	Anomalous enriched mid-ocean ridge and lithospheric mantle granulites
Eastern Cordillera Frontal				
Los Metales belt	Middle Ordovician to Devonian	Amphibolites, biotite + garnet and quartz + feldspar schists. Greenschists to amphibolite metamorphic facies	Serpentinized peridotites, gabbros, basaltic dikes, basaltic lavas	Enriched mid-ocean ridge basalts and oceanic island basalts
Central belt	Middle Ordovician to Devonian	Actinolite + tremolite schists, muscovite + biotite + garnet schists. Amphibolite-epidote to amphibolite metamorphic facies	Serpentinites. Scarce wherlites, lherzolites, dunite, gabbros, amphibolites. Associated with basaltic dike swarms	Enriched mid-ocean ridge to normal mid-ocean ridge
Western belt	Middle Ordovician to Devonian	Quartz + biotite + feldspar schists, biotite + garnet schists, amphibolite schists. Amphibolite metamorphic facies	Dunites, harzburgites, wehrlites, orthopyroxenites, associated with ophicalcites	Cumulate peridotite-gabbro section of amid-ocean ridge

ACKNOWLEDGMENTS

We appreciate guidance to the geology of these ophiolite assemblages from many colleagues and especially from Robert Coleman, Eldridge Moores, Miguel Haller, Stephen Davis, and Horacio Díaz, during several field visits to different areas. We also thank S. DeBari, E. Moores, and S.M. Kay for their critical reviews. The field research was partially covered by PID-CONICET 4162.

REFERENCES CITED

Astini, R., Ramos, V.A., Benedetto, J.L., Vaccari, N.E., and Cañas, F.L., 1996, La Precordillera: Un terreno exótico a Gondwana: 13° Congreso Geológico Argentino y 3° Congreso Exploración de Hidrocarburos (Buenos Aires), Actas, v. 5, p. 293–324.

Basei, M., Ramos, V.A., Vujovich, G.I., and Poma, S., 1998, El Basamento metamórfico de la Cordillera Frontal de Mendoza: Nuevos datos geocronológicos e isotópicos: 10° Congreso Latinoamericano de Geología (Buenos Aires), Actas, v. 2, p. 412–417.

Bjerg, E.A., Kostadinoff, J., Mogessie, A., Hoinkes, G., Stumplf, E.F., and Hauzenberger, C., 1996, La faja de rocas máficas y ultramáficas de las Sierras de San Luis: Nuevos hallazgos de minerales del grupo del platino: 2° Jornadas de Mineralogía, Petrografía y Metalogénesis de rocas máficas y ultramáficas, Publicación del Instituto de Recursos Minerales, Universidad Nacional de La Plata, v. 5, p. 303–310.

Blasco, G., and Ramos, V.A., 1976, Graptolitos caradocianos de la Formación Yerba Loca y del Cerro La Chilca, Departamento Jáchal, Provincia de San Juan: Ameghiniana, v. 13, p. 312–329.

Bonalumi, A.A., and Gigena, A.A., 1987, Relación entre las metamorfitas de alto grado y las rocas básicas y ultrabásicas en el departamento Calamuchita, Córdoba: Asociación Geológica Argentina, Revista, v. 42, p. 73–81.

Borrello, A.V., 1963, Elementos del magmatismo simaico en la evolución de la secuencia geosinclinal de la Precordillera: Instituto Nacional de Investigación de las Ciencias Naturales, Museo Argentino de Ciencias Naturales "Bernardino Rivadavia", Revista, Serie Ciencias Geológicas, v. 1, p. 1–19.

Borrello, A.V., 1969, Los geosinclinales de la Argentina: Dirección Nacional de Geología y Minería, Anales, v. 14, p. 1–188.

Boudier, F., and Nicolas, A., 1985, Harzburgite and lherzolite subtypes in ophiolitic and oceanic environments: Earth Planetary and Science Letters, v. 76, p. 84–92.

Buggish, W., von Gosen, W., Henjes-Kunst, F., and Krumm, S., 1994, The age of early Paleozoic deformation and metamorphism in the Argentine Precordillera: Evidence from K-Ar data: Zentralblatt für Geologie und Paläontologie, teil 1, p. 275–286.

Caminos, R., 1965, Geología de la vertiente oriental del Cordón del Plata, Cordillera Frontal de Mendoza: Asociación Geológica Argentina, Revista, v. 20, p. 351–392.

Caminos, R., Cordani, U.G., and Linares, E., 1979, Geología y geocronología de las rocas metamórficas y eruptivas de la Precordillera y Cordillera Frontal de Mendoza, República Argentina: 2° Congreso Geológico Chileno (Arica), Actas, v. 1(F), p. 43–61.

Caminos, R., Cingolani, C.A., Hervé, F., and Linares, E., 1982, Geochronology of the pre-Andean metamorphism and magmatism in the Andean Cordillera between latitudes 30° and 36° S: Earth Science Review, v. 18, p. 332–352.

Castro de Machuca, B., and Conte-Grand, A., 1994, Génesis de las anfibolitas de las Sierras Pampeanas Occidentales, provincia de San Juan, Argentina: 7° Congreso Geológico Chileno (Concepción), Actas, v. 2, p. 989–994.

Castro de Machuca, B., Conte-Grand, A., Meissi, E., Pontoriero, S., and Sumay, C., 1996, Petrología de las asociaciones máficas-ultramáficas de la Sierra de La Huerta: 13° Congreso Geológico Argentino y 3° Congreso de Exploración de Hidrocarburos (Buenos Aires), Actas, v. 3, p. 439–452.

Chincarini, A., Escayola, M., and Villar, L.M., 1996, Caracterización geoquímica del gabro estratificado del cerro San Lorenzo, Sierra de Comechingones, Córdoba, Argentina: 13° Congreso Geológico Argentino y 3° Congreso de Exploración de Hidrocarburos (Buenos Aires), Actas, v. 3, p. 387–398.

Cingolani, C., and Varela, R., 1975, Geocronología Rubidio-Estroncio de rocas ígneas y metamórficas de las Sierras Chica y Grande de Córdoba, República Argentina: 2° Congreso Iberoamericano de Geología Económica (Buenos Aires), Actas, v. 1, p. 9–35.

Coleman, R.G., 1977, Ophiolites, ancient oceanic lithosphere?: Berlin, Springer Verlag, 229 p.

Cortés, J.M., 1992, Lavas almohadilladas en el Grupo Ciénaga del Medio, extremo noroccidental de la Precordillera mendocina: Asociación Geológica Argentina, Revista, v. 47, p. 115–117.

Cortés, J.M., and Kay, S.M., 1994, Una dorsal oceánica como origen de las lavas almohadilladas del Grupo Ciénaga del Medio (Silúrico-Devónico) de la Precordillera de Mendoza: 7° Geológico Chileno (Concepción), Actas, v. 2, p. 1005–1009.

Cucchi, R.J., 1971, Edades radimétricas y correlación de metamorfitas de la Precordillera de San Juan y Mendoza, República Argentina: Asociación Geológica Argentina, Revista, v. 26, p. 503–515.

Dalla Salda, L.H., 1984, La estructura íntima de las sierras de Córdoba: Asociación Geológica Argentina, Revista, v. 39, p. 38–51.

Dalla Salda, L.H., and Varela, R., 1984, El metamorfismo en el tercio sur de la Sierra de Pié de Palo, San Juan: Asociación Geológica Argentina, Revista, v. 39, p. 68–93.

Davis, J.S., Roeske, S.M., and Moores, E.M., 1994, A nearly complete pre-Carboniferous ophiolite in the SW Precordillera, Argentina, requires new tectonic models: Geological Society of America Abstracts with Programs, v. 26, n. 7, p. 503.

Davis, J.S., Moores, E.M., Kay, S.M., and McClelland, W.C., 1995a, New information sheds light on the nature and emplacement history of Precordillera ophiolite, western Argentina: Geological Society of America Abstracts with Programs, v. 27, n. 6, p. 124.

Davis, J.S., Moores, E.M., Roeske, S.M., Kay, S.M., McClelland, W.C., and Snee, L.W., 1995b, New data from the western margin of Precordillera terrane, Argentina, constrain scenarios for the middle Paleozoic tectonics of western South America: Laurentian-Gondwanan Connections before Pangea Field Conference, Jujuy, Argentina, I.G.C.P. 376, UNESCO, Program with Abstracts, p. 15–16.

Davis, J.S., Roeske, S.M., and McClelland, W.C., 1999, Closing the ocean between the Precordillera terrane and Chilenia: Early Devonian ophiolite emplacement and deformation in the southwest Precordillera, *in* Ramos, V.A., and Keppie, J.D., eds., Laurentia-Gondwana Connections before Pangea: Geological Scoiety of America Special Paper 336, p. 115–138.

Dilek, Y., Moores, E.M., and Furnes, H., 1998, Structure of modern oceanic crust and ophiolites and implications for faulting and magmatism at oceanic spreading centers, *in* Buck, R., Karson, J., Delaney, P., and Lagabrielle, Y., eds., Faulting and magmatism at mid-ocean ridges: American Geophysical Union Geophysical Monograph 106, p. 219–266.

Escayola, M., 1994, Faja ultramáfica de la Sierra Grande de Córdoba entre Intiyaco y Estancia San Miguel, Calamuchita, Córdoba, Argentina [Ph.D. thesis]: Facultad de Ciencias Exactas, Físicas y Naturales, Universidad Nacional de Córdoba, Córdoba, 300 p.

Escayola, M.P., Ramé, G.A., and Kraemer, P.E., 1996, Caracterización y significado geotectónico de las fajas ultramáficas de Córdoba: 13° Congreso Geológico Argentino y 3° Congreso Exploración de Hidrocarburos (Buenos Aires), Actas, v. 3, p. 421–438.

Fernández Gianotti, J., 1979, Geología de la faja de rocas ultrabásicas de la ladera oriental de la Sierra de Comechingones [Ph. D. thesis]: Facultad de Ciencias Exactas y Naturales, Universidad de Buenos Aires, 87 p.

Ferreira da Fonseca, P.E.T., 1995, Estudo da sutura varisca no SW Ibérico nas regiões de Serpa-Beja-Torrão e alvito-Viana do Alentejo [Ph.D. thesis]: Facultade de Ciências da Universidade de Lisboa, Lisboa, 294 p.

Gervilla, F.J., Sabalua, C., Carrillo, R., Fenoll, P., and Acevedo, D., 1992, Mineralogy and mineral chemistry of the Las Aguilas Ni-Cu deposit, province of San Luis, *in* Fenoll, P., Torres Ruiz, H.A., and Gervilla, F.J., eds., Current research in geology applied to ore deposits: Rotterdam, A.A. Balkema Publishers, p. 461–464.

Giménez, M.E., 1997, Estudio genético y evolutivo de la cuenca del Bermejo (provincia de San Juan) a partir de datos de gravedad [Ph.D. thesis]: Facultad de Ciencias Exactas, Ingeniería y Agrimensura, Universidad Nacional de Rosario, Rosario, 142 p.

González Bonorino, F., 1962, Petrología de algunos cuerpos básicos de San Luis y las granulitas asociadas: Asociación Geológica Argentina, Revista, v. 16, p. 61–106.

Gregori, D.A., and Bjerg, E.A., 1997, New evidence on the nature of the Frontal Cordillera ophiolitic belt, Argentina: Journal of South American Earth Sciences, v. 10, p. 157–178.

Haller, M.A., 1995, A plume-influenced ridge-segment origin for the Precordillera Ophiolite: Laurentian-Gondwanan connections before Pangea Field Conference, Jujuy, Argentina, Program with Abstracts, p. 21.

Haller, M.A., and Ramos, V.A., 1984, Las ofiolitas famatinianas (Eopaleozoico) de las provincias de San Juan y Mendoza: 9° Congreso Geológico Argentino (Bariloche), Actas, v. 2, p. 66–83.

Haller, M.J., and Ramos, V.A., 1993, Las ofiolitas y otras rocas afines, *in* Ramos, V.A., ed., Geología y recursos naturales de Mendoza: 12° Congreso Geológico Argentino y 2° Congreso de Exploración de Hidrocarburos (Mendoza), Relatorio, v. 1, no. 4, p. 31–40.

Hawkins, J.W., 1995a, Evolution of the Lau Basin: Insights from ODP Leg 135, *in* Taylor, B., and Natland, J., eds., Active margins and marginal basins of the Western Pacific: New York, American Geophysical Union, Geophysical Monograph 88, p. 125–169.

Hawkins, J.W., 1995b, The geology of the Lau Basin, *in* Taylor, B., ed., Backarc basins: Tectonics and magmatism: New York, Plenum Press, p. 63–129.

Jacobsen, S.B., and Wasserburg, J.G., 1979, Nd and Sr isotopic study of the Bay of Islands ophiolite complex and the evolution of the source of mid-ocean ridge basalts: Journal of Geophysical Research, v. 84, p. 7429–7445.

Kay, S.M., Ramos, V.A., and Kay, R., 1984, Elementos mayoritarios y trazas de las vulcanitas ordovícicas en la Precordillera Occidental: Basaltos de rift oceánicos tempranos (?) próximos al margen continental: 9° Congreso Geológico Argentino (Bariloche), Actas, v. 2, p. 48–65.

Kay, S.M., Orrell, S., and Abbruzzi, J.M., 1996, Zircon and whole rock Nd-Pb isotopic evidence for a Grenville age and a Laurentian origin for the basement of the Precordilleran terrane in Argentina: Journal of Geology, v. 104, p. 637–648.

Keidel, J., 1939, Las estructuras de corrimientos paleozoicos de la Sierra de Uspallata, provincia de Mendoza: Physis, v. 14, no. 46, p. 1–96.

Kostadinoff, J., Bjerg, E.A., Delpino, S., Dimieri, L., Aberra, A., Mogessie, A., Hoinkes, G., Hauzenberger, C., and Felfernig, A., 1999, Gravimetric and magnetometric anomalies in the Sierras Pampeanas of San Luis: Asociación Geológica Argentina, Revista, v. 53, no. 4, p. 549–552.

Kraemer, P.E., Escayola, M.P., and Martino, R.D., 1995, Hipótesis sobre la evolución tectónica neoproterozoica de las Sierras Pampeanas de Córdoba (30°40′–32°40′), Argentina: Asociación Geológica Argentina, Revista, v. 50, p. 47–59.

Maiza, P.J., Bjerg, E.A., and Gregori, D.A., 1982, Mineralogía y quimismo de la zona de alteración en un perfil de Mina Salamanca, departamento Tupungato, provincia de Mendoza, Argentina: 5° Congreso Latinoamericano de Geología (Buenos Aires), Actas, v. 1, p. 665–674.

Martínez, M.P., 1997, Estudio del levantamiento de la Sierra de Valle Fértil (provincia de San Juan) [Ph.D. thesis]: Facultad de Ciencias Exactas, Ingeniería y Agrimensura, Universidad Nacional de Rosario, Rosario, 153 p.

Martino, R., Kraemer, P., Escayola, M., Giambastiani, M., and Arnosio, M., 1995, Transecta de las Sierras Pampeanas de Córdoba a los 32° S: Asociación Geológica Argentina, Revista, v. 50, p. 60–77.

Massabie, A., Mutti, D.I., and Fernández, S., 1994, Control tectónico y petrológico del emplazamiento y localización de las rocas ultramáficas en el basamento precámbrico de la Sierra de Comechingones, Córdoba, Argentina: São Paulo, 38th Congresso Brasileiro de Geologia, Actas, v. 1, p. 123–129.

McDonough, M.R., Ramos, V.A., Isachsen, C.E., Bowring, S.A., and Vujovich, G.I., 1993, Nuevas edades de circones del basamento de la Sierra de Pie de Palo, Sierras Pampeanas Occidentales de San Juan: Sus implicancias para los modelos del supercontinente proterozoico de Rodinia: 12° Congreso Geológico Argentino y 2° Congreso de Exploración de Hidrocarburos (Mendoza), Actas, v. 3, p. 340–342.

Mirré, J.C., 1976, Descripción geológica de la Hoja 19-c Valle Fértil, provincias de San Juan y La Rioja: Servicio Geológico Nacional, Buenos Aires, Boletín, v. 147, p. 1–70.

Mogessie, A., Hoinkes, G., Stumpfl, E., Berg, E., and Kostadinoff, J., 1995, Occurrence of platinum group minerals in the Las Aguilas ultramafic unit within a granulite facies basement, San Luis, Central Argentina, *in* Pasava, J., Krivek, B., and Zak, K., eds., Mineral deposits: From their origin to their environmental impacts: Rotterdam, A.A. Balkema Publishers, p. 897–900.

Moores, E.M., 1969, Petrology and structure of the Vourinos ophiolitic complex of northern Greece: Geological Society of America Special Paper 118, p. 174–189.

Moores, E.M., 1982, Origin and emplacement of ophiolites: Review of Geophysics and Space Physics, v. 2, p. 735–760.

Mutti, D.I., 1987, Estudio geológico del Complejo gabro-peridotítico del área de Bosque Alegre, provincia de Córdoba [Ph.D. thesis]: Universidad de Buenos Aires, 189 p.

Mutti, D.I., 1992, El complejo gabro-peridotítico de Bosque Alegre, provincia de Córdoba: Asociación Geológica Argentina, Revista, v. 47, p. 153–167.

Mutti, D.I., 1994, Los cromoespinelos del centro sur de las Sierras de Córdoba: Metalogénesis e implicancias geotectónicas: Publicación del Instituto de Recursos Minerales, v. 3, p. 545–570.

Mutti, D.I., 1995, The chromian spinels from Córdoba, *in* Pasava, J., Krivek, B., and Zak, K., eds., Mineral deposits: From their origin to their environmental impacts: Rotterdam, A.A. Balkema Publishers, p. 897–900.

Mutti, D.I., 1997, La secuencia ofiolítica basal desmembrada de las sierras de Córdoba: Asociación Geológica Argentina, Revista, v. 52, p. 275–285.

Mutti, D.I., and Di Marco, A.C., 1998, Discriminación litogeoquímica entre ambientes de arco y de retroarco en el centro y sur de las Sierras de Córdoba, Argentina. 10° Congreso Latinoamericano de Geología y 6° Congreso Nacional de Geología Económica (Buenos Aires), Actas, v. 2, p. 238–244.

Mutti, D.I., and Fernández, S.V., 1999, Alpine type chromite deposits in the basal ophiolitic sequence of Córdoba, Argentina: A suprasubduction zone?: Mineralium Deposita (in press).

Mutti, D.I., Di Marco, A.C., and Fernández, S.V., 1997, Neoproterozoic metalogenesis of the central south Sierras Pampeanas de Córdoba, Argentina, *in* Papunen, H., ed., Mineral deposits: Research and exploration: Where do they meet?: Rotterdam, A.A. Balkema Publishers, p. 863–866.

Nicolas, A., 1989, Structures of ophiolites and dynamics of ocean lithosphere: Dordrecht, The Netherlands, Kluwer Academic Publishers, Petrology and Structural Geology, v. 4, 367 p.

Oxburgh, E.R., 1972, Flake tectonics and continental collision: Nature, v. 239, p. 202–205.

Pearce, J.A., Lippard, S.J., and Robert, S., 1984, Characteristics and tectonic significance of suprasubduction zone ophiolite, *in* Kokelaar, B.P., and Hawells, M.F., eds., Marginal basin geology: Geological Society of London Special Publication 16, p. 77–96.

Pearce, J.A., Baker, P.E., Harvey, P.K., and Luff, I.W., 1995, Geochemical evidence for subduction fluxes, mantle melting, and fractional crystallization beneath the South Sandwich island arc: Journal of Petrology, v. 36, p. 1073–1109.

Polanski, J., 1964, Descripción geológica de la Hoja 25a Volcán San José, provincia de Mendoza: Dirección Nacional de Geología y Minería, Boletín, v. 98, p. 1–94.

Ramos, V.A., 1988, Tectonics of the Late Proterozoic-Early Paleozoic: A collisional history of southern South America: Episodes, v. 11, p. 168–174.

Ramos, V.A., 1994, Terranes of southern Gondwanaland and their control in the Andean structure (30–33°S lat.), *in* Reutter, K.J., Scheuber, E., and Wigger, P.J., eds., Tectonics of the Southern Central Andes: Structure and evolution of an active continental margin: Berlin, Springer Verlag, p. 249–261.

Ramos, V.A., 1995, Sudamérica: Un mosaico de continentes y océanos: Ciencia Hoy, v. 6, no. 32, p. 24–29.

Ramos, V.A., and Basei, M., 1997, The basement of Chilenia: An exotic continental terrane to Gondwana during the Early Paleozoic: Symposium on Terrane Dynamics '97 (New Zealand), p. 140–143.

Ramos, V.A., and Vujovich, G.I., 1995, Descripción geológica de la Hoja San Juan, provincias de San Juan y Mendoza: Servicio Geológico y Minero de Argentina, Open file report, p. 1–230.

Ramos, V.A., Jordan, T.E., Allmendinger, R.W., Kay, S.M., Cortés, J.M., and Palma, M.A., 1984, Chilenia: Un terreno alóctono en la evolución paleozoica de los Andes Centrales: 9° Congreso Geológico Argentino (Bariloche), Actas, v. 2, p. 84–106.

Ramos, V.A., Jordan, T.E., Allmendinger, R.W., Mpodozis, C., Kay, S., Cortés, J.M., and Palma, M.A., 1986, Paleozoic terranes of the Central Argentine-Chilean Andes: Tectonics, v. 5, p. 855–880.

Ramos, V.A., Vujovich, G.I., Kay, S.M., and McDonough, M.R., 1993, La orogénesis de Grenville en las Sierras Pampeanas Occidentales: La Sierra de Pie de Palo y su integración al supercontinente proterozoico: 12° Congreso Geológico Argentino y 2° Congreso de Exploración de Hidrocarburos (Mendoza), Actas, v. 3, p. 343–357.

Ramos, V.A., Vujovich, G.I., and Dallmeyer, R.D., 1996, Los klippes y ventanas tectónicas de la estructura preándica de la Sierra de Pie de Palo (San Juan): Edad e implicaciones tectónicas: 13° Congreso Geológico Argentino y 3° Congreso Exploración de Hidrocarburos (Buenos Aires), Actas, v. 5, p. 377–392.

Ramos, V.A., Dallmeyer, D., and Vujovich, G.I., 1998, Time constraints on the Early Paleozoic docking of the Precordillera, Central Argentina, *in* Pankhurst, R., and Rapela, C.W., eds., The Proto-Andean Margin of Gondwana: Geological Society of London Special Publication 142, p. 143–158.

Rapela, C.W., Pankhurst, R., Casquet, C., Baldo, E., Saavedra, J., and Galindo, C., 1998, Early evolution of the Proto-Andean margin of South America: Geology, v. 26, p. 707–710.

Ruviños, M.A., Gregori, D.A., and Bjerg, E.A., 1997, Condiciones de P y T del basamento metamórfico de la Cordillera Frontal de Mendoza, Argentina: 8° Congreso Geológico Chileno (Antofagasta), Actas, v. 2, p. 1512–1516.

Schmidt, C.J., Astini, R.A., Costa, C.H., Gardini, C.E., and Kraemer, P.E., 1995, Cretaceous rifting, alluvial fan sedimentation, and Neogene inversion, southern Sierras Pampeanas, Argentina, *in* Tankard, A., Suárez S., R., and Helsink, H.J., eds., Petroleum basins of South America: American Association of Petroleum Geologists Memoir 62, p. 341–358.

Sims, J., Stuart Smith, P., Lyons, P., and Skirrow, R., 1997, Informe geológico y metalogénico de las Sierras de San Luis y Comechingones, provincias de San Luis y Córdoba: Servicio Geológico Minero Argentino, Anales, v. 28, p. 1–148.

Sun, S.S., and McDonough, W.F., 1989, Chemical and isotopic systematics of oceanic basalts: Implications for mantle composition and processes, in Saunders, A.D., and Norry, M.J., eds., Magmatism in the ocean basins: Geological Society of London Special Publication 42, p. 313–345.

Taylor, S.R., and McLennan, S.M., 1985, The continental crust: Its composition and evolution: Oxford, Blackwell, 312 p.

Varela, R., and Dalla Salda, L., 1993, Geocronología Rb-Sr de metamorfitas y granitoides del extremo sur de la Sierra de Pie de Palo, San Juan: Asociación Geológica Argentina, Revista, v. 47, p. 271–276.

Villar, L.M., 1969, El Complejo ultrabásico de Novillo Muerto, Cordillera Frontal, Mendoza: Asociación Geológica Argentina, Revista, v. 24, p. 223–238.

Villar, L.M., 1975, Las fajas y otras manifestaciones ultrabásicas en la República Argentina y su significado metalogenético: 2° Congreso Ibero-Americano de Geología Económica (Buenos Aires), Actas, v. 3, p. 135–155.

Villar, L.M., 1985, Las fajas ultrabásicas argentinas, tipos de ultramáficas. Metalogenia: 4° Congreso Geológico Chileno (Antofagasta), Actas, v. 3, no. 4, p. 610–633.

Villar, L.M., 1998, Los gabros de fondo oceánico de la ofiolita de la Cordillera Frontal de Mendoza, Argentina: 10° Congreso Latinoamericano de Geología y 6° Congreso Nacional de Geología Económica (Buenos Aires), Actas, v. 2, p. 405.

Villar, L.M., and Donnari, E.I., 1989a, Geología y metalogénesis del Complejo máfico ultramáfico de Los Gateados, Departamento Tupungato, Mendoza: Facultad de Ciencias Naturales, Universidad Nacional de Tucumán, Correlación Geológica Serie, v. 3, p. 45–50.

Villar, L.M., and Donnari, E.I., 1989b, Asociación peridotítica-gabro estratificado de la sección septentrional de la faja ultramáfica de la Cordillera Frontal de Mendoza: Reunión sobre Geotransectas de América del Sur (Mar del Plata), Actas, p. 45–49.

Von Gosen, W., 1995, Polyphase structural evolution of the southwestern Argentine Precordillera: Journal of South American Earth Sciences, v. 8, p. 377–404.

Vujovich, G.I., 1993, Arcos magmáticos islándicos en el sector occidental del terreno Pampeano: 12° Congreso Geológico Argentino y 2° Congreso de Exploración de Hidrocarburos (Mendoza), Actas, v. 4, p. 17–22.

Vujovich, G., and Kay, S.M., 1996, Evidencias geoquímicas del origen y ambiente geológico de las rocas metamórficas de composición máfica e intermedia de las Sierras Pampeanas Occidentales: 13° Congreso Geológico Argentino y 3° Congreso Exploración de Hidrocarburos (Buenos Aires), Actas, v. 5, p. 273–291.

Vujovich, G., and Kay, S.M., 1998, A Laurentian? Grenville-age oceanic arc/backarc terrane in the Sierra de Pie de Palo, Western Sierras Pampeanas, Argentina, *in* Pankhurst, R., and Rapela, C.W., eds., The Proto-Andean Margin of Gondwana: Geological Society of London Special Publication 142, p. 159–180.

Vujovich, G., Ramos, V.A., and Miller, H., 1994, Proterozoic metavolcanics from Western Sierras Pampeanas Terrane, Argentina: Journal of South American Earth Sciences, v. 7, p. 309–323.

Wood, D.A., Joron, J.L., and Treuil, M., 1979a, A re-appraisal of the use of trace elements to classify and discriminate between magma series erupted in different tectonic settings: Earth and Planetary Science Letters, v. 45, p. 326–336.

Wood, D.A., Joron, J.L., Treuil, M., Norry, M., and Tarney, J., 1979b, Elemental and Sr isotope variations in basic lavas from Iceland and the surrounding ocean floor: Contributions to Mineralogy and Petrology, v. 70, p. 319–339.

MANUSCRIPT ACCEPTED BY THE SOCIETY FEBRUARY 23, 2000

Printed in the U.S.A.

Geological Society of America
Special Paper 349
2000

Prospecting for ophiolites along the California continental margin

Robert G. Coleman
*Department of Geological and Environmental Sciences, Stanford University, Stanford, California 94305, USA;
e-mail: coleman@pangea.stanford.edu*

ABSTRACT

This paper consolidates new data on ophiolite studies in California and presents alternative tectonic settings in light of new ideas. Recent geophysical studies using seismic refraction and reflection complemented by gravity and magnetic measurements have provided new facts concerning the subsurface distribution of ocean crust (ophiolites) in convergent and passive margins. The geophysical signature of ocean crust and its magmatic formation are well known as a result of marine oceanographic studies. Recent geophysical traverses across the California continental margin have been successful in tracking ocean crust as it enters the zone of convergence. The consistent geophysical signature of ocean crust and its sharp transition into underlying mantle is routinely imaged and traced along the base of accretionary wedges and along passive margins. For the purpose of this paper, the ophiolites of the California margin have been placed into five separate groups, each having an apparent uniqueness based on tectonic setting, stratigraphic relationships, petrochemical nature, geophysical parameters, and igneous age of formation. This diversity of ophiolites within the accreting California margin has produced new concepts concerning their emplacement. At least five separate tectonic settings are recognized for ophiolite accretion within the California margin: (1) Oceanic-crust slabs tectonically underplated the base of the subduction wedge along the edge of the margin. (2) Infrequently, wedges of abyssal peridotite within oceanic fracture zones detached during orthogonal subduction into the trench wedge. Presubduction oceanic detachment of peridotite wedges and their subsequent movement high into fracture zones allowed tectonic emplacement of monolithic slabs into trench mélange zones. (3) Supra-subduction zone ophiolites formed above forearc or backarc spreading centers associated with offshore oceanic arcs. These ophiolites became imbricated into accreting continental margins during subduction. (4) Magmatic underplating occurred along the slow-spreading, west-facing "infant" arc of the Sierra Nevada. The slow spreading rate allowed thickening of the crust along the continental margin. (5) Allochthonous ophiolites and seamounts formed in remote oceanic spreading centers were accreted into zones of convergence during the Mesozoic and Tertiary. These new ideas on the origin of ophiolites along the California margin suggest that oceanic crust may be added to continental margins by tectonic and magmatic underplating or wedging as well as by obduction.

INTRODUCTION

The seminal paper by Bailey et al. (1970) on the presence of oceanic crust as an integral part of crustal accretion along the Jurassic California continental margin led to a virtual Pandora's box of papers since then (Dickinson et al., 1996a). The first Penrose Conference on ophiolites, convened in the summer of 1972, produced a consensus definition for the term *ophiolite* (Penrose Conference Participants, 1972). Even though the Penrose Conference definition of ophiolite was imperfect, it provided guidelines for the rock associations and their sequence. During a subsequent Penrose conference on ophiolites in September 1998, participants reaffirmed the continued usefulness of the 1972 Penrose conference definition.

The new ideas on ophiolite formation have been fueled by abundant chemical data acquired by instrumental analytical methods that produced precise measurements of minor and trace elements. Since 1972, National Science Foundation and National Aeronautics and Space Administration research grants lavished support on petrologic study of basalts returned from the Moon and from the deep oceans by dredging and drilling. These new petrochemical data on rocks from the Moon and the deep-ocean crust soon overwhelmed all the previous geochemical and petrologic data on ophiolites from the Earth's orogenic belts.

Synthesis of these new data by Pearce and Cann (1971) and Pearce et al. (1981) brought into being a new method to establish the paleotectonic setting of ancient oceanic crust by using discriminant diagrams for selected minor and major elements. Spider diagrams were a natural extension of the Pearce discriminant diagrams in which a selected set of rare earth and other elements are normalized to abundances of these elements in chondrites (primitive Earth) or to abundances in other primary rocks.

These rapid graphic comparisons were ideal for evaluating multiple analyses generated by the Deep Sea Drilling Project (DSDP) and the successor Ocean Drilling Program (ODP). Shipboard analysis produced abundant documentation of known tectonic settings for oceanic crust and thus enhanced the use of discriminant diagrams to establish the tectonic setting of ophiolites.

Coleman, R.G., 2000, Prospecting for ophiolites along the California continental margin, *in* Dilek, Y., Moores, E.M., Elthon, D., and Nicolas, A., eds., Ophiolites and Oceanic Crust: New Insights from Field Studies and the Ocean Drilling Program: Boulder, Colorado, Geological Society of America Special Paper 349, p. 351–364.

The oceanographic community has routinely included petrologists on many DSDP and ODP cruises. Drilling into basement (i.e., basalt crust) soon became routine, and drilling techniques improved rapidly. As a result, a deep hole (ODP Hole 504B) was proposed to penetrate a complete oceanic section. The drilling into crust was done in small increments by using reentry techniques and in-hole monitoring between each drilling session. The goal was to establish a well-documented ocean-crust section so as to compare it with the gross stratigraphy of ophiolites in orogenic belts. Many return visits to Hole 504B by numerous ODP legs have extended the hole through the basaltic layer and into the sheeted-dike sequence for a total distance of 2111 m, confirming the gross stratigraphy indicated by on-land ophiolites (Anderson et al., 1982; Erzinger, 1992).

Black smokers discovered by deep submersible dives along the axis of active mid-ocean ridges and their massive sulfide deposits provided definitive evidence of the huge hydrothermal systems associated with newly forming oceanic crust. This knowledge led to the exciting conclusion that it was possible to relate these systems to massive sulfide deposits in on-land ophiolites (Coleman et al., 1979; Haymon et al., 1989). The connection led to mineral exploration within ophiolites of orogenic zones and resulted in numerous new discoveries of massive sulfide deposits throughout the world (Koski and Derkey, 1981).

Recent geophysical studies using seismic refraction and reflection complemented by gravity and magnetic measurements have provided new facts concerning the subsurface distribution of ocean crust (ophiolites) in convergent and passive margins (Brocher et al., 1994; Godfrey et al., 1997; Godfrey and Klemperer, 1998; Trehu et al., 1994; Gettings et al., 1986; Mooney et al., 1985).

The purpose of this paper is to consolidate the new data on ophiolite studies in California and to evaluate alternative tectonic settings in light of evolving concepts.

CALIFORNIA OPHIOLITES

Even though the standing definition of *ophiolite* (Penrose Conference Participants, 1972) does not encompass all the modes of ophiolite formation that I present herein, I consider all the occurrences discussed to be ophiolites in the broadest sense. Future workers may want to improve the definition and thereby give different statuses to the individual ophiolite types that I describe.

Ophiolite methodology

The early geologic maps of California represented nearly all mafic and ultramafic rocks as in situ intrusive rocks. Following the plate-tectonic revolution in the 1970s, many of the ultramafic rocks and their associates were cast into the ophiolite paradigm (Fig. 1). These rock complexes were considered to be allochthonous ophiolite fragments of ancient oceanic crust (Bailey et al., 1970; Page, 1972; Hopson and Frano, 1977; Evarts, 1977; Hopson et al., 1981). Since that time, many investigators have used structural, petrologic, and stratigraphic arguments to establish the oceanic setting in which these ophiolite assemblages formed (see Dickinson et al., 1996a). Up to this moment, nearly all ophiolite investigators have agreed that these complexes are not in situ intrusive rocks and that nearly all have been tectonically transported to their present location after forming in some sort of oceanic spreading center.

The criteria used to establish the tectonic setting of each ophiolite are manifold and inconsistent. Perhaps the most popular criterion now in use and widely accepted is the application of discriminant or spider diagrams. The use of spider diagrams is primarily based on petrochemical data obtained from suites of unaltered mafic volcanic rocks collected from known modern tectonic settings (Pearce et al., 1984). The usefulness of these diagrams becomes immediately apparent because magmas formed in mid-ocean spreading centers have a unique rare earth and trace element assemblage in comparison to island-arc lavas formed over subduction zones; however, even these element distributions may produce arbitrary distinctions (Keleman et al., 1990). Complications arise when magmas form in forearc or backarc settings where magmas become contaminated by mixing with materials related to both subduction and spreading. Thus the term *supra-subduction zone* (SSZ) was invented to describe various degrees of contamination of igneous rocks formed there without proscribing an exact tectonic setting (Pearce et al., 1984). In California, the geochemical arguments debating the formation of oceanic crust in forearc, backarc, or nascent arcs have disenchanted structural geologists looking for reasoning that did not conflict with their on-site geologic interpretations.

Other investigators have focused on the stratigraphy of the pelagic and other sediments contained within and directly overlying the pillow lavas. It is extremely critical to establish whether the sediments are interlayered within the extrusive lavas or are resting disconformably on top of the lavas (Cliff Hopson, 1998, personal communication). Some exposed chert sequences record time-stratigraphic sequences produced by preserved radiolarian tests. These studies produced strong evidence that selected sequences may have formed great distances from where they are now exposed (Hopson et al., 1996; Pessagno, 1973; Hopson and Frano, 1977; Robertson, 1989, 1990).

Radiometric dating using U-Pb in zircon from small leucocratic intrusions within the hypabyssal parts of the ophiolite have proved rewarding, and now the actual magmatic age of formation of some ophiolites is well known (Saleeby, 1996; Mattinson and Hopson, 1992; Harper et al., 1994; Edelman et al., 1989).

Paleomagnetic studies provide erratic results because in some cases, all of the primary magnetite has been destroyed by oceanic hydrothermal alteration (Frei and Blake, 1987). Other studies have had remarkable success in establishing the far-traveled paleomagnetic signatures where track lines of plate movements have been established (Beebe, 1986; Hopson et al., 1996; Hull et al., 1993).

Ad hoc plate-tectonic reconstructions have been used to combine stratigraphic evidence with ages of the Sierra Nevada arc and ophiolite sequences to form time-sequence scenarios related to plate movements. Some reconstructions using a combination of stratigraphic age and radiometric data provide convincing evidence of a far-traveled history of certain ophiolites (Dickinson et al., 1996a). The small geographic extent of the studied areas makes it difficult to extrapolate the results to encompass all of the Coast Range ophiolite.

For the purposes of this paper, the ophiolites of the California margin have been placed into four separate groups each having an apparent uniqueness based on tectonic setting, stratigraphic relationships, petrochemical nature, geophysical parameters, and igneous age of formation (Table 1).

Coast Range ophiolite

Most of the studies on California-Oregon ophiolites have centered on the Middle Jurassic Coast Range ophiolite (170–164 Ma), which is considered basement for the Great Valley Sequence that contains forearc sedimentary rocks (Fig. 2). The ophiolitic section rarely has a stratigraphic thickness exceeding 4–5 km and contains intrusive rocks that vary from tholeiite to calc-alkalic in composition

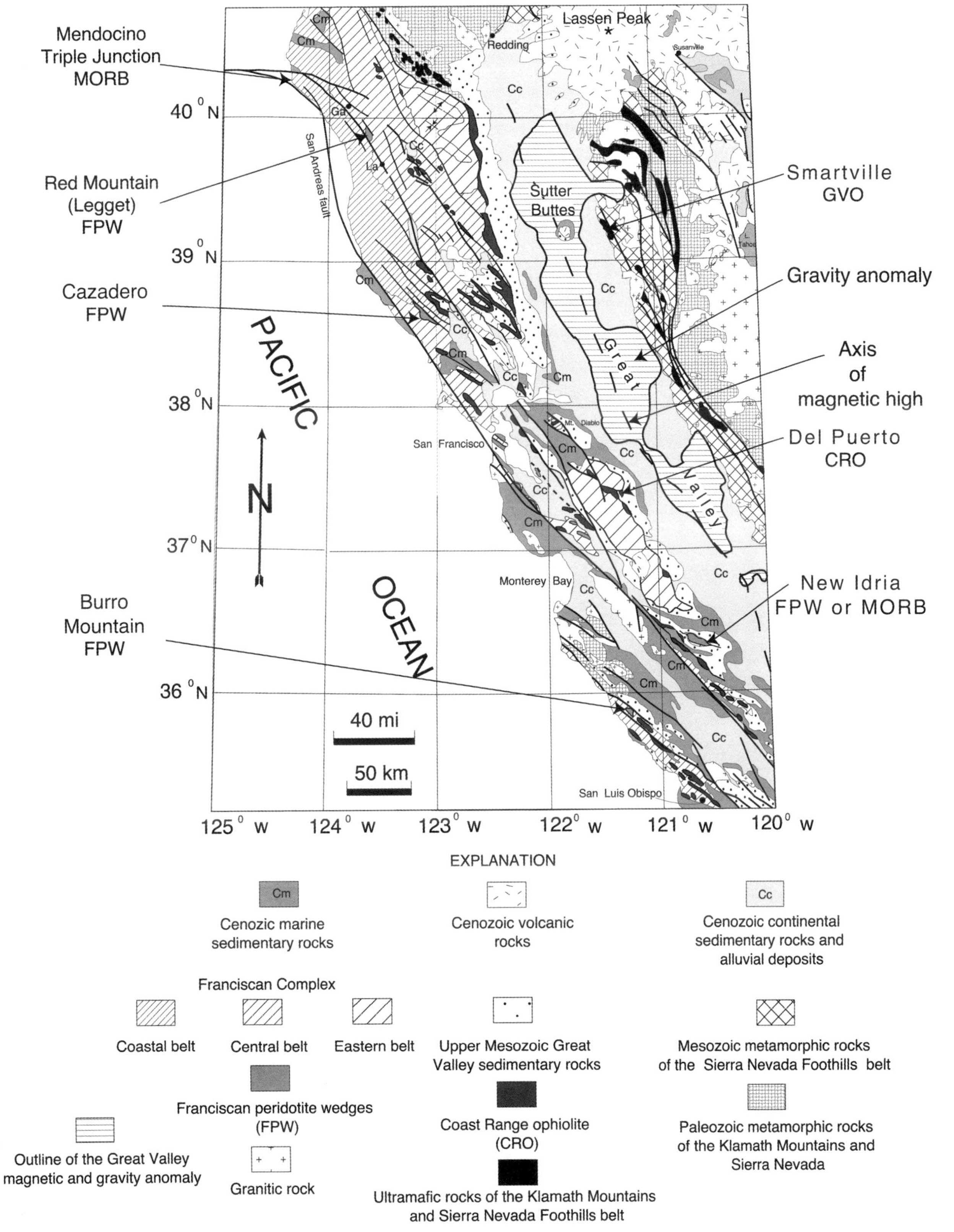

Figure 1. Geologic map of the western part of California showing the location of the ophiolites discussed in the text. On this map, the ophiolites are divided into Coast Range ophiolite (CRO), Franciscan peridotite wedges (FPW) and serpentine, and Great Valley ophiolite (GVO). The amount of magmatic underplating of the California margin can be estimated from the Great Valley ophiolite gravity anomaly and magnetic axis. Horizontal lines indicate area of Great Valley gravity anomaly.

TABLE 1. OPHIOLITE TYPES WITHIN ACCRETIONARY BELTS OF THE CALIFORNIA CONTINENTAL MARGIN

	MORB Mid-oceanic ridge basalt Pacific or Farallon plates	FPW Franciscan peridotite wedges	CRO Coast Range ophiolite	GVO Great Valley ophiolite
Tectonic setting of formation	Mid-ocean spreading center	Tectonically emplaced during orthogonal subduction of a fracture zone	Suprasubduction zone or accretion of suspect terranes	Suprasubduction zone and extension of the Jurassic ensimatic arc within the western Sierra Nevada
Age of formation	Jurassic to Holocene	Middle Jurassic or older	Middle Jurassic, 170–164 Ma	Middle Jurassic 165–155 Ma Smartville ophioite cut by calc-alkalic plutons 150–140 Ma.
Continental accretion	Related to faulting and tectonic underplating of oceanic crust during subduction	Detachment of abyssal periodotite from fracture zones	Authochthonous or accretion of ophiolite suspect terranes	Authochthonus continental edge extension and underplating
Configuration	Stalled MORB slabs at the base of subduction wedges	Detached tabular slabs imbricated in Franciscan accretionary wedge mélange	Base of western Great Valley sedimentary rocks. Thrust over Jurassic–Cretaceous Franciscan accretionary wedge	Basement for Great Valley forearc medial and eastern sedimentary rocks
Thickness	Uniform 6–7 km of mafic oceanic crust	~1–2 km, no overlying mafic layers or pelagic sedimentary rocks	<2 km to maximum 4–5 km, extremely irregular	Maximum thickness by geophysical measurements ~20 km; wedge-shaped
Structural history	East-directed subduction under the continental margin	Faulting in fracture zones elevates peridotite to the ocean floor. Subduction of irregular fracture zone detaches peridotite	Wedging and west-directed thrusting over Franciscan accretionary wedge	Extension and magmatic underplating with later, east-directed thrusting over Sierra Foothill belt
Geophysical properties	Uniform V_p, density, and magnetic properties	Variable V_p and density related to serpentinization	Variable V_p and gravity, strong magnetic anomalies associated with serpentinization	Strong positive magnetic and gravity signatures that produce Great Valley geophysical anomaly
Cover rocks	Pelagic and trench sedimentary rocks	Not known	Pelagic, forearc, and calc-alkalic pyroclastic sedimentary rocks	Great Valley sedimentary rocks and calc-alkalic pyroclastic rocks
Petrologic nature	Presumably with MORB affinities	Depleted harzburgite and dunite tectonite has geochemistry identical to abyssal peridotite from fracture zones. No associated mafic rocks	Calc-alkalic to MORB mafic rocks, pyroclastic rocks, cumulate gabbro, plagiogranite, and depleted harzburgite and dunite tectonites	Calc-alkalic and mafic pillow lava, sheeted dikes, thick gabbro complex sequences

disconformably overlain by pyroclastic rocks containing Tithonian radiolaria (90–80 Ma). Hopson et al. (1996) suggested that all the pyroclastic rocks interpreted formerly as part of the Coast Range ophiolite actually postdate the underlying ophiolite sequence (Fig. 3). The Coast Range ophiolite displays incomplete sequences and in some places is dismembered and included in serpentinite mélanges. Serpentinite mélanges may underlie some of the ophiolite, but it is difficult to determine whether a specific mélange is part of the Franciscan subduction complex or was formed from the basal Coast Range ophiolite peridotite member (Page et al., 1999).

The Coast Range ophiolite magmas appear to have been generated in a supra-subduction zone setting (Shervais and Kimbrough, 1985; Robertson, 1989; Shervais, 1990; Stern and Bloomer, 1992; Saleeby, 1992), i.e., during spreading above a subduction zone. As pointed out previously, magmas generated above a subduction zone become contaminated by fluids in the subduction trench or mantle wedge just above the descending plate. A basic conflict arises in evaluating the literature concerning the origin of the Coast Range ophiolite. Establishment of the forearc (i.e., SSZ) setting is based on geochemical evidence (Pearce and Cann, 1971) whereas other alternatives utilize stratigraphy, radiometric geochronology, and paleomagnetic studies. If all these data are synthesized for a single area, conflicts invariably arise, and there seems to be no sure-fire way of establishing the tectonic setting of ophiolites. Some of the Coast Range ophiolite occurrences contain radiolarian cherts whose affinities indicate deposition south of the paleoequator in Jurassic time (Hopson et al., 1996). Some of the stratigraphic variability found in Coast Range ophiolite is displayed in Figure 3.

Dickinson et al. (1996a) summarized the following arguments for three distinct tectonic settings advocated for the Coast Range ophiolite by different authors: (1) Backarc ocean crust formed behind an east-facing island arc and between that arc and the west-facing Sierra arc, all of which was accreted, imbricated, and partially subducted. (2) Far-traveled ocean-ridge crust docked along the continental margin prior to deposition of the Great Valley Sequence or the trench-related rocks of the Franciscan Complex. (3) Transtensional deformation along the margin induced forearc seafloor spreading along the west-facing Sierran-Klamath arc; rollback (i.e., western migration) of subduction during oblique convergence produced ophiolite accretion to the continental margin. All three models require the Coast Range ophiolite and Great Valley ophiolite (as defined in the next paragraph) to have formed in similar tectonic settings.

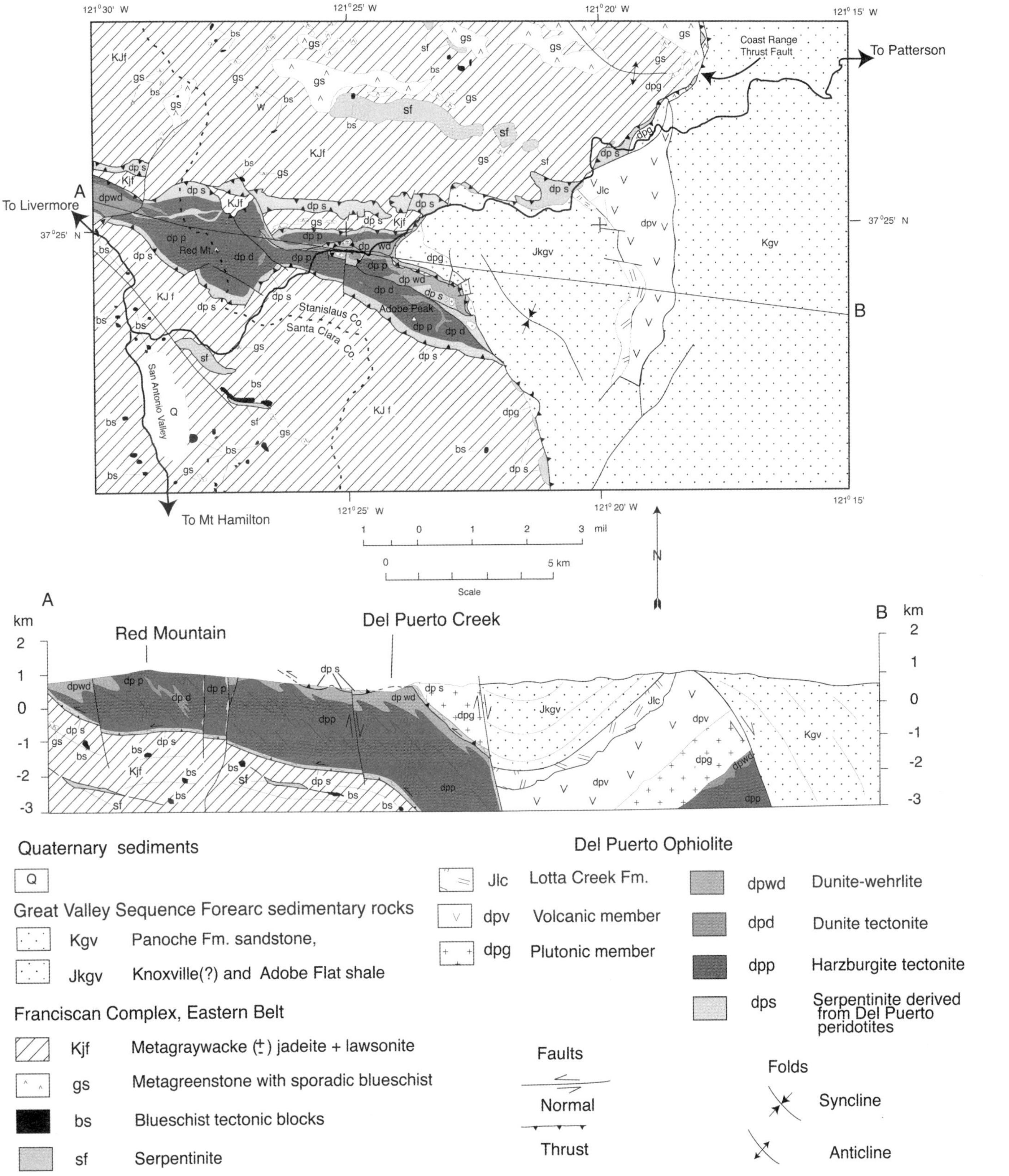

Figure 2. Geologic map of the Del Puerto ophiolite after Maddock (1964) and Evarts (1977), modified by R.G. Coleman (for location, see Fig. 1). The contact between the Franciscan Complex and the Great Valley Sequence is marked by normal faulting. The Coast Range thrust fault marks the tectonic contact between the Del Puerto ophiolite (part of the overall Coast Range ophiolite) and the Franciscan Complex, which was folded and faulted in the Late Cretaceous. The igneous stratigraphy of the Del Puerto ophiolite is shown in Figure 3. The reconstruction of the stratigraphy is obscured by folding and faulting. It is obvious that the Del Puerto ophiolite is the basement for the forearc sedimentary rocks of the Great Valley Sequence where the oldest unit, the Knoxville (Tithonian), is resting directly on the Lotta Creek Tuff. The Lotta Creek Tuff is a calc-alkalic pyroclastic tuff related to the igneous formation of the Del Puerto ophiolite.

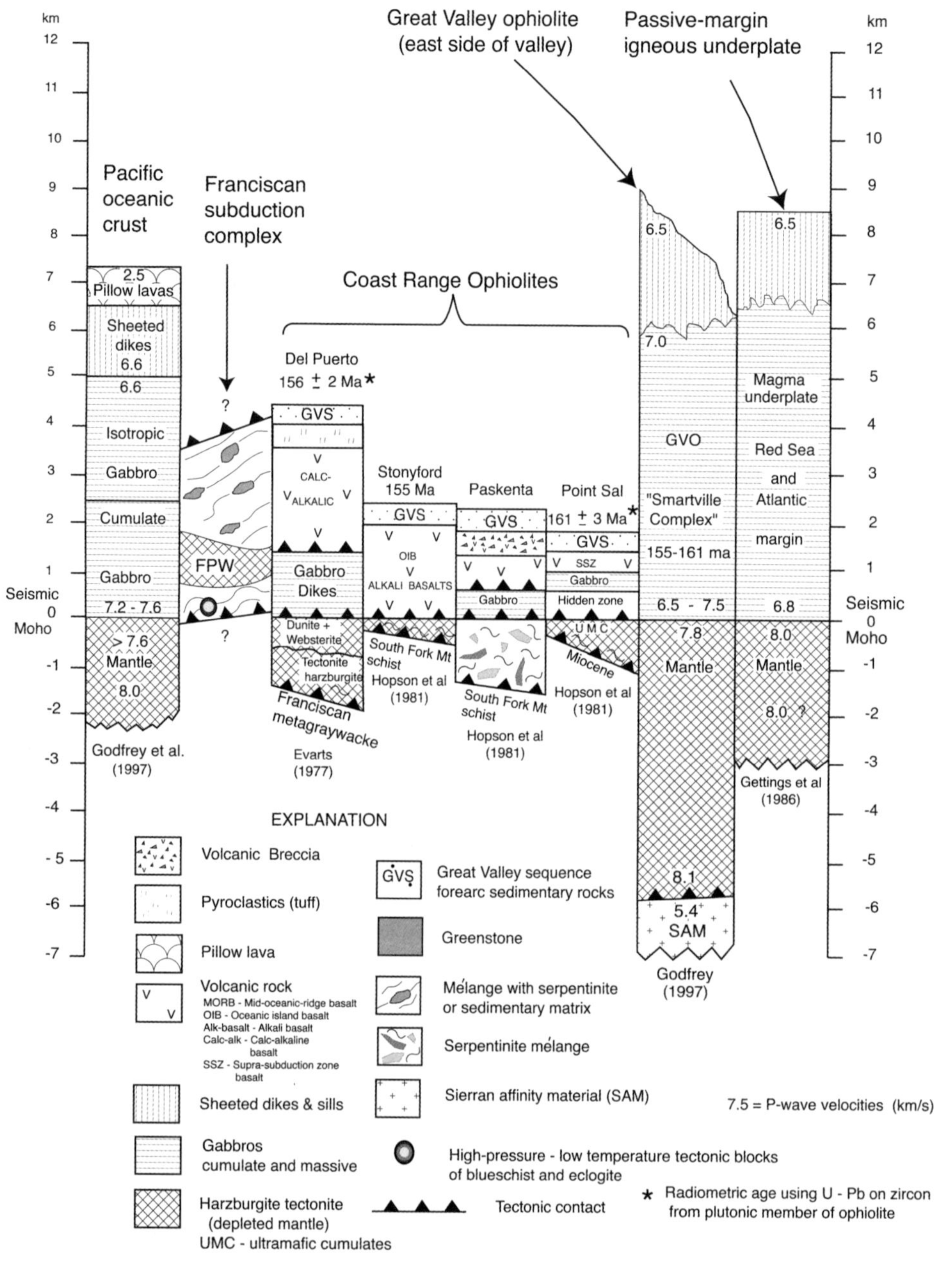

Figure 3. Relative comparison of oceanic-crust and ophiolite sequences of the California margin (for locations, see Fig. 1). Data for the Pacific Ocean crust and the Red Sea and Atlantic Ocean passive-margin crust are taken from various geophysical studies (Gettings et al., 1986; Godfrey et al., 1997; Holbrook and Kelemen, 1993; Shervais et al., 1999). The reconstructed ophiolite sections of the Coast Range ophiolite display a marked variation in thickness and in most cases fall short of the thickness of MORB crust.

Great Valley ophiolite

The Coast Range ophiolite is correlated with the basement of the Great Valley Sequence forearc sedimentary rocks. Both the Coast Range ophiolite and the basement are considered to be oceanic crust of Jurassic age formed in either a forearc or backarc setting (Dickinson et al., 1996b). This paper refers to the basement rocks underlying the physiographic Great Valley as the Great Valley ophiolite rather than as part of the Coast Range ophiolite.

The geophysical signature of the Great Valley ophiolite underlying the Great Valley is remarkable; the rocks produce continuous, strong, positive magnetic and gravity anomalies that extend the full length of this sedimentary basin (Cady, 1975) (Figs. 1 and 4). The Great Valley geophysical anomaly is nearly 500 km in length and ~50 km in width and has an average thickness of ~15–20 km along the axis of the magnetic high (Godfrey and Klemperer, 1998) (Fig. 4). Simple comparisons to one of the largest accreted masses of ophiolite—the Semail ophiolite, Oman—show that the the Great Valley ophiolite is nearly 1.5 times larger (by volume). Exploratory oil drilling has penetrated the top surface of the buried Great Valley ophiolite in many locations, but the drilling has penetrated only a few feet into the basement. Thus, only fragmentary evidence as to the petrologic nature of the Great Valley ophiolite is available. Publications have chronicled these basement drill cores as verifying the ophiolitic nature of the basement (Godfrey et al., 1997; May and Hewitt 1948; Godfrey and Klemperer, 1998; Wentworth et al., 1995). Close examination of these reports belies the superficial nature of these petrologic descriptions. These reports describe mainly calc-alkalic intrusive rocks with only a few descriptions of mafic oceanic rocks (mostly basalt), and deep-water oceanic sedimentary rocks are only rarely reported (Ross, 1979; Godfrey et al., 1997; Wentworth et al., 1995).

Geophysical modeling of the Great Valley magnetic and gravity highs reveals the presence of a continuous, large wedge-shaped body

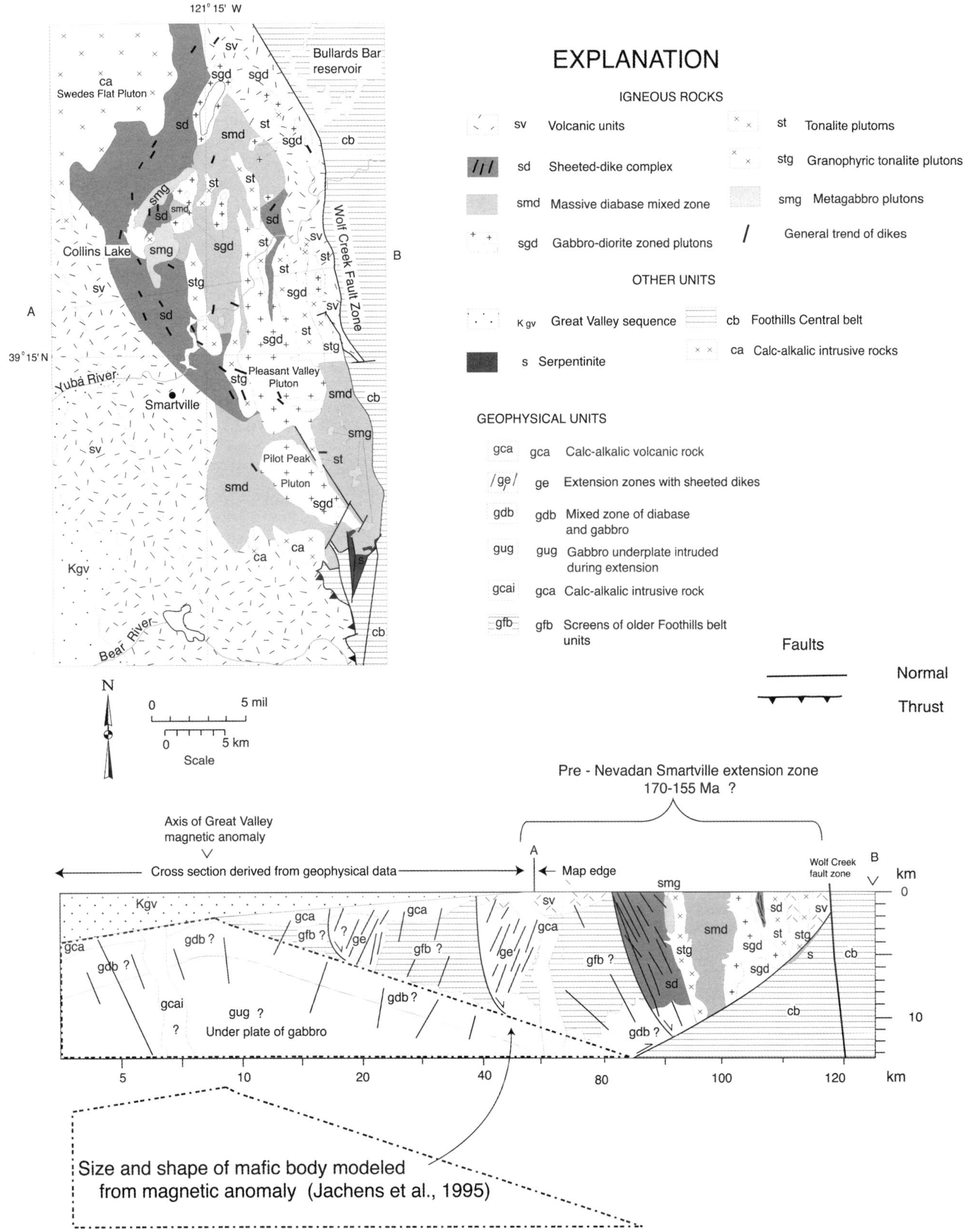

Figure 4. Geologic map of the Smartville rifted arc complex (Beard and Day, 1987; for location, see Fig. 1), considered to be a typical example of a part of the Great Valley ophiolite. Cross section shows size of the Great Valley mafic body occupying the medial and eastern parts of the Great Valley. Cross section beneath the Great Valley Sequence is controlled by magnetic modeling of the ophiolite body and extrapolation of the surface geology in and around the Smartville Complex. The Smartville Complex is considered to be autochthonous; it formed in the early stages of slow rifting on the western edge of the Sierra arc. Map modified by R.G. Coleman in 1996, after Beard and Day (1987).

(Jachens et al., 1995; Godfrey and Klemperer, 1998) (Fig. 4). If it is assumed that the modeling is correct, a maximum thickness of 20 km of gabbro-diabase extending horizontally more than 400 km is difficult to reconcile as oceanic crust whose maximum thickness rarely exceeds 12 km. Similar-sized, wedge-shaped mafic bodies underlie the eastern edge of the Red Sea and the Atlantic seaboard (Holbrook and Kelemen, 1993; Coleman and McGuire, 1988) (Fig. 5). These analogues have been related to underplating of mafic magma in the early stages of opening of the Red Sea and Atlantic Ocean. Middle to Late Jurassic extension has been well documented on the western edge of the Sierran arc (Beard and Day, 1987; Busby-Spera, 1990; Dilek et al., 1991; Saleeby, 1992; Dilek and Moores, 1992; Stern and Bloomer, 1992). This same extensional event occurred in the Klamath Mountains and is manifested by the Josephine ophiolite and Western Hayfork volcanic rocks (Hacker and Ernst, 1993; Harper et al., 1994). Careful studies of the Smartville complex situated on the eastern edge of the Great Valley reveal sheeted dikes associated with massive gabbro and diabase overlain by mafic volcanic rocks with pillow structures (Beard and Day, 1987) (Fig. 4). Surface exposures of the dike swarm indicate that a minimum of 7 km of extension is represented by the Smartville complex (Beard and Day, 1987).

Calc-alkalic plutons (150–140 Ma) cut the 165–155 Ma Smartville complex and restrict the duration of the possible rifting and extension event to ~14–25 m.y. (Beard and Day, 1987). The initial stages of the Great Valley ophiolite's formation may be related to SSZ extension along the very edge of the Jurassic Sierra arc (Dilek and Moores, 1992). Crustal extension preceded the actual rifting and may have been located to the west at the axis of the Great Valley anomaly (Fig. 4). The timing between actual rifting (that formed the Smartville dike swarm) and earlier extension is hard to estimate. Assuming that extension was slow (<1 cm/yr) during the initial period of magmatic underplating, such underplating could have produced a zone 10 km wide in ~10 m.y. As the extension rate increased, dike swarms and lavas formed and breached the surface of the proto–Pacific Ocean to the west, and new ocean crust began to form. An increase in spreading rate to 5 cm/yr would produce 50 km of new crust in the next 10 m.y. This scenario explains the wedge shape of the Great Valley ophiolite mass and can be compared to similar situations in the Red Sea and Atlantic Ocean where wedges of mafic intrusions thin oceanward from a maximum thickness of 20 km at the continental margins to the thickness of normal ocean crust (7–8 km). The thick part of each wedge was formed by massive infusions of mafic magma that underplated the continent during the initially slow opening of the rift. The Smartville dike swarm represents only a part of the huge Great Valley ophiolite wedge that slowly thinned westward as extension rates increased (Fig. 4). A similar suggestion for early extension in the formation of the Sierran forearc has been made by Stern and Bloomer (1992).

Franciscan ophiolite, in the broadest sense

Franciscan ophiolitic rocks. Complete ophiolite sequences have not been reported from the Franciscan Complex, but there are numerous blocks of greenstone (altered basalts, some pillowed) ranging in size up to 1 km in diameter (Bailey et al., 1964; Shervais, 1990; Shervais et al., 1999). Associated with these greenstones are serpentinite bodies often containing numerous tectonic blocks of blueschist and eclogite or larger coherent wedges of peridotite with serpentine selvages (Coleman, 1980).

Shervais has demonstrated that the Franciscan greenstones are fragments of the uppermost oceanic crust with variable geochemical signatures. MORB (mid-oceanic-ridge basalt) is the most common; OIB (oceanic-island basalt) alkalic basalts to transitional basalts represent seamounts (Shervais, 1990; Shervais and Kimbrough, 1985, 1987; Shervais and Hanan, 1989; MacPherson et al., 1990). The Franciscan greenstones with their MORB and seamount geochemical signatures are distinct from the supra-subduction zone and arc Coast Range ophiolite basalts. The nearly complete lack of gabbros and sheeted-dike swarms within the Franciscan Complex provides evidence that only thin slabs of upper oceanic crust detached during subduction and became incorporated into Franciscan trench sediments (Shervais et al., 1999).

Franciscan peridotite wedges. The geophysical signature of ocean crust as well as its magmatic generation in oceanic spreading centers is now quite well understood. Numerous geophysical transects across the western margin of North America inland to the Great Valley have revealed the presence of stalled slabs of oceanic crust at the base of the continental crust (Holbrook et al., 1996; Trehu et al., 1994; Howie et al., 1993; Page and Brocher, 1993; Brocher et al., 1994, 1998; Godfrey et al., 1997; Hole et al., 1998) (Fig. 6). Detached and stranded segments of oceanic crust are imaged at the base of the modern accretionary wedge south of the Mendocino triple junction at the latitude of San Francisco (Page and Brocher, 1993; Brocher et al., 1994; Hole et al., 1998). The existence of stranded ocean crust at the base of continental crust after passage of the Mendocino triple junction provides evidence that some segments of oceanic crust have accreted onto the base of the North American margin rather than all being recycled into the mantle as suggested by Atwater (1970) (Fig. 6).

It is obvious that tectonic underplating of ocean crust under the California continental margin opens further speculation concerning the occurrence of tectonic blocks of oceanic crust (ophiolite) in the Franciscan subduction complex. These stranded oceanic slabs extend eastward along the ductile transition zone beneath and across the San Andreas fault zone (Brocher et al., 1998). The stranded slabs are now the youngest ophiolites accreted to the California margin, and their distinctive layering with ~6 km of mafic material overlying mantle rocks can be easily traced by seismic refraction all the way to the western margin of the Great Valley (Brocher et al., 1998; Hole et al., 1998).

Exposures of partially serpentinized harzburgite tectonite wedges up to 4 km long and less than 1 km thick are detached fragments of oceanic crust basement also present within the Franciscan Complex (Bailey et al., 1964; Loney et al., 1971). The Burro Mountain (Fig. 7) and Cazadero harzburgite tectonite bodies are representative of other similar bodies and are referred to here as Franciscan peridotite wedges. They occur as tabular-like slabs within shaly-matrix mélange associated with tectonic blocks of metagraywacke, glaucophane schist, radiolarian cherts, and greenstones. Characteristically, gabbro and diabase are not found associated with the Franciscan peridotite wedges, nor is there field evidence that the wedges are related to dismembered ophiolites within the mélange. The borders of the wedges are sheared serpentinites forming high-angle faults with the surrounding mélange and may contain tectonic blocks of the Franciscan Complex (Bailey et al., 1964; Coleman and Keith, 1971; Burch, 1968) (Fig. 7). Geophysical studies on Franciscan peridotite wedges confirm that they are wedge shaped and allochthonous, clearly discrediting earlier claims that these bodies were igneous intrusions (Thompson and Robinson, 1975). A petrologic and structural study of the Burro Mountain Franciscan peridotite wedge demonstrated that it is a depleted harzburgite with a distinct tectonite fabric (Loney et al., 1971). Detailed geochemical work on the Burro Mountain (Loney et al., 1971) and Cazadero Franciscan peridotite wedges (Flanagan,

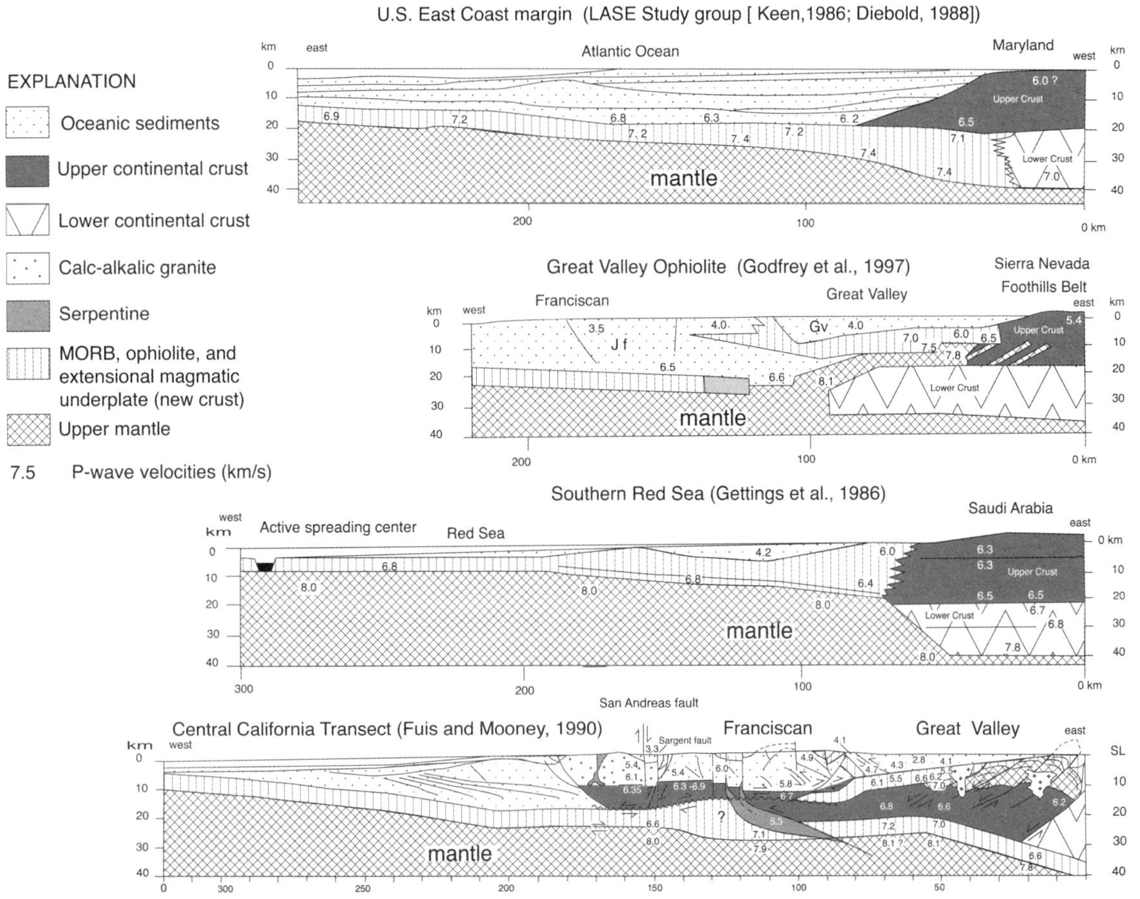

Figure 5. Comparison of magmatic underplating evolving into ocean crust in the Red Sea and beneath the Atlantic coast of North America with sections across the Great Valley by Godfrey et al. (1997) and Fuis and Mooney (1990). All cross sections are drawn at the same scale and reveal thick magmatic underplating at the continental margins during initial crustal thinning. Continued rifting with higher rates of extension leads to development of oceanic crust with normal thickness.

1969) demonstrates their similarity in chemistry and petrologic nature to abyssal peridotites dredged mainly from oceanic fracture zones (Dick, 1989).

The lack of gabbro and diabase associated with the peridotite wedges within the Franciscan subduction mélange is puzzling but may be explained if the peridotite wedges represent detached slabs of abyssal peridotite from oceanic fracture zones. Dick (1989) observed that extreme crustal thinning within oceanic fracture zones exposes mantle peridotites or leaves them with only a thin serpentinized crust and sporadically covered by pillow lavas. Orthogonal impingement of major fracture zones into subduction zones along the accretionary margin of North America since Jurassic time is a possible tectonic setting for emplacement of the Franciscan peridotite wedges. Other scenarios for emplacement of the wedges will undoubtedly be developed as future studies concentrate on this problem.

Tectonic implications

The understanding of ophiolites is at a new threshold because it is possible to use geophysical imaging from reflection, refraction, gravity, and magnetic surveys to locate ophiolites (ocean crust) within Earth's upper crust. Crossing this threshold will require integration of tectonics, petrology, and geophysics into multidisciplinary studies.

The geophysical signature of ocean crust and its magmatic formation are well known as a result of oceanographic studies. Recent geophysical traverses across the continental margin of western North America have successfully tracked ocean crust as it enters the zone of convergence (Fuis, 1998; Brocher et al., 1998). The consistent geophysical signature of ocean crust and the sharp transition from crust into underlying mantle can now routinely be imaged.

The layered nature of the ocean crust is a result of its spreading-ridge history (see Fig. 3). The thickness of the crust generated at oceanic spreading centers averages about 6–7 km; the crust shows a range in V_p velocities from ~3.0 to 6.5 km/s for layer 2 and 6.86 to 7.2 km/s for layer 3 (Christensen, 1978). The transition to upper-mantle peridotite, in which velocities range from 7.8 to 8.0 km/s, is sharp (Christensen and Salisbury, 1975). Stranded segments of such oceanic crust have been reported (Brocher et al., 1998; Page and Brocher, 1993) at the base of the modern accretionary wedge south of the Mendocino triple junction and extending eastward along the ductile transition zone across the San Andreas fault. Farther south, in the New Idria–Coalinga area, large masses of serpentinite in the fold-and-thrust belt have been transported to the surface by tectonic wedging (Wentworth et al., 1984) initiated by the passage of the triple junction at this latitude. The serpentinite bodies are derived from depleted mantle harzburgite (Franciscan peridotite wedges) whose age is unknown. These mantle peridotites are thought to be tectonically detached fracture-zone peridotites (Dick, 1989) that have been completely serpentinized; in the process, they acquired a lower density and plastic rheology that enhanced their upward tectonic mobil-

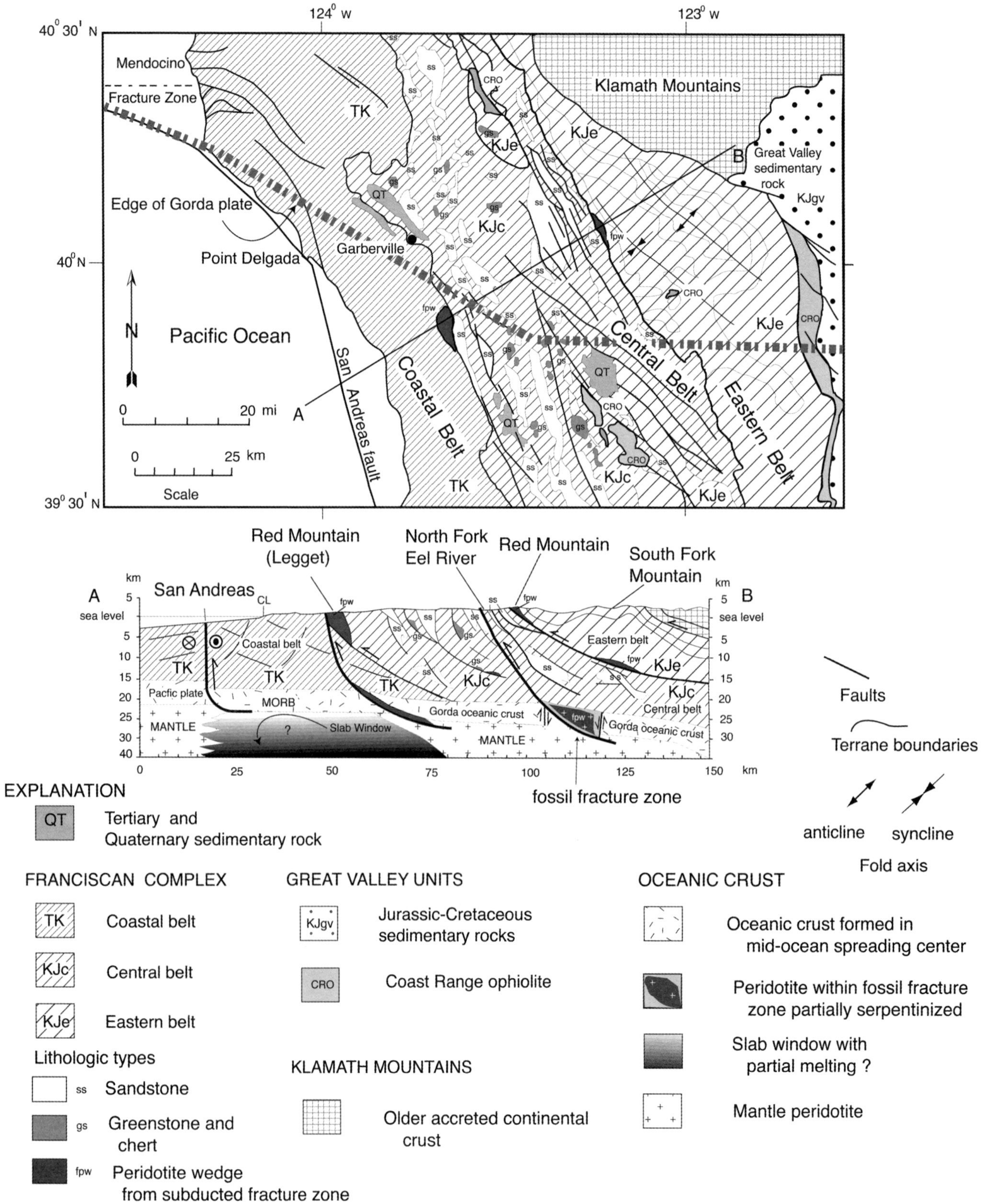

Figure 6. Geologic map of the Mendocino triple junction (Jayko and Blake, 1987), showing distribution of peridotite wedges and serpentine in the Franciscan Complex (for location, see Fig. 1). Continued eastward subduction has imbricated and uplifted Franciscan peridotite wedges into the central and eastern belts of the Franciscan Complex. Geophysical transects (Beaudoin et al., 1996; Godfrey et al., 1997; Henstock et al., 1997) have clearly tracked oceanic crust into the subduction zone and eastward under the accreting-margin basement. Vertical faults cutting the ocean crust, discovered in the transects, provide a mechanism to emplace fragments of oceanic crust into the subduction wedge (Hole et al., 1998). Subducted abyssal peridotites (now Franciscan peridotite wedges) within fracture zones are also immersed in or tectonically underplated below trench mélange. Subsequent serpentinization of the Franciscan peridotite wedges accelerates their tectonic elevation toward the surface. Geologic map, constructed by R.G. Coleman in October 1996, was modified after Jayko and Blake (1987); cross section modified after Beaudoin et al. (1996), Henstock et al. (1997), and Hole et al. (1998).

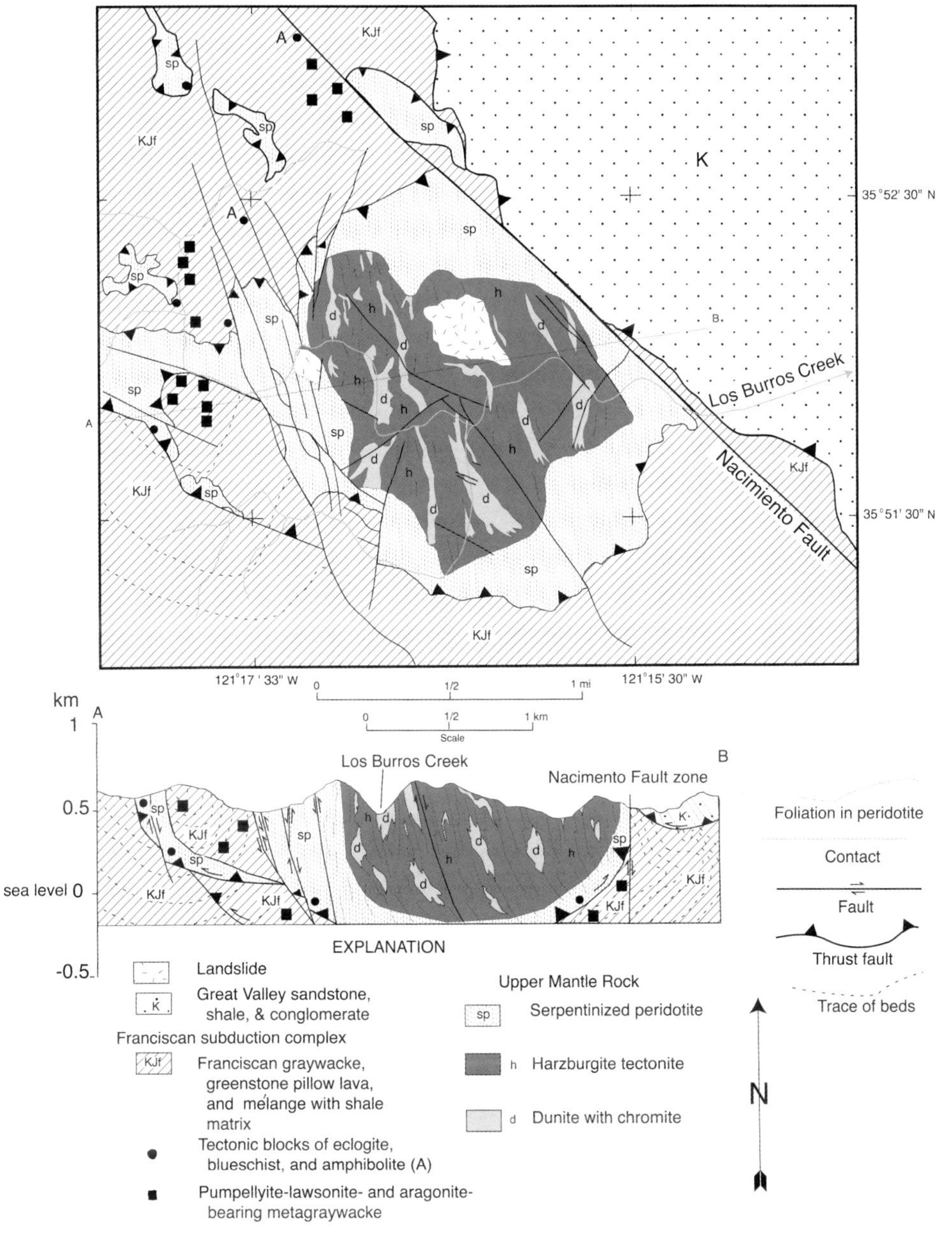

Figure 7. Geologic map (Loney et al., 1971) of the Burro Mountain peridotite slab (one of the Franciscan peridotite wedges; for location, see Fig. 1). The cross section illustrates shape and folding related to its emplacement into the Franciscan trench units. High-pressure–low-temperature metagraywackes surrounding the Burro Mountain peridotite wedge formed prior to the peridotite's emplacement and indicate that it was probably detached at depths greater than 20 km. Smaller tectonic blocks of amphibolite and higher-grade blueschist within the graywackes indicate even greater depths of emplacement. Geologic map by R.G. Coleman, modified after Loney et al. (1971).

ity. At least some of the detached or tectonically underplated mafic and ultramafic fragments are younger than similar ophiolite fragments embedded in the accretionary wedges (Franciscan Complex) of western North America.

The wedge-shaped mafic underplating that marks initial rifting of the Atlantic Ocean and the Red Sea were discovered by geophysical transects (Mooney et al., 1985; Gettings et al., 1986; Keen, 1986; Diebold, 1988). At these sites, the mafic wedges thicken to 20 km toward the continental margin and continuously taper seaward to the thickness of 7–8 km considered to be normal for oceanic crust (Fig. 5). In the early stages of attenuation, the crustal thinning rate is so slow that the rising magma reaches a neutral buoyancy at the base of the extended crust, underplating and thickening it (Coleman, 1993). As the spreading rate increases, the older, attenuated continental crust becomes very thin or disappears, and the magmas of the rift zone begin to form the basement of the expanding oceanic basin (as exemplified by the Great Valley ophiolite).

The Great Valley ophiolite's early stages of development are considered to have been within a SSZ on the western edge of the Sierra arc. The older, accreting margin was thinned by oblique strike-slip (transtensional) faulting, which initiated crustal thinning and mantle upwelling to produce a mafic magma that underplated and thickened the attenuated crust. As shown by geophysical studies, the Great Valley ophiolite is wedged shaped and thins to the west (Jachens et al., 1995). A comparison of its size and shape to that of the Red Sea and East Coast magmatically underplated mafic bodies is striking and provides a link between continental rifting and rifting within arcs (Fig. 5).

Two-dimensional geophysical imaging and modeling of oceanic crust (ophiolites) along continental margins reveal new and complex tectonic scenarios related to ophiolite accretion and subduction-angle changes along continental margins. From this new information, it is clear that not all ophiolites are allochthonous fragments of oceanic crust tectonically emplaced into suture zones along continental margins. Underplating by mafic magmas during initial rifting is a significant and widespread process. Stranding of slabs of ocean crust at the base of the continental margin, as imaged by geophysical methods, may also be an important tectonic process within subduction zones. Obduction of ophiolites is probably a rare event when compared to the stranding of oceanic-crust slabs in subduction zones along convergent margins. The possibility of magmatic underplating to form new mafic crust within arc systems has to be considered in light of the recently acquired geophysical data on the Great Valley ophiolite (Godfrey et al., 1997; Hole et al., 1998; Jachens et al., 1995; Godfrey and Dilek, this volume).

CONCLUSIONS

The diversity of the ophiolites within the California accreting margin is now obvious and leads to new ideas concerning their emplacement. At least five separate tectonic settings are recognized for ophiolites accreted within the California margin: (1) Stranded oceanic-crust slabs are tectonically underplated to the base of the continental margin in subduction zones. (2) Infrequently, abyssal peridotite detaches during orthogonal subduction of oceanic fracture zones. (3) SSZ ophiolites developed above forearc spreading centers offshore of the California continental margin. (4) Magmatic underplating filled rifts opened during extension of the west-facing "infant" arc of the Sierra Nevada. (5) Allochthonous ophiolites and seamounts that formed in remote oceanic spreading centers were accreted to North America in the zone of convergence during the Mesozoic and Tertiary.

New studies on the origin of ophiolites along the California margin should employ a multidisciplinary approach using stratigraphy, structure, petrology-geochemistry, and geophysics. Detailed unilateral studies on ophiolites by experts from each of these disciplines are no longer bearing scientific fruit. The reason may be that the early restrictive plate-tectonic concepts regarding oceanic-crust formation and its tectonic emplacement can no longer be used to explain these newly discovered settings.

ACKNOWLEDGMENTS

This paper is dedicated to Cliff Hopson, University of California at Santa Barbara, whose careful review of the first draft revealed gaps in my logic and lack of understanding concerning the complex history of the California ophiolites. Much of the knowledge of the Coast Range ophiolites is based on Hopson's pioneering and elegant studies, as will be evident to anyone reading this volume. John Shervais, University of South Carolina, provided an important second review that produced a better petrotectonic summary of the California ophiolites. Robert Jachens, U.S. Geological Survey, Menlo Park, gave patient guidance in proper use of the geophysical parameters utilized in this study.

REFERENCES CITED

Anderson, R.N., Honnerz, J., Becker, K., Adamson, A.C., Alt, J.C., Emmermann, R., Kempton, P.D., Kinoshita, H., Laverne, C., Mottl, M.J., and Newmark, R.L., 1982, DSDP Hole 504B, the first reference section over 1 km through Layer 2 of the oceanic crust: Nature, v. 300, no. 5893, p. 589–594.

Atwater, T., 1970, Implications of plate tectonics for the Cenozoic tectonic evolution of western North America: Geological Society of America Bulletin, v. 81, p. 3513–3536.

Bailey, E.H., Irwin, W.P., and Jones, D.L., 1964, Franciscan and related rocks: California Division of Mines and Geology, v. 183, p. 1–177.

Bailey, E.H., Blake, M.C., Jr., and Jones, D.L, 1970, On-land Mesozoic oceanic crust in California Coast Ranges: U.S. Geological Survey Professional Paper, v. 700C, p. 70C–81C.

Beard, J.S., and Day, H.W., 1987, The Smartville intrusive complex, Sierra Nevada, California: The core of a rifted arc: Geological Society of America Bulletin, v. 99, p. 779–791.

Beaudoin, B.C., Godfrey, N.A., Klemperer, S.L., Lendi, C., Trehu, A.M., Henstock, T.J., Levander, A., Holl, J.E., Meltzer, A.S., Luetgert, J.H., and Mooney, W.D., 1996, Transition from slab to slabless: Results from the 1993 Mendocino triple junction seismic experiment: Geology, v. 24, p. 195–199.

Beebe, W.J., 1986, A paleomagnetic study of the southern Coast Range ophiolite, California, and tectonic implications [MA. thesis]: University of California, Santa Barbara, 148 p.

Brocher, T.M., McCarthy, J., Hart, P.E., Holbrook, W.S., Furlong, K.P., McEvilly, T.V., Hole, J.A., and Klemperer, S.L., 1994, Seismic evidence for a lower-crustal detachment beneath San Francisco Bay, California: Science, v. 265, p. 1436–1441.

Brocher, T.M., Brink, U.-T., and Abramovitz, T., 1998, Synthesis of the deep crustal stucture in the San Francisco Bay area: International Geology Review, v. 41, no. 3, p. 263–274.

Burch, S.H., 1968, Tectonic emplacement of the Burro Mountain ultramafic body, Santa Lucia Range, California; Geological Society of America Bulletin, v. 60, p. 527–544.

Busby-Spera, C.J., 1990, Speculative tectonic model for the Lower Mesozoic arc of southwest Cordillera United States: Geology, v. 16, p. 1121–1125.

Cady, J.W., 1975, Magnetic and gravity anomalies in the Great Valley and western Sierra Nevada metamorphic belt, California: Geological Society of America Special Paper 168, 56 p.

Christensen, N.I., 1978, Ophiolites, seismic velocities and oceanic crustal structure: Tectonophysics, v. 47, p. 131–157.

Christensen, N.I., and Salisbury, M.H., 1975, Structure and constitution of the lower oceanic crust, Review of Geophysics and Space Physics, v. 13, no. 1, p. 57–86.

Coleman, R.G., 1980, Tectonic inclusions in serpentinites: Geneve, Archives Science, v. 33, p. 89–102.

Coleman, R.G., 1993, Geologic evolution of the Red Sea: New York, Oxford University Press, 186 p.

Coleman, R.G., and Keith, T.E.C., 1971, A chemical study of serpentinization: Burro Mountain, California: Journal of Petrology, v. 12, p. 311–328.

Coleman, R.G., and McGuire, A.V., 1988, Magma systems related to the Red Sea opening: Tectonophysics, v. 150, p. 77–100.

Coleman, R.G., Huston, C.C., El-Boushi, I.M., Al-Hinai, K.M., and Bailey, E.H., 1979, The Semail ophiolite and associated massive sulfide deposits, Jeddah, Kingdom of Saudi Arabia: Sultanate of Oman, Institute of Applied Geology Bulletin, King Abdullaziz University, v. 2, p. 179–192.

Dick, H.J.B., 1989, Abyssal peridotites, very slow spreading ridges and ocean ridge magmatism, *in* Saunders, A.D., and Norry, M.J., eds., Magmatism in the ocean basins: Geological Society of London Special Publication 42, p. 71–105.

Dickinson, W.R., Hopson, C.A., and Saleeby, J.R., 1996a, Alternate origins of the Coast Range Ophiolite (California): Introduction and implications: GSA Today, v. 6, no. 2, p. 1–10.

Dickinson, W.R., Schweikert, R.A., and Ingersoll, R.V., 1996b, 1. Coast Range ophiolite as back-arc-inter-arc basin lithosphere: GSA Today, v. 6, no. 2, p. 2–4.

Diebold, J.B., 1988, A large aperture seismic experiment in the Baltimore Canyon Trough, *in* Sheriden, R.E., and Grow, J.S., eds., The geology of North America: The Atlantic Continental Margin: Boulder, Colorado, Geological Society of America, v. I-2, p. 387–398.

Dilek, Y., and Moores, E.M., 1992, Island-arc evolution and fracture-zone tectonics in the Mesozoic Sierra Nevada, California, and implications for transform offset of the Sierran/Klamath convergent margin, *in* Bartholomew, M.J., Hyndman, D.W., Mogk, D.W., and Mason, R., eds., Basement Tectonics 8: Characterization and comparison of ancient and Mesozoic continental margins, Proceedings of the 8th International Conference on Basement Tectonics (Butte Montana, 1988): Dordrecht, Netherlands, Kluwer Academic Publishers, p. 179–196.

Dilek, Y., Thy, P., and Moores, E.M., 1991, Episodic dike intrusions in the northwestern Sierra Nevada, California: Implications for multistage evolution of a Jurassic arc terrane: Geology, v. 19, p. 180–184.

Edelman, S.H., Day, H.W., and Bickford, M.E., 1989, Implications of U-Pb zircon ages for the tectonic settings of the Smartville and Slate Creek complexes, northern Sierra Nevada, California: Geology, v. 17, p. 1032–1035.

Erzinger, J., 1992, Ocean Drilling Program cored the deepest hole in the ocean crust: 29th International Geologic Congress, Osaka, abstracts, v. 19, p. 19.

Evarts, R., 1977, The geology and petrology of the Del Puerto ophiolite, Diablo Range, central California: Oregon Department of Geology and Mineral Industries Bulletin, v. 95, p. 161–139.

Flanagan, F.J., 1969, U.S. Geological Survey standards. 2. First compilation of data for the new USGS rocks: Geochimica et Cosmochimica Acta, v. 33, no. 1, p. 81–120.

Frei, L.S., and Blake, C., 1987, Remagnetization of the Coast Range Ophiolite and lower part of the Great Valley Sequence in Northern California and Southwest Oregon: Journal of Geophysical Research, B, Solid Earth and Planets, v. 92, no. 5, p. 3487–3499.

Fuis, G.S., 1998, West margin of North America: A synthesis of recent seismic transects: Tectonophysics, v. 288, p. 265–292.

Fuis, G.S., and Mooney,W.D., 1990, Lithospheric structure and tectonics from seismic-refraction and other data, Wallace, R.E., ed, *in* The San Andreas Fault system, California: Washington, D.C., U.S. Geological Survey, Professional Paper 1515, p. 207–236.

Gettings, M.D., Blank, H.R., Jr., and Mooney, W.D., 1986, Crustal structure of southwestern Saudi Arabia: Journal of Geophysical Research, v. 91, p. 6491–6512.

Godfrey, N.J., and Klemperer, S.L., 1998, Ophiolitic basement to a forearc basin and implications for continental growth: The Coast Range/Great Valley ophiolite, California: Tectonics, v. 17, no.4, p. 558–570.

Godfrey, N.J., Beaudoin, B.C., Klemperer, S.L., Levander, A., Luetgert, J., Meltzer, A., Mooney, W., and Trehu, A., 1997, Ophiolitic basement to the Great Valley forearc basin, California, from seismic and gravity data: Implications for crustal growth at the North American continental margin: Geological Society of America Bulletin, v. 109, no. 12, p. 1536–1562.

Hacker, B.R., and Ernst, W.G., 1993, Jurassic Orogeny in the Klamath Mountains: A geochronological analysis, *in* Dunne, G., and McDougall, K. eds., Mesozoic paleogeography of the western United States—II: Pacific Section, SEPM Book 71, p. 37–60.

Harper, G.D., Saleeby, J.B., and Heizler, M., 1994, Formation and emplacement of the Josephine ophiolite and the Nevadan orogeny in the Klamath Mountains, California-Oregon: U/Pb and $^{40}Ar/^{39}Ar$ geochronology: Journal of Geophysical Research, v. 99, p. 4293–4321.

Haymon, R.M., Koski, R.A., and Abrams, M.J., 1989, Hydothermal discharge zones beneath massive sulfide deposits mapped in the Oman ophiolite: Geology, v. 17, p. 531–535.

Henstock, T.J., Levander, A., and Hole, J.A., 1997, Deformation in the lower crust of the San Andreas fault system in northern California, Science, v. 278, p. 650–653.

Holbrook, W.S., and Kelemen, P.B., 1993, Large igneous provinces on the US Atlantic margin and implications for magmatism during continental breakup: Nature, v. 364, p. 432–436.

Holbrook, W.S., Brocher, T.M., ten Brink, U.S., and Hole, J.A., 1996, Crustal structure beneath the San Francisco Bay block and the central California continental margin: Journal of Geophysical Research, v. 101, p. 22311–22334.

Hole, J.A., Beaudoin, B.C., and Henstock, T.J., 1998, Wide-angle seismic constraints on the evolution of the deep San Andreas plate boundary by Mendocino triple junction migration: Tectonics, v. 17, no.5, p. 802–918.

Hopson, C.A., and Frano, C.J., 1977, Igneous history of the Point Sal ophiolite, southern California: Oregon Department of Geology and Mineral Industries Bulletin, v. 95, p. 161–183.

Hopson, C.A., Mattinson, J.M., and Pessagno, E.A., 1981, Coast Range ophiolite, western California: Englewood Cliffs, New Jersey: Prentice-Hall. Rubey Colloquim Series, Ernst, W.G., ed., v. I, p. 418–510.

Hopson, C.A., Pessagno, E.A., Mattinson, J.M., Luyendyk, R.P., Beebe, W., Hull, D.M., Muñoz, I.M., and Blome, C.D., 1996, 2. Coast Range ophiolite as paleoequatorial mid-ocean lithosphere: GSA Today, v. 6, no. 2, p. 4–6.

Howie, J.M., Miller, K.C., and Savage, W.U., 1993, Integrated crustal structure across the south central California margin: Santa Lucia escarpment to the San Andreas fault: Journal of Geophysical Research, v. 98, p. 8173–8196.

Hull, D.M., Pessagno, E.A., Jr., Hopson, C.A., Blome, C.D., and Muñoz, I.M., 1993. Chronostratigraphic assignment of volcanopelagic strata above the Coast Range Ophiolite *in* Dunne, G., and McDougall, K., eds., Mesozoic paleogeography of the western United States—II: Pacific Section, SEPM Book 71, p. 151–170.

Jachens, R.C., Griscom, A., and Roberts, C.W., 1995, Regional extent of Great Valley basement west of the Great Valley, California: Implications for extensive tectonic wedging in the California Coast Ranges: Journal of Geophysical Research, v. 100, p. 12769–12790.

Jayko, A.S., and Blake, M.C., Jr., 1987, Geologic terranes of coastal Northern California and southern Oregon, *in* Schymiczek, H., and Sucksland, R.I., eds., Tectonics, sedimentation, and evolution of the Eel River and other coastal basins of northern California: Bakersfield, California, San Joaquin Geological Society, p. 1–12.

Keen, C.E., 1986, Deep structure of the US East Coast passive margin from large aperture seismic experiments (LASE): Marine and Petroleum Geology, v. 3, no. 3, p. 234–242.

Kelemen, P.B., Johnson, K.T.M., Kinzler, R.J., Irving, A.J., 1990, High-field-strength element depletions in arc basalts due to mantle-magma interaction: Nature, v. 345, p. 521–524.

Koski, R.A., and Derkey, R.E., 1981, Massive sulfide deposits in oceanic-crust and island-arc terranes of southwestern Oregon: Oregon Geology, v. 43, no. 9, p. 19–125.

Loney, R.A., Himmelberg, G.R., and Coleman, R.G., 1971, Structure and petrology of the Alpine-type peridotite at Burro Mountain, California, U.S.A.: Journal of Petrology, v. 12, p. 245–309.

MacPherson, G.J., Phipps, S.P., and Grossman, J.N., 1990, Diverse sources for igneous blocks in Franciscan mélanges, California Coast Ranges: Journal of Geology, v. 98, p. 845–862.

Maddock, M., 1964, Geologic map and sections of the Mount Boardman quadrangle, Santa Clara and Stanislaus Counties, California: California Division of Mines and Geology, Map Sheet 3, scale 1:62 500.

Mattinson, E.A., and Hopson, C.A., 1992, U/Pb ages of the Coast Range ophiolite: A critical reevaluation based on new high-precision Pb/Pb ages: American Association of Petroleum Geologists Bulletin, v. 7, p. 425.

May, J.C., and Hewitt, R.L., 1948, The basement complex in well samples from the Sacramento and San Joaquin Valleys, California: California Division of Mines and Geology Journal, v. 44, p. 129–158.

Mooney, W.D., Gettings, M.E., Blank, H.R., and Healy, J.H., 1985, Saudi Arabian seismic deep refraction profile: A travel time interpretation of crustal and upper mantle structure: Tectonophysics, v. 111, p. 173–246.

Page, B.M., 1972, Oceanic crust and mantle fragment in subduction complex near San Obispo, California: Geological Society of America Bulletin, v. 83, p. 957–972.

Page, B.M., and Brocher, T.M., 1993, Thrusting of the central California margin over the edge of the Pacific plate during the transform regime: Geology, v. 21, p. 635–638.

Page, B.M., DeVito, L.A., and Coleman, R.G., 1999, Tectonic emplacement of serpentinite southeast of San Jose, California: International Geology Review, v. 41, no. 6, p. 494–505.

Pearce, J.A., and Cann, J.R., 1973, Tectonic setting of basic volcanic rocks determined using trace element analyses: Earth and Planetary Science Letters, v. 19, p. 290–300.

Pearce, J.A., Alabaster, T., Shelton, A.W., and Searle, M.P., 1981, The Oman ophiolite as a Cretaceous arc-basin complex: Evidence and implications: Philosophical Transactions of the Royal Society London, v. A300, p. 299–317.

Pearce, J.A., Lippard, S.J., and Roberts, S., 1984, Characteristics and tectonic significance of supra-subduction zone ophiolites, *in* Kokelaar, B.P., and Howells, M.F., eds., Marginal basin geology: Geological Society of London Special Publication 16, p. 77–94.

Penrose Conference Participants, 1972, Penrose field conference on ophiolites: Geotimes, v. 17, no. 12, p. 24–25.

Pessagno, E.A., 1973, Age and geologic significance of radiolarian cherts in the California Coast Ranges: Geology, v. 1, p. 153–156.

Robertson, A.H.F., 1989, Paleoceanography and tectonic setting of the Jurassic Coast Range ophiolite, central California: Evidence from extrusive rocks and the volcaniclastic sediment cover: Marine and Petroleum Geology, v. 6, p. 194–219.

Robertson, A.H.F., 1990, Sedimentary and tectonic implications of ophiolite-derived clastics overlying the Jurassic Coast Range ophiolite, northern California: American Journal of Science, v. 290, p. 109–163.

Ross, D.C., 1979, Summary of petrographic data of basement rock samples from oil wells in the southeastern San Joaquin Valley, California: U.S. Geological Survey Open-File Report 79–400, map scale 1:125 000.

Saleeby, J.B., 1992, Petrotectonic and paleogeographic settings of U.S. Cordilleran ophiolites, *in* Burchfiel, B.C., et al., eds., The Cordilleran orogen: Conterminous U.S.: Boulder, Colorado: Geological Society of America, Geology of North America, v. G-3, p. 632–682.

Saleeby, J.R., 1996, 3. Coast Range ophiolite as parautochthonous fore-arc lithosphere: GSA Today, v. 6, no. 2, p. 7–9.

Shervais, J.W., 1990, Island arc and oceanic crust ophiolites: Contrasts in the petrology, geochemistry, and tectonic style of ophiolite assemblages in the California Coast Ranges, *in* Malpas, J., et al., eds., Ophiolites, oceanic crust analogues: Nicosia, Cyprus, Cyprus Geological Survey Department, p. 507–520.

Shervais, J.W., and Hanan, B., 1989, Jurassic volcanic glass from Stonyford volcanic complex, Franciscan assemblage, northern California Coast Ranges: Geology, v. 17, p. 510–514.

Shervais, J.W., and Kimbrough, D.L., 1985, Geochemical evidence for the tectonic setting of the Coast Range ophiolite: A composite island arc-oceanic crust terrane in western California: Geology, v. 13, p. 35–38.

Shervais, J.W., and Kimbrough, D.L., 1987, Alkaline and transitional subalkaline metabasalts in the Franciscan Complex mélange, California, *in* Morris, E.M., and Pasteris, J.D., eds., Mantle metasomatism and alkaline magmatism: Geological Society of America Special Paper 215, p. 165–182.

Stern, R.J., and Bloomer, S.H., 1992, Subduction zone infancy: Examples from the Eocene Izu-Bonin-Mariana and Jurassic California arcs: Geological Society of America Bulletin, v. 104, p. 1621–1636.

Thompson, G.A., and Robinson, R., 1975, Gravity and magnetic investigation of the Twin Sisters dunite, northern Washington, Geological Society of America Bulletin, v. 86, p. 1413–1422.

Trehu, A.M., Asudeh, I., Brocher, T.M., Luetgart, J.H., Mooney, W.D., Nabelek, J.L., and Nakamura, Y., 1994, Crustal architecture of the Cascadia Forearc: Science, v. 266, p. 237–243.

Wentworth, C.M., Blake, M.C., Jr., Jones, D.L., and Walter, A.W., 1984, Tectonic wedging associated with emplacement of the Franciscan assemblage, California Coast Ranges, *in* Blake, M.C., Jr., ed., Franciscan geology of Northern California: Los Angeles, Pacific Section, SEPM, v. 43, p. 163–173.

Wentworth, C.M., Fisher, G.R., Levine, P., and Jachens, R.C., 1995, The surface of crystalline basement, Great Valley and Sierra Nevada, California: A digital map database, U.S. Geological Survey Open-File Report 95-96, p. 1–16.

MANUSCRIPT ACCEPTED BY THE SOCIETY FEBRUARY 23, 2000

Printed in the U.S.A.

Geological Society of America
Special Paper 349
2000

Mesozoic assimilation of oceanic crust and island arc into the North American continental margin in California and Nevada: Insights from geophysical data

Nicola J. Godfrey
*Earth Science Department, University of Southern California, Los Angeles, California 90089, USA;
e-mail: nicola@jasper.usc.edu*
Yildirim Dilek
*Department of Geology, Miami University of Ohio, Oxford, Ohio 45056, USA;
e-mail: dileky@muohio.edu*

ABSTRACT

We present an updated, crustal-scale, geologic transect across northern California and Nevada that includes constraints from geologic data and geophysical models. We use this new model to explore the tectonic history of the region and conclude that prior to 175 Ma, oppositely dipping subduction zones were active along the North American continental margin at Sierra Nevada latitudes. An ensimatic arc terrane composed of the Jurassic Smartville Complex and its equivalents in the Sierra Nevada foothills developed on and across a disrupted ophiolitic basement in the upper plate of a west-dipping (in the present coordinate system) subduction zone. The east-dipping subduction zone was responsible for the formation of a Jurassic continental-margin arc that evolved on previously accreted terranes and a Permian–Triassic magmatic arc. As subduction continued at both convergent margins, the consumption of the intervening oceanic lithosphere resulted in the juxtaposition of the ensimatic arc terrane with the North American continental-margin arc at ca. 170 Ma. A short period of extension then followed during which the Coast Range ophiolite formed by backarc spreading. Almost immediately after the formation of the Coast Range ophiolite, tectonic underplating of the North American continental margin beneath the ensimatic arc terrane caused eastward transport of the ophiolites and the culmination of the Late Jurassic Nevadan orogeny. A new east-dipping subduction zone (Franciscan subduction) was then initiated west of the Coast Range ophiolite, following this collisional event. Franciscan subduction and Sierra Nevada magmatism continued along with the development of the Great Valley forearc basin until the Late Cretaceous. Intracontinental shearing within and/or along the eastern edge of the Sierra Nevada batholith (the Mojave–Snow Lake and Walker Lane shear zones) at ca. 140 Ma (Early Cretaceous) resulted in up to 400 km of dextral translation of a crustal block, causing a doubling of tectonic components along our east-west transect. In our interpretation, the Coast Range and (subsurface) Great Valley ophiolites, the Smartville arc complex and its equivalents in the Sierra Nevada foothills, and the late Paleozoic–early Mesozoic disrupted ophiolitic basement collectively form an intraoceanic terrane consisting of island arcs, intra-arc or backarc basins, and a multiply deformed oceanic substratum on which the island arc(s) were built. The Great Valley ophiolite is a major spatial and temporal link between the Coast Range ophiolite on the west and the Sierra Nevada foothills ophiolite belt on the east and is likely to represent the oceanic basement of this ensimatic Jurassic arc terrane. The accretion of the oceanic terrane to the North American continental margin in the Middle Jurassic marks a major orogenic event in the tectonic history of the North American Cordillera. We suggest that one of the most critical terrane boundaries in the western U.S. Cordillera occurs in the Sierra Nevada Foothills metamorphic belt, where the Jurassic intraoceanic terrane is juxtaposed with the Jurassic continental-margin arc and its basement.

INTRODUCTION

Much of western North America, particularly California and Nevada, formed as a result of volcanic-arc, oceanic-crust, and subduction-terrane accretion as well as by magmatic addition of granitic and tonalitic material during the Mesozoic, when the dominant process at the North American continental margin was subduction (Ernst, 1981). California and Nevada are, therefore, an ideal laboratory for investigating continental accretionary growth. The mode and timing of incorporation of the Jurassic ophiolites and island-arc complexes into

Godfrey, N.J., and Dilek, Y., 2000, Mesozoic assimilation of oceanic crust and island arc into the North American continental margin in California and Nevada: Insights from geophysical data, *in* Dilek, Y., Moores, E.M., Elthon, D., and Nicolas, A., eds., Ophiolites and Oceanic Crust: New Insights from Field Studies and the Ocean Drilling Program: Boulder, Colorado, Geological Society of America Special Paper 349, p. 365–382.

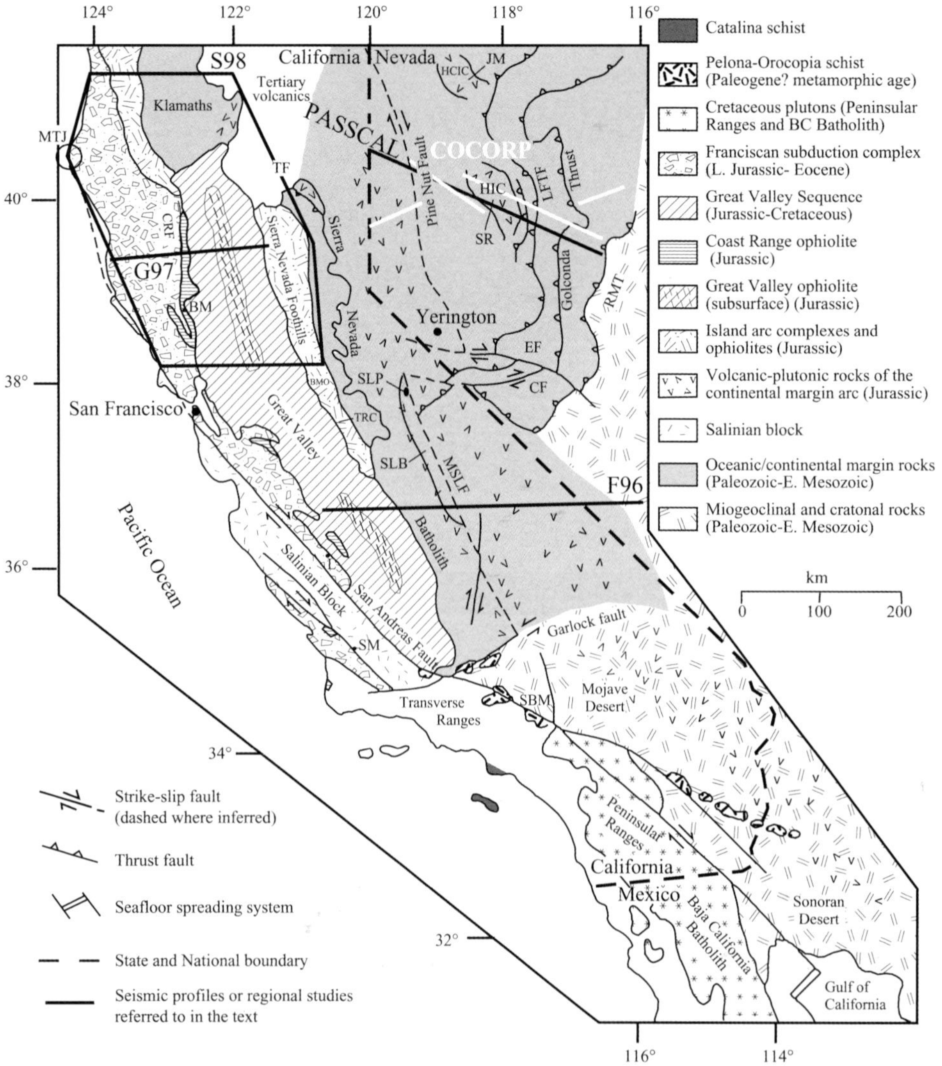

Figure 1. Simplified tectonic map of the western U.S. Cordillera (modified from Dilek and Moores, 1993). Key to labeling: BM—Black Mountain, BMO—Bear Mountains ophiolitic mélange, CA—California, CF—Coaldale fault, CRF—Coast Range fault, EF—Excelsior fault, HCIC—Happy Creek igneous complex, HIC—Humboldt igneous complex, JM—Jackson Mountains, L—Llanada, LFTF—Luning-Fencemaker thrust fault, MSLF—Mojave–Snow Lake fault, MTJ—Mendocino triple junction, NV—Nevada, RMT—Roberts Mountains thrust, SBM—San Bernardino Mountains, SLB—Snow Lake block, SLP—Snow Lake pendant, SM—Stanley Mountain, SR—Stillwater Range, TF—Taylorsville fault, TRC—Tuolumne River Complex. COCORP—line segments of Klemperer et al. (1986) and Hauge et al. (1987). F96—transect of Fliedner et al. (1996). G97—transect of Godfrey et al. (1997). PASSCAL—transect of Benz et al. (1990). S98—region of model by Stanley et al. (1998).

the North American continental margin have been controversial topics in Cordilleran geology; nevertheless, the resulting interpretations are widely applied to refine the tectonic models for the evolution of the western United States.

Beneath the Great Valley forearc basin, recent geophysical investigations have shown the existence of a slab of mafic and ultramafic material interpreted as oceanic crust and mantle that has been named the "Great Valley ophiolite" (Godfrey et al., 1997; Godfrey and Klemperer, 1998). The occurrence of this ophiolite and the nature of its contact with the bounding tectonic units have been the missing link in tectonic and paleogeographic models for the Mesozoic evolution of the western United States. The identification and presence of the Great Valley ophiolite plays a major role in creating and interpreting a coherent model for the tectonic evolution, including terrane accretion, of Mesozoic western North America. This paper brings together the geophysical and geologic evidence for the occurrence of the Great Valley ophiolite as well as its possible genetic and tectonic relationships with the Sierra Nevada Foothills metamorphic belt and the Coast Range ophiolite. We discuss the viability of current tectonic models based on these data and interpretations and present a geologic model and paleogeographic reconstruction for the Mesozoic tectonics of the western United States. We suggest that the Coast Range and Great Valley ophiolites, the Smartville arc complex and its equivalents in the Sierra Nevada foothills, and the multiply deformed ophiolitic basement beneath these arc fragments constitute different components of an intraoceanic arc terrane that included Jurassic volcanic archipelagoes (fringing island arcs), intra-arc or back-arc basins, and an older oceanic substratum. The accretion of this terrane into the North American continent in the Middle Jurassic marks a major orogenic event in Cordilleran geology.

REGIONAL GEOLOGY

During the Mesozoic and Paleogene (ca. 210–29 Ma), the Pacific and earlier oceanic plates converged with the North American continental plate. At ca. 29 Ma, the mid-ocean ridge producing the subducting oceanic crust began interacting with the subduction zone along the North American continental margin, creating two triple junctions (Atwater, 1970). These triple junctions migrated away from one another (McKenzie and Morgan, 1969), separated by the lengthening San Andreas transform system that replaced the subduction margin. East of the San Andreas fault system, outcrop geology reflects the Mesozoic and Paleogene subduction setting.

From west to east, at about lat 40°N, the following units are

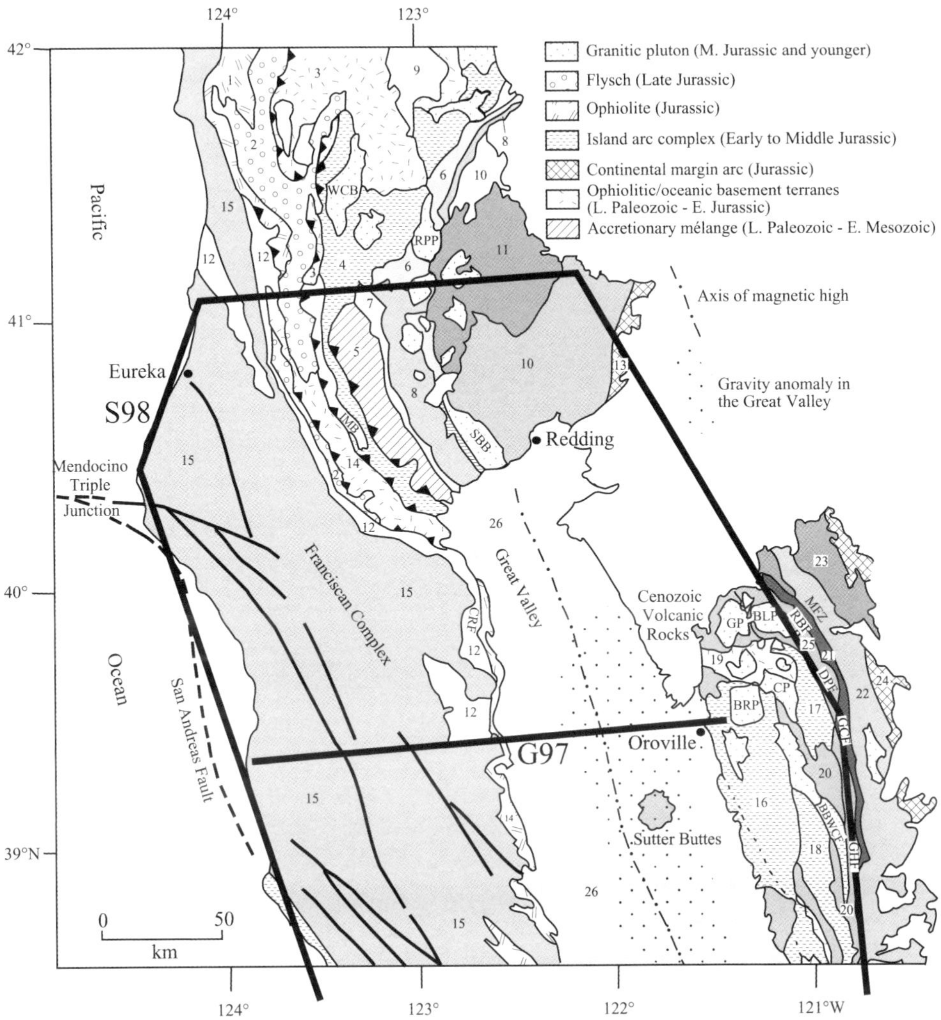

Figure 2. Simplified geologic map showing the distribution of ophiolites, island-arc complexes, and major tectonic units in the northern Sierra Nevada foothills and Klamath Mountains provinces of California (modified from Dilek and Moores, 1992). BLP—Bucks Lake pluton, BRP—Bald Rock pluton, CP—Cascade pluton, CRF—Coast Range fault, DPF—Dogwood Peak fault, GCF—Goodyear's Creek fault, GHF—Gillis Hill fault, GP—Grizzly pluton, IMB—Ironside Mountain batholith, MFZ—Melones fault zone, RBF—Rich Bar fault, RPP—Russian Peak pluton, SBB—Shasta Bally batholith, BBWCF—Big Bend-Wolf Creek fault, WCB—Wooley Creek batholith. Key to numbers: 1—Josephine ophiolite, 2—Galice Formation, 3—Rattlesnake Creek terrane, 4—Western Hayfork terrane, 5—Eastern Hayfork terrane, 6—Stuart Fork–Fort Jones terrane, 7—North Fork of the Salmon River terrane, 8—Central Metamorphic belt, 9—Condrey Mountain Schist, 10—Yreka-Callahan terrane (Paleozoic), 11—Trinity Complex, 12—South Fork Mountain Schist, 13—Eastern Klamath arc, 14—Coast Range ophiolite, 15—Franciscan Complex, 16—Smartville Complex, 17—Slate Creek Complex, 18—Lake Combie Complex, 19—Jarbo Gap ophiolite, 20—chert-argillite unit, 21—Feather River Peridotite, 22—Shoo Fly Complex, 23—Paleozoic volcanic arc complexes, 24—Jurassic continental-margin arc, 25—Calaveras Complex, 26—Great Valley Sequence. G97—transect of Godfrey et al. (1997). S98—region of model by Stanley et al. (1998).

exposed (Figs. 1 and 2): the late Mesozoic Franciscan Complex (accretionary prism), the Jurassic Coast Range ophiolite, the Jurassic to Cretaceous Great Valley forearc basin, a Jurassic island-arc complex and its early Paleozoic to early Mesozoic oceanic basement (the Sierra Nevada Foothills metamorphic belt), the Paleozoic to early Mesozoic Calaveras Complex (accretionary prism) (25 in Fig. 2), the Feather River Peridotite of the northern Sierra Nevada (21 in Fig. 2), the Sierra Nevada batholith, a Jurassic continental-margin arc (24 in Fig. 2), and the Paleozoic to lower Mesozoic passive-margin sequence of western North America (Dickinson and Rich, 1972; Dickinson, 1981; Ernst, 1981).

Franciscan Complex. The Franciscan Complex represents the Late Jurassic to Miocene accretionary prism of the subduction system that dipped eastward beneath the North American continent (McLaughlin et al., 1982; Blake et al., 1988). The accretionary prism displays in part a mélange character and consists of graywacke, shale, metasedimentary rocks, mafic volcanic rocks, chert, limestone, and ophiolite fragments (Bailey et al., 1964; Blake and Jones, 1974; Blake et al., 1988). The Franciscan Complex is subdivided into several terranes that become younger and exhibit a decrease in metamorphic grade from east (Late Jurassic to middle Cretaceous blueschist facies in the Eastern belt) to west (Late Cretaceous to Miocene prehnite-pumpellyite facies in the Coastal belt) (Irwin, 1960; Blake et al., 1985, 1988; McLaughlin et al., 1982, 1993). This change in metamorphic grade may correspond to the westward migration of the subduction zone, may record changing thermal conditions within the subduction zone, or may be the result of strike-slip faulting. East-west shortening across the Coast Ranges caused eastward wedging of Franciscan material beneath the Great Valley forearc basin (Wentworth et al., 1984; Wentworth and Zoback, 1990; Phipps and Unruh, 1992). This wedging, which might have begun in the Late Cretaceous as a result of the initiation of Franciscan subduction (Wentworth et al., 1984), is ongoing today because of compression across the San Andreas fault (Unruh and Moores, 1992) that resulted from a change in plate motions at ca. 5 Ma (Cox and Engebretson, 1985).

Coast Range ophiolite. The Coast Range ophiolite crops out discontinuously along the western margin of the Great Valley, north of lat 36°N. South of lat 36.5°N, there are also Coast Range ophiolite remnants that crop out west of the San Andreas fault (Fig. 1). Coast Range ophiolite outcrops are also found in southwest Oregon (Blake et al., 1985; Kosanke and Harper, 1997). The Coast Range fault is the present boundary between the Franciscan Complex and the Coast

Range ophiolite (or Great Valley Sequence where there is no Coast Range ophiolite) north of lat 36.5°N (Jayko et al., 1987; Harms et al., 1992) (Fig. 1). Coast Range ophiolite outcrops are generally highly sheared, dismembered, and highly serpentinized (Bailey et al., 1970; Hopson et al., 1981); a reconstructed ophiolite pseudostratigraphy consists, from the base upward, of variably serpentinized harzburgite, a plutonic member ranging from dunite cumulate through gabbroic cumulate to nonlayered gabbro, a basaltic sheeted-dike and -sill complex, and a volcanic member including pillowed and massive submarine basaltic lavas (Bailey et al., 1970; Hopson et al., 1981). Basal Great Valley Sequence sedimentary rocks, including the volcanic-pelagic sequences deposited between 157 and 147 Ma, were deposited conformably on the Coast Range ophiolite (Bailey et al., 1970; Evarts, 1977; Hopson et al., 1981), which formed during the Middle to Late Jurassic (ca. 172–158 Ma) (Hopson et al., 1981; Saleeby et al., 1984; Mattinson and Hopson, 1992) (Fig. 3).

Great Valley forearc basin. Cenozoic and Mesozoic sedimentary rocks of the Great Valley forearc basin crop out from the latitude of the Mendocino triple junction (the northern termination of the San Andreas fault) to the "Big Bend" in the San Andreas fault close to the Garlock fault (Jennings, 1977) (Fig. 1). The Great Valley Sequence consists of Upper Jurassic (deep-ocean and slope environment) to Paleogene (slope, shelf, and nonmarine environment) mudstones, sandstones, and conglomerates (Bailey et al., 1970; Ingersoll, 1982). These strata were deposited on oceanic crust that is now represented by the Coast Range ophiolite (Bailey et al., 1970; Evarts, 1977; Hopson et al., 1981), the subsurface Great Valley ophiolite (Godfrey et al., 1997; Godfrey and Klemperer, 1998), and, in the east north of 37°N, on the Smartville arc complex along an angular unconformity (Bailey and Blake, 1969; Harwood and Helley, 1987); in the east south of 37°N, the strata rest on the Sierra Nevada batholith itself.

Great Valley ophiolite. The Great Valley ophiolite is interpreted from geophysical data to underlie the Great Valley basin (Godfrey et al., 1997; Godfrey and Klemperer, 1998). The model on which this interpretation is based is discussed in the next section.

Jurassic island-arc terrane and its oceanic basement. Exposures of the ensimatic Jurassic island-arc terrane in the northern Sierra Nevada Foothills metamorphic belt include the Smartville, Slate Creek, and Lake Combie Complexes (units 16, 17, and 18 in Fig. 2) that are tectonically juxtaposed against ophiolitic and marine rocks of a late Paleozoic–early Mesozoic oceanic basement (the Jarbo Gap ophiolite and the overlying chert-argillite unit; units 19 and 20 in Fig. 2) (Day et al., 1985; Dilek et al., 1990).

The most extensive of the Jurassic island-arc fragments in the region, the Smartville arc complex, is composed of volcanic, plutonic, and hypabyssal rocks and sheeted-dike swarms (Moores and Day, 1984; Day et al., 1985; Dilek, 1989a) and is overlain unconformably to the west by Upper Cretaceous and younger rocks of the Great Valley Sequence. It is juxtaposed to the east and north against upper-greenschist–facies to amphibolite-facies ophiolitic and sedimentary rocks along the Big Bend–Wolf Creek fault zone (BBWCF in Fig. 2) (Dilek et al., 1990). The Smartville arc complex represents a composite island arc with a volcanic-volcaniclastic carapace and a plutonic and hypabyssal core. The hypabyssal core of the arc complex is Early Jurassic (205 ± 5 Ma; Bickford and Day, 1988) (Fig. 3). The sheeted dikes and tabular plutons intruding through the hypabyssal core and the lower volcanic unit have igneous ages of ca. 162 ± 5 Ma (Bickford and Day, 1988; Saleeby et al., 1989) (Fig. 3). Extrusion of mafic to felsic volcanic rocks in the upper volcanic unit may correspond to ca. 160 Ma dike intrusions that feed into different volcanic layers (Dilek, 1989c).

A late Paleozoic to early Mesozoic disrupted oceanic basement occurs north and east of the Smartville arc complex and includes dismembered ophiolite complexes (e.g., the Jarbo Gap ophiolite) and

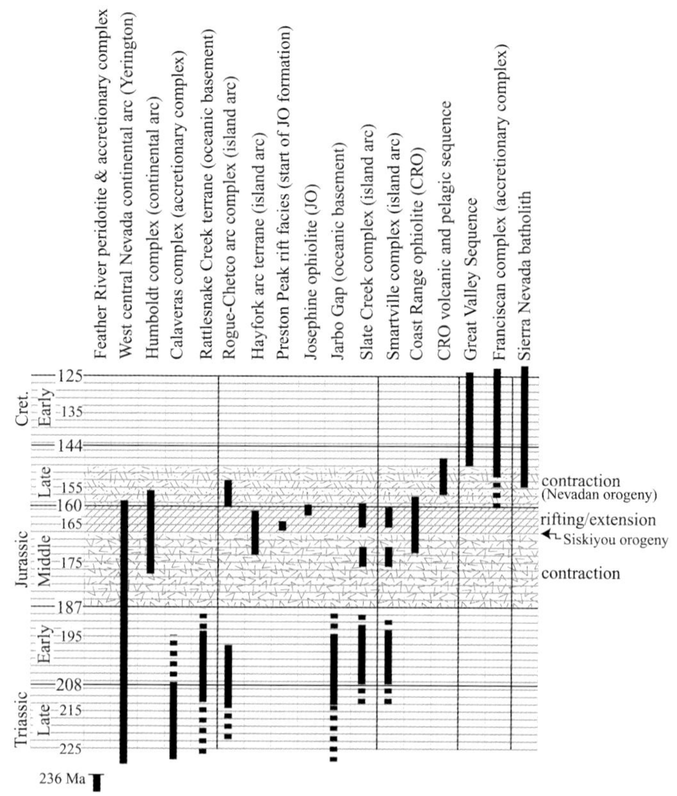

Figure 3. Timeline of major events discussed in the text. Time scale based on Harland et al. (1982) and Pessagno and Blome (1990). Solid line indicates well-determined ages, and dashed line indicates less-well-determined ages. Sources are as follows: Sierra Nevada magmatism—Schweickert and Cowan (1975); Franciscan Complex—Ingersoll and Schweickert (1986), Jayko and Blake (1986), Blake et al. (1988, 1989), McLaughlin et al. (1988), Moxon (1990); Coast Range ophiolite—Hopson et al. (1981), Phipps (1984), Ingersoll and Schweickert (1986), McLaughlin et al. (1988), Robertson (1989, 1990), Mattinson and Hopson (1992); Great Valley sequence—Ingersoll and Schweickert (1986), Jayko and Blake (1986); Rogue-Chetco arc complex—Yule (1996); Josephine ophiolite—Harper et al. (1994), Wyld and Wright (1988); Hayfork arc terrane—Harper and Wright (1984), Wright and Fahan (1988); Humboldt Complex—Willden and Speed (1974), Dilek and Moores (1991, 1995); Slate Creek Complex—Bickford and Day (1988), Saleeby et al. (1989), Dilek and Moores (1993); Smartville Complex—Harper and Wright (1984), Bickford and Day (1988), Wright and Fahan (1988), Saleeby et al. (1989), Dilek (1989b), Dilek and Moores (1992, 1993)—Jarbo Gap ophiolite, Saleeby et al. (1989), Dilek (1989b), Dilek et al. (1990), Dilek and Moores (1992); Rattlesnake Creek terrane—Wright and Fahan (1988), Wright and Wyld (1994); Calaveras Complex—Dilek and Moores (1993); Feather River Peridotite—Schweickert et al. (1980), Edelman et al. (1989), Dilek and Moores (1993); west-central Nevada volcanic arc—Schweickert et al. (1991); Nevadan orogeny—Schweickert et al. (1984), Ingersoll and Schweickert (1986), McLaughlin et al. (1988), Blake et al. (1989), Robertson (1989, 1990), Dilek and Moores (1992); Siskiyou orogeny—Coleman et al. (1988); Middle Jurassic compressional event—Wright and Fahan (1988), Dilek and Moores (1992).

a chert-argillite unit (units 19 and 20 in Fig. 2). Stratigraphic and structural relationships suggest that the ophiolitic units might have constituted a fragment of heterogeneous and multiply deformed oceanic crust onto which the chert-argillite unit was deposited during late Paleozoic to early Mesozoic time (Dilek et al., 1990). The chert-argillite unit comprises blocks of turbiditic sandstone, conglomerate, pebbly mudstone, fossiliferous limestone, graywacke, volcaniclastic rocks, massive volcanic rocks, gabbro, and serpentinite in a fine-grained argillite matrix. The existence of blocks of metagabbro, serpentinite, metavolcanic rocks, and deformed chert, enveloped by a sandstone-mudstone sequence in the chert-argillite unit, suggests a provenance of older, deformed oceanic crust exposed on the seafloor during the deposition of the chert-argillite unit in the Early Jurassic (Dilek et al., 1990).

On the basis of structural, geochemical, and geochronological data, we suggest that the disrupted oceanic basement in the northern Sierra Nevada foothills was accreted to the hanging wall of a subduction zone in Early Jurassic time and that it was subsequently intruded by subduction-zone magmas forming an ensimatic arc complex represented by the Smartville, Slate Creek, and Lake Combie Complexes. Similar intrusive relationships between the Early and Middle Jurassic arc plutons and an older ophiolitic basement, such as the Bear Mountains ophiolitic mélange in the Consumnes River area (BMO, Fig. 1) and the Tuolumne River ultramafic complex in the Tuolumne River area (TRC, Fig. 1), are documented farther south in the central and southern Sierra Nevada foothills (Sharp, 1988). Therefore, the ensimatic arc terrane must occupy much of the western Sierra Nevada Foothills metamorphic belt.

Calaveras Complex. The Calaveras Complex (unit 25 in Fig. 2) overlies the disrupted oceanic basement along the steeply east-dipping Dogwood Peak–Gillis Hill fault system (DPF and GHF in Fig. 2) and contains phyllite-chert, scattered exposures of volcanic rocks, and limestone blocks (Schweickert et al., 1988; Edelman et al., 1989). Late Paleozoic to Triassic fossils in various lithologic units within the Calaveras Complex suggest its evolution during late Paleozoic to early Mesozoic time, coeval with the disrupted oceanic basement to the west (Fig. 3). The Calaveras Complex displays a mélange character and is interpreted as an accretionary-prism assemblage (Sharp, 1988; Edelman et al., 1989).

Feather River Peridotite. The Feather River Peridotite (unit 21 in Fig. 2) occurs as a 150-km-long belt of metamorphosed mantle tectonite, ultramafic cumulate, mafic plutonic rocks, dikes, and pillow lavas between the Rich Bar fault on the west and the Downieville or "Melones" fault on the east in the northern Sierra Nevada foothills (RBF and MFZ in Fig. 2) (Hietanen, 1981; Edelman et al., 1989). Ages ranging from 387 to 236 Ma (Fig. 3) for dikes and some metamorphosed protoliths (Edelman et al., 1989; Saleeby et al., 1989) suggest a prolonged history of igneous activity and metamorphism within the Feather River Peridotite during much of middle to late Paleozoic time. Metasedimentary and metavolcanic rocks with blueschist-facies mineral assemblages within the Feather River Peridotite indicate that part of the peridotite massif represents a subduction complex metamorphosed at blueschist facies before Middle Jurassic time (pre–174 to 190 Ma) (Schweickert et al., 1980). Shear-sense indicators and minor folds within the Melones fault zone (MFZ in Fig. 2) indicate that both the Feather River Peridotite and the continental-margin rocks in the hanging wall of the Melones fault were situated in an oblique convergence zone (Gefell et al., 1989).

Sierra Nevada batholith. The Sierra Nevada batholith represents a magmatic arc that formed along the active western margin of North America between 155 Ma and 80 Ma (Fig. 3) (Evernden and Kistler, 1970; Kistler et al., 1971; Chen and Moore, 1982; Saleeby et al., 1987a, 1987b). The batholith consists of numerous plutons that are composed mainly of tonalitic and granodioritic to granitic rocks with rare monzonites and gabbros at upper- to middle-crustal levels. Recent studies of mafic-ultramafic xenoliths in Miocene volcanic rocks, which erupted through the central part of the Sierra Nevada batholith, as well as geophysical models, suggest that the granitoid rocks may be underlain by granulite- to eclogite-facies rocks forming a mafic-ultramafic keel beneath the batholith, a region transitional between crust and mantle (Fliedner, 1997; Ducea and Saleeby, 1998).

Jurassic continental-margin arc. A continental arc was active from the Permian through the Jurassic in present-day western Nevada. We refer to the Jurassic continental arc as the part of the arc formed during the Jurassic related to subduction that is also associated with the coeval ensimatic island arc described above. The continental-margin arc is represented by Early to Middle Jurassic plutonic and coeval metavolcanic rocks extending for hundreds of kilometers along the east side of the Sierra Nevada Foothills metamorphic belt. It includes formations at Mount Jura, Big Bend, Bowman Lake–Cisco Grove, and the North Fork of the American River areas and the Klamath Mountains as well as the Mineral King and Ritter Range pendants in the southern Sierra Nevada mountains (Figs. 1 and 2) (Burchfiel and Davis, 1981; Schweickert, 1978). The continental-margin arc is constructed over a composite and mainly continental basement consisting of previously accreted arc sequences and continental-margin rocks ranging in age from early Paleozoic through Permian–Triassic, including the Shoo Fly Complex, Grizzly and Sierra Buttes Formations, Taylor and Peale Formations, and Arlington Formation (Schweickert, 1978, 1981; Harwood, 1988, 1992). Metavolcanic rocks include basaltic to rhyolitic lavas, tuffs, and volcaniclastic rocks interbedded commonly with graywacke, conglomerate, siltstone, and limestone and locally with evaporite and eolian sandstone (Schweickert, 1978, and references therein). These sedimentary relationships together with fossil evidence suggest that the continental arc was mostly marine with locally temporal subaerial conditions (Schweickert, 1978; Busby-Spera, 1984; Tobisch et al., 1986; Fisher, 1990; Harwood, 1993; Saleeby and Busby, 1993).

The continental-margin arc in west-central Nevada contains a northwest-trending belt of plutonic, volcanic, volcanogenic, and carbonate rocks (Fig. 1) ranging in age from Middle Triassic(?) to Middle Jurassic (Speed, 1978; Oldow, 1984; Dilles and Wright, 1988). In the Yerington area (Fig. 1), Middle Triassic quartz porphyry and metadiorite and comagmatic andesitic to rhyolitic volcanic rocks, including the Northern Wassuk volcanic rocks, McConnell Canyon volcanic rocks, and Yerington batholith, are overlain by an Upper Triassic (Norian) to Lower Jurassic stratigraphic sequence consisting of volcanogenic rocks, siltstone, quartzose sandstone, and limestone, including the Gardnerville Formation and the Preachers Formation. This stratigraphic sequence is intruded and overlain, respectively, by monzonitic to granitic plutons (the Shamrock batholith) and andesitic to dacitic volcanic and volcaniclastic rocks (Artesia and Fulstone volcanic rocks) of Middle Jurassic age (Dilles and Wright, 1988).

Farther east, in the Stillwater Range (SR in Fig. 1) in west-central Nevada, the Jurassic Humboldt igneous complex (HIC in Fig. 1), which tectonically overlies the early Mesozoic shelf province, represents part of the continental-margin arc (Dilek and Moores, 1991). The Humboldt igneous complex consists of mafic to silicic plutonic rocks that are intruded by mafic to intermediate dikes and dike swarms and overlain by mafic to intermediate volcanic, volcanogenic, and pyroclastic rocks. Radiometric ages from various components of the complex range from ca. 179 to 157 Ma (Fig. 3) (Speed, 1978; Dilek and Moores, 1991). In the Jackson Mountains area (JM in Fig. 1), the continental-margin arc extends into the Late Triassic to Middle Jurassic Happy Creek igneous complex (HCIC in Fig. 1), which stratigraphically overlies a basinal assemblage of argillite, turbiditic calcarenite, and quartzose sandstone of Late Triassic age (Russell, 1984; Maher and Saleeby, 1986). In the Black Rock Desert area in northwestern Nevada, granitic to dioritic Jurassic plutons (Theodore, Big Creek, and Lone Tree plutons) that are intrusive into the Paleozoic to Triassic sedimentary and volcanic-volcaniclastic formations constitute northern components of the continental-margin arc (Wyld, 1996). Widespread Jurassic volcanic and volcaniclastic rocks and associated subvolcanic silicic plutons in the Mojave Desert and the Sonoran Desert (southeastern California, southern Arizona, and northernmost Sonora, Mexico) constitute the southward continuation of the continental-margin arc (Fig. 1).

These Jurassic arc sequences and structures associated with the continental-margin arc extending from the Sonoran Desert region to the northern Sierra Nevada foothills suggest that this greater than 1000-km-long arc might have occupied an extensional graben depression in Late Triassic to early Middle Jurassic time (Busby-Spera, 1988; Tosdal et al., 1989; Fisher, 1990; Busby-Spera et al., 1990). Volcanic and volcaniclastic rocks that make up the Early Jurassic continental-margin arc commonly form thick sequences and display evidence for syndepositional normal faulting, suggesting high subsidence rates synchronous with magmatic activity and deposition (Busby-Spera, 1988). The interplay of alkalic and calc-alkalic volcanism in the subsiding arc suggests that the arc was undergoing rifting during its evolution. The existence of mature quartz arenites that are associated with supracrustal rocks along the length of the continental-margin arc further indicates that detrital material derived from the craton had easy access to depocenters within the arc environment. Although some of its segments might have been locally and temporarily subaerial, the continental-margin arc apparently remained a submarine feature during much of its life span in Early to Middle Jurassic time, as indicated by the presence of interstratified marine rocks (e.g., limestone, siltstone, and mudstone). These lithologic relationships suggest that in general the upper plate of the convergent margin of western North America was characterized by a rifting and subsiding submarine arc, which bounded the craton and its miogeoclinal units to the east during the Jurassic.

Lower Mesozoic passive-margin sequence of western North America. Much of west-central Nevada is composed of an early Mesozoic marine province, which from east to west consists of Lower to Upper Triassic platform carbonates formed in a shelf environment and Triassic to Lower Jurassic basinal siliciclastic and carbonate rocks locally overlain by shallow-marine and subaerial sedimentary rocks (Fig. 1) (Speed, 1978; Oldow, 1984; Speed et al., 1988; Elison et al., 1990). The contacts between the shelf sequence, the basinal succession, and the continental-margin arc to the west are commonly tectonic, although primary stratigraphic relationships between the latter two are locally preserved, and rocks within all terranes are multiply folded and faulted (Speed, 1978; Oldow, 1984; Speed et al., 1988). The basinal succession and the continental-margin arc together constitute the Luning-Fencemaker allochthon of Oldow (1984) that rests on the Triassic shelf strata along the generally east-vergent Luning-Fencemaker thrust (LFTF in Fig. 1).

STRIKE-SLIP FAULTING AND TERRANE TRANSLATION

Recent studies in western Nevada and along the axis of the Sierra Nevada batholith suggest the existence of a tectonic discontinuity within and/or immediately east of the Sierra Nevada batholith (Stewart, 1988; Lahren and Schweickert, 1989; Schweickert and Lahren, 1990; Saleeby, 1991; Schweickert et al., 1991; Saleeby and Busby, 1993). Correlations of lithologic units and structures within the Snow Lake pendant (SLP in Fig. 1), a block in the central Sierra Nevada, with those in the Mojave Desert and the San Bernardino Mountains suggest nearly 400 km of dextral displacement and northward translation of the Snow Lake block from the western Mojave Desert along the Mojave–Snow Lake fault (MSLF in Fig. 1) during the Early Cretaceous (between 150 Ma and 100 Ma) (Lahren and Schweickert, 1989; Schweickert and Lahren, 1990). These correlations and the magnitude of the inferred dextral displacement along the Mojave–Snow Lake fault, however, have recently been disputed on the basis of batholithic petrochemical patterns and stratigraphic relationships discerned in the relatively well preserved lower Mesozoic rocks in the southern Sierra Nevada (Saleeby, 1991; Saleeby and Busby, 1993). Saleeby (1991) and Saleeby and Busby (1993) suggested lateral distension of the crust, with as much as 250 km displacement by dextral shearing and faulting as a result of the en echelon emplacement of large elongate plutons in the Cretaceous Sierra Nevada batholith. Regardless of its origin, the structural and stratigraphic evidence for strike-slip faulting along and on the eastern side of the Sierra Nevada batholith during the Early Cretaceous indicates that a major crustal block, containing the Sierra Nevada batholith and its country rocks both in the Sierra Nevada foothills and in western Nevada, was translated northwestward for as much as 400 km during that time. Tectonic models for the late Mesozoic paleogeography of the western United States should, therefore, consider the implications of this terrane translation in palinspastic reconstructions.

CORRELATION WITH THE WESTERN KLAMATH MOUNTAINS

The Klamath Mountains region includes Paleozoic and Mesozoic ophiolites and island-arc complexes in an imbricate stack of tectonic units bounded by generally east-dipping regional faults (Figs. 1 and 2). These tectonic units in general become progressively younger westward, with the Franciscan Complex occupying the lowest structural position. The island-arc and ophiolite complexes in the western Klamath Mountains show strong similarities to their counterparts in the Sierra Nevada foothills in terms of their lithostratigraphy and chronostratigraphy.

The Rattlesnake Creek terrane (unit 3 in Fig. 2; Fig. 4) consists of a serpentinite matrix mélange (ophiolitic basement) unconformably overlain by volcanic, hemipelagic, and clastic sedimentary rocks of Late Triassic to Early Jurassic age (Wright and Wyld, 1994). It correlates well with the dismembered ophiolites (e.g., Jarbo Gap ophiolite) and the chert-argillite cover in the northern Sierra Nevada foothills. The Middle Jurassic western Hayfork terrane (units 4 and 5 in Fig. 2; Fig. 4) includes mafic to felsic volcanic and volcaniclastic rocks and coeval intrusive rocks and is interpreted to have developed on and across an ophiolitic basement represented by the Rattlesnake Creek terrane (Wright and Fahan, 1988; Dilek and Moores, 1992). These lithologic relationships are similar to those of the Middle Jurassic island-arc terrane (Smartville arc complex and its equivalents) in the northern Sierra Nevada foothills. The Rogue-Chetco arc complexes are probably part of the same ensimatic Jurassic arc terrane in the western Klamath Mountains (Harper, 1994). The Josephine ophiolite (164–162 Ma; unit 1 in Fig. 2), which displays a complete ophiolite pseudostratigraphy, is interpreted to have developed as a result of intra-arc rifting of the western Hayfork and Rogue-Chetco

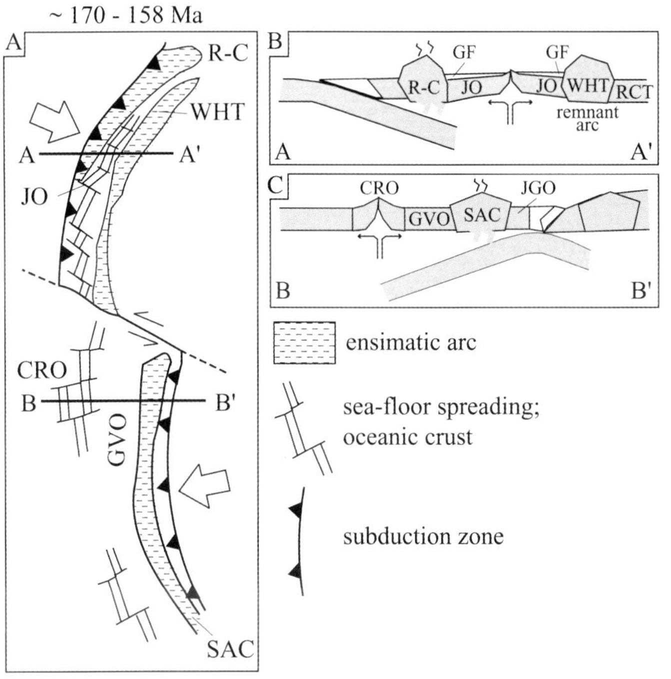

Figure 4. A, Map view of the Klamath and Sierra Nevada foothills region showing the oppositely dipping subduction zones connected by a strike-slip fault. B, Schematic section along A–A′ (Fig. 4A) showing the relationships among the Hayfork arc terrane, the Josephine ophiolite, the Rogue-Chetco arc complex and the Galice Formation in the Klamath Mountains at 170–158 Ma (after Harper and Wright, 1984). C, Schematic section along B–B′ (Fig. 4A) showing the relationships among the Coast Range, Great Valley, Jarbo Gap, and Smartville ophiolites at the latitude of the Sierra Nevada foothills at 170–158 Ma. CRO—Coast Range ophiolite, GF—Galice Formation, GVO—Great Valley ophiolite, JGO—Jarbo Gap ophiolite, WHT—West Hayfork arc terrane, JO—Josephine ophiolite, R-C—Rogue-Chetco arc complex, RCT—Rattlesnake Creek terrane, SAC—Smartville arc complex.

arc terrane (Harper, 1994; Harper et al., 1994). The Josephine ophiolite thus correlates with the late-stage rift component (sheeted-dike complex and zoned plutons dated at 162–157 Ma) of the Smartville arc complex in the Sierra Nevada foothills (Dilek, 1989c). The Galice Formation (unit 2 in Fig. 2), which was deposited on the Josephine ophiolite, is interpreted as a flysch sequence comparable to the Great Valley Sequence (Harper and Wright, 1984; Harper, 1994).

GEOPHYSICS OF THE GREAT VALLEY OPHIOLITE

Large magnetic and gravity anomalies (Figs. 2 and 5) run along the entire length of the Great Valley forearc basin (Cady, 1975). These anomalies were the first evidence presented for the existence of mafic and/or partially serpentinized ultramafic material beneath the Great Valley. Since then, many other analyses of geophysical data have concluded that the 700-km-long, 100-km-wide Great Valley basin is directly underlain by mafic and ultramafic (serpentinized and unserpentinized) material (Bailey et al., 1970; Griscom and Jachens, 1990; Jachens et al., 1995; Fliedner et al., 1996; Godfrey et al., 1997; Godfrey and Klemperer, 1998). Godfrey et al. (1997) named the interpreted mafic, serpentinized ultramafic, and unserpentinized ultramafic material beneath the Great Valley forearc basin the *Great Valley ophiolite.* No part of the Great Valley ophiolite crops out at the surface; therefore, it is impossible to verify that any of the specific geologic units included in the original Penrose definition (Penrose Conference Participants, 1972) of an ophiolite are present beneath the Great Valley. We follow Godfrey et al. (1997) and use the term *ophiolite* to denote an oceanic crustal and mantle sequence that is now part of the continent. Velocity and density models derived from seismic-refraction and gravity data are consistent with the occurrence of a complete ophiolite section beneath the northern Great Valley (Godfrey et al., 1997). These models also require material of lower velocity and density than unserpentinized peridotite structurally beneath the ophiolite (Fig. 5) (Fliedner et al., 1996; Godfrey et al., 1997; Godfrey and Klemperer, 1998; Stanley et al., 1998).

Two-dimensional velocity, density, and magnetic models derived from seismic-refraction data (collected as part of the Mendocino Triple Junction Seismic Experiment (Tréhu, 1995; Beaudoin et al., 1996), gravity and magnetic data along the same profile in northern California (Godfrey et al., 1997), and a three-dimensional velocity model derived from tomographic inversion of earthquake data (Stanley et al., 1998) provide the most complete analysis of the Great Valley subsurface to date. The most striking result of these models is a high-velocity region immediately beneath the Great Valley Sequence that extends from 4 to ~17 km depth beneath the central Great Valley and from 7 to ~19 km depth beneath the western Great Valley (Fig. 5, A and B). This high-velocity region has velocities of 5.5–6.5 km/s at the top and 8.1 km/s at the base. These velocities are interpreted as the mafic component (crustal) of an ophiolite sequence by comparison with velocities measured in the laboratory from samples of Coast Range ophiolite (Christensen, 1978; Nichols et al., 1980), from samples of other ophiolites around the world (Christensen, 1978), and from samples of in situ oceanic crust from ODP (Ocean Drilling Program) Hole 735B (Iturrino et al., 1991), as well as by comparison with velocity measurements of in situ oceanic crust from marine seismic-refraction experiments (White et al., 1992). The high 8.1 km/s velocities in this layer can only reasonably represent unserpentinized mantle rocks at unusually shallow crustal levels (15 km depth). Completely serpentinized peridotites have velocities of ~5.2 km/s (Nichols et al., 1980), whereas 20% serpentinization produces velocities of ~7.5 km/s and 5% serpentinization produces velocities of ~8.0 km/s (Clague and Straley, 1977). The 5.5–6.5 km/s material directly beneath the Great Valley basin could be interpreted as mafic volcanic arc rocks or highly serpentinized peridotites or a mixture of these. No significant S-wave data were generated in this seismic experiment, so Poisson's ratios cannot be calculated to aid differentiation of lithologies.

The three-dimensional velocity model (Stanley et al., 1998) requires a velocity inversion beneath the high-velocity region (the interpreted ophiolite sequence) down to the Moho at ~35 km depth. The two-dimensional velocity model by Godfrey et al. (1997) only defines the upper crust down to ~15 km depth (Fig. 5, A and B). However, a two-dimensional density model defined in the upper crust by the Godfrey et al. (1997) velocity model extends throughout the crust and requires a 16–19 km thickness of relatively low-density material between the base of the Great Valley ophiolite mantle (high velocity and high density) and the 34-km-deep Moho, as determined by regional earthquake traveltimes (Oppenheimer and Eaton, 1984) (Fig. 5D). This low-density (2.85–3.0 g/cm^3) region corresponds to the low-velocity region in the Stanley et al. (1998) model. The low-velocity and low-density material beneath the Great Valley ophiolite may be the older continental margin, now represented at the surface by the Sierra Nevada Foothills metamorphic belt, onto which the Great Valley ophiolite may have been thrust (Fig. 5G) (see later dis-

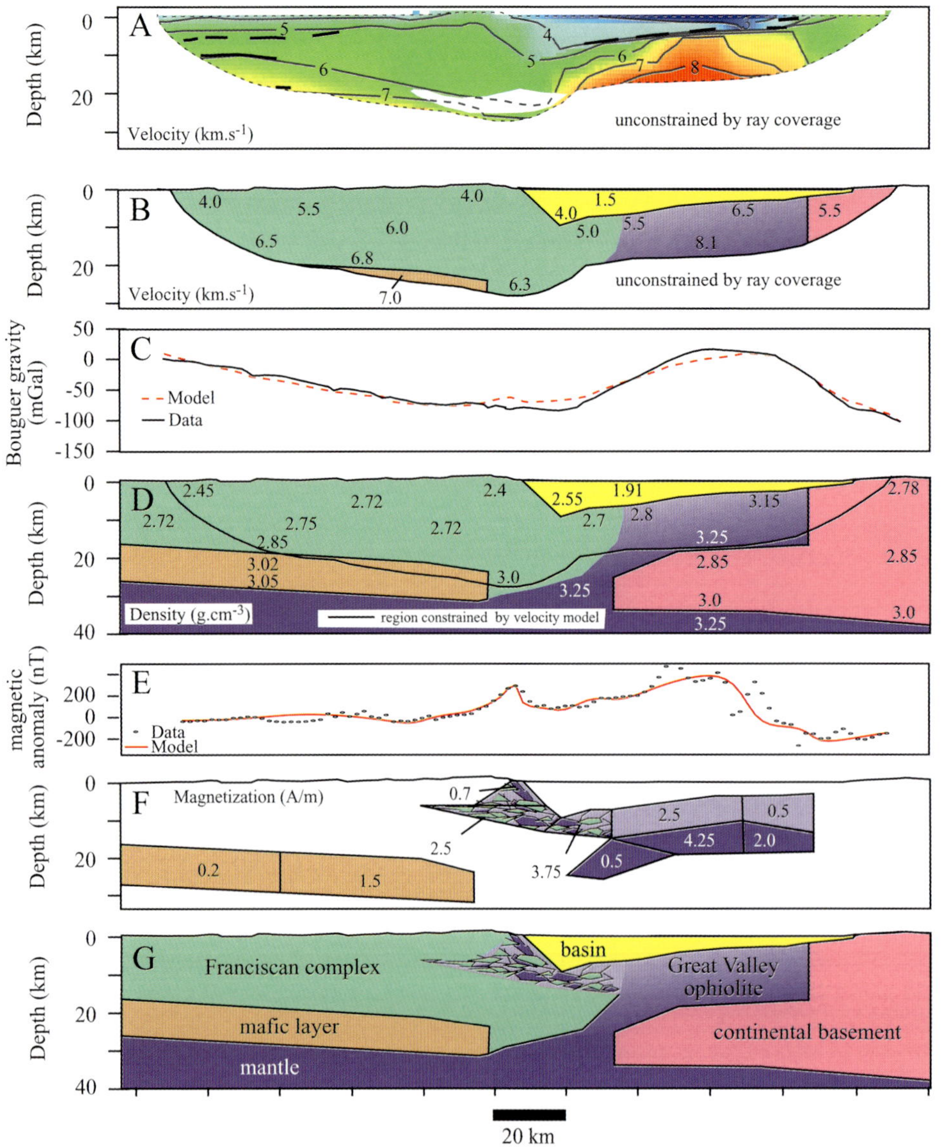

Figure 5. A, Velocity model along transect G97 (see Fig. 1). Velocities are contoured with contour values given in km/s. White regions are regions with no ray coverage and that are therefore not defined with regard to velocity. B, Block diagram of velocity model shown in A. C, Gravity data along transect G97 (solid line) and gravity calculated for the density model shown in D (dashed line). D, Density model along transect G97. The region above the solid line is fixed from the velocity model in A and B by an appropriate nonlinear velocity-density conversion (Godfrey et al., 1997). E, Magnetic data along transect G97 (circles) and magnetic intensity calculated for the magnetic model shown in F (solid line). F, Magnetic model along transect G97. White regions have zero magnetization. G, Interpreted model based on A through F. All models are shown at true scale.

cussion on tectonic scenarios). Alternatively, the low-velocity and low-density material may be old continental-margin material that was moved into its present position by strike-slip faulting, or it may represent magmatically underplated mafic plutons. The Godfrey et al. (1997) velocity and density model (Fig. 5) shows a near-vertical eastern boundary to the high-velocity and high-density Great Valley ophiolite; the model cannot, however, distinguish between a sharp or transitional eastern boundary. This vertical boundary could represent a pre–Great Valley Sequence (Late Cretaceous in this area) transform fault, in which case, the continental rocks beneath the Great Valley ophiolite could be from any age and place and do not need to be related to the continental crust east of the Great Valley.

There is no velocity or density expression of the Coast Range ophiolite beneath the western margin of the Great Valley in the velocity or density models (Fig. 5, A and D). Several alternative scenarios may explain this phenomenon: (1) The Coast Range ophiolite beneath the western margin of the Great Valley may consist of highly serpentinized ultramafic and mafic material whose velocity is indistinguishable from Franciscan material or Great Valley Sequence rocks, (2) the extent of the ophiolite may not be large enough to be detected by these data, or (3) the Coast Range ophiolite may not continue at depth beneath the western margin of the Great Valley.

The velocity and density models are consistent with magnetic data along the transect (Fig. 5, E and F) (Godfrey et al., 1997). The long-wavelength features of the magnetic signature are well matched by using "blocks" of material with constant magnetizations that have the same geometry as determined from the velocity and density models. A better fit to the shorter wavelengths could be obtained by using varying degrees of magnetizations within each block. The sources of magnetic signature in the magnetic model are mafic crust and serpentinized mantle of the Great Valley ophiolite and the highly serpentinized Coast Range ophiolite at the surface. The magnetic data require continuous magnetic material beneath the western margin of the Great Valley, and we interpret this material as a physical link between the Great Valley and Coast Range ophiolites. This interpretation does not require, however, that the Great Valley and Coast Range ophiolites formed adjacent to one another, only that they are currently physically connected.

A three-dimensional velocity model derived from seismic-refraction data from the southern Great Valley and Sierra Nevada

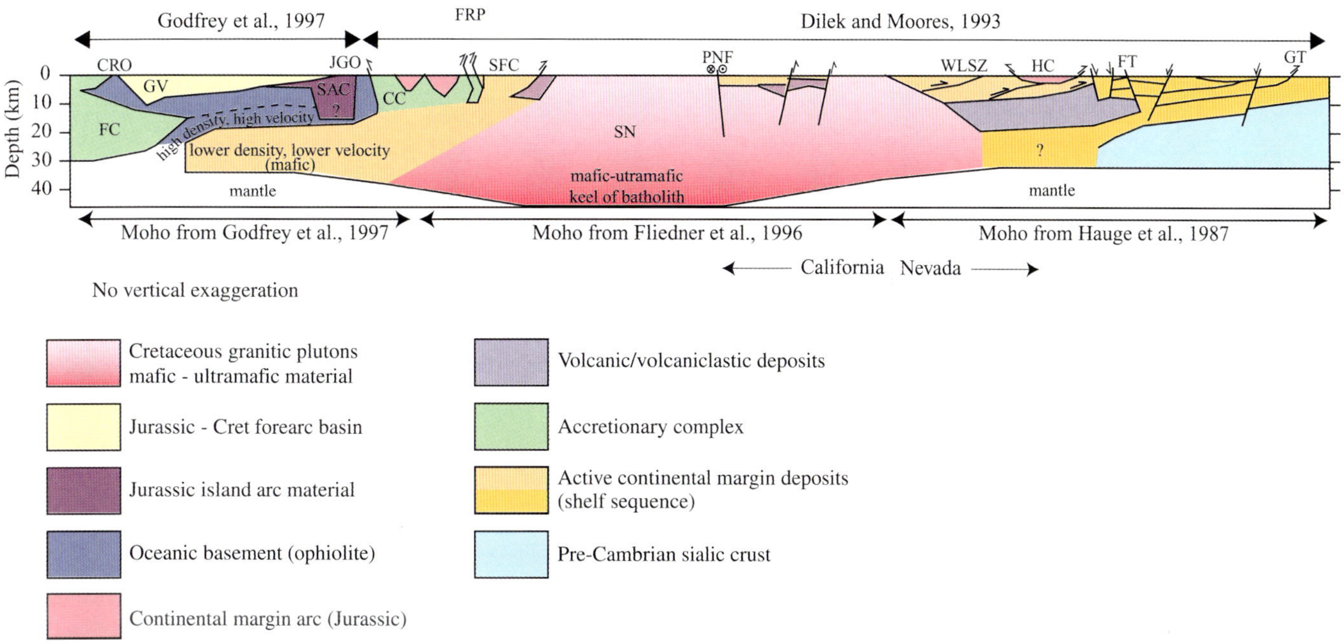

Figure 6. Geologic model along northern California–Nevada transect at lat 40°N. Abbreviations are as follows: CC—Calaveras Complex, CRO—Coast Range ophiolite, FC—Franciscan Complex, FRP—Feather River Peridotite, FT—Fencemaker thrust, GT—Golconda thrust, GV—Great Valley forearc basin, HC—Humboldt Complex, JGO—Jarbo Gap ophiolite, PNF—Pine Nut fault, SAC—Smartville arc complex, SFC—Shoo Fly Complex, SN—Sierra Nevada batholith, WLSZ—Walker Lane shear zone.

(Fliedner et al., 1996) strongly suggests the existence of a mafic and/or partially serpentinized ultramafic slab with a maximum velocity of 7.2 km/s overlying lower-velocity material. This geometry is consistent with oceanic crust and partially serpentinized mantle (herein called the Great Valley ophiolite) overlying older continental material, which extends down to the Moho at ~27 km depth. The unserpentinized Great Valley ophiolite mantle section that is present beneath the whole of the northern Great Valley is missing beneath the southern Great Valley. We suggest that the mantle peridotites here have been partially serpentinized (Godfrey and Klemperer, 1998). It is not possible to determine, however, how much of the material beneath the southern Great Valley is mafic rock and how much is serpentinized ultramafic rock.

In summary, the Great Valley forearc basin is most likely underlain by a gently west-dipping ophiolite that consists of 7–8-km-thick oceanic crust and 5–7-km-thick mantle. Much of the 7–8-km-thick mafic section could be volcanic arc rocks as observed locally in the Coast Range ophiolite (e.g., Llanada). Beneath the westernmost Great Valley, the fossil "ophiolite mantle" dips steeply west to become continuous with the present-day continental mantle. Beneath the Great Valley ophiolite there is 16–18-km-thick, lower-velocity and lower-density material that may represent older continental-margin material onto which the ophiolite was thrust.

Most of the Great Valley ophiolite is buried too deep beneath the sedimentary deposits of the Great Valley forearc basin to test more than the uppermost parts of the body by drilling. It may be possible to drill the easternmost parts of the Great Valley ophiolite where the base of the Great Valley basin is only 1–2 km deep. Seismic-reflection data across the eastern margin of the northern Great Valley may also provide more information about the relationship of the Great Valley ophiolite with the Sierra Nevada Foothills metamorphic belt and the proposed emplacement thrust surface.

NEW GEOLOGIC MODEL FOR A NORTHERN CALIFORNIA–NEVADA TRANSECT AT 40°N

We have created an updated geologic model along a transect at 40°N across northern California and Nevada that includes interpretations of subsurface geology from geophysical models (Fig. 6). Interpretation of the surface geology is taken from Dilek and Moores (1993, 1995). We include the Great Valley ophiolite and Moho beneath the Great Valley forearc basin based on the geophysical data and models already described herein (Godfrey et al., 1997; Godfrey and Klemperer, 1998; Stanley et al., 1998) and the depth to the Moho beneath the Sierra Nevada and Basin and Range from work by Klemperer et al. (1986), Hauge et al. (1987), Benz et al. (1990), and Fliedner et al. (1996).

Fliedner et al. (1996) presented a model derived from a seismic-refraction and wide-angle-reflection data set collected ~350 km south of our transect (F96 in Fig. 1). Their model is a three-dimensional velocity model with the depth to the Moho defined by wide-angle Moho reflections. An east-west section through their model extends from the Great Valley forearc basin, across the Sierra Nevada batholith, and east into the Basin and Range province. Their model shows the Moho at 30–34 km depth beneath the southern Great Valley and southwestern Sierra Nevada, reaching a maximum of ~43 km depth 40 km west of the highest topography. The Moho shallows again to 30–34 km depth beneath the southeastern Sierra Nevada and is less than 30 km deep beneath the Basin and Range province. We have extrapolated this Moho depth north to include it within our northern California–Nevada transect (Fig. 6). We point out that extrapolating the depth to the Moho 350 km north to the latitude of our transect invokes assumptions that may not be valid. The Bouguer gravity anomaly associated with the Sierra Nevada batholith becomes more pronounced from the location of the Fliedner et al. (1996) transect to the north, which could imply that either the crustal thickness or the degree of asthenospheric support is changing. Data on the depth to the Moho beneath the Basin and Range province in Nevada are based on seismic-reflection data from the COCORP 40°N transect (Fig. 1) (Klemperer et al., 1986; Hauge et al., 1987) and a refraction and wide-angle-reflection data set in the same region of Nevada (the PASSCAL experiment) (Fig. 1) (Benz et al., 1990). The seismic-reflection data set images a relatively flat reflective region in the lower crust, the base of which is interpreted as the reflection Moho (Klemperer et al., 1986; Hauge et al., 1987). The reflection Moho coincides with the top of a 1–5-km-thick layer at ~30 km depth that has velocities of 7.4–7.6 km/s, transitional between crust and mantle (Benz et al., 1990).

PALEOGEOGRAPHIC RECONSTRUCTIONS, OPHIOLITE TECTONICS, AND GEODYNAMIC EVOLUTION OF MESOZOIC WESTERN NORTH AMERICA

Paleogeographic reconstructions

The existence of the dextral shear zone along the eastern side of the Sierra Nevada requires palinspastic reconstruction and retrodeformation of the region to aid our understanding of the early Mesozoic paleogeography of the Cordilleran system. The removal of nearly 400 km of dextral translation along the inferred Mojave–Snow Lake fault and retrodeformation of other Mesozoic tectonic features provide a new tectonic configuration of the western U.S. Cordillera at ca. 170 Ma (Middle Jurassic) (Fig. 7). In our restored configuration, the marine province of western Nevada aligns with and becomes the northward continuation of the rifting and subsiding Jurassic continental-margin arc (Fig. 7). In the present-day configuration, the continental-margin arc seems to have been doubled near lat 40°N because of dextral faulting and shearing within and along the Sierra Nevada batholith. An important aspect of the reconstructed Middle Jurassic (170 Ma) paleogeography is the relative tectonic setting of the ensimatic or island-arc terrane in the west with respect to the continental-margin arc (Fig. 7). The ensimatic arc terrane and its forearc environment were apparently undergoing contractional deformation and thus internal telescoping and shortening in Middle and Late Jurassic time, as reported from the northern Sierra Nevada foothills (Bickford and Day, 1988; Dilek, 1989b, 1989c; Dilek et al., 1990) and the Klamath Mountains (Wright and Fahan, 1988), whereas the continental-margin arc was evolving in an extensional environment in the east (Tobisch et al., 1986; Busby-Spera, 1988; Tosdal et al., 1989; Schermer, 1993).

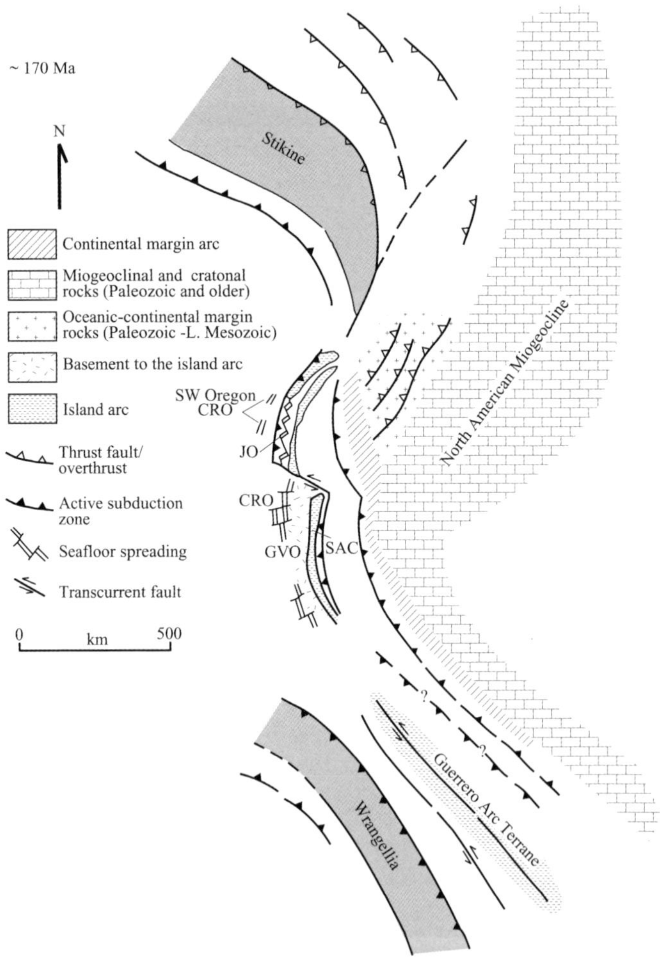

Figure 7. Inferred paleogeography of the western U.S. Cordillera for ca. 170 Ma (Middle Jurassic). See text for discussion. CRO—Coast Range ophiolite, GVO—Great Valley ophiolite, JO—Josephine ophiolite, SAC—Smartville arc complex.

Ophiolite tectonics and geodynamic evolution

The contact between the Great Valley ophiolite and the Sierra Nevada Foothills metamorphic belt underlies the Great Valley forearc-basin strata, although the nature of the contact is enigmatic. In the northern Great Valley basin (also known as the Sacramento basin), the contact underlying the Great Valley Sequence is associated with east-dipping faults (on the basis of borehole data and surface geology) (Godfrey and Klemperer, 1998) that offset the Great Valley Sequence and therefore postdate (at least in part) the Great Valley ophiolite and Sierran basement contact. The basement beneath the eastern margin of the Sacramento basin has not, to our knowledge, been imaged with seismic-reflection data. West-dipping reflectors imaged in seismic data from the southern Great Valley (San Joaquin) basin are the strongest evidence for shear zones that might be associated with ophiolite emplacement (Miller and Mooney, 1995).

There are very few constraints on the age of the high-velocity basement or underlying continental rocks. The high-velocity basement must be older than Tithonian (Late Jurassic) on the basis of the age of the basal Great Valley Sequence. There are no constraints on the maximum age of the high-velocity basement. One interpretation is that the basement comes to the surface in the Klamath Mountains, in which case the Trinity ophiolite may correspond to the ultramafic slab (G. Harper, 1999, personal communication). Continental crust ~20 km thick underlies the Trinity ophiolite, which could then correspond to the continental material beneath the high-velocity Great Valley basement. The Trinity ophiolite, however, is ca. 430 Ma, which is substantially older than the ca. 165 Ma Coast Range ophiolite. The age discrepancy makes this correlation implausible.

If, however, the Great Valley and Coast Range ophiolites are spatially related (Godfrey et al., 1997), we can use information about the origin of the Coast Range ophiolite, for which geologic, petrologic, and geochemical data are available, in conjunction with scarce petrologic and lithologic information from boreholes within the Great Valley, to address the origin of the Great Valley ophiolite.

The three main schools of thought on the origin of the Coast

Range ophiolite and Great Valley ophiolite are summarized by Dickinson et al. (1996).

Model I, after Dickinson et al. (Dickinson et al., 1996), suggests that the Coast Range ophiolite formed by seafloor spreading in a backarc setting behind an east-facing intraoceanic island arc (west-dipping subduction) during the Middle and Late Jurassic (172–158 Ma) (Fig. 8C). This island arc collided with the west-facing continental-margin arc after the intervening oceanic lithosphere was consumed during a Middle Jurassic collisional event (Fig. 8D1). Geophysical models suggest that the Great Valley ophiolite overlies mafic material that may represent older continental basement (Fig. 5; Godfrey et al., 1997; Godfrey and Klemperer, 1998). For this case we modify model I by requiring that the Late Jurassic Nevadan orogeny not only dock the ophiolite against the older continental margin but also emplace the ophiolite onto the older continental margin (Fig. 8D1). In this case, the velocity and density inversion that occurs at ~20 km depth in the geophysical models (Fig. 5) represents a major thrust surface. It should be noted that this scenario does not apply to the Josephine ophiolite, for which the Nevadan orogeny was a contractional but not a collisional event (Fig. 8D2). A new subduction zone (Franciscan subduction) was initiated to the west (Fig. 8, D1 and D2), which left the Coast Range and Great Valley ophiolites in a forearc setting. Subduction continued at the new trench, forming the Franciscan Complex (accretionary prism), with the subducting slab dipping east beneath the ophiolitic thrust sheets and the continent in the upper plate.

Model II, after Hopson et al. (Dickinson et al., 1996) proposes that the Coast Range ophiolite formed at a paleoequatorial mid-ocean ridge and was transported north to the latitude of the North American continental margin along strike-slip faults. The ophiolite was subsequently incorporated into the continental margin during the Nevadan orogeny. Again, in order to obtain the geometry seen in the geophysical models from the Sacramento Valley, we need to modify this model by emplacing the Coast Range and Great Valley ophiolite duo onto the older continental margin during the Nevadan orogeny. As in Model I, Franciscan subduction was initiated when the subduction zone jumped west and left the Coast Range and Great Valley ophiolites in a forearc setting. Hagstrum and Murchey (1996) and Hagstrum and Jones (1998) revisited the paleomagnetic data from the Coast Range ophiolite and concluded that previous paleolatitude determinations were based on a remagnetized component. Analysis of the primary magnetization, which had previously been overlooked, suggests that the Coast Range ophiolite formed at North American paleolatitudes (~35°N), rather than at paleoequatorial latitudes. A revision to model II therefore no longer requires a large northward translation component. Geochemical analyses suggest that the southern Coast Range ophiolite and parts of the northern Coast Range ophiolite are island-arc tholeiitic to calc-alkalic rocks comparable with an origin in a supra-subduction zone setting (Shervais and Kimbrough, 1985), whereas samples from the Stanley Mountain locality (a fossil seamount) of the Coast Range ophiolite compare well with enriched mid-oceanic-ridge basalts (MORBs) or ocean-island tholeiites (Shervais and Kimbrough, 1985) and samples from the Black Mountain locality with MORB affinity (Giaramita et al., 1998). Geochemical and petrologic data generally appear to support a supra-subduction zone environment rather than a mid-ocean ridge setting, although a complex melt and crustal evolution at a mid-ocean ridge setting for some segments of the Coast Range ophiolite (e.g., the Point Sal ophiolite) cannot be completely ruled out (C.A. Hopson, 1998, Penrose Conference presentation). In the light of the current data, we prefer a supra-subduction zone origin for the Coast Range and Great Valley ophiolites.

Model III, after Saleeby (Dickinson et al., 1996) suggests that the ophiolite formed in situ by seafloor spreading in a forearc setting in response to transtension caused by slab rollback, rather than by emplacement resulting from a collision. In this model, the Nevadan orogeny, which resulted in significant crustal shortening in the Sierra Nevada (Schweickert et al., 1984), is viewed as a highly contractional event as a result of oblique convergence along the active margin. To fit the regional geologic data, the extension that allowed the Coast Range and Josephine ophiolites to form by seafloor spreading must have been relatively local and related to slab rollback, whereas contraction associated with the Nevadan orogeny was regional. In this case, to place the high-velocity material beneath the proposed Great Valley ophiolite, as suggested by the geophysical models, requires mafic plutons associated with Sierran magmatism to be emplaced not only beneath the axis of the Sierra Nevada magmatic arc but also westward beneath the newly formed ophiolite (J. Saleeby, 1996, personal communication) (Fig. 8E2). Intruding plutons to the west of the main magmatic arc may be plausible beneath the southern Great Valley basin, which formed adjacent to the Cretaceous granitic and tonalitic intrusive bodies of the Sierra Nevada. Model III is less likely for the northern Great Valley basin, which formed adjacent to the Sierra Nevada Foothills metamorphic belt, since this scenario would require significant westward intrusion of Sierran plutons to result in tonalites beneath the Great Valley ophiolite.

Blake et al. (1989) suggested that the southern Coast Range ophiolite represents an island arc that formed above oceanic basement, consistent with the thicker mafic sections observed in the southern Great Valley, such as at the Llanada locality (Hopson et al., 1981; Godfrey and Klemperer, 1998), (Fig. 1) whereas the northern Coast Range ophiolite has a more oceanic character (forearc, backarc, or mid-ocean ridge), consistent with the 7- to 8-km-thick mafic section above unserpentinized mantle interpreted beneath the northern Great Valley (Godfrey and Klemperer, 1998).

For the northern Great Valley, we contend that (1) the Nevadan orogeny was an intense contractional, probably collisional event, (2) plutons of the main Sierra Nevada batholith occur only east of the eastern margin of the Great Valley basin, (3) the crust of the inferred Great Valley ophiolite is ~7 to 8 km thick (interpreted from geophysical models), and (4) the Coast Range ophiolite had a supra-subduction zone origin according to petrologic and geochemical data, although we cannot distinguish among backarc, arc, or forearc. For the southern Great Valley, we contend that (1) the Nevadan orogeny may have been a less intense contractional event, (2) Sierra Nevada plutons occur adjacent to and beneath the eastern margin of the Great Valley, (3) the crustal part of the Great Valley ophiolite may be up to 10 to 12 km thick (as suggested by geophysical models), if it is the Great Valley ophiolite and not some other mafic or serpentinized material, and (4) the Coast Range ophiolite originated in an island-arc setting, rather than a forearc or backarc environment, on the basis of petrologic and geochemical data.

Our favored model for the formation and emplacement of the Great Valley ophiolite is seafloor spreading in a backarc environment associated with an east-facing intra-oceanic arc (west-dipping subduction) built on a 200 Ma oceanic basement (similar to model I). However, it is possible that there were different modes of emplacement for the ophiolite from north to south. In the north, we suggest that the east-facing arc and backarc crust collided with, and were thrust onto, the older continental margin during the Nevadan orogeny (Fig. 8D1). The crust formed by backarc spreading became the base-

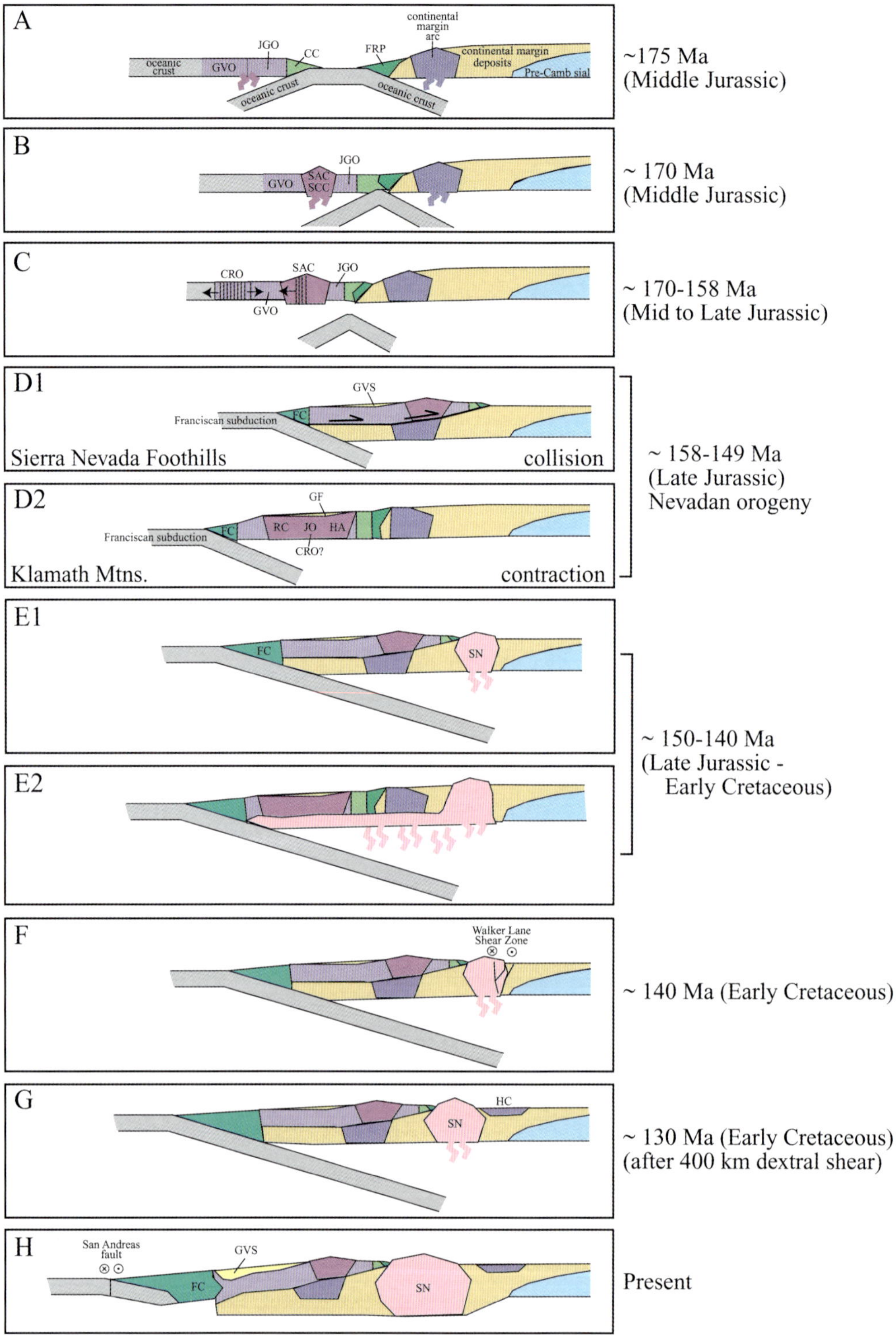

Figure 8. Cartoons depicting the tectonic development of the western margin of North America. A, Ca. 175 Ma: Oppositely dipping subduction zones. B, Ca. 170 Ma: Island arc and continental arc collide as intervening oceanic crust is consumed. C, 172–158 Ma: Coast Range ophiolite forms by backarc spreading and Smartville arc complex forms by intra-arc rifting. D1, 158–149 Ma: Coast Range and Great Valley ophiolites emplaced onto the continental crust. D2, 158–149 Ma: Nevadan orogeny docks Coast Range and Great Valley ophiolites and/or Josephine ophiolite, Hayfork arc terrane, and Rogue-Chetco arc complex with continental crust. E1, 150–140 Ma: Franciscan subduction and Sierra Nevada magmatism underway. Plutons intruded beneath axis of the Sierra Nevada batholith. E2, 150–140 Ma: Franciscan subduction and Sierra Nevada magmatism underway. Plutons intruded beneath axis of Sierra Nevada and west of the Sierra Nevada batholith. F, Ca. 140 Ma: Movement along Walker Lane shear zone results in up to 400 km dextral translation of material. G, Ca. 130 Ma: Franciscan–Sierra Nevadan system continues to develop. H, Present-day: San Andreas transform system well established. Abbreviations are as follows: CC—Calaveras Complex, CRO—Coast Range ophiolite, FC—Franciscan Complex, FRP—Feather River Peridotite, GF—Galice Formation, GVO—Great Valley ophiolite, GVS—Great Valley Sequence, HA—Hayfork arc terrane, HC—Humboldt Complex, JGO—Jarbo Gap ophiolite, JO—Josephine ophiolite, SN—Sierra Nevada batholith, RC—Rogue-Chetco arc complex, SAC—Smartville arc complex, SCC—Slate Creek Complex. Vertical lines with arrows represent rifting.

ment to the newly forming Great Valley basin (Fig. 8D1). After the Nevadan orogeny, Franciscan subduction was initiated west (outboard) of the Coast Range and Great Valley ophiolites (Fig. 8D1), and the associated Sierra Nevada magmatism evolved east of the emplaced ophiolite duo (Fig. 8E1). In the south, there are two possible models. (1) The same model invoked for the northern Great Valley may also be valid for the southern Great Valley. The only difference would be that in the southern Great Valley, serpentinization of ultramafic material was highly advanced prior to ophiolite emplacement. (2) The Nevadan orogeny was less intense in the south, and the east-facing arc and backarc crust collided with the older continental margin but were not thrust onto it. In the south, the Coast Range ophiolite arc itself became the basement to the newly forming Great Valley basin (Fig. 8D2). When Franciscan subduction began west of the Coast Range and Great Valley ophiolites, plutons of the associated Sierra Nevada magmatism may have intruded beneath the newly formed ophiolite in the southern Great Valley (Fig. 8E2). The east-dipping reflectors imaged beneath the southern Great Valley by Miller and Mooney (1995) are the strongest evidence for thrusts associated with ophiolite emplacement. By the Early Cretaceous, the entire Coast Range and Great Valley ophiolites had become part of the North American continent. North America continued to grow by the accretion of Franciscan terranes.

DISCUSSION AND CONCLUSIONS

We have developed a tectonic model for the development of the geologic transect presented in Figure 6 (Fig. 8).

We propose that during the Middle Jurassic (prior to 175 Ma) oppositely dipping subduction zones were active off the North American continent at the latitude of the Sierra Nevada foothills (Fig. 8A). The Jurassic continental-margin arc developed on and across a composite, but mainly continental basement in the eastern part of the Sierra Nevada Foothills metamorphic belt in the upper plate of an east-dipping subduction zone (Fig. 8, A and B). The basement of this continental-margin arc included the early Paleozoic Shoo Fly Complex successively overlain by Devonian and Mississippian and middle Permian arc sequences (Schweickert, 1981), suggesting a period of prolonged active-margin tectonics and magmatic arc construction along the western U.S. continental edge during much of the Paleozoic and early Mesozoic. The Feather River Peridotite and the Calaveras Complex west of the Paleozoic through Jurassic magmatic arc systems represent a subducted and subsequently uplifted and exhumed Paleozoic oceanic lithosphere and an accretionary prism, respectively, of the west-facing continental-margin arc and east-dipping subduction-zone system. The evolution of the ensimatic arc terrane containing the Smartville, Slate Creek, and Lake Combie Complexes in the north and the Tuolumne River Complex in the south was associated with a west-dipping subduction zone between 175 and 160 Ma (Fig. 8, A–C). Multiply deformed ophiolitic fragments and the chert-argillite unit of the late Paleozoic–early Mesozoic disrupted oceanic basement (i.e., Jarbo Gap and Great Valley ophiolites and the Rattlesnake Creek terrane) constituted the ensimatic basement of this Jurassic island-arc terrane (Figs. 4 and 8, A and B).

By ca. 170 Ma, the intervening ocean floor between the oppositely dipping subduction zones had been consumed, causing the continental-margin arc to collide with the ensimatic island-arc terrane in the Sierra Nevada foothills (Fig. 8B). Decoupling of the volcanic arc from its disrupted oceanic basement along east-directed thrust faults and imbrication of the island-arc fragments (Smartville, Slate Creek, and Lake Combie Complexes) occurred during this collisional event. We infer that this arc-continent collision caused a major and widespread contractional deformation across the Sierra Nevada Foothills metamorphic belt and continental-margin arc in the east (Wright and Fahan, 1988; Dilek, 1989c; Dilek and Moores, 1995). In the western Klamath Mountains, the Jurassic arc terrane (western Hayfork and Rogue-Chetco Complexes) developed above an east-dipping subduction zone (Harper and Wright, 1984; Harper et al., 1994) (Fig. 4). The Middle Jurassic deformation (ca. 169–165 Ma) in the Klamath Mountains is characterized by a major episode of contraction that caused structural imbrication across the eastern and western Hayfork, Rattlesnake Creek, and North Fork of the Salmon River terranes and regional metamorphism (decreasing in grade both toward the south and west in the present geographic reference frame; Coleman et al., 1988). This low-pressure, low-temperature deformation in the Klamath Mountains during the Middle Jurassic is interpreted as an intra-arc deformation and is known as the Siskiyou orogenic event (Coleman et al., 1988). The 160–167 Ma calc-alkalic plutons intruding across the thrust faults of the Siskiyou orogen provide a minimum age constraint of pre–167 Ma for the timing of the intra-arc deformation in the Klamath Mountains.

Between ca. 165 and 158 Ma, an extensional episode caused intra-arc rifting in the accreted and imbricated island-arc terranes (represented by the Smartville sheeted-dike complex and the Preston Peak rift facies of the Josephine ophiolite) (Fig. 2) (Harper et al., 1994; Yule, 1996) and the formation of the Coast Range ophiolite by backarc spreading of the Great Valley ophiolitic basement (Fig. 8C). We think that the backarc ophiolite was almost immediately incorporated into the North American continent during the Nevadan orogeny. In the northern Sierra Nevada, collision placed a relatively intact piece of backarc ophiolite crust and mantle (some of which was not serpentinized) onto the older continental crust (Fig. 8D1). In the southern Sierra Nevada, there are two possibilities. Either collision placed a relatively intact piece of arc crust and its serpentinized mantle onto the older continental crust (Fig. 8D1), or there was a significant amount of strike-slip faulting associated with strain partitioning due to oblique convergence that juxtaposed the oceanic arc and backarc ophiolite with the older continental margin (Fig. 8D2). In the Klamath Mountains, further shortening in and across the ensimatic arc system caused major imbrication along the Orleans and Madstone thrusts; this deformation emplaced the Josephine ophiolite beneath and over the eastern Hayfork (remnant arc) and the Rogue-Chetco arc (active arc) terranes, respectively (Harper et al., 1994; Fig. 8D2).

The apparent transport directions of island-arc terranes in the Klamath Mountains and the Sierra Nevada Foothills metamorphic belt are most likely related to differences in subduction-zone polarities. Our data and tectonic model suggest an east-facing island arc (west-dipping subduction zone) for the evolution of the Jurassic ensimatic arc terrane in the Sierra Nevada foothills, whereas the current tectonic models and the extant data from the western Klamath Mountains support a west-facing island-arc system (east-dipping subduction zone) for the evolution of the Jurassic arc terrane there (Fig. 4). We think that these oppositely dipping subduction zones and their arc-trench systems were separated by a sinistral oceanic transform fault or fracture zone outboard of the continental margin of western North America (Figs. 4 and 7; Dilek and Moores, 1992). Several lines of evidence support this interpretation: (1) the abrupt bend of the north-trending structures and tectonic boundaries to a conspicuous east trend along the Big Bend fault zone just to the north of the Smartville Complex; (2) oceanic fracture-zone characteristics of the disrupted ophiolitic basement within this fault zone (Dilek, 1989b); (3) occurrence of oceanic fracture zone–related rocks along the same latitude farther west in the northern Coast Range ophiolite (Harper et al., 1985; Shervais and Kimbrough, 1985; Dilek and Moores, 1992); (4) slight westward displacement of the magnetic anomaly over the strata of the Great Valley near lat 40°N along a west-northwest–trending strike-slip fault zone for as much as 40 km (Hinze and Hood, 1989). North of the Great Valley, the magnetic anomaly pattern follows the Josephine ophiolite in the western Klamath Mountains. There is no strong evidence suggesting that these ensimatic arc-trench systems (with opposite polarities) in the Sierran-Klamath regions developed adjacent to the North American continent prior to Middle Jurassic time. We consider the convergent-margin system in the southwestern Pacific Ocean, where the east-facing Tonga-Kermadec and the west-facing New Hebrides volcanic islands are offset by the sinistral fault zone containing the Fiji and Hunter fracture zones, as a modern analogue for the evolution of the Jurassic ensimatic arcs in the Sierra Nevada and Klamath regions.

After the Nevadan orogeny (ca. 149 Ma), the Franciscan subduction zone was initiated west of the Coast Range and Great Valley ophiolites, and Sierra Nevada continental arc magmatism associated with Franciscan subduction began (Fig. 8, D1 and D2). In the northern Sierra Nevada and possibly the southern Sierra Nevada, plutons were emplaced east of the Great Valley ophiolite (Fig. 8E1). An alternative model for the south is that plutons were emplaced both east of and beneath the ophiolite (Fig. 8E2). At ca. 140 Ma, the development of the Mojave–Snow Lake and the Walker Lane shear zones caused up to 400 km of dextral displacement along the eastern margin on the continental-margin arc (Fig. 8F). This displacement caused an apparent doubling of the continental-margin arc sequences at northern (~40°N) latitudes (Fig. 8G). A simple retrodeformation of these shear zones suggests that the Sierra Nevada and Klamath Mountains provinces were not laterally contiguous with the present Great Basin in Nevada during much of the Mesozoic, and that early to middle Mesozoic Nevada did not necessarily constitute a backarc environment to the Jurassic continental-margin arc. The Great Valley forearc basin developed on top of the stranded oceanic basement during the remainder of the Jurassic and Cretaceous as part of the continental-margin arc–Franciscan subduction system (Ingersoll and Schweickert, 1986) (Fig. 8, D–G). Sierra Nevada magmatism ceased at ca. 80 Ma (Late Cretaceous), and, finally, the San Andreas transform fault system was initiated at ca. 29 Ma (Fig. 8H).

The tectonic model presented in this synthesis suggests that the continental growth along the Mesozoic continental margin of western North America occurred mainly by accretion of oceanic crust and island-arc terranes and possibly by mafic igneous underplating through complex tectonic and magmatic events (Coleman, this volume). Accretion of island arcs and ophiolites was facilitated by an arc-continent collision, intra-arc and/or backarc rifting, and subsequent contractional deformation (Nevadan orogeny) at the present latitude of 40°N. The Great Valley ophiolite beneath the Great Valley forearc basin is a critical component in this assimilation history and represents a spatial link between the Coast Range ophiolite and the ensimatic arc terrane in the Sierra Nevada Foothills metamorphic belt. In our interpretation, the Coast Range and Great Valley ophiolites, the Smartville arc complex and its equivalents in the Sierra Nevada foothills, and the late Paleozoic–early Mesozoic disrupted ophiolitic basement collectively form an intraoceanic terrane. This intraoceanic terrane consists of volcanic archipelagoes (island arcs), intra-arc or backarc basins, and a multiply deformed oceanic substratum on which island arcs were built. The accretion of this intraoceanic terrane into the North American continental margin in the Middle Jurassic marks a major orogenic event in the tectonic history of the North American Cordillera (Moores, 1998). We therefore suggest that one of the most critical terrane boundaries in the western U.S. Cordillera occurs in the Sierra Nevada Foothills metamorphic belt, where the Jurassic intraoceanic terrane is juxtaposed against the Jurassic continental-margin arc and its basement (Paleozoic arc sequences and subduction-zone and accretionary-prism assemblages).

ACKNOWLEDGMENTS

We thank Darrell Cowan, Gregory Harper, Kate Miller, and George Thompson for critical reviews that greatly improved this manuscript.

REFERENCES CITED

Atwater, T., 1970, Implications of plate tectonics for the Cenozoic tectonic history of western North America: Geological Society of America Bulletin, v. 81, p. 3513–3536.

Bailey, E.H., and Blake, M.C., Jr., 1969, Tectonic development of western California during the late Mesozoic: Geotectonics, v. 3, p. 148–154.

Bailey, E.H., Irwin, W.P., and Jones, D.L., 1964, Franciscan and related rocks and their significance in the geology of western California: Bulletin of the California Division of Mines and Geology, v. 183, 177 p.

Bailey, E.H., Blake, M.C., Jr., and Jones, D.L., 1970, On-land Mesozoic oceanic crust in the California Coast Ranges: U.S. Geological Survey Professional Paper 700-C, p. C70–C81.

Beaudoin, B.C., Godfrey, N.J., Klemperer, S.L., Lendl, C., Tréhu, A.M., Henstock, T.J., Levander, A., Holl, J.E., Meltzer, A.S., Luetgert, J.H., and Mooney, W.D., 1996, The transition from slab to slabless: Results from the 1993 Mendocino Triple Junction Seismic Experiment: Geology, v. 24, p. 195–199.

Benz, H.M., Smith, R.B., and Mooney, W.D., 1990, Crustal structure of the northwestern Basin and Range province from the 1986 Program for Array Seismic Studies of the Continental Lithosphere seismic experiment: Journal of Geophysical Research, v. 95, p. 21823–21842.

Bickford, M.E., and Day, H.W., 1988, Jurassic ages of arc-ophiolite complexes, northern Sierra Nevada: Implications for duration of the Nevadan "Orogeny": Geological Society of America Abstracts with Programs, v. 20, p. A274.

Blake, M.C., Jr., and Jones, D.L., 1974, Origin of Franciscan mélanges in northern California, *in* Dott, R.H., Jr., and Shaver, R.H., eds., Modern and ancient geosynclinal sedimentation: Problems of palinspastic restoration: Society of Economic Paleontologists and Mineralogists Special Publication, v. 19, p. 345–357.

Blake, M.C., Jr., Bruhn, R.L., Miller, E.L., Moores, E.M., Smithson, S.B. and Speed, R.C., 1985, C-1 Mendocino triple junction to North American Craton: Centennial Continent/Ocean Transect no. 12: Boulder, Colorado, Geological Society of America, 30 p.

Blake, M.C., Jr., Jayko, A.S., McLaughlin, R.J., and Underwood, M.B., 1988, Metamorphic and tectonic evolution of the Franciscan Complex, northern California, *in* Ernst, W.G., ed., Metamorphic and crustal evolution of the western United States (Rubey Volume 7): Englewood Cliffs, New Jersey, Prentice-Hall, p. 1035–1060.

Blake, M.C., Jr., McLaughlin, R.J., and Jones, D.L., 1989, Terranes of the northern Coast Ranges, *in* Blake, M.C., Jr., ed., Tectonic evolution of Northern California: Field Trip Guidebook Volume T108: Washington D.C., American Geophysical Union, p. 3–18.

Burchfiel, B.C., and Davis, G.A., 1981, Triassic and Jurassic tectonic evolution of the Klamath Mountains-Sierra Nevada geologic terrane, *in* Ernst, W.G., ed., The geotectonic development of California (Rubey Volume 1): Englewood Cliffs, New Jersey, Prentice-Hall, p. 50–70.

Busby-Spera, C.J., 1984, The lower Mesozoic continental margin and marine intra-arc sedimentation at Mineral King, California, *in* Crouch, J.K., and Bachman, S.B., eds., Tectonics and sedimentation along the California margin: Los Angeles, Pacific Section, Society of Economic Paleontologists and Mineralogists, v. 38, p. 135–156.

Busby-Spera, C.J., 1988, Speculative tectonic model for the early Mesozoic arc of the southwest Cordilleran United States: Geology, v. 16, p. 1121–1125.

Busby-Spera, C.J., Mattinson, J.M., Riggs, N.R., and Schermer, E.R., 1990, The Triassic-Jurassic magmatic arc in the Mojave-Sonoran deserts and the Sierra-Klamath region: Similarities and differences in paleogeographic evolution, *in* Harwood, D.S., and Miller, M.M., eds., Paleozoic and early Mesozoic paleogeographic relations: Sierra Nevada, Klamath Mountains, and related terranes: Geological Society of America Special Paper 255, p. 325–337.

Cady, J.W., 1975, Magnetic and gravity anomalies in the Great Valley and western Sierra Nevada metamorphic belt, California: Geological Society of America Special Paper 168, 54 p.

Chen, J.H., and Moore, J.G., 1982, Uranium-lead isotopic ages from the Sierra Nevada Batholith, California: Journal of Geophysical Research, v. 87, p. 4761–4784.

Christensen, N.I., 1978, Ophiolites, seismic velocities, and oceanic crustal structure: Tectonophysics, v. 47, p. 131–157.

Clague, D.A., and Straley, P.F., 1977, Petrologic nature of the oceanic Moho: Geology, v. 5, p. 133–136.

Coleman, R.G., Manning, C.E., Donato, M.M., Mortimer, N., and Hill, L.B., 1988, Tectonic and regional metamorphic framework of the Klamath Mountains and adjacent Coast Ranges, California and Oregon, *in* Ernst, W.G., ed., Metamorphic and crustal evolution of the western United States (Rubey Volume 7): Englewood Cliffs, New Jersey, Prentice-Hall, p. 1061–1097.

Cox, A., and Engebretson, D.C., 1985, Change in motion of Pacific Plate at 5 Myr BP: Nature, v. 313, p. 472–474.

Day, H.W., Moores, E.M., and Tuminas, A.C., 1985, Structure and tectonics of the northern Sierra Nevada: Geological Society of America Bulletin, v. 96, p. 436–450.

Dickinson, W.R., 1981, Plate tectonics and the continental margin of California, *in* Ernst, W.G., ed., The geotectonic development of California (Rubey Volume 1): Englewood Cliffs, New Jersey, Prentice-Hall, p. 1–28.

Dickinson, W.R., and Rich, E.I., 1972, Petrologic intervals and petrofacies in the Great Valley Sequence, Sacramento Valley, California: Geological Society of America Bulletin, v. 83, p. 3007–3024.

Dickinson, W.R., Hopson, C.A., and Saleeby, J.B., 1996, Alternate origins of the Coast Range Ophiolite (California): Introduction and implications: GSA Today, v. 6, p. 1–10.

Dilek, Y., 1989a, Structure and petrology of selected Mesozoic ophiolite complexes in western North America and eastern Mediterranean: Tectonic significance of ophiolites in the Cordilleran and Alpine orogenic belts [Ph.D. dissert.]: Davis, University of California, 228 p.

Dilek, Y., 1989b, Structure and tectonics of an Early Mesozoic oceanic basement in the northern Sierra Nevada metamorphic belt, California: Evidence for transform faulting and ensimatic arc evolution: Tectonics, v. 8, p. 999–1014.

Dilek, Y., 1989c, Tectonic significance of post-accretion rifting of a Mesozoic island-arc terrane in the northern Sierra Nevada, California: Journal of Geology, v. 97, p. 503–518.

Dilek, Y., and Moores, E.M., 1991, Humboldt Complex, Nevada, as an allochthonous fragment of a Jurassic arc terrane: Geological Society of America Abstracts with Programs, v. 23, p. A192.

Dilek, Y., and Moores, E.M., 1992, Island-arc evolution and fracture-zone tectonics in the Mesozoic Sierra Nevada, California, and implications for transform offset of the Sierran/Klamath convergent margin, *in* Hyndman, D.W., Mogk, D.W., and Mason, R., eds., Basement tectonics 8: Characterization and comparison of ancient and Mesozoic continental margins—Proceedings of the 8th international conference on basement terranes (Butte, Montana): Dordrecht, The Netherlands, Kluwer Academic Publishers, p. 179–196.

Dilek, Y., and Moores, E.M., 1993, Across-strike anatomy of the Cordilleran orogen at 40°N latitude: Implications for the Mesozoic paleogeography of the western United States, *in* Dunne, G., and McDougall, K., eds., Mesozoic paleogeography of the western United States—II: Pacific Section, Society of Economic Paleontologists and Mineralogists Book 71, p. 333–346.

Dilek, Y., and Moores, E.M., 1995, Geology of the Humboldt igneous complex, Nevada, and its tectonic implications for the Jurassic magmatism in the Cordilleran orogen, *in* Miller, D.M., and Busby, C., eds., Jurassic magmatism and tectonics of the North American Cordillera: Geological Society of America Special Paper 299, p. 229–248.

Dilek, Y., Thy, P., Moores, E.M., and Grundvig, S., 1990, Late-Paleozoic–early Mesozoic oceanic basement of a Jurassic arc terrane in the northwestern Sierra Nevada, California, *in* Harwood, D.S., and Miller, M.M., eds., Paleozoic and early Mesozoic paleogeographic relations; Sierra Nevada, Klamath Mountains, and related terranes: Geological Society of America Special Paper 255, p. 351–369.

Dilek, Y., Thy, P., and Moores, E.M., 1991, Episodic dike intrusions in the northwestern Sierra Nevada, California: Implications for multistage evolution of a Jurassic arc terrane: Geology, v. 19, p. 180–184.

Dilles, J.H., and Wright, J.E., 1988, The chronology of Early Mesozoic arc magmatism in the Yerington district of western Nevada and its regional implications: Geological Society of America Bulletin, v. 100, p. 644–652.

Ducea, M., and Saleeby, J.B., 1998, Crustal recycling beneath continental arcs; silica-rich glass inclusions in ultramafic xenoliths from the Sierra Nevada, California: Earth and Planetary Science Letters, v. 156, p. 101–116.

Edelman, S.H., Day, H.W., Moores, E.M., Zigan, S.M., Murphy, T.P., and Hacker, B.R., 1989, Structure across a Mesozoic ocean-continent suture zone in the northern Sierra Nevada, California: Geological Society of America Special Paper 224, 56 p.

Elison, M.W., Speed, R.C., and Kistler, R.W., 1990, Geologic and isotopic constraints on the crustal structure of the northern Great Basin: Geological Society of America Bulletin, v. 102, p. 1077–1092.

Ernst, W.G., 1981, Summary of the geotectonic development of California, *in* Ernst, W.G., ed., The geotectonic development of California (Rubey Volume 1): Englewood Cliffs, New Jersey, Prentice-Hall, p. 601–613.

Evarts, R.C., 1977, The geology and petrology of the Del Puerto Ophiolite, Diablo Range, central California Coast Ranges: Oregon Department of Geology and Mineralogy Industries Bulletin 95, p. 121–139.

Evernden, J.F., and Kistler, R.W., 1970, Chronology of emplacement of Mesozoic batholithic complexes in California and western Nevada: U.S. Geological Survey Professional Paper 623, 42 p.

Fisher, G.R., 1990, Middle Jurassic syntectonic conglomerate in the Mt. Tallac roof pendant, northern Sierra Nevada, California, *in* Harwood, D.S., and Miller, M.M., eds., Paleozoic and early Mesozoic paleogeographic relations; Sierra Nevada, Klamath Mountains, and related terranes: Geological Society of America Special Paper 255, p. 339–350.

Fliedner, M.M., 1997, Making of continental crust by arc magmatism: Wide-angle seismic and gravity studies in the southern Sierra Nevada of California and the eastern Aleutian arc [Ph.D. dissert.]: Stanford University, California, 166 p.

Fliedner, M.M., Ruppert, S., and the Southern Sierra Continental Dynamics Working Group, 1996, Three-dimensional crustal structure of the southern Sierra Nevada from seismic fan profiles and gravity modeling: Geology, v. 4, p. 367–370.

Gefell, M.J., Twiss, R.J., and Moores, E.M., 1989, Ductile and brittle shear sense for the "Melones Fault Zone", northern Sierra Nevada, California: Geological Society of America Abstracts with Programs, v. 21, p. 83.

Giaramita, M., MacPherson, G.J., and Phipps, S.P., 1998, Petrologically diverse basalts from a fossil oceanic forearc in California: The Llanada and Black Mountain remnants of the Coast Range Ophiolite: Geological Society of America Bulletin, v. 110, p. 553–571.

Godfrey, N.J., and Klemperer, S.L., 1998, Ophiolitic basement to a forearc basin and implications for continental growth: The Coast Range/Great Valley ophiolite, California: Tectonics, v. 17, p. 558–570.

Godfrey, N.J., Beaudoin, B.C., Klemperer, S.L., and the Mendocino Working Group, 1997, Ophiolitic basement to the Great Valley forearc basin, northern California, from seismic and gravity data: Implications for crustal growth at the North American continental margin: Geological Society of America Bulletin, v. 109, p. 1536–1562.

Griscom, A., and Jachens, R.C., 1990, Tectonic implications of gravity and magnetic models along east-west seismic profiles across the Great Valley near Coalinga, *in* Rymer, M.J., and Ellsworth, W.L., eds., The Coalinga earthquake of May 2 1983: U.S. Geological Survey Professional Paper 1487, p. 69–78.

Hagstrum, J.T., and Jones, D.L., 1998, Paleomagnetism, paleogeographic origins, and uplift history of the Coast Range ophiolite at Mount Diablo, California: Journal of Geophysical Research, v. 103, p. 597–603.

Hagstrum, J.T., and Murchey, B.L., 1996, Paleomagnetism of Jurassic radiolarian chert above the Coast Range ophiolite at Stanley Mountain, California, and implications for its paleogeographic origins: Geological Society of America Bulletin, v. 108, p. 643–652.

Harland, W.B., Cox, A.V., Llewellyn, P.G., Pickton, C., Smith, A.G., and Walters, R., 1982, A geologic time scale: New York, Cambridge University Press, 131 p.

Harms, T.A., Jayko, A.S., and Blake, M.C., Jr., 1992, Kinematic evidence for extensional unroofing of the Franciscan Complex along the Coast Range fault, northern Diablo Range, California: Tectonics, v. 11, p. 228–241.

Harper, G.D., 1994, A review of hemipelagic and flysch sedimentation associated with the Josephine ophiolite, California: Ofioliti, v. 19, p. 397–411.

Harper, G.D., and Wright, J.E., 1984, Middle to Late Jurassic tectonic evolution of the Klamath Mountains, California-Oregon: Tectonics, v. 3, p. 759–772.

Harper, G.D., Saleeby, J.B., and Norman, A.S., 1985, Geometry and tectonic setting of sea-floor spreading for the Josephine ophiolite, and implications for Jurassic accretionary events along the California margin, *in* Howell, D.G., ed., Tectonostratigraphic terranes of the Circum-Pacific region: Circum-Pacific Council for the Energy and Mineral Resources, Earth Science Series 1, p. 239–257.

Harper, G.D., Saleeby, J.B., and Heizler, M., 1994, Formation and emplacement of the Josephine Ophiolite and the Nevadan Orogeny in the Klamath Mountains, California-Oregon; U/Pb zircon and $^{40}Ar/^{39}Ar$ geochronology: Journal of Geophysical Research, v. 99, p. 4293–4321.

Harwood, D.S., 1988, Tectonism and metamorphism in the northern Sierra terrane, *in* Ernst, W.G., ed., Metamorphic and crustal evolution of the western United States (Rubey Volume 7): Englewood Cliffs, New Jersey, Prentice-Hall, p. 764–788.

Harwood, D.S., 1992, Stratigraphy of Paleozoic and lower Mesozoic rocks in the northern Sierra terrane: U.S. Geological Survey Bulletin 1957, 78 p.

Harwood, D.S., 1993, Mesozoic geology of Mt. Jura, northern Sierra Nevada, California: A progress report, *in* Dunne, G., and McDougall, K., eds., Mesozoic paleogeography of the western United States—II: Pacific Section, Society of Economic Paleontologists and Mineralogists Book 71, p. 263–287.

Harwood, D.S., and Helley, E.J., 1987, Late Cenozoic tectonism of the Sacramento Valley, California: U.S. Geological Survey Professional Paper 1359, 45 p.

Hauge, T.A., Allmendinger, R.W., Caruso, C., Hauser, E.C., Klemperer, S.L., Opdyke, S., Potter, C.J., Sanford, W., Brown, L., Kaufman, S., and Oliver, J., 1987, Crustal structure of western Nevada from COCORP deep seismic-reflection data: Geological Society of America Bulletin, v. 98, p. 320–329.

Hietanen, A., 1981, Petrologic and structural studies in the northwestern Sierra Nevada, California: U.S. Geological Survey Professional Paper 1226A-C, 59 p.

Hinze, W.J., and Hood, P.J., 1989, The magnetic anomaly map of North America; A new tool for regional geologic mapping, *in* Bally, A.W., and Palmer, A.R., eds., The geology of North America: An overview: Boulder, Colorado, Geological Society of America DNAG Series, v. A, p. 29–38.

Hopson, C.A., Mattinson, J.M., and Pessagno, E.A., 1981, Coast Range ophiolite, western California, *in* Ernst, W.G., ed., The geotectonic development of California (Rubey Volume 1): Englewood Cliffs, New Jersey, Prentice-Hall, p. 418–510.

Ingersoll, R.V., 1982, Initiation and evolution of the Great Valley forearc basin of northern and central California, U.S.A.: Geological Society of London Special Publication 10, p. 459–467.

Ingersoll, R.V., and Schweickert, R.A., 1986, A plate-tectonic model for Late Jurassic ophiolite genesis, Nevadan Orogeny and forearc initiation, northern California: Tectonics, v. 5, p. 901–912.

Irwin, W.P., 1960, Geological reconnaissance of the northern Coast Ranges and Klamath Mountains, California, with a summary of mineral resources: California Division of Mines Bulletin 179, 80 p.

Iturrino, G.J., Christensen, N.I., Kirby, S., and Salisbury, M.H., 1991, Seismic velocities and elastic properties of oceanic gabbroic rocks from Hole 735B, Proceedings of the Ocean Drilling Program Scientific Results, v. 118: College Station, Texas, Ocean Drilling Program, p. 227–237.

Jachens, R.C., Griscom, A., and Roberts, C.W., 1995, Regional extent of Great Valley basement west of the Great Valley, California: Implications for extensive tectonic wedging in the California Coast Ranges: Journal of Geophysical Research, v. 100, p. 12769–12790.

Jayko, A.S., and Blake, M.C., Jr., 1986, Significance of Klamath rocks between the Franciscan Complex and Coast Range ophiolite, northern California: Tectonics, v. 5, p. 1055–1071.

Jayko, A.S., Blake, M.C., Jr., and Harms, T.A., 1987, Attenuation of the Coast Range ophiolite by extensional faulting, and nature of the Coast Range "thrust", California: Tectonics, v. 6, p. 475–488.

Jennings, C.W., 1977, Geologic map of California: California Division of Mines and Geology, scale 1:750 000.

Kistler, R.W., Evernden, J.F., and Shaw, H.R., 1971, Sierra Nevada plutonic cycle: 1. Origin of composite granitic batholiths: Geological Society of America Bulletin, v. 82, p. 853–868.

Klemperer, S.L., Hauge, T.A., Hauser, E.C., Oliver, J.E., and Potter, C.J., 1986, The Moho in the northern Basin and Range province, Nevada, along the COCORP 40°N seismic-reflection transect: Geological Society of America Bulletin, v. 97, p. 603–618.

Kosanke, S.B., and Harper, G.D., 1997, Geochemical variations in volcanic rocks from a fragment of the Jurassic Coast Range Ophiolite, Oregon; possible indication of fore-arc rifting: Geological Society of America Abstracts with Programs, v. 29, p. 159.

Lahren, M.M., and Schweickert, R.A., 1989, Proterozoic and Lower Cambrian miogeoclinal rocks of Snow Lake pendant, Yosemite-Emigrant Wilderness, Sierra Nevada, California: Evidence for major Early Cretaceous dextral translation: Geology, v. 17, p. 156–160.

Maher, K.A., and Saleeby, J.B., 1986, Geology of the Jackson Mountains, NW Nevada: Geological Society of America Abstracts with Programs, v. 18, p. 679.

Mattinson, J., and Hopson, C., 1992, U/Pb ages of the Coast Range Ophiolite: A critical re-evaluation based on new high-precision Pb/Pb ages [abs.]: American Association of Petroleum Geologists Bulletin, v. 76, p. 425.

McKenzie, D.P., and Morgan, W.J., 1969, Evolution of triple junctions: Nature, v. 224, p. 125–133.

McLaughlin, R.J., Kling, S.A., Poore, R.Z., McDougall, K., and Beutner, E.C., 1982, Post-middle Miocene accretion of Franciscan rocks, northwestern California: Geological Society of America Bulletin, v. 93, p. 595–605.

McLaughlin, R.J., Blake, M.C., Jr., Griscom, A., Blome, C.D., and Murchey, B., 1988, Tectonics of formation, translation, and dispersal of the Coast Range Ophiolite of California: Tectonics, v. 7, p. 1033–1056.

McLaughlin, R.J., Sliter, W.V., Frederiksen, N.O., Harbert, W.P., and McCulloch, D.S., 1993, Plate motions recorded in tectonostratigraphic terranes of the Franciscan Complex and evolution of the Mendocino triple junction, northwestern California: U.S. Geological Survey Bulletin 1997, 60 p.

Miller, K.C., and Mooney, W.D., 1995, Crustal structure and composition of the southern Foothills metamorphic belt, Sierra Nevada, California, from seismic data: Journal of Geophysical Research, v. 99, p. 6865–6880.

Moores, E.M., 1998, Ophiolites, the Sierra Nevada, "Cordillera", and orogeny along the Pacific and Caribbean margins of North and South America, *in* Ernst, W.G., and Nelson, C.A., eds., Integrated earth and environmental evolution of the southwestern United States: The Clarence A. Hall, Jr. Volume, Bellwether Publishing, p. 44–58.

Moores, E.M., and Day, H.W., 1984, An overthrust model for the Sierra Nevada: Geology, v. 12, p. 416–419.

Moxon, I.W., 1990, Stratigraphic and structural architecture of the San Joaquin-Sacramento basin [Ph.D. dissert.]: Stanford University, California, 371 p.

Nichols, J., Warren, N., Luyendyk, B.P., and Spudich, P., 1980, Seismic velocity structure of the ophiolite at Point Sal, southern California, determined from laboratory measurements: Geophysical Journal of the Royal Astronomical Society, v. 63, p. 165–185.

Oldow, J.S., 1984, Evolution of a late Mesozoic back-arc fold and thrust belt, northwestern Great Basin, U.S.A.: Tectonophysics, v. 102, p. 245–274.

Oppenheimer, D.H., and Eaton, J.P., 1984, Moho orientation beneath central California from regional earthquake traveltimes: Journal of Geophysical Research, v. 89, p. 10267–10282.

Penrose Conference Participants, 1972, Penrose field conference: Ophiolites: Geotimes, v. 17, p. 24–25.

Pessagno, E.A., Jr., and Blome, C.D., 1990, Implications of new Jurassic stratigraphic, geochronometric, and paleolatitudinal data from the western Klamath Terrane (Smith River and Rogue Valley subterranes): Geology, v. 18, p. 665–668.

Phipps, S.P., 1984, Ophiolitic olistostromes in the basal Great Valley sequence, Napa County, northern California Coast Ranges: Geological Society of America Special Paper 198, p. 103–125.

Phipps, S.P., and Unruh, J.R., 1992, Crustal-scale wedging beneath an imbricate roof-thrust system: Geology of a transect across the western Sacramento Valley and Northern Coast Ranges, California, *in* Erskine, M.C., ed., Field guide to the tectonics of the boundary between the California Coast Ranges and the Great Valley of California, p. 117–140.

Robertson, A.H.F., 1989, Palaeoceanography and tectonic setting of the Jurassic Coast Range ophiolite, central California: Evidence from the extrusive rocks and the volcaniclastic sediment cover: Marine and Petroleum Geology, v. 6, p. 194–220.

Robertson, A.H.F., 1990, Sedimentology and tectonic implications of ophiolite-derived clastics overlying the Jurassic Coast Range ophiolite, northern California: American Journal of Science, v. 290, p. 109–263.

Russell, B.J., 1984, Mesozoic geology of the Jackson Mountains, northwestern Nevada: Geological Society of America Bulletin, v. 95, p. 313–323.

Saleeby, J.B., 1991, The Cretaceous Sierra Nevada: A transtitching batholithic belt: Geological Society of America Abstracts with Programs, v. 23, p. 94.

Saleeby, J.B., and Busby, C., 1993, Paleogeographic and tectonic setting of axial and western metamorphic framework rocks of the southern Sierra Nevada, California, *in* Dunne, G., and McDougall, K., eds., Mesozoic paleogeography of the western United States—II: Pacific Section, Society of Economic Paleontologists and Mineralogists Book 71, p. 197–225.

Saleeby, J.B., Blake, M.C., and Coleman, R.G., 1984, Pb/U zircon ages on thrust plates of the western central Klamath Mountains and Coast Ranges, northern California and southern Oregon [abs.]: Eos, Transactions of the American Geophysical Union, v. 65, p. 1147.

Saleeby, J.B., Hannah, J.L., and Varga, R.J., 1987a, Isotopic age constraints on middle Paleozoic deformation in the northern Sierra Nevada, California: Geology, v. 15, p. 757–760

Saleeby, J.B., Sams, D.B. and Kistler, R.W., 1987b, U/Pb zircon, strontium, and oxygen isotopic and geochronological study of the southernmost Sierra Nevada Batholith, California: Journal of Geophysical Research, v. 92, p. 10443–10466

Saleeby, J.B., Shaw, H.F., Niemeyer, S., Moores, E.M., and Edelman, D.H., 1989, U/Pb, Sm/Nd, and Rb/Sr geochronological and isotopic study of northern Sierra Nevada ophiolitic assemblages, California: Contributions to Mineralogy and Petrology, v. 102, p. 205–220.

Schermer, E.R., 1993, Mesozoic structural evolution of the west-central Mojave Desert, *in* Dunne, G., and McDougall, K., eds., Mesozoic paleogeography of the western United States—II: Pacific Section, Society of Economic Paleontologists and Mineralogists Book 71, p. 307–322.

Schweickert, R.A., 1978, Triassic and Jurassic paleogeography of the Sierra Nevada ad adjacent regions, California and western Nevada, *in* Howell, D.G., and McDougall, K.A., eds., Mesozoic paleogeography of the western United States, Pacific Coast Paleogeography Symposium 2: Los Angeles, Society of Economic Paleontologists and Mineralogists, p. 362–384.

Schweickert, R.A., 1981, Tectonic evolution of the Sierra Nevada Range, *in* Ernst, W.G., ed., The geotectonic development of California (Rubey Volume 1): Englewood Cliffs, New Jersey, Prentice-Hall, p. 87–131.

Schweickert, R.A., and Cowan, D.S., 1975, Early Mesozoic tectonic evolution of the western Sierra Nevada, California: Geological Society of America Bulletin, v. 86, p. 1329–1336.

Schweickert, R.A., and Lahren, M.M., 1990, Speculative reconstruction of the Mojave–Snow Lake fault: Implications for Paleozoic and Mesozoic orogenesis in the western United States: Tectonics, v. 9, p. 1609–1629.

Schweickert, R.A., Armstrong, R.L., and Harkal, J.E., 1980, Lawsonite blueschist in the northern Sierra Nevada, California: Geology, v. 8, p. 27–31.

Schweickert, R.A., Bogen, N.L., Girty, G.H., Hanson, R.E., and Merguerian, C., 1984, Timing and structural expression of the Nevadan orogeny, Sierra Nevada, California: Geological Society of America Bulletin, v. 95, p. 967–979.

Schweickert, R.A., Merguerian, C., and Bogen, N.L., 1988, Deformational and metamorphic history of Paleozoic and Mesozoic basement terranes in the western Sierra Nevada metamorphic belt, *in* Ernst, W.G., ed., Metamorphic and crustal evolution of the western United States (Rubey Volume 7): Englewood Cliffs, New Jersey, Prentice-Hall, p. 789–822.

Schweickert, R.A., Fisher, G.R., Lahren, M.M., and Greene, D.C., 1991, Triassic-Jurassic magmatic arc in eastern California and western Nevada: Evolution, correlation, and evidence for the Mojave–Snow Lake fault: Geological Society of America Abstracts with Programs, v. 23, p. A192.

Sharp, W.D., 1988, Pre-Cretaceous crustal evolution in the Sierra Nevada region, California, *in* Ernst, W.G., ed., Metamorphic and crustal evolution of the western United States (Rubey Volume 7): Englewood Cliffs, New Jersey, Prentice-Hall, p. 823–864.

Shervais, J.W., and Kimbrough, D.L., 1985, Geochemical evidence for the tectonic setting of the Coast Range ophiolite: A composite island arc-oceanic crust terrane in western California: Geology, v. 13, p. 35–38.

Speed, R.C., 1978, Paleogeographic and plate tectonic evolution of the early Mesozoic marine province of the western Great Basin, *in* Howell, D.G., and McDougall, K.A., eds., Mesozoic paleogeography of the western United States, Pacific Coast Paleogeography Symposium 2: Los Angeles, Society of Economic Paleontologists and Mineralogists, p. 253–270.

Speed, R.C., Elison, M.W., and Heck, F.R., 1988, Phanerozoic tectonic evolution of the Great Basin, *in* Ernst, W.G., ed., Metamorphic and crustal evolution of the western United States (Rubey Volume 7): Englewood Cliffs, New Jersey, Prentice-Hall, p. 573–605.

Stanley, W.D., Benz, H.M., Walters, M.A., Villaseñor, A., and Rodriguez, B.D., 1998, Tectonic controls on magmatism in the Geysers-Clear Lake region: Evidence from new geophysical methods: Geological Society of America Bulletin, v. 110, p. 1193–1207.

Stewart, J.H., 1988, Tectonics of the Walker Lane Belt, western Great Basin: Mesozoic and Cenozoic deformation in a zone of shear, *in* Ernst, W.G., ed., Metamorphic and crustal evolution of the western United States (Rubey Volume 7): Englewood Cliffs, New Jersey, Prentice-Hall, p. 683–713.

Tobisch, O.T., Saleeby, J.B., and Fiske, R.S., 1986, Structural history of continental volcanic arc rocks along part of the eastern Sierra Nevada, California: A case for extensional tectonics: Tectonics, v. 5, p. 65–94.

Tosdal, R.M., Haxel, G.B., and Wright J.E., 1989, Jurassic geology of the Sonoran desert region, southern Arizona, southeastern California and northernmost Sonora: Construction of a continental-margin magmatic arc, *in* Jenney, J.P., and Reynolds, S.J., eds., Geological evolution of Arizona: Arizona Geological Society Digest, v. 17, p. 397–434.

Tréhu, A., 1995, Pulling the rug out from under California: Seismic images of the Mendocino triple junction region: Eos, Transactions, American Geophysical Union, v. 76, p. 369.

Unruh, J.R., and Moores, E.M., 1992, Quaternary blind thrusting in the southwestern Sacramento Valley, California: Tectonics, v. 11, p. 192–203.

Wentworth, C.M., and Zoback, M.D., 1990, Structure of the Coalinga area and thrust origin of the earthquake, *in* Rymer, M.J., and Ellsworth, W.L., eds., The Coalinga earthquake of May 2, 1983: U.S. Geological Survey Professional Paper 1487, p. 41–68.

Wentworth, C.M., Blake, M.C., Jr., Jones, D.L., Walter, A.W., and Zoback, M.D., 1984, Tectonic wedging associated with emplacement of the Franciscan Assemblage, California Coast Ranges, *in* Blake, M.C., Jr., ed., Franciscan geology of Northern California: Santa Barbara, California, Pacific Section, Society of Economic Paleontologists and Mineralogists, p. 163–173.

White, R.S., McKenzie, D. and O'Nions, R.K., 1992, Oceanic crustal thickness from seismic measurements and rare earth element inversions: Journal of Geophysical Research, v. 97, p. 19683–19715.

Willden, R., and Speed, R.C., 1974, The geology and mineral deposits of Churchill County, Nevada: Nevada Bureau of Mines and Geology Bulletin, v. 83, p. 95.

Wright, J.E., and Fahan, R., 1988, An expanded view of Jurassic orogenesis in the western U.S. Cordillera: Middle Jurassic (pre-Nevadan) regional metamorphism and thrust faulting within an active arc environment, Klamath Mountains, California: Geological Society of America Bulletin, v. 100, p. 859–876.

Wright, J.E., and Wyld, S.J., 1994, The Rattlesnake Creek Terrane, Klamath Mountains, California: An early Mesozoic volcanic arc and its basement of tectonically disrupted oceanic crust: Geological Society of America Bulletin, v. 106, p. 1033–1056.

Wyld, S.J., 1996, Early Jurassic deformation in the Pine Forest Range, northwest Nevada, and implications for Cordilleran tectonics: Tectonics, v. 15, p. 566–583.

Wyld, S.J., and Wright, J.E., 1988, The Devils Elbow Ophiolite remnant and overlying Galice Formation: New constraints on the Middle to Late Jurassic evolution of the Klamath Mountains, California: Geological Society of America Bulletin, v. 100, p. 29–44.

Yule, J.D., 1996, Geologic and tectonic evolution of Jurassic marginal ocean basin lithosphere, Klamath Mountains, Oregon [Ph.D. dissert.]: California Institute of Technology, California, 308 p.

MANUSCRIPT ACCEPTED BY THE SOCIETY FEBRUARY 23, 2000

Printed in the U.S.A.

Geological Society of America
Special Paper 349
2000

Tectonostratigraphic significance of sedimentary strata occurring within and above the Coast Range ophiolite (California Coast Ranges) and the Josephine ophiolite (Klamath Mountains), northwestern California

Emile A. Pessagno, Jr., and Donna M. Hull
Department of Geosciences, University of Texas at Dallas, P.O. Box 830688, Richardson, Texas 75083-0688, USA; e-mail: pessagno@utdallas.edu
Clifford A. Hopson
Department of Geological Sciences, University of California, Santa Barbara, California 93106, USA

ABSTRACT

In the study of ophiolites, an analysis of sedimentary rock that formed as interpillow sediment and as strata that rest in depositional contact above the lava can supply valuable data pertaining to ophiolite tectonogenesis. This paper focuses on two North American ophiolites: The Coast Range ophiolite in the California Coast Ranges and the Josephine ophiolite in the Klamath Mountains. The Coast Range ophiolite's lithosphere formed at an open-ocean spreading center where calcareous pelagic sedimentation and ferruginous and siliceous hydrothermal deposition accompanied volcanism. On the basis of an analysis of the physical stratigraphy, the geochronometry, and the radiolarian biostratigraphy, we conclude that a disconformity representing an 8 to 11 m.y. hiatus occurs between the Coast Range ophiolite and the overlying volcano-pelagic succession. This unconformity is correlative chronostratigraphically with that occurring in the San Pedro del Gallo terrane southwest of the Walper megashear in Mexico and in western Cuba. Therefore, we postulate that the regional unconformity indirectly reflects the titanic forces at work during the final breakup of Pangea and the opening up of the North Atlantic.

Radiolarian faunal and paleomagnetic data show that all remnants of the Coast Range ophiolite originated at near-equatorial paleolatitudes (Central Tethyan province) during the latest Middle Jurassic and were rapidly displaced northward to higher paleolatitudes (Northern Tethyan and Southern Boreal provinces) during the Late Jurassic. During northward tectonic transport, the Coast Range ophiolite "plate" subsided to abyssal depths and traversed a region of nondeposition (starved sedimentation, sediment bypassing, erosion) during the late Bathonian to early Oxfordian. It then moved through a region of siliceous (radiolarian) pelagic and tuffaceous volcaniclastic sedimentation—the depositional apron from an adjacent active volcanic arc—during the middle Oxfordian, Kimmeridgian, and Tithonian; the resulting deposits are termed "volcanopelagic." The contact between the volcanopelagic sedimentary rocks and the Great Valley Supergroup marks a period in which waning volcanogenic sedimentation was overwhelmed by a sudden influx of voluminous siliciclastic turbidite that originated from the continental margin (the uplifted Jurassic volcanoplutonic arc) at the beginning of the Nevadan orogeny (late Tithonian: lower Subzone 4α [based on Radiolaria]). A period of turbidite deposition occurred at precisely the same time and biohorizon in the distal backarc domain (San Pedro del Gallo terrane) in Mexico (age substantiated by ammonites, *Buchia,* and Radiolaria). Moreover, paleolatitudinal (faunal and paleomagnetic) data place these San Pedro del Gallo terrane remnants at approximately the same latitude as the Jurassic volcanoplutonic arc (Sierra Nevada segment) during the Late Jurassic (late Tithonian).

The Josephine ophiolite formed at a backarc or forearc spreading center close to a Jurassic volcanoplutonic arc where calc-alkalic interpillow tuff and tuffaceous chert occur within the volcanic member of the ophiolite. In the Smith River subterrane, the Josephine ophiolite (162 ± 1 Ma: late Callovian) is conformably overlain by volcanopelagic strata. Continuous volcanogenic sedimentation occurred during the late Callovian to early Oxfordian (? early middle Oxfordian). A low-latitude (Central Tethyan province) origin, coupled with northward transport, is also indicated by paleontological data (Radiolaria, ammonites, and *Buchia*) for the Jurassic Josephine ophiolite (Western Klamath terrane). The disconformable

Pessagno, Jr., E.A., Hull, D.M., and Hopson, C.A., 2000, Tectonostratigraphic significance of sedimentary strata occurring within and above the Coast Range ophiolite (California Coast Ranges) and the Josephine ophiolite (Klamath Mountains), northwestern California, *in* Dilek, Y., Moores, E.M., Elthon, D., and Nicolas, A., eds., Ophiolites and Oceanic Crust: New Insights from Field Studies and the Ocean Drilling Program: Boulder, Colorado, Geological Society of America Special Paper 349, p. 383–394.

contact between the volcanopelagic and the Galice Formation (s.l.) reflects a sudden influx of siliciclastic turbidite from a Jurassic volcanoplutonic arc source area and from older rocks of the accreted continental margin that lay inboard of the arc. The same event is represented in the Foothills terrane (Sierra Nevada) by the deposition of the siliciclastic turbidite of the Monte del Oro and Mariposa Formations. We suggest here that the event might also indirectly reflect the sudden deepening (inner neritic to upper abyssal or lower bathyal) of the paleo–Gulf of Mexico between the middle Oxfordian and early Kimmeridgian, which we demonstrate to occur in remnants of the San Pedro del Gallo terrane in Mexico and in Cuba. Faunal and paleomagnetic data from Mexican San Pedro del Gallo terrane remnants (distal backarc) indicate that these remnants were situated at Southern Boreal (>30°N) latitudes during the Middle and early Late Jurassic. Middle to Late Jurassic megafossils (ammonites and bivalves) from the Sierra Nevada indicate Southern Boreal paleolatitudes.

INTRODUCTION

The western part of the Cordilleran region is a collage of terranes, many of which have been displaced hundreds or even thousands of kilometers from their points of origin during the course of geologic time. The recognition of the first of these displaced terranes, Wrangellia, by Jones et al. (1977) revolutionized the geologic interpretation of the Circum-Pacific margin.

Since 1990, radiolarian biostratigraphy has had a great impact on the tectonostratigraphy (sensu Howell, 1995) of structurally complex eugeoclinal terranes in western North America and elsewhere in the world (e.g., Hopson et al., 1981; Pessagno and Blome, 1986; Pessagno et al., 1987a, 1987b, 1993, 1999; Montgomery et al., 1992, 1994a, 1994b; Cordey, 1998). Considerable progress has been made in developing detailed radiolarian zonations for much of the Paleozoic and Mesozoic. Moreover, it is now apparent from our analysis of radiolarian faunas from North America and elsewhere that Radiolaria can be utilized in paleolatitudinal investigations. As a result, radiolarian paleolatitudinal data can be used to document the presence of displaced terranes and chart their course through time (Pessagno and Blome, 1986; Pessagno et al., 1984, 1986, 1987a, 1993, 1999).

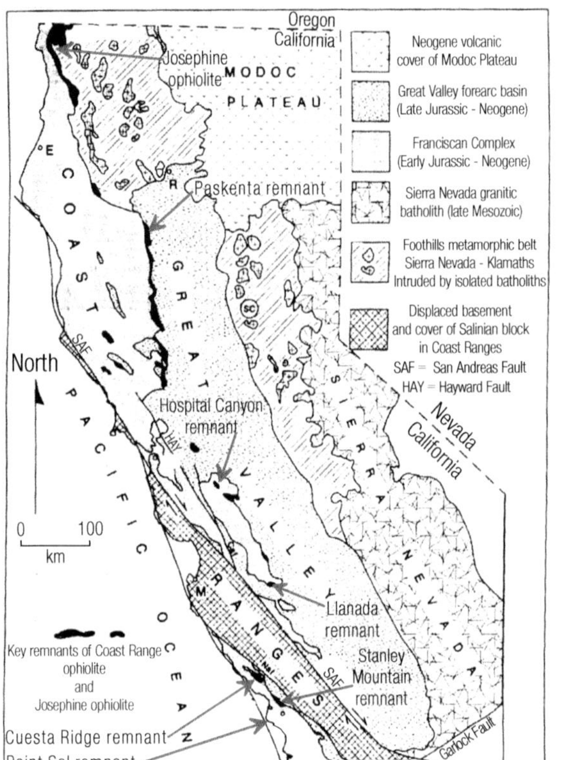

Figure 1. Index map showing important Coast Range ophiolite remnants, Josephine ophiolite, the Franciscan Complex, the Great Valley forearc strata, the Foothills metamorphic belt (Sierra Nevada and Klamath Mountains), Sierra Nevada batholith (granitic), etc. Modified from Dickinson et al. (1996). For complete inventory of Coast Range ophiolite localities, see Hopson et al. (1981, Fig. 14-1). E—Eureka, R—Redding, SC—Sacramento, SAF—San Andreas fault, HAY—Hayward fault.

COAST RANGE OPHIOLITE

Remnants of the Coast Range ophiolite occur at more than two dozen localities along the California Coast Ranges (Fig. 1). The Coast Range ophiolite consists of the Jurassic oceanic crustal basement that generally underlies Upper Jurassic volcanopelagic strata (Fig. 2). That succession is overlain in turn by the thick (composite thickness in Coast Ranges = ~17 500 m, W.R. Dickinson, 1999, personal communication), monotonous, siliciclastic-turbidite succession of the Great Valley Supergroup. The Coast Range ophiolite and its sedimentary cover are structurally juxtaposed against the Jurassic–Tertiary Franciscan Complex (Fig. 2) along the Coast Range fault.

Interpillow sedimentary rock within the ophiolite pillow lava provides clues to the tectonic setting in which the ophiolite formed. These sedimentary rocks are exclusively of three types: (1) light gray to cream-colored pelagic limestone (Fig. 3) (commonly metamorphosed deeper in the section to marble and calc-silicate rock), (2) red jasper (Fig. 6), and (3) rare red radiolarian chert. The jasper was

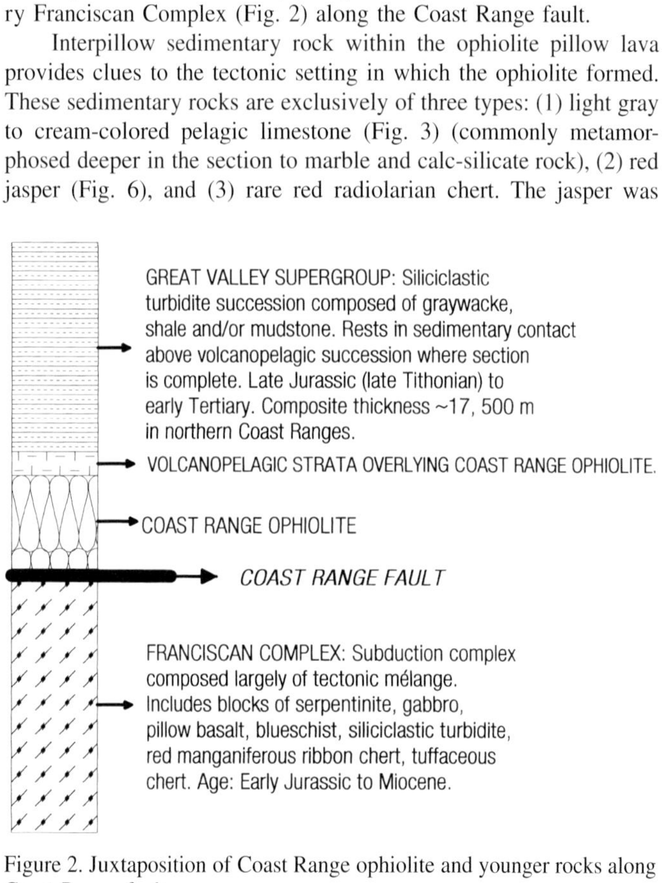

Figure 2. Juxtaposition of Coast Range ophiolite and younger rocks along Coast Range fault.

Figure 3. Interpillow limestone in volcanic member of Coast Range ophiolite just below unconformable contact with volcanopelagic succession. Point Sal remnant of Coast Range ophiolite, Santa Barbara County, California.

originally deposited on the sea floor as ferruginous umber by hydrothermal exhalations. The ophiolite lava is virtually devoid of volcaniclastic or terrigenous sedimentary strata. Hence, an open-ocean setting well beyond the reach of continental or volcanic arc-derived sediments is inferred (Hopson et al., 1981, 1986, 1992, 1996).

Directly overlying the Middle Jurassic Coast Range ophiolite are Upper Jurassic pelagic and volcanogenic strata (Fig. 2). This volcanopelagic succession displays a wide range of lithofacies that are included under two genetically related groups (described next).

Group 1: Distal lithofacies (30 to 130 m thick)

Most of the Coast Range ophiolite remnants (e.g., Point Sal, Stanley Mountain, and Cuesta Ridge; see Fig. 1 herein and Fig. 14-1 of Hopson et al., 1981) are covered by a thin sequence of mostly fine-grained marine sedimentary strata assignable to a *distal lithofacies.* The distal lithofacies consists of green to gray (commonly white-weathering) tuffaceous radiolarian chert (Figs. 4, 5), tuffaceous radiolarian mudstone, thin-bedded flinty tuff, coarse tuff including some zeolitized pumiceous tuff, and, at some localities, nodules, lenses and thin beds of gray- to cream-colored pelagic limestone (e.g., Stanley Mountain, Point Sal [Fig. 1]; described subsequently). These strata are inferred to represent the distal depositional fringe of fallout ash (tephra) and admixed pelagic ooze that accumulated on the seafloor adjacent to an active volcanic arc (Jurassic volcanoplutonic arc). The thickness of the distal facies is 30 to 130 m in most Coast Range ophiolite and volcanopelagic strata remnants.

It should be noted that our use of "Jurassic volcanoplutonic arc" corresponds closely to that of "Sierran-Klamath arc system" of Dickinson et al. (1996, p. 3). The Jurassic volcanoplutonic arc formed on oceanic crust relatively close to the continental margin—or sialized Paleozoic and early Mesozoic eugeoclinal rocks (approximately equal to the "continental margin arc" of Dilek and Moores, 1993). This volcanoplutonic arc complex would be continuous with that of the Klamath Mountains as well as that in the Blue Mountains of east-central Oregon, eastern Washington, and western Idaho. In California, the axis of the continental-margin arc corresponds to the axis of the granitic batholithic core of the Sierra Nevada. Granitic batholiths (largely quartz monzonite and quartz diorite) were emplaced during and after the Nevadan orogeny (Late Jurassic [late Tithonian] to Late Cretaceous).

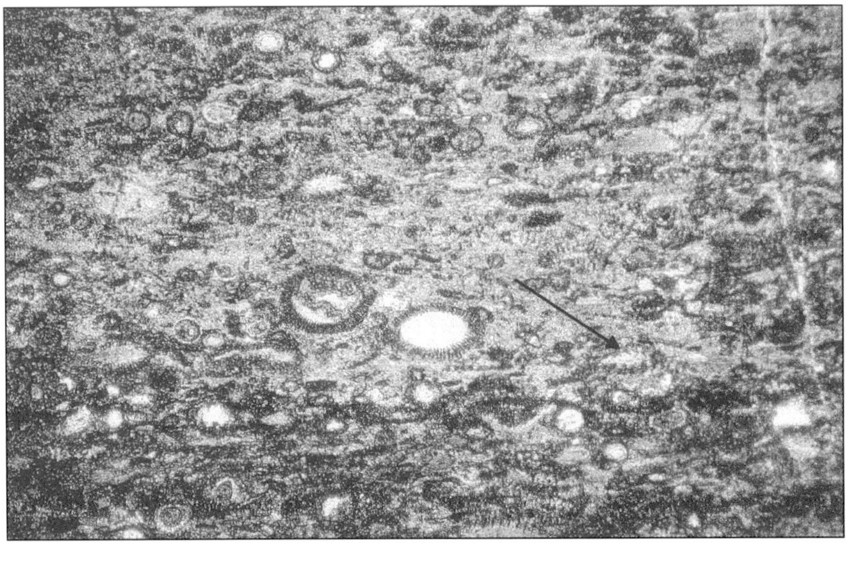

Figure 4. Photomicrograph of tuffaceous radiolarian chert from upper part of volcanopelagic succession, Coast Range ophiolite at Point Sal. Arrow points to *Parvicingula* sp. (0.7 mm).

Figure 5. Altered green and black tuffaceous chert in volcanopelagic succession above Coast Range ophiolite at Point Sal.

Figure 6. Red jasper in volcanic member of Coast Range ophiolite at Point Sal.

Figure 7. Andesite submarine laharic breccia at Llanada. See Figure 1 for location of Llanada remnant.

Group 2: Proximal lithofacies (250 to 900 m thick)

The *proximal lithofacies* includes a thicker succession composed of volcaniclastic sandstone (commonly turbidite), pumiceous and tuff-breccia beds (Fig. 7), and local volcaniclastic conglomerate, all interbedded with tuffaceous chert and tuffaceous radiolarian mudstone beds. The proximal lithofacies occurs at several Diablo Range localities: Llanada, Del Puerto, and Hospital Canyon (see Fig. 1) (Evarts, 1977; Hopson et al., 1981; Robertson, 1989). At Llanada, these strata grade up into ~500 m of crudely bedded to massive, cobbly to bouldery, andesitic diamictites, emplaced as submarine debris flows (Hull et al., 1993). This lithofacies represents a thick local variant of the base of the volcanopelagic proximal lithofacies. At Llanada, Del Puerto, and Hospital Creek, the basal volcanopelagic succession consists of a thin distal facies that grades up into the coarser, thicker proximal facies. At Llanada, ~10 m of green to gray tuffaceous radiolarian chert overlies the andesitic diamicrite at the top of the proximal lithofacies, separating it from the overlying basal Great Valley Supergroup (Hopson et al., 1981; Pessagno et al., 1984; Hull et al., 1993).

Prior to tectonic disruption and dispersion of Coast Range ophiolite oceanic crust into its present configuration of allochthonous remnants, the volcanopelagic succession formed a broad apron of deep-sea sediment that flanked the Jurassic volcanoplutonic arc (Hopson et al., 1981, 1986, 1992, 1996). At this site, the distal facies represented the outer fringe of water-laid tephra (airborne volcanic ash and dust) mixed with radiolarian ooze. The proximal facies comprises parts of a volcaniclastic turbidite and submarine debris-flow apron that lay closer to and thickened toward the volcanic arc source (Hopson et al., 1996). The trench formed a barrier to all arc-derived flowage deposits through most of the Late Jurassic; only the airborne fine ash was carried beyond it. But the trench became filled by the middle to late Tithonian (in response to Nevadan orogeny?), much like the filled trench of the modern Cascadian subduction zone. Thus, for the first time, the arc-derived volcaniclastic turbidites and debris flows formed a "proximal submarine apron" that extended across and beyond the trench, out over the deep-ocean floor to lay down the sediments that now constitute the volcanopelagic remnants at Llan-

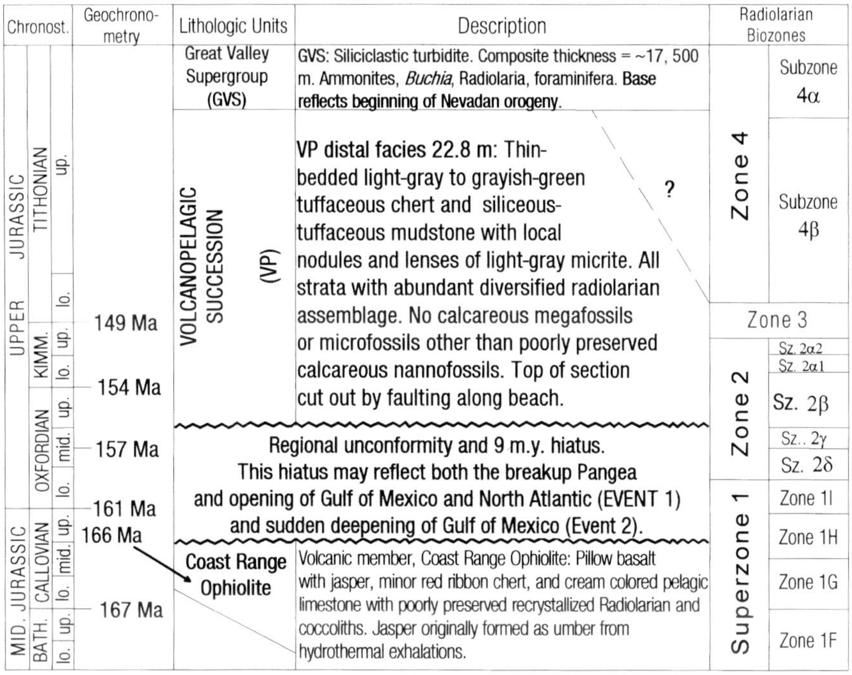

Figure 8. Stratigraphic summary for Point Sal remnant of Coast Range ophiolite. Time scale follows that of Gradstein et al. (1994). These workers relied on chronostratigraphic data from Pessagno et al. (1993) for Kimmeridgian and Oxfordian.

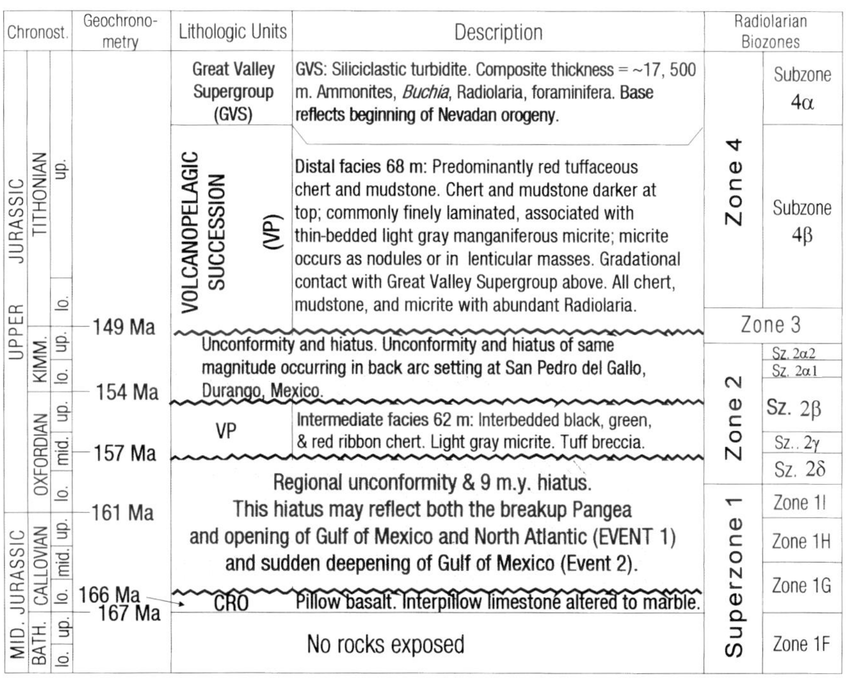

Figure 9. Stratigraphic summary for Stanley Mountain remnant of Coast Range ophiolite. Time scale follows that of Gradstein et al. (1994). These workers relied on data from Pessagno et al. (1993) for Kimmeridgian and Oxfordian.

ada, Del Puerto, and Hospital Creek. By the latest Tithonian, mud and silty-sandy turbidite (Great Valley Supergroup), derived from the uplift of the Nevadan orogen, spread out onto the oceanic crust and covered the volcanopelagic sediment (see Hopson et al., 1996, p. 6, Figs. 2 and 3).

The volcanopelagic succession rests disconformably upon the volcanic member of the Coast Range ophiolite (Figs. 8 and 9). Volcanopelagic strata were deposited onto the original upper surface of the ophiolite pillow lava, but only after a significant hiatus (Hopson et al., 1986, 1996). This hiatus is indicated both by the marked contrast in depositional environments represented by the Coast Range ophiolite interpillow sedimentary rocks and those of the overlying volcanopelagic succession and the marked differences in the age of the two units. The principal differences in depositional environment are as follows:

1. Ferruginous *hydrothermal* sedimentation (now red jasper: cf. Figs. 4 and 8) from submarine hot springs accompanied ophiolite volcanism, but terminated before the beginning of volcanopelagic sedimentation. Poorly preserved Radiolaria (often oxidized) compose a minor pelagic component of the hydrothermal sediment.

2. Calcareous ooze containing sparse, poorly preserved nannofossils (coccoliths) and Radiolaria and rare siliceous (radiolarian)

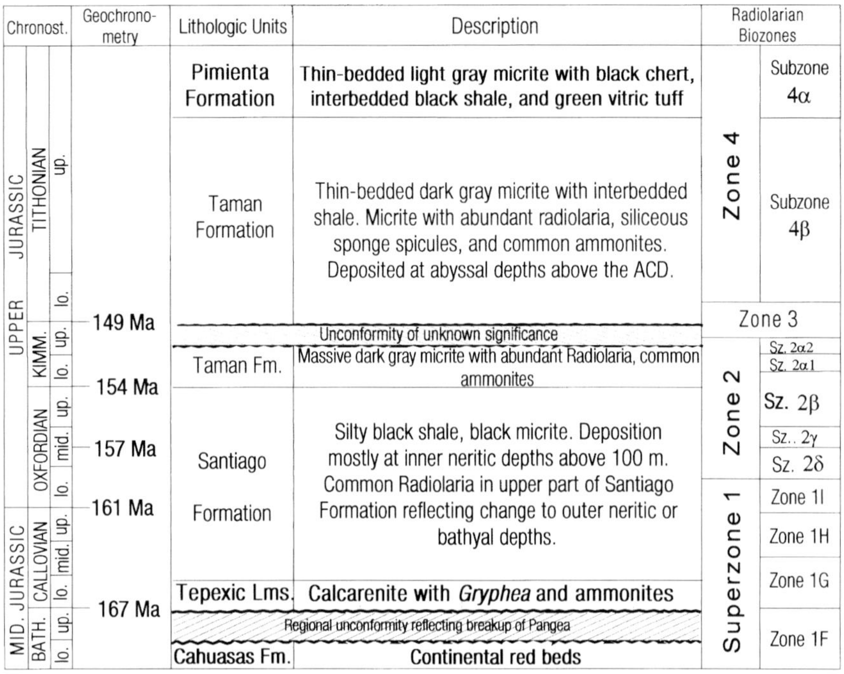

Figure 10. Huayacocotla remnant of San Pedro del Gallo terrane (Pessagno et al., 1999). Time scale follows that of Gradstein et al. (1994). These workers relied on data from Pessagno et al. (1993) for Kimmeridgian and Oxfordian.

ooze represents the only pelagic sediments deposited during ophiolite volcanism (see Fig. 3). These sediments were subsequently lithified to form interlava (including interpillow) pelagic limestone.

3. Arc-derived volcaniclastic sediments—even distal tephras—did *not* reach the Coast Range ophiolite oceanic crust during the period of igneous crust formation. However, they are the principal component of the overlying volcanopelagic succession, including its basal beds.

The sharp contrast in Coast Range ophiolite and volcanopelagic sedimentation can hardly be a coincidence; it clearly reflects the unconformable surface between these two units. The hiatus associated with this unconformity is substantiated by the 8 to 11 m.y. gap indicated by both geochronometric data from the ophiolite itself and radiolarian chronostratigraphic data from the base of the overlying volcanopelagic strata (Hull et al., 1993; Hull, 1991, 1995, 1997). At Stanley Mountain (Fig. 9), for example, middle Oxfordian (157 Ma) volcanopelagic strata rest disconformably on late Bathonian or early Callovian Coast Range ophiolite dated as 166.3 ± 1 Ma (U-Pb date on hornblende-quartz gabbro; see Gradstein et al., 1994).

That the unconformity separating the Coast Range ophiolite and volcanopelagic rocks would be accompanied by a significant hiatus is indicated by the following:

Interpillow pelagic limestone and red radiolarian chert within the Coast Range ophiolite volcanic member were formed from external pelagic ooze in an open-ocean environment (coccolith ooze and radiolarian ooze) with virtually no trace of arc-derived pyroclastic material. However, the overlying volcanopelagic unit, including its basal beds immediately above the contact with the Great Valley Supergroup, consists of a mixture of arc-derived volcanic ash and radiolarian ooze. Clearly, the externally derived sediment did not change so drastically immediately after the last submarine lava was erupted; it must have taken time for the Coast Range ophiolite ocean crust to move from an open-ocean environment to an environment near an active island arc. That time interval is marked by the Coast Range ophiolite–volcanopelagic strata unconformity and its associated hiatus. The Coast Range ophiolite U-Pb ages and the radiolarian biostratigraphy *quantify* the duration of the hiatus. However, they do not uniquely reveal it. This scenario is also explained in Hopson et al. (1996, p. 5, column 3).

It is postulated herein that the unconformity present between the volcanopelagic strata and the Coast Range ophiolite in the California Coast Ranges is indirectly related to that occurring in the San Pedro del Gallo terrane of west-central Mexico, east-central Mexico, and western Cuba (Fig. 10). In Mexico and Cuba, an unconformity of approximately the same age reflects the breakup of Pangea, the initiation of marine sedimentation, and the opening of the Gulf of Mexico and the North Atlantic during the late Bathonian (Pessagno et al., 1999). Our interpretation of biostratigraphic data from ODP Site 801 suggests that an unconformity and hiatus of identical age and magnitude can be recognized in the western Pacific (Pessagno and Hull, 1996). Moreover, our interpretation of European radiolarian biostratigraphic data presented by Baumgartner (1984, 1987) and Baumgartner et al. (1995) suggests that this unconformity and an accompanying hiatus of variable magnitude can be recognized throughout the Middle Jurassic of the southern Mediterranean. (The failure of Baumgartner [1984, 1987] and Baumgartner et al. [1995] to recognize reworking associated with this unconformity has corrupted chronostratigraphic correlation between southern Europe and North America [see discussions in Pessagno et al., 1987, 1993].)

Surely, a tectonic event of such titanic magnitude as the breakup of Pangea should be reflected in the Jurassic stratigraphic record. Powerful convection in the mantle not only may have led to the breakup of Pangea, but it also may have resulted in more rapid seafloor spreading along mid-ocean ridges associated with increased seismic and tectonic activity. With this scenario, the unconformity and hiatus between the Coast Range ophiolite and the volcanopelagic strata could be explained via more vigorous current action due to numerous changes in ocean-bottom topography. The result would be the erosion and bypassing of pelagic sediment.

No megafossils have been found in the volcanopelagic succession (s.s.). Moreover, no megafossils have been observed in the meager interpillow pelagic sedimentary rocks that occur within the Coast

Range ophiolite. Common to abundant Radiolaria and sparse, poorly preserved calcareous nannoplankton occur in the pelagic limestones that occur in both the Coast Range ophiolite (volcanic member) and the overlying volcanopelagic succession. No other calcareous microfossils are present. The sparse, poorly preserved nature of the calcareous nannofossil assemblage of both the interpillow and overlying volcanopelagic limestone may be due to the destruction of the microfossils from diagenetic alteration or from deposition at or below the CCD (calcite compensation depth). It is certain, however, that volcanopelagic strata formed at abyssal depths both below the ACD (aragonite compensation depth; absence of ammonites because of their aragonitic shells) and CCD (absence of pectenacids, *Buchia,* and calcareous foraminifera because of their calcite shells). The Jurassic strata of the overlying Great Valley Supergroup, however, contain common ammonites, abundant *Buchia,* calcareous foraminifera, calcareous nannoplankton, and abundant Radiolaria. The Jurassic Great Valley Supergroup appears to have been deposited for the most part above the ACD in waters of upper abyssal or bathyal depths.

Where the volcanopelagic strata–Great Valley Supergroup contact is exposed and not disrupted by faulting, it is conformable and in some cases, gradational. Two different types of gradational contacts can be observed:

1. Distal volcanopelagic strata grade upward into the Great Valley Supergroup. This type of contact can be observed at Llanada (Fig. 1). At this locality, greenish-gray to dark gray siliceous, tuffaceous mudstone and radiolarian chert (distal volcanopelagic facies) grade into the dark gray siliceous mudstone of the Great Valley Supergroup.

2. Distal volcanopelagic strata alternate with Great Valley Supergroup strata. This type of contact is well exposed at Stanley Mountain (Figs. 1, 9). Here black tuffaceous chert of the distal volcanopelagic facies alternates with the graywacke and dark gray siliceous mudstone of the Great Valley Supergroup over an interval of 30 m or more.

The volcanopelagic strata–Great Valley Supergroup contact marks a period in which waning volcanogenic sedimentation was overwhelmed by a sudden influx of voluminous siliciclastic turbidite that originated from the continental margin (uplifted Jurassic volcanoplutonic arc) at the beginning of the Nevadan orogeny (sensu King, 1959; Stanley, 1986) (late Tithonian: lower Subzone 4α [Radiolaria]). A period of turbidite deposition occurred at precisely the same time and biohorizon in the distal backarc domain in Mexico (San Pedro del Gallo terrane: Pessagno et al., 1999). We have been able to recognize this event by utilizing high-resolution radiolarian and ammonite biostratigraphy in remnants of the San Pedro del Gallo terrane (e.g., San Pedro del Gallo and Sierra Santa Rosa remnants of Pessagno et al., 1999, their Figs. 12 and 13). Moreover, paleolatitudinal (faunal and paleomagnetic) data place these terrane remnants at approximately the same latitude as the Jurassic volcanoplutonic arc (Sierra Nevada segment) during the Late Jurassic (Pessagno and Blome, 1986; Pessagno et al., 1999).

Radiolarian faunal data and paleomagnetic data show that the Point Sal and Llanada remnants of the Coast Range ophiolite originated at near-equatorial paleolatitudes (Central Tethyan province) during the latest Middle Jurassic. Subsequently, these remnants were rapidly displaced northward to higher paleolatitudes (Northern Tethyan and Southern Boreal provinces) during the Late Jurassic (Hopson et al., 1986, 1992, 1996; Hull, 1991, 1995, 1997; Hull et al., 1993; Pessagno et al., 1984, 1986, 1999). Radiolarian paleolatitudinal data indicate that the Stanley Mountain remnant of the Coast Range ophiolite likely originated at Central Tethyan paleolatitudes in the Middle Jurassic. Both radiolarian and paleomagnetic data indicate that the Stanley Mountain remnant of the Coast Range ophiolite was transported northward to Southern Boreal paleolatitudes by the Late Jurassic (Tithonian) (Hagstrum and Murchey, 1996; Hull, 1991, 1995, 1997; Hull et al., 1993, 1997; Hopson et al., 1996). A low-latitude (Central Tethyan province) origin, coupled with northward transport, is also indicated for the Jurassic Josephine ophiolite (Western Klamath terrane; see the subsequent section on the Josephine ophiolite).

U-Pb geochronometric data for the ophiolite remnants and biostratigraphic data for the volcanopelagic succession and basal Great Valley Supergroup, coupled with paleomagnetic, and faunal paleolatitudinal data, allow estimates of rates of northward displacement of volcanopelagic strata–Coast Range ophiolite oceanic lithosphere during the latest Middle Jurassic and Late Jurassic (Hopson et al., 1986, 1992, 1996; Hull, 1991, 1995, 1997; Hull et al., 1993, 1997; Pessagno et al., 1984, 1986, 1987a, 1999). The Point Sal Coast Range ophiolite remnant moved northward from between 1.7° ± 3.3°N and 1.7° ± 3.3°S at ~15–25 cm/yr. Arrival times of Coast Range ophiolite remnants at Southern Boreal paleolatitudes (>30°N) vary: e.g., Point Sal: late Kimmeridgian; Stanley Mountain: latest Kimmeridgian; Llanada: late Tithonian; and Paskenta: late Kimmeridgian (Hopson et al., 1986, 1992, 1996). The late arrival of the Llanada Coast Range ophiolite may suggest either a slower rate of transport or an origin farther south than that of the Point Sal remnant.

Various objections were raised by Hagstrum and Murchey (1996) to the paleolatitudinal displacement of the volcanopelagic strata–Coast Range ophiolite as exposed at Stanley Mountain. These objections were fully aired by Hull et al. (1997, p. 1633–1639) and Pessagno et al. (1999) and will not be reiterated herein.

JOSEPHINE OPHIOLITE

Western Klamath terrane

The Western Klamath terrane of Blake et al. (1985) includes Middle and Upper Jurassic rocks that crop out along the western margin of the Klamath Mountains and in a large outlier along the coast (northwestern California; southwestern Oregon). Blake et al. subdivided the Western Klamath terrane into five distinct subterranes that were each characterized by a different substrate (Fig. 11). Previous investigations by Pessagno et al. (1993) focused on two of these subterranes: the Rogue Valley subterrane and the Smith River subterrane (Fig. 11). In this paper, we focus on the Smith River subterrane because this subterrane contains the Josephine ophiolite.

Smith River subterrane

The Smith River subterrane (southwestern Oregon; northwestern California) structurally overlies the Rogue Valley subterrane and the Dry Butte subterrane along the Madstone Cabin thrust (Blake et al., 1985; Irwin, 1989) (Fig. 11). Our studies (Pessagno et al., 1993) of the Smith River subterrane focused primarily on the area mapped by Harper (1983, 1984) in northwestern California and in an adjacent area bridging the Oregon-California border. According to Harper (1983, 1984), the rocks of the Smith River subterrane were exposed to prehnite-pumpellyite to lower-greenschist metamorphism.

Volcanic member of the Josephine ophiolite. At the Turner Albright mine along the Oregon-California border, pillow lavas of the volcanic member of the Josephine ophiolite commonly are interbedded with masses of light green to gray tuffaceous radiolarian chert. Along the Middle Fork of the Smith River, Saleeby (1987) obtained a concordant U-Pb age of 162 ± 1 Ma (zircon in plagiogranite) from

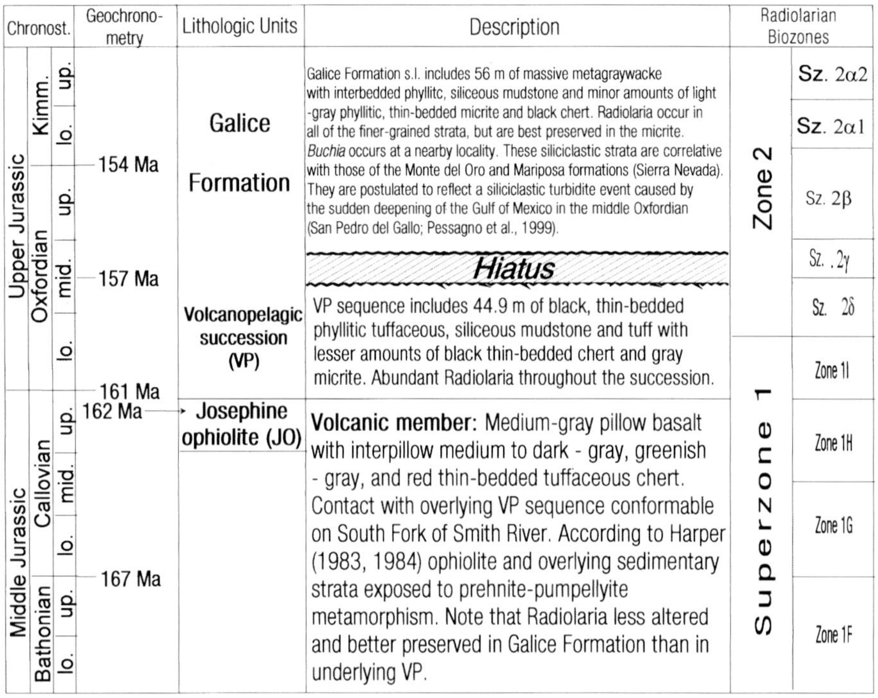

Figure 11. Josephine ophiolite, Smith River subterrane, Klamath Mountains, northwestern California. Time scale follows that of Gradstein et al. (1994). These workers relied on data from Pessagno et al. (1993) for Kimmeridgian and Oxfordian.

the ophiolite. Middle Jurassic (upper Callovian) Radiolaria assignable to Zone 1H were recovered from chert in the volcanic member (Fig. 11).

The presence of volcanogenic (calk-alkalic) interpillow sedimentary strata (greenish-gray tuff and tuffaceous chert) within the volcanic member of the Josephine ophiolite contrasts strongly with the lack of such strata within the volcanic member of the Coast Range ophiolite; the volcanogenic material suggests that the Josephine ophiolite formed within or near a magmatic arc. Moreover, the presence of an abundant, relatively diverse radiolarian assemblage, taken together with the lack of calcareous megafossils or microfossils, suggests deposition below the CCD at abyssal depths.

The depositional contact between the volcanic member of the Josephine ophiolite and overlying sedimentary strata is well exposed along the Middle Fork of the Smith River just upstream from the mouth of Little Jones Creek. Here, the sedimentary succession (100.9 m) consists of two units (Fig. 11): a lower volcanopelagic succession and an upper siliciclastic turbidite succession (Galice s.l.: see Pessagno et al., 1993, for the distinction between Galice s.l. and Galice s.s.).

Volcanopelagic succession. The volcanopelagic succession includes 44.9 m of black, thin-bedded, phyllitic, tuffaceous, siliceous mudstone (tuffaceous "argillite") and tuff, which are interbedded with lesser black, thin-bedded chert and thin-bedded, lenticular masses of phyllitic, medium gray pelagic limestone. These strata conformably overlie the volcanic member of the Josephine ophiolite. Radiolaria occur throughout the volcanopelagic succession, but are only well preserved in the pelagic limestone. No megafossils are known from the volcanopelagic succession at this locality or, for that matter, at any other locality in the Smith River subterrane. The presence of Radiolaria and the lack of calcareous microfossils and megafossils within the volcanopelagic succession suggest deposition below the CCD at abyssal depths. Radiolarian biostratigraphic data from the volcanopelagic succession indicate that these strata are assignable to Zone 1I and Zone 2, Subzone 2δ (upper Callovian to middle middle Oxfordian) (see Pessagno et al., 1993).

Galice Formation s.l. The Galice s.l. consists of 56 m of massive metagraywacke with interbedded, thick intervals of black, phyllitic, siliceous mudstone and minor amounts of light gray, phyllitic, thin-bedded pelagic limestone and thin-bedded black chert (Fig. 11). The siliceous mudstone beds in the upper 15.2 m of the siliciclastic turbidite succession contain common, small rip-up clasts of light gray pelagic limestone. Radiolaria occur in all of the finer-grained strata. However, they are only well preserved in the pelagic limestone. *Buchia concentrica* (middle Oxfordian to upper Kimmeridgian) was noted from Galice s.l. strata by Diller (1907) and Harper (1983) at nearby localities along Shelly Creek and Monkey Creek to the north. The presence of calcareous megafossils and microfossils (e.g., foraminifera in pelagic limestone) suggests deposition above the CCD at upper-abyssal depths. Radiolaria assignable to Zone 2, Subzone 2γ (upper part: middle middle Oxfordian) were recovered from the Galice Formation s.l. in the Rogue Valley subterrane (Pessagno et al., 1993).

Harper (1983, 1984) included both the volcanopelagic succession and the Galice Formation s.l. in the Galice Formation. Pessagno et al. (1993) excluded the volcanopelagic succession from the Galice in that it is neither lithologically nor genetically related to the type Galice Formation. The volcanopelagic succession was formed from an admixture of radiolarian ooze and fine pyroclastic material that were deposited in relative proximity to the Rogue island arc (= Nevadan island arc herein) (Saleeby and Harper, 1993). Pessagno et al. (1993) suggested that the volcanopelagic succession represents the distal equivalent of the Rogue Formation and thus restricted the term "Galice" along the Middle Fork of the Smith River to the Galice Formation s.l. The Galice Formation s.s. and Galice Formation s.l. consist of siliciclastic turbidite and are lithologically distinct from the volcanopelagic succession.

Both the volcanopelagic succession and the overlying Galice s.l. contain common thin-bedded layers or lenticular masses of light gray pelagic limestone and thin-bedded black chert. Although these rocks occur in the type Galice, they are rare and are only present in the lower part of the unit near its contact with the Rogue Formation (Pessagno et al., 1993). It should be noted that the Galice s.l. occur-

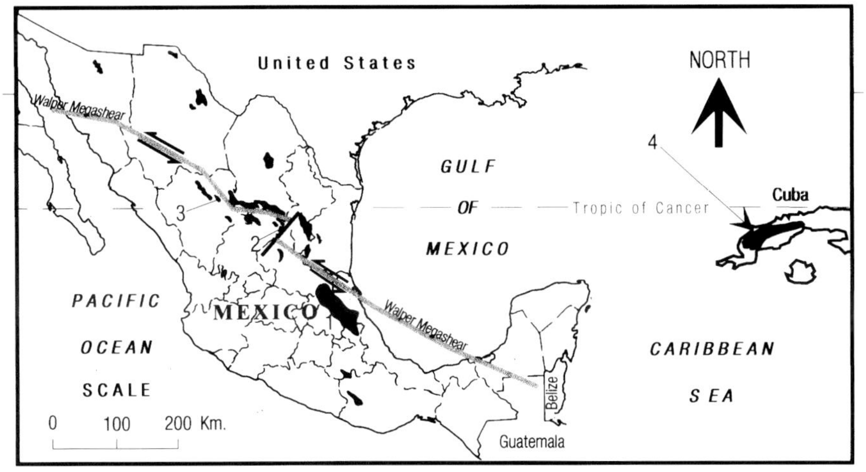

Figure 12. Index map showing important Jurassic localities (black areas) in Mexico and Cuba. The most important localities for this paper are numbered 1–4. A complete listing of all localities adjacent to the Walper megashear was presented by Pessagno et al. (1999). Key to localities herein: 1—Huayacocotla anticlinorium: Taman-Tamazunchale, San Luis Potosí; Huayacocotla, Veracruz; Huachinango, Puebla. 2—Sierra Santa Rosa, Zacatecas. 3—San Pedro del Gallo, Durango. 4—Cordillera de Guaniguanico, Cuba. Base map partly derived from that in Salvador et al. (1992). From Pessagno et al. (1999, Fig. 2).

ring along both the Middle Fork of the Smith River and Little Jones Creek is identical to the lower part of the Monte del Oro Formation at its type area near Oroville, California (Foothills terrane, Sierra Nevada; Pessagno et al., 1993, p. 111). The lower part of both units contain thin-bedded gray micritic limestone with common to abundant Radiolaria.

A dramatic difference in preservation occurs between radiolarian specimens extracted from pelagic limestone within the volcanopelagic succession and the overlying Galice s.l. The Radiolaria from the volcanopelagic succession tend to be considerably more recrystallized and deformed, whereas those from the overlying Galice are not recrystallized and are less deformed. These easily observable differences in the preservation of the Radiolaria (see Pessagno et al., 1993) suggest that the Josephine ophiolite and the volcanopelagic succession had been exposed to a period of diagenetic change (metamorphism?) prior to the deposition of the Galice. The dramatic difference in radiolarian preservation above and below the volcanopelagic strata–Galice s.l. contact taken together with the sudden change from hemipelagic (including calc-alkalic tuff and tuffaceous chert) sedimentation (volcanopelagic) to siliciclastic turbidite sedimentation (Galice s.l.) suggest that this contact is unconformable (Fig. 11).

The interface between the volcanopelagic succession and the Galice marks the onset of the deposition of siliciclastic strata (turbidite) both in the Klamath Mountains and in the Sierra Nevada. Regionally, there appears to be a genetic relationship between the Galice Formation s.s. and the Galice s.l. (Western Klamath terrane) and the Monte del Oro and Mariposa Formations (Foothills terrane; Sierra Nevada) (Imlay, 1980, p. 62; Pessagno et al., 1993). All of these lithostratigraphic units are of middle Oxfordian to early Kimmeridgian age and consist of similar siliciclastic turbidite successions. Pessagno et al. (1993) indicated that this siliciclastic turbidite event could conceivably represent an earlier pulse of the Nevadan orogeny. However, it might also reflect the sudden deepening (inner neritic to upper abyssal or lower bathyal) of the paleo–Gulf of Mexico between the middle Oxfordian and early Kimmeridgian. In the San Pedro del Gallo terrane (west-central and east-central Mexico), Pessagno et al. (1999) utilized high-resolution radiolarian and ammonite biostratigraphy and chronostratigraphy to document a regional middle Oxfordian to early Kimmeridgian disconformity, which can be found from the Huayacocotla anticlinorium to the southeast to San Pedro del Gallo, ~1000 km to the northwest (Figs. 10 and 12). Stratigraphic data (microfacies analysis) from numerous localities indicate that a shift in deposition occurred from inner neritic (<100 m) to outer neritic or upper abyssal (~3000 m) during the middle Oxfordian to early Kimmeridgian. Moreover, the faunal data (megafossils, i.e., ammonites and *Buchia*; Radiolaria) indicate that the San Pedro del Gallo terrane (depending on which remnant) was at Northern Tethyan to Southern Boreal paleolatitudes (>23°N paleolatitude). The San Pedro del Gallo remnant (Fig. 12: Durango, Mexico) was situated at Southern Boreal paleolatitudes (>30°N). This paleogeographic interpretation is substantiated by the ammonite, *Buchia*, radiolarian, and paleomagnetic data. J.G. Ogg's preliminary paleomagnetic data (in Pessagno, 1995, and Pessagno et al., 1999) for San Pedro del Gallo actually indicate 40°. The radiolarian and megafossil faunal data suggest that this remnant was situated at approximately the same Southern Boreal paleolatitude as the Smith River subterrane during the middle Oxfordian to early Kimmeridgian. Pessagno et al. (1999) regarded the San Pedro del Gallo terrane as being in the backarc domain of the Jurassic volcanoplutonic arc. Given the paleolatitudinal data, would not this tectonic event be reflected in the stratigraphic record of the Jurassic volcanoplutonic arc (Sierra Nevada and Klamath Mountains)? The Late Jurassic megafossil assemblage from the Sierra Nevada and Klamath Mountains is clearly Southern Boreal (>30°N) (Imlay, 1980; Pessagno and Blome, 1986).

Faunal data from strata within the volcanic member of the Josephine ophiolite and above the ophiolite indicate that the ophiolite originated at Central Tethyan paleolatitudes during the latest Callovian (162 Ma) and was carried northward to Northern Tethyan and Southern Boreal paleolatitudes during the early and middle Oxfordian (see Pessagno et al., 1993). Whether the Josephine ophiolite formed at a backarc spreading center as suggested by Saleeby et al. (1982) and Dickinson et al. (1996) or a forearc spreading center as suggested by Saleeby (1996), it is likely that the rate of movement northward was not much greater than that of the continent. It is conceivable that the ophiolite originated near the boundary between the Central and Northern Tethyan provinces (i.e., ~23°N: see Pessagno et al., 1987a, 1993). The presence of identical tuffaceous cherts both within the volcanic member of the ophiolite and within the overlying volcanopelagic succession strongly supports a backarc or forearc origin rather than an open-ocean origin for the Josephine ophiolite.

CONCLUSIONS AND COMPARISONS

1. The Coast Range ophiolite lithosphere formed at an open-ocean spreading center near the Jurassic paleoequator where calcareous pelagic sedimentation and ferruginous and siliceous hydrothermal deposition accompanied volcanism (Hopson et al., 1996). In

contrast, the Josephine ophiolite formed at a backarc or forearc spreading center (probably at the boundary between the Central Tethyan and Northern Tethyan provinces: ~23°N) close to the Jurassic volcanoplutonic arc where calc-alkalic tuff and calc-alkalic tuffaceous radiolarian ooze were deposited both within the basaltic pillow lava of the Josephine ophiolite as well as in the volcanopelagic succession above the Josephine ophiolite (Pessagno et al., 1993).

2. Both the Coast Range ophiolite and Josephine ophiolite lithosphere formed at abyssal depths below the CCD.

3. The Coast Range ophiolite "plate" subsided to abyssal depths, passing through a region of starved sedimentation (nondeposition; see the next point) during the late Bathonian to early Oxfordian and then through a region of siliceous (radiolarian) pelagic and airborne-tuffaceous sedimentation—the depositional apron from an adjacent active volcanic arc—during the middle Oxfordian, Kimmeridgian, and Tithonian.

4. High-resolution biostratigraphy and chronostratigraphy (ammonites and Radiolaria) indicate that the unconformable contact between the Coast Range ophiolite and overlying volcanopelagic succession can be correlated chronostratigraphically with that occurring in the San Pedro del Gallo terrane southwest of the Walper megashear in Mexico and in the San Pedro del Gallo terrane in western Cuba (Pessagno et al., 1987a, 1999; Meng, 1997). It is postulated here that the volcanopelagic strata–Coast Range ophiolite unconformity and its associated hiatus *indirectly* reflects the breakup of Pangea. An acceleration of powerful convection in the mantle not only may have led to the Pangean breakup, but also may have generated more rapid seafloor spreading along mid-ocean ridges and increased seismic and tectonic activity both along the ridge and ocean floor. Given such a scenario, the volcanopelagic strata–Coast Range ophiolite unconformity and hiatus could be explained by more vigorous current action due to numerous changes in bottom topography on and away from the Coast Range ophiolite spreading center. The end product would be bypassing and erosion of pelagic sediment—leaving a hiatus in the stratigraphic record.

Our interpretation of radiolarian biostratigraphic data presented by Baumgartner (1984, 1987) and Baumgartner et al. (1995) suggests that an unconformity and hiatus of similar magnitude can be recognized in southern Europe and north Africa. We propose that it likewise reflects the breakup of Pangea and the opening up of the North Atlantic and Gulf of Mexico. It should be noted that reworking associated with this unconformity has resulted in the misinterpretation of radiolarian biostratigraphic and chronostratigraphic data by Baumgartner (1984, 1987) and Baumgartner et al. (1995) and has resulted in invalid chronostratigraphic correlations between southern Europe and North America.

5. The younger Josephine ophiolite (162 ± 1 Ma: late Callovian) is conformably overlain by volcanopelagic strata in the Smith River subterrane. Continuous volcanogenic sedimentation occurred during the late Callovian to early Oxfordian (? early middle Oxfordian). The disconformable contact between the volcanopelagic and the Galice s.l. (Fig. 12) reflects the sudden influx of siliciclastic turbidite from a source area in the Jurassic volcanoplutonic arc; the same event is represented in the Foothills terrane (Sierra Nevada) by the deposition of the siliciclastic turbidite of the Monte del Oro and Mariposa Formations (Imlay, 1980; Pessagno et al., 1993). This unconformity reflects a middle Oxfordian tectonic event that can be recognized by high-resolution biostratigraphy (Radiolaria and ammonites) in the San Pedro del Gallo terrane for ~1000 km along the southwest side of the Walper megashear as well as in western Cuba (Pessagno et al., 1999). In the San Pedro del Gallo terrane, it reflects a sudden deepening of the basin of deposition from inner neritic (<200 m) to upper abyssal (~3000 m) depths.

6. In Coast Range ophiolite remnants, the interface between the volcanopelagic and Great Valley Supergroup reflects the sudden influx of Great Valley Supergroup siliciclastic turbidite during the Nevadan orogeny at the North American continental margin. High-resolution radiolarian biostratigraphic data from the Stanley Mountain, Cuesta Ridge, and Llanada remnants of the Coast Range ophiolite indicate that the volcanopelagic strata–Great Valley Supergroup contact is latest Tithonian (Pessagno et al., 1984, 1987a, 1987b; Hull 1991, 1995, 1997; Hull et al., 1993). High-resolution radiolarian and ammonite biostratigraphy and chronostratigraphy indicate that a turbidite event of precisely the same age (late Tithonian: lower part of radiolarian Subzone 4α) occurs in the Sierra Santa Rosa and San Pedro del Gallo remnants of the San Pedro del Gallo terrane of west-central Mexico (Fig. 12; Pessagno et al., 1999). Both of these San Pedro del Gallo terrane remnants are considered to be distal backarc by Pessagno et al. (1999). Moreover, both remnants were displaced from higher (Southern Boreal) paleolatitudes to lower (Tethyan) paleolatitudes during the Late Jurassic and the earliest Cretaceous. Given the faunal data (not just Radiolaria, but megafossils such *Buchia* and ammonites as well) and some paleomagnetic data from J.G. Ogg (in Pessagno, 1995, and Pessagno et al., 1999), it is probable that these San Pedro del Gallo terrane remnants might represent some of the missing Upper Jurassic backarc from farther north (Nevada?).

7. Radiolarian faunal data and paleomagnetic data show that the Point Sal, Stanley Mountain, and Llanada remnants of the Coast Range ophiolite originated at near-equatorial paleolatitudes (Central Tethyan province) during the latest Middle Jurassic and were rapidly displaced northward to higher paleolatitudes (Northern Tethyan and Southern Boreal provinces) during the Late Jurassic (Hopson et al., 1996; Hull, 1991, 1995; Hull and Pessagno, 1995; Hull et al., 1993; Pessagno et al., 1984, 1987a, 1999). A low-latitude (Central Tethyan province) origin, coupled with northward transport, is also indicated for the Josephine ophiolite (Pessagno et al., 1993).

ACKNOWLEDGMENTS

Pessagno's and Hull's investigations were supported by grants from the National Science Foundation (DES-72-01528, DES-72-01528-A01, GA35094, EAR-8615790, EAR-8816601, EAR-9418194, EAR-9304459). We thank William R. Dickinson (University of Arizona), Raymond Ingersoll (University of California, Los Angeles) and Cristopher A. Suczek (Western Washington University) for their reviews of the manuscript. We absolve the reviewers from our final interpretation of the data. Department of Geological Sciences, University of California, Santa Barbara, Contribution no. 910.

REFERENCES CITED

Baumgartner, P.O., 1984, Middle Jurassic–Early Cretaceous low latitude radiolarian zonation based on unitary associations and age of Tethyan radiolarites: Eclogae Geologicae Helvetiae, v. 77, p. 729–837.

Baumgartner, P.O., 1987, Age and genesis of Tethyan Jurassic radiolarites: Eclogae Geologicae Helvetiae, v. 80, p. 831–879.

Baumgartner, P.O., et al., 1995, Middle Jurassic to Lower Cretaceous radiolaria of Tethys: Occurrences, systematics, biochronology: Mémoires de Géologie (Lausanne): p. 1–1172.

Blake, M.C., Jr., Engebretson, D.C., Jayko, A.S., and Jones, D.L., 1985, Tectonostratigraphic terranes in southwest Oregon, *in* Howell, G.D., ed., Tectonostratigraphic terranes of Circum-Pacific region: Circum-Pacific Council for Energy and Mineral Resources Earth Sciences Series, v. 1, p. 147–157.

Cordey, F., 1998, Radiolaires des complexes d'Accrétion de La Cordillère Canadienne (Combie-Fritannique): Commission Géologique du Canada Bulletin 509, p. 1–209.

Dickinson, W.R., Hopson, C.A., and Saleeby, J.B., 1996, Alternate origins of the Coast Range ophiolite (California): Introduction and implications: GSA Today, v. 6, p. 1–10.

Dickinson, W.R., Schweickert, R.A., and Ingersoll, R.V., 1996, Coast Range ophiolite as back-arc inter-arc basin lithosphere, *in* Dickinson, W.R., Hopson, C.A., and Saleeby, J.B., Alternate origins of the Coast Range ophiolite (California): Introduction and implications: GSA Today, v. 6, p. 2–5.

Dilek, Y., and Moores, E.M., 1993, Across-strike anatomy of the Cordilleran orogen at 40°N latitude: Implications for the Mesozoic paleogeography of the western United States, *in* Dunne, G., and McDougall, K., eds., 1993, Mesozoic paleogeography of the western United States–II: Pacific Section, SEPM Book 71, p. 333–346.

Diller, J.S., 1907, The Mesozoic sediments of southwestern Oregon: American Journal of Science, v. 3, p. 401–421.

Evarts, R.C., 1977, The geology and petrology of the Del Puerto ophiolite, Diablo Range, central California Coast Ranges, *in* Coleman, R.G., and Irwin, W.P., eds., North American ophiolites: Oregon Department of Geology and Mineral Industries Bulletin 95, p. 121–140.

Gradstein, F.M., Agterberg, F.P., Ogg, J.G., Hardenbol, J., van Veen, P., Thierry, J., and Huang, Z., 1994, A Mesozoic time scale: Journal of Geophysical Research, v. 99, no. B12, p. 24051–24074.

Hagstrum, J.T., and Murchey, B.L., 1996, Paleomagnetism of Jurassic radiolarian chert above the Coast Range Ophiolite at Stanley Mountain, California, and its implications for its paleogeographic origins: Geological Society of America Bulletin, v. 108, p. 642–643.

Harper, G.D., 1983, A depositional contact between the Galice Formation and a Late Jurassic ophiolite in northwestern California and southwestern Oregon: Oregon Geology, v. 45, p. 3–9.

Harper, G.D., 1984, Josephine ophiolite, northwestern California: Geological Society of America Bulletin, v. 95, p. 1009–1026.

Hopson, C.A., Mattinson, J.M., and Pessagno, E.A., Jr., 1981, Coast Range ophiolite, western California, *in* Ernst, W.G., ed., The geotectonic development of California (Rubey Volume I): Prentice-Hall, Englewood Cliffs, New Jersey, p. 419–510.

Hopson, C., Beebe, W., Mattinson, J., Pessagno, E., and Blome, C., 1986, California Coast Range ophiolite, Jurassic tectonics: Eos, Transactions of the American Geophysical Union, v. 67, p. 1232.

Hopson, C.A., Mattinson, J., Luyendyk, B.P., Beebe, W., Pessagno, E.A., Jr., Hull, D.M., and Blome, C.D., 1992, California Coast Range ophiolite: Jurassic tectonic history: American Association of Petroleum Geologists Bulletin, v. 76, p. 422.

Hopson, C.A., Pessagno, E.A., Jr., Mattinson, J.M., Luyendyk, B.P., Beebe, W., Hull, D.M., Muñoz, I.M., and Blome, C.D., 1996, Coast Range ophiolite as paleoequatorial mid-ocean lithosphere, *in* Dickinson, W.R., Hopson, C.A., and Saleeby, J.B., Alternate origins of the Coast Range ophiolite (California): Introduction and implications: GSA Today, v. 6, p. 1–10.

Howell, D.G., 1995, Principles of terrane analysis: New applications for global tectonics: Topics in the Earth Science 8 (Second Edition), London, Chapman and Hall, 245 p.

Hull, D.M., 1991, Upper Jurassic radiolarian biostratigraphy of the lower member of the Taman Formation, east-central Mexico and of volcanopelagic strata overlying the Coast Range ophiolite, Stanley Mountain, southern California Coast Ranges [Ph.D. dissert.]: University of Texas at Dallas, 696 p.

Hull, D.M., 1995, Morphologic diversity and paleogeographic significance of the Family Parvicingulidae (Radiolaria): Micropaleontology, v. 41, p. 1–48.

Hull, D.M., 1997, Upper Jurassic Tethyan and Southern Boreal radiolarians from western North America: Micropaleontology, v. 43, supplement 2, p. 1–202.

Hull, D.M., Pessagno, E.A., Jr., Blome, C.D., Hopson, C.A., and Muñoz, I.M., 1993, Chronostratigraphic assignment of volcanopelagic strata above the Coast Range ophiolite, *in* Dunne, G., and McDougall, K., eds., 1993, Mesozoic paleogeography of the western United States–II: Pacific Section, SEPM Book 71, p. 157–170.

Hull, D.M., Blome, C.D., and Pessagno, E.A., Jr., 1997, Paleomagnetism of Jurassic radiolarian chert above the Coast Range Ophiolite at Stanley Mountain, California, and its implications for its paleogeographic origins: Discussion and reply: Geological Society of America Bulletin, v. 109, p. 1633–1639.

Imlay, R.W., 1980, Jurassic paleobiogeography of the conterminous United States in its continental setting, U.S. Geological Survey Professional Paper 1062, 134 p.

Irwin, W.P., 1989, Terranes of the Klamath Mountains, California and Oregon, *in* International Geological Congress Guidebook, Field Trip T108, Tectonic Evolution of Northern California, p. 19–32.

Jones, D.L., Silberling, N.J., and Hillhouse, J., 1977, Wrangellia: A displaced terrane in northwestern North America: Canadian Journal of Earth Sciences, v. 14, p. 2565–2577.

King, P.B., 1959, The evolution of North America: Princeton, New Jersey, Princeton University Press, p. 147.

Meng, X., 1997, Radiolarian biostratigraphy of the Upper Jurassic of San Pedro del Gallo terrane, north–central Mexico, and the Lower Cretaceous of Nooksack Group, Nooksack terrane, Northwestern Washington [Ph.D. dissertation]: University of Texas at Dallas, Graduate Program in Geosciences, p. 1–352.

Montgomery, H.A., Pessagno, E.A., Jr., and Muñoz, I., 1992, Jurassic (Tithonian) radiolaria from La Désirade (Lesser Antilles): Preliminary paleontological and tectonic implications: Tectonics, v. 11, p. 1426–1432.

Montgomery, H., Pessagno, E.A., Jr., Lewis, J.A., and Schellekens, J.H., 1994a, Paleogeography of the Jurassic fragments in the Caribbean: Tectonics, v. 13, p. 725–732.

Montgomery, H.A., Pessagno, E.A., Jr., and Pindell, J.L., 1994b, A 195 Ma terrane in a 165 Ma sea: Pacific origin of Caribbean Plate: GSA Today, v. 4, p. 3–6.

Pessagno, E.A., Jr., 1995, Stratigraphic evidence for northwest to southeast movement along the west side of the Walper Lineament: Geological Society of America Abstracts with Programs, v. 27, p. A75.

Pessagno, E.A., Jr., and Blome, C.D., 1986, Faunal affinities and tectonogenesis of Mesozoic rocks in the Blue Mountains Province of eastern Oregon and western Idaho, *in* Vallier, T.L., and Brooks, H.C., eds., Geology of the Blue Mountains Region of Oregon, Idaho, and Washington: Geologic implications of Paleozoic and Mesozoic Paleontology and Biostratigraphy, U.S. Geological Survey Professional Paper 1435, p. 65–78.

Pessagno, E.A., Jr., and Hull, D.M., 1996, Once upon a time in the Pacific: Chronostratigraphic misinterpretation of basal strata at ODP Site 801 (Central Pacific) and its impact on geochronology and plate tectonics models, *in* Riccardi, A.C., ed., Advances in Jurassic research: GeoResearch Forum, v. 1–2, p. 79–92.

Pessagno, E.A., Jr., Blome, C.D., and Longoria, J.F., 1984. A revised radiolarian zonation from Upper Jurassic of western North America: Bulletins of American Paleontology, v. 87, p. 1–51.

Pessagno, E.A., Jr., Whalen, P.A., and Yeh, K.-Y., 1986, Jurassic Nassellariina (Radiolaria) from North American geologic terranes: Bulletins of American Paleontology, v. 91, p. 1–75.

Pessagno, E.A., Jr., Longoria, J.F., MacLeod, N., and Six, W.M., 1987a, Upper Jurassic (Kimmeridgian-upper Tithonian) Pantanelliidae from the Taman Formation, east-central Mexico: Tectonostratigraphic, chronostratigraphic, and phylogenetic implications, *in* Culver, S.J., ed., Studies of North American Jurassic radiolaria, Part. I: Cushman Foundation Foraminiferal Research Special Publication 23, p. 1–51.

Pessagno, E.A., Jr., Blome, C.D., Carter, E.S., MacLeod, N., Whalen, P.A., and Yeh, K., 1987b, Preliminary radiolarian zonation for the Jurassic of North America, *in* Culver, S.J., ed., Studies of North American Jurassic radiolaria, Part II: Cushman Foundation Foraminiferal Research Special Publication 23, p. 1–18.

Pessagno, E.A., Jr., Blome, C.D., Hull, D., and Six, W.M., Jr., 1993, Middle and Upper Jurassic radiolaria from the Western Klamath terrane, Smith River subterrane, northwestern California: Their biostratigraphic, chronostratigraphic, geochronologic, and paleolatitudinal significance: Micropaleontology, v. 39, p. 93–166.

Pessagno, E.A., Jr., Cantú Chapa, A., Hull, D.M., Kelldorf, M., Longoria, J.F., Martin, C.B., Meng, X., Montgomery, H., Ogg, J.G., and Urrutia-Fucugauchi, J., 1999, Stratigraphic evidence for northwest to southeast tectonic transport of Jurassic terranes in Central Mexico and the Caribbean (western

Cuba), *in* Mann, P., ed. (Series Editor: Hsü, K.J.), Caribbean basins, Sedimentary basins of the world 4, Amsterdam, Elsevier Science B.V., p. 123–149.

Robertson, A.H.F., 1989, Paleoceanography and tectonic setting of the Jurassic Coast Range ophiolite, central California: Evidence from the extrusive rocks and the volcaniclastic sediment cover: Marine and Petroleum Geology, v. 6, p. 194–220.

Saleeby, J.B., 1987. Discordance patterns in Pb/U zircon ages of the Sierra Nevada and Klamath Mountains: Eos, Transactions of the American Geophysical Union, v. 68, p. 1514–1515.

Saleeby, J.B., 1996, Coast Range ophiolite as parautochthonous forearc lithosphere, *in* Alternate origins of the Coast Range ophiolite (California): Introduction and implications: GSA Today, v. 6, p. 7–8.

Saleeby, J.B., and Harper, G.D., 1993, Tectonic relations between the Galice Formation and the Condrey Mountain Schist, Klamath Mountains, Northern California, Mesozoic Paleogeography of the Western United States–II, Pacific Section SEPM, Book 71, p. 61–80.

Saleeby, J.B., Harper, G.D., Snoke, A.W., and Sharp, W.D., 1982. Time relations and structural–stratigraphic patterns in ophiolite accretion, west central Klamath Mountains, California. Journal of Geophysical Research, v. 87, p. 3831–3848.

Salvador, A., Westermann, G.E.G., Olóritz, F., Gordon, M.B., and Gursky, H.J., 1992, Chapter 5: Meso–America, *in* Westermann, G.E.G., ed., The Jurassic of the Circum–Pacific: Cambridge University Press, 676 p.

Stanley, S.M., 1986, Earth and life through time: W.H. Freeman and Company, p. 470.

Manuscript Accepted by the Society February 23, 2000

Printed in the U.S.A.

Geological Society of America
Special Paper 349
2000

Models for origin and emplacement of Jurassic ophiolites of northern California

Raymond V. Ingersoll
Department of Earth and Space Sciences, University of California, Los Angeles, California 90095-1567, USA; e-mail: ringer@zephyr.ess.ucla.edu

ABSTRACT

Four Jurassic ophiolite complexes in northern California have crystallization ages of 170–160 Ma; the Coast Range (including Great Valley) ophiolite is oldest (170–165 Ma), the Smartville ophiolite, intermediate (164–160 Ma), and the Josephine ophiolite, youngest (162 Ma). The Smartville ophiolite was obducted during the Sierran phase of the Nevadan orogeny (162–155 Ma), and the Josephine ophiolite was obducted during the Klamath phase of the Nevadan orogeny (153–150 Ma). The Great Valley ophiolite was not highly deformed during the Nevadan orogeny and became oceanic basement for the post-Nevadan forearc basin. Three conflicting models for origin of the Coast Range (including Great Valley) ophiolite have been proposed: (1) Formation by intra-arc and backarc spreading related to an east-facing intraoceanic arc, which collided with a west-facing continental-margin arc during the Nevadan orogeny (Sierran phase). (2) Formation by open-ocean seafloor spreading and incorporation into the continental margin during trench initiation outboard of an existing continental-margin trench. (3) Formation by forearc oblique rifting along the continental margin, followed by partial closure.

Lithospheric rifting is favored where there is either thick continental crust or thin mantle lithosphere because continental crust is weaker than oceanic crust and both are weaker than mantle lithosphere. Favorable sites for rifting are extant rifts, intra-arc settings, suture belts, and areas where mantle lithosphere has been delaminated from overlying crust. Upon cessation of magmatism related to rifting or subduction, mantle lithosphere rapidly strengthens, so that by 20 m.y., lithosphere strongly resists extensional stresses. Forearcs cool and strengthen even faster because of refrigeration by subducted slabs. Mature forearcs are strong and unlikely to yield to stresses, especially when weaker intra-arc and backarc settings are nearby.

Trench initiation is favored by (1) the presence of a preexisting lithospheric boundary, (2) the presence of thin oceanic lithosphere, (3) attempted subduction of buoyant crust, and (4) plate reorganization. Trench initiation in intact oceanic lithosphere older than 20 m.y. is implausible.

The following facts render models for Jurassic trench initiation (model 2) or rifting (model 3) in forearc settings implausible: (1) Subduction began along the California margin in the Triassic, so that forearc lithosphere was considerably older than 20 m.y. by the Late Jurassic. (2) The Triassic–Jurassic magmatic arc related to this earlier subduction zone died in eastern California at the same time that the Jurassic–Cretaceous magmatic arc was initiated in the westernmost Sierra Nevada. (3) The colliding buoyant crust that forced this plate reorganization is found in the Sierran foothills.

The most plausible model for formation of the Great Valley, Smartville, and Josephine ophiolites is by lithospheric rifting within and/or behind magmatic arcs. The Great Valley and Smartville ophiolites formed behind an east-facing intraoceanic arc. The west-facing continental-margin forearc of California was partially subducted beneath the east-facing intraoceanic arc, and this attempted subduction of buoyant crust resulted in the Sierran phase of the Nevadan orogeny, during which the Great Valley and Smartville ophiolites were obducted. North of this suture zone, no collision occurred, and the continental-margin arc expanded westward, resulting in intra-arc and/or backarc rifting to form the Josephine ophiolite. The Josephine ophiolite was then obducted over its arc during the Klamath phase of the Nevadan orogeny, as the Klamath trench propagated southward to trap the Great Valley ophiolite in the post-Nevadan forearc.

INTRODUCTION

Following recognition that mafic and ultramafic complexes and overlying sedimentary and volcanic strata of northern California represent on-land fragments of ancient ocean crust (e.g., Bailey et al., 1970), debate has been vigorous concerning the paleogeography and paleotectonics of ophiolite formation and obduction (e.g., Moores, 1970; Schweickert and Cowan, 1975; Hopson et al., 1981; Harper

Ingersoll, R.V., 2000, Models for origin and emplacement of Jurassic ophiolites of northern California, *in* Dilek, Y., Moores, E.M., Elthon, D., and Nicolas, A., eds., Ophiolites and Oceanic Crust: New Insights from Field Studies and the Ocean Drilling Program: Boulder, Colorado, Geological Society of America Special Paper 349, p. 395–402.

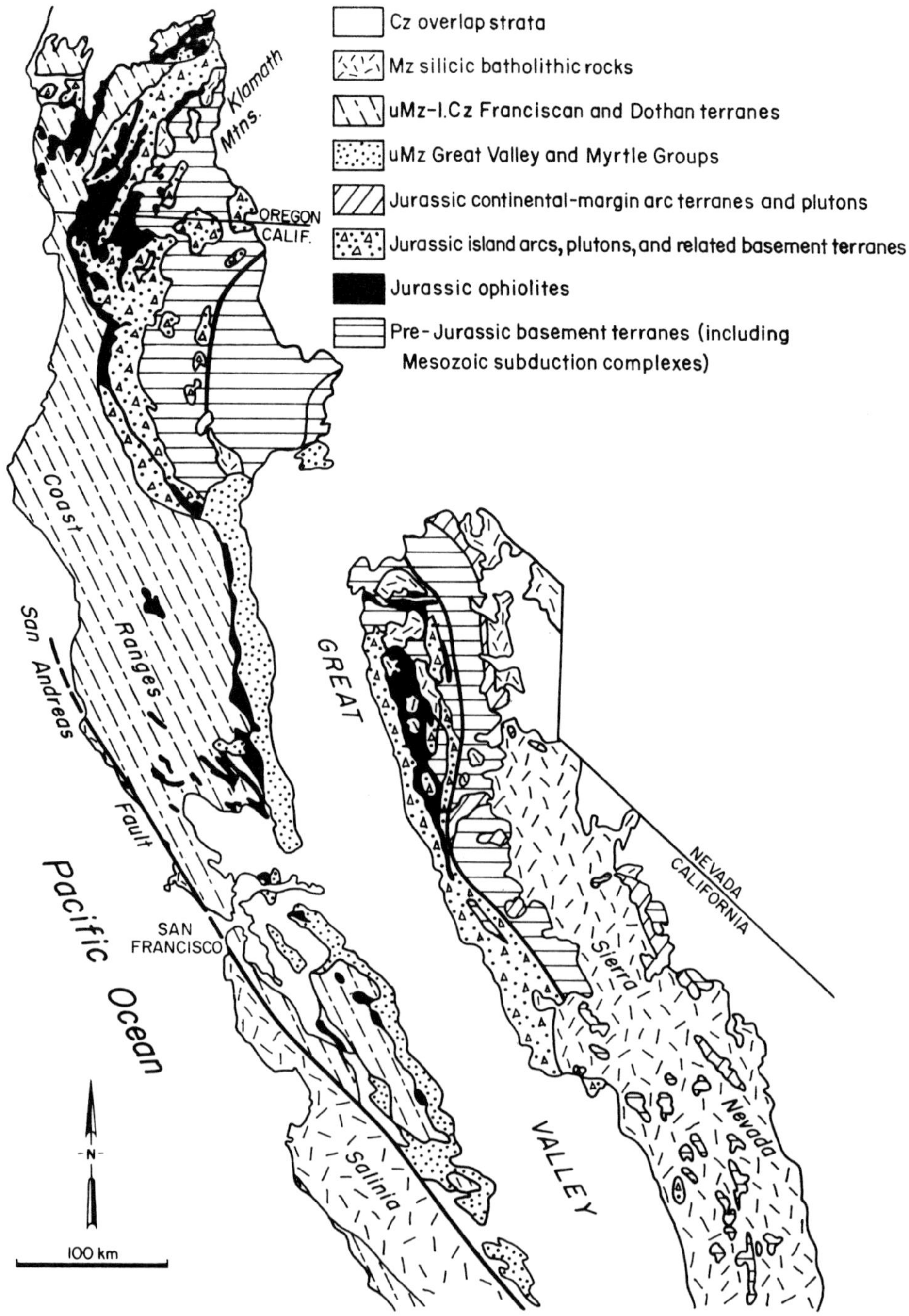

Figure 1. Location map of northern California, showing four belts of Jurassic ophiolites and their arc basements: Josephine ophiolite of the Klamath Mountains, Smartville ophiolite of northern Sierra Nevada, Great Valley ophiolite along the west side of the Great Valley, and other fragments of the Coast Range ophiolite in Coast Ranges. Also shown are Franciscan and Dothan terranes, Great Valley and Myrtle Groups, crystalline basement complexes and Cenozoic overlap strata. Modified from Ingersoll and Schweickert (1986).

and Wright, 1984; Ingersoll and Schweickert, 1986; Robertson, 1989, 1990; Saleeby, 1992; Saleeby and Busby-Spera, 1992; Dickinson et al., 1996a). Conflicting models may be evaluated by using plausibility arguments based on consistency with both theoretical models for physical and chemical lithospheric processes and actualistic analogue models for modern examples of lithospheric interactions. The construction of actualistic forward analogue models (e.g., Ingersoll, 1997) illustrates a plausible scenario by which ophiolites of northern California formed and were obducted.

Four different, but closely related, Late Jurassic ophiolite complexes occur in northern California (Fig. 1). All of the complexes have magmatic crystallization ages of 170–160 Ma, although, in detail, the Coast Range (including Great Valley) ophiolite is oldest (170–165 Ma), the Smartville ophiolite, intermediate (164–160 Ma), and the Josephine ophiolite, youngest (162 Ma) (Ingersoll and Schweickert, 1986; Dilek et al., 1990; Dilek and Moores, 1992; Saleeby and Busby-Spera, 1992; Harper et al., 1994). Stratigraphic ages of overlying pelagic and volcaniclastic strata are as young as 155 Ma, not including younger forearc deposits of the Great Valley Group that overlie the Great Valley ophiolite (Ingersoll, 1979, 1982; Robertson, 1989, 1990; Hopson et al., 1996). The Smartville ophiolite and its associated older intraoceanic arc were obducted and deformed during the Sierran phase of the Nevadan orogeny (162–155 Ma), and the Josephine ophiolite was obducted and deformed during the Klamath phase of the Nevadan orogeny (153–150 Ma) (Schweickert and Cowan, 1975; Harper and Wright, 1984; Ingersoll and Schweickert, 1986;

Saleeby and Busby-Spera, 1992). In contrast, the Great Valley ophiolite was generally not deformed during the Nevadan orogeny, as it remained at oceanic depths and formed the basement for the post-Nevadan Great Valley forearc basin (Ingersoll, 1979, 1982). Fragments of the Coast Range ophiolite are scattered through the Coast Ranges, on both sides of the San Andreas fault (Fig. 1). Although ages and stratigraphic sequences of other exposures of the Coast Range ophiolite are similar to those of the Great Valley ophiolite (Hopson et al., 1981, 1996), there has been too much late Mesozoic and Cenozoic tectonic disruption of these fragments to link them directly to the Great Valley ophiolite; because of their ambiguous genetic relationships to the Great Valley ophiolite, these other exposures of the Coast Range ophiolite are not considered further herein.

MODELS

Northern California ophiolites

Dickinson et al. (1996a) reviewed three conflicting models for origin of the Coast Range (including the Great Valley) ophiolite: (1) formation by intra-arc and backarc spreading related to an east-facing intraoceanic arc, which collided with a west-facing continental-margin arc during the Nevadan orogeny (Sierran phase) (Schweickert and Cowan, 1975; Dickinson et al., 1996b); (2) formation by open-ocean seafloor spreading and incorporation into the continental margin during trench initiation outboard of an existing continental-margin trench (Hopson et al., 1981, 1996); and (3) formation by forearc oblique rifting along the continental margin, followed by partial closure (Saleeby, 1992, 1996; Saleeby and Busby-Spera, 1992).

Model 1 is compatible with formation of the Josephine ophiolite by intra-arc and backarc spreading of the Klamath west-facing continental-margin arc during the collision of an intraoceanic arc in the Sierran region (Fig. 2; Harper and Wright, 1984; Ingersoll and Schweickert, 1986; Harper et al., 1994). Model 1 is also compatible with formation of first the Great Valley ophiolite and then the Smartville ophiolite by intra-arc and backarc spreading of an east-facing intraoceanic arc. The Copper Hill–Gopher Ridge complex is interpreted as a remnant arc, and the Logtown Ridge–Peñon Blanco complex constitutes the intraoceanic arc in this model (Ingersoll and Schweickert, 1986) (Figs. 2 and 3). North-to-south suturing and closure of the remnant ocean basin, within which the Mariposa Formation accumulated, occurred concurrently with rifting of the continental-margin arc in the Klamath region north of the suture zone (Fig. 2).

Intra-arc rifting north of the Sierran suture is analogous to recent opening of the Okinawa Trough north of the Taiwan suture (e.g., Suppe, 1984; Letouzey and Kimura, 1986; Viallon et al., 1986; Sibuet et al., 1998), with the exception that no active continental-margin arc is directly involved in the Taiwan suture (Fig. 4). Hsu and Sibuet (1995) and Sibuet and Hsu (1997) have inferred an extinct continental-margin arc west of and beneath Taiwan. Letouzey and Kimura (1986) and Viallon et al. (1986) proposed that opening of the Okinawa Trough was the direct result of the Taiwan collision, with lateral extrusion and/or trench rollback being important contributing factors. These authors demonstrated that backarc spreading could not be maintained for long without at least one of these additional factors. I suggest, by analogy, that the Josephine backarc basin opened only during the Sierran phase of the Nevadan orogeny; as the collision ended and a continuous trench was reestablished along the continental margin (Fig. 2), extensional forces resulting from indentation and

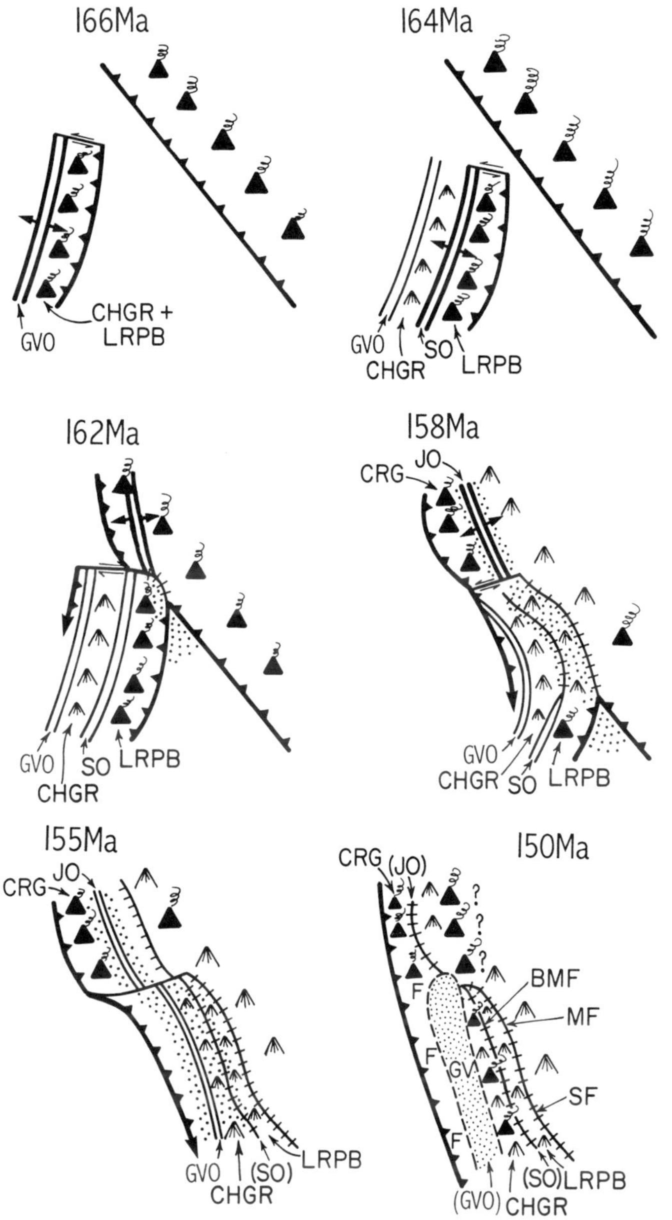

Figure 2. Sequential paleotectonic diagrams for Middle to Late Jurassic in northern California. Active magmatic arcs are shown with smoke; inactive, without. Active subduction zones are shown by barbed symbols; suture zones are shown by suture pattern. Rifted continental margin is shown by hachured line. Active spreading centers are shown by divergent arrows on double lines, without implication of exact spreading orientation; inactive spreading centers are shown by double lines without arrows. Transforms are shown by thin arrows. Southward-propagating trench is shown by large arrow. Dot pattern shows sites of deposition of the Mariposa and Galice Formations and the Great Valley (GV) forearc basin. Abbreviations: GVO—Great Valley ophiolite, SO—Smartville ophiolite, JO—Josephine ophiolite, CRG—Chetco-Rogue-Galice arc complex, F—Franciscan Complex, LRPB—Logtown Ridge arc complex and 200 Ma Peñon Blanco arc complex, CHGR—Copper Hill–Gopher Ridge arc complex, BMF—Bear Mountain fault, MF—Melones fault, SF—Sonora fault. Parentheses around ophiolite names in the diagrams indicate partial preservation within fault zones. Modified from Ingersoll and Schweickert (1986); based in part on Harper et al. (1994).

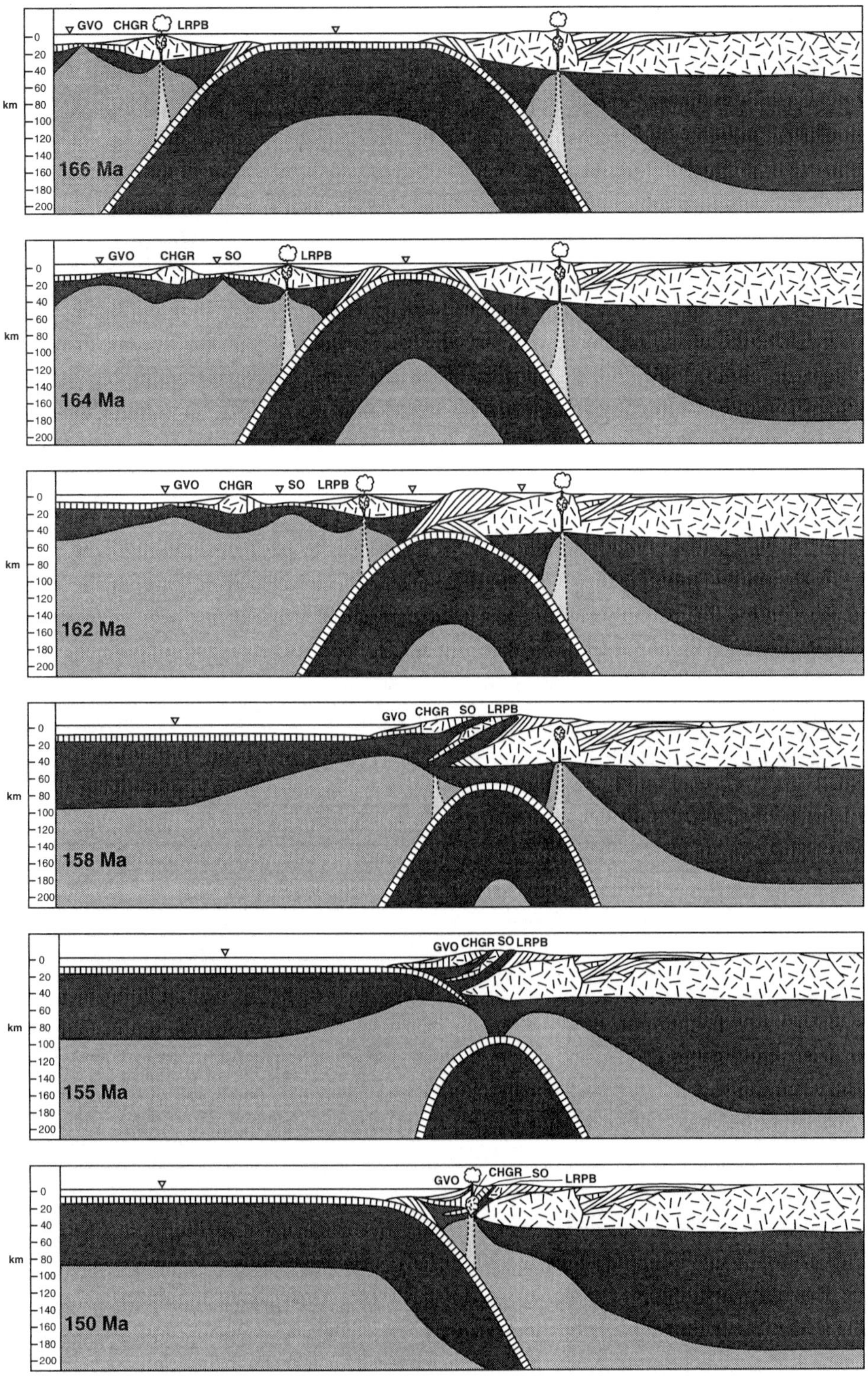

Figure 3. True-scale actualistic analogue sequential cross sections for the northern Coast Ranges, Great Valley, and Sierra Nevada, consistent with map relationships of Figure 2, geophysical cross sections of Godfrey et al. (1997), and analogue forward models of Ingersoll (1997). Timing relationships shown in cross sections are consistent with relationships along a line of latitude in the middle of the Franciscan–Great Valley–Sierra Nevada areas shown in Figure 2. Similar cross sections, but with different ages, could be drawn along the southward-propagating suture zone south of the Klamath Mountains. Abbreviations same as for Figure 2. Triangles show sea surface.

lateral extrusion decayed, and the Josephine backarc basin collapsed during the Klamath phase of the Nevadan orogeny. The Josephine ophiolite was then obducted westward over the Chetco-Rogue-Galice intraoceanic-arc complex soon after the southern intraoceanic arc and related remnant arc and backarc basins were obducted eastward over the continental-margin arc (Fig. 2).

The Great Valley ophiolite was trapped in the new forearc basin south of the Klamath Mountains as Franciscan subduction was initiated west of the collided arcs of the Sierra Nevada (Schweickert and Cowan, 1975) (Figs. 2 and 3). The pre-Nevadan continental-margin arc in California and western Nevada (Schweickert, 1976; Busby-Spera, 1988; Dilek and Moores, 1995) was extinguished by the collision; the post-Nevadan continental-margin magmatic arc was initiated farther west (e.g., Guadalupe Igneous Complex, 151 Ma; Saleeby et al., 1989) as Franciscan subduction began (Hamilton, 1969; Ernst, 1970).

Model 2 does not consider the Smartville or Josephine ophiolites, and it considers all exposures of the Coast Range ophiolite as a

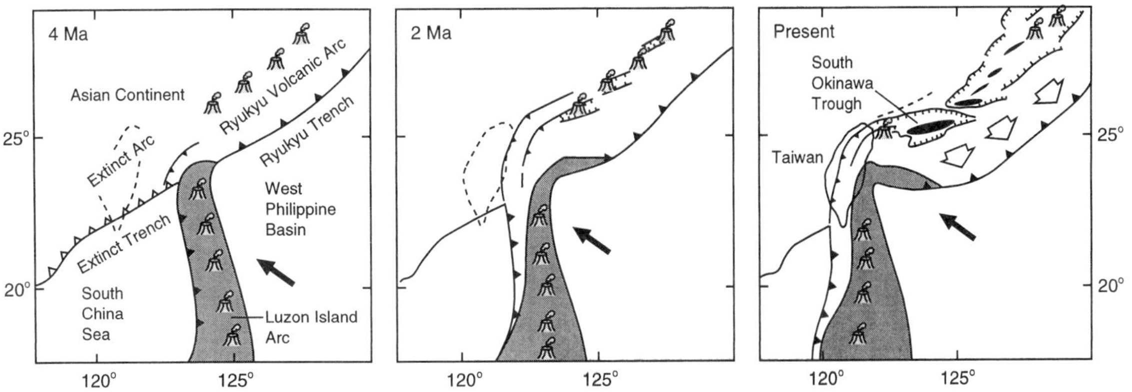

Figure 4. Structural evolution of the Taiwan collision zone, the Ryukyu intraoceanic arc, and the Okinawa Trough. Ryukyu magmatic arc constitutes a weak zone along the continental margin. Indentation due to collision of the north end of the Luzon intraoceanic arc with the continental margin has resulted in contraction in the Taiwan area; lateral extrusion north of the suture, coupled with slab rollback, has formed the Okinawa Trough, initially in an intra-arc setting and evolving into a backarc setting. See text for discussion. Modified from Letouzey and Kimura (1986), based partly on Sibuet and Hsu (1997).

single entity. The model utilizes geochemical, paleontological, paleoenvironmental, and paleomagnetic data to argue that the Coast Range ophiolite (sensu lato) formed far to the southwest of the present location of California and distal from any arc activity (Hopson et al., 1981, 1996). Subduction along the California margin resulted in northeastward transport of the future Coast Range ophiolite. A reorganization of plate motions resulted in initiation of the Franciscan trench outboard of the former continental-margin trench, resulting in trapping of the oceanic crust in a forearc setting. The mechanism for this trench relocation is unspecified in model 2.

Model 3 invokes oblique subduction as a cause for transtensional rifting in the continental-margin forearc to form all four Middle Jurassic ophiolite complexes (Saleeby, 1992, 1996; Saleeby and Busby-Spera, 1992). In this model, subduction was continuous through the Jurassic, with no Nevadan orogeny in the sense of Schweickert and Cowan (1975) and Ingersoll and Schweickert (1986). Obduction of the ophiolites resulted from a change in subduction parameters that resulted in transpressional deformation that produced diverse structures.

Lithospheric rifting

Rifting on Earth's surface is favored where there is either thick continental crust or thin mantle lithosphere (Kusznir and Park, 1987; Lynch and Morgan, 1987) because continental crust is weaker than oceanic crust, and both are weaker than mantle lithosphere (Vink et al., 1984; Steckler and ten Brink, 1986). Thus, favorable sites for rifting are extant rifts, intra-arc settings, suture belts, and areas where mantle lithosphere has been delaminated from overlying crust. Upon cessation of magmatism related to rifting or subduction, mantle lithosphere rapidly strengthens, so that by 20 m.y., the lithosphere strongly resists extensional stresses (Mueller and Phillips, 1991). Forearcs cool and strengthen even faster due to refrigeration as subducted slabs sink below forearcs (Dumitru, 1988; Bohannon and Parsons, 1995). Thus, mature forearcs are among the strongest plate settings on Earth's surface and are unlikely to yield to stresses, especially when weaker intra-arc and backarc settings are nearby.

This discussion refers to lithospheric rifting and rupture to form new oceanic crust, which subsequently might become ophiolite upon obduction. There is no doubt that upper-crustal extension is common in accretionary complexes (e.g., Platt, 1986; Jayko et al., 1987; Harms et al., 1992). However, this type of dynamic adjustment to changing subduction parameters has not been shown to lead to lithospheric rifting and rupture. Thus, the mere presence of normal faults in forearcs cannot be equated with forearc rifting.

Published examples of supposed late Cenozoic rifting of forearcs in the western Pacific (e.g., Taylor and Karner, 1983; Taylor, 1992; Hawkins, 1995) are rendered ambiguous because of rapid shifting of the location of arc magmatism in these examples, resulting in juxtaposition of backarc, intra-arc, and forearc positions through time. The only clear example of creation of completely new oceanic lithosphere directly adjacent to a trench appears to have resulted from the initiation of the Izu-Bonin-Mariana arc-trench system during the Eocene (e.g., Stern and Bloomer, 1992). This model for forearc rifting is applicable only to the initiation of new subduction zones; as Stern and Bloomer (1992) illustrated, as soon as true subduction is underway, forearc refrigeration leads to the development of strong, thick forearc lithosphere and localization of magmatism along a narrow arc axis. Thus, this model is not useful in explaining Jurassic ophiolites of California because subduction was active along the continental margin for tens of millions of years prior to their formation (Schweickert, 1976, 1981; Dickinson, 1981a, 1981b; Saleeby and Busby-Spera, 1992; Dilek and Moores, 1995; Ingersoll, 1997). The preponderance of lithospheric rifting in western Pacific arc-trench systems has been within or behind active arcs (e.g., Karig, 1972; Carey and Sigurdsson, 1984; Klaus et al., 1992; Marsaglia, 1995).

Saleeby (1981, 1992) suggested that oblique forearc rifting created the Jurassic ophiolites of California. He suggested the Andaman Sea of Southeast Asia as a modern analogue for such rifting. The failure of this model is that there is no active arc east of the Andaman Sea (e.g., Hamilton, 1979); the Andaman Sea is not even a forearc. Rather, it is a transtensional transform boundary. In contrast, a continental-margin magmatic arc was active along the California margin throughout the period of ophiolite formation (Saleeby and Busby-Spera, 1992; Dilek and Moores, 1995; Ingersoll, 1997). The useful modern analogue for California is the Sumatra arc-trench system (Hamilton, 1979), where strike-slip motion on the Barisan fault zone *within the magmatic arc* decouples the forearc from the backarc (Hamilton, 1979; Dewey, 1980; Saleeby, 1981; Dickinson, 1995), as probably occurred in California during the Cretaceous (Lahren and Schweickert, 1989; Schweickert and Lahren, 1990; Dilek and Moores, 1995). In spite of strongly oblique subduction under Sumatra, the forearc remains coherent, even as it slides northwestward relative to the remainder of Indonesia (Hamilton, 1979; Dickinson, 1995). There is no modern example of oblique forearc rifting, and the arguments presented herein render such a process implausible for mature forearcs.

Trench initiation

Trench initiation is favored by (1) the presence of a preexisting lithospheric boundary (e.g., fracture zone, transform, or extinct ridge); (2) the presence of thin (hot, young) oceanic lithosphere; (3) attempted subduction of buoyant (continental or arc) crust (proximal collision); and (4) plate reorganization (usually resulting from a distal collision) (Cloetingh et al., 1982; Casey and Dewey, 1984; Mueller and Phillips, 1991; Cloos, 1993; Dickinson, 1995). Trench initiation in intact oceanic lithosphere older than 20 m.y., therefore, is implausible.

PLAUSIBILITY ARGUMENTS

The following facts, therefore, render models for Jurassic trench initiation (e.g., Hopson et al., 1981, 1996) or rifting (Saleeby, 1992; Saleeby and Busby-Spera, 1992) in forearc settings implausible explanations for generation of the Great Valley ophiolite and initiation of the Great Valley forearc: (1) Subduction was initiated along the California margin in the Triassic following the Sonoma collisional orogeny (Hamilton, 1969; Dickinson, 1981a, 1981b; Ingersoll, 1997), so that forearc lithosphere was considerably older than 20 m.y. by the Late Jurassic. (2) The Triassic–Jurassic magmatic arc related to this earlier subduction zone (Schweickert, 1976; Busby-Spera, 1988; Dilek and Moores, 1995) died in eastern California at the same time the latest Jurassic–Cretaceous magmatic arc was initiated in the westernmost Sierra Nevada (now partially covered by Upper Cretaceous strata) (Ingersoll, 1979; Moxon and Graham, 1987; Fuis and Mooney, 1990; Godfrey et al., 1997). (3) The colliding buoyant crust that forced this proximal plate reorganization is found in the Sierran foothills (LRPB and CHGR, Figs. 2 and 3); attached to this intraoceanic magmatic-arc complex are the Smartville ophiolite (164–160 Ma) and the Great Valley ophiolite (170–165 Ma) beneath and along the west side of the Great Valley (Schweickert and Cowan, 1975; Schweickert, 1981; Ingersoll, 1982; Moores and Day, 1984; Dilek and Moores, 1992; Godfrey et al., 1997).

SEQUENTIAL DEVELOPMENT

These observations are consistent with formation of the Great Valley and Smartville ophiolites in intra-arc–backarc intraoceanic settings prior to collision with the continental-margin arc-trench system (Schweickert and Cowan, 1975; Ingersoll and Schweickert, 1986). The Franciscan trench and Great Valley forearc were initiated in the young, hot backarc lithosphere as the two facing arcs collided and were sutured to North America from north to south (Figs. 2 and 3). The slightly younger Josephine ophiolite of the Klamath Mountains formed by intra-arc–backarc spreading of the continental-margin arc north of the collision zone (Harper and Wright, 1984; Wyld and Wright, 1988; Harper et al., 1994). Opening of the Okinawa Trough north of the Taiwan collision zone (e.g., Letouzey and Kimura, 1986; Viallon et al., 1986) provides a mirror-image modern analogue for opening of the Josephine oceanic basin, although geometric details differ (compare Figs. 2 and 4). This model provides few constraints for origins of fragments of the Coast Range (other than Great Valley) ophiolite scattered through the Franciscan Complex and/or west of the Cenozoic San Andreas fault (Fig. 1).

The consequences of the collisional model of Schweickert and Cowan (1975) are illustrated and defined by the construction of actualistic analogue forward models (i.e., Ingersoll, 1997). Figure 3 illustrates the closing of the remnant ocean basin (e.g., Ingersoll et al., 1995) between the intraoceanic arc and the continental-margin arc during the Late Jurassic. As suturing occurred, the intraoceanic arc and its related backarcs and remnant arc overrode the continental-margin arc, following which the Franciscan subduction zone was initiated in the young and weak lithosphere of the westernmost backarc basin, which was thereby converted into the Great Valley forearc basin, underlain by the Great Valley ophiolite. As in the modern Taiwan (e.g., Suppe, 1984; Dorsey, 1988) and New Guinea (Weiland and Cloos, 1996) suture zones, uplift and erosion rates were extremely high, so that most of the subduction complex and forearc of the obducting arc-trench system were destroyed. The overridden subduction complex and forearc of the continental-margin arc-trench system are deeply buried and deformed (e.g., Godfrey et al., 1997). Figures 2 and 3 illustrate the four-dimensional development of the Nevadan orogeny and initiation of the Franciscan trench and the Great Valley forearc.

CONCLUSIONS

The most plausible model for creation of the Great Valley, Smartville, and Josephine ophiolites of northern California is that proposed by Ingersoll and Schweickert (1986), which is a hybrid of the models of Schweickert and Cowan (1975) and Harper and Wright (1984). All three ophiolites formed by lithospheric rifting within and/or behind magmatic arcs. The Great Valley and Smartville ophiolites formed behind an east-facing intraoceanic arc, with a remnant arc between them. The west-facing continental-margin forearc of California was partially subducted beneath the east-facing intraoceanic arc, and this attempted subduction of buoyant crust resulted in the Sierran phase of the Nevadan orogeny, during which the Great Valley and Smartville ophiolites were obducted onto the edge of the continent. North of this suture zone, no collision occurred, and the continental-margin arc expanded westward, resulting in intra-arc–backarc lithospheric rifting to form the Josephine ophiolite. Soon thereafter, as collision terminated in the Sierran region, the Josephine ophiolite was obducted over its arc during the Klamath phase of the Nevadan orogeny, as the Klamath trench propagated southward to trap the Great Valley ophiolite in the forearc of the soon-to-develop post-Nevadan continental-margin arc of the western Sierra Nevada. All aspects of this model are plausible and consistent with known geologic, geochemical, and geophysical relationships.

ACKNOWLEDGMENTS

I thank Yildirim Dilek and Eldridge Moores for convening the Penrose Conference on Ophiolites and for encouraging me to write this paper. I thank W.R. Dickinson, Y. Dilek, G.D. Harper, and E.M. Moores for reviewing the manuscript.

REFERENCES CITED

Bailey, E.H., Blake, M.C., Jr., and Jones, D.L., 1970, On-land Mesozoic oceanic crust in California Coast Ranges: U.S. Geological Survey Professional Paper 700-C, p. C70–C81.

Bohannon, R.G., and Parsons, T., 1995, Tectonic implications of post-30 Ma Pacific and North American relative plate motions: Geological Society of America Bulletin, v. 107, p. 937–959.

Busby-Spera, C.J., 1988, Speculative tectonic model for the early Mesozoic arc of the southwest Cordilleran United States: Geology, v. 16, p. 1121–1125.

Carey, S.N., and Sigurdsson, H., 1984, A model of volcanogenic sedimentation in marginal basins: Geological Society of London Special Publication 16, p. 37–58.

Casey, J.F., and Dewey, J.F., 1984, Initiation of subduction zones along transform and accreting plate boundaries, triple-junction evolution, and forearc spreading centres: Implications for ophiolite geology and obduction: Geological Society of London Special Publication 13, p. 269–290.

Cloetingh, S.A.P.L., Wortel, M.J.R., and Vlaar, N.J., 1982, Evolution of passive continental margins and initiation of subduction zones: Nature, v. 297, p. 139–142.

Cloos, M., 1993, Lithospheric buoyancy and collisional orogenesis: Subduction of oceanic plateaus, continental margins, island arcs, spreading ridges, and seamounts: Geological Society of America Bulletin, v. 105, p. 715–737.

Dewey, J.F., 1980, Episodicity, sequence, and style at convergent plate boundaries: Geological Association of Canada Special Paper 20, p. 553–573.

Dickinson, W.R., 1981a, Plate tectonic evolution of the southern Cordillera: Arizona Geological Society Digest, v. 14, p. 113–135.

Dickinson, W.R., 1981b, Plate tectonics and the continental margin of California, *in* Ernst, W.G., ed., The geotectonic development of California: Englewood Cliffs, New Jersey, Prentice-Hall, p. 1–28.

Dickinson, W.R., 1995, Forearc basins, *in* Busby, C.J., and Ingersoll, R.V., eds., Tectonics of sedimentary basins: Cambridge, Blackwell, p. 221–261.

Dickinson, W.R., Hopson, C.A., and Saleeby, J.B., 1996a, Alternate origins of the Coast Range ophiolite (California): Introduction and implications: GSA Today, v. 6, p. 1–2.

Dickinson, W.R., Schweickert, R.A., and Ingersoll, R.V., 1996b, 1. Coast Range ophiolite as back-arc–inter-arc basin lithosphere: GSA Today, v. 6, p. 2–4.

Dilek, Y., and Moores, E.M., 1992, Island-arc evolution and fracture-zone tectonics in the Mesozoic Sierra Nevada, California, and implications for transform offset of the Sierran/Klamath convergent margins, *in* Bartholomew, M.J., Hyndman, D.W., Mogk, D.W., and Mason, R., eds., Basement tectonics 8: Characterization and comparison of ancient and Mesozoic continental margins: Proceedings of the 8th International Conference on Basement Tectonics (Butte, Montana, 1988): Dordrecht, Kluwer Academic Publishers, p. 179–196.

Dilek, Y., and Moores, E.M., 1995, Geology of the Humboldt igneous complex, Nevada, and tectonic implications for the Jurassic magmatism in the Cordilleran orogen, *in* Miller, D.M., and Busby, C., eds., Jurassic magmatism and tectonics of the North American Cordillera: Geological Society of America Special Paper 299, p. 229–248.

Dilek, Y., Thy, P., Moores, E.M., and Grundvig, S., 1990, Late Paleozoic-early Mesozoic oceanic basement of a Jurassic arc terrane in the northwestern Sierra Nevada, California, *in* Harwood, D.S., and Miller, M.M., eds., Paleozoic and Mesozoic paleogeographic relations: Sierra Nevada, Klamath Mountains, and related terranes: Geological Society of America Special Paper 255, p. 351–369.

Dorsey, R.J., 1988, Provenance evolution and unroofing history of a modern arc-continent collision: Evidence from petrography of Plio-Pleistocene sandstones, eastern Taiwan: Journal of Sedimentary Petrology, v. 58, p. 208–218.

Dumitru, T.A., 1988, Subnormal geothermal gradients in the Great Valley forearc basin, California, during Franciscan subduction: A fission track study: Tectonics, v. 7, p. 1201–1221.

Ernst, W.G., 1970, Tectonic contact between the Franciscan mclange and the Great Valley sequence, crustal expression of a late Mesozoic Benioff zone: Journal of Geophysical Research, v. 75, p. 886–902.

Fuis, G.S., and Mooney, W.D., 1990, Lithospheric structure and tectonics from seismic-refraction and other data: U.S. Geological Survey Professional Paper 1515, p. 207–236.

Godfrey, N.J., Beaudoin, B.C., Klemperer, S.L., and Mendocino Working Group, 1997, Ophiolitic basement to the Great Valley forearc basin, California, from seismic and gravity data: Implications for crustal growth at the North American continental margin: Geological Society of America Bulletin, v. 108, p. 1536–1562.

Hamilton, W., 1969, Mesozoic California and the underflow of Pacific mantle: Geological Society of America Bulletin, v. 80, p. 2409–2430.

Hamilton, W., 1979, Tectonics of the Indonesian region: U.S. Geological Survey Professional Paper 1078, 345 p.

Harms, T.A., Jayko, A.S., and Blake, M.C., Jr., 1992, Kinematic evidence for extensional unroofing of the Franciscan Complex along the Coast Range fault, northern Diablo Range, California: Tectonics, v. 11, p. 228–241.

Harper, G.D., and Wright, J.E., 1984, Middle to Late Jurassic tectonic evolution of the Klamath Mountains, California-Oregon: Tectonics, v. 3, p. 759–772.

Harper, G.D., Saleeby, J.B., and Heizler, M., 1994, Formation and emplacement of the Josephine ophiolite and the Nevadan orogeny in the Klamath Mountains, California-Oregon: U/Pb zircon and $^{40}Ar/^{39}Ar$ geochronology: Journal of Geophysical Research, v. 99, p. 4293–4321.

Hawkins, J.W., 1995, Evolution of the Lau Basin: Insights from ODP Leg 135: American Geophysical Union Monograph 88, p. 125–173.

Hopson, C.A., and 7 others, 1996, 2. Coast Range ophiolite as paleoequatorial mid-ocean lithosphere: GSA Today, v. 6, p. 4–6.

Hopson, C.A., Mattinson, J.M., and Pessagno, E.A., Jr., 1981, Coast Range ophiolite, western California, *in* Ernst, W.G., ed., The geotectonic development of California [Rubey Volume I]: Englewood Cliffs, New Jersey, Prentice-Hall, p. 418–510.

Hsu, J.-C., and Sibuet, J.-C., 1995, Is Taiwan the result of arc-continent or arc-arc collision?: Earth and Planetary Science Letters, v. 136, p. 315–324.

Ingersoll, R.V., 1979, Evolution of the Late Cretaceous forearc basin, northern and central California: Geological Society of America Bulletin, v. 90, part I, p. 813–826.

Ingersoll, R.V., 1982, Initiation and evolution of the Great Valley forearc basin of northern and central California, *in* Leggett, J.K., ed., Trench-forearc geology: Sedimentation and tectonics on modern and ancient active plate margins: Geological Society of London Special Publication 10, p. 459–467.

Ingersoll, R.V., 1997, Phanerozoic tectonic evolution of central California and environs: International Geology Review, v. 39, p. 957–972.

Ingersoll, R.V., and Schweickert, R.A., 1986, A plate-tectonic model for Late Jurassic ophiolite genesis, Nevadan orogeny and forearc initiation, northern California: Tectonics, v. 5, p. 901–912.

Ingersoll, R.V., Graham, S.A., and Dickinson, W.R., 1995, Remnant ocean basins, *in* Busby, C.J., and Ingersoll, R.V., eds., Tectonics of sedimentary basins: Cambridge, Blackwell Science, p. 363–391.

Jayko, A.S., Blake, M.C., Jr., and Harms, T., 1987, Attenuation of the Coast Range ophiolite by extensional faulting, and nature of the Coast Range "thrust", California: Tectonics, v. 6, p. 475–488.

Karig, D.E., 1972, Remnant arcs: Geological Society of America Bulletin, v. 83, p. 1057–1068.

Klaus, A., and 6 coauthors, 1992, Structural and stratigraphic evolution of Sumisu Rift, Izu-Bonin Arc: Proceedings of the Ocean Drilling Program, Scientific Results, v. 126: College Station, Texas, Ocean Drilling Program, p. 555–574.

Kusznir, N.J., and Park, R.G., 1987, The extensional strength of the continental lithosphere: Its dependence on geothermal gradient, and crustal composition and thickness: Geological Society of London Special Publication 28, p. 35–52.

Lahren, M.M., and Schweickert, R.A., 1989, Proterozoic and Lower Cambrian miogeoclinal rocks of Snow Lake pendant, Yosemite-Emigrant Wilderness, Sierra Nevada, California: Evidence for major Early Cretaceous dextral translation: Geology, v. 17, p. 156–160.

Letouzey, J., and Kimura, M., 1986, The Okinawa Trough: Genesis of a back-arc basin developing along a continental margin: Tectonophysics, v. 125, p. 209–230.

Lynch, H.D., and Morgan, P., 1987, The tensile strength of the lithosphere and the localization of extension: Geological Society of London Special Publication 28, p. 53–65.

Marsaglia, K.M., 1995, Interarc and backarc basins, *in* Busby, C.J., and Ingersoll, R.V., eds., Tectonics of sedimentary basins: Cambridge, Blackwell Science, p. 299–329.

Moores, E., 1970, Ultramafics and orogeny, with models of the US Cordillera and the Tethys: Nature, v. 228, p. 837–842.

Moores, E.M., and Day, H.W., 1984, Overthrust model for the Sierra Nevada: Geology, v. 12, p. 416–419.

Moxon, I.W., and Graham, S.A., 1987, History and controls of subsidence in the Late Cretaceous-Tertiary Great Valley forearc basin, California: Geology, v. 15, p. 626–629.

Mueller, S., and Phillips, R.J., 1991, On the initiation of subduction: Journal of Geophysical Research, v. 96, p. 651–665.

Platt, J.P., 1986, Dynamics of orogenic wedges and uplift of high-pressure metamorphic rocks: Geological Society of America Bulletin, v. 97, p. 1037–1053.

Robertson, A.H.F., 1989, Palaeoceanography and tectonic setting of the Jurassic Coast Range ophiolite, central California: Evidence from the extrusive rocks and the volcaniclastic sediment cover: Marine and Petroleum Geology, v. 6, p. 194–220.

Robertson, A.H.F., 1990, Sedimentology and tectonic implications of ophiolite-derived clastics overlying the Jurassic Coast Range ophiolite, northern California: American Journal of Science, v. 290, p. 109–163.

Saleeby, J., 1981, Ocean floor accretion and volcanoplutonic arc evolution of the Mesozoic Sierra Nevada, *in* Ernst, W.G., ed., The geotectonic development of California: Englewood Cliffs, New Jersey, Prentice-Hall, p. 132–181.

Saleeby, J.B., 1992, Petrotectonic and paleogeographic settings of U.S. Cordilleran ophiolites, *in* Burchfiel, B.C., Lipman, P.W., and Zoback, M.L., eds., The Cordilleran orogen: Conterminous U.S.: Boulder, Colorado, Geological Society of America, Geology of North America, v. G-3, p. 653–682.

Saleeby, J.B., 1996, 3. Coast Range ophiolite as parautochthonous forearc lithosphere: GSA Today, v. 6, p. 6–8.

Saleeby, J.B., and Busby-Spera, C.J., 1992, Early Mesozoic tectonic evolution of the western U.S. Cordillera, *in* Burchfiel, B.C., Lipman, P.W., and Zoback, M.L., eds., The Cordilleran orogen: conterminous U.S.: Boulder, Colorado, Geological Society of America, Geology of North America, v. G-3, p. 107–168.

Saleeby, J.B., Geary, E.E., Paterson, S.R., and Tobisch, O.T., 1989, Isotopic systematics of Pb/U (zircon) and $^{40}Ar/^{39}Ar$ (biotite-hornblende) from rocks of the central Foothills terrane, Sierra Nevada, California: Geological Society of America Bulletin, v. 101, p. 1481–1492.

Schweickert, R.A., 1976, Shallow-level plutonic complexes in the eastern Sierra Nevada, California and their tectonic implications: Geological Society of America Special Paper 176, 58 p.

Schweickert, R.A., 1981, Tectonic evolution of the Sierra Nevada range, *in* Ernst, W.G., ed., The geotectonic development of California: Englewood Cliffs, New Jersey, Prentice-Hall, p. 87–131.

Schweickert, R.A., and Cowan, D.S., 1975, Early Mesozoic tectonic evolution of the western Sierra Nevada, California: Geological Society of America Bulletin, v. 86, p. 1329–1336.

Schweickert, R.A., and Lahren, M.M., 1990, Speculative reconstruction of the Mojave-Snow Lake fault: Implications for Paleozoic and Mesozoic orogenesis in the western United States: Tectonics, v. 9, p. 1609–1629.

Sibuet, J.-C., and Hsu, S.-K., 1997, Geodynamics of the Taiwan arc-arc collision: Tectonophysics, v. 274, p. 221–251.

Sibuet, J.-C., and 6 others, 1998, Okinawa trough backarc basin: Early tectonic and magmatic evolution: Journal of Geophysical Research, v. 103, p. 30245–30267.

Steckler, M.S., and ten Brink, U.S., 1986, Lithospheric strength variations as a control on new plate boundaries: Examples from the northern Red Sea region: Earth and Planetary Science Letters, v. 79, p. 120–132.

Stern, R.J., and Bloomer, S.H., 1992, Subduction zone infancy: Examples from the Eocene Izu-Bonin-Mariana and Jurassic California arcs: Geological Society of America Bulletin, v. 104, p. 1621–1636.

Suppe, J., 1984, Kinematics of arc-continent collision, flipping of subduction, and back-arc spreading near Taiwan: Geological Society of China Memoir 6, p. 21–33.

Taylor, B., 1992, Rifting and the volcanic-tectonic evolution of the Izu-Bonin-Mariana arc: Proceedings of the Ocean Drilling Program, Scientific Results, v. 126: College Station, Texas, Ocean Drilling Program, p. 627–651.

Taylor, B., and Karner, G.D., 1983, On the evolution of marginal basins: Reviews of Geophysics and Space Physics, v. 21, p. 1727–1741.

Viallon, C., Huchon, P., and Barrier, E., 1986, Opening of the Okinawa basin and collision in Taiwan: A retreating trench model with lateral anchoring: Earth and Planetary Science Letters, v. 80, p. 145–155.

Vink, G.E., Morgan, W.J., and Zhao, W.-L., 1984, Preferential rifting of continents: A source of displaced terranes: Journal of Geophysical Research, v. 89, p. 10072–10076.

Weiland, R.J., and Cloos, M., 1996, Pliocene-Pleistocene asymmetric unroofing of the Irian fold belt, Irian Jaya, Indonesia: Apatite fission-track thermochronology: Geological Society of America Bulletin, v. 108, p. 1438–1449.

Wyld, S.J., and Wright, J.E., 1988, The Devils Elbow ophiolite remnant and overlying Galice Formation: New constraints on the Middle to Late Jurassic evolution of the Klamath Mountains, California: Geological Society of America Bulletin, v. 100, p. 29–44.

MANUSCRIPT ACCEPTED BY THE SOCIETY FEBRUARY 23, 2000

Printed in the U.S.A.

Geological Society of America
Special Paper 349
2000

Geology and geochemistry of the ophiolitic Trinity terrane, California: Evidence of middle Paleozoic depleted supra-subduction zone magmatism in a proto-arc setting

Rodney V. Metcalf,* E. Timothy Wallin, Keith R. Willse, and Edward R. Muller
Department of Geoscience, University of Nevada, Las Vegas, Nevada 89154-4010, USA

ABSTRACT

As the largest exposure of ophiolitic rock in North America, the Trinity terrane constitutes an important part of the ophiolite record worldwide. It has been interpreted variously as a fragment of a mid-ocean ridge crust, a mantle diapir, the roots of a volcanic arc, and a backarc basin. A recent compilation of existing field, compositional, and geochronologic data from the Trinity terrane and the adjacent Eastern Klamath, Central Metamorphic, and Yreka terranes has led to the hypothesis that the Trinity terrane is a polygenetic supra-subduction zone (SSZ) ophiolite that formed in a trench-proximal extensional environment. New major and trace element geochemical data for mafic crustal rocks of the Trinity terrane are reported here and used to test the SSZ ophiolite hypothesis. Gabbro and mafic dike samples are depleted in high field strength elements relative to normal mid-oceanic-ridge basalt (MORB) and have high Th/Nb and Pb/Ce ratios, negative Ce anomalies, and positive Eu anomalies. These data are consistent with partial melting of a depleted, residual, MORB-type, mantle source enriched by subduction-derived fluids that included a sedimentary input. A model is presented for the development of the Trinity as a proto-arc formed during the inception of a single, east-dipping, middle Paleozoic subduction zone, analogous to the Eocene proto-arc of the Izu-Bonin-Mariana intraoceanic arc system.

INTRODUCTION

The emergence of the plate-tectonic paradigm led to interpretation of ophiolites as obducted slices of oceanic lithosphere formed by sea-floor spreading at mid-oceanic ridges. Subsequent research recognized that many ophiolites (e.g., Troodos) have a geochemical signature indicative of subduction-zone magmatism—specifically, enrichments in large ion lithophile elements (LILEs) relative to high field strength elements (HFSEs) (Pearce et al., 1984). These latter ophiolites were designated supra-subduction zone (SSZ) ophiolites by Pearce et al. (1984), signifying formation above an active subduction zone, in contrast to ophiolites formed at mid-oceanic ridges (MORB ophiolites).

The Trinity terrane in California has been recently interpreted as a polygenetic SSZ ophiolite that formed during the inception of an intraoceanic subduction zone (Wallin and Metcalf, 1998). Such SSZ ophiolites are an important component of the geologic record (Pearce et al., 1984; Bloomer et al., 1995), and hence the Trinity terrane provides an important three-dimensional observatory in which to examine the rocks and petrogenetic characteristics of a SSZ tectonic setting whose modern analogues are largely inaccessible. Here we report new geochemical data based on detailed field mapping that constitute the first comprehensive trace element data set for crustal rocks of the Trinity terrane. These data bear on both the paleotectonic setting of mafic intrusive complexes and reinforce our earlier interpretation of the Trinity terrane as an SSZ ophiolite formed in a proto-arc setting (Wallin and Metcalf, 1998).

The term "forearc" has been used widely to refer to lithosphere formed by magmatism in an extensional, trench-proximal setting above a nascent subduction zone (Stern and Bloomer, 1992; Bloomer et al., 1995). However, the crustal thickness in such environments is approximately half an order of magnitude less than that in typical intraoceanic magmatic arcs (cf. Gill, 1981). Although the term "forearc" adequately connotes the present tectonic setting of the crustal fragments, it is technically incorrect as a description of the site of nascent arc magmatism because no well-developed edifice of magmatic crust (i.e., $\geq \sim 20$ km thick) exists during such magmatism. In other words, one cannot have a forearc without an arc. Thus we prefer use of the term "proto-arc" (Woodhead et al., 1998) to refer to a geologic circumstance that predates and is fundamentally different from the geologic setting of a forearc. The reader should be aware that although we make this distinction here, these terms have been used interchangeably in the literature to refer to the same tectonic

* E-mail: metcalfr@nevada.edu

Metcalf, R.V., et al., 2000, Geology and geochemistry of the ophiolitic Trinity terrane, California: Evidence of middle Paleozoic depleted supra-subduction zone magmatism in a proto-arc setting, *in* Dilek, Y., Moores, E.M., Elthon, D., and Nicolas, A., eds., Ophiolites and Oceanic Crust: New Insights from Field Studies and the Ocean Drilling Program: Boulder, Colorado, Geological Society of America Special Paper 349, p. 403–418.

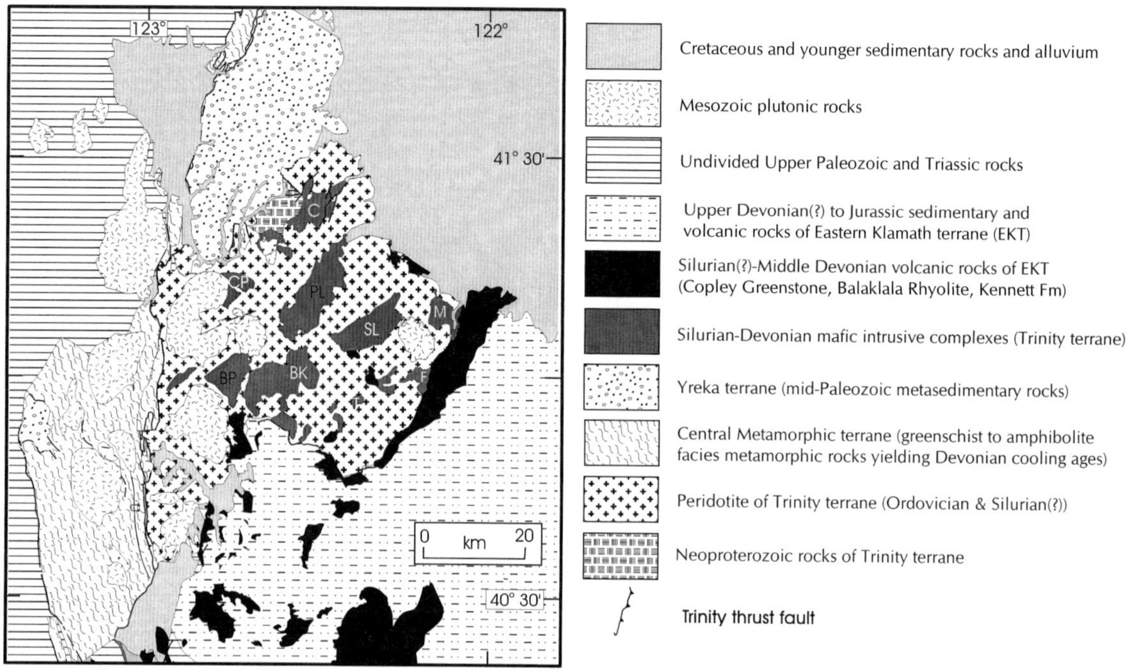

Figure 1. Geologic map of the eastern Klamath Mountains (modified from Irwin, 1994). Mafic intrusive complexes of Trinity terrane include the following: BK—Bonanza King, BP—Billy's Peak, C—China Mountain, CP—Craggy Peak, F—Flume Creek, M—Mount Bradley, PL—Porcupine Lake, SL—Seven Lakes, T—Tamarack Lake.

setting, namely intraoceanic, subduction-related, trench-proximal magmatism that occurs prior to the formation of a subduction-related magmatic edifice of crustal dimensions.

REGIONAL GEOLOGIC SETTING OF THE TRINITY TERRANE

The Klamath Mountains consist of east-dipping lithotectonic belts that record a history of episodic, subduction-related tectonic underplating, metamorphism, magmatism, and sedimentation ranging from middle Paleozoic in the east to Jurassic–Cretaceous in the west (Burchfiel and Davis, 1975; Irwin, 1989). The oldest rocks in the region occur in the eastern Klamath Mountains and are thought to preserve elements of a middle Paleozoic arc-trench system, with subduction-related volcanism (Eastern Klamath terrane), trench-forearc sedimentation (Yreka terrane), and trench-related metamorphism (Central Metamorphic terrane) (Fig. 1).

Metavolcanic rocks of the Eastern Klamath terrane

The middle Paleozoic volcanic succession includes flows and breccias of the basal Copley Greenstone and pyroclastic rocks of the overlying Balaklala Rhyolite (Kinkel et al., 1956) (Fig. 1). The Copley and lower Balaklala were intruded by the cogenetic Mule Mountain stock (Bence and Taylor 1985), which yields a concordant U-Pb zircon age of 400 $\pm$ 3 Ma (Albers et al., 1981). The Copley is composed primarily of island-arc tholeiite and boninite. Mafic rocks exhibit enrichments in LILEs (Sr, K, Rb, Ba, Th) relative to HFSEs (Nb, Ti, Zr Y) that are common in subduction-related magmas. The Copley–Balaklala–Mule Mountain system has been interpreted as a Silurian(?) to Early Devonian intraoceanic volcanic arc (Doe et al., 1985; Lapierre et al., 1985).

Sedimentary successions of the Yreka terrane

Metasedimentary strata of the Yreka terrane (Fig. 1) occur as nappes above a thin accretionary prism (Gregg Ranch Complex) preserved at the boundary between the Yreka and Trinity terranes. The Gazelle Formation contains volcaniclastic hemipelagic mudstones and volcanic turbidites and is interpreted to represent a trench-slope basin that was deposited unconformably on the Gregg Ranch Complex. The Moffett Creek Formation appears to record deposition in a trench (Wallin and Trabert, 1994). Concordant Late Silurian U-Pb ages (410–420 Ma) for detrital zircons in basal sedimentary rocks of the Yreka terrane, coupled with Devonian isotopic ages of metamorphism, reveal that virtually all sedimentation in the Yreka terrane occurred during the Early Devonian in the hanging wall of an intraoceanic convergent margin (Wallin and Gehrels, 1995).

Central Metamorphic terrane

The Central Metamorphic terrane (Fig. 1) comprises greenschist- to amphibolite-facies metabasite and siliceous pelite with minor metacarbonate and calc-silicate rocks (Davis et al., 1965; Peacock and Norris, 1989). The Central Metamorphic and Trinity terranes are juxtaposed along the east-dipping (~45°) Trinity thrust (Davis et al., 1965). The Central Metamorphic terrane has been interpreted as the lower plate of an east-dipping Devonian subduction zone (Peacock, 1987). The principal lines of evidence are (1) geophysical data, which indicate that the Central Metamorphic terrane continues east of the Trinity thrust at depth beneath the Trinity and Eastern Klamath terranes (Fuis et al., 1987), (2) metasomatism of the Trinity terrane adjacent to the Trinity thrust, which is interpreted to record subduction-related dewatering of the Central Metamorphic terrane (Peacock, 1987), (3) an inverted metamorphic gradient that formed in the Central Metamorphic terrane where the west end of the Trinity terrane was thrust over it (Peacock and Norris 1989), and (4) rocks of the Central Metamorphic terrane that yielded Devonian isotopic ages of metamorphism (Lanphere et al., 1968).

GEOLOGY OF THE TRINITY TERRANE

The Trinity terrane comprises three major lithologic assemblages: (1) a Neoproterozoic mafic tectonite assemblage, (2) a mantle assemblage of probable Ordovician and Silurian ages, and (3) Silurian and Devonian mafic intrusive complexes (Fig. 1). Undated pillow basalts crop out at numerous locations and appear to overlie the terrane. Several Mesozoic dioritic and granitic plutons intrude the Trinity terrane (Fig. 1).

The contact between the Trinity and Yreka terranes is a broad, structurally and lithologically complex zone whose tectonostratigraphic dip approximates the present topographic slope, thus producing a complicated map pattern. The boundary occurs principally within the accretionary prism (Gregg Ranch Complex) at the base of the Yreka terrane, which appears to be faulted everywhere. Contacts between the Trinity and Eastern Klamath terranes have been reported as both structural and depositional and the exact origin remains unclear (Schweickert and Irwin, 1989; Charvet et al., 1990). The amount of displacement among the Trinity, Yreka, and Eastern Klamath terranes is uncertain but is unlikely to be significant because of clear provenance links (Wallin and Metcalf, 1998).

Neoproterozoic mafic tectonite assemblage

The Neoproterozoic tectonite assemblage is restricted to the northwestern part of the Trinity terrane along the boundary with the Yreka terrane (Fig. 1). This assemblage consists of variably tectonized, amphibolite-facies hornblende metagabbro and metatonalite. Wallin et al. (1995) obtained slightly discordant $^{207}Pb/^{206}Pb$ ages of 556–579 Ma for zircon from deformed metagabbro and speculated that all metaplutonic rocks on the northwestern margin of the Trinity terrane constitute a coherent Neoproterozoic crustal fragment. The contact between the Trinity mantle peridotite and the Neoproterozoic mafic tectonite is a ductile shear zone that was intruded by the 415 Ma China Mountain gabbro pluton (Fig. 1) (Wallin et al., 1995). The Neoproterozoic tectonite assemblage has been interpreted as a block of oceanic crust that was emplaced tectonically into the margin of the Trinity terrane prior to 415 Ma (Wallin and Metcalf, 1998).

Ordovician–Silurian(?) mantle assemblage

A mantle assemblage consisting of harzburgite, dunite, and minor clinopyroxene-poor plagioclase lherzolite constitutes the bulk of the Trinity terrane (Quick, 1981; Boudier et al., 1989; Lindsley-Griffin, 1991) (Fig. 1). The peridotite exhibits high-temperature ductile deformational fabrics and varies from very weakly to strongly serpentinized. Plagioclase lherzolite from the mantle assemblage near China Mountain yielded an Ordovician Sm-Nd mineral isochron age of 472 ± 32 Ma (Jacobsen et al., 1984) and has been interpreted as a mantle diapir derived from a minimum depth of 30 km (Quick, 1981).

Various depleted Nd isotopic compositions have been recognized in mantle peridotite of the Trinity terrane. Plagioclase lherzolite has an Ordovician $\epsilon_{Nd(472)}$ signature of +10.4 (Jacobsen et al., 1984). $\epsilon_{Nd(423-402)}$ values of +7.1, +6.9, +2.0, and +0.5 have been reported for samples of undated lherzolite and harzburgite near the Porcupine Lake intrusive complex and east of the Seven Lakes intrusive complex (Gruau and Lécuyer, 1991; Gruau et al., 1995).

Silurian and Devonian mafic intrusive complexes

Mafic intrusive complexes represent the major crustal component in the Trinity terrane (Fig. 1). Eight major and several smaller complexes have been recognized and were mapped provisionally as Ordovician gabbro on the Weed 2° geologic map (Wagner and Saucedo, 1987). Previous workers have mapped at various scales all or parts of the China Mountain, Seven Lakes, Billy's Peak, and Tamarack Lake complexes (Quick, 1981; Cannat and Lécuyer, 1991; Lindsley-Griffin, 1991; Peterson et al., 1991). We have completed mapping (1:24,000) of two of these complexes, Bonanza King and Porcupine Lake (Fig. 1) and have examined all of the other major complexes. Three major units have been recognized in the mafic intrusive complexes and their relative ages have been determined by crosscutting and xenolith relationships. These units are (1) intrusive pyroxenite, (2) gabbroic plutons, and (3) aphanitic mafic dike swarms.

Intrusive pyroxenite. The oldest unit is an intrusive pyroxenite that is recognized in the Bonanza King, Porcupine Lake, Seven Lakes, and Tamarack Lake complexes. Intrusive pyroxenite occurs along the margins of the complexes where it intruded peridotite as large dikes and small stocks. Xenoliths of serpentinized peridotite are common in the pyroxenite (Fig. 2A); along many intrusive margins, peridotite forms a xenolithic breccia within an intrusive pyroxenite host. Peridotite xenoliths (± serpentinite) in pyroxenite typically exhibit a metamorphic rim (as wide as a few centimeters) of tremolite ± talc ± brucite ± chlorite (Schwindinger and Anderson, 1987; this study) (Fig. 2A).

Gabbroic plutons. Large gabbroic plutons are the dominant features of the mafic intrusive complexes. Xenolith relationships demonstrate that gabbroic plutons intruded both peridotite and intrusive pyroxenite (Fig. 2C). Considerable lithologic variation exists within gabbroic plutons. Several complexes (China Mountain, Seven Lakes, Porcupine Lake) exhibit basal layered-cumulate sequences (Quick, 1981; Schwindinger and Anderson, 1987; this study). A well-exposed southwest-dipping (45°–55°) sequence lies in the northern part of the Porcupine Lake complex. The base is composed of ~1200 m of dunite and wehrlite-pyroxenite interlayered at the centimeter to meter scale (Fig. 2B). Overlying this is a 1400 m section of coarse pyroxenite, which is in turn overlain by pyroxene gabbro. In the field, some outcrops of pyroxene gabbro exhibit centimeter-scale modal layering of pyroxene and plagioclase, but most outcrops are made of isotropic gabbro. The Craggy Peak complex appears to be dominated by pyroxene gabbro. At the structurally highest levels, gabbroic plutons are composed of hornblende (± clinopyroxene) gabbro and hornblende-quartz gabbro (e.g., Bonanza King, Billy's Peak, Flume Creek, Mount Bradley, northern part of China Mountain, parts of Seven Lakes). Hornblende gabbro and hornblende-quartz gabbro commonly exhibit isotropic textures, but modal layering of hornblende and plagioclase has been observed. Late segregations of pegmatitic pyroxene and hornblende-quartz gabbro and late trondhjemite dikes are ubiquitous features of the gabbroic plutons.

Aphanitic dike swarms. Kilometer-scale dike swarms intrude gabbro plutons in the Bonanza King, Billy's Peak (Peterson et al., 1991), Porcupine Lake, Flume Creek (the "Conant sheeted dikes" of Brouxel et al., 1988), and Mount Bradley complexes (Figs. 1 and 2D). Dike swarms consist of aphanitic dikes exhibiting both aphyric and porphyritic textures as well as chilled margins. Dike swarms are bimodal in composition, dominated by basalt and basaltic andesite

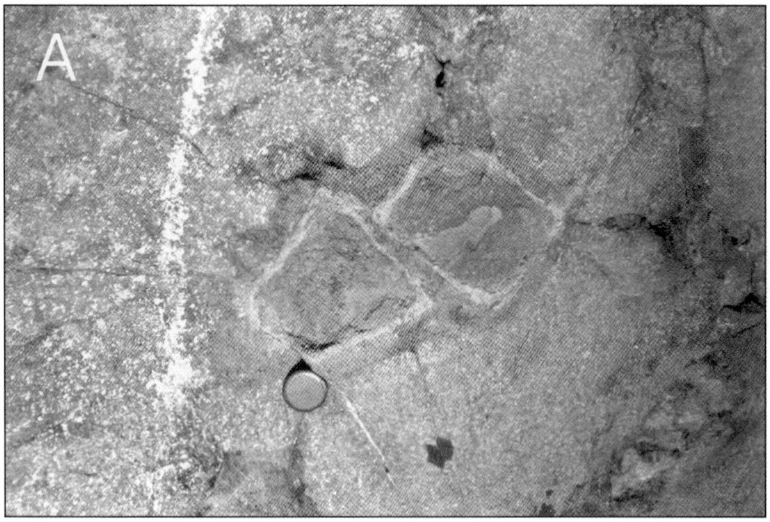

Figure 2 (on this and on facing page). A, View of peridotite xenoliths enclosed in an intrusive pyroxenite host from Seven Lakes mafic intrusive complex. Note the metamorphic rims on the xenoliths composed primarily of tremolite, talc, and chlorite. B, Interlayered cumulate dunite and pyroxenite (more resistant layers) from the basal section of the Porcupine Lake mafic intrusive complex. C, Xenoliths of pyroxenite enclosed in a gabbro host from Bonanza King mafic intrusive complex. D, Mafic dikes intruded into gabbroic pluton in the Billy's Peak mafic intrusive complex.

with lesser amounts of sodic rhyolite and trondhjemite. Mafic-felsic commingling features are common in the dikes. Dike swarms primarily intruded gabbro, although dikes cutting pyroxenite and peridotite are common. In the Bonanza King complex, two map units were used to indicate the areal extent of dike emplacement (Willse, 1999). The dike swarm unit constitutes areas where dikes, 0.5 to 3 m wide, make up >80% of outcrops, and dikes are separated by centimeter- to meter-scale screens of gabbro (Fig. 2D). The gabbro and dikes unit surrounds the dike swarm unit and constitutes areas where dikes make up 30%–80% of gabbro outcrops. A small percentage of aphanitic dikes is present in all of the mafic intrusive complexes, including those that lack major dike swarms. Throughout the Trinity terrane, aphanitic dikes strike approximately due east, dip steeply north and south, and have a dike-on-dike geometry (Boudier et al., 1989; Peterson et al., 1991; Willse, 1999). Collectively, dike swarms document a period of synmagmatic north-south extension late in the magmatic history of the mafic intrusive complexes.

Geochronology. Imprecise Silurian–Early Devonian Sm-Nd whole-rock isochron ages of 435–402 Ma have been reported for both gabbro plutons and dikes of microgabbro and pyroxenite (Jacobsen et al., 1984; Gruau and Lécuyer, 1991; Gruau et al., 1995). Pegmatitic gabbro from the China Mountain mafic intrusive complex, which intruded the shear zone separating the Neoproterozoic tectonite and Ordovician peridotite, yielded a concordant U-Pb zircon age of 415 ± 3 Ma (Late Silurian) (Wallin et al., 1995). New U-Pb zircon age data for pegmatitic gabbro from two mafic intrusive complexes clearly indicate a spectrum of Early Silurian (Bonanza King at 431 ± 3 Ma) to Early Devonian (Porcupine Lake at 404 ± 3 Ma) crystallization ages (Wallin and Metcalf, 1998).

Intrusive structure and emplacement history. The majority of contacts between peridotite and mafic intrusive complexes are intrusive, although some margins are faulted (e.g., China Mountain, Lindsley-Griffin, 1991) (Fig. 1). Some intrusive margins are steep, indicating pluton side walls (e.g., south side of Porcupine Lake complex; Cannat and Lécuyer 1991; this study), whereas others dip moderately (~45°) away from the intrusive complex (e.g., north side of Bonanza King), indicating transition to a roof zone. Topographically higher parts of the Bonanza King complex are rich in peridotite xenoliths, with map-scale septa of peridotite, indicating a roof zone. The Tamarack Lake complex may represent a roof zone or cupola (Cannat and Lécuyer, 1991).

Trinity dike swarms are unlike classical sheeted-dike complexes in two ways. First, dike swarms everywhere exhibit abundant screens and stoped blocks of country rock (either gabbro, peridotite, or both). Second, Trinity dike swarms are not laterally extensive sheets, but rather concentrated centers of dike emplacement within the gabbroic plutons. The Bonanza King dike swarm cuts at least 1000 vertical meters of gabbroic pluton. We infer that Trinity dike swarms represent the eroded remnants of subvolcanic feeder conduits.

Mafic intrusive complexes are composite plutonic features formed by the sequential emplacement of intrusive pyroxenite, gabbroic plutons, and dike swarms. Given that isotopic ages are available only for the gabbroic plutons, the duration of magmatism in a single intrusive complex is unknown. In addition, it is not known whether the emplacement of dike swarms occurred as a series of individual events over a protracted interval, temporally related to individual gabbroic plutons, or as a single event in the Early Devonian as might be suggested by their uniform due east strike.

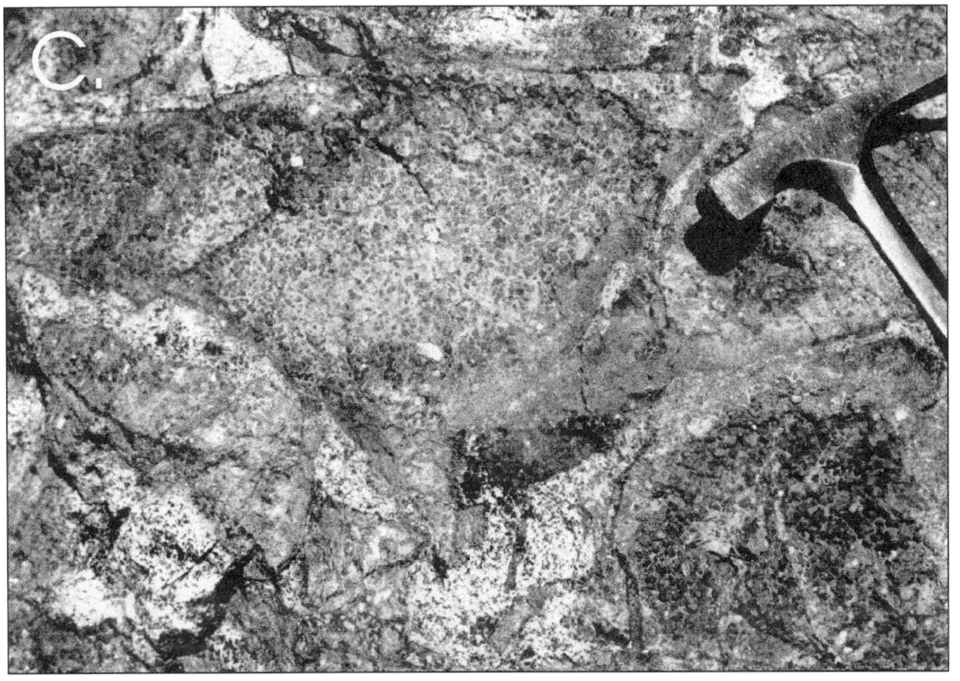

Greenschist-facies metamorphism. Metamorphism has partially altered primary ferromagnesian minerals to serpentine, actinolite, and/or chlorite and plagioclase to albite and epidote (Peterson et al., 1991). The effects of metamorphism are not ubiquitous; many rocks in the mafic intrusive complexes retain primary mineral assemblages and primary igneous textures (Quick, 1981; Peterson et al., 1991). A lack of deformational fabrics suggests metamorphism under static conditions subsequent to ductile deformation of the ultramafic country rock.

Pillow basalts

Undated, undeformed, mafic metabasalts, metadacite, and associated feeder dikes occur as isolated outcrops throughout much of the Trinity terrane and appear to overlie all other units. Many of the basalts exhibit pillow structures. These metavolcanic rocks are provisionally mapped as Copley Greenstone (Wagner and Saucedo, 1987) (Fig. 1).

IS THE TRINITY TERRANE OPHIOLITIC?

The Trinity terrane has been referred to as a mafic intrusion (Lipman, 1964), a mantle diapir (Quick, 1981), the roots of an island arc (Gruau and Lécuyer, 1991; Gruau et al., 1995; Wallin et al., 1995), an ophiolite (Moores, 1982; Saleeby, 1990), simply as a terrane (Silberling et al., 1992), and as a polygenetic SSZ ophiolite formed during the inception of intraoceanic subduction (Wallin and Metcalf, 1998). Lindsley-Griffin (1991) used the term "Trinity ophiolite" in referring to the Neoproterozoic tectonite unit of this paper, and the term "Trinity Complex" in referring collectively to the peridotite, mafic intrusive complexes, and pillow basalts. There is sufficient disparity in the literature that elucidation is in order. Much of the terminological confusion in the literature is the result of a failure to recognize the polygenetic nature of rocks in the Trinity terrane that resulted from a dearth of precise isotopic ages, relatively small study areas, and emphasis on particular lithologies rather than consideration of the evolution of the entire terrane. Previous workers noted the marked deviation of the Trinity terrane from the ideal Penrose ophiolite (Penrose Conference Participants, 1972); however, attempts to identify a monogenetic origin forced workers to discount evidence pointing to a more complex explanation (e.g., Boudier et al., 1989).

Some have questioned whether the Trinity terrane is an ophiolite; there is little merit in arguing this point, however, because it is clearly ophiolitic. The Trinity terrane contains the most essential elements of an ophiolite (Penrose Conference Participants, 1972), including, for example, tectonized mantle peridotite, plutonic sequences of layered ultramafic cumulates overlain by layered and isotropic gabbro, mafic dike complexes, and mafic volcanic rocks including pillow lavas. Although the pseudostratigraphy of the ideal Penrose ophiolite is absent, except internally in the mafic intrusive complexes, over 25 years of work has shown that such structure is not always present in ophiolites and is even absent along some segments of modern mid-oceanic ridges (e.g., the MARK area at the intersection of the Mid-Atlantic Ridge and the Kane Fracture Zone; Karson and Winters, 1992). Another feature of ophiolites, a basal thrust-fault system underlain by an inverted metamorphic sole (Moores, 1982), occurs where the Trinity terrane overlies the Central Metamorphic terrane along the Trinity thrust. Last, the general consensus is that ophiolites represent fossil oceanic lithosphere formed in an extensional environment prior to their tectonic emplacement onto a continental margin. The mafic and ultramafic lithologies of the Trinity terrane are consistent with those of modern oceanic lithosphere, and the strongly preferred orientation of mafic dikes is evidence of magmatism in an extensional environment.

Of greater concern are the paleogeographic and paleotectonic settings of these rocks, as well as what they reveal about tectonic and petrogenetic processes at those settings. Crustal rocks of the Trinity terrane are highly depleted, and a picture is emerging worldwide in which significant quantities of similarly depleted juvenile crust are generated in nascent intraoceanic subduction zones (Bloomer et al., 1995; Pearce et al., 1995; Taylor et al., 1992; Woodhead et al., 1998). The Trinity terrane constitutes an important three-dimensional observatory for the products of a tectonic setting whose modern analogues are largely inaccessible.

COMPOSITION OF MAFIC INTRUSIVE COMPLEXES

We present new major and trace element chemical data for a suite of 16 gabbro and mafic dike samples (Table 1) representing three of the mafic intrusive complexes, Bonanza King (three gabbros, three dikes), Porcupine Lake (three gabbros, three dikes), and Craggy Peak (three gabbros, one dike) (Fig. 1). We are concerned here primarily with the application of geochemical data to evaluating magma source and, by inference, the paleotectonic setting of formation. To avoid the effects of fractional crystallization, we excluded from the data set probable cumulate lithologies, specifically dunite and pyroxenite with obvious cumulate textures, high MgO (17–24 wt%), Cr (425–2083 ppm), and Ni (235–761 ppm) and mineralogically and compositionally layered gabbro, as well as obviously more evolved lithologies (>60 wt% SiO_2 and <7 wt% MgO) (Willse, 1999; Muller, 2000).

Petrography

Samples in the Trinity data set show the effects of secondary hydrothermal alteration to varying degrees from essentially fresh (<10% secondary minerals) (Fig. 3, A, B, D) to moderately altered (10%–50% secondary minerals) (Fig. 3C) to extensively altered samples (>50% secondary minerals) (Fig. 3E). Table 1 includes an indication of the degree of alteration for each sample. All of the samples have similar assemblages of secondary greenschist-facies minerals regardless of the degree (percent) of alteration. Alteration of plagioclase produced an aggregate of zeolite and epidote-group minerals (saussurite). Alteration of primary ferromagnesian minerals, clinopyroxene and hornblende, produced actinolite-tremolite with little or no chlorite. Secondary minerals largely formed pseudomorphs after the igneous minerals they replaced, and thus original igneous textures were preserved. The following discussion focuses on the primary igneous mineral assemblage and texture and degree of alteration for samples from each of the three mafic complexes.

The three Bonanza King pluton samples are hornblende gabbro composed of plagioclase (45%–55%) and brownish-green hornblende (40%–45%) with accessory iron-titanium oxide (1%–2%) or titanite (1%–2%). Quartz (~5%) is present in one sample (978-2). Samples are fine to medium grained and have subhedral granular textures. Plagioclase occurs as subhedral laths showing albite twinning; hornblende is subhedral to euhedral and shows twinning (Fig. 3A). Where present, quartz is a late interstitial mineral. The three Bonanza King dikes are composed of subequal amounts of plagioclase and hornblende with accessory iron-titanium oxide or titanite (Fig. 3B). Dike samples are porphyritic with 2%–10% phenocrysts of plagioclase and hornblende in aphanitic to fine-grained matrix. Plagioclase occurs as subhedral laths, and hornblende occurs as either small equant grains or as small acicular crystals (Fig. 3B). The degree of secondary alteration of Bonanza King samples varies from relatively fresh to extensively altered (Table 1).

The three Craggy Peak pluton samples are clinopyroxene gabbro composed of plagioclase (45%–55%) and clinopyroxene (40%–45%) with minor brown hornblende (5%–10%) and accessory iron-titanium oxide (trace). Samples are medium grained and exhibit subhedral granular texture. Plagioclase occurs as subhedral laths showing albite twinning; clinopyroxene is subhedral to euhedral and shows twinning (Fig. 3C). Hornblende occurs as a late interstitial mineral and as overgrowths on embayed (resorbed?) clinopyroxene crystals. The single Craggy Peak dike sample is composed of plagioclase (45%), clinopyroxene (30%), and brown hornblende (23%) with accessory iron-titanium oxide (1%–2%) (Fig. 3D). Some of clinopyroxene is rimmed by hornblende. The dike is very fine grained with aphyric texture. The degree of secondary alteration of Craggy Peak samples varies from a relatively fresh dike to moderately altered gabbro (Table 1).

The three Porcupine Lake samples are medium-grained clinopyroxene gabbro composed of plagioclase (45%–55%) and clinopyroxene (40%–45%) with accessory iron-titanium oxide (trace) (Fig. 3E). Where not affected by secondary alteration, plagioclase occurs as subhedral laths. The three Porcupine Lake dike samples are composed of subequal amounts of plagioclase and hornblende with accessory iron-titanium oxides. Dike samples are aphanitic and contain less than 5% phenocrysts of plagioclase and hornblende. Plagioclase occurs as subhedral laths and hornblende as small interstitial grains. Trachytic texture was observed in one sample. Both gabbro and dike

TABLE 1. WHOLE-ROCK GEOCHEMISTRY OF TRINITY TERRANE GABBRO AND MAFIC DIKES

Sample	977-4	978-2	978-42	978-9	978-11	978-50	P987-36	P987-40	P987-45	P987-47	P987-50	P988-20	986-5	986-8	986-7	986-6
Complex	BK	BK	BK	BK	BK	BK	PL	PL	PL	PL	PL	Pl	CP	CP	CP	CP
Lithology	gabbro	gabbro	gabbro	dike	dike	dike	gabbro	gabbro	gabbro	dike	dike	dike	dike	gabbro	gabbro	gabbro
Alteration	m	m	f	e	m	m	m	e	e	e	m	e	f	m	m	m
SiO_2	54.44	54.30	54.57	51.10	50.09	53.11	49.05	50.42	48.28	54.57	53.33	53.03	50.53	49.01	49.48	50.27
TiO_2	0.36	0.38	0.67	0.37	0.45	0.50	0.08	0.10	0.23	0.51	0.67	0.83	0.35	0.13	0.19	0.14
Al_2O_2	17.60	18.42	15.19	16.74	16.98	15.88	18.70	15.88	20.22	17.27	15.84	18.12	14.41	20.54	18.68	15.81
Fe_2O_3	7.63	5.33	8.94	6.18	6.74	7.09	3.96	5.62	7.69	6.15	7.54	7.94	7.03	5.12	5.90	5.56
MnO	0.13	0.10	0.17	0.09	0.13	0.10	0.08	0.12	0.15	0.10	0.14	0.16	0.13	0.10	0.12	0.12
MgO	7.82	6.47	6.52	9.15	7.81	7.21	11.83	15.13	10.04	8.09	8.32	7.44	13.97	10.20	10.67	13.16
CaO	11.08	10.70	6.82	12.11	9.61	7.64	15.49	12.42	12.36	9.41	7.99	7.20	13.35	14.69	14.63	14.79
Na_2O	1.41	2.65	5.19	2.14	3.02	3.35	0.50	0.01	0.23	3.92	4.14	4.97	0.35	0.30	0.29	0.16
K_2O	0.04	0.22	0.05	0.50	0.02	0.10	0.03	0.02	0.03	0.40	0.55	0.04	0.02	0.21	0.03	0.02
Total	100.51	98.57	98.12	98.38	94.84	94.99	99.72	99.73	99.22	100.43	98.51	99.73	100.13	100.30	99.99	100.02
Ba	9	17	17	49	6	29	18	6	5	24	110	13	4	2	4	1
Th	0.09	0.16	0.12	0.06	0.09	0.13	0.62	0.19	0.11	0.07	0.12	0.19	0.32	0.09	0.06	0.18
Cs	0.04	0.07	0.04	0.09	0.04	0.12	0.06	0.02	0.03	0.38	0.18	0.02	0.02	0.01	0.03	0.01
Rb	0.8	1.7	0.5	3.6	0.7	1.9	2.3	1	0.9	7.8	6.6	0.6	0.6	0.8	1	0.6
U	0.03	0.05	0.04	0.02	0.03	0.04	0.09	0.03	0.02	0.01	0.04	0.07	0.01	0.01	0.01	0.01
Pb	0.23	0.32	0.21	0.21	0.22	0.3	0.82	0.5	0.62	0.39	2.16	0.87	0.51	0.14	0.01	0.02
Ta	0.02	0.02	0.02	0.01	0.02	0.02	0.02	0.01	0.01	0.01	0.03	0.03	0.01	0.01	0.01	0.01
Nb	0.24	0.25	0.3	0.16	0.17	0.22	0.31	0.12	0.11	0.12	0.34	0.48	0.09	0.09	0.11	0.08
Zr	26	31	34	27	27	36	b.d.	b.d.	b.d.	16	34	36	b.d.	b.d.	b.d.	b.d.
Hf	0.54	0.76	0.86	0.56	0.7	0.86	0.09	0.05	0.08	0.71	1.07	1.32	0.3	0.07	0.1	0.07
Y	8.85	17.38	15.18	11.87	14.1	16.88	2.56	2.45	4.06	15.46	18.78	22.94	13.04	4	531	3.85
Sr	90	86	56	111	38	89	34	19	55	58	158	81	24	43	54	24
La	0.69	0.7	1.22	0.4	0.54	0.81	1.25	0.65	0.52	0.54	0.95	1.63	0.34	0.34	0.52	0.3
Ce	1.85	2.29	3.44	1.8	1.7	2.46	1.73	0.66	0.58	1.4	3.02	4.47	0.8	0.31	0.37	0.26
Pr	0.3	0.46	0.55	0.26	0.34	0.46	0.3	0.16	0.17	0.37	0.63	0.83	0.26	0.12	0.15	0.1
Nd	1.69	2.96	3.19	1.65	2.18	2.8	0.84	0.44	0.57	2.22	3.64	4.75	1.73	0.43	0.62	0.39
Sm	0.73	1.41	1.3	0.84	1.05	1.33	0.23	0.18	0.3	1.15	1.67	2.06	0.93	0.25	0.33	0.26
Eu	0.36	0.57	0.43	0.39	0.46	0.52	0.12	0.11	0.28	0.47	0.63	0.69	0.42	0.2	0.26	0.16
Gd	1.07	2.07	1.93	1.4	1.66	2.04	0.31	0.26	0.48	1.87	2.47	3.02	1.61	0.44	0.6	0.46
Tb	0.22	0.42	0.38	0.28	0.33	0.41	0.07	0.06	0.1	0.37	0.47	0.55	0.33	0.1	0.14	0.1
Dy	1.53	2.86	2.48	1.95	2.34	2.83	0.43	0.41	0.74	2.66	3.17	3.79	2.25	0.68	0.95	0.72
Ho	0.33	0.64	0.57	0.43	0.53	0.63	0.09	0.1	0.17	0.59	0.71	0.86	0.5	0.15	0.22	0.16
Er	0.95	1.83	1.6	1.27	1.51	1.81	0.28	0.29	0.48	1.73	2.08	2.43	1.43	0.44	0.62	0.45
Tm	0.14	0.27	0.23	0.19	0.23	0.27	0.04	0.04	0.07	0.26	0.3	0.35	0.2	0.07	0.09	0.06
Yb	0.93	1.75	1.5	1.22	1.41	1.67	0.26	0.29	0.49	1.67	1.95	2.3	1.28	0.43	0.54	0.44
Lu	0.15	0.28	0.24	0.19	0.22	0.27	0.04	0.04	0.08	0.27	0.31	0.35	0.21	0.07	0.09	0.07
Cr	158	168	177	188	236	174	84	260	42	179	310	50	783	182	336	285
Ni	120	76	71	76	84	85	98	101	83	95	138	75	256	85	121	89
Sc	54	40	44	47	40	38	46	46	51	39	39	46	59	43	51	55

Note: b.d.—below detection, BK—Bonanza King, PL—Porcupine Lake, CP—Craggy Peak, f—fresh, m—moderately altered (<50%), e—extensively altered (>50%). Oxides are in weight percent. Trace elements are in parts per million.

samples from Porcupine Lake are moderately to extensively altered (Table 1).

Geochemistry

Major elements and selected trace elements (Zr, Cr, Ni) were analyzed by X-ray fluorescence spectrometry at the University of Nevada, Las Vegas, using glass disks (analytical methods are reported in Willse, 1999). Samples were prepared by using tungsten carbide crushing equipment. Additional trace elements were analyzed by inductively coupled plasma–mass spectrometry (ICP-MS) in the GeoAnalytical Laboratory at the Washington State University; a discussion of the analytical methods is reported on the WSU web page (www.wsu.edu:8080/~geology/Pages/Services/Geolab.html). Samples were prepared for ICP-MS by using steel crushing equipment to avoid tungsten carbide contamination, which can interfere with analysis of Ta and Nb.

All of the samples in the data set are mafic in composition (SiO_2 48–55 wt%; MgO 7–15 wt%; Table 1). Pluton and mafic dike samples are subalkalic gabbro, basalt, and basaltic andesite with low TiO_2 (0.1–0.8 wt%) and low K_2O (0.1–0.55 wt%) (Table 1) similar to island-arc tholeiite. The suite exhibits high Mg# [molar MgO/(MgO + FeO)] (0.59–0.86; Fig. 4, Table 1) and shares a number of characteristics with boninites, although they do match well with one of the boninite subdivisions (e.g., high-Ca boninites vs. low-Ca boninites; Crawford et al., 1989). Low SiO_2 (<56 wt%, Fig. 4A), high CaO (7.6–14.7 wt%, Table 1), and high CaO/Al_2O_3 (0.4–0.94, Table 1, Fig. 4B) are more similar to high-Ca boninites. Total alkali concentrations (~0.05–5.2 wt%) and FeO concentrations (3.2–8 wt%), however, are more similar to low-Ca boninites (Crawford et al., 1989).

Figure 5 shows chondrite-normalized rare earth element (REE) plots for gabbro and dike samples. The sample set exhibits variously depleted light rare earth element (LREE) patterns. The heavy rare earth element (HREE) patterns are flat and commonly more depleted (~2–15 times chondrite) than average normal mid-ocean ridge basalt (N-MORB) (~20 times chondrite; Sun and McDonough, 1989). Pyroxene gabbro samples (Porcupine Lake and Craggy Peak, Fig. 5, C

Figure 3 (on this and facing page). Photomicrographs of Trinity rocks. A, Fresh (relatively unaltered) Bonanza King hornblende gabbro, sample 978-42. B, Moderately altered Bonanza King mafic dike, sample 978-11. C, Moderately altered (<50%) Craggy Peak pyroxene gabbro, sample 987-6. Dark patches within plagioclase (areas without albite twinning) are saussurite. D, Fresh Craggy Peak hornblende pyroxene diabase dike, sample 986-5. E, Extensively altered (>50%) Porcupine Lake pyroxene gabbro, sample 987-36. Abbreviations: P—plagioclase, H—hornblende, CPX—clinopyroxene, ACT—actinolite, S—saussurite.

and E) are more depleted than the mafic dike (all complexes, Fig. 5, B, D, and F) and hornblende gabbro (Bonanza King, Fig. 5A) samples.

Of particular note in the REE plots are large negative Ce and positive Eu anomalies displayed by the pyroxene gabbro samples (Fig. 5E). Less prominent negative Ce anomalies are evident in several dike samples (Fig. 5, B, D, and F). The magnitudes of the negative Ce and positive Eu anomalies are weakly correlated with one another; both anomalies are correlated with degree of depletion as indicated by decreasing HREE values (Fig. 5).

N-MORB–normalized spider diagrams for gabbro and mafic dike samples illustrate several important features (Fig. 6). Incompatible elements typically ascribed to a mantle component (e.g., Nb, Zr, Hf, Ti, Y, Yb) are depleted relative to N-MORB. Very highly incompatible elements (e.g., Nb, Zr $\approx$ 0.04–0.6 times N-MORB values) are more depleted than highly incompatible elements (e.g., Ti, Y, Yb $\approx$ 0.1–0.9 times N-MORB values). LILEs (e.g., Sr, K, Rb, Ba, Th), normally attributed to a subduction component, are enriched relative to HFSEs (e.g., Nb, Zr, Hf, Ti). Pyroxene gabbro samples are more depleted in HFSEs compared to samples of hornblende gabbro and mafic dikes (Fig. 6). Positive Pb and Sr anomalies of varying magnitude are present in most samples (Fig. 6).

Compositional data compiled from the literature for other mafic intrusive complexes and pillow basalts of the Trinity terrane show evidence of depletion consistent with that which we describe for the present samples (Gruau and Lécuyer, 1991; Gruau et al., 1995; Peterson et al., 1991). ϵ_{Nd} data have been published for the China Mountain (ϵ_{Nd} +7.3, Sm/Nd model age of 414 Ma; Jacobsen et al.,

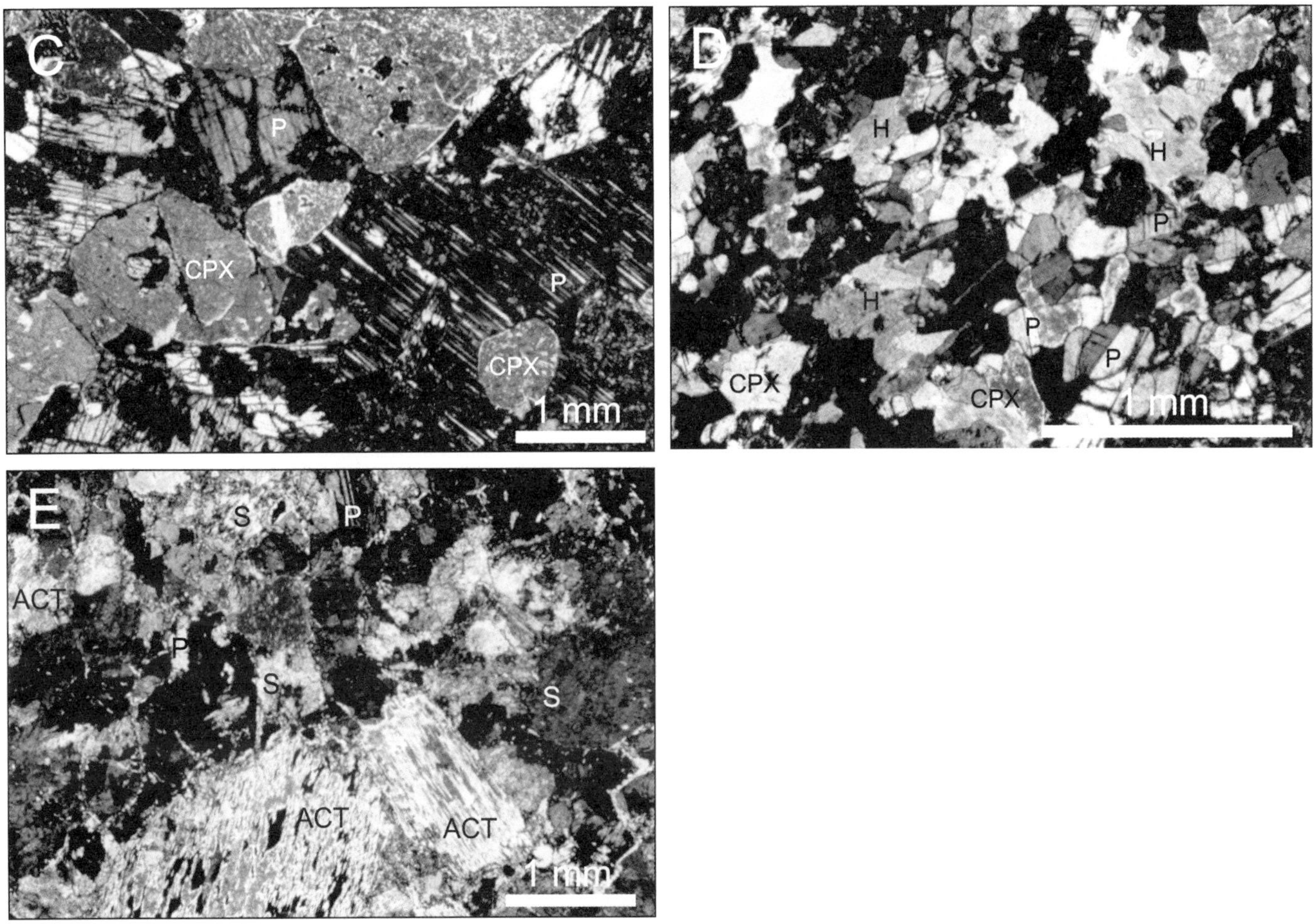

1984; Wallin et al., 1995), Porcupine Lake (ϵ_{Nd} +6.7, +6.4; Gruau and Lécuyer, 1991), Seven Lakes (ϵ_{Nd} +2.9, +2.5, −0.9; Gruau et al., 1995), and Bonanza King (ϵ_{Nd} +6.3 to +10.8; Metcalf et al., 1999) complexes.

PETROGENETIC INTERPRETATIONS

Trinity gabbro and mafic dikes as proxies for primary magmas

Application of whole-rock chemical data to questions of magma source and paleotectonic setting makes the assumption that whole-rock compositions approximate those of primary magmas. With regard to the Trinity data, two questions relate to the validity of this assumption. First, to what extent were whole-rock compositions affected by alteration? Second, to what degree do the whole-rock compositions record fractionation effects, particularly cumulate effects in the gabbro samples?

Two major types of alteration are recognized from ophiolite and deep-sea studies, both types presumably related to hydrothermal circulation of seawater near mid-ocean spreading centers (Humphris and Thompson, 1978; Harper et al., 1988; Valsami and Cann, 1992). Downward flowing cold seawater depletes crustal rocks in CaO while producing enrichments in MgO. This compositional shift results in chlorite-rich, amphibole- and epidote-poor assemblages (Humphris and Thompson, 1978). Hot, upwelling, Ca-rich fluids produce pipes of epidosite, a Ca-rich, Mg-poor rock composed essentially of nothing but epidote and quartz. Both types of alteration change whole-rock compositions and typically destroy primary igneous textures. In the Trinity samples, the lack of chlorite-rich or epidote + quartz mineral assemblages, the preservation of primary igneous textures, and magmatic concentrations of CaO and MgO preclude a significant amount of alteration of either type. Given that fresh rocks and extensively altered rocks have essentially identical compositions (e.g., Bonanza King gabbro and dike samples, Table 1, Figs. 5 and 6), there is no clear evidence that trace element compositions have been affected by alteration.

The Trinity dike rocks with fine-grained to aphanitic igneous textures, low phenocryst percentages, and chilled margins clearly have a noncumulate origin and are the best candidates for magmatic compositions. The mafic composition of the dikes (50.5–55.6 wt% SiO_2, 7.2–14 wt% MgO) precludes significant evolution by fractional crystallization, suggesting that the dikes may serve as suitable proxies for parental magmas. In general, petrographic evidence of a cumulate origin is lacking in the gabbro samples. Isotropic gabbro samples were collected specifically to avoid obvious cumulate features, and geochemical data are often used to screen data sets for cumulate samples. Specifically, samples with Sc greater than 50 ppm are considered to contain cumulate clinopyroxene and/or hornblende, whereas samples with Al_2O_3 greater than 20 wt% are considered to contain cumulate plagioclase (Pearce, 1996). Comparison of the Trinity data with these criteria shows 10 of the 16 samples clearly in the noncumulate field (Fig. 7). Only 3 samples—2 gabbro and 1 dike—are significantly above 50 ppm Sc (55–59 ppm). Two other samples both

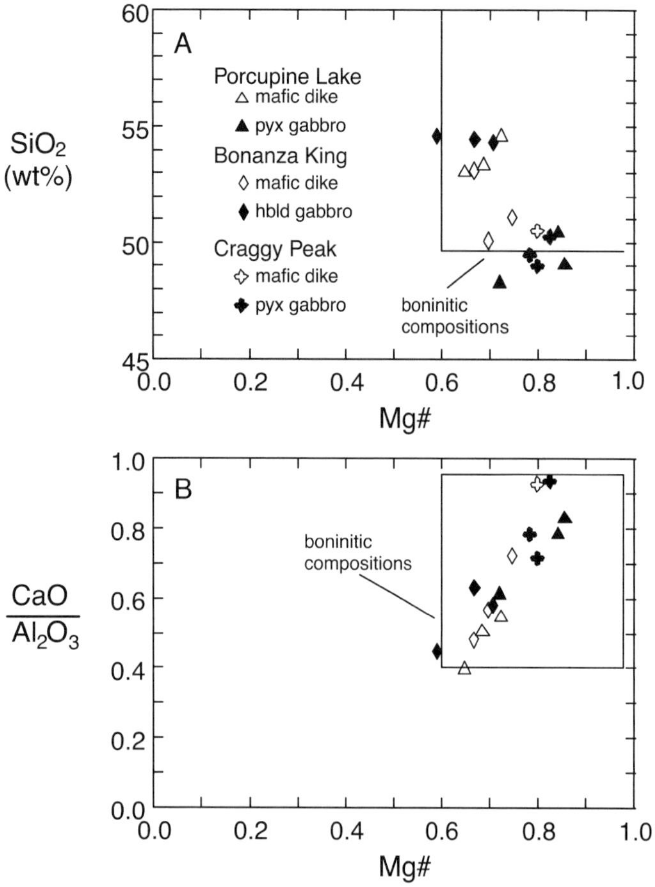

Figure 4. Plots illustrating similarities between Trinity samples and boninites. A, Mg# [molar Mg/(Mg + Fe)] vs. SiO_2. B, Mg# vs. CaO/Al_2O_3. Boninite fields from Crawford et al. (1989).

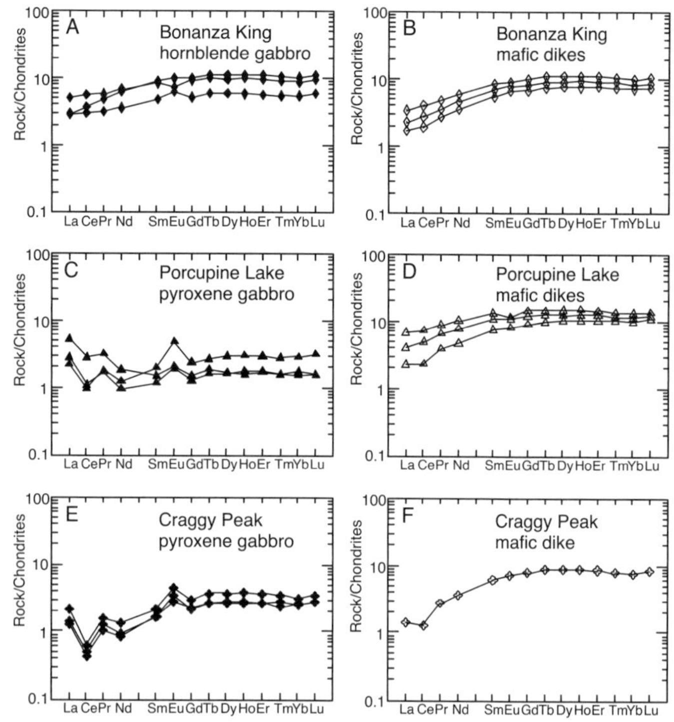

Figure 5. Chondrite-normalized rare earth element plots showing gabbro and dike samples from three mafic intrusive complexes. Normalizing values are from Sun and McDonough (1989). Symbols are the same as in Figure 4.

have 51 ppm Sc (Fig. 7). Two samples, both gabbro, have Al_2O_3 concentrations only slightly greater than 20 wt% (20.1, 20.5 wt%).

In summary, the dike samples appear to represent the most reasonable proxies for primary magmas based on field, petrographic, and chemical data. This interpretation is extended to the Bonanza King gabbro samples on the basis of compositional similarities with the dikes (Figs. 5 and 6). We refer collectively to the dike and Bonanza King gabbro samples as the "primary suite." We should point out that the Craggy Peak dike sample, a hornblende-bearing pyroxene basalt, plots in the pyroxene cumulate field (Fig. 7) and typically plots separate from the primary suite on subsequent diagrams. The pyroxene gabbro samples (Porcupine Lake and Craggy Peak) have similar, rather unique compositions, specifically the Ce and Eu anomalies and strongly depleted HFSEs and HREEs (Figs. 5 and 6). Petrogenetic interpretations are only slightly modified if pyroxene gabbro samples are included in the analysis, as discussed subsequently. We return later to the question of the anomalous composition of the pyroxene gabbro samples.

Depleted SSZ ophiolite

Ophiolites are subdivided into MORB and SSZ types primarily on the basis of geochemical data. The main criterion for recognizing MORB ophiolites, by definition, is the presence of crustal rocks with a MORB geochemical signature. It is clear that mafic rocks of the

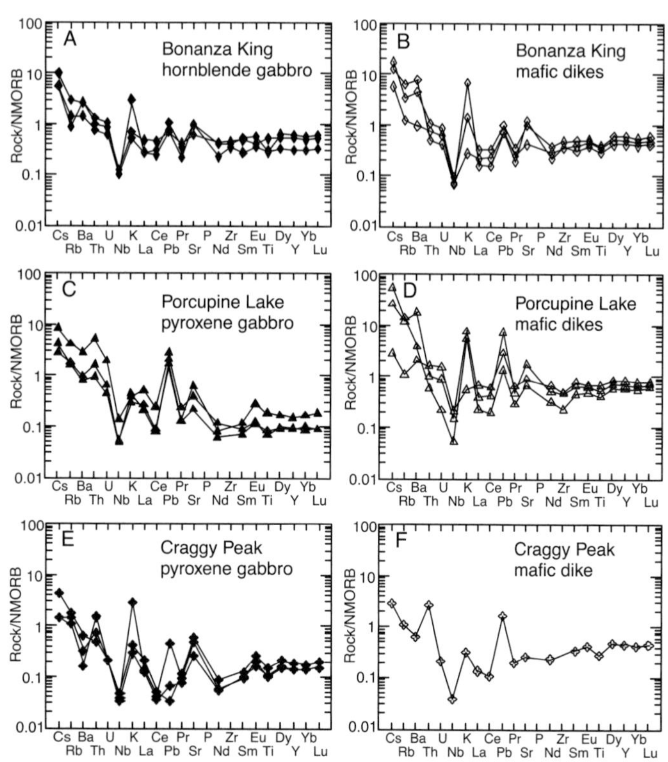

Figure 6. Trace element spider diagrams showing gabbro and dike data from three mafic intrusive complexes. Data are normalized to N-MORB by using values from Sun and McDonough (1989). Symbols are the same as in Figure 4.

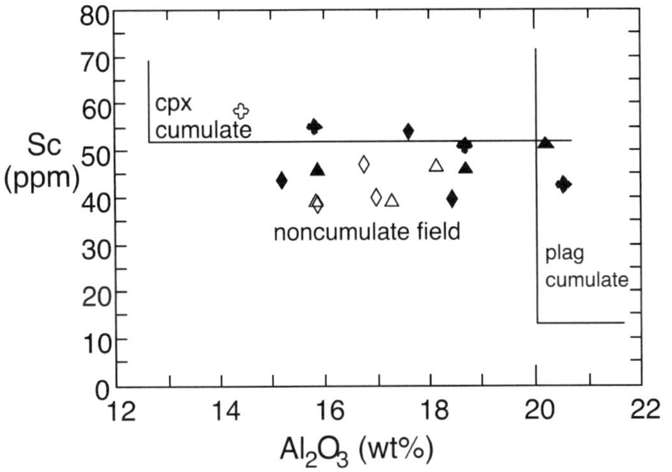

Figure 7. Plot of Sc vs. Al_2O_3 used to test for a cumulate origin. Samples with greater than 50 ppm Sc potentially contain cumulate clinopyroxene (cpx). Samples with greater than 20 wt% Al_2O_3 potentially contain cumulate plagioclase. Symbols are the same as in Figure 4.

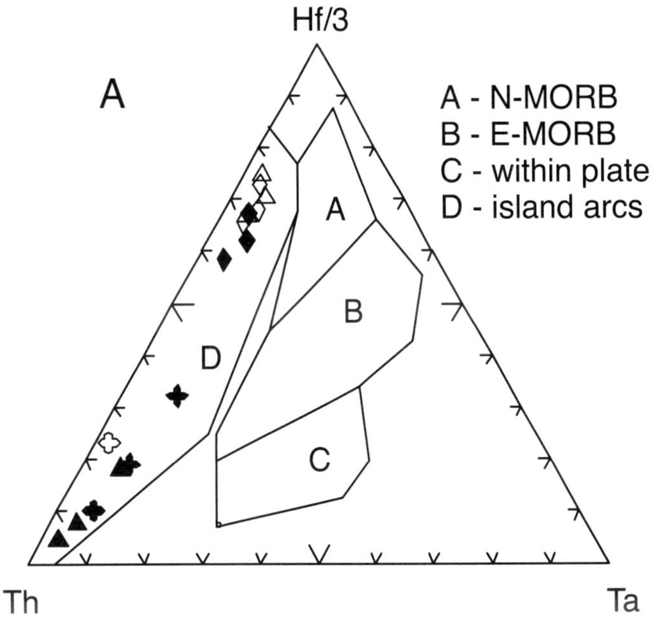

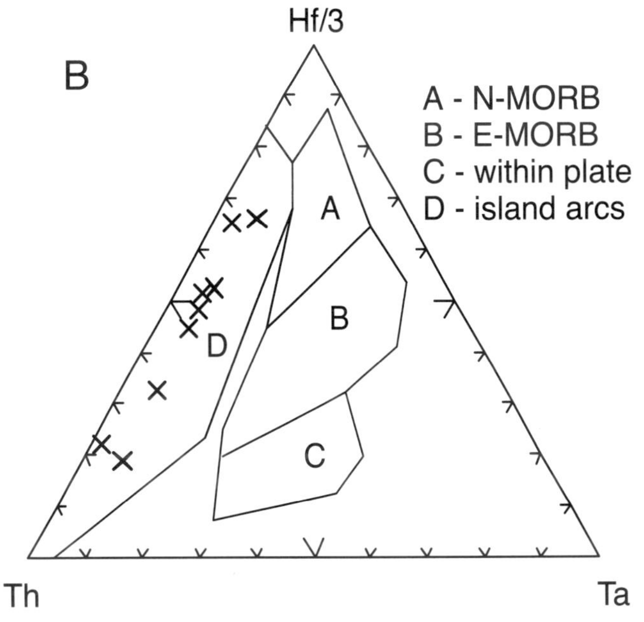

Figure 8. Th-Hf-Ta tectonic discrimination diagram after Wood (1980). A, Trinity data. Symbols are the same as in Figure 4. B, Data from the Copley Greenstone, Eastern Klamath Mountains. Copley data sources are Lapierre et al. (1985) and Brouxel et al. (1987, 1988).

Trinity terrane depart significantly from MORB compositions (Fig. 6).

SSZ magmas are characterized by high LILE/HFSE ratios. SSZ models hold that fluids derived from the subducting slab carry high concentrations of LILEs (± LREEs), whereas the HFSEs and HREEs are retained in the slab. These slab fluids metasomatize the overlying mantle wedge and trigger melting. In addition to contributions from the slab itself (i.e., metamorphosed basaltic crust), subduction fluids may carry elemental contributions from subducted pelagic sediments. HFSE and HREE concentrations in SSZ magmas are controlled primarily by the composition of the mantle. Consequently, the trace element signature of subduction-related magmas is apparently derived from three main sources, the slab-derived fluid, the subducted sediment, and the overlying mantle wedge (Perfit et al., 1980; Pearce, 1982; Arculus and Powell, 1986).

The Th-Hf-Ta discrimination diagram (Wood, 1980) has been used widely to discriminate between SSZ and MORB ophiolites, in part because these elements are less susceptible to mobility during low-grade metamorphism. On the Th-Hf-Ta diagram, the primary suite of the Trinity and Copley samples plots clearly in the island-arc (SSZ) field (Fig. 8). Pyroxene gabbro samples plot as a separate group in the island-arc field, near the Th corner. The characteristic subduction-related enrichment of LILEs (e.g., Cs, Rb, Ba Th, U, K, Sr) relative to HFSEs (e.g., Nb) is apparent in both the primary suite and the pyroxene gabbro group (Fig. 6). On the basis of these geochemical features alone, the Trinity terrane can be classified as a SSZ ophiolite.

The highly depleted HFSE and REE concentrations point to a residual, MORB-type, mantle source for the Trinity magmas. Ratio versus ratio plots such as the Nb/Yb vs. Th/Yb plot (Fig. 9) can be used to illustrate this point (Pearce, 1982; Pearce et al., 1995). Previous partial melting in the mantle source produces a decrease in both ratios, each of which consists of the concentration of a more incompatible element divided by the concentration of a less incompatible element, resulting in the source-depletion vector shown (Fig. 9). Addition of subduction-derived fluid increases the Th/Yb ratio, but not the Nb/Yb ratio, in the resulting SSZ melt and produces the subduction-component vector (Fig. 9). Liquid- and cumulate-fractionation trends also are shown in Figure 9. Extrapolation back along the subduction-component vector provides an indication of mantle-source composition. The Trinity data, including both the primary suite and the pyroxene gabbro samples, clearly point toward a highly depleted mantle source, more depleted than that which produces N-MORB (Fig. 9). Primary-suite data form a linear array consistent with some fractional crystallization (Fig. 9). Nd isotope data for the Trinity terrane, which lies mostly between +6.4 to +10.8, also supports the interpretation of a depleted-mantle source region. Modern analogues for depleted mafic rocks of the Trinity terrane occur in Eocene forearc basalts of the Izu-Bonin-Mariana system and Quaternary outer-arc basalts of the New Britain arc (Bloomer et al., 1995; Woodhead et al., 1998) (Fig. 10A).

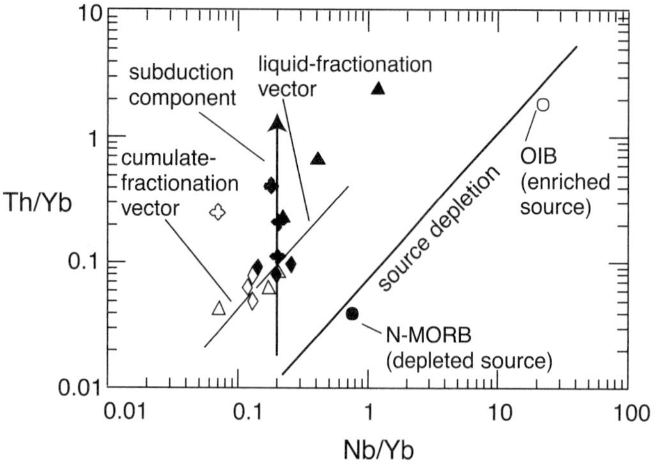

Figure 9. Trinity terrane data plotted on a Nb/Yb vs. Th/Yb diagram (modified from Pearce et al., 1995). The source-depletion vector illustrates the effect on magma chemistry of increasing degrees of depletion in the mantle-source region produced by prior melt extraction. The subduction-component vector illustrates the compositional effect of metasomatism by slab-derived subduction fluids. Fractional crystallization typically produces liquid and cumulate compositional vectors that parallel the source-depletion trend as shown. Trinity data (plotted) suggest subduction-related metasomatism of an ultradepleted, residual, MORB-type, mantle source. Symbols as in Figure 4. N-MORB and OIB (oceanic-island basalt) data from Sun and McDonough (1989).

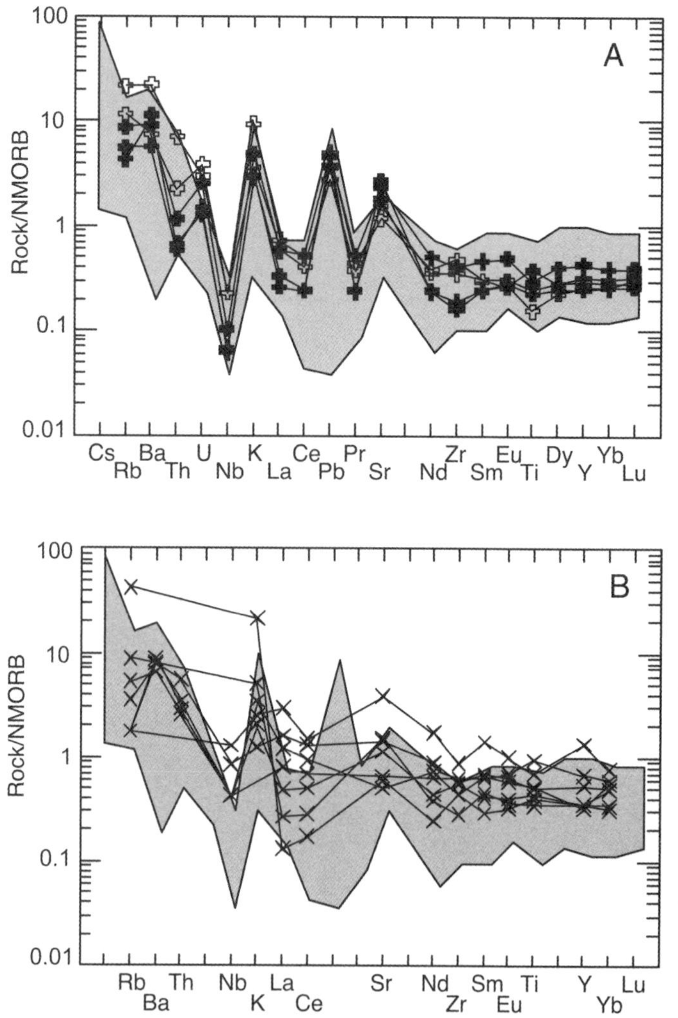

Figure 10. Trace element spider diagrams comparing Trinity data (shaded region) with data from other areas. A, Comparison with Eocene basalts of the Izu-Bonin-Mariana forearc (open symbols) and Quaternary basalts from the outer-arc region of the New Britain arc. Data are normalized to N-MORB by using normalizing values from Sun and McDonough (1989) Data sources are Bloomer et al. (1995) and Woodhead et al. (1998). B, Comparison with data from Copley Greenstone (X), Eastern Klamath terrane. Data sources: Lapierre et al. (1985) and Brouxel et al. (1987, 1988).

Subducted sediment and seawater components

Trace element data point to sediment-derived and seawater-derived components in the Trinity suite. Pb enrichment is typically thought to reflect the involvement of sediment in the subduction process. In the Trinity suite, Pb enrichment is seen in the positive Pb anomalies on the N-MORB–normalized spider diagrams (Fig. 6) or as elevated, though variable, Pb/Ce (0.03–1.09; mean = 0.35) (Fig. 11) and Pb/Nd (0.02–1.09; mean = 0.32) ratios (Table 1; compared to average N-MORB at Pb/Ce = 0.04 and Pb/Nd = 0.04; Sun and McDonough, 1989). Sr enrichment may reflect the involvement of seawater, either in the slab-derived subduction component or via hydrothermal circulation of seawater through oceanic crust. In the Trinity primary suite, Sr enrichment is seen in the positive Sr anomalies on the N-MORB–normalized spider diagrams (Fig. 6) or as elevated, though variable, Sr/Nd (7–100; mean = 42) (Fig. 11) ratios (Table 1; compared to average N-MORB at Sr/Nd = 12; Sun and McDonough, 1989).

Rare earth element anomalies

The pyroxene gabbro samples (Porcupine Lake and Craggy Peak) are notable for their negative Ce and positive Eu anomalies as well as their highly depleted compositions. Potential interpretations as to the origin of these REE anomalies include (1) cumulate processes, (2) element mobility during hydrothermal alteration, and (3) the involvement of sediment in the subduction process.

It is difficult to accept a cumulate origin for the REE anomalies. Only about half of the pyroxene gabbro samples plot clearly in either of the two cumulate fields on the Al_2O_3 vs. Sc plot (Fig. 7). Although plagioclase accumulation is a potential mechanism for production of positive Eu anomalies, only two of the six samples in question plot in the plagioclase field in Figure 10. Production of negative Ce anomalies would require fractionation of Ce^{4+} from trivalent LREEs. This, in turn, would require a high oxygen fugacity more typical of near-surface environments than that found in magmatic systems. Although a cumulate process could explain the depleted compositions of the pyroxene gabbro samples, in the case of the Trinity rocks, it does not appear to be a viable mechanism for production of the REE anomalies.

In a study of hydrothermal alteration in the Pindos ophiolite, Valsami and Cann (1992) found both negative Ce and positive Eu anomalies in epidosite pipes. In their study, a comparison of REE concentrations from rocks distal to the pipes suggests that epidosite anomalies were due largely to addition of La, Pr, and Eu by metasomatizing hydrothermal fluids. Although the Trinity pyroxene gabbro samples have been subjected to moderate to extensive alteration (Table 1), there is no evidence of the extreme alteration necessary to produce epidosites (simple epidote + quartz assemblages) and, con-

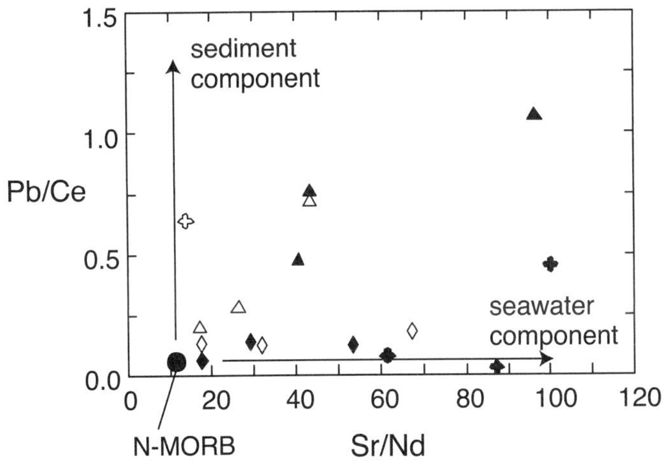

Figure 11. Plot of Pb/Ce vs. Sr/Nd. N-MORB data point from Sun and McDonough (1989).

sequently, hydrothermally generated Ce and Eu anomalies. Additionally, some samples from the primary suite are equally or more altered than the pyroxene gabbro samples, yet lack the REE anomalies. Although hydrothermal alteration and metasomatism are capable of generating Ce and Eu anomalies, such an interpretation is not consistent with the level of alteration in the Trinity samples.

Both negative Ce and positive Eu anomalies are observed in a number of modern, subduction-related island-arc suites (Elliott et al., 1997; Pearce et al., 1995; Woodhead et al., 1998). It is interesting that these anomalies are found primarily in highly depleted island-arc tholeiites (Woodhead et al., 1998). Ce anomalies in modern arcs are typically attributed to the subduction of pelagic sediments (Elliott et al., 1997; Woodhead et al., 1998), which are known to exhibit similar negative Ce anomalies (Hole et al., 1984). In cases where it is clear that positive Eu anomalies are not related to plagioclase accumulation or resorption, they appear to be related to preferential transport by slab-derived fluid of Eu^{2+} over trivalent REEs (Bau and Knittel, 1993; Woodhead et al., 1998). The outer-arc rocks of the New Britain arc (Woodhead et al., 1998) are an example of depleted island-arc tholeiites that carry both negative Ce and positive Eu anomalies. It should be noted that the magnitudes of the Ce and Eu anomalies in the Trinity rocks are larger than those reported from the New Britain arc.

DISCUSSION

Crustal rocks of the Trinity terrane carry a clear subduction-related geochemical signature (Figs. 6, 8, and 9). Depletion patterns in the HFSEs (Fig. 6) indicate a residual, MORB-type, mantle source, i.e., a source more depleted than that which produces N-MORB. Elevated Pb/Ce ratios suggest a subducted sediment component. These conclusions are valid for the primary-suite data and are based on only the less mobile elements (Th, Hf, Ta, Nb, Yb; Figs. 8 and 9). Consideration of the pyroxene gabbro samples in the interpretation requires a greater degree of source depletion and provides additional evidence of a subducted sediment component (negative Ce anomalies). These new data clearly point to a SSZ origin for Trinity crustal rocks.

Suggested tectonic settings for the generation of SSZ ophiolites include backarc basins, rifted island arcs (the initiation of backarc basin formation), and magmatic proto-arcs (elsewhere sometimes called forearcs) that form above nascent, retreating intraoceanic subduction zones. Results from the Deep Sea Drilling Project and Ocean Drilling Program that focused primarily, although not exclusively, on the Izu-Bonin-Mariana forearc-arc-backarc system provide data for testing various proposed origins of SSZ ophiolites (Stern and Bloomer, 1992; Taylor et al., 1992; Bloomer et al., 1995; Hawkins, 1995; Fryer, 1995; Taylor, 1995). Backarc-arc-forearc geochemical transects reveal variations in the composition of subduction-related magmatic rocks that are thought to reflect (1) a diminished effect of slab fluid in backarc regions and (2) across-strike variations in the composition of the mantle wedge (Taylor et al., 1992). Within the Izu-Bonin-Mariana system, only magmatic rocks of the forearc region have depleted compositions comparable to the those of the Trinity (Fig. 10).

Proto-arc (forearc) models for generation of SSZ ophiolites

Samples recovered from the Izu-Bonin-Mariana forearc consist of volcanic flows, pillow lavas, breccia, diabase, plutonic rocks, and serpentinized peridotite (Bloomer et al., 1995, and earlier references therein). These igneous rocks are island-arc tholeiites and boninites and related silicic differentiates, which carry SSZ geochemical signatures. Among the SSZ analogue regions of the Izu-Bonin-Mariana system, the forearc rocks are the most highly depleted in HFSEs. The Izu-Bonin-Mariana forearc (now 300 km wide by 1000 km long) developed by proto-arc magmatic processes in the Eocene immediately prior to the development of a narrow, well-defined volcanic arc in the Oligocene. In the Izu-Bonin-Mariana system, magmatism occurred for about 15 m.y. until vertical sinking of oceanic lithosphere gave way to downdip descent and true subduction began (Stern and Bloomer, 1992; Bloomer et al., 1995). This change marked the beginning of magmatism in a narrow, well-defined, volcanic arc. Collectively, work in the Izu-Bonin-Mariana system suggests that magmatic intraoceanic proto-arcs (intraoceanic forearcs) may serve as a model for the generation of many SSZ ophiolites.

According to the proto-arc model, subduction is initiated by density instability across an oceanic transform boundary (Stern and Bloomer, 1992, and earlier references therein). The older, denser lithosphere initially sinks, which produces a rollback of the newly forming subduction hinge. This rollback results in extension of the upper plate. Release of volatiles and/or siliceous partial melts by the descending slab, combined with extensional decompression, results in partial melting of the overlying mantle wedge. During extension, the overlying lithosphere becomes dismembered by faulting and is replaced by new SSZ-type magmatic rocks (lavas, dike swarms, gabbroic plutons) and mantle diapirs (highly depleted and/or serpentinized harzburgite). Preservation of blocks of older oceanic lithosphere in developing SSZ ophiolites may be favored by slower rates of hanging-wall extension; at higher rates of extension, evidence of older oceanic lithosphere may be absent (Bloomer et al., 1995). The Neoproterozoic assemblage may represent such a preserved block, suggesting slow extension rates.

Forearc paleotectonic setting for the Trinity terrane ophiolite

In a previous paper (Wallin and Metcalf, 1998), we cited geochronologic and genetic links between the Trinity terrane and the other terranes in the eastern Klamath Mountains—in particular, its juxtaposition with trench-related rocks of the Central Metamorphic and Yreka terranes and the sparse geochemical data suggesting its depleted SSZ geochemical signatures—as evidence favoring a proto-arc origin for the Trinity terrane. Our new geochemical data strengthens that interpretation by providing evidence of an SSZ origin and a depleted, residual, MORB-type, mantle source for magmas of the Trinity terrane.

This proto-arc model for the Trinity terrane holds that all four terranes in the eastern Klamath Mountains (Central Metamorphic, Yreka, Trinity, and Eastern Klamath) evolved principally during the inception of a single, east-dipping, middle Paleozoic convergent margin, analogous to the Eocene Izu-Bonin-Mariana intraoceanic arc system (Fig. 12). Ordovician peridotite (ϵ_{Nd} = +10.4) represents fragments of older oceanic lithosphere trapped in the hanging wall of the Silurian–Devonian subduction zone. Peridotite with ϵ_{Nd} values below +7.5 likely represents Silurian–Devonian mantle diapirs emplaced into Ordovician peridotite during proto-arc extension. Intrusive pyroxenite, gabbroic plutons, and dike swarms represent new crust formed during Silurian and Devonian forearc extensional magmatism and were emplaced into peridotite. The fact that the duration of middle Paleozoic magmatism in the eastern Klamaths (~30 m.y.) is longer than that recorded by the Izu-Bonin-Mariana forearc (~15 m.y.) suggests a slower spreading rate for the Trinity terrane proto-arc.

According to this model, Silurian and Devonian magmatism in the Trinity terrane was coeval and broadly cogenetic with middle Paleozoic metavolcanic rocks in the Eastern Klamath terrane such as the Copley Greenstone (compare Fig. 8, A and B, and compare Fig. 10, A and B). Geochemical and isotopic similarities between Trinity crustal rocks and volcanic rocks of the Eastern Klamath terrane are consistent with their derivation from a similar source (Figs. 8 and 10). Data from the Trinity terrane indicate a depleted mantle source enriched by fluids derived from a subducting slab that included a component of subducted sedimentary material. It is worth noting that volcanic rocks from the Izu-Bonin-Mariana forearc and the outer lavas of the New Britain arc also carry elevated Pb/Ce ratios (0.3–1.2; Bloomer et al., 1995; Woodhead et al., 1998); this similarity suggests sediment subduction in these modern intra-oceanic arc systems. The less depleted nature of Eastern Klamath terrane metavolcanic rocks is consistent with greater distance from the trench and tapping of a deeper, less depleted mantle source (Taylor et al., 1992; Wallin and Metcalf, 1998). Pb isotope data indicate a sedimentary component in the metavolcanic rocks of the Eastern Klamath terrane (Doe et al., 1985). The Yreka and Central Metamorphic terranes represent trench-forearc sedimentary basins and downgoing slab, respectively, that developed during proto-arc extension as the subducted slab made the transition from initial subsidence to true downdip subduction.

The Copley Greenstone and Balaklala Rhyolite in the Eastern Klamath terrane have been interpreted as an intraoceanic volcanic island arc (Doe et al., 1985; Lapierre et al., 1985). Such an interpretation, which would place the Trinity terrane in an arc-trench gap, is broadly consistent with our model. An alternative interpretation of the entire Copley–Balaklala–Mule Mountain system, however, is that it represents the thin volcanic cover of Silurian–Devonian forearc magmatism, rather than a fragment of an island arc as is generally believed. Depleted HFSE concentrations in the Copley Greenstone are consistent with a forearc origin (Wallin and Metcalf, 1998). Given that the Trinity terrane extends beneath the Eastern Klamath terrane

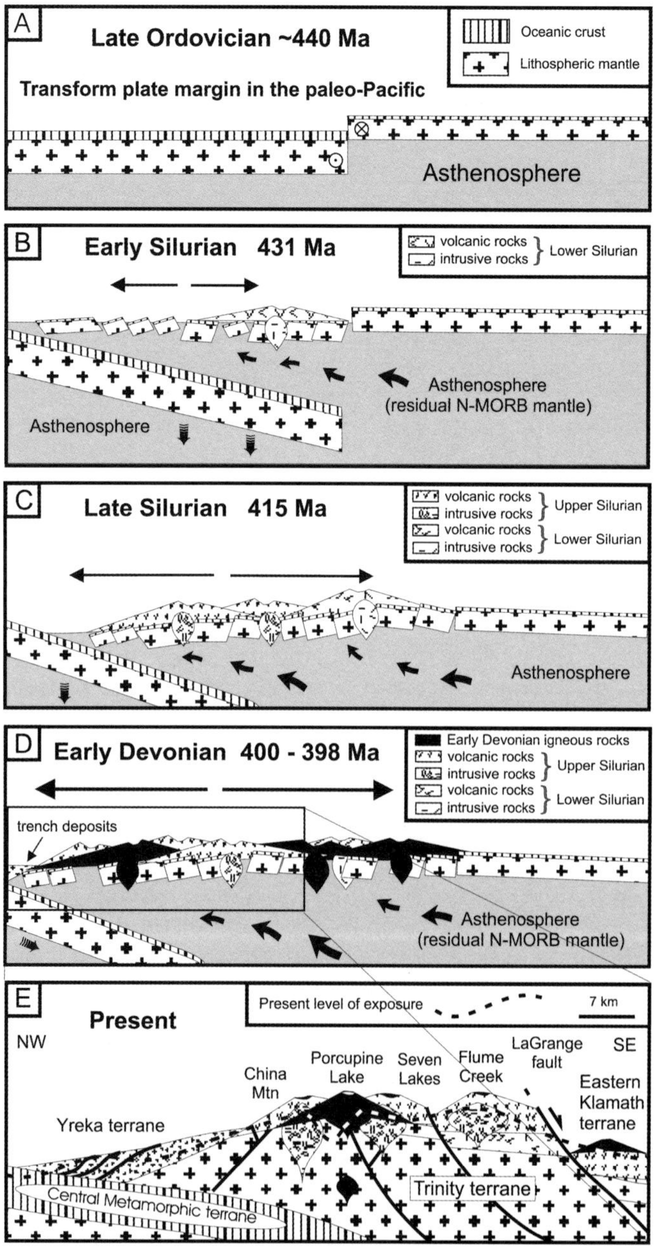

Figure 12. Schematic diagram illustrating the middle Paleozoic tectonic setting and evolution of the eastern Klamath Mountains. A, Intraoceanic subduction is initiated because of a density instability along a fracture zone. B, Vertical sinking of the denser slab causes asthenosphere to flow over the sinking slab with concomitant extension of the hanging-wall lithosphere. Adiabatic melting of asthenosphere (depleted, residual, MORB-type mantle), coupled with a flux of slab-derived fluids, results in the emplacement of primitive, depleted mafic intrusive complexes overlain by a thin, cogenetic, volcanic carapace. C and D, Continued trench rollback and associated hanging-wall extension produce mafic igneous complexes over a 30 m.y. period. E, Present configuration of middle Paleozoic rocks of the eastern Klamath Mountains, which exhibit little evidence of post-Devonian deformation. The Central Metamorphic terrane is a remnant of a downgoing oceanic slab. The Yreka terrane consists of mildly telescoped trench-related basins. The Trinity terrane comprises trapped Ordovician lithospheric mantle intruded by mafic intrusive complexes ranging from Early Silurian to Early Devonian (431–404 Ma) in age. Middle Paleozoic rocks of the Eastern Klamath terrane are a downthrown remnant of the volcanic cover of the eastern Klamath ophiolite preserved in the hanging wall of a normal fault having several kilometers of displacement. All these middle Paleozoic subduction-related igneous complexes (Trinity and Eastern Klamath terranes) formed above a subducting slab, but prior to the development of any stabilized volcanic-arc edifice of crustal dimensions (cf. Stern and Bloomer, 1992; Bloomer et al., 1995) (modified from Wallin and Metcalf, 1998).

(Fuis et al., 1987), we interpret that metavolcanic rocks in the Eastern Klamath terrane are lithostratigraphically equivalent to a thin volcanic carapace that once overlay the Trinity terrane and as such represent a part of the ophiolite. According to this interpretation, normal slip on terrane-bounding faults has preserved this carapace on the downthrown side, and the Trinity terrane is simply a deeper level of exposure than that of the Eastern Klamath terrane. Moreover, undated mafic volcanic rocks lying nonconformably on the Trinity terrane that are mapped provisionally as Devonian Copley Greenstone (Wagner and Saucedo, 1987) are probably remnants of that eroded volcanic carapace.

Collectively, observations indicate strongly that the Central Metamorphic, Yreka, Trinity, and Eastern Klamath terranes evolved together during the inception of a single, east-dipping, middle Paleozoic intraoceanic convergent margin (Fig. 12). This interpretation provides a unifying explanation for the seemingly disparate geology of these terranes.

CONCLUSIONS

Mafic intrusive complexes in the Trinity terrane are composite plutonic features that intruded peridotite by the sequential emplacement of intrusive pyroxenite, gabbroic plutons, and dike swarms during a period of lithospheric extension.

Gabbros and mafic dikes are depleted in HFSEs relative to N-MORB and have high LILE/HFSE ratios, high Pb/Ce ratios, negative Ce anomalies, and positive Eu anomalies. These data are consistent with partial melting of a depleted, residual, MORB-type, mantle source enriched by subduction-derived fluids that included a sedimentary input.

All four terranes in the eastern Klamath Mountains evolved principally as a proto-arc during the inception of a single, east-dipping, middle Paleozoic subduction zone, analogous to the Eocene Izu-Bonin-Mariana intraoceanic arc system.

ACKNOWLEDGMENTS

Research supported by National Science Foundation grant EAR 96-28552 to Metcalf and Wallin. We thank Y. Dilek, E. Moores, D. Elthon, and A. Nicolas for organizing the 1998 Penrose Conference on Ophiolites and Oceanic Crust and all of the attendees for stimulating discussions. This paper benefited greatly from careful and constructive reviews provided by Greg Harper and Sherman Bloomer. Alex Sanchez, Monica Camin, Clay Crow, and Jason Smith are thanked for their assistance in the rock-chemistry laboratory. We are particularly grateful to J. Garcia and the Other Ones for their assistance in the field.

REFERENCES CITED

Albers, J.P., Kistler, R.W., and Kwak, L., 1981, The Mule Mountain stock, an early Middle Devonian pluton in northern California: Isochron West, v. 31, p. 17.

Arculus, R.J., and Powell, R., 1986, Source component mixing in the regions of arc magma generation: Journal of Geophysical Research, v. 91, p. 5913–5926.

Bau, M., and Knittel, U., 1993, Significance of slab-derived partial melts and aqueous fluids for the genesis of tholeiitic and calc-alkaline island-arc basalts: Evidence from Mt. Arayat, Philippines: Chemical Geology, v. 105, p. 233–251.

Bence, A.E., and Taylor, B.E., 1985, Rare-earth element systematics of West Shasta metavolcanic rocks: Petrogenesis and hydrothermal alteration: Economic Geology, v. 80, p. 2164–2176.

Bloomer, S.H., Taylor, B., MacLeod, C.J., Stern, R.J., Fryer, P., Hawkins, J.W., and Johnson, L., 1995, Early arc volcanism and the ophiolite problem: A perspective from drilling in the western Pacific, *in* Taylor, B., and Natland, J., eds., Active margins and marginal basins of the western Pacific: Washington, D.C., American Geophysical Union Geophysical Monograph 88, p. 1–30.

Boudier, F., Le Sueur, E., and Nicolas, A., 1989, Structure of an atypical ophiolite: The Trinity Complex, eastern Klamath Mountains, California: Geological Society of America Bulletin, v. 101, p. 820–833.

Brouxel, M., Lapierre, H., Michard, A., and Albarede, F., 1987, The deep layers of a Paleozoic arc: Geochemistry of the Copley-Balaklala series, northern California: Earth and Planetary Science Letters, v. 85, p. 386–400.

Brouxel, M., Lapierre, H., Michard, A., and Albarede, F., 1988, Geochemical study of an early Paleozoic island-arc-back-arc basin system. Part 2. Eastern Klamath, early to middle Paleozoic island-arc volcanic rocks (northern California): Geological Society of America Bulletin, v. 100, p. 1120–1130.

Burchfiel, B.C., and Davis, G.A., 1975, Nature and controls of Cordilleran orogenesis, western United States: Extension of an earlier synthesis: American Journal of Science, v. 275A, p. 97–118.

Cannat, M., and Lécuyer, C., 1991, Ephemeral magma chambers in the Trinity peridotite, northern California: Tectonophysics, v. 186, p. 313–328.

Charvet, J., Lapierre, H., Rouer, O., Coulon, C., Campos, C., Martin, P., and Lécuyer, C., 1990, Tectono-magmatic evolution of Paleozoic and early Mesozoic rocks in the eastern Klamath Mountains, California, and the Blue River Mountains, eastern Oregon-western Idaho: Geological Society of America Special Paper 255, p. 255–276.

Crawford, A.J., Fallon, T.J., and Green, D.H., 1989, Classification, petrogenesis, and tectonic setting of boninites, *in* Crawford, A.J., ed., Boninites and related rocks: London, Unwin Hyman, p. 1–49.

Davis, G.A., Holdaway, M.J., Lipman, P.W., and Romey, W.D., 1965, Structure, metamorphism, and plutonism in the south-central Klamath Mountains, California: Geological Society of America Bulletin, v. 76, p. 933–966.

Doe, B.R., Delevaux, M.H., and Albers, J.P., 1985, The plumbotectonics of the West Shasta mining district, eastern Klamath Mountains, California: Economic Geology, v. 80, p. 2136–2148.

Elliott, T., Plank, T., Zindler, A., White, W., and Bourdon, B., 1997, Element transport from slab to volcanic front at Mariana arc: Journal of Geophysical Research, v. 102, p. 14991–15019.

Fryer, P., 1995, Geology of the Mariana trough, *in* Taylor, B., ed., Backarc basins: Tectonics and magmatism: New York, Plenum Press, p. 237–279.

Fuis, G.S., Zucca, J.J., Mooney, W.D., and Milkereit, B., 1987, A geologic interpretation of seismic-refraction results in northeastern California: Geological Society of America Bulletin, v. 98, p. 53–65.

Gill, J., 1981, Orogenic andesites and plate tectonics: Berlin, Springer-Verlag, 390 p.

Gruau, G., and Lécuyer, C., 1991, Origin and petrogenesis of the Trinity Ophiolite Complex (California): New constraints from REE and Nd isotope data, *in* Menzies, M.A., Dupuy, C., and Nicolas, A., eds., Orogenic lherzolites and mantle processes: Oxford, Oxford University Press, p. 229–242.

Gruau, G., Bernard-Griffiths, J., Lécuyer, C., Henin, O., Macé, J., and Cannat, M., 1995, Extreme Nd isotopic variation in the Trinity Ophiolite Complex and the role of melt/rock reactions in the oceanic lithosphere: Contributions to Mineralogy and Petrology, v. 121, p. 337–350.

Harper, G.D., Bowman, J.R., and Kuhns, R., 1988, Field, chemical and isotopic aspects of submarine hydrothermal metamorphism of the Josephine ophiolite, Klamath Mountains, California-Oregon: Journal of Geophysical Research, v. 93, p. 4625–4657.

Hawkins, J.W., Jr., 1995, The geology of the Lau basin, *in* Taylor, B., ed., Backarc basins: Tectonics and magmatism: New York, Plenum Press, p. 63–138.

Hole, M.J., Saunders, A.D., Marriner, G.F., and Tarney, J., 1984, Subduction of pelagic sediments: Implications for the origin of Ce-anomalous basalts from the Mariana Islands: Journal of the Geological Society of London, v. 141, p. 453–472.

Humphris, S.E., and Thompson, G., 1978, Hydrothermal alteration of oceanic basalts by seawater: Geochemica et Cosmochemica Acta, v. 42, p. 107–125.

Irwin, W.P., 1989, Terranes of the Klamath Mountains, California and Oregon, *in* Blake, M.C., and Harwood, D.S., eds., Tectonic evolution of northern California: International Geological Congress, 28th, Field Trip Guidebook T108, v. 108, p. 19–32.

Irwin, W.P., 1994, Geologic map of the Klamath Mountains, California and Oregon: U.S. Geological Survey, Miscellaneous Investigation Series Map I-2148, scale 1:500 000.

Jacobsen, S.B., Quick, J.E., and Wasserburg, G.J., 1984, A Nd and Sr isotopic study of the Trinity peridotite: Implications for mantle evolution: Earth and Planetary Science Letters, v. 68, p. 361–378.

Karson, J.A., and Winters, A.T., 1992, Along-axis variations in tectonic extension and accommodation zones in the MARK area, Mid-Atlantic Ridge 23°N latitude, *in* Parson, L.M., Murton, B.J., and Browning, P., eds., Ophiolites and their modern oceanic analogues: Geological Society of London Special Publication 60, Blackwell Scientific Publishers, p. 107–116.

Kinkel, A.R., Hall, W.E., and Albers, J.P., 1956, Geology and base-metal deposits of West Shasta copper-zinc district, Shasta County, California: U.S. Geological Survey Professional Paper, v. 285, p. 156.

Lanphere, M.A., Irwin, W.P., and Hotz, P.E., 1968, Isotopic age of the Nevadan orogeny and older plutonic and metamorphic events in the Klamath Mountains, California: Geological Society of America Bulletin, v. 79, p. 1027–1052.

Lapierre, H., Albarede, F., Albers, J.P., Cabanis, B., and Coulon, C., 1985, Early Devonian volcanism in the eastern Klamath Mountains, California: Evidence for an immature island arc: Canadian Journal of Earth Science, v. 22, p. 214–227.

Lindsley-Griffin, N., 1991, The Trinity Complex: A polygenetic ophiolitic assemblage, *in* Cooper, J.D., and Stevens, C.H., eds., Paleozoic paleogeography of the western United States: v. 2, Pacific Section, Society for Sedimentary Geology, p. 589–608.

Lipman, P.W., 1964, Structure and origin of an ultramafic pluton in the Klamath Mountains, California: American Journal of Science, v. 262, p. 199–222.

Metcalf, R.V., Willse, K.R., and Wallin, E.T., 1999, Trace element and Nd-Sr-Pb isotopic data from the Silurian Bonanza King mafic intrusive complex, Trinity terrane, CA: Evidence of supra-subduction zone magmatism in a proto-arc paleotectonic setting: Geological Society of America Abstracts with Programs, v. 31, no. 6, p. A-79.

Moores, E.M., 1982, Origin and emplacement of ophiolites: Reviews of Geophysics and Space Physics, v. 40, p. 735–760.

Muller, E.R., 2000, Petrogenetic study of the Porcupine Lake mafic intrusive complex, Trinity terrane, northern California [M.S. thesis]: University of Nevada–Las Vegas, 136 p.

Peacock, S.M., 1987, Serpentinization and infiltration metasomatism in the Trinity peridotite, Klamath province, northern California: Implications for subduction zones: Contributions to Mineralogy and Petrology, v. 95, p. 55–70.

Peacock, S.M., and Norris, P.J., 1989, Metamorphic evolution of the Central Metamorphic Belt, Klamath Province, California: An inverted metamorphic gradient beneath the Trinity peridotite: Journal of Metamorphic Geology, v. 7, p. 191–209.

Pearce, J.A., 1982, Trace element characteristics of lavas from destructive plate boundaries, *in* Thorpe, R.S., ed., Orogenic andesites and related rocks: Chichester, UK, John Wiley and Sons, p. 525–548.

Pearce, J.A., 1996, A user's guide to basalt discrimination diagrams, *in* Wyman, D.A., ed., Trace element geochemistry of volcanic rocks: Applications for massive sulphide exploration: Geological Association of Canada, Short Course Notes, v. 12, p. 79–113.

Pearce, J.A., Lippard, S.J., and Roberts, S., 1984, Characteristics and tectonic significance of supra-subduction zone ophiolites, *in* Kokelaar, B.P., and Howells, M.F., eds., Marginal basin geology: London, Blackwell Scientific, p. 77–94.

Pearce, J.A., Baker, P.E., Harvey, P.K., and Luff, I.W., 1995, Geochemical evidence for subduction fluxes, mantle melting and fractional crystallization beneath the South Sandwich island arc: Journal of Petrology, v. 36, p. 1073–1104.

Penrose Conference Participants, 1972, Penrose field conference on ophiolites: Geotimes, v. 17, no. 12, p. 24–25.

Perfit, M.R., Gust, D.A., Bence, A.E., Arculus, R.J., and Taylor, S.R., 1980, Chemical characteristics of island-arc basalts: Implications for mantle sources: Chemical Geology, v. 30, p. 227–256.

Peterson, S.W., Barnes, C.G., and Hoover, J.D., 1991, The Billy's Peak mafic complex of the Trinity sheet, California: Roots of a Paleozoic island arc, *in* Cooper, J.D., and Stevens, C.H., eds., Paleozoic paleogeography of the western United States: v. 2, Pacific Section, Society for Sedimentary Geology, p. 625–633.

Quick, J.E., 1981, Petrology and petrogenesis of the Trinity peridotite, an upper mantle diapir in the eastern Klamath Mountains, California: Journal of Geophysical Research, v. 86, p. 11837–11863.

Saleeby, J.B., 1990, Geochronological and tectonostratigraphic framework of Sierran-Klamath ophiolitic assemblages: Geological Society of America Special Paper, v. 255, p. 93–114.

Schweickert, R.A., and Irwin, W.P., 1989, Extensional faulting in the southern Klamath Mountains, California: Tectonics, v. 8, p. 135–149.

Schwindinger, K.R., and Anderson, A.T., Jr., 1987, Probable low-pressure intrusion of gabbro into serpentinized peridotite, northern California: Geological Society of America Bulletin, v. 98, p. 364–372.

Silberling, N.J., Jones, D.L., Monger, J.W.H., and Coney, P.J., 1992, Lithotectonic terrane map of the North American Cordillera: U.S. Geological Survey, Miscellaneous Geologic Investigations Map 2176, 1:500 000.

Stern, R.J., and Bloomer, S.H., 1992, Subduction zone infancy: Examples from the Eocene Izu-Bonin-Mariana and Jurassic California arcs: Geological Society of America Bulletin, v. 104, p. 1621–1636.

Sun, S.S., and McDonough, W.F., 1989, Chemical and isotopic systematics of oceanic basalts: Implications for mantle composition and processes, *in* Saunders, A.D., and Norris, M.J., eds., Magmatism in the oceanic basins, Geological Society of London Special Publication 42, p. 313–325

Taylor, B., ed., 1995, Backarc basins: Tectonics and magmatism: New York, Plenum Press, 524 p.

Taylor, R.N., Murton, B.J., and Nesbitt, R.W., 1992, Chemical transects across intra-oceanic arcs: Implications for the tectonic setting of ophiolites: Geological Society of London Special Publication 60, p. 117–132.

Valsami, E., and Cann, J.R., 1992, Mobility of rare earth elements in zones of intense hydrothermal alteration in the Pindos ophiolite, Greece, *in* Parson, L.M., Murton, B.J. and Browning, P., eds., Ophiolites and their modern oceanic analogues: Geological Society of London Special Publication 60, Blackwell Scientific Publishers, p. 219–232.

Wagner, D.L., and Saucedo, G.J., 1987, Geologic map of the Weed quadrangle: California Division of Mines and Geology, Regional Geologic Map Series 4A, scale 1:250 000.

Wallin, E.T., and Gehrels, G.E., 1995, Sedimentary record of Silurian volcanism reveals denudation of the Trinity terrane and the timing of siliciclastic sedimentation in the Yreka terrane, northern California: Geological Society of America Abstracts with Programs, v. 27, p. 83.

Wallin, E.T., and Metcalf, R.V., 1998, Supra-subduction zone ophiolite formed in an extensional forearc: Trinity terrane, Klamath Mountains, California: Journal of Geology, v. 106, p. 591–608.

Wallin, E.T., and Trabert, D.W., 1994, Eruption-controlled epiclastic sedimentation in a Devonian trench-slope basin: Evidence from sandstone petrofacies, Klamath Mountains, California: Journal Sedimentary Research, v. A64, p. 373–385.

Wallin, E.T., Coleman, D.S., Lindsley-Griffin, N., and Potter, A.W., 1995, Silurian plutonism in the Trinity terrane (Neoproterozoic and Ordovician), Klamath Mountains, California, USA: Tectonics, v. 14, p. 1007–1013.

Willse, K.R., 1999, Geochemistry and petrogenesis of the Bonanza King Complex, Trinity terrane, California [M.S. thesis]: University of Nevada–Las Vegas, 130 p.

Wood, D.A., 1980, The application of a Th-Hf-Ta diagram to problems of tectonomagmatic classification and to establishing the nature of crustal contamination of basaltic lavas of the British Tertiary volcanic province: Earth and Planetary Science Letters, v. 50, p. 11–30.

Woodhead, J.D., Eggins, S.M., and Johnson, R.W., 1998, Magma genesis in the New Britain island arc: Further insights into melting and mass transfer processes: Journal of Petrology, v. 39, p. 1641–1668.

Manuscript Accepted by the Society February 23, 2000

Printed in the U.S.A.

Geological Society of America
Special Paper 349
2000

Sulfide minerals as an indicator for petrogenesis and serpentinization of peridotites: An example from the Hayama-Mineoka belt, central Japan

Hiroshi Sato*
Doctoral Program in Geoscience, University of Tsukuba, Tsukuba, 305-8571 Japan
Yujiro Ogawa
Institute of Geoscience, University of Tsukuba, Tsukuba, 305-8571 Japan

ABSTRACT

Assemblages of sulfide minerals + awaruite + native copper are used to estimate conditions for petrogenesis and serpentinization of their host peridotites in the Hayama-Mineoka belt, central Japan. The presence of primary sulfide grains (pentlandite) suggests that they are residual after moderate degrees of partial melting. This idea is supported by spinel and bulk-rock compositions of the peridotites, which indicate 15% to 30% partial melting. Assemblages of secondary sulfide minerals + awaruite + native copper that replace the primary sulfide indicate that the serpentinization of peridotites in the Hayama-Mineoka belt occurred under low f_{O_2} and low f_{S_2} conditions.

The suggestion that the ophiolitic rocks in the Hayama-Mineoka belt might be equivalents of those in the Izu-Bonin-Mariana forearcs on the Philippine Sea plate is not supported by the results of our study. As the peridotites of the forearcs are lacking in sulfide minerals because of higher degrees of partial melting and as the characteristics of the peridotites in the two occurrences differ, we conclude that the peridotites in the Hayama-Mineoka belt were originally part of another oceanic plate.

INTRODUCTION

It is well known that the chemical composition of spinels in peridotites is useful for deducing the tectonic setting of peridotites, because their composition depends on the petrogenetic conditions such as temperatures, pressure, fugacities, and degree of partial melting (e.g., Dick and Bullen, 1984). The chemical composition of sulfide minerals in peridotites also reflects the petrogenetic conditions (e.g., Garuti et al., 1984; Lorand, 1987, 1988). Furthermore, the fact that the sulfide minerals are altered to secondary sulfide and native-metal mineral assemblages during serpentinization can be used to determine the conditions of serpentinization (e.g., Eckstrand, 1975; Shiga, 1983; Frost, 1985). Such assemblages have been reported from serpentinized mantle peridotites exposed on the ocean floor (Prichard et al., 1996; Sato et al., 1997, 1998; Alt and Shanks, 1998). Thus, beside spinels, sulfide minerals could provide an additional line of evidence for petrogenetic and serpentinization conditions of occurrences of mantle peridotite on land.

The ophiolitic rocks in the Hayama-Mineoka belt in central Japan are considered to be the eastern extension of the ophiolites within the Tertiary Shimanto belt in southwest Japan, and therefore they represent one of the youngest ophiolite occurrences in the world. Because they are situated between the northern terminus of the Izu-Bonin-Mariana arc system and the Honshu arc, the tectonic evolution of these Tertiary ophiolites is important for deciphering the tectonic development of the central Honshu arc. Geologic and petrologic studies of the Hayama-Mineoka belt have produced new interpretations and models for the tectonic evolution of this belt (e.g., Ogawa and Taniguchi, 1988; Taniguchi and Ogawa, 1990; Mohiuddin and Ogawa, 1998). However, petrologic and tectonic studies of the peridotites exposed in the Hayama-Mineoka belt have been limited (e.g., Uchida and Arai, 1978; Arai et al., 1983; Arai, 1991; Arai and Takahashi, 1988).

In this study of the ultramafic rocks in the Hayama-Mineoka belt, we present the characteristics of the peridotite and some of the included trace minerals—i.e., sulfides and associated awaruite and native copper—that are either primary minerals left in the residual rocks after partial melting or secondary minerals that formed during serpentinization. We then discuss utilization of the trace minerals for revealing the origin of ophiolitic rocks in the Hayama-Mineoka belt.

*Present address: Ocean Research Institute, University of Tokyo, Nakano, Tokyo 164-8639 Japan; e-mail: satohiro@ori.u-tokyo.ac.jp

Sato, H., and Ogawa, Y., 2000, Sulfide minerals as an indicator for petrogenesis and serpentinization of peridotites: An example from the Hayama-Mineoka belt, central Japan, *in* Dilek, Y., Moores, E.M., Elthon, D., and Nicolas, A., eds., Ophiolites and Oceanic Crust: New Insights from Field Studies and the Ocean Drilling Program: Boulder, Colorado, Geological Society of America Special Paper 349, p. 419–429.

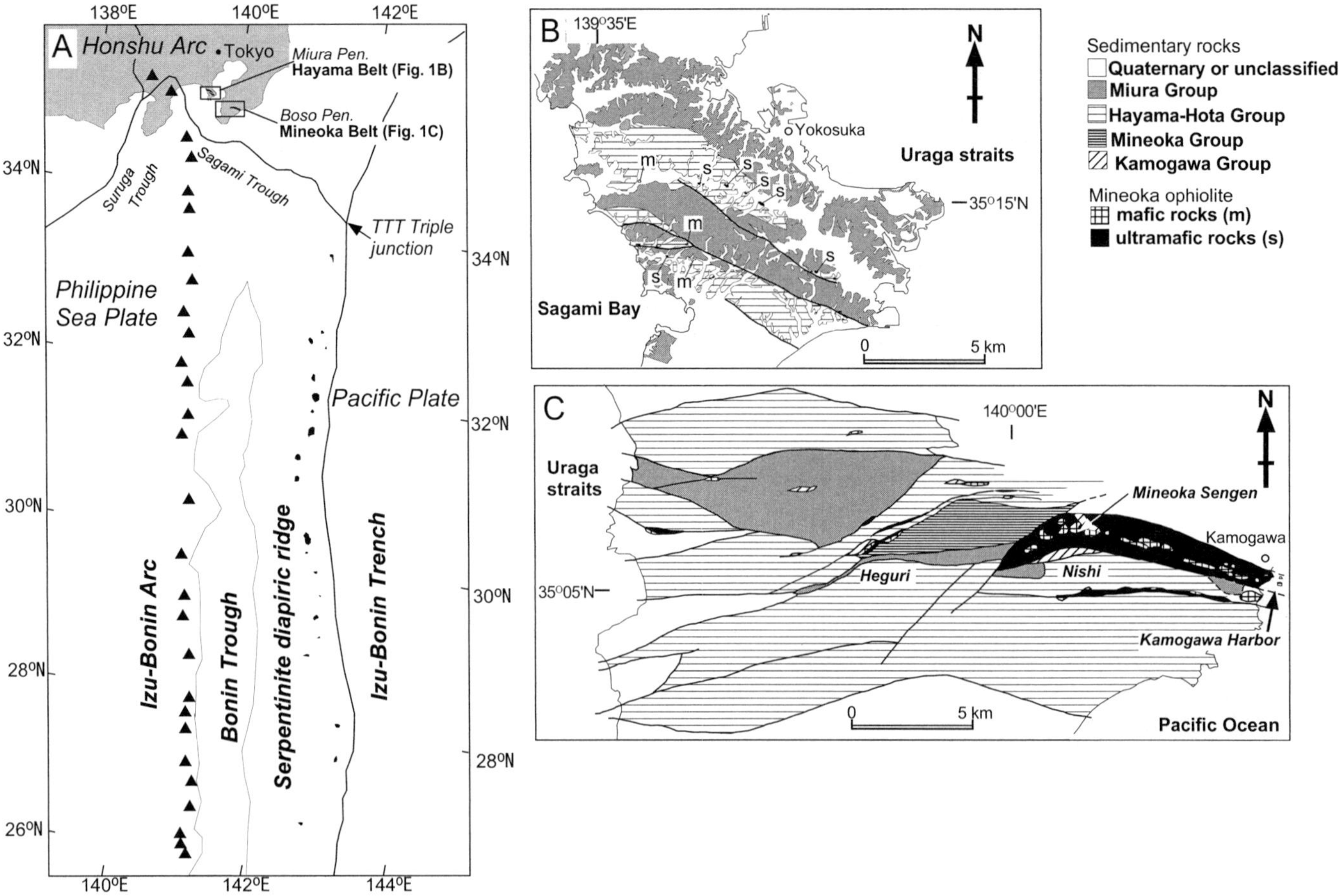

Figure 1. A, Index map showing the present plate setting around the Hayama and Mineoka belts. Simplified after Taylor (1992). Solid triangles: active volcanoes; black areas: diapiric serpentinite seamounts. B, Simplified geologic map of the Miura Peninsula (after Yokosuka City Museum, 1991). C, Simplified geologic map of the Boso Peninsula (after Mohiuddin and Ogawa, 1998).

GEOLOGIC SETTING

The Izu-Bonin-Mariana arc system is characterized by a "forearc ophiolite" (Ishii, 1985) exposed at the very toe of the inner slope of the trench as a ridge composed mainly of serpentinized mantle peridotites (Fryer and Fryer, 1987; Fryer et al., 1995) (Fig. 1). The northern extension of parts of this forearc ophiolite appears on land in the Hayama-Mineoka belt of central Japan (Fig. 1). Disrupted ophiolitic rocks in this belt were previously interpreted as an obducted or accreted forearc ophiolite fragment (Ogawa and Naka, 1984; Fujioka et al., 1995). However, an alternative model based on geologic and petrologic data suggests a mid-ocean ridge affinity for the origin of these ophiolitic rocks (e.g., Ogawa and Taniguchi, 1988; Taniguchi and Ogawa, 1990; Mohiuddin and Ogawa, 1998; Sato et al., 1999).

As reviewed and summarized by Sato et al. (1999), the ophiolitic rocks in the Miura Peninsula and western part of the Boso Peninsula of Japan are considered to be in fault contact with each other or with the Miocene forearc Hayama-Hota Groups (Ogawa and Taniguchi, 1988). Similar rocks are associated with the Paleocene to middle Miocene, pelagic to hemipelagic Kamogawa Group in the central to eastern parts of the Boso Peninsula (Mohiuddin and Ogawa, 1998). The former area is defined as the Hayama belt, and the latter, as the Mineoka belt in this paper. The basaltic rocks in the Hayama belt are all derivatives of an alkalic suite, have a hotspot origin, and have been dated at 36–37 Ma (Taniguchi and Ogawa, 1990), whereas most of the basaltic rocks in the Mineoka belt are composed of MORB (mid-oceanic-ridge basalt) or island-arc tholeiite (Ogawa and Taniguchi, 1988; Taniguchi, 1991) and have been dated at 50–40 Ma (Kaneoka et al., 1981).

Peridotites in both the Hayama and Mineoka belts are mostly serpentinized harzburgite with subordinate dunite. They are basically massive, but strongly sheared or brecciated particularly near basaltic blocks. Most of these harzburgites consist of olivine (Fo_{90-93}), orthopyroxene (En_{89-91}), and small amounts of clinopyroxene and Cr-rich spinel. Rare plagioclase has a composition of An_{88} (Arai and Takahashi, 1988). The chemical compositions of these minerals and the lack of cumulate textures in these peridotites suggest that these rocks are residual after partial melting. Harzburgites in general include sulfide and native-metal mineral assemblages (Kanehira et al., 1975), whereas dunite does only in limited places. According to detailed petrologic studies (Uchida and Arai, 1978), the peridotites in the Hayama belt equilibrated at ~700 °C, whereas those in the Mineoka belt did so at higher temperatures of ~800 to 900 °C. This difference is considered to be due to heterogeneity of the upper mantle (Arai et al., 1990).

TABLE 1. PETROGRAPHIC AND PETROLOGIC DATA OF REPRESENTATIVE SAMPLES

	Hayama belt				Mineoka belt								
Area sample	95 MR2	97 MN24	97 MN25	97 MN26	96 MN8	96 MN8-2	96 MN12	97 MN15-3	97 MN2-6	97 MN3-1	97 MN3-2	96 MN10	97 MN5
Modal composition (vol%)													
Altered products													
Serpentine	50.5	96.7	99.1	89.3	98.3	94.9	68.1	56.0	96.9	89.4	97.2	49.2	40.1
Altered minerals*	2.9	0.0	0.0	0.0	0.0	3.1	0.0	0.6	0.0	4.2	0.0	2.3	6.6
Relict minerals													
Olivine	14.4	1.5	0.0	3.4	0.0	0.7	5.4	11.4	3.0	2.6	0.3	10.6	10.9
Opx	25.4	0.0	0.0	4.1	0.0	0.0	25.8	31.3	0.0	1.4	0.0	36.3	37.7
Cpx	6.1	0.0	0.0	2.3	0.0	0.0	0.4	0.7	0.0	0.6	0.0	0.9	4.3
Plagioclase	0.0	0.0	0.0	0.0	0.0	0.0	0.0	0.0	0.0	0.0	0.0	0.3	0.0
Magnetite	0.0	1.4	0.5	0.4	1.0	0.7	0.0	0.0	0.0	0.8	0.6	0.0	0.1
Spinel	0.7	0.4	0.4	0.4	0.7	0.6	0.4	0.1	0.1	0.9	1.9	0.4	0.3
Sulfide minerals	0.05	0.03	0.00	0.03	0.00	0.02	0.01	0.02	0.00	0.01	0.00	0.02	0.02
Primary composition													
Olivine	67.4	75.7	100.0	76.3	100.0	75.3	73.7	67.8	100.0	83.5	100.0	61.6	54.8
Opx	26.3	24.3	0.0	21.5	0.0	24.7	25.9	31.5	0.0	15.8	0.0	37.4	40.6
Cpx	6.3	0.0	0.0	2.3	0.0	0.0	0.4	0.7	0.0	0.7	0.0	1.0	4.7
Original rock type†	lh	hz	du	hz	du	hz	hz	hz	du	hz	du	hz	pl-hz
Bulk-rock composition (wt%)													
SiO_2	45.58	43.76	41.15	43.99	45.83	44.13	44.31	44.17	40.99	45.13	42.74	44.08	45.81
TiO_2	0.04	0.03	0.02	0.03	0.02	0.05	0.04	0.03	0.03	0.03	0.03	0.04	0.04
Al_2O_3	1.09	0.75	0.28	1.21	0.37	1.63	0.87	0.98	0.30	1.08	1.04	1.48	2.40
Fe_2O_3	9.86	8.24	5.58	8.83	8.36	9.00	8.12	8.00	7.89	9.32	6.70	8.53	8.46
MnO	0.13	0.12	0.07	0.13	0.10	0.13	0.12	0.12	0.10	0.14	0.09	0.13	0.13
MgO	42.21	46.39	52.29	44.24	43.74	43.91	45.02	45.52	50.96	43.82	46.69	43.68	41.26
CaO	0.67	0.20	0.05	0.70	0.15	0.17	0.67	0.81	0.21	0.11	0.25	1.57	1.94
Total	99.59	99.50	99.44	99.13	98.57	99.03	99.16	99.63	100.49	99.65	97.56	99.51	100.06
H_2O	1.60	1.50	0.98	1.46	1.04	1.35	0.87	0.96	1.10	2.03	2.43	0.78	0.87
IGLS	12.02	12.91	14.86	12.40	12.46	12.52	10.43	11.02	14.61	12.92	14.14	10.42	8.33
Bead loss	0.94	0.85	0.85	0.91	1.06	0.88	0.88	0.85	0.89	0.96	0.82	0.88	0.82

* Most altered minerals are enriched in Al, and they are considered to be plagioclase primarily.
† Abbreviations: du—dunite, hz—plagioclase-poor harzburgite, lh—lherzolite, pl-hz—plagioclase-rich harzburgite.

PETROGRAPHIC AND PETROLOGIC CHARACTERISTICS OF PERIDOTITES

Peridotite masses are small in size (mostly less than 100 m in length) in the Hayama belt, and they are only sporadically distributed in the Miocene Hayama-Hota Groups (Fig. 1B). On the other hand, peridotite is considered to be widely distributed in the Mineoka belt (Fig. 1C). Samples were collected from more than 20 masses and outcrops, and up to 50 polished sections were studied. From these, 14 samples were selected to be analyzed by electron microprobe and X-ray fluorescence (XRF).

Modal and bulk-rock compositions

Modal compositions were calculated by the method of chemical-mapping analysis using an electron microprobe, JEOL JXA-8900R. Analyses were performed under the conditions of 15 kV accelerating voltage, 1×10^{-7} A beam current, and <10 μm beam diameter. Mg, Al, Ca, Cr, and S were detected by wavelength-dispersive X-ray spectrometer, and Si and Fe were detected by energy-dispersive spectrometer. Then, mineral-distribution maps were made from these data processed with application software Scion Image (developed at Scion Corporation, Frederick, Maryland, and available on the internet at http://www.scioncorp.com/), and modal compositions were obtained from these maps. Modal compositions of peridotites in the Hayama-Mineoka belt calculated in this way are listed in Table 1.

Obtained modal compositions are concordant with the bulk-rock chemistries. For example, modal percentages of orthopyroxene decrease with decreasing bulk-rock CaO and Al_2O_3 contents (Fig. 2, A and B). Furthermore, contents of H_2O^- and loss on ignition, which are believed to result from the presence of altered and/or hydrous minerals, correspond to the degree of serpentinization (Fig. 2C). Thus, the analytical method described in the previous paragraph is useful for estimating modal compositions.

The majority of peridotites are made of harzburgite or lherzolite with subordinate amounts of dunite, and they are commonly serpentinized up to 50%. It is noteworthy that most of the harzburgite includes nonnegligible amounts of sulfide minerals—from 0.01 to 0.05 vol%—whereas the dunite has less than 0.01 vol%. We use this distinction to classify the peridotites in the Hayama-Mineoka belt into two types: sulfide-bearing and sulfide-free peridotites. Generally harzburgite belongs to the former type, and dunite to the latter.

Chemical compositions of major minerals

Chemical compositions of the major minerals such as olivine, pyroxene, spinel, and plagioclase have already been described by Uchida and Arai (1978) and Arai and Takahashi (1988). Representative results of electron-microprobe analyses in this study are listed in Tables 2 and 3.

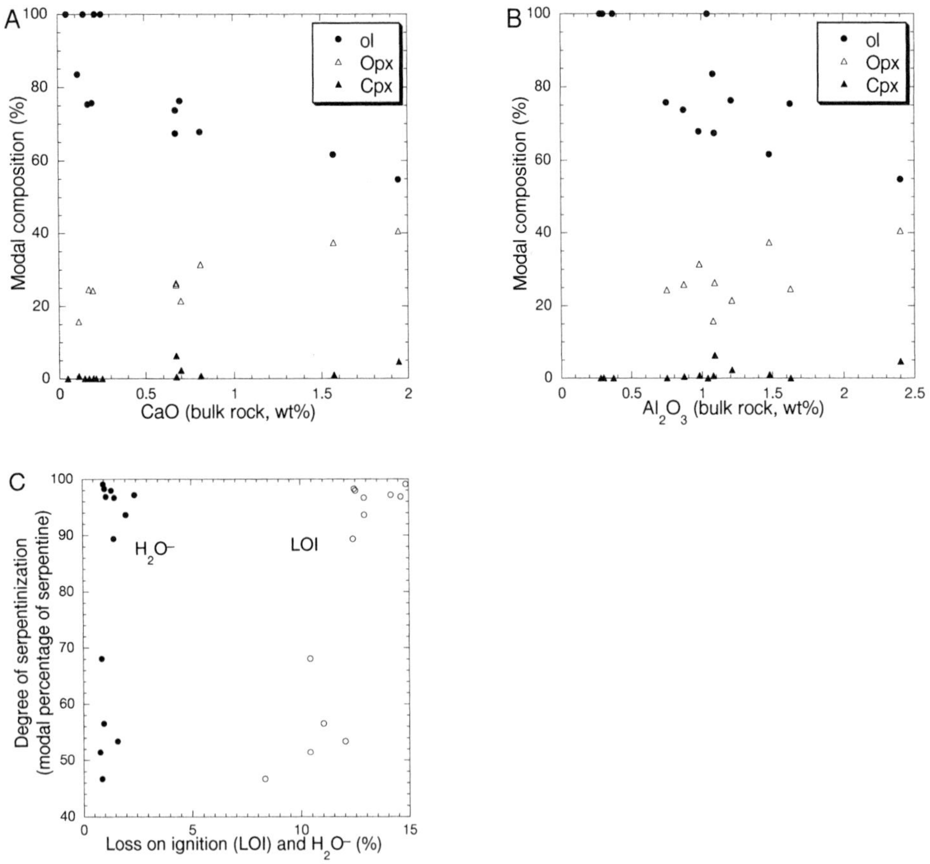

Figure 2. Plots showing relationships between chemical compositions determined by XRF and modal compositions determined by electron-microprobe analyses of peridotites from the Hayama-Mineoka belt. A, CaO (wt%) vs. modal compositions of olivine (ol), orthopyroxene (Opx), and clinopyroxene (Cpx). B, Al_2O_3 (wt%) vs. modal compositions of ol, Opx, and Cpx. C, Loss on ignition (LOI) and H_2O (%) of peridotites vs. degree of serpentinization diagram.

TABLE 2. REPRESENTATIVE COMPOSITIONS OF OLIVINE IN PERIDOTITES FROM HAYAMA-MINEOKA BELTS

	Sulfide-bearing peridotite									Sulfide-free peridotite	
Area	Hayama belt			Mineoka belt						Mineoka belt	
Sample	95MR2	97MN24	97MN26	96MN8-1	96MN5C	96MN12	97MN15-3	96MN10	97MN5	97MN2-6	96MN11-2
Rock type*	hz	hz	hz	hz	hz	hz	hz	hz	pl-hz	du	hz†
Number	117	77	10	18	99	160	11	59	B-1C	3	71
SiO_2	40.51	40.79	40.81	39.35	40.93	40.85	39.85	40.79	40.25	41.14	40.06
FeO	9.23	7.77	8.37	8.49	7.91	8.00	7.67	8.84	9.05	6.57	8.66
MnO	0.15	0.12	0.16	0.12	0.13	0.12	0.11	0.14	0.13	0.15	0.13
MgO	49.58	51.83	50.24	49.76	51.26	50.59	51.39	50.62	49.34	51.93	49.86
NiO	0.37	0.37	0.45	0.42	0.38	0.49	0.40	0.36	0.37	0.45	0.47
Total	99.93	100.93	100.18	98.23	100.78	100.13	99.55	100.86	99.31	100.35	99.28
O	8	8	8	8	8	8	8	8	8	8	8
Si	1.99	1.97	1.99	1.96	1.98	1.99	1.95	1.98	1.99	1.99	1.98
Fe	0.38	0.31	0.34	0.35	0.32	0.33	0.31	0.36	0.37	0.27	0.36
Mn	0.01	0.01	0.01	0.01	0.01	0.00	0.00	0.01	0.01	0.01	0.01
Mg	3.62	3.73	3.65	3.70	3.69	3.67	3.75	3.66	3.36	3.73	3.66
Ni	0.01	0.01	0.02	0.02	0.01	0.02	0.02	0.01	0.01	0.02	0.02
Range in each sample											
Fo max	91.17	93.40	91.78	91.67	92.06	92.72	92.31	91.51	91.11	93.48	92.77
Fo min	90.42	91.81	91.10	91.12	91.84	91.47	91.88	90.87	90.22	92.93	90.47
No. of grains	9	40	26	8	5	35	8	13	4	7	12

* Abbreviations: du—dunite; hz—plagioclase-poor harzburgite; pl-hz—plagioclase-rich harzburgite.
† Harzburgite with pyroxenite vein (2 cm wide).

TABLE 3. REPRESENTATIVE COMPOSITIONS OF SPINEL IN PERIDOTITES FROM HAYAMA-MINEOKA BELTS

	Sulfide-bearing peridotite												Sulfide-free peridotite					
Area	Hayama belt				Mineoka belt								Hayama belt		Mineoka belt			
Rock type*	lh	hz	hz	hz	hz	hz	hz	hz	hz	hz	hz	pl-hz	du	du	du	du	du	hz
Sample	95MR2	95MR10	97MN24	97MN26	96MN8-1	95MN23	93MN5	96MN12	97MN3-1	97MN15-3	96MN10	97MN5	95MR5	97MN25	96MN8	97MN2-6	97MN3-2	96MN11-2
No.	3C1	L1	141	22	5	C2-G83	C-A95	191	1	1	44	B-1	A-1	61	1	1	1	38
TiO_2	0.15	0.26	0.13	0.03	0.25	0.08	0.05	0.12	0.26	0.09	0.16	0.21	0.19	0.09	0.23	0.30	0.25	0.16
Al_2O_3	27.55	30.22	24.82	37.62	2.16	15.15	24.77	25.47	24.15	29.15	24.43	25.13	21.85	7.89	11.73	21.88	21.44	25.74
FeO	22.60	20.67	16.40	15.26	20.96	18.92	18.05	15.02	18.12	16.05	18.02	18.97	27.20	14.13	18.67	15.64	11.63	16.40
MnO	0.27	0.29	0.13	0.12	0.34	0.26	0.09	0.24	0.31	0.19	0.27	0.23	0.30	0.12	0.19	0.26	0.21	0.20
MgO	13.90	13.51	14.61	16.31	12.18	11.03	13.90	14.69	14.13	14.84	13.80	12.99	10.70	13.50	11.05	14.58	16.60	14.07
Cr_2O_3	33.74	34.57	44.22	30.63	42.05	54.51	44.15	44.15	42.69	39.70	42.87	41.87	39.24	64.12	56.85	45.76	48.92	42.61
NiO	0.18	0.17	0.12	0.17	0.00	0.03	0.04	0.09	0.23	0.09	0.16	0.17	0.16	0.03	0.05	0.16	0.09	0.09
Total	100.08	99.70	100.64	100.29	98.25	100.17	101.19	100.04	99.89	100.34	99.94	99.65	99.64	99.96	98.81	98.67	99.23	99.59
O	32	32	32	32	32	32	32	32	32	32	32	32	32	32	32	32	32	32
Fe^{3+}†	1.15	0.94	0.53	0.45	0.99	0.37	0.64	0.35	0.83	0.43	0.70	0.62	1.73	0.34	0.45	0.62	0.36	0.45
Fe^{2+}†	3.34	3.17	2.76	2.44	3.40	3.69	2.98	2.66	2.84	2.73	2.95	3.24	3.95	2.73	3.66	2.60	2.00	2.86
Ti	0.03	0.05	0.02	0.01	0.05	0.01	0.01	0.02	0.05	0.02	0.03	0.04	0.04	0.02	0.05	0.06	0.04	0.03
Al	7.72	8.47	7.01	10.04	6.54	4.57	6.99	7.21	6.90	8.10	6.98	7.22	6.44	2.42	3.63	6.35	6.13	7.33
Mn	0.05	0.06	0.03	0.02	0.07	0.06	0.02	0.05	0.06	0.04	0.06	0.05	0.06	0.03	0.04	0.05	0.04	0.04
Mg	4.92	4.79	5.21	5.50	4.55	4.21	4.96	5.25	5.10	5.21	4.98	4.71	3.99	5.24	4.33	5.35	6.00	5.06
Cr	6.34	6.50	8.37	5.48	8.33	11.03	8.36	8.38	8.18	7.40	8.22	8.06	7.76	13.19	11.82	8.91	9.38	8.14
Ni	0.04	0.03	0.02	0.03	0.00	0.01	0.01	0.02	0.04	0.02	0.03	0.03	0.03	0.01	0.01	0.03	0.02	0.02
Mg#	0.60	0.60	0.65	0.69	0.57	0.53	0.62	0.66	0.65	0.66	0.63	0.59	0.50	0.66	0.54	0.67	0.75	0.64
Cr#	0.45	0.43	0.54	0.35	0.56	0.71	0.54	0.54	0.54	0.48	0.54	0.53	0.55	0.84	0.76	0.58	0.60	0.53
Range in each sample																		
Mg# Max	0.60	0.60	0.66	0.73	0.67	0.58	0.62	0.70	0.66	0.76	0.66	0.64	0.53	0.66	0.54		0.76	0.69
Mg# Min	0.51	0.60	0.62	0.69	0.56	0.47	0.61	0.64	0.63	0.66	0.50	0.53	0.42	0.59	0.44		0.74	0.59
Cr# Max	0.46	0.43	0.55	0.36	0.62	0.71	0.57	0.56	0.55	0.48	0.54	0.55	0.62	0.86	0.77		0.61	0.55
Cr# Min	0.42	0.43	0.51	0.32	0.52	0.69	0.54	0.51	0.54	0.43	0.52	0.52	0.52	0.82	0.76		0.60	0.51
No. of grains	8	3	12	4	13	14	2	17	3	5	16	6	13	9	5	1	10	9

Abbreviations: * du—dunite, hz—plagioclase-poor harzburgite, lh—lherzolite, pl-hz—plagioclase-rich harzburgite.
† Fe^{2+} and Fe^{3+} are calculated by the method of Droop (1987).

TABLE 4. REPRESENTATIVE COMPOSITIONS OF SULFIDES AND NATIVE COPPER IN PERIDOTITES FROM HAYAMA-MINEOKA BELTS

Rock type	Hayama belt							Mineoka belt							
	Plagioclase-poor harzburgite							Plagioclase-poor harzburgite							
Mineral*	pn	pn	pn	pn	hz	pn	hz	aw	aw	aw	aw	aw	aw	hz	hz
Sample No.	95MR2 1A1	95MR2 1A2	95MR2 1A3	95MR10 L2	97MN24 1001	97MN24 1014	97MN26 1028	95MN23 C1B	95MN23 C1C	96MN12 1B	96MN12 1	96MN8-1 B2E	96MN8-1 B2H	95MN23 C2A	95MN23 C2B
Weight percent															
S	32.9	33.6	34.3	32.2	27.45	31.97	27.92	0.0	0.0	0.0	0.0	0.0	0.9	26.9	26.6
Fe	33.0	33.8	32.8	30.7	7.33	24.01	3.12	25.1	26.0	25.3	28.1	32.9	30.9	2.7	2.6
Co	0.6	0.6	0.5	2.6	0.45	6.98	0.40	0.8	0.6	0.2	0.6	0.7	0.6	2.2	1.9
Ni	27.5	31.0	25.4	34.1	67.75	37.21	70.93	74.3	73.9	73.0	68.6	65.1	60.4	68.3	69.1
Cu	5.6	1.9	8.6	0.0	0.09	0.00	0.26	0.0	0.0	1.8	2.6	0.7	5.6	0.0	0.0
Total	99.6	101.0	101.6	99.6	103.07	100.16	102.63	100.3	100.5	100.3	99.9	99.4	98.5	100.1	100.3
Atomic percent															
S	47.0	47.2	47.9	46.1	39.81	45.75	40.58	0.0	0.1	0.0	0.0	0.1	1.7	40.2	39.8
Fe	27.1	27.2	26.3	25.2	6.10	19.73	2.60	26.0	26.8	26.2	29.2	34.2	32.4	2.3	2.2
Co	0.5	0.5	0.4	2.0	0.36	5.43	0.32	0.8	0.6	0.2	0.6	0.7	0.6	1.8	1.5
Ni	21.4	23.8	19.4	26.6	53.66	29.08	56.31	73.2	72.5	71.9	67.8	64.4	60.2	55.7	56.4
Cu	4.1	1.3	6.0	0.0	0.07	0.00	0.19	0.0	0.0	1.7	2.4	0.7	5.2	0.0	0.0

Abbreviations: * aw—awaruite, hz—heazlewoodite, pn—pentlandite, n-Cu—native copper.

TABLE 4. REPRESENTATIVE COMPOSITIONS OF SULFIDES AND NATIVE COPPER IN PERIDOTITES FROM HAYAMA-MINEOKA BELTS (continued)

Rock type	Mineoka belt										
	Plagioclase-poor harzburgite					Plagioclase-bearing harzburgite					
Mineral*	hz	hz	hz	n-Cu	n-Cu	pn	pn	aw	aw	n-Cu	n-Cu
Sample No.	96MN12 2B	96MN8-1 B1C	96MN8-1 B1D	96MN8-2 C3B1	96MN8-2 C3B2	97MN5 A-1a	97MN5 A-1b	97MN5 A-4b3	97MN5 B-2a2	97MN5 A-2	97MN5 A-2b
Weight percent											
S	27.0	26.3	25.8	0.0	0.0	32.8	32.7	0.1	0.1	0.6	0.0
Fe	3.8	3.8	3.1	1.0	1.0	24.0	24.4	28.6	26.5	1.7	2.2
Co	0.1	0.2	0.1	0.0	0.0	0.7	0.8	0.5	0.4	0.0	0.0
Ni	68.2	69.6	70.6	2.0	1.6	41.9	41.1	68.7	72.0	1.9	1.4
Cu	0.0	0.1	0.1	97.8	98.3	0.1	0.1	0.4	0.3	95.2	98.0
Total	99.0	100.0	99.7	100.9	100.9	99.5	99.1	98.3	99.3	99.4	101.6
Atomic %											
S	40.6	39.5	39.0	0.0	0.0	46.9	47.0	0.2	0.2	1.2	0.0
Fe	3.3	3.3	2.7	1.1	1.2	19.7	20.1	30.1	27.7	1.9	2.5
Co	0.0	0.2	0.1	0.0	0.0	0.5	0.6	0.5	0.4	0.0	0.0
Ni	56.1	57.0	58.2	2.2	1.7	32.7	32.2	68.8	71.5	2.0	1.5
Cu	0.0	0.0	0.1	96.7	97.1	0.1	0.1	0.4	0.3	94.8	96.1

Fo contents of the olivines in the sulfide-free peridotite tend to be higher than those in sulfide-bearing peridotite (Table 2), and chemical compositions of spinel differ in the two peridotite types. Spinel grains in the sulfide-bearing peridotite have lower values of the Cr# [= Cr/(Cr + Al)] than those in the sulfide-free peridotite. Because it is well known that Cr# in spinel from residual peridotite increases with increasing degree of partial melting of the mantle source (e.g., Dick and Bullen, 1984), the sulfide-bearing peridotite, mostly harzburgite, appears to have undergone lower degrees of partial melting than the sulfide-free peridotites, commonly dunite in the present case.

MODE OF OCCURRENCE AND CHEMICAL COMPOSITION OF SULFIDES AND NATIVE COPPER

In the harzburgite, sulfide minerals and native copper are either scattered at the grain boundaries of orthopyroxene, olivine, and spinel in particular, or enclosed by these minerals and by pseudomorphs after them. The sulfide minerals are different in the Hayama and Mineoka belts. In the Hayama belt, pentlandite is dominant and is associated with heazlewoodite, magnetite, and chalcopyrite. Chalcopyrite replaces the peripheral zones or cracks of pentlandite or occurs as veins in spinel grains. Rarely, pentlandite is replaced by awaruite. On the other hand, in the Mineoka belt, the plagioclase-bearing harzburgite has an assemblage of pentlandite + awaruite + native copper + magnetite (Fig. 3A), whereas the plagioclase-free harzburgite contains an assemblage of awaruite + heazlewoodite + native copper + magnetite (Fig. 3B). Dunite contains small grains of pentlandite and awaruite, which are scattered and less common than in the peridotites.

Some sulfides and native copper are considered to have been formed by Ni liberated from olivine during serpentinization (e.g., Kanehira et al., 1975). However, in the case of the Hayama-Mineoka belt, the peridotites' volume percentage of sulfide minerals is not correlative with the degree of serpentinization, but rather with vari-

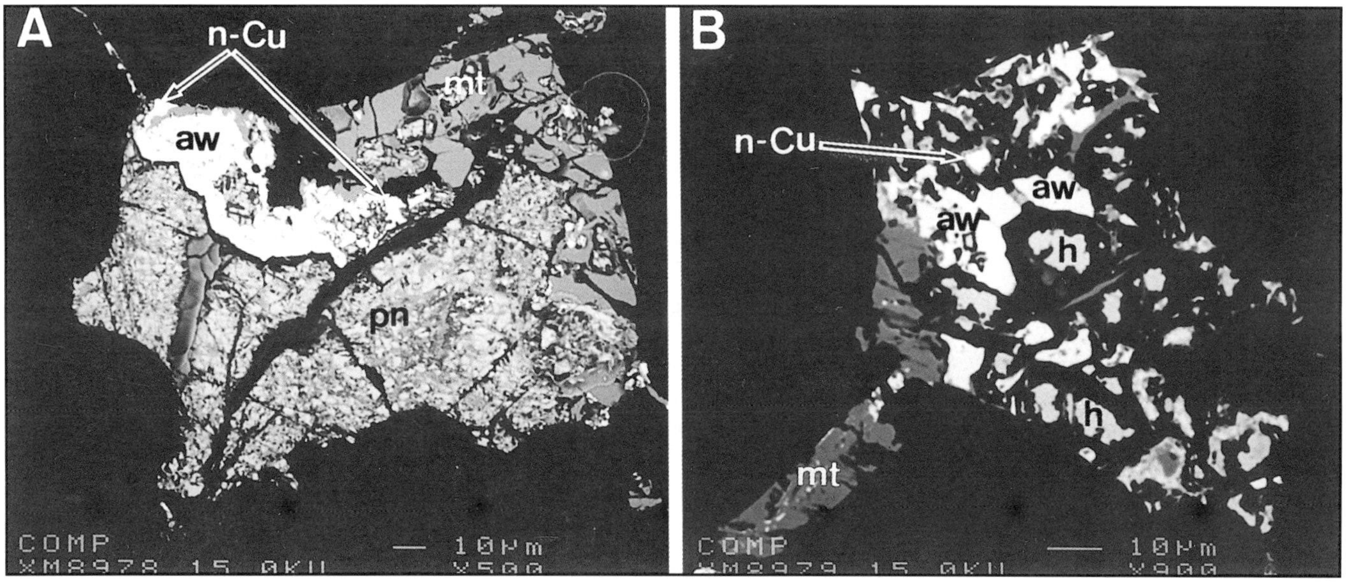

Figure 3. Compositional image by electron microprobe of sulfide minerals and native-copper assemblage. A, Assemblage of pentlandite (pn) + awaruite (aw) + native copper (n-Cu) enclosed by orthopyroxene and serpentine in plagioclase-rich harzburgite (97MN5) of the Mineoka belt. Scale bar is 10 μm. B, Assemblage of heazlewoodite (h) + awaruite (aw) + native copper (n-Cu) + magnetite (mt) in plagioclase-poor harzburgite of the Mineoka belt. Scale bar is 10 μm.

ations of the modal compositions of olivine and pyroxene: amounts of sulfide minerals increase with increasing modal compositions of pyroxene and with decreasing modal compositions of olivine (Fig. 4). Because peridotites are mostly serpentinized, sulfide minerals are not completely enclosed by silicate minerals or spinel. However, they are locally in direct contact with silicate minerals or spinel, a fact that suggests that primary sulfides might have formed before serpentinization. Pentlandite and heazlewoodite are common in the center of composite sulfide clusters, whereas awaruite, native copper, and magnetite occur on the edges. These observations indicate that pentlandite in the peridotites from the Hayama-Mineoka belt formed as a primary sulfide mineral by igneous processes, but that it was subsequently partially replaced by other sulfides and native copper during serpentinization in a later stage.

Representative chemical compositions determined by electron microprobe are listed in Table 4, and they are shown on the S-Fe-(Ni + Co) ternary diagram (Fig. 5).

THE ORIGIN OF SULFIDE MINERALS IN PERIDOTITES

Lorand (1988) proposed two processes for the origin of sulfides within oceanic upper-mantle peridotite: (1) The sulfide-mineral components in a fertile-mantle source survive within the depleted residual-mantle peridotite after the partial-melting event. In this case, sulfide minerals are closely associated with orthopyroxene. (2) The sulfide-mineral components are formed by sulfur-saturated melts extracted from harzburgite. In this case, sulfide minerals are associated with minerals that are crystallized in melt pockets trapped in the residual rocks, such as orthopyroxene, clinopyroxene, and spinel.

In the Hayama-Mineoka belt, sulfide-bearing peridotites are mostly classified as harzburgite in that they contain abundant orthopyroxene rather than clinopyroxene, and the primary pentlandite is hence thought to have originated by the first process proposed by Lorand (1988). According to his calculations, 23% partial melting is

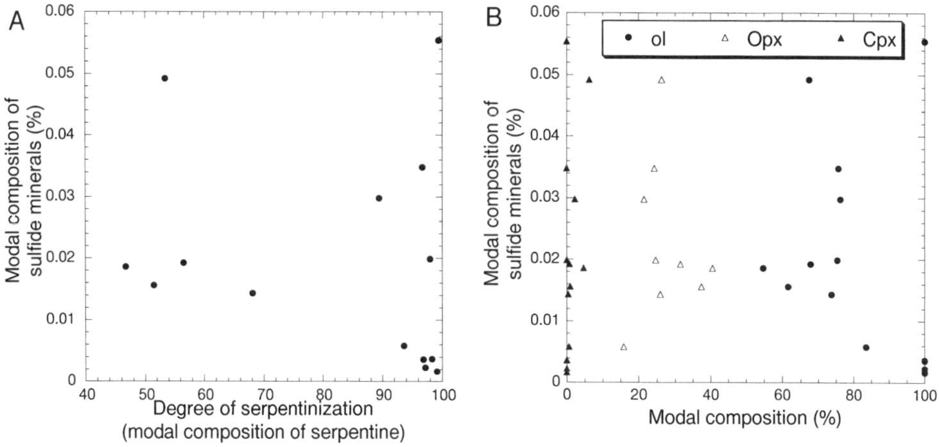

Figure 4. A, Relationships between degree of serpentinization and modal compositions of sulfide minerals. B, Relationships between modal compositions of ol, Opx, and Cpx and those of sulfide minerals.

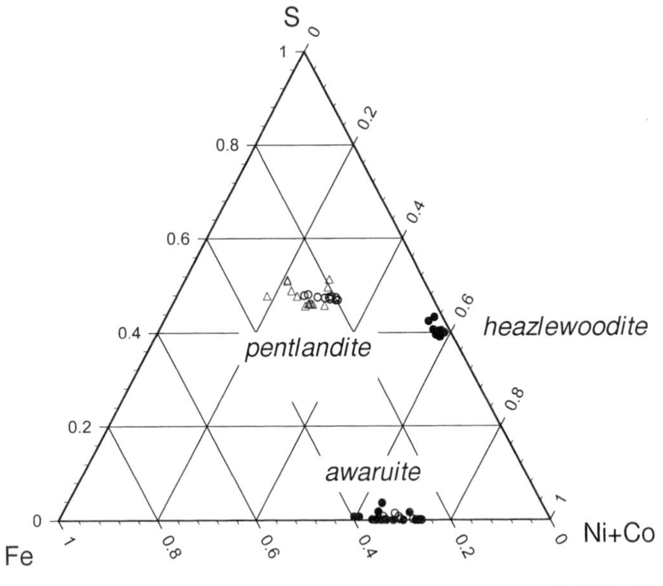

Figure 5. Ternary plot of sulfide minerals and native copper in S-Fe-(Ni + Co) system. Triangles—peridotites from the Hayama belt, open circles—plagioclase-rich peridotites from the Mineoka belt, solid circles—plagioclase-poor peridotites from the Mineoka belt.

necessary to remove all the sulfur from the mantle source. Furthermore, Garuti et al. (1984) concluded that iron sulfide (pyrrhotite) occurs only in relatively primitive peridotite (lherzolite), and sulfides in residual peridotite tend to be poorer in Fe and to be enriched in Ni and Cu with increasing degrees of partial melting. Thus, sulfides may change their compositions from pyrrhotite to heazlewoodite through pentlandite. In summary, because pentlandite is regarded as the primary sulfide in the Hayama-Mineoka belt peridotites, the sulfide-bearing mantle peridotite in the belt is considered to have undergone less than 23% partial melting.

This idea is supported by the chemical compositions of spinels in the peridotites. The chemical compositions of spinels from the Hayama-Mineoka belt peridotites are plotted on the Mg# versus Cr# diagram in Figure 6A. Cr# values in spinels from sulfide-bearing mantle peridotite (mostly harzburgite) range mostly from 0.4 to 0.6, values that are lower than those of sulfide-free peridotites (Fig. 6A). Because it is known that Cr# increases with increasing degrees of partial melting (Dick and Bullen, 1984), it is indicated that the sulfide-bearing peridotites underwent degrees of partial melting that are lower than those affecting the sulfide-free peridotites. Furthermore, the degree of partial melting of peridotite estimated by bulk-rock composition (Fig. 6B) based on the approach of Ishiwatari (1985) also shows that peridotites having the most sulfide underwent lower degrees of partial melting (between 15% and 30%) than sulfide-free peridotites. These spinel and bulk compositions support the hypoth-

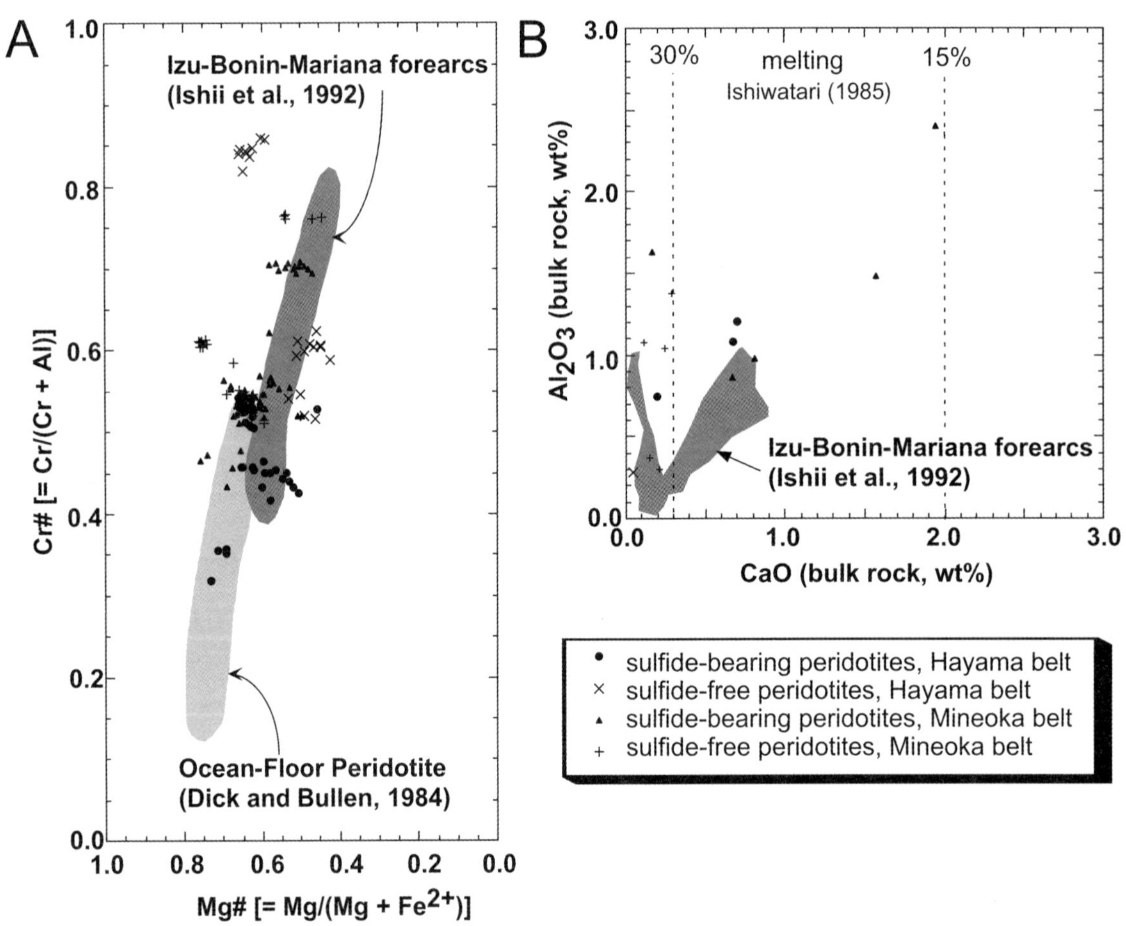

Figure 6. A, Chemical compositions of spinels in peridotites from the Hayama-Mineoka belt plotted on a Mg# vs. Cr# diagram. B, Bulk-rock compositions of peridotites in the Hayama-Mineoka belt. Degree of melting is inferred according to the method of Ishiwatari (1985).

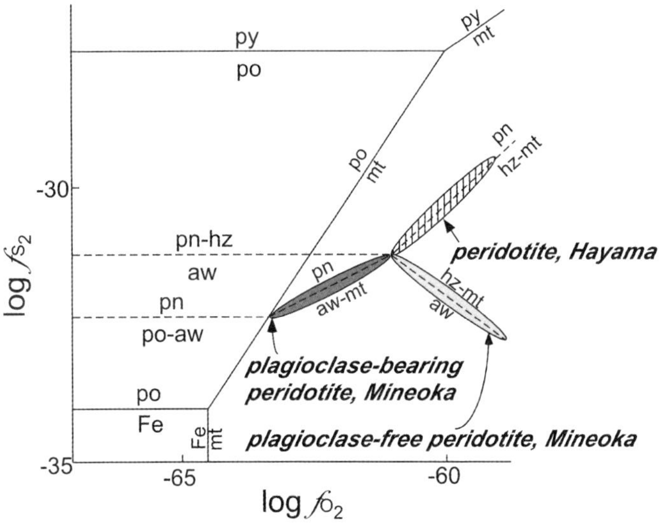

Figure 7. A log f_{O_2} vs. log f_{S_2} phase diagram at 127 °C (after Eckstrand, 1975), showing conditions of serpentinization in the Hayama-Mineoka belt. Abbreviations: aw—awaruite, hz—heazlewoodite, mt—magnetite, pn—pentlandite, po—pyrrhotite.

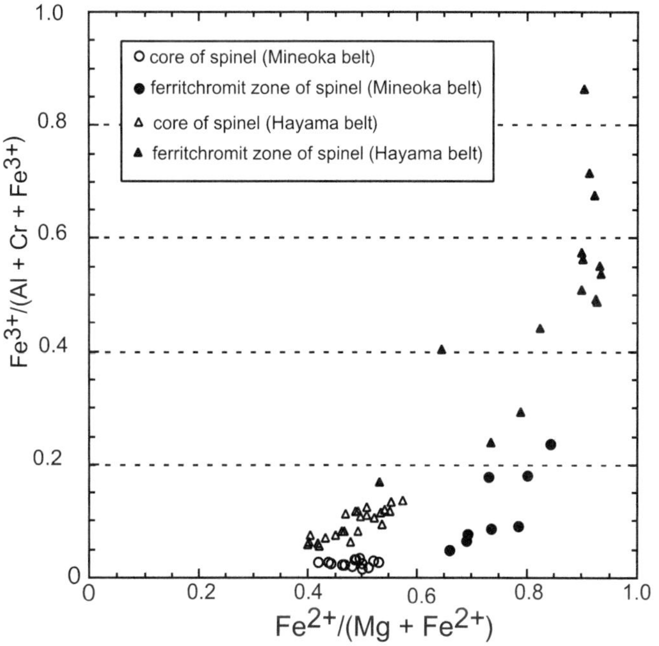

Figure 8. A plot of $Fe^{2+}/(Mg + Fe)^{2+}$ vs. $Fe^{3+}/(Cr + Al + Fe)^{3+}$, showing chemical compositions of spinel core and "ferritchromit."

esis that the primary sulfide minerals are residual after partial melting under the physical conditions already mentioned.

A high degree of partial melting (~30% or greater) is estimated for the mantle peridotites in diapiric serpentinite seamounts from the Izu-Bonin-Mariana forearcs (Ishii et al., 1992), whereas low to moderate degrees of partial melting are estimated for mantle peridotite at mid-ocean ridges (e.g., from 15% to 20% in the Mid-Atlantic Ridge; Casey, 1997). The results of these studies, therefore, indicate that most peridotites in the Hayama-Mineoka belt should be considered as residues of melting beneath a mid-ocean ridge, rather than in the mantle wedge associated with an island-arc mantle. Most sulfide-mineral components were removed through higher degrees of partial melting of the source-mantle peridotites in the Izu-Bonin-Mariana forearcs. This statement is consistent with the observations that most opaque minerals in the residual depleted peridotites in the Izu-Bonin-Mariana forearcs consist of magnetite (Stokking et al., 1992).

It is known that original sulfide minerals in peridotites are commonly replaced by aggregates of sulfide and native-metal minerals during serpentinization (Eckstrand, 1975; Shiga, 1983). Eckstrand (1975) discussed the physico-chemical conditions of these serpentinization products based on the thermodynamic data of the reactions involved and illustrated an equilibrium in the log f_{O_2} versus log f_{S_2} diagram for the Fe-Ni-S-O system at 127 °C (Fig. 7). Because serpentine in the Hayama-Mineoka belt consists of lizardite and/or chrysotile, considered to be formed under relatively low temperatures, the diagram can be used to estimate the conditions of formation of the sulfide and native-metal mineral assemblages. Figure 7 shows that (1) serpentinization in the Mineoka belt occurred under low f_{O_2} and low f_{S_2} conditions and (2) serpentinization in the Hayama belt also took place under similar, but slightly more oxidizing conditions. The "ferritchromit" zone around spinels may have formed by hydrothermal alteration of spinel (Spandenberg, 1943; Kimball, 1990), and the ratios of $Fe^{2+}/(Mg + Fe^{2+})$ and $Fe^{3+}/(Cr + Al + Fe^{3+})$ in this zone from the Hayama belt are higher than those of the Mineoka belt (Fig. 8). This result also suggests that serpentinization in the Hayama belt occurred under more oxidizing conditions than those in the Mineoka belt.

Assemblages of sulfide and native-metal minerals formed during serpentinization under low f_{O_2} and low f_{S_2} conditions are reported from many on-land ophiolite bodies (e.g., Panayiotou, 1980; Lorand, 1987, 1988). Similar mineral assemblages have been reported from ultramafic rocks exposed on the ocean floor (Prichard et al., 1996; Sato et al., 1997, 1998; Alt and Shanks, 1998).

Therefore, we infer that residual sulfide-mineral components in the Hayama-Mineoka belt peridotite were crystallized as pentlandite after medium degrees of partial melting of a fertile-mantle source in a mid-oceanic ridge environment rather than an island-arc environment. Then they were replaced by awaruite, heazlewoodite, and magnetite during serpentinization under low f_{O_2} and low f_{S_2} conditions.

TECTONIC IMPLICATION FOR OPHIOLITIC ROCKS IN THE HAYAMA-MINEOKA BELT

Several workers have suggested that the ophiolitic rocks in the Hayama-Mineoka belt represent remnants of a forearc ophiolite originated in the Izu-Bonin-Mariana forearc system, which was subsequently accreted to the Honshu arc (Ogawa and Naka, 1984; Fujioka et al., 1995). However, petrogenetic conditions and characteristics of the Hayama-Mineoka belt peridotites differ from those of the Izu-Bonin-Mariana peridotites. We suggest that the peridotites in the Hayama-Mineoka belt were originally part of a separate oceanic plate. On the basis of the new biostratigraphy of pelagic sedimentary units in the belt (Mohiuddin and Ogawa, 1998) and the chemistry and age determination of the basaltic rocks (Ogawa and Taniguchi, 1988; Taniguchi and Ogawa, 1990) from the Hayama-Mineoka belt exposures, we infer that the oceanic plate in which the belt formed existed in the northwest Pacific at least since the Eocene (Sato et al., 1999) (Fig. 9). Ogawa and Taniguchi (1988) called this oceanic plate the "Mineoka plate" as a mid-ocean ridge counterpart to the North New Guinea plate, which was moving south (Seno and Maruyama, 1984). Oceanic lithosphere of the Mineoka plate was emplaced on land as an ophiolite in the Hayama-Mineoka belt during the Miocene after

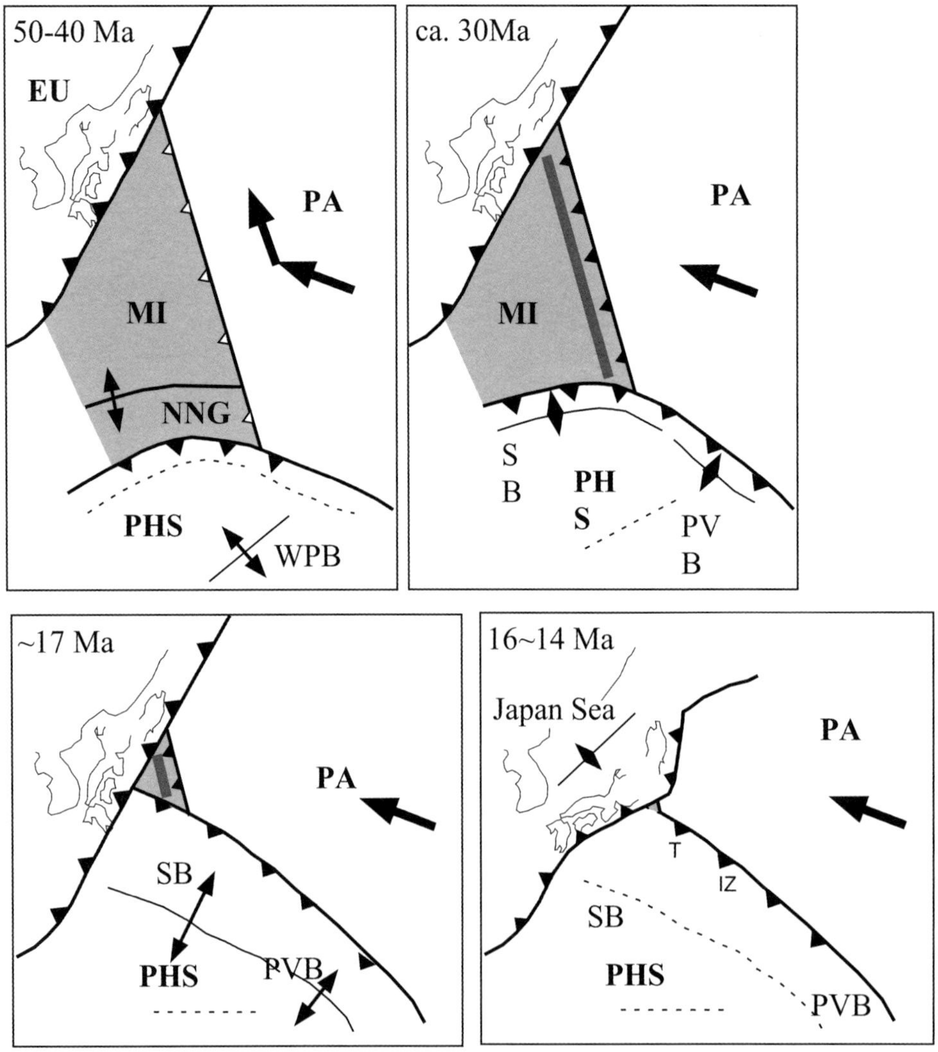

Figure 9. Simplified plate-tectonic model showing the Mineoka ophiolite (the Mineoka plate) and the adjacent area, modified after Sato et al. (1999). Abbreviations: EU—Eurasian, PA—Pacific, PHS—Philippine Sea, NNG—North New Guinea (Seno and Maruyama, 1984), and MI—Mineoka plates; WPB—West Philippine Basin; SB—Shikoku Basin; PVB—Parece-Vela Basin; T—Tanzawa block; IZ—Izu Peninsula.

the jump of the triple junction to its present location (Ogawa and Taniguchi, 1988). A southward-tilted high-density layer below the Hayama-Mineoka belt, inferred from gravity and magnetic anomalies (Tonouchi and Kobayashi, 1983), also supports the obducted plate model.

ACKNOWLEDGMENTS

We are grateful to Anthony Barber for his critical reviews of an early version of the manuscript and for fruitful advice. We deeply thank Mamoru Enjoji, Teruaki Ishii, and Hidetsugu Taniguchi for many valuable suggestions. Thanks are also extended to Norimasa Nishida, Shinji Kanayama, and Satoru Haraguchi for their help during chemical analysis. The manuscript was reviewed and revised by Patricia Fryer and two anonymous referees, to whom we are grateful.

REFERENCES CITED

Alt, J.C., and Shanks, W.C., 1998, Sulfur in serpentinized oceanic peridotites: Serpentinization processes and microbial sulfate reaction: Journal of Geophysical Research, v. 103, p. 9917–9929.

Arai, S., 1991, The Circum-Izu massif peridotite, central Japan as back-arc mantle fragments of the Izu-Bonin Arc system, *in* Peters, Tj., et al., eds., Ophiolite genesis and evolution of the oceanic lithosphere: Ministry of Petroleum and Minerals, Sultanate of Oman, p. 801–816.

Arai, S., and Takahashi, N., 1988, Relic plagioclase-bearing harzburgite from the Mineoka belt, Boso Peninsula, central Japan: Journal of Japan Association of Mineralogy, Petrology and Economic Geology, v. 83, p. 210–214 (in Japanese with English abstract).

Arai, S., Ito, T., and Ozawa, K., 1983, Ultramafic-mafic clastic rocks from the Mineoka belt, central Japan: The Journal of the Geological Society of Japan, v. 89, p. 287–297 (in Japanese with English abstract).

Arai, S., Ito, M., Nakayama, N., and Masuda, F., 1990, A suspect serpentinite mass in the Tokyo Bay area: Petrology and provenance of serpentinite pebbles in the Upper Cenozoic system in the Boso Peninsula, central Japan: The Journal of the Geological Society of Japan, v. 96, p. 171–179 (in Japanese with English abstract).

Casey, J.F., 1997, Comparison of major- and trace-element geochemistry of abyssal peridotites and mafic plutonic rocks with basalts from MARK region of the mid-Atlantic ridge, *in* Karson, J.A., Canat, M., Miller, D.J., and Elthon, D., eds., Proceedings of the Ocean Drilling Program, Scientific Results, vol. 153: College Station, Texas, Ocean Drilling Program, p. 181–241.

Dick, H.J.B., and Bullen, T., 1984, Chromian spinel as a petrogenetic indicator in abyssal and alpine-type peridotites and spatially associated lavas: Contributions to Mineralogy and Petrology, v. 86, p. 54–76.

Droop, G.T.R., 1987, A general equation for estimating Fe^{3+} concentrations in ferromagnesian silicates and oxides from microprobe analyses, using stoichiometric criteria: Mineralogical Magazine, v. 51, p. 431–435.

Eckstrand, O.R., 1975, The Dumount serpentinite: A model for control of nickeliferous opaque mineral assemblages by alteration reactions in ultramafic rocks: Economic Geology, v. 70, p. 183–201.

Frost, B.R., 1985, On the stability of sulfide, oxides, and native metals in serpentinite: Journal of Petrology, v. 26, p. 31–63.

Fryer, P., and Fryer, G.J., 1987, Origins of nonvolcanic seamounts in a forearc environment, *in* Keating, B.H., Fryer, P., Batiza, R., and Boehlert, G.W., eds., Seamounts, islands, and atolls: American Geophysical Union Geophysical Monograph 43, p. 61–69.

Fryer, P., Mottl, M., Johnson, L., Haggerty, J., Phipps, S., and Maekawa, H., 1995, Serpentine bodies in the forearcs of western Pacific convergent margins: Origin and associated fluids, *in* Taylor, B., and Natland, J., eds., Active margins and marginal basins of the western Pacific: American Geophysical Union Geophysical Monograph 88, p. 259–279.

Fujioka, K., Tanaka, T., and Aoike, K., 1995, Serpentine seamount in Izu-Bonin and Mariana forearcs: Observation by a submersible and its relation to onland serpentinite belt: Tokyo Geographical Society, Journal of Geography, v. 104, p. 473–494 (in Japanese with English abstract).

Garuti, G., Gorgoni, C., and Sighinolfi, G.P., 1984, Sulfide mineralogy and chalcophile and siderophile element abundances in the Ivrea-Verbano mantle peridotites (Western Italian Alps): Earth and Planetary Science Letters, v. 70, p. 69–87.

Ishii, T., 1985, Dredged samples from the Ogasawara fore-arc seamounts or "Ogasawara Paleoland": Fore-arc ophiolite, *in* Nasu, N., Kobayashi, K., Uyeda, S., Kushiro, I., and Kagami, H., eds., Formation of active ocean margins: Tokyo, TERRAPUB, p. 307–342.

Ishii, T., Robinson, P.T., Maekawa, H., and Fiske, R., 1992, Petrological studies of peridotites from diapiric serpentinite seamounts in the Izu-Ogasawara-Mariana Forearc, Leg 125, *in* Fryer, P., Pearce, J.A., Stokking, L.B., et al., eds., Proceedings of the Ocean Drilling Program, Scientific Results, vol. 125: College Station, Texas, Ocean Drilling Program, p. 445–485.

Ishiwatari, A., 1985, Igneous petrogenesis of the Yakuno ophiolites: Contributions to Mineralogy and Petrology, v. 89, p. 155–167.

Kanehira, K., Banno, S., and Yui, S., 1975, Awaruite, heazlewoodite, and native copper in serpentinized peridotite from the Mineoka district, southern Boso Peninsula: Journal of Japan Association of Mineralogy, Petrology and Economic Geology, v. 70, p. 338–394.

Kaneoka, I., Takigami, Y., Tonouchi, S., Furuta, T., Nakamura, Y., and Hirana M., 1981, Pre-Neogene volcanism in the central Japan based on K-Ar and Ar-Ar analysis [abs.]: Tokyo and Hakone, International Association of Volcanology and Chemistry of the Earth's Interior Symposium, p. 166.

Kimball, K.L., 1990, Effects of hydrothermal alteration on the compositions of chromian spinel: Contributions to Mineralogy and Petrology, v. 105, p. 337–346.

Lorand, J.P., 1987, Cu-Fe-Ni-S mineral assemblages in upper-mantle peridotites from the Table Mountain and Blow-Me-Down Mountain ophiolite massifs (Bay of Island area, Newfoundland): Lithos, v. 20, p. 59–76.

Lorand, J.P., 1988, Fe-Ni-Cu sulfide in tectonic peridotites from Maqsad district, Sumail Ophiolite, Southern Oman: Implications for the origin of the sulfide component in the oceanic upper mantle: Tectonophysics, v. 151, p. 57–73.

Mohiuddin, M.M., and Ogawa, Y., 1998, Late Paleocene-middle Miocene pelagic sequences in the Boso Peninsula, Japan: New light on northwest Pacific tectonics: The Island Arc, v. 7, p. 301–314.

Ogawa, Y., and Naka, J., 1984, Emplacement of ophiolitic rocks in forearc areas: Examples from central Japan and Izu-Mariana-Yap island arc system, *in* Gass, I.G., Lippard, S.J, and Shelton, A.W., eds., Ophiolite and oceanic lithosphere: Geological Society [London] Special Publication 13, p. 291–301.

Ogawa, Y., and Taniguchi, H., 1988, Geology and tectonics of the Miura-Boso Peninsulas and the adjacent area: Modern Geology, v. 12, p. 147–168.

Panayiotou, A., 1980, Cu-Ni-Co-Fe sulphide mineralization, Limassol Forest, Cyprus, *in* Panayiotou, A., ed., Ophiolite Proceedings, International Ophiolite Symposium, Cyprus 1979: Republic of Cyprus, Geological Survey Department, Ministry of Agriculture and Natural Resouces, p. 102–116.

Prichard, H.M., Puchelt, H., Eckhardt, J.D, and Fisher, P.C., 1996, Platinum-Group-Element concentrations in mafic and ultramafic lithologies drilled from Hess Deep, *in* Mével, C., Gillis, K.M., Allan, J.F., and Meyer, P.S., eds., Proceedings of the Ocean Drilling Program, Scientific Results, vol. 147: College Station, Texas, Ocean Drilling Program, p. 77–90.

Sato, H., Enjoji, M., and Ogawa, Y., 1997, The condition of serpentinization estimated by native alloys and metals [abs.]: Japan Earth and Planetary Science Joint Meeting, Nagoya, p. 622 (in Japanese).

Sato, H., Ishii, T., Taniguchi, H., and Ogawa, Y., 1998, Sulfide minerals as a petrogenetic and alteration indicator of ultramafic rocks in ocean floor [abs.]: Eos, Transactions of the American Geophysical Union, v. 79, No. 45, p. F398.

Sato, H., Taniguchi, H., Takahashi N., Mohiuddin, M.M., Hirano, N., and Ogawa, Y., 1999, The origin of the Mineoka ophiolite: Journal of Geography, Tokyo Geographical Society, v. 108, p. 177–189 (Japanese with English abstract).

Seno, T., and Maruyama, S., 1984, Paleogeographic reconstruction and origin of the Philippine Sea: Tectonophysics, v. 102, p. 53–84.

Shiga, Y., 1983, Fe-Ni (-Co)-S minerals in the Hayachine ultramafic rocks of the Kamaishi mining district, Northern Japan: Physico-chemical conditions during serpentinization: Mining Geology, v. 33, p. 23–38 (in Japanese with English abstract).

Spandenberg, K., 1943, Die chromitlaagerstatte von tampedal in Zobten: Zeitschrift fur Praktische Geologie, v. 51, p. 13–35.

Stokking, L.B., Merrill, D.L., Haston, R.B., Ali, J.R., and Saboda, K.L., 1992, Rock magnetic studies of serpentinite seamounts in the Mariana and Izu-Bonin regions, *in* Fryer, P., Pearce, J.A., Stokking, L.B., et al., eds., Proceedings of the Ocean Drilling Program, Scientific Results, vol. 125: College Station, Texas, Ocean Drilling Program, p. 561–579.

Taniguchi, H., 1991, The geological and geochemical study of the igneous rocks in the Hayama-Mineoka belt [Ph.D. thesis]: Tokyo, Nihon University, 108 p. (in Japanese with English abstract).

Taniguchi, H., and Ogawa, Y., 1990, Occurrence, chemistry and tectonic significance of alkali basaltic rocks in the Miura Peninsula, central Japan: The Journal of the Geological Society of Japan, v. 96, p. 101–116 (in Japanese with English abstract).

Taylor, B., 1992, Rifting and the volcanic-tectonic evolution of the Izu-Bonin-Mariana arc, *in* Taylor, B., Fujioka, K., et al., eds., Proceedings of the Ocean Drilling Program, Scientific Results, vol. 126: College Station, Texas, Ocean Drilling Program, p. 627–651.

Tonouchi, S., and Kobayashi, K., 1983, Paleomagnetic and geotectonic investigation of ophiolite suite and surrounding rocks in south-central Honshu, Japan, *in* Hashimoto, M., and Uyeda, S., eds., Accretion tectonics in the Circum-Pacific regions: Tokyo, TERRAPUB, p. 261–288.

Uchida, T., and Arai, S., 1978, Petrology of ultramafic rocks from the Boso Peninsula and the Miura Peninsula: The Journal of the Geological Society of Japan, v. 84, p. 561–570.

Yokosuka City Museum, 1991, Geological Map of the Miura Peninsula: Yokosuka City Museum, Kanagawa, Japan (in Japanese), scale 1:50 000.

MANUSCRIPT ACCEPTED BY THE SOCIETY FEBRUARY 23, 2000

Printed in the U.S.A.

Geological Society of America
Special Paper 349
2000

Section VI
Ophiolites from the Iapetus, Rheic-Pleionic, Neotethyan, and Indian Oceans

The Iapetus, Rheic-Pleionic, and Neotethyan Oceans closed during development of the Caledonian-Appalachian, Hercynian-Uralian, and Alpine-Himalayan orogens, respectively. Ophiolitic bodies in the Caledonian-Appalachian orogenic belt developed as the Iapetus Ocean evolved between North America and Baltica-Avalonia during much of the Paleozoic. The Rheic Ocean evolved between the Baltica-Avalonia and Gondwana continental masses, and the Hercynian ophiolites that are exposed in western and central Europe and in northwestern Africa developed in this complex ocean basin. A series of collisions during the Carboniferous between Baltica and south Europe and between Africa and south Europe caused the closure of the Rheic Ocean and its seaways, which in turn resulted in the emergence of the Hercynian orogenic belt. The Hercynian orogenic belt extends westward into the Alleghany-Ouachita belt in the southern coast of the United States. The Pleionic Ocean and its oceanic lithosphere evolved between Baltica–Eastern Europe and Kazakhstan-Siberia during the early Paleozoic through several episodes of seafloor spreading and active-margin tectonics. The gradual closure of the Pleionic Ocean through the collapse of its marginal basins, island arc–continent collisions, and finally continent-continent collision between Baltica and Kazakhstan-Siberia developed the Uralian orogeny during the Late Permian. The Neotethys was a wedge-shaped, east-trending (in present coordinates) ocean basin that evolved between Eurasia and Gondwana and had a complex paleogeography with several major seaways separated by continental fragments and/or carbonate platforms. The Neotethys closed progressively from west to east in a zipper fashion, as collisions between different continental fragments and eventually between Eurasia and Afro-Arabia and between Eurasia and India obliterated the Neotethyan seaways during the late Mesozoic (in the west) and Cenozoic (in the east).

Two papers in this section deal with the Iapetus Ocean. Suhr and Edwards report on data from the Lewis Hills massif of the Bay of Islands ophiolite, Newfoundland, one of the best-preserved and best-studied ophiolites. They argue for the existence of an oceanic fracture zone separating arc-type crust from backarc crust in the Lewis Hills massif, and they discuss that the mantle sequences (Springer Hill mantle) with boninitic pyroxenite dikes reflect development in an infant-arc environment, analogous to the peridotites exposed in the Mariana-Bonin forearc systems. Furnes et al. describe the Solund-Stavfjord ophiolite in the western Norwegian Caledonides as a complex formed in a backarc setting through two episodes of seafloor spreading. They present structural and geochemical evidence for a rift system that was spreading at intermediate rates and that propagated into a preexisting oceanic crust in this backarc basin, which might have evolved in proximity to a continental landmass bordering the Iapetus Ocean.

From the Rheic-Pleionic Ocean domains, Spadea and Scarrow describe Early Devonian boninites from an intraoceanic arc, the Magnitogorsk arc, that collided with the East European craton during the late Paleozoic Uralian orogeny. The geologic setting and composition of these arc rocks resemble the Izu-Bonin or Tonga arcs of the western Pacific. Progressive change from boninitic effusive rocks to explosive tholeiitic and calc-alkalic volcanic rocks in the arc stratigraphy and age data suggest that development of the Magnitogorsk island arc occurred in a 14–40 m.y. time period prior to its collision with the continent.

From the Neotethyan Ocean remnants, Rassios and Smith summarize information on the western Greek ophiolites (Vourinos, Pindos, and Othris). These ophiolites are thought to have been part of an originally mantle-rooted continuous thrust sheet emplaced from the southwest (present coordinates) onto the margin of the Pelagonian microcontinent. The ophiolites display marked compositional variation from supra-subduction zone (Vourinos) to mid-oceanic-ridge basalt–island-arc basalt (Pindos) to mid-oceanic-ridge basalt (Othris) types, which must have been originally within a relatively short distance in the same backarc basin (few kilometers to tens of kilometers). Ghazi and Hassanipak describe the Shahr-Babak ophiolite, thought to have been formed in an island arc–within plate (intraoceanic island) environment and emplaced onto the western margin (present coordinates) of the Central Iranian microcontinent. Nicolas et al. describe the Aswad massif, United Arab Emirates as one of the best-preserved ophiolite bodies in the Oman–United Arab Emirates ophiolite belt that has a horizontal paleo-Moho and a vertical sheeted-dike complex. They argue that the ~400-m-thick Moho transition zone in the central part of the massif probably originated from a mantle diapir, which was still active during the emplacement of the ophiolite nappe onto the Arabian continental margin. A north- to north-northwest–trending, paleo–fast-spreading ridge system is inferred for the tectonic setting of the igneous accretion of the Aswad massif. Gray et al. argue that emplacement of the Semail ophiolite was a two-stage subduction process involving partial subduction of a continental slab (microplate) beneath the Arabian continental margin, followed by the initiation of intraoceanic thrusting that resulted in the tectonic transport of the ophiolite nappe over the exhumed microplate and then onto Arabia as a gravity-driven thin sheet. This interpretation suggests that ophiolite emplacement in Oman was a complex tectonic process involving underthrusting toward the continental margin rather than away from it, as previous models have suggested.

Finally, Peters describes new information from the Late Jurassic Masirah ophiolite south of the Arabian peninsula, suggesting that it originated as a part of the Somali Basin during the evolution of the western Indian Ocean. A complex history of slow spreading and mid-plate volcanism is recorded in the complex, which was emplaced on the eastern margin of Arabia during latest Cretaceous–earliest Tertiary time, as the Carlsberg Ridge started operating between India and the Seychelles.

Geological Society of America
Special Paper 349
2000

Contrasting mantle sequences exposed in the Lewis Hills massif: Evidence for the early, arc-related history of the Bay of Islands ophiolite

Günter Suhr*
Max Planck Institut für Chemie, Postfach 3060, 55020 Mainz, Germany; e-mail: gsuhr@mpch-mainz.mpg.de
Stephen John Edwards
Lithospheric Processes and Resources Group, School of Earth and Environmental Sciences, University of Greenwich, Chatham Maritime, Kent ME4 4TB, England; e-mail: s.j.edwards@greenwich.ac.uk

ABSTRACT

A major fracture zone is preserved within the Lewis Hills massif (Bay of Islands ophiolite). It separates arc-type crust in the west from 17 m.y. younger, backarc-related oceanic crust in the east. We link mantle peridotites occurring in the Springers Hill locality to the arc crust and use their petrologic characteristics to define the early history of the arc. The Springers Hill mantle displays a weak shape fabric, high Cr/(Cr + Al) ratios in spinel (>60), very low, chondrite-normalized, heavy rare earth element abundances in clinopyroxene ($HREE_N$ < 1), and scattered calcic amphibole. It contains two important chromitite deposits and numerous pyroxenite dikes, some of which have a boninitic composition; and it is intruded by a late, large ultramafic body. The Springers Hill mantle peridotites are modeled by 26% melting of a primitive mantle fluxed by a slab melt or sediment (open-system melting with <1% total addition). The melting was outlasted by melt and fluid migration that caused middle rare earth element enrichment through chromatographic effects. This late infiltration event was fabric-destroying by causing orthopyroxene breakdown and recrystallization.

In the absence of a clearly identifiable old basement for the arc, and given the extreme degree of depletion of the mantle peridotites underlying the arc, we interpret the Springers Hill mantle as related to upwelling and extension in an infant arc setting. Such a primitive arc setting is supported by strong geochemical similarities of the Springers Hill mantle to peridotites exposed in the Mariana and Bonin forearc systems. In a later event, the infant arc was rifted to give way to mantle upwelling, backarc formation, and associated final exhumation of the Springers Hill mantle within an extensional fracture-zone setting. The exhumation was probably associated with renewed melting at depth and melt and fluid migration in the Springers Hill mantle.

INTRODUCTION

There is no doubt that among the well-preserved ophiolites, the Lewis Hills massif (Bay of Islands ophiolite, Newfoundland, Fig. 1A) ranks among the atypical bodies (Karson and Dewey, 1978). A steeply east-dipping, north-trending, transtensional strike-slip zone separates a mafic sequence in the west from a mainly ultramafic sequence in the east. It represents an oceanic fracture zone (Karson and Dewey, 1978). Isotopic evidence (Kurth et al., 1998) has confirmed an earlier geologic model (Cawood and Suhr, 1992) that the fracture zone separates a 17-m.y.-younger backarc lithosphere east of the fracture zone (the Bay of Islands Complex, 485 Ma) from an older arc-type lithosphere west of it (the Little Port Complex, 502 Ma). The fracture zone itself (Mount Barren Complex) was associated with highly extensional deformation, as demonstrated by the numerous magmatic bodies spatially associated with it (Karson and Dewey, 1978; Suhr and Cawood, 1999). More recent evidence suggests that another highly extensional strike-slip zone occurs farther east (Fig. 2).

The juxtaposition of arc and backarc lithosphere within the Lewis Hills offers a rather unique opportunity to study the extent of melting near a fracture zone (Suhr and Batanova, 1998) and the nature of the arc basement. For example, a major question is whether the mantle under an arc originates by decompression melting and ascent similar to that under ocean ridges (Plank and Langmuir, 1988). Also of interest is whether an arc is built on—and has preserved—a former, older oceanic basement with the source of the arc melts being located

*Present address: Mineralogisch-Geochemisches Institut, Universität zu Köln, Zülpicher Str. 49b, 50674 Köln, Germany; e-mail: suhr@min.uni-koeln.de

Suhr, G., and Edwards, S.J., 2000, Contrasting mantle sequences exposed in the Lewis Hills massif: Evidence for the early, arc-related history of the Bay of Islands ophiolite, *in* Dilek, Y., Moores, E.M., Elthon, D., and Nicolas, A., eds., Ophiolites and Oceanic Crust: New Insights from Field Studies and the Ocean Drilling Program: Boulder, Colorado, Geological Society of America Special Paper 349, p. 433–442.

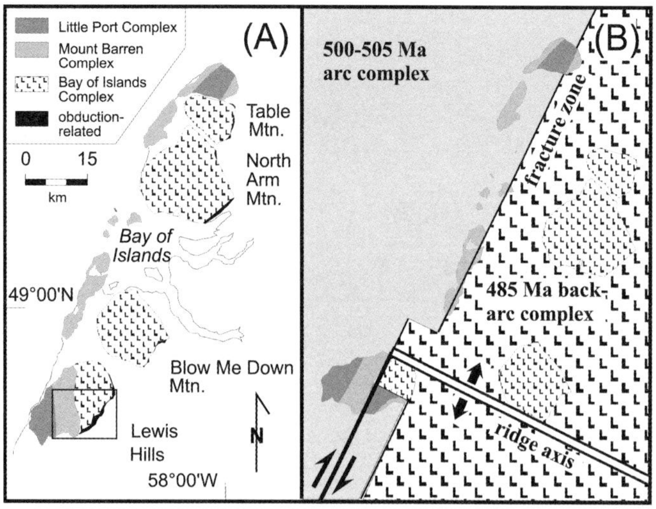

Figure 1. A, Location of Bay of Islands ophiolite and fourfold division into the backarc-related Bay of Islands Complex, the transform-related Mount Barren Complex, the arc-related Little Port Complex, and the obduction-related units (following Karson, 1984). B, Proposed paleotectonic arrangement of massifs (Suhr and Cawood, 1999). Note that southeastern Lewis Hills is considered to be part of the arc complex.

at greater depth (e.g., Davies and Stevenson, 1992; Iwamori, 1998) or whether the (relatively stationary) mantle located under arcs has a chemical signature acquired by pervasive interaction with melts generated at greater depth (Kelemen et al., 1995; Quick and Gregory, 1995).

Unfortunately, until recently, there was no evidence of exposed mantle in the western, arc-related, sequence. Mantle exposures were only reported from two, spatially separated locations in the eastern, presumably backarc-related, side of the Lewis Hills: the Wheelers Brook mantle in the northeastern Lewis Hills and the Springers Hill mantle in the southeastern Lewis Hills (Fig. 2).

In this paper, we report a new occurrence of mantle rocks exposed in the western, arc-related Lewis Hills (Western Lewis Hills mantle), and we suggest that it relates genetically to the Springers Hill mantle. We suggest that *both* occurrences are arc related because they are geochemically similar but radically different from the backarc-related Wheelers Brook mantle. The large exposure of the Springers Hill mantle thus offers an excellent window into the arc basement. We propose that the arc mantle formed by upwelling in a young mantle wedge, creating a juvenile basement for the contemporaneous and subsequent arc formation. In a later event, final exhumation of the subarc mantle occurred within an extensional fracture-zone setting involving limited ridge propagation into the old arc and inside-corner tectonic denudation of the arc lithosphere (Fig. 1B; Suhr and Cawood, 1999). The associated melt and fluid migration caused an overprint of both chemistry and microstructure of the Springers Hill mantle.

GEOLOGIC SITUATION

Mantle rocks are exposed as coherent bodies on the eastern side of all four massifs of the Bay of Islands ophiolite. In the Lewis Hills, the eastern mantle peridotites (Wheelers Brook mantle) are anhydrous clinopyroxene harzburgites to lherzolites with a strong, largely lineated shape fabric. As in the three northern massifs, wehrlite or gabbro bodies are absent within the Wheelers Brook mantle. Structurally below the Wheelers Brook mantle, but above the obduction-related metamorphic sole, a body of "exotic peridotite" several hundred meters thick was accreted to the base of the ophiolite during early detachment, and it was presumably derived from the overridden Little Port Complex (Suhr and Cawood, 1993). Until now, this sliver was the only suspected arc-derived mantle rock in the Bay of Islands Complex.

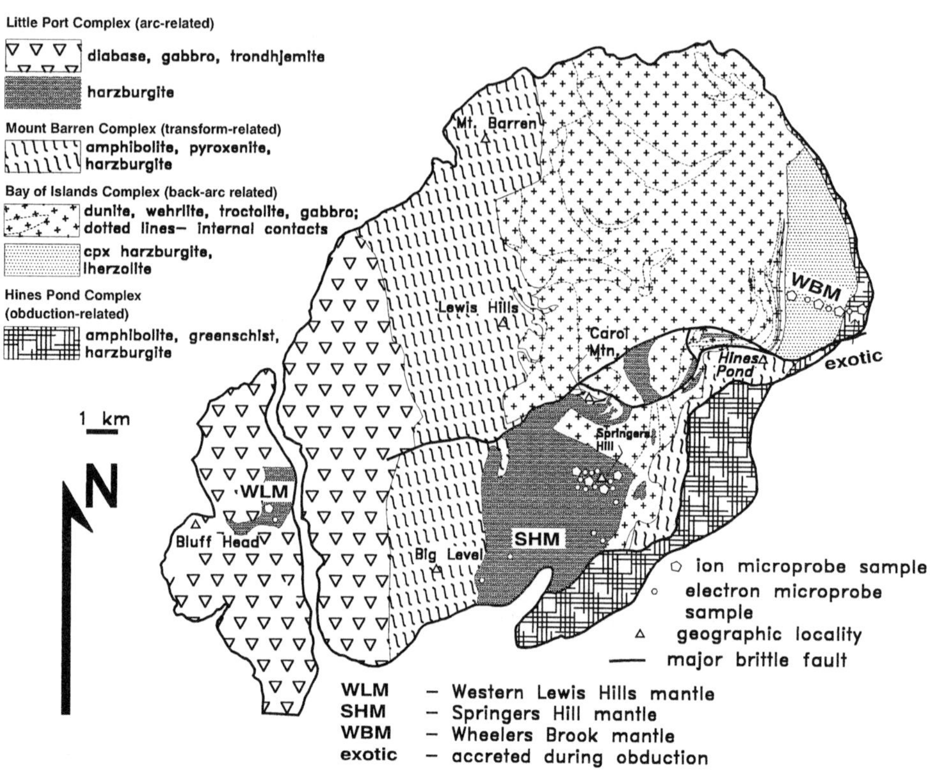

Figure 2. Simplified geologic map of the Lewis Hills massif. Geology from Karson and Dewey (1978) and Suhr and Cawood (1999). The Springers Hill mantle (SHM) represents the largest coherent exposure of mantle peridotite in the Lewis Hills. It is conventionally correlated with the Wheelers Brook mantle (WBM), and both units are considered to be part of the Bay of Islands Complex. We suggest here that the Springers Hill mantle correlates with a mantle exposure in the western Lewis Hills and that both the Springers Hill mantle and Western Lewis Hills mantle (WLM) represent basement to the Little Port Complex arc sequence.

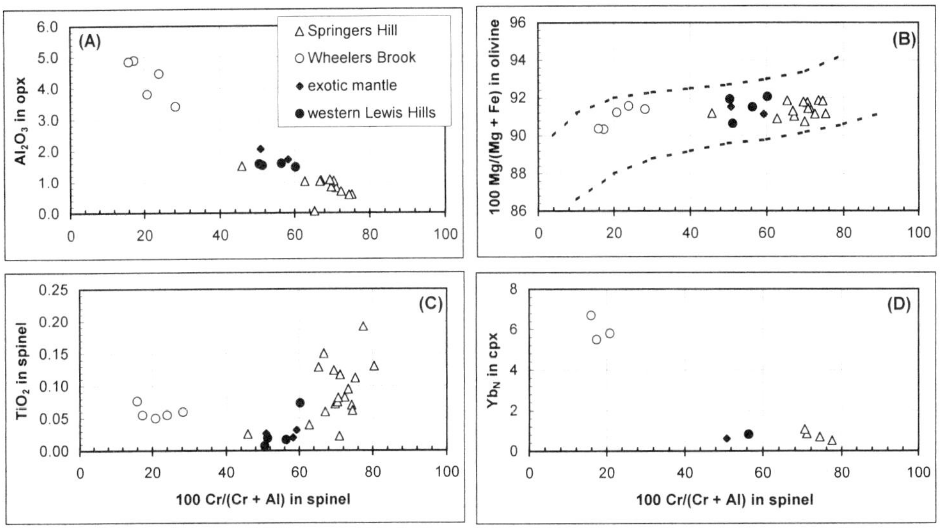

Figure 3. Mineral chemistry of harzburgites and lherzolites from mantle occurrences. Exotic mantle unit is presumably derived from Western Lewis Hills mantle. Mantle array from Arai (1994). Springers Hill sample with lowest Cr/(Cr + Al) ratio contains abundant calcic amphibole and is derived from the adjacent Mount Barren Complex.

Unlike in the northern massifs of the Bay of Islands Complex, the Lewis Hills exposes a second, large body of mantle rocks, located in the southeastern part of the massif (Fig. 2). It hosts a chromitite ore body known as Springers Hill (Smith, 1958; Dahl and Watkinson, 1986). The peridotites from this occurrence display complex structures and fabrics (Suhr et al., 1991), and their mineral chemistry on Springers Hill itself is very depleted (Suhr and Robinson, 1994). As pointed out by Edwards (1995), this very depleted character is not compatible with a cold-wall effect near an oceanic fracture zone (Fox and Gallo, 1984), which will tend to suppress shallow-level melting in the mantle and thus preserve below-average degrees of depletion (as in the Wheelers Brook mantle, Suhr and Batanova, 1998). The Springers Hill mantle hosts numerous magmatic bodies such as pyroxenitic dikes (Suhr et al., 1991), which range from tholeiitic to boninitic in character (Edwards, 1990, 1995), networks of gabbroic dikes (Suhr and Cawood, 1993), and a late, large dunite-wehrlite body that itself is multiply intrusive and weakly deformed (Dunsworth et al., 1986; Edwards and Malpas, 1995; Suhr and Cawood, 1999). During this late magmatic history of the Springers Hill mantle, the peridotites behaved largely as a lithospheric body. Examples for this behavior are the inefficient melt extraction as seen by the simple presence of the mantle-hosted magmatic bodies, the very heterogeneous strain contrasting with kilometer-scale unpartitioned strain during asthenospheric flow, and the mylonitic nature of pyroxenitic dikes within a matrix of low-strain harzburgite. This last feature represents ductile-strain localization, nucleating along fractures. The Springers Hill mantle is overlain by a major ductile normal fault. This shallowly north-dipping mylonitic shear zone is analogous to oceanic detachment faults within inside-corner ridge and transform intersections (Suhr and Cawood, 1999).

The low-strain Springers Hill harzburgites display low amounts of clinopyroxene (<1%) and scattered, mainly tremolitic, amphiboles. Amphibole is synkinematic in high-strain zones but late- to postkinematic in low-strain zones. The formation of amphibole occurs at the expense of orthopyroxene. Orthopyroxene forms pull-apart–like morphologies grading into wispy stringers and networks of dikes with partly tension-gash–like geometry, referred to as type II orthopyroxenites (Suhr et al., 1991; Edwards and Malpas, 1995). They obviously postdate the main fabric-forming event in the Springers Hill mantle. Spinel grains are large and rich in inclusions (Edwards and Malpas, 1995).

The regional mapping of Karson (1984) shows large intrusive peridotites also in the western Lewis Hills. Although we have not mapped the western Lewis Hills in detail, at least the southern exposures of the intrusive peridotites are, in our view, mantle peridotites. Their orthopyroxene to olivine ratio is typical for the mantle and homogeneous on the scale of the exposure. It is similar to the Springers Hill mantle in terms of the coarse, commonly mesh-like appearance of orthopyroxene. Numerous mafic dikes occur, and partial hydration to amphibole peridotite is common, as already mentioned herein for the Springers Hill mantle. Our local mapping shows that the mantle peridotites grade into dunite hosting the Bluff Head (Fig. 2) chromitite, which gives this orebody a typical mantle-crust transition-zone setting and may link it genetically to the Springers Hill chromitite occurrence.

MINERAL MAJOR ELEMENT CHEMISTRY

New mineral-chemistry data reported here are from harzburgitic samples taken south and west of Springers Hill, from the Western Lewis Hills mantle, and from the Wheelers Brook mantle. They are plotted together with data from Springers Hill itself as reported in Suhr and Robinson (1994). The sample distribution is shown in Figure 2.

Mineral-chemical data plot within two distinct domains (Fig. 3). The Wheelers Brook mantle shows low Cr# [= 100Cr/(Cr + Al)] values in spinel (15–30), whereas the Springers Hill harzburgites have Cr# values of 60–80. The Western Lewis Hills mantle and the exotic harzburgite have Cr# values between 50 and 60. A significantly amphibolitized sample of Springers Hill harzburgite located adjacent to the Mount Barren Complex has a somewhat lower Cr# of 45. The amphibole in this sample is calcic, with a low Al_2O_3 content of 2.5 wt%. Regionally, the scattered amphiboles present in the Springers Hill harzburgite have between 2 and 8 wt% Al_2O_3. The high Al_2O_3 trend of the Wheelers Brook mantle is also expressed in the pyroxene chemistry. Al_2O_3 is between 3% and 5% in the Wheelers Brook pyroxenes, whereas for the Springers Hill harzburgites, Al_2O_3 is below 1% and for Western Lewis Hills mantle and exotic harzburgites, Al_2O_3 is between 1% and 2% (Fig. 3A). Given that the Cr# in spinel and the Al_2O_3 in pyroxenes is an expression of the degree of depletion of a residual peridotite (Jaques and Green, 1980; Dick and Fisher, 1984), the Wheelers Brook mantle is clearly less refractory than the

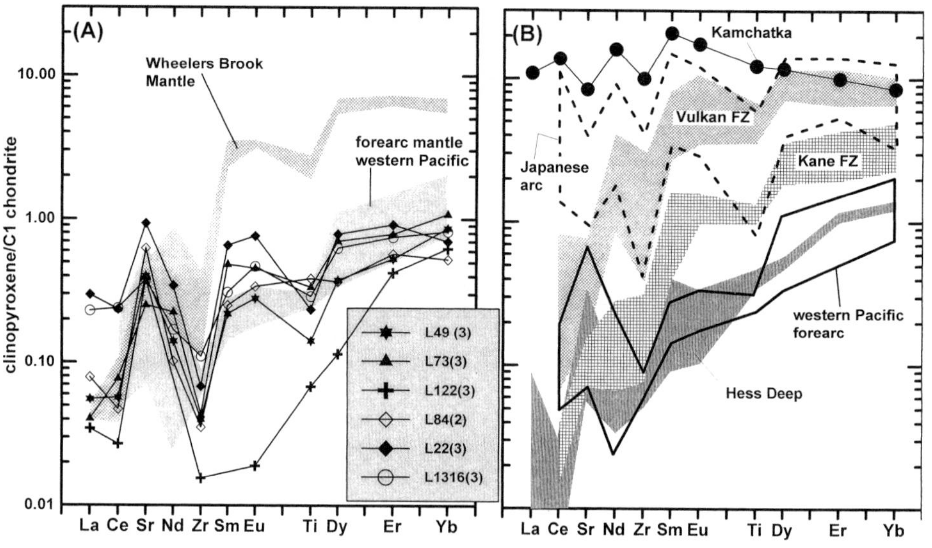

Figure 4. A, C1 chondrite–normalized REE, Ti, Zr, and Sr concentrations in clinopyroxene from Springers Hill harzburgite (L49, L73, L84, L2), Western Lewis Hills mantle (L1316), exotic harzburgite L122 (data in Table 1), Wheelers Brook mantle (Suhr and Batanova, 1998), and the western Pacific forearc peridotites (Parkinson et al., 1992). Number of analyses given in parentheses. Note the dramatic difference in HREEs between Wheelers Brook and Springers Hill mantle rocks and the similarity between pattern of L1316 and the Springers Hill mantle samples. B, Ranges of the same elements as in A for peridotites of the Vulcan (Johnson et al., 1990) and Kane (Ross and Elthon, 1997; sample 92-98 excluded) fracture zones, and the Hess Deep (Dick and Natland, 1996). Also shown are the ranges of the averages of several mantle xenoliths from the Japanese arc (Abe et al., 1998) and several from the Kamchatka arc (Kepezhinskas et al., 1996). Note the relatively enriched nature of the arc xenoliths compared to mantle exposed in forearcs. C1 chondrite data are from Sun and McDonough (1989).

Springers Hill mantle, Western Lewis Hills mantle, and exotic harzburgites. Although all olivine-spinel pairs fall into the mantle array of Arai (1994), the expected weak increase of the Mg/(Mg + Fe) ratio in olivine with increasing Cr# is not observed (Fig. 3B). The depletion trend for Al_2O_3 is not reflected in the TiO_2 content of spinel (Fig. 3C). This result is presumably related to a larger spinel/melt distribution coefficient for Ti in Cr-rich versus Al-rich spinels (Suhr and Robinson, 1994).

REEs, Ti, Zr, AND Sr IN CLINOPYROXENE

Trace element concentrations were determined by ion microprobe at the Yaroslavl Institute for Microelectronics by S. Simakin, following the methods described in Batanova et al. (1998) and Suhr (1999). In Figure 4A, the trace element distribution in clinopyroxene is shown for four samples from the Springers Hill mantle, one exotic, accreted harzburgite, one harzburgite from the Western Lewis Hills mantle, and three from the Wheelers Brook mantle (Table 1, sample locations in Fig. 2). The patterns readily fall into two groups on the basis of the heavy rare earth element (HREE) distribution, and these also correlate with the mineral chemistry (Fig. 3D). The concentrations of the middle rare earth elements (MREEs), HREEs, and Ti of the Wheelers Brook mantle are nearly one order of magnitude higher than in the Springers Hill clinopyroxenes. For the light rare earth elements (LREEs), both groups converge. Some Springers Hill and Western Lewis Hills mantle rocks display even higher LREE concentrations than Wheelers Brook mantle rocks. All patterns show negative Ti and Zr anomalies, but only non–Wheelers Brook mantle samples display a positive Sr anomaly. What is most important, the pattern from the Western Lewis Hills mantle shows the same characteristics as the Springers Hill mantle, i.e., very low HREEs, relatively high LREEs, and a positive Sr anomaly. In summary, mineral major and trace element data provide a strong link between peridotites from the Western Lewis Hills mantle and Springers Hill mantle.

Compared to other mantle-rock clinopyroxene data, the Springers Hill clinopyroxenes plot at the very depleted end of the spectrum (Fig. 4B). Both arc and abyssal peridotites (Johnson et al., 1990; Ross and Elthon, 1997; Abe et al., 1998; Kepezhinskas et al., 1996) are significantly more enriched. Only the puzzling Hess Deep peridotites (Dick and Natland, 1996) and, in particular, the peridotites now exposed in the Mariana and Bonin forearc (Parkinson et al., 1992; Parkinson and Pearce, 1998) plot at a similar level of depletion as the Springers Hill mantle rocks.

MODELING

The Wheelers Brook mantle can be modeled as 7% to 10% near-fractional melting of a typical MORB-source mantle (Suhr and Batanova, 1998). This value is among the least depleted for the spectrum of mantle rocks in the Bay of Islands Complex (Batanova et al., 1998) and in the mid range for the mid-ocean ridge system (Johnson et al., 1990). The rather low degree of depletion with respect to the Bay of Islands Complex reference frame may reflect thermal conditions in the vicinity of a transform fault (Suhr and Batanova, 1998). Characteristic for the Wheelers Brook mantle is a good fit to a simple melting model, i.e., there is no obvious need to call for mixing and reaction with migrating melts or fluids.

In contrast, a simple melting model cannot be used to explain the Springers Hill harzburgites. Qualitatively, (1) the strong HREE depletion in the clinopyroxene argues for very high degrees of de-

TABLE 1. MAJOR AND TRACE ELEMENT CHEMISTRY OF CLINOPYROXENES

Sample	Unit*	N†	SiO_2 (%)	TiO_2 (%)	Al_2O_3 (%)	Cr_2O_3 (%)	FeO (%)	MgO (%)	MnO (%)	CaO (%)	NiO (%)	Na_2O (%)	SUM (%)	Mg#
L22	SHM	2	54.59	0.03	0.89	0.62	2.02	18.79	0.09	23.34	0.05	0.13	100.51	94.3
L49	SHM	9	54.80	0.01	0.91	0.51	1.83	18.84	0.06	23.04	0.06	0.13	100.20	94.8
L73	SHM	4	54.55	0.01	0.85	0.61	1.79	18.47	0.07	23.84	0.01	0.10	100.30	94.8
L84	SHM	4	55.48	0.02	0.54	0.42	1.82	18.18	0.05	24.01	0.04	0.19	100.74	94.7
L122	exotic	8	53.73	0.00	0.99	0.41	1.82	18.51	0.08	23.88	0.05	0.04	99.59	94.8
L1316	WLM	2	53.67	0.02	1.11	0.44	1.63	17.50		24.80	0.08	0.11	99.42	95.0
L110	WBM	9	50.76	0.00	5.54	1.11	2.68	15.93	0.10	22.22	0.04	0.52	99.17	91.4
L113	WBM	2	51.89	0.00	3.49	0.65	2.14	16.83	0.08	23.70	0.03	0.33	99.54	93.3
L119	WBM	1	51.48	0.00	4.75	1.17	2.09	16.02	0.06	22.96	0.04	0.75	99.64	93.2

Sample	Unit*	N†		La (ppm)	Ce (ppm)	Sr (ppm)	Nd (ppm)	Zr (ppm)	Sm (ppm)	Eu (ppm)	Ti (ppm)	Dy (ppm)	Er (ppm)	Yb (ppm)
L49	SHM	3	avg†	0.0131	0.0349	2.9139	0.0656	0.1548	0.0336	0.0163	63	0.0955	0.0877	0.1469
			std†	0.0077	0.0077	0.4461	0.0141	0.0759	0.0093	0.0033	8	0.0155	0.0186	0.0219
L73	SHM	3	avg†	0.0096	0.047	1.8281	0.1049	0.1679	0.0741	0.0264	150	0.1785	0.13	0.1835
			std†	0.0022	0.0065	0.2088	0.0246	0.0084	0.0101	0.0026	9	0.0158	0.0218	0.0181
L122	exotic	3	avg†	0.0082	0.0165	2.774		0.0599		0.0033	30	0.0291	0.0703	0.1062
			std†	0.0007	0.0008	1.6281		0.0145		0	0	0.0059	0.0071	0.0249
L84	SHM	2	avg†	0.0187	0.0288	4.5448	0.0475	0.1365	0.0385	0.0197	172	0.0934	0.0947	0.0892
			std†	0.0062	0.0024	0.0773	0.0015	0.0572	0.0121	0.0045	19	0.0212	0.0204	0.0059
L22	SHM	3	avg†	0.0705	0.144	6.7728	0.1608	0.2629	0.1004	0.0443	105	0.2	0.1515	0.119
			std†	0.0175	0.0242	0.2363	0.0089	0.0171	0.0145	0.0143	3	0.0282	0.0734	0.0098
L1316	WLM	3	avg†	0.0556	0.1478	2.7472	0.079	0.4162	0.0476	0.0272	127	0.1609	0.1247	0.1397
			std†	0.0112	0.0385	0.3295	0.017	0.1174	0.0103	0.0108	19	0.0161	0.0381	0.0029
Mineral/melt distribution coefficients§														
Olivine				7×10^{-6}	1×10^{-5}	3×10^{-5}	7×10^{-5}	0.004	0.0007	0.001	0.015	0.004	0.009	0.014
Clinopyroxene				0.06	0.1	0.128	0.2	0.123	0.3	0.37	0.25	0.44	0.43	0.41
Orthopyroxene				0.0025	0.005	0.007	0.01	0.024	0.02	0.03	0.1	0.05	0.07	0.09
Spinel				0.0006	0.0006	0.0006	0.0006	0.07	0.0006	0.0006	0.15	0.0015	0.003	0.0045

* SHM—Springers Hill mantle; WLM—Western Lewis Hills mantle; WBM—Wheelers Brook mantle.
† *N*—number of analyses; avg—average; std—standard deviation.
§ From Suhr et al. (1998) except Ti between cpx and melt, which is adopted from Ozawa and Shimizu (1995).

pletion, (2) the lack of significant LREE depletion (and possibly the Sr enrichment) favors an enrichment event, and (3) the negative Zr and Ti anomalies favor a model in which the enrichment event did not include the high field strength elements Zr and Ti (cf. Ozawa and Shimizu, 1995). These criteria would typically indicate the involvement of a fluid as opposed to a melt (Pearce and Peate, 1995). Alternatively, the Zr-Ti anomalies arose during a late chromatographic overprint of a solid dominated by olivine and orthopyroxene (Kelemen et al., 1990).

We present first a single-stage, open-system melting model to explain the clinopyroxene trace element data. This single-stage model is remarkably successful in reproducing most features of the Springers Hill clinopyroxene chemistry, and it may therefore capture the chemically most relevant period. It fails, however, to account for negative Ti anomalies found in the Springers Hill samples. We therefore also present a multistage model. We demonstrate that the negative Ti anomalies may be generated by a late chromatographic overprint. This model also involves the melting of an already depleted residue. Both models are unable to explain the positive Sr anomalies present in the Springers Hill clinopyroxenes.

Single-Stage Model

This model simulates nonmodal melting (e.g., Johnson et al., 1990) of a continuously fluxed mantle above a slab. The protolith is a primitive, 2.5 times chondritic mantle with 15% to 18% initial clinopyroxene, and it is melted until the clinopyroxene-out condition is reached (see caption to Fig. 5 for exact modes). We use the unfractionated trace element concentration of the "global subducting sediment," or GLOSS (Plank and Langmuir, 1998), as composition of the flux. Such a flux, presumably carried as a fluid or melt, accounts for enhanced melting (compared to abyssal peridotites) due to the addition of water present in the fluid or melt (Hirose and Kawamoto, 1995; Gaetani and Grove, 1998). It also carries slab-derived incompatible trace elements that can explain the lack of LREE depletion. GLOSS has $LREE_N/HREE_N \approx 10$ (subscript N = C1 chondrite–normalized), $HREE_N \approx 20$, and weak negative Zr_N and Ti_N anomalies. The fluxing is modeled by the open-system melting model of Ozawa and Shimizu (1995). In this model, there is a constant ratio β between mass of melt added to mass of melt extracted (melt-fluxing rate). Melting is fractional beyond a retained melt fraction α [$\alpha = \phi/(1 - \phi)$, ϕ = melt fraction]. Because of fluxing, a trace element buffering is obtained, starting at lower degrees of melting with the lightest REEs (Fig. 5A). Relatively low values of β are sufficient to buffer the LREEs, whereas much higher fluxing rates are necessary to affect the HREEs and Ti. HREEs and Ti are thus excellent indicators of the degree of depletion (Ozawa and Shimizu, 1995). About 22%–26% melting is required to explain the observed HREE depletion in Springers Hill harzburgites (Fig. 5B). The fluxing rate is varied from 0.006 to 0.03. Multiplied with the extent of melting (0.26), the mass of sediment added to the residue (via a fluid or melt), relative to the original solid, is obtained, and it is <1%.

The main problem with this model is to explain the low Ti values in clinopyroxene that are coupled with high Sm and Eu concentrations. In the modeling, the samples do not develop a negative Ti anomaly since Sm and Eu are still governed by the melting term

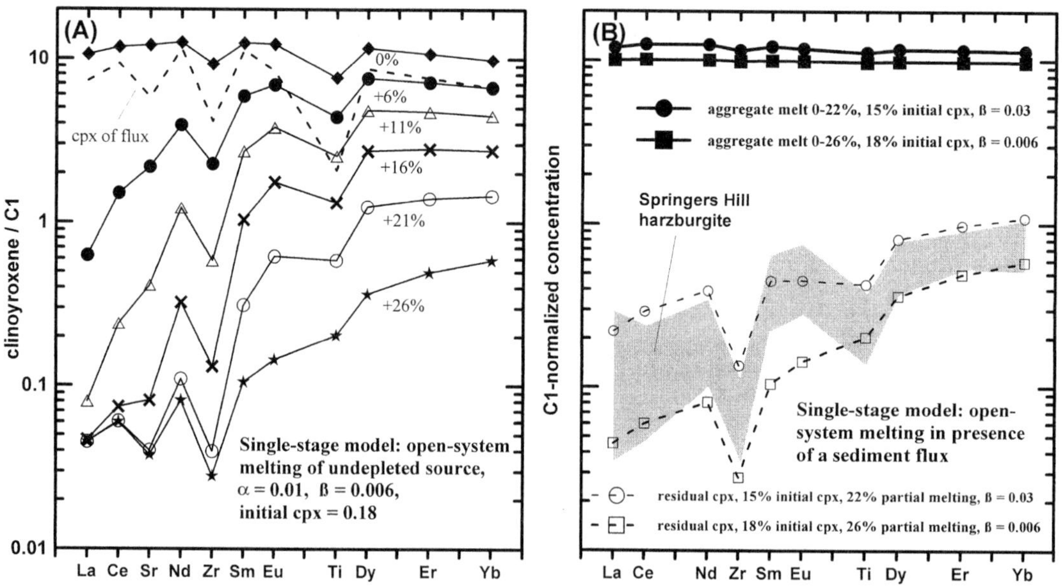

Figure 5. A, Progressive open-system melting (Ozawa and Shimizu, 1995) of a 2.5 times chondritic peridotite with initial modal fractions of 0.55 olivine, 0.18 clinopyroxene, 0.25 orthopyroxene, and 0.02 spinel. Melt modes are −0.28 olivine, 0.68 clinopyroxene, 0.54 orthopyroxene, 0.06 spinel. The retained melt fraction is 1%; the influx rate β is 0.006. The flux (GLOSS, plotted as equilibrium clinopyroxene) is from Plank and Langmuir (1998). B, Range of Springers Hill clinopyroxenes compared to modeled clinopyroxene in the single-stage model. More-depleted clinopyroxene is a residue of 26% melting, initial mode of clinopyroxene 18%, $\beta = 0.006$; less-depleted clinopyroxene is residue of 22% melting, initial mode of clinopyroxene is 15%, $\beta = 0.03$. Fit is good except for low modeled Sm and Eu. Sr could not be fitted and is omitted for clarity. Aggregated melt compositions from the models are also shown.

and not by the flux term. Increasing the fluxing rate can separate Sm and Eu from Ti, but it is not a satisfactory solution since, with the given composition of GLOSS, the LREE concentrations become too high. Another problem is that the aggregated melts from the melting event have HREE concentrations of >10 times C1 chondrites, whereas known basaltic melts from the Little Port Complex have HREE concentrations from 3 to 10 times C1 chondrites (Jenner et al., 1991). Therefore, an already somewhat depleted peridotitic source was possibly involved.

Multistage Model

This more complex model involves two additional features compared to the single-stage model: a late chromatographic overprint of the harzburgitic residue and an early depletion event. Conceptually, this model reflects the following three stages: (1) fractional melting with a retained porosity of 0.1% ("critical melting") of a largely primitive mantle peridotite in a ridge setting, (2) open-system melting of the depleted residue of stage 1 in the mantle wedge above a slab, (3) chromatographic exchange between the residue of stage 2 and a pooled melt from stage 2, probably occurring during the late exhumation history of the Springers Hill mantle (Fig. 6; Table 2). Type II orthopyroxenites present on Springers Hill are geologic evidence for an overprint by a late melt and fluid migration.

The relative extents of melting between stages 1 and 2 are poorly determined with our data, but they could potentially be determined with melt compositions from the western Lewis Hills and the degree of melting required to generate them. In the absence of better constraints, we use the Wheelers Brook mantle as the residue after stage 1, but any moderately depleted abyssal peridotite would suffice. The Wheelers Brook mantle can be modeled as 9% nonmodal, critical melting of an undepleted, 2.5 times chondrite source. Critical melting

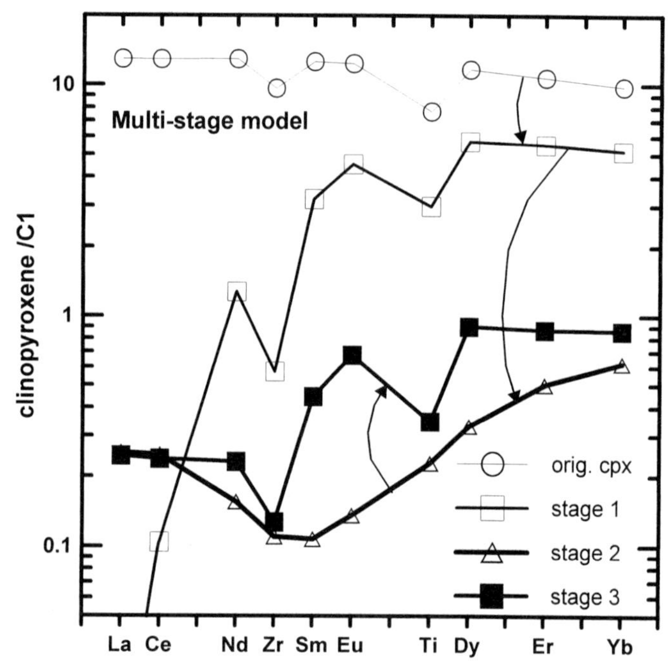

Figure 6. Proposed evolution of clinopyroxene trace element pattern in the Springers Hill mantle in the multistage model. Stage 1 reflects 9% critical melting (Sobolev and Shimizu, 1993) of a primitive-mantle source. Subsequently (stage 2), the residue from stage 1 undergoes 17% open-system melting (Ozawa and Shimizu, 1995) in an arc setting with a slab-melt flux (composition from Schiano et al., 1995). Finally (stage 3), the pooled melt of stage 2 interacts chromatographically with the residue of stage 2 and causes MREE and some HREE enrichment (see Fig. 7).

TABLE 2. INPUT PARAMETERS FOR MULTISTAGE MODEL

	Olivine	Cpx	Opx	Spinel
Start mode, stage 1	0.55	0.18	0.25	0.02
Melt mode, stage 1*	−0.3	0.82	0.42	0.06
Start mode, stage 2	0.634	0.117	0.233	0.016
Melt mode, stage 2†	−0.28	0.68	0.54	0.06
Start mode, stage 3	0.821	0.001	0.17	0.007
Grain size, stage 3 (cm)	0.3	0.02	0.3	0.2
Stage 1 (critical melting, Sobolev and Shimizu, 1993)				
Retained, critical porosity	0.001			
Stage 2 (open-system melting, Ozawa and Shimizu, 1995)				
Melt-fluxing rate β	0.02 or 0.004			
Retained porosity	0.01			
Stage 3 (numerical calculation)				
Percolation velocity	10 cm/yr			
Porosity	0.01			
Diffusion coefficient (solid)	10^{-3} $cm^2 \cdot s^{-1}$			
Density of melt; solid	2.8; 3.3 $g \cdot cm^{-3}$			
Total time	2000 yr			

* Kinzler and Grove (1992).
† Following trend for melting coefficients from anhydrous to hydrous (Gitani and Grove, 1998).

is fractional melting in the presence of a retained porosity (Sobolev and Shimizu, 1993), here taken as 0.1%.

The second stage is similar to the single-stage model, but the protolith is the residual mantle from stage 1. For the fluxing agent, we use a slab melt as obtained by direct measurement of melt inclusions in arc mantle xenoliths from the Luzon arc, Philippines (Schiano et al., 1995). This slab melt has $LREE_N/HREE_N$ ratios of ~100, its $HREE_N$ concentrations are 1–3, and it has no negative Zr_N and Ti_N anomalies. For 17% melting (therefore 26% total melting), fluxing rates β of 0.004 to 0.02 are suggested.

The third stage is an interaction between the residue of stage 2 and the linearly aggregated (pooled) melt of stage 2 and is modeled by using the physical parameters given in Table 2. Even though melt and residue of stage 2 are separated after each melting increment in stage 2, equilibrium between the final residue and the pooled melt is present for the LREEs because both La and Ce are dominated by the fluxing melt. For the HREEs, Ti, and Zr, the residue is moderately buffered by the presence of orthopyroxene, and chromatographic exchange is only noticeable for more enhanced interaction. The most pronounced and immediate effect of the chromatographic exchange is therefore in the concentrations of the MREEs (Fig. 7).

The multistage model contains robust and speculative components. The presence of an early depletion event (stage 1) is speculative, but the aforementioned HREE-depleted nature of the Little Port Complex mafic lavas can support such an interpretation. A two-stage melting history was inferred for the chemically similar Mariana-Bonin forearc mantle peridotites (Pearce et al., 1992). High degrees of melting totaling 26% are derived from the extreme HREE and Ti depletion in the clinopyroxene data. They are a robust feature because they correlate with other mineral-chemistry indicators such as very low Al_2O_3 in the pyroxenes, a high Cr# in spinel, and the scarcity of clinopyroxene. HREE concentrations are not readily overprinted during either open-system melting (Ozawa and Shimizu, 1995) or chromatographic exchange. Finally, we infer that a late fluid or melt chromatographic overprint is likely because of the difficulty in generating negative Ti anomalies in another way and because of the presence of the type II orthopyroxenites and scattered amphiboles. Specific assumptions about melt compositions, relative contributions of the two melting events, and numerical data of the chromatographic event are subject to debate. They definitely need to be better determined in the future.

Equilibrium Melts

Another way to evaluate the mantle rocks is to calculate the final melts they equilibrated with by using mineral/melt distribution coefficients (Fig. 8). Such equilibrium melts of Springers Hill harzburgites are depleted with respect to boninites, but enriched with respect to equilibrium melts of type I orthopyroxenite dikes occurring on Springers Hill (Edwards, 1995). The extremely depleted nature of these dikes suggests that even more depleted residual mantle rocks must be present below the Springers Hill mantle. Exotic peridotite sample L122 (Fig. 4) is evidence that such even more refractory mantle peridotites exist somewhere along the Little Port Complex. The amount of 26% total melting derived from the Springers Hill mantle may thus be a conservative estimate in the wider geologic context.

DISCUSSION

There appears to be little doubt that within the Lewis Hills, two different mantle domains are juxtaposed. Lithology and structure of

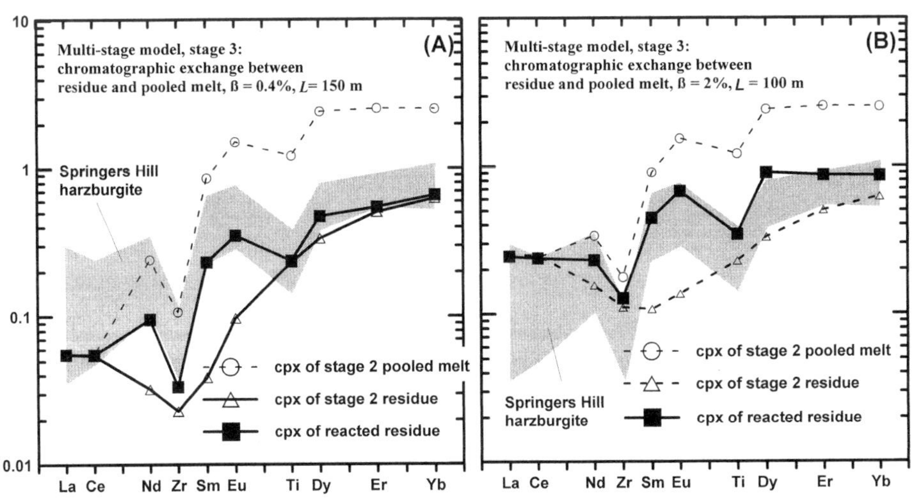

Figure 7. Chromatographic exchange between a residue of 9% + 17% open-system melting and the pooled melt of the 17% melting interval in stage 3 of the multistage model (numerical calculation). The range of Springers Hill harzburgites (shaded) is covered by a lower bound (in A) and an upper bound (in B). The flux is a slab melt (composition from Schiano et al., 1995). Note the development of negative Zr and Ti anomalies in the modified residue after chromatographic exchange. Chromatographic column parameters: time = 2000 yr; melt velocity = 10 cm/yr; column length L = 100–150 m (see Table 2).

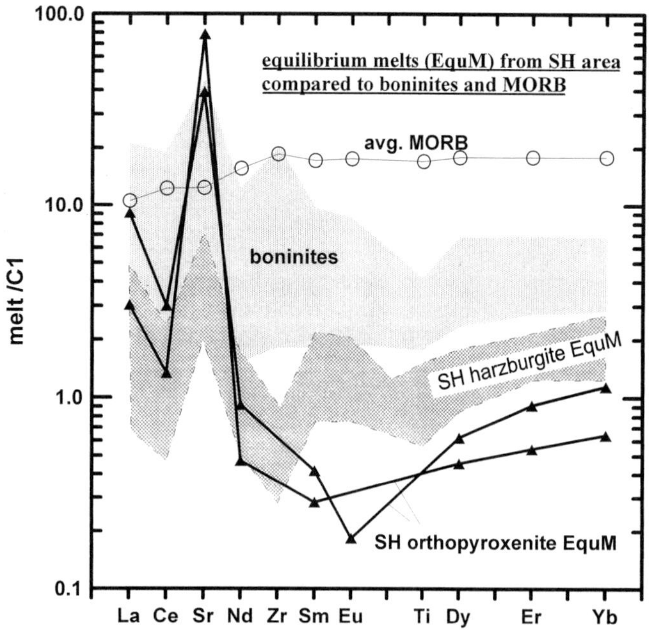

Figure 8. Trace element abundance patterns in a range of boninites (Cameron et al., 1983) compared to average mid-oceanic-ridge basalt (MORB; Sun and McDonough, 1989) and melts calculated to be in equilibrium with minerals from the Springers Hill (SH) area by using distribution coefficients as given in Table 1. Equilibrium melts of Springers Hill harzburgites are depleted with respect to boninites but enriched with respect to equilibrium melts of type I orthopyroxenites (from Edwards, 1995). Note that the extreme depletion of type I orthopyroxenites argues for the presence of a source even more depleted than the Springers Hill harzburgites.

the Wheelers Brook mantle are akin to the mantle exposed in the three northern massifs of the Bay of Islands Complex. The Wheelers Brook mantle records a strong flow history reflecting ductile mantle flow under a ridge near a fracture zone. The reduced amount of depletion relative to mantle rocks elsewhere in the Bay of Islands Complex (Batanova et al., 1998) has been related to the thermal cold-wall effect adjacent to a fracture zone (Suhr and Batanova, 1998). In contrast, the Springers Hill mantle shows heterogeneous and mostly low strain, brittle behavior, and intrusion, fractionation, and solidification of melt in the mantle, all capped by a ductile normal fault. Suhr and Cawood (1999) have interpreted the Springers Hill harzburgite as mantle that was originally located on the western side of the Mount Barren Complex fracture zone and that was exhumed during finite westward propagation of the Bay of Islands Complex ridge. The remarkable feature is that this mantle is extremely depleted; it is, therefore, inconsistent with being either a simple, offset Bay of Islands Complex–type mantle or a peridotite that was less depleted owing to a cold-wall effect.

Given (1) the isotopic evidence for an arc-origin of the Western Lewis Hills mantle (Kurth et al., 1998), (2) the similarity of the arc-hosted Western Lewis Hills mantle with the Springers Hill mantle, and (3) the rather unique trace element patterns in clinopyroxene involving extreme degrees of depletion coupled with LREE enrichment, we next briefly evaluate three possible interpretations for the origin of the Springers Hill mantle.

1. The Springers Hill harzburgite represents mantle onto which the 500–505 Ma Little Port Complex arc was built. There is, however, no evidence that the Little Port Complex arc was built onto an existing oceanic basement. Regional studies in the Little Port Complex and western Lewis Hills led Karson (1984) to conclude that the Little Port Complex is a deformed ophiolite with local low-strain inliers preserving sheeted dikes. This conclusion argues for an extensional environment. The crustal thickness cannot be unusually high, given that mantle rocks are exposed in the western Lewis Hills and farther north in the Little Port Complex (Suhr, personal observation). In summary, we do not see evidence that the Little Port Complex arc was built onto an existing oceanic basement.

2. The Springers Hill mantle represents reacted MORB or conventional arc-type mantle that got transformed into a very depleted mantle by the percolation of refractory melts. This interpretation is supported by the known occurrence of suitable refractory melts within the Springers Hill mantle, i.e., boninitic orthopyroxenite dikes (Edwards, 1995). It is clear that in this interpretation, a highly refractory mantle source is called for in order to generate the boninitic parental melts. The Springers Hill mantle is such a potential source. Therefore, in principle, not much is gained by calling for a pervasive melt and rock reaction model since a Springers Hill type mantle is required at depth anyway. The model can, however, elegantly explain the shallow occurrence of extremely depleted mantle without requiring upwelling and extension, i.e., it is fitting for a conventional arc setting. However, we have just argued that extension is rather obvious in the Little Port Complex such that upwelling of a depleted mantle is not at variance with field evidence. In addition, there is a lack of documented relicts of mantle rocks that would represent the unreacted protoliths prior to the pervasive melt and rock reaction. In summary, the model involving melt and rock reaction cannot be ruled out, but we have found no good evidence to support it either.

3. The Springers Hill harzburgite formed during upwelling in a hot mantle wedge above a downgoing slab.

Model 3 is preferred and depicted in Figure 9A. It shows the upwelling of hot, fluxed mantle above an older, sinking and dehydrating slab (compares to stage 2 of modeling), as suggested for the early Mariana-Bonin arc system (Stern and Bloomer, 1992; Pearce et al., 1992; Bloomer et al., 1995; Parkinson and Pearce, 1998). Upwelling explains both the high degrees of depletion of the Springers Hill harzburgites due to decompression melting and fluxing from the slab and its shallow-level position in the Lewis Hills. Stern and Bloomer (1992) derived half–spreading rates of a few centimeters per year for an infant-arc setting, i.e., rates comparable to slow-spreading centers. The occurrence of intermediate-composition to felsic magmatic rocks, as reflected in the abundant Little Port Complex trondhjemites, is also known from the Mariana-Bonin forearc (Bloomer et al., 1995). At least in the early spreading history, upwelling would have involved shallow-mantle material that has undergone a previous period of depletion (compares to stage 1 of modeling). We are unable to state whether a more mature arc did ever develop above this subduction zone. If so, the onset of corner flow in the mantle wedge could only have affected the mantle located *below* the Springers Hill mantle (Fig. 9B). Otherwise, the Springers Hill mantle would have been convectively removed.

Following what was probably a hiatus in magmatism, the Western Lewis Hills mantle and Little Port Complex were juxtaposed with the Bay of Islands Complex spreading center along a fracture zone at 485 Ma during formation of the Bay of Islands Complex (Fig. 9C). We conclude that the forearc basement must have been rifted before 485 Ma. Renewed melting of the Springers Hill mantle at depth probably occurred at this stage (Kurth et al., 1998).

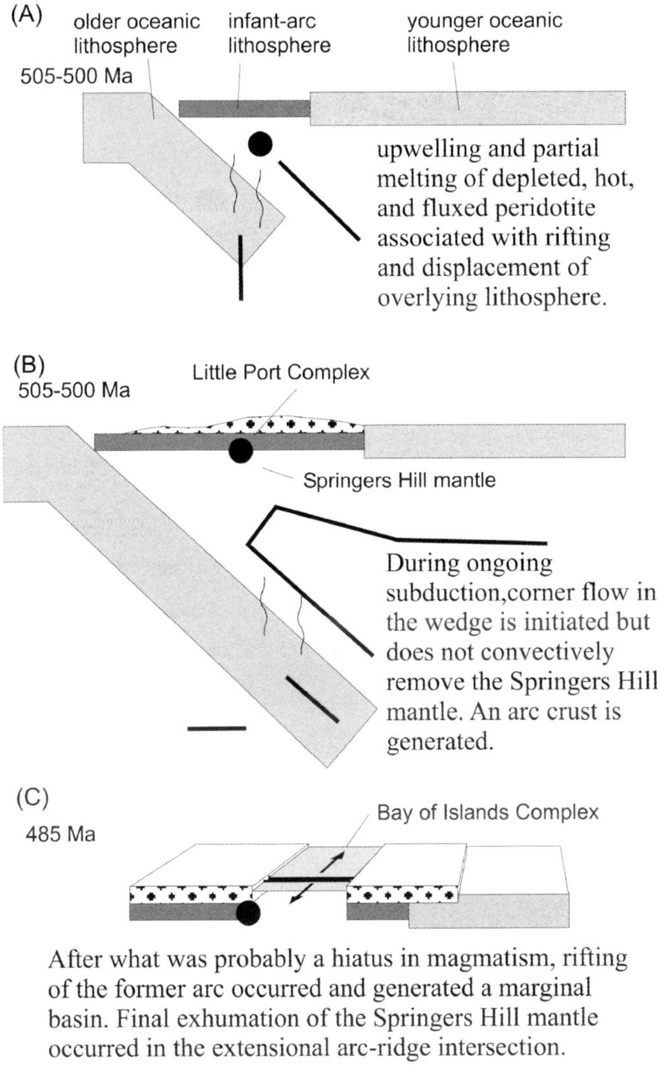

Figure 9. Model for the origin of the Springers Hill mantle, inspired by the infant-arc model of Stern and Bloomer (1992) and the Mariana-Bonin situation (Pearce et al., 1992). A, Upwelling of hot, formerly depleted, and fluxed mantle occurs during subduction-zone initiation. B, During a subsequent arc stage, corner flow is established. The Springers Hill mantle is located above the corner flow and remains at a shallow level. C, During a significantly later event, the arc is rifted along transtensional faults along which the Bay of Islands Complex spreading center forms. Along the transtensional arc-backarc boundary, final exhumation of the Springers Hill mantle occurs, coupled with reaction among melt, fluid, and rocks.

CONCLUSIONS

The Lewis Hills juxtaposes two remarkably different mantle sections. We relate the Springers Hill mantle to the arc-related Little Port Complex and use its petrologic characteristics to define the tectonic evolution of the arc. The lack of a basement for the Little Port Complex, the extremely HREE-depleted, but LREE-enriched mantle clinopyroxenes, and the nearly identical chemical characteristics of known forearc peridotites to the Springers Hill mantle support a primitive arc setting for the Western Lewis Hills peridotites and Little Port Complex. The final exhumation of the Springers Hill mantle occurred in a ridge-transform intersection setting. It was associated with melt and fluid movement that modified the chemical signature of the Springers Hill mantle.

ACKNOWLEDGMENTS

Sergei Simakin (Yaroslavl) is thanked for meticulous work on the ion probe. Don Elthon and Pavel Kepezhinskas are thanked for their constructive reviews. Yildirim Dilek's fine editorial handling is greatly appreciated. We wish to acknowledge generous financial support of this study by the Deutsche Forschungsgemeinschaft and the Max Planck Gesellschaft to G. Suhr and the School of Earth and Environmental Sciences to S.J. Edwards.

REFERENCES CITED

Abe, N., Arai, S., and Yurimoto, H., 1998, Geochemical characteristics of the uppermost mantle beneath the Japan island arcs: Implications for upper mantle evolution: Physics of the Earth and Planetary Interiors, v. 107, p. 233–248.

Arai, S., 1994, Characterization of spinel peridotites by olivine-spinel compositional relationships: Review and interpretation: Chemical Geology, v. 113, p. 191–204.

Batanova, V., Suhr, G., and Sobolev, A., 1998, Origin of geochemical heterogeneity in the mantle peridotites from the Bay of Islands ophiolite (Newfoundland, Canada): Ion probe study of clinopyroxenes: Geochimica et Cosmochimica Acta, v. 62, p. 853–866.

Bloomer, S.H., Taylor, B., MacLeod, C.J., Stern, R.J., Fryer, P., Hawkins, J.W., and Johnson, L., 1995, Early arc volcanism and the ophiolite problem: A perspective from drilling in the western Pacific, *in* Taylor, B., and Natland, J., eds., Active margins and marginal basins of the western Pacific: American Geophysical Union Geophysical Monograph 88, p. 1–30.

Cameron, W.E., McCulloch, M.T., and Walker, D.A., 1983, Boninite petrogenesis: Chemical and Nd-Sr isotopic constraints: Earth and Planetary Science Letters, v. 65, p. 75–89.

Cawood, P.A., and Suhr, G., 1992, Generation and obduction of ophiolites: Constraints from the Bay of Islands Complex, western Newfoundland: Tectonics, v. 11, p. 884–897.

Dahl, R., and Watkinson, D.H., 1986, Structural control of podiform chromitite in Bay of Islands ophiolite, Springer Hill area, Newfoundland: Current Research, Part B, Geological Survey of Canada, v. 86-1B, p. 757–766.

Davies, J.H., and Stevenson, D.J., 1992, Physical model of source region of subduction zone volcanics: Journal of Geophysical Research, v. 97, p. 2037–2070.

Dick, H.J.B., and Fisher, R.L., 1984, Mineralogical studies of the residues of mantle melting: Abyssal and alpine-type peridotites, *in* Kornprobst, J., ed., Kimberlite. II. The mantle and crust-mantle relationships: Amsterdam, Elsevier, p. 295–308.

Dick, H.J.B., and Natland, J.H., 1996, Late stage melt evolution and transport in the shallow mantle beneath the East Pacific Rise, *in* Mével, C., Gillis, K.M., Allan, J.F., and Meyer, P.S., eds., Proceedings of the Ocean Drilling Program, Scientific Results, v. 147: College Station, Texas, Ocean Drilling Program, p. 103–134.

Dunsworth, S., Calon, T., and Malpas, J., 1986, Structural and magmatic controls on the internal geometry of the plutonic complex and its chromite occurrences in the Bay of Islands ophiolite, Newfoundland: Current Research, Part A, Geological Survey of Canada, v. 86-1B, p. 471–482.

Edwards, S.J., 1990, Harzburgites and refractory melts in the Lewis Hills massif, Bay of Islands Ophiolite Complex: The base-metals and precious-metals story: Canadian Mineralogist, v. 28, p. 537–552.

Edwards, S.J., 1995, Boninitic and tholeiitic dikes in the Lewis Hills mantle section of the Bay of Islands ophiolite: Implications for magmatism adjacent to a fracture zone in a back-arc spreading environment: Canadian Journal of Earth Sciences, v. 32, p. 2128–2146.

Edwards, S.J., and Malpas, J., 1995, Multiple origins for mantle harzburgites: Examples from the Lewis Hills, Bay of Islands ophiolite, Newfoundland: Canadian Journal of Earth Sciences, v. 32, p. 1046–1057.

Fox, P.J., and Gallo, D.G., 1984, A tectonic model for ridge-transform-ridge plate boundaries: Implications for the structure of oceanic lithosphere: Tectonophysics, v. 104, p. 205–242.

Gaetani, G.A., and Grove, T.L., 1998, The influence of water on melting of mantle peridotite: Contributions to Mineralogy and Petrology, v. 131, p. 323–346.

Hirose, K., and Kawamoto, T., 1995, Hydrous partial melting of lherzolite at 1 GPa: The effect of H_2O on the genesis of basaltic magmas: Earth and Planetary Science Letters, v. 133, p. 463–473.

Iwamori, H., 1998, Transportation of H_2O and melting in subduction zones: Earth and Planetary Science Letters, v. 160, p. 65–80.

Jaques, A.L., and Green, D.H., 1980, Anhydrous melting of peridotite at 0–15Kb pressure and the genesis of tholeiitic basalts: Contributions to Mineralogy and Petrology, v. 73, p. 287–310.

Jenner, G.A., Dunning, G.R., Malpas, J., Brown, M., and Brace, T., 1991, Bay of Islands and Little Port Complexes, revisited: Age, geochemical and isotopic evidence confirm supra-subduction zone origin: Canadian Journal of Earth Sciences, v. 28, p. 1635–1652.

Johnson, K.T., Dick, H.J.B., and Shimizu, N., 1990, Melting in the oceanic upper mantle: An ion microprobe study of diopsides in abyssal peridotites: Journal of Geophysical Research, v. 95, p. 2661–2678.

Karson, J.A., 1984, Variations in structure and petrology in the Coastal Complex, Newfoundland: Anatomy of an oceanic fracture zone, *in* Lippard, S.J., Shelton, A.W., and Gass, I.G., eds., Ophiolites and oceanic lithosphere: Geological Society of London Special Publication 13, p. 131–144.

Karson, J., and Dewey, J.F., 1978, Coastal Complex, western Newfoundland: An early Ordovician oceanic fracture zone: Geological Society of America Bulletin, v. 89, p. 1037–1049.

Kelemen, P.B., Johnson, K.T.M., Kinzler, R.J., and Irving, A.J., 1990, High-field strength element depletions in arc basalts due to mantle-magma interaction: Nature, v. 345, p. 521–524.

Kelemen, P.B., Whitehead, J.A., Aharonov, E., and Jordahl, K.A., 1995, Experiments on flow focusing in soluble porous media, with applications to melt extraction from the mantle: Journal of Geophysical Research, v. 100, p. 475–496.

Kepezhinskas, P., Defant, M.J., and Drummond, M.S., 1996, Progressive enrichment of island arc mantle by melt-peridotite interaction inferred from Kamchatka xenoliths: Geochemica et Cosmochimica Acta, v. 60, p. 1217–1229.

Kinzler, R.J., and Grove, T.L., 1992, Primary magmas of mid-ocean ridge basalts. 1. Experiments and methods: Journal of Geophysical Research, v. 97, p. 6885–6906.

Kurth, M., Sassen, A., Suhr, G., and Mezger, K., 1998, Precise ages and isotopic constraints for the Lewis Hills (Bay of Islands ophiolite): Preservation of an arc-spreading ridge intersection: Geology, v. 26, p. 1127–1130.

Ozawa, K., and Shimizu, N., 1995, Open-system melting in the upper mantle: Constraints from the Hayachine-Miyamori ophiolite, northeastern Japan: Journal of Geophysical Research, v. 100, p. 22315–22335.

Parkinson, I.J., and Pearce, J.A., 1998, Peridotites from the Izu-Bonin-Mariana forearc (ODP Leg 125): Evidence for mantle melting and melt-mantle interaction in a supra-subduction zone setting: Journal of Petrology, v. 39, p. 1577–1618.

Parkinson, I.J., Pearce, J.A., Thirlwall, M.F., Johnson, K.T.M., and Ingram, G., 1992. Trace element geochemistry of peridotites from the Izu-Bonin-Mariana forearc, Leg 125, *in* Fryer, P., Pearce, J.A., and Stokking, L.B., et al., eds., Proceedings of the Ocean Drilling Program, Scientific Results, v. 125: College Station, Texas, Ocean Drilling Program, p. 487–506.

Pearce, J.A., and Peate, D.W., 1995, Tectonic implications of the composition of volcanic arc magmas: Annual Review of the Earth and Planetary Sciences, v. 23, p. 251–285.

Pearce, J.A., van der Laan, S.R., Arculus, R.J., Murton, B.J., Ishii, T., Peate, D.W., and Parkinson, I.J., 1992, Boninite and harzburgite from Leg 125 (Bonin-Mariana Forearc): A case study of magma genesis during the initial stages of subduction, *in* Fryer, P., Pearce, J.A., Stokking, L.B., et al., eds., Proceedings of the Ocean Drilling Program, Scientific Results, v. 125, College Station, Texas, Ocean Drilling Program, p. 623–659.

Plank, T., and Langmuir, C.H., 1988, An evaluation of the global variations in the major element chemistry of arc basalts: Earth and Planetary Science Letters, v. 90, p. 349–370.

Plank, T., and Langmuir, C.H., 1998, The chemical composition of subducting sediment and its consequences for the crust and mantle: Chemical Geology v. 145, p. 325–394.

Quick, J.E., and Gregory, R.T., 1995, Significance of melt-wall rock reaction: A comparative anatomy of three ophiolites: Journal of Geology, v. 103, p. 187–198.

Ross, K., and Elthon, D., 1997, Extreme incompatible trace-element depletion of diopside in residual mantle from south of the Kane fracture zone, *in* Karson, J.A., Cannat, M., Miller, D.J., and Elthon, D., eds., Proceedings of the Ocean Drilling Program, Scientific Results, v. 153: College Station, Texas, Ocean Drilling Program, p. 277–284.

Schiano, P., Clocchiatti, R., Shimizu, N., Maury, R.C., Jochum, K.P., and Hofmann, A.W., 1995, Hydrous, silica-rich melts in the sub-arc mantle and their relationship with erupted lavas: Nature, v. 377, p. 595–600.

Smith, C.H., 1958, Bay of Islands Igneous Complex, western Newfoundland: Geological Survey of Canada, Memoir 290, 129 p.

Sobolev, A.W., and Shimizu, N., 1993, Superdepleted melts and ocean mantle permeability: Transactions (Doklady) of the Russian Academy of Sciences, v. 328, p. 182–188.

Stern, R.J., and Bloomer, S.H.,1992, Subduction zone infancy: Examples from the Eocene Izu-Bonin-Mariana and Jurassic California arcs: Geological Society of America Bulletin, v. 104, p. 1621–1636.

Suhr, G., 1999, Significance of upper mantle hosted dunite bodies for melt migration and extraction under oceanic spreading centres: Inferences from reactive transport modeling: Journal of Petrology, v. 40, p. 575–599.

Suhr, G., and Batanova, V., 1998, Basal lherzolites in the Bay of Islands ophiolite: Origin by detachment-related telescoping of a ridge-parallel melting gradient: Terra Nova, v. 10, p. 1–5.

Suhr, G., and Cawood, P.A., 1993, Structural history of ophiolite obduction, Bay of Islands, Newfoundland: Geological Society of America Bulletin, v. 105, p. 399–410.

Suhr, G., and Cawood, P.A., 1999, The southeastern Lewis Hills (Bay of Islands ophiolite): Geology of a deeply eroded, inside-corner, ridge-transform intersection: Eos, Transactions of the American Geophysical Union, v. 80, p. S343.

Suhr, G., and Robinson, P.T., 1994, Origin of mineral chemical stratification in the mantle section of the Table Mountain massif (Bay of Islands ophiolite): Lithos, v. 31, p. 81–102.

Suhr, G., Calon, T., and Dunsworth, S.M., 1991, Origin of complex upper mantle structures in the southern Lewis Hills (Bay of Islands ophiolite, Newfoundland): Canadian Journal of Earth Sciences, v. 28, p. 774–787.

Suhr, G., Seck, H.A., Shimizu, N., Günther, D., and Jenner, G., 1998, Circulation of refractory melts in the lowermost oceanic crust: Evidence from a trace element study of dunite-hosted clinopyroxenes in the Bay of Islands ophiolite: Contributions to Mineralogy and Petrology, v. 131, p. 136–154.

Sun, S., and McDonough, W.F., 1989, Chemical and isotopic systematics of oceanic basalts: Implications for mantle composition and processes, *in* Saunders, A.D., and Norry, M.J., eds., Magmatism in the ocean basins, Geological Society of London Special Publication 42, p. 313–345.

Manuscript Accepted by the Society February 23, 2000

Printed in the U.S.A.

Geological Society of America
Special Paper 349
2000

Petrology, tectonics, and hydrothermal alteration of a fossil backarc oceanic crust: Solund-Stavfjord ophiolite complex of the western Norwegian Caledonides—A review

Harald Furnes
Geological Institute, University of Bergen, Allegaten 41, 5007 Bergen, Norway; e-mail: Harald.Furnes@geol.uib.no
Kjell Petter Skjerlie
Department of Geology, University of Tromsø, 9037 Tromsø, Norway; e-mail: Kjell.Skjerlie@ibg.uit.no
Yildirim Dilek
Department of Geology, Miami University, Oxford, Ohio 45056, USA; e-mail: Dileky@muohio.edu

ABSTRACT

The Late Ordovician Solund-Stavfjord ophiolite complex in southwest Norway developed in a backarc basin through at least two episodes of seafloor spreading. The youngest phase is a structural domain (domain 1) in oceanic crust that formed along a backarc rift system that opened at an intermediate spreading rate and propagated north-northeastward into preexisting oceanic crust (domain 2). These two domains are separated by a 1-km-wide zone of anomalous oceanic crust, with fracture-zone affinity (domain 3), that is composed of strongly sheared gabbros and metabasalt dikes hosting abundant serpentinite bodies. Domains 1 and 2 contain high-level gabbro, sheeted dikes, a transition zone of dikes and volcanic rocks, and a volcanic suite consisting of various proportions of pillow lavas, massive lava flows, and hyaloclastite breccias. The metabasalts of both domains have MORB (mid-oceanic-ridge basalt) composition, with minor evidence of an island-arc component, and those of domain 1 are dominantly Fe-Ti basalts. Their contents of FeO_t (total Fe as FeO), MgO, Al_2O_3, Cr, Ni, and Zr indicate that differences in depth of melting (30 to 12 kbar), crystal fractionation of olivine, plagioclase, and clinopyroxene at different levels in the mantle and crust (<8 kbar), and magma mixing controlled the composition of the metabasalts. The high proportion of massive lava flows, the lack of pervasive seafloor extensional tectonic deformation, and the hydrothermal history of the fossil oceanic crust in domain 1 are best comparable to the features of young oceanic crust of the Costa Rica Rift.

The Solund-Stavfjord ophiolite complex is conformably overlain by (1) a heterogeneous suite (Stavenes Group) of quartz-dominated metasediments hosting pillow lavas, volcaniclastic rocks, and minor intrusions of island-arc tholeiite–like composition with a clear subduction signature (the Heggøy Formation) and (2) the supposedly later calc-alkalic and alkalic rocks of the Hersvik and Smelvær units, respectively. This progressive change in chemical composition of the igneous rocks stratigraphically upward within the sedimentary cover indicates the development, through time, of a mature arc in the subduction-zone environment represented by the Solund-Stavfjord ophiolite complex. The mode and nature of magmatic, tectonic, and hydrothermal processes that operated during the evolution of the complex in a backarc basin are similar to those observed from modern mid-ocean ridge settings, particularly at intermediate-spreading ridge axes. The geologic features of the sedimentary cover of the Solund-Stavfjord ophiolite complex and the underlying Sunnfjord mélange suggest that the inferred Caledonian backarc basin evolved in proximity to a continental landmass, reminiscent of the tectonics of the modern Andaman Sea region.

INTRODUCTION

The global mid-ocean ridge system is responsible for nearly nine-tenths of the annual production of modern oceanic crust and thus contributes significantly to crustal evolution of the Earth (Hawkins, 1995). The seafloor-spreading rate, magma budget, and thermal regime along and beneath mid-ocean ridge axes strongly influence the morphotectonic evolution of oceanic crust (Phipps Morgan et al., 1994; Dilek et al., 1998), whereas the petrogenesis of basaltic rocks forming in this setting is controlled mainly by partial melting of a mantle source and subsequent modification by crystal-fractionation and magma-mixing processes. These processes collectively result in a wide range of chemical compositions of mid-ocean ridge basalts (e.g., Langmuir et al., 1992). A uniformitarian approach to plate tec-

Furnes, H., Skjerlie, K.P., and Dilek, Y., 2000, Petrology, tectonics, and hydrothermal alteration of a fossil backarc oceanic crust: Solund-Stavfjord ophiolite complex of the western Norwegian Caledonides—A review, *in* Dilek, Y., Moores, E.M., Elthon, D., and Nicolas, A., eds., Ophiolites and Oceanic Crust: New Insights from Field Studies and the Ocean Drilling Program: Boulder, Colorado, Geological Society of America Special Paper 349, p. 443–460.

tonics suggests that a great majority of oceanic crust in the geologic past should also have developed at mid-ocean ridge systems in the former ocean basins (e.g., Iapetus, Neotethys). However, many ophiolites, which are interpreted to represent the remnants of paleo–oceanic crust (e.g., Coleman, 1971; Moores and Vine, 1971), differ significantly from modern oceanic crust in that their crustal rock assemblages display geochemical and mineral fingerprints of a subduction-zone setting (e.g., Pearce et al., 1984). Both geochemical observations from ophiolites and detailed surveys from the convergent-margin settings in the southwest Pacific Ocean indicate that oceanic crust is also generated in supra-subduction zone environments, where mantle-derived melts produce arc, forearc, and backarc crust with highly heterogeneous chemical compositions (e.g., Taylor and Natland, 1995). Indeed, Bédard et al. (1998) recently showed that seafloor spreading can occur in a forearc region with magmas of boninitic composition.

Some of the well-developed backarc basins of the western Pacific Ocean have magnetic lineations that have been correlated with the geomagnetic time scale (i.e., Japan Sea, Coral Sea Basin, South China Basin, Woodlark Basin); therefore, the way in which oceanic crust forms in these basins may in fact be very similar to that occurring at mid-ocean ridges in major ocean basins. What is the nature of the magmatic, tectonic, and hydrothermal processes at backarc spreading centers, what is the interplay between these processes, and how do they differ from processes operating at mid-ocean ridges? The results of recent geologic and geophysical surveys and drilling in some of the modern backarc basins in the western Pacific Ocean by the Ocean Drilling Program (ODP) have provided us with new insights into the evolution of oceanic crust in backarc basins (Taylor and Natland, 1995, and references therein). For example, the evolution of the Lau Basin behind the Tonga Trench is recognized to have occurred first by crustal extension and rifting that created small subbasins partly filled with basaltic flows and sediments and then by opening via seafloor spreading through propagating rifts (Hawkins, 1995; Taylor et al., 1996). Indirect observations and limited sampling of the rock record in the Lau Basin make it difficult, however, to further define this two-stage evolution model and to determine and better understand the magmatic, tectonic, and hydrothermal processes responsible for the evolution of backarc oceanic crust.

Many ophiolites represent fossil oceanic crust with supra-subduction zone affinity; the Late Ordovician Solund-Stavfjord ophiolite complex in the western Norwegian Caledonides provides one example. Its paleobackarc tectonic setting is well determined owing to the characteristics of its sedimentary cover, geochemical signatures of its magmatic rocks, and its regional geology (Furnes et al., 1990). The results of our systematic interdisciplinary studies of this ophiolite complex are in accord with the tectonic model proposed for the evolution of the backarc-like Lau Basin and provide a four-dimensional view into oceanic-crust generation in a backarc tectonic setting. Thus, attention is focused on the spatial and temporal relationships among magmatism, faulting, and hydrothermal alteration, feedback systems between magma supply and seafloor spreading rate, and melt evolution and migration along a propagating rift in a backarc basin. In this paper, we present a review of the structure, petrology, and alteration history of the Late Ordovician backarc crust preserved in the Solund-Stavfjord ophiolite complex, discuss the mode and nature of magmatic and tectonic processes responsible for its characteristic features, and present a petrogenetic and tectonic model for its evolution that is internally coherent with its regional geology.

REGIONAL GEOLOGY

Ophiolitic and island-arc complexes of an Iapetus origin occur along the entire length of the Scandinavian Caledonides and can be subdivided into two different age groups. The oldest and most extensive group yields ages ranging from ca. 500 to 470 Ma (Dunning and Pedersen, 1988; Pedersen and Dunning, 1997). Among the younger magmatic complexes, only the Solund-Stavfjord ophiolite complex (dated to 443 $\pm$ 3 Ma; Dunning and Pedersen, 1988) can be classified as an ophiolite (Fig. 1); igneous complexes of the same age group farther to the north, e.g., Sulitjelma (Pedersen et al., 1991), represent continental rift–related intrusions, rather than true oceanic crust.

The Solund-Stavfjord ophiolite complex occurs in the Solund and Stavfjorden areas on the west coast of Norway and constitutes the uppermost allochthonous tectonic unit in a nappe package overlying Precambrian basement gneisses (Fig. 1A). The ophiolite complex rests tectonically on the Sunnfjord mélange, which consists of sedimentary and magmatic rocks and which tectonically overlies quartzites and schists, again resting on mangeritic gneisses of the Dalsfjord Suite (Fig. 1; Alsaker and Furnes, 1994). The presence of continentally derived detrital material in the sedimentary cover conformably overlying the ophiolite complex (the Stavenes Group; Fig. 1) further indicates that the basin in which the ophiolite evolved was proximal to a continental margin (Furnes et al., 1990).

INTERNAL STRUCTURE AND ARCHITECTURE

The Solund-Stavfjord ophiolite complex crops out in the area between Solund and Bremanger (Fig. 1A). On the basis of well-exposed structural and magmatic relationships on the island of Tviberg (Skjerlie and Furnes, 1990) (Fig. 1B), three structural domains are defined (Fig. 2A). The domains have different types of crustal architecture reflecting the mode and nature of magmatic and tectonic processes that operated during multistage seafloor spreading evolution of the ophiolite complex (Dilek et al., 1997).

Domain 1 has a north-northeast–trending structural grain defined by the consistent orientation of dike swarms on Tviberg and in the sheeted-dike complex of Oldra (Fig. 2, B and C, respectively). In the northernmost part on the island of Tviberg, domain 1 consists mainly of discrete dikes, whereas on the islands of Alden, Værlandet, and Oldra to the southwest (Fig. 1B), it contains an extensive volcanic sequence composed of pillow lava, pillow breccias, massive sheet flows, and fossil lava lakes stratigraphically overlying a well-developed sheeted-dike complex and high-level isotropic gabbros (Fig. 2C) (Furnes, 1972, 1974; Skjerlie et al., 1989; Furnes et al., 1992).

Domain 2 represents the oldest preserved crust in the domain-classified part of the ophiolite and contains a northwest-trending sheeted-dike complex and underlying isotropic gabbro, exposed on the island of Tviberg (Fig. 1B). Apart from the occurrence of domain 2 on the island of Tviberg, anticipated segments of domain 2 crop out on Lågøy and Strand in the south (Fig. 1B). In the Strand area, domain 2 is represented by a coherent section of volcanic rocks, sheeted dikes, and high-level gabbro (Hellevang, 1999).

In the central part of the island of Tviberg, where domains 1 and 2 approach each other, the sheeted dikes of domain 2 curve from a consistent northwest trend into an east-northeast trend, indicating a significant shift in the stress regime in the paleo–seafloor spreading system. This change in dike orientation from domain 1 to domain 2 occurs along and within a northeast-trending shear zone, which is composed mainly of anomalous oceanic crust with serpentinite breccias and fault-bounded serpentinite slivers in a deformed isotropic

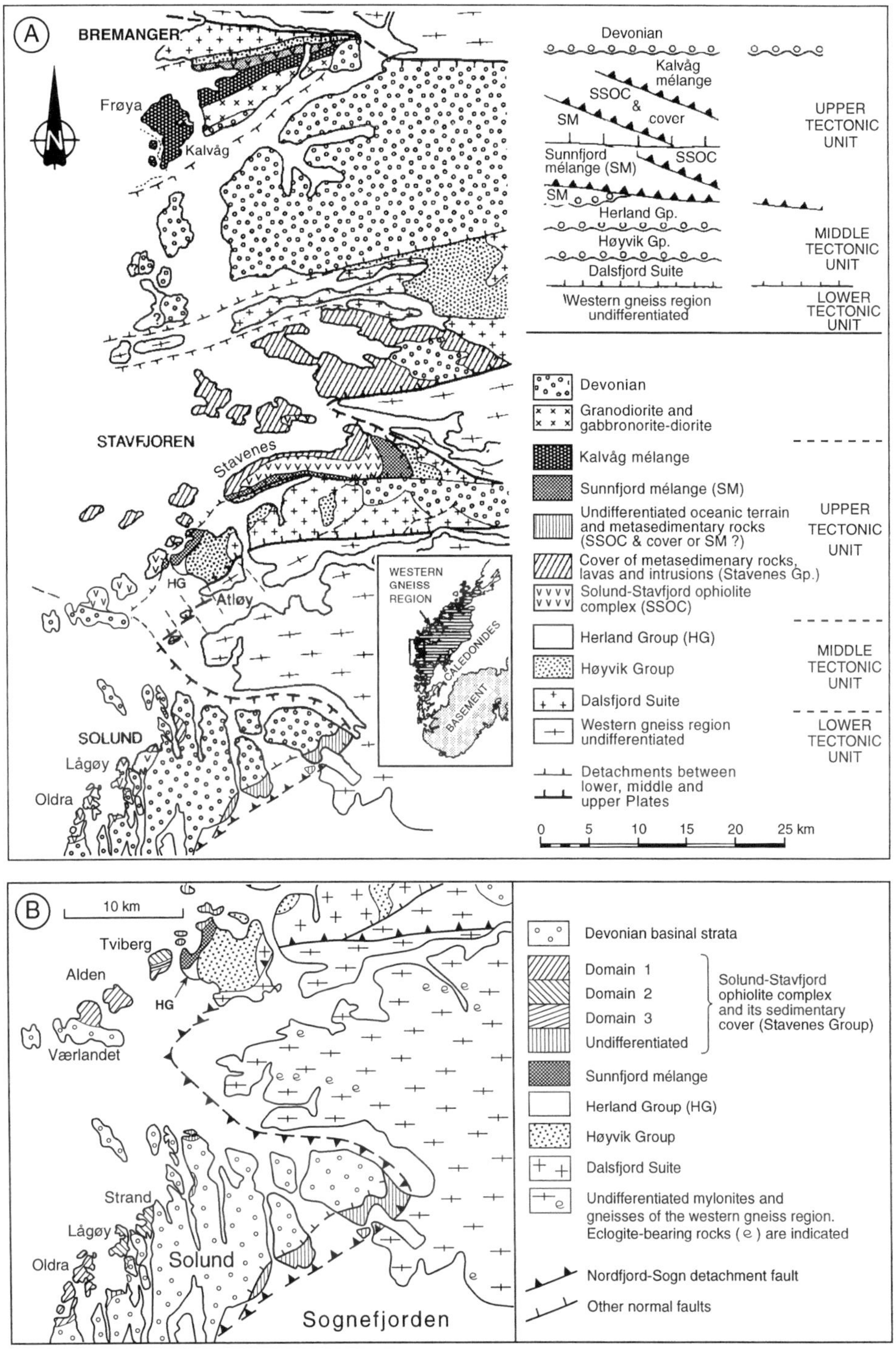

Figure 1. A, Simplified geologic map and tectonostratigraphy of the Solund-Bremanger area, compiled from Furnes et al. (1990), Hartz et al. (1994), Osmundsen and Andersen (1994), and Ravnås and Furnes (1995). The Solund-Stavfjord ophiolite complex (SSOC) is underlain by the Sunnfjord mélange (SM) and overlain by a thick sequence of volcanic and quartz-rich metasedimentary rocks of the Stavenes Group, which is divided into the Heggøy Formation and the Hersvik and Smelvær units (see Fig. 12 for stratigraphy). B, Simplified tectonic map of the Solund-Stavfjord ophiolite complex, modified from Dilek et al. (1997).

gabbro. This 1-km-wide shear zone (Fig. 2B) is interpreted as part of a fossil oceanic fracture zone (Skjerlie et al., 1989; Skjerlie and Furnes, 1990), which represents domain 3 in the Solund-Stavfjord ophiolite complex (Dilek et al., 1997). On the basis of the crosscutting relationships of these three structural domains, characteristic features of their volcanic sequence and internal architecture, and the geochemistry of the volcanic rocks and sheeted dikes, Dilek et al. (1997) proposed that the north-northeast–trending domain 1 represents the remnant of a rift system spreading at intermediate rates; this rift system propagated northeastward into preexisting oceanic crust developed along the northwest-trending doomed rift of domain 2 (Fig. 2A).

The ophiolitic lithostratigraphy in domains 1 and 2 and the undifferentiated part of the Solund-Stavfjord ophiolite complex in the Stavenes Peninsula (Fig. 1A) contain gabbro, sheeted dikes, volcanic rocks including pillow lavas, pillow breccias, and massive flows and are overlain by siliciclastic rocks. The state of preservation of magmatic and primary volcanic structures and of tectonic features of certain ocean-floor origin vary considerably from place to place. As previously stated, the three domains were defined on the island of

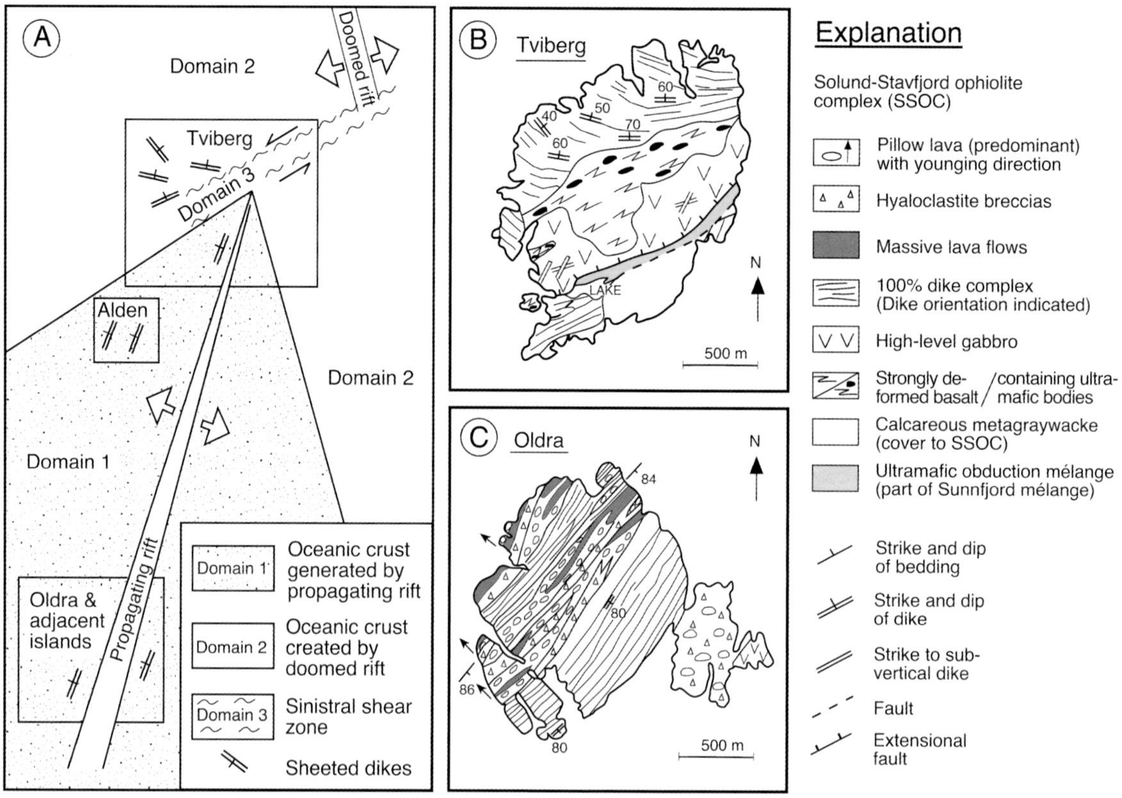

Figure 2. A, Reconstructed paleogeography of the Caledonian (Late Ordovician) marginal basin, as represented by the Solund-Stavfjord ophiolite complex. Modified from Dilek et al. (1997) and Furnes et al. (1998). B, Detailed geologic map of the island of Tviberg, on which the three domains have been defined. Modified from Skjerlie et al. (1989). C, Detailed geologic map of Oldra, defining part of domain 1. Modified from Dilek et al. (1997).

Tviberg, and domains 1 and 2 extend into the Solund area (Figs. 1 and 2). In general, the rocks between Tviberg and Oldra are little affected by Caledonian deformation, and hence, the following descriptions and illustrations of the volcanic rocks, the sheeted-dike complex, and the high-level plutonic rocks of the ophiolite complex are from domain 1 (Fig. 1B).

Domain 1

Extrusive sequence. The extrusive sequence consists of pillow lavas, hyaloclastite breccias, massive lava flows, and a fossil lava lake (Furnes 1972, 1974; Furnes et al., 1992). The cuspate shape of the pillow bodies (Fig. 3A) and well-defined drain-out structures in them (Fig. 3B) can be used effectively to determine the younging direction in the volcanic sequence. The pillows vary in size significantly from tube-like megapillows several meters long to small spherical bodies a few decimeters in diameter. Megapillows most commonly alternate with the massive flow units, marking the basal part of the pillow-lava layers. Pillows commonly decrease in size upward in a layer and may eventually grade into hyaloclastite breccias (Fig. 3C). The proportion of massive lava units to pillow lavas varies considerably along the axis of domain 1 (Fig. 4). The massive lava units and megapillows are most abundant near the top of the volcanic sequence, whereas pillow lavas and hyaloclastite breccias are more abundant at the lower stratigraphic levels. A fossil lava lake, consisting predominantly of massive lava units, occurs on the northwest part of Alden, in the northern part of domain 1 (Fig. 1B; Furnes et al., 1992).

Sheeted dikes. A mixed zone of dikes and screens of extrusive rocks represents a transition zone between the sheeted-dike complex and the extrusive sequence. The sheeted-dike complex (Fig. 3D) is dominated by parallel to subparallel regular, aphyric dikes. The majority of the dikes (65%) are nonsplit and range in thickness from 1.5 to 190 cm, averaging 71 cm (Ryttvad et al., 2000). Where younger dikes split older dikes, these intrusive relationships provide a way to study the evolution of magma through time (e.g., Robson and Cann, 1982). At least six magmatic intrusions, revealed by a series of dike-splittings, can be observed within the sheeted-dike complex. This sequence could represent six generations of magma input into the magma chamber or, alternatively, a continuous magmatic event if the system remained open for some time.

High-level plutonic rocks. On the basis of textural variations, two types of plutonic rocks are distinguished: fine- to medium-grained gabbro and variably textured gabbro (Fig. 3E). The aphyric, fine- to medium-grained gabbro is represented by irregular gabbro bodies intruded by 10–40-cm-wide regular, and 0.5–15-cm-wide irregular, mafic dikes. Locally, some of these dikes are rooted in fine-grained gabbro (Fig. 3F), defining a dike-gabbro transition zone, which is interpreted as a root zone of the sheeted-dike complex (Pedersen, 1986; Nicolas and Boudier, 1991; Skjerlie and Furnes, 1996). The variably textured gabbro includes medium-grained to pegmatitic gabbro with faint local lamination.

ANALYTICAL METHODS

Major and trace element analyses were performed on an X-ray fluorescence spectrometer (XRF) at the University of Bergen. The

Figure 3. Photographs of magmatic rocks from domain 1 of the Solund-Stavfjord ophiolite complex. A, Pillow lavas showing well-preserved primary shapes, chilled margins, and variolitic textures adjacent to the chilled zones. B, Pillow showing several drain-out structures, now filled with mainly epidote and quartz. C, Hyaloclastite breccia. D, Sheeted-dike complex with dikes running approximately top to bottom of photograph. E, Variably textured gabbro. F, Irregular aphanitic dikes (green) rooted in amoeboid-like diabase bodies (dark gray).

glass-bead technique of Padfield and Gray (1971) was used for the major elements, and pressed-powder pellets were used for the trace elements; international basalt standards with recommended or certified values from Govindaraju (1994) were used for calibration. The rare earth elements (REEs), Hf, Ta, and Th were analyzed by instrumental neutron activation analysis (INAA) at the University of Bergen, using a coaxial HP Ge detector, and the analytical methods were described by Brunfelt and Steinnes (1969).

Sr and Nd isotopes were analyzed on a Finnegan 262 mass spectrometer at the University of Bergen. All chemical processing was carried out in a clean-room environment with HEPA-filtered air supply and positive pressure, and the reagents were purified in two-bottle Teflon stills. Samples were dissolved in a mixture of HF and HNO_3. Sr and Rb were separated by specific extraction chromatography according to the method described by Pin et al. (1994). Sr was loaded on a double filament and analyzed in static mode. Sr isotope ratios

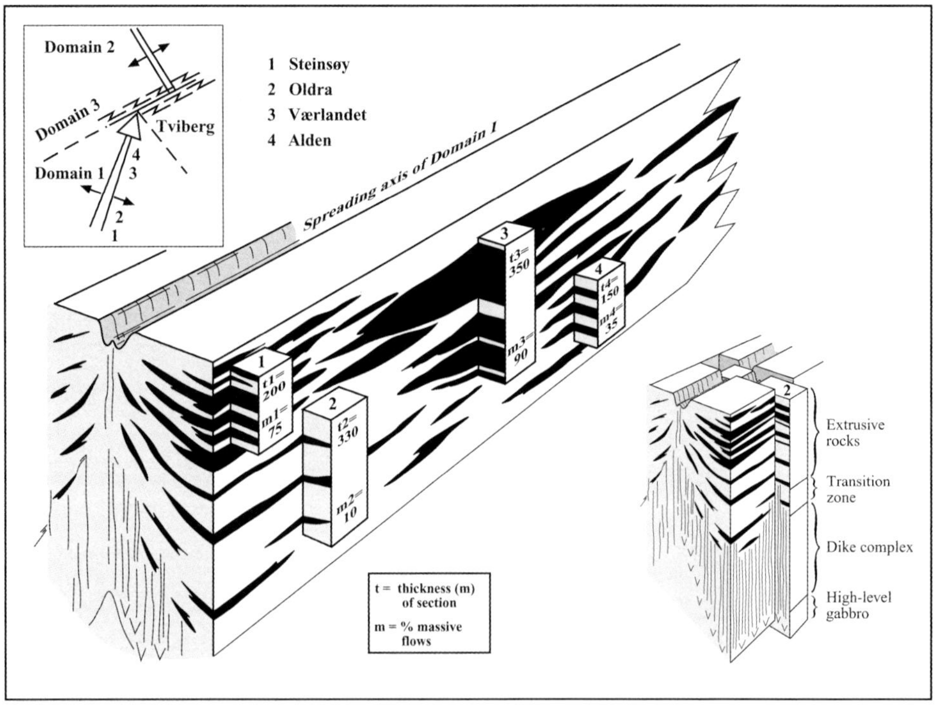

Figure 4. Cartoon illustrating the volcanic construction of domain 1 of the Solund-Stavfjord ophiolite complex. Black: massive sheet flows; white: predominantly pillow lava and minor hyaloclastite breccias. Positions of numbered columns are shown in inset that gives domain relationships. Modified from Fonneland (1997).

were corrected for mass fractionation by using a $^{88}Sr/^{86}Sr$ value of 8.375209. Repeated measurements of the NBS 987 Sr standard have yielded an average of 0.710251 ± 19 (2σ error shown on the least significant digits) ($n = 27$).

The oxygen isotope analyses were conducted on a Finnegan 252 mass spectrometer at the Department of Geology, University of Alberta, Canada. The oxygen isotope data are presented as the per mil enrichment of the heavy isotope in the sample, compared to the per mil concentration in the standard, i.e., $\delta = [(R_{sample}/R_{standard}) - 1] \times 1000$, where $R = {}^{18}O/{}^{16}O$. The data are reported in the usual delta notation of standard mean ocean water (SMOW) (Craig, 1961).

GEOCHEMISTRY

The focus of our geochemical studies and petrogenetic modeling is placed on domain 1 because (1) the geochemical composition of the metabasalts changes systematically along strike of the sheeted dikes within the ophiolite and (2) we have the comprehensive and systematic data from this domain in order to test our model for its magmatic development in relationship to rift propagation (Furnes et al., 1998).

Figure 5A shows a multielement diagram of three representative samples, normalized to MORB (mid-oceanic-ridge basalt): two extrusive rocks (SO-3 and EK3) and one sample from the sheeted-dike complex (D70) from domain 1 (Oldra; see Fig. 1B). The patterns defined by these samples are flat, and the element concentrations of Th, Ta, Nb, Ce, P_2O_5, Zr, Hf, TiO_2, Y, and Yb are generally two times their concentrations in normal MORB (N-MORB). Sample SO-3 displays a slight negative Ta anomaly. Most of the Nb and Zr concentrations are in the ranges of 8–11 ppm and 150–200 ppm, respectively (Fig. 5B), and the Cr concentrations are between 100 and 300 ppm (Fig. 5C). Thus, the metabasaltic rocks of the Solund-Stavfjord ophiolite complex may be regarded as representatives of average Cr (= 250 ppm) to moderately fractionated N-MORB magmas, with a possible minor signature of subduction-zone overprint, marked by the slight negative Ta anomaly. The geochemical characteristics of the metabasaltic dikes of domain 2 (as defined on the island of Tviberg; Fig. 2, A and B) are shown in Figure 5, D–F. The MORB-normalized multielement diagram of four representative samples shows a pattern of increasing rock/MORB ratios from Y through Nb (Fig. 5D). The ranges in the Zr, Nb, and Cr concentrations are ~100–280 ppm, 6–16 ppm, and 50–450 ppm, respectively (Fig. 5, E and F). These element relationships show that the metabasalts of domains 1 and 2 are different and that those of the latter contained more primitive as well as more extensively differentiated basalts than those of domain 1.

A petrogenetic model for the metabasalts of domain 1

Langmuir et al. (1992) showed that differences in the extent and depth of fractional melting cause variations in the geochemical compositions of primitive melts. Magmas generated by fractional melting at progressively higher pressures get increasingly richer in FeO and MgO (Langmuir et al., 1992). We compared the compositions of metabasaltic rocks from domain 1 with those of pooled melts generated within the range of 12 to 40 kbar (Langmuir et al., 1992) in order to determine the pressure conditions under which the metabasalts of domain 1 were generated. Figure 6 shows the MgO and FeO_t concentrations of the Oldra metabasalts superimposed on the diagrams of Langmuir et al. (1992), which show liquid lines of descent for melts generated at 40 to 12 kbar and fractionated under conditions of 8 kbar to 1 atm. These comparisons show that the Oldra metabasalts are best comparable with melts generated at ~12 to 30 kbar and that they were subsequently modified, predominantly by plagioclase and clinopyroxene fractionation at pressures below 8 kbar (Ryttvad et al., 2000). Thus the Oldra metabasalts may represent melts that were generated at different pressures in the mantle and that were subsequently modified by crystal fractionation at different levels in the mantle and within the crust.

According to Grove et al. (1992), MORB genesis may occur as

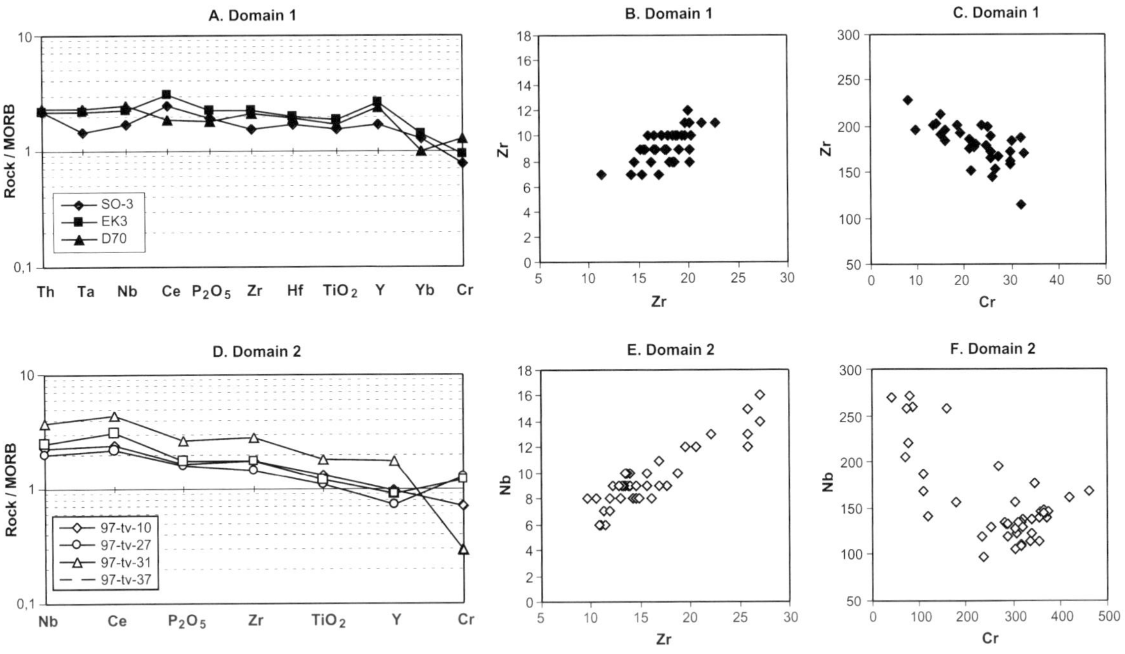

Figure 5. Representative geochemical data of metabasalts from domain 1 (plots A, B, C) and domain 2 (plots D, E, F) of the Solund-Stavfjord ophiolite complex. A and D, MORB-normalized multielement diagrams of representative metabasalt samples. Data taken from Furnes et al. (1990), Ryttvad et al. (2000), and Hellevang (1999). MORB values (Ta through Cr taken from Pearce [1980] and Th taken from Tarney et al. [1980]) are as follows (in ppm, unless otherwise indicated): Th = 0.2, Ta = 0.18, Nb = 4, Ce = 10, P_2O_5 = 0.12%, Zr = 90, Hf = 2.4, TiO_2 = 1.5%. Y = 30, Yb = 3.4, and Cr = 250. B, E and C, F, Nb vs. Zr, and Zr vs. Cr plots. Data from domains 1 and 2 are taken from Ryttvad et al. (2000) and Hellevang (1999), respectively.

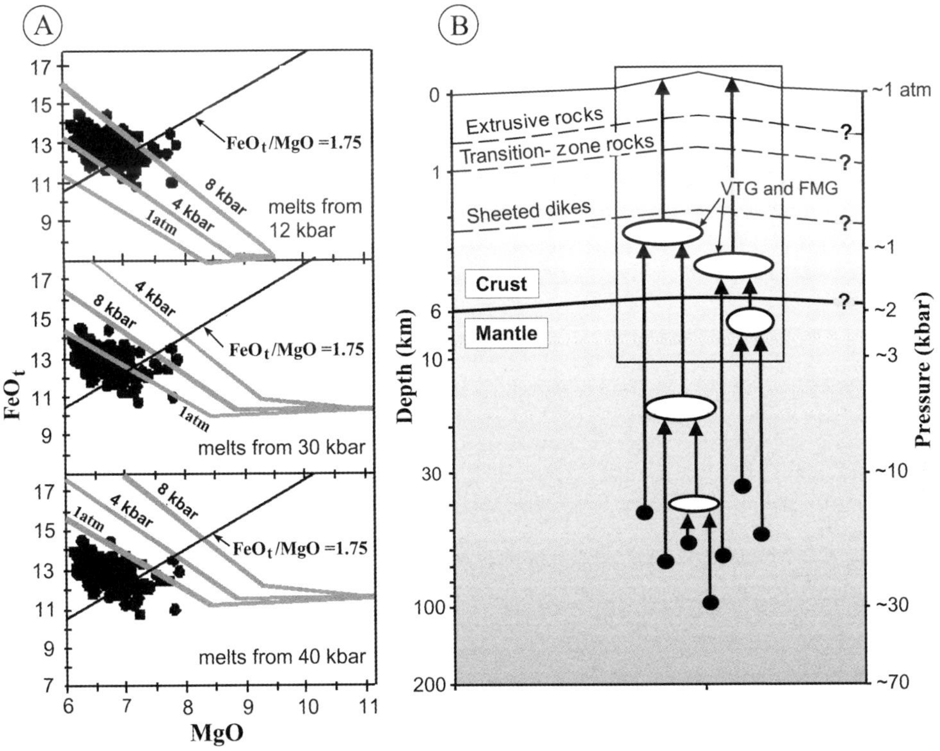

Figure 6. A, MgO vs. FeO_t data for the Oldra metabasalts and the calculated liquid lines of descent for melts generated by fractional melting at 12 kbar, 30 kbar, and 40 kbar and subsequently modified by fractional crystallization of olivine, plagioclase, and clinopyroxene at 1 atm, 4, and 8 kbar (modified from Langmuir et al., 1992). The diagonal line FeO_t/MgO = 1.75 separates the ferrobasalts (above, $FeO_t/MgO > 1.75$) from N-MORBs (below, $FeO_t/MgO < 1.75$). B, Cartoon depicting a petrogenetic model for the Oldra metabasalts. Solid circles are melts generated by fractional melting at pressures from 30 to 12 kbar. Solid arrows indicate upward melt transport and crystal fractionation of olivine, plagioclase, and clinopyroxene out of melts at pressures of less than 8 kbar. During ascent, individual batches of magma mix, as indicated by the ovals. VTG and FMG: variably textured gabbro and fine- to medium-grained gabbro, respectively. The boxed area shows the location of Figure 7C. Modified from Ryttvad et al. (2000).

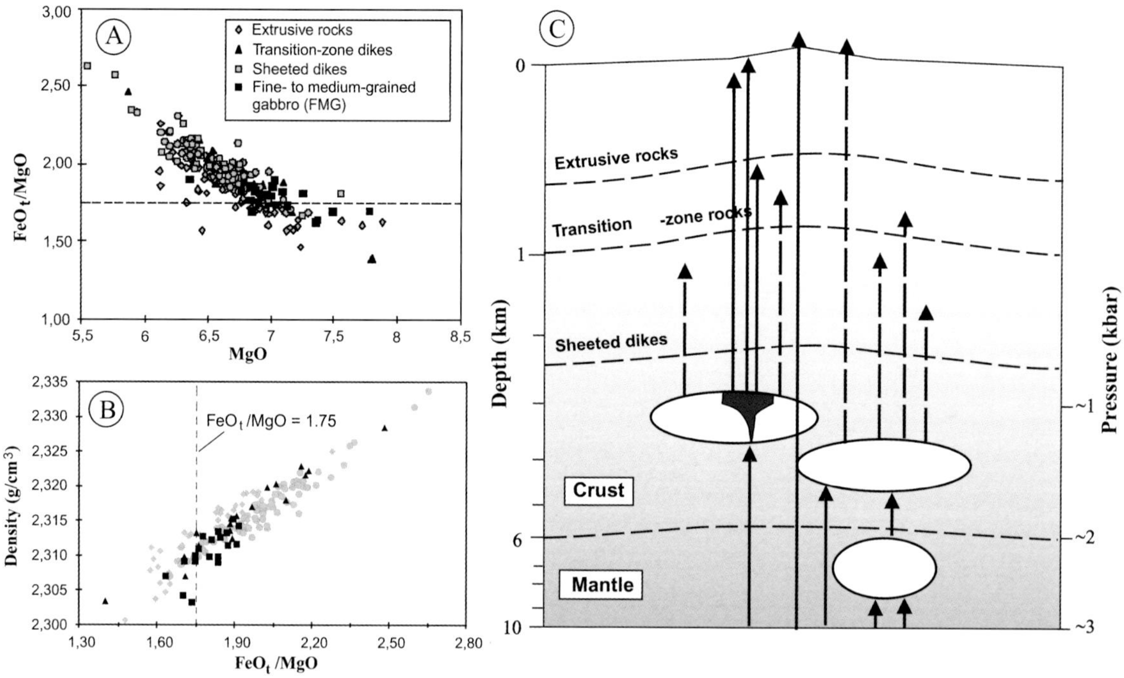

Figure 7. A, MgO vs. FeO_t/MgO for the metabasalts of Oldra. The horizontal, dotted line at $FeO_t/MgO = 1.75$ separates ferrobasalts (above) from N-MORBs (below) (from Sinton et al., 1983). B, FeO_t/MgO ratios vs. density for the extrusive rocks (gray diamonds), transition-zone dikes (filled triangles), sheeted dikes (gray circles), and fine- to medium-grained gabbro (filled squares). The vertical, dotted line $FeO_t/MgO = 1.75$ separates ferrobasalts (right) from N-MORBs (left). C, Enlargement of part of Figure 6B. Melts of Fe-Ti basalt (dashed arrows) and N-MORB (solid arrows). Ovals indicate areas of magma mixing and crystal fractionation. Low-density N-MORBs (solid diapir in oval) may ascend rapidly, without substantial mixing with high-density Fe-Ti basalts. Modified from Ryttvad et al. (2000).

a result of (1) partial melting and later fractional crystallization and mixing events that are separated by an aggregation and mixing step; (2) fractional melts are moved to shallower, cooler mantle, where they are modified by differentiation before aggregation and mixing beneath the spreading center; and (3) fractional melting, aggregation and mixing, and differentiation occur throughout the upper mantle, and the pressure ranges of aggregation, mixing, and differentiation overlap. Aggregation of melts may thus occur at many levels during the melt production and modification process, such as in the regions of melt production and melt modification. Our petrogenetic model (Fig. 6B) for the evolution of metabasaltic rocks in domain 1 favors the third alternative. Nd isotope data (Ryttvad et al., 2000) show only minor variations in the ϵ_{Nd} (T = 443 Ma) values (+7.8 to +8.4), which suggests a common depleted source for the Oldra magmas. The model (Fig. 6B) assumes crustal to subcrustal magma chambers in which new, incoming batches of less fractionated magma would only rarely bring the composition of the mixed liquids out of the Fe-Ti basaltic field.

The geochemical differences between the extrusive and intrusive rocks aid in further refining the petrogenetic model. By using the criteria $TiO_2 > 2$ wt%, $FeO_t > 12$ wt%, and $FeO_t/MgO > 1.75$, which divides ferrobasalts ($FeO_t/MgO > 1.75$) from N-MORBs ($FeO_t/MgO < 1.75$) (Melson et al., 1976; Sinton et al., 1983), we classify 80% of the metabasalts as Fe-Ti basalts (Fig. 7A); the remaining 20% are of N-MORB composition. There is, however, a significant difference in the distribution of the Fe-Ti basalts within the established ophiolite pseudostratigraphy exposed on Oldra. Nearly 51% of the extrusive rocks are made of Fe-Ti basalts, whereas 83% of the transition-zone dikes, 92% of the sheeted dikes, and 79% of the gabbroic rocks are of Fe-Ti basalt composition (Ryttvad et al., 2000).

Sparks et al. (1980) showed that primitive magmas with high MgO content have lower densities than MgO-poor, evolved Fe-Ti–rich basalts. Thus, primitive magmas injected into a magma chamber are buoyant and can physically mix with more fractionated magma. The ability of two magma batches to mix is controlled, however, by the density gaps between the melts. If a magma batch is relatively primitive, then substantial mixing with an extensively evolved liquid would be minimal (Huppert et al., 1986; Brooks et al., 1991), and a relatively primitive magma batch may ascend rapidly to erupt first, followed by more evolved, denser magma.

We have calculated (anhydrous) liquid densities at 1300 °C for the Oldra metabasalts by using partial molar volumes (Bottinga et al., 1983; Lange and Carmichael, 1987), and our findings show good correlation with the FeO_t/MgO ratios (Fig. 7B). The relatively low-density, buoyant N-MORB magmas would erupt more easily than the high-density Fe-Ti basalts. This density difference may explain why only 50% of the extrusive rocks are made of Fe-Ti basalts, whereas the majority of the transition-zone dikes and the sheeted dikes are made of Fe-Ti basalts (Fig. 7C).

Domain 1, a propagated rift segment of the Solund-Stavfjord ophiolite complex

Studies by Sinton et al. (1983) have demonstrated pronounced geochemical changes in basalt compositions along propagating rifts, ranging from a large variety of primitive to highly evolved basalts and even rhyodacites at the propagating tip. In their model, the com-

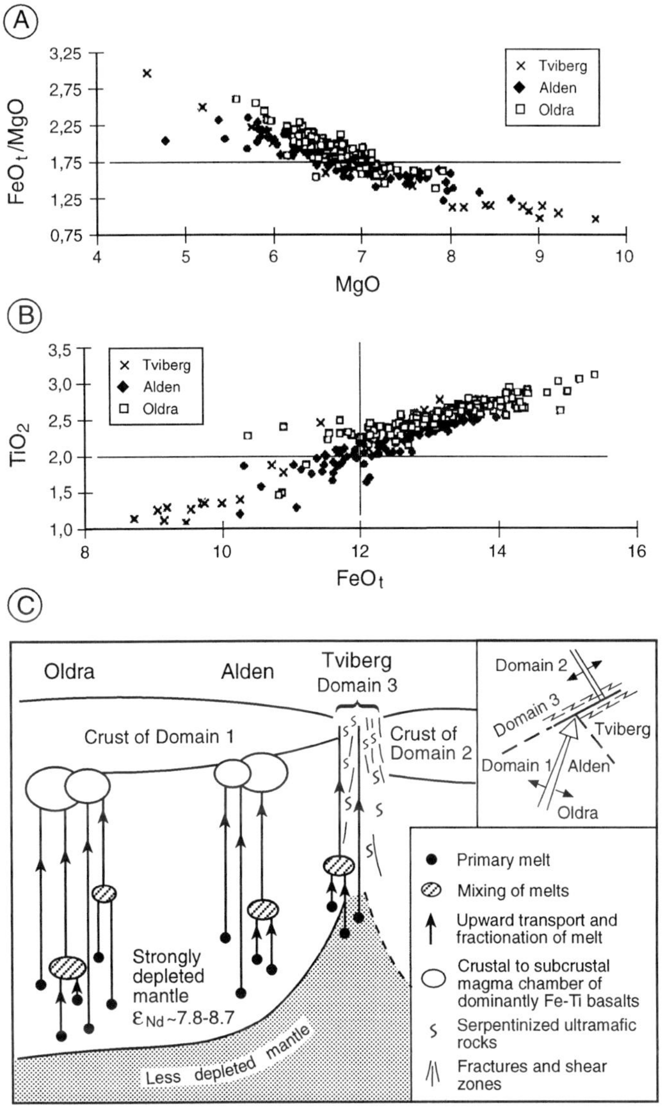

Figure 8. Relationships for metabasalts from Tviberg and Oldra. A, MgO vs. FeO_t/MgO. B, FeO_t vs. TiO_2. The definition of Fe-Ti basalts ($FeO_t/MgO > 1.75$) has been taken from Sinton et al. (1983). C, Petrogenetic cartoon depicting a generalized model for the magmatic development of domain 1 from the tip of the propagating rift in the north, where basaltic dikes intrude the sheared gabbroic rocks of domain 3 (on Tviberg), to Oldra situated 30 km south of domain 3. Modified from Furnes et al. (1998).

positional ranges reach a maximum at the propagating tip or a short distance behind it (~20–40 km), and the basalts gradually approach N-MORB compositions farther behind the tip (~60–200 km). In order to compare the magmatic evolution of domain 1 with that of modern propagating rifts, we examined the geochemical characteristics (in particular, the FeO_t/MgO ratios) of metabasaltic rocks in relationship to distance from the assumed propagating tip (Fig. 2A).

The MgO versus FeO_t/MgO relationships for the domain 1 dikes and lavas from Tviberg, Alden, and Oldra are shown in Figure 8A. The largest variations are displayed by the dikes from Tviberg, which show pronounced ranges in the MgO contents (4.5–9 wt%) and FeO_t/MgO ratios (1 to 3). The metabasaltic dikes and lavas from Oldra in the southern part of the ophiolite (Fig. 1) define a narrower range in the MgO content (5.6–8 wt%) and in the FeO_t/MgO ratios (1.40–2.65). At the same MgO content, the metabasalts from Oldra show higher FeO_t/MgO ratios than those from Tviberg. The metabasalts from Alden have compositions intermediate between those of Oldra and Tviberg. However, since Fe may be leached whereas Mg becomes enriched under seafloor hydrothermal conditions (Mottl and Holland, 1978; Seyfried and Bischoff, 1979), the original FeO_t/MgO ratios may be variably reduced. Ti, on the other hand, is reported to be a rather stable element in all low-temperature alteration studies of basalts. The FeO_t versus TiO_2 relationships (Fig. 8B) also show that the Tviberg samples are generally lowest in the FeO_t and TiO_2 contents and that the Oldra samples contain the highest contents of the same elements; the Alden samples again are intermediate between the two. The metabasaltic rocks from Tviberg are hence classified as primitive, high-MgO basalts to highly fractionated Fe-Ti basalts and quartz diorites (Skjerlie, 1988), whereas those from Oldra predominantly are Fe-Ti basalts.

The data presented in Figure 8, A and B, show a progressive change in the proportion of Fe-Ti basalts to N-MORB metabasalts of domain 1 with respect to an increasing distance from domain 3, in which the tip of the rift propagator, represented by north-northeast–trending dike swarms in the fossil fracture zone, is situated (Fig. 2A). There is an apparent shift southward along domain 1, from less representation of Fe-Ti basalts relative to N-MORB on Tviberg (43%) to a majority of Fe-Ti basalts on Alden (67%), and finally to a strong predominance of Fe-Ti basalts (80%) on Oldra.

The FeO_t/MgO ratios of the metabasalts of Oldra, Alden, and Tviberg show features that are comparable with those of modern propagating rifts. From Tviberg to Oldra, via Alden, the proportion of metabasalts with FeO_t/MgO ratios less than 1.75 decreases progressively (Furnes et al., 1998), a feature of modern propagating rifts as observed farther away (~50 to 100 km) from the rift tip (Sinton et al., 1983). The predominance of Fe-Ti basalts on Oldra, at the present distance of ~30 km south of the inferred propagated tip (Fig. 2A), is quite compatible with the pattern observed at some modern propagating rifts (e.g., Sinton et al., 1983).

A large part of the metabasalts from Alden are compositionally similar to those from Oldra, both in terms of FeO_t/MgO ratios (Fig. 8A) and Nd isotopes (ϵ_{Nd} values of +7.8 to +8.7; Furnes et al., 1992); hence, a petrogenetic model similar to that for Solund-Stavfjord ophiolite complex is suggested for the Alden segment of domain 1 (Fig. 8C). In the FeO_t/MgO versus MgO diagram, however, some of the Alden metabasalts show less FeO_t enrichment than those from Oldra and plot in an intermediate position between the Oldra and Tviberg samples (Fig. 8A). This compositional detail may indicate that, in general, melting occurred at somewhat lower pressures and/or that the liquids did not reach the same evolved level during fractional crystallization as did the liquids from the Oldra segment.

Of the investigated metabasalts from domain 1, those from Tviberg show the largest MgO variations and the lowest FeO_t/MgO ratio at a given MgO content (Fig. 8A). These features may be explained as a result of partial melting and fractional crystallization at a shallower depth beneath Tviberg compared to Alden and Oldra. The north-northeast–striking dike swarms on Tviberg might have originated from isolated, ephemeral magma pods, causing high-MgO liquids to solidify when magma supply was low. Intermittent, higher magma-supply rates, combined with long residence time to promote more extensive crystal fractionation, would yield a spectrum of liquids ranging from primitive basalts, through Fe-Ti basalts, to silicic differentiates (Christie and Sinton, 1981). Dikes and the gabbro from Tviberg are more enriched in the most incompatible trace elements (e.g., Th and LREEs [light REEs]) than those from Solund (Furnes

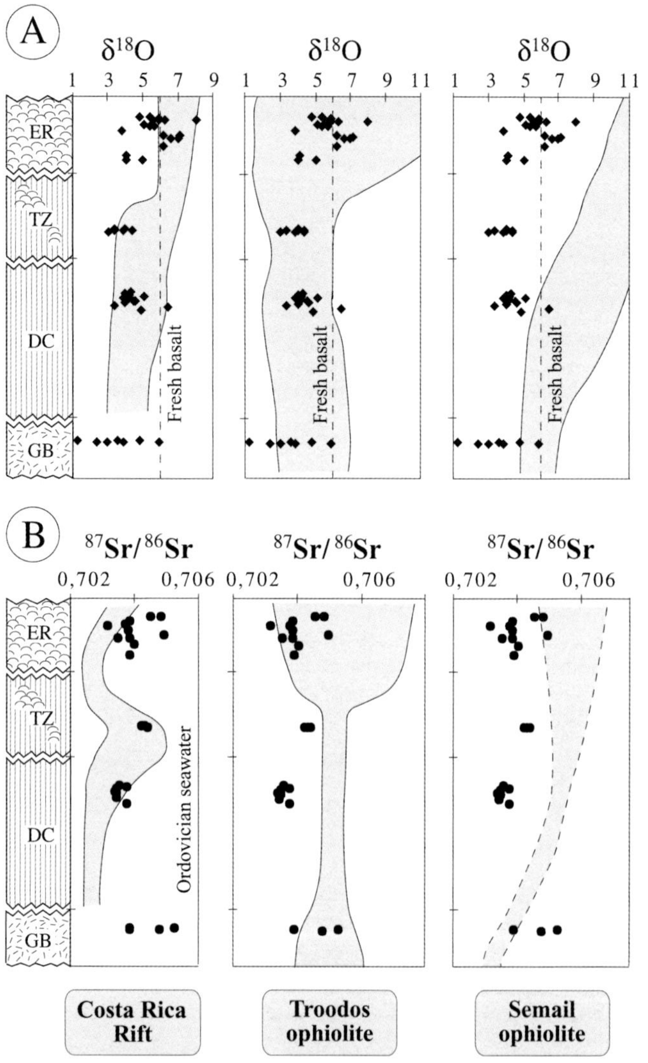

Figure 9. Isotope analyses. A, $\delta^{18}O$. B, $^{87}Sr/^{86}Sr$. Patterns of whole-rock results (points) from Oldra (data from Fonneland, 1997) compared to (shaded fields) metabasalts and gabbros from the Costa Rica Rift (Alt et al., 1996), Troodos ophiolite (Bickle and Teagle, 1992), and Semail ophiolite (McCulloch et al., 1981). Abbreviations: ER—extrusive rocks, TZ—transition-zone rocks, DC—dike complex, GB—high-level gabbro.

et al., 1982). This difference may reflect generation of melts from a less depleted mantle source for Tviberg, as indicated in Figure 8C.

Proximal to the thick, cold wall of transform-fault boundaries and/or fracture zones, lateral conductive heat flow reduces the ambient temperature and disrupts the upward migration of the hot upwelling asthenosphere. This process diminishes the volume of basaltic melt generated and impairs the ability of the melt that is segregated to reach the surface. At the initial stage of magmatism along a spreading ridge that is propagating into the fracture-zone crust (domain 3 on Tviberg), small blobs of primary magma would freeze and produce high-MgO, low–Fe-Ti basalts (Christie and Sinton, 1981). As a result, the total magmatic budget supplied to shallow-level magma chambers would decrease as the fracture zone with thin, anomalous oceanic crust is approached (Fox and Gallo, 1984). The thin and relatively cold crust proximal to fracture zone(s) would allow deeper seawater penetration, which could result in hydration of ultramafic rocks and subsequent mobilization of serpentinite diapirs along shear zones. The proposed magmatic and tectonic model (Fig. 8C) for the origin of domain 3 as a fossil oceanic fracture zone separating domains 1 and 2 is hence consistent with the thermal conditions for a ridge-transform intersection.

HYDROTHERMAL HISTORY

A systematic study of O and Sr isotopes of the volcanic rocks, sheeted dikes, and metagabbros from the southern part of domain 1 has been carried out on the island of Oldra (Fonneland, 1997), and the main results of this study are summarized in this section.

The ^{18}O values of the whole-rock samples show the largest variations in the volcanic and gabbroic rocks, whereas the metabasaltic rocks of the transition zone and dike complex show more narrow ranges (Fig. 9A). Temperature calculations from the ^{18}O values of epidote-quartz mineral pairs give 260–290 °C for the volcanic rocks, 420 °C for the transition zone, 280–350 °C for the sheeted-dike complex, and 290–470 °C for the gabbros (Fonneland, 1997).

Most of the whole-rock samples from Oldra are mainly depleted in $\delta^{18}O$ compared to the $\delta^{18}O$ value of fresh basalt (Fig. 9A). The downward decrease to more restricted ranges in $\delta^{18}O$ values is similar to that observed in several other ophiolites (e.g., Bickle and Teagle, 1992; Gregory and Taylor, 1981) and in the 5.9 Ma oceanic crust at the Costa Rica Rift (Alt et al., 1996). The increase in $\delta^{18}O$ values of the pillow lavas results from variable degrees of low-temperature alteration (Muehlenbachs and Clayton, 1972). This pattern, together with the estimated increase in temperatures, is consistent with interaction of isotopically heavy fluid (seawater or modified seawater) at progressively higher temperatures with depth.

The highest $^{87}Sr/^{86}Sr$ value from the Oldra whole-rock samples is ~0.705 (Fig. 9B), still far from being in equilibrium with Ordovician seawater (0.708). The $^{87}Sr/^{86}Sr$ data allow an estimate that the water/rock ratio due to circulating fluids was generally <10 throughout the Oldra sequence (Fonneland, 1997). The highest values are thought to have resulted from alteration by relatively unmodified Ordovician seawater. Even though the $^{87}Sr/^{86}Sr$ ratios are generally low, a change in the $^{87}Sr/^{86}Sr$ ratios with depth is detectable. The narrowness of the $^{87}Sr/^{86}Sr$ range in the extrusive rocks, transition-zone rocks, and dike complex may indicate an approach toward circulating fluid-rock equilibrium. In addition, a narrow $^{87}Sr/^{86}Sr$ range may be typical of a conventional layered oceanic crust in which channeled circulation in the extrusive rocks changes to more restricted circulation downward. The high $^{87}Sr/^{86}Sr$ ratios of the gabbros may display a channeled circulation whose origin is not certain; however, these ratios may be a result of permeability differences within the unit caused by textural differences, fractures initiated by cooling, or a combination of both. Larger changes in the $^{87}Sr/^{86}Sr$ ratios of the rocks are most likely the result of low-temperature alteration processes (Kawahata et al., 1987), whereas high-temperature oceanic fluids commonly display low $^{87}Sr/^{86}Sr$ ratios (0.703 for the East Pacific Rise at 21°N, 0.704 for the East Pacific Rise at 13°N; Albarede et al., 1981; Michard et al., 1984).

The isotope studies of the Oldra rocks show a depth-dependent variation, consistent with that observed from other ophiolites representing metamorphic grades ranging from zeolite to amphibolite facies, similar to that of modern ocean floor. Possible effects of orogen-related, pervasive greenschist-facies metamorphism could be a slight shift toward lower ^{18}O values (e.g., Valley, 1986). However, Sr and O isotopes show variations with depth that are not consistent with isotopic resetting into uniformity by orogenic deformation and meta-

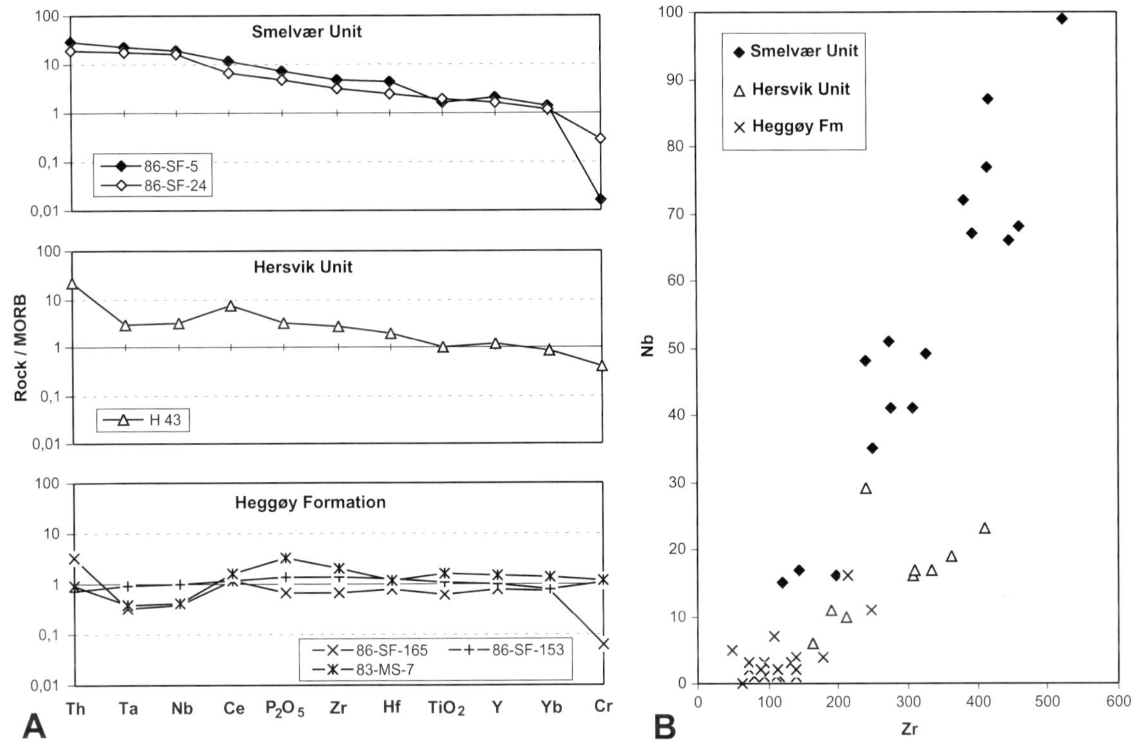

Figure 10. A, MORB-normalized multielement diagrams of representative samples from the Smelvær unit, Hersvik unit, and Heggøy Formation of the Stavenes Group. Data taken from Furnes et al. (1990) and Ryttvad et al. (2000). MORB values as in Figure 5. B, Nb vs. Zr relationships for the same rocks.

morphism. Hence, we interpret the isotope patterns of the Oldra metabasalts to reflect hydrothermal alteration at an oceanic stage.

Figure 9A shows the $\delta^{18}O$ patterns from Oldra in comparison to the patterns defined by the Troodos ophiolite in Cyprus, inferred to have formed at a slow spreading rate (Bickle and Teagle, 1992), the Semail ophiolite in Oman, inferred to have formed at a fast spreading rate (McCulloch et al., 1981), and the young ocean crust from the Costa Rica Rift, inferred to have formed at an intermediate spreading rate (Alt et al., 1996). The samples from Oldra plot within a range comparable to both the Troodos and the Costa Rica Rift. The Oldra rocks differ from those of the Costa Rica Rift by showing a much wider $\delta^{18}O$ range in the extrusive rocks, whereas the extrusive rocks from Troodos show a wider range than the samples from Oldra.

Figure 9B shows the $^{87}Sr/^{86}Sr$ ratios defined by the Oldra rocks compared to those of other ophiolites (Troodos, Bickle and Teagle, 1992, Semail; McCulloch et al., 1981) and to the 5.9 Ma oceanic crust of the Costa Rica Rift (Alt et al., 1996). If the $^{87}Sr/^{86}Sr$ patterns of the three mentioned examples are representative of hydrothermal alteration at the three examples' spreading rates, the hydrothermal history of Oldra is probably most comparable to that of the Costa Rica Rift.

GEOLOGY AND GEOCHEMISTRY OF THE VOLCANIC AND SEDIMENTARY COVER

The extrusive sequence of the Solund-Stavfjord ophiolite complex is overlain by a thick sequence of volcanic and quartz-rich metasedimentary rocks of the Stavenes Group, which is divided into the Heggøy Formation and the Hersvik and Smelvær units (Fig. 1A). The Heggøy Formation, conformably overlying the Solund-Stavfjord ophiolite complex, comprises siliciclastic rocks that are intruded by gabbros and basaltic dikes and sills, and/or are intercalated with pillow lavas and minor volcaniclastic rocks (Furnes et al., 1990). The volcanic rocks of the Hersvik unit consist of metabasaltic massive lavas and volcaniclastic rocks, whereas those of the Smelvær units are metabasalts, predominantly of volcaniclastic origin. Their precise stratigraphic position within the Stavenes Group is, however, unknown.

Similar to the Solund-Stavfjord ophiolite complex, the metabasaltic rocks of the Heggøy Formation produce a rather flat pattern in the MORB-normalized multielement diagram (Fig. 10A). The element concentrations are similar to those of MORB, some slightly lower and others slightly higher, but two of the samples (one sill intrusion, 83-MS-7; and one volcaniclastic rock, 86-SF-165) show significant negative Ta and Nb anomalies. The majority of the Heggøy metabasalts (mostly sills, but also pillow lavas) have lower concentrations of Nb and Zr than those of the Solund-Stavfjord ophiolite complex (compare Figs. 5 and 10B). The massive lava sample (H-43) from the Hersvik unit shows a progressive increase in the MORB-normalized values from Y through Th, except for Ta and Nb, which both show very low values and negative anomalies (Fig. 10A). Hence, the rocks of the Hersvik unit, mostly greenschists interpreted to represent metabasaltic tuffs, show high Zr/Nb ratios (Fig. 10B). The two samples from the Smelvær unit—one pillow lava (86-SF-5) and one gabbro (86-SF-24)—are similar to the pattern of the Hersvik unit, except for the lack of Ta and Nb anomalies (Fig. 10A) and consequently lower Zr/Nb ratios (Fig. 10B).

In summary, the rocks of the Heggøy Formation are rather similar to MORB, but their very low Ta and Nb concentrations are indicative of an initial island-arc tholeiite (IAT) affinity (e.g., Holm, 1985). The rocks of the Hersvik and Smelvær units show geochemical features typical of calc-alkalic and alkalic magmas, respectively (e.g., Thompson et al., 1984). In comparison to the igneous pseudostratigraphy of other well-investigated ophiolite and island-arc complexes in the Norwegian Caledonides (e.g., Pedersen and Hertogen, 1990), we infer that the calc-alkalic and alkalic lithologies of the Hersvik and Smelvær units are younger than the tholeiitic magmatic rocks of the Heggøy Formation.

GEOLOGY AND GEOCHEMISTRY OF THE UNDERLYING MÉLANGE

Suture zones mark the sites of obliteration of oceanic lithosphere by subduction and commonly include mélange units separating allochthonous thrust sheets of ophiolites in the upper plate from the autochthonous continental basement in the lower plate (Dewey, 1977). Obduction mélanges contain tectonic and olistostromal mixtures of ophiolitic material in serpentinite or pelagic and flyschoid matrix (Gansser, 1974). As the ophiolitic nappe systems override the continental basement, igneous, metamorphic, and sedimentary rocks derived from both the allochthonous units and the continental lower plate are incorporated into accretionary wedges, which may commonly be intensely deformed but locally well preserved in coherent stratigraphic packages. These accretionary mélanges can also provide important terrane linkage information that fixes the timing of ophiolite emplacement and suturing. In addition, geochemical characteristics of siliciclastic sedimentary rocks in accretionary mélanges may record signatures of the source-rock characteristics and hence may be useful in reconstructing the tectonostratigraphic history of the mélanges and in deciphering distinct tectonic settings of the provenance of mélange material. In this context, the Sunnfjord mélange beneath the Solund-Stavfjord ophiolite complex provides critical information on the ophiolite emplacement history.

The Sunnfjord mélange rests tectonically on basement gneisses and their supracrustal cover and is tectonically overlain by the Solund-Stavfjord ophiolite complex (Fig. 1). The mélange comprises a number of tectonic slices and lenses that were assembled during the emplacement of the overriding ophiolite (Andersen et al., 1990). Locally well-preserved stratigraphic relationships between the mélange and the underlying Lower Silurian rocks provide a terrane link between the ophiolite complex in the west and the continental margin in the east and define the timing of ophiolite emplacement as Early to Middle Silurian (Andersen et al., 1990). The maximum time span between the formation and obduction of the ophiolite complex is inferred to be 20 m.y. on the basis of U-Pb dating of zircon separates from quartz diorites (443 ± 3 Ma) in domain 2 (Dunning and Pedersen, 1988) and the occurrence of fossils (pentamerides) of Wenlockian age (ca. 425 Ma) within the mélange (Andersen et al., 1990).

The Sunnfjord mélange consists of two major units, the lower Øyravatn unit of predominantly fine-grained, calcareous, quartz-bearing, chlorite-muscovite schists and the upper Markavatn unit composed mainly of metagraywackes (Alsaker and Furnes, 1994). Both units contain blocks of metabasalt, marble, and meta-arkose; blocks of serpentinite occur only in the upper Markavatn unit.

Geochemical investigation of the metasedimentary rocks from the Sunnfjord mélange shows that the Markavatn unit on the average contains appreciably higher contents of Cr and Ni than the lower Øyravatn unit; the higher Cr and Ni are thought to reflect the presence of an ultramafic component (Fig. 11). Thus, the geochemistry of the metasedimentary rocks from both units is compatible with mixing of detritus from continental and ophiolite sources. Despite the strong and pervasive deformation of the mélange, its geochemical composition from the highest to the lowest tectonostratigraphic levels reflects that of the Solund-Stavfjord ophiolite complex pseudostratigraphy. Thus, only metabasalts derived from the ophiolite complex and continental meta-arkoses yielded detritus to the schists of the lower Øyravatn unit, whereas the geochemical composition of the metasedimentary rocks from the upper Markavatn unit shows that detritus derived from mafic and ultramafic rocks of the ophiolite complex mixed with detritus from continental meta-arkoses (Alsaker and Furnes, 1994).

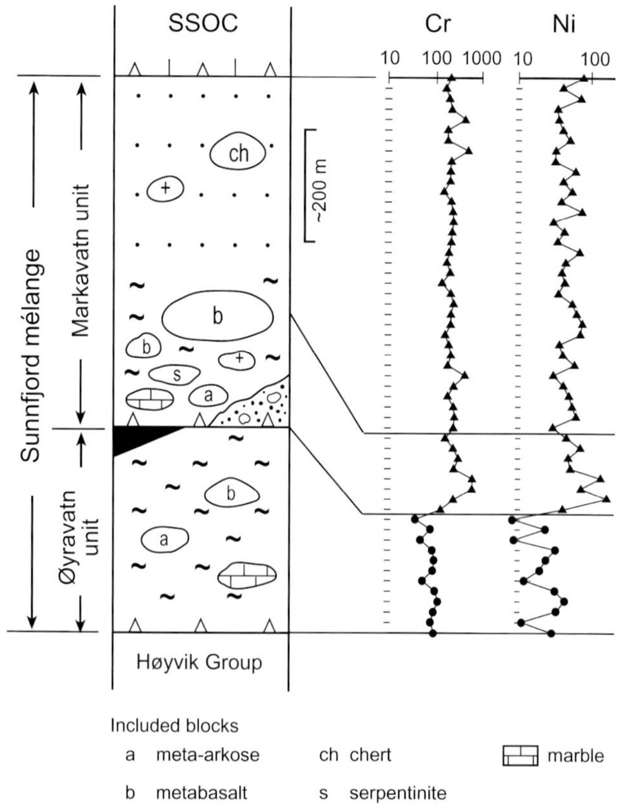

Figure 11. Cr and Ni values of systematically sampled metasedimentary rocks through the Sunnfjord mélange below the Solund-Stavfjord ophiolite complex (SSOC). The Øyravatn unit and lower part of the Markavatn unit have a matrix of calcareous quartz-bearing, chlorite-muscovite schist. Locally, at the top of the unit, bituminous schists occur. The upper part of the Markavatn unit has a matrix of calcareous metagraywacke (indicated by dots). Locally, at the bottom of the unit, matrix-supported lenses of polymictic conglomerate (containing pebbles of meta-arkose, quartz schist, chert, jasper, metabasalt) occur. The blocks "floating" in the matrix of both units are a—meta-arkose, b—metabasalt, ch—chert, s—serpentinite, brick pattern—marble, + pattern—granodiorite intrusions. The black wedge at the top of the Øyravatn unit consists of bituminous schist. The multipatterned wedge in the bottom part of the Markavatn unit consists of calcareous quartz-bearing, chlorite-muscovite schist. Inverted v shapes indicate main shear zones. Modified from Alsaker and Furnes (1994).

SEAFLOOR-SPREADING ORIGIN OF THE SOLUND-STAVFJORD OPHIOLITE COMPLEX

The internal structure and crustal architecture of oceanic lithosphere at modern spreading centers are a manifestation of the spatial and temporal relationships among the spreading rate, magma supply, tectonic extension rate, and thermal regime beneath and along the spreading axis (e.g., Macdonald, 1982, 1983; Phipps Morgan et al., 1994; Dilek et al., 1998; Karson, 1998). Thus, construction of oceanic crust at spreading centers with different spreading rates (slow vs. fast) results in variability in the mode and occurrence of hydrothermal, magmatic, and tectonic activities and thus in variations in volcanic buildup and geochemical and isotopic signatures. Along slow-spreading ridges, the highly variable magma budget causes a wide range of morphologic and geologic features along the rift valleys, and tectonic extension in the absence of a steady-state magma supply may result in the exposure of lower-crustal and upper-mantle rocks in contact with extrusive rocks on the seafloor (Karson, 1998). At ridges spreading at intermediate to fast rates, the magma budget is consistently higher and hence magmatic construction keeps pace with seafloor spreading without major tectonic deformation.

Intermediate-temperature alteration (<300 °C) as locally indicated in the high-level gabbros on Oldra (Fonneland, 1997) suggests that deep-seated normal faults were present within the plutonic sequence. The faults may have acted as fluid pathways to cool the crust and locally to disrupt and tilt the idealized layered crustal structure in the Solund-Stavfjord ophiolite complex. Tilting of fault blocks at slow-spreading centers could be explained by the episodic solidification of magma chambers (Harper, 1985). Terminal solidification of a magma chamber was probably not the case during the construction of domain 1 in the ophiolite complex because neither the most primitive nor the most fractionated rocks occur. The entire range in MgO content varies from 8% to 5.5% (Fig. 7A), which may suggest periodic decrease in the magma supply. During the periods of low magma supply, the upper and lower crust within the ophiolite complex might have been stretched tectonically.

In an attempt to reveal the effect of extensional forces during construction of domain 1 on Oldra, drain-out structures in pillow lavas and the orientation of the dikes were measured. The consistent orientation of the drain-out horizons shows that only minor tilting could have occurred during construction of the extrusive sequence (Fonneland, 1997). Furthermore, both the lack of pervasive extensional faulting and the presence of subvertical sheeted dikes indicate that tectonic extension was not dominant during construction of the ophiolite section in domain 1 (Dilek et al., 1997). Therefore, we infer that the displacement along localized shear zones was most likely of minor magnitude (on the order of some tens to a few hundreds of meters) at the ocean-floor stage, similar to the structural history of the oceanic crust along the Costa Rica Rift, which is spreading at an intermediate rate (Dilek, 1998).

Detailed examination of the volcanic stratigraphy of modern in situ oceanic crust in seafloor-drilling holes (i.e., ODP Hole 504B and DSDP [Deep Sea Drilling Project] Hole 418A) shows the occurrence of multiple extrusion cycles during construction of the volcanic sequence. The 600-m-thick volcanic sequence at the Costa Rica Rift (Hole 504B) appears to have been constructed during two main on-axis volcanic cycles and a later off-axis cycle (Ayadi et al., 1998), whereas the >550-m-thick volcanic pile of the old Atlantic Ocean crust (Hole 418A) seems to have been constructed during four or five main volcanic cycles (Staudigel et al., 1996). Our observations suggest that the volcanic buildup of the upper ~300 m of domain 1 of the Solund-Stavfjord ophiolite complex (Fig. 4) may have occurred during two main magmatic cycles, of which the pillow lavas at the lower stratigraphic levels define the upper part of one cycle and the predominance of massive lavas at the higher stratigraphic levels mark the lower part of another cycle. Studies by Hellevang (1999) show that the geochemical composition of the basalts varies through a cycle and that the most differentiated component is emplaced at the beginning stage of the cycle.

The constructional morphology of slow- to intermediate-rate-spreading mid-oceanic ridges is dominated by pillow lavas and lobate, sheet, and ponded lava flows erupted at high effusion rates and more subdued tectonic features (e.g., Bonatti and Harrison, 1988). The volcanic construction of domain 1 (Fig. 4) shows pronounced variability within relatively small distances. Locally, the lava pile may be completely dominated by massive lava flows (of which a substantial part may represent ponded lava lakes), as seen on the island of Værlandet (section 3 of Fig. 4), or by pillow lava (section 2 of Fig. 4). Averaging the findings of the four investigated sections (Steinsøy, Oldra, Værlandet, and Alden; Fig. 4), we find that massive lava flows constitute ~50%, whereas the rest is composed of pillow lavas and hyaloclastites. This distribution of massive lava flows is of the same order as, or slightly higher than, that which has been documented from the Costa Rica Rift of intermediate spreading rate (e.g., Alt et al., 1996). These comparisons suggest that domain 1 of the Solund-Stavfjord ophiolite complex represents a crustal section that developed at an axis having an intermediate spreading rate. This conclusion would also be in agreement with the proportion of massive flows to pillow lavas from the global mid-ocean ridge system, as proposed by Bonatti and Harrison (1988).

TECTONIC ENVIRONMENT OF FORMATION AND EMPLACEMENT OF THE OPHIOLITE COMPLEX AND ADJACENT ROCKS

The Early Ordovician ophiolites and island-arc complexes of the Norwegian Caledonides supposedly collided with, and were subsequently emplaced onto, the continental margin of the Laurentian Shield during the Caledonian orogeny (Pedersen et al., 1988, 1992). From field relationships in the Norwegian Caledonides, the precise dating of the accretion event of the island-arc and ophiolite system cannot be achieved. However, at ca. 470^{+9}_{-5} Ma, a mature stage of an ensimatic island arc can be documented, whereas the Solund-Stavfjord ophiolite complex represented a developed backarc basin at ca. 443 ± 3 Ma (Dunning and Pedersen, 1988). Hence, the emplacement of the Early Ordovician island arc and ophiolites must have taken place within the time interval of ca. 470–443 Ma (see Pedersen et al., 1992, and references therein). Development of the ophiolite complex started by attenuation and subsequent rifting of a basement consisting of the Early Ordovician ophiolite complexes and the Proterozoic to Archean gneisses beneath these ophiolitic nappes (Pedersen and Furnes, 1991; Pedersen et al., 1992).

The spatial and temporal relationships between continentally derived sediments and the volcanic carapace of the Solund-Stavfjord ophiolite complex indicate that the seafloor-spreading center was in proximity to a continental source such that volcanism and synchronous sedimentation of continentally derived quartzofeldspathic detritus were taking place on the seafloor, comparable to the tectonomagmatic setting of the Gulf of California (e.g., Saunders, 1983). However, the geochemical affinity of some of the Solund-Stavfjord ophiolite complex metabasalts and the volcaniclastic rocks requires a subduction-zone component in the tectonic setting of the inferred basin. Therefore, we envisage a marginal-basin setting (close to a continent) that was located in the upper plate of an active subduction zone, analogous to the Andaman Sea region of the Indonesian arc system. In this region, oblique subduction of the Indian plate oceanic lithosphere beneath the Burma plate has resulted in a pull-apart opening of the Andaman Sea since the middle Miocene, and oceanic crust is being generated along a highly oblique seafloor-spreading system (Curray et al., 1979, 1982). The oceanic crust of the Andaman Sea is hence developed in a supra-subduction zone setting adjacent to the continental margin of the Malay Peninsula to the east. The subduc-

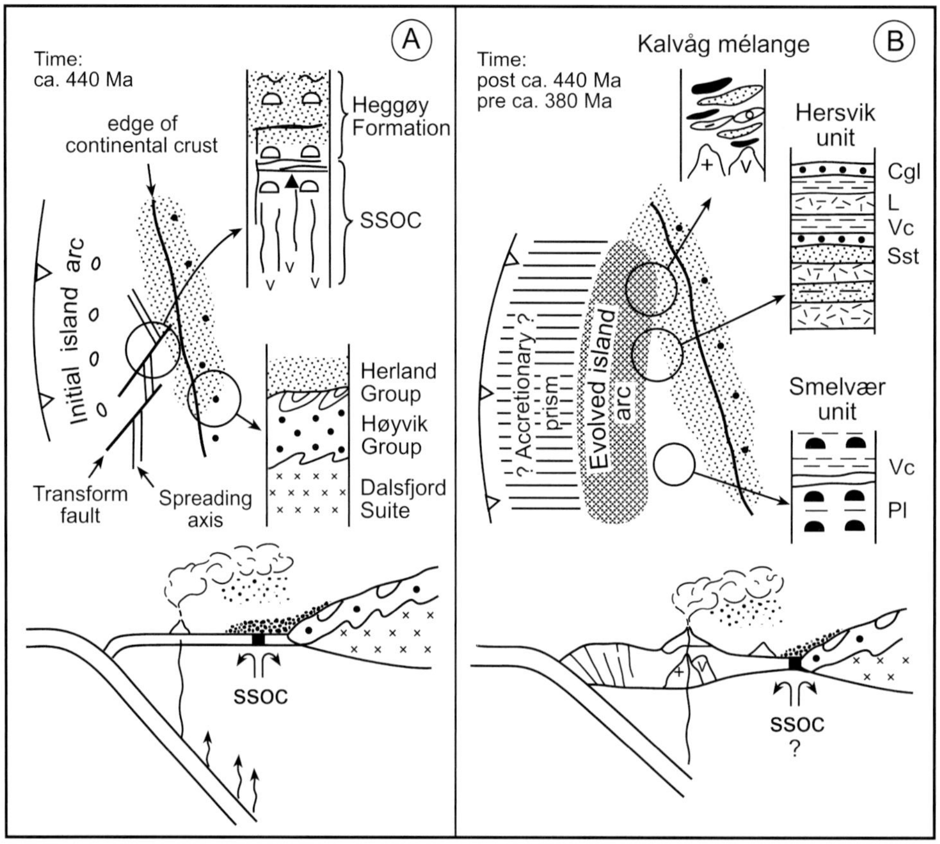

Figure 12. Proposed tectonomagmatic evolution of the Solund-Stavfjord ophiolite complex (SSOC) and associated rocks using the evolution of the present-day Andaman Sea as a model. Modified from Furnes et al. (1990). A, The geochemical effects of subduction upon the metabasalts of the ophiolite complex (indicated by arrows rooted in the subduction zone), as well as the presence of island-arc tholeiitic volcaniclastic rocks intercalated with the quartz-dominated sediments (assumed to have been derived from a nearby continent) of the Heggøy Formation, are indicated by dots at the SSOC-Høyvik Group boundary. (In B, the Dalsfjord Suite and the Høyvik and Herland Groups are described.) B, Presentation of a speculative model, for the evolution of a mature island arc, inferred by the presence of calc-alkalic lavas and volcaniclastic rocks of the Hersvik Group and the deposition of the Kalvåg mélange by progressive cannibalization of the volcanic apron in a backarc basin (Ravnås and Furnes, 1995). The Smelvær unit is assumed to have developed as an oceanic-island complex, distant enough from the continental margin and island arc not to receive quartz-dominated sediments and for the magmatic products not to be influenced by subduction processes, respectively. Explanation of lithologies: Heggøy Formation—metasandstone with intercalated metabasaltic pillow lavas and sills (see Furnes et al., 1990). Solund-Stavfjord ophiolite complex—pillow lavas, massive lavas, dikes, and metagabbro. Herland Group—various metasedimentary rocks (see Andersen et al., 1990). Høyvik Group—mainly quartzites (see Brekke and Solberg, 1987). Dalsfjord Suite—mainly mangeritic gneisses (see Skjerlie, 1969). Kalvåg mélange—olistoliths of shallow-marine deposits, deep-marine turbidites, bedded cherts, pebbly mudstones, calcareous rocks, lava-flow basalts and andesites, rhyolite flows, and rhyolitic ignimbrites embedded in a sheared, olistostromal "groundmass" (see Ravnås and Furnes, 1995). Hersvik unit—metagraywackes (Sst), conglomerates (Cgl), mafic volcaniclastic rocks (Vc), and massive lavas (L) (see Furnes et al., 1990). Smelvær unit—mainly mafic volcaniclastic rocks (Vc) and pillow lavas (Pl) (see Furnes et al., 1990).

tion-zone environment here consists of an inner active volcanic arc (Sumatra), a forearc region with active volcanoes (extending into Burma), and an outer ridge representing a well-developed accretionary prism (Ray et al., 1988). In the Andaman Sea model, the MORB chemical affinity of modern oceanic crust may or may not have a subduction-related geochemical fingerprint, and the volcaniclastic rocks intercalated with the sediments may represent pyroclastic debris associated with subduction-generated volcanism (Fig. 12A). At an initial stage of seafloor spreading in the Solund-Stavfjord ophiolite complex marginal basin (ca. 440 Ma), the quartz-rich metasedimentary rocks of the Herland Group (see Fig. 1) were deposited on the older, deformed quartz sandstones of the Høyvik Group (see Fig. 1) near the continental edge. Farther offshore, similar quartz-rich sediments of the Heggøy Formation would be deposited on the newly formed oceanic crust. These sediments were intercalated with MORB-type pillow lavas and intruded by sill intrusions (Fig. 12A). Later (between ca. 440 and 380 Ma), at a stage when the island arc had become fully developed and mature, subaerial calc-alkalic lavas and intrusions formed (the Hersvik unit), while gravity-driven debris flows and olistoliths, exemplified by the Kalvåg mélange, developed on the steep and unstable slopes of this arc (Figs. 1A and 12B). The alkalic metabasaltic lavas and volcaniclastic rocks of the Smelvær unit supposedly developed at the latest stage within this backarc basin, sufficiently far away from the subduction zone to avoid its chemical influence (Fig. 12B). Alkalic rocks are generally of minor abundance in island-arc systems, but may be represented at a late stage of their development. Thus, alkalic basaltic rocks represented in the Smelvær unit show incompatible trace element abundances and patterns similar to those of alkalic basalts from the Lau Basin (Gill, 1976) and west Mariana Ridge (Wood et al., 1982).

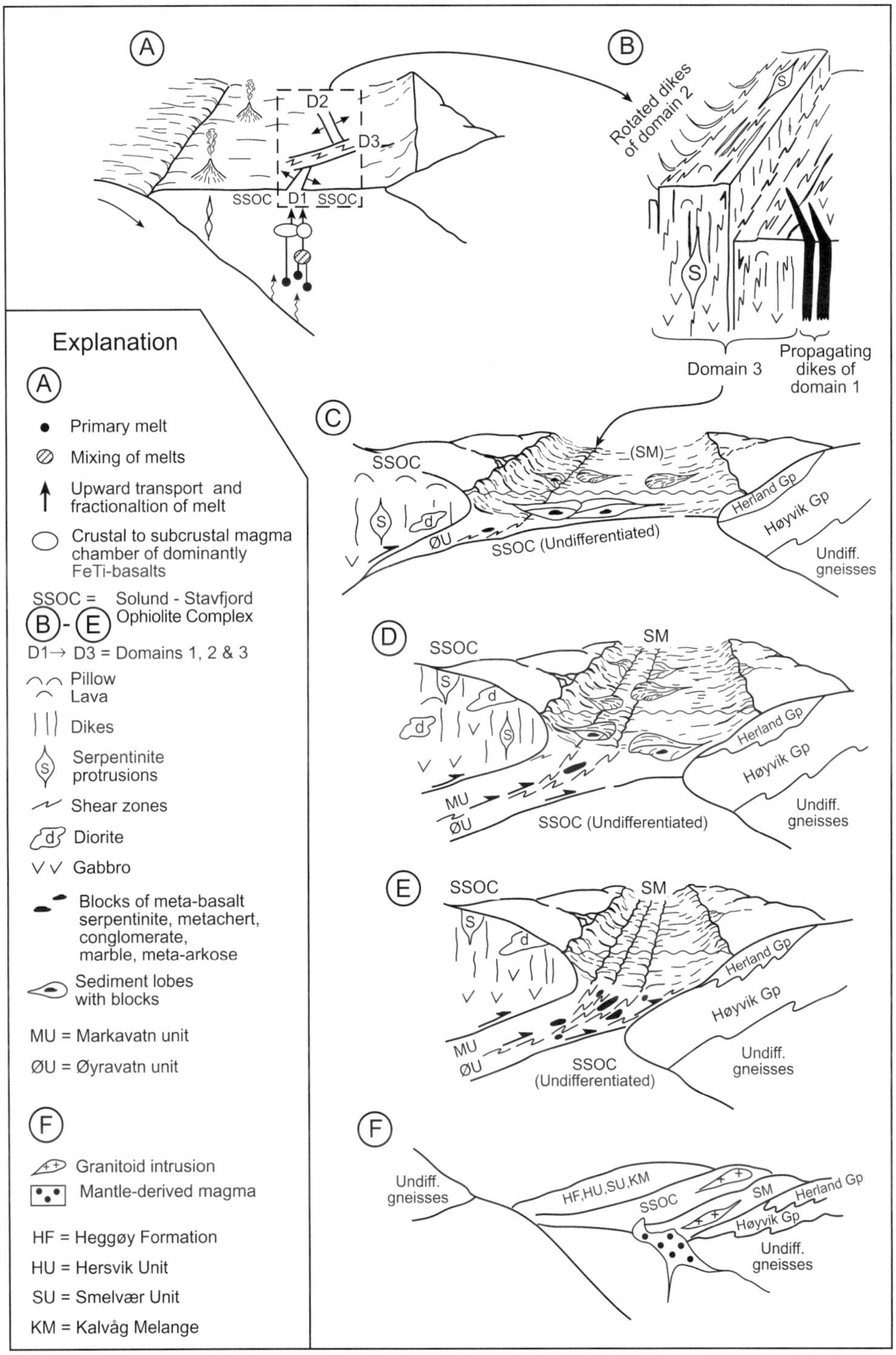

Figure 13. Evolution of the Sunnfjord mélange (SM) and the closure of the marginal basin in which the Solund-Stavfjord ophiolite complex (SSOC) and associated rocks developed. A, General tectonic setting of the ophiolite complex. D1, D2, and D3—domains 1, 2, and 3. Modified from Furnes et al. (1990, 1998). B, Detail of the ophiolite complex, showing the tectonic situation as deduced from the geologic relationships on Tviberg (Figs. 1 and 2), where serpentine diapirism (S) in a transform-fault complex causes uplift. Based on data from Skjerlie et al. (1989), Skjerlie and Furnes (1990), Dilek et al. (1997), and Furnes et al. (1998). C, Development of the Øyravatn unit, by mixing of continental detritus from the Høyvik Group and greenstone detritus from the Solund-Stavfjord ophiolite complex. The transform fault (domain 3) indicated in B now acts as the main shear zone of the Sunnfjord mélange. The sediments of the Wenlockian Herland Group (see Andersen et al., 1990) are at this stage undeformed. D, Development of the Markavatn unit, by mixing of continental detritus from the Høyvik Group and varied detritus consisting predominantly of metabasalts and serpentinites from the uplifted part of the ophiolite complex that was related to the transform fault (domain 3). The rocks of the Herland Group and its basement (the Høyvik Group) are deformed at this stage. E, Final configuration of the Sunnfjord mélange. Illustrations B through E are mainly modified from Alsaker and Furnes (1994). F, Closing of the marginal basin, final emplacement of the Solund-Stavfjord ophiolite complex and related island-arc–generated rocks, and associated granitoid intrusive activity. Modified from Furnes et al. (1990) and Skjerlie (1992). Gp—Group; undiff.—undifferentiated.

The emplacement of the Solund-Stavfjord ophiolite complex occurred as the Andaman-type marginal basin of Iapetus origin collapsed during a regional Middle to Late Silurian collisional event. This collisional event between Baltica and Laurentia, including the southeastward transport of the Early and Late Ordovician ophiolites and island arcs, is referred to as the Scandian phase of the Caledonian orogeny (see Andersen, 1998, and references therein). Initial displacement of the ocean floor most likely took place along an oceanic fracture zone as this plane of inherent weakness was converted to a mantle-rooted thrust fault during regional collision (Fig. 13). The Sunnfjord mélange developed in front of and beneath the ophiolitic nappe, which was displaced toward the continent (Fig. 13, C and D). Generally, the mélange is in tectonic contact with adjacent rocks. On Atløy, however, Andersen et al. (1990) demonstrated evidence for a stratigraphic contact (low-angle unconformity) with the underlying, fossiliferous Herland Group (Fig. 1A), thus suggesting that the maximum age of the mélange is Wenlockian. This fact indicates that the erosion of the ophiolite complex and development of the mélange, representing a terrane disrupted during ophiolite obduction, occurred some 20 m.y. later than the generation of domain 2 of the ophiolite complex (443 ± 3 Ma). Progressive unroofing of the ophiolite in the hanging wall provided gabbroic and peridotitic detrital material into the depocenter(s) of the mélange. Obduction of the ophiolite complex onto the continental margin was followed by the intrusion of granitoid magmas into the mélange and the overlying ophiolite (Fig. 13F; Skjerlie, 1992).

SUMMARY

The Late Ordovician Solund-Stavfjord ophiolite complex in the western Norwegian Caledonides represents a remnant of fossil oceanic crust developed in a backarc basin. Field relationships point to the existence of three structural domains with different tectonomagmatic histories within the ophiolite. The north-northeast–trending domain 1 has an internal structure and stratigraphy that show features characteristic of modern oceanic crust spreading at an intermediate rate. Dikes belonging to this domain penetrate laterally into the ophiolitic rocks in domains 2 and 3, which represent preexisting oceanic crust evolved along a northwest-trending spreading center and an east-northeast–trending fracture zone, respectively. Domain 1 is interpreted to represent a rift that propagated north-northeastward (in the present coordinate system). The metabasaltic rocks of domain 1 typically represent N-MORB magmas with a minor overprint of subduction-zone chemical affinity. Trace-element modeling of their chemistry suggests that these rocks might have formed from melts that were generated at different pressures (ranging from 12 to 30 kbar) in the mantle and that they were subsequently modified by crystal fractionation at different levels in the mantle and within the crust. Nearly 50% of the extrusive rocks on the island of Oldra are made of Fe-Ti basalts, whereas the majority of the transition-zone dikes and the sheeted dikes stratigraphically down section are composed of Fe-Ti basalts. A systematic evaluation of the rock chemistry in domain 1 shows a progressive increase in the proportion of Fe-Ti basalts to N-MORB metabasalts with increasing along-strike distance away from the tip of the rift propagator (in domain 3) in the north. These relationships are compatible with observations at modern propagating rifts at seafloor spreading centers.

The majority of the whole-rock samples from domain 1 on Oldra are depleted in ^{18}O compared to fresh basalt, and the observed changes in $\delta^{18}O$ values down section in the ophiolite pseudostratigraphy are similar to those documented from other ophiolites. The $^{87}Sr/^{86}Sr$ patterns defined by the representative metabasaltic rocks in domain 1 show similarities to those of the rocks from the Costa Rica Rift; therefore, we suggest that the hydrothermal alteration history of the Solund-Stavfjord ophiolite complex may result from oceanic-crust formation at an intermediate spreading rate.

The sedimentary cover of the Solund-Stavfjord ophiolite complex includes a thick sequence of volcanic and quartzofeldspathic metasedimentary rocks that are intruded by basaltic dikes and sills and that are intercalated with pillow lavas and volcaniclastic rocks. These igneous rocks display a chemical signature typical of an island-arc tholeiite within the lower part of the sedimentary cover, but the igneous rocks contained at progressively higher stratigraphic levels in the cover show geochemical features characteristic of calc-alkalic and alkalic affinity. We interpret this change as an indication of the establishment of a mature island-arc system through time.

The Middle Silurian Sunnfjord mélange beneath the Solund-Stavfjord ophiolite complex consists of both ophiolites and continental material and represents a typical obduction mélange. Higher Cr and Ni contents in the upper stratigraphic units of the mélange indicate more contribution from peridotites of the ophiolite as the mantle section was unroofed in the upper plate of the thrust fault cutting down section in the backarc crust within the advancing ophiolite nappe.

The structural, petrologic, hydrothermal, and tectonic features of the Solund-Stavfjord ophiolite complex suggest a complex evolutionary path in a narrow backarc basin, which developed in proximity to a continental landmass in the Iapetus Ocean. We interpret the tectonic setting of the Andaman Sea as a modern analogue for this backarc basin. In addition to the structure and geochemistry of its igneous rocks, the sedimentary cover and the underlying mélange of the Solund-Stavfjord ophiolite complex provide significant information in further defining its tectonic affinity and evolution.

ACKNOWLEDGMENTS

The field work carried out by Harald Furnes has been supported by grants from the Norwegian Research Council. Kjell Petter Skjerlie has mainly been supported by the University of Bergen, and Yildirim Dilek by Vassar College and Miami University (USA). Field work conducted by a large number of graduate students in the Solund-Stavfjord area from 1985 through 1996 has been of immense help for the present understanding and interpretation of the evolution of the Solund-Stavfjord ophiolite complex. Jean H. Bédard, Stephen J. Edwards, and James W. Hawkins gave constructive reviews of an early version of the manuscript. Jane Ellingsen helped with the illustrations.

REFERENCES CITED

Albarede, F., Michard, A., Minster, J.F., and Michard, G., 1981, $^{87}Sr/^{86}Sr$ ratios in hydrothermal waters and deposits from East Pacific Rise at 21°N: Earth and Planetary Science Letters, v. 55, p. 229–236.

Alsaker, E., and Furnes, H., 1994, The geochemistry of the Sunnfjord Mélange: Sediment mixing from different sources during obduction of the Solund–Stavfjord Ophiolite Complex, Norwegian Caledonides: Geological Magazine, v. 131, p. 105–121.

Alt, J.C., Teagle, D.A.H., Laverne, C., Vanko, D.A., Bach, W., Honnorez, J., Becker, K., Ayadi, M., and Pezard, P.A., 1996, Ridge flank alteration of upper ocean crust in the eastern Pacific: Synthesis of results for volcanic rocks of Holes 504B and 896A: Proceedings of the Ocean Drilling Program, Scientific Results, v. 148: College Station, Texas, Ocean Drilling Program, p. 435–450.

Andersen, T.B., 1998, Extensional tectonics in the Caledonides of southern Norway: An overview: Tectonophysics, v. 285, p. 333–351.

Andersen, T.B., Skjerlie, K.P., and Furnes, H., 1990, The Sunnfjord Mélange, evidence of Silurian ophiolite accretion in the west Norwegian Caledonides: Journal of the Geological Society, London, v. 147, p. 59–68.

Ayadi, M., Pezard, P.A., Laverne, C., and Bronner, G., 1998, Multi-scalar structure at DSDP/ODP Site 504, Costa Rica Rift. I. Stratigraphy of eruptive products and accretion processes, *in* Harvey, P.K., and Lovell, M.A., eds., Core-log integration: Geological Society of London Special Publication 136, p. 297–310.

Bédard, J.H., Lanziere, K., Tremblay, A., and Sangster, A., 1998, Evidence for forearc-spreading from the Betts Cove ophiolite, Newfoundland: Oceanic crust of boninitic affinity: Tectonophysics, v. 284, p. 233–245.

Bickle, M.J., and Teagle, D.A.H., 1992, Strontium alteration in the Troodos ophiolite: Implications for fluid fluxes and geochemical transport in mid-ocean ridge hydrothermal systems: Earth and Planetary Science Letters, v. 113, p. 219–237.

Bonatti, E., and Harrison, C.G.A., 1988, Eruption styles of basalt in oceanic spreading ridges and seamounts: Effect of magma temperature and viscosity: Journal of Geophysical Research, v. 93, p. 2967–2980.

Bottinga, Y., Richet, P., and Weill, D.F., 1983, Calculation of density and thermal expansion coefficient of silicate liquids: Bulletin de Mineralogie, v. 106, p. 129–138.

Brekke, H., and Solberg, P.O., 1987, The geology of Atløy, Sunnfjord, western Norway: Norges Geologiske Undersøkelse, Bulletin, v. 410, p. 73–94.

Brooks, C.K., Larsen, L.M., and Nielsen, T.F.D., 1991, Importance of iron-rich tholeiitic magmas at divergent plate margins: A reappraisal: Geology, v. 19, p. 269–272.

Brunfelt, A.O., and Steinnes, E., 1969, Instrumental activation analysis of silicate rocks with epithermal neutrons: Annal of Chemical Acta, v. 48, p. 13–24.

Christie, D.M., and Sinton, J.M., 1981, Evolution of abyssal lavas along propagating segments of the Galapagos spreading center: Earth and Planetary Science Letters, v. 56, p. 321–335.

Coleman, R.G., 1971, Plate tectonic emplacement of upper mantle peridotites along continental edges: Journal of Geophysical Research, v. 76, p. 1212–1222.

Craig, H., 1961, Standards for reporting concentrations of deuterium and ^{18}O in natural waters: Science, v. 133, p. 1702–1703.

Curray, J.R., Moore, D.G., Lawver, L.A., Emmel, F.J., Raitt, R.W., Henry, M., and Kieckhefner, R., 1979, Tectonics of the Andaman Sea and Burma, *in* Watkins, J., Montadert, L., and Dickerson, P., eds., Geological and geophysical investigations of continental margins: American Association of Petroleum Geologists Memoir 29, p. 189–198.

Curray, J.R., Emmel, F.J., Moore, D.G., and Raitt, R.W., 1982, Structure, tectonics, and geological history of the northeastern Indian Ocean, *in* Nairn, A.E.M., and Stehli, F.G., eds., The Indian Ocean: The ocean basins and margins 6: New York: Plenum Press, p. 339–450.

Dewey, J.F., 1977, Suture zone complexities: A review: Tectonophysics, v. 40, p. 53–67.

Dilek, Y., 1998, Structure and tectonics of intermediate-spread oceanic crust drilled at DSDP/ODP Holes 504B and 896A, Costa Rica Rift, *in* Cramp, A., MacLeod, C.J., Lee, S.V., and Jones, E.J.W., eds., Geological evolution of oceanic basins: Results from the Ocean Drilling Program: Geological Society of London Special Publication 131, p. 179–197.

Dilek, Y., Furnes, H., and Skjerlie, K.P., 1997, Propagating rift tectonics of a Caledonian marginal basin: Multi-stage seafloor spreading history of the Solund–Stavfjord ophiolite in Western Norway: Tectonophysics, v. 280, p. 213–238.

Dilek, Y., Moores, E.M., and Furnes, H., 1998, Structure of oceanic crust and ophiolites and implications for faulting and magmatism at oceanic spreading centers, *in* Buck, W.R., Delaney, P.T., Karson, J.A., and Lagabrielle, Y., eds., Faulting and magmatism at mid-ocean ridges: American Geophysical Union Geophysical Monograph 106, p. 219–265.

Dunning, G.R., and Pedersen, R.B., 1988, U/Pb ages of ophiolites and arc-related plutons of the Norwegian Caledonides: Implications for the development of Iapetus: Contributions to Mineralogy and Petrology, v. 98, p. 13–23.

Fonneland, H.C., 1997, The relationships between hydrothermal history, tectonism, and volcanic products in construction of a part of the Solund-Stavfjord ophiolite complex, Western Norway [Candidatus Scientiarium thesis]: University of Bergen, 84 p.

Fox, P.J., and Gallo, D.G., 1984, A tectonic model for ridge-transform-ridge plate boundaries: Implications for the structure of oceanic lithosphere: Tectonophysics, v. 104, p. 205–242.

Furnes, H., 1972, Meta-hyaloclastite breccias associated with Ordovician pillow lavas in the Solund area, west Norway: Norsk Geologisk Tidsskrift, v. 52, p. 385–407.

Furnes, H., 1974, Structural and metamorphic history of the Lower Paleozoic metavolcanics and associated sediments in the Solund area, Sogn: Norges Geologiske Undersøkelse, v. 302, p. 33–74.

Furnes, H., Thon, A., Nordås, J., and Garmann, L.B., 1982, Geochemistry of Caledonian metabasalts from some Norwegian ophiolite fragments: Contributions to Mineralogy and Petrology, v. 79, p. 295–307.

Furnes, H., Skjerlie, K.P., Pedersen, R.B., Anderson, T.B., Stillman, C.J., Suthren, R., Tysseland, M., and Garmann, L.B., 1990, The Solund–Stavfjord Ophiolite Complex and associated rocks, west Norwegian Caledonides: Geology, geochemistry, and tectonic environment: Geological Magazine, v. 127, p. 209–289.

Furnes, H., Johansen, R.J., and Skjerlie, K.P., 1992, FeTi-poor and FeTi-rich basalts in the Solund–Stavfjord ophiolite complex, west Norwegian Caledonides: Relationships and genesis: Neues Jahrbuch für Mineralogie, Monatsheft, v. 4, p. 153–168.

Furnes, H., Dilek, Y., Skjerlie, K.P., Ryttvad, H.L., and Fonneland, H.C., 1998, Variations in basaltic geochemistry along a propagating rift of the late Ordovician marginal basin of the West Norwegian Caledonides: Terra Nova, v. 10, p. 21–26.

Gansser, A., 1974, The ophiolitic mélange, a world-wide problem on Tethyan examples: Eclogae Geologicae Helvetiae, v. 67, p. 479–507.

Gill, J.B., 1976, Composition and age of Lau Basin and Ridge volcanic rocks: Implications for evolution of an interarc basin and remnant arc: Geological Society of America Bulletin, v. 87, p. 1384–1395.

Govindaraju, K., 1994, 1994, Compilation of working values and sample description for 383 geostandards: Geostandards Newsletter, v. 18, Special Issue, 158 p.

Gregory, R.T., and Taylor, H.P., Jr., 1981, An oxygen isotope profile in a section of Cretaceous oceanic crust, Samail ophiolite, Oman: Evidence for ^{18}O buffering of the oceans by deep (>5 km) seawater-hydrothermal circulation at mid-ocean ridges: Journal of Geophysical Research, v. 86, p. 2737–2755.

Grove, T.L., Kinzler, R.J., and Bryan, W.B., 1992, Fractionation of Mid-Ocean Ridge Basalts (MORB), *in* Morgan, J.P., Blackman, D.K., and Sinton, J. F., eds., Mantle flow and melt generation at mid-ocean ridges: American Geophysical Union, Geophysical Monograph 71, p. 281–310.

Harper, G.D., 1985, Tectonics of slow spreading mid-ocean ridges and consequences of a variable depth to the brittle/ductile transition: Tectonics, v. 4, p. 395–409.

Hartz, E., Andresen, A., and Andersen, T.B., 1994, Structural observations adjacent to a large scale extensional detachment zone in the hinterland of the Norwegian Caledonides: Tectonophysics, v. 231, p. 123–127.

Hawkins, J.W., 1995, Evolution of the Lau Basin: Insights from ODP Leg 135, *in* Taylor, B., and Natland, J., eds., Active margins and marginal basins of the western Pacific: American Geophysical Union Geophysical Monograph 88, p. 125–173.

Hellevang, B., 1999, Volcanic and geochemical cyclic development of Late Ordovician oceanic crust, represented by the Solund–Stavfjord Ophiolite Complex [Candidatus Scientiarium thesis]: University of Bergen, 114 p.

Holm, P.E., 1985, The geochemical fingerprints of different tectonomagmatic environments using hygromagmatophile element abundances of tholeiitic basalts and basaltic andesites: Chemical Geology, v. 51, p. 303–312.

Huppert, H.E., Sparks, R.S.J., Whitehead, J.A., and Hallworth, J., 1986, Replenishment of magma chambers with light inputs: Journal of Geophysical Research, v. 91, p. 6113–6122.

Karson, J., 1998, Internal structure of oceanic lithosphere: A perspective from tectonic windows, *in* Buck, W.R., Delaney, P.T., Karson, J.A., and Lagabrielle, Y., eds., Faulting and magmatism at mid-ocean ridges: American Geophysical Union Geophysical Monograph 106, p. 177–218.

Kawahata, H., Kusakabe, M., and Kikuchi, Y., 1987, Strontium, oxygen, and hydrogen isotope geochemistry of hydrothermally altered and weathered rocks in DSDP Hole 504B, Costa Rica Rift: Earth and Planetary Science Letters, v. 85, p. 343–355.

Lange, R.A., and Carmichael, I.S.E., 1987, Densities of Na_2O-K_2O-CaO-MgO-FeO-Fe_2O_3-Al_2O_3-TiO_2-SiO_2 liquids: New measurements and derived partial molar properties: Geochimica et Cosmochimica Acta, v. 51, p. 2931–2946.

Langmuir, C.H., Klein, E.M., and Plank, T., 1992, Petrological systematics of mid-ocean ridge basalts: Constraints on melt generation beneath ocean ridges, *in* Morgan, J.P., Blackman, D.K., and Sinton, J.F., eds., Mantle flow and melt generation at mid-ocean ridges: American Geophysical Union Geophysical Monograph 71, p. 183–280.

Macdonald, K.C., 1982, Mid-ocean ridges: Fine scale tectonic, volcanic, and hydrothermal processes within the plate boundary zone: Annual Review of Earth and Planetary Science Letters, v. 10, p. 155–190.

Macdonald, K.C., 1983, Crustal processes at spreading centers: Reviews of Geophysics and Space Physics, v. 21, p. 1441–1454.

McCulloch, M.T., Gregory, R.T., Wasserburg, G.J., and Taylor, H.P., Jr., 1981, Sm-Nd, Rb-Sr, and $^{18}O/^{16}O$ isotopic systematics in an oceanic crustal section: Evidence from the Semail ophiolite: Journal of Geophysical Research, v. 86, p. 2721–2735.

Melson, W.G., Vallier, T.L., Wright, T.L., Bryerly, G., and Nelson, J., 1976, Chemical diversity of abyssal volcanic glass erupted along the Pacific, Atlantic, and Indian Ocean seafloor spreading centers, *in* Sutton, G., Manghari, M.H., and Moberly, R., eds., The geophysics of the Pacific Ocean Basin and its margins: Geophysical Monograph Series, v. 12, p. 351–367.

Michard, G., Albarede, F., Michard, A., Minster, J.F., Charlou, J.L., and Tan, N., 1984, Chemistry of solutions from the 13°N East Pacific Rise hydrothermal site: Earth and Planetary Science Letters, v. 67, p. 297–307.

Moores, E.M., and Vine, F.J., 1971, The Troodos Massif, Cyprus and other ophiolites as ocean crust: Evaluation and implications: Philosophical Transactions of the Royal Society of London, series A, v. 268, p. 443–466.

Mottl, M.J., and Holland, H.D., 1978, Chemical exchange during hydrothermal alteration of basalt by seawater. I. Experimental results for major and minor components of seawater: Geochimica et Cosmochimica Acta, v. 42, p. 1103–1115.

Muehlenbachs, K., and Clayton, R.N., 1972, Oxygen isotope studies of fresh and weathered submarine basalts: Canadian Journal of Earth Science, v. 9, p. 172–184.

Nicolas, A., and Boudier, F., 1991, Rooting of the sheeted dike complex in the Oman ophiolite, *in* Peters, T., Nicolas, A., and Coleman, R.G., eds., Ophiolite genesis and evolution of the oceanic lithosphere: Series in Petrology and Structural Geology 5, p. 39–54.

Osmundsen, P.T., and Andersen, T.B., 1994, Caledonian compressional and late-orogenic extensional deformation in the Staveneset area, Sunnfjord: Journal of Structural Geology, v. 16, p. 1385–1401.

Padfield, T., and Gray, A., 1971, Major element rock analyses by X-ray fluorescence: A simple fusion method: Eindhoven, NV Philips, Analytical Equipment FS 35.

Pearce, J.A., 1980, Geochemical evidence for the genesis and eruptive setting of lavas from Tethyan ophiolites: Proceedings of the International Ophiolite Symposium, Nicosia, 1979, Ministry of Agriculture and Natural Resources, Geological Survey Department, Cyprus, p. 261–272.

Pearce, J.A., Lippard, S.J., and Roberts, S., 1984, Characteristics and tectonic significance of supra-subduction zone ophiolites, *in* Kokelaar, B. P., and Howells, M.F., eds., Marginal basin geology: Volcanic and associated sedimentary and tectonic processes in modern and ancient marginal basins: Geological Society of London Special Publication 16, p. 77–94.

Pedersen, R.B., 1986, The nature and significance of magma chamber margins in ophiolites: Examples from the Norwegian Caledonides: Earth and Planetary Science Letters, v. 77, p. 100–112.

Pedersen, R.B., and Dunning, G.R., 1997, Evolution of arc crust and relations between contrasting sources: U-Pb (age), Nd and Sr isotope systematics of the ophiolite terrain of SW Norway: Contributions to Mineralogy and Petrology, v. 128, p. 1–15.

Pedersen, R.B., and Furnes, H., 1991, Geology, magmatic affinity and geotectonic environment of some Caledonian ophiolites in Norway: Journal of Geodynamics, v. 13, p. 183–203.

Pedersen, R.B., and Hertogen, J., 1990, Magmatic evolution of the Karmøy Ophiolite Complex, SW Norway: Relationships between MORB-IAT-boniniticcalc-alkaline and alkaline magmatism: Contributions to Mineralogy and Petrology, v. 104, p. 277–293.

Pedersen, R.B., Furnes, H., and Dunning, G.R., 1988, Some Norwegian ophiolite complexes reconsidered: Norges Geologiske Undersøkelse, Special Publications, v. 3, p. 80–85.

Pedersen, R.B., Furnes, H., and Dunning, G.R., 1991, The U/Pb age of the Sulitjelma Gabbro, north Norway: Further evidence for the development of a Caledonian marginal basin in Ashgillian/Llandoverian time: Geological Magazine, v. 128, p. 141–153.

Pedersen, R.B., Bruton, D.L., and Furnes, H., 1992, Ordovician faunas, island arcs and ophiolites in the Scandinavian Caledonides: Terra Nova, v. 4, p. 217–222.

Phipps Morgan, J., Harding, A., Orcutt, J., Kent, G. and Chen, Y.J., 1994, An observational and theoretical synthesis of magma chamber geometry and crustal genesis along a mid-ocean ridge spreading center, *in* Ryan, M.P., ed., Magmatic systems: San Diego, Academic Press, Inc., p. 139–178.

Pin, C., Briot, D., Bassin, C., and Poitrasson, F., 1994, Concomitant separation of strontium and samarium-neodymium for isotope analyses in silicate samples, based on specific extraction chromatography: Analytica Chimica Acta, v. 298, p. 209–217.

Ravnås, R., and Furnes, H., 1995, The use of chemical data in determining the provenance and tectonic setting of ancient sedimentary successions: The Kalvåg Mélange, western Norwegian Caledonides, *in* Plint, G., ed., Sedimentary facies analysis: Special Publication of the International Association of Sedimentologists 22, p. 237–264.

Ray, K.K., Segupta, S., and Van den Hul, H.J., 1988, Chemical character of volcanic rocks from Andaman ophiolite, India: Journal of the Geological Society, London, v. 145, p. 393–400.

Robson, D., and Cann, J.R., 1982, A geochemical model of mid-ocean ridge magma chambers: Earth and Planetary Science Letters, v. 60, p. 93–104.

Ryttvad, H.L., Furnes, H., Skjerlie, K.P., and Rolfsen, R., 2000, Geochemistry and petrogenesis of extrusive rocks, dykes and high-level plutonic rocks on the island of Oldra, Solund–Stavfjord Ophiolite Complex, western Norway: Norsk Geologisk Tidsskrift, v. 80, p. 97–110.

Saunders, A.D., 1983, Geochemistry of basalts recovered from the Gulf of California during Leg 65 of the Deep Sea Drilling Project, *in* Lewis, B.T.R., Robinson, P., et al., eds., Initial Reports of the Deep Sea Drilling Project Leg 65: Washington D.C., U.S. Government Printing Office, p. 591–621.

Seyfried, W.E., Jr., and Bischoff, J.L., 1979, Low temperature basalt alteration by seawater: An experimental study at 70°C and 150°C: Geochimica et Cosmochimica Acta, v. 43, p. 1937–1947.

Sinton, J.M., Wilson, D.S., Christie, D.M., Hey, R.L., and Delaney, J.R., 1983, Petrologic consequences of rift propagation on oceanic spreading ridges: Earth and Planetary Science Letters, v. 62, p. 193–207.

Skjerlie, F.J., 1969, The pre-Devonian rocks in the Askvoll–Gaular area and adjacent districts, western Norway: Norges Geologiske Undersøkelse, Bulletin, v. 258, p. 325–359.

Skjerlie, K.P., 1988, Geologiske relasjoner mellom gabbro og gangkompleks, petrogenese og tektonomagmatisk modellutvikling i området Tviberg-NV-Atløy, Solund–Stavfjord ofiolittkompleks [Candidatus Scientiarium thesis]: University of Bergen, 179 p.

Skjerlie, K.P., 1992, Petrogenesis and significance of late Caledonian granitoid magmatism in western Norway: Contributions to Mineralogy and Petrology, v. 110, p. 473–487.

Skjerlie, K.P., and Furnes, H., 1990, Evidence for a fossil transform fault in the Solund–Stavfjord Ophiolite Complex: West Norwegian Caledonides: Tectonics, v. 9, p. 1631–1648.

Skjerlie, K.P., and Furnes, H., 1996, The gabbro-dyke transition zone demonstrated on Tviberg, Solund–Stavfjord Ophiolite Complex: Geological Magazine, v. 133, p. 573–582.

Skjerlie, K.P, Furnes, H., and Johansen, R.J., 1989, Magmatic development and tectonomagmatic models for the Solund–Stavfjord Ophiolite Complex, West Norwegian Caledonides: Lithos, v. 23, p. 137–151.

Sparks, R.S.J., Meyer, P., and Sigurdsson, H., 1980, Density variation amongst mid-ocean ridge basalts: Implications for magma mixing and the scarcity of primitive lavas: Earth and Planetary Science Letters, v. 46, p. 419–430.

Staudigel, H., Plank, T., White, B., and Schmincke, H-U., 1996, Geochemical fluxes during seafloor alteration of the basaltic upper oceanic crust: DSDP Sites 417 and 418: American Geophysical Union Geophysical Monograph 96, p. 19–38.

Tarney, J., Wood, D.A., Saunders, A.D., Cann, J.R., and Varet, J., 1980, Nature of mantle heterogeneity in the North Atlantic: Evidence from deep sea drilling: Philosophical Transactions of the Royal Society of London, A 297, p. 179–202.

Taylor, B., and Natland, J., eds., 1995, Active margins and marginal basins of the western Pacific: American Geophysical Union Geophysical Monograph 88, 417 p.

Taylor, B., Zellmer, K., Martiez, F., and Goodliffe, A., 1996, Sea-floor spreading in the Lau back-arc basin: Earth and Planetary Science Letters, v. 144, p. 33–40.

Thompson, R.N., Morrison, M.A., Hendry, G.L., and Parry, S.J., 1984, An assessment of the relative roles of crust and mantle in magma genesis: An elemental approach: Philosophical Transactions of the Royal Society of London, A 310, p. 549–590.

Valley, W., 1986, Stable isotope geochemistry of metamorphic rocks: Reviews in Mineralogy, v. 16, p. 445–489.

Wood, D.A., Marsh, N.G., Tarney, J., Joron, J-L., Fryer, P., and Treuil, M., 1982, Geochemistry of igneous rocks recovered from a transect across the Mariana trough, arc, fore-arc and trench, sites 453 through 461, *in* Lee, M., and Powell, R., eds., Initial Reports of the Deep Sea Drilling Project 60: Washington, D.C., U.S. Government Printing Office, p. 611–645.

Manuscript Accepted by the Society February 23, 2000

Printed in the U.S.A.

Geological Society of America
Special Paper 349
2000

Early Devonian boninites from the Magnitogorsk arc, southern Urals (Russia): Implications for early development of a collisional orogen

Piera Spadea and Jane H. Scarrow
Department Georisorse Territorio, University of Udine, Via Cotonificio 114, I-33100 Udine, Italy; e-mail: spadea@uniud.it

ABSTRACT

In the Paleozoic rocks of the Uralian orogen, boninites are well exposed in the Baimak-Buribai area. They are part of the arc complex of the Devonian Magnitogorsk zone that collided with the East European craton during the Uralian orogeny. The boninite lavas and dikes and shallow, high-Mg diorite intrusions compose the lowest stratigraphic level of a sequence that progresses to boninitic andesite lavas and then to explosive tholeiitic and calc-alkalic products (lavas and tuffs). Relict petrographic characteristics serve to identify the boninites: glassy texture, abundant olivine and Cr-rich spinel, and the absence of early-crystallized plagioclase. Compositional similarities to Tertiary Western Pacific boninites—particularly to the intermediate- and high-Ca boninites from the Izu-Bonin forearc region drilled during ODP (Ocean Drilling Program) Leg 125 and the high-Ca boninites from the northern Tonga Ridge—suggest that the Baimak-Buribai boninites and related rocks were erupted in the forearc region early in the development of an island arc. They give paleogeodynamic information about timing and location of the initiation of the subduction in the southern pre-Uralian ocean. The geologic setting of the lower Baimak-Buribai Complex was an extensional, non-accretionary forearc, but by the time the tholeiitic and calc-alkalic rocks erupted, the setting had changed to an established accretionary forearc. Comparison of an Emsian stratigraphic age for the boninites with radiometric dates from collisional metamorphic rocks indicates an interval of 14 to 40 m.y. between initiation of oceanic subduction and arc-continent collision. The inferred duration for arc development during the Paleozoic Uralian orogeny is similar to, or shorter than, the duration of present-day examples of ongoing orogeny in the Western Pacific region.

INTRODUCTION

Boninites, first defined as a new class of rocks in the Izu-Mariana arc (Shiraki et al., 1978), have been identified in many Western Pacific intraoceanic arcs (Sharaskin et al., 1980; Crawford et al., 1981; Natland, 1981; Natland and Tarney, 1981; Hickey and Frey, 1982; Umino, 1986; Bloomer and Hawkins, 1987; Crawford, 1989, and references therein; Falloon and Crawford, 1991; Stern and Bloomer, 1992; Sobolev and Danyushevsky, 1994; Danyushevsky et al., 1995). Petrographically, the Tertiary boninitic lavas of the Western Pacific arcs are glassy or contain olivine and orthopyroxene phenocrysts. Typically, they have high MgO contents, low abundances of high field strength elements (HFSEs) relative to large ion lithophile elements (LILEs—K, Rb, Sr, and Ba), and low alkalis and TiO_2 relative to basaltic and andesitic rocks with equivalent SiO_2 contents (Meijer, 1980; Natland and Tarney, 1981; Crawford et al., 1989). Overall, their petrochemical features suggest magma temperatures greater than 1300–1350 °C (Duncan and Green, 1987; Crawford et al., 1989). Their primitive composition combined with high LILE/HFSE ratios indicates a strongly depleted source with an added fluid or melt component.

Models proposed for generation of boninites invoke melting of depleted mantle at shallow depths (Crawford et al., 1989; van der Laan et al., 1989; Tatsumi and Maruyama, 1989; Pearce et al., 1992). The common denominator in these models is atypical subduction as plate motions change, either as subduction is initiated (Tatsumi, 1982; Hawkins et al., 1984; Tatsumi and Maruyama, 1989; Crawford et al., 1986; Pearce et al., 1992), e.g., at a transform fault (Casey and Dewey, 1984), or as a ridge is subducted (van der Laan et al., 1989; Crawford et al., 1989). On the basis of the work just cited, the eruptive setting of Tertiary boninites has been restricted to forearc regions of island arcs mostly associated with early arc development (Pearce et al., 1992; Bloomer et al., 1995). Therefore, boninites can be considered as important, unambiguous, paleogeodynamic markers of incipient subduction in a forearc setting.

Despite their importance in defining paleo–plate configurations and motions, relatively few examples of pre-Tertiary boninites have been described in the literature. Ancient boninites are interpreted to have been generated by processes similar to those in modern settings including the following:

- Supra-subduction zones—e.g., the Cambrian Heathcote greenstone belt, Victoria, Australia (Nelson et al., 1984; Crawford et al., 1984; Crawford and Cameron, 1985); Cambrian Mount Wellington

Spadea, P., and Scarrow, J.H., 2000, Early Devonian boninites from the Magnitogorsk arc, southern Urals (Russia): Implications for early development of a collisional orogen, *in* Dilek, Y., Moores, E.M., Elthon, D., and Nicolas, A., eds., Ophiolites and Oceanic Crust: New Insights from Field Studies and the Ocean Drilling Program: Boulder, Colorado, Geological Society of America Special Paper 349, p. 461–472.

greenstone belt, Victoria, Australia (Nelson et al., 1984); Ordovician Thetford Mines ophiolite, Quebec (Shaw and Wasserburg, 1984; Coish, 1989; Melancon et al., 1997); central Newfoundland ophiolites (MacLachlan and Dunning, 1998); Ordovician Ballantrae ophiolite, Scotland (Smellie et al., 1995); and Cretaceous Troodos ophiolite (Cameron et al., 1983; McCulloch and Cameron, 1983; Thy, 1984; Thy et al., 1985; Cameron, 1985; Flower and Levine, 1987; Rogers et al., 1989).

- Forearcs—e.g., the pre–Middle Cambrian boninite of western Tasmania (Brown and Jenner, 1989; Crawford and Berry, 1992); Ordovician Betts Cove ophiolite, Newfoundland (Coish, 1989; Bédard et al., 1998); and Ordovician Bay of Islands ophiolite (Varfalvy et al., 1996), Newfoundland.
- Areas undergoing initiation of subduction—the Late Jurassic Migdhalitsa ophiolite, Greece (Clift and Dixon, 1998).
- Arcs in early in stages of development—the Ordovician Southern Pacquet Harbour Group volcanic rocks, Newfoundland (Coish, 1989; Piercy, 1996).

In the Paleozoic rocks of the Uralian mountain chain, boninitic rocks were first described by Rumianzeva et al. (1989) within Early Silurian volcanic units of the Central Urals and later by Kuz'min and Kabanova (1991) within Early Devonian volcanic units of the Southern Urals. These rocks are particularly important in light of recent tectonic models for the Paleozoic Uralian mountain chain (Perez-Estaùn et al., 1997; Brown et al., 1996, 1998) that interpret the Central and Southern Urals as an orogen formed by the collision of the European paleo–continental margin and the Silurian–Devonian Tagil (Central Urals)–Magnitogorsk (Southern Urals) island arc. The original setting and evolution of the arc-related magmatism are, therefore, important for inferring the timing, geodynamic conditions of convergence, and collisional processes in the Urals orogen.

Further to the work of Kuz'min and Kabanova (1991), detailed data on stratigraphic relationships, petrography, mineral chemistry, and chemistry of boninitic and associated tholeiitic and calc-alkalic rocks have been recently reported (Spadea et al., 1998) for the boninites and associated arc lavas in the Baimak-Buribai area from the Magnitogorsk island arc, where the volcanic rocks are well exposed and their original structures are well preserved. In this paper, we present new geochemical and Sr and Nd isotope data. Notwithstanding the alteration of the volcanic rocks, their petrology and geochemistry allow us to infer the petrogenesis of the volcanic suites and to make a comparison with in situ Tertiary boninites. We conclude with a discussion of the significance and early evolution of the magmatism of the Magnitogorsk arc in the collisional history of the Southern Urals.

GEOLOGIC BACKGROUND

Regional geologic setting

The Urals are a 2500-km-long, north-trending mountain chain extending from the Arctic Ocean, 68°N, to just north of the Aral Sea, 48°N (Fig. 1). To their east is a collage of terranes making up eastern Asia, and to their west is the East European craton.

From west to east in the present-day Urals, the following units crop out (Fig. 1): the pre-Uralian marginal depression, west Uralian zone, and the Ural-Tau zone; these units consist of Paleozoic sedimentary and Precambrian basement rocks and form part of the former passive margin of the East European craton developed by Late Cambrian to Early Ordovician rifting (McKerrow, 1994; Puchkov, 1997; Dalziel, 1997; Smethurst et al., 1998). The Ural-Tau zone is bounded to the east by the principal suture of the orogen, the Main Uralian fault, which contains lherzolitic ophiolite massifs and serpentinite mélange (Savelieva, 1987; Savelieva and Nesbitt, 1996; Savelieva et al., 1997; Gaggero et al., 1997). Farther to the east, the Tagil and Tagil-Magnitogorsk zones are composed of allochthonous oceanic material. Finally, the Paleozoic sedimentary and Precambrian metamorphic rocks of the East Uralian zone form part of the Siberian craton continental crust.

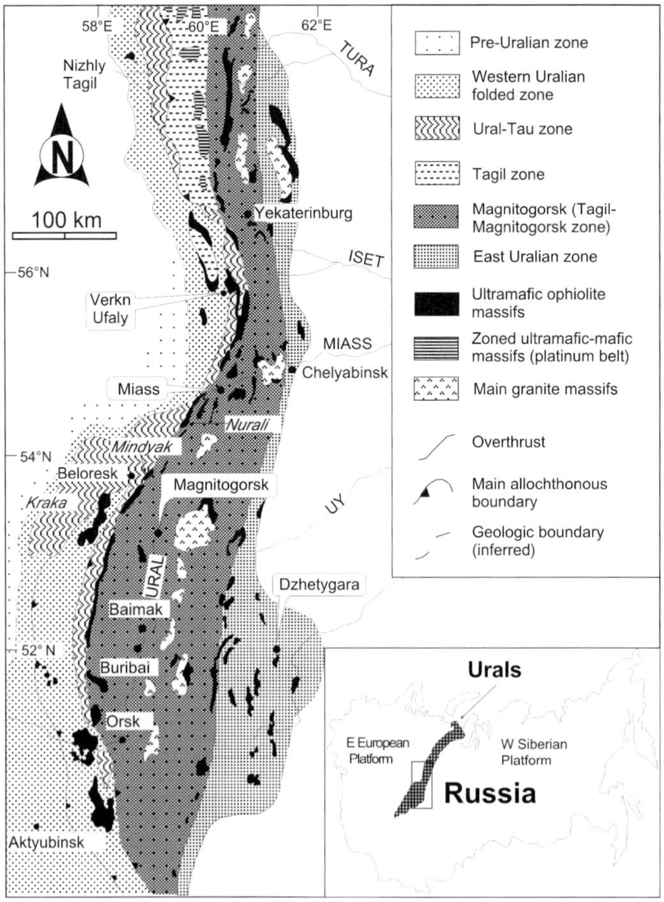

Figure 1. Geologic map of the Urals showing the location of Baimak.

Recent work has shown that the Urals developed by island-arc accretion to the East European craton continental margin followed by ocean closure and terminal collision. In particular, in the southern Urals, the main arc-continent collisional event occurred during the Late Devonian to Early Carboniferous, when continental lithosphere of the East European craton was subducted beneath the Magnitogorsk arc (Chemenda et al., 1997; Puchkov, 1997; Brown et al., 1998; Brown and Spadea, 1999). In Late Carboniferous to Permian time there was diachronous collision between the East European craton and the Siberian craton (Zonenshain et al., 1984; Matte, 1995; Puchkov, 1997).

Field and stratigraphic setting

The rocks in the present study come from the Baimak-Buribai area within the West Magnitogorsk zone (Fig. 1). This zone exposes a structurally well-preserved sequence comprising pillow lavas, hyaloclastites, dikes, volcanic-terrigenous deposits, and pelagic sedi-

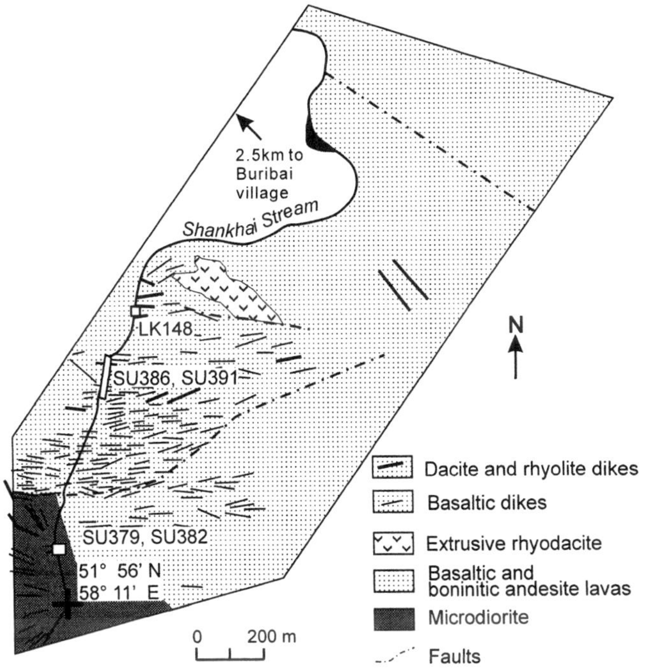

Figure 2. Schematic geologic map showing the Shankhai sample localities (redrawn from Kuz'min and Kabanova, 1991).

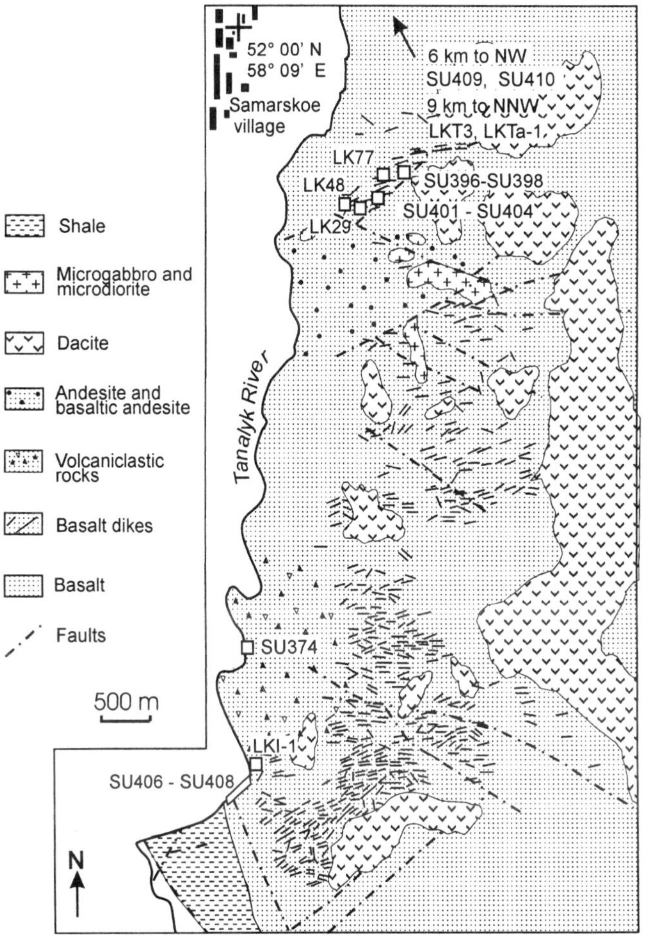

Figure 3. Schematic geologic map showing the Tanalyk sample localities (redrawn from Kuz'min and Kabanova, 1991).

ments, i.e., subvolcanic and volcanic materials extruded through and intruded into preexisting diorite that may represent arc plutonic rocks from the upper part of the middle crust. This sequence is here named the Baimak-Buribai Complex; it includes the mostly sedimentary and volcaniclastic Baimak-Buribai Complex of Emsian age (determined with conodonts by Ivanov et al., 1989) and the underlying extrusive and shallow intrusive rocks described in this paper. The Emsian deposits (410–394 Ma, according to the new time scale by Tucker et al., 1998) are directly overlain by the Turakta Formation shale and chert (Kuz'min and Kabanova, 1991) of Emsian–early Eifelian (394–388 Ma, according to Tucker et al., 1998) age (Maslov et al., 1993).

The lowest part of the Baimak-Buribai Complex exposed along the 2.5-km-long Shankhai stream bed (Fig. 2) consists of diorite cropping out at the base of the sequence extensively intruded by boninitic basalt to andesite dikes and dacite stocks and, up sequence, by a dike swarm with 2.5–3-m-wide northwest-trending dikes. Farther up section in the dike sequence dikes have chilled margins on southwest-dipping upper surfaces, indicating that the spreading center was to the northeast. The stratigraphically lowest lavas are boninitic andesite pillow- and pipe-shaped, 0.3–0.8-m-diameter, interdike and intersill screens. Higher up section are normal-position boninite pillow lavas, cut by boninitic, originally glassy, sills and dikes; at the top of the section these are overlain by hyaloclastite matrix pillow breccias and coarse breccias. All these rock types are cut by east-southeast–trending dacite and rhyolite intrusions.

Stratigraphically above the Shankhai stream sequence, 2–3 km to the northeast, is the Tanalyk sequence at Samarskoe (Figs. 3, 4). Here, over an area of several square kilometers, numerous complex dike swarms of boninitic andesite, 0.5–1.5 km in width, cut pillow lavas of similar composition; some lava screens are present at the swarm centers, but elsewhere dike-in-dike injections are common. The dikes, from decimeters to meters in width, strike between east-northeast and southeast and between north-northwest and north-northeast.

The structurally highest part of the Baimak-Buribai Complex crops out in a well-exposed sequence near Samarskoe. Here, steeply southwest-dipping volcaniclastic rocks and lavas crop out along an ~0.5-km-long section; they overlie boninitic pillow lavas and dike swarms of the Tanalyk sequence and underlie black shales, arenites, and cherts of the Turakta Formation (Emsian–early Eifelian). A 0.5-m-thick basalt dike cuts interlayered debris flow breccias, turbiditic sandstones and siltstones, and chert, at the base of this sequence. To the southwest, up section, are rhythmically interlayered evolved tuffs and packets of upward-fining volcaniclastic sandstones to siltstones and monogenetic silicified tuffs. These volcaniclastic rocks are overlain by a vesicular dacitic to andesitic lava and lava breccia. At the top of the sequence are debris flows: in the lowest, dacite clasts are dominant, whereas in the highest, the clasts are interlayered with red or green cherts, diverse volcanic rocks, and unsorted blocks. This sequence is an accretionary-prism deposit with background sedimentation of volcanic sandstones and siltstones interrupted by episodic, higher-energy, pulses of coarse sediment and debris flows from proximal volcanoes.

ANALYTICAL TECHNIQUES

Chemical data reported in Table 1 were obtained by XRF (X-ray fluorescence) spectrometry (major elements and some trace elements: Sc, V, Ni, Cu, Zn, Rb, Y, Sr, Zr, Nb, and Ba), ICP-AES (in-

TABLE 1. MAJOR AND TRACE ELEMENT DATA, Sr AND Nd ISOTOPE DATA FOR THE BAIMAK-BURIBAI MAGMATIC ROCKS

	Plutonic rocks		Boninitic basalt to andesite volcanic rocks									Boninites	
Sample	SU379	SU382	LK148-4a	LK148-5	LKTa1	LK148-1	SU398	SU401	SU386	LK148-3	LKI1	LK48-3	LK29-2b
Group	diorite	diorite	pic bas	bas and	bas and	bas and	bas and	bas and	bas and	and	and	bon	bon
Texture	granular	granular	ol phyr	aphyr	aphyr	ol sp phyr	ol phyr	ol px sp phyr	ol px sp phyr	ol sp phyr	pl phyr	aphyr	aphyr
SiO_2	54.25*	56.27	51.78	54.09	54.80	53.15	53.89	54.02	55.32	56.26	56.16	57.00	57.22
TiO_2	0.69	0.54	0.26	0.50	0.41	0.45	0.52	0.26	0.53	0.51	0.45	0.52	0.47
Al_2O_3	16.67	16.81	9.34	14.21	11.78	12.59	12.88	9.53	15.36	14.99	12.12	16.15	13.37
Fe_2O_3	13.10	9.32	9.20	9.44	9.55	10.17	8.47	8.50	9.28	8.68	9.41	6.76	7.42
MnO	0.37	0.26	0.18	0.28	0.17	0.23	0.13	0.24	0.37	0.26	0.17	0.12	0.17
MgO	9.34	8.20	18.65	12.92	10.35	12.65	14.13	14.76	11.81	12.31	9.74	8.46	9.68
CaO	2.00	4.20	8.88	4.18	9.05	6.92	5.89	10.20	4.21	2.52	7.32	4.16	5.95
Na_2O	4.00	4.28	1.63	4.09	3.89	4.00	3.68	2.41	3.53	4.14	4.51	2.75	5.71
K_2O	0.29	0.43	0.04	0.04	0.17	0.05	0.39	0.08	0.06	0.03	0.11	4.26	0.08
P_2O_5	0.08	0.07	0.08	0.11	0.10	0.09	0.05	0.04	0.07	0.11	0.10	0.10	0.09
Total	100.79	100.38	100.04	99.86	100.27	100.30	100.03	100.04	100.54	99.80	100.09	100.28	100.16
LOI	6.32	4.88	4.43	4.36	2.52	4.48	4.21	3.20	5.72	4.41	2.14	4.28	2.36
Mg#	0.59	0.64	0.80	0.73	0.68	0.71	0.77	0.77	0.72	0.74	0.67	0.71	0.72
Li	n.d.	7.03†	n.d.	n.d.	n.d.	n.d.	7.83	n.d.	6.33	n.d.	n.d.	n.d.	n.d.
Rb	3*	5.73	<1	2	4	<2	2.20	2	0.42	<2	<2	29	2
Cs	n.d.	0.14	n.d.	n.d.	n.d.	n.d.	0.13	n.d.	0.10	n.d.	n.d.	n.d.	n.d.
Be	n.d.	0.62	n.d.	n.d.	n.d.	n.d.	0.47	n.d.	0.55	n.d.	n.d.	n.d.	n.d.
Sr	49	121	27	33	109	28	66.0	91	71.5	23	55	185	68
Ba	n.d.	69.4	33	34	58	26	26.0	n.d.	18.6	26	44	385	42
Sc	n.d.	35.3	37	44	45	41	31.4	n.d.	30.5	35	44	41	34
V	307	156	136	205	196	222	151	n.d.	242	148	190	219	183
Cr	174	147	1450	580	382	490	1011	857	426	650	395	44	147
Co	n.d.	23.8	n.d.	n.d.	n.d.	n.d.	44.4	n.d.	33.1	n.d.	n.d.	n.d.	n.d.
Ni	61	57.7	445	144	127	150	381	352	228	188	108	31	82
Cu	144	13.8	3	<2	9	<2	16.3	2	57.1	110	3	443	<2
Zn	454	98.7	54	116	60	105	38.0	85	472	124	54	28	57
Ga	n.d.	13.9	n.d.	n.d.	n.d.	n.d.	9.25	n.d.	14.5	n.d.	n.d.	n.d.	n.d.
Y	15	15.9	6	11	12	11	11.1	7	12.9	13	12	14	11
Nb	1	1.06	<2	<2	3	2	0.63	1	0.81	3	<2	2	3
Ta	n.d.	0.07	n.d.	n.d.	n.d.	n.d.	0.04	n.d.	0.06	n.d.	n.d.	n.d.	n.d.
Zr	50	42.3	23	37	32	35	24.8	28	34.5	36	35	40	47
Hf	n.d.	1.38	n.d.	n.d.	n.d.	n.d.	0.82	n.d.	1.10	n.d.	n.d.	n.d.	n.d.
Mo	n.d.	1.13	n.d.	n.d.	n.d.	n.d.	1.11	n.d.	0.76	n.d.	n.d.	n.d.	n.d.
Sn	n.d.	0.67	n.d.	n.d.	n.d.	n.d.	2.04	n.d.	2.30	n.d.	n.d.	n.d.	n.d.
Tl	n.d.	0.01	n.d.	n.d.	n.d.	n.d.	0.00	n.d.	0.00	n.d.	n.d.	n.d.	n.d.
Pb	n.d.	1.46	n.d.	n.d.	n.d.	n.d.	0.57	n.d.	2.04	n.d.	n.d.	n.d.	n.d.
U	n.d.	0.16	n.d.	n.d.	n.d.	n.d.	0.10	n.d.	0.17	n.d.	n.d.	n.d.	n.d.
Th	n.d.	0.43	n.d.	n.d.	n.d.	n.d.	0.22	n.d.	0.41	n.d.	n.d.	n.d.	n.d.
La	n.d.	2.36	0.85§	1.03	1.26	1.15	1.01	n.d.	2.53	1.57	1.42	2.68	1.75
Ce	n.d.	6.27	1.43	2.69	3.49	3.29	2.83	n.d.	6.56	4.07	3.88	6.00	4.64
Pr	n.d.	0.96	n.d.	n.d.	n.d.	n.d.	0.48	n.d.	0.98	n.d.	n.d.	n.d.	n.d.
Nd	n.d.	4.91	1.71	2.33	2.91	2.76	2.64	n.d.	4.83	3.25	3.14	4.34	3.72
Sm	n.d.	1.69	0.44	0.70	0.94	0.94	0.97	n.d.	1.49	1.05	0.97	1.28	1.13
Eu	n.d.	0.61	0.15	0.29	0.35	0.36	0.29	n.d.	0.59	0.35	0.37	0.49	0.36
Gd	n.d.	2.15	0.75	1.12	1.39	1.33	1.39	n.d.	1.82	1.49	1.45	1.61	1.55
Tb	n.d.	0.39	n.d.	n.d.	n.d.	n.d.	0.25	n.d.	0.32	n.d.	n.d.	n.d.	n.d.
Dy	n.d.	2.72	0.95	1.44	1.84	1.73	1.86	n.d.	2.21	1.86	1.89	2.00	1.90
Ho	n.d.	0.63	n.d.	n.d.	n.d.	n.d.	0.43	n.d.	0.49	n.d.	n.d.	n.d.	n.d.
Er	n.d.	1.83	0.58	0.96	1.21	1.10	1.29	n.d.	1.37	1.18	1.02	1.27	1.21
Tm	n.d.	0.29	n.d.	n.d.	n.d.	n.d.	0.20	n.d.	0.21	n.d.	n.d.	n.d.	n.d.
Yb	n.d.	1.76	0.65	1.10	1.26	1.27	1.26	n.d.	1.32	1.23	1.26	1.33	1.27
Lu	n.d.	0.27	0.08	0.18	0.19	0.20	0.19	n.d.	0.21	0.18	0.20	0.20	0.20
$^{87}Sr/^{86}Sr$**	n.d.	0.70534	0.70694	n.d.	n.d.	n.d.	n.d.	n.d.	n.d.	n.d.	n.d.	n.d.	n.d.
$^{87}Sr/^{86}Sr_{leach}$††	n.d.	0.70523	0.70582	n.d.	n.d.	n.d.	n.d.	n.d.	n.d.	n.d.	n.d.	n.d.	n.d.
$^{143}Nd/^{144}Nd$**	n.d.	0.51300	0.51287	n.d.	n.d.	n.d.	n.d.	n.d.	n.d.	n.d.	n.d.	n.d.	n.d.

Note: Mg# = mol [Mg/(Mg + FeO_{tot})]

Abbreviations: pic—picrite, bas—basalt, bas and—basaltic andesite, and—andesite, bon—boninite, dac—dacite, aphyr—aphyric, ol—olivine, sp = spinel, phyr = porphyritic, px = pyroxene, pl = plagioclase, n.d. = not determined.

* Analyses by XRF at Udine University

† Analyses by ICP-MS at Granada University

§ Analyses by ICP-AES at Udine University

** Analyses by mass spectrometry at Granada University

†† Analyses by mass spectrometry at Napoli University

ductively-coupled-plasma atomic emission spectrometry) (rare earth elements [REEs]: La, Ce, Nd, Sm, Eu, Gd, Dy, Er, Yb, Lu), and ICP-MS (inductively-coupled-plasma mass spectrometry) (all REEs, except Pm, and 25 trace elements: Li, Rb, Cs, Be, Sr, Ba, Sc, V, Cr, Co, Ni, Cu, Zn, Ga, Y, Nb, Ta, Zr, Hf, Mo, Sn, Tl, Pb, U, and Th). Details on analytical techniques, also used for the newly analyzed samples, are reported in Spadea et al. (1998).

Samples (SU382, SU407, LK29–2, LK148–4a, AKT5, and AKT48) were prepared for $^{87}Sr/^{86}Sr$ and $^{143}Nd/^{144}Nd$ analysis by conventional ion-exchange resin separation techniques. Measurements were made by TIMS (thermal ionization mass spectrometry) at the University of Granada on a Finnigan MAT 262 RPQ spectrometer. Instrumental errors for determinations of $^{87}Sr/^{86}Sr$ and $^{143}Nd/^{144}Nd$ were ±0.002–0.004 rel% (relative percent) and ±0.0013–

TABLE 1. MAJOR AND TRACE ELEMENT DATA, Sr AND Nd ISOTOPE DATA FOR THE BAIMAK-BURIBAI MAGMATIC ROCKS (continued)

	Boninites				Tholeiitic and calc-alkalic volcanic rocks								
Sample	LK29-2	LK29-2c	LK148-4	LK148-2	SU407	AKT5	SU406	LK77	SU403	LKT3	AKT48	SU374	SU409
Group	bon	bon	bon	bon	bas	bas	bas and	bas and	and	and	and	and	dac
Texture	aphyr	aphyr	aphyr	ol sp phyr	ol pl phyr	aphyr	ol pyx pl phyr	px sp phyr	ol px sp phyr	ol sp phyr	px phyr	px sp phyr	pl pax phyr
SiO_2	57.70	58.37	62.31	58.36	48.36	52.52	56.16	56.43	58.68	59.30	62.26	62.88	66.89
TiO_2	0.46	0.47	0.39	0.49	0.63	0.92	1.00	0.43	0.79	0.39	0.63	0.44	0.47
Al_2O_3	12.17	14.77	13.70	13.02	17.91	19.21	17.34	16.89	15.89	17.00	14.26	15.56	16.39
Fe_2O_3	7.95	5.94	6.80	8.64	13.67	12.17	10.90	7.03	10.23	4.60	10.87	7.33	4.47
MnO	0.17	0.16	0.14	0.23	0.17	0.12	0.20	0.15	0.15	0.10	0.14	0.08	0.08
MgO	9.63	9.16	7.48	11.13	4.61	2.69	4.39	6.60	5.20	4.50	3.91	5.57	2.48
CaO	6.56	4.98	4.92	4.11	12.86	5.12	2.72	6.35	4.18	7.21	2.06	2.30	1.81
Na_2O	5.50	6.30	4.33	3.80	1.76	6.50	6.68	5.85	4.56	6.34	4.93	5.43	6.70
K_2O	0.10	0.11	0.03	0.03	0.03	0.06	0.13	0.40	0.55	0.80	0.32	0.27	0.83
P_2O_5	0.09	0.09	0.07	0.14	0.11	0.10	0.43	0.08	0.11	0.09	0.15	0.09	0.14
Total	100.33	100.35	100.17	99.93	100.11	99.41	99.95	100.21	100.34	100.33	99.53	99.95	100.26
LOI	1.89	3.05	5.09	4.28	7.29	2.46	4.80	4.06	3.38	2.46	2.96	3.23	2.32
Mg#	0.71	0.75	0.69	0.72	0.40	0.30	0.44	0.65	0.50	0.66	0.42	0.60	0.52
Li	n.d.	n.d.	n.d.	n.d.	13.3	1.76	8.20	n.d.	3.05	n.d.	3.16	23.4	12.3
Rb	3	<2	<2	<2	0.69	1.21	2.75	3	10.6	4	2.47	3.91	12.3
Cs	n.d.	n.d.	n.d.	n.d.	0.45	0.12	0.25	n.d.	0.17	n.d.	0.25	0.25	0.27
Be	n.d.	n.d.	n.d.	n.d.	0.91	0.72	0.75	n.d.	0.55	n.d.	0.62	0.48	1.02
Sr	61	79	66	47	40.9	150	106	57	170	58	40.7	140	66.8
Ba	44	48	31	29	22.0	16.4	88.8	66	96.8	98	18.9	34.8	63.5
Sc	39	36	33	43	45.4	31.1	41.8	41	33.8	24	33.2	30.8	14.9
V	203	146	175	160	493	196	232	193	326	137	274	169	80.4
Cr	133	142	100	550	37.8	20.8	17.5	142	28.6	54	42.2	88.3	74.4
Co	n.d.	n.d.	n.d.	n.d.	26.8	26.3	23.9	n.d.	21.1	n.d.	16.5	23.5	9.35
Ni	91	88	45	142	21.2	39.3	9.69	40	24.5	23	16.3	45.4	32.3
Cu	<1	<1	318	38	142	13.5	29.5	<1	82.7	<2	107	22.8	20.1
Zn	59	61	88	103	94.5	35.8	112	75	57.0	44	54	46.0	42.2
Ga	n.d.	n.d.	n.d.	n.d.	21.3	19.7	14.0	n.d.	15.1	n.d.	16.8	10.6	14.6
Y	11	13	9	11	18.2	19.3	19.8	20	15.8	17	13.1	11.4	13.5
Nb	4	2	2	3	0.86	2.12	2.13	2	1.21	3	1.01	0.92	2.10
Ta	n.d.	n.d.	n.d.	n.d.	0.05	0.07	0.12	n.d.	0.09	n.d.	0.00	0.06	0.14
Zr	48	48	37	36	33.2	68.8	41.9	40	42.2	39	19.7	34.6	84.3
Hf	n.d.	n.d.	n.d.	n.d.	1.13	2.11	1.44	n.d.	1.35	n.d.	0.59	1.13	2.43
Mo	n.d.	n.d.	n.d.	n.d.	2.39	0.09	0.54	n.d.	0.69	n.d.	0.26	0.64	1.70
Sn	n.d.	n.d.	n.d.	n.d.	1.18	0.00	1.03	n.d.	1.03	n.d.	0.00	1.00	3.20
Tl	n.d.	n.d.	n.d.	n.d.	0.03	0.08	0.01	n.d.	0.01	n.d.	0.02	0.00	0.03
Pb	n.d.	n.d.	n.d.	n.d.	6.35	1.14	1.65	n.d.	1.65	n.d.	1.96	5.27	1.03
U	n.d.	n.d.	n.d.	n.d.	0.58	0.15	0.25	n.d.	0.25	n.d.	0.26	0.48	0.23
Th	n.d.	n.d.	n.d.	n.d.	0.67	0.91	0.53	n.d.	0.53	n.d.	0.23	0.40	0.60
La	1.67	1.86	1.99	1.52	3.84	5.00	2.91	3.59	2.91	3.36	1.75	2.26	5.10
Ce	4.60	4.77	4.77	3.67	9.17	11.80	7.47	4.61	7.47	3.96	4.36	5.54	12.91
Pr	n.d.	n.d.	n.d.	n.d.	1.34	1.78	1.11	n.d.	1.11	n.d.	0.73	0.81	1.79
Nd	3.77	3.67	3.38	3.26	6.65	8.57	5.45	5.01	5.45	4.55	3.47	3.74	8.20
Sm	0.74	1.07	0.89	0.95	1.93	2.62	1.75	1.57	1.75	1.31	1.07	1.21	2.13
Eu	0.38	0.31	0.37	0.44	0.81	1.00	0.63	0.48	0.63	0.34	0.44	0.40	0.63
Gd	1.59	1.51	1.20	1.50	2.50	3.13	2.31	2.37	2.31	2.20	1.60	1.57	2.27
Tb	n.d.	n.d.	n.d.	n.d.	0.43	0.53	0.40	n.d.	0.40	n.d.	0.29	0.28	0.37
Dy	1.95	1.84	1.40	1.90	3.05	3.40	2.88	2.62	2.88	2.36	1.97	2.06	2.40
Ho	n.d.	n.d.	n.d.	n.d.	0.69	0.74	0.63	n.d.	0.63	n.d.	0.45	0.46	0.53
Er	1.20	1.15	0.89	1.19	2.02	2.08	1.79	1.58	1.79	1.49	1.44	1.30	1.53
Tm	n.d.	n.d.	n.d.	n.d.	0.31	0.33	0.27	n.d.	0.27	n.d.	0.22	0.20	0.23
Yb	1.32	1.17	0.90	1.25	1.95	2.11	1.67	1.26	1.67	1.22	1.46	1.29	1.48
Lu	0.24	0.16	0.12	0.20	0.30	0.32	0.27	0.19	0.27	0.19	0.23	0.20	0.24
$^{87}Sr/^{86}Sr$**	0.70576	n.d.	n.d.	n.d.	0.70595	0.70483	n.d.	n.d.	n.d.	n.d.	0.70585	n.d.	n.d.
$^{87}Sr/^{86}Sr_{leach}$††	0.70561	n.d.	n.d.	n.d.	0.70535	0.70479	n.d.	n.d.	n.d.	n.d.	n.d.	n.d.	n.d.
$^{143}Nd/^{144}Nd$**	0.51295	n.d.	n.d.	n.d.	0.51282	0.51291	n.d.	n.d.	n.d.	n.d.	0.51295	n.d.	n.d.

0.003 rel%, respectively. External precision (2σ) for $^{87}Sr/^{86}Sr$ from successive replicate measurements of standards was ±0.0028 rel% for standard WS-E (Whin Sill dolerite; Govindaruju et al., 1994) and was better than 0.0007% for the international standard for Sr (NBS-987). For $^{143}Nd/^{144}Nd$, the external precision (2σ) for WSE was ±0.0016 rel% and that for La Jolla was better than 0.0014 rel%. The normalization value for $^{143}Nd/^{144}Nd$ was $^{146}Nd/^{144}Nd = 0.7219$. The values of $^{87}Rb/^{86}Sr$ and $^{147}Sm/^{144}Nd$ were calculated from direct ICP-MS measurements (technique as already described herein) of $^{85}Rb/^{88}Sr$ and $^{147}Sm/^{145}Sm$. The spectrometer was calibrated from pure dissolutions of Rb and Sr ($^{87}Rb/^{86}Sr = 0.707265$). The precision of the method calculated on 10 replicates of standard WSE, analyzed over 2 months, was better than 1.2% (2σ). For Sm and Nd, the spectrometer was also calibrated from pure dissolutions ($^{147}Sm/^{144}Nd = 0.512240$). For this ratio, the precision of the method, also calculated on 10 replicates of standard WSE analyzed over 2 months, was better than 0.9% (2σ). In order to evaluate the extent of $^{87}Sr/^{86}Sr$ changes related to alteration, Sr isotope determinations on leached samples SU382, SU407, LK29-2, LK148-4a, and AKT5 were made at the Dipartimento di Geofisica e Vulcanologia (University Federico II of Naples). For these determinations, 0.300 mg of powder were leached with warm 6*N* HCl for 30 min, then rinsed thoroughly in pure sub-boiling double-distilled water, and finally dissolved with high purity $HF-HNO_3-HCl$ mixtures. Sr was extracted by conventional ion-exchange chromatographic techniques. Measurements were made by using a VG 354 double-collector thermal-ionization mass spectrometer

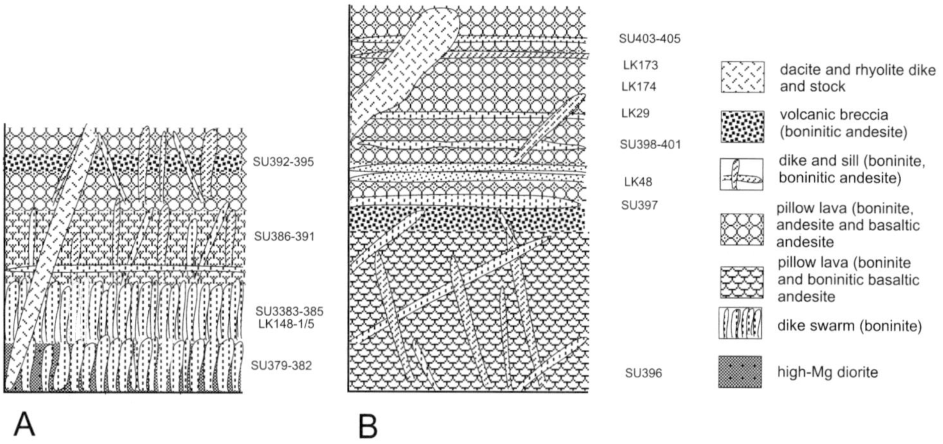

Figure 4. Schematic stratigraphic sections (after Spadea et al., 1998). A, Shankhai River succession. B, Tanalyk River succession (near Samarskoe). Not to scale.

running in jumping mode, by normalizing to $^{87}Sr/^{86}Sr = 0.1194$ for mass fractionation effects. The error given is the standard deviation of the mean ($2\sigma_m$) and refers to the last digit. Repeated analyses of the international reference standard NBS-987 yielded a mean value of $^{87}Sr/^{86}Sr = 0.71024 \pm 1$ ($N = 50$). The total blank was on the order of 6 ng during the period of measurements and thus was negligible for the measured samples.

SAMPLE CLASSIFICATION

Identification of typical boninitic minerals, e.g., enstatite, in the Baimak-Buribai rocks is not possible because low-grade hydrothermal metamorphism has pervasively transformed their original minerals and textures. Only some magmatic Cr-rich spinel and clinopyroxene are preserved, whereas olivine, orthopyroxene, and glass are completely replaced by pseudomorphs consisting of polyphase alteration-mineral assemblages. Low-grade metamorphic minerals are sub–greenschist-facies (prehnite, pumpellyite, albite, and chlorite) and greenschist-facies (actinolite, albite, epidote, chlorite, and titanite) assemblages related to seafloor water–rock interactions (cf. Alt et al., 1986). In view of the already-mentioned compositional changes, we do not use certain chemical classification criteria, e.g., LILE concentrations, that are typically applied to boninitic suites. Instead, initial classification was based on relict mineral constituents, including pseudomorphs after phenocrysts, and relict, glassy, texture; whole-rock major and trace element data were then used to identify significant intergroup compositional differences. From these considerations, we suggest that rocks from the Baimak-Buribai Complex are comparable to other boninitic suites worldwide. Some of the rocks are true boninites having glassy textures and some olivine- and Cr-rich spinel (with Cr# [Cr/(Cr + Al)] = 0.73–0.78) bearing lavas with boninitic REE concentrations; some are boninitic basalt to andesite (with Ca-rich pyroxene, but without plagioclase, Mg-rich pyroxene, and Cr-rich spinel [Cr# = 0.57–0.78] or boninitic trace element abundances), and others are arc tholeiitic and/or calc-alkalic basalts to dacite (containing relict plagioclase and having elemental characteristics transitional between depleted arc tholeiitic rocks and calc-alkalic rocks).

PETROGRAPHY

High-Mg plutonic rocks: Diorite

The high-Mg diorites (SU379, SU382) had a magmatic assemblage of plagioclase, clinopyroxene, hornblende, and iron-titanium oxides; these minerals have been replaced by a metamorphic assemblage of albite, actinolite, chlorite, epidote, quartz, magnetite, and titanite.

High-Mg volcanic rocks: Boninitic basalt to andesite

The high-Mg extrusive rocks (pillow lavas) and dikes include a picritic basalt and boninitic andesites. The picritic basalt is aphyric, whereas the boninitic basaltic andesites have relict olivine and/or orthopyroxene and, in one sample, plagioclase phenocrysts. Fresh clinopyroxene phenocrysts are enstatite-rich diopside or augite; within a single sample, clinopyroxenes have a range of Mg# values [e.g., 0.87–0.93; Mg# = Mg/(Mg + Fe)], suggesting that the magmas underwent fractional crystallization in a magma chamber prior to eruption; this interpretation is also indicated by a positive correlation between $Mg\#_{rock}$ and $Mg\#_{clinopyroxene}$ (Spadea et al., 1998). Cr-rich spinel microphenocrysts within pseudomorphs after olivine have Cr# values of 0.57–0.78 with Mg# values of 0.57–0.78. This diversity is seen even within a single sample.

Boninites

The boninites are aphyric, with spherulitic to dendritic and microlitic, and commonly amygdaloidal textures. Cr-rich spinel microphenocrysts from these rocks have Cr# values of 0.73–0.78, like those from modern boninites (Crawford et al., 1986) and Mg# values between 0.52 and 0.59. Their source was more refractory than that of MORB (mid-oceanic-ridge basalt) or primitive arc magmas.

Tholeiitic and calc-alkalic volcanic rocks: Basalt to dacite

The extrusive rocks (pillow lavas, dikes, and sills) exposed in the upper Tanalyk sequence and the Samarskoe type section are depleted tholeiitic and calc-alkalic in composition and become more evolved up sequence. The analyzed samples include a moderately phyric basalt with plagioclase, clinopyroxene (ferrosilite-rich augite), and (relict) olivine phenocrysts and a basaltic andesite. Clinopyroxene phenocrysts in these rocks, similar to the clinopyroxenes in the boninitic andesite lavas, show Mg# variations (70–80) and correlation with $Mg\#_{rock}$ that indicate fractional crystallization prior to eruption. The andesites contain clinopyroxene, fresh Cr-rich spinel, and, in two samples, titaniferous magnetite phenocrysts. The most evolved rock in this group is a dacite sampled from near the top of the Baimak-Buribai sequence. It is moderately porphyritic with plagioclase, minor

Ca-rich pyroxene, and titaniferous magnetite phenocrysts in an altered hyalopilitic groundmass.

GEOCHEMISTRY

Whole-rock data for the Baimak-Buribai Complex rocks are presented in Table 1. As noted above, there is a general trend from boninitic to arc-tholeiitic to calc-alkalic compositions up sequence as well as a general trend to more evolved compositions.

High-Mg plutonic rocks: Diorite

The diorite SiO_2 contents range from 54.25 wt% to 56.27 wt%. Compositionally, they are comparable to the extrusive boninitic andesites and, relative to the boninites, have higher Al_2O_3, FeO_{total}, TiO_2, and Zr and lower MgO and CaO (Fig. 5). The only analyzed high-Mg diorite has a flat chondrite-normalized REE pattern (Fig. 6). Initial Sr and Nd isotope ratios for the high-Mg diorite (Table 1, Fig. 7) suggest that, like the boninites and arc rocks, they were derived from a depleted source.

High-Mg volcanic rocks: Boninitic basalt to andesite

The most primitive sample in this group, and in the Baimak-Buribai Complex, is a picritic basalt that has the highest FeO_{total}, MgO, and CaO values and the lowest TiO_2 values. The rocks with intermediate silica contents are characterized by relatively high MgO contents (9.7–14.7 wt%). The picritic basalt has the highest values of Cr, 1450 ppm, Ni, 445 ppm, and Sr, and the lowest values of Zr; Cr and Ni are also high in the basaltic andesites (382–1011 ppm and 108–403 ppm, respectively). Intragroup trends are evident; SiO_2 correlates positively with Al_2O_3, total alkalis, TiO_2, and Zr and negatively with FeO_{total}, MgO, CaO, Cr, and Ni; these trends reflect fractional crystallization of olivine and/or pyroxene without accompanying plagioclase fractionation.

The boninitic basaltic-andesitic rocks are depleted tholeiites; they have the flat to slightly LREE-depleted chondrite-normalized REE patterns (La_N/Yb_N = 0.53–0.84, $Sm_N/Yb_N \leq 1$, and $La_N/Sm_N \leq 1$) typical of primitive arc tholeiites (Figs. 5 and 6).

Boninites

The SiO_2 and MgO contents of the boninites range from 57.0 to 62.3 wt% and from 7.5 to 11.1 wt%, respectively; nevertheless, there are few clear trends in the data set. The trace elements are also quite scattered; most elements, e.g., Cr, Ni, Zr, and Sr, show a range of values at approximately constant SiO_2. Al_2O_3 varies from 12 wt% to 16 wt% at similar values of SiO_2. For this reason, we suggest that, in addition to the ferromagnesian assemblage, plagioclase was fractionating from the more evolved samples in this group; this conclusion is supported by similar behavior in CaO and Sr.

Chondrite-normalized REE patterns of these rocks are flat to slightly concave upward, similar to typical boninites: $Sm_N/Yb_N \leq 1$ (0.61–1.07) and $La_N/Sm_N \geq 1$ (0.94–1.38). The suggestion of plagioclase fractionation is corroborated by weak, negative and positive, Eu anomalies (Fig. 6).

The Sr and Nd isotope ratios (Table 1, Fig. 7) are recalculated to 400 Ma, in accord with the Emsian age determined from conodonts in intralava sedimentary rocks (Ivanov et al., 1989) and the time scale by Tucker et al. (1998). From the spread in $^{87}Sr/^{86}Sr_{400}$ at restricted $^{143}Nd/^{144}Nd_{400}$ (Fig. 7), we suggest that the boninites were derived from a depleted source, i.e., a source with low time-integrated Rb/Sr and high Sm/Nd ratios relative to bulk Earth values. The Sr isotope ratios determined on unleached and leached samples (Table 1) show significant differences, mostly attributable to the carbonates present in the unleached sample. Thus this result indicates that the rocks were affected by seawater hydrothermal alteration. It should be noted, however, that the values obtained for the leached samples do not necessarily reflect primary values because some secondary phases, like epidote or titanite, are not affected by acid leaching. Therefore, the effect of their presence is not removed by this process.

Tholeiitic and calc-alkalic volcanic rocks: Basalt to dacite

An olivine- and plagioclase-phyric basalt, SU407, with the lowest SiO_2 contents, 48.4 wt%, has high loss-on-ignition and different element concentrations compared with other Baimak-Buribai Complex rocks. For these reasons, it is excluded from discussion in terms of major and trace elements; only HFSEs and REEs are used in the petrogenetic interpretation of this sample.

Tholeiitic and calc-alkalic extrusive rocks can be distinguished from the boninitic rocks by lower MgO contents and, generally, higher Al_2O_3 contents, at a given value of SiO_2. A clinopyroxene- and plagioclase-phyric basaltic andesite, SU406, also has characteristic high total alkalis, FeO_{total}, and TiO_2. Aphyric and clinopyroxene- and plagioclase-phyric andesites have a range of SiO_2 similar to that of the boninites; however, the andesites differ from the boninites in having higher FeO_{total} and lower MgO contents. The most evolved Baimak-Buribai Complex dacite has SiO_2 contents of 66.9 wt% and MgO contents of 2.5 wt%. In terms of their trace elements, compared with the boninitic rocks, the basaltic andesites have higher Sr and much lower Cr and Ni, whereas the andesites, which are obviously different in major element contents, are not clearly distinguished by trace elements; for example, Cr and Ni overlap in the two groups.

For the andesites, the fact that SiO_2 and Al_2O_3 correlate negatively with CaO and MgO indicates that plagioclase and ferromagnesian mineral(s) were fractionating during the evolution of this group. This conclusion is in accord with observed phenocryst assemblages. The dacite lies on an extension of the MgO and CaO andesite trends. Departure from the andesite trends in Al_2O_3, FeO_{total}, and TiO_2 probably reflects accumulation of plagioclase and titaniferous magnetite phenocrysts. Further evidence of plagioclase accumulation is, possibly, the particularly high Sr, 178 ppm, in the dacite.

This rock group has slight to more pronounced increases in chondrite-normalized LREE/HREE ratios (Fig. 6). The basaltic andesite has La_N/Yb_N = 2.01, the andesites have $La_N/Yb_N \approx 1.85$, and the dacite has La_N/Yb_N = 2.27. Negative Eu anomalies in the andesites and dacite ($Eu/Eu^* \approx 0.62$ and $Eu/Eu^* = 0.87$, respectively) imply involvement of plagioclase in the fractionating assemblages.

The initial $^{87}Sr/^{86}Sr$ and $^{143}Nd/^{144}Nd$ ratios for the porphyritic basalt (Table 1, Fig. 7) suggest that, like the boninites, the arc rocks were derived from a depleted source, i.e., low time-integrated Rb/Sr and high Sm/Nd ratios relative to bulk Earth ratios. Given the altered nature of the basalt, it is also likely, as we suggested for the boninites, that the $^{87}Sr/^{86}Sr$ ratio has been modified by alteration.

DISCUSSION

Comparison with Western Pacific Tertiary boninites

Petrographic and mineralogic comparisons between Baimak-Buribai Complex rocks and Western Pacific Tertiary boninites are

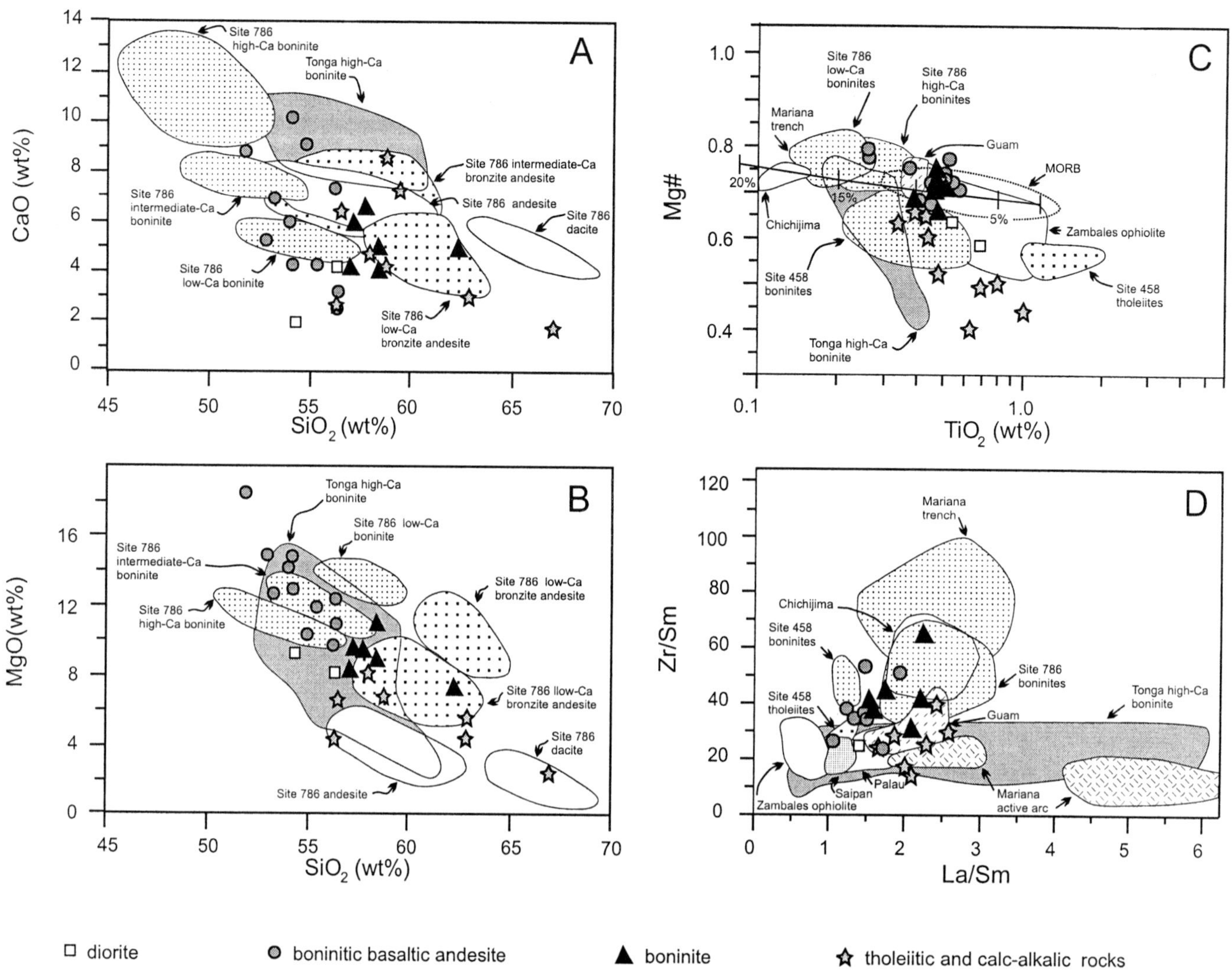

Figure 5. A, CaO vs. SiO_2 for Baimak-Buribai Complex rocks. B, MgO vs. SiO_2 for Baimak-Buribai Complex rocks; fields (in A and B) are for Tertiary rocks from West Pacific ODP (Ocean Drilling Program) Leg 125, Site 786, after Arculus et al. (1992), and for high-Ca boninites from the northern Tonga Ridge, after Falloon and Crawford (1991). C, Mg# vs. TiO_2 for Baimak-Buribai complex rocks; fields are for Tertiary rocks from the West Pacific after Pearce et al. (1992), and for high-Ca boninites from the northern Tonga Ridge, after Falloon and Crawford (1991). The annotated line is the trend for pooled melts from variably depleted residues, as defined by Figure 16 of Pearce et al. (1992). D, Zr/Sm vs. La/Sm for Baimak-Buribai complex rocks; fields are for Tertiary rocks from the West Pacific after Pearce et al. (1992) and for high-Ca boninites from the northern Tonga Ridge, after Falloon and Crawford (1991).

somewhat difficult because pervasive alteration has affected the former. Notwithstanding this feature, a significant, apparent, mineralogical difference between our rocks and typical low- and intermediate-Ca boninites is identified, i.e., the rarity of enstatite-clinoenstatite-bronzite phenocrysts. Furthermore, the common occurrence of clinopyroxene indicates that the magmas were Ca rich relative to typical boninitic magmas. On the other hand, primary, Cr-rich spinel in the Baimak-Buribai Complex boninites has high Cr# values comparable with Western Pacific Tertiary boninites (cf. Crawford et al., 1986). In addition, as in Western Pacific boninites, the more mafic boninitic rocks do not contain plagioclase. In summary, the boninites and boninitic basaltic andesite rocks have petrographic and mineralogic features, similar to those of intermediate- to high-Ca boninites (cf. van der Laan et al., 1992), which indicate that most of them crystallized from refractory melts with the following sequence: Cr-rich spinel, olivine, clinopyroxene, and, finally, plagioclase; orthopyroxene possibly preceded clinopyroxene in rare boninites that are lower in Ca.

Compositionally, the Baimak-Buribai Complex rocks can be compared to the geographically extensive Western Pacific Tertiary boninites and associated rocks, in particular the Izu-Bonin (Hickey and Frey, 1982; Bloomer and Hawkins, 1987; Crawford et al., 1986, 1989; Pearce et al., 1992; Taylor and Nesbitt, 1995) and Tonga arcs (Falloon and Crawford, 1991; Stern and Bloomer, 1992; Sobolev and Danyushevsky, 1994). In the CaO versus SiO_2 classification diagram of Arculus et al. (1992), the Baimak-Buribai boninites fall in the intermediate- to low-Ca boninitic andesite field; the boninitic basaltic andesite rocks fall in the intermediate-Ca boninite field; and, the tholeiitic and calc-alkalic rocks fall in the normal andesite and dacite fields (Fig. 5, A and B). Further comparisons can be made by using classification schemes that employ HFSEs and REEs; for example, the Baimak-Buribai boninitic and boninitic basaltic andesite have typical boninitic low TiO_2 contents and high Mg# values (cf. Pearce et al., 1992) whereas, again, the tholeiitic and calc-alkalic rocks have normal andesitic and dacitic values (Fig. 5C). In addition, our bon-

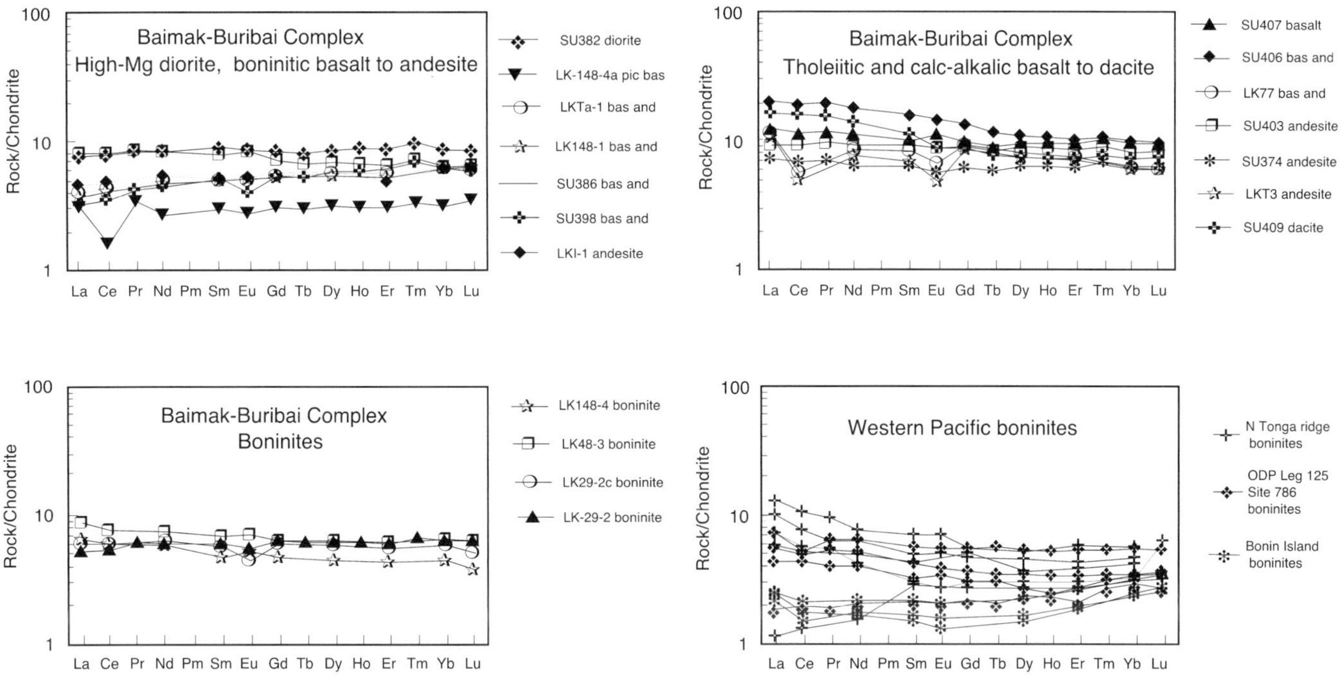

Figure 6. Chondrite-normalized REE diagrams for representative Baimak-Buribai Complex rocks. Normalization factors are those of Nakamura (1974). Shown for comparison in the lower right panel are the patterns for typical Bonin Island boninites (data from Shimizu et al., 1992); ODP Leg 125, Site 786, boninites (data from Murton et al., 1992); and northern Tonga Ridge boninites (data from Falloon and Crawford, 1991). Abbreviations: pic bas—picritic basalt, bas and—basaltic andesite.

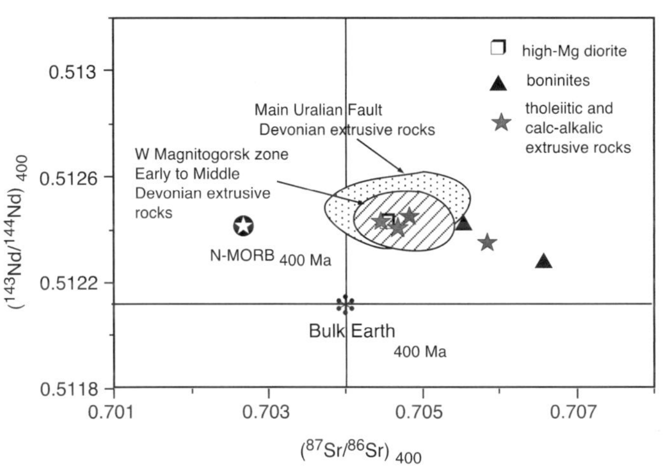

Figure 7. $^{143}Nd/^{144}Nd_{400}$ vs. $^{87}Sr/^{86}Sr_{400}$ for representative Baimak-Buribai Complex rocks, western Magnitogorsk extrusive rocks of Early to Middle Devonian age, and Devonian extrusive rocks along the Main Uralian fault described by Gaggero et al. (1997). Data source: this study and Spadea et al. (1999).

initic rocks show Zr enrichment (Fig. 5D) typical of Tertiary Western Pacific boninites (Taylor et al., 1992; Beccaluva and Serri, 1988).

Evolution of volcanism

Even though the Baimak-Buribai Complex is temporally and spatially limited, there is a clear progression from the lowest extended, preexisting, forearc crust, represented by the high-Mg diorites at the base of the Shankhai stream section through the Shankhai and Tanalyk boninites and boninitic rocks, to the overlying calc-alkalic rocks of the Baimak-Buribai sequence at Samarskoe. This up-sequence change is also reflected in the nature of the eruptive products that shift from effusive, high-temperature (~1150–1300°C, cf. Crawford et al., 1989), low-viscosity lavas to explosive, water-rich, higher-viscosity tuffs. They, thus, fit the petrogenetic models proposed for the Izu-Bonin-Mariana forearc system (Crawford et al., 1989; Pearce et al., 1992; Bloomer et al., 1995) in which generation of boninite magmas is ascribed to high-temperature melting of shallow, depleted mantle associated with the initiation of subduction of young, buoyant ocean crust; subsequently, calc-alkalic magmas are generated when older, denser oceanic crust is subducted at a greater angle in an established subduction zone, resulting in melting of deeper, less depleted mantle.

Stratigraphically above the Baimak-Buribai Complex, geographically to the north and west, are the main units of the Magnitogorsk arc: the Irendyk Formation (composed of basalts to rhyolites with tholeiitic to marked calc-alkalic compositions) and the Karamalytash Formation (composed of basalts to dacites also of tholeiitic to calc-alkalic affinity) (Sheravkin, 1986; Fershtater et al., 1997; Brown et al., 1998). These formations mark the full development of the Magnitogorsk arc during late Early to Middle Devonian time after initiation of arc activity in the Early Devonian.

Southern Uralian precollisional geodynamic evolution

The Baimak-Buribai Complex high-Mg diorites are, apparently, remnants of preexisting plutonic basement crystallized at upper middle-crustal depths into which the dikes and sills intruded and through which the volcanic rocks erupted. The high-Mg composition of the diorites suggests that the Shankhai and Tanalyk sequences represent a second-stage boninitic magmatism intruded into extended forearc crust. The association of the diorites and dikes exposed at the surface

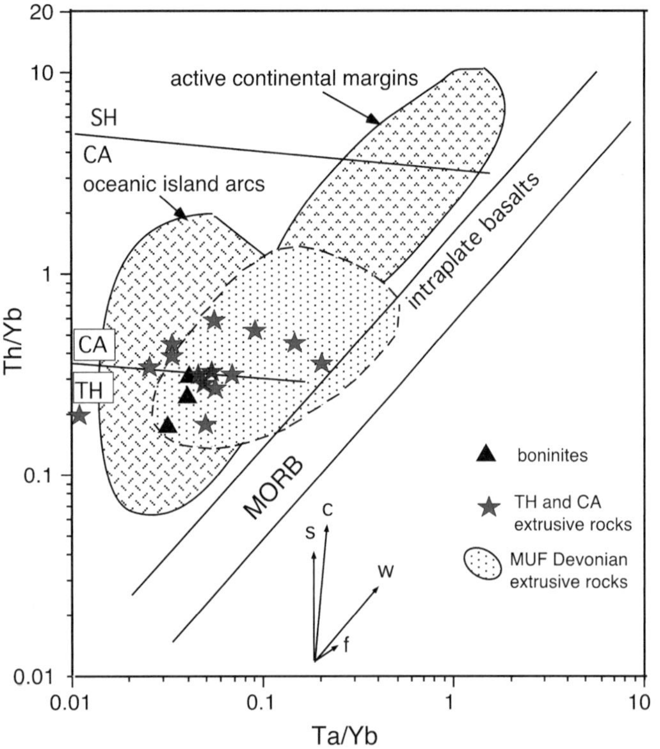

Figure 8. Th/Yb vs. Ta/Yb discriminant diagram (Pearce, 1983) for representative Baimak-Buribai Complex rocks, Devonian extrusive rocks of the western Magnitogorsk zone, and Devonian extrusive rocks along the Main Uralian fault (MUF). Data source: this study, Gaggero et al. (1997), and Spadea et al. (1999). Dashed lines indicate the boundaries of the tholeiitic (TH), calc-alkalic (CA), and shoshonitic (SH) fields; vectors indicate the influence of subduction components (s), within-plate enrichment (w), crustal contamination (c), and fractional crystallization (f).

with the lavas indicates that within the pre-Uralian ocean there was preorogenic active tectonic movement.

The Baimak-Buribai lower stratigraphic units, i.e., the Shankhai and Tanalyk pillow lavas and hyaloclastites, are products of subaqueous eruption. Moreover, in the lower part of the succession, there are shallow intrusive rocks devoid of associated arc-derived volcaniclastic or terrigenous interlayered sedimentary material such as would be expected in a marginal or backarc basin. So, we conclude that the geologic setting of the lower Baimak-Buribai Complex was an extensional, nonaccretionary, forearc, but, by the time the calc-alkalic Baimak-Buribai type sequence erupted, the geologic setting had evolved to an established accretionary forearc.

The supra-subduction setting of Baimak-Buribai Complex rocks confirms that there was intraoceanic convergence and subduction in the southern pre-Uralian ocean during the Emsian, 410–394 Ma. However, from the petrologic and geochemical evidence provided by this study and preliminary data on the volcanic units from the Magnitogorsk arc and dismembered volcanic units present in the Main Uralian fault (Figs. 7 and 8), we do not see any clear evidence of involvement of continental material in the Devonian magmatism. From this conclusion, we suggest that the arc in which the boninites were generated was far from continental input from either the East European continental margin or the continental block to the east of the pre-Uralian ocean. At 380 to 370 Ma, the intraoceanic convergence without involvement of continental material ended, as recorded by the radiometric ages of the Maksyutov Complex eclogites that formed during the subduction of the East European craton margin (Matte et al., 1993; Lennykh et al., 1995, Hetzel et al., 1998). Therefore, the time between initiation of subduction and arc-continent collision was 14 to 40 m.y.; this time constraint allows us to evaluate the rate of orogenic processes in the Paleozoic Uralian mountain chain and to conclude that the rate was similar to, or faster than, the rate of ongoing orogeny in the Western Pacific region.

CONCLUDING REMARKS

The Baimak-Buribai Complex of the southern Magnitogorsk oceanic arc, southern Urals, contains boninitic and depleted tholeiitic rocks intruded into, and erupted through, a preexisting upper part of middle crust with high-Mg diorite composition.

The volcanism progressed from mostly boninitic effusive activity to explosive activity with tholeiitic and calc-alkalic eruptive products.

In the Southern Urals, well-preserved oceanic crust gives insights into the processes and timing of precollisional geodynamic evolution of a typical collisional mountain belt.

ACKNOWLEDGMENTS

We are grateful to Fernando Bea for trace element and Sr and Nd isotope determinations, to Massimo D'Antonio for Sr isotope determinations, and to Larisa Kabanova, Alexei Pertsev, Galina Savelieva, and Alexander Saveliev for help with field work. This paper has benefited from engaged and encouraging reviews by Anthony Crawford and Martin Flower. To them go our warmest thanks. We acknowledge financial support from the European Commission (TMR-URO Programme) and the Italian Ministry of University and Scientific Research (Progetto Nazionale 40%). The Italian National Research Council is acknowledged for making possible the microprobe work in the Department of Mineralogy and Petrology of the University of Padua.

This is EUROPROBE publication number 295.

REFERENCES CITED

Alt, J.C., Honnorez, J., Laverne, C., and Emermann, R., 1986, Hydrothermal alteration of a 1 km section through the upper oceanic crust, Deep Sea Drilling Project Hole 504B: Mineralogy, chemistry, and evolution of seawater-basalt interactions: Journal of Geophysical Research, v. 91 B10, p. 309–335.

Arculus, R.J., Pearce, J.A., and van der Laan, S., 1992, Igneous stratigraphy and major element geochemistry of Holes 786A and 786B: Proceedings of ODP Scientific Research, v. 125: College Station, Texas, Ocean Drilling Program, p. 143–169.

Bédard, J.H., Lauziere, K., Tremblay, A., and Sangster, A., 1998, Evidence for forearc seafloor-spreading from the Betts Cove Ophiolite, Newfoundland: Oceanic crust of boninitic affinity: Tectonophysics, v. 284, p. 233–245.

Beccaluva, L., and Serri, G., 1988, Boninitic and low-Ti subduction-related lavas from intraoceanic arc-backarc systems and low-Ti ophiolites: A reappraisal of their petrogenesis and original tectonic setting: Tectonophysics, v. 146, p. 291–315.

Bloomer, S.H., and Hawkins, J.W., 1987, Petrology and geochemistry of boninite series volcanic rocks from the Mariana trench: Contributions to Mineralogy and Petrology, v. 89, p. 256–262.

Bloomer, S.H., Taylor, B., MacLeod, C.J., Stern, R.J., Fryer, P., Hawkins, J.W., and Johnson, L., 1995, Early arc volcanism and the ophiolite problem: A perspective from drilling in the Western Pacific, *in* Taylor, B., and Natland, J., eds, Active margins and marginal basins of the Western Pacific: American Geophysical Union Geophysical Monograph 88, p. 1–30.

Brown, A.V., and Jenner, G.A., 1989, Geological setting, petrology and chemistry of Cambrian boninite and low-Ti tholeiite lavas in western Tasmania, *in* Crawford, A.J., ed., Boninites: London, Unwin Hyman, p. 233–263.

Brown, D., Puchkov, V., Alvarez-Marron, J., and Perez-Estaùn, A., 1996, The structural architecture of the footwall to the Main Uralian Fault, southern Urals: Earth Science Reviews, v. 40, p. 125–147.

Brown, D., Juhlin, C., Alvarez-Marron, J., Perez-Estaun, A., and Oslinanski, A., 1998, Crustal-scale structure and evolution of an arc-continent collision zone in the southern Urals, Russia: Tectonophysics, v. 17, p. 158–171.

Brown, D., and Spadea, P., 1999, Processes of forearc and accretionary complex formation during arc-continent collision in the southern Urals: Geology, v. 27, p. 649–652.

Cameron, W.E., 1985, Petrology and origin of primitive lavas from the Troodos ophiolite, Cyprus: Contributions to Mineralogy and Petrology, v. 97, p. 361–377.

Cameron, W.E., McCulloch, M.T., and Walker, D.A, 1983, Boninite petrogenesis: Chemical and Nd-Sr isotopic constraints: Earth and Planetary Science Letters, v. 65, p. 75–89.

Casey, J.F., and Dewey, J.F., 1984, Initiation of subduction along transform and accreting plate boundaries, triple-junction evolution and forearc spreading centres: Implications for ophiolite geology and obduction: Geological Society of London Special Publication 13, p. 269–290.

Chemenda, A., Matte, P., and Sokolov, V., 1997, A model for Paleozoic obduction and exhumation of high-pressure/low-temperature rocks in the southern Urals: Tectonophysics, v. 276, p. 217–227.

Clift, P.D., and Dixon, J.E., 1998, Jurassic ridge collapse, subduction initiation, and ophiolite obduction in the southern Greek Tethys: Eclogae Geologicae Helvetiae, v. 91, p. 123–138.

Coish, R.A., 1989, Boninitic lavas in Appalachian ophiolites: A review, *in* Crawford, A.J., ed., Boninites: London, Unwin Hyman, p. 264–287.

Crawford, A.J., ed., 1989, Boninites: London, Unwin Hyman, 465 p.

Crawford, A.J., and Berry, R.F., 1992, Tectonic implications of late Proterozoic-early Paleozoic igneous rock associations in western Tasmania, *in* Fergusson, C.L., and Glen, R.A., eds., The Palaeozoic eastern margin of Gondwanaland: Tectonics of the Lachlan fold belt, southeastern Australia and related orogens: Tectonophysics, v. 214, p. 37–56.

Crawford, A.J., and Cameron, W.E., 1985, Petrology and geochemistry of Cambrian boninites and low-Ti andesites from Heathcote, Victoria: Contributions to Mineralogy and Petrology, v. 91, p. 93–104.

Crawford, A.J., Beccaluva, L., and Serri, G., 1981, Tectono-magmatic evolution of the West Philippine-Mariana region and the origin of boninites: Earth and Planetary Science Letters, v. 54, p. 346–356.

Crawford, A.J., Cameron, W.E., and Keays-Reid, R., 1984, The association of boninite low-Ti andesite-tholeiite in the Heathcote Greenstone Belt, Victoria: Ensimatic setting for the early Lachlan fold belt: Australian Journal of Earth Sciences, v. 31, p. 161–175.

Crawford, A.J., Beccaluva, L., Serri, G., and Dostal, J., 1986, Petrology, geochemistry, and tectonic implications of volcanics dredged from the intersection of the Yap and Mariana trenches: Earth and Planetary Science Letters, v. 80, p. 265–280.

Crawford, A.J., Falloon, T.J., and Green, D.H., 1989, Classification, petrogenesis, and tectonic setting of boninites, *in* Crawford, A.J., ed., Boninites: London, Unwin Hyman, p. 2–49.

Dalziel, W.D., 1997, Neoproterozoic-Paleozoic geography and tectonics: Review, hypothesis, environmental speculation: Geological Society of America Bulletin, v. 109, p. 16–42.

Danyushevsky, L.V., Sovbolev, A.V., and Falloon, T.J., 1995, North Tongan high-Ca boninite petrogenesis: The role of Samoan plume and subduction zone-transform fault transition: Journal of Geodynamics, v. 20, p. 219–241.

Duncan, R.A., and Green, D.H., 1987, The genesis of refractory melts in the formation of oceanic crust: Contributions to Mineralogy and Petrology, v. 96, p. 326–342.

Falloon, T.J., and Crawford, A.J., 1991, The petrogenesis of high-calcium boninite lavas dredged from the northern Tonga ridge: Earth and Planetary Science Letters, v. 102, p. 375–394.

Fershtater, G.B., Montero, P., Borodina, N.S., Pushkarev, E.V., Smirnov, V.N., and Bea, F., 1997, Uralian magmatism: An overview: Tectonophysics, v. 276, p. 87–102.

Flower, M.F.J., and Levine, H., 1987, Petrogenesis of a tholeiite-boninite sequence from Ayos Mamas, Troodos Ophiolite: Evidence for splitting of a volcanic arc: Contributions to Mineralogy and Petrology, v. 97, p. 509–524.

Gaggero, L., Spadea, P., Cortesogno, L., Savelieva, G.N., and Pertsev, A.N., 1997, Geochemical investigations of the igneous rocks from the Nurali Ophiolite mélange Zone, Southern Urals: Tectonophysics, v. 276, p. 139–161.

Govindaruju, K., Potts, P.J., Webb, P.C., and Watson, J.S., 1994, 1994 Report on Whin Sill dolerite WS-E from England and Pitscurrie microgabbro PM-S from Scotland: Assessment by one hundred and four international laboratories: Geostandards Newsletter, v. 18, p. 211–300.

Hawkins, J.W., Bloomer, S.H., Evans, C.A., and Melchior, J.T., 1984, Evolution of intra-oceanic, arc-trench systems: Tectonophysics, v. 102, p. 175–205.

Hetzel, R., Hechtler, H.P., Seifert, W., Schulte, A.B., and Ivanov, K.S., 1998, Subduction- and exhumation-related fabrics in the Paleozoic high-pressure/low-temperature Maksyutov complex, Antigan area, southern Urals, Russia: Geological Society of America Bulletin, v. 110, p. 916–930.

Hickey, R., and Frey, F.A., 1982, Geochemical characteristics of boninite series volcanics: Implications for their source: Geochimica et Cosmochimica Acta, v. 46, p. 2099–2115.

Ivanov, K.S., Puchkov, V.N., Nasedkina, V.A., and Pelevin, I.A., 1989, First results on revised stratigraphy of Polyakovka Formation based on conodonts: Ezhegodnik-88, Institute of Geology and Geochemistry, Uralian Department, Russian Academy of Science, Sverdlovsk, p. 11–12 (in Russian).

Kuz'min, M.I., and Kabanova, L.Ya., 1991, Boninite series of the Southern Urals: Geologic and petrographic description, compositional peculiarities and problem of origin: Nauka, Miass, p. 156–173 (in Russian).

Lennykh, V.I., Valiser, P.M., Beane, R., Leech, M., and Ernst, W.G., 1995, Petrotectonic evolution of the Makysutov Complex, southern Ural Mountains, Russia: Implications for ultrahigh-pressure metamorphism: International Geology Review, v. 37, p. 584–600.

MacLachlan, K., and Dunning, G., 1998, U-Pb ages and tectonomagmatic relationships of early Ordovician low-Ti tholeiites, boninites, and related plutonic rocks in central Newfoundland, Canada: Contributions to Mineralogy and Petrology, v. 133, p. 235–258.

Maslov, V.A., Cherkasov, V.L., Tischchenko, V.T., Smirnova, I.A., Artyushkova, O.V., and Pavlov, V.V., 1993, On the stratigraphy and correlation of the Middle Paleozoic complexes of the main copper-pyritic areas of the Southern Urals: Ufimsky Nauka Tsentr, Ufa, 217 p. (in Russian).

Matte, P., 1995, Southern Urals and Variscides: Comparison of their anatomies and evolutions: Geologie en Mijnbouw, v. 74, p. 151–166.

Matte, P., Maluski, H., Caby, R., Nicolas, A., Kepeshinskas, P., and Sobolev, S., 1993, Geodynamic model and $^{39}Ar/^{40}Ar$ dating for the generation and emplacement of the high pressure (HP) metamorphic rocks in SW Urals: Comptes Rendus de l'Academie de Science Series II, v. 317, p. 1667–1674.

McCulloch, M.T., and Cameron, W.E., 1983, Nd-Sr isotopic study of primitive lavas from the Troodos ophiolite, Cyprus: Evidence for a subduction-related setting: Geology, v. 11, p. 727–731.

McKerrow, W.S., 1994, Terrane assembly in the Variscan belt of Europe: Europrobe News, v. 5, p. 4–5.

Meijer, A., 1980, Primitive arc volcanism and a boninitic series: Examples from Western Pacific island arcs, *in* Hayes, D.E., ed., The tectonic and geologic evolution of Southeast Asian seas and islands: American Geophysical Union Geophysical Monograph 23, p. 269–282.

Melancon, B., Hebert, R., Laurent, R., and Dostal, J., 1997, Petrological and geochemical characteristics of the Bolton Igneous Group, southern Quebec Appalachians: American Journal of Science, v. 297, p. 527–549.

Murton, B.J., Peate, D.W., Arculus, R.J., Pearce, J.A., and van der Laan, S., 1992, Trace-element geochemistry of volcanic rocks from Site 786: The Izu-Bonin forearc: Proceedings of the Ocean Drilling Program, Scientific Results, v. 125: College Station, Texas, Ocean Drilling Program, p. 211–235.

Nakamura, N., 1974, Determination of REE, BA, Fe, Mg, Na, and K in carbonaceous and ordinary chondrites: Geochimica et Cosmochimica Acta, v. 38, p. 757–775.

Natland, J.H., 1981, Crystal morphologies and pyroxene compositions in boninites and tholeiitic basalts from Deep Sea Drilling Project Holes 458 and 459 in the Mariana forearc region: Initial Reports DSDP Leg 60: Washington, D.C., U.S. Government Printing Office, p. 681–709.

Natland, J.H., and Tarney, J., 1981, Petrological evolution of the Mariana arc and back-arc basin system: A synthesis of drilling results in the South Philippine Sea: Initial Reports DSDP Leg 60: Washington, D.C., U.S. Government Printing Office, p. 877–908.

Nelson, D.R., Crawford, A.J., and McCulloch, M.T., 1984, Nd-Sr isotopic and geochemical systematics in Cambrian boninites and tholeiites from Victoria, Australia: Contributions to Mineralogy and Petrology, v. 88, p. 164–172.

Pearce, J.A., 1983, The role of sub-continental lithosphere in magma genesis at destructive plate margins, *in* Hawkesworth, C.J., and Norry, M.J., eds, Continental basalts and mantle xenoliths: Nantwich, Shiva, p. 230–249.

Pearce, J.A., van der Laan, S.R., Arculus, R.J., Murton, B.J., Ishii, T., Peate, D.W., and Parkinson, I.J., 1992, Boninite and harzburgite from Leg 125 (Bonin-Marian forearc): A case study of magma genesis during the initial stage of subduction: Proceedings of the Ocean Drilling Program, Scientific Results, v. 125: College Station, Texas, Ocean Drilling Program, p. 623–659.

Perez-Estaùn, A., Alvarez-Marron, J., Brown, D., Puchkov, V., Gorozhanina, Y., and Baryshev, V., 1997, Along-strike structural variations in the foreland thrust and fold belt of the southern Urals: Tectonophysics, v. 276, p. 265–280.

Piercy, S.J., 1996, The geology and geochemistry of the southern Pacquet Harbour Group volcanics, Baie Verte Peninsula, Newfoundland [M.S. thesis]: Memorial University of Newfoundland, Canada, 123 p.

Puchkov, V.N., 1997, Structure and geodynamics of the Uralian orogen, *in* Burg, J.-P., and Ford, M., eds, Orogeny through time: Geological Society of London Special Publication 121, p. 201–236.

Rogers, N.W., MacCleod, C.J., and Murton, B.J., 1989, Petrogenesis of boninitic lavas from the Limassol Forest Complex, Cyprus, *in* Crawford, A.J., ed., Boninites: London, Unwin Hyman, p. 288–313.

Rumianzeva, N.A., Ushkova, G.A., Shmeleva, K.L., and Kukui, A.A., 1989, Silurian boninitic unit of the Urals: Reports of the Academy of Science USSR, v. 304, p. 947–951 (in Russian).

Savelieva, G.N., 1987, Gabbro-ultrabasite assemblages of the Urals ophiolites and their analogues in modern oceanic crust: Nauka, Moscow, 242 p. (in Russian).

Savelieva, G.N., and Nesbitt, R.W., 1996, A synthesis of the stratigraphy and tectonic setting of the Uralian ophiolites: Journal of the Geological Society of London, v. 153, p. 525–537.

Savelieva, G.N., Sharaskin, A., Saveliev, A.A., Spadea, P., and Gaggero, L., 1997, Ophiolites of Southern Uralides adjacent to the East European Continental Margin: Tectonophysics, v. 276, p. 117–137.

Sharaskin, A.Ya., Dobretsov, N.L., and Sobolev, N.V., 1980, Marianites: The clinoenstatite-bearing pillow lavas associated with the ophiolite assemblage of the Mariana Trench, *in* Panayoyou, A., ed., Proceedings of the International Ophiolite Symposium, Cyprus 1979: Nicosia, Geological Survey Department and Ministry of Agriculture and Natural Resources, Republic of Cyprus, p. 473–479.

Shaw, H.F., and Wasserburg, G.J., 1984, Isotopic constraints on the origin of Appalachian mafic complexes: American Journal of Science, v. 284, p. 319–349.

Sheravkin, I.B., 1986, Volcanism and massive sulfide deposits of the Southern Urals: Nauka, Moscow, 268 p. (in Russian).

Shimizu, H., Sawatari, H., Kawata, Y., Dunkley, P.N., and Masuda, A., 1992, Ce and Nd isotope geochemistry of island arc volcanic rocks with negative Ce anomaly: Existence of sources with concave REE patterns in the mantle beneath the Solomon and Bonin island arcs: Contributions to Mineralogy and Petrology, v. 110, p. 242–252.

Shiraki, K., Kuroda, N., Maruyama, S., and Urano, H., 1978, Evolution of the Tertiary volcanic rocks in the Izu Mariana Arc: Bulletin of Vulcanology, v. 41, p. 548–562.

Smellie, J.L., Stone, P., and Evans, J., 1995, Petrogenesis of boninites in the Ordovician Ballantrae Complex ophiolite, southwestern Scotland: Journal of Volcanology and Geothermal Research, v. 69, p. 323–342.

Smethurst, M.A., Khramov, A.N., and Torsvik, T.H., 1998, The Neoproterozoic and Palaeozoic palaeomagnetic data for the Siberian Platform: From Rodinia to Pangea: Earth Science Review, v. 43, p. 1–24.

Sobolev, A.V., and Danyushevsky, L.V., 1994, Petrology and geochemistry of boninites from the north termination of the Tonga trench: Constraints on the generation conditions of primary high-Ca boninite magmas: Journal of Petrology, v. 35, p. 1183–1211.

Spadea, P., Kabanova, L.Ya., and Scarrow J.H., 1998, Petrology and geodynamic significance of Mid-Devonian boninitic and calc-alkaline lavas from the Baimak-Buribai area (Magnitogorsk Zone, Southern Urals): Ofioliti, v. 23, p. 17–36.

Spadea, P., Scarrow, J.H., Brown, D., Gorozhanina, Y., and Kosarev, A., 1999, Oceanic island arc geochemical signature of volcanic suites (Early-Middle Devonian) of the Magnitogorsk Zone, Southern Urals: Journal of Conference Abstracts, v. 4, p. 398.

Stern, R.J., and Bloomer, S.H., 1992, Subduction zone infancy: Examples from the Eocene Izu-Bonin-Mariana and Jurassic California Arcs: Geological Society of America Bulletin, v. 104, p. 1621–1636.

Tatsumi, Y., 1982, Origin of high-magnesian andesite in the Setouchi volcanic belt, southwest Japan. II. Melting phase relations at high pressure: Earth and Planetary Science Letters, v. 60, p. 305–317.

Tatsumi, Y., and Maruyama, S., 1989, Boninites and high-Mg andesites: Tectonics and petrogenesis, *in* Crawford, A.J., ed., Boninites: London, Unwin Hyman, p. 112–147.

Taylor, R.N., and Nesbitt, R.W., 1995, Arc volcanism in an extensional regime at the initiation of subduction: A geochemical study of Hawajima, Bonin Islands, Japan, *in* Smellie, J.L., ed., Volcanism associated with extension at consuming plate margins: Geological Society of London Special Publication 81, p. 115–134.

Taylor, R.N., Murton, B.J., and Nesbitt, R.W, 1992, Chemical transects across intra-oceanic arcs: Implications for tectonic setting of ophiolites, *in* Parson, L.M., Murton, B.J., and Browning, P., eds., Ophiolites and their modern analogues: Geological Society of London Special Publication 60, p. 117–132.

Thy, P., 1984, On the nature of Troodos boninites: Ofioliti, v. 9, p. 555–568.

Thy, P., Brooks, C.K., and Walsh, J.N., 1985, Tectonic and petrogenetic implications of major and rare earth element geochemistry of Troodos glasses, Cyprus: Lithos, v. 18, p. 165–178.

Tucker, R.D., Bradley, D.C., Ver Straeten, C.A., Harris, A.G., Ebert, J.R., and McCutcheon, S.R., 1998, New U-Pb zircon ages and the duration and division of Devonian time: Earth and Planetary Science Letters, v. 158, p. 175–186.

Umino, S., 1986, Magma mixing in boninite sequence of Chichijima, the Bonin Islands: Journal of Volcanology and Geothermal Research, v. 29, p. 125–157.

van der Laan, S.R., Flower, M.F.J., and van Groos, A.F.K., 1989, Experimental evidence for the origin of boninites: Near-liquidus phase relations to 7.5 kbar, *in* Crawford, A.J., ed., Boninites: London, Unwin Hyman, p. 112–147.

van der Laan, S.R., Arculus, R.J., Pearce, J.A., and Murton, B.J., 1992, Petrography, mineral chemistry, and phase relations of the basement boninite series of Site 786, Izu-Bonin forearc: Proceedings of the Ocean Drilling Program, Scientific Results, v. 125: College Station, Texas, Ocean Drilling Program, p. 171–201.

Varfalvy, V., Hebert, R., and Bédard, J.H., 1996, Interactions between melt and upper-mantle peridotites in the North Arm Mountain massif, Bay of Islands Ophiolite, Newfoundland, Canada: Implications for the genesis of boninitic and related magmas: Chemical Geology, v. 129, p. 71–90.

Zonenshain, L.P., Korinevsky, V.G., Kazmin, V.G., Pechersky, D.M., Khain, V.V., and Matveenkov, V.V., 1984, Plate tectonic model of the South Urals development, *in* Zwart, H.J., Behr, H.J., and Oliver, J.E., eds., Appalachian and Hercynian fold belts: Tectonophysics, v. 109, p. 95–135.

MANUSCRIPT ACCEPTED BY THE SOCIETY FEBRUARY 23, 2000

Printed in the U.S.A.

Geological Society of America
Special Paper 349
2000

Constraints on the formation and emplacement age of western Greek ophiolites (Vourinos, Pindos, and Othris) inferred from deformation structures in peridotites

Annie Rassios
Institute of Geological and Mineral Exploration, Kozani Branch, Lefkovrisi, Kozani, Greece; e-mail: arassios@compulink.gr
Alan G. Smith
Department of Earth Sciences, Downing Street, Cambridge CB2 3EQ, England; e-mail: ags1@esc.cam.ac.uk

ABSTRACT

The age of ductile structures within ophiolites provides important constraints on models of ophiolite emplacement. In the peridotites of the western Hellenic ophiolite belt of Greece (Vourinos, Pindos, and Othris), *late* ductile structures appear to be contemporaneous with initial emplacement. The temperatures and depths at which the ductile structures formed were estimated and matched with similar conditions for cooling ocean floor. From this comparison, we determined that the structures must have formed less than 1–4 m.y. after the peridotites first formed at a spreading center. If shear heating is not an important factor, similar time-interval estimates are obtained from the *P-T* conditions for the metamorphic sole. The age of the sole is relatively well dated at 171 Ma (late Bajocian), implying that spreading ceased no earlier than early Bajocian time.

Mapping of the regional variation of peridotite fabrics shows that Othris peridotites are intensely mylonitized; Pindos peridotites are less mylonitized, and Vourinos are the least mylonitized. Compositional transitions accompany these fabric variations: from a supra-subduction zone setting in the Vourinos Complex, a harzburgite-lherzolite unit in the Pindos ophiolite that includes MORB (mid-oceanic-ridge basalt) to IAT (island-arc tholeiite) and BSV (boninite-series volcanic rock) chemical affinities and that links up, through the Koziakas ophiolite, with the Othris ophiolite, which is dominated by a plagioclase lherzolite-harzburgite unit with MORB-like lavas.

We speculate that the lateral variations in strain and composition are linked to the evolution of the mantle wedge above the subducting slab.

INTRODUCTION

The Jurassic Dinaric-Hellenic ophiolite belt (Fig. 1) crops out for a distance of at least 1000 km along the length of former Yugoslavia, Albania, and much of Greece. Smith (1993) summarized most of the earlier work on Greek ophiolites. This paper highlights more recent work, some of it published in literature that is not readily accessed, on the lithologic variations and emplacement structures in the peridotites of the major western Greek ophiolite complexes (Vourinos, Pindos, and Othris) (Figs. 2–5). These ophiolites crystallized, probably in Middle Jurassic time, and were emplaced shortly after crystallization (as discussed subsequently in the section "Conditions During Emplacement"). Structural, trace element, and metallogenic evidence suggests that all three major complexes were once linked together (Smith, 1979; Rassios et al., 1999), but have been separated tectonically by the formation of the Meso-Hellenic trough (IGRS-IFP, 1966).

Scattered dismembered ophiolite outcrops in the Kallidromon Range, in Beotia, in Evvia, and in Argolis are also probably continuations of the same ophiolite zone (Fig. 2). Their precise relationship to the major complexes is unclear because of tectonic dislocations and large rotations (Kissel and Laj, 1988; Morris, 1995), but detailed work in Argolis (Clift and Dixon, 1998) provides important geochemical evidence for their origin. The vergence of structures in the eastern Hellenic ophiolites (Fig. 2) (Axios or Vardar zone ophiolites) differs from that of the western Hellenic ophiolites: their relationship to the western ophiolites is controversial, and they will not be discussed further in this paper.

Much of the data reported here was collected during the compilation of technical reports on exploration for chromite and copper ore in the western Hellenides. Chromium and copper mineralization reflect both the structural and the compositional variations in the ophiolites (Rassios, 1989, 1990, 1994). Most of the structural data has not been previously published in readily accessible literature (it is in Greek or in European Union reports), or is nonconfidential information contained in confidential mining reports. It includes ma-

Rassios, A., and Smith, A.G., 2000, Constraints on the formation and emplacement age of western Greek ophiolites (Vourinos, Pindos, and Othris) inferred from deformation structures in peridotites, *in* Dilek, Y., Moores, E.M., Elthon, D., and Nicolas, A., eds., Ophiolites and Oceanic Crust: New Insights from Field Studies and the Ocean Drilling Program: Boulder, Colorado, Geological Society of America Special Paper 349, p. 473–483.

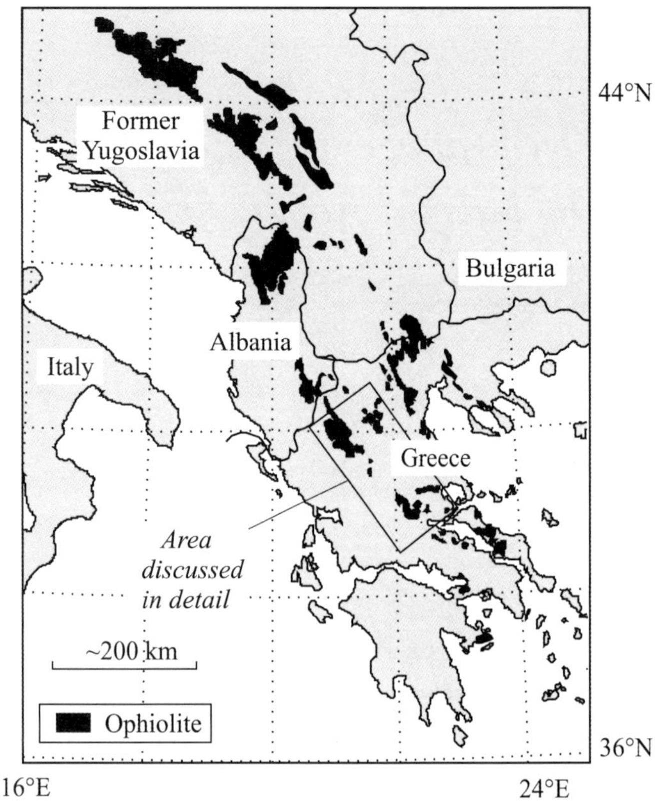

Figure 1. Map of Hellenic-Dinaric ophiolite outcrops.

terial collected during a 10-year EU-IGME (European Union–Institute of Geological and Mineral Exploration of Greece) development project to improve chromite exploration methodology (Rassios et al., 1986, 1991; Rassios, 1994). Greece is the only state within the European Union to possess such reserves. The western Hellenic ophiolites were mapped in detail on scales of 1:50 000 (Othris), 1:20 000 (Pindos), and from 1: 10 000 down to 1:1000 for individual ore districts in Vourinos.

Concentrations of chromite occur within dunite, but its abundance as an economic mineral varies: In Vourinos (Fig. 3), over 800 chromite occurrences have been mapped (Vrahatis and Grivas, 1980; Rassios et al., 1983, 1990) whereas in the Pindos ophiolite (Fig. 4), chromite shows are common, but nearly all subeconomic (Rassios et al., 1990); in Othris, chromite shows are very rare, although several chromite deposits have been mined near the Domokos massif and near Eretria (Fig. 5). As discussed subsequently, these differences reflect variations both in the initial tectonic and petrologic setting of the ophiolites and in the effect of subsequent deformation on chromite concentrations.

In the initial studies on chromite localization, an unexpected correlation was observed in Vourinos between the position of chromite orebodies and of peridotites that showed evidence of late ductile and early brittle structures and fabrics overprinting earlier high-temperature (T) ductile structures and fabrics (Grivas et al., 1986; Rassios et al., 1986). Initial attempts to interpret these features in terms of currently available structural models of oceanic spreading centers completely failed to locate new chromite bodies. However, their interpretation as structures and fabrics created in a thrust setting caused by emplacement led to a highly successful exploration model that was tested by drilling in eight ore districts (Grivas et al., 1993; Rassios, 1994). Orebodies not exposed at the surface were discovered at depths of more than 200 m.

These results guided the mapping program so that the field expression of the complete range of high-T to low-T structural fabrics was recorded for all the western Hellenic peridotites. It should be emphasized that this paper is not a review of ultramafic microfabrics, deformation mechanisms, and the like: several studies of this sort have already been undertaken for the Vourinos in particular (e.g., Courtin, 1979; Frison, 1987; Jackson et al., 1975; Ross et al., 1980). Instead, this paper summarizes for the first time the distribution and variations of peridotite fabrics and other structures observed in the field during a large mapping program. We speculate subsequently that processes in the mantle wedge above a subduction zone may have caused some of the observed fabric variations.

GENERAL LITHOLOGY AND STRUCTURE

Harzburgite, with associated dunite, is the dominant ultramafic component of all the western ophiolites, although a few massifs in Othris are predominantly lherzolite (Economou et al., 1986; Rassios et al., 1990). The dunite bodies range from tabular or podiform to irregularly shaped and deformed and in largest dimension from 1 m to more than 1 km. Folds are common, ranging in wavelength from a few centimeters (for example, within chromite deposits) to more than several kilometers, but *no* kilometer-scale recumbent folds (i.e., nappes) have been observed.

Vourinos is the least disrupted massif and includes a "complete" ophiolite section from an ultramafic tectonite base through a cumulate ultramafic sequence into gabbro, sheeted dikes, and minor pillow lavas (Moores, 1969; Rassios et al., 1983). The Pindos and Othris Complexes lack a continuous section, though several thrust units include overlapping stratigraphic elements that display all the structural and petrologic characteristics regarded as the hallmark of an ophiolite (Penrose Conference Participants, 1972).

Vourinos

The mantle rocks of the Vourinos Complex are exposed over an area of ~150 km^2, displaying 7–10 km of harzburgitic restite lying on a metamorphic sole (Zimmerman, 1972). The restite includes pods and lenses of dunite. Associated chromite occurrences define a metalliferous zone (Fig. 3): 85% of all chromite occurrences mapped by Vrahatis and Grivas (1980) and *all* economic deposits lie in this zone. The metalliferous zone is 0.5–1 km thick and is characterized by abundant dunites and by incipient imbrication structures (Grivas et al., 1986). The general form of the zone is parallel to high-T spreading-center fabrics mapped in the harzburgite (Frison, 1987). The high-T fabrics are overprinted by mylonitic deformation (Ross and Lewis, 1989; Grivas et al., 1993; Rassios et al., 1994). Ti-Y and Zr-Y plots of the Vourinos peridotites place them in the supra-subduction zone field (J. Pearce in Rassios, 1994).

The "petrologic Moho" is exposed in Vourinos (Jackson et al., 1975) and consists of a sharp intrusive contact between harzburgite restite and cumulate dunite (Harkins et al., 1980). All the rocks above this contact are of magmatic origin. They include ultramafic through mafic cyclic units; a massive, continuous gabbronorite unit (Jackson et al., 1975; Rassios et al., 1983); chromite-bearing dunite bodies; a 1-km-thick gabbroic diorite–diorite sequence and a sheeted-dike complex up to 1.5 km thick (Rassios et al., 1990); and minor pillow-lava screens, flows, and metalliferous sedimentary deposits that crop out

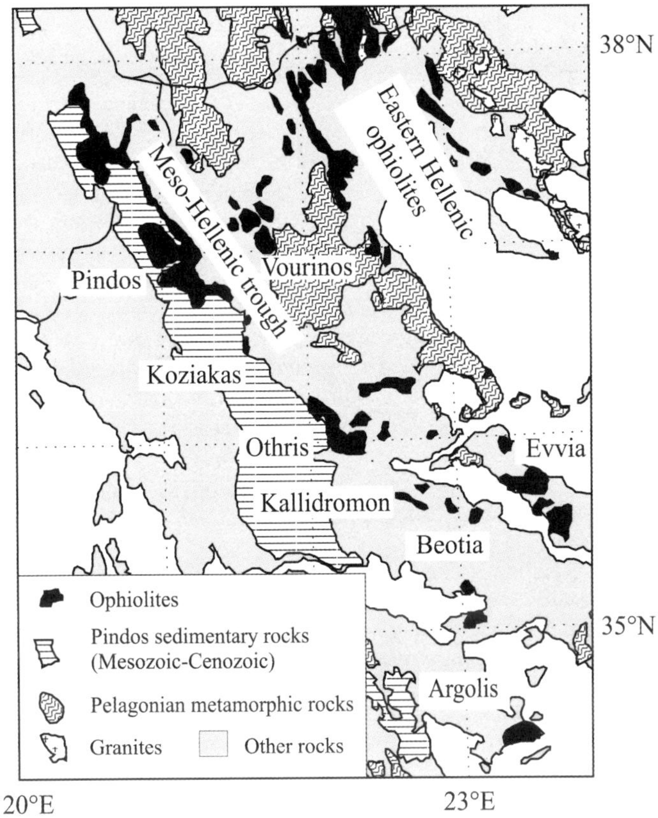

Figure 2. Map showing the ophiolite outcrops (black) in Greece and adjacent areas. The Hellenic ophiolites can be subdivided into the western Hellenic ophiolites (discussed in this paper) and the eastern Hellenic ophiolites of the Axios (Vardar) zone. The two zones are separated from one another by the metamorphic rocks of the Pelagonian zone and the associated Mesozoic carbonate platform. The western Hellenic ophiolites are thrust southwestward over the Triassic to Eocene Pindos sedimentary rocks. All the major western Hellenic ophiolites were originally emplaced to the northeast in later Jurassic time over the southwest margin of the Pelagonian continent. The Pindos and Vourinos ophiolites are now separated from one another by the Cenozoic Meso-Hellenic trough, but together with the Koziakis and Othris ophiolites are believed to have formed a continuous ophiolite zone after initial emplacement in Jurassic time. This zone may also have been linked with the scattered ophiolite outcrops farther to the south and east (Kallidromon, Beotia, Evvia, and Argolis).

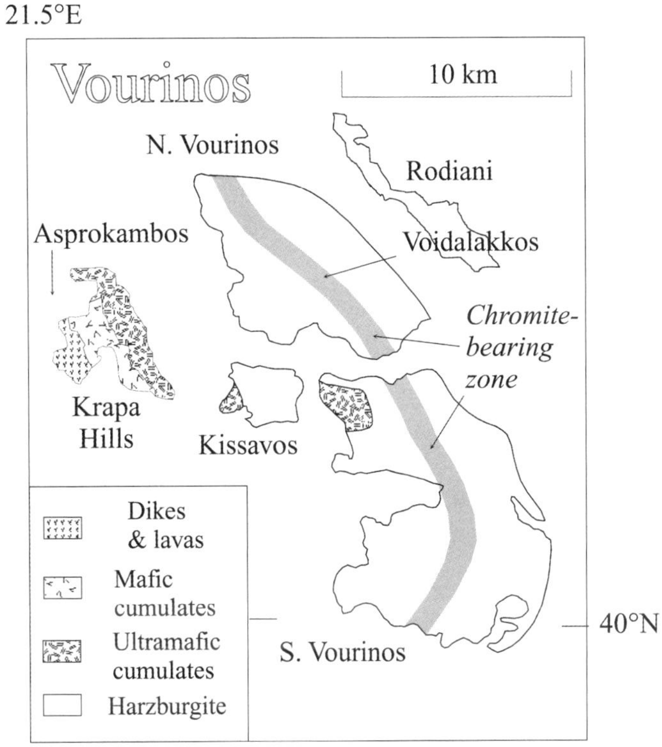

Figure 3. Outcrop map of Vourinos ophiolite. There is a broadly continuous igneous stratigraphy from the dikes and lavas, mafic rocks, and ultramafic cumulates in the Krapa Hills section to the tectonized harzburgite of Kissavos.

at the highest stratigraphic levels of the dike complex. A pillow-lava crustal section has not been preserved (Rassios et al., 1983).

The Rodiani ophiolite unit, 4 km northeast of Vourinos (Fig. 3), includes representative portions of all parts of a complete ophiolite, but its structural and petrogenetic relationships with Vourinos are obscure. A. Fotiades (in Rassios, 1994) mapped imbrications within the ophiolite and other structures indicating thrusting toward the northeast.

Pindos

The outcrop area of the Pindos ophiolite is >1300 km^2 (Fig. 4). The Dramala massif (Fig. 4; Rassios, 1991) is composed of a massive harzburgite unit lying structurally above a metamorphic sole composed mainly of amphibolites, with the following upward stratigraphic transition to the crustal sequence: 200 m of basal mylonitic

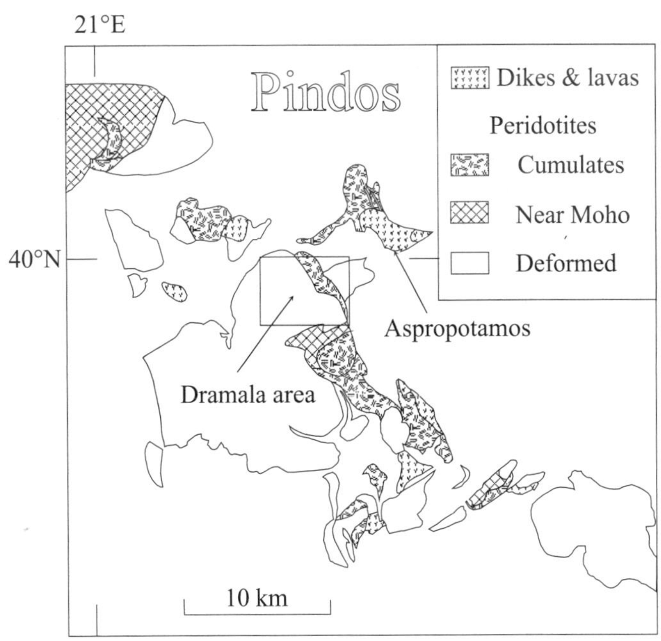

Figure 4. Outcrop map of the peridotites and associated igneous rocks in part of the Pindos ophiolite. With few exceptions, the lithologic boundaries are low-angle thrust faults.

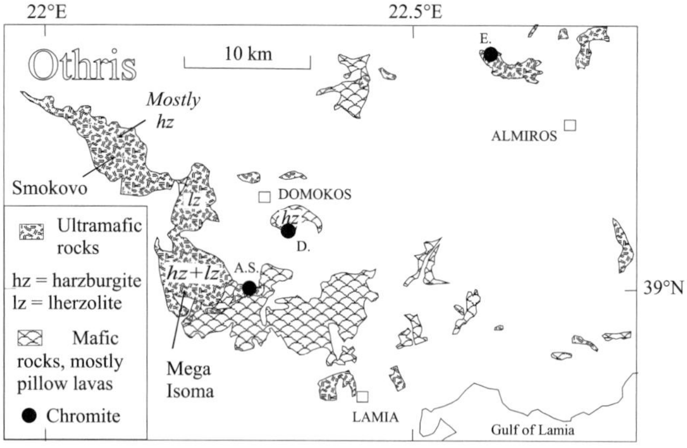

Figure 5. Outcrop map of Othris ophiolite. As in the case of the Pindos ophiolite, the structure is complex, and tectonic boundaries separate the lithologic units. The main chromite occurrences have been actively mined in the past (A.S.—Aghios Stephanos; D.—Domokos; E.—Eretria).

harzburgite is succeeded by harzburgite intruded by and codeformed with pyroxenite dikes; this zone is followed by harzburgite intruded by both pyroxenite (as dikes and massive bodies) and gabbro. A zone of harzburgite with both pyroxenite and gabbroic dikes and abundant dunite immediately underlies the transitional Moho unit.

Compared to the meter-scale petrologic Moho of Vourinos (Harkins et al., 1980), the transition in the Pindos ophiolite is much thicker, taking place over an interval of 0.5–1.5 km. The lowest unit in this interval lies on harzburgite restite and consists of a zone with both cumulate sills (plagioclase dunite and troctolite) and screens of restite, passing up into troctolite-gabbro cumulates. Locally, the cumulates are intruded by coarse-grained boninitic rocks and by boninitic dikes. Dunite bodies in both the cumulate and restite zones of the Pindos ophiolite are smaller than those of Vourinos: none is >5000 m^2 in area and most are <100 m^2.

Ti-Y and Zr-Y plots of the Pindos peridotites place them in the supra-subduction zone field and in the field transitional to mid-oceanic-ridge basalts (MORBs) (J. Pearce in Rassios, 1994). An evolution from MORBs to boninitic rocks suggested by the chemistry and layering of the cumulates in the Dramala complex has been proposed by Jones and Robertson (1991). Cr-Y and Zr-Y-Zr discrimination diagrams for the lavas and sheeted dikes of the eastern Aspropotamos Complex (Beccaluva et al., 1984; Fig. 2) show an evolution from N-MORB (normal MORB) to MORB-IAT (island-arc tholeiite), then IAT, and finally to BSV (boninite-series volcanic rock) units that crosscut preexisting rocks (Beccaluva et al., 1979).

Koziakas

The Koziakas ophiolite is a narrow north- to northwest-trending imbricated strip ~5 km wide between the Pindos and Othris ophiolites (Fig. 2). The northern and central part is mostly peridotite, with spinel harzburgites in the west and plagioclase lherzolites and lherzolites in the east, locally cut by dikes of gabbro. Each peridotite is a separate imbricate unit with pillow lavas, chert, amphibolite, and/or serpentine along the thrust contacts. Individual imbricates range from a few meters to ~400 m in thickness (G. Konstantopoulou in Rassios, 1994).

Othris

The Othris ophiolite has an outcrop area of 900 km^2 (Fig. 5) with most of the large peridotite thrust sheets cropping out in the western 450 km^2. Othris is generally recognized as a MORB-type ophiolite dominated by the presence of only slightly depleted (i.e., fertile) mantle rocks (Hynes, 1972; Menzies, 1974; Menzies and Allen, 1974; Courtin, 1979; Ferrière, 1982). Harzburgite occurs as a contiguous thrust sheet generally below the lherzolite and includes several highly strained dunite bodies with rare chromite shows. Chromite ores in the mines of Domokos and Eretria (Fig. 5) were probably associated with the harzburgite thrust sheet, but are today tectonically hosted by intensely sheared serpentinite. Imbrication has disrupted any petrologic Moho that may have once been present, although a "mixed zone," or near-Moho unit similar in appearance to that of the Pindos ophiolite, can be recognized among the thrust sheets overridden by the larger peridotite sheets. The petrologic Moho in Othris has apparently been sheared so intensely that it is no longer recognizable.

The ophiolitic units of Othris are stacked in reverse order, but not inverted; cherts and pillow lavas at the base have been overthrust by sheeted dikes and gabbro cumulates and then by ultramafic cumulates, a "near-Moho" section (as already described), a harzburgite thrust sheet, and finally a lherzolite–plagioclase lherzolite thrust sheet on top. Ophiolitic units were first emplaced in Jurassic time from the west or southwest toward a west- or southwest-facing passive continental margin (Smith et al., 1979). Early imbrications and the geometry of the deformation within the lavas in central and eastern Othris also confirm emplacement to the east or northeast (Rassios, 1989, 1990; Konstantopoulou et al., 1988).

Ti-Y and Zr-Y diagrams confirm that, apart from the supra-subduction zone character of the Domokos massif (Fig. 5), the Othris peridotites are almost exclusively confined to the abyssal tholeiite field (J. Pearce in Rassios, 1994).

BRITTLE AND DUCTILE STRAIN IN THE PERIDOTITES

Strain in the peridotites is recorded by a variety of structures. Ductile strain is expressed by the shape and orientation of the minerals, particularly orthopyroxene, giving rise to a characteristic sequence of pervasive fabric changes that ultimately lead to mylonites. Folds and ductile faults, including ductile imbricate structures, are also found. The transition to brittle structures such as brittle faults, ramps, and brittle shear zones is commonly associated with the development of discrete mylonite zones.

The structural history of all the peridotites is complex (Rassios et al., 1994) though much is only indirectly relevant to the present paper. Strain features that are relevant are (1) the regional distribution and intensity of the ductile strain, (2) the orientation of the earliest peridotite fabrics, and (3) the orientation and relationship of the late ductile structures to the early brittle emplacement features.

Peridotite deformation and chromite distribution

As already noted, chromite orebodies are found in the western Hellenic ophiolites only where earlier high-T peridotite fabrics have been overprinted by late ductile and early brittle fabrics (Grivas et al., 1986; Rassios et al., 1986). Chromitite is always more competent than harzburgite, which in turn is more competent than dunite. These relationships hold from a scale of a few millimeters to large outcrops of tens of meters or more. Chromitite is generally hosted by dunite interleaved with harzburgite. When such a sequence is deformed, the strain appears to concentrate in the dunite around the chromite ores, which eventually become mylonitized. These relationships led the field mapping program to concentrate on ductile structures and on those straddling the ductile to brittle transition and to develop a fabric classification scheme (Rassios et al., 1994).

Field classification of ductile fabrics

Ophiolitic peridotites show progressive changes in microstructure and fabrics with increasing strain (Mercier and Nicolas, 1975; Nicolas, 1989, Fig. 2.7). A similar sequence was recognized in hand specimens of peridotite during the mapping of the Pindos ophiolite, from which a nongenetic fabric-classification scheme was devised. The peridotites can be subdivided into five categories (Rassios, 1991), and the same scheme applied to all the ophiolites. Fabric class 0 shows essentially no discernible fabric in the field; class 1 shows a weak blocky alignment of orthopyroxene; class 2 shows a strong alignment of pyroxene; class 3 contains elongated orthopyroxenes (and neoblastic olivine visible in thin section), and class 4 is a totally mylonitic fabric (elongation ratios of orthopyroxenes in excess of 5: 1). The structural sequence at a given location among these structural classes is assumed to correspond to a sequence characteristic of falling temperatures (Rassios et al., 1994).

Thin-section examination shows that this nongenetic fabric classification generally does represent a gradation from relatively high-T conditions (in classes 0 and 1) through the lower-T conditions of classes 2 and 3 to the ductilely (mylonitically) deformed peridotites of class 4. Petrofabric studies of Vourinos show that, despite their lack of a macroscopic fabric in the field, class 0 and 1 fabrics have a high preferred lattice orientation in olivine, implying considerable lattice strain, attributed to high-T intracrystalline deformation (Ross et al., 1980). The macroscopic strain observed in the other classes represents strain by lower-T intercrystalline deformation, mostly along grain boundaries. In the mylonites, both olivine and orthopyroxene are deformed (Ross et al., 1980). The lowest-temperature structures (class 4) are mylonitic peridotites formed just above the peridotite ductile to brittle boundary. All these structures are gradational: that is, higher-T fabrics grade into lower-T fabrics along and/or across strike and younger (cooler) structures overprint older (hotter) structures by folding and/or crosscutting them (Rassios, 1991). The temperature of cessation of ductile deformation of peridotite corresponds with temperatures inferred independently from the metamorphic soles of ophiolites (Cabanes and Briqueu, 1987).

Regional variations in ductile fabrics

According to single-pyroxene thermobarometry, the earliest Vourinos structures (already described as belonging to fabric classes 0 and 1) were created at temperatures of 900–1200 °C and at depths of 40–100 km (Ross et al., 1980). These fabrics reflect strain imposed on the peridotite shortly after, and possibly contemporaneous with, partial melting in the asthenosphere (Nicolas, 1989) and have been used to determine strain below the presumed spreading center in Oman (Nicolas, 1989). In Vourinos, the affected rocks are coarse-grained harzburgites and lherzolites with blocky pyroxenes. Petrofabric orientations are quite strong (Ross et al., 1980; Jackson et al., 1975), but mineral foliations are poorly defined in hand specimen. This old structural fabric is found in all the ophiolites: in north Vourinos, it strikes northwest and dips southwest, but trends northeast in the more complexly deformed south Vourinos (Ross et al., 1980, Fig. 3). The fabric also trends northwest in Pindos and Othris (Rassios et al., 1994).

Mylonitic structures form between ~1000 and ~800 °C. Peridotite in this interval deforms through intercrystalline mechanisms, imparting a dominant shape-defined mineral foliation. Annealing of new crystals formed in the solid state under these conditions is greatly aided by the presence of fluids. The presence of amphiboles within mylonitic peridotites of the Pindos ophiolite and in the mylonite zones of Vourinos suggests that there were local variations in fluid concentrations. Some mylonitic samples from the Pindos ophiolite contain large, undeformed euhedral chromium-rich spinels that appear to have undergone metasomatic growth into and along mylonitic foliation planes.

In the Dramala massif of the Pindos ophiolite (Fig. 4), a pervasive mylonitic fabric affects peridotites immediately above the basal sole. This fabric is presumed to be a result of shearing during sole formation and detachment (Rassios, 1991) as described for other similar mylonitic peridotites (Nicolas, 1989). Elsewhere in the Dramala massif and in contiguous peridotite massifs, large zones of pervasive mylonitic material have also been identified.

The mylonite zones of Vourinos (Grivas et al., 1993; Ross et al., 1980) capture the ductile to brittle transition precisely. A series of mylonite zones striking northwest with southwest dip cuts across regional harzburgite fabric (northwest striking with less steep southwest dip); Ross et al. (1980) evaluated the T-P (temperature and pressure) conditions of the mylonitic fabrics to be ~900 °C and <40 km depth and those of the host harzburgite to be ~1200 °C and 80–100 km depth. Large-scale mapping of the mylonite zones (1:500 scale, Grivas et al., 1993) traced them into contiguous dunite bodies where the mylonitic zones changed in nature to brittle shear zones. Thus, while harzburgite was still undergoing ductile deformation within discrete mylonite zones, dunite was deforming as a brittle material. The offsets of orebodies within the mylonite zones demonstrate that they acted as ductile ramping structures, verging toward the northeast.

The relative proportion of the strain types is independent of the strain orientation and varies from ophiolite to ophiolite (Fig. 6). In Vourinos rocks, class 2 ductile fabric (strong orientation of blocky orthopyroxene) predominates, and there is an almost total absence of mylonitized fabrics outside of discrete mylonite zones (Grivas et al., 1986; Ross et al., 1980; Moores, 1969). Class 3 predominates in the Pindos ophiolite, but class 2 and class 4 structures are also significant, both as pervasive mylonitic rock fabric and as discrete mylonitic zones. The Othris ophiolite is the most intensely mylonitized, with classes 3 and 4 together making up over 80% of pervasive rock fabrics. The Smokovo, Makriraki, and Mega Isoma massifs of western Othris are almost entirely class 3 and 4 mylonites, as are the peridotites of southern Pindos.

Elsewhere, class 3 and 4 rock fabrics are generally confined to the three settings: (1) within discrete mylonite zones cutting preexisting fabrics in all ophiolites, (2) along the base of the peridotite immediately overriding the metamorphic sole in the Pindos ophiolite (but the corresponding peridotites of Vourinos and Othris are *not* mylonitic), and (3) at the base of "hot" imbricates, that is, internal to a major peridotite sheet (Pindos). In Othris, some minor amphibolitic units have been entrapped between imbricate units (Rassios and Konstantopoulou, 1992).

In short, Vourinos has not undergone much pervasive low-temperature ductile strain (mylonitic rocks occur in discrete mylonite zones); it was emplaced with a minimal imprint of transitional struc-

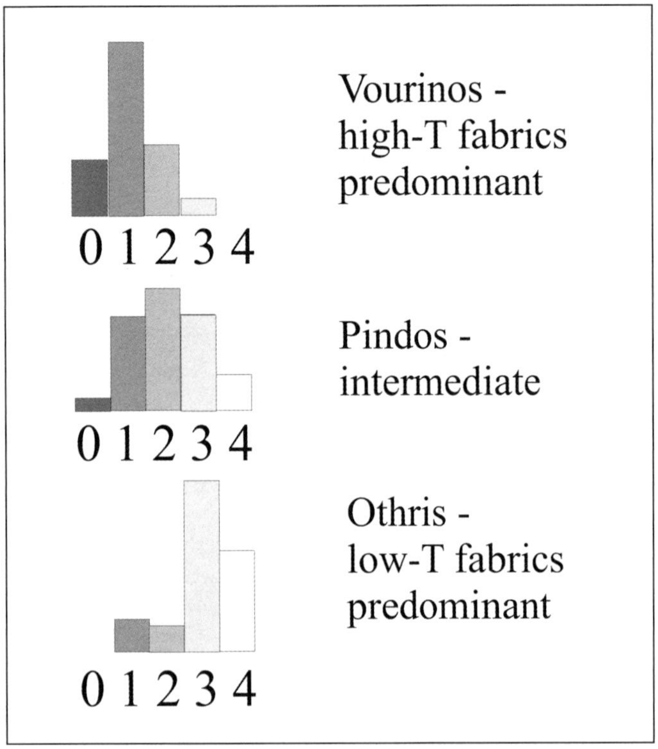

Figure 6. Relative abundances of fabric classes in Vourinos, Pindos (Dramala), and Othris (see text for definition of classes 0 to 4). The abundances are determined as follows: Vourinos and Pindos data are from the relative percentage of different peridotites exposed in each area so that they sum to 100%; Pindos data are the relative percentages of field observations of each type in the Dramala massif (377 observations in all). The fabrics show a clear trend from Vourinos (few mylonitized fabrics) to Othris (with pervasive fabrics, including the complete mylonization of class 4). This trend also corresponds to the geographic trend, with Vourinos in the northeast and Othris in the southwest.

tures. Pindos has undergone an intermediate degree of pervasive mylonitic deformation, and western Othris rocks show the most (Fig. 6). This regional tectonic variation is also partly mirrored in the chromite-ore distribution in the western Hellenic ophiolites, but the apparent correlation may reflect two quite different effects, as discussed next.

Fabric variation and chromite distribution

Highly depleted dunites are generally regarded as potentially favorable hosts to chromite ores (Pearce et al., 1984; Rassios et al., 1986; Roberts, 1988). The harzburgites and dunites of Pindos and western Othris (Smokovo massif) are depleted to highly depleted mantle but lack economic chromite deposits. This absence could be due to pervasive ductile cataclasis that has disseminated otherwise economic deposits throughout a greater volume of rock. However, contrary to this expectation, the highly sheared mantle sections of Domokos and Eretria in Othris host economic chromite deposits.

An alternative explanation is based on the observation that economic chromite deposits are abundant in ophiolites believed to have formed above a subduction zone, but less common or absent in ophiolites with a MORB signature (Roberts, 1988). Robertson et al. (1991) speculated that chromite was originally absent in the Pindos ophiolite because it was closer in composition to MORB than was Vourinos. However, the prevalence of mylonitic ductile strain in the Pindos dunites (Dramala massif) may also have contributed to the absence of economic chromite deposits.

The Othris ophiolite has a generally MORB-like character and lacks economic chromite. Only where there is evidence of a subduction zone—such as in the depleted harzburgites of Domokos and Eretria—do the associated dunites carry economic chromite. Presumably the Othris mafic rocks were derived mostly from the adiabatic decompressional partial melting of "normal" mantle, whereas the mafic rocks of Vourinos are the products of partial melting of a hydrated mantle wedge lying above a subduction zone. Metasomatic fluids from a mantle wedge have also been suggested as playing a role in other chromite deposits associated with ophiolites (Melcher et al., 1997; Proenza et al., 1999).

The regional differences in peridotite fabrics suggest that the Vourinos Complex cooled most rapidly through the ductile field and/or was in a low-strain setting, whereas western Othris spent the longest time in the ductile field and/or was in a high-strain setting. A simplified view of this apparent inverse correlation between chromite deposits and intensity of mylonitization is that the transition from a MORB setting (no economic chromite deposits) to a setting above a subduction zone (economic chromite deposits) was associated with a decrease in strain gradient.

Orientation of structures formed across the ductile to brittle transition

In Vourinos, the only strongly developed brittle shear set is geometrically similar to the oldest shear set of the Pindos and Othris ophiolites. Ductile-brittle shears display characteristics typical of both mylonitic and brittle fabrics (Ross and Lewis, 1989). We can thus assume that these structures were initiated in the ductile field. Except where they have been affected by tear faults, northeast-verging ductile structures of Vourinos such as mylonitization, sheath folds, and imbrication are identical in strain orientation to the strain field of brittle structures interpreted as emplacement structures. The transition between the two appears gradual rather than abrupt (Rassios et al., 1994), implying that parts of the ophiolite were still in the ductile field when emplacement to the northeast was initiated. A similar conclusion was reached by Ross et al. (1980). The structural setting is very similar to that inferred from (1) structural studies on the Vourinos metamorphic sole (Naylor and Harle, 1976) and (2) an analysis of thrusts in Vourinos (Wright, 1986).

Brittle shears are arrayed in shear zones, in ramps, or imposed upon older brittle structures. The relative ages of brittle systems are readily determined from crosscutting relationships in outcrops with multiple shear sets. For example, in the Mega Isoma massif of Othris (Rassios and Konstantopoulou, 1992) and in Dramala (Rassios, 1991), three distinct shear sets can be recognized. These three shear sets are the youngest structures *pervasively* affecting the peridotite. The oldest consists of ramps and duplexes indicating movement toward the northeast. The set of intermediate age verges to the southwest and is attributed to Eocene thrusting of the ophiolite over Tertiary flysch. The youngest set of shears again facilitated compression oriented toward the northeast and is a response to the most recent period of compression in the western Hellenides (Doutsos and Poulimeno, 1992). That the late ductile structures are the result of northeast thrusting related to initial emplacement is independently supported by the success of the chromite ore exploration program.

The apparent gradual transition from the ductile structures to the northeast-vergent brittle emplacement structures is central to the argument advanced here that the transition from spreading to initial emplacement took place in a short time interval. Numerical estimates of this interval can readily be made by considering the temperature changes in cooling ocean floor.

CONDITIONS DURING EMPLACEMENT

Metamorphic soles

During initial emplacement, metamorphic soles were created at the base of all three major ophiolites (Spray and Roddick, 1980). The sole formations consist of metamorphosed sedimentary rocks (Naylor and Harle, 1976; Jones, 1990) generally of amphibolitic facies but locally ranging into garnet amphibolite facies in small outcrops of the Othris sole near Lamia (Spray and Roddick, 1980). The compositions of the bulk of the amphibolites suggest an origin from a MORB source. Presumably the fracture along which motion was taking place was a low-angle thrust within the ocean floor along which the ophiolite-to-be was overriding other ocean floor (Spray and Roddick, 1980). This phase of intraoceanic thrusting is usually distinguished from the later phase of ophiolite emplacement onto a continent, a process generally referred to as obduction.

Within the Mega Isoma massif of Othris, the lower margins of the imbricates within the thrust sheets show layers of lower-amphibolite to greenschist facies (Rassios and Konstantopoulou, 1992). As discussed subsequently, these sole formations consistently give Middle Jurassic Ar-Ar dates (Spray and Roddick, 1980) that have not been reset by any shear heating associated with Tertiary thrusting. The amphibolitic sole rocks are ductilely deformed, but the sense of vergence is difficult to determine. Naylor and Harle (1976) inferred a northeast sense of emplacement from the geometry of folds in the sole units.

Ductile to brittle transition in cooling ocean floor

The rate at which temperatures change in newly formed ocean floor is readily estimated from the error function solution for the conduction of heat in a semi-infinite cooling slab (Carlslaw and Jaeger, 1959). For ocean floor younger than ~70 m.y., it is unnecessary to treat the ophiolite as the upper part of a plate of finite thickness as opposed to the upper part of a semi-infinite slab. For the purposes of this paper, the cooling effects of geothermal circulation on the upper oceanic crust can also be ignored. The temperatures and depths at which the ductile features in the ophiolites formed can be estimated. Thus by comparing the two estimates, limits can be placed on how old the ocean floor was at the time of ductile deformation (Fig. 7).

During its creation at a spreading center, the mantle section of an ophiolite is assumed here to have been at temperatures of ~1250 °C at the spreading center and to have cooled by conduction as it moved away. Peridotites deform in brittle fashion at temperatures below ~700 °C (Calmant, 1987; Watts et al., 1980), although in detail the ductile to brittle transition is complex (Kohlstedt et al., 1995).

The top of the ocean floor is always in the brittle field; the ductile to brittle transition for dry peridotite will correspond *approximately* to the 700 °C geotherm, but will be lower if volatiles are present. Below that level, the section will be in the ductile field. Thus, during spreading, ductile and brittle structures form contemporaneously at different levels and have the same strain orientation. As cooling takes place, lower-T brittle deformation penetrates to deeper levels and may be superposed on earlier higher-T ductile structures.

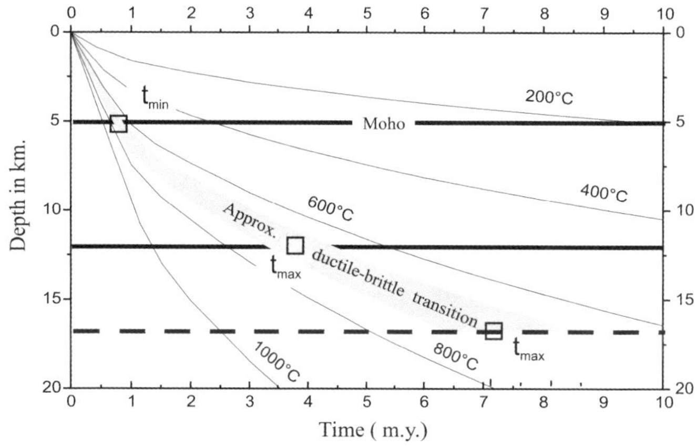

Figure 7. Isotherms for selected temperatures (200–1000 °C at 200 °C intervals) in cooling ocean floor. The values are calculated from the semi-infinite slab model because the interval of interest (0–10 m.y.) has solutions very close to the plate model for such a short interval. The solution is given by the error function (Carlslaw and Jaeger, 1959). The ductile to brittle transition for mantle rocks is approximated here by the region of the 700 °C isotherm (shaded). Two extreme estimates (t_{min} and t_{max}) for the time lag between crystallization and the beginning of emplacement are shown for the 12-km-thick Vourinos Complex (shown by the bold line). To determine the longest time lag, the entire slab is assumed to be in the brittle field (contrary to observation) before any emplacement structures are created. This assumption yields a maximum time lag t_{max} of ~4 m.y. The smallest time lag is given by the assumption that some ductile emplacement structures formed just below the Moho, giving a minimum time lag t_{min} of ~1 m.y. Also shown is a dashed line for a 17-km-thick ophiolite; though its t_{min} value would be the same as for the thinner Vourinos ophiolite, its t_{max} value would be ~7 m.y.

Because the depth of the ductile to brittle transition can be estimated, it is possible to link field observations of the position of brittle and ductile structures in an ophiolite slab to the maximum time that can have elapsed since the slab crystallized at a spreading center.

Vourinos provides the best data. The present minimum thickness of the Vourinos Complex is ~12 km, made up of 5 km of crust and 7 km of ultramafic rocks, the bottom 2 km of which is variably serpentinized (Moores, 1982). The temperatures of formation of the ductile structures are estimated as ~1200 °C for class 1 structures (high-T orthopyroxene fabric with a weak foliation) and ~1100–900 °C (Nicolas, 1989) for the remaining structures, with the ductile to brittle transition taking place at ~700 °C.

The western Hellenic ophiolites are presumed to be fragments of one or more thrust sheets. The thrust sheets currently have a length to thickness ratio of at least 5:1 or greater. It is assumed here that the most likely control on the location of the fracture is the position of the ductile to brittle transition (Nicolas, 1989). The thrust sheets are now cool and wholly in the brittle zone of deformation. However, the field evidence suggests that the ductile structures, which are now part of the thrust sheet, have the same geometry as the brittle emplacement structures and were formed contemporaneously with them.

The maximum time difference (t_{max}) between crystallization of a given piece of ophiolite at the spreading center and the beginning of its emplacement is the time taken for the base of the slab to pass through the ductile to brittle transition. Clearly, if no emplacement movement takes place prior to the transition through the ductile to the brittle regime, then the slab will show no ductile emplacement structures: they will be at a deeper level and not exposed to view. The data plotted in Figure 7 suggest that the maximum time difference (t_{max}) for a 12-km-thick slab such as in the Vourinos Complex is ~4 m.y. The smallest time difference (t_{min})is the time taken for those ductile emplacement structures that are shallowest to reach the ductile to brittle transition. If these structures are just below the Moho, the time is ~1 m.y. (Fig. 7). If the Vourinos is 17 km thick, then the slab cools to the temperature of the midpoint of the ductile-brittle transition in about 7 m.y. (Fig. 7).

That the time difference between crystallization and emplacement is short for many ophiolites is well known from isotopic studies. The time difference between crystallization and emplacement is less than 10 m.y. for many ophiolites and for one of the best-studied, the Semail ophiolite of Oman, it is less than 2 m.y. (Hacker et al., 1996). However, the estimates made here are entirely independent of any isotopic measurements. These estimates are not greatly affected by the range of uncertainty in the temperature of the ductile to brittle transition but depend on the assumption that the ductile structures are due to emplacement. If it is assumed instead that they are spreading features, then none of the currently available models of spreading centers correctly predicts the positions of chromite ores, whereas a model that assumes they are thrust-related emplacement features locates chromite ores, even when they are "blind," i.e., hidden at depths of 200 m or so. It is therefore concluded that the ocean floor must have been undergoing emplacement no more than 1–7 m.y. after its initial formation at the spreading center. Though crustal thicknesses are not known in Pindos and Othris, it is highly probable that the same range of time estimates apply to these ophiolites.

Similar inferences can be made from the *P-T* conditions inferred for the metamorphic sole at the base of the ophiolite, though all these inferences assume that shear heating is negligible. Sole conditions ranged up to 860 °C and 7 kbar in Othris (Spray and Roddick, 1980). The overall situation in Vourinos is very similar to that in Oman, where the metamorphic sole reached at least 825 °C and an isotopically determined interval of less than 2 m.y. exists between the crystallization of the Semail ophiolite and the age of the metamorphic sole (Hacker et al., 1996).

There are no well-defined isotopic dates for the crystallization of any of the western Hellenic ophiolites (Smith, 1993). The only relevant isotopic dates are Ar-Ar dates from amphibolites forming the metamorphic soles (Fig. 2) in Vourinos (171 ± 4 Ma [Spray et al., 1984]), Pindos (176 ± 5 Ma [Thuizat et al., 1981]; 167 ± 4 Ma [Spray et al., 1984]), and Othris (169 ± 4 Ma [Spray et al., 1984]). The mean age of 171 ± 4 Ma is late Bajocian, according to the time scale of Gradstein et al. (1995). If the formation of the metamorphic soles overlapped the period during which the ductile structures were active, then the formation of the ophiolite can be placed in the range 172 ± 4 Ma to 178 ± 4 Ma, i.e., middle Bajocian to latest Aalenian time or, most probably, early Bajocian time.

In Argolis and Evvia (Fig. 2), radiolarite lying on the pillow lavas ranges in age from late Bathonian (ca. 166 Ma) to early Oxfordian (ca. 158 Ma) (Jones et al., 1992). Emplacement of the Pindos ophiolite seems to have been underway by early Bathonian (ca. 168 Ma) to early Callovian (ca. 163 Ma) time as radiolarites of this age are interbedded with ophiolite clasts that include chromite-rich harzburgite debris (Jones et al., 1992, p. 395). Although there are still uncertainties of 3–4 m.y. in the Jurassic time scale (Gradstein et al., 1995), these data are entirely consistent with early Bajocian spreading followed by initial emplacement of the ophiolite in late Bajocian time.

Thermal considerations currently provide the best high-resolution tool for estimating relative time between spreading and initial emplacement, though high-precision isotopic dating and biostratigraphy may eventually provide "absolute" time estimates as well.

Emplacement models

Several models have attempted to place the western Hellenic ophiolites into a regional geologic setting, but they will not be discussed here. These models are generally derived from emplacement models of the Oman ophiolite (Coleman, 1981), as modified by Lippard et al. (1986). Instead, the backarc-basin setting is taken as given, and a model emphasizing the possible changes in the nature of the mantle wedge during subduction and their effects on ophiolite creation and strain are outlined. This speculative model may account for some of the field observations highlighted in this paper: (1) ductile deformation in the peridotites passing smoothly into brittle deformation associated with emplacement and (2) the intense mylonitization of the Othris peridotites compared with the much weaker deformation in the peridotites of Vourinos.

Associated with these strain differences are compositional differences: the rocks are largely of MORB chemical affinity in Othris but of IAT and BSV chemical affinity in Pindos and Vourinos; Vourinos rocks are the most highly boninitic. These remarkable lateral and horizontal differences appear to be transitional rather than abrupt. They are interpreted here as reflecting the influence of mantle wedge–derived volatiles, particularly water, on the ocean-floor rocks arising from the mantle wedge, rather than reflecting preexisting inhomogeneities in the mantle or the tectonic juxtaposition of ophiolites from widely separated source areas. Cartoons that may account for some of these differences are illustrated in Figure 8.

CONCLUSIONS

1. Vourinos is chiefly a supra-subduction zone ophiolite, lacking lherzolite and MORB, but with significant IAT and BSV additions. Pindos includes lherzolite and harzburgite and shows early MORB lavas overlain by lavas with trace element trends similar to those of Vourinos. Othris includes lherzolite and harzburgite as well as MORB lavas. These represent primary variations within a single oceanic (backarc) slab.

2. These variations reflect the influence of mantle wedge–derived volatiles, particularly water, on the ocean-floor rocks arising from the mantle wedge, rather than reflecting preexisting inhomogeneities in the mantle or the tectonic juxtaposition of ophiolites from widely separated source areas.

3. Emplacement of this unit resulted in a series of thrust sheets in Jurassic time; they were affected by Cenozoic thrusting and normal faulting, which has not significantly disrupted the original Jurassic geometry.

4. The stratigraphic thickness of the Vourinos is ~12 km. Significant tectonic thickening by thrust repetition of the ultramafic section has not been observed (Grivas et al., 1986). The amount of tectonic thinning, if present, is unknown. The thicknesses of the Pindos and Othris ophiolites are probably similar, but less well defined.

5. The gradual change in the ductile structures in Vourinos into northeast-vergent brittle emplacement structures indicates that the transition from spreading to initial emplacement was essentially a continuous process. The temperatures at which the ductile structures formed, together with the inferred *P-T* conditions of sole metamorphism, imply that the initial emplacement of the ophiolite took place less than 4 m.y.—and possibly as little as 1 m.y.—after crystallization.

6. Late ductile structures are penecontemporaneous with the formation of the metamorphic soles. The late Bajocian isotopic dates together with the ≤4 m.y. interval between initial emplacement and

spreading imply an early Bajocian spreading age for the western Hellenic ophiolites.

7. The oldest ductile structures in the peridotites, presumed to have formed below a Jurassic spreading center, have a northwest trend in north Vourinos, Pindos, and Othris.

8. Peridotite fabrics of Othris show intense mylonitization; Pindos shows less mylonitization, and Vourinos shows little mylonitization.

9. Economic chromite deposits are most abundant in Vourinos; they are rare in Pindos, and in Othris they occur only in association with harzburgites.

10. The lateral variations in strain and composition may reflect the variations in the geometry and composition of the mantle wedge associated with subduction.

ACKNOWLEDGMENTS

A. Rassios gives special thanks to E. Grivas, I. Vacondios, and G. Konstantopoulou. Y. Dilek, E.M. Moores, and J.G. Spray provided helpful reviews. A.G. Smith thanks the organizers of the Penrose Conference for the invitation to attend and St. John's College, Cambridge, for financial support. This is Earth Sciences Cambridge contribution no. 5670.

REFERENCES CITED

Beccaluva, L., Ohnenstetter, D., and Ohnenstetter, M., 1979, Geochemical discrimination between ocean floor and island arc tholeiites: Application to some ophiolites: Canadian Journal of Earth Sciences, v. 16, p. 1874–1882.

Beccaluva, L., Ohnenstetter, D., Ohnenstetter, M., and Paupy, A., 1984, Two magmatic series with island arc affinities within the Vourinos ophiolite: Contributions to Mineralogy and Petrology, v. 85, p. 253–271.

Cabanes, N., and Briqueu, L., 1987, Hydration of an active shear zone: Interaction between deformation, metasomatism, and magmatism: The spinel lherzolites from the Montferrier (southern France) Oligocene basalts: Earth and Planetary Science Letters, v. 81, p. 233–244.

Calmant, S., 1987, The elastic thickness of the lithosphere in the Pacific Ocean: Earth and Planetary Science Letters, v. 85, p. 277–288.

Carlslaw, H.S., and Jaeger, J.C., 1959, Conduction of heat in solids: Oxford, Oxford University Press, 510 p.

Clift, P.D., and Dixon, J.E., 1998, Jurassic ridge collapse, subduction initiation, and ophiolite obduction in the southern Greek Tethys: Eclogae Geologicae Helvetiae, v. 91, p. 128–138.

Coleman, R.G., 1981, Tectonic setting for ophiolite obduction in Oman: Journal of Geophysical Research, v. 86, p. 2497–2508.

Courtin, B., 1979, Étude géologique de la region du Domokos (Grèce): Le font des zones internes et les massifs ophiolitiques d'Othris occidentale [3rd cycle thesis]: University of Lille, 295 p.

Doutsos, T., and Poulimeno, G., 1992, Geometry and kinematics of active faults

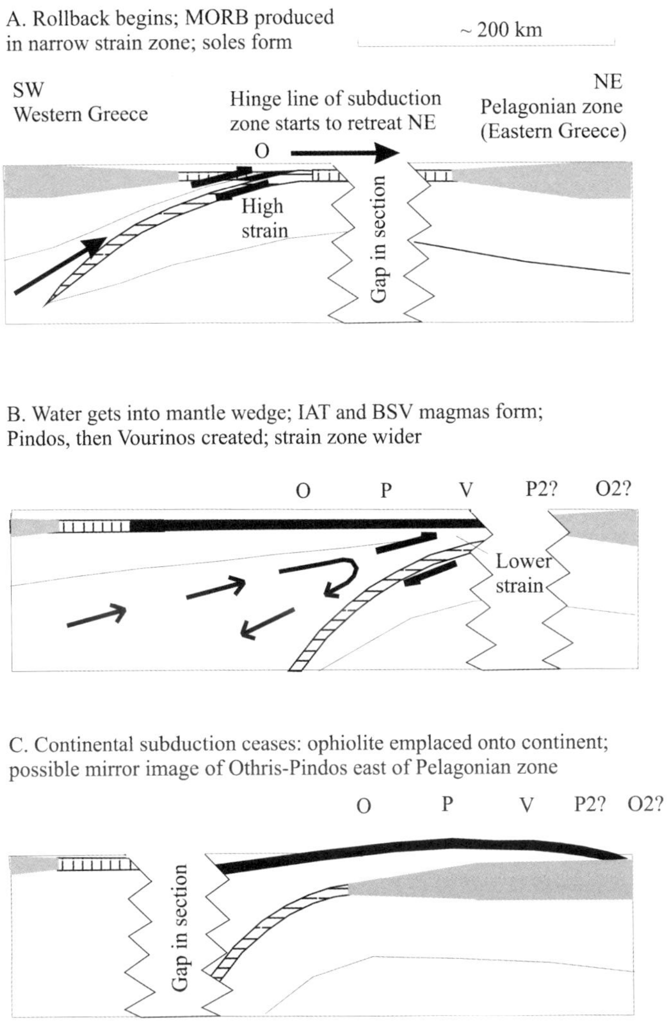

Figure 8. Cartoons showing an emplacement model that might account for mantle deformation and sole formation penecontemporaneous with spreading and for differences in strain and composition of the Othris, Pindos, and Vourinos ophiolites. All cartoons show a southwest-northeast cross section across the ophiolite belt from the Triassic passive margin of western Greece to that of the Pelagonian zone. The gray zones are continental crust; the vertically striped zones are Triassic(?) oceanic crust; the horizontally striped zones are subducted ocean crust; and the black strip is "backarc" ocean crust of the ophiolite generated during the retreat of the hinge line of the subduction zone relative to the southwest passive margin. The thin lines beneath these features are schematic lower boundaries of the lithosphere. A, In Middle Jurassic time, a narrow conduit is available for any partially melted asthenosphere that exists. Water is assumed not to be available, and the resulting magma composition is MORB. When the magma nears the surface, it is in the field of brittle deformation and in extension: sheeted dikes typical of an ophiolite are created. However, its base is hot and ductile and is simultaneously sheared, causing high strain, as in Othris (O). The northeastward-moving ophiolite also overrides the subducted ocean floor, creating sole rocks more or less contemporaneously with spreading and crystallization. B, No more than a few million years later, the circulating mantle wedge picks up water from the subducted slab, creating the IAT and BSV magmas found in the Pindos (P) and Vourinos (V) ophiolites. The resulting strain zone is much wider, causing progressively less strain in the Pindos and Vourinos peridotites. It is unclear whether magma is injected asymmetrically in such a setting, giving rise to a single lateral change from MORB to IAT to BSV, or whether it is injected symmetrically to produce a counterpart to the Pindos (P2?) and Othris (O2?) ophiolites to the east of Vourinos. C, In Late Jurassic time, the northeast passive continental margin (southwest Pelagonian zone) reaches the subduction zone. Subduction ceases, leaving the originally continuous Othris-Pindos-Vourinos (O-P-V) ophiolite on the southwest edge of the Pelagonian continental crust. It is unclear whether an eastern counterpart to the Pindos and Othris ophiolites existed at this time (or had been subducted). In western Vermion, ~70 km east of Vourinos, Sharp (1994, p. 139) mapped a mélange containing boninitic blocks, derived from the advancing margin of the Vourinos ophiolite. Farther east still, MORB ophiolites occur in the Axios (Vardar) zone of the Eastern Hellenic ophiolites (Fig. 2), but their relationship to the Vourinos Complex is unclear.

and their seismotectonic significance in the western Corinth-Patras rift: Journal of Structural Geology, v. 14, no. 6, p. 689–699.

Economou, M., Dimou, E., Economou, G., Migiros, G., Vacondios, I., Grivas, E., Rassios, A., and Dabitzias, S., 1986, Chromite deposits of Greece: Athens, Greece, Theophrastus Publications, S.A., p. 129–159.

Ferrière, J., 1982, Palaeogéographie et tectoniques superposées dans les Héllenides internes: Les massif de l'Othrys et du Pelion (Grèce septentrional): Societé Géologique du Nord, Lille Publications 8, p. 1–970.

Frison, J., 1987, Les péridotites du massif ophiolitique du Vourinos (Grèce): Étude petro-structurale mise en evidence d'une structure diapirique [Ph.D. thesis]: University of Paris, 240 p.

Gradstein, F.M., Agterberg, F.P., Ogg, J.G., Hardenbohl, J., van Veen, P., Thierry, T., and Huang, Z., 1995, A Triassic, Jurassic, and Cretaceous time scale, *in* Berggren, W.A., Kent, D.V., Aubry, M.-P., and Hardenbohl, J., eds., Geochronology, time scales and global stratigraphic correlation: Tulsa, Oklahoma, SEPM (Society for Sedimentary Geology), Special Publication 54, p. 95–128.

Grivas, E., Rassios, A., Vacondios, I., and Vrahatis, G., 1986, Structural traverses across Vourinos, *in* Rassios, A., Roberts, S., and Vacondios, I., eds., The application of a multidisciplinary concept for chromite exploration in the Vourinos Complex (N. Greece): Athens, Institute for Geological and Mineral Exploration, p. 258–283.

Grivas, E., Rassios, A., Konstantopoulou, G., Vacondios, I., and Vrahatis, G., 1993, Drilling for "blind" podiform chrome orebodies at Voidolakkos in the Vourinos ophiolite complex: Economic Geology, v. 88, p. 461–468.

Hacker, B.R., Mosenfelder, J.L., and Gnos, E., 1996, Rapid emplacement of the Oman ophiolite: Thermal and geochronologic constraints: Tectonics, v. 15, p. 1230–1247.

Harkins, E., Green, H.W., and Moores, E.M., 1980, Multiple intrusive events documented from the Vourinos ophiolite complex, northern Greece: American Journal of Science, v. 280A, p. 284–295.

Hynes, A.J., 1972, The geology of part of the western Othris Mountains [Ph.D. thesis]: Cambridge, University of Cambridge, p.

IGRS-IFP (Institut de Géologie et Recherches du sous-sol [Grèce] et Institut Français du pétrole. Mission Grèce), 1966, Etude géologique de l'Epire (Grèce nord-occidentale): Paris, Technipresse edition, 306 p.

Jackson, E.D., Green, H.W., and Moores, E.M., 1975, The Vourinos ophiolite, Greece: Cyclic units of lineated cumulates overlying harzburgite tectonite: Geological Society of America Bulletin, v. 86, p. 390–398.

Jones, G., 1990, Tectono-stratigraphy and evolution of the Mesozoic Pindos ophiolite and associated units, northwest Greece [Ph.D. thesis]: Edinburgh, University of Edinburgh, 394 p.

Jones, G., and Robertson, A.H.F., 1991, Tectono-stratigraphy and evolution of the Pindos ophiolite and associated units: Journal of the Geological Society (London), v. 148, p. 267–288.

Jones, G., De Wever, P., and Robertson, A.H.F., 1992, Significance of radiolarian age data to the Mesozoic tectonic and sedimentary evolution of the northern Pindos Mountains, Greece: Geological Magazine, v. 129, no. 4, p. 385–400.

Kissel, C., and Laj, C., 1988, The Tertiary geodynamical evolution of Greece: A paleomagnetic reconstruction: Tectonophysics, v. 146, p. 183–201.

Kohlstedt, D.L., Evans, B., and Mackwell, S.J., 1995, Strength of the lithosphere: Constraints imposed by experiment: Journal of Geophysical Research, v. 100, p. 17587–17602.

Konstantopoulou, G., Vacondios, I., and Vonisakou, M., 1988, Geology and copper mineralisation of the Ag. Ekaterini district of the Othris ophiolite: Athens, Institute of Geology and Mineral Exploration, 35 p.

Lippard, S.J., Shelton, A.W., and Gass, I.G., 1986, The ophiolite of northern Oman: Geological Society of London Memoir, v. 11, p. 178.

Melcher, F., Grum, W., Simon, G., Thalhammer, T., and Stumpfl, E., 1997, Petrogenesis of the ophiolitic giant chromite deposits of Kempirsai, Kazakhstan: A study of solid and fluid inclusions in chromite: Journal of Petrology, v. 38, no. 10, p. 1419–1458.

Menzies, M., 1974, Petrogenesis of the Makrirachi Ultramafic Complex [Ph.D. thesis]: Cambridge, University of Cambridge, 199 p.

Menzies, M., and Allen, C., 1974, Plagioclase lherzolite-residual mantle: Relationships within two eastern Mediterranean ophiolites: Contributions to Mineralogy and Petrology, v. 46, no. 2, p. 454–487.

Mercier, J., and Nicolas, A., 1975, Textures and fabrics of upper mantle peridotites as illustrated by basalt xenoliths: Journal of Petrology, v. 16, p. 454–487.

Moores, E.M., 1969, Petrology and structure of the Vourinos ophiolitic complex of northern Greece: Geological Society of America Special Paper 118, 74 p.

Moores, E.M., 1982, Origin and emplacement of ophiolites: Reviews of Geophysics and Space Physics, v. 20, p. 735–760.

Morris, A., 1995, Rotational deformation during Palaeogene thrusting and basin closure in eastern central Greece: Palaeomagnetic evidence from Mesozoic carbonates: Geophysical Journal International, v. 121, p. 827–847.

Naylor, M.A., and Harle, T.J., 1976, Palaeogeographic significance of rocks and structures beneath the Vourinos ophiolite, northern Greece: Journal of the Geological Society of London, v. 132, p. 667–675.

Nicolas, A., 1989, Structures of ophiolites and dynamics of oceanic lithosphere: London, Kluwer Academic Publishers, 367 p.

Pearce, J., Lippard, S.J., and Roberts, S., 1984, Characteristics and tectonic significance of supra-subduction zone ophiolites, *in* Kokelaar, B.P., and Howells, M.F., eds., Marginal basin geology: Geological Society of London Special Publication 16: Oxford, Blackwell Scientific Publications, p. 77–94.

Penrose Conference Participants, 1972, Penrose Field Conference on ophiolites: Geotimes, v. 17, no. 12, p. 24–25.

Proenza, J., Gervilla, F., Melgarejo, J., and Bodinier, J., 1999, Al- and Cr-rich chromitites from the Mayari-Baracoa ophiolitic belt (eastern Cuba): Consequence of interaction between volatile-rich melts and peridotites in suprasubduction mantle: Economic Geology, v. 94, no. 4, p. 547–566.

Rassios, A., 1989, The geology and ore environment of the Limogardi copper deposit, Othris ophiolite: Athens, Greece, Institute of Geology and Mining Exploration, 169 p.

Rassios, A., 1990, Geology and copper mineralization of the Vrinena area, east Othris ophiolite, Greece: Ofioliti, v. 15, no. 2, p. 287–304.

Rassios, A., 1991, Internal structure and pseudo-stratigraphy of the Dramala peridotite massif, Pindos Mountains, Greece: Bulletin of the Geological Society of Greece, v. 25 no. 1, p. 293–305.

Rassios, A., 1994, ed., Tectonic and geochemical methods for chrome exploration in ophiolites: Athens, Institute for Mining and Mineral Exploration, 98 p.

Rassios, A., and Konstantopoulou, D., 1992, Emplacement tectonics and the position of chrome ores in the Mega Isoma Peridotites, S.W. Othris, Greece: Bulletin, Geological Society of Greece, v. 28 no. 2, p. 463–474.

Rassios, A., Beccaluva, L., Bortolotti, V., Mavrides, A., and Moores, E.M., 1983, The Vourinos ophiolitic complex: A field excursion guidebook: Ofioliti, v. 8, no. 3, p. 275–292.

Rassios, A., Roberts, S., and Vacondios, I., 1986, The application of a multidisciplinary concept for chromite exploration in the Vourinos Complex (N. Greece): Athens, Institute of Geology and Mineral Exploration, p. 409.

Rassios, A., Konstantopoulou, G., Vacondios, I., and Grivas, E., 1990, Part 1, Synthesis, *in* Rassios, A., Konstantopoulou, G., and Vacondios, I., eds., Tectonic controls on chrome ore localisation in ophiolites, Greece: Athens, Greece, Institute of Geology and Mineral Exploration, p. 1–35.

Rassios, A., Grivas, E., Konstantopoulou, G., and Vacondios, I., 1994, The geometry of structures forming around the ductile-brittle transition in the Vourinos-Pindos-Othris oceanic slab: Bulletin of the Geological Society of Greece, v. 30 no. 2, p. 109–121.

Rassios, A., Grivas, E., and Vacondios, I., 1999, The structural-geochemical-metallogenic unification of the Pindos and Vourinos ophiolites: Journal of Conference Abstracts, European Union of Geosciences 10 Abstract volume, Strasbourg, France, v. 4, p. 407.

Roberts, S., 1988, Ophiolite chromite formation: A marginal basin phenomenon: Economic Geology, v. 83, p. 1034–1036.

Robertson, A.H.F., Clift, P.D., Degnan, P.J., and Jones, G., 1991, Palaeogeographic and palaeotectonic evolution of the Eastern Mediterranean Neotethys: Palaeogeography, Palaeoclimatology, Palaeoecology, v. 87, p. 289–343.

Ross, J., and Lewis, P., 1989, Brittle-ductile transition: Semi-brittle behaviour: Tectonophysics, v. 167, p. 63–81.

Ross, R.V., Mercier, J.C., Avé Lallemant, H.G., Carter, N.L., and Zimmerman, J., 1980, The Vourinos ophiolite complex, Greece: The tectonite suite: Tectonophysics, v. 70, p. 63–83.

Sharp, I., 1994, The Triassic to Tertiary sedimentary, tectonic, and magmatic evolution of the Pelagonian and Vardar (Axios) zones, Macedonia, Northern Greece [Ph.D. thesis]: University of Edinburgh, 815 p.

Smith, A.G., 1979, Othris, Pindos and Vourinos ophiolites and the Pelagonian zone, 6th Colloquium on the geology of the Aegean region: Athens, Greece, Proceedings of the Colloquium on the Geology of the Aegean Region 6, v. 3, p. 1369–1374.

Smith, A.G., 1993, The tectonic significance of the Hellenic-Dinaric ophiolites: Geological Society of London Special Publication 76, p. 213–243.

Smith, A.G., Woodcock, N.H., and Naylor, M.A., 1979, The structural evolution of a Mesozoic continental margin, Othris Mountains, Greece: Journal of the Geological Society of London, v. 136, p. 589–603.

Spray, J.G., and Roddick, J.C., 1980, Petrology and $^{40}Ar/^{39}Ar$ geochronology of some Hellenic sub-ophiolitic metamorphic rocks: Contributions to Mineralogy and Petrology, v. 72, p. 43–55.

Spray, J.G., Bebien, J., Rex, D.C., and Roddick, J.C., 1984, Age constraints on the igneous and metamorphic evolution of the Hellenic-Dinaric ophiolites, *in* Dixon, J.E., and Robertson, A.H.F., eds., The geological evolution of the Eastern Mediterranean: London, Geological Society, p. 619–627.

Thuizat, R., Whitechurch, H., Montigny, R., and Juteau, T., 1981, K-Ar dating of some infra-ophiolitic metamorphic soles from the Eastern Mediterranean: New evidence for oceanic thrustings before obduction: Earth and Planetary Science Letters, v. 52, p. 301–310.

Vrahatis, G., and Grivas, E., 1980, Geological mapping in Vourinos, 1:10 000 scale: Athens, Institute of Geological and Mineral Exploration, 80 p.

Watts, A., Bodine, J., and Steckler, M., 1980, Observation of flexure and the state of stress in the oceanic lithosphere: Journal of Geophysical Research, v. 85, p. 6369–6376.

Wright, L., 1986, The effect of deformation of the Vourinos ophiolite, *in* Rassios, A., Roberts, S., and Vacondios, I., eds., The application of a multidisciplinary concept for chromite exploration in the Vourinos Complex (N. Greece): Athens, Institute of Geology and Mineral Exploration, p. 156–216.

Zimmerman, J., 1972, Emplacement of the Vourinos ophiolitic complex, northern Greece, *in* Shagam, R., et al., eds., Studies in earth and space sciences: Geological Society of America Memoir 132, p. 225–239.

Manuscript Accepted by the Society February 23, 2000

Printed in the U.S.A.

Geological Society of America
Special Paper 349
2000

Petrology and geochemistry of the Shahr-Babak ophiolite, central Iran

A. Mohamad Ghazi
Department of Geology, Georgia State University, Atlanta, Georgia 30303, USA; e-mail: mghazi@gsu.edu
A.A. Hassanipak
Department of Mining Engineering, University of Tehran, Tehran, Iran

ABSTRACT

The Shahr-Babak ophiolite is a highly dismembered ophiolite complex located along the Nain-Baft fault zone in central Iran and is considered one of the complexes of the internal Iranian group of ophiolites. The igneous rocks of this complex consist of lherzolite, harzburgite, dunite, serpentinite, pyroxenite, gabbro, diorite, and a volcanic sequence that exhibits a wide range in composition from basaltic andesite to rhyodacite-rhyolite and trachyandesite. Associated with the ophiolite is a variety of shallow-marine rocks including pelagic fossiliferous carbonates. Also present are extensive units of radiolarian chert that are interbedded within the basaltic andesite sequence. A combination of petrographic observations and geochemical analyses indicates the presence of at least three different types of extrusive rocks in the Shahr-Babak ophiolite. The data show that the extrusive rocks were formed from two distinct magmatic sources: (1) the melt that produced the basaltic andesite and rhyodacite was generated in an island-arc environment, and (2) the melt that produced the trachyandesite was produced in a within-plate environment (e.g., an intraplate–oceanic-island environment). The geochemical data suggest that the basaltic andesite melt was generated either by a high degree of partial melting (~40%) of a mantle lherzolite source or by a lesser degree of partial melting (15%–20%) of a depleted mantle source. The data also suggest that the gabbro, diorite, basaltic andesite, and rhyodacite-rhyolite are petrogenetically related. The presence of compositionally diverse extrusive rocks intercalated with a variety of sedimentary rocks indicates that the Shahr-Babak ophiolite has both island-arc and intraoceanic components, which were brought together during the northwestward subduction of a narrow oceanic lithosphere under the Central Iranian microcontinent and Eurasia. This seaway was part of the southern Neotethyan Ocean that once surrounded the Central Iranian microcontinent.

INTRODUCTION

Iranian ophiolites are part of the Tethyan ophiolite belt of the Middle East. These ophiolites link the Middle Eastern ophiolites to other Asian ophiolites (e.g., Pakistani and Tibetan) as well as to the ophiolites in the Hellenides-Dinarides and the Mediterranean region (e.g., Turkish, Troodos, Greek, and East European). Main tectonic elements of Iran and the locations of the major Iranian ophiolites are depicted in Figure 1. On the basis of their age and abundance, the Iranian ophiolites may be classified into two groups: the less abundant Paleozoic ophiolites and more abundant Mesozoic ophiolites (Arvin and Robinson, 1994). However, more specifically, the Iranian ophiolites have been divided into four groups (Takin, 1972; Stocklin, 1974; McCall, 1997, 1998, written communication): (1) ophiolites of northern Iran along the Alborz Range, (2) ophiolites of the Zagros suture zone, including the Neyriz and the Kermanshah ophiolites, which appear to be extensions of the Oman ophiolite emplaced onto the Arabian continental margin, (3) ophiolites and colored mélanges of the Makran region, which are located to the south of the Sanandaj-Sirjan microcontinental block, including unfragmented complexes such as Sorkhband and Rudan, and (4) ophiolites and colored mélanges that mark the boundaries of the internal Iranian microcontinental block, including some of those in the Makran region (e.g., Band-e-Zeyarat, Dar Anar, Ganj) and those inside of the Sanandaj-Sirjan microcontinental block (Shahr-Babak, Nain, Baft, Sabsevar, and Tchehel Kureh).

Alavi (1991) used emplacement ages derived from field relationships to classify the Iranian ophiolites into three groups: (1) Proterozoic ophiolites are present as isolated outcrops on the western edge of the Lut block in central Iran, (2) pre-Jurassic ophiolites are located within the Alborz Range in the north, and (3) post-Jurassic ophiolites are the most abundant. Lippard et al. (1986) suggested emplacement ages of pre-Paleocene for all of the Mesozoic ophiolite complexes. However, more recent studies have shown that the volcanic rocks of the Makran ophiolites are intercalated with sedimentary deposits that contain Paleocene microfossils (McCall, 1997). Table 1 shows a summary of distribution, ages, chemical affinity, and major structural and tectonic characteristics of the Iranian ophiolites.

In this paper, we present results on the petrology and geochemistry of the Shahr-Babak ophiolite. The goal of this paper is to use field data, petrographic analyses, and geochemical data, including REE (rare earth element) and other incompatible trace element data,

Ghazi, A.M., and Hassanipak, A.A., 2000, Petrology and geochemistry of the Shahr-Babak ophiolite, central Iran, *in* Dilek, Y., Moores, E.M., Elthon, D., and Nicolas, A., eds., Ophiolites and Oceanic Crust: New Insights from Field Studies and the Ocean Drilling Program: Boulder, Colorado, Geological Society of America Special Paper 349, p. 485–497.

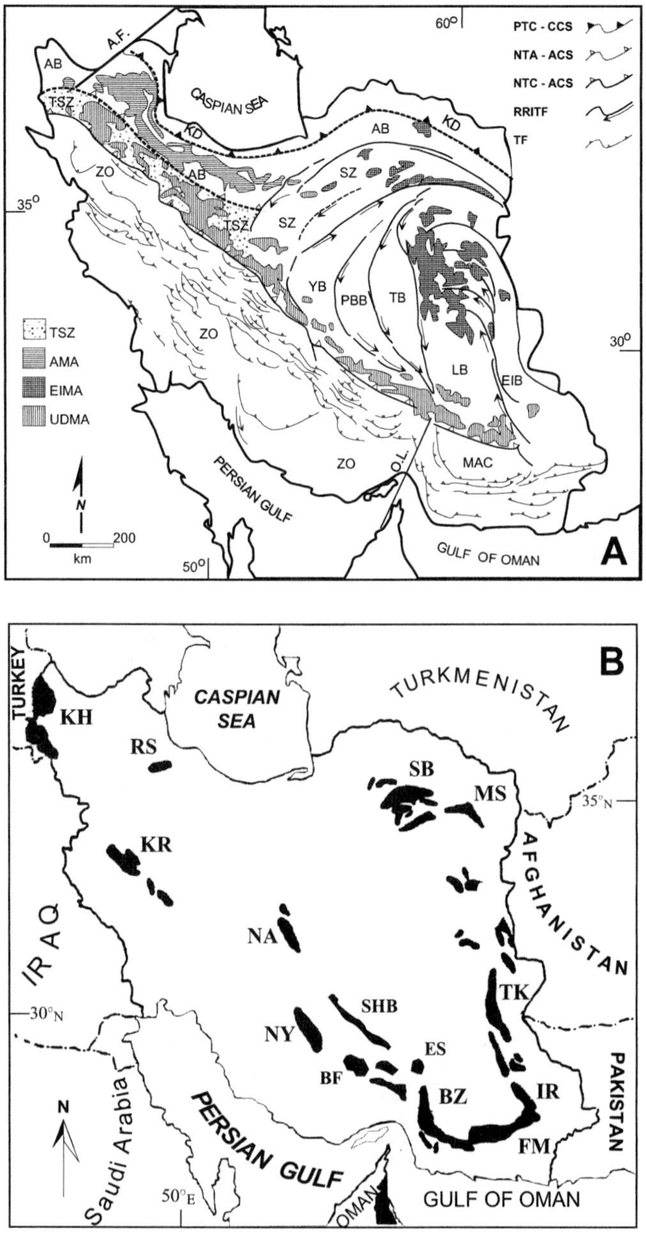

Figure 1. A, Index map showing main tectonic elements of Iran (after Alavi et al., 1997). Abbreviations: AB—Alborz belt, A.F.—Aras fault, AMA—Alborz magmatic belt, EIMA—East Iranian magmatic belt, KD—Kopeh Dagh zone, LB—Lut block, MAC—Makran accretionary complex, NTA-ACS—Neotethyan arc–arc collisional suture, O.L.—Oman Line, PBB—Poasht Badam block, RRITF—rotation-related intercontinental transfer fault, SZ—Sabzevar zone, TB—Tabas block, TSZ—Tabriz-Saveh zone, UDMA—Urumieh-Dokhtar magmatic arc, YB—Yazd block, ZO—Zagros orogen. B, Locations of major Iranian ophiolites. KH—Khoy, KR—Kermanshah, NY—Neyriz, BZ—Band-e-Zeyarat, NA—Nain, SHB—Shahr-Babak, BF—Baft, ES—Esphandagheh, FM—Fanuj-Maskutan, IR—Iranshahr, TK—Tchehel Kureh, MS—Mashad, SB—Sabzevar, RS—Rasht. PTC-CCS—Paleo-Tethyan continent-continent collisional suture, NTC-ACS—Neo-Tethyan continent-arc collisional suture, TF—thrust fault, EIB—East Iran block.

to (1) identify different lithologic units of this complex, (2) shed light on the geochemistry and petrogenetic processes of its formation, and (3) suggest a possible tectonic environment of formation for this ophiolite in the context of the Neotethyan tectonic reconstruction models of Iran and the Middle Eastern region.

REGIONAL GEOLOGY

The Shahr-Babak ophiolite is located in central Iran, and it is considered as one of the complexes of the internal Iranian group of ophiolites and colored mélanges. It covers an area of ~1500 km^2 and is located along the Nain-Baft fault zone, which marks the western border of the Central Iranian microcontinent. The Nain-Baft fault zone is characterized by a northwest-trending fault that is associated with the Tertiary volcanic rocks in central Iran. Although not confirmed by surface geology, this fault is probably the root zone from which several scattered ophiolites and mélanges (i.e., Nain, Shahr-Babak, Baft, and Esphandagheh) were emplaced along the eastern flank of the Sanandaj-Sirjan continental block; thus the fault probably separates this block from the Central Iranian microcontinent (Fig. 2) (Glennie et al., 1990). The fault-bounded area encircled by the central Iranian ophiolitic belt has been conceived by Takin (1972) as a large fragment of the Central Iranian continental plate, which for some time in the geologic past was separated from the rest of Iran by a narrow oceanic trough (Şengör, 1990; McCall, 1997).

OPHIOLITE STRATIGRAPHY

The ophiolite and colored mélange complex in the area of the Shahr-Babak consists of a mixture of ophiolitic and sedimentary rocks. Although the ophiolite stratigraphy in this area is largely dismembered, it contains all the rocks that are representative of an ophiolite suite except a sheeted-dike complex. The ultramafic rocks of the Shahr-Babak ophiolite are mainly exposed in the middle and northern sections of the complex (Fig. 2). The main mass of the Shahr-Babak peridotite is composed of plagioclase lherzolite, lherzolite, harzburgite, and dunite. Pyroxenite and isolated zones of two-phase olivine + Cr-rich spinel cumulate rocks are less abundant and form dikes or intrusive bodies (Fig. 2A). The serpentinite and the serpentinized harzburgite and lherzolite are surficially altered to magnesite and hydromagnesite. The less abundant gabbroic rocks largely occur in the southern part of the complex. They also occur as dikes intruding the ultramafic rocks. Two types of gabbros present in the area include pyroxene and the pegmatoid hornblende gabbros. In addition, there is a body of anorthositic gabbro in the central part of the area that contains both isotropic and layered outcrops consisting of clinopyroxene and plagioclase (An_{85-90}) (Fig. 2B). Other plutonic rocks of the complex include diorite and quartz diorite intrusions in the massive gabbros.

Felsic rocks of the Shahr-Babak ophiolite include plagiogranite and trondhjemite. They are small in volume and are largely associated with the gabbro, diorite, and colored mélange sections as diffused segregations or as discontinuous networks of veins with small dike-like bodies, several square meters in area. In general, the felsic bodies are moderately altered and have gradational contacts with each other.

Volcanic rocks of the Shahr-Babak ophiolite are more abundant in the southern parts of the complex (Fig. 2B) and include a few isolated outcrops of pillow lava and massive lava flows in blocky outcrops. These rocks have compositions ranging from basaltic andesite, andesite and dacite to rhyodacite and rhyolite as well as some trachyandesite.

TABLE 1. SUMMARY OF CHARACTERISTICS OF THE IRANIAN OPHIOLITES

Region	Ophiolite complex	Formation age (Ma)	Emplacement age (Ma)	Remarks
North-Northwest	Khoy	98-103 (Ar/Ar)[1,2]	95-96 (Ar/Ar)[1,2]	Complete sequence exposed, harzburgite E-MORB, T-MORB[3]
	Rasht	Pre-Jurassic[4]	(?)	Paleo tethys remnant[4]
Zagros	Kermanshah	(?)	Post-Campanian[5] Pre-Maastrichtian	Highly dismembered harzburgite, IAT, OIB[6]
	Sahneh	86.3[7] Ma	(?)	Highly dismembered harzburgite[7]
	Neyriz	93-95 Ma (Ar/Ar)[8]	89 Ma (Ar/Ar)[10]	Harzburgite, MORB, IAT[12]
		96-98 Ma (Ar/Ar)[9]	Post-Campanian[5] Pre-Maastrichtian	
South Central Iran	Nain	Cretaceous[5] (?)	Pre-Paleocene	Harzburgite-lherzolite, small ocean basin around Lut block[12]
	Shahr-Babak	Cretaceous[13]	Pre-Paleocene	Harzburgite-lherzolite, small ocean basin around Lut block, IAT[13,14]
	Baft Esphandagheh	Upper Cretaceous[5]	Pre-Paleocene	Harzburgite, small ocean basin around Lut block, MORB, IAT[15]
Makran	Band-e-Zeyarat	144-166 (Ar/Ar)[16]	Early-Paleocene	Harzburgite, back-arc marginal basin around Lut block, E-MORB and IAT[17,18,19,20]
	Ganj	135 (K/Ar)[17]		
	Dar-Anar (?)	170 (Ar/Ar)[9]		
	Iransharhr	Upper Cretaceous	Pre-Paleocene	Backarc marginal basin
	Fanuj-Maskutan	Upper Cretaceous	Pre-Paleocene	Backarc marginal basin
East Northeast	Tchehel-Kureh	Upper Cretaceous[5,7]	Pre-Paleocene	Mainly harzburgite, small ocean basin around Lut block MORB-IAT[18]
	Sabzevar	Upper Cretaceous	Pre-Paleocene	Harzburgite, small ocean basin between Lut and Asian blocks, IAT chemistry[21,22,23](?)
	Mashad	Pre-Jurassic[4,24]	(?)	

References: 1. Ghazi et al. (1997a); 2. Hassanipak and Ghazi (1999); 3. Ghazi et al. (1999); 4. Alavi (1996); 5. Lippard et al. (1986); 6. Ghazi and Hassanipak (1999); 7. Delaloye and Desmons (1980); 8. Ghazi et al. (1999); 9. Hayens and Reynolds (1980); 10. Lanphere and Pamic (1983); 11. Sarkarinejad (1994); 12. Davoudzadeh (1972); 13. Campbell et al. (1999); 14. Aftabi and Fathi (1992); 15. Arvin and Robinson (1994); 16. Ghazi et al. (1997b); 17. McCall (1985); 18. Desmons and Beccaluva (1983); 19. McCall (1997); 20. Hassanipak et al. (1996); 21. Ghazi et al. (1997); 22. Lensch et al. (1977); 23. Lensch (1980); 24. Alavi (1992).

The sedimentary rocks of the Shahr-Babak ophiolite consist of pelagic fossiliferous limestone and radiolarian chert. The limestone is intercalated with basalt and contains Upper Cretaceous (Campanian–Maastrichtian) fossils, suggesting a Late Cretaceous age for the youngest basalt. The radiolarite-bearing chert occurs as small bands within the basalt and basaltic andesite.

PETROGRAPHY

Plutonic rocks

Plutonic rocks of the Shahr-Babak ophiolite include both mafic and ultramafic rocks, and ultramafic rocks comprise nearly one third of the entire ophiolite complex. Among the peridotites studied in the area, harzburgite predominates. Plagioclase lherzolite and harzburgite have porphyroclastic texture defined by large grains of olivine (10%–15%), orthopyroxene (10%–15%), and clinopyroxene (5%–10%) set in a finer-grained matrix of olivine (50%–60%), orthopyroxene (10%–15%), clinopyroxene (2%–3%), plagioclase (3%–10%), and trace amphibole. Orthopyroxene porphyroclasts contain exsolution lamellae of high-Ca pyroxene in their interior, suggesting local re-equilibration. The absence of the lamellae at the porphyroclast rims suggests that the orthopyroxene is compositionally zoned. Plagioclase, amphibole, and some of the orthopyroxene and clinopyroxene occur as isolated grains or as clusters of grains that are interstitial to olivine. Lherzolite and harzburgite are massive and are mineralogically and texturally similar to the plagioclase lherzolite except that plagioclase is totally absent, amphibole is rare to absent, and the spinel is more equant and euhedral. The minerals of a typical lherzolite-harzburgite consist of porphyroclasts of olivine (5%–10%), orthopyroxene (15%–20%), and clinopyroxene (0%–5%) in a finer-grained matrix of olivine (70%–80%), clinopyroxene (1%–5%), and spinel (1%–2%). The composition of olivine in virtually all of the mantle rocks is approximately Fo_{90}. Although dunite is not very abundant in the area, it is more equigranular than the lherzolite and harzburgite and exhibits an interlocking granular texture composed mainly of olivine (95%–99%) with rare orthopyroxene as interstitial grains and traces of magnetite. Most of these rocks contain some degree of alteration in the form of serpentinization, which is locally pervasive and includes lizardite, chrysotile, and brucite with minor magnetite. Pyroxenite is associated with peridotite and occurs in the northern parts of the area and as small bodies within the colored mélange zone. Pyroxenite is generally fresh and massive; it displays a granular texture consisting of subhedral-euhedral grains of clinopyroxene and orthopyroxene with minor olivine and opaque minerals. Chrysotile has a honeycomb texture and appears to have formed entirely from Mg-rich olivine. Lizardite, on the contrary, was formed from serpentinization of orthopyroxene and as pseudomorphs after enstatite and bronzite.

The crustal sequence includes gabbro, diorite, quartz diorite, and plagiogranite (leucotonalite). Gabbro and diorite are more abundant than the plagiogranite and occur largely in the southern part of the area (Fig. 2). Isotropic gabbro is rare and consists of plagioclase (60%–70%; An_{65-75}), clinopyroxene (20%–25%), hornblende (8%–

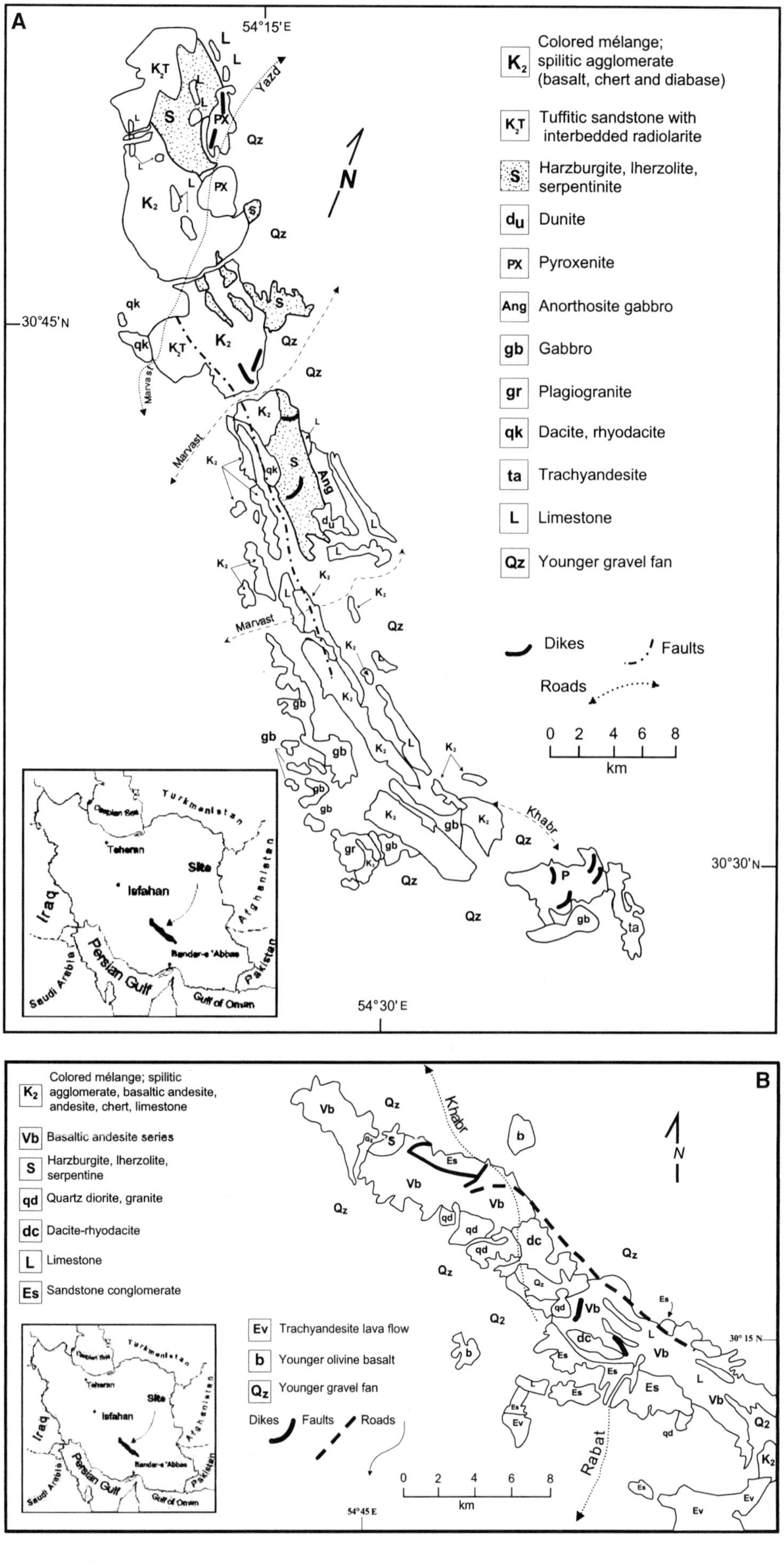

Figure 2. Geologic map of the Shahr-Babak ophiolite (modified from the geologic map of Iran, series 1:100 000). A, Kamrud Quadrangle (Antonovic, 1974). B, Robat Quadrangle (Petrovic and Isailovic, 1974).

10%), and olivine (1%–2%). Coarse-grained gabbro commonly contains large pegmatoid segregations and veins having grain sizes of up to 3–5 cm and diffuse and gradational or sharp and crosscutting contacts with the finer-grained host. Macroscopically, large pyroxene and amphibole crystals can be seen, up to 10 cm long. The pegmatoid segregations contain patches of leucogabbro and diorite, which have been mostly altered to a greenschist-facies assemblages of uralite, saussuritized plagioclase, epidote, and chlorite. The minerals in unaltered diorite consist of plagioclase (65%) with minor potassium feldspar, biotite, quartz, and hornblende. Anorthositic gabbro bodies are composed of plagioclase (85%–90%; An_{70-80}) and clinopyroxene (10%–15%).

Plagiogranite (leucotonalite) is equigranular and medium to fine grained; it is composed of 60%–90% zoned, subhedral plagioclase (An_{30-40}) and 10%–40% quartz. The texture ranges from hypidiomorphic granular, in which quartz forms a granular mosaic partly enclosing earlier-formed feldspars, to granophyric, in which a graphic intergrowth predominates. Also present in the plagiogranites are prismatic to fibrous amphiboles, which probably have replaced the original clinopyroxene. Plagiogranites are commonly altered. The secondary minerals include epidote, sphene, prehnite, and quartz.

Volcanic rocks

Volcanic rocks of the Shahr-Babak ophiolite occur mainly as massive flows and blocky outcrops and rarely as pillow lavas. Basaltic andesite pillow lavas are altered to some extent, but relict igneous texture is well preserved. In hand specimen, these rocks are largely aphyric and contain microvesicles that are filled with zeolite and carbonate. One group of basaltic andesite samples has trachytic texture with phenocrysts of plagioclase and pyroxene in a fluidal groundmass of plagioclase microlites. Porphyritic basaltic andesite contains phenocrysts of plagioclase occurring alone or as clusters of closely packed crystals (glomeroporphyritic) in a matrix of plagioclase microlites. Dacite, rhyodacite, and rhyolite are present as small lava flows consisting mainly of gray or pink felsites. The majority of these rocks are porphyritic, containing phenocrysts of plagioclase and clinopyroxene and, to a lesser extent, hornblende in a fine-grained matrix. The trachyandesite rocks are massive felsic lavas containing small (1–3 mm sized), white or pink feldspar phenocrysts, some of which have rounded shapes suggesting partial resorption during eruption. The groundmass is a fine-grained cryptocrystalline felsite composed largely of altered feldspar, quartz, carbonate, hematite, and chlorite.

ANALYTICAL METHODS

For major oxide determination, fused disks were prepared by fusing five parts of prepared flux (Spectroflux 105) and one part powdered rock sample. The mixtures were fused in crucibles of 95% Pt and 5% Au at ~950 to 1200 °C for 10 to 15 min to form a homogeneous melt. Each melt was then poured into a preheated mold to chill to a thick glass disk. A Rigaku 3070 X-ray fluorescence spectrometer (XRF) was used for the analysis. All the sample analyses were compared against standard calibration curves, which were made from a set of U.S. Geological Survey reference standards.

Samples for trace element determination, including rare earth elements, were prepared by a standard Teflon vial acid-digestion procedure using a mixture of HF-$HClO_4$-HNO_3. All samples were spiked to 50 ng/mL with indium to serve as an internal standard. REE determinations for this study were calibrated against a set of multielement working-standard solutions. The working standards were produced by mixing single-element SPEX standard solutions (SPEX Industries, USA). All solutions were introduced via a peristaltic pump to a SOLA inductively-coupled-plasma mass spectrometer (ICP-MS) for trace element determination. At the 0.5–2 ppm abundance level in the rocks, the relative precision for most elements determined by ICP-MS is 4.5% or better (range is 1.0% to 6.0%). An indication of accuracy and relative precision of determination for most elements is given in Table 2.

GEOCHEMISTRY

Major and trace elements

Twenty-seven samples of igneous rocks from the Shahr-Babak ophiolite have been analyzed for major elements (Table 2). The mantle sequence, represented by lherzolite, plagioclase lherzolite, and harzburgite, has a narrow compositional range; SiO_2 contents vary only from 40% to 42% (Table 2). Lherzolite is typically fresh and has a Mg# [Mg = Mg/(Mg + Fe^{2+})] of 90–91. The Mg# values for these lherzolites are similar to the average values of 90 for the plagioclase lherzolite from the Trinity ophiolite (Quick, 1981) and the Lanzo peridotite massif (Boudier and Nicolas, 1985). The pyroxenite (websterite) is characterized by higher CaO contents ranging between 11% and 14% and lower Mg# values (78–80). Diorite shows a wide variety of key chemical compositional values; SiO_2 ranges between 56.3% and 70.2% and Mg# values range from 0.4 to 0.6. Further, the diorite from the Shahr-Babak contains characteristically less TiO_2 (0.38%–0.87%) than diorites from the Kermanshah, Khoy, and Samail ophiolites (1.1%–1.5%) (Ghazi and Hassanipak, 1999; Hassanipak and Ghazi, 1999; Lippard et al., 1986, respectively).

The extrusive rocks of the Shahr-Babak ophiolite in general underwent extensive secondary changes due to sea-floor alteration and low-grade hydrothermal metamorphism. This alteration typically resulted in losses or gains of most of the major elements. Thus some of the discrimination diagrams, such as the alkali-silica and AFM diagrams (Middlemost, 1977; Irvine and Baragar, 1971), which use major element oxides for characterizing chemical types, are not useful. The effects of alteration may even extend to some geochemically important trace elements, especially large ion lithophile elements such as Ba, Sr, and Rb. Nevertheless, on the basis of a diagram that plots SiO_2 versus Zr/Ti (Winchester and Floyd, 1977), the extrusive rocks from the Shahr-Babak can be divided into at least three different types (Fig. 3A): (1) rhyodacite and rhyolite, characterized by highest SiO_2 contents (69%–75%); (2) basaltic andesite with SiO_2 contents ranging from 53% to 55%, and (3) trachyandesite with SiO_2 contents ranging from 50% to 56%. The latter two groups can be separated by the lower Zr/Ti ratios of the basaltic andesite.

Selected minor and trace elements (e.g., Ti, Zr, Y, and Nb) thought to be relatively immobile under conditions of metasomatism and low-grade hydrothermal metamorphism are used to characterize extrusive rocks with respect to their original composition and possible tectonic environment of formation (e.g., Pearce and Cann, 1973; Winchester and Floyd, 1977; Pearce, 1996; Jenner, 1996). The Nb/Y versus Zr/Ti discrimination diagram (Winchester and Floyd, 1977) confirms the presence of three different types of extrusive rocks (Fig. 3B). Group 1 extrusive rocks plot in the andesite and rhyodacite fields, group 2 extrusive rocks plot in the basaltic andesite field, and group 3 extrusive rocks show trachyandesite chemical affinity (Fig. 3B). In a Zr-Nb-Y discrimination diagram (Meschede, 1986), although the results from all the extrusive rocks plot in two different

TABLE 2. BULK-ROCK ELEMENTAL ANALYSES OF PLUTONIC AND VOLCANIC ROCKS FROM SHAHR-BABAK OPHIOLITE, IRAN

	Basaltic andesite						Rhyodacite-rhyolite						BHVO	
Sample	SHB12-2	SHB13-2	SHB17-1	SHB18-3	ASHB12-2	ASHB17-1	SHB14-2	SHB15-1	SHB15-1b	SHB16-2	SHB16-3	SHB7-2	Mea.	Rec.
SiO_2	54.5	53.5	53.1	55.1	54.5	53.1	71.9	75.1	75.6	69.7	71.1	69.3	49.9	49.9
TiO_2	0.33	0.59	0.43	0.63	0.33	0.43	0.48	0.25	0.25	0.37	0.38	0.57	2.69	2.71
Al_2O_3	15.9	16.1	14.9	14.9	15.9	14.9	12.7	11.9	12	13.5	13.3	13.6	13.8	13.8
Fe_2O_3	8.71	7.39	9.2	7.9	8.71	8.2	4.49	3.14	2.59	4.54	3.17	5.14	12.1	13.2
MnO	0.14	0.15	0.12	0.14	0.14	0.12	0.05	0.04	0.03	0.09	0.04	0.09	0.17	0.17
MgO	5.13	3.52	6.1	3.93	5.13	6.1	1.55	0.51	0.5	1.08	1.1	1.27	7.31	7.23
CaO	7.5	6.9	8.2	7.4	7.5	8.2	0.66	0.48	0.73	1.36	1.2	0.94	11.3	11.4
Na_2O	4.83	5.53	4.51	4.9	4.83	4.51	4.98	5.26	5.07	5.61	5.47	6.12	2.29	2.26
K2O	0.65	0.39	0.55	0.5	0.65	0.55	1.21	1.21	1.15	0.58	1.79	0.49	0.54	0.52
P_2O_5	0.04	0.19	0.08	0.12	0.04	0.08	0.09	0.04	0.05	0.08	0.08	0.13	0.28	0.27
LOI	2.15	2.77	1.87	1.98	2.15	1.87	1.48	1.12	1.49	2.55	1.93	1.56		
Total	97.77	95.23	99	97.5	97.7	99	98.09	97.88	97.52	96.95	96.6	97.71		
MG#	54	48	55	50	65	55	40	24	28	32	41	33		
Rb	14.72	6.29	6.3	7.3	10.5	6.9	13	10.3	11.31	8.16	28.45	5.72	9.5	9.7
Sr	603	490	559	498	546	526	60	104	85	127	116	58	411	420
Y	9.6	15.3	8.5	19.3	12.4	13.1	34.2	43.5	38.9	18.3	26.3	36.9	26.5	25.8
Zr	28.5	48.1	25.3	45.2	38.2	30.1	123	103	97.5	68	91.9	97.6	167	181
Nb	1.45	2.93	1.8	3.9	2.09	3.25	4.2	4.09	4.17	3.18	3.68	3.83	18.5	19
Ba	206	112	191	131	159	161	184	184	193	137	260	141	135	133
Hf	0.57	1.28	0.77	2.1	0.92	1.48	5.1	4.42	3.6	2.92	3.09	3.26	4.51	4.38
Ta	1.1	1.04	1	1.14	1.01	1	2.4	3.82	2.21	1.91	0.7	2.08	1.22	1.16
Pb	0.58	3.01	0.67	3.52	1.79	2.09	1.42	4.13	2.52	2	2.65	1.63	2.21	2.05
Th	0.18	0.63	0.15	0.53	0.18	0.34	2.78	3.11	2.93	4.05	3.01	1.95	1.05	1.23
U	0.17	0.3	0.13	0.4	0.15	0.18	0.5	0.83	1.21	1.2	1.43	0.46	0.49	0.41
Sc	5.27	1.15	6.3	2.18	3.21	4.24	2.86	0.38	0.26	4.75	13.7	5.8	32.7	31.8
V	217	33.9	222	40.8	125	131	12.4			19.9	19.8	31.6	319	317
Cr	27.4	19.6	32.6	15.2	25.4	27.4							289	291
Co	60	24.3	70.4	35.4	42.1	69.2	67.3	87.1	85.1	60.3	64.7	138	46	45
Ni	30.4	14.1	35.4	18.9	22.2	27.1	1.44	0.2	1.63	2.24	2.17	4	118	121
Cu	85.1	40.4	92.1	45.8	20.7	68.9	226	7.92	8.12	13.6	10.9	27.5	144	136
Zn	65.3	71.4	67.8	79.5	68.3	73.6	48.8	66.3	70.3	66.4	63	90.5	101	105
La	2.4	7.4	2.1	7.9	5.4	4.2	10.2	12.5	11.5	10	7.7	10.7	15.5	15.8
Ce	5.55	13.5	5.21	14.6	11.5	10.1	23.1	26.7	23.1	23.3	20.4	22.8	38	37.8
Nd	5.06	8.59	4.72	9.8	6.89	5.89	16	17.9	15.8	13.8	12.6	15.5	24.5	24.8
Sm	1.46	2.27	1.21	2.6	1.96	1.81	4.86	5.17	4.6	4.02	3.35	4.11	6.26	6.1
Eu	0.63	0.96	0.51	1	0.59	0.65	1.25	1.6	1.1	1.06	1.07	1.09	2.18	1.98
Gd	1.96	2.48	1.72	2.8	2.22	2.26	5.1	6.8	5.9	4.8	4.21	5.21	6.78	6.4
Dy	1.93	2.59	1.82	2.79	2.22	2.3	5.19	9.91	5.86	4.88	4.53	6.15	4.78	5.2
Er	1.07	1.67	0.82	1.91	1.67	1.36	2.77	6.93	4.17	2.7	2.85	3.52	2.43	2.4
Yb	1.25	1.98	1	2.1	1.61	1.51	3.48	6.28	3.74	2.51	3.18	2.78	2.27	2.02
Lu	0.09	0.16	0.09	0.23	0.12	0.2	0.74	1.19	0.53	0.44	0.57	0.53	0.29	0.29
Eu/Eu*	1.15	1.24	1.09	1.14	0.87	0.99	0.77	0.83	0.65	0.74	0.88	0.73		
$(La/Sm)_N$	0.99	1.97	1.05	1.85	1.66	1.41	1.28	1.47	1.52	1.51	1.4	1.59		
$(La/Yb)_N$	1.26	2.45	1.37	2.48	2.19	1.83	1.94	1.31	2.03	2.63	1.59	2.54		

fields, the basaltic andesite and rhyodacite-rhyolite groups show volcanic arc chemical affinity, whereas the trachyandesite shows within-plate–type chemical affinity (Fig. 3C). In a set of bivariant trace element diagrams using Zr, Nb, and Y, the three types of extrusive rocks are clearly distinguishable as three different clusters (Fig. 4). Both diagrams reveal a positive trend between the basaltic andesite and rhyodacite and rhyolite, whereas the trachyandesite samples plot as an isolated cluster.

Primitive-mantle–normalized incompatible element patterns for all the extrusive rocks from the Shahr-Babak ophiolite are illustrated in Figure 5. In general, the patterns for these extrusive rocks form three recognizably different semiparallel envelopes. All patterns are

TABLE 2. BULK-ROCK ELEMENTAL ANALYSES OF PLUTONIC AND VOLCANIC ROCKS FROM SHAHR-BABAK OPHIOLITE, IRAN (continued)

	Websterite-pyroxenite				Diorite				Trachyandesite					Lherzolites	
Sample	SHB10-3a	SHB10-3b	SHB11-1	SHB11-2	SHB10-1	SHB9-1	SHB9-2	SHB9-3	SHB15-4	SHB16-1	SHB16-4	SHB2-3	SHB5-4	SHB-L1	SHB-L2
SiO_2	48.7	50.3	45.3	47.5	58.5	60.0	56.3	70.2	54.0	56.3	55.3	52.1	49.8	43.5	42.7
TiO_2	0.25	0.25	0.14	0.13	0.38	0.75	0.87	0.47	0.79	0.68	0.71	1.09	0.79	0.08	0.05
Al_2O_3	3.9	4.2	9.6	16.4	15.3	15.4	16.6	14.2	18.9	19.1	18.6	16.6	16.3	2.57	1.74
Fe_2O_3	10.8	11.1	7.46	3.49	5.41	8.1	9.17	4.08	5.65	6.16	6.55	10.8	11.33	8.81	8.81
MnO	0.21	0.25	0.13	0.08	0.12	0.17	0.18	0.06	0.15	0.13	0.18	0.15	0.21	0.12	0.12
MgO	21.8	22.3	19.3	11.1	5.33	3.5	3.75	1.41	1.68	1.48	2.32	3.45	4.95	40.3	41.3
CaO	11.5	11.8	14.3	17.2	10.7	6.89	7.47	4.73	4.51	5.1	4.64	6.66	4.91	1.95	2.12
Na_2O	0.04	0.05	0.15	0.7	2.12	2.92	3.05	2.98	3.23	4.46	4.54	3.28	5.99	0.07	0.1
K_2O				0.37	0.63	0.51	0.95	0.31	7.17	4.4	4.98	2.28	0.27		
P_2O_5	0.01	0.01	0.02	0.02	0.07	0.14	0.28	0.11	0.5	0.36	0.37	0.33	0.41		
LOI	0.51	0.55	3.49	3.5	1.28	1.4	2.78	1.17	3.46	1.61	2.49	3.25	5.94	1.08	1.98
Total	97.70	100.80	96.43	96.95	98.66	98.36	98.70	98.51	96.61	98.19	98.11	96.76	94.91	98.5	98.9
MG#	80.0	80.0	84.0	86.0	66.0	46.0	45.0	40.0	37.0	32.0	41.0	39.0	46.0	90.0	90.0
Rb	1.5	1.4	1.71	5.59	6.75	6.52	21.65	4.57	158.4	92.27	98.7	36.058	5.42		
Sr	61	64	89	303	254	142	220	208	992	747	801	319	381		
Y	5.14	5.4	4.51	4.58	10.9	21.8	41.4	15.2	25.2	25.3	25.9	24.3	19.3		
Zr	6.95	6.78	5.93	6.15	16.1	93.8	16.1	8.69	193	178	162	148	156		
Nb	1.94	1.89	2.03	2.01	2.75	3.76	4.03	3.43	14.2	8.77	8.7	6.63	5.6		
Ba	18.6	18.9	20.5	83.1	172	298	249	177	581	520	554	598	343		
Hf						0.43			6.86	5.7	4.91	3.58	2.28		
Ta	0.89	1.25	0.97	1.06	1.35	1.99	1.56	1.71	1.4	1.02	0.93	1.13	1.02		
Pb					0.64		0.36		39.2	15.4	13.7	4.56	3.89		
Th					0.72	0.98		0.75	29.3	8.85	9.73	4.99	7.15		
U					0.21				6.9	3.06	2.34	1.74	1.83		
Sc	52.3	51.8	14.19	13.5	24.1	7.97	5.15	17.9	17.1	11.8	3.05				
V	416	411	108	107	154	290	243	85.8	134	149	102				
Cr	2589	2476	998	389	74.83	17.8	6.07								
Co	199	196	72	52	102	105	132	93.5	36.6	48.7	43.2				
Ni	385	393	242	133	43.9	15.1	25.1	9.12	7.27	10.8	7.56				
Cu			78.3	52.9	4.9	10.7	17.2	16.5	38.7	47.7	93.5				
Zn			51.1	40.5	53.2	78.3	131	39.8	150	98.7	92.4				
La	0.7	0.7	0.7	0.7	4.3	7.5	11.4	6.0	60.4	28.9	29.3	16.9	21.6		
Ce	2.98	3.06	2.57	2.9	8.97	18.7	27.9	12.2	117	56.6	56.3	38.0	43.7		
Nd	3.06	3.91	2.79	2.51	6.99	11.5	16.4	7.82	50.9	27.3	23.9	18.4	20.6		
Sm	0.94	1.1	0.88	0.61	2.26	3.71	5.41	2.72	11.2	5.93	5.96	5.55	5.5		
Eu					0.87	1.2	1.37	1.05	4.35	1.82	1.54	1.66	1.65		
Gd	1.29	1.4	1.18	1.15	2.98	5.16	6.36	3.65	9.77	4.06	4.86	4.21	4.49		
Dy	1.25	0.82	1.12	0.97	3.41	6.05	7.54	3.71	9.4	4.61	4.07	3.58	4.02		
Er	0.43	0.33	0.32	0.41	1.54	3.07	4.55	1.99	3.66	2.24	2.39	2.62	1.86		
Yb	0.37	0.36	0.34	0.3	1.76	3.15	3.97	2.24	3.2	2.46	2.65	2.15	1.92		
Lu	0.06	0.04	0.05	0.03	0.15	0.69	0.48	0.22	0.8	0.58	0.57	0.41	0.29		
Eu/Eu*					1.04	0.85	0.72	1.03	1.26	1.08	0.86	1.02	0.99		
$(La/Sm)_N$	0.43	0.37	0.46	0.74	1.15	1.23	1.28	1.34	3.3	2.97	3	1.85	2.39		
$(La/Yb)_N$	1.2	1.21	1.3	1.63	1.6	1.56	1.89	1.76	12.46	7.77	7.3	5.18	7.43		

characterized by a sizable trough at Nb. Patterns for the trachyandesite group show the highest relative abundances and most scatter for the more mobile elements (e.g., Rb, Sr). These patterns also have a steeper slope to the right and thus overlap the rhyodacite and rhyolite patterns in the middle to heavy REE (i.e., MREE to HREE) region. This crosscutting relationship between the patterns for trachyandesite and the rhyodacite and rhyolite suggests that they are petrogenetically unrelated and were most likely derived from melts of different initial composition or perhaps formed in different tectonic settings.

For comparison, the pattern for a typical island-arc tholeiite (IAT; Sun, 1980) is plotted along with the data for each of these sets. Although the patterns for all the Shahr-Babak samples are more enriched in all of the elements, there are some similarities with a typical

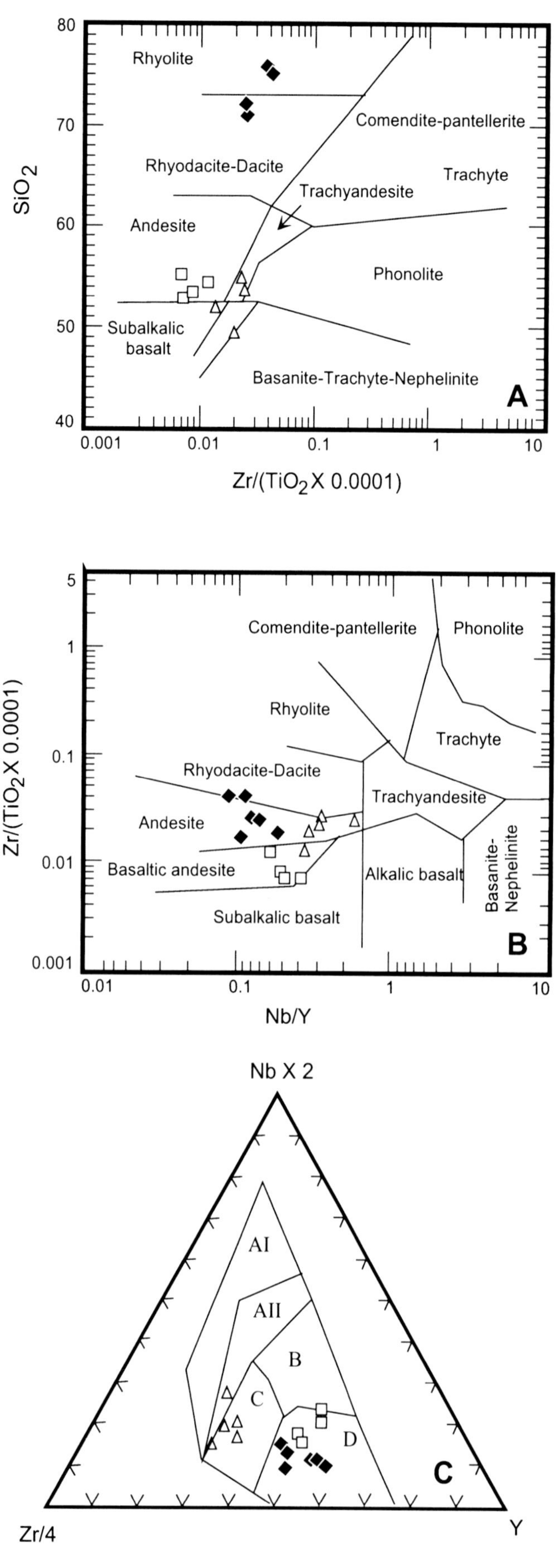

Figure 3. A, SiO_2 vs. Zr/Ti geochemical discrimination diagram (after Winchester and Floyd, 1977) showing two types of basaltic rocks. Solid diamonds—rhyodacite and rhyolite, open squares—basaltic andesite, and open triangles—trachyandesite. B, Zr/Ti vs. Nb/Y geochemical discrimination diagram (after Winchester and Floyd, 1977) showing three types of extrusive rocks. Symbols as in A. C, Geochemical discrimination diagram (Meschede, 1986) showing three types of extrusive rocks. Fields: AI—WPA (within-plate alkalic basalt), AII—WPT (within-plate tholeiite), B—P-MORB (plume-type mid-oceanic-ridge basalt), D—N-MORB (mid-oceanic-ridge basalt), C–D—VAB (volcanic-arc basalt).

IAT pattern particularly in the case of the basaltic andesite samples. The resemblance of the basaltic andesite patterns from the Shahr-Babak ophiolite to the IAT pattern (Fig. 5) suggests that they were derived from magma sources with IAT characteristics. The large trough at Nb is a characteristic feature of extrusive rocks influenced by continental lithosphere or related to volcanic arc environments, which has been explained in terms of fractionation and retention of this element in the source during partial melting of the subducting plate (Pearce, 1996). Thus, from the evidence of the primitive-mantle–normalized incompatible trace element diagram (Fig. 5) and the discrimination plots (Fig. 4), it appears that these extrusive rocks were derived from melts generated in an island-arc environment, similar to some of the transitional IAT basalt reported for the Neyriz ophiolite (Sarkarinejad, 1994) and subalkalic basalt reported for the Kermanshah ophiolite (Ghazi and Hassanipak, 1999).

Rare earth element patterns

Figure 6 shows the chondrite-normalized REE patterns for whole-rock samples of the entire sequence of igneous rocks of the Shahr-Babak ophiolite. The patterns for the mantle rocks are convex upward and are characterized by a strong depletion in light REEs [LREEs; $(La_N/Sm_N)_{avg}$ = 0.33] (Fig. 6A). These REE patterns are significantly different from the patterns obtained for the mantle rocks from other ophiolites in the region. For example, Ghazi and Hassanipak (1999) and Hassanipak and Ghazi (1999) reported LREE-enriched, convex-downward (v-shaped) patterns for the harzburgite of Kermanshah and Khoy ophiolites, respectively. Pallister and Knight (1981) observed similar LREE-enriched, convex-downward patterns for the mantle rocks from the Samail ophiolite and concluded that the Samail's peridotite and its crustal-suite rocks are not related as parent and daughter in a simple partial-melting relationship. The REE concentrations for the diorite-series rocks from the Shahr-Babak ophiolite are 20–30 times chondritic values, and their REE patterns are virtually parallel. The patterns for some of the diorites have distinct negative Eu anomalies $(Eu/Eu^*)_{avg}$ = 0.91.

Chondrite-normalized REE patterns for the extrusive rocks are shown in Figure 6B. In general, all types of the extrusive rocks are characterized by an enrichment in LREEs. However, they have different REE patterns, i.e., the trachyandesite samples are more LREE enriched [$(La_N/Yb_N)_{avg}$ = 8.16] than the basaltic andesite and the rhyodacite-rhyolite [$(La_N/Yb_N)_{avg}$ = 1.72 and $(La_N/Yb_N)_{avg}$ = 1.46], respectively. Furthermore, the fact that the patterns for samples within each suite are largely parallel to one another and form narrow envelopes suggests a potential cogenetic relationship among the samples. The patterns for the trachyandesite cut across those of the rhyodacite and rhyolite, especially the MREEs and HREEs. This relationship further supports the earlier interpretation that there are two compositionally different suites of extrusive rocks in the Shahr-Babak

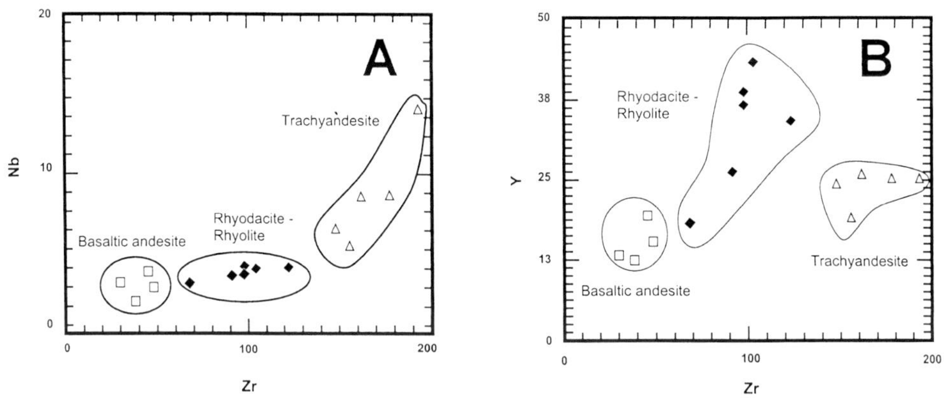

Figure 4. Bivariant trace element diagram showing three chemically distinct groups of extrusive rocks from the Shahr-Babak ophiolite. A, Zr vs. Nb. B, Y vs. Zr. Note the possible positive linear correlation between the basaltic andesite and the rhyodacite and rhyolite.

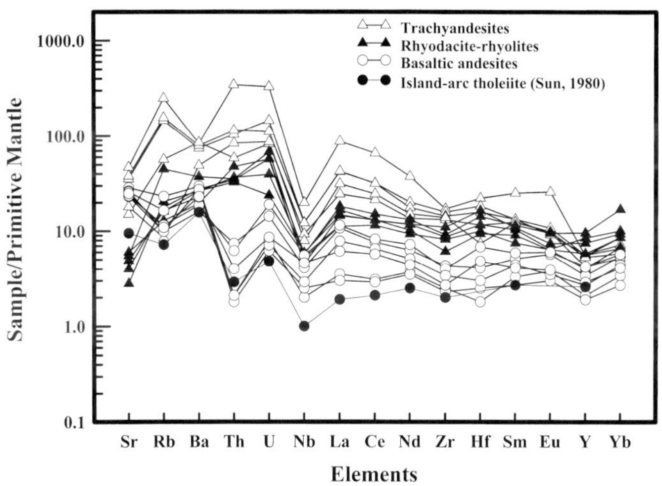

Figure 5. Incompatible element patterns of rhyodacite and rhyolite, basaltic andesite, and trachyandesite normalized to primitive mantle. Shown for comparison is a pattern for a typical island-arc tholeiite (IAT; Sun, 1980). Normalizing values for primitive mantle are from Sun and McDonough (1989).

ophiolite and that they were derived from melts with different geochemical signatures that most likely were generated in different tectonic settings. The patterns for the rhyodacite-rhyolite and the basaltic andesite do not have a crosscutting relationship, and the rhyodacite-rhyolite patterns show overall enrichments in all of the REEs. The only major difference between the REE patterns for the latter suite of extrusive rocks is that the rhyodacite-rhyolite samples have a pronounced negative Eu anomaly (Eu/Eu* = 0.77).

PETROGENESIS

The chondrite-normalized REE patterns for all the volcanic rocks, which include the basaltic andesite, rhyodacite-rhyolite, and trachyandesite (late extrusive rocks) are LREE enriched. These patterns clearly distinguish three different groups of extrusive rocks within the Shahr-Babak ophiolite. For example, the patterns for both basaltic andesite and rhyodacite-rhyolite are only slightly LREE enriched—i.e., $(La_N/Sm_N)_{avg} = 1.49$ and $(La_N/Yb_N)_{avg} = 1.72$ for the former and $(La_N/Sm_N)_{avg} = 1.46$ and $(La_N/Yb_N)_{avg} = 2.01$ for the latter. Furthermore, the patterns for the rhyodacite-rhyolite have slight negative Eu anomalies $[(Eu/Eu^*)_{avg} = 0.77]$, whereas most of the basaltic andesite rocks show slight positive Eu anomalies $[(Eu/Eu^*)_{avg} = 1.09]$.

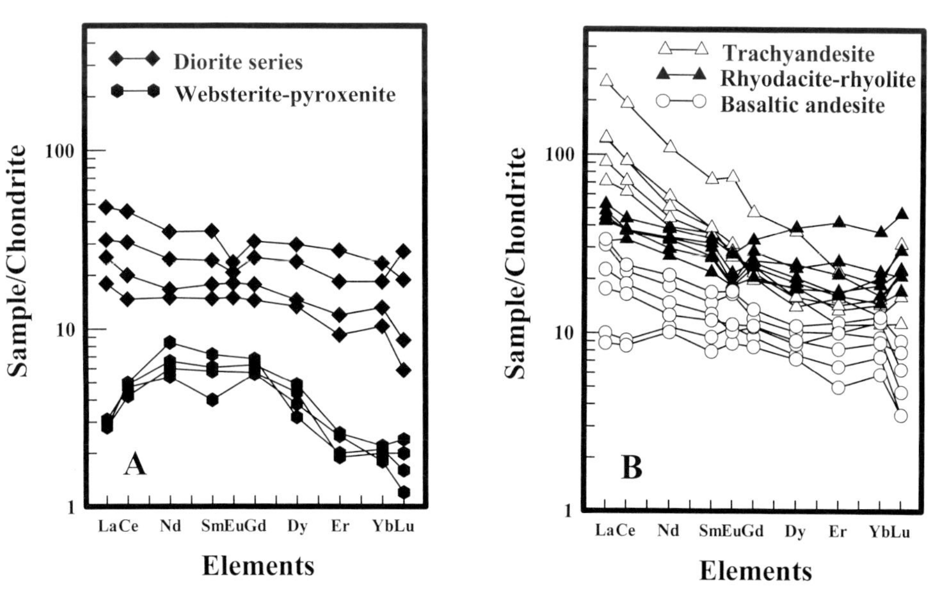

Figure 6. A, Chondrite-normalized REE patterns for the diorite-series rocks and the ultramafic rocks from the Shahr-Babak ophiolite. B, Chondrite-normalized REE patterns for the extrusive rocks from the Shahr-Babak ophiolite.

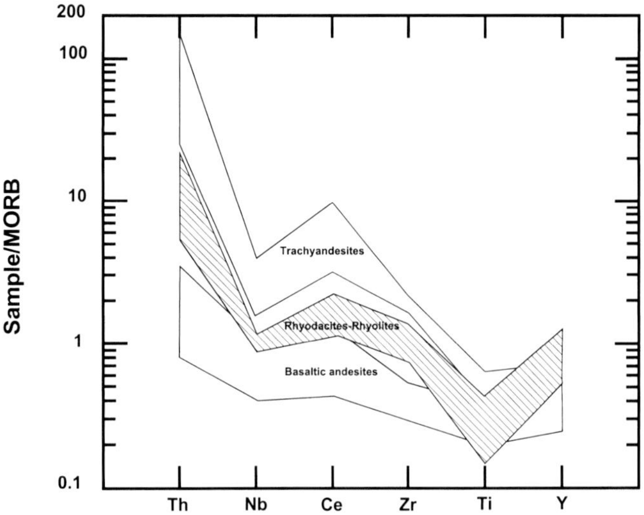

Figure 7. MORB-normalized incompatible trace element diagram for the extrusive rocks from the Shahr-Babak ophiolite (Pearce, 1996). Normalizing values for MORB are from Sun and McDonough (1989).

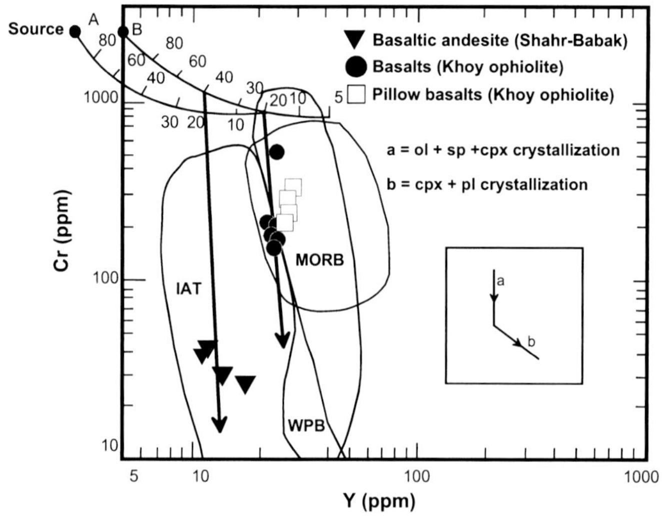

Figure 8. Cr (ppm) vs. Y semilog plot for the basaltic andesite rocks from the Shahr-Babak ophiolite, showing higher-degree partial melting of a lherzolitic mantle source or lower degree of partial melting of more depleted mantle source. For comparison, the data from pillow and massive basalt from the Khoy ophiolite are included, showing partial-melting trends of a lherzolitic mantle source and fractional-crystallization paths for all the basalt. MORB—mid-oceanic ridge basalt, WPB—within-plate basalt, IAT—island-arc tholeiite. (Figure modified after Pearce, 1982, and Malpas et al., 1994.)

The volume of rhyodacite-rhyolite is small relative to the basaltic andesite so that the negative Eu anomaly for the rhyolite can be attributed to the cumulative effects of plagioclase crystallization and the removal of Eu from the melt under low oxygen fugacity (Drake, 1975) and the relatively high abundance of alkali feldspar. Plagioclase crystallization has typically only minor effects on the REE abundances of the melt, except that the large distribution coefficient for Eu will contribute to a negative Eu anomaly in the melt (Hanson, 1989). Thus, the rhyodacite-rhyolite may be the end product of a fractional-crystallization process from a parent magma with a basaltic to andesitic composition where plagioclase was removed in the early-formed rocks. The exceptional parallelism between the REE patterns for these two groups of extrusive rocks also strongly suggests a cogenetic relationship (Fig. 6A). The higher concentrations of REEs in rhyodacite-rhyodacite are understandable if these rocks were the end products of extensive fractionation of a magma with composition similar to an average basaltic andesite (e.g., similar to SHB12-2; see Table 2).

Figure 7 shows a plot of selected, highly incompatible trace elements normalized to MORB (mid-oceanic-ridge basalt) for all the extrusive rocks (Pearce, 1996). It is apparent that parallelism among the patterns exists and that the patterns' shape is retained throughout the entire sequence. However, there are some exceptions that should be considered. One of the key features among the patterns for these extrusive rocks is the significant negative Nb anomaly with respect to Th and Ce. Petrogenetically, this negative Nb anomaly is characteristic of volcanic arc magmas and can be attributed to fractionation of Nb relative to Ce and Th caused by preferential retention of Nb within the amphibole and minor phases such as titanite and rutile during dehydration and/or partial melting of the subducting plate (Pearce and Peate, 1995; Pearce, 1996). Another feature supporting the volcanic arc petrogenetic origin for the rocks is that virtually all of the patterns for basaltic andesite and rhyodacite show absolute depletion in Nb, Zr, Ti, and Y relative to MORB. However, the patterns for the trachyandesite do not show the latter feature; they show depletion for Ti + Y, but not the other two. The petrogenetic relationship between the basaltic andesite and the rhyolite is further demonstrated by the increased size of the negative Ti anomaly from basaltic andesite through rhyodacite and rhyolite. This distinct behavior has been attributed to continuous removal of Ti (as Ti-bearing oxides) causing magma to become depleted in Ti while Zr continues to be enriched (Pearce, 1996). Again, the lack of such a relationship between the trachyandesite (and rhyodacite) samples and the basaltic andesite samples suggests that either no petrogenetic relationship exists or that some secondary magmatic processes have affected the probable genetic relationships. The latter argument is further supported by the crosscutting behavior of the chondrite-normalized REE patterns for the trachyandesite. These patterns show differences in slope from the other two group of extrusive rocks [i.e., $(La_N/Sm_N)_{avg} = 2.78$ and $(La_N/Yb_N)_{avg} = 8.16$], resulting in crossing REE patterns.

Figure 8, from Pearce (1982), shows a simple equilibrium partial-melting diagram adapted to model the primary-melt composition from a mantle lherzolite having about 2000 ppm Cr (Pearce, 1982; Malpas et al., 1994; Lippard et al., 1986, p. 122). The two crystal fractionation trends that can be modeled here include (1) fractionation of olivine, spinel, and clinopyroxene and (2) fractionation of clinopyroxene and plagioclase. When applied to the data from the Shahr-Babak ophiolite, this model clearly shows that the basaltic andesite rocks are extremely depleted in Cr and Y and plot in the IAT field. According to this model, the most compositionally primitive volcanic rocks (basaltic andesite) from the Shahr-Babak ophiolite were derived either from a large degree of melting (~40%) of a mantle lherzolite source or a lesser degree of melting (15%–20%) of a previously depleted mantle source. The model also suggests that the basaltic andesite from the Shahr-Babak ophiolite lies along the fractional-crystallization trend that is controlled by removal of olivine, spinel, clinopyroxene, and plagioclase. In comparison, both types of basalt from the Khoy ophiolite lie along the fractional-crystallization trend from a melt that was produced from 20% to 25% partial melting of a primary mantle lherzolite (Hassanipak and Ghazi, 1999). Also, Ar-

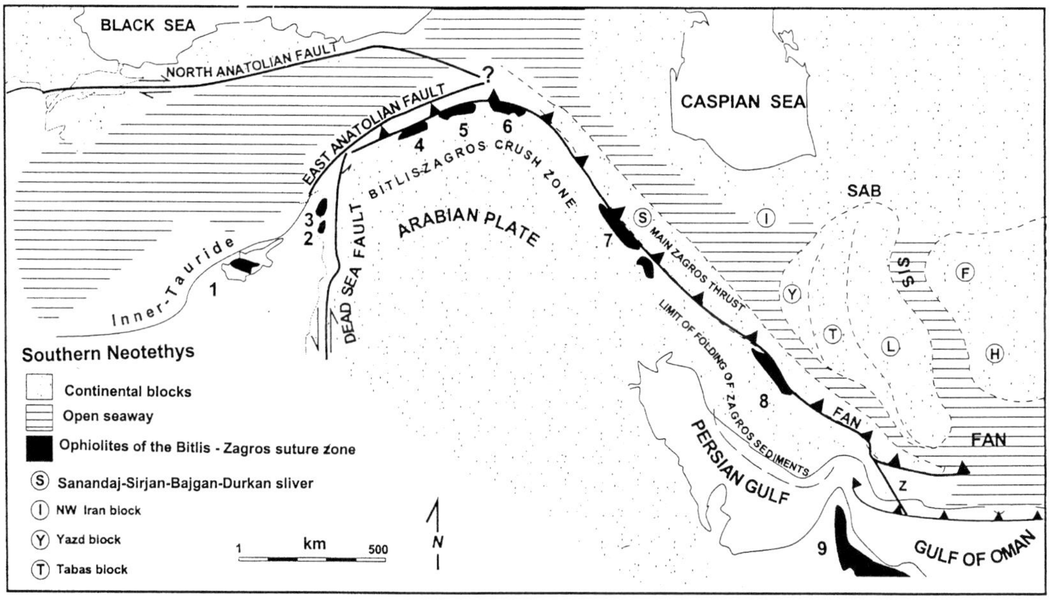

Figure 9. Simplified paleogeographic reconstruction of the Late Cretaceous Neotethys in Iran and the adjacent parts of Turkey, Arabia, and Pakistan, showing the various sectors of the postulated Inner Mesozoic Ocean, which includes the Sabsevar (SAB) and Sistan (SIS) Oceans of Şengör (1990), as well as the Fanuj Ocean (FAN) of McCall (1997). These postulated seaways are now represented by ophiolites around the Central Iranian microcontinent, including the Shahr-Babak ophiolite. Also shown is the distribution of Cretaceous Neotethyan ophiolites around the Arabian promontory and along the Bitlis-Zagros suture zone (modified after Dilek and Delaloye, 1992, and Dilek et al., 1999). Key to the numbers: 1—Troodos, 2—Baer-Bassit, 3—Kizildag, 4—Ispendere-Komurhan, 5—Guleman, 6—Cilo, 7—Kermanshah, 8—Neyriz, 9—Samail. Other elements in this diagram include S—Sanandaj-Sirjan–Bajgan-Durkan microcontinental block, Z—Zendan fault, part of the Oman Line fault system (Coleman, 1981), L—Lut block, F—Farah block, and H—Helmand block.

vin and Robinson (1994) determined 20%–25% partial melting of a source with similar chemical composition to produce the basaltic lavas in the Baft ophiolite, which is located to the southeast of the Shahr-Babak ophiolite. Similarly, Lippard et al. (1986) suggested 20%–30% partial melting to produce melt for the clinopyroxene-phyric basalt unit of the Samail ophiolite.

SUMMARY AND CONCLUSION

The main mass of the Shahr-Babak peridotite is composed of plagioclase lherzolite, lherzolite, harzburgite, and dunite. Pyroxenite and isolated zones of two-phase olivine + chrome spinel cumulates are less abundant and form dikes or cumulate intrusive bodies. The gabbroic rocks are not very abundant and largely occur as normal and pegmatoid hornblende gabbros. Felsic rocks (plagiogranites) of the Shahr-Babak ophiolite are associated with gabbro and diorite and occur as low- and high-K_2O types.

There is no sheeted-dike complex, and there are only a few isolated outcrops of pillow and massive lava flows. Petrologically, the volcanic rocks of the Shahr-Babak exhibit a wide compositional range including basaltic andesite, rhyodacite to rhyolite, and trachyandesite. Sedimentary rocks of the Shahr-Babak ophiolite consist of pelagic fossiliferous limestone and radiolarian chert. The limestones are intercalated with basalt and contain Upper Cretaceous fossils (Campanian–Maastrichtian), suggesting a Late Cretaceous age for the formation of the youngest basalt during the final phase of the formation of oceanic crust.

The chemical data on extrusive rocks of the Shahr-Babak ophiolite indicate their generation from two different magmatic sources: (1) a source that produced the basaltic andesite and rhyodacite from a magma that was generated in an island-arc environment and (2) a source that produced the trachyandesite from a melt generated in a within-plate environment. The basaltic andesite rocks have incompatible trace element patterns virtually identical to typical IAT (Sun, 1980) and are chemically similar to the subalkalic basalt from the Kermanshah ophiolite (Ghazi and Hassanipak, 1999) and transitional IAT basalt reported for the Neyriz ophiolite (Sarkarinejad, 1994). Compositionally, the initial magma for the Shahr-Babak ophiolites was generated in an island arc and was derived either from a large degree of melting (~40%) of a mantle lherzolite source or a lesser degree of melting (15–20%) from a previously depleted mantle source. The gabbro, diorite and quartz diorite, basaltic andesite, rhyodacite, and rhyolite are petrogenetically related and were most likely derived from a basaltic tholeiite magma whose original REE concentrations were ~10 times chondritic values. Furthermore, rhyodacite and rhyolite are the end products of fractional crystallization of the same magma. The trachyandesite rocks are not petrogenetically related to the latter groups, or if so, the composition of the initial magma was affected by secondary processes. Geochemically, the trachyandesite samples from the Shahr-Babak ophiolite are similar to the trachytes from the Haybi volcanic rocks of the Oman ophiolite (Lippard et al., 1986) and the trachytes from the Kermanshah ophiolite, Iran (Ghazi and Hassanipak, 1999). On the basis of the abundances and ratios of immobile elements Ti, Zr, and Y, the trachyandesites are typical of within-plate magmas that are found in several tectonic environments, including intraplate oceanic islands (Pearce and Cann, 1973). Oceanic-island basalts often show a considerable range in total alkali contents from mildly alkalic to strongly alkalic or potassic type, which is the case with the trachyandesite samples from the Shahr-Babak ophiolite. Rocks from the potassic end member include basalt, trachybasalt, and trachyandesite, which are often typified by a more evolved basaltic composition (trachybasalt, hawaiite) than true basalt (Floyd, 1991).

In terms of the tectonics of the Neotethyan oceanic lithosphere and the Mesozoic–Cenozoic geologic evolution of the eastern Mediterranean region, the data from this work suggest that the ultramafic-mafic complex at the Shahr-Babak is a dismembered ophiolite in which the sheeted-dike complex is absent. The ophiolite is located along a northwest-trending fault zone (Nain-Baft) that contains several relatively intact fragments of Neotethyan ophiolite massifs. Other major complexes along this fault zone include the Nain ophiolite (Davoudzadeh, 1972), the Khoy ophiolite to the northwest (Hassanipak and Ghazi, 1999), and the Baft complex to the southeast (Arvin and Robinson, 1994). Glennie et al. (1990) suggested that the Nain-Baft fault zone was probably the root zone from which the scattered ophiolites and mélange complexes were emplaced along the east flank of the Sanandaj-Sirjan microcontinental block. Thus, the presence of the ophiolites and mélange complexes along this fault zone are strong indications of the previously existing seaway that separated the Central Iranian and the Sanandaj-Sirjan microcontinents. Şengör (1990) suggested that such a seaway surrounded the entire Central Iranian microcontinent. McCall (1997) proposed a similar scenario for the geotectonic evolution of the Makran region and suggested that the emplacement of some of the southern Iranian ophiolites in the Makran region (e.g., Band-e-Zeyarat, Ganj, Remshk, and Talkhab) was the result of closure of the southern branch of this narrow Mesozoic seaway (Fig. 9).

Geochemically, unlike other studied Iranian ophiolites (e.g., Khoy, Band-e-Zeyarat, Neyriz, Kermanshah, and Baft) that contain the relicts of MORB-like signatures, the results from the Shahr-Babak ophiolite show only IAT and possibly intraoceanic signatures. The presence of MORB-like signatures in the extrusive units of Neotethyan ophiolites suggests that the fossil oceanic lithosphere preserved in these ophiolites developed at an oceanic spreading center (Hassanipak and Ghazi, 1999; Tankut et al., 1998). The occurrence of the MORB-like volcanic rocks in the Baft area to the southeast of the Shahr-Babak (Arvin and Robinson, 1994) can be best explained in the context of tectonic setting of the area. The Baft ophiolite is located at a relatively complex junction where three major tectonic elements meet (i.e., the Lut block, the Makran accretionary prism, and the Zagros suture zone). These tectonic elements were separated at one time by southern Tethyan oceanic crust, and it is likely that this oceanic crust may have included a MORB-like component. On the contrary, the Shahr-Babak ophiolite has a simpler tectonic setting (located along the Nain-Baft fault zone), and it is basically a segment of a narrow seaway that did not contain a MORB-like volcanic component. Therefore, considering the previous interpretation for tectonic reconstruction of this region together with geochemical results from this study, we propose that volcanic rocks from the Shahr-Babak ophiolite represent the island-arc and intraoceanic components that were emplaced during northeast-dipping subduction of a narrow oceanic lithosphere that was part of the southern Tethyan Ocean. This inferred ocean separated the Afro-Arabian plate from the Eurasian plate, surrounded the Central Iranian microcontinent, and extended in the northwest direction into Turkey.

ACKNOWLEDGMENTS

We thank Marion Wampler for his review of the original manuscript and R.T. Gregory, R.V. Metcalf, and an anonymous reviewer for constructive comments. We also thank Michael Latta for his assistant in drafting and cartography. The XRF and the laser-ablation ICP-MS were purchased with funds from National Science Foundation grants (EAR-86-17682) to T.E. LaTour and (EAR-94-05716) to A.M. Ghazi, respectively.

REFERENCES CITED

Aftabi, A., and Fathi, G., 1992, Geology and mining potential of the ophiolitic belt of Kamroud and Shahr-Babak (in Farsi): Mining and Metals, v. 50, p. 60–74.

Alavi, M., 1991, Tectonic map of the Middle East: Geological Survey of Iran, 1 sheet, scale 1:5,000,000.

Alavi, M., 1992, Thrust tectonics of the Binaload region, NE Iran: Tectonics, v. 11, p. 360–370.

Alavi, M., 1996, Tectonostratigraphic synthesis and structural style of the Alborz Mountain system in northern Iran: Journal of Geodynamics, v. 21, p. 1–33.

Alavi, M., Vaziri, H., Seyed-Emami, K., and Lasemi, Y., 1997, Triassic and associated rocks of the Nakhlak and Aghdarband areas in central and northeastern Iran as remnants of the southern Turanian active continental margin: Geological Society of America Bulletin, v. 109, 1563–1575.

Antonovic, A., 1974, Geological maps of Kamrud Quadrangle: Geological Survey of Iran, 1 sheet, scale 1:100 000.

Arvin, M., and Robinson, P.T., 1994, The petrogenesis and tectonic setting of lavas from the Baft Ophiolitic Mélange, southwest of Kerman, Iran: Canadian Journal of Earth Sciences, v. 31, p. 824–834.

Boudier, F. and Nicolas, A., 1985, Harzburgite and lherzolite subtype in ophiolitic and oceanic environments: Earth and Planetary Science Letters, v. 76, p. 84–92.

Campbell, K., Ghazi, A. Mohamad, LaTour, T., and Hassanipak, A.A., 1999, Geochemistry, petrology, and tectonics of the Shahr-Babak ophiolite, SE Iran, Geological Society of America Southeastern Section, Abstracts with Programs, v. 31, p. 9.

Coleman, R.J., 1981, Tectonic setting for ophiolite obduction in Oman: Journal of Geophysical Research, v. 86, p. 2495–2782.

Davoudzadeh, M., 1972, Geology and petrography of the area north of Nain, central Iran: Geological Survey of Iran, Report No. 39, 89 p.

Delaloye, M., and Desmons, J., 1980, Ophiolites and mélange terranes in Iran: A geochronological study and its paleotectonic implications: Tectonophysics, v. 68, p. 83–111.

Desmons, J., and Beccaluva, L., 1983, Mid-oceanic ridge and island arc affinities in ophiolites from Iran: Paleogeographic implication: Chemical Geology, v. 39, p. 39–63.

Dilek, Y., and Delaloye, M., 1992, Structure of the Kizildag ophiolite, a slow-spread Cretaceous ridge segment north of the Arabian promontory: Geology, v. 20, p. 19–22.

Dilek, Y., Thy, P., Hacker, B., and Grundvig, S., 1999, Structure and petrology of Tauride ophiolites and mafic dike intrusions (Turkey): Implications for the Neotethyan Ocean: Geological Society of America Bulletin, v. 111, p. 1192–1216.

Drake, M.J., 1975, The oxidation state of europium as an indicator of oxygen fugacity: Geochimica et Cosmochimica Acta, v. 39, p. 55–64.

Floyd, P., 1991, Ocean islands and seamounts, *in* P.A. Floyd, ed., Oceanic basalts: New York, Blackie, p. 174–218.

Ghazi, A.M., and Hassanipak, A.A., 1999, Geochemistry and petrology of subalkaline and alkaline extrusives of Kermanshah ophiolite, Zagros Suture Zone, SW Iran: Journal Asian Earth Sciences, v. 17, p. 319–332.

Ghazi, A.M., Hassanipak, A.A., Duncan, R.A., Hogen, L.G., and Mahoney, J.J., 1997a, Geochemistry, $^{40}Ar/^{39}Ar$ ages and preliminary isotopic analyses of the Khoy ophiolite, northwestern Iran [abs.]: Eos, Transactions of the American Geophysical Union, v. 79, p. F654.

Ghazi, A.M., Hassanipak, A.A., and Wallace, K., 1997b, Geochemistry, petrology, and geology of the Sabzevar ophiolite, northeastern Iran: Implication on Tethyan tectonics: Geological Society of America Abstracts with Programs, v. 29, p. A-229.

Ghazi, A.M., Chatham, B., Hassanipak, A.A., Mahoney, J.J., and Duncan, R.A., 1999, A petrogenetic investigation of the Khoy ophiolite, NW Iran: Implications for Tethyan magmatism and ophiolite genesis: RIDGE Field School: The Troodos Ophiolite and Mid-Ocean Ridge Processes: Abstract and Field School Notes, Larnaca, Cyprus, p. 13.

Gladney, E.S., and Roelandts, I., 1988, 1987 compilation of elemental concentration data for USGS BHVO, MAG, QLO-1, RGM-1, Sco-1, SDC-1, SGR-1 and STM-1: Geostandards Newsletter, v. 12, p. 253–362.

Glennie, K.W., Hughes Clarke, M.W., Boeuf, M.G.A., Pilaar, W.F.H., and Reinhardt, B.M., 1990, Inter-relationship of Makran-Oman Mountains: Belts of convergence, *in* Robertson, A.H.F., Searle, M.P., and Ries, A.C., eds., The geology and tectonics of the Oman region: Geological Society of London Special Publication 49, p. 773–786.

Govindaraju, K., 1989, Compilation of working values and sample descriptions for 272 Geostandards: Geostandards Newsletter, v. 13, p. 1–113.

Hanson, G.N., 1989, An approach to trace element modeling using a simple igneous system as an example: Mineralogical Society of America Reviews in Mineralogy, v. 21, p. 79–97.

Hassanipak, A.A., and Ghazi, A.M., 1999, Petrology, geochemistry, and tectonic setting of the Khoy ophiolite, Northwest Iran: Journal of Asian Earth Sciences, v. 18, p. 43–55.

Hassanipak, A.A., Ghazi, A.M., and Wampler, J.M., 1996, REE characteristics and K/Ar ages of the Band Ziarat ophiolite complex, Southeastern Iran: Canadian Journal of Earth Sciences, v. 33, p. 1534–1542.

Haynes, S.J., and Reynolds, P.H., 1980, Early development of Tethys and Jurrasic ophiolite displacement: Nature, v. 283, p. 561–563.

Irvine, T.N., and Baragar, W.R.A., 1971, A guide to chemical classification of common volcanic rocks: Canadian Journal of Earth Sciences, v. 8, p. 523–547.

Jenner, G.J., 1996, Trace element geochemistry of igneous rocks: Geochemical and nomenclature and analytical geochemistry, *in* Wyman, D.A., ed., Trace element geochemistry of volcanic rocks: Applications for massive sulphide exploration: Geological Association of Canada, Short Course Notes, no. 12, p. 51–78.

Lanphere, M.A., and Pamic, T., 1983, $^{40}Ar/^{39}Ar$ Ages and tectonic setting of ophiolites from Neyriz area, south-east Zagros Range, Iran: Tectonophysics, v. 96, p. 245–256.

Lensch, G., 1980, Major element geochemistry of the ophiolites in north-eastern Iran, *in* Panayiotou A., ed., Ophiolites: Proceedings of the international ophiolite symposium, Ministry of Agriculture and Natural Resources, Geological Survey Department, Republic of Cyprus, p. 398–401.

Lensch, G., Mihm, A., and Alavi-Tehrani, N., 1977, Petrography and geology of the ophiolite belt north of Sabzevar/Khorasan (Iran): Neues Jahrbuch fuer Mineralogie, Abhandlungen v. 131, p. 56–178.

Lippard, S.J., Shelton, A.W., and Gass, I.G., 1986, The ophiolite of Northern Oman: Geological Society of London Memoir, 11, 178 p.

Malpas, J., Calon, T.J., and MacDonald, R.W.J., 1994, The Shulaps ophiolite complex of British Columbia, Canada: A Palaeozoic/Mesozoic arc-related microterrane, *in* Ishiwatari, A., et al., eds., Circum-Pacific Ophiolites: Proceedings, 29th International Geological Congress, Part D, p. 69–87.

McCall, G.J.H., 1985, Explanatory text of the Minab Quadrangle Map, J13, Geological Survey of Iran, 530 p., scale 1:250 000.

McCall, G.J.H., 1997, The geotectonic history of the Makran and adjacent areas of southern Iran: Journal of Asian Earth Sciences, v. 15, p. 517–531.

Meschede, M., 1986, A method of discriminating between different types of mid-ocean ridge basalts and continental tholeiites with the Nb, Zr, Y diagram: Chemical Geology, v. 56, p. 207–218.

Middlemost, E.A.K., 1977, The basalt clan: Earth Science Review, v. 1, p. 51–57.

Pallister, J.S., and Knight, R.J., 1981, Rare earth element geochemistry of the Samail ophiolite near Ibra, Oman: Journal of Geophysical Research, v. 86, p. 2673–2697.

Pearce, J.A., 1982, Trace element characteristics of lava from destructive plate boundaries, *in* Thrope, R.S., ed., Andesite: London, John Wiley and Sons, p. 525–548.

Pearce, J.A., 1996, A user's guide to basalt discrimination diagrams, *in* Wyman, D.A., ed., Trace element geochemistry of volcanic rocks: Applications for massive sulphide exploration: Geological Association of Canada, Short Course Notes, no.12, p. 79–113.

Pearce, J.A., and Cann, J.R., 1973, Tectonic setting of basic volcanic rocks determined using trace element analyses: Earth and Planetary Science Letters, v. 19, p. 290–300.

Pearce, J.A., and Peate, D.W., 1995, Tectonic implications of composition of volcanic arc magmas: Annual Review of Earth Planetary Science, v. 23, p. 251–285.

Petrovic, B., and Isailovic, S., 1974, Geological maps of Robat Quadrangle: Geological Survey of Iran, 1 sheet, scale 1:100 000.

Quick, J.E., 1981, Petrology and petrogenesis of the Trinity peridotite, an upper mantle diapir in the Eastern Klamath Mountains, Northern California: Journal of Geophysical Research, v. 86, p. 11837–11863.

Sarkarinejad, K., 1994, Petrology and tectonic setting of the Neyriz ophiolite, southeast Iran, *in* Ishiwatari, A., et al., eds., Circum-Pacific ophiolites: Proceedings, 29th International Geological Congress, Part D, p. 221–234.

Şengör, A.M.C., 1990, A new model for the late Paleozoic-Mesozoic tectonic evolution of Iran and implications for Oman, *in* Robertson, A.H.F., Searle, M.P., and Ries, A.C., eds., The geology and tectonics of the Oman region: Geological Society of London Special Publication 49, p. 797–831.

Stocklin, J., 1974, Possible ancient continental margins in Iran, *in* Burke, C.A., and Drake, C.L., eds., The geology of continental margins: New York, Springer-Verlag, p. 873–887.

Sun, S.S., 1980, Lead isotopic study of young volcanic rocks from mid-ocean ridges, oceanic islands and island arcs: Philosophical Transaction of Royal Society of London, v. A297, p. 409–445.

Sun, S.S., and McDonough, W.F., 1989, Chemical and isotopic systematics of ocean basalts: Implication for mantle composition and processes, *in* Saunders, A.D., and Norry, M.J., eds., Magmatism in ocean basins: Geological Society of London Special Publication 42, p. 313–345.

Takin, M., 1972, Iranian geology and continental drift in the Middle East: Nature, v. 235, p. 147–150.

Tankut, A., Dilek, Y., and Önen, P., 1998, Petrology and geochemistry of Neo-Tethyan volcanism as revealed in the Ankara mélange, Turkey: Journal of Volcanology and Geothermal Research, v. 85, p. 265–284.

Winchester, J.A., and Floyd, P.A., 1977, Geochemical discrimination of different magma series and their differentiation products using immobile elements: Chemical Geology, v. 20, p. 325–342.

Manuscript Accepted by the Society February 23, 2000

Printed in the U.S.A.

Geological Society of America
Special Paper 349
2000

Aswad massif (United Arab Emirates): Archetype of the Oman-UAE ophiolite belt

Adolphe Nicolas, Françoise Boudier, K. Michibayashi,* and L. Gerbert-Gaillard
Laboratoire de Tectonophysique CNRS-UM2, place Eugène Bataillon, cc049, 34095 Montpellier cedex 05, France; e-mail: tectono@dstu.univ-montp2.fr

ABSTRACT

Among the 12 massifs constituting the Oman-United Arab Emirates (UAE) ophiolite belt, the Aswad massif has the simplest internal structure and is the best analogue for a fast-spreading oceanic ridge. Typically the Moho is horizontal along the western margin of the massif for more than 50 km. The vertical sheeted dike complex strikes north, at a right angle to the average trend of both the high temperature (high-*T*) plastic lineations in the mantle and the magmatic lineations in the crust. Below the ridge of origin, the high-*T* flow in the uppermost mantle was on average at a right angle to the ridge axis, diverging somewhat from a central north-trending axis centered on a mantle window. This axis could have been the axis of the paleoridge of origin, which was frozen in situ before the ophiolite obduction. The Moho Transition Zone (MTZ) between the mantle harzburgites and crustal gabbros is on average 50 m thick, but attains a thickness of 400 m in the central part of the massif. Thick MTZ is regarded as being issued from a mantle diapir, which is possibly located in the mantle window along the presumed ridge axis, within an area of steep mantle foliations and lineations, below a MTZ that is thick and rich in chromite deposits. The Aswad massif is separated to the south from the Fizh massif by a northwest-trending major shear domain. This could represent the transform domain between a propagating ridge segment (Aswad axis) and a doomed ridge segment (located to the west of Fizh).

INTRODUCTION

The Oman and United Arab Emirates (UAE) ophiolite has received much attention over the past 20 years. Many results are contained in four volumes devoted to this ophiolite (Coleman, 1981; Lippard et al., 1986; Boudier and Nicolas, 1988; Peters et al., 1991). These volumes and more recent publications (Boudier and Juteau, 2001) are oriented toward thematic issues, with selective information from various massifs. An alternative view of this ophiolite is to present a complete description of a single massif. The choice of the Aswad massif, located in the UAE part of the belt, is justified by the fact that among the 12 principal massifs composing the belt, Aswad stands as one of the largest, covering over 2000 km^2 and as the one with the simplest internal structure. Along the western margin of the massif for a north to south length of 50 km, the Moho is flat lying and even completely horizontal; the rocks containing the Moho form a topography of cuestas and buttes. The horizontal domain covers 800 km^2, but it is known to extend westward another 15 km beneath the Quaternary dunes of the UAE (Fig. 1).

*Present address: Institute of Geosciences, Shizuoka University, Shizuoka 422-8529, Japan; e-mail: sekmich@ipc.shizuoka.ac.jp

There is a broad consensus that the Oman-UAE ophiolite belt was being generated at a fast-spreading ridge system (Ildefonse et al., 1993). The structural mapping has, however, revealed that the internal structures of most massifs are complex, and that the most intense deformations, such as steep Moho-bearing sections, major folds, and shear zones, were not related to the obduction but to the ridge activity, namely ridge propagation (Nicolas and Boudier, 1995; Nicolas et al., 2001). Large and rapid rotations resulting in local thrusts may reflect the activity of a microplate such as the Easter or Juan Fernandez microplates along the East Pacific Rise (EPR) or the Manus microplate in the Bismark Sea (Boudier et al., 1997). The only noteworthy deformations recorded in the Aswad massif were induced by shearing between the Fizh massif to the south and the Khawr Fakkan massif to the north. This latter massif is also thrust over the northeast corner of Aswad (Fig. 1). These deformations occurred while the Aswad ridge was still active, similar to the early tectonic activity described in other parts of the belt (Boudier et al., 1997). The effects of the thrusting are intense but local. Otherwise this massif is, to our knowledge, the best analogue to a ridge segment of a fast-spreading center such as the East Pacific Rise or the Valu Fa Ridge in the Lau Basin. In a more general study (Nicolas et al., 2001) we interpret the Aswad ridge segment as belonging to a propagating ridge system, a model accounting for the regional data and the tectonic activity observed in Aswad.

Nicolas, A., et al., 2000, Aswad massif (United Arab Emirates): Archetype of the Oman-UAE ophiolite belt, *in* Dilek, Y., Moores, E.M., Elthon, D., and Nicolas, A., eds., Ophiolites and Oceanic Crust: New Insights from Field Studies and the Ocean Drilling Program: Boulder, Colorado, Geological Society of America Special Paper 349, p. 499–512.

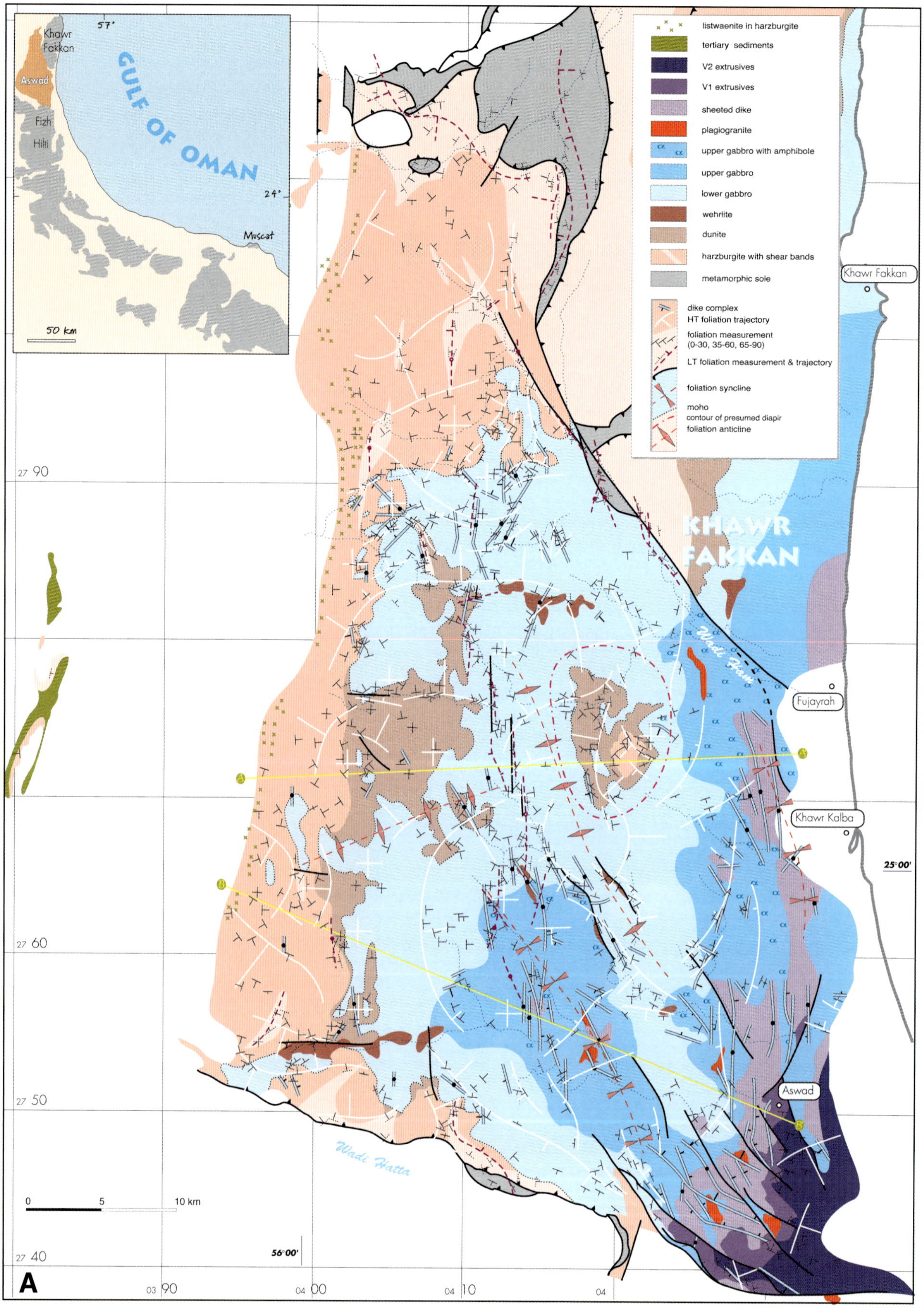

Figure 1 (on this and next two pages). Structural maps of the Aswad massif (inset: location of the Aswad massif in the ophiolite belt). A, Planar structures: foliation in peridotites and in upper gabbros, layering in lower gabbros, and bedding in lavas and associated sedimentary rocks. B, Linear structures: plastic flow–related mineral lineations in peridotites, magmatic flow–related mineral lineations in gabbros, and mineral lineations in metamorphic shear zones. C, Dikes and veins: diabase, plagiogranite, gabbro, and pyroxenite dikes and hydrothermal and dunite veins; numbers indicate the thickness of the MTZ. HT—high-temperature, LT—low-temperature.

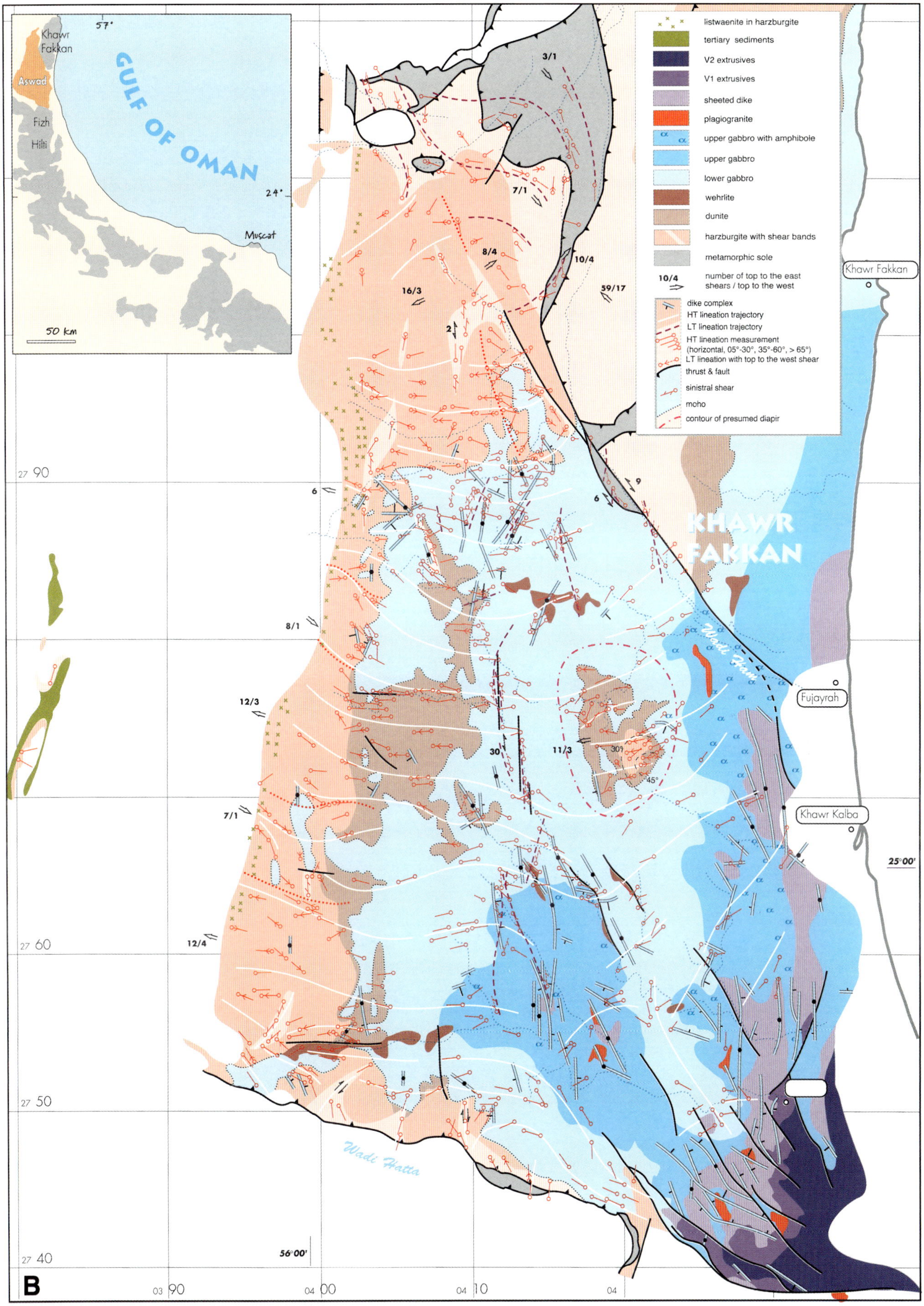

Figure 1. Continued.

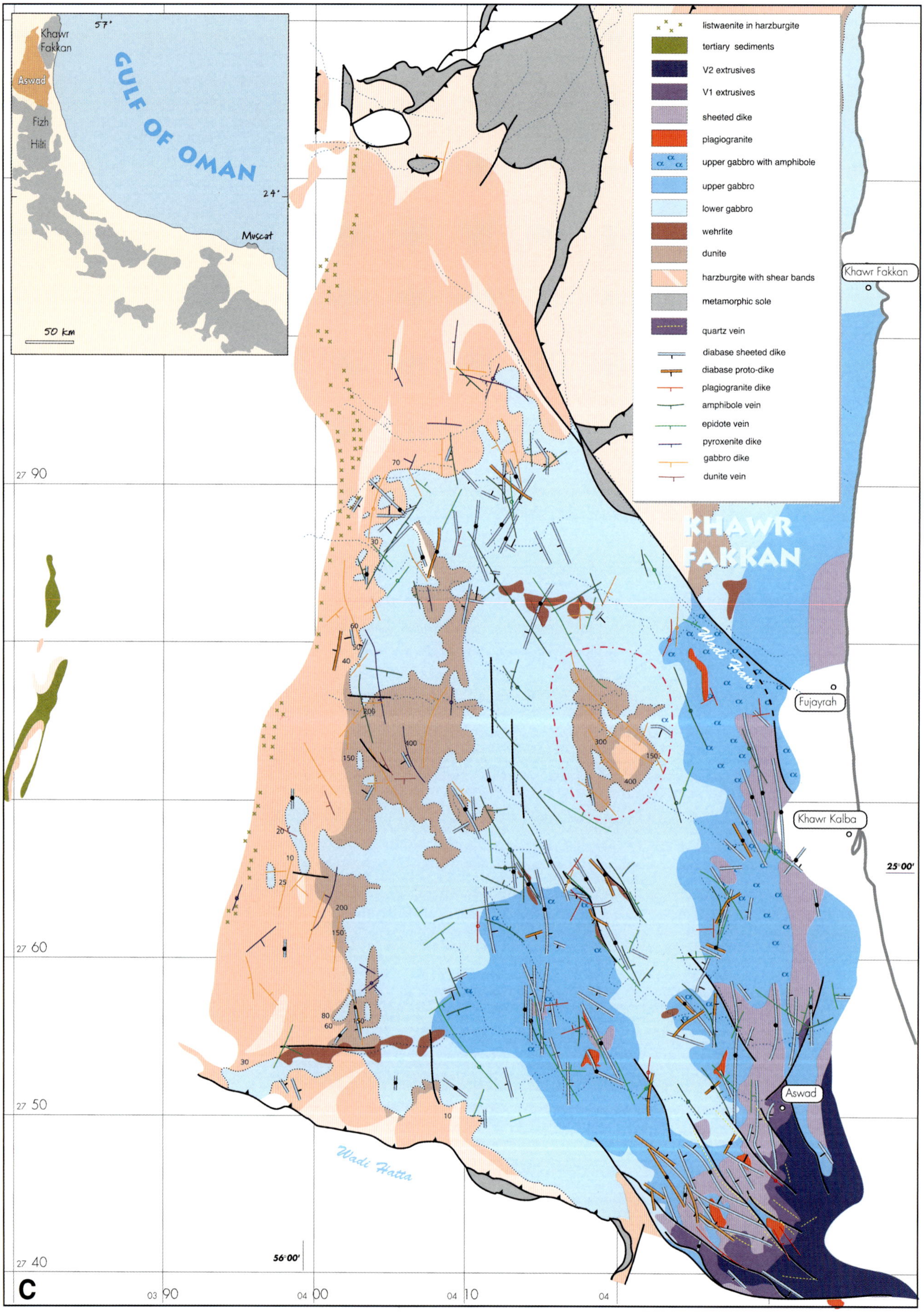

Figure 1. Continued.

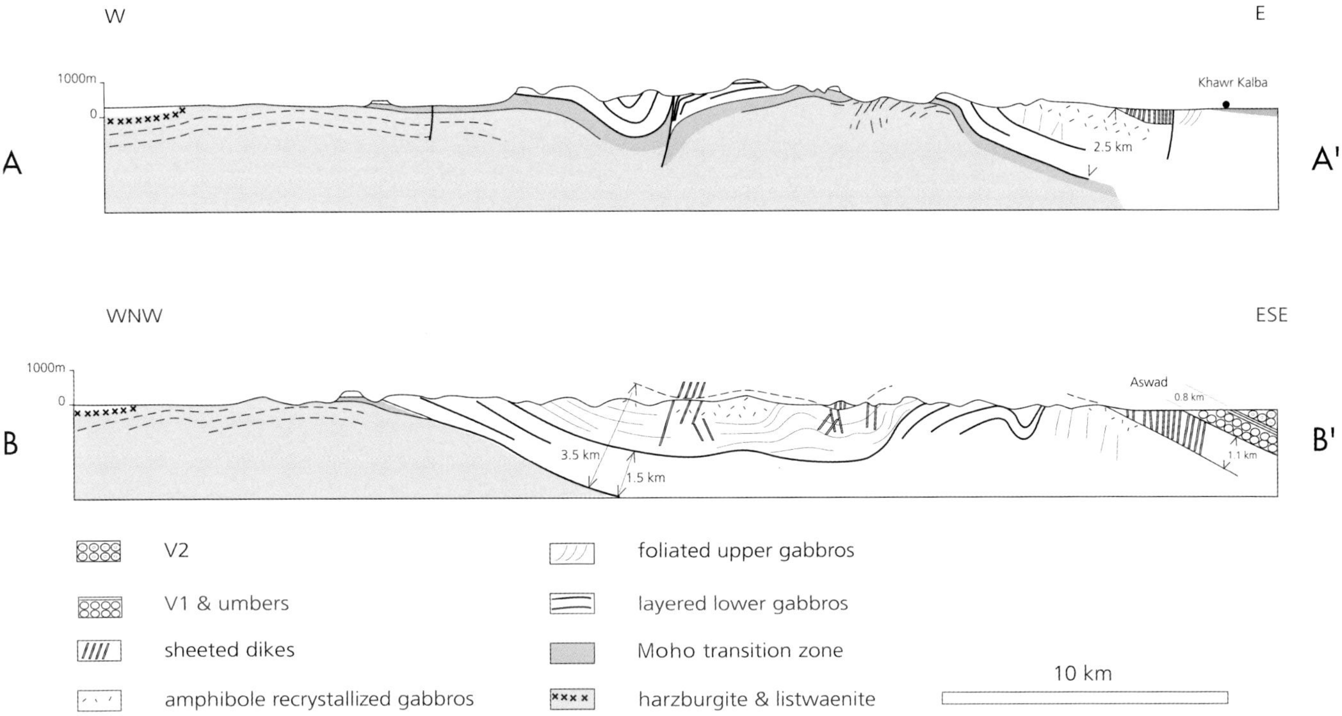

Figure 2. Cross sections through the Aswad massif (location in Fig. 1A).

GEOLOGIC SETTING

The Aswad massif is located in the northern part of the ophiolite belt where the Bouguer gravity anomaly exceeds 150 mgal (Ravaut et al., 1997); this value points to an ophiolite thickness of some ≥ 10 km. Aswad shares the highest gravity anomalies of the entire belt with the Khawr Fakkan massif. This latter massif was thrust over the northeastern part of Aswad and sheared against it, along the northeast-trending Wadi Ham suture. Both massifs are separated to the south from the Fizh massif and from the rest of the belt by a major sinistral shear zone following Wadi Hatta, which corresponds more or less to the border between the UAE and Oman. In contrast with most other massifs of the belt, which resulted from the dispersal of parts of a continuous nappe during and after obduction, the Khawr Fakkan, Aswad, and Fizh massifs are still continuous, being affected only by shear bands initiated at high temperature and thus reflecting tectonic events that occurred close to the ridge system of origin.

Wadi Hatta suture zone

As in most other sutures between individual massifs of the belt, the Aswad and Fizh massifs are separated in their western part by the Hawasina Formation, which consists of marine sedimentary rocks that have been thrust into and below the ophiolite nappe (Fig. 1). Locally, slivers of metamorphic aureoles from the soles of these massifs are also exposed. In contrast, their eastern contact is the termination of a major sinistral shear band that splays into the Aswad massif and extends more than 20 km southward into the Fizh massif. In the Fizh massif, mylonite bands were deformed at 900–1000 °C in the mantle section and to amphibolite-facies conditions in the crustal gabbros.

Wadi Ham suture zone

This steep, sinistral shear zone, several hundred meters thick, evolves to the west into a northwest-directed thrust, carrying the Khawr Fakkan massif over the Aswad massif (Fig. 1). Mylonitic peridotites and flaser gabbros developed in the steeply dipping shear zone. They reflect initial temperatures of deformation as high as 800–900 °C, followed by greenschist-facies conditions (~400 °C). Within the shear zone, lineations plunge gently southwest in accordance with the northwest-directed overthrusting by the Khawr Fakkan massif. This geometry has been confirmed by a structural study of the Khawr Fakkan massif (Gnos and Nicolas, 1995) showing, in particular, northwest-directed, low-T plastic-flow deformation in the basal peridotites of this massif. The underlying and exceptionally well-developed metamorphic aureole displays an inverse metamorphic gradient that is in continuity with the basal, low-T peridotites. It grades from amphibolite-facies, quartz-rich gneisses to greenschist-facies quartz-rich phyllites (Lippard, 1983). This scheme may be too simple because west of the metamorphic aureole, low-T peridotites that have been affected by dominantly southeast-oriented thrusting are also present (Fig. 1B).

Between these two sheared limits, the Aswad massif has a generally eastward dip: lavas are exposed along the Indian Ocean coastal plain, and high-T peridotites crop out along the western part of the massif. Cross sections (Fig. 2) show that the massif is remarkably flat lying along its western border and western extension below the sand dunes; to the east, the massif is gently flexed eastward and dips 20°–30°E only in its easternmost parts.

DESCRIPTION OF THE UNITS

Peridotites

With the exception of a small window in the center of the massif, the peridotites from the mantle unit crop out along the western border of the massif. They are composed of harzburgites with minor dunites, either as discordant veins or as centimeter-wide bands, interlayered with pyroxenites (mainly orthopyroxenites). From the structural and petrologic points of view, they are completely similar to those observed in the rest of the belt.

West of the massif, the peridotites disappear below the sand dunes, reappearing locally as small hills. They extend at least as far as Dayd to the west, where they constitute small ridges covered with Maastrichtian limestones. There, the high-T harzburgites, largely serpentinized, were exposed on the seafloor as deduced from their occurrence in a sedimentary breccia belonging to the basal level of the Maastrichtian limestones. Along the western margin of the massif, where they disappear below the dunes, the peridotites are transformed to listwaenite, a silicified peridotite, by a progressive silica invasion. The serpentinite network can be traced into totally silicified peridotites. The listwaenite, which has been described as birbirite by Glennie et al. (1974), constitutes a flat horizon, a few meters thick, that grades upward into continental conglomerates. These authors regarded the listwaenite as the product of tropical alteration during late Tertiary to recent time.

Eastward, the peridotites are covered by the lower gabbros of the crustal section, along a Moho that is incredibly flat for over 50 km from north to south (Fig. 1). Just below the Moho, the Moho transition zone (MTZ) dominantly composed of dunite, grades downward into harzburgite. As in the rest of the belt (Boudier and Nicolas, 1995), the MTZ dunites are variously impregnated by gabbros and pyroxenites, ranging from disseminated plagioclase and clinopyroxene grains within an olivine framework, to gabbros or websterite sills tens of meters thick that are generally well layered. Chromite pods are locally present. The MTZ is a few hundred meters thick in the central part of the massif (Fig. 1). As dunites erode more easily than harzburgites, the gabbro cuestas stand 400 m above the dunitic lowlands, where small hills represent layered gabbros sills, or more exceptionally, clinopyroxenite sills. North and south of the central area, the MTZ thins to 10–80 m. The thinning is sharp, occurring over 1–2 km, as in other parts of the belt (Nicolas and Boudier, this volume).

Lower gabbros

The lower gabbros are dominantly clinopyroxene-olivine-anorthite gabbros, commonly troctolitic by olivine enrichment. They display a magmatic foliation and a layering defined either by discrete, modal, centimeter-sized mineral segregations or by distinct and streaky layers of pegmatitic and fine-grained gabbros grading into doleritic norites and anorthosites, websterites, and wehrlites. In contrast with the gabbros from the Oman part of the belt, the modally graded bedding is nearly absent in this massif, both in the layered gabbros from the lower crust and in lenses within the MTZ. Because the streaky layers are locally slightly oblique with respect to the discrete layering and to the magmatic foliation, they are believed to represent sills or tectonically transposed dikes. The gabbro layers show an internal magmatic foliation and internal structures that are locally symmetrical. Anorthosite layers, for example, underlie both walls or only the upper wall. This geometry indicates that the gabbros and anorthosites represent early injections into the crystal mush of the magma chamber; in contrast, the fine-grained and doleritic norites were injected outside the chamber, within gabbros below the solidus. Although the evidence is not as good as at Aswad, the intrusive origin of such layers has already been suggested for the Oman part of the belt (Nicolas et al., 1988; Boudier and Nicolas, 1995).

Compared to lower gabbros in the Oman part of the belt, the Aswad lower gabbros present a few other dissimilarities. They are, on average, much less deformed by magmatic flow. As shown in Figure 3, traces of magmatic flow are nearly absent in a central north-northwest-trending domain centered on the mantle window. This lack of strain may explain why the analysis of the origin of different types of layering is simpler here than in the southern part of the belt where the large flow resulted in a smoothing out of all structures. Lack of strain also accounts for the near-total absence of the small-scale synmagmatic tectonic structures that are so common in the rest of the belt. In contrast, traces of plastic deformation (mechanical twinning, lattice bending, and fine-grained recrystallization) are common at all levels in these gabbros, whereas in the Oman part they are restricted to the lowermost gabbros. Another difference is the local development of norites in place of olivine gabbros (Fig. 3). Although the norites are uniformly distributed in the upper-gabbro domains, in the lower gabbros, the norites occur mainly along a north-trending band rich in amphibolite-facies shear zones. In contrast, norites are absent in the central domain of the massif where the lower gabbros are weakly deformed by magmatic flow. In norites, the orthopyroxene forms elongated prisms that crystallized late as shown by their inclusion of plagioclase and clinopyroxene and their association with magmatic hornblende. Orthopyroxene coexists with relicts of olivine in some rocks. This assemblage indicates that late during the crystallization of the olivine gabbros and after the end of magmatic flow, water penetrated the system (Boudier et al., 2001). This water augmentation favored the growth of orthopyroxene and amphibole (Lachize et al., 1996).

Upper gabbros

As in the Oman part of the belt, the upper gabbros have a magmatic foliation but tend to have lost any layering and thus tend to be very massive. Their foliation also steepens upsection. The magmatic deformation is also weaker than in the Oman part of the belt, and the dominant facies are not clinopyroxene-olivine gabbros but clinopyroxene-orthopyroxene-hornblende norites (Fig. 3). In the eastern and southeastern part of the massif, mainly near the root zone of the diabase sheeted dikes, the upper-level rocks have been recrystallized into amphibole gabbros. Petrologic facies are highly complex in the amphibole gabbros as a result of successive injections of (1) high-T diabase dikes (i.e., the protodikes of Nicolas and Boudier, 1991), (2) diorite to plagiogranite dikes rich in hornblende and (locally) pistacite, plagiogranites forming irregular dikes and, more commonly, magmatic breccia together with diabase, (3) diabase dikes with chilled margins, and (4) hydrothermal veins. The high-T diabase dikes are always orthopyroxene-hornblende bearing, in upper as well as in lower gabbros, where they are a component of the layering. They display various degrees of recrystallization and magmatic deformation, which generally increase down section. Mylonitic deformation at amphibolite- to greenschist-facies conditions is ubiquitous, mostly as decimeter- to meter-thick steeply east- or west-dipping shear bands.

Zone of transition to the sheeted dike complex

The zone of transition from the upper gabbros to the sheeted-dike complex, also called the root zone of the sheeted-dike complex (Nicolas and Boudier, 1991), is especially well exposed in the Aswad massif, mainly because large areas of this massif are flat laying and have been eroded to this level. We place the boundary between amphibole gabbros and this root zone where the ratio of diabase dikes with chilled margins to other rocks attains 50%, and we place the boundary between the root zone and the diabase sheeted-dike complex where this ratio exceeds 90%. The petrologic facies and their

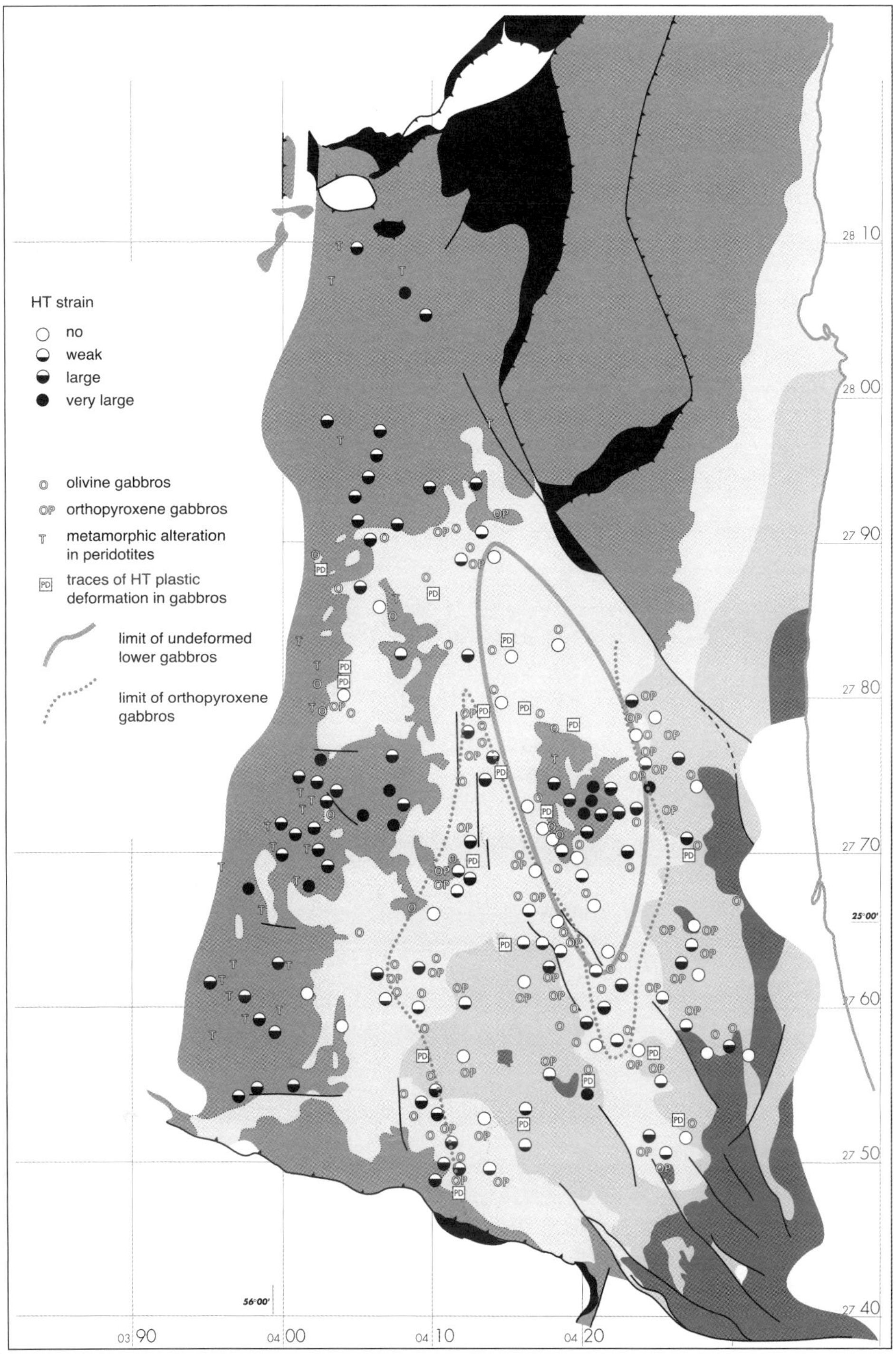

Figure 3. Map of tectonic and petrologic facies. High-*T* (HT) plastic strain in the mantle section and magmatic strain in the gabbro sections have been estimated qualitatively. Only significant, very large strain is noted in the mantle section. Note the presence of a central domain in lower gabbros where no magmatic strain is observed. As the petrology of the peridotites is very homogeneous, the only conspicuous feature is, locally, the presence of greenschist-facies alteration marked by the presence of talc, chlorite, actinolite, and serpentine. In gabbros, orthopyroxene-bearing norites are distinguished from olivine gabbros. For lithology, refer to Fig. 1.

mutual relationships are similar to the upper-level amphibole gabbro unit. There is, however, a major difference: the amphibole gabbros were partly derived from granular and foliated gabbros by hydrous recrystallization and partial melting, as deduced from the observation of ubiquitous relicts of variously assimilated foliated gabbros inside the amphibole gabbros. Such relicts are never observed within the root zone; in contrast, a new doleritic foliation appears locally in the otherwise equant amphibole gabbros. All these relationships are similar to those described in the Oman part of the belt (Rothery, 1983; Nehlig and Juteau, 1988; Nicolas and Boudier, 1991; MacLeod and Rothery, 1992).

Because of the dominantly horizontal attitude of the gabbro unit, the thickness of the diabase dike root zone cannot be estimated in most areas, and it is possible that this critical zone is, on average, thicker here than in the rest of the belt, where it does not exceed 200 m. In contrast, in a few places, the transition from granular and foliated gabbros devoid of any diabase dikes to the diabase sheeted-dike complex is abrupt, occurring over 10–20 m or less, in the ab-

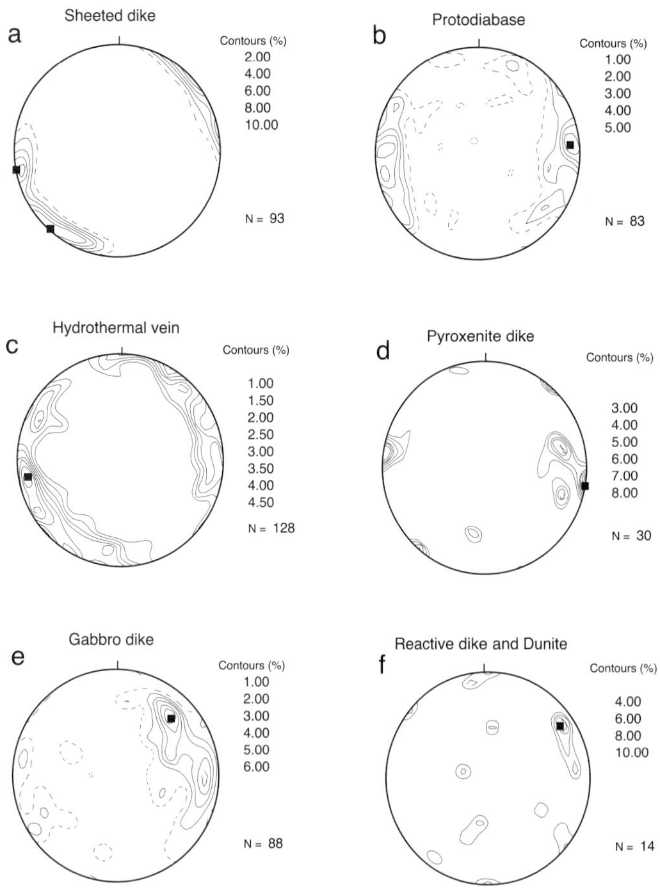

Figure 4. Stereonets of dikes and veins. A, Diabase dikes from the diabase sheeted-dike complex; each of the 93 reported measurements represents a field station with an average of 10 individual measurements. B, Diabase dikes without chilled margins and protodiabase. C, Hornblende and epidote + prehnite hydrothermal veins in mantle section. D, Pyroxenite dikes in mantle section. E, Gabbro dikes in mantle section. F, Dunite veins and gabbro reactive dikes in mantle section.

sence of any fault. Such situations do not represent a normal stratification of the sheeted-dike unit on top of the upper gabbros. They rather suggest intrusions of sheeted dikes into cooled gabbros, presumably occurring off-axis.

Diabase sheeted-dike complex (DSDC)

Near its lower transition to gabbro, the diabase sheeted-dike complex still contains a few gabbro screens that disappear upsection; epidote-rich hydrothermal alteration is also important, and some dikes are brecciated and tilted. Otherwise, the diabase sheeted-dike complex contains on average remarkably vertical dikes (Fig. 4A); dike azimuth is more variable, with local sets of dikes a few tens to a few hundreds of meters thick, departing by several tens of degrees from the dominant azimuth. Hydrothermal, epidote-rich veins commonly mark the chilled margins of dikes as already observed in Haylayn massif (Nehlig and Juteau, 1988). Diabase in the diabase sheeted-dike complex has a fine-grained doleritic texture with interstitial green amphibole, an opaque phase, and locally quartz between laths of plagioclase. Exceptionally, clinopyroxene is still visible inside amphibole. Orthopyroxene has not been observed, in contrast with the protodiabase intruding the gabbros, which has abundant orthopyroxene.

Lavas and associated sedimentary rocks

These uppermost levels of the ophiolite complex are represented only in the southeast corner of the massif east and south of the Aswad locality. The V_2 (Lasail) upper lavas are much developed compared to the V_1 ("Geotimes") lower lavas. The V_1 lavas are in direct contact with the sheeted dikes of the ophiolite and separated from the V_2 lavas by a few meters of umbers and metalliferous sedimentary rocks. Lava flows and sedimentary beds are regularly oriented with an average 30°-SE dip. To our knowledge, there are no chemical analyses of lavas from the Aswad massif. In the Oman part of the belt, the V_1 is regarded as having MORB affinities, and the V_2 is thought to have calc-alkalic affinity (Lippard et al., 1986; Ernewein et al., 1988).

Wehrlites

The duality of wehrlite occurrences as it has been described in Oman (Benn et al., 1988; Juteau et al., 1988) is also verified in Aswad. As already mentioned, wehrlite sills, a few tens of centimeters to a few meters thick, are a ubiquitous component of the layering in the layered gabbros; in addition, they form large sills and discordant intrusions through the gabbro unit. In the central part of the massif, sills 100 m thick and kilometers wide predominate. They intrude the lower levels of the gabbro unit and are difficult to distinguish from the locally thick MTZ. Both the small and large wehrlite sills were emplaced inside or near the boundary of the magma chamber as demonstrated by their internal magmatic foliation. These wehrlite sills are themselves injected in layer-by-layer fashion by gabbro sills; in contrast, the gabbro roof is discordantly intruded by these wehrlites.

North and south of the central part of the massif, the wehrlites constitute discordant intrusions, several kilometers wide. Along the contact, the gabbros are plastically folded, demonstrating that they intruded the section outside the magma chamber. Wehrlite intrusions also follow north-northwest-trending discontinuities that are hydrothermal fracture zones parallel to the overlying sheeted dikes. (Because the serpentinized wehrlites and the altered gabbros erode faster than the fresh gabbros, the wadis preferentially follow the north-northwest trend.) The relationship between the intrusions and the hydrothermal fractures suggests that the wehrlitic mush was injected during hydrothermal alteration when the gabbro unit was at 400–500 °C, i.e., still close to the ridge axis.

Dikes and veins

In the peridotite section, the nature and contact relationships of the gabbro and/or pyroxenite dikes and the dunite veins are similar to those of the rest of the belt (Nicolas and Boudier, 1995). The dunite veins are the oldest. They formed from pyroxene reaction with a circulating melt, as reviewed by Nicolas (1989, p. 21). Where this melt has been trapped, it has crystallized as gabbro dikes and chromite pods that are separated from the harzburgite matrix by a dunite wall, generally a few centimeters to a few tens of centimeters wide (Fig. 4F). The other dikes are gabbros (Fig. 4E) and pyroxenites (Fig. 4D) with sharp and non-reactive margins and were clearly emplaced in a cooling peridotite.

In the gabbro section, excluding the sills and transposed dikes, (wehrlites, pyroxenites, pegmatitic gabbros, anorthosites, protodikes, microgabbros, and dolerites) that contribute to the streaky layering of the lower gabbros, the intrusive dikes are mainly protodikes that represent diabase dikes emplaced into a gabbro still at high temperatures, as shown by their microgranular margins and local internal recrystallization to granular and foliated microgabbros (Nicolas and Boudier, 1991) (Fig. 4B). These dikes and sills are remarkably abundant in the Aswad massif. Hydrothermal veins are also exceptionally abundant in this massif. The earliest veins are fissures along which a dark green hornblende developed. Juteau et al. (1988) regarded these early veins as having resulted from downward penetration of seawater into the still hot gabbro, along a "cracking front"; fluid inclusions yield temperatures of 400–530 °C. Water circulating at greenschist-facies conditions bleached the gabbros and formed epidote + prehnite veins (temperature of 200–400 °C). In a given locality, they commonly split into one set that follows the orientation of the sheeted dikes and another set at a high angle (Fig. 4C). The upper gabbros, mainly in the root zone of the diabase sheeted-dike complex, are intruded by hydrous dioritic dikes related to plagiogranite dikes; both are poorly oriented. Plagiogranites are commonly associated with diabase dikes, forming a magmatic breccia. It is interesting that a comparable magmatic breccia, involving protodikes and having a dioritic eutectic composition, has been locally observed, resulting from the same mixing of basaltic and felsic eutectic melts, but at a higher temperature.

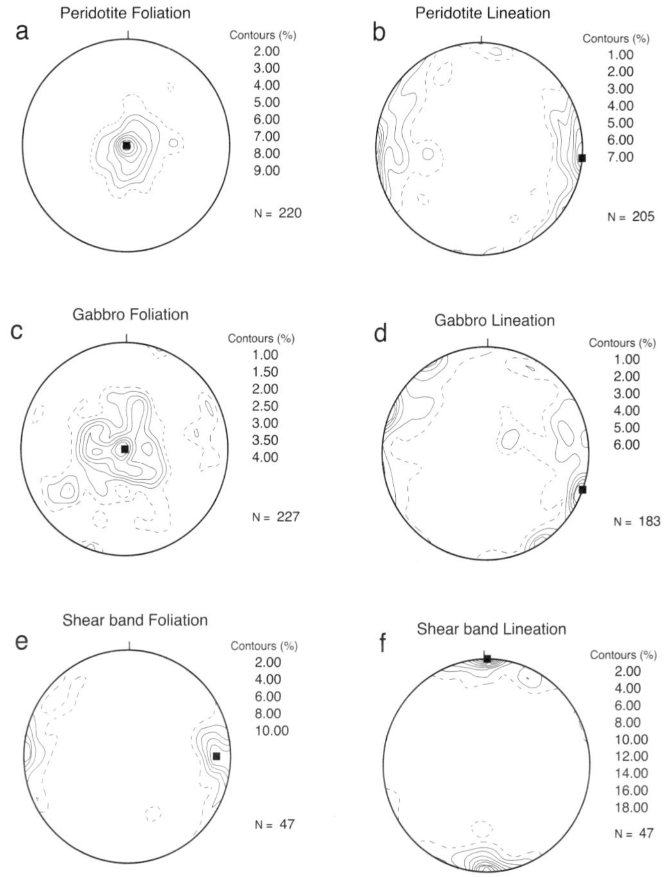

Figure 5. Stereonets of planar and linear structures for the massif. A, Peridotite high-*T* foliation. B, Peridotite lineation. C, Gabbro magmatic foliation. D, Gabbro magmatic lineation. E, Foliation in mylonitic peridotite and in amphibolite-facies shear bands in gabbro. F, Stretching lineation in mylonite and shear band. Lower hemisphere projections; black squares show maximum density.

STRUCTURES

Mantle unit

The main part of the mantle unit is represented by the high-*T* harzburgites and dunites of the western part of the massif. Characterized by coarse grains (2 mm) and a well-recovered substructure, these high-*T* peridotites are ascribed to hypersolidus to solidus conditions (~1200 °C) below the ridge of origin, and their deformation reflects the asthenospheric flow below this ridge (Nicolas, 1989, p. 27). In contrast, rocks with low-temperature (low *T*) microstructures have small neoblasts (≤0.5 mm) and tight substructures. They reflect deformations at lithospheric conditions and temperatures in the 1100 °C-800 °C range. In Aswad, the low-*T* peridotites are restricted to a few north-trending sinistral shear bands within the western peridotite; these are more developed in the northern part of the massif close to the sheared and thrust-faulted contacts with the Khawr Fakkan massif (Fig. 1). Along the southern, Wadi Hatta boundary, low-*T* peridotites constitute a few north-trending shear zones and a discontinuous basal aureole-related thrust band.

The high-*T* foliation and associated mineral banding in Aswad are horizontal or moderately dipping (Figs. 1A and 5A). The small mantle window present in the center of the massif displays, within the MTZ located below horizontal gabbros, a flat-lying foliation that steepens downward. Unfortunately, the window is too small and the level of erosion insufficient to rule out that the steepening of foliations is local and insignificant.

The high-*T* lineations in peridotites have a strong east trend (Figs. 1B and 5B). A few steep lineations are recorded within the central window into the mantle rocks. The shear sense (Fig. 1B) along the western front is dominantly top to the west (48 top-to-the-west shears vs. 25 top-to-the-east shears). The general ratio of 2/1 westward:eastward shears is abnormally low compared to the southern part of the ophiolite belt. Shears in the central part of the domain of thick MTZ rocks are dominantly west directed, but this domain is flanked by two bands, 5 km wide, displaying clear east-directed measurements (Fig. 1B). North and south of these central domains, the shear is dominantly west directed. In the northern part of the massif, it is possible to draw a north-northwest-trending line between dominantly west-directed (16/3) and east-directed shear senses (8/4). However, the analysis is complicated by the importance of parallel, low-*T* shear bands related to the nearby Khawr Fakkan massif. A somewhat comparable situation prevails along the southern boundary of the Aswad massif where high-*T* and low-*T* deformations interfere.

A qualitative estimate of strain in the high-*T* peridotites shows that the most strained facies cluster in the two domains of thick MTZ rocks—the mantle window in the center of the massif and its western extension (Fig. 3).

Gabbro unit

The conspicuous layering of the lower gabbros is systematically parallel to the Moho and to the foliation in the uppermost mantle peridotites. The magmatic foliation, defined by planar plagioclase and streaky olivine aggregates, is parallel to this layering (Fig. 1A and 5C). In upper gabbros, the layering disappears and the magmatic foliation steepens, trending toward the attitude of the diabase sheeted dikes as in other massifs of the belt, except in the central synform where the layering is locally transverse with respect to the sheeted dikes. Steep and sheeted-dike parallel foliations in the upper gabbros have been interpreted as having been induced by magmatic flow parallel to the magma chamber walls (Nicolas and Boudier, 1995). Therefore, the locally transverse foliations are not understood. Mineral lineations carried by the magmatic foliation are parallel to the plastic flow-related lineations in the underlying mantle peridotites, at least in the lower part of the gabbro section (Figs. 1B and 5D). This behavior of magmatic foliations and lineations is the same as in the rest of the belt.

Shear zones

In the foregoing, we have given a general description of the major sinistral shear zones bounding the Aswad massif to the north (see also Gnos and Nicolas, 1995) and to the south (see also Reuber, 1988). Much smaller and more widespread shear bands affect the central part of the Aswad massif. They correspond to similar deformation conditions (~900 °C in the peridotites and 600 °C in the gabbros). They are mainly concentrated along a north-trending corridor that is also heavily hydrothermally altered (Fig. 1). In this corridor, the small shear bands are dominantly vertical, trend due north (Fig. 5E), and have a horizontal-stretching lineation (Fig. 5F), and all shear indicators are sinistral.

Dikes and veins

Plots of the preferred orientations of the various dikes and veins (Fig 4), in conjunction with the map (Fig. 1C), clearly show that all dikes and veins in a given domain tend to be parallel. In the most intact domains of the massif—the western domain exposing the uppermost mantle and the lower gabbros and the eastern domain exposing the upper level gabbros and diabase sheeted-dike complex—the dikes and veins are preferentially aligned due north, and they are steep. They constitute the north-trending subarea in the stereonet of Figure 4A. This subarea is accompanied by a subsidiary, north-northeast–trending alignment. In the central part of the massif, east of the north-trending thin corridor of intense shearing and hydrothermal activity, dikes and veins strike dominantly north-northwest. This 20-km-wide domain extends southward into the Wadi Hatta shear zone where the dikes have been rotated farther into a northwest strike. Together with the diabase sheeted-dike complex of this southeastern end of the massif, they are responsible for the northwest-trending subarea of Figure 4A. The general relationship between the local dominant orientation of the dikes and shear zones is presented in Figure 6.

DISCUSSION

Ridge structure

Ridge reference frame. As shown in Figures 1 and 3, in large parts of the massif, the Moho is horizontal and there is no need to rotate the orientation data to restore them to the ridge reference frame. The diabase sheeted-dike complex is vertical and trends due north in undisturbed areas, so the ridge of origin was also oriented due north. As a consequence, the maps of Figure 1 can be viewed as directly representing a piece of oceanic lithosphere, issued from a due north-trending ridge.

Possible diapir and ridge location. Following observations of diapirs made in other parts of the belt (Jousselin et al., 1998), there are several indications suggesting that a fossil mantle diapir is nestled below the mantle window in the center of the massif:

- A few steep, high-*T* foliations and lineations have been measured within this window. It is the only area in Aswad where steep, high-*T* structures are seen in the mantle unit. The location of the steep structures in the eastern part of the window and the dominant west-directed shear sense suggest that the window corresponds to the western flank of a possible diapir (Fig. 1B). As already mentioned, however, the window is too small and insufficiently eroded to fully support this interpretation.
- High-*T* lineations that trend nearly due west at the scale of the massif display some divergence away from the window: they have a bias toward west-northwest trends west of the window and east-northeast trends east of the window. In the northern part of the massif, a line might also separate west-directed shear senses to the west from mixed senses to the east. The band of mixed senses, together with a number of north-northwest–trending lineations, may correspond to the uppermost mantle just below the paleoridge axis.
- MTZ rocks thicker than 100 m and associated with chromite deposits of economic size are observed within the window. West of the window, a thick MTZ occurs in the wake of the presumed diapir, and sharply grades into an MTZ thinner than 100 m to the north and to the south. As predicted by Jousselin and Nicolas (2001), the large wehrlite bodies intrusive into the crust just outside the limit of the thick MTZ rocks could represent the products of the compaction of the thick and melt-impregnated MTZ.
- As already mentioned, on the basis of shear senses, the central part of the thick MTZ rocks with dominant west-directed shears can be distinguished on the basis of shear senses from two bordering corridors with opposite shear sense. In our models of active mantle flow (Nicolas et al., 1994), the mantle flow moving away from the diapir is particularly intense within the thick MTZ, as confirmed here by high-*T* plastic-strain estimations (Fig. 3). We speculate that the corridors may record some return flow toward the central MTZ.
- The crustal section is thinner above the mantle window than away from it (Fig. 2) in conformity with other diapirs in the belt (Nicolas and Boudier, 1995).
- Magmatic strain is absent or very weak in the lower gabbros from a north-northwest–trending central domain (Fig. 3), a situation exceptional throughout the entire belt. It is tempting to interpret this lack of strain as a consequence of the cessation of the chamber's magmatic flow because of the dying ridge activity. However, it must be noted that just below, the mantle window reveals that a large plastic flow regime was still very active at that time.

Spreading rate

There are a few indications from the gabbros suggesting that the Aswad ridge segment was spreading more slowly than the southern part of the belt. The main indication is the significantly weaker magmatic flow features in these gabbros compared to those in the southern part of the belt. This finding points to a less active and possibly smaller magma chamber, which in turn points to smaller melt budget and slower spreading. A thicker lid (diabase sheeted-dike complex + V_1 lavas) (2 km compared to 1.8 km on average in the south: Nicolas and Boudier, this volume), and a thicker diabase sheeted-dike complex root zone (~0.4 km vs. 0.2 km) suggest that the melt lens on top of the magma chamber could have been 0.4 km deeper than in the southern part of the belt. Therefore, we conclude that the spreading rate was smaller in Aswad, because marine data show an inverse

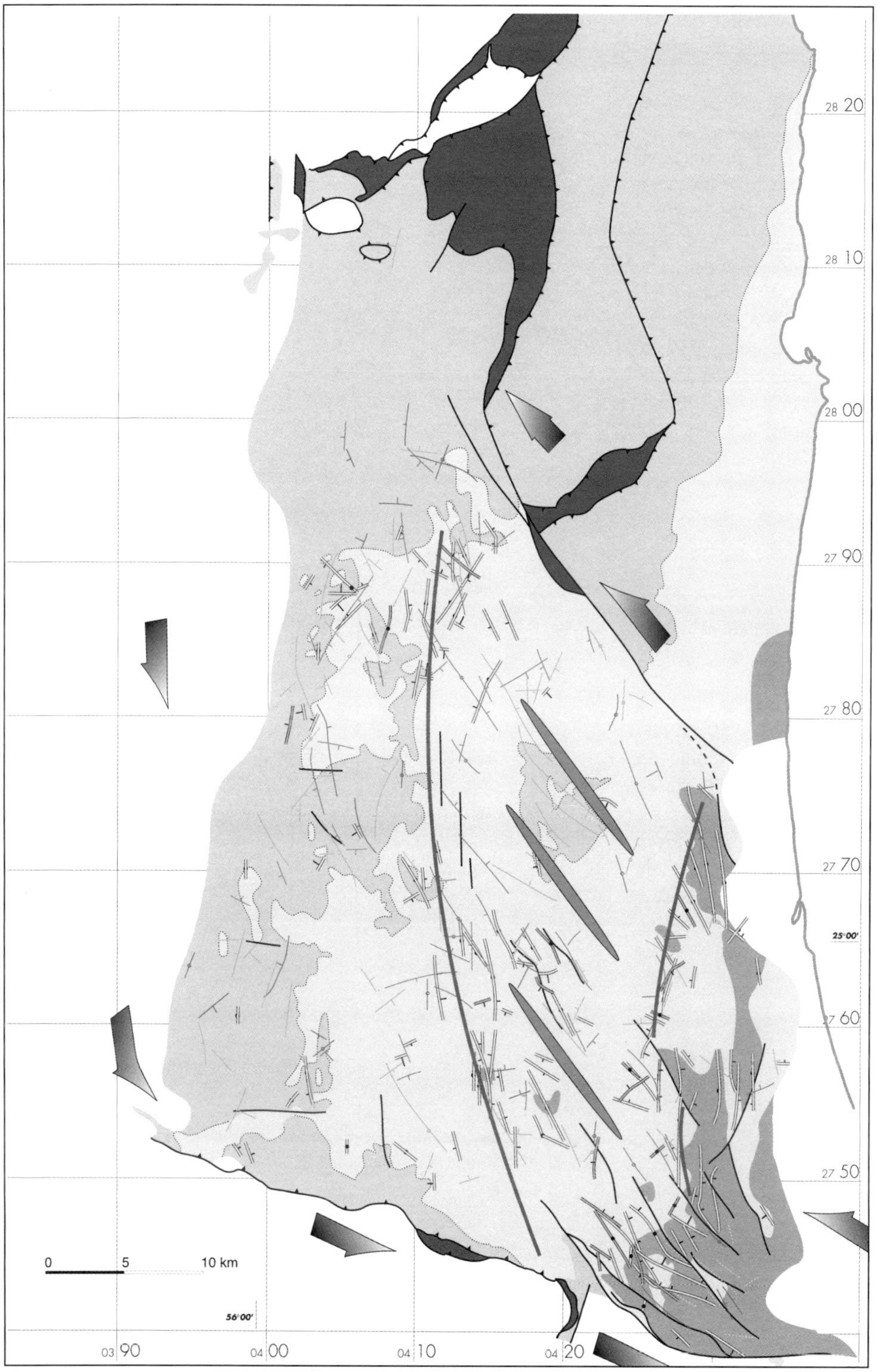

Figure 6. Map illustrating the relationship between the dominant directions of dikes and veins (see also Fig. 1C) and those of the main shear zones. Dark gray lenses represent en echelon extension bands. Arrows indicate the main shear (single-barbed arrow) and thrust (double-barbed arrow) motions. For lithology, refer to Fig. 1.

relationship between depth of the melt lens and spreading rate (Purdy et al., 1992; Chen and Phipps Morgan, 1996). Finally, from a mass-balance and heat-budget study at fast-spreading ridges, it has been proposed that faster spreading rates cause the injection of melt in the lower crust to form sills with graded-bedding (Chenevez and Nicolas, 1997). The absence of graded-bedding sequences in Aswad, in contrast with their presence in the rest of the belt, would imply slow spreading during the formation of the Aswad massif rocks.

Finally, we can indirectly estimate the spreading rate by comparing the size of the magma chamber shown by our mapping and modeling with the size deduced from seismic and thermal modeling of fast-spreading ridges. As already mentioned, the thick MTZ above the diapirs was sharply compacted sideways presumably when the soft constituent units came into contact with the more rigid lithosphere at the site of accretion (Jousselin and Nicolas, 2001). In this case we assume that the lithosphere is defined by the 1185 °C isotherm, which is the eutectic temperature of the gabbros and thus the temperature of the magma-chamber walls (Chenevez et al., 1998; Nicolas and Boudier, 1995). In Figure 1C, the sharp limit between the thick MTZ and the thin MTZ along the western front of the massif follows the NS0400 Lambert coordinate. It is 20 km west of the mantle window considered to belong to the ridge's axial diapir. Thus in Aswad, the magma chamber would extend laterally 20 km away from the axis. Marine geophysical investigations across the East Pacific Rise near 9°30′N, which is spreading at 11 cm/yr, have suggested the presence of melt in a domain that, depending on the technique used, extends between 4–8 km to 20 km on each side of the axis near the Moho (Garmany, 1989; Vera et al., 1990; Wilcock et al., 1995; Dunn and Toomey, 1997; Crawford et al., 1998). By reference to ridge thermal models that include a spreading-rate dependence (Morton and Sleep, 1985; Pelayo et al., 1994), the 20 km distance suggests that the Aswad ridge full-spreading rate was ~12 cm/yr. Spreading in the southern part of the belt should have been even faster.

Detachment-related tectonics

We ascribe detachment structures such as thrusts, shear zones, and associated dike and vein formation (distribution shown in Fig. 6) to ridge-related tectonics, for the following reasons:

- Shear zones were formed in the peridotites between 1000 and 800 °C and in the gabbros between 600 and 200 °C (amphibolite- to greenschist-facies conditions). In the light of cooling models for an oceanic ridge (Cormier et al., 1995), these temperatures correspond to a lithosphere ~1 m.y. old that has either drifted ~50 km from the axis, or cooled in situ for this period of time. The tectonic activity may have started even earlier, as the presence of magmatic hornblende and orthopyroxene in these normally olivine-bearing gabbros has been explained as a result of the introduction of water into the crystallizing lower gabbros in the central area (see the description of these rocks and Fig. 3).
- Sheeted dikes and hydrothermal veins were injected along a new northwest strike in the vicinity of shear zones (Fig. 6), suggesting that accretion was still occurring in the new tectonic context leading to emplacement. We locate the paleoridge axis above the mantle window, possibly along the north-northwest–trending extension of the Khawr Fakkan shear zone, and speculate that the shear zone formed along the weak ridge axis. The Fizh shear zone along the southern end of Aswad massif cuts this axis. It is thus possible that the thrusting and shearing between the Khawr Fakkan and Fizh massifs were initiated while the Aswad ridge was still active or dying.

The internal and more diffuse north-trending sinistral shearing may be a consequence of squeezing of the Aswad massif between the Khawr Fakkan and Fizh massifs. The locations of shear bands dominantly in the central part of the massif might also be related to proximity to the ridge axis. The dominantly northwest trend of all dikes, from mantle to diabase sheeted-dike complex in the central area—instead of due north as in the rest of the massif—suggests that the dying ridge activity occurred as the horizontal minimum principal stress rotated from east-west to northeast-southwest. The intrusion of obliquely oriented packets of diabase sheeted dikes in sharp intrusive contacts with upper gabbros is possibly also related to this emplacement-related and perturbed tectonic environment.

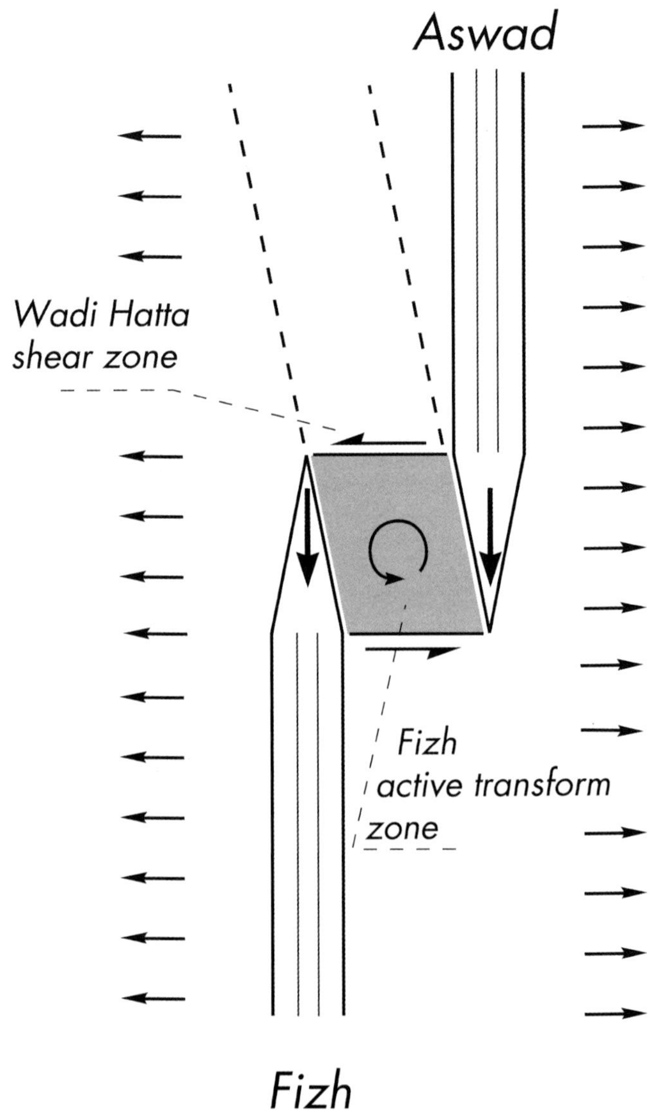

Figure 7. Model for right-stepped propagating and receding ridges (arrows point to tip displacement with time). Active shearing occurs in the overlapping domain. The track of the tips of the receding and propagating ridges (bold dashed lines) limit a transform-zone domain of past shearing. The shaded domain represents the active transform zone (modified from Hey et al., 1980).

Ridge propagation model

Although the location of the Aswad paleoridge in the frame of the Oman-UAE paleoridge system is discussed elsewhere (Nicolas et al., 2001), we speculate here on the possible relationship between the Aswad and Fizh ridge segments in order to derive a ridge model inspired by present-day fast-spreading systems.

From shear-sense indicators in the connected Hilti-Fizh massifs, which also trend due north, we conclude that the ridge axis of their origin was to the west, possibly along the front of the ophiolite nappe (Nicolas et al., 2001). The present data suggest that the Aswad ridge axis is located inside the massif, that is, east of the Fizh paleo–spreading axis. This discrepancy suggests that the Wadi Hatta and northern Fizh shear zones were initiated by nontransform ridge offset (Hey et al., 1980). Previous workers have already noted the abundance in the northern Fizh crust of amphibolite facies and sinistral shear zones, intense hydrothermal alteration, orthopyroxene-rich intrusions, and successive injections of diabase dikes accompanied by large rotations. This area has been successively interpreted as a leaky transform (Smewing, 1980), a large-offset overlapping spreading center (Reuber, 1988), and/or a propagating rift structure (MacLeod and Rothery, 1992). Our analysis complements these studies on the crust of the Fizh massif by tracing the crustal shear zones into kilometer-wide, low-*T* shear zones in the mantle section, also showing a dominant sinistral sense. We conclude that the domain extending mainly south of the Wadi Hatta shear zone and covering the northern Fizh massif could represent the transform zone related to the southward propagation of the Aswad segment at the expense of the Fizh doomed segment (Fig. 7).

CONCLUDING REMARKS

The Aswad massif in the UAE part of the Oman ophiolite belt has been described in detail because of its exemplary nature. We regard it as the best ophiolite model for a fast-spreading ridge segment. This massif differs only in a few ways from the other massifs of the UAE-Oman ophiolite:

- absence of graded-bedding sequences in the layered gabbros and in the MTZ at the expense of gabbroic and wehrlitic sills,
- minor imprint of magmatic deformation in the gabbros,
- abnormally large development of recrystallized amphibole gabbros at the top of the gabbro section, associated with a similarly exceptional development of the root zone of the sheeted-dike complex; it cannot be excluded that this development is accentuated by the remarkably flat attitude of this horizon inside the massif.

These differences are ascribed to the fact that the spreading rate in Aswad, though fast—estimated at ~12 cm/yr—was smaller than the spreading rate in the southern part of the Oman ophiolite belt.

The absolutely horizontal attitude of the Moho for 50 km from north to south is the most striking feature of this massif. It points to the rigidity of the thrust-fault–bounded ophiolite nappe, and confirms earlier conclusions about the limited tectonic deformation accompanying and postdating the UAE-Oman ophiolite obduction (Nicolas and Boudier, 1995; Nicolas et al., 2001).

Finally, we suggest that the Aswad massif contains a north-trending to north-northwest–trending ridge axis and an underlying mantle diapir. The Aswad ridge segment was possibly propagating southward and being separated by a west-northwest–trending transform zone from a doomed ridge segment farther west that was responsible for the accretion of the Fizh-Hilti massifs. Now limited by the Wadi Hatta fault and shear zone, the active transform zone was responsible during its southern migration for the dismembering by shearing and hydrothermal alteration of the northern part of the Fizh massif.

We extend this conclusion to the tectonic events that affected both the Khawr Fakkan and the Fizh massifs including the thrusting of the Khawr Fakkan massif over the northeast corner of the Aswad massif. On the basis of our analysis of the last high-*T* deformations accompanying the dike injections that are recorded in the center of the Aswad massif, we show that tectonic rotation and shearing occurred before the ridge segment accreting the Aswad massif died.

ACKNOWLEDGMENTS

This work has benefited from regular financial support by the Centre National de la Recherche Scientifique (CNRS) and, in Oman and the United Arab Emirates, from the willingness of the authorities and the hospitality of the people. Technical help in the laboratory is also acknowledged, including Emmanuel Ball for the drawings, Christophe Nevado for the thin sections, and Bernadette Marie for preparing the manuscript. We thank B. Hacker, Y. Dilek, and two anonymous referees for their help in improving the manuscript.

REFERENCES CITED

Benn, K., Nicolas, A., and Reuber, I., 1988, Mantle-crust transition zone and origin of wehrlitic magmas: Evidence from the Oman ophiolite: Tectonophysics, v. 151, p. 75–85.

Boudier, F., and Nicolas, A., 1988, The ophiolites of Oman: Tectonophysics, v. 151, p. 1–401.

Boudier, F., and Nicolas, A., 1995, Nature of the Moho transition zone in the Oman ophiolite: Journal of Petrology, v. 36, p. 777–796.

Boudier, F., Nicolas, A., Ildefonse, B., and Jousselin, D., 1997, EPR microplates, a model for the Oman Ophiolite: Terra Nova, v. 9, no. 2, p. 79–82.

Boudier, F., and Juteau, T., 2001, The ophiolite of Oman and United Arab Emirates, Marine Geophysical Researches, special issue, in press.

Boudier, F., Godard, M., and Armbruster, C., 2001, Significance of gabbronorite occurrence in the crustal section of the Semail ophiolite, Marine Geophysical Researches, special issue, in press.

Chen, Y.J., and Phipps Morgan, J., 1996, The effect of spreading rate, the magma budget, and the geometry of magma emplacement on the axial heat flux at mid-ocean ridges: Journal of Geophysical Research, v. 101, p. 11475–11482.

Chenevez, J., and Nicolas, A., 1997, Crustal feeding in the Oman ophiolite: From top and from bottom? A thermal and mass balance model: Geophysical Research Letters, v. 24, p. 1811–1814.

Chenevez, J., Machetel, P., and Nicolas, A., 1998, Numerical models of magma chamber in the Oman ophiolite: Journal of Geophysical Research, v. 103, p. 15443–15455.

Coleman, R.G., 1981, Tectonic setting for ophiolite obduction in Oman: Journal of Geophysical Research, v. 86, p. 2545–2556.

Cormier, M.H., Macdonald, K.C., and Wilson, D.S., 1995, A three-dimensional gravity analysis of the East Pacific Rise from 18° to 21°30′S: Journal of Geophysical Research, v. 100, p. 8063–8082.

Crawford, W.C., Webb, S.C., and Hildebrand, J.A., 1998, Estimating shear velocities in the oceanic crust from compliance measurements by two-dimensional finite difference modeling: Journal of Geophysical Research, v. 103, p. 9895–9916.

Dunn, R.S., and Toomey, D.R., 1997, Seismological evidence for three-dimensional melt migration beneath the East Pacific Rise: Nature, v. 388, p. 259–262.

Ernewein, M., Pflumio, C., and Whitechurch, H., 1988, The death of an accretion zone as evidenced by the magmatic history of the Sumail ophiolite (Oman): Tectonophysics, v. 151, p. 247–274.

Garmany, J., 1989, Accumulation of melt at the base of young oceanic crust: Nature, v. 340, p. 628–632.

Glennie, K.W., Boeuf, M.G.A., Hughes Clarke, M.W., Moody-Stuart, M., Pilaar, W.F.H., and Reinhardt, B.M., 1974, Geological map of the Oman Mountains, scale 1:500 000.

Gnos, E., and Nicolas, A., 1995, Structural evolution of the northern end of the Oman Ophiolite and enclosed granulites: Tectonophysics, v. 254, p. 111–137.

Hey, R.N., Duennebier, F.K., and Morgan, W.J., 1980, Propagating rifts on mid-ocean ridges: Journal of Geophysical Research, v. 85, p. 3647–3658.

Ildefonse, B., Nicolas, F., and Boudier, F., 1993, Evidence from the Oman ophiolite for sudden stress changes during melt injection at oceanic spreading centres: Nature, v. 366, p. 673–675.

Jousselin, D., and Nicolas, A., 2001, The Moho transition zone in Oman ophiolite-relation with wehrlites in the crust and dunites in the mantle: Marine Geophysical Researches, special issue, in press.

Jousselin, D., Nicolas, A., and Boudier, F., 1998, Detailed mapping of mantle diapir below a paleo-spreading center in the Oman ophiolite: Journal of Geophysical Research, v. 103, p. 153–170.

Juteau, T., Ernewein, M., Reuber, I., Whitechurch, H., and Dahl, R., 1988, Duality of magmatism in the plutonic sequence of the Sumail nappe, Oman: Tectonophysics, v. 151, no. 1/4, p. 107–135.

Lachize, M., Lorand, J.P., and Juteau, T., 1996, Calc-alkaline differentiation trend in the plutonic sequence of the Wadi Haymilyiah section, Haylayn massif, Semail ophiolite, Oman: Lithos, v. 38, p. 207–232.

Lippard, S.J., 1983, Cretaceous high pressure metamorphism in SE Oman and its relationship to subduction and ophiolite nappe emplacement: Journal of the Geological Society of London, v. 140, p. 97–104.

Lippard, S.J., Shelton, A.W., and Gass, I.G., 1986, The ophiolite of Northern Oman: Geological Society of London Memoir 11, 178 p.

MacLeod, C.J., and Rothery, D.A., 1992, Ridge axial segmentation in the Oman ophiolite: Evidence from along-strike variations in the sheeted dyke complex, *in* Parson, L.M., Murton, B.J., and Browning, P., eds., Ophiolites and their modern oceanic analogues: Geological Society of London Special Publication 60, p. 39–63.

Morton, J.L., and Sleep, N.H., 1985, A mid-ocean ridge thermal model: Constraints on the volume of axial hydrothermal heat flux: Journal of Geophysical Research, v. 90, p. 11345–11353.

Nehlig, P., and Juteau, T., 1988, Flow porosities, permeabilities, and preliminary data on fluid inclusions and fossil thermal gradients in the crustal sequence of the Sumail ophiolite (Oman): Tectonophysics, v. 151, p. 199–221.

Nicolas, A., 1989, Structures of ophiolites and dynamics of oceanic lithosphere: Kluwer Academic Publishers, Petrology and Structural Geology Series 4, 367 p.

Nicolas, A., and Boudier, F., 1991, Rooting of the sheeted dike complex in Oman ophiolite, *in* Peters, Tj., Nicolas, A., and Coleman, R.G., eds., Ophiolite genesis and evolution of the oceanic lithosphere: Norwell, Massachusetts, Kluwer Academic Publishers, Petrology and Structural Geology Series 5, p. 39–54.

Nicolas, A., and Boudier, F., 1995, Mapping oceanic ridge segments in Oman ophiolites: Journal of Geophysical Research, v. 100, p. 6179–6197.

Nicolas, A., and Boudier, F., 2001, Large mantle upwellings and related variations in crustal thickness in the Oman ophiolite: Marine Geophysical Researches, special issue, in press.

Nicolas, A., Reuber, I., and Benn, K., 1988, A new magma chamber model based on structural studies in the Oman ophiolite: Tectonophysics, v. 151, p. 87–105.

Nicolas, A., Boudier, F., and Ildefonse, B., 1994, Evidence from the Oman ophiolite for active mantle upwelling beneath a fast-spreading ridge: Nature, v. 370, p. 51–53.

Nicolas, A., Boudier, F., Ildefonse, B., and Ball, E., 2001, Accretion of Oman and United Arab Emirates ophiolite: Discussion of a new structural map: Marine Geophysical Researches, special issue, in press.

Pelayo, A.M., Stein, S.S., and Stein, C.A., 1994, Estimation of oceanic hydrothermal heat flux from heat flow and depths of midocean ridge seismicity and magma chambers: Geophysical Research Letters, v. 21, p. 713–716.

Penrose Conference Participants, 1972, Penrose field conference on ophiolites: Geotimes, v. 17, no. 12, p. 24–25.

Peters, Tj., Nicolas, A., and Coleman, R.G., 1991, Ophiolite genesis and evolution of oceanic lithosphere: Petrology and structural geology: Kluwer Academic Publishers, Petrology and Structural Geology Series 5, 903 p.

Purdy, J.M., Kong, L.S.L., Christeson, G.S., and Solomon, S.C., 1992, Relationship between spreading rate and the seismic structure of mid-ocean ridges: Nature, v. 355, p. 137–165.

Ravaut, P., Bayer, R., Hassani, R., Al Yahya'eh, A., and Rousset, D., 1997, Structure and evolution of the northern Oman margin: Gravity and seismic constraints over the Zagros-Makran-Oman collision zone: Tectonophysics, v. 279, p. 253–280.

Reuber, I., 1988, Complexity of the crustal sequence in northern Oman ophiolite (Fizh and southern Aswad block): The effect of early slicing?: Tectonophysics, v. 151, p. 137–165.

Rothery, D.A., 1983, The base of a sheeted dyke complex, Oman ophiolite: Implications for magma chambers at oceanic spreading axes: Journal of the Geological Society of London, v. 140, p. 287–296.

Smewing, J.D., 1980, An upper cretaceous ridge-transform intersection in the Oman ophiolite, *in* Panayioutou, A., ed., Ophiolites: Cyprus, Ministry of Agriculture and Natural Resources Geological Survey, p. 407–413.

Vera, E.E., Buhl, P., Mutter, J.C., Harding, A.J., Orcutt, J.A., and Detrick, R.S., 1990, The structure of 0–0.2 My old oceanic crust at 9°N in the Pacific Rise from expanded spread profiles: Journal of Geophysical Research, v. 95, p. 15529–15556.

Wilcock, S.D., Solomon, S.C., Purdy, G.M., and Toomey, D.R., 1995, Seismic attenuation structure of the East Pacific Rise near 9°30′N: Journal of Geophysical Research, v. 100, p. 24147–24165.

Manuscript Accepted by the Society February 23, 2000

Printed in the U.S.A.

Geological Society of America
Special Paper 349
2000

A new structural profile along the Muscat-Ibra transect, Oman: Implications for emplacement of the Samail ophiolite

David R. Gray
Department of Earth Sciences, Monash University, Melbourne, Victoria 3168, Australia; e-mail: dgray@artemis.earth.monash.edu.au
Robert T. Gregory
Stable Isotope Laboratory, Department of Geological Science, Southern Methodist University, Dallas, Texas 75275, USA; e-mail: bgregory@mail.smu.edu
John McL. Miller*
Department of Earth Sciences, Monash University, Melbourne, Victoria 3168, Australia

ABSTRACT

The Oman Mountains preserve evidence for a long-lived, kinematically coherent, tectonic "event" linked to the emplacement of the Oman ophiolite. Three windows (Hawasina, Jabal Akhdar, and Saih Hatat) hold the keys to the puzzle of the ophiolite obduction and the tectonic development of this part of the Arabian Peninsula. In the southeasternmost Saih Hatat window, where continental-shelf sequences have been subjected to blueschist- to eclogite-facies metamorphism, a major fold nappe and two major crustal-scale shear zones reveal the pre-emplacement geometry of the Arabian margin. They necessitate initial underthrusting toward the margin, rather than away from it as in most other published models. The stretching lineation is consistent in all of the units including the ?Pre-Cambrian Hatat Schist in the core of a regional fold nappe, the major shear zones and schistosity of the fold nappes, the high-pressure rocks beneath the fold nappe, and the high-temperature sole (intraoceanic thrust) of the uppermost nappe (Samail ophiolite). This consistency suggests that the deformation and metamorphism displayed in the Saih Hatat window are part of this long-lived event. The tectonic environment for ophiolite obduction involved initial subduction and high-pressure metamorphism of a continental slab (microplate) toward the former Arabian margin followed by neutral buoyancy of the slab, causing intraoceanic thrusting, and then buoyant rise and internal deformation of the slab along the former subduction interface. Slab rise led to gravitational instability of the partially obducted ophiolite, thinning along normal faults, and eventual gravity-driven emplacement of the Samail nappe onto the Arabian margin.

INTRODUCTION

Ever since Coleman (1971, 1977) introduced the term "obduction," the Oman ophiolite has been recognized as perhaps the best example of the emplacement of allochthonous ocean lithosphere onto the continental crust. In the Oman Mountains, former Tethyan oceanic lithosphere now sits on the Arabian continental margin (Fig. 1). It overlies Permian–Cretaceous shelf carbonates and allochthonous deep-water basinal sedimentary rocks that were "bulldozed" in front of and then overridden by the allochthonous ophiolite sheet. In 1981, Coleman suggested that the ophiolite formed near a spreading center and was emplaced in a two-stage process onto the Arabian margin (see Fig. 6 in Coleman, 1981). The existence of high-pressure (*P*) metamorphic rocks in the Saih Hatat window is an important constraint on this overall process (see Lippard, 1983; Lippard et al., 1986; Goffé et al., 1988; El Shazly and Coleman, 1990; El Shazly et al., 1990; Searle et al., 1994; Gregory et al., 1998; Miller et al., 1998).

Several workers (e.g., Searle, 1985; Mann and Hanna, 1990; Hanna, 1990) interpreted the structure of the Oman Mountains with a classic thrust-belt geometry at a time when thrusting was a popular paradigm. The style of emplacement of the ophiolite has therefore been linked to thrusting scenarios. It is interesting that the Oman and Bay of Islands ophiolites share some common elements including (1) present geometry as a thin sheet (<5 km thickness) bounded by a relatively planar basal fault, (2) structural thickening of ophiolite sheet by folding, (3) ophiolitic folds are truncated by a basal fault in a manner more typical of extensional fault systems (cf. Gregory et

*Present address: School of Earth Sciences, University of Melbourne, Melbourne, Victoria 3052, Australia; e-mail: jomiller@unimelb.edu.au

Gray, D.R., Gregory, R.T., and Miller, J.McL., 2000, A new structural profile along the Muscat-Ibra transect, Oman: Implications for emplacement of the Samail ophiolite, *in* Dilek, Y., Moores, E.M., Elthon, D., and Nicolas, A., eds., Ophiolites and Oceanic Crust: New Insights from Field Studies and the Ocean Drilling Program: Boulder, Colorado, Geological Society of America Special Paper 349, p. 513–523.

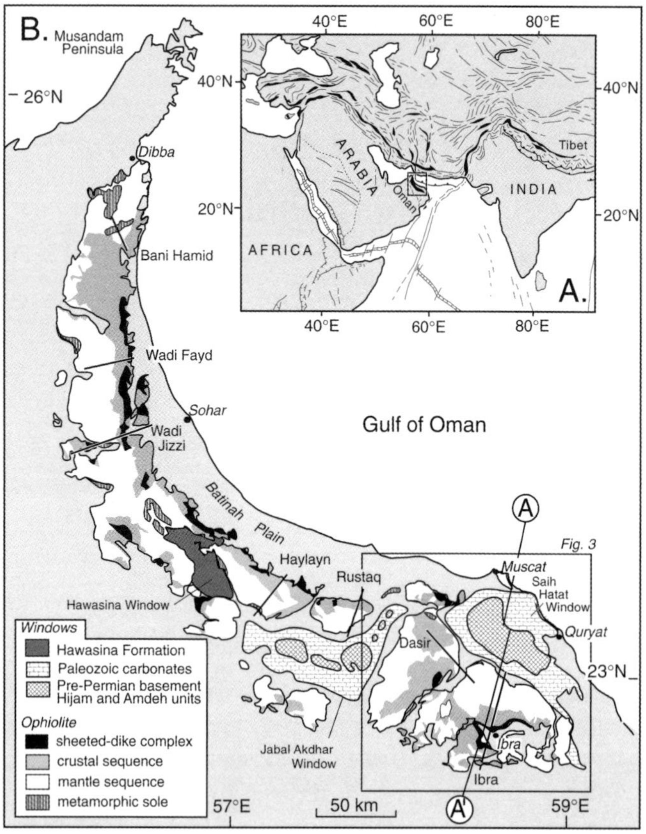

Figure 1. Map of the Samail ophiolite showing internal stratigraphy of the ophiolite and locations of the metamorphic sole and the Hawasina, Jabal Akhdar, and Saih Hatat windows (modified from Fig. 1 of Boudier et al., 1988, and Fig. 1 of Hacker and Gnos, 1997). Location of the regional structural profile A–A′ of Figure 2 is shown, as well as the section lines across the ophiolite of Figures 3 and 5. Inset shows the location of Oman in the tectonic setting of other Tethyan ophiolites (black) and the structural grain (lines) (modified from Fig. 1 of Coleman, 1981).

al., 1998, for Oman with Williams and Smyth, 1973, for the Bay of Islands), and (4) intraoceanic thrusting within young (<2 m.y.), hot oceanic lithosphere indicated by the contact-aureole assemblages.

In Oman, high-*P* metamorphism of the underlying shelf rocks provides new constraints for the obduction history. Oman also provides an unparalleled opportunity to investigate the obduction process because of the exceptional exposure and the fact that the rocks beneath the ophiolite are exposed in windows that give access to lower structural levels beneath the ophiolite nappe. The windows expose a complex of variously deformed and metamorphosed oceanic and continental sequences. They show different parts of the tectonostratigraphy and preserve different segments of the tectonic history.

Mechanisms of obduction of oceanic lithosphere, the nature of the processes involved, and the tectonic setting remain controversial (e.g., Coleman, 1971, Moores, 1982, Nicolas, 1989). Most obduction models for Oman involve overthrusting of oceanic lithosphere due to subduction dipping away from the continental margin (e.g., Gealey, 1977, Lippard et al., 1986, Chemenda et al., 1996, Searle and Cox, 1999). Evidence is based on structural geometry (e.g., Mattauer and Ritz, 1996; Fig. 8 of Chemenda et al., 1996) as well as supra-subduction affinities of the ophiolite (e.g., Fig. 2 of Searle and Cox, 1999). New detailed structural mapping in the Saih Hatat window requires at least two major, southwest-dipping, crustal-scale shear zones that potentially root into the Moho. The presence of these shear zones necessitates convergence, involving *underthrusting* directed toward the Arabian continental shield for at least part of the geologic evolution of the Oman Mountains (see Gregory et al., 1998). The significance of this southwestward-directed underthrusting for the tectonic evolution of this part of the Arabian peninsula is the focus of this paper.

THE OMAN MOUNTAINS

The Samail nappe is draped over the Oman Mountains as the highest allochthonous sheet in the structural pile (Fig. 1). The Samail nappe now consists of a segmented sheet ~600 km in length, as much as 150 km in width, and between 5 and 10 km in thickness (Nicolas et al., 1988; Boudier et al., 1988; Shelton, 1990). In general, the inferred original oceanic lithospheric thickness of the ophiolite is greater than the present-day nappe (e.g., Dasir and Ibra blocks, Fig. 1, where it may be <5 km in thickness). Major geologic units of the Oman Mountains include *allochthonous units*—the Samail nappe (remnants of oceanic lithosphere), Hawasina nappe (remnants of ocean-basin sedimentary rocks), and Sumeini nappe (remnants of continental shelf); *the para-autochthonous units*—the Hajar Supergroup consisting of middle Permian–Cenomanian continental-shelf units (dominated by carbonate) of the former Arabian margin and pre-Permian basement units of the Arabian Shield (exposed in windows associated with Jabal Akhdar and Saih Hatat); and *onlapping Maastrichtian to Miocene sedimentary rocks* dominated by shallow-water carbonates locally conglomeratic at the base (post–ophiolite emplacement).

Topographically, the Oman Mountains (Fig. 1) are dominated by a series of en echelon antiformal domes, the largest of that are the Saih Hatat and Jebal Akhdar domes. These domal structures are now eroded to form the series of windows (Glennie et al., 1974; Searle, 1985). The domal culminations are interpreted to be ramp-related anticlines (Searle, 1985; Mann and Hanna, 1990; Cawood et al., 1990; Mount et al., 1998). In a southeastward direction, the windows in turn expose deeper structural levels of the Oman Mountains. The northern Hawasina window features structures within the Hawasina nappe immediately beneath the ophiolite. The central part the Jabal Akhdar window exposes the upper limb of a regional nappe structure that is fully exposed within the upper part of the Saih Hatat window (see Le Métour et al., 1990). The southernmost Saih Hatat window preserves the deepest structural levels within the Oman Mountains (Fig. 2). A major detachment surface is exposed within the northern part of the Saih Hatat window; the detachment separates the higher-*P* blueschist and eclogite of As Sifah (located in Fig. 3) from less metamorphosed upper-plate rocks involving nappes developed in the pre-Permian basement and Hajar Supergroup (see Gregory et al., 1998; Miller et al., 1998).

The uplift and present-day morphology of the Oman Mountains is due to Tertiary (post-Maastrichtian) Arabian Shield–directed overthrusting and folding (Glennie et al., 1974; Mount et al., 1998). Glennie (1995) proposed 30–35 m.y. of uplift and erosion during the later Cenozoic. Apatite fission-track data from a cobble from the Precambrian Mistal Formation limits the uplift history of the Oman Mountains to the Oligocene (Mount et al., 1998). Modeling of the data suggests that the most likely thermal history consists of two stages: an initial rapid-cooling phase at ca. 30 Ma (from 120 to 60 °C) followed by a second phase of slow cooling from 25 Ma to the present. Recent uplift is shown by several levels of terraces in the

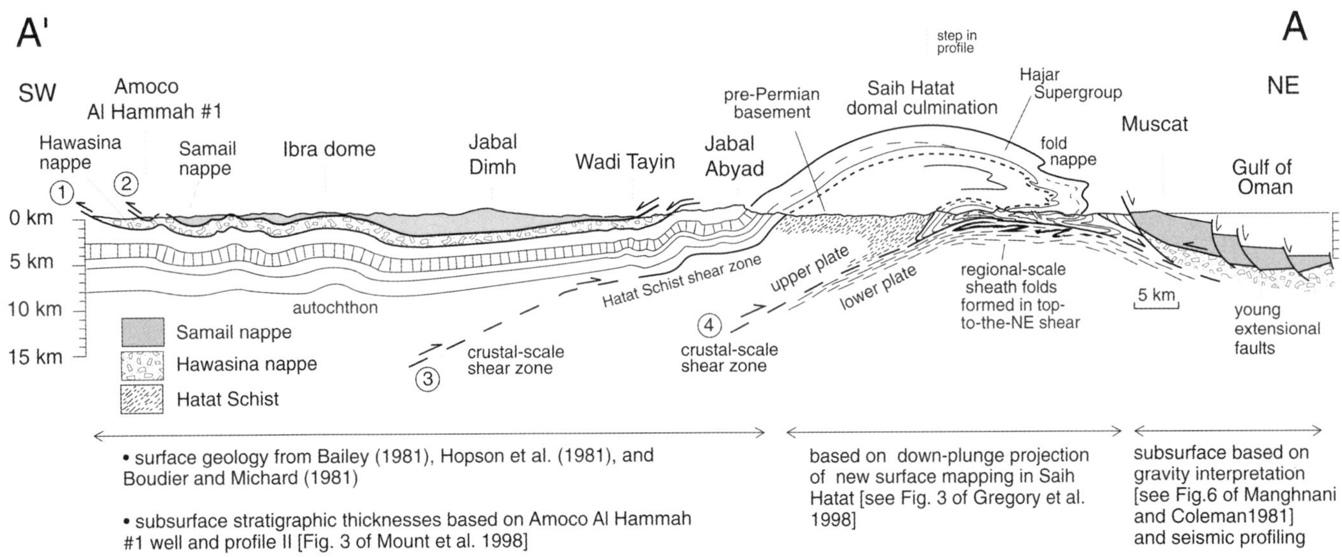

Figure 2. Regional structural profile A–A′ across the Saih Hatat window and Ibra block of the Samail ophiolite (see Fig. 1 for location). The section is modified from Bailey (1981), Hopson et al. (1981, Fig. 1 therein), Boudier and Michard (1981), and Gregory et al. (1998, Fig. 7 therein). Key elements are the (1) Hawasina and (2) Samail nappes, the Ibra dome, the Saih Hatat domal culmination and regional-scale fold nappe, (3) and (4) the crustal-scale shear zones, and the normal faults off the Oman coast.

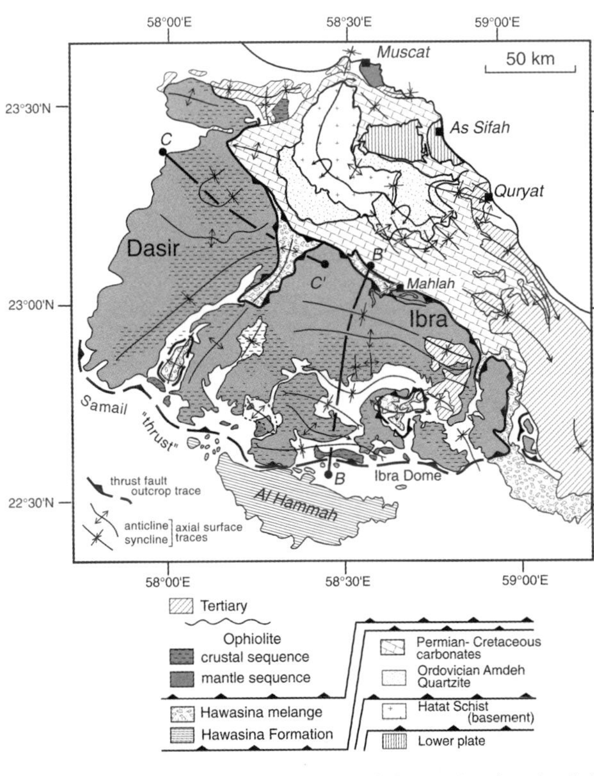

Figure 3. Map of the northeast corner of Oman showing the Saih Hatat window and the Dasir, Ibra, and Muscat ophiolite blocks. Axial-surface traces of Tertiary folds and the locations of the ophiolite structural profiles of Figure 5 are shown. Prominent structures are the Saih Hatat window (northeast), the Ibra dome (south), and refolded folds both in the Dasir (northwest) and Ibra (south) blocks. The ophiolite shows northeast-trending folds overprinting northeast-trending folds. Note that the fold axial-surface traces in Saih Hatat are collinear with those in the Tertiary rocks south of Quryat. Fold interference also occurs in the Tertiary rocks west of Muscat where east-trending folds are overprinted by north-trending folds.

interior of the Saih Hatat dome and by raised strandlines particularly along the coast east of Muscat, between Sur and Quryat.

THE OPHIOLITE NAPPE

Geometry

The ophiolite is now preserved as a thin sheet bounded by a flat-lying basal fault (Shelton, 1984, 1990; Gregory et al., 1998). Even though the pseudostratigraphy generalized for ophiolites indicates a lithospheric structure at least 10 and possibly 20 km thick, the present-day Samail nappe (Figs. 3, 4) is much thinner (see also Table 1 of Shelton, 1990). This discrepancy requires the structure of the ancient oceanic lithosphere to be truncated by the Samail "thrust" (B.M. Rheinhardt in Glennie et al., 1974; Hopson et al., 1981; Shelton, 1990). The present basal fault crosscuts the phyllitic contact-aureole blocks (metamorphic sole vestiges) that are preserved sporadically as isolated remnants along the contact between the ophiolite and the underlying units (Fig. 2).

The ophiolite thrust nappe is broken into approximately one dozen blocks separated by cross-strike faults or by imbricate thrusts (Fig. 1). There is marked structural thinning and imbrication along the western and southern edges of these blocks (Boudier et al., 1988). Gravity modeling suggests that they are tapered at their leading and trailing edges (Shelton, 1990). In the northern part of the Oman Mountains, the ophiolite nappe, as shown by the intact oceanic-crust pseudostratigraphy, is a northeast-dipping and northeast-younging, homoclinally dipping sheet that is folded in part (Fig. 4).

In the southeastern part, the ophiolite nappe contains a series of folds that themselves are refolded into dome and basin structures (Fig. 3). The folds generally tend to have east-west axes, inclined axial surfaces, and overturned and attenuated limbs to the south. What is most important is that the earliest generation of folds appears to be truncated by the Samail "thrust," which does not appear to be folded to the same degree (Fig. 5).

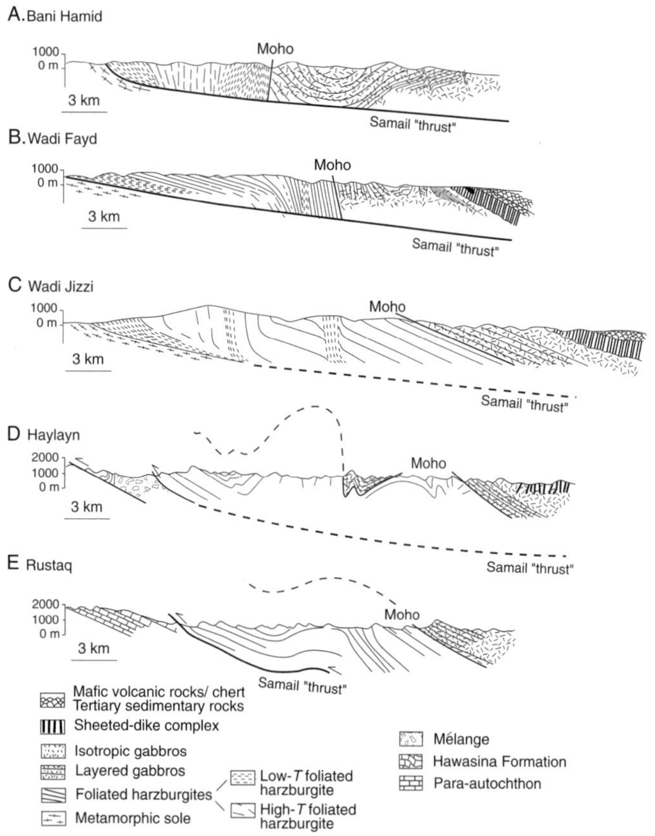

Figure 4. Structural profiles across the Oman ophiolite. A, Bani Hamid. B, Wadi Fayd. C, Wadi Jizzi. D, Haylayn. E, Rustaq. Profiles A–C are from Boudier et al. (1988, Fig. 9, A and B, therein) and profiles D and E are from Nicolas et al. (1988, Fig. 9, A and B, therein). See Figure 1 for locations of profiles.

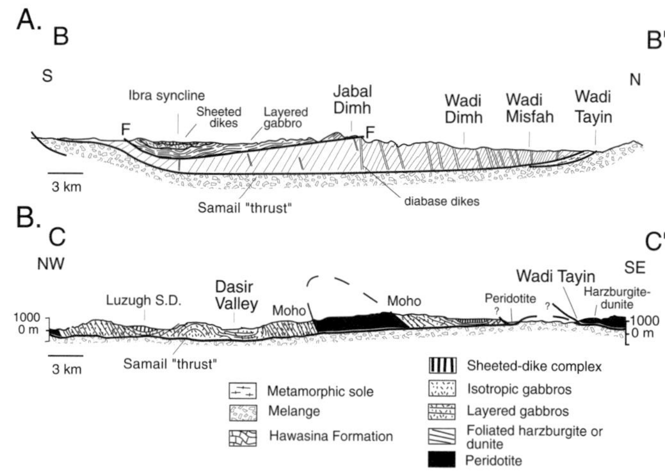

Figure 5. Structural profiles across (A) the Ibra and (B) Dasir-Ibra blocks of the Samail ophiolite. See Figure 3 for locations. Profile B–B′ from Boudier and Michard (1988) and profile C–C′ is from Gregory et al. (1998, Fig. 3 therein).

Metamorphism

There is a general lack of any high-*P*, low-*T* (temperature) metamorphism in any part of the preserved Samail nappe. Changes in secondary-mineral assemblages are correlated with oxygen isotope profiles (e.g., Gregory and Taylor, 1981) resulting from seafloor hydrothermal alteration. These alteration assemblages are correlated with original depth below the seafloor and predate the folding. The ophiolite pseudostratigraphy provides some insights into the conditions that might have been present within the lithosphere shortly before the ophiolite's emplacement.

Along the Ibra transect (B–B′, Figs. 3 and 5) south of Saih Hatat, the Jabal Dima section is the most complete section of the ophiolite. Here it consists of a 5–8-km-thick crustal section of pillow lava, sheeted dikes, and gabbro bounded by by a 9–12-km-thick peridotite section (Hopson et al., 1981; Boudier and Coleman, 1981; Gregory, 1983). The deepest levels in the lithosphere are inferred to be a harzburgite-dunite zone consisting of isoclinally folded layers (1–5 m scale) of harzburgite and dunite. The minimum estimate of ~10 km for the lithospheric thickness requires that the 5 kbar isobar is reached at the base of the section (Gregory, 1983). The contact-aureole rocks welded to the harzburgite-dunite zone at Wadi Tayin provide insights into the temperature state of the lithosphere at the time the ophiolite became allochthonous.

Contact-metamorphic sole: Conditions of metamorphism

The metamorphic sole at the base of the ophiolite consists of small, tectonically bounded blocks of high-*T* garnet amphibolite–facies metamorphic rocks (Searle and Malpas, 1980; Ghent and Stout, 1981; Montigny et al., 1988; Hacker, 1990, 1991; Hacker et al., 1996; Hacker and Gnos, 1997; Hacker and Mosenfelder, 1997; Gnos, 1998). The conditions of metamorphism are not precisely defined within the high-*T* amphibolites. Temperature and pressure estimates range from 450 to 850 °C and from 4 to 15 kbar, respectively (Ghent and Stout, 1981; Gnos, 1998; Searle and Cox, 1999). Gnos (1998) identified kyanite in the metamorphic sole at Asimah (northern Oman) for which he inferred pressures of >11 kbar associated with peak metamorphic temperatures of 450–550 °C on the basis of magnesio-hornblende + albite assemblages (Bucher, 1991).

Gnos (1998) calculated a pressure of 11 ± 2 kbar from the sole at Tayin. This pressure is approximately a factor of two higher than the pressure that could be produced by the full thickness of the preserved ophiolite pseudostratigraphy. Gregory (1983) estimated that the 5 kbar isobar was exceeded at the base of the ophiolite at Wadi Tayin on the basis of lithologic zonations within the peridotite section. The basal harzburgite-dunite zone represents the deepest level of exposure, and these original mantle structures are crosscut by the second-stage Samail "thrust". The foliation in the metamorphic sole at Wadi Tayin is roughly conformable with the overlying basal harzburgite-dunite zone.

Structure of the Wadi Tayin metamorphic sole

The metamorphic sole is an amphibolite-facies high-strain zone that represents the early intra-oceanic thrusting phase of ophiolite obduction (Boudier et al., 1982; Michard et al., 1984). It is now preserved as a series of relict slices along the basal fault of the ophiolite. West of Mahlah (Fig. 3) at the Green Pool locality, a ~230m thick metamorphic slice consists of structurally intercalated hornblende + clinopyroxene + garnet gneiss and schistose hornblende + plagioclase gneiss that are isoclinally folded with calc-silicate and quartzite (metachert) layers locally containing piedmontite (R.G. Coleman and R.T. Gregory in Ghent and Stout, 1981; Hacker and Mosenfelder, 1996). Mineral lattice preferred orientations of fabrics within the amphibolitic rocks (Michard et al., 1984; Boudier et al., 1988) indicate that the metamorphic sole developed while the ophiolite was still hot enough to produce pyroxene, garnet, and amphibole in the mafic schists (i.e. during the intraoceanic thrusting stage). This interpretation is supported by lattice preferred orientations of quartz in the intercalated quartzites that indicate a dominantly simple-shear regime with a sense of east-northeast over west-southwest, consistent with that determined from orthopyroxene in the basal, mylonitic peridotite (see Fig. 4 of Boudier et al., 1982).

The amphibolite-facies rocks are underlain by an ~150-m-thick lower zone of greenschist and phyllite. These rocks appear infolded with the quartzite-amphibolite package by a large, north-closing isoclinal fold best shown in the metachert layering (Fig. 8 of Gray and Gregory, in press). Structural breaks within this strongly foliated package are marked by discordances in foliation and truncation of layering. Foliation in the amphibolites adjacent to the peridotite contact is approximately parallel to the contact, but is locally truncated by it. The rodingitic assemblages present as boudins along the contact indicate later movement.

Within the Mahlah slice there appear to be two stretching-lineation directions (Fig. 3 of Boudier et al., 1988). An E-NE lineation in the quartzite and amphibolitic gneisses had been considered the oldest and the highest-temperature fabric that dates to older than 90 Ma. A more NE lineation occurs in the lower grade phyllitic rocks and was therefore considered a younger fabric element. Recent mapping of the Mahlah slice (Gray and Gregory, in press) suggests, however, that the changing lineation is a consequence of early, isoclinal folding within the amphibolitic gneisses combined with later younger, NW-trending folding that folds the ophiolite, the metamorphic sole, and the Hawasina complex. The change from a more NE to ENE-trending lineation appears to coincide with the axial surface trace of the major, north-closing isoclinal fold closure within the quartzite units, suggesting that the lineation is folded.

BENEATH THE OPHIOLITE: NEW INSIGHTS FROM THE SAIH HATAT WINDOW

Mapping along the northern part of the Ibra-Muscat transect (Gregory et al., 1998; Miller et al., 1998) and Ar-Ar geochronology (Miller et al., 1999) provide new insights into the evolution of the high-*P* rocks of Saih Hatat as well as implications for the evolution of the mountains. In contrast to other parts of the autochthonous continental margin exposed in the Oman Mountains, the northeastern part of the Saih Hatat window shows evidence for high-*P* blueschist- to eclogite-facies metamorphism accompanied by intense deformation involving nappe development.

Structure

A marked southwest to northeast deformation gradient exists across the Saih Hatat structural dome (Le Métour et al., 1990). On the southwestern side of the dome, limestones and dolomites of the Hajar Supergroup are homoclinally dipping and almost undeformed (Figs. 2, 3). The deepening of the structural level toward the northeast is accompanied by a strong increase in fabric development, culminating in a pervasive schistosity and regional fold nappes. These changes occur because the Saih Hatat domal culmination is cored by an asymmetric east-facing fold nappe whose upper limb is pinned to the relatively undeformed strata of the autochthon and whose lower limb is isoclinally folded and intensely deformed (Fig. 2).

Fold kinematics require top-to-the-northeast–directed shear, opposite to the emplacement direction of the ophiolite. The strain gradient in the fold nappe increases toward a major shear zone that defines distinct upper and lower plates within the window (Gregory et al., 1998; Miller et al., 1998). Kinematic indicators (shear bands, tails on porphyroclasts, and pressure shadows around porphyroblasts) require top-to-the-northeast–directed shear. This is accompanied by increasing strain, increasing pressure of metamorphism, and a change to isoclinal, sheath-like fold geometry with markedly attenuated fold limbs (see Fig. 3B of Miller et al., 1998).

At the structurally lowest levels, in the lower plate, megaboudins of eclogite-facies metabasalts are enveloped by the regional schistosity that is parallel to the axial surfaces of regional-scale sheath nappes. The megaboudins contain early east-trending lineations associated with a west-over-east sense of shear. The lower-plate structure is a pervasive schistosity and stretching lineation associated with the formation of regional nappes of sheath-like form (see Fig. 2, A and B, of Miller et al., 1998). The earlier, higher-grade assemblages are folded around these nappes, whereas the lowest-grade assemblages are synchronous with the formation of regional-scale sheath-like nappes in the As Sifah region (Figs. 3 and 4 of Miller et al., 1999). In the Saih Hatat window, the stretching lineation trends consistently south-southwestward within the upper and lower plates, the Hatat Schist (the basement), and the metamorphic sole of the ophiolite.

A marked deformation gradient between the Hatat Schist, which cores the antiformal closure, and the relatively undeformed Permian–Cretaceous shelf carbonates on the northwest side of the dome (i.e., the upper limb of the fold nappe) requires a fault break at this level in the structure (Fig. 2). This detachment or break must either have predated or been synchronous with the development of the fold nappe, as it is clearly folded by it (see Fig. 7 of Gregory et al., 1998). The Hatat Schist is folded into the core of the fold nappe and the schistosity, enclosing lineation, and degree of strain match those of the intensely deformed rocks of the nappe's lower limb. The intensity of deformation in the Hatat Schist is relatively uniform across the window, particularly in the northern half of the dome. This break either roots into an undeformed basement-cover contact to the southwest (i.e., toward the craton that forms the Arabian Shield) or is part of a throughgoing crustal-scale shear that appears to deepen and intersect the Moho to the southwest (see Fig. 2). This shear system also includes the upper-plate–lower-plate shear-zone discontinuity.

Metamorphism

The Saih Hatat window had formerly been subdivided into three regions of different metamorphism (see Fig. 2 of El Shazly and Coleman, 1990). The definition of a major crustal shear zone within the window (see Gregory et al., 1998), however, has simplified the interpretation of the metamorphic rocks into two major regions—defined as the upper (formerly regions I and II) and lower (formerly region III) plates—that have different tectono-metamorphic histories (Miller et al., 1998) (see Fig. 3). Carpholite-bearing assemblages in the upper plate versus glaucophane-bearing eclogites within mafic megaboudins in the lower plate reflect a possible pressure difference of at least 13 kbar. Metamorphic temperature and pressure ranges for the upper plate are 315–435 °C and 6.8–9 kbar (El Shazly, 1994, 1995), whereas for the lower plate, they are 500–580 °C and >12 kbar (El Shazly et al., 1990); $P > 20$ kbar for the lower plate has been suggested by Searle et al. (1994). Field relationships, as well as new Ar-Ar geochronology on white mica separates from fabric elements within the two plates, have shown that the upper and lower

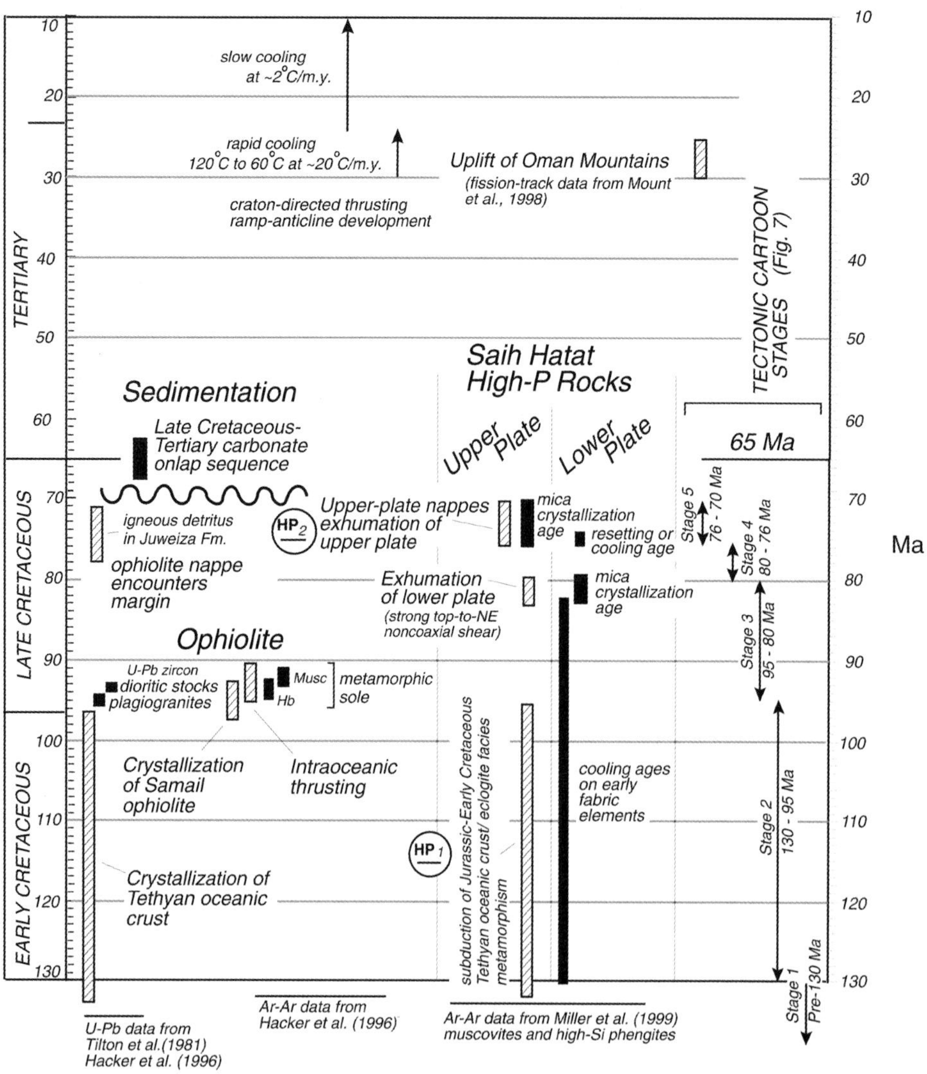

Figure 6. Time and space diagram illustrating the major timing constraints for ophiolite emplacement, sedimentation, and the high-*P* metamorphism.

plates have undergone metamorphism at different times and at different structural levels during the margin evolution (Miller et al., 1998, 1999; see HP_1 and HP_2, Fig. 6; the specifics of the obduction model shown in Fig. 7 are discussed later).

TIMING RELATIONSHIPS

The ophiolite

The igneous crystallization age of the Samail ophiolite is ca. 95 Ma on the basis of U-Pb zircon ages from plagiogranites (Tilton et al., 1981); biostratigraphic data (Tippit et al., 1981) suggest that it formed at a Tethyan spreading center between 94 and 97 Ma (Fig. 6). Dated plagiogranites mostly have crystallization ages between 95.4 and 94.5 ± 0.5 Ma (Tilton et al., 1981; Hacker et al., 1996, p. 1241). These ages are concordant with fossil ages from the lower and middle volcanic rocks. Radiolaria in the lower volcanic rocks (V_1) are Cenomanian (98.9–93.4 Ma), whereas those in the middle volcanic member (V_2) are Turonian (93.5–89.0 Ma).

The ophiolite was detached via intraoceanic thrusting along the metamorphic sole at ca. 94–93 Ma ($^{40}Ar/^{39}Ar$ on hornblende from amphibolite-facies tectonite; Hacker et al., 1996, and references therein). Lower-grade phyllitic, mica-bearing cherts and calc-silicates from the sole (Fig. 6) have Ar-Ar muscovite ages of 90.9 ± 1.5 to 93.6 ± 1.5 ($^{40}Ar/^{39}Ar$; Hacker et al., 1996). Movement on the "sole" was originally considered to have lasted at least into the late Campanian and perhaps into the Maastrichtian, but Hacker et al. (1996) have shown that a weakly deformed, greenschist-facies diabase dike that cuts the metamorphic sole has an Ar-Ar age of 93.7 ± 0.8 Ma. This clearly postdates the amphibolite-facies metamorphism and the major deformation of the metamorphic sole. The age also indicates that amphibolite-facies deformation ended within 1–2 m.y. of plagiogranite crystallization (Hacker et al., 1996).

Igneous detritus within the middle to late Campanian Juweiza Formation signals arrival of the ophiolite nappe onto the margin in the period 78–71 Ma (Warburton et al., 1990), with final emplacement by ca. 68 Ma (Fig. 6). Upper Maastrichtian (ca. 68–67 Ma) autochthonous sedimentary rocks unconformably overlie all sequences, indicating a minimum age of emplacement (Coleman, 1981).

Blueschist and eclogite metamorphism in Saih Hatat

$^{40}Ar/^{39}Ar$ geochronology on white mica populations from various fabric elements at different structural levels within Saih Hatat (Miller et al., 1999) has shown the following: (1) Phengites and muscovites from older fabrics preserved in lower-plate eclogite boudins

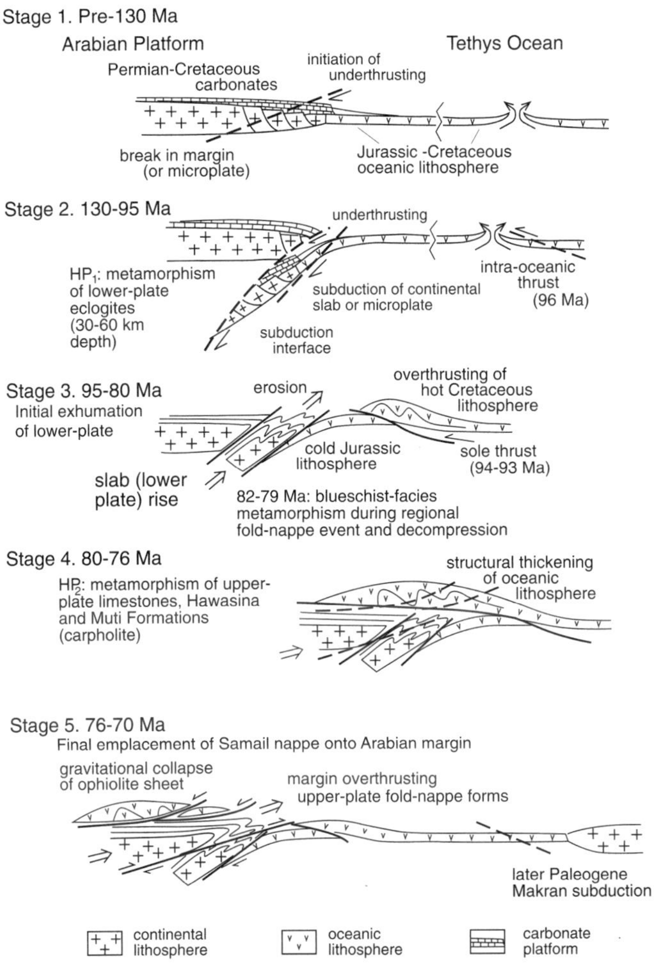

Figure 7. Tectonic cartoons showing the evolution of the Arabian continental margin (modified from Fig. 8 of Gregory et al., 1998). Steps include the initial closure of the ocean by margin-directed subduction, subduction of a part of the margin to depths of 35–50 km with subsequent exhumation, oceanic thrusting of the ophiolite with convergence toward the craton forming the Arabian Shield, rise of the "buoyant" high-*P* slab, and eventual gravitational emplacement of the ophiolite onto the margin by pseudo-extensional faulting (Samail basal fault).

show older cooling ages (131–82 Ma), some of which are not due to excess argon. (2) Micas from late, superimposed fabrics associated with lower-*P* metamorphism in the lower plate (e.g., calc-schists with shear bands that enclose the eclogite megaboudins) have tightly determined cooling ages of 82–79 Ma. (3) Micas from fabric elements in the upper plate associated with fold nappes show crystallization ages of 76–70 Ma.

These data require the blueschist and eclogite metamorphism (HP_1) of the lower plate to predate ophiolite emplacement (see Fig. 6). Initial cooling of the lower-plate eclogites below ~400 °C from temperatures of 550–580 °C is related to underthrusting of colder units in an inferred subduction zone possibly as early as 119–115 Ma. Continued underthrusting led to cooling of the lower plate between 95 and 82 Ma. Partial exhumation of the lower plate, recorded by cooling ages of 82–79 Ma, is highlighted by a progressive drop in metamorphic grade around a consistent southwest-trending stretching lineation related to folding and transposition of the lower-plate, high-*P* fabrics.

The geochronologic data also indicate that lower-*P* (HP_2) metamorphism of the upper plate is associated with emplacement of the Samail ophiolite and in particular with development of regional-scale nappes that evolved from 76 to 70 Ma during a northeast-directed shearing event that occurred while the overall lithostatic pressure was decreasing. It resulted in the juxtaposition of the higher-grade lower-plate units against an upper plate consisting of nappes in mainly carbonate platform rocks. This upper-plate, nappe-forming event was synchronous with structurally higher tectonic thinning and rapid erosion of the nappe pile prior to deposition of Maastrichtian autochthonous units at ca. 68–67 Ma.

STRUCTURAL ARGUMENTS FOR GEOMETRY OF THE ARABIAN MARGIN DURING OPHIOLITE OBDUCTION

Shear zones and a regional fold nappe in the core of the Saih Hatat domal culmination not only have important consequences for the distribution and evolution of the high-*P* metamorphic rocks (see Gregory et al., 1998; Miller et al., 1998), but also for the pre-emplacement structural evolution of the mountains. Most structural and/or tectonic interpretations have either not realized or ignored the presence of the fold nappes in Saih Hatat and the shear zones that separate them (e.g., Fig. 12 of Hanna, 1990; Searle et al., 1994; Searle and Cox, 1999). The direction that these shear zones root into the Moho is crucial to any tectonic reconstruction.

In removing the effects of the Tertiary doming across Saih Hatat, the shear zones within the window must root into the Moho to the southwest (compare with Fig. 17B of Goffé et al., 1988, and Fig. 4, frame 2, of Montigny et al., 1988). Some previous scenarios have hypothesized the existence of shear zones equivalent to the upper plate–lower plate discontinuity that are truncated by a younger fault underlying the domal culmination in a regime where the dominant shear zones dip to the northeast (see Fig. 3 of Mattauer and Ritz, 1996). These interpretations for the Saih Hatat metamorphic and structural zonation link such shear zones into a system involving underthrusting of the Arabian margin to the northeast with shear zones dipping to the northeast (e.g., Michard et al., 1994; Searle et al., 1994; Hacker and Gnos, 1997; Searle and Cox, 1999).

These models are mainly based on the premises that, first, sense of shear is top to the north and, second, the high-grade rocks underlie the lower-grade rocks. As a result, the relative motion of the high-grade rocks is to come up from the north toward the south (effectively crustal extension; see Fig. 9 of Michard et al., 1994, and Fig. 3B of Mattauer and Ritz, 1996). Top-to-the-northeast (dextral) shearing within the structurally lowest rocks has also been attributed to north-northeast–directed subduction (e.g., Fig. 7 of Le Métour et al., 1990).

The overall geometry of the mountains and the thrusting of the ophiolite to the west and southwest have predisposed many models to have the margin underthrust to the northeast. To accommodate thrusting toward the margin, the simplest interpretation of the Oman Mountains is to have thrust faults dipping eastward. These faults root in the east or northeast and climb stratigraphic section in the direction of tectonic transport to the west or southwest. Most interpretive, regional cross sections have footwall ramps beneath the domal culminations of the mountains (e.g., Fig. 2, Goffé et al., 1988; Figs. 10 and 11 of Hanna, 1990; Figs. 11 and 12 of Cawood et al., 1990; and Fig. 3 of Michard et al., 1991).

The timing of domal culmination development is, however, somewhat controversial. Glennie et al. (1974, 1995) and Mount et al. (1998) argued that these are Tertiary structures. Other workers have the domal structures forming as a consequence of the Late Cretaceous obduction event (e.g., most notably Hanna, 1990, Mann and Hanna, 1990, and Cawood et al., 1990). Fission-track data, however, support the original interpretation of Glennie et al. (1974) in which an Oligocene (ca. 30 Ma) compressional event was held responsible for the uplift and present geometry of the mountains (Mount et al., 1998). Although extensional faults are common along the northeast flank of the ophiolite, particularly along the Batinah coast (see Mann et al., 1990), the Tertiary sequence from Muscat south to Quryat is clearly folded, is involved in ramp-anticline development, and has dips as steep as 60° in places. Flexing of the existing fold nappe at ~10%–20% shortening will produce the current geometry (see Fig. 2).

Significance of the crustal-scale stacking

In Saih Hatat, the crustal-scale stacking provides important clues to the tectonic evolution of the margin. The timing is crucial as is whether the various units behaved actively or passively during their emplacement. This latter aspect is implicit in the term "overthrusting," where the hanging wall is active relative to a "fixed" footwall, versus "underthrusting," where the footwall is active relative to a "fixed" hanging wall.

The crustal structure in Saih Hatat consists of a three-tiered system involving the uppermost Samail and Hawasina nappes, where the hanging wall was active (i.e., emplaced as an overthrusting event from 95 to 73 Ma). The middle tier involves the Saih Hatat fold nappe—or upper plate of Gregory et al. (1998) and Miller et al. (1998)—where the hanging wall is essentially passive (i.e., deformed as part of an underthrusting event from 76 to 70 Ma). In the upper plate, strain increases toward and is partitioned into the upper plate–lower plate shear zone (see strain data on structural profiles in Fig. 2, B and C, of Miller et al., 1998).

The third and structurally lowest member is the lower plate, which exhibits high internal strain and deformation via regional-scale, sheath folding during exhumation (decompression) involving top-to-the-northeast shear from 82 to 79 Ma. An earlier high-*P* eclogite-facies metamorphism (ca. 130–100 Ma) is preserved in megaboudins in the structurally lowest, easternmost part of the lower plate at As Sifah (Fig. 3). The train of megaboudins is folded by the regional-scale sheath-nappe folding event in the lower plate (see Fig. 6 of Gregory et al., 1998).

All of these data suggest that the lower plate was "active" during its initial exhumation. For lower-plate exhumation (i.e., footwall up), then the simplest geometry requires the early history of the lower-plate shear zone to have a normal-movement sense associated with a possible southwest-dipping shear zone. However, there is no evidence for an earlier top-to-the-southwest or -west sense of shear. This lack of evidence means either that the shear zone dipped to the northeast, with exhumation by top-to-the-northeast shear (i.e., normal sense) or that the original fabrics have been completely obliterated.

OBDUCTION OF THE OPHIOLITE

Obduction of the ophiolite involved initial intraoceanic thrusting, overthrusting, and structural thickening of the ophiolite nappe (hanging wall), and final emplacement onto the margin as a thin sheet (Fig. 7). The orientation of the ridge system, reconstructed by using the present-day attitudes of the sheeted dikes (Pallister, 1981), required the pre-emplacement geometry of the ophiolite to have a north-trending spreading center with possible east-trending transforms. Lineations in the metamorphic sole, as well as shape fabrics in basal peridotite mylonites that overlie the metamorphic sole, reflect the direction of intraoceanic thrusting (Boudier et al., 1982, 1988; Michard et al., 1984).

Thrusting was largely south directed, although it was west directed in the southernmost massifs. These directions have been interpreted (see Boudier et al., 1988) to reflect early south-directed (ridge-parallel) thrusting dated at 95.7 to 93.5 Ma (Hacker et al., 1996), followed by west-directed (ridge-normal) thrusting dated at 94.9–92.6 Ma (Hacker et al., 1996). An elapsed time of ~1 m.y. between igneous crystallization and intraoceanic thrusting (Hacker et al., 1996) requires rapid thrusting of the Samail ophiolite over old, cold lithosphere at this time (Hacker, 1990; Hacker et al., 1996).

Emplacement of oceanic thrust sheets has been related to overthrusting and/or gravity-induced emplacement mechanisms. With arguments against thrust emplacement (e.g., Elliott, 1976), most models have involved gravity-driven emplacement of the ophiolite related to structural thickening with "extensional collapse" (e.g., Cawood, 1991). Other scenarios have involved (1) a rapidly rising arc producing a regional surface slope to induce down-surface slope stresses to emplace ophiolite thrust sheets (Elliott, 1976, p. 960), (2) crustal loading, flexure, and "isostatic rebound" (Shelton and Egan, 1991), and (3) buoyancy-induced collapse (Chemenda et al., 1995, 1996; Gregory et al., 1998).

In this paper, we have adopted a "paddle board" (see also Gregory et al., 1998), analogous to the buoyancy-driven "rebound-type" model of Chemenda et al. (1995, 1996), to explain the exhumation of the high-*P* metamorphic rocks and the emplacement of the ophiolite (Fig. 7). In this scenario, the ophiolite undergoes final gravity-driven emplacement along pseudo-extensional faults due to the rising slab (see Fig. 8C in Chemenda et al., 1996). Differences between the Chemenda et al. (1996) model and our model relate primarily to the crustal structure beneath Saih Hatat (compare Fig. 8E of Chemenda et al., 1996, with Fig. 2 of this paper) driven by the geometry of the upper-plate fold nappe and the shear zones associated with the Hatat Schist and the upper plate–lower plate discontinuity.

The differences in interpretation on deep-crustal structure have led to different interpretations in the direction of subduction (toward or away from the margin), the timing of high-*P* metamorphism (later or earlier), and the timing of the Saih Hatat domal culmination (Cretaceous or Tertiary). In our model, the geometry of Saih Hatat is controlled by the regional, northeast-facing, antiformal fold nappe, and the shear zones that are associated with it (intervals 3 and 4 in Fig. 7). These record a period of underthrusting at the deepest levels, when the upper-plate fold nappe was forming in top-to-the-northeast shear. At the same time, the ophiolite was largely being emplaced to the southwest, and the upper plate was being juxtaposed against the lower plate with its prior history. Internal deformation within each of the present crustal elements (e.g., Samail and Hawasina nappes and the upper and lower plates of Saih Hatat) indicate that the lower crust at this time was not behaving as a series of rigid blocks sliding past each other (cf. Chemenda et al., 1996). Each was deforming internally relative to the adjacent footwall and hanging-wall blocks, and the positions and nature of these blocks have changed through time.

TECTONIC EVOLUTION OF THE ARABIAN MARGIN

In the Oman Mountains, the new data from the Saih Hatat window have enabled us to put together a tectonic scenario to explain the evolution and present geometry of the mountains (Fig. 7). This evolutionary sequence is based on the following key points and observations:

- High-*P* metamorphism and exhumation relate to a long-lived, kinematically coherent, tectonic "event" linked to the emplacement of the Oman ophiolite (as shown by the consistency of the stretching lineations in all units).
- A major fold nappe and two major crustal-scale shear zones act as constraints on interpretations of the pre-emplacement geometry of the Arabian margin, necessitating an interpretation of underthrusting toward the margin, rather than away from it, as in most other published models.
- Metamorphism at different structural levels at different times (cf. El Shazly and Coleman, 1990) in two stages—HP_1 metamorphism predates ophiolite obduction, and HP_2 metamorphism is due to ophiolite obduction loading.
- The ophiolite is now preserved as a thin sheet bounded by a flat-lying basal fault.
- Exhumation is a multistage phenomenon and certainly not linked to one process.
- Buoyant rise and internal deformation of high-*P* slabs are controlled by the former subduction interface.
- Slab rise led to gravitational instability and normal faulting of the partially obducted ophiolite and eventual gravity-driven emplacement of the Samail nappe onto the Arabian margin.
- The uplift and present-day morphology of the Oman Mountains is due to Tertiary (post-Maastrichtian) southwestward-directed overthrusting and folding (Glennie et al., 1974; Mount et al., 1998).

In this model, the high-*P* metamorphism (HP_1) predates or is at least contemporaneous with ophiolite formation (see Fig. 6) and is therefore not due to thrust loading (structural thickening) by the ophiolite. The sequence of fabric development in the eclogite megaboudins at As Sifah shows early east-trending fabrics overprinted by north-northeast–trending fabrics. This change requires subduction of older (Jurassic?) oceanic crust (e.g., the age of Masirah Island oceanic crust) , where eclogite-facies metamorphism accompanied subduction of part of the former Arabian margin, or part of a rifted piece of the margin, i.e., a microplate (stages 1 and 2, Fig. 7). The subduction interface must have been at a sufficiently low angle or sufficiently short lived to dampen magma generation.

Neutral buoyancy of the slab or microplate at ca. 96 Ma choked the subduction zone and caused a break in the oceanic lithosphere near the spreading center (stage 2, Fig. 7). The tectonic setting in the Hawasina Ocean changed from extension to compression, which led to overthrusting at this time. Slab rise led to initial exhumation of the Saih Hatat lower plate accompanied by blueschist metamorphism and development of regional-scale sheath folds (stage 3, Fig. 7). Preservation of the early eclogite- and blueschist-facies metamorphic evidence requires continued underthrusting of cold lithosphere during slab ascent. Overthrusting and structural thickening of the oceanic Samail thrust nappe (internally folded) ended with lock-up along the basal fault as the oceanic nappe encountered the former Arabian margin (stages 3 and 4, Fig. 7). At this time, emplacement of the overlying oceanic thrust nappe caused metamorphism (HP_2) at the leading edge of the Arabian margin (Stage 4, Fig. 7).

At depth, slab rise continued, largely owing to buoyancy of the less dense, now deformed and metamorphosed, continental slab. The slab rise took place along the former subduction interface. Gravitational instability within the ophiolite thrust nappe resulted in the development of extensional faults that allowed thinning of the ophiolite nappe during its eventual final emplacement (stage 5, Fig. 7). At the same time, the underlying margin was being deformed into a regional-scale fold nappe (now preserved within the Saih Hatat domal culmination).

Discussion of the model

There are two competing scenarios for the origin and emplacement of the Samail ophiolite complex. These scenarios are influenced by the choice of the tectonic setting for the ophiolite: whether the ophiolite formed at some type of mid-ocean ridge or in the hanging wall of a subduction zone. The evidence for the ophiolite's forming in the hanging wall of a subduction zone is largely based upon geochemical arguments (Lippard et al., 1986; Searle and Cox, 1999). Interpreting the geometry of the emplacement under this scenario starts with the assumption that the subduction zone dipped away from the Arabian margin (e.g., Gealey, 1977).

The mid-ocean ridge scenario has the ophiolite obduction occurring on some intraoceanic thrust system whose shear zones dip away from the continental margin (e.g., Coleman, 1971, 1981; Boudier et al., 1988). The conditions of the metamorphism within the contact-aureole rocks are critical to the latter scenario because they provide insights into the condition of the lithosphere at the time of thrusting. The presence of eclogites and blueschists of continental-margin affinity complicates the history, because the metamorphism of the continental margin appears to be coeval with the formation of the ophiolite at the spreading center and the intraoceanic thrusting event. This timing relationship may require the existence of two subduction zones (Hacker, 1991) or a reversal in the polarity of the convergence (Gregory et al., 1998).

For the "away-from-the-margin"–directed subduction, the high-*P*, low-*T* rocks were exhumed and returned to the surface from the north to the south. The Saih Hatat lower-plate rocks must have either overridden the trailing continental-margin rocks or extruded to a free surface and shed continental-shelf debris in front of the advancing ophiolite. The lower-plate rocks themselves must then be thrust under the upper plate that is tied to the Arabian margin.

In this scenario, the timing of the Hatat Schist shear zone is difficult to reconcile with its incorporation into the core of the upper-plate nappe. The intensity of deformation in the Hatat Schist is comparable to that of the lower plate, and the stretching lineations are consistent with the development of both shear zones as part of a single progressive event. In contrast, if the original underthrusting of the rifted and attenuated continental-margin rocks that now comprise the lower plate was directed toward the margin, the Hatat Schist shear zone (structure 3 in Fig. 2) can be interpreted as part of the original subduction interface. The lower-plate rocks were exhumed along this interface and are themselves underlain by oceanic lithosphere.

There are three key points in our model: (1) the Hatat Schist was involved in the Cretaceous deformation, (2) the HP_2 (i.e., metamorphism of the upper plate) was driven by the load of the ophiolite and the development of the lower limb of the upper-plate nappe, and (3) these upper-plate units were emplaced by northeast-directed transport over the lower plate (or underthrusting of the lower plate in the fixed-upper-plate frame of reference). The presence of the shear zones beneath Saih Hatat (structures 3 and 4, Fig. 2) record the former underthrusting event (i.e., nascent subduction toward the margin). The lack of any calc-alkalic or constructional-pile arc volcanism along the Arabian margin or in the former Hawasina Ocean is a problem common to both subduction polarities (toward or away from the margin). Subduction toward the margin, however, may have produced conditions favorable to the formation one of the world's largest oil-bearing basins (the Rub Al Khali basin), as part of extension in the overriding plate, a possible consequence of slab rollback. The Rub Al Khali basin and the high-*P* rocks of Saih Hatat are therefore the only direct evidence for the pre–100 Ma convergent margin.

CONCLUSIONS

Obduction of the Samail ophiolite is a consequence of failed, margin-directed subduction along the former Arabian continental margin. New data suggest that this process was protracted over a time period of ~35–40 m.y. Kinematic consistency shown by the stretching lineation at all levels in the pile and at different stages (time periods) indicates a sequence of progressive events culminating in the emplacement of the ophiolite. New structural and geological mapping in the Saih Hatat window identifies two major shear zones linked to a major, northeast-facing, regional-scale, sheath-like, fold nappe whose upper limb is pinned to the margin. These require underthrusting toward the margin rather than away from it.

The structurally lowest shear zone exposed in Saih Hatat represents the upper contact with a buoyant slab (lower plate), but this slab has deformed internally via sheath-nappe folding during ascent (exhumation) and shows a consistent top-to-the-northeast shear sense. This slab is linked to eventual obduction of the Samail ophiolite nappe as a gravity-driven thin sheet, in that the gravitational instability was created by the rise of the slab. Inherent in the model is the concept that the former subduction interface, preserved as major shear zones within the now continental-margin crust, controls both the descent and ascent of the buoyant slab. The shear zones therefore record a prior history of the margin geometry.

ACKNOWLEDGMENTS

Funding for the research was from Australian Research Council Grants ARCSM95 and A39601548 and National Science Foundation Grant EAR91-06016. We thank M. Kassim and Hilal Al Azry of the Ministry of Commerce of Oman for sponsorship and support during the field work. We also thank Jean-Paul Breton (BRGM, Oman) for logistical and moral support, Gavin Graham (Petroleum Development Oman) for satellite imagery and base maps when we needed them most, and Samir Hannah for an overview field trip into Jabal Akhdar. Our knowledge of the geology of Oman has benefited from discussions with Bob Coleman. Comments from Eldridge Moores, Tjerk Peters, and John Wakabayashi on an earlier draft of the manuscript were very helpful.

REFERENCES CITED

Bailey, E.H., 1981, Geologic map of the Muscat-Ibra area, Sultanate of Oman: Journal of Geophysical Research, v. 86, pocket map.

Boudier, F., and Coleman, R.G., 1981, Cross-section through the peridotite in the Samail Ophiolite, southeastern Oman: Journal of Geophysical Research, v. 86, p. 2573–2592.

Boudier, F., and Michard, A., 1981, Oman ophiolite: The quiet obduction of oceanic crust: Terra Cognita, v. 1, p. 109–118.

Boudier, F., Nicolas, A., and Bouchez, J.L., 1982, Kinematics of oceanic thrusting and subduction from basal sections of ophiolites: Nature, v. 296, p. 825–828.

Boudier, F., Ceuleneer, G., and Nicolas, A., 1988, Shear zones, thrusts, and related magmatism in the Oman ophiolite: Initiation of thrusting on an oceanic ridge: Tectonophysics, v. 151, p. 275–296.

Bucher, M., 1991, The metamorphic series associated with the Semail Ophiolite nappe of the Oman Mountains in the United Arab Emirates [M.Sc. thesis]: University of Berne, Switzerland, 178 p.

Cawood, P.A., 1991, Processes of ophiolite emplacement in Oman and western Newfoundland, *in* Peters, Tj., Nicolas, A., and Coleman, R., eds., Ophiolite genesis and evolution of the ocean lithosphere: Boston, Kluwer Academic Publishers, p. 501–516.

Cawood, P.A., Green, F.K., and Calon, T.J., 1990, Origin of culminations within the southeast Oman Mountains at Jebel Ma Jhool and Ibra Dome, *in* Robertson, A.H.F., Searle, M.P., and Ries, A.C., eds., The geology and tectonics of the Oman region: Geological Society of London Special Publication 49, p. 429–446.

Chemenda, A.I., Mattauer, M., Malaveille, J., and Bokun, A.N., 1995, A mechanism for syn-collisional deep-rock exhumation and associated normal faulting: Results from physical modelling: Earth and Planetary Science Letters, v. 132, p. 225–232.

Chemenda, A.I., Mattauer, M., and Bokun, A.N., 1996, Continental subduction and a mechanism for exhumation of high-pressure metamorphic rocks: New modelling and field data from Oman: Earth and Planetary Science Letters, v. 143, p. 173–182.

Coleman, R.G., 1971, Plate tectonic emplacement of upper mantle peridotites along continental edges: Journal of Geophysical Research, v. 76, p. 1212–1222.

Coleman, R.G., 1977, Ophiolites: Ancient oceanic lithosphere: New York, Springer, 229 p.

Coleman, R.G., 1981, Tectonic setting for ophiolite obduction in Oman: Journal of Geophysical Research, v. 86, p. 2497–2508.

Elliott, D., 1976, The motion of thrust sheets: Journal of Geophysical Research, v. 81, p. 949–963.

El-Shazly, A.K., 1994, Petrology of lawsonite-, pumpellyite- and sodic amphibole-bearing metabasites from NE Oman: Journal of Metamorphic Geology, v. 12, p. 23–48.

El-Shazly, A.K., 1995, Petrology of Fe-Mg-carpholite-bearing metasediments from NE Oman: Journal of Metamorphic Geology, v. 13, p. 379–396.

El-Shazly, A.K., and Coleman, R.G., 1990, Metamorphism in the Oman Mountains in relation to the Semail ophiolite emplacement, *in* Robertson, A.H.F., Searle, M.P., and Ries, A.C., The geology and tectonics of the Oman region: Geological Society of London Special Publication 49, p. 473–493.

El-Shazly, A.K., Coleman, R.G., and Liou, J.G., 1990, Eclogites and blueschists from northeast Oman: Petrology and P-T evolution: Journal of Petrology, v. 31, p. 629–666.

Gealey, W.K., 1977, Ophiolite obduction and geologic evolution of the Oman Mountains and adjacent areas: Geological Society of America Bulletin, v. 88, p. 1183–1191.

Ghent, E.D., and Stout, M.Z., 1981, Metamorphism at the base of the Semail Ophiolite, southeastern Oman Mountains: Journal of Geophysical Research, v. 86, p. 2557–2572.

Glennie, K.W., 1995, The geology of the Oman Mountains: An outline of their origin: United Kingdom, Scientific Press Limited, 92 p.

Glennie, K.W., Hughes Clarke, M.W., Boeuf, M.G., Pilaar, W.F., and Reinhardt, B.M., 1974, Geology of the Oman Mountains: Verhandelingen van het Koninklinjk Nederlands geolgisch minjbouwkundig Genootsschap, Amsterdam, v. 31, 423 p.

Gnos, E., 1998, Peak metamorphic conditions of garnet amphibolites beneath the Semail Ophiolite: Implications for an inverted pressure gradient: International Geology Review, v. 40, p. 281–304.

Goffé, B., Michard, A., Kienast, J.R., and le Mer, O., 1988, A case of obduction-related high pressure, low temperature metamorphism in upper crustal nappes, Arabian continental margin, Oman: Tectonophysics, v. 151, p. 363–386.

Gray, D.R., and Gregory, R.T., 2001, Implications of the structure of the Wadi Tayin metamorphic sole, the Ibra-Dasir block of the Samail Ophiolite, and the Saih Hatat window for late stage extensional ophiolite emplacement, Oman, *in* Boudier, F., and Juteau, T., eds., The ophiolite of Oman and the United Arab Emirates, Marine Geophysical Researches Special Issue, in press.

Gregory, R.T., 1983, Melt percolation beneath a spreading ridge: Evidence from the Semail peridotite, Oman, *in* Gass, I.G., Lippard, S.J., and Shelton, A.W., eds., Ophiolites and oceanic lithosphere: Geological Society of London, Blackwell Scientific, p. 55–62.

Gregory, R.T., and Taylor, H.P., Jr., 1981, An oxygen isotope profile in a section of Cretaceous oceanic crust, Samail ophiolite, Oman: Evidence for $\delta^{18}O$ buffering of the oceans by deep (>5km) seawater-hydrothermal circulation at mid-ocean ridges: Journal of Geophysical Research, v. 86, p. 2737–2755.

Gregory, R.T., Gray, D.R., and Miller, J. Mcl., 1998, Tectonics of the Arabian Margin associated with the emplacement of the Oman Margin along the

Ibra Transect: New evidence from northeast Saih Hatat: Tectonics, v. 17, p. 657–670.

Hacker, B.R., 1990, Simulation of the metamorphic and deformational history of the metamorphic sole of the Oman ophiolite: Journal of Geophysical Research, v. 95, p. 4895–4907.

Hacker, B.R., 1991, The role of deformation in the formation of metamorphic gradients: Ridge subduction beneath the Oman ophiolite: Tectonics, v. 10, p. 455–473.

Hacker, B.R., and Gnos, E., 1997, The conundrum of Samail: Explaining the metamorphic history: Tectonophysics, v. 279, p. 215–226.

Hacker, B.R., and Mosenfelder, J.L., 1996, Metamorphism and deformation of the sole beneath the Samail Ophiolite: Tectonics, v. 10, p. 455–473.

Hacker, B.R., Mosenfelder, J.L., and Gnos, E., 1996, Rapid emplacement of the Oman Ophiolite: Thermal and geochronological constraints: Tectonics, v. 15, p. 1230–1247.

Hanna, S., 1990, The Alpine deformation of the central Oman Mountains, *in* Robertson, A.H.F., Searle, M.P., and Ries, A.C., eds., The geology and tectonics of the Oman region: Geological Society of London Special Publication 49, p. 341–359.

Hopson, C.A., Coleman, R.G., Gregory, R.T., Pallister, J.S., and Bailey, E.H., 1981, Geological section through the Samail Ophiolite and associated rocks along a Muscat-Ibra transect, SE Oman mountains: Journal of Geophysical Research, v. 86, p. 2527–2544.

Le Métour, J., Rabu, D., Tegyey, M., Béchennec, F., Beurrier, M., and Villey, M., 1990, Subduction and obduction: Two stages in the Eo-Alpine tectonometamorphic evolution of the Oman Mountains, *in* Robertson, A.H.F., Searle, M.P., and Ries, A.C., eds., The geology and tectonics of the Oman region: Geological Society of London Special Publication 49, p. 327–339.

Lippard, S.J., 1983, Cretaceous high-pressure metamorphism in northeast Oman and its relationship to subduction and ophiolite nappe emplacement: Journal of the Geological Society of London, v. 140, p. 97–104.

Lippard, S.J., Shelton, A.W., and Gass, I.G., 1986, The ophiolite of northern Oman, Geological Society of London Memoir No. 11, 178 p.

Manghnani, M.H., and Coleman, R.G., 1981, Gravity profiles across the Samail Ophiolite, Oman: Journal of Geophysical Research, v. 86, p. 2509–2525.

Mann, A., and Hanna, S., 1990, The tectonic evolution of Pre-Permian rocks, central and southeastern Oman Mountains, *in* Robertson, A.H.F., Searle, M.P., and Ries, A.C., eds., The geology and tectonics of the Oman region: Geological Society of London Special Publication 49, p. 307–325.

Mann, A., Hanna, S., and Nolan, S.C., 1990, The post-Campanian tectonic evolution of the central Oman Mountains: Tertiary extension of the eastern Arabian margin, *in* Robertson, A.H.F., Searle, M.P., and Ries, A.C., eds., The geology and tectonics of the Oman region: Geological Society of London Special Publication 49, p. 549–563.

Mattauer, M., and Ritz, J-F., 1996, Arguments géologiques en faveur d'un modèle de subduction continentale pour l'exhumation du metamorphisme hautepression d'Oman: Centre Rendue Acadamie des Sciences Paris, série IIa, v. 322, p. 869–876.

Michard, A., Bouchez, J.L., and Ouazziani-Touhami, O., 1984, Obduction-related planar and linear fabrics in Oman: Journal of Structural Geology, v. 6, p. 39–49.

Michard, A., Boudier, F., and Goffé, B., 1991, Obduction versus subduction and collision in the Oman case and other Tethyan settings, *in* Peters, Tj., et al., eds., Ophiolite genesis and evolution of oceanic lithosphere: Ministry of Petroleum and Minerals, Sultanate of Oman, p. 447–467.

Michard, A., Goffé, B., Saddiqi, O., Oberhänsli, R., and Wendt, A.S., 1994, Late Cretaceous exhumation of the Oman blueschists and eclogites: A two-stage extensional mechanism: Terra Nova, v. 6, p. 404–413.

Miller, J. McL., Gray, D.R., and Gregory, R.T., 1998, Exhumation of high-pressure rocks in northeastern Oman: Geology, v. 26, p. 235–238.

Miller, J. McL., Gregory, R.T., Gray, D.R., and Foster, D.A., 1999, Geological and geochronological constraints on the exhumation of a high-pressure metamorphic terrane, Oman, *in* Ring, U., Brandon, M.T., Lister, G.S., and Willett, S.D., eds., Exhumation processes: Normal faulting, ductile flow and erosion: Geological Society of London Special Publication 154, p. 241–260.

Montigny, R., Le Mer, O., and Whitechurch, H., 1988, K-Ar and Ar^{40}/Ar^{39} study of metamorphic rocks associated with the Oman ophiolite: Tectonic implications: Tectonophysics, v. 151, p. 345–362.

Moores, E.M., 1982, Origin and emplacement of ophiolites: Reviews of Geophysics, v. 20, p. 735–760.

Mount, V.S., Crawford, R.I.S., and Bergman, S.C., 1998, Regional structural style of the central and southern Oman Mountains: Jabal Akhdar, Saih Hatat and the northern Ghaba Basin: GeoArabia, v. 3, p. 475–490.

Nicolas, A., 1989, Structures of ophiolites and dynamics of oceanic lithosphere: Dordrecht, The Netherlands, Kluwer Academic Publishers, 367 p.

Nicolas, A., Ceuleneer, G., Boudier, F., and Misseri, M., 1988, Structural mapping in the Oman ophiolites: Mantle diapirism along an oceanic ridge: Tectonophysics, v. 151, p. 27–56.

Pallister, J.S., 1981, Structure of the sheeted dike complex of the Samail Ophiolite near Ibra, Oman: Journal of Geophysical Research, v. 86, p. 2661–2672.

Searle, M.P., 1985, Sequence of thrusting and origin of culminations in the northern and central Oman Mountains: Journal of Structural Geology, v. 7, p. 129–143.

Searle, M.P., and Cox, J., 1999, Tectonic setting, origin, and obduction of the Oman ophiolite: Geological Society of America Bulletin, v. 111, p. 104–122.

Searle, M.P., and Malpas, J., 1980, The structure and metamorphism of rocks beneath the Samail Ophiolite of the Oman Mountains and their significance in ophiolite obduction: Transactions of the Royal Society of Edinburgh, Earth Sciences, v. 71, p. 213–248.

Searle, M.P., Waters, D.J., Martin, H.N., and Rex, D.C., 1994, Structure and metamorphism of blueschist-eclogite facies rocks from the northeast Oman Mountains: Journal of the Geological Society of London, v. 151, p. 555–576.

Shelton, A.W., 1984, Geophysical studies on the northern Oman ophiolite [Ph.D. thesis]: Milton Keynes, Open University, 353 p.

Shelton, A.W., 1990, The interpretation of gravity data in Oman: Constraints on the ophiolite emplacement mechanism, *in* Robertson, A.H.F., Searle, M.P., and Ries, A.C., eds., The geology and tectonics of the Oman region: Geological Society of London Special Publication 49, p. 459–471.

Shelton, A.W., and Egan, S.S., 1991, The obduction of the northern Oman ophiolite: Crustal loading and flexure, *in* Peters, Tj., et al., eds., Ophiolite genesis and evolution of the oceanic lithosphere: Ministry of Petroleum and Minerals, Sultanate of Oman, p. 469–483.

Tilton, G.R., Hopson, C.A., and Wright, J.E., 1981, Uranium-Lead isotopic ages of the Samail Ophiolite, Oman, with application to Tethyan ridge tectonics: Journal of Geophysical Research, v. 86, p. 2763–2776.

Tippit, P.R., Pessagno, E.A., and Smewing, J.D., 1981, The biostratigraphy of sediments in the volcanic unit of the Samail Ophiolite: Journal of Geophysical Research, v. 86, p. 2756–2762.

Warburton, J., Burnhill, T.J., Graham, R.H., and Isaac, K.P., 1990, The evolution of the Oman Mountains foreland basin, *in* Robertson, A.H.F., Searle, M.P., and Ries, A.C., eds., The geology and tectonics of the Oman region: Geological Society of London Special Publication 49, p. 419–427.

Williams, H., and Smyth, W.R., 1973, Metamorphic aureoles beneath ophiolite suites and alpine peridotites: Tectonic implications with west Newfoundland examples: American Journal of Science, v. 273, p. 594–621.

MANUSCRIPT ACCEPTED BY THE SOCIETY FEBRUARY 23, 2000

Printed in the U.S.A.

Geological Society of America
Special Paper 349
2000

Formation and evolution of the western Indian Ocean as evidenced by the Masirah ophiolite: A review

Tjerk Peters
Mineralogisch-petrologisches Institut, Baltzerstrasse 1, CH-3012, Berne, Switzerland; e-mail: tjerk@mpi.unibe.ch

ABSTRACT

This paper describes the results of the investigations of the Late Jurassic Masirah ophiolite and its sedimentary autochthonous cover. In the Masirah ophiolite, all members of a "classic" ophiolite sequence are present, although the gabbroic lower crust measures only 200 to 500 m in thickness, making the Masirah ophiolite a good example of thin oceanic crust formed under low magma supply. Generation of the Masirah ophiolite is explained by interaction of a ridge and a transform fault where they intersected. The basaltic rocks forming the Late Jurassic crust show a compositional variation between E-MORB and N-MORB.

After the Late Jurassic ridge formation, the sedimentary record shows a continuous deepening below the calcite compensation depth (CCD) until the late Hauterivian, when oceanic crust was uplifted above sea level and platform carbonates were deposited. Simultaneously, a second magmatic event involved the formation of seamount structures and the intrusion of isotropic gabbro, alkalic dolerite, and K-feldspar–rich granite. This middle Cretaceous magmatism is interpreted as the result of oceanic rifting and/or hotspot activity at the proto–Owen Fracture Zone in response to the rearrangement of the plates in the Southern Hemisphere. The intraoceanic thrusting—resulting in the nappe structure of the Masirah ophiolite—and its obduction between late Maastrichtian and Paleocene time coincide with the opening of the Carlsberg Ridge. The middle Miocene block faulting registered on the island of Masirah and the Batain coast is correlated with the extensional tectonics caused by the opening of the Gulf of Aden since magnetic anomaly A5.

INTRODUCTION

Most of the evidence for the formation and evolution of the oceanic lithosphere of the Indian Ocean since the breakup of Gondwana originates from Deep Sea Drilling Project, Ocean Drilling Program, and industrial drilling or from geophysical methods such as magnetic anomaly patterns or apparent polar-wander paths. Compared to the numerous studies on Tethyan ophiolites, there are only few studies (Gnos et al., 1997) on parts of Indian Ocean lithosphere that have been emplaced onto the Arabian or the Eurasian plate. The Masirah ophiolite with its sedimentary cover has offered the opportunity to obtain data on the formation of part of the oceanic crust of the Mesozoic western Indian Ocean. The autochthonous sedimentary cover of the ophiolite enables the study of its subsequent history as part of the ocean floor during a time span of 85 m.y. from the latest Jurassic up to the early Paleocene. Furthermore, the Masirah ophiolite represents a good example for a thin oceanic lithosphere (Peters and Mercolli, 1998), where, in spite of its reduced thickness, all members of a model ophiolite are present in a coherent sequence. In addition, the effects of a later stage of hotspot- and/or rift-related magmatism can be studied. The ridge and rift magmatism, intraoceanic thrusting, and obduction were accompanied by faulting, which provides insight into the opening of pathways for hydrothermal alteration.

This paper summarizes the results of a research project, parts of which have been presented elsewhere (Peters et al., 1995; Marquer et al., 1995; Gnos and Perrin, 1995; Immenhauser, 1996; Meyer et al., 1996; Gnos et al., 1997; Nägler and Frey, 1997; Peters and Mercolli, 1997; Peters and Mercolli, 1998; Marquer et al., 1998.

Geologic setting

The Masirah ophiolite, which constitutes a large part of the island of Masirah, is the largest of the remnants of oceanic crust exposed on the southeast coast of Oman (Fig. 1). On Masirah, the uppermost two of a stack of ophiolite nappes, overlying the autochthonous cover of the Arabian continental platform, are exposed (Fig. 2), as inferred from the geophysical studies of Barton et al. (1990) and Beauchamp et al. (1995). Peters et al. (1995) and Peters and Mercolli (1997) have shown that the Masirah ophiolite is not related to the Semail ophiolite (Glennie et al., 1974) nor is it an up-thrust zone of oceanic crust, separated from the passive margin by the so-called Masirah fault or line (Abbotts, 1979; Smewing et al., 1991). Paleomagnetic studies (Gnos and Perrin, 1995) indicate formation at a latitude of 48°S ± 10° (owing to a calculation error, the original paper gives 38°S!) on the southern side of an east-trending ridge and point to a polar-wandering path for the ophiolite similar to the Indian plate.

Peters, Tj., 2000, Formation and evolution of the western Indian Ocean as evidenced by the Masirah ophiolite: A review, *in* Dilek, Y., Moores, E.M., Elthon, D., and Nicolas, A., eds., Ophiolites and Oceanic Crust: New Insights from Field Studies and the Ocean Drilling Program: Boulder, Colorado, Geological Society of America Special Paper 349, p. 525–536.

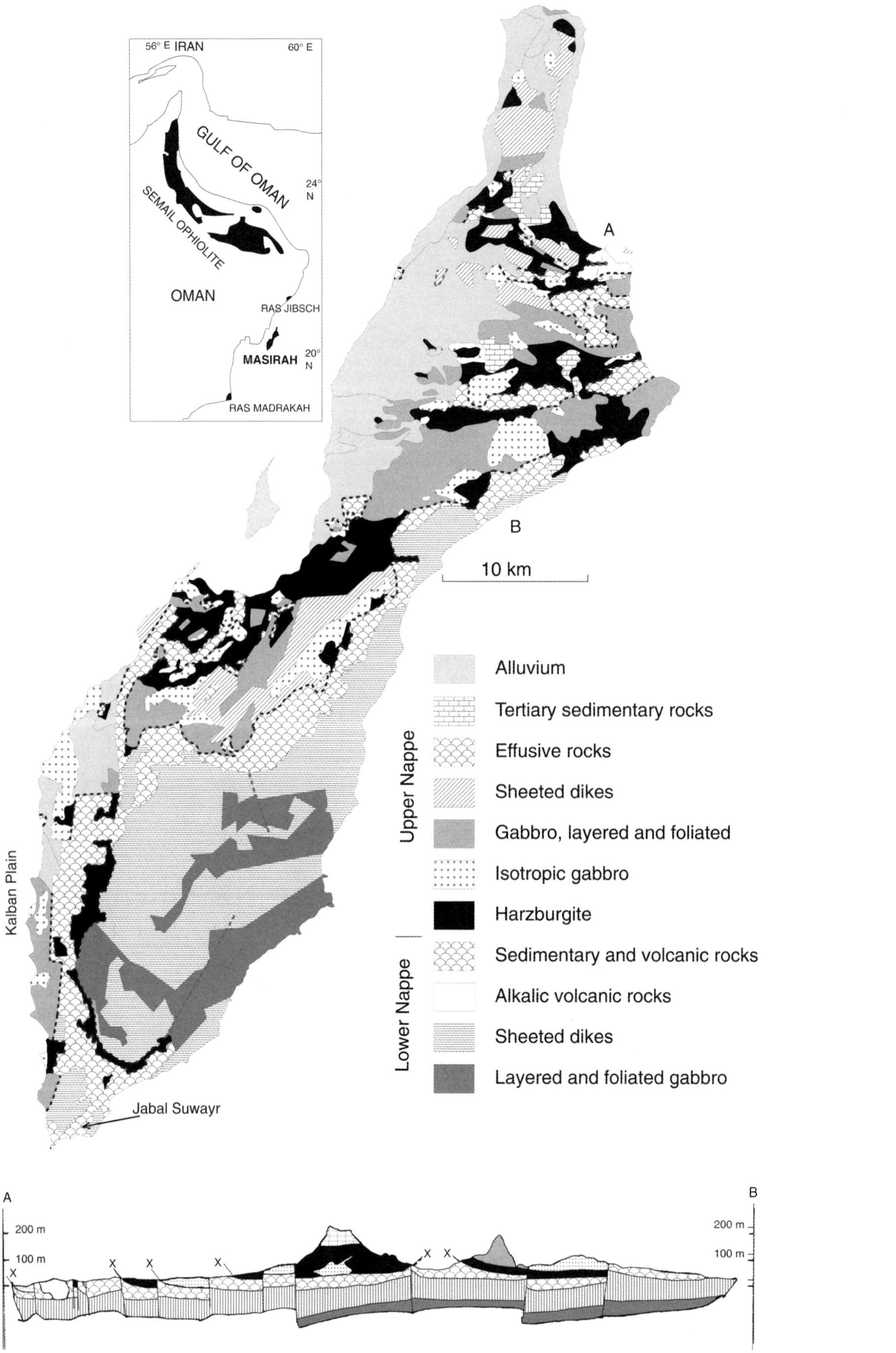

Figure 1. A, Geologic-tectonic sketch map of Masirah Island (Sultanate of Oman). B, Cross section through the northeastern part of Masirah Island along A–B in part A.

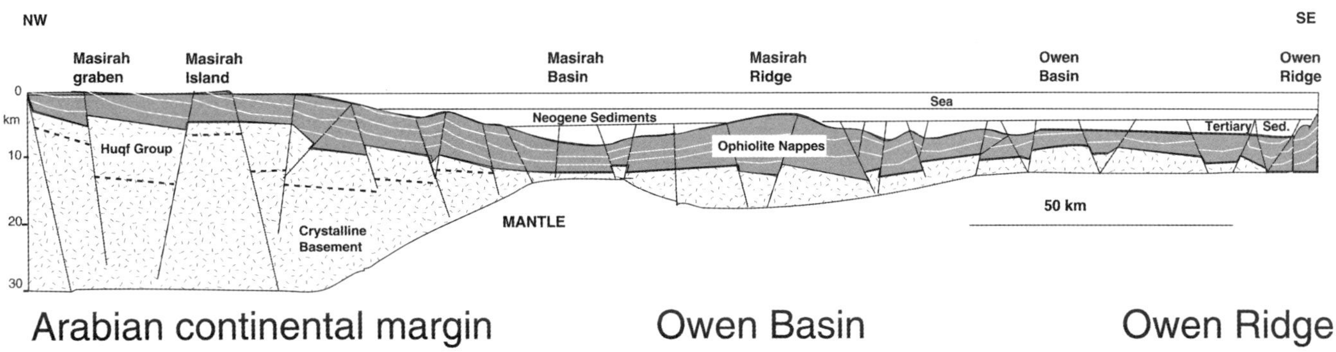

Figure 2. Section through the Owen Basin showing oceanic crust as thrust sheets on the Arabian continental margin. The continental crust and the ophiolite nappes have been thinned by a series of normal faults as a result of the opening of the Gulf of Aden since the middle Miocene. Seismic information from Barton et al. (1990).

The position must have been between Somalia–Madagascar–southeast Arabia and the northeast corner of greater India, far away from the Tethyan Ocean. The ophiolite sequence has been concordantly dated at 150 Ma by radiolarian micrites of Tithonian age overlying and interlayered with pillow lavas, by U-Pb ages on zircons from two gabbro pods in the mantle section, and by Ar-Ar patterns on plagioclase from these gabbros (R. Frey, 1999, personal communication). After formation, the oceanic crust subsided as shown by the deposits of ribbon cherts (Immenhauser, 1996). During the middle Cretaceous, differential uplift of crustal blocks, alkalic magmatism (Meyer et al., 1996), and the intrusion of a suite of gabbro, dolerite, and granite occurred. The platform carbonates on the uplifted blocks are stratigraphically dated as upper Hauterivian to lower Barremian (Immenhauser, 1996). This age is confirmed by the U-Pb ages of 123 ± 3 Ma on the zircons in the associated alkalic volcanic rocks. From Aptian to Santonian, the uplifted oceanic crust subsided again with deposition of radiolarian cherts. They partly interfinger with and are overlain by flysch-like sediments of Coniacian to Maastrichtian age. During or after the late Maastrichtian, an intraoceanic, south-to-north–directed, and almost horizontal thrusting led to the formation of both the lower and upper Masirah ophiolite nappes. The westward-directed obduction of the nappe stacks onto the Arabian plate took place before formation of the laterite within the Paleocene unconformity (Peters et al., 1995). Eocene and lower Oligocene platform carbonates form the postnappe autochthonous cover.

Oceanic basement

In the lower nappe (Fig. 3), the crustal rocks were formed exclusively by Late Jurassic ridge magmatism; in the upper nappe, in contrast, a major part of the rocks are the products of the middle Cretaceous magmatism.

PETROGRAPHY

Late Jurassic ridge magmatism

Harzburgite with orthopyroxene contents ranging between 10% and 30%, picotite (1%–3%), and minor clinopyroxene make up the main part of the mantle section. Dunite is scarce. Foliation defined by deformed spinel is not everywhere parallel to the locally observed layering marked by orthopyroxene enrichment. The foliation indicates a high-temperature deformation during mantle flow (Nicolas, 1989). Near the petrologic Moho, defined by the upper limit of harzburgite,

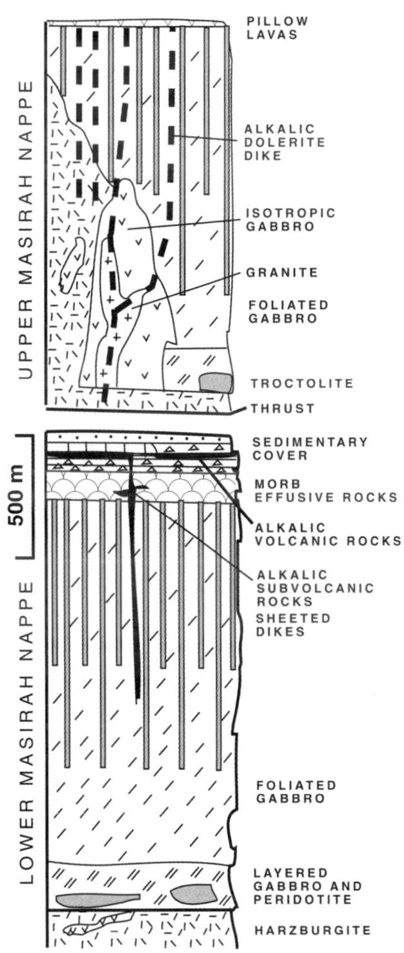

Figure 3. Composite lithostratigraphic section through the lower and upper nappes of the Masirah ophiolite. The lower nappe is composed of an Upper Jurassic "classic" ophiolite sequence with a thin lower-crustal section. In the upper nappe, the ophiolite sequence is intruded by middle Cretaceous isotropic gabbro, K-feldspar granite, and alkalic dolerite dikes. An autochthonous sedimentary cover ranging from Tithonian to late Maastrichtian is exposed only in the lower nappe.

subhorizontally and vertically layered clinopyroxene gabbro increases to form up to 50% of the outcrop.

The lower-crustal section above the petrographic Moho (Fig. 3) comprises interlayered olivine gabbro, plagioclase dunite, plagioclase-bearing wehrlite, and melatroctolite. The lower-crustal rocks grade upward into layered olivine gabbro in which compositional layering is defined by grain size and abundance of olivine. The uppermost part of the lower-crustal section consists of foliated gabbro, where the foliation is the result of preferred orientation and abundance of diopsidic clinopyroxene and plagioclase. In the lower part of the section, this foliation is subhorizontal, parallel to the gabbro layering, but turns subvertical parallel to the dikes in the transition to the upper-crustal section. This structural pattern is very similar to that of the Semail ophiolite (Nicolas et al., 1988). Within layered as well as foliated gabbro, small (3–30 cm) veins and pods of anorthosite with less than 5% clinopyroxene are present.

The transition of the lower-crustal section into the sheeted-dike complex of the upper-crustal section is gradational over a few hundreds of meters with a continuous increase in the dike to gabbro-screen volume ratio. In contrast, the transition of the sheeted-dike complex into overlying pillow lavas takes place within some tens of meters. On the basis of macroscopic textural criteria and crosscutting relationships, three types of dikes can be recognized: microgabbroic dolerite, porphyritic dolerite (plagioclase crystals of up to 2 cm), and aphyric dolerite. Statistically, the dikes are east striking over the whole island, with a dip of ~90°. Locally, a northeast or north-northeast trend is present, which must have resulted from the rotation of 5–10 km^2 blocks of oceanic crust during ridge formation. Tonalite (plagiogranite) is found in places as 10–50-cm-thick dikes in the roots of the sheeted-dike complex. The lower extrusive rocks consist of meter-sized pillows of lava without interpillow material. They grade upward into smaller-sized (0.5 m) pillow and tube lavas with few intercalated sheet flows. Under the microscope, the lavas show porphyritic texture with plagioclase, olivine, clinopyroxene, or Cr-rich spinel phenocrysts in a subophitic to intersertal texture.

The mineral compositions in the lower-crustal section show little variation: olivine with 86 mol% Fo, plagioclase with 82–85 mol% An, and clinopyroxene with 6 mol% $FeSiO_3$ and 5–6 mol% Al-Tsk components. Dolerite is composed of 48 ± 7 vol% plagioclase (65%–70% An), 37 vol% diopsidic clinopyroxene, 13 ± 5 vol% pseudomorphs after olivine, and 3 vol% ilmenite as well as accessory Cr-rich spinel and apatite. Pillow lavas are generally richer in clinopyroxene than intrusive rocks.

Middle Cretaceous alkalic magmatism

Middle Cretaceous magmatism yielded two rock suites: a gabbro, diorite, and granite suite in the upper nappe and an alkalic volcanic and subvolcanic suite in the lower nappe. The age of the alkalic volcanic suite is stratigraphically defined as early Barremian. For the gabbro, diorite, and granite suite, U-Pb ages on zircon and Ar/Ar ages on hornblende and biotite range between 125 and 130 Ma, overlapping with the radiometric ages of 123 to 130 Ma from the volcanic suite (R. Frey, 1999, personal communication).

The gabbros of the gabbro, diorite, and granite suite (Fig. 3) crop out as oval-shaped, 0.5–3.5-km-wide intrusions into preexisting mantle and lower-crustal sections. The gabbro is isotropic and pegmatitic with a heterogeneous grain-size distribution. Coarse-grained gabbros exhibit extensive replacement of clinopyroxene by brown magnesian hastingsitic amphibole. Gabbros also contain some olivine (0–1%) or biotite (0–2%) as well as 30%–60% bytownitic plagioclase (An_{80-70}). In fine- to medium-grained, isotropic gabbro, only rarely is more than 5% clinopyroxene preserved, and the amount of biotite can reach up to 10 vol%. In mylonitic shear zones, green hornblende is the main mafic phase. This deformation must have occurred during a late stage emplacement of the gabbro bodies, as the radiometric ages of this green hornblende overlap with those of the magmatic brown hornblende (R. Frey, 1999, personal communication).

Dolerite and gabbro are partly contemporaneous, and they show incomplete mixing structures. Other dolerite crosscuts the gabbro, but may be associated with granite with mingling and remelting structures. The mineral contents of the different dolerites is similar; they have brown hornblende and varying amounts of biotite but no clinopyroxene.

Granite sills and dikes of variable lengths and thicknesses (1–200 m) crop out mainly in one large isotropic gabbro body and in surrounding harzburgite in the northeastern part of Masirah Island (Fig. 1). Scattered, 0.5–1-m-thick granite dikes are present in harzburgite in other parts of the island. They strike northeast, similar to the associated dolerite dikes and the dolerites of the alkalic subvolcanic suite. As described by Abbotts (1978), mineralogically they are granites (quartz, 40%–44%; K-feldspar—30%–38%; plagioclase [An_{10-20}], 20%–25%; and biotite, 1%–3% along with accessory zircon, monazite, and titanomagnetite).

The middle Cretaceous alkalic volcanic and subvolcanic suite described by Meyer et al. (1996) comprises (1) intercalations of pillow lava, pillow breccia, and pyroclastic as well as epiclastic rocks within ribbon cherts of the Maghilah unit, (2) pillow lava and pillow breccia deposited on denuded Late Jurassic ridge sequences, (3) seamount structures, and (4) alkalic dikes intruded into the Late Jurassic ridge sequences. In the upper ophiolite nappe, only alkalic dikes are found. The mineral assemblage of the alkalic pillow lavas, dolerites, and microgabbro differs from the assemblages of the dolerites and pillow lavas of the Late Jurassic ridge magmatism by a higher plagioclase (An_{75-30}) and a lower clinopyroxene content. The alkalic pillow lavas are richer in apatite and ilmenite and contain brown hornblende and biotite. Clinopyroxene is Ti rich (2%–4% TiO_2). Trachyandesite, trachyte, and rhyolite are mainly encountered in the seamount structures. In these, brown hornblende, biotite, and sanidine become predominant. Brown hornblende is hastingsitic to kaersutitic, and biotite is Fe rich (Mg/Fe = 0.4–0.45).

GEOCHEMISTRY

Late Jurassic ridge magmatism

Dolerite and pillow lava are uniformly of basaltic composition. Apart from some tonalite, no more-evolved rocks are found in the Late Jurassic igneous sequence, as indicated in the analyses presented by Peters and Mercolli (1998). Mg# values [Mg/(Mg + Fe)] range between 0.7 and 0.6, near the value for melts in equilibrium with mantle olivine. The MgO versus Ni variation (Fig. 4) suggests a late-stage, low-pressure olivine fractionation of ~10% starting from a melt with 12% MgO. Fractionation of plagioclase and clinopyroxene is difficult to estimate because accumulation of phenocrysts masks fractionation effects on elements such as Ca, Na, and Al. The linear correlations between Ti, Zr, Y, and Nb indicate an olivine-plagioclase–dominated fractionation (Pearce and Norry, 1979). The flat to slightly light rare earth element (LREE)–enriched pattern (Fig. 5) suggests an affinity with E-MORB (enriched mid-oceanic-ridge basalt). The absolute Ti, Zr, and Y contents (Fig. 6) show lower con-

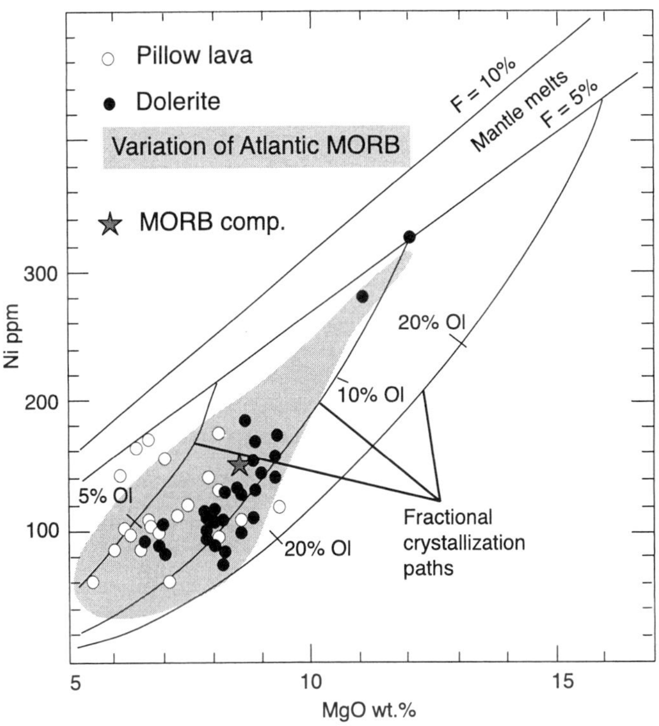

Figure 4. Ni vs. MgO variation diagram of the Late Jurassic pillow lava and dolerite. Fractional-crystallization paths after Hart and Davies (1988). F—percentage of melting.

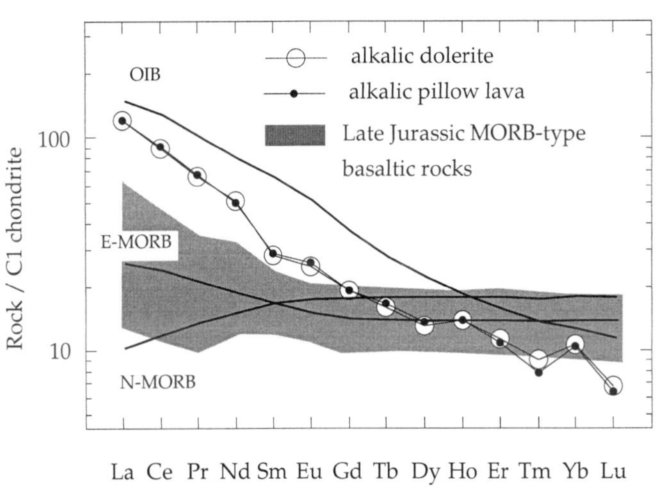

Figure 5. Chondrite (C1)–normalized REE patterns of the Late Jurassic basaltic rocks with N-MORB to E-MORB chemical affinity and the middle Cretaceous pillow lavas and dolerites with OIB chemical affinity. C1, N-MORB, E-MORB, and OIB after Sun and McDonough (1989).

centrations than is typical for N-MORB (normal MORB). The incompatible-element ratios Zr/Nb, Y/Nb, Zr/Y, and La/Ce (Fig. 6) scatter between OIB (oceanic-island basalt) and N-MORB. The moderate degree of low-pressure fractionation indicates that this variation is too large to have been generated in melts by fractional-crystallization processes. The variation is present in the dolerite of the sheeted-dike complex at the scale of a few meters as well as tens of kilometers perpendicular to strike, implying a few million years in time. A significant part of the variation in incompatible-element ratios must have been inherited from heterogeneities in the mantle source. Heterogeneities in Pb and Nd isotopes seem to be characteristic of the mantle domain of the Indian Ocean (Mahoney et al., 1998). In addition, influences of subcontinental mantle from the African as well as the Indian continental plate could also have been involved as the Masirah Ridge formed only shortly after the breakup of Gondwana, before an evolved oceanic mantle could develop.

In conclusion, Late Jurassic basaltic rocks are the product of moderate fractionation of olivine, plagioclase, and clinopyroxene from basaltic melt formed out of a heterogeneous mantle source with a low degree of melting at pressures of 10–15 kbar.

Middle Cretaceous magmatism

Geochemically, the alkalic volcanic and subvolcanic suite defines a coherent sodic alkalic magmatic suite. Mafic members fall into the trachybasalt field of the total alkali versus silica (TAS) diagram after Le Maître (1984) or the alkali basalt field in the discrimination diagram (Fig. 7) by Winchester and Floyd (1977). In the MORB-normalized spider and chondrite-normalized REE diagrams (Fig. 5), the affinity to OIB is evident. Less mafic members of this suite define a complete differentiation trend toward trachytic or rhyolitic rocks. This trend can be explained by two steps of differentiation (Meyer et al., 1996). An initial step with mainly olivine was followed by hornblende as a fractionation phase. The slightly different fractionation trends in the three different seamount structures (Fig. 7) could be the result of fractionation in local magma chambers below individual seamounts. The primitive melts of the alkalic volcanic and subvolcanic suite were produced at great depth by low degrees of partial melting from a nondepleted mantle.

In the gabbro, diorite, and granite suite, the geochemical relationships are more complex. The composition of the dolerite varies from E-MORB toward OIB affinities, as shown in the discrimination diagram (Fig. 7) by Winchester and Floyd (1977) or in the spider and REE diagrams (Fig. 5). Between the dolerites and the granites there are no intermediate compositions as in the seamount trends. As the Pb and Nd isotope signatures (Nägler and Frey, 1997) of the granites indicate a pure mantle origin, the assumption of amphibole fractionation, unmixing of a basaltic and a granitic melt, or partial melting of a K-rich, amphibole-bearing metasomatized ultramafic mantle source rock is required. The original mafic melt could have originated from a higher degree of melting from a more depleted mantle source than that of the alkalic volcanic suite.

The middle Cretaceous magmatism could be the result of hotspot activity (Marion?), an intraoceanic rift, or a contribution of both.

Sedimentary cover

Autochthonous sedimentary rocks on the Masirah ophiolite are divisible into three lithostratigraphic and biostratigraphic units (Peters et al., 1995; Immenhauser, 1996): the Upper Jurassic and Lower Cretaceous Maghilah unit, comprising pelagic radiolarian shale and ribbon shale; the middle to Upper Cretaceous Kalban unit, composed of shallow-marine to hemipelagic limestone and radiolarian shale and chert; and the Coniacian to upper Maastrichtian Fayah unit, characterized by flysch-like conglomerate, sandstone, and shale with exotic blocks. The small overlaps in age among these three units enable reconstruction of the geologic history of the ophiolite as part of the ocean floor.

The Maghilah unit (Fig. 8) has been divided into three members (Immenhauser, 1996). The basal member (member 1) comprises red

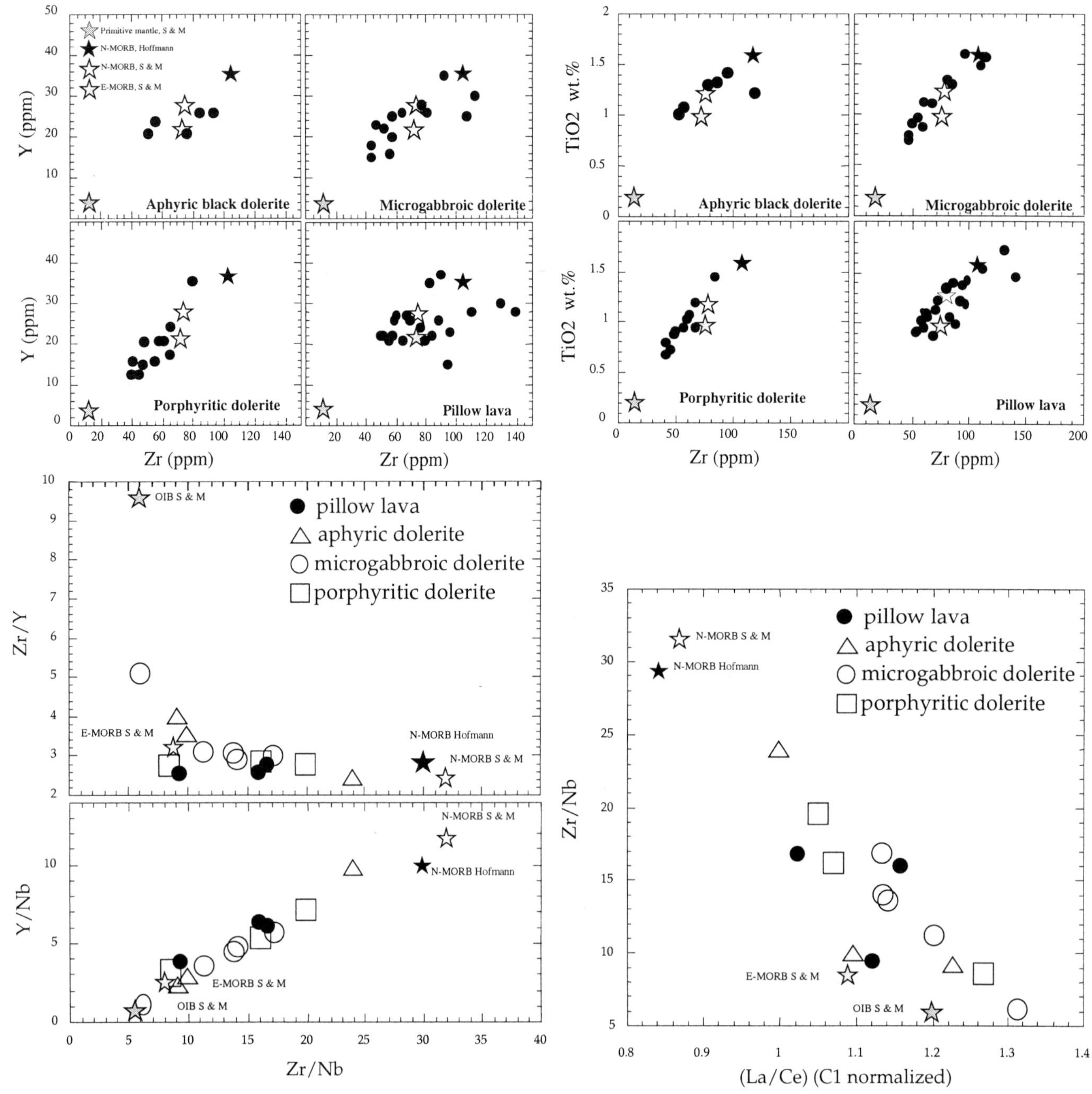

Figure 6. Zr/Nb, Y/Nb, Zr/Y, and La/Ce ratios and Zr, TiO_2, and Y contents of Late Jurassic pillow lavas and dolerites. The contents are lower and the ratios are higher than in N-MORB, indicating a heterogeneous mantle source. Compositions: S & M—Sun and McDonough (1989), Hofmann—Hofmann (1988).

and green radiolarian micrites and turbidites composed of ophiolitic material with micritic clasts. It ranges in age from Tithonian to Valanginian and varies in thickness between 7 and 15 m. Member 2 comprises dominantly brown, red, and green ribbon cherts. It has an age ranging from Valanginian to upper Hauterivian and reaches ~12 to 15 m in thickness. The first appearance of dark brown-green, alkalic volcanic breccia marks the base of member 3, which reaches thicknesses of ~20 m. Its age is late Hauterivian to Aptian. Member 3 consists of alkalic volcanic breccia and associated graded sandstone turbidites (mainly ophiolite material), shaly interbeds, and nodular cherts.

The Kalban unit comprises Barremian or Aptian limestones and Albian–Coniacian cherts. Kalban limestone beds are generally white but also brownish, reddish, or greenish. A variety of facies types was recognized: (1) coarse-grained bioclastic limestones with rudists, corals, oysters, and red algae, (2) echinoderm breccia, (3) peloidal micrites with oncoids and worn, coated bioclasts, (4) fine-grained micrites with pelagic foraminifera and ostracode shells, and (5) radiolarian limestones. No ooids were found. Many of the bioclasts have been strongly altered and underwent later silicification. The majority of these limestones are breccias embedded in a fine-grained, biomicritic matrix. Small clasts of all rock types of the ophiolite are com-

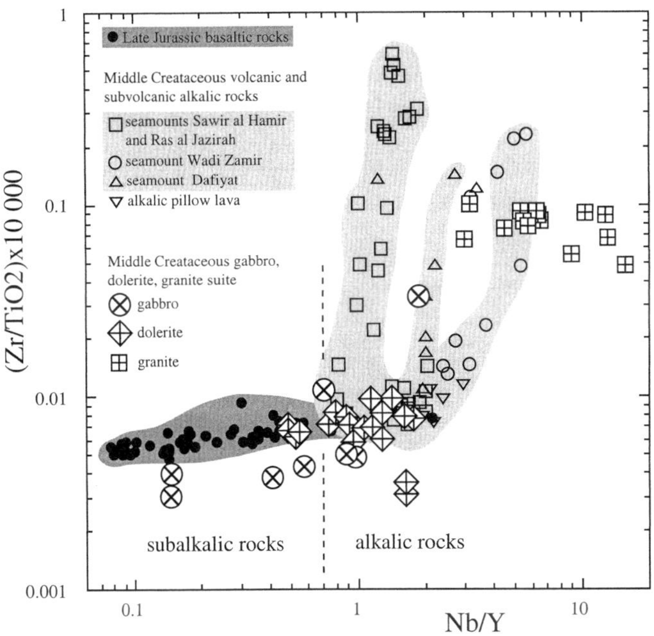

Figure 7. Discrimination diagram after Winchester and Floyd (1977). Different volcanic centers of the middle Cretaceous magmatism show individualized differentiation trends. The granites of the isotropic gabbro, dolerite, and granite suite show a more variable evolution, indicating more than one episode of differentiation.

mon components. A maximum thickness of ~80 m was measured at the northern end of Jabal Suwayr; smaller outcrops in the Kalban Plain show less than 30 m and often even less than 10 m thickness (Fig. 8). The Jabal Suwayr, ~1.8 km in outcrop length and reaching 100 m above its surroundings, is the largest outcrop of Kalban limestone on Masirah Island.

The Fayah unit, named after the type locality of the Fayah Formation (Béchennec et al., 1992) in Jabal Fayah on the Batain coast, is characterized by quartz-rich lithologies (Fig. 8). Three facies are recognized: (1) Fayah sandstones, shales, and claystones form centimeter- to meter-thick saffron-yellow beds alternating with brown and red shales. Flute and tool marks and sinuous current ripples are common. Maximum thicknesses range from 10 to 20 m. The marly intercalations are Coniacian–Campanian in age. (2) Fayah conglomerates and claystones form the second facies; the conglomerates have components from all ophiolitic rock types and from limestone and chert beds of the Kalban unit. During deposition of the conglomerates and claystones, the underlying marls and sandstones were eroded. (3) The Fayah exotic facies comprises exotic blocks of crystalline continental basement as well as 10-m-sized blocks of sedimentary rocks of a different Mesozoic age in a shaly matrix. The youngest component is determined to be of latest Maastrichtian age, comparable with the results obtained by Shackleton and Ries (1990).

By using these three sedimentary units, Immenhauser (1996) reconstructed the bathymetric evolution after the formation of the Late Jurassic ridge until its obduction onto the Arabian continent (Fig. 9). Thermal subsidence (1.5×10^{-2} cm/yr) of oceanic crust, formed at an initial depth of 2.5 km during Tithonian time, would result in the crust's reaching a depth of 4 km in middle Hauterivian time. In middle Barremian time, a horst of crystalline oceanic basement reached sea level, and the formation of carbonate platforms ensued. Renewed subsidence starting in middle Aptian time led to drowning of the platforms, and the CCD was reached in early Albian time. The calculated subsidence rate of 5×10^{-2} cm/yr is much higher than a normal thermal subsidence rate (1.5×10^{-2} cm/yr), which may indicate involvement of additional mechanisms such as normal faulting. Deep-sea deposition prevailed until the Campanian, after which time continental influx became more important. Deposition of marls in the late Maastrichtian indicates that uplift most likely correlated with the episodes of intraoceanic thrusting and obduction.

Structural evolution

Structural geology, deduced from observations of continuous exposures in three dimensions, is one of the subjects in which field studies on ophiolites have an advantage over studies of recent ocean floor. The structures revealed by the Masirah ophiolite are the result of several magmatic and tectonic events: (1) Tithonian ridge magmatism formed the ophiolite sequence with magmatic and high-temperature deformation structures and textures (Peters et al., 1995). (2) Intensive small- and large-scale normal faulting disrupted the section during its spreading away from the active ridge. (3) Middle Cretaceous normal faulting led to brittle extension of the Late Jurassic crust and syn- to postmagmatic ductile extension in the newly emplaced isotropic gabbro and granite intrusions (Marquer et al., 1998). (4) Late Cretaceous normal faulting disrupted the ocean floor. (5) Intraoceanic thrusting emplaced the upper Masirah nappe over the lower Masirah nappe between late Maastrichtian and Eocene time (Marquer et al., 1995). (6) The Masirah ophiolite nappes and associated rocks were obducted onto the Arabian continental margin. As the base of the lower Masirah nappe is not exposed, however, there is no direct structural evidence for this obduction. (7) Post-Oligocene normal faulting displaced and slightly folded the Tertiary platform carbonates (Peters et al., 1995).

FAULTING PATTERN

Faulting opens pathways enabling seawater penetration into the oceanic crust where, through water and rock interaction, hydrothermal fluids are enriched in solutes. Further along in the cycle, these solutes can be precipitated.

Intensive faulting is one of the most striking features in the Masirah ophiolite. The "patchwork" appearance of the geologic map (Fig. 1) is the result of faulting on a hectometer scale. Statistically, north-, north-northeast–, and east-trending fault directions predominate in the Mesozoic rocks, and the same holds true for the Tertiary rocks. One might conclude that all the observed faulting in the ophiolite is of post-Tertiary age. This age is only apparent because field relationships indicate that most of the faulting occurred earlier and that post-Tertiary faulting has merely reactivated preexisting structures. Direct evidence for old, ridge-related faulting is present in the sheeted-dike complex, where faults filled with hydrothermal products are cut by middle Cretaceous alkali basaltic dolerites. Furthermore, the effects of these early oceanic tectono-hydrothermal events were not found in either the overlying autochthonous sedimentary rocks nor in the middle Cretaceous gabbro, diorite, and granite suite. In order to obtain more quantitative information on the intensity of faulting during the different deformation phases, the fault densities in rocks formed at a known time were measured statistically. The ophiolite is very suitable for such a study since different rock types behave similarly under brittle deformation. The dolerite dikes that are present in all units and that formed during the two different magmatic events are very good markers for horizontal displacements as they are nearly vertical. The layering in the gabbro and the nearly hori-

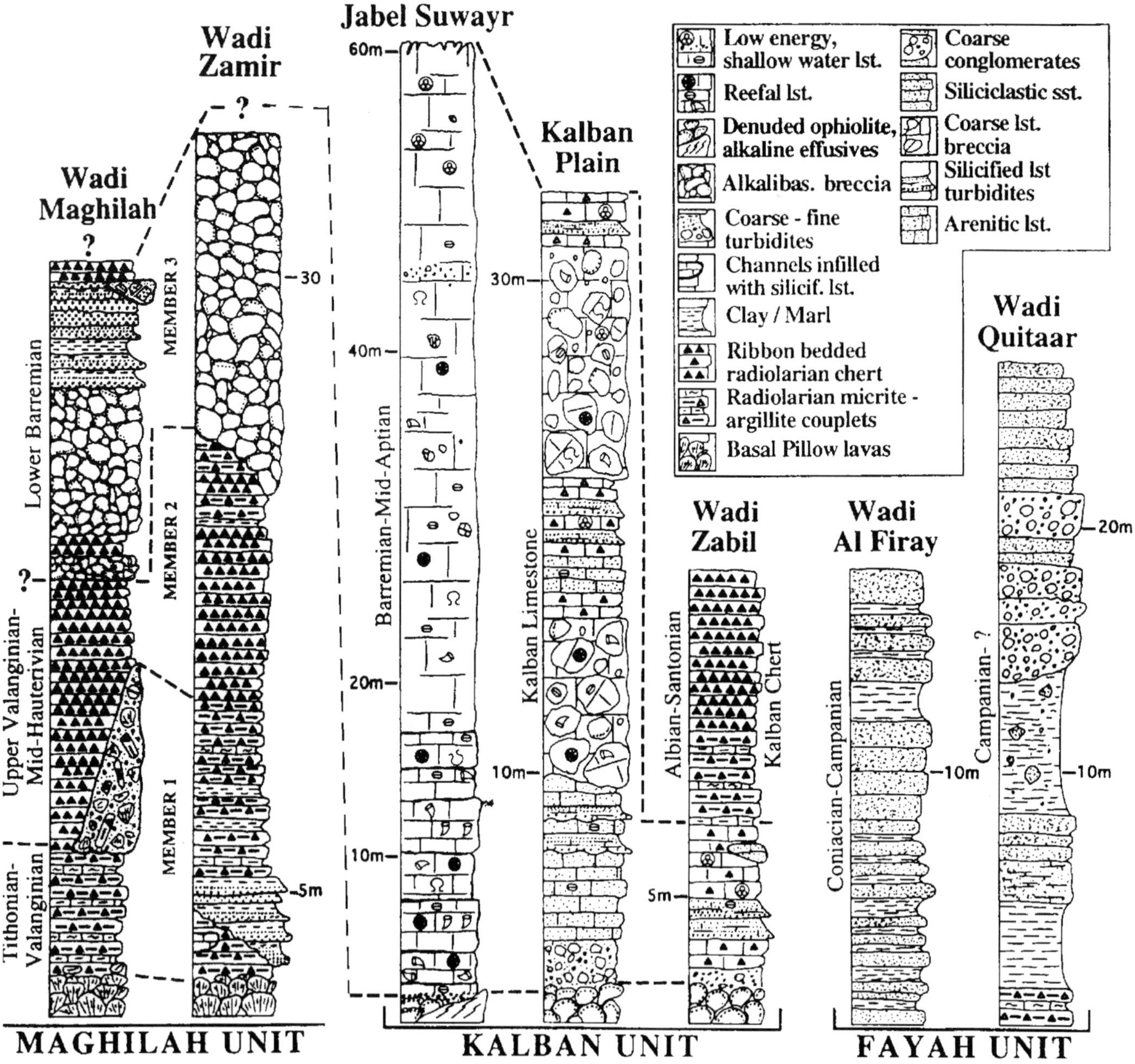

Figure 8. Stratigraphic sections of the autochthonous sedimentary cover of the Masirah ophiolite after Immenhauser (1996). The Maghilah unit comprises the pelagic sedimentary rocks deposited subsequent to the Late Jurassic ridge magmatism. The Kalban unit, which slightly overlaps in time with the Maghilah unit, starts with Barremian platform carbonates followed by slope debris and indicates renewed subsidence. The flysch-like deposits of the Fayah unit indicate a strong continental influx.

zontal granite veins are suited to determine vertical displacements. Abundances of minor faults (Table 1) were measured in the middle of blocks to avoid the effects of major faults.

In the Tertiary strata, the mean distance between north-striking faults ranges from 200 to 600 m, and the mean distance between east-striking faults ranges from 300 to 350 m. The faults are normal faults with displacements ranging from 15 to 20 m to several hundreds of meters. No minor faults were detected in either the higher Tertiary units or at the base of the laterite. In the alkalic volcanic units, the mean distance (Table 1) between minor faults (displacements of 5–50 cm) is 2 m; the mean distance between major faults (displacements of 5–20 m) is 20 to 30 m. The apparent horizontal displacements are sinistral as well as dextral, corresponding to larger normal faults. In the Late Jurassic ophiolitic rocks, measurements were made in the harzburgites of the mantle section (pyroxenite layers as markers), the layered gabbro and peridotite section, the sheeted dikes, and the tholeiitic pillow lavas. In the pillow lavas, measurements were difficult to carry out because of the lack of good markers. There were very many small faults draped in the hyaloclastites around the dense pillows. Here, some intercalated sheet flows were also used as markers. In the volcanic rocks, the mean distance between minor faults (mainly north trending) ranges from 20 to 80 cm (Table 1). In the sheeted dikes, the mean distance between minor faults is 25 cm. In the layered gabbro and peridotite section, the mean distance was 22 cm. In the mantle section, orthopyroxene layers showed a minor-fault spacing of 30 cm. In the whole Late Jurassic

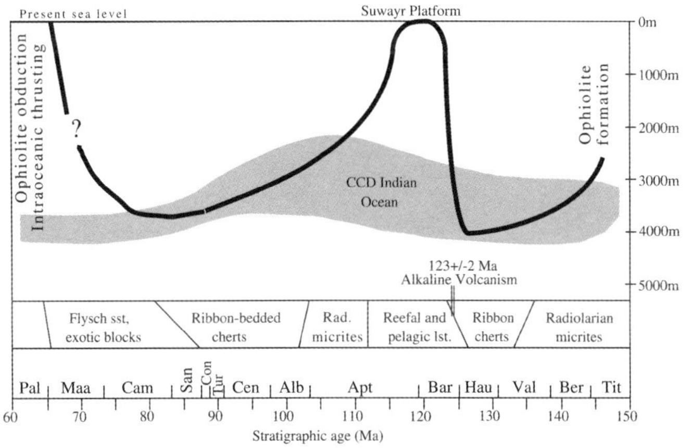

Figure 9. Bathymetric evolution of the Masirah ophiolite derived from the sedimentary cover. Modified after Immenhauser (1996). The continuous deepening after the Late Jurassic ridge formation is interpreted to be the result of strong uplift due to block faulting. This uplift was related to middle Cretaceous alkalic volcanism (hotspot activity?). The renewed subsidence lasted until the ophiolite's obduction onto the Arabian continent. Shortly before Paleocene obduction, influx of continentally derived material from the Arabian plate is registered. Stratigraphic boundaries after Harland et al. (1982).

TABLE 1. FREQUENCIES OF MINOR NORMAL FAULTS RELATED TO THE LATE JURASSIC RIDGE MAGMATISM AND MIDDLE CRETACEOUS OCEANIC RIFT AND/OR HOTSPOT MAGMATISM

Unit	Marker	Faults/Metre
Syn- and post Late Jurassic ridge magmatism		
Mantle sequence	opx layer	3.3
Layered gabbro–peridotite	Layering	4
Layered gabbro–peridotite	Gabbro dike	6
Layered gabbro–peridotite	Gabbro dike	4.8
Layered gabbro–peridotite	Dunite dike	4.5
Foliated gabbro	Dolerite dike	7.1
Sheeted dike	Chilled margin	4
Pillow lavas	Hyaloclastite margin	2
Syn– and post–middle Cretaceous magmatism		
Mantle peridotite	Alk. dolerite dike	0.5
Mantle peridotite	Granite dikes	0.5
Isotropic gabbro	Dolerite dike	0.5
Subvolc. microgabbro	Rhyolite dike	0.52

ophiolite section, an average distance between major faults of 15–20 m was found.

In both magmatic sequences, major faults similar to those observed displacing the Tertiary section are also encountered. Their abundance and displacements are often difficult to assess directly. They can, however, be inferred indirectly from the map. The distance between them ranges from 200 to 500 m. It is difficult to assess how many of these major faults could also be pre-Tertiary. The observed contemporaneous deposition of middle Cretaceous alkalic lavas and sedimentary rocks on different levels of the ophiolite sequence (harzburgite, gabbro, sheeted dikes, and effusive rocks) indicates that major block faulting on a 500 m to kilometer scale took place before or during the middle Cretaceous magmatic event.

DISCUSSION

The Late Jurassic ridge formation

The Masirah ophiolite is characterized by two features that merit discussion: (1) its reduced thickness of 2 km compared to the normal 6 km, caused by the extremely reduced thickness (only 100–500 m) of the gabbroic lower-crustal layer, and (2) the heterogeneity of the mantle source for the basaltic melts.

A reduced thickness of oceanic crust is common in slow-spreading centers. The only indication for the spreading rate of the Late Jurassic ridge segment where the Masirah ophiolite formed comes from a comparison with the northern Somali Basin. The paleolatitude of 48°S ± 12° during formation (Gnos and Perrin, 1995) in the Late Jurassic approximately corresponds to the latitude of the northern Somali Basin (Scotese et al., 1988). The spreading rate there is occurring at a rate of ~2 cm/yr; consequently, the crust should be ~6 km thick (Lin and Phipps Morgan, 1992). Hence some other factors must have played a role in causing the thin crustal section of the Masirah ophiolite. Since the Masirah ophiolite formed only 5–10 m.y. after the beginning of spreading in the Somali Basin, the ridge was less than 200 km away from the continental margins of the African-Arabian and Indian plates. According to White et al. (1992), this position could have caused additional cooling and thus could have led to a thin crust. Yet another factor could be the location of Masirah at the inside corner of a ridge-transform intersection. The following evidence supports this interpretation: (1) With east-trending ridges that offset the north-northeast–trending continental margin, a large number of north-trending transform faults are needed, resulting in ridge-segment lengths shorter than the normal 40–80 km (White et al., 1984). The probability of the 20-km-wide Masirah ophiolite having been formed near such a transform fault therefore would be very high. (2) The layering in the lower-crustal sections dips irregularly as much as 50°, in comparison to the almost horizontal layering in the Semail ophiolite (Nicolas and Boudier, 1995). (3) Kilometer-sized blocks of the sheeted-dike complex with rotated trends are reminiscent of those described from the Arakapas fault zone on Cyprus (Macleod and Murton, 1993). (4) Even before emergence of the ophiolite, its topographic relief was complex, as indicated by differences in sedimentation in the north-trending small basins (Immenhauser, 1996).

The cause of the mantle heterogeneity can also be deduced from the peculiar tectonic setting. The opening of the Somali and Masirah Basins separated two cratons (Africa and India) that were assembled during the late Proterozoic. Therefore, remnants of this old subcontinental mantle could be one of the components in the fertile source of the Masirah basalts. Furthermore, a number of currently active hotspots in the western Indian Ocean are arranged subparallel to the opening direction of the Somali and Masirah Basins (Duncan and Richards, 1991). Thus it is realistic to infer a contribution of such deep-mantle anomalies to the breakup of Africa and India and consequently the input of a new source component to the Masirah melting system.

The middle Cretaceous alkalic magmatism

The gabbro, diorite, and granite suite and the alkalic volcanic and subvolcanic suite of the middle Cretaceous magmatism have many similarities, but also some differences: They are coeval within a time interval of 123–130 Ma, the northeast strike of the dikes is common to both suites, their mineral assemblages are very similar, and both developed evolved silicic magmas from melts of pure mantle origin. The gabbro, diorite, and granite suite is intrusive into the mantle and lower-crustal sections, whereas the volcanic and subvolcanic suite is mainly extrusive. The gabbro-dolerite-granite is of bimodal (dolerite-granite) composition, whereas the volcanic suite shows a gradual transition from mafic to silicic members. The dolerites from the gabbro, diorite, and granite suite range from MORB to OIB, whereas the volcanic suite is of OIB chemical affinity.

Although the relationships between the two suites are not quite clear, they are likely a product of a common tectonic event. The middle Cretaceous magmatism coincides with the general plate reorganization that involved the Indian, African, and Australian plates and Antarctica at this time. The northeast trend of the dike structures together with the northwest lineations of the shear zones indicate northwest-southeast extensional tectonics (Marquer et al., 1998). This extension can be related to the beginning of the relative motion along a transform fault between Madagascar and the Indian plate. The fault would have extended into the Late Jurassic–Early Cretaceous oceanic crust as an intraoceanic rift. The alkalic magmatism might have been related to this oceanic rift system and/or to a hotspot. In such a scenario, the volcanic and subvolcanic suite would be the immediate products of melts originating deep in the mantle, whereas the gabbro, diorite, and granite suite would originate from such melts that reacted with the overlying depleted mantle.

Mesozoic evolution of the western Indian Ocean

The breakup of Gondwana at the end of the time of the Jurassic Magnetic Quiet Zone resulted in the opening of basins such as the Mozambique and west Somali Basins, connected by transform faults. In the west Somali Basin, the ridge segments between magnetic anomalies M25 and M0 trend east and are connected by north-trending transform faults (Cochran, 1988). The east-trending sheeted dikes of Masirah would fit in this picture. The possible paleogeographic situation of the Masirah ophiolite at M20 time (150 Ma) during formation of the ridge segment is indicated in Figure 10A.

In the western Somali Basin, spreading took place until anomaly M0 time, while Madagascar, relative to Africa, moved south along the Davie Fracture Zone. As indicated by the presence of anomalies M6 to M0 between Antarctica and Africa and between India and Australia (these anomalies are missing between Antarctica and India), the movement of India could have been decoupled from the movement of Madagascar. A translation must have taken place between India and Madagascar that might have had a continuation to the north as a precursor of the Chain and Owen Ridges (Fig. 10B). The Hauterivian-Barremian extensional regime as registered in the Masirah ophiolite has been related to this translation (Peters et al., 1995). Whether magmatism in the middle Cretaceous was also influenced by hotspot activity that could have been a precursor of the Marion hotspot remains an open question.

According to the reconstruction of Gnos et al. (1997), oceanic crust formed north of greater India from 130 Ma onward. Masirah could have either stayed with India or moved with the northward side of the ridge (Fig. 10C). At ca. 100 Ma, the Semail ophiolite formed north of Oman. The Kabul block either stayed attached to Pakistan with the Indian plate or moved north toward Eurasia.

On Masirah, the sedimentation patterns indicate a continuous deepening from Barremian (125 Ma) to the Santonian-Coniacian boundary (86 Ma). The influx of flysch-like sandstones at this boundary indicates a drastic change in sedimentation. It could also indicate that Masirah did not stay with the Indian plate. This sudden change in sedimentation coincides (A34) with the counterclockwise movement of India together with the Seychelles relative to Madagascar

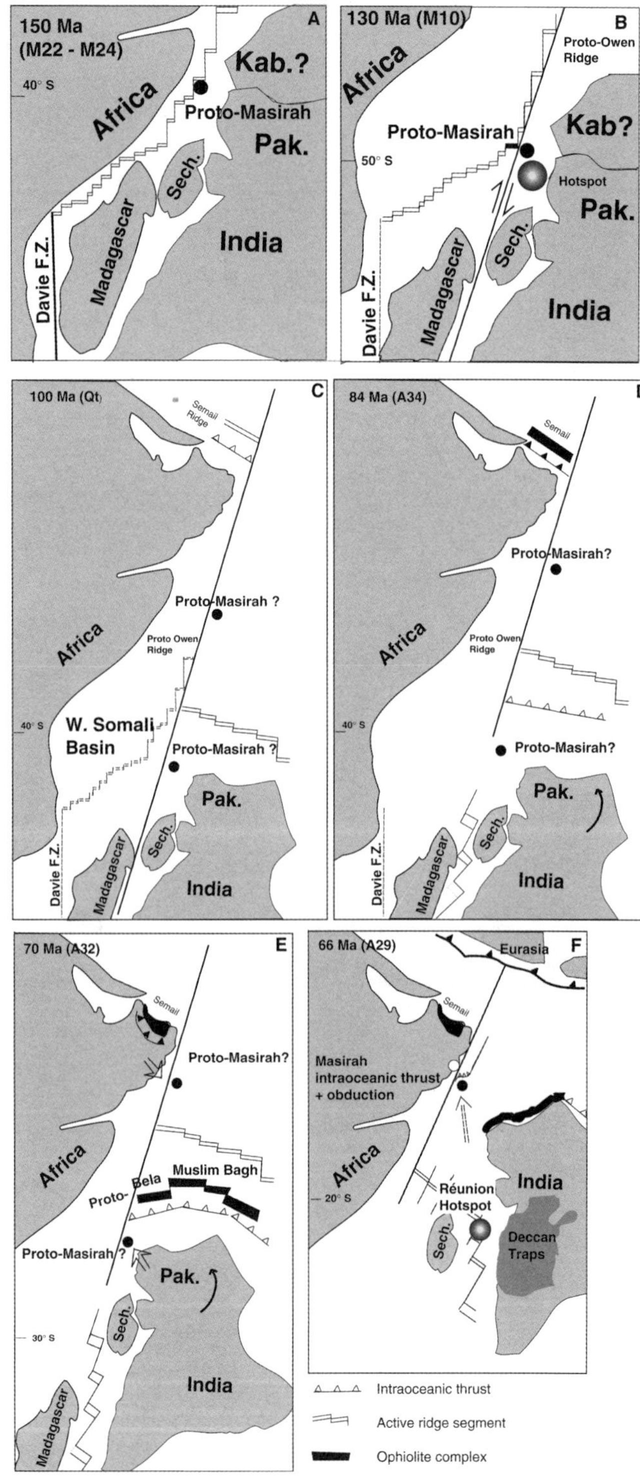

Figure 10. Tentative plate reconstructions for the evolution of the western Indian Ocean since the Late Jurassic as discussed in the text. Pak.—Pakistan, Sech.—Seychelles, F.Z.—Fracture Zone, Qt—Jurassic Magnetic Quiet Zone.

(Fig. 10D) and the obduction of the Semail ophiolite onto the northern Arabian continental margin.

Up to the early Maastrichtian, the ocean around what would become the Masirah ophiolite stayed deep (Fig. 9) during sedimentation of the flysch sandstones, as shown by the fact that they interfinger with radiolarian cherts. At this time, Masirah must have been in the neighborhood of the eastern Arabian continent, as large crystalline boulders in the sandstone originate from this area (Fig. 10E). The metamorphic sole of the 70 Ma Bela, Muslim Bagh, and Khost ophiolites in Pakistan (Mahmood et al., 1995) could have formed at this stage in an intraoceanic-ridge situation when the Indian plate started its rapid northward movement.

At the Cretaceous-Tertiary boundary (ca. 66 Ma), intraoceanic thrusting of the upper Masirah nappe over the northward-moving lower Masirah nappe (Fig. 10F) took place. The thrusting may have been caused by north-south compression from frictional resistance to the movement of the Indian plate along the Owen Fracture Zone or by friction of the subducting Indian plate below Eurasia. At the same time, the Carlsberg Ridge opened between the Seychelles and India, and the Deccan Traps formed over the Réunion hotspot. The counterclockwise component of the northward movement of India must have caused transpression along the Owen Fracture Zone that could have resulted in the obduction of the Masirah ophiolite nappes onto the margin of the eastern Arabian platform.

The normal block faulting on Masirah during the middle Miocene can be related to the onset of the opening of the Gulf of Aden since magnetic anomaly 5 (ca. 10 Ma). On the other hand, the formation of the Owen Basin could have been the result of the counterclockwise rotation of Arabia, whereby the continental margin with the overthrust ophiolite nappes (Fig. 3) was thinned by extensional tectonics.

ACKNOWLEDGMENTS

The research was financed by the Swiss National Funds (project No. 22-33559.92). Mohammed Bin Hussein Bin Kassim and Hilal Al Azry are thanked for their logistical support. We thank R.G. Coleman, R. Harris, and A. Nicolas for reviews and helpful suggestions.

REFERENCES CITED

Abbotts, I.L., 1978, High-potassium granite in the Masirah ophiolite of Oman: Geological Magazine, v. 115, p. 415–425.

Abbotts, I.L., 1979, A field petrological and geochemical study of the Masirah ophiolite, Oman [Ph.D. thesis]: University of Birmingham, 287 p.

Barton, P.J., Owen, T.R.E., and White, R.S., 1990, The deep structure of the east Oman continental margin: Preliminary results and interpretation: Tectonophysics v. 173, p. 319–331.

Beauchamp, W.H., Ries, A.C., Coward, M.P., and Miles, J.A., 1995, Masirah Graben, Oman: A hidden Cretaceous rift basin?: American Association of Petroleum Geologists Bulletin, v. 79, p. 864–879.

Béchennec, F., Roger, J., Chevrel, S., and Le Métour, J., 1992, Geological map of Al Ashkharah, Sheet NF 40–12, with explanatory notes: Directorate General of Minerals, Oman Ministry of Petroleum and Minerals, scale 1: 250 000.

Cochran, J.R., 1988, Somali basin, Chain Ridge, and origin of the Northern Somali Basin gravity and geoid low: Journal of Geophysical Research v. 93, p. 11985–12008.

Duncan, R.A., and Richards, M.A., 1991, Hotspots, mantle plumes, flood basalts and true polar wander: Reviews in Geophysics, v. 29, no. 1, p. 31–50.

Glennie, K.W., Boeuf, M.G.A., Hughes Clarke, M.W., Moody-Stuart, M., Pilaar, W.F.H., and Reinhardt, B.M., 1974, Geology of the Oman Mountains: Verhandelingen van het Koninklijk Nederlands geologisch mijnbouwkundig Genootschap, Rijswijk, v. 31, 423 p.

Gnos, E., and Perrin, M., 1995, Formation and evolution of the Masirah ophiolite constrained by palaeomagnetic study of volcanic rocks: Tectonophysics, v. 253, p. 53–64.

Gnos, E., Immenhauser, A., and Peters, Tj., 1997, Late Cretaceous/Early Tertiary convergence between the Indian and Arabian plates recorded in ophiolites and related sediments: Tectonophysics, v. 271, p. 1–19.

Harland, W.B., Cox, A.V., Llewellyn, P.G., Pickton, C.A.G., Smith, A.G. and Walters, S., 1982, A geologic timescale: Cambridge, Cambrige University Press, 128 p.

Hart, S.R., and Davies, K.E., 1988, Nickel partitioning between olivine and silicate melt: Earth and Planetary Science Letters, v. 40, p. 203–219.

Hofmann, A.W., 1988, Chemical differences of the Earth: The relationship between mantle, continental crust and oceanic crust: Earth and Planetary Science Letters, v. 90, p 297–314.

Immenhauser, A., 1996, Cretaceous sediments on the Masirah Island ophiolite (Sultanate of Oman): Evidence for an unusual bathymetric history: Journal of the Geological Society London, v. 153, p. 539–551.

Le Maître, R.W., 1984, A proposal by the IUGS subcommission on the systematics of igneous rocks for a chemical classification of volcanic rocks based on the total alkali silica (TAS) diagram: Australian Journal of Earth Sciences, v. 31, p. 243–255.

Lin, Y., and Phipps Morgan, J., 1992, The spreading dependence of three dimensional mid-ocean ridge gravity structure: Geophysical Research Letters, v. 19, p. 13–16.

MacLeod, C.J., and Murton, B.J., 1993, Structure and tectonic evolution of the southern Troodos transform fault zone, Cyprus, *in* Prichard, H.M., Alabaster, T., Harris, N.B.W., and Neary, C.R., eds., Magmatic processes and plate tectonics: Geological Society of London Special Publication 76, p. 141–176.

Mahmood, K., Boudier, F.E.G., Monié, P., and Nicolas A., 1995, The emplacement of the Muslim Bagh ophiolite, Pakistan: Tectonophysics, v. 250, p. 169–181.

Mahoney, J.J., Frei, R., Tejada, M.L.G., Mo, X.X., Leat, P.T., and Nägler, T.F., 1998, Tracing the Indian Ocean mantle domain through time: Isotopic results from old West Indian, East Tethyan, and South Pacific seafloor: Journal of Petrology, v. 39/7, p. 1285–1306.

Marquer, D., Peters, T., and Gnos, E., 1995, A new structural interpretation for the emplacement of the Masirah ophiolites (Oman): A main Paleocene intraoceanic thrust: Geodynamica Acta, v. 8, p. 13–19.

Marquer, D., Mercolli, I., and Peters, Tj., 1998, Early Cretaceous intra-oceanic rifting in the Proto-Indian Ocean recorded in the Masirah ophiolite, Sultanate of Oman: Tectonophysics, v. 292, p. 1–16.

Meyer, J., Mercolli, I., and Immenhauser, A., 1996, Off-ridge alkaline magmatism and seamount volcanoes in the Masirah Island ophiolite, Oman: Tectonophysics, v. 267, p. 187–208.

Nägler, T.F., and Frey, R., 1997, True K-feldspar granites in oceanic crust (Masirah ophiolite, Sultanate of Oman): A U-Pb and Sm-Nd isotope study: Chemical Geology, v. 138, p. 119–126.

Nicolas, A., 1989, Structures of ophiolites and dynamics of oceanic lithosphere: Dordrecht, Netherlands, Kluwer Academic Publishers, 367 p.

Nicolas, A., and Boudier, F., 1995, Mapping oceanic ridge segments in Oman ophiolite: Journal of Geophysical Research, v. 100, p. 6179–6197.

Nicolas, A., Reuber, I., and Benn, K., 1988, A new magma chamber model based on structural studies in the Oman ophiolite: Tectonophysics, v. 151, p. 87–105.

Pearce, J.A., and Norry, M.J., 1979, Petrogenetic implications of Ti, Zr, Y and Nb variations in volcanic rocks: Contributions to Mineralogy and Petrology, v. 69, p. 33–47.

Peters, Tj., and Mercolli, I., 1997, Formation and evolution of the Masirah ophiolite (Sultanate of Oman): Ofioliti, v. 22/1, p. 15–34.

Peters, Tj., and Mercolli, I., 1998, Extremely thin oceanic crust in the Proto-Indian Ocean: Evidence from the Masirah ophiolite, Sultanate of Oman: Journal of Geophysical Research, v. 103/B1, p. 677–689.

Peters, Tj., Immenhauser, A., Mercolli, I., and Meyer, J., 1995, Geological map of Masirah north and Masirah south with explanatory notes, sheet K768-north and sheet K768-south: Directorate General of Minerals, Ministry of Petroleum and Minerals, Oman, Muscat, scale 1:50 000.

Scotese, C.R., Gahagan, L.M., and Larson, R., 1988, Plate tectonic reconstruction of the Cretaceous and Cenozoic ocean basins: Tectonophysics, v. 155, p. 27–48.

Shackleton, R.M., and Ries, A.C., 1990, Tectonics of the Masirah Fault Zone and eastern Oman, *in* Robertson, A.H.F., Searle, M.P., and Ries, A.C. eds., The geology and tectonics of the Oman region: Geological Society of London Special Publication 49, p. 715–724.

Smewing, J.D., Abbotts, I.L., Dunne, L.A., and Rex, D.C., 1991, Formation and emplacement ages of the Masirah ophiolite, Sultanate of Oman: Geology, v. 19, p. 453–456.

Sun, S.S., and McDonough, W.F., 1989, Chemical and isotope systematics of oceanic basalts: Implications for mantle composition and process, *in* Sanders, A.D., and Norry, M.J.,, eds., Magmatism in the ocean basins: Cambridge, Massachusetts, Blackwell Science., p. 313–346.

White, R.S., Detrick, R.S., Sinha, M.C., and Cormier, M.H., 1984, Anomalous seismic crustal structure of oceanic fracture zones: Geophysical Journal of the Royal Astronomical Society, v. 79, p. 779–798.

White, R.S., Mckenzie, D., and O'Nions, R.K., 1992, Oceanic crustal thickness from seismic measurements and rare earth element inversions: Journal of Geophysical Research, v. 97, p. 19683–19715.

Winchester, J.A., and Floyd, P.A., 1977, Geochemical discrimination of different magma series and their differentiation products using immobile elements: Chemical Geology, v. 20, p. 325–343.

Manuscript Accepted by the Society February 23, 2000

Printed in the U.S.A.

Index

A

abyssal peridotites
  Al_2O_3 and CaO vs. MgO plots (whole rock), 23
  Cr vs. Yb plot, 143
Adria plate
  continental margin, 23, 29
  and Europe, conjugate margins formed by, 28
age of oceanic crust, 3
ages of modern supra-subduction zone samples
  Izu-Bonin forearc area, 285
  Tonga forearc, 285
ages of ophiolites
  Bay of Islands ophiolite, 441
  Betts Cove, 285
  California continental margin ophiolites, 356
  Coast Range ophiolite, 387
  Del Puerto, 285
  Iranian ophiolites, 487
  Josephine ophiolite, 285
  Karmøy, 285
  Kizildag ophiolite, 285
  Leka, 285
  Macquarie Island, 308–309
  of mantle source, southern Chile Ridge, 15
  Masirah ophiolite, 527
  of metamorphic soles of ophiolites, 53, 59–60
  Northern Apennine ophiolites, 28
  of ophiolites along California continental margin, 354
  of origin and emplacement of northern California ophiolites, 397
  Pindos, 285
  of San Pedro del Gallo terrane, Huayacocotla remnant of, 388
  Semail ophiolite, 285, 518
  southern Andean ophiolites, 331–346
  southern Turkey, Cyprus, and farther east, 5
  TAG (Trans-Atlantic Geotraverse) hydrothermal field mounds, 216
  and their metamorphic soles, 56–60
  Trinity ophiolite, 285, 406
  Troodos ophiolite, 285
  Tuolumne Complex, California, 61
  young relative to timing of subduction, 16
  Zambales, 285
ages of ophiolite-sole pairs
  Bay of Islands ophiolite, 57
  Brooks Range ophiolite, 57
  Coast Range ophiolite, 57, 61
  Semail ophiolite, 57, 59
  Trinity ophiolite, 57
Ainyn terrane, 48
Al_2O_3 and CaO vs. MgO plots (whole rock)
  abyssal peridotites, 23
  Northern Apennine ophiolites, 23
  primordial mantle estimate, 23
Al_2O_3 vs. 100 Cr/(Cr + Al), plot of, 435
Al_2O_3 vs. CaO plot (whole rock), Hayama-Mineoka belt(s), 426
Al_2O_3 vs. Sc plot, Trinity ophiolite, 413
Alexander terrane, western North America, 8–9
Alpine-Apennine belt, characteristics of, 23
Alpine-Apennine belts and MARK (Mid-Atlantic Ridge–Kane Fracture Zone intersection) area, 89
Alpine-type serpentinite masses
  Izu-Bonin forearc area and, 46
  Mariana forearc and, 46
alteration (hydrothermal) of oceanic crust
  and differences in ophiolites, 279–280
  and fluid fluxes, 273–280
  and ophiolites, 274–278
alteration (hydrothermal) patterns in supra-subduction zone ophiolites, 283–293
  Betts Cove, 288, 291
  Del Puerto, 289, 291
  epidosites, 290
  hydrothermal characteristics of, 284, 286–290
  Izu-Bonin forearc area, 289
  Josephine ophiolite, 289, 291
  Karmøy, 288, 291
  Kizildag ophiolite, 288, 291
  Leka, 288, 291
  metalliferous sedimentary rocks, 290
  ore deposits, 290
  Pindos, 288, 291
  plutonic sequences, 287, 290
  Semail ophiolite, 288, 291
  sheeted dikes, 287
  stockworks, 290
  Tonga forearc, 289
  Trinity ophiolite, 289, 291
  Troodos ophiolite, 288, 291
  volcanic rocks, 284, 286–287
  volcanic sequence–sheeted dike transition, 287
  Zambales, 289, 291
alteration intensity, Ocean Drilling Program Hole 735B, 79
alteration of basalt
  chlorite, 220
  paragonite, 220
  in TAG (Trans-Atlantic Geotraverse) up-flow zone, 226–230
alteration of gabbro, in cracking front, 263
Alvin (or Double) Mound, TAG hydrothermal deposits, 216
amphibole
  analyses of, from Hess Deep, 265
  analyses of, from Oman ophiolite belt, 266
  compositional parameters, Hess Deep and Oman ophiolite belt, 267
  micrographs of, 263
  Northern Apennine ophiolites, rare earth element plot, 24
  and plagioclase, equilibrium temperature vs. depth below dike-gabbro transition, 268
  and plagioclase, metamorphic temperatures based on, 264, 266–268
  veins of, associated with cracking front, 263
  Zabargad Island, Red Sea, Sm-Nd isotope plot, 30
amplification curves, definition of, 155
analogue models, Rayleigh-Taylor instability experiment, 125
analogues to ophiolites, modern, 4, 29–31
  Galicia margin of western Spain, 29
  Mariana convergent margin, 54
  Red Sea, 29
anatexis and heat transport, 110–111
Andaman Sea, mantle source, 7
Andean ophiolites of southern South America
  ages of, 346
  chemical compositions of, 335, 339
  Cordillera Frontal ophiolites, 343–345
  country rocks of, 346
  Eastern Córdoba ophiolite belt, 333–335
  location of (map), 332
  Pie de Palo belt, 337–339
  rare earth element plot, 336
  tectonic settings of, 346
  ultramafic and mafic belts and ophiolite assemblages, 331–333
  Western Córdoba ophiolite belt, 334–337
  Western Precordillera ophiolite, 339–343
Andean Proterozoic–early Paleozoic ophiolites, 331–346
anhydrite
  model for formation of TAG (Trans-Atlantic Geotraverse) hydrothermal field, active mound, 232
  up-flow zone, TAG (Trans-Atlantic Geotraverse), 222–224
Ankara mélange, location of (map), 5
anorthosite boudins (photograph), Bay of Islands ophiolite, 107
anorthosite "event," 10
Antarctic plate, 14
Appalachian ultramafic belt, sedimentary serpentinites and, 47
Apulian continental block, 23
Arabian continental margin, cross section of, 527
Arabian group I xenoliths, Sm-Nd isotope plot, 30
architecture
  Costa Rica Rift, extrusive rocks of, 186–188, 191
  East Pacific Rise, 184
  East Pacific Rise, extrusive rocks of, 183
  Troodos ophiolite, extrusive rocks of, 184–186, 191
  upper oceanic crust, 188, 191
arc-type chemical signature. *See* serpentine mud volcanoes; supra-subduction zone
Argolis ophiolite, location of (map), 475
Aspropotamos Complex, lithology and structure, 475–476
associated rock types, 3–4
Aswad massif
  cross section of, 503
  description of ophiolitic units, 503–507
  dike and vein orientations, 502, 506, 508
  fast-spreading ridge analogue, 499
  geologic setting, 503
  linear structures, 501, 507
  planar structures, 500, 507
  structural maps of, 500–502
  structures, 507–508
  tectonic and petrologic facies (map of), 505
  Wadi Ham, 503
  Wadi Hatta, 503, 510
Ataq xenoliths, Sm-Nd isotope plot, 30
Atlantica composite continent, 10
Atlantic coast of North America and Great Valley, 359
Atlantis Bank, Southwest Indian Ridge, 76
Atlantis Fracture Zone
  geochemistry, 97
  MARK (Mid-Atlantic Ridge–Kane Fracture Zone intersection) area, comparison with, 89
  on Na8 vs. Fe8 diagram, 116
Atlantis II Fracture Zone, Southwest Indian Ridge, 76
aureole, metamorphic. *See* metamorphic soles of ophiolites
Australian-Antarctic discordance
  depth to top of magma lens as a function of spreading rate, 163
  location of (map), 162
  and steady-state magma lens, 175
Austroalpine unit, 29
awaruite, chemical composition, Hayama-Mineoka belt(s), 424–427
axial morphology, 175
Axios (Vardar) zone of eastern Hellenic ophiolites, 475, 481

B

backarc basins, and mantle compositional domains, 7
backarc setting for ophiolite generation, 54
  forearc provenance, comparison with, 60–61
  of Solund-Stavfjord ophiolite complex, 443–458
  of western Hellenic ophiolites, 481
  See also forearc provenance
Baër-Bassit ophiolite, location of (map), 5
Baër-Bassit ophiolite, regional paleogeographic reconstruction, 495

Baft ophiolite
characteristics of, 487
tectonic setting of, 486
Bahia Barrientos ophiolite, location of (map), 14
Baimak-Buribai Complex. *See* Magnitogorsk arc
Ballantrae, 60
Balleny Fracture Zone, location of (map), 302
Baltica, 10
Band-e-Zeyarat ophiolite
characteristics of, 487
tectonic setting of, 486
Bay of Islands ophiolite
age of, 441
ages of ophiolite and sole, 57
anorthosite boudins (photograph), 107
Bay of Islands Complex (backarc-related), 434, 441
Brooks Range ophiolites analogous to, 58
cumulates (photograph), 107
division into four complexes, 434
evolution of, 441
geologic setting (and map), 434–435
Lewis Hills massif, 433–441
Little Port Complex (arc-related), 434, 441
major geographic components of (map), 106
Mariana-Bonin arc system, comparison with, 441
metamorphic conditions of sole formation, 57, 60
Mount Barren Complex (transform-related), 434–435
obduction-related units, 434
ophiolite-sole pair ages inconsistent with forearc provenance, 57–58
petrogenetic provenance of, 13
size, 57
spreading center–trench interaction, 17
Springers Hill mantle, 434–439, 441
tectonic setting, 57
Tethyan type, 57
Th-Hf-Ta discrimination diagram, 15
trace elements, 17, 435–440
Western Lewis Hills mantle, 434–437
Wheelers Brook mantle, 434, 436–437
Bear Mountains ophiolitic mélange, geologic map of, 366
Beotia ophiolite, location of (map), 475
Betts Cove
age of, 285
alteration (hydrothermal) patterns in supra-subduction zone ophiolites, 291
background alteration, 288
chemical composition of lavas, 285
epidosite, 288
hydrothermal characteristics of (table), 288
key geologic and hydrothermal characteristics of, 291
location of (map), 286
metamorphic overprint, 288
ore deposits, 288
sedimentary cover, 285
stockworks, 288
sulfide deposit in, 291
Big Blue Seamount, 36–37, 49
Kuyul ophiolite terrane as an analogue for, 48
location of (bathymetric map), 38
modes of origin of, 45
and New Idria serpentinite, 46
Bily's Peak mafic intrusive complex, Trinity ophiolite, 407
biological specimens collected during *Shinkai* 6500 submersible dives, 44
Bitlis-Zagros suture zone
location of (map), 5
regional paleogeographic reconstruction, 495
black smokers
S isotopes, 277
Sr isotopes, 275
TAG (Trans-Atlantic Geotraverse) hydrothermal field, 215–216, 218–222, 231
Blue Moon Seamount, 49
location of (bathymetric map), 38
blueschist-grade mud-volcano materials, and Mariana convergent margin, 37
blueschist-grade mud-volcano pebbles, pressure and temperature of formation of, 41
Bonanza King mafic intrusive complex, Trinity ophiolite, 407, 409–410, 412–415
Bonin Islands
Mg# vs. TiO_2 plot, and Baimak-Buribai Complex of Magnitogorsk arc, 468
rare earth element plot, 469
boninites
from Magnitogorsk arc, southern Urals, 461–470
rare earth element plot, 469
western Hellenic ophiolites, 481
Bonin-Mariana inner-trench-wall samples, Cr vs. Yb plot, 143
Bonin Trough, tectonic settings, 420
borehole-enlargement frequency vs. azimuth, Ocean Drilling Program Hole 504B, 199
borehole-televiewer (BHTV) images
Ocean Drilling Program Hole 504B, 199–200
transit-time and amplitude data, 200
Boso Peninsula, geologic map of, 420
brittle and ductile strain, western Hellenic ophiolites, 476–481
Brooks Range ophiolites
ages of ophiolite and sole, 57–58
compatible with mid-ocean ridge provenance, 61
incompatible with backarc provenance, 61
metamorphic conditions of sole formation, 58
ophiolite-sole pair ages inconsistent with forearc provenance, 58
size, 58
tectonic setting, 58
Tethyan vs. Cordilleran type, 58

C

Cadomian–Pan-African, distribution of ophiolites, 5
Calaveras Complex
description of, 369
geologic map of, 367
geologic model along northern California–Nevada transect, 373
tectonic development of western North American continental margin, 376
timeline plot of, 368
Caledonian orogeny, 10
Caledonides of Scandinavia, sedimentary serpentinites in, 47
California active-plate-margin development, subduction polarity flip, 61
California continental margin ophiolites, 351–362
ages of, 356
tectonic implications of, 359, 361–362
types of, 354
Campbell Plateau, location of (map), 303
CaO and Al_2O_3 vs. MgO plots (whole rock)
abyssal peridotites, 23
Northern Apennines ophiolites, 23
primordial mantle estimate, 23
CaO vs. Al_2O_3 plot (whole rock), Hayama-Mineoka belt(s), 426
CaO vs. SiO_2 plot
Baimak-Buribai Complex of Magnitogorsk arc, 468
Ocean Drilling Program Leg 125, Site 786, 468
Tonga Ridge, 468
Catalina schists, 59
Cayman Rise, 246
Celestial Seamount, 37, 49
description of, 39
location of (bathymetric map), 38, 42
modes of origin of, 45
perspective view of, 44
Central Alps, 29
Central Dinaric ophiolite belt, rare earth elements plot, 24
Central Metamorphic terrane
pressure and temperature of metamorphism of, 58
remnant of downgoing slab, 416
Trinity ophiolite, 404–405
Ce/Pb vs. $^{206}Pb/^{204}Pb$ discrimination diagram, southern Chile Ridge, 15
Ce/Pb vs. Nb/U discrimination diagram, southern Chile Ridge, 15
$CH_4 + H_2O \pm H_2$–bearing fluids, 248–249
Chamorro Seamounts, 37
chamosite
phase diagrams, 229
TAG (Trans-Atlantic Geotraverse) active mound, 227
channel created by lithospheric spreading, 16
characteristics of Alpine-Apennine ophiolites, 23
chemical composition
amphibole and plagioclase, from Hess Deep, 265
amphibole and plagioclase, from Oman ophiolite belt, 266
Andean ophiolites of southern South America, 335
anorthite in plagioclase vs. Mg# in clinopyroxene, 132
Bay of Islands ophiolite, minerals in, 435, 437
Betts Cove (lavas of), 285
boninites and other rocks of Magnitogorsk arc, 464–465
clinopyroxene, Northern Apennines ophiolites, 131
Cordillera Frontal ophiolites, 344–345
Cr vs. MgO plot, 143
Cr vs. Yb plot, 143
Del Puerto (lavas of), 285
evolution of primary magma, 105
External Liguride peridotites, 23–26
Gorda Ridge eruption plume, 254
Hayama-Mineoka belt(s), 421–427
Hess Deep and Oman ophiolite belt, amphibole and plagioclase, 267
Izu-Bonin forearc area (lavas of), 285
Josephine ophiolite (lavas of), 285
Karmøy (lavas of), 285
Kizildag ophiolite (lavas of), 285
Leka (lavas of), 285
Macquarie Island glass end members, 316
major elements and igneous petrogenetic constraints, 131–132
and mantle domains, 6
MARK (Mid-Atlantic Ridge–Kane Fracture Zone intersection) area, 132
Mid-Cayman Rise, 132
modern analogues to ophiolites, 30
modern ridges, 8
Northern Apennines ophiolites, 132
Ocean Drilling Program Hole 735B, 81
Ocean Drilling Program Hole 735B, samples measured, 245
of parental liquids, 134–135
Pindos (lavas of), 285
richterite-edenite exchange, 267
Semail ophiolite (lavas of), 285
Shahr-Babak ophiolite, 490–491
Sierra de Pie de Palo ophiolites, 339–340
Sierras de Córdoba ophiolites, 335–336
Solund-Stavfjord ophiolite complex, 449–451
Southwest Indian Ridge, 132, 246
subduction component in mantle-source region, 13
sulfide minerals and native copper, 424–427
TAG (Trans-Atlantic Geotraverse) fresh and altered basalts, 228
Tethyan ophiolites, 4–5
Timor ophiolite, 324
Tonga forearc (lavas of), 285
Trinity ophiolite, 412
Trinity ophiolite (lavas of), 285
Trinity terrane, mafic intrusive complexes of, 408–415
Troodos ophiolite (lavas of), 285
and various tectonic settings, 13

Western Precordillera ophiolite, 341–342
Zambales (lavas of), 285
See also cryptic variation; geochemistry
Chetco-Rogue-Galice arc complex, paleotectonic sequence diagrams for, 397
Chile Trench, location of (map), 14
Chile triple junction, 14
chlorite
alteration of basalt, 220
TAG (Trans-Atlantic Geotraverse) active mound, 227–229
chromatographic exchange, Bay of Islands ophiolite, melting model for, 439
chromian spinel, crystallization of, 144
chromite
distribution in western Hellenic ophiolites, 476–477
distribution of, and fabric variation, 478
fracture properties, 152
gradients affecting deformation of, 157
model for fracture of, 150–157
texture of, 151
Vourinos Complex, 475
chromitite. *See* podiform chromitite
chrysotile, in New Idria serpentinite, 46
Cilo ophiolite, regional paleogeographic reconstruction, 495
Cimmeride island arcs, 8–9
clathrate melting temperatures, MARK (Mid-Atlantic Ridge–Kane Fracture Zone intersection) area, fluid inclusion from, 242
clinochlore
phase diagrams, 229
TAG (Trans-Atlantic Geotraverse) active mound, 227
clinopyroxene, Bay of Islands ophiolite, 435
chemical composition, 437
trace elements, 436
clinopyroxene, Northern Apennine ophiolites
liquids in equilibrium with, 134
major element composition of, 131
rare earth element plot, 24
trace elements, 132–134, 136
clinopyroxene, Zabargad Island, Red Sea, Sm-Nd isotope plot, 30
CO_2 + CH_4 + H_2O–bearing fluids, and evolution of magmatic fluids, 244–246
CO_2 + H_2O + NaCl–bearing fluids, and evolution of magmatic fluids, 242
Coast Range ophiolite
ages of ophiolite and sole, 57–59
ages of ophiolite-sole pairs, 61
andesite submarine laharic breccia at Llanada (photograph of), 386
backarc and forearc provenance compared, 60–61
chert in Point Sal remnant (photographs of), 386
Cordilleran type, 59
cross section of formation and emplacement of, 56
cross section showing, 371
description of, 367–368
dismembered metamorphic sole, 59
exposures of, 58
forearc provenance, 47
forearc provenance of associated igneous rocks, 60
Franciscan Formation, 59
geologic map and cross section of Del Puerto ophiolite, 355
geologic map of, 366–367, 384
geologic model along northern California–Nevada transect, 373
interpillow limestone in Point Sal remnant (photograph of), 385
location of (map), 353, 396
metamorphic conditions of sole formation, 59–60
oceanic crust, comparison with, 356
ophiolite-sole pair ages mainly inconsistent with forearc provenance, 59
paleogeographic reconstruction of western U.S. Cordillera, 374
pressure-temperature-time path, 59
red jasper in Point Sal remnant (photograph of), 385
size, 58–59
stratigraphic summary, Point Sal remnant of, 387
stratigraphic summary, Stanley Mountain remnant of, 387
studies of, 352–355
tectonic development of western North American continental margin, 376
tectonostratigraphic significance of associated sedimentary strata, 383–389
timeline plot of, 368
Coast Ranges, California, sequential cross sections of, 398
cogenetic relationship between mantle peridotites and associated magmatic rocks, 21
lack of, 28, 31
C-O-H-S speciation, and evolution of magmatic fluids, 240–242
cold-mantle anomaly (cold spot), Australian-Antarctic discordance, 162–163
conductive heating, fluid-flow rates, distances, and, 226
Conical Seamount, 37, 48
location of (bathymetric map), 38
modes of origin of, 45–46
and Mysterious Valley Formation (California), 47
conjugate margins, Adria and Europe, 28
contact relationships, Bay of Islands ophiolite, 110, 112
contamination of mantle source, southern Chile Ridge, 15–16
continental accretion, sedimentary serpentinites and, 46
continental-assembly episodes and highly magmatic rifted margins, 9–10
continental convergence and dispersal history, 10
continental-margin sequences, 9
control on ridge-axis thermal structure, 164
conundrum. *See* ophiolite conundrum
convection model, 7
convergent-margin geochemical signature, 53
convergent margins, serpentine mud volcanism and, 35–49
convergent-margin setting, exposure of forearc mantle and, 48
Copley Greenstone, Eastern Klamath Mountains
Th-Hf-Ta discrimination diagram, 413
trace elements, 414
Copper Hill–Gopher Ridge arc complex, paleotectonic sequence diagrams for, 397
Coppermine River basalts, northwestern Canada, 10
Cordillera de Guaniguanico, Cuba, location of (map), 391
Cordillera Frontal ophiolites
Central belt, 343–344
Eastern belt, 344
geologic map of, 343
Los Metales belt, 343
summary of, 346
tectonic settings, 344
Cordilleran-type ophiolites
cross sections of emplacement of Coast Range ophiolite, 56
definition of, 3, 54
core-mantle boundary, 6
Costa Rica Rift
deformation decoupling horizon, 191
depth to top of magma lens as a function of spreading rate, 163
isostatic compensation for lava-flow loading, 189
layer 2–layer 3 boundary, 195–200
lithology of samples, 238, 240
location of (map), 196
postdepositional rotation (about horizontal axis) of lava flows, 188, 190–191
rate of lava deposition, 187
rate of spreading, 187
Solund-Stavfjord ophiolite complex, comparison with, 452
See also architecture; Ocean Drilling Program Hole 504B
cracking front
fast-spreading ridges and, 261–270
in lower oceanic crust, 261–270
pore-fluid phase separation and, 243
propagation of, 269–270
temperature of, 267–269
Craggy Peak mafic intrusive complex, Trinity ophiolite, 409, 411–415
Cr compatibility during partial melting, 142–144
Cr/(Cr + Al) vs. Al_2O_3, Mg/(Mg+ Fe), and TiO_2 plots, 435
Cretaceous superplume epoch, 8–10
Cr solubility in basaltic melt, 142
crustal accretion and ridge-axis topography
dependence on spreading rate, mantle temperature, and hydrothermal cooling, 161–176
crustal-genesis model of mid-ocean ridges, 164–176
boundary between steady-state magma lens and no crustal magma lens, 175
boundary conditions, 166
brittle failure, 166
conceptual model, 164–165
controlling parameters, 165
crustal thermal structure and flow vectors, 169
crustal thickness as a function of mantle temperature, 171–172
depth to melt lens as a function of mantle temperature, 174
effects of mantle temperature, 170–172
future improvements to, 176
governing equations, 165–166
hydrothermal cooling, 166
hydrothermal cooling, influence of, 172–173
melt injection, 166
melt injection and crustal flow, 165
melt migration, 165
melt-migration efficiency factor, 164–165
melt production, 164–165, 167
model formulation, 165–166
model parameters, 166, 168
numerical procedure and solution algorithm, 167
partial-melting region, melt fraction in, 170–171
results of, 168–174
sensitivity to mantle temperature and hydrothermal cooling, 173
spreading-rate dependence, 168–170
crustal-scale stacking, Semail ophiolite, 520
crustal thickness as a function of mantle temperature, 172
crustal-thickness variations and mantle upwellings, Oman ophiolite belt, 67–73
crust-mantle tectonic history, 8
Cr vs. Al_2O_3 plot, Troodos peridotite, 143
Cr vs. MgO plot
Ronda peridotite, 143
Tinaquillo, 143
Western Alps, 143
xenoliths, 143
Zabargad Island, 143
Cr vs. Yb plot
abyssal peridotites, 143
Bonin-Mariana inner-trench-wall samples, 143
Cr vs. Y discrimination diagram, Shahr-Babak ophiolite, 494
cryptic variation
Ocean Drilling Program Hole 735B, 91
Skaergaard intrusion, 95
Troodos ophiolite, 93
crystallization order, in oceanic magma chambers, 96

crystal-plastic deformation, Ocean Drilling Program Hole 735B, 82
Cuba and Mexico, Jurassic localities in (map of), 391
Cuban sedimentary serpentinite, 49
  lithologies present in, 47
cumulate-descent model for lower-crust formation, 121–126
cumulate-origin test, Trinity ophiolite, 413
cumulates
  Bay of Islands ophiolite (photograph), 107
  disaggretion of, 113
  flow of, 123–124
  formation in shallow magma lens, 123
  modal ratios in, cotectic or syntectic?, 115
  slurry of, 121–126
Cyprus Crustal Study Project (CCSP) CY-1 and CY-1A, Troodos ophiolite, 181–191
Cyprus Crustal Study Project (CCSP) CY-4, Sr isotopes, 275

D

Dar-Anar (?) ophiolite, characteristics of, 487
decompression-driven melting, 167
decoupling of deformation between lava pile and sheeted dikes
  Costa Rica Rift, 191
  Troodos ophiolite, 186, 191
Deep Sea Drilling Project Hole 334, mantle source, 7
Deep Sea Drilling Project Hole 458, Mg# vs. TiO_2 plot and Baimak-Buribai Complex of Magnitogorsk arc, 468
Deep Sea Drilling Project Hole 504B. *See* Ocean Drilling Program Hole 504B
Deep Sea Drilling Project Legs 37, 45, and 83, layer 3 rocks sampled, 76
Deep Sea Drilling Project Sites 278 and 279, location of (map), 303
deformation
  of chromite grains, 152–157
  Costa Rica Rift, 181–192
  due to lava-flow loading, 191
  as a function of strain, in chromitite pods, 155
  mantle, 481
  Northern Apennine ophiolites (photomicrograph), 135
  Ocean Drilling Program Hole 504B, 181–192
  Ocean Drilling Program Hole 735B, 90, 98
  structures in peridotites, and chromite distribution, 476–477
  structures in peridotites, and Vourinos Complex, 479
  Troodos ophiolite, 92, 181–192
Del Puerto ophiolite
  age of, 285
  alteration (hydrothermal) patterns in supra-subduction zone ophiolites, 291
  background alteration, 289
  chemical composition of lavas, 285
  epidosite, 289
  geologic map and cross section of, 355
  hydrothermal characteristics of (table), 289
  key geologic and hydrothermal characteristics of, 291
  location of (map), 286, 353
  Lotta Creek Tuff and, 355
  metamorphic overprint, 289
  oceanic crust, comparison with, 356
  ore deposits, 289
  sedimentary cover, 285
  stockworks, 289
depleted mantle composition. *See* mantle compositions
depth of metamorphism
  Bay of Islands ophiolite, 60
  Semail ophiolite and sole, 60
depth to melt lens as a function of mantle temperature, 172–174
Diablo Range, California
  New Idria serpentinite, 46
  protrusive serpentinite piercement structures in, 47
differentiation, in oceanic magma chambers, 96–98
dike orientation, rapid change in, characteristic of upper oceanic crust and ophiolites, 190
dikes and syntexis, 111–113
dike swarms
  Tethyan ophiolites, 5
  on Zabargad Island, 30
disaggregation of cumulates, 113
discrimination diagrams
  Baimak-Buribai Complex of Magnitogorsk arc, 470
  Bay of Islands ophiolite, 15
  Ce/Pb vs. $^{206}Pb/^{204}Pb$, 15
  Ce/Pb vs. Nb/U, 15
  Copley Greenstone, Eastern Klamath Mountains, 413
  Cr vs. Y discrimination diagram, 494
  incompatible elements, 493–494
  La/Ce vs. Zr/Nb, 530
  Masirah ophiolite, 529–531
  Nazca plate sediment, 15
  Nb/U vs. Ce/Pb, 15
  Nb vs. Zr, 493
  Nb/Y–Zr–Y, 492
  Nb/Y vs. Zr/Ti discrimination diagram, 492, 531
  Ni vs. MgO, 529
  $^{206}Pb/^{204}Pb$, 15
  Pie de Palo belt, southern Andes, 340
  provenance discrimination through trace-element signatures, misleading conclusions from, 16
  Semail massif, 15
  Shahr-Babak ophiolite, 492–494
  southern Chile Ridge, 15
  Th-Hf-Ta, 15, 340, 413
  Th/Yb vs. Ta/Yb, 470
  Timor ophiolite, 325
  Trinity ophiolite, 413
  Uralian fault extrusive rocks, 470
  whole-rock, 325
  Y vs. Zr, 493
  Zr/Nb vs. Y/Nb, 530
  Zr/Nb vs. Zr/Y, 530
  Zr vs. Nb, 493
  Zr vs. TiO_2, 530
  Zr vs. Y, 530
dissolution or reaction?, 113–114
distribution coefficients (mineral/melt), 437
Double (or Alvin) Mound, TAG hydrothermal deposits, 216
Dramala massif, lithology and structure, 475
DSDP. *See* Deep Sea Drilling Project
ductile fabrics
  field classification of, 477
  regional variations in, 477–478
  ductile to brittle transition
  in cooling ocean floor, 479–480
  orientation of structures formed across, 478–479

E

earthquakes, focal mechanisms of, in Mariana forearc, 37, 40
Eastern Córdoba ophiolite belt, summary of, 346
Eastern Klamath terrane
  metavolcanic rocks of, 404
  volcanic cover of easter Klamath ophiolite, 416
East Pacific Rise
  architecture based on geophysical data, 183–184
  cracking front, 261–270
  depth to melt lens as a function of mantle temperature, 174
  depth to top of magma lens as a function of spreading rate, 163
  dip of isochrons in layer 2A in, 184
  geophysical results, 183
  lithology of samples, 239
  location of (map), 14, 238
  model for magma chamber at, 88
  paleomagnetic data, cross section based on, 183
  seismic data, cross section based on, 183
  thickness of layer 2A in, 184
East Pacific Rise, lat 9° N, and Oman ophiolite belt, 73
embryonic oceans, 21
Emerald Basin, location of (map), 303
emplacement
  conditions during, 479
  of Cordilleran-type ophiolites, definition of, 54
  Jurassic ophiolites, northern California, 395–400
  models for, 480–481
  of ophiolites in forearc setting, 54
  Semail ophiolite, 519
  Solund-Stavfjord ophiolite complex, 455–458
  of supra-subduction zone Cordilleran-type ophiolites, 56
  of supra-subduction zone Tethyan-type ophiolites, 55
  timing of, for ophiolites of southern Turkey, Cyprus, and farther east, 5
  Trinity ophiolite, history of, 406
enriched mantle composition. *See* mantle compositions
environmental setting for ophiolite formation. *See* ophiolite conundrum
epidosite
  Betts Cove, 288
  Del Puerto, 289
  Izu-Bonin forearc area, 289
  Josephine ophiolite, 289
  Karmøy, 288
  Kizildag ophiolite, 288
  Leka, 288
  Pindos, 288
  Semail ophiolite, 288
  Tonga forearc, 289
  Trinity ophiolite, 289
  Troodos ophiolite, 288
  Zambales, 289
Esphandagheh ophiolite
  characteristics of, 487
  tectonic setting of, 486
evolution
  of Arabian margin, 521–522
  bathymetric, Masirah ophiolite, 533
  of Bay of Islands ophiolite, 441
  of gabbros of Northern Apennine ophiolites, 136
  of Luzon intraoceanic arc, 399
  of magma chambers at slow-spreading ridges, 87–101
  of magmatic volatiles in mid-ocean ridges, 240–246
  of Masirah ophiolite, 534
  of Masirah ophiolite structure, 531
  of MORB and continental-flood-basalt geochemistry, 115–116
  of northern California ophiolites and arc complexes, 397–398, 400
  of Okinawa Trough, 399
  of ophiolite tectonics and geodynamics of Mesozoic western North America, 374–377
  primary magma chemistry, 105
  of Ryukyu intraoceanic arc, 399
  of Semail ophiolite, 519
  of Solund-Stavfjord ophiolite complex, 456–457
  southern Uralian precollisional geodynamic, 469–470
  of Taiwan collision zone, 399
  of Timor ophiolite, 328–329
  of volcanism, Baimak-Buribai Complex of Magnitogorsk arc, 469
Evvia ophiolite, location of (map), 475

experimental design, diagram showing boundary conditions, 152
experimental design for determining stress in chromitite pods
finite-element method, 152–153
results, 154–155
stress loading in composites, 150
exposure of forearc mantle, model for, 48
extent of fractionation, southern Chile Ridge, 14
External Ligurides. *See* Northern Apennine ophiolites
definition of, 23

F

Fanuj-Maskutan ophiolite
characteristics of, 487
Fanuj Ocean, paleogeographic reconstruction of, 495
tectonic setting of, 486
Farallon-Izanagi spreading center, Panthalassa, 8–9
fast-spreading ridges
active or passive mantle upwelling, 73
cracking front, 261–270
fluid inclusions, 241
Hess Deep, 4, 241
models for magma chambers at, 88
Ocean Drilling Program Hole 504B, 241
processes at slow-spreading ridges, comparison with, 87
production of lower oceanic crust by dense cumulate slurry, 121–126
Semail ophiolite, 4, 57, 241
sensitivity of magma-lens depth to mantle temperature, 173
western Mediterranean ophiolites, 4
fault gouge
flushing of, 48
mobilized, 36
Feather River Peridotite
description of, 369
geologic map of, 367
geologic model along northern California–Nevada transect, 373
tectonic development of western North American continental margin, 376
timeline plot of, 368
FeO_t/MgO vs. density and MgO vs. FeO_t plots, Solund-Stavfjord ophiolite complex, 450
FeO_t/MgO vs. MgO plot, Solund-Stavfjord ophiolite complex, 451
FeO_t vs. TiO_2 plot, Solund-Stavfjord ophiolite complex, 451
"ferritchromit," chemical composition, Hayama-Mineoka belt(s), 427
ferrobasaltic liquids, replenishment with, 100
ferrobustamite, in Skaergaard intrusion, 95
finite-difference modeling, seismic data, 206, 208–209
finite-element method, experimental design for determining stress in chromitite pods, 152–153
fluid contribution to melting, 5
fluid-flow and mixing processes, TAG (Trans-Atlantic Geotraverse) hydrothermal field, 213–232
fluid fluxes
and hydrothermal alteration of oceanic crust, 273–280
in oceanic crust and ophiolites, 278–279
fluid inclusions
clathrate melting temperatures, 249
Costa Rica Rift, 251
and evolution of volatiles in mid-ocean ridges, 240
first-melting temperatures, 251
halite-dissolution temperatures, 244
Hess Deep, 241–242, 244, 246–247, 252
homogenization temperatures, 242, 244, 246–247, 249, 251–252
MARK (Mid-Atlantic Ridge–Kane Fracture Zone intersection) area, 241–242, 244, 246–247, 249, 252
Ocean Drilling Program Hole 504B, 241, 251
Ocean Drilling Program Hole 735B, 245, 249
salinities, 242, 246–247, 252
Semail ophiolite, 241
Southwest Indian Ridge, 241–242, 245–247, 249
temperature vs. depth, plot of, 223
Troodos ophiolite, 241–242, 244, 246–247
up-flow zone, TAG (Trans-Atlantic Geotraverse), 223
fluid phase in shear zones, origin of, 135
fluids discharging from TAG (Trans-Atlantic Geotraverse) hydrothermal field, 221–222
fluids in the shallow oceanic crust
metabasalts and, 250
sheeted dikes and, 249–250
up-flow zones and, 250
flux composition, Bay of Islands ophiolite, model for, 438
focal mechanisms of earthquakes, in Mariana forearc, 37, 40
forearc-basin fill, Great Valley Group, 47
forearc genesis of ophiolites. *See* forearc provenance
forearc mantle, exposure of, 48
forearc provenance
age relationships of ophiolite-sole pairs, 56
backarc setting for ophiolite generation, comparison with, 60–61
Coast Range ophiolite, 47
cross sections of formation of supra-subduction zone ophiolites, 55
felsic rocks associated with Coast Range ophiolite, 61
implausibility of forearc setting for Great Valley ophiolite, 400
metamorphic soles of ophiolites and, 53–61
of ophiolites, 54
paleotectonic setting, Trinity ophiolite, 415–416
of sedimentary serpentinites, 49
Semail ophiolite, 57
structural and geochronologic relationships hard to reconcile with, 60
tectonics of forearc setting, 48
formation micro scanner (FMS), Ocean Drilling Program Hole 504B, 199
fractionation, extent of, southern Chile Ridge, 14
fractionation of liquid, 99–100
fracture systems, heat transport and, 270
Franciscan Formation, California
Burro Mountain peridotite wedge, geologic map and cross section of, 361
Coast Range ophiolite, 59–60
cross section of, 355, 359–360
description of, 367
geologic model along northern California–Nevada transect, 373
gravity model across northern California, 372
location of (map), 353, 396
magnetic anomaly model across northern California, 372
map of, 355, 360, 367, 384
model cross section, 56
oceanic crust, comparison with, 356
ophiolitic rocks of, 358
paleotectonic sequence diagrams for, 397
protrusive serpentinite piercement structures in, 47
serpentine mud volcanoes and, 49
studies of, 358–359
subduction complex, 47
tectonic development of western North American continental margin, 376
timeline plot of, 368
velocity model across northern California, 372
Franciscan peridotite wedges and serpentine, location of (map), 353
Franciscan subduction complex, California
Coast Range ophiolite, 61
geologic map and cross section of, 361
geologic map of, 366
model cross section, 56

G

Galice Formation, sites of deposition of, 397
Galicia margin of western Spain, 31
cross section, 29
modern analogue to Alpine-Apennine ophiolites, 29, 31
Northern Apennine ophiolites, comparison with, 21–22
Ganj ophiolite, characteristics of, 487
garnet-clinopyroxene geothermometer, 59–60
geochemical-reservoir volume, 8
geochemistry
alteration of up-flow zone basalt, TAG (Trans-Atlantic Geotraverse) hydrothermal field, 226–230
arc-like, 4
Atlantis Fracture Zone, 97
boninites and other rocks of Magnitogorsk arc, 467
conditions of serpentinization, 427
degree of serpentinization vs. modal sulfides, 425
Eastern Córdoba ophiolite belt, 334–336
fluid inclusions, 240–249
Late Jurassic ridge magmatism, Masirah ophiolite, 528–529
Macquarie Island, N-MORB to E-MORB magmatism, 301–318
mafic intrusive complexes of Trinity terrane, 409–415
Mg# vs. Cr#, 426
modal silicates vs. sulfides, 425
Ocean Drilling Program Hole 735B, 80–81, 91, 97
oxygen fugacity vs. sulfur fugacity, 427
partial-melting estimates from whole-rock compositions, 326, 328
Pie de Palo belt, southern Andes, 337–340
Semail ophiolite, 4
Shahr-Babak ophiolite, 489–495
Skaergaard intrusion, 95, 97
Solund-Stavfjord ophiolite complex, 448–454
Southwest Indian Ridge, 97
TAG (Trans-Atlantic Geotraverse) active mound, stockwork (or up-flow) zone, 218
TAG (Trans-Atlantic Geotraverse) black and white smokers, 221
TAG (Trans-Atlantic Geotraverse) fresh and altered basalts, 228
TAG (Trans-Atlantic Geotraverse) Sr and O isotopes in anhydrite, 224–225
Troodos ophiolite, 4
Western Córdoba ophiolite belt, 335–336
Western Precordillera ophiolite, 341–343
See also chemical composition; major elements; rare earth elements; trace elements
geochronologic tests, of forearc provenance of ophiolites, 56–58
geologic model along northern California–Nevada transect
cross section of, 373
description of, 373–374
geology, of Macquarie Island, 310–315
geometry of lava flows, model for, 182
geometry of rift valley
Troodos ophiolite, 187
geophysical data
Mesozoic terrane assimilation into North American continental margin, 365–378
ophiolites along California continental margin, 354
geophysical modeling, seismic data, 203–209
geophysical results, East Pacific Rise, 183
global mid-ocean ridge system and oceanic hotspots, map of, 162
Gondwana, 8–10
and Masirah ophiolite, 534

Gorda Ridge eruption plume
temperature anomalies, 253
thermophiles and, 253
volatiles in, 254
gravity model across northern California, cross section of, 372
Great Valley, California, sequential cross sections of, 398
Great Valley Group
cross section of, 359
description of, 368
forearc-basin fill, 47
geologic map and cross section of, 357, 360–361
geologic map of, 366–367, 384
location of (map), 353, 396
map and cross section showing, 355
model cross section, 56
sites of deposition of, 397
suture zone possibly buried by, 61
tectonic development of western North American continental margin, 376
timeline plot of, 368
Great Valley ophiolite
Atlantic coast of North America, comparison with, 359
cross section showing, 371
description of, 368
geologic map of, 357, 366
geophysics of, 371–373
gravity model across northern California, 372
implausibility of forearc setting for, 400
location of (map), 353, 396
magnetic anomaly model across northern California, 372
magnetic modeling, 357
oceanic crust, comparison with, 356
paleotectonic sequence diagrams for, 397
Red Sea, comparison with, 359
studies of, 356–359
tectonic development of western North American continental margin, 376
velocity model across northern California, 372
greenstone belts, 3
Grenville orogeny, 10
Guajira Peninsula of Columbia, sedimentary serpentinites in, 48
Guam, Mg# vs. TiO_2 plot, and Baimak-Buribai Complex of Magnitogorsk arc, 468
Guleman ophiolite, regional paleogeographic reconstruction, 495

H

Happy Creek igneous complex, geologic map of, 366
Hawasina Formation, 72, 514, 519
Hayama-Mineoka belt(s)
geologic setting and map of, 420
petrogenesis and serpentinization, 419, 425–427
sulfides, 419, 424–427
heat-flow budget, 6, 8
Oman ophiolite belt, 72
heat transport and anatexis, 110–111
heat transport and fracture systems, 270
heazlewoodite, chemical composition, Hayama-Mineoka belt(s), 424–427
Hellenic ophiolite belt, western
brittle and ductile strain, 476–479
emplacement conditions, 479–480
lithology and structure, 474–476
location of (map), 474
sole formation, 481
subduction-zone rollback, 481
Hercynian, post-, gabbroic bodies, 29
Hercynian-Appalachian-Altaid orogeny, 8–10
Hercynian continental crust, Northern Apennine ophiolites and, 21, 23
Hercynian orogeny, 4, 31
distribution of ophiolites, 5
Hess Deep
amphibole and plagioclase analyses, 265
amphibole and plagioclase compositional parameters, 267
amphibole-plagioclase equilibration temperature vs. depth, 268
cracking front, 261–270
fluid inclusions, 241–242, 244, 246–247, 252
lithology of samples, 238–239
location of (map), 238
O isotopes, 276
Sr isotopes, 275
temperature of amphibole-plagioclase equilibration in microcracks, 268
trace elements, 436
heterogeneous mantle, 6–8
Masirah ophiolite, 530
highly magmatic rifted margins and continental-assembly episodes, 9–10
historical approaches to California ophiolites, 352
historical-contingency model, and mantle sources, 8, 10
hornblende
ages of ophiolite-sole pairs, 57–59
hazards of age dating ophiolitic rocks with, 56
hot abyssal layer, 8
hotspot, Reykjanes Ridge, 163
Huayacocotla anticlinorium, Mexico, location of (map), 391
Humboldt igneous complex
geologic map of, 366
geologic model along northern California–Nevada transect, 373
tectonic development of western North American continental margin, 376
timeline plot of, 368
hybridization, 112
by syntexis, 113
hydration metamorphism
serpentine mud volcanoes and, 48
supra-subduction zone mantle and, 48
hydrothermal alteration. *See* alteration (hydrothermal) of oceanic crust
hydrothermal characteristics of (table)
Betts Cove, 288
Del Puerto, 289
Izu-Bonin forearc area, 289
Josephine ophiolite, 289
Karmøy, 288
Kizildag ophiolite, 288
Leka, 288
Pindos, 288
Semail ophiolite, 288
Tonga forearc, 289
Trinity ophiolite, 289
Troodos ophiolite, 288
Zambales, 289
hydrothermal cooling, mantle temperature, and spreading rate
dependence of crustal accretion and ridge-axis topography on, 161–176
hydrothermal deposits, TAG (Trans-Atlantic Geotraverse) hydrothermal field, 216
hydrothermal history, Solund-Stavfjord ophiolite complex, 452–453
hydrothermal metamorphism, hazards of age dating ophiolitic rocks, 56
hydrothermal plumbing system
fluids bearing $CH_4 + H_2O \pm H_2$, 248–249
low- to moderate-salinity fluids, 246–248
hydrothermal system, TAG (Trans-Atlantic Geotraverse) hydrothermal field, 213–232

I

Iapetus Ocean, 9
Iceland hotspot, location of (map), 162
incipient arcs, Lau Basin, 4
incipient rifts
Gulf of California, 4
Red Sea, 4
Indian Ocean, mantle source, 7
infant-arc model, 441
Inner Mesozoic Ocean, paleogeographic reconstruction of, 495
intercratonic collision, sedimentary serpentinites and, 46
intermediate-rate spreading ridges
Chile Ridge, 14
fluid inclusions, 241
Hole 504B (south of Costa Rica Rift), 4
Ocean Drilling Program Hole 504B, 241
Ordovician ophiolites of Norway, 4
sensitivity of magma-lens depth to mantle temperature, 173
thermal structure, 172
Internal Ligurides. *See* Northern Apennine ophiolites
definition of, 23
internal structures
Solund-Stavfjord ophiolite complex, 444–446
and spreading rate, 4
intra-arc setting for ophiolite generation, 54
See also forearc provenance
Iranian ophiolites
ages of, 487
geologic map of Shahr-Babak ophiolite, 488
geologic setting, 486
Iranshahr ophiolite
characteristics of, 487
tectonic setting of, 486
island-arc tholeiites, western Hellenic ophiolites, 481
isochrons
at Costa Rica Rift, 188
in East Pacific Rise layer 2A, dip of, 184
through East Pacific Rise based on lava-deposition distribution, 184
through Troodos ophiolite based on paleomagnetic inclinations, 185
isostatic compensation for lava-flow loading
at Akaki River section of Troodos ophiolite, 189
based on postdepositional rotation of lava flows, 189
at Costa Rica Rift, 189
isotherms in cooling ocean floor, Vourinos Complex, 479
isotherms offset upward, Oman ophiolite belt, 72
isotope exchange, in anhydrite from TAG (Trans-Atlantic Geotraverse) hydrothermal field, 225
isotopic anomalies (Hf and Nd), in arc-basin systems, 45
Ispendere-Komurhan ophiolite, regional paleogeographic reconstruction, 495
Ivrea-Verbano unit, 29
Izanagi-Pacific spreading center, 8–9
Izmir-Ankara-Erzincan suture zone, 4
Izu-Bonin forearc area
age of, 285
and Alpine-type serpentinite masses, 46
background alteration, 289
bathymetric and geologic map of, 36
chemical composition of lavas, 285
epidosite, 289
hydrothermal characteristics of (table), 289
location of (map), 286
metamorphic overprint, 289
ore deposits, 289
ridge of serpentinized peridotite in, 45
sedimentary cover, 285
stockworks, 289
Izu-Bonin-Mariana forearc, trace elements, 414
Izu-Bonin-Mariana forearc area
Al_2O_3 vs. CaO plot (whole rock), 426
Mg# vs. Cr#, 426
Izu-Bonin Trench, tectonic settings, 420

J

Jabal Akhdar window, Semail ophiolite, 514
Japan Sea, mantle source, 7
Jarbo Gap ophiolite
cross section showing, 371
geologic map of, 367

geologic model along northern California–Nevada transect, 373
tectonic development of western North American continental margin, 376
timeline plot of, 368
Josephine ophiolite
age of, 285
alteration (hydrothermal) patterns in supra-subduction zone ophiolites, 291
background alteration, 289
chemical composition of lavas, 285
cross section showing, 371
epidosite, 289
Galice Formation, 390–391
geologic map of, 367
hydrothermal characteristics of (table), 289
key geologic and hydrothermal characteristics of, 291
location of (map), 286, 396
metamorphic overprint, 289
ore deposits, 289
paleogeographic reconstruction of western U.S. Cordillera, 374
paleotectonic sequence diagrams for, 397
sedimentary cover, 285
Smith River subterrane, 389
stockworks, 289
stratigraphic summary, Smith River subterrane of, 391
sulfide deposit in, 291
tectonic development of western North American continental margin, 376
tectonostratigraphic significance of associated sedimentary strata, 383, 389–391
timeline plot of, 368
volcanic member of, 389–390
volcanopelagic succession, 390
Western Klamath terrane, 389
Juan de Fuca Ridge
depth to top of magma lens as a function of spreading rate, 163
mantle source, 7
Jurassic (Middle to Late), northern California, paleotectonic sequence diagrams for, 397
Jurassic continental-margin arc, description of, 369–370
Jurassic island-arc terrane and its oceanic basement, description of, 368
Jurassic localities in Mexico and Cuba (map of), 391
Jurassic ophiolites, northern California
location of (map), 396
models for origin and emplacement of, 395–400

K

Kallidromon ophiolite, location of (map), 475
Kane Fracture Zone
cross section of Mid-Atlantic Ridge, 89
on Na8 vs. Fe8 diagram, 116
trace elements, 436
See also MARK (Mid-Atlantic Ridge–Kane Fracture Zone intersection) area
Karmøy
age of, 285
alteration (hydrothermal) patterns in supra-subduction zone ophiolites, 291
background alteration, 288
chemical composition of lavas, 285
epidosite, 288
hydrothermal characteristics of (table), 288
key geologic and hydrothermal characteristics of, 291
location of (map), 286
metamorphic overprint, 288
ore deposits, 288
sedimentary cover, 285
stockworks, 288
sulfide deposit in, 291
Kermanshah ophiolite
characteristics of, 487
regional paleogeographic reconstruction, 495
tectonic setting of, 486
Khoy ophiolite
characteristics of, 487
tectonic setting of, 486
kinematic histories, contrasting between lava pile and sheeted dikes, 191
Kissavos, lithology and structure, 475
Kizildag ophiolite
age of, 285
alteration (hydrothermal) patterns in supra-subduction zone ophiolites, 291
background alteration, 288
chemical composition of lavas, 285
epidosite, 288
hydrothermal characteristics of (table), 288
key geologic and hydrothermal characteristics of, 291
location of (map), 5, 286
metamorphic overprint, 288
ore deposits, 288
regional paleogeographic reconstruction, 495
sedimentary cover, 285
stockworks, 288
Klamath Mountains
correlation with Sierra Nevada foothills, 370
geologic map and cross section of, 360
geologic map of, 367, 404
Josephine ophiolite, 383, 389–391
location of (map), 396
oppositely dipping subduction zones, 371
ultramafic rocks and associated rock units, 353
Knoxville Formation, California, sedimentary serpentinites in, 47
Koryak fold-and-thrust belt, Kamchatka, sedimentary serpentinites, 48
Koziakas ophiolite
lithology and structure, 476
map of, 475
Kremlin area, TAG (Trans-Atlantic Geotraverse) hydrothermal field, 220, 229
Kuyul ophiolite terrane as an analogue
for Big Blue Seamount, 48
for Mariana forearc, 48
for Pacman Seamount, 48

L

La/Ce vs. Zr/Nb discrimination diagram, Masirah ophiolite, 530
Lake Combie Complex, geologic map of, 367
Lake Sevan, Armenia, protrusive serpentinite piercement structures in, 47
Lanzo North and South (Western Alps)
rare earth element plot, 24, 27
Sm-Nd isotope plot, 26
Sr-Nd isotope plot, 24
large igneous provinces (LIPs), 7–8, 10
La/Sm vs. Zr/Sm plot, Baimak-Buribai Complex of Magnitogorsk arc, 468
Lau Basin, depth to top of magma lens as a function of spreading rate, 163
Laurasia, 8–10
Laurentia, 10
Laurussia, 10
lava-deposition distribution
Costa Rica Rift, 187
model for, 184
Troodos ophiolite, 187
layer 2–layer 3 boundary
Costa Rica Rift, 195–200
a stress change?, 195–200
layer 3 rocks dredged
Mid-Atlantic Ridge, 75
Southwest Indian Ridge, 75
layer 3 rocks sampled, Deep Sea Drilling Project Legs 37, 45, and 83, 76
layered gabbro, Bay of Islands ophiolite, 107–108
layered intrusions and ophiolites, magma-chamber evolution, 87–101
layering, in oceanic magma chambers, 96
leaky-transform rifts, Neotethys, 8
Leka
age of, 285
alteration (hydrothermal) patterns in supra-subduction zone ophiolites, 291
background alteration, 288
chemical composition of lavas, 285
epidosite, 288
hydrothermal characteristics of (table), 288
key geologic and hydrothermal characteristics of, 291
location of (map), 286
metamorphic overprint, 288
ore deposits, 288
sedimentary cover, 285
stockworks, 288
leopard texture (photograph), Bay of Islands ophiolite, 109
Lesser Caucasus, sedimentary serpentinites in, 47
lherzolite, foliated (photograph), 111
Ligurian ophiolites, 21–31, 137
location of (map), 5
See also Northern Apennine ophiolites
Ligurian Sea sector of the Tethys, 21, 23, 31
cross section of, 136
diagrammatic model for evolution of, 29
liquid lines of descent, Solund-Stavfjord ophiolite complex, 449
Lister model, the
and constraints on the cracking front in the lower crust, 267–268
temperature distribution vs. depth below dike-gabbro transition, 269
lithology, of Macquarie Island, 311–312
lithospheric rifting, 399–400
lithostratigraphy
Iranian ophiolites, 486–487
Masirah ophiolite, 527
Ocean Drilling Program Hole 735B, 90
sedimentary cover, Masirah ophiolite, 532
lit-par-lit intrusions (photograph), Bay of Islands ophiolite, 109
Little Port Complex (arc-related), Bay of Islands ophiolite, 434
lizardite, in sedimentary serpentinites, 46
Logtown Ridge arc complex, paleotectonic sequence diagrams for, 397
Lotta Creek Tuff, and Del Puerto ophiolite, 355
lower oceanic crust
constraints on cracking front in, 267
cracking front, 261–270
fast-spreading ridges, 121–126
generation by syntexis, 105–116
melt migration across mushy, 110
models for formation of, 122
Ocean Drilling Program Hole 735B, 75–85
production of by dense cumulate slurry, 121–126
as a reactive filter, 115
Southwest Indian Ridge, 75–85
viscosity of magma lens and upper gabbros, 124–125
low-temperature zone, TAG hydrothermal deposits, summary of, 216
Luzon intraoceanic arc, structural evolution of, 399

M

MacKenzie dike swarm, northwestern Canada, 10
Macquarie Island
basaltic and sedimentary rocks, 311
experimentally determined cotectics and origin of glass end members, 316–318
gabbroic rocks, 311–312
geodynamic history of, 307–308
geologic map of, 304–305
geology of, 310–315
glass end members, analyses of, 316
harzburgite association, 312–313
Macquarie Ridge Complex, location of (map), 302–303

magmatism, summary of, 318
magmatism in a waning system, 318
metamorphism, 313–314
mixed lithologies, 312
model ages, 308–309
N-MORB to E-MORB magmatism, 301–318
petrogenesis of N-MORB to E-MORB, 315–318
plate-boundary geometry at 11 Ma, 308–309
postspreading history, 309
"primitive" glass compositions, significance of, 316
sheeted dolerite dikes, 311
spreading and postspreading events, 314
structural map of, 306–307
transpression, 314–315
troctolite, 312
mafic and ultramafic cumulates, Bay of Islands ophiolite, 108–110
magma chambers
axial, model for, 88
depth to top of magma lens as a function of spreading rate, 163
tectonism and evolution of, 100–101
zoning in, 99–100
magma chambers at fast-spreading ridges, models for, 88, 123
magma chambers at slow-spreading ridges
evolution of, 87–101
magmatic and tectonic controls on, 87–101
magmatic foliation variations, Ocean Drilling Program Hole 735B, 82
magmatism, N-MORB to E-MORB. *See* Macquarie Island
magnetic anomaly model across northern California, cross section of, 372
magnetic modeling, Great Valley area, California, 357
Magnitogorsk arc
Baimak-Buribai Complex, major and trace element data, 464–465
geologic map of Urals, 462
location of Baimak-Buribai Complex, 462
rare earth element plot, 469
Shankhai and Tanalyk sample localities, 463
Sr-Nd isotope plot, 469
stratigraphic sections, Shankhai and Tanalyk, 466
stratigraphic setting, 462
Western Pacific Tertiary boninites, comparison with, 467–469
Magnitogorsk arc, southern Urals, Early Devonian boninites from, 461–470
major elements
igneous petrogenetic constraints and, 131–132
Ocean Drilling Program Hole 735B, 81
mantle compositional domains and origins, 6
mantle compositions, 6
mantle flow rates, Oman ophiolite belt, 71
mantle heterogeneity, 6–7
Masirah ophiolite, 530
mantle mixing models, 6
mantle source
depleted, 4
historical-contingency model, 8, 10
non-MORB (mid-oceanic-ridge basalt), 4
mantle-source regions, changes in, 10
mantle structure, model for, 8
mantle temperature, hydrothermal cooling, and spreading rate
dependence of crustal accretion and ridge-axis topography on, 161–176
"threshold" mantle temperature and changes in axial thermal structure, 172
mantle upwellings and crustal-thickness variations, Oman ophiolite belt, 67–73
mantle variations beneath modern oceans, 6–7
Maqsad area
chromite fracture properties, 152
Oman ophiolite belt, 151
Mariana active arc, Mg# vs. TiO_2 plot, and Baimak-Buribai Complex of Magnitogorsk arc, 468
Mariana convergent margin
Alpine-type serpentinite masses and, 46
analogues to ophiolites, modern, 54
bathymetric and geologic map of, 36
bathymetric map of, 38, 41–43
and blueschist-grade mud-volcano materials, 37
focal mechanisms of earthquakes in, 37, 40
Kuyul ophiolite terrane as an analogue to, 48
model for emplacement of serpentine masses and, 48–49
serpentine mud volcanism and, 35–36
Mariana Trench, Mg# vs. TiO_2 plot, and Baimak-Buribai Complex of Magnitogorsk arc, 468
Mariposa Formation, sites of deposition of, 397
MARK (Mid-Atlantic Ridge–Kane Fracture Zone intersection) area
chemical composition, 132
clathrate melting temperatures from fluid inclusion, 242
cross section of, 89
fluid inclusions, 241–242, 244, 246–247, 249, 252
lithology of samples, 238–239
location of (map), 238
mantle source, 7
Raman (micro-) analyses from fluid inclusion, 242
Sr isotopes, 275
trace elements, 134
Mashad ophiolite
characteristics of, 487
tectonic setting of, 486
Masirah ophiolite, 7
bathymetric evolution of, 533
cross section of, 526
faulting pattern of, 531–533
faults related to magmatic events, 533
geologic setting, 525–527
geologic-tectonic map of, 526
Late Jurassic ridge formation, 533
lithostratigraphy, 527
location of (map), 5, 526
Mesozoic ridge magmatism, 527
middle Cretaceous alkalic magmatism, 528–529, 533–534
Ni vs. MgO discrimination diagram, 529
paleogeographic reconstruction of, 534
rare earth elements, 529
sedimentary cover, 529–530
massive sulfides
model for formation of TAG (Trans-Atlantic Geotraverse) hydrothermal field, active mound, 232
TAG (Trans-Atlantic Geotraverse) active mound, 218–220
TAG (Trans-Atlantic Geotraverse) hydrothermal field, 216
mass spectrometry, Ocean Drilling Program Hole 735B, samples measured, 245
megamullions, formation of seamounts and, 48
melt and peridotite reaction, 144
melt circulation and reaction zones, Oman ophiolite belt, 70
melt focusing below spreading axis, and formation of podiform chromitite, 144–145
melting
Cr compatibility during, 142–144
decompression driven, 167
depth to melt lens as a function of mantle temperature, 172
melting, degree of, Hayama-Mineoka belt(s), 426
melting, open-system modeling of, Bay of Islands ophiolite, 438–439
melting without separation, 167
melt migration
across a mushy lower crust, 110
in oceanic magma chambers, 98–99
melt-migration efficiency factor, crustal-genesis model of mid-ocean ridges, 164–165
Mendocino triple junction, geologic map and cross section of, 360
Meso-Hellenic trough, location of (map), 475
mesospheric boundary layer, 6
mesostructural cross section, of Northern Apennine ophiolites, 130
Mesozoic terrane assimilation into North American continental margin, geophysical data and, 365–378
metamorphic aureole. *See* metamorphic soles of ophiolites
metamorphic conditions of sole formation, 59–60
Bay of Islands ophiolite, 57, 60
Brooks Range ophiolites, 58
Coast Range ophiolite, 59–60
Semail ophiolite, 57, 60
Trinity ophiolite, 58
metamorphic expansion of supra-subduction zone ultramafic rocks, exposure of forearc mantle and, 48
metamorphic overprint
Betts Cove, 288
Del Puerto, 289
Izu-Bonin forearc area, 289
Josephine ophiolite, 289
Karmøy, 288
Kizildag ophiolite, 288
Leka, 288
Pindos, 288
Semail ophiolite, 288
Tonga forearc, 289
Trinity ophiolite, 289
Troodos ophiolite, 288
Zambales, 289
metamorphic soles of ophiolites, 53–61
age approximates time of subduction, 53
age relationships of ophiolite-sole pairs, 56–57, 59–60
cross sections showing forearc setting, 54
definition of, 53
in fault contact with ophiolite, 60
inverted metamorphic gradients in, 6, 60
metamorphic conditions of formation, 53
oceanic protoliths, 53
ophiolite-sole ages, 56–57
Semail ophiolite, 514, 516, 518
tectonic settings, 60–61
timing of metamorphism, 53
vertical tectonic transport, 60
welded to ophiolite base, 60
western Hellenic ophiolites, 480–481
metamorphic temperatures, from amphibole-plagioclase pairs, 264, 266
metamorphism and alteration
at greenschist facies, Trinity terrane, 407
Ocean Drilling Program Hole 735B, 78–80
metamorphism and microfractures, 262–263
metamorphism by hydration
hazards of age dating ophiolitic rocks, 56
serpentine mud volcanoes and, 48
metasomatized subarc mantle, 16
methodology for ophiolite studies, 352
Mexico and Cuba, Jurassic localities in (map of), 391
Mg/(Mg+ Fe) vs. Cr/(Cr + Al) plot, 435
Mg# vs. TiO_2 plot, Baimak-Buribai Complex of Magnitogorsk arc, 468
MgO vs. Al_2O_3 and CaO plots (whole rock)
abyssal peridotites, 23
Northern Apennine ophiolites, 23
primordial mantle estimate, 23
MgO vs. Al_2O_3 plot, Troodos peridotite, 143
MgO vs. CaO and Al_2O_3 plots (whole rock)
abyssal peridotites, 23
Northern Apennine ophiolites, 23
primordial mantle estimate, 23
MgO vs. FeO, Solund-Stavfjord ophiolite complex, 449
MgO vs. FeO_t and FeO_t/MgO vs. density plots, Solund-Stavfjord ophiolite complex, 450
MgO vs. FeO_t/MgO plot, Solund-Stavfjord ophiolite complex, 451

MgO vs. SiO_2 plot
Baimak-Buribai Complex of Magnitogorsk arc, 468
Ocean Drilling Program Leg 125, Site 786, 468
Tonga Ridge, 468
microcracking temperatures and depth, 266
microcracks
cracking temperature, 268–269
origin of by semibrittle failure, 266
propagation of cracking front, 269–270
scale of, 268
microfracturing, in cracking front, 262–263
microthermometry, Ocean Drilling Program Hole 735B, samples measured, 245
Mid-Atlantic Ridge
cross section of, 89
depth to magma-lens top as a function of spreading rate, 163
location of (map), 238
and steady-state magma lens, 175
See also TAG (Trans-Atlantic Geotraverse) hydrothermal field
Mid-Cayman Rise, chemical composition, 132
Midcontinent rift, North America, 10
mid-oceanic ridges
Coast Range ophiolite, 61
compatible setting for some ophiolites, 61
forearc provenance, comparison with, 61
and mantle compositional domains, 7
new observations of remote, 162–164
supra-subduction zone ophiolites, comparison with crust of, 292–293
volatiles in, 237–255
See also crustal-genesis model of mid-ocean ridges
Mineoka ophiolite. *See* Hayama-Mineoka belt(s)
mineral assemblages
oceanic crust, 274–276
ophiolites, 274
TAG (Trans-Atlantic Geotraverse) active mound, stockwork (or up-flow) zone, 218
mineral chemistry
Bay of Islands ophiolite, 435–436
Timor ophiolite, 327
mineral compositions
Hess Deep, 263–264
Oman ophiolite belt, 263–264
Mir zone, TAG hydrothermal deposits, summary of, 216
Mitsero graben, Troodos ophiolite
as an abandoned spreading center, 190
deformation decoupling horizon, 186
Miura Peninsula, geologic map of, 420
mixing of continental and greenstone detritus, and Solund-Stavfjord ophiolite complex, 457
modal compositions
in cumulates, cotectic or syntectic?, 115
Hayama-Mineoka belt(s), 421–422
Ocean Drilling Program Hole 735B, 78
in Timor ophiolite, 323
model for
Atlantis Bank, Southwest Indian Ridge, 76
axial magma chambers, 88
Bay of Islands ophiolite, 436–439
boundary conditions for crustal genesis, 166
chromite-grain fracture, 150–157
convection, 7
crustal-genesis model of mid-ocean ridges, 164–168
cumulate descent to form lower crust, 121–126
emplacement of serpentine masses and, 48–49
emplacement of serpentine mud volcanoes, 48–49
emplacement of western Hellenic ophiolites, 480–481
exposure of forearc mantle, 48
formation of lower oceanic crust, 122
geometry of lava flows, 182
governing equations for crustal genesis, 165–166
historical contingency bearing on mantle sources, 8, 10
hydrothermal cooling at mid-ocean ridges, 166
lava-deposition distribution, 184
Ligurian Sea sector of the Tethys, evolution of, 29
magma chamber at East Pacific Rise, 88
magma chambers at fast-spreading ridges, 88, 123
magmatic development from propagating-rift tip, Solund-Stavfjord ophiolite complex, 451
magmatic fluid evolution, 245
mantle mixing, 6
mantle structure, 8
melting, parameters for, 439
melt injection and crustal flow at mid-ocean ridges, 165
melt injection at mid-ocean ridges, 166
melt migration at mid-ocean ridges, 165
melt production at mid-ocean ridges, 164–165, 167
numerical procedure and solution algorithm for crustal genesis, 167
oceanization process in Red Sea, 31
ophiolite generation, 54
origin and emplacement of northern California Jurassic ophiolites, 395–400
plate tectonics and Mineoka ophiolite, 428
Rayleigh-Taylor instability, 125
RIDGE2M, 125
sedimentary serpentinites (model of Lockwood), 35, 48
seismic data, 203–209
Semail ophiolite magma chamber, 88
stress in chromitite pods, 154–155
subducted slabs in lower mantle, 6
subduction initiation, 48
TAG (Trans-Atlantic Geotraverse) hydrothermal field, active mound, 230–232
tectonic evolution of Arabian margin, 521–522
Moho bulge, mantle gravity anomalies above East Pacific Rise and, 151
Moho transition zone
map of, 69
Oman ophiolite belt, 67–73
thickness variations of, 70–71
MORB signatures
southern Chile Ridge, 15–16
western Hellenic ophiolites, 481
Mount Barren Complex (transform-related), Bay of Islands ophiolite, 434
mud volcanoes not related to convergent margin, 37
mud volcanoes related to convergent margin. *See* serpentine mud volcanoes
Muskox intrusion, northwestern Canada, 10
Myrtle Group, California, location of (map), 396
Mysterious Valley Formation
Conical Seamount, compared with, 47
possible serpentine mud volcano and, 47, 49

N

Na8 vs. Fe8 diagram, 116
Na8 vs. Fe8 diagram, FAMOUS area on, 116
NaCl + H_2O–bearing fluids and magmatic fluid evolution, 243–244
Nain ophiolite
characteristics of, 487
tectonic setting of, 486
nascent forearc setting for ophiolite generation. *See* forearc setting
native copper and sulfides, chemical composition, 424–426
Nazca plate, 14
Nazca plate sediment, Ce/Pb vs. $^{206}Pb/^{204}Pb$ discrimination diagram, 15
Nb/U vs. Ce/Pb discrimination diagram, southern Chile Ridge, 15
Nb vs. Zr discrimination diagram, Shahr-Babak ophiolite, 493
Nb/Yb vs. Th/Yb plot, Trinity ophiolite, 414
Nb/Y–Zr–Y discrimination diagram, Shahr-Babak ophiolite, 492
Nb/Y vs. Zr/Ti discrimination diagram
Masirah ophiolite, 531
Shahr-Babak ophiolite, 492
Nd-Sr isotope plot
Baimak-Buribai Complex of Magnitogorsk arc, 469
Uralian fault extrusive rocks, 469
Nena composite continent, 10
Neotethyan ophiolites
forearc provenance of associated igneous rocks, 60
MARK (Mid-Atlantic Ridge–Kane Fracture Zone intersection) area, 89
Neotethys, 4
distribution of ophiolites, 5, 7
leaky-transform rifts, 8
paleogeographic reconstruction of, 495
rifts of, 10
Nevada. *See* California continental margin ophiolites; North American continental margin
New (or Shikai) Mound, TAG hydrothermal deposits, 216
New Britain arc, trace elements, 414
"New Global Tectonics," 3
New Guinea (eastern), subduction polarity flip, 61
New Idria serpentinite
and Big Blue Seamount, 46
location of (map), 353
and Pacman Seamount, 46
New Zealand ultramafic belt, sedimentary serpentinites and, 47
Neyriz ophiolite
characteristics of, 487
regional paleogeographic reconstruction, 495
tectonic setting of, 486
Ni vs. Al_2O_3 plot, Troodos peridotite, 143
Ni vs. MgO discrimination diagram, Masirah ophiolite, 529
N-MORB (average), trace elements in, 134
nonuniformitarian relationship between ophiolite composition and tectonic environment, 8
North American continental margin
Mesozoic terrane assimilation into, 365–378
regional geology, 366–370
See also California continental margin ophiolites
North Arm Mountain massif, Bay of Islands ophiolite
contact relationships, 110
layered gabbro, 107–108
leopard texture (photograph), 109
lit-par-lit intrusions (photograph), 109
mafic and ultramafic cumulates, 108–110
on Na8 vs. Fe8 diagram, 116
See also Bay of Islands ophiolite
North Chamorro Seamount, 49
description of, 44
location of (bathymetric map), 38, 43
perspective view of, 45
Northern Apennines, Italy, ophiolite–oceanic lithosphere analogue, 21–31
Northern Apennine ophiolites
CaO and Al_2O_3 vs. MgO, plots of, 23
chemical composition, 132
evolution of gabbroic rocks of, 129–137
fluid phase in shear zones, origin of, 135
geodynamic implications for, 136–137
geologic setting, 130
location of (map), 23, 130
lower oceanic crust (modern) and, 129–137
mesostructural cross section of, 130
$^{187}Os/^{188}Os$ vs. Al_2O_3 plot, 25
petrology of, 23–27
rare earth elements in, 24, 27
slow-spreading ridges (modern) and, 129–137
Sm-Nd isotope plot, 26

Sr-Nd isotope plot, 24
structure of, 22–23
tectonic denudation, 28
tectonic setting of, 27–29
trace elements, 134
transform-zone tectonic setting and, 28
Northern Mound, TAG hydrothermal deposits, 216
Nusselt number, crustal-genesis model of mid-ocean ridges, 166, 172, 174

O

obduction events, serpentine mud volcanoes and, 49
obduction of Semail nappe and Arabian margin geometry, 519–520
obduction-related units, Bay of Islands ophiolite, 434
Ocean Drilling Program, Mariana convergent margin, 37
Ocean Drilling Program Hole 504B
azimuth of borehole elongation, 197
borehole cross sections, 200
borehole-enlargement frequency vs. azimuth, 199
borehole-televiewer (BHTV) images, data from, 199
Costa Rica Rift, extrusive rocks of, 186–187
density, 205
downhole geophysical data, 204–205
downhole measurements and images, 196–198
downhole stress analysis, 198–200
drilling history, 196–197
electrical resistivity, 197
fluid inclusions, 241
formation micro scanner (FMS), data from, 199
images, electrical, 197
intermediate-rate spreading ridges, 4
layer 2–layer 3 boundary, 75, 195–200
lithology of samples, 238
lithostratigraphy, 197
location of (map), 196
O isotopes, 276
paleomagnetic inclinations of lavas and dikes in, 181–191
porosity, 197, 205
sediment properties at, 205
seismic data, 205
seismic energy loss and attenuation, 204
seismic energy loss vs. two-way traveltime, 206–208
seismic reflection coefficient series, 207
S isotopes, 277
Sr isotopes, 275
stress-regime change in, 197
synthetic results compared with real seismic data, 206
velocity, 205
Ocean Drilling Program Hole 735B
average primary modal compositions, 78
chemical composition, 81
cryptic variation, 91
deformation, 90, 98
downhole profile, 78
fluid inclusions, 249
geochemistry, 80–81
lithology of lower crust, 77–78
lithology of samples, 238
lithostratigraphic column, 77, 90
location of (map), 76
lower oceanic crust, 75–85
major oxide compositions, 81
MARK (Mid-Atlantic Ridge–Kane Fracture Zone intersection) area, comparison with, 89
metamorphism and alteration, 78–80
O isotopes, 276
paleomagnetism, 82–83
processes in the magma chamber, 95–100
S isotopes, 277
Southwest Indian Ridge, 75–85
Sr isotopes, 275
structural geology, 81–82
tectonic settings, 76–77
Ocean Drilling Program Hole 735B, samples measured, 245
Ocean Drilling Program Hole 895, lithology of samples, 238
Ocean Drilling Program Hole 896A
O isotopes, 276
Sr isotopes, 275
Ocean Drilling Program Holes 417D and 418A
O isotopes, 276
Sr isotopes, 275
Ocean Drilling Program Leg 118. *See* Ocean Drilling Program Hole 735B
Ocean Drilling Program Leg 125, Site 786
CaO vs. SiO_2 plot, and Baimak-Buribai Complex of Magnitogorsk arc, 468
MgO vs. SiO_2 plot, and Baimak-Buribai Complex of Magnitogorsk arc, 468
rare earth element plot, 469
Ocean Drilling Program Leg 125, and Torishima Forearc Seamount, 45
Ocean Drilling Program Leg 141, Site 862, 14
Ocean Drilling Program Leg 153
trace elements, 134
See also MARK (Mid-Atlantic Ridge–Kane Fracture Zone intersection) area
Ocean Drilling Program Leg 158, geochemical and isotopic data, 227
Ocean Drilling Program Sites 920–924, lithology of samples, 238
Ocean Drilling Program Sites 921–924, mantle source, 7
oceanic crust
age of, 3
in California continental margin ophiolites, 360
fluid fluxes and hydrothermal alteration, 273–280
isotopic profiles, 275–278
lithology of samples, 238
mineral assemblages, 274–276
O isotopes, 276
S isotopes, 277
Sr isotopes, 275
submarine hydrothermal systems, 274
See also lower oceanic crust; the specific ophiolite and Ocean Drilling Program entries; upper oceanic crust
oceanic hotspots and global mid-ocean ridge system, map of, 162
oceanization
early stages of, Northern Apennine ophiolites and, 22
process in Red Sea, model for, 31
ODP. *See* Ocean Drilling Program
O isotopes
oceanic crust, 276
Semail ophiolite, 276
Troodos ophiolite, 276
Okinawa Trough, structural evolution of, 399
olistoliths
Northern Apennine ophiolite occurrences as, 23–24
and Solund-Stavfjord ophiolite complex, 456
olistrostromes containing serpentinite, in California Coast Ranges, 47
olivine
Bay of Islands ophiolite, 435
chemical composition, Hayama-Mineoka belt(s), 422
Oman Mountains, Semail ophiolite, 514–515
Oman ophiolite, location of (map), 5
Oman ophiolite belt
active or passive mantle upwelling, 73
amphibole and plagioclase analyses, 266
amphibole and plagioclase compositional parameters, 267
amphibole-plagioclase equilibration temperature vs. depth, 268
area that drifted away from ridge, 72
block diagrams of, 151
cracking front, 261–270
cross section of, 122
crustal-thickness variations and mantle upwellings, 67–73
deformation of, 71
mantle upwellings and crustal-thickness variations, 67–73
map of, 68, 515
maps and cross sections of, 151
Moho transition zone, 67–73
olivine-melt suspension, 67
projection of cross sections, 70
propagating ridge segment, 67, 70
restoration of gravity sliding, 69
size of mantle upwellings, 73
structural signature of mantle diapirs, 67
temperature of amphibole-plagioclase equilibration in microcracks, 268
viscosity, 67
See also Semail ophiolite
ophiolite conundrum, 4, 7–8
ophiolite–oceanic lithosphere analogue
Ligurian ophiolites, 21–31
Northern Apennines, Italy, 21–31
ophiolites and layered intrusions, magma-chamber evolution, 87–101
ophiolite-sole pair ages inconsistent with forearc provenance
Bay of Islands ophiolite, 57–58
Brooks Range ophiolites, 58
Coast Range ophiolite, 59
Semail ophiolite, 57, 59
Trinity ophiolite (not definitive for), 58
ophiolite tectonics and geodynamic evolution, Mesozoic western North America, 374
ophiolitic sulfide deposits, TAG (Trans-Atlantic Geotraverse) hydrothermal field, comparison with, 217
orebodies of podiform chromitite, formation of, 140
ore deposits
alteration (hydrothermal) patterns in supra-subduction zone ophiolites, 290
Betts Cove, 288
Del Puerto, 289
Izu-Bonin forearc area, 289
Josephine ophiolite, 289
Karmøy, 288
Kizildag ophiolite, 288
Leka, 288
Pindos, 288
Semail ophiolite, 288
Tonga forearc, 289
Trinity ophiolite, 289
Troodos ophiolite, 288
Zambales, 289
origin of anomalous MORB signatures, southern Chile Ridge, 15–16
origin of mantle compositional domains, 6
orthopyroxene, Bay of Islands ophiolite, 435
orthopyroxene porphyroclasts (photograph), 111
$^{187}Os/^{188}Os$ vs. Al_2O_3 plot
Northern Apennine ophiolites and, 25
Pyrenees peridotites, 25
Ronda peridotite, 25
Os isotope data, Zabargad Island, 30
Othris ophiolite
deformational fabrics in, 478
maps of, 475–476
strain in, compared to other western Hellenic ophiolites, 481
Owen Basin, cross section of, 527
Owen Ridge
cross section of, 527
paleogeographic reconstruction of, 534
oxygen fugacity vs. sulfur fugacity, conditions of serpentination, 427

P

Pacific-Antarctic Ridge, location of (map), 14
Pacific-Farallon spreading center, 8–9
Pacman Seamount, 37, 49

Kuyul ophiolite terrane as an analogue for, 48
location of (bathymetric map), 38
modes of origin of, 45
and New Idria serpentinite, 46
Palau, Mg# vs. TiO_2 plot, and Baimak-Buribai Complex of Magnitogorsk arc, 468
paleogeographic reconstructions
Late Cretaceous Neotethys (Iran and parts of Turkey, Arabia, and Pakistan), 495
Masirah ophiolite, 534
Mesozoic western North America, 374
Middle Jurassic, western U.S. Cordillera, 374
Solund-Stavfjord ophiolite complex, 446
western Indian Ocean, 534
paleomagnetic data, cross section based on, East Pacific Rise, 183
paleomagnetic inclinations of lavas and dikes
in Costa Rica Rift, 188, 191
in Ocean Drilling Program Hole 504B, 181–191
in Troodos drill core, 191
in Troodos ophiolite, 181–191
paleomagnetism, Ocean Drilling Program Hole 735B, 82–83
Paleotethyan subduction, 5, 8–9
Pan-African–Brasiliano orogeny, 10
Pan-African metamorphism, on Zabargad Island, 30
Panthalassa (Paleotethys), 8–9
paragonite
alteration of basalt, 220
alteration of chloritized basalt, 228–229
TAG (Trans-Atlantic Geotraverse) active mound, 227
parental liquids, 134–135
See also primary magma proxies
pargasite (titanian), Northern Apennine ophiolites, trace elements, 136
Paskenta, oceanic crust, comparison with, 356
passive continental margins, 21
formation of, 28
lower Mesozoic sequence of western North America, 370
$^{206}Pb/^{204}Pb$ vs. Ce/Pb discrimination diagram, southern Chile Ridge, 15
Pb/Ce vs. Sr/Nd plot, N-MORB, 415
Peacock Seamount, 49
description of, 39
location of (bathymetric map), 38, 42
modes of origin of, 45
perspective view of, 44
Peñon Blanco arc complex, paleotectonic sequence diagrams for, 397
Penshinskaya Gulf, Kamchatka, sedimentary serpentinites, 48
pentlandite, chemical composition, Hayama-Mineoka belt(s), 424–427
peridotite ridge (supra-subduction zone mantle) in Izu-Bonin forearc area, 45
Permian subcontinental lithosphere, Northern Apennine ophiolites and, 22, 28, 31
petrogenesis
Bay of Islands ophiolite, 436–439
of N-MORB to E-MORB, Macquarie Island, 315–318
and serpentinization, Hayama-Mineoka belt(s), 419, 425–427
Shahr-Babak ophiolite, 493–495
Solund-Stavfjord ophiolite complex, 448–449
petrogenetic implications, Trinity primary magma and, 411–412
petrography
boninites and other rocks of Magnitogorsk arc, 466–467
Hayama-Mineoka belt(s), 421
Iranian ophiolites, 487, 489
Late Jurassic ridge magmatism, 527
Trinity ophiolite, 408
petrology
blueschist-grade mud-volcano materials, 41, 44
Hayama-Mineoka belt(s), 421
of metamorphic soles of ophiolites, 53
Northern Apennine ophiolites, 23–27
ophiolites along California continental margin, 354
of podiform chromitite, 140
of Solund-Stavfjord ophiolite complex, 443–458
phase diagrams
bearing on chloritization followed by paragonization, 229
experimentally determined cotectics and origin of Macquarie Island glass end members, 316–318
Fo-An-Di at 1 atm, 112
Fo-An-Qtz at 1 atm, 112
NaCl-H_2O system, 248
oxygen fugacity vs. sulfur fugacity, 427
pressure vs. temperature, and hypothetical paths of seawater circulation, 247
S-Fe-(Ni + Co) ternary plot, 426
SiO_2-olivine-Cr_2O_3, 114
sulfide minerals and native copper, 426
two-phase liquid + vapor curve at various temperatures, 243
phase-equilibrium topology, influence of during magma mixing, 114–115
phase separation of magmatic fluids, and evolution of magmatic fluids, 242–243
physical properties, Ocean Drilling Program Hole 735B, 83
Pie de Palo belt, southern Andes
chemical compositions of, 339
geochemistry, 337–339
geologic map and photograph of, 338
summary of, 346
tectonic setting of, 339
Th-Hf-Ta discrimination diagram, 340
Pindos
age of, 285
alteration (hydrothermal) patterns in supra-subduction zone ophiolites, 291
background alteration, 288
chemical composition of lavas, 285
epidosite, 288
hydrothermal characteristics of (table), 288
key geologic and hydrothermal characteristics of, 291
location of (map), 286
metamorphic overprint, 288
ore deposits, 288
sedimentary cover, 285
stockworks, 288
sulfide deposit in, 291
Pindos ophiolite
deformational fabrics in, 479
lithology and structure, 475–476
maps of, 475
strain in, compared to other western Hellenic ophiolites, 481
plagioclase
and amphibole, equilibrium temperature vs. depth below dike-gabbro transition, 268
and amphibole, metamorphic temperatures based on, 264, 266–268
analyses of, from Hess Deep, 265
analyses of, from Oman ophiolite belt, 266
compositional parameters, Hess Deep and Oman ophiolite belt, 267
plume, from eruption event, 253
plume-generated rifting, 9
plume production, 7
podiform chromitite
deformation of, 150
formation of, 140–141, 149–150
gauging stress from, 149–157
texture of chromite in, 151
water and formation of, 139–145
Point Sal, oceanic crust, comparison with, 356
Porcupine Lake mafic intrusive complex, Trinity ophiolite, 406, 409, 411–415
pore-fluid pressure vs. depth, plot of, and evolution of magmatic fluids, 243
Precordillera fold-and-thrust belt, geologic map of, 340
pressure and temperature of metamorphism
Bay of Islands ophiolite, sole of, 57
of blueschist-grade mud-volcano pebbles, 41
ophiolite sole, 60
of South Chamorro Seamount, 46
Trinity ophiolite, sole of (Central Metamorphic belt), 58
See also individual ophiolites
pressure-temperature-time path, Coast Range ophiolite, 59
primary magma proxies
Trinity ophiolite, 411–412
See also parental liquids
primordial mantle estimate
Al_2O_3 and CaO vs. MgO plots (whole rock), 23
processes
generation of lower oceanic crust by syntexis, 105–116
in oceanic magma chambers, 95–100, 122–124
See also evolution
propagated rift segment, Solund-Stavfjord ophiolite complex, 450
propagating suture zone, northern California, 398
Proterozoic–early Paleozoic southern Andean ophiolites, 331–346
Proterozoic subcontinental lithosphere, Northern Apennine ophiolites and, 21–24, 28, 31
proto-arc setting, Trinity ophiolite and, 403–404, 415
protrusive serpentinite piercement structures, triggered by seismic activity, 48
provenance discrimination through trace-element signatures, misleading conclusions from, 16
provenance of ophiolites
See mantle source; tectonic settings
pulsation of subducting plate, 16
"punctuations" on uniform process, 10
Pyrenees peridotites
$^{187}Os/^{188}Os$ vs. Al_2O_3 plot, 25
Sm-Nd isotope plot, 26
Sr-Nd isotope plot, 24
pyrite
alteration of chloritized basalt, 228–229
TAG (Trans-Atlantic Geotraverse) active mound, 227

Q

quartz
alteration of chloritized basalt, 228–229
fluid-inclusion trapping temperatures from, TAG (Trans-Atlantic Geotraverse) up-flow zone, 223–225
TAG (Trans-Atlantic Geotraverse) active mound, 227

R

radiolarian chert, Coast Range ophiolite, 386
Raman (micro-) analyses, MARK (Mid-Atlantic Ridge–Kane Fracture Zone intersection) area, fluid inclusion from, 242
Raman (micro-) analyses from fluid inclusion, Ocean Drilling Program Hole 735B, samples measured, 245
rare earth elements
Baimak-Buribai Complex of Magnitogorsk arc, 469
Bay of Islands ophiolite, 436, 438
Bonin Islands, 469
boninites, 469
Central Dinaric peridotites, 24
Lanzo North (Western Alps) gabbros, 27
Lanzo South ophiolites, 24
Masirah ophiolite, 529
Northern Apennine ophiolites, 24, 27, 132
Ocean Drilling Program Leg 125, Site 786, 469

Pie de Palo belt, southern Andes, 340
Pyrenees peridotites, 24
Ronda peridotite, 24
Shahr-Babak ophiolite, 490–495
Sierras de Córdoba ophiolites, 336
Tonga Ridge, 469
Trinity ophiolite, 412, 414–415
Rasht ophiolite
characteristics of, 487
tectonic setting of, 486
rate of lava deposition
Costa Rica Rift, 187
Troodos ophiolite, 187
rate of spreading
Costa Rica Rift, 187
Troodos ophiolite, 187
Rattlesnake Creek terrane
cross section showing, 371
geologic map of, 367
timeline plot of, 368
Rayleigh-Taylor instability, experimental model of, 125
reaction or dissolution?, 113–114
reactive filter, lower oceanic crust as, 105, 115
reconstruction of Earth's surface and interior, 8
recrystallization during decreasing *P* and *T* conditions
External Ligurides, 24, 29
Galicia peridotites of western Spain, 29
Internal Ligurides, 24–25
Red Sea
formation of, 31
and Great Valley, comparison with, 359
modern analogue to Alpine-Apennine ophiolites, 29, 31
Northern Apennine ophiolites, comparison with, 21–22
Sm-Nd isotope plot, 30
replenishment with ferrobasaltic liquids, 100
Resolution Ridge, location of (map), 302
Resolution Rift margin, location of (map), 303
retrograde metamorphism, 29
Northern Apennine ophiolites and, 21, 24, 31
on Zabargad Island, 30
Reykjanes peninsula lavas, on Na8 vs. Fe8 diagram, 116
Reykjanes Ridge
depth to melt lens as a function of mantle temperature, 172
depth to top of magma lens as a function of spreading rate, 163
location of (map), 162
and presence of steady-state magma lens, 175
rheology
of "chromite" inclusion, 153
of dunite, 153
podiform chromite, 151
richterite-edenite exchange, Hess Deep and Oman ophiolite belt, 267
RIDGE2M, crustal-genesis model of mid-ocean ridges, 167–168
ridge-axis topography and crustal accretion
dependence on spreading rate, mantle temperature, and hydrothermal cooling, 161–176
ridge-propagation model, Aswad massif, 510
ridge-trench-trench triple junctions, 13
Rodinia, 10
Rogue-Chetco arc complex
cross section showing, 371
tectonic development of western North American continental margin, 376
timeline plot of, 368
Ronda peridotite
Cr vs. MgO plot, 143
$^{187}Os/^{188}Os$ vs. Al_2O_3 plot, 25
Sm-Nd isotope plot, 26
Sr-Nd isotope plot, 24
rotation (postdepositional) about horizontal axis
based on paleomagnetic inclinations, 191
Costa Rica Rift, 188, 190–191
in Troodos ophiolite, 185–186, 190–191
in upper oceanic crust, 182–184
Ryukyu intraoceanic arc, structural evolution of, 399

S

Sabzevar ophiolite
characteristics of, 487
Sabsevar Ocean, paleogeographic reconstruction of, 495
tectonic setting of, 486
Sahneh ophiolite, characteristics of, 487
Saih Hatat window, Semail ophiolite
location of (map), 514–515
metamorphism in, 517–519
structure of, 517
Saipan, Mg# vs. TiO_2 plot, and Baimak-Buribai Complex of Magnitogorsk arc, 468
Salinian block, 59
geologic map of, 366
Samail ophiolite. *See* Semail ophiolite
Scandinavian Caledonides, sedimentary serpentinites in, 47
Sc vs. Al_2O_3 plot, Trinity ophiolite, 413
seafloor piercements of serpentinite, sources of sedimentary serpentinite, 47
seafloor-spreading origin, of Solund-Stavfjord ophiolite complex, 454–455
seafloor topography, global, map of, 162
seamounts
megamullions, and formation of, 48
subduction of, 48
See also individual seamounts; serpentine mud volcanoes
secondary minerals, Ocean Drilling Program Hole 735B, 79
sedimentary cover
Betts Cove, 285
Del Puerto, 285
Izu-Bonin forearc area, 285
Josephine ophiolite, 285
Karmøy, 285
Kizildag ophiolite, 285
Leka, 285
Masirah ophiolite, 529–532
Pindos, 285
Semail ophiolite, 285
Tonga forearc, 285
Trinity ophiolite, 285
Troodos ophiolite, 285
Zambales, 285
sedimentary serpentinites
along Pambak River in the Lesser Caucasus, 47
continental accretion and, 46
in Cuba, 48–49
difficulty of recognizing in the geologic record, 49
in Guajira Peninsula of Columbia, 48
in Koryak fold-and-thrust belt, Kamchatka, 48
lithologies and minerals present in, 46
model of Lockwood about, 35, 48
on-land examples of, 46–48
along Penshinskaya Gulf, Kamchatka, 48
sedimentary structures in, 49
from Wilbur Springs, northern California, 47
seismic data, cross section based on, East Pacific Rise, 183–184
seismic discontinuity at 670 km depth, 8
seismic events, protrusive serpentinite and, 48
seismic modeling
multiple back-scattering (MBS) convolution model, 208
seismic energy loss, 209
two-dimensional finite-difference model (FDM), 208
seismic velocity model, Atlantis Bank, Southwest Indian Ridge, 76
seismological models, subducted slabs in lower mantle, 6
Semail massif, 13, 15, 17, 68, 151
Semail ophiolite, 7, 68, 72
age of, 285
ages of ophiolite and sole, 57, 59
alteration (hydrothermal) patterns in suprasubduction zone ophiolites, 291
background alteration, 288
Brooks Range ophiolites analogous to, 58
chemical composition of lavas, 285
compatible with mid-ocean ridge provenance, 61
epidosite, 288
fluid inclusions, 241
geochronology of, 518–519
geologic map of, 514
hydrothermal characteristics of (table), 288
incompatible with backarc provenance, 61
key geologic and hydrothermal characteristics of, 291
location of (map), 286
maps and cross sections of, 151
metamorphic conditions of sole formation, 57, 60
metamorphic overprint, 288
models for magma chamber at, 88
Muscat-Ibra structural profile, 513–516
O isotopes, 276
ophiolite nappe, 515–517
ophiolite-sole pair ages inconsistent with forearc provenance, 57
ore deposits, 288
petrogenetic provenance of, 13
regional paleogeographic reconstruction, 495
sedimentary cover, 285
size, 57
Solund-Stavfjord ophiolite complex, comparison with, 452
spreading center–trench interaction, 17
Sr isotopes, 275
stockworks, 288
structural profile, 515
structure of Wadi Tayin metamorphic sole, 516–517
sulfide deposit in, 291
tectonic setting, 57
Th-Hf-Ta discrimination diagram, 15
Semail sole, 60, 514
semibrittle failure, origin of microcracks by, 266
serpentine diapirism, and Solund-Stavfjord ophiolite complex, 457
serpentine mud volcanism
convergent margins and, 35–49
distribution of, 37
fault zones or faulted serpentized blocks and, 48
structural setting of, 37, 41
serpentine mud volcanoes
Big Blue Seamount, 36–38, 45–46, 49
Blue Moon Seamount, 38, 49
Celestial Seamount, 37–39, 42, 44–45, 49
Chamorro Seamounts, 37
conduit for, preserved, 46, 49
Conical Seamount, 37–38, 45–46, 48
deformation of, 48
depositional form of, 48
description of, 48
long-lived, 36
model for emplacement of, 48–49
modes of origin of, 45–46
Mysterious Valley Formation, 47, 49
North Chamorro Seamount, 38, 43–45, 49
obduction events and, 49
Pacman Seamount, 37–38, 45–46, 49
Peacock Seamount, 38–39, 42, 44–45, 49
preservation of, 49
pressure and temperature of formation of, 46
South Chamorro Seamount, 38–39, 41, 43–45, 48–49
stages of formation, 37
Turquoise Seamount, 38–39, 42, 44, 48–49
See also sedimentary serpentinites
serpentinization and petrogenesis, Hayama-Mineoka belt(s), 419, 425–427
Sesia-Lanzo unit, 29
Seven Lakes mafic intrusive complex, Trinity ophiolite, 406
Shahr-Babak ophiolite

characteristics of, 487
chemical composition of, 490–491
geologic map of, 488
petrogenesis of, 493–495
petrography of, 487, 489
stratigraphy of, 486–487
tectonic setting of, 486
trace elements, 490–494
shear zones, origin of fluid phase in, 135
Shikai (or New) Mound, TAG hydrothermal deposits, summary of, 216
Shimmering Mound, TAG hydrothermal deposits, 216
Shinkai 6500 submersible dives
biological specimens collected during, 44
Mariana convergent margin, 37
mud volcano samples collected, 41, 44
to South Chamorro Seamount summit, 44
Shuksan suite, 59
Siberia, 10
sidescan sonar backscatter image, active mound, TAG (Trans-Atlantic Geotraverse) hydrothermal field, 215
Sierra de Pie de Palo, geologic map of, 338
Sierra Nevada
geologic map of northern Foothills, 367
ophiolitic rocks within, 61
sequential cross sections of, 398
suture zone possibly within, 61
ultramafic rocks and associated rock units, 353
Sierra Nevada batholith
description of, 369
timeline plot of, 368
Sierra Nevada foothills
correlation with Klamath Mountains, 370–371
oppositely dipping subduction zones, 371
Sierra Santa Rosa, Zacatecas, location of (map), 391
Sierras de Córdoba
Eastern Córdoba ophiolite belt, 333–335
geologic map of, 333
Western Córdoba ophiolite belt, 334–337
silicate inclusions in podiform chromite, and evidence for water in melt, 142
S isotopes
black smokers, 277
oceanic crust, 277
Troodos ophiolite, 277
Sistan Ocean, paleogeographic reconstruction of, 495
Skaergaard intrusion, 93–95
slab-derived fluids, 36, 48, 438
Hellenic ophiolite belt, western, 481
slab rollback, western Pacific, 399
slab-window development, 16
Slate Creek Complex
geologic map of, 367
tectonic development of western North American continental margin, 376
timeline plot of, 368
slow-spreading ridges
crustal-genesis model and, 170
crustal thickness as a function of mantle temperature, 171–172
depth to melt lens as a function of mantle temperature, 172
fluid inclusions, 241
Jurassic Ligurian Tethys cross section, 136
magma chambers at, 87–101
MARK (Mid-Atlantic Ridge–Kane Fracture Zone intersection) area, 4, 241
Northern Apennine ophiolites, 136
Southwest Indian Ridge, 241
Troodos ophiolite, 241
slurry, gravitationally unstable. *See* serpentine mud volcanism
slurry of dense cumulate, and lower-gabbro production, 121–126
Smartville Complex
cross section showing, 371
geologic map of, 357, 367
geologic model along northern California–Nevada transect, 373
location of (map), 353, 396
oceanic crust, comparison with, 356
paleogeographic reconstruction of western U.S. Cordillera, 374
paleotectonic sequence diagrams for, 397
tectonic development of western North American continental margin, 376
timeline plot of, 368
Sm-Nd isotope plots
Lanzo North peridotites, 26
Northern Apennine ophiolites, 26
Pyrenees peridotites, 26
Red Sea area, 30
Ronda peridotite, 26
Zabargad Island, 30
Society Islands, 16
Solund-Stavfjord ophiolite complex
Costa Rica Rift, comparison with, 452
emplacement of, 455–458
evolution of, 456–457
geochemistry, 448–451, 453–454
geologic map of, 445
geologic map of Tviberg and Oldra, 446
hydrothermal history, 452–453
internal structure and architecture, 444–446
magmatic rocks (photographs), 447
model for magmatic development from propagating-rift tip, 451
O and Sr isotope analyses, 452
paleogeographic reconstruction of, 446
petrogenetic model for, 448–449
propagated rift segment of, 450
regional geology, 443–444
seafloor-spreading origin of, 454–455
Semail ophiolite, comparison with, 452
tectonic settings, 455–458
Troodos ophiolite, comparison with, 452
volcanic construction model, 448
sonar image. *See* sidescan sonar backscatter image
Sonomia, western United States, 8–9
source-depletion vector, in Nb/Yb vs. Th/Yb plot, 414
sources of sedimentary serpentinite, seafloor piercements of serpentinite, 47
South Alpine unit, 29
South American plate, 14
South Chamorro Seamount, 48–49
description of, 39, 41
location of (bathymetric map), 38, 43
perspective view of, 45
pressure and temperature of formation of, 46
still actively undergoing protrusion, 44
Southeast Indian Ridge
depth to top of magma lens as a function of spreading rate, 163
location of (map), 302
and steady-state magma lens, 175
southern Chile Ridge
geochemistry, 14–15
mantle source, 7
spreading-ridge subduction, 16
subduction-related contamination of mantle beneath, 13–17
tectonic setting, 13–14
Southern Cross Expedition, southern Chile Ridge, 14
Southern Mound, TAG hydrothermal, 216
South Tasman Rise, location of (map), 302
Southwest Indian Ridge
C concentrations vs. $\delta^{13}C$, 246
chemical composition, 132
fluid inclusions, 241–242, 245–247, 249
gabbroic complexes, 90–91
geochemistry, 97
lithology of samples, 238–239
location of (map), 238
lower oceanic crust, 75–85
magma-chamber development at, 87
MARK (Mid-Atlantic Ridge–Kane Fracture Zone intersection) area, comparison with, 89
See also Ocean Drilling Program Hole 735B
spinel
Bay of Islands ophiolite, 435
chemical composition, Hayama-Mineoka belt(s), 423, 426–427
spreading center–trench interaction
Bay of Islands ophiolite, 17
Semail massif, 17
southern Chile Ridge, 17
spreading environment, 250–252
spreading rate, 4
Aswad massif, 508–509
spreading rate, mantle temperature, and hydrothermal cooling
dependence of crustal accretion and ridge-axis topography on, 161–176
spreading-ridge subduction, southern Chile Ridge, 16
Sr isotopes
black smokers, 275
Cyprus Crustal Study Project (CCSP) CY-4, 275
oceanic crust, 275
Semail ophiolite, 275
Troodos ophiolite, 275
Sr-Nd isotope plot
Lanzo (Western Alps) peridotites, 24
Northern Apennine ophiolites, 24
Pyrenees peridotites, 24
Ronda peridotite, 24
St. Anthony Complex, 60
Stavenes Group and Solund-Stavfjord ophiolite complex, 445, 453
Stikinia terrane, western North America, 8–9
stockwork
TAG (Trans-Atlantic Geotraverse) hydrothermal field, 218
See also up-flow zone
stockworks
alteration (hydrothermal) patterns in suprasubduction zone ophiolites, 290
Betts Cove, 288
Del Puerto, 289
Izu-Bonin forearc area, 289
Josephine ophiolite, 289
Karmøy, 288
Kizildag ophiolite, 288
Leka, 288
Pindos, 288
Semail ophiolite, 288
Tonga forearc, 289
Trinity ophiolite, 289
Troodos ophiolite, 288
Zambales, 289
Stonyford, oceanic crust, comparison with, 356
stoping, 113
strain differences in western Hellenic ophiolites, 481
stress in chromitite pods
amplification curves, 155
deformation as a function of strain, 155
experimental design for determining, 150
gradients in, 156
Oman ophiolite belt, 149–157
plastic yield in chromite inclusion, 154–155
results of modeling experiment, 154–155
strain rate during experiment, 154
stress-regime change from compressional to strike slip, Ocean Drilling Program Hole 504B, 197
strike-slip faulting and terrane translation, western North America, 370
structural geology, Ocean Drilling Program Hole 735B, 81–82
structural history
Macquarie Island, 301–318
ophiolites along California continental margin, 354
structural settings
oceanic magma chambers and, 95–96

serpentine and blueschist mud volcanoes, 37, 41
structure
of Aswad massif, 500–501, 506–508, 509
faulting pattern of Masirah ophiolite, 532–533
faults and Masirah ophiolite, 531, 533
of Northern Apennine ophiolites, 22–23, 130–131
of ridge segment (based on Oman ophiolite belt), 71
of Semail nappe, 515–517
of TAG (Trans-Atlantic Geotraverse) active mound and stockwork (or up-flow) zone, 219
subcontinental lithospheric mantle
Northern Apennine ophiolites and, 21, 31
northern Red Sea and, 30–31
subducted sediment and seawater components, Trinity ophiolite, 414
subducted slabs in lower mantle, 6, 8, 10
subduction
of ridges, 48
of seamounts, 48
subduction-component vector, in Nb/Yb vs. Th/Yb plot, 414
subduction erosion, 54
subduction initiation, model for, 48
subduction polarity flips
California active-plate-margin development, 61
eastern New Guinea, 61
southwest Pacific, 61
subophiolitic sole generation, 61
Sulawesi, 61
subduction-related contamination of mantle beneath southern Chile Ridge, 13–17
subduction-related geochemical signature, Trinity ophiolite, 415–417
subduction zones
creation of ophiolites nearby, 16
oppositely dipping, 371, 376
rollback and western Hellenic ophiolites, 481
subophiolitic soles
definition of, 53
See also metamorphic soles of ophiolites
Sulawesi, subduction polarity flip, 61
sulfide deposit
in Betts Cove, 291
in Josephine ophiolite, 291
in Karmøy, 291
in Pindos, 291
in Semail ophiolite, 291
in Troodos ophiolite, 291
in Zambales, 291
sulfides
composition of, Hayama-Mineoka belt(s), 424–427
origin of in peridotites, 425–427
TAG (Trans-Atlantic Geotraverse) hydrothermal field, 225
sulfur fugacity vs. oxygen fugacity, conditions of serpentination, 427
Sumail massif. *See* Oman ophiolite belt
Sunnfjord mélange, and Solund-Stavfjord ophiolite complex, 445, 454, 457
supra-subduction zone
backarc setting for ophiolite generation, comparison with, 60–61
chemical signature of, 8, 13
fluid inclusions, 241
forearc provenance of ophiolites hard to reconcile, 60
ophiolite generation in, 54
Tethyan ophiolites, 5
Trinity ophiolite, 403–404, 415–417
Troodos ophiolite, 241
See also individual ophiolites
supra-subduction zone chemistry, depleted, Trinity ophiolite, 412–413
supra-subduction zone mantle
hydration metamorphism and, 48
isotopic nature, 7
peridotite ridge in Izu-Bonin forearc area, 45
supra-subduction zone ophiolites
alteration (hydrothermal) patterns in, 283–293
geologic characteristics of, 285
mid-oceanic ridge crust, comparison with, 292–293
and modern supra-subduction zone settings, 290, 292
suture zone
possibly buried by Great Valley Group, 61
possibly buried by Sierra Nevada, 61
syntectic assimilation reactions, 105
syntexis, 105–116
definition of, 105–107
dikes and, 111–113

T

Table Mountain, California, protrusive serpentinite piercement structures in, 47
TAG (Trans-Atlantic Geotraverse) hydrothermal field, 213–232
active mound, 214–215
active mound, bathymetry of, 220
active mound, cross section of, 219–220
active mound, fluid flow within, 225
active mound, fluids discharging from, 221
active mound, hydrothermal deposits of, 216
active mound, implications of fluid flow within, 226
active mound, interior of, 219–220
active mound, mineral assemblages in, 217–219
active mound, model for formation of, 230–232
active mound, stockwork (or up-flow) zone, 218–219, 221–222
active mound, surface of, 217, 219
active mound, trapping temperatures of fluid inclusions from, 223
active-mound morphology, 217
active mound, Sr and O isotope data, 224
alteration of basalt, 227–230
alteration of up-flow zone basalt, 226–230
Alvin zone, 214–215
anhydrite, 222–224
black smokers, 215, 218–222, 231
chloritization of fresh basalt, 227–228
geologic setting, 213–217
low-temperature zone, 215
Mir zone, 215
ophiolitic sulfide deposits, comparison with, 217
outstanding problems, 232
paragonitization of chloritized basalt, 228–229
pyritiztation of chloritized basalt, 228–229
quartz, 224
role of faulting in development of, 215, 217
silicification of chloritized basalt, 228
Sr and O isotopes and temperatures of formation, 224, 225
sulfides, 225
up-flow zone, implications of alteration within, 230
white smokers, 216, 218–219, 221–222, 231
TAG segment of Mid-Atlantic Ridge
bathymetric map, 214
geologic setting, 213–214
Taitao Peninsula, 13, 16
location of (map), 14
trace elements, 17
Taitao Ridge (submarine morphologic feature), 16
location of (map), 14
trace elements, 17
Taiwan collision zone, structural evolution of, 399
Tasman Ridge, location of (map), 302
Tchehel Kureh ophiolite
characteristics of, 487
tectonic setting of, 486
tectonic controls
on magma chambers at slow-spreading ridges, 87–101
over mud-volcano distribution, 37
tectonic denudation, 21
Northern Apennine ophiolites and, 22, 28–29, 31
tectonic development
Arabian margin, 520–522
cross sections of western North American continental margin, 376
Semail nappe, 519
tectonic implications
of California continental margin ophiolites, 359, 361–362
of Hayama-Mineoka ophiolites, 427–428
tectonic model for ophiolite generation, requirements for, 54
tectonic settings
Andean ophiolites of southern South America, 346
backarc and forearc provenance compared, 60–61
Bay of Islands ophiolite, 17, 57
Betts Cove, 285
Bonin Trough, 420
Brooks Range ophiolites, 58
central Anatolia, 4
chemical composition, 13
Coast Range ophiolite, 58–61
Cordillera Frontal ophiolites, 344
cross sections of formation of supra-subduction zone ophiolites, 54–55
cross sections of ophiolites and metamorphic soles, 54–55
Del Puerto, 285
Eastern Córdoba ophiolite belt, 334
forearc, 48
forearc provenance of ophiolites hard to reconcile, 60
indicators of, 13
Iranian ophiolites, 486
Izu-Bonin forearc area, 285
Izu-Bonin Trench, 420
Josephine ophiolite, 285
Karmøy, 285
Kizildag ophiolite, 285
Leka, 285
Macquarie Island, 301–318
most compatible with all data, 61
Neotethys, 4
Northern Apennine ophiolites, 27–29
Ocean Drilling Program Hole 735B, 76–77
of ophiolite generation, 54, 54–56
ophiolites along California continental margin, 351, 354, 371, 373–374, 376, 397–398
Pie de Palo belt, southern Andes, 339
Pindos, 285
possibly an artifact of postmagmatic emplacement, 54
Semail ophiolite, 17, 57, 285
serpentine mud volcanism, 48–49
Solund-Stavfjord ophiolite complex, 455–458
southern Chile Ridge, 14
southern Turkey, Cyprus, and farther east, 5
summary of and examples, 461–462
TAG (Trans-Atlantic Geotraverse) hydrothermal field, 216
Tethyan ophiolites, 514
Tonga forearc, 285
trench rollback and, 48
Trinity ophiolite, 58, 285, 403–404, 415–417
Troodos ophiolite, 285
western Balkan complexes, 4
Western Córdoba ophiolite belt, 337
western Pacific, 399
Zambales, 285
See also mantle source
tectonic transport, vertical, metamorphic soles of ophiolites, 60
tectonism and magma-chamber evolution, 100–101
tectonostratigraphic significance of associated sedimentary strata
Coast Range ophiolite, 383–389, 391–392
Josephine ophiolite, 383, 389–392

temperature
of amphibole-plagioclase equilibratin in microcracks, 268
of anhydrite formation, TAG (Trans-Atlantic Geotraverse) hydrothermal field, 222–225
of cracking, 268–269
of cracking front, variation with depth and distance from ridge axis, 269
distribution vs. depth below dike-gabbro transition, 269
of eruption event plume, Gorda Ridge, 253
of microcracking, and its variation with depth, 266
of quartz formation, TAG (Trans-Atlantic Geotraverse) hydrothermal field, 223
and richterite-edenite exchange in amphiboles, Hess Deep and Oman ophiolite belt, 267
during uplift of Oman Mountains, 518
temperature and pressure of metamorphism
of blueschist-grade mud-volcano pebbles, 41
ophiolite sole, 59–60
See also individual ophiolites
temperature of initial cracking
Hess Deep, 267
Oman ophiolite belt, 267
temperature vs. depth below dike-gabbro transition, Hess Deep and Oman ophiolite belt, 268
Tethyan and Tethyan-type ophiolites
Alpine-Apennine belt, 4
central Anatolia, 4
compatible with mid-ocean ridge provenance, 61
cross sections of emplacement of supra-subduction zone ophiolites, 55
definition of, 3, 53
Dinaride-Hellenide complexes, 4
eastern Greece and central Turkey, 4
and Hercynian-Appalachian orogeny, 8
incompatible with backarc provenance, 61
and Pan-African orogeny, 8
tectonic settings, 5–6, 8, 10, 514
western Balkan complexes, 4
Tethys, Ligurian Sea sector of, 21
texture
of microscopic amphibole veins, 263
of podiform chromitite, 140
thermal boundary layer in mantle, 7
thermal structure
flow vectors and, 172
"threshold" mantle temperature and changes in, 172
thermophiles, and Gorda Ridge eruption plume, 253
Th-Hf-Ta discrimination diagram
Bay of Islands ophiolite, 15
Copley Greenstone, Eastern Klamath Mountains, 413
Semail massif, 15
southern Chile Ridge, 15
Trinity ophiolite, 413
thickness
of crust in various ridge segments, 175
of layer 2A, East Pacific Rise, 184
of Oman crust, 72
of Oman lithologic units, 72
of Oman Moho transition zone and crust, 70–71
thin lithosphere of ophiolites, 16
"threshold" mantle temperature, and changes in axial thermal structure, 172
Th/Yb vs. Ta/Yb discrimination diagram
Baimak-Buribai Complex of Magnitogorsk arc, 470
Uralian fault extrusive rocks, 470
Tibetan ophiolites, 7
Tihama Asir Magmatic Complex, 31
Timor ophiolite, 321–329
chemical composition, 324
covariation diagrams, 328
geologic map of, 322
geologic setting, 322–323
lithologic units of, 323–324
location of (map), 322
mineral chemistry, 327–328
modal abundances, 323
origin and tectonic evolution of, 328–329
partial-melting estimates from whole-rock compositions, 326
petrography, 327
structural section across West Timor, 323
whole-rock chemistry, 327–328
whole-rock discrimination diagrams, 325
Tinaquillo, Cr vs. MgO plot, 143
TiO_2 vs. 100 Cr/(Cr + Al) plot of, 435
TiO_2 vs. FeO_t plot, Solund-Stavfjord ophiolite complex, 451
TiO_2 vs. Mg# plot, Baimak-Buribai Complex of Magnitogorsk arc, 468
TiO_2 vs. Zr discrimination diagram, Masirah ophiolite, 530
Tonga forearc
age of, 285
background alteration, 289
chemical composition of lavas, 285
epidosite, 289
hydrothermal characteristics of (table), 289
location of (map), 286
metamorphic overprint, 289
ore deposits, 289
sedimentary cover, 285
stockworks, 289
Tonga Ridge
CaO vs. SiO_2 plot, and Baimak-Buribai Complex of Magnitogorsk arc, 468
Mg# vs. TiO_2 plot, and Baimak-Buribai Complex of Magnitogorsk arc, 468
MgO vs. SiO_2 plot, and Baimak-Buribai Complex of Magnitogorsk arc, 468
rare earth element plot, 469
topography, global seafloor, map of, 162
Torishima Forearc Seamount, sampled during Ocean Drilling Program Leg 125, 45
trace elements
average N-MORB, 134
Bay of Islands ophiolite, 17, 436, 438–439
Bay of Islands ophiolite (Springers Hll mantle), compared with boninites and MORB, 440
clinopyroxene, Northern Apennine ophiolites, 136
Copley Greenstone, Eastern Klamath Mountains, 414
incompatible, Shahr-Babak ophiolite, 493–494
Izu-Bonin-Mariana forearc, 414
MARK (Mid-Atlantic Ridge–Kane Fracture Zone intersection) area, 134
New Britain arc, 414
Northern Apennine ophiolites, 134
Ocean Drilling Program Leg 153, 134
pargasite (titanian), Northern Apennine ophiolites, 136
Shahr-Babak ophiolite, 490–495
Solund-Stavfjord ophiolite complex, 449
southern Chile Ridge, 17
Stavenes Group, Solund-Stavfjord ophiolite complex, 453
Sunnfjord mélange, Solund-Stavfjord ophiolite complex, 454
Taitao Peninsula, 17
Taitao Ridge (submarine morphologic feature), 17
Trinity ophiolite, 412, 414
trace elements vs. Zr plot, Northern Apennine ophiolites, 133
transform faults, subduction initiation and, 48
transform-zone tectonic setting, Northern Apennine ophiolites and, 28
transpression, Macquarie Island, 314–315
trench initiation, 400
trench rollback, 48
Trinity ophiolite
age of, 285, 406
ages of ophiolite and sole, 57–58
alteration (hydrothermal) patterns in supra-subduction zone ophiolites, 291
background alteration, 289
Bily's Peak mafic intrusive complex, 407
Bonanza King mafic intrusive complex, 407, 409–410, 412–415
Central Metamorphic belt, 58, 404–405
chemical composition of lavas, 285
Cordilleran type, 58
Craggy Peak mafic intrusive complex, 409, 411–415
epidosite, 289
geochemistry of, 408–415
geologic map of, 367, 404
geology of, 403–408
hydrothermal characteristics of (table), 289
key geologic and hydrothermal characteristics of, 291
location of (map), 286
metamorphic conditions of sole formation, 58
metamorphic overprint, 289
metavolcanic rocks of Eastern Klamath terrane, 404
Neoproterozoic mafic tectonite assemblage, 405
ophiolite-sole pair ages possibly inconsistent with forearc provenance, 58
Ordovician–Silurian(?) mantle assemblage, 405
ore deposits, 289
origin of, 416
petrography, 408
Porcupine Lake mafic intrusive complex, 409, 411–415
sedimentary cover, 285
sedimentary successions of Yreka terrane, 404
Seven Lakes mafic intrusive complex, 406
Silurian and Devonian mafic intrusive complexes, 405–407
size, 58
stockworks, 289
supra-subduction zone magmatism, 403–404, 415–417
tectonic setting, 58
Th-Hf-Ta discrimination diagram, 413
trace elements, 414
xenoliths in (photographs), 406–407
Trinity terrane
ophiolitic?, 407–408
trapped Ordovician lithospheric mantle, 416
triple junctions, 13–14
Troodos ophiolite
age of, 285
Akaki River Canyon, lava-flow orientations, 187
alteration (hydrothermal) patterns in supra-subduction zone ophiolites, 291
background alteration, 288
chemical composition of lavas, 285
cross section of, 92
Cr vs. Al_2O_3 plot, 143
cryptic variation, 93
Cyprus Crustal Study Project (CCSP) CY-1 and CY-1A, 181–191
deformation decoupling horizon, 186, 191
deformation of, 92
East Pacific Rise, comparison with, 187
epidosite, 288
fluid inclusions, 241–242, 244, 246–247
geologic map of Akaki and Maroulena River area, 186
geometry of rift valley, 187
hydrothermal characteristics of (table), 288
isostatic compensation for lava-flow loading, 189
Kamara Potamos, lava-flow orientations, 187
key geologic and hydrothermal characteristics of, 291
Klirou bridge area, lava-flow orientations, 187
lithology of samples, 238, 240
lithology of samples, Cyprus Crustal Study Project (CCSP) CY-4, 238
location of (map), 5, 286

location of paleo–ridge crest, 185
metamorphic overprint, 288
MgO vs. Al_2O_3 plot, 143
Mitsero graben, 186
Ni vs. Al_2O_3 plot, 143
O isotopes, 276
ore deposits, 288
paleomagnetic inclinations of lavas and dikes in, 181–191
postdepositional rotation (about horizontal axis) of lava flows, 185–187, 190–191
rate of lava deposition, 187
rate of spreading, 187
regional paleogeographic reconstruction, 495
sedimentary cover, 285
S isotopes, 277
Solund-Stavfjord ophiolite complex, comparison with, 452
Sr isotopes, 275
stockworks, 288
sulfide deposit in, 291
supra-subduction zone, 17
Tuolumne Complex, California, age of, 61
Turquoise Seamount, 48–49
description of, 39
location of (bathymetric map), 38, 42
perspective view of, 44

U

ultra-slow–spreading ridges. *See* slow-spreading ridges
up-flow zone, TAG (Trans-Atlantic Geotraverse)
alteration of basalt in, 226–230
anhydrite, 222–224
fluid inclusions, 223
quartz, 223–225
vent fluids and minerals, 222–226
uplift, evidence for, retrograde metamorphism on Zabargad Island, 30
upper oceanic crust
deformation of, 181–192
dike orientation, rapid change in, 190
rotation (postdepositional) about horizontal axis, 182–184
seismic amplitude modeling, 203–209
seismic energy loss and attenuation, 204
seismic energy loss vs. two-way traveltime, 206
Uralian fault extrusive rocks, Nd-Sr isotopes, 469
Uralian orogeny, 10, 469–470

V

vein intensity variations, Ocean Drilling Program Hole 735B, 82
velocity model across northern California, cross section showing, 372
vent fluids and minerals, up-flow zone, TAG (Trans-Atlantic Geotraverse) area, 222–226
viscosity of magma lens and upper gabbros, 124–125
volatiles
Gorda Ridge eruption plume, 254
mid-oceanic ridges, 237–255
released during dike intrusion and eruptive events, 252–255
from subducting plate, 36
volcanic construction model, Solund-Stavfjord ophiolite complex, 448
Vourinos Complex
chromite-bearing zone, 475
deformational fabrics in, 479
isotherms in cooling ocean floor, 479
lithology and structure, 474–475
maps of, 475
strain in, compared to other western Hellenic ophiolites, 481
Vulcan Fracture Zone, trace elements, 436

W

Wadi Ham, Aswad massif, 503
Wadi Hatta, Aswad massif, 503, 510
Wadi Tayin metamorphic sole, structure of, 516–517
Walper megashear, Mexico, location of (map), 391
water
dissolved in basaltic and andesitic melts, 141–142
and formation of podiform chromitite, 139–145
welding of metamorphic sole to ophiolite, 60
Western Alps, 29
Cr vs. MgO plot, 143
rare earth element plot, 24
Western Córdoba ophiolite belt, summary of, 346
western Pacific Tertiary rocks, and Magnitogorsk arc, Urals
Mg# vs. TiO_2 plot, 468
rare earth element plot, 469
Western Precordillera ophiolite
geochemistry of, 341–343
geologic map of, 340
summary of, 346
western U.S., Cordillera, tectonic map of, 366
white smokers, TAG (Trans-Atlantic Geotraverse) hydrothermal field, 218–219, 221–222, 231
Wrangellia terrane, western North America, 8–9

X

xenoliths, 113–114
from arcs, trace elements, 436
Cr vs. MgO plot, 143
metabasaltic, 100
in Trinity ophiolite (photographs), 406–407
xenoliths of western Arabian peninsula, Sm-Nd isotope plot, 30

Y

Yb_N vs. Cr/(Cr + Al), plot of, 435
Yerington, west-central Nevada continental arc, timeline plot of, 368
Y/Nb vs. Zr/Nb discrimination diagram, Masirah ophiolite, 530
Yreka terrane
mildly telescoped trench-related basins, 416
sedimentary successions of, 404
Y vs. Zr discrimination diagram
Masirah ophiolite, 530
Shahr-Babak ophiolite, 493

Z

Zabargad Island, 21
Cr vs. MgO plot, 143
Sm-Nd isotope plot, 30
Zambales
age of, 285
alteration (hydrothermal) patterns in supra-subduction zone ophiolites, 291
background alteration, 289
chemical composition of lavas, 285
epidosite, 289
hydrothermal characteristics of (table), 289
key geologic and hydrothermal characteristics of, 291
location of (map), 286
metamorphic overprint, 289
Mg# vs. TiO_2 plot, and Baimak-Buribai Complex of Magnitogorsk arc, 468
ore deposits, 289
sedimentary cover, 285
stockworks, 289
sulfide deposit in, 291
zircon
age dating plagiogranites with, 56
ages of ophiolite-sole pairs, 57–59
Zr/Nb vs. La/Ce discrimination diagram, Masirah ophiolite, 530
Zr/Nb vs. Y/Nb discrimination diagram, Masirah ophiolite, 530
Zr/Nb vs. Zr/Y discrimination diagram, Masirah ophiolite, 530
Zr/Sm vs. La/Sm plot, Baimak-Buribai Complex of Magnitogorsk arc, 468
Zr/Ti vs. Nb/Y discrimination diagram
Masirah ophiolite, 531
Shahr-Babak ophiolite, 492
Zr/Ti vs. SiO_2 discrimination diagram, Shahr-Babak ophiolite, 492
Zr vs. Nb discrimination diagram, Shahr-Babak ophiolite, 493
Zr vs. TiO_2 discrimination diagram, Masirah ophiolite, 530
Zr vs. Y discrimination diagram, Masirah ophiolite, 530
Zr/Y vs. Zr/Nb discrimination diagram
Masirah ophiolite, 530

Printed in the U.S.A.